Physics Galaxy

22 Years'

2002-2023 (Offline & Online)

JEE MAIN

Chapter-Wise Solutions

Physics

Ashish Arora
Founder & Chief Mentor

PHYSICS GALAXY

World's largest encyclopedia of online video courses for High School Students preparing for

* JEE Main * JEE Advanced * NEET * NSEP/INPhO/IPhO

First Edition : December, 2000

Second Edition : May, 2019

Revised Edition : November 24, 2023

Title : Physics Galaxy : Physics 22 Years' (2002-2023) JEE Main Chatper-Wise Solutions

Language : English

Author's Name : Ashish Arora

Typeset & Published by :

Career Launcher Infrastructure (P) Ltd.

A-45, Mohan Cooperative Industrial Area, Near Mohan Estate Metro Station, New Delhi - 110044

Marketed by :

G.K. Publications (P) Ltd.

Plot No. 9A, Sector-27A, Mathura Road, Faridabad, Haryana-121003

ISBN : **978-93-56817-89-0**

Printer's Details : Printed in India, New Delhi.

For product information :

Visit ***www.gkpublications.com*** or email to ***gkp@gkpublications.com***

Dedicated

to

My Parents, Son, Daughter

and

My beloved wife

In his teaching career since 1992 Ashish Arora personally mentored more than 10000 IITians and students who reached global heights in various career and profession chosen. It is his helping attitude toward students with which all his students remember him in life for his contribution in their success and keep connections with him live. Below is the list of some of the successful students in International Olympiad personally taught by him.

NAVNEET LOIWAL	***International GOLD Medal in IPhO-2000 at LONDON***, Also secured **AIR-4** in **IIT JEE 2000** PROUD FOR INDIA : Navneet Loiwal was the first Indian Student who won first International GOLD Medal for our country in International Physics Olympiad.
DUNGRA RAM CHOUDHARY	**AIR-1** in **IIT JEE 2002**
HARSHIT CHOPRA	***National Gold Medal in INPhO-2002*** and got **AIR-2** in **IIT JEE-2002**
KUNTAL LOYA	A Girl Student got position **AIR-8** in **IIT JEE 2002**
LUV KUMAR	***National Gold Medal in INPhO-2003*** and got **AIR-3** in **IIT JEE-2003**
RAJHANS SAMDANI	***National Gold Medal in INPhO-2003*** and got **AIR-5** in **IIT JEE-2003**
SHANTANU BHARDWAJ	***International SILVER Medal in IPhO-2002 at INDONESIA***
SHALEEN HARLALKA	***International GOLD Medal in IPhO-2003 at CHINA*** and got **AIR-46** in **IIT JEE-2003**
TARUN GUPTA	***National GOLD Medal in INPhO-2005***
APEKSHA KHANDELWAL	***National GOLD Medal in INPhO-2005***
ABHINAV SINHA	***Hon'ble Mension Award in APhO-2006* at *KAZAKHSTAN***
RAMAN SHARMA	***International GOLD Medal in IPhO-2007* at *IRAN*** and got **AIR-20** in **IIT JEE-2007**
PRATYUSH PANDEY	***International SILVER Medal in IPhO-2007* at *IRAN*** and got **AIR-85 in IIT JEE-2007**
GARVIT JUNIWAL	***International GOLD Medal in IPhO-2008* at *VIETNAM*** and got **AIR-10** in **IIT JEE-2008**
ANKIT PARASHAR	***National GOLD Medal in INPhO-2008***
HEMANT NOVAL	***National GOLD Medal in INPhO-2008*** and got **AIR-25** in **IIT JEE-2008**
ABHISHEK MITRUKA	***National GOLD Medal in INPhO-2009***
SARTHAK KALANI	***National GOLD Medal in INPhO-2009***
ASTHA AGARWAL	***International SILVER Medal in IJSO-2009* at *AZERBAIJAN***
RAHUL GURNANI	***International SILVER Medal in IJSO-2009* at *AZERBAIJAN***
AYUSH SINGHAL	***International SILVER Medal in IJSO-2009* at *AZERBAIJAN***
MEHUL KUMAR	***International SILVER Medal in IPhO-2010* at *CROATIA*** and got **AIR-19** in **IIT JEE-2010**
ABHIROOP BHATNAGAR	***National GOLD Medal in INPhO-2010***
AYUSH SHARMA	***International Double GOLD Medal in IJSO-2010*** at ***NIGERIA***
AASTHA AGRAWAL	***Hon'ble Mension Award in APhO-2011*** at ***ISRAEL*** and got **AIR-93** in **IIT JEE 2011**
ABHISHEK BANSAL	***National GOLD Medal in INPhO-2011***
SAMYAK DAGA	***National GOLD Medal in INPhO-2011***
SHREY GOYAL	***National GOLD Medal in INPhO-2012*** and secured **AIR-24** in **IIT JEE 2012**
RAHUL GURNANI	***National GOLD Medal in INPhO-2012***
JASPREET SINGH JHEETA	***National GOLD Medal in INPhO-2012***
DIVYANSHU MUND	***National GOLD Medal in INPhO-2012***
SHESHANSH AGARWAL	***International SILVER Medal in IAO-2012 at KOREA***
SWATI GUPTA	***International SILVER Medal in IJSO-2012 at IRAN***
PRATYUSH RAJPUT	***International SILVER Medal in IJSO-2012 at IRAN***
SHESHANSH AGARWAL	***International BRONZE Medal in IOAA-2013 at GREECE***
SHESHANSH AGARWAL	***International GOLD Medal in IOAA-2014 at ROMANIA***
SHESHANSH AGARWAL	***International SILVER Medal in IPhO-2015 at INDIA*** and secured **AIR-58** in **JEE(Advanced)-2015**
VIDUSHI VARSHNEY	***International SILVER Medal in IJSO-2015 at SOUTH KOREA***
AMAN BANSAL	**AIR-1 in JEE Advanced 2016**
KUNAL GOYAL	**AIR-3 in JEE Advanced 2016**
GOURAV DIDWANIA	**AIR-9 in JEE Advanced 2016**
DIVYANSH GARG	***International SILVER Medal in IPhO-2016 at SWITZERLAND***
NALIN KHANDELWAL	**AIR-1 in NEET 2019**
MRIDUL AGARWAL	**AIR-1 in JEE Advanced 2021**

ABOUT THE AUTHOR

The complexities of Physics have given nightmares to many, but the homegrown genius of Jaipur- Ashish Arora has helped several students to live their dreams by decoding it.

Newton Law of Gravitation and Faraday's Magnetic force of attraction apply perfectly well with this unassuming genius. A Pied Piper of students, his webportal https://www.physicsgalaxy.com, The world's largest encyclopedia of video lectures on high school Physics possesses strong gravitational pull and magnetic attraction for students who want to make it big in life.

Ashish Arora, gifted with rare ability to train masterminds, has mentored over 10,000 IITians in his past 24 years of teaching sojourn including lots of students made it to Top 100 in IIT-JEE/ JEE(Advance) including AIR-1 and many in Top-10. Apart from that, he has also groomed hundreds of students for cracking International Physics Olympiad. No wonder his student Navneet Loiwal brought laurel to the country by becoming the first Indian to win a Gold medal at the 2000 - International Physics Olympiad in London (UK).

His special ability to simplify the toughest of the Physics theorems and applications rates him as one among the best Physics teachers in the world. With this, Arora simply defies the logic that perfection comes with age. Even at 18 when he started teaching Physics while pursuing engineering, he was as engaging as he is now. Experience, besides graying his hair, has just widened his horizon.

Now after encountering all tribes of students - some brilliant and some not-so-intelligent - this celebrated teacher has embarked upon a noble mission to make the entire galaxy of Physics inform of his webportal PHYSICSGALAXY.COM to serve and help global students in the subject. Today students from 183 countries are connected with this webportal. On any topic of physics students can post their queries in comment section in Physics Galaxy Youtube channel on which many Physics experts with Ashish Arora reply to several queries posted online by students.

Dedicated to global students of middle and high school level, his website *www.physicsgalaxy.com* also has teaching sessions dubbed in American accent and subtitles in 87 languages. For students in India preparing for JEE & NEET, his online courses will be available soon on PHYSICSGALAXY.COM.

In 2007 he started authoring Physics Galaxy Youtube Channel which has now become World's Largest Youtube Channel for online physics lectures on high school physics and preparation of JEE & NEET. Daily about 1.5 lakh+ video lectures are being watched on the channel by students preparing for JEE, NEET & High school Physics.

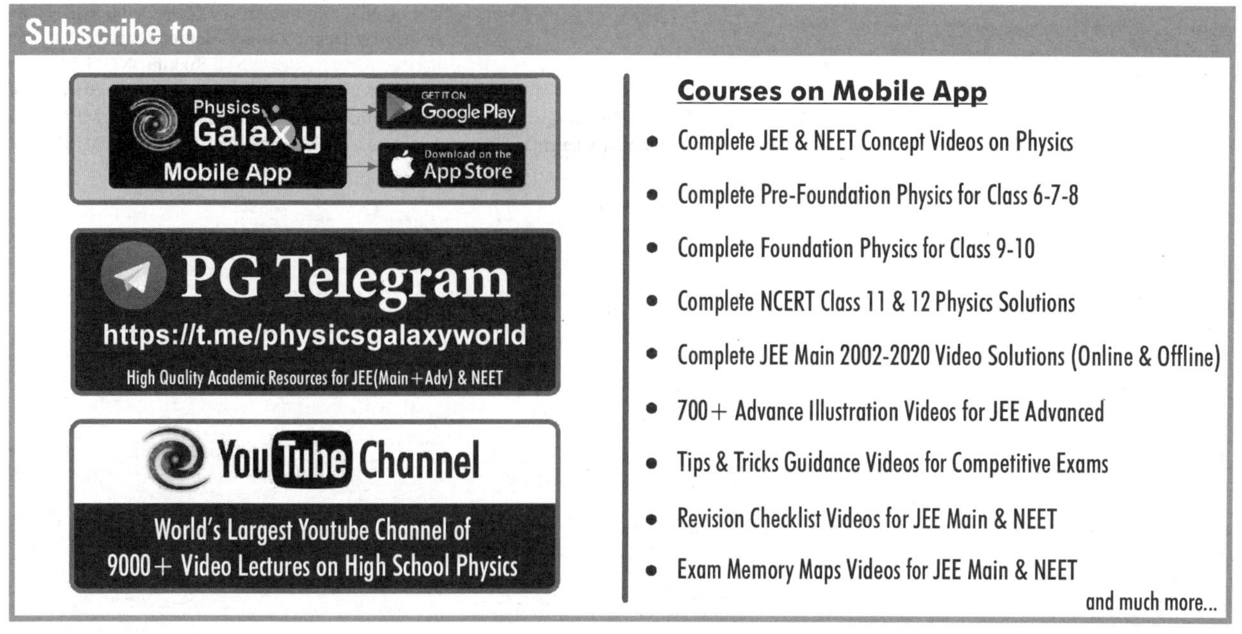

FOREWORD

I am extremely happy to share my views about this book written by Er. Ashish Arora. He has contributed immensely to the teaching-learning of Physics though many innovative ways. His approach has always been student centric and this contribution is yet another example of the same.

As the title states "Physics Galaxy JEE Main Solutions" the book provides valuable information and covers the necessary components by offering solution to all questions asked in old JEE Main (all Online & Offline 2002-2023) with detailed explanations. All questions are categorized into 30 topics based on weightage of questions asked in the past 22 years (2002-2023) however in this book total 22 papers of all online and offline shifts of JEE Main are included with detailed comprehensive solutions and upto 2023 all solutions to questions are linked with video solutions of most of questions available on YouTube channel of Physics Galaxy. This is to indicate and promote understanding of the methods or reasons for specific/particular solutions. Aspirants can also scan the respective QR code given in the book to see its video explanation as and when required. All solutions of the questions are based on the conceptual analysis covered in the popular Physics Galaxy book series (5 volumes) so that aspirants following the book series will be able to relate and recall all the explanations already done.

I am sure the book promotes the concept of deeper learning in the domain of Physics. My best wishes to the many aspirants who will certainly find this book a very valuable tool to enhance thorough understanding of the subject.

Date : November 24, 2023

Prof. Sandeep Sancheti
Ph. D. (U.K.), B.Tech. FIETE, FIE (I), SMIEEE

Vice-Chancellor (Provost) Marwadi University, Rajkot
Member Executive Council, Association of Commonwealth Universities, London
Former President, AIU, New Delhi
Former Vice-Chancellor, SRMIST, Chennai
Former President, Manipal University, Jaipur
Former-Director NIT Surathkal, NIT New Delhi, NIT Trichy,
NIT Calicut, SPA Delhi; Former-Mentor Director NIT Goa,
NIT Sikkim, NIT Puducherry
Former-Member JEE Apex Board (JAB), MHRD, GoI
Former-Chairman, Central Counselling Board (CCB), AIEEE, MHRD, GoI
Former-Chairman, Direct Admission of Students Abroad (DASA), MHRD, GoI

PREFACE

For a science student, preparing for JEE Main or NEET, Physics is the most important subject, unlike to other subjects it requires logical reasoning and high imagination of brain. Without improving the level of physics it is very difficult to achieve a goal in the present age of competitions. To score better, one does not require hard working at least in physics. It just requires a simple understanding and approach to think a physical situation. Actually physics is the surrounding of our everyday life. If you wish to make the concepts of physics strong, you should try to understand core concepts of physics in practical approach rather than theoretical.

This book "*PHYSICS GALAXY JEE Main Solutions*" is designed in a totally different and friendly approach to develop the physics concepts psychologically. The book covers detailed solutions to all the questions previously asked in JEE Main all online and offline papers from 2002 to 2023. From the analysis of 2021 all four attempts papers it is clearly seen that all these papers held were designed on the basis of past 22 year papers. More than 60% questions are based on concepts asked in previous years which were matching in style with the questions of these old papers, called derived PYQs.

All the questions of these online and offline papers are divided in 30 topics for topic level understanding of questions asked in JEE Main. In my classroom teaching also I found that topic level understanding of previous year papers of exam helps students a lot in concept visualization from the point of view of preparation of final exam as well as for the purpose of last moment revision of all topics.

For every question asked in these old JEE Main papers, video solutions are recorded and made available for students on YouTube channel of Physics Galaxy as well as these video solutions can be accessed through Physics Galaxy mobile app available on iOS and Android play store. While following the book if students face any doubt in understanding the solution of the specific question then they can scan the QR codes given at the end of the book for all the questions to directly start the video solution of that question.

I don't have words for my best friend my wife Anuja for always being together with me to complete this book in the unique style and format along with the whole series of Physics Galaxy book series and lecture notes series.

I would like to pay my gratitude to Sh. Dayashankar Prajapati in assisting me to complete the task in Design Labs of PHYSICSGALAXY.COM and presenting the book in totally new format of second edition.

At last but the most important person, my father who has devoted his valuable time to finally present the book in such a format and a simple language, thanks is a very small word for his dedication in this book.

In this first edition of book I have tried my best to make this book error free but owing to the nature of work, inadvertently, there is possibility of errors left untouched. I shall be grateful to the readers, if they point out me regarding errors and oblige me by giving their valuable and constructive suggestions via emails for further improvement of the book.

Date : November 24, 2023

Ashish Arora

PHYSICSGALAXY.COM
B-80, Model Town, Malviya Nagar, Jaipur-302017
e-mails: ashisharora@physicsgalaxy.com

CONTENTS

* * * * *

CHAPTER

1 UNITS AND DIMENSIONS

1.1 MEASUREMENTS AND DIMENSIONAL ANALYSIS

1. Identify the pair whose dimensions are equal : **[2002]**
(A) Torque and work (B) Stress and energy
(C) Force and stress (D) Force and work

2. The physical quantities not having same dimensions are : **[2003]**
(A) torque and work
(B) momentum and planck's constant
(C) stress and young's modulus
(D) speed and $(\mu_0\varepsilon_0)^{-1/2}$

3. Dimensions of $\frac{1}{\mu_0\varepsilon_0}$, where symbols have their usual meaning, are : **[2003]**
(A) $[L^{-1} T]$ (B) $[L^{-2} T^2]$
(C) $[L^2 T^{-2}]$ (D) $[L T^{-1}]$

4. Which one of the following represents the correct dimensions of the coefficient of viscosity ? **[2004]**
(A) $[ML^{-1}T^{-1}]$ (B) $[MLT^{-1}]$
(C) $[ML^{-1}T^{-2}]$ (D) $[ML^{-2}T^{-2}]$

5. Out of the following pair, which one does NOT have identical dimensions : **[2005]**
(A) Impulse and momentum
(B) Angular momentum and planck's constant
(C) Work and torque
(D) Moment of inertia and moment of a force

6. Which of the following units denotes the dimensions $\frac{ML^2}{Q^2}$, where Q denotes the electric charge ? **[2006]**
(A) Wb/m^2 (B) Henry (H)
(C) H/m^2 (D) Weber (Wb)

7. The dimensions of magnetic field in M, L, T and C (coulomb) is given as : **[2008]**
(A) $[MLT^{-1}C^{-1}]$ (B) $[MT^2C^{-2}]$
(C) $[MT^{-1}C^{-1}]$ (D) $[MT^{-2}C^{-1}]$

8. Given that K = energy, V = velocity, T = time. If they are chosen as the fundamental units, then what is dimensional formula for surface tension ? **[2012 (Online)]**
(A) $[K V^{-2} T^{-2}]$ (B) $[K^2 V^2 T^{-2}]$
(C) $[K^2 V^{-2} T^{-2}]$ (D) $[K V^2 T^2]$

9. Let $[\epsilon_0]$ denote the dimensional formula of the permittivity of vacuum. If M = mass, L = length, T = time and A = electric current, then : **[2013]**
(A) $\epsilon_0 = [M^{-1} L^{-3} T^2 A]$ (B) $\epsilon_0 = [M^{-1} L^{-3} T^4 A^2]$
(C) $\epsilon_0 = [M^1 L^2 T^1 A^2]$ (D) $\epsilon_0 = [M^1 L^2 T^1 A]$

10. The dimensions of angular momentum, latent heat and capacitance are, respectively : **[2013 (Online)]**
(A) $ML^2T^1A^2, L^2T^{-2}, M^{-1}L^{-2}T^2$
(B) $ML^2T^{-2}, L^2T^2, M^{-1}L^{-2}T^4A^2$
(C) $ML^2T^{-1}, L^2T^{-2}, ML^2TA^2$
(D) $ML^2T^{-1}, L^2T^{-2}, M^{-1}L^{-2}T^4A^2$

11. If the time period t of the oscillation of a drop of liquid of density d, radius r, vibrating under surface tension s is given by the formula $t = \sqrt{r^{2b} s^c d^{a/2}}$. It is observed that the time period is directly proportional to $\sqrt{\frac{d}{s}}$. The value of b should therefore be : **[2013 (Online)]**
(A) $\frac{3}{4}$ (B) $\sqrt{3}$
(C) $\frac{3}{2}$ (D) $\frac{2}{3}$

12. In terms of resistance R and time T, the dimensions of ratio $\frac{\mu}{\varepsilon}$ of the permeability μ and permittivity ε is : **[2014 (Online)]**
(A) $[RT^{-2}]$ (B) $[R^2T^{-1}]$
(C) $[R^2]$ (D) $[R^2T^2]$

13. From the following combinations of physical constants (expressed through their usual symbols) the only combination, that would have the same value in different systems of units, is : **[2014 (Online)]**
(A) $\frac{ch}{2\pi \epsilon_e^2}$
(B) $\frac{e^2}{2\pi \epsilon_0 Gm_e^2}$ (m_e = mass of electron)
(C) $\frac{\mu_0 \epsilon_0}{c^2} \frac{G}{he^2}$
(D) $\frac{2\pi\sqrt{\mu_0 \epsilon_0}}{ce^2} \frac{h}{G}$

14. If the capacitance of a nanocapacitor is measured in terms of a unit 'u' made by combining the electric charge 'e', Bohr radius 'a_0', Planck's constant 'h' and speed of light 'c' then : **[2015 (Online)]**

(A) $u = \dfrac{e^2 h}{a_0}$ (B) $u = \dfrac{hc}{e^2 a_0}$

(C) $u = \dfrac{e^2 c}{h a_0}$ (D) $u = \dfrac{e^2 a_0}{hc}$

15. If electronic charge e, electron mass m, speed of light in vacuum c and Planck's constant h are taken as fundamental quantities, the permeability of vacuum μ_0 can be expressed in units of : **[2015 (Online)]**

(A) $\left(\dfrac{h}{me^2}\right)$ (B) $\left(\dfrac{hc}{me^2}\right)$

(C) $\left(\dfrac{h}{ce^2}\right)$ (D) $\left(\dfrac{mc^2}{he^2}\right)$

16. In the following 'I' refers to current and other symbols have their usual meaning, Choose the option that corresponds to the dimensions of electrical conductivity : **[2016 (Online)]**

(A) $M^{-1} L^{-3} T^3 I$ (B) $M^{-1} L^{-3} T^3 I^2$

(C) $M^{-1} L^3 T^3 I$ (D) $M L^{-3} T^3 I^2$

17. A, B, C and D are four different physical quantities having different dimensions. None of them is dimensionless. But we know that the equation $AD = C \ln (BD)$ holds true. Then which of the combination is not a meaningful quantity ? **[2016 (Online)]**

(A) $\dfrac{C}{BD} - \dfrac{AD^2}{C}$ (B) $A^2 - B^2C^2$

(C) $\dfrac{A}{B} - C$ (D) $\dfrac{(A-C)}{D}$

18. Time (T), velocity (C) and angular momentum (h) are chosen as fundamental quantities instead of mass, length and time. In terms of these, the dimensions of mass would be : **[2017 (Online)]**

(A) $[M] = [T^{-1} C^{-2} h]$ (B) $[M] = [T^{-1} C^2 h]$

(C) $[M] = [T^{-1} C^{-2} h^{-1}]$ (D) $[M] = [T C^{-2} h]$

19. The characteristic distance at which quantum gravitational effects are significant, the Planck length, can be determined from a suitable combination of the fundamental physical constants G, h and c. Which of the following correctly gives the Planck length ? **[2018 (Online)]**

(A) G^2hc (B) $\left(\dfrac{Gh}{c^3}\right)^{\frac{1}{2}}$

(C) $G^{\frac{1}{2}} h^2 c$ (D) Gh^2c^3

20. Expression for time in terms of G (universal gravitational constant), h (Planck constant) and c (speed of light) is proportional to : **[9 Jan. 2019 (Evening)]**

(A) $\sqrt{\dfrac{Gh}{c^3}}$ (B) $\sqrt{\dfrac{hc^5}{G}}$

(C) $\sqrt{\dfrac{c^3}{Gh}}$ (D) $\sqrt{\dfrac{Gh}{c^5}}$

21. The density of a material in SI units is 128 kg m^{-3}. In certain units in which the unit of length is 25 cm and the unit of mass is 50 g, the numerical value of density of the material is : **[10 Jan. 2019 (Morning)]**

(A) 410 (B) 640

(C) 16 (D) 40

22. The force of interaction between two atoms is given by $F = \alpha\beta \exp\left(-\dfrac{x^2}{\alpha kt}\right)$; where x is the distance, k is the Boltzmann constant and T is temperature and α and β are two constants. The dimension of β is : **[11 Jan. 2019 (Morning)]**

(A) $M^2L^2T^{-2}$ (B) M^2LT^{-4}

(C) $M^0L^2T^{-4}$ (D) MLT^{-2}

23. If speed (V), acceleration (A) and force (F) are considered as fundamental units, the dimension of Young's modulus will be : **[11 Jan. 2019 (Evening)]**

(A) $V^{-2} A^2F^2$ (B) $V^{-4}A^2F$

(C) $V^{-4}A^{-2}F$ (D) $V^{-2}A^2F^{-2}$

24. Let l, r, c and v represent inductance, resistance, capacitance and voltage, respectively. The dimension of $\dfrac{l}{rcv}$ in SI units will be : **[12 Jan. 2019 (Evening)]**

(A) [LTA] (B) $[LA^{-2}]$

(C) $[A^{-1}]$ (D) $[LT^2]$

25. If surface tension (S), Moment of inertia (I) and Planck's constant (h), were to be taken as the fundamental units, the dimensional formula for linear momentum would be : **[8 April 2019 (Evening)]**

(A) $S^{3/2}I^{1/2}h^0$ (B) $S^{1/2}I^{1/2}h^0$

(C) $S^{1/2}I^{1/2}h^{-1}$ (D) $S^{1/2}I^{3/2}h^{-1}$

26. In the formula $X = 5YZ^2$, X and Z have dimensions of capacitance and magnetic field, respectively. What are the dimensions of Y in SI units ? **[10 April 2019 (Evening)]**

(A) $[M^{-2} L^{-2} T^6 A^3]$ (B) $[M^{-1} L^{-2} T^4 A^2]$
(C) $[M^{-3} L^{-2} T^8 A^4]$ (D) $[M^{-2} L^0 T^{-4} A^{-2}]$

27. A quantity f is given by $f=\sqrt{\dfrac{hc^5}{G}}$ where c is speed of light, G universal gravitational constant and h is the Planck's constant. Dimension of f is that of: **[9 Jan. 2020 (Morning)]**
(A) Momentum (B) Area
(C) Energy (D) Volume

28. If speed V, area A and force F are chosen as fundamental units, then the dimension of Young's modulus will be : **[2 Sept. 2020 (Morning)]**
(A) $FA^{-1}V^0$ (B) FA^2V^{-1}
(C) FA^2V^{-3} (D) FA^2V^{-2}

29. If momentum (P), area (A) and time (T) are taken to be the fundamental quantities then the dimensional formula for energy is : **[2 Sept. 2020 (Evening)]**
(A) $[PA^{-1} T^{-2}]$ (B) $[PA_1/2T^{-1}]$
(C) $[P_2AT^{-2}]$ (D) $[P_1/2AT^{-1}]$

30. Dimensional formula for thermal conductivity is (here K denotes the temperature) : **[4 Sept. 2020 (Morning)]**
(A) $MLT^{-3}K$ (B) $MLT^{-2}K$
(C) $MLT^{-2}K^{-2}$ (D) $MLT^{-3}K^{-1}$

31. A quantity x is given by (IFv^2/WL^4) in terms of moment of inertia I, force F, velocity v, work W and Length L. The dimensional formula for x is same as that of: **[4 Sept. 2020 (Evening)]**
(A) Planck's constant (B) Force constant
(C) Energy density (D) Coefficient of viscosity

32. The quantities $x=\dfrac{1}{\sqrt{\mu_0 \epsilon_0}}$, $y=\dfrac{E}{B}$ and $z=\dfrac{l}{CR}$ are defined where C-capacitance, R-Resistance, l-length, E-Electric field, B-magnetic field and ϵ_0 & μ_0 free are permittivity and permeability of free space respectively. Then : **[5 Sept. 2020 (Evening)]**
(A) Only x and y have the same dimension
(B) x, y and z have the same dimension
(C) Only x and z have the same dimension
(D) Only y and z have the same dimension

33. The workdone by a gas molecule in an isolated system is given by, $W=\alpha\beta^2 e^{-\frac{x^2}{\alpha kT}}$, where x is the displacement, k is the Boltzmann constant and T is the temperature. α and β are constants. Then the dimensions of β will be : **[24 Feb. 2021 (Morning)]**
(A) $[M L T^{-2}]$ (B) $[M^2 L T^2]$
(C) $[M^0 L T^0]$ (D) $[M L^2 T^{-2}]$

34. Match List-I with List-II : **[25 Feb. 2021 (Morning)]**

List-I	List-II
(a) h (Planck's constant)	(i) $[M L T^{-1}]$
(b) E (kinetic energy)	(ii) $[M L^2 T^{-1}]$
(c) V (electric potential)	(iii) $[M L^2 T^{-2}]$
(d) P (linear momentum)	(iv) $[M L^2 I^{-1} T^{-3}]$

(A) (a)-(ii), (b)-(iii), (c)-(iv), (d)-(i)
(B) (a)-(iii), (b)-(ii), (c)-(iv), (d)-(i)
(C) (a)-(i), (b)-(ii), (c)-(iv), (d)-(iii)
(D) (a)-(iii), (b)-(iv), (c)-(ii), (d)-(i)

35. If e is the electronic charge, c is the speed of light in free space and h is Planck's constant, the quantity $\dfrac{1}{4\pi\varepsilon_0}\dfrac{|e|^2}{\hbar c}$ has dimensions of : **[25 Feb. 2021 (Evening)]**
(A) $[L C^{-1}]$ (B) $[M L T^0]$
(C) $[M^0 L^0 T^0]$ (D) $[M L T^{-1}]$

36. In a typical combustion engine the workdone by a gas molecule is given by $W=\alpha^2\beta e^{\frac{-\beta x^2}{kT}}$, where x is the displacement, k is the Boltzmann constant and T is the temperature. If α and β are constants, dimensions of α will be : **[26 Feb. 2021 (Morning)]**
(A) $[M^2 L T^{-1}]$ (B) $[M L T^{-2}]$
(C) $[M L T^{-1}]$ (D) $[M^0 L T^0]$

37. If C and V represent capacity and voltage respectively then what are the dimensions of λ where $C/V=\lambda$? **[26 Feb. 2021 (Evening)]**
(A) $[M^{-2} L^{-3} I^2 T^6]$ (B) $[M^{-1} L^{-3} I^{-2} T^{-7}]$
(C) $[M^{-3} L^{-4} I^3 T^7]$ (D) $[M^{-2} L^{-4} I^3 T^7]$

38. One main scale division of a vernier callipers is 'a' cm and n^{th} division of the vernier scale coincide with $(n-1)^{th}$ division of the main scale. The least count of the callipers in mm is : **[16 March 2021 (Morning)]**
(A) $\dfrac{10na}{(n-1)}$ (B) $\dfrac{10a}{(n-1)}$
(C) $\left(\dfrac{n-1}{10n}\right)a$ (D) $\dfrac{10a}{n}$

39. Suppose you have taken a dilute solution of oleic acid in such a way that its concentration becomes 0.01 cm^3 of oleic

acid per cm^3 of the solution. Then you make a thin film of this solution (monomolecular thickness) of area 4 cm^2 by considering 100 spherical drops of radius $\left(\frac{3}{40\pi}\right)^{\frac{1}{3}} \times 10^{-3}$ cm. Then the thickness of oleic acid layer will be $x \times 10^{-14}$ m.
Where x is _____. **[17 March 2021 (Evening)]**

40. The entropy of any system is given by

$$S = \alpha^2 \beta \ln \left[\frac{\mu k R}{J\beta^2} + 3\right]$$

where α and β are the constants. μ, J. k and R are no. of moles, mechanical equivalent of heat, Boltzmann constant and gas constant respectively. [Take $S = \frac{dQ}{T}$] Choose the incorrect options from the following : **[20 July 2021 (Morning)]**
(A) S and α have different dimensions
(B) S, β, k and μR have the same dimensions
(C) α and J have the same dimensions
(D) α and k have the same dimensions

41. If time (t), velocity (v), and angular momentum (l) are taken as the fundamental units. Then the dimension of mass (m) in terms of t, v, and l is : **[20 July 2021 (Evening)]**
(A) $[t^{-1} v^{-2} l^1]$ (B) $[t^{-2} v^{-2} l^1]$
(C) $[t^{-1} v^2 l^1]$ (D) $[t^{-1} v^1 l^{-2}]$

42. The force is given in terms of time t and displacement x by the equation $F = A \cos Bx + C \sin Dt$
The dimensional formula of $\frac{AD}{B}$ is : **[25 July 2021 (Evening)]**
(A) $[ML^2T^{-3}]$ (B) $[M^2L^2T^{-3}]$
(C) $[M^0LT^{-1}]$ (D) $[M^1L^1T^{-2}]$

43. Match List-I with List-II : **[27 July 2021 (Evening)]**

List-I	List-II
(a) Capacitance, C	(i) $M^1L^1T^{-3}A^{-1}$
(b) Permittivity of free space, ε_0	(ii) $M^{-1}L^{-3}T^4A^2$
(c) Permeability of free space, μ_0	(iii) $M^{-1}L^{-2}T^4A^2$
(d) Electric field, E	(iv) $M_1L^1T^{-2}A^{-2}$

Choose the correct answer from the options given below :
(A) (a)-(iii), (b)-(iv), (c)-(ii), (d)-(i)
(B) (a)-(iii), (b)-(ii), (c)-(iv), (d)-(i)
(C) (a)-(iv), (b)-(ii), (c)-(iii), (d)-(i)
(D) (a)-(iv), (b)-(iii), (c)-(ii), (d)-(i)

44. In a Screw Gauge, fifth division of the circular scale coincides with the reference line when the ratchet is closed. There are 50 divisions on the circular scale, and the main scale moves by 0.5 mm on a complete rotation. For a particular observation the reading on the main scale is 5 mm and the 20th division of the circular scale coincides with reference line. Calculate the true reading : **[26 Aug. 2021 (Morning)]**
(A) 5.00 mm (B) 5.20 mm
(C) 5.25 mm (D) 5.15 mm

45. If E, L, M and G denote the quantities as energy, angular momentum, mass and constant of gravitation respectively, then the dimension of P in the formula $P = EL^2M^{-5}G^{-2}$ are : **[26 Aug. 2021 (Morning)]**
(A) $[M^1L^1T^{-2}]$ (B) $[M^{-1}L^{-1}T^2]$
(C) $[M^0L^1T^0]$ (D) $[M^0L^0T^0]$

46. Match List-I with List-II :

List-I	List-II
(a) Magnetic Induction	(i) $ML^2T^{-2}A^{-1}$
(b) Magnetic Flux	(ii) $M^0L^{-1}A$
(c) Magnetic Permeability	(iii) $MT^{-2}A^{-1}$
(d) Magnetization	(iv) $MLT^{-2}A^{-2}$

Choose the most appropriate answer form the options given below : **[26 Aug. 2021 (Evening)]**
(A) (a)-(iii), (b)-(ii), (c)-(iv), (d)-(i)
(B) (a)-(iii), (b)-(i), (c)-(iv), (d)-(ii)
(C) (a)-(ii), (b)-(iv), (c)-(i), (d)-(iii)
(D) (a)-(ii), (b)-(i), (c)-(iv), (d)-(iii)

47. Which of the following is not a dimensionless quantity ? **[27 Aug. 2021 (Morning)]**
(A) Permeability of free space (μ_0)
(B) Relative magnetic permeability (μ_r)
(C) Power factor
(D) Quality factor

48. If E and H represents the intensity of electric field and magnetising field respectively, then the unit of E/H will be : **[27 Aug. 2021 (Morning)]**
(A) Newton (B) mho
(C) ohm (D) Joule

49. Match List-I with List-II :

List-I	List-II
(a) R_H (Rydberg constant)	(i) kg m^{-1} s^{-1}
(b) h (Planck's constant)	(ii) kg m^{-2} s^{-1}
(c) μ_B (Magnetic filed energy density	(iii) m^{-1}
(d) η (coefficient of viscosity)	(iv) kg m^{-1} s^{-2}

Choose the most appropriate answer from the options given below : **[27 Aug. 2021 (Evening)]**
(A) (a)-(iv), (b)-(ii), (c)-(i), (d)-(iii)
(B) (a)-(ii), (b)-(iii), (c)-(iv), (d)-(i)
(C) (a)-(iii), (b)-(ii), (c)-(iv), (d)-(i)
(D) (a)-(iii), (b)-(ii), (c)-(i), (d)-(iv)

50. If force (F), length (L) and time (T) are taken as the fundamental quantities. Then what will be the dimension of density : **[27 Aug. 2021 (Evening)]**

(A) $[FL^{-4}T^2]$ (B) $[FL^{-3}T^3]$
(C) $[FL^{-3}T^2]$ (D) $[FL^{-5}T^2]$

51. Match List-I with List-II : **[31 Aug. 2021 (Morning)]**

List-I	**List-II**
(a) Torque	(i) MLT^{-1}
(b) Impulse	(ii) MT^{-2}
(c) Tension	(iii) ML^2T^{-2}
(d) Surface Tension	(iv) MLT^{-2}

Choose the most appropriate answer from the option given below :

(A) (a)-(iii), (b)-(iv), (c)-(i), (d)-(ii)
(B) (a)-(ii), (b)-(i), (c)-(iv), (d)-(iii)
(C) (a)-(i), (b)-(iii), (c)-(iv), (d)-(ii)
(D) (a)-(iii), (b)-(i), (c)-(iv), (d)-(ii)

52. Which of the following equations is dimensionally incorrect ?
Where t = time, h = height, s = surface tension, θ = angle, ρ = density, a, r = radius, g = acceleration due to gravity, v = volume, p = pressure, W = work done, Γ = torque, $\in$ = permittivity, E = electric field, J = current density, L = length : **[31 Aug. 2021 (Morning)]**

(A) $J = \in \dfrac{\partial E}{\partial t}$ (B) $v = \dfrac{\pi p a^4}{8\eta L}$

(C) $h = \dfrac{2s\cos\theta}{\rho r g}$ (D) $W = \Gamma\theta$

53. If velocity $[V]$ time $[T]$ and force $[F]$ are chosen as the base quantities, the dimensions of the mass will be : **[31 Aug. 2021 (Evening)]**

(A) $[FVT^{-1}]$ (B) $[FT^{-1}V^{-1}]$
(C) $[FT^2V]$ (D) $[FTV^{-1}]$

54. The diameter of a spherical bob is measured using a vernier callipers 9 divisions of the main scale, in the vernier callipers, are equal to 10 divisions of vernier scale. One main scale division is 1 mm. The main scale reading is 10 mm and 8th division of vernier scale was found to coincide exactly with one of the main scale division. If the given vernier callipers has positive zero error of 0.004 cm, then the radius of the bob is _____ $\times 10^{-2}$ cm : **[31 Aug. 2021 (Evening)]**

55. Identify the pair of physical quantities which have different dimensions : **[24 June 2022 (Morning)]**

(A) Wave number and Rydberg's constant
(B) Stress and Coefficient of elasticity
(C) Coercivity and Magnetisation
(D) Specific heat capacity and Latent heat

56. Identify the pair of physical quantities that have same dimensions : **[24 June 2022 (Evening)]**

(A) Velocity gradient and decay constant
(B) Wien's constant and Stefan constant
(C) Angular frequency and angular momentum
(D) Wave number and Avogadro number

57. Two buses P and Q start from a point at the same time and move in a straight line and their positions are represented by $X_P(t) = \alpha t + \beta t^2$ and $X_Q(t) = ft - t^2$. At what time, both the buses have same velocity ? **[25 June 2022 (Evening)]**

(A) $\dfrac{\alpha - f}{1+\beta}$ (B) $\dfrac{\alpha + f}{2(\beta - 1)}$

(C) $\dfrac{\alpha + f}{2(1+\beta)}$ (D) $\dfrac{f - \alpha}{2(1+\beta)}$

58. An expression for a dimensionless quantity P is given by $P = \dfrac{\alpha}{\beta}\log_e\left(\dfrac{kt}{\beta x}\right)$; where α and β are constants, x is distance; k is Boltzmann constant and t is the temperature. Then the dimensions of α will be : **[26 June 2022 (Morning)]**

(A) $[M^0L^{-1}T^0]$ (B) $[ML^0T^{-2}]$
(C) $[MLT^{-2}]$ (D) $[ML^2T^{-2}]$

59. The dimension of mutual inductance is : **[26 June 2022 (Evening)]**

(A) $[ML^2T^{-2}A^{-1}]$ (B) $[ML^2T^{-3}A^{-1}]$
(C) $[ML^2T^{-2}A^{-2}]$ (D) $[ML^2T^{-3}A^{-2}]$

60. The SI unit of a physical quantity is pascal-second. The dimensional formula of this quantity will be : **[27 June 2022 (Evening)]**

(A) $[ML^{-1}T^{-1}]$ (B) $[ML^{-1}T^{-2}]$
(C) $[ML^2T^{-1}]$ (D) $[M^{-1}L^3T^0]$

61. If L, C and R are the self inductance, capacitance and resistance respectively, which of the following does not have the dimension of time ? **[27 June 2022 (Evening)]**

(A) RC (B) $\dfrac{L}{R}$

(C) $\sqrt{LC}$ (D) $\dfrac{L}{C}$

62. If momentum [P], area [A] and time [T] are taken as fundamental quantities, then the dimensional formula for coefficient of viscosity is : **[25 July 2022 (Morning)]**

(A) $[P A^{-1} T^0]$ (B) $[P A T^{-1}]$
(C) $[P A^{-1} T]$ (D) $[P A^{-1} T^{-1}]$

63. An expression of energy density is given by $u = \frac{\alpha}{\beta}\sin\left(\frac{\alpha x}{kt}\right)$, where α, β are constants, x is displacement, k is Boltzmann constant and t is the temperature. The dimensions of β will be : **[27 July 2022 (Evening)]**

(A) $[ML^2T^{-2}\theta^{-1}]$ (B) $[M^0L^2T^{-2}]$
(C) $[M^0L^0T^0]$ (D) $[M^0L^2T^0]$

64. The dimensions of $\left(\frac{B^2}{\mu_0}\right)$ will be :
(If μ_0 : permeability of free space and B : magnetic field) **[28 July 2022 (Morning)]**

(A) $[ML^2T^{-2}]$ (B) $[MLT^{-2}]$
(C) $[ML^{-1}T^{-2}]$ (D) $[ML^2T^{-2}A^{-1}]$

65. Consider the efficiency of Carnot's engine is given by

$$\eta = \frac{\alpha\beta}{\sin\theta}\log_e\frac{\beta x}{kT}$$, where α and β are constants. If T is temperature, k is Boltzman constant, θ is angular displacement and x has the dimensions of length. Then, choose the **incorrect** option : **[28 July 2022 (Evening)]**

(A) Dimensions of β is same as that of force
(B) Dimensions of $\alpha^{-1}x$ is same as that of energy
(C) Dimensions of $\eta^{-1}\sin\theta$ is same as that of αβ
(D) Dimensions of α is same as that of β

66. Given below are two statements : One is labelled as **Assertion-A** and other is labelled as **Reason-R**.

Assertion-A: Time period of oscillation of a liquid drop depends on surface tension (S), if density of the liquid is p and radius of the drop is r, then $T = k\sqrt{pr^3/s^{3/2}}$ is dimensionally correct, where K is dimensionless.

Reason-R: Using dimensional analysis we get RHS having different dimension than that of time period.

In the light of above statements, choose the correct answer from the options given below : **[29 July 2022 (Morning)]**

(A) Both **A** and **R** are true and **R** is the correct explanation of **A**
(B) Both **A** and **R** are true but **R** is not the correct explanation of **A**
(C) **A** is true but **R** is false
(D) **A** is false but **R** is true

67. Match List-I with List-II : **[29 July 2022 (Evening)]**

List-I	List-II
(a) Torque	(i) Nms^{-1}
(b) Stress	(ii) $J\,kg^{-1}$
(c) Latent Heat	(iii) Nm
(d) Power	(iv) Nm^{-2}

Choose the correct answer from the options given below :

(A) (a)-(iii), (b)-(ii), (c)-(i), (d)-(iv)
(B) (a)-(iii), (b)-(iv), (c)-(ii), (d)-(i)
(C) (a)-(iv), (b)-(i), (c)-(iii), (d)-(ii)
(D) (a)-(ii), (b)-(iii), (c)-(i), (d)-(iv)

68. The frequency (ν) of an oscillating liquid drop may depend upon radius (r) of the drop, density (ρ) of liquid and the surface tension (s) of the liquid as $\nu = r^a\rho^b s^c$. The values of a, b and c respectively are : **[24 Jan. 2023 (Evening)]**

(A) $\left(-\frac{3}{2}, -\frac{1}{2}, \frac{1}{2}\right)$ (B) $\left(\frac{3}{2}, -\frac{1}{2}, \frac{1}{2}\right)$

(C) $\left(\frac{3}{2}, \frac{1}{2}, -\frac{1}{2}\right)$ (D) $\left(-\frac{3}{2}, \frac{1}{2}, \frac{1}{2}\right)$

69. Match List-I with List-II : **[25 Jan. 2023 (Evening)]**

List-I	List-II
(a) Young's Modulus (Y)	(i) $[ML^{-1}T^{-1}]$
(b) Co-efficient of Viscosity (η)	(ii) $[ML^2T^{-1}]$
(c) Planck's Constant (h)	(iii) $[ML^{-1}T^{-2}]$
(d) Work Function (ϕ)	(iv) $[ML^2T^{-2}]$

Choose the correct answer from the options given below :

(A) (a)-(ii), (b)-(iii), (c)-(iv), (d)-(i)
(B) (a)-(iii), (b)-(i), (c)-(ii), (d)-(iv)
(C) (a)-(i), (b)-(iii), (c)-(iv), (d)-(ii)
(D) (a)-(i), (b)-(ii), (c)-(iii), (d)-(iv)

70. Match List-I with List-II : **[29 Jan. 2023 (Morning)]**

List-I (Physical Quantity)	List-II (Dimensional Formula)
(a) Pressure gradient	(i) $[M^0L^2T^{-2}]$
(b) Energy density	(ii) $[M^1L^{-1}T^{-2}]$
(c) Electric Field	(iii) $[M^1L^{-2}T^{-2}]$
(d) Latent heat	(iv) $[M^1L^1T^{-3}A^{-1}]$

Choose the correct answer from the options given below :

(A) (a)-(iii), (b)-(ii), (c)-(i), (d)-(iv)
(B) (a)-(ii), (b)-(iii), (c)-(iv), (d)-(i)
(C) (a)-(iii), (b)-(ii), (c)-(iv), (d)-(i)
(D) (a)-(ii), (b)-(iii), (c)-(i), (d)-(iv)

71. The equation of a circle is given by $x^2 + y^2 = a^2$, where a is the radius. If the equation is modified to change the origin other than (0, 0), then find out the correct dimensions of A and B in a new equation : $(x - At)^2 + \left(y - \frac{t}{B}\right)^2 = a^2$. The dimensions of t is given as $[T^{-1}]$: **[29 Jan. 2023 (Evening)]**

(A) $A = [L^{-1}T], B = [LT^{-1}]$ (B) $A = [LT], B = [L^{-1}T^{-1}]$
(C) $A = [L^{-1}T^{-1}], B = [LT^{-1}]$ (D) $A = [L^{-1}T^{-1}], B = [LT]$

72. Electric field in a certain region is given by $\vec{E} = \left(\frac{A}{x^2}\hat{i} + \frac{B}{y^3}\hat{j}\right)$. The SI unit of A and B are :

[30 Jan. 2023 (Morning)]

(A) $Nm^3\,C^{-1}$; $Nm^2\,C^{-1}$ (B) $Nm^2\,C^{-1}$; $Nm^3\,C^{-1}$
(C) $Nm^3\,C$; $Nm^2\,C$ (D) $Nm^2\,C$; $Nm^3\,C$

73. Match List-I with List-II : **[30 Jan. 2023 (Evening)]**

List-I	**List-II**
(a) Torque	(i) $kg\,m^{-1}\,s^{-2}$
(b) Energy density	(ii) $kg\,ms^{-1}$
(c) Pressure gradient	(iii) $kg\,m^{-2}\,s^{-2}$
(d) Impulse	(iv) $kg\,m^2\,s^{-2}$

Choose the correct answer from the options given below :
(A) (a)-(iv), (b)-(iii), (c)-(i), (d)-(ii)
(B) (a)-(i), (b)-(iv), (c)-(iii), (d)-(ii)
(C) (a)-(iv), (b)-(i), (c)-(ii), (d)-(iii)
(D) (a)-(iv), (b)-(i), (c)-(iii), (d)-(ii)

74. Match List-I with List-II : **[31 Jan. 2023 (Evening)]**

List-I	**List-II**
(a) Angular momentum	(i) $[ML^2T^{-2}]$
(b) Torque	(ii) $[ML^{-2}T^{-2}]$
(c) Stress	(iii) $[ML^2T^{-1}]$
(d) Pressure gradient	(iv) $[ML^{-1}T^{-2}]$

Choose the correct answer from the options given below :
(A) (a)-(i), (b)-(iv), (c)-(iii), (d)-(ii)
(B) (a)-(iii), (b)-(i), (c)-(iv), (d)-(ii)
(C) (a)-(ii), (b)-(iii), (c)-(iv), (d)-(i)
(D) (a)-(iv), (b)-(ii), (c)-(i), (d)-(iii)

75. If the velocity of light c, universal gravitational constant G and planck's constant h are chosen as fundamental quantities. The dimensions of mass in the new system is :

[1 Feb. 2023 (Evening)]

(A) $[h^{1/2}c^{1/2}G^1]$ (B) $[h^1c^1G^{-1}]$
(C) $[h^{-1/2}c^{1/2}G^{1/2}]$ (D) $[h^{1/2}c^{1/2}G^{-1/2}]$

76. Dimension of $\frac{1}{\mu_0\,\epsilon_0}$ should be equal to :

[8 April 2023 (Morning)]

(A) T^2/L^2 (B) L/T
(C) L^2/T^2 (D) T/L

77. Match List-I with List-II : **[8 April 2023 (Evening)]**

List-I	**List-II**
(a) Torque	(i) $ML^{-2}T^{-2}$
(b) Stress	(ii) ML^2T^{-2}
(c) Pressure gradient	(iii) $ML^{-1}T^{-1}$
(d) Coefficient of viscosity	(iv) $ML^{-1}T^{-2}$

Choose the correct answer from the options given below :
(A) (a)-(iii), (b)-(iv), (c)-(i), (d)-(ii)
(B) (a)-(iv), (b)-(ii), (c)-(iii), (d)-(i)
(C) (a)-(ii), (b)-(iv), (c)-(i), (d)-(iii)
(D) (a)-(ii), (b)-(i), (c)-(iv), (d)-(iii)

78. Given below are two statements :
Statements-I : Astronomical unit (Au). Parsec (Pc) and Light year (ly) are units for measuring astronomical distances.
Statements-II : Au < Parsec (Pc) < *ly*
In the light of the above statements. Choose the most appropriate answer from the options given below :

[11 April 2023 (Morning)]

(A) Both Statements-I and Statements-II are correct
(B) Statements-I is correct but Statements-II is incorrect
(C) Both Statements-I and Statements-II are incorrect
(D) Statements-I is incorrect but statements-II is correct

79. If force (F), velocity (V) and time (T) are considered as fundamental physical quantity, then dimensional formula of density will be : **[11 April 2023 (Evening)]**

(A) $FV^{-2}T^2$ (B) FV^4T^2
(C) FV^4T^6 (D) $F^2V^2T^6$

80. In the equation $\left[X + \frac{a}{Y^2}\right][Y - b] = RT$, X is pressure, Y is volume, R is universal gas constant and T is temperature. The physical quantity equivalent to the ratio $\frac{a}{b}$ is:

[13 April 2023 (Evening)]

(A) Energy (B) Impulse
(C) Pressure gradient (D) Coefficient of viscosity

81. The speed of a wave produced in water is given by $\upsilon = \lambda^a g^b \rho^c$. Where λ, g and ρ are wavelength of wave, acceleration due to gravity and density of water respectively. The values of a, b and c respectively, are :

[15 April 2023 (Morning)]

(A) $\frac{1}{2}, \frac{1}{2}, 0$ (B) 1, 1, 0
(C) 1, –1, 0 (D) $\frac{1}{2}, 0, \frac{1}{2}$

* * * * *

CHAPTER

2 MOTION IN A STRAIGHT LINE

2.1 MOTION IN ONE DIMENSION

1. If a body loses half of its velocity on penetrating 3 cm in a wooden block, then how much will it penetrate more before coming to rest? **[2002]**

(A) 1 cm (B) 2 cm
(C) 3 cm (D) 4 cm

2. From a building two balls A and B are thrown with equal speeds such that A is thrown upwards and B downwards (both vertically). If v_A and v_B are their respective velocities on reaching the ground, then : **[2002]**

(A) $v_B > v_A$
(B) $v_A = v_B$
(C) $v_A > v_B$
(D) their velocities depends on their masses

3. Speeds of two identical cars are u and $4u$ at a specific instant. At constant retardation the ratio of the respective distances at which the two cars are stopped from that instant is : **[2002]**

(A) 1 : 1 (B) 1 : 4
(C) 1 : 8 (D) 1 : 16

4. A car moving with a speed of 50 km/hr, can be stopped by brakes after at least 6 m. If the same car is moving at a speed of 100 km/hr, the minimum stopping distance is : **[2003]**

(A) 12m (B) 18m
(C) 24m (D) 6m

5. A ball is released from the top of a tower of height h metres. It takes T seconds to reach the ground. What is the position of the ball after $T/3$ seconds? **[2004]**

(A) $h/9$ metre from the ground
(B) $7h/9$ metre from the ground
(C) $8h/9$ metre from the ground
(D) $17h/9$ metre from the ground

6. An automobile travelling with a speed of 60 km/h, can brake to stop within a distance of 20 m. If the car is going twice as fast, i.e. 120 km/h, the stopping distance will be : **[2004]**

(A) 20m (B) 40m
(C) 60m (D) 80m

7. The relation between time t and distance x is $t = ax^2 + bx$, where a and b are constants. The acceleration is : **[2005]**

(A) $-2abv^2$ (B) $2bv^2$
(C) $-2av^3$ (D) $2av^3$

8. A car, starting from rest, accelerates at the rate f through a distance S, then continues at constant speed for time t and then decelerates at the rate $f/2$ to come to rest. If the total distance travelled is 15 S, then : **[2005]**

(A) $S = ft$ (B) $S = \frac{1}{6}ft^2$
(C) $S = \frac{1}{72}ft^2$ (D) $S = \frac{1}{4}ft^2$

9. A particle is moving eastwards with a velocity of 5 ms^{-1}. In 10 second the velocity changes to 5 ms^{-1} northwards. The average acceleration in this time is : **[2005]**

(A) $\frac{1}{\sqrt{2}}$ ms^{-1} towards north-west
(B) $\frac{1}{2}$ ms^{-2} towards north
(C) Zero
(D) $\frac{1}{2}$ ms^{-2} towards north-west

10. A parachutist after bailing out falls 50 m without friction. When parachute opens, it decelerates at 2 m/s^2. He reaches the ground with a speed of 3 m/s. At what height, did he bail out? **[2005]**

(A) 91 m (B) 182 m
(C) 293 m (D) 111 m

11. A particle located at $x = 0$ at time $t = 0$, starts moving along the positive x-direction with a velocity v that varies as $v = \alpha\sqrt{x}$. The displacement of the particle varies with time as : **[2006]**

(A) $t^{1/2}$ (B) t^3
(C) t^2 (D) t

12. The velocity of a particle is $v = v_0 + gt + ft^2$. If its position is $x = 0$ at $t = 0$, then its displacement after unit time ($t = 1$) is : **[2007]**

(A) $v_0 + 2g + 3f$ (B) $v_0 + \frac{g}{2} + \frac{f}{3}$
(C) $v_0 + g + f$ (D) $v_0 + \frac{g}{2} + f$

13. An object, moving with a speed of 6.25 m/s, is decelerated at a rate given by $\frac{dv}{dt} = -2.5\sqrt{v}$ where v is the instantaneous speed. The time taken by the object, to come to rest, would be: **[2011]**

(A) 1s (B) 2s
(C) 4s (D) 8s

14. A good train accelerating uniformly on a straight railway track, approaches an electric pole standing on the side of track. Its engine passes the pole with velocity u and the guard's room passes with velocity v. The middle wagon of the train passes the pole with a velocity : **[2012 (Online)]**

(A) $\frac{u+v}{2}$ (B) $\frac{1}{2}\sqrt{u^2+v^2}$
(C) $\sqrt{uv}$ (D) $\sqrt{\left(\frac{u^2+v^2}{2}\right)}$

15. The graph of an object's motion (along the x-axis) is shown in the figure. The instantaneous velocity of the object at point A and B are v_A and v_B respectively. Then : **[2012 (Online)]**

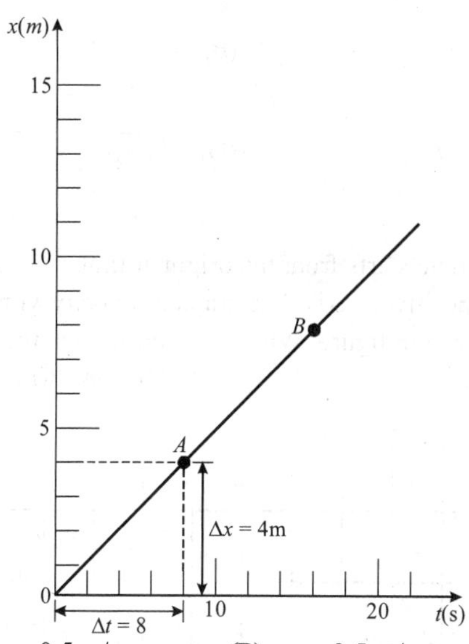

(A) $v_A = v_B = 0.5$ m/s (B) $v_A = 0.5$ m/s $< v_B$
(C) $v_A = 0.5$ m/s $> v_B$ (D) $v_A = v_B = 2$ m/s

16. The distance travelled by a body moving along a line in time t is proportional to t^3. The acceleration-time (a, t) graph for the motion of the body will be : **[2012 (Online)]**

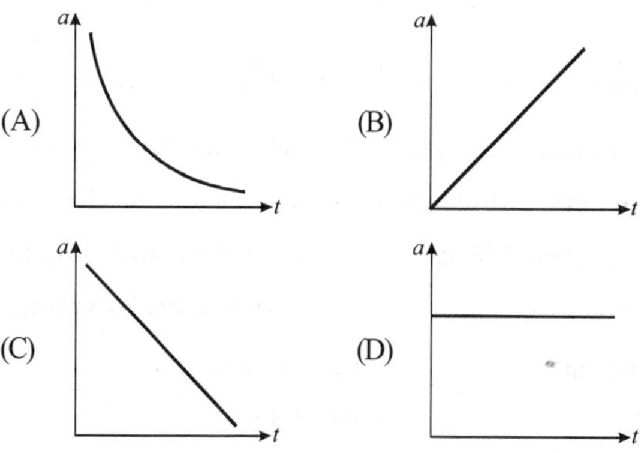

17. A car covers the first half of the distance between two places at 40 km/h and other half at 60 km/h. The average speed of the car is : **[2012 (Online)]**

(A) 40 km/h (B) 45 km/h
(C) 48 km/h (D) 60 km/h

18. From a tower of height H, a particle is thrown vertically upwards with a speed u. The time taken by the particle, to hit the ground, is n times that taken by it to reach the highest point of its path. The relation between H, u and n is : **[2014]**

(A) $2g\,H = n^2u^2$ (B) $g\,H = (n-2)^2\,u^2$
(C) $2g\,H = nu^2(n-2)$ (D) $g\,H = (n-2)\,u^2$

19. A person climbs up a stalled escalator in 60 s. If standing on the same but escalator running with constant velocity he takes 40 s. How much time is taken by the person to walk up the moving escalator ? **[2014 (Online)]**

(A) 37 s (B) 27 s
(C) 24 s (D) 45 s

20. A body is thrown vertically upwards. Which one of the following graphs correctly represent the velocity *vs* time ? **[2017]**

(A)
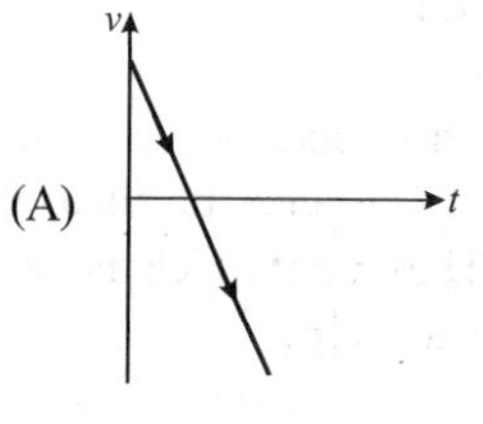

(B)
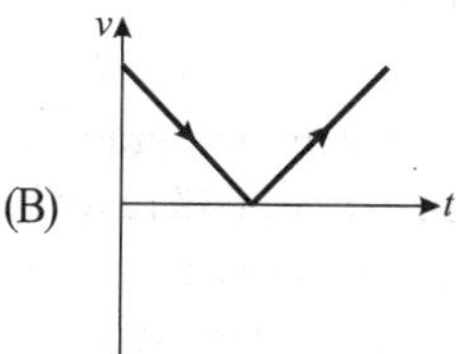

(C)
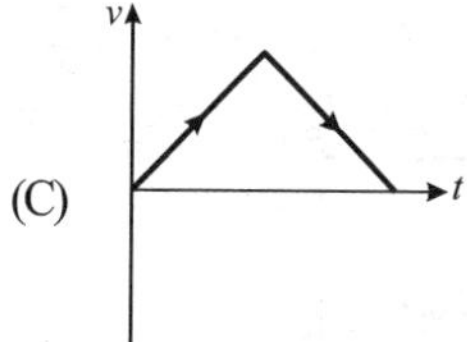

(D)
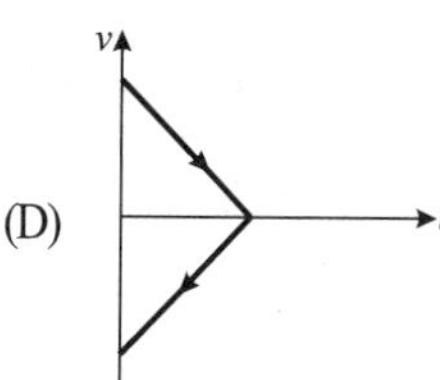

21. Which graph corresponds to an object moving with a constant negative acceleration and a positive velocity ? **[2017 (Online)]**

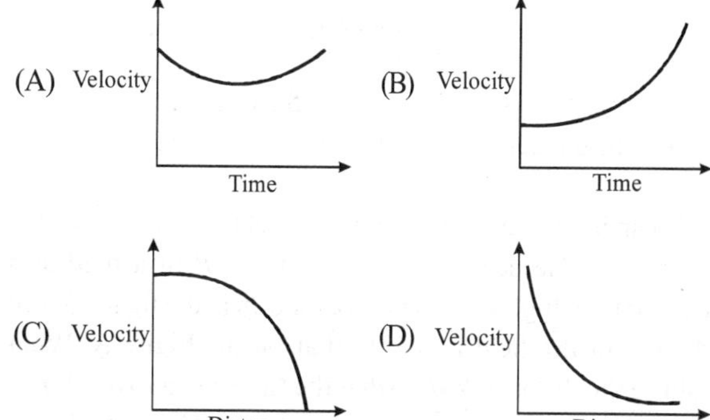

22. All the graphs below are intended to represent the same motion. One of them does it incorrectly. Pick it up : **[2018]**

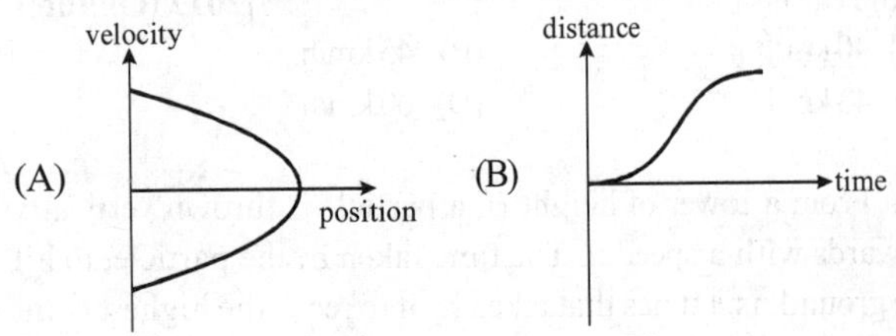

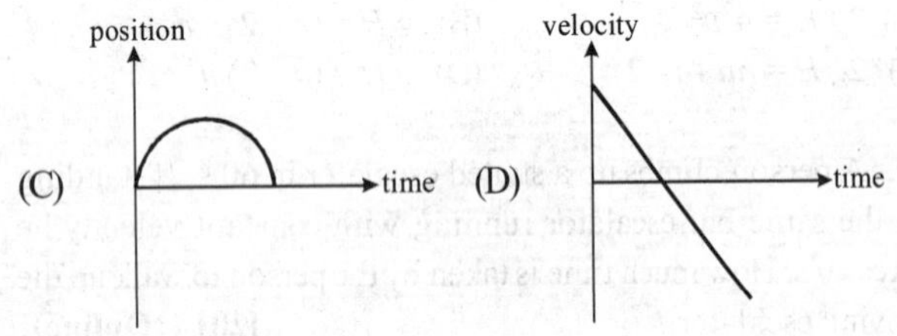

23. An automobile, travelling at 40 km/h, can be stopped at a distance of 40 m by applying brakes. If the same automobile is travelling at 80 km/h, the minimum stopping distance, in metres, is (assume no skidding) : **[2018 (Online)]**

(A) 75m (B) 160m
(C) 100m (D) 150m

24. The velocity-time graph of a car and a scooter are shown in the figure. (i) The difference between the distance travelled by the car and the scooter in 15 s and (ii) The time at which the car will catch up with the scooter are, respectively : **[2018 (Online)]**

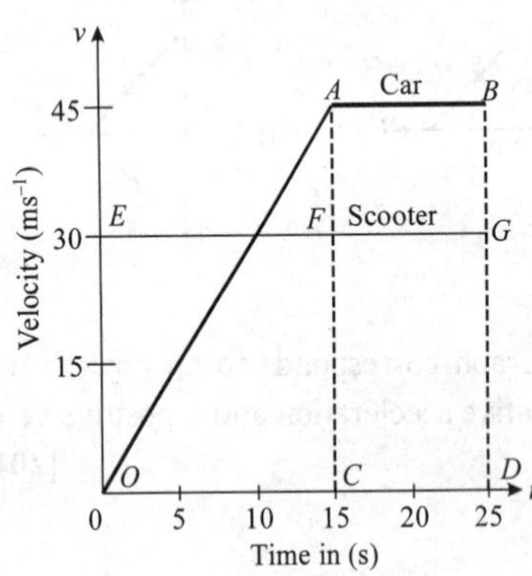

(A) 337.5 m and 25 s (B) 225.5 m and 10 s
(C) 112.5 m and 22.5 s (D) 112.5 m and 15 s

25. A man in a car at location Q on a straight highway is moving with speed v. He decides to reach a point P in a field at a distance d from highway (point M) as shown in the figure. Speed of the car in the field is half to that on the highway. What should be the distance RM, so that the time taken to reach P is minimum? **[2018 (Online)]**

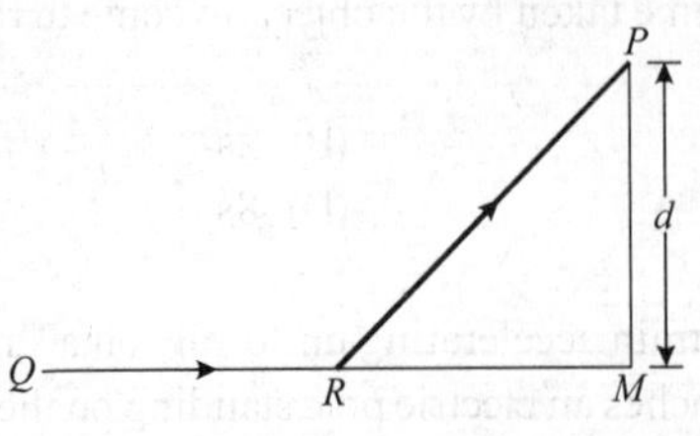

(A) $\frac{d}{\sqrt{3}}$ (B) $\frac{d}{2}$

(C) $\frac{d}{\sqrt{2}}$ (D) d

26. In a car race on straight road, car A takes a time t less than car B at the finish and passes finishing point with a speed 'v' more than that of car B. Both the cars start from rest and travel with constant acceleration a_1 and a_2 respectively. Then 'v' is equal to : **[9 Jan. 2019 (Evening)]**

(A) $\frac{a_1 + a_2}{2}t$ (B) $\sqrt{2a_1a_2}\,t$

(C) $\frac{2a_1a_2}{a_1 + a_2}t$ (D) $\sqrt{a_1a_2}\,t$

27. A particle starts from the origin at time $t = 0$ and moves along the positive x-axis. The graph of velocity with respect to time is shown in figure. What is the position of the particle at time $t = 5$s ? **[10 Jan. 2019 (Evening)]**

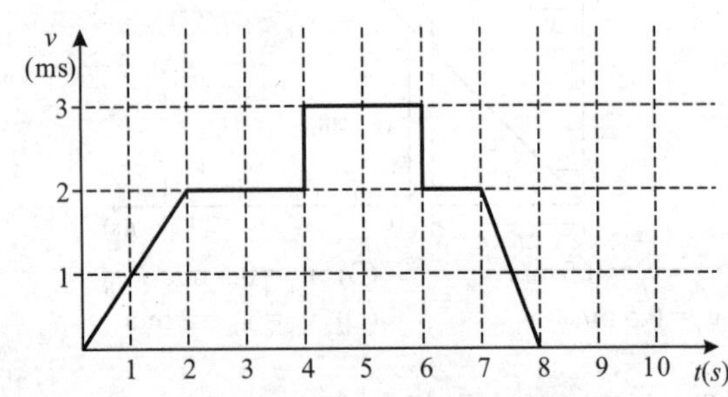

(A) 6m (B) 9m
(C) 3m (D) 10m

28. A particle moves from the point $(2.0\hat{i} + 4.0\hat{j})\,m$, at $t = 0$, with an initial velocity $(5.0\hat{i} + 4.0\hat{j})\,\text{ms}^{-1}$. It is acted upon by a constant force which produces a constant acceleration $(4.0\hat{i} + 4.0\hat{j})\,\text{ms}^{-2}$. What is the distance of the particle from the origin at time 2s ? **[11 Jan. 2019 (Evening)]**

(A) $20\sqrt{2}$ m (B) $10\sqrt{2}$ m
(C) 5m (D) 15m

29. A passenger train of length 60 m travels at a speed of 80 km/hr. Another freight train of length 120 m travels at a speed of 30 km/hr. The ratio of times taken by the passenger train to completely cross the freight train when : (i) they are moving in the same direction, and (ii) in the opposite directions is :

[12 Jan. 2019 (Morning)]

(A) $\frac{5}{2}$ (B) $\frac{25}{11}$

(C) $\frac{3}{2}$ (D) $\frac{11}{5}$

30. A particle starts from origin O from rest and moves with a uniform acceleration along the positive x-axis. Identify all figures that correctly represent the motion qualitatively. (a = acceleration, v = velocity, x = displacement, t = time) :

[8 April 2019 (Evening)]

(1) 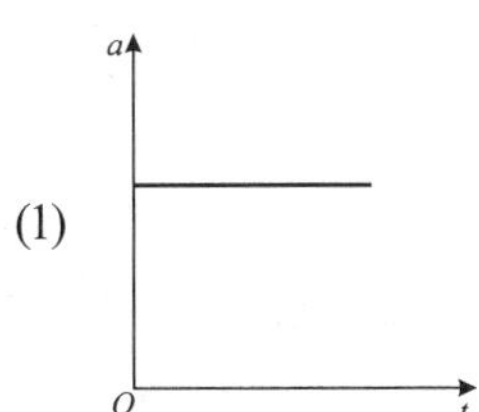

(2) 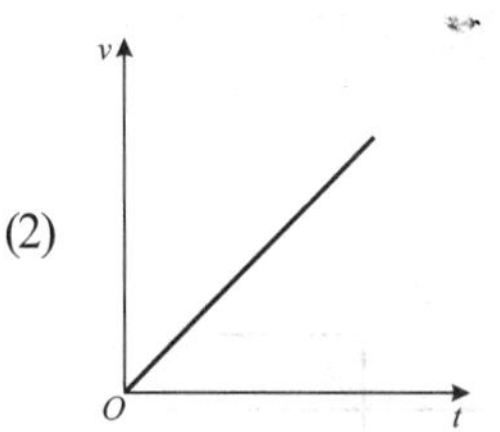

(3) 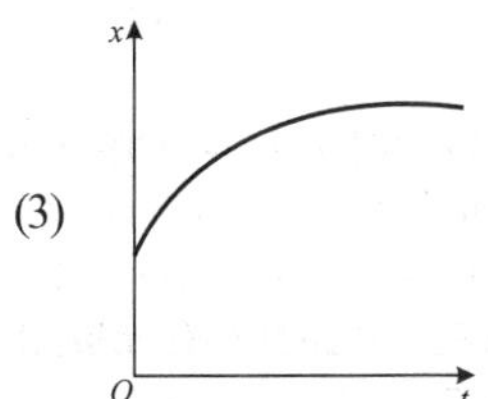

(4) 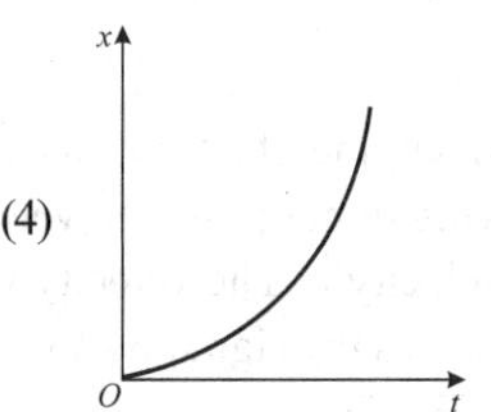

(A) (1), (2), (3) (B) (1)

(C) (1), (2), (4) (D) (2), (3)

31. Let $|\vec{A}_1| = 3$, $|\vec{A}_2| = 5$ and $|\vec{A}_1 + \vec{A}_2| = 5$. The value of $(2\vec{A}_1 + 3\vec{A}_2)\cdot(3\vec{A}_1 - 2\vec{A}_2)$ is : **[8 April 2019 (Evening)]**

(A) -112.5 (B) -106.5

(C) -118.5 (D) -99.5

32. A ball is thrown vertically up (taken as + z-axis) from the ground. The correct momentum-height (p-h) diagram is :

[9 April 2019 (Morning)]

(A) 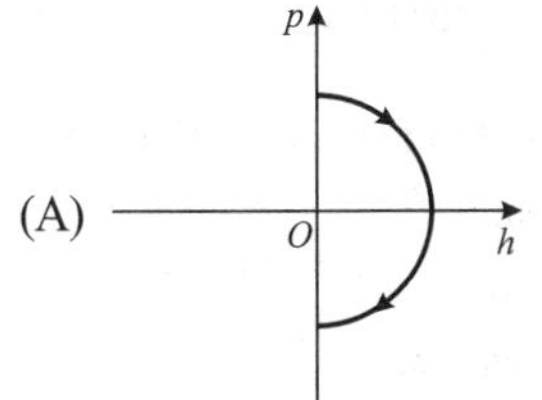

(B) 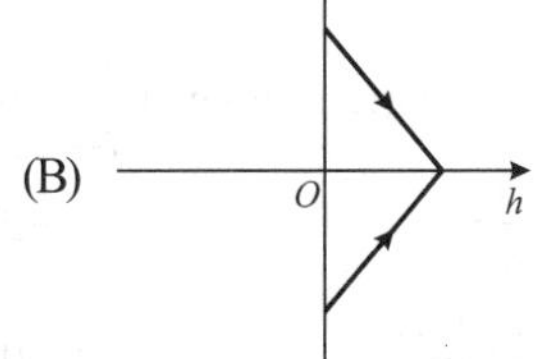

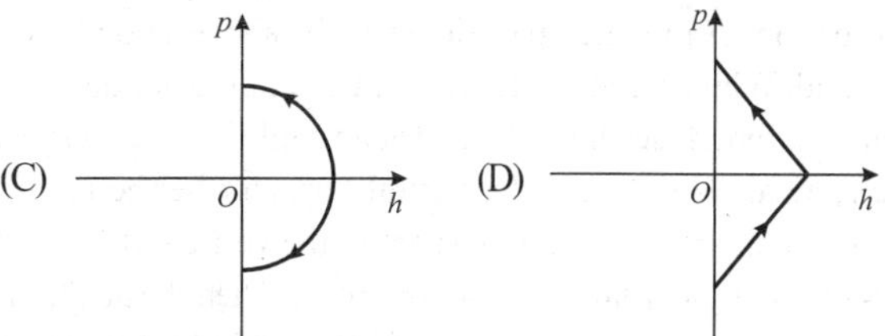

33. The position of a particle as a function of time t, is given by $x(t) = at + bt^2 - ct^3$ where a, b and c are constants. When the particle attains zero acceleration, then its velocity will be :

[9 April 2019 (Evening)]

(A) $a + \frac{b^2}{4c}$ (B) $a + \frac{b^2}{c}$

(C) $a + \frac{b^2}{2c}$ (D) $a + \frac{b^2}{3c}$

34. A particle is moving with speed $v = b\sqrt{x}$ along positive x-axis. Calculate the speed of the particle at time $t = \tau$: (Assume that the particle is at origin at $t = 0$) **[12 April 2019 (Evening)]**

(A) $\frac{b^2\tau}{4}$ (B) $\frac{b^2\tau}{2}$

(C) $b^2\tau$ (D) $\frac{b^2\tau}{\sqrt{2}}$

35. A ball is dropped from the top of a 100 m high tower on a planet. In the last $\frac{1}{2}s$ before hitting the ground, it covers a distance of 19 m. Acceleration due to gravity (in ms^{-2}) near the surface on that planet is _____. **[8 Jan. 2020 (Evening)]**

36. The distance x covered by a particle in one dimensional motion varies with time t as $x^2 = at^2 + 2bt + c$. If the acceleration of the particle depends on x as x^{-n}, where n is an integer, the value of n is _____. **[9 Jan. 2020 (Morning)]**

37. The speed verses time graph for a particle is shown in the figure. The distance travelled (in m) by the particle during the time interval $t = 0$ to $t = 5$ s will be _____ :

[4 Sept. 2020 (Evening)]

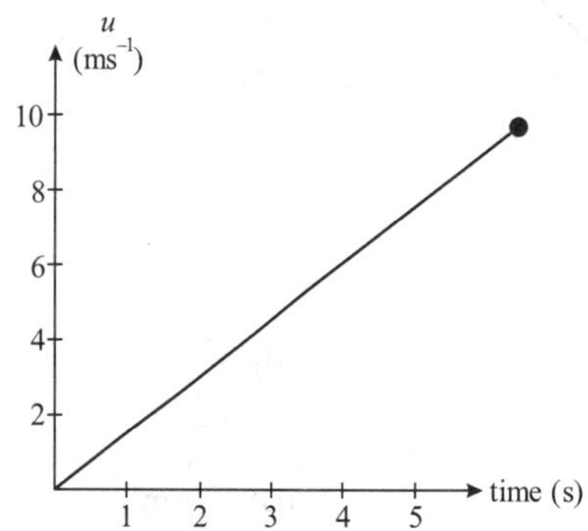

38. A balloon is moving up in air vertically above a point A on the ground. When it is at a height h_1, a girl standing at a distance d (point B) from A (see figure) sees it at an angle 45° with respect to the vertical. When the balloon climbs up a further height h_2, it is seen at an angle 60° with respect to the vertical if the girl moves further by a distance 2.464 d (point C). Then the height h_2 is (given tan 30° = 0.5774) : **[5 Sept. 2020 (Morning)]**

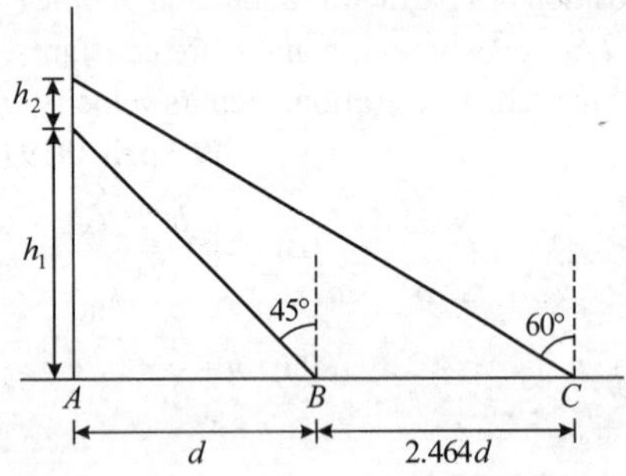

(A) d (B) $0.732\,d$

(C) $1.464\,d$ (D) $0.464\,d$

39. A helicopter rises from rest on the ground vertically upwards with a constant acceleration g. A food packet is dropped from the helicopter when it is a height h. The time taken by the packet to reach the ground is close to [g is the acceleration due to gravity] : **[5 Sept. 2020 (Morning)]**

(A) $t=\sqrt{\dfrac{2h}{3g}}$ (B) $t=1.8\sqrt{\dfrac{h}{g}}$

(C) $t=3.4\sqrt{\left(\dfrac{h}{g}\right)}$ (D) $t=\dfrac{2}{3}\sqrt{\left(\dfrac{h}{g}\right)}$

40. The velocity (v) and time (t) graph of a body in a straight line motion is shown in the figure. The point S is at 4.333 seconds. The total distance covered by the body in 6s is : **[5 Sept. 2020 (Evening)]**

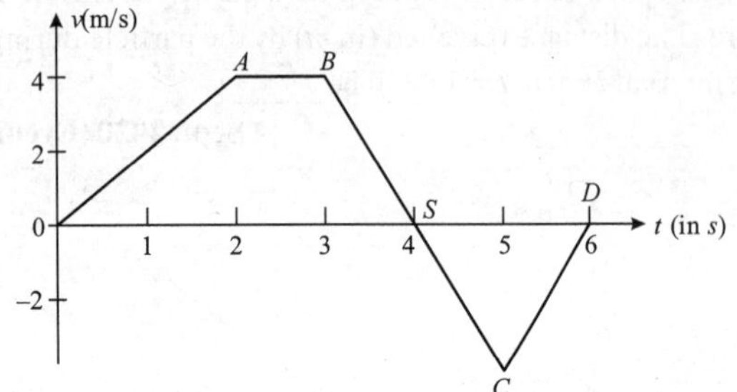

(A) 12m (B) $\dfrac{49}{4}$ m

(C) 11m (D) $\dfrac{37}{3}$ m

41. If the velocity-time graph has the shape AMB, what would be the shape of the corresponding acceleration-time graph ? **[24 Feb. 2021 (Morning)]**

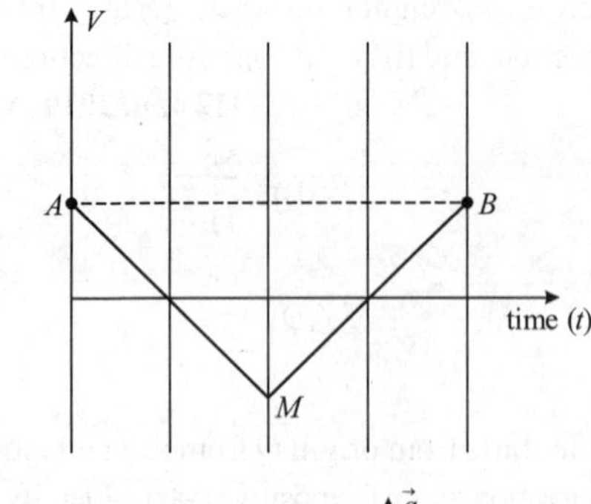

(A) 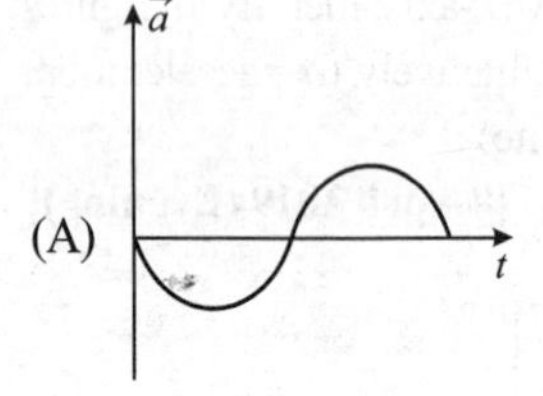 (B) 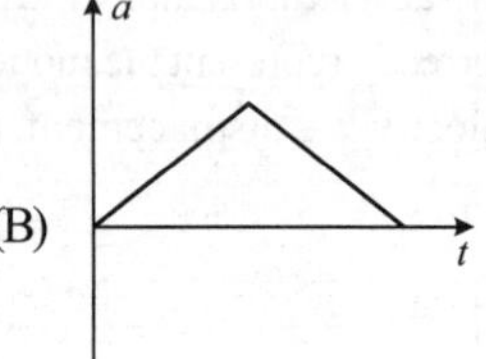

(C) 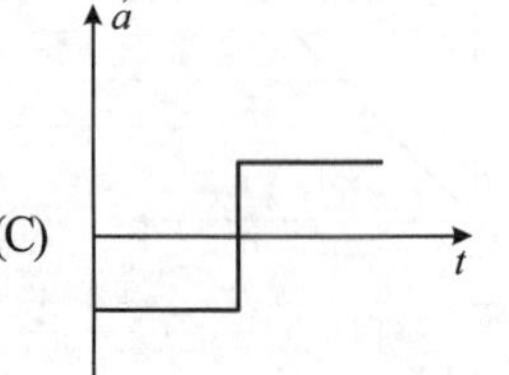 (D) 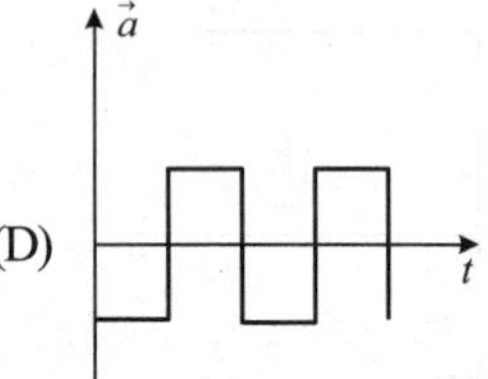

42. An engine of a train, moving with uniform acceleration, passes the signal-post with velocity u and the last compartment with velocity v. The velocity with which middle point of the train passes the signal post is : **[25 Feb. 2021 (Morning)]**

(A) $\sqrt{\dfrac{v^2+u^2}{2}}$ (B) $\sqrt{\dfrac{v^2-u^2}{2}}$

(C) $\dfrac{v-u}{2}$ (D) $\dfrac{u+v}{2}$

43. A stone is dropped from the top of a building. When it crosses a point 5 m below the top, another stone starts to fall from a point 25 m below the top. Both stones reach the bottom of building simultaneously. The height of the building is : **[25 Feb. 2021 (Evening)]**

(A) 35m (B) 45m

(C) 50m (D) 25m

44. A scooter accelerates from rest for time t_1 at constant rate a_1 and then retards at constant rate a_2 for time t_2 and comes to rest. The correct value of $\dfrac{t_1}{t_2}$ will be : **[26 Feb. 2021 (Evening)]**

(A) $\dfrac{a_1}{a_2}$ (B) $\dfrac{a_2}{a_1}$

(C) $\dfrac{a_1+a_2}{a_2}$ (D) $\dfrac{a_1+a_2}{a_1}$

45. The velocity-displacement graph describing the motion of a bicycle is shown in the figure. **[16 March 2021 (Morning)]**

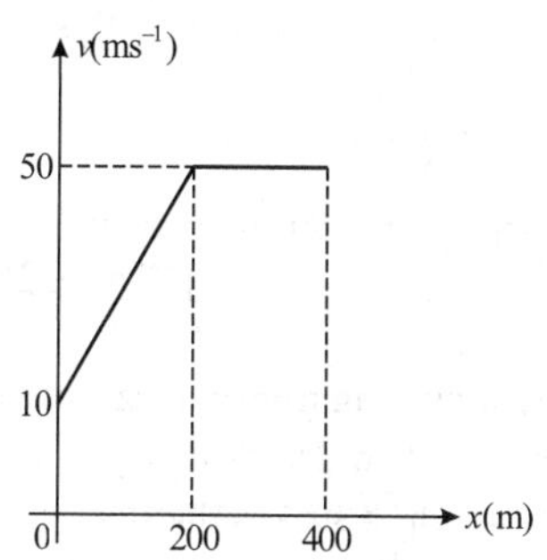

The acceleration-displacement graph of the bicycle's motion is best described by :

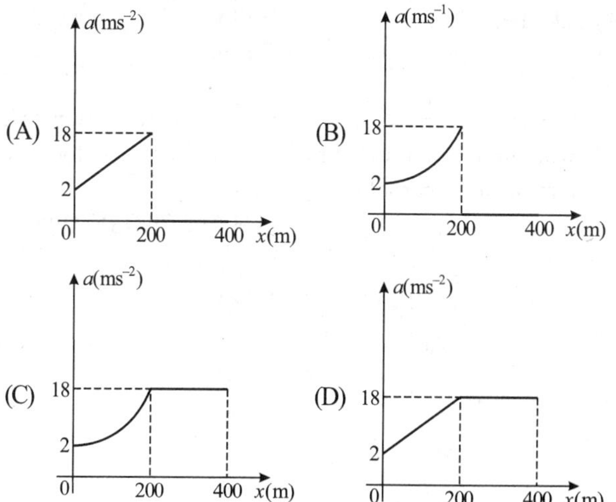

46. A body of mass 2 kg moves under a force of $(2\hat{i}+3\hat{j}+5\hat{k})$N. It starts from rest and was at the origin initially. After 4 s, its new coordinates are (8, *b*, 20). The value of *b* is _____. (Round off to the Nearest Integer) **[16 March 2021 (Evening)]**

47. A car accelerates from rest at a constant rate α for some time after which it decelerates at a constant rate β to come to rest. If the total time elapsed is *t* seconds, the total distance travelled is : **[17 March 2021 (Morning)]**

(A) $\frac{4\alpha\beta}{(\alpha+\beta)}t^2$ (B) $\frac{2\alpha\beta}{(\alpha+\beta)}t^2$

(C) $\frac{\alpha\beta}{2(\alpha+\beta)}t^2$ (D) $\frac{\alpha\beta}{4(\alpha+\beta)}t^2$

48. The velocity of a particle is $v=v_0+gt+Ft^2$. Its position is $x=0$ at $t=0$; then its displacement after time $(t=1)$ is : **[17 March 2021 (Evening)]**

(A) v_0+g+F (B) $v_0+\frac{g}{2}+\frac{F}{3}$

(C) $v_0+\frac{g}{2}+F$ (D) $v_0+2g+3F$

49. The position, velocity and acceleration of a particle moving with a constant acceleration can be represented by : **[18 March 2021 (Morning)]**

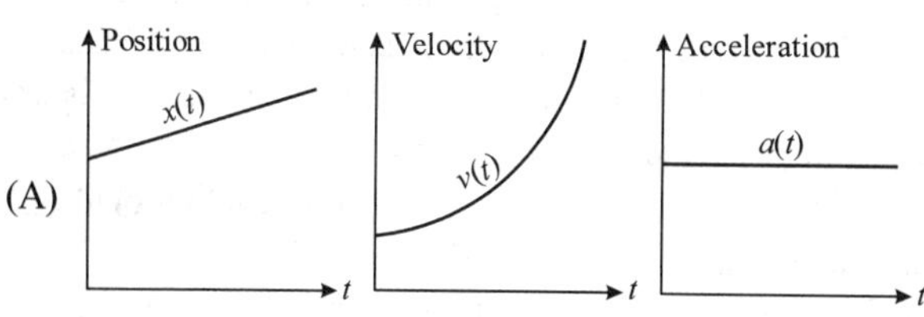

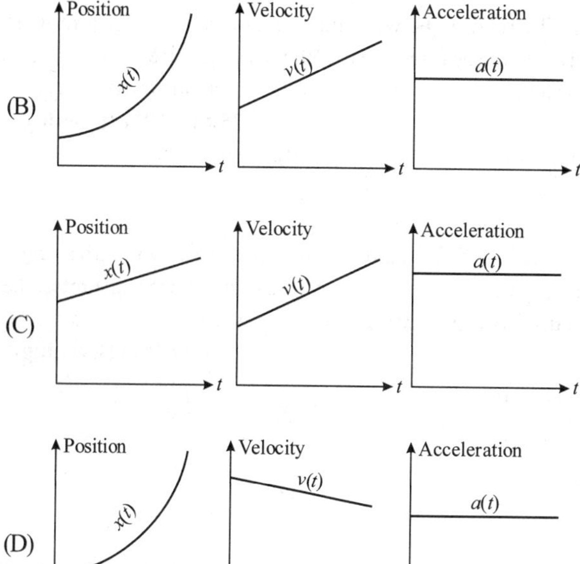

50. The velocity-displacement graph of a particle is shown in the figure.

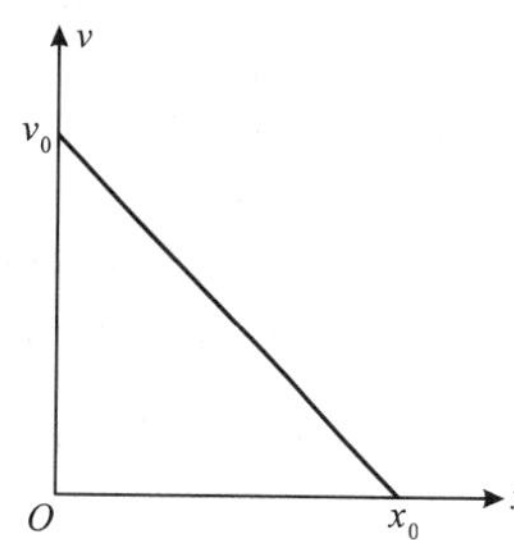

The acceleration-displacement graph of the same particle is represented by : **[18 March 2021 (Evening)]**

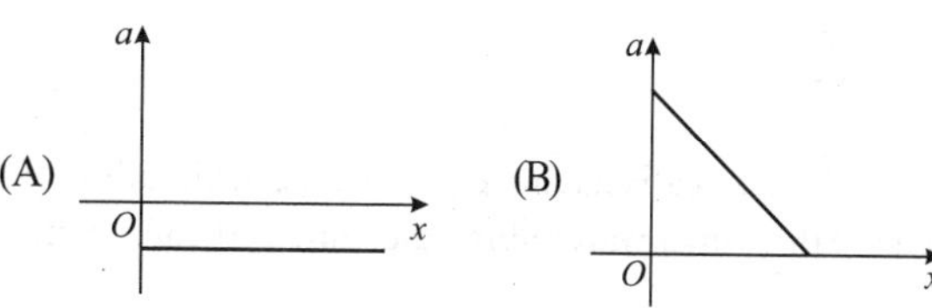

(C) (D)

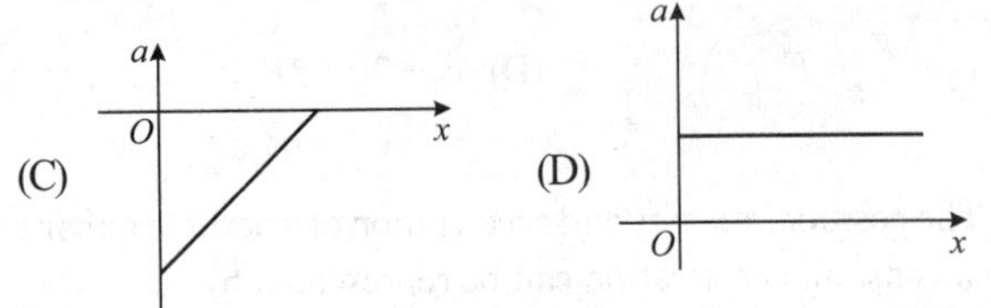

51. Water droplets are coming from an open tap at a particular rate. The spacing between a droplets observed at 4th second after its fall to the next droplet is 34.3 m. At what rate the droplets are coming from the tap ? (Take $g = 9.8$ m/s^2)

[25 July 2021 (Morning)]

(A) 1 drop/second (B) 1 drop/7 seconds

(C) 2 drops/second (D) 3 drops/2 seconds

52. The relation between time t and distance x for a moving body is given as $t = mx^2 + nx$, where m and n are constants. The retardation of the motion is : (Where v stands for velocity)

[25 July 2021 (Evening)]

(A) $2\,mnv^3$ (B) $2\,nv^3$

(C) $2\,mv^3$ (D) $2\,n^2v^3$

53. The instantaneous velocity of a particle moving in a straight line is given as $v = \alpha t + \beta t^2$, where α and β are constants. The distance travelled by the particle between 1 s and 2 s is :

[25 July 2021 (Evening)]

(A) $\frac{\alpha}{2}+\frac{\beta}{3}$ (B) $\frac{3}{2}\alpha+\frac{7}{3}\beta$

(C) $\frac{3}{2}\alpha+\frac{7}{2}\beta$ (D) $3\alpha+7\beta$

54. A balloon was moving upwards with a uniform velocity of 10 m/s. An object of finite mass is dropped from the balloon when it was at a height of 75 m from the ground level. The height of the balloon from the ground when object strikes the ground was around. (Takes the value of g as 10 m/s^2)

[25 July 2021 (Evening)]

(A) 125 m (B) 200 m

(C) 300 m (D) 250 m

55. A ball is thrown up with a certain velocity so that it reaches a height h. Find the ratio of the two different times of the ball reaching $\frac{h}{3}$ in both the directions : **[27 July 2021 (Morning)]**

(A) $\frac{\sqrt{2}-1}{\sqrt{2}+1}$ (B) $\frac{1}{3}$

(C) $\frac{\sqrt{3}-\sqrt{2}}{\sqrt{3}+\sqrt{2}}$ (D) $\frac{\sqrt{3}-1}{\sqrt{3}+1}$

56. Two spherical balls having equal masses with radius of 5 cm each are thrown upwards along the same vertical direction at an interval of 3s with the same initial velocity of 35 m/s, then these balls collide at a height of ______ m. (Take g = 10 m/s^2)

[26 Aug. 2021 (Morning)]

57. If the velocity of a body related to displacement x is given by $v = \sqrt{5000 + 24x}$ m/s, then calculate the acceleration of the body in m/s^2. **[27 Aug. 2021 (Morning)]**

58. Water drops are falling from a nozzle of a shower onto the floor from a height of 9.8 m. The drops fall at a regular interval of time. When the first drop strikes the floor, at that instant, the third drop begins to fall. Locate the position of second drop from the floor when the first drop strikes the floor :

[27 Aug. 2021 (Evening)]

(A) 2.94 m (B) 4.18 m

(C) 2.45 m (D) 7.35 m

59. A particle is moving with constant acceleration 'a'. Following graph shows v^2 versus x (displacement) plot. The acceleration of the particle is ______ m/s^2.

[31 Aug. 2021 (Evening)]

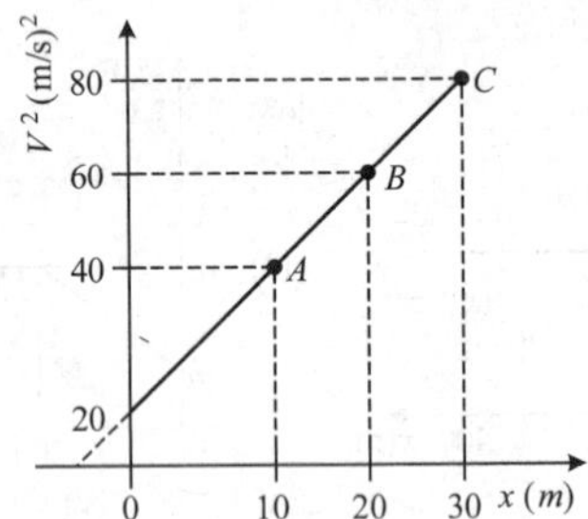

60. From the top of a tower, a ball is thrown vertically upward which reaches the ground in 6 s. A second ball thrown vertically downward from the same position with the same speed reaches the ground in 1.5 s. A third ball released, from the rest from the same location, will reach the ground in ______ s.

[24 June 2022 (Morning)]

61. $\vec{A}$ is a vector quantity such that $|\vec{A}|$ = non-zero constant. Which of the following expressions is true for $\vec{A}$?

[25 June 2022 (Morning)]

(A) $\vec{A}\cdot\vec{A} = 0$ (B) $\vec{A}\times\vec{A} < 0$

(C) $\vec{A}\times\vec{A} = 0$ (D) $\vec{A}\times\vec{A} > 0$

62. Which of the following relations is true for two unit vectors $\hat{A}$ and $\hat{B}$ making an angle θ to each other ?

[25 June 2022 (Morning)]

(A) $|\hat{A}+\hat{B}| = |\hat{A}-\hat{B}|\tan\frac{\theta}{2}$

(B) $|\hat{A}-\hat{B}| = |\hat{A}+\hat{B}| \tan \frac{\theta}{2}$

(C) $|\hat{A}+\hat{B}| = |\hat{A}-\hat{B}| \cos \frac{\theta}{2}$

(D) $|\hat{A}-\hat{B}| = |\hat{A}+\hat{B}| \cos \frac{\theta}{2}$

63. Given below are two statements. One is labelled as **Assertion-A** and the other is labelled as **Reason-R**

Assertion-A: Two identical balls A and B thrown with same velocity u at two different angles with horizontal attained the same range R. If A and B reached the maximum height h_1 and h_2 respectively, then $R = 4\sqrt{h_1 h_2}$

Reason-R: Product of said heights

$$h_1 h_2 = \left(\frac{u^2 \sin^2 \theta}{2g}\right) \cdot \left(\frac{u^2 \cos^2 \theta}{2g}\right)$$

Choose the CORRECT answer : **[25 June 2022 (Evening)]**

(A) Both **A** and **R** are true and **R** is the correct explanation of **A**

(B) Both **A** and **R** are true but **R** is **NOT** the correct explanation of **A**

(C) **A** is true but **R** is false

(D) **A** is false but **R** is true

64. A ball of mass 0.5 kg is dropped from the height of 10 m. The height, at which the magnitude of velocity becomes equal to the magnitude of acceleration due to gravity, is _____ m. (Use $g = 10$ m/s^2) **[26 June 2022 (Morning)]**

65. A ball is projected vertically upward with an initial velocity of 50 ms^{-1} at $t = 0$ s. At $t = 2$ s. Another ball is projected vertically upward with same velocity. At $t =$ _____ s, second ball will meet the first ball. ($g = 10$ ms^{-2}) **[26 June 2022 (Evening)]**

66. When a ball is dropped into a lake from a height 4.9 m above the water level, it hits the water with a velocity v and then sinks to the bottom with the constant velocity v. It reaches the bottom of the lake 4.0 s after it is dropped. The approximate depth of the lake is : **[27 June 2022 (Evening)]**

(A) 19.6m (B) 29.4m

(C) 39.2m (D) 73.5m

67. Velocity (v) and acceleration (a) in two systems of units 1 and 2 are related as $v_2 = \frac{n}{m^2} v_1$ and $a_2 = \frac{a_1}{mn}$ respectively. Here m and n are constants. The relations for distance and time in two systems respectively are : **[28 June 2022 (Evening)]**

(A) $\frac{n^3}{m^3} L_1 = L_2$ and $\frac{n^2}{m} T_1 = T_2$

(B) $L_1 = \frac{n^4}{m^2} L_2$ and $T_1 = \frac{n^2}{m} T_2$

(C) $L_1 = \frac{n^2}{m} L_2$ and $T_1 = \frac{n^4}{m^2} T_2$

(D) $\frac{n^2}{m} L_1 = L_2$ and $\frac{n^4}{m^2} T_1 = T_2$

68. A car covers AB distance with first one-third at velocity v_1 ms^{-1}, second one-third at v_2 ms^{-1} and last one-third at v_3 ms^{-1}. If $v_3 = 3v_1$, $v_2 = 2v_1$ and $v_1 = 11$ ms^{-1} then the average velocity of the car is _____ ms^{-1}. **[28 June 2022 (Evening)]**

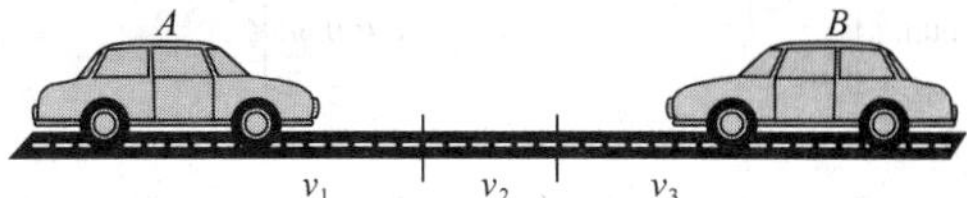

69. Two balls A and B are placed at the top of 180 m tall tower. Ball A is released from the top at $t = 0$ s. Ball B is thrown vertically down with an initial velocity 'u' at $t = 2$ s. After a certain time, both balls meet 100 m above the ground. Find the value of 'u' in ms^{-1}: [use $g = 10$ ms^{-2}] **[29 June 2022 (Morning)]**

(A) 10 (B) 15

(C) 20 (D) 30

70. A positive charge particle of 100 mg is thrown in opposite direction to a uniform electric field of strength 1×10^5 NC^{-1}. If the charge on the particle is 40 μC and the initial velocity is 200 ms^{-1}, how much distance it will travel before coming to the rest momentarily : **[29 June 2022 (Morning)]**

(A) 1m (B) 5m

(C) 10m (D) 0.5m

71. A small toy starts moving from the position of rest under a constant acceleration. If it travels a distance of 10 m in t s,. The distance travelled by the toy in the next t s will be : **[29 June 2022 (Evening)]**

(A) 10m (B) 20m

(C) 30m (D) 40m

72. A car is moving with speed of 150 km/h and after applying the brake it will move 27 m before it stops. If the same car is moving with a speed of one third the reported speed then it will stop after travelling _____ m distance. **[25 July 2022 (Morning)]**

73. A particle is moving in a straight line such that its velocity is increasing at 5 ms^{-1} per meter. The acceleration of the particle is _____ ms^{-2} at a point where its velocity is 20 ms^{-1}. **[25 July 2022 (Evening)]**

74. A bullet is shot vertically downwards with an initial velocity of 100 m/s from a certain height. Within 10 s, the bullet reaches the ground and instantaneously comes to rest due to the perfectly inelastic collision. The velocity-time curve for total time $t = 20$ s will be : (Take $g = 10$ m/s^2) **[27 July 2022 (Morning)]**

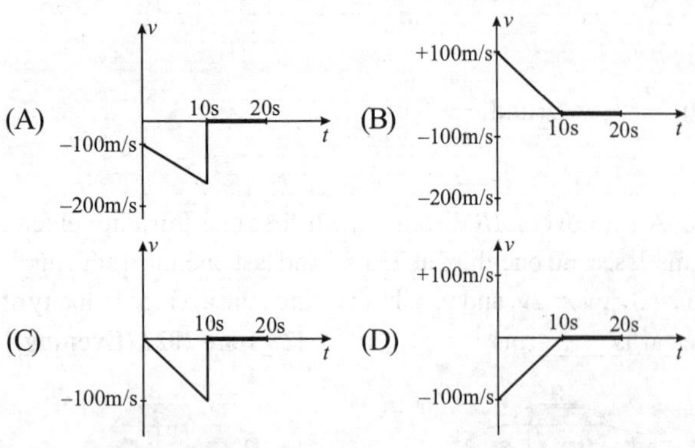

75. A *NCC* parade is going at a uniform speed of 9 km/h under a mango tree on which a monkey is sitting at a height of 19.6 m. At any particular instant, the monkey drops a mango. A cadet will receive the mango whose distance from the tree at time of drop is : (Given $g = 9.8$ m/s^2) **[28 July 2022 (Morning)]**

(A) 5 m (B) 10 m
(C) 19.8 m (D) 24.5 m

76. At time $t = 0$ a particle starts travelling from a height $7\hat{z}$ cm in a plane keeping z coordinate constant. At any instant of time it's position along the $\hat{x}$ and $\hat{y}$ directions are defined as $3t$ and $5t^3$ respectively. At $t = 1$ s acceleration of the particle will be : **[28 July 2022 (Evening)]**

(A) $-30\,\hat{y}$ (B) $30\,\hat{y}$
(C) $3\,\hat{x} + 15\,\hat{y}$ (D) $3\,\hat{x} + 15\,\hat{y} + 7\hat{z}$

77. A ball is thrown vertically upwards with a velocity of 19.6 ms^{-1} from the top of a tower. The ball strikes the ground after 6 s. The height from the ground up to which the ball can rise will be $\left(\frac{k}{5}\right)$ m. The value of k is ______. (use $g = 9.8$ m/s^2) **[28 July 2022 (Evening)]**

78. A ball is thrown up vertically with a certain velocity so that, it reaches a maximum height h. Find the ratio of the times in which it is at height $\frac{h}{3}$ while going up and coming down respectively : **[29 July 2022 (Morning)]**

(A) $\frac{\sqrt{2}-1}{\sqrt{2}+1}$ (B) $\frac{\sqrt{3}-\sqrt{2}}{\sqrt{3}+\sqrt{2}}$
(C) $\frac{\sqrt{3}-1}{\sqrt{3}+1}$ (D) $\frac{1}{3}$

79. If $t = \sqrt{x} + 4$, then $\left(\frac{dx}{dt}\right)_{t=4}$ is : **[29 July 2022 (Morning)]**

(A) 4 (B) Zero
(C) 8 (D) 16

80. A juggler throws balls vertically upwards with same initial velocity in air. When the first ball reaches its highest position, he throws the next ball. Assuming the juggler throws n balls per second, the maximum height the balls can reach is : **[29 July 2022 (Evening)]**

(A) $g/2n$ (B) g/n
(C) $2gn$ (D) $g/2n^2$

81. A ball is released from a height h. If t_1 and t_2 be the time required to complete first half and second half of the distance respectively. Then, choose the correct relation between t_1 and t_2 : **[29 July 2022 (Evening)]**

(A) $t_1 = (\sqrt{2})t_2$ (B) $t_1 = (\sqrt{2}-1)t_2$
(C) $t_2 = (\sqrt{2}+1)t_1$ (D) $t_2 = (\sqrt{2}-1)t_1$

82. The velocity time graph of a body moving in a straight line is shown in figure. **[24 Jan. 2023 (Evening)]**

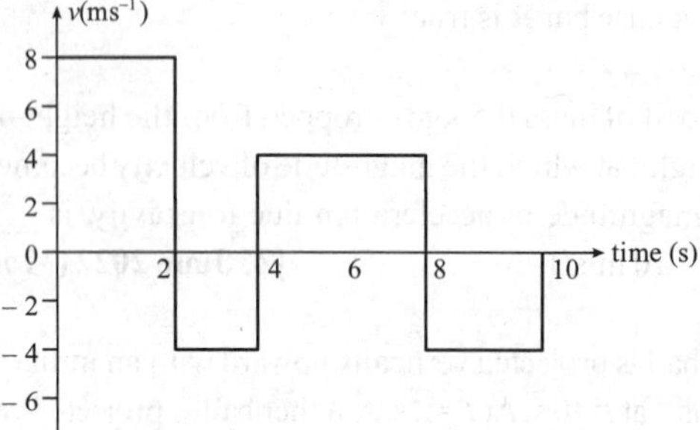

The ratio of displacement to distance travelled by the body in time 0 to 10 s is :

(A) 1 : 1 (B) 1 : 4
(C) 1 : 2 (D) 1 : 3

83. A car travels a distance of x with speed V_1 and then same distance x with speed V_2 in the same direction. The average speed of the car is : **[25 Jan. 2023 (Morning)]**

(A) $\frac{v_1 v_2}{2(v_1 + v_2)}$ (B) $\frac{v_1 + v_2}{2}$
(C) $\frac{2x}{v_1 + v_2}$ (D) $\frac{2v_1 v_2}{v_1 + v_2}$

84. The distance travelled by a particle is related to time t as $x = 4t^2$. The velocity of the particle at $t = 5$ s is. **[25 Jan. 2023 (Evening)]**

(A) 40 ms^{-1} (B) 25 ms^{-1}
(C) 20 ms^{-1} (D) 8 ms^{-1}

85. Match Column-I with Column-II:

[30 Jan. 2023 (Morning)]

Column-I (**x-t graphs**)	**Column-II** (**v-t graphs**)
(a)	(i)
(b)	(ii)
(c)	(iii)
(d)	(iv)

Choose the correct answer from the options given below :

(A) (a)-(ii), (b)-(iv), (c)-(iii), (d)-(i)

(B) (a)-(i), (b)-(ii), (c)-(iii), (d)-(iv)

(C) (a)-(ii), (b)-(iii), (c)-(iv), (d)-(i)

(D) (a)-(i), (b)-(iii), (c)-(iv), (d)-(ii)

86. A horse rider covers half the distance with 5 m/s speed. The remaining part of the distance was travelled with speed 10 m/s for half the time and with speed 15 m/s for other half of the time. The mean speed of the rider averaged over the whole time of motion is $x/7$ m/s. The value of x is.

[30 Jan. 2023 (Morning)]

87. A vehicle travels 4 km with speed of 3 km/h and another 4 km with speed of 5 km/h, then its average speed is :

[30 Jan. 2023 (Evening)]

(A) 4.25 km/h (B) 3.50 km/h

(C) 4.00 km/h (D) 3.75 km/h

88. An object moves with speed v_1, v_2 and v_3 along a line segment AB, BC and CD respectively as shown in figure. Where $AB = BC$ and $AD = 3\ AB$, then average speed of the object will be : **[1 Feb. 2023 (Morning)]**

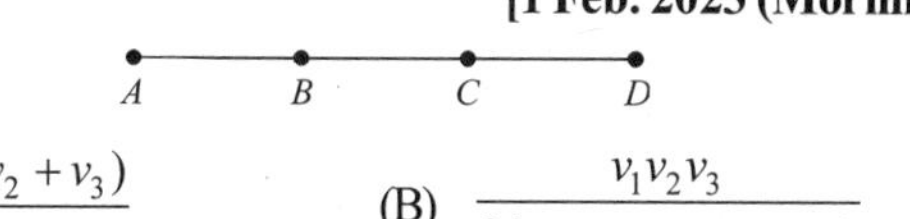

(A) $\frac{(v_1 + v_2 + v_3)}{3}$ (B) $\frac{v_1 v_2 v_3}{3(v_1 v_2 + v_2 v_3 + v_3 v_1)}$

(C) $\frac{3 v_1 v_2 v_3}{v_1 v_2 + v_2 v_3 + v_3 v_1}$ (D) $\frac{(v_1 + v_2 + v_3)}{3 v_1 v_2 v_3}$

89. For a train engine moving with speed of 20 ms^{-1}. The driver must apply brakes at a distance of 500 m before the station for the train to come to rest at the station. If the brakes were applied at half of this distance, the train engine would cross the station with speed $\sqrt{x}$ ms^{-1}. The value of x is _____ (Assuming same retardation is produced by brakes). **[1 Feb. 2023 (Evening)]**

90. A particle starts with an initial velocity of 10.0 ms^{-1} along x-direction and accelerates uniformly at the rate of 2.0 ms^{-2}. The time taken by the particle to reach the velocity of 60.0 ms^{-1} is _____. **[6 April 2023 (Evening)]**

(A) 6 s (B) 3 s

(C) 30 s (D) 25 s

91. The trajectory of projectile, projected from the ground is given by $y = x - \frac{x^2}{20}$. Where x and y are measured in meter. The maximum height attained by the projectile will be :

[8 April 2023 (Evening)]

(A) 5m (B) $10\sqrt{2}$m

(C) 200m (D) 10m

92. Given below are two statements :

Statement-I: Area under velocity-time graph gives the distance travelled by the body in a given time

Statement-II: Area under acceleration-time graph is equal to the change in velocity-in the given time

In the light of given statements, choose the correct answer from the options given below : **[8 April 2023 (Evening)]**

(A) Both Statement-I and Statement-II are true

(B) Statement-I is correct but Statement-II is false

(C) Statement-I is incorrect but Statement-II is true

(D) Both Statement-I and Statement-II are False

93. The range of the projectile projected at an angle of 15° with horizontal is 50 m. If the projectile is projected with same velocity at an angle of 45° with horizontal, then its range will be:

[10 April 2023 (Morning)]

(A) 50 m (B) $50\sqrt{2}$ m

(C) 100m (D) $100\sqrt{2}$ m

94. The position-time graphs for two students A and B returning from the school to their homes are shown in figure:

[10 April 2023 (Morning)]

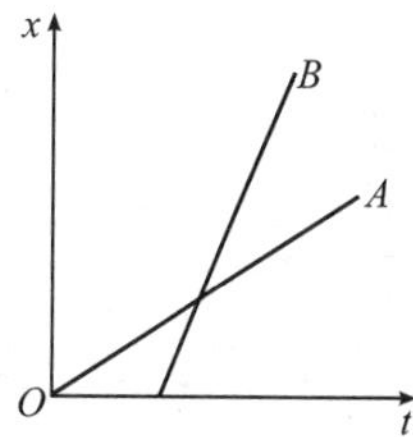

(a) A lives closer to the school
(b) B lives closer to the school
(c) A takes lesser time to reach home
(d) A travels faster than B
(e) B travels faster than A
Choose the correct answer from the options given below:
(A) (a) and (e) only (B) (b) and (e) only
(C) (a), (c) and (e) only (D) (a), (c) and (d) only

95. A person travels x distance with velocity v_1 and then x distance with velocity v_2 in the same direction. The average velocity of the person is v, then the relation between v, v_1 and v_2 will be: **[10 April 2023 (Evening)]**

(A) $v = v_1 + v_2$ (B) $v = \frac{v_1 + v_2}{2}$
(C) $\frac{2}{v} = \frac{1}{v_1} + \frac{1}{v_2}$ (D) $\frac{1}{v} = \frac{1}{v_1} + \frac{1}{v_2}$

96. From the $v - t$ graph shown. The ratio of distance to displacement in 25 s of motion: **[11 April 2023 (Morning)]**

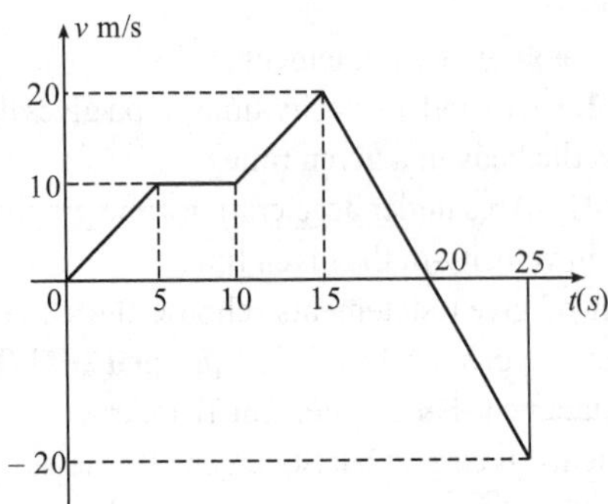

(A) $\frac{3}{5}$ (B) $\frac{1}{2}$
(C) $\frac{5}{3}$ (D) 1

97. A body of mass 500 g moves along x-axis such that it's velocity varies with displacement x according to the relation $v = 10\sqrt{x}$ m/s the force acting on the body is: **[11 April 2023 (Evening)]**
(A) 166 N (B) 25 N
(C) 125 N (D) 5 N

98. A ball is thrown vertically upward with an initial velocity of 150 m/s. The ratio of velocity after 3 s and 5 s is $\frac{x+1}{x}$. The value of x is _____. (Take: $g = 10$ m/s^2) **[12 April 2023 (Morning)]**
(A) 6 (B) 5
(C) –5 (D) 10

99. Two trains A and B of length l and $4l$ are travelling into a tunnel of length L in parallel tracks from opposite directions with velocities 108 km/h and 72 km/h, respectively. If train A takes 35 s less time than train 'B' to cross the tunnel then, length L of tunnel is : (Given $L = 60l$) **[13 April 2023 (Morning)]**
(A) 1200 m (B) 2700 m
(C) 1800 m (D) 900 m

100. A passenger sitting in a train A moving at 90 km/h observes another train B moving in the opposite direction for 8 s. If the velocity of the train B is 54 km/h, then length of train B is: **[13 April 2023 (Evening)]**
(A) 80 m (B) 200 m
(C) 120 m (D) 320 m

101. The distance travelled by an object in time t is given by $s = (2.5)t^2$. The instantaneous speed of the object at $t = 5$ s will be: **[13 April 2023 (Evening)]**
(A) 12.5 ms^{-1} (B) 62.5 ms^{-1}
(C) 5 ms^{-1} (D) 25 ms^{-1}

102. The position of a particle related to time is given by $x = (5t^2 - 4t + 5)$ m. The magnitude of velocity of the particle at $t = 2$ s will be : **[15 April 2023 (Morning)]**
(A) 10 ms^{-1} (B) 14 ms^{-1}
(C) 16 ms^{-1} (D) 6 ms^{-1}

* * * * *

CHAPTER

3 MOTION IN A PLANE

3.1 MOTION IN TWO DIMENSIONS

1. A ball whose kinetic energy is E, is projected at an angle of 45° to the horizontal. The kinetic energy of the ball at the highest point of its flight will be : **[2002]**

(A) E (B) $E/\sqrt{2}$

(C) $E/2$ (D) Zero

2. A boy playing on the roof of a 10 m high building throws a ball with a speed of 10 m/s at an angle of 30° with the horizontal. How far from the throwing point will the ball be at the height of 10 m from the ground? **[2002]**

[$g = 10$ m/s^2, $\sin 30° = 1/2$, $\cos 30° = \frac{\sqrt{3}}{2}$]

(A) 5.20m (B) 4.33m

(C) 2.60m (D) 8.66m

3. The coordinates of a moving particle at any time t are given by $x = \alpha t^3$ and $y = \beta t^3$. The speed of the particle at time t is given by : **[2003]**

(A) $\sqrt{\alpha^2 + \beta^2}$ (B) $3t^2\sqrt{\alpha^2 + \beta^2}$

(C) $t^2\sqrt{\alpha^2 + \beta^2}$ (D) $\sqrt{\alpha^2 + \beta^2}$

4. A ball is thrown from a point with a speed v_0 at angle of projection θ. From the same point and at the same instant, a person starts running with a constant speed $v_0/2$ to catch the ball? If yes, what should be the angle of projection? **[2004]**

(A) Yes, 60° (B) Yes, 30°

(C) No (D) Yes, 45°

5. If $\vec{A} \times \vec{B} = \vec{B} \times \vec{A}$, then the angle between A and B is : **[2004]**

(A) $\frac{\pi}{2}$ (B) $\frac{\pi}{3}$

(C) π (D) $\frac{\pi}{4}$

6. A projectile can have the same range 'R' for two angles of projection. If 'T_1' and 'T_2' to be time of flights in the two cases, then the product of the two time of flights is directly proportional to : **[2004]**

(A) R (B) $\frac{1}{R}$

(C) $\frac{1}{R^2}$ (D) R^2

7. A particle is projected at 60° to the horizontal with a kinetic energy K. The kinetic energy at the highest point is : **[2007]**

(A) K (B) Zero

(C) $K/4$ (D) $K/2$

8. A particle has an initial velocity of $3\hat{i} + 4\hat{j}$ and an acceleration of $0.4\hat{i} + 0.3\hat{j}$. Its speed after 10 s is : **[2009]**

(A) $7\sqrt{2}$ units (B) 7 units

(C) 8.5 units (D) 10 units

9. A particle is moving with velocity $\vec{v} = K(y\hat{i} + x\hat{j})$, where K is a constant. The general equation for its path is : **[2010]**

(A) $y = x^2$ + constant (B) $y^2 = x$ + constant

(C) xy = constant (D) $y^2 = x^2$ + constant

10. A water fountain on the ground sprinkles water all around it. If the speed of water coming out of the fountain is v, the total area around the fountain that gets wet is : **[2011]**

(A) $\pi \frac{v^4}{g^2}$ (B) $\frac{\pi}{2}\frac{v^4}{g^2}$

(C) $\pi \frac{v^2}{g^2}$ (D) $\pi \frac{v^4}{g}$

11. A boy can throw a stone up to a maximum height of 10 m. The maximum horizontal distance that the boy can throw the same stone up to, will be : **[2012]**

(A) $20\sqrt{2}$m (B) 10m

(C) $10\sqrt{2}$m (D) 20m

12. A projectile is given an initial velocity of $(\hat{i} + 2\hat{j})$ m/s, where $\hat{i}$ is along the ground and $\hat{j}$ is along the vertical. If $g = 10$ m/s^2, the equation of its trajectory is : **[2013]**

(A) $y = 2x - 5x^2$ (B) $4y = 2x - 5x^2$

(C) $4y = 2x - 25x^2$ (D) $y = x - 5x^2$

13. The maximum range of a bullet fired from a toy pistol mounted on a car at rest is $R_0 = 40$ m. What will be the acute angle of inclination of the pistol for maximum range when the car is moving in the direction of firing with uniform velocity $v = 20$ m/s, on a horizontal surface ? ($g = 10$ m/s^2) **[2013 (Online)]**

(A) 30° (B) 60°

(C) 75° (D) 45°

14. A ball projected from ground at an angle of 45° just clears a wall in front. If point of projection is 4 m from the foot of wall and ball strikes the ground at a distance of 6 m on the other side of the wall, the height of the wall is : **[2013 (Online)]**

(A) 4.4m (B) 2.4m
(C) 3.6m (D) 1.6m

15. The initial speed of a bullet fired from a rifle is 630 m/s. The rifle is fired at the centre of a target 700 m away at the same level as the target. How far above the centre of the target ? **[2014 (Online)]**

(A) 1.0m (B) 4.2m
(C) 6.1m (D) 9.8m

16. The position of a projectile launched from the origin at $t=0$ is given by $\vec{r} = (40\hat{i}+50\hat{j})$ m at $t = 2$ s. If the projectile was launched at an angle θ from the horizontal, then θ is : (take $g = 10\ \text{ms}^{-2}$) **[2014 (Online)]**

(A) $\tan^{-1}\frac{2}{3}$ (B) $\tan^{-1}\frac{3}{2}$
(C) $\tan^{-1}\frac{7}{4}$ (D) $\tan^{-1}\frac{4}{5}$

17. A vector $\vec{A}$ is rotated by a small angle $\Delta\theta$ radian ($\Delta\theta << 1$) to get a new vector $\vec{B}$. In that case $|\vec{B}-\vec{A}|$ is : **[2015 (Online)]**

(A) $|\vec{A}|\Delta\theta$ (B) $|\vec{B}|\Delta\theta - |\vec{A}|$
(C) $|\vec{A}|\left(1-\frac{\Delta\theta^2}{2}\right)$ (D) 0

18. Let $\vec{A} = (\hat{i}+\hat{j})$ and $\vec{B} = (2\hat{i}-\hat{j})$. The magnitude of a coplanar vector $\vec{C}$ such that $\vec{A}\cdot\vec{C} = \vec{B}\cdot\vec{C} = \vec{A}\cdot\vec{B}$ is given by: **[2018 (Online)]**

(A) $\sqrt{\frac{5}{9}}$ (B) $\sqrt{\frac{10}{9}}$
(C) $\sqrt{\frac{20}{9}}$ (D) $\sqrt{\frac{9}{12}}$

19. A particle is moving with a velocity $\vec{v} = K(y\hat{i}+x\hat{j})$, where K is a constant. The general equation for its path is : **[9 Jan. 2019 (Morning)]**

(A) xy = constant (B) $y^2 = x^2$ + constant
(C) $y = x^2$ + constant (D) $y^2 = x$ + constant

20. The position co-ordinates of a particle moving in a 3-D coordinate system is given by

$x = a\cos\omega t$

$y = a\sin\omega t$

and $z = a\omega t$

The speed of the particle is : **[9 Jan. 2019 (Evening)]**

(A) $a\omega$ (B) $\sqrt{3}\,a\omega$
(C) $\sqrt{2}\,a\omega$ (D) $2a\omega$

21. Two guns A and B can fire bullets at speeds 1 km/s and 2 km/s respectively. From a point on a horizontal ground, they are fired in all possible directions. The ratio of maximum areas covered by the bullets fired by the two guns, on the ground is : **[10 Jan. 2019 (Morning)]**

(A) 1 : 2 (B) 1 : 4
(C) 1 : 8 (D) 1 : 16

22. In the cube of side 'a' shown in the figure, the vector from the central point of the face $ABOD$ to the central point of the face $BEFO$ will be : **[10 Jan. 2019 (Morning)]**

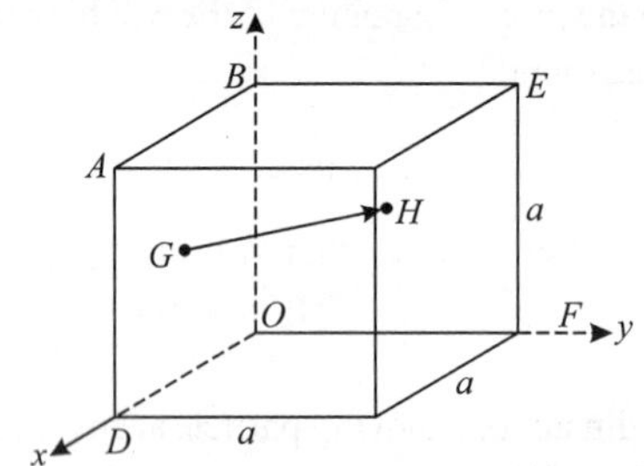

(A) $\frac{1}{2}a(\hat{i}-\hat{k})$ (B) $\frac{1}{2}a(\hat{j}-\hat{i})$
(C) $\frac{1}{2}a(\hat{k}-\hat{i})$ (D) $\frac{1}{2}a(\hat{j}-\hat{k})$

23. Two forces P and Q of magnitude $2F$ and $3F$, respectively, are at an angle θ with each other. If the force Q is doubled, then their resultant also gets doubled. Then, the angle is : **[10 Jan. 2019 (Evening)]**

(A) 30° (B) 60°
(C) 90° (D) 120°

24. Two vectors $\vec{A}$ and $\vec{B}$ have equal magnitudes. The magnitude of $(\vec{A}+\vec{B})$ is 'n' times the magnitude of $(\vec{A}-\vec{B})$. The angle between $\vec{A}$ and $\vec{B}$ is : **[10 Jan. 2019 (Evening)]**

(A) $\sin^{-1}\left[\frac{n^2-1}{n^2+1}\right]$ (B) $\cos^{-1}\left[\frac{n-1}{n+1}\right]$
(C) $\cos^{-1}\left[\frac{n^2-1}{n^2+1}\right]$ (D) $\sin^{-1}\left[\frac{n-1}{n+1}\right]$

25. A body is projected at $t = 0$ with a velocity 10 ms^{-1} at an angle of 60° with the horizontal. The radius of curvature of its trajectory at $t = 1$ s is R. Neglecting air resistance and taking acceleration due to gravity $g = 10$ ms^{-2}, the value of R is : **[11 Jan. 2019 (Morning)]**

(A) 2.5 m (B) 10.3 m
(C) 2.8 m (D) 5.1 m

26. The position vector of a particle changes with time according to the relation $\vec{r}(t) = 15t^2\hat{i} + (4 - 20t^2)\hat{j}$. What is the magnitude of the acceleration at $t = 1$?

[9 April 2019 (Evening)]

(A) 40 (B) 100
(C) 25 (D) 50

27. A plane is inclined at an angle $\alpha = 30°$ with a respect to the horizontal. A particle is projected with a speed $u = 2$ ms^{-1} from the base of the plane, making an angle $\theta = 15°$ with respect to the plane as shown in the figure. The distance from the base, at which the particle hits the plane is close to : (Take $g = 10$ ms^{-2})

[10 April 2019 (Evening)]

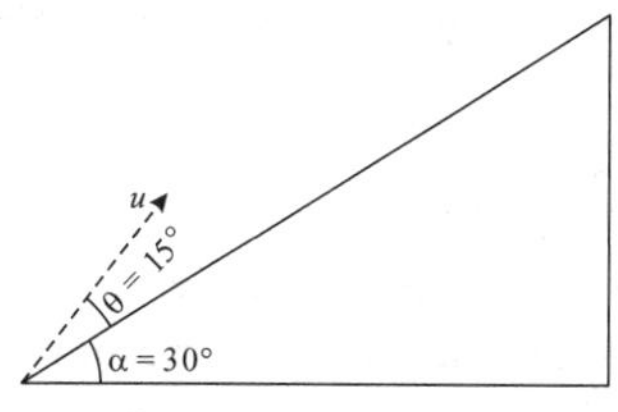

(A) 14 cm (B) 20 cm
(C) 18 cm (D) 26 cm

28. The trajectory of a projectile near the surface of the earth is given as $y = 2x - 9x^2$. If it were launched at an angle θ_0 with speed v_0 then ($g = 10$ ms^{-2}) : **[12 April 2019 (Morning)]**

(A) $\theta_0 = \cos^{-1}\left(\frac{1}{\sqrt{5}}\right)$ and $v_0 = \frac{5}{3}$ ms^{-1}

(B) $\theta_0 = \sin^{-1}\left(\frac{1}{\sqrt{5}}\right)$ and $v_0 = \frac{5}{3}$ ms^{-1}

(C) $\theta_0 = \sin^{-1}\left(\frac{2}{\sqrt{5}}\right)$ and $v_0 = \frac{3}{5}$ ms^{-1}

(D) $\theta_0 = \cos^{-1}\left(\frac{2}{\sqrt{5}}\right)$ and $v_0 = \frac{3}{5}$ ms^{-1}

29. A shell is fired from a fixed artillery gun with an initial speed u such that it hits the target on the ground at a distance R from it. If t_1 and t_2 are the values of the time taken by it to hit the target in two possible ways, the product $t_1 t_2$ is :

[12 April 2019 (Morning)]

(A) R/g (B) $R/4g$
(C) $2R/g$ (D) $R/2g$

30. Two particles are projected from the same point with the same speed u such that they have the same range R, but different maximum heights, h_1 and h_2. Which of the following is correct?

[12 April 2019 (Evening)]

(A) $R^2 = 2h_1h_2$ (B) $R^2 = 16h_1h_2$
(C) $R^2 = 4h_1h_2$ (D) $R^2 = h_1h_2$

31. A particle is moving along the x-axis with its coordinate with the time 't' given be $x(t) = 10 + 8t - 3t^2$. Another particle is moving the y-axis with its coordinate as a function of time given by $y(t) = 5 - 8t^3$. At $t = 1$ s, the speed of the second particle as measured in the frame of the first particle is given as $\sqrt{v}$. Then v (in m/s) is _____. **[8 Jan. 2020 (Morning)]**

32. A particle moves such that its position vector $\vec{r}(t) = \cos\omega t\hat{i} + \sin\omega t\hat{j}$ where ω is a constant and t is time. Then which of the following statements is true for the velocity $\vec{v}(t)$ and acceleration $\vec{a}(t)$ of the particle : **[8 Jan. 2020 (Evening)]**

(A) $\vec{v}$ is perpendicular to $\vec{r}$ and $\vec{a}$ is directed towards the origin

(B) $\vec{v}$ and $\vec{a}$ both are parallel to $\vec{r}$

(C) $\vec{v}$ and $\vec{a}$ both are perpendicular to $\vec{r}$

(D) $\vec{v}$ is perpendicular to $\vec{r}$ and $\vec{a}$ is directed away from the origin

33. A particle starts from the origin at $t = 0$ with an initial velocity of $3.0\,\hat{i}$ m/s and moves in the x-y plane with a constant acceleration $(6.0\,\hat{i} + 4.0\,\hat{j})$ m/s^2. The x-coordinate of the particle at the instant when its y-coordinate is 32 m is D meters. The value of D is : **[9 Jan. 2020 (Evening)]**

(A) 50 (B) 32
(C) 60 (D) 40

34. Starting from the origin at time $t = 0$, with initial velocity $5\hat{j}$ ms^{-1}, a particle moves in the x-y plane with a constant acceleration of $(10\hat{i} + 4\hat{j})$ ms^{-2}. At time t, its coordinates are (20 m, y_0m). The values of t and y_0, are respectively :

[4 Sept. 2020 (Morning)]

(A) 4 s and 52 m (B) 2 s and 24 m
(C) 2 s and 18 m (D) 5 s and 25 m

35. In an octagon $ABCDEFGH$ of equal side, what is the sum of $\overrightarrow{AB} + \overrightarrow{AC} + \overrightarrow{AD} + \overrightarrow{AE} + \overrightarrow{AF} + \overrightarrow{AG} + \overrightarrow{AH}$, if, $\overrightarrow{AO} = 2\hat{i} + 3\hat{j} - 4\hat{k}$: **[25 Feb. 2021 (Morning)]**

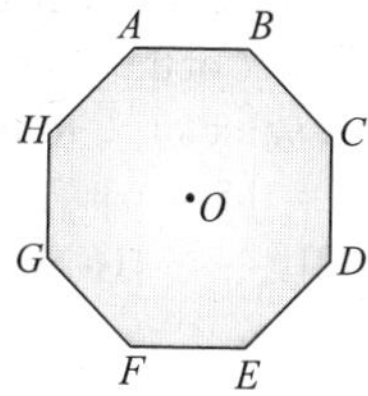

(A) $-16\hat{i} - 24\hat{j} + 32\hat{k}$ (B) $16\hat{i} + 24\hat{j} - 32\hat{k}$
(C) $16\hat{i} + 24\hat{j} + 32\hat{k}$ (D) $16\hat{i} - 24\hat{j} + 32\hat{k}$

36. An electron with kinetic energy K_1 enters between parallel plates of a capacitor at an angle 'α' with the plates. It leaves the plates at angle 'β' with kinetic energy K_2. Then the ratio of kinetic energies $K_1 : K_2$ will be : **[25 Feb. 2021 (Evening)]**

(A) $\frac{\cos\beta}{\sin\alpha}$ (B) $\frac{\cos\beta}{\cos\alpha}$

(C) $\frac{\sin^2\beta}{\cos^2\alpha}$ (D) $\frac{\cos^2\beta}{\cos^2\alpha}$

37. If $\vec{P}\times\vec{Q} = \vec{Q}\times\vec{P}$, the angle between $\vec{P}$ and $\vec{Q}$ is $\theta(0° < \theta < 360°)$. The value of 'θ' will be ______°. **[25 Feb. 2021 (Evening)]**

38. The trajectory of a projectile in a vertical plane is $y = \alpha x - \beta x^2$, where α and β are constants and x & y are respectively the horizontal and vertical distances of the projectile from the point of projection. The angle of projection θ and the maximum height attained H are respectively given by : **[26 Feb. 2021 (Evening)]**

(A) $\tan^{-1}\alpha, \frac{4\alpha^2}{\beta}$ (B) $\tan^{-1}\left(\frac{\beta}{\alpha}\right), \frac{\alpha^2}{\beta}$

(C) $\tan^{-1}\alpha, \frac{\alpha^2}{4\beta}$ (D) $\tan^{-1}\beta, \frac{\alpha^2}{2\beta}$

39. Consider a frame that is made up of two thin massless rods AB and AC as shown in the figure. A vertical force $\vec{P}$ of magnitude 100 N is applied at point A of the frame.

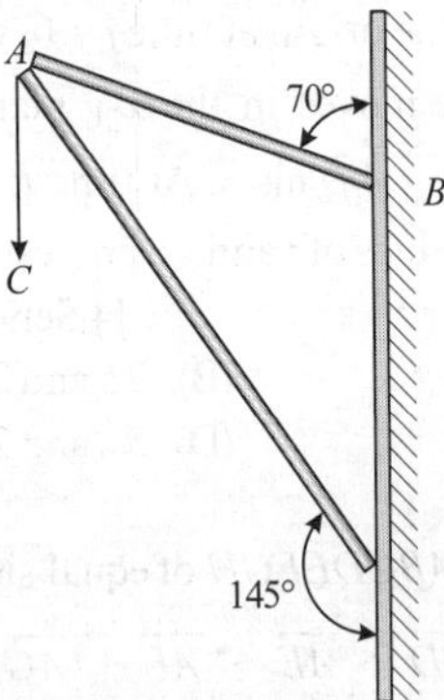

Suppose the force is $\vec{P}$ resolved parallel to the arms AB and AC of the frame. The magnitude of the resolved component along the arm AC is xN. The value of x, to the nearest integer, is ______. [Given : sin(35°) = 0.573, cos(35°) = 0.819; sin(110°) = 0.939, cos(110°) = –0.342] **[16 March 2021 (Morning)]**

40. A mosquito is moving with a velocity $\vec{v} = 0.5t^2\hat{i} + 3t\hat{j} + 9\hat{k}$ m/s and accelerating in uniform conditions. What will be the direction of mosquito after 2s ? **[16 March 2021 (Evening)]**

(A) $\tan^{-1}\left(\frac{2}{3}\right)$ from x-axis (B) $\tan^{-1}\left(\frac{2}{3}\right)$ from y-axis

(C) $\tan^{-1}\left(\frac{5}{2}\right)$ from y-axis (D) $\tan^{-1}\left(\frac{5}{2}\right)$ from x-axis

41. A person is swimming with a speed of 10 m/s at an angle of 120° with the flow and reaches to a point directly opposite on the other side of the river. The speed of the flow is 'x' m/s. The value of 'x' to the nearest integer is ______. **[18 March 2021 (Morning)]**

42. The projectile motion of a particle of mass 5 g is shown in the figure.

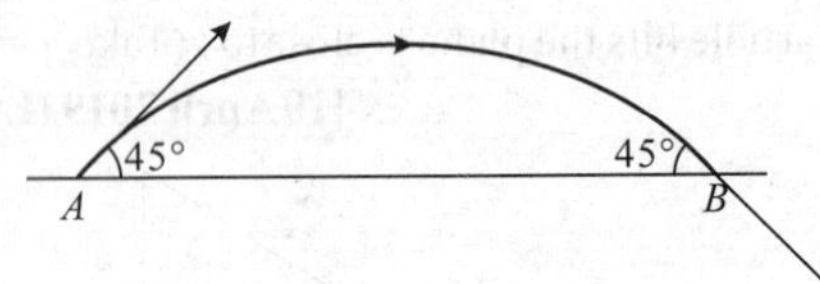

The initial velocity of the particle is $5\sqrt{2}\,\text{ms}^{-1}$ and the air resistance is assumed to be negligible. The magnitude of the change in momentum between the points A and B is $x \times 10^{-2}\,\text{kgms}^{-1}$. The value of x, to the nearest integer, is ______. **[18 March 2021 (Evening)]**

43. If $\vec{A}$ and $\vec{B}$ are two vectors satisfying the relation

$$\vec{A}\cdot\vec{B} = |\vec{A}\times\vec{B}|$$

Then the value of $|\vec{A}-\vec{B}|$ will be : **[20 July 2021 (Morning)]**

(A) $\sqrt{A^2+B^2+2AB}$ (B) $\sqrt{A^2+B^2-\sqrt{2}AB}$

(C) $\sqrt{A^2+B^2}$ (D) $\sqrt{A^2+B^2+\sqrt{2}AB}$

44. Two vectors $\vec{P}$ and $\vec{Q}$ have equal magnitudes. If the magnitude of $\vec{P}+\vec{Q}$ is n times the magnitude of $\vec{P}-\vec{Q}$, then angle between $\vec{P}$ and $\vec{Q}$ is : **[20 July 2021 (Evening)]**

(A) $\cos^{-1}\left(\frac{n-1}{n+1}\right)$ (B) $\sin^{-1}\left(\frac{n-1}{n+1}\right)$

(C) $\cos^{-1}\left(\frac{n^2-1}{n^2+1}\right)$ (D) $\sin^{-1}\left(\frac{n^2-1}{n^2+1}\right)$

45. What will be the projection of vector $\vec{A} = \hat{i}+\hat{j}+\hat{k}$ on vector $\vec{B} = \hat{i}+\hat{j}$? **[22 July 2021 (Evening)]**

(A) $2(\hat{i}+\hat{j}+\hat{k})$ (B) $\sqrt{2}(\hat{i}+\hat{j})$

(C) $(\hat{i}+\hat{j})$ (D) $\sqrt{2}(\hat{i}+\hat{j}+\hat{k})$

46. Three particles P, Q and R moving along the vectors $\vec{A}=\hat{i}+\hat{j}$, $\vec{B}=\hat{j}+\hat{k}$ and $\vec{C}=-\hat{i}+\hat{j}$ respectively. They strike on a point and start to move in different directions. Now particle P is moving normal to the plane which contains vector $\vec{A}$ and $\vec{B}$. Similarly particle Q is moving normal to the plane which contains vector $\vec{A}$ and $\vec{C}$. The angle between the direction of motion of P and Q is $\cos^{-1}\left(\frac{1}{\sqrt{x}}\right)$. Then the value of x is _____ .

[22 July 2021 (Evening)]

47. Match List-I with List-II : **[25 July 2021 (Morning)]**

(a) $\vec{C}-\vec{A}-\vec{B}=0$ (i)

(b) $\vec{A}-\vec{C}-\vec{B}=0$ (ii)

(c) $\vec{B}-\vec{A}-\vec{C}=0$ (iii)

(d) $\vec{A}+\vec{B}=-\vec{C}$ (iv)

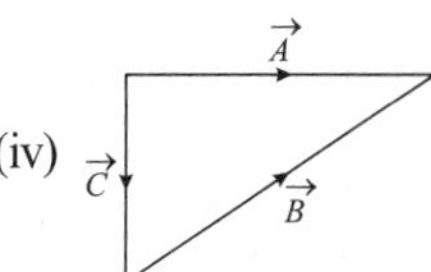

Choose the correct answer from the options given below :

(A) (a)-(iii), (b)-(ii), (c)-(iv), (d)-(i)
(B) (a)-(iv), (b)-(iii), (c)-(i), (d)-(ii)
(C) (a)-(i), (b)-(iv), (c)-(ii), (d)-(iii)
(D) (a)-(iv), (b)-(i), (c)-(iii), (d)-(ii)

48. Two vectors $\vec{X}$ and $\vec{Y}$ have equal magnitude. The magnitude of $(\vec{X}+\vec{Y})$ is n times the magnitude of $(\vec{X}+\vec{Y})$. The angle between $\vec{X}$ and $\vec{Y}$ is: **[25 July 2021 (Evening)]**

(A) $\cos^{-1}\left(\frac{n^2+1}{-n^2-1}\right)$ (B) $\cos^{-1}\left(\frac{n^2-1}{-n^2-1}\right)$

(C) $\cos^{-1}\left(\frac{-n^2-1}{n^2-1}\right)$ (D) $\cos^{-1}\left(\frac{n^2+1}{n^2-1}\right)$

49. A force $\vec{F}=(40\hat{i}+10\hat{j})N$ acts N acts on a body of mass 5 kg. If the body starts from rest, its position vector $\vec{r}$ at time $t=10$ s, will be : **[25 July 2021 (Evening)]**

(A) $(100\hat{i}+100\hat{j})$ (B) $(400\hat{i}+400\hat{j})$m

(C) $(100\hat{i}+400\hat{j})$m (D) $(400\hat{i}+100\hat{j})$m

50. Assertion-A: If A, B, C, D are four points on a semi-circular arc with centre at 'O' such that $|\overrightarrow{AB}|=|\overrightarrow{BC}|=|\overrightarrow{CD}|$, then

$$\overrightarrow{AB}+\overrightarrow{AC}+\overrightarrow{AD}=4\overrightarrow{AO}+\overrightarrow{OB}+\overrightarrow{OC}$$

Reason-R: Polygon law of vector addition yields

$$\overrightarrow{AB}+\overrightarrow{BC}+\overrightarrow{CD}=\overrightarrow{AD}=2\overrightarrow{AO}$$

[27 July 2021 (Morning)]

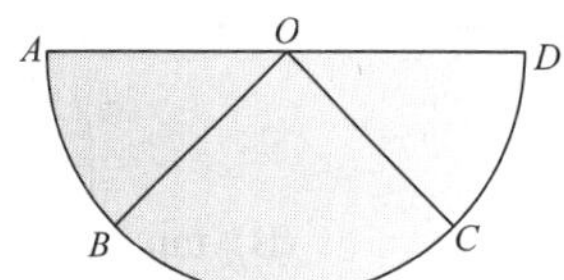

In the light of the above statements, choose the most appropriate answer from the options given below :

(A) A is not correct but R is correct
(B) Both A and R are correct but R is not the correct explanation of A
(C) A is correct but R is not correct
(D) Both A and R are correct and R is the correct explanation of A

51. A swimmer wants to cross a river from point A to point B. Line AB makes an angle of 30° with the flow of river. Magnitude of velocity of the swimmer is same as that of the river. The angle θ with the line AB should be _____°, so that the swimmer reaches point B. **[27 July 2021 (Evening)]**

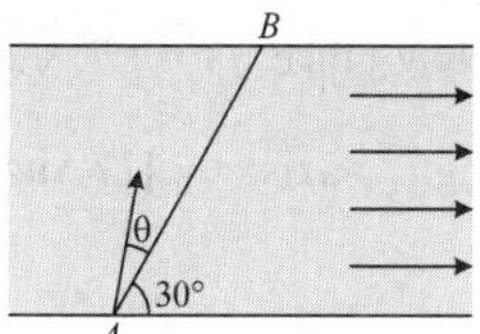

52. The magnitude of vector $\overrightarrow{OA}$, $\overrightarrow{OB}$ and $\overrightarrow{OC}$ in the given figure are equal. The direction of $\overrightarrow{OA}+\overrightarrow{OB}-\overrightarrow{OC}$ with x-axis will be : **[26 Aug. 2021 (Morning)]**

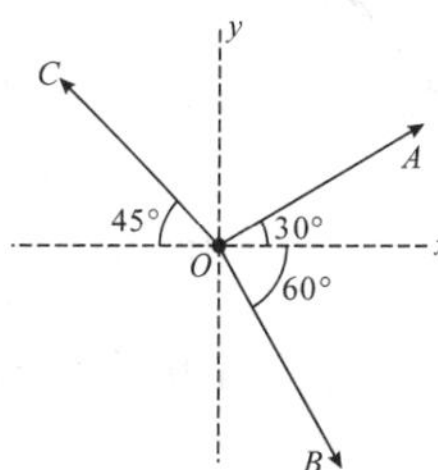

(A) $\tan^{-1}\frac{(1+\sqrt{3}-\sqrt{2})}{(1-\sqrt{3}-\sqrt{2})}$ (B) $\tan^{-1}\frac{(1-\sqrt{3}-\sqrt{2})}{(1+\sqrt{3}+\sqrt{2})}$

(C) $\tan^{-1}\dfrac{(\sqrt{3}-1-\sqrt{2})}{(1+\sqrt{3}+\sqrt{2})}$ (D) $\tan^{-1}\dfrac{(\sqrt{3}-1+\sqrt{2})}{(1+\sqrt{3}-\sqrt{2})}$

53. The angle between vector $(\vec{A})$ and $(\vec{A}-\vec{B})$ is : **[26 Aug. 2021 (Evening)]**

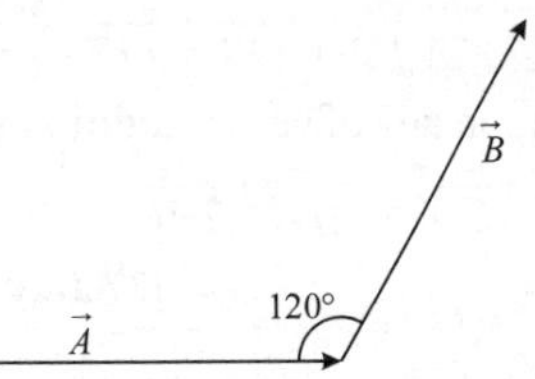

(A) $\tan^{-1}\left(\dfrac{A}{0.7B}\right)$ (B) $\tan^{-1}\left(\dfrac{-\frac{B}{2}}{A-B\frac{\sqrt{3}}{2}}\right)$

(C) $\tan^{-1}\left(\dfrac{\sqrt{3}B}{2A-B}\right)$ (D) $\tan^{-1}\left(\dfrac{B\cos\theta}{A-B\sin\theta}\right)$

54. A bomb is dropped by a fighter plane flying horizontally. To an observer sitting in the plane, the trajectory of the bomb is a : **[26 Aug. 2021 (Evening)]**

(A) Hyperbola
(B) Straight line vertically down the plane
(C) Parabola in a direction opposite to the motion of plane
(D) Parabola in the direction of motion of plane

55. The resultant of these forces $\overline{OP}$, $\overline{OQ}$, $\overline{OR}$, $\overline{OS}$ and $\overline{OT}$ is approximately N. [Take $\sqrt{3}=1.7$, $\sqrt{2}=1.4$, Given $\hat{i}$ and $\hat{j}$ unit vectors along x, y-axis] **[27 Aug. 2021 (Morning)]**

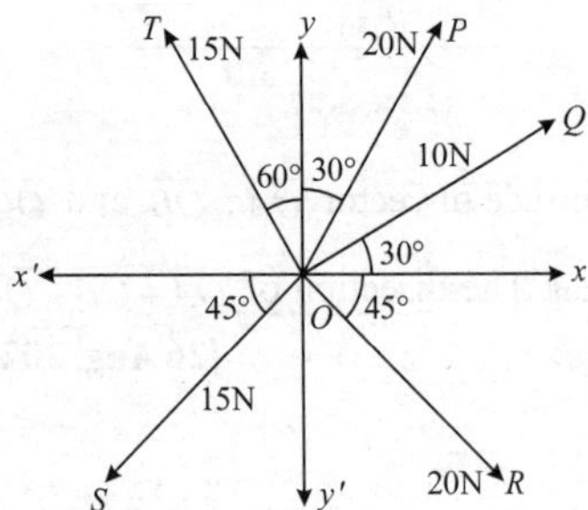

(A) $9.25\hat{i}+5\hat{j}$ (B) $3\hat{i}+15\hat{j}$

(C) $-1.5\hat{i}-15.5\hat{j}$ (D) $2.5\hat{i}-14.5\hat{j}$

56. A player kicks a football with an initial speed of $25\ ms^{-1}$ at an angle of 45° from the ground. What are the maximum height and the time taken by the football to reach at the highest point during motion? (Take $g = 10\ ms^{-2}$) **[27 Aug. 2021 (Evening)]**

(A) $h_{max} = 15.625$ m T = 1.77 s (B) $h_{max} = 3.54$ m T = 0.125 s
(C) $h_{max} = 10$ m T = 2.5 s (D) $h_{max} = 15.625$ m T = 3.54 s

57. A helicopter is flying horizontally with a speed v at an altitude h has to drop a food packet for a man on the ground. What is the distance of helicopter from the man when the food packet is dropped ? **[31 Aug. 2021 (Morning)]**

(A) $\sqrt{\dfrac{2ghv^2+1}{h^2}}$ (B) $\sqrt{2ghv^2+h^2}$

(C) $\sqrt{\dfrac{2gh}{v^2}+h^2}$ (D) $\tan^{-1}\dfrac{(\sqrt{3}-1+\sqrt{2})}{(1+\sqrt{3}-\sqrt{2})}$

58. A car is moving on a plane inclined at 30° to be horizontal with an acceleration of $10\ ms^{-2}$ parallel to the plane upward. A bob is suspended by a string from the roof of car. The angle in degrees which the string makes with the vertical is _____ . **[31 Aug. 2021 (Morning)]**

59. Statement-I: If three forces $\vec{F}_1, \vec{F}_2$ and $\vec{F}_3$ are represented by three sides of triangle and $\vec{F}_1+\vec{F}_2 = -\vec{F}_3$, then three forces are concurrent forces and satisfy the condition for equilibrium.
Statement-II: A triangle made up of three forces $\vec{F}_1, \vec{F}_2$ and $\vec{F}_3$ as its sides taken in the same order, satisfy the condition for translatory equilibrium. In the light of the above statements, choose the most appropriate answer from the options given below. **[31 Aug. 2021 (Evening)]**

(A) Both Statement-I and Statement-II are false
(B) Both Statement-I and Statement-II are true
(C) Statement-I is false but Statement-II is true
(D) Statement-I is true but Statement-II is false

60. Statement-I: Two forces $(\vec{P}+\vec{Q})$ and $(\vec{P}-\vec{Q})$ where $\vec{P}\perp\vec{Q}$, when act at an angle θ_1 each other, the magnitude of their resultant is $\sqrt{3(P^2+Q^2)}$, when they act at an angle θ_2, then magnitude of their resultant becomes $\sqrt{2(P^2+Q^2)}$. This is possible only when $\theta_1 < \theta_2$.
Statement-II: In the situation given above.
$\theta_1 = 60°$ and $\theta_2 = 90°$
In the light of the above statement, choose the most appropriate answer from the options given below : **[31 Aug. 2021 (Evening)]**

(A) Statement-I is true but Statement-II is false
(B) Both Statement-I and Statement-II are true
(C) Statement-I is false but Statement-II is true
(D) Both Statement-I and Statement-II are false

61. The ranges and heights for two projectiles projected with the same initial velocity at angles 42° and 48° with the horizontal are R_1, R_2 and H_1, H_2 respectively. Choose the correct option : **[01 Sep. 2021 (Evening)]**

(A) $R_1 > R_2$ and $H_1 = H_2$ (B) $R_1 = R_2$ and $H_1 < H_2$
(C) $R_1 < R_2$ and $H_1 < H_2$ (D) $R_1 = R_2$ and $H_1 = H_2$

62. A projectile is projected with velocity of 25 m/s at an angle θ with the horizontal. After t seconds its inclination with horizontal becomes zero. If R represents horizontal range of the projectile, the value of θ will be : [use $g = 10$ m/s^2] **[24 June 2022 (Morning)]**

(A) $\frac{1}{2}\sin^{-1}\left(\frac{5t^2}{4R}\right)$ (B) $\frac{1}{2}\sin^{-1}\left(\frac{4R}{5t^2}\right)$

(C) $\tan^{-1}\left(\frac{4t^2}{5R}\right)$ (D) $\cot^{-1}\left(\frac{R}{20t^2}\right)$

63. A body is projected from the ground at an angle of 45° with the horizontal. Its velocity after 2 s is 20 ms^{-1}. The maximum height reached by the body during its motion is ______ m. (Use $g = 10$ ms^{-2}) **[24 June 2022 (Evening)]**

64. A fighter jet is flying horizontally at a certain altitude with a speed of 200 ms^{-1}. When it passes directly overhead an antiaircraft gun, a bullet is fired from the gun, at an angle θ with the horizontal, to hit the jet. If the bullet speed is 400 m/s, the value of θ will be ______°. **[26 June 2022 (Morning)]**

65. A projectile is launched at an angle α with the horizontal with a velocity 20 ms^{-1}. After 10 s, its inclination with horizontal is β. The value of tan β will be : ($g = 10$ ms^{-2}) **[27 June 2022 (Morning)]**

(A) tan α + 5 sec α (B) tan α − 5 sec α
(C) 2 tan α − 5 sec α (D) 2 tan α + 5 sec α

66. Two vectors $\vec{A}$ and $\vec{B}$ have equal magnitudes. If magnitude of $\vec{A}+\vec{B}$ is equal to two times the magnitude of $\vec{A}-\vec{B}$, then the angle between $\vec{A}$ and $\vec{B}$ will be : **[29 June 2022 (Morning)]**

(A) $\sin^{-1}\left(\frac{3}{5}\right)$ (B) $\sin^{-1}\left(\frac{1}{3}\right)$

(C) $\cos^{-1}\left(\frac{3}{5}\right)$ (D) $\cos^{-1}\left(\frac{1}{3}\right)$

67. A person can throw a ball upto a maximum range of 100 m. How high above the ground he can throw the same ball? **[29 June 2022 (Evening)]**

(A) 25 m (B) 50 m
(C) 100 m (D) 200 m

68. A ball is projected from the ground with a speed 15 ms^{-1} at an angle θ with horizontal so that its range and maximum height are equal, then tan θ will be equal to : **[25 July 2022 (Evening)]**

(A) $\frac{1}{4}$ (B) $\frac{1}{2}$
(C) 2 (D) 4

69. Two projectile thrown at 30° and 45° with the horizontal respectively, reach the maximum height in same time. The ratio of their initial velocities is : **[26 July 2022 (Morning)]**

(A) $1:\sqrt{2}$ (B) $2:1$
(C) $\sqrt{2}:1$ (D) $1:2$

70. If the initial velocity in horizontal direction of a projectile is unit vector $\hat{i}$ and the equation of trajectory is $y = 5x(1-x)$. The y component vector of the initial velocity is ______ $\hat{j}$. (Take $g = 10$ m/s^2) **[26 July 2022 (Morning)]**

71. Two projectiles are thrown with same initial velocity making an angle of 45° and 30° with the horizontal respectively. The ratio of their respective ranges will be : **[26 July 2022 (Evening)]**

(A) $1:\sqrt{2}$ (B) $\sqrt{2}:1$
(C) $2:\sqrt{3}$ (D) $\sqrt{3}:2$

72. If $\vec{A} = (2\hat{i}+3\hat{j}-\hat{k})m$ and $\vec{B} = (\hat{i}+2\hat{j}+2\hat{k})m$. The magnitude of component of vector $\vec{A}$ along vector $\vec{B}$ will be ______ m. **[26 July 2022 (Evening)]**

73. A ball of mass m is thrown vertically upward. Another ball of mass $2m$ is thrown at an angle θ with the vertical. Both the balls stay in air for the same period of time. The ratio of the heights attained by the two balls respectively is $\frac{1}{x}$. The value of x is ______. **[27 July 2022 (Morning)]**

74. If the projection of $2\hat{i}+4\hat{j}-2\hat{k}$ on $\hat{i}+2\hat{j}+\alpha\hat{k}$ is zero. Then, the value of α will be. **[28 July 2022 (Morning)]**

75. A ball is projected with kinetic energy E, at an angle of 60° to the horizontal. The kinetic energy of this ball at the highest point of its flight will become : **[29 July 2022 (Morning)]**

(A) Zero (B) $\frac{E}{2}$

(C) $\frac{E}{4}$ (D) E

76. An object is projected in the air with initial velocity u at an angle θ. The projectile motion is such that the horizontal range R, is maximum. Another object is projected in the air with a horizontal range half of the range of first object. The initial velocity remains same in both the case. The value of the angle of projection, at which the second object is projected, will be _____ degree. **[29 July 2022 (Morning)]**

77. The maximum vertical height to which a man can throw a ball is 136 m. The maximum horizontal distance upto which he can throw the same ball is : **[24 Jan. 2023 (Morning)]**

(A) 192 m (B) 136 m
(C) 272 m (D) 68 m

78. Vectors $a\hat{i}+b\hat{j}+\hat{k}$ and $2\hat{i}-3\hat{j}+4\hat{k}$ are perpendicular to each other when $3a+2b=7$, then ratio of a to b is $\frac{x}{2}$. The value of x is _____. **[24 Jan. 2023 (Morning)]**

79. If two vectors $\vec{P}=\hat{i}+2m\hat{j}+m\hat{k}$ and $\vec{Q}=4\hat{i}-2\hat{j}+m\hat{k}$ are perpendicular to each other. Then, the value of m will be: **[24 Jan. 2023 (Evening)]**

(A) 1 (B) −1
(C) −3 (D) 2

80. If $\vec{P}=3\hat{i}+\sqrt{3}\hat{j}+2\hat{k}$ and $\vec{Q}=4\hat{i}+\sqrt{3}\hat{j}+2.5\hat{k}$ then. The unit vector in the direction of $\vec{P}\times\vec{Q}$ is $\frac{1}{x}(\sqrt{3}\hat{i}+\hat{j}-2\sqrt{3}\hat{k})$. The value of x is. **[25 Jan. 2023 (Morning)]**

81. Two objects are projected with same velocity u however at different angles α and β with the horizontal. If $\alpha+\beta=90°$, the ratio of horizontal range of the first object to the 2nd object will be : **[25 Jan. 2023 (Evening)]**

(A) 4 : 1 (B) 2 : 1
(C) 1 : 2 (D) 1 : 1

82. A stone is projected at angle 30° to the horizontal. The ratio of kinetic energy of the stone at point of projection to its kinetic energy at the highest point of flight will be : **[29 Jan. 2023 (Morning)]**

(A) 1 : 2 (B) 1 : 4
(C) 4 : 1 (D) 4 : 3

83. A ball of mass 200 g rests on a vertical post of height 20 m. A bullet of mass 10 g, travelling in horizontal direction, hits the centre of the ball. After collision both travels independently. The ball hits the ground at a distance 30 m and the bullet at a distance of 120 m from the foot of the post. The value of initial velocity of the bullet will be : (If $g=10\ m/s^2$) **[30 Jan. 2023 (Morning)]**

(A) 120 m/s (B) 60 m/s
(C) 400 m/s (D) 360 m/s

84. Two bodies are projected from ground with same speeds 40 ms^{-1} at two different angles with respect to horizontal. The bodies were found to have same range. If one of the body was projected at an angle of 60°, with horizontal then sum of the maximum heights, attained by the two projectiles, is _____ m. (Given $g=10\ ms^{-2}$) **[31 Jan. 2023 (Evening)]**

85. A child stands on the edge of the cliff 10 m above the ground and throws a stone horizontally with an initial speed of 5 ms^{-1}. Neglecting the air resistance, the speed with which the stone hits the ground will be _____ ms^{-1} (given, $g=10\ ms^{-2}$) : **[1 Feb. 2023 (Morning)]**

(A) 20 (B) 15
(C) 30 (D) 25

86. For a body projected at an angle with the horizontal from the ground, choose the correct statement : **[1 Feb. 2023 (Evening)]**

(A) Gravitational potential energy is maximum at the highest point
(B) The horizontal component of velocity is zero at highest point
(C) The vertical component of momentum is maximum at the highest point
(D) The kinetic energy ($K.E.$) is zero at the highest point of projectile motion

87. Given below are two statements: one is labelled as **Assertion-A** and the other is labelled as **Reason-R**.

Assertion-A: When a body is projected at an angle 45°, it's range is maximum.

Reason-R : For maximum range, the value of sin 2θ should be equal to one.

In the light of the above statements, choose the **correct** answer from the options given below : **[6 April 2023 (Morning)]**

(A) Both **A** and **R** are correct but **R** is **NOT** the correct explanation of **A**.
(B) Both **A** and **R** are correct **R** is the correct explanation of **A**
(C) **A** is true but **R** is false
(D) **A** is false but **R** is true

88. Two projectiles A and B are thrown with initial velocities of

40 m/s and 60 m/s at angles 30° and 60° with the horizontal respectively. The ratio of their ranges respectively is:
($g = 10$ m/s^2) **[8 April 2023 (Morning)]**

(A) $\sqrt{3}:2$ (B) $2:\sqrt{3}$
(C) 1 : 1 (D) 4 : 9

89. At any instant the velocity of a particle of mass 500g is $(2t\hat{i}+3t^2\hat{j})$ ms^{-1}. If the force acting on the particle at $t = 1$ s is $(\hat{i}+x\hat{j})$ N. Then the value of x will be :
[8 April 2023 (Morning)]

(A) 3 (B) 4
(C) 6 (D) 2

90. Two projectiles are projected at 30° and 60° with the horizontal with the same speed. The ratio of the maximum height attained by the two projectiles respectively is:
[10 April 2023 (Evening)]

(A) $2:\sqrt{3}$ (B) $\sqrt{3}:1$
(C) 1:3 (D) $1:\sqrt{3}$

91. A projectile fired at 30° to the ground is observed to be at same height at time 3 s and 5 s after projection, during its flight. The speed of projection of the projectile is______ ms^{-1}.
[11 April 2023 (Morning)]

92. A projectile is projected at 30° from horizontal with initial velocity 40 ms^{-1}. The velocity of the projectile at $t = 2$ s from the start will be : (Given $g = 10$ m/s^2) **[11 April 2023 (Evening)]**

(A) $20\sqrt{3}$ ms^{-1} (B) $40\sqrt{3}$ ms^{-1}
(C) 20 ms^{-1} (D) Zero

93. When vector $\vec{A} = 2\hat{i}+3\hat{j}+2\hat{k}$ is subtracted from vector $\vec{B}$, it gives a vector equal to $2\hat{j}$. Then the magnitude of vector $\vec{B}$ will be : **[11 April 2023 (Evening)]**

(A) $\sqrt{13}$ (B) 3
(C) $\sqrt{6}$ (D) $\sqrt{5}$

94. The position vector of a particle related to time t is given by $\vec{r} = (10t\hat{i}+15t^2\hat{j}+7\hat{k})$ m. The direction of net force experienced by the particle is : **[15 April 2023 (Morning)]**

(A) Positive y-axis (B) Positive x-axis
(C) Positive z-axis (D) In x-y plane

95. A vector in x-y plane makes an angle of 30° with y-axis The magnitude of y-component of vector is $2\sqrt{3}$. The magnitude of x-component of the vector will be : **[15 April 2023 (Morning)]**

(A) $\frac{1}{\sqrt{3}}$ (B) 6
(C) $\sqrt{3}$ (D) 2

* * * * *

CHAPTER 3

MOTION IN A PLANE

3.2 RELATIVE MOTION

1. A body is at rest at $x = 0$. At $t = 0$, it starts moving in the positive x-direction with a constant acceleration. At the same instant another body passes through $x = 0$ moving in the positive x-direction with a constant speed. The position of the first body is given by $x_1(t)$ after time 't' and that of second body by $x_2(t)$ after the same time interval. Which of the following graphs correctly describes $(x_1 - x_2)$ as a function of time 't' ? **[2008]**

(A) 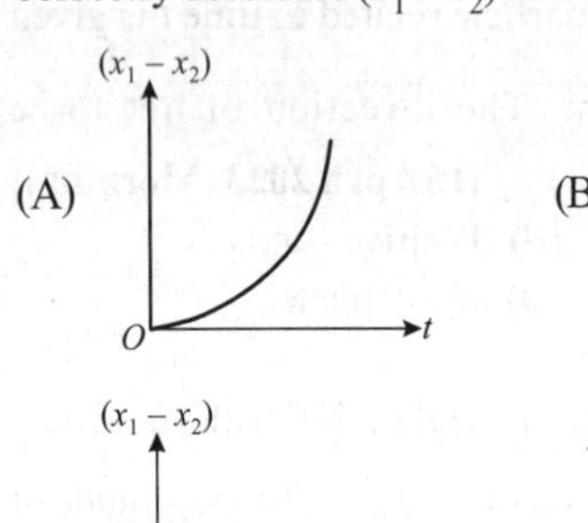

(B) 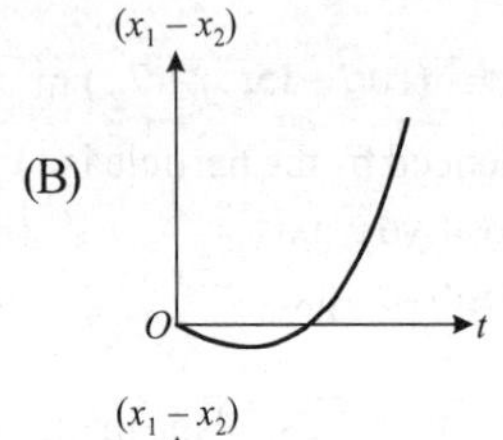

(C) 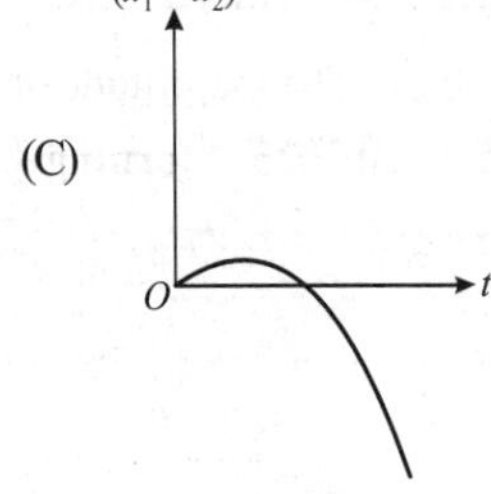

(D) 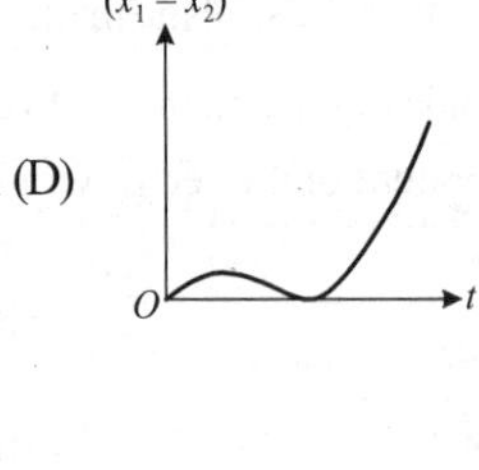

2. Two fixed frictionless inclined planes making an angle 30° and 60° with the vertical are shown in the figure. Two blocks A and B are placed on the two planes. What is the relative vertical acceleration of A with respect to B? **[2010]**

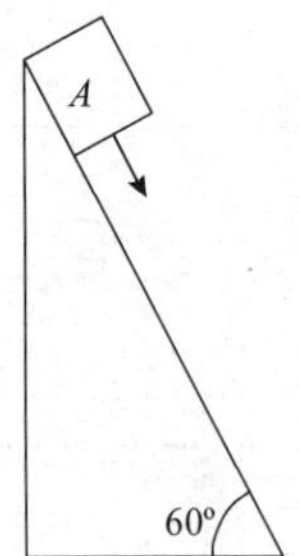

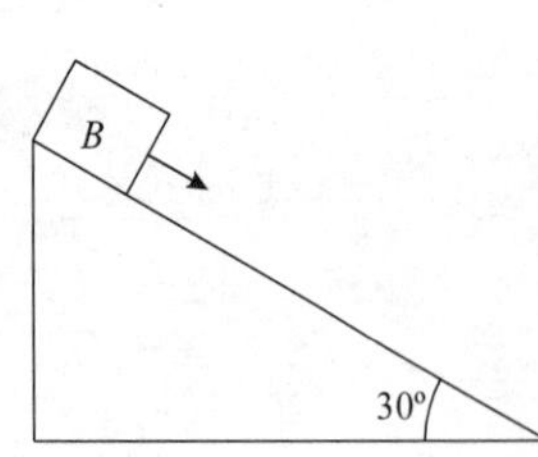

(A) 4.9 ms^{-2} in horizontal direction
(B) 9.8 ms^{-2} in vertical direction
(C) Zero
(D) 4.9 ms^{-2} in vertical direction

3. Two stones are thrown up simultaneously from the edge of a cliff 240 m high with initial speed of 10 m/s and 40 m/s respectively. Which of the following graph best represents the time variation of relative position of the second stone with respect of the first ?

(Assume stones do not rebound after hitting the ground and neglect air resistance, take $g = 10$ m/s^2)

(The figure are schematic and not drawn to scale) **[2015]**

(A) 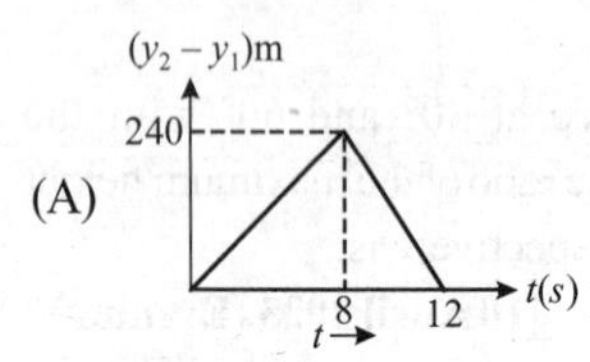

(B) 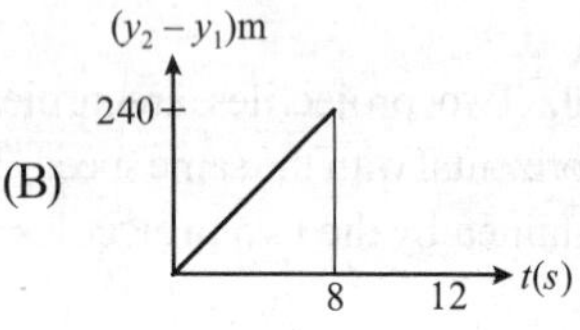

(C) 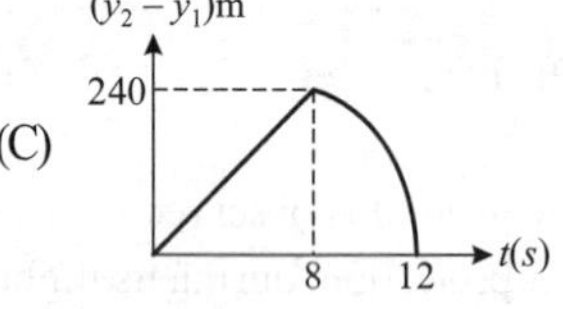

(D) 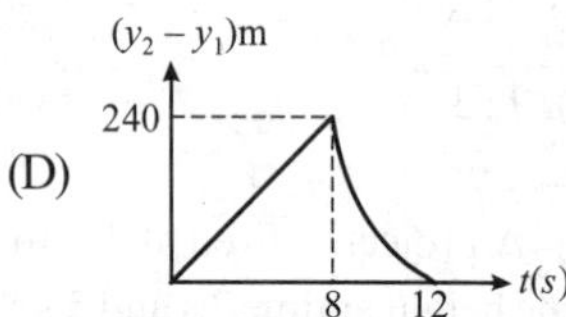

4. A car is standing 200 m behind a bus, which is also at rest. The two start moving at the same instant but with different forward accelerations. The bus has acceleration 2 m/s^2 and the car has acceleration 4 m/s^2. The car will catch up with the bus after a time of: **[2017 (Online)]**

(A) $\sqrt{100}$ s (B) $\sqrt{120}$ s
(C) $10\sqrt{2}$ s (D) 15 s

5. Ship A is sailing towards north-east with velocity $\vec{v} = 30\hat{i} + 50\hat{j}$ km/hr where $\hat{i}$ points east and $\hat{j}$, north. Ship B is at a distance of 80 km east and 150 km north of Ship A and is sailing towards west at 10 km/hr. A will be at minimum distance from B in : **[8 April 2019 (Morning)]**

(A) 4.2 hrs. (B) 2.2 hrs.
(C) 3.2 hrs. (D) 2.6 hrs.

6. The stream of a river is flowing with a speed of 2 km/h. A swimmer can swim at a speed of 4 km/h. What should be the direction of the swimmer with respect to the flow of the river to cross the river straight ? **[9 April 2019 (Morning)]**

(A) 60° (B) 150°
(C) 90° (D) 120°

7. Train A and train B are running on parallel tracks in the opposite directions with speeds of 36 km/hour and 72 km/hour,

respectively. A person is walking in train A in the direction opposite to its motion with a speed of 1.8 km/hr. Speed (in ms^{-1}) of this person as observed from train B will be close to : (take the distance between the tracks as negligible) :

[2 Sept. 2020 (Morning)]

(A) 30.5 ms^{-1} (B) 29.5 ms^{-1}
(C) 31.5 ms^{-1} (D) 28.5 ms^{-1}

8. When a car is at rest, its driver sees rain drops falling on it vertically. When driving the car with speed v, he sees that rain drops are coming at an angle 60° from the horizontal. On further increasing the speed of the car to $(1 + \beta)v$, this angle changes to 45°. The value of b is close to : **[6 Sept. 2020 (Evening)]**

(A) 0.41 (B) 0.50
(C) 0.37 (D) 0.73

9. A swimmer can swim with velocity of 12 km/h in still water. Water flowing in a river has velocity 6 km/h. The direction with respect to the direction of flow of river water he should swim in order to reach the point on the other bank just opposite to his starting point is _____°. (Round off to the Nearest Integer) (find the angle in degree) **[16 March 2021 (Evening)]**

10. A butterfly is flying with a velocity $4\sqrt{2}$ m/s in North-East direction. Wind is slowly blowing at 1 m/s from North to South. The resultant displacement of the butterfly in 3 seconds is :

[20 July 2021 (Morning)]

(A) $12\sqrt{2}$ m (B) 15 m
(C) 3 m (D) 20 m

11. A boy reaches the airport and finds that the escalator is not working. He walks up the stationary escalator in time t_1. If he remains stationary on a moving escalator then the escalator takes him up in time t_2. The time taken by him to walk up on the moving escalator will be : **[20 July 2021 (Evening)]**

(A) $t_2 - t_1$ (B) $\frac{t_1 t_2}{t_2 - t_1}$

(C) $\frac{t_1 t_2}{t_2 + t_1}$ (D) $\frac{t_1 + t_2}{2}$

12. A girl standing on road holds her umbrella at 45° with the vertical to keep the rain away. If she starts running without umbrella with a speed of $15\sqrt{2}$ km h^{-1}, the rain drops hit her head vertically. The speed of rain drops with respect to the moving girl is : **[27 June 2022 (Morning)]**

(A) 30 km h^{-1} (B) $\frac{25}{\sqrt{2}}$ km h^{-1}

(C) $\frac{30}{\sqrt{2}}$ km h^{-1} (D) 25 km h^{-1}

13. The initial speed of a projectile fired from ground is u. At the highest point during its motion, the speed of projectile is $\frac{\sqrt{3}}{2}u$. The time of flight of the projectile is :

[31 Jan. 2023 (Morning)]

(A) $\frac{u}{2g}$ (B) $\frac{u}{g}$

(C) $\frac{2u}{g}$ (D) $\frac{\sqrt{3}u}{g}$

14. The speed of a swimmer is 4 km h^{-1} in still water. If the swimmer makes his strokes normal to the flow of river of width 1 km, he reaches a point 750 m down the stream on the opposite bank. The speed of the river water is _____ km h^{-1}.

[31 Jan. 2023 (Morning)]

* * * * *

CHAPTER

4 LAWS OF MOTION

4.1 NEWTON'S LAWS OF MOTION

1. When forces F_1, F_2, F_3 are acting on a particle of mass m such that F_2 and F_3 are mutually perpendicular, then the particle remains stationary. If the force F_1 is now removed then the acceleration of the particle is : **[2002]**

(A) F_1/m (B) F_2F_3/mF_1

(C) $(F_2 - F_3)/m$ (D) F_2/m

2. Three identical blocks of masses $m = 2$ kg are drawn by a force F with an acceleration of $0.6\ ms^{-2}$ on a frictionless surface, then what is the tension (in N) in the string between the blocks B and C : **[2002]**

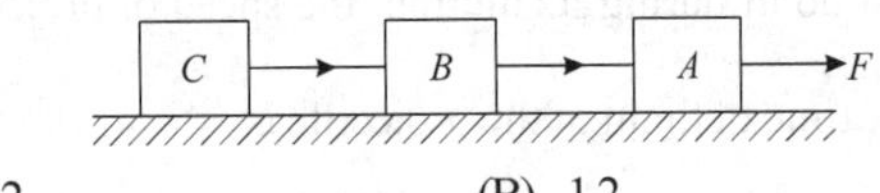

(A) 9.2 (B) 1.2

(C) 4 (D) 9.8

3. One end of massless rope, which passes over a massless and frictionless pulley P is tied to a hook C while the other end is free. Maximum tension that the rope can bear is 840 N. With what value of maximum safe acceleration (in ms^{-2}) can a man of 60 kg climb on the rope? **[2002]**

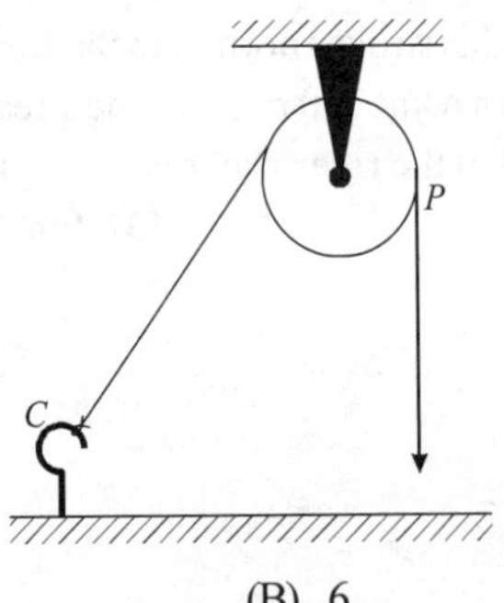

(A) 16 (B) 6

(C) 4 (D) 8

4. A solid sphere, a hollow sphere and a ring are released from top of an inclined plane (frictionless) so that they slide down the plane. Then maximum acceleration down the plane is for (no rolling) : **[2002]**

(A) Solid sphere (B) Hollows sphere

(C) Ring (D) All same

5. A lift is moving down with acceleration a. A man in the lift drops a ball inside the lift. The acceleration of the ball as observed by the man in the lift and a man standing stationary on the ground are respectively : **[2002]**

(A) g, g (B) $g - a, g - a$

(C) $g - a, g$ (D) a, g

6. Two forces are such that the sum of their magnitudes is 18 N and their resultant is 12 N which is perpendicular to the smaller force. Then the magnitudes of the forces are : **[2002]**

(A) 12 N, 6 N (B) 13 N, 5 N

(C) 10 N, 8 N (D) 16 N, 2 N

7. A light string passing over a smooth light pulley connects two blocks of masses m_1 and m_2 (vertically). If the acceleration of the system is $g/8$, then the ratio of the masses is : **[2002]**

(A) 8 : 1 (B) 9 : 7

(C) 4 : 3 (D) 5 : 3

8. A light spring balance hangs from the hook of the other light spring balance and a block of mass M kg hangs from the former one. Then the true statement about the scale reading is: **[2003]**

(A) Both the scales read M kg each

(B) The scale of the lower one reads M kg and of the upper one zero

(C) The reading of the two scales can be anything but the sum of the readings will be M kg

(D) Both the scales read $M/2$ kg

9. A spring balance is attached to the ceiling of a lift. A man hangs his bag on the spring and the spring reads 49 N, when the lift is stationary. If the lift moves downward with an acceleration of $5\ m/s^2$, the reading of the spring balance will be : **[2003]**

(A) 24 N (B) 74 N

(C) 15 N (D) 49 N

10. A rocket which has a mass of 3.5×10^4 kg is blasted upwards with an initial acceleration of $10\ m/s^2$. Then the initial thrust of the blast is : **[2003]**

(A) 3.5×10^5 N (B) 7.0×10^5 N

(C) 14.0×10^5 N (D) 1.75×10^5 N

11. Three forces start acting simultaneously on a particle moving with velocity $\vec{v}$. These forces are represented in magnitude and direction by the three sides of a triangle ABC (as shown). The particle will now move with velocity : **[2003]**

(A) Less than $\vec{v}$

(B) Greater than $\vec{v}$

(C) $|v|$ in the direction of largest force BC

(D) $\vec{v}$, remaining unchanged

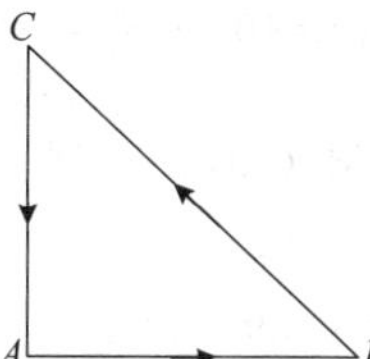

12. A block of mass M is pulled along a horizontal frictionless surface by a rope of mass m. If a force P is applied at the free end of the rope, the force exerted by the rope on the block is : **[2003]**

(A) $\frac{Pm}{M+m}$ (B) $\frac{Pm}{M-m}$

(C) P (D) $\frac{PM}{M+m}$

13. Two masses $m_1 = 5$ kg and $m_2 = 4.8$ kg tied to a string are hanging over a light frictionless pulley. What is the acceleration of the masses when they are free to move ? ($g = 9.8$ m/s^2) **[2004]**

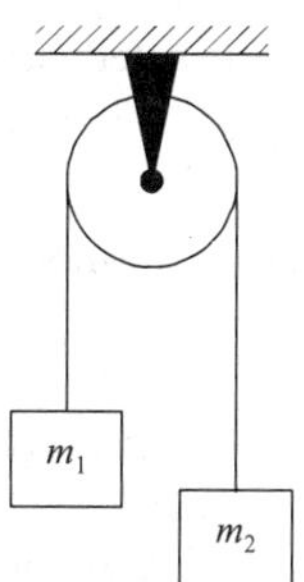

(A) 0.2 m/s^2 (B) 9.8 m/s^2

(C) 5 m/s^2 (D) 4.8 m/s^2

14. A machine gun fires a bullet of mass 40 g with a velocity 1200 ms^{-1}. The man holding it, can exert maximum force of 144 N on the gun. How many bullets can he fire per second at the most? **[2004]**

(A) One (B) Four

(C) Two (D) Three

15. A block is kept on a frictionless inclined surface with angle of inclination α. The incline is given an acceleration a to keep the block stationary. Then a is equal to : **[2005]**

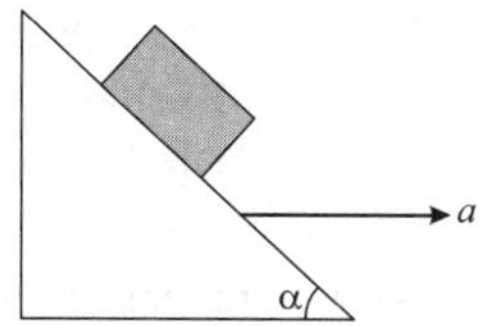

(A) $g/\tan\alpha$ (B) $g \operatorname{cosec} \alpha$

(C) g (D) $g \tan\alpha$

16. A player caught a cricket ball of mass 150 g moving at a rate of 20 m/s. If the catching process is completed in 0.1 s, the force of the blow exerted by the ball on the hand of the player is equal to : **[2006]**

(A) 150 N (B) 3 N

(C) 30 N (D) 300 N

17. A ball of mass 0.2 kg is thrown vertically upwards by applying a force by hand. If the hand moves 0.2 m while applying the force and the ball goes upto 2 m height further, find the magnitude of the force. (Consider $g = 10$ m/s^2) **[2006]**

(A) 4 N (B) 16 N

(C) 20 N (D) 22 N

18. A block of mass m is connected to another block of mass M by a spring (massless) of spring constant k. The blocks are kept on a smooth horizontal plane. Initially the blocks are at rest and the spring is unstretched. Then a constant force F starts acting on the block of mass M to pull it. Find the force on the block of mass m : **[2007]**

(A) $\frac{mF}{M}$ (B) $\frac{(M+m)F}{m}$

(C) $\frac{mF}{(m+M)}$ (D) $\frac{MF}{(m+M)}$

19. If a spring of stiffness 'k' is cut into two parts 'A' and 'B' of length $l_A : l_B = 2:3$, then the stiffness of spring 'A' is given by : **[2011]**

(A) $\frac{5}{2}k$ (B) $\frac{3k}{5}$

(C) $\frac{2k}{5}$ (D) k

20. A particle of mass m is at rest at the origin at time $t = 0$. It is subjected to a force $F(t) = F_0 e^{-bt}$ in the x-direction. Its speed $v(t)$ is depicted by which of the following curves ? **[2012]**

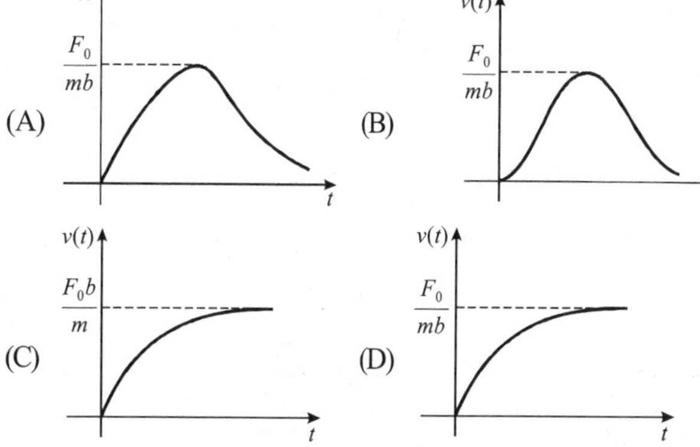

21. This question has Statement-I and Statement-II. Of the four choices given after the Statements, choose the one that best describes the two Statements. **[2012 (Online)]**

Statement-I: If you push on a cart being pulled by a horse so that it does not move, the cart pushes you back with an equal and opposite force :

Statement-II: The cart does not move because the force described in Statement-I cancel each other.

(A) Statement-I is true, Statement-II is true, Statement-II is the correct explanation of Statement-I

(B) Statement-I is false, Statement-II is true

(C) Statement-I is true, Statement-II is false

(D) Statement-I is true, Statement-II is true, Statement-II is not the correct explanation of Statement-I.

22. Two blocks of mass $M_1 = 20$ kg and $M_2 = 12$ kg are connected by a metal rod of mass 8 kg. The system is pulled vertically up by applying a force of 480 N as shown. The tension at the mid-point of the rod is : **[2013 (Online)]**

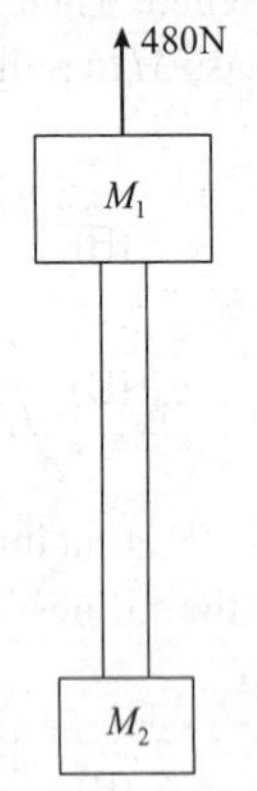

(A) 144 N (B) 96 N
(C) 240 N (D) 192 N

23. A uniform sphere of weight W and radius 5 cm is being held by a string as shown in the figure. The tension in the string will be : **[2013 (Online)]**

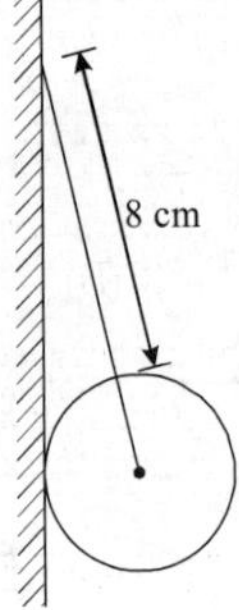

(A) $12\frac{W}{5}$ (B) $5\frac{W}{12}$
(C) $13\frac{W}{5}$ (D) $13\frac{W}{12}$

24. A body of mass 5 kg under the action of constant force $\vec{F} = F_x\hat{i} + F_y\hat{j}$ has velocity at $t = 0$ s as $\vec{v} = (6\hat{i} - 2\hat{j})$ m/s and at $t = 10$ s as $\vec{v} = +6\hat{j}$ m/s. The force $\vec{F}$ is : **[2014 (Online)]**

(A) $(-3\hat{j} + 4\hat{j})$ N (B) $\left(-\frac{3}{5}\hat{i} + \frac{4}{5}\hat{j}\right)$N

(C) $(3\hat{i} - 4\hat{j})$ N (D) $\left(\frac{3}{5}\hat{i} - \frac{4}{5}\hat{j}\right)$N

25. A heavy box is to dragged along a rough horizontal floor. To do so, person A pushes it at an angle 30° from the horizontal and requires a minimum force F_A, while person B pulls the box at an angle 60° from the horizontal and needs minimum force F_B. If the coefficient of friction between the box and the floor is $\frac{\sqrt{3}}{5}$, the ratio $\frac{F_A}{F_B}$ is : **[2014 (Online)]**

(A) $\sqrt{3}$ (B) $\frac{5}{\sqrt{3}}$

(C) $\sqrt{\frac{3}{2}}$ (D) $\frac{2}{\sqrt{3}}$

26. A particle of mass m is acted upon by a force F given by the empirical law $F = \frac{R}{t^2} v(t)$. If this law is to be tested experimentally by observing the motion starting from rest, the best way is to plot : **[2016 (Online)]**

(A) log $v(t)$ against $\frac{1}{t}$ (B) $v(t)$ against t^2

(C) log $v(t)$ against $\frac{1}{t^2}$ (D) log $v(t)$ against t

27. A rocket is fired vertically from the earth with an acceleration of 2 g, where g is the gravitational acceleration. On an inclined plane inside the rocket, making an angle θ with the horizontal, a point object of mass m is kept. The minimum coefficient of friction μ_{min} between the mass and the inclined surface such that the mass does not move is : **[2016 (Online)]**

(A) tan 2θ (B) tan θ
(C) 3 tan θ (D) 2 tan θ

28. A mass of 10 kg is suspended vertically by a rope from the roof. When a horizontal force is applied on the rope at some point, the rope deviated at an angle of 45° at the roof point. If the suspended mass is at equilibrium, the magnitude of the force applied is ($g = 10$ ms^{-2}) : **[9 Jan. 2019 (Evening)]**

(A) 200 N (B) 100 N
(C) 140 N (D) 70 N

29. A particle of mass m is moving in a straight line with momentum p. Starting at time $t = 0$, a force $F = kt$ acts in the same direction on the moving particle during time interval T so that its momentum changes from p to $3p$. Here k is a constant.

The value of T is : **[11 Jan. 2019 (Evening)]**

(A) $2\sqrt{\frac{p}{k}}$ (B) $\sqrt{\frac{2p}{k}}$

(C) $\sqrt{\frac{2k}{p}}$ (D) $2\sqrt{\frac{k}{p}}$

30. A mass of 10 kg is suspended by a rope of length 4 m, from the ceiling. A force F is applied horizontally at the mid-point of the rope such that the top half of the rope makes an angle of 45° with the vertical. Then F equals: (Take $g = 10\text{ ms}^{-2}$ and the rope to be massless) **[7 Jan. 2020 (Evening)]**

(A) 100 N (B) 90 N
(C) 75 N (D) 70 N

31. The sum of two forces $\vec{P}$ and $\vec{Q}$ is $\vec{R}$ such that $|\vec{R}| = |\vec{P}|$. The angle θ (in degrees) that the resultant of $2\vec{P}$ and $\vec{Q}$ will make with $\vec{Q}$ is, _____. **[7 Jan. 2020 (Evening)]**

32. A small ball of mass is thrown upward with velocity u from the ground. The ball experiences a resistive force mkv^2 where v is its speed. The maximum height attained by the ball is : **[4 Sept. 2020 (Evening)]**

(A) $\frac{1}{2k}\tan^{-1}\frac{ku^2}{g}en$ (B) $\frac{1}{2k}\ln\left(1+\frac{ku^2}{g}\right)$

(C) $\frac{1}{k}\tan^{-1}\frac{ku^2}{2g}$ (D) $\frac{1}{k}\ln\left(1+\frac{ku^2}{2g}\right)$

33. A spaceship in space sweeps stationary interplanetary dust. As a result, its mass increases at a rate $\frac{dM(t)}{dt} = bv^2(t)$, where $v(t)$ is its instantaneous velocity. The instantaneous acceleration of the satellite is : **[5 Sept. 2020 (Evening)]**

(A) $-\frac{2bv^3}{M(t)}$ (B) $-\frac{bv^3}{2M(t)}$

(C) $-bv^3(t)$ (D) $-\frac{bv^3}{M(t)}$

34. A particle is projected with velocity v_0 along x-axis. A damping force is acting on the particle which is proportional to the square of the distance from the origin i.e. $ma = -\alpha x^2$. The distance at which the particle stops : **[24 Feb. 2021 (Evening)]**

(A) $\left(\frac{2v_0^2}{3\alpha}\right)^{1/2}$ (B) $\left(\frac{3v_0^2}{2\alpha}\right)^{1/3}$

(C) $\left(\frac{2v_0}{3\alpha}\right)^{1/3}$ (D) $\left(\frac{3v_0^2}{2\alpha}\right)^{1/2}$

35. A person standing on a spring balance inside a stationary lift measures 60 kg. The weight of that person if the lift descends with uniform downward acceleration of 1.8 m/s^2 will be _____ N. [$g = 10$ m/s^2]. **[26 Feb. 2021 (Morning)]**

36. A boy pushes a box of mass 2 kg with a force $\vec{F} = (20\hat{i} + 10\hat{j})$N on a frictionless surface. If the box was initially at rest, then _____ m is displacement along the x-axis after 10 s. **[26 Feb. 2021 (Morning)]**

37. A modern grand-prix racing car of mass m is travelling on a flat track in a circular arc of radius R with a speed v. If the coefficient of static friction between the tyres and the track is μ_s, then the magnitude of negative lift F_L acting downwards on the car is : (Assume forces on the four tyres are identical and g = acceleration due to gravity) **[17 March 2021 (Morning)]**

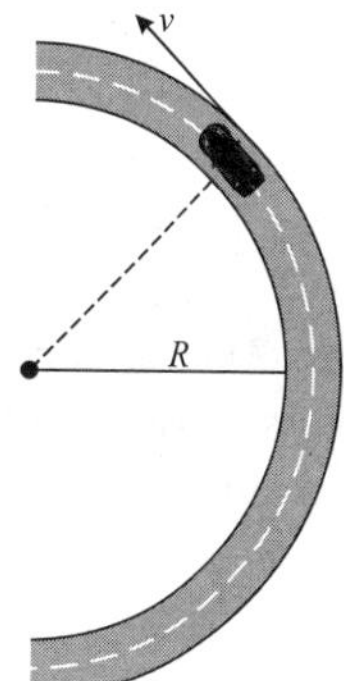

(A) $m\left(\frac{v^2}{\mu_s R} + g\right)$ (B) $m\left(\frac{v^2}{\mu_s R} - g\right)$

(C) $m\left(g - \frac{v^2}{\mu_s R}\right)$ (D) $-m\left(g + \frac{v^2}{\mu_s R}\right)$

38. A boy of mass 4 kg is standing on a piece of wood having mass 5 kg. If the coefficient of friction between the wood and the floor is 0.5, the maximum force that the boy can exert on the rope so that the piece of wood does not move from its place is _____ N. (Round off to the Nearest Integer) [Take $g = 10\text{ ms}^{-2}$] **[17 March 2021 (Evening)]**

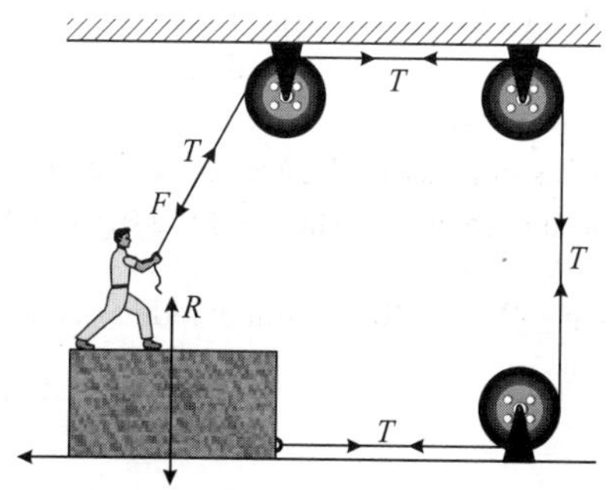

39. A bullet of mass 0.1 kg is fired on a wooden block to pierce through it, but it stops after moving a distance of 50 cm into it. If the velocity of bullet before hitting the wood is 10 m/s and it slows down with uniform deceleration, then the magnitude of effective retarding force on the bullet is 'x' N. The value of 'x' to the nearest integer is_____. **[18 March 2021 (Morning)]**

40. A steel block of 10 kg rests on horizontal floor as shown. When three iron cylinders are placed on its as shown, the block and cylinders go down with an acceleration 0.2 m/s^2. Then normal reaction R' by the floor if mass of the iron cylinders are equal and of 20 kg each, is _____ N. (Take $g = 10$ m/s^2 and $\mu_s = 0.2$) **[20 July 2021 (Morning)]**

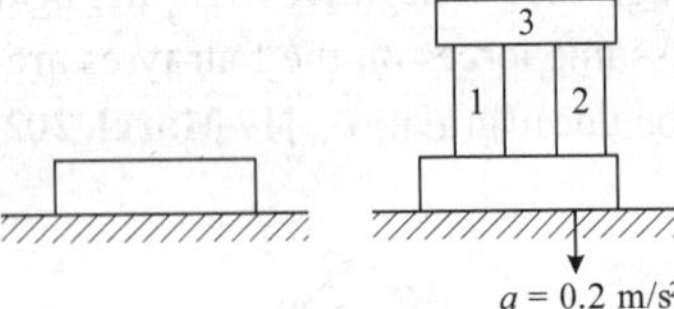

(A) 686 (B) 714
(C) 684 (D) 716

41. A particle of mass M originally at rest is subjected to a force whose direction is constant but magnitude varies with time according to the relation

$$F = F_0\left[1 - \left(\frac{t-T}{T}\right)^2\right]$$

Where F_0 and T are constants. The force acts only for the time interval $2T$. The velocity v of the particle after time $2T$ is : **[27 July 2021 (Evening)]**

(A) $\frac{2F_0T}{M}$ (B) $\frac{F_0T}{2M}$
(C) $\frac{4F_0T}{3M}$ (D) $\frac{F_0T}{3M}$

42. The initial mass of a rocket is 1000 kg. Calculate at what rate the fuel should be burnt so that rocket is given an acceleration of 20 ms^{-2}. The gases come out at a relative speed of 500 ms^{-1} with respect to the rocket : [Use $g = 10$ m/s^2] **[26 Aug. 2021 (Morning)]**

(A) 6.0×10^2 kg s^{-1} (B) 500 kg s^{-1}
(C) 10 kg s^{-1} (D) 60 kg s^{-1}

43. Two blocks of masses 3 kg are connected by a metal wire going over a smooth pulley. The breaking stress of the metal is $\frac{24}{\pi} \times 10^2$ Nm^{-2}. What is the minimum radius of the wire ? (Take $g = 10$ ms^{-2}) **[26 Aug. 2021 (Evening)]**

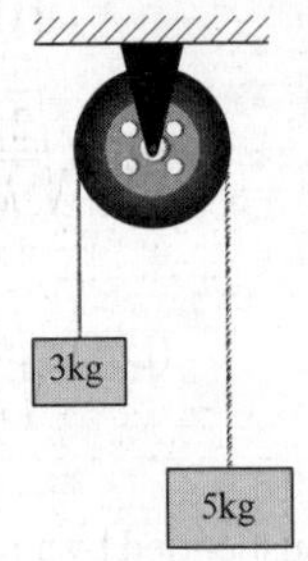

(A) 1250 cm (B) 1.25 cm
(C) 125 cm (D) 12.5 cm

44. The boxes of masses 2 kg and 8 kg are connected by a massless string passing over smooth pulleys. Calculate the time taken by box of mass 8 kg to strike the ground starting from rest. (Use $g = 10$ m/s^2) : **[27 Aug. 2021 (Evening)]**

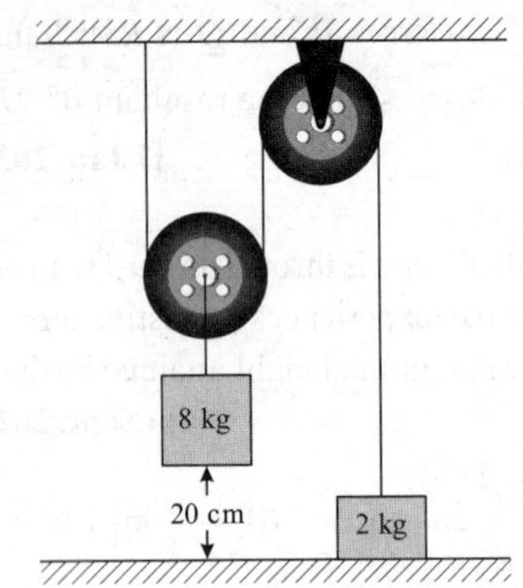

(A) 0.25 s (B) 0.34 s
(C) 0.2 s (D) 0.4 s

45. An object of mass m is being moved with a constant velocity under the action of an applied force of 2 N along a frictionless surface with following surface profile. The correct applied force vs distance graph will be : **[01 Sep. 2021 (Evening)]**

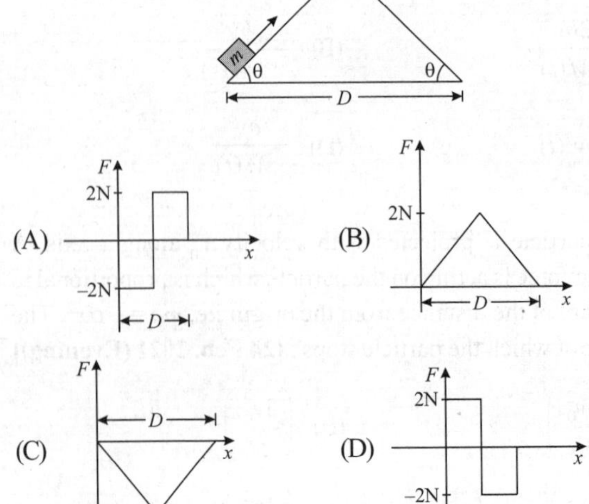

46. A block of mass m slides on the wooden wedge, which in turn slides backwards on the horizontal surface. The acceleration of the block with respect to the wedge is : Given $m = 8$ kg, $M = 16$ kg Assume all the surfaces shown in the figure to be frictionless : **[01 Sep. 2021 (Evening)]**

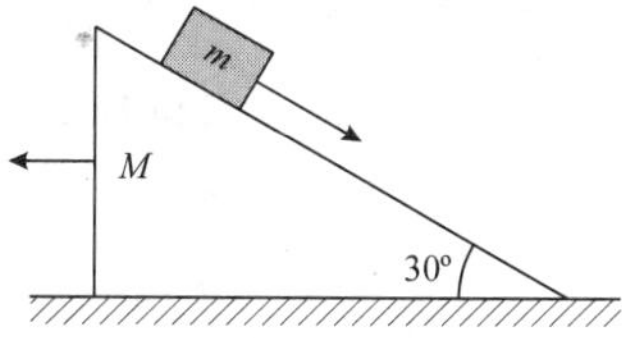

(A) $\frac{4}{3}g$ (B) $\frac{6}{5}g$

(C) $\frac{3}{5}g$ (D) $\frac{2}{3}g$

47. If force $\vec{F} = 3\hat{i} + 4\hat{j} - 2\hat{k}$ acts on a particle having position vector $2\hat{i} + \hat{j} + 2\hat{k}$ then, the torque about the origin will be : **[25 June 2022 (Morning)]**

(A) $3\hat{i} + 4\hat{j} - 2\hat{k}$ (B) $-10\hat{i} + 10\hat{j} + 5\hat{k}$

(C) $10\hat{i} + 5\hat{j} - 10\hat{k}$ (D) $10\hat{i} + \hat{j} - 5\hat{k}$

48. A force on an object of mass 100 g is $(10\hat{i} + 5\hat{j})N$. The position of that object at $t = 2$ s is $(a\hat{i} + b\hat{j})m$ after starting from rest. The value of $\frac{a}{b}$ will be _____. **[25 June 2022 (Morning)]**

49. A block of mass 200 g is kept stationary on a smooth inclined plane by applying a minimum horizontal force $F = \sqrt{x}N$ as shown in figure. The value of x = _____. **[25 June 2022 (Evening)]**

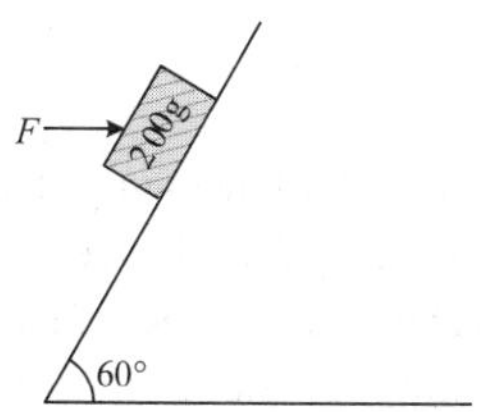

50. A person is standing in an elevator. In which situation, he experiences weight loss ? **[26 June 2022 (Morning)]**

(A) When the elevator moves upward with constant acceleration

(B) When the elevator moves downward with constant acceleration

(C) When the elevator moves upward with uniform velocity

(D) When the elevator moves downward with uniform velocity

51. In the arrangement shown in figure a_1, a_2, a_3 and a_4 are the accelerations of masses m_1, m_2, m_3 and m_4 respectively. Which of the following relation is true for this arrangement? **[26 June 2022 (Evening)]**

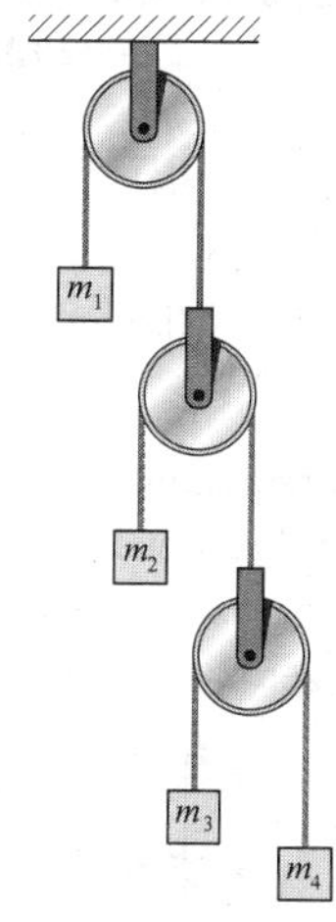

(A) $4a_1 + 2a_2 + a_3 + a_4 = 0$ (B) $a_1 + 4a_2 + 3a_3 + a_4 = 0$

(C) $a_1 + 4a_2 + 3a_3 + 2a_4 = 0$ (D) $2a_1 + 2a_2 + 3a_3 + a_4 = 0$

52. A system to 10 balls each of mass 2 kg are connected via massless and unstretchable string. The system is allowed to slip over the edge of a smooth table as shown in figure. Tension on the string between the 7th and 8th ball is _____ N when 6th ball just leaves the table. **[26 June 2022 (Evening)]**

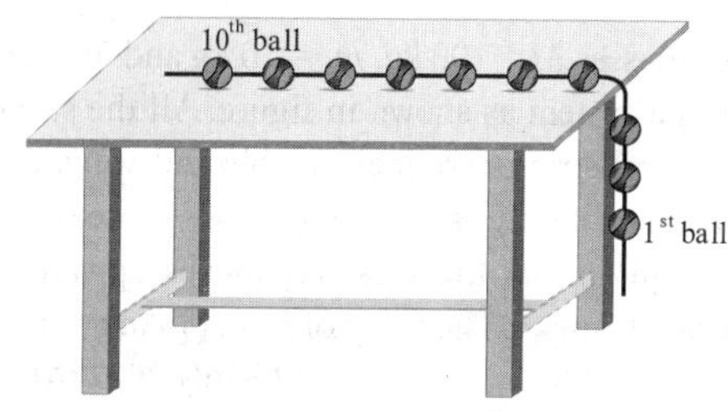

53. A mass of 10 kg is suspended vertically by a rope of length 5 m from the roof. A force of 30 N is applied at the middle point of rope in horizontal direction. The angle made by upper half of the rope with vertical is $\theta = \tan^{-1}(x \times 10^{-1})$. The value of x is _____. (Given $g = 10$ m/s^2) **[27 June 2022 (Evening)]**

54. A hanging mass M is connected to a four times bigger mass by using a string-pulley arrangement as shown in the figure. The bigger mass is placed on a horizontal ice-slab and being pulled by 2 Mg force. In this situation tension in the string is $\frac{x}{5}Mg$ for x = _____. Neglect mass of the string and friction of the block (bigger mass) with ice slab. [Given g = acceleration due to gravity] **[28 June 2022 (Morning)]**

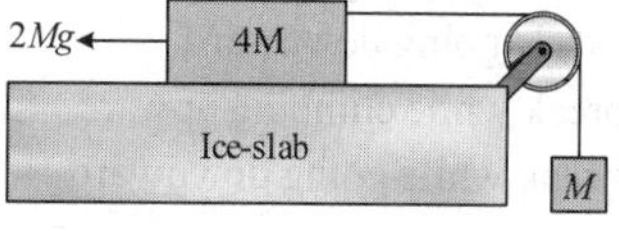

55. A block of mass m placed inside a box descends vertically with acceleration a. The block exerts a force equal to one-fourth of its weight on the floor of the box. The value of a will be :

[29 June 2022 (Evening)]

(A) $\frac{g}{4}$ (B) $\frac{g}{2}$

(C) $\frac{3g}{4}$ (D) g

56. For a free body diagram shown in the figure, the four forces are applied in the x and y directions. What additional force must be applied and at what angle with positive x-axis so that the net acceleration of body is zero? **[25 July 2022 (Evening)]**

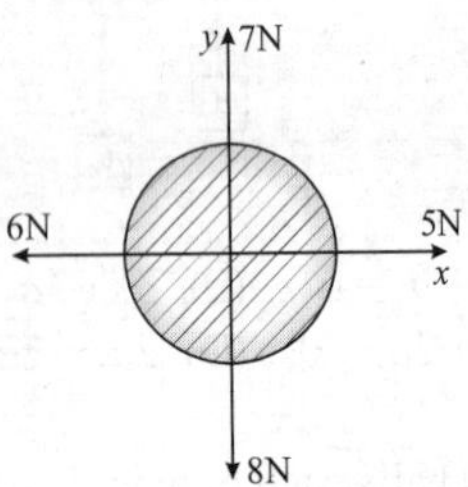

(A) $\sqrt{2}$ N, 45° (B) $\sqrt{2}$ N, 135°

(C) $\frac{2}{\sqrt{3}}$ N, 30° (D) 2 N, 45°

57. Three masses $M = 100$ kg, $m_1 = 10$ kg and $m_2 = 20$ kg are arranged in a system as shown in figure. All the surfaces are frictionless and strings are inextensible and weightless. The pulleys are also weightless and frictionless. A force F is applied on the system so that the mass m_2 moves upward with an acceleration of 2 ms^{-2}. The value of F is : (Take $g = 10$ ms^{-2})

[26 July 2022 (Morning)]

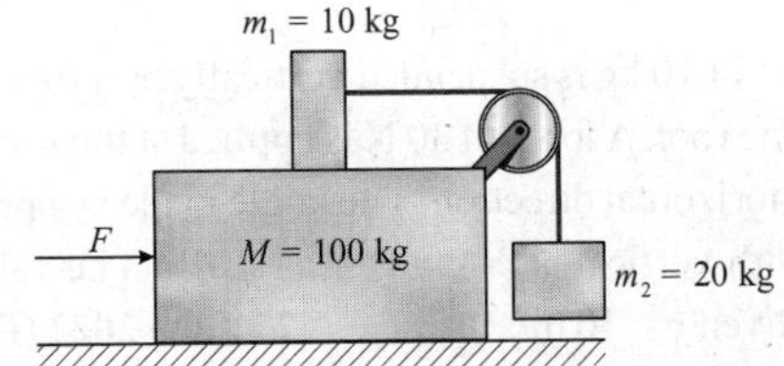

(A) 3360 N (B) 3380 N

(C) 3120 N (D) 3240 N

58. A monkey of mass 50 kg climbs on a rope which can withstand the tension (T) of 350 N. If monkey initially climbs down with an acceleration of 4 m/s^2 and then climbs up with an acceleration of 5 m/s^2. Choose the correct option : ($g = 10$ m/s^2)

[26 July 2022 (Morning)]

(A) $T = 700$ N while climbing upward

(B) $T = 350$ N while going downward

(C) Rope will break while climbing upward

(D) Rope will break while going downward

59. Two masses M_1 and M_2 are tied together at the two ends of a light inextensible string that passes over a frictionless pulley. When the mass M_2 is twice that of M_1, the acceleration of the system is a_1. When the mass M_2 is thrice that of M_1, the acceleration of the system is a_2. The ratio $\frac{a_1}{a_2}$ will be :

[26 July 2022 (Evening)]

(A) $\frac{1}{3}$ (B) $\frac{2}{3}$

(C) $\frac{3}{2}$ (D) $\frac{1}{2}$

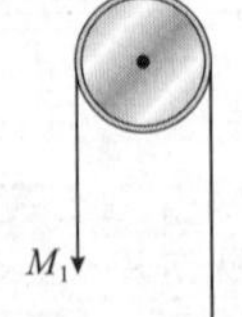

60. A block A takes 2 s to slide down a frictionless incline of 30° and length l, kept inside a lift going up with uniform velocity v. If the incline is changed to 45°, the time taken by the block, to slide down the incline, will be approximately :

[27 July 2022 (Evening)]

(A) 2.66 s (B) 0.83 s

(C) 1.68 s (D) 0.70 s

61. Two inclined planes are placed as shown in figure. A block is projected from the Point A of inclined plane AB along its surface with a velocity just sufficient to carry it to the top Point B at a height 10 m. After reaching the Point B the block slides down on inclined plane BC. Time it takes to reach to the point C from point A is $t(\sqrt{2}+1)$ s. The value of t is _____.

(Use $g = 10$ m/s^2) **[27 July 2022 (Evening)]**

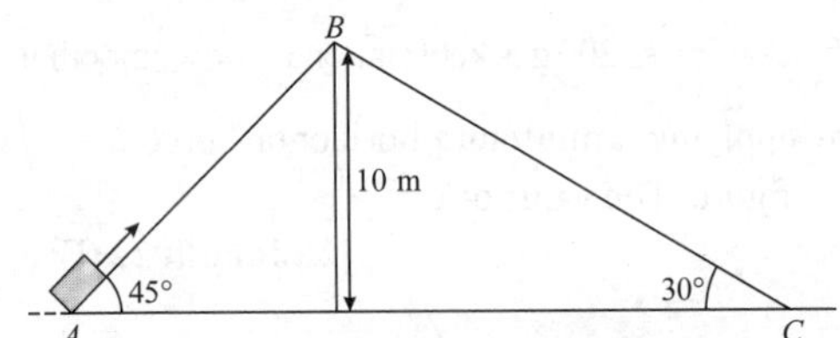

62. A uniform metal chain of mass m and length L passes over a massless and frictionless pulley. It is released from rest with a part of its length l is hanging on one side and rest of its length $L - l$ is hanging on the other side of the pulley.

At a certain point of time, when $l = \frac{L}{x}$, the acceleration of the chain is $\frac{g}{2}$. The value of x is _____. **[28 July 2022 (Evening)]**

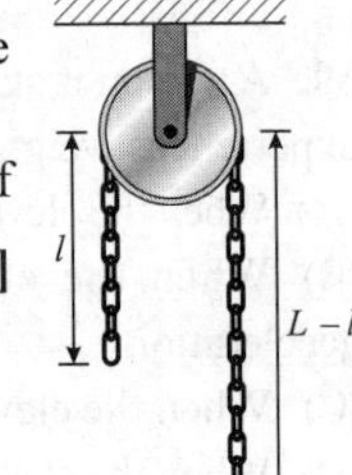

(A) 6

(B) 2

(C) 1.5

(D) 4

63. Two bodies of masses $m_1 = 5$ kg and $m_2 = 3$ kg are connected by a light string going over a smooth light pulley on a smooth inclined plane as shown in the figure. The system is at rest. The force exerted by the inclined plane on the body of mass m_1 will be : [Take $g = 10\ \text{ms}^{-2}$] **[29 July 2022 (Evening)]**

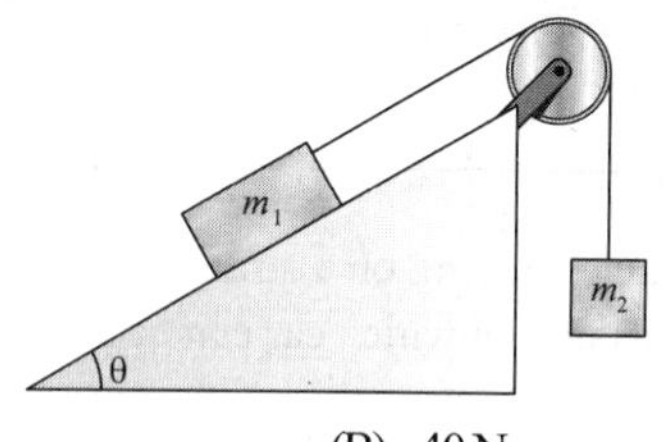

(A) 30 N (B) 40 N
(C) 50 N (D) 60 N

64. Given below are two statements :

Statement-I : An elevator can go up or down with uniform speed when its weight is balanced with the tension of its cable.

Statement-II : Force exerted by the floor of an elevator on the foot of a person standing on it is more than his/her weight when the elevator goes down with increasing speed.

In the light of the above statements, choose the correct answer from the options given below : **[24 Jan. 2023 (Morning)]**

(A) Both statement-I and statement-II are false
(B) Statement-I is true but Statement-II is false
(C) Both Statement-I and Statement-II are true
(D) Statement-I is false but Statement-II is true

65. As per given figure, a weightless pulley P is attached on a double inclined frictionless surface. The tension in the string (massless) will be : (If $g = 10\ \text{m/s}^2$) **[24 Jan. 2023 (Morning)]**

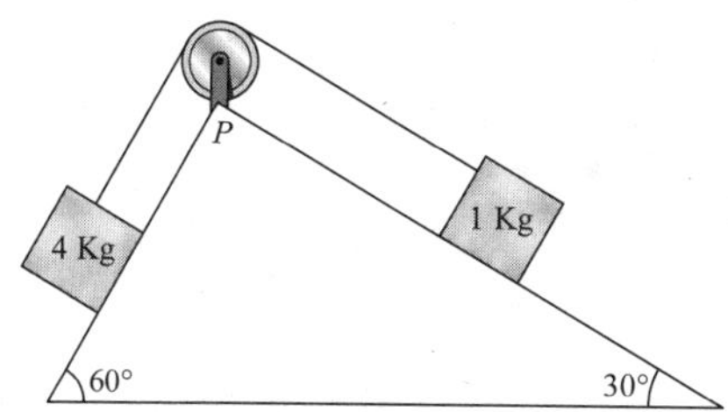

(A) $(4\sqrt{3}+1)$ N (B) $4(\sqrt{3}+1)$ N
(C) $4(\sqrt{3}-1)$ N (D) $(4\sqrt{3}-1)$ N

66. Force acts for 20 s on a body of mass 20 kg, starting from rest, after which the force ceases and then body describes 50 m in the next 10 s. The value of force will be :
[29 Jan. 2023 (Evening)]

(A) 40 N (B) 5 N
(C) 20 N (D) 10 N

67. The figure represents the momentum time (p-t) curve for a particle moving along an axis under the influence of the force. Identify the regions on the graph where the magnitude of the force is maximum and minimum respectively ? If $(t_3 - t_2) < t_1$.
[30 Jan. 2023 (Morning)]

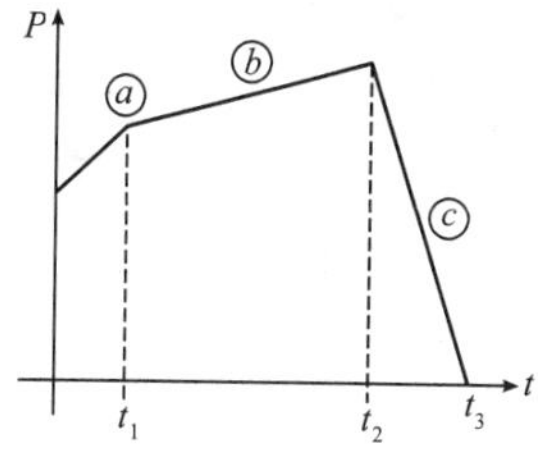

(A) c and a (B) b and c
(C) c and b (D) a and b

68. A block of $\sqrt{3}$ kg is attached to a string whose other end is attached to the wall. An unknown force F is applied so that the string makes an angle of 30° with the wall. The tension T is : (Given $g = 10\ \text{ms}^{-2}$) **[30 Jan. 2023 (Evening)]**

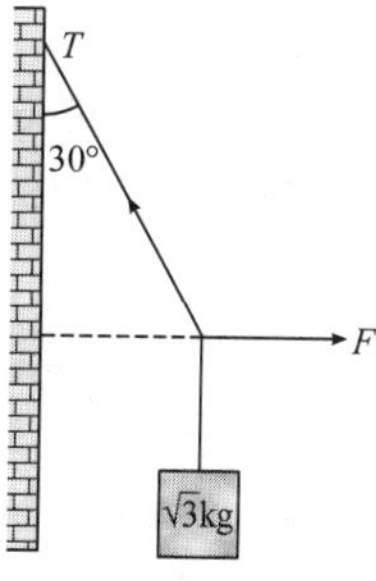

(A) 20 N (B) 25 N
(C) 10 N (D) 15 N

69. Two forces having magnitude A and $\frac{A}{2}$ are perpendicular to each other. The magnitude of their resultant is :
[8 April 2023 (Morning)]

(A) $\frac{\sqrt{5}A}{4}$ (B) $\frac{5A}{2}$
(C) $\frac{\sqrt{5}A^2}{2}$ (D) $\frac{\sqrt{5}A}{2}$

70. Three force $F_1 = 10$ N, $F_2 = 8$ N, $F_3 = 6$ N are acting on a particle of mass 5 kg. The forces F_2 and F_3 are applied perpendicular so that particle remains at rest. If the force F_1 is removed, then the acceleration of the particle is:
[12 April 2023 (Morning)]

(A) $2\ \text{ms}^{-2}$ (B) $0.5\ \text{ms}^{-2}$
(C) $4.8\ \text{ms}^{-2}$ (D) $7\ \text{ms}^{-2}$

* * * * *

CHAPTER

4 LAWS OF MOTION

4.2 FRICTION

1. A horizontal force of 10 N is necessary to just hold a block stationary against a wall. The coefficient of friction between the block and the wall is 0.2. The weight of the block is : **[2003]**

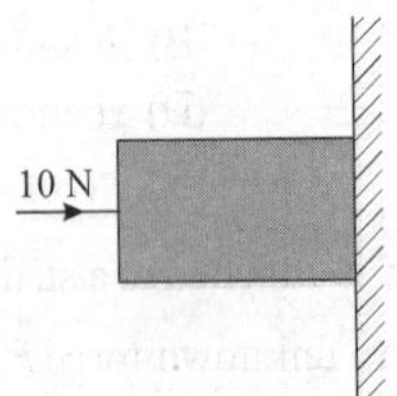

(A) 20 N (B) 50 N
(C) 100 N (D) 2 N

2. A marble block of mass 2 kg lying on ice when given a velocity of 6 m/s is stopped by friction in 10 s. Then the coefficient of friction is : **[2003]**
(A) 0.02 (B) 0.03
(C) 0.06 (D) 0.01

3. A block rests on a rough inclined plane making an angle of 30° with the horizontal. The coefficient of static friction between the block and the plane is 0.8. If the frictional force on the block is 10 N, the mass of the block (in kg) is : (taken $g = 10\ m/s^2$) **[2004]**
(A) 2.0 (B) 4.0
(C) 1.6 (D) 2.5

4. A smooth block is released at rest on a 45° incline and then slides a distance d. The time taken to slide is n times as much to slide on rough incline than on a smooth incline. The coefficient of friction is : **[2005]**

(A) $\mu_k = 1 - \frac{1}{n^2}$ (B) $\mu_k = \sqrt{1 - \frac{1}{n^2}}$

(C) $\mu_s = 1 - \frac{1}{n^2}$ (D) $\mu_s = \sqrt{1 - \frac{1}{n^2}}$

5. The upper half of an inclined plane with inclination ϕ is perfectly smooth, while the lower half is rough. A body starting from rest at the top will again come to rest at the bottom, if the coefficient of friction for the lower half is given by : **[2005]**
(A) $2 \sin \phi$ (B) $2 \cos \phi$
(C) $2 \tan \phi$ (D) $\tan \phi$

6. Consider a car moving on a straight road with a speed of 100 m/s. The distance at which car can be stopped, is : $[\mu_k = 0.5]$ **[2005]**
(A) 800 m (B) 1000 m
(C) 100 m (D) 400 m

7. The minimum force required to start pushing a body up a rough (frictional coefficient μ) inclined plane is F_1 while the minimum force needed to prevent it from sliding down is F_2. If the inclined plane makes an angle θ from the horizontal such that $\tan \theta = 2\mu$ then the ratio $\frac{F_1}{F_2}$ is : **[2011]**
(A) 4 (B) 1
(C) 2 (D) 3

8. An insect crawls up a hemispherical surface very slowly. The coefficient of friction between the insect and the surface is 1/3. If the line joining the centre of the hemispherical surface to the insect makes an angle α with the vertical, the maximum possible value of α so that the insect does not slip is given by: **[2012 (Online)]**

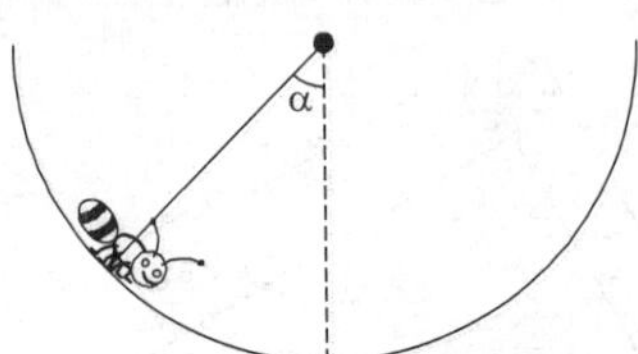

(A) $\cot \alpha = 3$ (B) $\sec \alpha = 3$
(C) $\operatorname{cosec} \alpha = 3$ (D) $\cos \alpha = 3$

9. A block of weight W rest on a horizontal floor with coefficient of static friction μ. It is desired to make the block move by applying minimum amount of force. The angle θ from the horizontal at which the force should be applied and magnitude of the force F are respectively : **[2012 (Online)]**

(A) $\theta = \tan^{-1}(\mu)$, $F = \frac{\mu W}{\sqrt{1+\mu^2}}$

(B) $\theta = \tan^{-1}\left(\frac{1}{\mu}\right)$, $F = \frac{\mu W}{\sqrt{1+\mu^2}}$

(C) $\theta = 0$, $F = \mu W$

(D) $\theta = \tan^{-1}\left(\frac{\mu}{1+\mu}\right)$, $F = \frac{\mu W}{1+\mu}$

10. A body starts from rest on a long inclined plane of slope 45°. The coefficient of friction between the body and the plane varies as $\mu = 0.3\,x$, where x is distance travelled down the plane. The body will have maximum speed : **[2013 (Online)]**

(A) 9.8 m (B) 27 m
(C) 12 m (D) 3.33 m

11. A block is placed on a rough horizontal plane. A time dependent horizontal force $F = kt$ acts on the block, where k is a positive constant. The acceleration-time graph of the block is: **[2013 (Online)]**

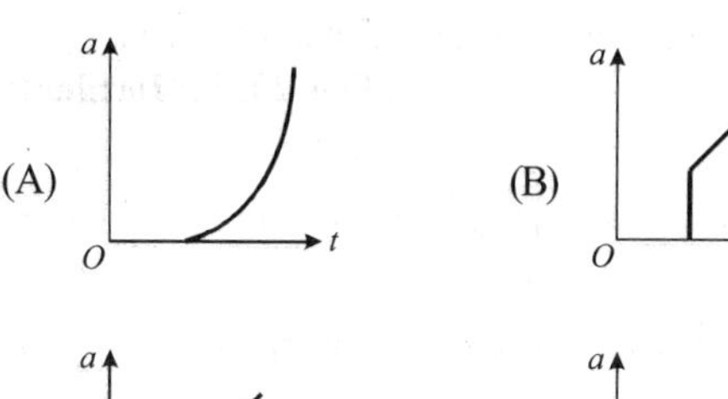

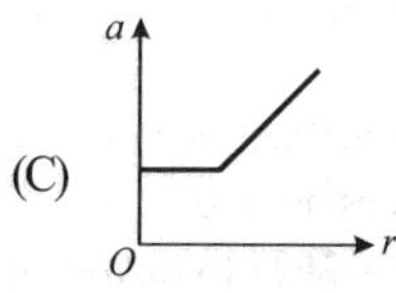

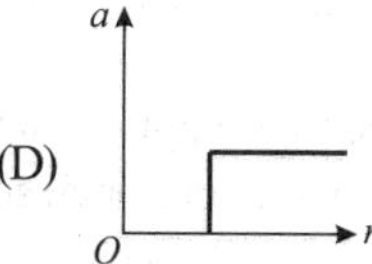

12. A block of mass m is placed on a surface with a vertical cross section given by $y = \frac{x^3}{6}$. If the coefficient of friction is 0.5, the maximum height above the ground at which the block can be placed without slipping is : **[2014]**

(A) $\frac{1}{6}m$ (B) $\frac{2}{3}m$
(C) $\frac{1}{3}m$ (D) $\frac{1}{2}m$

13. Given in the figure are two blocks A and B of weight 20 N and 100 N, respectively. These are being pressed against a wall by a force F as shown. If the coefficient of friction between the blocks is 0.1 and between block B and the wall is 0.15, the frictional force applied by the wall on block B is : **[2015]**

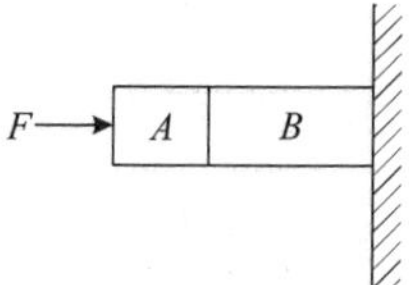

(A) 100 N (B) 80 N
(C) 120 N (D) 150 N

14. The masses m_1 = 5 kg and m_2 = 10 kg, connected by an inextensible string over a friction less pulley, are moving as shown in the figure. The coefficient of friction of horizontal surface is 0.15. The minimum weight m that should be put on top of m_2 to stop the motion is : **[2018]**

m
m_2
T
T
m_1
m_1g

(A) 18.3 kg (B) 27.3 kg
(C) 43.3 kg (D) 10.3 kg

15. A given object taken n times more time to slide down a 45° rough inclined plane as it takes to slide down a perfectly smooth 45° incline. The coefficient of kinetic friction between the object and the incline is : **[2018 (Online)]**

(A) $\sqrt{1-\frac{1}{n^2}}$ (B) $1-\frac{1}{n^2}$
(C) $\frac{1}{2-n^2}$ (D) $\sqrt{\frac{1}{1-n^2}}$

16. A body of mass 2 kg slide down with an acceleration of 3 m/s^2 on a rough inclined plane having a slope of 30°. The external force required to take the same body up the plane with the same acceleration will be : (g = 10 m/s^2) **[2018 (Online)]**

(A) 4 N (B) 14 N
(C) 6 N (D) 20 N

17. A block of mass 10 kg is kept on a rough inclined plane as shown in the figure. A force of 3 N is applied on the block. The coefficient of static friction between the plane and the block is 0.6. What should be the minimum value of force P, such that the block does not move downward ? [Take g = 10 ms^{-2}] **[9 Jan. 2019 (Morning)]**

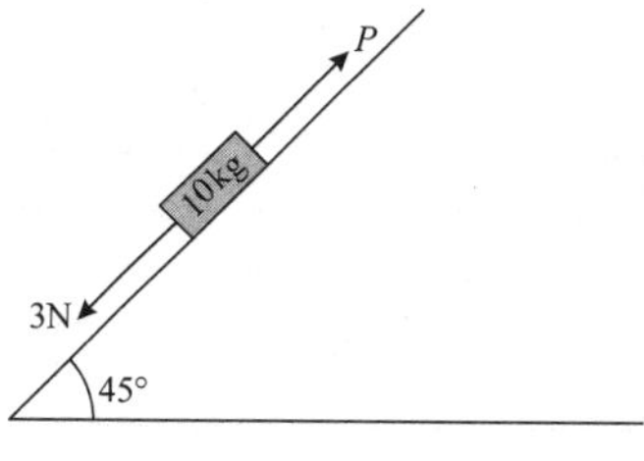

(A) 32 N (B) 25 N
(C) 23 N (D) 18 N

18. A block kept on a rough inclined plane, as shown in the figure, remains at rest upto a maximum force 2 N down the inclined plane. The maximum external force up the inclined plane that does not move the block is 10 N. The coefficient of static friction between the block and the plane is : [Take g = 10 m/s^2] **[12 Jan. 2019 (Evening)]**

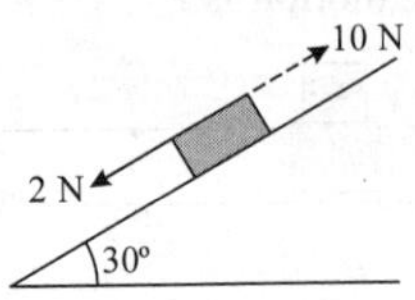

(A) $\frac{2}{3}$ (B) $\frac{\sqrt{3}}{2}$

(C) $\frac{\sqrt{3}}{4}$ (D) $\frac{1}{2}$

19. Two blocks A and B of masses $m_A = 1$ kg and $m_B = 3$ kg are kept on the table as shown in figure. The coefficient of friction between A and B is 0.2 and between B and the surface of the table is also 0.2. The maximum force F that can be applied on B horizontally, so that the block A does not slide over the block B is : [Take $g = 10$ m/s^2] **[10 April 2019 (Evening)]**

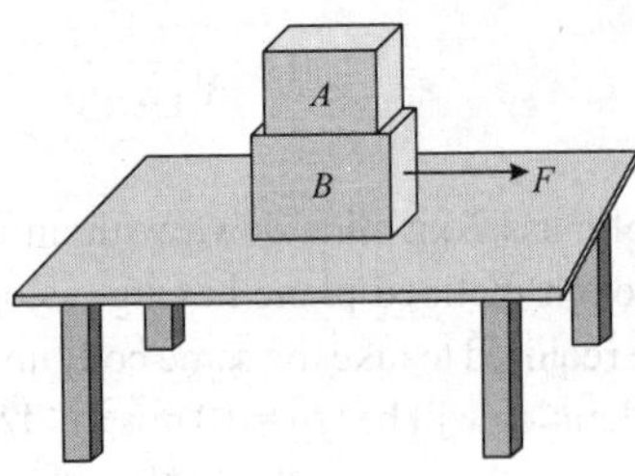

(A) 16 N (B) 40 N
(C) 12 N (D) 8 N

20. A block of mass 5 kg is (i) pushed in case (A) and (ii) pulled in case (B), by a force $F = 20$ N, making an angle of 30° with the horizontal, as shown in the figures. The coefficient of friction between the block and floor is $\mu = 0.2$. The difference between the accelerations of the block, in case (B) and case (A) will be : ($g = 10$ ms^{-2}) **[12 April 2019 (Evening)]**

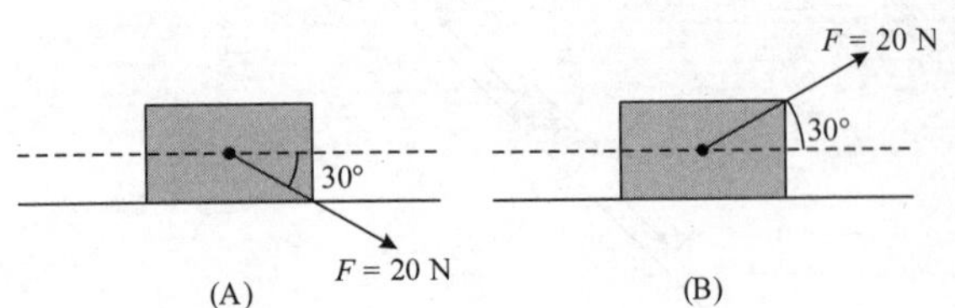

(A) 0 ms^{-2} (B) 0.8 ms^{-2}
(C) 0.4 ms^{-2} (D) 3.2 ms^{-2}

21. An insect is at the bottom of a hemispherical ditch of radius 1 m. It crawls up the ditch but starts slipping after it is at height h from the bottom. If the coefficient of friction between the ground and the insect is 0.75, then h is : ($g = 10$ ms^{-2}) **[6 Sept. 2020 (Morning)]**

(A) 0.80 m (B) 0.60 m
(C) 0.45 m (D) 0.20 m

22. The coefficient of static friction between a wooden block of mass 0.5 kg and a vertical rough wall is 0.2. The magnitude of horizontal force that should be applied on the block to keep it adhere to the wall will be ______ N. [$g = 10$ ms^{-2}] **[24 Feb. 2021 (Morning)]**

23. An inclined plane is bent in such a way that the vertical cross-section is given by $y = \frac{x^2}{4}$ where y is in vertical and x in horizontal direction. If the upper surface of this curved plane is rough with coefficient of friction $\mu = 0.5$, the maximum height in cm at which a stationary block will not slip downward is ______ cm. **[24 Feb. 2021 (Morning)]**

24. As shown in the figure, a block of mass $\sqrt{3}$ kg is kept on a horizontal rough surface of coefficient of friction $\frac{1}{3\sqrt{3}}$. The critical force to be applied on the vertical surface as shown at an angle 60° with horizontal such that it does not move, will be x. The value of x will be ______. **[26 Feb. 2021 (Morning)]**

[$g = 10$ m/s^2; $\sin 60° = \frac{\sqrt{3}}{2}$; $\cos 60° = \frac{1}{2}$]

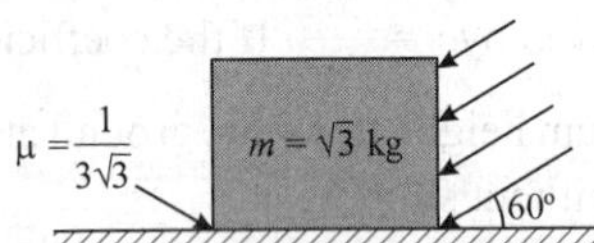

25. A block of mass m slides along a floor while a force of magnitude F is applied to it at an angle θ as shown in figure. The coefficient of kinetic friction is μ_k. Then, the block's acceleration 'a' is given by : (g is acceleration due to gravity) **[16 March 2021 (Morning)]**

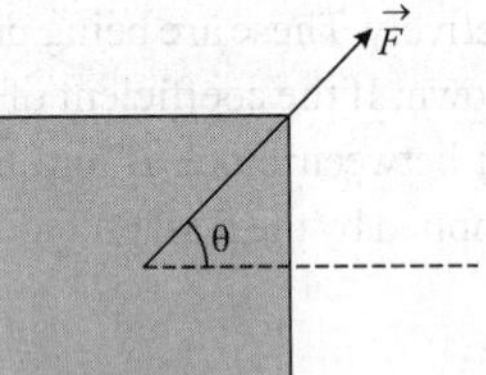

(A) $-\frac{F}{m}\cos\theta - \mu_k\left(g - \frac{F}{m}\sin\theta\right)$

(B) $\frac{F}{m}\cos\theta - \mu_k\left(g - \frac{F}{m}\sin\theta\right)$

(C) $\frac{F}{m}\cos\theta - \mu_k\left(g + \frac{F}{m}\sin\theta\right)$

(D) $\frac{F}{m}\cos\theta - \mu_k\left(g - \frac{F}{m}\sin\theta\right)$

26. Two blocks ($m = 0.5$ kg and $M = 4.5$ kg) are arranged on a horizontal frictionless table as shown in figure. The coefficient of static friction between the two blocks is $\frac{3}{7}$. Then the maximum horizontal force that can be applied on the larger block so that the blocks move together is ______ N. (Round off to the Nearest Integer) [Take g as 9.8 ms^{-2}] **[17 March 2021 (Morning)]**

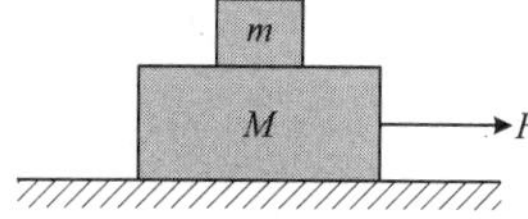

27. A body of mass 1 kg rests on a horizontal floor with which it has a coefficient of static friction $\frac{1}{\sqrt{3}}$. It is desired to make the body move by applying the minimum possible force F N. The value of F will be ______. (Round off to the Nearest Integer) [Take $g = 10$ ms^{-2}] **[17 March 2021 (Evening)]**

28. The normal reaction 'N' for a vehicle of 800 kg mass, negotiating a turn on a 30° banked road at maximum possible speed without skidding is ______ $\times 10^3$ kg m/s^2. (Given cos 30° = 0.87, $\mu s = 0.2$) **[20 July 2021 (Morning)]**

(A) 7.2 (B) 12.4
(C) 10.2 (D) 6.98

29. A body of mass 'm' is launched up on a rough inclined plane making and angle of 30° with the horizontal. The coefficient of friction between the body and plane is $\frac{\sqrt{x}}{5}$ if the time of ascent is half of the time of descent. The value of x is ______. **[20 July 2021 (Evening)]**

30. The coefficient of static friction between two blocks is 0.5 and the table is smooth. The maximum horizontal force that can be applied to move the blocks together is ______ N. (take $g = 10$ ms^{-2}) **[26 Aug. 2021 (Evening)]**

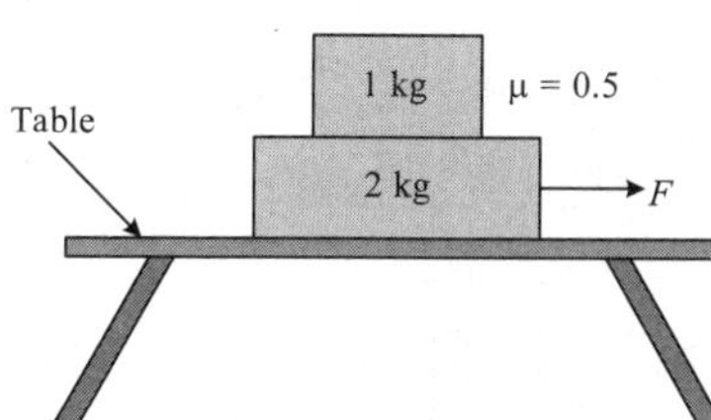

31. When a body slides down from rest along a smooth incline plane making an angle of 30° with the horizontal, it takes time T. When the same body slides down the rest along a rough inclined plane making the same angle and through the same distance, it takes time αT, where α is a constant greater than 1. The co-efficient of friction between the body and the rough plane is $\frac{1}{\sqrt{x}}\left(\frac{\alpha^2 - 1}{\alpha^2}\right)$ where $x =$ ______. **[01 Sep. 2021 (Evening)]**

32. Consider a block kept on an inclined plane (inclined at 45°) as shown in the figure. If the force required to just push it up the incline is 2 times the force required to just prevent it from sliding down, the coefficient of friction between the block and inclined plane (μ) is equal to : **[25 Jan. 2023 (Evening)]**

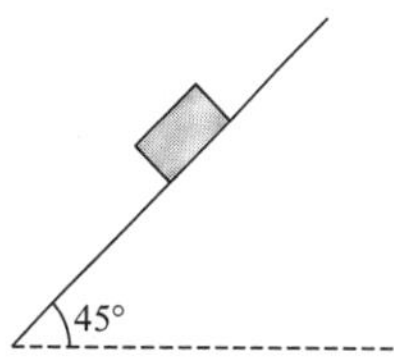

(A) 0.33 (B) 0.60
(C) 0.25 (D) 0.50

33. A block of mass m slides down the plane inclined at angle 30° with an acceleration $\frac{g}{4}$. The value of coefficient of kinetic friction will be : **[29 Jan. 2023 (Morning)]**

(A) $\frac{2\sqrt{3}+1}{2}$ (B) $\frac{1}{2\sqrt{3}}$

(C) $\frac{\sqrt{3}}{2}$ (D) $\frac{2\sqrt{3}-1}{2}$

34. The time taken by an object to slide down 45° rough inclined plane is n times as it takes to slide down a perfectly smooth 45° incline plane. The coefficient of kinetic friction between the object and the incline plane is : **[29 Jan. 2023 (Evening)]**

(A) $\sqrt{\frac{1}{1-n^2}}$ (B) $\sqrt{1-\frac{1}{n^2}}$

(C) $1+\frac{1}{n^2}$ (D) $1-\frac{1}{n^2}$

35. A body of mass 10 kg is moving with an initial speed of 20 m/s. The body stops after 5 s due to friction between body and the floor. The value of the coefficient of friction is : (Take acceleration due to gravity $g = 10$ ms^{-2}) **[31 Jan. 2023 (Evening)]**

(A) 0.2 (B) 0.3
(C) 0.5 (D) 0.4

36. A block of mass 5 kg is placed at rest on a table of rough surface. Now, if a force of 30 N is applied in the direction parallel to surface of the table, the block slides through a distance of 50 m in an interval of time 10 s. Coefficient of kinetic friction is

(Given, $g = 10\ ms^{-2}$) **[1 Feb. 2023 (Morning)]**

(A) 0.60 (B) 0.75

(C) 0.50 (D) 0.25

37. As shown in the figure a block of mass 10 kg lying on a horizontal surface is pulled by a force F acting at an angle 30°, with horizontal. For $\mu_s = 0.25$, the block will just start to move for the value of F : [Given $g = 10\ ms^{-2}$] **[1 Feb. 2023 (Evening)]**

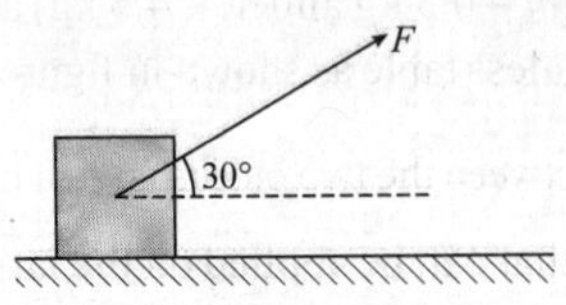

(A) 33.3 N (B) 25.2 N

(C) 20 N (D) 35.7 N

* * * * *

CHAPTER 5

WORK ENERGY POWER & APPLICATIONS

5.1 WORK ENERGY POWER

1. A spring of force constant 800 N/m has an extension of 5 cm. The work done in extending it from 5 cm to 15 cm is : **[2002]**

(A) 16 J (B) 8 J
(C) 32 J (D) 24 J

2. A spring of spring constant 5×10^3 N/m is stretched initially by 5 cm from the unstretched position. Then the work required to stretch it further by another 5 cm is : **[2003]**

(A) 12.50 N-m (B) 18.75 N-m
(C) 25.00 N-m (D) 6.25 N-m

3. A body is moved along a straight line by a machine delivering a constant power. The distance moved by the body in time t is proportional to : **[2003]**

(A) $t^{3/4}$ (B) $t^{3/2}$
(C) $t^{1/4}$ (D) $t^{1/2}$

4. A particle moves in a straight line with retardation proportional to its displacement. Its loss of kinetic energy for any displacement x is proportional to : **[2004]**

(A) x^2 (B) e^x
(C) x (D) $\log_e x$

5. A body of mass m accelerates uniformly from rest to v_1 in time t_1. The instantaneous power delivered to the body as a function of time t is : **[2004]**

(A) $\frac{mv_1t}{t_1}$ (B) $\frac{mv_1^2t}{t_1^2}$

(C) $\frac{mv_1t^2}{t_1}$ (D) $\frac{mv_1^2t}{t_1}$

6. A uniform chain of length 2 m is kept on a table such that a length of 60 cm hangs freely from the edge of the table. The total mass of the chain is 4 kg. What is the work done in pulling the entire chain on the table : **[2004]**

(A) 7.2 J (B) 3.6 J
(C) 120 J (D) 1200 J

7. A force $\vec{F} = (5\vec{i} + 3\vec{j} + 2\vec{k})N$ is applied over a particle which displaces it from its origin to the point $\vec{r} = (2\vec{i} - \vec{j})m$. The work done on the particle in joule is : **[2004]**

(A) +10 (B) +7
(C) −7 (D) +13

8. A bullet fired into a fixed target loses half of its velocity after penetrating 3 cm. How much further it will penetrate before coming to rest, assuming that it faces constant resistance to motion ? **[2005]**

(A) 3.0m (B) 2.0m
(C) 1.5 cm (D) 1.0 cm

9. The block of mass M moving on the frictionless horizontal surface collides with the spring of spring constant k and compresses it by length L. The maximum momentum of the block after collision is : **[2005]**

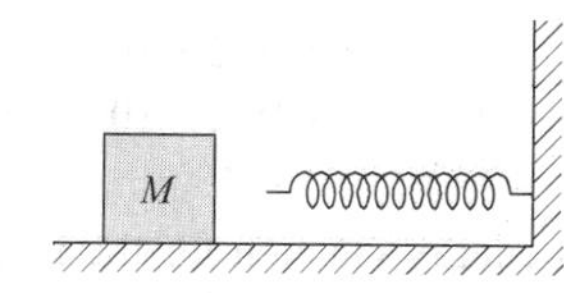

(A) $\sqrt{Mk}\,L$ (B) $\frac{kL^2}{2M}$

(C) Zero (D) $\frac{ML^2}{k}$

10. A spherical ball of mass 20 kg is stationary at the top of a hill of height 100 m. It rolls down a smooth surface to the ground, then climbs up another hill of height 30 m and finally rolls down to a horizontal base at a height of 20 m above the ground. The velocity attained by the ball is : **[2005]**

(A) 20 m/s (B) 40 m/s
(C) $10\sqrt{30}$ m/s (D) 10 m/s

11. A particle of mass 100 g is thrown vertically upwards with a speed of 5 m/s. The work done by the force of gravity during the time the particle goes up is : **[2006]**

(A) −0.5 J (B) −1.25 J
(C) 1.25 J (D) 0.5 J

12. The potential energy of a 1 kg particle free to move along the x-axis is given by $V(x) = \left(\frac{x^4}{4} - \frac{x^2}{2}\right)$ J. The total mechanical energy of the particle is 2 J. Then, the maximum speed (in m/s) is: **[2006]**

(A) $3/\sqrt{2}$ (B) $\sqrt{2}$
(C) $1/\sqrt{2}$ (D) 2

13. A mass of M kg is suspended by a weightless string. The horizontal force that is required to displace it until the string makes an angle of 45° with the initial vertical direction is : **[2006]**

(A) $Mg(\sqrt{2}+1)$ (B) $Mg\sqrt{2}$

(C) $\frac{Mg}{\sqrt{2}}$ (D) $Mg(\sqrt{2}-1)$

14. A 2 kg block slides on a horizontal floor with a speed of 4 m/s. It strikes an uncompressed spring, and compresses it till the block is motionless. The kinetic friction force is 15 N and spring constant is 10,000 N/m. The spring compresses by : **[2007]**

(A) 5.5 cm (B) 2.5 cm

(C) 11.0 cm (D) 8.5 cm

15. An athlete in the Olympic games covers distance of 100 m in 10 s. His kinetic energy can be estimated to be in the range : **[2008]**

(A) 200 J – 500 J (B) 2×10^5 J – 3×10^5 J

(C) 2×10^4 J – 3×10^4 J (D) 2×10^3 J – 5×10^3 J

16. The potential energy function for the force between two atoms in a diatomic molecule is approximately given by $U(x)=\frac{a}{x^{12}}-\frac{b}{x^6}$, where a and b are constant and x is the distance between the atoms. If the dissociation energy of the molecule is $D=[U(x=\infty)-U_{\text{at equilibrium}}]$, D is : **[2010]**

(A) $\frac{b^2}{6a}$ (B) $\frac{b^2}{2a}$

(C) $\frac{b^2}{12a}$ (D) $\frac{b^2}{4a}$

17. At time $t=0$s particle starts moving along the x-axis. If its kinetic energy increases uniformly with time 't', the net force acting on it must be proportional to : **[2011]**

(A) $\sqrt{t}$ (B) Constant

(C) t (D) $\frac{1}{\sqrt{t}}$

18. This question has Statement-I and Statement-II. Of the four choices given after the statements, choose the one that best describes the two statements.

If two springs S_1 and S_2 of force constants k_1 and k_2, respectively, are stretched by the same force, it is found that more work is done on spring S_1 than on spring S_2.

Statement-I: If stretched by the same amount, work done on S_1, will be more than that on S_2.

Statement-II: $k_1 < k_2$. **[2012]**

(A) Statement-I is true, Statement-II is true and Statement-II is not the correct explanation of Statement-I.

(B) Statement-I is false, Statement-II is true

(C) Statement-I is true, Statement-II is false

(D) Statement-I is true, Statement-II is true and Statement-2 is the correct explanation of Statement-I.

19. The force $\vec{F}=F\hat{i}$ on a particle of mass 2 kg, moving along the x-axis is given in the figure as a function of its position x. The particle is moving with a velocity of 5 m/s along the x-axis at $x=0$. What is the kinetic energy of the particle at $x=8$ m? **[2012 (Online)]**

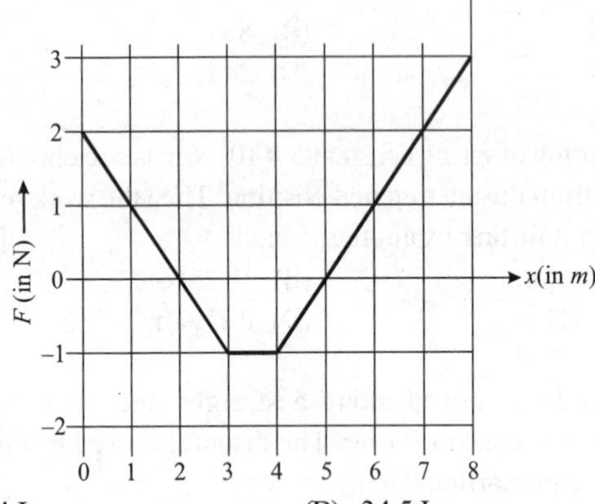

(A) 34 J (B) 34.5 J

(C) 4.5 J (D) 29.4 J

20. A particle gets displaced by $\Delta\vec{r}=(2\hat{i}+3\hat{j}+4\hat{k})$ m under the action of a force $\vec{F}\,(7\hat{i}+4\hat{j}+3\hat{k})$. The change in its kinetic energy is : **[2012 (Online)]**

(A) 38 J (B) 70 J

(C) 52.5 J (D) 126 J

21. The springs of force constants 300 N/m (Spring A) and 400 N/m (Spring B) are joined together in series. The combination is compressed by 8.75 cm. The ratio of energy stored in A and B is $\frac{E_A}{E_B}$. Then $\frac{E_A}{E_B}$ is equal : **[2013 (Online)]**

(A) $\frac{4}{3}$ (B) $\frac{16}{9}$

(C) $\frac{3}{4}$ (D) $\frac{9}{16}$

22. A 70 kg man leaps vertically into the air from a crouching position. To take the leap the man pushes the ground with a constant force F to raise himself. The center of gravity rises by 0.5 m before he leaps. After the leap the c.g. rises by another 1 m. The maximum power delivered by the muscles is :

[Take $g=10$ ms^{-2}] **[2013 (Online)]**

(A) 6.26×10^3 Watts at the start

(B) 6.26×10^3 Watts at take off

(C) 6.26×10^4 Watts at the start

(D) 6.26×10^4 Watts at take off

23. When a rubber-band is stretched by a distance x, it exerts a restoring force of magnitude $F = ax + bx^2$ where a and b are constants. The work done in stretching the unstretched rubber-band by L is : **[2014]**

(A) $aL^2 + bL^3$ (B) $\frac{1}{2}(aL^2 + bL^3)$

(C) $\frac{aL^2}{2} + \frac{bL^3}{3}$ (D) $\frac{1}{2}\left(\frac{aL^2}{2} + \frac{bL^3}{3}\right)$

24. A spring of unstretched length 1 has a mass m with one end fixed to a rigid support. Assuming spring to be made of a uniform wire, the kinetic energy possessed by it if its free end is pulled with uniform velocity v is : **[2014 (Online)]**

(A) $\frac{1}{2}mv^2$ (B) mv^2

(C) $\frac{1}{3}mv^2$ (D) $\frac{1}{6}mv^2$

25. A wind-powered generator converts wind energy into electrical energy. Assume that the generator converts a fixed fraction of the wind energy intercepted by its blades into electrical energy. For wind speed v, the electrical power output will be most likely proportional to : **[2014 (Online)]**

(A) v^4 (B) v^2

(C) v (D) v^3

26. A bullet looses $\left(\frac{1}{n}\right)^{th}$ of its velocity passing through one plank. The number of such planks that are required to stop the bullet can be : **[2014 (Online)]**

(A) $\frac{n^2}{2n-1}$ (B) $\frac{2n^2}{n-1}$

(C) Infinite (D) n

27. A small ball of mass m starts at a point A with speed v_0 and moves along a frictionless track AB as shown. The track BC has coefficient of friction μ. The ball comes to stop at C after travelling a distance L which is : **[2014 (Online)]**

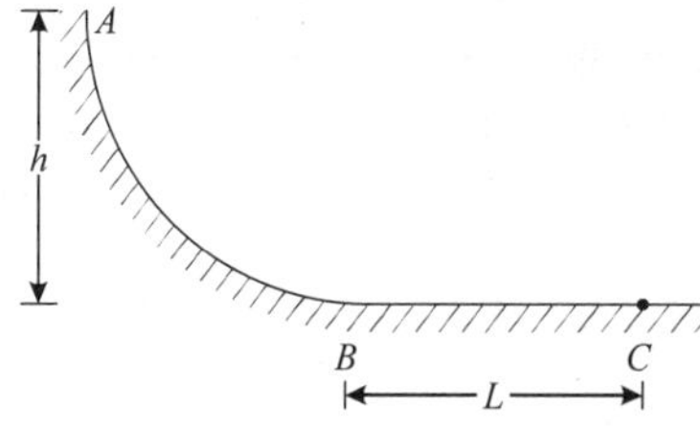

(A) $\frac{2h}{\mu} + \frac{v_0^2}{2\mu g}$ (B) $\frac{h}{\mu} + \frac{v_0^2}{2\mu g}$

(C) $\frac{h}{2\mu} + \frac{v_0^2}{\mu g}$ (D) $\frac{h}{2\mu} + \frac{v_0^2}{2\mu g}$

28. A point particle of mass m, moves along the uniformly rough track PQR as shown in the figure. The coefficient of friction, between the particle and the rough track equals m. The particle released, from rest, from the point P and it comes to rest at a point R. The energies, lost by the ball, over the parts, PQ and QR, of the track, are equal to each other, and no energy is lost when particle changes direction from PQ to QR The values of the coefficient of friction m and the distance $x(= QR)$, are, respectively close to : **[2016]**

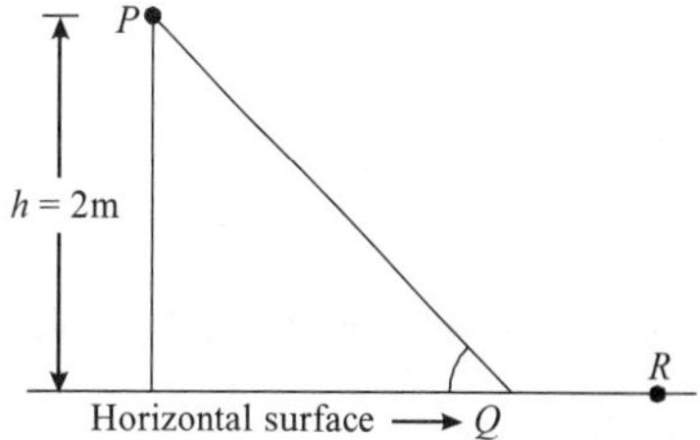

(A) 0.2 and 6.5 m (B) 0.2 and 3.5 m

(C) 0.29 and 3.5 m (D) 0.29 and 6.5 m

29. A person trying to lose weight by burning fat lifts a mass of 10 kg upto a height of 1 m 1000 times. Assume that the potential energy lost each time he lowers the mass is dissipated. How much fat will he use up considering the work done only when the weight is lifted up? Fat supplies 3.8×10^7 J of energy per kg which is converted to mechanical energy with a 20% efficiency rate. Take $g = 9.8\ ms^{-2}$: **[2016]**

(A) 2.45×10^{-3} kg (B) 6.45×10^{-3} kg

(C) 9.89×10^{-3} kg (D) 12.89×10^{-3} kg

30. Velocity-time graph for a body of mass 10 kg is shown in figure. Work-done on the body in first two seconds of the motion is : **[2016 (Online)]**

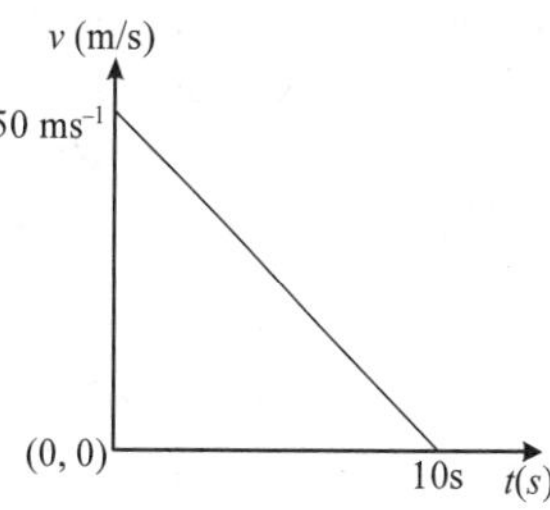

(A) –9300 J (B) 12000 J

(C) –4500 J (D) – 12000 J

31. A car of weight W is on an inclined road that rises by 100 m over a distance of 1 Km and applies a constant frictional force $\frac{W}{20}$ on the car. While moving uphill on the road at a speed of 10 ms^{-1}, the car needs power P. If it needs power $\frac{P}{2}$ while moving downhill at speed v then value of v is : **[2016 (Online)]**

(A) 20 ms^{-1} (B) 5 ms^{-1}

(C) 15 ms^{-1} (D) 10 ms^{-1}

32. A time dependent force $F = 6t$ acts on a particle of mass 1 kg. If the particle starts from rest, the work done by the force during the first 1 second will be : **[2017]**

(A) 9 J (B) 18 J
(C) 4.5 J (D) 22 J

33. A body of mass $m = 10^{-2}$ kg is moving in a medium and experiences a frictional force $F = -kv^2$. Its initial speed is $v_0 = 10\ ms^{-1}$. If, after 10 s, its energy is $\frac{1}{8}mv_0^2$, the value of k will be : **[2017]**

(A) 10^{-4} kg m^{-1} (B) 10^{-1} kg m^{-1} s^{-1}
(C) 10^{-3} kg m^{-1} (D) 10^{-3} kg s^{-1}

34. A particle is moving in a circular path of radius a under the action of an attractive potential $U = -\frac{k}{2r^2}$. Its total energy is : **[2018]**

(A) $-\frac{k}{4a^2}$ (B) $\frac{k}{2a^2}$
(C) Zero (D) $-\frac{3}{2}\frac{k}{a^2}$

35. Two particles of the same mass m are moving in circular orbits because of force, given by $F(r) = \frac{-16}{r} - r^3$.
The first particle is at a distance $r = 1$, and the second, at $r = 4$. The best estimate for the ratio of kinetic energies of the first and the second particle is closest to : **[2018 (Online)]**

(A) 10^{-1} (B) 6×10^{-2}
(C) 6×10^2 (D) 3×10^{-3}

36. A body of mass m starts moving from rest along x-axis so that its velocity varies as $v = a\sqrt{s}$ where a is a constant and s is the distance covered by the body. The total work done by all the forces acting on the body in the first t second after the start of the motion is : **[2018 (Online)]**

(A) $\frac{1}{8}ma^4t^2$ (B) $4\ ma^4t^2$
(C) $8\ ma^4t^2$ (D) $\frac{1}{4}ma^4t^2$

37. A block of mass m, lying on a smooth horizontal surface, is attached to a spring (of negligible mass) of spring constant k. The other end of the spring is fixed, as shown in the figure. The block is initially at rest in its equilibrium position. If now the block is pulled with a constant force F, the maximum speed of the block is : **[9 Jan. 2019 (Morning)]**

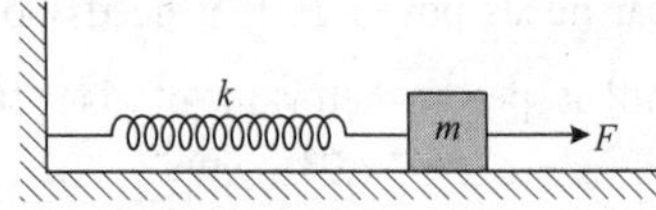

(A) $\frac{\pi F}{\sqrt{mk}}$ (B) $\frac{2F}{\sqrt{mk}}$
(C) $\frac{F}{\sqrt{mk}}$ (D) $\frac{F}{\pi\sqrt{mk}}$

38. A force acts on a 2 kg object so that its position is given as a function of time as $x = 3t^2 + 5$. What is the work done by this force in first 5 seconds ? **[9 Jan. 2019 (Evening)]**

(A) 850 J (B) 900 J
(C) 950 J (D) 875 J

39. A block of mass m is kept on a platform which starts from rest with constant acceleration $g/2$ upward, as shown in figure. Work done bv normal reaction on block in time t is : **[10 Jan. 2019 (Morning)]**

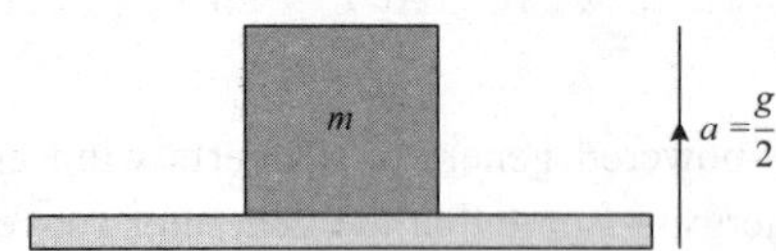

(A) 0 (B) $\frac{3mg^2t^2}{8}$
(C) $-\frac{mg^2t^2}{8}$ (D) $\frac{mg^2t^2}{8}$

40. A particle which is experiencing a force, given by $\vec{F} = 3\vec{i} - 12\vec{j}$, undergoes a displacement of $\vec{d} = 4\vec{i}$. If the particle had a kinetic energy of 3 J at the beginning of the displacement, what is its kinetic energy at the end of the displacement? **[10 Jan. 2019 (Evening)]**

(A) 15 J (B) 10 J
(C) 12 J (D) 9 J

41. A body of mass 1 kg falls freely from a height of 100 m on a platform of mass 3 kg which is mounted on a spring having spring constant $k = 1.25 \times 10^6$ N/m. The body sticks to the platform and the spring's maximum compression is found to be x. Given that $g = 10\ ms^{-2}$, the value of x will be close to : **[11 Jan. 2019 (Morning)]**

(A) 4 cm (B) 8 cm
(C) 80 cm (D) 40 cm

42. A particle moves in one dimension from rest under the influence of a force that varies with the distance travelled by the particle as shown in the figure. The kinetic energy of the particle after it has travelled 3 m is : **[8 April 2019 (Morning)]**

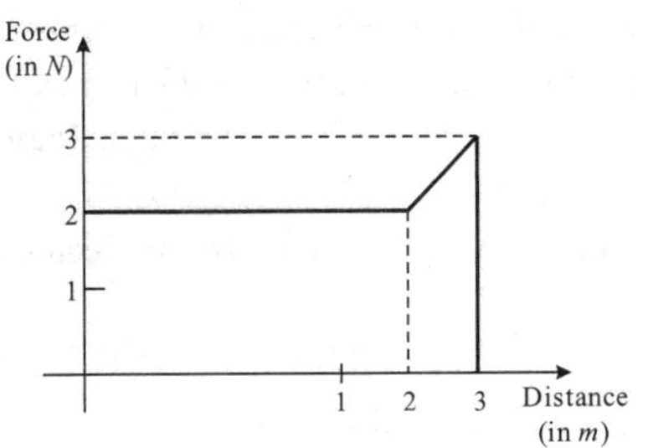

(A) 6.5 J (B) 2.5 J
(C) 4 J (D) 5 J

43. A uniform cable of mass 'M' and length 'L' is placed on a horizontal surface such that its $\left(\frac{1}{n}\right)^{th}$ part is hanging below the edge of the surface. To lift the hanging part of the cable upto the surface, the work done should be :

[9 April 2019 (Morning)]

(A) $\frac{MgL}{n^2}$ (B) $\frac{MgL}{2n^2}$

(C) $\frac{2MgL}{n^2}$ (D) $nMgL$

44. A ball is thrown upward with an initial velocity V_0 from the surface of the earth. The motion of the ball is affected by a drag force equal to $m\gamma\upsilon^2$ (where m is mass of the ball, υ is its instantaneous velocity and γ is a constant). Time taken by the ball to rise to its zenith is : **[10 April 2019 (Morning)]**

(A) $\frac{1}{\sqrt{\gamma g}}\sin^{-1}\left(\sqrt{\frac{\gamma}{g}}V_0\right)$ (B) $\frac{1}{\sqrt{\gamma g}}\tan^{-1}\left(\sqrt{\frac{\gamma}{g}}V_0\right)$

(C) $\frac{1}{\sqrt{2\gamma g}}\tan^{-1}\left(\sqrt{\frac{2\gamma}{g}}V_0\right)$ (D) $\frac{1}{\sqrt{\gamma g}}\ln\left(1+\sqrt{\frac{\gamma}{g}}V_0\right)$

45. A spring whose unstretched length is l has a force constant k. The spring is cut into two pieces of unstretched lengths l_1 and l_2 where, $l_1 = nl_2$ and n is an integer. The ratio k_1/k_2 of the corresponding force constants, will be :

[12 April 2019 (Evening)]

(A) $\frac{1}{n^2}$ (B) n^2

(C) $\frac{1}{n}$ (D) n

46. A 60 HP electric motor lifts an elevator having a maximum total load capacity of 2000 kg. If the frictional force on the elevator is 4000 N, the speed of the elevator at full load is close to : (1 HP = 746 W, $g = 10\ ms^{-2}$) **[7 Jan. 2020 (Morning)]**

(A) $1.7\ ms^{-1}$ (B) $2.0\ ms^{-1}$
(C) $1.9\ ms^{-1}$ (D) $1.5\ ms^{-1}$

47. A particle ($m = 1$ kg) slides down a frictionless track (AOC) starting from rest at a point A (height 2 m). After reaching C, the particle continues to move freely in air as a projectile. When it reaching its highest point P (height 1 m), the kinetic energy of the particle (in J) is : (Figure drawn is schematic and not to scale; take $g = 10\ ms^{-2}$)_____ : **[7 Jan. 2020 (Morning)]**

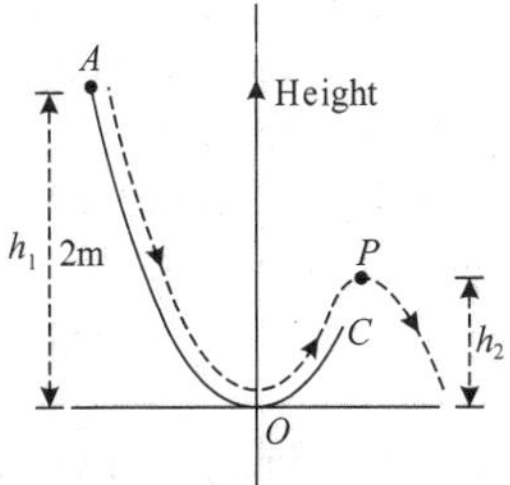

48. An elevator in a building can carry a maximum of 10 persons, with the average mass of each person being 68 kg. The mass of the elevator itself is 920 kg and it moves with a constant speed 3 m/s. The frictional force opposing the motion is 6000 N. If the elevator is moving up with its full capacity, the power delivered by the motor to the elevator ($g = 10\ m/s^2$) must be at least :

[7 Jan. 2020 (Evening)]

(A) 56300 W (B) 48000 W
(C) 66000 W (D) 62360 W

49. Consider a force $\vec{F} = -x\hat{i} + y\hat{j}$. The work done by this force in moving a particle from point $A(1, 0)$ to $B(0, 1)$ along the line segment is : (all quantities are in SI units)

[9 Jan. 2020 (Morning)]

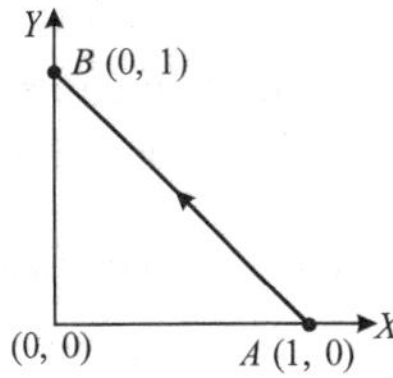

(A) $\frac{3}{2}$ (B) 1

(C) 2 (D) $\frac{1}{2}$

50. A small block starts slipping down from a point B on an inclined plane AB, which is making an angle θ with the horizontal section BC is smooth and the remaining section CA is rough with a coefficient of friction μ. It is found that the block comes to rest as it reaches the bottom (point A) of the inclined plane. If $BC = 2AC$, the coefficient of friction is given by $\mu = k\tan\theta$). The value of k is : **[2 Sept. 2020 (Morning)]**

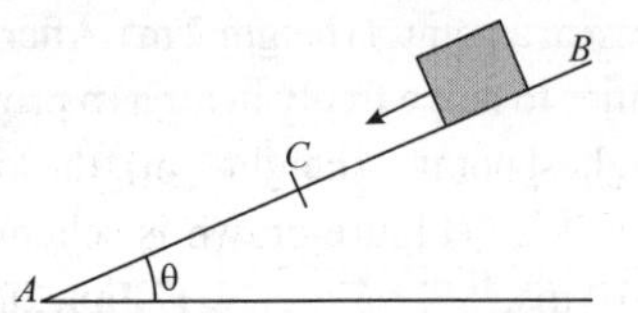

51. A cricket ball of mass 0.15 kg is thrown vertically up by a bowling machine so that it rises to a maximum height of 20 m after leaving the machine. If the part pushing the ball applies a constant force F on the ball and moves horizontally a distance of 0.2 m while launching the ball, the value of F(in N) is ($g = 10\ \text{ms}^{-2}$) _____. **[3 Sept. 2020 (Morning)]**

52. A particle is moving unidirectionally on a horizontal plane under the action of a constant power supplying energy source. The displacement (s)-time (t) graph that describes the motion of the particle is: (Graphs are drawn schematically and are not to scale) **[3 Sept. 2020 (Evening)]**

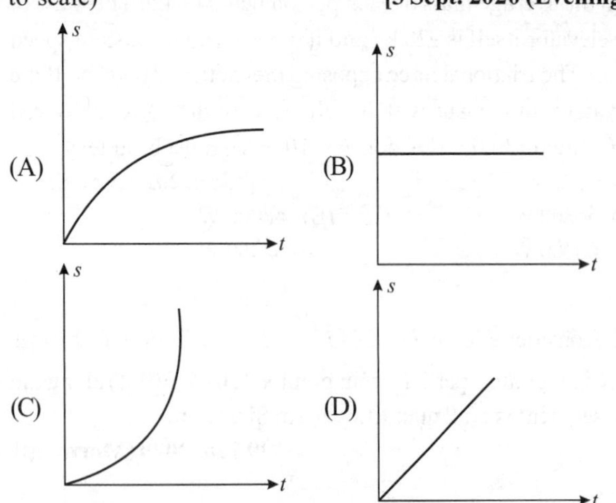

53. A block starts moving up an inclined plane of inclination 30° with an initial velocity of v_0. It comes back to its initial position with velocity $\frac{v_0}{2}$. The value of the coefficient of kinetic friction between the block and the inclined plane is close to $\frac{I}{1000}$. The nearest integer to I is _____. **[3 Sept. 2020 (Evening)]**

54. A person pushes a box on a rough horizontal platform surface. He applies a force of 200 N over a distance of 15 m. Thereafter, he gets progressively tired and his applied force reduces linearly with distance of 100 N. The total distance through which the box has been moved is 30 m. What is the work done by the person during the total movement of the box ? **[4 Sept. 2020 (Evening)]**

(A) 5690 J (B) 5250 J
(C) 3280 J (D) 2780 J

55. A body of mass 2 kg is driven by an engine delivering a constant power 1 J/s. The body starts from rest and moves in a straight line. After 9 seconds, the body has moved a distance (in m) _____. **[5 Sept. 2020 (Evening)]**

56. If the potential energy between two molecules is given by $U = -\frac{A}{r^6} + \frac{B}{r^{12}}$, then at equilibrium, separation between molecules, and the potential energy are : **[6 Sept. 2020 (Morning)]**

(A) $\left(\frac{B}{A}\right)^{1/6}, 0$ (B) $\left(\frac{B}{2A}\right)^{1/6}, -\frac{A^2}{2B}$

(C) $\left(\frac{2B}{A}\right)^{1/6}, -\frac{A^2}{4B}$ (D) $\left(\frac{2B}{A}\right)^{1/6}, -\frac{A^2}{2B}$

57. The potential energy (U) of a diatomic molecule is a function dependent on r (interatomic distance) as

$$U = \frac{\alpha}{r^{10}} - \frac{\beta}{r^5} - 3$$

where, α and β are positive constants. The equilibrium distance between two atoms will be $\left(\frac{2\alpha}{\beta}\right)^{a/b}$, where $a =$ _____. **[25 Feb. 2021 (Morning)]**

58. A boy is rolling a 0.5 kg ball on the frictionless floor with the speed of 20 ms^{-1}. The ball gets deflected by an obstacle on the way. After deflection it moves with 5% of its initial kinetic energy. What is the speed of the ball now ? **[17 March 2021 (Morning)]**

(A) 19.0 ms^{-1} (B) 4.47 ms^{-1}
(C) 14.41 ms^{-1} (D) 1.00 ms^{-1}

59. Two identical blocks A and B each of mass m resting on the smooth horizontal floor are connected by a light spring of natural length L and spring constant K. A third block C of mass m moving with a speed v along the line joining A and B collides with A. The maximum compression in the spring is : **[17 March 2021 (Evening)]**

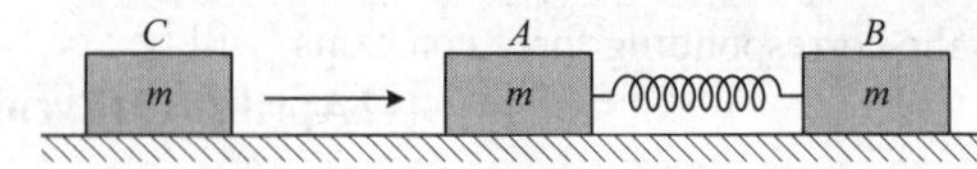

(A) $v\sqrt{\frac{M}{2K}}$ (B) $\sqrt{\frac{mv}{2K}}$

(C) $\sqrt{\frac{mv}{K}}$ (D) $\sqrt{\frac{m}{2K}}$

60. A particle of mass m moves in a circular orbit in a central potential field $U(r) = U_0 r^4$. If Bohr's quantization conditions are applied, radii of possible orbitals r_n vary with $n^{1/\alpha}$, where α is _____. **[17 March 2021 (Evening)]**

61. A constant power delivering machine has towed a box, which was initially at rest, along a horizontal straight line. The distance moved by the box in time 't' is proportional to : **[18 March 2021 (Morning)]**

(A) $t^{2/3}$ (B) $t^{3/2}$

(3) t (D) $t^{1/2}$

62. As shown in the figure, a particle of mass 10 kg is placed at a point A. When the particle is slightly displaced to its right, it starts moving and reaches the point B. The speed of the particle at B is x m/s. (Take $g = 10\text{ m/s}^2$)

The value of 'x' to the nearest integer is_____. **[18 March 2021 (Morning)]**

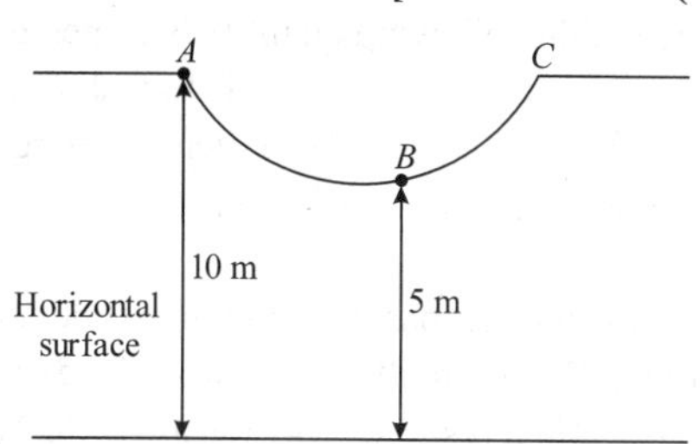

63. A ball of mass 4 kg, moving with a velocity of 10 ms^{-1}, collides with a spring of length 8 m and force constant 100 Nm^{-1}. The length of the compressed spring is x m. The value of x, to the nearest integer, is_____. **[18 March 2021 (Evening)]**

64. If the Kinetic energy of a moving body becomes four times its initial Kinetic energy, then the percentage change in its momentum will be : **[20 July 2021 (Evening)]**

(A) 400% (B) 100%

(C) 300% (D) 200%

65. A body at rest is moved along a horizontal straight line by a machine delivering a constant power. The distance moved by the body in time 't' is proportional to : **[20 July 2021 (Evening)]**

(A) $t^{1/2}$ (B) $t^{1/4}$

(C) $t^{3/2}$ (D) $t^{3/4}$

66. A porter lifts a suitcase of mass 80 kg and at the destination lowers it down by a distance of 80 cm with a constant velocity. Calculate the work done by the porter in lowering the suitcase. (Take $g = 9.8\text{ ms}^{-2}$) **[22 July 2021 (Evening)]**

(A) 784.0 J (B) −62700.0 J

(C) 627.2 J (D) −627.2 J

67. A force of $F = (5y + 20)\hat{j}$ N acts on a particle. The work done by this force when the particle is moved from $y = 10$ m to $y = 10$ m is_____. **[25 July 2021 (Evening)]**

68. An automobile of mass 'm' accelerates starting from origin and initially at rest, while the engine supplies constant power P. The position is given as a function of time by : **[27 July 2021 (Evening)]**

(A) $\left(\frac{8P}{9m}\right)^{1/2} t^{3/2}$ (B) $\left(\frac{9P}{8m}\right)^{1/2} t^{3/2}$

(C) $\left(\frac{9m}{8P}\right)^{1/2} t^{3/2}$ (D) $\left(\frac{8P}{9m}\right)^{1/2} t^{2/3}$

69. Given below is the plot of a potential energy function $U(x)$ for a system, in which a particle is in one dimensional motion, while a conservative force $F(x)$ acts on it. Suppose that **[27 July 2021 (Evening)]**

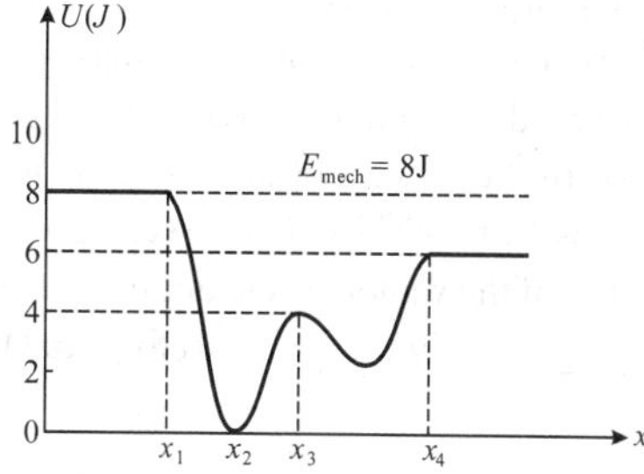

(A) At $x = x_3$, $K.E. = 4$

(B) At $x > x_4$, $K.E.$ is constant throughout the region

(C) At $x < x_1$, $K.E.$ is smallest and the particle is moving at the slowest speed

(D) At $x = x_2$, $K.E.$ is greatest and the particle is moving at the fastest speed

70. A uniform chain of length 3 meter and mass 3 kg overhangs a smooth table with 2 meter laying on the table. If k is the kinetic energy of the chain in joule as it completely slips off the table, then the value of k is _____. **[26 Aug. 2021 (Morning)]**

71. Two persons A and B perform same amount of work in moving a body through a certain distance d with application of forces acting at angles 45° and 60° with the direction of displacement respectively. The ratio of force applied by person A to the force applied by person B is $\frac{1}{\sqrt{x}}$. The value of x is _____. **[27 Aug. 2021 (Morning)]**

72. A block moving horizontally on a smooth surface with a speed of 40 ms^{-1} splits into two equal parts. If one of the parts moves at 60 ms^{-1} in the same direction, them the fractional change in the kinetic energy will be $x : 4$ where $x =$_____. **[31 Aug. 2021 (Morning)]**

73. A block moving horizontally on a smooth surface with a speed of 40 m/s splits into two parts with masses in the ratio of 1 : 2. If the smaller part moves at 60 m/s in the same direction, then the fractional change in kinetic energy is : **[31 Aug. 2021 (Evening)]**

(A) $\frac{1}{8}$ (B) $\frac{2}{3}$

(C) $\frac{1}{3}$ (D) $\frac{1}{4}$

74. A body of mass m dropped from a height h reaches the ground with a speed of $0.8\sqrt{gh}$. The value of work done by the air - friction is : **[01 Sep. 2021 (Evening)]**

(A) -0.68 mgh (B) 0.64 mgh

(C) mgh (D) 1.64 mgh

75. An engine is attached to a wagon through a shock absorber of length 1.5 m. The system with a total mass of 40000 kg is moving with a speed of 72 kmh^{-1} when the brakes are applied to bring it to rest. In the process of the system being brought to rest, the spring of the shock absorber gest compressed by 1.0 m. If 90% of energy of the wagon is lost due of friction the spring constant is _____ $\times 10^5$ N/m. **[01 Sep. 2021 (Evening)]**

76. A particle experiences a variable force $\vec{F} = (4x\hat{i} + 3y^2\hat{j})$ in a horizontal x-y plane. Assume distance in meters and force is newton. If the particle moves from point (1, 2) to point (2, 3) in the x-y plane, the Kinetic Energy changes by :

[24 June 2022 (Morning)]

(A) 50.0 J (B) 12.5 J

(C) 25.0 J (D) 0 J

77. A ball of mass 100 g is dropped from a height $h = 10$ cm on a platform fixed at the top of vertical spring (as shown in figure). The ball stays on the platform and the platform is depressed by a distance $h/2$. The spring constant is _____ Nm.

[Use $g = 10$ ms^{-1}] **[24 June 2022 (Morning)]**

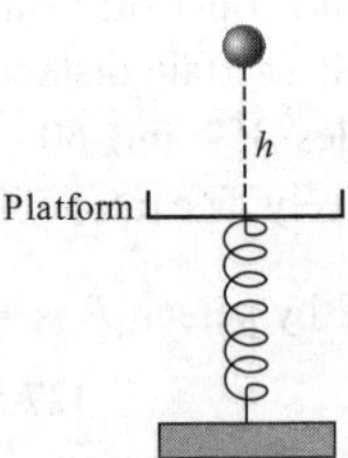

78. Potential energy as a function of r is given by $U = \frac{A}{r^{10}} - \frac{b}{r^5}$, where r is the interatomic distance, A and B are positive constants. The equilibrium distance between the two atoms will be : **[24 June 2022 (Evening)]**

(A) $\left(\frac{A}{B}\right)^{1/5}$ (B) $\left(\frac{B}{A}\right)^{1/5}$

(C) $\left(\frac{2A}{B}\right)^{1/5}$ (D) $\left(\frac{B}{2A}\right)^{1/5}$

79. A uniform chain of 6 m length is placed on a table such that a part of its length is hanging over the edge of the table. The system is at rest. The co-efficient of static friction between the chain and the surface of the table is 0.5, the maximum length of the chain hanging from the table is _____ m.

[25 June 2022 (Morning)]

80. A 0.5 kg block moving at a speed of 12 ms^{-1} compresses a spring through a distance 30 cm when its speed is halved. The spring constant of the spring will be _____ Nm^{-1}.

[25 June 2022 (Morning)]

81. An object is thrown vertically upwards. At its maximum height, which of the following quantity becomes zero?

[26 June 2022 (Morning)]

(A) Momentum (B) Potential energy

(C) Acceleration (D) Force

82. Arrange the four graphs in descending order of total work done; where W_1, W_2, W_3 and W_4 are the work done corresponding to figure a, b, c and d respectively : **[26 June 2022 (Evening)]**

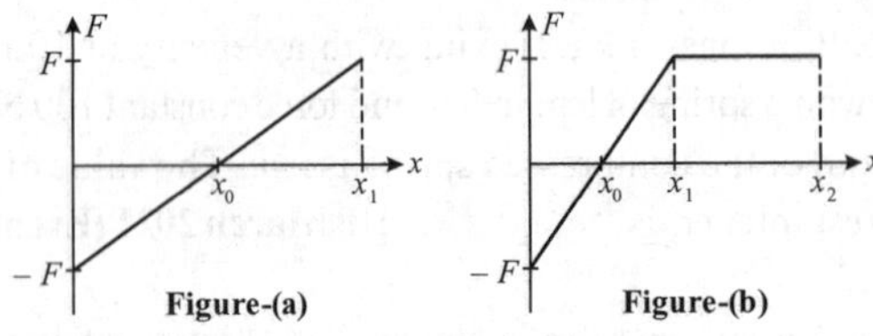

Figure-(a) Figure-(b)

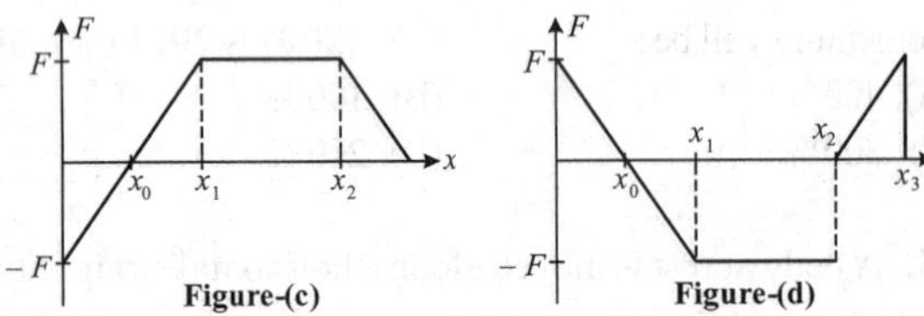

Figure-(c) Figure-(d)

(A) $W_3 > W_2 > W_1 > W_4$ (B) $W_3 > W_2 > W_4 > W_1$

(C) $W_2 > W_3 > W_4 > W_1$ (D) $W_2 > W_3 > W_1 > W_4$

83. One end of a massless spring of spring constant k and natural length l_0 is fixed while the other end is connected to a small object of mass m lying on a frictionless table. The spring remains horizontal on the table. If the object is made to rotate at an angular velocity ω about an axis passing through fixed end, then the elongation of the spring will be :

[27 June 2022 (Evening)]

(A) $\frac{k - m\omega^2 l_0}{m\omega^2}$ (B) $\frac{m\omega^2 l_0}{k + m\omega^2}$

(C) $\frac{m\omega^2 l_0}{k - m\omega^2}$ (D) $\frac{k + m\omega^2 l_0}{m\omega^2}$

84. A block of mass 2 kg moving on a horizontal surface with speed of 4 ms^{-1} enters a rough surface ranging from $x = 0.5$ m to $x = 1.5$ m. The retarding force in this range of rough surface is related to distance by $F = -kx$ where $k = 12$ Nm^{-1}. The speed of the block as it just crosses the rough surface will be :

[28 June 2022 (Evening)]

(A) Zero (B) 1.5 ms^{-1}
(C) 2.0 ms^{-1} (D) 2.5 ms^{-1}

85. Water fall from a 40 m high dam at the rate of 9×10^4 kg per hour. Fifty percentage of gravitational potential energy can be converted into electrical energy. Using this hydroelectric energy number of 100 W lamps, that can be lit, is : [Take $g = 10$ ms^{-2}]

[28 June 2022 (Evening)]

(A) 25 (B) 50
(C) 100 (D) 18

86. A particle of mass 500 gm is moving in a straight line with velocity $v = bx^{5/2}$. The work done by the net force during its displacement from $x = 0$ to $x = 4$ m is : (Take $b = 0.25$ $m^{-3/2}$ s^{-1})

[29 June 2022 (Morning)]

(A) 2 J (B) 4 J
(C) 8 J (D) 16 J

87. In the given figure, the block of mass m is dropped from the point A. The expression for kinetic energy of block when it reaches point B is : **[29 June 2022 (Evening)]**

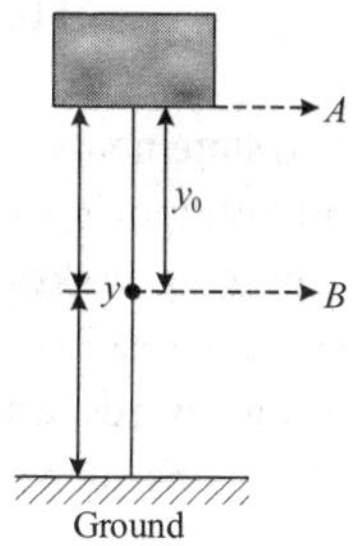

(A) $\frac{1}{2}mg\,y_0^2$ (B) $\frac{1}{2}mg\,y^2$
(C) $mg(y - y_0)$ (D) $mg\,y_0$

88. A body of mass 0.5 kg travels on straight line path with velocity $v = (3x^2 + 4)$ m/s. The net workdone by the force during its displacement from $x = 0$ to $x = 2$ m is :

[25 July 2022 (Morning)]

(A) 64 J (B) 60 J
(C) 120 J (D) 128 J

89. As per the given figure, two blocks each of mass 250 g are connected to a spring of spring constant 2 Nm^{-1}. If both are given velocity v in opposite directions, then maximum elongation of the spring is : **[26 July 2022 (Morning)]**

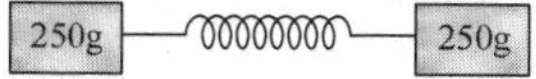

(A) $\frac{v}{2\sqrt{2}}$ (B) $\frac{v}{2}$
(C) $\frac{v}{4}$ (D) $\frac{v}{\sqrt{2}}$

90. A body of mass 8 kg and another of mass 2 kg are moving with equal kinetic energy. The ratio of their respective momenta will be : **[26 July 2022 (Evening)]**

(A) 1 : 1 (B) 2 : 1
(C) 1 : 4 (D) 4 : 1

91. Sand is being dropped from a stationary dropper at a rate of 0.5 kgs^{-1} on a conveyor belt moving with a velocity of 5 ms^{-1}. The power needed to keep belt moving with the same velocity will be : **[27 July 2022 (Morning)]**

(A) 1.25 W (B) 2.5 W
(C) 6.25 W (D) 12.5 W

92. A block of mass m (as shown in figure) moving with kinetic energy E compresses a spring through a distance 25 cm when, its speed is halved. The value of spring constant of used spring will be nE Nm^{-1} for n = _____. **[28 July 2022 (Morning)]**

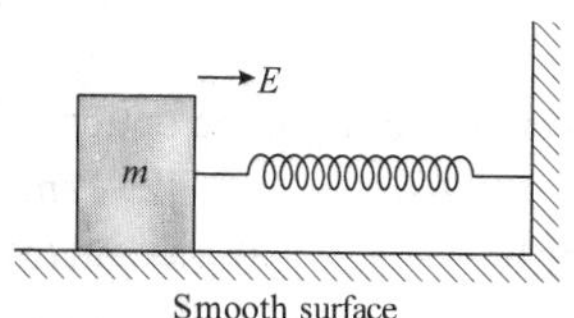

93. If momentum of a body is increased by 20%, then its kinetic energy increases by : **[29 July 2022 (Evening)]**

(A) 36% (B) 40%
(C) 44% (D) 48%

94. A spherical body of mass 2 kg starting from rest acquires a kinetic energy of 10000 J at the end of 5th second. The force acted on the body is _____ N. **[24 Jan. 2023 (Morning)]**

95. A body of mass 1 kg begins to move under the action of a time dependent force $\vec{F} = (t\hat{i} + 3t^2\hat{j})$ N. Where $\hat{i}$ and $\hat{j}$ are the unit vectors along x and y axis. The power developed by above force, at the time $t = 2$ s. Will be _____ W.

[24 Jan. 2023 (Evening)]

96. An object of mass m initially at rest on a smooth horizontal plane starts moving under the action of force $F = 2$ N. In the process of its linear motion, the angle θ (as shown in figure) between the direction of force and horizontal varies as $\theta = kx$, where k is a constant and x is the distance covered by the object from its initial position. The expression of kinetic energy of the object will be $E = \frac{n}{k}\sin\theta$. The value of n is _____.

[25 Jan. 2023 (Morning)]

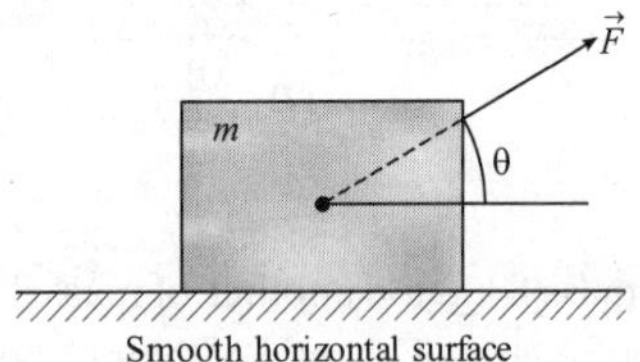

97. A 0.4 kg mass takes 8 s to reach ground when dropped from a certain height P above surface of earth. The loss of potential energy in the last second of fall is ______ J. [Take $g = 10\ m/s^2$] **[29 Jan. 2023 (Morning)]**

98. Identify the correct statements from the following :

(a) Work done by a man in lifting a bucket out of a well by means of a rope tied to the bucket is negative

(b) Work done by gravitational force in lifting a bucket out of a well by a rope tied to the bucket is negative

(c) Work done by friction on a body sliding down an inclined plane is positive

(d) Work done by an applied force on a body moving on a rough horizontal plane with uniform velocity in zero

(e) Work done by the air resistance on an oscillating pendulum in negative

Choose the correct answer from the options given below :

[29 Jan. 2023 (Evening)]

(A) (b) and (e) only (B) (a) and (c) only

(C) (b), (d) and (e) only (D) (b) and (d) only

99. A lift of mass $M = 500$ kg is descending with speed of $2\ ms^{-1}$. Its supporting cable begins to slip thus allowing it to fall with a constant acceleration of $2\ ms^{-2}$. The kinetic energy of the lift at the end of fall through a distance of 6 m will be _____ kJ.

[31 Jan. 2023 (Morning)]

100. A small particle moves to position $5\hat{i} - 2\hat{j} + \hat{k}$ from its initial position $2\hat{i} + 3\hat{j} - 4\hat{k}$ under the action of force $5\hat{i} + 2\hat{j} + 7\hat{k}$ N. The value of work done will be _____ J.

[1 Feb. 2023 (Morning)]

101. A force $F = (5 + 3y^2)$ acts on a particle in the y-direction, where F is newton and y is in meter. The work done by the force during a displacement from $y = 2$ m to $y = 5$ m is _____ J.

[1 Feb. 2023 (Evening)]

102. A block is fastened to a horizontal spring. The block is pulled to a distance $x = 10$ cm from its equilibrium position (at $x = 0$) on a frictionless surface from rest. The energy of the block at $x = 5$ cm is 0.25 J. The spring constant of the spring is _____ Nm^{-1}. **[1 Feb. 2023 (Evening)]**

103. A particle of mass 10 g moves in a straight line with retardation $2x$, where x is the displacement in SI units. Its loss of kinetic energy for above displacement is $\left(\frac{10}{x}\right)^{-n}$ J. The value of n will be _____. **[6 April 2023 (Morning)]**

104. If the maximum load carried by an elevator is 1400 kg (600 kg – Passenger + 800 kg – elevator), which is moving up with a uniform speed of $3\ ms^{-1}$ and the frictional force acting on it is 2000 N, then the maximum power used by the motor is _____ kW. ($g = 10\ m/s^2$) **[10 April 2023 (Evening)]**

105. A force $\vec{F} = (2 + 3x)\hat{i}$ acts on a particle in the x direction where F is in newton and x is in meter. The work done by this force during a displacement from $x = 0$ to $x = 4$ m, is_____ J.

[11 April 2023 (Morning)]

106. A block of mass 5 kg starting from rest pulled up on a smooth incline plane making an angle of 30° with horizontal with an affective acceleration of $1\ ms^{-2}$. The power delivered by the pulling force at $t = 10$ s from the start is_____ W. [Use $g = 10\ ms^{-2}$] (Calculate the nearest integer value)

[11 April 2023 (Evening)]

107. Given below are two statements :

Statement-I: A truck and a car moving with same kinetic energy are brought to rest by applying brakes which provide equal retarding forces. Both come to rest in equal distance.

Statement-II: A car moving towards east takes a turn and moves towards north, the speed remains unchanged. The acceleration of the car is zero.

In the light of given statements, choose the most appropriate answer from the options given below.

[12 April 2023 (Morning)]

(A) Statement-I is correct but Statement-II is incorrect

(B) Statement-I is incorrect but Statement-II is correct

(C) Both Statement-I is correct but Statement-II are incorrect

(D) Both Statement-I is correct but Statement-II are correct

108. To maintain a speed of 80 km/h by a bus of mass 500 kg on a plane rough road for 4 km distance, the work done by the engine of the bus will be _____ KJ. [The coefficient of friction between tyre of bus and road is 0.04].

[12 April 2023 (Morning)]

109. The ratio of powers of two motors is $\frac{3\sqrt{x}}{\sqrt{x}+1}$, that are capable of raising 300 kg water in 5 minutes and 50 kg water in 2 minutes respectively from a well of 100 m deep. The value of x will be : **[13 April 2023 (Morning)]**

(A) 2 (B) 4
(C) 2.4 (D) 16

110. A body of mass (5 ± 0.5) kg is moving with a velocity of (20 ± 0.4) m/s. Its kinetic energy will be : **[13 April 2023 (Morning)]**

(A) (1000 ± 140) J (B) (1000 ± 0.14) J
(C) (500 ± 0.14) J (D) (500 ± 140) J

111. Two bodies are having kinetic energies in the ratio 16 : 9. If they have same linear momentum, the ratio of their masses respectively is : **[13 April 2023 (Morning)]**

(A) 4 : 3 (B) 3 : 4
(C) 16 : 9 (D) 9 : 16

112. A car accelerates from rest to u m/s. The energy spent in this process is E J. The energy required to accelerate the car from u m/s to $2u$ m/s is nE J. The value of n is_____. **[13 April 2023 (Evening)]**

113. A block of mass 10 kg is moving along x-axis under the action of force $F = 5x$ N. The work done by the force in moving the block from $x = 2$ m to 4 m will be_____ J. **[15 April 2023 (Morning)]**

* * * * *

CHAPTER **5**

WORK ENERGY POWER & APPLICATIONS

5.2 CIRCULAR MOTION

1. The minimum velocity (in ms^{-1}) with which a car driver must traverse a flat curve of radius 150 m and coefficient of friction 0.6 to avoid skidding is : **[2002]**

(A) 60 (B) 30

(C) 15 (D) 25

2. A particle is acted upon by a force of constant magnitude which is always perpendicular to the velocity of the particle. The motion of the particle takes place in a plane, it follows that: **[2004]**

(A) Its velocity is constant

(B) Its acceleration is constant

(C) Its kinetic energy is constant

(D) It moves in a straight line

3. Which of the following statements is false for a particle moving in a circle with a constant angular speed? **[2004]**

(A) The velocity vector is tangent to the circle

(B) The acceleration vector is tangent to the circle

(C) The acceleration vector point to the center of the circle

(D) The velocity and acceleration vectors are perpendicular to each other

4. For a particle in uniform circular motion in anticlockwise sense the acceleration at $\vec{a}$ point $P(R, \theta)$ on the circle of radius R is (Here θ is measured from the x-axis) : **[2010]**

(A) $-\frac{v^2}{R}\cos\theta\,\hat{i}+\frac{v^2}{R}\sin\theta\,\hat{j}$ (B) $-\frac{v^2}{R}\sin\theta\,\hat{i}+\frac{v^2}{R}\cos\theta\,\hat{j}$

(C) $-\frac{v^2}{R}\cos\theta\,\hat{i}-\frac{v^2}{R}\sin\theta\,\hat{j}$ (D) $\frac{v^2}{R}\hat{i}+\frac{v^2}{R}\hat{j}$

5. A point P moves in counter-clockwise direction on a circular path as shown in the figure. The movement of 'P' is such that it sweeps out a length $s = t^3 + 5$, where s is in metres and t is in seconds. The radius of the path is 20 m. The acceleration of 'P' when $t = 2$ s is nearly : **[2010]**

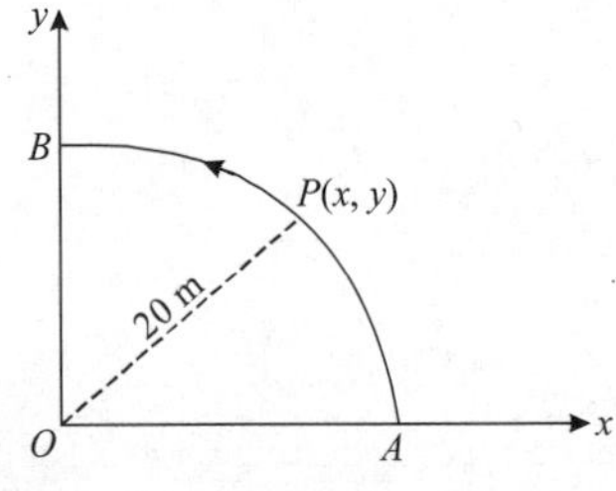

(A) 13 m/s^2 (B) 12 m/s^2

(C) 7.2 m/s^2 (D) 14 m/s^2

6. Two cars of masses m_1 and m_2 are moving in circles of radii r_1 and r_2, respectively. Their speeds are such that they make complete circles in the same time t. The ratio of their centripetal acceleration is : **[2012]**

(A) $m_1 r_1 : m_2 r_2$ (B) $m_1 : m_2$

(C) $r_1 : r_2$ (D) $1 : 1$

7. A body of mass 'm' is tied to one end of a spring and whirled round in a horizontal plane with a constant angular velocity. The elongation in the spring is 1 cm. If the angular velocity is doubled, the elongation in the spring is 5 cm. The original length of the spring is : **[2013 (Online)]**

(A) 15 cm (B) 12 cm

(C) 16 cm (D) 10 cm

8. A particle is released on a vertical smooth semicircular track from point X so that OX makes angle θ from the vertical (see figure). The normal reaction of the track on the particle vanishes at point Y where OY makes angle ϕ with the horizontal. Then : **[2014 (Online)]**

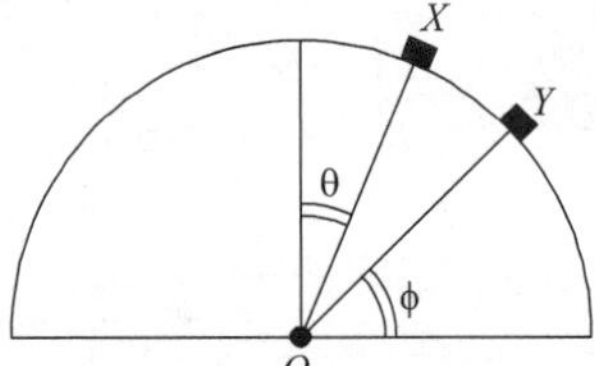

(A) $\sin\phi = \cos\phi$ (B) $\sin\phi = \frac{1}{2}\cos\theta$

(C) $\sin\phi = \frac{2}{3}\cos\theta$ (D) $\sin\phi = \frac{3}{4}\cos\theta$

9. If a body moving in circular path maintains constant speed of 10 ms^{-1}, then which of the following correctly describes relation between acceleration and radius ? **[2015 (Online)]**

(A) 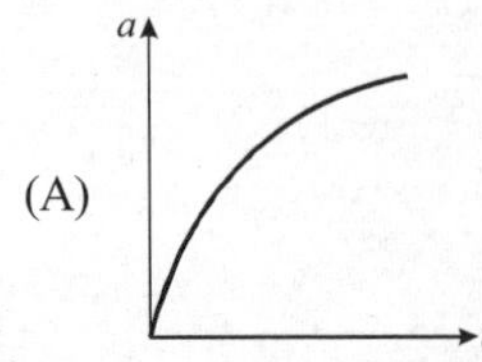

(B) 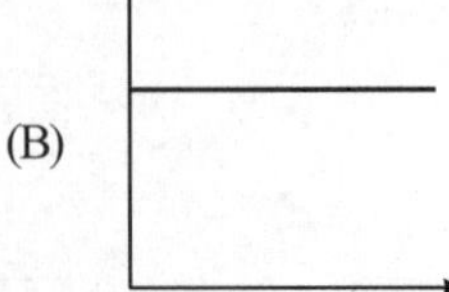

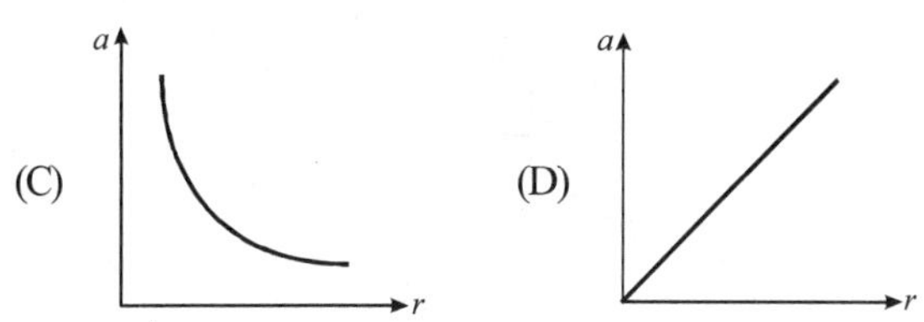

10. A particle is moving in a circle of radius r under the action of a force $F = \alpha r^2$ which is directed towards centre of the circle. Total mechanical energy (kinetic energy + potential energy) of the particle is: (Take potential energy = 0 for $r = 0$) **[2015 (Online)]**

(A) $\frac{1}{2}\alpha r^3$ (B) $\frac{5}{6}\alpha r^3$

(C) $\frac{4}{3}\alpha r^3$ (D) αr^3

11. A particle of mass M is moving in a circle of fixed radius R in such a way that its centripetal acceleration at time t is given by $n^2 R t^2$ where n is a constant. The power delivered to the particle by the force acting on it, is : **[2016 (Online)]**

(A) $\frac{1}{2}M n^2 R^2 t^2$ (B) $M n^2 R^2 t$

(C) $M n R^2 t^2$ (D) $M n R^2 t$

12. Concrete mixture is made by mixing cement, stone and sand in a rotating cylinder drum. If the drum rotates too fast, the ingredients remain stuck to the wall of the drum and proper mixing of ingredients does not take place. The maximum rotational speed of the drum in revolutions per minute (rpm) to ensure proper mixing is close to : [Take the radius of the drum to be 1.25 m and its axle to be horizontal] **[2016 (Online)]**

(A) 27.0 (B) 0.4

(C) 1.3 (D) 8.0

13. A conical pendulum of length 1 m making an angle $\theta = 45^\circ$ w.r.t. Z-axis and moves in a circle in the XY plane. The radius of the circle is 0.4 m and its centre is vertically below O. The speed of the pendulum, in its circular path, will be : [Take $g = 10\ ms^{-2}$] **[2017 (Online)]**

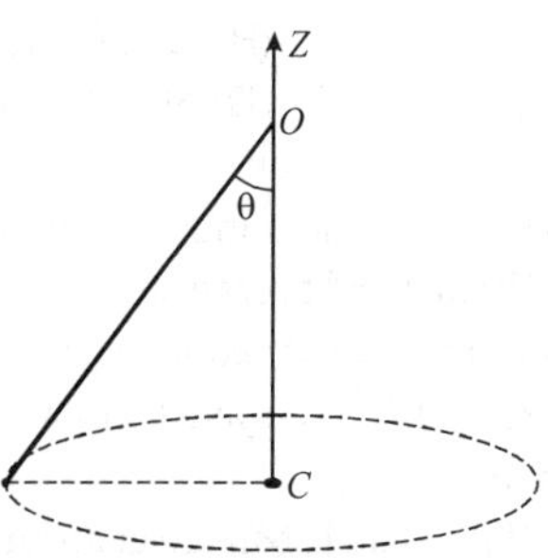

(A) 0.4 m/s (B) 4 m/s

(C) 0.2 m/s (D) 2 m/s

14. A disc rotates about its axis of symmetry in a horizontal plane at a steady rate of 3.5 revolutions per second. A coin placed at a distance of 1.25 cm from the axis of rotation remains at rest on the disc. The coefficient of friction between the coin and the disc is: ($g = 10\ m/s^2$) **[2018 (Online)]**

(A) 0.5 (B) 0.7

(C) 0.3 (D) 0.6

15. A particle is moving along a circular path with a constant speed of 10 ms^{-1}. What is the magnitude of the change in velocity of the particle, when it moves through an angle of 60° around the centre of the circle? **[11 Jan. 2019 (Morning)]**

(A) Zero (B) 10 m/s

(C) $10\sqrt{3}$ m/s (D) $10\sqrt{2}$ m/s

16. Two particles A, B are moving on two concentric circles of radii R_1 and R_2 with equal angular speed ω. At $t = 0$, their positions and direction of motion are shown in the figure. The relative velocity $\vec{\upsilon}_A - \vec{\upsilon}_B$ at $t = \pi/2\omega$ is given by : **[12 Jan. 2019 (Evening)]**

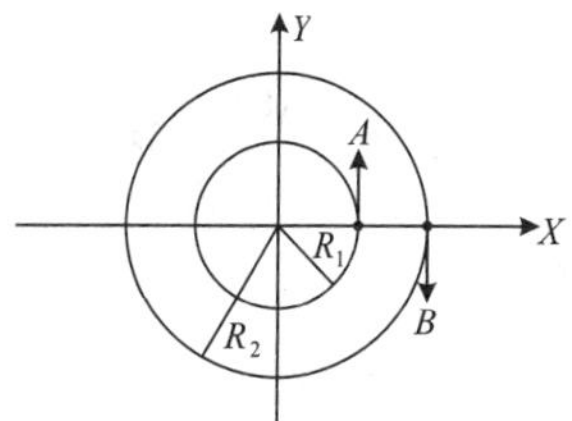

(A) $-\omega(R_1 + R_2)\hat{i}$ (B) $\omega(R_1 + R_2)\hat{i}$

(C) $\omega(R_1 - R_2)\hat{i}$ (D) $\omega(R_2 - R_1)\hat{i}$

17. A smooth wire of length $2\pi r$ is bent into a circle and kept in a vertical plane. A bead can slide smoothly on the wire. When the circle is rotating with angular speed ω about the vertical diameter AB, as shown in figure, the bead is at rest with respect to the circular ring at position P as shown. Then the value of ω^2 is equal to : **[12 April 2019 (Evening)]**

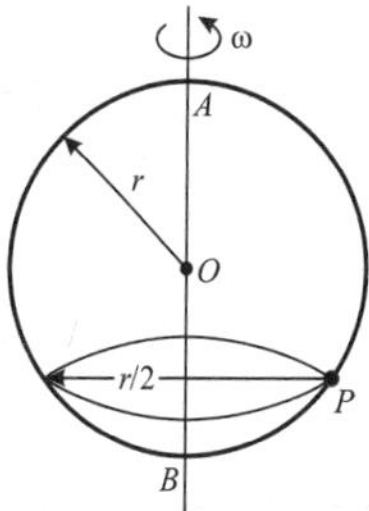

(A) $(g\sqrt{3})/r$ (B) $\frac{\sqrt{3}g}{2r}$

(C) $2g/r$ (D) $2g/(r\sqrt{3})$

18. A particle of mass m is fixed to one end of a light spring having force constant k and unstretched length l. The other end is fixed. The system is given an angular speed ω about the fixed end of the spring such that it rotates in a circle in gravity free space. Then the stretch in the spring is :

[8 Jan. 2020 (Morning)]

(A) $\dfrac{ml\omega^2}{k+m\omega^2}$ (B) $\dfrac{ml\omega^2}{k-m\omega^2}$

(C) $\dfrac{ml\omega^2}{k-\omega m}$ (D) $\dfrac{ml\omega^2}{k+m\omega}$

19. A spring mass system (mass m, spring constant k and natural length l) rest in equilibrium on a horizontal disc. The free end of the spring is fixed at the centre of the disc. If the disc together with spring mass system, rotates about it's axis with an angular velocity ω, ($k >> m\omega^2$) the relative change in the length of the spring is best given by the option :

[9 Jan. 2020 (Evening)]

(A) $\dfrac{2m\omega^2}{k}$ (B) $\dfrac{m\omega^2}{3k}$

(C) $\sqrt{\dfrac{2}{3}}\left(\dfrac{m\omega^2}{k}\right)$ (D) $\dfrac{m\omega^2}{k}$

20. A bead of mass m stays at point $P(a, b)$ on a wire bent in the shape of a parabola $y = 4Cx^2$ and rotating with angular speed ω (see figure). The value of ω is (neglect friction) :

[2 Sept. 2020 (Morning)]

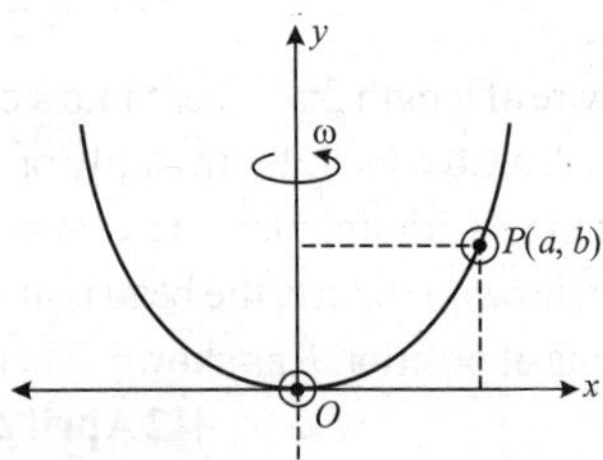

(A) $\sqrt{\dfrac{2gC}{ab}}$ (B) $2\sqrt{2gC}$

(C) $\sqrt{\dfrac{2g}{C}}$ (D) $2\sqrt{gC}$

21. A cylindrical vessel containing a liquid is rotated about its axis so that the liquid rises at its sides as shown in the figure. The radius of vessel is 5 cm and the angular speed of rotation is ω rad s^{-1}. The difference in the height, h (in cm) of liquid at the centre of vessel and at the side will be :

[2 Sept. 2020 (Morning)]

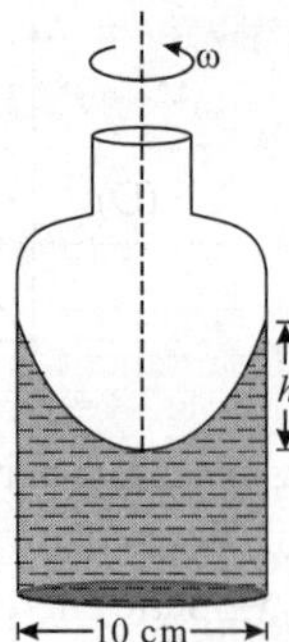

(A) $\dfrac{25\omega^2}{2g}$ (B) $\dfrac{2\omega^2}{5g}$

(C) $\dfrac{5\omega^2}{2g}$ (D) $\dfrac{2\omega^2}{25g}$

22. A clock has a continuously moving second's hand of 0.1 m length. The average acceleration of the tip of the hand (in units of ms^{-2}) is of the order of : **[6 Sept. 2020 (Morning)]**

(A) 10^{-3} (B) 10^{-2}

(C) 10^{-4} (D) 10^{-1}

23. A small bob tied at one end of a thin string of length 1 m is describing a vertical circle so that the maximum and minimum tension in the string are in the ratio 5 : 1. The velocity of the bob at the highest position is _____ m/s. (Take $g = 10$ m/s^2).

[25 Feb. 2021 (Morning)]

24. A particle is moving with uniform speed along the circumference of a circle of radius R under the action of a central fictitious force F which is inversely proportional to R^3. Its time period of revolution will be given by : **[26 Feb. 2021 (Morning)]**

(A) $T \propto R^2$ (B) $T \propto R^{3/2}$

(C) $T \propto R^{5/2}$ (D) $T \propto R^{4/3}$

25. A block of 200 g mass moves with a uniform speed in a horizontal circular groove, with vertical side walls of radius 20 cm. If the block takes 40 s to complete one round, the normal force by the side walls of the groove is :

[16 March 2021 (Morning)]

(A) 0.0314 N (B) 9.859×10^{-2} N

(C) 6.28×10^{-3} N (D) 9.859×10^{-4} N

26. Statement-I : A cyclist is moving on an unbanked road with a speed of 7 kmh^{-1} and takes a sharp circular turn along a path of radius of 2m without reducing the speed. The static friction coefficient is 0.2. The cyclist will not slip and pass the curve ($g = 9.8$ m/s^2)

Statement-II : If the road is banked at an angle of 45°, cyclist can cross the curve of 2m radius with the speed of 18.5 kmh^{-1} without slipping.

In the light of the above statements, choose the correct answer from the options given below. **[16 March 2021 (Evening)]**

(A) Statement-I is incorrect and statement-II is correct
(B) Statement-I is correct and statement-II is incorrect
(C) Both statement-I and statement-II are false
(D) Both statement-I and statement-II are true

27. The angular speed of truck wheel is increased from 900 rpm to 2460 rpm in 26 seconds. The number of revolutions by the truck engine during this time is _____.
(Assuming the acceleration to be uniform).
[17 March 2021 (Morning)]

28. A particle of mass m moves in a circular orbit under the central potential field, $U(r) = \frac{-C}{r}$, where C is a positive constant. The correct radius – velocity graph of the particle's motion is :
[18 March 2021 (Evening)]

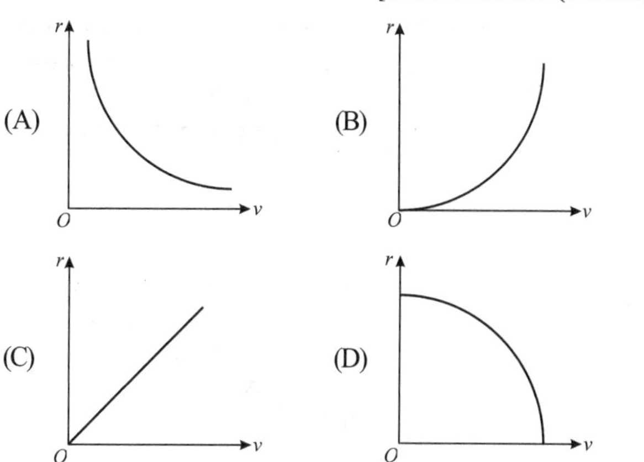

29. A body rotating with an angular speed of 600 rpm is uniformly accelerated to 1800 rpm in 10 s. The number of rotations made in the process is _____ .
[20 July 2021 (Evening)]

30. A pendulum bob has a speed of 3 m/s at its lowest position. The pendulum is 50 cm long. The speed of bob, when the length makes an angle of 60° to the vertical will be ($g = 10$ m/s^2) _____ m/s : **[25 July 2021 (Morning)]**

31. A small block slides down from the top of hemisphere of radius $R = 3$ m as shown in the figure. The height h at which the block will lose contact with the surface of the sphere is _____ (Assume there is no friction between the block and the hemisphere) : **[27 July 2021 (Evening)]**

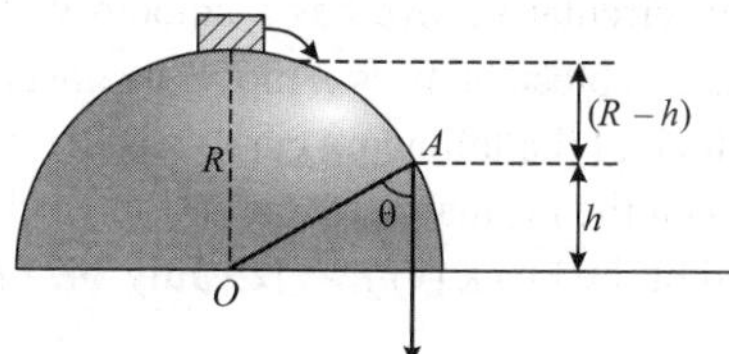

32. Particle of mass m is suspended from a celling through a string of length L. The particle moves in a horizontal circle of radius r such that $r = \frac{L}{\sqrt{2}}$. The speed of particle will be :
[26 Aug. 2021 (Evening)]

(A) $\sqrt{\frac{rg}{2}}$ (B) $2\sqrt{rg}$
(C) $\sqrt{2rg}$ (D) $\sqrt{rg}$

33. A huge circular arc of length 4.4 LY subtends an angle 4s at the centre of circle. How long it would take for a body to complete 4 revolution if its speed is 8 AU per second ?
(Given: 1 light year = 9.46×10^{15} m) **[27 Aug. 2021 (Morning)]**

(A) 4.1×10^8 s (B) 4.5×10^{10} s
(C) 7.2×10^8 s (D) 3.5×10^6 s

34. A boy ties a stone of mass 100 g to the end of a 2 m long string and whirls it around in a horizontal plane. The string can withstand the maximum tension of 80 N. If the maximum speed with which the stone can revolve is $\frac{K}{\pi}$ rev/min. The value of K is: [Assume the string is massless and unstretchable]
[24 June 2022 (Morning)]

(A) 400 (B) 300
(C) 600 (D) 800

35. A stone of mass m, tied to a string is being whirled in a vertical circle with a uniform speed. The tension in the string is :
[24 June 2022 (Evening)]

(A) The same throughout the motion
(B) Minimum at the highest position of the circular path
(C) Minimum at the lowest position of the circular path
(D) Minimum when the rope is in the horizontal position

36. A disc with a flat small bottom beaker placed on it at a distance R from its center is revolving about an axis passing through the center and perpendicular to its plane with an angular velocity ω. The coefficient of static friction between the bottom of the beaker and the surface of the disc is μ. The beaker will revolve with the disc if : **[25 June 2022 (Evening)]**

(A) $R \le \frac{\mu g}{2\omega^2}$ (B) $R \le \frac{\mu g}{\omega^2}$

(C) $R \ge \frac{\mu g}{2\omega^2}$ (D) $R \ge \frac{\mu g}{\omega^2}$

37. For a particle in uniform circular motion, the acceleration a at any point $P(R, \theta)$ on the circular path of radius R is (when θ is measured from the positive x-axis and v is uniform speed) :
[25 June 2022 (Evening)]

(A) $-\frac{v^2}{R}\sin\theta\hat{i} + \frac{v^2}{R}\cos\theta\hat{j}$ (B) $-\frac{v^2}{R}\cos\theta\hat{i} + \frac{v^2}{R}\sin\theta\hat{j}$

(C) $-\frac{v^2}{R}\cos\theta\hat{i} - \frac{v^2}{R}\sin\theta\hat{j}$ (D) $-\frac{v^2}{R}\hat{i} + \frac{v^2}{R}\hat{j}$

38. A curved in a level road has a radius 75 m. The maximum speed of a car turning this curved road can be 30 m/s without skidding. If radius of curved road is changed to 48 m and the coefficient of friction between the tyres and the road remains same, then maximum allowed speed would be _____ m/s.

[25 June 2022 (Evening)]

39. A ball is released from rest from point P of a smooth semi-spherical vessel as shown in figure. The ratio of the centripetal force and normal reaction on the ball at point Q is A while angular position of point Q is α with respect to point P. Which of the following graphs represent the correct relation between A and α when ball goes from Q to R?

[26 June 2022 (Morning)]

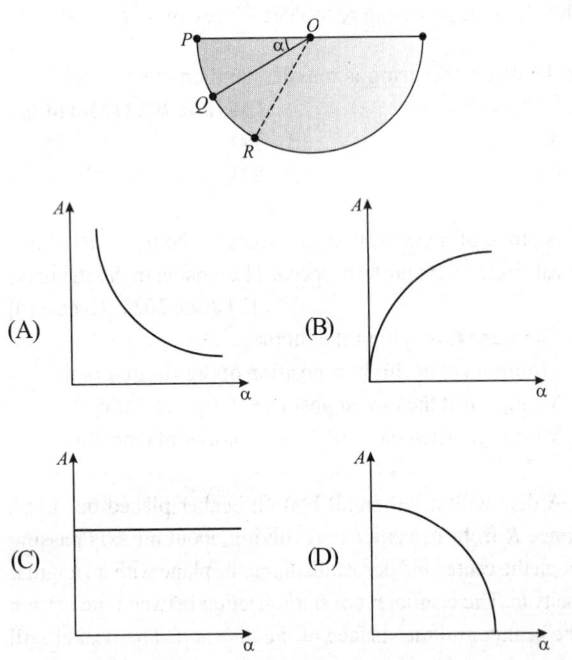

40. A stone tide to a string of length L is whirled in a vertical circle with the other end of the string at the centre. At a certain instant of time, the stone is at its lowest position and has a speed u. The magnitude of change in its velocity, as it reaches a position where the string is horizontal, is $\sqrt{x(u^2 - gL)}$. The value of x is:

[27 June 2022 (Evening)]

(A) 3 (B) 2
(C) 1 (D) 5

41. A particle of mass m is moving in a circular path of constant radius r such that its centripetal acceleration (a) is varying with time t as $a = k^2rt^2$. Where k is a constant. The power delivered to the particle by the force acting on it is given as:

[28 June 2022 (Morning)]

(A) Zero (B) $mk^2r^2t^2$
(C) mk^2r^2t (D) mk^2rt

42. Motion of a particle in x-y plane is described by a set of following equations $x = 4\sin\left(\frac{\pi}{2} - \omega t\right)$ m and $y = 4\sin(\omega t)$ m. The path of particle will be:

[28 June 2022 (Morning)]

(A) Circular (B) Helical
(C) Parabolic (D) Elliptical

43. A pendulum is suspended by a string of length 250 cm. The mass of the bob of the pendulum is 200 g. The bob is pulled aside until the string is at 60° with vertical as shown in the figure. After releasing the bob, the maximum velocity attained by the bob will be _____ ms^{-1}. [If $g = 10$ m/s^2]

[28 June 2022 (Morning)]

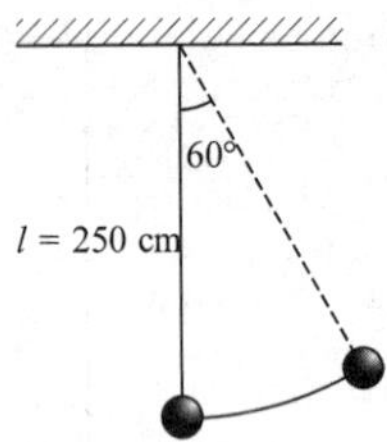

44. A person moved from A to B on a circular path as shown in figure. If the distance travelled by him is 60 m, then the magnitude of displacement would be: [Given cos 135° = – 0.7]

[25 July 2022 (Morning)]

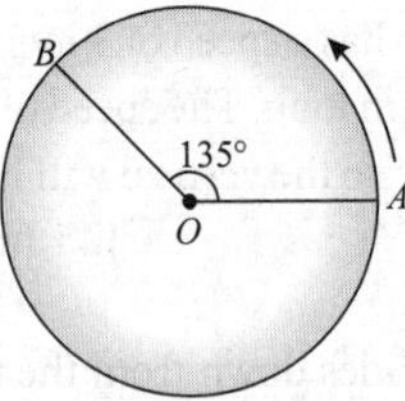

(A) 42 m (B) 47 m
(C) 19 m (D) 40 m

45. A smooth circular groove has a smooth vertical wall as shown in figure. A block of mass m moves against the wall with a speed v. Which of the following curve represents the correct relation between the normal reaction on the block by the wall (N) and speed of the block (v)? **[29 July 2022 (Morning)]**

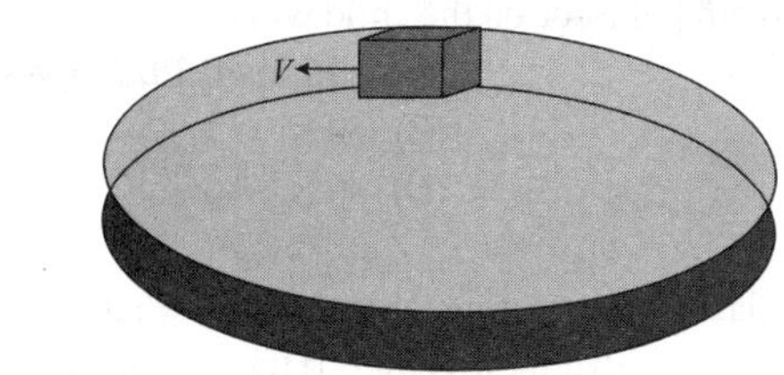

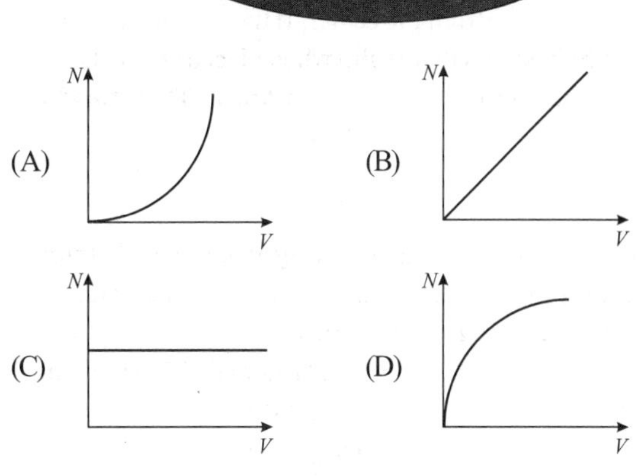

46. A body of mass 200 g is tied to a spring of spring constant 12.5 N/m, while the other end of spring is fixed at point O. If the body moves about O in a circular path on a smooth horizontal surface with constant angular speed 5 rad/s, then the ratio of extension in the spring to its natural length will be :

[24 Jan. 2023 (Evening)]

(A) 1 : 2 (B) 1 : 1
(C) 2 : 3 (D) 2 : 5

47. A car is moving with a constant speed of 20 m/s in a circular horizontal track of radius 40 m. A bob is suspended from the roof of the car by a massless string. The angle made by the string with the vertical will be : (Take $g = 10$ m/s^2)

[25 Jan. 2023 (Morning)]

(A) $\frac{\pi}{6}$ (B) $\frac{\pi}{2}$

(C) $\frac{\pi}{4}$ (D) $\frac{\pi}{3}$

48. A car is moving on a horizontal curved road with radius 50 m. The approximate maximum speed of car will be, if friction between tyres and road is 0.34. (Take $g = 10$ ms^{-2})

[29 Jan. 2023 (Morning)]

(A) 3.4 ms^{-1} (B) 22.4 ms^{-1}
(C) 13 ms^{-1} (D) 17 ms^{-1}

49. An object moves at a constant speed along a circular path in a horizontal plane with centre at the origin. When the object is at $x = +2$ m, its velocity is $-4\hat{j}$ m/s. The object's velocity (v) and acceleration (a) at $x = -2$ m will be :

[29 Jan. 2023 (Evening)]

(A) $v = 4\hat{i}$ m/s, $a = 8\hat{j}$ m/s^2

(B) $v = 4\hat{j}$ m/s, $a = 8\hat{i}$ m/s^2

(C) $v = -4\hat{j}$ m/s, $a = 8\hat{i}$ m/s^2

(D) $v = -4\hat{i}$ m/s, $a = -8\hat{j}$ m/s^2

50. A car is moving on a circular path of radius 600 m such that the magnitudes of the tangential acceleration and centripetal acceleration are equal. The time taken by the car to complete first quarter of revolution, if it is moving with an initial speed of 54 km/hr is $t(1 - e^{-\pi/2})$s. The value of t is _____.

[29 Jan. 2023 (Evening)]

51. A body of mass 2 kg is initially at rest. It starts moving unidirectionally under the influence of a source of constant power P. Its displacement in 4s is $\frac{1}{3}\alpha^2\sqrt{Pm}$. The value of α will be _____.

[30 Jan. 2023 (Evening)]

52. A stone tied to 180 cm long string at its end is making 28 revolutions in horizontal circle in every minute. The magnitude of acceleration of stone is $\frac{1936}{x}$ ms^{-2}. The value of x $\left(\text{Take } \pi = \frac{22}{7}\right)$

[30 Jan. 2023 (Evening)]

53. A body is moving with constant speed, in a circle of radius 10 m. The body completes one revolution in 4 s. At the end of 3rd second, the displacement of body (in m) from its starting point is :

[31 Jan. 2023 (Evening)]

(A) 30 (B) 15π

(C) 5π (D) $10\sqrt{2}$

54. A particle is moving with constant speed in a circular path. When the particle turns by an angle 90°, the ratio of instantaneous velocity to its average velocity is $\pi : x\sqrt{2}$. The value of x will be:

[6 April 2023 (Morning)]

(A) 2 (B) 5
(C) 1 (D) 7

55. A small block of mass 100 g is tied to a spring of spring constant 7.5 N/m and length 20 cm. The other end of spring is fixed at a particular point A. If the block moves in a circular path on a smooth horizontal surface with constant angular velocity 5 rad/s about point A, then tension in the spring is:

[6 April 2023 (Morning)]

(A) 1.5 N (B) 0.75 N
(C) 0.25 N (D) 0.50 N

56. As shown in the figure, a particle is moving with constant speed π m/s. Considering its motion from A to B, the magnitude of the average velocity is: **[6 April 2023 (Evening)]**

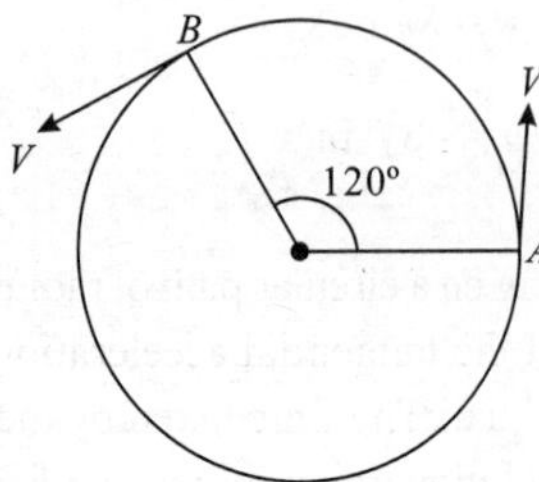

(A) π m/s (B) $\sqrt{3}$ m/s

(C) $2\sqrt{3}$ m/s (D) $1.5\sqrt{3}$ m/s

57. A child of mass 5 kg is going round a merry-go-round that makes 1 rotation in 3.14 s. The radius of the merry-go-round is 2 m. The centrifugal force on the child will be: **[6 April 2023 (Evening)]**

(A) 80 N (B) 50 N

(C) 100 N (D) 40 N

58. A coin placed on a rotating table just slips when it is placed at a distance of 1 cm from the center. If the angular velocity of the table in halved, it will just slip when placed at a distance of _____ from the centre: **[11 April 2023 (Morning)]**

(A) 2 cm (B) 1 cm

(C) 8 cm (D) 4 cm

59. A vehicle of mass 200 kg is moving along a levelled curved road of radius 70 m with angular velocity of 0.2 rad/s. The centripetal force acting on the vehicle is: **[13 April 2023 (Evening)]**

(A) 560 N (B) 2800 N

(C) 14 N (D) 2240 N

* * * * *

CHAPTER 6

SYSTEM OF PARTICLES & ROTATIONAL MOTION

6.1 SYSTEM OF PARTICLES

1. Two identical particles move towards each other with velocity $2v$ and v respectively. The velocity of centre of mass is: **[2002]**

(A) v (B) $v/3$
(C) $v/2$ (D) Zero

2. Consider the following two statements :
A : Linear momentum of a system of particles is zero
B : Kinetic energy of a system of particles is zero
Then : **[2003]**
(A) A does not imply B and B does not imply A
(B) A implies B but B does not imply A
(C) A does not imply B but B implies A
(D) A implies B and B implies A

3. Two spherical bodies of mass M and $5M$ and radii R and $2R$ respectively are released in free space with initial separation between their centres equal to $12R$. It they attract each other due to gravitational force only, then the distance covered by the smaller body just before collision is : **[2003]**
(A) $2.5R$ (B) $4.5R$
(C) $7.5R$ (D) $1.5R$

4. A body A of mass M while falling vertically downwards under gravity breaks into two parts; a body B of mass $1/3M$ and, a body C of mass $2/3M$. The centre of mass of bodies B and C taken together shifts compared to that of body A towards : **[2005]**
(A) Depends on height of breaking
(B) Does not shift
(C) Body C
(D) Body B

5. A mass m moves with a velocity v and collides inelastically with another identical mass. After collision the I[st] mass moves with velocity $v/\sqrt{3}$ in a direction perpendicular to the initial direction of motion. Find the speed of the second mass after collision : **[2005]**
(A) v (B) $\sqrt{3}\,v$
(C) $\frac{2}{\sqrt{3}}v$ (D) $\frac{v}{\sqrt{3}}$

6. A T shaped object with dimensions shown in the figure, is lying on a smooth floor. A force $\vec{F}$ is applied at the point P parallel to AB, such that the object has only the translational motion without rotation. Find the location of P with respect to C : **[2005]**

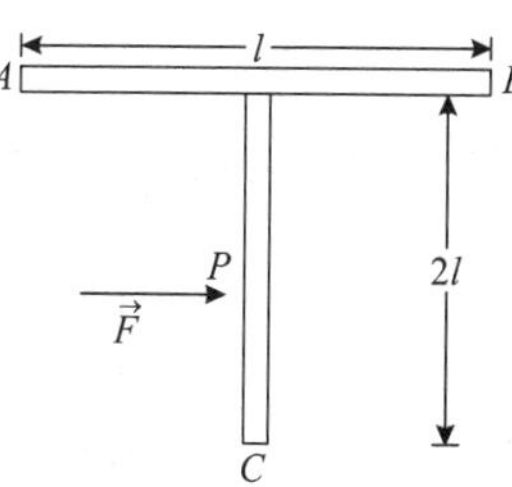

(A) $\frac{2}{3}l$ (B) $\frac{3}{2}l$
(C) $\frac{4}{3}l$ (D) l

7. A bomb of mass 16 kg at rest explodes into two pieces of masses 4 kg and 12 kg. The velocity of the 12 kg mass is 4 ms^{-1}. The kinetic energy of the other mass is : **[2006]**
(A) 144 J (B) 288 J
(C) 192 J (D) 96 J

8. Consider a two particle system with particles having masses m_1 and m_2. If the first particle is pushed towards the centre of mass through a distance d, by what distance should the second particle be moved, so as to keep the centre of mass at the same position ? **[2006]**
(A) $\frac{m_2}{m_1}d$ (B) $\frac{m_1}{m_1+m_2}d$
(C) $\frac{m_1}{m_2}d$ (D) d

9. A block of mass 0.50 kg is moving with a speed of 2.00 ms^{-1} on a smooth surface. It strikes another mass of 1.00 kg and then they move together as a single body. The energy loss during the collision is : **[2008]**
(A) 0.16 J (B) 1.00 J
(C) 0.67 J (D) 0.34 J

10. A thin rod of length 'L' is lying along the x-axis with its ends at $x = 0$ and $x = L$. It linear density (mass/length) varies with x as $k\left(\frac{x}{L}\right)^n$ where n can be zero or any positive number. If the position x_{CM} of the centre of mass of the rod is plotted against 'n', which of the following graphs best approximates the dependence of x_{CM} on n ? **[2008]**

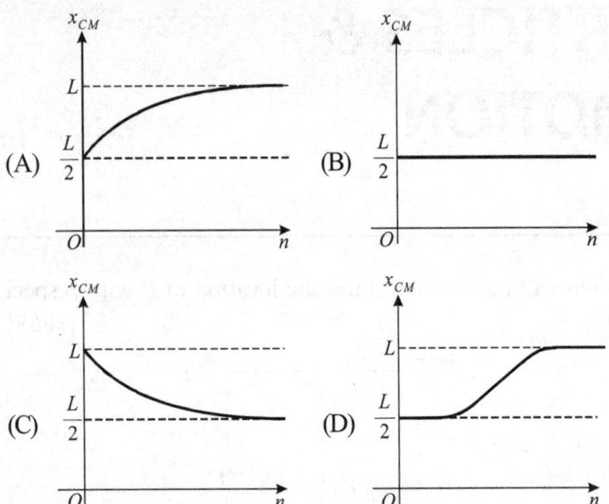

11. Consider a rubber ball freely falling from a height $h = 4.9$ m onto a horizontal elastic plate. Assume that the duration of collision is negligible and the collision with the plate is totally elastic. Then the velocity as a function of time and the height as a function of time will be : **[2009]**

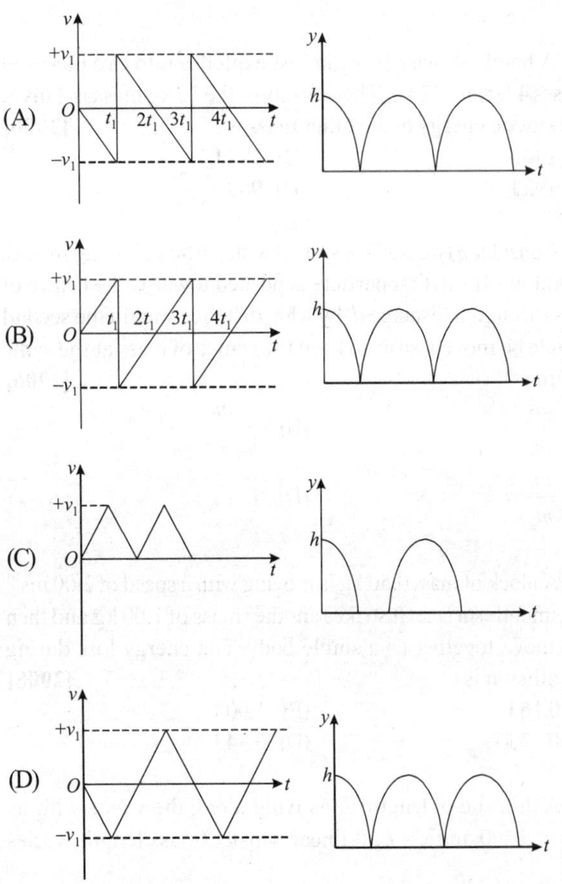

12. The figure shows the position-time (x-t) graph of one dimensional motion of a body of mass 0.4 kg. The magnitude of each impulse is : **[2010]**

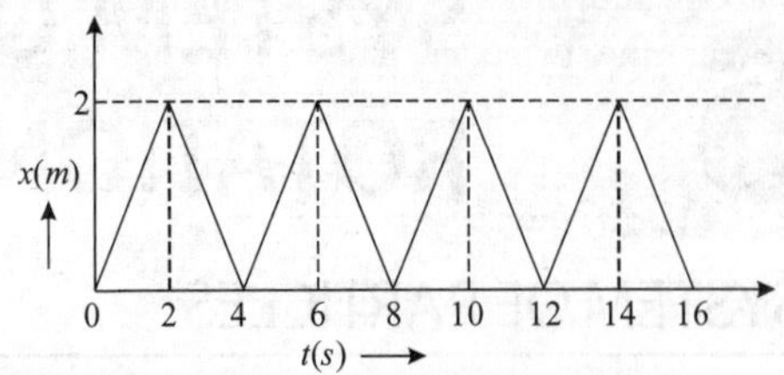

(A) 0.4 Ns (B) 0.8 Ns
(C) 1.6 Ns (D) 0.2 Ns

Directions : This Questions contain Statement-I and Statement-II.

Of the four choices given after the statements, choose the one that best describes the two statements.

13. Statement-I : Two particles moving in the same direction do not lose all their energy in a completely inelastic collision.
Statement-II : Principle of conservation of momentum holds true for all kinds of collisions. **[2010]**

(A) Statement-I is true, Statement-II is false
(B) Statement-I is true, Statement-II is true; Statement-II is the correct explanation of Statement-I
(C) Statement-I is true, Statement-II is true; Statement-II is not the correct explanation of Statement-I
(D) Statement-I is false, Statement-II is true

14. A particle of mass 'm' is projected with a velocity v making an angle of 30° with the horizontal. The magnitude of angular momentum of the projectile about the point of projection when the particle is at its maximum height 'h' is: **[2011]**

(A) $\frac{\sqrt{3}}{2}\frac{mv^2}{g}$ (B) Zero
(C) $\frac{mv^3}{\sqrt{2}g}$ (D) $\frac{\sqrt{3}}{16}\frac{mv^3}{g}$

15. A spring is compressed between two blocks of masses m_1 and m_2 placed on a horizontal frictionless surface as shown in the figure. When the blocks are released, they have initial velocity of v_1 and v_2 as shown. The block travel distances x_1 and x_2 respectively before coming to rest. The ratio (x_1/x_2) is : **[2012 (Online)]**

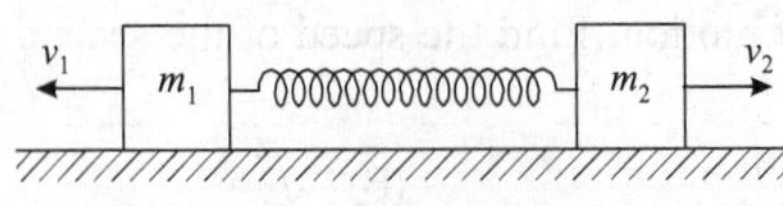

(A) $\frac{m_2}{m_1}$ (B) $\frac{m_1}{m_2}$
(C) $\sqrt{\frac{m_2}{m_1}}$ (D) $\sqrt{\frac{m_1}{m_2}}$

16. A projectile moving vertically upwards with a velocity of 200 ms^{-1} breaks into two equal parts at a height of 490 m. One part starts moving vertically upwards with a velocity of 400 ms^{-1}. How much time it will take, after the break up with the other part to hit the ground ? **[2012 (Online)]**

(A) $2\sqrt{2}$ s (B) 5s
(C) 10s (D) $\sqrt{10}$ s

17. Two bodies A and B of mass m and $2m$ respectively are placed on a smooth floor. They are connected by a spring of negligible mass. A third body C of mass m is placed on the floor. The body C moves with a velocity v_0 along the line joining A and B and collides elastically with A. At a certain time after the collision it is found that the instantaneous velocities of A and B are same and the compression of the spring is x_0. The spring constant k will be : **[2012 (Online)]**

(A) $m\dfrac{v_0^2}{x_0^2}$ (B) $m\dfrac{v_0}{2x_0}$

(C) $2m\dfrac{v_0}{x_0}$ (D) $\dfrac{2}{3}m\left(\dfrac{v_0}{x_0}\right)^2$

18. A moving particle of mass m, makes a head on elastic collision with another particle of mass $2m$, which is initially at rest. The percentage loss in energy of the colliding particle on collision, is close to : **[2012 (Online)]**

(A) 33% (B) 67%
(C) 90% (D) 10%

Directions : This Questions contain Statement-I and Statement-II.

Of the four choices given after the statements, choose the one that best describes the two statements.

19. This question has statement-I and statement-II. Of the four choices given after the statements, choose the one that best describes the two statements.

Statement-I : A point particle of mass m moving with speed v collides with stationary point particle of mass M. If the maximum energy loss possible is given as $f\left(\dfrac{1}{2}mv^2\right)$ then $f=\left(\dfrac{m}{M+m}\right)$.

Statement-II : Maximum energy loss occurs when the particles get stuck together as a result of the collision. **[2013]**

(A) Statement-I is true, Statement-II is false; Statement-II is the correct explanation of Statement-I.
(B) Statement-I is true, Statement-II is true; Statement-II is not the correct explanation of Statement-II
(C) Statement-I is true, Statement-II is false
(D) Statement-I is false, Statement-II is true

20. A projectile of mass M is fired so that the horizontal range is 4 km. At the highest point the projectile explodes in two parts of masses $M/4$ and $3M/4$ respectively and the heavier part starts falling down vertically with zero initial speed. The horizontal range (distance from point of firing) of the lighter part is : **[2013 (Online)]**

(A) 16km (B) 1km
(C) 10km (D) 2km

21. A boy of mass 20 kg is standing on a 80 kg free to move on a long cart. There is negligible friction between cart and ground. Initially, the boy is standing 25 m from a wall. If he walks 10 m on the cart towards the wall, then the final distance of the boy from the wall will be : **[2013 (Online)]**

(A) 15m (B) 12.5m
(C) 15.5m (D) 17m

22. Three masses m, $2m$ and $3m$ are moving in x-y plane with speed $3u$, $2u$ and u respectively as shown in figure. The three masses collide at the same point at P and stick together. The velocity of resulting mass will be : **[2014 (Online)]**

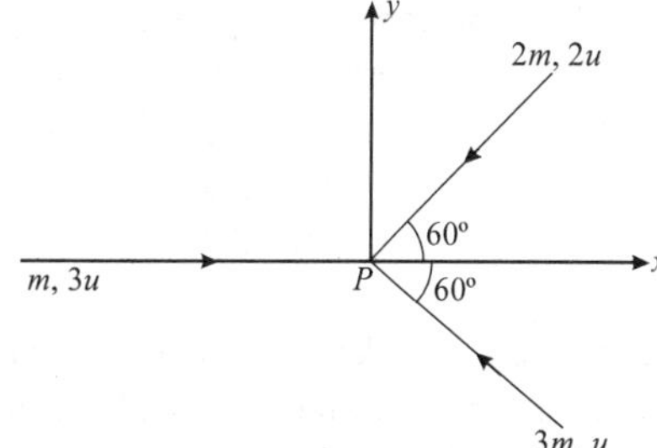

(A) $\dfrac{u}{12}(\hat{i}+\sqrt{3}\hat{j})$ (B) $\dfrac{u}{12}(\hat{i}-\sqrt{3}\hat{j})$

(C) $\dfrac{u}{12}(-\hat{i}+\sqrt{3}\hat{j})$ (D) $\dfrac{u}{12}(-\hat{i}-\sqrt{3}\hat{j})$

23. A thin bar of length L has a mass per unit length λ, that increases linearly with distance from one end. If its total mass is M and its mass per unit length at the lighter end is λ_0, then the distance of the centre of mass from the lighter end is : **[2014 (Online)]**

(A) $\dfrac{L}{2}-\dfrac{\lambda_0L^2}{4M}$ (B) $\dfrac{L}{3}+\dfrac{\lambda_0L^2}{8M}$

(C) $\dfrac{L}{3}+\dfrac{\lambda_0L^2}{4M}$ (D) $\dfrac{2L}{3}-\dfrac{\lambda_0L^2}{6M}$

24. Distance of the centre of mass of a solid uniform cone from its vertex is z_0. If the radius of its base is R and its height is h then z_0 is equal to : **[2015]**

(A) $\dfrac{h^2}{4R}$ (B) $\dfrac{3h}{4}$

(C) $\dfrac{5h}{8}$ (D) $\dfrac{3h^2}{8R}$

25. A particle of mass m moving in the x direction with speed $2v$ is hit by another particle of mass $2m$ moving in the y direction with speed v. If the collision is perfectly inelastic, the percentage loss in the energy during the collision is close to : **[2015]**

(A) 44% (B) 50%
(C) 56% (D) 62%

26. A bullet of mass 4 g is fired horizontally with a speed of 300 m/s into 0.8 kg block of wood at rest on a table. If the coefficient of friction between the block and the table is 0.3, how far will the block slide approximately? **[2015 (Online)]**

(A) 0.19m (B) 0.379m
(C) 0.569m (D) 0.758m

27. A block of mass $m = 0.1$ kg is connected to a spring of unknown spring constant k. It is compressed to a distance x from its equilibrium position and released from rest. After approaching half the distance $\left(\frac{x}{2}\right)$ from equilibrium position, it hits another block and comes to rest momentarily, while the other block moves with a velocity 3 ms^{-1}. The total initial energy of the spring is : **[2015 (Online)]**

(A) 0.3 J (B) 0.6 J
(C) 0.8 J (D) 1.5 J

28. A block of mass $m = 10$ kg rest on a horizontal table. The coefficient of friction between the block and the table is 0.05. When hit by a bullet of mass 50 g moving with speed v, that gets embedded in it, the block moves and comes to stop after moving a distance of 2 m on the table. If a freely falling object were to acquire speed $\frac{v}{10}$ after being dropped from height H, then neglecting energy losses and taking $g = 10$ ms^{-2}, the value of H is close to : **[2015 (Online)]**

(A) 0.05km (B) 0.02km
(C) 0.03km (D) 0.04km

29. A large number (n) of identical beads, each of mass m and radius r are strung on a thin smooth rigid horizontal rod of length L ($L >> r$) and are at rest at random positions. The rod is mounted between two rigid supports (see figure). If one of the beads is now given a speed v, the average force experienced by each support after a long time is (assume all collisions are elastic): **[2015 (Online)]**

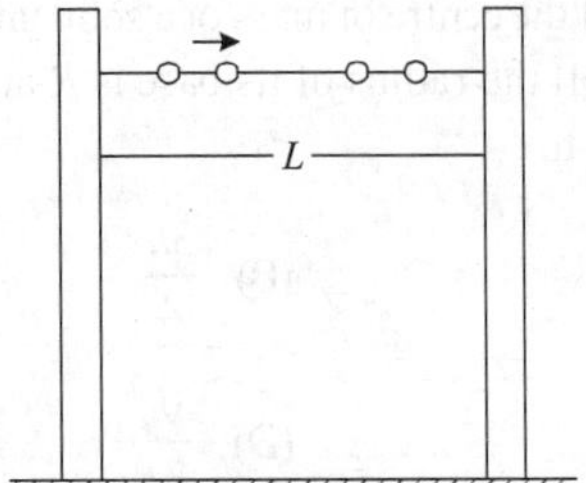

(A) $\frac{mv^2}{2(L-nr)}$ (B) $\frac{mv^2}{L-2nr}$

(C) $\frac{mv^2}{L-nr}$ (D) Zero

30. A uniform thin rod AB of length L has linear mass density $\mu(x) = a + \frac{bx}{L}$, where x is measured from A. If the CM of the rod lies at a distance of $\left(\frac{7}{12}\right) L$ from A, then a and b are related as: **[2015 (Online)]**

(A) $a = 2b$ (B) $2a = b$
(C) $a = b$ (D) $3a = 2b$

31. In the figure shown ABC is a uniform wire. If centre of mass of wire lies vertically below point A, then $\frac{BC}{AB}$ is close to : **[2016 (Online)]**

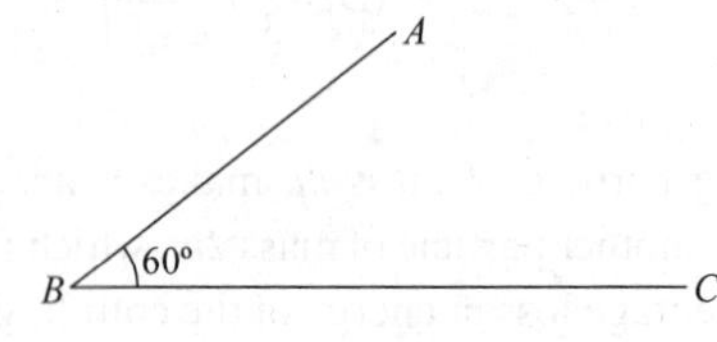

(A) 1.85 (B) 1.5
(C) 1.37 (D) 3

32. An object is dropped from a height h from the ground. Every time it hits the ground it looses 50% of its kinetic energy. The total distance covered as $t \to \infty$ is : **[2017 (Online)]**

(A) $3h$ (B) ∞
(C) $\frac{5}{3}h$ (D) $\frac{8}{3}h$

33. Two particles A and B of equal mass M are moving with the same speed v as shown in the figure. They collide completely inelastically and move as a single particle C. The angle θ that the path of C makes with the X-axis is given by : **[2017 (Online)]**

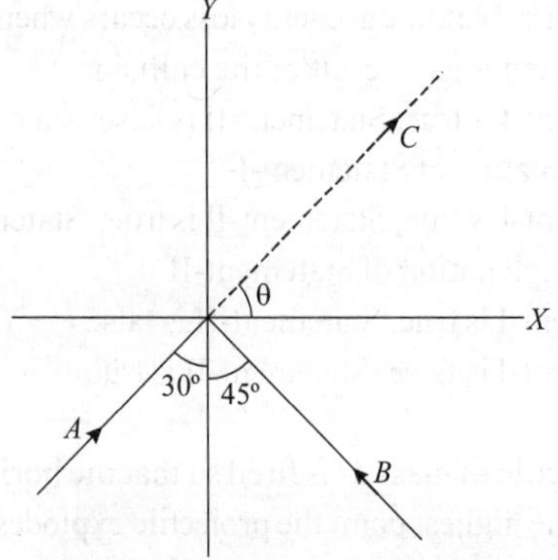

(A) $\tan\theta = \frac{\sqrt{3}+\sqrt{2}}{1-\sqrt{2}}$ (B) $\tan\theta = \frac{\sqrt{3}-\sqrt{2}}{1-\sqrt{2}}$

(C) $\tan\theta = \frac{1-\sqrt{2}}{\sqrt{2}(1+\sqrt{3})}$ (D) $\tan\theta = \frac{1-\sqrt{3}}{1+\sqrt{2}}$

34. In a collinear collision, a particle with an initial speed v_0 strikes a stationary particle of the same mass. If the final total kinetic energy is 50% greater than the original kinetic energy, the magnitude of the relative velocity between the two particles, after collision, is : **[2018]**

(A) $\frac{v_0}{4}$ (B) $\sqrt{2}v_0$

(C) $\frac{v_0}{2}$ (D) $\frac{v_0}{\sqrt{2}}$

35. The mass of a hydrogen molecule is 3.32×10^{-27} kg. If 10^{23} hydrogen molecules strike, per second, a fixed wall of area 2 cm^2 at an angle of 45° to the normal, and rebound elastically with a speed of 10^3 m/s, then the pressure on the wall is nearly : **[2018]**

(A) 2.35×10^3 N/m^2 (B) 4.70×10^3 N/m^2
(C) 2.35×10^2 N/m^2 (D) 4.70×10^2 N/m^2

36. It is found that if a neutron suffers an elastic collinear collision with deuterium at rest, fractional loss of its energy is P_d; while for its similar collision with carbon nucleus at rest, fractional loss of energy is P_c. The values of P_d and P_c are respectively : **[2018]**

(A) (·89, ·28) (B) (·28, ·89)
(C) (0, 0) (D) (0, 1)

37. A proton of mass m collides elastically with a particle of unknown mass at rest. After the collision, the proton and the unknown particle are seen moving at an angle of 90° with respect to each other. The mass of unknown particle is : **[2018 (Online)]**

(A) $\frac{m}{\sqrt{3}}$ (B) $\frac{m}{2}$
(C) 2m (D) m

38. Three blocks A, B and C are lying on a smooth horizontal surface, as shown in the figure. A and B have equal masses, m while C has mass M. Block A is given an initial speed v towards B due to which it collides with B perfectly inelastically. The combined mass collides with C, also perfectly inelastically $\frac{5}{6}$th of the initial kinetic energy is lost in whole process. What is value of M/m ? **[9 Jan. 2019 (Morning)]**

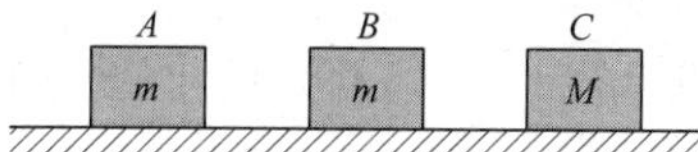

(A) 4 (B) 5
(C) 3 (D) 2

39. A piece of wood of mass 0.03 kg is dropped from the top of a 100 m height building. At the same time, a bullet of mass 0.02 kg is fired vertically upward, with a velocity 100 ms^{-1}, from the ground. The bullet gets embedded in the wood. Then the maximum height to which the combined system reaches above the top of the building before falling below is : (g = 10 ms^{-2}) **[10 Jan. 2019 (Morning)]**

(A) 30m (B) 10m
(C) 40m (D) 20m

40. A simple pendulum, made of a string of length l and a bob of mass m, is released from a small angle θ_0. It strikes a block of mass M, kept on a horizontal surface at its lowest point of oscillations, elastically. It bounces back and goes up to an angle θ_1. Then M is given by : **[12 Jan. 2019 (Morning)]**

(A) $\frac{m}{2}\left(\frac{\theta_0-\theta_1}{\theta_0+\theta_1}\right)$ (B) $\frac{m}{2}\left(\frac{\theta_0+\theta_1}{\theta_0-\theta_1}\right)$

(C) $m\left(\frac{\theta_0+\theta_1}{\theta_0-\theta_1}\right)$ (D) $m\left(\frac{\theta_0-\theta_1}{\theta_0+\theta_1}\right)$

41. The position vector of the centre of mass $\vec{r}$ cm of an symmetric uniform bar of negligible area of cross-section as shown in figure is : **[12 Jan. 2019 (Morning)]**

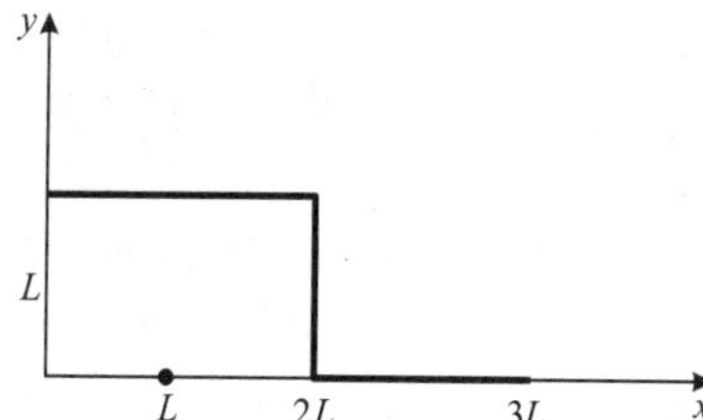

(A) $\vec{r}$ cm $= \frac{13}{8}L\hat{x}+\frac{5}{8}L\hat{y}$ (B) $\vec{r}$ cm $= \frac{11}{8}L\,\hat{x}+\frac{3}{8}L\hat{y}$

(C) $\vec{r}$ cm $= \frac{3}{8}L\,\hat{x}+\frac{11}{8}L\hat{y}$ (D) $\vec{r}$ cm $= \frac{5}{8}L\,\hat{x}+\frac{13}{8}L\hat{y}$

42. An alpha-particle of mass m suffers 1-dimensional elastic collision with a nucleus at rest of unknown mass. It is scattered directly backwards losing, 64% of its initial kinetic energy. The mass of the nucleus is : **[12 Jan. 2019 (Evening)]**

(A) 4m (B) 3.5m
(C) 2m (D) 1.5m

43. Four particles A, B, C and D with masses $m_A = m$, $m_B = 2m$, $m_C = 3m$ and $m_D = 4m$ are at the corners of a square. They have accelerations of equal magnitude with directions as shown.

The acceleration of the centre of mass of the particles is : **[8 April 2019 (Morning)]**

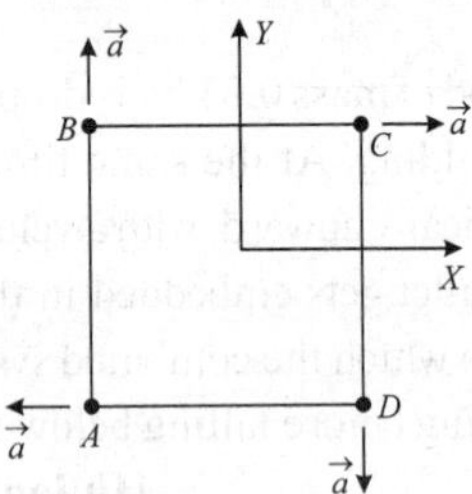

(A) $\frac{a}{5}(\hat{i}-\hat{j})$ (B) $\frac{a}{5}(\hat{i}+\hat{j})$

(C) Zero (D) $a(\hat{i}+\hat{j})$

44. A uniform rectangular thin sheet $ABCD$ of mass M has length a and breadth b, as shown in the figure. If the shaded portion $HBGO$ is cut-off, the coordinates of the centre of mass of the remaining portion will be : **[8 April 2019 (Evening)]**

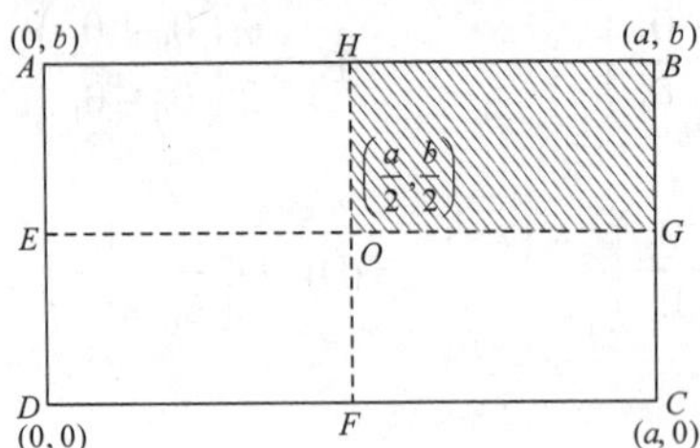

(A) $\left(\frac{2a}{3},\frac{2b}{3}\right)$ (B) $\left(\frac{5a}{3},\frac{5b}{3}\right)$

(C) $\left(\frac{3a}{4},\frac{3b}{4}\right)$ (D) $\left(\frac{5a}{12},\frac{5b}{12}\right)$

45. A body of mass m_1 moving with an unknown velocity of $v_1\hat{i}$, undergoes a collinear collision with a body of mass m_2 moving with a velocity $v_2\hat{i}$. After collision, m_1 and m_2 move with velocities of $v_3\hat{i}$ and $v_4\hat{i}$, respectively. If $m_2 = 0.5\, m_1$ and $v_3 = 0.5v_1$, then v_1 is : **[8 April 2019 (Evening)]**

(A) $v_4 - \frac{v_2}{4}$ (B) $v_4 - \frac{v_2}{2}$

(C) $v_4 - v_2$ (D) $v_4 + v_2$

46. A body of mass 2 kg makes an elastic collision with a second body at rest and continues to move in the original direction but with one fourth of its original speed. What is the mass of the second body ? **[9 April 2019 (Morning)]**

(A) 1.8 kg (B) 1.2 kg

(C) 1.5 kg (D) 1.0 kg

47. A particle of mass 'm' is moving with speed '$2v$' and collides with a mass '$2m$' moving with speed 'v' in the same direction. After collision, the first mass is stopped completely while the second one splits into two particles each of mass 'm', which move at angle 45° with respect to the original direction. The speed of each of the moving particle will be : **[9 April 2019 (Evening)]**

(A) $v/(2\sqrt{2})$ (B) $2\sqrt{2}v$

(C) $\sqrt{2}v$ (D) $v/\sqrt{2}$

48. A wedge of mass $M = 4m$ lies on a frictionless plane. A particle of mass m approaches the wedge with speed v. There is no friction between the particle and the plane or between the particle and the wedge. The maximum height climbed by the particle on the wedge is given by : **[9 April 2019 (Evening)]**

(A) $\frac{2v^2}{7g}$ (B) $\frac{v^2}{g}$

(C) $\frac{2v^2}{5g}$ (D) $\frac{v^2}{2g}$

49. Two particles, of masses M and $2M$, moving, as shown, with speeds of 10 m/s and 5 m/s, collide elastically at the origin. After the collision, they move along the indicated directions with speeds υ_1 and υ_2, respectively. The values of υ_1 and υ_2 are nearly : **[10 April 2019 (Morning)]**

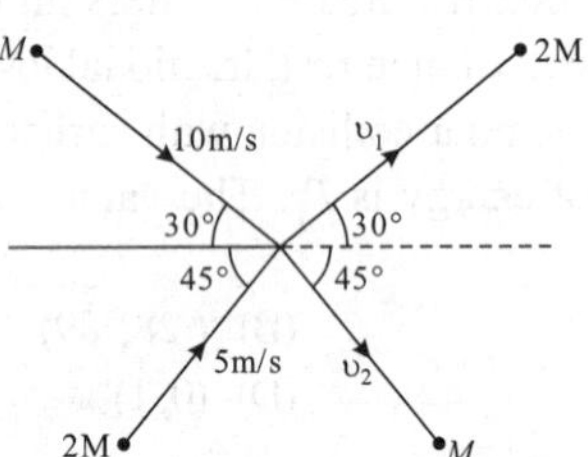

(A) 3.2 m/s and 6.3 m/s (B) 3.2 m/s and 12.6 m/s

(C) 6.5 m/s and 6.3 m/s (D) 6.5 m/s and 3.2 m/s

50. A bullet of mass 20 g has an initial speed of 1 ms^{-1}, just before it starts penetrating a mud wall of thickness 20 cm. If the wall offers a mean resistance of 2.5×10^{-2} N, the speed of the bullet after emerging from the other side of the wall is close to: **[10 April 2019 (Evening)]**

(A) 0.4 ms^{-1} (B) 0.1 ms^{-1}

(3) 0.3 ms^{-1} (D) 0.7 ms^{-1}

51. A person of mass M is, sitting on a swing of length L and swinging with an angular amplitude θ_0. If the person stands up when the swing passes through its lowest point, the work done by him, assuming that his centre of mass moves by a distance $l(l << L)$, is close to : **[12 April 2019 (Morning)]**

(A) Mgl (B) $Mgl(1+\theta_0^2)$

(C) $Mgl(1-\theta_0^2)$ (D) $Mgl\left(1+\frac{\theta_0^2}{2}\right)$

52. A man (mass = 50 kg) and his son (mass = 20 kg) are standing on a frictionless surface facing each other. The man pushes his son so that he starts moving at a speed of $0.70\ ms^{-1}$ with respect to the man. The speed of the man with respect to the surface is: **[12 April 2019 (Morning)]**

(A) $0.20\ ms^{-1}$ (B) $0.14\ ms^{-1}$
(C) $0.47\ ms^{-1}$ (D) $0.28\ ms^{-1}$

53. Three particles of masses 50 g, 100 g and 150 g are placed at the vertices of an equilateral triangle of side 1m (as shown in the figure). The (x, y) coordinates of the centre of mass will be: **[12 April 2019 (Evening)]**

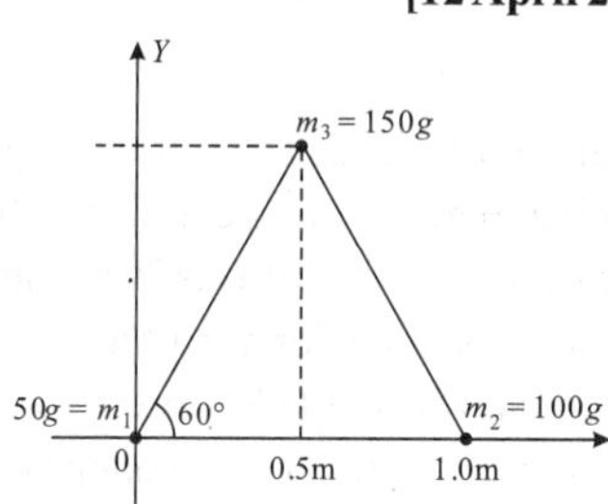

(A) $\left(\frac{7}{12}m, \frac{\sqrt{3}}{8}m\right)$ (B) $\left(\frac{\sqrt{3}}{4}m, \frac{5}{12}m\right)$
(C) $\left(\frac{7}{12}m, \frac{\sqrt{3}}{4}m\right)$ (D) $\left(\frac{\sqrt{3}}{8}m, \frac{7}{12}m\right)$

54. Three point particles of masses 1.0 kg, 1.5 kg and 2.5 kg are placed at three corners of a right angle triangle of sides 4.0 cm, 3.0 cm and 5.0 cm as shown in the figure. The center of mass of the system is at a point : **[7 Jan. 2020 (Morning)]**

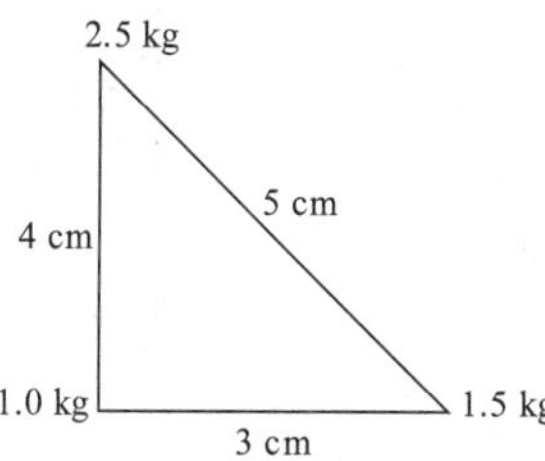

(A) 1.5 cm right and 1.2 cm above 1 kg mass
(B) 0.9 cm right and 2.0 cm above 1 kg mass
(C) 0.6 cm right and 2.0 cm above 1 kg mass
(D) 2.0 cm right and 0.9 cm above 1 kg mass

55. The coordinates of centre of mass of a uniform flag shaped lamina (thin flat plate) of mass 4 kg. (The coordinates of the same are shown in figure) are : **[8 Jan. 2020 (Morning)]**

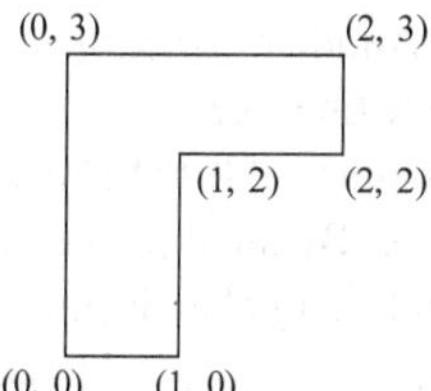

(A) (1.25 m, 1.50 m) (B) (1 m, 1.75 m)
(C) (0.75 m, 0.75 m) (D) (0.75 m, 1.75 m)

56. A body A, of mass m = 0.1 kg has an initial velocity of $3\hat{i}ms^{-1}$. It collides elastically with another body, B of the same mass which has an initial velocity of $5\hat{j}ms^{-1}$. After collision, A moves with a velocity $\vec{v} = 4(\hat{i}+\hat{j})$. The energy of B after collision is written as $\frac{x}{10}J$. The value of x is _____. **[8 Jan. 2020 (Morning)]**

57. As shown in figure, when a spherical cavity (centred at O) of radius 1 is cut out of a uniform sphere of radius R (centred at C), the centre of mass of remaining (shaded) part of sphere is at G, i.e, on the surface of the cavity. R can be determined by the equation : **[8 Jan. 2020 (Evening)]**

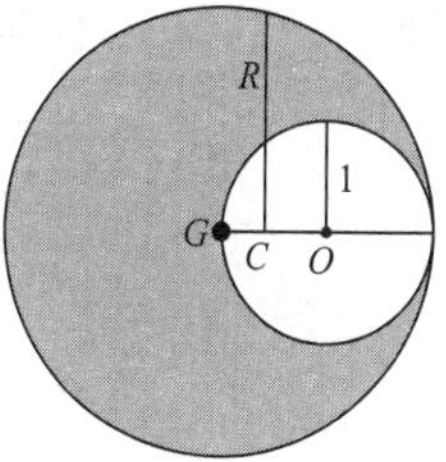

(A) $(R^2 - R + 1)(2 - R) = 1$ (B) $(R^2 + R - 1)(2 - R) = 1$
(C) $(R^2 + R + 1)(2 - R) = 1$ (D) $(R^2 - R - 1)(2 - R) = 1$

58. A particle of mass m is dropped from a height h above the ground. At the same time another particle of the same mass is thrown vertically upwards from the ground with a speed of $\sqrt{2gh}$. If they collide head-on completely inelastically, the time taken for the combined mass to reach the ground, in units of $\sqrt{\frac{h}{g}}$ is : **[8 Jan. 2020 (Evening)]**

(A) $\frac{1}{2}$ (B) $\sqrt{\frac{1}{2}}$
(C) $\sqrt{\frac{3}{4}}$ (D) $\sqrt{\frac{3}{2}}$

59. Two particles of equal mass m have respective initial velocities $u\hat{i}$ and $u\left(\frac{\hat{i}+\hat{j}}{2}\right)$. They collide completely inelastically. The energy lost in the process is : **[9 Jan. 2020 (Morning)]**

(A) $\frac{3}{4}mu^2$ (B) $\frac{1}{8}mu^2$
(C) $\sqrt{\frac{2}{3}}mu^2$ (D) $\frac{1}{3}mu^2$

60. A rod of length L has non-uniform linear mass density given by $\rho(x) = a + b\left(\frac{x}{L}\right)^2$, where a and b are constants and $0 \le x \le L$. The value of x for the centre of mass of the rod is at : **[9 Jan. 2020 (Evening)]**

(A) $\frac{4}{3}\left(\frac{a+b}{2a+3b}\right)L$ (B) $\frac{3}{2}\left(\frac{a+b}{2a+b}\right)L$

(C) $\frac{3}{2}\left(\frac{2a+b}{3a+b}\right)L$ (D) $\frac{3}{4}\left(\frac{2a+b}{3a+b}\right)L$

61. A particle of mass m is projected with a speed u from the ground at an angle $\theta = \frac{\pi}{3}$ w.r.t. horizontal (x-axis). When it has reached its maximum height, it collides completely inelastically with another particle of the same mass and velocity $u\hat{i}$. The horizontal distance covered by the combined mass before reaching the ground is : **[9 Jan. 2020 (Evening)]**

(A) $\frac{3\sqrt{2}}{4}\frac{u^2}{g}$ (B) $2\sqrt{2}\frac{u^2}{g}$

(C) $\frac{3\sqrt{3}}{8}\frac{u^2}{g}$ (D) $\frac{5}{8}\frac{u^2}{g}$

62. A particle of mass m with an initial velocity $u\hat{i}$ collides perfectly elastically with a mass $3m$ at rest. It moves with a velocity $v\hat{j}$ after collision, then, v is given by : **[2 Sept. 2020 (Morning)]**

(A) $v = \sqrt{\frac{2}{3}}u$ (B) $v = \frac{1}{\sqrt{6}}u$

(C) $v = \frac{u}{\sqrt{3}}$ (D) $v = \frac{u}{\sqrt{2}}$

63. A particle of mass m is moving along the x-axis with initial velocity $u\hat{i}$. It collides elastically with a particle of mass 10 m at rest and then moves with half its initial kinetic energy (see figure). If $\sin\theta_1 = \sqrt{n}\sin\theta_2$ then value of n is _____. **[2 Sept. 2020 (Evening)]**

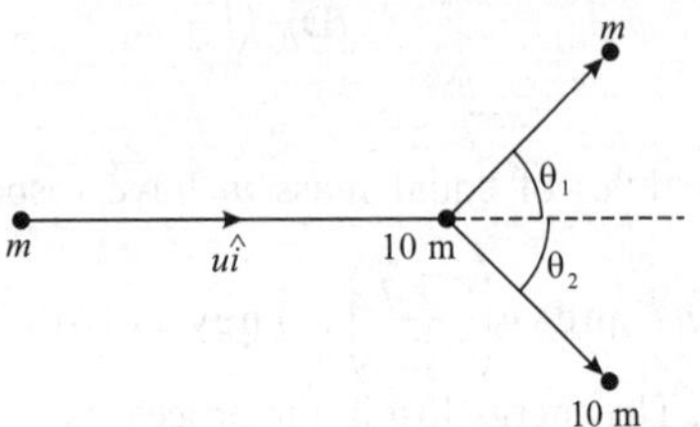

64. A square shaped hole of side $l = \frac{a}{2}$ is carved out at a distance $d = \frac{a}{2}$ from the centre 'O' of a uniform circular disk of radius a. If the distance of the centre of mass of the remaining portion from O is $-\frac{a}{X}$, value of X (to the nearest integer) is _____. **[2 Sept. 2020 (Evening)]**

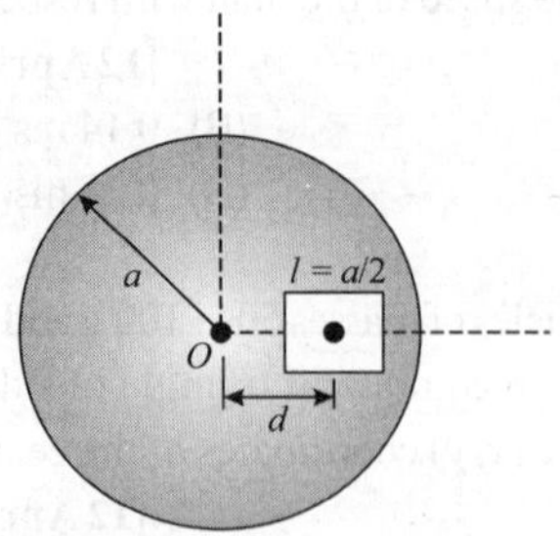

65. A block of mass 1.9 kg is at rest at the edge of a table, of height 1 m. A bullet of mass 0.1 kg collides with the block and sticks to it. If the velocity of the bullet is 20 m/s in the horizontal direction just before the collision then the kinetic energy just before the combined system strikes the floor, is: [Take g = 10 m/s^2. Assume there is no rotational motion and loss of energy after the collision is negligible.] **[3 Sept. 2020 (Evening)]**

(A) 21 J (B) 23 J

(C) 19 J (D) 20 J

66. A Tennis ball is released from a height h and after freely falling on a wooden floor it rebounds and reaches height $\frac{h}{2}$. The velocity versus height of the ball during its motion may be represented graphically by: (Graph are drawn schematically and on not to scale) **[4 Sept. 2020 (Morning)]**

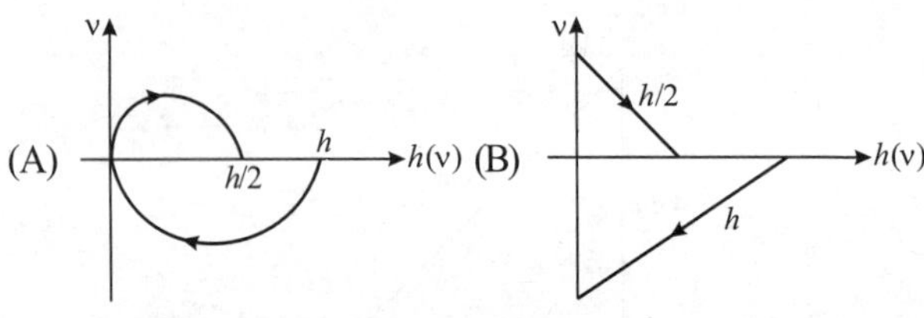

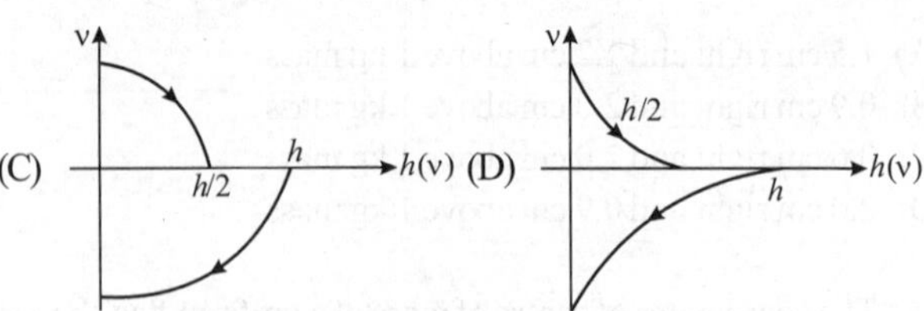

67. Blocks of masses m, $2m$, $4m$ and $8m$ are arranged in a line on a frictionless floor. Another block of mass m, moving with speed v along the same line (see figure) collides with mass m in perfectly inelastic manner. All the subsequent collisions are also perfectly inelastic. By the time the last block of mass $8m$ starts moving the total energy loss is p% of the original energy. Value of 'p' is close to : **[4 Sept. 2020 (Morning)]**

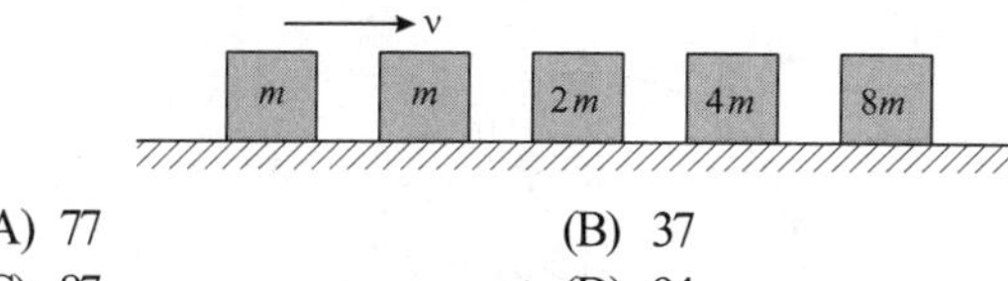

(A) 77 (B) 37
(C) 87 (D) 94

68. Two bodies of the same mass are moving with the same speed, but in different directions in a plane. They have a completely inelastic collision and move together thereafter with a final speed which is half of their initial speed. The angle between the initial velocities of the two bodies (in degree) is _____. **[6 Sept. 2020 (Morning)]**

69. Particle A of mass m_1 moving with velocity $(\sqrt{3}\hat{i}+\hat{j})$ ms^{-1} collides with another particle B of mass m_2 which is at rest initially. Let $\vec{V_1}$ and $\vec{V_2}$ be the velocities of particles A and B after collision respectively. If $m_1 = 2m_2$ and after collision $\vec{V_1} = (\hat{i}+\sqrt{3}\hat{j})\text{ms}^{-1}$, the angle between $\vec{V_1}$ and $\vec{V_2}$ is: **[6 Sept. 2020 (Evening)]**

(A) 60° (B) 15°
(C) –45° (D) 105°

70. The centre of mass of a solid hemisphere of radius 8 cm is X cm from the centre of the flat surface. Then value of x is. **[6 Sept. 2020 (Evening)]**

71. A ball with a speed of 9 m/s collides with another identical ball at rest. After the collision, the direction of each ball makes an angle of 30° with the original direction. The ratio of velocities of the balls after collision is $x : y$, where x is _____. **[24 Feb. 2021 (Morning)]**

72. A circular hole of radius $\left(\frac{a}{2}\right)$ is cut out of a circular disc of radius 'a' as shown in figure. The centroid of the remaining circular portion with respect to point 'O' will be : **[24 Feb. 2021 (Evening)]**

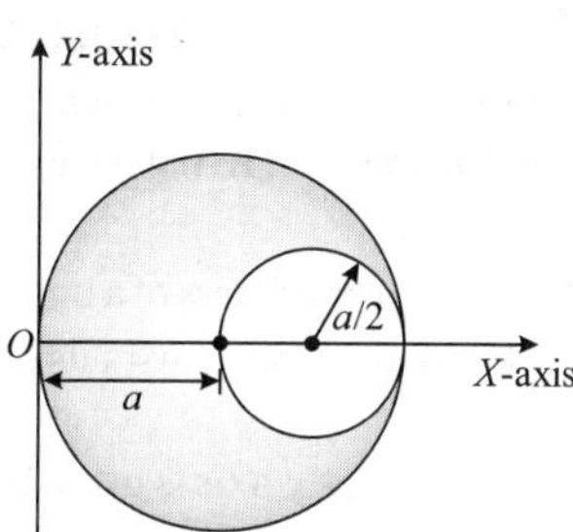

(A) $\frac{5}{6}a$ (B) $\frac{2}{3}a$

(C) $\frac{1}{6}a$ (D) $\frac{10}{11}a$

73. Two solids A and B of mass 1 kg and 2 kg respectively are moving with equal linear momentum. The ratio of their kinetic energies (K.E.)$_A$: (K.E.)$_B$ will be $\frac{A}{1}$, so the value of A will be _____. **[24 Feb. 2021 (Evening)]**

74. Two particles having masses 4 g and 16 g respectively are moving with equal kinetic energies. The ratio of the magnitudes of their linear momentum is n : 2. The value of n will be _____. **[25 Feb. 2021 (Evening)]**

75. Given below are two statements : one is labelled as **Assertion A** and the other is labelled as **Reason R**

Assertion-A : Body 'P' having mass M moving with speed 'u' has head-on collision elastically with another body 'Q' having mass 'm' initially at rest. If $m << M$, body 'Q' will have a maximum speed equal to '$2u$' after collision.

Reason-R : During elastic collision, the momentum and kinetic energy are both conserved

In the light of the above statement, choose the most appropriate answer from the options given below : **[26 Feb. 2021 (Morning)]**

(A) **A** is not correct but **R** is correct
(B) Both **A** and **R** are correct and **R** is the correct explanation of **A**
(C) **A** is correct but **R** is not correct
(D) Both **A** and **R** are correct but **R** is NOT the correct explanation of **A**

76. Two masses A and B, each of mass M are fixed together by a massless spring. A force acts on the mass B as shown in figure. If the mass A starts moving away from mass B with acceleration 'a', then the acceleration of mass B will be : **[26 Feb. 2021 (Evening)]**

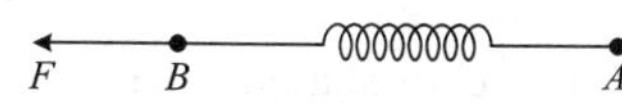

(A) $\frac{MF}{F+Ma}$ (B) $\frac{F+Ma}{M}$

(C) $\frac{F-Ma}{M}$ (D) $\frac{Ma-F}{M}$

77. A ball of mass 10 kg moving with a velocity $10\sqrt{3}$ ms^{-1} along X-axis, hits another ball of mass 20 kg which is at rest. After collision, the first ball comes to rest and the second one disintegrates into two equal pieces. One of the pieces starts moving along Y-axis at a speed of 10 m/s. The second piece starts moving at a speed of 20 m/s at an angle θ (degree) with respect to the X-axis.

The configuration of pieces after collision is shown in the figure. The value of θ to the nearest integer is _____.

[16 March 2021 (Morning)]

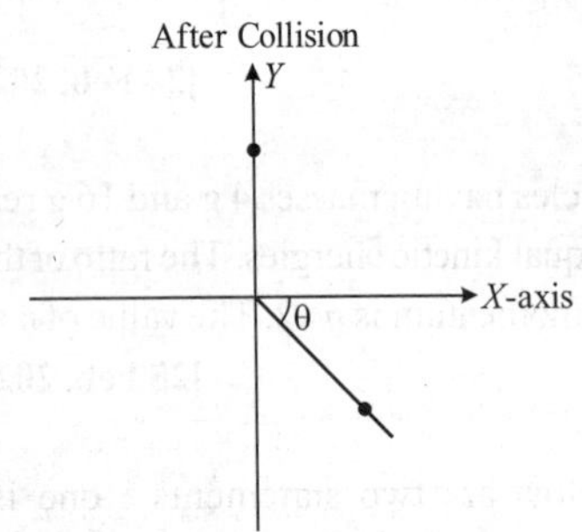

78. A large block of wood of mass $M = 5.99$ kg is hanging from two long massless cords. A bullet of mass $m = 10$g is fired into the block and gets embedded in it. The (block + bullet) then swing upwards, their centre of mass rising a vertical distance $h = 9.8$ cm before the (block + bullet) pendulum comes momentarily to rest at the end of its arc. The speed of the bullet just before collision is : (Take $g = 9.8\ ms^{-2}$)

[16 March 2021 (Evening)]

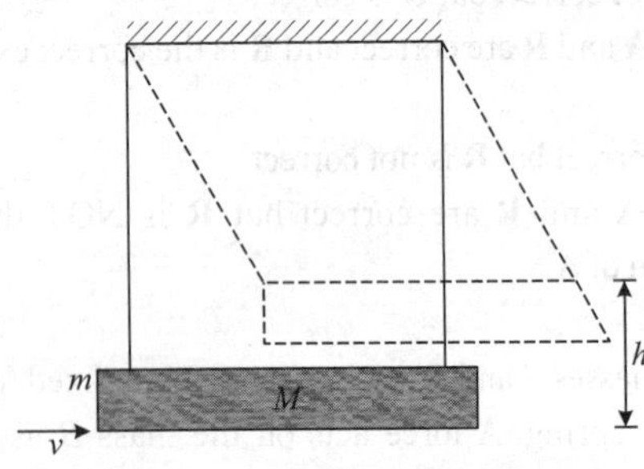

(A) 841.4 m/s (B) 811.4 m/s
(C) 831.4 m/s (D) 821.4 m/s

79. A rubber ball is released from a height of 5 m above the floor. It bounces back repeatedly, always rising to $\frac{81}{100}$ of the height through which it falls. Find the average speed of the ball. (Take $g = 10\ ms^{-2}$) **[17 March, 2021 (Evening)]**

(A) $3.0\ ms^{-1}$ (B) $3.50\ ms^{-1}$
(C) $2.0\ ms^{-1}$ (D) $2.50\ ms^{-1}$

80. The disc of mass M with uniform surface mass density σ is shown in the figure. The centre of mass of the quarter disc (the shaded area) is at the position $\frac{x}{3}\frac{a}{\pi}, \frac{x}{3}\frac{a}{\pi}$ where x is _____. (Round off to the Nearest Integer) [a is an area as shown in the figure] **[17 March 2021 (Evening)]**

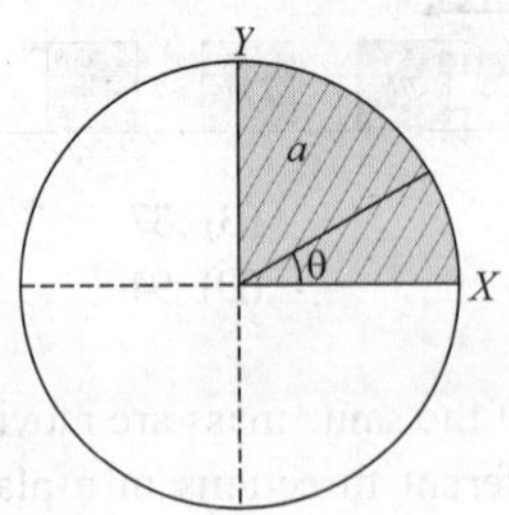

81. A ball of mass 10 kg moving with a velocity $10\sqrt{3}$ m/s along the x-axis, hits another ball of mass 20 kg which is at rest. After the collision, first ball comes to rest while the second ball disintegrates into two equal pieces. One piece starts moving along y-axis with a speed of 10 m/s. The second piece starts moving at an angle of 30° with respect to the x-axis. The velocity of the ball moving at 30° with x-axis is x m/s. The configuration of pieces after collision is shown in the figure below. The value of x to the nearest integer is _____.

[18 March 2021 (Morning)]

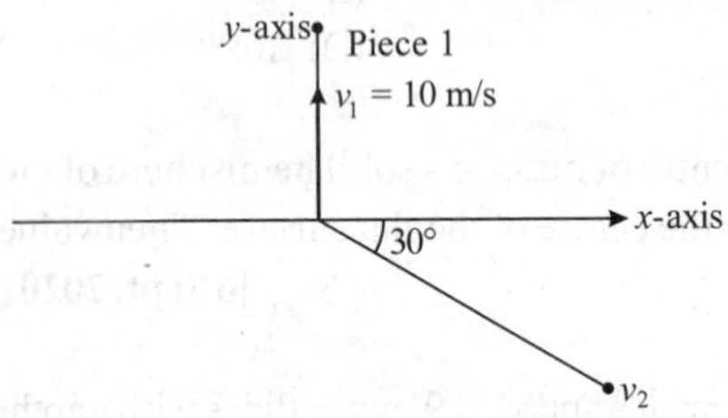

82. An object of mass m_1 collides with another object of mass m_2, which is at rest. After the collision the objects move with equal speeds in opposite direction. The ratio of the masses $m_2 : m_1$ is : **[18 March 2021 (Evening)]**

(A) 3 : 1 (B) 2 : 1
(C) 1 : 2 (D) 1 : 1

83. A bullet of 4 g mass is fired from a gun of mass 4 kg. If the bullet moves the muzzle speed of $50\ ms^{-1}$, the impulse imparted to the gun and velocity of recoil of gun are :

[22 July 2021 (Evening)]

(A) $0.4\ kg\ ms^{-1}, 0.1\ ms^{-1}$ (B) $0.2\ kg\ ms^{-1}, 0.1\ ms^{-1}$
(C) $0.2\ kg\ ms^{-1}, 0.05\ ms^{-1}$ (D) $0.4\ kg\ ms^{-1}, 0.05\ ms^{-1}$

84. The position of the centre of mass of a uniform semi-circular wire of radius 'R' placed in x-y plane with its centre at the origin and the line joining its ends as x-axis is given by $\left(0, \frac{xR}{\pi}\right)$. Then, the value of $|x|$ is _____. **[22 July 2021 (Evening)]**

85. Three objects A, B and C are kept in a straight line on a frictionless horizontal surface. The masses of A, B and C are m, $2m$ and $2m$ respectively. A moves towards B with a speed of

9 m/s and makes an elastic collision with it. Thereafter B makes a completely inelastic collision with C. All motions occur along same straight line. The final speed of C is :

[27 July 2021 (Morning)]

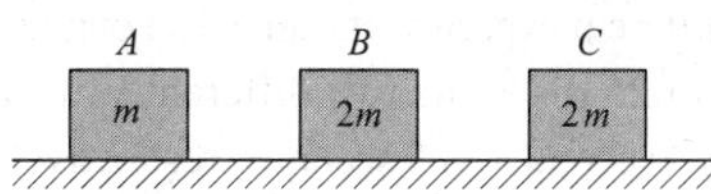

(A) 3 m/s (B) 9 m/s
(C) 4 m/s (D) 6 m/s

86. A bullet of 10g, moving with velocity v, collides head-on with the stationary bob of a pendulum and recoils with velocity 100 m/s. The length of the pendulum is 0.5 m and mass of the bob is 1 kg. The minimum value of v = _____ m/s so that the pendulum describes a circle. (Assume the string to be inextensible and g = 10 m/s^2). **[27 Aug. 2021 (Evening)]**

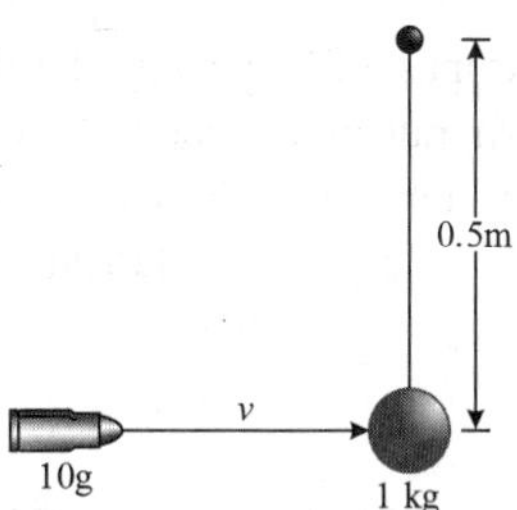

87. A body of mass M moving at speed V_0 collides elastically with another body of same mass m at rest. After the collision, the two masses move at angles θ_1 and θ_2 with respect to the initial direction of motion of the body of mass M. The largest possible value of the ratio M/m, for which the angles θ_1 and θ_2 will be equal, is : **[31 Aug. 2021 (Morning)]**

(A) 1 (B) 4
(C) 2 (D) 3

88. A metre scale is balanced on a knife edge at its centre. When two coins, each of mass 10 g are put one on the top of the other at the 10.0 cm mark the scale is found to be balanced at 40.0 cm mark. The mass of the metre scale is found to be $x \times 10^{-2}$ kg. The value of x is : **[24 June 2022 (Morning)]**

89. A batsman hits back a ball of mass 0.4 kg straight in the direction of the bowler without changing its initial speed of 15 ms^{-1}. The impulse imparted to the ball is _____ Ns.

[26 June 2022 (Evening)]

90. Two blocks of masses 10 kg and 30 kg are placed on the same straight line with coordinates (0, 0) cm and (x, 0) cm respectively. The block of 10 kg is moved on the same line through a distance of 6 cm towards the other block. The distance through which the block of 30 kg must be moved to keep the position of centre of mass of the system unchanged is :

[27 June 2022 (Morning)]

(A) 4 cm towards the 10 kg block
(B) 2 cm away from the 10 kg block
(C) 2 cm towards the 10 kg block
(D) 4 cm away from the 10 kg block

91. What percentage of kinetic energy of a moving particle is transferred to a stationary particle when it strikes the stationary particle of 5 times its mass ? [Assume the collision to be head-on elastic collision] **[27 June 2022 (Morning)]**

(A) 50.0% (B) 66.6%
(C) 55.6% (D) 33.3%

92. A pendulum of length 2 m consists of a wooden bob of mass 50 g. A bullet of mass 75 g is fired towards the stationary bob with a speed v. The bullet emerges out of the bob with a speed $\frac{v}{3}$ and the bob just completes the vertical circle. The value of v is _____ ms^{-1}. (If g = 10 m/s^2)

[27 June 2022 (Morning)]

93. A lead bullet penetrates into a solid object and melts. Assuming that 40% of its kinetic energy is used to heat it, the initial speed of bullet is :

[Given, initial temperature of the bullet = 127°C,
Melting point of the bullet = 327°C,
Latent heat of fusion of lead = 2.5×10^4 J Kg^{-1},
Specific heat capacity of lead = 125J/kg K]

[27 June 2022 (Evening)]

(A) 125 ms^{-1} (B) 500 ms^{-1}
(C) 250 ms^{-1} (D) 600 ms^{-1}

94. A man of 60 kg is running on the road and suddenly jumps into a stationary trolly car of mass 120 kg. Then the trolly car starts moving with velocity 2 ms^{-1}. The velocity of the running man was _____ ms^{-1}, when he jumps into the car.

[28 June 2022 (Morning)]

95. A body of mass M at rest explodes into three pieces, in the ratio of masses 1 : 1 : 2. Two smaller pieces fly off perpendicular to each other with velocities of 30 ms^{-1} and 40 ms^{-1} respectively. The velocity of the third piece will be :

[29 June 2022 (Morning)]

(A) 15 ms^{-1} (B) 25 ms^{-1}
(C) 35 ms^{-1} (D) 50 ms^{-1}

96. A block of metal weighing 2 kg is resting on a frictionless plane (as shown in figure). It is struck by a jet releasing water at a rate of 1 kgs^{-1} and at a speed of 10 ms^{-1}. Then, the initial acceleration of the block, in ms^{-2}, will be :

[29 June 2022 (Morning)]

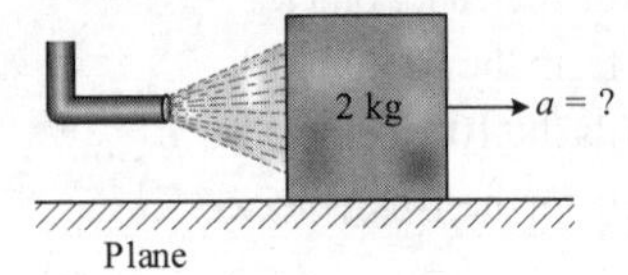

(A) 3 (B) 6
(C) 5 (D) 4

97. Two billiard balls of mass 0.05 kg each moving in opposite directions with 10 ms^{-1} collide and rebound with the same speed. If the time duration of contact is $t = 0.005$ s, then what is the force exerted on the ball due to each other ? **[25 July 2022 (Evening)]**

(A) 100 N (B) 200 N
(C) 300 N (D) 400 N

98. A bag of sand of mass 9.8 kg is suspended by a rope. A bullet of 200 g travelling with speed 10 ms^{-1} gets embedded in it, then loss of kinetic energy will be : **[25 July 2022 (Evening)]**

(A) 4.9 J (B) 9.8 J
(C) 14.7 J (D) 19.6 J

99. Three identical spheres each of mass M are placed at the corners of a right angled triangle with mutually perpendicular sides equal to 3 m each. Taking point of intersection of mutually perpendicular sides as origin, the magnitude of position vector of centre of mass of the system will be $\sqrt{x}m$. The value of x is. **[25 July 2022 (Evening)]**

100. A ball of mass 0.15 kg hits the wall with its initial speed of 12 ms^{-1} and bounces back without changing its initial speed. If the force applied by the wall on the ball during the contact is 100 N, calculate the time duration of the contact of ball with the wall : **[26 July 2022 (Evening)]**

(A) 0.018 s (B) 0.036 s
(C) 0.009 s (D) 0.072 s

101. A body of mass 10 kg is projected at an angle of 45° with the horizontal. The trajectory of the body is observed to pass through a point (20, 10). If T is the time of flight, then its momentum vector, at time $t = \frac{T}{\sqrt{2}}$, is :_____ [Take $g = 10$ m/s^2] **[27 July 2022 (Evening)]**

(A) $100\hat{i} + (100\sqrt{2} - 200)\hat{j}$ (B) $100\sqrt{2}\hat{i} + (100 - 200\sqrt{2})\hat{j}$
(C) $100\hat{i} + (100 - 200\sqrt{2})\hat{j}$ (D) $100\sqrt{2}\hat{i} + (100\sqrt{2} - 200)\hat{j}$

102. The velocity of the bullet becomes one third after it penetrates 4 cm in a wooden block. Assuming that bullet is facing a constant resistance during its motion in the block. The bullet stops completely after travelling at $(4 + x)$ cm inside the block. The value of x is : **[27 July 2022 (Evening)]**

(A) 2.0 (B) 1.0
(C) 0.5 (D) 1.5

103. In two different experiments, an object of mass 5 kg moving with a speed of 25 ms^{-1} hits two different walls and comes to rest within (i) 3 second, (ii) 5 seconds, respectively. Choose the correct option out of the following : **[28 July 2022 (Morning)]**

(A) Impulse and average force acting on the object will be same for both the cases
(B) Impulse will be same for both the cases but the average force will be different
(C) Average force will be same for both the cases but the impulse will be different
(D) Average force and impulse will be different for both the cases

104. A balloon has mass of 10 g in air. The air escapes from the balloon at a uniform rate with velocity 4.5 cm/s. If the balloon shrinks in 5 s completely. Then, the average force acting on that balloon will be : [in dyne] **[28 July 2022 (Morning)]**

(A) 3 (B) 9
(C) 12 (D) 18

105. A bullet of mass 200 g having initial kinetic energy 90 J is shot inside a long swimming pool as shown in the figure. If it's kinetic energy reduces to 40 J within 1 s, the minimum length of the pool, the bullet has a to travel so that it completely comes to rest is : **[28 July 2022 (Evening)]**

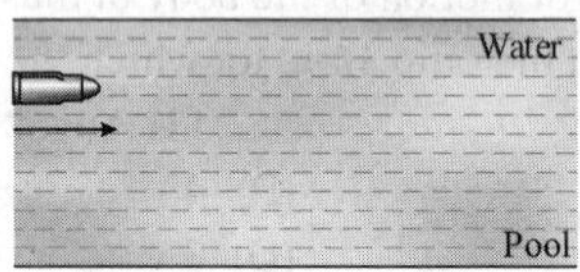

(A) 45 m (B) 90 m
(C) 125 m (D) 25 m

106. The distance of centre of mass from end A of a one dimensional rod (AB) having mass density $\rho = \rho_0\left(1 - \frac{x^2}{L^2}\right)$ kg/m and length L (in meter) is $\frac{3L}{\alpha}$. The value of α is _____ [where x is the distance form end A] **[28 July 2022 (Evening)]**

107. Two bodies of mass 1 kg and 3 kg have position vectors $\hat{i} + 2\hat{j} + \hat{k}$ and $-3\hat{i} - 2\hat{j} + \hat{k}$ respectively. The magnitude of position vector of centre of mass of this system will be similar to the magnitude of vector : **[29 July 2022 (Morning)]**

(A) $\hat{i} - 2\hat{j} + \hat{k}$ (B) $-3\hat{i} - 2\hat{j} + \hat{k}$
(C) $-2\hat{i} + 2\hat{k}$ (D) $-2\hat{i} - \hat{j} + 2\hat{k}$

108. A body of mass 1 kg collides head on elastically with a stationary body of mass 3 kg. After collision, the smaller body reverses its direction of motion and moves with a speed of 2 m/s. The initial speed of the smaller body before collision is _____ ms^{-1}. **[25 Jan. 2023 (Evening)]**

109. A tennis ball is dropped on to the floor from a height of 9.8 m. It rebounds to a height 5.0 m. Ball comes in contact with the floor for 0.2 s. The average acceleration during contact is _____ ms^{-2}. [Given $g = 10\ ms^{-2}$] **[29 Jan. 2023 (Morning)]**

110. As per the given figure, a small ball P slides down the quadrant of a circle and hits the other ball Q of equal mass which is initially at rest. Neglecting the effect of friction and assume the collision to be elastic, the velocity of ball Q after collision will be : ($g = 10\ m/s^2$) **[30 Jan. 2023 (Morning)]**

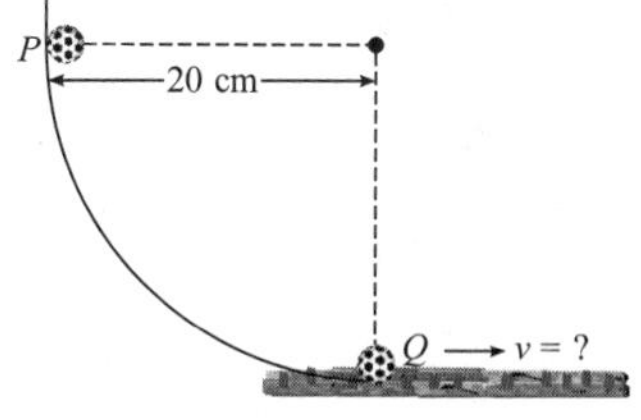

(A) 0 (B) 0.25 m/s
(C) 2 m/s (D) 4 m/s

111. A machine gun of mass 10 kg fires 20 g bullets at the rate of 180 bullets per minute with a speed of 100 ms^{-1} each. The recoil velocity of the gun is : **[30 Jan. 2023 (Evening)]**

(A) 0.02 m/s (B) 2.5 m/s
(C) 1.5 m/s (D) 0.6 m/s

112. 100 balls each of mass m moving with speed v simultaneously strike a wall normally and reflected back with same speed, in time t s. The total force exerted by the balls on the wall is : **[31 Jan. 2023 (Morning)]**

(A) $\frac{100mv}{t}$ (B) $\frac{200mv}{t}$
(C) $200\,mvt$ (D) $\frac{mv}{100t}$

113. A ball is dropped from a height of 20 m. If the coefficient of restitution for the collision between ball and floor is 0.5, after hitting the floor, the ball rebounds to a height of_____ m. **[31 Jan. 2023 (Evening)]**

114. Figures (a), (b), (c) and (d) show variation of force with time : **[1 Feb. 2023 (Evening)]**

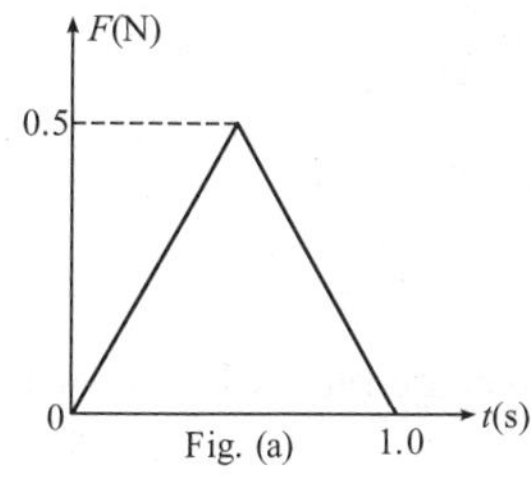

Fig. (a)

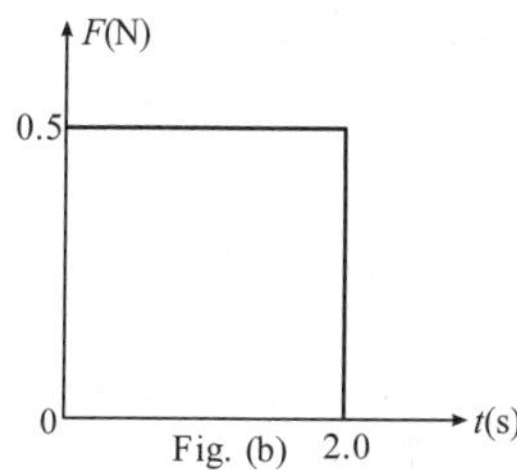

Fig. (b)

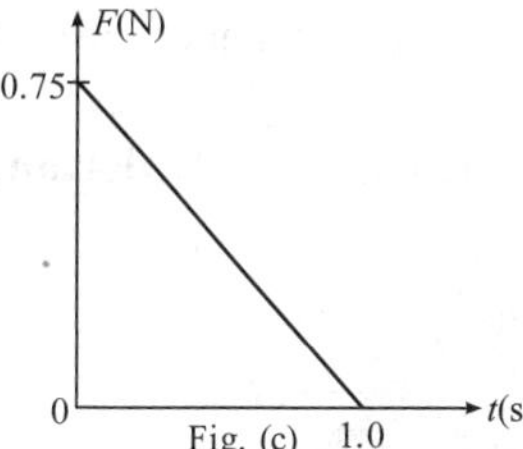

Fig. (c)

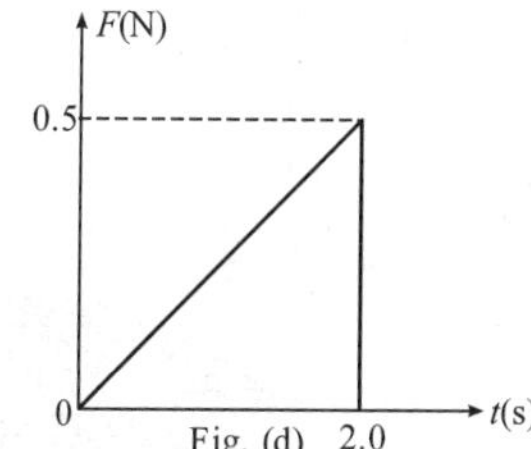

Fig. (d)

The impulse is highest in figure :

(A) Fig. (c) (B) Fig. (b)
(C) Fig. (a) (D) Fig. (d)

115. A body is dropped on ground from a height h_1 and after hitting the ground, it rebounds to a height h_2. If the ratio of velocities of the body just before and after hitting ground is 4, then percentage loss in kinetic energy of the body is $\frac{x}{4}$. The value of x is_____. **[6 April 2023 (Evening)]**

116. The momentum of a body is increased by 50%. The percentage increase in the kinetic energy of the body is_____% **[8 April 2023 (Morning)]**

117. A bullet of mass 0.1 kg moving horizontally with speed 400 ms^{-1} hits a wooden block of mass 3.9 kg kept on a horizontal rough surface. The bullet gets embedded into the block and moves 20 m before coming to rest. The coefficient of friction between the block and the surface is_____. (Given $g = 10\ ms^2$) **[8 April 2023 (Evening)]**

(A) 0.50 (B) 0.90
(C) 0.65 (D) 0.25

118. A body of mass 5 kg is moving with a momentum of 10 kg ms^{-1}. Now a force of 2 N acts on the body in the direction of its motion for 5 s. The increase in the Kinetic energy of the body is : **[8 April 2023 (Evening)]**

119. A particle of mass m moving with velocity v collides with a stationary particle of mass 2 m. After collision, they stick together and continue to move together with velocity: **[10 April 2023 (Morning)]**

(A) v
(B) $\frac{v}{2}$

(C) $\frac{v}{3}$
(D) $\frac{v}{4}$

120. A closed circular tube of average radius 15 cm, whose inner walls are rough, is kept in vertical plane. A block of mass 1 kg just fit inside the tube. The speed of block is 22 m/s, when it is introduced at the top of tube. After completing five oscillations, the block stops at the bottom region of tube. The work done by the tube on the block is ______ J.
[Given $g = 10\ m/s^2$] **[10 April 2023 (Morning)]**

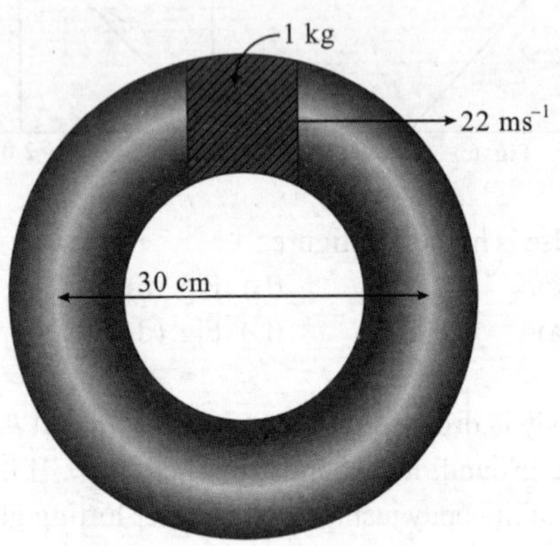

121. An average force of 125 N is applied on a machine gun firing bullets each of mass 10 g at the speed of 250 m/s to keep it in position. The number of bullets fired per second by the machine gun is : **[11 April 2023 (Morning)]**

(A) 5
(B) 50
(C) 100
(D) 25

122. A bullet 10 g leaves the barrel of gun with a velocity of 600 m/s. If the barrel of gun is 50 cm long and mass of gun is 3 kg, then value of impulse supplied to the gun will be : **[13 April 2023 (Morning)]**

(A) 12 Ns
(B) 6 Ns
(C) 36 Ns
(D) 3 Ns

* * * * *

CHAPTER 6

SYSTEM OF PARTICLES & ROTATIONAL MOTION

6.2 RIGID BODY DYNAMICS

1. Initial angular velocity of a circular disc of mass M is ω_1. Then two small spheres of mass m are attached gently to two diametrically opposite points on the edge of the disc. What is the final angular velocity of the disc ? **[2002]**

(A) $\left(\frac{M+m}{M}\right)\omega_1$ (B) $\left(\frac{M+m}{m}\right)\omega_1$

(C) $\left(\frac{M}{M+4m}\right)\omega_1$ (D) $\left(\frac{M}{M+2m}\right)\omega_1$

2. Moment of inertia of a circular wire of mass M and radius R about its diameter is : **[2002]**

(A) $MR^2/2$ (B) MR^2

(C) $2MR^2$ (D) $MR^2/4$

3. A particle of mass m moves along line PC with velocity v as shown. What is the angular momentum of the particle about O? **[2002]**

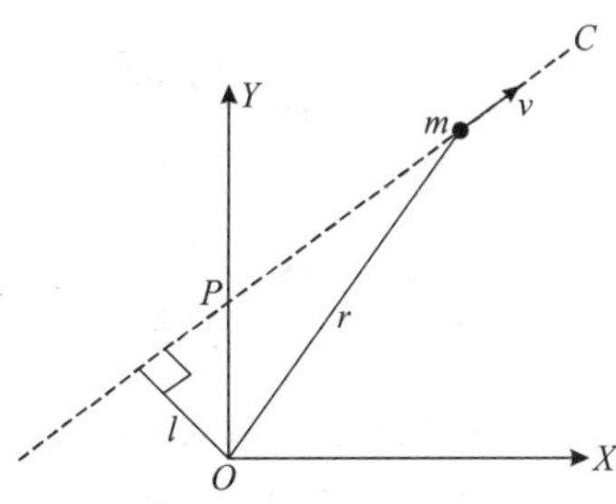

(A) mvL (B) mvl

(C) mvr (D) Zero

4. A circular disc X of radius R is made from an iron plate of thickness t and another disc Y of radius $4R$ is made from an iron plate of thickness $t/4$. Then the relation between the moment of inertia I_X and I_Y is : **[2003]**

(A) $I_Y = 32\, I_X$ (B) $I_Y = 16\, I_X$

(C) $I_Y = I_X$ (D) $I_Y = 64\, I_X$

5. A particle performing uniform circular motion has angular momentum L. If its angular frequency is doubled and its kinetic energy halved, then the new angular momentum is : **[2003]**

(A) $\frac{L}{4}$ (B) $2L$

(C) $4L$ (D) $\frac{L}{2}$

6. Let $\vec{F}$ be the force acting on particle having position vector $\vec{r}$ and $\vec{\tau}$ be the torque of this force about the origin. Then : **[2003]**

(A) $\vec{r}\cdot\vec{\tau} = 0$ and $\vec{F}\cdot\vec{\tau} \neq 0$ (B) $\vec{r}\cdot\vec{\tau} \neq 0$ and $\vec{F}\cdot\vec{\tau} = 0$

(C) $\vec{r}\cdot\vec{\tau} \neq 0$ and $\vec{F}\cdot\vec{\tau} \neq 0$ (D) $\vec{r}\cdot\vec{\tau} = 0$ and $\vec{F}\cdot\vec{\tau} = 0$

7. A solid sphere is rotating in free space. If the radius of the sphere is increased keeping mass same which one of the following will not be affected ? **[2004]**

(A) Moment of inertia (B) Angular momentum

(C) Angular velocity (D) Rotational kinetic energy

8. One solid sphere A and another hollow sphere B are of same mass and same outer radii. Their moment of inertia about their diameters are respectively I_A and I_B such that : **[2004]**

(A) $I_A = I_B$ (B) $I_A > I_B$

(C) $I_A < I_B$ (D) $\frac{I_A}{I_B} = \frac{d_A}{d_B}$

Where d_A and d_B are their densities.

9. An annular ring with inner and outer radii R_1 and R_2 is rolling without slipping with a uniform angular speed. The ratio of the forces experienced by the two particles situated on the inner and outer parts of the ring F_1/F_2 is : **[2005]**

(A) $\frac{R_2}{R_1}$ (B) $\left(\frac{R_1}{R_2}\right)^2$

(C) 1 (D) $\frac{R_1}{R_2}$

10. The moment of inertia of uniform semicircular disc of mass M and radius r about a line perpendicular to the plane of the disc through the centre is : **[2005]**

(A) $\frac{1}{4}Mr^2$ (B) $\frac{2}{5}Mr^2$

(C) Mr^2 (D) $\frac{1}{2}Mr^2$

11. Four point masses, each of value m, are placed at the corners of a square $ABCD$ of side l. The moment of inertia of this system about an axis passing through A and parallel to BD is : **[2006]**

(A) $2ml^2$ (B) $\sqrt{3}\, ml^2$

(C) $3ml^2$ (D) ml^2

12. A force of $-F\hat{k}$ acts on O, the origin of the co-ordinate system. The torque about the point (1, –1) is : **[2006]**

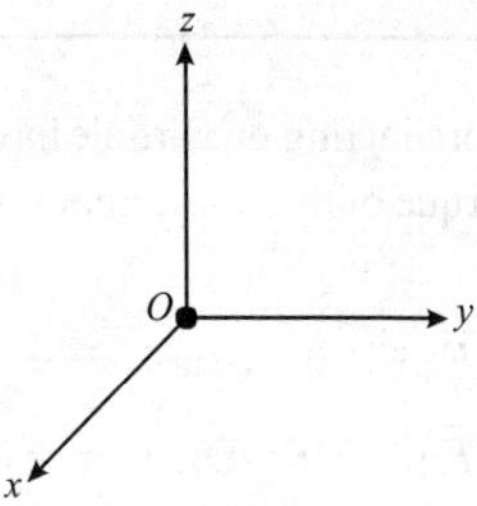

(A) $F(\hat{i}-\hat{j})$ (B) $-F(\hat{i}+\hat{j})$

(C) $F(\hat{i}+\hat{j})$ (D) $-F(\hat{i}-\hat{j})$

13. A thin circular ring of mass m and radius R is rotating about its axis with a constant angular velocity ω. Two objects each of mass M are attached gently to the opposite ends of a diameter of the ring. The ring now rotates with an angular velocity ω'= :

[2006]

(A) $\dfrac{\omega(m+2m)}{M}$ (B) $\dfrac{\omega(m-2m)}{(m+2M)}$

(C) $\dfrac{\omega m}{(m+M)}$ (D) $\dfrac{\omega m}{(m+2M)}$

14. For the given uniform square lamina $ABCD$, whose centre is O: **[2007]**

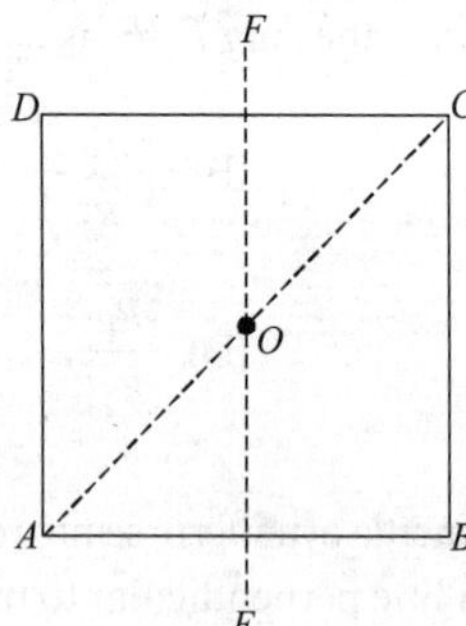

(A) $\sqrt{2}\, I_{AC} = I_{EF}$ (B) $I_{AD} = 3I_{EF}$

(C) $I_{AC} = I_{EF}$ (D) $I_{AC} = \sqrt{2}\, I_{EF}$

15. A circular disc of radius R is removed from a bigger circular disc of radius 2 R such that the circumferences of the discs coincide. The centre of mass of the new disc is αR from the centre of the bigger disc. The value of α is : **[2007]**

(A) 1/3 (B) 1/2

(C) 1/6 (D) 1/4

16. A round uniform body of radius R, mass M and moment of inertia I, rolls down (without slipping) an inclined plane making an angle θ with the horizontal. Then its acceleration is : **[2007]**

(A) $\dfrac{g\sin\theta}{1+I/MR^2}$ (B) $\dfrac{g\sin\theta}{1+MR^2/I}$

(C) $\dfrac{g\sin\theta}{1-I/MR^2}$ (D) $\dfrac{g\sin\theta}{1-MR^2/I}$

17. Angular momentum of the particle rotating with a central force is constant due to : **[2007]**

(A) Constant force (B) Constant linear momentum

(C) Zero torque (D) Constant torque

18. Consider a uniform square plate of side 'a' and mass 'm' the moment of inertia of this plate about an axis perpendicular to its plane and passing through one of its corners is : **[2008]**

(A) $\dfrac{5}{6}ma^2$ (B) $\dfrac{1}{12}ma^2$

(C) $\dfrac{7}{12}ma^2$ (D) $\dfrac{2}{3}ma^2$

19. A thin uniform rod of length l and mass m is swinging freely about a horizontal axis passing through its end. Its maximum angular speed is ω. Its centre of mass rises to a maximum height of : **[2009]**

(A) $\dfrac{1}{6}\dfrac{l\omega}{g}$ (B) $\dfrac{1}{2}\dfrac{l^2\omega^2}{g}$

(C) $\dfrac{1}{6}\dfrac{l^2\omega^2}{g}$ (D) $\dfrac{1}{3}\dfrac{l^2\omega^2}{g}$

20. A small particle of mass m is projected at an angle θ with the x-axis with an initial velocity v_0 in the x-y plane as shown in the figure. At a time $t < \dfrac{v_0 \sin\theta}{g}$, the angular momentum of the particle is : **[2010]**

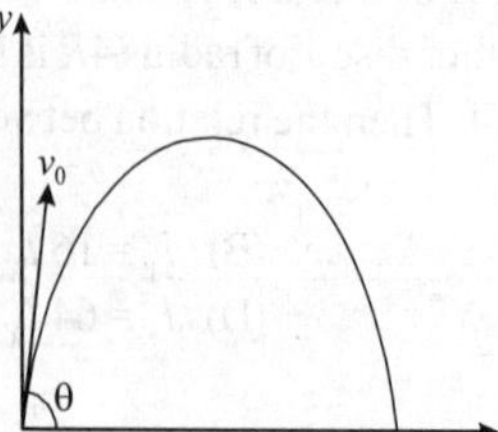

(A) $\dfrac{1}{2}mg\, v_0\, t^2 \cos\theta\, \hat{i}$ (B) $-mg\, v_0\, t^2 \cos\theta\, \hat{j}$

(C) $mg\, v_0\, t \cos\theta\, \hat{k}$ (D $-\dfrac{1}{2}mg\, v_0\, t^2 \cos\theta\, \hat{k}$

Where $\hat{i}, \hat{j}$ and $\hat{k}$ are unit vectors along x, y and z-axis respectively.

21. A mass m hangs with the help of a string wrapped around a pulley on a frictionless bearing. The pulley has mass m and radius R. Assuming pulley to be a perfect uniform circular disc, the acceleration of the mass m, if the string does not slip on the pulley, is : **[2011]**

(A) g (B) $\frac{2}{3}g$

(C) $\frac{g}{3}$ (D) $\frac{3}{2}g$

22. A thin horizontal circular disc is rotating about a vertical axis passing through its centre. An insect is at rest at a point near the rim of the disc. The insect now moves along a diameter of the disc to reach its other end. During the journey of the insect, then angular speed of the disc : **[2011]**

(A) Continuously increases
(B) First increases and then decreases
(C) Remains unchanged
(D) Continuously decreases

23. A pulley of radius $2m$ is rotated about its axis by a force $F = (20t - 5t^2)$ newton (where t is measured in seconds) applied tangentially. If the moment of inertia of the pulley about its axis of rotation is 10 kg-m^2, the number of rotations made by the pulley before its direction of motion it reversed, is : **[2011]**

(A) More than 6 but less than 9
(B) More than 9
(C) Less than 3
(D) More than 3 but less than 6

24. A diameter R is cut from a disc of mass M and radius R; the circumference of the cut passes through the centre of the disc. The moment of inertia of the remaining portion of the disc about an axis perpendicular to the disc and passing through its centre is : **[2012 (Online)]**

(A) $\left(\frac{15}{32}\right)MR^2$ (B) $\left(\frac{1}{8}\right)MR^2$

(C) $\left(\frac{3}{8}\right)MR^2$ (D) $\left(\frac{13}{32}\right)MR^2$

25. A solid sphere having mass m and radius r rolls down an inclined plane. Then its kinetic energy is : **[2012 (Online)]**

(A) $\frac{5}{7}$ rotational and $\frac{2}{7}$ translational

(B) $\frac{2}{7}$ rotational and $\frac{5}{7}$ translational

(C) $\frac{2}{5}$ rotational and $\frac{3}{5}$ translational

(D) $\frac{1}{2}$ rotational and $\frac{1}{2}$ translational

26. This question has Statement-I and Statement-II. Of the four choices given after the Statements, choose the one the best describes the two Statements. **[2012 (Online)]**

Statement-I : When moment of inertia I of a body rotating about an axis with angular speed ω increases, its angular momentum L is unchanged but the kinetic energy K increases if there is no torque applied on it.

Statement-II : $L = I\omega$, kinetic energy of rotation $= \frac{1}{2}I\omega^2$.

(A) Statement-I is true, Statement-II is true, Statement-II is not the correct explanation of Statement-I
(B) Statement-I is false, Statement-II is true
(C) Statement-I is true, Statement-II is true, Statement-II is correct explanation of the Statement-I
(D) Statement-I is true, Statement-II is false

27. A stone of mass m, tied to the end of a string, is whirled around in a circle on a horizontal frictionless table. The length of the string is reduced gradually keeping the angular momentum of the stone about the centre of the circle constant. Then, the tension in the string is given by $T = Ar^n$, where A is a constant, r is the instantaneous radius of the circle. The value of n is equal to : **[2012 (Online)]**

(A) -1 (B) -2
(C) -4 (D) -3

28. A solid sphere is rolling on a surface as shown in figure, with a translational velocity v ms^{-1}. If it is to climb the inclined surface continuing to roll without slipping, then minimum velocity for this to happen is : **[2012 (Online)]**

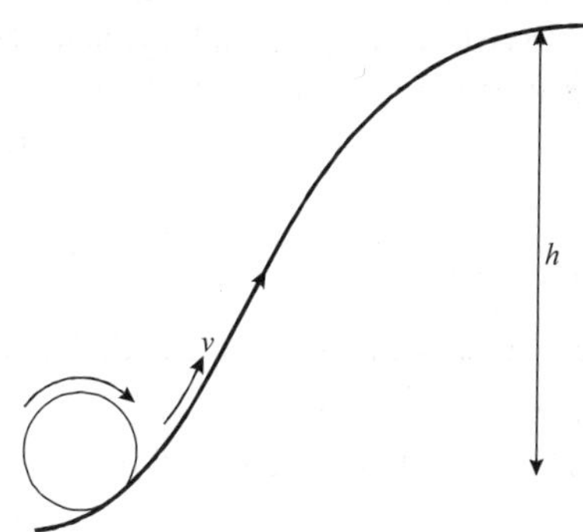

(A) $\sqrt{2gh}$ (B) $\sqrt{\frac{7}{5}gh}$

(C) $\sqrt{\frac{7}{2}gh}$ (D) $\sqrt{\frac{10}{7}gh}$

29. A thick-walled hollow sphere has outside radius R_0. It rolls down an incline without slipping and its speed at the bottom is v_0. Now the incline is waxed, so that it is practically frictionless and the sphere is observed to slide down (without any rolling). Its speed at the bottom is observed to be $5v_0/4$. The radius of gyration of the hollow sphere about an axis through its centre is : **[2012 (Online)]**

(A) $3R_0/2$ (B) $3R_0/4$

(C) $9R_0/16$ (D) $3R_0$

30. Two point masses of mass $m_1 = fM$ and $m_2 = (1-f)M$ $(f < 1)$ are in outer space (far from gravitational influence of other objects) at a distance R from each other. They move in circular orbits about their centre of mass with angular velocities ω_1 for m_1 and ω_2 for m_2. In that case : **[2012 (Online)]**

(A) $(1-f)\,\omega_1 = f\omega$

(B) $\omega_1 = \omega_2$ and independent of f

(C) $f\omega_1 = (1-f)\,\omega_2$

(D) $\omega_1 = \omega_2$ and depend on f

31. A hoop of radius r and mass m rotating with an angular velocity ω_0 is placed on a rough horizontal surface. The initial velocity of the centre of the hoop is zero. What will be the velocity of the centre of the hoop when it ceases to slip ? **[2013]**

(A) $\dfrac{r\omega_0}{4}$ (B) $\dfrac{r\omega_0}{3}$

(C) $\dfrac{r\omega_0}{2}$ (D) $r\omega_0$

32. A bullet of mass 10 g and speed 500 m/s is fired into a door and gets embedded exactly at the centre of the door. The door is 1.0 m wide and weighs 12 kg. It is hinged at one end and rotates about a vertical axis practically without friction. The angular speed of the door just after the bullet embeds into it will be : **[2013 (Online)]**

(A) 6.25 rad/sec (B) 0.625 rad/sec

(C) 3.35 rad/sec (D) 0.335 rad/sec

33. A particle of mass 2 kg is moving such that at time t, its position, in meter, is given by $\vec{r}\,(t) = 5\hat{i} - 2t^2\hat{j}$. The angular momentum of the particle at $t = 2$s about the origin in kg m^{-2} s^{-1} is : **[2013 (Online)]**

(A) $-80\,\hat{k}$ (B) $(10\hat{i} - 16\hat{j})$

(C) $-40\,\hat{k}$ (D) $40\,\hat{k}$

34. A tennis ball (treated as hollow spherical shell) starting from O rolls down a hill. At point A the ball becomes air borne leaving at an angle of 30° with the horizontal. The ball strikes the ground at B. What is the value of the distance AB ? (Moment of inertia of a spherical shell of mass m and radius R about its diameter $= 2/3mR^2$) : **[2013 (Online)]**

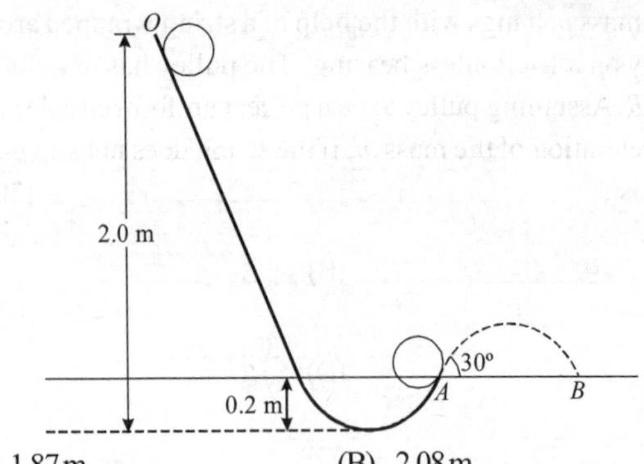

(A) 1.87 m (B) 2.08 m

(C) 1.57 m (D) 1.77 m

35. A ring of mass M and radius R is rotating about its axis with angular velocity ω. Two identical bodies each of mass m are now gently attached at the two ends of a diameter of the ring. Because of this, the kinetic energy loss will be : **[2013 (Online)]**

(A) $\dfrac{m(M+2m)}{M}\omega^2R^2$ (B) $\dfrac{Mm}{(M+m)}\omega^2R^2$

(C) $\dfrac{Mm}{(M+2m)}\omega^2R^2$ (D) $\dfrac{(M+m)M}{(M+2m)}\omega^2R^2$

36. A mass 'm' is supported by a massless string wound around a uniform hollow cylinder of mass m and radius R. If the string does not slip on the cylinder, with what acceleration will the mass fall on release ? **[2014]**

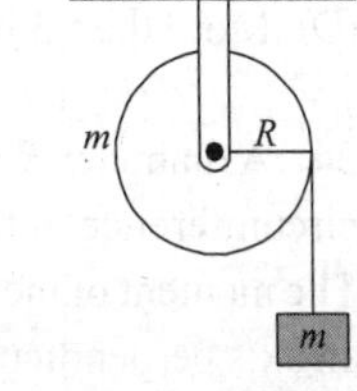

(A) $\dfrac{2g}{3}$

(B) $\dfrac{g}{2}$

(C) $\dfrac{5g}{6}$

(D) g

37. A bob of mass m attached to an inextensible string of length l is suspended from a vertical support. The bob rotates in a horizontal circle with an angular speed ω rad/s about the vertical. About, the point of suspension : **[2014]**

(A) Angular momentum is conserved

(B) Angular momentum changes in magnitude but not in direction

(C) Angular momentum changes in direction but not in magnitude

(D) Angular momentum changes both in direction and magnitude

38. Consider a cylinder of mass M resting on a rough horizontal rug that is pulled out from under it with acceleration 'a' perpendicular to the axis of the cylinder. What is $F_{friction}$ at point P ? It is assumed that the cylinder does not slip : **[2014 (Online)]**

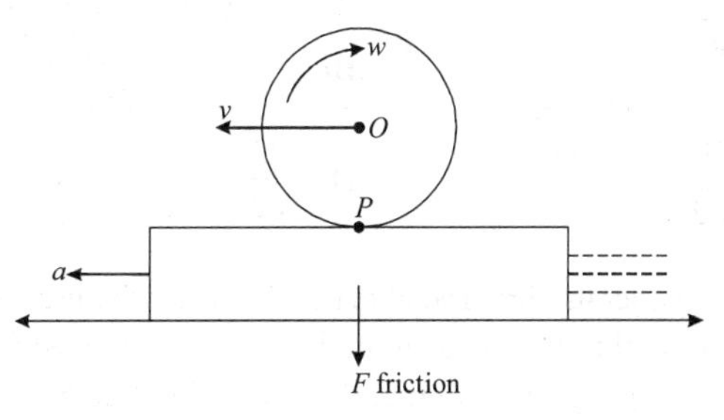

(A) Mg (B) Ma

(C) $\frac{Ma}{2}$ (D) $\frac{Ma}{3}$

39. A cylinder of mass M_C and sphere of mass M_S are placed at point A and B of two inclines, respectively (See figure). If they roll on the incline without slipping such that their acceleration are the same, then the ratio $\frac{\sin\theta_C}{\sin\theta_S}$ is : **[2014 (Online)]**

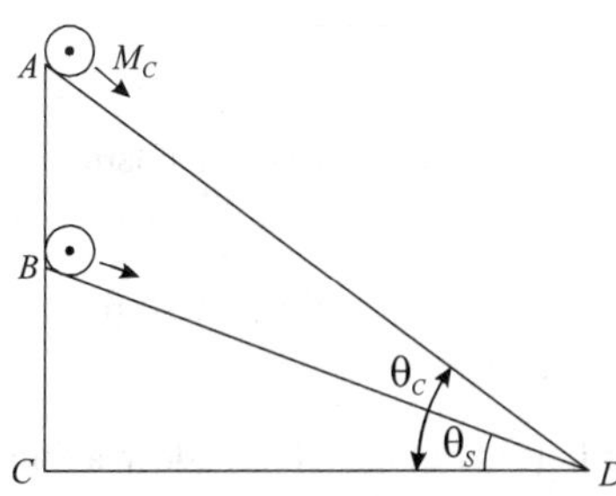

(A) $\sqrt{\frac{8}{7}}$ (B) $\sqrt{\frac{15}{14}}$

(C) $\frac{8}{7}$ (D) $\frac{15}{14}$

40. A ball of mass 160 g is thrown up at an angle of 60° to the horizontal at a speed of 10 ms^{-1}. The angular momentum of the ball at the highest point of the trajectory with respect to the point from which the ball is thrown is nearly : ($g = 10$ ms^{-2}) **[2014 (Online)]**

(A) 1.73 kg m^2/s (B) 3.0 kg m^2/s

(C) 3.46 kg m^2/s (D) 6.9 kg m^2/s

41. A particle of mass m is moving in a circular path of radius a, with a constant velocity v as shown in the figure. The centre of circle is marked by 'C'. The angular momentum from the origin O can be written as : **[2014 (Online)]**

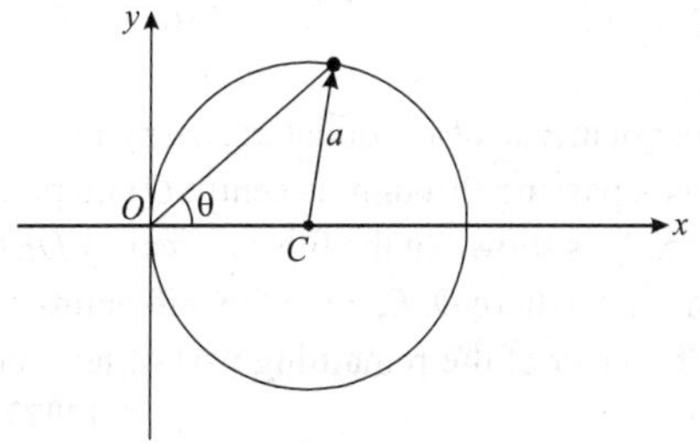

(A) $mva\,(1 + \cos 2\theta)$ (B) $mva\,(1 + \cos\theta)$

(C) $mva \cos 2\theta$ (D) mva

42. From a solid sphere of mass M and radius R a cube of maximum possible volume is cut. Moment of inertia of cube about an axis passing through its center and perpendicular to one of its faces is : **[2015]**

(A) $\frac{MR^2}{32\sqrt{2}\pi}$ (B) $\frac{MR^2}{16\sqrt{2}\pi}$

(C) $\frac{4MR^2}{9\sqrt{3}\pi}$ (D) $\frac{4MR^2}{3\sqrt{3}\pi}$

43. A uniform solid cylindrical roller of mass 'm' is being pulled on a horizontal surface with force F parallel to the surface and applied at its centre. If the acceleration of the cylinder is 'a' and it rolling without slipping then the value of 'F' is : **[2015 (Online)]**

(A) ma (B) $\frac{5}{3}ma$

(C) $\frac{3}{2}ma$ (D) $2ma$

44. A particle of mass 2 kg is on a smooth horizontal table and moves in a circular path of radius 0.6 m. The height of the table from the ground is 0.8 m. If the angular speed of the particle is 12 rad s^{-1}, the magnitude of its angular momentum about a point on the ground right under the centre of the circle is : **[2015 (Online)]**

(A) 14.4 kg m^2 s^{-1} (B) 8.64 kg m^2 s^{-1}

(C) 0.16 kg m^2 s^{-1} (D) 11.52 kg m^2 s^{-1}

45. Consider a thin uniform square sheet made of a rigid material. If its side is 'a' mass m and moment of inertia I about one of its diagonals, then : **[2015 (Online)]**

(A) $I > \frac{ma^2}{12}$ (B) $\frac{ma^2}{24} < I < \frac{ma^2}{12}$

(C) $I = \frac{ma^2}{24}$ (D) $I = \frac{ma^2}{12}$

46. A particle of mass m is moving along the side of a square of side 'a', with a uniform speed v in x-y plane as shown in the figure : **[2016]**

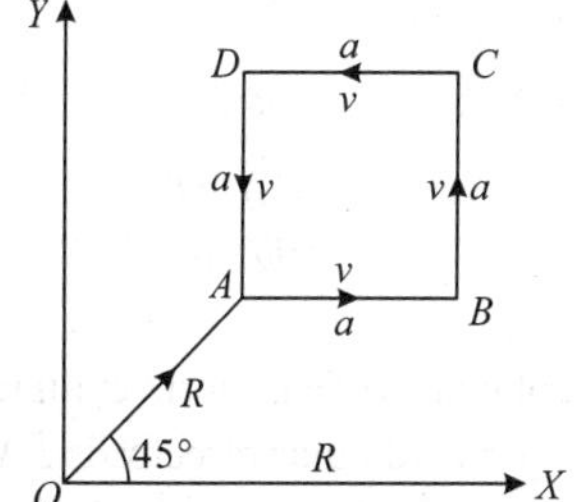

Which of the following statements is false for the angular momentum $\vec{L}$ about the origin ?

(A) $\vec{L} = mv\left[\frac{R}{\sqrt{2}}+a\right]\hat{k}$ when the particle is moving from B to C

(B) $\vec{L} = \frac{mv}{\sqrt{2}}R\hat{k}$ when the particle is moving from D to A

(C) $\vec{L} = -\frac{mv}{\sqrt{2}}R\hat{k}$ when the particle is moving from A to B

(D) $\vec{L} = mv\left[\frac{R}{\sqrt{2}}-a\right]\hat{k}$ when the particle is moving from C to D

47. A roller is made by joining together two cones at their vertices O. It is kept on two rails AB and CD, which are placed asymmetrically (see figure), with its axis perpendicular to CD and its centre O at the centre of line joining AB and CD (see figure). It is given a light push so that its starts rolling with its centre O moving parallel to CD in the direction shown. As it moves, the roller will tend to : **[2016]**

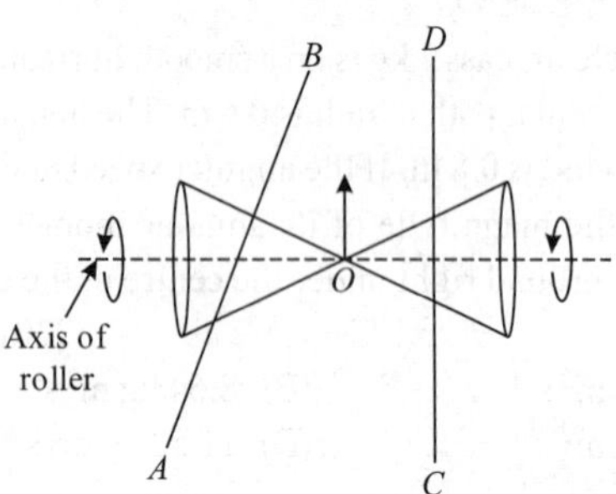

(A) Go straight
(B) Turn left and right alternately
(C) Turn left
(D) Turn right

48. A cubical block of side 30 cm is moving with velocity 2 ms^{-1} on a smooth horizontal surface. The surface has a bump at a point O as shown in figure. The angular velocities (in rad/s) of the block immediately after it hits the bump, is : **[2016 (Online)]**

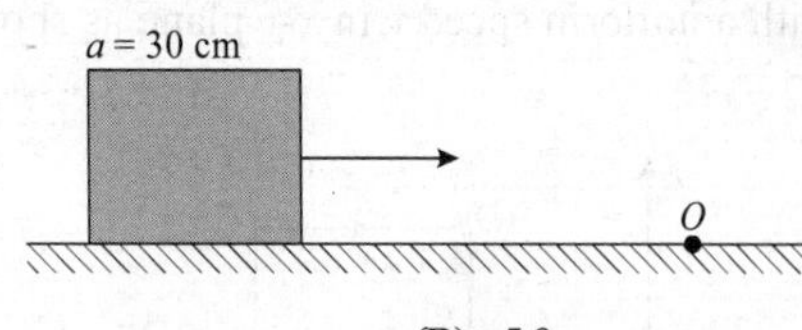

(A) 13.3 (B) 5.0
(C) 9.4 (D) 6.7

49. The moment of inertia of a uniform cylinder of length l and radius R about its perpendicular bisector is I. What is the ratio l/R such that the moment of inertia is minimum ? **[2017]**

(A) 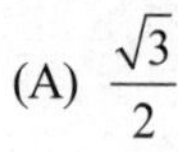 $\frac{\sqrt{3}}{2}$ (B) 1

(C) $\frac{3}{\sqrt{2}}$ (D) $\sqrt{\frac{3}{2}}$

50. A slender uniform rod of mass M and length l is pivoted at one end so that it can rotate in a vertical plane (see figure.) There is negligible friction at the pivot. The free end is held vertically above the pivot and then released. The angular acceleration of the rod when it make an angle θ with the vertical is : **[2017]**

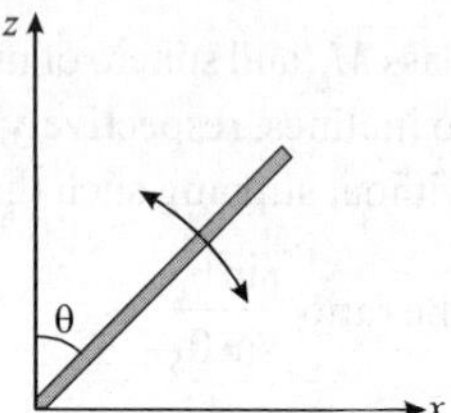

(A) $\frac{3g}{3l}\sin\theta$ (B) $\frac{3g}{3l}\cos\theta$

(C) $\frac{2g}{3l}\cos\theta$ (D) $\frac{3g}{2l}\sin\theta$

51. A circular hole of radius $R/4$ is made in a thin uniform disc having mass M and radius R, as shown in figure. The moment of inertia of the remaining portion of the disc about an axis passing through the point O and perpendicular to the plane of the disc is : **[2017 (Online)]**

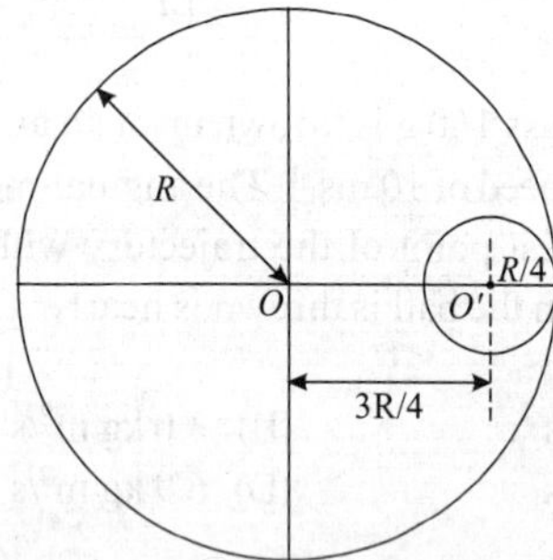

(A) $\frac{219MR^2}{256}$ (B) $\frac{237MR^2}{512}$

(C) $\frac{19MR^2}{512}$ (D) $\frac{197MR^2}{256}$

52. Moment of inertia of an equilateral triangular lamina ABC, about the axis passing through its centre O and perpendicular to its plane is I_0 as shown in the figure. A cavity DEF is cut out from the lamina, where D, E, F are the mid points of the sides. Moment of inertia of the remaining part of lamina about the same axis is : **[2017 (Online)]**

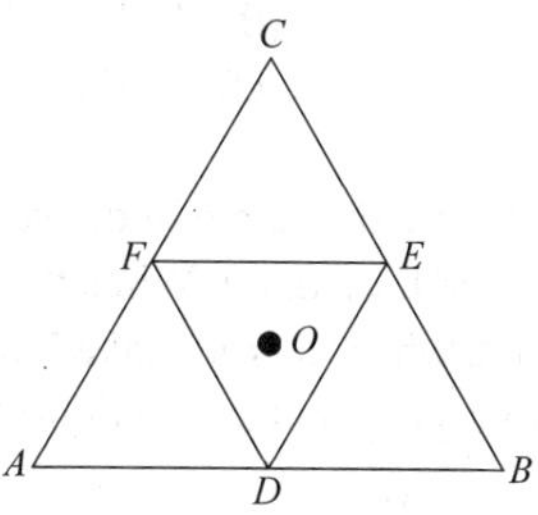

(A) $\frac{7}{8}I_0$ (B) $\frac{15}{16}I_0$

(C) $\frac{3I_0}{4}$ (D) $\frac{31I_0}{32}$

53. The machine as shown has 2 rods of length 1 m connected by a pivot at the top. The end of one rod is connected to the floor by a stationary pivot and the end of the other rod has a roller that rolls along the floor in a slot. As the roller goes back and forth, a 2 kg weight moves up and down. If the roller is moving towards right at a constant speed, the weight moves up with a : **[2017 (Online)]**

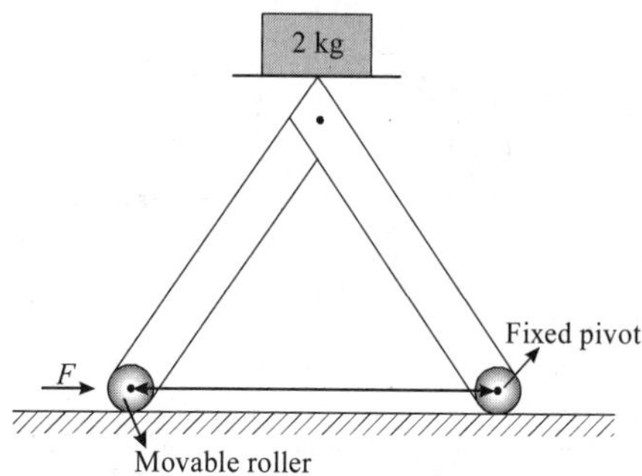

(A) Constant speed (B) Decreasing speed
(C) Increasing speed
(D) Speed which is 3/4th of that of the roller when the weight is 0.4 m above the ground

54. In a physical balance working on the principle of moments, when 5 mg weight is placed on the left pan, the beam becomes horizontal. Both the empty pans of the balance are of equal mass. Which of the following statements is correct ? **[2017 (Online)]**

(A) Left arm is longer than the right arm
(B) Both the arms are of same length
(C) Left arm is shorter than the right arm
(D) Every object that is weighed using this balance appears lighter than its actual weight

55. A particle is moving with a uniform speed in a circular orbit of radius R in a central force inversely proportional to the nth power of R. If the period of rotation of the particle is T, then : **[2018]**

(A) $T \propto R^{3/2}$ for any n (B) $T \propto R^{n/2+1}$
(C) $T \propto R^{(n+1)/2}$ (D) $T \propto R^{n/2}$

56. Seven identical circular planar disks, each of mass M and radius R are welded symmetrically as shown. The moment of inertia of the arrangement about the axis normal to the plane and passing through the point P is : **[2018]**

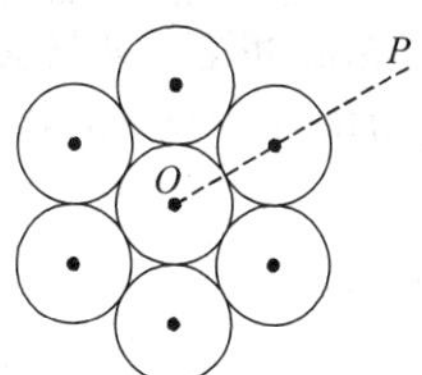

(A) $\frac{19}{2}$MR^2 (B) $\frac{55}{2}MR^2$
(C) $\frac{73}{2}MR^2$ (D) $\frac{181}{2}MR^2$

57. From a uniform circular disc of radius R and mass $9M$, a small disc of radius $R/3$ is removed as shown in the figure. The moment of inertia of the remaining disc about an axis perpendicular to the plane of the disc and passing, through centre of disc is : **[2018]**

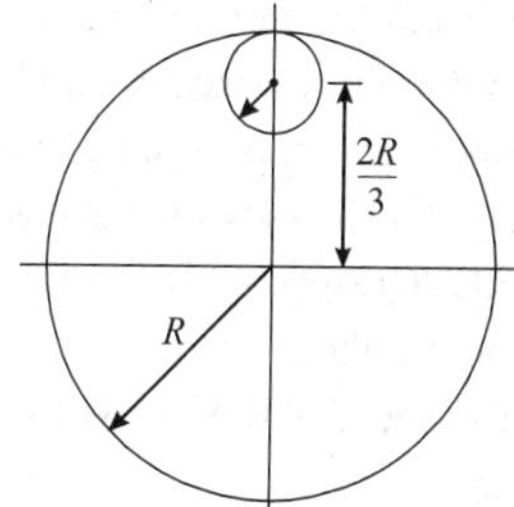

(A) 4 MR^2 (B) $\frac{40}{9}$MR^2

(C) 10 MR^2 (D) $\frac{37}{9}$MR^2

58. A force of 40 N acts on a point B a the end of an L-shaped object, as shown in the figure. The angle θ that will produce maximum moment of the force about point A is given by : **[2018 (Online)]**

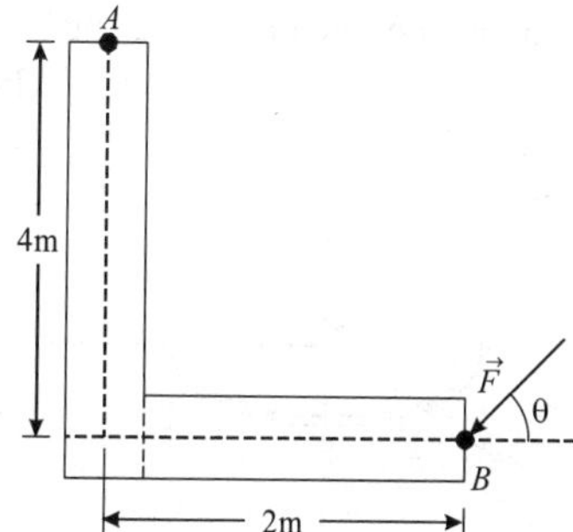

(A) $\tan\theta = \frac{1}{4}$ (B) $\tan\theta = 2$

(C) $\tan\theta = \frac{1}{2}$ (D) $\tan\theta = 4$

59. A uniform rod AB is suspended from a point X, at a variable distance from x from A, as shown. To make the rod horizontal, a mass m is suspended from its end A. A set of (m, x) values is recorded. The appropriate variable that give a straight line, when plotted, are : **[2018 (Online)]**

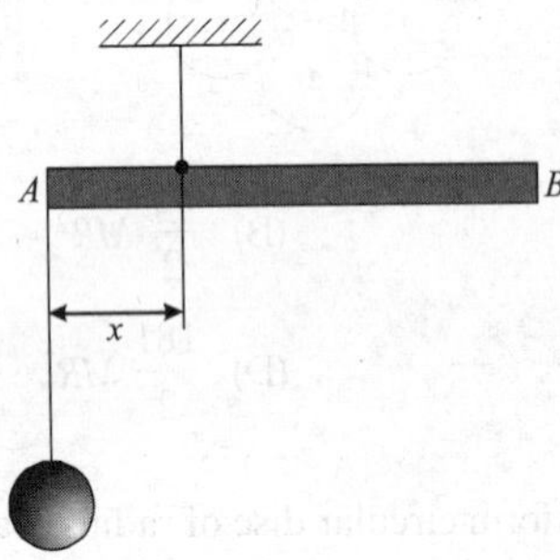

(A) $m, \frac{1}{x}$ (B) $m, \frac{1}{x^2}$

(C) m, x (D) m, x^2

60. A thin uniform bar of length L and mass $8m$ lies on a smooth horizontal table. Two point masses m and $2m$ moving in the same horizontal plane from opposite sides of the bar with speeds $2v$ and v respectively. The masses stick to the bar after collision at a distance $L/3$ and $L/6$ respectively from the centre of the bar. If the bar starts rotating about its center of mass as a result of collision, the angular speed of the bar will be : **[2018 (Online)]**

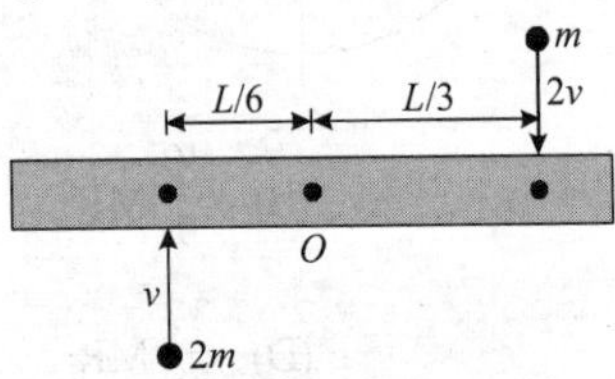

(A) $\frac{v}{6L}$ (B) $\frac{6v}{5L}$

(C) $\frac{3v}{5L}$ (D) $\frac{v}{5L}$

61. A thin circular disc is in the xy plane as shown in the figure. The ratio of its moment of inertia about z and z' axes will be : **[2018 (Online)]**

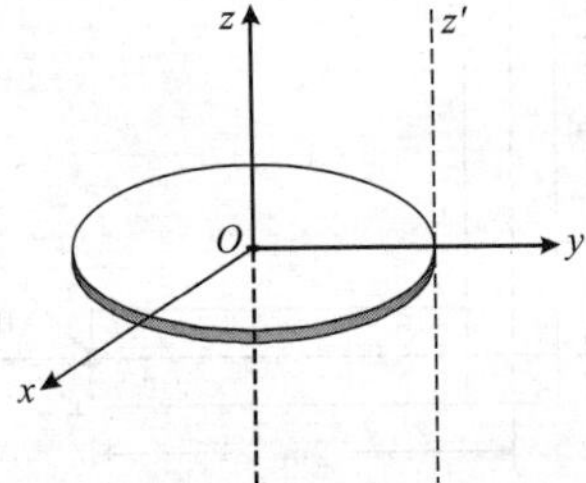

(A) 1 : 2 (B) 1 : 4

(C) 1 : 3 (D) 1 : 5

62. A thin rod MN, free the rotate in the vertical plane about the fixed end N, is held horizontal. When the end M is released the speed of this end, when the rod makes an angle α with the horizontal, will be proportional to (see figure) : **[2018 (Online)]**

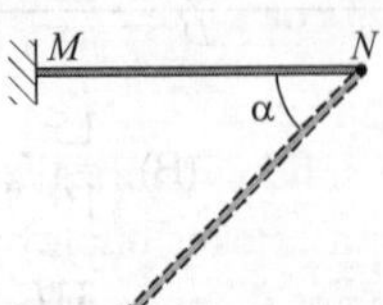

(A) $\sqrt{\cos\alpha}$ (B) $\cos\alpha$

(C) $\sin\alpha$ (D) $\sqrt{\sin\alpha}$

63. An L-shaped object, made of thin rods of uniform mass density, is suspended with a string as shown in figure. If $AB = BC$, and the angle made by AB with downward vertical is θ, then : **[9 Jan. 2019 (Morning)]**

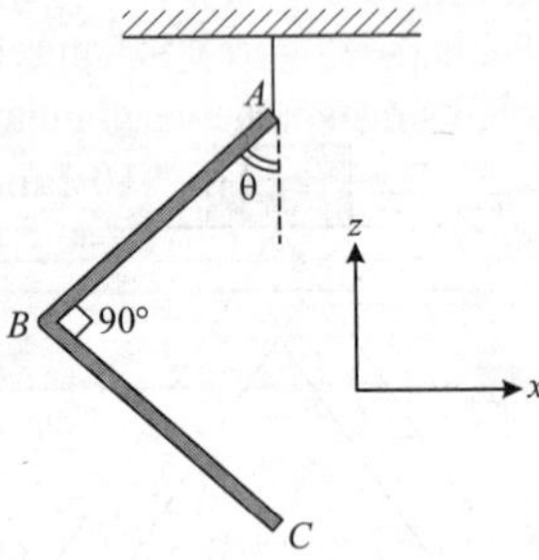

(A) $\tan\theta = \frac{2}{\sqrt{3}}$ (B) $\tan\theta = \frac{1}{3}$

(C) $\tan\theta = \frac{1}{2}$ (D) $\tan\theta = \frac{1}{2\sqrt{3}}$

64. A rod of length 50 cm is pivoted at one end. It is raised such that it makes an angle of 30° from the horizontal as shown and released from rest. Its angular speed when it passes through the horizontal (in rad s^{-1}) will be ($g = 10\ ms^{-2}$) : **[9 Jan. 2019 (Evening)]**

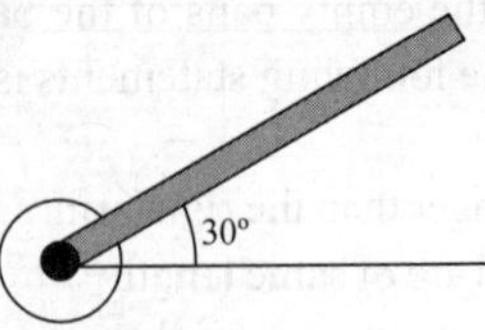

(A) $\sqrt{30}$ (B) $\sqrt{\frac{30}{2}}$

(C) $\frac{\sqrt{30}}{2}$ (D) $\frac{\sqrt{20}}{3}$

65. To mop-clean a floor, a cleaning machine presses a circular mop of radius R vertically down with a total force F and rotates it with a constant angular speed about its axis. If the force F is distributed uniformly over the mop and if coefficient of friction between the mop and the floor is μ, the torque, applied by the machine on the mop is : **[10 Jan. 2019 (Morning)]**

(A) $\frac{2}{3}\mu FR$ (B) $\mu FR/3$
(C) $\mu FR/2$ (D) $\mu FR/6$

66. A homogeneous solid cylindrical roller of radius R and mass M is pulled on a cricket pitch by a horizontal force. Assuming rolling without slipping, angular acceleration of the cylinder is: **[10 Jan. 2019 (Morning)]**

(A) $\frac{3F}{2mR}$ (B) $\frac{F}{3mR}$
(C) $\frac{2F}{3mR}$ (D) $\frac{F}{2mR}$

67. A rigid massless rod of length $3l$ has two masses attached at each end as shown in the figure. The rod is pivoted at point P on the horizontal axis (see figure). When released from initial horizontal position, its instantaneous angular acceleration will be : **[10 Jan. 2019 (Evening)]**

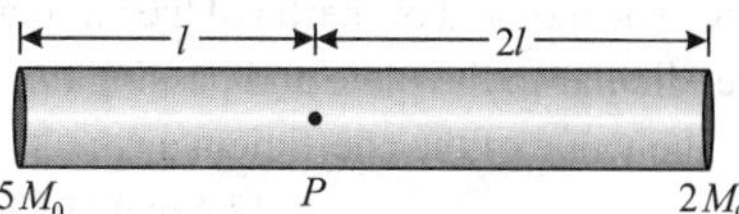

(A) $\frac{g}{2l}$ (B) $\frac{7g}{3l}$
(C) $\frac{g}{13l}$ (D) $\frac{g}{3l}$

68. Two identical spherical balls of mass M and radius R each are stuck on two ends of a rod of length $2R$ and mass M (see figure). The moment of inertia of the system about the axis passing perpendicularly through the centre of the rod is : **[10 Jan. 2019 (Evening)]**

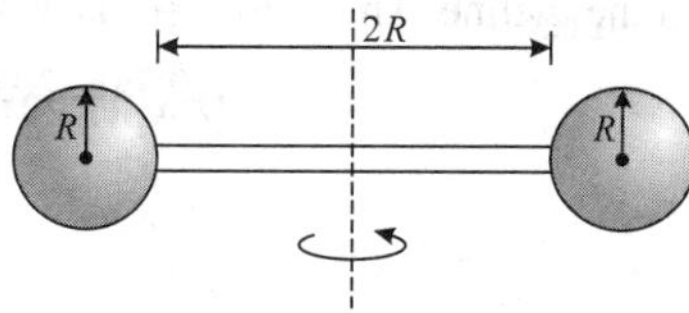

(A) $\frac{152}{15}MR^2$ (B) $\frac{17}{15}MR^2$
(C) $\frac{137}{15}MR^2$ (D) $\frac{209}{15}MR^2$

69. An equilateral triangle ABC is cut from a thin solid sheet of wood. (see figure) D, E and F are the mid-points of its sides as shown and G is the centre of the triangle. The moment of inertia of the triangle about an axis passing through G and perpendicular to the plane of the triangle is I_0. It the smaller triangle DEF is removed from ABC, the moment of inertia of the remaining figure about the same axis is I. Then : **[11 Jan. 2019 (Morning)]**

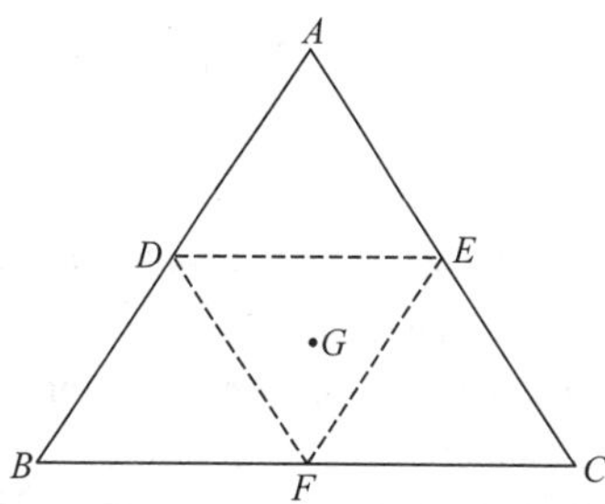

(A) $I = \frac{9}{16}I_0$ (B) $I = \frac{3}{4}I_0$
(C) $I = \frac{I_0}{4}$ (D) $I = \frac{15}{16}I_0$

70. A slab is subjected to two forces $\vec{F_1}$ and $\vec{F_2}$ of same magnitude F as shown in the figure. Force $\vec{F_2}$ is in XY-plane while force F_1 acts along z-axis at the point $(2\vec{i} + 3\vec{j})$. The moment of these forces about point O will be : **[11 Jan. 2019 (Morning)]**

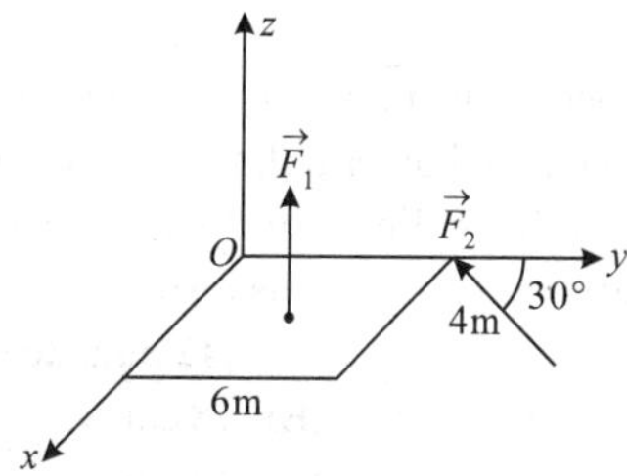

(A) $(3\hat{i} - 2\hat{j} - 3\hat{k})F$ (B) $(3\hat{i} + 2\hat{j} + 3\hat{k})F$
(C) $(3\hat{i} + 2\hat{j} - 3\hat{k})F$ (D) $(3\hat{i} - 2\hat{j} + 3\hat{k})F$

71. A string is wound around a hollow cylinder of mass 5 kg and radius 0.5 m. If the string is now pulled with a horizontal force of 40 N, and the cylinder is rolling without slipping on a horizontal surface (see figure), then the angular acceleration of the cylinder will be (Neglect the mass and thickness of the string) : **[11 Jan. 2019 (Evening)]**

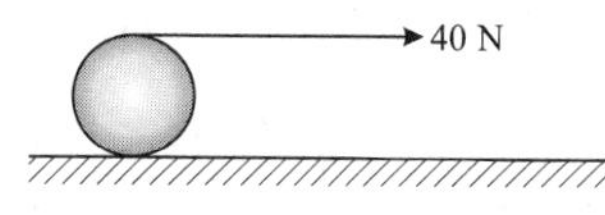

(A) 12 rad/s^2 (B) 16 rad/s^2
(C) 10 rad/s^2 (D) 20 rad/s^2

72. The magnitude of torque on a particle of mass 1 kg is 2.5 Nm about the origin. If the force acting on it is 1 N, and the distance of the particle from the origin is 5 m, the angle between the force and the position vector is : (In radians) **[11 Jan. 2019 (Evening)]**

(A) $\frac{\pi}{8}$ (B) $\frac{\pi}{6}$

(C) $\frac{\pi}{4}$ (D) $\frac{\pi}{3}$

73. A circular disc D_1 of mass M and radius R has two identical discs D_2 and D_3 of the same mass M and radius R attached rigidly at its opposite ends (see figure). The moment of inertia of the system about the axis OO', passing through the centre of D_1, as shown in the figure, will be: **[11 Jan. 2019 (Evening)]**

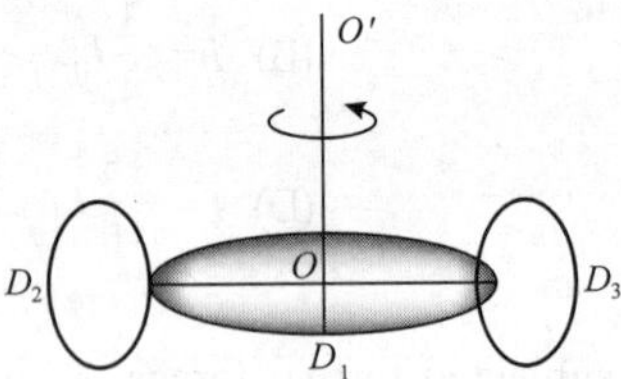

(A) $3MR^2$ (B) $\frac{2}{3}MR^2$

(C) MR^2 (D) $\frac{4}{5}MR^2$

74. Let the moment of inertia of a hollow cylinder of length 30 cm (inner radius 10 cm and outer radius 20 cm), about its axis be I. The radius of a thin cylinder of the same mass such that its moment of inertia about its axis is also I, is : **[12 Jan. 2019 (Morning)]**

(A) 12 cm (B) 18 cm
(C) 16 cm (D) 14 cm

75. A particle of mass 20 g is released with an initial velocity 5 m/s along the curve from the point A, as shown in the figure. The point A is at height h from point B. The particle slides along the frictionless surface. When the particle reaches point B, its angular momentum about O will be : [Take $g = 10$ m/s^2] **[12 Jan. 2019 (Evening)]**

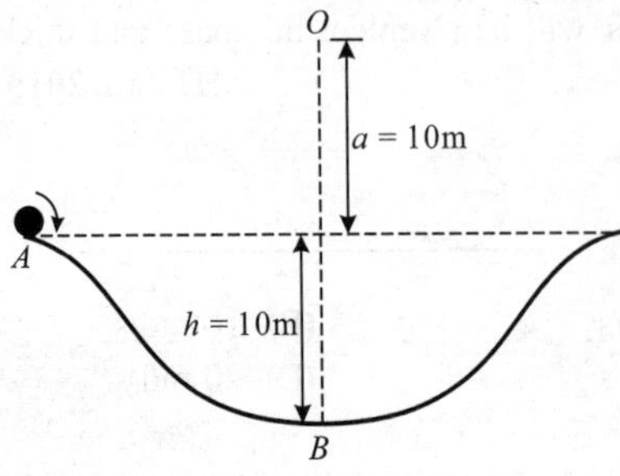

(A) 8 kg-m^2/s (B) 6 kg-m^2/s
(C) 3 kg-m^2/s (D) 2 kg-m^2/s

76. The moment of inertia of a solid sphere, about an axis parallel to its diameter and at a distance of x from it, is $I(x)'$. Which one of the graphs represents the variation of $I(x)$ with x correctly? **[12 Jan. 2019 (Evening)]**

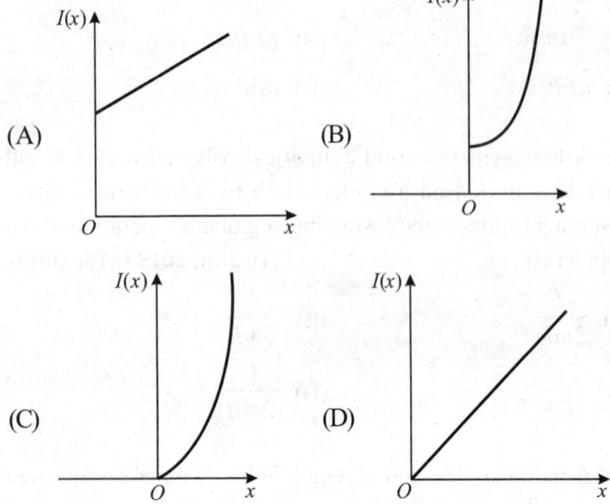

77. A thin circular plate of mass M and radius R has its density varying as $\rho(r) = \rho_0 r$ with ρ_0 as constant and r is the distance from its centre. The moment of Inertia of the circular plate about an axis perpendicular to the plate and passing through its edge is $I = aMR^2$. The value of the coefficient a is : **[8 April 2019 (Morning)]**

(A) $\frac{3}{2}$ (B) $\frac{1}{2}$

(C) $\frac{3}{5}$ (D) $\frac{8}{5}$

78. A solid sphere and solid cylinder of identical radii approach an incline with the same linear velocity (see figure). Both roll without slipping all throughout. The two climb maximum heights h_{sph} and h_{cyl} on the incline. The ratio $\frac{h_{sph}}{h_{cyl}}$ is given by : **[8 April 2019 (Evening)]**

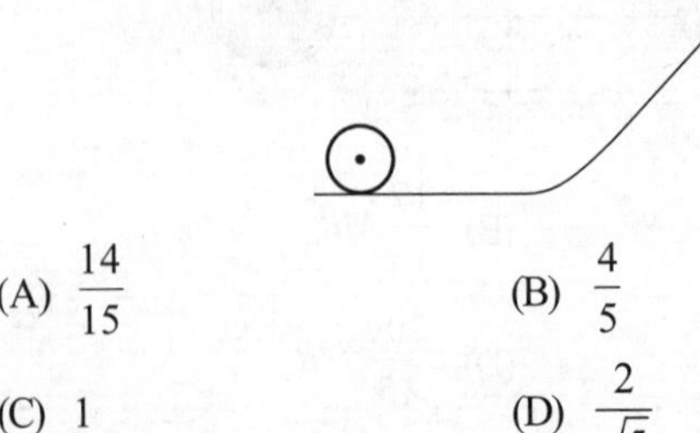

(A) $\frac{14}{15}$ (B) $\frac{4}{5}$

(C) 1 (D) $\frac{2}{\sqrt{5}}$

79. A rectangular solid box of length 0.3 m is held horizontally, with one of its sides on the edge of a platform of height 5 m. When released, it slips off the table in a very short time $\tau = 0.01$ s, remaining essentially horizontal. The angle by which it would rotate when it hits the ground will be (in radians) close to : **[8 April 2019 (Evening)]**

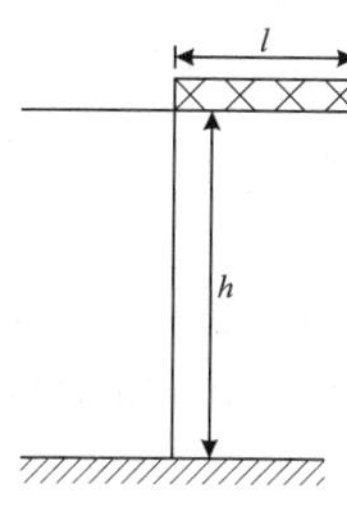

(A) 0.02 (B) 0.28
(C) 0.5 (D) 0.3

80. The following bodies are made to roll up (without slipping) the same inclined plane from a horizontal plane : (i) a ring of radius R, (ii) a solid cylinder of radius $\frac{R}{2}$ and (iii) a solid sphere of radius $\frac{R}{4}$. If in each case, the speed of the centre of mass at the bottom of the incline is same, the ratio of the maximum heights they climb is : **[9 April 2019 (Morning)]**

(A) 4 : 3 : 2 (B) 14 : 15 : 20
(C) 10 : 15 : 7 (D) 2 : 3 : 4

81. A stationary horizontal disc is free to rotate about its axis. When a torque is applied on it, its kinetic energy as a function of θ, where θ is the angle by which it has rotated, is given as $k\theta^2$. If its moment of inertia is I then the angular acceleration of the disc is : **[9 April 2019 (Morning)]**

(A) $\frac{k}{2I}\theta$ (B) $\frac{k}{I}\theta$
(C) $\frac{k}{4I}\theta$ (D) $\frac{2k}{I}\theta$

82. Moment of inertia of a body about a given axis is 1.5 kg m^2. Initially the body is at rest. In order to produce a rotational kinetic energy of 1200 J, the angular acceleration of 20 rad/s^2 must be applied about the axis for a duration of : **[9 April 2019 (Evening)]**

(A) 2s (B) 5s
(C) 2.5s (D) 3s

83. A thin smooth rod of length L and mass M is rotating freely with angular speed ω_0 about an axis perpendicular to the rod and passing through its center. Two beads of mass m and negligible size are at the center of the rod initially. The beads are free to slide along the rod. The angular speed of the system, when the beads reach the opposite ends of the rod, will be : **[9 April 2019 (Evening)]**

(A) $\frac{M\omega_0}{M+3m}$ (B) $\frac{M\omega_0}{M+m}$
(C) $\frac{M\omega_0}{M+2m}$ (D) $\frac{M\omega_0}{M+6m}$

84. A thin disc of mass M and radius R has mass per unit area $\sigma(r) = kr^2$ where r is the distance from its centre. Its moment of inertia about an axis going through its centre of mass and perpendicular to its plane is : **[10 April 2019 (Morning)]**

(A) $\frac{MR^2}{6}$ (B) $\frac{MR^2}{3}$
(C) $\frac{2MR^2}{3}$ (D) $\frac{MR^2}{2}$

85. Two coaxial discs, having moments of inertia I_1 and $\frac{I_1}{2}$, are rotating with respective angular velocities ω_1 and $\frac{\omega_1}{2}$, about their common axis. They are brought in contact with each other and thereafter they rotate with a common angular velocity. If E_f and E_i are the final and initial total energies, then $(E_f - E_i)$ is : **[10 April 2019 (Morning)]**

(A) $\frac{I_1\omega_1^2}{12}$ (B) $\frac{3}{8}I_1\omega_1^2$
(C) $\frac{I_1\omega_1^2}{6}$ (D) $\frac{I_1\omega_1^2}{24}$

86. A particle of mass m is moving along a trajectory given by
$x = x_0 + a\cos\omega_1 t$
$y = y_0 + b\sin\omega_2 t$
The torque, acting on the particle about the origin, at $t = 0$ is : **[10 April 2019 (Morning)]**

(A) $m(-x_0 b + y_0 a)\,\omega_1^2\hat{k}$ (B) $+my_0 a\,\omega_1^2\hat{k}$
(C) $-m(x_0 b\omega_2^2 - y_0 a\omega_1^2)\hat{k}$ (D) Zero

87. A metal coin of mass 5g and radius 1 cm is fixed to a thin stick AB of negligible mass as shown in the figure. The system is initially at rest. The constant torque, that will make the system rotate about AB at 25 rotations per second in 5 s, is close to : **[10 April 2019 (Evening)]**

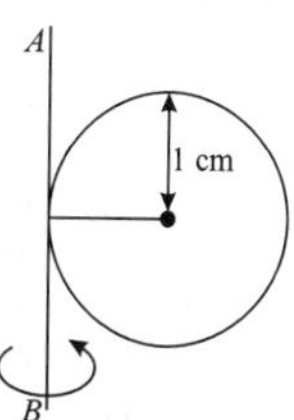

(A) 4.0×10^{-6} Nm (B) 2.0×10^{-5} Nm
(C) 1.6×10^{-5} Nm (D) 7.9×10^{-6} Nm

88. The time dependence of the position of a particle of mass $m = 2$ is given by $\vec{r}(t) = 2t\hat{i} - 3t^2\hat{j}$. Its angular momentum, with respect to the origin, at time $t = 2$ is : **[10 April 2019 (Evening)]**

(A) $36\hat{k}$ (B) $-34(\hat{k} - \hat{i})$

(C) $48(\hat{i} + \hat{j})$ (D) $-48\hat{k}$

89. A solid sphere of mass M and radius R is divided into two unequal parts. The first part has a mass of $\frac{7M}{8}$ and is converted into a uniform disc of radius $2R$. The second part is converted into a uniform solid sphere. Let I_1 be the moment of inertia of the disc about its axis and I_2 be the moment of inertia of the new sphere about its axis. The ratio I_1/I_2 is given by :

[10 April 2019 (Evening)]

(A) 185 (B) 65

(C) 285 (D) 140

90. A circular disc of radius b has a hole of radius a at its centre (see figure). If the mass per unit area of the disc varies as $\left(\frac{\sigma_0}{r}\right)$, then the radius of gyration of the disc about its axis passing through the centre is : **[12 April 2019 (Morning)]**

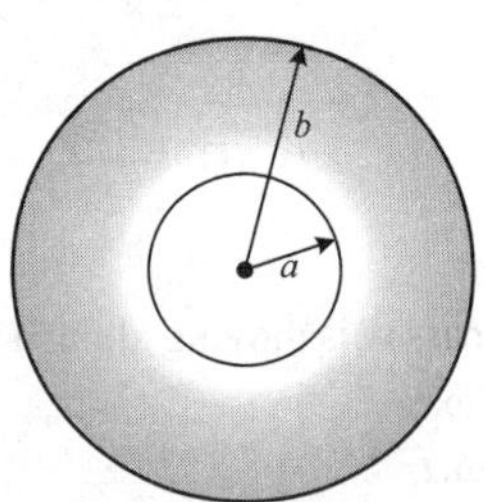

(A) $\frac{a+b}{2}$ (B) $\frac{a+b}{3}$

(C) $\sqrt{\frac{a^2+b^2+ab}{2}}$ (D) $\sqrt{\frac{a^2+b^2+ab}{3}}$

91. A uniform rod of length l is being rotated in a horizontal plane with a constant angular speed about an axis passing through one of its ends. If the tension generated in the rod due to rotation is $T(x)$ at a distance x from the axis, then which of the following graphs depicts it most closely ?

[12 April 2019 (Morning)]

(A) 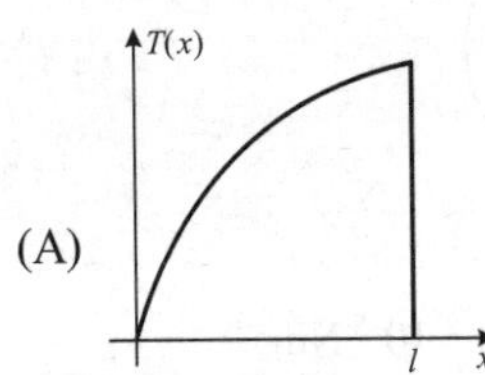

(B) 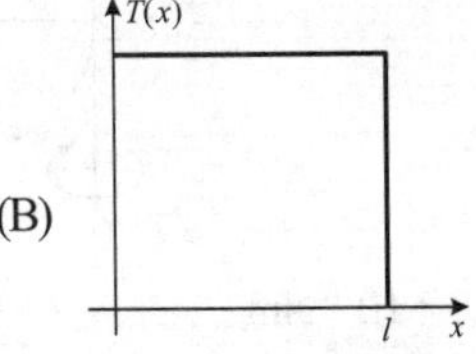

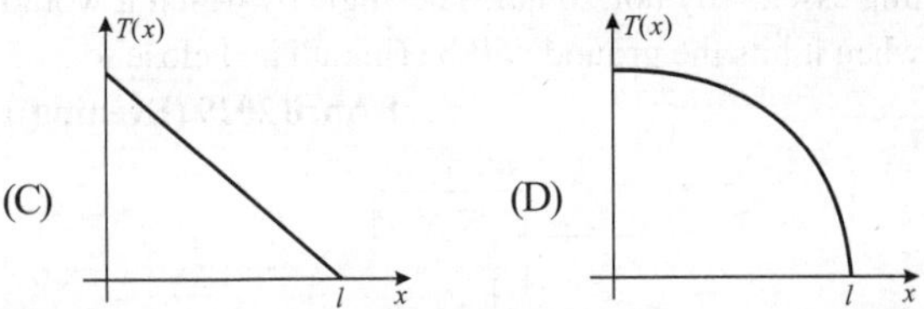

92. As shown in the figure, a bob of mass m is tied by a massless string whose other end portion is wound on a fly wheel (disc) of radius r and mass m. When released from rest the bob starts falling vertically. When it has covered a distance of h, the angular speed of the wheel will be : **[7 Jan. 2020 (Morning)]**

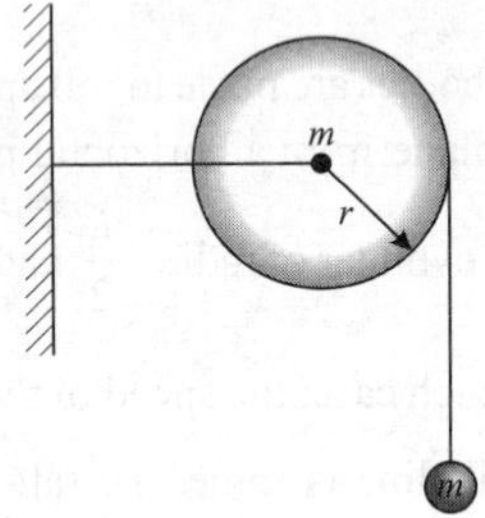

(A) $\frac{1}{r}\sqrt{\frac{2gh}{3}}$ (B) $r\sqrt{\frac{3}{4gh}}$

(C) $\frac{1}{r}\sqrt{\frac{4gh}{3}}$ (D) $r\sqrt{\frac{3}{2gh}}$

93. The radius of gyration of a uniform rod of length l, about an axis passing through a point $l/4$ away from the centre of the rod, and perpendicular to it, is : **[7 Jan. 2020 (Morning)]**

(A) $\frac{1}{8}l$ (B) $\sqrt{\frac{7}{48}}l$

(C) $\sqrt{\frac{3}{8}}l$ (D) $\frac{1}{4}l$

94. Mass per unit area of a circular disc of radius a depends on the distance r from its centre as $\sigma(r) = A + Br$. The moment of inertia of the disc about the axis, perpendicular to the plane and passing through its centre is : **[7 Jan. 2020 (Evening)]**

(A) $2\pi a^4\left(\frac{A}{4} + \frac{aB}{5}\right)$ (B) $\pi a^4\left(\frac{A}{4} + \frac{aB}{5}\right)$

(C) $2\pi a^4\left(\frac{aA}{4} + \frac{B}{5}\right)$ (D) $2\pi a^4\left(\frac{A}{4} + \frac{B}{5}\right)$

95. Consider a uniform cubical box of side a on a rough floor that is to be moved by applying minimum possible force F at a point b above its centre of mass (see figure). If the coefficient of friction is $\mu = 0.4$, the maximum possible value of $100 \times \frac{b}{a}$ for a box not to topple before moving is _____. **[7 Jan. 2020 (Evening)]**

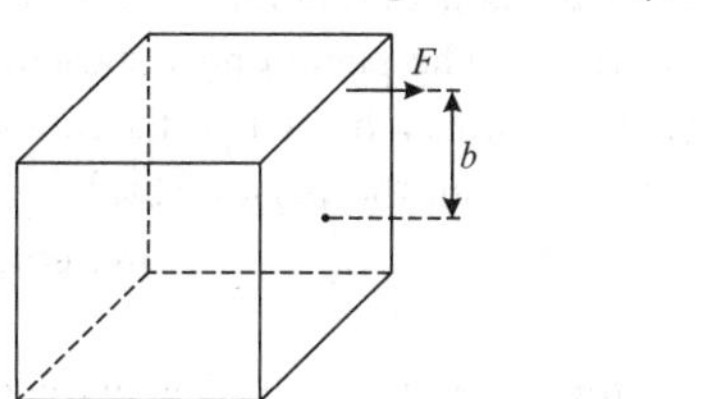

96. Consider a uniform rod of mass $M = 4m$ and length l pivoted about its centre. A mass m moving with velocity v making angle $\theta = \pi/4$ to the rod's long axis collides with one end of the rod and sticks to it. The angular speed of the rod-mass system just after the collision is : **[8 Jan. 2020 (Morning)]**

(A) $\frac{3}{7\sqrt{2}}\frac{v}{l}$ (B) $\frac{3\sqrt{2}}{7}\frac{v}{l}$

(C) $\frac{4}{7}\frac{v}{l}$ (D) $\frac{3}{7}\frac{v}{l}$

97. A uniform sphere of mass 500 g rolls without slipping on a plane horizontal surface with its centre moving at a speed of 5.00 cm/s. Its kinetic energy is : **[8 Jan. 2020 (Evening)]**

(A) 8.75×10^{-4} J (B) 8.75×10^{-3} J

(C) 6.25×10^{-4} J (D) 1.13×10^{-3} J

98. Three solid spheres each of mass m and diameter d are stuck together such that the lines connecting the centres form an equilateral triangle of side of length d. The ratio I_0/I_A of moment of inertia I_0 of the system about an axis passing the centroid and about center of any of the spheres I_A and perpendicular to the plane of the triangle is : **[9 Jan. 2020 (Morning)]**

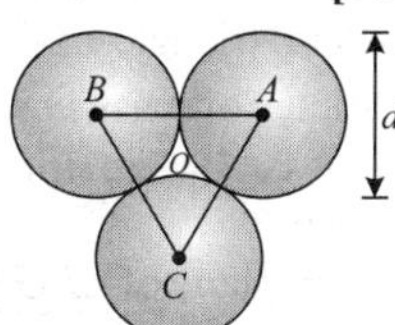

(A) $\frac{13}{23}$ (B) $\frac{15}{13}$

(C) $\frac{23}{13}$ (D) $\frac{13}{15}$

99. One end of a straight uniform 1 m long bar is pivoted on horizontal table. It is released from rest when it makes an angle 30° from the horizontal (see figure). Its angular speed when it hits the table is given as $\sqrt{n}\,s^{-1}$, where n is an integer. The value of n is _____ . **[9 Jan. 2020 (Morning)]**

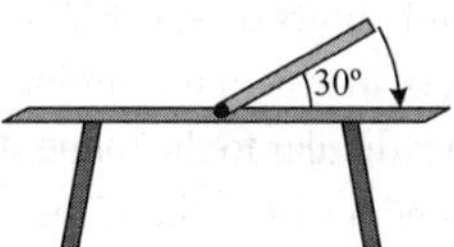

100. A uniformly thick wheel with moment of inertia I and radius R is free to rotate about its centre of mass (see figure). A massless string is wrapped over its rim and two blocks of masses m_1 and m_2 $(m_1 > m_2)$ are attached to the ends of the string. The system is released from rest. The angular speed of the wheel when m_1 descents by a distance h is : **[9 Jan. 2020 (Evening)]**

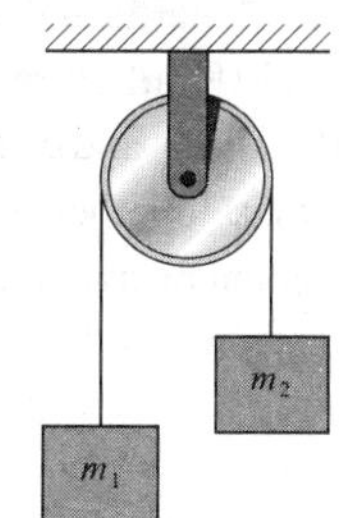

(A) $\left[\frac{m_1 + m_2}{(m_1 + m_2)R^2 + I}\right]^{\frac{1}{2}} gh$ (B) $\left[\frac{2(m_1 - m_2)gh}{(m_1 + m_2)R^2 + I}\right]^{\frac{1}{2}}$

(C) $\left[\frac{2(m_1 + m_2)gh}{(m_1 + m_2)R^2 + I}\right]^{\frac{1}{2}}$ (D) $\left[\frac{(m_1 - m_2)}{(m_1 + m_2)R^2 + I}\right]^{\frac{1}{2}} gh$

101. Shown in the figure is rigid and uniform one meter long rod AB held in horizontal position by two strings tied to its ends and attached to the ceiling. The rod is of mass 'm' and has another weight of mass $2m$ hung at a distance of 75 cm from A. The tension in the string at A is : **[2 Sept. 2020 (Morning)]**

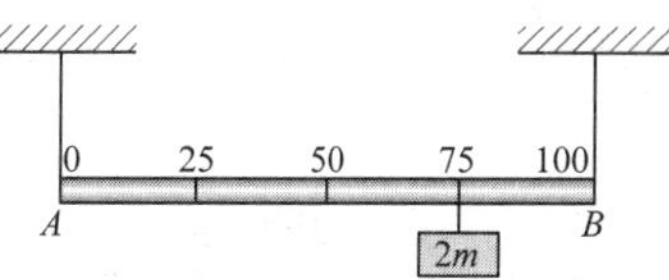

(A) $2mg$ (B) $0.5mg$

(C) $0.75mg$ (D) $1mg$

102. A uniform cylinder of mass M and radius R is to be pulled over a step of height a $(a < R)$ by applying a force F at its centre 'O' perpendicular to the plane through the axes of the cylinder on the edge of the step (see figure). The minimum value of F required is : **[2 Sept. 2020 (Morning)]**

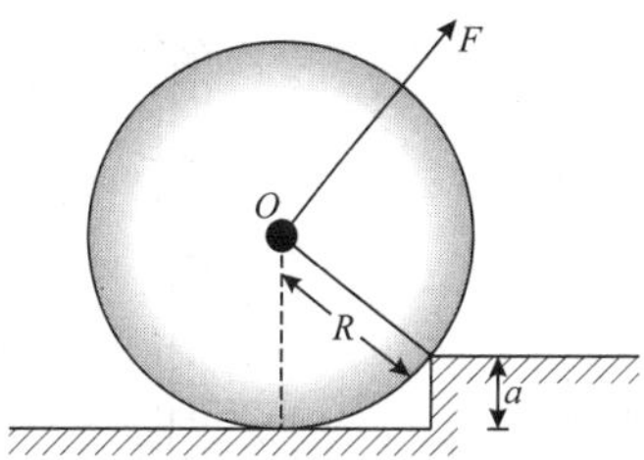

(A) $Mg\sqrt{1 - \frac{a^2}{R^2}}$ (B) $Mg\sqrt{\left(\frac{R}{R-a}\right)^2 - 1}$

(C) $Mg\frac{a}{R}$ (D) $Mg\sqrt{1-\left(\frac{R-a}{R}\right)^2}$

103. Two uniform circular discs are rotating independently in the same direction around their common axis passing through their centres. The moment of inertia and angular velocity of the first disc are 0.1 kg-m^2 and 10 rad s^{-1} respectively while those for the second one are 0.2 kg-m^2 and 5 rad s^{-1} respectively. At some instant they get stuck together and start rotating as a single system about their common axis with some angular speed. The Kinetic energy of the combined system is :

[2 Sept. 2020 (Evening)]

(A) $\frac{10}{3}$ J (B) $\frac{2}{3}$ J

(C) $\frac{5}{3}$ J (D) $\frac{20}{3}$ J

104. Moment of inertia of a cylinder of mass M, length L and radius R about an axis passing through its centre and perpendicular to the axis of the cylinder is given as

$$I = M\left(\frac{R^2}{4} + \frac{L^2}{12}\right)$$

If such a cylinder is to be made for a given mass of material, the ratio L/R for it to have minimum possible I is :

[3 Sept. 2020 (Morning)]

(A) $\sqrt{\frac{2}{3}}$ (B) $\frac{3}{2}$

(C) $\sqrt{\frac{3}{2}}$ (D) $\frac{2}{3}$

105. A block of mass m = 1 kg slides with velocity v = 6 m/s on a frictionless horizontal surface and collides with a uniform vertical rod and sticks to it as shown. The rod is pivoted about O and swings as a result of the collision making angle θ before momentarily coming to rest. If the rod has mass M = 2 kg, and length l = 1 m, the value of θ is approximately :

(Take g = 10 m/s^2) : **[3 Sept. 2020 (Morning)]**

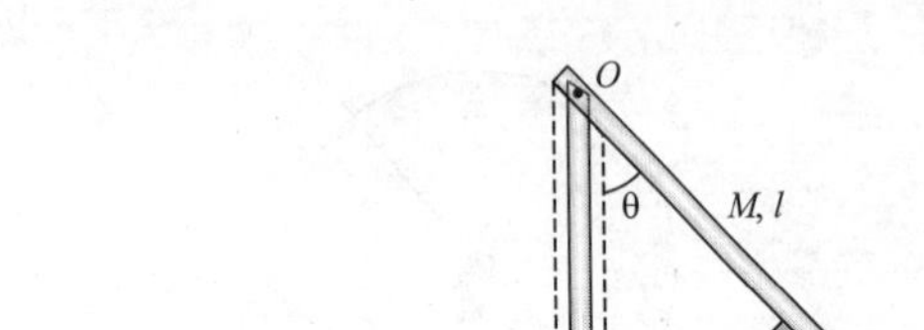

(A) 69° (B) 63°

(C) 55° (D) 49°

106. A person of 80 kg mass is standing on the rim of a circular platform of mass 200 kg rotating about its axis as 5 revolutions per minute (rpm). The person now starts moving towards the centre of the platform. What will be the rotational speed (in rpm) of the platform when the person reaches its centre _____.

[3 Sept. 2020 (Morning)]

107. A uniform rod of length 'l' is pivoted at one of its ends on a vertical shaft of negligible radius. When the shaft rotates at angular speed ω the rod makes an angle θ with it (see figure). To find θ equate the rate of change of angular momentum (direction going into the paper) $\frac{ml^2}{12}\omega^2 \sin\theta \cos\theta$ about the centre of mass (CM) to the torque provided by the horizontal and vertical forces F_H and F_V about the CM. The value of θ is then such that : **[3 Sept. 2020 (Evening)]**

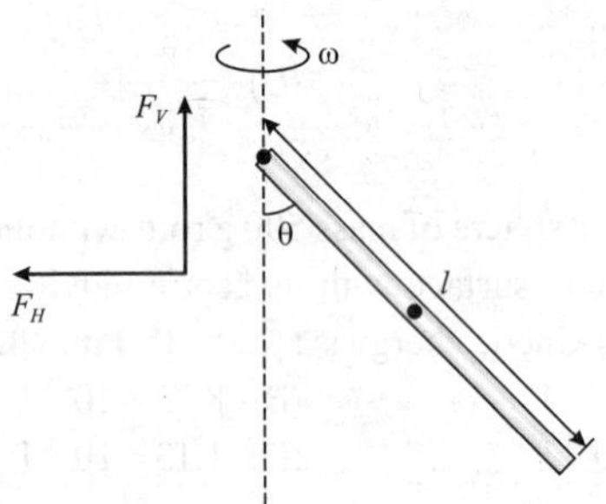

(A) $\cos\theta = \frac{g}{2l\omega^2}$ (B) $\cos\theta = \frac{3g}{2l\omega^2}$

(C) $\cos\theta = \frac{2g}{3l\omega^2}$ (D) $\cos\theta = \frac{g}{l\omega^2}$

108. A massless equilateral triangle EFG of side 'a' (As shown in figure) has three particles of mass m situated at its vertices. The moment of inertia of the system about the line EX perpendicular to EG in the plane of EFG is $\frac{N}{20}ma^2$ where N is an integer. The value of N is _____.

[3 Sept. 2020 (Evening)]

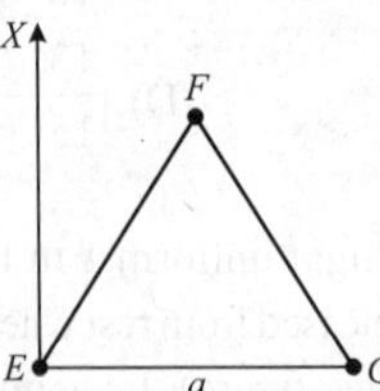

109. ABC is a plane lamina of the shape of an equilateral triangle. D, E are mid points of AB, AC and G is the centroid of the lamina. Moment of inertia of the lamina about an axis passing through G and perpendicular to the plane ABC is I_0. If part ADE is removed, the moment of inertia of the remaining part about the same axis is $\frac{NI_0}{16}$ where N is an integer. Value of N is _____.

[4 Sept. 2020 (Morning)]

110. A circular disc of mass M and radius R is rotating about its axis with angular speed ω_1. If another stationary disc having radius $\frac{R}{2}$ and same mass M is dropped co-axially on to the rotating disc. Gradually both discs attain constant angular speed ω_2. The energy lost in the process is $p\%$ of the initial energy. Value of p is _____. **[4 Sept. 2020 (Morning)]**

111. Consider two uniform discs of the same thickness and different radii $R_1 = R$ and $R_2 = \alpha R$ made of the same material. If the ratio of their moments of inertia I_1 and I_2, respectively, about their axes is $I_1 : I_2 = 1 : 16$ then the value of α is :

[4 Sept. 2020 (Evening)]

(A) $\sqrt{2}$ (B) 2

(C) 4 (D) $2\sqrt{2}$

112. For a uniform rectangular sheet shown in the figure, the ratio of moments of inertia about the axes perpendicular to the sheet and passing through O (the centre of mass) and O' (corner point) is : **[4 Sept. 2020 (Evening)]**

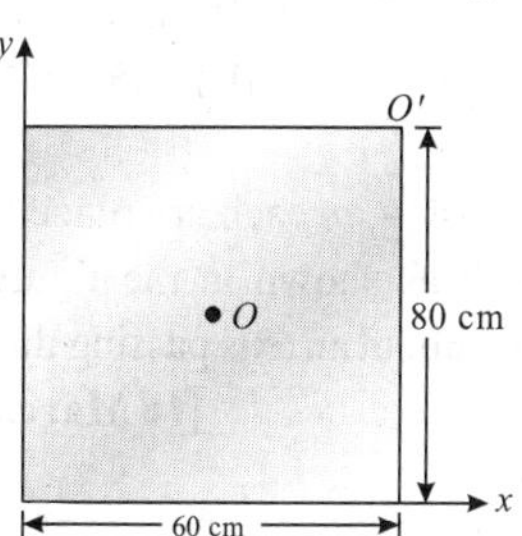

(A) 1/2 (B) 2/3

(C) 1/8 (D) 1/4

113. A wheel is rotating freely with an angular speed ω on a shaft. The moment of inertia of the wheel is I and the moment of inertia of the shaft is negligible. Another wheel of moment of inertia $3I$ initially at rest is suddenly coupled to the same shaft. The resultant fractional loss in the kinetic energy of the system is : **[5 Sept. 2020 (Morning)]**

(A) 0 (B) $\frac{1}{4}$

(C) $\frac{3}{4}$ (D) $\frac{5}{6}$

114. A force $\vec{F} = (\hat{i} + 2\hat{j} + 3\hat{k})$ N acts at a point $(4\hat{i} + 3\hat{j} - \hat{k})m$. Then the magnitude of torque about the point $(\hat{i} + 2\hat{j} + \hat{k})m$ will be $\sqrt{x}$ N-m. The value of x is _____.

[5 Sept. 2020 (Morning)]

115. A thin rod of mass 0.9 kg and length 1 m is suspended, at rest, from one end so that it can freely oscillate in the vertical plane. A particle of move 0.1 kg moving in a straight line with velocity 80 m/s hits the rod at its bottom most point and sticks to it (see figure). The angular speed (in rad/s) of the rod immediately after the collision will be _____.

[5 Sept. 2020 (Evening)]

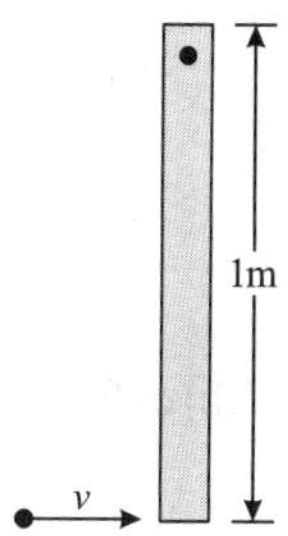

116. Shown in the figure is a hollow ice-cream cone (it is open at the top). If its mass is M, radius of its top, R and height, H, then its moment of inertia about its axis is :

[6 Sept. 2020 (Morning)]

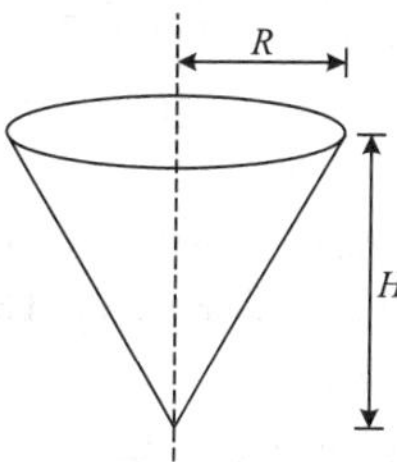

(A) $\frac{MR^2}{2}$ (B) $\frac{MH^2}{3}$

(C) $\frac{MR^2}{3}$ (D) $\frac{M(R^2 + H^2)}{4}$

117. Four point masses, each of mass m, are fixed at the corners of a square of side l. The square is rotating with angular frequency ω, about an axis passing through one of the corners of the square and parallel to its diagonal, as shown in the figure. The angular momentum of the square about this axis is :

[6 Sept. 2020 (Morning)]

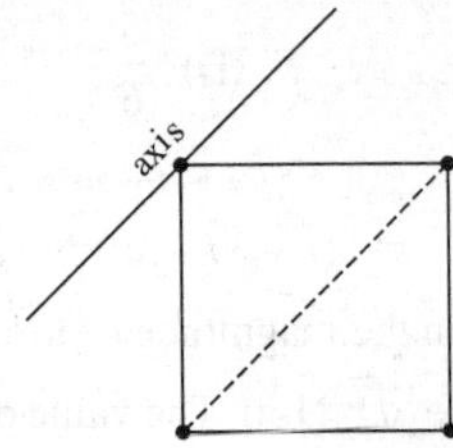

(A) $2ml^2\omega$ (B) $3ml^2\omega$
(C) $ml^2\omega$ (D) $4ml^2\omega$

118. The linear mass density of a thin rod AB of length L varies from A to B as $\lambda(x)=\lambda_0\left(1+\frac{x}{L}\right)$, where x is the distance from A. If M is the mass of the rod then its moment of inertia about an axis passing through A and perpendicular to the rod is : **[6 Sept. 2020 (Evening)]**

(A) $\frac{5}{12}ML^2$ (B) $\frac{3}{7}ML^2$
(C) $\frac{2}{5}ML^2$ (D) $\frac{7}{18}ML^2$

119. Moment of inertia (M.I.) of four bodies, having same mass and radius, are reported as;
I_1 = M.I. of thin circular ring about its diameter,
I_2 = M.I. of circular disc about an axis perpendicular to disc and going through the centre,
I_3 = M.I. of solid cylinder about its axis and
I_4 = M.I. of solid sphere about its diameter
Then : **[24 Feb. 2021 (Morning)]**

(A) $I_1+I_3<I_2+I_4$ (B) $I_1+I_2=I_3+\frac{5}{2}I_4$
(C) $I_1=I_2=I_3>I_4$ (D) $I_1=I_2=I_3<I_4$

120. A uniform thin bar of mass 6 kg and length 24 meter is bent to make an equilateral hexagon. The moment of inertia about an axis passing through the centre of mass and perpendicular to the plane of hexagon is _____ $\times 10^{-1}$ kg-m^2. **[24 Feb. 2021 (Evening)]**

121. A sphere of radius 'a' and mass 'm' rolls along a horizontal plane with constant speed v_0. It encounters an inclined plane at angle θ and climbs upward. Assuming that it rolls without slipping, how far up the sphere will travel ? **[25 Feb. 2021 (Evening)]**

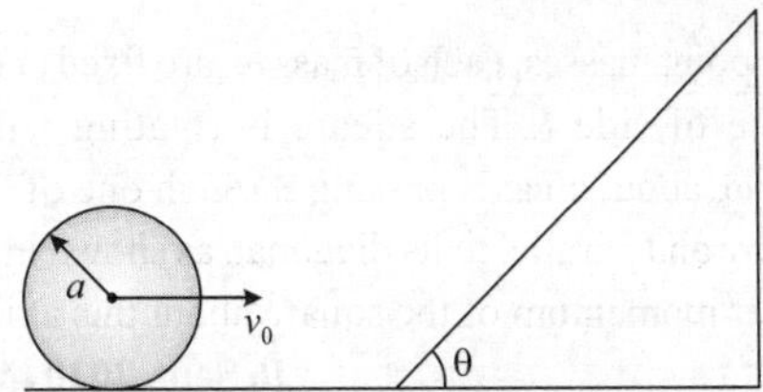

(A) $\frac{v_0^2}{5g\sin\theta}$ (B) $\frac{v_0^2}{2g\sin\theta}$
(C) $\frac{2}{5}\frac{v_0^2}{g\sin\theta}$ (D) $\frac{10v_0^2}{7g\sin\theta}$

122. Four identical solid spheres each of mass 'm' and radius 'a' are placed with their centres on the four corners of a square of side 'b'. The moment of inertia of the system about one side of square where the axis of rotation is parallel to the plane of the square is : **[26 Feb. 2021 (Morning)]**

(A) $\frac{4}{5}ma^2$ (B) $\frac{4}{5}ma^2+2mb^2$
(C) $\frac{8}{5}ma^2+2mb^2$ (D) $\frac{8}{5}ma^2+mb^2$

123. A cord is wound round the circumference of wheel of radius r. The axis of the wheel is horizontal and the moment of inertia about it is I. A weight mg is attached to the cord at the end. The weight falls from rest. After falling through a distance 'h', the square of angular velocity of wheel will be : **[26 Feb. 2021 (Evening)]**

(A) $2gh$ (B) $\frac{2mgh}{I+2mr^2}$
(C) $\frac{2gh}{I+mr^2}$ (D) $\frac{2mgh}{I+mr^2}$

124. The maximum and minimum distances of a comet from the Sun are 1.6×10^{12} m and 8.0×10^{10} m respectively. If the speed of the comet at the nearest point is 6×10^4 ms^{-1}, the speed at the farthest point is : **[16 March 2021 (Morning)]**
(A) 1.5×10^3 m/s (B) 6.0×10^3 m/s
(C) 3.0×10^3 m/s (D) 4.5×10^3 m/s

125. Four equal masses, m each are placed at the corners of a square of length (l) as shown in the figure. The moment of inertia of the system about an axis passing through A and parallel to DB would be : **[16 March 2021 (Morning)]**

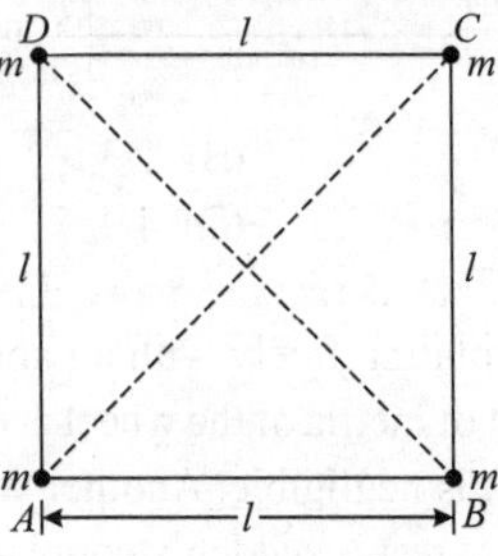

(A) ml^2 (B) $2ml^2$
(C) $3ml^2$ (D) $\sqrt{3}ml^2$

126. Consider a 20 kg uniform circular disk of radius 0.2 m. It is pin supported at its center and is at rest initially. The disk is acted upon by a constant force $F = 20$ N through a massless string wrapped around its periphery as shown in the figure.

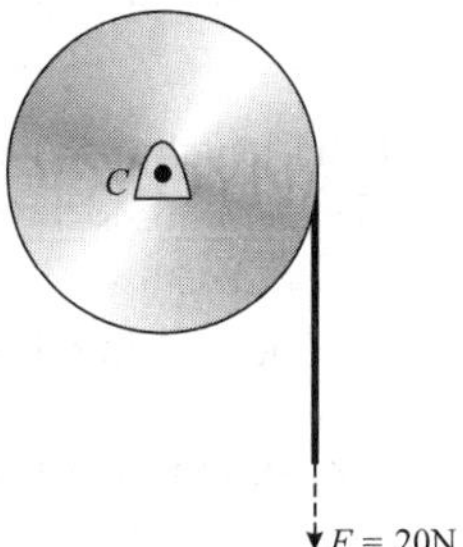

Suppose the disk makes n number of revolutions to attain an angular speed of 50 rad s^{-1}. The value of n, to the nearest integer, is _____ . [Given : In one complete revolution, the disk rotates by 6.28 rad] **[16 March 2021 (Morning)]**

127. A force $\vec{F} = 4\hat{i} + 3\hat{j} + 4\hat{k}$ is applied on an intersection point of $x = 2$ plane and x-axis. The magnitude of torque of this force about a point (2, 3, 4) is _____ . (Round off to the Nearest Integer) **[16 March 2021 (Evening)]**

128. A solid disc of radius 'a' and mass 'm' rolls down without slipping on an inclined plane making an angle θ with the horizontal. The acceleration of the disc will be $\frac{2}{b} g \sin\theta$ where b is _____ . (Round off to the Nearest Integer)
(g = acceleration due to gravity) (θ = angle as shown in figure) **[16 March, 2021 (Evening)]**

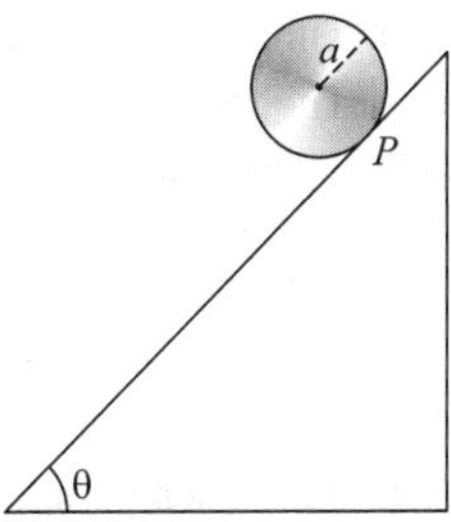

129. A triangular plate is shown. A force $\vec{F} = 4\hat{i} - 3\hat{j}$ is applied at point P. The torque at point P with respect to point 'O' and 'Q' are : **[17 March 2021 (Morning)]**

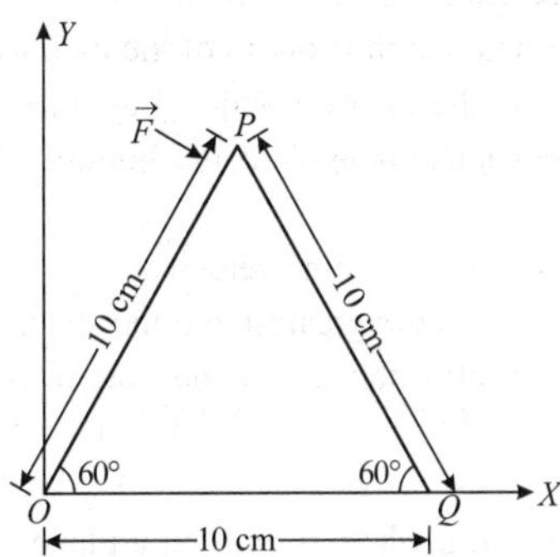

(A) $-15-20\sqrt{3}, 15-20\sqrt{3}$ (B) $15+20\sqrt{3}, 15-20\sqrt{3}$
(C) $15-20\sqrt{3}, 15+20\sqrt{3}$ (D) $-15+20\sqrt{3}, 15+20\sqrt{3}$

130. A mass M hangs on a massless rod of length l which rotates at a constant angular frequency. The mass M moves with steady speed in a circular path of constant radius. Assume that the system is in steady circular motion with constant angular velocity ω. The angular momentum of M about point A is L_A which lies in the positive z direction and the angular momentum of M about B is L_B. The correct statement for this system is : **[17 March 2021 (Morning)]**

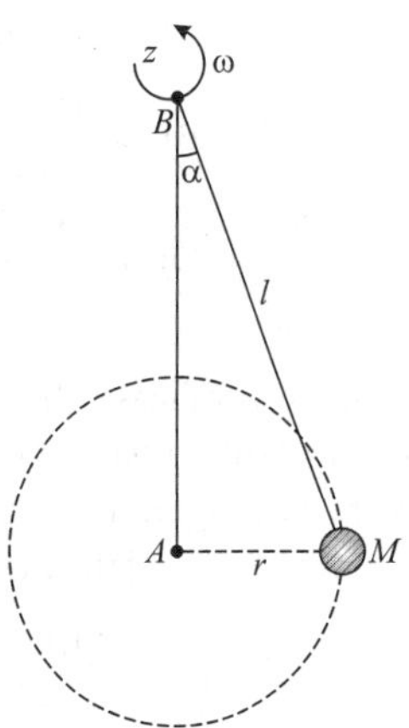

(A) L_A and L_B are both constant in magnitude and direction
(B) L_B is constant in direction with varying magnitude
(C) L_B is constant, both in magnitude and direction
(D) L_A is constant, both in magnitude and direction

131. The following bodies,
(1) a ring (2) a disc
(3) a solid cylinder
(4) a solid sphere, of same mass 'm' and radius 'R' are allowed to roll down without slipping simultaneously from the top of the inclined plane. The body which will reach first at the bottom of the inclined plane is _____ .
[Mark the body as per their respective numbering given in the question] **[17 March 2021 (Morning)]**

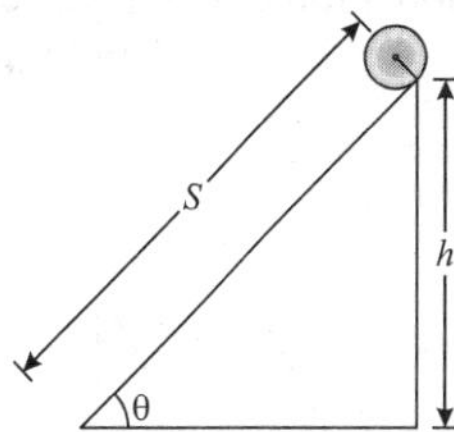

132. A sphere of mass 2 kg and radius 0.5 m is rolling with an initial speed of 1 ms^{-1} goes up an inclined plane which makes an angle of 30° with the horizontal plane, without slipping. How long will the sphere take to return to the starting point A ? **[17 March 2021 (Evening)]**

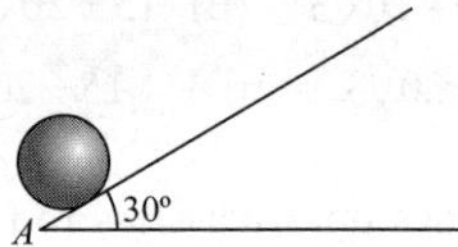

(A) 0.60 s (B) 0.52 s
(C) 0.57 s (D) 0.80 s

133. A thin circular ring of mass M and radius r is rotating about its axis with an angular speed ω. Two particles having mass m each are now attached at diametrically opposite points. The angular speed of the ring will become :

[18 March 2021 (Morning)]

(A) $\omega \frac{M}{M+m}$ (B) $\omega \frac{M+2m}{M}$

(C) $\omega \frac{M}{M+2m}$ (D) $\omega \frac{M-2m}{M+2m}$

134. Consider a uniform wire of mass M and length L. It is bent into a semicircle. Its moment of inertia about a line perpendicular to the plane of the wire passing through the centre is :

[18 March 2021 (Evening)]

(A) $\frac{1}{4}\frac{ML^2}{\pi^2}$ (B) $\frac{2}{5}\frac{ML^2}{\pi^2}$

(C) $\frac{ML^2}{\pi^2}$ (D) $\frac{1}{2}\frac{ML^2}{\pi^2}$

135. The angular momentum of a planet of mass M moving around the sun in an elliptical orbit is $\vec{L}$. The magnitude of the areal velocity of the planet is : **[18 March 2021 (Evening)]**

(A) $\frac{4L}{M}$ (B) $\frac{L}{M}$

(C) $\frac{2L}{M}$ (4) $\frac{L}{2M}$

136. A solid cylinder of mass m is wrapped with an inextensible light string and, is placed on a rough inclined plane as shown in the figure. The frictional force acting between the cylinder and the inclined plane is : **[18 March 2021 (Evening)]**

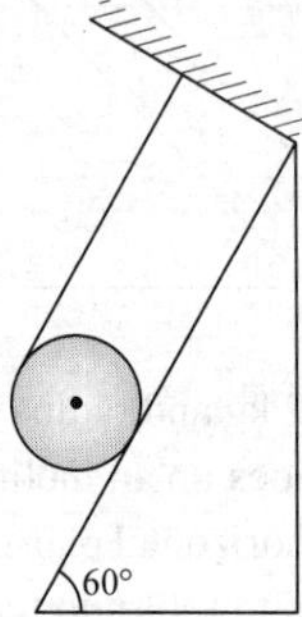

[The coefficient of static friction, μ_s, is 0.4]

(A) $\frac{7}{2}$mg (B) 5 mg

(C) $\frac{mg}{5}$ (D) 0

137. A rod of mass M and length L is lying on a horizontal frictionless surface. A particle of mass 'm' travelling along the surface hits at one end of the rod with a velocity 'u' in a direction perpendicular to the rod. The collision is completely elastic. After collision, particle comes to rest. The ratio masses $\left(\frac{m}{M}\right)$ is $\frac{1}{x}$. The value of 'x' will be _____. **[20 July 2021 (Morning)]**

138. A circular disc reaches from top to bottom of an inclined plane of length L. When it slips down the plane, it takes time t_1. When it rolls down the plane, it takes time t_2. The value of $\frac{t_2}{t_1}$ is $\sqrt{\frac{3}{x}}$. Then value of x will be _____.

[20 July 2021 (Morning)]

139. A body rolls down an inclined plane without slipping. The kinetic energy of rotation is 50% of its translational kinetic energy. The body is : **[20 July 2021 (Evening)]**

(A) Hollow cylinder (B) Solid sphere
(C) Sold cylinder (D) Ring

140. Two bodies, a ring and a solid cylinder of same material are rolling down without slipping and inclined plane. The radii of the bodies are same. The ratio of velocity of the centre of mass at the bottom of the inclined plane of the ring to that of the cylinder is $\frac{\sqrt{x}}{2}$. Then, the value of x is _____.

[20 July 2021 (Evening)]

141. Consider a situation in which a ring, a solid cylinder and a solid sphere roll down on the same inclined plane without slipping, Assume that they start rolling from rest and having identical diameter. The correct statement for this situation is :

[22 July 2021 (Evening)]

(A) The sphere has the greatest and the ring has least velocity of the centre of mass at the bottom of the inclined plane.
(B) The ring has the greatest and the cylinder has the least velocity of the centre of mass at the bottom of the inclined plane.
(C) All of them will have same velocity
(D) The cylinder has the greatest and the sphere has the least velocity of the centre of mass at the bottom of the inclined plane

142. The centre of a wheel rolling on a plane surface moves with a speed V_0. A particle on the rim of the wheel at the same level as the centre will be moving at speed $\sqrt{x}V_0$. Then the value of x is ______ . **[22 July 2021 (Evening)]**

143. Given below are two statements : one is labelled as **Assertion A** and the other is labelled as **Reason R**.

Assertion-A: Moment of inertia of a circular disc of mass 'M' and radius 'R' about X, Y axes (passing through its plane) and Z-axis which is perpendicular to its plane were found to be I_x, I_Y & I_Z respectively. The respective radii of gyration about all the three axes will be the same.

Reason-R: A rigid body making rotational motion has fixed mass and shape.

In the light of the above statements, choose the most appropriate answer from the options given below :

[25 July 2021 (Morning)]

(A) Both **A** and **R** are correct and **R** is the correct explanation of **A**.

(B) Both **A** and **R** are correct but **R** is *NOT* the correct explanation of **A**.

(C) **A** is not correct but **R** is correct.

(D) **A** is correct hut **R** is not correct.

144. The minimum and maximum distances of a planet revolving around the Sun are x_1 and x_2. If the minimum speed of the planet on its trajectory is v_0 then its maximum speed will be :

[25 July 2021 (Morning)]

(A) $\frac{v_0 x_2^2}{x_1^2}$ (B) $\frac{v_0 x_2}{x_1}$

(C) $\frac{v_0 x_1}{x_2}$ (D) $\frac{v_0 x_1^2}{x_2^2}$

145. Two billiard balls of equal mass 30 g strike a wall with normally same speed of 108 kmph but at different angles. If the balls get reflected with the same speed then the ratio of the magnitude of impulses imparted to ball A and ball B by the wall along 'X' direction is : **[25 July 2021 (Morning)]**

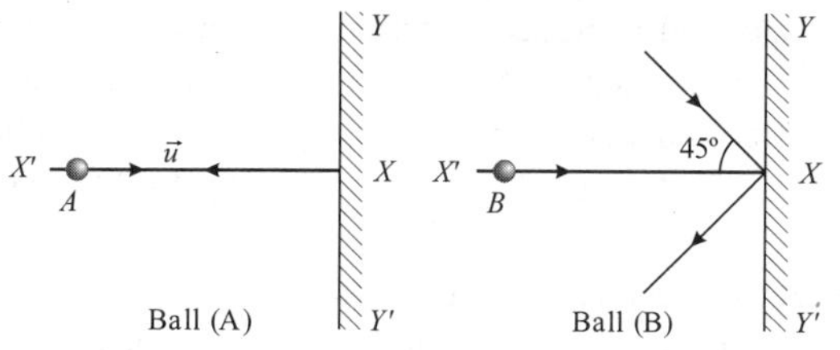

(A) $\sqrt{2}:1$ (B) $1:\sqrt{2}$

(C) $2:1$ (D) $1:1$

146. A body of mass 2 kg moving with a speed of 4 m/s, makes an elastic collision with another body at rest and continues to move in the original direction but with one fourth of its initial speed. The speed of the two body centre of mass is $\frac{x}{10}$ m/s. Then the value of x is ______. **[25 July 2021 (Morning)]**

147. A particle of mass m is moving in time t on a trajectory given by $\vec{r} = 10\alpha t^2\hat{i} + 5\beta(t-5)\hat{j}$ Where a and b are dimensional constants. The angular momentum of the particle about origin becomes the same as it was for $t = 0$ at time $t =$ ______ seconds.

[25 July 2021 (Morning)]

148. A solid disc of radius 20 cm and mass 10 kg is rotating with an angular velocity of 600 rpm, about an axis normal to its circular plane and passing through its centre of mass. The retarding torque required to bring the disc at rest in 10 s is ______ $\pi \times 10^{-1}$ Nm : **[25 July 2021 (Evening)]**

149. The figure shows two solid discs with radius R and r respectively. If mass per unit area is same for both, what is the ratio of *MI* of bigger disc around axis AB (which is ⊥ to the plane) of the disc and passing through its centre) to *MI* of smaller disc around one of its diameters lying on its plane ? Given M is the mass of the larger disc. (*MI* stands for moment of inertia) **[27 July 2021 (Morning)]**

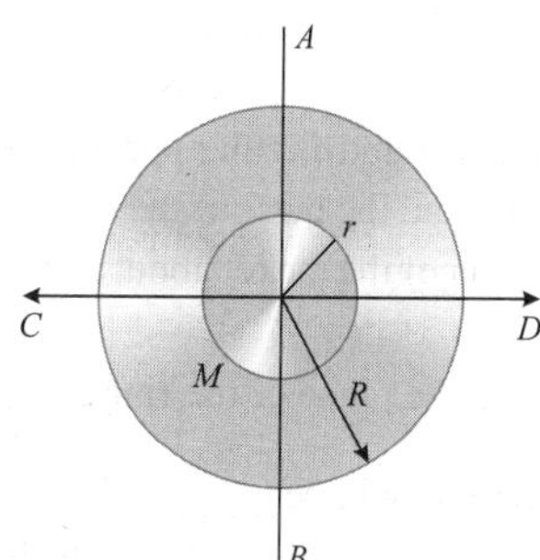

(A) $2R^4 : r^4$ (B) $2R^2 : r^2$

(C) $R^2 : r^2$ (D) $2r^4 : R^4$

150.

List-I	**List-II**
(a) *MI* of the rod (length L, Mass M, about an axis ⊥ to the rod passing through the mid point)	(i) $8\,ML^2/3$
(b) *MI* of the rod (length L, Mass $2M$, about an axis ⊥ to the rod passing through one of its end)	(ii) $ML^2/3$
(c) *MI* of the rod (length $2L$, Mass M, about an axis ⊥ to the rod passing through its midpoint)	(iii) $ML^2/12$
(d) *MI* of the rod (length $2L$, mass $2M$, about an axis ⊥ to the rod passing through one of its end	(iv) $2\,ML^2/3$

[27 July 2021 (Morning)]

(A) (a)-(iii), (b)-(iv), (c)-(i), (d)-(ii)
(B) (a)-(ii), (b)-(i), (c)-(iii), (d)-(iv)
(C) (a)-(ii), (b)-(i), (c)-(iii), (d)-(iv)
(D) (a)-(iii), (b)-(iv), (c)-(ii), (d)-(i)

151. In the given figure, two wheels P and Q are connected by a belt B. The radius of P is three times as that of Q. In case of same rotational kinetic energy, the ratio of rotational inertias $\left(\frac{I_1}{I_2}\right)$ will be $x : 1$. The value of x will be _____. **[27 July 2021 (Evening)]**

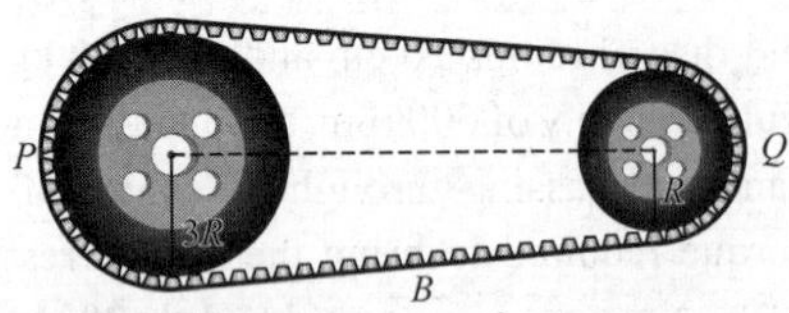

152. Consider a badminton racket with length scales as shown in the figure.

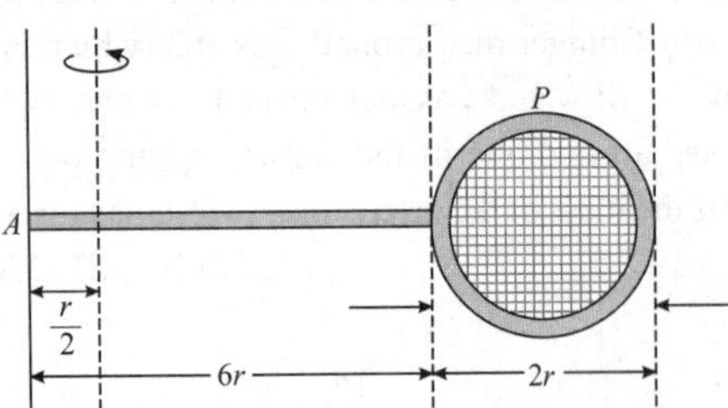

If the mass of the linear and circular portions of the badminton racket are same (M) and the mass of the threads are negligible, the moment of inertia of the racket about an axis perpendicular to the handle and in the plane of the ring at, $\frac{r}{2}$ distance from the end A of the handle will be _____ Mr^2. **[26 Aug. 2021 (Morning)]**

153. The solid cylinder of length 80 cm and mass M has a radius of 20 cm. Calculate the density of the material used if the moment of inertia of the cylinder about an axis CD parallel of AB as shown in figure is 2.7 kg m^2 : **[26 Aug. 2021 (Evening)]**

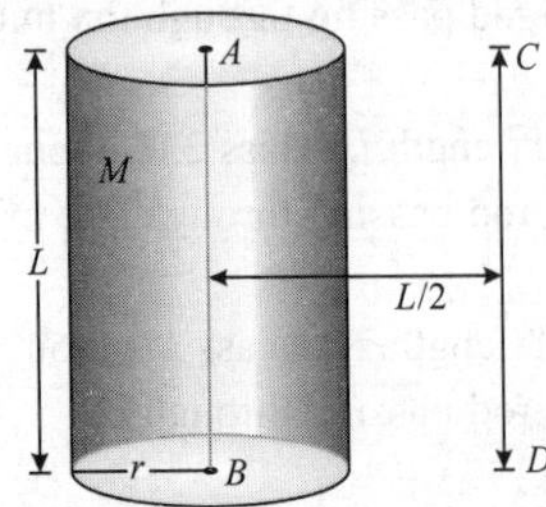

(A) 7.5×10^2 kg/m^3 (B) 7.5×10^1 kg/m^3
(C) 14.9 kg/m^3 (D) 1.49×10^2 kg/m^3

154. Moment of inertia of a square plate of side l about the axis passing through one of the corner and perpendicular to the plane of square plate is given by : **[27 Aug. 2021 (Morning)]**

(A) $\frac{2}{3}Ml^2$ (B) $\frac{Ml^2}{12}$
(C) $\frac{Ml^2}{6}$ (D) Ml^2

155. Two discs have moments of inertia I_1 and I_2 about their respective axes perpendicular to the plane and passing through the centre. They are rotating with angular speeds, ω_1 and ω_2 respectively and are brought into contact face to face with their axes of rotation coaxial. The loss in kinetic energy of the system in the process is given by : **[27 Aug. 2021 (Evening)]**

(A) $\frac{I_1 I_2}{(I_1+I_2)}(\omega_1-\omega_2)^2$ (B) $\frac{(\omega_1-\omega_2)^2}{2(I_1+I_2)}$
(C) $\frac{I_1 I_2}{2(I_1+I_2)}(\omega_1-\omega_2)^2$ (D) $\frac{(I_1-I_2)^2\omega_1\omega_2}{2(I_1+I_2)}$

156. Angular momentum of a single particle moving with constant speed along circular path : **[31 Aug. 2021 (Morning)]**
(A) Changes in magnitude but remains same in the direction
(B) Is zero
(C) Remains same in magnitude and direction
(D) Remains same in magnitude but changes in the direction

157. A system consists of two identical spheres each of mass 1.5 kg and radius 50 cm at the ends of a light rod. The distance between the centres of the two spheres is 5 m. What will be the moment of inertia of the system about an axis perpendicular to the passing through its midpoint ? **[31 Aug. 2021 (Evening)]**
(A) 1.905×10^5 kgm^2 (B) 19.05 kgm^2
(C) 18.75 kgm^2 (D) 18.45×10^5 kgm^2

158. A 2 kg steel rod of length 0.6 m is clamped on a table vertically at its lower end and is free to rotate in vertical plane. The upper end is pushed so that the rod falls under gravity. Ignoring the friction due to clamping at its lower end, the speed of the free end when it passes through its lowest position is _____ ms^{-1}. (Take $g = 10$ ms^{-2}) **[01 Sep. 2021 (Evening)]**

159. A fly wheel is accelerated uniformly from rest and rotates through 5 rad in the first second. The angle rotated by the fly wheel in the next second, will be : **[24 June 2022 (Evening)]**
(A) 7.5 rad (B) 15 rad
(C) 20 rad (D) 30 rad

160. Moment of Inertia of four bodies having same mass M and radius $2R$ are as follows :

I_1 = M.I. of solid sphere about its diameter
I_2 = M.I. of solid cylinder about its axis
I_3 = M.I. of solid circular disc about its diameter
I_4 = M.I. of thin circular ring about its diameter
If $2(I_2 + I_3) + I_4 = x. I_1$ then the value of x will be_____.

[25 June 2022 (Evening)]

161. A thin circular ring of mass M and radius R is rotating with a constant angular velocity 2 rad s^{-1} in a horizontal plane about an axis vertical to its plane and passing through the center of the ring. If two objects each of mass m be attached gently to the opposite ends of a diameter of ring, the ring will then rotate with an angular velocity : (in rad s^{-1})

[26 June 2022 (Morning)]

(A) $\frac{M}{(M+m)}$ (B) $\frac{(M+2m)}{2M}$

(C) $\frac{2M}{(M+2m)}$ (D) $\frac{2(M+2m)}{M}$

162. Solid spherical ball is rolling on a frictionless horizontal plane surface about its axis of symmetry. The ratio of rotational kinetic energy of the ball to its total kinetic energy is :

[26 June 2022 (Evening)]

(A) $\frac{2}{5}$ (B) $\frac{2}{7}$

(C) $\frac{1}{5}$ (D) $\frac{7}{10}$

163. A rolling wheel of 12 kg is on an inclined plane at position P and connected to a mass of 3 kg through a string of fixed length and pulley as shown in figure. Consider PR as friction free surface. The velocity of centre of mass of the wheel when it reaches at the bottom Q of the inclined plane PQ will be $\frac{1}{2}\sqrt{xgh}$ m/s. The value of x is_____.

[27 June 2022 (Evening)]

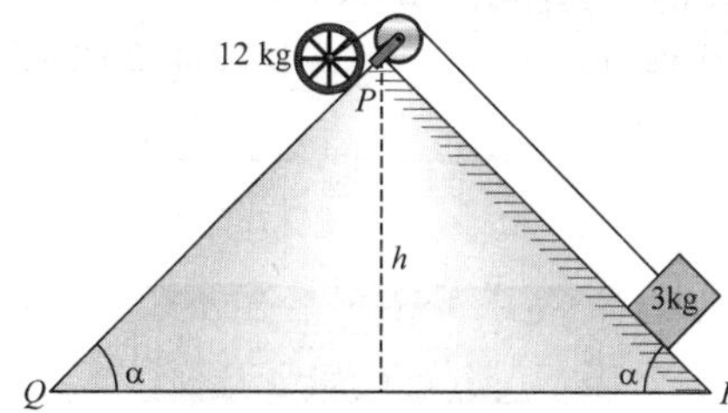

164. Match List-I with List-II : **[28 June 2022 (Morning)]**

List-I	**List-II**
(a) Moment of inertia of solid sphere of radius R about any tangent	(i) $\frac{5}{3}MR^2$
(b) Moment of inertia of hollow sphere of radius (R) about any tangent	(ii) $\frac{7}{5}MR^2$
(c) Moment of inertia of circular ring of radius (R) about its diameter	(iii) $\frac{1}{4}MR^2$
(d) Moment of inertia of circular disc of radius (R) about any diameter	(iv) $\frac{1}{2}MR^2$

Choose the correct answer from the options given below :

(A) (a)-(ii), (b)-(ii), (c)-(iv), (d)-(iii)
(B) (a)-(i), (b)-(ii), (c)-(iv), (d)-(iii)
(C) (a)-(ii), (b)-(i), (c)-(iii), (d)-(iv)
(D) (a)-(i), (b)-(ii), (c)-(iii), (d)-(iv)

165. The position vector of 1 kg object is $\vec{r} = (3\hat{i} - \hat{j})m$ and its velocity $\vec{v} = (3\hat{j} + \hat{k})$ ms^{-1}. The magnitude of its angular momentum is $\sqrt{x}$ Nm where x is_____.

[28 June 2022 (Morning)]

166. A ball is spun with angular acceleration $\alpha = 6t^2 - 2t$ where t is in second and α is in rad s^{-2}. At $t = 0$, the ball has angular velocity of 10 rad s^{-1} and angular position of 4 rad. The most appropriate expression for the angular position of the ball is :

[28 June 2022 (Evening)]

(A) $\frac{3}{2}t^4 - t^2 + 10t$ (B) $\frac{t^4}{2} - \frac{t^3}{3} + 10t + 4$

(C) $\frac{2t^4}{3} - \frac{t^3}{6} + 10t + 12$ (D) $2t^4 - \frac{t^3}{2} + 5t + 4$

167. A $\sqrt{34}m$ long ladder weighing 10 kg leans on a frictionless wall. Its feet rest on the floor 3 m away from the wall as shown in the figure. If F_f and F_w are the reaction forces of the floor and the wall, then ratio of F_w/F_f will be : [Use $g = 10$ m/s^2]

[28 June 2022 (Evening)]

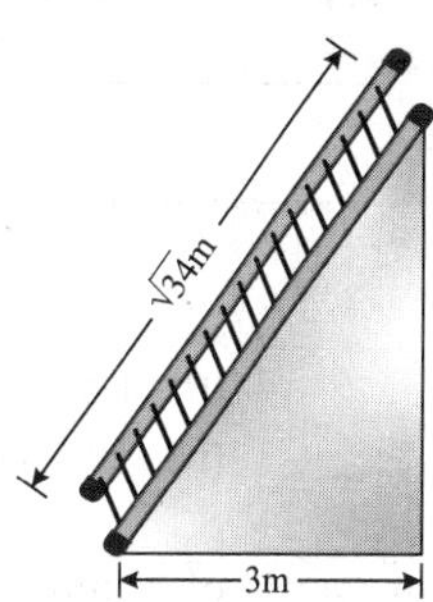

(A) $\frac{6}{\sqrt{110}}$ (B) $\frac{3}{\sqrt{113}}$

(C) $\frac{3}{\sqrt{109}}$ (D) $\frac{2}{\sqrt{109}}$

168. A uniform disc with mass $M = 4$ kg and radius $R = 10$ cm is mounted on a fixed horizontal axle as shown in figure. A block with mass $m = 2$ kg hangs from a massless cord that is wrapped around the rim of the disc. During the fall of the block, the cord does not slip and there is no friction at the axle. The tension in the cord is _____ N. [Take $g = 10$ ms^{-2}]

[28 June 2022 (Evening)]

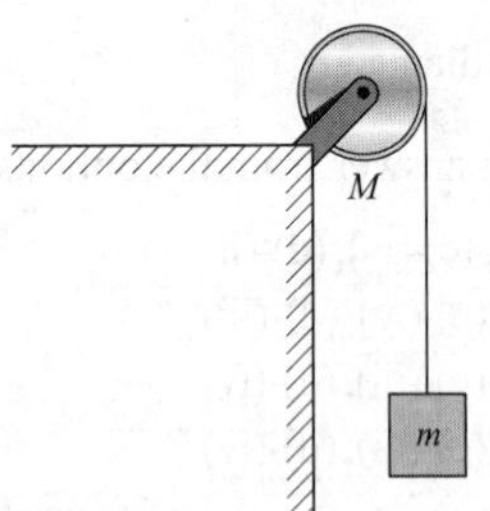

169. A spherical shell of 1 kg mass and radius R is rolling with angular speed ω on horizontal plane (as shown in figure). The magnitude of angular momentum of the shell about the origin O is $\frac{a}{3}R^2\omega$. The value of a will be : **[29 June 2022 (Morning)]**

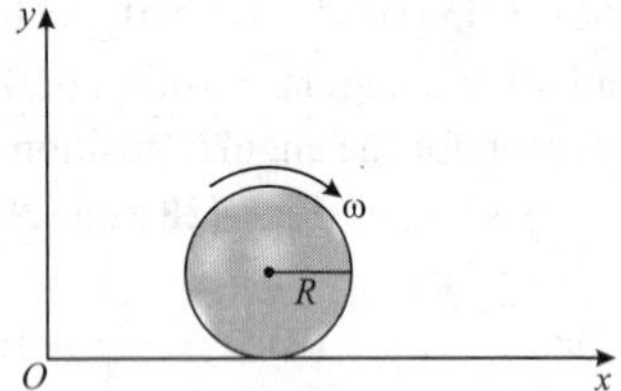

(A) 2 (B) 3
(C) 5 (D) 4

170. A block of mass 40 kg slides over a surface, when a mass of 4 kg is suspended through an inextensible massless string passing over frictionless pulley as shown below. The coefficient of kinetic friction between the surface and block is 0.02. The acceleration of block is. [Given $g = 10$ ms^{-2}]

[29 June 2022 (Evening)]

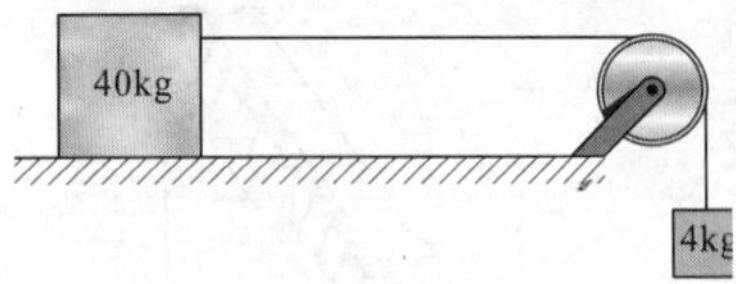

(A) 1 ms^{-1} (B) 1/5 ms^{-2}
(C) 4/5 ms^{-2} (D) 8/11 ms^{-2}

171. The moment of inertia of a uniform thin rod about a perpendicular axis passing through one end is I_1. The same rod is bent into a ring and its moment of inertia about a diameter is I_2. If $\frac{I_1}{I_2}$ is $\frac{x\pi^2}{3}$, then the value of x will be _____.

[29 June 2022 (Evening)]

172. A solid cylinder and a solid sphere, having same mass M and radius R, roll down the same inclined plane from top without slipping. They start from rest. The ratio of velocity of the solid cylinder to that of the solid sphere, with which they reach the ground, will be : **[25 July 2022 (Morning)]**

(A) $\sqrt{\frac{5}{3}}$ (B) $\sqrt{\frac{4}{5}}$
(C) $\sqrt{\frac{3}{5}}$ (D) $\sqrt{\frac{14}{15}}$

173. A disc of mass 1 kg and radius R is free of rotate about a horizontal axis passing through its centre and perpendicular to the plane of disc. A body of same mass as that of disc is fixed at the highest point of the disc. Now the system is released, when the body comes to the lowest position, its angular speed will be $4\sqrt{\frac{x}{3R}}$ rad s^{-1} where $x =$ _____ [$g = 10$ ms^{-2}]

[26 July 2022 (Morning)]

174. The radius of gyration of a cylindrical rod about an axis of rotation perpendicular to its length and passing through the center will be _____ m. Given, the length of the rod is $10\sqrt{3}$ m.

[26 July 2022 (Evening)]

175. A pulley of radius 1.5 m is rotated about its axis by a force $F = (12t - 3t^2)$ N applied tangentially (while t is measured in seconds). If moment of inertia of the pulley about its axis of rotation is 4.5 kg m^2, the number of rotations made by the pulley before its direction of motion is reversed, will be $\frac{K}{\pi}$. The value of K is _____. **[27 July 2022 (Morning)]**

176. A solid cylinder length is suspended symmetrically through two massless strings, as shown in the figure. The distance from the initial rest position, the cylinder should by unbinding the strings to achieve a speed of 4 ms^{-1}, is _____ cm.
[Take $g = 10$ ms^{-2}] **[27 July 2022 (Evening)]**

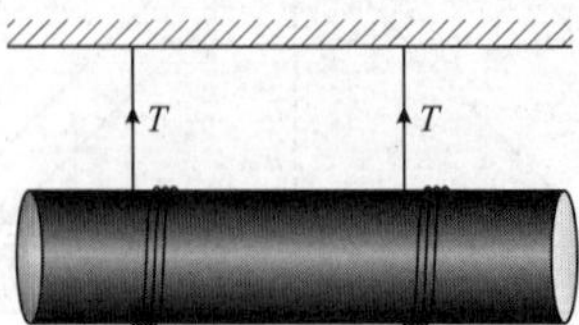

177. Four identical discs each of mass M and diameter a are arranged in a small plane as shown in figure. If the moment of inertia of the system about OO' is $\frac{x}{4}Ma^2$. Then, the value of x will be _____. **[28 July 2022 (Morning)]**

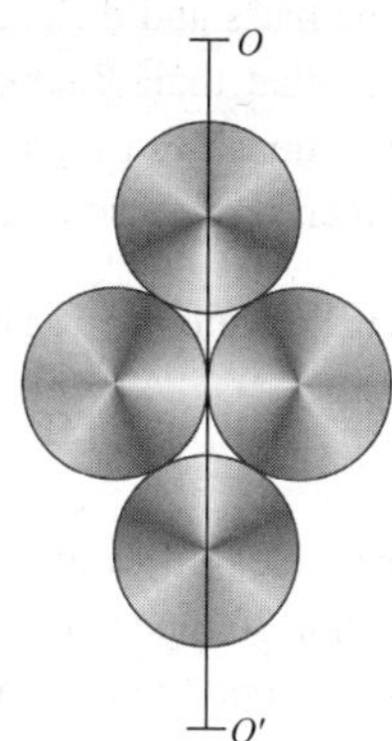

178. A string of area of cross-section 4 mm^2 and length 0.5 is connected with a rigid body of mass 2 kg. The body is rotated in a vertical circular path of radius 0.5 m. The body acquires a speed of 5 m/s at the bottom of the circular path. Strain produced in the string when the body is at the bottom of the circle is _____ $\times 10^{-5}$.
[Use Young's modulus 10^{11} N/m^2 and $g = 10$ m/s^2]
[28 July 2022 (Evening)]

179. The torque of a force $5\hat{i}+3\hat{j}-7\hat{k}$ about the origin is τ. If the force acts on a particle whose position vector is $2\hat{i}+2\hat{j}+\hat{k}$, then the value of τ will be : **[29 July 2022 (Evening)]**

(A) $11\hat{i}+19\hat{j}-4\hat{k}$
(B) $-11\hat{i}+19\hat{j}-16\hat{k}$
(C) $-17\hat{i}+19\hat{j}-4\hat{k}$
(D) $17\hat{i}+9\hat{j}+16\hat{k}$

180. Solid sphere A is rotating about an axis PQ. If the radius of the sphere is 5 cm then its radius of gyration about PQ will be $\sqrt{x}$ cm. The value of x is _____. **[24 Jan. 2023 (Morning)]**

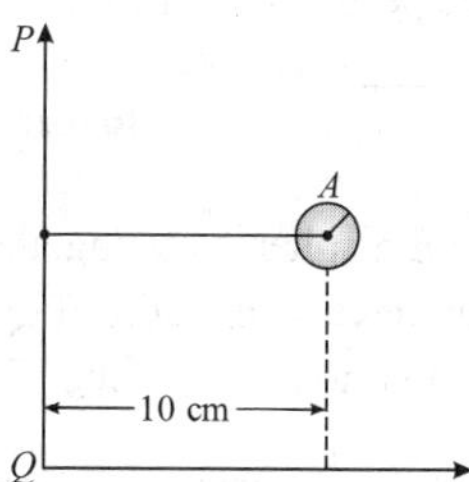

181. A uniform solid cylinder with radius R and length L has moment of inertia I_1, about the axis of cylinder. A concentric solid cylinder of radius $R' = \dfrac{R}{2}$ and length $L' = \dfrac{L}{2}$ original cylinder. If I_2 is the moment of inertia of the carved out portion to the cylinder then $\dfrac{I_1}{I_2} =$ _____ (Both I_1 and I_2 are about the axis of the cylinder). **[24 Jan. 2023 (Evening)]**

182. An object of mass 8 kg is hanging from one end of a uniform rod CD of mass 2 kg and length 1 m pivoted at its end C on a vertical wall as shown in figure. It is supported by a cable AB such that the system is in equilibrium. The tension in the cable is : (Take $g = 10$ m/s^2) **[25 Jan. 2023 (Morning)]**

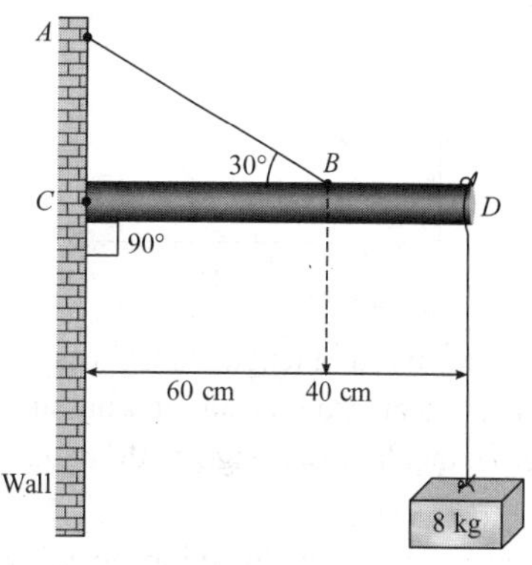

(A) 240 N
(B) 90 N
(C) 300 N
(D) 30 N

183. I_{CM} is moment of inertia of a circular disc about an axis (CM) passing through its center and perpendicular to the plane of disc. I_{AB} is it's moment of inertia about an axis AB perpendicular to plane and parallel to axis CM at a distance $\dfrac{2}{3}R$ from center. Where R is the radius of the disc. The ratio of I_{AB} and I_{CM} is $x : 9$. The value of x is _____.
[25 Jan. 2023 (Morning)]

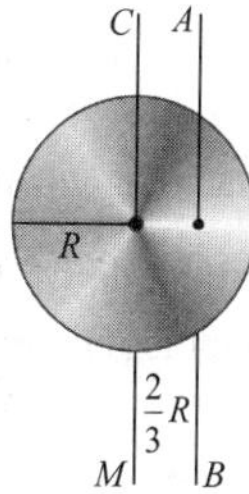

184. If a solid sphere of mass 5 kg and a disc of mass 4 kg have the same radius. Then the ratio of moment of inertia of the disc about a tangent in its plane to the moment of inertia of the sphere about its tangent will be $\dfrac{x}{7}$. The value of x is _____.
[25 Jan. 2023 (Evening)]

185. A solid sphere of mass 2 kg is making pure rolling on a horizontal surface with kinetic energy 2240 J. The velocity of centre of mass of the sphere will be _____ ms^{-1}
[29 Jan. 2023 (Morning)]

186. A particle of mass 100 g is projected at time $t = 0$ with a speed 20 ms^{-1} at an angle 45° to the horizontal as given in the figure. The magnitude of the angular momentum of the particle about the starting point at time $t = 2$s is found to be $\sqrt{K}$ kg m²/s. The value of K is ______. (Take $g = 10$ ms^{-2})

[29 Jan. 2023 (Evening)]

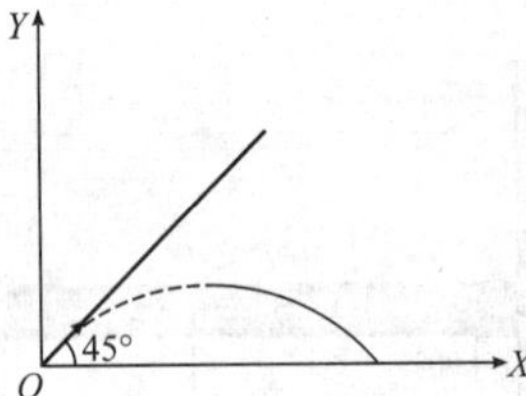

187. A thin uniform rod of length 2 m. Cross sectional area A and density d is rotated about an axis passing through the centre and perpendicular to its length with angular velocity ω. If value of ω in terms of its rotational kinetic energy E is $\sqrt{\frac{\alpha E}{Ad}}$ then the value of α is ______. **[30 Jan. 2023 (Morning)]**

188. A uniform disc of mass 0.5 kg and radius r is projected with velocity 18 m/s at $t = 0$ s on a rough horizontal surface. It starts off with a purely sliding motion at $t = 0$ s. After 2 s it acquires a purely rolling motion (see figure). The total kinetic energy of the disc after 2s will be ______ J. (given, coefficient of friction is 0.3 and $g = 10$ m/s²) **[30 Jan. 2023 (Evening)]**

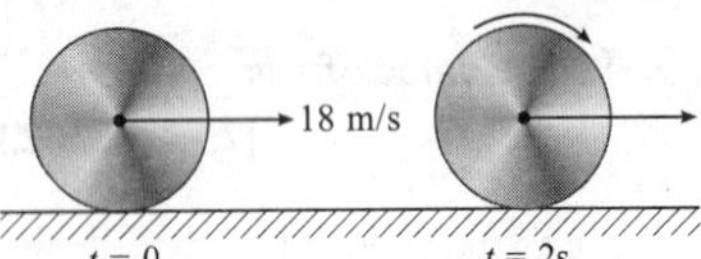

189. As shown in figure, a 70 kg garden roller is pushed with a force of $\vec{F} = 200$ N at an angle of 30° with horizontal. The normal reaction on the roller is (Given $g = 10$ ms^{-2}) :

[31 Jan. 2023 (Morning)]

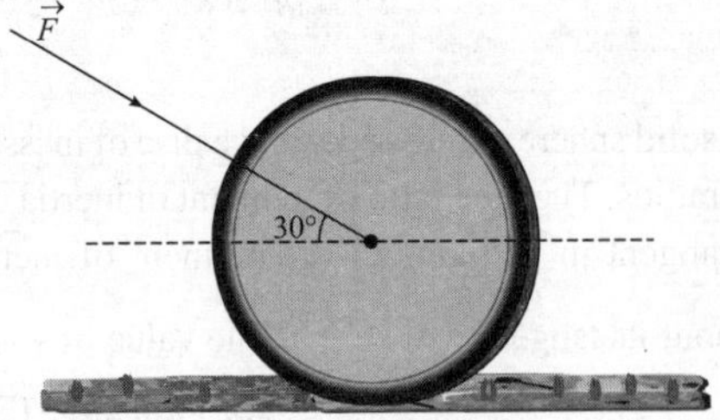

(A) $800\sqrt{2}$ N (B) 600 N

(C) 800 N (D) $200\sqrt{3}$ N

190. A solid sphere of mass 1 kg rolls without slipping on a plane surface. Its kinetic energy is 7×10^{-3} J. The speed of the centre of mass of the sphere is ______ cm s^{-1}.

[31 Jan. 2023 (Morning)]

191. Two discs of same mass and different radii are made of different materials such that their thicknesses are 1 cm and 0.5 cm respectively. The densities of materials are in the ratio 3 : 5. The moment of inertia of these discs respectively about their diameters will be in the ratio of $\frac{x}{6}$. The value of x is______.

[31 Jan. 2023 (Evening)]

192. A solid cylinder is released from rest from the top of an inclined plane of inclination 30° and length 60 cm. If the cylinder rolls without slipping, its speed upon reaching the bottom of the inclined plane is ______ ms^{-1}. (Given $g = 10$ ms^{-2})

[1 Feb. 2023 (Morning)]

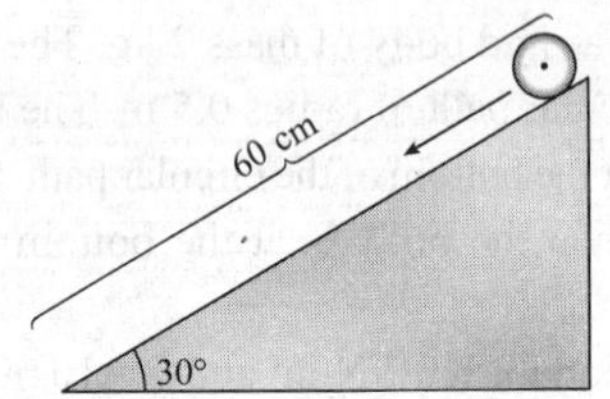

193. Moment of inertia of a disc of mass M and radius R about any of its diameter is $\frac{MR^2}{4}$. The moment of inertia of this disc about an axis normal to the disc and passing through a point on its edge will be, $\frac{x}{2}MR^2$. The value of x is ______.

[1 Feb. 2023 (Evening)]

194. Two identical solid spheres each of mass 2 kg and radii 10 cm are fixed at the ends of a light rod. The separation between the centres of the spheres is 40 cm. The moment of inertia of the system about an axis perpendicular to the rod passing through its middle point is ______ $\times 10^{-3}$ kg-m²

[6 April 2023 (Morning)]

195. A ring and a solid sphere rotating about an axis passing through their centers have same radii of gyration. The axis of rotation is perpendicular to plane of ring. The ratio of radius of ring to that of sphere is $\sqrt{\frac{2}{x}}$. The value of x is ______.

[6 April 2023 (Evening)]

196. The moment of inertia of semicircular ring about an axis, passing through the center and perpendicular to the plane of ring, is $\frac{1}{x}MR^2$, where R is the radius and M is the mass of semicircular ring. The value of x will be.

[8 April 2023 (Morning)]

197. A hollow spherical ball of uniform density rolls up a curved surface with an initial velocity 3 m/s (as shown in figure). Maximum height with respect to the initial position covered by it will be _____ cm. **[8 April 2023 (Evening)]**

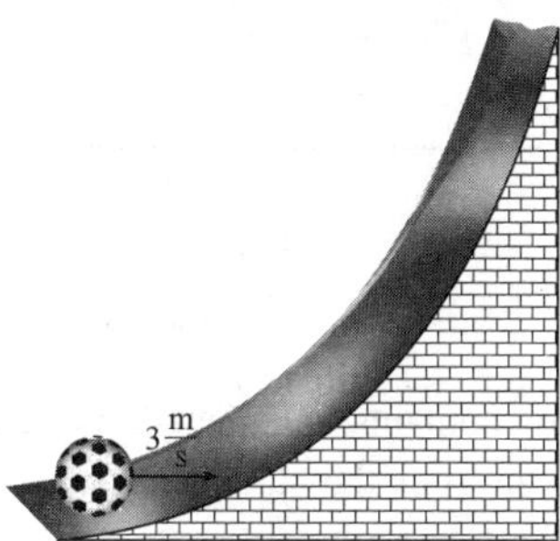

198. Given below are two Statements: one is labelled as Assertion-A and the other is labelled as Reason-R.

Assertion-A: An electric fan continous to rotate for some time after the current is switched off.

Reason-R: Fan continuous to rotate due to inertia of motion.

In the light of above Statements, choose the most appropriate answer from the options given below: **[10 April 2023 (Evening)]**

(A) **A** is correct but **R** is not correct.

(B) Both **A** and **R** are correct and **R** is the correct explanation of **A**.

(C) **A** is not correct but **R** is correct.

(D) Both **A** and **R** are correct but **R** is NOT the correct explanation of **A**.

199. A solid sphere of mass 500 g and radius 5 cm is rotated about one of its diameter with angular speed of 10 rad s^{-1}. If the moment of inertia of the sphere about its tangent is $x \times 10^{-2}$ times its angular momentum about the diameter. Then the value of x will be_____. **[11 April 2023 (Morning)]**

200. Match **List-I** with **List-II** :

	List-I		List-II
(a)	Spring constant	(i)	(T^{-1})
(b)	Angular speed	(ii)	(MT^{-2})
(c)	Angular momentum	(iii)	(ML^2)
(d)	Moment of Inertia	(iv)	(ML^2T^{-1})

Choose the correct answer from the options given below : **[12 April 2023 (Morning)]**

(A) (a)-(ii), (b)-(i), (c)-(iv), (d)-(iii)

(B) (a)-(iv), (b)-(i), (c)-(iii), (d)-(ii)

(C) (a)-(ii), (b)-(iii), (c)-(i), (d)-(iv)

(D) (a)-(i), (b)-(iii), (c)-(ii), (d)-(iv)

201. For a rolling spherical shell, the ratio of rotational kinetic energy and total kinetic energy is $\frac{x}{5}$. The value of x is_____. **[12 April 2023 (Morning)]**

202. A disc is rolling without slipping on a surface. The radius of the disc is R. At $t = 0$, the top most point on the disc is A as shown in figure. When the disc completes half of its rotation, the displacement of point A from its initial position is : **[13 April 2023 (Morning)]**

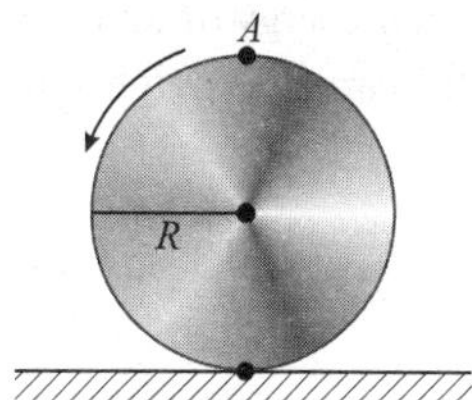

(A) $R\sqrt{(\pi^2+4)}$ (B) $R\sqrt{(\pi^2+1)}$

(C) $2R$ (D) $2R\sqrt{(1+4\pi^2)}$

203. A solid sphere is rolling on a horizontal plane without slipping. If the ratio of angular momentum about axis of rotation of the sphere to the total energy of moving sphere is $\pi : 22$ then, the value of its angular speed will be _____ rad/s. **[13 April 2023 (Morning)]**

204. A light rope is wound around a hollow cylinder of mass 5 kg and radius 70 cm. The rope is pulled with a force of 52.5 N. The angular acceleration of the cylinder will be _____ rad s^{-2}. **[13 April 2023 (Evening)]**

205. A solid sphere and a solid cylinder of same mass and radius are rolling on a horizontal surface without slipping. The ratio of their radius of gyrations respectively ($k_{sph} : k_{cyl}$) is $2 : \sqrt{x}$, then value of x is _____. **[15 April 2023 (Morning)]**

* * * * *

CHAPTER 7 GRAVITATION

7.1 GRAVITATIONAL FIELD

1. The kinetic energy needed to project a body of mass m from the earth's surface (radius R) to infinity is : **[2002]**

(A) $\frac{mgR}{2}$ (B) $2mgR$

(C) mgR (D) $\frac{mgR}{4}$

2. If g is the acceleration due to gravity on the earth's surface, the gain in the potential energy of an object of mass m raised from the surface of the earth to a height equal to the radius R of the earth, is : **[2004]**

(A) $2mgR$ (B) $\frac{1}{2}mgR$

(C) $\frac{1}{4}mgR$ (D) mgR

3. A particle of mass 10 g is kept on the surface of a uniform sphere of mass 100 kg and radius 10 cm. Find the work to be done against the gravitational force between them, to take the particle far away from the sphere : [you may take $G = 6.67 \times 10^{-11}$ Nm^2/kg^2] **[2005]**

(A) 13.34×10^{-10} J (B) 3.33×10^{-10} J

(C) 6.67×10^{-9} J (D) 6.67×10^{-10} J

4. The change in the value of g at a height h above the surface of the earth is the same as at a depth d below the surface of earth. When both d and h are much smaller than the radius of earth, then which one of the following is correct ? **[2005]**

(A) $d = \frac{h}{2}$ (B) $d = \frac{3h}{2}$

(C) $d = 2h$ (D) $d = h$

5. Average density of the earth : **[2005]**

(A) Does not depend on g

(B) Is a complex function of g

(C) Is directly proportional to g

(D) Is inversely proportional to g

6. If g_E and g_M are the accelerations due to gravity on the surfaces of the earth and the moon respectively and if Millikan's oil drop experiment could be performed on the two surfaces, one will find the ratio $\frac{\text{electronic charge on the moon}}{\text{electronic charge on the earth}}$ to be : **[2007]**

(A) 1 (B) Zero

(C) g_E/g_M (D) g_M/g_E

7. This question contains statement-1 and statement-2 of the four choices given after the statements, choose the one that best describes the two statements. **[2008]**

Statement-I : For a mass M kept at the centre of a cube of side 'a', the flux of gravitational field passing through its sides is 4π GM.

Statement-I : If the direction of a field due to a point source is radial and its dependence on the distance 'r' from the source is given as l/r^2, its flux through a closed surface depends only on the strength of the source enclosed by the surface and not on the size or shape of the surface.

(A) Statement-1 is false, Statement-2 is true

(B) Statement-1 is true, Statement-2 is true; Statement-2 is a correct explanation for Statement-1

(C) Statement-1 is true, Statement-2 is true; Statement-2 is not a correct explanation for Statement-1

(D) Statement-1 is true, Statement-2 is false

8. The height at which the acceleration due to gravity becomes $\frac{g}{9}$ (where g = the acceleration due to gravity on the surface of the earth) in terms of R, the radius of the earth, is : **[2009]**

(A) $\frac{R}{2}$ (B) $\sqrt{2}R$

(C) $2R$ (D) $\frac{R}{\sqrt{2}}$

9. Two bodies of masses m and $4m$ are placed at a distance r. The gravitational potential at a point on the line joining them where the gravitational field is zero is : **[2011]**

(A) $-\frac{6Gm}{r}$ (B) $-\frac{9Gm}{r}$

(C) Zero (D) $-\frac{4Gm}{r}$

10. The mass of a spaceship is 1000 kg. It is to be launched from the earth's surface out into free space. The value of 'g' and 'R' (radius of earth) are 10 m/s^2 and 6400 km respectively. The required energy for this work will be : [2012]

(A) 6.4×10^{10} J (B) 6.4×10^{11} J

(C) 6.4×10^{8} J (D) 6.4×10^{9} J

11. Assuming the earth to be a sphere of uniform density, the acceleration due to gravity inside the earth at a distance of r from the centre is proportional to : [2012 (Online)]

(A) r (B) r^{-1}

(C) r^2 (D) r^{-2}

12. The variation of acceleration due to gravity g with distance d from centre of the earth is best represented by (R = Earth's radius) : [2012 (Online)]

(A) 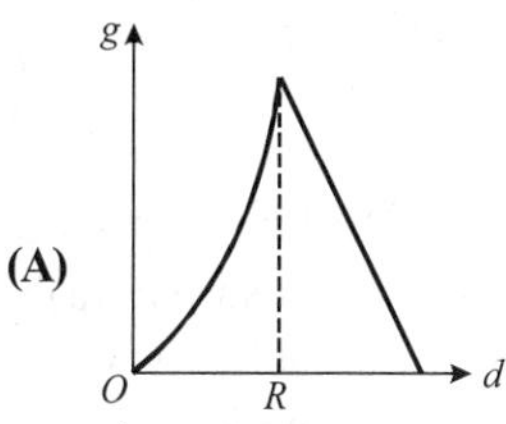

(B) 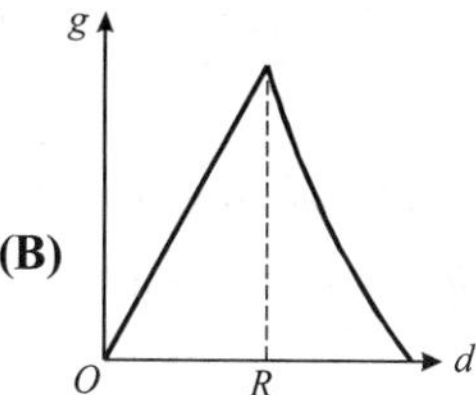

(C) 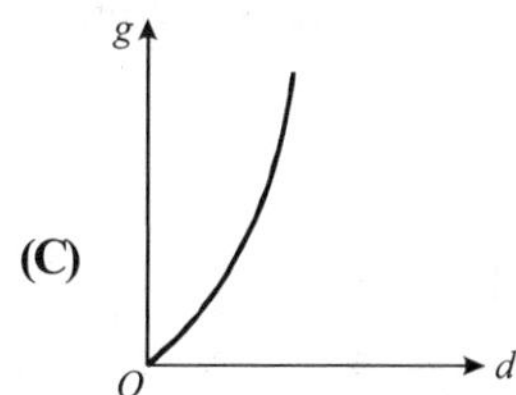

(D) 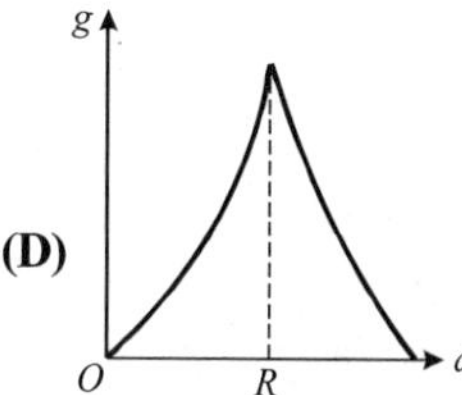

13. The gravitational field, due to the 'left over part' of a uniform sphere (from which a part as shown, has been 'removed out'), at a very far off point P, located as shown, would be (nearly) : [2013 (Online)]

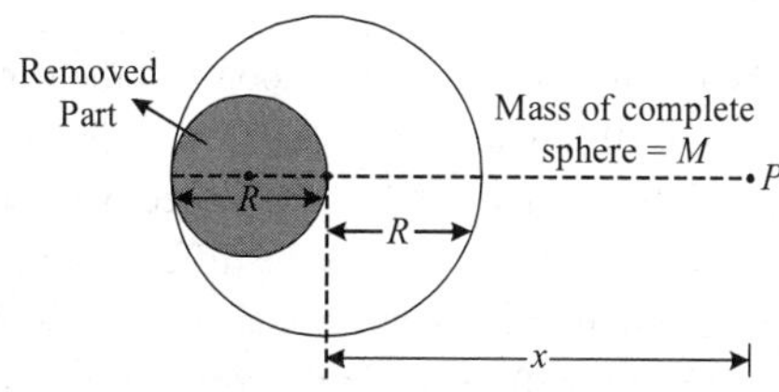

(A) $\frac{5}{6}\frac{GM}{x^2}$ (B) $\frac{8}{9}\frac{GM}{x^2}$

(C) $\frac{7}{8}\frac{GM}{x^2}$ (D) $\frac{6}{7}\frac{GM}{x^2}$

14. The change in the value of acceleration of earth towards sun, when the moon comes from the position of solar eclipse to the position on the other side of earth in line with sun is : (mass of the moon = 7.36×10^{22} kg, radius of the moon's orbit = 3.8×10^{8} m) [2013 (Online)]

(A) 6.73×10^{-5} m/s^2 (B) 6.73×10^{-3} m/s^2

(C) 6.73×10^{-2} m/s^2 (D) 6.73×10^{-4} m/s^2

15. Four particles, each of mass M and equidistant from each other, move along a circle of radius R under the action of their mutual gravitational attraction. The speed of each particle is : [2014]

(A) $\sqrt{\frac{GM}{R}}$ (B) $\sqrt{2\sqrt{2}\frac{GM}{R}}$

(C) $\sqrt{\frac{GM}{R}(1+2\sqrt{2})}$ (D) $\frac{1}{2}\sqrt{\frac{GM}{R}(1+2\sqrt{2})}$

16. A capillary tube is immersed vertically in water and the height of the water column is x. When this arrangement is taken into a mine of depth d, the height of the water column is y. If R is the radius of earth, the ratio $\frac{x}{y}$ is : [2014 (Online)]

(A) $\left(1-\frac{d}{R}\right)$ (B) $\left(1-\frac{2d}{R}\right)$

(C) $\left(\frac{R-d}{R+d}\right)$ (D) $\left(\frac{R+d}{R-d}\right)$

17. The gravitational field in a region is given by $\vec{g}$ = 5 N/kg $\hat{i}$ + 12 N/kg $\hat{j}$. The change in the gravitational potential energy of a particle of mass 1 kg when it is taken from the origin to a point (7 m, –3 m) is : [2014 (Online)]

(A) 71 J (B) $13\sqrt{58}$ J

(C) – 71 J (D) 1 J

18. From a sphere of mass M and radius R, a smaller sphere of radius $\frac{R}{2}$ is carved out such that the cavity made in the original sphere is between its centre and the periphery (See figure). For the configuration in the figure where the distance between the centre of the original sphere and the removed sphere is $3R$, the gravitational force between the two sphere is : [2014 (Online)]

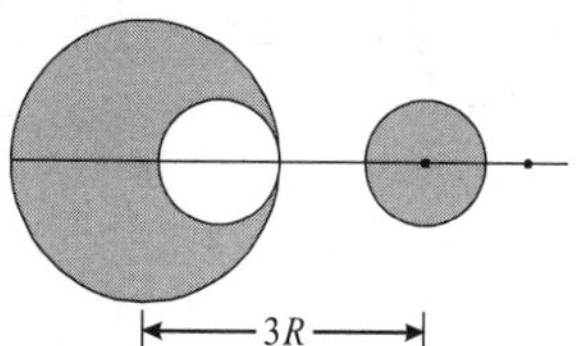

(A) $\frac{41GM^2}{3600R^2}$ (B) $\frac{41GM^2}{450R^2}$

(C) $\frac{59GM^2}{450R^2}$ (D) $\frac{GM^2}{225R^2}$

19. Two hypothetical planets of masses m_1 and m_2 are at rest when they are infinite distance apart. Because of the gravitational force they move towards each other along the line joining their centres. What is their speed when their separation is 'd' ? (Speed of m_1 is v_1 and that of m_2 is v_2) **[2014 (Online)]**

(A) $v_1 = v_2$

(B) $v_1 = m_2\sqrt{\frac{2G}{(d(m_1+m_2)}}$ $\quad v_2 = m_1\sqrt{\frac{2G}{(d(m_1+m_2)}}$

(C) $v_1 = m_1\sqrt{\frac{2G}{(d(m_1+m_2)}}$ $\quad v_2 = m_2\sqrt{\frac{2G}{(d(m_1+m_2)}}$

(D) $v_1 = m_2\sqrt{\frac{2G}{m_1}}$ $\quad v_2 = m_2\sqrt{\frac{2G}{m_2}}$

20. From a solid sphere of mass M and radius R, a spherical portion of radius $\frac{R}{2}$ is removed, as shown in figure. Taking gravitational potential $V = 0$ at $r = \infty$, the potential at the centre of the cavity thus formed is : **[2015]**
(G = gravitational constant)

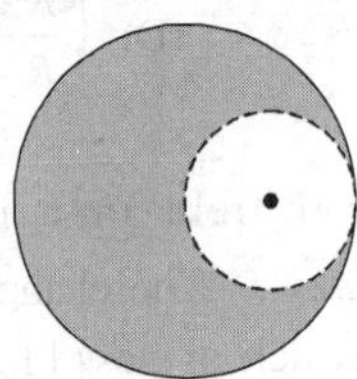

(A) $\frac{-GM}{2R}$ (B) $\frac{-GM}{R}$

(C) $\frac{-2GM}{3R}$ (D) $\frac{-2GM}{R}$

21. Which of the following most closely depicts the correct variation of the gravitational potential $V(r)$ due to a large planet of radius R and uniform mass density ? (Figures are not drawn to scale) **[2015 (Online)]**

(A) 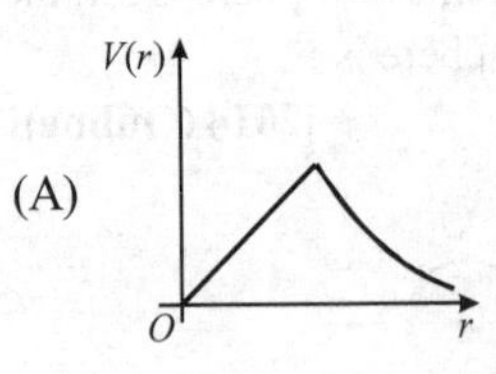

(B) 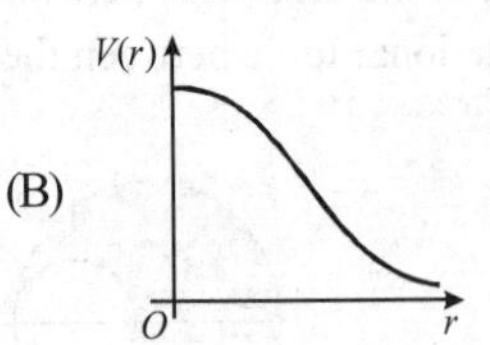

(C) 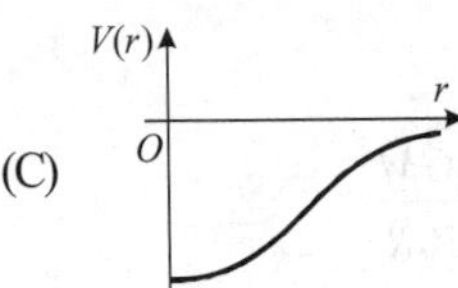

(D) 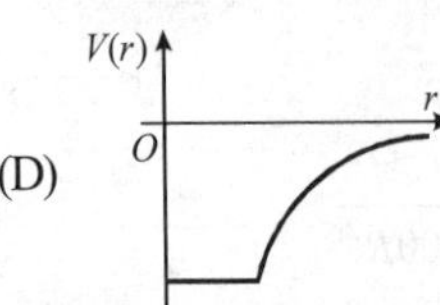

22. A point particle is held on the axis of a ring of mass m and radius r at a distance r from it's centre C. When released, it reaches C under the gravitational attraction of the ring. Its speed at C will be : **[2016 (Online)]**

(A) $\sqrt{\frac{2Gm}{r}(\sqrt{2}-1)}$ (B) $\sqrt{\frac{Gm}{r}}$

(C) $\sqrt{\frac{2Gm}{r}\left(1-\frac{1}{\sqrt{2}}\right)}$ (D) $\sqrt{\frac{2Gm}{r}}$

23. If the Earth has no rotational motion, the weight of a person on the equator is W. Determine the speed with which the earth would have to rotate about its axis so that the person at the equator will weight 3/4 W. Radius of the Earth is 6400 km and $g = 10$ m\s^2 : **[2017 (Online)]**

(A) 1.1×10^{-3} rad/s (B) 0.83×10^{-3} rad/s
(C) 0.63×10^{-3} rad/s (D) 0.28×10^{-3} rad/s

24. The mass density of a spherical body is given by $\rho(r) = \frac{k}{r}$ for $r \le R$ and $\rho(r) = 0$ for $r > R$, where r is the distance from the centre. The correct graph that describes qualitatively the acceleration a, of a test particle as a function of r is : **[2017 (Online)]**

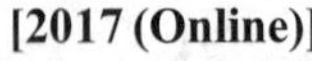

(A) 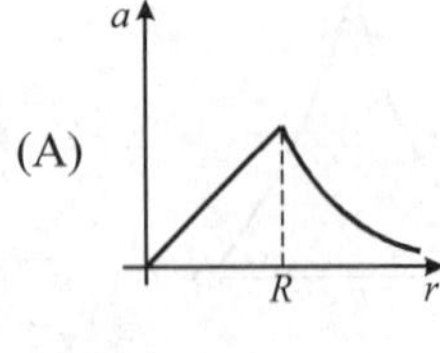

(B) 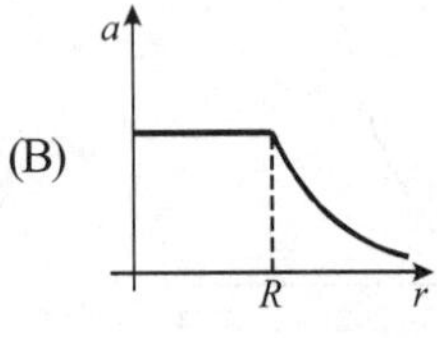

(C) 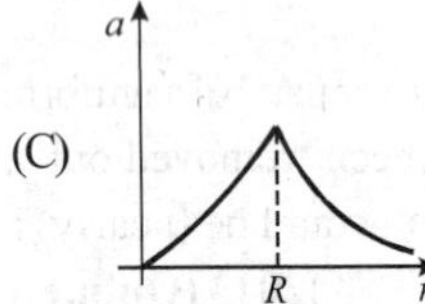

(D) 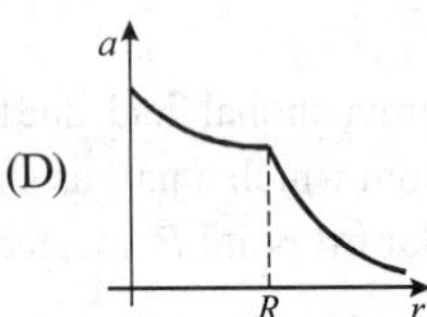

25 Take the mean distance of the moon and the sun from the earth to be 0.4×10^6 km and 150×10^6 km respectively. Their masses are 8×10^{22} kg and 2×10^{30} kg respectively. The radius of the earth is 6400 km. Let ΔF_1 be the difference in the forces exerted by the moon at the nearest and farthest points on the earth and ΔF_2 be the difference in the force exerted by the sun at the nearest and farthest points on the earth. Then, the number closest to $\frac{\Delta F_1}{\Delta F_2}$ is : **[2018 (Online)]**

(A) 2 (B) 6
(C) 10^{-2} (D) 0.6

26. Suppose that the angular velocity of rotation of earth is increased. Then, as a consequence : **[2018 (Online)]**

(A) There will be no change in weight anywhere on the earth
(B) Weight of the object, everywhere on the earth, will decrease
(C) Weight of the object, everywhere on the earth, will increase
(D) Except at poles, weight of the object on the earth will decrease

27. The mass and the diameter of a planet are three times the respective values for the Earth. The period of oscillation of a simple pendulum on the Earth is 2 s. The period of oscillation of the same pendulum on the planet would be :

[11 Jan. 2019 (Evening)]

(A) $\frac{2}{\sqrt{3}}$ s (B) $2\sqrt{3}$ s

(C) $\frac{\sqrt{3}}{2}$ s (D) $\frac{3}{2}$ s

28. A straight rod of length L extends from $x = a$ to $x = L + a$. The gravitational force exerted on a point mass 'm' at $x = 0$, if the mass per unit length of the rod is $A + Bx^2$, is given by :

[12 Jan. 2019 (Morning)]

(A) $Gm\left[A\left(\frac{1}{a+L}-\frac{1}{a}\right)-BL\right]$

(B) $Gm\left[A\left(\frac{1}{a}-\frac{1}{a+L}\right)+BL\right]$

(C) $Gm\left[A\left(\frac{1}{a+L}-\frac{1}{a}\right)+BL\right]$

(D) $Gm\left[A\left(\frac{1}{a}-\frac{1}{a+L}\right)-BL\right]$

29. Four identical particles of mass M are located at the corners of a square of side a. What should be their speed if each of them revolves under the influence of other's gravitational field in a circular orbit circumscribing the square ?

[8 April 2019 (Morning)]

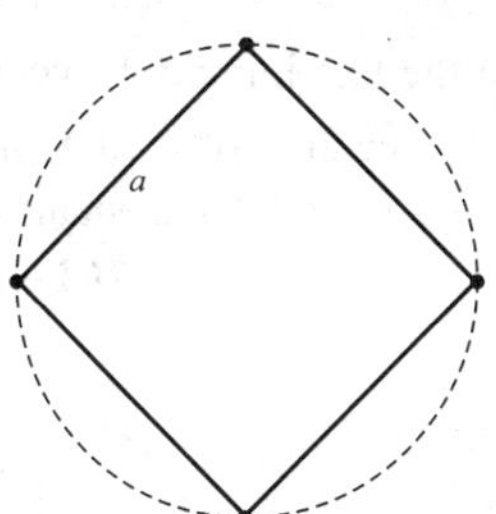

(A) $1.21\sqrt{\frac{GM}{a}}$ (B) $1.41\sqrt{\frac{GM}{a}}$

(C) $1.16\sqrt{\frac{GM}{a}}$ (D) $1.35\sqrt{\frac{GM}{a}}$

30. A rocket has to be launched from earth in such a way that it never returns. If E is the minimum energy delivered by the rocket launcher, what should be the minimum energy that the launcher should have if the same rocket is to be launched from the surface of the moon ? Assume that the density of the earth and the moon are equal and that the earth's volume is 64 times the volume of the moon :

[8 April 2019 (Evening)]

(A) $\frac{E}{4}$ (B) $\frac{E}{16}$

(C) $\frac{E}{32}$ (D) $\frac{E}{64}$

31. A solid sphere of mass 'M' and radius 'a' is surrounded by a uniform concentric spherical shell of thickness $2a$ and mass $2M$. The gravitational field at distance '$3a$' from the centre will be :

[9 April 2019 (Morning)]

(A) $\frac{2GM}{9a^2}$ (B) $\frac{GM}{3a^2}$

(C) $\frac{GM}{9a^2}$ (D) $\frac{2GM}{3a^2}$

32. The value of acceleration due to gravity at Earth's surface is 9.8 ms^{-2}. The altitude above its surface at which the acceleration due to gravity decreases to 4.9 ms^{-2}, is close to : (Radius of earth = 6.4×10^6 m) **[10 April 2019 (Morning)]**

(A) 1.6×10^6 m (B) 6.4×10^6 m

(C) 9.0×10^6 m (D) 2.6×10^6 m

33. A box weighs 196 N on a spring balance at the north pole. Its weight recorded on the same balance if it is shifted to the equator is close to (Take $g = 10$ ms^{-2} at the north pole and the radius of the earth = 6400 km) : **[7 Jan. 2020 (Evening)]**

(A) 195.66 N (B) 194.66 N

(C) 194.32 N (D) 195.32 N

34. Consider two solid spheres of radii $R_1 = 1$ m, $R_2 = 2$ m and masses M_1 and M_2, respectively. The gravitational field due to sphere (1) and (2) are shown. The value of $\frac{M_1}{M_2}$ is :

[8 Jan. 2020 (Morning)]

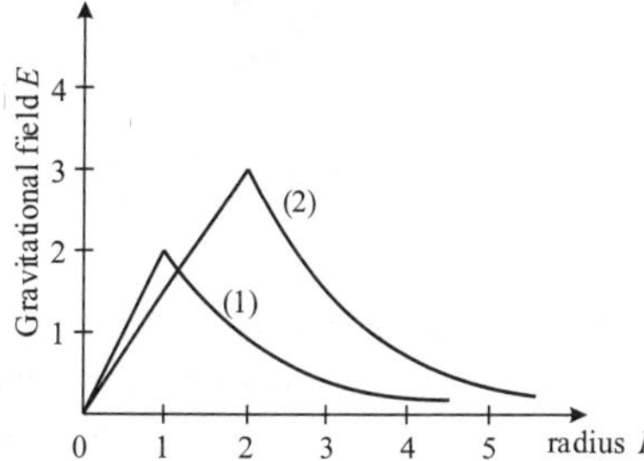

(A) $\frac{1}{2}$ (B) $\frac{2}{3}$

(C) $\frac{1}{3}$ (D) $\frac{1}{6}$

35. The height 'h' at which the weight of a body will be the same as that at the same depth 'h' from the surface of the earth is (Radius of the earth is R and effect of the rotation of the earth is neglected) : **[2 Sept. 2020 (Evening)]**

(A) $\frac{\sqrt{5}R-R}{2}$ (B) $\frac{\sqrt{5}}{2}R-R$

(C) $\frac{R}{2}$ (D) $\frac{\sqrt{3}R-R}{2}$

36. The mass density of a planet of radius R varies with the distance r from its centre as $\rho(r)=\rho_0\left(1-\frac{r^2}{R^2}\right)$. Then the gravitational field is maximum at : **[3 Sept. 2020 (Evening)]**

(A) $r=\frac{1}{\sqrt{3}}R$ (B) $r=\sqrt{\frac{5}{9}}R$

(C) $r=\sqrt{\frac{3}{4}}R$ (D) $r=R$

37. On the x-axis and a distance x from the origin, the gravitational field due to a mass distribution is given by $\frac{Ax}{(x^2+a^2)^{3/2}}$ in the x-direction. The magnitude of gravitational potential on the x-axis at a distance x, taking its value to be zero at infinity, is : **[4 Sept. 2020 (Morning)]**

(A) $\frac{A}{(x^2+a^2)^{1/2}}$ (B) $\frac{A}{(x^2+a^2)^{3/2}}$

(C) $A(x^2+a^2)^{3/2}$ (D) $A(x^2+a^2)^{1/2}$

38. The value of the acceleration due to gravity is g_1 at a height $h=R/2$ (R = radius of the earth) from the surface of the earth. It is again equal to g_1 at a depth d below the surface of the earth. The ratio (d/R) equals : **[5 Sept. 2020 (Morning)]**

(A) $\frac{7}{9}$ (B) $\frac{4}{9}$

(C) $\frac{1}{3}$ (D) $\frac{5}{9}$

39. The acceleration due to gravity on the earth's surface at the poles is g and angular velocity of the earth about the axis passing through the pole is ω. An object is weighed at the equator and at a height h above the poles by using a spring balance. If the weights are found to be same, then h is : ($h<<R$, where R is the radius of the earth) : **[5 Sept. 2020 (Evening)]**

(A) $\frac{R^2\omega^2}{8g}$ (B) $\frac{R^2\omega^2}{4g}$

(C) $\frac{R^2\omega^2}{g}$ (D) $\frac{R^2\omega^2}{2g}$

40. A body weighs 49 N on a spring balance at the north pole. What will be its weight recorded on the same weighing machine, if it is shifted to the equator ? **[24 Feb. 2021 (Evening)]**

[Use $g=\frac{GM}{R^2}=9.8\text{ ms}^{-2}$ and radius of earth, $R=6400$ km.]

(A) 49.17 N (B) 49.83 N

(C) 49 N (D) 48.83 N

41. A solid sphere of radius R gravitationally attracts a particle placed at $3R$ from its centre with a force F_1. Now a spherical cavity of radius $\left(\frac{R}{2}\right)$ is made in the sphere (as shown in figure) and the force becomes F_2. The value of $F_1 : F_2$ is : **[25 Feb. 2021 (Morning)]**

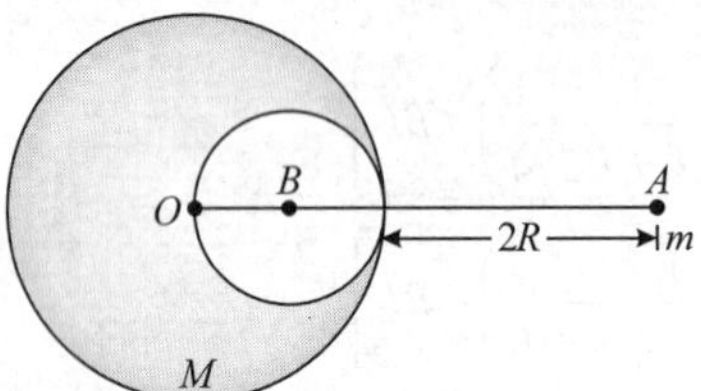

(A) 25 : 36 (B) 50 : 41

(C) 41 : 50 (D) 36 : 36

42. Find the gravitational force of attraction between the ring and sphere as shown in the diagram, where the plane of the ring is perpendicular to the line joining the centres. If $\sqrt{8}R$ is the distance between the centres of a ring (of mass 'm') and a sphere (mass 'M') where both have equal radius 'R'. **[26 Feb. 2021 (Morning)]**

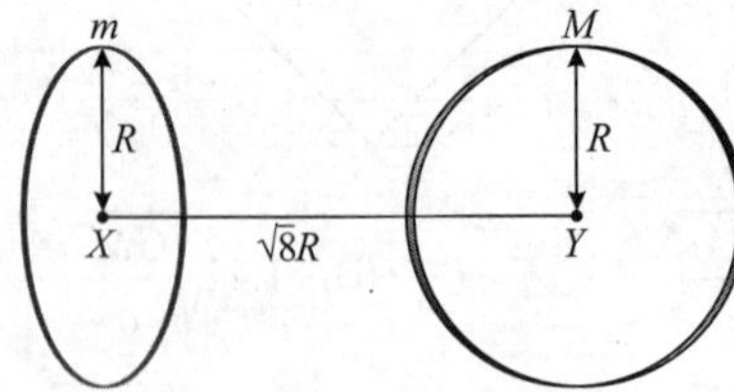

(A) $\frac{1}{3\sqrt{8}}\cdot\frac{GMm}{R^2}$ (B) $\frac{\sqrt{8}}{9}\cdot\frac{GmM}{R}$

(C) $\frac{\sqrt{8}}{27}\cdot\frac{GmM}{R^2}$ (D) $\frac{2\sqrt{2}}{3}\cdot\frac{GMm}{R^2}$

43. In the reported figure of earth, the value of acceleration due to gravity is same at point A and C but it is smaller than that of its value at point B (surface of the earth). The value of $OA : AB$ will be $x : y$. The value of x is _____. **[26 Feb. 2021 (Evening)]**

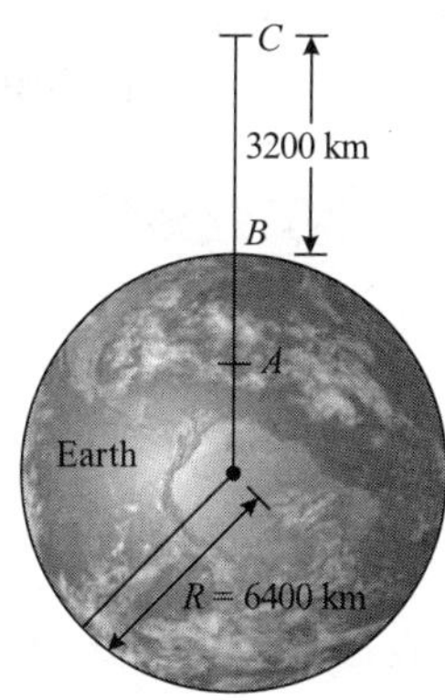

44. If one wants to remove all the mass of the earth to infinity in order to break it up completely. The amount of energy that needs to be supplied will be $\frac{x}{5}\frac{GM^2}{R}$ where x is _____ (Round off to the Nearest Integer)
(M is the mass of earth, R is the radius of earth, G is the gravitational constant) **[16 March 2021 (Evening)]**

45. A person whose mass is 100 kg travels from Earth to Mars in a spaceship. Neglect all other objects in sky and take acceleration due to gravity on the surface of the Earth and Mars as 10 m/s^2 and 4 m/s^2 respectively. Identify from the below figures, the curve that fits best for the weight of the passenger as a function of time : **[20 July 2021 (Morning)]**

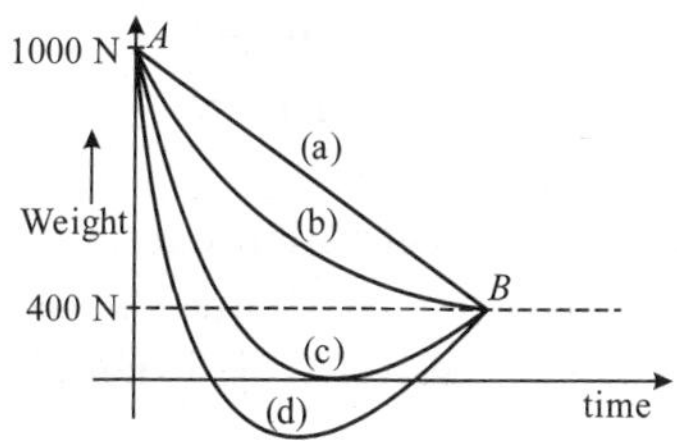

(A) (c) (B) (d)
(C) (b) (D) (a)

46. Consider a binary star system of star A and star B with masses m_A and m_B revolving in a circular orbit of radii r_A and r_B, respectively. If T_A and T_B are the time period of star A and star B, respectively, then : **[20 July 2021 (Evening)]**

(A) $T_A = T_B$ (B) $T_A > T_B$ (if $r_A > r_B$)

(C) $\frac{T_A}{T_B} = \left(\frac{r_A}{r_B}\right)^{3/2}$ (D) $T_A > T_B$ (if $m_A > m_B$)

47. A body is projected vertically upwards from the surface of earth with a velocity sufficient enough to carry to infinity. The time taken by it to reach height h is_____s :

[22 July 2021 (Evening)]

(A) $\sqrt{\frac{2R_e}{g}}\left[\left(1+\frac{h}{R_e}\right)^{\frac{3}{2}}-1\right]$ (B) $\sqrt{\frac{R_e}{2g}}\left[\left(1+\frac{h}{R_e}\right)^{\frac{3}{2}}-1\right]$

(C) $\frac{1}{3}\sqrt{\frac{R_e}{2g}}\left[\left(1+\frac{h}{R_e}\right)^{\frac{3}{2}}-1\right]$ (D) $\frac{1}{3}\sqrt{\frac{2R_e}{g}}\left[\left(1+\frac{h}{R_e}\right)^{\frac{3}{2}}-1\right]$

48. Consider a planet in some solar system which has a mass double the mass of earth and density equal to the average density of earth. If the weight of an object on earth is W, the weight of the same object on that planet will be :

[25 July 2021 (Evening)]

(A) 2 W (B) $\sqrt{2}$ W

(C) W (D) $2^{\frac{1}{3}}$ W

49. Suppose two planets (spherical in shape) of radii R and $2R$, but mass M and $9M$ respectively have a centre to centre separation $8R$ as shown in the figure. A satellite of mass 'm' is projected from the surface of the planet of mass 'M' directly towards the centre of the second planet. The minimum speed 'v' required for the satellite to reach the surface of the second planet is $\sqrt{\frac{a}{7}\frac{GM}{R}}$ then the value of 'a' is _____ .

[Given : the two planets are fixed in their position]

[27 July 2021 (Morning)]

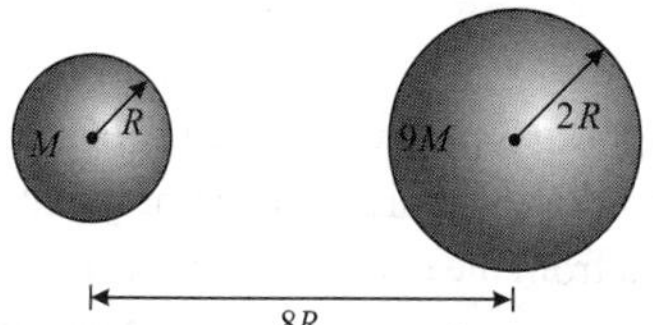

50. Two identical particles of mass 1 kg each to round a circle of radius R, under the action of their mutual gravitational attraction. The angular speed of each particle is :

[27 July 2021 (Evening)]

(A) $\frac{1}{2R}\sqrt{\frac{1}{G}}$ (B) $\sqrt{\frac{2G}{R^3}}$

(C) $\frac{1}{2}\sqrt{\frac{G}{R^3}}$ (D) $\sqrt{\frac{G}{2R^3}}$

51. The planet Mars has two moons, if one of them has a period 7 hours, 30 minutes and an orbital radius of 9.0×10^3 km.

Find the mass of Mars: (Given $\frac{4\pi^2}{G} = 6 \times 10^{11}\ N^{-1}\ m^{-2}\ kg^2$)

[27 July 2021 (Evening)]

(A) 7.02×10^{25} kg (B) 2.25×10^{21} kg

(C) 5.96×10^{19} kg (D) 6.00×10^{23} kg

52. Inside a uniform spherical shell :

[26 Aug. 2021 (Morning)]

(a) The gravitational field is zero
(b) The gravitational field is non-zero but uniform
(c) The gravitational field is same everywhere
(d) The gravitational potential is same everywhere
(e) All of the above

Choose the most appropriate answer from the options given below :

(A) (b), (c), and (d) only (B) (e) only

(C) (a), (c) and (d) only (D) (a), (b) and (c) only

53. A body of mass $(2M)$ splits into four masses $[m, M-m, m, M-m]$ which are rearranged to form a square as shown in the figure. The ratio of M/m for which, the gravitational potential energy of the system becomes maximum is $x : 1$. The value of x is _____ .

[27 Aug. 2021 (Morning)]

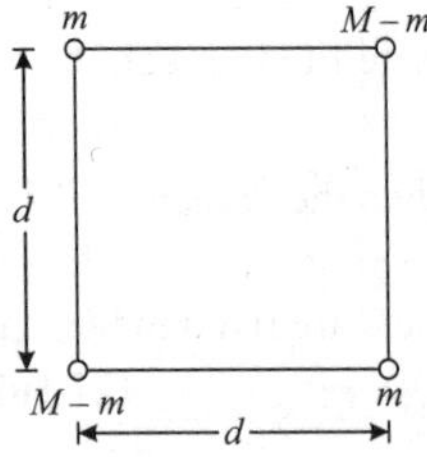

54. A mass of 50 kg is placed at the centre of a uniform spherical shell of mass 100 kg and radius 50 m. If the gravitational potential at a point, 25 m from the centre is V kg/m. The value of V is :

[27 Aug. 2021 (Evening)]

(A) $+2G$ (B) $-20G$

(C) $-4G$ (D) $-60G$

55. If R_E be the radius of Earth, then the ratio between the acceleration due to gravity at a depth r below and a height r above the earth surface is : (Given : $r < R_E$)

[31 Aug. 2021 (Evening)]

(A) $1+\frac{r}{R_E}-\frac{r^2}{R_E^2}-\frac{r^3}{R_E^3}$ (B) $1+\frac{r}{R_E}-\frac{r^2}{R_E^2}+\frac{r^3}{R_E^3}$

(C) $1-\frac{r}{R_E}-\frac{r^2}{R_E^2}-\frac{r^3}{R_E^3}$ (D) $1+\frac{r}{R_E}+\frac{r^2}{R_E^2}+\frac{r^3}{R_E^3}$

56. Four particle of mass M, move along a circle of radius R under the action of their mutual gravitational attraction as shown in figure. The speed of each particle is :

[01 Sep. 2021 (Evening)]

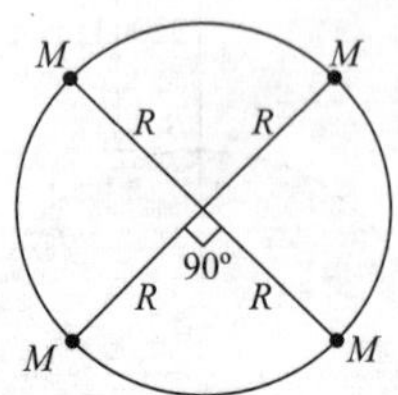

(A) $\frac{1}{2}\sqrt{\frac{GM}{R(2\sqrt{2}+1)}}$ (B) $\frac{1}{2}\sqrt{\frac{GM}{R}}(2\sqrt{2}-1)$

(C) $\frac{1}{2}\sqrt{\frac{GM}{R}}(2\sqrt{2}+1)$ (D) $\sqrt{\frac{GM}{R}}$

57. The approximate height from the surface of earth at which the weight of the body becomes $\frac{1}{3}$ of its weight on the surface of earth is : [Radius of earth $R = 6400$ km and $\sqrt{3} = 1.732$]

[24 June 2022 (Morning)]

(A) 3840 km (B) 4685 km

(C) 2133 km (D) 4267 km

58. The height of any point P above the surface of earth is equal to diameter of earth. The value of acceleration due to gravity at point P will be : [Given g = acceleration due to gravity at the surface of earth]

[25 June 2022 (Morning)]

(A) $g/2$ (B) $g/4$

(C) $g/3$ (D) $g/9$

59. The variation of acceleration due to gravity (g) with distance (r) from the center of the earth is correctly represented by : [Given R = radius of earth]

[26 June 2022 (Morning)]

(A)

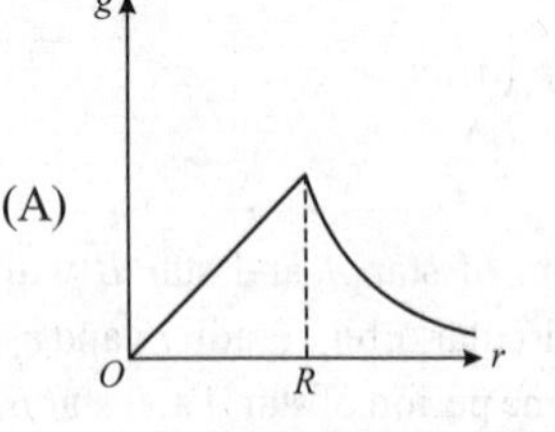

(B)

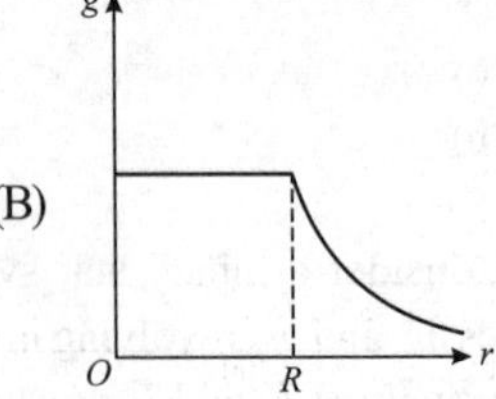

(C)

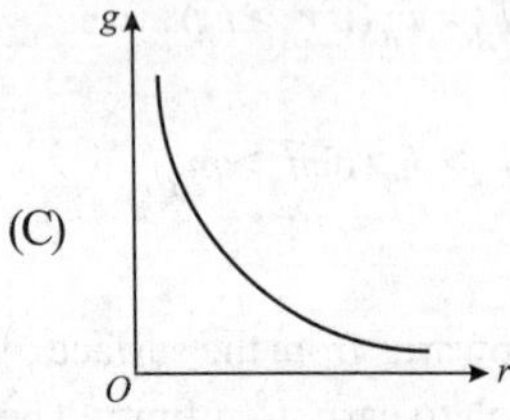

(D)

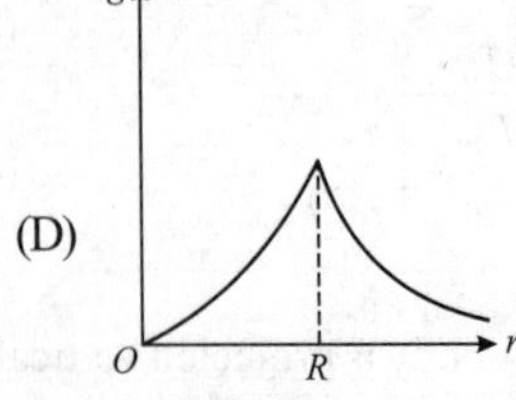

60. The elongation of a wire on the surface of the earth is 10^{-4} m. The same wire of same dimensions is elongated by 6×10^{-5} m on another planet. The acceleration due to gravity on the planet will be _____ ms^{-2}. [Take acceleration due to gravity on the surface of earth = 10 m/s^{-2}] **[26 June 2022 (Morning)]**

61. Given below are two statements : One is labelled as Assertion-**A** and the other is labelled as Reason-**R**.

Assertion-A: If we move from poles to equator, the direction of acceleration due to gravity of earth always points towards the center of earth without any variation in its magnitude.

Reason-R: At equator, the direction of acceleration due to the gravity is towards the center of earth.

In the light of above statements, choose the correct answer from the options given below : **[26 June 2022 (Evening)]**

(A) Both **A** and **R** are true and **R** is the correct explanation of **A**.
(B) Both **A** and **R** are true but **R** is **NOT** the correct explanation of **A**.
(C) **A** is true but **R** is false.
(D) **A** is false but **R** is true.

62. Given below are two statements : **[27 June 2022 (Morning)]**

Statement-I : The law of gravitation holds good for any pair of bodies in the universe

Statement-II : The weight of any person becomes zero when the person is at the centre of the earth

In the light of the above statements, choose the correct answer from the options given below :

(A) Both statement-I and Statement-II are true
(B) Both statement-I and Statement-II are false
(C) Statement-I is true but Statement-II are false
(D) Statement-I is false but Statement-II is true

63. Four spheres each of mass m form a square of side d (as shown in figure). A fifth sphere of mass M is situated at the centre of square. The total gravitational potential energy of the system is : **[27 June 2022 (Evening)]**

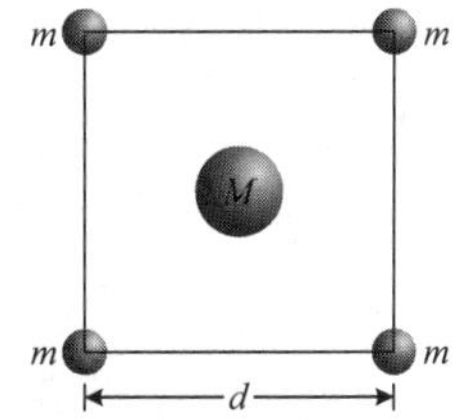

(A) $-\frac{Gm}{d}[(4+\sqrt{2})m+4\sqrt{2}M]$

(B) $-\frac{Gm}{d}[(4+\sqrt{2})M+4\sqrt{2}m]$

(C) $-\frac{Gm}{d}[3m^2+4\sqrt{2}M]$

(D) $-\frac{Gm}{d}[6m^2+4\sqrt{2}M]$

64. Two objects of equal masses placed at certain distance from each other attracts each other with a force of F. If one-third mass of one object is transferred to the other object, then the new force will be : **[28 June 2022 (Evening)]**

(A) $\frac{2}{9}F$ (B) $\frac{16}{9}F$
(C) $\frac{8}{9}F$ (D) F

65. Three identical particle A, B and C of mass 100 kg each are placed in a straight line with $AB = BC = 13$ m. The gravitational force on a fourth particle P of the same mass is F, when placed at a distance 13 m from the particle B on the perpendicular bisector of the line AC. The value of F will be approximately : **[25 July 2022 (Morning)]**

(A) 21G (B) 100G
(C) 59G (D) 42G

66. An object is taken to a height above the surface of earth at a distance $\frac{5}{4}R$ from the centre of the earth. Where radius of earth, $R = 6400$ km. The percentage decrease in the weight of the object will be : **[25 July 2022 (Evening)]**

(A) 36% (B) 50%
(C) 64% (D) 25%

67. The percentage decrease in the weight of a rocket, when taken to a height of 32 km above the surface of earth will, be : [Radius of earth = 6400 km] **[26 July 2022 (Morning)]**

(A) 1% (B) 3%
(C) 4% (D) 0.5%

68. A body is projected vertically upwards from the surface of earth with a velocity equal to one third of escape velocity. The maximum height attained by the body will be : [Take radius of earth = 6400 km and $g = 10\ ms^{-2}$] **[26 July 2022 (Evening)]**

(A) 800 km (B) 1600 km
(C) 2133 km (D) 4800 km

69. If the radius of earth shrinks by 2% while its mass remains same. The acceleration due to gravity on the earth's surface will approximately : **[28 July 2022 (Morning)]**

(A) decrease by 2% (B) decrease by 4%
(C) increase by 2% (D) increase by 4%

70. Assume there are two identical simple pendulum Clocks-1 is placed on the earth and Clock-2 is placed on a space station located at a height h above the earth surface. Clock-1 and Clock-2 operate at time periods 4 s and 6 s respectively. Then the value of h is : [Consider radius of earth $R_E = 6400$ km and g on earth 10 m/s^2] **[28 July 2022 (Evening)]**

(A) 1200 km (B) 1600 km
(C) 3200 km (D) 4800 km

71. If the acceleration due to gravity experienced by a point mass at a height h above the surface of earth is same as that of the acceleration due to gravity at a depth $\alpha h (h << R_e)$ from the earth surface. The value of α will be _____ .
[Use $R_e = 6400$ km] **[29 July 2022 (Morning)]**

72. An object of mass 1 kg is taken to a height from the surface of earth which is equal to three times the radius of earth. The gain in potential energy of the object will be :
[If, $g = 10$ ms^{-2} and radius of earth = 6400 km]
[29 July 2022 (Evening)]

(A) 48 MJ (B) 24 MJ
(C) 36 MJ (D) 12 MJ

73. The weight of a body at the surface of earth is 18 N. The weight of the body at an altitude of 3200 km above the earth's surface is : (Given, radius of earth $R_e = 6400$ km)
[24 Jan. 2023 (Morning)]

(A) 9.8 N (B) 4.9 N
(C) 19.6 N (D) 8 N

74. Given below are two statements:
Statement-I: Acceleration due to earth's gravity decreases as you go 'up' or 'down' from earth's surface.
Statement-II: Acceleration due to earth's gravity is same at a height h and depth d from earth's surface, if $h = d$
In the light of above statements, choose the most appropriate answer form the options given below.
[24 Jan. 2023 (Evening)]

(A) Statement-I is incorrect but statement-II is correct
(B) Both Statement-I and Statement-II are incorrect
(C) Statement-I is correct but statement-II is incorrect
(D) Both Statement-I and II are correct

75. Given below are two statements:
One is labelled as **Assertion-A** and the other is labelled as **Reason-R**.
Assertion-A: A pendulum clock when taken to Mount Everest becomes fast.
Reason-R: The value of g (acceleration due to gravity) is less at Mount Everest than its value on the surface of earth.
In the light of the above statements, choose the most appropriate answer from the options given below :
[24 Jan. 2023 (Evening)]

(A) Both **A** and **R** are correct but **R** is NOT the correct explanation of **A**.
(B) Both **A** and **R** are correct and **R** is the correct explanation of **A**.
(C) **A** is not correct but **R** is correct.
(D) **A** is correct but **R** is not correct.

76. Assume that the earth is a solid sphere of uniform density and a tunnel is dug along its diameter throughout the earth. It is found that when a particle is released in this tunnel, it executes a simple harmonic motion. The mass of the particle is 100 g. The time period of the motion of the particle will be (approximately)
(Take $g = 10$ ms^{-2}, radius of earth = 6400 km)
[25 Jan. 2023 (Morning)]

(A) 24 hours (B) 1 hour 24 minutes
(C) 1 hour 40 minutes (D) 12 hours

77. A body of mass is taken from earth surface to the height h equal to twice the radius of earth (R_e), the increase in potential energy will be : (g = acceleration due to gravity on the surface of Earth) **[25 Jan. 2023 (Evening)]**

(A) $3\,mgR_e$ (B) $\frac{1}{3}mgR_e$
(C) $\frac{2}{3}mgR_e$ (D) $\frac{1}{2}mgR_e$

78. Two particles of equal mass m move in a circle of radius r under the action of their mutual gravitational attraction. The speed of each particle will be : **[29 Jan. 2023 (Morning)]**

(A) $\sqrt{\frac{GM}{2r}}$ (B) $\sqrt{\frac{4GM}{r}}$
(C) $\sqrt{\frac{GM}{r}}$ (D) $\sqrt{\frac{GM}{4r}}$

79. If the gravitational field in the space is given as $\left(-\frac{K}{r^2}\right)$. Taking the reference point to be at $r = 2$ cm with gravitational potential $V = 10$ J/kg. Find the gravitational potential at $r = 3$ cm in SI unit : (Given, that $K = 6$ J cm/kg)
[30 Jan. 2023 (Morning)]

(A) 9 (B) 11
(C) 12 (D) 10

80. An object is allowed to fall from a height R above the earth, where R is the radius of earth. Its velocity when it strikes the earth's surface, ignoring air resistance, will be :
[30 Jan. 2023 (Evening)]

(A) $2\sqrt{gR}$ (B) $\sqrt{gR}$
(C) $\sqrt{\frac{gR}{2}}$ (D) $\sqrt{2gR}$

81. At a certain depth d below surface of earth. Value of acceleration due to gravity becomes four times that of its value at a height $3R$ above earth surface. Where R is Radius of earth (Take $R = 6400$ km). The depth d is equal to :
[31 Jan. 2023 (Morning)]

(A) 5260 km (B) 640 km
(C) 2560 km (D) 4800 km

82. A body weight W, is projected vertically upwards from earth's surface to reach a height above the earth which is equal to nine times the radius of earth. The weight of the body at that height will be : **[31 Jan. 2023 (Evening)]**

(A) $\frac{W}{91}$ (B) $\frac{W}{100}$

(C) $\frac{W}{9}$ (D) $\frac{W}{3}$

83. Given below are two statements :

Statement-I : Acceleration due to gravity is different at different places on the surface of earth

Statement-II : Acceleration due to gravity increases as we go down below the earth's surface

In the light of the above statements, choose the correct answer from the options given below : **[1 Feb. 2023 (Morning)]**

(A) Both Statement-I and Statement-II are true

(B) Both Statement-I and Statement-II are false

(C) Statement-I is true but Statement-II is false

(D) Statement-I is false but Statement-II is true

84. A planet has double the mass of the earth. Its average density is equal to that of the earth. An object weighing W on earth will weigh on that planet : **[6 April 2023 (Morning)]**

(A) $2^{2/3} W$ (B) W

(C) $2^{1/3} W$ (D) $2W$

85. The weight of a body on the surface of the earth is 100 N. The gravitational force on it when taken at a height, from the surface of earth, equal to one-fourth the radius of the earth is: **[6 April 2023 (Evening)]**

(A) 100 N (B) 64 N

(C) 50 N (D) 25 N

86. The weight of a body on the earth is 400 N. Then weight of the body when taken to a depth half of the radius of the earth will be : **[8 April 2023 (Morning)]**

(A) Zero (B) 300 N

(C) 100 N (D) 200 N

87. Graphical variation of electric field due to a uniformly charged insulating solid sphere of radius R, with distance r from the centre O is represented by : **[8 April 2023 (Morning)]**

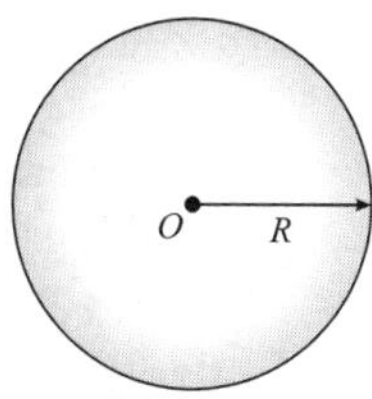

(A)

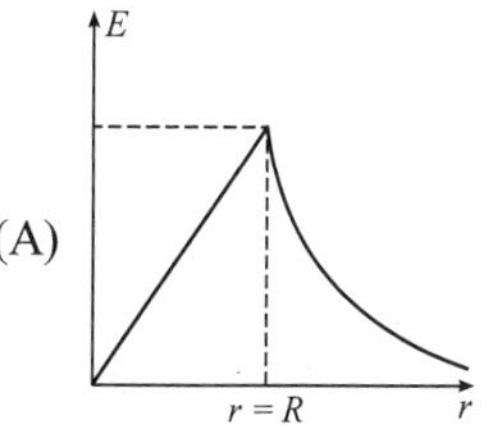

(B)

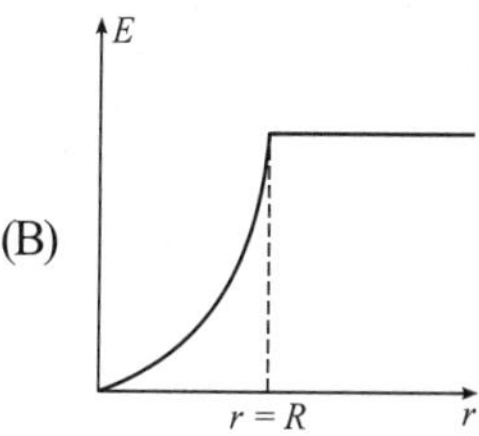

(C)

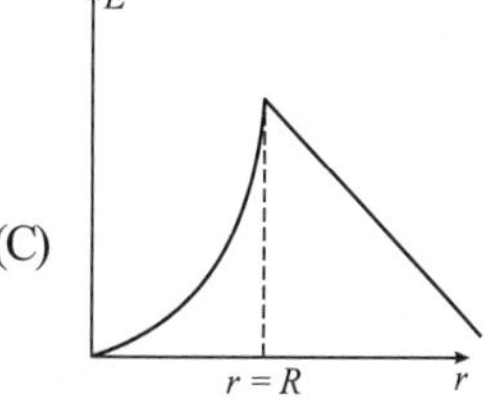

(D)

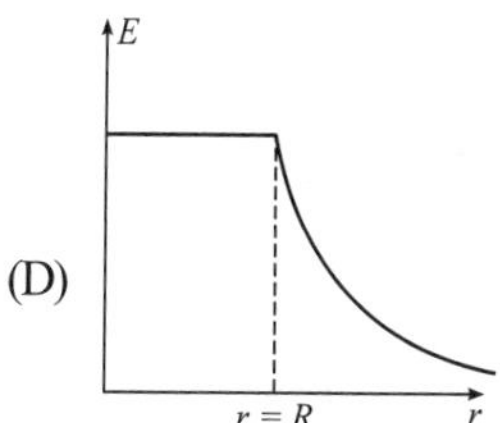

88. The acceleration due to gravity at height h above the earth if $h << R$ (radius of earth) is given by : **[8 April 2023 (Evening)]**

(A) $g' = g\left(1-\frac{2h}{R}\right)$ (B) $g' = g\left(1-\frac{2h^2}{R^2}\right)$

(C) $g' = g\left(1-\frac{h}{2R}\right)$ (D) $g' = g\left(1-\frac{h^2}{2R^2}\right)$

89. Assuming the earth to be a sphere of uniform mass density, the weight of a body at a depth $d = \frac{R}{2}$ from the surface of earth, if its weight on the surface of earth is 200 N, will be:

(Given R = Radius of earth) **[10 April 2023 (Morning)]**

(A) 400 N (B) 500 N

(C) 300 N (D) 100 N

90. If the earth suddenly shrinks to $\frac{1}{64}$ th of its original volume with its mass remaining the same, the period of rotation of earth becomes $\frac{24}{x} h$. The value of x is _____.

[10 April 2023 (Morning)]

91. Given below are two Statements:

Statement-I: Rotation of the earth shows effect on the value of acceleration due to gravity (g).

Statement-II: The effect of rotation of the earth on the value of g at the equator is minimum and that at the pole is maximum.

In the light of the above Statements, choose the correct answer from the options given below: **[10 April 2023 (Evening)]**

(A) Statement-I is false but Statement-II is true
(B) Statement-I is true but Statement-II are false
(C) Both Statement-I and Statement-II are true
(D) Both Statement-I and Statement-II are false

92. The radii of two planets A and B are R and $4R$ and their densities are ρ and $\rho/3$ respectively. The ratio of acceleration due to gravity at their surfaces ($g_A : g_B$) will be : **[11 April 2023 (Morning)]**

(A) 1:16 (B) 3:16
(C) 3:4 (D) 4:3

93. Two planets A and B of radii R and $1.5R$ have densities ρ and $\rho/2$ respectively. The ratio of acceleration due to gravity at the surface of B to A is: **[13 April 2023 (Evening)]**

(A) 2:3 (B) 2:1
(C) 3:4 (D) 4:3

94. Two identical particles each of mass m go round a circle of radius a under the action of their mutual gravitational attraction. The angular speed of each particle will be : **[15 April 2023 (Morning)]**

(A) $\sqrt{\frac{Gm}{2a^3}}$ (B) $\sqrt{\frac{Gm}{8a^3}}$
(C) $\sqrt{\frac{Gm}{4a^3}}$ (D) $\sqrt{\frac{Gm}{a^3}}$

95. A body is released from a height equal to the radius R of the earth. The velocity of the body when it strikes the surface of the earth will be : (Given : g = acceleration due to gravity on the earth) **[15 April 2023 (Morning)]**

(A) $\sqrt{gR}$ (B) $\sqrt{4gR}$
(C) $\sqrt{2gR}$ (D) $\sqrt{\frac{gR}{2}}$

* * * * *

7.2 SATELLITE MOTION

1. If suddenly the gravitational force of attraction between earth and a satellite revolving around it becomes zero, then the satellite will : **[2002]**

(A) Continue to move in its orbit with same velocity
(B) Move tangentially to the original orbit with same velocity
(C) Become stationary in its orbit
(D) Move towards the earth

2. Energy required to move a body of mass m from an orbit of radius $2R$ to $3R$ is : **[2002]**

(A) $GMm/12R$ (B) $GMm/3R$
(C) $GMm/8R$ (D) $GMm/6R$

3. The escape velocity of a body depends upon mass as : **[2002]**

(A) m^0 (B) m^1
(C) m^2 (D) m^3

4. The time period of a satellite of earth is 5 hours. If the separation between the centre of earth and the satellite is increased to 4 times the previous value, the new time period will become : **[2003]**

(A) 10 h (B) 80 h
(C) 40 h (D) 20 h

5. The escape velocity for a body projected vertically upwards from the surface of earth is 11 km/h. If the body is projected at an angle of 45° with the vertical, the escape velocity will be : **[2003]**

(A) $11\sqrt{2}$ km/s (B) 22 km/s
(C) 11 km/s (D) $\frac{11}{\sqrt{2}}$ km/s

6. Suppose the gravitational force varies inversely as the nth power of distance. Then the time period of a planet in circular orbit of radius R around the sun will be proportional to : **[2004]**

(A) $R^{\left(\frac{n+1}{2}\right)}$ (B) $R^{\left(\frac{n-1}{2}\right)}$
(C) R^n (D) $R^{\left(\frac{n-2}{2}\right)}$

7. The time period of an earth satellite in circular orbit is independent of : **[2004]**

(A) The mass of the satellite
(B) Radius of its orbit
(C) Both the mass and radius of the orbit
(D) Neither the mass of the satellite nor the radius of its orbit

8. A satellite of mass m revolves around the earth of radius R at a height x from its surface. If g is the acceleration due to gravity on the surface of the earth, the orbital speed of the satellite is : **[2004]**

(A) gx (B) $\frac{gR}{R-x}$
(C) $\frac{gR^2}{R+x}$ (D) $\left(\frac{gR^2}{R+x}\right)^{1/2}$

9. A planet in a distant solar system is 10 times more massive than the earth and its radius is 10 times smaller. Given that the escape velocity from the earth is 11 km s^{-1}, the escape velocity from the surface of the planet would be : **[2008]**

(A) 1.1 km s^{-1} (B) 11 km s^{-1}
(C) 110 km s^{-1} (D) 0.11 km s^{-1}

10. Two particles of equal mass 'm' go around a circle of radius R under the action of their mutual gravitational attraction. The speed of each particle with respect to their centre of mass is : **[2011]**

(A) $\sqrt{\frac{Gm}{R}}$ (B) $\sqrt{\frac{Gm}{4R}}$
(C) $\sqrt{\frac{Gm}{3R}}$ (D) $\sqrt{\frac{Gm}{2R}}$

11. India's Mangalyan was sent to the Mars by launching it into a transfer orbit EOM around the sun. It leaves the earth at E and meets Mars at M. If the semi-major axis of Earth's orbit is $a_e = 1.5 \times 10^{11}$ m, that of Mars orbit $a_m = 2.28 \times 10^{11}$ m, taken Kepler's laws give the estimate of time for Mangalyan to reach Mars from Earth to be close to : **[2014 (Online)]**

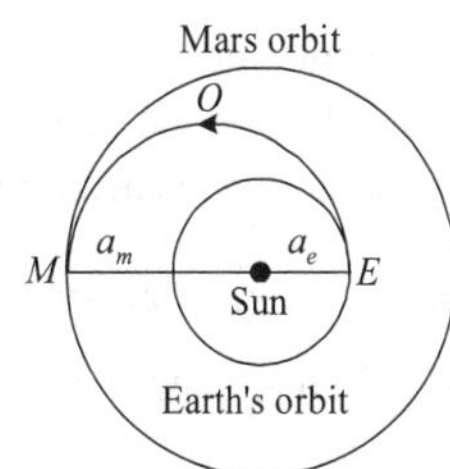

(A) 500 days (B) 320 days
(C) 260 days (D) 220 days

12. A very long (length L) cylindrical galaxy is made of uniformly distributed mass and has radius R ($R << L$). A star outside the galaxy is orbiting the galaxy in a plane perpendicular to the galaxy and passing through its centre. If the time period of star is T and its distance from the galaxy's axis is r, then : **[2015 (Online)]**

(A) $T \propto r$ (B) $T \propto \sqrt{r}$
(C) $T \propto r^2$ (D) $T^2 \propto r^3$

13. A satellite is revolving in a circular orbit at a height 'h' from the earth's surface (radius of earth R; $h << R$). The minimum increase in its orbital velocity required, so that the satellite could escape from the earth's gravitational field, is close to : (Neglect the effect of atmosphere.) **[2016]**

(A) $\sqrt{2gR}$ (B) $\sqrt{gR}$
(C) $\sqrt{gR/2}$ (D) $\sqrt{gR}(\sqrt{2}-1)$

14. Figure shows elliptical path $abcd$ of a planet around the sun S such that the area of triangle csa is 1/4 the area of the ellipse. (See figure) With db as the semimajor axis and ca as the semiminor axis. If t_1 is the time taken for planet to go over path abc and t_2 for path taken over cda then : **[2016 (Online)]**

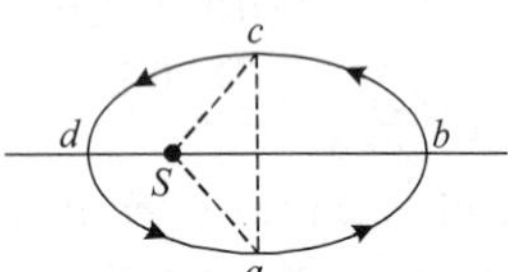

(A) $t_1 = 4t_2$ (B) $t_1 = 2t_2$
(C) $t_1 = 3t_2$ (D) $t_1 = t_2$

15. An astronaut of mass m is working on a satellite orbiting the earth at a distance h from the earth's surface. The radius of the earth is R, while its mass is M. The gravitational pull F_G on the astronaut is : **[2016 (Online)]**

(A) Zero since astronaut feels weightless

(B) $\frac{GMm}{(R+h)^2} < F_G < \frac{GMm}{R^2}$

(C) $F_G = \frac{GMm}{(R+h)^2}$ (D) $0 < F_G < \frac{GMm}{R^2}$

16. A body of mass m is moving in a circular orbit of radius R about a planet of mass M. At some instant, it splits into two equal masses. The first mass moves in a circular orbit of radius $\frac{R}{2}$, and the other mass, in a circular orbit of radius $\frac{3R}{2}$. The difference between the final and initial total energies is : **[2018 (Online)]**

(A) $-\frac{GMm}{2R}$ (B) $+\frac{GMm}{6R}$
(C) $-\frac{GMm}{6R}$ (D) $\frac{GMm}{2R}$

17. If the angular momentum of a planet of mass m, moving around the Sun in a circular orbit is L, about the center of the Sun, its areal velocity is : **[9 Jan. 2019 (Morning)]**

(A) $\frac{4L}{m}$ (B) $\frac{L}{m}$
(C) $\frac{L}{2m}$ (D) $\frac{2L}{m}$

18. The energy required to take a satellite to a height 'h' above Earth surface (radius of Earth = 6.4×10^3 km) is E_1 and kinetic energy required for the satellite to be in a circular orbit at this height is E_2. The value of h for which E_1 and E_2 are equal, is : **[9 Jan. 2019 (Evening)]**

(A) 1.28×10^4 km (B) 6.4×10^3 km
(C) 3.2×10^3 km (D) 1.6×10^3 km

19. A satellite is moving with a constant speed v in circular orbit around the earth. An object of mass 'm' is ejected from the satellite such that it just escapes from the gravitational pull of the earth. At the time of ejection, the kinetic energy of the object is : **[10 Jan. 2019 (Morning)]**

(A) $\frac{3}{2}mv^2$ (B) mv^2
(C) $2mv^2$ (D) $\frac{1}{2}mv^2$

20. Two stars of masses 3×10^{31} kg each, and at distance 2×10^{11} m rotate in a plane about their common centre of mass O. A meteorite passes through O moving perpendicular to the star's rotation plane. In order to escape from the gravitational field of this double star, the minimum speed that meteorite should have at O is :
(Take Gravitational constant $G = 6.67 \times 10^{-11}$ Nm^2 kg^{-2}) **[10 Jan. 2019 (Evening)]**

(A) 1.4×10^5 m/s (B) 24×10^4 m/s
(C) 3.8×10^4 m/s (D) 2.8×10^5 m/s

21. A satellite is revolving in a circular orbit at a height h from the earth surface, such that $h << R$ where R is the radius of the earth. Assuming that the effect of earth's atmosphere can be neglected then minimum increase in the speed required so that the satellite could escape from the gravitational field of earth is : **[11 Jan. 2019 (Morning)]**

(A) $\sqrt{gR}(\sqrt{2}-1)$ (B) $\sqrt{2gR}$
(C) $\sqrt{gR}$ (D) $\sqrt{\frac{gR}{2}}$

22. A satellite of mass M is in a circular orbit of radius R about the centre of the earth. A meteorite of the same mass, falling towards the earth, collides with the satellite completely inelastically. The speeds of the satellite and the meteorite are the same, just before the collision. The subsequent motion of the combined body will be : **[12 Jan. 2019 (Morning)]**

(A) In a circular orbit of a different radius
(B) In the same circular orbit of radius R
(C) In an elliptical orbit
(D) Such that it escapes to infinity

23. Two satellites, A and B, have masses m and $2m$ respectively. A is in a circular orbit of radius R, and B is in a circular orbit of radius $2R$ around the earth. The ratio of their kinetic energies, T_A/T_B, is : **[12 Jan. 2019 (Evening)]**

(A) 2 (B) $\sqrt{\frac{1}{2}}$
(C) 1 (D) $\frac{1}{2}$

24. A test particle is moving in a circular orbit in the gravitational field produced by a mass density $\rho(r) = \frac{K}{r^2}$. Identify the correct relation between the radius R of the particle's orbit and its period T: **[9 April 2019 (Evening)]**

(A) T/R^2 is a constant (B) TR is a constant
(C) T^2/R^3 is a constant (D) T/R is a constant

25. A spaceship orbits around a planet at a height of 20 km from its surface. Assuming that only gravitational field of the planet acts on the spaceship, what will be the number of complete revolutions made by the spaceship in 24 hours around the planet ?

[Given : Mass of planet $= 8 \times 10^{22}$ kg;
Radius of planet $= 2 \times 10^6$ m,
Gravitational constant $G = 6.67 \times 10^{-11}$ Nm^2/kg^2]
[10 April 2019 (Evening)]

(A) 9 (B) 11
(C) 13 (D) 17

26. The ratio of the weights of a body on the Earth's surface to that on the surface of a planet is 9 : 4. The mass of the planet is 1/9th of that of the Earth. If 'R' is the radius of the Earth, what is the radius of the planet ? (Take the planets to have the same mass density) : **[12 April 2019 (Evening)]**

(A) $\frac{R}{3}$ (B) $\frac{R}{2}$
(C) $\frac{R}{4}$ (D) $\frac{R}{9}$

27. A satellite of mass m is launched vertically upwards with an initial speed u from the surface of the earth. After it reaches height R (R = radius of the earth), it ejects a rocket of mass $m/10$ so that subsequently the satellite moves in a circular orbit. The kinetic energy of the rocket is (G is the gravitational constant; M is the mass of the earth) : **[7 Jan. 2020 (Morning)]**

(A) $\frac{m}{20}\left(u-\sqrt{\frac{2GM}{3R}}\right)^2$ (B) $5m\left(u^2-\frac{119}{200}\frac{GM}{R}\right)$

(C) $\frac{3m}{8}\left(u+\sqrt{\frac{5GM}{6R}}\right)^2$ (D) $\frac{m}{20}\left(u^2+\frac{113}{200}\frac{GM}{R}\right)$

28. An asteroid is moving directly towards the centre of the earth. When at a distance of $10R$ (R is the radius of the earth) from the earths centre, it has a speed of 12 km/s. Neglecting the effect of earths atmosphere, what will be the speed of the asteroid when it hits the surface of the earth (escape velocity from the earth is 11.2 km/s) ? Give your answer to the nearest integer in kilometer/s _____. **[8 Jan. 2020 (Evening)]**

29. A body A of mass m is moving in a circular orbit of radius R about a planet. Another body B of mass $\frac{m}{2}$ collides with A with a velocity which is half $\left(\frac{\vec{v}}{2}\right)$ the instantaneous velocity $\vec{v}$ of A. The collision is completely inelastic. Then, the combined body : **[9 Jan. 2020 (Morning)]**

(A) Starts moving in an elliptical orbit around the planet
(B) Continues to move in a circular orbit
(C) Falls vertically downwards towards the planet
(D) Escapes from the Planet's Gravitational field.

30. Planet A has mass M and radius R. Planet B has half the mass and half the radius of Planet A. If the escape velocities from the Planets A and B are v_A and v_B, respectively, then $\frac{v_A}{v_B} = \frac{n}{4}$. The value of n is : **[9 Jan. 2020 (Evening)]**

(A) 4 (B) 1
(C) 2 (D) 3

31. The mass density of a spherical galaxy varies as $\frac{K}{r}$ over a large distance 'r' from its centre. In that region, a small star is in a circular orbit of radius R. Then the period of revolution, T depends on R as : **[2 Sept. 2020 (Morning)]**

(A) $T \propto R$ (B) $T^2 \propto \frac{1}{R^3}$
(C) $T^2 \propto R$ (D) $T^2 \propto R^3$

32. A satellite is moving in a low nearly circular orbit around the earth. Its radius is roughly equal to that of the earth's radius R_e. By firing rockets attached to it, its speed is instantaneously increased in the direction of its motion so that is become $\sqrt{\frac{3}{2}}$ times larger. Due to this the farthest distance from the centre of the earth that the satellite reaches is R, value of R is : **[3 Sept. 2020 (Morning)]**

(A) $4R_e$ (B) $3R_e$
(C) $2R_e$ (D) $2.5R_e$

33. A body is moving in a low circular orbit about a planet of mass M and radius R. The radius of the orbit can be taken to be R itself. Then the ratio of the speed of this body in the orbit to the escape velocity from the planet is : **[4 Sept. 2020 (Evening)]**

(A) 1 (B) 2
(C) $\frac{1}{\sqrt{2}}$ (D) $\sqrt{2}$

34. A satellite is in an elliptical orbit around a planet P. It is observed that the velocity of the satellite when it is farthest from the planet is 6 times less than that when it is closest to the planet. The ratio of distances between the satellite and the planet at closest and farthest points is : **[6 Sept. 2020 (Morning)]**

(A) $1:6$ (B) $3:4$
(C) $1:3$ (D) $1:2$

35. Two planets have masses M and $16\,M$ and their radii are a and $2a$, respectively. The separation between the centres of the planets is $10a$. A body of mass m is fired from the surface of the larger planet towards the smaller planet along the line joining their centres. For the body to be able to reach at the surface of smaller planet, the minimum firing speed needed is : **[6 Sept. 2020 (Evening)]**

(A) $\sqrt{\frac{GM^2}{ma}}$ (B) $\frac{3}{2}\sqrt{\frac{5GM}{a}}$
(C) $4\sqrt{\frac{GM}{a}}$ (D) $2\sqrt{\frac{GM}{a}}$

36. Two stars of masses m and $2m$ at a distance d rotate about their common centre of mass in free space. The period of revolution is : **[24 Feb. 2021 (Morning)]**

(A) $\frac{1}{2\pi}\sqrt{\frac{3Gm}{d^3}}$ (B) $2\pi\sqrt{\frac{d^3}{3Gm}}$
(C) $2\pi\sqrt{\frac{3Gm}{d^3}}$ (D) $\frac{1}{2\pi}\sqrt{\frac{d^3}{3Gm}}$

37. Consider two satellites S_1 and S_2 with periods of revolution 1 hr and 8 hr respectively revolving around a planet in circular orbits. The ratio of angular velocity of satellite S_1 to the angular velocity of satellite S_2 is : **[24 Feb. 2021 (Morning)]**

(A) $2:1$ (B) $1:4$
(C) $1:8$ (D) $8:1$

38 Four identical particles of equal masses 1 kg made to move the circumference of a circle of radius 1 m under the action of their own mutual gravitational attraction. The speed of each particle will be : **[24 Feb. 2021 (Morning)]**

(A) $\sqrt{G(1+2\sqrt{2})}$ (B) $\sqrt{\frac{G}{2}(1+2\sqrt{2})}$
(C) $\sqrt{\frac{(1+2\sqrt{2})G}{2}}$ (D) $\sqrt{\frac{G}{2}(2\sqrt{2}-1)}$

39. Given below are two statements : one is labelled as **Assertion-A** and the other is labelled as **Reason-R**.

Assertion-A : The escape velocities of planet A and B are same. But A and B are of unequal mass.

Reason-R : The product of their mass and radius must be same. $M_1R_1 = M_2R_2$.

In the light of the above statements, choose the most appropriate answer from the options given below : **[25 Feb. 2021 (Morning)]**

(A) Both **A** and **R** are correct but **R** is NOT the correct explanation of **A**.
(B) Both **A** and **R** are correct and **R** is the correct explanation of **A**.
(C) **A** is correct but **R** is not correct.
(D) **A** is not correct but **R** is correct.

40. Two satellites A and B of masses 200 kg and 400 kg are revolving round the earth at height of 600 km and 1600 km respectively. If T_A and T_B are the time periods of A and B respectively then the value of $T_B - T_A$: **[25 Feb. 2021 (Morning)]**

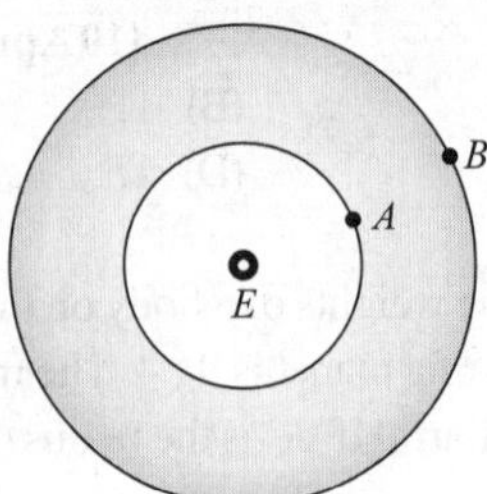

[Given : radius of earth = 6400 km, mass of earth = 6×10^{24} kg]

(A) 4.24×10^2 s (B) 3.33×10^2 s
(C) 1.33×10^3 s (D) 4.24×10^3 s

41. The initial velocity v_i required to project a body vertically upward from the surface of the earth to reach a height of $10R$, where R is the radius of the earth, may be described in terms of escape velocity v_e such that $v_i = \sqrt{\frac{x}{y}} \times v_e$. The value of x will be _____. **[25 Feb. 2021 (Evening)]**

42. A planet revolving in elliptical orbit has : **[26 Feb. 2021 (Morning)]**

(1) A constant velocity of revolution
(2) Has the least velocity when it is nearest to the sun
(3) Its areal velocity is directly proportional to its velocity
(4) Areal velocity is inversely proportional to its velocity
(5) To follow a trajectory such that the areal velocity is constant

Choose the correct answer from the options given below :

(A) 3 only (B) 5 only
(C) 4 only (D) 1 only

43. Assume that a tunnel is dug along a chord of the earth, at a perpendicular distance ($R/2$) from the earth's centre, where 'R' is the radius of the Earth. The wall of the tunnel is frictionless. If a particle is released in this tunnel, it will execute a simple harmonic motion with a time period :**[26 Feb. 2021 (Morning)]**

(A) $2\pi\sqrt{\frac{R}{g}}$ (B) $\frac{2\pi R}{g}$

(C) $\frac{g}{2\pi R}$ (D) $\frac{1}{2\pi}\sqrt{\frac{g}{R}}$

44. The radius in kilometer to which the present radius of earth ($R = 6400$ km) to be compressed so that the escape velocity is increased 10 time is _____. **[17 March 2021 (Morning)]**

45. A geostationary satellite is orbiting around an arbitrary planet 'P' at a height of 11 R above the surface of 'P', R being the radius of 'P'. The time period of another satellite in hours at a height of $2R$ from the surface of 'P' is _____. 'P' has the time period of 24 hours. **[17 March 2021 (Evening)]**

(A) $6\sqrt{2}$ (B) $\frac{6}{\sqrt{2}}$

(C) 3 (D) 5

46. The time period of a satellite in a circular orbit of radius R is T. The period of another satellite in a circular orbit of radius $9R$ is : **[18 March 2021 (Morning)]**

(A) 9 T (B) 27 T
(C) 12 T (D) 3 T

47. A satellite is launched into a circular orbit of radius R around earth, while a second satellite is launched into a circular orbit of radius $1.02R$. The percentage difference in the time periods of the two satellites is : **[20 July 2021 (Evening)]**

(A) 1.5 (B) 0.7
(C) 3.0 (D) 2.0

48. The masses and radii of the earth and moon are (M_1, R_1) and (M_2, R_2) respectively. Their centers are at a distance r apart. Find the minimum escape velocity for a particle of mass m to be projected from the middle of these two masses : **[31 Aug. 2021 (Morning)]**

(A) $V = \frac{1}{2}\sqrt{\frac{2GM(M_1+M_2)}{r}}$ (B) $V = \sqrt{\frac{4G(M_1+M_2)}{r}}$

(C) $V = \frac{\sqrt{2G}(M_1+M_2)}{r}$ (D) $V = \frac{1}{2}\sqrt{\frac{4GM(M_1+M_2)}{r}}$

49. Two satellites revolve around a planet in coplanar circular orbits in anticlockwise direction Their period of revolutions are 1 hour and 8 hours respectively. The radius of the orbit of nearer satellite is 2×10^3 km. The angular speed of the farther satellite as observed from the nearer satellite at the instant when both the satellites are closest is $\frac{\pi}{x}$ rad h^{-1} where x is _____. **[01 Sep. 2021 (Evening)]**

50. The distance between Sun and Earth is R. The duration of year if the distance between Sun and Earth becomes $3R$ will be : **[24 June 2022 (Evening)]**

(A) $\sqrt{3}$ years (B) 3 years

(C) 9 years (D) $3\sqrt{3}$ years

51. Two satellites S_1 and S_2 are revolving in circular orbits around a planet with radius $R_1 = 3200$ km and $R_2 = 800$ km respectively. The ratio of speed of satellite S_1 to the speed of satellite S_2 in their respective orbits would be $\frac{1}{x}$ where x =_____. **[25 June 2022 (Evening)]**

52. The distance of the Sun from earth is 1.5×10^{11} m and its angular diameter is (2000) s when observed from the earth. The diameter of the Sun will be : **[27 June 2022 (Evening)]**

(A) 2.45×10^{10} m (B) 1.45×10^{10} m
(C) 2.45×10^{9} m (D) 0.14×10^{9} m

53. Two planets A and B of equal mass are having their period of revolutions T_A and T_B such that $T_A = 2T_B$. These planets are revolving in the circular orbits of radii r_A and r_B respectively. Which out of the following would be the correct relationship of their orbits ? **[28 June 2022 (Morning)]**

(A) $2r_A^2 = r_B^2$ (B) $r_A^3 = 2r_B^3$

(C) $r_A^3 = 4r_B^3$ (D) $T_A^2 - T_B^2 = \frac{\pi^2}{GM}(r_B^3 - 4r_A^3)$

54. The escape velocity of a body on a planet A is 12 kms^{-1}. The escape velocity of the body on another planet B, whose density is four times and radius is half of the planet A, is :

[29 June 2022 (Morning)]

(A) 12 kms^{-1} (B) 24 kms^{-1}

(C) 36 kms^{-1} (D) 6 kms^{-1}

55. The time period of a satellite revolving around earth in a given orbit is 7 hours. If the radius of orbit is increased to three times its previous value, then approximate new time period of the satellite will be : **[29 June 2022 (Evening)]**

(A) 40 hours (B) 36 hours

(C) 30 hours (D) 25 hours

56. Two satellites A and B having masses in the ratio 4 : 3 are revolving in circular orbits of radii $3r$ and $4r$ respectively around the earth. The ratio of total mechanical energy of A to B is :

[27 July 2022 (Morning)]

(A) 9:16 (B) 16:9

(C) 1:1 (D) 4:3

57. A body of mass m is projected with velocity λv_e in vertically upward direction from the surface of the earth into space. It is given that v_e is escape velocity and $\lambda < 1$. If air resistance is considered to the negligible, then the maximum height from the centre of earth, to which the body can go, will be :

[R : Radius of earth] **[27 July 2022 (Evening)]**

(A) $\dfrac{R}{1+\lambda^2}$ (B) $\dfrac{R}{1-\lambda^2}$

(C) $\dfrac{R}{1-\lambda}$ (D) $\dfrac{\lambda^2 R}{1-\lambda^2}$

58. If the distance of the earth from Sun is 1.5×10^6 km. Then the distance of an imaginary planet from Sun, if its period of revolution is 2.83 years is : **[24 Jan. 2023 (Evening)]**

(A) 6×10^7 km (B) 6×10^6 km

(C) 3×10^6 km (D) 3×10^7 km

59. Every planet revolves around the sun in an elliptical orbit :

(a) The force acting on a planet is inversely proportional to square of distance from sun

(b) Force acting on planet is inversely proportional to product of the masses of the planet and the sun

(c) The centripetal force acting on the planet is directed away from the sun

(d) The square of time period of revolution of planet around sun is directly proportional to cube of semi-major axis of elliptical orbit

Choose the correct answer from the options given below :

[25 Jan. 2023 (Evening)]

(A) (a) and (d) only (B) (c) and (d) only

(C) (b) and (c) only (D) (a) and (c) only

60. The time period of a satellite of earth is 24 hours. If the separation between the earth and the satellite is decreased to one fourth of the previous value, then its new time period will become : **[29 Jan. 2023 (Evening)]**

(A) 4 hours (B) 6 hours

(C) 12 hours (D) 3 hours

61. If earth has a mass nine times and radius twice to the of a planet P. Then $\frac{v_e}{3}\sqrt{x}$ ms^{-1} will be the minimum velocity required by a rocket to pull out of gravitational force of P, where v_e is escape velocity on earth. The value of x is :

[1 Feb. 2023 (Morning)]

(A) 2 (B) 3

(C) 18 (D) 1

62. The escape velocities of two planets A and B are in the ratio 1 : 2. If the ratio of their radii respectively is 1 : 3, then the ratio of acceleration due to gravity of planet A to the acceleration of gravity of planet B will be : **[1 Feb. 2023 (Evening)]**

(A) $\dfrac{4}{3}$ (B) $\dfrac{3}{2}$

(C) $\dfrac{2}{3}$ (D) $\dfrac{3}{4}$

63. Given below are two statements: one is labelled as **Assertion-A** and the other is labelled as **Reason-R.**

Assertion-A: Earth has atmosphere whereas moon doesn't have any atmosphere.

Reason-R: The escape velocity on moon is very small as compared to that on earth.

In the light of the above statement, choose the correct answer from the options given below : **[6 April 2023 (Morning)]**

(A) **A** is true but **R** is false

(B) **A** is true but **R** is false

(C) Both **A** and **R** are correct but **R** is NOT the correct explanation of **A**

(D) Both **A** and **R** are correct but **R** is correct explanation of **A**

64. Choose the incorrect statement from the following:

[6 April 2023 (Evening)]

(A) The speed of satellite in a given circular orbit remains constant.

(B) For a planet revolving around the sun in an elliptical orbit, the total energy of the planet remains constant.

(C) When a body fall towards earth, the displacement of earth towards the body is negligible.

(D) The linear speed of a planet revolving around the sun remains constant.

65. Given below are two statements :

Statement-I : If E be the total energy of a satellite moving around the earth, then its potential energy will be $\frac{E}{2}$.

Statement-II : The kinetic energy of a satellite revolving in an orbit is equal to the half the magnitude of total energy E.

In the light of the above statements, choose the most appropriate answer from the options given below :

[8 April 2023 (Morning)]

(A) Both Statement-I and Statement-II are correct
(B) Both Statement-I and Statement-II are incorrect
(C) Statement-I is incorrect but Statement-II is correct
(D) Statement-I is correct but Statement-II is incorrect

66. The orbital angular momentum of a satellite is L, when it is revolving in a circular orbit at height h from earth surface. If the distance of satellite from the earth centre is increased by eight times to its initial value, then the new angular momentum will be : **[8 April 2023 (Evening)]**

(A) $8L$ (B) $4L$
(C) $9L$ (D) $3L$

67. Two satellites of masses m and $3m$ revolve around the earth in circular orbits of radii r and $3r$ respectively. The ratio of orbital speeds of the satellites respectively is :

[10 April 2023 (Morning)]

(A) $1:1$ (B) $3:1$
(C) $\sqrt{3}:1$ (D) $9:1$

68. The time period of a satellite, revolving above earth's surface at a height equal to R will be.
(Given $g = \pi^2$ m/s^2, R = radius of earth)

[10 April 2023 (Evening)]

(A) $\sqrt{4R}$ (B) $\sqrt{8R}$
(C) $\sqrt{32R}$ (D) $\sqrt{2R}$

69. A spaceship of mass 2×10^4 kg is launched into a circular orbit close to the earth surface. The additional velocity to be imparted to the spaceship in the orbit to overcome the gravitational pull will be : (If $g = 10$ m/s^2 and radius of earth = 6400 km) **[11 April 2023 (Evening)]**

(A) $11.2(\sqrt{2}-1)$ km/s (B) $7.9(\sqrt{2}-1)$ km/s
(C) $8(\sqrt{2}-1)$ km/s (D) $7.4(\sqrt{2}-1)$ km/s

70. Two satellites A and B move round the earth in the same orbit. The mass of A is twice the mass of B. The quantity which is same for the two satellites will be :

[12 April 2023 (Morning)]

(A) Potential Energy (B) Total energy
(C) Kinetic energy (D) Speed

71. The ratio of escape velocity of a planet to the escape velocity of earth will be :

Given: Mass of the planet is 16 times mass of earth and radius of the planet is 4 times the radius of earth.

[12 April 2023 (Morning)]

(A) $4:1$ (B) $2:1$
(C) $1:\sqrt{2}$ (D) $1:4$

72. A planet having mass $9M_e$ and radius $4R_e$, where M_e and R_e are mass and radius of earth respectively, has escape velocity in km/s given by :
(Given escape velocity on earth $V_e = 11.2 \times 10^3$ m/s)

[13 April 2023 (Morning)]

(A) 67.2 (B) 16.8
(C) 33.6 (D) 11.2

73. Given below are two statements:

Statement-I : For a planet, if the ratio of mass of the planet to its radius increases, the escape velocity from the planet also increases.

Statement-II : Escape velocity is independent of the radius of the planet.

In the light of above statements, choose the most appropriate answer from the options given below:

[13 April 2023 (Evening)]

(A) Both Statement-I and Statement-II are incorrect
(B) Statement-I is correct but statement-II is incorrect
(C) Statement-I is incorrect but statement-II is correct
(D) Both Statement-I and Statement-II are correct

* * * * *

CHAPTER 8

MECHANICAL PROPERTIES OF SOLIDS

8.1 ELASTICITY

1. A wire suspended vertically from one of its ends is stretched by attaching a weight of 200 N to the lower ends. The weight stretches the wire by 1 mm. Then the elastic energy stored in the wire is : **[2003]**

(A) 0.2 J (B) 10 J
(C) 20 J (D) 0.1 J

2. A wire fixed at the upper end stretches by length l by applying a force F. The work done in stretching is : **[2004]**

(A) $\frac{F}{2l}$ (B) Fl
(C) $2Fl$ (D) $\frac{Fl}{2}$

3. If S is stress and Y is Young's modulus of material of a wire, the energy stored in the wire per unit volume is : **[2005]**

(A) $2S^2Y$ (B) $\frac{S^2}{2Y}$
(C) $\frac{2Y}{S^2}$ (D) $\frac{S}{2Y}$

4. A wire elongates by l mm when a load W is hanged from it. If the wire goes over a pulley and two weights W each are hung at the two ends, the elongation of the wire will be (in *mm*) : **[2006]**

(A) l (B) $2l$
(C) Zero (D) $l/2$

5. Two wires are made of the same material and have the same volume. However wire 1 has cross-sectional area A and wire 2 has cross-sectional area $3A$. If the length of wire 1 increases by Δx on applying force F, how much force is needed to stretch wire 2 by the same amount ? **[2009]**

(A) $6F$ (B) $9F$
(C) F (D) $4F$

6. The load versus elongation graphs for four wires of same length and made of the same material are shown in the figure. The thinnest wire is represented by the line : **[2012 (Online)]**

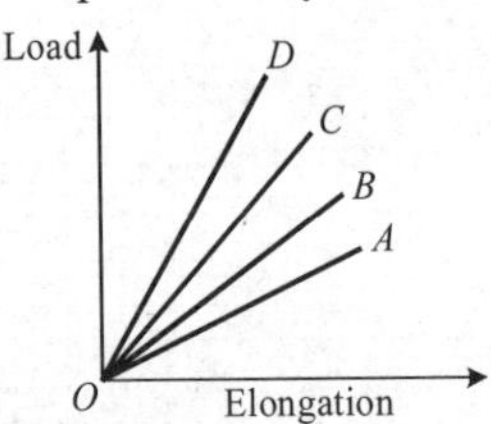

(A) OA (B) OC
(C) OD (D) OB

7. A structural steel rod has a radius of 10 mm and length of 1.0 m. A 100 kN for stretches it along its length. Young's modulus of structural steel is 2×10^{11} Nm^{-2}. The percentage strain is about : **[2012 (Online)]**

(A) 0.16% (B) 0.32%
(C) 0.08% (D) 0.24%

8. A steel wire can sustain 100 kg weight without breaking. If the wire is cut into two equal parts, each can sustain a weight of : **[2012 (Online)]**

(A) 50 kg (B) 400 kg
(C) 100 kg (D) 200 kg

9. Two blocks of masses m and M are connected by means of a metal wire of cross-sectional area A passing over a frictionless fixed pulley as shown in the figure. The system is then released. If M = 2m, then the stress produced in the wire is : **[2013 (Online)]**

(A) $\frac{2mg}{3A}$

(B) $\frac{4mg}{3A}$

(C) $\frac{mg}{A}$

(D) $\frac{3mg}{4A}$

10. If the ratio of lengths, radii and Young's moduli of steel and brass wires in the figure are a, b and c respectively, then the corresponding ratio of increase in their lengths is : **[2013 (Online)]**

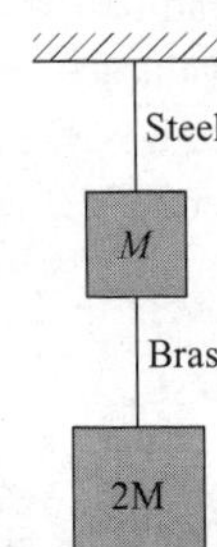

(A) $\frac{3c}{2ab^2}$

(B) $\frac{2a^2c}{b}$

(C) $\frac{3a}{2b^2c}$

(D) $\frac{2ac}{b^2}$

11. A uniform wire (Young's modulus 2×10^{11} Nm^{-2}) is subjected to longitudinal tensile stress of 5×10^7 Nm^{-2}. If the overall volume change in the wire is 0.02%, the fractional decrease in the radius of the wire is close to : **[2013 (Online)]**

(A) 1.0×10^{-4} (B) 1.5×10^{-4}
(C) 0.25×10^{-4} (D) 5×10^{-4}

12. A copper wire of length 1.0 m and a steel wire of length 0.5 m having equal cross-sectional areas are joined end to end. The composite wire is stretched by a certain load which stretches the copper wire by 1 mm. If Young's moduli of copper and steel are respectively 1.0×10^{11} Nm^{-2} and 2.0×10^{11} Nm^{-2}, the total extension of the composite wire is : **[2013 (Online)]**

(A) 1.75 mm (B) 2.0 mm
(C) 1.50 mm (D) 1.25 mm

13. In materials like aluminium and copper, the correct order of magnitude of various elastic moduli is : **[2014 (Online)]**

(A) Young's modulus < shear modulus < bulk modulus
(B) Bulk modulus < shear modulus < Young's modulus
(C) Shear modulus < Young's modulus < bulk modulus
(D) Bulk modulus < Young's modulus < shear modulus

14. The bulk moduli of ethanol, mercury and water are given as 0.9, 25 and 2.2 respectively in units of 10^9 Nm^{-2}. For a given value of pressure, the fractional compression in volume is $\frac{\Delta V}{V}$. Which of the following statements about $\frac{\Delta V}{V}$ for these three liquids is correct ? **[2014 (Online)]**

(A) Ethanol > Water > Mercury
(B) Water > Ethanol > Mercury
(C) Mercury > Ethanol > Water
(D) Ethanol > Mercury > Water

15. Steel ruptures when a shear of 3.5×10^8 Nm^{-2} is applied. The force needed to punch a 1 cm diameter hole in a steel sheet 0.3 cm thick is nearly : **[2014 (Online)]**

(A) 1.4×10^4 N (B) 2.7×10^4 N
(C) 3.3×10^4 N (D) 1.1×10^4 N

16. A uniform tapering conical wire is made from a material of Young's modulus Y and has a normal, unextended length L. The radii, at the upper and lower ends of this conical wire, have values R and $3R$, respectively. The upper end of the wire is fixed to a rigid support and a mass M is suspended from its lower end. The equilibrium extended length, of this wire, would equal: **[2016 (Online)]**

(A) $L\left(1+\frac{2}{9}\frac{Mg}{\pi YR^2}\right)$ (B) $L\left(1+\frac{1}{9}\frac{Mg}{\pi YR^2}\right)$
(C) $L\left(1+\frac{1}{3}\frac{Mg}{\pi YR^2}\right)$ (D) $L\left(1+\frac{2}{3}\frac{Mg}{\pi YR^2}\right)$

17. A bottle has an opening of radius a and length b. A cork of length b and radius $(a + \Delta a)$ where $(\Delta a << a)$ is compressed to fit into the opening completely (see figure). If the bulk modulus of cork is B and frictional coefficient between the bottle and cork is μ then the force needed to push the cork into the bottle is : **[2016 (Online)]**

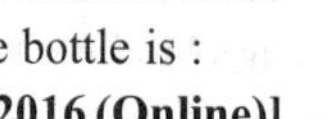

(A) $(\pi\mu Bb)a$ (B) $(2\pi\mu Bb)\Delta a$
(C) $(\mu\pi Bb)\Delta a$ (D) $(4\pi\mu Bb)\Delta a$

18. A load of mass M kg is suspended from a steel wire of length 2 m and radius 1.0 mm in Searle's apparatus experiment. The increase in length produced in the wire is 4.0 mm. Now the load is fully immersed in a liquid of relative density 2. The relative density of the material of load is 8. The new value of increase in length of the steel wire is : **[12 Jan. 2019 (Evening)]**

(A) 4.0 mm (B) 3.0 mm
(C) 5.0 mm (D) Zero

19. A boy's catapult is made of rubber cord which is 42 cm long, with 6 mm diameter of cross-section and of negligible mass. The boy keeps a stone weighing 0.02 kg on it and stretches the cord by 20 cm by applying a constant force. When released, the stone flies off with a velocity of 20 ms^{-1}. Neglect the change in the area of cross-section of the cord while stretched. The Young's modulus of rubber is closest to : **[8 April 2019 (Morning)]**

(A) 10^4 Nm^{-2} (B) 10^8 Nm^{-2}
(C) 10^6 Nm^{-2} (D) 10^3 Nm^{-2}

20. A steel wire having a radius of 2.0 mm, carrying a load of 4 kg, is hanging from a ceiling. Given that $g = 3.1\,\pi$ ms^{-2}, what will be the tensile stress that would be developed in the wire ? **[8 April 2019 (Morning)]**

(A) 4.8×10^6 Nm^{-2} (B) 5.2×10^6 Nm^{-2}
(C) 6.2×10^6 Nm^{-2} (D) 3.1×10^6 Nm^{-2}

21. Young's moduli of two wires A and B are in the ratio 7 : 4. Wire A is 2 m long and has radius R. Wire B is 1.5 m long and has radius 2 mm. If the two wires stretch by the same length for a given load, then the value of R is close to : **[8 April 2019 (Evening)]**

(A) 1.9 mm (B) 1.7 mm
(C) 1.5 mm (D) 1.3 mm

22. The elastic limit of brass is 379 MPa. What should be the minimum diameter of a brass rod if it is to support a 400 N load without exceeding its elastic limit ? **[10 April 2019 (Evening)]**

(A) 1.16 mm (B) 0.90 mm
(C) 1.36 mm (D) 1.00 mm

23. In an experiment, bras and steel wires of length 1 m each with areas of cross section 1 mm^2 are used. The wires are connected in series and one end of the combined wire is connected to a rigid support and other end is subjected to elongation. The stress required to produce a net elongation of 0.2 mm is:
(Given, the Young's Modulus for steel and brass are respectively, 120×10^9 N/m^2 and 60×10^9 N/m^2) **[10 April 2019 (Evening)]**

(A) 0.2×10^6 N/m^2 (B) 4.0×10^6 N/m^2
(C) 1.8×10^6 N/m^2 (D) 1.2×10^6 N/m^2

24. At 40°C, a brass wire of 1 mm radius is hung from the ceiling. A small mass, M is hung from the free end of the wire. When the wire is cooled down from 40°C to 20°C it regains its original length of 0.2 m. The value of M is close to : (Coefficient of linear expansion and Young's modulus of brass are 10^{-5}/°C and 10^{11} N/m^2, respectively; $g = 10$ ms^{-2}) **[12 April 2019 (Morning)]**

(A) 1.5 kg (B) 9 kg
(C) 0.9 kg (D) 0.5 kg

25. A uniform cylindrical rod of length L and radius r, is made from a material whose Young's modulus of Elasticity equals Y. When this rod is heated by temperature T and simultaneously subjected to a net longitudinal compressional force F, its length remains unchanged. The coefficient of volume expansion, of the material of the rod, is (nearly) equals to : **[12 April 2019 (Evening)]**

(A) $F/(3\pi r^2 YT)$ (B) $3F/(\pi r^2 YT)$
(C) $6F/(\pi r^2 YT)$ (D) $9F/(\pi r^2 YT)$

26. A body of mass $m = 10$ kg is attached to one end of a wire of length 0.3 m. The maximum angular speed (in rad s^{-1}) with which it can be rotated about its other end in space station is (Breaking stress of wire $= 4.8 \times 10^7$ Nm^{-2} and area of cross-section of the wire $= 10^{-2}$ cm^2) is : **[9 Jan. 2020 (Morning)]**

27. Two steel wires having same length are suspended from a ceiling under the same load. If the ratio of their energy stored per unit volume is 1 : 4, the ratio of their diameters is : **[9 Jan. 2020 (Evening)]**

(A) $1:\sqrt{2}$ (B) $1:2$
(C) $2:1$ (D) $\sqrt{2}:1$

28. A wire of density 9×10^3 kg cm^{-3} is stretched between two clamps 1 m apart. The resulting strain in the wire is 4.9×10^{-4}. The lowest frequency of the transverse vibrations in the wire is (Young's modulus of wire $Y = 9 \times 10^{10}$ Nm^{-2}), (to the nearest integer). **[2 Sept. 2020 (Evening)]**

29. A cube of metal is subjected to a hydrostatic pressure of $4GPa$. The percentage change in the length of the side of the cube is close to : (Given bulk modulus of metal, $B = 8 \times 10^{10}$ Pa) : **[4 Sept. 2020 (Evening)]**

(A) 0.6 (B) 1.67
(C) 5 (D) 20

30. An object of mass m is suspended at the end of a massless wire of length L and area of cross-section, A. Young modulus of the material of the wire is Y. If the mass is pulled down slightly its frequency of oscillation along the vertical direction is : **[6 Sept. 2020 (Morning)]**

(A) $f = \frac{1}{2\pi}\sqrt{\frac{YA}{mL}}$ (B) $f = \frac{1}{2\pi}\sqrt{\frac{YL}{mA}}$

(C) $f = \frac{1}{2\pi}\sqrt{\frac{mA}{YL}}$ (D) $f = \frac{1}{2\pi}\sqrt{\frac{mL}{YA}}$

31. If Y, K and η are the values of Young's modulus, bulk modulus and modulus of rigidity of any material respectively. Choose the correct relation for these parameters : **[24 Feb. 2021 (Morning)]**

(A) $K = \frac{Y\eta}{9\eta - 3Y}$ N/m^2 (B) $Y = \frac{9Y\eta}{9K - \eta}$ N/m^2

(C) $\eta = \frac{3YK}{9K + Y}$ N/m^2 (D) $Y = \frac{9K\eta}{2\eta + 3K}$ N/m^2

32. A uniform metallic wire is elongated by 0.04 m when subjected to a linear force F. The elongation, if its length and diameter is doubled and subjected to the same force will be _____ cm. **[24 Feb. 2021 (Evening)]**

33. The normal density of a material is ρ and its bulk modulus of elasticity is K. The magnitude of increase in density of material, when a pressure P is applied uniformly on all sides, will be : **[26 Feb. 2021 (Morning)]**

(A) $\frac{\rho P}{K}$ (B) $\frac{K}{\rho P}$

(C) $\frac{\rho K}{P}$ (D) $\frac{PK}{\rho}$

34. The length of metallic wire is l_1 when tension in it is T_1. It is l_2 when the tension is T_2. The original length of the wire will be : **[26 Feb. 2021 (Evening)]**

(A) $\frac{l_1 + l_2}{2}$ (B) $\frac{T_2 l_1 + T_1 l_2}{T_1 + T_2}$

(C) $\frac{T_1 l_1 - T_2 l_2}{T_2 - T_1}$ (D) $\frac{T_2 l_1 - T_1 l_2}{T_2 - T_1}$

35. In order to determine the Young's Modulus of a wire of radius 0.2 cm (measured using a scale of least count = 0.001 cm) and length 1m (measured using a scale of least count = 1 mm), a weight of mass 1 kg (measured using a scale of least count = 1 g) was hanged to get the elongation of 0.5 cm (measured using a scale of least count 0.001 cm). What will be the fractional error in the value of Young's Modulus determined by this experiment ?

[16 March 2021 (Evening)]

(A) 0.14% (B) 0.9%
(C) 9% (D) 1.4%

36. An object is located at 2 km beneath the surface of the water. If the fractional compression $\frac{\Delta V}{V}$ is 1.36% , the ratio of hydraulic stress to the corresponding hydraulic strain will be _____.

[17 March 2021 (Evening)]

[Given : density of water is 1000 kg m^{-3} and $g = 9.8$ ms^{-2}]

(A) 1.96×10^7 Nm^{-2} (B) 1.44×10^7 Nm^{-2}
(C) 2.26×10^9 Nm^{-2} (D) 1.44×10^9 Nm^{-2}

37. In Young's double slit arrangement, slits are separated by a gap of 0.5 mm, and the screen is placed at a distance of 0.5 m from them. The distance between the first and the third bright fringe formed when the slits are illuminated by a monochromatic light of 5890 Å is : **[18 March 2021 (Morning)]**

(A) 1178×10^{-9} m (B) 1178×10^{-6} m
(C) 1178×10^{-12} m (D) 5890×10^{-7} m

38. Two separate wires A and B are stretched by 2 mm and 4 mm respectively, when they are subjected to a force of 2 N. Assume that both the wires are made up of same material and the radius of wire B is 4 times that of the radius of wire A. The length of the wires A and B are in the ratio of $a : b$. Then a/b can be expressed as $1/x$ where x is _____ .

[18 March 2021 (Morning)]

39. For an adiabatic expansion of an ideal gas, the fractional change in its pressure is equal to (where γ is the ratio of specific heats) : **[18 March 2021 (Evening)]**

(A) $-\gamma\frac{dV}{V}$ (B) $-\gamma\frac{V}{dV}$
(C) $-\frac{1}{\gamma}\frac{dV}{V}$ (D) $\frac{dV}{V}$

40. The value of tension in a long thin metal wire has been changed from T_1 to T_2. The lengths of the metal wire at two different values of tension T_1 and T_2 are l_1 and l_2 respectively. The actual length of the metal wire is : **[20 July 2021 (Morning)]**

(A) $\frac{T_1 l_1 - T_2 l_2}{T_1 - T_2}$ (B) $\frac{l_1 + l_2}{2}$
(C) $\frac{T_1 l_2 + T_2 l_1}{T_1 - T_2}$ (D) $\sqrt{T_1 T_2 l_1 l_2}$

41. The length of a metal wire is l_1, when the tension in it is T_1 and is l_2 when the tension is T_2. The natural length of the wire is : **[20 July 2021 (Evening)]**

(A) $\sqrt{l_1 l_2}$ (B) $\frac{l_1 + l_2}{2}$
(C) $\frac{l_1 T_2 - l_2 T_1}{T_2 - T_1}$ (D) $\frac{l_1 T_2 + l_2 T_1}{T_2 + T_1}$

42. The area of cross-section of a railway track is 0.01 m^2. The temperature variation is 10°C. Coefficient of linear expansion of material of track 10^{-5}/°C. The energy stored per meter in the track is_____J/m. (Young's modulus of material of track is 10^{11} Nm^{-2}) **[22 July 2021 (Evening)]**

43. Two wires of same length and radius are joined end to end and loaded. The Young's modulii of the materials of the two wires are Y_1 and Y_2. The combination behaves as a single wire then its Young's modulus is : **[25 July 2021 (Morning)]**

(A) $Y = \frac{Y_1 Y_2}{Y_1 + Y_2}$ (B) $Y = \frac{Y_1 Y_2}{2(Y_1 + Y_2)}$
(C) $Y = \frac{2Y_1 Y_2}{3(Y_1 + Y_2)}$ (D) $Y = \frac{2Y_1 Y_2}{Y_1 + Y_2}$

44. In Young's double slit experiment, if the source of light changes from orange to blue then : **[27 July 2021 (Morning)]**

(A) The distance between consecutive fringes will decrease
(B) The distance between consecutive fringes will increase
(C) The intensity of the minima will increase
(D) The central bright fringe will become a dark fringe

45. A stone of mass 20 g is projected from a rubber catapult of length 0.1 m and area of cross section 10^{-6} m^2 stretched by an amount 0.04 m. The velocity of the projected stone is _____ m/s. (Young's modulus of rubber = 0.5×10^9 N/m^2)

[27 July 2021 (Morning)]

46. Wires W_1 and W_2 are made of same material having the breaking stress of 1.25×10^9 N/m^2. W_1 and W_2 have cross-sectional area of 8×10^{-7} m^2 and 4×10^{-7} m^2, respectively. Masses of 20 kg and 10 kg hang from them as shown in the figure. The maximum mass that can be placed in the pan without breaking the wires is_____kg. (Use $g = 10$ m/s^2)

[27 Aug. 2021 (Evening)]

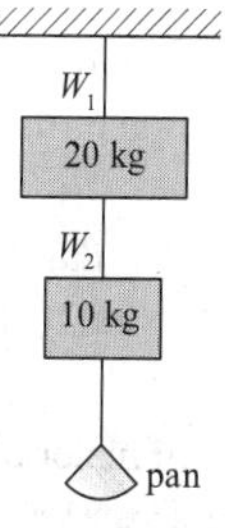

47. A uniform heavy rod of weight 10 kg ms^{-2}, cross-sectional area 100 cm^2 and length 20 cm is hanging from a fixed support.

Young modulus of the material of the rod is 2×10^{11} Nm^{-2}. Neglecting the lateral contraction, find the elongation of rod due to its own weight : **[31 Aug. 2021 (Morning)]**

(A) 4×10^{-8} m (B) 5×10^{-10} m
(C) 2×10^{-9} m (D) 5×10^{-8} m

48. When a rubber ball is taken to a depth of_____ m in deep sea, its volume decrease by 0.5%. (The bulk modulus of rubber $= 9.8 \times 10^8$ Nm^{-2}, Density of sea water $= 10^3$ kgm^{-3}, $g = 9.8$ m/s^2) **[31 Aug. 2021 (Morning)]**

49. Four identical hollow cylindrical columns of mild steel support a big structure of mass 50×10^3 kg. The inner and outer radii of each column are 50 cm and 100 cm respectively. Assuming uniform local distribution, calculate the compression strain of each column. (Use $Y = 2.0 \times 10^{11}$ Pa, $g = 9.8$ m/s^2) : **[31 Aug. 2021 (Evening)]**

(A) 1.87×10^{-3} (B) 60×10^{-8}
(C) 7.07×10^{-4} (D) 2.60×10^{-7}

50. In a Young's double slit experiment, the slits are separated by 0.3 mm and the screen is 1.5 m away from the plane of slits. Distance between fourth bright fringes on both sides of central bright fringe is 2.4 cm. The frequency of light used is _____ $\times 10^{14}$ Hz: **[31 Aug. 2021 (Evening)]**

51. A student determined Young's Modulus of elasticity using the formula $Y = \frac{MgL^3}{4bd^3\delta}$. The value of g is taken to be 9.8 m/s^2, without any significant error, his observation are as following.

Physical Quantity	Least count of the Equipment Used for measurement	Observed Value
Mass (M)	1 g	2 kg
Length of bar (L)	1 mm	1 m
Breadth of bar (b)	0.1 mm	4 cm
Thickness of bar (d)	0.01 mm	0.4 cm
Depression (δ)	0.01 mm	5 mm

Then the fractional error in the measurement of Y is : **[01 Sep. 2021 (Evening)]**

(A) 0.083 (B) 0.0155
(C) 0.155 (D) 0.0083

52. The bulk modulus of a liquid is 3×10^{10} Nm^{-2}. The pressure required to reduce the volume of liquid by 2% is : **[24 June 2022 (Morning)]**

(A) 3×10^8 Nm^{-2} (B) 9×10^8 Nm^{-2}
(C) 6×10^8 Nm^{-2} (D) 12×10^8 Nm^{-2}

53. The elastic behaviour of material for linear stress and linear strain, is shown in the figure. The energy density for a linear strain of 5×10^{-4} is_____ kJ/m^3. Assume that material is elastic upto the linear strain of 5×10^{-4}. **[26 June 2022 (Morning)]**

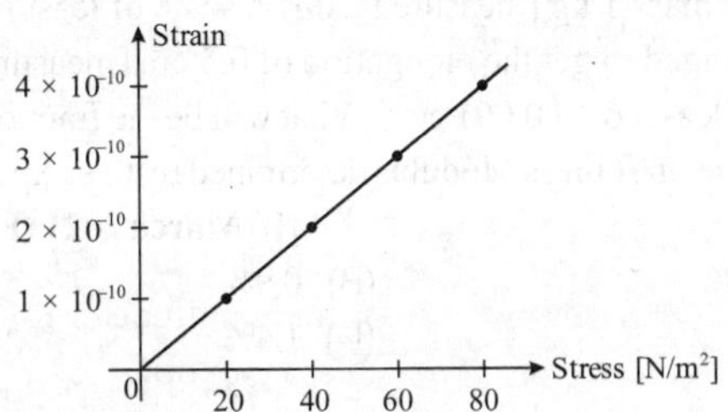

54. If ρ is the density and η is coefficient of viscosity of fluid which flows with a speed v in the pipe of diameter d, the correct formula for Reynolds number R_e is : **[26 June 2022 (Evening)]**

(A) $R_e = \frac{\eta d}{\rho v}$ (B) $R_e = \frac{\rho v}{\eta d}$

(C) $R_e = \frac{\rho v d}{\eta}$ (D) $R_e = \frac{\eta}{\rho v d}$

55. Given below are two statements : One is labelled as Assertion-**A** and the other is labelled as Reason-**R**.

Assertion-A: Product of Pressure (P) and time (t) has the same dimension as that of coefficient of viscosity.

Reason-R: Coefficient of viscosity $= \frac{\text{Force}}{\text{Velocity gradient}}$.

Choose the correct answer from the options given below : **[28 June 2022 (Morning)]**

(A) Both **A** and **R** true, and **R** is correct explanation of **A**.
(B) Both **A** and **R** are true but **R** is NOT the correct explanation of **A**.
(C) **A** is true but **R** is false.
(D) **A** is false but **R** is true.

56. A wire of length L is hanging from a fixed support. The length changes to L_1 and L_2 when masses 1 kg and 2 kg are suspended respectively from its free end. Then the value of L is equal to : **[29 June 2022 (Morning)]**

(A) $\sqrt{L_1 L_2}$ (B) $\frac{L_1 + L_2}{2}$
(C) $2L_1 - L_2$ (D) $3L_1 - 2L_2$

57. A wire of length L and radius r is clamped rigidly at one end. When the other end of the wire is pulled by a force F, its length increases by 5 cm. Another wire of the same material of length $4L$ and radius $4r$ is pulled by a force $4F$ under same conditions. The increase in length of this wire is _____ cm. **[25 July 2022 (Morning)]**

58. In an experiment of determine the Young's modulus of wire of a length exactly 1 m, the extension in the length of the wire is measured as 0.4 mm with an uncertainty of $\pm$ 0.02 mm when a load of 1 kg is applied. The diameter of the wire is measured as

0.4 mm with an uncertainty of $\pm$ 0.01 mm. The error in the measurement of Young's modulus (ΔY) is found to be $x \times 10^{10}$ Nm^{-2}. The value of x is _____. [Take $g = 10$ m/s^2]

[26 July 2022 (Morning)]

59. The area of cross section of the rope used to lift a load by a crane is 2.5×10^{-4} m^2. The maximum lifting capacity of the crane is 10 metric tons. To increase the lifting capacity of the crane to 25 metric tons, the required area of cross section of the rope should be : [Take g =10 ms^{-2}] **[26 July 2022 (Evening)]**

(A) 6.25×10^{-4} m^2 (B) 10×10^{-4} m^2

(C) 1×10^{-4} m^2 (D) 1.67×10^{-4} m^2

60. A uniform heavy rod of mass 20 kg, cross sectional area 0.4 m^2 and length 20 m is hanging from a fixed support. Neglecting the lateral contraction, the elongation in the rod due to its own weight is $x \times 10^{-9}$ m. The value of x is _____.

[Given : Young's modulus $Y = 2 \times 10^{11}$ Nm^{-2} and $g = 10$ ms^{-2}]

[26 July 2022 (Evening)]

61. A square aluminium (shear modulus is 25×10^9 Nm^{-2}) slab of side 60 cm and thickness 15 cm is subjected to a shearing force (on its narrow face) of 18.0×10^4 N. The lower edge is riveted to the floor. The displacement of the upper edge is _____ μm. **[27 July 2022 (Morning)]**

62. A steel wire of length 3.2 m ($Y_S = 2.0 \times 10^{11}$ Nm^{-2}) and a copper wire of length 4.4 m ($Y_C = 1.1 \times 10^{11}$ Nm^{-2}), both of radius 1.4 mm are connected end to end. When stretched by a load, the net elongation is found to be 1.4 mm. The load applied, in Newton, will be : [Given $\pi = \frac{22}{7}$] **[27 July 2022 (Evening)]**

(A) 360 (B) 180

(C) 1080 (D) 154

63. In an experiment to determine the Young's modulus, steel wires of five different lengths (1, 2, 3, 4 and 5 m) but of same cross section (2 mm^2) were taken and curves between extension and load were obtained. The slope (extension/load) of the curves were plotted with the wire length and the following graph is obtained. If the Young's modulus of given steel wires is $x \times 10^{11}$ Nm^{-2}, then the value of x is _____.

[27 July 2022 (Evening)]

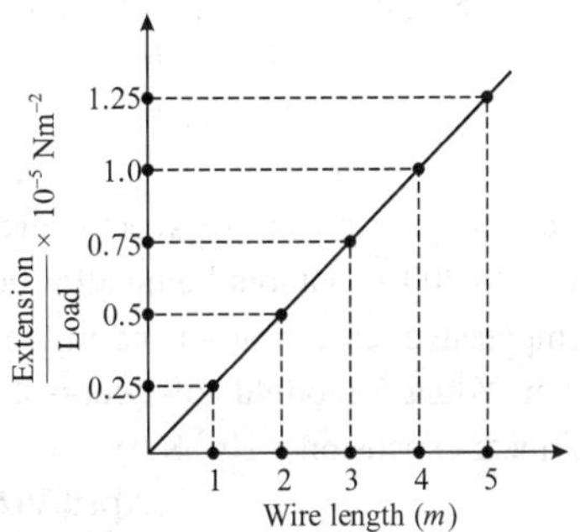

64. The force required to stretch a wire of cross-section 1 cm^2 to double its length will be :

[Given Yong's modulus of the wire = 2×10^{11} N/m^2]

[28 July 2022 (Morning)]

(A) 1×10^7 N (B) 1.5×10^7 N

(C) 2×10^7 N (D) 2.5×10^7 N

65. If the length of a wire is made double and radius is halved of its respective values. Then, the Young's modules of the material of the wire will : **[29 July 2022 (Morning)]**

(A) Remains same

(B) Become 8 times its initial value

(C) Become $\frac{1}{4}^{th}$ of its initial value

(D) Become 4 times its initial value

66. A metal wire of length 0.5 m and cross-sectional area 10^{-4} m^2 has breaking stress 5×10^8 Nm^{-2}. A block of 10 kg is attached at one end of the string and is rotating in a horizontal circle. The maximum linear velocity of block will be _____ ms^{-1}.

[29 July 2022 (Evening)]

67. A 100 m long wire having cross-sectional area 6.25×10^{-4} m^2 and Young's modulus is 10^{10} Nm^{-2} is subjected to a load of 250 N, then the elongation in the wire will be :

[24 Jan. 2023 (Morning)]

(A) 6.25×10^{-3} m (B) 4×10^{-4} m

(C) 6.25×10^{-6} m (D) 4×10^{-3} m

68. Given below are two statements:

One is labelled as **Assertion-A** and the other is labelled as **Reason-R**.

Assertion-A: Steel is used in the construction of buildings and bridges

Reason-R: Steel is more elastic and its elastic limit is high.

In the light of above statements, choose the most appropriate answer from the options given below:

[24 Jan. 2023 (Evening)]

(A) Both **A** and **R** are correct but **R** is NOT the correct explanation of **A**.

(B) **A** is not correct but **R** is correct.

(C) Both **A** and **R** are correct and **R** is the correct explanation of **A**.

(D) **A** is correct but **R** is not correct.

69. As shown in the figure, in an experiment to determine Young's modulus of a wire, the extension-load curve is plotted. The curve is a straight line passing through the origin and makes an angle of 45° with the load axis. The length of wire is 62.8 cm and its diameter is 4 mm. The Young's modulus is found to be $x \times 10^4$ Nm^{-2}. The value of x is _____.

[25 Jan. 2023 (Morning)]

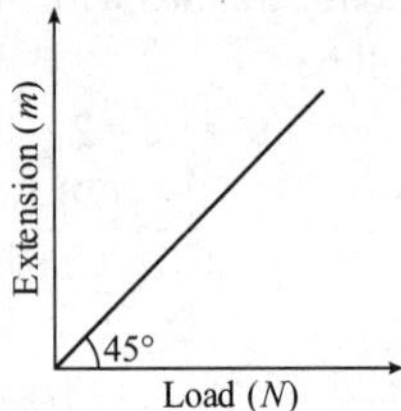

70. Choose the correct relationship between Poisson ratio (σ). Bulk modulus (K) and modulus of rigidity (η) of a given solid object : **[30 Jan. 2023 (Morning)]**

(A) $\sigma = \frac{3K - 2\eta}{6K + 2\eta}$ (B) $\sigma = \frac{6K + 2\eta}{3K - 2\eta}$

(C) $\sigma = \frac{3K + 2\eta}{6K + 2\eta}$ (D) $\sigma = \frac{6K - 2\eta}{3K - 2\eta}$

71. A force is applied to a steel wire A, rigidly clamped at one end. As a result elongation in the wire is 0.2 mm. If same force is applied to another steel wire B of double the length and a diameter 2.4 times that of the wire A, the elongation in the wire B will be : (Wires having uniform circular cross sections) **[30 Jan. 2023 (Evening)]**

(A) 6.06×10^{-2} mm (B) 2.77×10^{-2} mm
(C) 3.0×10^{-2} mm (D) 6.9×10^{-2} mm

72. A thin rod having a length of 1 m and area of cross-section 3×10^{-6} m^2 is suspended vertically from one end. The rod is cooled from 210°C to 160°C. After cooling, a mass M is attached at the lower end of the rod such that the length of rod again becomes 1 m. Young's modulus and coefficient of linear expansion of the rod are 2×10^{11} Nm^{-2} and 2×10^{-5} K^{-1}, respectively. The value of M is_____ kg. (Take $g = 10$ ms^{-2}) **[31 Jan. 2023 (Morning)]**

73. For a solid rod, the Young's modulus of elasticity is 3.2×10^{11} Nm^{-2} and density is 8×10^3 kg m^{-3}. The velocity of longitudinal wave in the rod will be : **[31 Jan. 2023 (Evening)]**

(A) 145.75×10^3 ms^{-1} (B) 3.65×10^3 ms^{-1}
(C) 18.96×10^3 ms^{-1} (D) 6.32×10^3 ms^{-1}

74. Under the same load, wire A having length 5.0 m and cross section 2.5×10^{-5} m^2 stretches uniformly by the same amount as another wire B of length 6.0 m and a cross section of 3.0×10^{-5} m^2 stretches. The ratio of the Young's modulus of wire A to that of wire B will be : **[31 Jan. 2023 (Evening)]**

(A) 1 : 4 (B) 1 : 1
(C) 1 : 10 (D) 1 : 2

75. $\left(P + \frac{a}{V^2}\right)(V - b) = RT$ represents the equation of state of some gases. Where P is the pressure, V is the volume, T is the temperature and a, b, R are the constants. The physical quantity, which has dimensional formula as that of $\frac{b^2}{a}$, will be : **[1 Feb. 2023 (Morning)]**

(A) Bulk modulus (B) Modulus of rigidity
(C) Compressibility (D) Energy density

76. A certain pressure P is applied to 1 litre of water and 2 litre of a liquid separately. Water gets compressed to 0.01% whereas the liquid gets compressed to 0.03%. The ratio of Bulk modulus of water to that of the liquid is $\frac{3}{x}$. The value of x is _____. **[1 Feb. 2023 (Morning)]**

77. The Young's modulus of a steel wire of length 6 m and cross-sectional area 3 mm^2, is 2×11^{11} N/m^2. The wire is suspended from its support on a given planet. A block of mass 4 kg is attached to the free end of the wire. The acceleration due to gravity on the planet is $\frac{1}{4}$ of its value on the earth. The elongation of wire is : (Take g on the earth = 10 m/s^2) **[1 Feb. 2023 (Evening)]**

(A) 1 cm (B) 1 mm
(C) 0.1 mm (D) 0.1 cm

78. A steel rod has a radius of 20 mm and a length of 2.0 m. A force of 62.8 kN stretches it along its length. Young's modulus of steel is 2.0×10^{11} N/m^2. The longitudinal strain produced in the wire is _____ $\times 10^{-5}$. **[6 April 2023 (Morning)]**

79. A metal block of mass m is suspended from a rigid support through a metal wire of diameter 14 mm. The tensile stress developed in the wire under equilibrium state is 7×10^5 Nm^{-2}. The value of mass m is _____ kg.
[Take, $g = 9.8$ ms^{-2} and $\pi = \frac{22}{7}$] **[6 April 2023 (Evening)]**

80. An aluminium rod with Young's modulus $Y = 7.0 \times 10^{10}$ N/m^2 undergoes elastic strain of 0.04%. The energy per unit volume stored in the rod in SI unit is : **[8 April 2023 (Morning)]**

(A) 5600 (B) 8400
(C) 2800 (D) 11200

81. A steel rod of length 1 m and cross sectional area 10^{-4} m^2 is heated from 0°C to 200°C without being allowed to extend or bend. The compressive tension produced in the rod is _____ $\times 10^4$ N. (Given Young's modulus of steel = 2×10^{11} Nm^{-2}, coefficient of linear expansion = 10^{-5} K^{-1}) **[8 April 2023 (Evening)]**

82. Two wires each of radius 0.2 cm and negligible mass, one made of steel and other made of brass are loaded as shown in the figure. The elongation of the steel wire is _____ $\times 10^{-6}$ m. [Young's modulus for steel = 2×10^{11} Nm^{-2} and $g = 10$ ms^{-2}].

[10 April 2023 (Morning)]

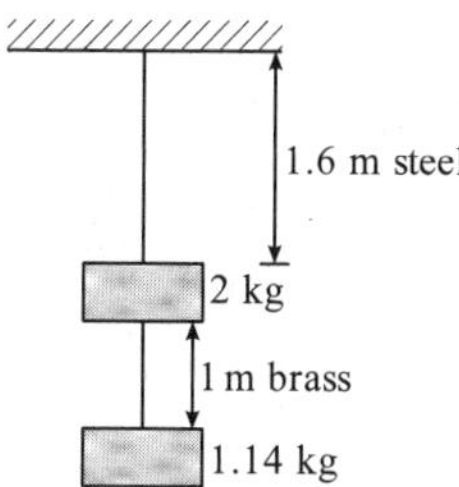

83. Young's moduli of the material of wires A and B are in the ratio of 1 : 4, while its area of cross sections are in the ratio of 1 : 3. If the same amount of load is applied to both the wires, the amount of elongation produced in the wires A and B will be in the ratio of: **[10 April 2023 (Evening)]**

(A) 36 : 1 (B) 12 : 1

(C) 1 : 36 (D) 1 : 12

84. The length of wire becomes l_1 and l_2 when 100 N and 120 N tensions are applied respectively. If $10\ l_2 = 11\ l_1$, the natural length of wire will be $\frac{1}{x} l_1$. Here the value of x is_____.

[11 April 2023 (Morning)]

85. Under isothermal condition, the pressure of a gas is given by $P = aV^{-3}$, where a is a constant and V is the volume of the gas. The bulk modulus at constant temperature is equal to :

[13 April 2023 (Morning)]

(A) $\frac{P}{2}$ (B) $3P$

(C) $2P$ (D) P

86. The elastic potential energy stored in a steel wire of length 20 m stretched through 2 cm is 80 J. The cross sectional area of the wire is _____ mm^2. (Given, $y = 2.0 \times 10^{11}$ Nm^{-2})

[13 April 2023 (Morning)]

87. A wire of length L and radius r is clamped rigidly at one end. When the other end of the wire is pulled by a force f, its length increases by l. Another wire of same material of length $2L$ and radius $2r$ is pulled by a force $2f$. Then the increase in its length will be : **[15 April 2023 (Morning)]**

(A) $2l$ (B) l

(C) $4l$ (D) $l/2$

* * * * *

CHAPTER 9

MECHANICAL PROPERTIES OF FLUIDS

9.1 FLUID STATICS

1. A jar is filled with two non-mixing liquids 1 and 2 having densities ρ_1 and ρ_2, respectively. A solid ball, made of a material of density ρ_3, is dropped in the jar. It comes to equilibrium in the position shown in the figure. Which of the following is true for ρ_1, ρ_2 and ρ_3 : **[2008]**

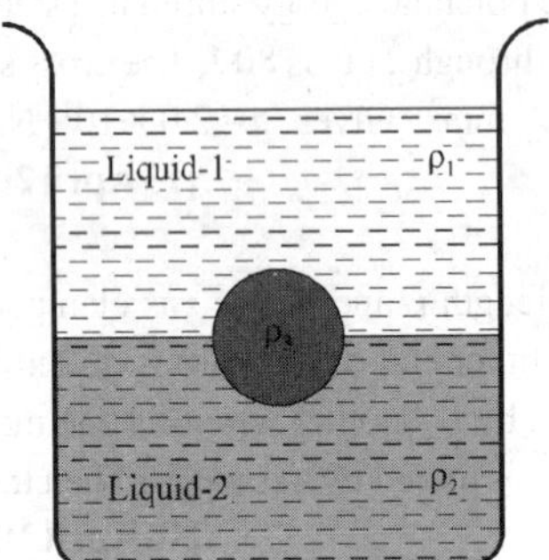

(A) $\rho_3 < \rho_1 < \rho_2$ (B) $\rho_3 < \rho_2 < \rho_1$
(C) $\rho_1 < \rho_2 < \rho_3$ (D) $\rho_1 < \rho_3 < \rho_2$

2. A ball is made of a material of density ρ where $\rho_{oil} < \rho < \rho_{water}$ with ρ_{oil} and ρ_{water} representing the densities of oil and water, respectively. The oil and water are immiscible. If the above ball is in equilibrium in a mixture of this oil and water, which of the following pictures represents its equilibrium position ? **[2010]**

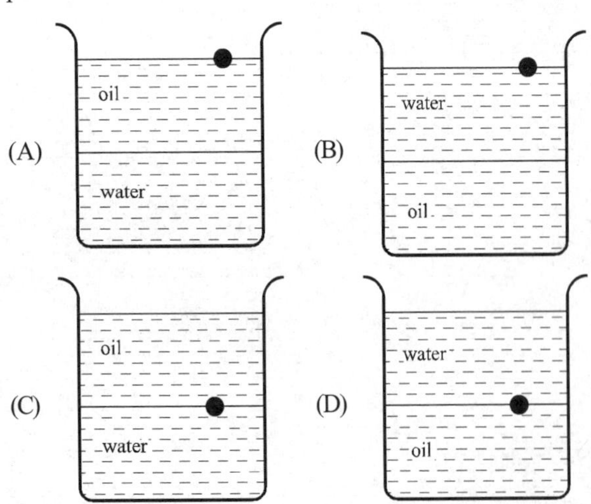

3. A uniform cylinder of length L and mass M having cross-sectional area A is suspended, with its length vertical, from a fixed point by a massless spring, such that it is half submerged in a liquid of density σ at equilibrium position. The extension x_0 of the spring when it is in equilibrium is : **[2013]**

(A) $\frac{Mg}{k}\left(1-\frac{LA\sigma}{M}\right)$ (B) $\frac{Mg}{k}\left(1-\frac{LA\sigma}{2M}\right)$

(C) $\frac{Mg}{k}\left(1+\frac{LA\sigma}{M}\right)$ (D) $\frac{Mg}{k}$

4. There is a circular tube in a vertical plane. Two liquids which do not mix and of densities d_1 and d_2 are filled in the tube. Each liquid subtends 90° angle at centre. Radius joining their interface makes an angle α with vertical. Ratio d_1/d_2 is : **[2014]**

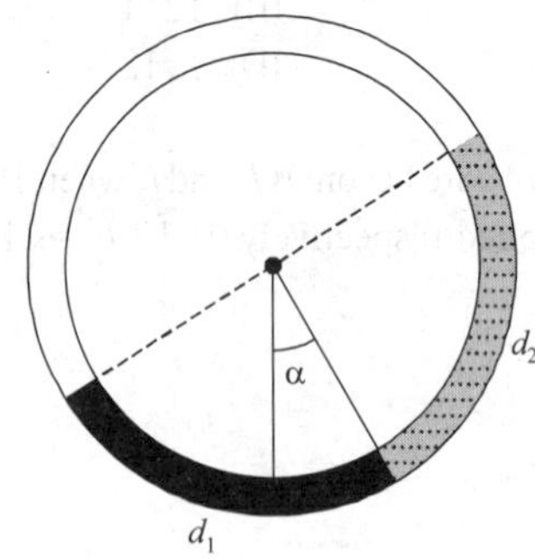

(A) $\frac{1+\sin\alpha}{1-\sin\alpha}$ (B) $\frac{1+\cos\alpha}{1-\cos\alpha}$

(C) $\frac{1+\tan\alpha}{1-\tan\alpha}$ (D) $\frac{1+\sin\alpha}{1-\cos\alpha}$

5. A wooden block floating in a bucket of water has 4/5 of its volume submerged. When certain amount of an oil is poured into the bucket, it is found that the block is just under the oil surface with half of its volume under water and half in oil. The density of oil relative to that of water is :
[9 April 2019 (Evening)]

(A) 0.5 (B) 0.7
(C) 0.6 (D) 0.8

6. A submarine experiences a pressure of 5.05×10^6 Pa at a depth of d_1 in a sea. When it goes further to a depth of d_2, it experiences a pressure of 8.08×10^6 Pa. Then $d_2 - d_1$ is approximately (density of water = 10^3 kg/m^3 and acceleration due to gravity = 10 ms^{-2}) : **[10 April 2019 (Evening)]**

(A) 500 m (B) 400 m
(C) 300 m (D) 600 m

7. A cubical block of side 0.5 m floats on water with 30% of its volume under water. What is the maximum weight that can be put on the block without fully submerging it under water? (Take density of water = 10^3 kg/m^3) **[10 April 2019 (Evening)]**

(A) 65.4 kg (B) 87.5 kg

(C) 30.1 kg (D) 46.3 kg

8. Consider a solid sphere of radius R and mass density $\rho(r) = \rho_0\left(1-\frac{r^2}{R^2}\right)$, $0 < r \le R$. The minimum density of a liquid in which it will float is : **[8 Jan. 2020 (Morning)]**

(A) $\frac{\rho_0}{5}$ (B) $\frac{\rho_0}{3}$

(C) $\frac{2\rho_0}{3}$ (D) $\frac{2\rho_0}{5}$

9. Two liquids of densities ρ_1 an $\rho_2(\rho_2 = 2\rho_1)$ are filled up behind a square wall of side 10 m as shown in figure. Each liquid has a height of 5 m. The ratio of the forces due to these liquids exerted on upper part MN to that at the lower part NO is: (Assume that the liquids are not mixing) **[8 Jan. 2020 (Evening)]**

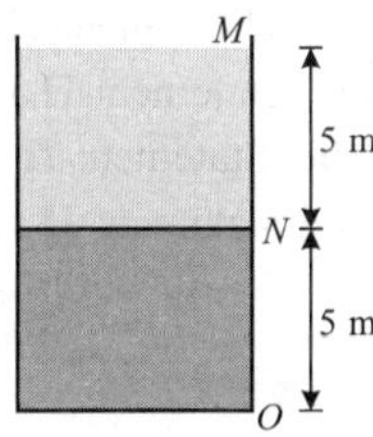

(A) 1/4 (B) 2/3

(C) 1/3 (D) 1/2

10. A air bubble of radius 1 cm in water has an upward acceleration 9.8 cm s^{-2}. The density of water is 1 gm cm^{-3} and water offers negligible drag force on the bubble. The mass of the bubble is : (g = 980 cm/s^2) **[4 Sept. 2020 (Morning)]**

(A) 3.15 gm (B) 4.51 gm

(C) 4.15 gm (D) 1.52 gm

11. Two identical cylindrical vessels are kept on the ground and each contain the same liquid of density d. The area of the base of both vessels is S but the height of liquid in one vessel is x_1 and in the other, x_2. When both cylinders are connected through a pipe of negligible volume very close to the bottom, the liquid flows from one vessel to the other until it comes to equilibrium at a new height. The change in energy of the system in the process is : **[4 Sept. 2020 (Evening)]**

(A) $gdS(x_2 + x_1)^2$ (B) $\frac{3}{4}gdS(x_2 - x_1)^2$

(C) $\frac{1}{4}gdS(x_2 - x_1)^2$ (D) $gdS(x_2^2 + x_1^2)$

12. A hollow spherical shell at outer radius R floats just submerged under the water surface. The inner radius of the shell is r. If the specific gravity of the shell material is $\frac{27}{8}$ w.r.t. water, the value of r is : **[5 Sept. 2020 Morning]**

(A) $\frac{4}{9}R$ (B) $\frac{8}{9}R$

(C) $\frac{1}{3}R$ (D) $\frac{2}{3}R$

13. A hydraulic press can lift 100 kg when a mass 'm' is placed on the smaller piston. It can lift _____ kg when the diameter of the larger piston is increased by 4 times and that of the smaller piston is decreased by 4 times keeping the same mass 'm' on the smaller piston. **[24 Feb. 2021 (Morning)]**

14. The pressure acting on a submarine is 3×10^5 Pa at a certain depth. If the depth is doubled, the percentage increase in the pressure acting on the submarine would be : (Assume that atmospheric pressure is 1×10^5 Pa density of water is 10^3 kg m^{-3}, g = 10 ms^{-2}) **[16 March 2021 (Morning)]**

(A) $\frac{200}{3}$% (B) $\frac{200}{5}$%

(C) $\frac{5}{200}$% (D) $\frac{3}{200}$%

15. A raindrop with radius R = 0.2 mm falls from a cloud at a height h = 2000 m above the ground. Assume that the drop is spherical throughout is fall and the force of buoyance may be neglected, then the terminal speed attained by the raindrop is : (Density of water f_w = 1000 kg m^{-3} and density of air f_a = 1.2 kg m^{-3}, g = 10 m/s^2, Coefficient of viscosity of air = 1.8×10^{-5} Nsm^{-2}) **[27 July 2021 (Evening)]**

(A) 4.94 ms^{-1} (B) 250.6 ms^{-1}

(C) 43.56 ms^{-1} (D) 14.4 ms^{-1}

16. Two narrow bores of diameter 5.0 mm and 8.0 mm are joined together to form a U-shaped tube open at both ends. If this U-tube contains water, what is the difference in level of two limbs of the tube? [Take surface tension of water $T = 7.3 \times 10^{-2}$ Nm^{-1}, angle of contact = 0, g = 10 ms^{-2} and density of water = 1.0×10^3 kg m^{-3}] **[26 Aug. 2021 (Morning)]**

(A) 4.97 mm (B) 2.19 mm

(C) 5.34 mm (D) 3.62 mm

17. In Millikan's oil drop experiment, what is viscous force acting on a uncharged drop of radius 2×10^{-5} m and density 1.2×10^3 kgm^{-3}? Take viscosity of liquid = 1.8×10^{-5} Nsm^{-2}: (Neglect buoyancy due to air) **[27 Aug. 2021 (Morning)]**

(A) 5.8×10^{-10} N (B) 1.8×10^{-10} N

(C) 3.8×10^{-11} N (D) 3.9×10^{-10} N

18. Two cylindrical vessels of equal cross-sectional area 16 cm^2 contain water upto heights 100 cm and 150 cm respectively. The vessels are interconnected so that the water levels in them become equal. The work done by the force of gravity during the process, is :
[Take density of water = 10^3 kg/m^3 and $g = 10$ ms^{-2}]
[27 July 2022 (Morning)]

(A) 0.25 J
(B) 1 J
(C) 8 J
(D) 12 J

19. A tube of length 50 cm is filled completely with an incompressible liquid of mass 250 g and closed at both ends. The tube is then rotated in horizontal plane about one of its ends with a uniform angular velocity $x\sqrt{F}$ rad s^{-1}. If F be the force exerted by the liquid at the other end then the value of x will be ______. **[29 July 2022 (Evening)]**

20. Given below are two statements: one is labelled as **Assertion-A** and the other is labelled as **Reason-R**.
Assertion-A: When you squeeze one end of a tube to get toothpaste out from the other end, Pascal's principle is observed.
Reason-R: A change in the pressure applied to an enclosed incompressible fluid is transmitted undiminished to every portion of the fluid and to the walls of its container.
In the light of the above statements, choose the most appropriate answer from the options given below :
[6 April 2023 (Evening)]

(A) **A** is not correct but **R** is correct.
(B) **A** is correct but **R** is not correct.
(C) Both **A** and **R** are correct and **R** is the correct explanation of **A**.
(D) Both **A** and **R** are correct but **R** is **NOT** the correct explanation of **A**.

21. A hydraulic automobile lift is designed to lift vehicles of mass 5000 kg. The area of cross section of the cylinder carrying the load is 250 cm^2. The maximum pressure the smaller piston would have to bear is : [Assume $g = 10$ m/s^2]
[8 April 2023 (Evening)]

(A) $200 \times 10^{+6}$ Pa
(B) $20 \times 10^{+6}$ Pa
(C) $2 \times 10^{+6}$ Pa
(D) $2 \times 10^{+5}$ Pa

22. Given below are two Statements:
Statement-I: Pressure in a reservoir of water is same at all points at the same level of water.
Statement-II: The pressure applied to enclosed water is transmitted in all directions equally.
In the light of the above Statements, choose the correct answer from the options given below: **[10 April 2023 (Morning)]**
(A) Statement-**I** is false but Statement-**II** is true
(B) Both Statement-**I** and Statement-**II** are true
(C) Statement-**I** is true but Statement-**II** is false
(D) Both Statement-**I** and Statement-**II** are false

* * * * *

CHAPTER 9

MECHANICAL PROPERTIES OF FLUIDS

9.2 FLUID DYNAMICS

1. A cylinder of height 20 m is completely filled with water. The velocity of efflux of water (in ms^{-1}) through a small hole on the side wall of the cylinder near its bottom, is : **[2002]**

(A) 10 (B) 20

(C) 25.5 (D) 5

2. Water is flowing continuously from a tap having an internal diameter 8×10^{-3} m. The water velocity as it leaves the tap is 0.4 ms^{-1}. The diameter of the water stream at a distance 2×10^{-1} m below the tap is close to : **[2011]**

(A) 9.6×10^{-3} m (B) 3.6×10^{-3} m

(C) 5.0×10^{-3} m (D) 7.5×10^{-3} m

3. A square hole of side length l is made at a depth of h and a circular hole of radius r is made at a depth of $4h$ from the surface of water in a water tank kept on a horizontal surface. If $l << h$, $r << h$ and the rate of water flow from the holes is the same, then r is equal to : **[2012]**

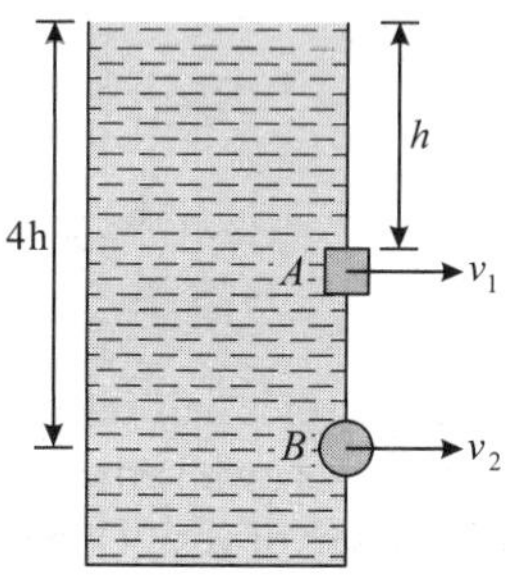

(A) $\frac{l}{\sqrt{2\pi}}$ (B) $\frac{l}{\sqrt{3\pi}}$

(C) $\frac{l}{3\pi}$ (D) $\frac{l}{2\pi}$

4. Water is flowing through a horizontal tube having cross-sectional area of its two ends being A and A' such that the ratio A/A' is 5. If the pressure difference of water between the two ends is 3×10^5 Nm^{-2}, the velocity of water with which it enters the tube will be (neglect gravity effects) : **[2012 (Online)]**

(A) 5 ms^{-1} (B) 10 ms^{-1}

(C) 25 ms^{-1} (D) $50\sqrt{10}$ ms^{-1}

5. In a cylindrical water tank, there are two small holes A and B on the wall at a depth of h_1, from the surface of water and at a height of h_2 from the bottom of water tank. Surface of water is at height H from the bottom of water tank. Water coming out from both holes strikes the ground at the same point S. Find the ratio of h_1 and h_2 : **[2012 (Online)]**

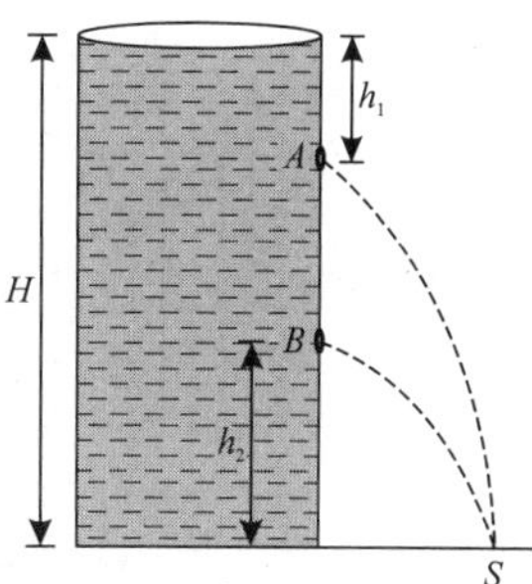

(A) Depends on H (B) 1 : 1

(C) 2 : 2 (D) 1 : 2

6. Air of density 1.2 kg m^{-3} is blowing across the horizontal wings of an aeroplane in such a way that its speeds above and below the wings are 150 ms^{-1} and 100 ms^{-1}, respectively. The pressure difference between the upper and lower sides of the wings, is : **[2013 (Online)]**

(A) 60 Nm^{-2} (B) 180 Nm^{-2}

(C) 7500 Nm^{-2} (D) 12500 Nm^{-2}

7. Water is flowing at a speed of 1.5 ms^{-1} through horizontal tube of cross-sectional area 10^{-2} m^2 and you are trying to stop the flow by your palm. Assuming that the water stops immediately after hitting the palm, the minimum force that you must exert should be : (density of water = 10^3 kgm^{-3}) **[2014 (Online)]**

(A) 15 N (B) 22.5 N

(C) 33.7 N (D) 45 N

8. A cylindrical vessel of cross-section A contains water to a height h. There is a hole in the bottom of radius 'a'. The time in which it will be emptied is : **[2014 (Online)]**

(A) $\frac{2A}{\pi a^2}\sqrt{\frac{h}{g}}$ (B) $\frac{\sqrt{2}A}{\pi a^2}\sqrt{\frac{h}{g}}$

(C) $\frac{2\sqrt{2}A}{\pi a^2}\sqrt{\frac{h}{g}}$ (D) $\frac{A}{\sqrt{2}\pi a^2}\sqrt{\frac{h}{g}}$

9. A tank with a small hole at the bottom has been filled with water and kerosene (specific gravity 0.8). The height of water is 3 m and that of kerosene 2 m. When the hole is opened the velocity of fluid coming out from it is nearly : (take $g = 10\ \text{ms}^{-2}$ and density of water $= 10^3\ \text{kg m}^{-3}$) : **[2014 (Online)]**

(A) $10.7\ \text{ms}^{-1}$ (B) $9.6\ \text{ms}^{-1}$
(C) $8.5\ \text{ms}^{-1}$ (D) $7.6\ \text{ms}^{-1}$

10. In the diagram shown, the difference in the two tubes of the manometer is 5 cm, the cross section of the tube at *A* and *B* is 6 mm^2 and 10 mm^2 respectively. The rate at which water flows through the tube is ($g = 10\ \text{ms}^{-2}$) : **[2014 (Online)]**

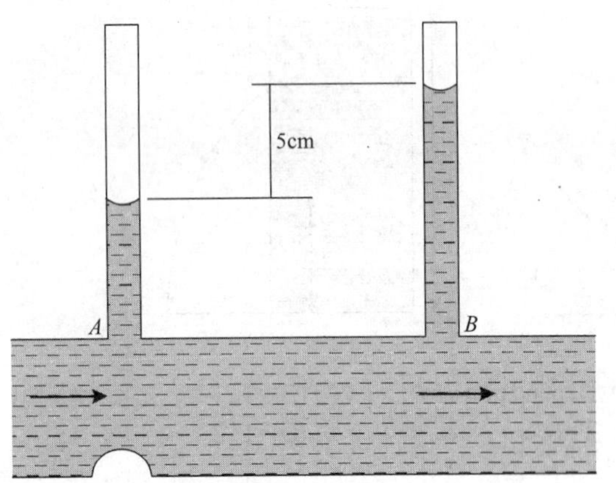

(A) 7.5 cc/s (B) 8.0 cc/s
(C) 10.0 cc/s (D) 12.5 cc/s

11. Consider a water jar of radius *R* that has water filled up to height *H* and is kept on a stand of height *h* (see figure). Through a hole of radius r ($r << R$) at its bottom, the water leaks out and the stream of water coming down towards the ground has a shape like a funnel as shown in the figure. If the radius of the cross-section of water stream when it hits the ground is x. Then: **[2016 (Online)]**

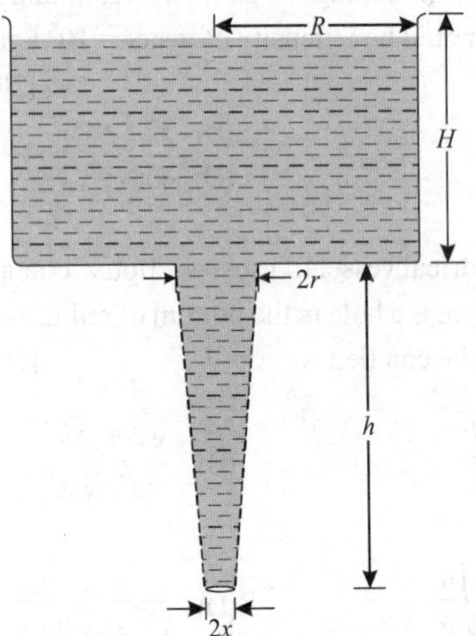

(A) $x = r\left(\dfrac{H}{H+h}\right)^{\frac{1}{4}}$ (B) $x = r\left(\dfrac{H}{H+h}\right)$

(C) $x = r\left(\dfrac{H}{H+h}\right)^{2}$ (D) $x = r\left(\dfrac{H}{H+h}\right)^{\frac{1}{2}}$

12. The top of a water tank is open to air and its water level is maintained. It is giving out 0.74 m^3 water per minute through a circular opening of 2 cm radius in its wall. The depth of the centre of the opening from the level of water in the tank is close to : **[9 Jan. 2019 (Evening)]**

(A) 9.6 m (B) 4.8 m
(C) 2.9 m (D) 6.0 m

13. Water flows into a large tank with flat bottom at the rate of $10^{-4}\ \text{m}^3\text{s}^{-1}$. Water is also leaking out of a hole of area 1 cm^2 at its bottom. If the height of the water in the tank remains steady, then this height is : **[10 Jan. 2019 (Morning)]**

(A) 4 cm (B) 2.9 cm
(C) 1.7 cm (D) 5.1 cm

14. A liquid of density ρ is coming out of a hose pipe of radius a with horizontal speed v and hits a mesh. 50% of the liquid passes through the mesh unaffected. 25% looses all of its momentum and 25% comes back with the same speed. The resultant pressure on the mesh will be : **[11 Jan. 2019 (Morning)]**

(A) ρv^2 (B) $\dfrac{3}{4}\rho v^2$

(C) $\dfrac{1}{2}\rho v^2$ (D) $\dfrac{1}{4}\rho v^2$

15. A long cylindrical vessel is half filled with a liquid. When the vessel is rotated about its own vertical axis, the liquid rises up near the wall. If the radius of vessel is 5 cm and its rotational speed is 2 rotations per second, then the difference in the heights between the centre and the sides, in cm, will be : **[12 Jan. 2019 (Evening)]**

(A) 1.2 (B) 0.1
(C) 2.0 (D) 0.4

16. Water from a tap emerges vertically downwards with an initial speed of $1.0\ \text{ms}^{-1}$. The cross-sectional area of the tap is $10^{-4}\ \text{m}^2$. Assume that the pressure is constant throughout the stream of water and that the flow is streamlined. The cross-sectional area of the stream, 0.15 m below the tap would be : (Take $g = 10\ \text{ms}^{-2}$) **[10 April 2019 (Evening)]**

(A) $1 \times 10^{-5}\ \text{m}^2$ (B) $5 \times 10^{-5}\ \text{m}^2$
(C) $2 \times 10^{-5}\ \text{m}^2$ (D) $5 \times 10^{-4}\ \text{m}^2$

17. An ideal fluid flows (laminar flow) through a pipe of non-uniform diameter. The maximum and minimum diameters of the pipes are 6.4 cm and 4.8 cm, respectively. The ratio of the minimum and the maximum velocities of fluid in this pipe is : **[7 Jan. 2020 (Evening)]**

(A) $\frac{\sqrt{3}}{2}$ (B) $\frac{3}{4}$

(C) $\frac{81}{256}$ (D) $\frac{9}{16}$

18. Water flows in a horizontal tube (see figure). The pressure of water changes by 700 Nm^{-2} between A and B where the area of cross section are 40 cm^2 and 20 cm^2, respectively. Find the rate of flow of water through the tube. (Density of water = 1000 kgm^{-3}) **[9 Jan. 2020 (Morning)]**

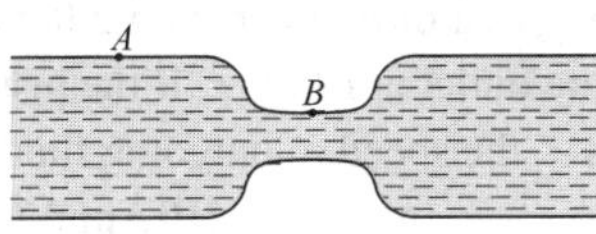

(A) 1810 cm^3/s (B) 3020 cm^3/s
(C) 2720 cm^3/s (D) 2420 cm^3/s

19. A fluid is flowing through a horizontal pipe of varying cross-section, with speed v ms^{-1} at a point where the pressure is P Pascal. At another point where pressure is $P/2$ Pascal its speed is V ms^{-1}. If the density of the fluid is ρ kgm^{-3} and the flow is streamline, then V is equal to : **[6 Sept. 2020 (Evening)]**

(A) $\sqrt{\frac{P}{2\rho}+v^2}$ (B) $\sqrt{\frac{P}{\rho}+v^2}$

(C) $\sqrt{\frac{2P}{\rho}+v^2}$ (D) $\sqrt{\frac{P}{\rho}+v}$

20. Consider a water tank as shown in the figure. It's cross-sectional area is 0.4 m^2. The tank has an opening B near the bottom whose cross-section area is 1 cm^2. A load of 24 kg is applied on the water at the top when the height of the water level is 40 cm above the bottom, the velocity of water coming out the opening B is v ms^{-1}. The value of v, to the nearest integer, is_____. [Take value of g to be 10 ms^{-2}] **[18 March 2021 (Evening)]**

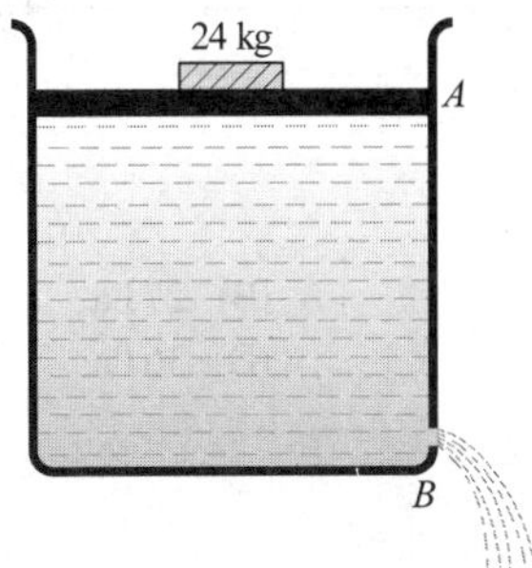

21. A light cylindrical vessel is kept on a horizontal surface. Area of base is A. A hole of cross sectional area 'a' is made just at the bottom side. The minimum coefficient of friction necessary to prevent sliding the vessel due to the impact force of the emerging liquid is ($a << A$) : **[27 July 2021 (Morning)]**

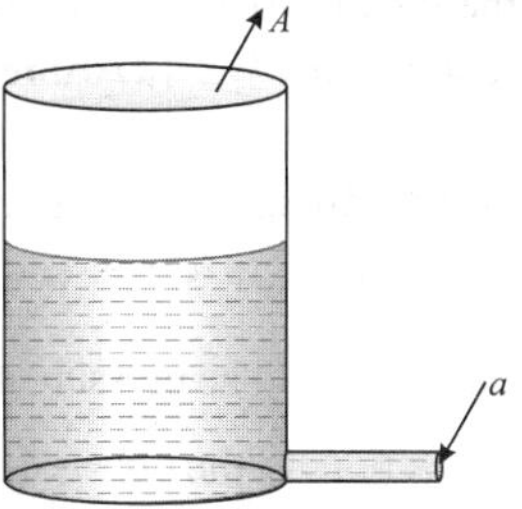

(A) $\frac{A}{2a}$ (B) $\frac{a}{A}$

(C) $\frac{2a}{A}$ (D) None of these

22. The water is filled upto height of 12 m in a tank having vertical side walls. A hole is made in one of the walls at a depth 'h' below the water level. The value of 'h' for which the emerging stream of water strikes the ground at the maximum range is _____ m. **[27 July 2021 (Evening)]**

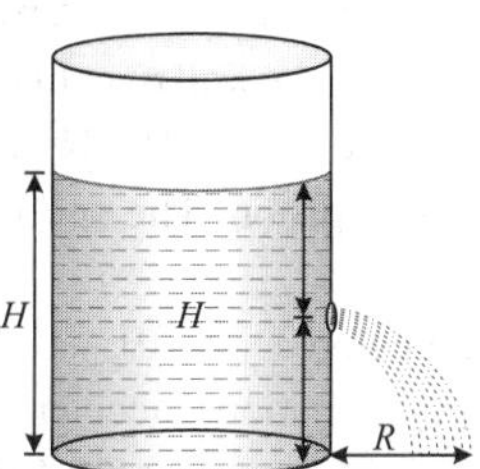

23. An ideal fluid of density 800 kgm^{-3}, flows smoothly through a bent pipe (as shown in figure) that tapers in cross-sectional area from a to $\frac{a}{2}$. The pressure difference between the wide and narrow sections of pipe is 4100 Pa. At wider section, the velocity of fluid is $\frac{\sqrt{x}}{6}$ ms^{-1} for x = _____.
[Given : $g = 10\ ms^{-2}$] **[26 June 2022 (Morning)]**

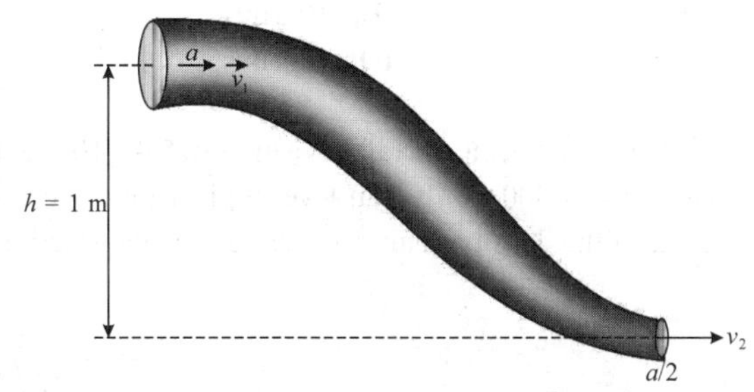

24. The area of cross-section of a large tank is 0.5 m^2. It has a narrow opening near the bottom having area of cross-section 1 cm^2. A load of 25 kg is applied on the water at the top in the tank. Neglecting the speed of water in the tank, the velocity of the water, coming out of the opening at the time when the height of water level in the tank is 40 cm above the bottom, will be _____ cms^{-1}. [Take $g = 10\ ms^{-2}$] **[27 June 2022 (Morning)]**

25. A liquid of density 750 kg m^{-3} flows smoothly through a horizontal pipe that tapers in cross-sectional area from $A_1 = 1.2 \times 10^{-2}\ m^2$ to $A_2 = \frac{A_1}{2}$. The pressure difference between the wide and narrow sections of the pipe is 4500 Pa. The rate of flow of liquid is _____ $\times 10^{-3}\ m^3\ s^{-1}$.

[28 June 2022 (Evening)]

26. A pressure-pump has a horizontal tube of cross-sectional area 10 cm^2 for the outflow of water at a speed of 20 m/s. The force exerted on the vertical wall just in front of the tube which stops water horizontally flowing out of the tube, is :
[Given : density of water = 1000 kg/m^3] **[28 July 2022 (Evening)]**

(A) 300 N (B) 500 N
(C) 250 N (D) 400 N

27. Consider a cylindrical tank of radius 1 m is filled with water. The top surface of water is at 15 m from the bottom of the cylinder. There is a hole on the wall of cylinder at a height of 5 m from the bottom. A force of 5×10^5 N is applied an the top surface of water using a piston. The speed of efflux from the hole will be :
[Given atmospheric pressure $P_A = 1.01 \times 10^5$ Pa, density of water $\rho_w = 1000\ kg/m^3$ and gravitational acceleration $g = 10\ m/s^2$]

[28 July 2022 (Evening)]

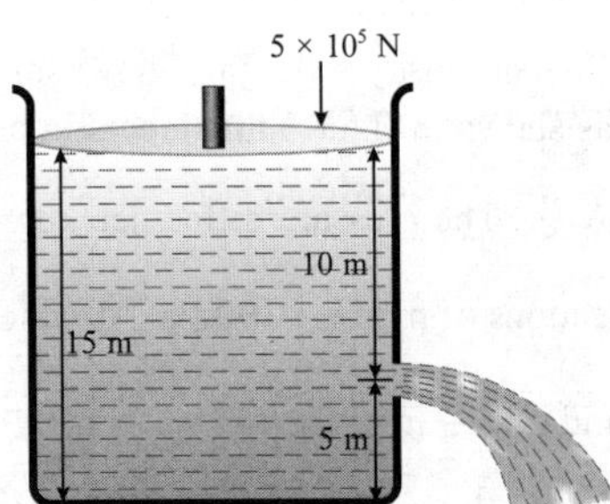

(A) 11.6 m/s (B) 10.8 m/s
(C) 17.8 m/s (D) 14.4 m/s

28. A fully loaded boeing aircraft has a mass of 5.4×10^5 kg. Its total wing area is 500 m^2. It is in level flight with a speed of 1080 km/h. If the density of air ρ is 1.2 kg m^{-3}, the fractional increase in the speed of the air on the upper surface of the wing relative to the lower surface in percentage will be: ($g = 10\ m/s^2$)

[29 Jan. 2023 (Evening)]

(A) 16 (B) 6
(C) 8 (D) 10

29. The surface of water in a water tank of cross section area 750 cm^2 on the top of a house is h m. Above the tap level. The speed of water coming out through the tap of cross section area 500 mm^2 is 30 cm/s. At that instant, $\frac{dh}{dt}$ is $x \times 10^{-3}$ m/s. The value of x will be_____. **[1 Feb. 2023 (Evening)]**

30. Figure below shows a liquid being pushed out of the tube by a piston having area of cross section 2.0 cm^2. The area of cross section at the outlet is 10 mm^2. If the piston is pushed at a speed of 4 cm s^{-1}, the speed of outgoing fluidis_____cm s^{-1}.

[10 April 2023 (Evening)]

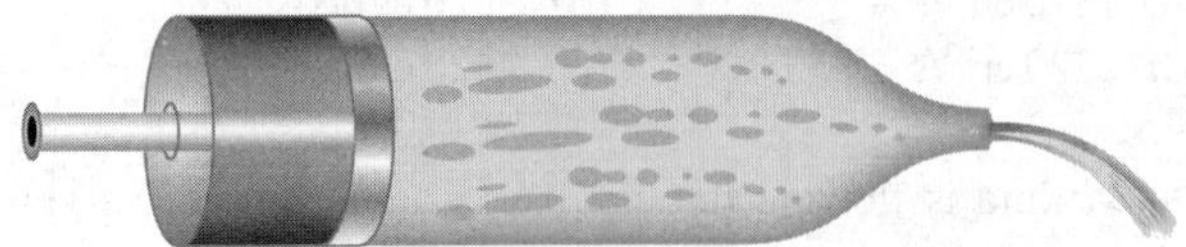

31. Glycerine of density 1.25×10^3 kg m^{-3} is flowing through the conical section of pipe. The area of cross-section of the pipe at its ends is 10 cm^2 and 5 cm^2 and pressure drop across its length is 3 Nm^{-2}. The rate of flow of glycerine through the pipe is $x \times 10^{-5}\ m^3\ s^{-1}$. The value of x is _____.

[12 April 2023 (Morning)]

32. The figure shows a liquid of given density flowing steadily in horizontal tube of varying cross-section. Cross sectional areas at A is 1.5 cm^2, and B is 25 mm^2, if the speed of liquid at B is 60 cm/s then $(P_A - P_B)$ is :
(Given P_A and P_B are liquid pressures at A and B points. Density $\rho = 1000$ kg m^{-3}, A and B are on the axis of tube)

[13 April 2023 (Morning)]

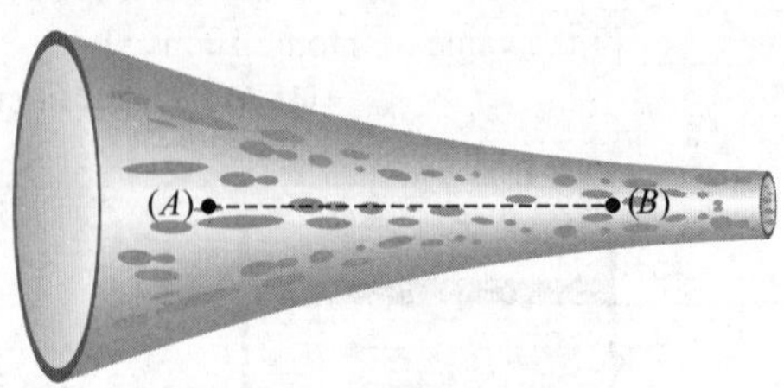

(A) 175 Pa (B) 27 Pa
(C) 135 Pa (D) 36 Pa

* * * * *

CHAPTER

9 MECHANICAL PROPERTIES OF FLUIDS

9.3 VISCOSITY

1. Spherical balls of radius R are falling in a viscous fluid of viscosity η with a velocity v. The retarding viscous force acting on the spherical ball is : **[2004]**

(A) Directly proportional to R but inversely proportional to v
(B) Directly proportional to both radius R and velocity v
(C) Inversely proportional to both radius R and velocity v
(D) Inversely proportional to R but directly proportional to velocity v

2. If the terminal speed of a sphere of gold (density = 19.5 kg/m^3) is 0.2 m/s in a viscous liquid (density = 1.5 kg/m^3), find the terminal speed of a sphere of silver (density = 10.5 kg/m^3) of the same size in the same liquid : **[2006]**

(A) 0.4 m/s (B) 0.133 m/s
(C) 0.1 m/s (D) 0.2 m/s

3. A spherical solid ball of volume V is made of a material of density ρ_1. It is falling through a liquid of density ρ_2 ($\rho_2 < \rho_1$). Assume that the liquid applies a viscous force on the ball that is proportional to the square of its speed v, i.e., $F_{viscous} = -kv^2$ ($k > 0$). Then terminal speed of the ball is : **[2008]**

(A) $\sqrt{\frac{Vg(\rho_1 - \rho_2)}{k}}$ (B) $\frac{Vg\rho_1}{k}$
(C) $\sqrt{\frac{Vg\rho_1}{k}}$ (D) $\frac{Vg(\rho_1 - \rho_2)}{k}$

4. If a ball of steel (density $\rho = 7.8$ g cm^{-3}) attains a terminal velocity of 10 cm s^{-1} when falling in a tank of water (coefficient of viscosity $\eta_{water} = 8.5 \times 10^{-4}$ Pa.s) then its terminal velocity in glycerine : **[2011]**
($\rho = 12$ g cm^{-3}, $\eta = 13.2$ Pa.s) would be nearly :

(A) 1.6×10^{-5} cm s^{-1} (B) 6.25×10^{-4} cm s^{-1}
(C) 6.45×10^{-4} cm s^{-1} (D) 1.5×10^{-5} cm s^{-1}

5. The terminal velocity of a small sphere of radius a in a viscous liquid is proportional to : **[2012 (Online)]**

(A) a^2 (B) a^3
(C) a (D) a^{-1}

6. In an experiment, a small steel ball falls through a liquid at a constant speed of 10 cm/s. If the steel ball is pulled upward with a force equal to twice its effective weight, how fast will it move upward ? **[2013 (Online)]**

(A) 5 cm/s (B) Zero
(C) 10 cm/s (D) 20 cm/s

7. The average mass of rain drops is 3.0×10^{-5} kg and their average terminal velocity is 9 m/s. Calculate the energy transferred by rain to each square metre of the surface at a place which receives 100 cm of rain in a year : **[2014 (Online)]**

(A) 3.5×10^5 J (B) 4.05×10^4 J
(C) 3.0×10^5 J (D) 9.0×10^4 J

8. The velocity of water in a river is 18 km/hr near the surface. If the river is 5 m deep, find the shearing stress between the horizontal layers of water. The co-efficient of viscosity of water = 10^{-2} poise : **[2014 (Online)]**

(A) 10^{-1} N/m^2 (B) 10^{-2} N/m^2
(C) 10^{-3} N/m^2 (D) 10^{-4} N/m^2

9. If it takes 5 minutes to fill a 15 litre bucket from a water tap of diameter $\frac{2}{\sqrt{\pi}}$ cm then the Reynolds number for the flows is (density of water = 10^3 kg/m^3) and viscosity of water = 10^{-3} Pa.s) close to : **[2015 (Online)]**

(A) 1100 (B) 11,000
(C) 550 (D) 5500

10. Which of the following option correctly describes the variation of the speed v and acceleration 'a' of a point mass falling vertically in a viscous medium that applies a force $F = -kv$, where 'k' is a constant, on the body ? (Graphs are schematic and not drawn to scale) **[2016 (Online)]**

(A) 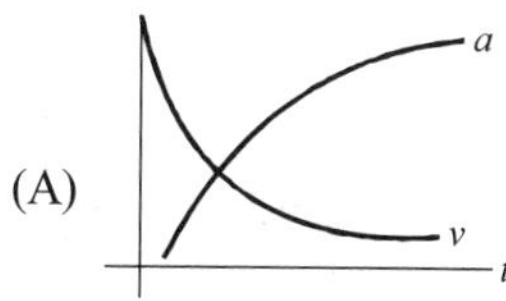

(B) 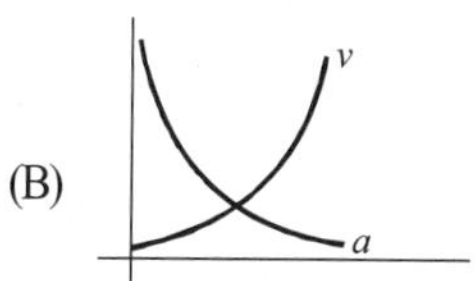

(C) 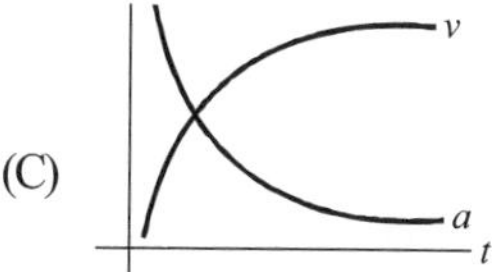

(D) 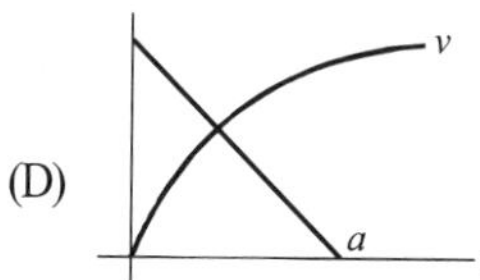

11. Two tubes of radii r_1 and r_2, and lengths l_1 and l_2, respectively, are connected in series and a liquid flows through each of them in streamline conditions. P_1 and P_2 are pressure differences across the two tubes. If P_2 is $4P_1$ and l_2 is $l_1/4$, then the radius r_2 will be equal to : **[2017 (Online)]**

(A) r_1 (B) $2r_1$

(C) $4r_1$ (D) $\frac{r_1}{2}$

12. Water from a pipe is coming at a rate of 100 litres per minute. If the radius of the pipe is 5 cm, the Reynolds number for the flow is of the order of : (density of water = 1000 kg/m^3, coefficient of viscosity of water = 1 mPas) : **[8 April 2019 (Morning)]**

(A) 10^6 (B) 10^3

(C) 10^4 (D) 10^2

13. A solid sphere, of radius R acquires a terminal velocity v_1 when falling (due to gravity) through a viscous fluid having a coefficient of viscosity η. The sphere is broken into 27 identical solid spheres. If each of these spheres acquires a terminal velocity v_2, when falling through the same fluid, the ratio (v_1/v_2) equals : **[12 April 2019 (Evening)]**

(A) 1/27 (B) 1/9

(C) 27 (D) 9

14. In an experiment to verify Stokes law, a small spherical ball of radius r and density ρ falls under gravity through a distance h in air before entering a tank of water. If the terminal velocity of the ball inside water is same as its velocity just before entering the water surface, then the value of h is proportional to (ignore viscosity of air) : **[5 Sept. 2020 (Evening)]**

(A) r (B) r^4

(C) r^3 (D) r^2

15. What will be the nature of flow of water from a circular tap, when its flow rate increased from 0.18 L/min to 0.48 L/min ? The radius of the tap and viscosity of water are 0.5 cm and 10^{-3} Pa s, respectively. (Density of water : 10^3 kg/m^3) **[16 March 2021 (Evening)]**

(A) Unsteady to steady flow

(B) Remains steady flow

(C) Remains turbulent flow

(D) Steady flow to unsteady flow

16. The terminal velocity (v_t) of the spherical rain drop depends on the radius (r) of the spherical rain drop as : **[25 June 2022 (Morning)]**

(A) $r^{1/2}$ (B) r

(C) r^2 (D) r^3

17. The velocity of upper layer of water in a river is 36 kmh^{-1}. Shearing stress between horizontal layers of water is 10^{-3} Nm^{-2}. Depth of the river is _____ m. (Co-efficiency of viscosity of water is 10^{-2} Pa.s) **[25 June 2022 (Morning)]**

18. The velocity of a small ball of mass m and density d_1, when dropped in a container filled with glycerine, becomes constant after some time. If the density of glycerine is d_2, then the viscous force acting on the ball, will be : **[27 June 2022 (Morning)]**

(A) $mg\left(1-\frac{d_1}{d_2}\right)$ (B) $mg\left(1-\frac{d_2}{d_1}\right)$

(C) $mg\left(\frac{d_1}{d_2}-1\right)$ (D) $mg\left(\frac{d_2}{d_1}-1\right)$

19. A water drop of radius 1 μm falls in a situation where the effect of buoyant force is negligible. Co-efficient of viscosity of air is 1.8×10^{-5} Nsm^{-2} and its density is negligible as compared to that of water 10^6 gm^{-3}. Terminal velocity of the water drop is : [Take acceleration due to gravity = 10 ms^{-2}] **[28 June 2022 (Evening)]**

(A) 145.4×10^{-6} ms^{-1} (B) 118.0×10^{-6} ms^{-1}

(C) 132.6×10^{-6} ms^{-1} (D) 123.4×10^{-6} ms^{-1}

20. A small spherical ball of radius 0.1 mm and density 10^4 kg m^{-3} falls freely under gravity through a distance h before entering a tank of water. If after entering the water the velocity of ball does not change and it continue to fall with same constant velocity inside water, then the value of h will be _____ m. [Given : $g = 10$ ms^{-2}, viscosity of water = 1.0×10^{-5} N-sm^{-2}] **[29 June 2022 (Evening)]**

21. The diameter of an air bubble which was initially 2 mm, rises steadily through a solution of density 1750 kg m^{-3} at the rate of 0.35 cms^{-1}. The coefficient of viscosity of the solution is _____ poise (in nearest integer). [The density of air is negligible] **[28 July 2022 (Morning)]**

22. The velocity of a small ball of mass 0.3 g and density 8 g/cc when dropped in a container filled with glycerine becomes constant after some time. If the density of glycerine is 1.3 g/cc, then the value of viscous force acting on the ball will be $x \times 10^{-4}$ N, the value of x is _____. [Use g = 10m/s^2] **[29 July 2022 (Evening)]**

23. A Spherical ball of radius 1 mm and density 10.5 g/cc is dropped in glycerine of coefficient of viscosity 9.8 poise and density 1.5 g/cc. Viscous force on the ball when it attains constant velocity is 3696×10^{-x} N. The value of x is. (Given, g = 9.8 m/s^2 and $\pi = \frac{22}{7}$) **[24 Jan. 2023 (Evening)]**

24. Match List-I with List-II : **[25 Jan. 2023 (Morning)]**

List-I	List-II
(a) Surface tension	(i) Kg m^{-1} s^{-1}
(b) Pressure	(ii) Kg ms^{-1}
(c) Viscosity	(iii) Kg m^{-1} s^{-2}
(d) Impulse	(iv) Kg s^{-2}

(A) (a)-(iv), (b)-(iii), (c)-(ii), (d)-(i)
(B) (a)-(iv), (b)-(iii), (c)-(i), (d)-(ii)
(C) (a)-(iii), (b)-(iv), (c)-(i), (d)-(ii)
(D) (a)-(ii), (b)-(i), (c)-(iii), (d)-(iv)

25. A metal block of base area 0.20 m^2 is placed on a table, as shown in figure. A liquid film of thickness 0.25 mm is inserted between the block and the table. The block is pushed by a horizontal force of 0.1 N and moves with a constant speed. If the viscosity of the liquid is 5.0×10^{-3} *Pl*, the speed of block is _____ $\times 10^{-3}$ m/s. **[29 Jan. 2023 (Evening)]**

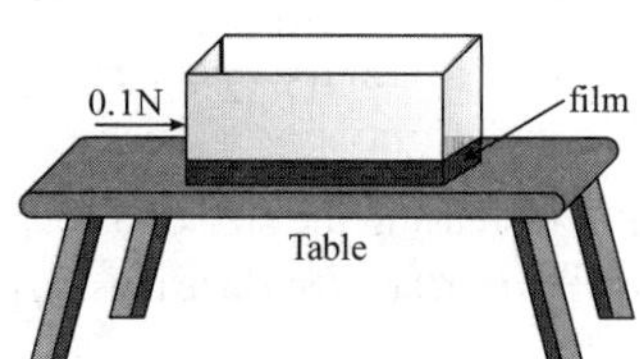

26. A small ball of mass *M* and density ρ is dropped in a viscous liquid of density ρ_0. After some time, the ball falls with a constant velocity. What is the viscous force on the ball ? **[6 April 2023 (Morning)]**

(A) $F = Mg\left(1 - \frac{\rho_0}{\rho}\right)$ (B) $F = Mg\left(1 + \frac{\rho}{\rho_0}\right)$

(C) $F = Mg\left(1 + \frac{\rho_0}{\rho}\right)$ (D) $F = Mg(1 \pm \rho\rho_0)$

27. An air bubble of diameter 6 mm rises steadily through a solution of density 1750 kg/m^3 at the rate of 0.35 cm/s. The co-efficient of viscosity of the solution (neglect density of air) is _____ Pa.s (Given, $g = 10$ ms^{-2}). **[8 April 2023 (Morning)]**

28. Eight equal drops of water are falling through air with a steady speed of 10 cm/s. If the drops coalesce, the new velocity is : **[11 April 2023 (Evening)]**

(A) 10 cm/s (B) 40 cm/s
(C) 16 cm/s (D) 5 cm/s

29. Given below are two statements : one is labelled as **Assertion-A** and the other is labelled as **Reason-R**.

Assertion-A : A spherical body of radius (5 ± 0.1) mm having a particular density is falling through a liquid of constant density is falling through a liquid of constant density. The percentage error in the calculation of its terminal velocity is 4%.

Reason-R : The terminal velocity of the spherical body falling through the liquid is inversely proportional to its radius.

In the light of the above statements, choose the correct answer form the options given below. **[13 April 2023 (Evening)]**

(A) Both **A** and **R** are true but **R** is Not the correct explanation of **A**.
(B) Both **A** and **R** are true and **R** is the correct explanation of **A**.
(C) **A** is false but **R** is true.
(D) **A** is true but **R** is false.

* * * * *

CHAPTER 9

MECHANICAL PROPERTIES OF FLUIDS

9.4 SURFACE TENSION

1. If two soap bubbles of different radii are connected by a tube : **[2004]**

(A) Air flows from the bigger bubble to the smaller bubble till the sizes become equal

(B) Air flows from bigger bubble to the smaller bubble till the sizes are interchanged

(C) Air flows from the smaller bubble to the bigger bubble

(D) There is no flow of air

2. A 20 cm long capillary tube is dipped in water. The water rises upto 8 cm. If the entire arrangement is put in a freely falling elevator, the length of water column in the capillary tube will be: **[2005]**

(A) 8 cm (B) 10 cm

(C) 4 cm (D) 20 cm

3. A capillary tube (A) is dipped in water. Another identical tube (B) is dipped in a soap-water solution. Which of the following shows the relative nature of the liquid columns in the two tubes : **[2008]**

(A)

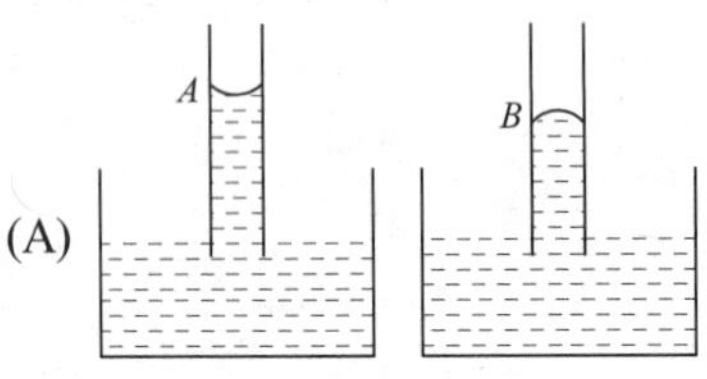

(B)

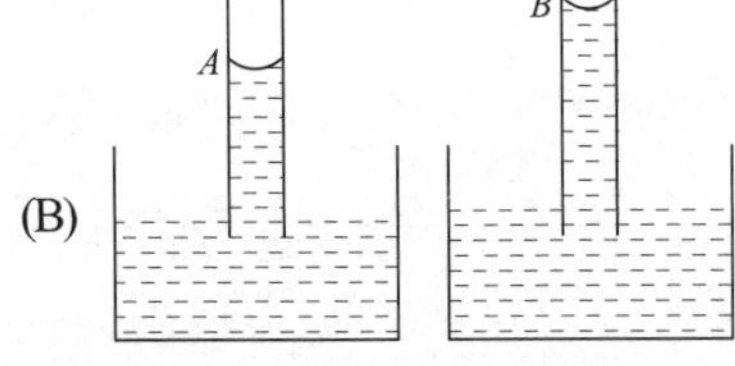

(C)

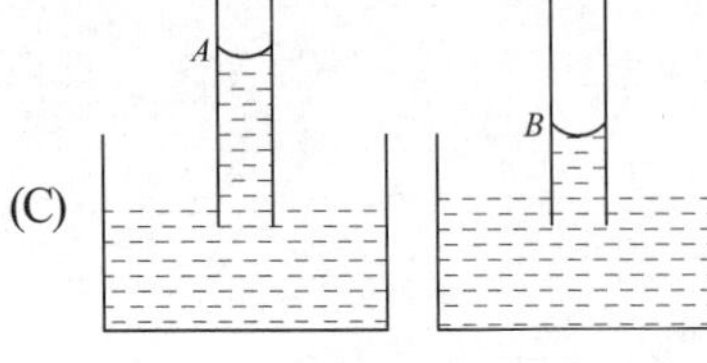

(D)

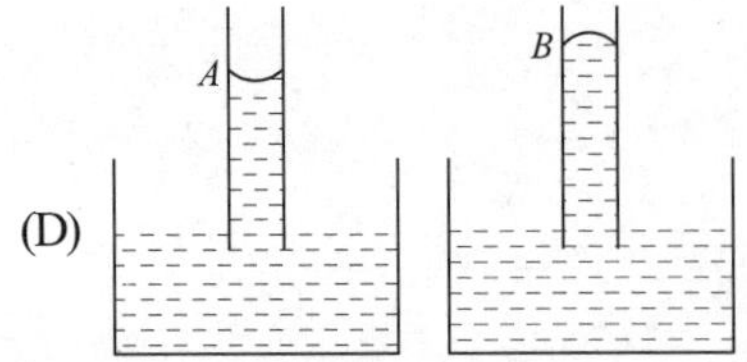

4. Two mercury drops (each of radius 'r') merge to form a bigger drop. The surface energy of the bigger drop, if T is the surface tension, is : **[2011]**

(A) $2^{5/3}\pi r^2 T$ (B) $4\pi r^2 T$

(C) $2\pi r^2 T$ (D) $2^{8/3}\pi r^2 T$

5. Work done in increasing the size of a soap bubble from a radius of 3 cm to 5 cm is nearly (Surface tension of soap solution $= 0.03\ Nm^{-1}$) : **[2011]**

(A) 2π mJ (B) 0.4π mJ

(C) 4π mJ (D) 0.2π mJ

6. A thin liquid film formed between a U-shaped wire and a light slider supports a weight of 1.5×10^{-2} N (see figure). The length of the slider is 30 cm and its weight negligible. The surface tension of the liquid film is : **[2012]**

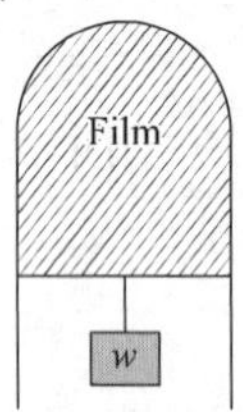

(A) $0.025\ Nm^{-1}$ (B) $0.012525\ Nm^{-1}$

(C) $0.125\ Nm^{-1}$ (D) $0.0525\ Nm^{-1}$

7. A large number of liquid drops each of radius r coalesce to from a single drop of radius R. The energy released in the process is converted into kinetic energy of the big drop so formed. The speed of the big drop is (given, surface tension of liquid T, density ρ) : **[2012 (Online)]**

(A) $\sqrt{\frac{T}{\rho}\left(\frac{1}{r}-\frac{1}{R}\right)}$ (B) $\sqrt{\frac{2T}{\rho}\left(\frac{1}{r}-\frac{1}{R}\right)}$

(C) $\sqrt{\frac{4T}{\rho}\left(\frac{1}{r}-\frac{1}{R}\right)}$ (D) $\sqrt{\frac{6T}{\rho}\left(\frac{1}{r}-\frac{1}{R}\right)}$

8. Assume that a drop of liquid evaporates by decrease in its surface energy, so that its temperature remains unchanged. What should be the minimum radius of the drop for this to be possible ? The surface tension is T, density of liquid is ρ and L is its latent heat of vaporization : **[2013]**

(A) $\sqrt{T/\rho L}$ (B) $T/\rho L$

(C) $2T/\rho L$ (D) $\rho L/T$

9. This question has Statement-1 and Statement-2. Of the four choices given after the Statements, choose the one that best describes the two Statements. **[2013 (Online)]**

Statement-1 : A capillary is dipped in a liquid and liquid rises to a height h in it. As the temperature of the liquid is raised, the height h increases. (If the density of the liquid and the angle of contact remain the same)

Statement-2 : Surface tension of a liquid decreases with the rise in its temperature.

(A) Statement-1 is true, Statement-2 is true; Statement-2 is not the correct explanation for Statement-1
(B) Statement-1 is false, Statement-2 is true
(C) Statement-1 is true, Statement-2 is false
(D) Statement-1 is true, Statement-2 is true; Statement-2 is the correct explanation for Statement-1

10. Wax is coated on the inner wall of a capillary tube and the tube is then dipped in water. Then, compared to the unwaxed capillary, the angle of contact θ and the height h upto which water rises change. These changes are : **[2013 (Online)]**

(A) θ increases and h also increases
(B) θ decreases h also decreases
(C) θ increases and h decreases
(D) θ decreases and h increases

11. On heating water, bubbles being formed at the bottom of the vessel detach and rise. Take the bubbles to be spheres of radius R and making a circular contact of radius r with the bottom of the vessel. If $r << R$ and the surface tension of water is T, value of r just before bubbles detach is : (density of water is ρ_w) : **[2014]**

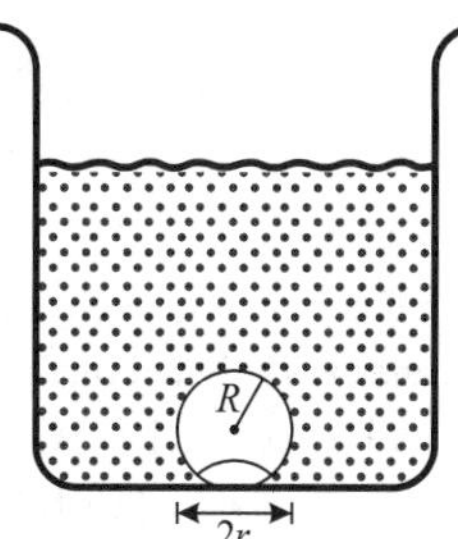

(A) $R^2\sqrt{\frac{2\rho_w g}{3T}}$ (B) $R^2\sqrt{\frac{\rho_w g}{6T}}$
(C) $R^2\sqrt{\frac{\rho_w g}{T}}$ (D) $R^2\sqrt{\frac{3\rho_w g}{T}}$

12. An air bubble of radius 0.1 cm is in a liquid having surface tension 0.06 N/m and density 10^3 kg/m^3. The pressure inside the bubble is 1100 Nm^{-2} greater than the atmospheric pressure. At what depth is the bubble below the surface of the liquid ? ($g = 9.8$ ms^{-2}) **[2014 (Online)]**

(A) 0.1 m (B) 0.15 m
(C) 0.20 m (D) 0.25 m

13. Two soap bubbles coalesce to form a single bubble. If V is the subsequent change in volume of contained air and S change in total surface are, T is the surface tension and P atmospheric pressure, then which of the following relation is correct ? **[2014 (Online)]**

(A) $4PV+3ST=0$ (B) $3PV+4ST=0$
(C) $2PV+3ST=0$ (D) $3PV+2ST=0$

14. If two glass plates have water between them and are separated by very small distance (see figure), it is very difficult to pull them apart. It is because the water in between form cylindrical surface on the side that gives rise to lower pressure in the water in comparison to atmosphere. If the radius of the cylindrical surface is R and surface tension of water is T then the pressure in water between the plates is lower by : **[2015 (Online)]**

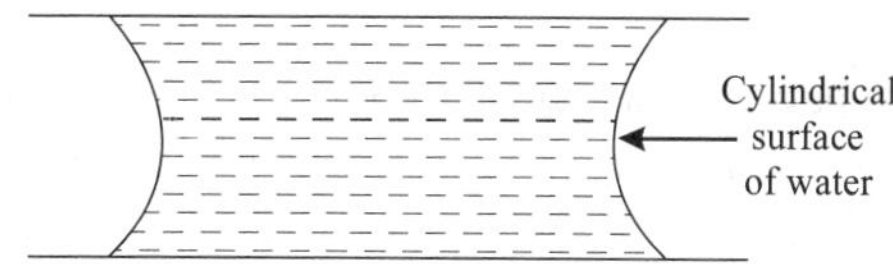

(A) $\frac{2T}{R}$ (B) $\frac{4T}{R}$
(C) $\frac{T}{4R}$ (D) $\frac{T}{R}$

15. A soap bubble, blown by a mechanical pump at the mouth of a tube, increases in volume, with time, at a constant rate. The graph that correctly depicts the time dependence of pressure inside the bubble is given by : **[12 Jan. 2019 (Evening)]**

(A)
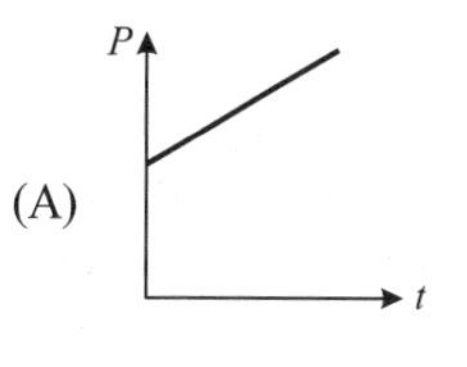

(B)
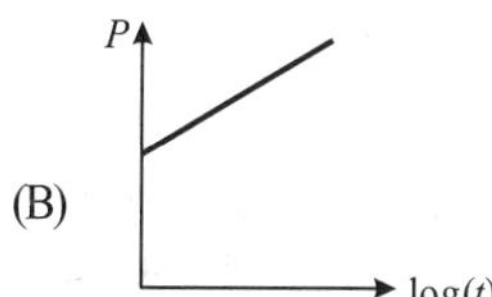

(C)
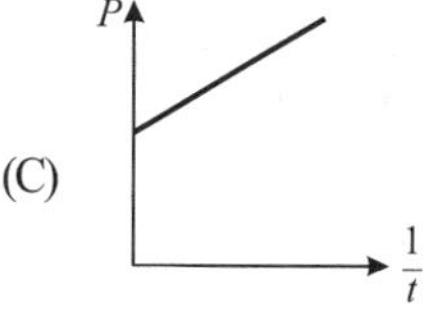

(D)
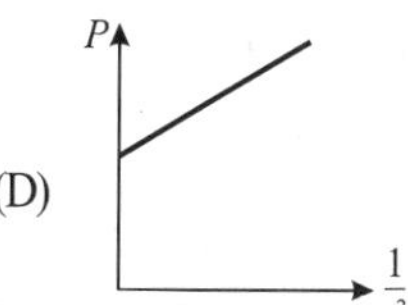

16. If 'M' is the mass of water that rises in a capillary tube of radius 'r', then mass of water which will rise in a capillary tube of radius '$2r$' is : **[9 April 2019 (Morning)]**

(A) $4M$ (B) M
(C) $2M$ (D) $M/2$

17. The ratio of surface tensions of mercury and water is given to be 7.5 while the ratio of their densities is 13.6. Their contact angles, with glass, are close to 135° and 0°, respectively. It is observed that mercury gets depressed by an amount h in a capillary tube of radius r_1, while water rises by the same amount h in a capillary tube of radius r_2. The ratio, (r_1/r_2), is then close to : **[10 April 2019 (Morning)]**

(A) 2/3 (B) 3/5
(C) 2/5 (D) 4/5

18. A small spherical droplet of density d is floating exactly half immersed in a liquid of density ρ and surface tension T. The radius of the droplet is: (Take note that the surface tension applies an upward force on the droplet) **[9 Jan. 2020 (Evening)]**

(A) $r=\sqrt{\frac{2T}{3(d+\rho)g}}$ (B) $r=\sqrt{\frac{3T}{(2d-\rho)g}}$

(C) $r=\sqrt{\frac{T}{(d-\rho)g}}$ (D) $r=\sqrt{\frac{T}{(d+\rho)g}}$

19. A capillary tube made of glass of radius 0.15 mm is dipped vertically in a beaker filled with methylene iodide surface tension $= 0.05\ \text{Nm}^{-1}$, density $= 667\ \text{kgm}^{-3}$) which rises to height h in the tube. It is observed that the two tangents drawn from liquid-glass interfaces (from opposite sides of the capillary) make an angle of 60° with one another. Then h is close to: $(g = 10\ \text{ms}^{-2})$ **[2 Sept. 2020 (Evening)]**

(A) 0.137 m (B) 0.172 m
(C) 0.087 m (D) 0.049 m

20. Pressure inside two soap bubbles are 1.01 and 1.02 atmosphere, respectively. The ratio of their volumes is : **[3 Sept. 2020 (Morning)]**

(A) 8 : 1 (B) 0.8 : 1
(C) 2 : 1 (D) 4 : 1

21. When a long glass capillary tube of radius 0.015 cm is dipped in a liquid, the liquid rises to a height of 15 cm within it. If the contact angle between the liquid and glass to close to 0°, the surface tension of the liquid, in milli Newton m^{-1}, is $[\rho_{(\text{liquid})} = 900\ \text{kgm}^{-3}, g = 10\ \text{ms}^{-2}]$ (Give answer in closest integer)_____. **[3 Sept. 2020 (Morning)]**

22. When two soap bubbles of radii a and b $(b > a)$ coalesce, the radius of curvature of common surface is : **[17 March 2021 (Morning)]**

(A) $\frac{ab}{b-a}$ (B) $\frac{a+b}{ab}$

(C) $\frac{b-a}{ab}$ (D) $\frac{ab}{a+b}$

23. Two small drops of mercury each of radius R coalesce to form a single large drop. The ratio of total surface energy before and after the change is : **[20 July 2021 (Evening)]**

(A) $1 : 2^{1/3}$ (B) $2^{3/1} : 1$
(C) 2 : 1 (D) 1 : 2

24. Two spherical soap bubbles of radii r_1 and r_2 in vacuum combine under isothermal conditions. The resulting bubble has a radius equal to : **[25 July 2021 (Evening)]**

(A) $\frac{r_1+r_2}{2}$ (B) $\sqrt{r_1 r_2}$

(C) $\sqrt{r_1^2+r_2^2}$ (D) $\frac{r_1 r_2}{r_1+r_2}$

25. A soap bubble of radius 3 cm is formed inside the another soap bubble of radius 6 cm. The radius of an equivalent soap bubble which has the same excess pressure as inside the smaller bubble with respect to the atmospheric pressure is_____cm. **[26 Aug. 2021 (Morning)]**

26. A water drop of diameter 2 cm is broken into 64 equal droplets. The surface tension of water is 0.075 N/m. In this process the gain in surface energy will be : **[28 June 2022 (Morning)]**

(A) 2.8×10^{-4} J (B) 1.5×10^{-3} J
(C) 1.9×10^{-4} J (D) 9.4×10^{-5} J

27. A drop of liquid of density ρ is floating half immersed in a liquid of density σ and surface tension $7.5 \times 10^{-4}\ \text{Ncm}^{-1}$. The radius of drop in cm will be : [Take : $g = 10\ \text{m/s}^2$] **[25 July 2022 (Evening)]**

(A) $\frac{15}{\sqrt{2\rho-\sigma}}$ (B) $\frac{15}{\sqrt{\rho-\sigma}}$

(C) $\frac{3}{2\sqrt{\rho-\sigma}}$ (D) $\frac{3}{20\sqrt{2\rho-\sigma}}$

28. A water drop of radius 1 cm is broken into 729 equal droplets. If surface tension of water is 75 dyne/cm, then the gain in surface energy upto first decimal place will be : [Given : $\pi = 3.14$] **[26 July 2022 (Morning)]**

(A) 8.5×10^{-4} J (B) 8.2×10^{-4} J
(C) 7.5×10^{-4} J (D) 5.3×10^{-4} J

29. A spherical soap bubble of radius 3 cm is formed inside another spherical soap bubble of radius 6 cm. If the internal pressure of the smaller bubble of radius 3 cm in the above system is equal to the internal pressure of the another single soap bubble of radius r cm. The value of r is _____. **[27 July 2022 (Evening)]**

30. Given below are two statements:
One is labelled as **Assertion-A** and the other is labelled as **Reason-R**.

Assertion-A: Clothes containing oil or grease stains cannot be cleaned by water wash.

Reason-R: Because the angle of contact between the oil/ grease and water is obtuse.

In the light of the above statements, choose the correct answer from the option given below : **[29 July 2022 (Morning)]**

(A) Both **A** and **R** are true and **R** is the correct explanation of **A**.

(B) Both **A** and **R** are true but **R** is not the correct explanation of **A**.

(C) **A** is true but **R** is false.

(D) **A** is true but **R** is true.

31. A spherical drop of liquid splits into 1000 identical spherical drops. If u_i is the surface energy of the original drop and u_f is the total surface energy of the resulting drops (ignoring evaporation). $\frac{u_f}{u_i} = \left(\frac{10}{x}\right)$ then value of x is _____.

[25 Jan. 2023 (Evening)]

32. Surface tension of a soap bubble is 2.0×10^{-2} Nm^{-1}. Work done to increase the radius of soap bubble from 3.5 cm to 7 cm will be : [Take $\pi = \frac{22}{7}$] **[29 Jan. 2023 (Morning)]**

(A) 0.72×10^{-4} J (B) 5.76×10^{-4} J

(C) 18.48×10^{-4} J (D) 9.24×10^{-4} J

33. The height of liquid column raised in a capillary tube of certain radius when dipped in liquid A vertically is, 5 cm. If the tube is dipped in a similar manner in another liquid B of surface tension and density double the values of liquid A, the height of liquid column raised in liquid B would be _____ m.

[30 Jan. 2023 (Morning)]

(A) 0.20 (B) 0.5

(C) 0.05 (D) 0.10

34. If 1000 droplets of water of surface tension 0.07 N/m. Having same radius 1 mm each, combine to from a single drop. In the process the released surface energy is : $\left(\text{Take } \pi = \frac{22}{7}\right)$

[31 Jan. 2023 (Morning)]

(A) 7.92×10^{-6} J (B) 7.92×10^{-4} J

(C) 9.68×10^{-4} J (D) 8.8×10^{-5} J

35. A mercury drop of radius 10^{-3} m is broken into 125 equal size droplets. Surface tension of mercury is 0.45 Nm^{-1}. The gain in surface energy is : **[1 Feb. 2023 (Morning)]**

(A) 2.26×10^{-5} J (B) 28×10^{-5} J

(C) 17.5×10^{-5} J (D) 5×10^{-5} J

36. The surface tension of soap solution is 3.5×10^{-2} Nm^{-1}. The amount of work done required to increase the radius of soap bubble from 10 cm to 20 cm is _____ $\times 10^{-4}$ J.

[11 April 2023 (Evening)]

37. There is an air bubble of radius 1.0 mm in a liquid of surface tension 0.075 Nm^{-1} and density 1000 kg m^{-3} at a depth of 10 cm below the free surface. The amount by which the pressure inside the bubble is greater than the atmospheric pressure is_____. Pa ($g = 10$ ms^{-2}) **[15 April 2023 (Morning)]**

* * * * *

CHAPTER

10 THERMAL PROPERTIES OF MATTER

10.1 THERMAL EXPANSION

1. A metal rod of Young's modulus Y and coefficient of thermal expansion α is held at its two ends such that its length remains invariant. If its temperature is raised by t°C, the linear stress developed in it is : **[2011]**

(A) $\frac{\alpha t}{Y}$ (B) $\frac{Y}{\alpha t}$

(C) $Y\alpha t$ (D) $\frac{1}{(Y\alpha t)}$

2. An aluminium sphere of 20 cm diameter is heated from 0°C to 100°C. Its volume changes by (given that coefficient of linear expansion for aluminium $\alpha_{Al} = 23 \times 10^{-6}$/°C) : **[2011]**

(A) 28.9 cc (B) 2.89 cc

(C) 9.28 cc (D) 49.8 cc

3. A wooden wheel of radius R is made of two semicircular parts (see figure). The two parts are held together by a ring made of a metal strip of cross sectional area S and Length L. L is slightly less than $2\pi R$. To fit the ring on the wheel, it is heated so that its temperature rises by ΔT and it just steps over the wheel. As it cools down to surrounding temperature, it presses the semicircular parts together. If the coefficient of linear expansion of the metal is α, and its Young's modulus is Y, the force that one part of the wheel applies on the other part is : **[2012]**

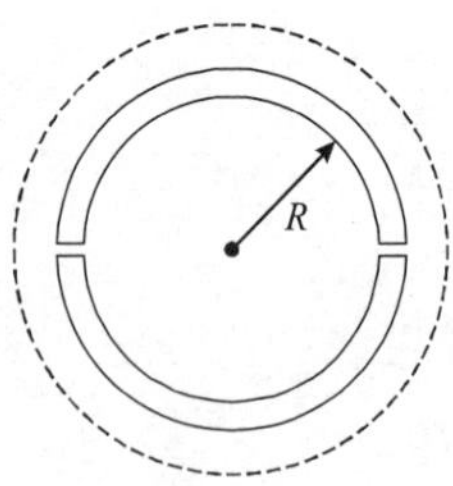

(A) $2SY\alpha\Delta T$ (B) $2\pi SY\alpha\Delta T$

(C) $SY\alpha\Delta T$ (D) $\pi SY\alpha\Delta T$

4. On a linear temperature scale Y, water freezes at – 160°Y and boils at – 50°Y. On this Y scale, a temperature of 340 K would be read as : (water freezes at 273 K and boils at 373 K) : **[2013 (Online)]**

(A) –73.7°Y (B) –233.7°Y

(C) –86.3°Y (D) –106.3°Y

5. The ratio of the coefficient of volume expansion of a glass container to that of a viscous liquid kept inside the container is 1 : 4. What fraction of the inner volume of the container should the liquid occupy so that the volume of the remaining vacant space will be same at all temperatures ? **[2013 (Online)]**

(A) 2 : 5 (B) 1 : 4

(C) 1 : 64 (D) 1 : 8

6. The pressure that has to be applied to the ends of a steel wire of length 10 cm to keep its length constant when its temperature is raised by 100°C is (For steel Young's modulus is 2×10^{11} Nm^{-2} and coefficient of thermal expansion is 1.1×10^{-5} K^{-1}) : **[2014]**

(A) 2.2×10^{8} Pa (B) 2.2×10^{9} Pa

(C) 2.2×10^{7} Pa (D) 2.2×10^{6} Pa

7. A pendulum clock loses 12 s a day if the temperature is 40°C and gains 4 s a day if the temperature is 20°C. The temperature at which the clock will show correct time, and the co-efficient of linear expansion (α) of the metal of the pendulum shaft are respectively : **[2016]**

(A) 25°C; $\alpha = 1.85 \times 10^{-5}$/°C (B) 60°C; $\alpha = 1.85 \times 10^{-4}$/°C

(C) 30°C; $\alpha = 1.85 \times 10^{-3}$/°C (D) 55°C; $\alpha = 1.85 \times 10^{-2}$/°C

8. An external pressure P is applied on a cube at 0°C so that it is equally compressed from all sides. K is the bulk modulus of the material of the cube and α is its coefficient of linear expansion. Suppose we want to bring the cube to its original size by heating. The temperature should be raised by : **[2017]**

(A) $\frac{3\alpha}{PK}$ (B) $3PK\alpha$

(C) $\frac{P}{3\alpha K}$ (D) $\frac{P}{\alpha K}$

9. A compressive force, F is applied at the two ends of a long thin steel rod. It is heated, simultaneously, such that its temperature increases by ΔT. The net change in its length is zero. Let l be the length of the rod, A its area of cross-section, Y its Young's modulus, and α its coefficient of linear expansion. Then, F is equal to : **[2017 (Online)]**

(A) $l^2 Y\alpha\Delta T$ (B) $lA\,Y\alpha\Delta T$

(C) $AY\alpha\Delta T$ (D) $\frac{AY}{\alpha\Delta T}$

10. A steel rail of length 5 m and area of cross-section 40 cm^2 is prevented from expanding along its length while the temperature rises by 10°C. If coefficient of linear expansion and Young's modulus of steel are 1.2×10^{-5} K^{-1} and 2×10^{11} Nm^{-2} respectively, the force developed in the rail is approximately : **[2017 (Online)]**

(A) 2×10^7 N (B) 1×10^5 N
(C) 2×10^9 N (D) 3×10^{-5} N

11. A rod, of length L at room temperature and uniform area of cross section A, is made of a metal having coefficient of linear expansion α/°C^{-1}. It is observed that an external compressive force F, is applied on each of its ends, prevents any change in the length of the rod, when its temperature rises by ΔT K. Young's modulus Y, for this metal is : **[9 Jan. 2019 (Morning)]**

(A) $\dfrac{F}{2A\alpha\Delta T}$ (B) $\dfrac{F}{A\alpha(\Delta T - 273)}$

(C) $\dfrac{F}{A\alpha\Delta T}$ (D) $\dfrac{2F}{A\alpha\Delta T}$

12. Two rods A and B of identical dimensions are at temperature 30°C. If A is heated upto 180°C and B upto T°C, then the new lengths are the same. If the ratio of the coefficients of linear expansion of A and B is 4 : 3, then the value of T is : **[11 Jan. 2019 (Evening)]**

(A) 270°C (B) 230°C
(C) 250°C (D) 200°C

13. A thermometer graduated according to a linear scale reads a value x_0 when in contact with boiling water, and $x_0/3$ when in contact with ice. What is the temperature of an object in °C, if this thermometer is in the contact with the object reads $x_0/2$? **[11 Jan. 2019 (Evening)]**

(A) 35 (B) 25
(C) 60 (D) 40

14. A non-isotropic solid metal cube has coefficients of linear expansion as :
5×10^{-5}/°C along the x-axis and 5×10^{-6}/°C along the y and the z-axis. If the coefficient of volume expansion of the solid is $C \times 10^{-16}$/°C then the value of C is ______. **[7 Jan. 2020 (Morning)]**

15. A leak proof cylinder of length 1 m, made of a metal which has very low coefficient of expansion is floating vertically in water at 0°C such that its height above the water surface is 20 cm. When the temperature of water is increased to 4°C, the height of the cylinder above the water surface becomes 21 cm. The density of water at T = 4°C, relative to the density at T = 0°C is close to : **[8 Jan. 2020 (Morning)]**

(A) 1.01 (B) 1.04
(C) 1.03 (D) 1.26

16. When the temperature of a metal wire is increased from 0°C to 10°C, its length increases by 0.02 %. The percentage change in its mass density will be closest to : **[2 Sept. 2020 (Evening)]**

(A) 0.008 (B) 0.06
(C) 0.8 (D) 2.3

17. A bakelite beaker has volume capacity of 500 cc at 30°C. When it is partially filled with V_m volume (at 30°) of mercury, it is found that the unfilled volume of the beaker remains constant as temperature is varied. If $\gamma_{(beaker)} = 6 \times 10^{-6}$ °C^{-1} and $\gamma_{(mercury)} = 1.5 \times 10^{-4}$ °C^{-1}, where γ is the coefficient of volume expansion, then V_m (in cc) is close to______. **[3 Sept. 2020 (Morning)]**

18. Two different wires having lengths L_1 and L_2, and respective temperature coefficient of linear expansion α_1 and α_2, are joined end-to-end. Then the effective temperature coefficient of linear expansion is : **[5 Sept. 2020 (Evening)]**

(A) $4\dfrac{\alpha_1\alpha_2}{\alpha_1+\alpha_2}\dfrac{L_2L_1}{(L_2+L_1)^2}$ (B) $2\sqrt{\alpha_1\alpha_2}$

(C) $\dfrac{\alpha_1+\alpha_2}{2}$ (D) $\dfrac{\alpha_1L_1+\alpha_2L_2}{L_1+L_2}$

19. Each side of a box made of metal sheet in cubic shape is 'a' at room temperature 'T', the coefficient of linear expansion of the metal sheet is 'α'. The metal sheet is heated uniformly, by a small temperature ΔT, so that its new temperature is $T + \Delta T$. Calculate the increase in the volume of the metal box : **[24 Feb. 2021 (Morning)]**

(A) $3a^3\alpha\Delta T$ (B) $4\pi a^3\alpha\Delta T$

(C) $4a^3\alpha\Delta T$ (D) $\dfrac{4}{3}\pi a^3\alpha\Delta T$

20. Given below are two statements : one is labelled as **Assertion-A** and the other is labelled as **Reason-R**.
Assertion-A : When a rod lying freely on a smooth surface is heated, no thermal stress is developed in it.
Reason-R : On heating, the length of the rod increases.
In the light of the above statements, choose the correct answer from the options given below : **[25 Feb. 2021 (Morning)]**

(A) Both **A** and **R** are true and **R** is the correct explanation of **A**.
(B) **A** is false but **R** is true.
(C) **A** is true but **R** is false.
(D) Both **A** and **R** are true but **R** is NOT the correct explanation of **A**.

21. A bimetallic strip consists of metals A and B. It is mounted rigidly as shown. The metal A has higher coefficient of expansion compared to that of metal B. When the bimetallic strip is placed in a cold both, it will : **[16 March 2021 (Evening)]**

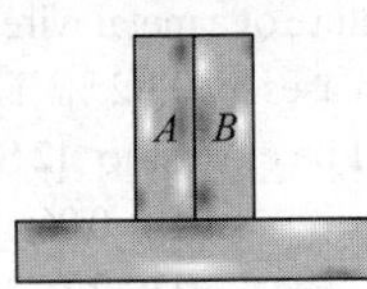

(A) Bend towards the right (B) Not bend but shrink
(C) Neither bend nor shrink (D) Bend towards the left

22. The resistance of a conductor at 15°C is 16 Ω and at 100°C is 20 Ω. What will be the temperature coefficient of resistance of the conductor ? **[27 July 2021 (Evening)]**

(A) $0.010°C^{-1}$ (B) $0.033°C^{-1}$
(C) $0.003°C^{-1}$ (D) $0.042°C^{-1}$

23. A steel rod with $y = 2.0 \times 10^{11}$ Nm^{-2} and $\alpha = 10^{-5}$ $°C^{-1}$ of length 4 m and area of cross-section 10 cm^2 is heated from 0°C to 400°C without being allowed to extend. The tension produced in the rod is $x \times 10^5$ N where the value of x is : **[01 Sep. 2021 (Evening)]**

24. A solid metallic cube having total surface area 24 m^2 is uniformly heated. If its temperature is increased by 10°C, calculate the increase in volume of the cube :
[Given : $\alpha = 5.0 \times 10^{-4}$ $°C^{-1}$] **[25 June 2022 (Evening)]**

(A) 2.4×10^6 cm^3 (B) 1.2×10^5 cm^3
(C) 6.0×10^4 cm^3 (D) 4.8×10^5 cm^3

25. At what temperature a gold ring of diameter 6.230 cm be heated so that it can be fitted on a wooden bangle of diameter 6.241 cm ? Both the diameters have been measured at room temperature (27°C): (Given – coefficient of linear thermal expansion of gold $\alpha_L = 1.4 \times 10^{-5}$ K^{-1}) **[29 June 2022 (Evening)]**

(A) 125.7°C (B) 91.7°C
(C) 425.7°C (D) 152.7°C

26. A unit scale is to be prepared whose length does not change with temperature and remains 20 cm, using a bimetallic strip made of brass and iron each of different length. The length of both components would change in such a way that difference between their lengths remains constant. If length of brass is 40 cm and length of iron will be ____ cm.
[$\alpha_{iron} = 1.2 \times 10^{-5}$ K^{-1} and $\alpha_{brass} = 1.8 \times 10^{-5}$ K^{-1}] **[25 July 2022 (Morning)]**

27. A hole is drilled in a metal sheet. At 27°C, the diameter of hole is 5 cm. When the sheet is heated to 177°C, the change in the diameter of hole is $d \times 10^{-3}$ cm. The value of d will be ____ if coefficient of linear expansion of the metal is 1.6×10^{-5}/°C. **[24 Jan. 2023 (Morning)]**

28. The graph between two temperature scales P and Q is shown in the figure. Between upper fixed point and lower fixed point there are 150 equal divisions of scale P and 100 divisions on scale Q. The relationship for conversion between the two scales is given by : **[25 Jan. 2023 (Evening)]**

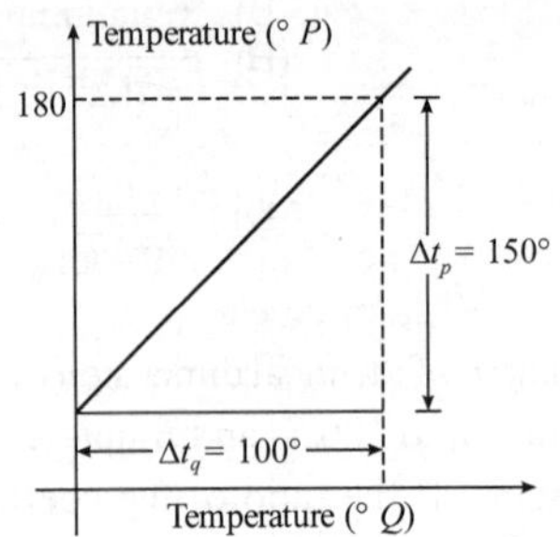

(A) $\frac{t_Q}{150} = \frac{t_P - 180}{100}$ (B) $\frac{t_Q}{100} = \frac{t_P - 30}{150}$

(C) $\frac{t_P}{180} = \frac{t_Q - 40}{100}$ (D) $\frac{t_P}{100} = \frac{t_Q - 180}{150}$

29. A faulty thermometer reads 5°C in melting ice and 95°C in steam. The correct temperature on absolute scale will be ____ K when the faulty thermometer reads 41°C. **[30 Jan. 2023 (Evening)]**

30. On a temperature scale 'X'. The boiling point of water is 65° X and the freezing point is – 15°X. Assume that the X scale is linear. The equivalent temperature corresponding to – 95°X on the Fahrenheit scale would be : **[11 April 2023 (Morning)]**

(A) –64°F (B) –112°F
(C) –48°F (D) –148°F

* * * * *

CHAPTER

10 THERMAL PROPERTIES OF MATTER

10.2 CALORIMETRY

1. Heat given to a body which raises its temperature by 1°C is: **[2002]**

(A) Water equivalent (B) Thermal capacity
(C) Specific heat (D) Temperature gradient

2. If mass-energy equivalence is taken into account, when water is cooled to form ice, the mass of water should : **[2002]**

(A) Increase (B) Remain unchanged
(C) Decrease (D) First increase then decrease

3. Given that 1 g of water in liquid phase has volume 1 cm^3 and in vapour phase 1671 cm^3 at atmospheric pressure and the latent heat of vaporization of water is 2256 J/g; the change in the internal energy in joules for 1 g of water at 373 K when it changes from liquid phase to vapour phase at the same temperature is : **[2013 (Online)]**

(A) 2256 (B) 167
(C) 2089 (D) 1

4. 500 g of water and 100 g of ice at 0°C are in a calorimeter whose water equivalent is 40 g. 10 g of steam at 100°C is added to it. Then water in the calorimeter is : (Latent heat of ice = 80 cal/g, Latent heat of steam = 540 cal/g) : **[2013 (Online)]**

(A) 580 g (B) 590 g
(C) 600 g (D) 610 g

5. Water of volume 2*L* in a closed container is heated with a coil of 1 kW. While water is heated, the container loses energy at a rate of 160 J/s. In how much time will the temperature of water rise from 27°C to 77°C ? (Specific heat of water is 4.2 kJ/kg and that of the container is negligible) : **[2014 (Online)]**

(A) 8 min 20 s (B) 6 min 2 s
(C) 7 min (D) 14 min

6. An experiment takes 10 minutes to raise the temperature of water in a container from 0°C to 100°C and another 55 minutes to convert it totally into steam by a heater supplying heat at a uniform rate. Neglecting the specific heat of the container and taking specific heat of water to be 1 cal/g °C, the heat of vapourization according to this experiment will come out to be: **[2015 (Online)]**

(A) 560 cal/g (B) 550 cal/g
(C) 540 cal/g (D) 530 cal/g

7. A copper ball of mass 100 gm is at a temperature *T*. It is dropped in a copper calorimeter of mass 100 gm, filled with 170 gm of water at room temperature. Subsequently, the temperature of the system is found to be 75°C. *T* is given by (Given : room temperature = 30°C, specific heat of copper = 0.1 cal/gm°C) : **[2017]**

(A) 1250°C (B) 825°C
(C) 800°C (D) 885°C

8. In an experiment a sphere of aluminium of mass 0.20 kg is heated upto 150°C. Immediately, it is put into water of volume 150 cc at 27°C kept in a calorimeter of water equivalent to 0.025 kg. Final temperature of the system is 40°C. The specific heat of aluminium is : [Take 4.2 Joule = 1 calorie] **[2017 (Online)]**

(A) 378 J/kg – °C (B) 315 J/kg – °C
(C) 476 J/kg – °C (D) 434 J/kg – °C

9. An unknown metal of mass 192 g heated to a temperature of 100°C was immersed into a brass calorimeter of mass 128 g containing 240 g of water a temperature of 8.4°C Calculate the specific heat of the unknown metal if water temperature stabilizes at 21.5°C : (Specific heat of brass is 394 J kg^{-1} K^{-1}) **[10 Jan. 2019 (Evening)]**

(A) 1232 J kg^{-1} K^{-1} (B) 458 J kg^{-1} K^{-1}
(C) 654 J kg^{-1} K^{-1} (D) 916 J kg^{-1} K^{-1}

10. Ice at –20°C is added to 50 g of water at 40°C. When the temperature of the mixture reaches 0°C, it is found that 20 g of ice is still unmelted. The amount of ice added to the water was close to : **[11 Jan. 2019 (Morning)]**

(Specific heat of water = 4.2 J/g/°C)
Specific heat of Ice = 2.1 J/g/°C
Heat of fusion of water at 0°C = 334 J/g)

(A) 50 g (B) 40 g
(C) 60 g (D) 100 g

11. When 100 g of a liquid *A* at 100°C is added to 50 g of a liquid *B* at temperature 75°C, the temperature of the mixture becomes 90°C. The temperature of the mixture, if 100 g of liquid *A* at 100°C is added to 50 g of liquid *B* at 50°C, will be : **[11 Jan. 2019 (Evening)]**

(A) 80°C (B) 60°C
(C) 70°C (D) 85°C

12. A metal ball of mass 0.1 kg is heated upto 500°C and dropped into a vessel of heat capacity 800 JK^{-1} and containing 0.5 kg water. The initial temperature of water and vessel is 30°C. What is the approximate percentage increment in the temperature of the water ? [Specific Heat Capacities of water and metal are, respectively, 4200 $JKg^{-1}K^{-1}$ and 400 $JKg^{-1}K^{-1}$]

[11 Jan. 2019 (Evening)]

(A) 30% (B) 20%
(C) 25% (D) 15%

13. A thermally insulated vessel contains 150 g of water at 0°C. Then the air from the vessel is pumped out adiabatically. A fraction of water turns into ice and the rest evaporates at 0°C itself. The mass of evaporated water will be closest to :
(Latent heat of vaporization of water = 2.10×10^6 J kg^{-1} and Latent heat of Fusion of water = 3.36×10^5 J kg^{-1})

[8 April 2019 (Morning)]

(A) 130 g (B) 35 g
(C) 20 g (D) 150 g

14. When M_1 gram of ice at –10°C (specific heat = 0.5 cal g^{-1} °C^{-1}) is added to M_2 gram of water at 50°C, finally no ice is left and the water is at 0°C. The value of latent heat of ice, in cal g^{-1} is : **[12 April 2019 (Morning)]**

(A) $\frac{5M_1}{M_2} - 50$ (B) $\frac{50M_2}{M_1}$

(C) $\frac{50M_2}{M_1} - 5$ (D) $\frac{5M_2}{M_1} - 5$

15. M grams of steam at 100°C is mixed with 200 g of ice at its melting point in a thermally insulated container. If it produces liquid water at 40°C [heat of vaporization of water is 540 cal/g and heat of fusion of ice is 80 cal/g], the value of M is______.

[7 Jan. 2020 (Evening)]

16. Three containers C_1, C_2 and C_3 have water at different temperatures. The table below shows the final temperature T when different amounts of water (given in litres) are taken from each containers and mixed (assume no loss of heat during the process). **[8 Jan. 2020 (Evening)]**

C_1	C_2	C_3	T
1l	2l	–	60°C
–	1l	2l	30°C
2l	–	1l	60°C
1l	1l	1l	θ

The value of θ (in °C to the nearest integer) is _____.

17. A calorimeter of water equivalent 20 g contains 180 g of water at 25°C. 'm' grams of steam at 100°C is mixed in it till the temperature of the mixture is 31°C. The value of 'm' is close to (Latent heat of water = 540 cal g^{-1}, specific heat of water = 1 cal g^{-1} °C^{-1}) : **[3 Sept. 2020 (Evening)]**

(A) 2.6 (B) 2
(C) 4 (D) 3.2

18. The specific heat of water = 4200 J kg^{-1} K^{-1} and the latent heat of ice = 3.4×10^5 J kg^{-1}. 100 grams of ice at 0°C is placed in 200 g of water at 25°C. The amount of ice that will melt as the temperature of water reaches 0°C is close to (in grams) :

[4 Sept. 2020 (Morning)]

(A) 61.7 (B) 63.8
(C) 69.3 (D) 64.6

19. A bullet of mass 5 g, travelling with a speed of 210 m/s, strikes a fixed wooden target. One half of its kinetic energy is converted into heat in the bullet while the other half is converted into heat in the wood. The rise of temperature of the bullet if the specific heat of its material is 0.030 cal/(g–°C) (1 cal = 4.2×10^7 ergs) close to : **[5 Sept. 2020 (Morning)]**

(A) 83.3 °C (B) 87.5 °C
(C) 119.2 °C (D) 38.4 °C

20. A large number of water drops, each of radius r, combine to have a drop of radius R. If the surface tension is T and mechanical equivalent of heat is J, the rise in heat energy per unit volume will be : **[26 Feb. 2021 (Morning)]**

(A) $\frac{2T}{rJ}$ (B) $\frac{3T}{rJ}$

(C) $\frac{3T}{J}\left(\frac{1}{r}-\frac{1}{R}\right)$ (D) $\frac{2T}{J}\left(\frac{1}{r}-\frac{1}{R}\right)$

21. The temperature of equal masses of three different liquids x, y and z are 10°C, 20°C and 30°C respectively. The temperature of mixture when x is mixed with y is 16°C and that when y is mixed with z is 26°C. The temperature of mixture when x and z are mixed will be : **[26 Aug. 2021 (Evening)]**

(A) 20.28°C (B) 25.62°C
(C) 23.84°C (D) 28.32°C

22. The height of Victoria falls is 63 m. What is the difference in temperature of water at the top and at the bottom of fall? [Given 1 cal = 4.2 J and specific heat of water = 1 cal g^{-1}° C^{-1}] :

[27 Aug. 2021 (Evening)]

(A) 14.76 °C (B) 1.476°C
(C) 0.147°C (D) 0.014°C

23. Due to cold weather a 1 m water pipe of cross-sectional area 1 cm^2 is filled with ice at – 10°C. Resistance heating is used to melt the ice. Current of 0.5 A is passed through 4 kΩ resistance. Assuming that all the heat produced is used for melting, what is the minimum time required ?

(Given latent heat of fusion for water/ice = 3.33×10^5 J kg^{-1}, specific heat of ice = 2×10^3 J kg^{-1} and density of ice = 10^3 kg/m^3)

[01 Sep 2021 Evening]

(A) 0.353 s (B) 3.53 s
(C) 70.6 s (D) 35.3 s

24. A 100 g of iron nail is hit by a 1.5 kg hammer striking at a velocity of 60 ms^{-1}. What will be the rise in the temperature of the nail if one fourth of energy of the hammer goes into heating the nail ? **[24 June 2022 (Evening)]**

[Specific heat capacity of iron = 0.42 Jg^{-1} °C^{-1}]

(A) 675°C (B) 1600°C
(C) 16.07°C (D) 6.75°C

25. A steam engine intakes 50 g of steam at 100°C per minute and cools it down to 20°C. If latent heat of vaporization of steam is 540 cal g^{-1}, then the heat rejected by the steam engine per minute is _____ $\times 10^3$ Cal. **[25 June 2022 (Morning)]**

26. A copper block of mass 5.0 kg is heated to a temperature of 500°C and is placed on a large ice block. What is the maximum amount of ice that can melt ?

[Specific heat of copper : 0.39 Jg^{-1} °C^{-1} and latent heat of fusion of water : 335 J g^{-1}] : **[25 June 2022 (Evening)]**

(A) 1.5 kg (B) 5.8 kg
(C) 2.9 kg (D) 3.8 kg

27. A geyser heats water flowing at a rate of 2.0 kg per minute from 30°C to 70°C. If geyser operates on a gas burner, the rate of combustion of fuel will be _____ g min^{-1}. [Heat of combustion = 8×10^3 Jg^{-1}, Specific heat of water = 4.2 Jg^{-1} °C^{-1}]

[26 June 2022 (Evening)]

28. A block of ice of mass 120g at temperature 0°C is put in 300 gm of water at 25°C. The xg of ice melts as the temperature of the water reaches 0°C. The value of x is.

[Use : Specific heat capacity of water = 4200 Jkg^{-1} K^{-1}, Latent heat of ice = 3.5×10^5 Jkg^{-1}] **[25 July 2022 (Evening)]**

29. Heat energy of 184 kJ is given to ice of mass 600 g at −12°C, Specific heat of ice is 2222.3 J kg^{-1}° C^{-1} and latent heat of ice in 336 kJ/kg^{-1}.

(a) Final temperature of system will be 0°C

(b) Final temperature of the system will be greater than 0°C

(c) The final system will have a mixture of ice and water in the ratio of 5 : 1

(d) The final system will have a mixture of ice and water in the ratio of 1 : 5

(e) The final system will have water only

Choose the correct answer from the options given below :

[29 Jan. 2023 (Evening)]

(A) (a) and (d) only (B) (b) and (d) only
(C) (a) and (e) only (D) (a) and (c) only

30. A water heater of power 2000 W is used to heat water. The specific heat capacity of water is 4200 J kg^{-1} K^{-1}. The efficiency of heater is 70%. Time required to heat 2 kg of water from 10°C to 60°C is _____ s.

(Assume that the specific heat capacity of water remains constant over the temperature range of the water)

[31 Jan. 2023 (Evening)]

* * * * *

CHAPTER

10 THERMAL PROPERTIES OF MATTER

10.3 HEAT TRANSFER

1. Two spheres of the same material have radii 1 m and 4 m and temperatures 4000 K and 2000 K respectively. The ratio of the energy radiated per second by the first sphere to that by the second is : **[2002]**

(A) 1 : 1 (B) 16 : 1

(C) 4 : 1 (D) 1 : 9

2. Infrared radiations are detected by : **[2002]**

(A) Spectrometer (B) Pyrometer

(C) Nanometer (D) Photometer

3. Which of the following is more close to a black body : **[2002]**

(A) Black board paint (B) Green leaves

(C) Black holes (D) Red roses

4. The earth radiates in the infrared region of the spectrum. The spectrum is correctly given by : **[2003]**

(A) Rayleigh Jeans law (B) Plank's law of radiation

(C) Stefan's law of radiation (D) Wien's law

5. According to Newton's law of cooling, the rate of cooling of a body is proportional to $(\Delta\theta)^n$, where $\Delta\theta$ is the difference of the temperature of the body and the surroundings, and n is equal to : **[2003]**

(A) Two (B) Three

(C) Four (D) One

6. If the temperature of the sun were to increase from T to $2T$ and its radius from R to $2R$, then the ratio of the radiant energy received on earth to what it was previously, will be : **[2004]**

(A) 4 (B) 16

(C) 32 (D) 64

7. The temperature of the two outer surfaces of a composite slab, consisting of two materials having coefficients of thermal conductivity K and $2K$ and thickness x and $4x$, respectively are T_2 and T_1 $(T_2 > T_1)$. The rate of heat transfer through the slab, in a steady state is $\left(\frac{A(T_2 - T_1)K}{x}\right) f$, with f equals to : (A = cross sectional area) **[2004]**

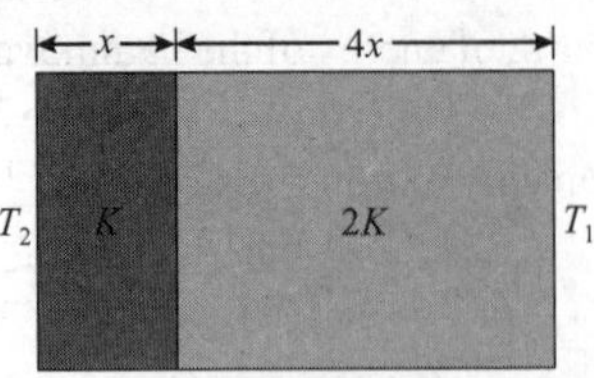

(A) 1 (B) 1/2

(C) 2/3 (D) 1/3

8. The figure shows a system of two concentric spheres of radii r_1 and r_2 and kept at temperatures T_1 and T_2, respectively. The radial rate of flow of heat in a substance between the two concentric spheres, is proportional to : **[2005]**

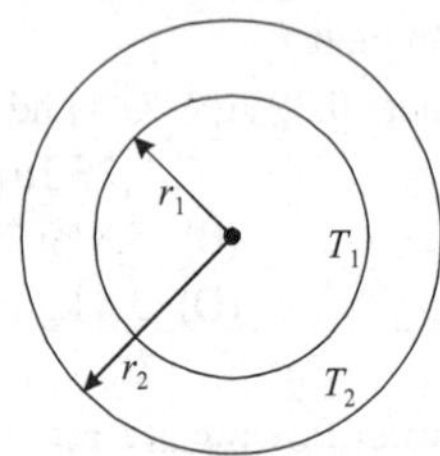

(A) $\frac{(r_2 - r_1)}{(r_1 r_2)}$ (B) $\ln\left(\frac{r_2}{r_1}\right)$

(C) $\frac{r_1 r_2}{(r_2 - r_1)}$ (D) $(r_2 - r_1)$

9. Assuming the sun to be a spherical body of radius R at a temperature of T K, evaluate the total radiant power, incident on earth, at a distance r from the sun-(when radius of earth is r_0) : **[2006]**

(A) $4\pi r_0^2 R^2 \sigma T^4 / r^2$ (B) $\pi r_0^2 R^2 \sigma T^4 / r^2$

(C) $r_0^2 R^2 \sigma T^4 / 4\pi r^2$ (D) $R^2 \sigma T^4 / r^2$

10. One end of a thermally insulated rod is kept at a temperature T_1 and the other at T_2. The rod is composed of two sections of lengths l_1 and l_2 and thermal conductivities K_1 and K_2 respectively. The temperature at the interface of the two sections is : **[2007]**

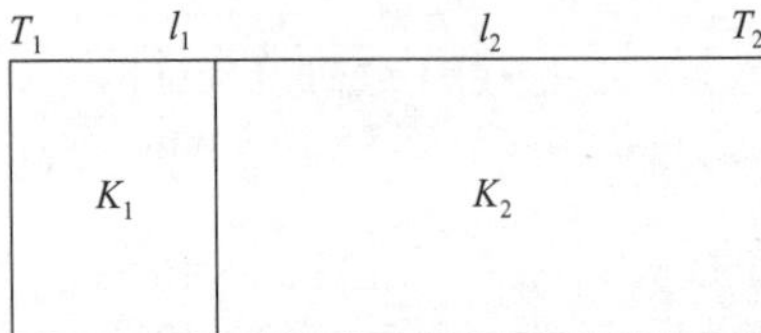

(A) $(K_2 l_2 T_1 + K_1 l_1 T_2)/(K_1 l_1 + K_2 l_2)$

(B) $(K_2 l_1 T_1 + K_1 l_2 T_2)/(K_2 l_1 + K_1 l_2)$

(C) $(K_1 l_2 T_1 + K_2 l_1 T_2)/(K_1 l_2 + K_2 l_1)$

(D) $(K_1 l_1 T_1 + K_2 l_2 T_2)/(K_1 l_1 + K_2 l_2)$

11. A long metallic bar is carrying heat from one of its ends to the other end under steady-state. The variation of temperature θ along the length x of the bar from its hot end is best described by which of the following figures? **[2009]**

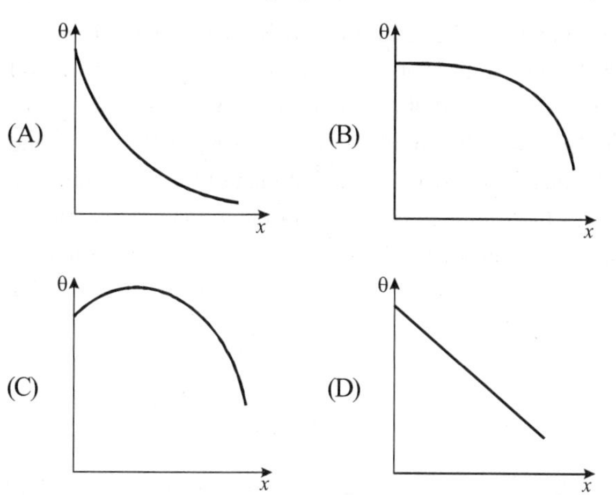

12. A liquid in a beaker has temperature $\theta(t)$ at time t and θ_0 is temperature of surroundings, then according to Newton's law of cooling the correct graph between $\log_e(\theta - \theta_0)$ and t is : **[2012]**

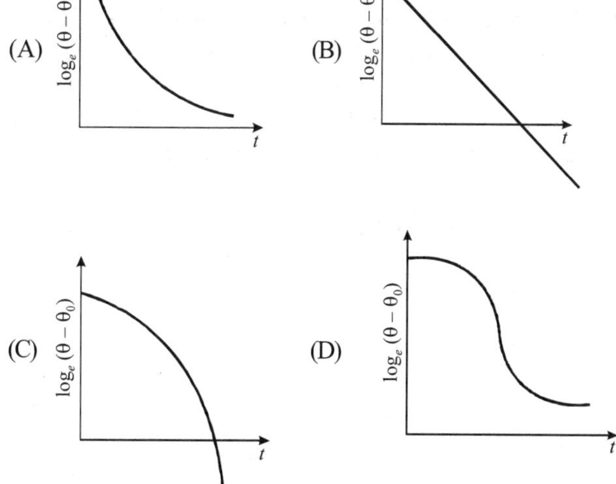

13. The heat radiated per unit area in 1 hour by a furnace whose temperature is 3000 K is ($\sigma = 5.7 \times 10^{-8}\ Wm^{-2}\ K^{-4}$) : **[2012 (Online)]**

(A) 1.7×10^{10} J (B) 1.1×10^{12} J

(C) 2.8×10^{8} J (D) 4.6×10^{6} J

14. A large cylindrical rod of length L is made by joining two identical rods of copper and steel of length $\left(\frac{L}{2}\right)$ each. The rods are completely insulated from the surroundings. If the free end of copper rod is maintained at 100°C and that of steel at 0°C then the temperature of junction is (Thermal conductivity of copper is 9 times that of steel) : **[2012 (Online)]**

(A) 90°C (B) 50°C

(C) 10°C (D) 67°C

15. If a piece of metal is heated to temperature θ and then allowed to cool in a room which is at temperature θ_0 the graph between the temperature T of the metal and time t will be closed to : **[2013]**

(A)

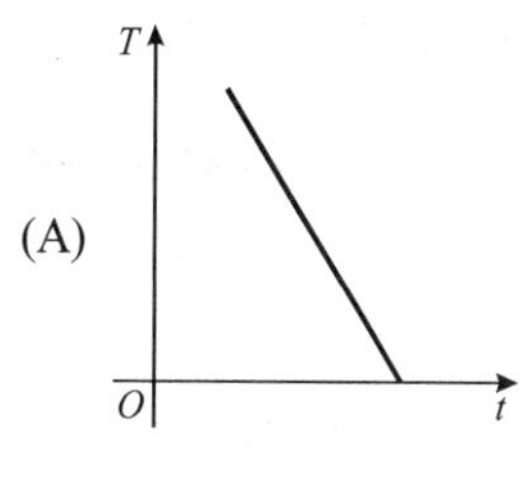

(B)

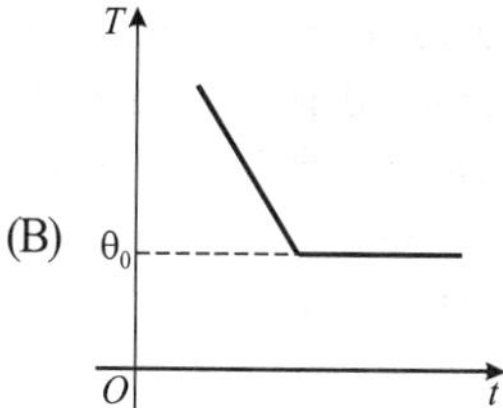

(C)

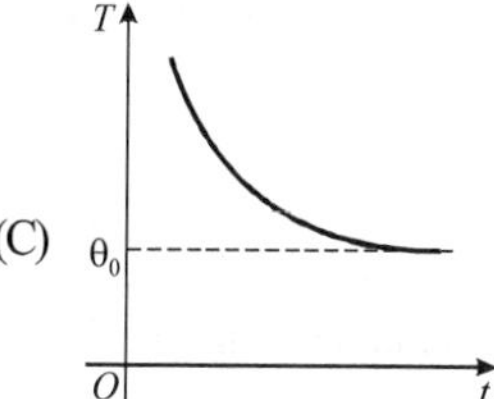

(D)

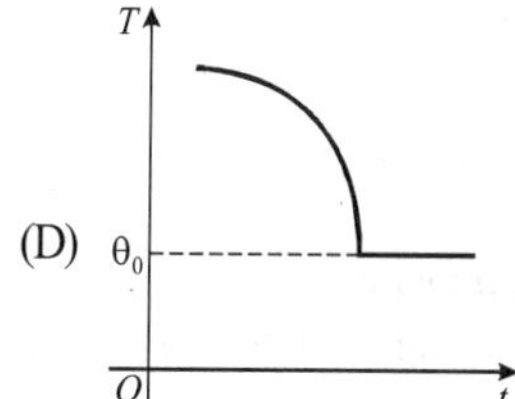

16. A mass of 50g of water in a closed vessel, with surroundings at a constant temperature takes 2 minutes to cool from 30°C to 25°C. A mass of 100 g of another liquid in an identical vessel with identical surroundings takes the same time to cool from 30° C to 25°C. The specific heat of the liquid is :
(The water equivalent of the vessel is 30 g) : **[2013 (Online)]**

(A) 2.0 kcal/kg (B) 7 kcal/kg

(C) 3 kcal/kg (D) 0.5 kcal/kg

17. Three rods of Copper, Brass and Steel are welded together to form a Y-shaped structure. Area of cross-section of each rod = 4 cm^2. End of copper rod is maintained at 100°C where as ends of brass and steel are kept at 0°C. Lengths of the copper, brass and steel rods are 46, 13 and 12 cms respectively. The rods are thermally insulated from surroundings except at ends. Thermal conductivities of copper, brass and steel are 0.92, 0.26 and 0.12 *CGS* units respectively. Rate of heat flow through copper rod is : **[2014]**

(A) 1.2 cal/s (B) 2.4 cal/s

(C) 4.8 cal/s (D) 6.0 cal/s

18. A black coloured solid sphere of radius R and mass M is inside a cavity with vacuum inside. The walls of the cavity are maintained at temperature T_0. The initial temperature of the sphere is $3T_0$. If the specific heat of the material of the sphere varies as αT^3 per unit mass with the temperature T of the sphere, where α is a constant, then the time taken for the sphere to cool down to temperature $2T_0$ will be (σ is Stefan Boltzmann constant): **[2014 (Online)]**

(A) $\frac{M\alpha}{4\pi R^2\sigma}\ln\left(\frac{3}{2}\right)$ (B) $\frac{M\alpha}{4\pi R^2\sigma}\ln\left(\frac{16}{3}\right)$

(C) $\frac{M\alpha}{16\pi R^2\sigma}\ln\left(\frac{16}{3}\right)$ (D) $\frac{M\alpha}{16\pi R^2\sigma}\ln\left(\frac{3}{2}\right)$

19. A hot body, obeying Newton's law of cooling is cooling down from its peak value 80°C to an ambient temperature of 30°C. it takes 5 minutes in cooling down from 80°C to 40°C. How much time will it take to cool down from 62°C to 32°C? (Given In 2 = 0.693, In 5 = 1.609): **[2014 (Online)]**

(A) 3.75 minutes (B) 8.6 minutes
(C) 9.6 minutes (D) 6.5 minutes

20. Hot water cools from 60°C to 50°C in the first 10 minutes and to 42°C in the next 10 minutes. The temperature of the surroundings is : **[2014 (Online)]**

(A) 25°C (B) 10°C
(C) 15°C (D) 20°C

21. Temperature difference of 120°C is maintained between two ends of a uniform rod AB of length $2L$. Another bent rod PQ, of same cross-section as AB and length $\frac{3L}{2}$, is connected across AB (See figure). In steady state, temperature difference between P and Q will be close to : **[9 Jan. 2019 (Morning)]**

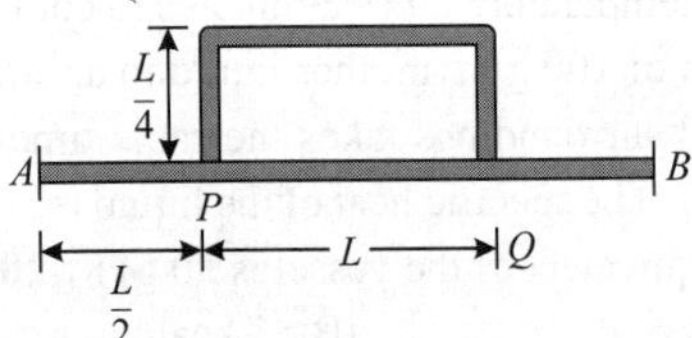

(A) 60°C (B) 75°C
(C) 35°C (D) 45°C

22. A heat source at $T = 10^3$ K is connected to another heat reservoir at $T = 10^2$ K by a copper slab which is 1 m thick. Given that the thermal conductivity of copper is 0.1 WK^{-1} m^{-1}, the energy flux through it in the steady state is : **[10 Jan. 2019 (Morning)]**

(A) 90 Wm^{-2} (B) 200 Wm^{-2}
(C) 65 Wm^{-2} (D) 120 Wm^{-2}

23. A cylinder of radius R is surrounded by a cylindrical shell of inner radius R and outer radius $2R$. The thermal conductivity of the material of the inner cylinder is K_1 and that of the outer cylinder is K_2. Assuming no loss of heat, the effective thermal conductivity of the system for heat flowing along the length of the cylinder is : **[12 Jan. 2019 (Morning)]**

(A) $K_1 + K_2$ (B) $\frac{K_1 + K_2}{2}$

(C) $\frac{2K_1 + 3K_2}{5}$ (D) $\frac{K_1 + 3K_2}{4}$

24. Two identical beakers A and B contain equal volumes of two different liquids at 60°C each and left to cool down. Liquid in A has density of 8×10^2 kg/m^3 and specific heat of 2000 J kg^{-1} K^{-1} while liquid in B has density of 10^3 kg m^{-3} and specific heat of 4000 J kg^{-1} K^{-1}. Which of the following best describes their temperature versus time graph schematically? (Assume the emissivity of both the beakers to be the same) **[8 April 2019 (Morning)]**

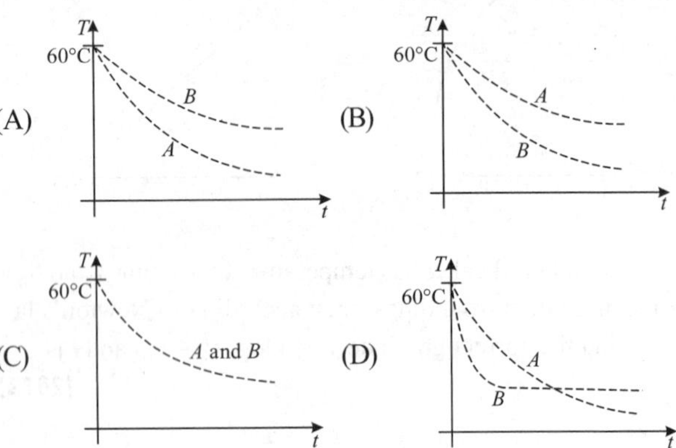

25. Two materials having coefficients of thermal conductivity '$3K$' and 'K' and thickness 'd' and '$3d$', respectively, are joined to form a slab as shown in the figure. The temperatures of the outer surfaces are 'θ_2' and 'θ_1' respectively, ($\theta_2 > \theta_1$). The temperature at the interface is : **[9 April 2019 (Evening)]**

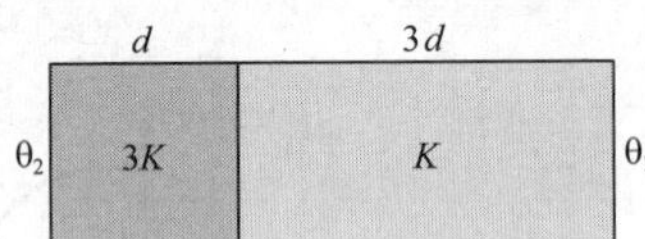

(A) $\frac{\theta_2 + \theta_1}{2}$ (B) $\frac{\theta_1}{10} + \frac{9\theta_2}{10}$

(C) $\frac{\theta_1}{3} + \frac{2\theta_2}{3}$ (D) $\frac{\theta_1}{6} + \frac{5\theta_2}{6}$

26. A metallic sphere cools from 50°C to 40°C in 300 s. If atmospheric temperature around is 20°C, then the sphere's temperature after the next 5 minutes will be close to : **[3 Sept. 2020 (Evening)]**

(A) 33°C (B) 35°C
(C) 31°C (D) 28°C

27. Amount of solar energy received on the earth's surface per unit area per unit time is defined a solar constant. Dimension of solar constant is : **[3 Sept. 2020 (Evening)]**

(A) ML^2T^{-2} (B) MLT^{-2}
(C) $M^2L^0T^{-1}$ (D) ML^0T^{-3}

28. Three rods of identical cross-section and lengths are made of three different materials of thermal conductivity K_1, K_2, and K_3, respectively. They are joined together at their ends to make a long rod (see figure). One end of the long rod is maintained at 100°C and the other at 0°C (see figure). If the joints of the rod are at 70°C and 20°C in steady state and there is no loss of energy from the surface of the rod, the correct relationship between K_1, K_2 and K_3 is : **[6 Sept. 2020 (Evening)]**

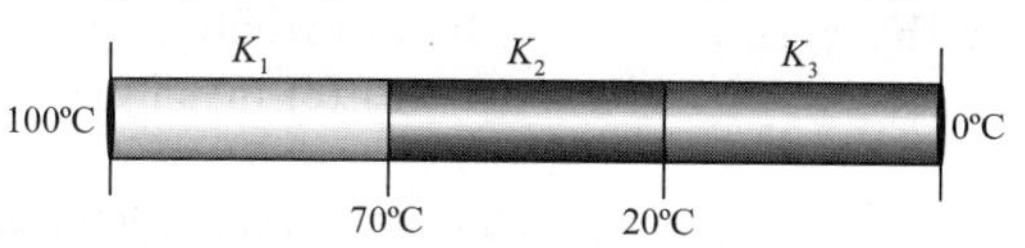

(A) $K_1 : K_3 = 2 : 3; K_2 : K_3 = 2 : 5$
(B) $K_1 < K_2 < K_3$
(C) $K_1 : K_2 = 5 : 2; K_1 : K_3 = 3 : 5$
(D) $K_1 > K_2 > K_3$

29. The temperature θ at the junction of two insulating sheets, having thermal resistances R_1 and R_2 as well as top and bottom temperatures θ_1 and θ_2 (as shown in figure) is given by : **[26 Feb. 2021 (Morning)]**

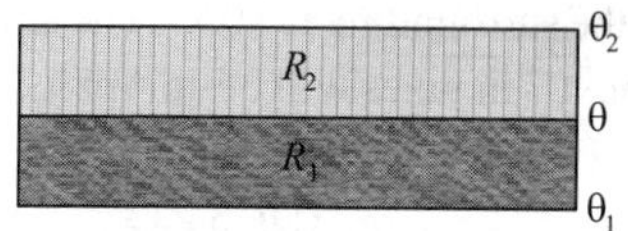

(A) $\frac{\theta_1 R_2 + \theta_2 R_1}{R_1 + R_2}$ (B) $\frac{\theta_2 R_2 - \theta_1 R_1}{R_2 - R_1}$

(C) $\frac{\theta_1 R_1 + \theta_2 R_2}{R_1 + R_2}$ (D) $\frac{\theta_1 R_1 - \theta_2 R_1}{R_2 - R_1}$

30. Two identical metal wires of thermal conductivities K_1 and K_2 respectively are connected in series. The effective thermal conductivity of the combination is : **[17 March 2021 (Morning)]**

(A) $\frac{2K_1 K_2}{K_1 + K_2}$ (B) $\frac{K_1 + K_2}{2K_1 K_2}$

(C) $\frac{K_1 + K_2}{K_1 K_2}$ (D) $\frac{K_1 K_2}{K_1 + K_2}$

31. In 5 minutes, a body cools from 75°C to 65°C at room temperature of 25°C. The temperature of body at the end of next 5 minutes is _____°C. **[22 July 2021 (Evening)]**

32. A body takes 4 min to cool from 61°C to 59°C. If the temperature of the surroundings is 30°C, the time taken by the body to cool from 51°C to 49°C is : **[27 July 2021 (Morning)]**

(A) 3 min (B) 8 min
(C) 6 min (D) 4 min

33. A rod *CD* of thermal resistance 10.0 KW^{-1} is joined at the middle of an identical rod *AB* as shown in figure. The end *A*, *B* and *D* are maintained at 200°C, 100°C and 125°C respectively. The heat current in *CD* is *P* watt. The value of *P* is _____. **[27 Aug. 2021 (Morning)]**

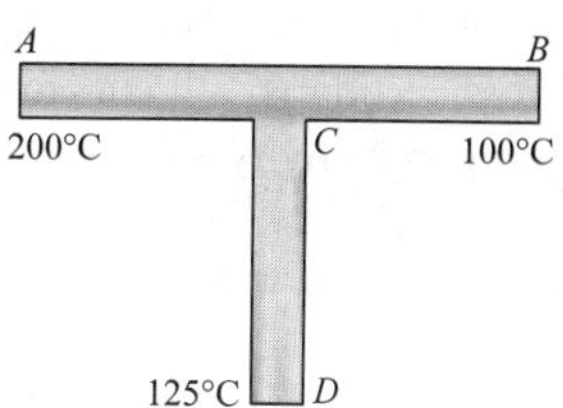

34. Two thin metallic spherical shells of radii r_1 and r_2 ($r_1 < r_2$) are placed their centres coinciding. A material of thermal conductivity *K* is filled in the space between the shells. The inner shell is maintained at temperature θ_1 and the outer shell at temperature θ_2 ($\theta_1 < \theta_2$). The rate at which heat flows radially through the material is : **[31 Aug. 2021 (Evening)]**

(A) $\frac{K(\theta_2 - \theta_1)(r_2 - r_1)}{4\pi r_1 r_2}$ (B) $\frac{\pi r_1 r_2 (\theta_2 - \theta_1)}{r_2 - r_1}$

(C) $\frac{K(\theta_2 - \theta_1)}{r_2 - r_1}$ (D) $\frac{4\pi K r_1 r_2 (\theta_2 - \theta_1)}{r_2 - r_1}$

35. Two metallic blocks M_1 and M_2 of same area of cross-section are connected to each other (as shown in figure). If the thermal conductivity of M_2 is *K* then the thermal conductivity of M_1 will be : [Assume steady state heat conduction] **[24 June 2022 (Morning)]**

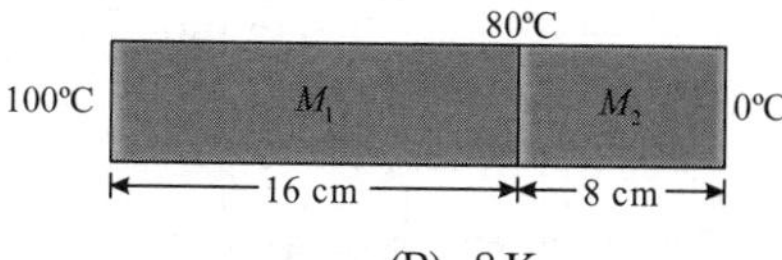

(A) 10K (B) 8K
(C) 12.5K (D) 2K

36. In an experiment to verify Newton's law of cooling, a graph is plotted between, the temperature difference (ΔT) of the water and surroundings and time as shown in figure. The initial temperature of water is taken as 80°C. The value of t_2 as mentioned in the graph will be _____. **[24 June 2022 (Evening)]**

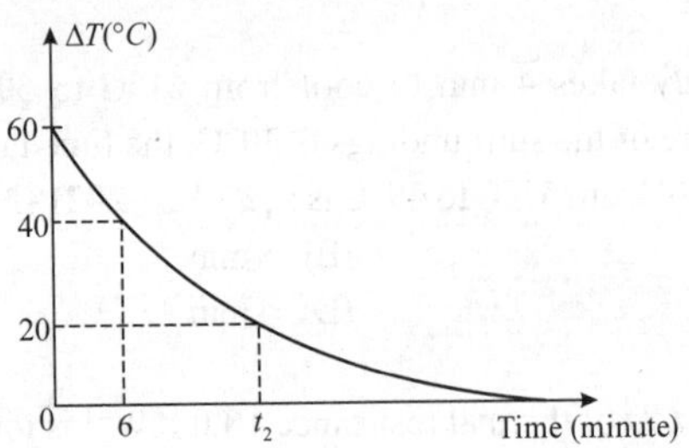

37. As per the given figure, two plates A and B of thermal conductivity K and $2K$ are joined together to form a compound plate. The thickness of plates are 4.0 cm and 2.5 cm respectively and the area of cross-section is 120 cm^2 for each plate. The equivalent thermal conductivity of the compound plate is $\left(1+\frac{5}{\alpha}\right)K$, then the value of α will be _____.

[29 June 2022 (Morning)]

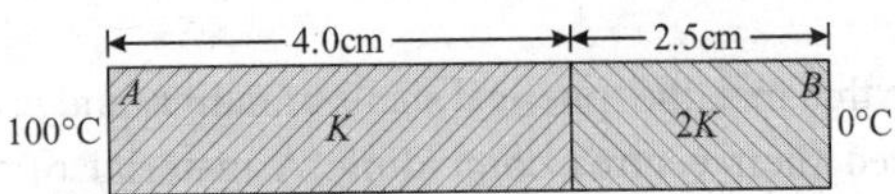

38. An ice cube of dimensions 60 cm × 50 cm × 20 cm is placed in an insulation box of wall thickness 1 cm. The box keeping the ice cube at 0°C of temperature is brought to a room of temperature 40°C. The rate of melting of ice is approximately : [Latent heat of fusion of ice is 3.4×10^5 J kg^{-1} and thermal conducting of insulation wall is 0.05 Wm^{-1}°C^{-1}] **[26 July 2022 (Evening)]**

(A) 61×10^{-1} kg s^{-1}
(B) 61×10^{-5} kg s^{-1}
(C) 208 kg s^{-1}
(D) 30×10^{-5} kg s^{-1}

39. If K_1 and K_2 are the thermal conductivities L_1 and L_2 are the lengths and A_1 and A_2 are the cross sectional areas of steel and copper rods respectively such that $\frac{K_2}{K_1}=9, \frac{A_1}{A_2}=2, \frac{L_1}{L_2}=2.$ Then, for the arrangement as shown in the figure, the value of temperature T of the steel-copper junction in the steady state will be : **[27 July 2022 (Morning)]**

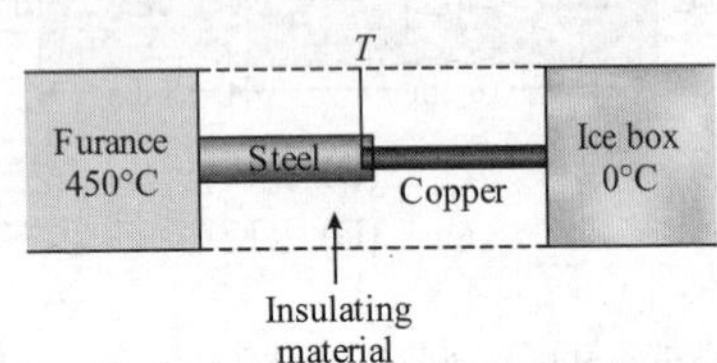

(A) 18°C
(B) 14°C
(C) 45°C
(D) 150°C

40. Read the following statements :

(a) When small temperature difference between a liquid and its surrounding is doubled the rate of loss of heat of the liquid becomes twice.

(b) Two bodies P and Q having equal surface areas are maintained at temperature 10°C and 20°C. The thermal radiation emitted in a given time by P and Q are in the ratio 1 : 1.15.

(c) A carnot Engine working between 100 K and 400 K has an efficiency of 75%.

(d) When small temperature difference between a liquid and its surrounding is quadrupled, the rate of loss of heat of the liquid becomes twice.

Choose the correct answer from the options given below : **[27 July 2022 (Morning)]**

(A) (a), (b), (c) only
(B) (a), (b) only
(C) (a), (c) only
(D) (b), (c), (d) only

41. A bowl filled with very hot soup cools from 98°C to 86°C in 2 minutes when the room temperature is 22°C. How long it will take to cool from 75°C to 69°C ? **[25 Jan. 2023 (Morning)]**

(A) 2 minutes
(B) 1.4 minutes
(C) 0.5 minute
(D) 1 minute

42. A body cools from 60°C to 40°C in 6 minutes. If, temperature of surroundings is 10°C. Then, after the next 6 minutes, its temperature will be _____°C. **[29 Jan. 2023 (Morning)]**

43. A body cools in 7 minutes from 60°C to 40°C. The temperature of the surrounding is 10°C. The temperature of the body after the next 7 minutes will be: **[6 April 2023 (Evening)]**

(A) 32°C
(B) 30°C
(C) 28°C
(D) 34°C

44. A body cools from 80°C to 60°C in 5 minutes. The temperature of the surrounding is 20°C. The time it takes to cool from 60°C to 40°C is: **[12 April 2023 (Morning)]**

(A) 500 s
(B) $\frac{25}{3}$ s
(C) 450 s
(D) 420 s

45. Two plates A and B have thermal conductivities 84 Wm^{-1}K^{-1} and 126 Wm^{-1} K^{-1} respectively. They have same surface area and same thickness. They are placed in contact along their surfaces. If the temperatures of the outer surfaces of A and B are kept at 100 °C and 0 °C respectively, then the temperature of the surface of contact in steady state is _____°C.

[13 April 2023 (Evening)]

* * * * *

CHAPTER 11 THERMODYNAMICS

11.1 LAWS OF THERMODYNAMICS

1. Which statement is incorrect ? **[2002]**

(A) All reversible cycles have same efficiency

(B) Reversible cycle has more efficiency than an irreversible one

(C) Carnot cycle is a reversible one

(D) Carnot cycle has the maximum efficiency in all cycles

2. Even carnot engine cannot give 100% efficiency because we cannot : **[2002]**

(A) Prevent radiation

(B) Find ideal sources

(C) Reach absolute zero temperature

(D) Eliminate friction

3. A carnot engine takes 3×10^6 cal of heat from a reservoir at 627 °C and gives it to a sink at 27 °C. The work done by the engine is : **[2003]**

(A) 4.2×10^6 J (B) 8.4×10^6 J

(C) 16.8×10^6 J (D) Zero

4. During an adiabatic process, the pressure of a gas is found to be proportional to the cube of its absolute temperature. The ratio C_P/C_V for the gas is : **[2003]**

(A) 4/3 (B) 2

(C) 5/3 (D) 3/2

5. Which of the following parameters does not characterise the thermodynamic state of matter ? **[2003]**

(A) Temperature (B) Pressure

(C) Work (D) Volume

6. "Heat cannot be itself flow from a body at lower temperature to a body at higher temperature" is a statement or consequence of : **[2003]**

(A) Second law of thermodynamics

(B) Conservation of momentum

(C) Conservation of mass

(D) First law of thermodynamics

7. Which of the following statements is correct for any thermodynamic system ? **[2004]**

(A) The internal energy changes in all processes

(B) Internal energy and entropy are state functions

(C) The change in entropy can never be zero

(D) The work done in an adiabatic process is always zero

8. A system goes from A to B via two processes I and II as shown in figure. If ΔU_1 and ΔU_2 are the changes in internal energies in the processes I and II respectively then : **[2005]**

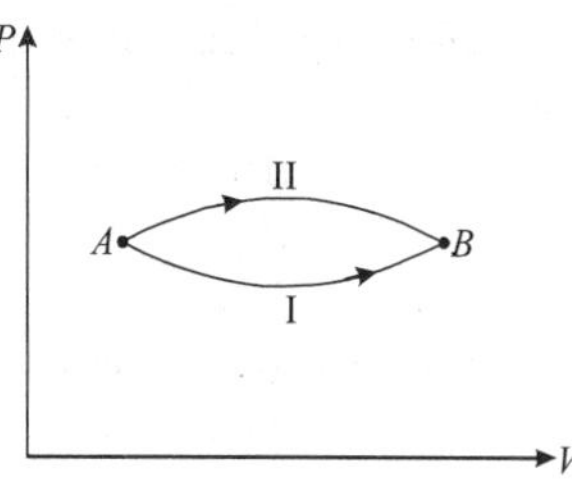

(A) $\Delta U_1 = \Delta U_2$

(B) Relation between ΔU_1 and ΔU_2 cannot be determined

(C) $\Delta U_2 > \Delta U_1$ (D) $\Delta U_2 < \Delta U_1$

9. The temperature-entropy diagram of a reversible engine cycle is given in the figure. Its efficiency is : **[2005]**

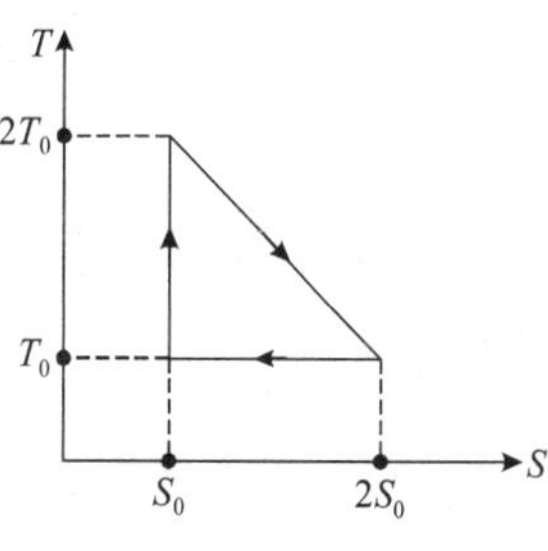

(A) 1/2 (B) 1/4

(C) 1/3 (D) 2/3

10. Which of the following is incorrect regarding the first law of thermodynamics ? **[2005]**

(A) It is applicable to any cyclic process

(B) It is a restatement of the principle of conservation of energy

(C) It introduces the concept of the internal energy

(D) It introduced the concept of the entropy

11. The work of 146 kJ is performed in order to compress one kilo mole of a gas adiabatically and in this process the temperature of the gas increases by 7°C. The gas is : $[R = 8.3 \text{ J mol}^{-1} \text{K}^{-1}]$ **[2006]**

(A) Diatomic (B) Triatomic

(C) A mixture of monoatomic and diatomic

(D) Monoatomic

12. When a system is taken from state i to state f along the path iaf, it is found that $Q = 50$ cal and $W = 20$ cal. Along the path ibf $Q = 36$ cal. W along the path ibf is : **[2007]**

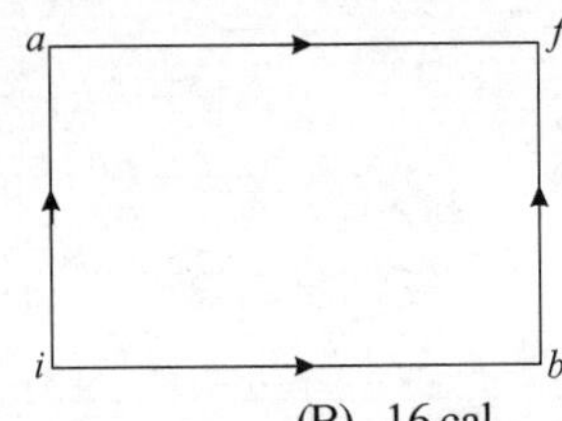

(A) 6 cal (B) 16 cal
(C) 66 cal (D) 14 cal

13. A carnot engine, having an efficiency of $\eta = 1/10$ as heat engine, is used as a refrigerator. If the work done on the system is 10 J, the amount of energy absorbed from the reservoir at lower temperature is : **[2007]**

(A) 99 J (B) 90 J
(C) 1 J (D) 100 J

14. Directions : This question is based on the following paragraph. Two moles of helium gas are taken over the cycle *ABCDA*, as shown in the *P-T* diagram.

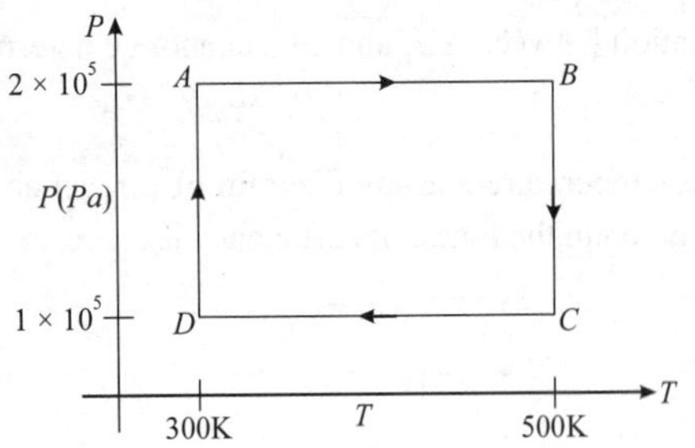

Assuming the gas to be ideal, the work done on the gas in taking it from *A* to *B* is : **[2009]**

(A) 400 R (B) 500 R
(C) 200 R (D) 300 R

15. Directions : This question is based on the following paragraph. Two moles of helium gas are taken over the cycle *ABCDA*, as shown in the *P-T* diagram. **[2009]**

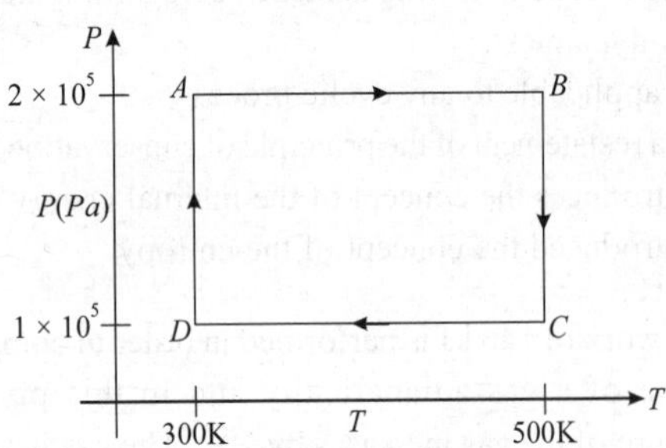

The work done on the gas in taking it from *D* to *A* is :

(A) –690 R (B) +690 R
(C) –414 R (D) +414 R

16. Directions : This question is based on the following paragraph.
Two moles of helium gas are taken over the cycle *ABCDA*, as shown in the *P-T* diagram.

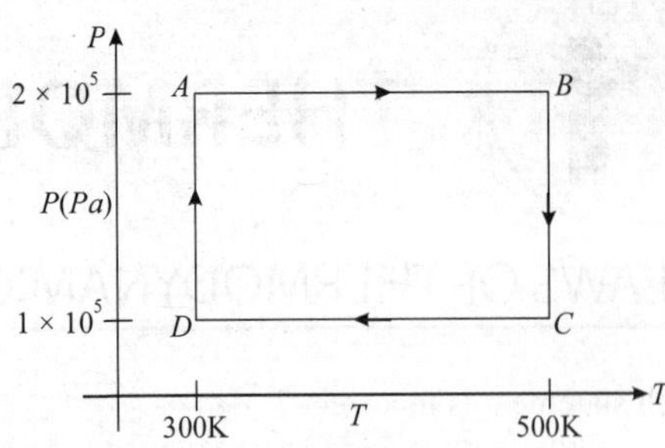

The net work done on the gas in the cycle *ABCDA* is : **[2009]**

(A) 1076 R (B) 1904 R
(C) Zero (D) 276 R

17. A diatomic ideal gas is used in a carnot engine as the working substance. If during the adiabatic expansion part of the cycle the volume of the gas increases from V to $32V$, the efficiency of the engine is : **[2010]**

(A) 0.25 (B) 0.5
(C) 0.75 (D) 0.99

18. 100 g of water is heated from 30°C to 50°C Ignoring the slight expansion of the water, the change in its internal energy is (specific heat of water is 4184 J/kg/K) : **[2011]**

(A) 84 kJ (B) 2.1 kJ
(C) 4.2 kJ (D) 8.4 kJ

19. A container with insulating walls is divided into two equal parts by a partition fitted with a valve. One part is filled with an ideal gas at a pressure P and temperature T, whereas the other part is completely evacuated. If the valve is suddenly opened, the pressure and temperature of the gas will be : **[2011]**

(A) $\frac{P}{2}, T$ (B) $\frac{P}{2}, \frac{T}{2}$

(C) P, T (D) $P, \frac{T}{2}$

20. The specific heat capacity of a metal at low temperature (T) is given as $C_p(\text{kJK}^{-1}\,\text{kg}^{-1}) = 32\left(\frac{T}{400}\right)^3$. A 100 gram vessel of this metal is to be cooled from 20 K to 4 K by a special refrigerator operating at room temperature (27°C). The amount of work required to cool the vessel is : **[2011]**

(A) Equal to 0.002 kJ
(B) Greater than 0.148 kJ
(C) Between 0.148 kJ and 0.028 kJ
(D) Less than 0.028 kJ

21. A Carnot engine operating between temperatures T_1 and T_2 has efficiency has efficiency 1/6. When T_2 is lowered by

62 K, its efficiency increases to 1/3. Then T_1 and T_2 are, respectively : **[2011]**

(A) 330 K and 268 K (B) 310 K and 248 K
(C) 372 K and 310 K (D) 372 K and 330 K

22. A thermally insulated vessel contains an ideal gas of molecular mass M and ratio of specific heats γ. It is moving with speed v and is suddenly brought to rest. Assuming no heat is lost to the surroundings, its temperature increases by : **[2011]**

(A) $\frac{\gamma M v^2}{2R}$ K (B) $\frac{(\gamma-1)}{2R} Mv^2$ K

(C) $\frac{(\gamma-1)}{2(\gamma+1)R} Mv^2$ K (D) $\frac{(\gamma-1)}{2\gamma R} Mv^2$ K

23. A Carnot engine, whose efficiency is 40% takes in heat from a source maintained at a temperature of 500 K. It is desired to have an engine of efficiency 60%. Then, the intake temperature for the same exhaust (sink) temperature must be : **[2012]**

(A) 600 K
(B) Efficiency of Carnot engine cannot be made larger than 50%
(C) 1200 K
(D) 750 K

24. Helium gas goes through a cycle $ABCDA$ (consisting of two isochoric and two isobaric lines) as shown in figure. Efficiency of this cycle is nearly (Assume the gas to be close to ideal gas) **[2012]**

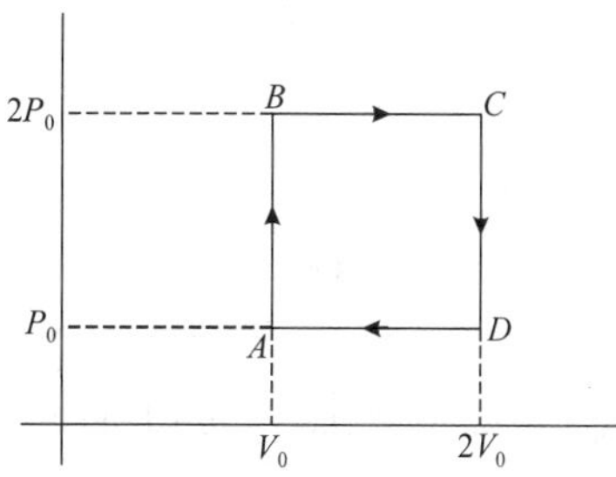

(A) 12.5% (B) 15.4%
(C) 9.1% (D) 10.5%

25. This question has Statement-1 and Statement-2. Of the four choices given after the Statements, choose the one that best describes the two Statements. **[2012 (Online)]**

Statement-1 : An inventor claims to have constructed an engine that has an efficiency of 30% when operated between the boiling and freezing points of water. This is not possible.

Statement-2 : The efficiency of a real engine is always less than the efficiency of a carnot engine operating between the same two temperatures.

(A) Statement-1 is true; Statement-2 is true, Statement-2 not the correct explanation of Statement-1
(B) Statement-1 is true, Statement-2 is false
(C) Statement-1 is false, Statement-2 is true
(D) Statement-1 is true, Statement-2 is true, Statement-2 is the correct explanation of Statement-1

26. This question has Statement-1 and Statement-2. Of the four choices given after the Statements, choose the one that best describes the two Statements. **[2012 (Online)]**

Statement-1 : In an adiabatic process, change in internal energy of a gas is equal to work done on/by the gas in the process.

Statement-2 : The temperature of a gas remains constant in an adiabatic process.

(A) Statement-1 is true; Statement-2 is true, Statement-2 is a correct explanation of Statement-1
(B) Statement-1 is true, Statement-2 is true
(C) Statement-1 is false, Statement-2 is true
(D) Statement-1 is false, Statement-2 is true, Statement-2 is not a correct explanation of Statement-1

27. n moles of an ideal gas undergo a process $A \rightarrow B$ as shown in the figure. Maximum temperature of the gas during the process is : **[2012 (Online)]**

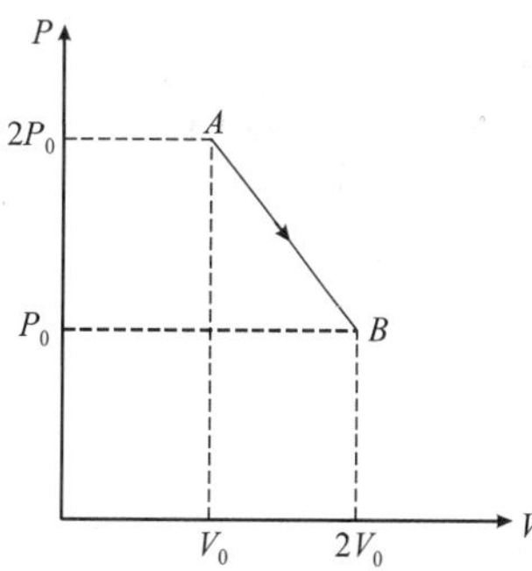

(A) $\frac{9P_0V_0}{nR}$ (B) $\frac{3P_0V_0}{2nR}$

(C) $\frac{9P_0V_0}{2nR}$ (D) $\frac{9P_0V_0}{4nR}$

28. The pressure of an ideal gas varies with volume as $P = \alpha V$, where α is a constant. One mole of the gas is allowed to undergo expansion such that its volume becomes 'm' times its initial volume. The work done by the gas in the process is : **[2012 (Online)]**

(A) $\frac{\alpha V}{2}(m^2-1)$ (B) $\frac{\alpha^2 V^2}{2}(m^2-1)$

(C) $\frac{\alpha}{2}(m^2-1)$ (D) $\frac{\alpha V^2}{2}(m^2-1)$

29. An ideal monatomic gas with pressure P, volume V and temperature T is expanded isothermally to a volume $2V$ and a final pressure P_i. If the same gas is expanded adiabatically to a volume $2V$, the final pressure is P_a. The ratio $\frac{P_a}{P_i}$ is : **[2012 (Online)]**

(A) $2^{-1/3}$ (B) $2^{1/3}$
(C) $2^{2/3}$ (D) $2^{-2/3}$

30. A given ideal gas with $\gamma = \frac{C_p}{C_v} = 1.5$ at a temperature T. If the gas is compressed adiabatically to one-fourth of its initial volume, the final temperature will be : **[2012 (Online)]**

(A) $2\sqrt{2}T$ (B) $4T$
(C) $2T$ (D) $8T$

31. The door of a working refrigerator is left open in a well insulated room. The temperature of air in the room will : **[2012 (Online)]**

(A) Decrease
(B) Increase in winters and decrease in summers
(C) Remains the same
(D) Increase

32. The above p-v diagram represents the thermodynamic cycle of an engine, operating with an ideal monoatomic gas. The amount of heat, extracted from the source in a single cycle is : **[2013]**

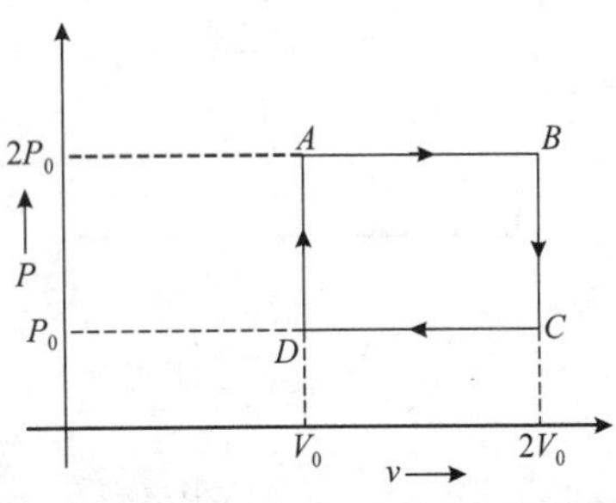

(A) p_0v_0 (B) $\left(\frac{13}{2}\right)p_0v_0$
(C) $\left(\frac{11}{2}\right)p_0v_0$ (D) $4p_0v_0$

33. An ideal gas at atmospheric pressure is adiabatically compressed so that its density becomes 32 times of its initial value. If the final pressure of gas is 128 atmospheres, the value of 'γ' of the gas is : **[2013 (Online)]**

(A) 1.5 (B) 1.4
(C) 1.3 (D) 1.6

34. A certain amount of gas is taken through a cyclic process (A, B, C, D, A) that has two isobars, one isochore and one isothermal. The cycle can be represented on a P-V indicator diagram as : **[2013 (Online)]**

(A) 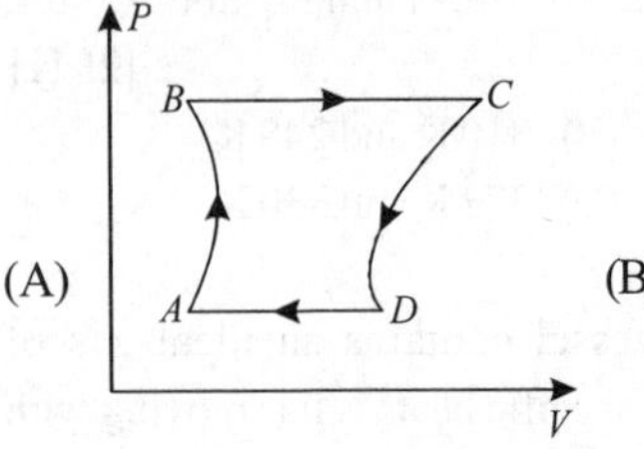

(B) 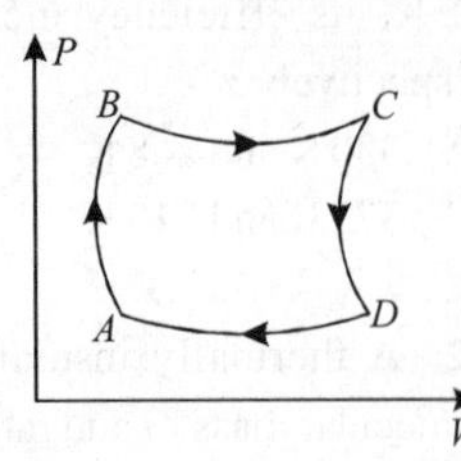

(C) 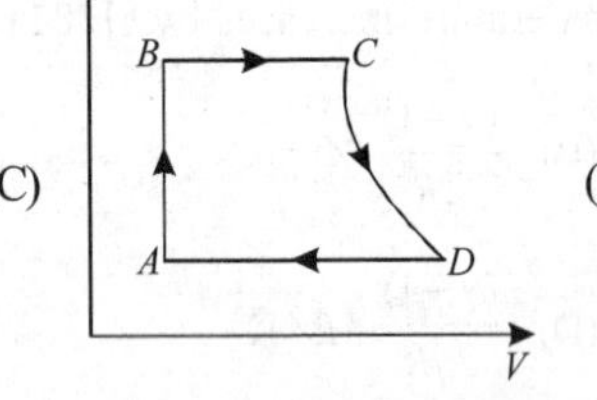

(D) 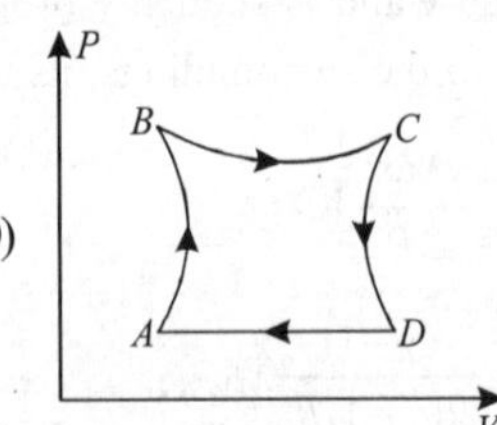

35. In the isothermal expansion of 10 g of gas from volume V to 2 V the work done by the gas is 575 J. What is the root mean square speed of the molecules of the gas at that temperature ? **[2013 (Online)]**

(A) 398 m/s (B) 520 m/s
(C) 499 m/s (D) 532 m/s

36. One mole of diatomic ideal gas undergoes a cyclic process ABC as shown in figure. The process BC is adiabatic. The temperatures at A, B and C are 400 K, 800 K and 600 K respectively. Choose the correct statement : **[2014]**

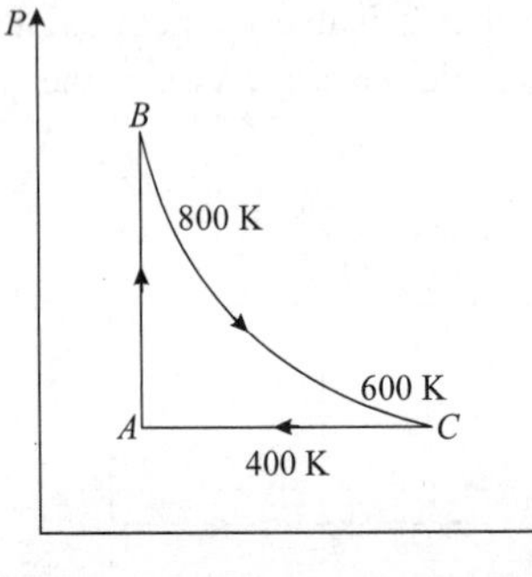

(A) The change in internal energy in whole cyclic process is 250 R
(B) The change in internal energy in the process CA is 700 R
(C) The change in internal energy in the process AB is –350 R
(D) The change in internal energy in the process BC is –500 R

37. The equation of state for a gas is given by $PV = nRT + \alpha V$, where n is the number of moles and α is a positive constant. The initial temperature and pressure of one mole of the gas contained in a cylinder are T_0 and P_0 respectively. The work done by the gas when its temperature doubles isobarically will be : **[2014 (Online)]**

(A) $\frac{P_0T_0R}{P_0 - \alpha}$ (B) $\frac{P_0T_0R}{P_0 + \alpha}$
(C) $P_0T_0R \ln 2$ (D) P_0T_0R

38. An ideal monoatomic gas is confined in a cylinder by a spring loaded piston of cross section 8.0×10^{-3} m^2. Initially the gas is at 300 K and occupies a volume of 2.4×10^{-3} m^3 and the spring is in its relaxed state as shown in figure. The gas is heated by a small heater until the piston moves out slowly by 0.1 m. The force constant of the spring is 8000 N/m and the atmospheric pressure is 1.0×10^5 N/m^2. The cylinder and the piston are thermally insulated. The piston and the spring are massless and there is no friction between the piston and the cylinder. The final temperature of the gas will be :
(Neglect the heat loss through the lead wires of the heater. The heat capacity of the heater coil is also negligible) : **[2014 (Online)]**

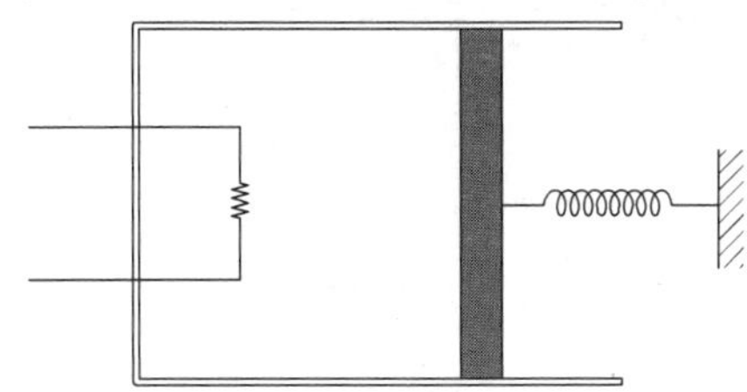

(A) 300 K (B) 800 K
(C) 500 K (D) 1000 K

39. During an adiabatic compression, 830 J of work is done on 2 moles of a diatomic ideal gas to reduce its volume by 50%. The change in its temperature is nearly : $[R = 8.3\ \text{JK}^{-1}\ \text{mol}^{-1}]$ **[2014 (Online)]**

(A) 40 K (B) 33 K
(C) 20 K (D) 14 K

40. A gas is compressed from a volume of 2 m^3 to a volume of 1 m^3 at a constant pressure of 100 N/m^2. Then it is heated at constant volume by supplying 150 J of energy. As a result, the internal energy of the gas : **[2014 (Online)]**

(A) Increases by 250 J (B) Decreases by 250 J
(C) Increases by 50 J (D) Decreases by 50 J

41. A Carnot engine absorbs 1000 J of heat energy from a reservoir at 127°C and rejects 600 J of heat energy during each cycle. The efficiency of engine and temperature of sink will be : **[2014 (Online)]**

(A) 20% and – 43°C (B) 40% and – 33°C
(C) 50% and – 20°C (D) 70% and – 10°C

42. Consider an ideal gas confined in an isolated closed chamber. As the gas undergoes an adiabatic expansion, the average time of collision between molecules increases as V^q, where V is the volume of the gas. The value of q is $\left(\gamma = \frac{C_p}{C_v}\right)$: **[2015]**

(A) $\frac{3\gamma + 5}{6}$ (B) $\frac{3\gamma - 5}{6}$

(C) $\frac{\gamma + 1}{2}$ (D) $\frac{\gamma - 1}{2}$

43. Consider a spherical shell of radius R at temperature T. The black body radiation inside it can be considered as an ideal gas of photons with internal energy per unit volume $u = \frac{U}{V} \propto T^4$ and pressure $p = \frac{1}{3}\left(\frac{U}{V}\right)$. If the shell now undergoes an adiabatic expansion the relation between T and R is : **[2015]**

(A) $T \propto \frac{1}{R}$ (B) $T \propto \frac{1}{R^3}$

(C) $T \propto e^{-R}$ (D) $T \propto e^{-3R}$

44. A solid body of constant heat capacity 1 J/°C is being heated by keeping it in contact with reservoirs in two ways : **[2015]**

(i) Sequentially keeping in contact with 2 reservoirs such that each reservoir supplied same amount of heat
(ii) Sequentially keeping in contact with 8 reservoirs such that each reservoir supplies same amount of heat
In both the cases body is brought from initial temperature 100°C to final temperature 200°C. Entropy change of the body in the two cases respectively is :

(A) ln2, 2 ln2 (B) 2 ln2, 8 ln2
(C) ln2, 4 ln2 (D) ln2, ln2

45. An ideal gas goes through a reversible cycle $a \to b \to c \to d$ has the V-T diagram shown below. Process $d \to a$ and $b \to c$ are adiabatic.

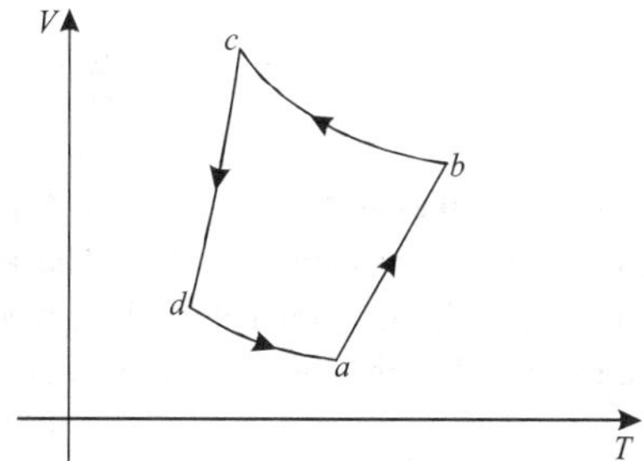

The corresponding P-V diagram for the process is (all figures are schematic and not drawn to scale) : **[2015 (Online)]**

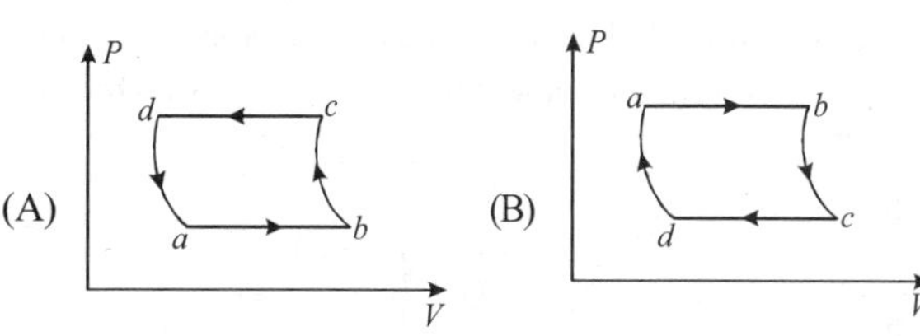

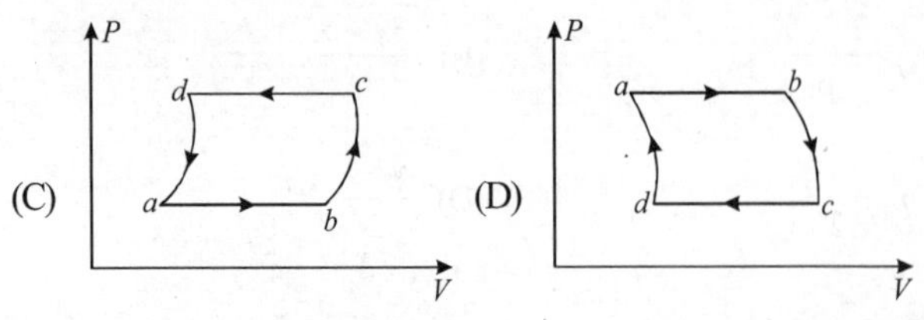

46. 'n' moles of an ideal gas undergoes a process $A \to B$ as shown in the figure. The maximum temperature of the gas during the process will be : **[2016]**

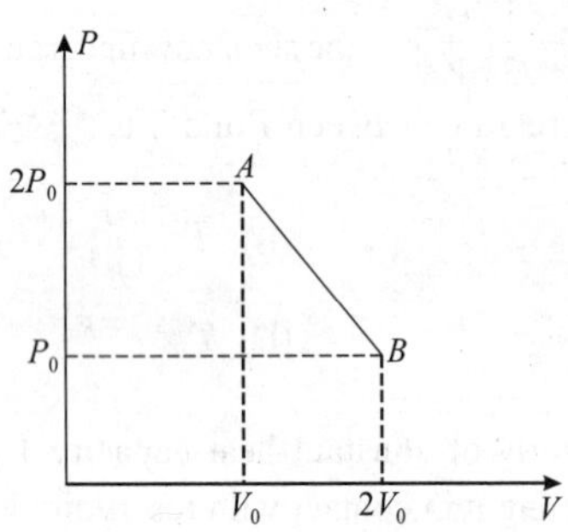

(A) $\frac{9P_0V_0}{4nR}$ (B) $\frac{3P_0V_0}{2nR}$

(C) $\frac{9P_0V_0}{2nR}$ (D) $\frac{9P_0V_0}{nR}$

47. An ideal gas undergoes a quasi static, reversible process in which its molar heat capacity C remains constant. If during this process the relation of pressure P and volume V is given by PV^n = constant, then n is given by (Here C_P and C_V are molar specific heat at constant pressure and constant volume, respectively) : **[2016]**

(A) $n = \frac{C_P}{C_V}$ (B) $n = \frac{C - C_P}{C - C_V}$

(C) $n = \frac{C_P - C}{C - C_V}$ (D) $n = \frac{C - C_V}{C - C_P}$

48. A Carnot freezer takes heat from water at 0°C inside it and rejects it to the room at a temperature of 27°C. The latent heat of ice is 336×10^3 J kg^{-1}. If 5 kg of water at 0°C is converted into ice at 0°C by the freezer, then the energy consumed by the freezer is closed to : **[2016 (Online)]**

(A) 1.51×10^5 J (B) 1.68×10^6 J

(C) 1.71×10^7 J (D) 1.67×10^5 J

49. The ratio of work done by an ideal monoatomic gas to the heat supplied to it in an isobaric process is : **[2016 (Online)]**

(A) $\frac{2}{5}$ (B) $\frac{3}{2}$

(C) $\frac{3}{5}$ (D) $\frac{2}{3}$

50. 200 g water is heated from 40°C to 60°C. Ignoring the slight expansion of water, the change in its internal energy is close to : (Given specific heat of water = 4184 J/kg K) **[2016 (Online)]**

(A) 167.4 kJ (B) 8.4 kJ

(C) 4.2 kJ (D) 16.7 kJ

51. An engine operates by taking n moles of an ideal gas through the cycle $ABCDA$ shown in figure. The thermal efficiency of the engine is : (Take C_v = 1.5 R, where R is gas constant) **[2017 (Online)]**

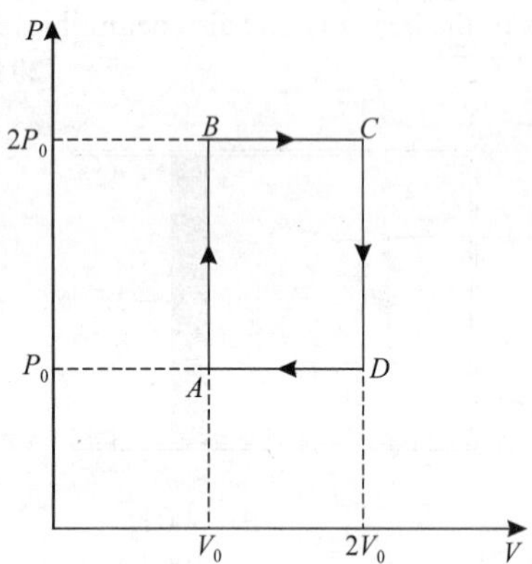

(A) 0.24 (B) 0.15

(C) 0.32 (D) 0.08

52. A gas can be taken from A to B via two different processes ACB and ADB. When path ACB is used 60 J of heat flows into the system and 30 J of work is done by the system. If path ADB is used work done by the system is 10 J. The heat Flow into the system in path ADB is : **[9 Jan. 2019 (Morning)]**

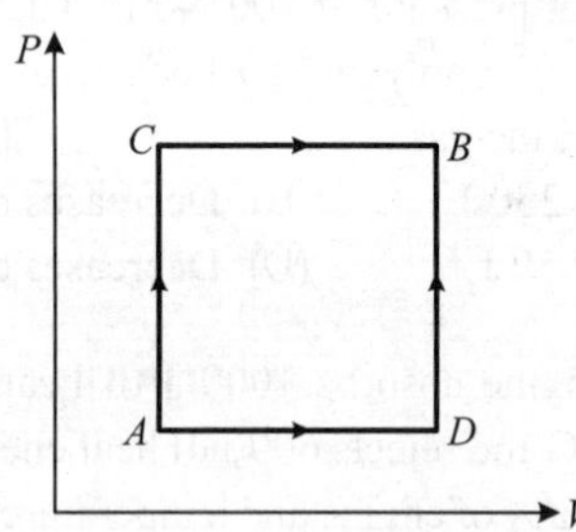

(A) 80 J (B) 20 J

(C) 100 J (D) 40 J

53. A 15 g mass of nitrogen gas is enclosed in a vessel at a temperature 27°C. Amount of heat transferred to the gas, so that rms velocity of molecules is doubled, is about : [Take R = 8.3 J/K mole] **[9 Jan. 2019 (Evening)]**

(A) 10 kJ (B) 0.9 kJ

(C) 6 kJ (D) 14 kJ

54. Two Carnot engines A and B are operated in series. The first one, A, receives heat at T_1 (= 600 K) and rejects to a reservoir at temperature T_2. The second engine B receives heat rejected by the first engine and, in turn, rejects to a heat reservoir at T_3 (= 400 K). Calculate the temperature T_2 if the work outputs of the two engines are equal : **[9 Jan. 2019 (Evening)]**

(A) 400 K (B) 600 K

(C) 500 K (D) 300 K

55. Three Carnot engines operate in series between a heat source at a temperature T_1 and a heat sink at temperature T_4 (see figure). There are two other reservoirs at temperature T_2, and T_3, as shown, with $T_2 > T_2 > T_3 > T_4$. The three engines are equally efficient if: **[10 Jan. 2019 (Morning)]**

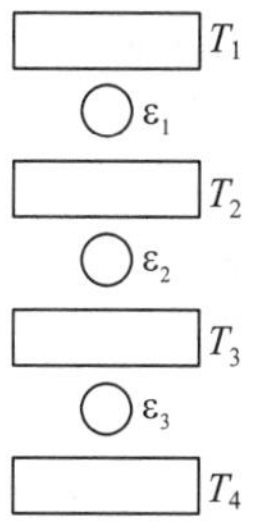

(A) $T_2 = (T_1^2 T_4)^{1/3}; T_3 = (T_1 T_4^2)^{1/3}$

(B) $T_2 = (T_1 T_4^2)^{1/3}; T_3 = (T_1^2 T_4)^{1/3}$

(C) $T_2 = (T_1^3 T_4)^{1/4}; T_3 = (T_1 T_4^3)^{1/4}$

(D) $T_2 = (T_1 T_4)^{1/2}; T_3 = (T_1^2 T_4)^{1/3}$

56. Half mole of an ideal monatomic gas is heated at constant pressure of 1 atm from 20 °C to 90°C. Work done by gas is close to : (Gas constant $R = 8.31$ J /mol.K) **[10 Jan. 2019 (Evening)]**

(A) 73 J (B) 291 J

(C) 581 J (D) 146 J

57. A rigid diatomic ideal gas undergoes an adiabatic process at room temperature. The relation between temperature and volume of this process is TV^x = constant, then x is : **[11 Jan. 2019 (Morning)]**

(A) $\frac{5}{3}$ (B) $\frac{2}{5}$

(C) $\frac{2}{3}$ (D) $\frac{3}{5}$.

58. In a process, temperature and volume of one mole of an ideal monoatomic gas are varied according to the relation $VT = K$, where K is a constant. In this process the temperature of the gas is increased by ΔT. The amount of heat absorbed by gas is (R is gas constant) : **[11 Jan. 2019 (Evening)]**

(A) $\frac{1}{2}R\Delta T$ (B) $\frac{3}{2}R\Delta T$

(C) $\frac{1}{2}KR\Delta T$ (D) $\frac{2K}{3}\Delta T$

59. For the given cyclic process CAB as shown for a gas, the work done is : **[12 Jan. 2019 (Morning)]**

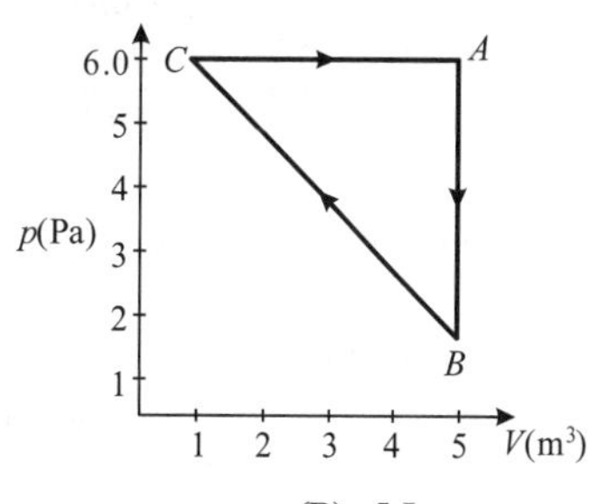

(A) 1 J (B) 5 J

(C) 10 J (D) 30 J

60. The given diagram shows four processes i.e., isochoric, isobaric, isothermal and adiabatic. The correct assignment of the processes, in the same order is given by : **[8 April 2019 (Evening)]**

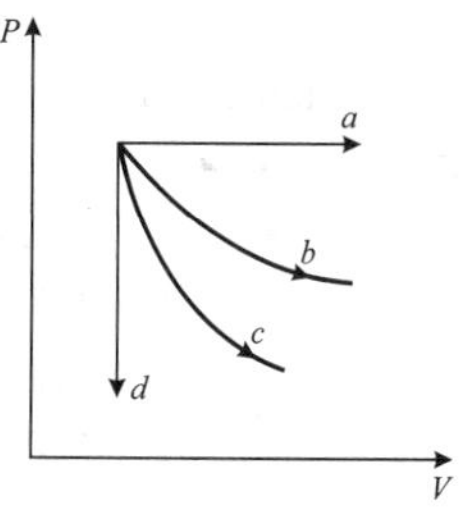

(A) $d\ a\ c\ b$ (B) $a\ d\ c\ b$

(C) $a\ d\ b\ c$ (D) $d\ a\ b\ c$

61. Following figure shows two processes A and B for a gas. If ΔQ_A and ΔQ_B are the amount of heat absorbed by the system in two cases, and ΔU_A and ΔU_B are changes in internal energies, respectively, then : **[9 April 2019 (Morning)]**

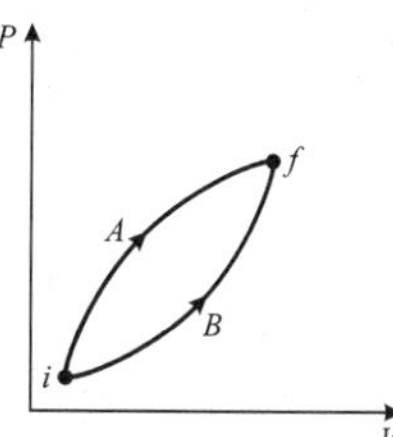

(A) $\Delta Q_A = \Delta Q_B; \Delta U_A = \Delta U_B$ (B) $\Delta Q_A > \Delta Q_B; \Delta U_A = \Delta U_B$

(C) $\Delta Q_A > \Delta Q_B; \Delta U_A > \Delta U_B$ (D) $\Delta Q_A < \Delta Q_B; \Delta U_A < \Delta U_B$

62. A cylinder with fixed capacity of 67.2 lit contains helium gas at STP. The amount of heat needed to raise the temperature of the gas by 20°C is : [Given that $R = 8.31$ J mol^{-1} K^{-1}] **[10 April 2019 (Morning)]**

(A) 748 J (B) 374 J

(C) 350 J (D) 700 J

63. n moles of an ideal gas with constant volume heat capacity C_V undergo an isobaric expansion by certain volume. The ratio of the work done in the process, to the heat supplied is : **[10 April 2019 (Morning)]**

(A) $\frac{4nR}{C_V - nR}$ (B) $\frac{nR}{C_V - nR}$

(C) $\frac{nR}{C_V + nR}$ (D) $\frac{4nR}{C_V + nR}$

64. When heat Q is supplied to a diatomic gas of rigid molecules, at constant volume its temperature increases by ΔT. The heat required to produce the same change in temperature, at a constant pressure is : **[10 April 2019 (Evening)]**

(A) $\frac{7}{5}Q$ (B) $\frac{3}{2}Q$

(C) $\frac{5}{3}Q$ (D) $\frac{2}{3}Q$

65. One mole of an ideal gas passes through a process where pressure and volume obey the relation $P = P_o\left[1-\frac{1}{2}\left(\frac{V_0}{V}\right)^2\right]$. Here P_o and V_o are constants. Calculate the change in the temperature of the gas if its volume changes from V_o to $2V_o$: **[10 April 2019 (Evening)]**

(A) $\frac{1}{2}\frac{P_oV_o}{R}$ (B) $\frac{3}{4}\frac{P_oV_o}{R}$

(C) $\frac{5}{4}\frac{P_oV_o}{R}$ (D) $\frac{1}{4}\frac{P_oV_o}{R}$

66. A sample of an ideal gas is taken through the cyclic process *abca* as shown in the figure. The change in the internal energy of the gas along the path *ca* is – 180 J. The gas absorbs 250 J of heat along the path ab and 60 J along the path *bc*. The work done by the gas along the path *abc* is : **[12 April 2019 (Morning)]**

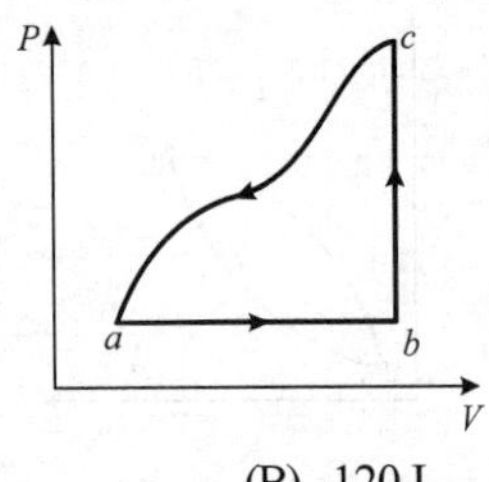

(A) 100 J (B) 120 J

(C) 140 J (D) 130 J

67. A Carnot engine has an efficiency of 1/6. When the temperature of the sink is reduced by 62°C, its efficiency is doubled. The temperatures of the source and the sink are, respectively : **[12 April 2019 (Evening)]**

(A) 124°C, 62°C (B) 37°C, 99°C

(C) 62°C, 124°C (D) 99°C, 37°C

68. A diatomic gas with rigid molecules does 10 J of work when expanded at constant pressure. What would be the heat energy absorbed by the gas, in this process ? **[12 April 2019 (Evening)]**

(A) 35 J (B) 40 J

(C) 25 J (D) 30 J

69. A litre of dry air at STP expands adiabatically to a volume of 3 litres. If $\gamma = 1.40$, the work done by air is : $(3^{1.4} = 4.6555)$ [Take air to be an ideal gas] : **[7 Jan. 2020 (Morning)]**

(A) 90.5 J (B) 48 J

(C) 60.7 J (D) 100.8 J

70. Two moles of an ideal gas with $\frac{C_P}{C_V} = \frac{5}{3}$ are mixed with 3 moles of another ideal gas with $\frac{C_P}{C_V} = \frac{4}{3}$. The value of $\frac{C_P}{C_V}$ for the mixture is : **[7 Jan. 2020 (Morning)]**

(A) 1.50 (B) 1.42

(C) 1.45 (D) 1.47

71. A Carnot engine operates between two reservoirs of temperatures 900 K and 300 K. The engine performs 1200 J of work per cycle. The heat energy (in J) delivered by the engine to the low temperature reservoir, in a cycle, is______. **[7 Jan. 2020 (Morning)]**

72. Two infinite planes each with uniform surface charge density $+\sigma$ are kept in such a way that the angle between them is 30°. The electric field in the region shown between them is given by : **[7 Jan. 2020 (Morning)]**

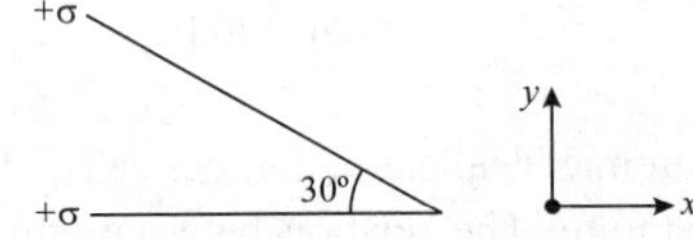

(A) $\frac{\sigma}{\varepsilon_0}\left[\left(1+\frac{\sqrt{3}}{2}\right)\hat{y}+\frac{\hat{x}}{2}\right]$ (B) $\frac{\sigma}{2\varepsilon_0}\left[\left(1-\frac{\sqrt{3}}{2}\right)\hat{y}-\frac{\hat{x}}{2}\right]$

(C) $\frac{\sigma}{2\varepsilon_0}\left[(1+\sqrt{3})\hat{y}+\frac{\hat{x}}{2}\right]$ (D) $\frac{\sigma}{2\varepsilon_0}\left[(1+\sqrt{3})\hat{y}-\frac{\hat{x}}{2}\right]$

73. Two ideal Carnot engines operate in cascade (all heat given up by one engine is used by the other engine to produce work) between temperatures, T_1 and T_2. The temperature of the hot reservoir of the first engine is T_1 and the temperature of the cold reservoir of the second engine is T_2. T is temperature of the sink of first engine which is also the source for the second engine. How is T related to T_1 and T_2, if both the engines perform equal amount of work ? **[7 Jan. 2020 (Evening)]**

(A) $T=\dfrac{2T_1T_2}{T_1+T_2}$ (B) $T=\sqrt{T_1T_2}$

(C) $T=\dfrac{T_1+T_2}{2}$ (D) $T=0$

74. Under an adiabatic process, the volume of an ideal gas gets doubled. Consequently the mean collision time between the gas molecule changes from τ_1 to τ_2. If $\dfrac{C_p}{C_v}=\gamma$ for this gas then a good estimate for $\dfrac{\tau_2}{\tau_1}$ is given by : **[7 Jan. 2020 (Evening)]**

(A) $\left(\dfrac{1}{2}\right)^{\frac{\gamma+1}{2}}$ (B) 2

(C) $\dfrac{1}{2}$ (D) $\left(\dfrac{1}{2}\right)^{\gamma}$

75. A thermodynamic cycle *xyzx* is shown on a *V-T* diagram. The *P-V* diagram that best describes this cycle is : (Diagrams are schematic and not to scale) **[8 Jan. 2020 (Morning)]**

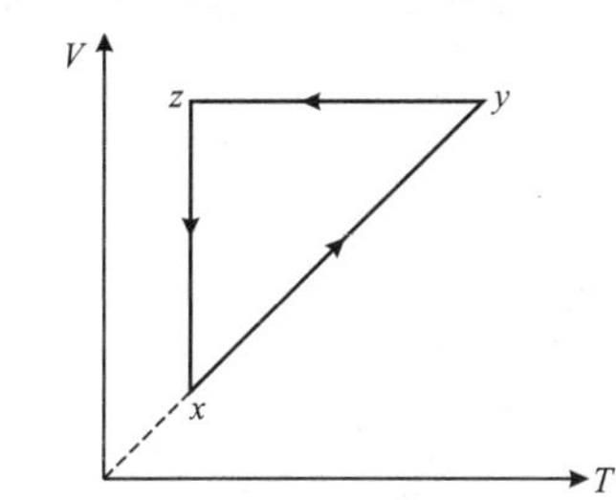

(A) 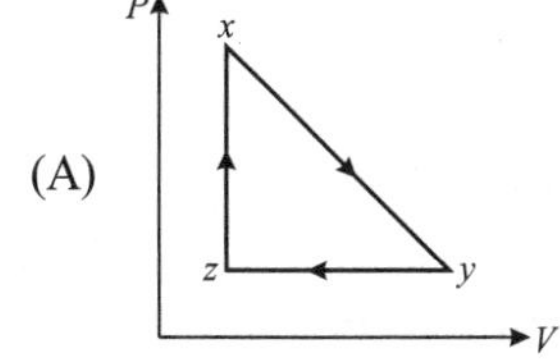

(B) 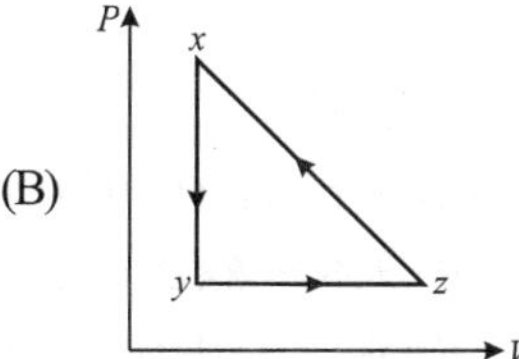

(C) 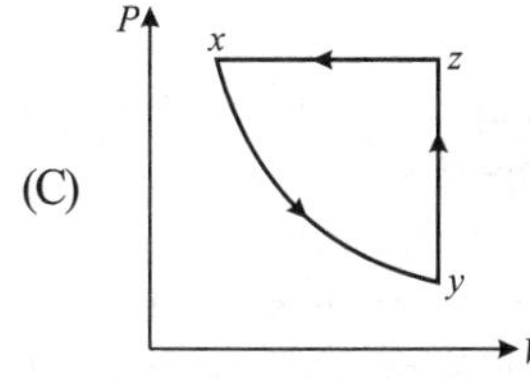

(D) 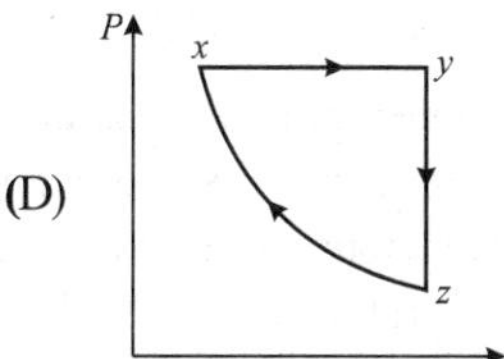

76. A Carnot engine having an efficiency of 1/10 is being used as a refrigerator. If the work done on the refrigerator is 10 J, the amount of heat absorbed from the reservoir at lower temperature is : **[8 Jan. 2020 (Evening)]**

(A) 99 J (B) 100 J

(C) 90 J (D) 1 J

77. Consider a mixture of *n* moles of helium gas and 2 *n* moles of oxygen gas (molecules taken to be rigid) as an ideal gas. Its C_P/C_V value will be : **[8 Jan. 2020 (Evening)]**

(A) 67/45 (B) 19/13

(C) 23/15 (D) 40/27

78. Consider two ideal diatomic gases *A* and *B* at some temperature *T*. Molecules of the gas *A* are rigid, and have a mass *m*. Molecules of the gas *B* have an additional vibrational mode, and have a mass *m*/4. The ratio of the specific heats (C_V^A and C_V^B) of gas *A* and *B*, respectively is :

[9 Jan. 2020 (Morning)]

(A) 7 : 9 (B) 5 : 7

(C) 3 : 5 (D) 5 : 9

79. Which of the following is an equivalent cyclic process corresponding to the thermodynamic cyclic given in the figure ? Where, $1 \rightarrow 2$ is adiabatic. (Graphs are schematic and are not to scale) **[9 Jan. 2020 (Morning)]**

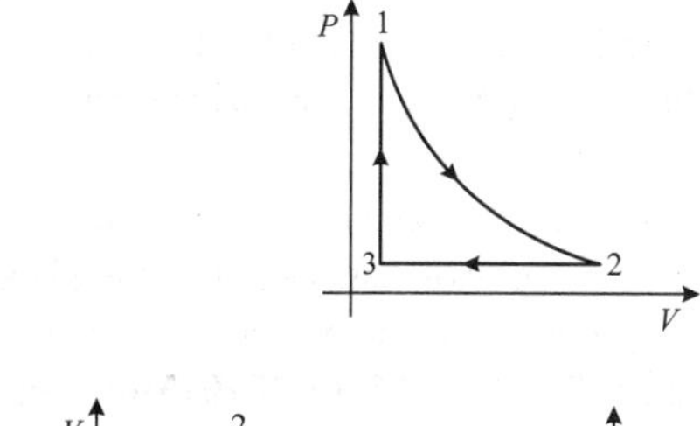

(A) 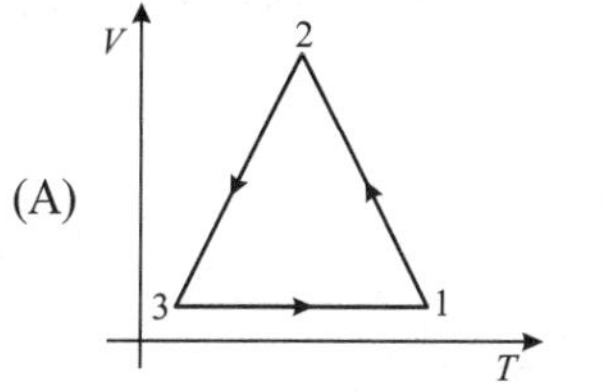

(B) 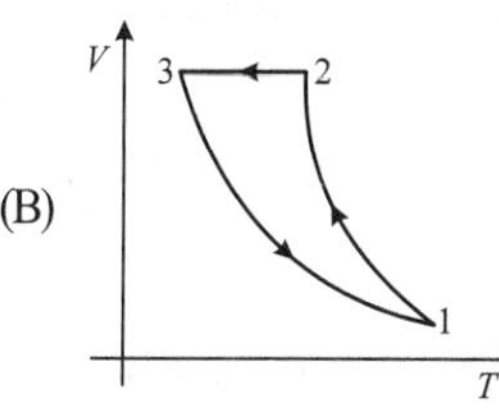

(C) 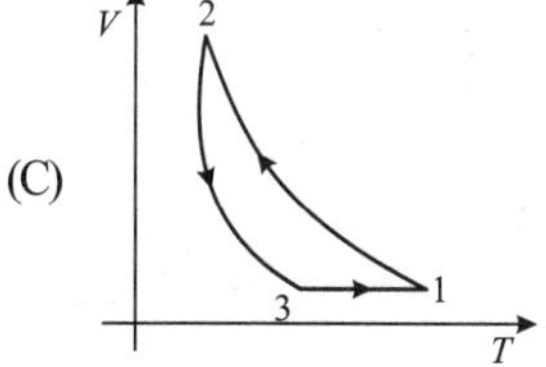

(D) 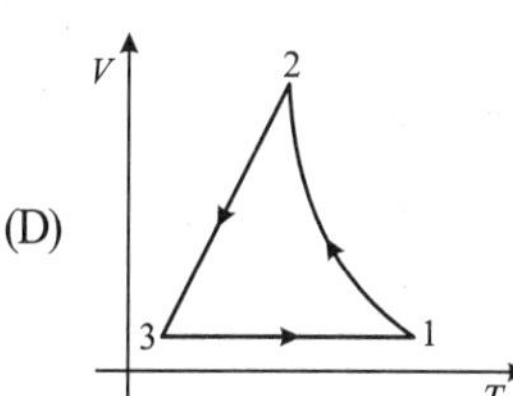

80. Starling at temperature 300 K, one mole of an ideal diatomic gas ($\gamma = 1.4$) is first compressed adiabatically from volume V_1 to $V_2=\dfrac{V_1}{16}$. It is then allowed to expand isobarically to volume $2V_2$. If all the processes are the quasi-static then the final temperature of the gas (in °K) is (to the nearest integer) _____.

[9 Jan. 2020 (Evening)]

81. A gas mixture consists of 3 moles of oxygen and 5 moles of argon at temperature *T*. Assuming the gases to be ideal and the oxygen bond to be rigid, the total internal energy (in units of

RT) of the mixture is : **[2 Sept. 2020 (Morning)]**
(A) 11 (B) 15
(C) 20 (D) 13

82. An engine takes in 5 moles of air at 20°C and 1 atm, and compresses it adiabatically to $1/10^{th}$ of the original volume. Assuming air to be a diatomic ideal gas made up of rigid molecules, the change in its internal energy during this process comes out to be X kJ. The value of X to the nearest integer is : **[2 Sept. 2020 (Morning)]**

83. A heat engine is involved with exchange of heat of 1915 J, –40 J, +125 J and QJ, during one cycle achieving an efficiency of 50.0%. The value of Q is : **[2 Sept. 2020 (Evening)]**
(A) 640 J (B) 400 J
(C) 980 J (D) 40 J

84. A balloon filled with helium (32°C and 1.7 atm.) bursts. Immediately afterwards the expansion of helium can be considered as : **[3 Sept. 2020 (Morning)]**
(A) Irreversible isothermal (B) Irreversible adiabatic
(C) Reversible adiabatic (D) Reversible isothermal

85. Consider a gas of triatomic molecules. The molecules are assumed to be triangular and made of massless rigid rods whose vertices are occupied by atoms. The internal energy of a mole of the gas at temperature T is : **[3 Sept. 2020 (Morning)]**

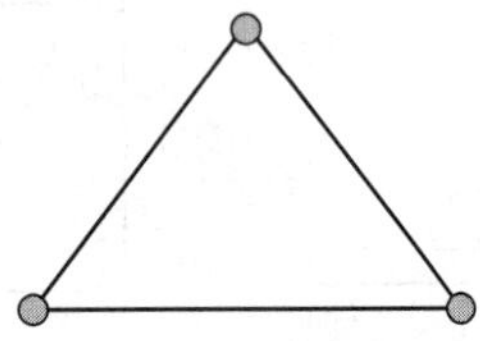

(A) $\frac{9}{2}RT$ (B) $\frac{3}{2}RT$

(C) $\frac{5}{2}RT$ (D) $3RT$

86. If minimum possible work is done by a refrigerator in converting 100 grams of water at 0°C to ice, how much heat (in calories) is released to the surrounding at temperature 27°C (Latent heat of ice = 80 Cal/gram) to the nearest integer ? **[3 Sept. 2020 (Evening)]**

87. A closed vessel contains 0.1 mole of a monoatomic ideal gas at 200 K. If 0.05 mole of the same gas at 400 K is added to it, the final equilibrium temperature (in K) of the gas in the vessel will be closed to _____. **[4 Sept. 2020 (Morning)]**

88. Match the thermodynamic processes taking place in a system with the correct conditions. In the table, ΔQ is the heat supplied, ΔW is the work done and ΔU is change in internal energy of the system : **[4 Sept. 2020 (Evening)]**

Process	Condition
(I) Adiabatic	(1) $\Delta W=0$
(II) Isothermal	(2) $\Delta Q=0$
(III) Isochoric	(3) $\Delta U\neq 0, \Delta W\neq 0, \Delta Q\neq 0$
(IV) Isobaric	(4) $\Delta U=0$

(A) I-2, II-4, III-1, IV-3 (B) I-2, II-1, III-4, IV-3
(C) I-4, II-1, III-2, IV-3 (D) I-1, II-2, III-3, IV-4

89. The change in the magnitude of the volume of an ideal gas when a small additional pressure ΔP is applied at a constant temperature, is the same as the change when the temperature is reduced by a small quantity ΔT at constant pressure. The initial temperature and pressure of the gas were 300 K and 2 atm respectively. If $|\Delta T| = C|\Delta P|$ then value of C in (K/atm) is _____. **[4 Sept. 2020 (Evening)]**

90. Three different processes that can occur in an ideal monoatomic gas are shown in the P vs V diagram. The paths are labelled as $A \rightarrow B, A \rightarrow C$ and $A \rightarrow D$. The change in internal energies during these process are taken as EAB, EAC and EAD and the workdone as WAB, WAC and WAD. The correct relation between these parameters are : **[5 Sept. 2020 (Morning)]**

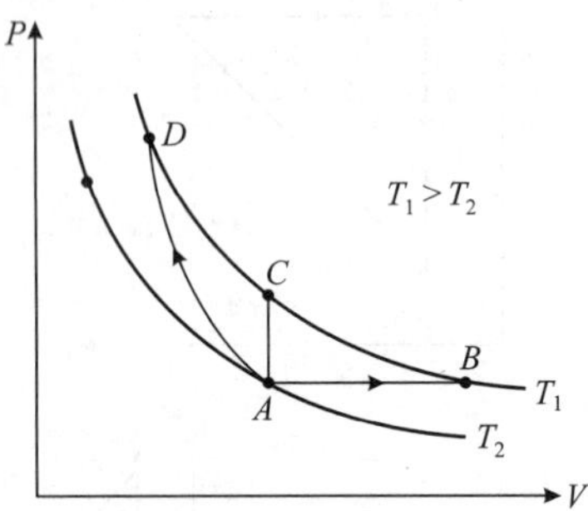

(A) $EAB = EAC = EAD, WAB > 0, WAC = 0, WAD > 0$
(B) $EAB < EAC < EAD, WAB > 0, WAC > WAD$
(C) $EAB = EAC < EAD, WAB > 0, WAC = 0, WAD < 0$
(D) $EAB > EAC > EAD, WAB < WAC < WAD$

91. Number of molecules in a volume of 4 cm^3 of a perfect monoatomic gas at some temperature T and at a pressure of 2 cm of mercury is close to ? (Given, mean kinetic energy of a molecule (at T) is 4×10^{-14} erg, $g = 980$ cm/s^2, density of mercury = 13.6 g/cm^3) **[5 Sept. 2020 (Morning)]**
(A) 5.8×10^{18} (B) 5.8×10^{16}
(C) 4.0×10^{18} (D) 4.0×10^{16}

92. In an adiabatic process, the density of a diatomic gas becomes 32 times its initial value. The final pressure of the gas is found to be n times the initial pressure. The value of n is : **[5 Sept. 2020 (Evening)]**

(A) 326 (B) $\frac{1}{32}$

(C) 32 (D) 128

93. Molecules of an ideal gas are known to have three translational degrees of freedom and two rotational degrees of freedom. The gas is maintained at a temperature of T. The total internal energy, U of a mole of this gas, and the value of $\gamma\left(=\frac{C_P}{C_v}\right)$ given, respectively, by : **[6 Sept. 2020 (Morning)]**

(A) $U=\frac{5}{2}RT$ and $\gamma=\frac{6}{5}$ (B) $U=5RT$ and $\gamma=\frac{7}{5}$

(C) $U=5RT$ and $\gamma=\frac{6}{5}$ (D) $U=\frac{5}{2}RT$ and $\gamma=\frac{7}{5}$

94. An engine operates by taking a monatomic ideal gas through the cycle shown in the figure. The percentage efficiency of the engine is close to : **[6 Sept. 2020 (Evening)]**

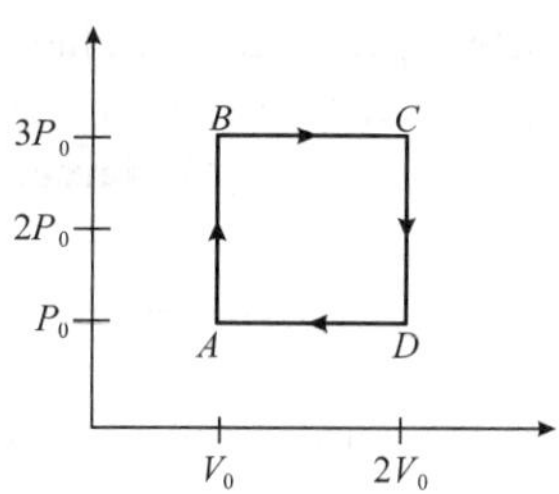

95. Match List-I with List-II. **[24 Feb. 2021 (Morning)]**

List-I	**List-II**
(a) Isothermal	(i) Pressure constant
(b) Isochoric	(ii) Temperature constant
(c) Adiabatic	(iii) Volume constant
(d) Isobaric	(iv) Heat content is constant

Choose the correct answer from the options given below :

(A) (a)-(ii), (b)-(iv), (c)-(iii), (d)-(i)
(B) (a)-(i), (b)-(iii), (c)-(ii), (d)-(iv)
(C) (a)-(iii), (b)-(ii), (c)-(i), (d)-(iv)
(D) (a)-(ii), (b)-(iii), (c)-(iv), (d)-(i)

96. n mole of a perfect gas undergoes a cyclic process $ABCA$ (see figure) consisting of the following processes. **[24 Feb. 2021 (Morning)]**

$A\to B$: Isothermal expansion at temperature T so that the volume is doubled from V_1 to $V_2=2V_1$ and pressure changes from P_1 to P_2

$B\to C$: Isobaric compression at pressure P_2 to initial volume V_1

$C\to A$: Isochoric change leading to change of pressure from P_2 to P_1

Total workdone in the complete cycle $ABCA$ is :

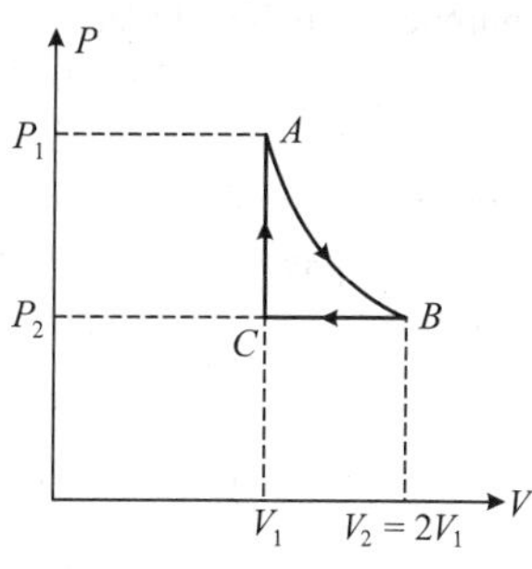

(A) $nRT\left(\ln 2-\frac{1}{2}\right)$ (B) $nRT\left(\ln 2+\frac{1}{2}\right)$

(C) $nRT\ln 2$ (D) 0

97. If one mole of an ideal gas at (P_1, V_1) is allowed to expand reversibly and isothermally (A to B) its pressure is reduced to one-half of the original pressure (see figure). This is followed by a constant volume cooling till its pressure is reduced to one-fourth of the initial value ($B\to C$). Then its is restored to its initial state by a reversible adiabatic compression (C to A). The net workdone by the gas is equal to : **[24 Feb. 2021 (Evening)]**

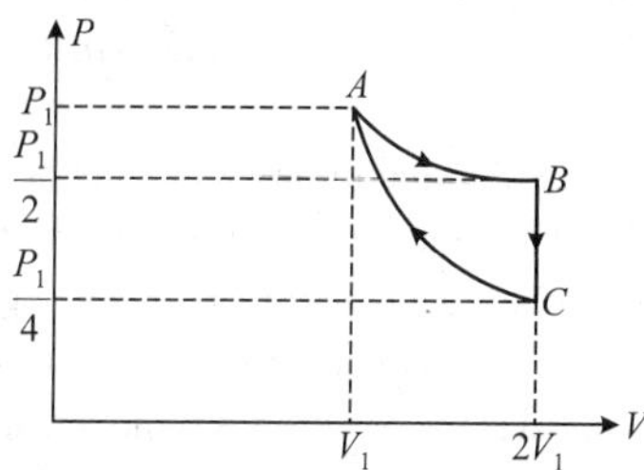

(A) $-\frac{RT}{2(\gamma-1)}$ (B) 0

(C) $RT\ln 2$ (D) $RT\left(\ln 2-\frac{1}{2(\gamma-1)}\right)$

98. A diatomic gas, having $C_P=\frac{7}{2}R$ and $C_V=\frac{5}{2}R$, is heated at constant pressure. The ratio $dU:dQ:dW$: **[25 Feb. 2021 (Morning)]**

(A) 3 : 5 : 2 (B) 5 : 7 : 2

(C) 3 : 7 : 2 (D) 5 : 7 : 3

99. A monoatomic gas of mass 4.0 u is kept in an insulated container. Container is moving with velocity 30 m/s. If container is suddenly stopped then change in temperature of the gas (R = gas constant) is $\frac{x}{3R}$. Value of x is ______.

[25 Feb. 2021 (Morning)]

100. In a certain thermodynamical process, the pressure of a gas depends on its volume as kV^3. The work done when the temperature changes from 100°C to 300°C will be _____ nR, where n denotes number of moles of a gas.

[25 Feb. 2021 (Morning)]

101. Thermodynamic process is shown below on a P-V diagram for one mole of an ideal gas. If $V_2 = 2V_1$ then the ratio of temperature T_2/T_1 is : **[25 Feb. 2021 (Evening)]**

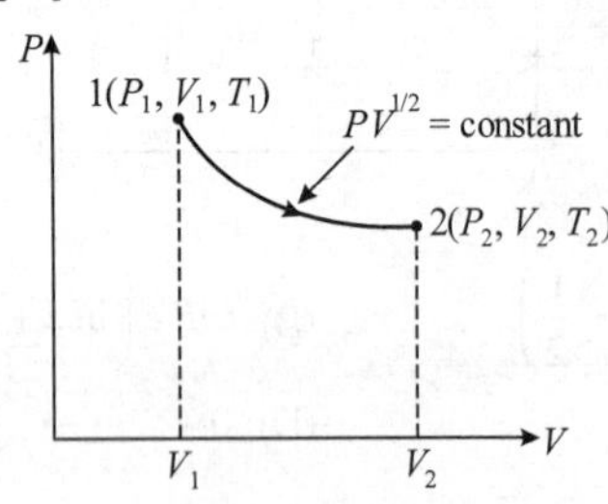

(A) 2 (B) $\sqrt{2}$

(C) $\frac{1}{2}$ (D) $\frac{1}{\sqrt{2}}$

102. A reversible heat engine converts one-fourth of the heat input into work. When the temperature of the sink is reduced by 52 K, its efficiency is doubled. The temperature in Kelvin of the source will be _____. **[25 Feb. 2021 (Evening)]**

103. A container is divided into two chambers by a partition. The volume of first chamber is 4.5 litre and second chamber is 5.5 litre. The first chamber contain 3.0 moles of gas at pressure 2.0 atom and second chamber contain 4.0 moles of gas at pressure 3.0 atm. After the partition is removed and the mixture attains equilibrium, then, the common equilibrium pressure existing in the mixture is $x \times 10^{-1}$ atm. Value of x is _____.

[26 Feb. 2021 (Morning)]

104. The internal energy (U), pressure (P) and volume (V) of an ideal gas are related as $U = 3PV + 4$. The gas is :

[26 Feb. 2021 (Evening)]

(A) Either monoatomic or diatomic

(B) Polyatomic only

(C) Diatomic only

(D) Monoatomic only

105. 1 mole of rigid diatomic gas performs a work of $\frac{Q}{5}$ when heat Q is supplied to it. The molar heat capacity of the gas during this transformation is $\frac{xR}{8}$. The value of x is _____.

[R = universal gas constant] **[26 Feb. 2021 (Evening)]**

106. The volume V of a given mass of monoatomic gas changes with temperature T according to the relation $V = KT^{2/3}$. The workdone when temperature changes by 90 K will be xR. The value of x is _____. **[26 Feb. 2021 (Evening)]**

[R = universal gas constant]

107. In thermodynamics, heat and work are :

[16 March 2021 (Morning)]

(A) Path functions

(B) Intensive thermodynamic state variables

(C) Extensive thermodynamic state variables

(D) Point functions

108. For an ideal heat engine, the temperature of the source is 127°C. In order to have 60% efficiency the temperature of the sink should be _____°C. (Round off to the Nearest Integer)

[16 March 2021 (Evening)]

109. A polyatomic ideal gas has 24 vibrational modes. What is the value of γ ? **[17 March 2021 (Morning)]**

(A) 1.03 (B) 1.30

(C) 1.37 (D) 10.3

110. A Carnot's engine working between 400 K and 800 K has a work output of 1200 J per cycle. The amount of heat energy supplied to the engine from the source in each cycle is :

[17 March 2021 (Morning)]

(A) 3200 J (B) 1800 J

(C) 1600 J (D) 2400 J

111. Which one is the correct option for the two different thermodynamic processes ? **[17 March 2021 (Evening)]**

(a)
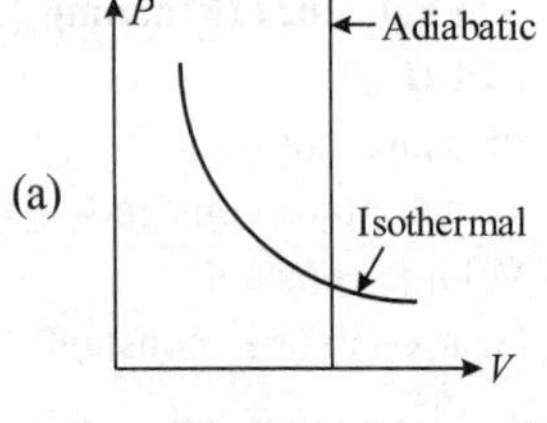

(b)
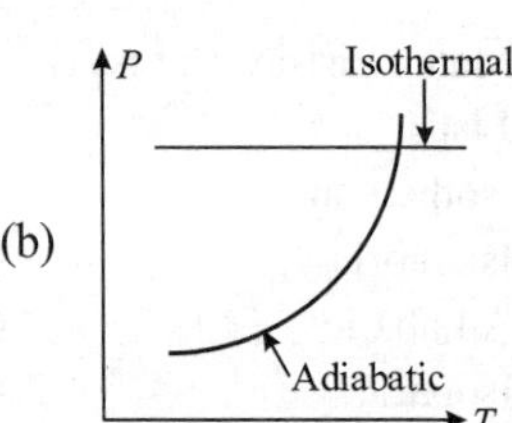

(c)
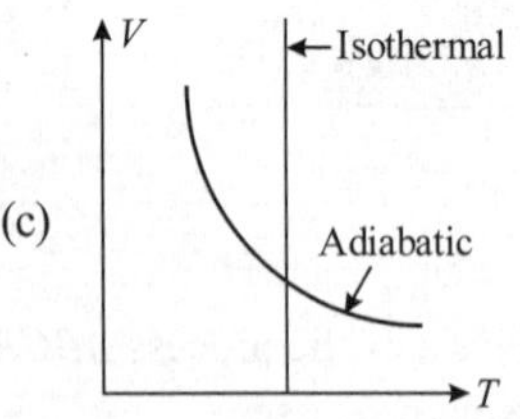

(d)
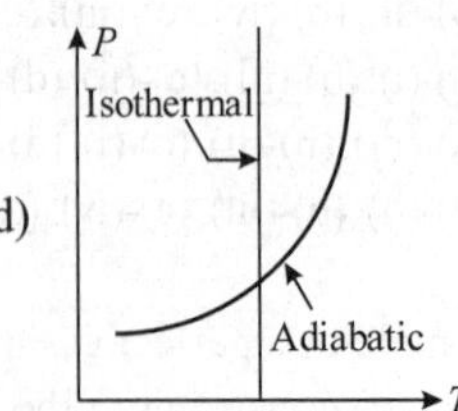

(A) (c) and (a) (B) (c) and (d)

(C) (a) only (D) (b) and (c)

112. The P-V diagram of a diatomic ideal gas system going under cyclic process as shown in figure. The work done during an adiabatic process CD is : (Use $\gamma = 1.4$)

[18 March 2021 (Morning)]

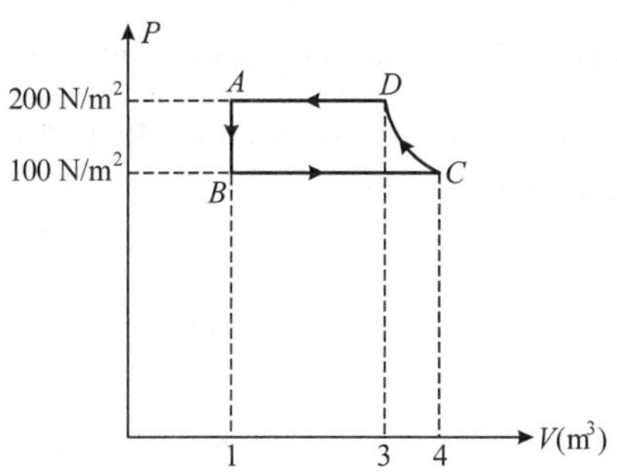

(A) – 500 J (B) – 400 J
(C) 400 J (D) 200 J

113. The amount of heat needed to raise the temperature of 4 moles of a rigid diatomic gas from 0°C to 50°C when no work is done is _____. (R is the universal gas constant)

[20 July 2021 (Morning)]

(A) $750R$ (B) $500R$
(C) $250R$ (D) $175R$

114. In the reported figure, heat energy absorbed by a system in going through a cyclic process is _____ πJ.

[20 July 2021 (Morning)]

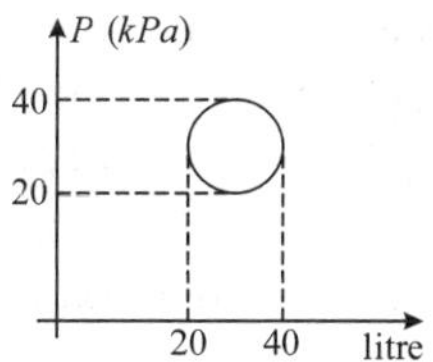

115. One mole of an ideal gas at 27°C is taken from A to B as shown in the given PV indicator diagram. The work done by the system will be _____ $\times 10^{-1}$ J. [Given : $R = 8.3$ J/mole K, ln2 = 0.6931] (Round off to the nearest integer)

[20 July 2021 (Evening)]

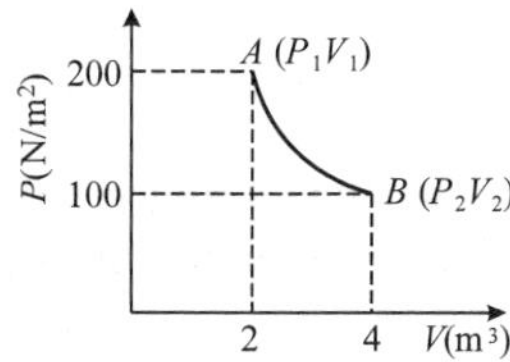

116. A monoatomic ideal gas, initially at temperature T_1 is enclosed in a cylinder fitted with a frictionless piston. The gas is allowed to expand adiabatically to temperature T_2 by releasing the piston suddenly. If l_1 and l_2 are the lengths of the gas column, before and after the expansion respectively, then the value of $\frac{T_2}{T_1}$ will be : **[25 July 2021 (Morning)]**

(A) $\frac{l_2}{l_1}$ (B) $\frac{l_1}{l_2}$
(C) $\left(\frac{l_2}{l_1}\right)^{\frac{2}{3}}$ (D) $\left(\frac{l_1}{l_2}\right)^{\frac{2}{3}}$

117. Two different metal bodies A and B of equal mass are heated at a uniform rate under similar conditions. The variation of temperature of the bodies is graphically represented as shown in the figure. The ratio of specific heat capacities is :

[25 July 2021 (Morning)]

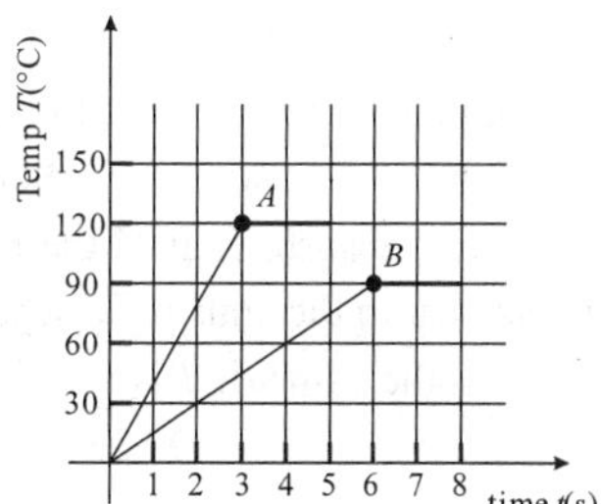

(A) $\frac{3}{4}$ (B) $\frac{4}{3}$
(C) $\frac{8}{3}$ (D) $\frac{3}{8}$

118. A heat engine has an efficiency of $\frac{1}{6}$. When the temperature of sink is reduced by 62°C, its efficiency get doubled. The temperature of the source is :

[25 July 2021 (Evening)]

(A) 99° (B) 124 °C
(C) 62 °C (D) 37 °C

119. In the reported figure, there is a cycle process $ABCDA$ on a sample of 1 mol of a diatomic gas. The temperature of the gas during the process $A \rightarrow B$ and $C \rightarrow D$ are T_1 and T_2 ($T_1 > T_2$) respectively.

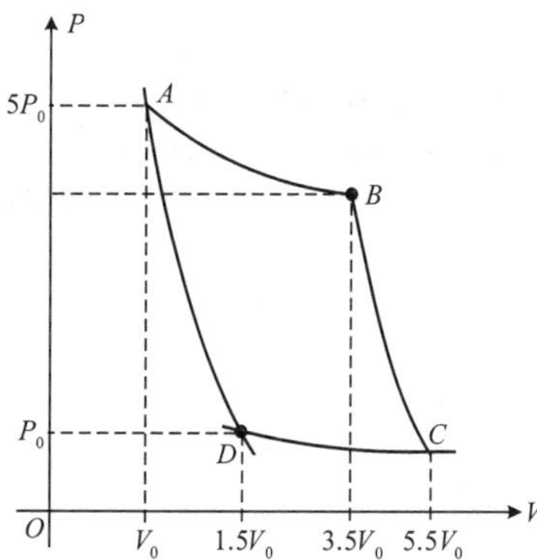

Choose the correct option out of the following for work done if processes BC and DA are adiabatic : **[27 July 2021 (Morning)]**

(A) $W_{BC} + W_{DA} > 0$ (B) $W_{AB} < W_{CD}$
(C) $W_{AB} = W_{DC}$ (D) $W_{AD} = W_{BC}$

120. One mole of an ideal gas is taken through an adiabatic process where the temperature rises from 27°C to 37°C. If the ideal gas is composed of polyatomic molecule that has 4 vibrational modes, which of the following is true ? ($R = 8.314$ J mol^{-1} K^{-1}) **[27 July 2021 (Evening)]**

(A) Work done by the gas is close to 582 J
(B) Work done on the gas is close to 332 J
(C) Work done on the gas is close to 582 J
(D) Work done by the gas is close to 332 J

121. Two Carnot engines *A* and *B* operate in series such that engine *A* absorbs heat at T_1 and rejects heat to a sink at temperature *T*. Engine *B* absorbs half of the heat rejected by Engine *A* and rejects heat to the sink at T_3, when workdone in both the cases is equal, the value of *T* is : **[27 July 2021 (Evening)]**

(A) $\frac{3}{2}T_1 + \frac{1}{3}T_3$ (B) $\frac{1}{3}T_1 + \frac{2}{3}T_3$
(C) $\frac{2}{3}T_1 + \frac{3}{2}T_3$ (D) $\frac{2}{3}T_1 + \frac{1}{3}T_3$

122. An electric appliance supplies 6000 J/min heat to the system. If the system delivers a power of 90 W. How long it would take to increase the internal energy by 2.5×10^3 J ? **[26 Aug. 2021 (Morning)]**

(A) 4.1×10^1 s (B) 2.4×10^3 s
(C) 2.5×10^2 s (D) 2.5×10^1 s

123. A refrigerator consumes an average 35 W power to operate between temperatures – 10ºC to 25ºC. If there is no loss of energy then how much average heat per second does it transfer? **[26 Aug. 2021 (Evening)]**

(A) 263 J/s (B) 298 J/s
(C) 350 J/s (D) 35 J/s

124. An ideal gas is expanding such that PT^3 = constant. The coefficient of volume expansion of the gas is : **[27 Aug. 2021 (Morning)]**

(A) $\frac{1}{T}$ (B) $\frac{3}{T}$
(C) $\frac{4}{T}$ (D) $\frac{2}{T}$

125. A heat engine operates between a cold reservoir at temperature $T_2 = 400$ K and a hot reservoir at temperature T_1. It takes 300 J of heat from the hot reservoir and delivers 240 J of heat to the cold reservoir in a cycle. The minimum temperature of the hot reservoir has to be_____ K. **[27 Aug. 2021 (Evening)]**

126. A reversible engine has an efficiency of $\frac{1}{4}$. If the temperature of the sink is reduced by 58°C, its efficiency becomes double. Calculate the temperature of the sink : **[31 Aug. 2021 (Morning)]**

(A) 180.4 °C (B) 382 °C
(C) 174 °C (D) 280 °C

127. A sample of gas with $\gamma = 1.5$ is taken through an adiabatic process in which the volume is compressed from 1200 cm^3 to 300 cm^3. If the initial pressure is 200 kPa. The absolute value of the work done by the gas in the process = _____ J. **[31 Aug. 2021 (Evening)]**

128. The temperature of 3.00 mol of an ideal diatomic gas is increased by 40.0ºC without changing the pressure of the gas. The molecules in the gas rotate but do not oscillate. If the ratio of change in internal energy of the gas to the amount of work done by the gas is $x/10$. Then the value of x (round off to the nearest) is_____. (Given $R = 8.31$ J mol^{-1} K^{-1}) **[01 Sep. 2021 (Evening)]**

129. A Carnot engine whose heat sinks at 27ºC, has an efficiency of 25%. By how many degrees should the temperature of the source be changed to increase the efficiency by 100% of the original efficiency ? **[24 June 2022 (Morning)]**

(A) Increase by 18ºC (B) Increase by 200ºC
(C) Increase by 120ºC (D) Increase by 73°

130. A Carnot engine take 5000 kcal of heat from a reservoir at 727°C and gives heat to a sink at 127°C. The work done by the engine is : **[24 June 2022 (Evening)]**

(A) 3×10^6 J (B) Zero
(C) 12.6×10^6 J (D) 8.4×10^6 J

131. A monoatomic gas performs a work of $\frac{Q}{4}$ where Q is the heat supplied to it. The molar heat capacity of the gas will be ______ *R* during this transformation. Where *R* is the gas constant. **[24 June 2022 (Evening)]**

132. The ratio of specific heats $\left(\frac{C_P}{C_V}\right)$ in terms of degree of freedom (f) is given by : **[25 June 2022 (Evening)]**

(A) $\left(1+\frac{f}{3}\right)$ (B) $\left(1+\frac{2}{f}\right)$
(C) $\left(1+\frac{f}{2}\right)$ (D) $\left(1+\frac{1}{f}\right)$

133. The efficiency of a Carnot's engine, working between steam point and ice point, will be : **[26 June 2022 (Morning)]**

(A) 26.81% (B) 37.81%
(C) 47.81% (D) 57.81%

134. A heat engine operates with the cold reservoir at temperature 324 K. The minimum temperature of the hot reservoir, if the heat engine takes 300 J heat from the hot reservoir and delivers 180 J heat to the cold reservoir per cycle, is _____ K. **[26 June 2022 (Evening)]**

135. In a carnot engine, the temperature of reservoir is 527°C and that of sink is 200 K. If the workdone by the engine when it transfers heat from reservoir to sink is 12000 kJ, the quantity of heat absorbed by the engine from reservoir is _____ $\times 10^6$ J. **[27 June 2022 (Morning)]**

136. A diatomic gas ($\gamma = 1.4$) does 400 J of work when it is expanded isobarically. The heat given to the gas in the process is _____ J. **[27 June 2022 (Evening)]**

137. Given below are two statement : **[28 June 2022 (Morning)]**
Statement-I : What μ amount of an ideal gas undergoes adiabatic change from state (P_1, V_1, T_1) to state (P_2, V_2, T_2), the work done is $W = \frac{\mu R(T_2 - T_1)}{1-\gamma}$, where $\gamma = \frac{C_P}{C_V}$ and R = universal gas constant,
Statement-II : In the above case, when work is done on the gas, the temperature of the gas would rise
Choose the correct answer from the options given below :
(A) Both statement-I and statement-II are true
(B) Both statement-I and statement-II are false
(C) Statement-I is true but statement-II is false
(D) Statement-I is false but statement-II is true

138. The total internal energy of two mole monoatomic ideal gas at temperature $T = 300$ K will be J. [Given $R = 8.31$ J/mol.K] **[28 June 2022 (Morning)]**

139. A sample of an ideal gas is taken through the cyclic process $ABCA$ as shown in figure. It absorbs, 40 J of heat during the part AB, no heat during BC and rejects 60 J of heat during CA. A work 50 J is done on the gas during the part BC. The internal energy of the gas at A is 1560 J. The work done by the gas during the part CA is : **[28 June 2022 (Evening)]**

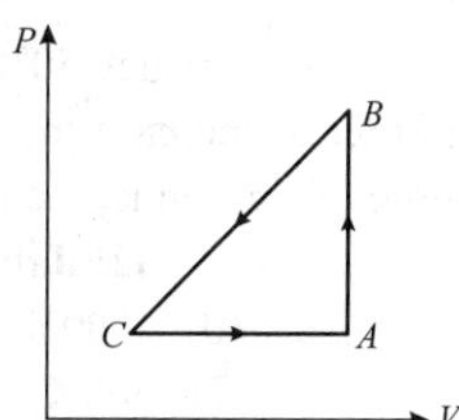

(A) 20 J (B) 30 J
(C) -30 J (D) -60 J

140. 300 calories of heat is given to a heat engine and it rejects 225 cal of heat. If source temperature is 227°C, then the temperature of sink will be _____ °C. **[29 June 2022 (Morning)]**

141. Starting with the same initial conditions, an ideal gas expands from volume V_1 to V_2 in three different ways. The work done by the gas is W_1 if the process is purely isothermal. W_2 if the process is purely adiabatic and W_3 if the process is purely isobaric. Then, choose the correct option : **[29 June 2022 (Evening)]**

(A) $W_1 < W_2 < W_3$ (B) $W_2 < W_3 < W_1$
(C) $W_3 < W_1 < W_2$ (D) $W_2 < W_1 < W_3$

142. A certain amount of gas of volume V at 27°C temperature and pressure 2×10^7 Nm^{-2} expands isothermally until its volume gets doubled. Later it expands adiabatically until its volume gets redoubled. The final pressure of the gas will be :
[Use $\gamma = 1.5$] **[25 July 2022 (Morning)]**

(A) 3.536×10^5 Pa (B) 3.536×10^6 Pa
(C) 1.25×10^6 Pa (D) 1.25×10^5 Pa

143. Let η_1 is the efficiency of an engine at $T_1 = 447$°C and $T_2 = 147$°C while η_2 is the efficiency at $T_1 = 947$°C and $T_2 = 47$°C. The ratio $\frac{\eta_1}{\eta_2}$ will be : **[25 July 2022 (Evening)]**

(A) 0.41 (B) 0.56
(C) 0.73 (D) 0.70

144. A monoatomic gas at pressure P and volume V is suddenly compressed to one eighth of its original volume. The final pressure at constant entropy will be : **[26 July 2022 (Morning)]**

(A) P (B) $8P$
(C) $32P$ (D) $64P$

145. A gas has n degrees of freedom. The ratio of specific heat of gas at constant volume to the specific heat of gas at constant pressure will be : **[26 July 2022 (Evening)]**

(A) $\frac{n}{n+2}$ (B) $\frac{n+2}{n}$

(C) $\frac{n}{2n+2}$ (D) $\frac{n}{n-2}$

146. In 1st case, Carnot engine operates between temperatures 300 K and 100 K. In 2nd case, as shown in the figure, a combination of two engines is used. The efficiency of this combination (in 2nd case) will be : **[27 July 2022 (Evening)]**

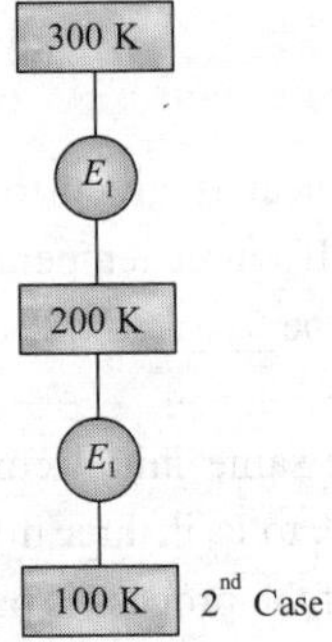

(A) Same as the 1st case
(B) Always greater than the 1st case
(C) Always less than the 1st case
(D) May increase or decrease with respect to the 1st case

147. A Carnot engine has efficiency of 50%. If the temperature of sink is reduced by 40°C, its efficiency increases by 30%. The temperature of the source will be : **[28 July 2022 (Morning)]**
(A) 166.7 K (B) 255.1 K
(C) 266.7 K (D) 367.7 K

148. At a certain temperature, the degrees of freedom per molecule for gas is 8. The gas performs 150 J of work when it expands under constant pressure. The amount of heat absorbed by the gas will be _____ J. **[28 July 2022 (Evening)]**

149. The pressure P_1 and density d_1 of diatomic gas $\left(\lambda = \frac{7}{5}\right)$ changes suddenly to $P_2 (> P_1)$ and d_2 respectively during an adiabatic process. The temperature of the gas increases and becomes _____ times of its initial temperature.

[Given $\frac{d_2}{d_1} = 32$] **[29 July 2022 (Morning)]**

150. One mole of a monoatomic gas is mixed with three moles of a diatomic gas. The molecular specific heat of mixture at constant volume is $\frac{\alpha^2}{4} R$ J/mol K; then the value of α will be _____. [Assume that the given diatomic gas has no vibrational mode] **[29 July 2022 (Morning)]**

151. A thermodynamic system is taken from an original state D to an intermediate state E by the linear process shown in the figure. Its volume is then reduced to the original volume from E to F by an isobaric process. The total work done by the gas from D to E to F will be : **[29 July 2022 (Evening)]**

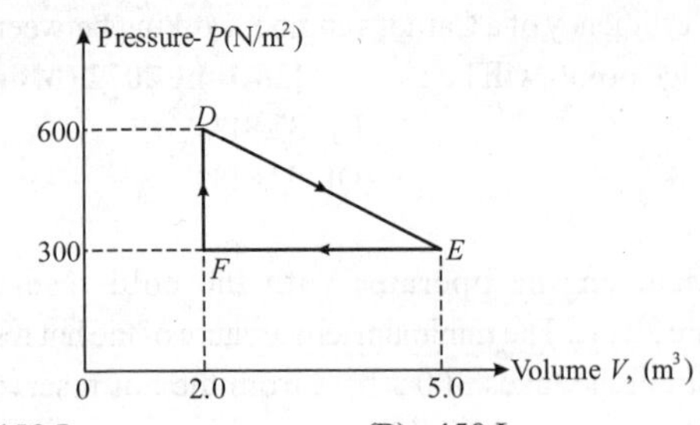

(A) −450 J (B) 450 J
(C) 900 J (D) 1350 J

152. 1 g of a liquid is converted to vapour at 3×10^5 Pa pressure. If 10% of the heat supplied is used for increasing the volume by 1600 cm^3 during this phase change, then the increase in internal energy in the process will be : **[24 Jan. 2023 (Morning)]**
(A) 4320 J (B) 432000 J
(C) 4800 J (D) 4.32×10^8 J

153. Let γ_1 be the ratio of molar specific heat at constant pressure and molar specific heat at constant volume of a monoatomic gas and γ_2 be the similar ratio of diatomic gas. Considering the diatomic gas molecule as a rigid rotator, the ratio, $\frac{\gamma_1}{\gamma_2}$ is : **[24 Jan. 2023 (Evening)]**
(A) $\frac{27}{35}$ (B) $\frac{35}{27}$
(C) $\frac{25}{21}$ (D) $\frac{21}{25}$

154. In an Isothermal change, the change in pressure and volume of a gas can be represented for three different temperature; $T_3 > T_2 > T_1$ as : **[24 Jan. 2023 (Evening)]**

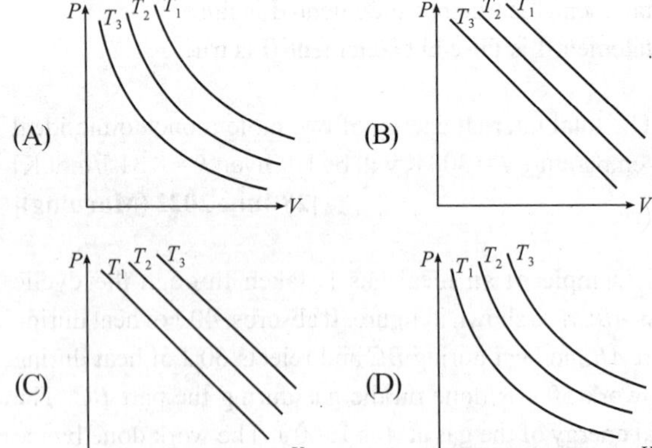

155. A Carnot engine with efficiency 50% takes heat from a source at 600 K. In order to increase the efficiency to 70%, keeping the temperature of sink same, the new temperature of the source will be : **[25 Jan. 2023 (Morning)]**
(A) 360 K (B) 1000 K
(C) 900 K (D) 300 K

156. Match List-I with List-II : **[25 Jan. 2023 (Evening)]**

List-I		List-II
(a) Isothermal Process	(i)	Work done by the gas decreases internal energy
(b) Adiabatic Process	(ii)	No change in internal energy
(c) Isochoric Process	(iii)	The heat absorbed goes partly to increase internal energy and partly to do work
(d) Isobaric Process	(iv)	No work is done on or by the gas

Choose the correct answer from the options given below :

(A) (a)-(ii), (b)-(i), (c)-(iii), (d)-(iv)
(B) (a)-(ii), (b)-(i), (c)-(iv), (d)-(iii)
(C) (a)-(i), (b)-(ii), (c)-(iv), (d)-(iii)
(D) (a)-(i), (b)-(ii), (c)-(iii), (d)-(iv)

157. Given below are two statements. One is labelled as Assertion-A and the other is labelled as Reason-R.

Assertion-A : If dQ and dW represent the heat supplied to the system and the work done on the system respectively. Then according to the first law of thermodynamics $dQ = dU - dW$.

Reason-R : First law of thermodynamics is based on law of conservation of energy.

In the light of the above statements, choose the correct answer from the option given below : **[29 Jan. 2023 (Morning)]**

(A) **A** is correct but **R** is not correct.
(B) **A** is not correct but **R** is correct.
(C) Both **A** and **R** are correct and **R** is the correct explanation of **A**.
(D) Both **A** and **R** are correct but **R** is not the correct explanation of **A**.

158. Heat is given to an ideal gas in an isothermal process.

(a) Internal energy of the gas will decrease
(b) Internal energy of the gas will increase
(c) Internal energy of the gas will not change
(d) The gas will do positive work
(e) The gas will do negative work

Choose the correct answer from the options given below : **[30 Jan. 2023 (Morning)]**

(A) (a) and (e) only (B) (b) and (d) only
(C) (c) and (e) only (D) (c) and (d) only

159. Given below are two statements : One is labelled as **Assertion-A** and the other is labelled as **Reason-R**.

Assertion-A: Efficiency of a reversible heat engine will be highest at –273°C temperature of cold reservoir

Reason-R: The efficiency of Carnot's engine depends not only on temperature of cold reservoir but it depends on the temperature of hot reservoir too and is given as $\eta = \left(1 - \frac{T_2}{T_1}\right)$.

In the light of the above statements, choose the correct answer from the options given below : **[30 Jan. 2023 (Evening)]**

(A) **A** is true but **R** is false.
(B) Both **A** and **R** are true but **R** is NOT the correct explanation of **A**.
(C) **A** is false but **R** is true.
(D) Both **A** and **R** are true and **R** is the correct explanation of **A**.

160. The pressure of a gas changes linearly with volume from A to B as shown in figure. If no heat is supplied to or extracted from the gas then change in the internal energy of the gas will be : **[31 Jan. 2023 (Morning)]**

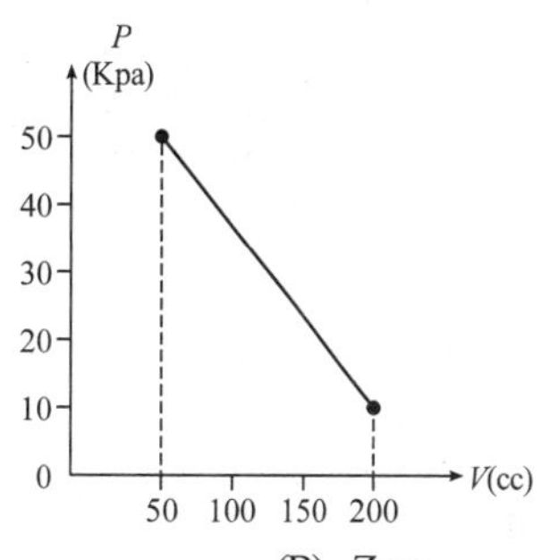

(A) 6 J (B) Zero
(C) –4.5 J (D) 4.5 J

161. The correct relation between $\gamma = \frac{C_p}{C_v}$ and temperature T is : **[31 Jan. 2023 (Morning)]**

(A) $\gamma \propto \frac{1}{\sqrt{T}}$ (B) $\gamma \propto T^o$
(C) $\gamma \propto \frac{1}{T}$ (D) $\gamma \propto T$

162. A hypothetical gas expands adiabatically such that its volume changes from 08 litres to 27 litres. If the ratio of final pressure of the gas to initial pressure of the gas is $\frac{16}{81}$. Then the ratio of $\frac{C_P}{C_V}$ will be : **[31 Jan. 2023 (Evening)]**

(A) $\frac{4}{3}$ (B) $\frac{3}{1}$
(C) $\frac{1}{2}$ (D) $\frac{3}{2}$

163. A sample of gas at temperature T is adiabatically expanded to double its volume. The work done by the gas in the process is : $\left(\text{given, } \lambda = \frac{3}{2}\right)$ **[1 Feb. 2023 (Morning)]**

(A) $W = TR[\sqrt{2} - 2]$ (B) $W = \frac{T}{R}[\sqrt{2} - 2]$
(C) $W = \frac{R}{T}[2 - \sqrt{2}]$ (D) $W = RT[2 - \sqrt{2}]$

164. A Carnot engine operating between two reservoirs has efficiency $\frac{1}{3}$. When the temperature of cold reservoir raised by x, its efficiency decreases to $\frac{1}{6}$. The value of x, if the temperature of hot reservoir is 99° C, will be : **[1 Feb. 2023 (Evening)]**

(A) 16.5 K (B) 33 K
(C) 66 K (D) 62 K

165. A source supplies heat to a system at the rate of 1000 W. If the system performs work at a rate of 200 W. The rate at which internal energy of the system increases :

[6 April 2023 (Morning)]

(A) 1200 W (B) 600 W
(C) 500 W (D) 800 W

166. Given below are two statements :

Statement-I : If heat is added to a system, its temperature must increase

Statement-II : If positive work is done by a system in a thermodynamic process, its volume must increase.

In the light of the above statements, choose the correct answer from the options given below : **[8 April 2023 (Morning)]**

(A) Statement-I is true but Statement-II is false
(B) Both Statement-I and Statement-II are true
(C) Both Statement-I and Statement-II are false
(D) Statement-I is false but Statement-II is true

167. Work done by a Carnot engine operating between temperatures 127°C and 27°C is 2 kJ. The amount of heat transferred to the engine by the reservoir is :

[8 April 2023 (Evening)]

(A) 4 kJ (B) 2 kJ
(C) 8 kJ (D) 2.67 kJ

168. Consider two containers A and B containing monoatomic gases at the same Pressure (P), Volume (V) and Temperature (T). The gas in A is compressed isothermally to $\frac{1}{8}$ of its original volume while the gas B is compressed adiabatically to $\frac{1}{8}$ of its original volume. The ratio of final pressure of gas in B to that of gas in A is: **[10 April 2023 (Morning)]**

(A) 8 (B) $8^{3/2}$
(C) $\frac{1}{8}$ (D) 4

169. A gas is compressed adiabatically, which one of the following statement is NOT true: **[10 April 2023 (Evening)]**

(A) There is no heat supplied to the system
(B) The temperature of the gas increases
(C) The change in the internal energy is equal to the work done on the gas
(D) There is no change in the internal energy

170. 1 kg of water at 100°C is converted into steam at 100°C by boiling at atmospheric pressure. The volume of water changes from 1.00×10^{-3} m^3 as a liquid to 1.671 m^3 as steam. The change in internal energy of the system during the process will be (Given latent heat of vaporisaiton = 2257 kJ/kg. Atmospheric pressure = 1 × 105 Pa) : **[11 April 2023 (Morning)]**

(A) +2090 kJ (B) –2090 kJ
(C) –2426 kJ (D) +2476 kJ

171. The thermodynamic process, in which internal energy of the system remains constant is : **[11 April 2023 (Evening)]**

(A) Isochoric (B) Isothermal
(C) Adiabatic (D) Isobaric

172. An engine operating between the boiling and freezing points of water will have :

(1) Efficiency more than 27%.
(2) Efficiency less than the efficiency of a Carnot engine operating between the same two temperatures.
(3) Efficiency equal to 27%.
(4) Efficiency less than 27%. **[12 April 2023 (Morning)]**

(A) (2), (3) and (4) only (B) (2) and (3) only
(C) (2) and (4) only (D) (1) and (2) only

173. The initial pressure and volume of an ideal gas are P_0 and V_0. The final pressure of the gas when the gas is suddenly compressed to volume $\frac{V_0}{4}$ will be:

(Given γ = ratio of specific heats at constant pressure and at constant volume) **[13 April 2023 (Evening)]**

(A) $P_0(4)^{1/\gamma}$ (B) $P_0(4)^{\gamma}$
(C) P_0 (D) $4P_0$

174. A thermodynamic system is taken through cyclic process. The total work done in the process is :

[15 April 2023 (Morning)]

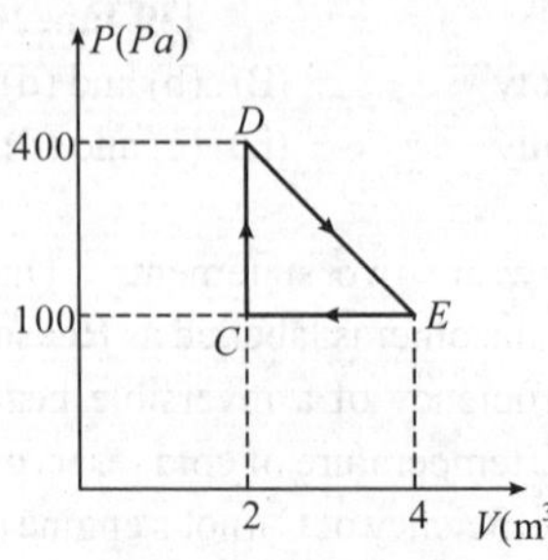

(A) 100 J (B) 300 J
(C) Zero (D) 200 J

* * * * *

CHAPTER

12 KINETIC THEORY

12.1 KINETIC THEORY OF GASES

1. Cooking gas containers are kept in a lorry moving with uniform speed. The temperature of the gas molecules inside will : **[2002]**

(A) Increase (B) Decrease

(C) Remains same

(D) Decrease for some, while increase for others

2. At what temperature is the rms velocity of a hydrogen molecule equal to that of an oxygen molecules at 47° C? **[2002]**

(A) 80 K (B) – 73 K

(C) 3 K (D) 20 K

3. 1 mole of a gas with $\gamma = 7/5$ is mixed with 1 mole of a gas with $\gamma = 5/3$, then the value of γ for the resulting mixture is : **[2002]**

(A) 7/5 (B) 2/5

(C) 24/16 (D) 12/7

4. Two thermally insulated vessels 1 and 2 are filled with air at temperatures (T_1, T_2), volume (V_1, V_2) and pressure (P_1, P_2) respectively. If the valve joining the two vessels is opened, the temperature inside the vessel at equilibrium will be : **[2004]**

(A) $T_1 + T_2$ (B) $\dfrac{(T_1 + T_2)}{2}$

(C) $\dfrac{T_1T_2(P_1V_1 + P_2V_2)}{P_1V_1T_2 + P_2V_2T_1}$ (D) $\dfrac{T_1T_2(P_1V_1 + P_2V_2)}{P_1V_1T_1 + P_2V_2T_2}$

5. One mole of ideal monoatomic gas ($\gamma = 5/3$) is mixed with one mole of diatomic gas ($\gamma = 7/5$). What is γ for the mixture ? γ denotes the ratio of specific heat at constant pressure, to that at constant volume : **[2004]**

(A) 3/2 (B) 23/15

(C) 35/23 (D) 4/3

6. A gaseous mixture consists of 16 g of helium and 16 g of oxygen. The ratio C_p/C_v of the mixture is : **[2005]**

(A) 1.59 (B) 1.62

(C) 1.4 (D) 1.54

7. Two rigid boxes containing different ideal gases are placed on a table. Box A contains one mole of nitrogen at temperature T_0, while box B contains one mole of helium at temperature (7/3) T_0. The boxes are then put into thermal contact with each other, and heat flows between them until the gases reach a common final temperature (Ignore the heat capacity of boxes). Then, the final temperature of gases, T_f in terms of T_0 is : **[2006]**

(A) $T_f = \dfrac{3}{7}T_0$ (B) $T_f = \dfrac{7}{3}T_0$

(C) $T_f = \dfrac{3}{2}T_0$ (D) $T_f = \dfrac{5}{2}T_0$

8. If C_P and C_V denote the specific heats of nitrogen per unit mass at constant pressure and constant volume respectively, then : **[2007]**

(A) $C_P - C_V = R/28$ (B) $C_P - C_V = R/14$

(C) $C_P - C_V = R$ (D) $C_P - C_V = 28R$

9. An insulated container of gas has two chambers separated by an insulating partition. One of the chambers has volume V_1 and contains ideal gas at pressure P_1 and temperature T_1. The other chamber has volume V_2 and contains ideal gas at pressure P_2 and temperature T_2. If the partition is removed without doing any work on the gas, the final equilibrium temperature of the gas in the container will be : **[2008]**

(A) $\dfrac{T_1T_2(P_1V_1 + P_2V_2)}{P_1V_1T_2 + P_2V_2T_1}$ (B) $\dfrac{P_1V_1T_1 + P_2V_2T_2}{P_1V_1 + P_2V_2}$

(C) $\dfrac{P_1V_1T_2 + P_2V_2T_1}{P_1V_1 + P_2V_2}$ (D) $\dfrac{T_1T_2(P_1V_1 + P_2V_2)}{P_1V_1T_1 + P_2V_2T_2}$

10. One kg of a diatomic gas is at a pressure of 8×10^4 N/m^2. The density of the gas is 4 kg/m^3. What is the energy of the gas due to its thermal motion ? **[2009]**

(A) 6×10^4 J (B) 7×10^4 J

(C) 3×10^4 J (D) 5×10^4 J

11. Three perfect gases at absolute temperatures T_1, T_2 and T_3 are mixed. The masses of molecules are m_1, m_2, and m_3 and the number of molecules are n_1, n_2 and n_3 respectively. Assuming no loss of energy, then final temperature of the mixture is : **[2011]**

(A) $\dfrac{n_1T_1^2 + n_2T_2^2 + n_3T_3^2}{n_1T_1 + n_2T_2 + n_3T_3}$ (B) $\dfrac{n_1^2T_1^2 + n_2^2T_2^2 + n_3^2T_3^2}{n_1T_1 + n_2T_2 + n_3T_3}$

(C) $\dfrac{T_1 + T_2 + T_3}{3}$ (D) $\dfrac{n_1T_1 + n_2T_2 + n_3T_3}{n_1 + n_2 + n_3}$

12. A perfect gas at 27°C is heated at constant pressure so as to double its volume. The final temperature of the gas will be, close to : [2012 (Online)]

(A) 327°C (B) 200°C

(C) 54°C (D) 300°C

13. A thin tube sealed at both ends in 100 cm long. It lies horizontally, the middle 20 cm containing mercury and two equal ends containing air at standard atmospheric pressure. If the tube is now turned to a vertical position, by what amount will the mercury be displaced ? (Given : cross-section of the tube can be assumed to be uniform) **[2013 (Online)]**

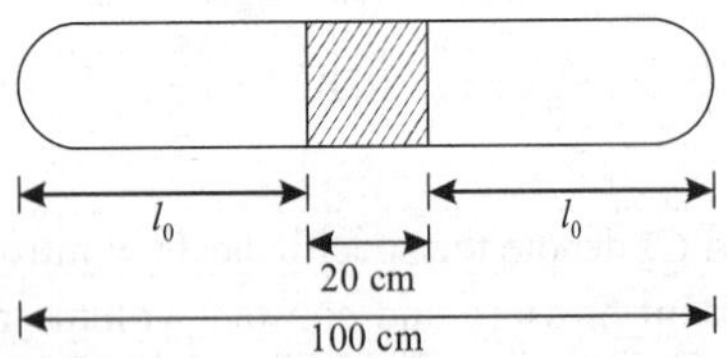

(A) 2.95 cm (B) 5.18 cm

(C) 8.65 cm (D) 0.0 cm

14. Figure shows the variation in temperature (ΔT) with the amount of heat supplied (Q) in an isobaric process corresponding to a monoatomic (M), diatomic (D) and a polyatomic (P) gas. The initial state of all the gases are the same and the scales for the two axes coincide. Ignoring vibrational degrees of freedom, the lines a, b and c respectively correspond to : **[2013 (Online)]**

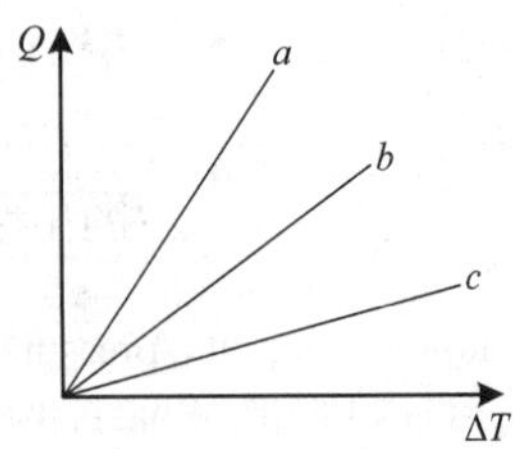

(A) P, M and D (B) M, D and P

(C) P, D and M (D) D, M and P

15. There are two identical chambers, completely thermally insulated from surroundings. Both chambers have a partition wall dividing the chambers in two compartments. Compartment 1 is filled with an ideal gas and Compartment 3 is filled with a real gas. Compartments 2 and 4 are vacuum. A small hole (orifice) is made in the partition walls and the gases are allowed to expand in vacuum. **[2013 (Online)]**

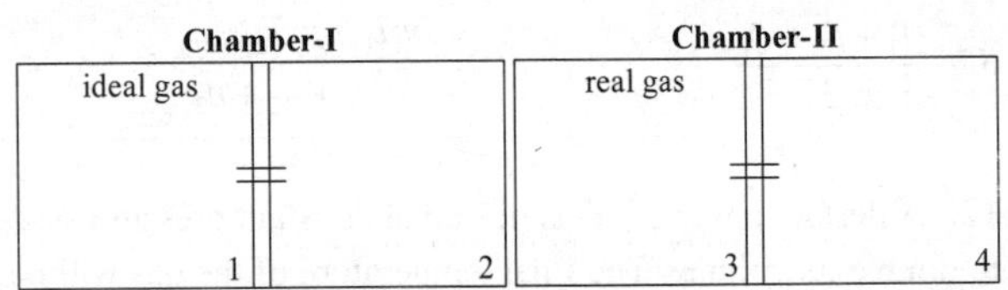

Statement-1 : No change in temperature of the gas takes place when ideal gas expands in vacuum. However, the temperature of real gas goes down (cooling) when it expands in vacuum.

Statement-2 : The internal energy of an ideal gas in only kinetic. The internal energy of a real gas is kinetic as well as potential.

(A) Statement-1 is false and Statement-2 is true

(B) Statement-1 and Statement-2 both are true. Statement-2 is the correct explanation of Statement-1

(C) Statement-1 is true and Statement-2 is false

(D) Statement-1 and Statement-2 both are true. Statement-2 is not correct explanation of Statement-1

16. An open glass tube is immersed in mercury in such a way that a length of 8 cm extends above the mercury level. The open end of the tube is then closed and sealed and the tube is raised vertically up by additional 46 cm. What will be length of the air column above mercury in the tube now ?
(Atmospheric pressure = 76 cm of Hg) **[2014]**

(A) 16 cm (B) 22 cm

(C) 38 cm (D) 6 cm

17. Modern vacuum pumps can evacuate a vessel down to a pressure of 4.0×10^{-15} atom. At room temperature (300 K). Taking $R = 8.0\ \text{JK}^{-1}\ \text{mole}^{-1}$, atom $= 10^5\ Pa$ and $N_{\text{Avogadro}} = 6 \times 10^{23}\ \text{mole}^{-1}$, the mean distance between molecules of gas in an evacuated vessel will be of the order of : **[2014 (Online)]**

(A) 0.2 μm (B) 0.2 mm

(C) 0.2 cm (D) 0.2 nm

18. At room temperature a diatomic gas is found to have an rms speed of 1930 ms^{-1}. The gas is : **[2014 (Online)]**

(A) H_2 (B) Cl_2

(C) O_2 (D) F_2

19. A gas molecule of mass M at the surface of the earth has kinetic energy equivalent to 0°C. If it were to go up straight without colliding with any other molecules, how high it would rise ? Assume that the height attained in much less than radius of the earth. (k_B is Boltzmann constant) : **[2014 (Online)]**

(A) 0 (B) $\dfrac{273k_B}{2Mg}$

(C) $\dfrac{546k_B}{3Mg}$ (D) $\dfrac{819k_B}{2Mg}$

20. Using equipartition of energy, the specific heat (in $\text{J kg}^{-1}\ \text{K}^{-1}$) of aluminium at room temperature can be estimated to be (atomic weight of aluminium = 27) : **[2015 (Online)]**

(A) 410 (B) 25

(C) 1850 (D) 925

21. In an ideal gas at temperature T, the average force that a molecule applies on the walls of a closed container depends on T as T^q. A good estimate for q is : **[2015 (Online)]**

(A) $\frac{1}{2}$ (B) 2

(C) 1 (D) $\frac{1}{4}$

22. C_p and C_v are specific heats at constant pressure and constant volume respectively. It is observed that

$C_p - C_v = a$ for hydrogen gas

$C_p - C_v = b$ for nitrogen gas

The correct relation between a and b is : **[2017]**

(A) $a = 14b$ (B) $a = 28b$

(C) $a = \frac{1}{14}b$ (D) $a = b$

23. The temperature of an open room of volume 30 m^3 increases from 17°C to 27°C due to sunshine. The atmospheric pressure in the room remains 1×10^5 *Pa*. If n_i and n_f are the number of molecules in the room before and after heating, the $n_f - n_i$ will be : **[2017]**

(A) 2.5×10^{25} (B) -2.5×10^{25}

(C) -1.61×10^{23} (D) 1.38×10^{23}

24. An ideal gas has molecules with 5 degrees of freedom. The ratio of specific heats at constant pressure (C_p) and at constant volume (C_v) is : **[2017 (Online)]**

(A) 6 (B) $\frac{7}{2}$

(C) $\frac{5}{2}$ (D) $\frac{7}{5}$

25. N moles of a diatomic gas in a cylinder are at a temperature T. Heat is supplied to the cylinder such that the temperature remains constant but n moles of the diatomic gas get converted into monoatomic gas. What is the change in the total kinetic energy of the gas ? **[2017 (Online)]**

(A) $\frac{1}{2}nRT$ (B) 0

(C) $\frac{3}{2}nRT$ (D) $\frac{5}{2}nRT$

26. For the *P-V* diagram given for an ideal gas, out of the following which one correctly represents the *T-P* diagram? **[2017 (Online)]**

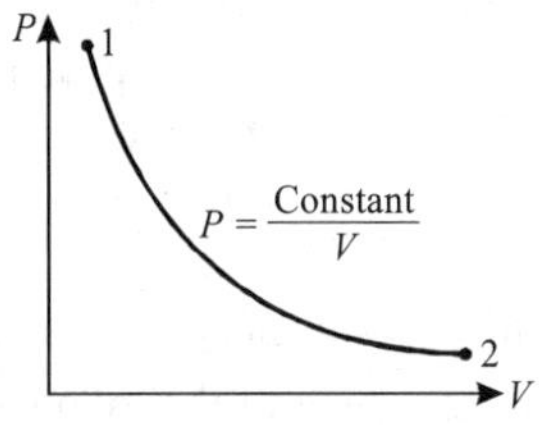

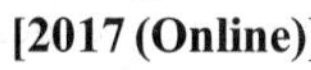

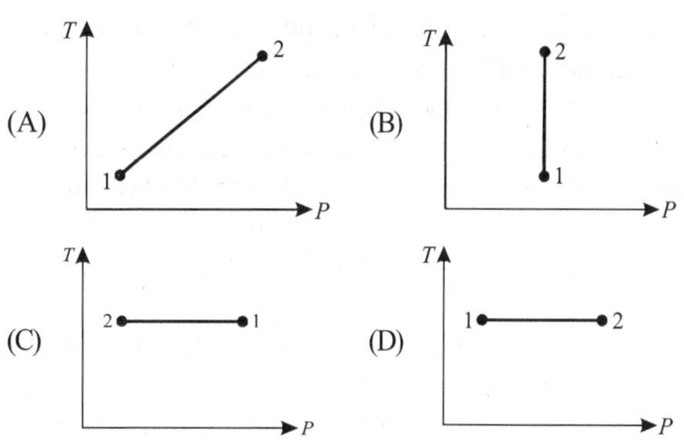

27. A mixture of 2 moles of helium gas (atomic mass = 4 u), and 1 mole of argon gas (atomic mass = 40u) is kept at 300 K in a container. The ratio of their rms speeds $\left[\frac{V_{rms}(\text{helium})}{V_{rms}(\text{argon})}\right]$, is close to : **[9 Jan. 2019 (Morning)]**

(A) 2.24 (B) 0.45

(C) 0.32 (D) 3.16

28. Two kg of a monoatomic gas is at a pressure of 4×10^4 N/m^2. The density of the gas is 8 kg/m^3. What is the order of energy of the gas due to its thermal motion ? **[10 Jan. 2019 (Evening)]**

(A) 10^3 J (B) 10^5 J

(C) 10^6 J (D) 10^4 J

29. The gas mixture consists of 3 moles of oxygen and 5 moles of argon at temperature T. Considering only translational and rotational modes, the total internal energy of the system is : **[11 Jan. 2019 (Morning)]**

(A) $12RT$ (B) $20RT$

(C) $15RT$ (D) $4RT$

30. An ideal gas occupies a volume of 2 m^3 at a pressure of 3×10^6 Pa. The energy of the gas is : **[12 Jan. 2019 (Morning)]**

(A) 3×10^2 (B) 10^8 J

(C) 6×10^4 J (D) 9×10^6 J

31. A vertical closed cylinder is separated into two parts by a frictionless piston of mass m and of negligible thickness. The piston is free to move along the length of the cylinder. The length of the cylinder above the piston is l_1, and that below the piston is l_2, such that $l_1 > l_2$. Each part of the cylinder contains n moles of an ideal gas at equal temperature T. If the piston is stationary, its mass, m, will be given by : (R is universal gas constant and g is the acceleration due to gravity) : **[12 Jan. 2019 (Evening)]**

(A) $\frac{nRT}{g}\left[\frac{1}{l_2}+\frac{1}{l_1}\right]$ (B) $\frac{nRT}{g}\left[\frac{l_1-l_2}{l_1 l_2}\right]$

(C) $\frac{RT}{g}\left[\frac{2l_1+l_2}{l_1 l_2}\right]$ (D) $\frac{RT}{ng}\left[\frac{l_1-3l_2}{l_1 l_2}\right]$

32. An ideal gas is enclosed in a cylinder at pressure of 2 atm and temperature, 300 K. The mean time between two successive collisions is 6×10^{-8} s. If the pressure is doubled and temperature is increased to 500 K, the mean time between two successive collisions will be close to : **[12 Jan. 2019 (Evening)]**

(A) 4×10^{-8} s (B) 3×10^{-6} s
(C) 2×10^{-7} s (D) 0.5×10^{-8} s

33. If 10^{22} gas molecules each of mass 10^{-26} kg collide with a surface (perpendicular to it) elastically per second over an area 1 m^2 with a speed 10^4 m/s, the pressure exerted by the gas molecules will be of the order of: **[8 April 2019 (Morning)]**

(A) 10^8 N/m^2 (B) 10^4 N/m^2
(C) 10^3 N/m^2 (D) 10^{16} N/m^2

34. The temperature, at which the root mean square velocity of hydrogen molecules equals their escape velocity from the earth, is closest to :

[Boltzmann Constant $k_B = 1.38 \times 10^{-23}$ J/K
Avogadro Number $N_A = 6.02 \times 10^{26}$ /kg
Radius of Earth : 6.4×10^6 m
Gravitational acceleration on Earth = 10 ms^{-2}]

[8 April 2019 (Evening)]

(A) 650 K (B) 3×10^5 K
(C) 10^4 K (D) 800 K

35. An HCl molecule has rotational, translational and vibrational motions. If the rms velocity of HCl molecules in its gaseous phase is $\bar{v}$, m is its mass and k_B is Boltzmann constant, then its temperature will be : **[9 April 2019 (Morning)]**

(A) $\dfrac{m\bar{v}^2}{6k_B}$ (B) $\dfrac{m\bar{v}^2}{5k_B}$
(C) $\dfrac{m\bar{v}^2}{3k_B}$ (D) $\dfrac{m\bar{v}^2}{7k_B}$

36. For a given gas at 1 atm pressure, rms speed of the molecule is 200 m/s at 127°C. At 2 atm pressure and at 227°C, the rms speed of the molecules will be : **[9 April 2019 (Morning)]**

(A) 80 m/s (B) $100\sqrt{5}$ m/s
(C) $80\sqrt{5}$ m/s (D) 100 m/s

37. The specific heats, C_P and C_V of a gas of diatomic molecules, A, are given (in units of J mol^{-1} K^{-1}) by 29 and 22, respectively. Another gas of diatomic molecules, B, has the corresponding values 30 and 21. If they are treated as ideal gases, then : **[9 April 2019 (Evening)]**

(A) A has one vibrational mode and B has two
(B) Both A and B have a vibrational mode each
(C) A is rigid but B has a vibrational mode
(D) A has a vibrational mode but B has none

38. A 25×10^{-3} m^3 volume cylinder is filled with 1 mol of O_2 gas at room temperature (300 K). The molecular diameter of O_2, and its root mean square speed, are found to be 0.3 nm, and 200 m/s, respectively. What is the average collision rate (per second) for an O_2 molecule ? **[10 April 2019 (Morning)]**

(A) $\sim 10^{11}$ (B) $\sim 10^{13}$
(C) $\sim 10^{10}$ (D) $\sim 10^{12}$

39. Two moles of helium gas is mixed with three moles of hydrogen molecules (taken to be rigid). What is the molar specific heat of mixture at constant volume ? ($R = 8.3$ J/mol K): **[12 April 2019 (Morning)]**

(A) 21.6 J/mol K (B) 19.7 J/mol K
(C) 17.4 J/mol K (D) 15.7 J/mol K

40. The number density of molecules of a gas depends on their distance r from the origin as, $n(r) = n_0 e^{-\alpha r^4}$. Then the total number of molecules is proportional to : **[12 April 2019 (Evening)]**

(A) $n_0 \alpha^{1/4}$ (B) $n_0 \alpha^{-3}$
(C) $n_0 \alpha^{-3/4}$ (D) $\sqrt{n_0}\, \alpha^{1/2}$

41. The plot that depicts the behavior of the mean free time t (time between two successive collisions) for the molecules of an ideal gas, as a function of temperature (T), qualitatively, is : (Graphs are schematic and not drawn to scale) : **[8 Jan. 2020 (Morning)]**

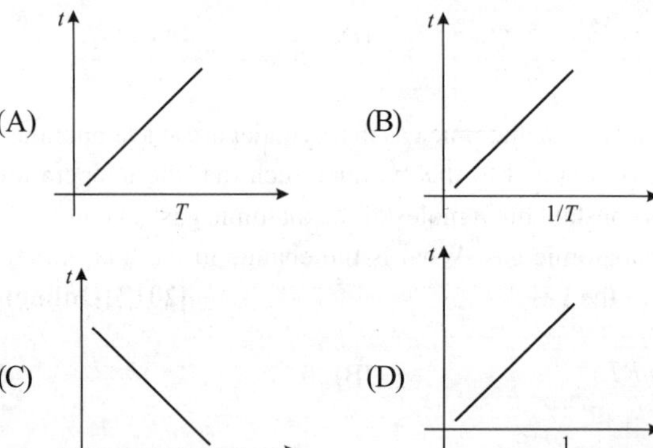

42. Two gases-argon (atomic radius 0.07 nm, atomic weight 40) and xenon (atomic radius 0.1 nm, atomic weight 140) have the same number density and are at the same temperature. The ratio of their respective mean free times is closest to : **[9 Jan. 2020 (Evening)]**

(A) 3.67 (B) 4.67
(C) 1.83 (D) 2.3

43. An ideal gas in a closed container is slowly heated. As its temperature increases, which of the following statements are true : [2 Sept. 2020 (Evening)]

(i) The mean free path of the molecules decreases

(ii) The mean collision time between the molecules decreases

(iii) The mean free path remains unchanged

(iv) The mean collision time remains unchanged

(A) (iii) and (iv) (B) (i) and (ii)

(C) (i) and (iv) (D) (ii) and (iii)

44. To raise the temperature of a certain mass of gas by 50ºC at a constant pressure, 160 calories of heat is required. When the same mass of gas is cooled by 100ºC at constant volume, 240 calories of heat is released. How many degrees of freedom does each molecule of this gas have (assume gas to be ideal) ? **[3 Sept. 2020 (Evening)]**

(A) 5 (B) 3

(C) 6 (D) 7

45. Match the C_P/C_V ratio for ideal gases with different type of molecules : **[4 Sept. 2020 (Morning)]**

Molecular type	C_P/C_V
(a) Monoatomic	(i) 7/5
(b) Diatomic rigid molecules	(ii) 9/7
(c) Diatomic non-rigid molecules	(iii) 4/3
(d) Triatomic rigid molecules	(iv) 5/3

(A) (a)-(iv), (b)-(i), (c)-(ii), (d)-(iii)

(B) (a)-(iv), (b)-(ii), (c)-(i), (d)-(iii)

(C) (a)-(iii), (b)-(iv), (c)-(ii), (d)-(i)

(D) (a)-(ii), (b)-(iii), (c)-(i), (d)-(iv)

46. Nitrogen gas is at 300°C temperature. The temperature (in K) at which the rms speed of a H_2 molecule would be equal to the rms speed of a nitrogen molecule, is______. (Molar mass of N_2 gas 28g) **[5 Sept. 2020 (Evening)]**

47. Initially a gas of diatomic molecules is contained in a cylinder of volume V_1 at a pressure P_1 and temperature 250 K. Assuming that 25% of the molecules get dissociated causing a change in number of moles. The pressure of the resulting gas at temperature 2000 K, when contained in a volume $2V_1$ is given by P_2. The ratio P_2/P_1 is : **[6 Sept. 2020 (Morning)]**

48. In a dilute gas at pressure P and temperature T, the mean time between successive collisions of a molecule varies with T as : **[6 Sept. 2020 (Evening)]**

(A) $\sqrt{T}$ (B) $\frac{1}{T}$

(C) $\frac{1}{\sqrt{T}}$ (D) T

49. On the basis of kinetic theory of gases, the gas exerts pressure because its molecules : **[24 Feb. 2021 (Evening)]**

(A) Continuously stick to the walls of container

(B) Are attracted by the walls of container

(C) Continuously lose their energy till it reaches wall

(D) Suffer change in momentum when impinge on the walls of container

50. The root mean square speed of molecules of a given mass of a gas at 27°C and 1 atmosphere pressure is 200 ms^{-1}. The root mean square speed of molecules of the gas at 127°C and 2 atmosphere pressure is $\frac{x}{\sqrt{3}}$ ms^{-1}. The value of x will be______.

[24 Feb. 2021 (Evening)]

51. Given below are two statements :**[25 Feb. 2021 (Evening)]**

Statement-I : In a diatomic molecule, the rotational energy at a given temperature obeys Maxwell's distribution

Statement-II : In a diatomic molecule, the rotational energy at a given temperature equals the translational kinetic energy for each molecule

In the light of the above statements, choose the correct answer from the options given below :

(A) Both Statement-I and Statement-II are false

(B) Statement-I is false but Statement-II is true

(C) Statement-I is true but Statement-II is false

(D) Both Statement-I and Statement-II are true

52. The volume V of an enclosure contains a mixture of three gases, 16 g of oxygen, 28 g of nitrogen and 44 g of carbon dioxide at absolute temperature T. Consider R as universal gas constant. The pressure of the mixture of gases is :

[16 March 2021 (Morning)]

(A) $\frac{88RT}{V}$ (B) $\frac{3RT}{V}$

(C) $\frac{5}{2}\frac{RT}{V}$ (D) $\frac{4RT}{V}$

53. Calculate the value of mean free path (λ) for oxygen molecules at temperature 27°C and pressure 1.01×10^5 Pa. Assume the molecular diameter 0.3 nm and the gas is ideal. ($k = 1.38 \times 10^{-23}$ JK^{-1}) **[16 March 2021 (Evening)]**

(A) 58 nm (B) 32 nm

(C) 86 nm (D) 102 nm

54. Two ideal polyatomic gases at temperatures T_1 and T_2 are mixed so that there is no loss of energy. If F_1 and F_2, m_1 and m_2, n_1 and n_2 be the degrees of freedom, masses, number of molecules of the first and second gas respectively, the temperature of mixture of these two gases is :

[17 March 2021 (Morning)]

(A) $\frac{n_1T_1 + n_2T_2}{n_1 + n_2}$ (B) $\frac{n_1F_1T_1 + n_2F_2T_2}{n_1F_1 + n_2F_2}$

(C) $\frac{n_1F_1T_1 + n_2F_2T_2}{F_1 + F_2}$ (D) $\frac{n_1F_1T_1 + n_2F_2T_2}{n_1 + n_2}$

55. If one mole of the polyatomic gas is having two vibrational modes and β is the ratio of molar specific heats for polyatomic gas $\left(\beta = \frac{C_P}{C_V}\right)$ then the value of β is : **[17 March 2021 (Evening)]**

(A) 1.02 (B) 1.2
(C) 1.25 (D) 1.35

56. What will be the average value of energy along one degree of freedom for an ideal gas in thermal equilibrium at a temperature T? (k_B is Boltzmann constant) **[18 March 2021 (Morning)]**

(A) $\frac{1}{2}k_BT$ (B) $\frac{2}{3}k_BT$
(C) $\frac{3}{2}k_BT$ (D) k_BT

57. An ideal gas in a cylinder is separated by a piston in such a way that the entropy of one part is S_1 and that of the other part is S_2. Given that $S_1 > S_2$. If the piston is removed then the total entropy of the system will be : **[18 March 2021 (Evening)]**

(A) $S_1 \times S_2$ (B) $S_1 - S_2$
(C) $\frac{S_1}{S_2}$ (D) $S_1 + S_2$

58. Consider a sample of oxygen behaving like an ideal gas. At 300 K, the ratio of root mean square (rms) velocity to the average velocity of gas molecule would be :
(Molecular weight of oxygen is 32 g/mol; $R = 8.3\ \text{J K}^{-1}\ \text{mol}^{-1}$)
[18 March 2021 (Evening)]

(A) $\sqrt{\frac{3}{3}}$ (B) $\sqrt{\frac{8}{3}}$
(C) $\sqrt{\frac{3\pi}{8}}$ (D) $\sqrt{\frac{8\pi}{3}}$

59. Consider a mixture of gas molecule of types A, B and C having masses $m_A < m_B < m_C$. The ratio of their root mean square speeds at normal temperature and pressure is :
[20 July 2021 (Morning)]

(A) $\frac{1}{v_A} > \frac{1}{v_B} > \frac{1}{v_C}$ (B) $v_A = v_B = v_C = 0$
(C) $v_A = v_B \neq v_C$ (D) $\frac{1}{v_A} < \frac{1}{v_B} < \frac{1}{v_C}$

60. The correct relation between the degrees of freedom f and the ratio of specific heat γ is : **[20 July 2021 (Evening)]**

(A) $f = \frac{\gamma+1}{2}$ (B) $f = \frac{1}{\gamma+1}$
(C) $f = \frac{2}{\gamma+1}$ (D) $f = \frac{2}{\gamma-1}$

61. Which of the following graphs represent the behaviour of an ideal gas ? Symbols have their usual meaning :
[20 July 2021 (Evening)]

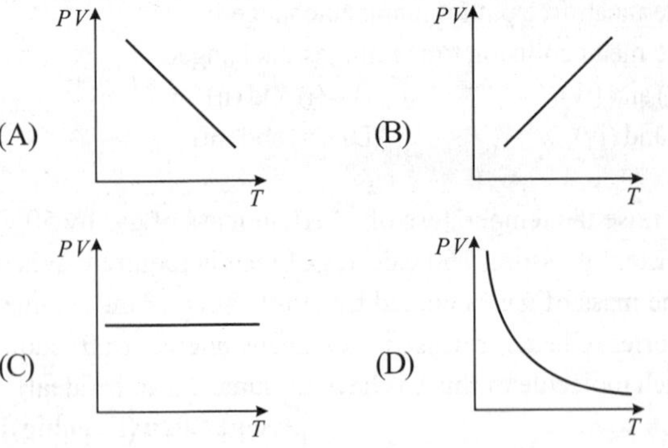

62. What will be the average kinetic energy for a monoatomic gas molecule in thermal equilibrium at temperature T?
[22 July 2021 (Evening)]

(A) $\frac{1}{2}k_BT$ (B) $\frac{3}{2}k_BT$
(C) $\frac{2}{3}k_BT$ (D) k_BT

63. For a gas $C_P - C_V = R$ in a state P and $C_P - C_V = 1.10\,R$ in a state Q, T_P and T_Q are the temperature in two different state P and Q respectively. Then : **[25 July 2021 (Morning)]**

(A) $T_P > T_Q$ (B) $T_P = T_Q$
(C) $T_P = 0.9\,T_Q$ (D) $T_P < T_Q$

64. A system consists of two types of gas molecules A and B having same number density $2 \times 10^{25}\ /\text{m}^3$. The diameter of A and B are 10 Å and 5 Å respectively. They suffer collision at room temperature. The ratio of average distance covered by the molecule A to that of B between two successive collision is ______ $\times 10^{-2}$: **[25 July 2021 (Evening)]**

65. The number of molecules in one litre of an ideal gas at 300 K and 2 atmospheric pressure with mean kinetic energy 2×10^{-9} J per molecule is : **[27 July 2021 (Morning)]**

(A) 3×10^{11} (B) 6×10^{11}
(C) 0.75×10^{11} (D) 1.5×10^{11}

66. The rms speed of the molecules of Hydrogen, Oxygen and Carbon dioxide at the same temperature are V_H, V_O and V_C respectively then : **[26 Aug. 2021 (Morning)]**

(A) $V_H > V_O > V_C$ (B) $V_H = V_O > V_C$
(C) $V_C > V_O > V_H$ (D) $V_H = V_O = V_C$

67. A cylindrical container of volume $4.0 \times 10^{-3}\ \text{m}^3$ contains one mole of hydrogen and two moles of carbon dioxide. Assume the temperature of the mixture is 400 K. The pressure of the mixture of gases is : (Take gas constant as 8.3 J mol^{-1} K^{-1}] :

[26 Aug. 2021 (Evening)]

(A) 249×10^1 Pa (B) 24.9 Pa

(C) 24.9×10^5 Pa (D) 24.9×10^3 Pa

68. A balloon carries a total load of 185 kg at normal pressure and temperature of 27°C. What load will the balloon carry on rising to a height at which the barometric pressure is 45 cm of Hg and the temperature is – 7°C. Assuming constant volume?

[27 Aug. 2021 (Morning)]

(A) 2.14.15 kg (B) 219.07 kg

(C) 181.46 kg (D) 123.54 kg

69. If the *RMS* speed of oxygen molecules at 0°C is 160 m/s. Find the rms speed of hydrogen molecules at 0°C.

[27 Aug. 2021 (Evening)]

(A) 40 m/s (B) 80 m/s

(C) 640 m/s (D) 332 m/s

70. For an ideal gas the instantaneous change in pressure p with volume V is given by the equation $dp/dV = -ap$. If $p = p_0$ at $V = 0$ is the given boundary condition, then the maximum temperature one mole of gas can attain is : (Here R is the gas constant) **[31 Aug. 2021 (Morning)]**

(A) $\frac{ap_0}{eR}$ (B) $\frac{p_0}{aeR}$

(C) Infinity (D) 0°C

71. A mixture of hydrogen and oxygen has volume 500 cm^3, temperature 300 K, pressure 400 kPa and mass 0.76 g. The ratio of masses of oxygen to hydrogen will be :

[31 Aug. 2021 (Evening)]

(A) 3 : 8 (B) 16 : 3

(C) 3 : 16 (D) 8 : 3

72. Consider two separate ideal gases of electrons and protons having same number of particles. The temperature of both the gases are same. The ratio of the uncertainty in determining the position of an electron to that of a proton is proportional to :

[31 Aug. 2021 (Evening)]

(A) $\sqrt{\frac{m_p}{m_e}}$ (B) $\sqrt{\frac{m_e}{m_p}}$

(C) $\frac{m_p}{m_e}$ (D) $\left(\frac{m_p}{m_e}\right)^{3/2}$

73. The average translational kinetic energy of N_2 gas molecules at______°C becomes equal to the kinetic energy of an electron accelerated from rest through a potential difference of 0.1 volt. (Given $k = 1.38 \times 10^{-23}$ J/K) (Fill the nearest integer)

[01 Sep. 2021 (Evening)]

74. 0.056 kg of Nitrogen is enclosed in a vessel at a temperature of 127°C. The amount of heat required to double the speed of its molecules is _____ k cal. [Take : $R = 2$ cal mole^{-1} K^{-1}]

[24 June 2022 (Morning)]

75. The relation between root mean square speed (v_{rms}) and most probable speed (v_p) for the molar mass M of oxygen gas molecule at the temperature of 300 K will be :

[25 June 2022 (Morning)]

(A) $v_{rms} = \sqrt{\frac{2}{3}} v_p$ (B) $v_{rms} = \sqrt{\frac{3}{2}} v_p$

(C) $v_{rms} = v_p$ (D) $v_{rms} = \sqrt{\frac{1}{3}} v_p$

76. When a gas filled in a closed vessel is heated by raising the temperature by 1°C, its pressure increase by 0.4%. The initial temperature of the gas is _____ K.

[25 June 2022 (Evening)]

77. A thermally insulated vessel contains an ideal gas of molecular mass M and ratio of specific heats 1.4. Vessel is moving with speed v and is suddenly brought to rest. Assuming no heat is lost to the surrounding and vessel temperature of the gas increases by : [R = universal gas constant]

[26 June 2022 (Morning)]

(A) $\frac{Mv^2}{7R}$ (B) $\frac{Mv^2}{5R}$

(C) $2\frac{Mv^2}{7R}$ (D) $7\frac{Mv^2}{5R}$

78. A flask contains argon and oxygen in the ratio of 3 : 2 in mass and the mixture is kept at 27°C. The ratio of their average kinetic energy per molecule respectively will be :

[26 June 2022 (Evening)]

(A) 3 : 2 (B) 9 : 4

(C) 2 : 3 (D) 1 : 1

79. A mixture of hydrogen and oxygen has volume 2000 cm^3, temperature 300 K, pressure 100 kPa and mass 0.76 g. The ratio of number of moles of hydrogen to number of moles of oxygen in the mixture will be : **[27 June 2022 (Morning)]**

(A) $\frac{1}{3}$ (B) $\frac{3}{1}$

(C) $\frac{1}{16}$ (D) $\frac{16}{1}$

80. For a perfect gas, two pressures P_1 and P_2 are shown in figure. The graph shows : **[27 June 2022 (Evening)]**

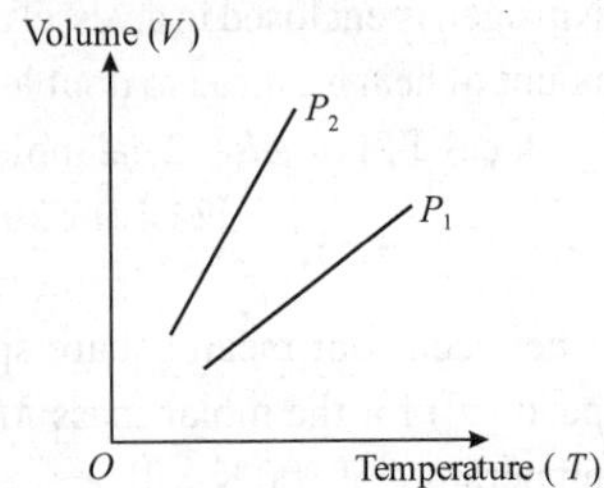

(A) $P_1 > P_2$ (B) $P_1 < P_2$
(C) $P_1 = P_2$
(D) Insufficient data to draw any conclusion

81. According to kinetic theory of gases,
(a) The motion of the gas molecules freezes at 0°C
(b) The mean free path of gas molecules decreases if the density of molecules is increased
(c) The mean free path of gas molecules increases if temperature is increased keeping pressure constant
(d) Average kinetic energy per molecule per degree of freedom is $\frac{3}{2}k_BT$ (for monoatomic gases)
Choose the most appropriate answer from the options given below: **[27 June 2022 (Evening)]**
(A) (a) and (c) only (B) (b) and (c) only
(C) (a) and (b) only (D) (c) and (d) only

82. What will be the effect on the root mean square velocity of oxygen molecules if the temperature is doubled and oxygen molecule dissociates into atomic oxygen? **[28 June 2022 (Evening)]**
(A) The velocity of atomic oxygen remains same
(B) The velocity of atomic oxygen doubles
(C) The velocity of atomic oxygen becomes half
(D) The velocity of atomic oxygen becomes four times

83. A cylinder of fixed capacity of 44.8 litres contains helium gas at standard temperature and pressure. The amount of heat needed to raise the temperature of gas in the cylinder by 20.0°C will be : [Given : Gas constant $R = 8.3\ \text{JK}^{-1}\ \text{mol}^{-1}$] **[29 June 2022 (Morning)]**
(A) 249 J (B) 415 J
(C) 498 J (D) 830 J

84. In Vander Waals equation $\left[P+\frac{a}{V^2}\right][V-b]=RT$; P is pressure, V is volume, R is universal gas constant and T is temperature. The ratio of constants $\frac{a}{b}$ is dimensionally equal to : **[29 June 2022 (Morning)]**
(A) $\frac{P}{V}$ (B) $\frac{V}{P}$
(C) PV (D) PV^3

85. A vessel contains 16 g of hydrogen and 128 g of oxygen at standard temperature and pressure. The volume of the vessel in cm^3 is : **[29 June 2022 (Evening)]**
(A) 72×10^5 (B) 32×10^5
(C) 27×10^4 (D) 54×10^4

86. Following statements are given : **[25 July 2022 (Morning)]**
(1) The average kinetic energy of a gas molecule decreases when the temperature is reduced
(2) The average kinetic energy of a gas molecule increases with increase in pressure at constant temperature
(3) The average kinetic energy of a gas molecule decreases with increases in volume
(4) Pressure of a gas increases with increase in temperature at constant pressure
(5) The volume of gas decreases with increase in temperature
Choose the correct answer from the options given below :
(A) (1) and (4) only (B) (1), (2) and (4) only
(C) (2) and (4) only (D) (1), (2) and (5) only

87. Sound travels in a mixture of two moles of helium and n moles of hydrogen. If rms speed of gas molecules in the mixture is $\sqrt{2}$ times the speed of sound, then the value of n will be : **[25 July 2022 (Evening)]**
(A) 1 (B) 2
(C) 3 (D) 4

88. 7 mole of certain monoatomic ideal gas undergoes a temperature increase of 40 K at constant pressure. The increase in the internal energy of the gas in this process is :
[Given : $R = 8.3\ \text{JK}^{-1}\ \text{mol}^{-1}$] **[26 July 2022 (Morning)]**
(A) 5810 J (B) 3486 J
(C) 11620 J (D) 6972 J

89. Same gas is filled in two vessels of the same volume at the same temperature. If the ratio of the number of molecules is 1 : 4, then :
(a) The rms velocity of gas molecules in two vessels will be the same
(b) The ratio of pressure in these vessels will be 1 : 4
(c) The ratio of pressure will be 1 : 1
(d) The rms velocity of gas molecules in two vessels will be in the ratio of 1 : 4 **[27 July 2022 (Morning)]**
(A) (a) and (c) only (B) (b) and (d) only
(C) (a) and (b) only (D) (c) and (d) only

90. Which statements are correct about degrees of freedom ?
(a) A molecule with n degrees of freedom has n^2 different ways of storing energy.
(b) Each degree of freedom is associated with $\frac{1}{2}RT$ average energy per mole.

(c) A monoatomic gas molecule has 1 rotational degree of freedom where as diatomic molecule has 2 rotational degrees of freedom.

(d) CH_4 has a total to 6 degrees of freedom.

Choose the correct answer from the option given below :

[27 July 2022 (Evening)]

(A) (b) and (c) only (B) (b) and (d) only

(C) (a) and (b) only (D) (c) and (d) only

91. Given below are two statements : **[28 July 2022 (Morning)]**

Statement-I : The average momentum of a molecule in a sample of an ideal gas depends on temperature.

Statement-II : The rms speed of oxygen molecules in a gas is v. If the temperature is doubled and the oxygen molecules dissociate into oxygen atoms, the rms speed will become $2v$.

In the light of the above statements, choose the correct answer from the options given below :

(A) Both Statement-I and Statement-II are true

(B) Both Statement-I and Statement-II are false

(C) Statement-I is true but Statement-II is false

(D) Statement-I is false but Statement-II is true

92. A vessel contains 14 g of nitrogen gas at a temperature of 27°C. The amount of heat to be transferred to the gas to double the rms speed of its molecules will be :

[Take : $R = 8.32\ \text{J mol}^{-1}\ \text{K}^{-1}$] **[28 July 2022 (Evening)]**

(A) 2229 J (B) 5616 J

(C) 9360 J (D) 13, 104 J

93. The root mean square speed of smoke particles of mass 5×10^{-17} kg in their Brownian motion in air at NTP is approximately : [Given : $k = 1.38 \times 10^{-23}\ \text{JK}^{-1}$]

[29 July 2022 (Evening)]

(A) $60\ \text{mm s}^{-1}$ (B) $12\ \text{mm s}^{-1}$

(C) $15\ \text{mm s}^{-1}$ (D) $36\ \text{mm s}^{-1}$

94. Given below are two statements :

Statements-I: The temperature of a gas is – 73°C. When the gas is heated to 527°C, the root mean square speed of the molecules is doubled

Statement-II: The product of pressure and volume of an ideal gas will be equal to translational kinetic energy of the molecules.

In the light of the above statements, choose the correct answer from the options given below : **[24 Jan. 2023 (Morning)]**

(A) Both statement-I and Statement-II are true

(B) Statement-I is true but Statement-II is false

(C) Both Statement-I and Statement-II are false

(D) Statement-I is false but Statement-II is true

95. The root mean square velocity of molecules of gas is :

[25 Jan. 2023 (Morning)]

(A) Proportional to square of temperature (T^2)

(B) Inversely proportional to square root of temperature $\sqrt{\frac{1}{T}}$

(C) Proportional to square root of temperature $\sqrt{T}$

(D) Proportional to temperature (T)

96. According to law of equipartition of energy the molar specific heat of a diatomic gas at constant volume where the molecule has one additional vibrational mode is :

[25 Jan. 2023 (Evening)]

(A) $\frac{9}{2}R$ (B) $\frac{5}{2}R$

(C) $\frac{3}{2}R$ (D) $\frac{7}{2}R$

97. A bicycle tyre is filled with air having pressure of 270 kPa at 27°C. The approximate pressure of the air in the tyre when the temperature increases to 36°C is : **[29 Jan. 2023 (Morning)]**

(A) 270 kPa (B) 262 kPa

(C) 278 kPa (D) 360 kPa

98. At 300 K, the rms speed of oxygen molecules is $\sqrt{\frac{\alpha+5}{\alpha}}$ times to that of its average speed in the gas. Then, the value of α will be (used $\pi = \frac{22}{7}$) **[29 Jan. 2023 (Evening)]**

(A) 32 (B) 28

(C) 24 (D) 24

99. The pressure (P) and temperature (T) relationship of an ideal gas obeys the equation PT^2 = constant. The volume expansion coefficient of the gas will be :

[30 Jan. 2023 (Morning)]

(A) $3T^2$ (B) $\frac{3}{T^2}$

(C) $\frac{3}{T^3}$ (D) $\frac{3}{T}$

100. A flask contains hydrogen and oxygen in the ratio of 2 : 1 by mass at temperature 27°C. The ratio of average kinetic energy per molecule of hydrogen and oxygen respectively is :

[30 Jan. 2023 (Evening)]

(A) 2 : 1 (B) 1 : 1

(C) 1 : 4 (D) 4 : 1

101. Heat energy of 735 J is given to a diatomic gas allowing the gas to expand at constant pressure. Each gas molecule rotates around an internal axis but do not oscillate. The increase in the internal energy of the gas will be :

[31 Jan. 2023 (Evening)]

(A) 525 J (B) 441 J
(C) 572 J (D) 735 J

102. The average kinetic energy of a molecule of the gas is :

[1 Feb. 2023 (Morning)]

(A) Proportional to absolute temperature
(B) Proportional to volume
(C) Proportional to pressure
(D) Dependent on the nature of the gas

103. For three low density gases A, B, C pressure versus temperature graphs are plotted while keeping them at constant volume, as shown in the figure.

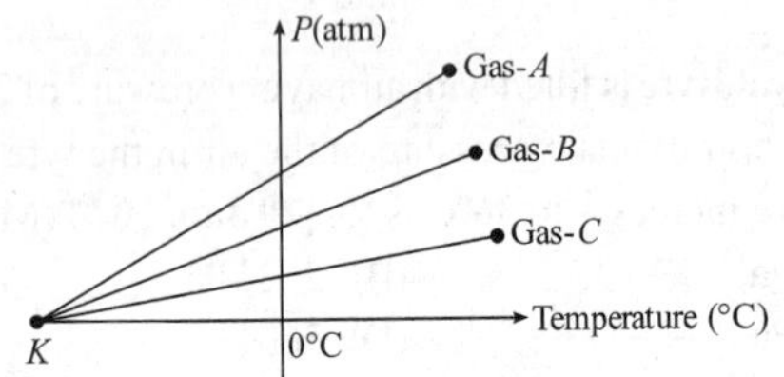

The temperature corresponding to the point 'K' is :

[1 Feb. 2023 (Evening)]

(A) −273°C (B) −100°C
(C) −373°C (D) −40°C

104. The number of air molecules per cm^3 increased from 3×10^{19} to 12×10^{19}. The ratio of collision frequency of air molecules before and after the increase in number respectively is **[6 April 2023 (Morning)]**

(A) 1.25 (B) 0.25
(C) 0.75 (D) 0.50

105. The ratio of speed of sound in hydrogen gas to the speed of sound in oxygen gas at the same temperature is :

[6 April 2023 (Evening)]

(A) 4 : 1 (B) 1 : 2
(C) 1 : 4 (D) 1 : 1

106. The temperature of an ideal gas is increased from 200 K to 800 K. If rms speed of gas at 200 K is v_0. Then, rms speed of the gas at 800 K will be: **[6 April 2023 (Evening)]**

(A) v_0 (B) $4v_0$

(C) $\frac{v_0}{4}$ (D) $2v_0$

107. An air bubble of volume 1 cm^3 rises from the bottom of a lake 40 m deep to the surface at a temperature of 12°C. The atmospheric pressure is 1×10^5 Pa, the density of water is 1000 kg/m^3 and g = 10 m/s^2. There is no difference of the temperature of water at the depth of 40 m and on the surface. The volume of air bubble when it reaches the surface will be :

[8 April 2023 (Morning)]

(A) 5 cm^3 (B) 2 cm^3
(C) 4 cm^3 (D) 3 cm^3

108. The temperature at which the kinetic energy of oxygen molecules becomes double than its value at 27°C is :

[8 April 2023 (Evening)]

(A) 1227°C (B) 927°C
(C) 327°C (D) 627°C

109. Match **List-I** with **List-II**:

List-I	**List-II**
(a) 3 Translational degrees of freedom	(i) Monoatomic gases
(b) 3 Translational, 2 rotational degrees of freedoms	(ii) Polyatomic gases
(c) 3 Translational, 2 rotational and 1 vibrational degrees of freedom	(iii) Rigid diatomic gases
(d) 3 Translational, 3 rotational and more than one vibrational degrees of freedom	(iv) Non-rigid diatomic gases

Choose the correct answer from the options given below:

[10 April 2023 (Morning)]

(A) (a)-(iv), (b)-(iii), (c)-(ii), (d)-(i)
(B) (a)-(iv), (b)-(ii), (c)-(i), (d)-(iii)
(C) (a)-(i), (b)-(iii), (c)-(iv), (d)-(ii)
(D) (a)-(i), (b)-(iv), (c)-(iii), (d)-(ii)

110. A gas mixture consists of 2 moles of oxygen and 4 moles of neon at temperature T. Neglecting all vibrational modes, the total internal energy of the system will be:

[10 April 2023 (Evening)]

(A) 8 RT (B) 16 RT
(C) 4 RT (D) 11 RT

111. Three vessels of equal volume contain gases at the same temperature and pressure. The first vessel contains neon (monoatomic), the second contains chlorine (diatomic) and third contains uranium hexafluoride (polyatomic). Arrange these on the

basis of their root mean square speed (v_{rms}) and choose the correct answer from the options given below :

[11 April 2023 (Morning)]

(A) v_{rms} (mono) = v_{rms} (dia) = v_{rms} (poly)
(B) v_{rms} (mono) > v_{rms} (dia) > v_{rms} (poly)
(C) v_{rms} (dia) < v_{rms} (poly) < v_{rms} (mono)
(D) v_{rms} (mono) < v_{rms} (dia) < v_{rms} (poly)

112. The root mean square speed of molecules of nitrogen gas at 27°C is approximately :
(Given mass of a nitrogen molecule = 4.6×10^{-26} kg and take Boltzmann constant $k_B = 1.4 \times 10^{-23}$ JK^{-1})

[11 April 2023 (Evening)]

(A) 523 m/s (B) 1262 m/s
(C) 91 m/s (D) 27.4 m/s

113. If the rms speed of chlorine molecule is 490 m/s at 27°C, the rms speed of argon molecules at the same temperature will be: (Atomic mass of argon = 39.9u, molecular mass of chlorine = 70.9u) **[12 April 2023 (Morning)]**

(A) 751.7 m/s (B) 451.7 m/s
(C) 651.7 m/s (D) 551.7 m/s

114. The rms speed of oxygen molecule in a vessel at particular temperature is $\left(1+\frac{5}{x}\right)^{1/2}$ υ, where υ is the average speed of the molecule. The value of x will be : (Take $\pi = \frac{22}{7}$)

[13 April 2023 (Morning)]

(A) 28 (B) 27
(C) 8 (D) 4

115. The mean free path of molecules of a certain gas at STP is 1500 d, where d is the diameter of the gas molecules. While maintaining the standard pressure, the mean free path of the molecules at 373 K is approximately: **[13 April 2023 (Evening)]**

(A) 1098 d (B) 2049 d
(C) 750 d (D) 1500 d

116. (A flask contains Hydrogen and Argon in the ratio 2 : 1 by mass. The temperature of the mixture is 30°C. The ratio of average kinetic energy per molecule of the two gases (K argon/ K hydrogen) is: (Given: Atomic Weight of Ar = 39.9)

[15 April 2023 (Morning)]

(A) 1 (B) 2
(C) $\frac{39.9}{2}$ (D) 39.9

* * * * *

CHAPTER

13 OSCILLATIONS

13.1 SIMPLE HARMONIC MOTION

1. If a spring has time period T, and is cut into n equal parts, then the time period of each part will be : **[2002]**

(A) $T\sqrt{n}$ (B) $\frac{T}{\sqrt{n}}$

(C) nT (D) T

2. In a simple harmonic oscillator, at the mean position : **[2002]**

(A) Kinetic energy is minimum, potential energy is maximum
(B) Both kinetic and potential energies are maximum
(C) Kinetic energy is maximum, potential energy is minimum
(D) Both kinetic and potential energies are minimum

3. A child swinging on a swing in sitting position, stands up, then the time period of the swing will : **[2002]**

(A) Increase
(B) Decrease
(C) Remain same
(D) Increase if the child is long and decrease if the child is short

4. A mass M is suspended from a spring of negligible mass. The spring is pulled a little and then released so that the mass executes SHM of time period T. If the mass is increased by m, the time period becomes $5T/3$, then the ratio of m/M is :**[2003]**

(A) $\frac{3}{5}$ (B) $\frac{25}{9}$

(C) $\frac{16}{9}$ (D) $\frac{5}{3}$

5. The displacement of a particle varies according to the relation $x = 4(\cos\pi t + \sin\pi t)$. The amplitude of the particle is : **[2003]**

(A) -4 (B) 4

(C) $4\sqrt{2}$ (D) 8

6. A body executes simple harmonic motion. The potential energy (PE), the kinetic energy (KE) and total energy (TE) are measured as function of displacement x. Which of the following statements is true ? **[2003]**

(A) KE is maximum when $x = 0$
(B) TE is zero when $x = 0$
(C) KE is maximum when x is maximum
(D) PE is maximum when $x = 0$

7. Two particles A and B of equal masses are suspended from two massless springs of spring constants k_1 and k_2, respectively. If the maximum velocities, during oscillations are equal, the ratio of amplitudes of A and B is : **[2003]**

(A) $\sqrt{k_1/k_2}$ (B) k_1/k_2

(C) $\sqrt{k_2/k_1}$ (D) k_2/k_1

8. The length of a simple pendulum executing simple harmonic motion is increased by 21%. The percentage increase in the time period of the pendulum of increased length is : **[2003]**

(A) 11% (B) 21%
(C) 42% (D) 10%

9. The bob of a simple pendulum executes simple harmonic motion in water with a period t, while the period of oscillation of the bob is t_0 in air. Neglecting frictional force of water and given that the density of the bob is $(4/3) \times 1000$ kg/m^3. What relationship between t and t_0 is true? **[2004]**

(A) $t = t_0$ (B) $t = t_0/2$
(C) $t = 2t_0$ (D) $t = 4t_0$

10. A particle at the end of a spring executes simple harmonic motion with a period t_1, while the corresponding period for another spring is t_2. If the period of oscillation with the two springs in series is T, then : **[2004]**

(A) $T = t_1 + t_2$ (B) $T^2 = t_1^2 + t_2^2$
(C) $T^{-1} = t_1^{-1} + t_2^{-1}$ (D) $T^{-2} = t_1^{-2} + t_1^{-2}$

11. The total energy of a particle, executing simple harmonic motion is : **[2004]**

(A) $\propto x$ (B) $\propto x^2$
(C) independent of x (D) $\propto x^{1/2}$

12. A particle of mass m is attached to a spring (of spring constant k) and has a natural angular frequency ω_0. An external force $F(t)$ proportional to $\cos\omega t$ ($\omega \neq \omega_0$) is applied to the oscillator. The time displacement of the oscillator will be proportional to : **[2004]**

(A) $\frac{m}{\omega_0^2 - \omega^2}$ (B) $\frac{1}{m(\omega_0^2 - \omega^2)}$

(C) $\frac{1}{m(\omega_0^2 + \omega^2)}$ (D) $\frac{m}{(\omega_0^2 + \omega^2)}$

13. In forced oscillation of a particle the amplitude is maximum for a frequency ω_1 of the force while the energy is maximum for a frequency ω_2 of the force; then : **[2004]**
(A) $\omega_1 < \omega_2$ when damping is small and $\omega_1 > \omega_2$ when damping is large
(B) $\omega_1 > \omega_2$
(C) $\omega_1 = \omega_2$
(D) $\omega_1 < \omega_2$

14. A particle of mass 0.3 kg is subjected to a force $F = -kx$ with $k = 15$ N/m. What will be its initial acceleration, if it is released from a point 20 cm away from the origin ? **[2005]**
(A) 3 m/s^2 (B) 15 m/s^2
(C) 5 m/s^2 (D) 10 m/s^2

15. The function $\sin^2(\omega t)$ represents : **[2005]**
(A) A periodic, but not simple harmonic, motion with a period $2\pi/\omega$
(B) A periodic, but not simple harmonic, motion with a period π/ω
(C) A simple harmonic motion with a period $2\pi/\omega$
(D) A simple harmonic motion with a period π/ω

16. If a simple harmonic motion is represented by $\frac{d^2x}{dt^2} + \alpha x = 0$, its time period is : **[2005]**
(A) $\frac{2\pi}{\alpha}$ (B) $\frac{2\pi}{\sqrt{\alpha}}$
(C) $2\pi\alpha$ (D) $2\pi\sqrt{\alpha}$

17. The bob of a simple pendulum is a spherical hollow ball filled with water. A plugged hole near the bottom of the oscillating bob gets suddenly unplugged. During observation, till water is coming out, the time period of oscillation would : **[2005]**
(A) First increase and then decrease to the original value
(B) First decrease and then increase to the original value
(C) Remain unchanged
(D) Increase towards a saturation value

18. Two simple harmonic motions are represented by the equations $y_1 = 0.1 \sin\left(100\pi t + \frac{\pi}{3}\right)$ and $y_2 = 0.1 \cos 100\,\pi t$. The phase difference of the velocity of particle 1, with respect to the velocity of particle 2 is : **[2005]**
(A) $\frac{-\pi}{6}$ (B) $\frac{\pi}{3}$
(C) $\frac{-\pi}{3}$ (D) $\frac{\pi}{6}$

19. The maximum velocity of a particle, executing simple harmonic motion with an amplitude 7 mm, is 4.4 m/s. The period of oscillation is : **[2006]**
(A) 0.01 s (B) 10 s
(C) 0.1 s (D) 100 s

20. Starting from the origin a body oscillates simple harmonically with a period of 2s. After what time will its kinetic energy be 75% of the total energy ? **[2006]**
(A) $\frac{1}{6}$ s (B) $\frac{1}{4}$ s
(C) $\frac{1}{3}$ s (D) $\frac{1}{12}$ s

21. A coin is placed on a horizontal platform which undergoes vertical simple harmonic motion of angular frequency ω. The amplitude of oscillation is gradually increased. The coin will leave contact with the platform for the first time : **[2006]**
(A) At the mean position of the platform
(B) For an amplitude of $\frac{g}{\omega^2}$
(C) For an amplitude of $\frac{g^2}{\omega^2}$
(D) At the highest position on the platform

22. The displacement of an object attached to a spring and executing simple harmonic motion is given by $x = 2 \times 10^{-2} \cos \pi t$ metre. The time at which the maximum speed first occurs is : **[2007]**
(A) 0.5 s (B) 0.75 s
(C) 0.125 s (D) 0.25 s

23. A point mass oscillates along the x-axis according to the law $x = x_0 \cos(\omega t - \pi/4)$ If the acceleration of the particle is written as- $a = A\cos(\omega t + \delta)$, then : **[2007]**
(A) $A = x_0, \delta = -\pi/4$ (B) $A = x_0\omega^2, \delta = \pi/4$
(C) $A = x_0\omega^2, \delta = -\pi/4$ (D) $A = x_0\omega^2, \delta = 3\pi/4$

24. Two springs, of force constants k_1 and k_2, are connected to a mass m as shown. The frequency of oscillation of the mass is f. If both k_1 and k_2 are made four times their original values, the frequency of oscillation becomes : **[2007]**

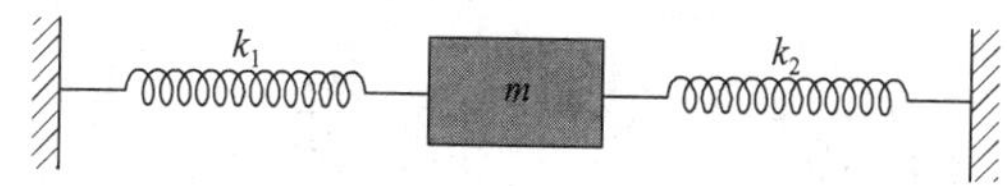

(A) $f/2$ (B) $f/4$
(C) $4f$ (D) $2f$

25. A particle of mass m executes simple harmonic motion with amplitude a and frequency ν. The average kinetic energy during its motion from the position of equilibrium to the end is : **[2007]**

(A) $\pi^2 ma^2\nu^2$ (B) $\frac{1}{4}ma^2\nu^2$

(C) $4\pi^2 ma^2\nu^2$ (D) $2\pi^2 ma^2\nu^2$

26. If x, v and a denote the displacement, the velocity and the acceleration of a particle executing simple harmonic motion of time period T, then, which of the following does not change with time ? **[2009]**

(A) $aT + 2\pi v$ (B) $\frac{aT}{v}$

(C) $a^2T^2 + 4\pi^2v^2$ (D) $\frac{aT}{x}$

27. A mass M, attached to a horizontal spring, executes SHM with amplitude A_1. When the mass M passes through its mean position then a smaller mass m is placed over it and both of them move together with amplitude A_2. The ratio of $\left(\frac{A_1}{A_2}\right)$ is : **[2011]**

(A) $\left(\frac{M}{M+m}\right)^{1/2}$ (B) $\left(\frac{M+m}{M}\right)^{1/2}$

(C) $\frac{M}{M+m}$ (D) $\frac{M+m}{M}$

28. Two particles are executing simple harmonic motion of the same amplitude A and frequency ω along the x-axis. Their mean position is separated by distance X_0 $(X_0 > A)$. If the maximum separation between them is $(X_0 + A)$, the phase difference between their motion is : **[2011]**

(A) $\frac{\pi}{4}$ (B) $\frac{\pi}{6}$

(C) $\frac{\pi}{2}$ (D) $\frac{\pi}{3}$

29. A wooden cube (density of wood 'd') of side 'l' floats in a liquid of density 'ρ' with its upper and lower surfaces horizontal. If the cube is pushed slightly down and released, it performs simple harmonic motion of period 'T'. Then, 'T' is equal to : **[2011]**

(A) $2\pi\sqrt{\frac{l\pi}{(\rho-d)g}}$ (B) $2\pi\sqrt{\frac{ld}{\rho g}}$

(C) $2\pi\sqrt{\frac{l\rho}{dg}}$ (D) $2\pi\sqrt{\frac{ld}{(\rho-d)g}}$

30. If a simple pendulum has significant amplitude (up to a factor of $1/e$ of original) only in the period between $t = 0s$ to $t = \tau s$, then τ may be called the average life of the pendulum. When the spherical bob of the pendulum suffers a retardation (due to viscous drag) proportional to its velocity, with 'b' as the constant of proportionality, the average life time of the pendulum is (assuming damping is small) in seconds : **[2012]**

(A) $\frac{2}{b}$ (B) $\frac{0.693}{b}$

(C) b (D) $\frac{1}{b}$

31. A ring is suspended from a point S on its rim as shown in the figure. When displaced from equilibrium, it oscillates with time period of 1 second. The radius of the ring is : (Take $g = \pi^2$) **[2012 (Online)]**

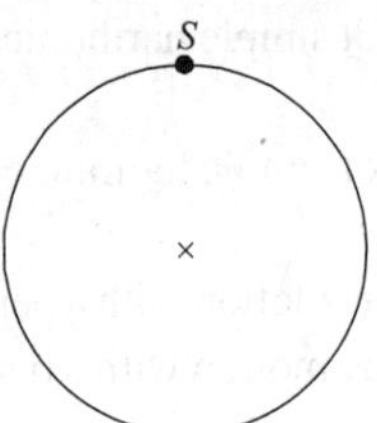

(A) 0.125 m (B) 1.5 m

(C) 1.0 m (D) 0.5 m

32. The displacement $y(t) = A\sin(\omega t + \phi)$ of a pendulum for $\phi = \frac{2\pi}{3}$ is correctly represented by : **[2012 (Online)]**

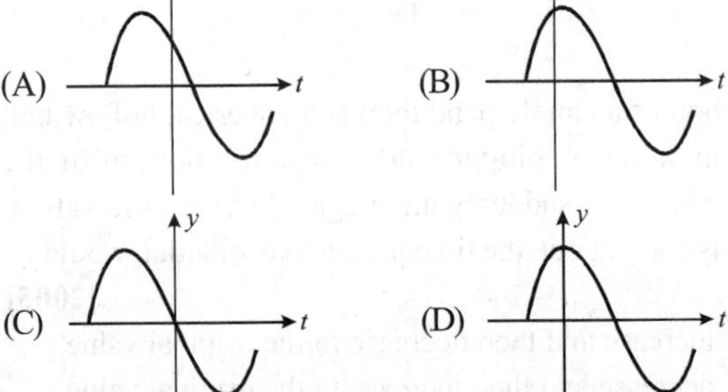

33. An ideal gas enclosed in a vertical cylindrical container supports a freely moving piston of mass M. The piston and the cylinder have equal cross sectional area A. When the piston is in equilibrium, the volume of the gas is V_0 and its pressure is P_0. The piston is slightly displaced from the equilibrium position and released. Assuming that the system is completely isolated from its surrounding, the piston executes a simple harmonic motion with frequency : **[2013]**

(A) $\frac{1}{2\pi}\frac{A\gamma P_0}{V_0 M}$ (B) $\frac{1}{2\pi}\frac{V_0 M P_0}{A^2\gamma}$

(C) $\frac{1}{2\pi}\sqrt{\frac{A^2\gamma P_0}{MV_0}}$ (D) $\frac{1}{2\pi}\sqrt{\frac{MV_0}{A\gamma P_0}}$

34. The amplitude of a damped oscillator decreases to 0.9 times its original magnitude in 5s. In another 10s it will decrease to α times its original magnitude, where α equals : **[2013]**

(A) 0.7 (B) 0.81
(C) 0.729 (D) 0.6

35. Two simple pendulums of length 1 m and 4 m respectively are both given small displacement in the same direction at the same instant. They will be again in phase after the shorter pendulum has completed number of oscillations equal to : **[2013 (Online)]**

(A) 2 (B) 7
(C) 5 (D) 3

36. A mass $m = 1.0$ kg is put on a flat pan attached to a vertical spring fixed on the ground. The mass of the spring and the pan is negligible. When pressed slightly and released, the mass executes simple harmonic motion. The spring constant is 500 N/m. What is the amplitude A of the motion, so that the mass m tends to get detached from the pan? (Take $g = 10$ m/s^2) The spring is stiff enough so that it does not get distorted during the motion : **[2013 (Online)]**

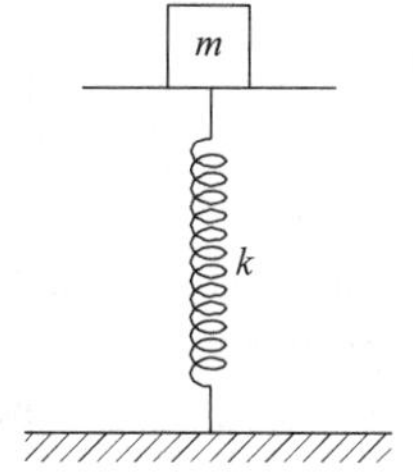

(A) $A > 2.0$ cm (B) $A = 2.0$ cm
(C) $A < 2.0$ cm (D) $A = 1.5$ cm

37. A uniform cylinder of length L and mass M having cross-sectional area A is suspended, with its length vertical, from a fixed point by a massless spring, such that it is half submerged in a liquid of density σ at equilibrium position. When the cylinder is given a downward push and released, it starts oscillating vertically with a small amplitude. The time period T of the oscillations of the cylinder will be : **[2013 (Online)]**

(A) Smaller than $2\pi\left[\frac{M}{(k + A\sigma g)}\right]^{1/2}$

(B) $2\pi\sqrt{\frac{M}{k}}$

(C) Larger than $2\pi\left[\frac{M}{(k + A\sigma g)}\right]^{1/2}$

(D) $2\pi\left[\frac{M}{(k + A\sigma g)}\right]^{1/2}$

38. Bob of a simple pendulum of length l is made of iron. The pendulum is oscillating over a horizontal coil carrying direct current. If the time period of the pendulum is T then : **[2013 (Online)]**

(A) $T < 2\pi\sqrt{\frac{l}{g}}$ and damping is smaller than in air alone

(B) $T = 2\pi\sqrt{\frac{l}{g}}$ and damping is larger than in air alone

(C) $T > 2\pi\sqrt{\frac{l}{g}}$ and damping is smaller than in air alone

(D) $T < 2\pi\sqrt{\frac{l}{g}}$ and damping is larger than in air alone

39. A particle moves with simple harmonic motion in a straight line. In first τ, s after starting from rest it travels a distance a, and in next τ s it travels $2a$, in same direction, then : **[2014]**

(A) Amplitude of motion is $3a$
(B) Time period of oscillations is 8τ
(C) Amplitude of motion is $4a$
(D) Time period of oscillations is 6τ

40. The amplitude of a simple pendulum, oscillating in air with a small spherical bob, decreases from 10 cm to 8 cm in 40 seconds. Assuming that Stokes law is valid, and ratio of the coefficient of viscosity of air to that of carbon dioxide is 1.3. The time in which amplitude of this pendulum will reduce from 10 cm to 5 cm in carbon dioxide will be close to :
(In 5 = 1.601, In 2 = 0.693) **[2014 (Online)]**

(A) 231 s (B) 208 s
(C) 161 s (D) 142 s

41. In an experiment for determining the gravitational acceleration g of a place with the help of a simple pendulum, the measured time period square is plotted against the string length of the pendulum in the figure. What is the value of g at the place ? **[2014 (Online)]**

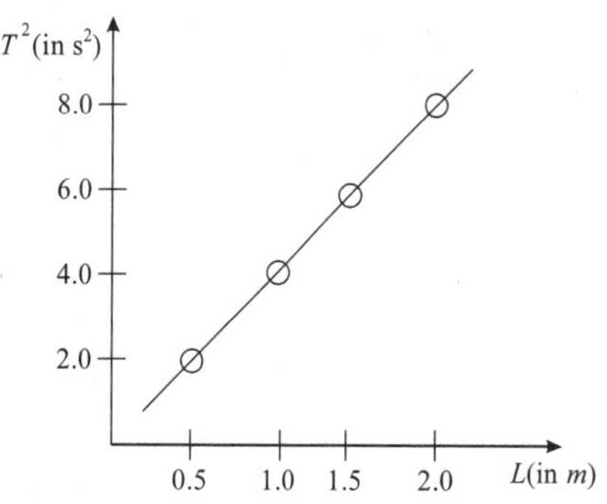

(A) 9.81 m/s^2 (B) 9.87 m/s^2
(C) 9.91 m/s^2 (D) 10.0 m/s^2

42. A particle which is simultaneously subjected to two perpendicular simple harmonic motions represented by; $x = a_1 \cos \omega t$ and $y = a_2 \cos 2\omega t$ traces a curve given by : **[2014 (Online)]**

(A) 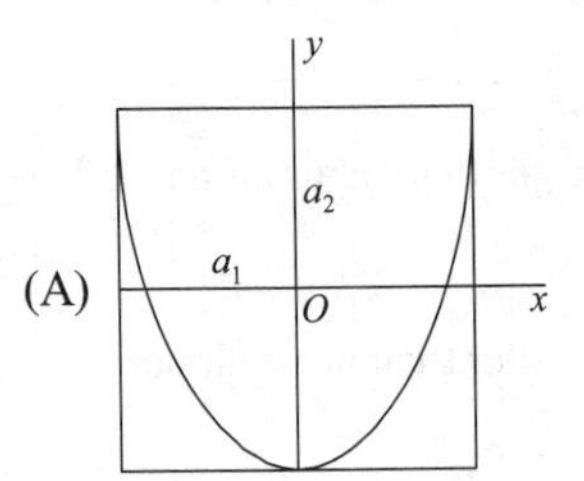

(B) 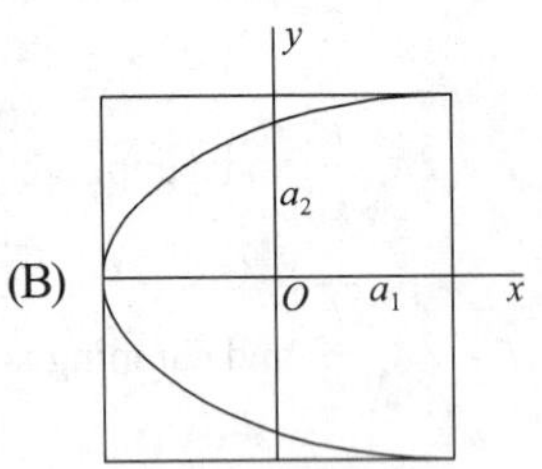

(C) 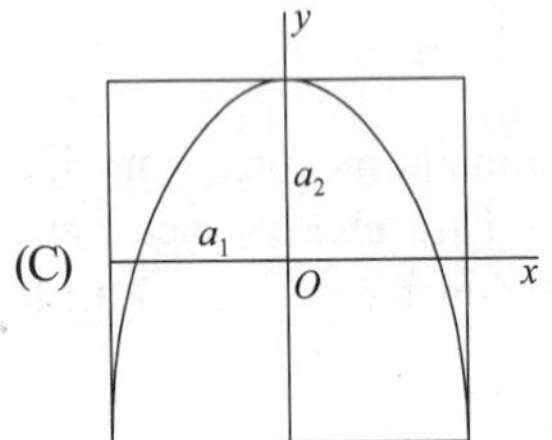

(D) 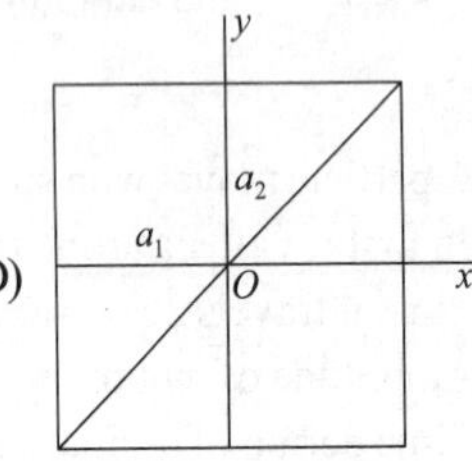

43. Which of the following expressions corresponds to simple harmonic motion along a straight line, where x is the displacement and a, b, c are positive constants ? **[2014 (Online)]**

(A) $a + bx - cx^2$ (B) bx^2
(C) $a - bx + cx^2$ (D) $-bx$

44. A body is in simple harmonic motion with time period half second ($T = 0.5$ s) and amplitude one cm ($A = 1$ cm). Find the average velocity in the interval in which it moves form equilibrium position to half of its amplitude : **[2014 (Online)]**

(A) 4 cm/s (B) 6 cm/s
(C) 12 cm/s (D) 16 cm/s

45. Two bodies of masses 1 kg and 4 kg are connected to a vertical spring, as shown in the figure. The smaller mass executes simple harmonic motion of angular frequency 25 rad/s, and amplitude 1.6 cm while the bigger mass remains stationary on the ground. The maximum force exerted by the system on the floor is (take $g = 10$ ms^{-2}) : **[2014 (Online)]**

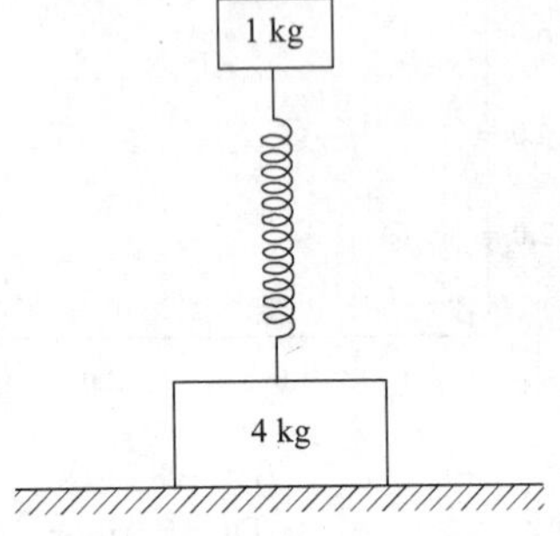

(A) 20 N (B) 10 N
(C) 60 N (D) 40 N

46. The angular frequency of the damped oscillator is given by, $\omega = \sqrt{\frac{k}{m} - \frac{r^2}{4m^2}}$ where k is the spring constant, m is the mass of the oscillator and r is the damping constant. If the ratio $\frac{r^2}{mk}$ is 8%, the change in time period compared to the undamped oscillator is approximately as follows : **[2014 (Online)]**

(A) Increases by 1% (B) Increases by 8%
(C) Decreases by 1% (D) Decreases by 8%

47. A pendulum made of a uniform wire of cross sectional area A has time period T. When an additional mass M is added to its bob, the time period changes to T_M. If the Young's modulus of the material of the wire is Y then $1/Y$ is equal to : (g = gravitational acceleration) **[2015]**

(A) $\left[\left(\frac{T_M}{T}\right)^2 - 1\right]\frac{A}{Mg}$ (B) $\left[\left(\frac{T_M}{T}\right)^2 - 1\right]\frac{Mg}{A}$

(C) $\left[1 - \left(\frac{T_M}{T}\right)^2\right]\frac{A}{Mg}$ (D) $\left[1 - \left(\frac{T}{T_M}\right)^2\right]\frac{A}{Mg}$

48. For a simple pendulum, a graph is plotted between its kinetic energy (KE) and potential energy (PE) against its displacement d. Which one of the following represents these correctly ? (Graph are schematic and not drawn to scale) **[2015]**

(A) 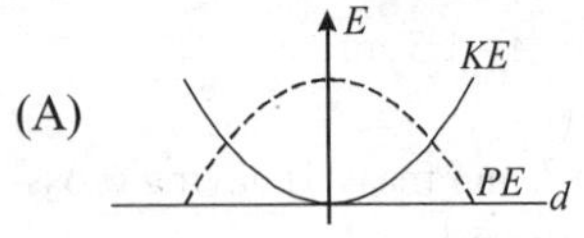

(B) 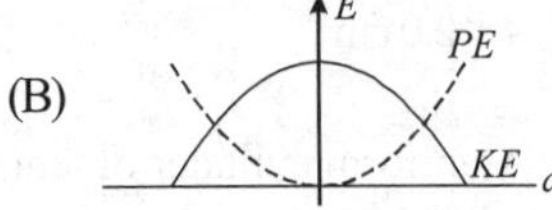

(C) 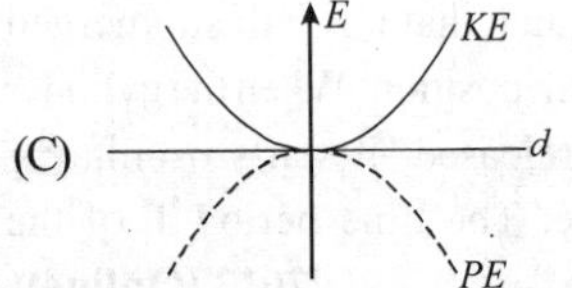

(D) 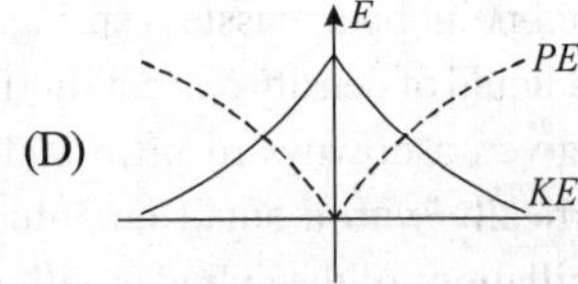

49. A pendulum with time period of 1s is losing energy. At certain time its energy is 45 J. If after completing 15 oscillations, its energy has become 15 J, its damping constant (in s^{-1}) is : **[2015 (Online)]**

(A) $\frac{1}{2}$ (B) $\frac{1}{30} \ln 3$

(C) 2 (D) $\frac{1}{15} \ln 3$

50. x and y displacements of a particle are given as $x(t) = a \sin \omega t$ and $y(t) = a \sin 2\omega t$. Its trajectory will look like : **[2015 (Online)]**

(A) 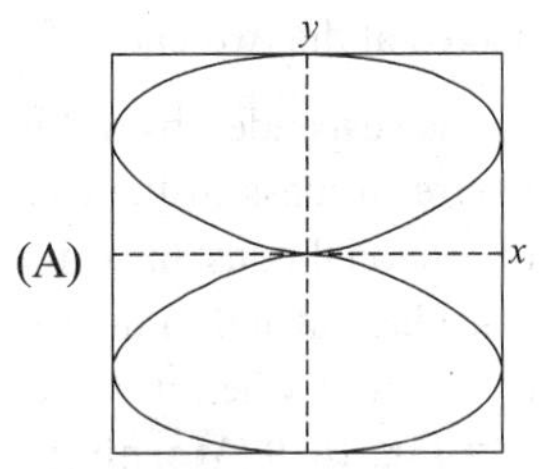

(B) 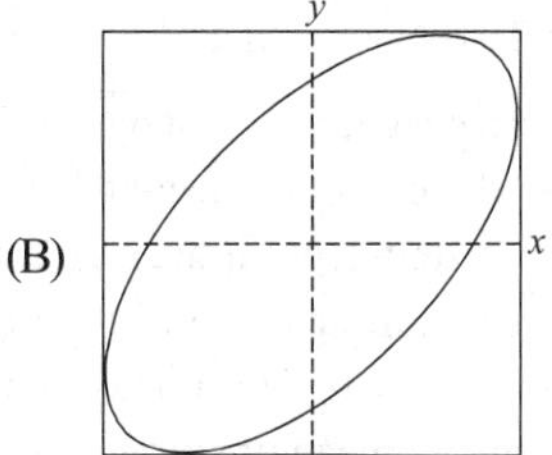

(C) 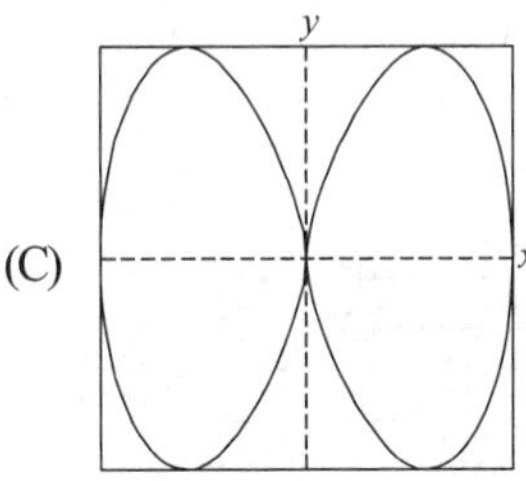

(D) 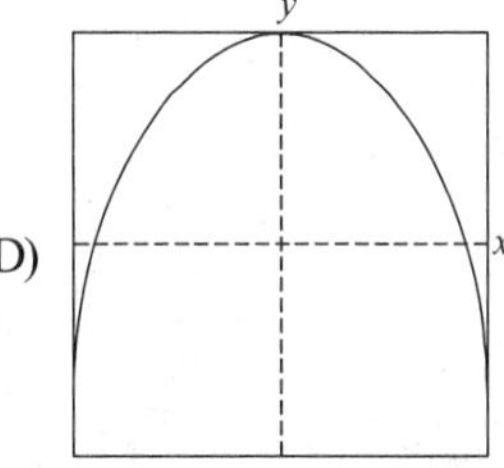

51. A simple harmonic oscillator of angular frequency 2 rad s^{-1} is acted upon by an external force $F = \sin t$ N. If the oscillator is at rest in its equilibrium position at $t = 0$, its position at later times is proportional to : **[2015 (Online)]**

(A) $\sin t + \frac{1}{2}\cos 2t$ (B) $\cot t - \frac{1}{2}\sin 2t$

(C) $\sin t - \frac{1}{2}\sin 2t$ (D) $\sin t + \frac{1}{2}\sin 2t$

52. A particle performs simple harmonic motion with amplitude A. Its speed is trebled at the instant that it is at a distance $\frac{2A}{3}$ from equilibrium position. The new amplitude of the motion is : **[2016]**

(A) $\frac{A}{3}\sqrt{41}$ (B) $3A$

(C) $A\sqrt{3}$ (D) $\frac{7A}{3}$

53. In an engine the piston undergoes vertical simple harmonic motion with amplitude 7 cm. A washer rests on top of the piston and moves with it. The motor speed is slowly increased. The frequency of the piston at which the washer no longer stays in contact with the piston, is close to : **[2016 (Online)]**

(A) 0.7 Hz (B) 1.9 Hz
(C) 1.2 Hz (D) 0.1 Hz

54. Two particles are performing simple harmonic motion in a straight line about the same equilibrium point. The amplitude and time period for both particles are same and equal to A and T, respectively. At time $t = 0$ one particle has displacement A while the other one has displacement $-A/2$ and they are moving towards each other. If they cross each other at time t, then t is : **[2016 (Online)]**

(A) $\frac{5T}{6}$ (B) $\frac{T}{3}$

(C) $\frac{T}{4}$ (D) $\frac{T}{6}$

55. A particle is executing simple harmonic motion with a time period T. At time $t = 0$, it is at its position of equilibrium. The kinetic energy-time graph of the particle will look like : **[2017]**

(A) 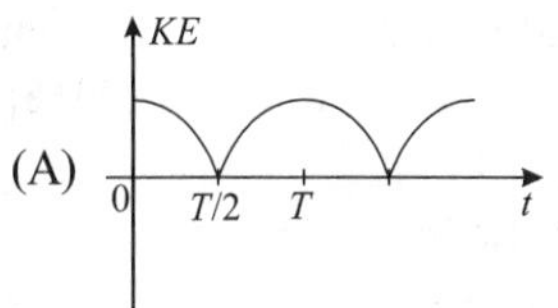

(B) 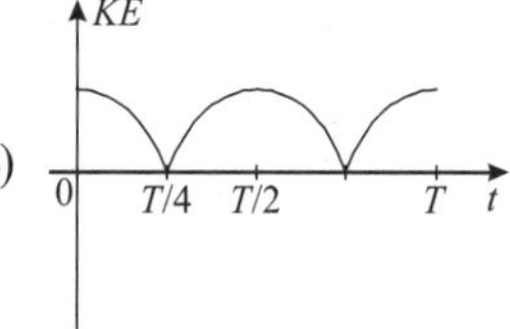

(C) 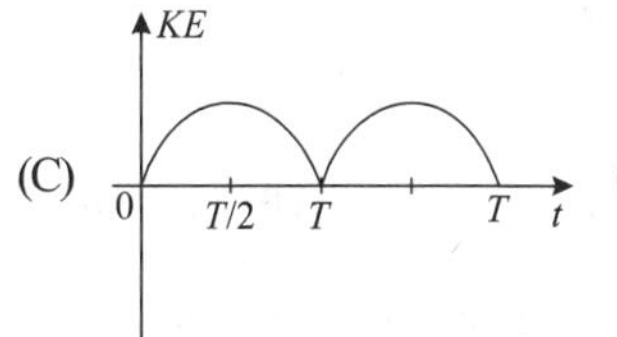

(D) 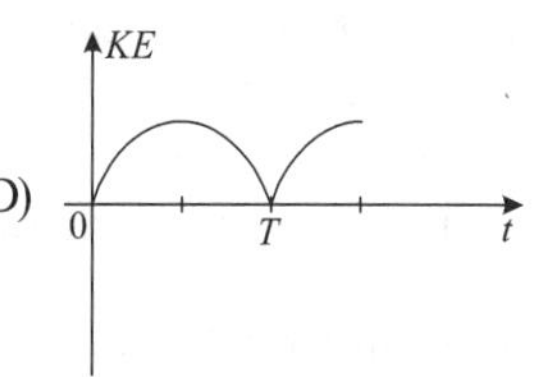

56. A 1 kg block attached to a spring vibrates with a frequency of 1 Hz on a frictionless horizontal table. Two springs identical to the original spring are attached in parallel to an 8 kg block placed on the same table. So, the frequency of vibration of the 8 kg block is : **[2017 (Online)]**

(A) $\frac{1}{4}$Hz (B) $\frac{1}{2\sqrt{2}}$Hz

(C) $\frac{1}{2}$Hz (D) 2 Hz

57. In an experiment to determine the period of a simple pendulum of length 1 m, it is attached to different spherical bobs of radii r_1 and r_2. The two spherical bobs have uniform mass distribution. If the relative difference in the periods, is found to be 5×10^{-4} s, the difference in radii, $|r_1 - r_2|$ is best given : **[2017 (Online)]**

(A) 1 cm (B) 0.1 cm
(C) 0.5 cm (D) 0.01 cm

58. A block of mass 0.1 kg is connected to an elastic spring of spring constant 640 Nm^{-1} and oscillates in a medium of constant 10^{-2} kg s^{-1}. The system dissipates its energy gradually. The time taken for its mechanical energy of vibration to drop to half of its initial value, is closest to : **[2017 (Online)]**

(A) 2 s (B) 3.5 s
(C) 5 s (D) 7 s

59. The ratio of maximum acceleration to maximum velocity in a simple harmonic motion is 10 s^{-1}. At, $t = 0$ the displacement is 5 m. What is the maximum acceleration ? The initial phase is $\pi/4$: **[2017 (Online)]**

(A) 500 m/s^2 (B) $500\sqrt{2}$ m/s^2

(C) 750 m/s^2 (D) $750\sqrt{2}$ m/s^2

60. A silver atom in a solid oscillates in simple harmonic motion in some direction with a frequency of 10^{12}/sec. What is the force constant of the bonds connecting one atom with the other? (Mole wt. of silver = 108 and Avogadro number = 6.02×10^{23} gm mole^{-1}) : **[2018]**

(A) 6.4 N/m (B) 7.1 N/m
(C) 2.2 N/m (D) 5.5 N/m

61. Two simple harmonic motions, as shown, are at right angles. They are combined to form Lissajous figures.

$x(t) = A\sin(at + \delta)$
$y(t) = B\sin(bt)$

Identify the correct match below **[2018 (Online)]**

(A) Parameters : $A = B, a = 2b; \delta = \frac{\pi}{2}$; Curve: Circle

(B) Parameters : $A = B, a = b; \delta = \frac{\pi}{2}$; Curve: Line

(C) Parameters : $A \neq B, a = b; \delta = \frac{\pi}{2}$; Curve: Ellipse

(D) Parameters : $A \neq B, a = b; \delta = 0$; Curve: Parabola

62. A particle executes simple harmonic motion and is located at $x = a$, b and c at times t_0, $2t_0$ and $3t_0$ respectively. The frequency of the oscillation is : **[2018 (Online)]**

(A) $\frac{1}{2\pi t_0}\cos^{-1}\left(\frac{a+b}{2c}\right)$ (B) $\frac{1}{2\pi t_0}\cos^{-1}\left(\frac{a+b}{3c}\right)$

(C) $\frac{1}{2\pi t_0}\cos^{-1}\left(\frac{2a+3c}{b}\right)$ (D) $\frac{1}{2\pi t_0}\cos^{-1}\left(\frac{a+c}{2b}\right)$

63. An oscillator of mass M is at rest in its equilibrium position in a potential $V = \frac{1}{2}k(x - X)^2$. A particle of mass m comes from right with speed u and collides completely inelastically with M and sticks to it. This process repeats every time the oscillator crosses its equilibrium position. The amplitude of oscillations after 13 collisions is : [$M = 10, m = 5, u = 1, k = 1$] **[2018 (Online)]**

(A) $\frac{1}{2}$ (B) $\frac{1}{\sqrt{3}}$

(C) $\frac{2}{3}$ (D) $\sqrt{\frac{3}{5}}$

64. Two masses m and $\frac{m}{2}$ are connected at the two ends of a massless rigid rod of length l. The rod is suspended by a thin wire of torsional constant k at the centre of mass of the rod-mass system (see figure). Because of torsional constant k, the restoring torque is $\tau = k\theta$ for angular displacement θ. If the rod is rotated by θ_0 and released, the tension in it when it passes through its mean position will be : **[9 Jan. 2019 (Morning)]**

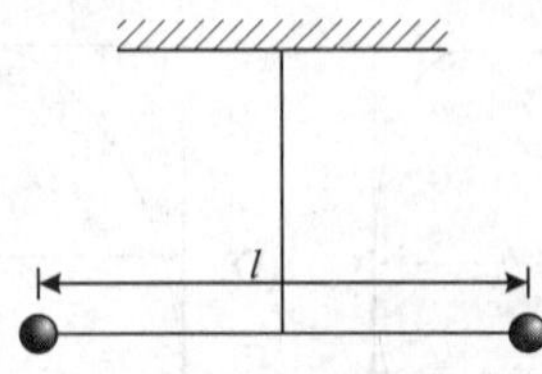

(A) $\frac{3k\theta_0^2}{l}$ (B) $\frac{k\theta_0^2}{2l}$

(C) $\frac{2k\theta_0^2}{l}$ (D) $\frac{k\theta_0^2}{l}$

65. A rod of mass M and length $2L$ is suspended at its middle by a wire. It exhibits torsional oscillations; If two masses each of m are attached at distance $L/2$ from its centre on both sides, it reduces the oscillation frequency by 20%. The value of ratio m/M is close to : **[9 Jan. 2019 (Evening)]**

(A) 0.17 (B) 0.37
(C) 0.57 (D) 0.77

66. A particle is executing simple harmonic motion (SHM) of amplitude A, along the x-axis, about $x = 0$. When its potential Energy (PE) equals kinetic energy (KE), the position of the particle will be : **[9 Jan. 2019 (Evening)]**

(A) $\frac{A}{2}$ (B) $\frac{A}{2\sqrt{2}}$

(C) $\frac{A}{\sqrt{2}}$ (D) A

67. A particle executes simple harmonic motion with an amplitude of 5 cm. When the particle is at 4 cm from the mean position, the magnitude of its velocity in SI units is equal to that of its acceleration. Then, its periodic time in seconds is : **[10 Jan. 2019 (Evening)]**

(A) $\frac{7}{3}\pi$ (B) $\frac{3}{8}\pi$

(C) $\frac{4\pi}{3}$ (D) $\frac{8\pi}{3}$

68. A cylindrical plastic bottle of negligible mass is filled with 310 ml of water and left floating in a pond with still water. If pressed downward slightly and released, it starts performing simple harmonic motion at angular frequency ω. If the radius of the bottle is 2.5 cm then ω close to :

(Density of water = 10^3 kg/m^3) **[10 Jan. 2019 (Evening)]**

(A) 5.00 rad s^{-1} (B) 1.25 rad s^{-1}

(C) 3.75 rad s^{-1} (D) 2.50 rad s^{-1}

69. A particle undergoing simple harmonic motion has time dependent displacement given by $x(t)\ A\sin\frac{\pi t}{90}$. The ratio of potential to kinetic energy of this particle at $t = 210$s will be :

[11 Jan. 2019 (Morning)]

(A) 2 (B) $\frac{1}{9}$

(C) 3 (D) 1

70. A simple pendulum of length 1 m is oscillating with an angular frequency 10 rad/s. The support of the pendulum starts oscillating up and down with a small angular frequency of 1 rad/s and an amplitude of 10^{-2} m. The relative change in the angular frequency of the pendulum is best given by :

[11 Jan. 2019 (Evening)]

(A) 10^{-3} rad/s (B) 10^{-1} rad/s

(C) 1 rad/s (D) 10^{-5} rad/s

71. A pendulum is executing simple harmonic motion and its maximum kinetic energy is K_1. If the length of the pendulum is doubled and it performs simple harmonic motion with the same amplitude as in the first case, its maximum kinetic energy is K_2. Then : **[11 Jan. 2019 (Evening)]**

(A) $K_2 = \frac{K_1}{4}$ (B) $K_2 = \frac{K_1}{2}$

(C) $K_2 = 2\,K_1$ (D) $K_2 = K_1$

72. Two light identical springs of spring constant k are attached horizontally at the two ends of a uniform horizontal rod AB of length l and mass m. The rod is pivoted at its centre O and can rotate freely in horizontal plane. The other ends of the two springs are fixed to rigid supports as shown in figure. The rod is gently pushed through a small angle and released. The frequency of resulting oscillation is : **[12 Jan. 2019 (Morning)]**

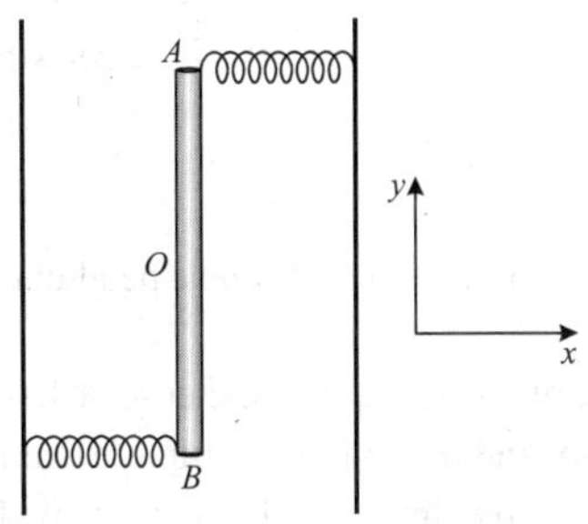

(A) $\frac{1}{2\pi}\sqrt{\frac{6k}{m}}$ (B) $\frac{1}{2\pi}\sqrt{\frac{2k}{m}}$

(C) $\frac{1}{2\pi}\sqrt{\frac{k}{m}}$ (D) $\frac{1}{2\pi}\sqrt{\frac{3k}{m}}$

73. A simple harmonic motion is represented by

$$y = 5\,(\sin 3\pi t + \sqrt{3}\cos 3\pi t)\text{ cm}$$

The amplitude and time period of the motion are :

[12 Jan. 2019 (Evening)]

(A) 5 cm, $\frac{3}{2}$ s (B) 5 cm, $\frac{2}{3}$ s

(C) 10 cm, $\frac{3}{2}$ s (D) 10 cm, $\frac{2}{3}$ s

74. A damped harmonic oscillator has a frequency of 5 oscillations per second. The amplitude drops to half its value for every 10 oscillations. The time it will take to drop to 1/1000 of the original amplitude is close to : **[8 April 2019 (Evening)]**

(A) 100 s (B) 20 s

(C) 10 s (D) 50 s

75. A simple pendulum oscillating in air has period T. The bob of the pendulum is completely immersed in a nonviscous liquid. The density of the liquid is $\frac{1}{16}$ th of the material of the bob. If the bob is inside liquid all the time, its period of oscillation in this liquid is : **[9 April 2019 (Morning)]**

(A) $4T\sqrt{\frac{1}{15}}$ (B) $2T\sqrt{\frac{1}{10}}$

(C) $4T\sqrt{\frac{1}{14}}$ (D) $2T\sqrt{\frac{1}{14}}$

76. A massless spring (k = 800 N/m), attached with a mass (500 g) is completely immersed in 1 kg of water. The spring is stretched by 2 cm and released so that it starts vibrating. What would be the order of magnitude of the change in the temperature of water when the vibrations stop completely ? (Assume that the water container and spring receive negligible heat and specific heat of mass = 400 J/kg K, specific heat of water = 4184 J/kg K) : **[9 April 2019 (Evening)]**

(A) 10^{-3} K (B) 10^{-4} K

(C) 10^{-1} K (D) 10^{-5} K

77. The displacement of a damped harmonic oscillator is given by $x(t) = e^{-0.1t}\cos(10\pi t + \phi)$. Here t is in seconds. The time taken for its amplitude of vibration to drop to half of its initial value is close to : **[10 April 2019 (Morning)]**

(A) 13 s (B) 7 s

(C) 27 s (D) 4 s

78. The displacement time graph of a particle executing SHM is given in figure. (Sketch is schematic and not to scale)

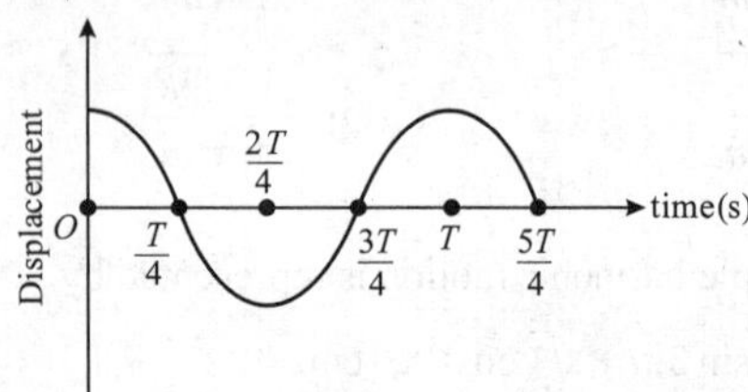

Which of the following statements is/are true for this motion : **[2 sept. 2020 (Evening)]**

(i) The force is zero at $t = \frac{3T}{4}$

(ii) The acceleration is maximum at $t = T$

(iii) The speed is maximum at $t = \frac{T}{4}$

(iv) The potential energy is equal to kinetic energy of the oscillation at $t = \frac{T}{2}$

(A) (i), (ii) and (iv) (B) (ii), (iii) and (iv)
(C) (i) and (iv) (D) (i), (ii) and (iii)

79. A block of mass m attached to massless spring is performing oscillatory motion of amplitude A on a frictionless horizontal plane. If half of the mass of the block breaks off when it is passing through its equilibrium point, the amplitude of oscillation for the remaining system become fA. The value of f is : **[3 Sept. 2020 (Evening)]**

(A) $\frac{1}{2}$ (B) $\sqrt{2}$

(C) 1 (D) $\frac{1}{\sqrt{2}}$

80. A ring is hung on a nail. It can oscillate, without slipping or sliding (i) in its plane with a time period T_1 and, (ii) back and forth in a direction perpendicular to its plane, with a period T_2. The ratio T_1/T_2 will be : **[5 Sept. 2020 (Evening)]**

(A) $\frac{2}{\sqrt{3}}$ (B) $\frac{\sqrt{2}}{3}$

(C) $\frac{2}{3}$ (D) $\frac{3}{\sqrt{2}}$

81. When a particle of mass m is attached to a vertical spring of spring constant k and released, its motion is described by $y(t) = y_0 \sin^2 \omega t$, where '$y$' is measured from the lower end of unstretched spring. Then ω is : **[6 Sept. 2020 (Evening)]**

(A) $\sqrt{\frac{g}{y_0}}$ (B) $\sqrt{\frac{g}{2y_0}}$

(C) $\frac{1}{2}\sqrt{\frac{g}{y_0}}$ (D) $\sqrt{\frac{2g}{y_0}}$

82. In the given figure, a mass M is attached to a horizontal spring which is fixed on one side to a rigid support. The spring constant of the spring is k. The mass oscillates on a frictionless surface with time period T and amplitude A. When the mass is in equilibrium position, as shown in the figure, another mass m is gently fixed upon it. The new amplitude of oscillation will be : **[24 Feb. 2021 (Morning)]**

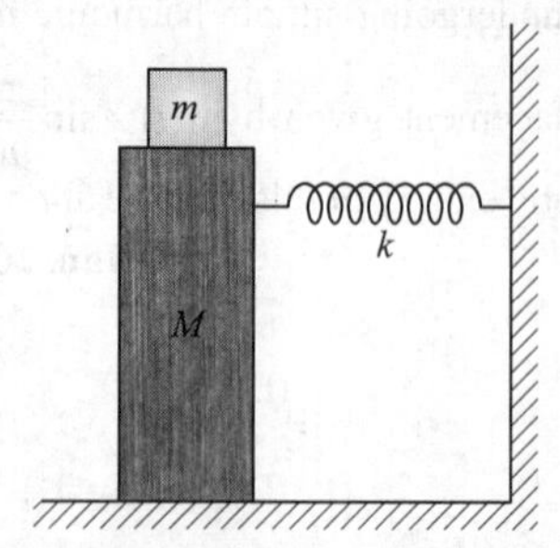

(A) $A\sqrt{\frac{M}{M+m}}$ (B) $A\sqrt{\frac{M-m}{M}}$

(C) $A\sqrt{\frac{M}{M-m}}$ (D) $A\sqrt{\frac{M+m}{M}}$

83. When a particle executes SHM, then nature of graphical representation of velocity as a function of displacement is : **[24 Feb. 2021 (Evening)]**

(A) Straight line (B) Circular
(C) Elliptical (D) Parabolic

84. In the given figure, a body of mass M is held between two massless springs, on a smooth inclined plane. The free ends of the springs are attached to firm supports. If each spring has spring constant k, the frequency of oscillation of given body is : **[24 Feb. 2021 (Evening)]**

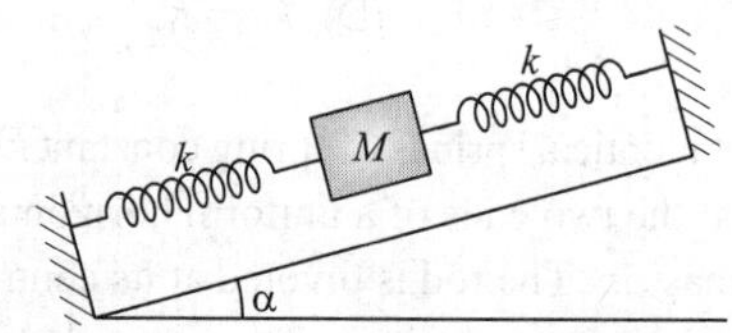

(A) $\frac{1}{2\pi}\sqrt{\frac{k}{Mg\sin\alpha}}$ (B) $\frac{1}{2\pi}\sqrt{\frac{k}{2M}}$

(C) $\frac{1}{2\pi}\sqrt{\frac{2k}{M}}$ (D) $\frac{1}{2\pi}\sqrt{\frac{2k}{Mg\sin\alpha}}$

85. The period of oscillation of a simple pendulum is $T = 2\pi\sqrt{\frac{L}{g}}$.

Measured value of L is 1.0 m from meter scale having a minimum division of 1 mm and time of one complete oscillation is 1.95 s measured from stopwatch of 0.01 s resolution. The percentage error in the determination of g will be :

[24 Feb. 2021 (Evening)]

(A) 1.03% (B) 1.33%
(C) 1.13% (D) 1.30%

86. If the time period of a two meter long simple pendulum is 2 s, the acceleration due to gravity at the place where pendulum is executing SHM is : **[25 Feb. 2021 (Morning)]**

(A) $\pi^2\ ms^{-2}$ (B) $9.8\ ms^{-2}$
(C) $16\ m/s^2$ (D) $2\pi^2\ ms^{-2}$

87. $Y = A\sin(\omega t + \phi_0)$ is the time-displacement equation of a SHM. At $t = 0$ the displacement of the particle is $Y = \frac{A}{2}$ and it is moving along negative x-direction. Then the initial phase angle ϕ_0 will be : **[25 Feb. 2021 (Evening)]**

(A) $\frac{2\pi}{3}$ (B) $\frac{\pi}{3}$
(C) $\frac{\pi}{6}$ (D) $\frac{5\pi}{6}$

88. Two identical springs of spring constant $2k$ are attached to a block of mass m and to fixed support (see figure). When the mass is displaced from equilibrium position on either side, it executes simple harmonic motion. The time period of oscillations of this system is : **[25 Feb. 2021 (Evening)]**

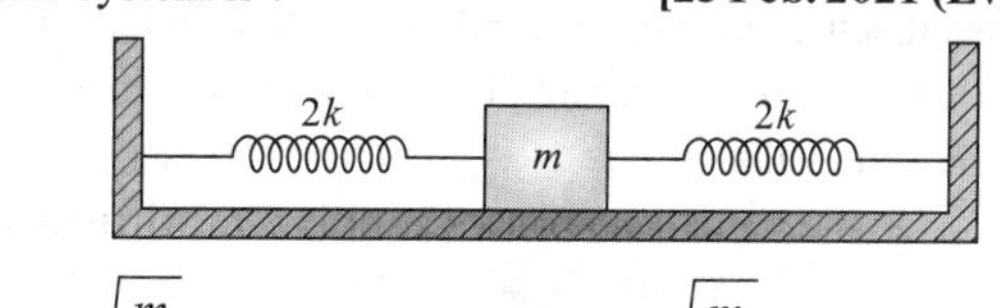

(A) $2\pi\sqrt{\frac{m}{2k}}$ (B) $\pi\sqrt{\frac{m}{2k}}$
(C) $\pi\sqrt{\frac{m}{k}}$ (D) $2\pi\sqrt{\frac{m}{k}}$

89. The point A moves with a uniform speed along the circumference of a circle of radius 0.36 m and covers 30° in 0.1 s. The perpendicular projection P from A on the diameter MN represents the simple harmonic motion of P. The restoration force per unit mass when P touches M will be :

[25 Feb. 2021 (Evening)]

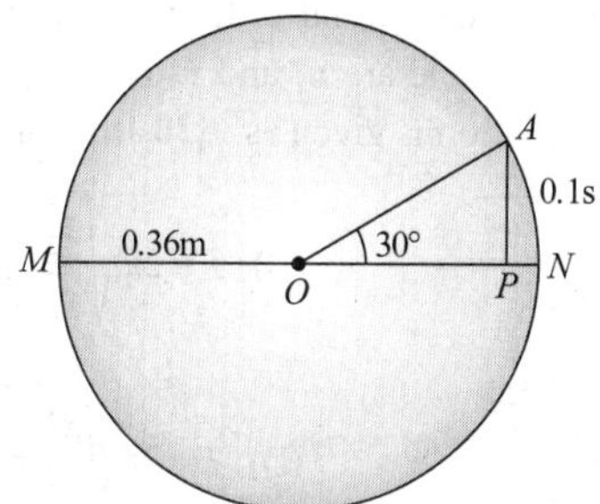

(A) 9.87 N (B) 0.49 N
(C) 100 N (D) 50 N

90. If two similar springs each of spring constant K_1 are joined in series, the new spring constant and time period would be changed by a factor : **[26 Feb. 2021 (Morning)]**

(A) $\frac{1}{4}, \sqrt{2}$ (B) $\frac{1}{2}, \sqrt{2}$
(C) $\frac{1}{4}, 2\sqrt{2}$ (D) $\frac{1}{2}, 2\sqrt{2}$

91. A particle executes SHM, the graph of velocity as a function of displacement is : **[26 Feb. 2021 (Evening)]**

(A) An ellipse (B) A parabola
(C) A helix (D) A circle

92. Given below are two statements :

Statement-I : A second's pendulum has a time period of 1 second

Statement-II : It takes precisely one second to move between the two extreme positions

In the light of the above statements, choose the correct answer from the options given below :

(A) Both Statement-I and Statement-II are false
(B) Statement-I is true but Statement-II is false
(C) Statement-I is false but Statement-II is true
(D) Both Statement-I and Statement-II are true

[26 Feb. 2021 (Evening)]

93. A particle executes SHM with amplitude a and time period T. The displacement of the particle when its speed is half of maximum speed is $\frac{\sqrt{x}a}{2}$. The value of x is ______.

[26 Feb. 2021 (Evening)]

94. Time period of a simple pendulum is T. The time taken to complete $\frac{5}{8}$ oscillations starting from mean position is $\frac{\alpha}{\beta}T$. The value of α is ______. **[26 Feb. 2021 (Evening)]**

95. Time period of a simple pendulum is T inside a lift when the lift is stationary. If the lift moves upwards with an acceleration $g/2$, the time period of pendulum will be :

[16 March 2021 (Morning)]

(A) $\sqrt{3}T$ (B) $\frac{T}{\sqrt{3}}$
(C) $\sqrt{\frac{3}{2}}T$ (D) $\sqrt{\frac{2}{3}}T$

96. Amplitude of a mass-spring system, which is executing simple harmonic motion decreases with time. If mass = 500 g,

Decay constant = 20 g/s then how much time is required for the amplitude of the system to drop to half of its initial value ? ($\ln 2 = 0.693$) **[16 March 2021 (Evening)]**

(A) 34.65 s (B) 17.32 s
(C) 0.034 s (D) 15.01 s

97. For what value of displacement the kinetic energy and potential energy of a simple harmonic oscillation become equal ? **[17 March 2021 (Morning)]**

(1) $x = 0$ (B) $x = \pm A$

(C) $x = \pm \frac{A}{\sqrt{2}}$ (D) $x = \frac{A}{2}$

98. Consider two identical springs each of spring constant k and negligible mass compared to the mass M as shown. Figure-1 shows one of them and figure-2 shows their series combination. The ratios of time period of oscillation of the two SHM is $\frac{T_b}{T_a} = \sqrt{x}$, where value of x is ______.

(Round off to the Nearest Integer) **[17 March 2021 (Morning)]**

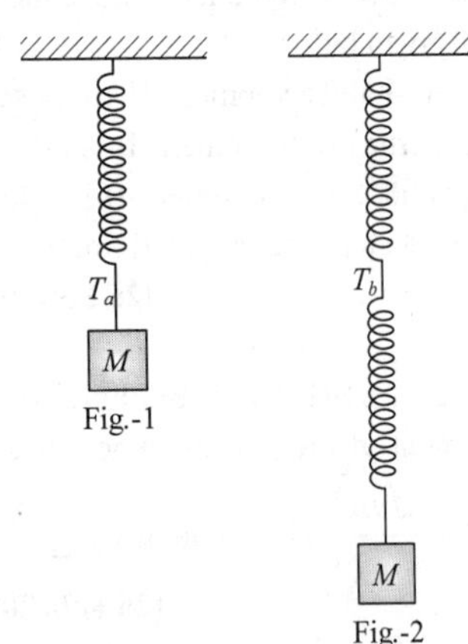

Fig.-1 Fig.-2

99. A block of mass 1 kg attached to a spring is made to oscillate with an initial amplitude of 12 cm. After 2 minutes the amplitude decreases to 6 cm. Determine the value of the damping constant for this motion. (Take $\ln 2 = 0.693$) **[17 March 2021 (Evening)]**

(A) 0.69×10^2 kg s^{-1} (B) 3.3×10^2 kg s^{-1}
(C) 1.16×10^2 kg s^{-1} (D) 5.7×10^{-3} kg s^{-1}

100. Two particles A and B of equal masses are suspended from two massless springs of spring constants K_1 and K_2 respectively. If the maximum velocities during oscillations are equal, the ratio of the amplitude of A and B is : **[17 March 2021 (Evening)]**

(A) $\frac{K_2}{K_1}$ (B) $\frac{K_1}{K_2}$

(C) $\sqrt{\frac{K_1}{K_2}}$ (D) $\sqrt{\frac{K_2}{K_1}}$

101. A particle performs simple harmonic motion with a period of 2 second. The time taken by the particle to cover a displacement equal to half of its amplitude from the mean position is $\frac{1}{a}$ s. The value of a to the nearest integer is ______. **[18 March 2021 (Morning)]**

102. If the angular velocity of earth's spin is increased such that the bodies at the equator start floating, the duration of the day would be approximately :
(Take : $g = 10$ ms^{-2}, the radius of earth, $R = 6400 \times 10^3$ m, Take $\pi = 3.14$) **[18 March 2021 (Evening)]**

(A) 60 minutes (B) Does not change
(C) 1200 minutes (D) 84 minutes

103. The function of time representing a simple harmonic motion with a period of π/ω is : **[18 March 2021 (Evening)]**

(A) $\sin(\omega t) + \cos(\omega)$
(B) $\cos(\omega t) + \cos(2\omega t) + \cos(3\omega t)$
(C) $\sin^2(\omega t)$
(D) $3\cos\left(\frac{\pi}{4} - 2\omega t\right)$

104. In a spring gun having spring constant 100 N/m a small ball B of mass 100 g is put in its barrel (as shown in figure) by compressing the spring through 0.05 m. There should be a box placed at a distance d on the ground so that the ball falls in it. If the ball leaves the gun horizontally at a height of 2 m above the ground. The value of d is ______ m. ($g = 10$ m/s^2) **[20 July 2021 (Morning)]**

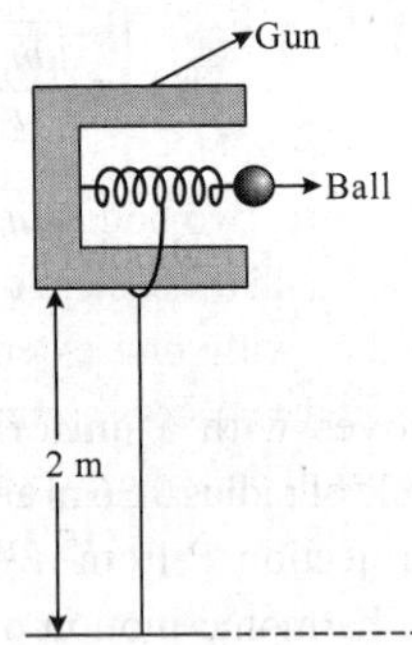

105. A particle is making simple harmonic motion along, the X-axis. If at a distances x_1 and x_2 from the mean position the velocities of the particle are v_1 and v_2 respectively. The time period of its oscillation is given as : **[20 July 2021 (Evening)]**

(A) $T = 2\pi\sqrt{\frac{x_2^2 + x_1^2}{v_1^2 - v_2^2}}$ (B) $T = 2\pi\sqrt{\frac{x_2^2 + x_1^2}{v_1^2 + v_2^2}}$

(C) $T = 2\pi\sqrt{\frac{x_2^2 - x_1^2}{v_1^2 + v_2^2}}$ (D) $T = 2\pi\sqrt{\frac{x_2^2 - x_1^2}{v_1^2 - v_2^2}}$

106. T_0 is the time period of a simple pendulum at a place. If the length of the pendulum is reduced to $\frac{1}{16}$ times of its initial value, the modified time period is : **[22 July 2021 (Evening)]**

(A) T_0 (B) $\frac{1}{4}T_0$

(C) $8\pi T_0$ (D) $4T_0$

107. The motion of a mass on a spring, with spring constant K is as shown in figure. The equation of motion is given by

$$x(t) = A\sin\omega t + B\cos\omega t \quad \omega = \sqrt{\frac{K}{m}}$$

Suppose that at time $t = 0$, the position of mass is $x(0)$ and velocity $v(0)$, then its displacement can also be represented as $x(t) = C\cos(\omega t - \phi)$, where C and ϕ are : **[22 July 2021 (Evening)]**

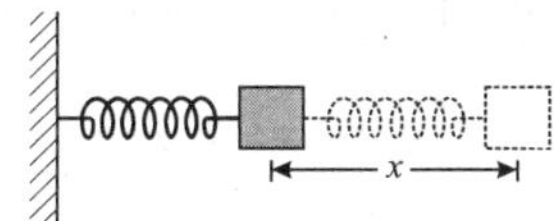

(A) $C = \sqrt{\frac{v(0)^2}{\omega^2} + x(0)^2}$, $\phi = \tan^{-1}\left(\frac{v(0)}{x(0)\omega}\right)$

(B) $C = \sqrt{\frac{2v(0)^2}{\omega^2} + x(0)^2}$, $\phi = \tan^{-1}\left(\frac{x(0)\omega}{2v(0)}\right)$

(C) $C = \sqrt{\frac{2v(0)^2}{\omega^2} + x(0)^2}$, $\phi = \tan^{-1}\left(\frac{v(0)}{x(0)\omega}\right)$

(D) $C = \sqrt{\frac{v(0)^2}{\omega^2} + x(0)^2}$, $\phi = \tan^{-1}\left(\frac{x(0)\omega}{v(0)}\right)$

108. In the reported figure, two bodies A and B of masses 200 g and 800 g are attached with the system of springs. Springs are kept in a stretched position with some extension when the system is released. The horizontal surface is assumed to be frictionless. The angular frequency will be ______ rad/s when $k = 20$ N/m : **[25 July 2021 (Morning)]**

109. In a simple harmonic oscillation, what fraction of total mechanical energy is in the form of kinetic energy, when the particle is midway between mean and extreme position : **[25 July 2021 (Evening)]**

(A) $\frac{3}{4}$ (B) $\frac{1}{4}$

(C) $\frac{1}{2}$ (D) $\frac{1}{3}$

110. A particle starts executing simple harmonic motion (SHM) of amplitude a and total energy E. At any instant, its kinetic energy is $\frac{3E}{4}$ then its displacement y is given by : **[27 July 2021 (Morning)]**

(A) $y = \frac{a}{2}$ (B) $y = a$

(C) $y = \frac{a\sqrt{3}}{2}$ (D) $y = \frac{a}{\sqrt{2}}$

111. In a uniform magnetic field, the magnetic needle has a magnetic moment 9.85×10^{-2} A/m^2 and moment of inertia 5×10^{-6} kg m^2. If it performs 10 complete oscillations in 5 seconds then the magnitude of the magnetic field is ______ mT. [Take π^2 as 9.85] **[27 July 2021 (Morning)]**

112. An object of mass 0.5 kg is executing simple harmonic motion. Its amplitude is 5 cm and time period T is 0.2 s. What will be the potential energy of the object at an instant $t = \frac{T}{4}$ s starting from mean position. Assume that the initial phase of the oscillation is zero : **[27 July 2021 (Evening)]**

(A) 6.2×10^3 J (B) 1.2×10^3 J

(C) 0.62 J (D) 6.2×10^{-3} J

113. A particle executes simple harmonic motion represented by displacement function as $x(t) = A\sin(\omega t + \phi)$. If the position and velocity of the particle at $t = 0$ s and $x = 2$ cm is 2ω cm s^{-1} respectively, then its amplitude is $x\sqrt{2}$ cm where the value of x is ______ **[27 July 2021 (Evening)]**

114. If the length of the pendulum in pendulum clock increases by 0.1%, then the error in time per day is : **[26 Aug. 2021 (Evening)]**

(A) 43.2 s (B) 8.64 s

(C) 4.32 s (D) 86.4 s

115. Two simple harmonic motions are represented by the equations $x_1 = 5\sin\left(2\pi t + \frac{\pi}{4}\right)$ and $x_2 = 5\sqrt{2}(\sin 2\pi t + \cos 2\pi t)$. The amplitude of second motion is ______ times the amplitude in first motion. **[26 Aug. 2021 (Evening)]**

116. The variation of displacement with time of a particle executing free simple harmonic motion is shown in the figure below

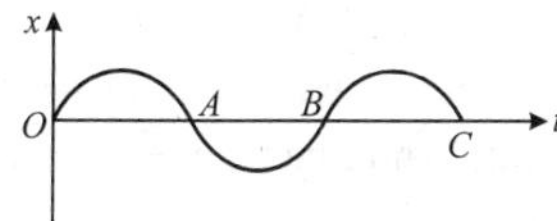

The potential energy curve of the particle with time is represented as: **[27 Aug. 2021 (Morning)]**

(A) 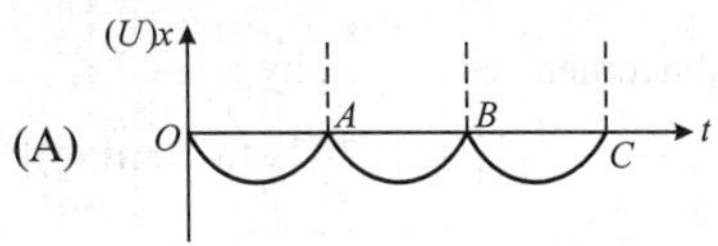

(B) 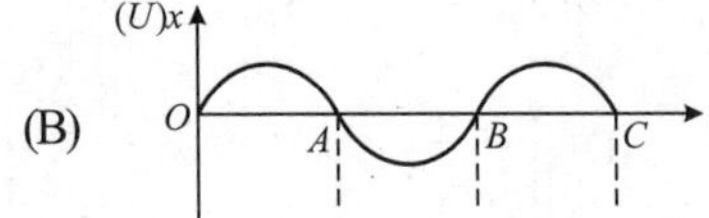

(C) 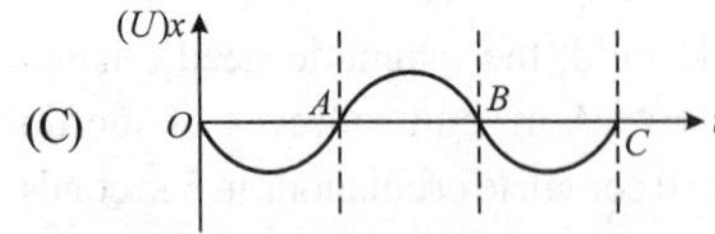

(D)

117. Two simple harmonic motion, are represented by the equations

$$y_1 = 10\sin\left(3\pi t + \frac{\pi}{3}\right); y_2 = 5(\sin 3\pi t + \sqrt{3}\cos 3\pi t)$$

Ratio of amplitude of y_1 to $y_2 = x : 1$. The value of x is ______.

[27 Aug. 2021 (Evening)]

118. A particles of mass 1 kg is hanging from a spring of force constant 100 Nm^{-1}. The mass is pulled slightly downward and released so that it executes free simple harmonic motion with time period T. The time when the kinetic energy and potential energy of the system will become equal, is $\frac{T}{x}$. The value of x is______. **[31 Aug. 2021 (Morning)]**

119. A bob of mass m suspended by a thread of length l undergoes simple harmonic oscillations with time period T. If the bob is immersed in a liquid that has density $\frac{1}{4}$ times that of the bob and the length of the thread is increased by 1/3rd of the original length, then the time period of the simple harmonic oscillations will be : **[31 Aug. 2021 (Evening)]**

(A) T (B) $\frac{4}{3}T$

(C) $\frac{3}{2}T$ (D) $\frac{3}{4}T$

120. For a body executing SHM :

(a) potential energy is always equal to its K.E.

(b) Average potential and kinetic energy over any given time interval are always equal

(c) Sum of the kinetic and potential energy at any point of time is constant

(d) Average K.E. in one time period is equal to average potential energy in one time period

Choose the most appropriate option from the options given below: **[31 Aug. 2021 (Evening)]**

(A) (c) and (d) (B) Only (b)

(C) Only (c) (D) (b) and (c)

121. A mass of 5 kg is connected to a spring. The potential energy curve of the simple harmonic motion executed by the system is shown in the figure. A simple pendulum of length 4 m has the same period of oscillation as the spring system. What is the value of acceleration due to gravity on the planet where these experiments are performed? **[01 Sep. 2021 (Evening)]**

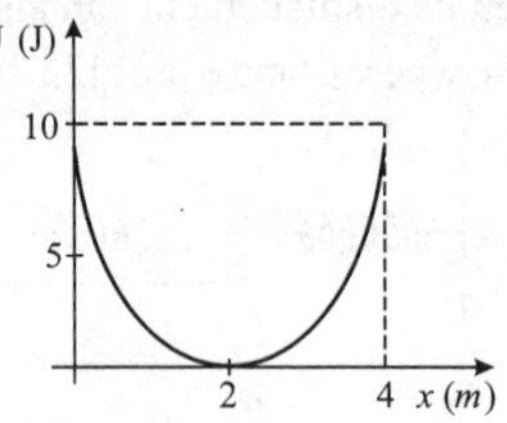

(A) 10 m/s^2 (B) 5 m/s^2

(C) 9.8 m/s^2 (D) 4 m/s^2

122. Two massless springs with spring constants $2k$ and $9k$, carry 50 g and 100 g masses at their free ends. These two masses oscillate vertically such that their maximum velocities are equal. Then, the ratio of their respective amplitudes will be : **[24 June 2022 (Evening)]**

(A) 1 : 2 (B) 3 : 2

(C) 3 : 1 (D) 2 : 3

123. Time period of a simple pendulum in a stationary lift is T. If the lift accelerates with $\frac{g}{6}$ vertically upwards then the time period will be : [Where g = acceleration due to gravity] **[26 June 2022 (Morning)]**

(A) $\sqrt{\frac{6}{5}}T$ (B) $\sqrt{\frac{5}{6}}T$

(C) $\sqrt{\frac{6}{7}}T$ (D) $\sqrt{\frac{7}{6}}T$

124. The displacement of simple harmonic oscillator after 3 seconds starting from its mean position is equal to half of its amplitude. The time period of harmonic motion is : **[27 June 2022 (Morning)]**

(A) 6 s (B) 8 s

(C) 12 s (D) 36 s

125. The equation of a particle executing simple harmonic motion is given by $x = \sin \pi\left(t+\frac{1}{3}\right)$ m. At $t = 1$ s, the speed of particle will be : [Given : $\pi = 3.14$] **[27 June 2022 (Evening)]**

(A) 0 cm s^{-1} (B) 157 cm s^{-1}

(C) 272 cm s^{-1} (D) 314 cm s^{-1}

126. A particle executes simple harmonic motion. Its amplitude is 8 cm and time period is 6 s. The time it will take to travel from its position of maximum displacement to the point corresponding to half of its amplitude, is _____ s.

[27 June 2022 (Evening)]

127. A body is performing simple harmonic motion with an amplitude of 10 cm. The velocity of the body was tripled by air Jet when it is at 5 cm from its mean position. The new amplitude of vibration is $\sqrt{x}$ cm. The value of x is _____.

[29 June 2022 (Morning)]

128. The motion of a simple pendulum executing SHM is represented by following equation. $Y = A \sin(\pi t + \phi)$, where time is measured in second. The length of pendulum is :

[29 June 2022 (Evening)]

(A) 97.23 cm (B) 25.3 cm

(C) 99.4 cm (D) 406.1 cm

129. In figure-(A), mass $2m$ is fixed on mass m which is attached to two springs of spring constant k. In figure-(B), mass m is attached to two spring of spring constant k and $2k$. If mass m in (A) and (B) are displaced by distance x horizontally and then released, then time period T_1 and T_2 corresponding to (A) and (B) respectively follow the relation :

[25 July 2022 (Morning)]

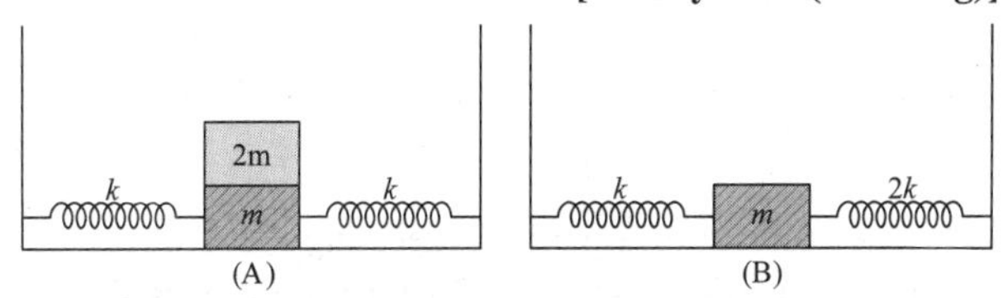

(A) $\frac{T_1}{T_2} = \frac{3}{\sqrt{2}}$ (B) $\frac{T_1}{T_2} = \sqrt{\frac{3}{2}}$

(C) $\frac{T_1}{T_2} = \sqrt{\frac{2}{3}}$ (D) $\frac{T_1}{T_2} = \frac{\sqrt{2}}{3}$

130. The length of a seconds pendulum at a height $h = 2R$ from earth surface will be :

[Given : R = Radius of earth and acceleration due to gravity at the surface of earth $g = \pi^2$ m/s^{-2}] **[25 July 2022 (Evening)]**

(A) $\frac{2}{9}$ m (B) $\frac{4}{9}$ m

(C) $\frac{8}{9}$ m (D) $\frac{1}{9}$ m

131. Two waves executing simple harmonic motion travelling in the same direction with same amplitude and frequency are superimposed. The resultant amplitude is equal to the $\sqrt{3}$ times of amplitude of individual motions. The phase difference between the two motions is _____ (degree).

[25 July 2022 (Evening)]

132. When a particle executes simple Harmonic motion, the nature of graph of velocity as function of displacement will be :

[26 July 2022 (Morning)]

(A) Circular (B) Elliptical

(C) Sinusoidal (D) Straight line

133. As per given figures, two springs of spring constants K and $2K$ are connected to mass m. If the period of oscillation in figure-(a) is 3s, then the period of oscillation in figure-(b) will be $\sqrt{x}$ s. The value of x is _____. **[26 July 2022 (Evening)]**

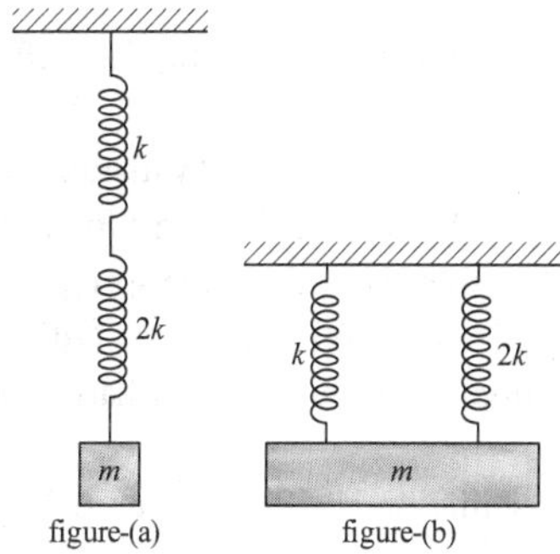

134. Two identical positive charges Q each are fixed at a distance of $2a$ apart from each other. Another point charge q_0 with mass m is placed at midpoint between two fixed charges. For a small displacement along the line joining the fixed charges, the charge q_0 executes SHM. The time period of oscillation of charge q_0 will be : **[27 July 2022 (Morning)]**

(A) $\sqrt{\frac{4\pi^3\varepsilon_0 ma^3}{q_0 Q}}$ (B) $\sqrt{\frac{q_0 Q}{4\pi^3\varepsilon_0 ma^3}}$

(C) $\sqrt{\frac{2\pi^3\varepsilon_0 ma^3}{q_0 Q}}$ (D) $\sqrt{\frac{8\pi^3\varepsilon_0 ma^3}{q_0 Q}}$

135. A mass 0.9 kg, attached a horizontal spring, executes SHM with an amplitude A_1. When this mass passes through its mean position, then a smaller mass of 124 g is placed over it and both masses move together with amplitude A_2. If the ratio $\frac{A_1}{A_2}$ is $\frac{\alpha}{\alpha - 1}$, then the value of α will be _____.

[27 July 2022 (Morning)]

136. The potential energy of a particle of mass 4 kg in motion along the x-axis is given by $U = 4(1 - \cos 4x)$ J. The time period of the particle for small oscillation ($\sin\theta \approx \theta$) is $\left(\frac{\pi}{K}\right)$ s. The value of K is _____. **[28 July 2022 (Evening)]**

137. The time period of oscillation of a simple pendulum of length L suspended from the roof of a vehicle, which moves without friction down an inclined plane of inclination α, is given by : **[29 July 2022 (Morning)]**

(A) $2\pi\sqrt{L/(g\cos\alpha)}$ (B) $2\pi\sqrt{L/(g\sin\alpha)}$

(C) $2\pi\sqrt{L/g}$ (D) $2\pi\sqrt{L/(g\tan\alpha)}$

138. The metallic bob of simple pendulum has the relative density 5. The time period of this pendulum is 10 s. If the metallic bob is immersed in water, then the new time period becomes $5\sqrt{x}$ s. The value of x will be _____. **[29 July 2022 (Evening)]**

139. A block of mass 2 kg is attached with two identical springs of spring constant 20 N/m each. The block is placed on a frictionless surface and the ends of the springs are attached to rigid supports (see figure). When the mass is displaced from its equilibrium position, it executes a simple harmonic motion. The time period of oscillation is $\frac{\pi}{\sqrt{x}}$ in SI unit. The value of x is _____. **[24 Jan. 2023 (Morning)]**

2 Kg

140. A mass m attached to free end of a spring executes SHM with a period of 1 s. If the mass is increased by 3 kg the period of oscillation increases by one second, the value of mass m is _____ kg. **[24 Jan. 2023 (Evening)]**

141. T is the time period of simple pendulum on the earth's surface. Its time period becomes xT when taken to a height R (equal to earth's radius) above the earth's surface. Then, the value of x will be : **[25 Jan. 2023 (Morning)]**

(A) 4 (B) 2

(C) $\frac{1}{2}$ (D) $\frac{1}{4}$

142. A particle executes simple harmonic motion between $x = -A$ and $x = +A$. If time taken by particle to go from $x = 0$ to $\frac{A}{2}$ is 2 s then time taken by particle in going from $x = \frac{A}{2}$ to A is : **[25 Jan. 2023 (Evening)]**

(A) 3 s (B) 2 s

(C) 1.5 s (D) 4 s

143. Two simple harmonic waves having equal amplitudes of 8 cm and equal frequency of 10 Hz are moving along the same direction. The resultant amplitude is also 8 cm. The phase difference between the individual waves is _____ degree. **[29 Jan. 2023 (Morning)]**

144. A particle of mass 250 g executes a simple harmonic motion under a periodic force $F = (-25x)$ N. The particle attains a maximum speed of 4 m/s during its oscillation. The amplitude of the motion is _____ cm. **[29 Jan. 2023 (Evening)]**

145. The general displacement of a simple harmonic oscillator is $x = A\sin\omega t$. Let T be its time period. The slope of its potential energy (U) – time (t) curve will be maximum when $t = \frac{t}{\beta}$. The value of β is _____. **[30 Jan. 2023 (Morning)]**

146. For a simple harmonic motion in a mass spring system shown, the surface is frictionless. When the mass of the block is 1 kg, the angular frequency is ω_1. When the mass block is 2 kg the angular frequency is ω_2. The ratio ω_2/ω_1 is : **[30 Jan. 2023 (Evening)]**

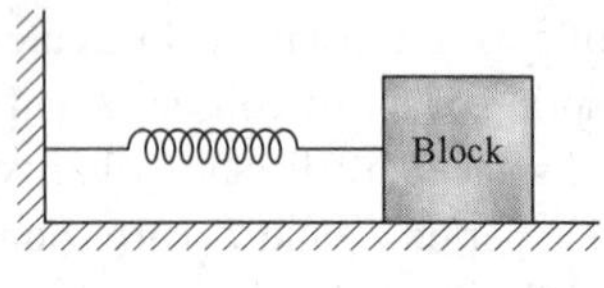

(A) $\sqrt{2}$ (B) $\frac{1}{\sqrt{2}}$

(C) 2 (D) $\frac{1}{2}$

147. The velocity of a particle executing SHM varies with displacement (x) as $4v^2 = 50 - x^2$. The time period of oscillations is $\frac{x}{7}$ s. The value of x is _____ $\left(\text{Take } \pi = \frac{22}{7}\right)$ **[30 Jan. 2023 (Evening)]**

148. The maximum potential energy of a block executing simple harmonic motion is 25 J. A is amplitude of oscillation. At $A/2$, the kinetic energy of the block is : **[31 Jan. 2023 (Morning)]**

(A) 37.5 J (B) 9.75 J

(C) 18.75 J (D) 12.5 J

149. In the figure given below, a block of mass $M = 490$ g placed on a frictionless table is connected with two springs having same spring constant ($K = 2\text{ N m}^{-1}$). If the block is horizontally displaced through X m then the number of complete oscillations it will make in 14π seconds will be_____.

[31 Jan. 2023 (Morning)]

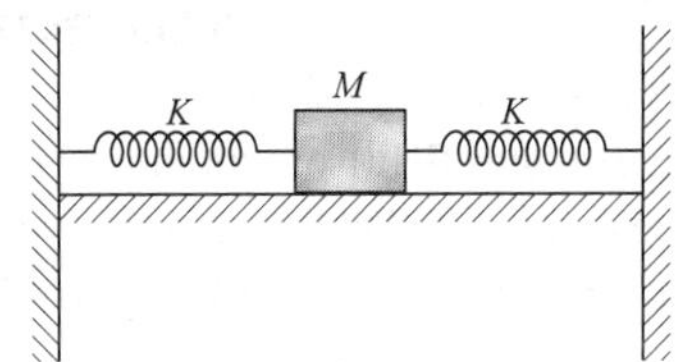

150. The amplitude of a particle executing SHM is 3 cm. The displacement at which its kinetic energy will be 25% more than the potential energy is _____ cm. **[1 Feb. 2023 (Morning)]**

151. Choose the correct length (L) versus square of time period (T^2) graph for a simple pendulum executing simple harmonic motion : **[1 Feb. 2023 (Evening)]**

(A) 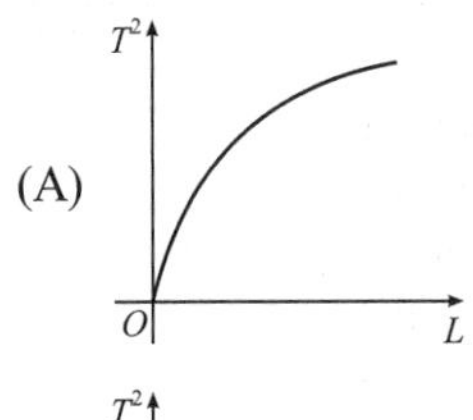

(B) 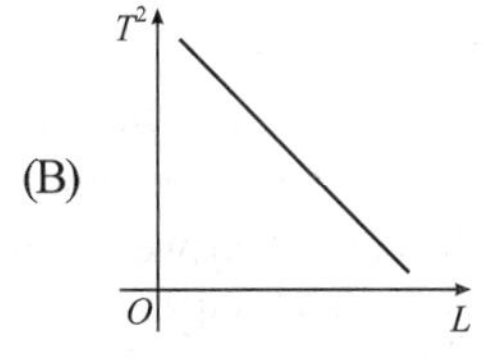

(C) 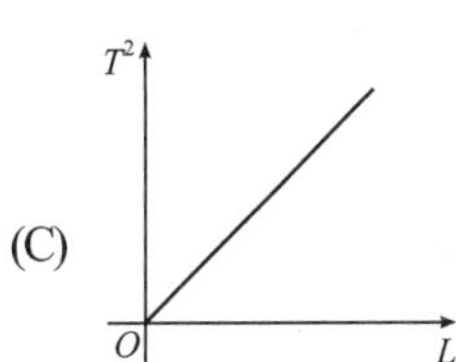

(D) 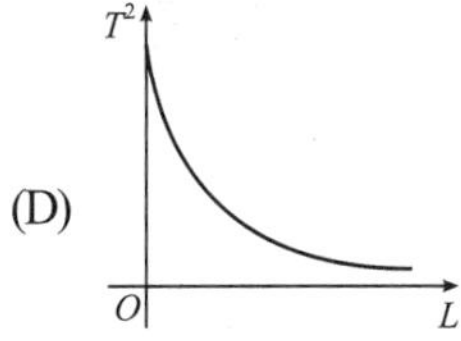

152. A mass m is attached to two springs as shown in figure. The spring constants of two springs are K_1 and K_2. For the frictionless surface, the time period of oscillation of mass m is

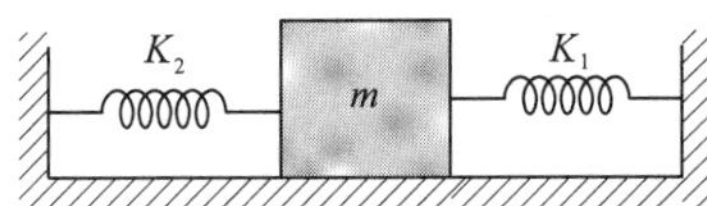

[06 April 2023 (Morning)]

(A) $\frac{1}{2\pi}\sqrt{\frac{K_1+K_2}{m}}$ (B) $\frac{1}{2\pi}\sqrt{\frac{K_1-K_2}{m}}$

(C) $2\pi\sqrt{\frac{m}{K_1+K_2}}$ (D) $2\pi\sqrt{\frac{m}{K_1-K_2}}$

153. A simple pendulum with length 100 cm and bob of mass 250 g is executing SHM of amplitude 10 cm. The maximum tension in the string is found to be $\frac{x}{40}$ N. The value of x is_____.

[6 April 2023 (Evening)]

154. For particle P revolving round the centre O with radius of circular path r and angular velocity ω, as shown in below figure, the projection of OP on the x-axis at time t is :

[8 April 2023 (Evening)]

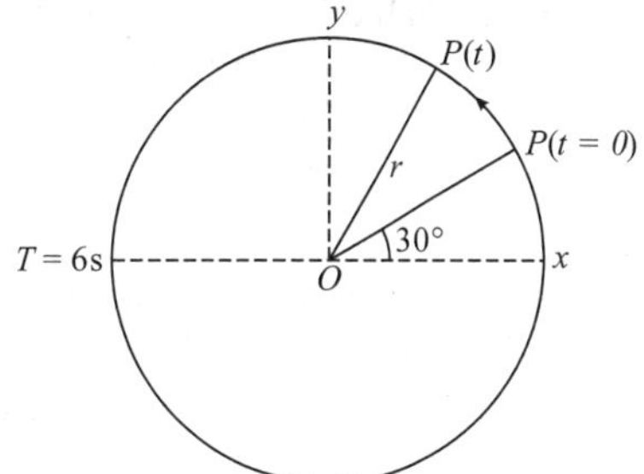

(A) $x(t) = r\cos\left(\omega t + \frac{\pi}{6}\right)$ (B) $x(t) = r\cos(\omega t)$

(C) $x(t) = r\sin\left(\omega t + \frac{\pi}{6}\right)$ (D) $x(t) = r\cos\left(\omega t - \frac{\pi}{6}\omega\right)$

155. A particle executes SHM of amplitude A along x-axis. At $t = 0$, the position of the particle is $x = \frac{A}{2}$ and it moves along positive x-axis the displacement of particle in time t $x = A\sin(\omega t + \delta)$, then the value δ will be:

[10 April 2023 (Morning)]

(A) $\frac{\pi}{6}$ (C) $\frac{\pi}{3}$

(C) $\frac{\pi}{4}$ (D) $\frac{\pi}{2}$

156. A rectangular block of mass 5 kg attached to a horizontal spiral spring executes simple harmonic motion of amplitude 1 m and time period 3.14 s. The maximum force exerted by spring on block is______ N. **[10 April 2023 (Evening)]**

157. The variation of kinetic energy (KE) of a particle executing simple harmonic motion with the displacement (x) starting from mean position to extreme position (A) is given by :

[11 April 2023 (Morning)]

(A) 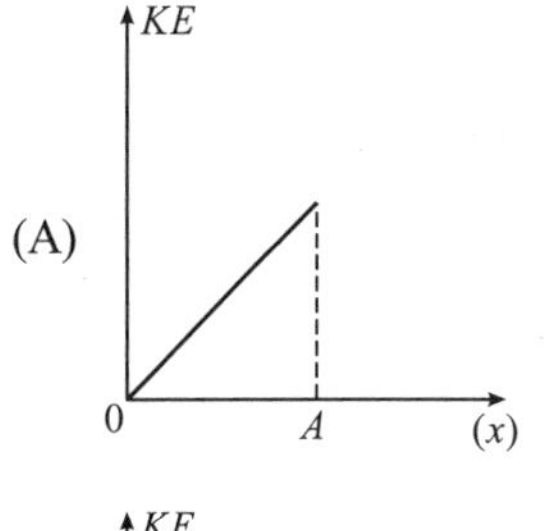

(B) 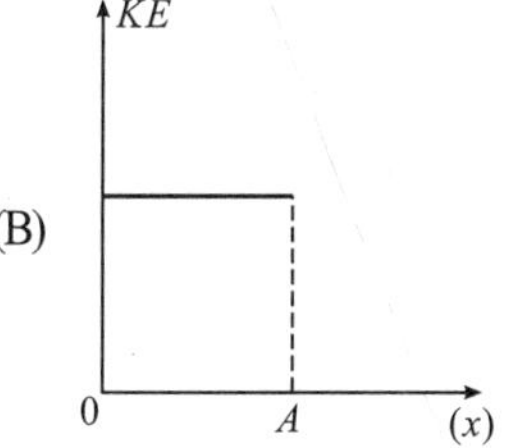

(C) 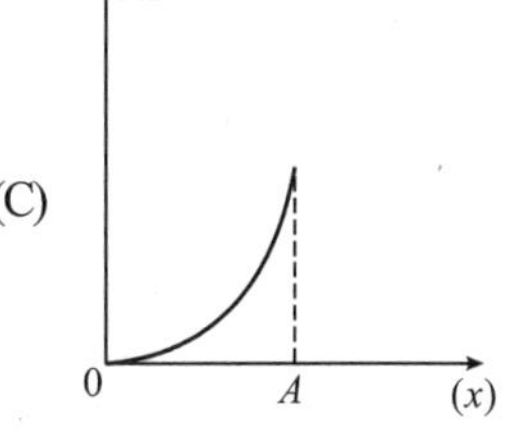

(D) 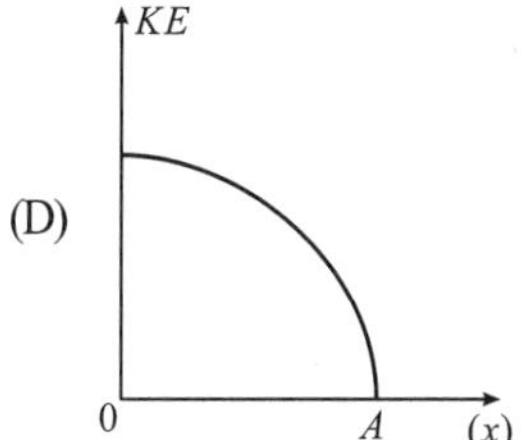

158. A particle is executing Simple Harmonic Motion (SHM). The ratio of potential energy and kinetic energy of the particle when its displacement is half of its amplitude will be: **[12 April 2023 (Morning)]**

(A) 1 : 1 (B) 2 : 1

(C) 1 : 4 (D) 1 : 3

159. Which graph represents the difference between total energy and potential energy of a particle executing SHM with it's distance from mean position ? **[13 April 2023 (Morning)]**

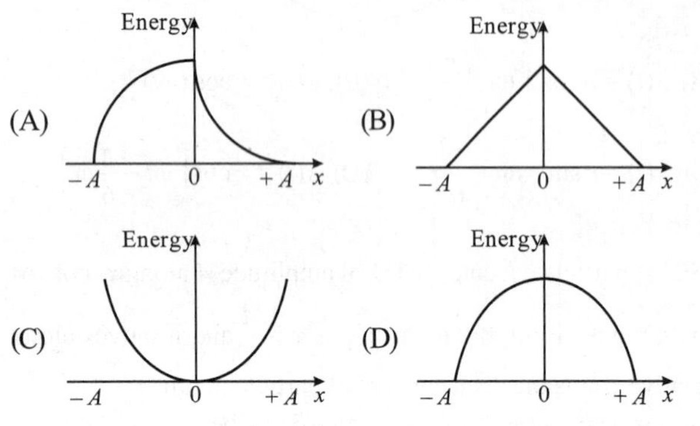

160. At a given point of time the value of displacement of a simple harmonic oscillator is given as $y = A\cos(30°)$. If amplitude is 40 cm and kinetic energy at that time is 200 J, the value of force constant is 1.0×10^x Nm^{-1}. The value of x is _____. **[13 April 2023 (Morning)]**

161. A particle executes SHM of amplitude A. The distance from the mean position when its's kinetic energy becomes equal to its potential energy is: **[13 April 2023 (Evening)]**

(A) $\sqrt{2}A$ (B) $2A$

(C) $\frac{1}{\sqrt{2}}A$ (D) $\frac{1}{2}A$

162. In a linear simple harmonic motion (SHM) : **[15 April 2023 (Morning)]**

(a) Restoring force is directly proportional to the displacement.

(b) The acceleration and displacement are opposite in direction.

(c) The velocity is maximum at mean position

(d) The acceleration is minimum at extreme points

Choose the correct answer from the options given below :

(A) (a), (b) and (c) only (B) (c) and (d) only

(C) (a), (b) and (d) only (D) (a), (c) and (d) only

* * * * *

CHAPTER

14 WAVES

14.1 WAVE MOTION

1. The displacement y of a wave travelling in the x-direction is given by $y = 10^{-4}\sin\left(600t - 2x + \frac{\pi}{3}\right)$ metre, where, x is expressed in metres and t in seconds. The speed of the wave-motion, in ms^{-1} is : **[2003]**

(A) 300 (B) 600
(C) 1200 (D) 200

2. A sound absorber attenuates the sound level by $20dB$. The intensity decreases by a factor of : **[2007]**

(A) 1000 (B) 10000
(C) 10 (D) 100

3. A wave travelling along the x-axis is described by the equation $y(x, t) = 0.005\cos(\alpha x - \beta t)$. If the wavelength and the time period of the wave in 0.08 m and 2.0 s respectively then α and β in appropriate units are : **[2008]**

(A) $\alpha = 25.00\,\pi, \beta = \pi$ (B) $\alpha = \frac{0.08}{\pi}, \beta = \frac{2.0}{\pi}$
(C) $\alpha = \frac{0.04}{\pi}, \beta = \frac{1.0}{\pi}$ (D) $\alpha = 12.50\,\pi, \beta = \frac{\pi}{2.0}$

4. The speed of sound in oxygen (O_2) at a certain temperature is 460 ms^{-1}. The speed of sound in helium (He) at the same temperature will be (assume both gases to be ideal) : **[2008]**

(A) $460\sqrt{\frac{200}{21}}\ ms^{-1}$ (B) $500\sqrt{\frac{200}{21}}\ ms^{-1}$
(C) $650\sqrt{2}\ ms^{-1}$ (D) $330\sqrt{2}\ ms^{-1}$

5. The equation of a wave on a string of linear mass density 0.04 kg m^{-1} is given by $y = 0.02\ (m)\ \sin\left[2\pi\left(\frac{t}{0.04(s)} - \frac{x}{0.50(m)}\right)\right]$. The tension in the string is : **[2010]**

(A) 6.25 N (B) 4.0 N
(C) 12.5 N (D) 0.5 N

6. **Statement-1 :** Two longitudinal waves given by equations: $y_1(x, t) = 2a\sin(\omega t - kx)$ and $y_2(x, t) = a\sin(2\omega t - 2kx)$ will have equal intensity.

Statement-2 : Intensity of waves of given frequency in same medium is proportional to square of amplitude only. **[2011]**

(A) Statement-1 is false, statement-2 is true
(B) Statement-1 is true, statement-2 is false
(C) Statement-1 is true, statement-2 true; statement-2 is the correct explanation of statement-1
(D) Statement-1 is true, statement-2 is true; statement-2 is not correct explanation of statement-1

7. The transverse displacement $y(x, t)$ of a wave on a string is given by $y(x, t) = e^{-(ax^2 + bt^2 + 2\sqrt{ab}xt)}$. This represents a : **[2011]**

(A) Standing wave of frequency $\sqrt{b}$
(B) Standing wave of frequency $\frac{1}{\sqrt{b}}$
(C) Wave moving in $+x$ direction with speed $\sqrt{\frac{a}{b}}$
(D) Wave moving in $-x$ direction with speed $\sqrt{\frac{b}{a}}$

8. The disturbance $y(x, t)$ of a wave propagating in the positive x-direction is given by $y = \frac{1}{1+x^2}$ at time $t = 0$ and by $y = \frac{1}{[1+(x-1)^2]}$ at $t = 2$s, where x and y are in meters. The shape of the wave disturbance does not change during the propagation. The velocity of wave in m/s is : **[2012 (Online)]**

(A) 2.0 (B) 4.0
(C) 0.5 (D) 1.0

9. When two sound waves travel in the same direction in a medium, the displacements of a particle located at 'x' at time 't' is given by :

$$y_1 = 0.05\cos(0.50\pi x - 100\pi t)$$
$$y_2 = 0.05\cos(0.45\pi x - 92\pi t)$$

where y_1, y_2 and x are in meters and t in seconds. The speed of sound in the medium is : **[2013 (Online)]**

(A) 92 m/s (B) 200 m/s
(C) 100 m/s (D) 332 m/s

10. In a transverse wave the distance between a crest and neighbouring trough at the same instant is 0.4 cm and the distance between a crest and trough at the same place is 1.0 cm. The next crest appears at the same place after a time interval of 0.4s. The maximum speed of the vibrating particles in the medium is : **[2013 (Online)]**

(A) $\frac{3\pi}{2}$ cm/s (B) $\frac{5\pi}{2}$ cm/s

(C) $\frac{\pi}{2}$ cm/s (D) 2π cm/s

11. An engine approaches a hill with a constant speed. When it is at a distance of 0.9 km, it blows a whistle whose echo is heard by the driver after 5 seconds. If the speed of sound in air is 330 m/s, then the speed of the engine is : **[2013 (Online)]**

(A) 32 m/s (B) 27.5 m/s

(C) 60 m/s (D) 30 m/s

12. A transverse wave is represented by

$$y = \frac{10}{\pi}\sin\left(\frac{2\pi}{T}t - \frac{2\pi}{\lambda}x\right)$$

For what value of the wavelength the wave velocity is twice the maximum particle velocity? **[2014 (Online)]**

(A) 40 cm (B) 20 cm

(C) 10 cm (D) 60 cm

13. A uniform string of length 20 m is suspended from a rigid support. A short wave pulse is introduced at its lowest end. It starts moving up the string. The time taken to reach the support is : (take $g = 10\ ms^{-2}$) : **[2016]**

(A) $2\pi\sqrt{2}$ s (B) 2s

(C) $2\sqrt{2}$ s (D) $\sqrt{2}$ s

14. A heavy ball of mass M is suspended from the ceiling of a car by a light string of mass m ($m << M$). When the car is at rest, the speed of transverse waves in the string is 60 ms^{-1}. When the car has acceleration a, the wave-speed increases to 60.5 ms^{-1}. The value of a, in terms of gravitational acceleration g, is closest to : **[9 Jan. 2019 (Morning)]**

(A) $\frac{g}{5}$ (B) $\frac{g}{20}$

(C) $\frac{g}{10}$ (D) $\frac{g}{30}$

15. Equation of travelling wave on a stretched string of linear density 5 g/m is $y = 0.03 \sin(450t - 9x)$ where distance and time are measured is SI units. The tension in the string is : **[11 Jan. 2019 (Morning)]**

(A) 10N (B) 12.5N

(C) 7.5N (D) 5N

16. A travelling harmonic wave is represented by the equation $y(x, t) = 10^{-3}\sin(50t + 2x)$, where x and y are in meter and t is in seconds. Which of the following is a correct statement about the wave? The wave is propagating along the : **[12 Jan. 2019 (Morning)]**

(A) Negative x-axis with speed 25 ms^{-1}

(B) The wave is propagating along the positive x-axis with speed 25 ms^{-1}

(C) The wave is propagating along the positive x-axis with speed 100 ms^{-1}

(D) The wave is propagating along the negative x-axis with speed 100 ms^{-1}

17. The pressure wave, $P = 0.01 \sin[1000t - 3x]\ Nm^{-2}$, corresponds to the sound produced by a vibrating blade on a day when atmospheric temperature is 0°C. On some other day, when temperature is T, the speed of sound produced by the same blade and at the same frequency is found to be 336 ms^{-1}. Approximate value of T is : **[9 April 2019 (Morning)]**

(A) 15°C (B) 12°C

(C) 4°C (D) 11°C

18. A progressive wave travelling along the positive x-direction is represented by $y(x, t) = A\sin(kx - \omega t + \phi)$. Its snapshot at $t = 0$ is given in the figure. For this wave, the phase ϕ is : **[12 April 2019 (Morning)]**

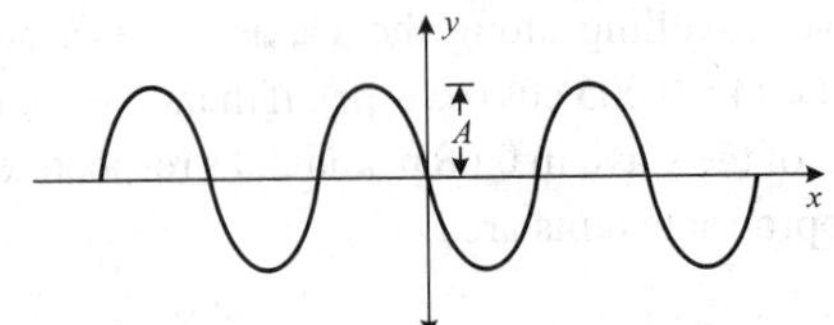

(A) 0 (B) $-\frac{\pi}{2}$

(C) π (D) $\frac{\pi}{2}$

19. A small speaker delivers 2 W of audio output. At what distance from the speaker will one detect 120 dB intensity sound? [Given reference intensity of sound as $10^{-12}\ W/m^2$] : **[12 April 2019 (Evening)]**

(A) 10 cm (B) 30 cm

(C) 40 cm (D) 20 cm

20. Speed of a transverse wave on a straight wire (mass 6.0 g, length 60 cm and area of cross-section 1.0 mm^2) is 90 ms^{-1}. If the Young's modulus of wire is $16 \times 10^{11}\ Nm^{-2}$, the extension of wire over its natural length is : **[7 Jan. 2020 (Morning)]**

(A) 0.02 mm (B) 0.04 mm

(C) 0.03 mm (D) 0.01 mm

21. A transverse wave travels on a taut steel wire with a velocity of v when tension in it is 2.06×10^4 N. When the tension is changed to T, the velocity changed to $v/2$. The value of T is close to : **[8 Jan. 2020 (Evening)]**

(A) 10.2×10^2 N (B) 5.15×10^3 N

(C) 2.50×10^4 N (D) 30.5×10^4 N

22. A uniform thin rope of length 12 m and mass 6 kg hangs vertically from a rigid support and a block of mass 2 kg is attached to its free end. A transverse short wavetrain of wavelength 6 cm is produced at the lower end of the rope. What is the wavelength of the wavetrain (in cm) when it reaches the top of the rope ? **[3 Sept. 2020 (Morning)]**

(A) 9 (B) 12
(C) 6 (D) 3

23. For a transverse wave travelling along a straight line, the distance between two peaks (crests) is 5 m, while the distance between one crest and one trough is 1.5 m. The possible wavelengths (in m) of the waves are : **[4 Sept. 2020 (Morning)]**

(A) $1, 2, 3, \ldots$ (B) $\frac{1}{2}, \frac{1}{4}, \frac{1}{6}, \ldots$

(C) $1, 3, 5, \ldots$ (D) $\frac{1}{1}, \frac{1}{3}, \frac{1}{5}, \ldots$

24. Assume that the displacement(s) of air is proportional to the pressure difference (Δp) created by a sound wave. Displacement(s) further depends on the speed of sound (v), density of air (r) and the frequency (f). If $\Delta p \sim 10 Pa$, $v \sim 300$ m/s, $p \sim 1$ kg/m^3 and $f \sim 1000$ Hz, then s will be the order of (take multiplicative constant to be 1) : **[5 Sept. 2020 (Morning)]**

(A) 10 mm (B) $\frac{3}{100}$ mm

(C) 1 mm (D) $\frac{1}{10}$ mm

25. Which of the following equations represents a travelling wave ? **[24 Feb. 2021 (Evening)]**

(A) $y = Ae^{-x^2}(vt + \theta)$ (B) $y = Ae^{x}\cos(\omega t - \theta)$
(C) $y = A\sin x \cos \omega t$ (D) $y = A\sin(15x - 2t)$

26. The percentage increase in the speed of transverse waves produced in a stretched string if the tension is increased by 4%, will be ______%. **[25 Feb. 2021 (Evening)]**

27. The mass per unit length of a uniform wire is 0.135 g/cm. A transverse wave of the form $y = -0.21 \sin(x + 30t)$ is produced in it, where x is in meter and t is in second. Then, the expected value of tension in the wire is $x \times 10^{-2}$ N. Value of x is ______. (Round-off to the nearest integer). **[26 Feb. 2021 (Morning)]**

28. A sound wave of frequency 245 Hz travels with the speed of 300 ms^{-1} along the positive x-axis. Each point of the wave moves to and fro through a total distance of 6 cm. What will be the mathematical expression of this travelling wave ? **[17 March 2021 (Evening)]**

(A) $Y(x, t) = 0.03\,[\sin 5.1\,x - (0.2 \times 10^3)t]$
(B) $Y(x, t) = 0.06\,[\sin 5.1\,x - (1.5 \times 10^3)t]$
(C) $Y(x, t) = 0.06\,[\sin 0.8\,x - (0.5 \times 10^3)t]$
(D) $Y(x, t) = 0.03\,[\sin 5.1\,x - (1.5 \times 10^3)t]$

29. The amplitude of wave disturbance propagating in the positive x-direction is given by $y = \frac{1}{(1+x)^2}$ at time $t = 0$ and $y = \frac{1}{1+(x-2)^2}$ at $t = 1$ s, where x and y are in metres. The shape of wave doesn't change during the propagation. The velocity of the wave will be ______ m/s. **[20 July 2021 (Morning)]**

30. Two travelling waves produces a standing wave represented by equation. $y = 1.0$ mm $\cos(1.57$ cm$^{-1})x$ $\sin(78.5$ s$^{-1})t$. The node closest to the origin in the region $x > 0$ will be at $x =$ ______ cm. **[26 Aug. 2021 (Morning)]**

31. A longitudinal wave is represented by

$$x = 10 \sin 2\pi\left(nt - \frac{x}{\lambda}\right) \text{cm}$$

The maximum particle velocity will be four times the wave velocity if the determined value of wavelength is equal to : **[29 June 2022 (Morning)]**

(A) 2π (B) 5π

(C) π (D) $\frac{5\pi}{2}$

32. A transverse wave is represented by $y = 2 \sin(\omega t - kx)$ cm. The value of wavelength (in cm) for which the wave velocity becomes equal to the maximum particle velocity, will be : **[26 July 2022 (Evening)]**

(A) 4π (B) 2π
(C) π (D) 2

33. In the wave equation $y = 0.5 \sin\frac{2\pi}{\lambda}(400t - x)\,m$ the velocity of the wave will be : **[28 July 2022 (Morning)]**

(A) 200 m/s (B) $200\sqrt{2}$ m/s

(C) 400 m/s (D) $400\sqrt{2}$ m/s

34. The speed of a transverse wave passing through a string of length 50 cm and mass 10 g is 60 ms^{-1}. The area of cross-section of the wire is 2.0 mm^2 and its Young's modulus is 1.2×10^{11} Nm^{-2}. The extension of the wire over its natural length due to its tension will be $x \times 10^{-5}$ m. The value of x is ______. **[29 July 2022 (Evening)]**

35. A travelling wave is described by the equation

$y(x, t) = [0.05 \sin (8x - 4t)]m$

The velocity of the wave is : [all the quantities are in SI unit]

[24 Jan. 2023 (Morning)]

(A) 4 ms^{-1} (B) 2 ms^{-1}

(C) 0.5 ms^{-1} (D) 8 ms^{-1}

36. The distance between two consecutive points with phase difference of 60° in a wave of frequency 500 Hz is 6.0 m. The velocity with which wave is traveling is _____ km/s.

[25 Jan. 2023 (Morning)]

37. A steel wire with mass per unit length 7.0×10^{-3} kg m^{-1} is under tension of 70 N. The speed of transverse waves in the wire will be : **[1 Feb. 2023 (Morning)]**

(A) 200π m/s (B) 100 m/s

(C) 10 m/s (D) 50 m/s

38. A transverse harmonic wave on a string is given by

$y(x, t) = 5 \sin (6t + 0.003x)$

where x and y are in cm and t in sec. The wave velocity is _____ ms^{-1}. **[10 April 2023 (Morning)]**

39. For a periodic motion represented by the equation

$Y = \sin \omega t + \cos \omega t.$

The amplitude of the motion is. **[10 April 2023 (Evening)]**

(A) 0.5 (B) $\sqrt{2}$

(C) 1 (D) 2

40. The equation of wave is given by

$$Y = 10^{-2} \sin 2\pi \left(160t - 0.5x + \frac{\pi}{4}\right)$$

Where x and Y are in m and t in s. The speed of the wave is_____ kmh^{-1}. **[11 April 2023 (Morning)]**

41. A wire of density 8×10^3 kg/m^3 is stretched between two clamps 0.5 m apart. The extension developed in the wire is 3.2×10^{-4} m. If $Y = 8 \times 10^{10}$ N/m^2, the fundamental frequency of vibration in the wire will be_____ Hz.

[11 April 2023 (Evening)]

* * * * *

CHAPTER

14 WAVES

14.2 STATIONARY WAVES AND BEATS

1. Tube A has both ends open while tube B has one end closed, otherwise they are identical. The ratio of fundamental frequency of tubes A and B is : **[2002]**

(A) 1 : 2 (B) 1 : 4
(C) 2 : 1 (D) 4 : 1

2. A tuning fork arrangement (pair) produces 4 beats/s with one fork of frequency 288 cps. A little wax is placed on the unknown fork and it then produces 2 beats/s. The frequency of the unknown fork is : **[2002]**

(A) 286 cps (B) 292 cps
(C) 294 cps (D) 288 cps

3. Length of a string tied to two rigid supports is 40 cm. Maximum length (wavelength in cm) of a stationary wave produced on it, is : **[2002]**

(A) 20 (B) 80
(C) 40 (D) 120

4. A wave $y = a \sin(\omega t - kx)$ on a string meets with another wave producing a node at $x = 0$. Then the equation of the unknown wave is : **[2002]**

(A) $y = a \sin(\omega t + kx)$ (B) $y = -a \sin(\omega t + kx)$
(C) $y = a \sin(\omega t - kx)$ (D) $y = -a \sin(\omega t - kx)$

5. A tuning fork of known frequency 256 Hz makes 5 beats per second with the vibrating string of a piano. The beat frequency decreases to 2 beats per second when the tension in the piano string is slightly increased. The frequency of the piano string before increasing the tension was : **[2003]**

(A) (256 + 2) Hz (B) (256 – 2) Hz
(C) (256 – 5) Hz (D) (256 + 5) Hz

6. When two tuning forks (fork 1 and fork 2) are sounded simultaneously, 4 beats per second are heard. Now, some tape is attached on the prong of the fork 2. When the tuning forks are sounded again, 6 beats per second are heard. If the frequency of fork 1 is 200 Hz, then what was the original frequency of fork 2? **[2005]**

(A) 200 Hz (B) 202 Hz
(C) 196 Hz (D) 204 Hz

7. A string is stretched between fixed points separated by 75.0 cm. It is observed to have resonant frequencies of 420 Hz and 315 Hz. Their are no other resonant frequencies between these two. Then, the lowest resonant frequency for this string is : **[2006]**

(A) 105 Hz (B) 1.05 Hz
(C) 1050 Hz (D) 10.5 Hz

8. While measuring the speed of sound by performing a resonance column experiment, a student gets the first resonance condition at column length of 18 cm during winter. Repeating the same experiment during summer, student measures the column length to be x cm for the second resonance. Then : **[2008]**

(A) $18 > x$ (B) $x > 54$
(C) $54 > x > 36$ (D) $36 > x > 18$

9. Three sound waves of equal amplitudes have frequencies $(\nu - 1)$, ν, $(\nu + 1)$. They superpose to give beats. The number of beats produced per second will be : **[2009]**

(A) 2 (B) 1
(C) 4 (D) 3

10. A travelling wave represented by $y = A \sin(\omega t - kx)$ is superimposed on another wave represented by $y = A \sin(\omega t + kx)$. The resultant is : **[2011]**

(A) A standing wave having nodes at $x = \left(n + \frac{1}{2}\right)\frac{\lambda}{2}, n = 0, 1, 2$

(B) A wave travelling along $+x$ direction

(C) A wave travelling along $-x$ direction

(D) A standing wave having nodes at $x = \frac{n\lambda}{2}; n = 0, 1, 2$

11. A cylindrical tube, open at both ends, has a fundamental frequency f, in air. The tube is dipped vertically in water so that half of it is in water. The fundamental frequency of the air-column is now : **[2012]**

(A) $2f$ (B) f
(C) $f/2$ (D) $3f/4$

12. Following are expressions for four plane simple harmonic waves : **[2012 (Online)]**

(i) $y_1 = A \cos 2\pi \left(n_1 t + \frac{x}{\lambda_1}\right)$

(ii) $y_2 = A \cos 2\pi \left(n_1 t + \frac{x}{\lambda_1} + \pi\right)$

(iii) $y_3 = A \cos 2\pi \left(n_2 t + \dfrac{x}{\lambda_2} \right)$

(iv) $y_4 = A \cos 2\pi \left(n_2 t - \dfrac{x}{\lambda_2} \right)$

The pair of waves which will produce destructive interference and stationary wave respectively in a medium, are :

(A) (iii, iv), (i, ii) (B) (i, iii), (ii, iv)
(C) (i, iv), (ii, iii) (D) (i, ii), (iii, iv)

13. A wave represented by the equation $y_1 = a \cos (kx - \omega t)$ is superimposed with another wave to form a stationary wave such that the point $x - 0$ is node. The equation for the other wave is : **[2012 (Online)]**

(A) $a \cos (kx - \omega t - \pi)$ (B) $a \cos (kx + \omega t + \pi)$

(C) $a \cos \left(kx + \omega t + \dfrac{\pi}{2} \right)$ (D) $a \cos \left(kx - \omega t + \dfrac{\pi}{2} \right)$

14. A uniform tube of length 60.5 cm is held vertically with its lower end dipped in water. A sound source of frequency 500 Hz sends sound waves into the tube. When the length of tube above water is 16 cm and again when it is 50 cm, the tube resonates with the source of sound. Two lowest frequencies (in Hz), to which tube will resonate when it is taken out of water, are (approximately) : **[2012 (Online)]**

(A) 281, 562 (B) 281, 843
(C) 276, 552 (D) 272, 544

15. An air column in a pipe, which is closed at one end, will be in resonance with a vibrating tuning fork of frequency 264 Hz if the length of the column in cm is (velocity of sound = 330 m/s): **[2012 (Online)]**

(A) 125.00 (B) 93.75
(C) 62.50 (D) 187.50

16. A sonometer wire of length 1.5 m is made of steel. The tension in it produces an elastic strain of 1%. What is the fundamental frequency of steel if density and elasticity of steel are 7.7×10^3 kg/m^3 and 2.2×10^{11} N/m^2 respectively? **[2013]**

(A) 188.5 Hz (B) 178.2 Hz
(C) 200.5 Hz (D) 770 Hz

17. A sonometer wire of length 114 cm is fixed at both the ends. Where should the two bridges be placed so as to divide the wire into three segments whose fundamental frequencies are in the ratio 1 : 3 : 4 ? **[2013 (Online)]**

(A) At 36 cm and 84 cm from one end
(B) At 24 cm and 72 cm from one end
(C) At 48 cm and 96 cm from one end
(D) At 72 cm and 96 cm from one end

18. A pipe of length 85 cm is closed from one end. Find the number of possible natural oscillations of air column in the pipe whose frequencies lie below 1250 Hz. The velocity of sound in air is 340 m/s : **[2014]**

(A) 12 (B) 8
(C) 6 (D) 4

19. The total length of a sonometer wire between fixed ends is 110 cm. Two bridges are placed to divide the length of wire in ratio 6 : 3 : 2. The tension in the wire is 400 N and the mass per unit length is 0.01 kg/m. What is the minimum common frequency with which three parts can vibrate ? **[2014 (Online)]**

(A) 1100 Hz (B) 1000 Hz
(C) 166 Hz (D) 100 Hz

20. A pipe open at both ends has a fundamental frequency f in air. The pipe is dipped vertically in water so that half of it is in water. The fundamental frequency of the air column is now : **[2016]**

(A) $\dfrac{f}{2}$ (B) $\dfrac{3f}{4}$
(C) $2f$ (D) f

21. A standing wave is formed by the superposition of two waves travelling in opposite directions. The transverse displacement is given by $y(x, t) = 0.5 \sin \left(\dfrac{5\pi}{4} x \right) \cos (200\, \pi t)$. What is the speed of the travelling wave moving in the positive x direction ? (x and t are in meter and second, respectively) **[2017 (Online)]**

(A) 160 m/s (B) 90 m/s
(C) 180 m/s (D) 120 m/s

22. Two wires W_1 and W_2 have the same radius r and respective densities ρ_1 and ρ_2 such that $\rho_2 = 4\rho_1$. They are joined together at the point O, as shown in the figure. The combination is used as a sonometer wire and kept under tension T. The point O is midway between the two bridges. When a stationary waves is set up in the composite wire, the joint is found to be a node. The ratio of the number of antinodes formed in W_1 to W_2 is : **[2017 (Online)]**

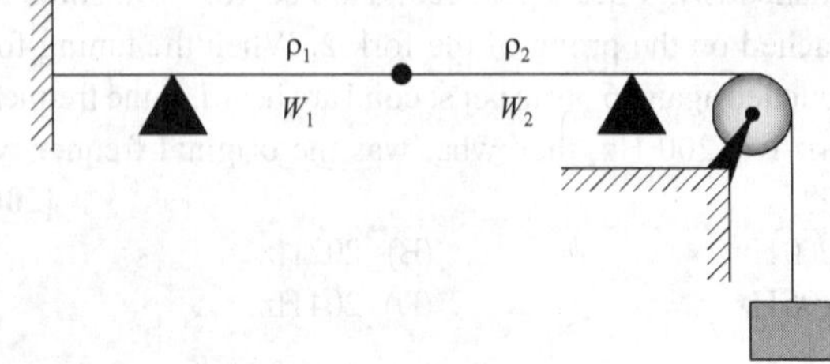

(A) 1 : 1 (B) 1 : 2
(C) 1 : 3 (D) 4 : 1

23. A granite rod of 60 cm length is clamped at its middle point and is set into longitudinal vibrations. The density of granite is 2.7×10^3 kg/m^3 and its Young's modulus is 9.27×10^{10} Pa. What will be the fundamental frequency of the longitudinal vibrations? **[2018]**

(A) 5 kHz (B) 2.5 kHz
(C) 10 kHz (D) 7.5 kHz

24. Two sitar strings, A and B, playing the note 'Dha' are slightly out of tune and produce beats and frequency 5 Hz. The tension of the string B is slightly increased and the beat frequency is found to decrease by 3 Hz. If the frequency of A is 425 Hz, the original frequency of B is : **[2018 (Online)]**

(A) 430 Hz (B) 428 Hz
(C) 422 Hz (D) 420 Hz

25. 5 beats/second are heard when a turning fork is sounded with a sonometer wire under tension, when the length of the sonometer wire is either 0.95 m or 1 m. The frequency of the fork will be : **[2018 (Online)]**

(A) 195 Hz (B) 251 Hz
(C) 150 Hz (D) 300 Hz

26. A tuning fork vibrates with frequency 256 Hz and gives one beat per second with the third normal mode of vibration of an open pipe. What is the length of the pipe ? (Speed of sound of air is 340 s^{-1}) **[2018 (Online)]**

(A) 190 cm (B) 180 cm
(C) 220 cm (D) 200 cm

27. The end correction of a resonance column is 1 cm. If the shortest length resonating with the tuning fork is 10 cm, the next resonating length should be : **[2018 (Online)]**

(A) 32 cm (B) 40 cm
(C) 28 cm (D) 36 cm

28. A musician using an open flute of length 50 cm produces second harmonic sound waves. A person runs towards the musician from another end of a hall at a speed of 10 km/h. If the wave speed is 330 m/s, the frequency heard by the running person shall be close to : **[9 Jan. 2019 (Evening)]**

(A) 753 Hz (B) 500 Hz
(3) 333 Hz (D) 666 Hz

29. A string of length 1 m and mass 5 g is fixed at both ends. The tension in the string is 8.0 N. The string is set into vibration using an external vibrator of frequency 100 Hz. The separation between successive nodes on the string is close to : **[10 Jan. 2019 (Morning)]**

(A) 16.6 cm (B) 20.0 cm
(C) 10.0 cm (D) 33.3 cm

30. A closed organ pipe has a fundamental frequency of 1.5 kHz. The number of overtones that can be distinctly heard by a person with this organ pipe will be : (Assume that the highest frequency a person can hear is 20,000 Hz) **[10 Jan. 2019 (Evening)]**

(A) 7 (B) 5
(C) 6 (D) 4

31. A resonance tube is old and has jagged end. It is still used in the laboratory to determine velocity of sound in air. A tuning fork of frequency 512 Hz produces first resonance when the tube is filled with water to a mark 11 cm below a reference mark, near the open end of the tube. The experiment is repeated with another fork of frequency 256 Hz which produces first resonance when water reaches a mark 27 cm below the reference mark. The velocity of sound in air, obtained in the experiment, is close to : **[12 Jan. 2019 (Evening)]**

(A) 328 ms^{-1} (B) 322 ms^{-1}
(C) 341 ms^{-1} (D) 335 ms^{-1}

32. A wire of length $2L$, is made by joining two wires A and B of same length but different radii r and $2r$ and made of the same material. It is vibrating at a frequency such that the joint of the two wires forms a node. If the number of antinodes in wire A is p and that in B is q then the ratio $p : q$ is : **[8 April 2019 (Morning)]**

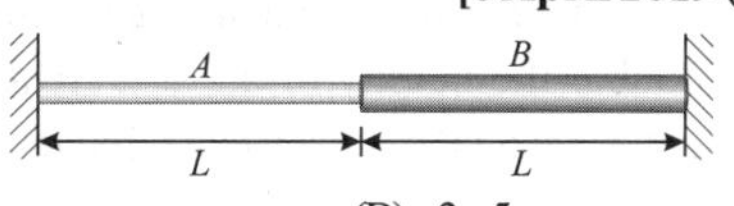

(A) 4 : 9 (B) 3 : 5
(C) 1 : 4 (D) 1 : 2

33. A string is clamped at both the ends and it is vibrating in its 4th harmonic. The equation of the stationary wave is $Y = 0.3 \sin(0.157x) \cos(200\pi t)$. The length of the string is : (All quantities are in SI units) **[9 April 2019 (Morning)]**

(A) 20 m (B) 80 m
(C) 60 m (D) 40 m

34. A string 2.0 m long and fixed at its ends is driven by a 240 Hz vibrator. The string vibrates in its third harmonic mode. The speed of the wave and its fundamental frequency is : **[9 April 2019 (Evening)]**

(A) 320 m/s, 120 Hz (B) 180 m/s, 80 Hz
(C) 180 m/s, 120 Hz (D) 320 m/s, 80 Hz

35. The correct figure that shows, schematically, the wave pattern produced by superposition of two waves of frequencies 9 Hz and 11 Hz is : **[10 April 2019 (Evening)]**

(A) 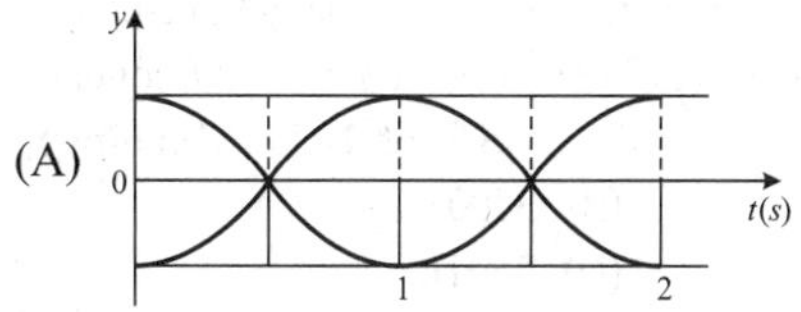

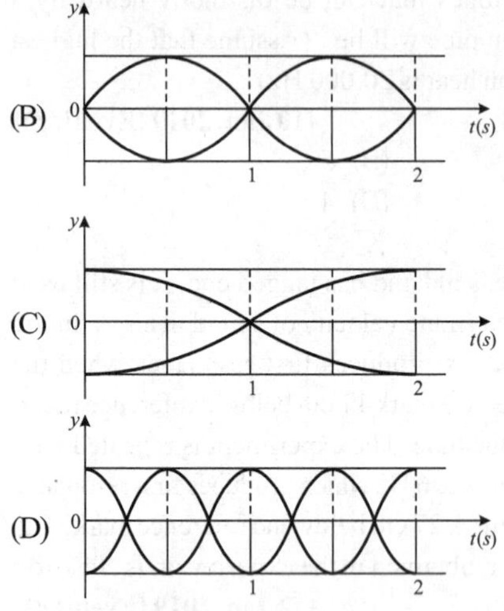

36. A tuning fork of frequency 480 Hz is used in an experiment for measuring speed of sound (ν) in air by resonance tube method. Resonance is observed to occur at two successive lengths of the air column, $l_1 = 30$ cm and $l_2 = 70$ cm. Then ν is equal to : **[12 April 2019 (Evening)]**

(A) 332 ms^{-1} (B) 379 ms^{-1}
(C) 384 ms^{-1} (D) 338 ms^{-1}

37. A one metre long (both ends open) organ pipe is kept in a gas that has double the density of air at STP. Assuming the speed of sound in air at STP is 300 m/s, the frequency difference between the fundamental and second harmonic of this pipe is ______ Hz. **[8 Jan. 2020 (Morning)]**

38. A wire of length L and mass per unit length 6.0×10^{-3} kgm^{-1} is put under tension of 540 N. Two consecutive frequencies that it resonates at are 420 Hz and 490 Hz. Then L in meters is : **[9 Jan. 2020 (Evening)]**

(A) 8.1 m (B) 5.1 m
(C) 1.1 m (D) 2.1 m

39. Two identical strings X and Z made of same material have tension T_X and T_Z in them. It their fundamental frequencies are 450 Hz and 300 Hz, respectively, then the ratio T_X/T_Z is : **[2 Sept. 2020 (Morning)]**

(A) 0.44 (B) 1.5
(C) 2.25 (D) 1.25

40. In a resonance tube experiment when the tube is filled with water up to height of 17.0 cm from bottom, it resonates with a given tuning fork. When the water level is raised the next resonance with the same tuning fork occurs at a height of 24.5 cm. If the velocity of sound in air is 330 m/s, the tuning fork frequency is : **[5 Sept. 2020 (Morning)]**

(A) 1100 Hz (B) 3300 Hz
(C) 2200 Hz (D) 550 Hz

41. A student is performing the experiment of resonance column. The diameter of the column tube is 6 cm. The frequency of the tuning fork is 504 Hz. Speed of the sound at the given temperature is 336 m/s. The zero of the metre scale coincides with the top end of the resonance column tube. The reading of the water level in the column when the first resonance occurs is : **[25 Feb. 2021 (Morning)]**

(A) 18.4 cm (B) 14.8 cm
(C) 16.6 cm (D) 13 cm

42. A tuning fork A of unknown frequency produces 5 beats/s with a fork of known frequency 340 Hz. When fork A is filed, the beat frequency decreases to 2 beats/s. What is the frequency of fork A ? **[26 Feb. 2021 (Evening)]**

(A) 342 Hz (B) 335 Hz
(C) 338 Hz (D) 345 Hz

43. A closed organ pipe of length L and an open organ pipe contain gases of densities ρ_1 and ρ_2 respectively. The compressibility of gases are equal in both the pipes. Both the pipes are vibrating in their first overtone with same frequency.

The length of the open pipe is $\frac{x}{3}L\sqrt{\frac{\rho_1}{\rho_2}}$ where x is ______.

(Round off to the Nearest Integer) **[16 March 2021 (Evening)]**

44. Two waves are simultaneously passing through a string and their equations are :

$$y_1 = A_1 \sin k(x - vt),\ y_2 = A_2 \sin k(x - vt + x_0)$$

Given amplitudes $A_1 = 12$ mm and $A_2 = 5$ mm, $x_0 = 3.5$ cm and wave number $k = 6.28$ cm^{-1}. The amplitude of resulting, wave will be ______ mm. **[26 Aug. 2021 (Evening)]**

45. A tuning fork is vibrating at 250 Hz. The length of the shortest closed organ pipe that will resonate with the tuning fork will be______cm. (Take speed of sound in air as 340 ms^{-1}). **[27 Aug. 2021 (Evening)]**

46. A wire having a linear mass density 9.0×10^{-4} kg/m is stretched between two rigid supports with a tension of 900 N. The wire resonates at a frequency of 500 Hz. The next higher frequency at which the same wire resonates is 550 Hz. The length of the wire is______m : **[31 Aug. 2021 (Morning)]**

47. The equations of two waves are given by :

$y_1 = 5 \sin 2\pi (x - vt)$ cm

$y_2 = 3 \sin 2\pi (x - vt + 1.5)$ cm

These waves are simultaneously passing through a string. The amplitude of the resulting wave is : **[24 June 2022 (Morning)]**

(A) 2 cm (B) 4 cm
(C) 5.8 cm (D) 8 cm

48. Two travelling waves of equal amplitudes and equal frequencies move in opposite directions along a string. They interfere to produce a stationary wave whose equation is given by

$$y = (10 \cos \pi x \sin \frac{2\pi t}{T}) \text{ cm}$$

The amplitude of the particle at $x = \frac{4}{3}$ cm will be _____ cm.

[24 June 2022 (Evening)]

49. The first overtone frequency of an open organ pipe is equal to the fundamental frequency of a closed organ pipe. If the length of the closed organ pipe is 20 cm. The length of the open organ pipe is _____ cm. **[25 June 2022 (Morning)]**

50. A set of 20 tuning forks are arranged in a series of increasing frequencies. If each fork gives 4 beats with respect to the preceding fork and the frequency of the last fork is twice the frequency of the first, then the frequency of last fork is _____ Hz. **[26 June 2022 (Evening)]**

51. The velocity of sound in a gas in which two wavelengths 4.08 m and 4.16 m produce 40 beats in 12 s, will be :

[28 June 2022 (Morning)]

(A) 282.8 ms^{-1}
(B) 175.5 ms^{-1}
(C) 353.6 ms^{-1}
(D) 707.2 ms^{-1}

52. A tuning fork of frequency 340 Hz resonates in the fundamental mode with an air column of length 125 cm in a cylindrical tube closed at one end. When water is slowly poured in it, the minimum height of water required for observing resonance once again is _____ cm.
[Velocity of sound in air is 340 ms^{-1}]**[28 June 2022 (Evening)]**

53. In an experiment to determine the velocity of sound in air at room temperature using a resonance tube, the first resonance is observed when the air column has a length of 20.0 cm for a tuning fork of frequency 400 Hz is used. The velocity of the sound at room temperature is 336 ms^{-1}. The third resonance is observed when the air column has a length of _____ cm.

[29 June 2022 (Evening)]

54. A wire of length 30 cm, stretched between rigid supports, has it's n^{th} and $(n + 1)^{th}$ harmonics at 400 Hz and 450 Hz, respectively. If tension in the string is 2700 N, it's linear mass density is _____ kg/m. **[27 July 2022 (Evening)]**

55. An organ pipe 40 cm long is open at both ends. The speed of sound in air is 360 ms^{-1}. The frequency of the second harmonic is _____ Hz. **[8 April 2023 (Morning)]**

56. A guitar string of length 90 cm vibrates with a fundamental frequency of 120 Hz. The length of the string producing a fundamental frequency of 180 Hz will be _____ cm.

[8 April 2023 (Evening)]

57. For a certain organ pipe, the first three resonance frequencies are in the ratio of 1 : 3 : 5 respectively. If the frequency of fifth harmonic is 405 Hz and the speed of sound in air is 324 ms^{-1} the length of the organ pipe is _____ m.

[12 April 2023 (Morning)]

58. In an experiment with sonometer when a mass of 180g is attached to the string, it vibrates with fundamental frequency of 30 Hz. When a mass m is attached, the string vibrates with fundamental frequency of 50 Hz. The value of m is _____ g.

[13 April 2023 (Evening)]

59. The fundamental frequency of vibration of a string stretched between two rigid support is 50 Hz. The mass of the string is 18 g and its linear mass density is 20 g/m. The speed of the transverse waves so produced in the string is _____ ms^{-1}.

[15 April 2023 (Morning)]

* * * * *

CHAPTER

14 WAVES

14.3 DOPPLER'S EFFECT

1. An observer moves towards a stationary source of sound, with a velocity one-fifth of the velocity of sound. What is the percentage increase in the apparent frequency ? **[2005]**

(A) Zero (B) 0.5%
(C) 5% (D) 20%

2. A whistle producing sound waves of frequencies 9500 Hz and above is approaching a stationary person with speed v ms^{-1}. The velocity of sound in air is 300 ms^{-1}. If the person can hear frequencies upto a maximum of 10,000 Hz, the maximum value of v upto which he can hear the whistle is : **[2006]**

(A) $15\sqrt{2}$ ms^{-1} (B) $15/\sqrt{2}$ ms^{-1}
(C) 15 ms^{-1} (D) 30 ms^{-1}

3. A motor cycle starts from rest and accelerates along a straight path at 2 m/s^2. At the starting point of the motor cycle there is a stationary electric siren. How far has the motor cycle gone when the driver hears the frequency of the siren at 94% of its value when the motor cycle was at rest ? (Speed of sound = 330 ms^{-1}) : **[2009]**

(A) 147 m (B) 196 m
(C) 49 m (D) 98 m

4. This question has Statement-1 and Statement-2. Of the four choices given after the Statements, choose the one that best describes the two Statements. **[2012 (Online)]**

Statement-1 : Bats emitting ultrasonic waves can detect the location of a prey by hearing the waves reflected from it.

Statement-2 : When the source and the detector are moving, the frequency of reflected waves is changed.

(A) Statement-1 is false, Statement-2 is true
(B) Statement-1 is true, Statement-2 is false
(C) Statement-1 is true, Statement-2 is true, Statement-2 is not the correct explanation of Statement-1
(D) Statement-1 is true, Statement-2 is true, Statement-2 is the correct explanation of Statement-1

5. A and B are two sources generating sound waves. A listener is situated at C. The frequency of the source at A is 500 Hz. A, now, moves towards C with a speed 4 m/s. The number of beats heard at C is 6. When A moves away from C with speed 4 m/s, the number of beats heard at C is 18. The speed of sound is 340 m/s. The frequency of the source at B is : **[2013 (Online)]**

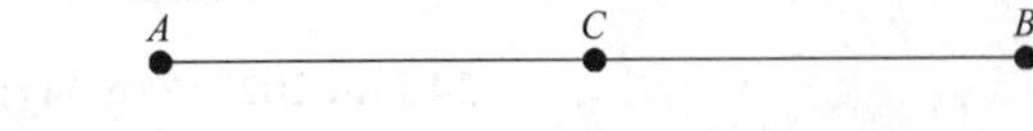

(A) 500 Hz (B) 506 Hz
(C) 512 Hz (D) 494 Hz

6. Two factories are sounding their sirens at 800 Hz. A man goes from one factory to other at a speed of 2 m/s. The velocity of sound is 320 m/s. The number of beats heard by the person in one second will be : **[2014 (Online)]**

(A) 2 (B) 4
(C) 8 (D) 10

7. A source of sound A emitting waves of frequency 1800 Hz is falling towards ground with a terminal speed v. The observer B on the ground directly beneath the source receives waves of frequency 2150 Hz. The source A receives waves, reflected from ground of frequency nearly : (Speed of sound = 343 m/s) **[2014 (Online)]**

(A) 2150 Hz (B) 2500 Hz
(C) 1800 Hz (D) 2400 Hz

8. A train is moving on a straight track with speed 20 ms^{-1}. It is blowing its whistle at the frequency of 1000 Hz. The percentage change in the frequency heard by a person standing near the track as the train passes him is (speed of sound = 320 ms^{-1}) close to : **[2015]**

(A) 6% (B) 12%
(C) 18% (D) 24%

9. A bat moving at 10 ms^{-1} towards a wall sends a sound signal of 8000 Hz towards it. On reflection it hears a sound of frequency f. The value of f in Hz is close to (speed of sound = 320 ms^{-1}) **[2015 (Online)]**

(A) 8516 (B) 8258
(C) 8424 (D) 800

10. A source of sound emits sound waves at frequency f_0. It is moving towards an observer with fixed speed vs ($vs < v$, where v is the speed of sound in air). If the observer were to move towards the source with speed v_0, one of the following two graphs (A and B) will given the correct variation of the frequency f heard by the observer as v_0 is changed : **[2015 (Online)]**

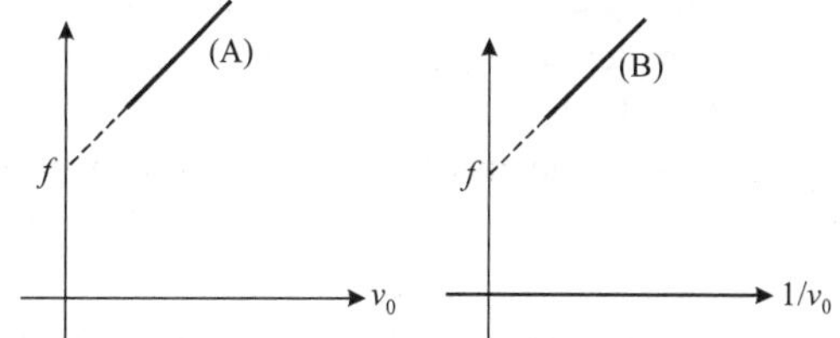

The variation of f with v_0 is given correctly by :

(A) Graph A with slope = $\dfrac{f_0}{(v+v_s)}$

(B) Graph B with slope = $\dfrac{f_0}{(v-v_s)}$

(C) Graph A with slope = $\dfrac{f_0}{(v-v_s)}$

(D) Graph B with slope = $\dfrac{f_0}{(v+v_s)}$

11. Two engines pass each other moving in opposite directions with uniform speed of 30 m/s. One of them is blowing a whistle of frequency 540 Hz. Calculate the frequency heard by driver of second engine before they pass each other. Speed of sound is 330 m/sec : **[2016 (Online)]**

(A) 450 Hz (B) 540 Hz
(C) 270 Hz (D) 648 Hz

12. A toy-car, blowing its horn, is moving with a steady speed of 5 m/s, away from a wall. An observer, towards whom the toy car is moving, is able to hear 5 beats per second. If the velocity of sound in air is 340 m/s, the frequency of the horn of the toy car is close to : **[2016 (Online)]**

(A) 680 Hz (B) 510 Hz
(C) 340 Hz (D) 170 Hz

13. An observer is moving with half the speed of light towards a stationary microwave source emitting waves at frequency 10 GHz. What is the frequency of the microwave measured by the observer ? (Speed of light = 3×10^8 ms^{-1}) : **[2017]**

(A) 17.3 GHz (B) 15.3 GHz
(C) 10.1 GHz (D) 12.1 GHz

14. A train moves towards a stationary observer with speed 34 m/s. The train sounds a whistle and its frequency registered by the observer is f_1. If the speed of the train is reduced to 17 m/s, the frequency registered is f_2. If speed of sound is 340 m/s, then the ratio f_1/f_2 is : **[10 Jan. 2019 (Morning)]**

(A) 18/17 (B) 19/18
(C) 20/19 (D) 21/20

15. A person standing on an open ground hears the sound of a jet aeroplane, coming from north at an angle 60° with ground level. But he finds the aeroplane right vertically above his position. If v is the speed of sound, speed of the plane is : **[12 Jan. 2019 (Morning)]**

(A) $\dfrac{2v}{\sqrt{3}}$ (B) v

(C) $\dfrac{v}{2}$ (D) $\dfrac{\sqrt{3}}{2}v$

16. Two cars A and B are moving away from each other in opposite directions. Both the cars are moving with a speed of 20 ms^{-1} with respect to the ground. If an observer in car A detects a frequency 2000 Hz of the sound coming from car B, what is the natural frequency of the sound source in car B ? (speed of sound in air = 340 ms^{-1}) **[9 April 2019 (Evening)]**

(A) 2250 Hz (B) 2060 Hz
(C) 2150 Hz (D) 2300 Hz

17. A stationary source emits sound waves of frequency 500 Hz. Two observers moving along a line passing through the source detect sound to be of frequencies 480 Hz and 530 Hz. Their respective speeds are, in ms^{-1}, (Given speed of sound = 300 m/s) : **[10 April 2019 (Morning)]**

(A) 16, 14 (B) 12, 18
(C) 12, 16 (D) 8, 18

18. A source of sound S is moving with a velocity of 50 m/s towards a stationary observer. The observer measures the frequency of the source as 1000 Hz. What will be the apparent frequency of the source when it is moving away from the observer after crossing him ? (Take velocity of sound in air is 350 m/s) **[10 April 2019 (Evening)]**

(A) 857 Hz (B) 807 Hz
(C) 750 Hz (D) 1143 Hz

19. A submarine (A) travelling at 18 km/hr is being chased along the line of its velocity by another submarine (B) travelling at 27 km/hr. B sends a sonar signal of 500 Hz to detect A and receives a reflected sound of frequency ν. The value of ν is close to : (Speed of sound in water = 1500 ms^{-1}) : **[12 April 2019 (Morning)]**

(A) 499 Hz (B) 502 Hz
(C) 507 Hz (D) 504 Hz

20. Two sources of sound S_1 and S_2 produce sound waves of same frequency 660 Hz. A listener is moving from source S_1 towards S_2 with a constant speed u m/s and he hears 10 beats/s. The velocity of sound is 330 m/s. Then, u equals : **[12 April 2019 (Evening)]**

(A) 2.5 m/s (B) 15.0 m/s
(C) 5.5 m/s (D) 10.0 m/s

21. A stationary observer receives sound from two identical tuning forks, one of which approaches and the other one recedes with the same speed (much less than the speed of sound). The observer hears 2 beats/sec. The oscillation frequency of each tuning fork is $v_0 = 1400$ Hz and the velocity of sound in air is 350 m/s. The speed of each tuning fork is close to :

[7 Jan. 2020 (Evening)]

(A) $\frac{1}{8}$ m/s (B) $\frac{1}{2}$ m/s

(C) 1 m/s (D) $\frac{1}{4}$ m/s

22. The driver of a bus approaching a big wall notices that the frequency of his bus's horn changes from 420 Hz to 490 Hz, when he hears it after it gets reflected from the wall. Find the speed of the bus if speed of the sound is 330 ms^{-1} :

[4 Sept. 2020 (Evening)]

(A) 91 kmh^{-1} (B) 71 kmh^{-1}

(C) 81 kmh^{-1} (D) 61 kmh^{-1}

23. A driver in a car, approaching a vertical wall notices that the frequency of his car horn, has changed from 440 Hz to 480 Hz, when it gets reflected from the wall. If the speed of sound in air is 345 m/s, then the speed of the car is :

[5 Sept. 2020 (Evening)]

(A) 36 km/hr (B) 24 km/hr

(C) 18 km/hr (D) 54 km/hr

24. A sound source S is moving along a straight track with speed v, and is emitting sound of frequency v_0 (see figure). An observer is standing at a finite distance, at the point O, from the track. The time variation of frequency heard by the observer is best represented by :

(t_0 represents the instant when the distance between the source and observer is minimum) **[6 Sept. 2020 (Morning)]**

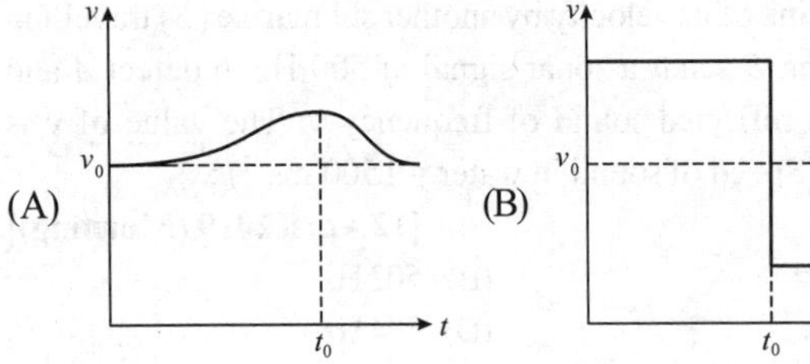

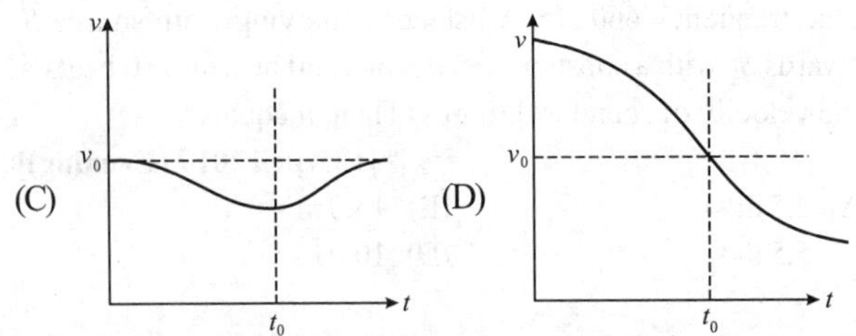

25. Two cars are approaching each other at an equal speed of 7.2 km/hr. When they see each other, both blow horns having frequency of 676 Hz. The beat frequency heard by each driver will be ______ Hz. [Velocity of sound in air is 340 m/s]

[24 Feb. 2021 (Evening)]

26. A galaxy is moving away from the earth at a speed of 286 kms^{-1}. The shift in the wavelength of a red line at 630 nm is $x \times 10^{-10}$ m. The value of x, to the nearest integer, is ______. [Take the value of speed of light c, as 3×10^8 ms^{-1}]

[18 March 2021 (Evening)]

27. The frequency of a car horn encountered a change from 400 Hz to 500 Hz, when the car approaches a vertical wall. If the speed of sound is 330 m/s. Then the speed of car is ______ km/h. **[20 July 2021 (Morning)]**

28. With what speed should a galaxy move outward, with respect to earth so that the sodium-D line at wavelength 5890 Å is observed at 5896 Å : **[20 July 2021 (Evening)]**

(A) 322 km/sec (B) 296 km/sec

(C) 306 km/sec (D) 336 km/sec

29. A source and a detector move away from each other in absence of wind with a speed of 20 m/s with respect to the ground. If the detector detects a frequency of 1800 Hz of the sound coming from the source, then the original frequency of source considering speed of sound in air 340 m/s will be ______Hz. **[26 Aug. 2021 (Morning)]**

30. Two cars X and Y are approaching each other with velocities 36 km/h and 72 km/h respectively. The frequency of a whistle sound as emitted by a passenger in car X, heard by the passenger in car Y is 1320 Hz. If the velocity of sound in air is 340 m/s, the actual frequency of the whistle sound produced is ______Hz.

[27 Aug. 2021 (Morning)]

31. For a specific wavelength 670 nm of light coming from a galaxy moving with velocity v, the observed wavelength is 670.7 nm. The value of v is : **[26 June 2022 (Evening)]**

(A) 3×10^8 ms^{-1} (B) 3×10^{10} ms^{-1}

(C) 3.13×10^5 ms^{-1} (D) 4.48×10^5 ms^{-1}

32. An observer moves towards a stationary source of sound with a velocity equal to one-fifth of the velocity of sound. The percentage change in the frequency will be :

[27 June 2022 (Morning)]

(A) 20% (B) 10%

(C) 5% (D) 0%

33. An observer is riding on a bicycle and moving towards a hill at 18 kmh^{-1}. He hears a sound from a source at some distance behind him directly as well as after its reflection from the hill. If the original frequency of the sound as emitted by source is 640 Hz and velocity of the sound in air is 320 m/s, the beat frequency between the two sounds heard by observer will be _____ Hz. **[25 July 2022 (Morning)]**

34. When a car is approaching the observer, the frequency of horn is 100 Hz. After passing the observer, it is 50 Hz. If the observer moves with the car, the frequency will be $\frac{x}{3}$ Hz where $x =$ _____. **[26 July 2022 (Morning)]**

35. The frequency of echo will be _____ Hz if the train blowing a whistle of frequency 320 Hz is moving with a velocity of 36 km/h towards a hill from which an echo is heard by the train driver. Velocity of sound in air is 330 m/s. **[28 July 2022 (Morning)]**

36. A train blowing a whistle of frequency 320 Hz approaches an observer standing on the platform at a speed of 66 m/s. The frequency observed by the observer will be (given speed of sound = 330 ms^{-1}) _____ Hz. **[25 Jan. 2023 (Evening)]**

37. A person observes two moving trains, *A* reaching the station and *B* leaving the station with equal speed of 30 m/s. If both trains emit sounds with frequency 300 Hz, (Speed of sound : 330 m/s) approximate difference of frequencies heard by the person will be : **[29 Jan. 2023 (Morning)]**

(A) 33 Hz (B) 55 Hz
(C) 80 Hz (D) 10 Hz

38. A person driving car at a constant speed of 15 m/s is approaching a vertical wall. The person notices a change of 40 Hz in the frequency of his car's horn upon reflection from the wall. The frequency of horn is _____ Hz.
[Given : Speed of sound : 330 m/s] **[6 April 2023 (Morning)]**

39. The engine of a train moving with speed 10 ms^{-1} towards a platform sounds a whistle at frequency 400 Hz. The frequency heard by a passenger inside the train is (neglect air speed. Speed of sound in air 330 ms^{-1}) **[8 April 2023 (Morning)]**

(A) 200 Hz (B) 400 Hz
(C) 412 Hz (D) 388 Hz

40. A car *P* travelling at 20 ms^{-1} sounds its horn at a frequency of 400 Hz. Another car *Q* is travelling behind the first car in the same direction with a velocity 40 ms^{-1}. The frequency heard by the passenger of the car *Q* is approximately: [Take, velocity of sound = 360 ms^{-1}] **[11 April 2023 (Evening)]**

(A) 514 Hz (B) 421 Hz
(C) 485 Hz (D) 471 Hz

* * * * *

CHAPTER

15 ELECTRIC CHARGES AND FIELDS

15.1 ELECTRIC FORCE AND ELECTRIC FIELD

1. If a charge q is placed at the centre of the line joining two equal charges Q such that the system is in equilibrium then the value of q is : **[2002]**

(A) $Q/2$ (B) $-Q/2$

(C) $Q/2$ (D) $-Q/4$

2. Three charges $-q_1$, $+q_2$ and $-q_3$ are place as shown in the figure. The x-component of the force on $-q_1$ is proportional to: **[2003]**

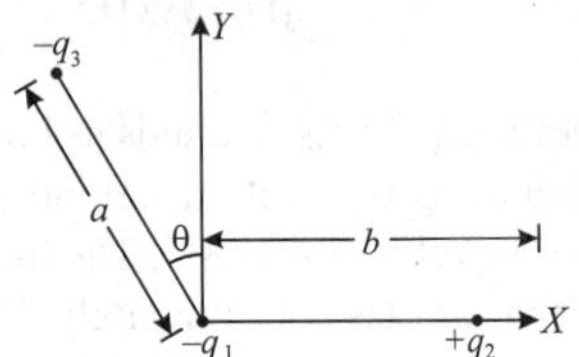

(A) $\frac{q_2}{b^2} - \frac{q_3}{a^2}\cos\theta$ (B) $\frac{q_2}{b^2} + \frac{q_3}{a^2}\sin\theta$

(C) $\frac{q_2}{b^2} + \frac{q_3}{a^2}\cos\theta$ (D) $\frac{q_2}{b^2} - \frac{q_3}{a^2}\sin\theta$

3. Four charges equal to $-Q$ are placed at the four corners of a square and a charge q is at its centre. If the system is in equilibrium, the value of q is : **[2004]**

(A) $-\frac{Q}{4}(1+2\sqrt{2})$ (B) $\frac{Q}{4}(1+2\sqrt{2})$

(C) $-\frac{Q}{2}(1+2\sqrt{2})$ (D) $\frac{Q}{2}(1+2\sqrt{2})$

4. Two spherical conductors B and C having equal radii and carrying equal charges in them, repel each other with a force F when kept apart at some distance. A third spherical conductor having same radius as that of B but uncharged, is brought in contact with B, then brought in contact with C and finally removed away from both. The new force of repulsion between B and C is : **[2004]**

(A) $\frac{F}{4}$ (B) $\frac{3F}{4}$

(C) $\frac{F}{8}$ (D) $\frac{3F}{8}$

5. A charged oil drop is suspended in a uniform field of 3×10^4 v/m so that it neither falls nor rises. The charge on the drop will be (Take the mass of the charge = 9.9×10^{-15} kg and $g = 10$ m/s^2) : **[2004]**

(A) 1.6×10^{-18} C (B) 3.2×10^{-18} C

(C) 3.3×10^{-18} C (D) 4.8×10^{-18} C

6. Two point charges $+8q$ and $-2q$ are located at $x = 0$ and $x = L$ respectively. The location of a point on the x-axis at which the net electric field due to these two point charges is zero is : **[2005]**

(A) $2L$ (B) $L/4$

(C) $8L$ (D) $4L$

7. A charged ball B hangs from a silk thread S, which makes an angle θ with a large charged conducting sheet P, as shown in the figure. The surface charge density σ of the sheet is proportional to : **[2005]**

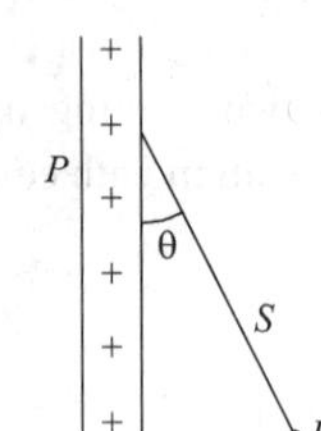

(A) $\cos\theta$ (B) $\cot\theta$

(C) $\sin\theta$ (D) $\tan\theta$

8. A thin spherical shell of radius R has a charge Q spread uniformly over its surface. Which of the following graphs most closely represents the electric field $E(r)$ produced by the shell in the range $0 \le r < \infty$, where r is the distance from the centre of the shell : **[2008]**

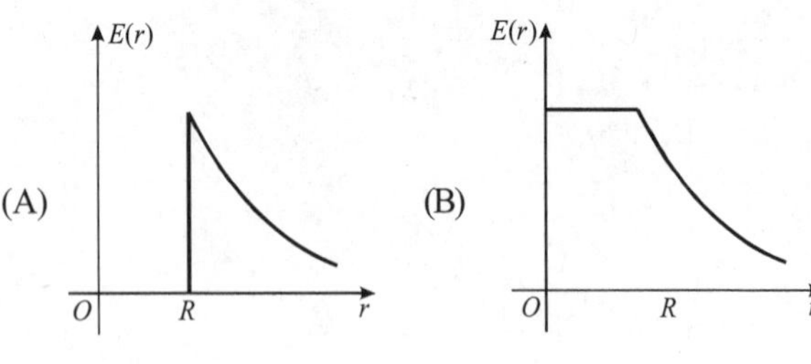

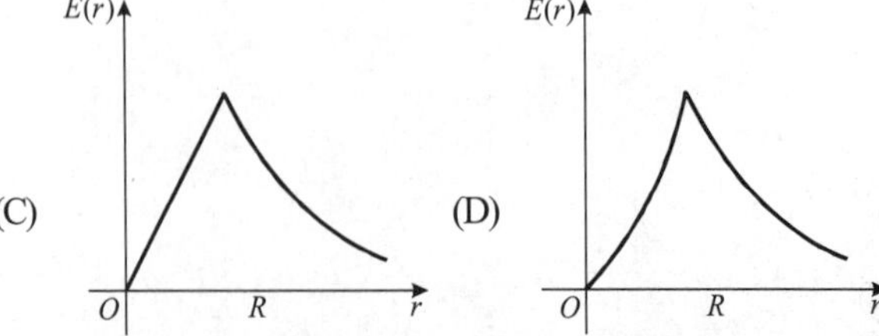

9. A charge Q is placed at each of the opposite corners of a square. A charge q is placed at each of the other two corners. If the net electrical force on Q is zero, then $\frac{Q}{q}$ equals : **[2009]**

(A) 1 (B) $-\frac{1}{\sqrt{2}}$

(C) $-2\sqrt{2}$ (D) -1

10. Let $\rho(r) = \frac{Q}{\pi R^4}r$ be the charge density distribution for a solid sphere of radius R and total charge Q. For a point 'p' inside the sphere at distance r_1 from the centre of the sphere, the magnitude of electric field is : **[2009]**

(A) $\frac{Qr_1^2}{4\pi \in_0 R^4}$ (B) $\frac{Qr_1^2}{3\pi \in_0 R^4}$

(C) 0 (D) $\frac{Q}{4\pi \in_0 r_1^2}$

11. Two identical charged spheres are suspended by strings of equal lengths. The strings make an angle of 30° with each other. When suspended in a liquid of density 0.8 g cm^{-3}, the angle remains the same. If density of the material of the sphere is 1.6 g cm^{-3}, the dielectric constant of the liquid is : **[2010]**

(A) 1 (B) 4

(C) 3 (D) 2

12. Let there be a spherically symmetric charge distribution with charge density varying as $\rho(r) = \rho_0\left(\frac{5}{4} - \frac{r}{R}\right)$ upto $r = R$, and $\rho(r) = 0$ for $r > R$, where r is the distance from the origin. The electric field at a distance r $(r < R)$ from the origin is given by : **[2010]**

(A) $\frac{\rho_0 r}{3\in_0}\left(\frac{5}{4} - \frac{r}{R}\right)$ (B) $\frac{4\pi\rho_0 r}{3\in_0}\left(\frac{5}{3} - \frac{r}{R}\right)$

(C) $\frac{\rho_0 r}{4\in_0}\left(\frac{5}{3} - \frac{r}{R}\right)$ (D) $\frac{4\rho_0 r}{3\in_0}\left(\frac{5}{4} - \frac{r}{R}\right)$

13. A thin semi-circular ring of radius r has a positive charge q distributed uniformly over it. The net field $\vec{E}$ at the centre O is: **[2010]**

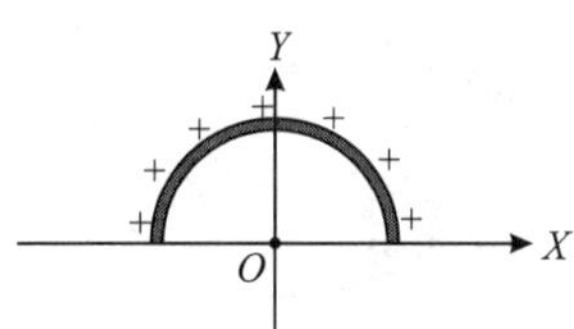

(A) $\frac{q}{2\pi^2 \in_0 r^2}\hat{j}$ (B) $\frac{q}{4\pi^2 \in_0 r^2}\hat{j}$

(C) $-\frac{q}{4\pi^2 \in_0 r^2}\hat{j}$ (D) $-\frac{q}{2\pi^2 \in_0 r^2}\hat{j}$

14. Two identical charged spheres suspended from a common point by two massless string of length l are initially a distance $d(d << l)$ apart because of their mutual repulsion. The charge begins to leak from both the spheres at a constant rate. As a result the charges approach each other with a velocity v. Then as a function of distance x between them : **[2011]**

(A) $v \propto x^{1/x}$ (B) $v \propto x$

(C) $v \propto x^{-1/2}$ (D) $v \propto x^{-1}$

15. In a uniformly charged sphere of total charge Q and radius R, the electric field E is plotted as a function of distance from the centre. The graph which would correspond to the above will be : **[2012]**

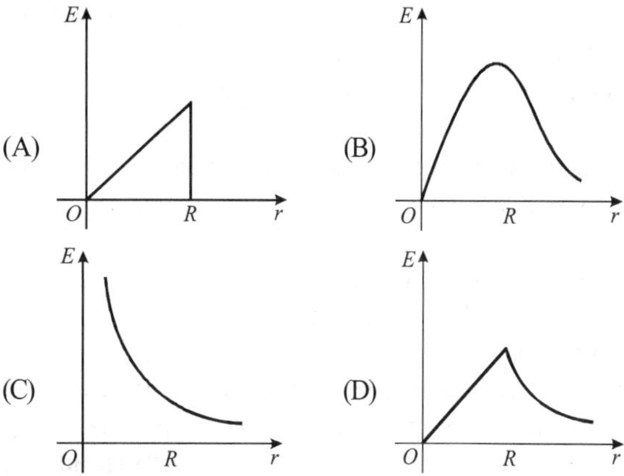

16. Two charges, each equal to q, are kept at $x = -a$ and $x = a$ on the x-axis. A particle of mass m and charge $q_0 = \frac{q}{2}$ is placed at the origin. If charge q_0 is given a small displacement $(y << a)$ along the y-axis, the net force acting on the particle is proportional to : **[2013]**

(A) y (B) $-y$

(C) $\frac{1}{y}$ (D) $-\frac{1}{y}$

17. A liquid drop having 6 excess electrons is kept stationary under a uniform electric field of 25.5 kVm^{-1}. The density of liquid is 1.26×10^3 kg m^{-3}. The radius of the drop is (neglect buoyancy) : **[2013 (Online)]**

(A) 4.3×10^{-7} m (B) 7.8×10^{-7} m

(C) 0.078×10^{-7} m (D) 3.4×10^{-7} m

18. The surface charge density of a thin charged disc of radius R is σ. The value of the electric field at the centre of the disc is $\frac{\sigma}{2\in_0}$. With respect to the field at the centre, the electric field along the axis at a distance R from the centre of the disc : **[2013 (Online)]**

(A) Reduces by 70.7% (B) Reduces by 29.3%

(C) Reduces by 9.7% (D) Reduces by 14.6%

19. Two balls of same mass and carrying equal charge are hung from a fixed support of length l. At electrostatic equilibrium, assuming that angles made by each thread is small, the separation, x between the balls is proportional to : **[2013 (Online)]**

(A) l (B) l^2
(C) $l^{2/3}$ (D) $l^{1/3}$

20. The magnitude of the average electric field normally present in the atmosphere just above the surface of the Earth is about 150 N/C, directed inward towards the center of the Earth. This gives the total net surface charge carried by the Earth to be :
[Given $\varepsilon_0 = 8.85 \times 10^{-12}$ C^2/N-m^2, $R_E = 6.37 \times 10^6$ m] **[2014 (Online)]**

(A) +670 kC (B) −670 kC
(C) −680 kC (D) +680 kC

21. A spherically symmetric charge distribution is characterised by a charge density having the following variations :

$$\rho(r) = \rho_0\left(1-\frac{r}{R}\right) \text{ for } r < R$$

$$\rho(r) = 0 \text{ for } r \ge R$$

Where r is the distance from the centre of the charge distribution ρ_0 is a constant. The electric field at an internal point ($r < R$) is: **[2014 (Online)]**

(A) $\frac{\rho_0}{4\varepsilon_0}\left(\frac{r}{3}-\frac{r^2}{4R}\right)$ (B) $\frac{\rho_0}{\varepsilon_0}\left(\frac{r}{3}-\frac{r^2}{4R}\right)$

(C) $\frac{\rho_0}{3\varepsilon_0}\left(\frac{r}{3}-\frac{r^2}{4R}\right)$ (D) $\frac{\rho_0}{12\varepsilon_0}\left(\frac{r}{3}-\frac{r^2}{4R}\right)$

22. A thin disc of radius $b = 2a$ has a concentric hole of radius 'a' in it (see figure). It carries uniform surface charge 'σ' on it. If the electric field on its axis at height 'h' ($h << a$) from its centre is given as 'Ch' then value of 'C' is : **[2015 (Online)]**

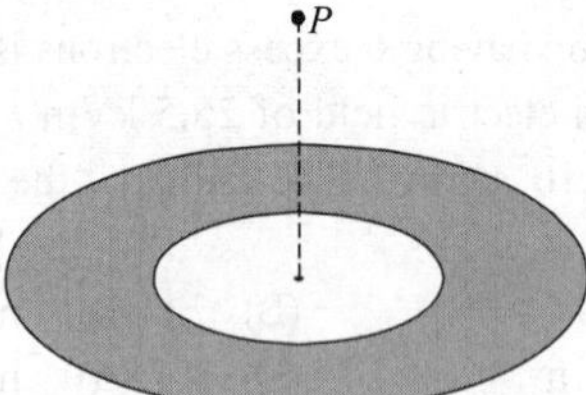

(A) $\frac{\sigma}{4a\,\epsilon_0}$ (B) $\frac{\sigma}{8a\,\epsilon_0}$

(C) $\frac{\sigma}{a\,\epsilon_0}$ (D) $\frac{\sigma}{2a\,\epsilon_0}$

23. A wire of length L (= 20 cm), is bent into a semicircular arc. If the two equal halves of the arc were each to be uniformly charged with charges $\pm Q$, [$|Q| = 10^3\varepsilon_0$ Coulomb where ε_0 is the permittivity (in SI units) of free space] the net electric field at the centre O of the semicircular arc would be : **[2015 (Online)]**

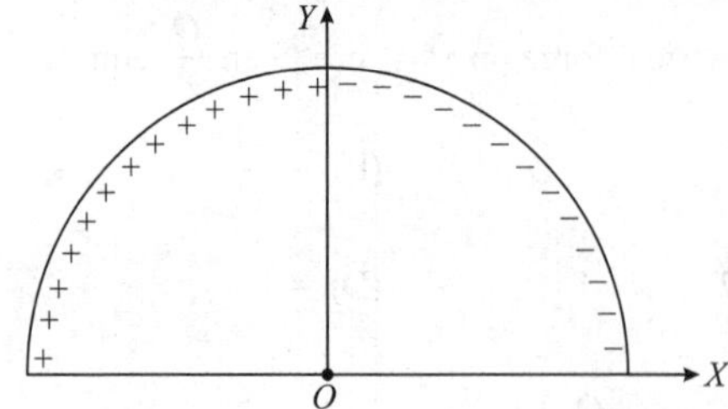

(A) $(50 \times 10^3 \text{ N/C})\,\hat{j}$ (B) $(50 \times 10^3 \text{ N/C})\,\hat{i}$
(C) $(25 \times 10^3 \text{ N/C})\,\hat{j}$ (D) $(25 \times 10^3 \text{ N/C})\,\hat{i}$

24. Shown in the figure are two point charges $+Q$ and $-Q$ inside the cavity of a spherical shell. The charges are kept near the surface of the cavity on opposite sides of the centre of the shell. If σ_1 is the surface charge on the inner surface and Q_1 net charge on it and σ_2 the surface charge on the outer surface and Q_2 net charge on it then : **[2015 (Online)]**

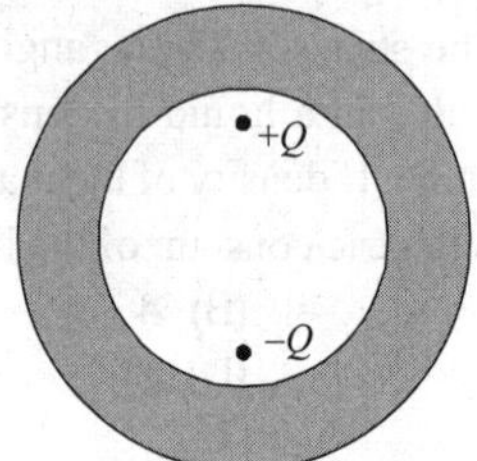

(A) $\sigma_1 \ne 0, Q_1 = 0$; $\sigma_2 \ne 0, Q_2 = 0$ (B) $\sigma_1 \ne 0, Q_1 = 0$; $\sigma_2 = 0, Q_2 = 0$
(C) $\sigma_1 = 0, Q_1 = 0$; $\sigma_2 = 0, Q_2 = 0$ (D) $\sigma_1 \ne 0, Q_1 \ne 0$; $\sigma_2 \ne 0, Q_2 \ne 0$

25. The region between two concentric spheres of radii 'a' and 'b' respectively (see figure), has volume charge density $\rho = \frac{A}{r}$, where A is a constant and r is the distance from the centre. At the centre of the spheres is a point charge Q. The value of A such that the electric field in the region between the spheres will be constant, is : **[2016]**

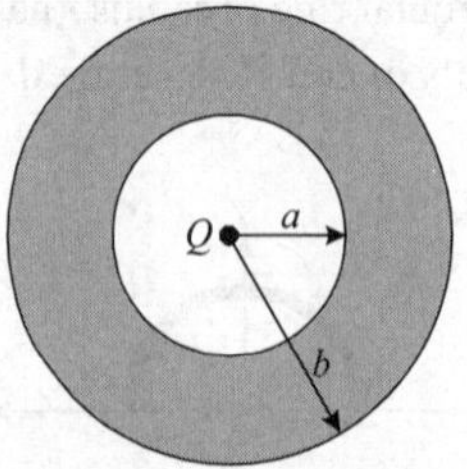

(A) $\frac{Q}{2\pi a^2}$ (B) $\frac{Q}{2\pi(b^2-a^2)}$

(C) $\frac{2Q}{\pi(a^2-b^2)}$ (D) $\frac{2Q}{\pi a^2}$

26. A solid ball of radius R has a charge density ρ given by $\rho = \rho_0\left(1-\frac{r}{R}\right)$ for $0 \le r \le R$. The electric field outside the ball is : **[2018 (Online)]**

(A) $\frac{\rho_0 R^3}{\epsilon_0 r^2}$ (B) $\frac{4\rho_0 R^3}{3\epsilon_0 r^2}$

(C) $\frac{3\rho_0 R^3}{4\epsilon_0 r^2}$ (D) $\frac{\rho_0 R^3}{12\epsilon_0 r^2}$

27. A body of mass M and charge q is connected to a spring of spring constant k. It is oscillating along x-direction about its equilibrium position, taken to be at $x = 0$, with an amplitude A. An electric field E is applied along the x-direction. Which of the following statements is correct ? **[2018 (Online)]**

(A) The total energy of the system is $\frac{1}{2}m\omega^2A^2 + \frac{1}{2}\frac{q^2E^2}{k}$

(B) The new equilibrium position is at a distance : $\frac{2qE}{k}$ from $x=0$

(C) The new equilibrium position is at a distance : $\frac{qE}{2k}$ from $x=0$

(D) The total energy of the system is $\frac{1}{2}m\omega^2A^2 - \frac{1}{2}\frac{q^2E^2}{k}$

28. Two identical conducting sphere A and B, carry equal charge. They are separated by a distance much larger than their diameter, and the force between them is F. A third identical conducting sphere, C, is uncharged. Sphere C is first touched to A, then to B, and then removed. As a result, the force between A and B would be equal to : **[2018 (Online)]**

(A) $\frac{3F}{4}$ (B) $\frac{F}{2}$

(C) F (D) $\frac{3F}{8}$

29. Three charges $+Q$, q, $+Q$ are placed respectively, at distance 0, $d/2$ and d from the origin, on the x-axis. If the net force experienced by $+Q$, placed at $x = 0$, is zero, then value of q is : **[9 Jan. 2019 (Morning)]**

(A) $+Q/2$ (B) $-Q/2$

(C) $-Q/4$ (D) $+Q/4$

30. For a uniformly charged ring of radius R, the electric field on its axis has the largest magnitude at a distance h from its centre. Then value of h is : **[9 Jan. 2019 (Morning)]**

(A) $\frac{R}{\sqrt{5}}$ (B) R

(C) $\frac{R}{\sqrt{2}}$ (D) $R\sqrt{2}$

31. Two point charges $q_1(\sqrt{10}\ \mu C)$ and $q_2(-25\ \mu C)$ are placed on the x-axis at $x = 1$ m and $x = 4$ m respectively. The electric field (in V/m) at a point $y = 3$ m on y-axis is :

$\left[\text{take } \frac{1}{4\pi\varepsilon_0} = 9\times10^9\ \text{Nm}^2\text{C}^{-2}\right]$ **[9 Jan. 2019 (Evening)]**

(A) $(-63\hat{i}+27\hat{j})\times10^2$ (B) $(81\hat{i}-81\hat{j})\times10^2$

(C) $(63\hat{i}-27\hat{j})\times10^2$ (D) $(-81\hat{i}+81\hat{j})\times10^2$

32. Charge is distributed within a sphere of radius R with a volume charge density $\rho(r) = \frac{A}{r^2}e^{-2r/a}$, where A and a are constants. If Q is the total charge of this charge distribution, the radius R is : **[9 Jan. 2019 (Evening)]**

(A) $\frac{a}{2}\log\left(1-\frac{Q}{2\pi aA}\right)$ (B) $a\log\left(1-\frac{Q}{2\pi aA}\right)$

(C) $a\log\left(\frac{1}{1-\frac{Q}{2\pi aA}}\right)$ (D) $\frac{a}{2}\log\left(\frac{1}{1-\frac{Q}{2\pi aA}}\right)$

33. The bob of a simple pendulum has mass 2 g and a charge of 5.0 μC. It is at rest in a uniform horizontal electric field of intensity 2000 V/m. At equilibrium, the angle that the pendulum makes with the vertical is : (take $g = 10$ m/s^2) **[8 April 2019 (Morning)]**

(A) $\tan^{-1}(5.0)$ (B) $\tan^{-1}(2.0)$

(C) $\tan^{-1}(0.5)$ (D) $\tan^{-1}(0.2)$

34. Four point charges $-q$, $+q$, $+q$ and $-q$ are placed on y-axis at $y = -2d$, $y = -d$, $y = +d$ and $y = +2d$, respectively. The magnitude of the electric field E at a point on the x-axis at $x = D$, with $D >> d$, will behave as : **[9 April 2019 (Evening)]**

(A) $E \propto \frac{1}{D}$ (B) $E \propto \frac{1}{D^3}$

(C) $E \propto \frac{1}{D^2}$ (D) $E \propto \frac{1}{D^4}$

35. Let a total charge $2Q$ be distributed in a sphere of radius R, with the charge density given by $\rho(r) = kr$, where r is the distance from the centre. Two charges A and B, of $-Q$ each, are placed on diametrically opposite points, at equal distance a from the centre. If A and B do not experience any force, then : **[12 April 2019 (Evening)]**

(A) $a = \frac{3R}{2^{1/4}}$ (B) $a = R/\sqrt{3}$

(C) $a = 8^{-1/4}R$ (D) $a = 2^{-1/4}R$

36. Three charged particle A, B and C with charges $-4q$, $2q$ and $-2q$ are present on the circumference of a circle of radius d. The charged particles A, C and centre O of the circle formed an equilateral triangle as shown in figure. Electric field at O along x-direction is : **[8 Jan. 2020 (Morning)]**

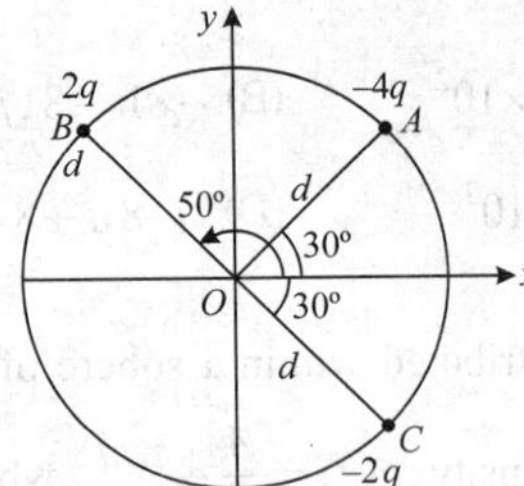

(A) $\dfrac{2\sqrt{3}q}{\pi\varepsilon_0 d^2}$ (B) $\dfrac{\sqrt{3}q}{4\pi\varepsilon_0 d^2}$

(C) $\dfrac{3\sqrt{3}q}{4\pi\varepsilon_0 d^2}$ (D) $\dfrac{\sqrt{3}q}{\pi\varepsilon_0 d^2}$

37. A particle of mass m and charge q is released from rest in a uniform electric field. If there is no other force on the particle, the dependence of its speed v on the distance x travelled by it is correctly given by (graphs are schematic and not drawn to scale) : **[8 Jan. 2020 (Evening)]**

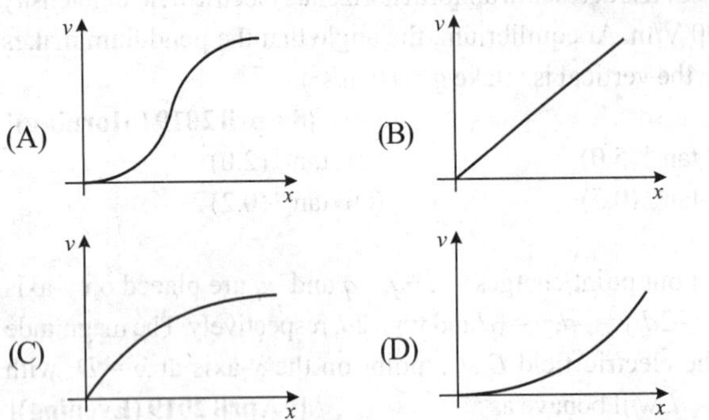

38. Consider a sphere of radius R which carries a uniform charge density ρ. If a sphere of radius $\frac{R}{2}$ is carved out of it, as shown, the ratio $\frac{|\vec{E}_A|}{|\vec{E}_B|}$ of magnitude of electric field $\vec{E}_A$ and $\vec{E}_B$, respectively, at points A and B due to the remaining portion is : **[9 Jan. 2020 (Morning)]**

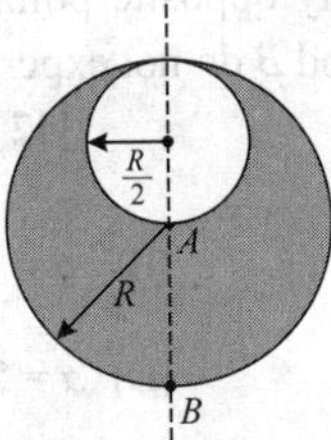

(A) $\frac{18}{54}$ (B) $\frac{21}{34}$

(C) $\frac{17}{54}$ (D) $\frac{18}{34}$

39. A charge Q is distributed over two concentric conducting thin spherical shells radii r and R $(R > r)$. If the surface charge densities on the two shells are equal, the electric potential at the common centre is : **[2 Sept. 2020 (Evening)]**

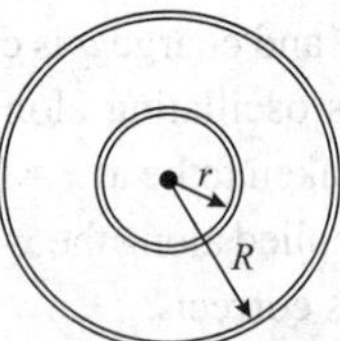

(A) $\dfrac{1}{4\pi\varepsilon_0}\dfrac{(R+2r)Q}{2(R^2+r^2)}$ (B) $\dfrac{1}{4\pi\varepsilon_0}\dfrac{(R+r)}{2(R^2+r^2)}Q$

(C) $\dfrac{1}{4\pi\varepsilon_0}\dfrac{(R+r)}{(R^2+r^2)}Q$ (D) $\dfrac{1}{4\pi\varepsilon_0}\dfrac{(2R+r)}{(R^2+r^2)}Q$

40. A small point mass carrying some positive charge on it, is released from the edge of a table. There is a uniform electric field in this region in the horizontal direction. Which of the following options then correctly describe the trajectory of the mass : (Curves are drawn schematically and are not to scale) **[2 Sept. 2020 (Evening)]**

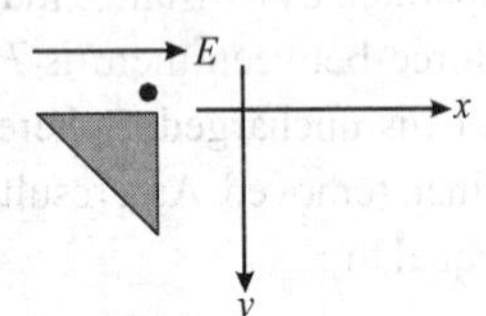

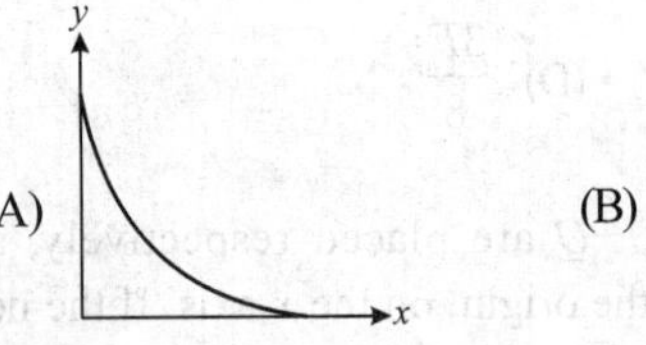

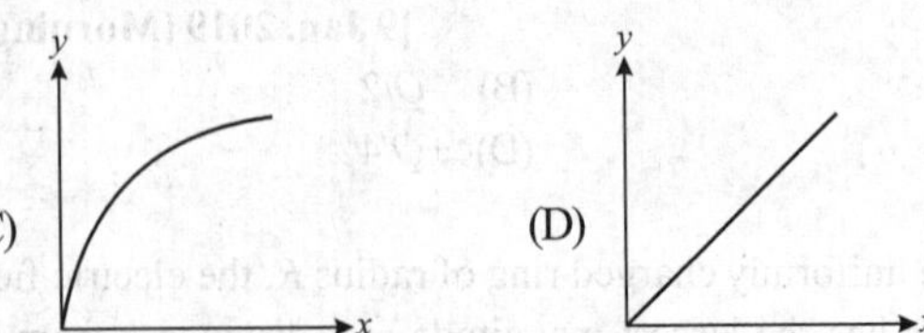

41. Charges Q_1 and Q_2 are at points A and B of a right angle triangle OAB (see figure). The resultant electric field at point O is perpendicular to the hypotenuse, then Q_1/Q_2 is proportional to : **[6 Sept. 2020 (Morning)]**

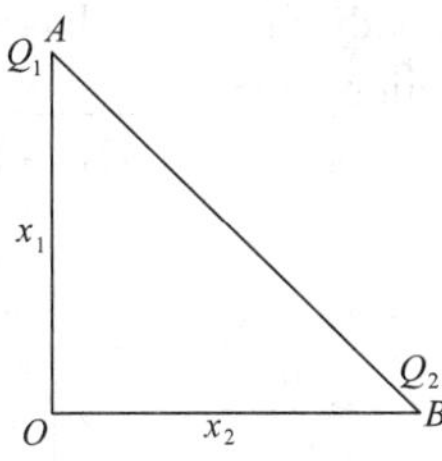

(A) $\frac{x_2^2}{x_1^2}$ (B) $\frac{x_1^3}{x_2^3}$

(C) $\frac{x_1}{x_2}$ (D) $\frac{x_2}{x_1}$

42. Consider the force F on a charge q due to a uniformly charged spherical shell of radius R carrying charge Q distributed uniformly over it. Which one of the following statements is true for F, if q is placed at distance r from the centre of the shell :

[6 Sept. 2020 (Evening)]

(A) $F = \frac{1}{4\pi\varepsilon_0}\frac{Qq}{r^2}$ for $r > R$

(B) $\frac{1}{4\pi\varepsilon_0}\frac{qQ}{R^2} > F > 0$ for $r < R$

(C) $F = \frac{1}{4\pi\varepsilon_0}\frac{Qq}{r^2}$ for all r

(D) $F = \frac{1}{4\pi\varepsilon_0}\frac{Qq}{R^2}$ for $r < R$

43. A cube of side a has point charges $+Q$ located at each of its vertices except at the origin where the charge is $-Q$. The electric field at the centre of cube is : **[24 Feb. 2021 (Morning)]**

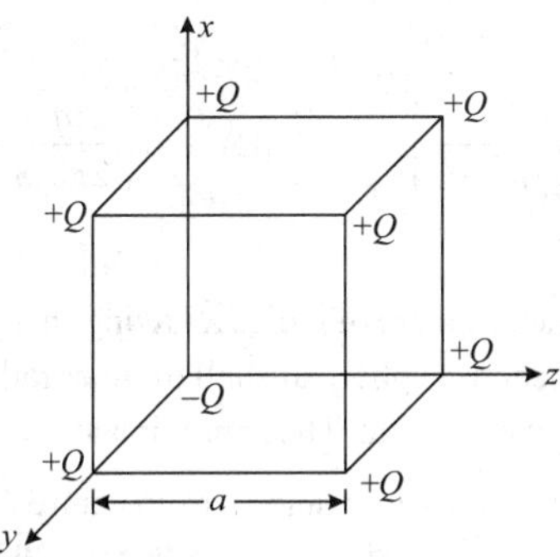

(A) $\frac{2Q}{3\sqrt{3}\pi\varepsilon_0 a^2}(\hat{x}+y+\hat{z})$ (B) $\frac{Q}{3\sqrt{3}\pi\varepsilon_0 a^2}(\hat{x}+y+\hat{z})$

(C) $\frac{-2Q}{3\sqrt{3}\pi\varepsilon_0 a^2}(\hat{x}+y+\hat{z})$ (D) $\frac{-Q}{3\sqrt{3}\pi\varepsilon_0 a^2}(\hat{x}+y+\hat{z})$

44. Two electrons each are fixed at a distance $2d$. A third charge proton placed at the midpoint is displaced slightly by a distance $x(x << d)$ perpendicular to the line joining the two fixed charges. Proton will execute simple harmonic motion having angular frequency : (m = mass of charged particle) :

[24 Feb. 2021 (Evening)]

(A) $\left(\frac{\pi\varepsilon_0 md^3}{2q^2}\right)^{1/2}$ (B) $\left(\frac{q^2}{2\pi\varepsilon_0 md^3}\right)^{1/2}$

(C) $\left(\frac{2q^2}{\pi\varepsilon_0 md^3}\right)^{1/2}$ (D) $\left(\frac{2\pi\varepsilon_0 md^3}{q^2}\right)^{1/2}$

45. Two identical conducting spheres with negligible volume have 2.1 nC and – 0.1 nC charges, respectively. They are brought into contact and then separated by a distance of 0.5 m. The electrostatic force acting between the spheres is _____ $\times 10^{-9}$ N.

[Given : $4\pi\varepsilon_0 = \frac{1}{9\times10^9}$ SI unit] **[25 Feb. 2021 (Evening)]**

46. Two small spheres each of mass 10 mg are suspended from a point by threads 0.5 m long. They are equally charged and repel each other to a distance of 0.20 m. The charge on each of the sphere is $\frac{a}{21} \times 10^{-8}$ C. The value of a will be ______.

[Given $g = 10\ \text{ms}^{-2}$] **[25 Feb. 2021 (Evening)]**

47. Find the electric field at point P (as shown in figure) on the perpendicular bisector of a uniformly charged thin wire of length L carrying a charge Q. The distance of the point P from the centre of the rod is $a = \frac{\sqrt{3}}{2}L$: **[26 Feb. 2021 (Morning)]**

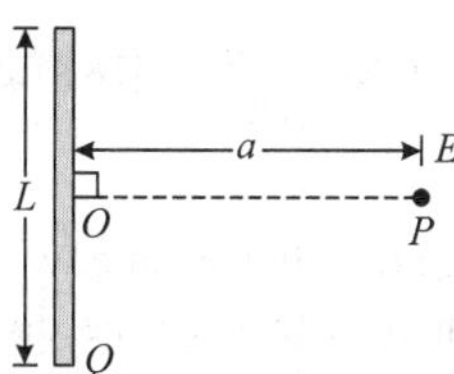

(A) $\frac{Q}{4\pi\varepsilon_0 L^2}$ (B) $\frac{\sqrt{3}Q}{4\pi\varepsilon_0 L^2}$

(C) $\frac{Q}{2\sqrt{3}\pi\varepsilon_0 L^2}$ (D) $\frac{Q}{3\pi\varepsilon_0 L^2}$

48. An inclined plane making an angle of 30° with the horizontal is placed in a uniform horizontal electric field $200\frac{N}{C}$ as shown in the figure. A body of mass 1 kg and charge 5 mC is allowed to slide down from rest at a height of 1 m. If the coefficient of friction is 0.2, find the time taken by the body to reach the bottom. **[26 Feb. 2021 (Evening)]**

$[g = 9.8\ m/s^2;\ \sin 30° = \frac{1}{2};\ \cos 30° = \frac{\sqrt{3}}{2}]$

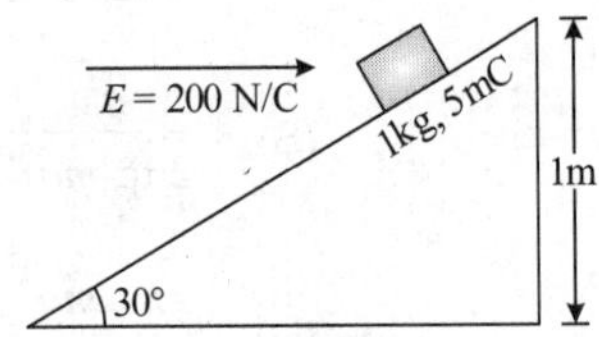

(A) 0.46 s (B) 2.3 s
(C) 0.92 s (D) 1.3 s

49. Find out the surface charge density at the intersection of point $x = 3$ m plane and x-axis, in the region of uniform line charge of 8 nC/m lying along the z-axis in free space. **[16 March 2021 (Evening)]**

(A) $0.424\ nC\ m^{-2}$ (B) 47.88 C/m
(C) $0.07\ nC\ m^{-2}$ (D) $4.0\ nC\ m^{-2}$

50. An oil drop of radius 2 mm with a density $3g\ cm^{-3}$ is held stationary under a constant electric field $3.55 \times 10^5\ V\ m^{-1}$ in the Millikan's oil drop experiment. What is the number of excess electrons that the oil drop will possess ?
(Consider $g = 9.81\ m/s^2$) **[18 March 2021 (Morning)]**

(A) 48.8×10^{11} (B) 1.73×10^{10}
(C) 17.3×10^{10} (D) 1.73×10^{12}

51. An infinite number of point charges, each carrying 1 μC charge, are placed along the y-axis at $y = 1$ m, 2 m, 4 m, 8 m .... The total force on a 1 C point charge, placed at the origin, is $x \times 10^3$ N. The value of x, to the nearest integer, is______.

[Take $\frac{1}{4\pi \epsilon_0} = 9 \times 10^9 Nm^2/C^2$] **[18 March 2021 (Evening)]**

52. A certain charge Q is divided into two parts q and $(Q - q)$. How should the charges Q and q be divided so that q and $(Q - q)$. Placed at a certain distance apart experience maximum electrostatics repulsion : **[20 July 2021 (Morning)]**

+q ——— r ——— Q − q

(A) $Q = 4q$ (B) $Q = 4q$
(C) $Q = \frac{q}{2}$ (D) $Q = 2q$

53. A body having specific charge 8 μC/g is resting on a frictionless plane at a distance 10 cm from the wall (as shown in the figure). It starts moving towards the wall when a uniform electric field of 100 V/m is applied horizontally towards the wall. If the collision of the body with the wall is perfectly elastic, the time period of the motion will be ______s. **[20 July 2021 (Morning)]**

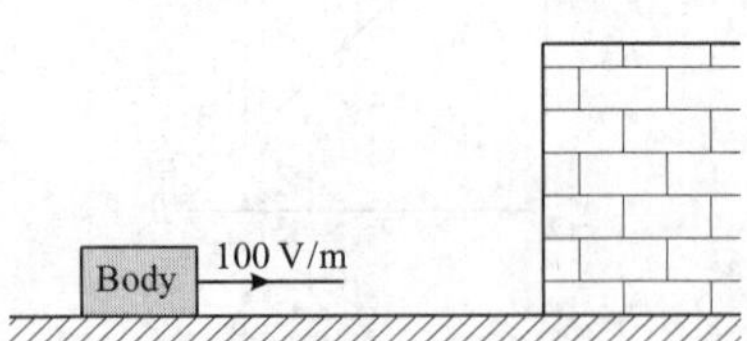

54. An electric dipole is placed on x-axis in proximity to a line charge of linear charge density 3.0×10^{-6} C/m. Line charge is placed on z-axis and positive and negative charge of dipole is at a distance of 10 mm and 12 mm from the origin respectively. If total force of 4 N is exerted on the dipole, find out the amount of positive or negative charge of the dipole : **[22 July 2021 (Evening)]**

(A) 0.485 mC (B) 4.44 μC
(C) 815.1 nC (D) 8.8 μC

55. A particle of mass 1 mg and charge q is lying at the mid-point of two stationary particle kept at a distance 2 m when each is carrying same charge q. If the free charged particle is displaced from its equilibrium position through distance x ($x << 1$ m). The particle executes SHM. Its angular frequency of oscillation will be ______ $\times 10^5$ rad/s if $q^2 = 10\ C^2$: **[25 July 2021 (Morning)]**

56. Two identical tennis balls each having mass 'm' and charge 'q' are suspended from a fixed point by threads of length 'l'. What is the equilibrium separation when each thread makes a small angle 'θ' with the vertical ? **[27 July 2021 (Morning)]**

(A) $x = \left(\frac{q^2 l^2}{2\pi\varepsilon_0 m^2 g}\right)^{1/3}$ (B) $x = \left(\frac{q^2 l}{2\pi\varepsilon_0 mg}\right)^{1/3}$

(C) $x = \left(\frac{q^2 l^2}{2\pi\varepsilon_0 m^2 g^2}\right)^{1/3}$ (D) $x = \left(\frac{q^2 l}{2\pi\varepsilon_0 mg}\right)^{1/3}$

57. A solid metal sphere of radius R having charge q is enclosed inside the concentric spherical shell of inner radius a and outer radius b as shown in figure. The approximate variation of electric field $\vec{E}$ as a function of distance r from centre O is given by :

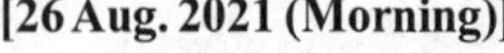

[26 Aug. 2021 (Morning)]

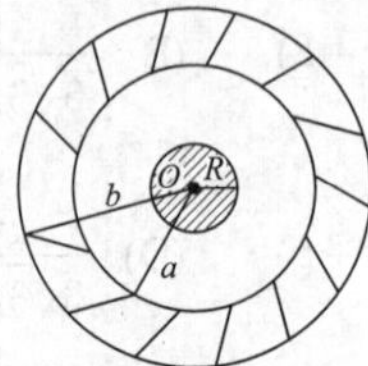

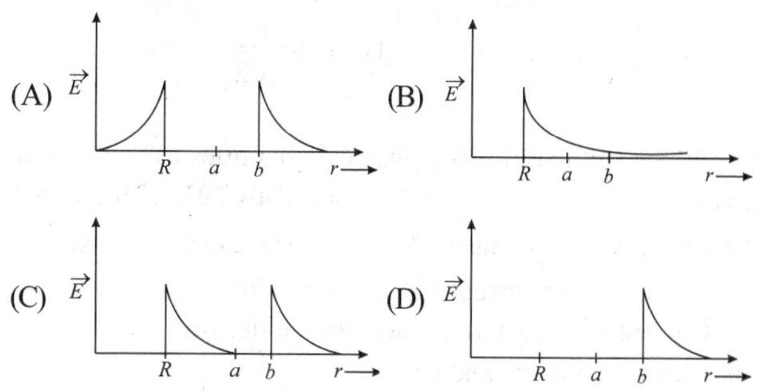

58. A uniformly charged disc of radius R having surface charge density σ is placed in the xy plane with its center at the origin. Find the electric field intensity along the z-axis at a distance Z from origin : **[27 Aug. 2021 (Morning)]**

(A) $E=\frac{\sigma}{2\varepsilon_0}\left(\frac{1}{Z^2+R^2}\right)+\frac{1}{Z^2}$

(B) $E=\frac{2\varepsilon_0}{2\sigma}\left(\frac{1}{(Z^2+R^2)^{1/2}}+Z\right)$

(C) $E=\frac{\sigma}{2\varepsilon_0}\left(1+\frac{Z}{(Z^2+R^2)^{1/2}}\right)$

(D) $E=\frac{\sigma}{2\varepsilon_0}\left(1-\frac{Z}{(Z^2+R^2)^{1/2}}\right)$

59. Figure shows a rod AB, which is bent in a 120° circular are of radius R. A charge $(-Q)$ is uniformly distributed over rod AB. What is the electric filed $\vec{E}$ at the centre of curvature O ? **[27 Aug. 2021 (Evening)]**

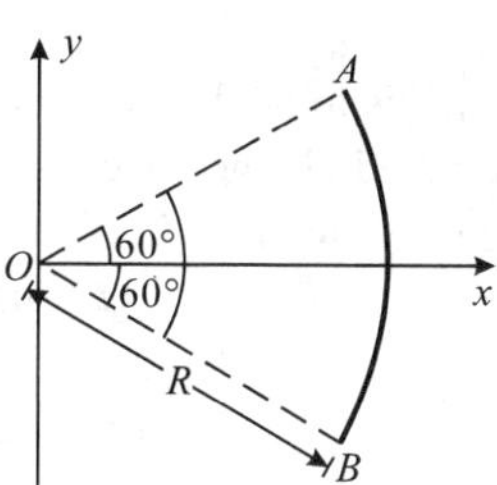

(A) $\frac{3\sqrt{3}Q}{8\pi^2\varepsilon_0 R^2}(\hat{i})$

(B) $\frac{3\sqrt{3}Q}{8\pi\varepsilon_0 R^2}(\hat{i})$

(C) $\frac{3\sqrt{3}Q}{8\pi^2\varepsilon_0 R^2}(-\hat{i})$

(D) $\frac{3\sqrt{3}Q}{16\pi^2\varepsilon_0 R^2}(\hat{i})$

60. Two particles A and B having charges 20 μC and –5 μC respectively are held fixed with a separation of 5 cm. At what position a third charged particle should be placed so that it does not experience a net electric force ? **[31 Aug. 2021 (Morning)]**

20μC ———— 5 cm ———— –5μC
A B

(A) At midpoint between two charges
(B) At 1.25 cm from a –5 μC between two charges
(C) At 5 cm from 20 μC on the left side of system
(D) At 5 cm from –5 μC on the right side

61. A vertical electric field of magnitude 4.9×10^5 N/C just prevents a water droplet of a mass 0.1 g from falling. The value of charge on the droplet will be : [Given : $g=9.8$ m/s^2] **[24 June 2022 (Morning)]**

(A) 1.6×10^{-9} C
(B) 2.0×10^{-9} C
(C) 3.2×10^{-9} C
(D) 0.5×10^{-9} C

62. Two identical charged particles each having a mass 10 g and charge 2.0×10^{-7} C area placed on a horizontal table with a separation of L between then such that they stay in limiting equilibrium. If the coefficient of friction between each particle and the table is 0.25, find the value of L. [Use g = 10 ms^{-2}] : **[24 June 2022 (Evening)]**

(A) 12 cm
(B) 10 cm
(C) 8 cm
(D) 5 cm

63. A long cylindrical volume contains a uniformly distributed charge of density ρ. The radius of cylindrical volume is R. A charge particle (q) revolves around the cylinder in a circular path. The kinetic of the particle is : **[24 June 2022 (Evening)]**

(A) $\frac{\rho qR^2}{4\varepsilon_0}$
(B) $\frac{\rho qR^2}{2\varepsilon_0}$
(C) $\frac{q\rho}{4\varepsilon_0 R^2}$
(D) $\frac{4\varepsilon_0 R^2}{q\rho}$

64. In the figure, a very large plane sheet of positive charge is shown. P_1 and P_2 are two points at distance l and $2l$ from the charge distribution. If σ is the surface charge density, then the magnitude of electric fields E_1 and E_2 at P_1 and P_2 respectively are : **[25 June 2022 (Morning)]**

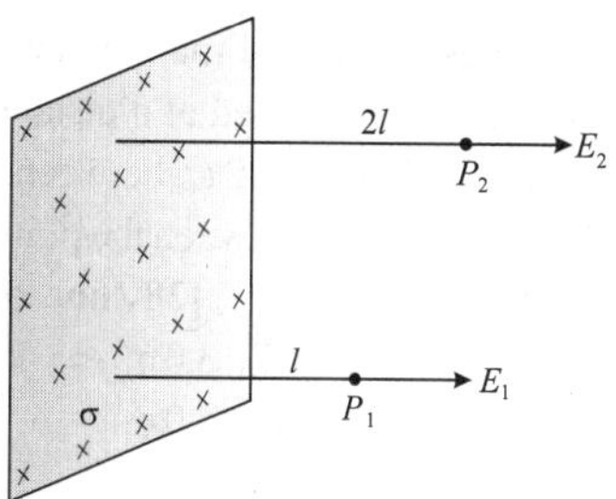

(A) $E_1=\sigma/\varepsilon_0, E_2=\sigma/2\varepsilon_0$
(B) $E_1=2\sigma/\varepsilon_0, E_2=\sigma/\varepsilon_0$
(C) $E_1=E_2=\sigma/2\varepsilon_0$
(D) $E_1=E_2=\sigma/\varepsilon_0$

65. Sixty four conducting drops each of radius 0.02 m and each carrying a charge of 5 μC are combined to form a bigger drop. The ratio of surface density of bigger drop to the smaller drop will be : **[26 June 2022 (Evening)]**

(A) 1:4 (B) 4:1
(C) 1:8 (D) 8:1

66. Three identical charged balls each of charge $2C$ are suspended from a common point P by silk threads of 2 m each (as shown in figure). They form an equilateral triangle of side 1 m. The ratio of net force on a charged ball to the force between any two charged balls will be : **[27 June 2022 (Evening)]**

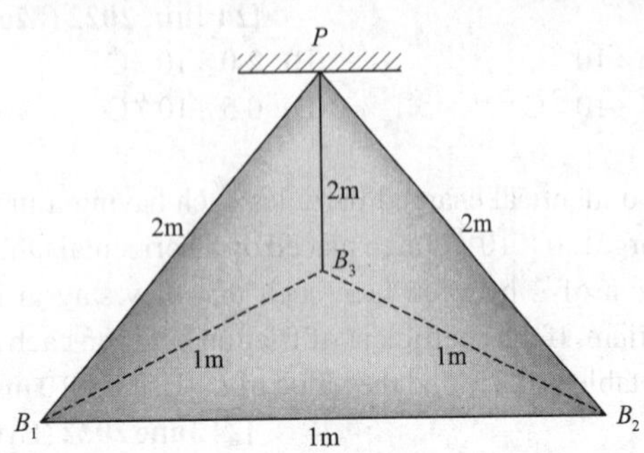

(A) 1:1 (B) 1:4
(C) $\sqrt{3}:2$ (D) $\sqrt{3}:1$

67. The three charges $q/2$, q and $q/2$ are placed at the corners A, B and C of a square of side a as shown in figure. The magnitude of electric field E at the corner D of the square, is : **[28 June 2022 (Morning)]**

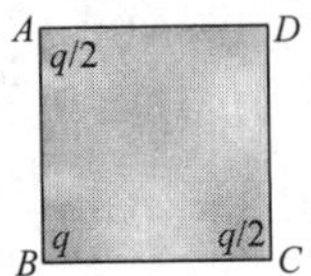

(A) $\frac{q}{4\pi \epsilon_0 a^2}\left(\frac{1}{\sqrt{2}}+\frac{1}{2}\right)$ (B) $\frac{q}{4\pi \epsilon_0 a^2}\left(1+\frac{1}{2}\right)$

(C) $\frac{q}{4\pi \epsilon_0 a^2}\left(1+\frac{1}{\sqrt{2}}\right)$ (D) $\frac{q}{4\pi \epsilon_0 a^2}\left(\frac{1}{\sqrt{2}}-\frac{1}{2}\right)$

68. Two point charges A and B of magnitude $+8 \times 10^{-6}$ C and -8×10^{-6} C respectively are placed at a distance d apart. The electric field at the middle point O between the charges is 6.4×10^4 NC^{-1}. The distance d between the point charges A and B is : **[28 June 2022 (Evening)]**

(A) 2.0 m (B) 3.0 m
(C) 1.0 m (D) 4.0 m

69. Two point charges Q each are placed at a distance d apart. A third point charge q is placed at a distance x from mid-point on the perpendicular bisector. The value of x at which charge q will experience the maximum Coulomb's force is : **[29 June 2022 (Evening)]**

(A) $x = d$ (B) $x = \frac{d}{2}$
(C) $x = \frac{d}{\sqrt{2}}$ (D) $x = \frac{d}{2\sqrt{2}}$

70. Which of the following physical quantities have the same dimensions? **[25 July 2022 (Morning)]**

(A) Electric displacement $(\vec{D})$ and surface charge density
(B) Displacement current and electric field
(C) Current density and surface charge density
(D) Electric potential and energy

71. Three point charges of magnitude 5 μC, 0.16 μC and 0.3 μC are located at the vertices A, B, C of a right angled triangle whose sides are $AB = 3$ cm, $BC = 3\sqrt{2}$ cm and $CA = 3$ cm and point A is the right angle corner. Charge at point A experiences _____ N of electrostatic force due to the other two charges. **[26 July 2022 (Evening)]**

72. A long cylindrical volume contains a uniformly distributed charge of density ρ Cm^{-3}. The electric field inside the cylindrical volume at a distance $x = \frac{2\varepsilon_0}{\rho}$ m from its axis is _____ Vm^{-1}. **[27 July 2022 (Morning)]**

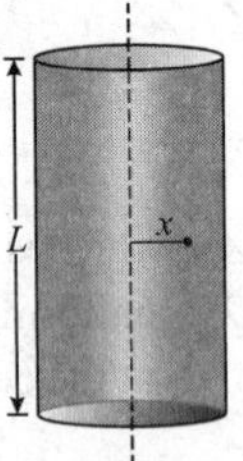

73. A charge of 4 μC is to be divided into two. The distance between the two divided charges is constant. The magnitude of the divided charges so that the force between them is maximum, will be : **[27 July 2022 (Evening)]**

(A) 1 μC and 3 μC (B) 2 μC and 2 μC
(C) 0 and 4 μC (D) 1.5 μC and 2.5 μC

74. A uniform electric field $E = (8m/e)$ V/m is created between two parallel plates of length 1 m as shown in figure, (where m = mass of electron and e = charge of electron). An electron enters the field symmetrically between the plates with a speed of 2 m/s. The angle of the deviation (θ) of the path of the electron as it comes out of the field will be _____. **[28 July 2022 (Evening)]**

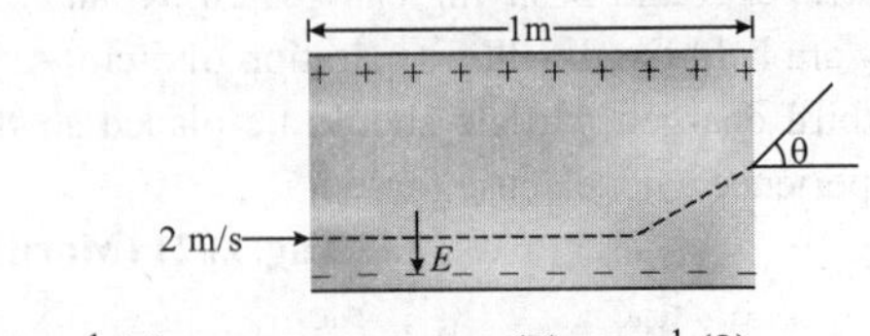

(A) $\tan^{-1}(4)$ (B) $\tan^{-1}(2)$

(C) $\tan^{-1}\left(\frac{1}{3}\right)$ (D) $\tan^{-1}(3)$

75. A spherically symmetric charge distribution is considered with charge density varying as

$$\rho(r) = \begin{cases} \rho_0\left(\frac{3}{4} - \frac{r}{R}\right) & \text{for } r \le R \\ \text{Zero} & \text{for } r > R \end{cases}$$

Where, $r(r < R)$ is the distance from the centre O (as shown in figure). The electric field at point P will be :

[29 July 2022 (Morning)]

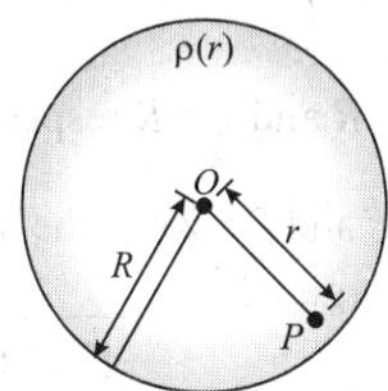

(A) $\frac{\rho_0 r}{4\varepsilon_0}\left(\frac{3}{4} - \frac{r}{R}\right)$ (B) $\frac{\rho_0 r}{3\varepsilon_0}\left(\frac{3}{4} - \frac{r}{R}\right)$

(C) $\frac{\rho_0 r}{4\varepsilon_0}\left(1 - \frac{r}{R}\right)$ (D) $\frac{\rho_0 r}{5\varepsilon_0}\left(1 - \frac{r}{R}\right)$

76. Two identical metallic spheres A and B when placed at certain distance in air repel each other with a force of F. Another identical uncharged sphere C is first placed in contact with A and then in contact with B and finally placed at midpoint between spheres A and B. The force experienced by sphere C will be :

[29 July 2022 (Evening)]

(A) $3F/2$ (B) $3F/4$
(C) F (D) $2F$

77. If two charges q_1 and q_2 are separated with distance d and placed in a medium of dielectric constant K. What will be the equivalent distance between charges in air for the same electrostatic force ? **[24 Jan. 2023 (Morning)]**

(A) $d\sqrt{k}$ (B) $k\sqrt{d}$
(C) $1.5d\sqrt{k}$ (D) $2d\sqrt{k}$

78. A stream of a positively charged particles having $\frac{q}{m} = 2 \times 10^{11} \frac{C}{\text{kg}}$ and velocity $\vec{v}_0 = 3 \times 10^7\ \hat{i}$ m/s is deflected by an electric field $1.8\hat{j}$ kV/m. The electric field exists in a region of 10 cm along x direction. Due to the electric field, the deflection of the charge particles in the y direction is _____ mm.

[24 Jan. 2023 (Morning)]

79. A point charge of 10 μC is placed at the origin. At what location on the X-axis should a point charge of 40 μC be placed so that the net electric field is zero at $x = 2$ cm on the X-axis ?

[25 Jan. 2023 (Evening)]

(A) $x = 6$ cm (B) $x = 4$ cm
(C) $x = 8$ cm (D) $x = -4$ cm

80. A point charge $q_1 = 4q_0$ is placed at origin. Another point charge $q_2 = -q_0$ is placed at $x = 12$ cm. Charge of proton is q_0. The proton is placed on x-axis so that the electrostatic force on the proton in zero. In this situation, the position of the proton from the origin is _____ cm. **[29 Jan. 2023 (Morning)]**

81. As shown in the figure, a point charge Q is placed at the centre of conducting spherical shell of inner radius a and outer radius b. The electric field due to charge Q in three different regions I, II and III is given by : (I : $r < a$, II : $a < r < b$, III : $r > b$)

[30 Jan. 2023 (Evening)]

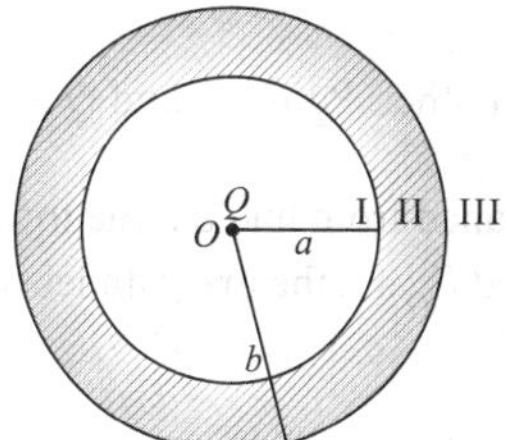

(A) $E_{\text{I}} = 0, E_{\text{II}} = 0, E_{\text{III}} \ne 0$ (B) $E_{\text{I}} \ne 0, E_{\text{II}} = 0, E_{\text{III}} \ne 0$
(C) $E_{\text{I}} = 0, E_{\text{II}} = 0, E_{\text{III}} = 0$ (D) $E_{\text{I}} = 0, E_{\text{II}} = 0, E_{\text{III}} = 0$

82. Two equal positive point charges are separated by a distance $2a$. The distance of a point from the centre of the line joining two charges on the equatorial line (perpendicular bisector) at which force experienced by a test charge q_0 becomes maximum is $\frac{a}{\sqrt{x}}$. The value of x is _____.

[1 Feb. 2023 (Morning)]

83. An electron revolves around an infinite cylindrical wire having uniform linear change density 2×10^{-8} cm^{-1} in circular path under the influence of attractive electrostatic field as shown in the figure. The velocity of electron with which it is revolving is_____ $\times 10^6$ ms^{-1}. Given mass of electron = 9×10^{-31} kg.

[10 April 2023 (Evening)]

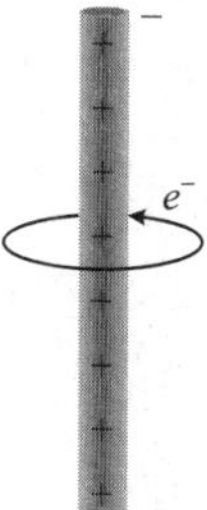

84. As shown in the figure a configuration of two equal point charges (q_0 = +2 μC) is placed on an inclined plane. Mass of each point charge is 20 g. Assume that there is no friction between charge and plane. For the system of two point charges to be in equilibrium (at rest) the height $h = x \times 10^{-3}$ m The value of x is______. (Take $\frac{1}{4\pi\varepsilon_0} = 9 \times 10^9 \text{ N m}^2\text{C}^{-2}, g = 10 \text{ ms}^{-1}$)

[11 April 2023 (Morning)]

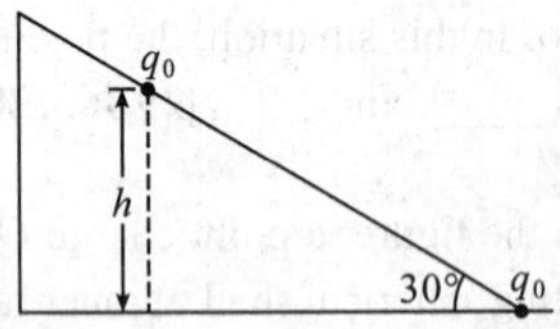

85. A thin infinite sheet charge and an infinite line charge of respective charge densities +σ and +λ are placed parallel at 5 m distance from each other. Points P and Q are at $\frac{3}{\pi}$ m and $\frac{4}{\pi}$ m perpendicular distance from line charge towards sheet charge, respectively. E_P and E_Q are the magnitudes of resultant electric field intensities at point P and Q, respectively. If $\frac{E_P}{E_Q} = \frac{4}{a}$ for $2|\sigma| = |\lambda|$. Then the value of a is______.

[13 April 2023 (Morning)]

86. A 10 μC charge is divided into two parts and placed at 1 cm distance so that the repulsive force between them is maximum. The charges of the two parts are: **[13 April 2023 (Evening)]**

(A) 9 μC, 1 μC
(B) 5 μC, 5μC
(C) 7 μC, 3 μC
(D) 8 μC, 2 μC

87. Three point charges q, $-2q$ and $2q$ are placed on x-axis at a distance $x = 0$, $x = \frac{3}{4}R$ and $x = R$ respectively from origin as shown. If $q = 2 \times 10^{-6}$ C and $R = 2$ cm, the magnitude of net force experienced by the charge $-2q$ is______ N.

[13 April 2023 (Evening)]

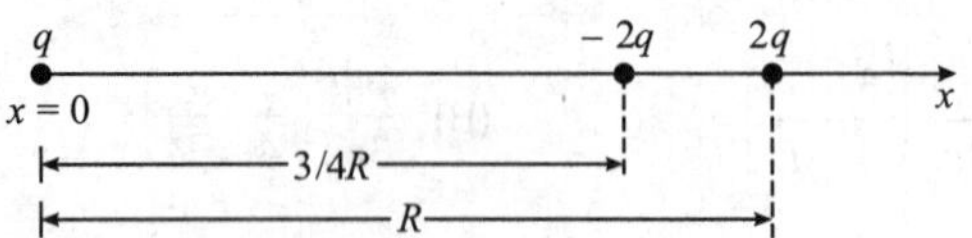

* * * * *

CHAPTER 15

ELECTRIC CHARGES AND FIELDS

15.2 ELECTRIC FLUX AND GAUSS'S LAW

1. A charged particle q is placed at the centre O of cube of length L ($ABCDEFGH$). Another same charge q is placed at a distance L from O. Then the electric flux through $ABCD$ is : **[2002]**

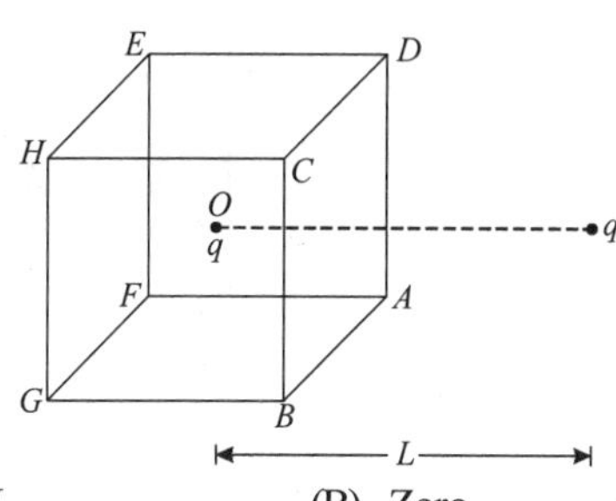

(A) $q/4\pi\in_0 L$ (B) Zero
(C) $q/2\pi\in_0 L$ (D) $q/3\pi\in_0 L$

2. If the electric flux entering and leaving an enclosed surface respectively is ϕ_1 and ϕ_2, the electric charge inside the surface will be : **[2003]**

(A) $(\phi_2-\phi_1)\varepsilon_0$ (B) $(\phi_1-\phi_2)/\varepsilon_0$
(C) $(\phi_2-\phi_1)/\varepsilon_0$ (D) $(\phi_1-\phi_2)\varepsilon_0$

3. The electrostatic potential inside a charged spherical ball is given by $\phi = ar^2 + b$ where r is the distance from the centre; a, b are constant. Then the charge density inside the ball is : **[2011]**

(A) $-24\pi a \in_0$ (B) $-6 a \in_0$
(C) $-24\pi a \in_0 r$ (D) $-6 a \in_0 r$

4. The flat base of a hemisphere of radius a with no charge inside it lies in a horizontal plane. A uniform electric field $\vec{E}$ is applied at an angle $\pi/4$ with the vertical direction. The electric flux through the curved surface of the hemisphere is : **[2012 (Online)]**

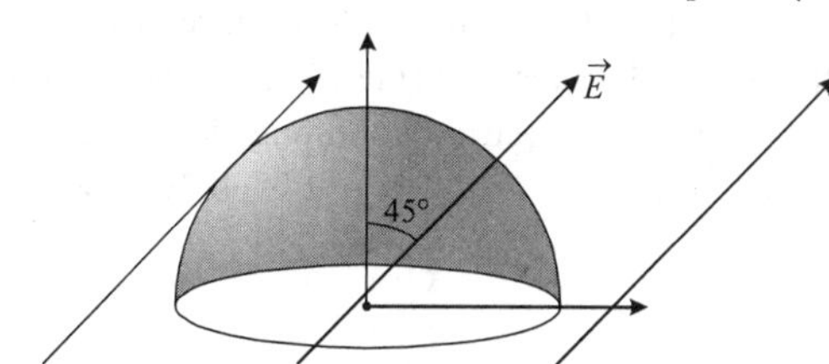

(A) $\pi a^2 E$ (B) $\dfrac{\pi a^2 E}{\sqrt{2}}$
(C) $\dfrac{\pi a^2 E}{2\sqrt{2}}$ (D) $\dfrac{(\pi+2a\pi)^2 E}{(2\sqrt{2}^2)}$

5. Three positive charges of equal value q are placed at vertices of an equilateral triangle. The resulting lines of force should be sketched as in : **[2012 (Online)]**

(A) (B)

(C) (D)

6. The electric potential $V(x)$ in a region around the origin is given by $V(x) = 4x^2$ volts. The electric charge enclosed in a cube of 1m side with its centre at the origin is : (In coulomb) **[2012 (Online)]**

(A) $8\varepsilon_0$ (B) $-4\varepsilon_0$
(C) 0 (D) $-8\varepsilon_0$

7. The electric field in a region of space is given by, $\vec{E} = E_0\hat{i} + 2E_0\hat{j}$ where E_0 = 100 N/C. The flux of the field through a circular surface of radius 0.02 m parallel to the Y-Z plane is nearly : **[2014 (Online)]**

(A) 0.125 Nm^2/C (B) 0.02 Nm^2/C
(C) 0.005 Nm^2/C (D) 3.14 Nm^2/C

8. A long cylindrical shell carries positive surface charge ρ in the upper half and negative surface charge $-\sigma$ in the lower half. The electric field lines around the cylinder will look like figure given in (figures are schematic and not drawn to scale) : **[2015]**

(A)
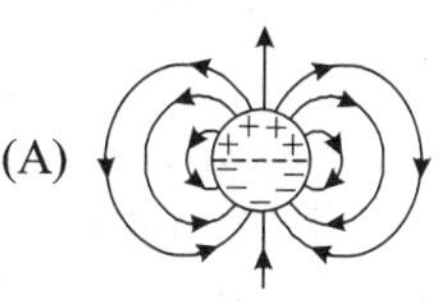

(B)
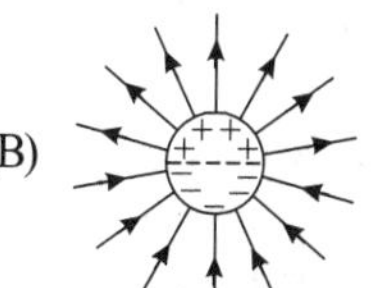

(C)
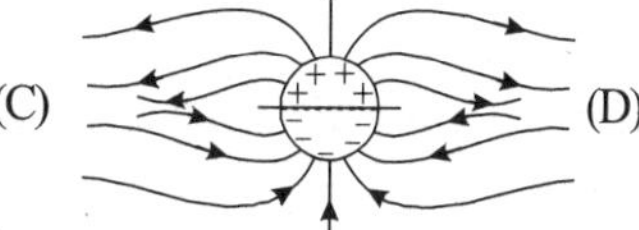

(D)
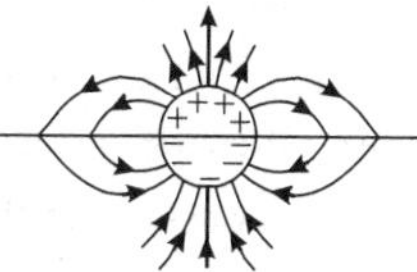

9. Four closed surfaces and corresponding charge distributions are shown below :

Let the respective electric fluxes through the surfaces be Φ_1, Φ_2, Φ_3 and Φ_4. Then : **[2017 (Online)]**

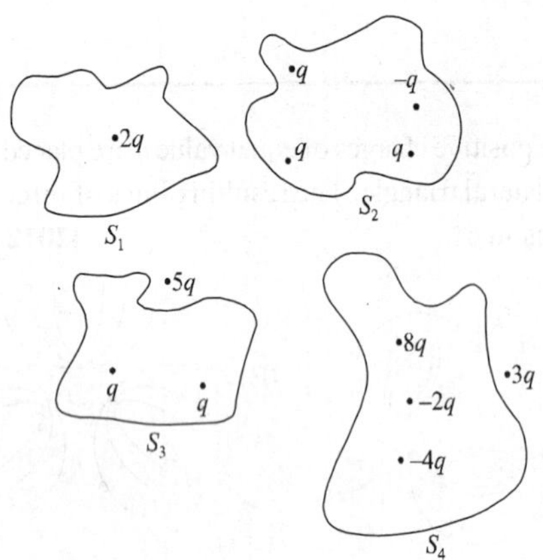

(A) $\Phi_1 < \Phi_2 = \Phi_3 > \Phi_4$ (B) $\Phi_1 > \Phi_2 > \Phi_3 > \Phi_4$
(C) $\Phi_1 = \Phi_2 = \Phi_3 = \Phi_4$ (D) $\Phi_1 > \Phi_3 ; \Phi_2 < \Phi_4$

10. A charge Q is placed at a distance $a/2$ above the centre of the square surface of edge a as shown in the figure. The electric flux through the square surface is : **[2018 (Online)]**

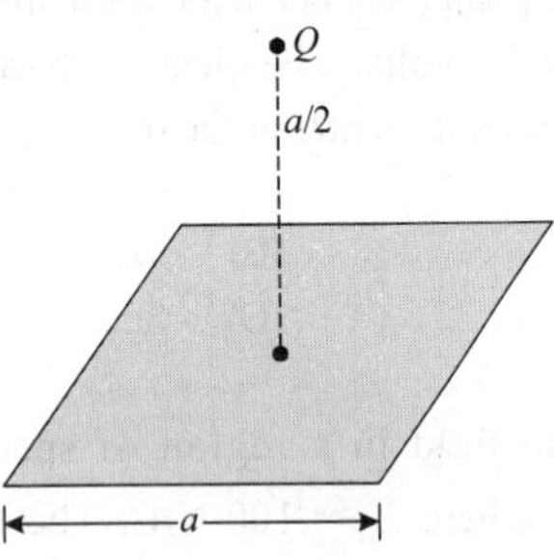

(A) $\frac{Q}{3\epsilon_0}$ (B) $\frac{Q}{6\epsilon_0}$

(C) $\frac{Q}{2\epsilon_0}$ (D) $\frac{Q}{\epsilon_0}$

11. In finding the electric field using Gauss Law the formula $|\vec{E}| = \frac{q_{enc}}{\varepsilon_0 |A|}$ is applicable. In the formula ε_0 is permittivity of free space, A is the area of Gaussian surface and q_{enc} is charge enclosed by the Gaussian surface. The equation can be used in which of the following situation ? **[8 Jan. 2020 (Morning)]**

(A) Only when the Gaussian surface is an equipotential surface
(B) Only when $|\vec{E}|$ = constant on the surface
(C) For any choice of Gaussian surface
(D) Only when the Gaussian surface is an equipotential surface and $|\vec{E}|$ is constant on the surface

12. An electric field $\vec{E} = 4x\hat{i} - (y^2 + 1)\hat{j}$ N/C passes through the box shown in figure. The flux of the electric field through surfaces $ABCD$ and $BCGF$ are marked as ϕ_I and ϕ_{II} respectively. The difference between $(\phi_I - \phi_{II})$ is (in Nm^2/C) ______. **[9 Jan. 2020 (Evening)]**

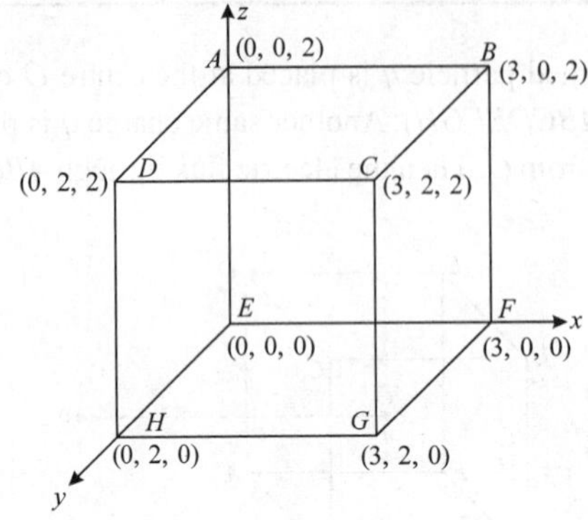

13. Two charged thin infinite plane sheets of uniform surface charge density σ_+ and σ_- where $|\sigma_+| > |\sigma_-|$ intersect at right angle. Which of the following best represents the electric field lines for this system : **[4 Sept. 2020 (Morning)]**

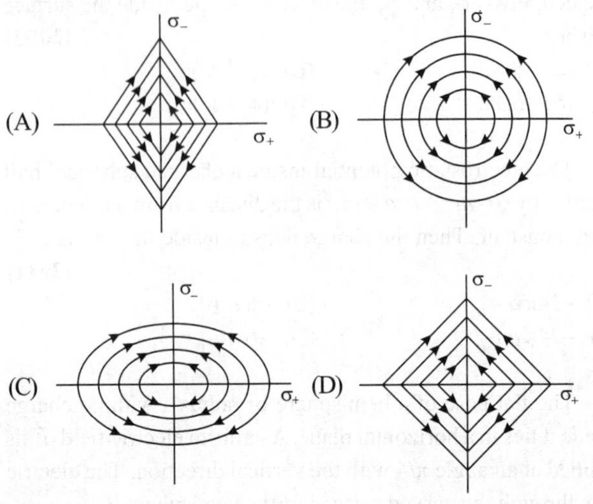

14. A point charge of + 12 μC is at a distance 6 cm vertically above the centre of a square of side 12 cm as shown in figure. The magnitude of the electric flux through the square will be ______ $\times 10^3$ Nm^2/C **[24 Feb. 2021 (Evening)]**

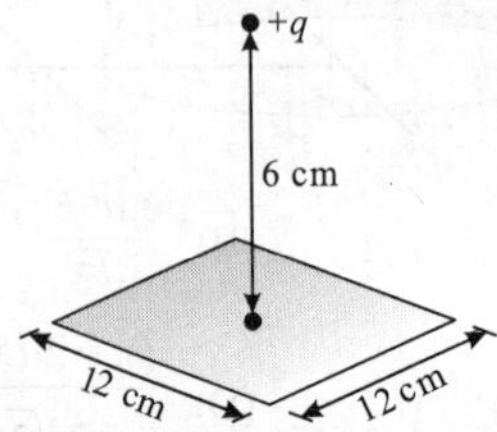

15. The electric field in a region is given by

$$\vec{E} = \left(\frac{3}{5}E_0\hat{i} + \frac{4}{5}E_0\hat{j}\right)\frac{N}{C}$$

The ratio of flux of reported field through the rectangular surface of area 0.2 m^2 (parallel to y-z plane) to that of the surface of area 0.3 m^2 (parallel to x-z plane) is $a : b$, where a = ______.

[Here $\hat{i}$, $\hat{j}$ and $\hat{k}$ are unit vectors along x, y and z-axes respectively] **[25 Feb. 2021 (Morning)]**

16. A charge q is placed at one corner of a cube as shown in figure. The flux of electrostatic field $\vec{E}$ through the shaded area is : **[25 Feb. 2021 (Evening)]**

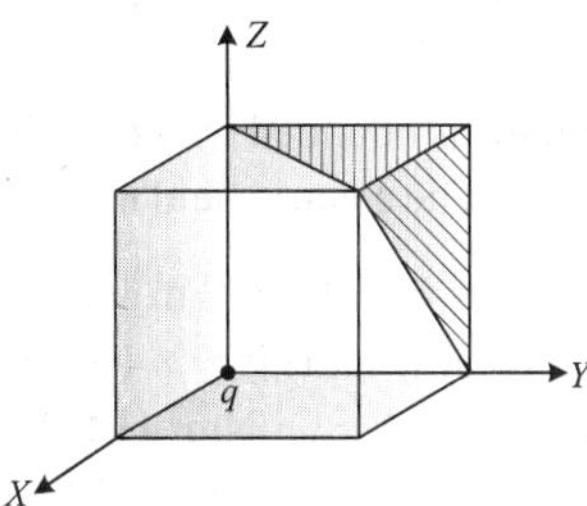

(A) $\frac{q}{4\varepsilon_0}$ (B) $\frac{q}{8\varepsilon_0}$

(C) $\frac{q}{48\varepsilon_0}$ (D) $\frac{q}{24\varepsilon_0}$

17. Given below are two statements :

Statement-I : An electric dipole is placed at the centre of a hollow sphere. The flux of electric field through the sphere is zero but the electric field is not zero anywhere in the sphere.

Statement-II : If R is the radius of a solid metallic sphere and Q be the total charge on it. The electric field at any point on the spherical surface of radius $r(< R)$ is zero but the electric flux passing through this closed spherical surface of radius r is not zero.

In the light of the above statements, choose the correct answer from the options given below : **[26 Feb. 2021 (Evening)]**

(A) Statement-I is true but Statement-II is false
(B) Both Statement-I and Statement-II are false
(C) Both Statement-I and Statement-II are true
(D) Statement-I is false but Statement-II is true

18. The electric field in a region is given by $\vec{E} = \frac{2}{5}E_0\hat{i} + \frac{3}{5}E_0\hat{j}$ with $E_0 = 4.0 \times 10^3 \frac{N}{C}$. The flux of this field through a rectangular surface area 0.4 m^2 parallel to the $Y-Z$ plane is _____Nm^{2}C^{-1}. **[17 March 2021 (Evening)]**

19. The total charge enclosed in an incremental volume of 2×10^{-9} m^3 located at the origin is ______ nC, if electric flux density of its field is found as $D = e^{-x}\sin y\,\hat{i} - e^{-x}\cos y\,\hat{j} + 2z\,\hat{k}$ C/m^2. **[22 July 2021 (Evening)]**

20. Choose the incorrect statement : **[31 Aug. 2021 (Evening)]**

(a) The electric lines of force entering into a Gaussian surface provide negative flux
(b) A charge q is placed at the centre of a cube. The flux through all the faces will be the same
(c) In a uniform electric field net flux through a closed Gaussian surface containing no net charge, is zero.
(d) When electric field is parallel to a Gaussian surface, it provides a finite non-zero flux

Choose the most appropriate answer from the options given below :

(A) (b) and (d) only (B) (a) and (c) Only
(C) (d) Only (D) (c) and (d) Only

21. A cube is placed inside an electric field, $\vec{E} = 150y^2\,\hat{j}$. The side of the cube is 0.5 m and is placed in the field as shown in the given figure. The charge inside the cube is : **[01 Sep. 2021 (Evening)]**

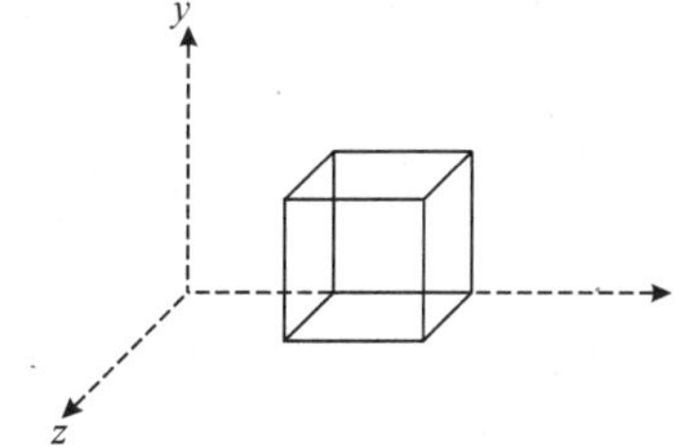

(A) 8.3×10^{-12} C (B) 8.3×10^{-11} C
(C) 3.8×10^{-12} C (D) 3.8×10^{-11} C

22. If a charge q is placed at the centre of a closed hemispherical non-conducting surface, the total flux passing through the flat surface would be : **[27 June 2022 (Evening)]**

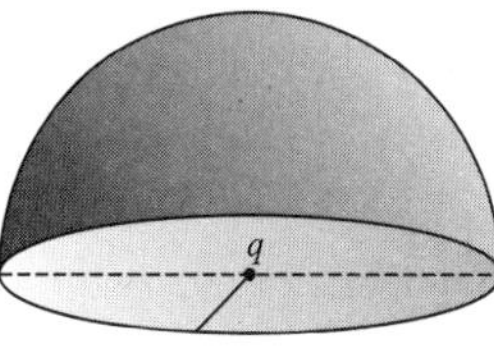

(A) $\frac{q}{\varepsilon_0}$ (B) $\frac{q}{2\varepsilon_0}$

(C) $\frac{q}{4\varepsilon_0}$ (D) $\frac{q}{2\varepsilon_0}$

23. The volume charge density of a sphere of radius 6 m is 2 μCcm^{-3}. The number of lines of force per unit surface area coming out from the surface of the sphere is _____ $\times 10^{10}\ NC^{-1}$. [Given : Permittivity of vacuum $\in_0 = 8.85 \times 10^{-12}\ C^2 N^{-1} - m^{-2}$].

[25 July 2022 (Morning)]

24. In a cuboid of dimension $2L \times 2L \times L$, a charge q is placed at the centre of the surface S having area of $4L^2$. The flux through the opposite surface to S is given by :

[29 Jan. 2023 (Morning)]

(A) $\frac{q}{12\varepsilon_0}$ (B) $\frac{q}{3\varepsilon_0}$

(C) $\frac{q}{2\varepsilon_0}$ (D) $\frac{q}{6\varepsilon_0}$

25. As shown in figure, a cuboid lies in a region with electric field $E = 2x^2\hat{i} - 4y\hat{j} + 6\hat{k}$ N/C. The magnitude of charge within the cuboid is $n \in_0 C$. The value of n is _____. (if dimension of cuboid is $1 \times 2 \times 3\ m^3$) **[30 Jan. 2023 (Evening)]**

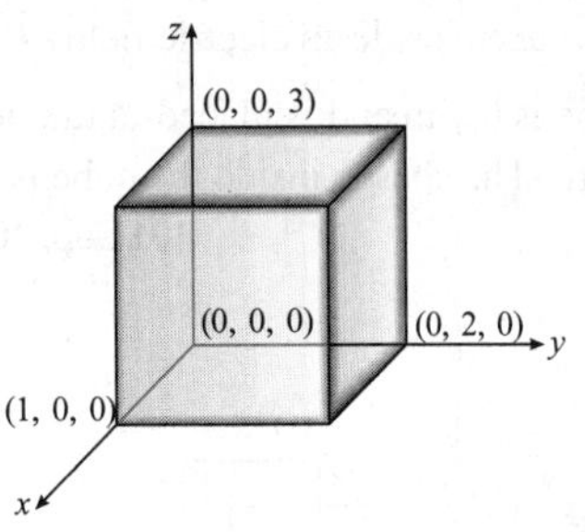

26. Expression for an electric field is given by $\vec{E} = 4000x^2\hat{i}\frac{V}{m}$. The electric flux through the cube of side 20 cm when placed in electric field (as shown in the figure) is _____ V cm.

[31 Jan. 2023 (Morning)]

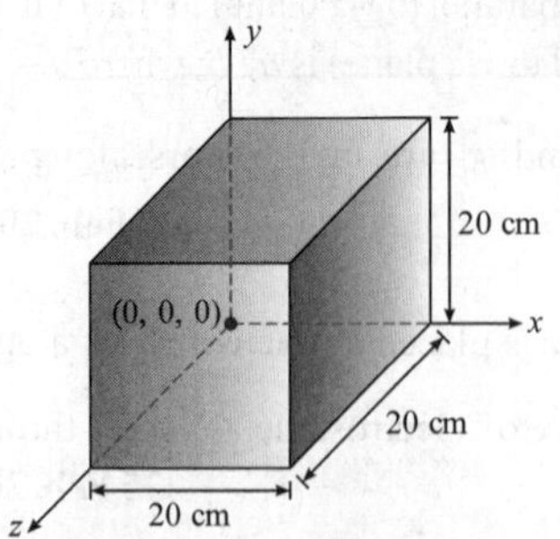

27. A cubical volume is bounded by the surfaces $x = 0, x = a, y = 0, y = a, z = 0, z = a$. The electric field in the region is given by $\vec{E} = E_0 x\hat{i}$. Where $E_0 = 4 \times 10^4\ NC^{-1}\ m^{-1}$. If $a = 2$ cm, the charge contained in the cubical volume is $Q \times 10^{-14}$ C. The value of Q is _____. (Take $\in_0 = 9 \times 10^{-12}\ C^2/Nm^2$) **[1 Feb. 2023 (Evening)]**

* * * * *

CHAPTER 16 ELECTRIC POTENTIAL AND CAPACITANCE

16.1 ELECTRIC POTENTIAL AND DIPOLE

1. On moving a charge of 20 C by 2 cm, 2 J of work is done, then the potential difference between the points is : **[2002]**

(A) 0.1 V (B) 8 V
(C) 2 V (D) 0.5 V

2. A thin spherical conducting shell of radius R has a charge q. Another charge Q is placed at the centre of the shell. The electrostatic potential at a point P at a distance $R/2$ from the centre of the shell is : **[2003]**

(A) $\frac{2Q}{4\pi \in_0 R}$ (B) $\frac{2Q}{4\pi \in_0 R} - \frac{2q}{4\pi \in_0 R}$

(C) $\frac{2Q}{4\pi \in_0 R} + \frac{q}{4\pi \in_0 R}$ (D) $\frac{(q+Q)}{4\pi \in_0} \frac{2}{R}$

3. A charged particle q is shot towards another charged particle Q which is fixed, with a speed v. It approaches Q upto a closest distance r and then returns. If q was given a speed $2v$, the closest distance of approach would be : **[2004]**

(A) r (B) $2r$
(C) $r/2$ (D) $r/4$

4. Two thin wire rings each having a radius R are placed at a distance d apart with their axes coinciding. The charges on the two rings are $+q$ and $-q$. The potential difference between the centres of the two rings is : **[2005]**

(A) $\frac{qR}{4\pi \in_0 d^2}$ (B) $\frac{q}{2\pi \in_0}\left[\frac{1}{R} - \frac{1}{\sqrt{R^2+d^2}}\right]$

(C) Zero (D) $\frac{q}{4\pi \in_0}\left[\frac{1}{R} - \frac{1}{\sqrt{R^2+d^2}}\right]$

5. Two spherical conductors A and B of radii 1 mm and 2 mm are separated by a distance of 5 cm and are uniformly charged. If the spheres are connected by a conducting wire then in equilibrium condition, the ratio of the magnitude of the electric fields at the surfaces of spheres A and B is : **[2006]**

(A) 4 : 1 (B) 1 : 2
(C) 2 : 1 (D) 1 : 4

6. An electric dipole is placed at an angle of 30° to a non-uniform electric field. The dipole will experience : **[2006]**

(A) A translational force only in the direction of the field
(B) A translational force only in a direction normal to the direction of the field
(C) A torque as well as a translational force
(D) A torque only

7. The potential at a point x (measured in μm) due to some charges situated on the x-axis is given by :
$V(x) = 20/(x^2 - 4)$ volt. The electric field E at $x = 4$ μm is given by: **[2007]**

(A) $\frac{5}{3}$ V/μm and in the $-ve$ x direction

(B) $\frac{5}{3}$ V/μm and in the $+ve$ x direction

(C) $\frac{10}{9}$ V/μm and in the $-ve$ x direction

(D) $\frac{10}{9}$ V/μm and in the $+ve$ x direction

8. Charges are placed on the vertices of a square as shown. Let $\vec{E}$ be the electric field and V the potential at the centre. If the charges on A and B are interchanged with those on D and C respectively, then : **[2007]**

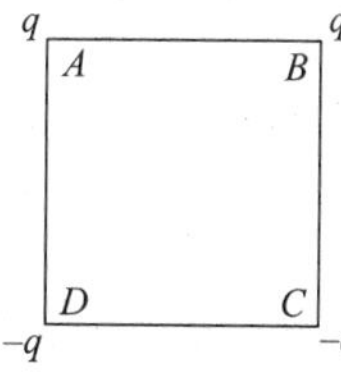

(A) $\vec{E}$ remains unchanged, V changes
(B) both $\vec{E}$ and V change
(C) $\vec{E}$ and V remain unchanged
(D) $\vec{E}$ changes and V remains unchanged

9. An electric charge 10^{-3} μC is placed at the origin (0, 0) of X-Y coordinate system. Two points A and B are situated at $(\sqrt{2}, \sqrt{2})$ and (2, 0) respectively. The potential difference between the points A and B will be : **[2007]**

(A) 9 V (B) Zero
(C) 2 V (D) 4.5 V

10. Statement-1 : For a charged particle moving from point P to point Q the net work done by an electrostatic field on the particle is independent of the path connecting point P to point Q.
Statement-2 : The net work done by a conservative force on an object moving along closed loop is zero. **[2009]**
(A) Statement-1 is true, Statement-2 is true; Statement-2 is not the correct explanation of Statement-1
(B) Statement-1 is false, Statement-2 is true
(C) Statement-1 is true, Statement-2 is false
(D) Statement-1 is true, Statement-2 is true; Statement-2 is the correct explanation of Statement-1

11. Two points P and Q are maintained at the potential of 10 V and –4 V, respectively. The work done in moving 100 electrons from P to Q is : **[2009]**
(A) -2.24×10^{-16} J (B) 2.24×10^{-16} J
(C) -9.60×10^{-17} J (D) 9.60×10^{-17} J

12. Two positive charges of magnitude 'q' are placed at the ends of a side (side-1) of a square of side '$2a$'. Two negative charges of the same magnitude are kept at the other corners. Starting from rest, if a charge Q moves from the middle of side-1 to the centre of square, its kinetic energy at the centre of square is : **[2011]**
(A) $\frac{1}{4\pi\epsilon_0}\frac{2qQ}{a}\left(1-\frac{1}{\sqrt{5}}\right)$ (B) Zero
(C) $\frac{1}{4\pi\epsilon_0}\frac{2qQ}{a}\left(1+\frac{1}{\sqrt{5}}\right)$ (D) $\frac{1}{4\pi\epsilon_0}\frac{2qQ}{a}\left(1-\frac{2}{\sqrt{5}}\right)$

13. An insulating solid sphere of radius R has a uniformly positive charge density ρ. As a result of this uniform charge distribution there is a finite value of electric potential at the centre of the sphere, at the surface of the sphere and also at a point out side the sphere. The electric potential at infinity is zero.
Statement-1 : When a charge 'q' is taken from the centre to the surface of the sphere, its potential energy changes by $\frac{q\rho}{3\epsilon_0}$.
Statement-2 : The electric field at a distance $r(r<R)$ from the centre of the sphere is $\frac{\rho r}{3\epsilon_0}$. **[2012]**
(A) Statement-1 is true, Statement-2 is true and Statement-2 is the correct explanation of Statement-1
(B) Statement-1 is true, Statement-2 is true and Statement-2 is not the correct explanation of Statement-1
(C) Statement-1 is true, Statement-2 is false
(D) Statement-1 is false, Statement-2 is true

14. A charge of total amount Q is distributed over two concentric hollow spheres of radii r and $R(R>r)$ such that the surface charge densities on the two spheres are equal. The electric potential at the common centre is : **[2012 (Online)]**
(A) $\frac{1}{4\pi\varepsilon_0}\frac{(R-r)Q}{(R^2+r^2)}$ (B) $\frac{1}{4\pi\varepsilon_0}\frac{(R+r)Q}{2(R^2+r^2)}$
(C) $\frac{1}{4\pi\varepsilon_0}\frac{(R+r)Q}{(R^2+r^2)}$ (D) $\frac{1}{4\pi\varepsilon_0}\frac{(R-r)Q}{(R^2+r^2)}$

15. A charge Q is uniformly distributed over a long rod AB of length L as shown in the figure. The electric potential at the point O lying at a distance L from the end A is : **[2013]**

O ←—L—→ A ▭▭▭ B ←—L—→

(A) $\frac{Q}{8\pi\epsilon_0 L}$ (B) $\frac{3Q}{4\pi\epsilon_0 L}$
(C) $\frac{Q}{4\pi\epsilon_0 L\ln 2}$ (D) $\frac{Q\ln 2}{4\pi\epsilon_0 L}$

16. Two point dipoles of dipole moment $\vec{P}_1$ and $\vec{P}_2$ are at a distance x from each other and $\vec{P}_1 \parallel \vec{P}_2$. The force between the dipoles is : **[2013 (Online)]**
(A) $\frac{1}{4\pi\varepsilon_0}\frac{4p_1p_2}{x^4}$ (B) $\frac{1}{4\pi\varepsilon_0}\frac{3p_1p_2}{x^3}$
(C) $\frac{1}{4\pi\varepsilon_0}\frac{6p_1p_2}{x^4}$ (D) $\frac{1}{4\pi\varepsilon_0}\frac{8p_1p_2}{x^4}$

17. This question has Statement-1 and Statement-2. Of the four choices given after the Statements, choose the one that best describes the two Statements. **[2013 (Online)]**
Statement-1 : No work is required to be done to move a test charge between any two points on an equipotential surface.
Statement-2 : Electric lines of force at the equipotential surfaces are mutually perpendicular to each other.
(A) Statement-1 is true, Statement-2 is true, Statement-2 is the correct explanation of Statement-1
(B) Statement-1 is true, Statement-2 is true, Statement-2 is not the correct explanation of Statement-1
(C) Statement-1 is true, Statement-2 is false
(D) Statement-1 is false, Statement-2 is true

18. A point charge of magnitude + 1 μC is fixed at (0, 0, 0). An isolated uncharged spherical conductor, is fixed with its center at (4, 0, 0). The potential and the induced electric field at the centre of the sphere is : **[2013 (Online)]**
(A) 1.8×10^5 V and -5.625×10^6 V/m
(B) 0 V and 0 V/m
(C) 2.25×10^5 V and -5.625×10^6 V/m
(D) 2.25×10^5 V and 0 V/m

19. Consider a finite insulated, uncharged conductor placed near a finite positively charged conductor. The uncharged body must have a potential : **[2013 (Online)]**

(A) Less than the charged conductor and more than at infinity
(B) More than the charged conductor and less than at infinity
(C) More than the charged conductor and more than at infinity
(D) Less than the charged conductor and less than at infinity

20. Two small equal point charges of magnitude q are suspended from a common point on the ceiling by insulating mass less strings of equal lengths. They come to equilibrium with each string making angle θ from the vertical. If the mass of each charge is m, then the electrostatic potential at the centre of line joining them will be $\left(\frac{1}{4\pi \in_0} = k\right)$: **[2013 (Online)]**

(A) $2\sqrt{kmg \tan\theta}$ (B) $\sqrt{kmg \tan\theta}$
(C) $4\sqrt{kmg / \tan\theta}$ (D) $4\sqrt{kmg \tan\theta}$

21. Assume that an electric field $\vec{E} = 30x^2\hat{i}$ exists in space. Then the potential difference $V_A - V_O$, where V_O is the potential at the origin and V_A the potential at $x = 2$ m is : **[2014]**

(A) 120 V (B) – 120 V
(C) – 80 V (D) 80 V

22. A uniformly charged solid sphere of radius R has potential V_0 (measured with respect to ∞) on its surface. For this sphere the equipotential surfaces with potentials $\frac{3V_0}{2}, \frac{5V_0}{4}, \frac{3V_0}{4}$ and $\frac{V_0}{4}$ have radius R_1, R_2, R_3, and R_4 respectively. Then : **[2015]**

(A) $R_1 = 0$ and $R_2 > (R_4 - R_3)$
(B) $R_1 \neq 0$ and $(R_2 - R_1) > (R_4 - R_3)$
(C) $R_1 = 0$ and $R_2 < (R_4 - R_3)$
(D) $2R < R_4$

23. An electric field $\vec{E} = (25\hat{i} + 30\hat{j})NC^{-1}$ exists in a region of space. If the potential at the origin is taken to be zero then the potential at $x = 2$m, $y = 2$m is : **[2015 (Online)]**

(A) – 110 J/C (B) – 140 J/C
(C) – 120 J/C (D) – 130 J/C

24. The potential (in volts) of a charge distribution is given by

$V(z) = 30 - 5z^2$ for $|z| \leq 1$m

$V(z) = 35 - 10|z|$ for $|z| \geq 1$m

$V(z)$ does not depend on x and y. If this potential is generated by a constant charge per unit volume ρ_0 (in units of ε_0) which is spread over a certain region, then choose the correct statement: **[2016 (Online)]**

(A) $\rho_0 = 20\,\varepsilon_0$ in the entire region
(B) $\rho_0 = 10\,\varepsilon_0$ for $|z| \leq 1$ m and $p_0 = 0$ elsewhere
(C) $\rho_0 = 20\,\varepsilon_0$ for $|z| \leq 1$ m and $p_0 = 0$ elsewhere
(D) $\rho_0 = 40\,\varepsilon_0$ in the entire region

25. Within a spherical charge distribution of charge density $\rho(r)$, N equipotential surfaces of potential $V_0, V_0 + \Delta V, V_0 + 2\Delta V, \ldots V_0 + N\Delta V (\Delta V > 0)$, are drawn and have increasing radii $r_0, r_1, r_2, \ldots r_N$, respectively. If the difference in the radii of the surfaces is constant for all values of V_0 and ΔV then : **[2016 (Online)]**

(A) $\rho(r)$ = constant (B) $\rho(r) \propto \frac{1}{r^2}$
(C) $\rho(r) \propto \frac{1}{r}$ (D) $\rho(r) \propto r$

26. There is a uniform electrostatic field in a region. The potential at various points on a small sphere centred at P, in the region, is found to vary between in the limits 589.0 V to 589.8 V. What is the potential at a point on the sphere whose radius vector makes an angle of 60° with the direction of the field ? **[2017 (Online)]**

(A) 589.5 V (B) 589.2 V
(C) 589.4 V (D) 589.6 V

27. Three concentric metal shells A, B and C of respective radii a, b and c $(a < b < c)$ have surface charge densities $+\sigma$, $-\sigma$ and $+\sigma$ respectively. The potential of shell B is : **[2018]**

(A) $\frac{\sigma}{\epsilon_0}\left[\frac{a^2 - b^2}{a} + c\right]$ (B) $\frac{\sigma}{\epsilon_0}\left[\frac{a^2 - b^2}{b} + c\right]$

(C) $\frac{\sigma}{\epsilon_0}\left[\frac{b^2 - c^2}{b} + a\right]$ (D) $\frac{\sigma}{\epsilon_0}\left[\frac{b^2 - c^2}{c} + a\right]$

28. A charge Q is distributed over three concentric spherical shells of radii a, b, c $(a < b < c)$ such that their surface charge densities are equal to one another. The total potential at a point at distance r from their common centre, where $r < a$, would be: **[10 Jan. 2019 (Morning)]**

(A) $\frac{Q}{4\pi\varepsilon_0(a + b + c)}$ (B) $\frac{Q(a + b + c)}{4\pi\varepsilon_0(a^2 + b^2 + c^2)}$

(C) $\frac{Q}{12\pi\varepsilon_0}\frac{ab + bc + ca}{abc}$ (D) $\frac{Q}{4\pi\varepsilon_0}\frac{(a^2 + b^2 + c^2)}{(a^3 + b^3 + c^3)}$

29. Two electric dipoles, A, B with respective dipole moments $\vec{d}_A = -4qa\hat{i}$ and $\vec{d}_B = -2qa\hat{i}$ placed on the x-axis with a separation R, as shown in the figure : **[10 Jan. 2019 (Morning)]**

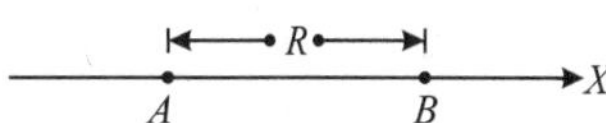

The distance from A at which both of them produce the same potential is :

(A) $\frac{\sqrt{2}R}{\sqrt{2}+1}$ (B) $\frac{R}{\sqrt{2}+1}$

(C) $\frac{\sqrt{2}\,R}{\sqrt{2}-1}$ (D) $\frac{R}{\sqrt{2}-1}$

30. Charges $-q$ and $+q$ located at A and B, respectively, constitute an electric dipole. Distance $AB = 2a$, O is the mid point of the dipole and OP is perpendicular to AB. A charge Q is placed at P where $OP = y$ and $y >> 2a$. The charge Q experiences and electrostatic force F. If Q is now moved along the equatorial line to P' such that $OP' = \left(\frac{y}{3}\right)$, the force on Q will be close to. $\left(\frac{y}{3} >> 2a\right)$: **[10 Jan. 2019 (Evening)]**

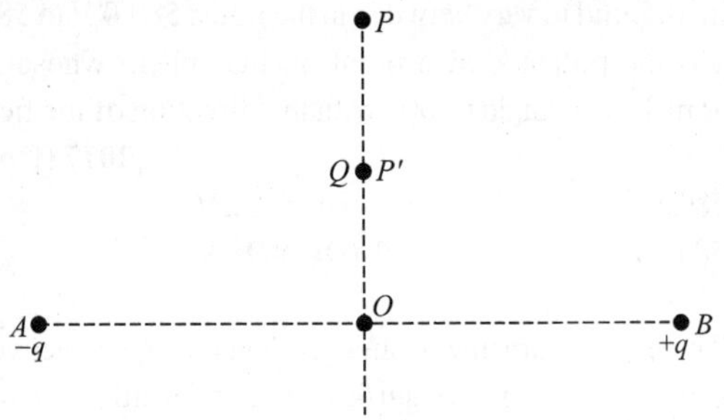

(A) $\frac{F}{3}$ (B) $3F$

(C) $9F$ (D) $27F$

31. Four equal point charges Q each are placed in the xy plane at (0, 2), (4, 2), (4, –2) and (0, –2). The work required to put a fifth charge Q at the origin of the coordinate system will be : **[10 Jan. 2019 (Evening)]**

(A) $\frac{Q^2}{2\sqrt{2}\pi\varepsilon_0}$ (B) $\frac{Q^2}{4\pi\varepsilon_0}\left(1+\frac{1}{\sqrt{5}}\right)$

(C) $\frac{Q^2}{4\pi\varepsilon_0}\left(1+\frac{1}{\sqrt{3}}\right)$ (D) $\frac{Q^2}{4\pi\varepsilon_0}$

32. The charges Q, $+q$ and $+q$ are placed at the vertices of a right-angle isosceles triangle as shown below. The net electrostatic energy of the configuration is zero, if the value of Q is: **[11 Jan. 2019 (Morning)]**

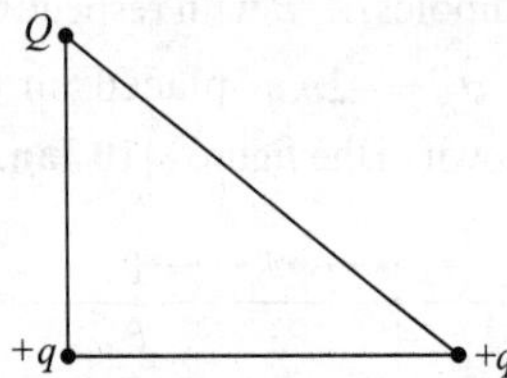

(A) $\frac{-\sqrt{2}q}{\sqrt{2}+1}$ (B) $-2q$

(C) $\frac{-q}{1+\sqrt{2}}$ (D) $+q$

33. The given graph shows variation (with distance r from centre) of: **[11 Jan. 2019 (Morning)]**

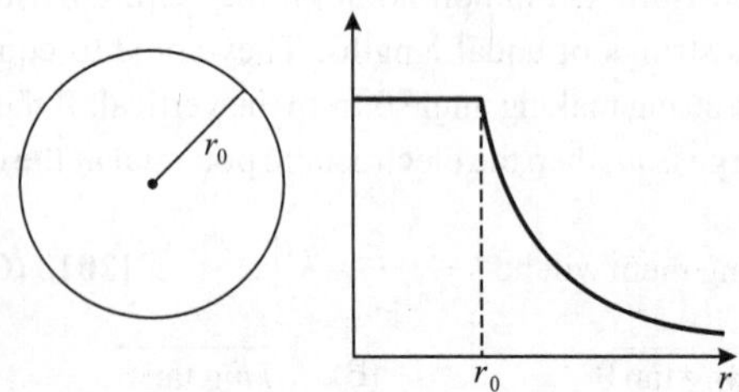

(A) Potential of a uniformly charged sphere
(B) Potential of a uniformly charged spherical shell
(C) Electric field of uniformly charged spherical shell
(D) Electric field of uniformly charged sphere

34. An electric field of 1000 V/m is applied to an electric dipole at angle of 45°. The value of electric dipole moment is 10^{-29} C.m. What is the potential energy of the electric dipole ? **[11 Jan. 2019 (Evening)]**

(A) -9×10^{-20} J (B) -7×10^{-27} J
(C) -10×10^{-29} J (D) -20×10^{-18} J

35. Determine the electric dipole moment of the system of three charges, placed on the vertices of an equilateral triangle, as shown in the figure : **[12 Jan. 2019 (Morning)]**

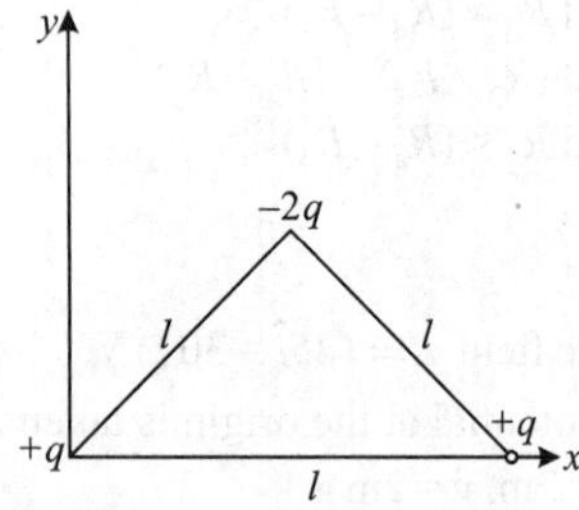

(A) $(ql)\frac{\hat{i}+\hat{j}}{\sqrt{2}}$ (B) $\sqrt{3}ql\frac{\hat{j}-\hat{i}}{\sqrt{2}}$

(C) $-\sqrt{3}ql\,\hat{j}$ (D) $2ql\,\hat{j}$

36. There is a uniform spherically symmetric surface charge density at a distance R_0 from the origin. The charge distribution is initially at rest and starts expanding because of mutual repulsion. The figure that represents best the speed $V(R(t))$ of the distribution as a function of its instantaneous radius $R(t)$ is : **[12 Jan. 2019 (Morning)]**

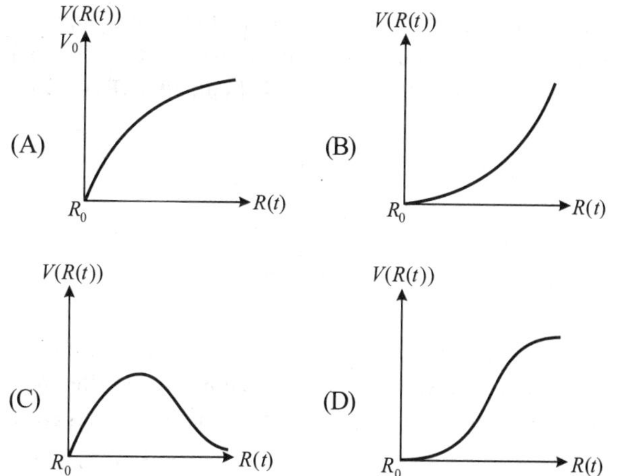

37. A solid conducting sphere, having a charge Q, is surrounded by an uncharged conducting hollow spherical shell. Let the potential difference between the surface of the solid sphere and that of the outer surface of the hollow shell be V. If the shell is now given a charge of $-4Q$, the new potential difference between the same two surfaces is :

[8 April 2019 (Morning)]

(A) V (B) 2V
(C) –2 V (D) 4V

38. An electric dipole is formed by two equal and opposite charges q with separation d. The charges have same mass m. It is kept in a uniform electric field E. If it is slightly rotated from its equilibrium orientation, then its angular frequency ω is :

[8 April 2019 (Evening)]

(A) $\sqrt{\frac{qE}{2md}}$ (B) $2\sqrt{\frac{qE}{md}}$
(C) $\sqrt{\frac{2qE}{md}}$ (D) $\sqrt{\frac{qE}{md}}$

39. A positive point charge is released from rest at a distance r_0 from a positive line charge with uniform density. The speed (v) of the point charge, as a function of instantaneous distance r from line charge, is proportional to : **[8 April 2019 (Evening)]**

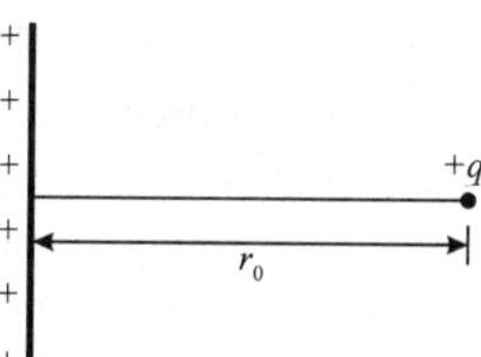

(A) $v \propto e^{+r/r_0}$ (B) $v \propto \ln\left(\frac{r}{r_0}\right)$
(C) $v \propto \left(\frac{r}{r_0}\right)$ (D) $v \propto \sqrt{\ln\left(\frac{r}{r_0}\right)}$

40. The electric field in a region is given by $\vec{E} = (Ax + B)\hat{i}$, where E is in NC^{-1} and x is in metres. The values of constants are $A = 20$ SI unit and $B = 10$ SI unit. If the potential at $x = 1$ is V_1 and that at $x = -5$ is V_2, then $V_1 - V_2$ is :

[8 April 2019 (Evening)]

(A) –48 V (B) –520 V
(C) 180 V (D) 320 V

41. A system of three charges are placed as shown in the figure.

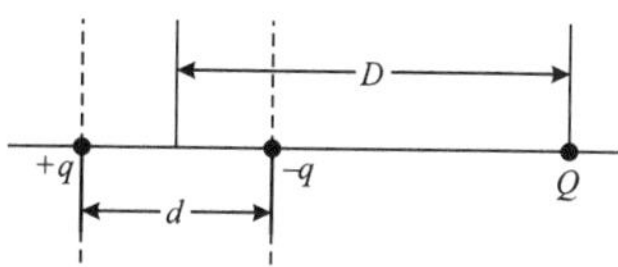

If $D >> d$, the potential energy of the system is best given by :

[9 April 2019 (Morning)]

(A) $\frac{1}{4\pi\varepsilon_0}\left[-\frac{q^2}{d} - \frac{qQd}{2D^2}\right]$ (B) $\frac{1}{4\pi\varepsilon_0}\left[+\frac{q^2}{d} + \frac{qQd}{D^2}\right]$

(C) $\frac{1}{4\pi\varepsilon_0}\left[-\frac{q^2}{d} + \frac{2qQd}{D^2}\right]$ (D) $\frac{1}{4\pi\varepsilon_0}\left[-\frac{q^2}{d} - \frac{qQd}{D^2}\right]$

42. A uniformly charged ring of radius $3a$ and total charge q is placed in xy-plane centred at origin. A point charge q is moving towards the ring along the z-axis and has speed u at $z = 4a$. The minimum value of u such that it crosses the origin is :

[10 April 2019 (Morning)]

(A) $\sqrt{\frac{2}{m}}\left(\frac{1}{15}\frac{q^2}{4\pi\varepsilon_0 a}\right)^{1/2}$ (B) $\sqrt{\frac{2}{m}}\left(\frac{2}{15}\frac{q^2}{4\pi\varepsilon_0 a}\right)^{1/2}$

(C) $\sqrt{\frac{2}{m}}\left(\frac{4}{15}\frac{q^2}{4\pi\varepsilon_0 a}\right)^{1/2}$ (D) $\sqrt{\frac{2}{m}}\left(\frac{1}{5}\frac{q^2}{4\pi\varepsilon_0 a}\right)^{1/2}$

43. In free space, a particle A of charge 1 μC is held fixed at a point P. Another particle B of the same charge and mass 4 μg is kept at a distance of 1 mm from P. If B is released, then its velocity at a distance of 9 mm from P is :

$\left[\text{Take } \frac{1}{4\pi\varepsilon_0} = 9 \times 10^9 \text{ Nm}^2\text{C}^{-2}\right]$ **[10 April 2019 (Evening)]**

(A) 2.0×10^3 m/s (B) 3.0×10^4 m/s
(C) 1.5×10^2 m/s (D) 1.0 m/s

44. Shown in the figure is a shell made of a conductor. It has inner radius a and outer radius b, and carries charge Q. At its centre is a dipole $\vec{p}$ as shown. In this case :

[12 April 2019 (Morning)]

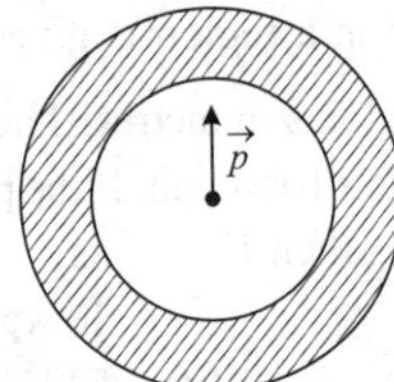

(A) Electric field outside the shell is the same as that of a point charge at the centre of the shell

(B) Surface charge density on the inner surface of the shell is zero everywhere

(C) Surface charge density on the inner surface is uniform and equal to $\frac{(Q/2)}{4\pi a^2}$

(D) Surface charge density on the outer surface depends on $|\vec{p}|$

45. A point dipole $\vec{p} = -p_0\hat{x}$ is kept at the origin. The potential and electric field due to this dipole on the *y*-axis at a distance *d* are, respectively : (Take $V = 0$ at infinity) : **[12 April 2019 (Morning)]**

(A) $\frac{|\vec{p}|}{4\pi\varepsilon_0 d^2}, \frac{-\vec{p}}{4\pi\varepsilon_0 d^3}$ (B) $0, \frac{\vec{p}}{4\pi\varepsilon_0 d^3}$

(C) $\frac{|\vec{p}|}{4\pi\varepsilon_0 d^2}, \frac{\vec{p}}{4\pi\varepsilon_0 d^3}$ (D) $0, \frac{-\vec{p}}{4\pi\varepsilon_0 d^3}$

46. Consider two charged metallic spheres S_1 and S_2 of radii R_1 and R_2, respectively. The electric fields E_1 (on S_1) and E_2 (on S_2) on their surfaces are such that $E_1/E_2 = R_1/R_2$. Then the ratio V_1 (on S_1)/V_2 (on S_2) of the electrostatic potentials on each sphere is : **[8 Jan. 2020 (Evening)]**

(A) (R_2/R_1) (B) $(R_1/R_2)^3$

(C) R_1/R_2 (D) $(R_1/R_2)^2$

47. An electric dipole of moment $\vec{p} = (-\hat{i} - 3\hat{j} + 2\hat{k}) \times 10^{-29}$ C.m is at the origin (0, 0, 0). The electric field due to this dipole at $\vec{r} = +\hat{i} + 3\hat{j} + 5\hat{k}$ (note that $\vec{r} \cdot \vec{p} = 0$) is parallel to : **[9 Jan. 2020 (Morning)]**

(A) $(-\hat{i} + 3\hat{j} - 2\hat{k})$ (B) $(+\hat{i} - 3\hat{j} - 2\hat{k})$

(C) $(+\hat{i} + 3\hat{j} - 2\hat{k})$ (D) $(-\hat{i} - 3\hat{j} + 2\hat{k})$

48. Two isolated conducting spheres S_1 and S_2 of radius $\frac{2}{3}R$ and $\frac{1}{3}R$ have 12 μC and – 3 μC charges, respectively, and are at a large distance from each other. They are now connected by a conducting wire. A long time after this is done the charges on S_1 and S_2 are respectively : **[3 Sept. 2020 (Morning)]**

(A) 6 μC and 3 μC (B) + 4.5 μC and – 4.5 μC

(C) 3 μC and 6 μC (D) 4.5 μC on both

49. Concentric metallic hollow spheres of radii *R* and 4*R* hold charges Q_1 and Q_2 respectively. Given that surface charge densities of the concentric spheres are equal, the potential difference $V(R) - V(4R)$ is : **[3 Sept. 2020 (Evening)]**

(A) $\frac{3Q_1}{16\pi\varepsilon_0 R}$ (B) $\frac{Q_2}{4\pi\varepsilon_0 R}$

(C) $\frac{3Q_1}{4\pi\varepsilon_0 R}$ (D) $\frac{3Q_2}{4\pi\varepsilon_0 R}$

50. Hydrogen ion and singly ionized helium atom are accelerated, from rest, through the same potential difference. The ratio of final speeds of hydrogen and helium ions is close to : **[3 Sept. 2020 (Evening)]**

(A) 5 : 7 (B) 1 : 2

(C) 10 : 7 (D) 2 : 1

51. A two point charges 4*q* and –*q* are fixed on the *x*-axis at $x = -\frac{d}{2}$ and $x = \frac{d}{2}$, respectively. If a third point charge '*q*' is taken from the origin to $x = d$ along the semicircle as shown in the figure, the energy of the charge will : **[4 Sept. 2020 (Morning)]**

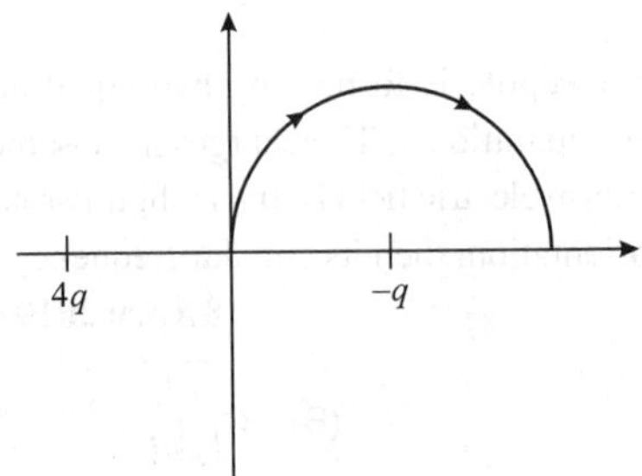

(A) Increase by $\frac{2q^2}{3\pi\varepsilon_0 d}$ (B) Increase by $\frac{3q^2}{4\pi\varepsilon_0 d}$

(C) Decrease by $\frac{4q^2}{3\pi\varepsilon_0 d}$ (D) Decrease by $\frac{q^2}{4\pi\varepsilon_0 d}$

52. A particle of charge *q* and mass m is subjected to an electric field $E = E_0(1 - ax^2)$ in the *x*-direction, where *a* and E_0 are constants. Initially the particle was at rest at $x = 0$. Other than the initial position the kinetic energy of the particle becomes zero when the distance of the particle from the origin is : **[4 Sept. 2020 (Evening)]**

(A) $\sqrt{\frac{2}{a}}$ (B) $\sqrt{\frac{1}{a}}$

(C) a (D) $\sqrt{\frac{3}{a}}$

53. A solid sphere of radius R carries a charge $(Q+q)$ distributed uniformly over its volume. A very small point like piece of it of mass m gets detached from the bottom of the sphere and falls down vertically under gravity. This piece carries charge q. If it acquires a speed v when it has fallen through a vertical height y (see figure), then : (Assume the remaining portion to be spherical) **[5 Sept. 2020 (Morning)]**

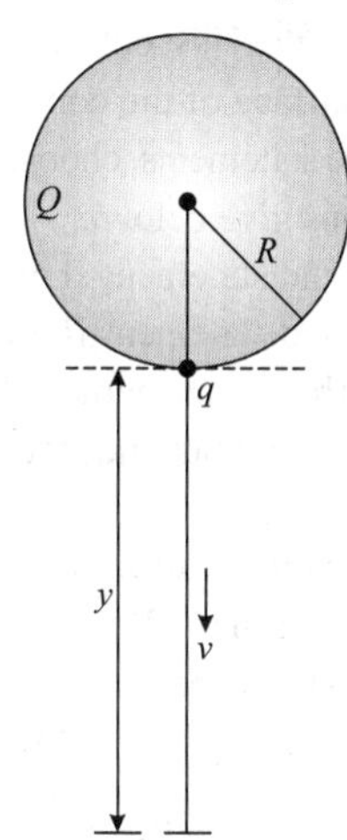

(A) $v^2 = 2y\left[\dfrac{qQ}{4\pi\in_0 R(R+y)m}+g\right]$

(B) $v^2 = y\left[\dfrac{qQ}{4\pi\in_0 R^2 ym}+g\right]$

(C) $v^2 = 2y\left[\dfrac{qQR}{4\pi\in_0 (R+y)^3 m}+g\right]$

(D) $v^2 = y\left[\dfrac{qQ}{4\pi\in_0 R(R+y)m}+g\right]$

54. Ten charges are placed on the circumference of a circle of radius R with constant angular separation between successive charges. Alternate charges 1, 3, 5, 7, 9 have charge $(+q)$ each, while 2, 4, 6, 8, 10 have charge $(-q)$ each. The potential V and the electric field E at the centre of the circle are respectively : (Take $V=0$ at infinity) **[5 Sept. 2020 (Evening)]**

(A) $V=\dfrac{10q}{4\pi\in_0 R}; E=\dfrac{10q}{4\pi\in_0 R^2}$

(B) $V=0, E=\dfrac{10q}{4\pi\in_0 R^2}$

(C) $V=0, E=0$

(D) $V=\dfrac{10q}{4\pi\in_0 R}; E=0$

55. Two identical electric point dipoles have dipole moments $\vec{p}_1 = p\hat{i}$ and $\vec{p}_2 = -p\hat{i}$ and are held on the x-axis at distance 'a' from each other. When released, they move along the x-axis with the direction of their dipole moments remaining unchanged. If the mass of each dipole is 'm', their speed when they arc infinitely far apart is : **[6 Sept. 2020 (Evening)]**

(A) $\dfrac{p}{a}\sqrt{\dfrac{1}{\pi\varepsilon_0 ma}}$ (B) $\dfrac{p}{a}\sqrt{\dfrac{3}{2\pi\varepsilon_0 ma}}$

(C) $\dfrac{p}{a}\sqrt{\dfrac{1}{2\pi\varepsilon_0 ma}}$ (D) $\dfrac{p}{a}\sqrt{\dfrac{2}{\pi\varepsilon_0 ma}}$

56. 512 identical drops of mercury are charged to a potential of 2 V each. The drops are joined to form a single drop. The potential of this drop is ______ V. **[25 Feb. 2021 (Morning)]**

57. 27 similar drops of mercury are maintained at 10 V each. All these spherical drops combine into a single big drop. The potential energy of the bigger drops is ______ times that of a smaller drop. **[26 Feb. 2021 (Evening)]**

58. Two ideal electric dipoles A and B, having their dipole moment p_1 and p_2 respectively are placed on a plane with their centres at O as shown in the figure. At point C on the axis of dipole A, the resultant electric field is making an angle of 37° with the axis. The ratio of the dipole moment of A and B, $\dfrac{P_1}{P_2}$ is :

(Take $\sin 37° = \dfrac{3}{5}$) **[25 July 2021 (Evening)]**

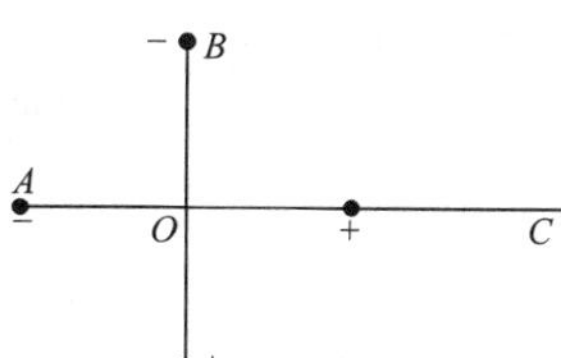

(A) $\dfrac{3}{2}$ (B) $\dfrac{2}{3}$

(C) $\dfrac{3}{8}$ (D) $\dfrac{4}{3}$

59. The two thin coaxial rings, each of radius a and having charges $+Q$ and $-Q$ respectively are separated by a distance of s. The potential difference between the centres of the two rings is : **[26 Aug. 2021 (Evening)]**

(A) $\dfrac{Q}{4\pi\varepsilon_0}\left[\dfrac{1}{a}-\dfrac{1}{\sqrt{s^2+a^2}}\right]$ (B) $\dfrac{Q}{4\pi\varepsilon_0}\left[\dfrac{1}{a}-\dfrac{1}{\sqrt{s^2+a^2}}\right]$

(C) $\dfrac{Q}{4\pi\varepsilon_0}\left[\dfrac{1}{a}+\dfrac{1}{\sqrt{s^2+a^2}}\right]$ (D) $\dfrac{Q}{2\pi\varepsilon_0}\left[\dfrac{1}{a}+\dfrac{1}{\sqrt{s^2+a^2}}\right]$

60. 27 identical drops are charged at 22 V each. They combine to form a bigger drop. The potential of the bigger drop will be _____ V. **[25 June 2022 (Evening)]**

70. Given below two statements:
One is labelled as Assertion-(A) and other is labelled as Reason-(R) **[26 June 2022 (Morning)]**
Assertion-(A) : Non-polar materials do not have my permanent dipole moment.
Reason-(R) : When an non-polar material is placed in a electric field. The centre of the positive charge distribution of it's individual atom or molecule coincides with the centre of the negative charge distribution.
In the light of above statements, choose the most appropriate answer from the options given below :
(A) Both (A) and (R) are correct and (R) is the correct explanation of (A)
(B) Both (A) and (R) are correct and (R) is not the correct explanation of (A)
(C) (A) is correct but (R) is not correct
(D) (A) is not correct but (R) is correct

71. If the electric potential at any point (x, y, z) m in space is given by $V = 3x^2$ volt. The electric field at the point $(1, 0, 3)m$ will be : **[29 June 2022 (Evening)]**
(A) 3 Vm^{-1}, directed along positive x-axis
(B) 3 Vm^{-1}, directed along negative x-axis
(C) 6 Vm^{-1}, directed along positive x-axis
(D) 6 Vm^{-1}, directed along negative x-axis

72. Capacitance of an isolated conducting sphere of radius R_1 becomes n times when it is enclosed by a concentric conducting sphere of radius R_2 connected to earth. The ratio of their radii $\left(\frac{R_2}{R_1}\right)$ is : **[25 July 2022 (Evening)]**
(A) $\frac{n}{n-1}$ (B) $\frac{2n}{2n+1}$
(C) $\frac{n+1}{n}$ (D) $\frac{2n+1}{n}$

73. Two uniformly charged spherical conductors A and B of radii 5 mm and 10 mm are separated by a distance of 2 cm. If the spheres are connected by a conducting wire, then in equilibrium condition, the ratio of the magnitudes of the electric fields at the surface of the sphere A and B will be : **[26 July 2022 (Evening)]**
(A) 1 : 2 (B) 2 : 1
(C) 1 : 1 (D) 1 : 4

74. Two electric dipoles of dipole moments 1.2×10^{-30} C-m and 2.4×10^{-30} C-m are placed in two different uniform electric fields of strength 5×10^4 NC^{-1} and 15×10^4 NC^{-1} respectively. The ratio of maximum torque experienced by the electric dipoles will be $\frac{1}{x}$. The value of x is _____. **[28 July 2022 (Morning)]**

75. Given below are two statements. **[29 July 2022 (Morning)]**
Statement-I : Electric potential is constant within and at the surface of each conductor
Statement-II : Electric field just outside a charged conductor is perpendicular to the surface of the conductor at every point.
In the light of the above statements, choose the most appropriate answer from the options give below :
(A) Both statement-I and statement-II are correct
(B) Both statement-I and statement-II are incorrect
(C) Statement-I is correct but statement-II is incorrect
(D) Statement-I is incorrect but and statement-II is correct

76. An α particle and a proton are accelerated from rest through the same potential difference. The ratio of linear momenta acquired by above two particles will be : **[29 July 2022 (Evening)]**
(A) $\sqrt{2}:1$ (B) $2\sqrt{2}:1$
(C) $4\sqrt{2}:1$ (D) $8:1$

77. The electric potential at the centre of two concentric half rings of radii R_1 and R_2, having same linear charge density λ is : **[24 Jan. 2023 (Evening)]**

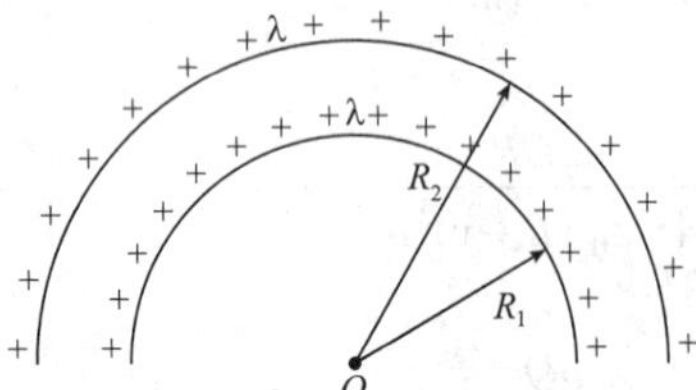

(A) $\frac{2\lambda}{\epsilon_0}$ (B) $\frac{\lambda}{2\epsilon_0}$
(C) $\frac{\lambda}{4\epsilon_0}$ (D) $\frac{\lambda}{\epsilon_0}$

78. A uniform electric field of 10 N/C is created between two parallel charged plates (as shown in figure). An electron enters the field symmetrically between the plates with a kinetic energy 0.5 eV. The length of each plate is 10 cm. The angle (θ) of deviation of the path of electron as it comes out of the field is _____. (In degree) **[25 Jan. 2023 (Morning)]**

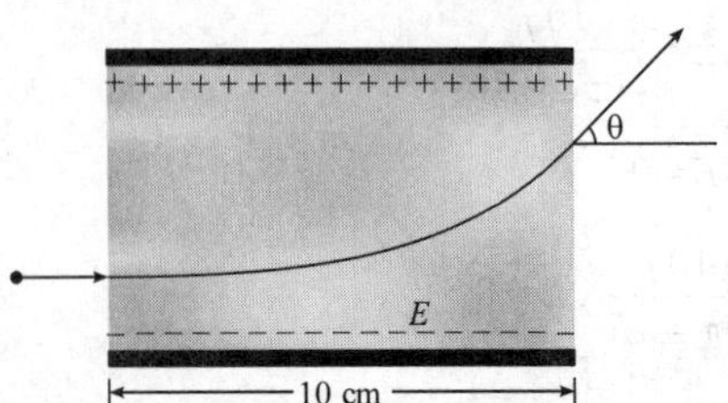

79. A point charge 2×10^{-2} C is moved from P to S in a uniform electric field of 30 NC^{-1} directed along positive x-axis. If coordinates of P and S are (1, 2, 0) m and (0, 0, 0) m respectively, the work done by electric field will be : **[29 Jan. 2023 (Evening)]**

(A) 1200 mJ (B) 600 mJ
(C) – 600 mJ (D) – 1200 mJ

80. For a charged spherical ball, electrostatic potential inside the ball varies with r as $V = 2ar^2 + b$. Here, a and b are constant and r is the distance from the center. The volume charge density inside the ball is $-\lambda a\varepsilon$. The value of λ is _____. ε = permittivity of medium. **[29 Jan. 2023 (Evening)]**

81. Two isolated metallic solid spheres of radii R and $2R$ are charged such that both have same charge density σ. The spheres are then connected by a thin conducting wire. If the new charge density of the bigger sphere is σ'. The ratio $\frac{\sigma'}{\sigma}$ is : **[30 Jan. 2023 (Morning)]**

(A) $\frac{9}{4}$ (B) $\frac{4}{3}$

(C) $\frac{5}{3}$ (D) $\frac{5}{6}$

82. Which of the following correctly represents the variation of electric potential (V) of a charged spherical conductor of radius (R) with radial distance (r) from the centre ? **[31 Jan. 2023 (Morning)]**

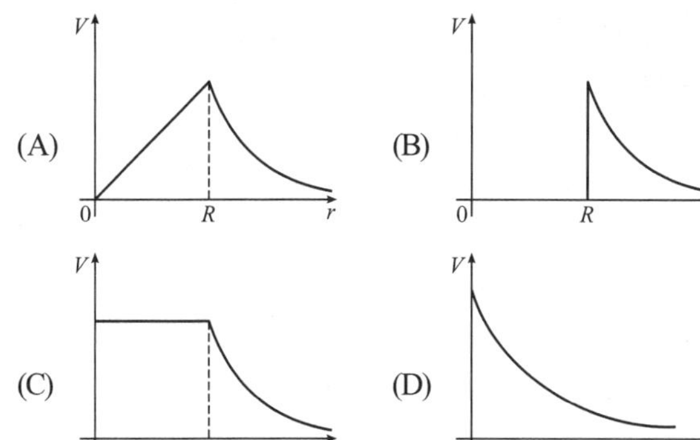

83. Considering a group of positive charges, which of the following statements is correct ? **[31 Jan. 2023 (Evening)]**

(A) Net potential of the system cannot be zero at a point but net electric field can be zero at that point

(B) Net potential of the system at a point can be zero but net electric field can't be zero at that point

(C) Both the net potential and the net field can be zero at a point

(D) Both the net potential and the net electric field cannot be zero at a point

84. For a uniformly charged thin spherical shell, the electric potential (V) radially away from the center (O) of shell can be graphically represented as **[06 April 2023 (Morning)]**

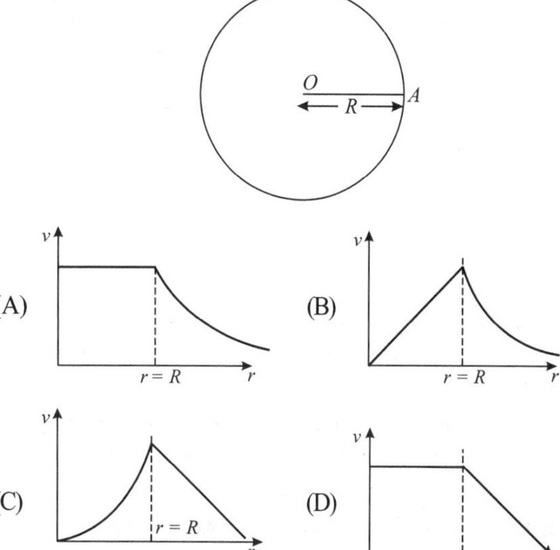

85. A dipole comprises of two charged particles of identical magnitude q and opposite in nature. The mass m of the positive charged particle is half of the mass of the negative charged particle. The two charges are separated by a distance l. If the dipole is placed in a uniform electric field $\vec{E}$; such a way that dipole axis makes a very small angle with the electric field, $\vec{E}$. The angular frequency of the oscillations of the dipole when released is given by: **[6 April 2023 (Evening)]**

(A) $\sqrt{\frac{8qE}{3ml}}$ (B) $\sqrt{\frac{4qE}{ml}}$

(C) $\sqrt{\frac{4qE}{3ml}}$ (D) $\sqrt{\frac{8qE}{ml}}$

86. An electric dipole of dipole moment is 6.0×10^{-6} cm placed in a uniform electric field of 1.5×10^3 NC^{-1} in such a way that dipole moment is along electric field. The work done in rotating dipole by 180° in this field will be _____ mJ. **[8 April 2023 (Morning)]**

87. Electric potential at a point P due to a point charge of 5×10^{-9} C is 50 V. The distance of P from the point charge is :

[Assume, $\frac{1}{4\pi\varepsilon_0} = 9 \times 10^{+9}$ $Nm^2 C^{-2}$] **[8 April 2023 (Evening)]**

(A) 3 cm (B) 9 cm
(C) 90 cm (D) 0.9 cm

88. Three concentric spherical metallic shells X, Y and Z of radius a, b and c respectively $[a < b < c]$ have surface charge densities σ, $-\sigma$ and σ, respectively. The shells X and Z are at same potential. If the radii of X and Y are 2 cm and 3 cm, respectively. The radius of shell Z is ______ cm.

[10 April 2023 (Morning)]

89. If V is the gravitational potential due to sphere of uniform density on it's surface, then it's value at the center of sphere will be : **[11 April 2023 (Evening)]**

(A) $\frac{3V}{2}$ (B) V

(C) $\frac{4}{3}V$ (D) $\frac{V}{2}$

90. Given below are two statement:

One is labelled as **Assertion-A** and the other is labelled as **Reason-R.**

Assertion-A: If an electric dipole of dipole moment 30×10^{-5} cm is enclosed by a closed surface, the net flux coming out of the surface will be zero.

Reason-R: Electric dipole consists of two equal and opposite charges.

In the light of above, statements, choose the correct answer from the options given below: **[12 April 2023 (Morning)]**

(A) Both **A** and **R** are true and **R** is the correct explanation of **A**.

(B) **A** is true but **R** is false.

(C) Both **A** and **R** true but **R** is NOT the correct explanation of **A**.

(D) **A** is false but **R** is true.

91. 64 identical drops each charged upto potential of 10 mV are combined to form a bigger drop. The potential of the bigger drop will be _____ mV. **[12 April 2023 (Morning)]**

92. Two charges each of magnitude 0.01 C and separated by a distance of 0.4 mm constitute an electric dipole. If the dipole is placed in an uniform electric field $\vec{E}$ of 10 dyne/C making 30° angle with $\vec{E}$, the magnitude of torque acting on dipole is : **[13 April 2023 (Morning)]**

(A) 4.0×10^{-10} Nm (B) 2.0×10^{-10} Nm

(C) 1.0×10^{-8} Nm (D) 1.5×10^{-9} Nm

93. The electric field due to a short electric dipole at a large distance (r) from center of dipole on the equatorial plane varies with distance as : **[15 April 2023 (Morning)]**

(A) r (B) $\frac{1}{r}$

(C) $\frac{1}{r^3}$ (D) $\frac{1}{r^2}$

* * * * *

CHAPTER 16

ELECTRIC POTENTIAL AND CAPACITANCE

16.2 CAPACITANCE

1. Capacitance (in F) of a spherical conductor having radius 1 m, is : **[2002]**

(A) 1.1×10^{-10} (B) 10^{-6}
(C) 9×10^{-9} (D) 10^{-3}

2. If there are n capacitors in parallel connected to V volt source, then the energy stored is equal to : **[2002]**

(A) CV (B) $\frac{1}{2}nCV^2$
(C) CV^2 (D) $\frac{1}{2n}CV^2$

3. The work done in placing a charge of 8×10^{-18} C on a condenser of capacity 100 micro-farad is : **[2003]**

(A) 16×10^{-32} J (B) 3.1×10^{-26} J
(C) 4×10^{-10} J (D) 32×10^{-32} J

4. A sheet of aluminium foil of negligible thickness is introduced between the plates of a capacitor. The capacitance of the capacitor : **[2003]**

(A) Decreases (B) Remains unchanged
(C) Becomes infinite (D) Increases

5. A parallel plate capacitor is made by stacking n equally spaced plates connected alternatively. If the capacitance between any two adjacent plates is C, then the resultant capacitance is : **[2005]**

(A) $(n-1)C$ (B) $(n+1)C$
(C) C (D) nC

6. A fully charged capacitor has a capacitance C. It is discharged through a small coil of resistance wire embedded in a thermally insulated block of specific heat capacity s and mass m. If the temperature of the block is raised by ΔT, the potential difference V across the capacitance is : **[2005]**

(A) $\sqrt{\frac{2mC\Delta T}{s}}$ (B) $\frac{mC\Delta T}{s}$
(C) $\frac{ms\Delta T}{C}$ (D) $\sqrt{\frac{2ms\Delta T}{C}}$

7. Two insulating plates are both uniformly charged in such a way that the potential difference between them is $V_2 - V_1 = 20$ V. (i.e., plate 2 is at a higher potential). The plates are separated by $d = 0.1$ m and can be treated as infinitely large. An electron is released from rest on the inner surface of plate 1. What is its speed when it hits plate 2? **[2006]**

($e = 1.6 \times 10^{-19}$ C, $m_0 = 9.11 \times 10^{-31}$ kg)

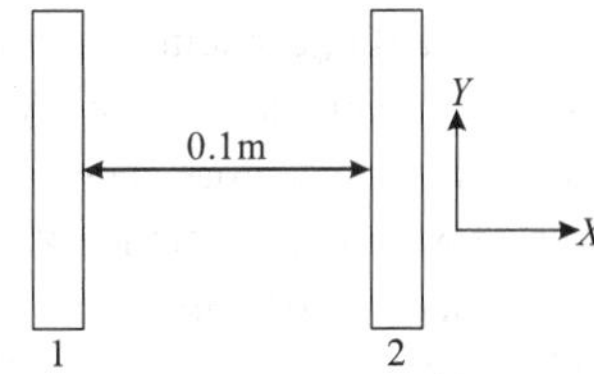

(A) 2.65×10^6 m/s (B) 7.02×10^{12} m/s
(C) 1.87×10^6 m/s (D) 32×10^{-19} m/s

8. A parallel plate condenser with a dielectric of dielectric constant K between the plates has a capacity C and is charged to a potential V volts. The dielectric slab is slowly removed from between the plates and then re-inserted. The net work done by the system in this process is : **[2007]**

(A) $\frac{1}{2}(K-1)CV^2$ (B) $CV^2(K-1)/K$
(C) $(K-1)CV^2$ (D) Zero

9. A battery is used to charge a parallel plate capacitor till the potential difference between the plates becomes equal to the electromotive force of the battery. The ratio of the energy stored in the capacitor and the work done by the battery will be : **[2007]**

(A) 1 (B) 2
(C) 1/4 (D) 1/2

10. A parallel plate capacitor with air between the plates has a capacitance of 9 pF. The separation between its plates is 'd'. The space between the plates is now filled with two dielectrics. One of the dielectric has dielectric constant $K_1 = 3$ and thickness $\frac{d}{3}$ while the other one has dielectric constant $K_2 = 6$ and thickness $\frac{2d}{3}$. Capacitance of the capacitor is now : **[2008]**

(A) 1.8 pF (B) 45 pF
(C) 40.5 pF (D) 20.25 pF

11. Two circuits (a) and (b) have charged capacitors of capacitance C, 2C and 3C with open switches. Charges on each of the capacitor are as shown in the figures. On closing the switches : **[2012 (Online)]**

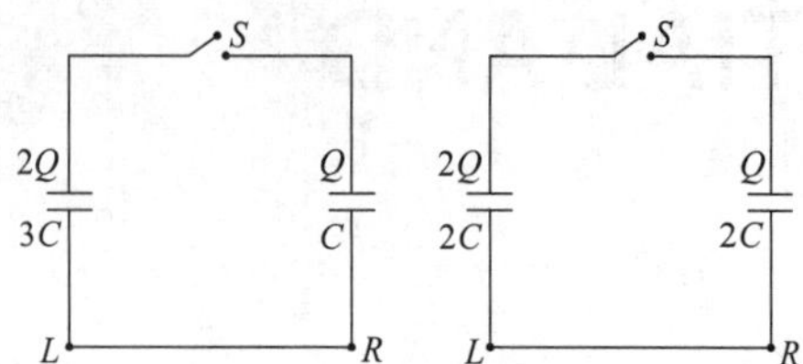

(A) No charge flows in (a) but charge flows from R to L in (b)
(B) Charges flow from L to R in both (a) and (b)
(C) Charges flows from R to L in (a) and from L to R in (b)
(D) No charge flows in (a) but charge flows from L to R in (b)

12. A series combination of n_1 capacitors, each of capacity C_1 is charged by source of potential difference 4 V. When another parallel combination of n_2 capacitors each of capacity C_2 is charged by a source of potential difference V, it has the same total energy stored in it as the first combination has. The value of C_2 in terms of C_1 is then : **[2012 (Online)]**

(A) $16\frac{n_2}{n_1}C_1$ (B) $\frac{2C_1}{n_1 n_2}$
(C) $2\frac{n_2}{n_1}C_1$ (D) $\frac{16C_1}{n_1 n_2}$

13. This question has Statement-1 and Statement-2. Of the four choices given after the Statements, choose the one that best describes the two Statements. **[2012 (Online)]**
Statement-1 : It is not possible to make a sphere of capacity 1 farad using a conducting material.
Statement-2 : It is possible for earth as its radius is 6.4×10^6 m.
(A) Statement-1 is true, Statement-2 is true, Statement-2 is the correct explanation of Statement-1
(B) Statement-1 is false, Statement-2 is true
(C) Statement-1 is true, Statement-2 is true, Statement-2 is not the correct explanation of Statement-1
(D) Statement-1 is true, Statement-2 is false

14. Two capacitors C_1 and C_2 are charged to 120 V and 200 V respectively. It is found that by connecting them together the potential on each one can be made zero. Then : **[2013]**
(A) $5C_1 = 3C_2$ (B) $3C_1 = 5C_2$
(C) $3C_1 + 5C_2 = 0$ (D) $9C_1 = 4C_2$

15. A uniform electric field $\vec{E}$ exists between the plates of a charged condenser. A charged particle enters the space between the plates and perpendicular to $\vec{E}$. The path of the particle between the plates is a : **[2013 (Online)]**
(A) Straight line (B) Hyperbola
(C) Parabola (D) Circle

16. To establish an instantaneous current of 2 A through a 1 μF capacitor; the potential difference across the capacitor plates should be changed at the rate of : **[2013 (Online)]**
(A) 2×10^4 V/s (B) 4×10^6 V/s
(C) 2×10^6 V/s (D) 4×10^4 V/s

17. A parallel plate capacitor having a separation between the plates d, plate are A and material with dielectric constant K has capacitance C_0. Now one-third of the material is replaced by another material with dielectric constant 2 K, so that effectively there are two capacitors one with area $\frac{1}{3}A$, dielectric constant 2K and another with area $\frac{2}{3}A$ and dielectric constant K. If the capacitance of this new capacitor is C then C/C_0 is : **[2013 (Online)]**
(A) 1 (B) 4/3
(C) 2/3 (D) 1/3

18. A parallel plate capacitor is made of two circular plates separated by a distance of 5 mm and with a dielectric of dielectric constant 2.2 between them. When the electric field in the dielectric is 3×10^4 V/m, the charge density of the positive plate will be closed to : **[2014]**
(A) 6×10^{-7} C/m^2 (B) 3×10^{-7} C/m^2
(C) 3×10^4 C/m^2 (D) 6×10^4 C/m^2

19. The gap between the plates of a parallel plate capacitor of area A and distance between plates d, is filled with a dielectric whose permittivity varies linearly from $\in_1$ at one plate to $\in_2$ at the other. The capacitance of capacitor is: **[2014 (Online)]**
(A) $\in_0(\in_1 + \in_2)A/d$ (B) $\in_0(\in_2 + \in_1)A/2d$
(C) $\in_0 A/[d\ln(\in_2/\in_1)]$ (D) $\in_0(\in_2 - \in_1)A/[d\ln(\in_2/\in_1)]$

20. Three capacitors, each of 3 μF, are provided these cannot be combined to provide the resultant capacitance of : **[2014 (Online)]**
(A) 1 μF (B) 2 μF
(C) 4.5 μF (D) 6 μF

21. A parallel plate capacitor is made of two plates of length l, width w and separated by distance d. A dielectric slab (dielectric constant K) that fits exactly between the plates is held near the edge of the plates. It is pulled into the capacitor by a force $F = -\frac{\partial U}{\partial x}$ where U is the energy of the capacitor when dielectric is inside the capacitor up to distance x (See figure). If the charge on the capacitor is Q then the force on the dielectric when it is near the edge is : **[2014 (Online)]**

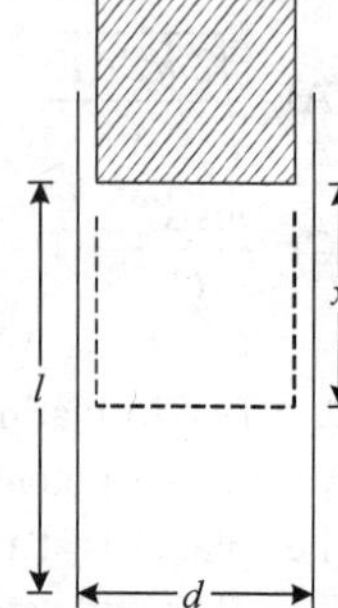

(A) $\frac{Q^2 d}{2\omega l^2 \varepsilon_0}K$

(B) $\frac{Q^2 w}{2\omega l^2 \varepsilon_0}(K-1)$

(C) $\frac{Q^2 d}{2wl^2 \varepsilon_0}(K-1)$

(D) $\frac{Q^2 w}{2dl^2 \varepsilon_0}K$

22. The space between the plates of a parallel plate capacitor is filled with a 'dielectric' whose 'dielectric constant' varies with distance as per the relation :

$K(x) = K_0 + \lambda x$ ($\lambda = a$ constant)

The capacitance C, of the capacitor, would be related to its vacuum capacitance C_0 for the relation : **[2014 (Online)]**

(A) $C = \frac{\lambda d}{\ln(1+K_0\lambda d)} C_0$ (B) $C = \frac{\lambda}{d \cdot \ln(1+K_0\lambda d)} C_0$

(C) $C = \frac{\lambda d}{\ln(1+\lambda d / K_0)} C_0$ (D) $C = \frac{\lambda}{d \cdot \ln(1+K_0 / \lambda d)} C_0$

23. In the given circuit, charge Q_2 on the 2 μF capacitor changes as C is varied from 1 μF to 3 μF. Q_2 as a function of 'C' is given properly by : (figures are drawn schematically and are not to scale) **[2015]**

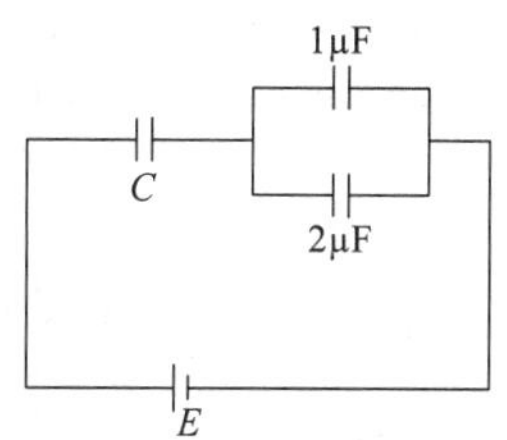

(A) 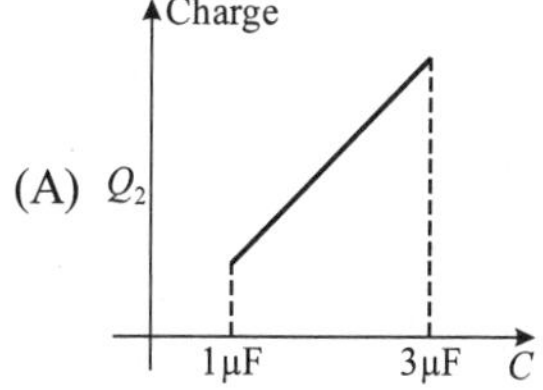

(B) 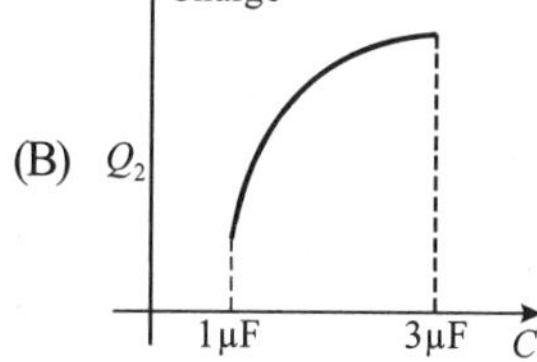

(C) 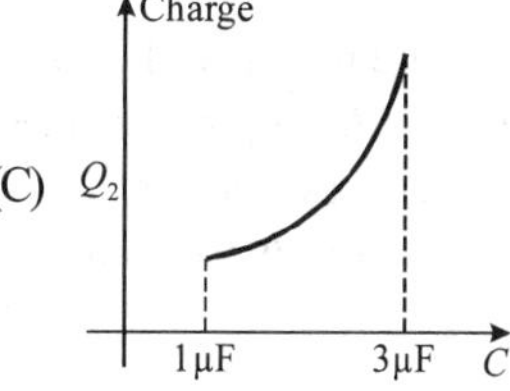

(D) 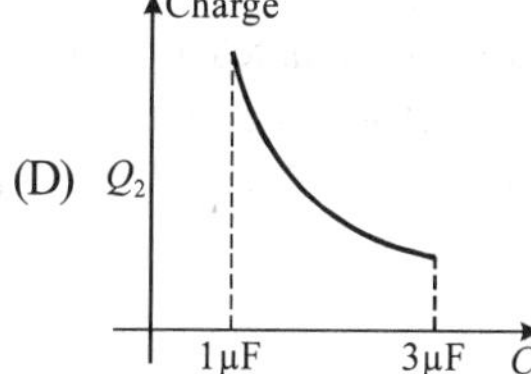

24. In figure a system of four capacitors connected across a 10 V battery is shown. Charge that will flow from switch S when it is closed is : **[2015 (Online)]**

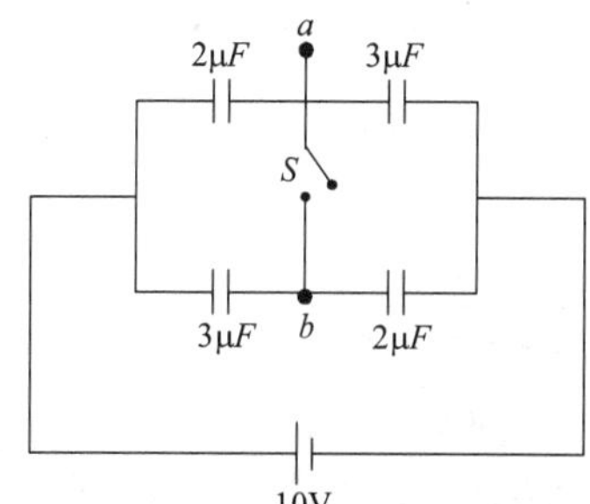

(A) 5 μC from b to a (B) 20 μC from a to b

(C) Zero (D) 5 μC from a to b

25. A combination of capacitors is set up as shown in the figure. The magnitude of the electric field, due to a point charge Q (having a charge equal to the sum of the charges on the 4 μF and 9 μF capacitors), at a point distant 30 m from it, would equal : **[2016]**

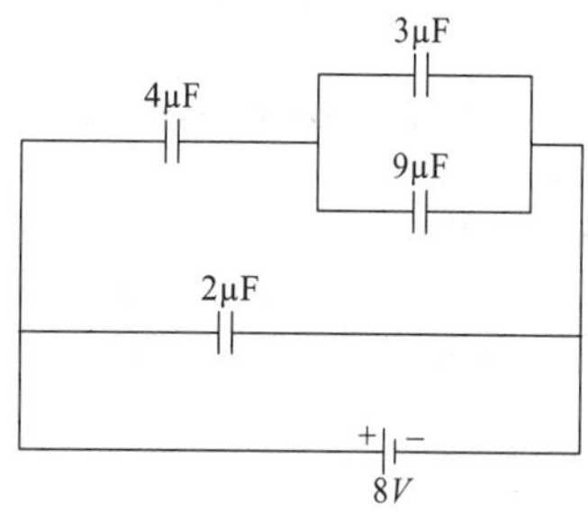

(A) 240 N/C (B) 360 N/C

(C) 420 N/C (D) 480 N/C

26. Three capacitors each of 4 μF are to be connected in such a way that the effective capacitance is 6 μF. This can be done by connecting them : **[2016 (Online)]**

(A) All in series

(B) All in parallel

(C) Two in parallel and one in series

(D) Two in series and one in parallel

27. Figure shows a network of capacitors where the numbers indicates capacitances in micro Farad. The value of capacitance C if the equivalent capacitance between point A and B is to be 1 μF is : **[2016 (Online)]**

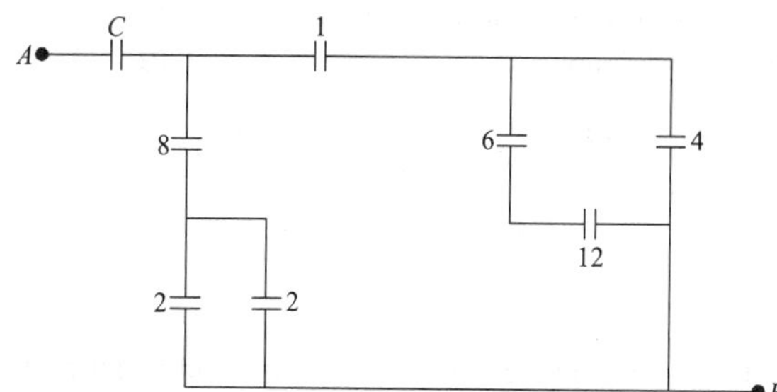

(A) $\frac{32}{23} \mu F$ (B) $\frac{31}{23} \mu F$

(C) $\frac{33}{23} \mu F$ (D) $\frac{34}{23} \mu F$

28. A capacitance of 2 μF is required in an electrical circuit across a potential difference of 1.0 kV. A large number of 1 μF capacitors are available which can withstand a potential difference of not more than 300 V. The minimum number of capacitors required to achieve this is : **[2017]**

(A) 24 (B) 32

(C) 2 (D) 16

29. The energy stored in the electric field produced by a metal sphere is 4.5 J. If the sphere contains 4 μC charge, its radius will be : [Take : $\frac{1}{4\pi\varepsilon_0} = 9 \times 10^9\ N-m^2/C^2$] **[2017 (Online)]**

(A) 20 mm (B) 32 mm
(C) 28 mm (D) 16 mm

30. A combination of parallel plate capacitors is maintained at a certain potential difference. When a 3 mm thick slab is introduced between all the plates, in order to maintain the same potential difference, the distance between the plates in increased by 2.4 mm. Find the dielectric constant of the slab : **[2017 (Online)]**

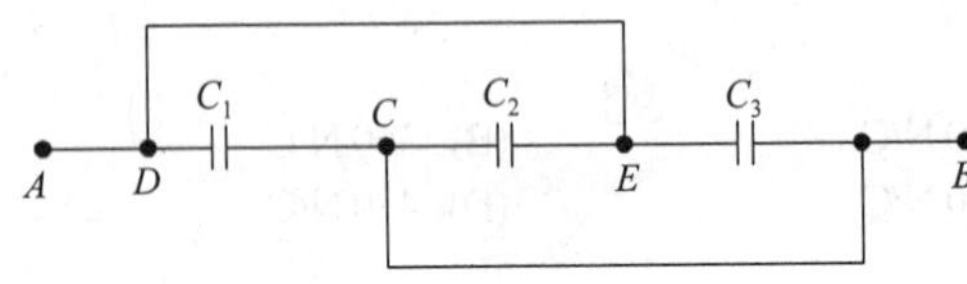

(A) 3 (B) 4
(C) 5 (D) 6

31. A parallel plate capacitor of capacitance 90 pF is connected to a battery of emf 20 V. If a dielectric material of dielectric constant $k = 5/3$ is inserted between the plates, the magnitude of the induced charge will be : **[2018]**

(A) 1.2 nC (B) 0.3 nC
(C) 2.4 nC (D) 0.9 nC

32. A parallel plate capacitor with area 200 cm^2 and separation between the plates 1.5 cm, is connected across a battery of emf V. If the force of attraction between the plates is 25×10^{-6} N, the value of V is approximately : **[2018 (Online)]**

$$\left(\epsilon_0 = 8.85 \times 10^{-12} \frac{C^2}{N.m^2}\right)$$

(A) 150 V (B) 100 V
(C) 250 V (D) 300 V

33. The equivalent capacitance between A and B in the circuit given below is : **[2018 (Online)]**

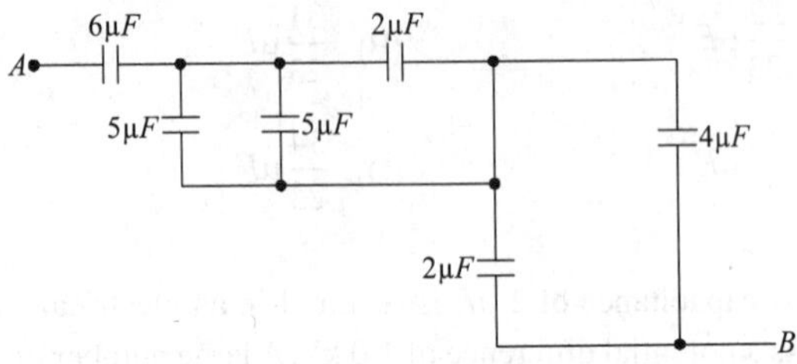

(A) 4.9 μF (B) 3.6 μF
(C) 5.4 μF (D) 2.4 μF

34. A capacitor $C_1 = 1$ μF is charged up to a voltage $V = 60$ V by connecting it to battery B through switch (1), Now C_1 is disconnected from battery and connected to a circuit consisting of two uncharged capacitors $C_2 = 3.0$ μF and $C_3 = 6.0$ μF through a switch (2) as shown in the figure. The sum of final charges on C_2 and C_3 is : **[2018 (Online)]**

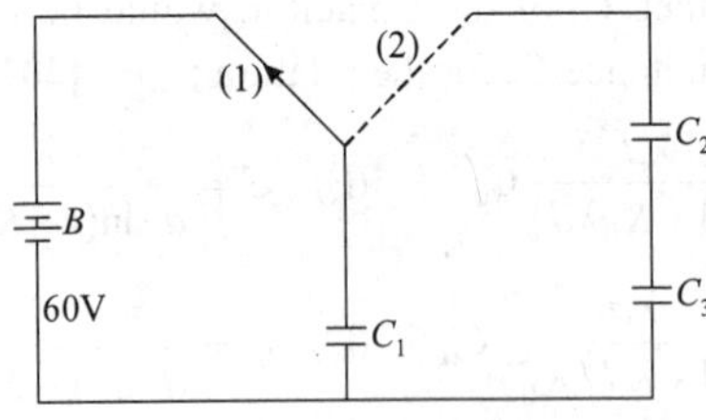

(A) 36 μC (B) 20 μC
(C) 54 μC (D) 40 μC

35. A parallel plate capacitor is made of two square plates of side 'a', separated by a distance d ($d << a$). The lower triangular portion is filled with a dielectric of dielectric constant K, as shown in the figure. Capacitance of this capacitor is : **[9 Jan. 2019 (Morning)]**

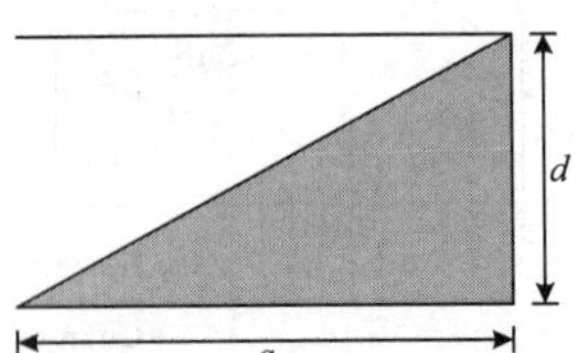

(A) $\frac{1}{2}\frac{k \in_0 a^2}{d}$ (B) $\frac{k \in_0 a^2}{d} \ln K$

(C) $\frac{k \in_0 a^2}{d(K-1)} \ln K$ (D) $\frac{k \in_0 a^2}{2d(K+1)}$

36. A parallel plate capacitor with square plates is filled with four dielectrics of dielectric constants K_1, K_2, K_3, K_4 arranged as shown in the figure. The effective dielectric constant K will be : **[9 Jan. 2019 (Evening)]**

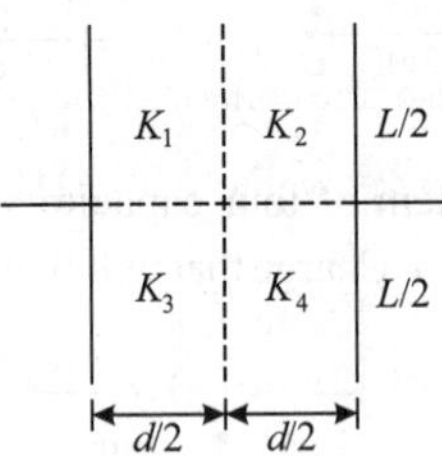

(A) $K = \frac{(K_1 + K_2)(K_3 + K_4)}{2(K_1 + K_2 + K_3 + K_4)}$

(B) $K = \frac{(K_1 + K_2)(K_3 + K_4)}{(K_1 + K_2 + K_3 + K_4)}$

(C) $K = \frac{(K_1 + K_4)(K_2 + K_3)}{2(K_1 + K_2 + K_3 + K_4)}$

(D) $K = \frac{(K_1 + K_3)(K_2 + K_4)}{K_1 + K_2 + K_3 + K_4}$

37. A parallel plate capacitor is of area 6 cm^2 and a separation 3 mm. The gap is filled with three dielectric materials of equal thickness (see figure) with dielectric constants $K_1 = 10$, $K_2 = 12$ and $K_3 = 14$. The dielectric constant of a material which when fully inserted in above capacitor, gives same capacitance would be : **[10 Jan. 2019 (Morning)]**

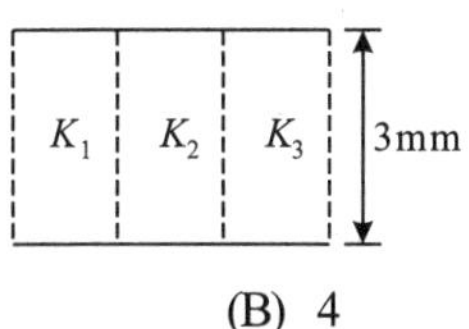

(A) 12 (B) 4
(C) 36 (D) 14

38. A parallel plate capacitor having capacitance 12 pF is charged by a battery to a potential difference of 10 V between its plates. The charging battery is now disconnected and a porcelain slab of dielectric constant 6.5 is slipped between the plates, the work done by the capacitor on the slab is : **[10 Jan. 2019 (Evening)]**

(A) 692 pJ (B) 60 pJ
(C) 508 pJ (D) 560 pJ

39. In the figure shown below, the charge on the left plate of the 10 μF capacitor is –30 μC. The charge on the right plate of the 6 μF capacitor is : **[11 Jan. 2019 (Morning)]**

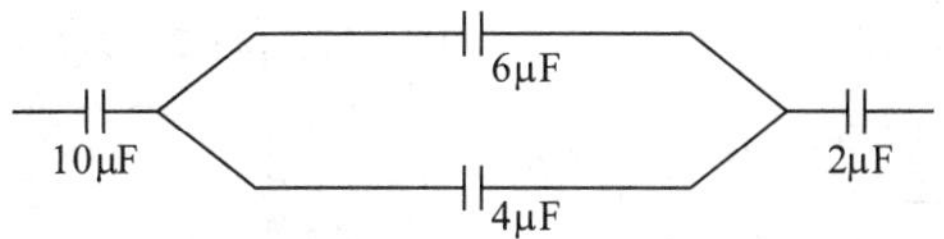

(A) –18 μC (B) –12 μC
(C) +12 μC (D) +18 μC

40. Seven capacitors, each of capacitance 2 μF, are to be connected in a configuration to obtain an effective capacitance of $\left(\frac{6}{13}\right)$ μF. Which of the combinations, shown in figures below, will achieve the desired value ? **[11 Jan. 2019 (Evening)]**

(A) 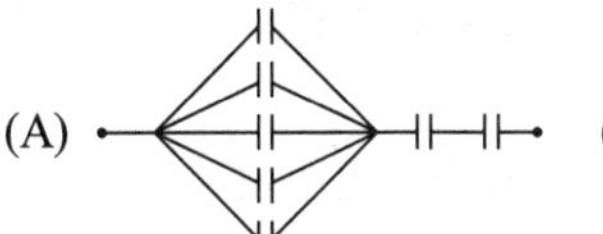 (B) 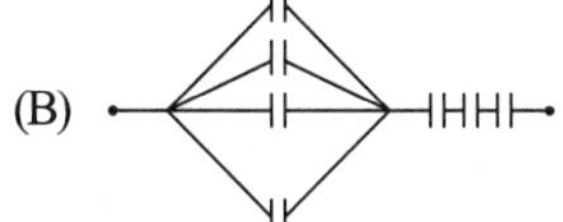

(C) 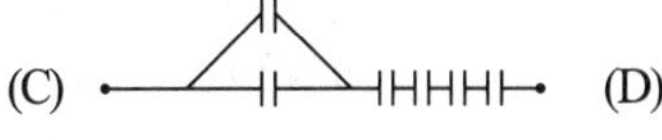 (D) 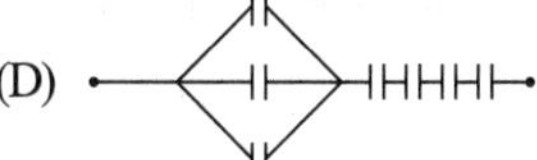

41. In the figure shown, after the switch 'S' is turned from position 'A' to position 'B', the energy dissipated in the circuit in terms of capacitance 'C' and total charge 'Q' is : **[12 Jan. 2019 (Morning)]**

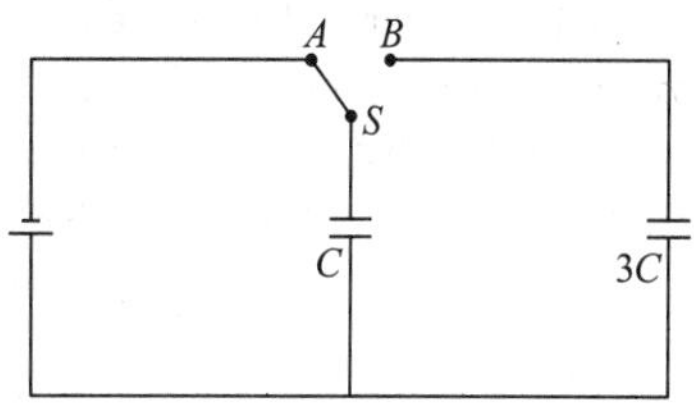

(A) $\frac{3}{8}\frac{Q^2}{C}$ (B) $\frac{3}{4}\frac{Q^2}{C}$

(C) $\frac{1}{8}\frac{Q^2}{C}$ (D) $\frac{5}{8}\frac{Q^2}{C}$

42. In the circuit shown, find C if the effective capacitance of the whole circuit is to be 0.5 μF. All values in the circuit are in μF. **[12 Jan. 2019 (Evening)]**

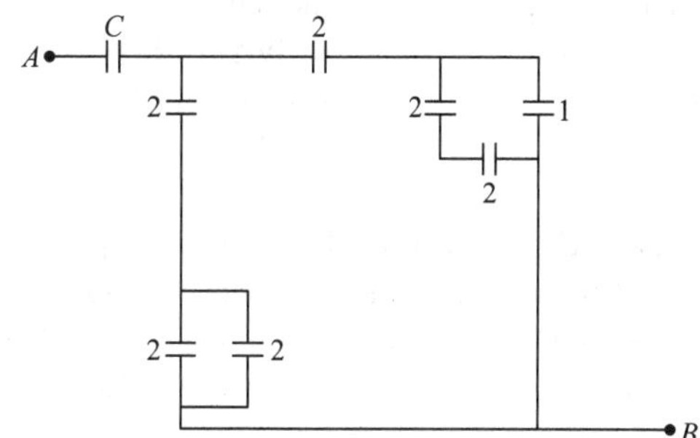

(A) $\frac{7}{10}\mu F$ (B) $\frac{7}{11}\mu F$

(C) $\frac{6}{5}\mu F$ (D) 4 μF

43. The charge on a capacitor plate in a circuit, as a function of time, is shown in the figure. What is the value of current at $t = 4$ s ? **[12 Jan. 2019 (Evening)]**

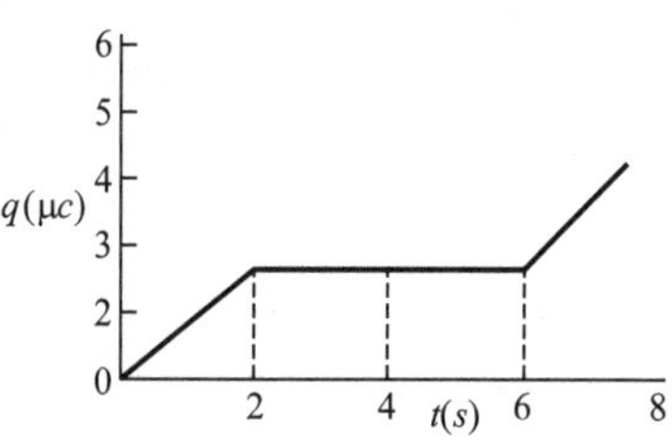

(A) 3 μA (B) 2 μA
(C) Zero (D) 1.5 μA

44. A parallel plate capacitor with plates of area 1 m^2 each, and at a separation of 0.1 m. If the electric field between the plates is 100 N/C, the magnitude of charge each plate is :

[Take $\varepsilon_0 = 8.85 \times 10^{-12} \frac{C^2}{N-m^2}$] **[12 Jan. 2019 (Evening)]**

(A) 7.85×10^{-10} C (B) 6.85×10^{-10} C
(C) 9.85×10^{-10} C (D) 8.85×10^{-10} C

45. Voltage rating of a parallel plate capacitor is 500 V. Its dielectric can withstand a maximum electric field of 10^6 V/m. The plate area is 10^{-4} m^2. What is the dielectric constant of the capacitance is 15 pF ? [Given $\epsilon_0 = 8.86 \times 10^{-12}$ C^2/Nm^2] **[8 April 2019 (Morning)]**

(A) 3.8 (B) 4.5
(C) 6.2 (D) 8.5

46. A parallel plate capacitor has 1 μF capacitance. One of its two plates is given +2 μC charge and the other plate, +4 μC charge. The potential difference developed across the capacitor is : **[8 April 2019 (Evening)]**

(A) 5V (B) 2V
(C) 3V (D) 1V

47. A capacitor with capacitance 5 μF is charged to 5 μC. If the plates are pulled apart to reduce the capacitance to 2 μF, how much work is done ? **[9 April 2019 (Morning)]**

(A) 3.75×10^{-6} J (B) 2.55×10^{-6} J
(C) 2.16×10^{-6} J (D) 6.25×10^{-6} J

48. The parallel combination of two air filled parallel plate capacitors of capacitance C and nC is connected to a battery of voltage, V. When the capacitors are fully charged, the battery is removed and after that a dielectric material of dielectric constant K is placed between the two plates of the first capacitor. The new potential difference of the combined system is : **[9 April 2019 (Evening)]**

(A) $\dfrac{V}{K+n}$ (B) V

(C) $\dfrac{(n+1)V}{(K+n)}$ (D) $\dfrac{nV}{K+n}$

49. Figure shows charge (q) versus voltage (V) graph for series and parallel combination of two given capacitors. The capacitances are : **[10 April 2019 (Morning)]**

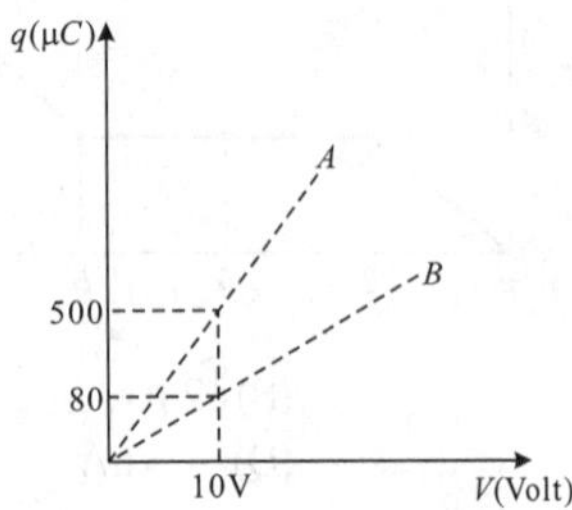

(A) 50 μF and 30 μF (B) 20 μF and 30 μF
(C) 60 μF and 40 μF (D) 40 μF and 10 μF

50. A simple pendulum of length L is placed between the plates of a parallel plate capacitor having electric field E, as shown in figure. Its bob has mass m and charge q. The time period of the pendulum is given by : **[10 April 2019 (Evening)]**

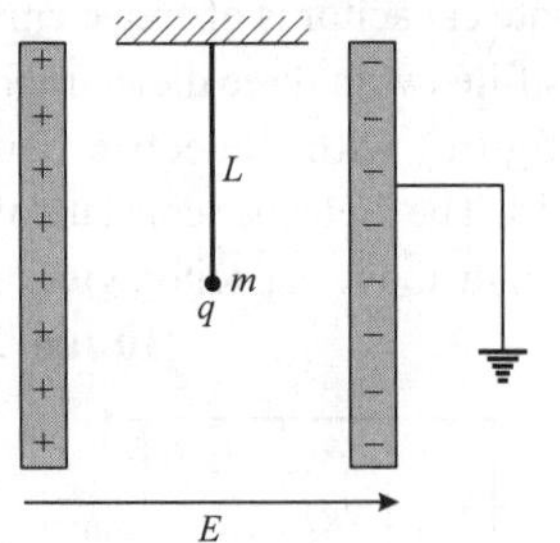

(A) $2\pi\sqrt{\dfrac{L}{\sqrt{g^2+\left(\dfrac{qE}{m}\right)^2}}}$ (B) $2\pi\sqrt{\dfrac{L}{\left(g+\dfrac{qE}{m}\right)}}$

(C) $2\pi\sqrt{\dfrac{L}{\left(g-\dfrac{qE}{m}\right)}}$ (D) $2\pi\sqrt{\dfrac{L}{\sqrt{g^2-\dfrac{q^2E^2}{m^2}}}}$

51. Two identical parallel plate capacitors, of capacitance C each, have plates of area A, separated by a distance d. The space between the plates of the two capacitors, is filled with three dielectrics, of equal thickness and dielectric constants K_1, K_2 and K_3. The first capacitor is filled as shown in figure-I, and the second one is filled as shown in figure-II. If these two modified capacitors are charged by the same potential V, the ratio of the energy stored in the two, would be (E_1 refers to capacitor (I) and E_2 to capacitor (II)) : **[12 April 2019 (Morning)]**

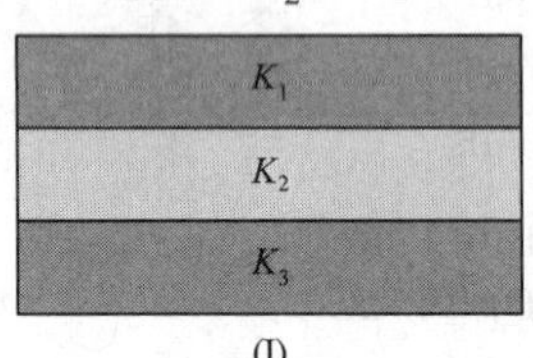

(I)

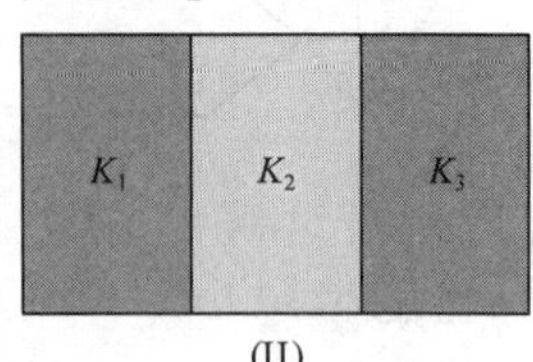

(II)

(A) $\dfrac{E_1}{E_2} = \dfrac{9K_1K_2K_3}{(K_1+K_2+K_3)(K_2K_3+K_3K_1+K_1K_2)}$

(B) $\dfrac{E_1}{E_2} = \dfrac{K_1K_2K_3}{(K_1+K_2+K_3)(K_2K_3+K_3K_1+K_1K_2)}$

(C) $\dfrac{E_1}{E_2} = \dfrac{(K_1+K_2+K_3)(K_2K_3+K_3K_1+K_1K_2)}{K_1K_2K_3}$

(D) $\dfrac{E_1}{E_2} = \dfrac{(K_1+K_2+K_3)(K_2K_3+K_3K_1+K_1K_2)}{9K_1K_2K_3}$

52. In the given circuit, the charge on 4 μF capacitor will be : **[12 April 2019 (Evening)]**

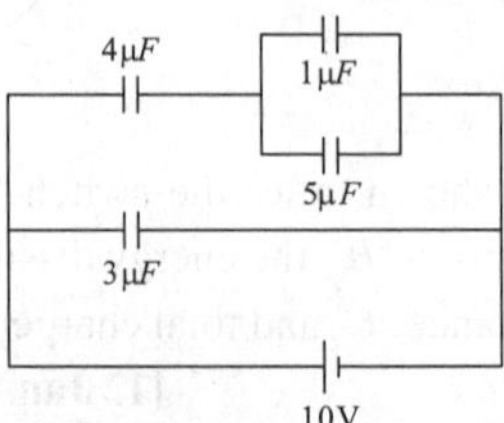

(A) $5.4\,\mu C$ (B) $24\,\mu C$
(C) $13.4\,\mu C$ (D) $9.6\,\mu C$

53. A parallel plate capacitor has plates of area A separated by distance 'd' between them. It is filled with a dielectric which has a dielectric constant that varies as $k(x) = K(1 + \alpha x)$ where 'x' is the distance measured from one of the plates. If $(\alpha d) << 1$, the total capacitance of the system is best given by the expression : **[7 Jan. 2020 (Morning)]**

(A) $\frac{AK\varepsilon_0}{d}\left(1+\frac{\alpha d}{2}\right)$ (B) $\frac{A\varepsilon_0 K}{d}\left(1+\left(\frac{\alpha d}{2}\right)^2\right)$

(C) $\frac{A\varepsilon_0 K}{d}\left(1+\frac{\alpha^2 d^2}{2}\right)$ (D) $\frac{AK\varepsilon_0}{d}(1+\alpha d)$

54. A 60 pF capacitor is fully charged by a 20 V supply. It is then disconnected from the supply and is connected to another uncharged 60 pF capacitor in parallel. The electrostatic energy that is lost in this process by the time the charge is redistributed between them is (in nJ) ______. **[7 Jan. 2020 (Evening)]**

55. Effective capacitance of parallel combination of two capacitors C_1 and C_2 is 10 μF. When these capacitors are individually connected to a voltage source of 1 V, the energy stored in the capacitor C_2 is 4 times that of C_1. If these capacitors are connected in series, their effective capacitance will be : **[8 Jan. 2020 (Morning)]**

(A) $3.2\,\mu F$ (B) $8.4\,\mu F$
(C) $1.6\,\mu F$ (D) $4.2\,\mu F$

56. A capacitor is made of two square plates each of side 'a' making a very small angle α between them, as shown in figure. The capacitance will be close to : **[8 Jan. 2020 (Evening)]**

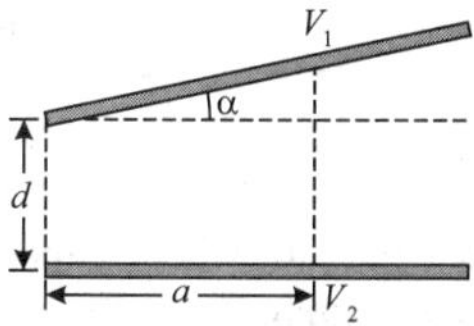

(A) $\frac{\varepsilon_0 a^2}{d}\left(1-\frac{3\alpha a}{2d}\right)$ (B) $\frac{\varepsilon_0 a^2}{d}\left(1-\frac{\alpha a}{4d}\right)$

(C) $\frac{\varepsilon_0 a^2}{d}\left(1+\frac{\alpha a}{d}\right)$ (D) $\frac{\varepsilon_0 a^2}{d}\left(1-\frac{\alpha a}{2d}\right)$

57. A 5 μF capacitor is charged fully by a 220 V supply. It is then disconnected from the supply and is connected in series to another uncharged 2.5 μF capacitor. If the energy change during the charge redistribution is $\frac{X}{100}$ J then value of X to the nearest integer is : **[2 Sept. 2020 (Morning)]**

58. A 10 μF capacitor is fully charged to a potential difference of 50 V. After removing the source voltage it is connected to an uncharged capacitor in parallel. Now the potential difference across them becomes 20 V. The capacitance of the second capacitor is : **[2 Sept. 2020 (Evening)]**

(A) $10\,\mu F$ (B) $15\,\mu F$
(C) $20\,\mu F$ (D) $30\,\mu F$

59. In the circuit shown in the figure, the total charge is 750 μC and the voltage across capacitor C_2 is 20 V. Then the charge on capacitor C_2 is : **[3 Sept. 2020 (Morning)]**

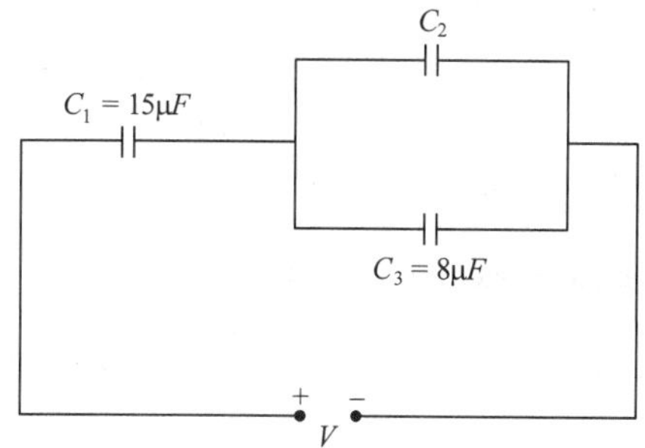

(A) $590\,\mu C$ (B) $450\,\mu C$
(C) $650\,\mu C$ (D) $160\,\mu C$

60. A capacitor C is fully charged with voltage V_0. After disconnecting the voltage source, it is connected in parallel with another uncharged capacitor of capacitance $\frac{C}{2}$. The energy loss in the process after the charge is distributed between the two capacitors is : **[4 Sept. 2020 (Evening)]**

(A) $\frac{1}{6}CV_0^2$ (B) $\frac{1}{2}CV_0^2$

(C) $\frac{1}{3}CV_0^2$ (D) $\frac{1}{4}CV_0^2$

61. Two capacitors of capacitances C and 2 C are charged to potential differences V and 2 V, respectively. These are then connected in parallel in such a manner that the positive terminal of one is connected to the negative terminal of the other. The final energy of this configuration is :**[5 Sept. 2020 (Morning)]**

(A) $\frac{9}{2}CV^2$ (B) $\frac{25}{6}CV^2$

(C) Zero (D) $\frac{3}{2}CV^2$

62. A parallel plate capacitor has plate of length 'l', width 'w' and separation of plates is 'd'. It is connected to a battery of *emf* V. A dielectric slab of the same thickness 'd' and of dielectric constant $k = 4$ is being inserted between the plates of the capacitor. At what length of the slab inside plates, will be energy stored in the capacitor be two times the initial energy stored? **[5 Sept. 2020 (Evening)]**

(A) $l/4$ (B) $l/2$
(C) $l/3$ (D) $2l/3$

63. In the circuit shown, charge on the 5 μF capacitor is : **[5 Sept. 2020 (Evening)]**

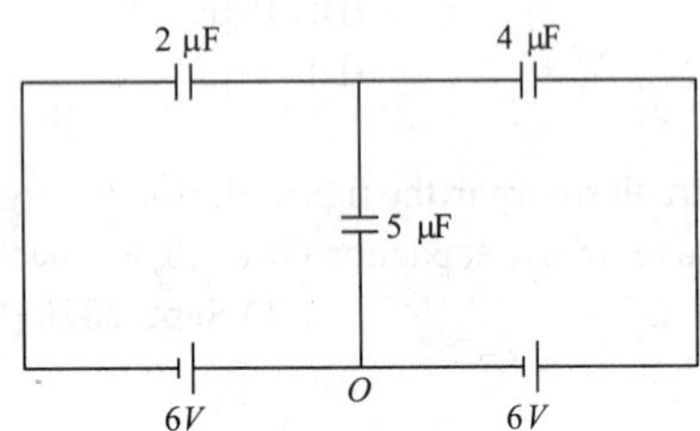

(A) 5.45 μC (B) 16.36 μC
(C) 10.90 μC (D) 18.00 μC

64. Two equal capacitors are first connected in series and then in parallel. The ratio of the equivalent capacities in the two cases will be : **[24 Feb. 2021 (Morning)]**

(A) 4 : 1 (B) 1 : 4
(C) 1 : 2 (D) 2 : 1

65. Consider the combination of 2 capacitors C_1 and C_2, with $C_2 > C_1$, when connected in parallel, the equivalent capacitance is $\frac{15}{4}$ times the equivalent capacitance of the same connected in series. Calculate the ratio of capacitors, $\frac{C_2}{C_1}$:

[26 Feb. 2021 (Morning)]

(A) $\frac{15}{4}$ (B) $\frac{29}{15}$

(C) $\frac{15}{11}$ (D) $\frac{111}{80}$

66. For changing the capacitance of a given parallel plate capacitor, a dielectric material of dielectric constant K is used, which has the same area as the plates of the capacitor. The thickness of the dielectric slab is $\frac{3}{4}d$, where 'd' is the separation between the plates of parallel plate capacitor. The new capacitance (C') in terms of original capacitance (C_0) is given by the following relation : **[16 March 2021 (Morning)]**

(A) $C' = \frac{3+K}{4K} C_0$ (B) $C' = \frac{4+K}{3} C_0$

(C) $C' = \frac{4K}{K+3} C_0$ (D) $C' = \frac{4}{3+K} C_0$

67. In a parallel plate capacitor set up, the plate area of capacitor is 2 m^2 and the plates are separated by 1 m. If the space between the plates are filled with a dielectric material of thickness 0.5 m and area 2 m^2 (see figure) the capacitance of the set-up will be ______ ε_0. (Dielectric constant of the material = 3.2)
(Round off to the Nearest Integer) **[16 March 2021 (Evening)]**

68. Four identical rectangular plates with length, $l = 2$ cm and breadth, $b = \frac{3}{2}$ cm are arranged as shown in figure. The equivalent capacitance between A and C is $\frac{x\varepsilon_0}{d}$. The value of x is _____. (Round off to the Nearest Integer)
[17 March 2021 (Morning)]

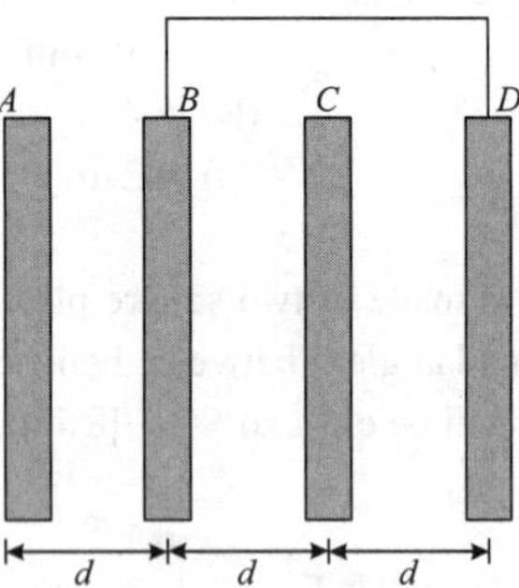

69. A parallel plate capacitor whose capacitance C is 14 pF is charged by a battery to a potential difference $V = 12$ V between its plates. The charging battery is now disconnected and a porcelain plate with $k = 7$ is inserted between the plates, then the plate would oscillate back and forth between the plates with a constant mechanical energy of ______ pJ.
(Assume no friction) **[17 March 2021 (Morning)]**

70. A 2 μF capacitor C_1 is first charged to a potential difference of 10 V using a battery. Then the battery is removed and the capacitor is connected to an uncharged capacitor C_2 of 8 μF. The charge in C_2 on equilibrium condition is______ μC. (Round off to the Nearest Integer) **[17 March 2021 (Evening)]**

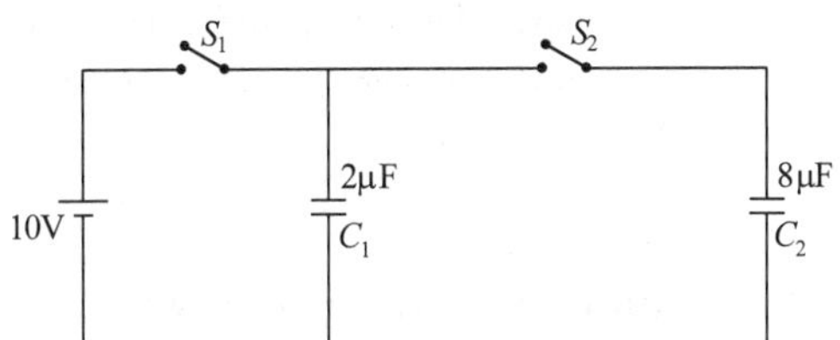

71. A parallel plate capacitor has plate area 100 m^2 and plate separation of 10 m. The space between the plates is filled up to a thickness 5 m with a material of dielectric constant of 10. The resultant capacitance of the system is 'x' pF.
The value of $\varepsilon_0 = 8.85 \times 10^{-12}$ F.m^{-1}.
The value of 'x' to the nearest integer is______. **[18 March 2021 (Morning)]**

72. A parallel plate capacitor with plate area 'A' and distance of separation d is filled with a dielectric. What is the capacitance of the capacitor when permittivity of the dielectric varies as : **[25 July 2021 (Morning)]**

$$\varepsilon(x) = \varepsilon_0 + kx, \text{ for } \left(0 < x \le \frac{d}{2}\right)$$

$$\varepsilon(x) = \varepsilon_0 + k(d - x), \text{ for } \left(\frac{d}{2} \le x \le d\right)$$

Where x is the distance from one plate of capacitor.

(A) $\dfrac{kA}{2\ln\left(\dfrac{2\varepsilon_0 + kd}{2\varepsilon_0}\right)}$ (B) $\left(\varepsilon_0 + \dfrac{kd}{2}\right)^{2/kA}$

(C) $\dfrac{kA}{2}\ln\left(\dfrac{2\varepsilon_0}{2\varepsilon_0 - kd}\right)$ (D) 0

73. If q_f is the free charge on the capacitor plates and q_b is the bound charge on the dielectric slab of dielectric constant k placed between the capacitor plates, then bound charge q_b can be expressed as : **[25 July 2021 (Evening)]**

(A) $q_b = q_f\left(1 + \dfrac{1}{k}\right)$ (B) $q_b = q_f\left(1 - \dfrac{1}{\sqrt{k}}\right)$

(C) $q_b = q_f\left(1 + \dfrac{1}{\sqrt{k}}\right)$ (D) $q_b = q_f\left(1 - \dfrac{1}{k}\right)$

74. In the reported figure, a capacitor is formed by placing a compound dielectric between the plates of parallel plate capacitor. The expression for the capacity of the said capacitor will be : (Given area of plate = A) : **[27 July 2021 (Morning)]**

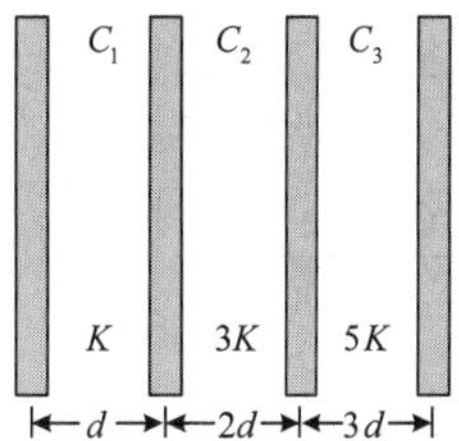

(A) $\dfrac{9}{6}\dfrac{K\varepsilon_0 A}{d}$ (B) $\dfrac{25}{6}\dfrac{K\varepsilon_0 A}{d}$

(C) $\dfrac{15}{6}\dfrac{K\varepsilon_0 A}{d}$ (D) $\dfrac{15}{34}\dfrac{K\varepsilon_0 A}{d}$

75. Two capacitors of capacities 2 C and C are joined in parallel and charged up to potential V. The battery is removed and the capacitor of capacity C is filled completely with a medium of dielectric constant K. The potential difference across the capacitors will now be : **[27 July 2021 (Morning)]**

(A) $\dfrac{V}{K}$ (B) $\dfrac{3V}{K+2}$

(C) $\dfrac{3V}{K}$ (D) $\dfrac{V}{K+2}$

76. A capacitor of capacitance C = 1 μF is suddenly connected to a battery of 100 volt through a resistance R = 100 Ω. The time taken for the capacitor to be charged to get 50 V is :
(Take ln 2 = 0.69) **[27 July 2021 (Morning)]**

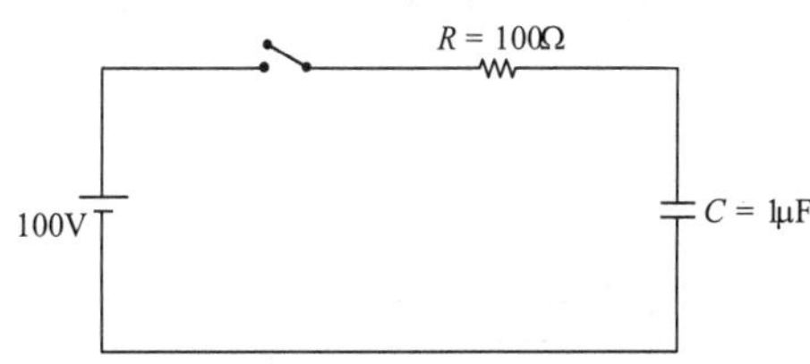

(A) 0.30×10^{-4} s (B) 0.69×10^{-4} s
(C) 1.44×10^{-4} s (D) 3.33×10^{-4} s

77. A simple pendulum of mass 'm', length 'l' and charge $+q$' suspended in the electric field produced by two conducting parallel plates as shown. The value of deflection of pendulum in equilibrium position will be : **[27 July 2021 (Evening)]**

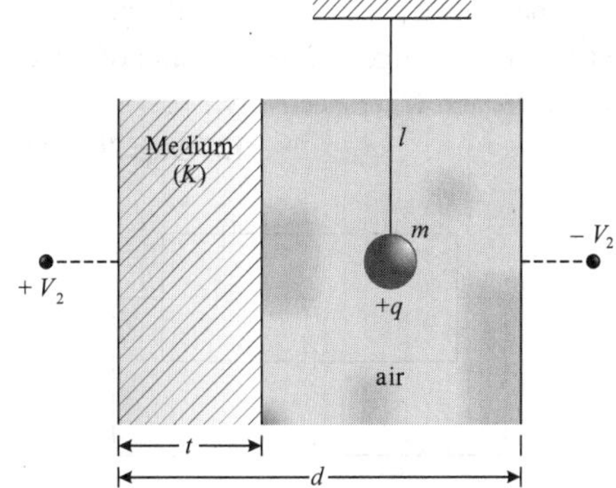

(A) $\tan^{-1}\left[\frac{q}{mg}+\frac{C_1(V_2-V_1)}{(C_1+C_2)(d-t)}\right]$

(B) $\tan^{-1}\left[\frac{q}{mg}+\frac{C_1(V_2+V_1)}{(C_1+C_2)(d-t)}\right]$

(C) $\tan^{-1}\left[\frac{q}{mg}+\frac{C_2(V_1+V_2)}{(C_1+C_2)(d-t)}\right]$

(D) $\tan^{-1}\left[\frac{q}{mg}+\frac{C_2(V_2-V_1)}{(C_1+C_2)(d-t)}\right]$

78. The material filled between the plates of a parallel plate capacitor has resistivity 200 Ωm. The value of capacitance of the capacitor is 2 pF. If a potential difference of 40 V is applied across the plates of the capacitor, then the value of leakage current following out of the capacitor is : (given the value of relative permittivity of material is 50) **[26 Aug. 2021 (Morning)]**

(A) 9.0 μA (B) 9.0 mA
(C) 0.9 mA (D) 0.9 μA

79. A parallel-plate capacitor with plate area A has separation d between the plates. Two dielectric slabs of dielectric constant K_1 and K_2 of same area $A/2$ and thickness $d/2$ are inserted in the space between the plates. The capacitance of the capacitor will be given by : **[26 Aug. 2021 (Evening)]**

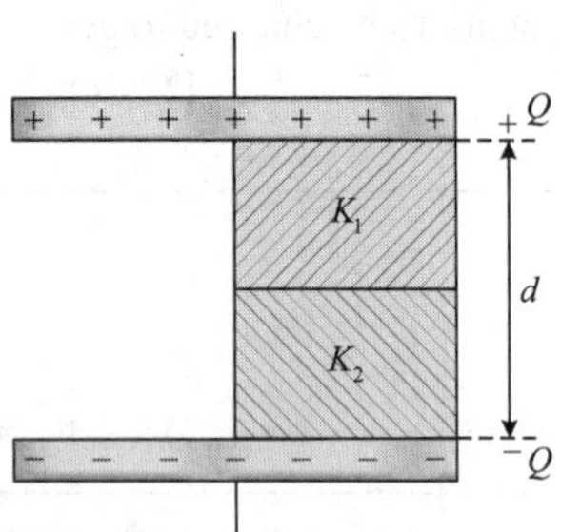

(A) $\frac{\varepsilon_0 A}{d}\left(\frac{1}{2}+\frac{2(K_1+K_2)}{K_1K_2}\right)$ (B) $\frac{\varepsilon_0 A}{d}\left(\frac{1}{2}+\frac{K_1+K_2}{K_1K_2}\right)$

(C) $\frac{\varepsilon_0 A}{d}\left(\frac{1}{2}+\frac{K_1K_2}{K_1+K_2}\right)$ (D) $\frac{\varepsilon_0 A}{d}\left(\frac{1}{2}+\frac{K_1K_2}{2(K_1+K_2)}\right)$

80. Three capacitors $C_1 = 2$ μF, $C_2 = 6$ μF and $C_3 = 12$ μF are connected as shown in figure. Find the ratio of the charges on capacitors C_1, C_2 and C_3 respectively :

[27 Aug. 2021 (Evening)]

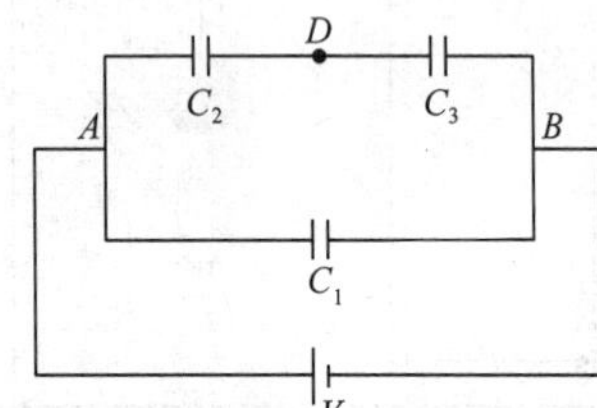

(A) 3 : 4 : 4 (B) 2 : 3 : 3
(C) 2 : 1 : 1 (D) 1 : 2 : 2

81. A parallel plate capacitor of capacitance 200 μF is connected to a battery of 200 V. A dielectric slab of dielectric constant 2 is now inserted into the space between plates of capacitor while the battery remain connected. The change in the electrostatic energy in the capacitor will be ______ ?

[31 Aug. 2021 (Evening)]

82. A parallel plate capacitor is formed by two plates each of area 30 π cm² separated by 1 mm. A material of dielectric strength 3.6×10^7 Vm^{-1} is filled between the plates. If the maximum charge that can be stored on the capacitor without causing any dielectric breakdown is 7×10^{-6} C, the value of dielectric constant of the material is :

$\left\{\text{Use} : \frac{1}{4\pi\varepsilon_0} = 9\times10^9\ \text{Nm}^2\text{C}^{-2}\right\}$ **[24 June 2022 (Morning)]**

(A) 1.66 (B) 1.75
(C) 2.25 (D) 2.33

83. If the charge on a capacitor is increased by 2 C, the energy stored in it increases by 44%. The original charge on the capacitor is (in C) : **[24 June 2022 (Evening)]**

(A) 10 (B) 20
(C) 30 (D) 40

84. The equivalent capacitance between points A and B in below shown figure will be _____ μF.

[25 June 2022 (Morning)]

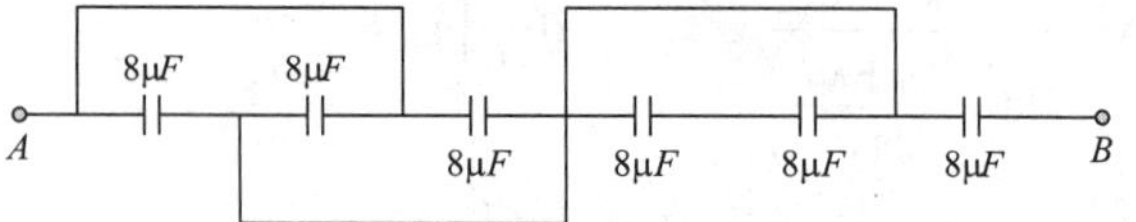

85. Two metallic plates form a parallel plate capacitor. The distance between the plates is d. A metal sheet of thickness $d/2$ and of area equal to area of each plate is introduced between the plates. What will be the ratio of the new capacitance to the original capacitance of the capacitor ? **[25 June 2022 (Evening)]**

(A) 2 : 1 (B) 1 : 2
(C) 1 : 4 (D) 4 : 1

86. Two capacitors having capacitance C_1 and C_2 respectively are connected as shown in figure. Initially, capacitor C_1 is charged to a potential difference V volt by a battery. The battery is then removed and the charged capacitor C_1 is now connected to uncharged capacitor C_2 by closing the switch S. The amount of charge on the capacitor C_2, after equilibrium is :

[26 June 2022 (Morning)]

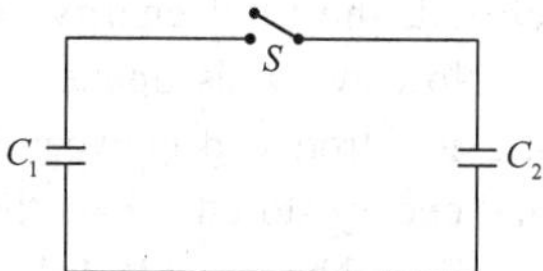

(A) $\frac{C_1C_2}{(C_1+C_2)}V$
(B) $\frac{(C_1+C_2)}{C_1C_2}V$
(C) $(C_1+C_2)V$
(D) $(C_1-C_2)V$

87. The charge on capacitor of capacitance 15 μF in the figure given below is : **[26 June 2022 (Evening)]**

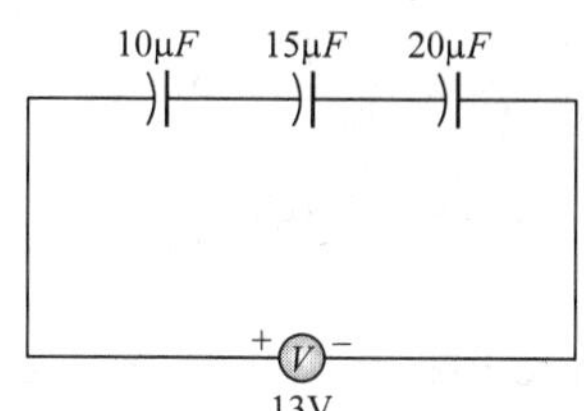

(A) 60 μC
(B) 130 μC
(C) 260 μC
(D) 585 μC

88. A parallel plate capacitor with plate area A and plate separation d = 2 m has a capacitance of 4 μF. The new capacitance of the system if half of the space between them is filled with a dielectric material of dielectric constant $K = 3$ (as shown in figure) will be : **[26 June 2022 (Evening)]**

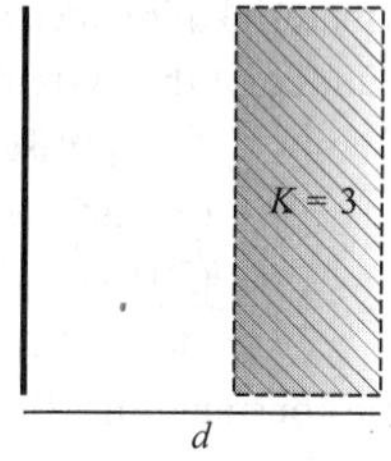

(A) 2 μF
(B) 32 μF
(C) 6 μF
(D) 8 μF

89. A force of 10 N acts on a charged particle placed between two plates of a charged capacitor. If one plate of capacitor is removed, then the force acting on that particle will be : **[27 June 2022 (Morning)]**

(A) 5 N
(B) 10 N
(C) 20 N
(D) Zero

90. A capacitor of capacitance 50 pF is charged by 100 V source. It is then connected to another uncharged identical capacitor. Electrostatic energy loss in the process is _____ nJ. **[27 June 2022 (Morning)]**

91. A parallel plate capacitor is made up of stair like structure with a plate area A of each stair and that is connected with a wire of length b, as shown in the figure. The capacitance of the arrangement is $\frac{x}{15}\frac{\varepsilon_0 A}{b}$. The value of x is _____. **[27 June 2022 (Evening)]**

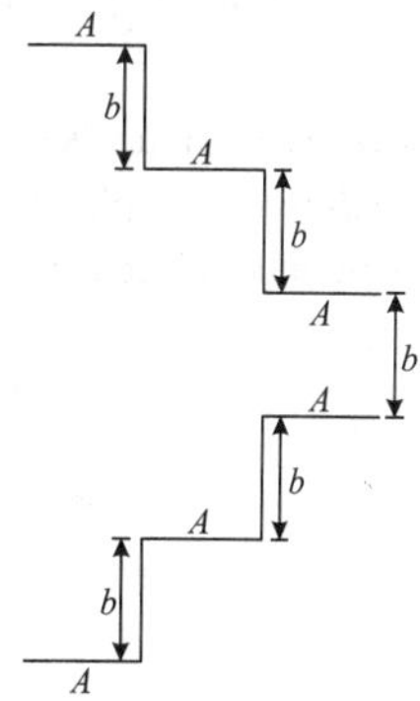

92. Given below are two statements : **[28 June 2022 (Morning)]**

Statement-I : A point charge is brought in an electric field. The value of electric field at a point near to the charge may increase if the charge is positive.

Statement-II : An electric dipole is placed in a non-uniform electric field. The net electric force on the dipole will not be zero.

Choose the correct answer from the options given below :

(A) Both statement-I and statement-II are true
(B) Both statement-I and statement-II are false
(C) Statement-I is true but statement-II is false
(D) Statement-I is false but statement-II is true

93. A capacitor C_1 of capacitance 5 μF is charged to a potential of 30 V using a battery. The battery is then removed and the charged capacitor is connected to an uncharged capacitor C_2 of capacitance 10 μF as shown in figure. When the switch is closed charge flows between the capacitors. At equilibrium, the charge on the capacitor C_2 is _____ μC. **[28 June 2022 (Evening)]**

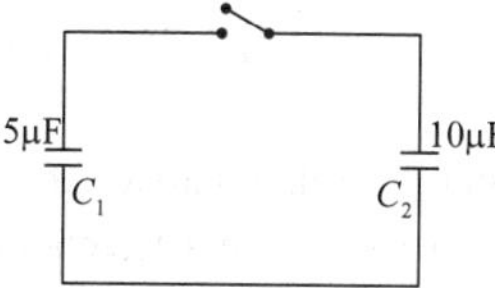

94. A parallel plate capacitor filled with a medium of dielectric constant 10, is connected across a battery and is charged. The dielectric slab is replaced by another slab of dielectric constant 15. Then the energy of capacitor will : **[29 June 2022 (Morning)]**

(A) Increase by 50%
(B) Decrease by 15%
(C) Increase by 25%
(D) Increase by 33%

95. The displacement current of 4.425 μA is developed in the space between the plates of parallel plate capacitor when voltage is changing at a rate of 10^6 Vs^{-1}. The area of each plate of the capacitor is 40 cm^2. The distance between each plate of the capacitor is $x \times 10^{-3}$ m. The value of x is, (Permittivity of free space, $\varepsilon_0 = 8.85 \times 10^{-12}$ $C^2 N^{-1} m^{-2}$). **[29 June 2022 (Evening)]**

96. A condenser of 2 μF capacitance is charged steadily from 0 to 5 C. Which of the following graph represents correctly the variation of potential difference (V) across it's plates with respect to the charge (Q) on the condenser ?

[25 July 2022 (Morning)]

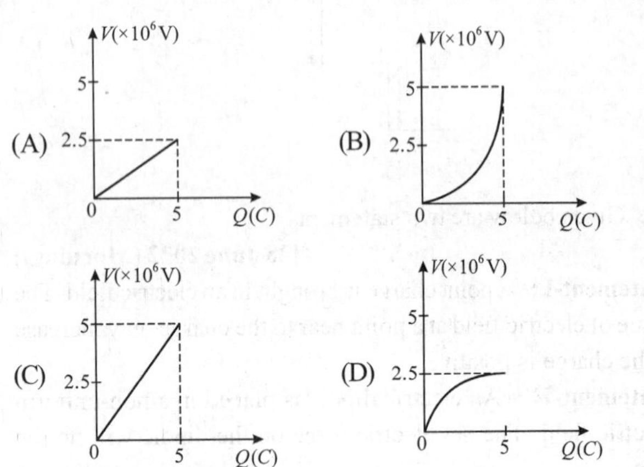

97. Two parallel plate capacitors of capacity C and 3C are connected in parallel combination and charged to a potential difference 18 V. The battery is then disconnected and the space between the plates of the capacitor of capacity C is completely filled with a material of dielectric constant 9. The final potential difference across the combination of capacitors will be ______ V. **[25 July 2022 (Evening)]**

98. The total charge on the system of capacitance $C_1 = 1$ μF, $C_2 = 2$ μF, $C_3 = 4$ μF and $C_4 = 3$ μF connected in parallel is (Assume a battery of 20 V is connected to the combination) :

[26 July 2022 (Morning)]

(A) 200 μC (B) 200 C

(C) 10 μC (D) 10 C

99. A composite parallel plate capacitor is made up of two different dielectric materials with different thickness (t_1 and t_2) as shown in figure. The two different dielectric material are separated by a conducting foil F. The voltage of the conducting foil is ______ V. **[26 July 2022 (Morning)]**

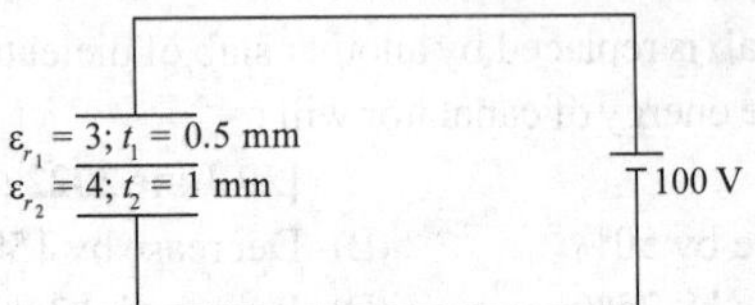

100. A source of potential difference V is connected to the combination of two identical capacitors as shown in the figure. When key K is closed, the total energy stored across the combination is E_1. Now key K is opened and dielectric of dielectric constant 5 is introduced between the plates of the capacitors. The total energy stored across the combination is now E_2. The ratio E_1/E_2 will be : **[26 July 2022 (Evening)]**

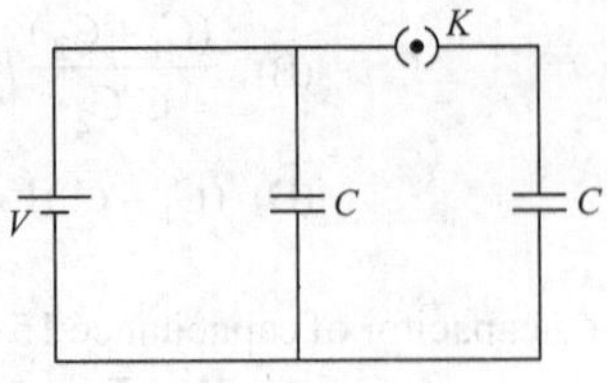

(A) $\frac{1}{10}$ (B) $\frac{2}{5}$

(C) $\frac{5}{13}$ (D) $\frac{5}{26}$

101. A parallel plate capacitor with width 4 cm, length 8 cm and separation between the plates of 4 mm is connected to a battery of 20 V. A dielectric slab of dielectric constant 5 having length 1 cm, width 4 cm and thickness 4 mm is inserted between the plates of parallel plate capacitor. The electrostatic energy of this system will be ______ ϵ_0 J. (Where ϵ_0 is the permittivity of free space). **[27 July 2022 (Evening)]**

102. Two capacitors, each having capacitance 40 μF are connected in series. The space between one of the capacitors is filled with dielectric material of dielectric constant k such that the equivalence capacitance of the system became 24 μF. The value of k will be : **[28 July 2022 (Morning)]**

(A) 1.5 (B) 2.5

(C) 1.2 (D) 3

103. A slab of dielectric constant K has the same cross-sectional area as the plates of a parallel plate capacitor and thickness $\frac{3}{4}d$, where d is the separation of the plates. The capacitance of the capacitor when the slab is inserted between the plates will be :

(Given C_0 = capacitance of capacitor with air as medium between plates) **[28 July 2022 (Evening)]**

(A) $\frac{4KC_0}{3+K}$ (B) $\frac{3KC_0}{3+K}$

(C) $\frac{3+K}{4KC_0}$ (D) $\frac{K}{4+K}$

104. Two identical thin metal plates have charge q_1 and q_2 respectively such that $q_1 > q_2$. The plates were brought close to each other to form a parallel plate capacitor of capacitance C. The potential difference between them is :

[29 July 2022 (Evening)]

(A) $\frac{(q_1+q_2)}{C}$ (B) $\frac{(q_1-q_2)}{C}$

(C) $\frac{(q_1-q_2)}{2C}$ (D) $\frac{2(q_1-q_2)}{C}$

105. A parallel plate capacitor with air between the plate has a capacitance of 15 pF. The separation between the plate becomes twice and the space between them is filled with a medium of dielectric constant 3.5. Then the capacitance becomes $\frac{x}{4}$ pF. The value of x is _____. **[24 Jan. 2023 (Evening)]**

106. A parallel plate capacitor has plate area 40 cm^2 and plates separation 2 mm. The space between the plates is filled with a dielectric medium of a thickness 1 mm and dielectric constant 5. The capacitance of the system is : **[25 Jan. 2023 (Morning)]**

(A) $24\,\varepsilon_0$ F (B) $\frac{3}{10}\varepsilon_0$ F

(C) $\frac{10}{3}\varepsilon_0$ F (D) $10\,\varepsilon_0$ F

107. A capacitor has capacitance 5 μF when it's parallel plates are separated by air medium of thickness d. A slab of material of dielectric constant 1.5 having area equal to that of plates but thickness $\frac{d}{2}$ is inserted between the plates. Capacitance of the capacitor in the presence of slab will be _____ μF.

[25 Jan. 2023 (Evening)]

108. A capacitor of capacitance 900 μF is charged by a 100 V battery. The capacitor is disconnected from the battery and connected to another uncharged identical capacitor such that one plate of uncharged capacitor connected to positive plate and another plate of uncharged capacitor connected to negative plate of the charged capacitor. The loss of energy in this process is measured as $x \times 10^{-2}$ J. The value of x is _____.

[30 Jan. 2023 (Morning)]

109. Two parallel plate capacitors C_1 and C_2 each having capacitance of 10 μF are individually charged by a 100 V D.C. source. Capacitor C_1 is kept connected to the source and a dielectric slab is inserted between it plates. Capacitor C_2 is disconnected from the source and then a dielectric slab is inserted in it. Afterwards the capacitor C_1 is also disconnected from the source and the two capacitors are finally connected in parallel combination. The common potential of the combination will be _____ V. (Assuming Dielectric constant = 10)

[31 Jan. 2023 (Evening)]

110. Let σ be the uniform surface charge density of two infinite thin plane sheets shown in figure. Then the electric fields in three different region E_I, E_{II} and E_{III} are :

[1 Feb. 2023 (Morning)]

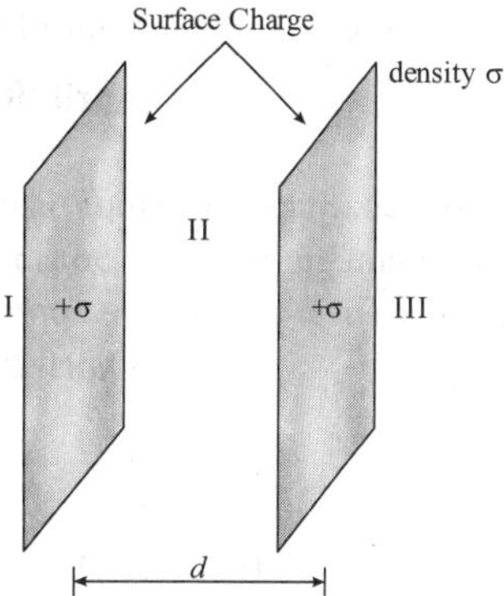

(A) $\vec{E}_I = \frac{2\sigma}{\epsilon_0}\hat{n}$, $\vec{E}_{II} = 0$, $\vec{E}_{III} = \frac{2\sigma}{\epsilon_0}\hat{n}$

(B) $\vec{E}_I = 0$, $\vec{E}_{II} = \frac{\sigma}{\epsilon_0}\hat{n}$, $\vec{E}_{III} = 0$

(C) $\vec{E}_I = \frac{\sigma}{2\epsilon_0}\hat{n}$, $\vec{E}_{II} = 0$, $\vec{E}_{III} = \frac{\sigma}{2\epsilon_0}\hat{n}$

(D) $\vec{E}_I = -\frac{\sigma}{2\epsilon_0}\hat{n}$, $\vec{E}_{II} = 0$, $\vec{E}_{III} = \frac{\sigma}{\epsilon_0}\hat{n}$

111. Given below are two statements : One is labelled as **Assertion-A** and the other is labelled as **Reason-R**.

Assertion-A : Two metallic spheres are charged to the same potential. One of them is hollow and another is solid, and both have the same radii. Solid sphere will have lower charge than the hollow one.

Reason-R : Capacitance of metallic spheres depend on the radii of spheres.

In the light of the above statements, choose the correct answer from the options given below: **[1 Feb. 2023 (Evening)]**

(A) **A** is false but **R** is true.

(B) Both **A** and **R** are true and **R** is the correct explanation of **A**.

(C) **A** is true but **R** is false.

(D) Both **A** and **R** are true but **R** is not the correct explanation of **A**.

112. A parallel plate capacitor with plate area A and plate separation d is filled with a dielectric material of dielectric constant $K = 4$. The thickness of the dielectric material is x, where $x < d$.

Let C_1 and C_2 be the capacitance of the system for $x = \frac{1}{3}d$ and $x = \frac{2d}{3}$, respectively. If $C_1 = 2\ \mu F$ the value of C_2 is ______ μF. **[6 April 2023 (Morning)]**

113. A capacitor of capacitance 150.0 μF is connected to an alternating source of emf given by $E = 36 \sin(120\ \pi t)$ V. The maximum value of current in the circuit is approximately equal to : **[6 April 2023 (Evening)]**

(A) 2 A (B) $\frac{1}{\sqrt{2}}$ A

(C) $\sqrt{2}$ A (D) $2\sqrt{2}$ A

114. As shown in the figure, two parallel plate capacitors having equal plate area of 200 cm^2 are joined in such a way that $a \neq b$. The equivalent capacitance of the combination is $x \in_0 F$. The value of x is_____. **[6 April 2023 (Evening)]**

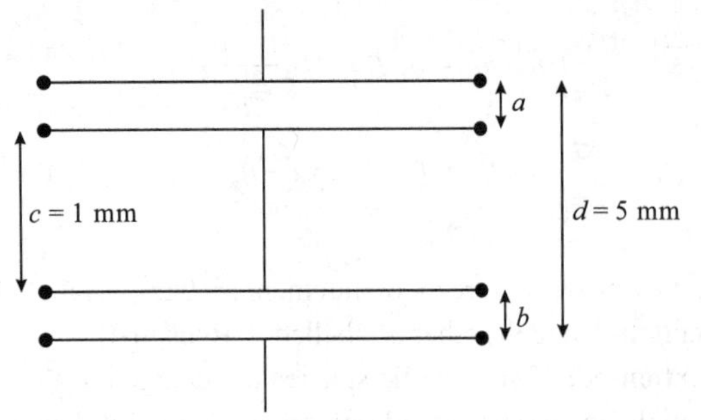

115. A 600 pF capacitor is charged by 200 V supply. It is then disconnected from the supply and is connected to another uncharged 600 pF capacitor. Electrostatic energy lost in the process is _____ μJ. **[8 April 2023 (Evening)]**

116. The equivalent capacitance of the combination shown is: **[10 April 2023 (Morning)]**

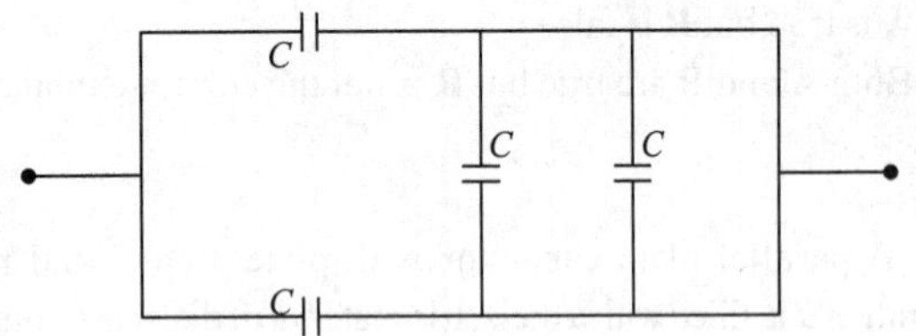

(A) $\frac{C}{2}$ (B) 4 C

(C) 2 C (D) $\frac{5}{3}C$

117. The distance between two plates of a capacitor is d and its capacitance is C_1, when air is the medium between the plates. If a metal sheet of thickness $\frac{2d}{3}$ and of same area as plate is introduced between the plates, the capacitance of the capacitor becomes C_2. The ratio $\frac{C_2}{C_1}$ is: **[10 April 2023 (Evening)]**

(A) 2 : 1 (B) 4 : 1

(C) 3 : 1 (D) 1 : 1

118. A force of $-P\hat{k}$. acts on the origin of the coordinate system. The torque about the point (2, – 3) is $P(a\hat{i} + b\hat{j})$, the ratio of $\frac{a}{b}$ is $\frac{x}{2}$. The value of x is. **[10 April 2023 (Evening)]**

119. A parallel plate capacitor of capacitance 2 F is charged to a potential V. The energy stored in the capacitor is E_1. The capacitor is now connected to another uncharged identical capacitor in parallel combination. The energy stored in the combination is E_2. The ratio E_2/E_1 is : **[11 April 2023 (Morning)]**

(A) 2 : 1 (B) 1 : 2

(C) 1 : 4 (D) 2 : 3

120. A capacitor of capacitance C is charged to a potential V. The flux of the electric field through a closed surface enclosing the positive plate of the capacitor is : **[11 April 2023 (Evening)]**

(A) $\frac{CV}{2\varepsilon_0}$ (B) $\frac{2CV}{\varepsilon_0}$

(C) $\frac{CV}{\varepsilon_0}$ (D) Zero

121. In the given circuit,
$C_1 = 2\ \mu F$, $C_2 = 0.2\ \mu F$, $C_3 = 2\ \mu F$, $C_4 = 4\ \mu F$, $C_5 = 2\ \mu F$, $C_5 = 2\ \mu F$, $C_6 = 2\ \mu F$, the charge stored on capacitor C_4 is_____μC. **[11 April 2023 (Evening)]**

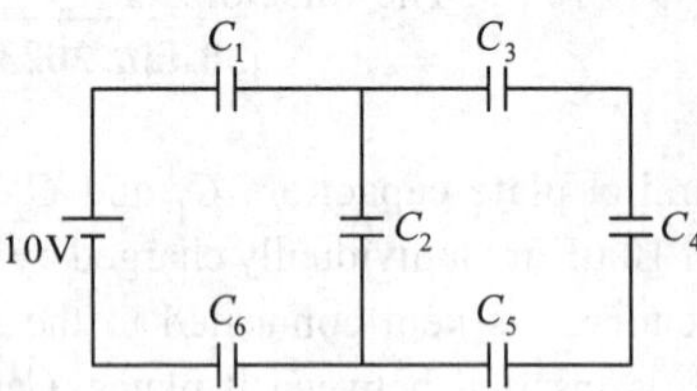

122. A circular plate is rotating in horizontal plane, about an axis passing through its center and perpendicular to the plate, with an angular velocity ω. A person sits at the center having two dumbbells in his hands. When he stretches out his hands, the moment of inertia of the system becomes triple. If E be the initial Kinetic energy of the system, then final Kinetic energy will be $\frac{E}{x}$. The value of x is. **[11 April 2023 (Evening)]**

123. In the given figure the total charge stored in the combined of capacitors is 100 μC. The value of 'x' is _____.

[15 April 2023 (Morning)]

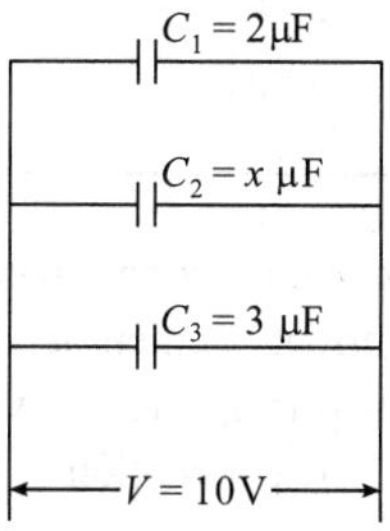

$C_1 = 2\mu F$

$C_2 = x\ \mu F$

$C_3 = 3\ \mu F$

$V = 10V$

* * * * *

CHAPTER

17 CURRENT ELECTRICITY

17.1 CURRENT ELECTRICITY

1. The length of a given cylindrical wire is increased by 100%. Due to the consequent decrease in diameter the change in the resistance of the wire will be : **[2003]**

(A) 200% (B) 100%

(C) 50% (D) 300%

2. The thermistor are usually made of : **[2004]**

(A) metals with low temperature coefficient of resistivity

(B) metals with high temperature coefficient of resistivity

(C) metal oxides with high temperature coefficient of resistivity

(D) semiconducting materials having low temperature coefficient of resistivity

3. An electric current is passed through a circuit containing two wires of the same material, connected in parallel. If the lengths and radii of the wires are in the ratio of 4/3 and 2/3, then the ratio of the currents passing through the wire will be : **[2004]**

(A) 3 (B) 1/3

(C) 8/9 (D) 2

4. The resistance of the series combination of two resistance is S. When they are joined in parallel, the total resistance is P. If $S = nP$, then the minimum possible value of n is : **[2004]**

(A) 4 (B) 3

(C) 2 (D) 1

5. The resistance of bulb filament is 100 Ω at a temperature of 100°C. If its temperature coefficient of resistance be 0.005 /°C, its resistance will become 200 Ω at a temperature of : **[2006]**

(A) 300 °C (B) 400 °C

(C) 500 °C (D) 200 °C

6. A material 'B' has twice the specific resistance of 'A'. A circular wire made of 'B' has twice the diameter of a wire made of 'A'. Then for the two wires to have the same resistance, the ratio l_B/l_A of their respective lengths must be: **[2006]**

(A) 1 (B) $\frac{1}{2}$

(C) $\frac{1}{4}$ (D) 2

7. The resistance of a wire is 5 ohm at 50°C and 6 ohm at 100°C. The resistance of the wire at 0°C will be : **[2007]**

(A) 3 ohm (B) 2 ohm

(C) 1 ohm (D) 4 ohm

8. Consider a block of conducting material of resistivity 'ρ' shown in the figure. Current 'I' enters at 'A' and leaves from 'D'. We apply superposition principle to find voltage 'ΔV' developed between 'B' and 'C'. The calculation is done in the following steps :

(i) Take current 'I' entering from 'A' and assume it to spread over a hemispherical surface in the block.

(ii) Calculate field $E(r)$ at distance 'r' from A by using Ohm's law $E = \rho j$, where j is the current per unit area at 'r'

(iii) From the 'r' dependence of $E(r)$, obtain the potential $V(r)$ at r.

(iv) Repeat (i), (ii) and (iii) for current 'I' leaving 'D' and superpose results for 'A' and 'D'

For current entering at A, the electric field at a distance 'r' from A is : **[2008]**

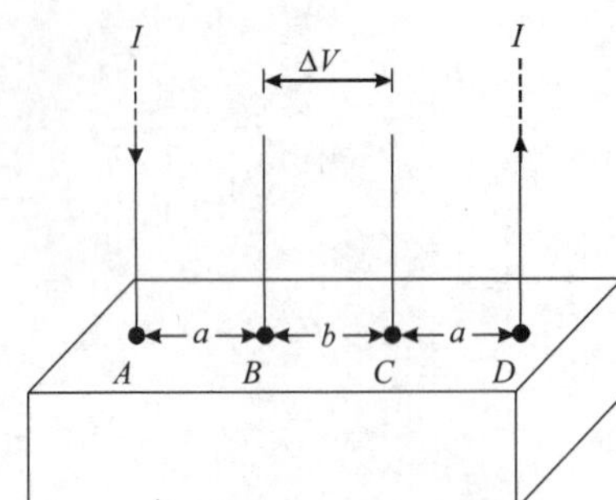

(A) $\frac{\rho I}{8\pi r^2}$ (B) $\frac{\rho I}{r^2}$

(C) $\frac{\rho I}{2\pi r^2}$ (D) $\frac{\rho I}{4\pi r^2}$

9. Consider a block of conducting material of resistivity 'ρ' shown in the figure. Current 'I' enters at 'A' and leaves from 'D'. We apply superposition principle to find voltage 'ΔV' developed between 'B' and 'C'. The calculation is done in the following steps :

(i) Take current 'I' entering from 'A' and assume it to spread over a hemispherical surface in the block.

(ii) Calculate field $E(r)$ at distance 'r' from A by using Ohm's law $E = \rho j$, where j is the current per unit area at 'r'

(iii) From the 'r' dependence of $E(r)$, obtain the potential $V(r)$ at r.

(iv) Repeat (i), (ii) and (iii) for current 'I' leaving 'D' and superpose results for 'A' and 'D'

ΔV measured between B and C is : **[2008]**

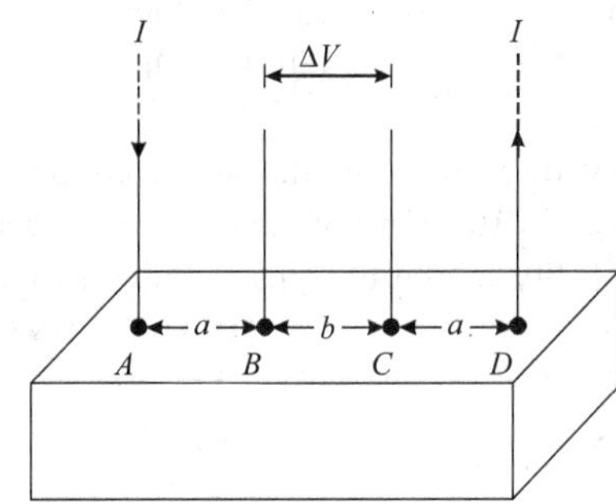

(A) $\frac{\rho I}{\pi a} - \frac{\rho I}{\pi(a+b)}$ (B) $\frac{\rho I}{a} - \frac{\rho I}{(a+b)}$

(C) $\frac{\rho I}{2\pi a} - \frac{\rho I}{2\pi(a+b)}$ (D) $\frac{\rho I}{2\pi(a+b)}$

10. Statement-1 : The temperature dependence of resistance is usually given as $R = R_0(1 + \alpha\Delta t)$. The resistance of a wire changes from 100 Ω to 150 Ω when its temperature is increased from 27°C to 227°C. This implies that $\alpha = 2.5 \times 10^{-3}$ /°C.

Statement-2 : $R = R_0(1 + \alpha\Delta t)$ is valid only when the change in the temperature ΔT is small and $\Delta R = (R - R_0) << R_0$. **[2009]**

(A) Statement-1 is true, Statement-2 is true; Statement-2 is not the correct explanation of Statement-1

(B) Statement-1 is false, Statement-2 is true

(C) Statement-1 is true, Statement-2 is false

(D) Statement-1 is true, Statement-2 is true; Statement-2 is the correct explanation of Statement-1

11. Two conductors have the same resistance at 0°C but their temperature coefficients of resistance are α_1 and α_2. The respective temperature coefficients of their series and parallel combinations are nearly : **[2010]**

(A) $\frac{\alpha_1+\alpha_2}{2}, \frac{\alpha_1+\alpha_2}{2}$ (B) $\frac{\alpha_1+\alpha_2}{2}, \alpha_1+\alpha_2$

(C) $\alpha_1+\alpha_2, \frac{\alpha_1+\alpha_2}{2}$ (D) $\alpha_1+\alpha_2, \frac{\alpha_1\alpha_2}{\alpha_1+\alpha_2}$

12. If 400 Ω of resistance is made by adding four 100 Ω resistance of tolerance 5%, then the tolerance of the combination is : **[2011]**

(A) 20% (B) 5%

(C) 10% (D) 15%

13. If a wire is stretched to make it 0.1 % longer, its resistance will : **[2011]**

(A) Decrease by 0.2% (B) Decrease by 0.05%

(C) Increase by 0.05% (D) Increase by 0.2%

14. The resistance of a wire is R. It is bent at the middle by 180° and both the ends are twisted together to make a shorter wire. The resistance of the new wire is : **[2012 (Online)]**

(A) $2R$ (B) $R/2$

(C) $R/4$ (D) $R/8$

15. When 5 V potential difference is applied across a wire of length 0.1 m, the drift speed of electrons is 2.5×10^{-4} ms^1. If the electron density in the wire is 8×10^{28} m^{-3}, the resistivity of the material is close to : **[2015]**

(A) 1.6×10^{-8} Ωm (B) 1.6×10^{-7} Ωm

(C) 1.6×10^{-6} Ωm (D) 1.6×10^{-5} Ωm

16. Suppose the drift velocity v_d in a material varied with the applied electric field E as $v_d \propto \sqrt{E}$. The $V - I$ graph for a wire made of such a material is best given by : **[2015 (Online)]**

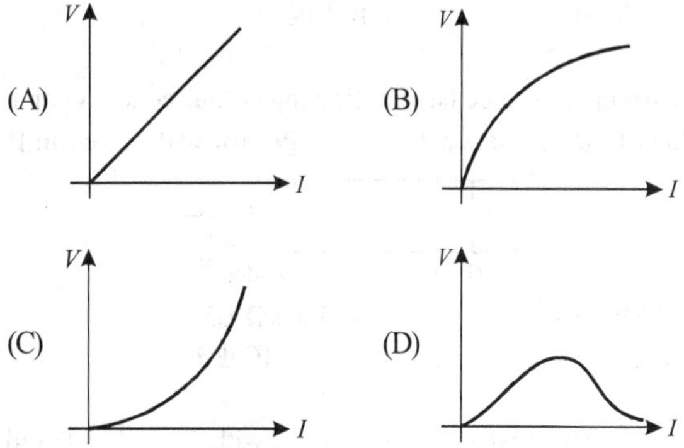

17. A uniform wire of length l and radius r has a resistance of 100 Ω. It is recast into a wire of radius $r/2$. The resistance of new wire will be : **[2017 (Online)]**

(A) 1600 Ω (B) 400 Ω

(C) 200 Ω (D) 100 Ω

18. A copper rod of cross-sectional area A carries a uniform current I through it. At temperature T, if the volume charge density of the rod is ρ, how long will the charges take to travel a distance d ? **[2018 (Online)]**

(A) $\frac{2\rho dA}{IT}$ (B) $\frac{2\rho dA}{I}$

(C) $\frac{\rho dA}{I}$ (D) $\frac{\rho dA}{IT}$

19. A heating element has a resistance of 100 Ω at room temperature. When it is connected to a supply of 220 V, a steady current of 2 A passes in it and temperature is 500ºC more than room temperature. What is the temperature coefficient of resistance of the heating element ? **[2018 (Online)]**

(A) 1×10^{-4} ºC^{-1} (B) 5×10^{-4} ºC^{-1}

(C) 2×10^{-4} ºC^{-1} (D) 0.5×10^{-4} ºC^{-1}

20. A resistance is shown in the figure. Its value and tolerance are given respectively by : **[9 Jan. 2019 (Morning)]**

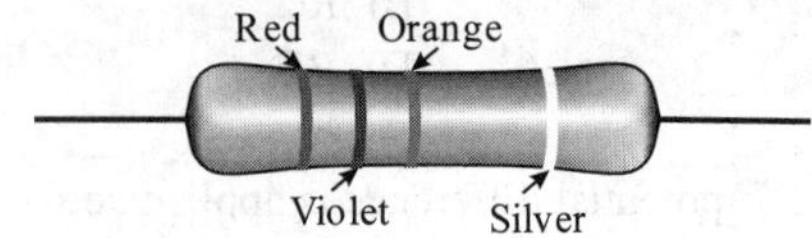

(A) 27 KΩ, 20% (B) 270 KΩ, 5%
(C) 270 KΩ, 10% (D) 27 KΩ, 10%

21. Drift speed of electrons, when 1.5 A of current flows in a copper wire of cross section 5 mm^2, is v. If the electron density in copper is 9×10^{28} /m^3 the value of v in mm/s is close to (Take charge of electron to be = 1.6×10^{-19} C)
[9 Jan. 2019 (Morning)]
(A) 0.2 (B) 3
(C) 2 (D) 0.02

22. A carbon resistance has a following colour code. What is the value of the resistance ? **[9 Jan. 2019 (Evening)]**

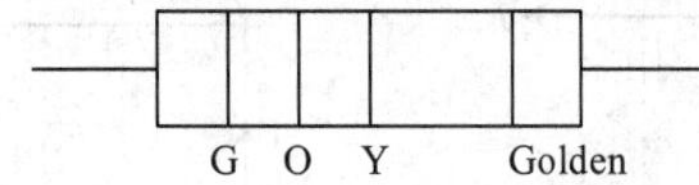

(A) 1.64 MΩ ± 5% (B) 530 kΩ ± 5%
(C) 64 kΩ ± 10% (D) 5.3 MΩ ± 5%

23. A 2 W carbon resistor is color coded with green, black, red and brown respectively. The maximum current which can be passed through this resistor is : **[10 Jan. 2019 (Morning)]**
(A) 63 mA (B) 0.4 mA
(C) 100 mA (D) 20 mA

24. A 200 Ω resistor has a certain color code. If one replaces the red color by green in the code, the new resistance will be :
[8 April 2019 (Morning)]
(A) 100 Ω (B) 400 Ω
(C) 500 Ω (D) 300 Ω

25. A metal wire of resistance 3 Ω is elongated to make a uniform wire of double its previous length. This new wire is now bent and the ends joined to make a circle. If two points on this circle make an angle 60° at the centre, the equivalent resistance between these two points will be :
[9 April 2019 (Evening)]
(A) $\frac{12}{5}\Omega$ (B) $\frac{5}{3}\Omega$
(C) $\frac{5}{2}\Omega$ (D) $\frac{7}{2}\Omega$

26. In a conductor, if the number of conduction electrons per unit volume is 8.5×10^{28} m^{-3} and mean free time is $25 fs$ (femtosecond), it's approximate resistivity is :
($m_e = 9.1 \times 10^{-31}$ kg) **[9 April 2019 (Evening)]**
(A) 10^{-5} Ωm (B) 10^{-6} Ωm
(C) 10^{-7} Ωm (D) 10^{-8} Ωm

27. A current of 5 A passes through a copper conductor (resistivity = 1.7×10^{-8} Ωm) of radius of cross-section 5 mm. Find the mobility of the charges if their drift velocity is 1.1×10^{-3} m/s : **[10 April 2019 (Morning)]**
(A) 1.3 m^2/Vs (B) 1.5 m^2/Vs
(C) 1.8 m^2/Vs (D) 1.0 m^2/Vs

28. In an experiment, the resistance of a material is plotted as a function of temperature (in some range). As shown in the figure, it is a straight line. One may conclude that :
[10 April 2019 (Morning)]

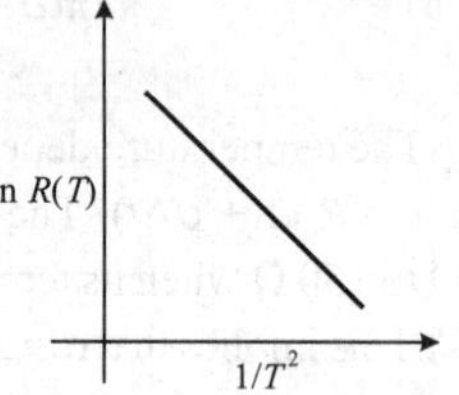

(A) $R(T) = \frac{R_0}{T^2}$ (B) $R(T) = R_0 e^{-T^2/T_0^2}$
(C) $R(T) = R_0 e^{-T_0^2/T^2}$ (D) $R(T) = R_0 e^{T^2/T_0^2}$

29. Space between two concentric conducting spheres of radii a and $b (b > a)$ is filled with a medium of resistivity ρ. The resistance between the two spheres will be :
[10 April 2019 (Evening)]
(A) $\frac{\rho}{4\pi}\left(\frac{1}{a} - \frac{1}{b}\right)$ (B) $\frac{\rho}{2\pi}\left(\frac{1}{a} - \frac{1}{b}\right)$
(C) $\frac{\rho}{2\pi}\left(\frac{1}{a} + \frac{1}{b}\right)$ (D) $\frac{\rho}{4\pi}\left(\frac{1}{a} + \frac{1}{b}\right)$

30. Consider four conducting materials copper, tungsten, mercury and aluminium with resistivity $\rho_C < \rho_T < \rho_M$ and ρ_A respectively. Then : **[2 Sept. 2020 (Morning)]**
(A) $\rho_A > \rho_T > \rho_C$ (B) $\rho_C > \rho_A > \rho_T$
(C) $\rho_A > \rho_M > \rho_C$ (D) $\rho_M > \rho_A > \rho_C$

31. A current through a wire depends on time as $i = \alpha_0 t + \beta t^2$ where $\alpha_0 = 20$ A/s and $\beta = 8$ As^{-2}. Find the charge crossed through a section of the wire in 15 s: **[24 Feb. 2021 (Morning)]**
(A) 260 C (B) 2250 C
(C) 2100 C (D) 11250 C

32. A cylindrical wire of radius 0.5 mm and conductivity 5×10^7 S/m is subjected to an electric field of 10 mV/m. The expected value of current in the wire will be $x^3\pi$ mA. The value of x is ______. **[24 Feb. 2021 (Evening)]**

33. A conducting wire of length 'l', area of cross-section A and electric resistivity ρ is connected between the terminals of a battery. A potential difference V is developed between its ends, causing an electric current. If the length of the wire of the same material is doubled and the area of cross-section is halved, the resultant current would be : **[16 March 2021 (Morning)]**

(A) $\frac{1}{4}\frac{VA}{\rho l}$ (B) $\frac{3}{4}\frac{VA}{\rho l}$

(C) $\frac{1}{4}\frac{\rho l}{VA}$ (D) $4\frac{VA}{\rho l}$

34. In the figure given, the electric current flowing through the 5 kΩ resistor is 'x' mA.

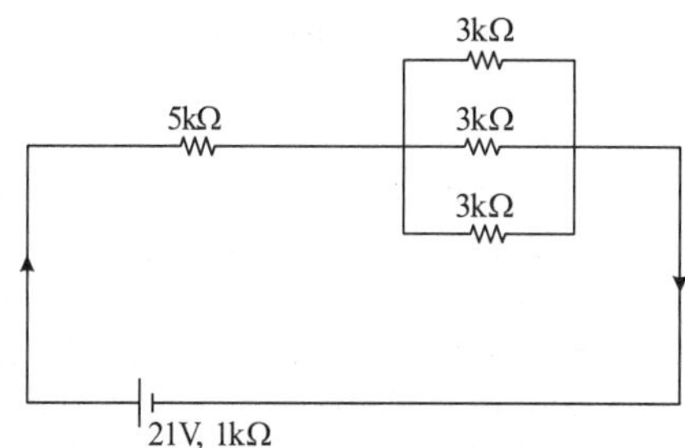

The value of x to the nearest integer is _______.

[16 March 2021 (Morning)]

35. A current of 10 A exists in a wire of cross-sectional area of 5 mm^2 with a drift velocity of 2×10^{-3} ms^{-1}. The number of free electrons in each cubic meter of the wire is _______.

[17 March 2021 (Morning)]

(A) 2×10^6 (B) 625×10^{25}

(C) 2×10^{25} (D) 1×10^{23}

36. In the experiment of Ohm's law, a potential difference of 5.0 V is applied across the end of a conductor of length 10.0 cm and diameter of 5.00 mm. The measured current in the conductor is 2.00 A. The maximum permissible percentage error in the resistivity of the conductor is : **[18 March 2021 (Morning)]**

(A) 3.9 (B) 8.4

(C) 7.5 (D) 3.0

37. A current of 5 A is a passing through a non-linear magnesium wire of cross-section 0.04 m^2. At every point the direction of current density is at an angle of 60° with the unit vector of area of cross-section. The magnitude of electric field at every point of the conductor is :

(Resistivity of magnesium $\rho = 44 \times 10^{-8}$ Ωm)

[20 July 2021 (Morning)]

(A) 11×10^{-7} V/m (B) 11×10^{-5} V/m

(C) 11×10^{-2} V/m (D) 11×10^{-3} V/m

38. In the given figure, a battery of emf E is connected across a conductor PQ of length 'l' and different area of cross-sections having radii r_1 and $r_2 (r_2 < r_1)$. Choose the correct option as one moves from P to Q : **[27 July 2021 (Morning)]**

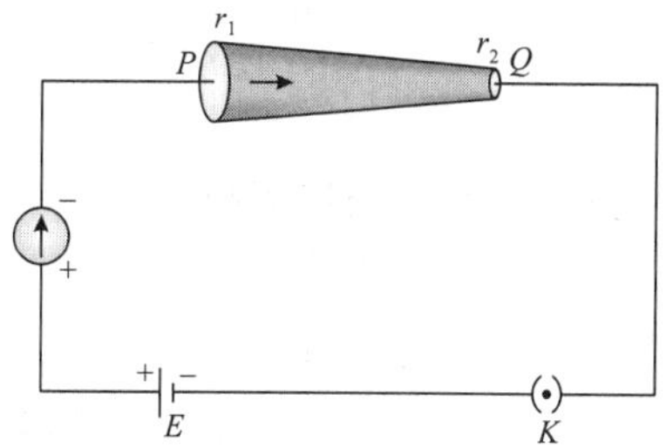

(A) Drift velocity of electron increases

(B) Electric field decreases

(C) All of these

(D) Electron current decreases

39. For the circuit shown, the value of current at time $t = 3.2$ s will be _______ A.

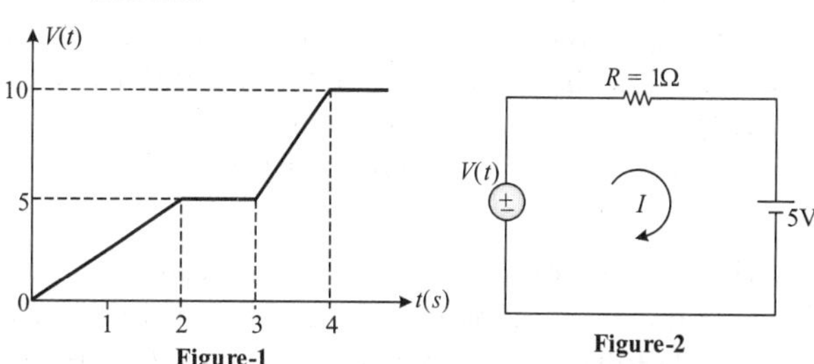

Figure-1 Figure-2

[Voltage distribution $V(t)$ is shown by figure-(1) and the circuit is shown in figure-(2)]. **[27 July 2021 (Evening)]**

40. What equal length of an iron wire a copper-nickel alloy wire, each of 2 mm diameter connected parallel to give an equivalent resistance of 3 Ω ? [Given resistivities of iron and copper-nickel alloy wire are 12 μΩcm and 51 μΩ cm respectively]

[26 Aug. 2021 (Morning)]

(A) 82 m (B) 97 m

(C) 110 m (D) 90 m

41. The colour coding on a carbon resistor is shown in the given figure. The resistance value of the given resistor is :

[27 Aug. 2021 (Evening)]

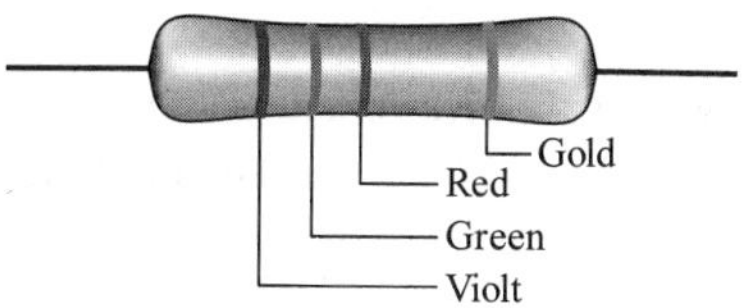

(A) (7500 ± 375)Ω (B) (5700 ± 285)Ω

(C) (7500 ± 755)Ω (D) (5700 ± 375)Ω

42. The length of a given cylindrical wire is increased to double of its original length. The percentage increase in the resistance of the wire will be _____ %. **[25 June 2022 (Evening)]**

43. An aluminium wire is stretched to make its length, 0.4% larger. Then percentage change in resistance is : **[26 June 2022 (Morning)]**

(A) 0.4% (B) 0.2%
(C) 0.8% (D) 0.6%

44. The current density in a cylindrical wire of radius 4 mm is $4 \times 10^6\ Am^{-2}$. The current through the outer portion of the wire between radial distance $\frac{R}{2}$ and R is _____ πA. **[27 June 2022 (Morning)]**

45. The current density in a cylindrical wire of radius $r = 4.0$ mm is $1.0 \times 10^6\ A/m^2$. The current through the outer portion of the wire between radial distances $r/2$ and r is $x\pi A$; where x is _____. **[27 June 2022 (Evening)]**

46. Resistance of the wire is measured as 2 Ω and 3 Ω at 10°C and 30°C respectively. Temperature co-coefficient of resistance of the material of the wire is : **[28 June 2022 (Evening)]**

(A) $0.033°C^{-1}$ (B) $-0.033°C^{-1}$
(C) $0.011°C^{-1}$ (D) $0.055°C^{-1}$

47. The variation of applied potential and current flowing through a given wire is shown in figure. The length of wire is 31.4 cm. The diameter of wire is measured as 2.4 cm. The resistivity of the given wire is measured as $x \times 10^{-3}$ Ω cm. The value of x is _____. [Take $\pi = 3.14$] **[29 June 2022 (Morning)]**

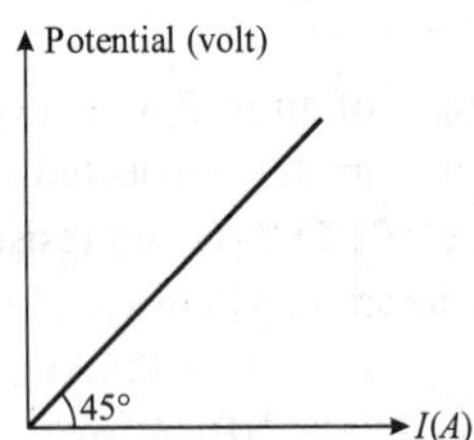

48. Eight copper wire of length l and diameter d are joined in parallel to form a single composite conductor of resistance R. If a single copper wire of length $2l$ have the same resistance (R) then its diameter will be _____ d. **[25 July 2022 (Morning)]**

49. The current I in the given circuit will be : **[26 July 2022 (Morning)]**

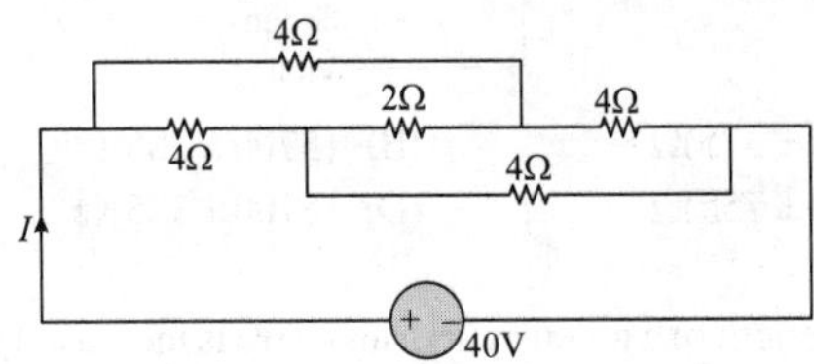

(A) 10 A (B) 20 A
(C) 4 A (D) 40 A

50. A battery of 6 V is connected to the circuit as shown below. The current I drawn from the battery is : **[26 July 2022 (Evening)]**

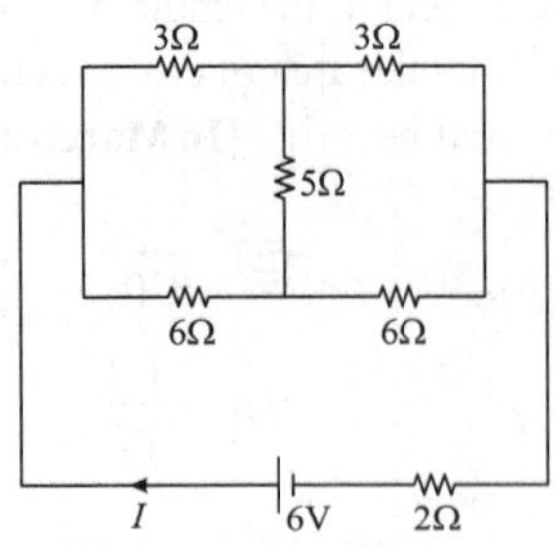

(A) 1 A (B) 2 A
(C) $\frac{6}{11}$ A (D) $\frac{4}{3}$ A

51. Two sources of equal emfs are connected in series. This combination is connected to an external resistance R. The internal resistances of the two sources are r_1 and r_2 $(r_1 > r_2)$. If the potential difference across the source of internal resistance r_1 is zero then the value of R will be : **[25 June (Evening) & 27 July (Morning)]**

(A) $r_1 - r_2$ (B) $\frac{r_1 r_2}{r_1 + r_2}$
(C) $\frac{r_1 + r_2}{2}$ (D) $r_2 - r_1$

52. A 1 m long copper wire carries a current of 1 A. If the cross section of the wire is 2.0 mm² and the resistivity of copper is 1.7×10^{-8} Ωm, the force experienced by moving electron in the wire is _____ $\times 10^{-23}$ N. [Charge on electron = 1.6×10^{-19} C] **[27 July 2022 (Morning)]**

53. Which of the following statement(s) is/are correct ? **[27 July 2022 (Evening)]**

(a) The drift velocity of electrons decreases with the increase in the temperature of conductor
(b) The drift velocity is inversely proportional to the area of cross-section of given conductor
(c) The drift velocity does not depend on the applied potential difference to the conductor
(d) The drift velocity of electron is inversely proportional to the length of the conductor
(e) The drift velocity increases with the increase in the temperature of conductor

(A) (a) and (b) only (B) (a) and (d) only
(C) (b) and (e) only (D) (b) and (c) only

54. As shown in the figure, in steady state, the charge stored in the capacitor is _____ $\times 10^{-6}$ C. **[27 July 2022 (Evening)]**

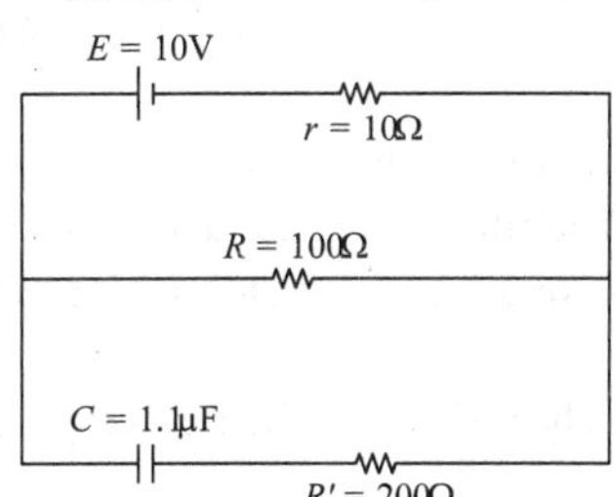

55. A wire of resistance R_1 is drawn out so that its length is increased by twice of its original length. The ratio of new resistance to original resistance is : **[28 July 2022 (Morning)]**

(A) 9 : 1 (B) 1 : 9
(C) 4 : 1 (D) 3 : 1

56. Two metallic wires of identical dimensions are connected in series. If σ_1 and σ_2 are the conductivities of the these wires respectively, the effective conductivity of the combination is : **[29 July 2022 (Morning)]**

(A) $\frac{\sigma_1\sigma_2}{\sigma_1+\sigma_2}$ (B) $\frac{2\sigma_1\sigma_2}{\sigma_1+\sigma_2}$

(C) $\frac{\sigma_1+\sigma_2}{2\sigma_1\sigma_2}$ (D) $\frac{\sigma_1+\sigma_2}{\sigma_1\sigma_2}$

57. Given below are two statements :
One is labelled as **Assertion-A** and the other is labelled as **Reason-R**.
Assertion-A : Alloys such as constantan and manganin are used in making standard resistance coils.
Reason-R : Constantan and manganin have very small value of temperature coefficient of resistance.
In the light of the above statements, choose the correct answer from the options given below: **[29 July 2022 (Evening)]**

(A) Both **A** and **R** are true and **R** is the correct explanation of **A**.
(B) Both **A** and **R** are true but **R** is NOT the correct explanation of **A**.
(C) **A** is true but **R** is false.
(D) **A** is false but **R** is true.

58. A 1 m long wire is broken into two unequal parts X and Y. The X part of the wire is stretched into another wire W. Length of W is twice the length of X and the resistance of W is twice that of Y. Find the ratio of length of X and Y. **[29 July 2022 (Evening)]**

(A) 1 : 4 (B) 1 : 2
(C) 4 : 1 (D) 2 : 1

59. A hollow cylindrical conductor has length of 3.14 m, while its inner and outer diameters are 4 mm and 8 mm respectively. The resistance of the conductor is $n \times 10^{-3}$ Ω. If the resistivity of the material is 2.4×10^{-8} Ωm. The value of n is _____. **[24 Jan. 2023 (Morning)]**

60. If a copper wire is stretched to increase its length by 20°%. The percentage increase in resistance of the wire is _____ %. **[24 Jan. 2023 (Evening)]**

61. A uniform metallic wire carries a current 2 A. When 3.4 V battery is connected across it. The mass of uniform metallic wire is 8.92×10^{-3} kg. Density is 8.92×10^{3} kg/m^3 and resistivity is 1.7×10^{-8} Ω – m. The length of wire is : **[25 Jan. 2023 (Morning)]**

(A) $l = 6.8$ m (B) $l = 10$ m
(C) $l = 5$ m (D) $l = 100$ m

62. The resistance of a wire is 5 Ω. It's new resistance in ohm if stretched to 5 times of it's original length will be : **[25 Jan. 2023 (Evening)]**

(A) 625 (B) 5
(C) 125 (D) 25

63. The charge flowing in a conductor changes with time as $Q(t) = \alpha t - \beta t^2 + \gamma t^3$. Where α, β and γ are constants. Minimum value of current is : **[30 Jan. 2023 (Morning)]**

(A) $\alpha - \frac{3\beta^2}{\gamma}$ (B) $\alpha - \frac{\gamma^2}{3\beta}$

(C) $\beta - \frac{\alpha^2}{3\gamma}$ (D) $\alpha - \frac{\beta^2}{3\gamma}$

64. The drift velocity of electrons for a conductor connected in an electrical circuit is V_d. The conductor in now replaced by another conductor with same material and same length but double the area of cross section. The applied voltage remains same. The new drift velocity of electrons will be : **[31 Jan. 2023 (Morning)]**

(A) V_d (B) $\frac{V_d}{2}$

(C) $\frac{V_d}{4}$ (D) $2V_d$

65. A current of 2 A flows through a wire of cross-sectional area 25.0 mm^2. The number of free electrons in a cubic meter are 2.0×10^{28}. The drift velocity of the electrons is _____ $\times 10^{-6}$ ms^{-1} (Given, charge on electron = 1.6×10^{-19} C) **[8 April 2023 (Morning)]**

66. The number density of free electrons in copper is nearly 8×10^{28} m^{-3}. A copper wire has its area of cross section $= 2 \times 10^{-6}$ m^2 and is carrying a current of 3.2 A. The drift speed of the electrons is _____ $\times 10^{-6}$ ms^{-1} **[8 April 2023 (Evening)]**

67. 10 resistors each of resistance 10 Ω can be connected in such as to get maximum and minimum equivalent resistance. The ratio of maximum and minimum equivalent resistance will be _____. **[10 April 2023 (Morning)]**

68. In a metallic conductor, under the effect of applied electric field, the free electrons of the conductor: **[10 April 2023 (Evening)]**

(A) Drift from higher potential to lower potential
(B) Move in the curved paths from lower potential to higher potential
(C) Move with the uniform velocity throughout from lower potential to higher potential
(D) Move in the straight line paths in the same direction

69. A rectangular parallelepiped is measured as 1 cm × 1 cm × 100 cm. If its specific resistance is 3×10^{-7} Ωm, then the resistance between its two opposite rectangular faces will be _____ $\times 10^{-7}$ Ω. **[10 April 2023 (Evening)]**

70. A wire of resistance 160 Ω is melted and drawn in wire of one-fourth of its length. The new resistance of the wire will be **[12 April 2023 (Morning)]**

(A) 10 Ω (B) 640 Ω
(C) 40 Ω (D) 16 Ω

71. The current flowing through a conductor connected across a source is 2 A and 1.2 A at 0°C and 100°C respectively. The current flowing through the conductor at 50°C will be _____ × 10^2 mA. **[12 April 2023 (Morning)]**

72. Given below are two statements:

Statement-I : The equivalent resistance of resistors in a series combination is smaller than least resistance used in the combination.

Statement-II : The resistivity of the material is independent of temperature.

In the light of the above statements, choose the correct answer from the options given below : **[15 April 2023 (Morning)]**

(A) Statement-I is false but Statement-II is true
(B) Both Statement-I and Statement-II are false
(C) Statement-I is true but Statement-II is false
(D) Both Statement-I and Statement-II are true

73. In the given circuit, the current (I) through the battery will be : **[15 April 2023 (Morning)]**

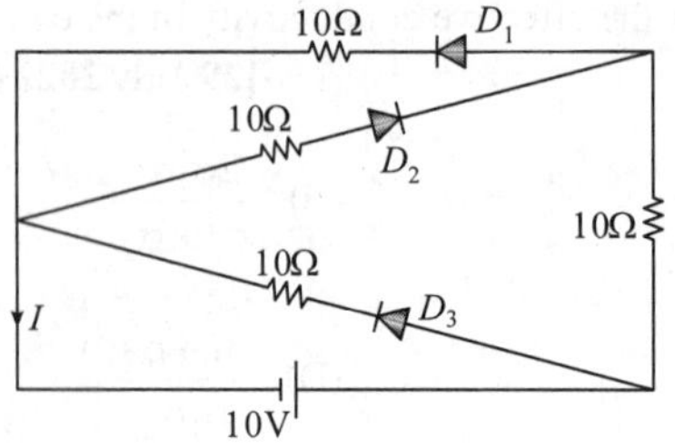

(A) 1.5 A (B) 1 A
(C) 2.5 A (D) 2 A

* * * * *

CHAPTER

17 CURRENT ELECTRICITY

17.2 CIRCUIT ANALYSIS

1. A 3 volt battery with negligible internal resistance is connected in a circuit as shown in the figure. The current I, in the circuit will be : **[2003]**

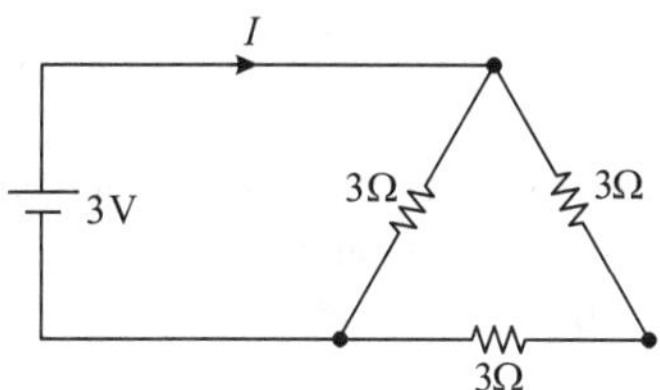

(A) 1 A (B) 1.5 A
(C) 2 A (D) 1/3 A

2. Two sources of equal emf are connected to an external resistance R. The internal resistances of the two sources are R_1 and R_2 ($R_2 > R_1$). If the potential difference across the source having internal resistance R_2, is zero, then : **[2005]**

(A) $R = \dfrac{R_2 \times (R_1 + R_2)}{(R_2 - R_1)}$ (B) $R = R_2 - R_1$

(C) $R = \dfrac{R_1 R_2}{(R_1 + R_2)}$ (D) $R = \dfrac{R_1 R_2}{(R_2 - R_1)}$

3. The current I drawn from the 5 volt source will be : **[2006]**

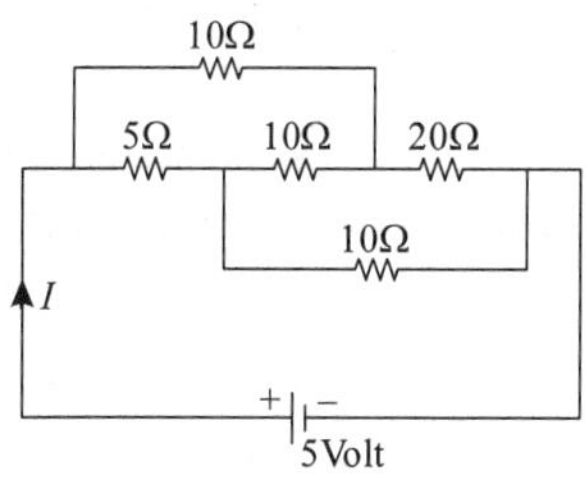

(A) 0.33 A (B) 0.5 A
(C) 0.67 A (D) 0.17 A

4. In a Wheatstone's bridge, three resistances P, Q and R are connected in the three arms and the fourth arm is formed by two resistances S_1 and S_2 connected in parallel. The condition for the bridge to be balanced will be : **[2006]**

(A) $\dfrac{P}{Q} = \dfrac{2R}{S_1 + S_2}$ (B) $\dfrac{P}{Q} = \dfrac{R(S_1 + S_2)}{S_1 S_2}$

(C) $\dfrac{P}{Q} = \dfrac{R(S_1 + S_2)}{2S_1 S_2}$ (D) $\dfrac{P}{Q} = \dfrac{R}{S_1 + S_2}$

5. The Kirchhoff's first law ($\Sigma i = 0$) and second law ($\Sigma iR = \Sigma E$), Where the symbols have their usual meanings, are respectively based on : **[2006]**

(A) conservation of charge, conservation of momentum
(B) conservation of energy, conservation of charge
(C) conservation of momentum, conservation of charge
(D) conservation of charge, conservation of energy

6. A 5 V battery with internal resistance 2 Ω and a 2 V battery with internal resistance 1 Ω are connected to a 10 Ω resistor as shown in the figure. The current in the 10 Ω resistor is :**[2008]**

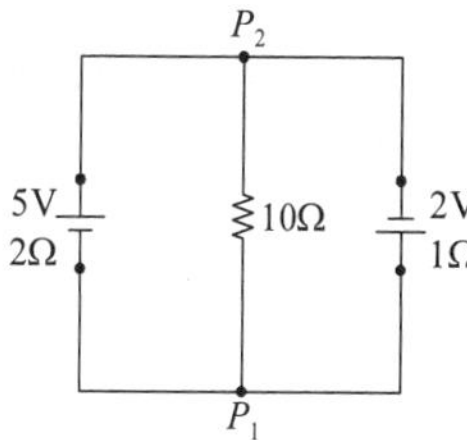

(A) 0.27 A P_2 to P_1 (B) 0.03 A P_1 to P_2
(C) 0.03 A P_2 to P_1 (D) 0.27 A P_1 to P_2

7. A d.c. source of emf E_1 = 100 V and internal resistance r = 0.5 Ω, a storage battery of emf E_2 = 90 V and an external resistance R are connected as shown in figure. For what value of R no current will pass through the battery ?**[2013 (Online)]**

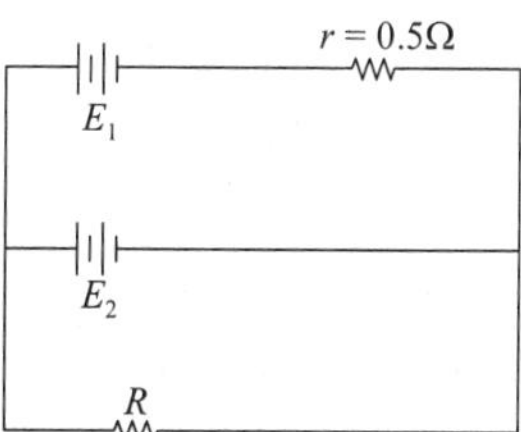

(A) 5.5 Ω (B) 3.5 Ω
(C) 4.5 Ω (D) 2.5 Ω

8. A letter 'A' is constructed of a uniform wire with resistance 1.0 Ω per cm. The sides of the letter are 20 cm and the cross piece in the middle is 10 cm long. The apex angle is 60°. The resistance between the ends of the legs is closed to : **[2013 (Online)]**

(A) 50.0 Ω (B) 10 Ω
(C) 36.7 Ω (D) 26.7 Ω

9. Six equal resistances are connected between points P, Q and R as shown in figure. Then net resistance will be maximum between : **[2013 (Online)]**

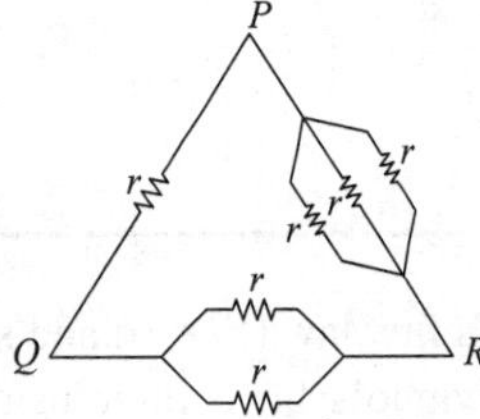

(A) P and R (B) P and Q
(C) Q and R (D) Any two points

10. A d.c. main supply of e.m.f. 220 V is connected across a storage battery of e.m.f. 200 V through a resistance of 1 Ω. The battery terminals are connected to an external resistance 'R'. The minimum value of 'R', so that a current passes through the battery to charge it is : **[2014 (Online)]**

(A) 7 Ω (B) 9 Ω
(C) 11 Ω (D) Zero

11. In the circuit shown, current (in A) through 50 V and 30 V batteries are, respectively : **[2014 (Online)]**

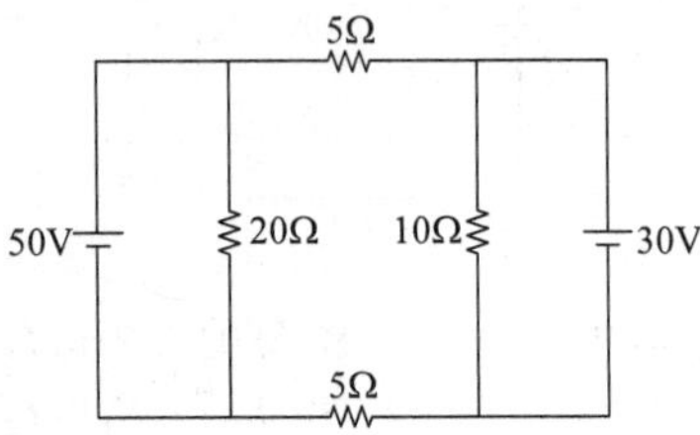

(A) 2.5 and 3 (B) 3.5 and 2
(C) 4.5 and 1 (D) 3 and 2.5

12. In the circuit shown, the current in the 1 Ω resistor is : **[2015]**

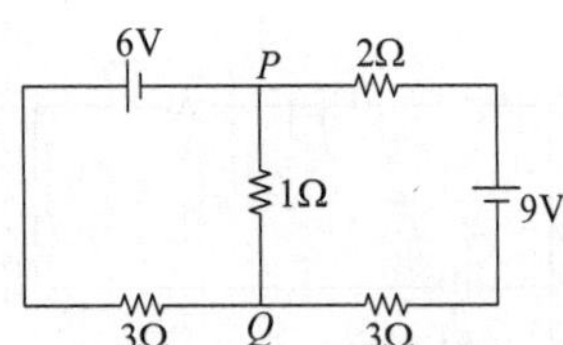

(A) 1.3 A, from P to Q (B) 0 A
(C) 0.13 A, from Q to P (D) 0.13 A, from P to Q

13. In the electric network shown, when no current flows through the 4 Ω resistor in the arm EB, the potential difference between the points A and D will be : **[2015 (Online)]**

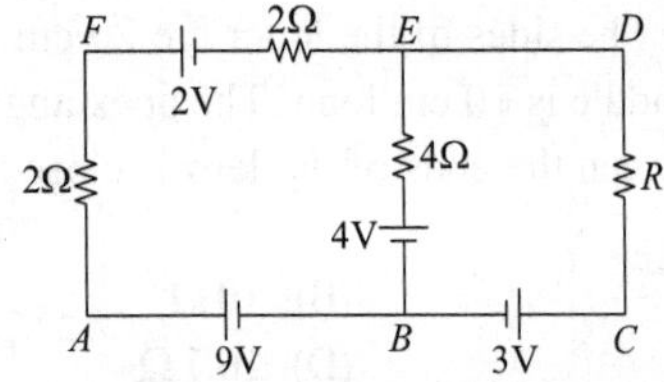

(A) 6V (B) 3V
(C) 5V (D) 4V

14.

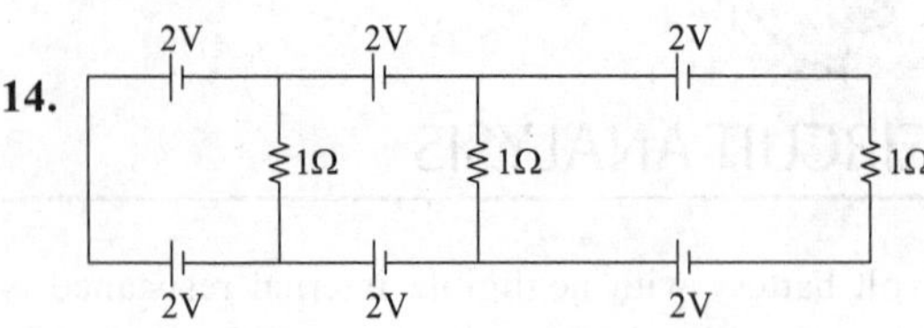

In the above circuit the current in each resistance is : **[2017]**

(A) 0.5 A (B) 0 A
(C) 1 A (D) 0.25 A

15. A 9 V battery with internal resistance of 0.5 Ω is connected across an infinite network as shown in the figure. All ammeters A_1, A_2, A_3 and voltmeter V are ideal : **[2017 (Online)]**

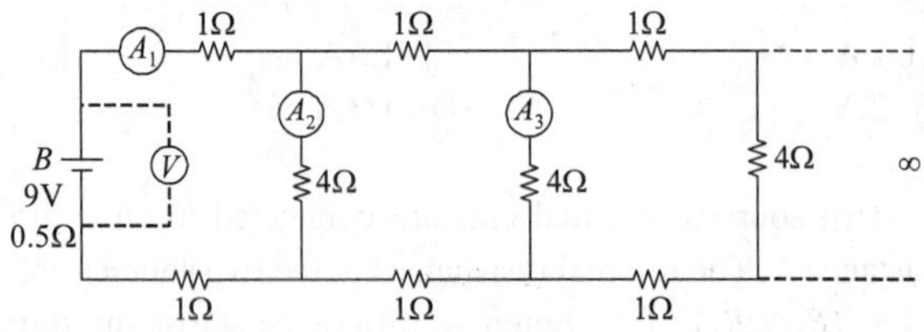

(A) Reading of A_1 is 2 A (B) Reading of A_1 is 18 A
(C) Reading of V is 9 V (D) Reading of V is 7 V

16. Two batteries with e.m.f. 12 V and 13 V are connected in parallel across a load resistor of 10 Ω. The internal resistance of the two batteries are 1 Ω and 2 Ω respectively. The voltage across the load lies between : **[2018]**

(A) 11.6 V and 11.7 V (B) 11.5 V and 11.6 V
(C) 11.4 and 11.5 V (D) 11.7 V and 11.8 V

17. In the given circuit all resistances are of volume R ohm each. The equivalent resistance between A and B is : **[2018 (Online)]**

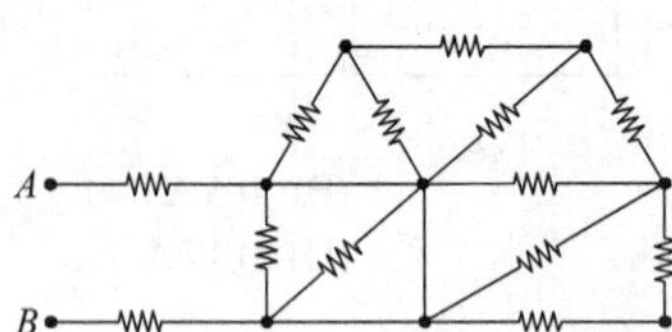

(A) $2R$ (B) $\frac{5R}{2}$

(C) $\frac{5R}{3}$ (D) $3R$

18. When the switch S, in the circuit shown, is closed, then the value of current i will be : **[9 Jan. 2019 (Morning)]**

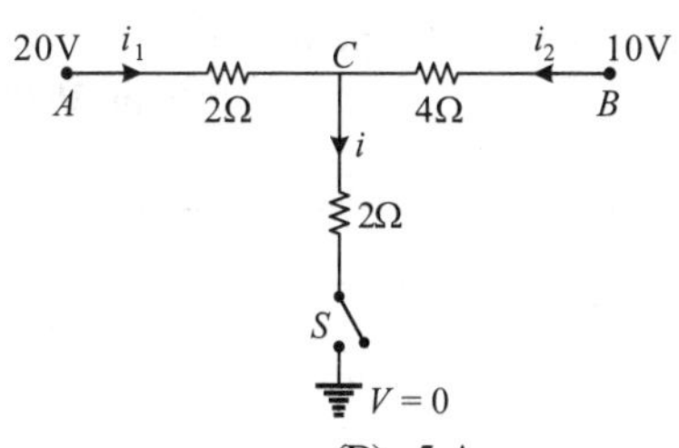

(A) 3 A (B) 5 A
(C) 4 A (D) 2 A

19. A uniform metallic wire has a resistance of 18 Ω and is bent into an equilateral triangle. Then, the resistance between any two vertices of the triangle is : **[10 Jan. 2019 (Morning)]**

(A) 8 Ω (B) 12 Ω
(C) 4 Ω (D) 2 Ω

20. In the given circuit the cells have zero internal resistance. The currents (in Amperes) passing through resistance R_1, and R_2 respectively, are : **[10 Jan. 2019 (Morning)]**

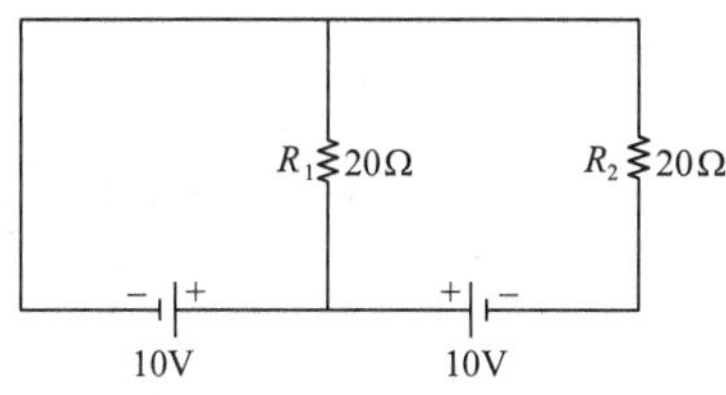

(A) 2, 2 (B) 0,1
(C) 1, 2 (D) 0.5, 0

21. In the circuit, the potential difference between A and B is: **[11 Jan. 2019 (Evening)]**

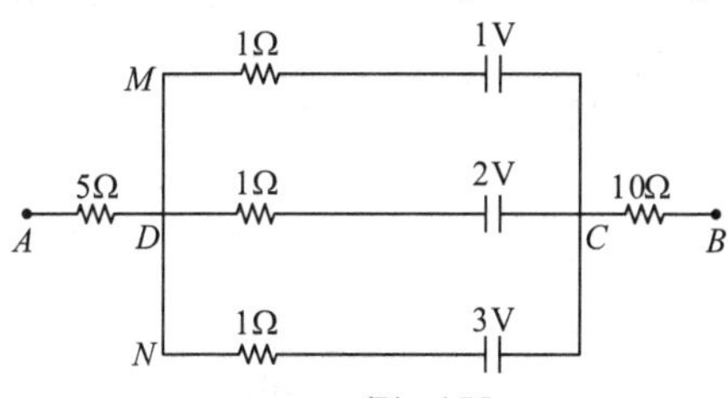

(A) 6 V (B) 1 V
(C) 3 V (D) 2 V

22. In the given circuit diagram, the currents, $I_1 = -0.3$ A, $I_4 = 0.8$ A and $I_5 = 0.4$ A, are flowing as shown. The currents I_2, I_3 and I_6, respectively, are : **[12 Jan. 2019 (Evening)]**

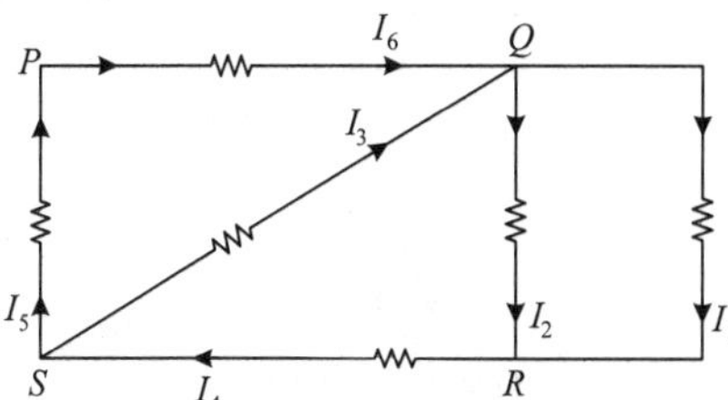

(A) 1.1 A, 0.4 A, 0.4 A (B) −0.4 A, 0.4 A, 1.1 A
(C) 0.4 A, 1.1 A, 0.4 A (D) 1.1 A, −0.4 A, 0.4 A

23. For the circuit shown, with $R_1 = 1.0\,\Omega$, $R_2 = 2.0\,\Omega$, $E_1 = 2$ V and $E_2 = E_3 = 4$ V, the potential difference between the points 'a' and 'b' is approximately (in V) : **[8 April 2019 (Morning)]**

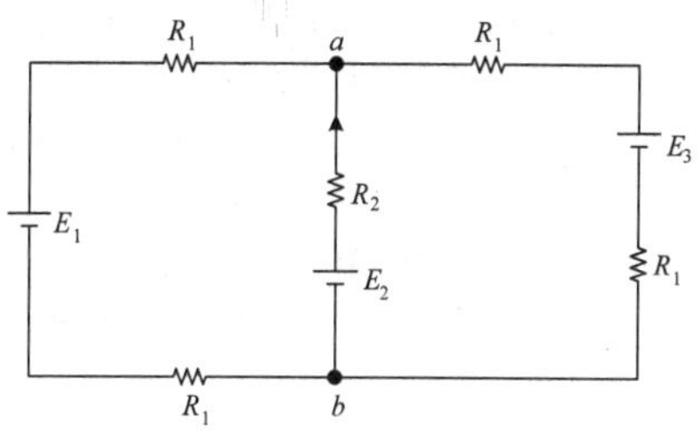

(A) 2.7 (B) 3.3
(C) 2.3 (D) 3.7

24. In the figure shown, what is the current (in Ampere) drawn from the battery ? You are given.

$R_1 = 15\,\Omega, R_2 = 10\,\Omega, R_3 = 20\,\Omega, R_4 = 5\,\Omega, R_5 = 25\,\Omega,$
$R_6 = 30\,\Omega, E = 15$ V **[8 April 2019 (Evening)]**

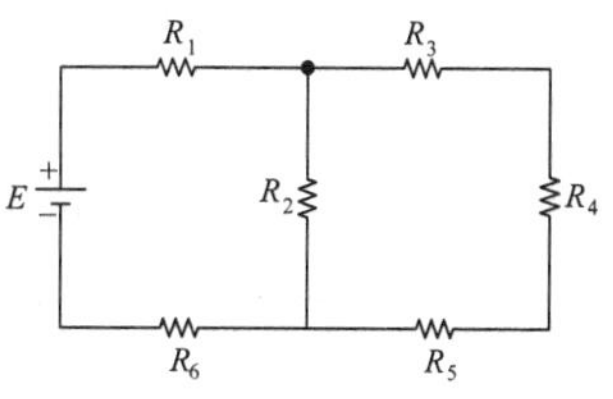

(A) 7/18 (B) 13/24
(C) 9/32 (D) 20/3

25. A wire of resistance R is bent to form a square $ABCD$ as shown in the figure. The effective resistance between E and C is : (E is mid-point of arm CD) : **[9 April 2019 (Morning)]**

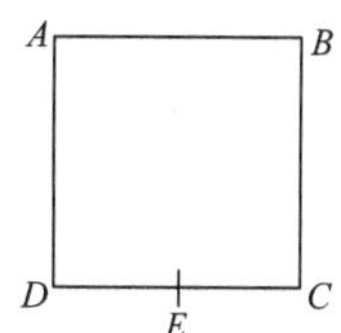

(A) R (B) $\frac{1}{16}R$
(C) $\frac{7}{64}R$ (D) $\frac{3}{4}R$

26. The current I_1(in A) flowing through 1 Ω resistor in the following circuit is : **[7 Jan. 2020 (Morning)]**

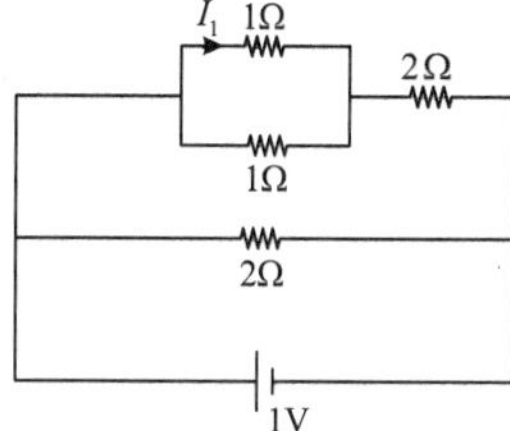

(A) 0.5 (B) 0.2
(C) 0.25 (D) 0.4

27. The series combination of two batteries, both of the same emf 10 V, but different internal resistance of 20 Ω and 5 Ω, is connected to the parallel combination of two resistors 30 Ω and $R\Omega$. The voltage difference across the battery of internal resistance 20 Ω is zero, the value of R (in Ω) is ______.

[8 Jan. 2020 (Evening)]

28. In the given circuit diagram, a wire is joining points B and D. The current in this wire is : **[9 Jan. 2020 (Morning)]**

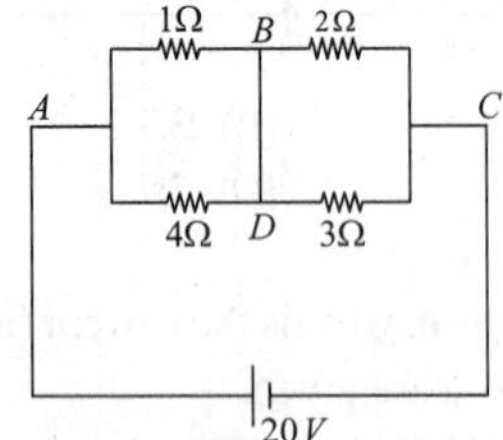

(A) 4 A (B) 2 A
(C) 0.4 A (D) Zero

29. The value of current i_1 flowing from A to C in the circuit diagram is : **[4 Sept. 2020 (Evening)]**

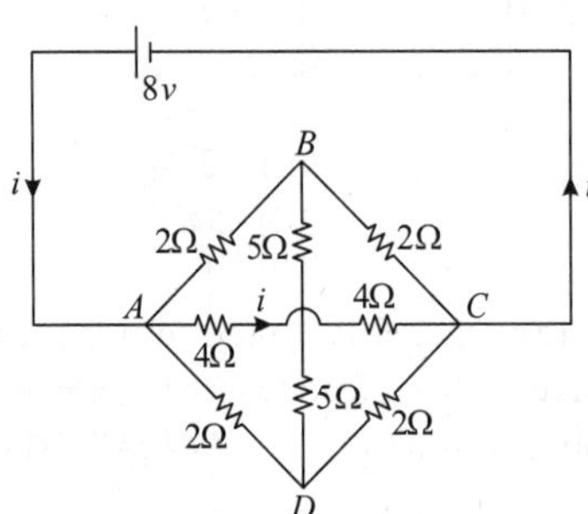

(A) 5 A (B) 2 A
(C) 4 A (D) 1 A

30. Four resistances 40 Ω, 60 Ω, 90 Ω and 110 Ω make the arms of a quadrilateral $ABCD$. Across AC is a battery of emf 40 V and internal resistance negligible. The potential difference across BD is V is ______. **[4 Sept. 2020 (Evening)]**

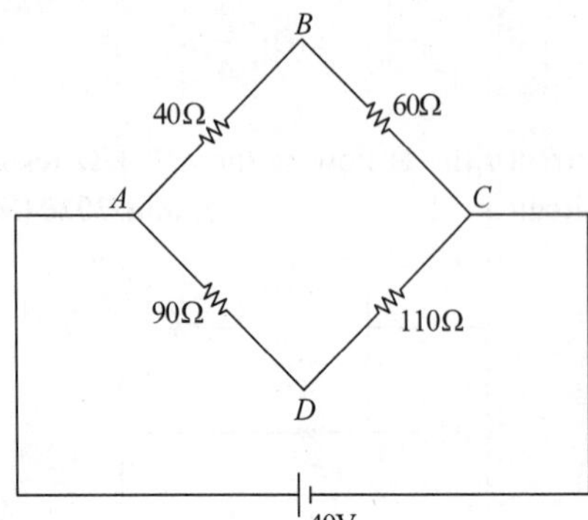

31. In the circuit, given in the figure currents in different branches and value of one resistor are shown. Then potential at point B with respect to the point A is : **[5 Sept. 2020 (Evening)]**

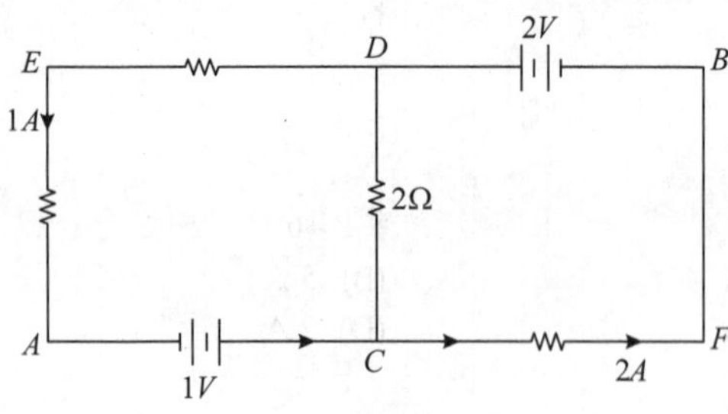

(A) +1 V (B) −1 V
(C) −2 V (D) +2 V

32. In the figure shown, the current in the 10 V battery is close to : **[6 Sept. 2020 (Evening)]**

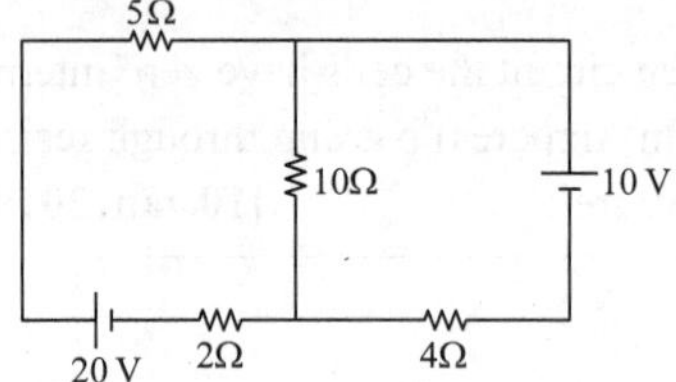

(A) 0.36 A from negative to positive terminal
(B) 0.71 A from positive to negative terminal
(C) 0.21 A from positive to negative terminal
(D) 0.42 A from positive to negative terminal

33. A cell E_1 of emf 6 V and internal resistance 2 Ω is connected with another cell E_2 of emf 4 V and internal resistance 8 Ω (as shown in the figure). The potential difference across points X and Y is : **[24 Feb. 2021 (Morning)]**

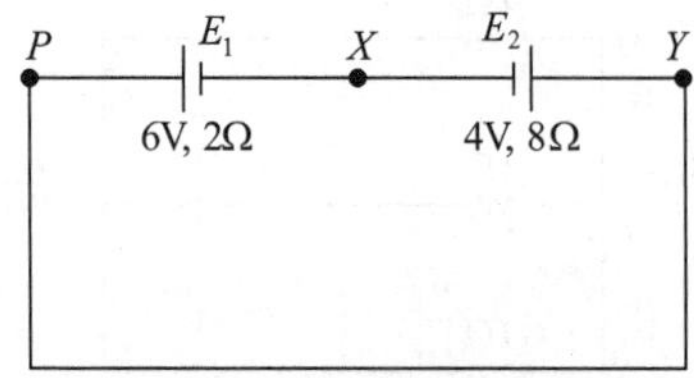

(A) 2.0 V (B) 10.0 V
(C) 5.6 V (D) 3.6 V

34. A current of 6 A enters one corner P of an equilateral triangle PQR having 3 wires of resistance 2 Ω each and leaves by the corner R. The currents i_1 in ampere is ______. **[25 Feb. 2021 (Evening)]**

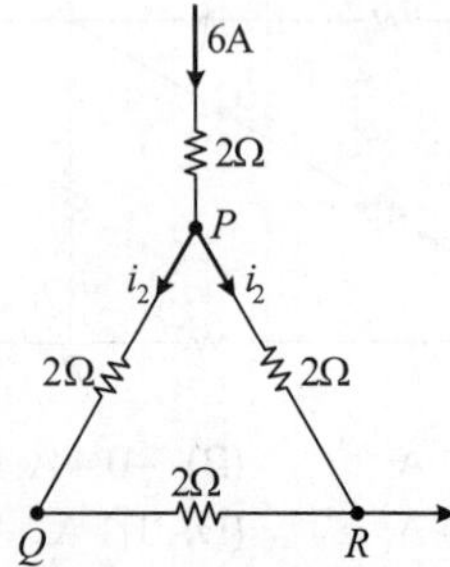

35. Five equal resistances are connected in a network as shown in figure. The net resistance between the points A and B is :

[26 Feb. 2021 (Morning)]

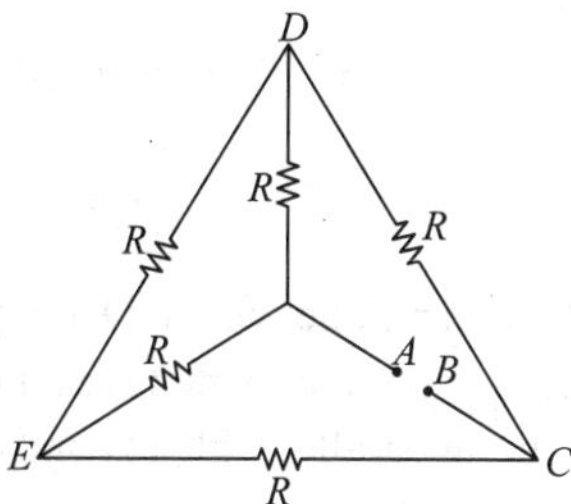

(A) $2R$ (B) $\frac{R}{2}$

(C) R (D) $\frac{3R}{2}$

36. In an electrical circuit, a battery is connected to pass 20 C of charge through it in a certain given time. The potential difference between two plates of the battery is maintained at 15 V. The workdone by the battery is ______ J.

[26 Feb. 2021 (Morning)]

37. The equivalent resistance of series combination of two resistors is 's'. When they are connected in parallel, the equivalent resistance is 'p'. If $s = np$, then the minimum value for n is ______. (Round off to the Nearest Integer)

[17 March 2021 (Morning)]

38. Two cells of emf $2E$ and E with internal resistance r_1 and r_2 respectively are connected in series to an external resistor R (see figure). The value of R, at which the potential difference across the terminals of the first cell becomes zero is :

[17 March 2021 (Evening)]

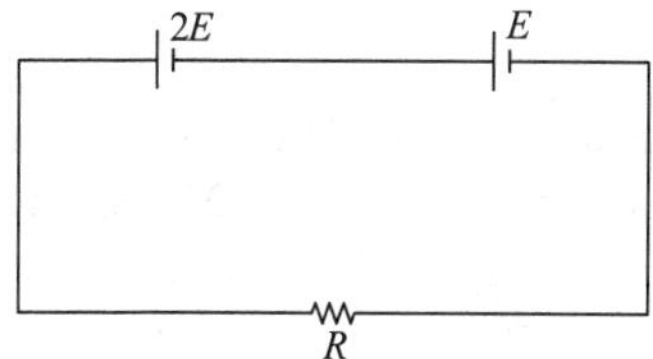

(A) $r_1 + r_2$ (B) $\frac{r_1}{2} - r_2$

(C) $\frac{r_1}{2} + r_2$ (D) $r_1 - r_2$

39. The four arms of a Wheatstone bridge have resistances as shown in the figure. A galvanometer of 15 Ω resistance is connected across BD. Calculate the current through the galvanometer when a potential difference of 10 V is maintained across AC.

[17 March 2021 (Evening)]

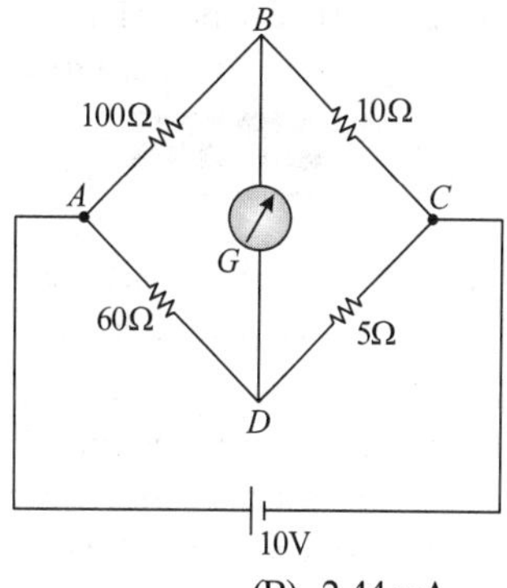

(A) 2.44 μA (B) 2.44 mA

(C) 4.87 mA (D) 4.87 μA

40. The voltage across the 10 Ω resistor in the given circuit is x volt.

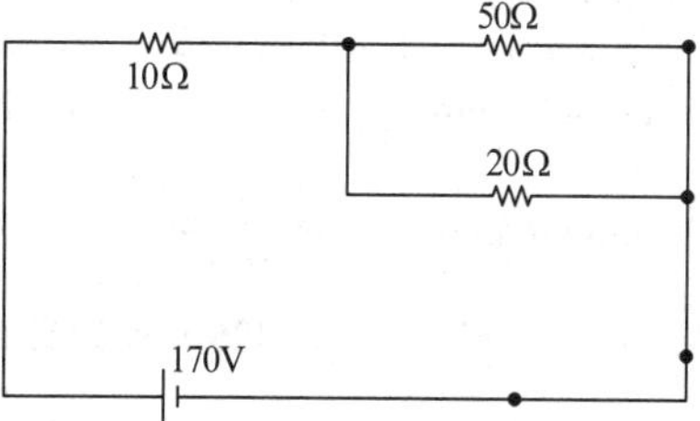

The value of 'x' to the nearest integer is ______.

[18 March 2021 (Morning)]

41. Two wires of same length and thickness having specific resistances 6 Ω cm and 3 Ω cm respectively are connected in parallel. The effective resistivity is ρΩ cm. The value of ρ, to the nearest integer, is______. **[18 March 2021 (Evening)]**

42. The value of current in the 6 Ω resistance is :

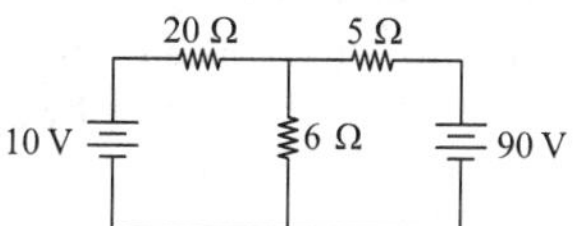

[20 July 2021 (Morning)]

(A) 4 A (B) 10 A

(C) 8 A (D) 6 A

43. In the given figure switches S_1 and S_2 are in open condition. The resistance across ab when the switches S_1 and S_2 are closed is ______Ω. **[20 July 2021 (Evening)]**

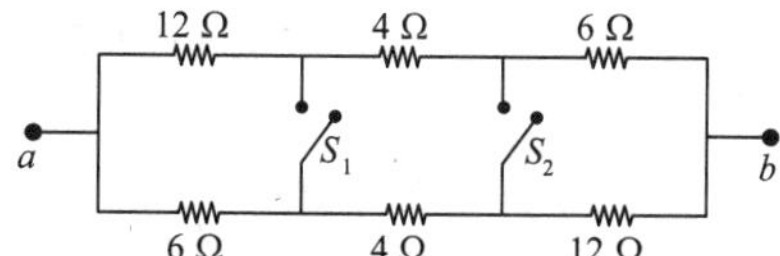

44. A copper (Cu) rod of length 25 cm and cross-sectional area 3 mm² is joined with a similar Aluminium (Al) rod as shown in figure. Find the resistance of the combination between the ends

A and B : (Take Resistivity of Copper $= 1.7 \times 10^{-8}\ \Omega$m) **[22 July 2021 (Evening)]**

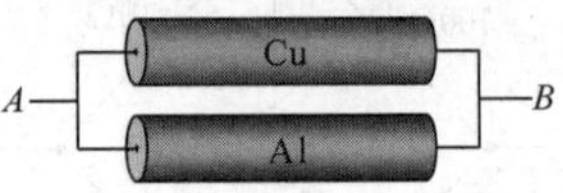

(A) $0.858\ m\Omega$ (B) $0.0858\ m\Omega$
(C) $2.170\ m\Omega$ (D) $1.420\ m\Omega$

45. In an electric circuit, a cell of certain emf provides a potential difference of 1.25 V across a load resistance of 5 Ω. However, it provides a potential difference of 1 V across a load resistance of 2 Ω. The emf of the cell is given by $\frac{x}{10}$ V. Then the value of x is ______. **[22 July 2021 (Evening)]**

46. An inductor of 10 mH is connected to a 20 V battery through a resistor of 10 Ω and a switch. After a long time, when maximum current is set up in the circuit, the current is switched off. The current in the circuit after 1 ms is $\frac{x}{100}$ mA. Then x is equal to ______. (Take $e^{-1} = 0.37$) **[25 July 2021 (Morning)]**

47. A 16 Ω wire is bend to from a square loop. A 9 V supply having internal resistance of 1 Ω is connected across one of its sides. The potential drop across the diagonals of the square loop is ______ $\times 10^{-1}$ V. **[25 July 2021 (Evening)]**

48. Consider an electrical circuit containing a two way switch S. Initially S is open and then T_1 is connected to T_2. As the current in $R = 6\ \Omega$ attains a maximum value of steady state level, T_1 is disconnected from T_2 and immediately connected to T_3. Potential drop across $r = 3\ \Omega$ resistor immediately after T_1 is connected to T_3 is ______ V. (Round off to the Nearest integer). **[27 July 2021 (Morning)]**

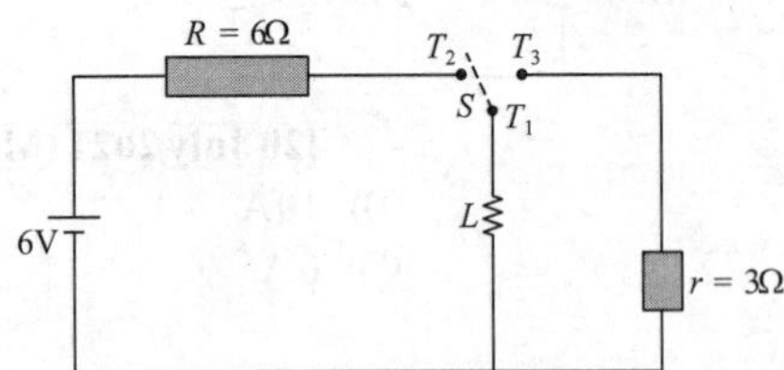

49. In the given figure, the emf of the cell is 2.2 V and if internal resistance is 0.6 Ω. Calculate the power dissipated in the whole circuit : **[26 Aug. 2021 (Morning)]**

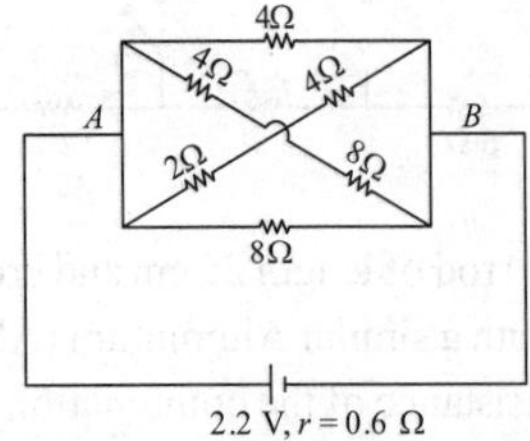

(A) 0.65 W (B) 4.4 W
(C) 1.32 W (D) 2.2 W

50. If you are provided a set of resistances 2 Ω, 4 Ω, and 8 Ω. Connect these resistances so as to obtain an equivalent resistance of $\frac{46}{3}\Omega$: **[26 Aug. 2021 (Evening)]**

(A) 2 Ω and 4 Ω are in parallel with 6 Ω and 8 Ω in series
(B) 6 Ω and 8 Ω are in parallel with 2 Ω and 4 Ω in series
(C) 2 Ω and 6 Ω are in parallel with 4 Ω and 8 Ω in series
(D) 4 Ω and 6 Ω are in parallel with 2 Ω and 8 Ω in series

51. Five identical cells each of internal resistance 1 Ω and emf 5 V are connected in series and in parallel with an external resistance R. For what value of R, current in series and parallel combination will remain the same ? **[27 Aug. 2021 (Morning)]**

(A) 10 Ω (B) 1 Ω
(C) 25 Ω (D) 5 Ω

52. First, a set of n equal resistors of 10 Ω each are connected in series to a battery of emf 20 V and internal resistance 10 Ω. A current I is observed to flow. Then, the n resistors are connected in parallel to the same battery. It is observed that the current is increased 20 times, then the value of n is ______. **[27 Aug. 2021 (Morning)]**

53. The ratio of the equivalent resistance of the network (shown in figure) between the points a and b when switch is open and switch is closed is $x : 8$. The value of x is______. **[27 Aug. 2021 (Evening)]**

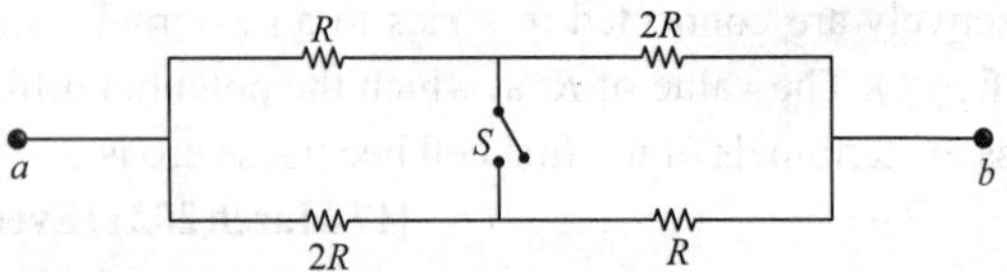

54. The voltage drop across 15 Ω resistance in the given figure will be ______ V. **[31 Aug. 2021 (Morning)]**

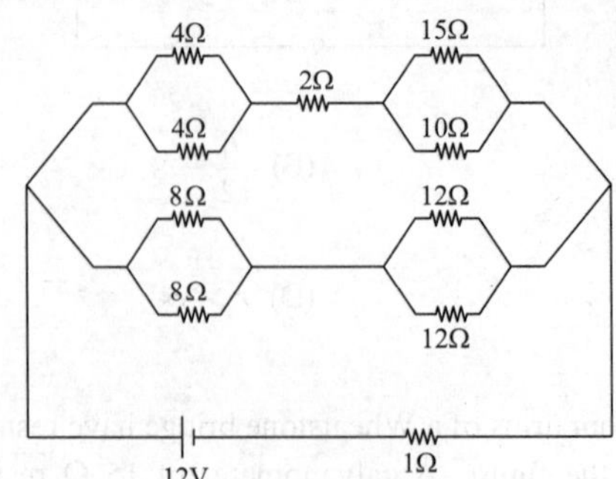

55. A square shaped wire with resistance of each side 3 Ω is bent to form a complete circle. The resistance between two diametrically opposite points of the circle in unit of Ω will be ______: **[31 Aug. 2021 (Morning)]**

56. The equivalent resistance of the given circuit between the terminals A and B is : **[31 Aug. 2021 (Evening)]**

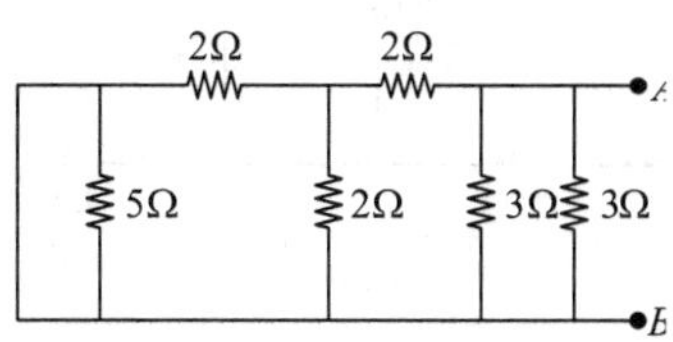

(A) $1\,\Omega$ (B) $3\,\Omega$

(C) $0\,\Omega$ (D) $\frac{9}{2}\Omega$

57. Two resistors $R_1 = (4 \pm 0.4)\,\Omega$ and $R_2 = (4 \pm 0.4)\,\Omega$ are connected in parallel. The equivalent resistance of their parallel combination will be : **[01 Sep. 2021 (Evening)]**

(A) $(4 \pm 0.3)\,\Omega$ (B) $(2 \pm 0.4)\,\Omega$

(C) $(4 \pm 0.4)\,\Omega$ (D) $(2 \pm 0.3)\,\Omega$

58. Two identical cells each of emf 1.5 V are connected in parallel across a parallel combination of two resistors each of resistance $20\,\Omega$. A voltmeter connected in the circuit measures 1.2 V. The internal resistance of each cell is : **[24 June 2022 (Morning)]**

(A) $2.5\,\Omega$ (B) $4\,\Omega$

(C) $5\,\Omega$ (D) $10\,\Omega$

59. What will be the most suitable combination of three resistors $A = 2\,\Omega, B = 4\,\Omega, C = 6\,\Omega$ so that $\left(\frac{22}{3}\right)\Omega$ is equivalent resistance of combination ? **[24 June 2022 (Evening)]**

(A) Parallel combination of A and C connected in series with B

(B) Parallel combination of A and B connected in series with C

(C) Series combination of A and C connected in parallel with B

(D) Series combination of B and C connected in parallel with A

60. The total current supplied to the circuit as shown in figure by the 5 V battery is _____ A. **[25 June 2022 (Morning)]**

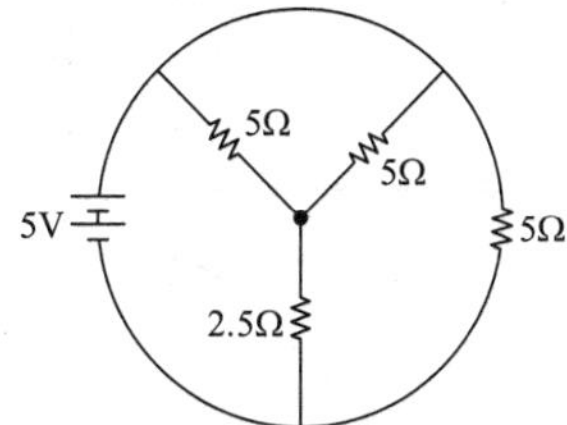

61. The equivalent resistance between points A and B in the given network is : **[26 June 2022 (Evening)]**

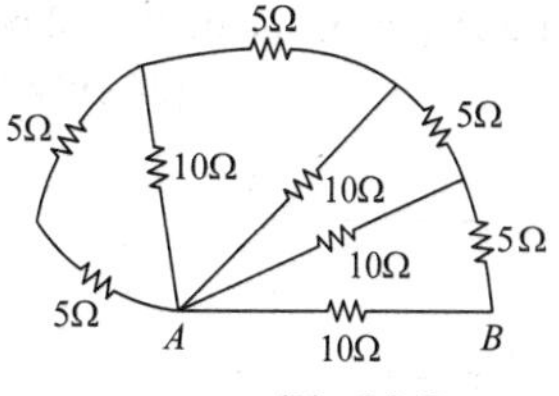

(A) $65\,\Omega$ (B) $20\,\Omega$

(C) $5\,\Omega$ (D) $2\,\Omega$

62. In the given circuit a is an arbitrary constant. The value of m for which the equivalent circuit resistance is minimum, will be

$\sqrt{\frac{x}{2}}$. The value of x is ______. **[27 June 2022 (Evening)]**

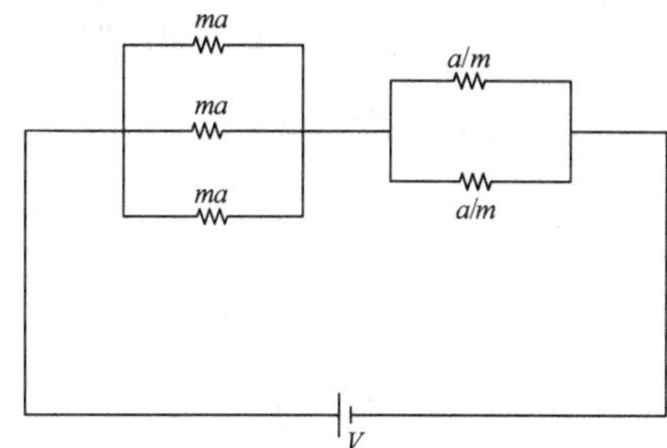

63. All resistances in figure are $1\,\Omega$ each. The value of current I is $\frac{a}{5}$ A. The value of a is ______. **[28 June 2022 (Evening)]**

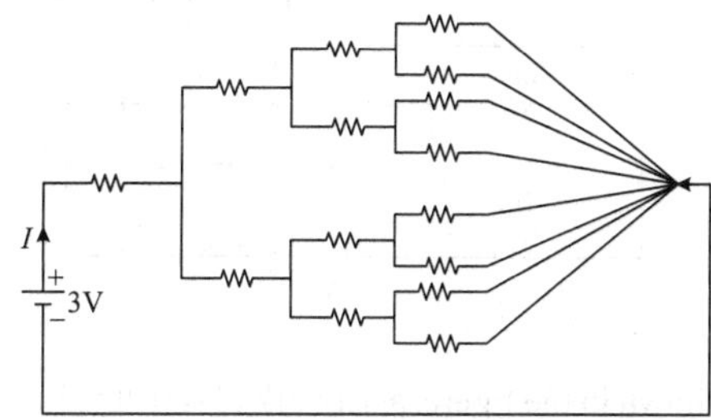

64. For the network shown below, the value $V_B - V_A$ is _____ V. **[29 June 2022 (Morning)]**

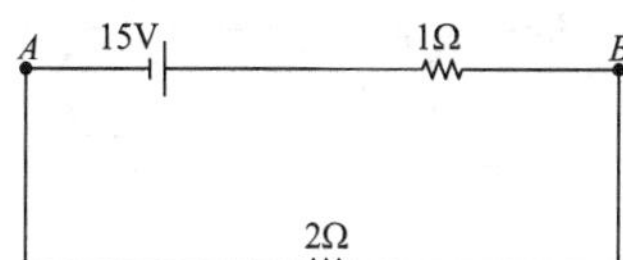

65. The combination of two identical cells, whether connected in series or parallel combination provides the same current through an external resistance of $2\,\Omega$. The value of internal resistance of each cell is : **[29 June 2022 (Evening)]**

(A) $2\,\Omega$ (B) $4\,\Omega$

(C) $6\,\Omega$ (D) $8\,\Omega$

66. In the given figure, the value of V_0 will be ______V. **[25 July 2022 (Morning)]**

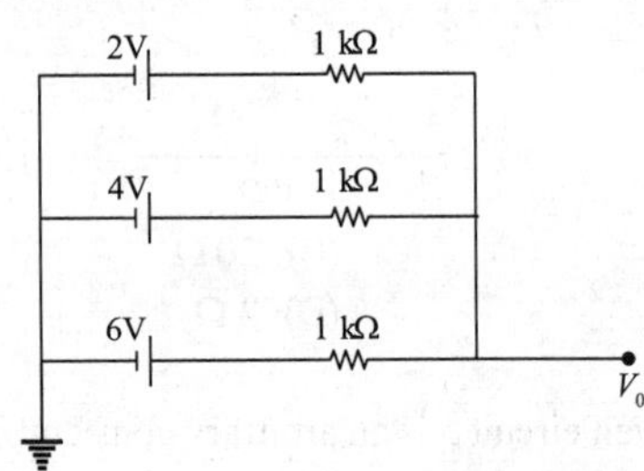

67. A current of 15 mA flows in the circuit as shown in figure. The value of potential difference between the points A and B will be : **[25 July 2022 (Evening)]**

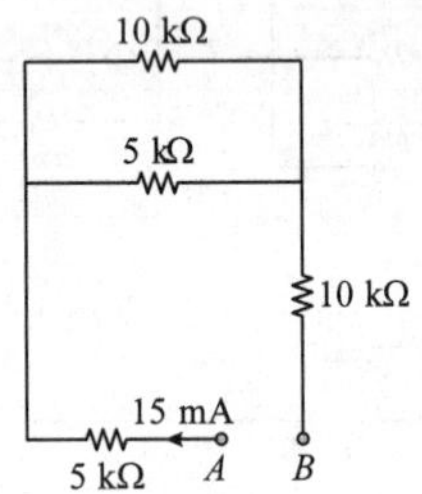

(A) 50 V (B) 75 V
(C) 150 V (D) 275 V

68. The current I flowing through the given circuit will be ______ A. **[29 July 2022 (Morning)]**

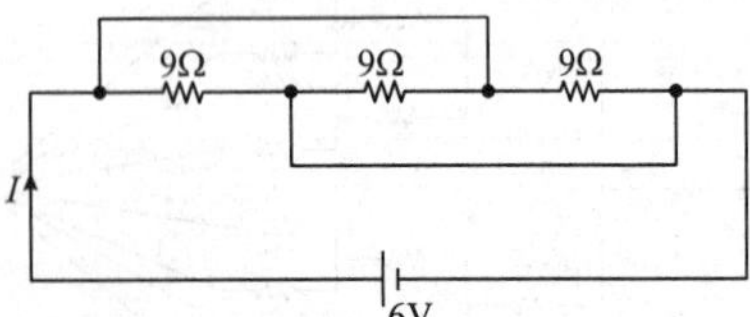

69. As shown in the figure, a network of resistors is connected to a battery of 24 V with an internal resistance of 3 Ω. The currents through the resistors R_4 and R_5 are I_4 and I_5 respectively. The values of I_4 and I_5 are : **[24 Jan. 2023 (Morning)]**

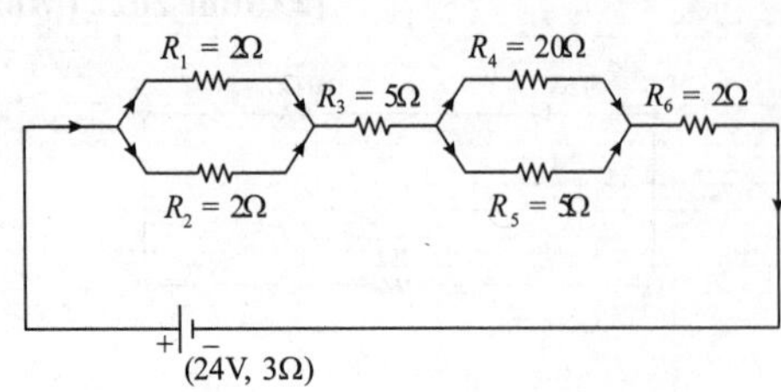

(A) $I_4 = \frac{8}{5}A$ and $I_5 = \frac{2}{5}A$ (B) $I_4 = \frac{24}{5}A$ and $I_5 = \frac{6}{5}A$

(C) $I_4 = \frac{6}{5}A$ and $I_5 = \frac{24}{5}A$ (D) $I_4 = \frac{2}{5}A$ and $I_5 = \frac{8}{5}A$

70. In the given circuit, the equivalent resistance between the terminal A and B is ______ Ω. **[25 Jan. 2023 (Morning)]**

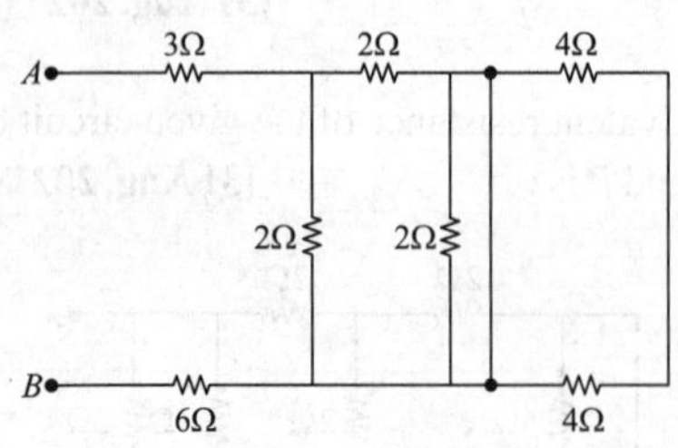

71. Two cells are connected between points A and B as shown. Cell 1 has emf of 12 V and internal resistance of 3 Ω. Cell 2 has emf of 6 V and internal resistance of 6 Ω. An external resistor R of 4 Ω is connected across A and B. The current flowing through R will be ______ A. **[25 Jan. 2023 (Evening)]**

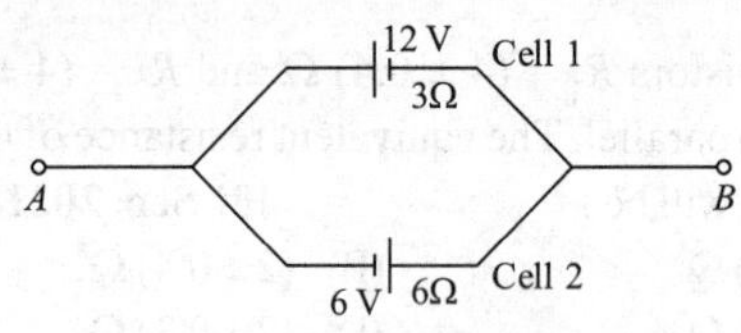

72. In the following circuit, the current I_1, is ______ A. **[30 Jan. 2023 (Morning)]**

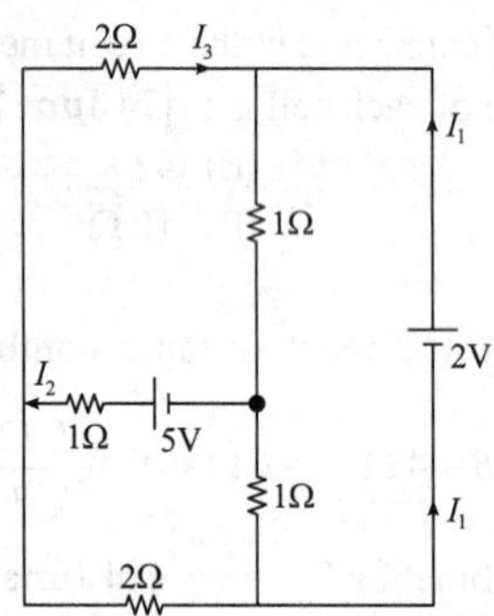

73. The equivalent resistance between A and B is … **[30 Jan. 2023 (Evening)]**

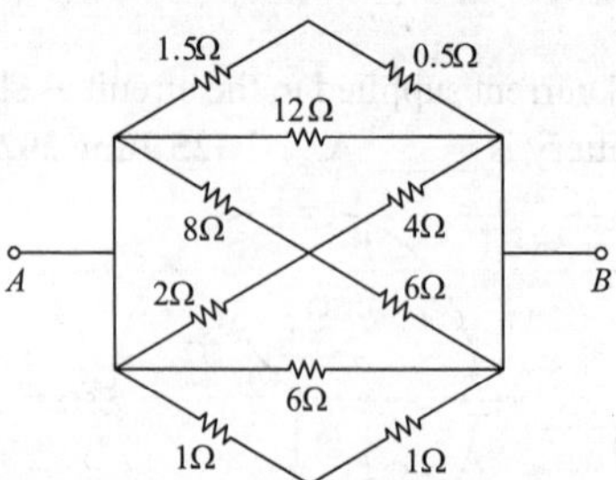

(A) $\frac{2}{3}\Omega$ (B) $\frac{1}{2}\Omega$

(C) $\frac{3}{2}\Omega$ (D) $\frac{1}{3}\Omega$

74. If the potential difference between B and D is zero, the value of x is $\frac{1}{n}\Omega$. The value of n is ______.

[30 Jan. 2023 (Evening)]

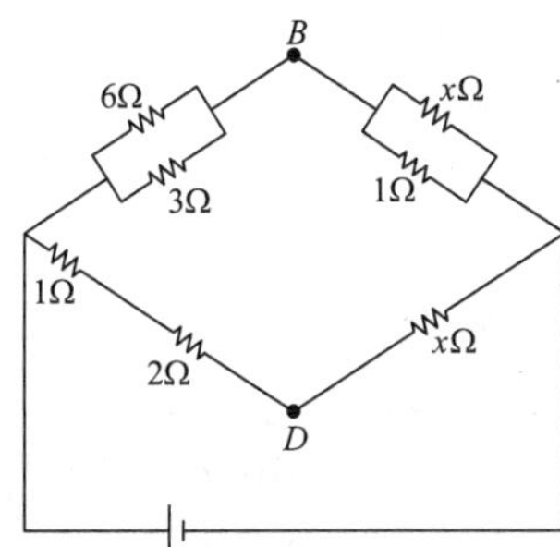

75. Two identical cells, when connected either in parallel or in series gives same current in an external resistance 5 Ω. The internal resistance of each cell will be ______ Ω.

[31 Jan. 2023 (Morning)]

76. The equivalent resistance between A and B of the network shown in figure : **[1 Feb. 2023 (Morning)]**

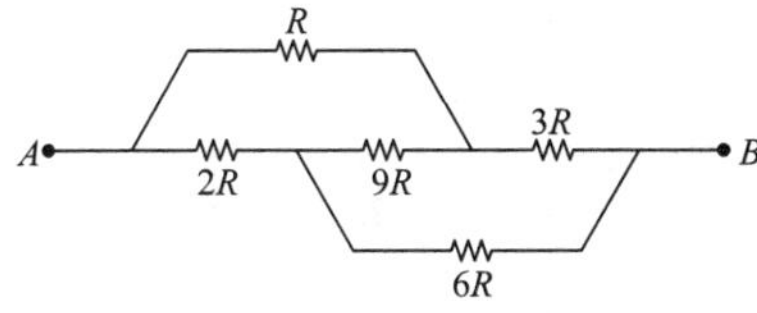

(A) $11\frac{2R}{3}$ (B) $14R$

(C) $21R$ (D) $\frac{8}{3}R$

77. Equivalent resistance between the adjacent corners of a regular n-sided polygon of uniform wire of resistance R would be : **[1 Feb. 2023 (Evening)]**

(A) $\frac{(n-1)R}{n^2}$ (B) $\frac{(n-1)R}{(2n-1)}$

(C) $\frac{n^2R}{n-1}$ (D) $\frac{(n-1)R}{n}$

78. In the given circuit the value of $\left|\frac{I_1+I_3}{I_2}\right|$ is :

[1 Feb. 2023 (Evening)]

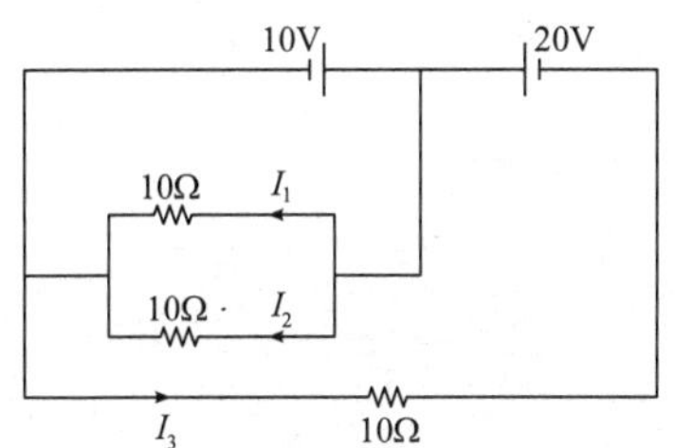

79. Figure shows a part of an electric circuit. The potentials at points a, b and c are 30 V, 12 V and 2 V respectively. The current through the 20 Ω resistor will be.

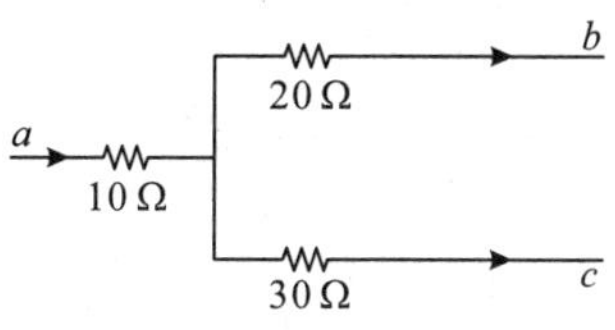

[6 April 2023 (Evening)]

(A) 0.4 A (B) 0.2 A

(C) 0.6 A (D) 1.0 A

80. The equivalent resistance between A and B as shown in figure is : **[8 April 2023 (Evening)]**

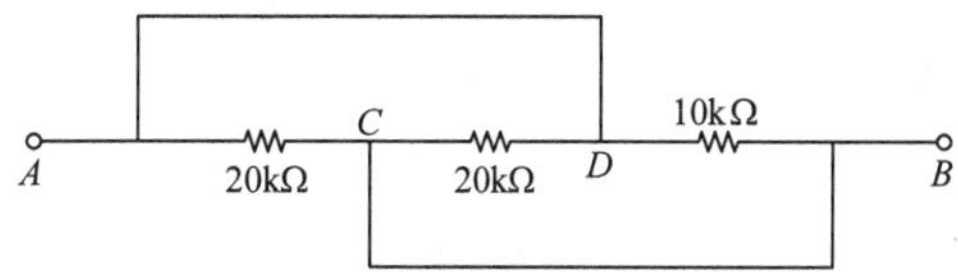

(A) 5 kΩ (B) 30 kΩ

(C) 10 kΩ (D) 20 kΩ

81. The equivalent resistance of the circuit shown below between points a and b is: **[10 April 2023 (Morning)]**

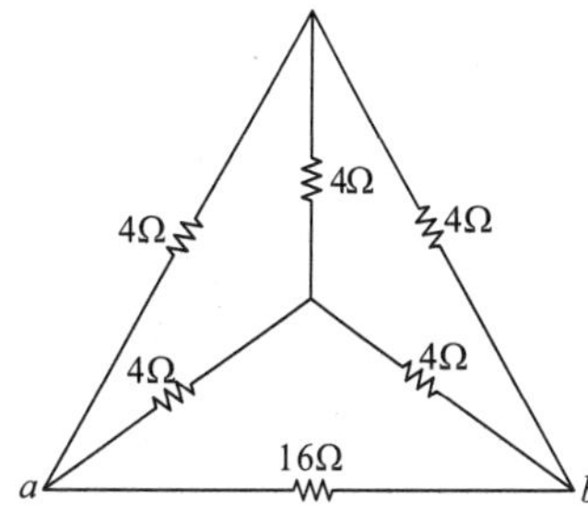

(A) 24 Ω (B) 3.2 Ω

(C) 20 Ω (D) 16 Ω

82. In the circuit diagram shown in figure given below, the current flowing through resistance 3 Ω is $\frac{x}{3}A$. The value of x is______. **[11 April 2023 (Morning)]**

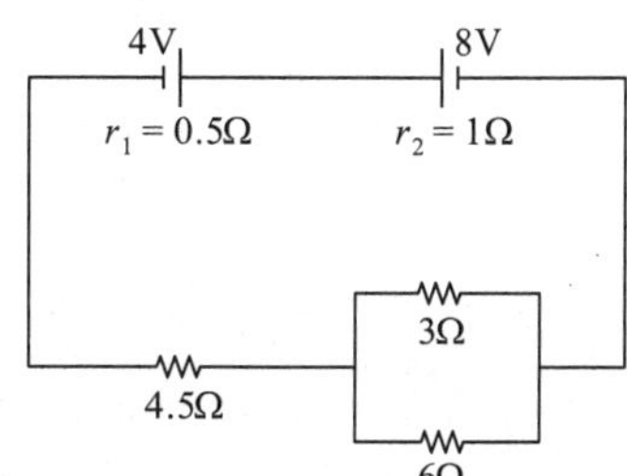

83. The current flowing through R_2 is : **[11 April 2023 (Evening)]**

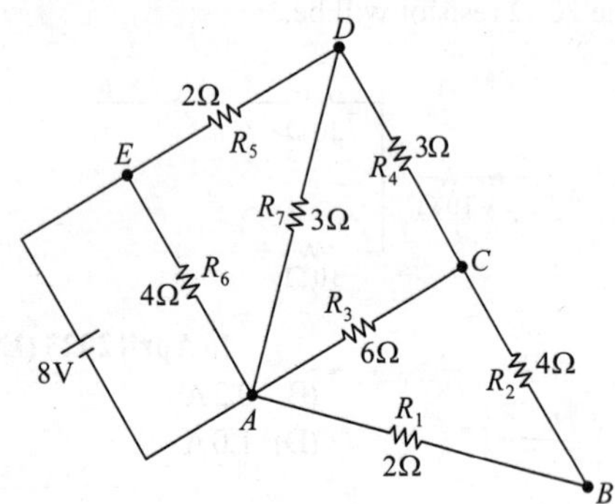

(A) $\frac{2}{3}$ A (B) $\frac{1}{4}$ A

(C) $\frac{1}{2}$ A (D) $\frac{1}{3}$

84. Two identical cells each of emf 1.5 V are connected in series across a 10 Ω resistance. An ideal voltmeter connected across 10 Ω resistance reads 1.5 V. The internal resistance of each cell is _____Ω. **[11 April 2023 (Evening)]**

85. Different combination of 3 resistors of equal resistance R are shown in the figures. The increasing order for power dissipation is : **[13 April 2023 (Morning)]**

(a)
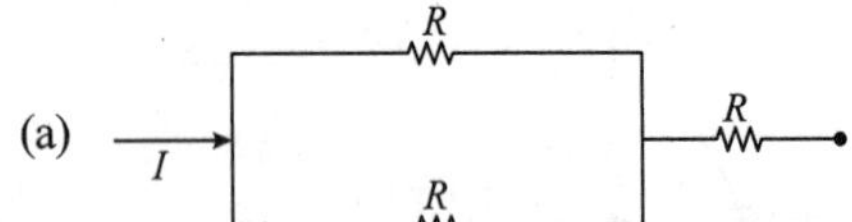

(b)
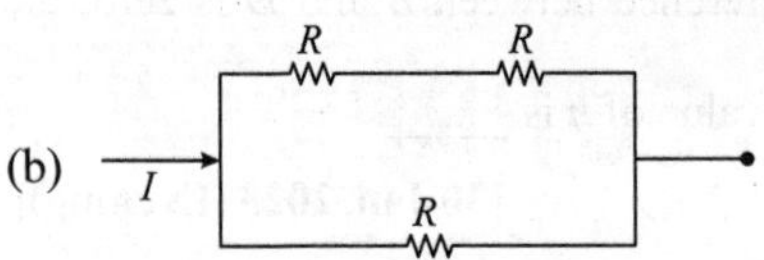

(c)
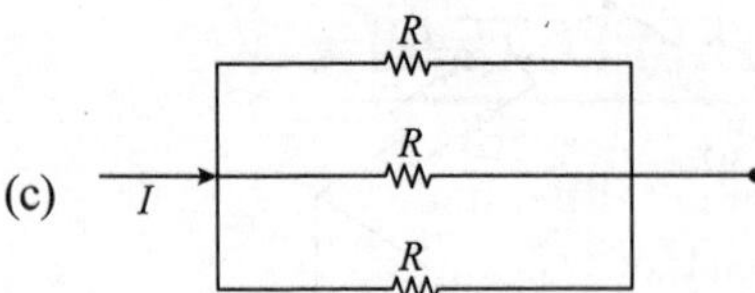

(d) I ⟶ R R R

(A) $P_a < P_b < P_c < P_d$ (B) $P_c < P_d < P_a < P_b$

(C) $P_b < P_c < P_d < P_a$ (D) $P_c < P_b < P_a < P_d$

86. A network of four resistances is connected to 9 V battery, as shown in figure. The magnitude of voltage difference between the points A and B is _____V. **[15 April 2023 (Morning)]**

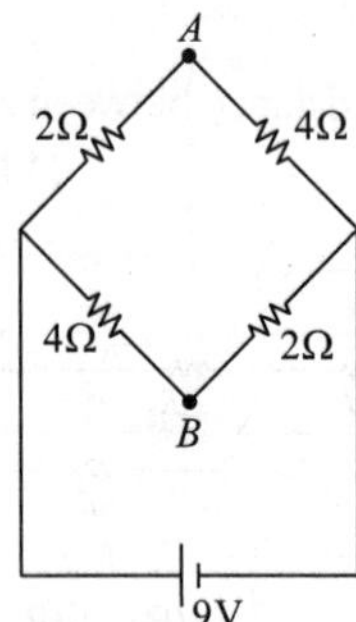

* * * * *

CHAPTER 17

CURRENT ELECTRICITY

17.3 THERMAL EFFECTS OF CURRENT & RC CIRCUITS

1. The mass of product liberated on anode in an electrochemical cell depends on : **[2002]**

(A) $(It)^{1/2}$ (B) It

(C) I/t (D) I^2t

2. A wire when connected to 220 V mains supply has power dissipation P_1. Now the wire is cut into two equal pieces which are connected in parallel to the same supply. Power dissipation in this case is P_2. Then $P_2 : P_1$ is: **[2002]**

(A) 1 (B) 4

(C) 2 (D) 3

3. If in the circuit, power dissipation is 150 W, then R is : **[2002]**

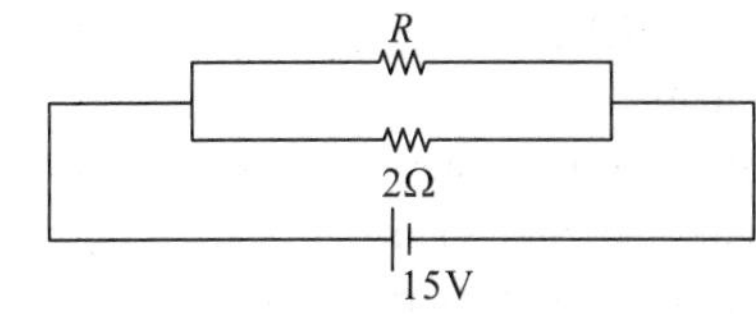

(A) 2 Ω (B) 6 Ω

(C) 5 Ω (D) 4 Ω

4. The thermo e.m.f. of a thermocouple is 25 μ V/°C at room temperature. A galvanometer of 40 ohm resistance, capable of detecting current as low as 10^{-5} A, is connected with the thermocouple. The smallest temperature difference that can be detected by this system is : **[2003]**

(A) 16°C (B) 12°C

(C) 8°C (D) 20°C

5. The negative Zn pole of a Daniell cell, sending a constant current through a circuit, decreases in mass by 0.13 g in 30 minutes. If the electrochemical equivalent of Zn and Cu are 32.5 and 31.5 respectively, the increase in the mass of the positive Cu pole in this time is : **[2003]**

(A) 0.180 g (B) 0.141 g

(C) 0.126 g (D) 0.242 g

6. A 220 volt, 1000 watt bulb is connected across a 110 volt mains supply. The power consumed will be : **[2003]**

(A) 750 watt (B) 500 watt

(C) 250 watt (D) 1000 watt

7. The thermo emf of a thermocouple varies with the temperature θ of the hot junction as $E = a\theta + b\theta^2$ in volts where the ratio a/b is 700°C. If the cold junction is kept at 0°C, then the neutral temperature is : **[2004]**

(A) 1400 °C

(B) 350 °C

(C) 700 °C

(D) No neutral temperature is possible for this thermocouple

8. The electrochemical equivalent of a metal is 3.35×10^{-7} kg per Coulomb. The mass of the metal liberated at the cathode when a 3A current is passed for 2 seconds will be : **[2004]**

(A) 6.6×10^{57} kg (B) 9.9×10^{-7} kg

(C) 19.8×10^{-7} kg (D) 1.1×10^{-7} kg

9. Time taken by a 836 W heater to heat one litre of water from 10°C to 40°C is : **[2004]**

(A) 150 s (B) 100 s

(C) 50 s (D) 200 s

10. An energy source will supply a constant current into the load, if its internal resistance is : **[2005]**

(A) Equal to the resistance of the load

(B) Very large as compared to the load resistance

(C) Zero

(D) Non-zero but less than the resistance of the load

11. Two voltameters, one of copper and another of silver, are joined in parallel. When a total charge q flows through the voltameters, equal amount of metals are deposited. If the electrochemical equivalents of copper and silver are Z_1 and Z_2 respectively the charge which flows through the silver voltameter is : **[2005]**

(A) $\dfrac{q}{1+\dfrac{Z_2}{Z_1}}$ (B) $\dfrac{q}{1+\dfrac{Z_1}{Z_2}}$

(C) $q\dfrac{Z_2}{Z_1}$ (D) $q\dfrac{Z_1}{Z_2}$

12. A heater coil is cut into two equal parts and only one part is now used in the heater. The heat generated will now be : **[2005]**

(A) Four times (B) Doubled

(C) Halved (D) One fourth

13. The resistance of hot tungsten filament is about 10 times the cold resistance. What will be the resistance of 100 W and 200 V lamp when not in use ? **[2005]**

(A) 20 Ω (B) 40 Ω
(C) 200 Ω (D) 400 Ω

14. A thermocouple is made from two metals, Antimony and Bismuth. If one junction of the couple is kept hot and the other is kept cold, then, an electric current will : **[2006]**

(A) Flow from Antimony to Bismuth at the hot junction
(B) Flow from Bismuth to Antimony at the cold junction
(C) Now flow through the thermocouple
(D) Flow from Antimony to Bismuth at the cold junction

15. An electric bulb is rated 220 V-100 W. The power consumed by it when operated on 110 V will be : **[2006]**

(A) 75 W (B) 40 W
(C) 25 W (D) 50 W

16. Let C be the capacitance of a capacitor discharging through a resistor R. Suppose t_1 is the time taken for the energy stored in the capacitor to reduce to half its initial value and t_2 is the time taken for the charge to reduce to one-fourth its initial value. Then the ratio t_1/t_2 will be : **[2010]**

(A) 2 (B) 1
(C) 1/2 (D) 1/4

17. Combination of two identical capacitors, a resistor R and a dc voltage source of voltage 6 V is used in an experiment on $(C-R)$ circuit. It is found that for a parallel combination of the capacitor the time in which the voltage of the fully charged combination reduces to half its original voltage is 10 second. For series combination the time needed for reducing the voltage of the fully charged series combination by half is : **[2011]**

(A) 20 second (B) 10 second
(C) 5 second (D) 2.5 second

18. A resistor 'R' and 2μF capacitor in series is connected through a switch to 200 V direct supply. Across the capacitor is a neon bulb that lights up at 120 V. Calculate the value of R to make the bulb light up 5s after the switch has been closed. $(\log_{10} 2.5 = 0.4)$: **[2011]**

(A) $2.7 \times 10^6\,\Omega$ (B) $3.3 \times 10^7\,\Omega$
(C) $1.3 \times 10^4\,\Omega$ (D) $1.7 \times 10^5\,\Omega$

19. Two electric bulbs marked 25W-220 V and 100 W-220 V are connected in series to a 440 V supply. Which of the bulbs will fuse ? **[2012]**

(A) Neither (B) Both
(C) 100 W (D) 25 W

20. The figure shows an experimental plot for discharging of a capacitor in an R–C circuit. The time constant τ of this circuit lies between : **[2012]**

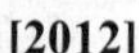

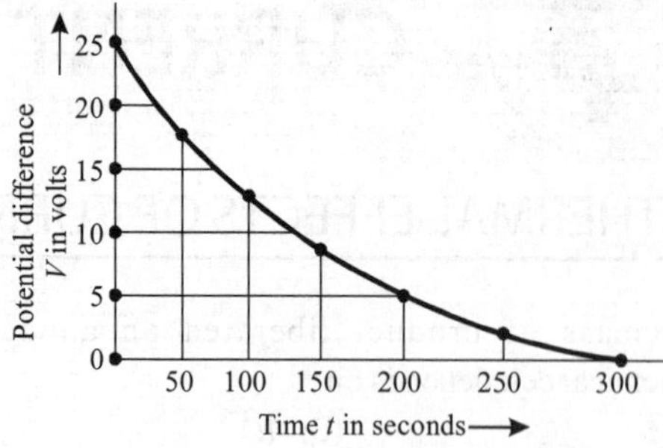

(A) 100 sec and 150 sec (B) 150 sec and 200 sec
(C) 0 and 50 sec (D) 50 sec and 100 sec

21. This question has Statement-1 and Statement-2. Of the four choice given after the Statements, choose the one that best describes the two Statements.

Statement-1 : The possibility of an electric bulb fusing is higher at the time of switching ON.

Statement-2 : Resistance of an electric bulb when it is not lit up is much smaller than when it is lit up. **[2012 (Online)]**

(A) Statement-1 is true, Statement-2 is false
(B) Statement-1 is false, Statement-2 is true, Statement-2 is not a correct explanation of Statement-1
(C) Statement-1 is true, Statement-2 is true, Statement-2 is a correct explanation of Statement-1
(D) Statement-1 is false, Statement-2 is true

22. Three resistors of 4 Ω, 6 Ω and 12 Ω are connected in parallel and the combination is connected in series with a 1.5 V battery of 1 Ω internal resistance. The rate of Joule heating in the 4 Ω resistor is : **[2012 (Online)]**

(A) 0.55 W (B) 0.33 W
(C) 0.25 W (D) 0.86 W

23. A 6.0 volt battery is connected to two light bulbs as shown in figure. Light bulb-1 has resistance 3 ohm while light bulb-2 has resistance 6 ohm. Battery has negligible internal resistance. Which bulb will glow brighter ? **[2012 (Online)]**

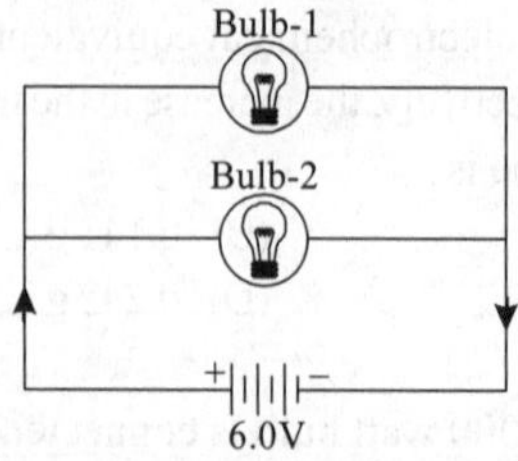

(A) Bulb-1 will glow more first and then its brightness will become less than bulb-2
(B) Bulb-1
(C) Bulb-2
(D) Both glow equally

24. A resistance R and a capacitance C are connected in series to a battery of negligible internal resistance through a key. The key is closed at $t = 0$. If after t sec the voltage across the capacitance was seven times the voltage across R, the value of t is : **[2012 (Online)]**

(A) 3 RC ln 2 (B) 2 RC ln 2
(C) 2 RC ln 7 (D) 3 RC ln 7

25. The supply voltage to a room is 120 V. The resistance of the lead wires is 6 Ω. A 60 W bulb is already switched on. What is the decrease of voltage across the bulb, when a 240 W heater is switched on in parallel to the bulb? **[2013]**

(A) Zero Volt (B) 2.9 Volt
(C) 13.3 Volt (D) 10.04 Volt

26. Which of the four resistances P, Q, R and S generate the greatest amount of heat when a current flows from A to B? **[2013 (Online)]**

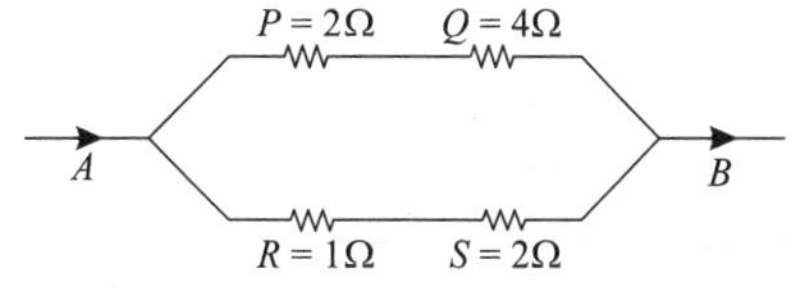

(A) Q (B) S
(C) P (D) R

27. In a large building, there are 15 bulbs of 40 W, 5 bulbs of 100 W, 5 fans of 80 W and 1 heater of 1 kW. The voltage of the electric mains is 220 V. The minimum capacity of the main fuse of the building will be : **[2014]**

(A) 8 A (B) 10 A
(C) 12 A (D) 14 A

28. Four bulbs B_1, B_2, B_3 and B_4 of 100 W each are connected to 220 V main as shown in the figure. The reading in an ideal ammeter will be : **[2014 (Online)]**

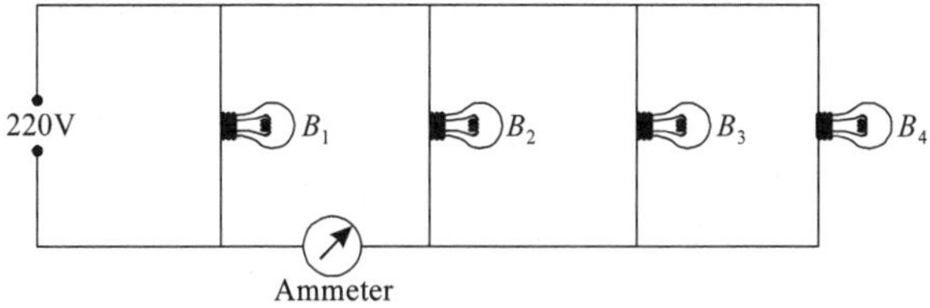

(A) 0.45 A (B) 0.90 A
(C) 1.35 A (D) 1.80 A

29. The circuit shown here has two batteries of 8.0 V and 16.0 V and three resistors 3 Ω, 9 Ω and 9 Ω and a capacitor of 5.0 μF. How much is the current I in the circuit in steady state ? **[2014 (Online)]**

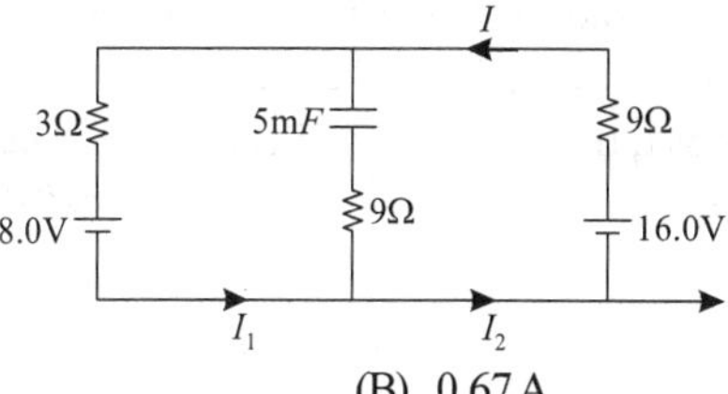

(A) 1.6 A (B) 0.67 A
(C) 2.5 A (D) 0.25 A

30. In the circuit shown, the resistance r is a variable resistance. If for $r = fR$, the heat generation in r is maximum then the value of f is : **[2016 (Online)]**

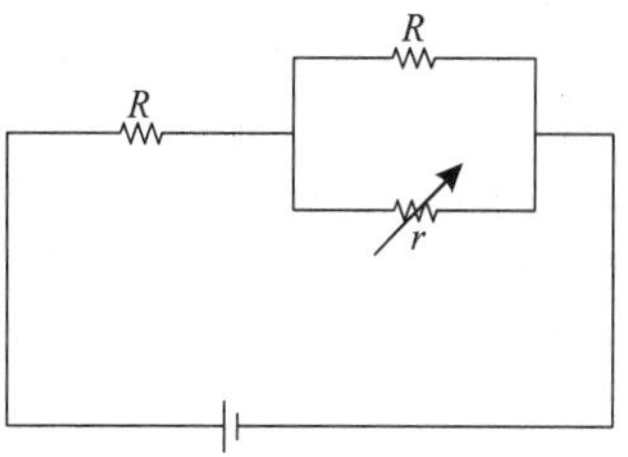

(A) $\frac{1}{2}$ (B) 1
(C) $\frac{1}{4}$ (D) $\frac{3}{4}$

31. The resistance of an electrical toaster has a temperature dependence given by $R(T) = R_0[1 + \alpha(T - T_0)]$ in its range of operation. At $T_0 = 300$ K, $R = 100\ \Omega$ and at $T = 500$ K, $R = 120\ \Omega$. The toaster is connected to a voltage source at 200 V and its temperature is raised at a constant rate from 300 to 500 K in 30 s. The total work done in raising the temperature is : **[2016 (Online)]**

(A) $400 \ln\frac{5}{6}$ J (B) $200 \ln\frac{2}{3}$ J
(C) 300 J (D) $400 \ln\frac{1.5}{1.3}$ J

32. In the given circuit diagram when the current reaches steady state in the circuit, the charge on the capacitor of capacitance C will be : **[2017]**

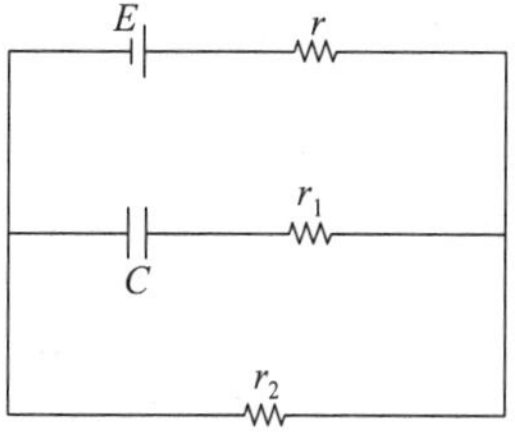

(A) $CE\frac{r_2}{(r + r_2)}$ (B) $CE\frac{r_1}{(r_1 + r)}$
(C) CE (D) $CE\frac{r_1}{(r_2 + r)}$

33. The figure shows three circuits I, II, and III which are connected to a 3 V battery. If the powers dissipated by the configurations I, II, and III are P_1, P_2 and P_3 respectively, then : **[2017 (Online)]**

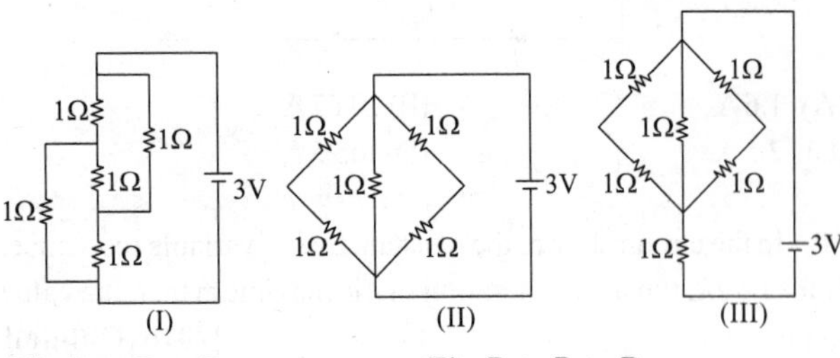

(A) $P_1 > P_2 > P_3$ (B) $P_1 > P_3 > P_2$
(C) $P_2 > P_1 > P_3$ (D) $P_3 > P_2 > P_1$

34. In the following circuit, the switch S is closed at $t = 0$. The charge on the capacitor C_1 as a function of time will be given by $\left(C_{eq} = \frac{C_2C_2}{C_1 + C_2}\right)$. **[2018 (Online)]**

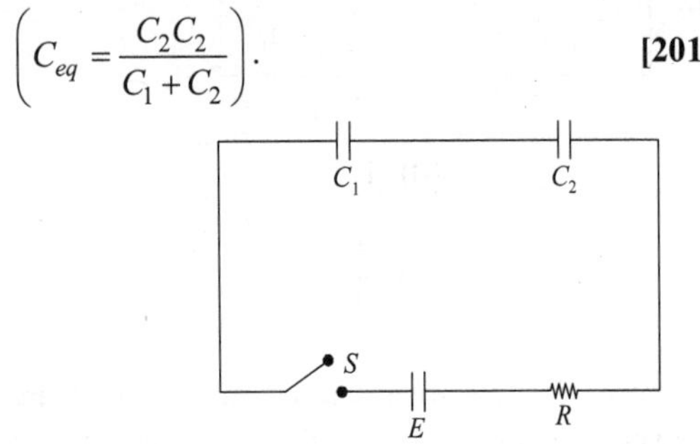

(A) $C_{eq}E[1 - \exp(-t/RC_{eq})]$ (B) $C_1E[1 - \exp(-tR/C_1)]$
(C) $C_2E[1 - \exp(-t/RC_2)]$ (D) $C_{eq}E\exp(-t/RC_{eq})$

35. A constant voltage is applied between two ends of a metallic wire. If the length is halved and the radius of the wire is doubled, the rate of heat developed in the wire will be : **[2018 (Online)]**
(A) Increased 8 times (B) Doubled
(C) Halved (D) Unchanged

36. A current of 2 mA was passed through an unknown resistor which dissipated a power of 4.4 W. Dissipated power when an ideal power supply of 11 V is connected across it is : **[10 Jan. 2019 (Evening)]**
(A) 11×10^{-5} W (B) 11×10^{-4} W
(C) 11×10^{5} W (D) 11×10^{-3} W

37. Two equal resistance when connected in series to a battery, consume electric power of 60 W. If these resistances are now connected in parallel combination to the same battery, the electric power consumed will be : **[11 Jan. 2019 (Morning)]**
(A) 60 W (B) 240 W
(C) 30 W (D) 120 W

38. Two electric bulbs, rated at (25 W, 220 V) and (100 W, 220 V), are connected in series across a 220 V voltage source. If the 25 W and 100 W bulbs draw powers P_1 and P_2 respectively, then : **[12 Jan. 2019 (Morning)]**
(A) $P_1 = 9$ W, $P_2 = 16$ W (B) $P_1 = 4$ W, $P_2 = 16$ W
(C) $P_1 = 16$ W, $P_2 = 4$ W (D) P_1 16 W, $P_2 = 9$ W

39. A 20 Henry inductor coil is connected to a 10 ohm resistance in series as shown in figure. The time at which rate of dissipation of energy (joule's heat) across resistance is equal to the rate at which magnetic energy is stored in the inductor is : **[8 April 2019 (Morning)]**

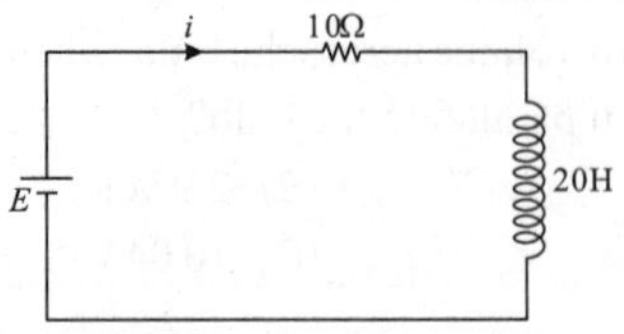

(A) $\frac{2}{\ln 2}$ (B) $\ln 2$
(C) $2 \ln 2$ (D) $\frac{1}{2}\ln 2$

40. A cell of internal resistance r drives current through an external resistance R. The power delivered by the cell to the external resistance will be maximum when : **[8 April 2019 (Evening)]**
(A) $R = 1000\,r$ (B) $R = 0.001\,r$
(C) $R = 2r$ (D) $R = r$

41. Determine the charge on the capacitor in the following circuit : **[9 April 2019 (Morning)]**

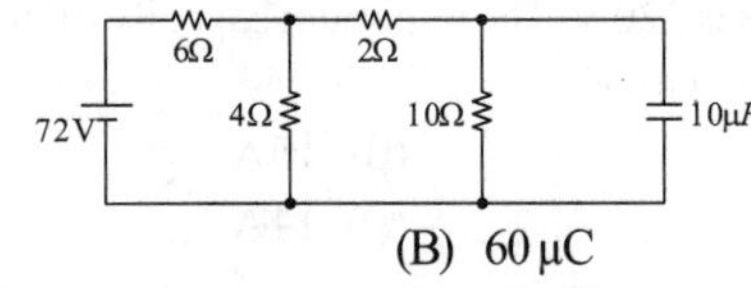

(A) 2 μC (B) 60 μC
(C) 200 μC (D) 10 μC

42. The resistive network shown below is connected to a D.C. source of 16 V. The power consumed by the network is 4 Watt. The value of R is : **[12 April 2019 (Morning)]**

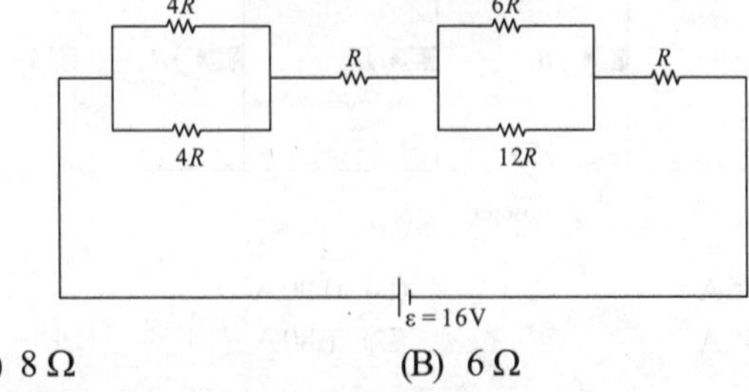

(A) 8 Ω (B) 6 Ω
(C) 1 Ω (D) 16 Ω

43. One kg of water, at 20°C, is heated in an electric kettle whose heating element has a mean (temperature averaged) resistance of 20 Ω. The rms voltage in the mains is 200 V. Ignoring heat loss from the kettle, time taken for water to evaporate fully, is close to : [Specific heat of water = 4200 J/kg°C). Latent heat of water = 2260 kJ/kg] **[12 April 2019 (Evening)]**

(A) 3 minutes (B) 22 minutes
(C) 10 minutes (D) 16 minutes

44. In a building there are 15 bulbs of 45 W, 15 bulbs of 100 W, 15 small fans of 10 W and 2 heaters of 1 kW. The voltage of electric main is 220 V. The minimum fuse capacity (rated value) of the building will be : **[7 Jan. 2020 (Evening)]**

(A) 10A (B) 25 A
(C) 15A (D) 20 A

45. An ideal cell of emf 10 V is connected in circuit shown in figure. Each resistance is 2 Ω. The potential difference (in *V*) across the capacitor when it is fully charged is : **[2 Sept. 2020 (Evening)]**

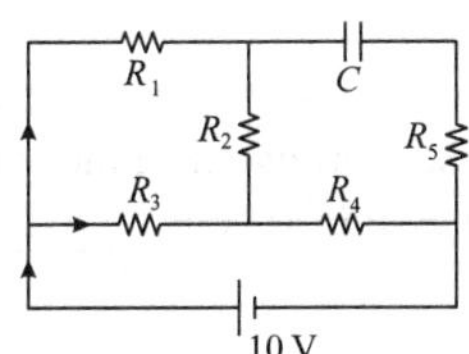

46. Model of a torch battery of length *l* to be made up of a thin cylindrical bar of radius *a* and a concentric thin cylindrical shell of radius *b* filled in between with an electrolyte of resistivity ρ (see figure). If the battery is connected to a resistance of value *R*, the maximum Joule heating in *R* will take place for : **[3 Sept. 2020 (Morning)]**

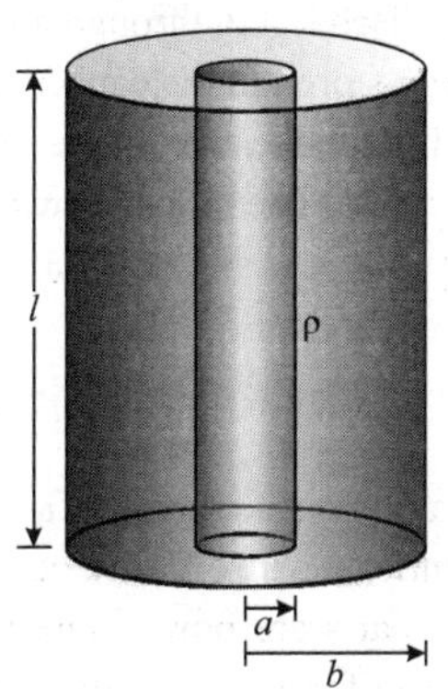

(A) $R = \frac{2\rho}{\pi l}\ln\left(\frac{b}{a}\right)$ (B) $R = \frac{\rho}{\pi l}\ln\left(\frac{b}{a}\right)$

(C) $R = \frac{\rho}{2\pi l}\left(\frac{b}{a}\right)$ (D) $R = \frac{\rho}{2\pi l}\ln\left(\frac{b}{a}\right)$

47. A battery of 3.0 V is connected to a resistor dissipating 0.5 W of power. If the terminal voltage of the battery is 2.5 V, the power dissipated within the internal resistance is : **[4 Sept. 2020 (Morning)]**

(A) 0.50 W (B) 0.125 W
(C) 0.072 W (D) 0.10 W

48. An electrical power line, having a total resistance of 2 W, delivers 1 kW at 220 V. The efficiency of the transmission line is approximately : **[5 Sept. 2020 (Morning)]**

(A) 72% (B) 96%
(C) 91% (D) 85%

49. An *RC* circuit as shown in the figure is driven by a *AC* source generating a square wave. The output wave pattern monitored by *CRO* would look close to : **[16 March 2021 (Morning)]**

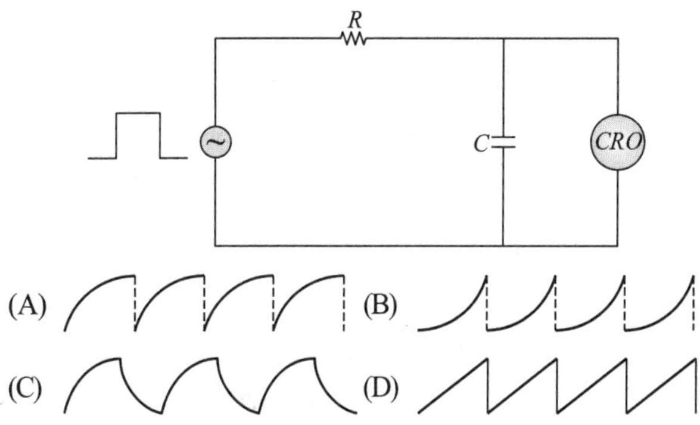

50. A resistor develops 500 J of thermal energy in 20 s when a current of 1.5 A is passed through it. If the current is increased from 1.5 A to 3 A, what will be the energy developed in 20 s. **[16 March 2021 (Evening)]**

(A) 1500 J (B) 1000 J
(C) 500 J (D) 2000 J

51. The energy dissipated by a resistor is 10 mJ in 1 s when an electric current of 2 mA flows through it. The resistance is ______ Ω. (Round off to the Nearest Integer) **[16 March 2021 (Evening)]**

52. The circuit shown in the figure consists of a charged capacitor of capacity 3 μF and a charge of 30 μC. At time $t = 0$, when the key is closed, the value of current flowing through the 5 MΩ resistor is '*x*' μ-A. The value of *x* to the nearest integer is ______. **[18 March 2021 (Morning)]**

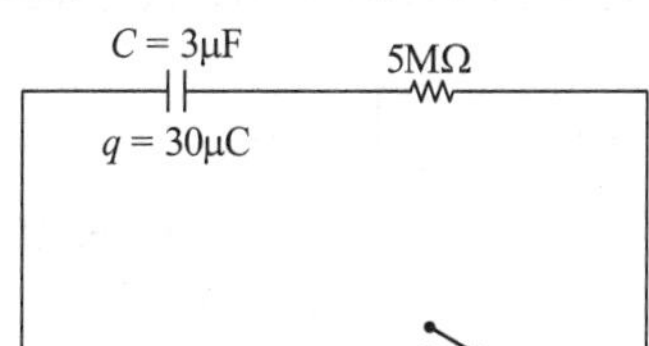

53. An electric bulb rated as 200 W at 100 V is used in a circuit having 200 V supply. The resistance *R* that must be put in series with the bulb so that the bulb delivers the same power is ______Ω. **[25 July 2021 (Morning)]**

54. An electric bulb of 500 W at 100 V is used in a circuit having a 200 V supply. Calculate the resistance R to be connected in series with the bulb so that the power delivered by the bulb is 500 W : **[26 Aug. 2021 (Evening)]**

(A) 10 Ω (B) 5 Ω
(C) 30 Ω (D) 20 Ω

55. Calculate the amount of charge on capacitor of 4 μF. The internal resistance of battery is 1 Ω : **[27 Aug. 2021 (Morning)]**

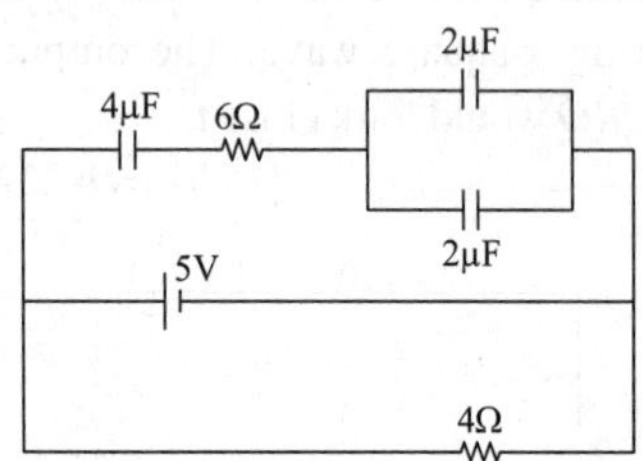

(A) Zero (B) 8 μC
(C) 16 μC (D) 4 μC

56. A capacitor of 50 μF is connected in a circuit as shown in figure. The charge on the upper plate of the capacitor is ______ μC. **[31 Aug. 2021 (Morning)]**

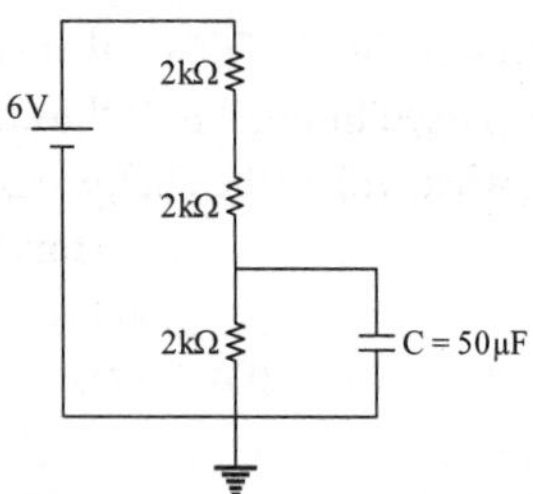

57. A resistor dissipated 192 J of energy in 1 s when a current of 4 A is passed through it. Now, when the current is doubled, the amount of thermal energy dissipated in 5 s is ______ J. **[31 Aug. 2021 (Evening)]**

58. A capacitor is connected to a 20 V battery through a resistance of 10 Ω. It is found that the potential difference across the capacitor rises to 2 V in 1 μs. The capacitance of the capacitor is ______ μF. Given $ln\left(\frac{10}{9}\right) = 0.105$.

[01 Sep. 2021 (Evening)]

(A) 1.85 (B) 9.52
(C) 0.95 (D) 0.105

59. For the given circuit the current i through the battery when the key is closed and the steady state has been reached is ______. **[01 Sep. 2021 (Evening)]**

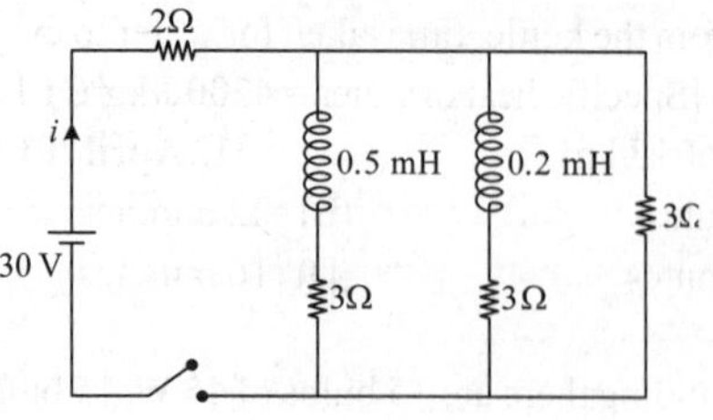

(A) 0 A (B) 25 A
(C) 10 A (D) 6 A

60. A uniform heating wire of resistance 36 Ω is connected across a potential difference of 240 V. The wire is then cut into half and a potential difference of 240 V is applied across each half separately. The ratio of power dissipation in first case to the total power dissipation in the second case would be 1 : x, where x is______. **[01 Sep. 2021 (Evening)]**

61. A resistor develops 300 J of thermal energy in 15 s, when a current of 2 A is passed through it. If the current increases to 3 A, the energy developed in 10 s is ______ J. **[25 June 2022 (Morning)]**

62. Two coils require 20 minutes and 60 minutes respectively to produce same amount of heat energy when connected separately to the same source. If they are connected in parallel arrangement to the same source; the time required to produce same amount of heat by the combination of coils, will be ______ min. **[29 June 2022 (Morning)]**

63. A capacitor is discharging through a resistor R. Consider in time t_1, the energy stored in the capacitor reduces to half of its initial value and in time t_2, the charge stored reduces to one eighth of its initial value. The ratio t_1/t_2 will be : **[29 June 2022 (Evening)]**

(A) 1/2 (B) 1/3
(C) 1/4 (D) 1/6

64. Given below are two statements : **[28 July 2022 (Evening)]**

Statement-I : A uniform wire of resistance 80 Ω is cut into four equal parts. These parts are now connected in parallel. The equivalent resistance of the combination will be 5 Ω

Statement-II : Two resistance $2R$ and $3R$ are connected in parallel in a electric circuit. The value of thermal energy developed in $3R$ and $2R$ will be in the ratio 3 : 2

In the light of the above statements, choose the most appropriate answer from the options given below :

(A) Both statement-I and statement-II are correct
(B) Both statement-I and statement-II are incorrect
(C) Statement-I is correct but statement-II is incorrect
(D) Statement-I is incorrect but statement-II is correct

65. An electrical bulb rated 220 V, 100 W, is connected in series with another bulb rated 220 V, 60 W. If the voltage across combination is 220 V, the power consumed by the 100 W bulb will be about _____ W. **[28 July 2022 (Evening)]**

66. Ratio of thermal energy released in two resistor R and $3R$ connected in parallel in an electric circuit is : **[29 Jan. 2023 (Morning)]**

(A) 3 : 1
(B) 1 : 1
(C) 1 : 3
(D) 1 : 27

67. The H amount of thermal energy is developed by a resistor in 10 s when a current of 4 A is passed through it. If the current is increased to 16 A, the thermal energy developed by the resistor in 10 s will be : **[31 Jan. 2023 (Evening)]**

(A) H
(B) 16H
(C) $\frac{H}{4}$
(D) 4H

68. For the given circuit, in the steady state, $|VB - VD| =$ _____ V. **[31 Jan. 2023 (Evening)]**

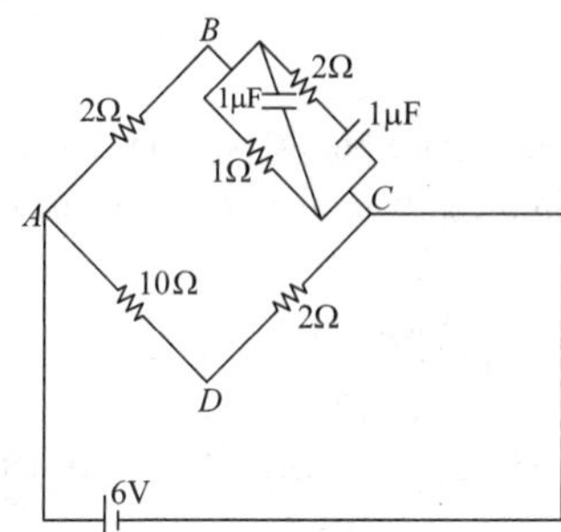

69. In this figure the resistance of the coil of galvanometer G is 2Ω. The emf of the cell is 4 V. The ratio of potential difference across C_1 and C_2 is : **[8 April 2023 (Morning)]**

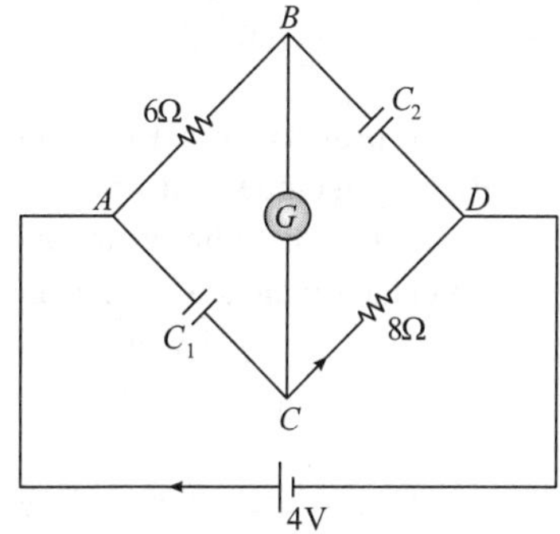

(A) 1
(B) $\frac{4}{5}$
(C) $\frac{3}{4}$
(D) $\frac{5}{4}$

70. Two identical heater filaments are connected first in parallel and then in series. At the same applied voltage, the ratio of heat produced in same time for parallel to series will be : **[11 April 2023 (Morning)]**

(A) 4 : 1
(B) 2 : 1
(C) 1 : 2
(D) 1 : 4

71. A potential V_0 is applied across a uniform wire of resistance R. The power dissipation is P_1. The wire is then cut into two equal halves and a potential of V_0 is applied across the length of each half. The total power dissipation across two wires is P_2. The ratio $P_2 : P_1$ is $\sqrt{x} : 1$. The value of x is _____. **[13 April 2023 (Morning)]**

72. In the network shown below, the charge accumulated in the capacitor in steady state will be: **[13 April 2023 (Evening)]**

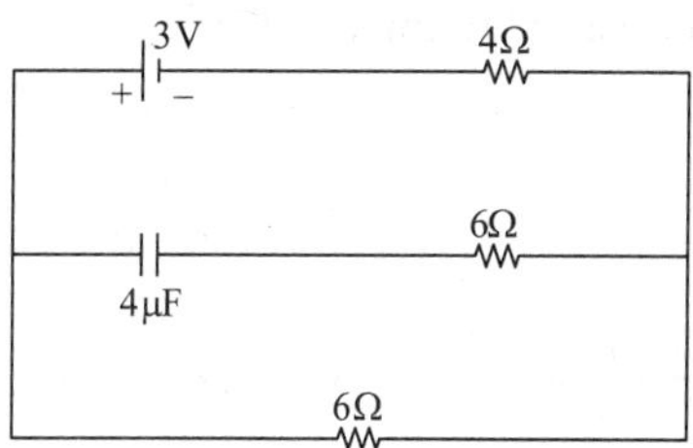

(A) 7.2 μC
(B) 4.8 μC
(C) 10.3 μC
(D) 12 μC

73. In the circuit shown, the energy stored in the capacitor is n μJ. The value of n is _____. **[13 April 2023 (Evening)]**

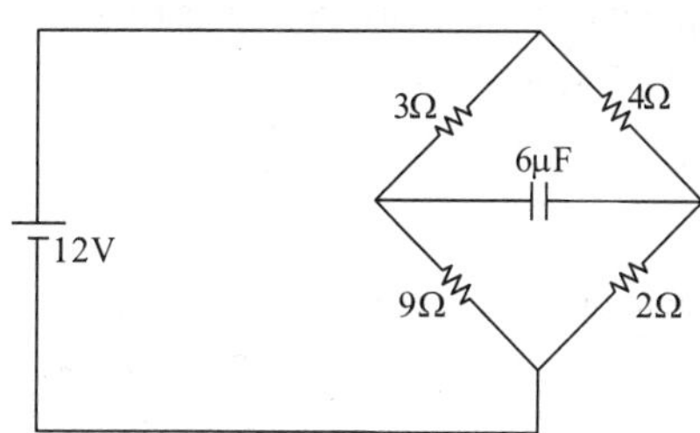

* * * * *

CURRENT ELECTRICITY

17.4 ELECTRICAL MEASURING INSTRUMENTS

1. If an ammeter is to be used in place of a voltmeter, then we must connect with the ammeter a : **[2002]**

(A) low resistance in parallel
(B) high resistance in parallel
(C) high resistance in series
(D) low resistance in series

2. An ammeter reads upto 1 A. Its internal resistance is 0.81 Ω. To increase the range to 10 A, the value of the required shunt is : **[2003]**

(A) 0.03 Ω (B) 0.3 Ω
(C) 0.9 Ω (D) 0.09 Ω

3. The length of a wire of a potentiometer is 100 cm, and the emf of its stand and cell is E volt. It is employed to measure the emf of a battery whose internal resistance is 0.5 Ω. If the balance point is obtained at $l = 30$ cm from the positive end, the emf of the battery is : **[2003]**

(A) $\frac{30E}{100.5}$ (B) $\frac{30E}{100-0.5}$

(C) $\frac{30(E-0.5i)}{100}$, where i is the current in the potentiometer wire

(D) $\frac{30E}{100}$

4. In a metre bridge experiment, null point is obtained at 20 cm from one end of the wire when resistance X is balanced against another resistance Y. If $X < Y$, then where will be the new position of the null point from the same end, if one decides to balance a resistance of $4X$ against Y? **[2004]**

(A) 50 cm (B) 80 cm
(C) 40 cm (D) 70 cm

5. In a potentiometer experiment the balancing with a cell is at length 240 cm. On shunting the cell with a resistance of 2 Ω, the balancing length becomes 120 cm. The internal resistance of the cell is : **[2005]**

(A) 1 Ω (B) 0.5 Ω
(C) 4 Ω (D) 2 Ω

6. In the circuit, the galvanometer G shows zero deflection. If the batteries A and B have negligible internal resistance, the value of the resistor R will be : **[2005]**

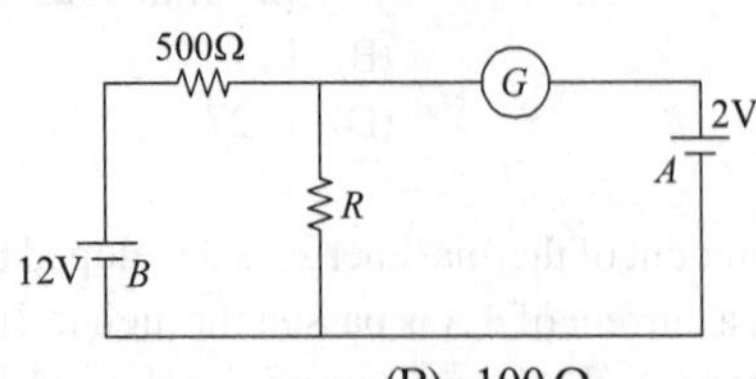

(A) 200 Ω (B) 100 Ω
(C) 500 Ω (D) 1000 Ω

7. A moving coil galvanometer has 150 equal divisions. Its current sensitivity is 10 divisions per milli ampere and voltage sensitivity is 2 divisions per millivolt. In order that each division reads 1 V, the resistance in Ohm's needed to be connected in series with the coil will be : **[2005]**

(A) 10^3 (B) 10^5
(C) 99995 (D) 9995

8. Shown in the figure below is a meter-bridge set up with null deflection in the galvanometer. The value of the unknown resistor R is : **[2008]**

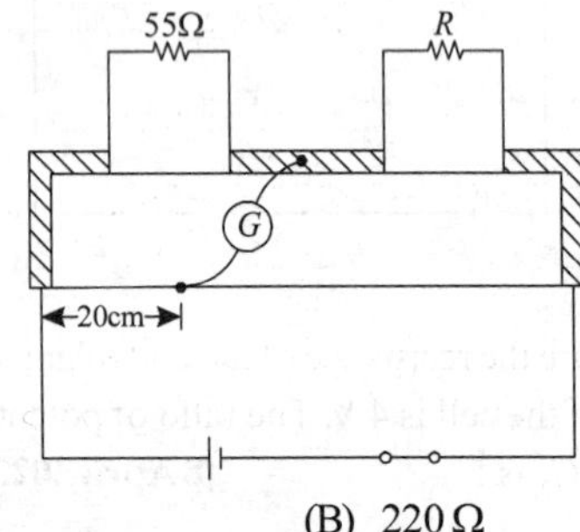

(A) 13.75 Ω (B) 220 Ω
(C) 110 Ω (D) 55 Ω

9. The current in the primary circuit of a potentiometer is 0.2 A. The specific resistance and cross-section of the potentiometer wire are 4×10^{-7} ohm metre and 8×10^{-7} m^2 respectively. The potential gradient will be equal to : **[2011]**

(A) 0.2 V/m (B) 1 V/m
(C) 0.5 V/m (D) 0.1 V/m

10. It is preferable to measure the emf of a cell by potentiometer than by a voltmeter because of the following possible reasons : **[2012 (Online)]**

(i) In case of potentiometer, no current flows through the cell
(ii) The length of the potentiometer allows greater precision
(iii) Measurement by the potentiometer is quicker
(iv) The sensitivity of the galvanometer, when using a

potentiometer is not relevant

(A) (i), (iii), (iv) (B) (i), (iii), (iv)
(C) (i), (ii) (D) (i), (ii), (iii), (iv)

11. In an experiment of potentiometer for measuring the internal resistance of primary cell a balancing length l is obtained on the potentiometer wire when the cell is open circuit. Now the cell is short circuited by a resistance R. If R is to be equal to the internal resistance of the cell the balancing length on the potentiometer wire will be : **[2012 (Online)]**

(A) l (B) $2l$
(C) $l/2$ (D) $l/4$

12. This question has Statement-1 and Statement-2. Of the four choice given after the Statements, choose the one that best describes the two Statements.

Statement-1 : Higher the range, greater is the resistance of ammeter.

Statement-2 : To increase the range of ammeter, additional shunt needs to be used across it **[2013]**

(A) Statement-1 is true, Statement-2 is true, Statement-2 is the correct explanation of Statement-1
(B) Statement-1 is true, Statement-2 is true, Statement-2 is not the correct explanation of Statement-1
(C) Statement-1 is true, Statement-2 is false
(D) Statement-1 is false, Statement-2 is true

13. Correct set up to verify Ohm's law is : **[2013 (Online)]**

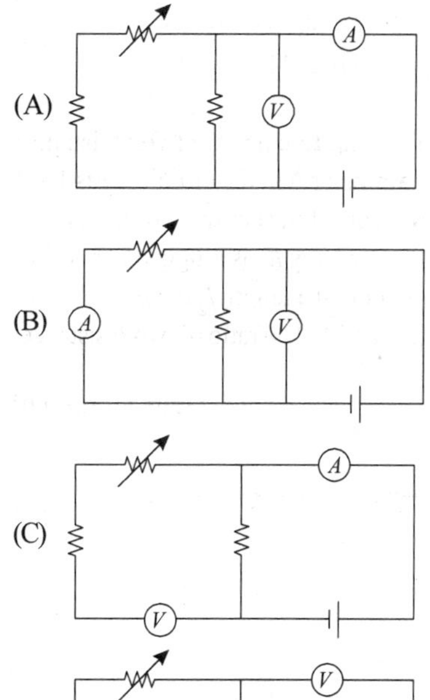

14. In a metre bridge experiment null point is obtained at 40 cm from one end of the wire when resistance X is balanced against another resistance Y. If $X < Y$, then the new position of the null point from the same end, if one decides to balance a resistance of $3X$ against Y, will be close to : **[2013 (Online)]**

(A) 80 cm (B) 75 cm
(C) 67 cm (D) 50 cm

15. A shunt of resistance 1 Ω is connected across a galvanometer of 120 Ω resistance. A current of 5.5 ampere gives full scale deflection in the galvanometer. The current that will give full scale deflection in the absence of the shunt is nearly : **[2013 (Online)]**

(A) 5.5 ampere (B) 0.5 ampere
(C) 0.004 ampere (D) 0.045 ampere

16. To find the resistance of a galvanometer by the half deflection method the following circuit is used with resistances R_1 = 9970 Ω, R_2 = 30 Ω and R_3 = 0. The deflection in the galvanometer is d. With R_3 = 107 Ω the deflection changed to $\frac{d}{2}$. The galvanometer resistance is approximately : **[2013 (Online)]**

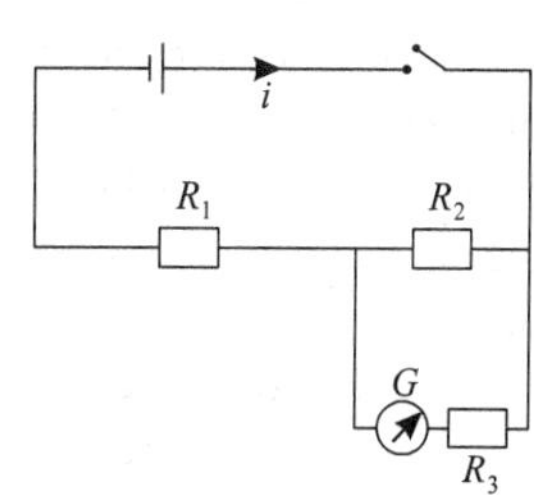

(A) 107 Ω (B) 137 Ω
(C) 107/2 Ω (D) 77 Ω

17. In the circuit diagrams (A, B, C and D) shown below, R is a high resistance and S is a resistance of the order of galvanometer resistance G. The correct circuit, corresponding to the half deflection method for finding the resistance and figure of merit of the galvanometer, is the circuit labelled as : **[2014 (Online)]**

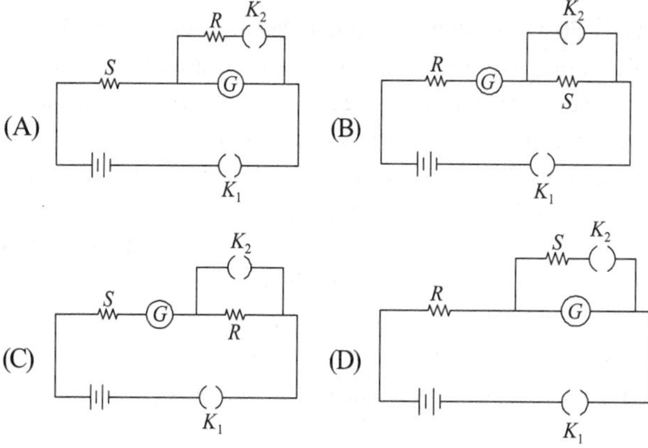

(A) Circuit A with $G = \dfrac{RS}{(R-S)}$

(B) Circuit B with $G=S$
(C) Circuit C with $G=S$
(D) Circuit D with $G=\dfrac{RS}{(R-S)}$

18. A 10 V battery with internal resistance 1 Ω and a 15 V battery with internal resistance 0.6 Ω are connected in parallel to a voltmeter (see figure). The reading in the voltmeter will be close to : **[2015 (Online)]**

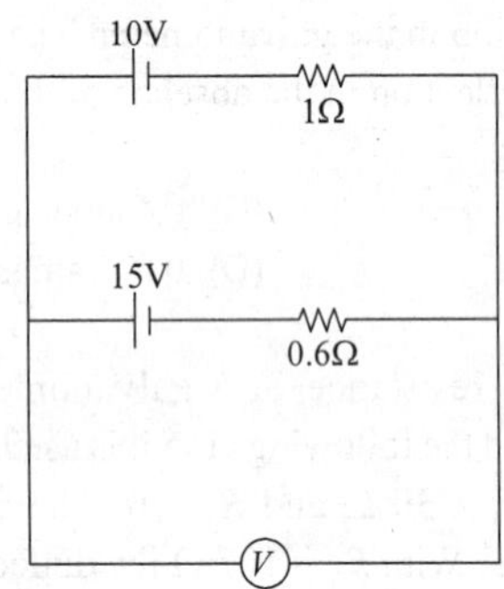

(A) 12.5 V (B) 24.5 V
(C) 13.1 V (D) 11.9 V

19. A galvanometer having a coil resistance of 100 Ω gives a full scale deflection, when a current of 1 mA is passed through it. The value of the resistance, which can convert this galvanometer into ammeter giving a full scale deflection for a current of 10 A, is : **[2016]**
(A) 0.01 Ω (B) 2 Ω
(C) 0.1 Ω (D) 3 Ω

20. To know the resistance G of a galvanometer by half deflection method, a battery of emf V_E and resistance R is used to deflect the galvanometer by angle θ. If a shunt of resistance S is needed to get half deflection then G, R and S related by the equation : **[2016 (Online)]**
(A) $S(R+G)=RG$ (B) $2S(R+G)=RG$
(C) $2G=S$ (D) $2S=G$

21. A 50 Ω resistance is connected to a battery of 5 V. A galvanometer of resistance 100 Ω is to be used as an ammeter to measure current through the resistance, for this a resistance r_s is connected to the galvanometer. Which of the following connections should be employed if the measured current is within 1% of the current without the ammeter in the circuit? **[2016 (Online)]**
(A) $r_s = 0.5$ Ω in series with the galvanometer
(B) $r_s = 1$ Ω in series with galvanometer
(C) $r_s = 1$ Ω in parallel with galvanometer
(D) $r_s = 0.5$ Ω in parallel with the galvanometer

22. A galvanometer has a 50 division scale. Battery has no internal resistance. It is found that there is deflection of 40 divisions when $R = 2400$ Ω. Deflection becomes 20 divisions when resistance taken from resistance box is 4900 Ω. Then we can conclude : **[2016 (Online)]**

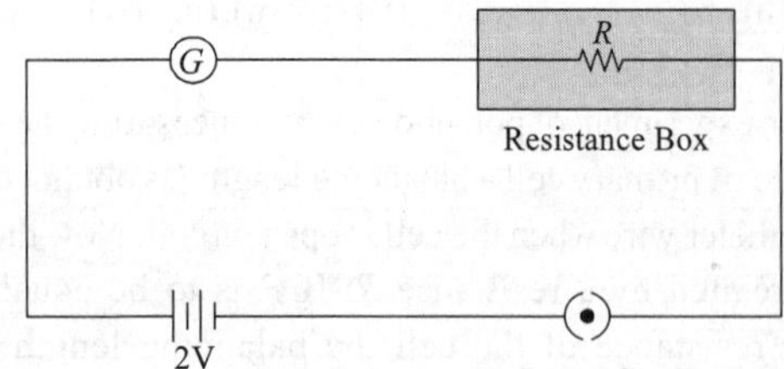

(A) Current sensitivity of galvanometer is 20 μA/division
(B) Resistance of galvanometer is 200 Ω
(C) Resistance required on $R.B.$ for a deflection of 10 divisions is 9800 Ω
(D) Full scale deflection current is 2 mA

23. Which of the following statements is false ? **[2017]**
(A) A rheostat can be used as a potential divider
(B) Kirchhoff's second law represents energy conservation
(C) Wheatstone bridge is the most sensitive when all the four resistances are of the same order of magnitude
(D) In a balanced Wheatstone bridge if the cell and the galvanometer are exchanged, the null point is disturbed

24. When a current of 5 mA is passed through a galvanometer having a coil of resistance 15 Ω, it shows full scale deflection. The value of the resistance to be put in series with the galvanometer to convert it into to voltmeter of range 0 – 10 V is: **[2017]**
(A) 2.535×10^3 Ω (B) 4.005×10^3 Ω
(C) 1.985×10^3 Ω (D) 2.045×10^3 Ω

25. A potentiometer PQ is set up to compare two resistances as shown in the figure. The ammeter A in the circuit reads 1.0 A when two way key K_3 is open. The balance point is at a length l_1 cm from P when two way key K_3 is plugged in between 2 and 1, while the balance point is at a length l_2 cm from P when key K_3 is plugged in between 3 and 1. The ratio of two resistances $\dfrac{R_1}{R_2}$, is found to be : **[2017 (Online)]**

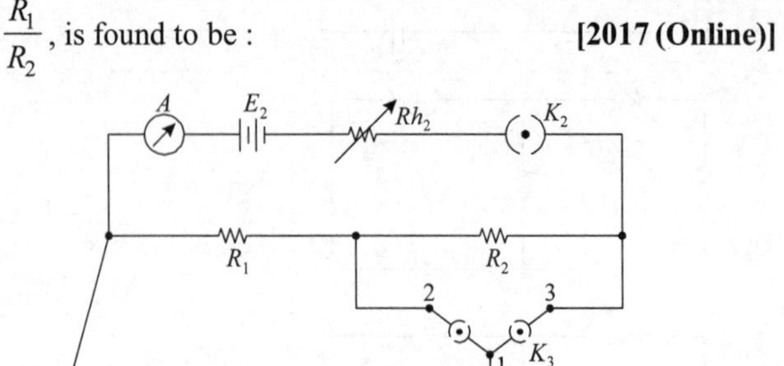

(A) $\frac{l_1}{l_1+l_2}$ (B) $\frac{l_2}{l_2-l_1}$

(C) $\frac{l_1}{l_1-l_2}$ (D) $\frac{l_1}{l_2-l_1}$

26. In a meter bridge experiment resistances are connected as shown in the figure. Initially resistance $P = 4\ \Omega$ and the neutral point N is at 60 cm from A. Now an unknown resistance R is connected in series to P and the new position of the neutral point is at 80 cm from A. The value of unknown resistance R is: **[2017 (Online)]**

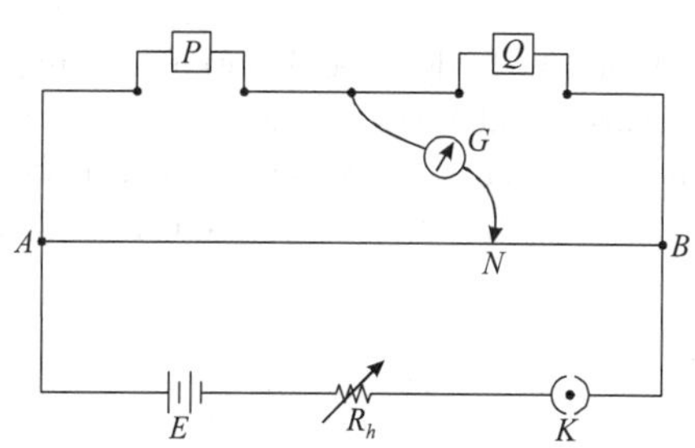

(A) $\frac{33}{5}\Omega$ (B) $6\ \Omega$

(C) $7\ \Omega$ (D) $\frac{20}{3}\Omega$

27. In a potentiometer experiment, it is found that no current passes through the galvanometer when the terminals of the cell are connected across 52 cm of the potentiometer wire. If the cell is shunted by a resistance of $5\ \Omega$, a balance is found when the cell is connected across 40 cm of the wire. Find the internal resistance of the cell : **[2018]**

(A) $1\ \Omega$ (B) $1.5\ \Omega$
(C) $2\ \Omega$ (D) $2.5\ \Omega$

28. On interchanging the resistances, the balance point of a meter bridge shifts to the left by 10 cm. The resistance of their series combination is $1\ k\Omega$. How much was the resistance on the left slot before interchanging the resistances? **[2018]**

(A) $990\ \Omega$ (B) $505\ \Omega$
(C) $550\ \Omega$ (D) $910\ \Omega$

29. In a meter bridge, as shown in the figure, it is given that resistance $Y = 12.5\ \Omega$ and that the balance is obtained at a distance 39.5 cm from end A (by jockey J). After interchanging the resistances X and Y, a new balance point is found at a distance l_1 from end A. What are the values of X and l_1 ? **[2018 (Online)]**

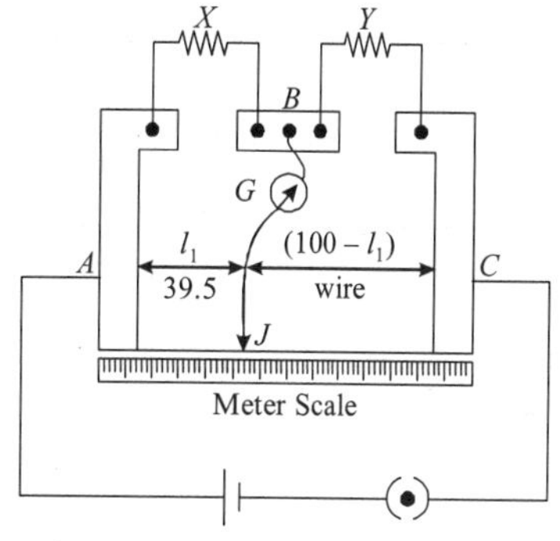

(A) $19.15\ \Omega$ and 39.5 cm (B) $8.16\ \Omega$ and 60.5 cm
(C) $19.15\ \Omega$ and 60.5 cm (D) $8.16\ \Omega$ and 39.5 cm

30. In a circuit for finding the resistance of a galvanometer by half deflection method, a 6 V battery and a high resistance of $11\ k\Omega$ are used. The figure of merit of the galvanometer $60\ \mu A$/division. In the absence of shunt resistance, the galvanometer produces a deflection of $\theta = 9$ divisions when current flows in the circuit. The value of the shunt resistance that can cause the deflection of $\theta/2$, is closest to : **[2018 (Online)]**

(A) $55\ \Omega$ (B) $110\ \Omega$
(C) $220\ \Omega$ (D) $550\ \Omega$

31. A galvanometer with its coil resistance $25\ \Omega$ requires a current of 1 mA for its full deflection. In order to construct an ammeter to read up to a current of 2 A, the approximate value of the shunt resistance should be : **[2018 (Online)]**

(A) $2.5 \times 10^{-2}\ \Omega$ (B) $1.25 \times 10^{-3}\ \Omega$
(C) $2.5 \times 10^{-3}\ \Omega$ (D) $1.25 \times 10^{-2}\ \Omega$

32. In the given circuit the internal resistance of the 18 V cell is negligible. If $R_1 = 400\ \Omega$, $R_3 = 100\ \Omega$ and $R_4 = 500\ \Omega$ and the reading of an ideal voltmeter across R_4 is 5 V, then the value R_2 will be : **[9 Jan. 2019 (Evening)]**

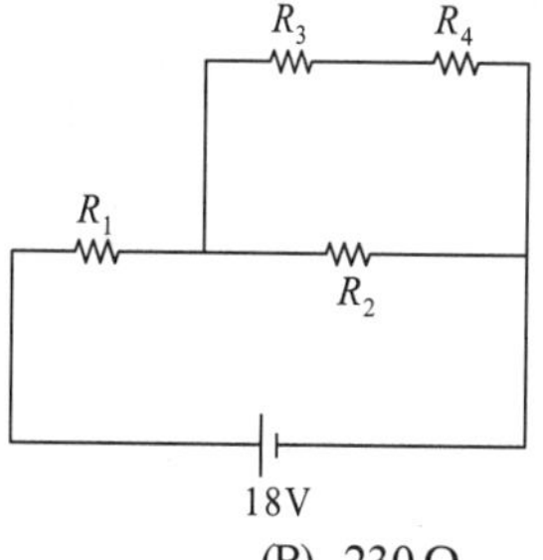

(A) $300\ \Omega$ (B) $230\ \Omega$
(C) $450\ \Omega$ (D) $550\ \Omega$

33. A potentiometer wire AB having length L and resistance $12r$ is joined to a cell D of emf ε and internal resistance r. A cell C having emf $\varepsilon/2$ and internal resistance $3r$ is connected. The length AJ at which the galvanometer as shown in figure shows no deflection is : **[10 Jan. 2019 (Morning)]**

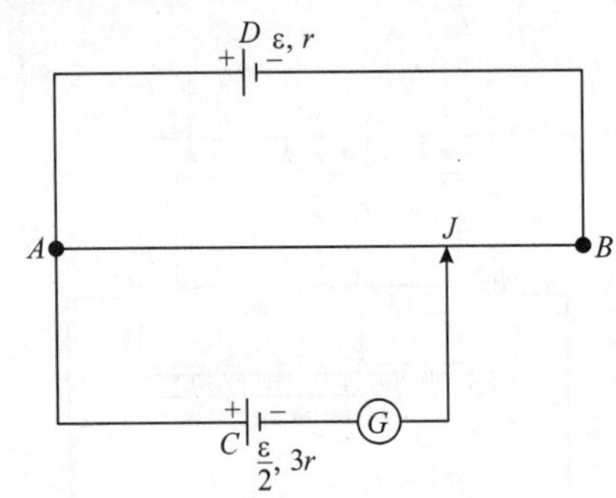

(A) $\frac{5}{12}L$ (B) $\frac{11}{24}L$

(C) $\frac{11}{12}L$ (D) $\frac{13}{24}L$

34. The actual value of resistance R, shown in the figure is 30 Ω. This is measured in an experiment as shown using the standard formula $R = \frac{V}{I}$, where V and I are the readings of the voltmeter and ammeter, respectively. If the measured value of R is 5% less, then the internal resistance of the voltmeter is : **[10 Jan. 2019 (Evening)]**

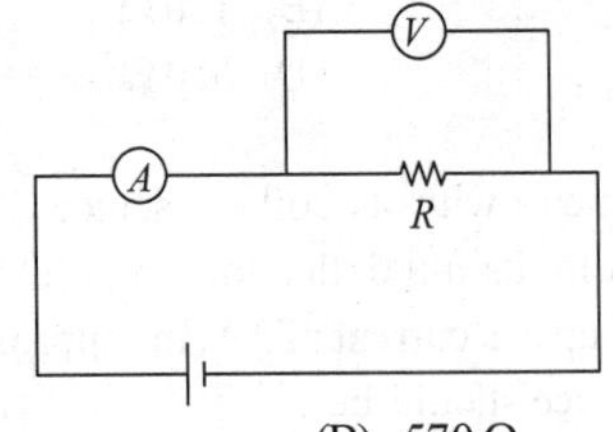

(A) 350 Ω (B) 570 Ω

(C) 35 Ω (D) 600 Ω

35. The Wheatstone bridge shown in Figure here, gets balanced when the carbon resistor used as R_1 has the colour code (Orange, Red, Brown). The resistors R_2 and R_4 are 80 Ω and 40 Ω, respectively.

Assuming that the colour code for the carbon resistors gives their accurate values, the colour code for the carbon resistor, used as R_3, would be : **[10 Jan. 2019 (Evening)]**

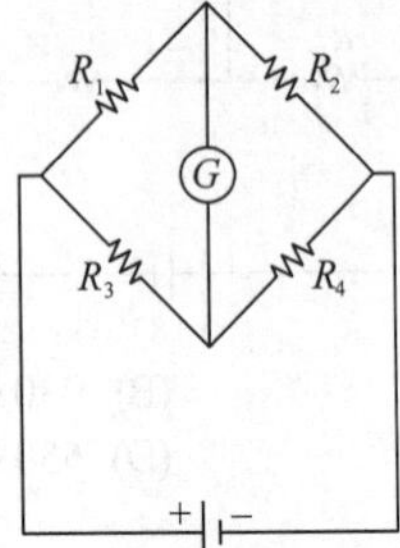

(A) Red, Green, Brown (B) Brown, Blue, Brown

(C) Grey, Black, Brown (D) Brown, Blue, Black

36. The resistance of the meter bridge AB in given figure is 4 Ω. With a cell of emf ε = 0.5 V and rheostat resistance R_h = 2 Ω the null point is obtained at some point J. When the cell is replaced by another one of emf ε = ε_2 the same null point J is found for R_h = 6 Ω. The emf ε_2 is : **[11 Jan. 2019 (Morning)]**

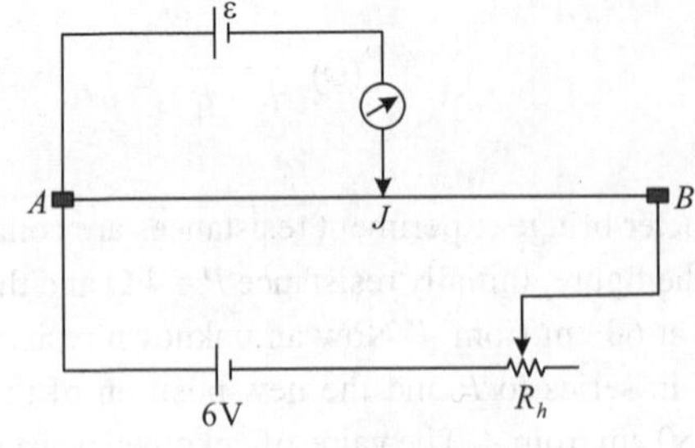

(A) 0.6 V (B) 0.5 V

(C) 0.3 V (D) 0.4 V

37. In a Wheatstone bridge (see figure). Resistances P and Q are approximately equal. When R = 400 Ω, the bridge is balanced. On inter-changing P and Q, the value of R, for balance, is 405 Ω. The value of X is close to : **[11 Jan. 2019 (Morning)]**

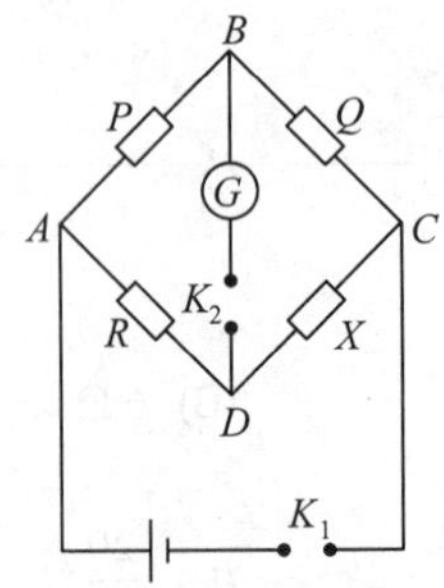

(A) 403.5 ohm (B) 404.5 ohm

(C) 401.5 ohm (D) 402.5 ohm

38. A galvanometer having a resistance of 20 Ω and 30 divisions on both sides has figure of merit 0.005 ampere/division. The resistance that should be connected in series such that it can be used as a voltmeter upto 15 volt, is : **[11 Jan. 2019 (Evening)]**

(A) 80 Ω (B) 120 Ω

(C) 125 Ω (D) 100 Ω

39. In the experimental set up of metre bridge shown in the figure, the null point is obtained at a distance of 40 cm from A. If a 10 Ω resistor is connected in series with R_1, the null point shifts by 10 cm. The resistance that should be connected in parallel with $(R_1 + 10)$ Ω such that the null point shifts back to its initial position is : **[11 Jan. 2019 (Evening)]**

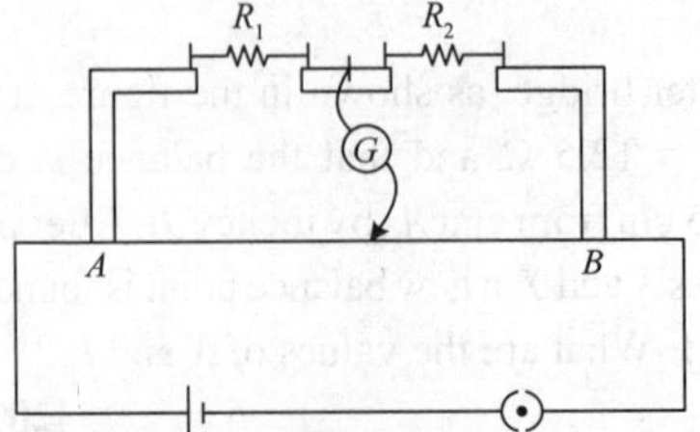

(A) 40 Ω (B) 60 Ω
(C) 20 Ω (D) 30 Ω

40. The galvanometer deflection, when key K_1 is closed but K_2 is open, equals θ_0 (see figure). On closing K_2 also and adjusting R_2 to 5 Ω, the deflection in galvanometer becomes $\frac{\theta_0}{5}$. The resistance of the galvanometer is, then, given by [Neglect the internal resistance of battery] :

[12 Jan. 2019 (Morning)]

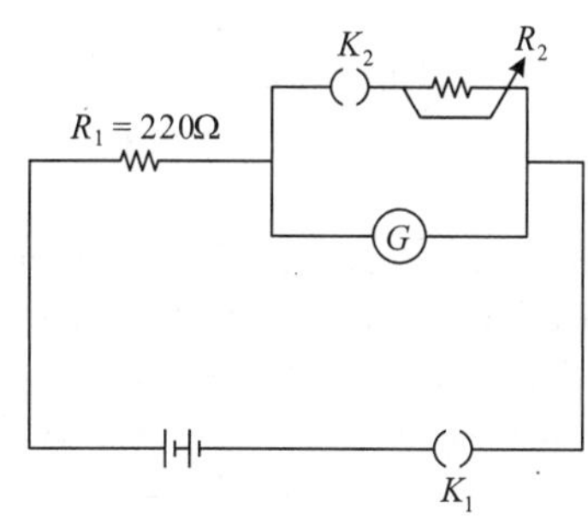

(A) 12 Ω (B) 25 Ω
(C) 5 Ω (D) 22 Ω

41. In a meter bridge, the wire of length 1 m has a non-uniform cross-section such that, the variation $\frac{dR}{dl}$ of its resistance R with length l is $\frac{dR}{dl} \propto \frac{1}{\sqrt{l}}$. Two equal resistances are connected as shown in the figure. The galvanometer has zero deflection when the jockey is at point P. What is the length AP ?

[12 Jan. 2019 (Morning)]

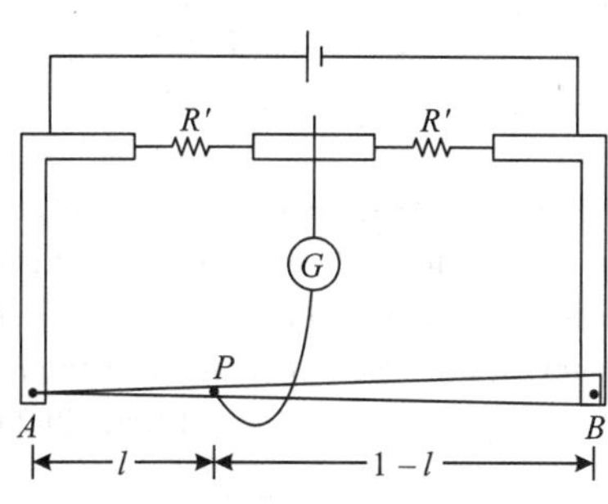

(A) 0.25 m (B) 0.3 m
(C) 0.35 m (D) 0.2 m

42. An ideal battery of 4 V and resistance R are connected in series in the primary circuit of a potentiometer of length 1 m and resistance 5 Ω. The value of R, to give a potential difference of 5 mV across 10 cm of potentiometer wire, is :

[12 Jan. 2019 (Morning)]

(A) 490 Ω (B) 480 Ω
(C) 395 Ω (D) 495 Ω

43. A galvanometer, whose resistance is 50 ohm, has 25 divisions in it. When a current of 4×10^{-4} A passes through it, its needle (pointer) deflects by one division. To use this galvanometer as a voltmeter of range 2.5 V, it should be connected to a resistance of : **[12 Jan. 2019 (Evening)]**

(A) 6250 ohm (B) 250 ohm
(C) 200 ohm (D) 6200 ohm

44. In the circuit shown, a four-wire potentiometer is made of a 400 cm long wire, which extends between A and B. The resistance per unit length of the potentiometer wire is $r = 0.01$ Ω/cm. If an ideal voltmeter is connected as shown with jockey J at 50 cm from end A, the expected reading of the voltmeter will be :

[8 April 2019 (Evening)]

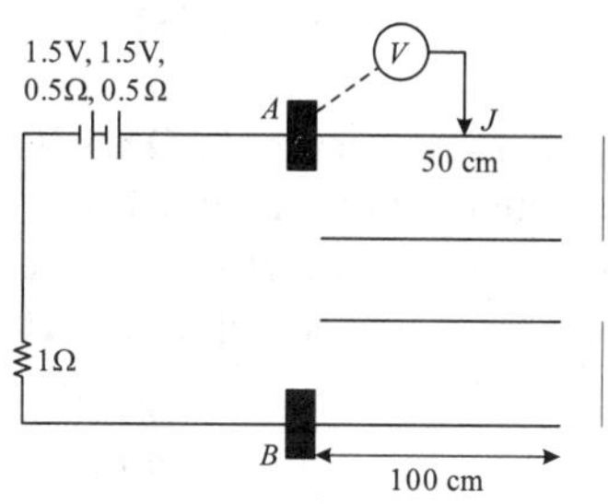

(A) 0.20 V (B) 0.25 V
(C) 0.75 V (D) 0.50 V

45. The resistance of a galvanometer is 50 ohm and the maximum current which can be passed through it is 0.002 A. What resistance must be connected to it in order to convert it into an ammeter of range 0 – 0.5 A ? **[9 April 2019 (Evening)]**

(A) 0.2 ohm (B) 0.002 ohm
(C) 0.02 ohm (D) 0.5 ohm

46. In a meter bridge experiment, the circuit diagram and the corresponding observation table are shown in figure

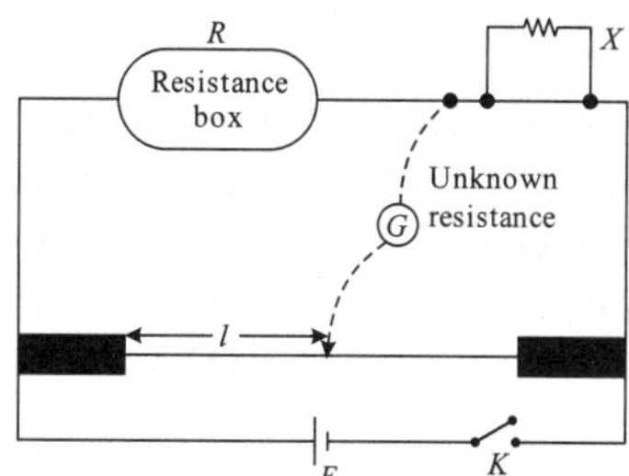

SI.No.	$R(\Omega)l$	l(cm)
1.	1000	60
2.	100	13
3.	10	1.5
4.	1	1.0

Which of the readings is inconsistent ?

[10 April 2019 (Morning)]

(A) 4 (B) 1
(C) 2 (D) 3

47. In the given circuit, an ideal voltmeter connected across the 10 Ω resistance reads 2 V. The internal resistance r, of each cell is : **[10 April 2019 (Morning)]**

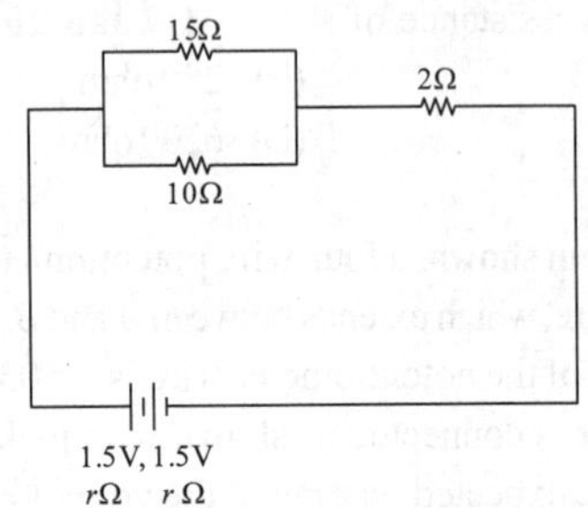

(A) 1 Ω (B) 1.5 Ω
(C) 0 Ω (D) 0.5 Ω

48. A galvanometer of resistance 100 Ω has 50 divisions on its scale and has sensitivity of 20 μA/division. It is to be converted to a voltmeter with three ranges, of 0 – 2 V, 0 – 10 V and 0 – 20 V. The appropriate circuit to do so is : **[12 April 2019 (Morning)]**

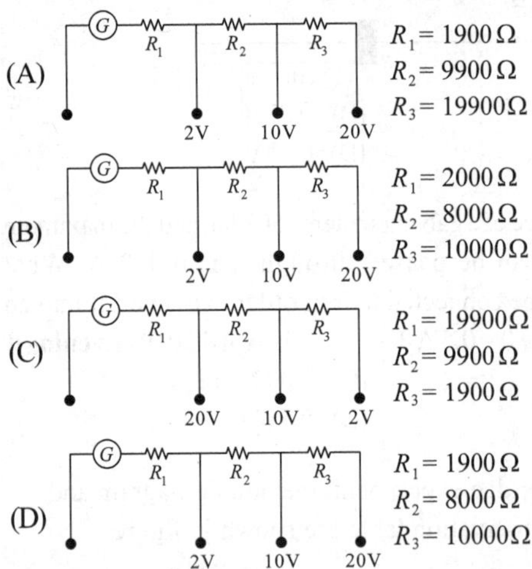

49. To verify Ohm's law, a student connects the voltmeter across the battery as, shown in the figure. The measured voltage is plotted as a function of the current, and the following graph is obtained. If V_0 is almost zero, identify the correct statement : **[12 April 2019 (Morning)]**

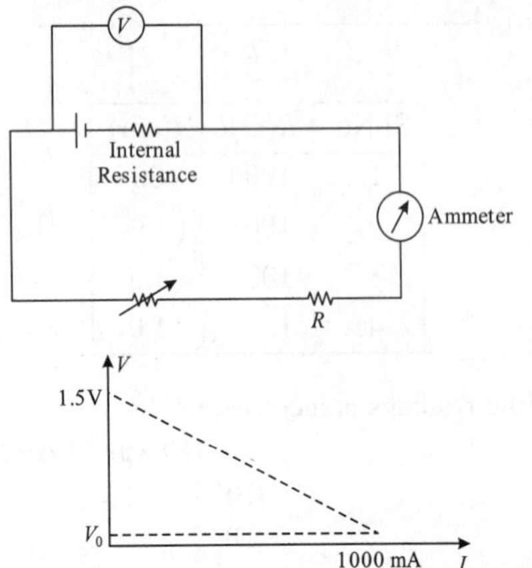

(A) The value of the resistance R is 1.5 Ω

(B) The emf of the battery is 1.5 V and the value of R is 1.5 Ω

(C) The emf of the battery is 1.5 V and its internal resistance is 1.5 Ω

(D) The potential difference across the battery is 1.5 V when it sends a current of 1000 mA

50. The balancing length for a cell is 560 cm in a potentiometer experiment. When an external resistance of 10 Ω is connected in parallel to the cell, the balancing length changes by 60 cm. If the internal resistance of the cell is $\frac{N}{10}\Omega$, where N is an integer then value of N is ______. **[7 Jan. 2020 (Evening)]**

51. The length of a potentiometer wire is 1200 cm and it carries a current of 60 mA. For a cell of emf 5 V and internal resistance of 20 Ω, the null point on it is found to be a 1000 cm. The resistance of whole wire is : **[8 Jan. 2020 (Morning)]**

(A) 120 Ω (B) 60 Ω
(C) 80 Ω (D) 100 Ω

52. Four resistances of 15 Ω, 12 Ω, 4 Ω and 10 Ω respectively in cyclic order to form. Wheatstone's network. The resistance that is to be connected in parallel with the resistance of 10 Ω to balance the network is ______ Ω. **[8 Jan. 2020 (Morning)]**

53. A galvanometer having a coil resistance 100 Ω gives a full scale deflection when a current of 1 mA is passed through it. What is the value of the resistance which can convert this galvanometer into a voltmeter giving full scale deflection for a potential difference of 10 V ? **[8 Jan. 2020 (Evening)]**

(A) 9.9 kΩ (B) 8.9 kΩ
(C) 7.9 kΩ (D) 10 kΩ

54. In a meter bridge experiment S is a standard resistance. R is a resistance wire. It Is found that balancing length is $l = 25$ cm. If R is replaced by a wire of half length and half diameter that of R of same material, then the balancing distance l' (in cm) will now be_______. **[9 Jan. 2020 (Evening)]**

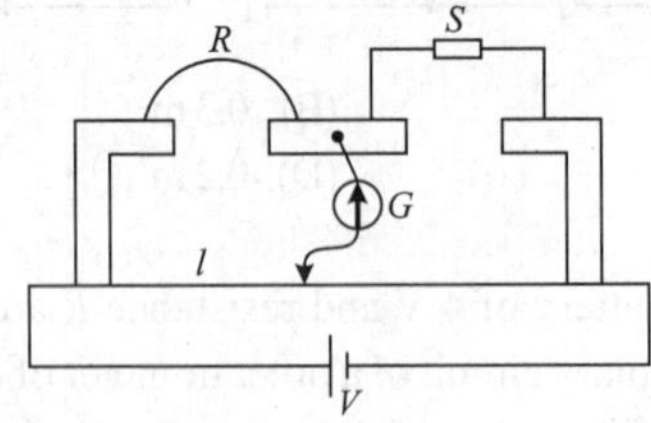

55. A potentiometer wire PQ of 1 m length is connected to a standard cell E_1. Another cell E_2 of emf 1.02 V is connected with a resistance 'r' and switch S (as shown in figure). With switch S open, the null position is obtained at a distance of 49 cm from Q. The potential gradient in the potentiometer wire is : **[2 Sept. 2020 (Evening)]**

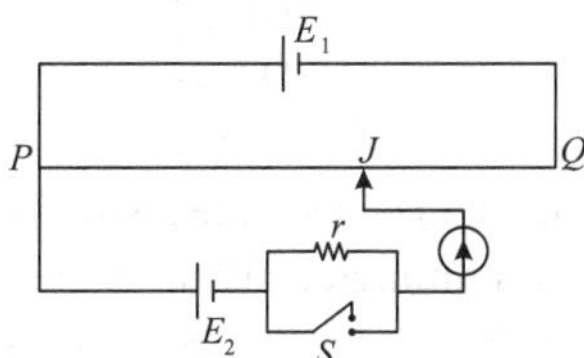

(A) 0.02 V/cm (B) 0.04 V/cm
(C) 0.01 V/cm (D) 0.03 V/cm

56. Which of the following will NOT be observed when a multimeter (operating in resistance measuring mode) probes connected across a component, are just reversed ?

[3 Sept. 2020 (Evening)]

(A) Multimeter shows NO deflection in both cases *i.e.* before and after reversing the probes if the chosen component is capacitor

(B) Multimeter shows a deflection, accompanied by a splash of light out of connected component in one direction and NO deflection on reversing the probes if the chosen component is LED

(C) Multimeter shows NO deflection in both cases *i.e.* before and after reversing the probes if the chosen component is metal wire

(D) Multimeter shows an equal deflection in both cases *i.e.* before and after reversing the probes if the chosen component is resistor

57. Two resistors 400 Ω and 800 Ω are connected in series across a 6 V battery. The potential difference measured by a voltmeter of 10 $k\Omega$ across 400 Ω resistor is close to :

[3 Sept. 2020 (Evening)]

(A) 2 V (B) 1.95 V
(C) 2.05 V (D) 1.8 V

58. A galvanometer of resistance G is converted into a voltmeter of range 0 – 1 V by connecting a resistance R_1 in series with it. The additional resistance that should be connected in series with R_1 to increase the range of the voltmeter to 0 – 2 V will be :

[5 Sept. 2020 (Morning)]

(A) R_1 (B) $R_1 + G$
(C) $R_1 - G$ (D) G

59. A galvanometer is used in laboratory for detecting the null point in electrical experiments. If, on passing a current of 6 mA it produces a deflection of 2°, its figure of merit is close to :

[5 Sept. 2020 (Evening)]

(A) 3×10^{-3} A/div (B) 333° A/div
(C) 6×10^{-3} A/div (D) 666° A/div

60. A circuit to verify Ohm's law uses ammeter and voltmeter in series or parallel connected correctly to the resistor. In the circuit : **[6 Sept. 2020 (Evening)]**

(A) Ammeter is always connected in series and voltmeter in parallel.

(B) Both, ammeter and voltmeter must be connected in series.

(C) Both ammeter and voltmeter must be connected in parallel.

(D) Ammeter is always used in parallel and voltmeter is series.

61. In the given circuit of potentiometer, the potential difference E across AB (10 m length) is larger than E_1 and E_2 as well. For key K_1 (closed), the jockey is adjusted to touch the wire at point J_1 so that there is no deflection in the galvanometer. Now the first battery (E_1) is replaced by second battery (E_2) for working by making K_1 open and K_2 closed. The galvanometer gives then null deflection at J_2. The value of $\frac{E_1}{E_2}$ is $\frac{a}{b}$, where a = ______.

[25 Feb. 2021 (Morning)]

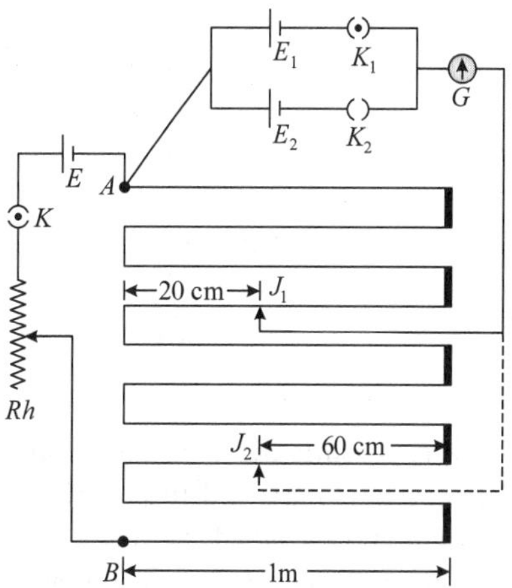

62. Consider a 72 cm long wire AB as shown in the figure. The galvanometer jockey is placed at P on AB at a distance x cm from A. The galvanometer shows zero deflection.

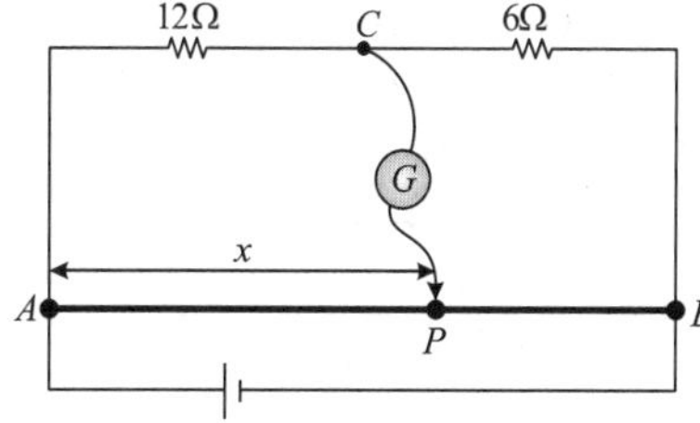

The value of x, to the nearest integer, is ?

[18 March 2021 (Evening)]

63. In the given figure, there is a circuit of potentiometer of length AB = 10 m. The resistance per unit length of 0.1 Ω per cm. Across AB, a battery of emf E and internal resistance r is connected. The maximum value of emf measured by this potentiometer is : **[25 July 2021 (Morning)]**

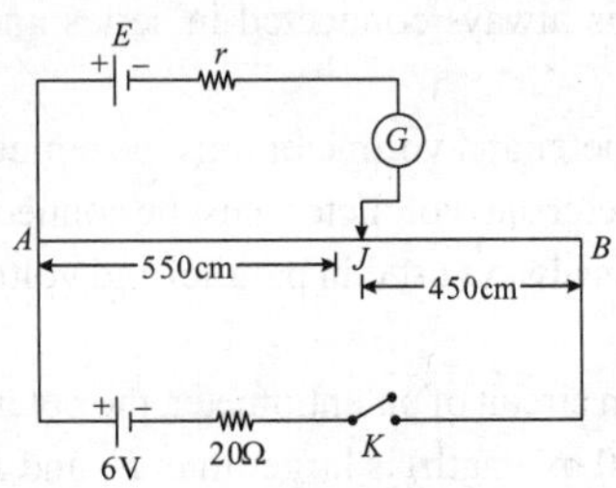

(A) 2..25 V (B) 6 V
(C) 5 V (D) 2.75 V

64. The given potentiometer has its wire of resistance 10 Ω When the sliding contact is in the middle of the potentiometer wire, the potential drop across 2 Ω resistor is : **[25 July 2021 (Evening)]**

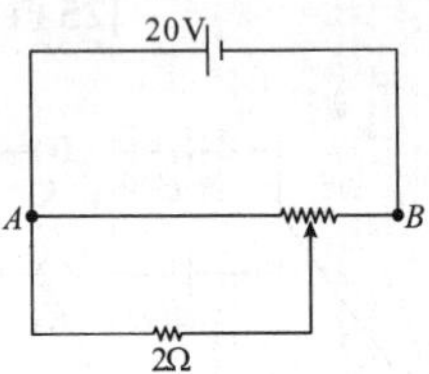

(A) $\frac{40}{9}$V (B) 5V

(C) $\frac{40}{11}$V (D) 10V

65. In the given potentiometer circuit arrangement, the balancing length AC is measured to be 250 cm. When the galvanometer connection is shifted from point (1) to point (2) in the given diagram, the balancing length becomes 400 cm. The ration of the emf of two cells, $\frac{\varepsilon_1}{\varepsilon_2}$ is : **[25 July 2021 (Evening)]**

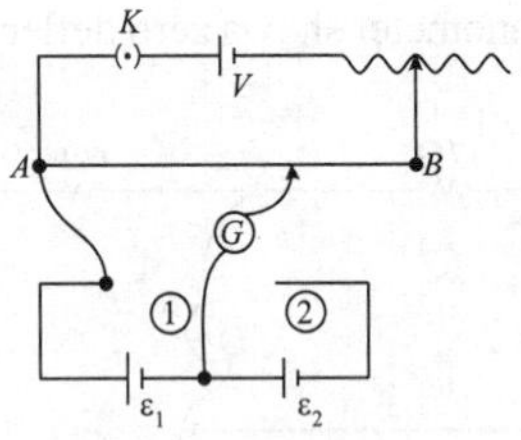

(A) $\frac{4}{3}$ (B) $\frac{5}{3}$

(C) $\frac{8}{5}$ (D) $\frac{3}{2}$

66. For full scale deflection of total 50 divisions, 50 mV voltage is required in galvanometer. The resistance of galvanometer if its current sensitivity is 2 div/mA will be : **[27 Aug. 2021 (Evening)]**

(A) 1 Ω (B) 2 Ω
(C) 5 Ω (D) 4 Ω

67. Consider a galvanometer shunted with 5 Ω resistance and 2% of current passes through it. What is the resistance of the given galvanometer ? **[31 Aug. 2021 (Morning)]**

(A) 226 Ω (B) 245 Ω
(C) 344 Ω (D) 300 Ω

68. In a potentiometer arrangement, a cell gives a balancing point at 75 cm length of wire. This cell is now replaced by another cell of unknown emf. If the ratio of the emf's of two cells respectively is 3 : 2, the difference in the balancing length of the potentiometer wire in above two cases will be _____ cm. **[24 June 2022 (Morning)]**

69. A potentiometer wire of length 10 m and resistance 20 Ω is connected in series with a 25 V battery and an external resistance 30 Ω. A cell of emf E in secondary circuit is balanced by 250 cm long potentiometer wire. The value of E (in volt) is $\frac{x}{10}$. The value of x is _____. **[24 June 2022 (Evening)]**

70. A teacher in his physics laboratory allotted an experiment to determine the resistance (G) of a galvanometer. Students took the observations for $\frac{1}{3}$ deflection in the galvanometer. Which of the below is **true** for measuring value of G ? **[25 June 2022 (Morning)]**

(A) $\frac{1}{3}$ deflection method cannot be used for determining the resistance of the galvanometer

(B) $\frac{1}{3}$ deflection method can be used and in this case the G equals to twice the value of shunt resistance(s)

(C) $\frac{1}{3}$ deflection method can be used and in this case, the G equals to three times the value of shunt resistance(s)

(D) $\frac{1}{3}$ deflection method can be used and in this case the G value equals to the shunt resistance(s)

71. If n represents the actual number of deflections in a converted galvanometer of resistance G and shunt resistance S. Then the total current I when its figure of merit is K will be : **[25 June 2022 (Evening)]**

(A) $\frac{KS}{(S+G)}$ (B) $\frac{(G+S)}{nKS}$

(C) $\frac{nKS}{(G+S)}$ (D) $\frac{nK(G+S)}{S}$

72. A 72 Ω galvanometer is shunted by a resistance of 8 Ω. The percentage of the total current which passes through the galvanometer is : **[27 June 2022 (Morning)]**

(A) 0.1 %
(B) 10 %
(C) 25 %
(D) 0.25 %

73. A cell, shunted by a 8 Ω resistance, is balanced across a potentiometer wire of length 3 m. The balancing length is 2 m when the cell is shunted by 4 Ω resistance. The value of internal resistance of the cell will be ________ Ω.

[27 June 2022 (Morning)]

74. A meter bridge setup is shown in the figure. It is used to determine an unknown resistance R using a given resistor of 15 Ω. The galvanometer (G) shows null deflection when tapping key is at 43 cm mark from end A. If the end correction for end A is 2 cm. Then the determined value of R will be _____ Ω.

[28 June 2022 (Morning)]

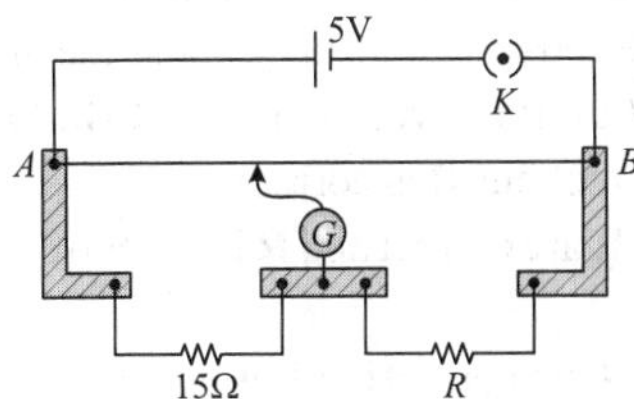

75. Current measured by the ammeter in the reported circuit when no current flows through 10 Ω resistance. Will be _____ A. **[28 June 2022 (Morning)]**

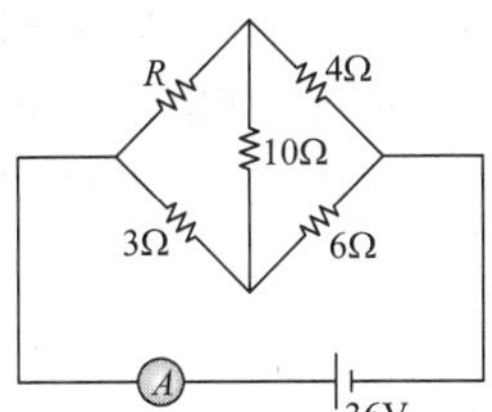

76. Two resistors are connected in series across a battery as shown in figure. If a voltmeter of resistance 2000 Ω is used to measure the potential difference across 500 Ω resistor, the reading of the voltmeter will be _____ V.

[29 June 2022 (Evening)]

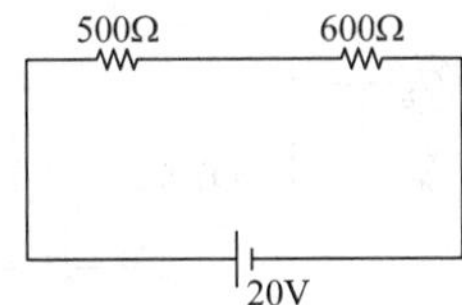

77. In a potentiometer arrangement, a cell of emf 1.20 V gives a balance point at 36 cm length of wire. This cell is now replaced by another cell of emf 1.80 V. The difference in balancing length of potentiometer wire in above conditions will be _____ cm.

[25 July 2022 (Evening)]

78. Resistance are connected in a meter bridge circuit as shown in the figure. The balancing length l_1 is 40 cm. Now an unknown resistance x is connected in series with P and new balancing length is found to be 80 cm measured from the same end. Then the value of x will be _____ Ω. **[26 July 2022 (Morning)]**

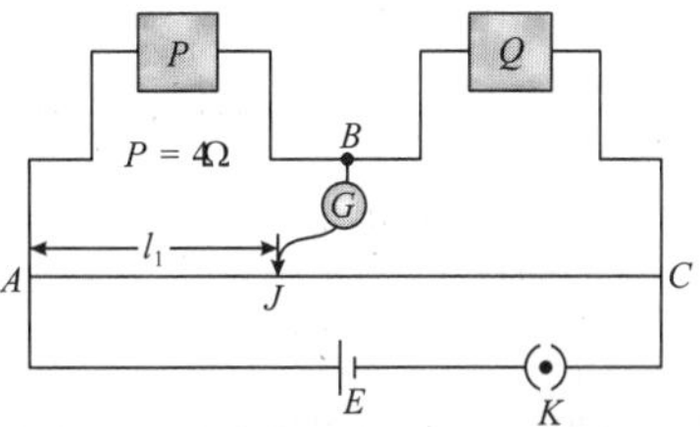

79. A potentiometer wire of length 300 cm is connected in series with a resistance 780 Ω and a standard cell of emf 4 V. A constant current flows through potentiometer wire. The length of the null point for cell of emf 20 mV is found to be 60 cm. The resistance of the potentiometer wire is _____ Ω.

[26 July 2022 (Evening)]

80. In meter bridge experiment for measuring unknown resistance S, the null point is obtained at a distance 30 cm from the left side as shown at point D. If R is 5.6 kΩ, then the value of unknown resistance S will be _____ Ω.

[27 July 2022 (Morning)]

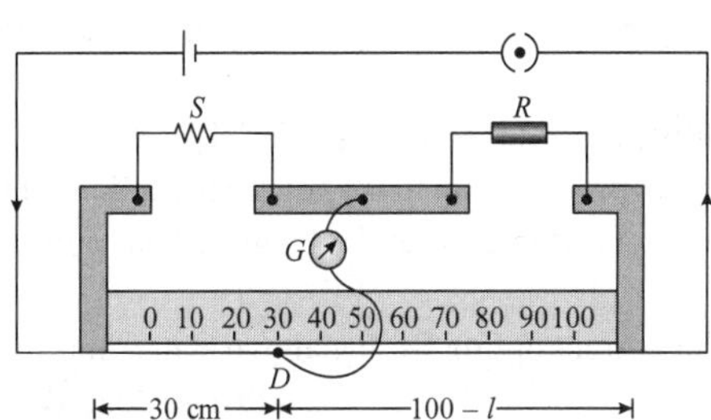

81. In the given figure of meter bridge experiment, the balancing length AC corresponding to null deflection of the galvanometer is 40 cm. The balancing length, if the radius of the wire AB is doubled, will be _____ cm. **[27 July 2022 (Evening)]**

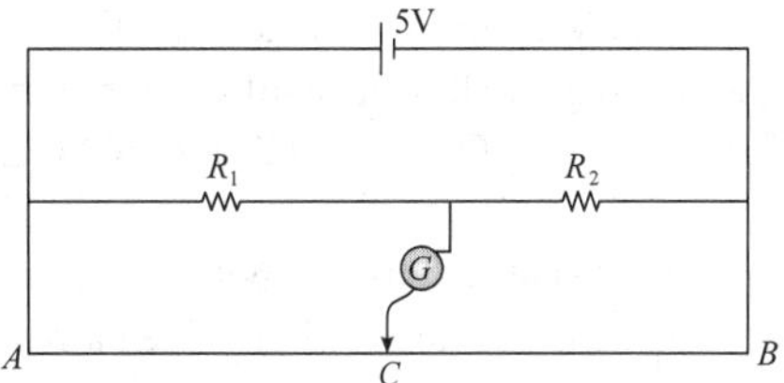

82. As shown in the figure, a potentiometer wire of resistance 20 Ω and length 300 cm is connected with resistance box (*R.B.*) and a standard cell of emf 4 V. For a resistance *R* of resistance box introduced into the circuit, the null point for a cell of 20 mV is found to be 60 cm. The value of *R* is ______ Ω.

[28 July 2022 (Morning)]

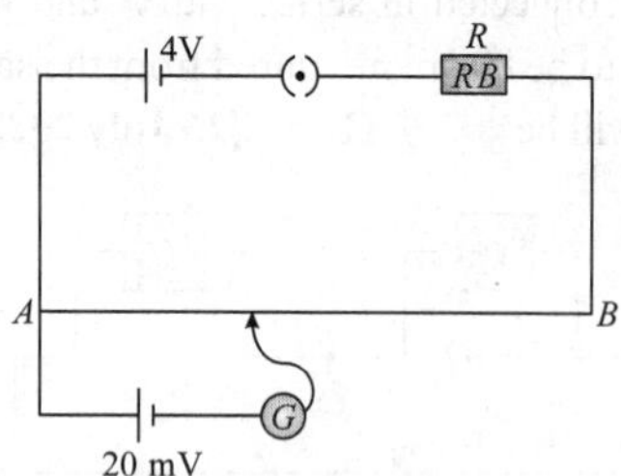

83. A cell of emf 90 V is connected across series combination of two resistors each of 100 Ω resistance. A voltmeter of resistance 400 Ω is used to measure the potential difference across each resistor. The reading of the voltmeter will be :

[24 Jan. 2023 (Evening)]

(A) 40V (B) 45V
(C) 80V (D) 90V

84. In a metre bridge experiment the balance point in obtained if the gaps are closed by 2 Ω and 3 Ω. A shunt of *X*Ω is added to 3 Ω resistor to shift the balancing point by 22.5 cm. The value of *X* is ______. **[29 Jan. 2023 (Morning)]**

85. With the help of potentiometer, we can determine the value of emf of a given cell. The sensitivity of the potentiometer is :

(a) Directly proportional to the length of the potentiometer wire
(b) Directly proportional to the potential gradient of the wire
(c) Inversely proportional to the potential gradient of the wire
(d) Inversely proportional to the length of the potentiometer wire

Choose the correct option for the above statements :

[29 Jan. 2023 (Evening)]

(A) (b) and (d) only (B) (a) and (c) only
(C) (a) only (D) (c) only

86. When two resistance R_1 and R_2 connected in series and introduced into the left gap of a meter bridge and a resistance of 10 Ω is introduced into the right gap, a null point is found at 60 cm from left side. When R_1 and R_2 are connected in parallel and introduced into the left gap, a resistance of 3 Ω is introduced into the right-gap to get null point at 40 cm from left end. The product of R_1R_2 is ______Ω^2. **[29 Jan. 2023 (Evening)]**

87. A null point is found at 200 cm in potentiometer when cell in secondary circuit is shunted by 5 Ω. When a resistance of 15 Ω is used for shunting null point moves to 300 cm. The internal resistance of the cell is ______ Ω.

[29 Jan. 2023 (Evening)]

88. In an experiment to find emf of a cell using potentiometer, the length of null point for a cell of emf 1.5 V is found to be 60 cm. If this cell is replaced by another cell of emf *E*. The length of null point increases by 40 cm. The value of *E* is $\frac{x}{10}$ *V*. The value of *x* is. **[1 Feb. 2023 (Morning)]**

89. Given below are two statements : One is labelled as **Assertion-A** and the other is labelled as **Reason-R**.

Assertion-A : For measuring the potential difference across a resistance of 600 Ω, the voltmeter with resistance 1000 Ω will be preferred over voltmeter with resistance 4000 Ω.

Reason-R : Voltmeter with higher resistance will draw smaller current than voltmeter with lower resistance.

In the light of the above statements, choose the most appropriate answer from the options given below: **[1 Feb. 2023 (Evening)]**

(A) **A** is not correct but **R** is correct.
(B) Both **A** and **R** are correct and **R** is the correct explanation of **A**.
(C) Both **A** and **R** are correct but **R** is not the correct explanation of **A**.
(D) **A** is correct but **R** is not correct.

90. A student is provided with a variable voltage source *V*, a test resistor $R_r = 10\ \Omega$, two identical galvanometers G_1 and G_2 and two additional resistors, $R_1 = 10\ M\Omega$ and $R_2 = 0.001\ \Omega$. For conducting an experiment to verify ohms law, the most suitable circuit is: **[6 April 2023 (Evening)]**

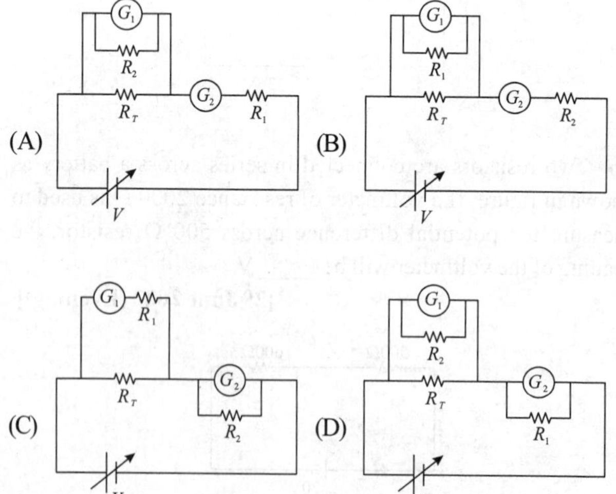

91. As shown in the figure, the voltmeter reads 2 V across 5 Ω resistor. The resistance of the voltmeter is _____ Ω.

[6 April 2023 (Evening)]

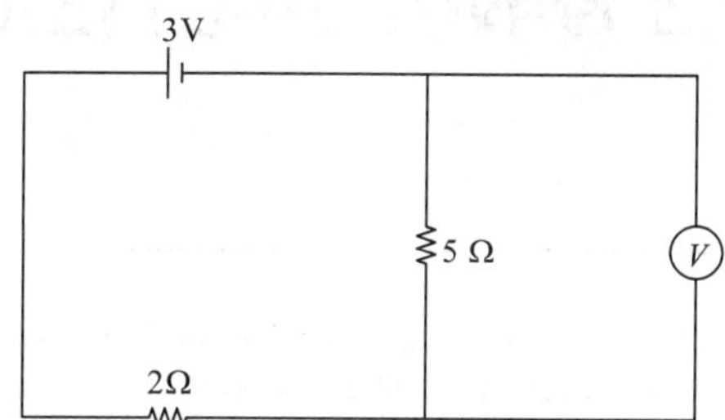

92. When a resistance of 5 Ω is shunted with a moving coil galvanometer, it shows a full scale deflection for a current of 250 mA, however when 1050 Ω resistance is connected with it in series, it gives full scale deflection for 25 volt. The resistance of galvanometer is_____Ω. **[13 April 2023 (Morning)]**

93. For designing a voltmeter of range 50 V and an ammeter of range 10 mA using a galvanometer which has a coil of resistance 54 Ω showing a full scale deflection for 1 mA as shown in figure below. **[15 April 2023 (Morning)]**

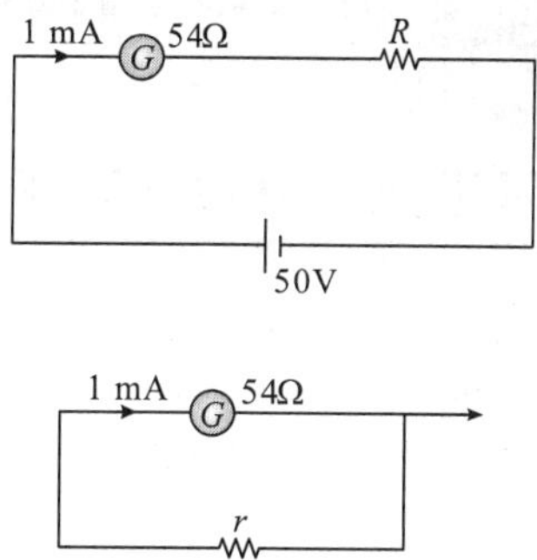

(a) For voltmeter $R \approx 50\ k\Omega$ (b) For ammeter $r \approx 0.2\ k\Omega$

(c) For ammeter $r \approx 6\ \Omega$ (d) For voltmeter $R \approx 5\ k\Omega$

(e) For voltmeter $R \approx 500\ \Omega$

Choose the correct answer from the options given below :

(A) (C) and (E) (B) (C) and (D)

(C) (A) and (C) (D) (A) and (B)

* * * * *

CHAPTER

18 MOVING CHARGES & MAGNETISM

18.1 MAGNETIC EFFECTS OF CURRENT

1. If in a circular coil A of radius R, current i is flowing and in another coil B of radius $2R$ a current $2i$ is flowing, then the ratio of the magnetic fields, B_A and B_B produced by them will be : **[2002]**

(A) 1 (B) 2
(C) 1/2 (D) 4

2. The magnetic field due to a current carrying circular loop of radius 3 cm at a point on the axis at a distance of 4 cm from the centre is 54 μT. What will be its value at the centre of the loop ? **[2004]**

(A) 250 μT (B) 150 μT
(C) 125 μT (D) 75 μT

3. A long wire carries a steady current. It is bent into a circle of one turn and the magnetic field at the centre of the coil is B. It is then bent into a circular loop of n turns. The magnetic field at the centre of the coil will be : **[2004]**

(A) nB (B) n^2B
(C) $2nB$ (D) $2n^2B$

4. A current i ampere flows along an infinitely long straight thin walled tube, then the magnetic induction at any point inside the tube is : **[2004]**

(A) $\frac{\mu_0}{4\pi} \cdot \frac{2i}{r}$ tesla (B) Zero

(C) Infinite (D) $\frac{2i}{r}$ tesla

5. Two concentric coils each of radius equal to 2π cm are placed at right angles to each other. 3 A and 4 A are the currents flowing in each coil respectively. The magnetic induction in Wb/m^2 at the centre of the coils will be : $[\mu_0 = 4\pi \times 10^{-7}$ Wb/Am$]$ **[2005]**

(A) 12×10^{-5} (B) 10^{-5}
(C) 5×10^{-5} (D) 7×10^{-5}

6. A long solenoid has 200 turns per cm and carries a current i. The magnetic field at its centre is 6.28×10^{-2} weber/m^2. Another long solenoid has 100 turns per cm and it carries a current $i/3$. The value of the magnetic field at its centre is : **[2006]**

(A) 1.05×10^{-2} weber/m^2 (B) 1.05×10^{-5} weber/m^2
(C) 1.05×10^{-3} weber/m^2 (D) 1.05×10^{-4} weber/m^2

7. Two identical conducting wires AOB and COD are placed at right angles to each other. The wire AOB carries an electric current I_1 and COD carries a current I_2. The magnetic field on a point lying at a distance d from O, in a direction perpendicular to the plane of the wires AOB and COD, will be given by : **[2007]**

(A) $\frac{\mu_0}{2\pi}\left(\frac{I_1+I_2}{d}\right)^{1/2}$ (B) $\frac{\mu_0}{2\pi d}(I_1^2+I_2^2)^{1/2}$

(C) $\frac{\mu_0}{2\pi d}(I_1+I_2)$ (D) $\frac{\mu_0}{2\pi d}(I_1^2+I_2^2)$

8. A long straight wire of radius a carries a steady current i. The current is uniformly distributed across its cross section. The ratio of the magnetic field at $a/2$ and $2a$ is : **[2007]**

(A) 1/2 (B) 1/4
(C) 4 (D) 1

9. A current I flows along the length of an infinitely long, straight, thin walled pipe. Then : **[2007]**

(A) The magnetic field at all points inside the pipe is the same, but not zero
(B) The magnetic field is zero only on the axis of the pipe
(C) The magnetic field is different at different points inside the pipe
(D) The magnetic field at any point inside the pipe is zero

10. A horizontal overhead power line is at a height of 4 m from the ground and carries a current of 100 A from east to west. The magnetic field directly below it on the ground is : $[\mu_0 = 4\pi \times 10^{-7}$ TmA$^{-1}]$: **[2008]**

(A) 2.5×10^{-7} T southward (B) 5×10^{-6} T northward
(C) 5×10^{-6} T southward (D) 2.5×10^{-7} T northward

11. The magnitude of the magnetic field (B) due to the loop $ABCD$ at the origin (O) is : **[2009]**

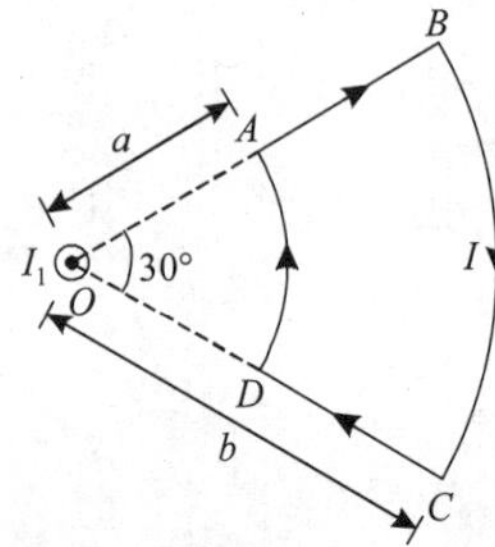

(A) $\frac{\mu_0 I}{4\pi}\left[\frac{b-a}{ab}\right]$ (B) $\frac{\mu_0 I}{4\pi}\left[2(b-a)+\frac{\pi}{3}(a+b)\right]$

(C) Zero (D) $\frac{\mu_0 I(b-a)}{24ab}$

12. Due to the presence of the current I_1 at the origin :**[2009]**

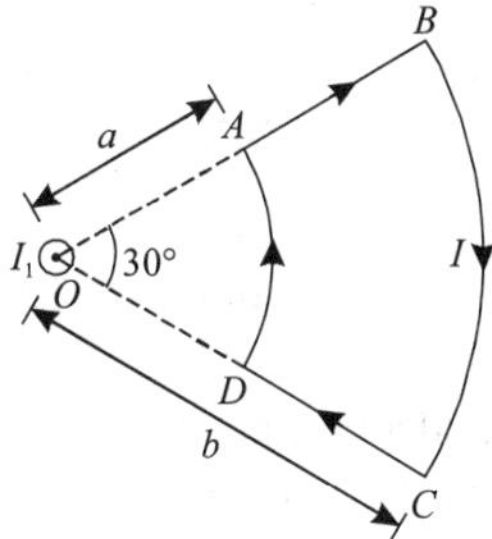

(A) The magnitude of the net force on the loop is given by $\frac{I_1 I}{4\pi}\mu_0\left[2(b-a)+\frac{\pi}{3}(a+b)\right]$

(B) The magnitude of the net force on the loop is given by $\frac{\mu_0 I I_1}{24ab}(b-a)$

(C) The forces on AB and DC are zero

(D) The forces on AD and BC are zero

13. Two long parallel wire are at a distance 2d apart. They carry steady equal currents flowing out of the plane of the paper as shown. The variation of the magnetic field B along the line XX' is given by : **[2010]**

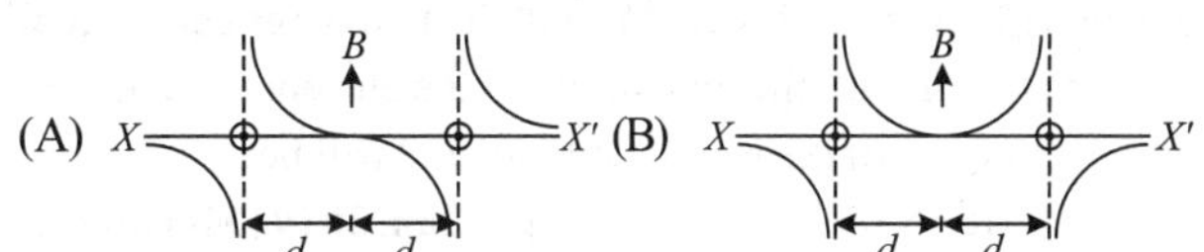

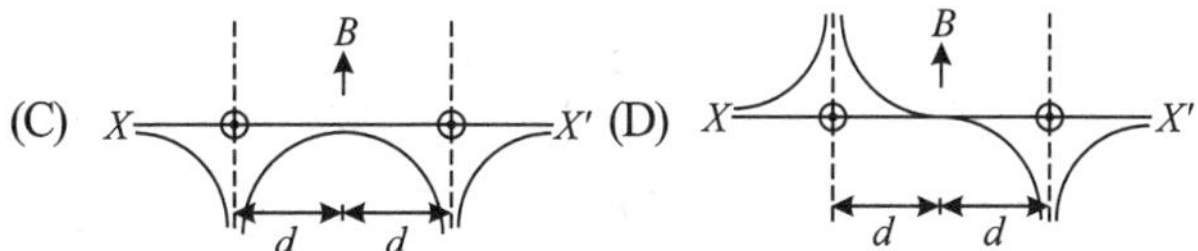

14. A current I flows in an infinity long wire with cross section in the from of a semicircular ring of radius R. The magnitude of the magnetic induction along its axis is : **[2011]**

(A) $\frac{\mu_0 I}{2\pi R}$ (B) $\frac{\mu_0 I}{4\pi R}$

(C) $\frac{\mu_0 I}{\pi^2 R}$ (D) $\frac{\mu_0 I}{2\pi^2 R}$

15. A charge Q is uniformly distributed over the surface of non-conducting disc of radius R. The disc rotates about an axis perpendicular to its plane and passing through its centre with an angular velocity ω. As a result of this rotation a magnetic field of induction B is obtained at the centre of the disc. If we keep both the amount of charge placed on the disc and its angular velocity to be constant and vary the radius of the disc then the variation of the magnetic induction at the centre of the disc will be represented by the figure : **[2012]**

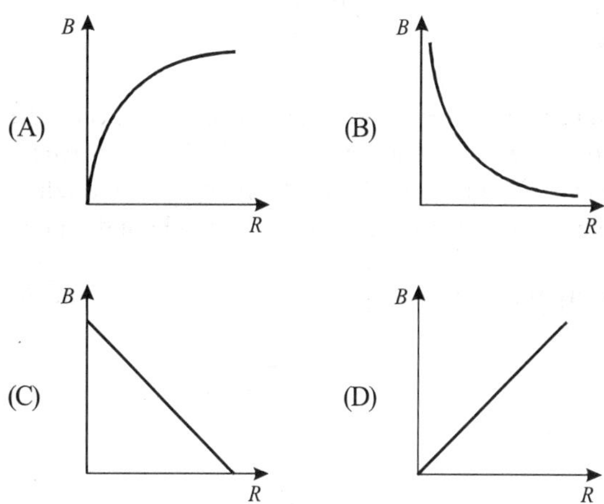

16. An electric current is flowing through a circular coil of radius R. The ratio of the magnetic field at the centre of the coil and that at a distance $2\sqrt{2}R$ from the centre of the coil and on its axis is : **[2013 (Online)]**

(A) $2\sqrt{2}$ (B) 27
(C) 36 (D) 8

17. A current i is flowing in a straight conductor of length L. The magnetic induction at a point on its axis at a distance $L/4$ from its centre will be : **[2013 (Online)]**

(A) Zero (B) $\frac{\mu_0 i}{2\pi L}$

(C) $\frac{\mu_0 i}{\sqrt{2}L}$ (D) $\frac{4\mu_0 i}{\sqrt{5}\mu L}$

18. Choose the correct sketch of the magnetic field lines of a circular current loop shown by the dot ⊙ and the cross ⊗ : **[2013 (Online)]**

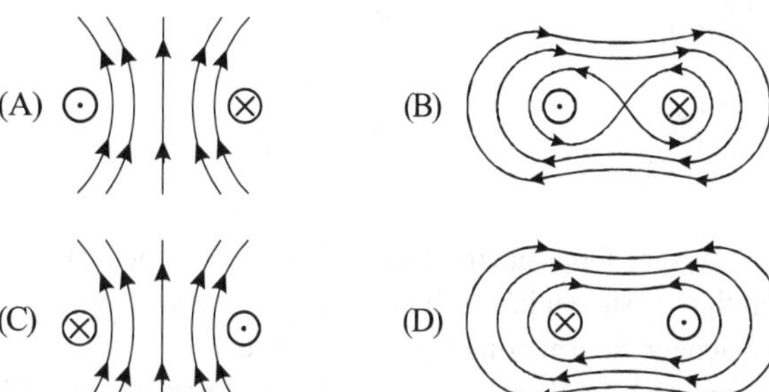

19. Consider two thin identical conducting wires covered with very thin insulating material. One of the wires is bent into a loop and produces magnetic field B_1, at its centre when a current I passes through it. The second wire is bent into a coil with three identical loops adjacent to each other and produces magnetic field B_2 at the centre of the loops when a current $I/3$ passes through it. Then ratio $B_1 : B_2$ is : **[2014 (Online)]**

(A) 1 : 1 (B) 1 : 3
(C) 1 : 9 (D) 9 : 1

20. Two identical wires A and B, each of length 'l', carry the same current I. Wire A is bent into a circle of radius R and wire B is bent to form a square of side 'a'. If B_A and B_B are the values of magnetic field at the centres of the circle and square respectively, then the ratio $\frac{B_A}{B_B}$ is : **[2016]**

(A) $\frac{\pi^2}{8}$ (B) $\frac{\pi^2}{16\sqrt{2}}$
(C) $\frac{\pi^2}{16}$ (D) $\frac{\pi^2}{8\sqrt{2}}$

21. A current of 1 A is flowing on the sides of an equilateral triangles of side 4.5×10^{-2} m. The magnetic field at the centre of the triangle will be : **[2018 (Online)]**

(A) 4×10^{-5} Wb/m^2 (B) Zero
(C) 2×10^{-5} Wb/m^2 (D) 8×10^{-5} Wb/m^2

22. A Helmholtz coil has pair of loops, each with N turns and radius R. They are placed coaxially at distance R and the same current I flows through the loops in the same direction. The magnitude of magnetic field at P, midway between the centres A and C, is given by (Refer to figure) : **[2018 (Online)]**

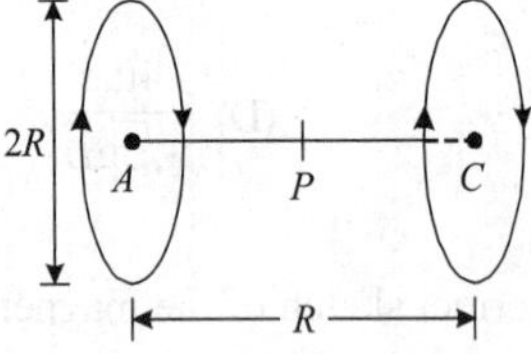

(A) $\frac{4N\mu_0 I}{5^{3/2} R}$ (B) $\frac{8N\mu_0 I}{5^{3/2} R}$
(C) $\frac{4N\mu_0 I}{5^{1/2} R}$ (D) $\frac{8N\mu_0 I}{5^{1/2} R}$

23. A current loop, having two circular arcs joined by two radial lines is shown in the figure. It carries a current of 10 A. The magnetic field at point O will be close to : **[9 Jan. 2019 (Morning)]**

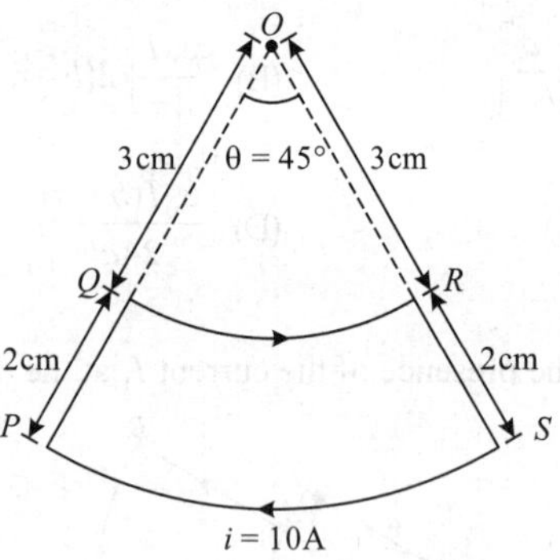

(A) 1.0×10^{-5} T (B) 1.5×10^{-5} T
(C) 1.0×10^{-7} T (D) 1.0×10^{-7} T

24. One of the two identical conducting wires of length L is bent in the form of a circular loop and the other one into a circular coil of N identical turns. If the same current is passed in both, the ratio of the magnetic field at the center of the loop (B_L) to that at the centre of the coil (B_C), then $\frac{B_L}{B_C}$ will be : **[9 Jan. 2019 (Evening)]**

(A) $\frac{1}{N}$ (B) N^2
(C) $\frac{1}{N^2}$ (D) N

25. As shown in the figure, two infinitely long, identical wires are bent by 90° and placed in such a way that the segments LP and QM are along the x-axis, while segments PS and QN are parallel to the y-axis. If $OP = OQ = 4$ cm, and the magnitude of the magnetic field at O is 10^{-4} T, and the two wires carry equal currents (see figure), the magnitude of the current in each wire and the direction of the magnetic field at O will be: ($\mu_0 = 4\pi \times 10^{-7}$ NA^{-2}) **[12 Jan. 2019 (Morning)]**

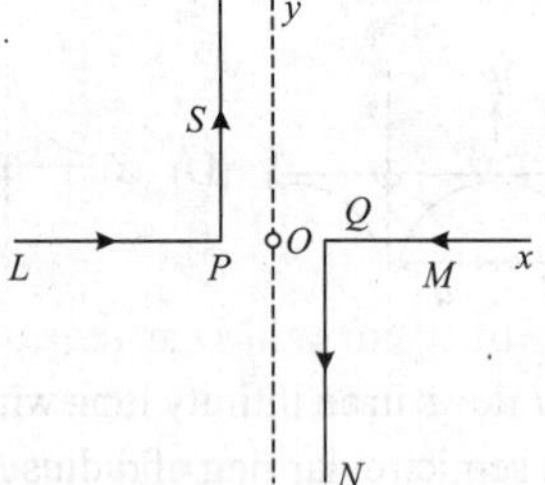

(A) 40 A, perpendicular into the page
(B) 40 A, perpendicular out of the page
(C) 20 A, perpendicular out of the page
(D) 20 A, perpendicular into the page

26. In SI units, the dimensions of $\sqrt{\frac{\epsilon_0}{\mu_0}}$ is : **[8 April 2019 (Morning)]**

(A) $A^{-1}\ TML^3$ (B) $A^2T^3M^{-1}L^{-2}$

(C) $AT^2M^{-1}L^{-1}$ (D) $AT^{-3}ML^{3/2}$

27. Two very long, straight, and insulated wires are kept at 90° angle from each other in xy-plane as shown in the figure. These wires carry currents of equal magnitude I, whose directions are shown in the figure. The net magnetic field at point P will be : **[8 April 2019 (Evening)]**

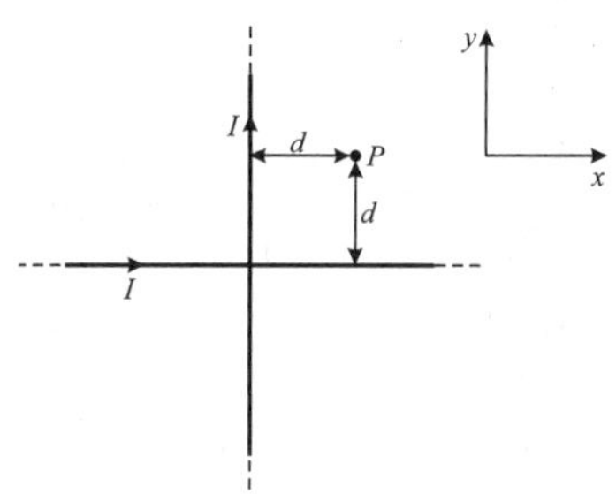

(A) Zero (B) $\frac{+\mu_0 I}{\pi d}(\hat{z})$

(C) $-\frac{\mu_0 I}{2\pi d}(\hat{x}+\hat{y})$ (D) $\frac{\mu_0 I}{2\pi d}(\hat{x}+\hat{y})$

28. The magnitude of the magnetic field at the center of an equilateral triangular loop of side 1 m which is carrying a current of 10 A is : [Take $\mu_0 = 4\pi \times 10^{-7}\ NA^{-2}$] **[10 April 2019 (Evening)]**

(A) 18 μT (B) 3 μT

(C) 1 μT (D) 9 μT

29. Which of the following combinations has the dimension of electrical resistance (ε_0 is the permittivity of vacuum and μ_0 is the permeability of vacuum) ? **[12 April 2019 (Morning)]**

(A) $\sqrt{\frac{\varepsilon_0}{\mu_0}}$ (B) $\frac{\mu_0}{\varepsilon_0}$

(C) $\sqrt{\frac{\mu_0}{\varepsilon_0}}$ (D) $\frac{\varepsilon_0}{\mu_0}$

30. A thin ring of 10 cm radius carries a uniformly distributed charge. The ring rotates at a constant angular speed of 40π rad s^{-1} about its axis, perpendicular to its plane. If the magnetic field at its centre is 3.8×10^{-9} T, then the charge carried by the ring is close to ($\mu_0 = 4\pi \times 10^{-7}\ N/A^2$) : **[12 April 2019 (Morning)]**

(A) 2×10^{-6} C (B) 3×10^{-5} C

(C) 4×10^{-5} C (D) 7×10^{-6} C

31. Find the magnetic field at point P due to a straight line segment AB of length 6 cm carrying a current of 5 A. (See figure) ($\mu_0 = 4\pi \times 10^{-7}$ N-A^{-2}) **[12 April 2019 (Evening)]**

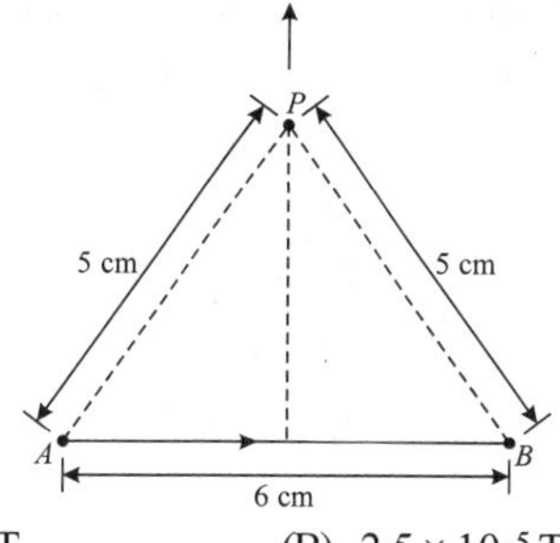

(A) 3.0×10^{-5} T (B) 2.5×10^{-5} T

(C) 2.0×10^{-5} T (D) 1.5×10^{-5} T

32. The dimension of $B^2/2\mu_0$, where B is magnetic field and μ_0 is the magnetic permeability of vacuum, is : **[7 Jan. 2020 (Evening)]**

(A) $ML^{-1}T^{-2}$ (B) ML^2T^{-1}

(C) MLT^{-2} (D) ML^2T^{-2}

33. A very long wire $ABDMNDC$ is shown in figure carrying current I. AB and BC parts are straight, long and at right angle. At D wire forms a circular turn $DMND$ of radius R. AB, BC parts are tangential to circular turn at N and D. Magnetic field at the centre of circle is : **[8 Jan. 2020 (Evening)]**

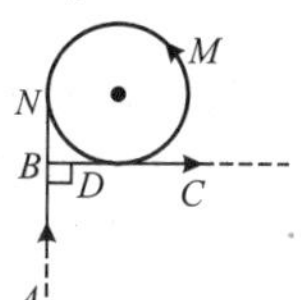

(A) $\frac{\mu_0 I}{2R}$

(B) $\frac{\mu_0 I}{2\pi R}(\pi+1)$

(C) $\frac{\mu_0 I}{2\pi R}\left(\pi+\frac{1}{\sqrt{2}}\right)$ (D) $\frac{\mu_0 I}{2\pi R}\left(\pi-\frac{1}{\sqrt{2}}\right)$

34. A long, straight wire of radius a carries a current distributed uniformly over its cross-section. The ratio of the magnetic fields due to the wire at distance $a/3$ and $2a$, respectively from the axis of the wire is : **[9 Jan. 2020 (Morning)]**

(A) $\frac{2}{3}$ (B) $\frac{3}{2}$

(C) $\frac{1}{2}$ (D) 2

35. Magnitude of magnetic field (in SI units) at the centre of a hexagonal shape coil of side 10 cm, 50 turns and carrying current I (Ampere) in units of $\mu_0 I/\pi$ is : **[3 Sept. 2020 (Morning)]**

(A) $250\sqrt{3}$ (B) $5\sqrt{3}$

(C) $500\sqrt{3}$ (D) $50\sqrt{3}$

36. A wire A, bent in the shape of an arc of a circle, carrying a current of $2A$ and having radius 2 cm and another wire B, also bent in the shape of arc of a circle, carrying a current of $3A$ and having radius of 4 cm, are placed as shown in the figure. The ratio of the magnetic fields due to the wires A and B at the common centre O is : **[4 Sept. 2020 (Morning)]**

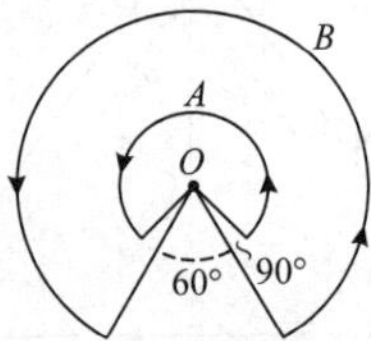

(A) 4:6 (B) 6:4
(C) 6:5 (D) 2:5

37. Magnetic fields at two points on the axis of a circular coil at a distance of 0.05 m and 0.2 m from the centre are in the ratio 8 : 1. The radius of coil is ______. **[25 Feb. 2021 (Morning)]**
(A) 0.1 m (B) 0.2 m
(C) 1.0 m (D) 0.15 m

38. A charge Q is moving $\vec{dl}$ distance in the magnetic field $\vec{B}$. Find the value of work done by $\vec{B}$: **[16 March 2021 (Evening)]**
(A) 1 (B) Infinite
(C) Zero (D) –1

39. A solenoid of 1000 turns per metre has a core with relative permeability 500. Insulated windings of the solenoid carry an electric current of 5 A. The magnetic flux density produced by the solenoid is : **[17 March 2021 (Morning)]**
(permeability of free space = $4\pi \times 10^{-7}$ H/m)
(A) πT (B) $2 \times 10^{-3}\,\pi$T
(C) $\frac{\pi}{5}$T (D) $10^{-4}\pi$T

40. A hairpin like shape as shown in figure is made by bending a long current carrying wire. What is the magnitude of a magnetic field at point P which lies on the centre of the semicircle ? **[17 March 2021 (Evening)]**

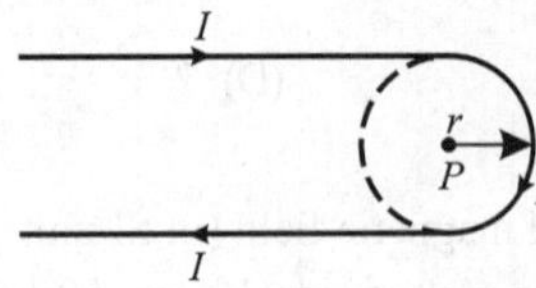

(A) $\frac{\mu_0 I}{4\pi r}(2-\pi)$ (B) $\frac{\mu_0 I}{4\pi r}(2+\pi)$
(C) $\frac{\mu_0 I}{2\pi r}(2+\pi)$ (D) $\frac{\mu_0 I}{2\pi r}(2-\pi)$

41. Figure-(A) and (B) show two long straight wires of circular cross-section (a and b with $a < b$), carrying current I which is uniformly distributed across the cross-section. The magnitude of magnetic field B varies with radius r and an be represented as : **[27 July 2021 (Evening)]**

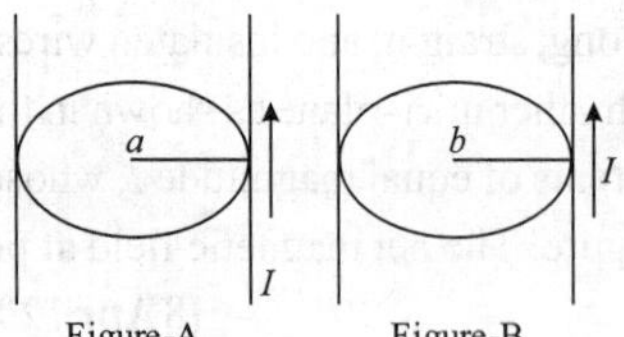

(A) (B)

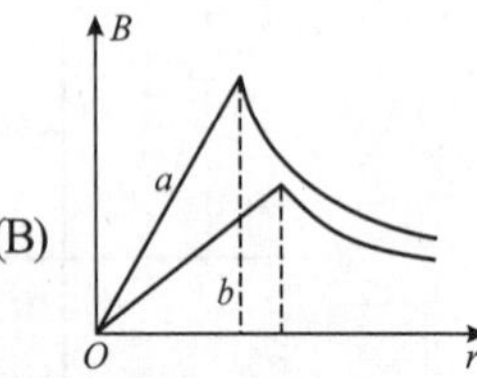

(C)

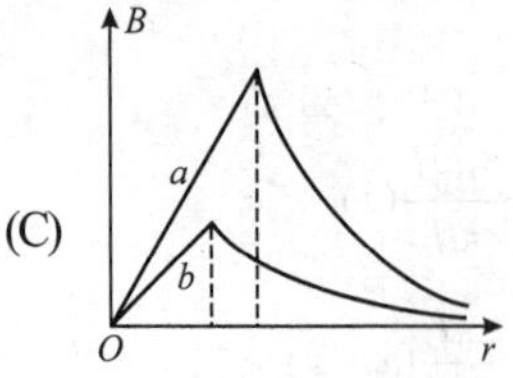

(D)

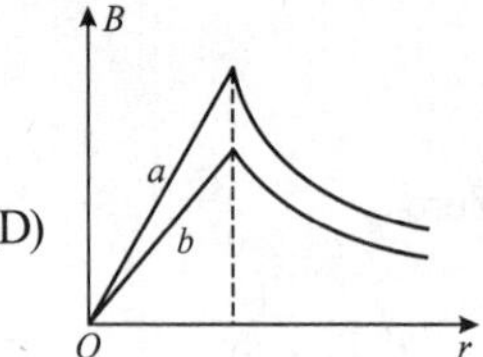

42. What will be the magnitude of electric field at point O as shown in figure ? Each side of the figure is l and perpendicular to each other ? **[27 July 2021 (Evening)]**

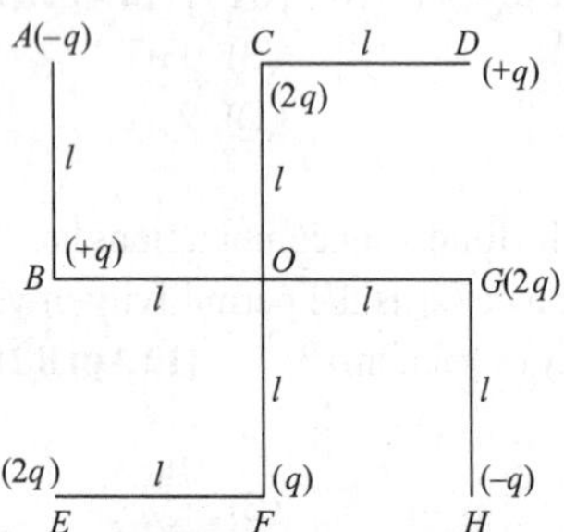

(A) $\frac{1}{4\pi\varepsilon_0}\frac{q}{(2l^2)}(2\sqrt{2}-1)$ (B) $\frac{1}{4\pi\varepsilon_0}\frac{q}{(2l^2)}(\sqrt{2})$
(C) $\frac{1}{4\pi\varepsilon_0}\frac{q}{l^2}$ (D) $\frac{1}{4\pi\varepsilon_0(2l)^2}$

43. The fractional change in the magnetic field intensity at distance 'r' from centre on the axis of current carrying coil of radius 'a' to the magnetic field intensity at centre of the same coil is : (Take $r << a$) **[26 Aug. 2021 (Morning)]**

(A) $\frac{2}{3}\frac{r^2}{a^2}$ (B) $\frac{3}{2}\frac{r^2}{a^2}$
(C) $\frac{2}{3}\frac{a^2}{r^2}$ (D) $\frac{3}{2}\frac{a^2}{r^2}$

44. A coaxial cable consists of an inner wire of radius a surrounded by an outer shell of inner and outer radii b and c respectively. The inner wire carries an electric current i_0, which is distributed uniformly across cross-sectional area. The outer shell carries an equal current in opposite direction and distributed uniformly. What will be the ratio of the magnetic field at a distance x from the axis when (i) $x < a$ and (ii) $a < x < b$?

[27 Aug. 2021 (Evening)]

(A) $\frac{a^2}{x^2}$ (B) $\frac{b^2 - a^2}{x^2}$

(C) $\frac{x^2}{b^2 - a^2}$ (D) $\frac{x^2}{a^2}$

45. A coil having N turns is wound tightly in the form of a spiral with inner and outer radii a and b respectively. Find the magnetic field at centre, when a current I passes through coil :

[31 Aug. 2021 (Morning)]

(A) $\frac{\mu_0 I}{a(a-b)}\left[\frac{1}{a} - \frac{1}{b}\right]$ (B) $\frac{\mu_0 IN}{2(b-a)}\log_e\left(\frac{b}{a}\right)$

(C) $\frac{\mu_0 I}{8}\left[\frac{a+b}{a-b}\right]$ (D) $\frac{\mu_0 I}{8}\left[\frac{a-b}{a-b}\right]$

46. A current of 1.5 A is flowing through a triangle, of side 9 cm each. The magnetic field at the centroid of the triangle is : (Assume that the current is flowing in the clockwise direction)

[31 Aug. 2021 (Evening)]

(A) $2\sqrt{3} \times 10^{-7}$ T, outside the plane of triangle

(B) 3×10^{-7} T, outside the plane of triangle

(C) $2\sqrt{3} \times 10^{-5}$ T, inside the plane of triangle

(D) 3×10^{-5} T, inside the plane of triangle

47. There are two infinitely long straight current carrying conductors and they are held at right angles to each other so that their common ends meet at the origin as shown in the figure given below. The ratio of current in both conductors is 1 : 1. The magnetic field at point P is ______.

[01 Sep. 2021 (Evening)]

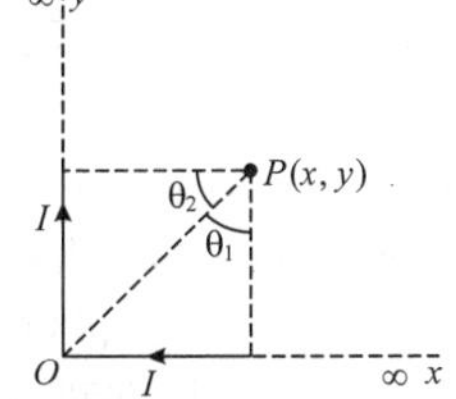

(A) $\frac{\mu_0 I}{4\pi xy}[\sqrt{x^2 + y^2} - (x + y)]$

(B) 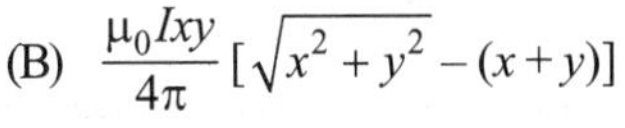 $\frac{\mu_0 Ixy}{4\pi}[\sqrt{x^2 + y^2} - (x + y)]$

(C) $\frac{\mu_0 Ixy}{4\pi}[\sqrt{x^2 + y^2} - (x + y)]$

(D) $\frac{\mu_0 I}{4\pi xy}[\sqrt{x^2 + y^2} + (x + y)]$

48. The magnetic field at the centre of a circular coil of radius r, due to current I flowing through it, is B. The magnetic field at a point along the axis at a distance $\frac{r}{2}$ from the centre is :

[24 June 2022 (Morning)]

(A) $B/2$ (B) $2B$

(C) $\left(\frac{2}{\sqrt{5}}\right)^3 B$ (D) $\left(\frac{2}{\sqrt{3}}\right)^3 B$

49. A long straight wire with a circular cross-section having radius R, is carrying a steady current I. The current I is uniformly distributed across this cross-section. Then the variation of magnetic field due to current I with distance $r (r < R)$ from its centre will be :

[25 June 2022 (Morning)]

(A) $B \propto r^2$ (B) $B \propto r$

(C) $B \propto \frac{1}{r^2}$ (D) $B \propto \frac{1}{r}$

50. A long solenoid carrying a current produces a magnetic field B along its axis. If the current is doubled and the number of turns per cm is halved, the new value of magnetic field will be equal to :

[25 June 2022 (Evening)]

(A) B (B) $2B$

(C) $4B$ (D) $\frac{B}{2}$

51. Two 10 cm long, straight wires, each carrying a current of 5 A are kept parallel to each other. If each wire experienced a force of 10^{-5} N, then separation between the wires is ______ cm.

[26 June 2022 (Evening)]

52. An infinitely long hollow conducting cylinder with radius R carries a uniform current along its surface. Choose the correct representation of magnetic field (B) as a function of radial distance (r) from the axis of cylinder : **[28 June 2022 (Morning)]**

(A) 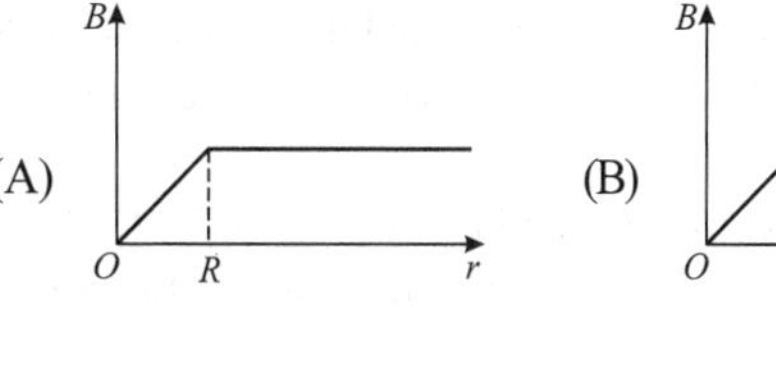

(B)

(C) 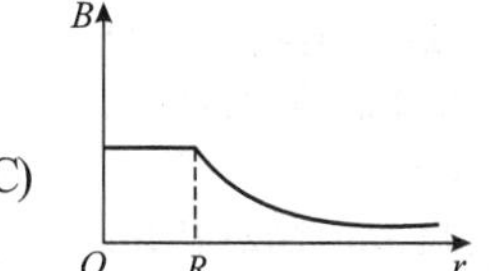

(D) 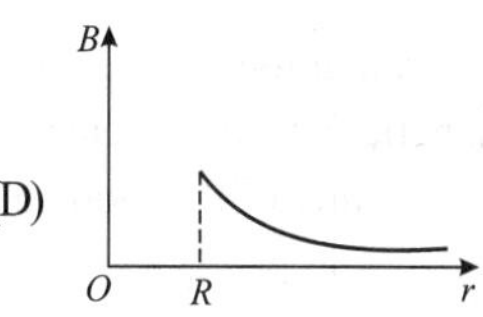

53. Two long current carrying conductors are placed parallel to each other at a distance of 8 cm between them. The magnitude of magnetic field produced at mid-point between the two conductors due to current flowing in them is 300 μT. The equal current flowing in the two conductors is :

[29 June 2022 (Evening)]

(A) 30 A in the same direction
(B) 30 A in the opposite direction
(C) 60 A in the opposite direction
(D) 300 A in the opposite direction

54. The electric current in a circular coil of 2 turns produces a magnetic induction B_1 at its centre. The coil is unwound and is rewound into a circular coil of 5 turns and the same current produces a magnetic induction B_2 at its centre. The ratio of $\frac{B_2}{B_1}$ is :

[25 July 2022 (Evening)]

(A) $\frac{5}{2}$ (B) $\frac{25}{4}$

(C) $\frac{5}{4}$ (D) $\frac{25}{2}$

55. B_X and B_Y are the magnetic field at the centre of two coils of two coils X and Y respectively, each carrying equal current. If coil X has 200 turns and 20 cm radius and coil Y has 400 turns and 20 cm radius, the ratio of B_X and B_Y is :

[26 July 2022 (Morning)]

(A) 1 : 1 (B) 1 : 2
(C) 2 : 1 (D) 4 : 1

56. The magnetic field at the center of current carrying circular loop is B_1. The magnetic field at a distance of $\sqrt{3}$ times radius of the given circular loop from the center on its axis is B_2. The value of B_1/B_2 will be :

[28 July 2022 (Evening)]

(A) 9 : 4 (B) $12\sqrt{5}$
(C) 8 : 1 (D) $5 : \sqrt{3}$

57. A closely wounded circular coil of radius 5 cm produces a magnetic field of 37.68×10^{-4} T at its center. The current through the coil is ______ A.
[Given : Number of turns in the coil is 100 and $\pi = 3.14$]

[29 July 2022 (Morning)]

58. A circular loop of radius r is carrying current I A. The ratio of magnetic field at the centre of circular loop and at a distance r from the center of the loop on its axis is :

[24 Jan. 2023 (Morning)]

(A) $1 : 3\sqrt{2}$ (B) $3\sqrt{2} : 2$

(C) $2\sqrt{2} : 1$ (D) $1 : \sqrt{2}$

59. A long solenoid is formed by winding 70 turns cm^{-1}. If 2.0 A current flows, then the magnetic field produced inside the solenoid is ______. ($\mu_0 = 4\pi \times 10^{-7}$ TmA^{-1})

[24 Jan. 2023 (Evening)]

(A) 1232×10^{-4} T (B) 176×10^{-4} T
(C) 352×10^{-4} T (D) 88×10^{-4} T

60. Match List-I with List-II : **[25 Jan. 2023 (Morning)]**

List-I (Current configuration)	**List-II (Magnetic field at point O)**
(a) 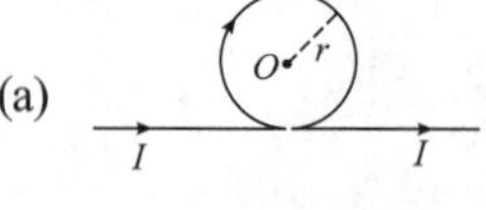	(i) $B_0 = \frac{\mu_0 I}{4\pi r}[\pi + 2]$
(b)	(ii) $B_0 = \frac{\mu_0}{4}\frac{I}{r}$
(c) 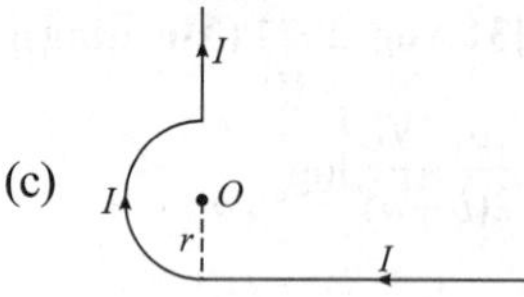	(iii) $B_0 = \frac{\mu_0 I}{2\pi r}[\pi - 1]$
(d)	(iv) $B_0 = \frac{\mu_0 I}{4\pi r}[\pi + 1]$

Choose the correct answer from the option given below :
(A) (a)-(iii), (b)-(iv), (c)-(i), (d)-(ii)
(B) (a)-(i), (b)-(iii), (c)-(iv), (d)-(ii)
(C) (a)-(iii), (b)-(i), (c)-(iv), (d)-(ii)
(D) (a)-(ii), (b)-(i), (c)-(iv), (d)-(iii)

61. A solenoid of 1200 turns is wound uniformly in a single layer on a glass tube 2 m long and 0.2 m in diameter. The magnetic intensity at the center of the solenoid when a current of 2 A flows through it is :

[25 Jan. 2023 (Morning)]

(A) 2.4×10^3 Am^{-1} (B) 1.2×10^3 Am^{-1}
(C) 1 Am^{-1} (D) 2.4×10^{-3} Am^{-1}

62. Two long parallel wires carrying currents 8 A and 15 A in opposite directions are placed at a distance of 7 cm from each other. A point P is at equidistant from both the wires such that the lines joining the point P to the wires are perpendicular to each other. The magnitude of magnetic field at P is ______ $\times 10^{-6}$ T. (Given : $\sqrt{2} = 1.4$)

[25 Jan. 2023 (Evening)]

63. A single current carrying loop of wire carrying current I flowing in anticlockwise direction seen from $+ve$ z-direction and lying in xy plane is shown in figure. The plot of $\hat{j}$ component of magnetic field (B_y) at a distance a (less than the radius of the coil) and on yz plane vs z coordinate look like :

[29 Jan. 2023 (Morning)]

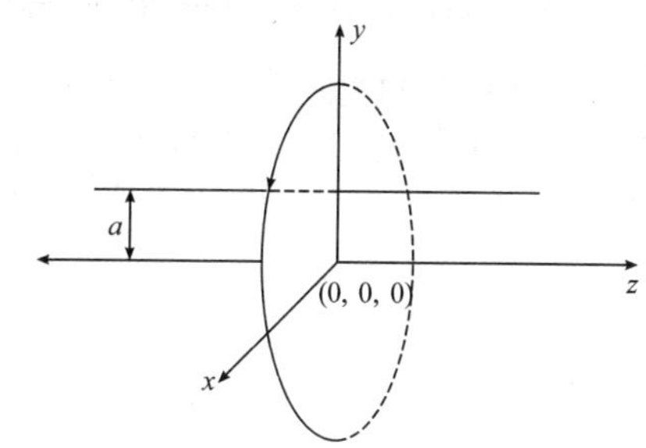

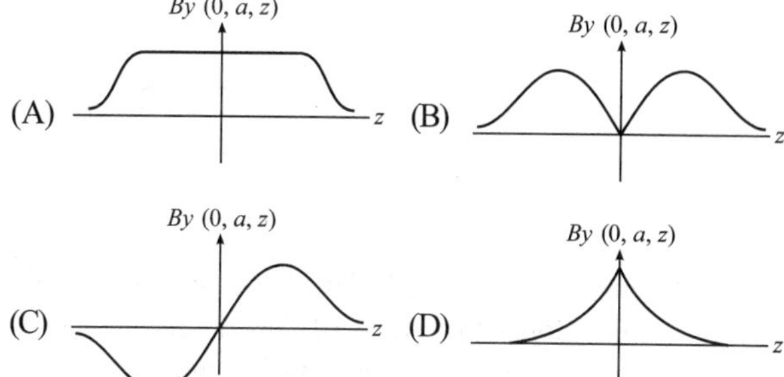

64. The magnitude of magnetic induction at midpoint O due to current arrangement as shown in figure will be :

[29 Jan. 2023 (Morning)]

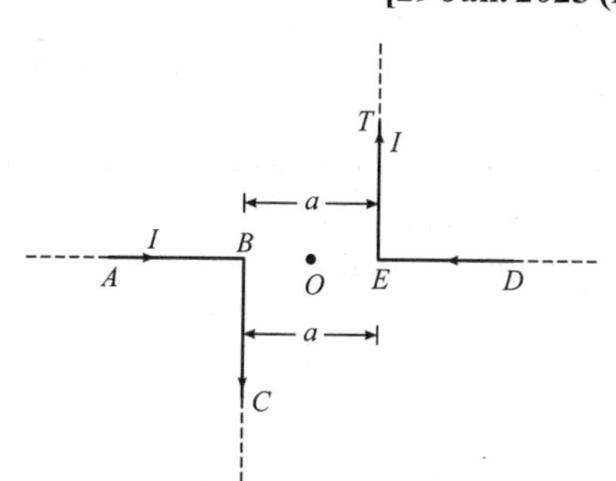

(A) $\frac{\mu_0 I}{2\pi a}$ (B) 0

(C) $\frac{\mu_0 I}{4\pi a}$ (D) $\frac{\mu_0 I}{\pi a}$

65. The electric current in a circular coil of four turns produces a magnetic induction 32 T at its centre. The coil is unwound and is rewound into a circular coil of single turn, the magnetic induction at the centre of the coil by the same current will be :

[29 Jan. 2023 (Evening)]

(A) 8 T (B) 4 T
(C) 2 T (D) 16 T

66. As shown in the figure, a current of 2 A flowing in an equilateral triangle of side $4\sqrt{3}$ cm. The magnetic field at the centroid O of the triangle is :

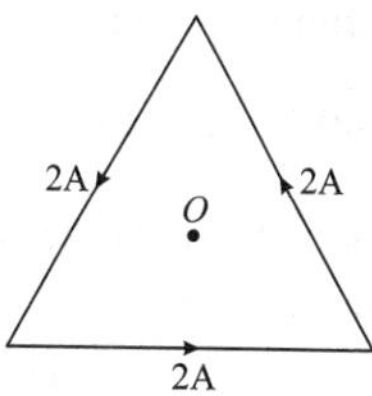

(Neglect the effect of earth's magnetic field)

[30 Jan. 2023 (Evening)]

(A) $4\sqrt{3} \times 10^{-4}$ T (B) $4\sqrt{3} \times 10^{-5}$ T

(C) $\sqrt{3} \times 10^{-4}$ T (D) $3\sqrt{3} \times 10^{-5}$ T

67. A rod with circular cross-section area 2 cm^2 and length 40 cm is wound uniformly with 400 turns of an insulated wire. If a current of 0.4 A flows in the wire windings, the total magnetic flux produced inside windings is $4\pi \times 10^{-6}$ Wb. The relative permeability of the rod is :

(Given : Permeability of vacuum $\mu_0 = 4\pi \times 10^{-7}$ Na^{-2})

[31 Jan. 2023 (Morning)]

(A) 12.5 (B) $\frac{32}{5}$

(C) 125 (D) $\frac{5}{16}$

68. A long conducting wire having a current I flowing through it, is bent into a circular coil of N turns. Then it is bent into a circular coil of n turns. The magnetic field is calculated at the centre of coils in both the cases. The ratio of the magnetic field in first case to that of second case is :

[31 Jan. 2023 (Evening)]

(A) $N : n$ (B) $n^2 : N^2$
(C) $N^2 : n^2$ (D) $n : N$

69. Find the magnetic field at the point P in figure. The curved portion is a semicircle connected to two long straight wires :

[1 Feb. 2023 (Morning)]

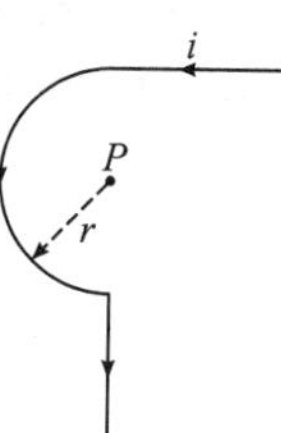

(A) $\frac{\mu_0 i}{2r}\left(1+\frac{2}{\pi}\right)$ (B) $\frac{\mu_0 i}{2r}\left(1+\frac{1}{\pi}\right)$

(C) $\frac{\mu_0 i}{2r}\left(\frac{1}{2}+\frac{1}{2\pi}\right)$ (D) $\frac{\mu_0 i}{2r}\left(\frac{1}{2}+\frac{1}{\pi}\right)$

70. As shown in the figure, a long straight conductor with semicircular arc of radius $\frac{\pi}{10} m$ is carrying current $I = 3$ A. The magnitude of the magnetic field. At the center O of the arc is : (The permeability of the vacuum $= 4\pi \times 10^{-7}$ NA^{-2})

[1 Feb. 2023 (Evening)]

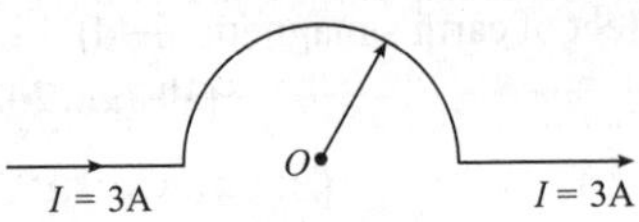

(A) 6 μT (B) 1 μT
(C) 4 μT (D) 3 μT

71. A coil is placed in magnetic field such that plane of coil is perpendicular to the direction of magnetic field. The magnetic flux through a coil can be changed :

(a) By changing the magnitude of the magnetic field within the coil

(b) By changing the area of coil within the magnetic field

(c) By changing the angle between the direction of magnetic field and the plane of the coil

(d) By reversing the magnetic field direction abruptly without changing its magnitude

Choose the most appropriate answer from the options given below : **[1 Feb. 2023 (Evening)]**

(A) (a) and (b) only (B) (a), (b) and (c) only
(C) (a), (b) and (d) only (D) (a) and (c) only

72. A long straight wire of circular cross-section (radius a) is carrying steady current I. The current I is uniformly distributed across this cross-section. The magnetic field is

[6 April 2023 (Morning)]

(A) Zero in the region $r < a$ and inversely proportional to r in the region $r > a$

(B) Inversely proportional to r in the region $r < a$ and uniform throughout in the region $r > a$

(C) Directly proportional to r in the region $r < a$ and inversely proportional to r in the region $r > a$

(D) Uniform in the region $r < a$ and inversely proportional to distance r from the axis, in the region $r > a$

73. Two identical circular wires of radius 20 cm and carrying current $\sqrt{2}$ A are placed in perpendicular planes as shown in figure. The net magnetic field at the centre of the circular wire is ______ $\times 10^{-8}$ T. (Take $\pi = 3.14$) **[6 April 2023 (Morning)]**

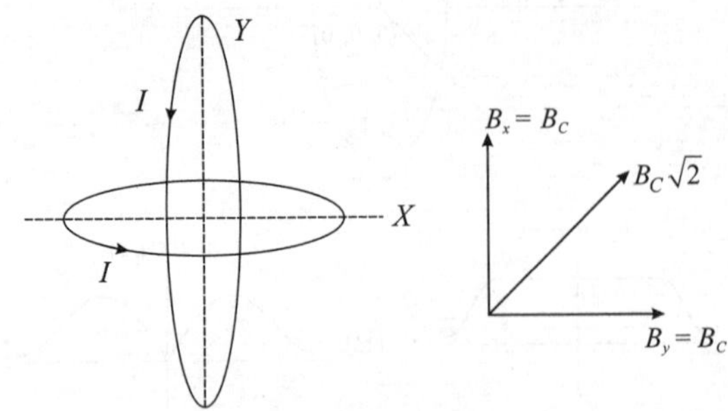

74. The magnetic intensity at the centre of a long current carrying solenoid is found to be 1.6×10^3 Am^{-1}. If the number of turns is 8 per cm, then the current flowing through the solenoid is______ A. **[8 April 2023 (Morning)]**

75. The ratio of magnetic field at the centre of a current carrying coil of radius r to the magnetic field at distance r from the centre of coil on its axis is $\sqrt{x}:1$. The value of x is ______.

[8 April 2023 (Evening)]

76. A straight wire carrying a current of 14 A is bent into a semicircular are of radius 2.2 cm as shown in the figure. The magnetic field produced by the current at the centre (O) of the arc is ______ $\times 10^{-4}$ T. **[10 April 2023 (Evening)]**

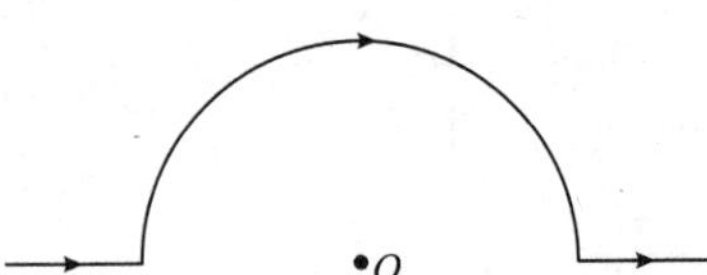

77. An electron in a hydrogen atom revolves around its nucleus with a speed of 6.76×10^6 ms^{-1} in an orbit of radius 0.52 A°. The magnetic field produced at the nucleus of the hydrogen atom is ______ T. **[15 April 2023 (Morning)]**

* * * * *

CHAPTER

18 MOVING CHARGES & MAGNETISM

18.2 MAGNETIC FORCE ON MOVING CHARGES

1. Wires 1 and 2 carrying currents i_1 and i_2 respectively are inclined at an angle θ to each other. What is the force on a small element dl of wire 2 at a distance r from wire 1 (as shown in figure) due to the magnetic field of wire 1 ? **[2002]**

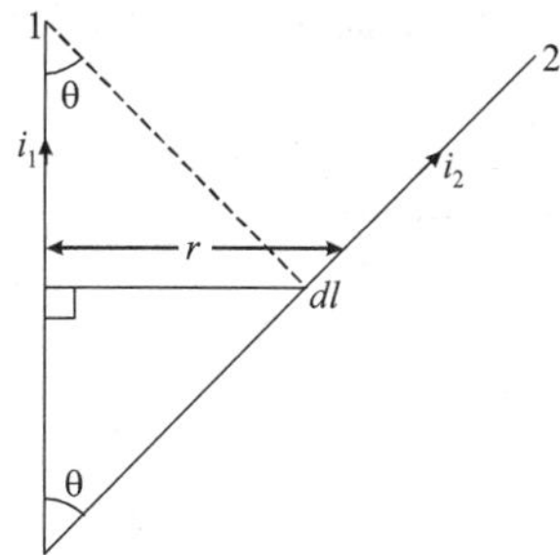

(A) $\frac{\mu_0}{2\pi r} i_1 i_2 \, dl \tan\theta$ (B) $\frac{\mu_0}{2\pi r} i_1 i_2 \, dl \sin\theta$

(C) $\frac{\mu_0}{2\pi r} i_1 i_2 \, dl \cos\theta$ (D) $\frac{\mu_0}{4\pi r} i_1 i_2 \, dl \sin\theta$

2. The time period of a charged particle undergoing a circular motion in a uniform magnetic field is independent of its : **[2002]**

(A) Speed (B) Mass

(C) Charge (D) Magnetic induction

3. If a current is passed through a spring then the spring will : **[2002]**

(A) Expand (B) Compress

(C) Remain same (D) None of these

4. If an electron and a proton having same momentum enter perpendicularly to a magnetic field, then : **[2002]**

(A) Curved path of electron and proton will be same (ignoring the sense of revolution)

(B) They will move undeflected

(C) Curved path of electron is more curved than that of proton

(D) Path of proton is more curved

5. A particle of charge -16×10^{-18} C moving with velocity 10 ms^{-1} along the x-axis enters a region where a magnetic field of induction B is along the y-axis and an electric field of magnitude 10^4 V/m is along the negative z-axis If the charged particle continues moving along the x-axis, the magnitude of B is : **[2003]**

(A) 10^3 Wb/m^2 (B) 10^5 Wb/m^2

(C) 10^{16} Wb/m^2 (D) 10^{-3} Wb/m^2

6. A magnetic needle laying parallel to a magnetic field requires W unit of work to turn it through 60°. The torque needed to maintain the needle in this position will be : **[2003]**

(A) $\sqrt{3}\,W$ (B) W

(C) $(\sqrt{3}/2)\,W$ (D) $2\,W$

7. A particle of mass M and charge Q moving with velocity $\vec{v}$ describes a circular path of radius R when subjected to a uniform transverse magnetic field of induction B. The work done by the field when the particle completes one full circle is : **[2003]**

(A) $\left(\frac{Mv^2}{R}\right) 2\pi R$ (B) Zero

(C) $BQ\,2\pi R$ (D) $BQv\,2\pi R$

8. Two long conductors, separated by a distance d carry currents I_1 and I_2 in the same direction. They exert a force F on each other. Now the current in one of them is increased to two times and its direction is reversed. The distance is also increased to $3d$. The new value of the force between them is : **[2004]**

(A) $-2F$ (B) $F/3$

(C) $-2F/3$ (D) $-F/3$

9. A charged particle of mass m and charge q travels on a circular path of radius r that is perpendicular to a magnetic field B. The time taken by the particle to complete one revolution is : **[2005]**

(A) $\frac{2\pi mq}{B}$ (B) $\frac{2\pi q^2 B}{m}$

(C) $\frac{2\pi qB}{m}$ (D) $\frac{2\pi m}{qB}$

10. Two thin, long, parallel wires, separated by a distance d carry a current of 'i' in the same direction. They will : **[2005]**

(A) attract each other with a force of $\frac{\mu_0 i^2}{(2\pi d)}$

(B) repel each other with a force of $\frac{\mu_0 i^2}{(2\pi d)}$

(C) attract each other with a force of $\frac{\mu_0 i^2}{(2\pi d^2)}$

(D) repel each other with a force of $\frac{\mu_0 i^2}{(2\pi d^2)}$

11. A uniform electric field and a uniform magnetic field are acting along the same direction in a certain region. If an electron is projected along the direction of the fields with a certain velocity then : **[2005]**

(A) Its velocity will increase
(B) Its velocity will decrease
(C) It will turn towards left of direction of motion
(D) It will turn towards right of direction of motion

12. In a region, steady and uniform electric and magnetic fields are present. These two fields are parallel to each other. A charged particle is released from rest in this region. The path of the particle will be a : **[2006]**

(A) helix (B) straight line
(C) ellipse (D) circle

13. A charged particle moves through a magnetic field perpendicular to its direction. Then : **[2007]**

(A) the momentum changes but the kinetic energy is constant
(B) both momentum and kinetic energy of the particle are not constant
(C) both momentum and kinetic energy of the particle are constant
(D) kinetic energy changes but the momentum is constant

14. A charged particle with charge q enters a region of constant, uniform and mutually orthogonal fields $\vec{E}$ and $\vec{B}$ with a velocity $\vec{V}$ perpendicular to both $\vec{E}$ and $\vec{B}$, and comes out without any change in magnitude or direction of $\vec{V}$. Then : **[2007]**

(A) $\vec{V} = \dfrac{\vec{E}\times\vec{B}}{B^2}$ (B) $\vec{V} = \dfrac{\vec{B}\times\vec{E}}{B^2}$
(C) $\vec{V} = \dfrac{\vec{E}\times\vec{B}}{E^2}$ (D) $\vec{V} = \dfrac{\vec{B}\times\vec{E}}{E^2}$

15. A thin circular disk of radius R is uniformly charged with density $\sigma > 0$ per unit area. The disk rotates about its axis with a uniform angular speed ω. The magnetic moment of the disk is : **[2011]**

(A) $2\pi R^4\sigma\omega$ (B) $\pi R^4\sigma\omega$
(C) $\dfrac{\pi R^4}{2}\sigma\omega$ (D) $\dfrac{\pi R^4}{4}\sigma\omega$

16. An electric charge $+q$ moves with velocity $\vec{V} = 3\hat{i} + 4\hat{j} + \hat{k}$, in an electromagnetic field given by : $\vec{E} = 3\hat{i} + \hat{j} + 2\hat{k}$. $\vec{B} = \hat{i} + \hat{j} - 3\hat{k}$. The y-component of the force experienced by $+q$ is : **[2011]**

(A) $2q$ (B) $11q$
(C) $5q$ (D) $3q$

17. Proton, Deuteron and alpha particle of the same kinetic energy are moving in circular trajectories in a constant magnetic field. The radii of proton, deuteron and alpha particle are respectively r_p, r_d and r_α. Which one of the following relations is correct ? **[2012]**

(A) $r_\alpha = r_d > r_p$ (B) $r_\alpha = r_d = r_p$
(C) $r_\alpha = r_p < r_d$ (D) $r_\alpha > r_d > r_p$

18. The circuit in figure consists of wires at the top and bottom and identical springs as the left and right sides. The wire at the bottom has a mass of 10 g and is 5 cm long. The wire is hanging as shown in the figure. The springs stretch 0.5 cm under the weight of the wire and the circuit has a total resistance of 12 Ω. When the lower wire is subjected to a static magnetic field, the springs, stretch an additional 0.3 cm. The magnetic field is : **[2012 (Online)]**

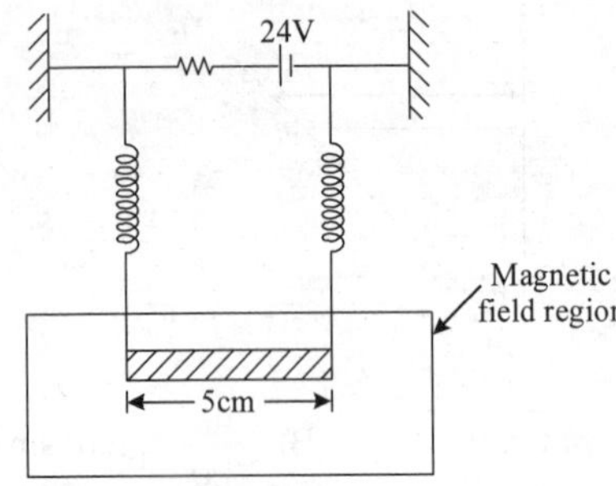

(A) 0.6 T and directed out of page
(B) 1.2 T and directed into the plane of page
(C) 0.6 T and directed into the plane of page
(D) 1.2 T and directed out of page

19. Currents of a 10 ampere and 2 ampere are passed through two parallel thin wires A and B respectively in opposite directions. Wire A is infinitely long and the length of the wire B is 2 m. The force acting on the conductor B, which is situated at 10 cm distance from A will be : **[2012 (Online)]**

(A) 8×10^{-5} N (B) 5×10^{-5} N
(C) $8\pi \times 10^{-7}$ N (D) $4\pi \times 10^{-7}$ N

20. The velocity of certain ions that pass undeflected through crossed electric field $E = 7.7$ k V/m and magnetic field $B = 0.14$ T is : **[2012 (Online)]**

(A) 18 km/s (B) 77 km/s
(C) 55 km/s (D) 1078 km/s.

21. The magnetic force acting on charged particle of charge 2 μC in magnetic field of 2T acting in y-direction, when the particle velocity is $(2\hat{i} + 3\hat{j}) \times 10^6$ ms^{-1} is : **[2012 (Online)]**

(A) 8 N in z-direction (B) 8 N in y-direction
(C) 4 N in y-direction (D) 4 N in z-direction

22. A proton and a deuteron are both accelerated through the same potential difference and enter in a magnetic field perpendicular to the direction of the field. If the deuteron follows a path of radius R, assuming the neutron and proton masses are nearly equal, the radius of the proton's path will be : **[2012 (Online)]**

(A) $\sqrt{2}R$ (B) $\frac{R}{\sqrt{2}}$

(C) $\frac{R}{2}$ (D) R

23. This question has Statement-1 and Statement-2. Of the four choices given after the Statements, choose the one that best describes the two Statements.

Statement-1 : A charged particle is moving at right angle to a static magnetic field. During the motion the kinetic energy of the charge remains unchanged.

Statement-2 : Static magnetic field exert force on a moving charge in the direction perpendicular to the magnetic field : **[2012 (Online)]**

(A) Statement-1 is false, Statement-2 is true

(B) Statement-1 is true, Statement-2 is true, Statement-2 is not the correct explanation of Statement-1

(C) Statement-1 is true, Statement-2 is false

(D) Statement-1 is true, Statement-2 is true, Statement-2 is the correct explanation of Statement-1

24. A rectangular loop of wire, supporting a mass m, hangs with one end in a uniform magnetic field B pointing out of the plane of the paper. A clockwise current is set up such that $i > mg/Ba$, where a is the width of the loop. Then : **[2013 (Online)]**

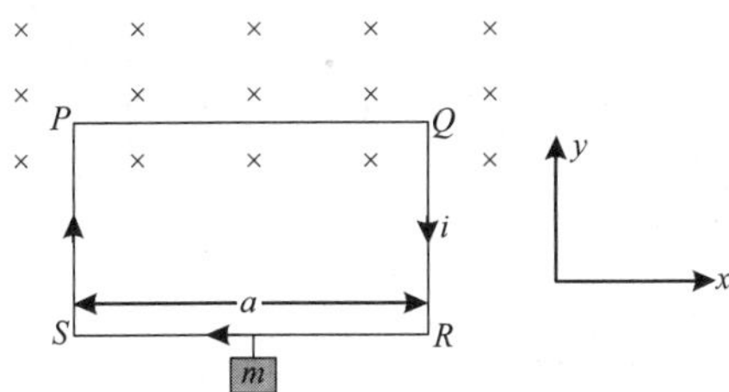

(A) The weight rises due to a vertical force caused by the magnetic field and work is done on the system

(B) The weight do not rise due to vertical for caused by the magnetic field and work is done on the system

(C) The weight rises due to a vertical force caused by the magnetic field but no work is done on the system

(D) The weight rises due to a vertical force caused by the magnetic field and work is extracted from the magnetic field

25. A particle of charge 16×10^{-16} C moving with velocity 10 ms^{-1} along x-axis enters a region where magnetic field of induction $\vec{B}$ is along the y-axis and an electric field of magnitude 10^4 Vm^{-1} is along the negative z-axis. If the charged particle continues moving along x-axis, the magnitude of $\vec{B}$ is : **[2013 (Online)]**

(A) 16×10^3 Wb m^{-2} (B) 2×10^3 Wb m^{-2}

(C) 1×10^3 Wb m^{-2} (D) 4×10^3 Wb m^{-2}

26. A metal sample carrying a current along X-axis with density J_X is subjected to magnetic field B_Z (along Z-axis). The electric field E_Y developed along Y-axis is directly proportional to J_X as well as B_Z. The constant of proportionality has SI unit : **[2013 (Online)]**

(A) $\frac{\text{m}^2}{\text{A}}$ (B) $\frac{\text{m}^3}{\text{As}}$

(C) $\frac{\text{m}^2}{\text{As}}$ (D) $\frac{\text{As}}{\text{m}^3}$

27. A conductor lies along the z-axis at $-1.5 \le z < 1.5$ m and carries a fixed current of 10.0 A in $-\hat{a}_Z$ direction (see figure). For a field $\vec{B} = 3.0 \times 10^{-4} e^{-0.2x} \hat{a}_y$ T, find the power required to move the conductor at constant speed to $x = 2.0$ m, $y = 0$ m in 5×10^{-3} s. Assume parallel motion along the x-axis: **[2014]**

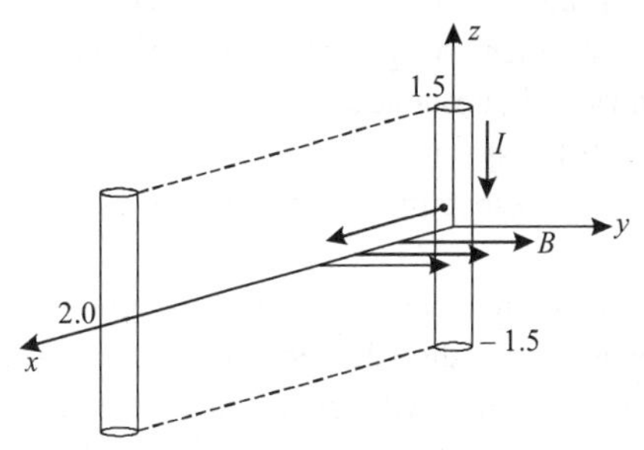

(A) 1.57 W (B) 2.97 W

(C) 14.85 W (D) 29.7 W

28. Three straight parallel current carrying conductors are shown in the figure. The force experienced by the middle conductor of length 25 cm is : **[2014 (Online)]**

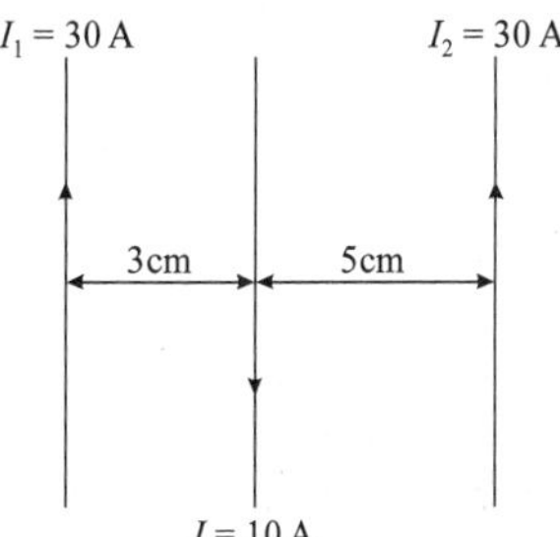

(A) 3×10^{-4} N toward right (B) 6×10^{-4} N toward right

(C) 9×10^{-4} N toward right (D) Zero

29. A positive charge 'q' of mass 'm' is moving along the $+x$ axis. We wish to apply a uniform magnetic field B for time Δt so that the charge reverses its direction crossing the y-axis at a distance d. Then : **[2014 (Online)]**

(A) $B = \frac{mv}{qd}$ and $\Delta t = \frac{\pi d}{v}$ (B) $B = \frac{mv}{2qd}$ and $\Delta t = \frac{\pi d}{2v}$

(C) $B = \frac{2mv}{qd}$ and $\Delta t = \frac{\pi d}{2v}$ (D) $B = \frac{2mv}{qd}$ and $\Delta t = \frac{\pi d}{v}$

30. A rectangular loop of sides 10 cm and 5 cm carrying a current I of 12 A is placed in different orientations as shown in the figures below :

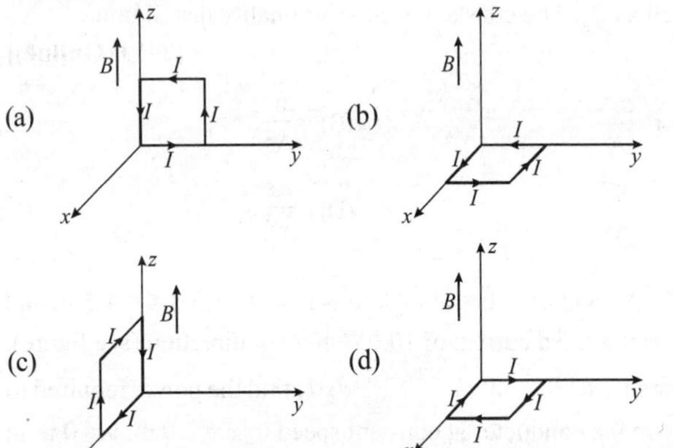

If there is a uniform magnetic field of 0.3 T in the positive z direction, in which orientations the loop would be in (i) stable equilibrium and (ii) unstable equilibrium? **[2015]**

(A) (a) and (b), respectively (B) (a) and (c), respectively
(C) (b) and (d), respectively (D) (b) and (c), respectively

31. Two long current carrying thin wires, both with current I, are held by insulating threads of length L and are in equilibrium as shown in the figure, with threads making an angle 'θ' with the vertical. If wires have mass λ per unit length then the value of I is : (g = gravitational acceleration) : **[2015]**

(A) $\sin\theta\sqrt{\dfrac{\pi\lambda gL}{\mu_0\cos\theta}}$

(B) $2\sin\theta\sqrt{\dfrac{\pi\lambda gL}{\mu_0\cos\theta}}$

(C) $2\sqrt{\dfrac{\pi gL}{\mu_0}\tan\theta}$

(D) $\sqrt{\dfrac{\pi\lambda gL}{\mu_0}\tan\theta}$

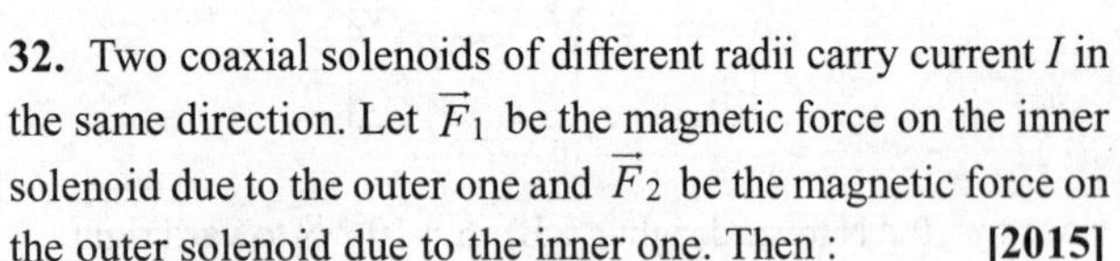

32. Two coaxial solenoids of different radii carry current I in the same direction. Let $\vec{F}_1$ be the magnetic force on the inner solenoid due to the outer one and $\vec{F}_2$ be the magnetic force on the outer solenoid due to the inner one. Then : **[2015]**

(A) $\vec{F}_1 = \vec{F}_2 = 0$

(B) $\vec{F}_1$ is radially inwards and $\vec{F}_2$ is radially outwards

(C) $\vec{F}_1$ is radially inwards and $\vec{F}_2 = 0$

(D) $\vec{F}_1$ is radially outwards and $\vec{F}_2 = 0$

33. A wire carrying current I is tied between points P and Q and is in the shape of a circular arc of radius R due to a uniform magnetic field B (perpendicular to the plane of the paper, shown by ×××) in the vicinity of the wire. If the wire subtends an angle $2\theta_0$ at the centre of the circle (of which it forms an arc) then the tension in the wire is : **[2015 (Online)]**

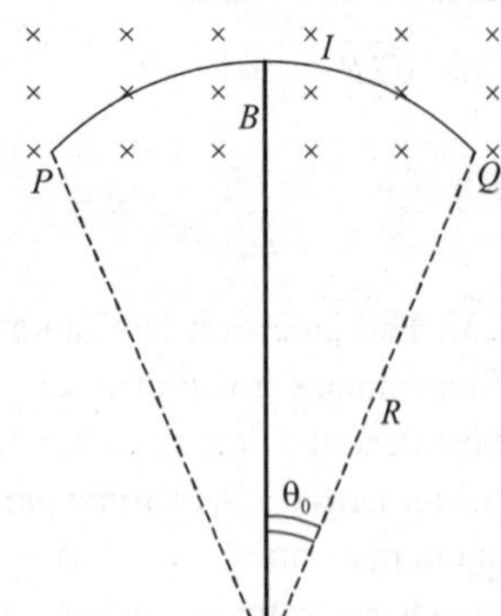

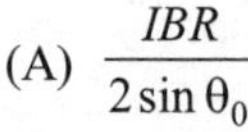

(A) $\dfrac{IBR}{2\sin\theta_0}$

(B) $\dfrac{IBR\theta_0}{\sin\theta_0}$

(C) IBR

(D) $\dfrac{IBR}{\sin\theta_0}$

34. Two long straight parallel wires, carrying (adjustable) current I_1 and I_2, are kept at a distance d apart. If the force 'F' between the two wires is taken as 'positive' when the wires repel each other and 'negative' when the wires attract each other, the graph showing the dependence of 'F', on the product $I_1 I_2$, would be : **[2015 (Online)]**

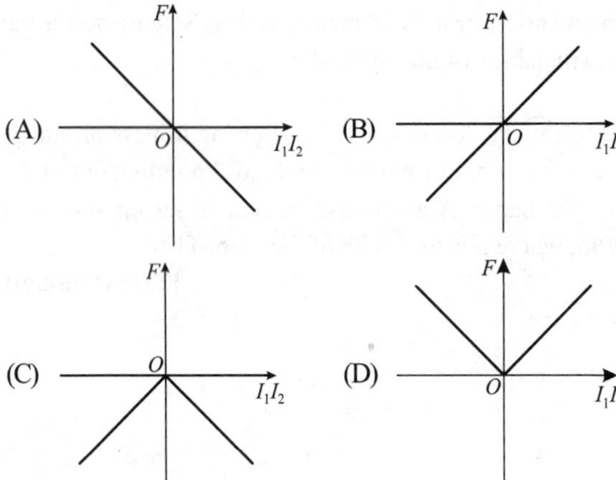

35. A proton (mass m) accelerated by a potential difference V flies through a uniform transverse magnetic field B. The field occupies a region of space by width 'd'. If α be the angle of deviation of proton from initial direction of motion (see figure), the value of $\sin\alpha$ will be : **[2015 (Online)]**

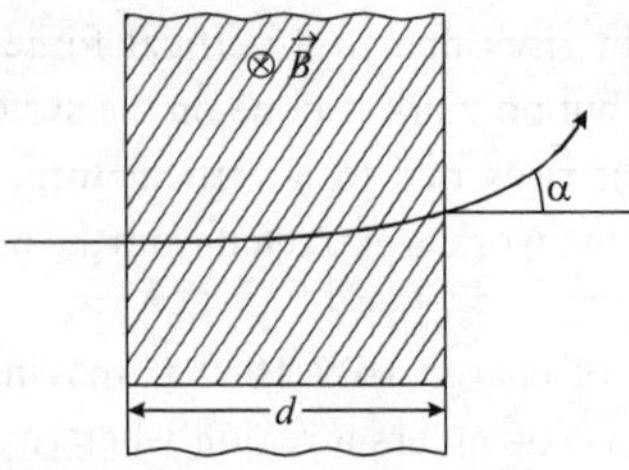

(A) $qV\sqrt{\dfrac{Bd}{2m}}$

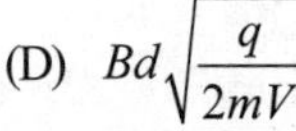

(B) $\dfrac{B}{2}\sqrt{\dfrac{qd}{mV}}$

(C) $\dfrac{B}{d}\sqrt{\dfrac{q}{2mV}}$

(D) $Bd\sqrt{\dfrac{q}{2mV}}$

36. A uniform magnetic field B of 0.3 T is along the positive Z-direction. A rectangular loop ($abcd$) of sides 10 cm × 5 cm carries a current I of 12 A. Out of the following different orientations which one corresponds to stable equilibrium? **[2017 (Online)]**

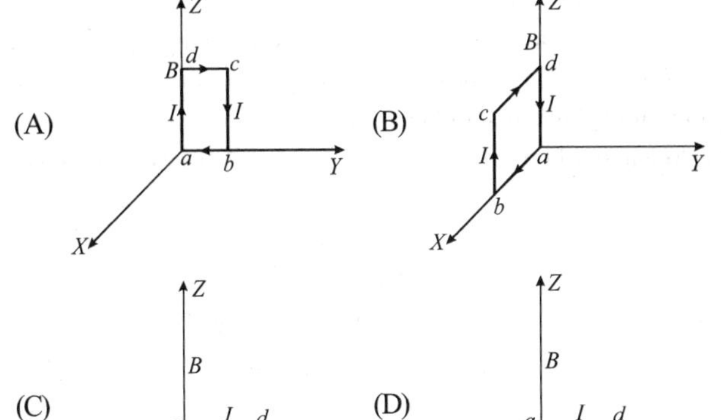

(A) (B) (C) (D)

37. In a certain region static electric and magnetic fields exist. The magnetic field is given by $\vec{B} = B_0(\hat{i} + 2\hat{j} - 4\hat{k})$. If a test charge moving with a velocity $\vec{v} = v_0(3\hat{i} - \hat{j} + 2\hat{k})$ experiences no force in that region, then the electric field in the region, in SI units, is : **[2017 (Online)]**

(A) $\vec{E} = -v_0B_0(3\hat{i} - 2\hat{j} - 4\hat{k})$

(B) $\vec{E} = -v_0B_0(\hat{i} + \hat{j} + 7\hat{k})$

(C) $\vec{E} = v_0B_0(14\hat{j} + 7\hat{k})$

(D) $\vec{E} = -v_0B_0(14\hat{j} + 7\hat{k})$

38. A negative test charge is moving near a long straight wire. Carrying a current. The force acting on the test charge is parallel to the direction of the current. The motion of the charge is : **[2017 (Online)]**

(A) Away from the wire

(B) Towards the wire

(C) Parallel to the wire along the current

(D) Parallel to the wire opposite to the current

39. The dipole moment of a circular loop carrying a current I, is m and the magnetic field at the centre of the loop is B_1. When the dipole moment is doubled by keeping the current constant, the magnetic field at the centre of the loop is B_2. The ratio $\frac{B_1}{B_2}$ is : **[2018]**

(A) 2 (B) $\sqrt{3}$

(C) $\sqrt{2}$ (D) $\frac{1}{\sqrt{2}}$

40. An electron, a proton and an alpha particle having the same kinetic energy are moving in circular orbits of radii r_e, r_p, r_α respectively in a uniform magnetic field B. The relation between r_e, r_p, r_α is : **[2018]**

(A) $r_e > r_p = r_\alpha$ (B) $r_e < r_p = r_\alpha$

(C) $r_e < r_p < r_\alpha$ (D) $r_e < r_\alpha < r_p$

41. A charge q is spread uniformly over an insulated loop of radius r. If it is rotated with an angular velocity ω with respect to normal axis then the magnetic moment of the loop is : **[2018 (Online)]**

(A) $\frac{1}{2}q\omega r^2$ (B) $\frac{4}{3}q\omega r^2$

(C) $\frac{3}{2}q\omega r^2$ (D) $q\omega r^2$

42. An infinitely long current carrying wire and a small current carrying loop are in the plane of the paper as shown. The radius of the loop is 'a' and distance of its centre from the wire is d ($d >> a$). If the loop applies a force F on the wire then : **[9 Jan. 2019 (Morning)]**

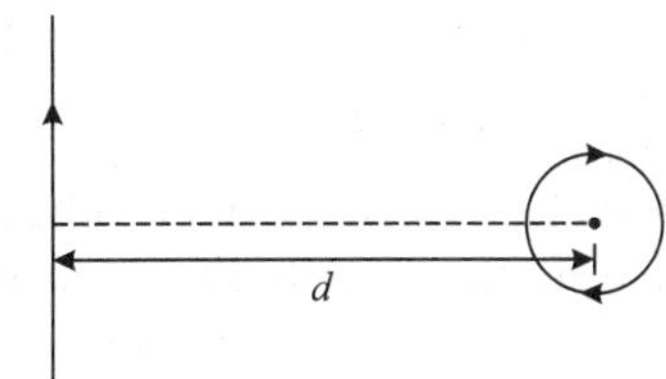

(A) $F \propto \left(\frac{a^2}{d^3}\right)$ (B) $F \propto \left(\frac{a}{d}\right)$

(C) $F \propto \left(\frac{a}{d}\right)^2$ (D) $F = 0$

43. A particle having the same charge as of electron moves in a circular path of radius 0.5 cm under the influence of a magnetic field of 0.5 T. If an electric field of 100 V/m makes it to move in a straight path, then the mass of the particle is : (Given charge of electron $= 1.6 \times 10^{-19}$ C) **[9 Jan. 2019 (Evening)]**

(A) 2.0×10^{-24} kg (B) 1.6×10^{-19} kg

(C) 1.6×10^{-27} kg (D) 9.1×10^{-31} kg

44. An insulating thin rod of length l has a linear charge density $\rho(x) = \rho_0 \frac{x}{l}$ on it. The rod is rotated about an axis passing through the origin ($x = 0$) and perpendicular to the rod. If the rod makes n rotation per second, then the time averaged magnetic moment of the rod is : **[10 Jan. 2019 (Morning)]**

(A) $\frac{\pi}{4}n\rho l^3$ (B) $n\rho l^3$

(C) $\pi n\rho l^3$ (D) $\frac{\pi}{3}n\rho l^3$

45. In an experiment electrons are accelerated, from rest, by applying a voltage of 500 V. Calculate the radius of the path if a magnetic field 100 mT is then applied.
[Charge of the electron = 1.6×10^{-19} C
Mass of the electron = 9.1×10^{-31} kg] : **[11 Jan. 2019 (Morning)]**
(A) 7.5×10^{-4} m (B) 7.5×10^{-3} m
(C) 7.5 m (D) 7.5×10^{-2} m

46. The region between $y = 0$ and $y = d$ contains a magnetic field $\vec{B} = B\hat{z}$. A particle of mass m and charge q enters the region with a velocity $\vec{v} = v\hat{i}$. If $d = \frac{mv}{2qB}$, the acceleration of the charged particle at the point of its emergence at the other side is : **[11 Jan. 2019 (Evening)]**
(A) $\frac{qvB}{m}\left(\frac{\hat{i}+\hat{j}}{\sqrt{2}}\right)$ (B) $\frac{qvB}{m}\left(\frac{1}{2}\hat{i}-\frac{\sqrt{3}}{\sqrt{2}}\hat{j}\right)$
(C) $\frac{qvB}{m}\left(\frac{-\hat{j}+\hat{i}}{\sqrt{2}}\right)$ (D) $\frac{qvB}{m}\left(\frac{\sqrt{3}}{2}\hat{i}+\frac{1}{2}\hat{j}\right)$

47. A particle of mass m and charge q is in an electric and magnetic field given by

$$\vec{E} = 2\hat{i} + 3\hat{j}\ ;\ \vec{B} = 4\hat{j} + 6\hat{k}.$$

The charged particle is shifted from the origin to the point $P(x = 1; y = 1)$ along a straight path. The magnitude of the total work done is : **[11 Jan. 2019 (Evening)]**
(A) $(0.35)q$ (B) $(0.15)q$
(C) $(2.5)q$ (D) $5q$

48. A proton and an α-particle (with their masses in the ratio of 1 : 4 and charges in the ratio of 1 : 2) are accelerated from rest through a potential difference V. If a uniform magnetic field (B) is set up perpendicular to their velocities, the ratio of the radii $r_p : r_\alpha$ of the circular paths described by them will be : **[12 Jan. 2019 (Morning)]**
(A) $1 : \sqrt{2}$ (B) 1 : 2
(C) 1 : 3 (D) $1 : \sqrt{3}$

49. A circular coil having N turns and radius r carries a current I. It is held in the XZ plane in a magnetic field $B\hat{i}$. The torque on the coil due to the magnetic field is : **[8 April 2019 (Morning)]**
(A) $B\pi r^2 IN$ (B) $\frac{Br^2 I}{\pi N}$
(C) Zero (D) $\frac{B\pi r^2 I}{N}$

50. A moving coil galvanometer has resistance 50 Ω and it indicates full deflection at 4 mA current. A voltmeter is made using this galvanometer and a 5 kΩ resistance. The maximum voltage, that can be measured using this voltmeter, will be close to : **[9 April 2019 (Morning)]**
(A) 10 V (B) 20 V
(C) 40 V (D) 15 V

51. A rigid square loop of side 'a' and carrying current I_2 is lying on a horizontal surface near a long current I_1 carrying wire in the same plane as shown in figure. The net force on the loop due to wire will be : **[9 April 2019 (Morning)]**

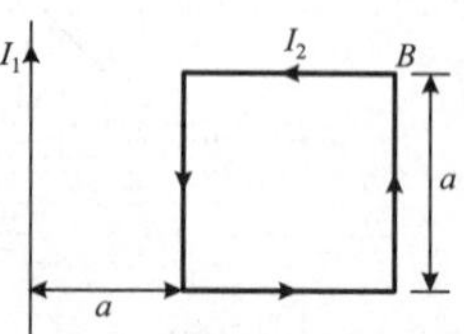

(A) Attractive and equal to $\frac{\mu_0 I_1 I_2}{3\pi}$
(B) Repulsive and equal to $\frac{\mu_0 I_1 I_2}{4\pi}$
(C) Repulsive and equal to $\frac{\mu_0 I_1 I_2}{2\pi}$
(D) Zero

52. A rectangular coil (Dimension 5 cm × 2.5 cm) with 100 turns, carrying a current of 3A in the clock-wise direction is kept centered at the origin and in the X-Z plane. A magnetic field of 1T is applied along X-axis. If the coil is tilted through 45° about Z-axis, then the torque on the coil is : **[9 April 2019 (Morning)]**
(A) 0.55 Nm (B) 0.27 Nm
(C) 0.38 Nm (D) 0.42 Nm

53. A moving coil galvanometer has a coil with 175 turns and area 1 cm^2. It uses a torsion band of torsion constant 10^{-6} N-m/rad. The coil is placed in a magnetic field B parallel to its plane. The coil deflects by 1° for a current of 1 mA. The value of B (in Tesla) is approximately : **[9 April 2019 (Evening)]**
(A) 10^{-3} (B) 10^{-1}
(C) 10^{-4} (D) 10^{-2}

54. A proton, an electron, and a Helium nucleus, have the same energy. They are in circular orbits in a plane due to magnetic field perpendicular to the plane. Let r_p, r_e and r_{He} be their respective radii, then : **[10 April 2019 (Morning)]**
(A) $r_e > r_p > r_{He}$ (B) $r_e < r_p < r_{He}$
(C) $r_e < r_p = r_{He}$ (D) $r_e > r_p = r_{He}$

55. Two wires A & B are carrying currents I_1 & I_2 as shown in the figure. The separation between them is d. A third wire C

carrying a current I is to be kept parallel to them at a distance x from A such that the net force acting on it is zero. The possible values of x are : **[10 April 2019 (Morning)]**

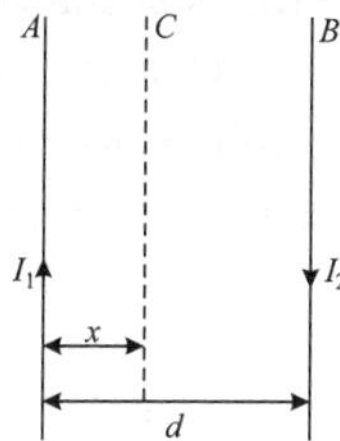

(A) $x=\left(\frac{I_1}{I_1-I_2}\right)d$ and $x=\frac{I_2}{(I_1+I_2)}d$

(B) $x=\pm\frac{I_1 d}{(I_1-I_2)}$

(C) $x=\left(\frac{I_1}{I_1+I_2}\right)d$ and $x=\frac{I_2}{(I_1-I_2)}d$

(D) $x=\left(\frac{I_2}{I_1+I_2}\right)d$ and $x=\left(\frac{I_2}{I_1-I_2}\right)d$

56. A moving coil galvanometer allows a full scale current of 10^{-4} A. A series resistance of 2 $M\Omega$ is required to convert the above galvanometer into a voltmeter of range 0-5 V. Therefore the value of shunt resistance required to convert the above galvanometer into an ammeter of range 0-10 mA is : **[10 April 2019 (Morning)]**

(A) 200 Ω (B) 100 Ω
(C) 10 Ω (D) 500 Ω

57. A moving coil galvanometer, having a resistance G, produces full scale deflection when a current I_g flows through it. This galvanometer can be converted into (i) an ammeter of range 0 to I_0 ($I_0 > I_g$) by connecting a shunt resistance R_A to it and (ii) into a voltmeter of range 0 to V($V = GI_0$) by connecting a series resistance R_V to it. Then : **[12 April 2019 (Evening)]**

(A) $R_A R_V = G^2\left(\frac{I_g}{I_0-I_g}\right)$ and $\frac{R_A}{R_V}=\left(\frac{I_0-I_g}{I_g}\right)^2$

(B) $R_A R_V = G^2$ and $\frac{R_A}{R_V}=\left(\frac{I_g}{I_0-I_g}\right)^2$

(C) $R_A R_V = G^2$ and $\frac{R_A}{R_V}=\frac{I_g}{(I_0-I_g)}$

(D) $R_A R_V = G^2\left(\frac{I_0-I_g}{I_g}\right)$ and $\frac{R_A}{R_V}=\left(\frac{I_g}{I_0-I_g}\right)^2$

58. An electron, moving along the x-axis with an initial energy of 100 eV, enters a region of magnetic field $\vec{B}=(1.5\times10^{-3}T)\hat{k}$ at S (See figure). The field extends between $x = 0$ and $x = 2$ cm. The electron is detected at the point Q on a screen placed 8 cm away from the point S. The distance d between P and Q (on the screen) is : (electron's charge = 1.6×10^{-19} C, mass of electron = 9.1×10^{-31} kg) : **[12 April 2019 (Evening)]**

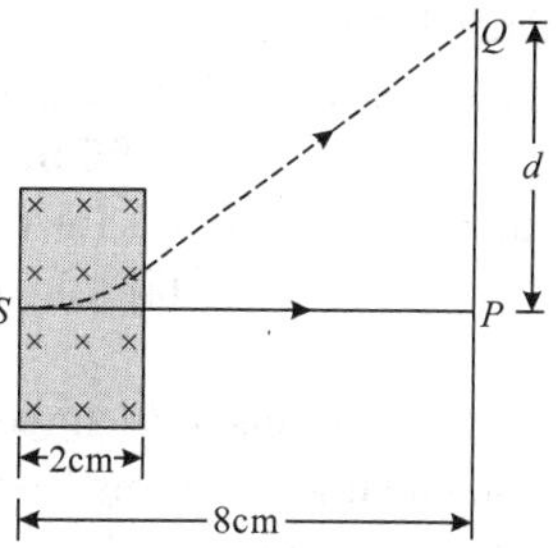

(A) 12.87 cm (B) 1.22 cm
(C) 11.65 cm (D) 2.25 cm

59. A particle of mass m and charge q has an initial velocity $\vec{v}=-v_0\hat{j}$. If an electric field $\vec{E}=E_0\hat{i}$ and magnetic field $\vec{B}=B_0\hat{i}$ act on the particle, its speed will double after a time : **[7 Jan. 2020 (Evening)]**

(A) $\frac{2mv_0}{qE_0}$ (B) $\frac{3mv_0}{qE_0}$

(C) $\frac{\sqrt{3}mv_0}{qE_0}$ (D) $\frac{\sqrt{2}mv_0}{qE_0}$

60. Protons with kinetic energy of 1 MeV moves from south to north. It gets an acceleration of 10^{12} m/s^2 by an applied magnetic field (west to east). The value of magnetic field : (Rest mass of proton is 1.6×10^{-27} kg) : **[8 Jan. 2020 (Morning)]**

(A) 71 mT (B) 7.1 mT
(C) 0.071 mT (D) 0.71 mT

61. A charged particle of mass 'm' and charge 'q' moving under the influence of uniform electric field $E\vec{i}$ and a uniform magnetic field $B\vec{k}$ follows a trajectory from point P to Q as shown in figure. The velocities at P and Q are respectively, $v\vec{i}$ and $-2v\vec{j}$. Then which of the following statements (a, b, c, d) are the correct ? (Trajectory shown is schematic and not to scale)

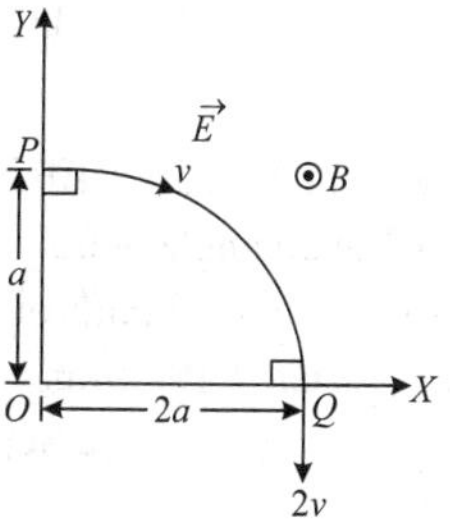

(a) $E=\frac{3}{4}\left(\frac{mv^2}{qa}\right)$

(b) Rate of work done by the electric field at P is $\frac{3}{4}\left(\frac{mv^3}{a}\right)$

(c) Rate of work done by both the fields at Q is zero

(d) The difference between the magnitude of angular momentum of the particle at P and Q is 2 mav.

[9 Jan. 2020 (Morning)]

(A) (a), (b), (c), (d) (B) (a), (b), (c)

(C) (b), (c), (d) (D) (a), (c), (d)

62. An electron gun is placed inside a long solenoid of radius R on its axis. The solenoid has n turns/length and carries a current I. The electron gun shoots an electron along the radius of the solenoid with speed v. If the electron does not hit the surface of the solenoid, maximum possible value of v is : (all symbols have their standard meaning) **[9 Jan. 2020 (Evening)]**

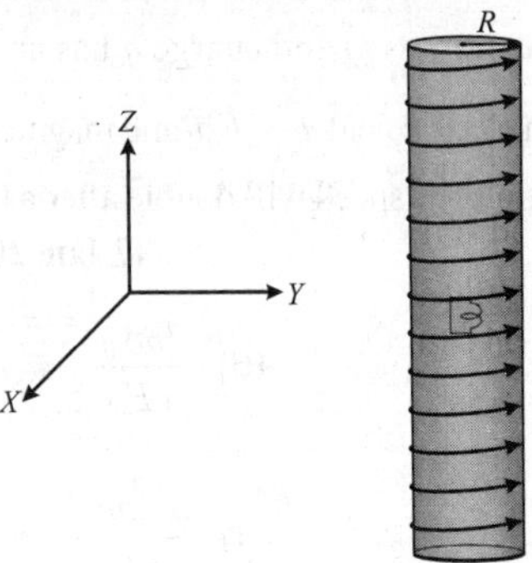

(A) $\frac{e\mu_0 nIR}{m}$ (B) $\frac{e\mu_0 nIR}{2m}$

(C) $\frac{2e\mu_0 nIR}{m}$ (D) $\frac{e\mu_0 nIR}{4m}$

63. A small circular loop of conducting wire has radius a and carries current I. It is placed in a uniform magnetic field B perpendicular to its plane such that when rotated slightly about its diameter and released, it starts performing simple harmonic motion of time period T. If the mass of the loop is m then :

[9 Jan. 2020 (Evening)]

(A) $T=\sqrt{\frac{\pi m}{2IB}}$ (B) $T=\sqrt{\frac{2\pi m}{IB}}$

(C) $T=\sqrt{\frac{\pi m}{IB}}$ (D) $T=\sqrt{\frac{2m}{IB}}$

64. A beam of protons with speed 4×10^5 ms^{-1} enters a uniform magnetic field of 0.3 T at an angle of 60° to the magnetic field. The pitch of the resulting helical path of protons is close to : (Mass of the proton = 1.67×10^{-27} kg, charge of the proton = 1.69×10^{-19} C) **[2 Sept. 2020 (Morning)]**

(A) 12 cm (B) 4 cm

(C) 5 cm (D) 2 cm

65. A charged particle (mass m and charge q) moves along X-axis with velocity V_0. When it passes through the origin it enters a region having uniform electric field $\vec{E}=-E\hat{j}$ which extends upto $x=d$. Equation of path of electron in the region $x>d$ is : **[2 Sept. 2020 (Morning)]**

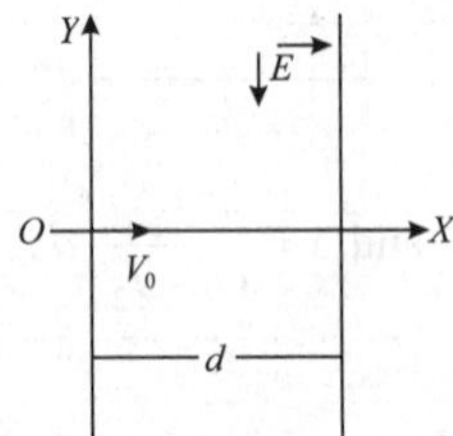

(A) $y=\frac{qEd}{mV_0^2}\left(\frac{d}{2}-x\right)$ (B) $y=\frac{qEd}{mV_0^2}(x-d)$

(C) $y=\frac{qEd}{mV_0^2}x$ (D) $y=\frac{qEd^2}{mV_0^2}x$

66. The figure shows a region of length 'l' with a uniform magnetic field of 0.3 T in it and a proton entering the region with velocity 4×10^5 ms^{-1} making an angle 60° with the field. If the proton completes 10 revolution by the time it cross the region shown, 'l' is close to (mass of proton = 1.67×10^{-27} kg, charge of the proton = 1.6×10^{-19} C) : **[2 Sept. 2020 (Evening)]**

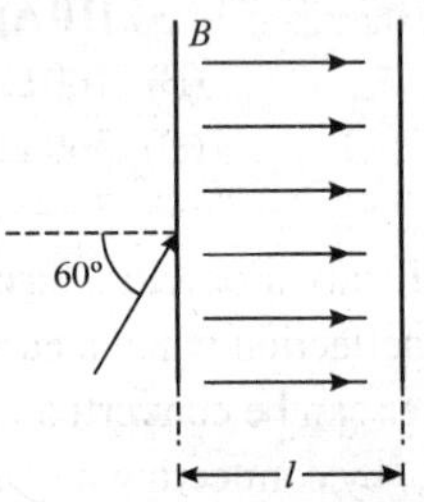

(A) 0.11 m (B) 0.22 m

(C) 0.44 m (D) 0.88 m

67. A wire carrying current I is bent in the shape $ABCDEFA$ as shown, where rectangle $ABCDA$ and $ADEFA$ are perpendicular to each other. If the sides of the rectangles are of lengths a and b, then the magnitude and direction of magnetic moment of the loop $ABCDEFA$ is : **[2 sept. 2020 Evening]**

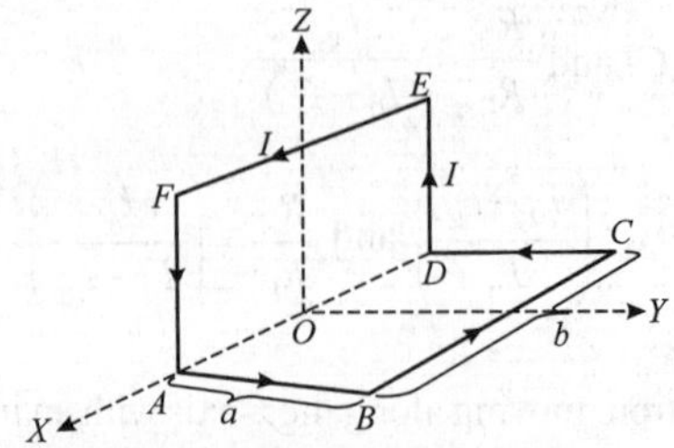

(A) $\sqrt{2}\,abI$, along $\left(\frac{\hat{j}}{\sqrt{2}}+\frac{\hat{k}}{\sqrt{2}}\right)$

(B) $\sqrt{2}\,abI$, along $\left(\frac{\hat{j}}{\sqrt{5}}+\frac{2\hat{k}}{\sqrt{5}}\right)$

(C) abI, along $\left(\frac{\hat{j}}{\sqrt{2}}+\frac{\hat{k}}{\sqrt{2}}\right)$

(D) abI, along $\left(\frac{\hat{j}}{\sqrt{5}}+\frac{2\hat{k}}{\sqrt{5}}\right)$

68. A charged particle carrying charge 1 μC is moving with velocity $(2\hat{i}+3\hat{j}+4\hat{k})$ ms^{-1}. If an external magnetic field of $(5\hat{i}+3\hat{j}-6\hat{k})\times10^{-3}$ T exists in the region where the particle is moving then the force on the particle is $\vec{F}\times10^{-9}$ N. The vector $\vec{F}$ is : **[3 Sept. 2020 (Morning)]**

(A) $-0.30\hat{i}+0.32\hat{j}-0.09\hat{k}$

(B) $-300\hat{i}+320\hat{j}-90\hat{k}$

(C) $-30\hat{i}+32\hat{j}+9\hat{k}$

(D) $-3.0\hat{i}+3.2\hat{j}-0.9\hat{k}$

69. A galvanometer coil has 500 turns and each turn has an average area of 3×10^{-4} m^2. If a torque of 1.5 Nm is required to keep this coil parallel to magnetic field when a current of 0.5 A is flowing through it, the strength of the field (in T) is______. **[3 Sept. 2020 (Evening)]**

70. A circular coil has moment of inertia 0.8 kg m^2 around any diameter and is carrying current to produce a magnetic moment of 20 Am^2. The coil is kept initially in a vertical position and it can rotate freely around a horizontal diameter. When a uniform magnetic field of 4 T is applied along the vertical, it starts rotating around its horizontal diameter. The angular speed the coil acquires after rotating by 60° will be : **[4 Sept. 2020 (Evening)]**

(A) 10 rad s^{-1} (B) 20 π rad s^{-1}

(C) 10 π rad s^{-1} (D) 20 rad s^{-1}

71. A square loop of side $2a$, and carrying current I, is kept in XZ plane with its centre at origin. A long wire carrying the same current I is placed parallel to the z-axis and passing through the point $(0, b, 0)$, $(b >> a)$. The magnitude of the torque on the loop about z-axis is given by : **[5 Sept. 2020 (Morning)]**

(A) $\frac{2\mu_0 I^2 a^2}{\pi b}$ (B) $\frac{\mu_0 I^2 a^3}{2\pi b^2}$

(C) $\frac{\mu_0 I^2 a^2}{2\pi b}$ (D) $\frac{2\mu_0 I^2 a^3}{\pi b^2}$

72. An iron rod of volume 10^{-3} m^3 and relative permeability 1000 is placed as core in a solenoid with 10 turns/cm. If a current of 0.5 A is passed through the solenoid, then the magnetic moment of the rod will be : **[5 Sept. 2020 (Evening)]**

(A) 0.5×10^2 Am^2 (B) 50×10^2 Am^2

(C) 500×10^2 Am^2 (D) 5×10^2 Am^2

73. A particle of charge q and mass m is moving with a velocity $-v\hat{i}$ $(v\neq0)$ towards a large screen placed in the $Y-Z$ plane at a distance d. If there is a magnetic field $\vec{B}=B_0\hat{k}$, the minimum value of v for which the particle will not hit the screen is : **[6 Sept. 2020 (Morning)]**

(A) $\frac{qdB_0}{2m}$ (B) $\frac{qdB_0}{m}$

(C) $\frac{2qdB_0}{m}$ (D) $\frac{qdB_0}{3m}$

74. An electron is moving along $+x$ direction with a velocity of 6×10^6 ms^{-1}. It enters a region of uniform electric field of 300 V/cm pointing along $+y$ direction. The magnitude and direction of the magnetic field set up in this region such that the electron keeps moving along the x direction will be : **[6 Sept. 2020 (Morning)]**

(A) 5×10^{-3} T, along $+z$ direction

(B) 3×10^{-4} T, along $-z$ direction

(C) 3×10^{-4} T, along $+z$ direction

(D) 5×10^{-3} T, along $-z$ direction

75. A particle moving in the xy plane experiences a velocity dependent force $\vec{F}=k(v_y\hat{i}+v_x\hat{j})$, where v_x and v_y are the x and y components of its velocity $\vec{v}$. If $\vec{a}$ is the acceleration of the particle, then which of the following statements is true for the particle : **[6 Sept. 2020 (Evening)]**

(A) Quantity $\vec{v}\cdot\vec{a}$ is constant in time

(B) Kinetic energy of particle is constant in time

(C) Quantity $\vec{v}\times\vec{a}$ is constant in time

(D) $\vec{F}$ arises due to a magnetic field

76. A charged particle going around in a circle can be considered to be a current loop. A particle of mass m carrying charge q is moving in a plane with speed v under the influence of magnetic field $\vec{B}$. The magnetic moment of this moving particle : **[6 Sept. 2020 (Evening)]**

(A) $-\frac{mv^2\vec{B}}{B^2}$ (B) $-\frac{mv^2\vec{B}}{2\pi B^2}$

(C) $-\frac{mv^2\vec{B}}{3B^2}$ (D) $-\frac{mv^2\vec{B}}{2B^2}$

77. A square loop of side $2a$ and carrying current I is kept in xz plane with its centre at origin. A long wire carrying the same current I is placed parallel to z-axis and passing through point $(0, b, 0)$, $(b >> a)$. The magnitude of torque on the loop about z-axis will be : **[6 Sept. 2020 (Evening)]**

(A) $\frac{2\mu_0 I^2 a^2 b}{\pi(a^2+b^2)}$ (B) $\frac{\mu_0 I^2 a^2 b}{2\pi(a^2+b^2)}$

(C) $\frac{\mu_0 I^2 a^2}{2\pi b}$ (D) $\frac{2\mu_0 I^2 a^2}{\pi b}$

78. A proton, a deuteron and an α particle are moving with same momentum in a uniform magnetic field. The ratio of magnetic forces acting on them is ______ and their speed is ______, in the ratio : **[25 Feb. 2021 (Morning)]**

(A) 1 : 2 : 4 and 1 : 1 : 2 (B) 2 : 1 : 1 and 4 : 2 : 1
(C) 4 : 2 : 1 and 2 : 1 : 1 (D) 1 : 2 : 4 and 2 : 1 : 1

79. A proton and an α-particle, having kinetic energies K_p and K_α, respectively, enter into a magnetic field at right angles. The ratio of the radii of trajectory of proton to that of α-particle is 2 : 1. The ratio of $K_p : K_\alpha$ is : **[18 March 2021 (Evening)]**

(A) 1 : 8 (B) 8 : 1
(C) 1 : 4 (D) 4 : 1

80. A deuteron and an alpha particle having equal kinetic energy enter perpendicularly into a magnetic filed. Let r_d and r_a be their respective radii of circular path. The value of $\frac{r_d}{r_a}$ is : **[20 July 2021 (Morning)]**

(A) $\frac{1}{\sqrt{2}}$ (B) $\sqrt{2}$

(C) 2 (D) 1

81. Two ions having same mass have charge in the ratio 1 : 2. They are projected normally in a uniform magnetic field with their speeds in the ration 2 : 3. The ratio of the radii of their circular trajectories is : **[25 July 2021 (Evening)]**

(A) 1 : 4 (B) 2 : 3
(C) 4 : 3 (D) 3 : 1

82. If the maximum value of acceleration potential provided by a radio frequency oscillator is 12 kV. The number of revolution made by a proton in cyclotron to achieve one sixth of the speed of light is ______.
[m_p = 1.67 × 10^{-27} kg, e = 1.6 × 10^{-19} C, Speed of light = 3 × 108 m/s]. **[26 Aug. 2021 (Evening)]**

83. Two ions of masses 4 amu and 16 amu have charges $+2e$ and $+3e$ respectively. These ions pass through the region of constant perpendicular magnetic field. The kinetic energy of both ions is same. Then : **[27 Aug. 2021 (Morning)]**

(A) No ion will be deflected
(B) Lighter ion will be deflected less than heavier ion
(C) Lighter ion will be deflected more than heavier ion
(D) Both ions will be deflected equally.

84. The magnetic field vector of an electromagnetic wave is given by $B = B_0 \frac{\hat{i}+\hat{j}}{\sqrt{2}} \cos(kz - \omega t)$ where $\hat{i}, \hat{j}$ represents unit vector x and y-axis respectively. At $t = 0$ s, two electric changes q_1 of 4π coulomb and q_2 of 2π coulomb located at $\left(0, 0, \frac{\pi}{k}\right)$ and $\left(0, 0, \frac{3\pi}{k}\right)$, respectively, have the same velocity of 0.5 $c\hat{i}$, (where c is the velocity of light). The ratio of the force acting on charge q_1 to q_2 is : **[31 Aug. 2021 (Evening)]**

(A) $1 : \sqrt{2}$ (B) $\sqrt{2} : 1$

(C) $2\sqrt{2} : 1$ (D) $2 : 1$

85. Given below are two statements: One is labelled as **Assertion-A** and the other is labelled as **Reason-R.**

Assertion-A: In an uniform magnetic field, speed and energy remains the same for a moving charged particle

Reason-R: Moving charged particle experiences magnetic force perpendicular to its direction of motion : **[24 June 2022 (Morning)]**

(A) Both **A** and **R** are true and **R** is the correct explanation of **A**.
(B) Both **A** and **R** are true but **R** is NOT the correct explanation of **A**.
(C) **A** is true but **R** is false.
(D) **A** is false but **R** is true.

86. A proton, a deuteron and an α-particle with same kinetic energy enter into a uniform magnetic field at right angle to magnetic field. The ratio of the radii of their respective circular paths is : **[24 June 2022 (Evening)]**

(A) $1 : \sqrt{2} : \sqrt{2}$ (B) $1 : 1 : \sqrt{2}$

(C) $\sqrt{2} : 1 : 1$ (D) $1 : \sqrt{2} : 1$

87. A proton and an alpha particle of the same speed enter in a uniform magnetic field which is acting perpendicular to their direction of motion. The ratio of the circular paths described by the alpha particle and proton is : **[26 June 2022 (Morning)]**

(A) 1 : 4 (B) 4 : 1
(C) 2 : 1 (D) 1 : 2

88. Two long parallel conductors S_1 and S_2 are separated by a distance 10 cm and carrying currents of 4 A and 2 A respectively. The conductors are placed along x-axis in X-Y plane. There is a point P located between the conductors (as shown in figure). A charge particle of 3π coulomb is passing through the point P with velocity $\vec{v} = (2\hat{i} + 3\hat{j})$ m/s; where $\hat{i}$ and $\hat{j}$ represents unit vector along x & y axis respectively. The force acting on the charge particle is $4\pi \times 10^{-5}(-x\hat{i} + 2\hat{j})N$. The value of x is : **[27 June 2022 (Evening)]**

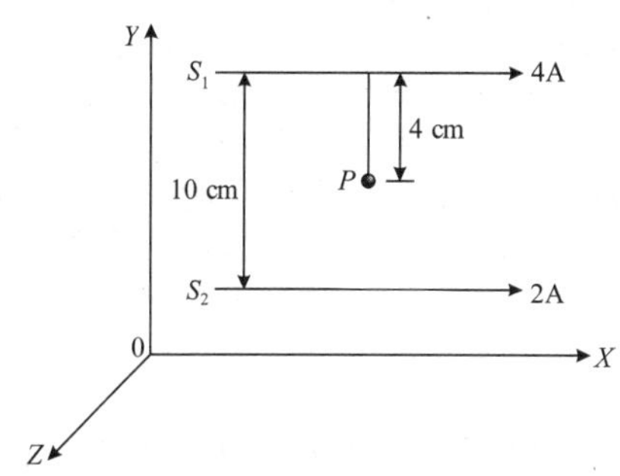

(A) 2
(B) 1
(C) 3
(D) – 3

89. A deuteron and a proton moving with equal kinetic energy enter into to a uniform magnetic field at right angle to the field. If r_d and r_p are the radii of their circular paths respectively, then the ratio $\frac{r_d}{r_p}$ will be $\sqrt{x}:1$ where x is ______. **[27 June 2022 (Evening)]**

90. A singly ionized magnesium atom ($A = 24$) ion is accelerated to kinetic energy 5 keV and is projected perpendicularly into a magnetic field B of the magnitude 0.5 T. The radius of path formed will be ______ cm. **[28 June 2022 (Morning)]**

91. Two parallel, long wires are kept 0.20 m apart in vacuum, each carrying current of x A in the same direction. If the force of attraction per meter of each wire is 2×10^{-6} N, then the value of x is approximately: **[28 June 2022 (Evening)]**

(A) 1
(B) 2.4
(C) 1.4
(D) 2

92. A charged particle moves along circular path in a uniform magnetic field in a cyclotron. The kinetic energy of the charged particle increases to 4 times its initial value. What will be the ratio of new radius to the original radius of circular path of the charged particle : **[29 June 2022 (Morning)]**

(A) 1 : 1
(B) 1 : 2
(C) 2 : 1
(D) 1 : 4

93. Given below are two statements : **[29 June 2022 (Evening)]**

Statement-I : The electric force changes the speed of the charged particle and hence changes its kinetic energy, whereas the magnetic force does not change the kinetic energy of the charged particle

Statement-II : The electric force accelerates the positively charged particle perpendicular to the direction of electric field. The magnetic force accelerates the moving charged particle along the direction of magnetic field. In the light of the above statements, choose the most appropriate answer from the options given below :

(A) Both Statement-I and Statement-II are correct
(B) Both Statement-I and Statement-II are incorrect
(C) Statement-I is correct but Statement-II is incorrect
(D) Statement-I is incorrect but Statement-II is correct

94. Two charged particles, having same kinetic energy, are allowed to pass through a uniform magnetic field perpendicular to the direction of motion. If the ratio of radii of their circular paths is 6 : 5 and their respective masses ratio is 9 : 4. Then, the ratio of their charges will be : **[25 July 2022 (Morning)]**

(A) 8 : 5
(B) 5 : 4
(C) 5 : 3
(D) 8 : 7

95. A charge particle is moving in a uniform magnetic field $(2\hat{i} + 3\hat{j})$ T. If it has an acceleration of $(\alpha\hat{i} - 4\hat{j})$ m/s^2, then the value of α will be : **[26 July 2022 (Morning)]**

(A) 3
(B) 6
(C) 12
(D) 2

96. Two concentric circular loops of radii $r_1 = 30$ cm and $r_2 = 50$ cm are placed in X-Y plane as shown in the figure. A current $I = 7$ A is flowing through them in the direction as shown in figure. The net magnetic moment of this system of two circular loops is approximately : **[26 July 2022 (Evening)]**

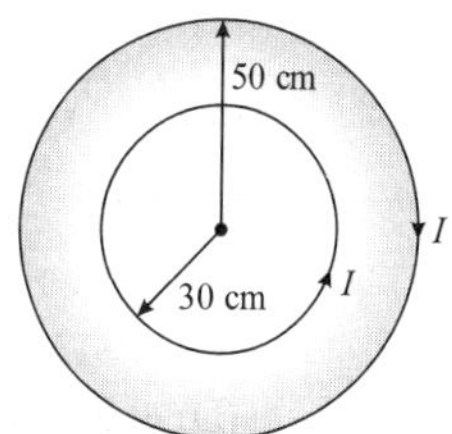

(A) $\frac{7}{2}\hat{k}$ Am^2
(B) $-\frac{7}{2}\hat{k}$ Am^2
(C) $7\hat{k}$ Am^2
(D) $-7\hat{k}$ Am^2

97. A velocity selector consists of electric field $\vec{E} = E\hat{k}$ and magnetic field $\vec{B} = B\hat{j}$ with $B = 12$ mT. The value E required for an electron of energy 728 eV moving along the positive x-axis to pass undeflected is : [Given, mass of electron $= 9.1 \times 10^{-31}$ kg] **[26 July 2022 (Evening)]**

(A) 192 kVm^{-1}
(B) 192 m Vm^{-1}
(C) 9600 kVm^{-1}
(D) 16 kVm^{-1}

98. A cyclotron is used to accelerate protons. If the operating magnetic field is 1.0 T and the radius of the cyclotron 'dees' is 60 cm, the kinetic energy of the accelerated protons in MeV will be : [Use : $m_p = 1.6 \times 10^{-27}$ kg, $e = 1.6 \times 10^{-19}$ C] **[27 July 2022 (Evening)]**

(A) 12 (B) 18
(C) 16 (D) 32

99. As shown in the figure, a metallic rod of linear density 0.45 kg m^{-1} is lying horizontally on a smooth incline plane which makes an angle of 45° with the horizontal. The minimum current flowing in the rod required to keep it stationary, when 0.15 T magnetic field is acting on it in the vertical upward direction, will be : [Use : $g = 10$ m/s^2] **[28 July 2022 (Morning)]**

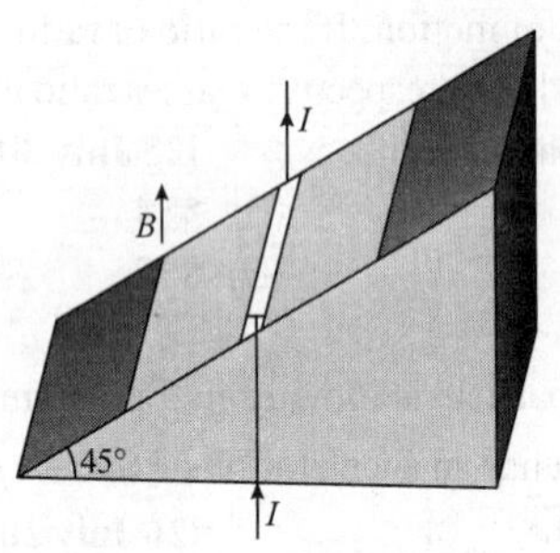

(A) 30 A (B) 15 A
(C) 10 A (D) 3 A

100. A triangular shaped wire carrying 10 A current is placed in a uniform magnetic field of 0.5 T, as shown in figure. The magnetic force on segment CD is :
[Given : $BC = CD = BD = 5$ cm] **[28 July 2022 (Evening)]**

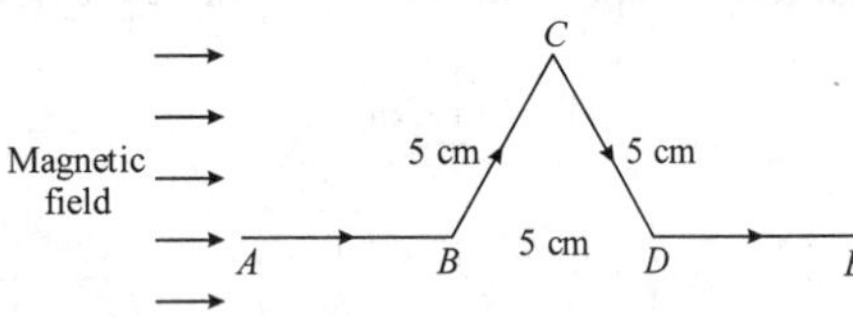

(A) 0.125 N (B) 0.312 N
(C) 0.216 N (D) 0.245 N

101. A wire of length 314 cm carrying current of 14 A is bent to form a circle. The magnetic moment of the coil is _____ A-m^2.
[Given : $\alpha = 3.14$] **[29 July 2022 (Morning)]**

102. A wire X of length 50 cm carrying a current of 2 A is placed parallel to a long wire Y of length 5 m. The wire Y carries a current of 3 A. The distance between two wires is 5 cm and currents flow in the same direc tion. The force acting on the wire Y is : **[29 July 2022 (Evening)]**

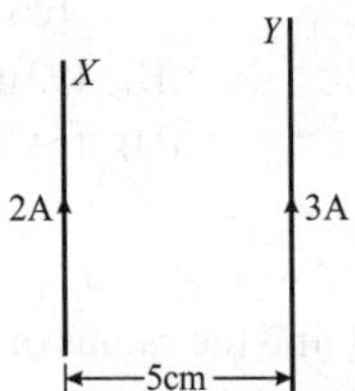

(A) 1.2×10^{-5} N directed towards wire X
(B) 1.2×10^{-4} N directed away from wire X
(C) 1.2×10^{-4} N directed towards wire X
(D) 2.4×10^{-5} N directed towards wire X

103. Two long straight wires P and Q carrying equal current 10 A each were kept parallel to each other at 5 cm distance. Magnitude of magnetic force experienced by 10 cm length of wire P is F_1. If distance between wires is halved and currents on them are doubled, force F_2 on 10 cm length of wire P will be : **[24 Jan. 2023 (Morning)]**

(A) $8F_1$ (B) $10F_1$
(C) $F_1/8$ (D) $F_1/10$

104. A single turn current loop in the shape of a right angle triangle with sides 5 cm, 12 cm, 13 cm is carrying a current of 2A. The loop is in a uniform magnetic field of magnitude 0.75 T whose direction is parallel to the current in the 13 cm side of the loop. The magnitude of the magnetic force on the 5 cm side will be $\frac{x}{130}$ N. The value of x is _____. **[24 Jan. 2023 (Evening)]**

105. For a moving coil galvanometer, the deflection in the coil is 0.05 rad when a current of 10 mA is passed through it. If the torsional constant of suspension wire is 4.0×10^{-5} Nm rad^{-1}, the magnetic field is 0.01 T and the number of turns in the coil is 200, the area of each turn (in cm^2) is : **[25 Jan. 2023 (Evening)]**

(A) 2.0 (B) 1.0
(C) 1.5 (D) 0.5

106. Find the mutual inductance in the arrangement, when a small circular loop of wire of radius 'R' is placed inside a large square loop of wire of side L ($L >> R$). The loops are coplanar and their centres coincide : **[29 Jan. 2023 (Morning)]**

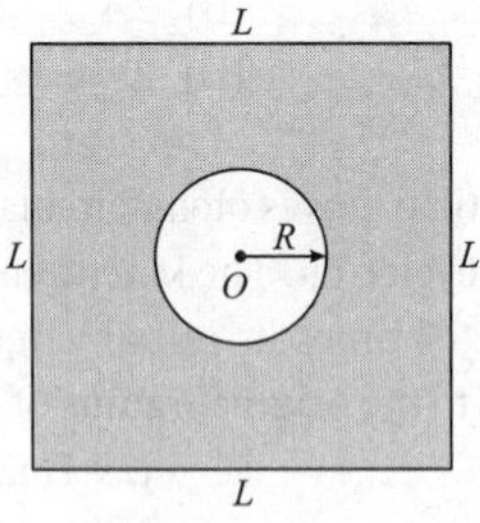

(A) $M = \frac{\sqrt{2}\mu_0 R^2}{L}$ (B) $M = \frac{2\sqrt{2}\mu_0 R}{L^2}$

(C) $M = \frac{2\sqrt{2}\mu_0 R^2}{L}$ (D) $M = \frac{\sqrt{2}\mu_0 R}{L^2}$

107. The magnetic moments associated with two closely wound circular coils A and B of radius r_A = 10 cm and r_B = 20 cm respectively are equal if : (Where N_A, I_A and N_B, I_B are number of turn and current of A and B respectively)

[30 Jan. 2023 (Morning)]

(A) $2N_AI_A = N_BI_B$
(B) $N_A = 2N_B$
(C) $N_AI_A = 4N_BI_B$
(D) $4N_AI_A = N_BI_B$

108. A massless square loop, of wire of resistance 10 Ω. Supporting a mass of *Ig*. Hangs vertically with one of its sides in a uniform magnetic field of 10^3 G, directed outwards in the shaded region. A dc voltage V is applied to the loop. For what value of V. The magnetic force will exactly balance the weight of the supporting mass of *Ig* ?

(If sides of the loop = 10 cm, $g = 10\ ms^{-2}$)

[30 Jan. 2023 (Morning)]

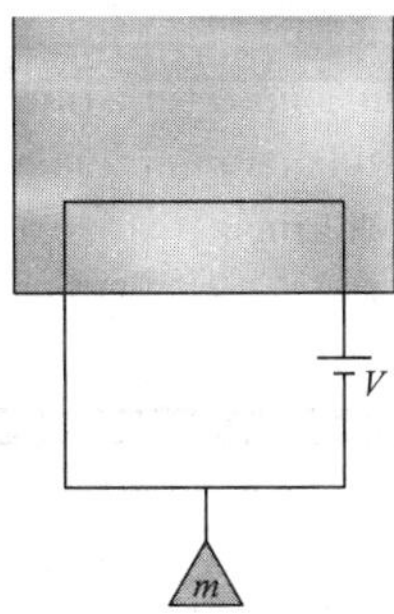

(A) $\frac{1}{10}$V
(B) 100V
(C) 1V
(D) 10V

109. A current carrying rectangular loop $PQRS$ is made of uniform wire. The length $PR = QS = 5$ cm and $PQ = RS = 100$ cm. If ammeter current reading changes from I to $2I$, the ratio of magnetic forces per unit length on the wire PQ due to wire RS in the two cases respectively $f_{PQ}^{1} : f_{PQ}^{2I}$ is :

[30 Jan. 2023 (Evening)]

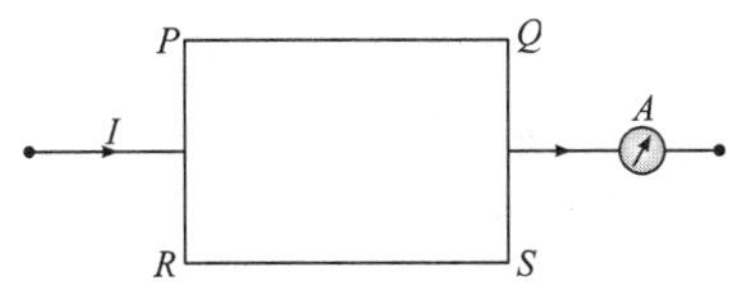

(A) 1 : 2
(B) 1 : 4
(C) 1 : 5
(D) 1 : 3

110. A bar magnet with a magnetic moment 5.0 Am^2 is placed in parallel position relative to a magnetic field of 0.4 T. The amount of required work done in turning the magnet form parallel to antiparallel position relative to the field direction is _____.

[31 Jan. 2023 (Morning)]

(A) 4 J
(B) 1 J
(C) 2 J
(D) Zero

111. The number of turns of the coil of a moving coil galvanometer is increased in order to increase current sensitivity by 50%. The percentage change in voltage sensitivity of the galvanometer will be :

[31 Jan. 2023 (Evening)]

(A) 100%
(B) 50%
(C) 75%
(D) 0%

112. A charge particle of 2 μC accelerated by a potential difference of 100 V enters a region of uniform magnetic field of magnitude 4 mT at right angle to the direction of field. The charge particle completes semicircle of radius 3 cm inside magnetic field. The mass of the charge particle is_____ × 10^{-18} kg.

[1 Feb. 2023 (Morning)]

113. A proton with a kinetic energy of 2.0 eV moves into a region of uniform magnetic field of magnitude $\frac{\pi}{2} \times 10^{-3}$ T. The angle between the direction of magnetic field and velocity of proton is 60°. The pitch of the helical path taken by the proton is_____cm. [Take, mass of proton = 1.6×10^{-27} kg and Charge on proton = 1.6×10^{-19} C]

[6 April 2023 (Evening)]

114. A charge particle moving in magnetic field B, has the components of velocity along B as well as perpendicular to B. The path of the charge particle will be :

[8 April 2023 (Morning)]

(A) Helical path with the axis perpendicular to the direction of magnetic field B
(B) Straight along the direction of magnetic field B
(C) Helical path with the axis along magnetic field B
(D) Circular path

115. Given below are two Statement:

Statement-I: If the number of turns in the coil of a moving coil galvanometer is doubled then the current sensitivity becomes double.

Statement-II: Increasing current sensitivity of a moving coil galvanometer by only increasing the number of turns in the coil will also increase its voltage sensitivity in the same ratio.

In the light of the above Statement, choose the correct answer from the options given below. **[10 April 2023 (Morning)]**

(A) Statement-**I** is false but Statement-**II** is true
(B) Both Statement-**I** and Statement-**II** are true
(C) Both Statement-**I** and Statement-**II** are false
(D) Statement-**I** is true but Statement-**II** is false

116. The current sensitivity of moving coil galvanometer is increased by 25%. This increase is achieved only by changing in the number of turns of coils and area of cross section of the wire while keeping the resistance of galvanometer coil constant. The percentage change in the voltage sensitivity will be :

[11 April 2023 (Morning)]

(A) +25%
(B) –50%
(C) Zero
(D) –25%

117. An electron is allowed to move with constant velocity along the axis of current carrying straight solenoid.

(a) The electron will experience magnetic force along the axis of the solenoid

(b) The electron will not experience magnetic force

(c) The electron will continue to move along the axis of the solenoid

(d) The electron will be accelerated along the axis of the solenoid

(e) The electron will follow parabolic path–inside the solenoid

Choose the correct answer from the options given below :

[11 April 2023 (Evening)]

(A) (b), (c) and (d) only (B) (b) and (c) only

(C) (a) and (d) only (D) (b) and (e) only

118. An electron is moving along the positive x-axis. If the uniform magnetic field is applied parallel to the negative z-zxis. then

(a) The electron will experience magnetic force along positive y-axis

(b) The electron will expperience magnetic force along negative y-axis

(c) The electron will not experience any force in magnetic field

(d) The electron will continue to move along the positive x-axis

(e) The electron will move along circular path in magnetic field

Choose the corrct ansver from the options given below:

[13 April 2023 (Evening)]

(A) (b) and (e) only (B) (a) to (e) only

(C) (c) and (d) only (D) (b) and (d) only

119. A straight wire AB of mass 40 g and length 50 cm is suspended by a pair of flexible leads in uniform magnetic field of magnitude 0.40 T as shown in the figure. The magnitude of the current required in the wire to remove the tension in the supporting leads is ____ A. (Take $g = 10\ \text{ms}^{-2}$).

[13 April 2023 (Evening)]

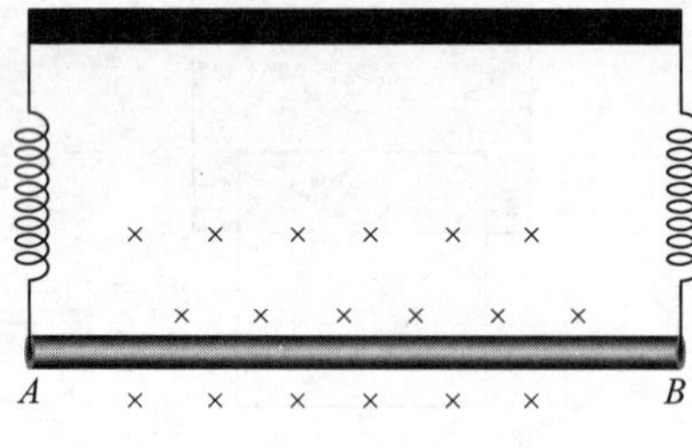

* * * * *

CHAPTER 19

MAGNETISM AND MATTER

19.1 CLASSICAL AND TERRESTRIAL MAGNETISM

1. The magnetic lines of force inside a bar magnet : **[2003]**
(A) are from north-pole to south-pole of the magnet
(B) do not exist
(C) depend upon the area of cross-section of the bar magnet
(D) are from south-pole to north-pole of the magnet

2. A thin rectangular magnet suspended freely has a period of oscillation equal to T. Now it is broken into two equal halves (each having half of the original length) and one piece is made to oscillate freely in the same field. If its period of oscillation is T', the ratio $\frac{T'}{T}$ is : **[2003]**

(A) $\frac{1}{2\sqrt{2}}$ (B) $\frac{1}{2}$
(C) 2 (D) $\frac{1}{4}$

3. The length of a magnet is large compared to its width and breadth. The time period of its oscillation in a vibration magnetometer is 2 s. The magnet is cut along its length into three equal parts and these parts are then placed on each other with their like poles together. The time period of this combination will be : **[2004]**

(A) $2\sqrt{3}$ s (B) $\frac{2}{3}$ s
(C) 2 s (D) $\frac{2}{\sqrt{3}}$ s

4. A magnetic needle is kept in a non-uniform magnetic field. It experiences : **[2005]**
(A) a torque but not a force (B) neither a force nor a torque
(C) a force and a torque (D) a force but not a torque

5. A bar magnet of length 6 cm has a magnetic moment of 4 JT^{-1}. Find the strength of magnetic field at a distance of 200 cm from the centre of the magnet along its equatorial line : **[2012 (Online)]**
(A) 4×10^{-8} tesla (B) 3.5×10^{-8} tesla
(C) 5×10^{-8} tesla (D) 3×10^{-8} tesla

6. Two short bar magnets of length 1 cm each have magnetic moments 1.20 Am^2 and 1.00 Am^2 respectively. They are placed on a horizontal table parallel to each other with their N poles pointing towards the South. They have a common magnetic equator and are separated by a distance of 20.0 cm. The value of the resultant horizontal magnetic induction at the mid-point O of the line joining their centres is close to : (Horizontal component of earth's magnetic induction is 3.6×10^{-5} Wb/m^2) : **[2013]**
(A) 3.6×10^{-5} Wb/m^2 (B) 2.56×10^{-4} Wb/m^2
(C) 3.50×10^{-4} Wb/m^2 (D) 5.80×10^{-4} Wb/m^2

7. The earth's magnetic field lines resemble that of a dipole at the centre of the earth. If the magnetic moment of this dipole at the centre of the earth. If the magnetic moment of this dipole is close to 8×10^{22} Am^2, the value of earth's magnetic field near the equator is close to (radius of the earth = 6.4×10^6 m) : **[2013 (Online)]**
(A) 0.6 Gauss (B) 1.2 Gauss
(C) 1.8 Gauss (D) 0.32 Gauss

8. The mid points of two small magnetic dipoles of length d in end-on positions, are separated by a distance x, ($x >> d$). The force between them is proportional to x^{-n} where n is : **[2014 (Online)]**

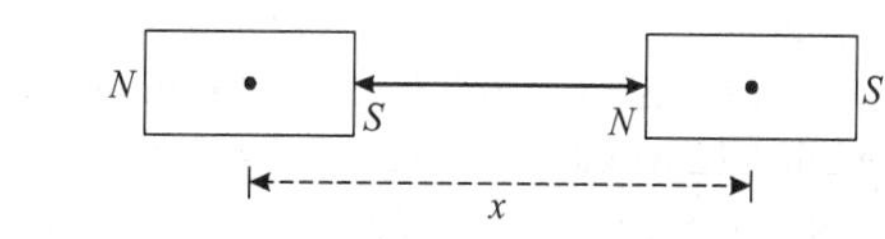

(A) 1 (B) 2
(C) 3 (D) 4

9. The magnetic field of earth at the equator is approximately 4×10^{-5} T. The radius of earth is 6.4×10^6 m. Then the dipole moment of the earth will be nearly of the order of : **[2014 (Online)]**
(A) 10^{23} Am^2 (B) 10^{20} Am^2
(C) 10^{16} Am^2 (D) 10^{10} Am^2

10. A short bar magnet is placed in the magnetic meridian of the earth with north pole pointing north. Neutral points are found at a distance of 30 cm from the magnet on the East-West line, drawn through the middle point of the magnet. The magnetic moment of the magnet in Am^2 is close to : (Given $\frac{\mu_0}{4\pi} = 10^{-7}$ in SI units and B_H = Horizontal component of earth's magnetic field = 3.6×10^{-5} tesla) **[2015 (Online)]**
(A) 14.6 (B) 19.4
(C) 9.7 (D) 4.9

11. A 25 cm long solenoid has radius 2 cm and 500 total number of turns. It carries a current of 15 A. If it is equivalent to a magnet of the same size and magnetization $\vec{M}$ (magnetic moment/volume), then $|\vec{M}|$ is : **[2015 (Online)]**

(A) $30000\,\pi\,Am^{-1}$ (B) $3\,\pi\,Am^{-1}$
(C) $30000\,Am^{-1}$ (D) $300\,Am^{-1}$

12. A magnetic dipole is acted upon by two magnetic fields which are inclined to each other at an angle of 75°. One of the fields has a magnitude of 15 mT. The dipole attains stable equilibrium at an angle of 30° with this field. The magnitude of the other field (in *mT*) is close to : **[2016 (Online)]**

(A) 1 (B) 11
(C) 36 (D) 1060

13. A magnetic needle of magnetic moment $6.7 \times 10^{-2}\,Am^2$ and moment of inertia $7.5 \times 10^{-6}\,kg\,m^2$ is performing simple harmonic oscillations in a magnetic field of 0.01 T. Time taken for 10 complete oscillations is : **[2017]**

(A) 6.98 s (B) 8.76 s
(C) 6.65 s (D) 8.89 s

14. A Magnetic dipole in a constant magnetic field has : **[2017 (Online)]**

(A) Maximum potential energy when the torque is maximum
(B) Zero potential energy when the torque is minimum
(C) Zero potential energy when the torque is maximum
(D) Minimum potential energy when the torque is maximum

15. At some location on earth the horizontal component of earth's magnetic field is 18×10^{-6} T. At this location, magnetic needle of length 0.12 m and pole strength 1.8 Am is suspended from its mid-point using a thread, it makes 45° angle with horizontal in equilibrium. To keep this needle horizontal, the vertical force that should be applied at one of its ends is : **[10 Jan. 2019 (Evening)]**

(A) 3.6×10^{-5} N (B) 6.5×10^{-5} N
(C) 1.3×10^{-5} N (D) 1.8×10^{-5} N

16. A hoop and a solid cylinder of same mass and radius are made of a permanent magnetic material with their magnetic moment parallel to their respective axes. But the magnetic moment of hoop is twice of solid cylinder. They are placed in a uniform magnetic field in such a manner that their magnetic moments make a small angle with the field. If the oscillation periods of hoop and cylinder are T_h and T_c respectively, then: **[10 Jan. 2019 (Evening)]**

(A) $T_h = 0.5\,T_c$ (B) $T_h = 2\,T_c$
(C) $T_h = 1.5\,T_c$ (D) $T_h = T_c$

17. A square loop is carrying a steady current *I* and the magnitude of its magnetic dipole moment is *m*. If this square loop is changed to a circular loop and it carries the same current, the magnitude of the magnetic dipole moment of circular loop will be : **[10 April 2019 (Evening)]**

(A) $\frac{3m}{\pi}$ (B) $\frac{4m}{\pi}$
(C) $\frac{2m}{\pi}$ (D) $\frac{m}{\pi}$

18. A magnetic compass needle oscillates 30 times per minute at a place where the dip is 45°, and 40 times per minute where the dip is 30°. If B_1 and B_2 are respectively the total magnetic field due to the earth at the two places, then the ratio B_1/B_2 is best given by : **[12 April 2019 (Morning)]**

(A) 2.2 (B) 1.8
(C) 0.7 (D) 3.6

19. A small bar magnet placed with its axis at 30° with an external field of 0.06 T experiences a torque of 0.018 Nm. The minimum work required to rotate it from its stable to unstable equilibrium position is : **[4 Sept. 2020 (Morning)]**

(A) 9.2×10^{-3} J (B) 6.4×10^{-2} J
(C) 11.7×10^{-3} J (D) 7.2×10^{-2} J

20. A bar magnet of length 14 cm is placed in the magnetic meridian with its north pole pointing towards the geographic north pole. A neutral point is obtained at a distance of 18 cm from the center of the magnet. If $B_H = 0.4$ G, the magnetic moment of the magnet is (1 G $= 10^{-4}$T) **[16 March 2021 (Morning)]**

(A) $2.880 \times 10^3\,J\,T^{-1}$ (B) $2.880 \times 10^2\,J\,T^{-1}$
(C) $2.880\,J\,T^{-1}$ (D) $28.80\,J\,T^{-1}$

21. At an angle of 30° to the magnetic meridian, the apparent dip is 45°. Find the true dip : **[20 July 2021 (Evening)]**

(A) $\tan^{-1}\frac{\sqrt{3}}{2}$ (B) $\tan^{-1}\frac{1}{\sqrt{3}}$

(C) $\tan^{-1}\sqrt{3}$ (D) $\tan^{-1}\frac{2}{\sqrt{3}}$

22. Choose the correct option : **[22 July 2021 (Evening)]**

(A) True dip is always equal to apparent dip
(B) True dip is less than the apparent dip
(C) True dip is always greater than the apparent dip
(D) True dip is not mathematically related to apparent dip

23. A bar magnet having a magnetic moment of $2.0 \times 10^5\,J\,T^{-1}$, is placed along the direction of uniform magnetic field of magnitude $B = 14 \times 10^{-5}$ T. The work done in rotating the magnet slowly through 60° from the direction of field is : **[26 June 2022 (Evening)]**

(A) 14 J (B) 8.4 J
(C) 4 J (D) 1.4 J

24. At a certain place the angle of dip is 30° and the horizontal component of earth's magnetic field is $0.5G$. The earth's total magnetic field (in G), at that certain place, is : **[29 June 2022 (Morning)]**

(A) $\frac{1}{\sqrt{3}}$ (B) $\frac{1}{2}$
(C) $\sqrt{3}$ (D) 1

25. Two bar magnets oscillate in a horizontal plane in earth's magnetic field with time periods of 3 s and 4 s respectively. If their moments of inertia are in the ratio of 3 : 2 then the ratio of their magnetic moments will be : **[27 July 2022 (Morning)]**

(A) 2 : 1 (B) 8 : 3
(C) 1 : 3 (D) 27 : 16

26. A magnet hung at 45° with magnetic meridian makes an angle of 60° with the horizontal. The actual value of the angle of dip is : **[27 July 2022 (Morning)]**

(A) $\tan^{-1}\left(\sqrt{\frac{3}{2}}\right)$ (B) $\tan^{-1}(\sqrt{6})$
(C) $\tan^{-1}\left(\sqrt{\frac{2}{3}}\right)$ (D) $\tan^{-1}\left(\sqrt{\frac{1}{2}}\right)$

27. A compass needle of oscillation magnetometer oscillates 20 times per minute at a place P of dip 30°. The number of oscillations per minute become 10 at another place Q of 60° dip. The ratio of the total magnetic field at the two places ($B_Q : B_P$) is : **[27 July 2022 (Evening)]**

(A) $\sqrt{3} : 4$ (B) $4 : \sqrt{3}$
(C) $\sqrt{3} : 2$ (D) $2 : \sqrt{3}$

28. The current sensitivity of a galvanometer can be increased by :

(a) Decreasing the number of turns
(b) Increasing the magnetic field
(c) Decreasing the area of the coil
(d) Decreasing the torsional constant of the spring

Choose the most appropriate answer from the options given below : **[28 July 2022 (Morning)]**

(A) (b) and (c) only (B) (c) and (d) only
(C) (a) and (c) only (D) (b) and (d) only

29. The vertical component of the earth's magnetic field is 6×10^{-5} T at any place where the angle of dip is 37°. The earth's resultant magnetic field at that place will be :

[Given : $\tan 37° = \frac{3}{4}$] **[29 July 2022 (Evening)]**

(A) 8×10^{-5} T (B) 6×10^{-5} T
(C) 5×10^{-4} T (D) 1×10^{-4} T

30. Spherical insulating ball and a spherical metallic ball of same size and mass are dropped from the same height. Choose the correct statement out of the following: [Assume negligible air friction] **[31 Jan. 2023 (Morning)]**

(A) Time taken by them to reach the earth's surface will be independent of the properties of their materials
(B) Insulating ball will reach the earth's surface earlier than the metal ball
(C) Both will reach the earth's surface simultaneously
(D) Metal ball will reach the earth's surface earlier than the insulating ball

31. Given below are two statements; one is labelled as **Assertion-A** and the other is labelled as **Reason-R**.

Assertion-A: Electromagnets are made of soft iron

Reason-R: Soft iron has high permeability and low retentivity.

In the light of above, statements, choose the most appropriate answer from the options given below : **[8 April 2023 (Evening)]**

(A) **A** is not correct but **R** is correct.
(B) Both **A** and **R** are correct and **R** is the correct explanation of **A**.
(C) Both **A** and **R** are correct but **R** is NOT the correct explanation of **A**.
(D) **A** is correct but **R** is not correct.

32. The current required to be passed through a solenoid of 15 cm length and 60 turns in order to demagnetise a bar magnet of magnetic intensity 2.4×10^3 Am^{-1} is ______ A. **[10 April 2023 (Morning)]**

33. A bar magnet is released from rest along the axis of a very long vertical copper tube. After some time the magnet will: **[10 April 2023 (Evening)]**

(A) Move down with almost constant speed
(B) Oscillate inside the tube
(C) Move down with an acceleration greater than g
(D) Move down with an acceleration equal to g

34. The free space inside a current carrying toroid is filled with a material of susceptibility 2×10^{-2}. The percentage increase in the value of magnetic field inside the toroid will be : **[11 April 2023 (Morning)]**

(A) 2% (B) 0.2%
(C) 0.1% (D) 1%

35. A compass needle oscillates 20 times per minute at a place where the dip is 30° and 30 times per minute where the dip is 60°. The ratio of total magnetic field due to the earth at two place respectively is $\frac{4}{\sqrt{x}}$. The value of x is. **[12 April 2023 (Morning)]**

* * * * *

MAGNETISM AND MATTER

19.2 MAGNETIC PROPERTIES OF MATERIALS

1. Curie temperature is the temperature above which : **[2003]**
(A) A ferromagnetic material becomes paramagnetic
(B) A paramagnetic material becomes diamagnetic
(C) A ferromagnetic material becomes diamagnetic
(D) A paramagnetic material becomes ferromagnetic

2. The materials suitable for making electromagnets should have : **[2004]**
(A) High retentivity and low coercivity
(B) Low retentivity and low coercivity
(C) High retentivity and high coercivity
(D) Low retentivity and high coercivity

3. Needles N_1, N_2 and N_3 are made of a ferromagnetic, a paramagnetic and a diamagnetic substance respectively. A magnet when brought close to them will : **[2006]**
(A) Attract N_1 and N_2 strongly but repel N_3
(B) Attract N_1 strongly, N_2 weakly and repel N_3 weakly
(C) Attract N_1 strongly, but repel N_2 and N_3 weakly
(D) Attract all three of them

4. Relative permittivity and permeability of a material ε_r and μ_r, respectively. Which of the following values of these quantities are allowed for a diamagnetic material ? **[2008]**
(A) $\varepsilon_r = 0.5, \mu_r = 1.5$ (B) $\varepsilon_r = 1.5, \mu_r = 0.5$
(C) $\varepsilon_r = 0.5, \mu_r = 0.5$ (D) $\varepsilon_r = 1.5, \mu_r = 1.5$

5. The coercivity of a small magnet where the ferromagnet gets demagnetized is 3×10^3 A m^{-1}. The current required to be passed in a solenoid of length 10 cm and number of turns 100, so that the magnet gets demagnetized when inside the solenoid, is : **[2014]**
(A) 30 mA (B) 60 mA
(C) 3 A (D) 6 A

6. Three identical bars A, B and C are made of different magnetic materials. When kept in a uniform magnetic field, the field lines around them look as follows :

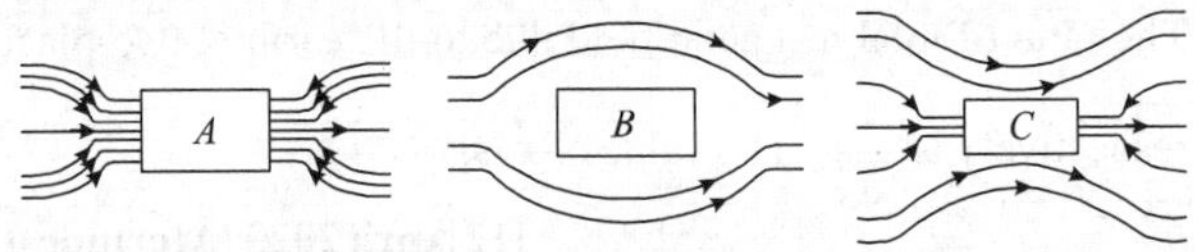

Make the correspondence of these bars with their material being diamagnetic (D), ferromagnetic (F) and paramagnetic (P) : **[2014 (Online)]**
(A) $A \leftrightarrow D, B \leftrightarrow P, C \leftrightarrow F$ (B) $A \leftrightarrow F, B \leftrightarrow D, C \leftrightarrow P$
(C) $A \leftrightarrow P, B \leftrightarrow F, C \leftrightarrow D$ (D) $A \leftrightarrow F, B \leftrightarrow P, C \leftrightarrow D$

7. An example of a perfect diamagnet is a superconductor. This implies that when a superconductor is put in a magnetic field of intensity B, the magnetic field B_s inside the superconductor will be such that : **[2014 (Online)]**
(A) $B_s = -B$ (B) $B_s = 0$
(C) $B_s = B$ (D) $B_s < B$ but $Bs \neq 0$

8. Hysteresis loops for two magnetic materials A and B are given below :

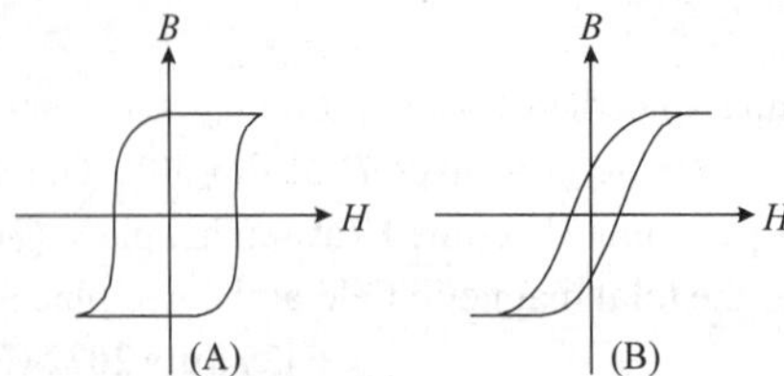

These materials are used to make magnets for electric generators, transformer core and electromagnet core. Then it is proper to use : **[2016]**
(A) A for electric generators and transformers
(B) A for electromagnets and B for electric generators
(C) A for transformers and B for electric generators
(D) B for electromagnets and transformers

9. The B-H curve for a ferromagnet is shown in the figure. The ferromagnet is placed inside a long solenoid with 1000 turns/cm. The current that should be passed in the solenoid to demagnetize the ferromagnet completely is : **[2018 (Online)]**

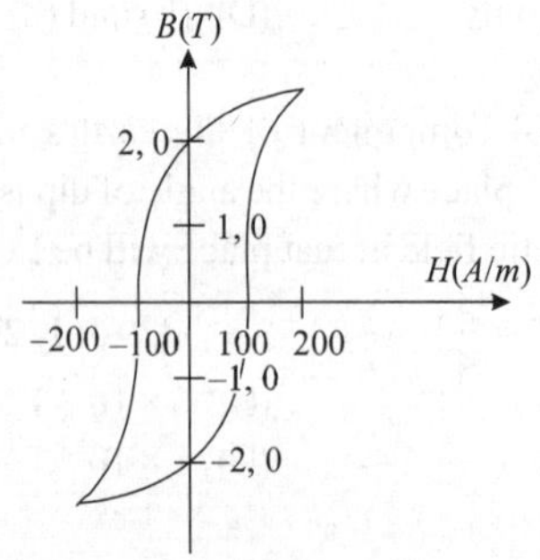

(A) 2 mA (B) 1 mA
(C) 40 μA (D) 20 μA

10. A bar magnet is demagnetized by inserting it inside a solenoid of length 0.2 m, 100 turns, and carrying a current of 5.2 A. The coercivity of the bar magnet is : **[9 Jan. 2019 (Morning)]**

(A) 1200 A/m (B) 2600 A/m
(C) 520 A/m (D) 285 A/m

11. A paramagnetic substance in the form of a cube with sides 1 cm has a magnetic dipole moment of 20×10^{-6} J/T when a magnetic intensity of 60×10^3 A/m is applied. Its magnetic susceptibility is : **[11 Jan. 2019 (Evening)]**

(A) 2.3×10^{-2} (B) 3.3×10^{-2}
(C) 3.3×10^{-4} (D) 4.3×10^{-2}

12. A paramagnetic material has 10^{28} atoms/m^3. Its magnetic susceptibility at temperature 350 K is 2.8×10^{-4}. Its susceptibility at 300 K is : **[12 Jan. 2019 (Evening)]**

(A) 3.672×10^{-4} (B) 3.726×10^{-4}
(C) 3.267×10^{-4} (D) 2.672×10^{-4}

13. Two magnetic dipoles X and Y are placed at a separation d, with their axes perpendicular to each other. The dipole moment of Y is twice that of X. A particle of charge q is passing, through their midpoint P, at angle $\theta = 45°$ with the horizontal line, as shown in figure. What would be the magnitude of force on the particle at that instant ? (d is much larger than the dimensions of the dipole) : **[8 April 2019 (Evening)]**

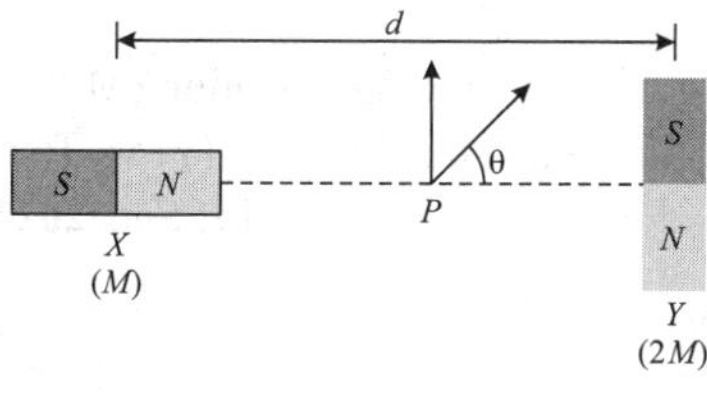

(A) $\sqrt{2}\left(\frac{\mu_0}{4\pi}\right)\frac{M}{(d/2)^3}\times qv$ (B) $\left(\frac{\mu_0}{4\pi}\right)\frac{2M}{(d/2)^3}\times qv$

(C) $\left(\frac{\mu_0}{4\pi}\right)\frac{M}{(d/2)^3}\times qv$ (D) 0

14. The figure gives experimentally measured B vs H variation in a ferromagnetic material. The retentivity, coercivity and saturation, respectively, of the material are : **[7 Jan. 2020 (Evening)]**

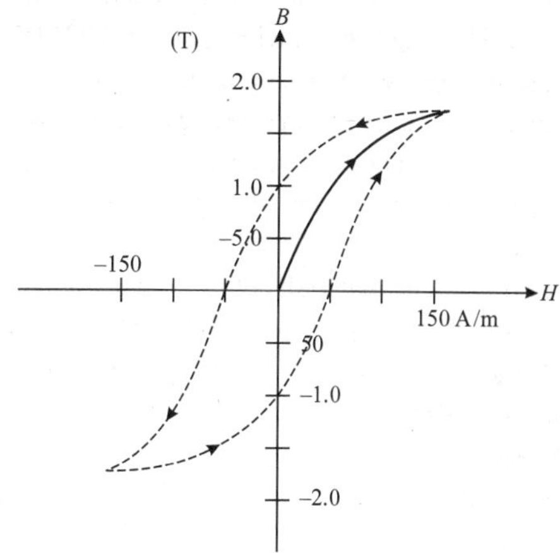

(A) 150 A/m, 1.0 T and 1.5 T (B) 1.0 T, 50 A/m and 1.5 T
(C) 1.5 T, 50 A/m and 1.0 T (D) 1.5 T, 50 A/m and 1.0 T

15. Magnetic materials used for making permanent magnets (P) and magnets in a transformer (T) have different properties of the following, which property best matches for the type of magnet required : **[2 Sept. 2020 (Morning)]**

(A) T : Large retentivity, small coercivity
(B) P : Small retentivity, large coercivity
(C) T : Large retentivity, large coercivity
(D) P : Large retentivity, large coercivity

16. A perfectly diamagnetic sphere has a small spherical cavity at its centre, which is filled with a paramagnetic substance. The whole system is placed in a uniform magnetic field $\vec{B}$. Then the field inside the paramagnetic substance is : **[3 Sept. 2020 (Evening)]**

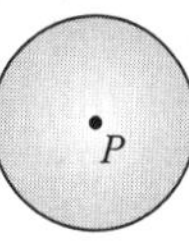

(A) Zero
(B) $\vec{B}$
(C) Much large than $|\vec{B}|$ but opposite to $\vec{B}$
(D) Much large than $|\vec{B}|$ and parallel to $\vec{B}$

17. A paramagnetic sample shows a net magnetisation of 6 A/m when it is placed in an external magnetic field of 0.4 T at a temperature of 4 K. When the sample is placed in an external magnetic field of 0.3 T at a temperature of 24 K, then the magnetisation will be : **[4 Sept. 2020 (Evening)]**

(A) 4 A/m (B) 0.75 A/m
(C) 2.25 A/m (D) 1 A/m

18. A soft ferromagnetic material is placed in an external magnetic field. The magnetic domains : **[24 Feb. 2021 (Evening)]**

(A) Increase in size but no change in orientation

(B) May increase or decrease in size and change its orientation

(C) Decrease in size and changes orientation

(D) Have no relation with external magnetic field

19. In a ferromagnetic material, below the curie temperature, a domain is defined as : **[25 Feb. 2021 (Evening)]**

(A) A macroscopic region with zero magnetization

(B) A macroscopic region with consecutive magnetic dipoles oriented in opposite direction

(C) A macroscopic region with randomly oriented magnetic dipoles

(D) A macroscopic region with saturation magnetization

20. Which of the following statements are correct?

(a) Electric monopoles do not exist whereas magnetic monopoles exist.

(b) Magnetic field lines due to a solenoid at its ends and outside cannot be completely straight and confined.

(c) Magnetic field lines are completely confined within a toroid.

(d) Magnetic field lines inside a bar magnet are not parallel.

(e) $\chi = -1$ is the condition for a perfect diamagnetic material, where χ is its magnetic susceptibility.

Choose the correct answer from the options given below :

[18 March 2021 (Evening)]

(A) (c) and (e) only (B) (b) and (d) only

(C) (a) and (b) only (D) (b) and (c) only

21. The magnetic susceptibility of a material of a rod is 499. Permeability in vacuum is $4\pi \times 10^{-7}$ H/m. Absolute permeability of the material of the rod is : **[20 July 2021 (Evening)]**

(A) $\pi \times 10^{-4}$ H/m (B) $3\pi \times 10^{-4}$ H/m

(C) $4\pi \times 10^{-4}$ H/m (D) $2\pi \times 10^{-4}$ H/m

22. Statement-I: The ferromagnetic property depends on temperature. At high temperature, ferromagnet becomes paramagnet.

Statement-II: At high temperature, the domain wall area of a ferromagnetic substance increases.

In the light of the above statements, choose the most appropriate answer from the option given below : **[22 July 2021 (Evening)]**

(A) Both Statement-I and Statement-II are true

(B) Statement-I is false but Statement-II is true

(C) Both Statement-I and Statement-II are false

(D) Statement-I is true but Statement-II is false

23. Two short magnetic dipoles m_1 and m_2 each having magnetic moment of 1 Am^2 are placed at point O and P respectively. The distance between OP is 1 meter. The torque experienced by the magnetic dipole m_2 due to the presence of m1 is ______ $\times 10^{-7}$ Nm : **[26 Aug. 2021 (Morning)]**

24. A coil in the shape of an equilateral triangle of side 10 cm lies in a vertical place between the pole pieces of permanent magnet producing a horizontal magnetic field 20 mT. The torque acting on the coil when a current of 0.2 A is passed through it and its plane becomes parallel to the magnetic field will be $\sqrt{x} \times 10^5$ Nm. The value of x is ______.

[26 Aug. 2021 (Evening)]

25. A uniform conducting wire of length $24a$, and resistance R is wound up as a current carrying coil in the shape of an equilateral triangle of side a and then in the form of a square of side a. The coil is connected to a voltage source V_0. The ratio of magnetic moment of the coils in case of equilateral triangle to that for square is $1 : y$ where y is ______.

[27 Aug. 2021 (Morning)]

26. A long solenoid with 1000 turns/m has a core material with relative permeability 500 and volume 10^3 cm^3. If the core material is replaced by another material having relative permeability of 750 with same volume maintain same current of 0.75 A in the solenoid. The fractional change in the magnetic moment of the core would be approximately $\left(\frac{x}{499}\right)$. Find the value of x :

[31 Aug. 2021 (Evening)]

27. Following plots show Magnetization (M) vs Magnetising field (H) and Magnetic susceptibility (χ) vs Temperature (T) graph : **[01 Sep. 2021 (Evening)]**

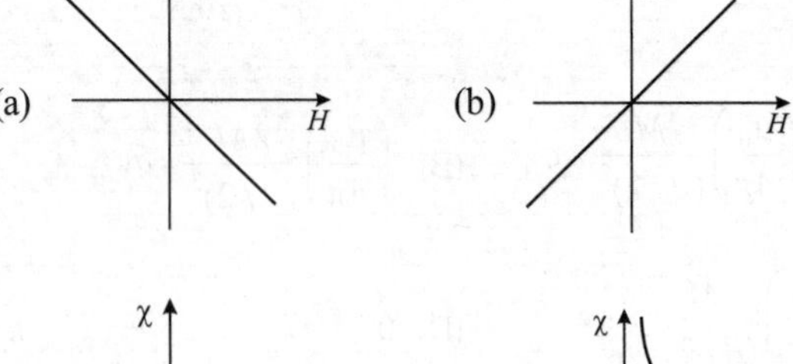

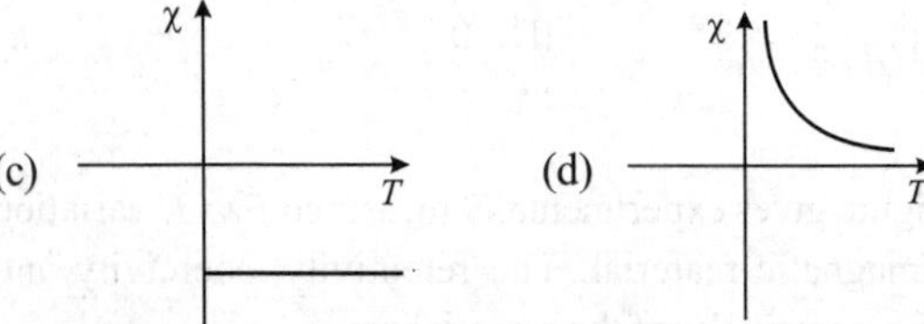

Which of the following combination will be represented by a diamagnetic material ?

(A) (b), (d) (B) (b), (c)

(C) (a), (d) (D) (a), (c)

28. The soft-iron is a suitable material for making an electromagnet. This is because soft-iron has :

[24 June 2022 (Evening)]

(A) Low coercively and high retentively
(B) Low coercively and low permeability
(C) High permeability and low retentively
(D) High permeability and high retentively

29. Given below are two statements : **[25 June 2022 (Evening)]**

Statement-I : Susceptibilities of paramagnetic and ferromagnetic substances increase with decrease in temperature.

Statement-II : Diamagnetism is a result of orbital motions of electrons developing magnetic moments opposite to the applied magnetic field.

Choose the **CORRECT** answer from the options given below :

(A) Both statement-I and statement-II are true
(B) Both statement-I and Statement-II are false
(C) Statement-I is true but statement-II is false
(D) Statement-I is false but Statement-II is true

30. The susceptibility of a paramagnetic material is 99. The permeability of the material in Wb/A-m is : [Permeability of free space $\mu_0 = 4\pi \times 10^{-7}$ Wb/A-m] **[27 June 2022 (Morning)]**

(A) $4\pi \times 10^{-7}$ (B) $4\pi \times 10^{-4}$
(C) $4\pi \times 10^{-5}$ (D) $4\pi \times 10^{-6}$

31. The space inside a straight current carrying solenoid is filled with a magnetic material having magnetic susceptibility equal to 1.2×10^{-5}. What is fractional increase in the magnetic field inside solenoid with respect to air as medium inside the solenoid? **[28 June 2022 (Evening)]**

(A) 1.2×10^{-5} (B) 1.2×10^{-3}
(C) 1.8×10^{-3} (D) 2.4×10^{-5}

32. Certain galvanometers have a fixed core made of non magnetic metallic material. The function of this metallic material is : **[8 April 2023 (Morning)]**

(A) To oscillate the coil in magnetic field for longer period of time
(B) To bring the coil to rest quickly
(C) To produce large deflecting torque on the coil
(D) To make the magnetic field radial

33. Given below are two Statements:

Statement-I: For diamagnetic substance, $-1 \le c < 0$, where c is the magnetic susceptibility.

Statement-II: Diamagnetic substances when placed in an external magnetic field, tend to move from stronger to weaker part of the field.

In the light of the above statements, choose the:

[10 April 2023 (Evening)]

(A) Both Statement-I and Statement-II are false
(B) Both Statement-I and Statement-II are true
(C) Statement-I is incorrect but Statement-II is true
(D) Statement-I is correct but Statement-II is false

34. Given below are two statements : one is labelled as **Assertion-A** and the other is labelled as **Reason-R**.

Assertion-A : A bar magnet dropped through a metallic cylindrical pipe takes more time to come down compared to a non-magnetic bar with same geometry and mass.

Reason-R : For the magnetic bar, Eddy currents are produced in the metallic pipe which oppose the motion of the magnetic bar.

In the light of the above statements, choose the correct answer from the options given below : **[11 April 2023 (Evening)]**

(A) Both **A** and **R** are true but **R** is NOT the correct explanation of **A**.
(B) **A** is true but **R** is false.
(C) Both **A** and **R** are true and **R** is the correct explanation of **A**.
(D) **A** is false but **R** is true.

35. Given below are two statements:

Statement-I: The diamagnetic property depends on temperature.

Statement-II: The included magnetic dipole moment in a diamagnetic sample is always opposite to the magnetizing field.

In the light of given statement, choose the correct answer from the options given below: **[12 April 2023 (Morning)]**

(A) Statement-I is incorrect but Statement-II is true
(B) Both Statement-I and Statement-II are true
(C) Both Statement-I and Statement-II are false
(D) Statement-I is correct but Statement-II is false

* * * * *

CHAPTER

20 ELECTROMAGNETIC INDUCTION

20.1 ELECTROMAGNETIC INDUCTION

1. A conducting square loop of side L and resistance R moves in its plane with a uniform velocity v perpendicular to one of its sides. A magnetic induction B constant in time and space, pointing perpendicular and into the plane at the loop exists everywhere with half the loop outside the field, as shown in figure. The induced emf is : **[2002]**

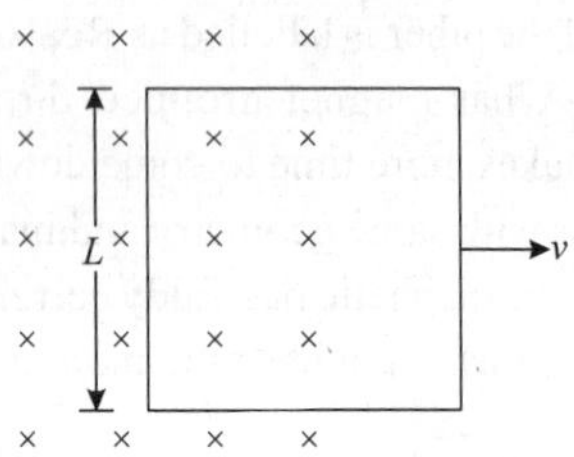

(A) Zero (B) RvB

(C) $\frac{vBL}{R}$ (D) vBL

2. A metal conductor of length 1 m rotates vertically about one of its ends at angular velocity 5 radians per second. If the horizontal component of earth's magnetic field is 0.2×10^{-4} T, then the emf developed between the two ends of the conductor is : **[2004]**

(A) $5 \mu V$ (B) $50 \mu V$

(C) 5mV (D) 50mV

3. In a uniform magnetic field of induction B, a wire in the form of semicircle of radius r rotates about the diameter of the circle with angular frequency ω. If the total resistance of the circuit is R, the mean power generated per period of rotation is: **[2004]**

(A) $\frac{B\pi r^2 \omega}{2R}$ (B) $\frac{(B\pi r^2 \omega)^2}{8R}$

(C) $\frac{(B\pi r \omega)^2}{2R}$ (D) $\frac{(B\pi r \omega^2)^2}{8R}$

4. A coil having n turns and resistance $R\,\Omega$ is connected with a galvanometer of resistance $4\,R\Omega$. This combination is moved in time t seconds from a magnetic field W_1 weber to W_2 weber. The induced current in the circuit is : **[2004]**

(A) $\frac{W_2 - W_1}{5Rnt}$ (B) $-\frac{n(W_2 - W_1)}{5Rt}$

(C) $-\frac{(W_2 - W_1)}{Rnt}$ (D) $-\frac{n(W_2 - W_1)}{Rt}$

5. One conducting U-tube can slide inside another as shown in figure, maintaining electrical contacts between the tubes. The magnetic field B is perpendicular to the plane of the figure. If each tube moves towards the other at a constant speed v, then the emf induced in the circuit in terms of B, l and v, where l is the width of each tube, will be : **[2005]**

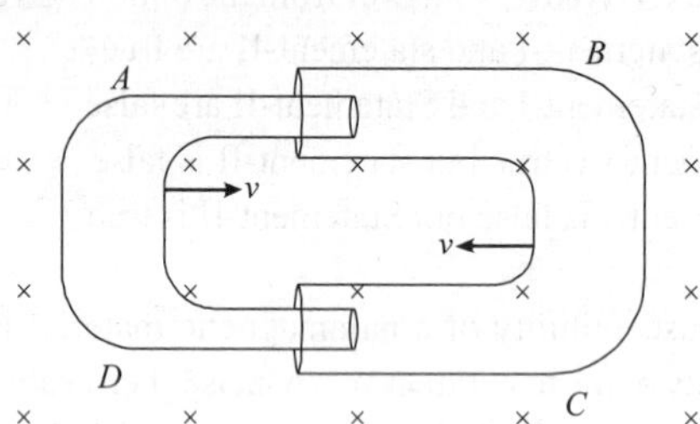

(A) Blv (B) $-Blv$

(C) zero (D) $2Blv$

6. A rectangular loop has a sliding connector PQ of length l and resistance $R\Omega$ and it is moving with a speed v as shown. The set-up is placed in a uniform magnetic field going into the plane of the paper. The three currents I_1, I_2 and I are : **[2010]**

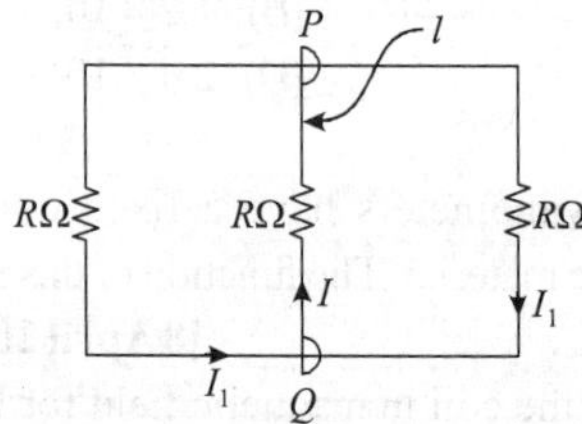

(A) $I_1 = I_2 = \frac{Blv}{6R}, I = \frac{Blv}{3R}$ (B) $I_1 = -I_2 = \frac{Blv}{R}, I = \frac{2Blv}{R}$

(C) $I_1 = I_2 = \frac{Blv}{3R}, I = \frac{2Blv}{3R}$ (D) $I_1 = I_2 = , I = \frac{Blv}{R}$

7. A horizontal straight wire 20 m long extending from east to west is falling with a speed of 5.0 m/s, at right angles to the horizontal component of the earth's magnetic field 0.30×10^{-4} Wb/m^2. The instantaneous value of the emf induced in the wire will be : **[2011]**

(A) 6.0mV (B) 3mV

(C) 4.5mV (D) 1.5mV

8. A boat is moving due east in a region where the earth's magnetic field is 5.0×10^{-5} NA^{-1} m^{-1} due north and horizontal. The boat carries a vertical aerial 2 m long. If the speed of the boat is 1.50 ms^{-1}, the magnitude of the induced emf in the wire of aerial is : **[2011]**

(A) 0.50 mV (B) 0.15 mV
(C) 1 mV (D) 0.75 mV

9. A coil is suspended in a uniform magnetic field, with the plane of the coil parallel to the magnetic lines of force. When a current is passed through the coil it starts oscillating; it is very difficult to stop. But if an aluminium plate is placed near to the coil, it stops. This is due to : **[2012]**

(A) Electromagnetic induction in the aluminium plate giving rise to electromagnetic damping
(B) Development of air current when the plate is placed
(C) Induction of electrical charge on the plate
(D) Shielding of magnetic lines of force as aluminium is a paramagnetic material

10. Magnetic flux through a coil of resistance 10 Ω is changed by $\Delta\phi$ in 0.1 s. The resulting current in the coil varies with time as shown in the figure. Then $|\Delta\phi|$ is equal to (in weber) : **[2012 (Online)]**

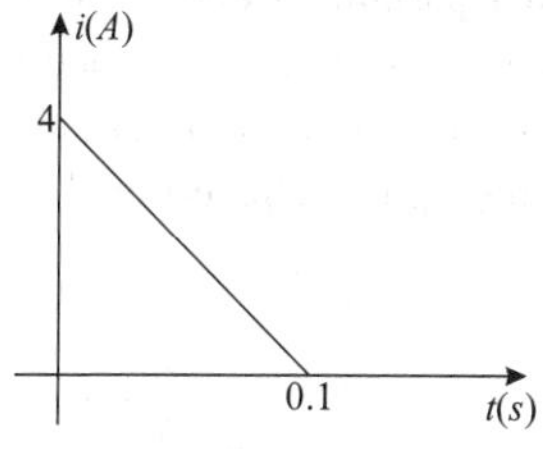

(A) 6 (B) 4
(C) 2 (D) 8

11. A circular loop of radius 0.3 cm lies parallel to a much bigger circular loop of radius 20 cm. The centre of the small loop is on the axis of the bigger loop. The distance between their centres is 15 cm. If a current of 2.0 A flows through the smaller loop, then the flux linked with bigger loop is : **[2013]**

(A) 9.1×10^{-11} weber (B) 6×10^{-11} weber
(C) 3.3×10^{-11} weber (D) 6.6×10^{-9} weber

12. A metallic rod of length 'l' is tied to a string of length $2l$ and made to rotate with angular speed ω on a horizontal table with one end of the string fixed. If there is a vertical magnetic field 'B' in the region, the emf induced across the ends of the rod is : **[2013]**

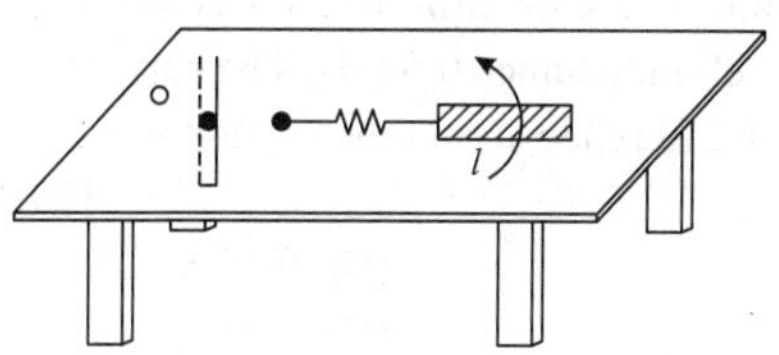

(A) $\frac{2B\omega l^2}{2}$ (B) $\frac{3B\omega l^2}{2}$
(C) $\frac{4B\omega l^2}{2}$ (D) $\frac{5B\omega l^2}{2}$

13. A square frame of side 10 cm and a long straight wire carrying current 1 A are in the plate of the paper. Starting from close to the wire, the frame moves towards the right with a constant speed of 10 ms^{-1} (see figure).
The emf induced at the time the left arm of the frame is at $x = 10$ cm from the wire is: **[2014 (Online)]**

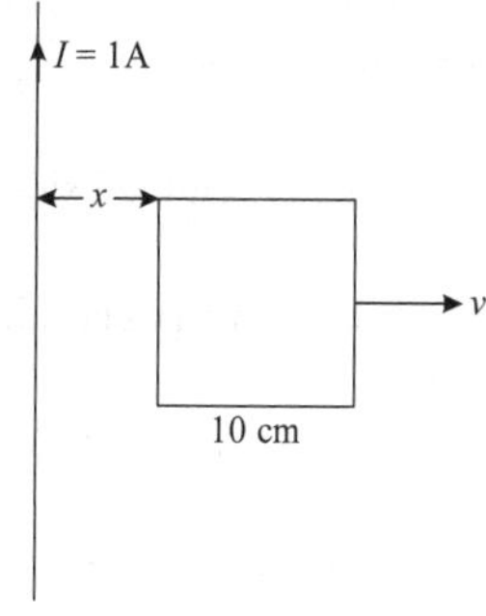

(A) 2 μV (B) 1 μV
(C) 0.75 μV (D) 0.5 μV

14. A coil of circular cross-section having 1000 turns and 4 cm^2 face area is placed with its axis parallel to a magnetic field which decreases by 10^{-2} Wb m^{-2} in 0.01 s. The emf induced in the coil is : **[2014 (Online)]**

(A) 400 mV (B) 200 mV
(C) 4 mV (D) 0.4 mV

15. Figure shows a circular area of radius R where a uniform magnetic field $\vec{B}$ is going into the plane of paper and increasing in magnitude at a constant rate.
In that case, which of the following graphs, drawn schematically, correctly shows the variation of the induced electric field $E(r)$? **[2014 (Online)]**

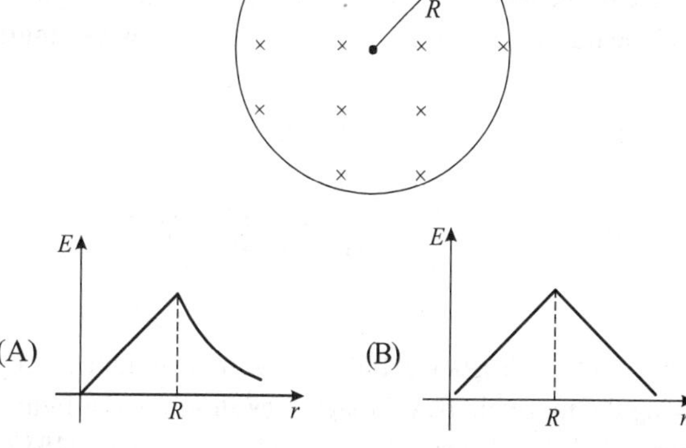

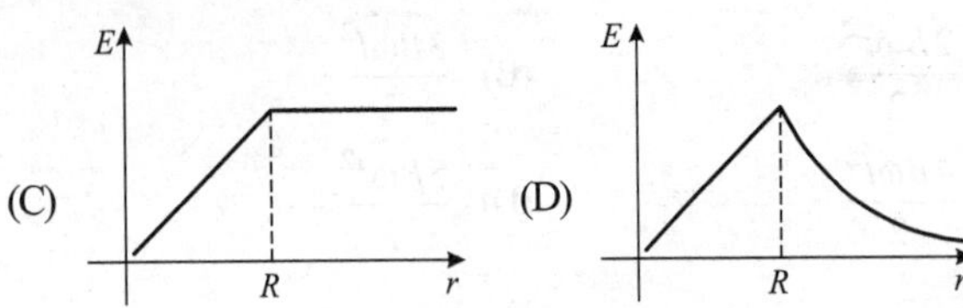

16. A fighter plane of length 20 m, wing span (distance from tip of one wing to the tip of the other wing) of 15 m and height 5 m is lying towards east over Delhi. Its speed is 240 ms^{-1}. The earth's magnetic field over Delhi is 5×10^{-5} T with the declination angle ~ 0° and dip of θ such that sin θ = 2/3. If the voltage developed is V_B between the lower and upper side of the plane of V_W between the tips of the wings then V_B and V_W are close to: **[2016 (Online)]**

(A) V_B = 40 mV; V_w = 120 mV with left side of pilot at higher voltage

(B) V_B = 45 mV; V_W = 120 mV with right side of pilot at higher voltage

(C) V_B = 40 mV; V_W = 135 mV with right side of pilot at higher voltage

(D) V_B = 45 mV; V_W = 120 mV with left side of pilot at higher voltage

17. Consider a thin metallic sheet perpendicular to the plane of the paper moving with speed 'v' in a uniform magnetic field B going into the plane of the paper (See figure). If charge densities σ_1 and σ_2 are induced on the left and right surfaces, respectively, of the sheet then (ignore fringe effects) : **[2016 (Online)]**

(A) $\sigma_1 = \frac{-\epsilon_0 vB}{2}, \sigma_2 = \frac{\epsilon_0 vB}{2}$

(B) $\sigma_1 = \epsilon_0 vB, \sigma_2 = -\epsilon_0 vB$

(C) $\sigma_1 = \frac{\epsilon_0 vB}{2}, \sigma_2 = \frac{-\epsilon_0 vB}{2}$

(D) $\sigma_1 = \sigma_2 = \epsilon_0 vB$

18. A conducting metal circular-wire-loop of radius r is placed perpendicular to a magnetic field which varies with time as $B = B_0 e^{-t/\tau}$, where B_0 and τ are constants, at time $t = 0$. If the resistance of the loop is R then the heat generated in the loop after a long time ($t \to \infty$) is : **[2016 (Online)]**

(A) $\frac{\pi^2 r^2 B_0^4}{2\tau R}$ (B) $\frac{\pi^2 r^4 B_0^2}{2\tau R}$

(C) $\frac{\pi^2 r^4 B_0^2 R}{\tau}$ (D) $\frac{\pi^2 r^4 B_0^2}{\tau R}$

19. In a coil of resistance 100 Ω, a current is induced by changing the magnetic flux through it as shown in the figure. The magnitude of change in flux through the coil is : **[2017]**

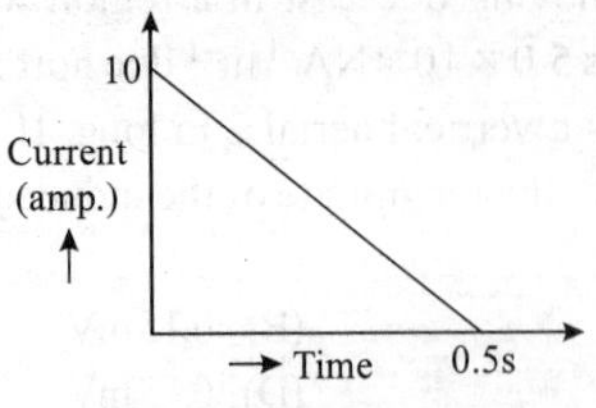

(A) 250 Wb (B) 275 Wb

(C) 200 Wb (D) 225 Wb

20. At the centre of a fixed large circular coil of radius R, a much smaller circular coil of radius r is placed. The two coils are concentric and are in the same plane. The larger coil carries a current I. The smaller coil is set to rotate with a constant angular velocity ω about an axis along their common diameter. Calculate the emf induced in the smaller coil after a time t of its start of rotation : **[2018 (Online)]**

(A) $\frac{\mu_0 I}{2R} \omega r^2 \sin \omega t$ (B) $\frac{\mu_0 I}{4R} \omega \pi r^2 \sin \omega t$

(C) $\frac{\mu_0 I}{2R} \omega \pi r^2 \sin \omega t$ (D) $\frac{\mu_0 I}{4R} \omega r^2 \sin \omega t$

21. A copper rod of mass m slides under gravity on two smooth parallel rails, with separation 1 and set at angle of θ with the horizontal. At the bottom, rails are joined by a resistance R. There is a uniform magnetic field B normal to the plane of the rails, as shown in the figure. The terminal speed of the copper rod is : **[2018 (Online)]**

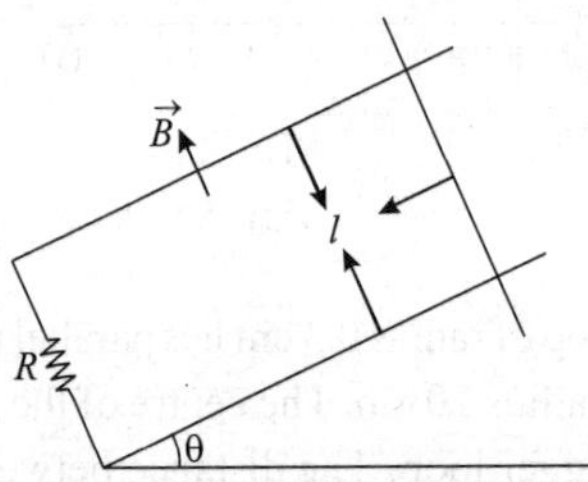

(A) $\frac{mgR \cos \theta}{B^2 l^2}$ (B) $\frac{mgR \sin \theta}{B^2 l^2}$

(C) $\frac{mgR \tan \theta}{B^2 l^2}$ (D) $\frac{mgR \cot \theta}{B^2 l^2}$

22. An ideal capacitor of capacitance 0.2 μF is charged to a potential difference of 10 V. The charging battery is then disconnected. The capacitor is then connected to an ideal inductor of self inductance 0.5 mH. The current at a time when the potential difference across the capacitor is 5 V, is : **[2018 (Online)]**

(A) 0.17 A (B) 0.15 A

(C) 0.34 A (D) 0.25 A

23. A coil of cross-sectional area A having n turns is placed in a uniform magnetic field B. When it is rotated with an angular velocity ω, the maximum emf induced in the coil will be : **[2018 (Online)]**

(A) $nBA\,\omega$ (B) $\frac{3}{2}\cdot nBA\,\omega$

(C) $3\,nBA\,\omega$ (D) $\frac{1}{2}\,nBA\,\omega$

24. A conducting circular loop made of a thin wire, has area 3.5×10^{-3} m^2 and resistance 10 Ω. It is placed perpendicular to a time dependent magnetic field $B(t) = (0.4\text{T}) \sin(50\pi t)$. The field is uniform in space. Then the net charge flowing through the loop during $t = 0$ s and $t = 10$ ms is close to : **[9 Jan. 2019 (Morning)]**

(A) 14 mC (B) 21 mC

(C) 6 mC (D) 7 mC

25. A solid metal cube of edge length 2 cm is moving in a positive y-direction at a constant speed of 6 m/s. There is a uniform magnetic field of 0.1 T in the positive z-direction. The potential difference between the two faces of the cube perpendicular to the x-axis, is : **[10 Jan. 2019 (Morning)]**

(A) 6 mV (B) 1 mV

(C) 12 mV (D) 2 mV

26. A magnet of total magnetic moment $10^{-2}\,\hat{i}$ Am^2 is placed in a time varying magnetic field, $B\,\hat{i}$ (cost ωt) where $B = 1$ Tesla and $\omega = 0.125$ rad/s. The work done for reversing the direction of the magnetic moment at $t = 1$ second, is : **[10 Jan. 2019 (Morning)]**

(A) 0.007 J (B) 0.014 J

(C) 0.01 J (D) 0.028 J

27. A 10 m long horizontal wire extends from North East to South West. It is falling with a speed of 5.0 ms^{-1}, at right angles to the horizontal component of the earth's magnetic field, of 0.3×10^{-4} Wb/m^2. The value of the induced emf in wire is : **[12 Jan. 2019 (Evening)]**

(A) 2.5×10^{-3} V (B) 1.1×10^{-3} V

(C) 0.3×10^{-3} V (D) 1.5×10^{-3} V

28. A thin strip 10 cm long is on a U shaped wire of negligible resistance and it is connected to a spring of spring constant 0.5 Nm^{-1} (see figure). The assembly is kept in a uniform magnetic field of 0.1 T. If the strip is pulled from its equilibrium position and released, the number of oscillation it performs before its amplitude decreases by a factor of e is N. If the mass of the strip is 50 grams, its resistance 10 Ω and air drag negligible, N will be close to : **[8 April 2019 (Morning)]**

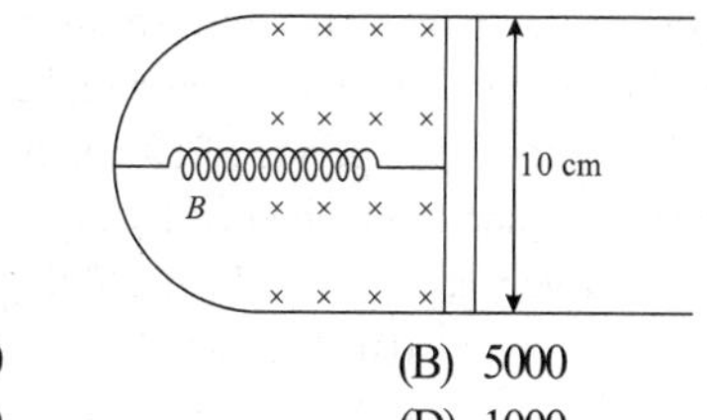

(A) 50000 (B) 5000

(C) 10000 (D) 1000

29. The figure shows a square loop L of side 5 cm which is connected to a network of resistances. The whole setup is moving towards right with a constant speed of 1 cms^{-1}. At some instant, a part of L is in a uniform magnetic field of 1 T, perpendicular to the plane of the loop. If the resistance of L is 1.7 Ω, the current in the loop at that instant will be close to : **[12 April 2019 (Morning)]**

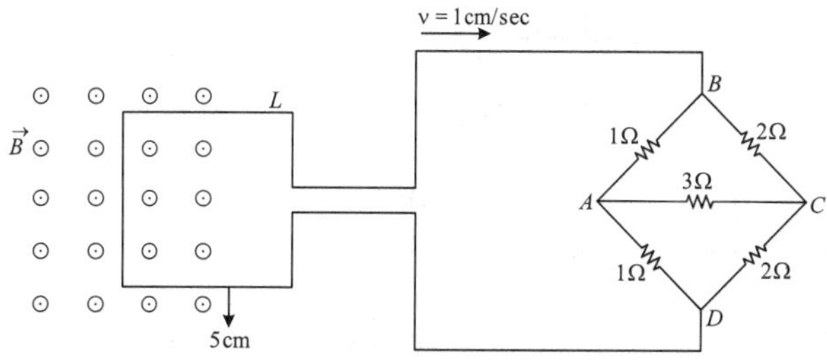

(A) 115 μA (B) 170 μA

(C) 60 μA (D) 150 μA

30. A long solenoid of radius R carries a time (t)-dependent current $I(t) = I_0 t(1 - t)$. A ring of radius $2R$ is placed coaxially near its middle. During the time interval $0 \le t \le 1$, the induced current (I_R) and the induced EMF(V_R) in the ring change as : **[7 Jan. 2020 (Morning)]**

(A) At $t = 0.5$ direction of I_R reverses and V_R is zero

(B) Direction of I_R remains unchanged and V_R is zero at $t = 0.25$

(C) Direction of I_R remains unchanged and V_R is maximum at $t = 0.5$

(D) At $t = 0.25$ direction of I_R reverses and V_R is maximum

31. Consider a circular coil of wire carrying constant current I, forming a magnetic dipole. The magnetic flux through an infinite plane that contains the circular coil and excluding the circular coil area is given by ϕ_i. The magnetic flux through the area of the circular coil area is given by ϕ_0. Which of the following option is correct ? **[7 Jan. 2020 (Morning)]**

(A) $\phi_i = -\phi_0$ (B) $\phi_i = \phi_0$

(C) $\phi_i < \phi_0$ (D) $\phi_i > \phi_0$

32. A loop $ABCDEFA$ of straight edges has six corner points $A(0,0,0)$, $B(5,0,0)$, $C(5,5,0)$, $D(0,5,0)$, $E(0,5,5)$ and $F(0,0,5)$. The magnetic field in this region is $\vec{B} = (3\hat{i} + 4\hat{k})T$. The quantity of flux through the loop $ABCDEFA$ (in Wb) is ______. **[7 Jan. 2020 (Morning)]**

33. A planar loop of wire rotates in a uniform magnetic field. Initially, at $t = 0$, the plane of the loop is perpendicular to the magnetic field. If it rotates with a period of 10 s about an axis in its plane then the magnitude of induced emf will be maximum and minimum, respectively at : **[7 Jan. 2020 (Evening)]**

(A) 2.5 s and 7.5 s (B) 5.0 s and 7.5 s
(C) 5.0 s and 10.0 s (D) 2.5 s and 5.0 s

34. At time $t = 0$ magnetic field of 1000 Gauss is passing perpendicularly through the area defined by the closed loop shown in the figure. If the magnetic field reduces linearly to 500 Gauss, in the next 5s, then induced EMF in the loop is : **[8 Jan. 2020 (Morning)]**

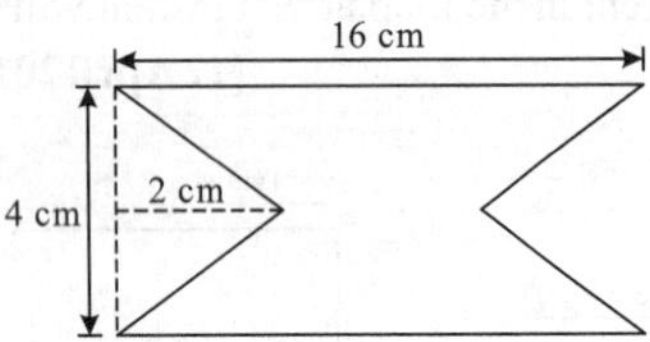

(A) 36 μV (B) 48 μV
(C) 56 μV (D) 28 μV

35. A circular coil of radius 10 cm is placed in a uniform magnetic field of 3.0×10^{-5} T with its plane perpendicular to the field initially. It is rotated at constant angular speed about an axis along the diameter of coil and perpendicular to magnetic field so that it undergoes half of rotation in 0.2 s. The maximum value of EMF induced (in μV) in the coil will be close to the integer. **[2 Sept. 2020 (Morning)]**

36. An elliptical loop having resistance R, of semi major axis a, and semi minor axis b is placed in a magnetic field as shown in the figure. If the loop is rotated about the x-axis with angular frequency ω, the average power loss in the loop due to Joule heating is : **[3 Sept. 2020 (Morning)]**

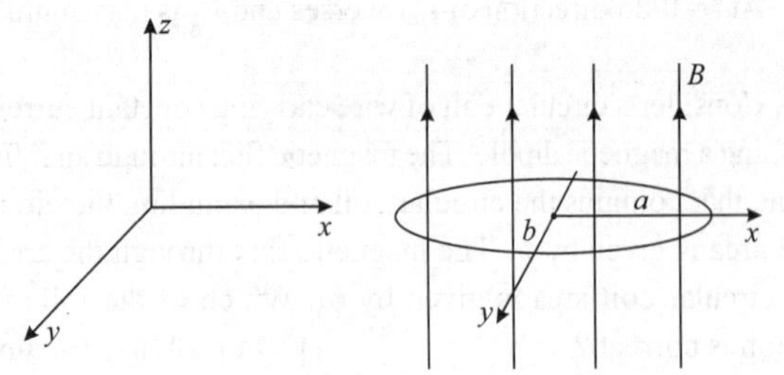

(A) $\dfrac{\pi^2 a^2 b^2 B^2 \omega^2}{2R}$ (B) Zero

(C) $\dfrac{\pi^2 a^2 b^2 B^2 \omega^2}{R}$ (D) $\dfrac{\pi abB\omega}{R}$

37. A uniform magnetic field B exists in a direction perpendicular to the plane of a square loop made of a metal wire. The wire has a diameter of 4 mm and a total length of 30 cm. The magnetic field changes with time at a steady rate $dB/dt = 0.032$ Ts^{-1}. The induced current in the loop is close to (Resistivity of the metal wire is 1.23×10^{-8} Wm) : **[3 Sept. 2020 (Evening)]**

(A) 0.61 A (B) 0.34 A
(C) 0.43 A (D) 0.53 A

38. A small bar magnet is moved through a coil at constant speed from one end to the other. Which of the following series of observations will be seen on the galvanometer G attached across the coil ? Three positions shown describe : (a) the magnet's entry (b) magnet is completely inside and (c) magnet's exit : **[4 Sept. 2020 (Morning)]**

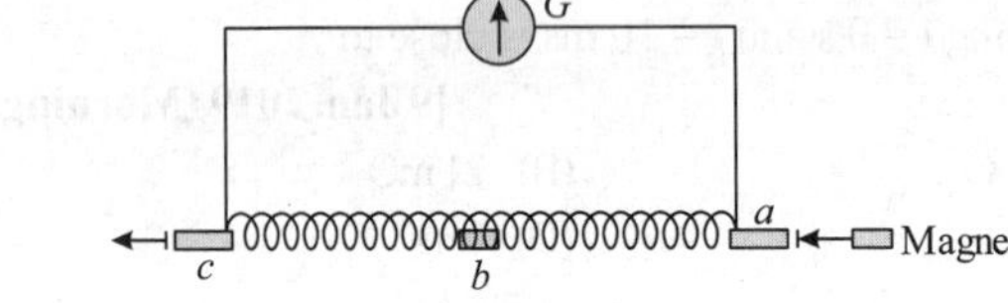

(A) 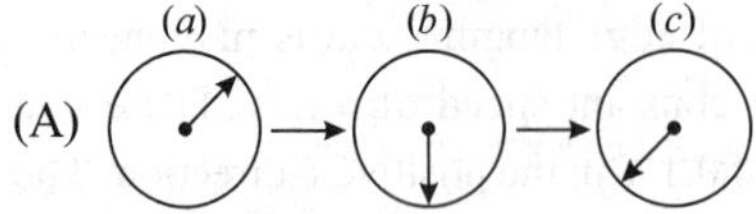

(B) 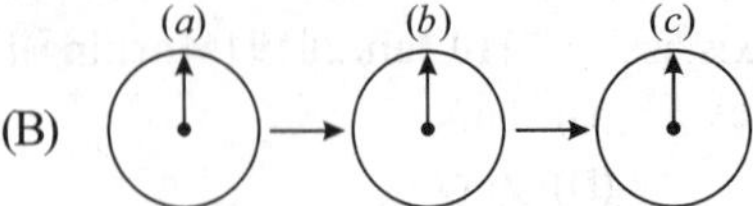

(C) 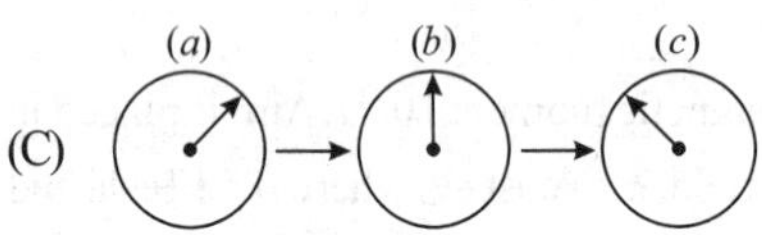

(D) 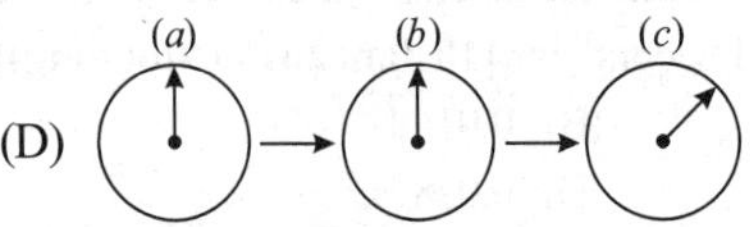

39. Two concentric circular coils, C_1 and C_2, are placed in the XY plane. C_1 has 500 turns, and a radius of 1 cm. C_2 has 200 turns and radius of 20 cm. C_2 carries a time dependent current $I(t) = (5t^2 - 2t + 3)$. A where t is in s. The emf induced in C_1 (in mV), at the instant $t = 1$s is $\dfrac{4}{x}$. The value of x is : **[5 Sept. 2020 (Morning)]**

40. An infinitely long straight wire carrying current I, one side opened rectangular loop and a conductor C with a sliding connector are located in the same plane, as shown in the figure. The connector has length l and resistance R. It slides to the right with a velocity v. The resistance of the conductor and the self inductance of the loop are negligible. The induced current in the loop, as a function of separation r, between the connector and the straight wire is : **[5 Sept. 2020 (Evening)]**

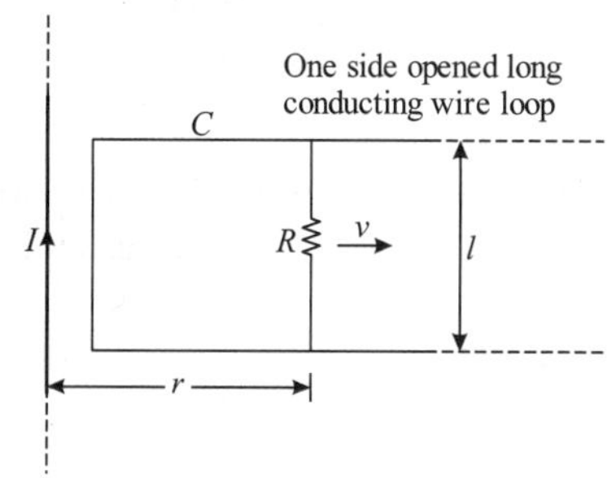

(A) 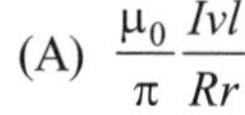 $\frac{\mu_0}{\pi}\frac{Ivl}{Rr}$

(B) $\frac{\mu_0}{2\pi}\frac{Ivl}{Rr}$

(C) $\frac{2\mu_0}{\pi}\frac{Ivl}{Rr}$

(D) $\frac{\mu_0}{4\pi}\frac{Ivl}{Rr}$

41. An aeroplane, with its wings spread 10 m, is flying at a speed of 180 km/h in a horizontal direction. The total intensity of earth's field at that part is 2.5×10^{-4} Wb/m^2 and the angle of dip is 60°. The emf induced between the tips of the plane wings will be ______. **[26 Feb. 2021 (Evening)]**

(A) 62.50 mV
(B) 54.125 mV
(C) 88.37 mV
(D) 108.25 mV

42. A conducting bar of length L is free to slide on two parallel conducting rails as shown in the figure.

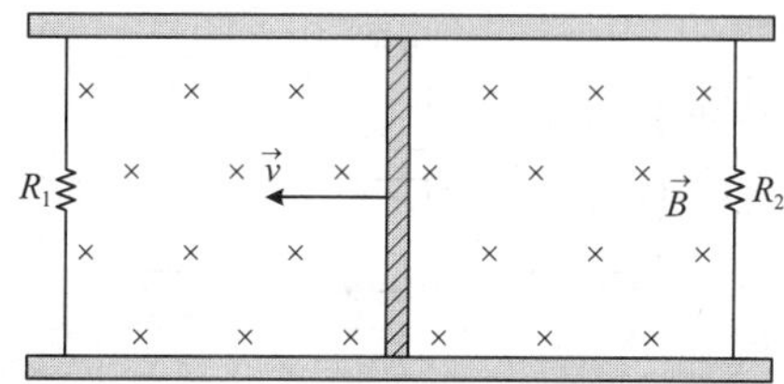

Two resistors R_1 and R_2 are connected across the ends of the rails. There is a uniform magnetic field $\vec{B}$ pointing into the page. An external agent pulls the bar to the left at a constant speed v. The correct statement about the directions of induced currents I_1 and I_2 flowing through R_1 and R_2 respectively is : **[16 March 2021 (Morning)]**

(A) Both I_1 and I_2 are in anticlockwise direction
(B) Both I_1 and I_2 are in clockwise direction
(C) I_1 is in clockwise direction and I_2 is in anticlockwise direction
(D) I_1 is in anticlockwise direction and I_2 is in clockwise direction

43. The magnetic field in a region is given by $\vec{B} = B_0\left(\frac{x}{a}\right)\hat{k}$. A square loop of side d is placed with its edges along the x and y axes. The loop is moved with a constant velocity $\vec{v} = v_0\hat{i}$. The emf induced in the loop is : **[16 March 2021 (Evening)]**

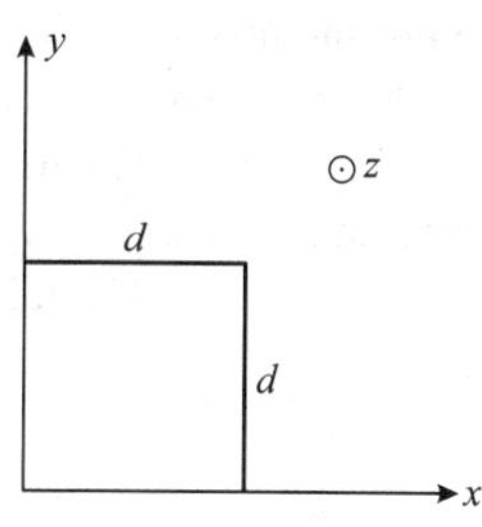

(A) $\frac{B_0 v_0^2 d}{2a}$

(B) $\frac{B_0 v_0 d}{2a}$

(C) $\frac{B_0 v_0 d^2}{a}$

(D) $\frac{B_0 v_0 d^2}{2a}$

44. A loop of flexible wire of irregular shape carrying current is placed in an external magnetic field. Identify the effect of the field on the wire. **[18 March 2021 (Morning)]**

(A) Loop assumes circular shape with its plane normal to the field.
(B) Loop assumes circular shape with its plane parallel to the field.
(C) Wire gets stretched to become straight.
(D) Shape of the loop remains unchanged.

45. The arm PQ of a rectangular conductor is moving from $x = 0$ to $x = 2b$ outwards and then inwards from $x = 2b$ to $x = 0$ as shown in the figure. A uniform magnetic field perpendicular to the plane is acting from $x = 0$ to $x = b$. Identify the graph showing the variation of different quantities with distance : **[20 July 2021 (Morning)]**

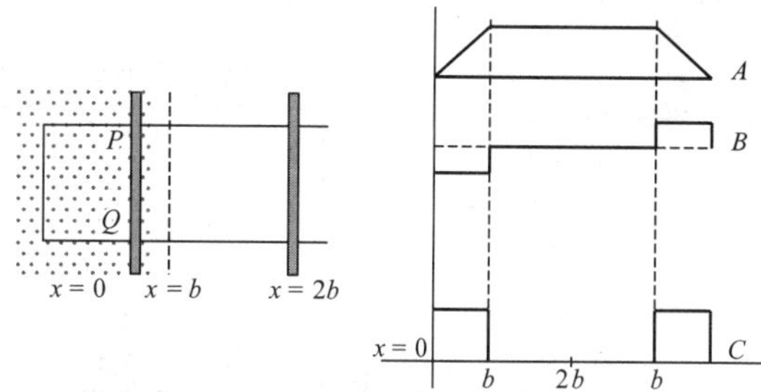

(A) A-Flux, B-power dissipated, C-EMF
(B) A-Flux, B-power dissipated, C-Flux
(C) A-Power dissipated, B-Flux, C-EMF
(D) A-Flux, B-EMF, C-Power dissipated

46. A circular conducting coil of radius 1 m is being heated by the change of magnetic field $\vec{B}$ passing perpendicular to the plane in which the coil is laid. The resistance of the coil is 2 μΩ. The magnetic field is slowly switched off such that its magnitude changes in time as $B = \frac{4}{\pi} \times 10^{-3}\left(1 - \frac{t}{100}\right)$ tesla. The energy dissipated by the coil before the magnetic field is switched off completely is $E =$ ______ mJ. **[25 July 2021 (Morning)]**

47. In the given figure the magnetic flux through the loop increases according to the relation $\phi_B(t) = 10t^2 + 20t$, where ϕ_B is in milliwebers and t is in seconds. The magnitude of current through $R = 2\Omega$ resistor at $t = 5$ s is _____ mA.

[27 July 2021 (Evening)]

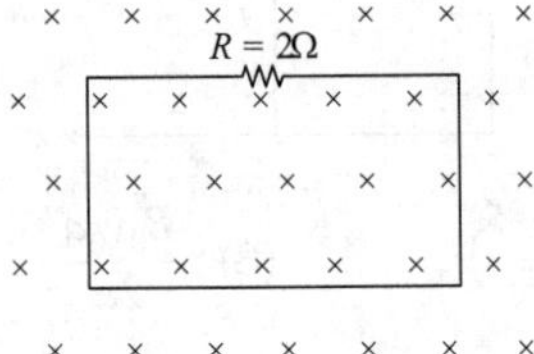

48. A circular coil of radius 8.0 cm and 20 turns is rotated about its vertical diameter with an angular speed of 50 rad s^{-1} in a uniform horizontal field of 3.0×10^{-2} T. The maximum emf induced the coil will be _____ $\times 10^{-2}$ volt. (rounded off to the nearest integer) **[26 Aug. 2021 (Evening)]**

49. A bar magnet is passing through a conducting loop of radius R with velocity v. The radius of the bar magnet is such that it just passes through the loop. The induced emf in the loop can be represented by the approximate curve :

[27 Aug. 2021 (Morning)]

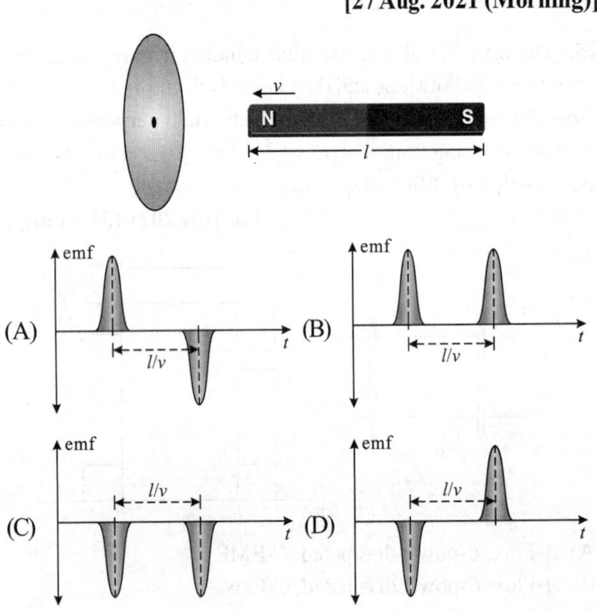

50. A constant magnetic field of 1 T is applied in the $x > 0$ region. A metallic circular ring of radius 1 m is moving with a constant velocity of 1 m/s along the x-axis. At $t = 0$ s, the centre O of the ring is at $x = -1$m. What will be the value of the induced emf in the ring at $t = 1$s ? (Assume the velocity of the ring does not change) **[27 Aug. 2021 (Evening)]**

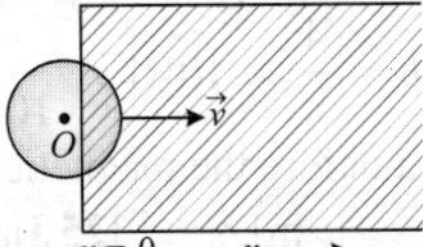

(A) 0V
(B) 2V
(C) 2πV
(D) 1V

51. A coil is placed in a magnetic field $\vec{B}$ as shown below.

[31 Aug. 2021 (Evening)]

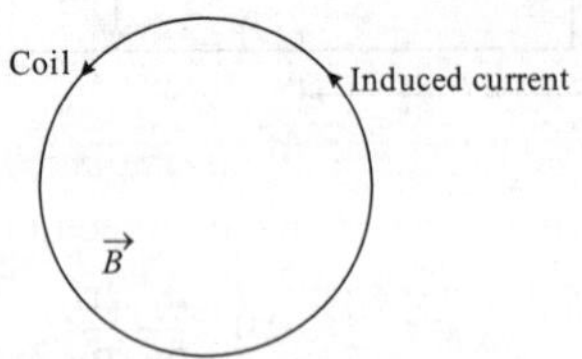

A current is induced in the coil because $\vec{B}$ is :

(A) parallel to the plane of coil and decreasing with time
(B) outward and increasing with time
(C) parallel to the plane of coil and increasing with time
(D) outward and decreasing with time

52. A square loop of side 20 cm and resistance 1 Ω is moved towards right a constant speed v_0. The right arm of the loop is in a uniform magnetic field of 5 T. The field is perpendicular to the plane of the plane of loop and is going into it. The loop is connected to a network of resistor each of value 4 Ω. What should be the value of v_0 so that a steady current of 2 mA flows in the loop ? **[01 Sep. 2021 (Evening)]**

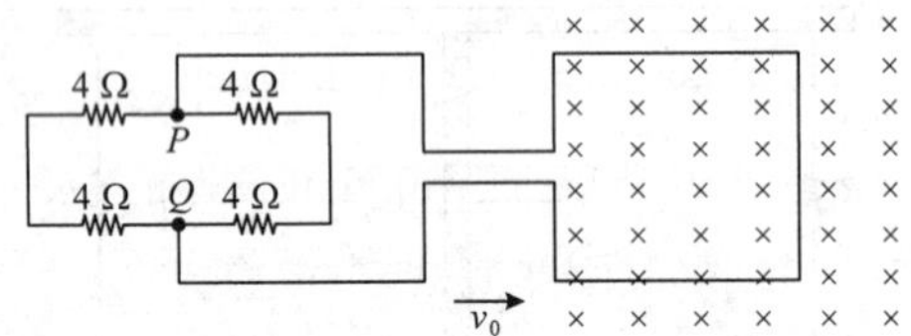

(A) 1 cm/s
(B) 10^{-2} cm/s
(C) 10^2 m/s
(D) 1 m/s

53. A circular coil of 1000 turns each with area 1 m^2 is rotated about its vertical diameter at the rate of one revolution per second in a uniform horizontal magnetic field of 0.07 T. The maximum voltage generation will be _____ V.

[24 June 2022 (Evening)]

54. The magnetic flux through a coil perpendicular to its plane is varying according to the relation $\phi = (5t^3 + 4t^2 + 2t - 5)$ Weber. If the resistance of the coil is 5 ohm, then the induced current through the coil at $t = 2$ s will be : **[26 June 2022 (Morning)]**

(A) 15.6A
(B) 16.6A
(C) 17.6A
(D) 18.6A

55. A 10 Ω, 20 mH coil carrying constant current is connected to a battery of 20 V through a switch. Now after switch is opened current becomes zero in 100 μs. The average emf induced in the coil is _____ V. **[26 June 2022 (Morning)]**

56. A metallic conductor of length 1m rotates in a vertical plane parallel to east-west direction about one of its end with angular velocity 5 rad/s. If the horizontal component of earth's magnetic field is 0.2×10^{-4} T, then emf induced between the two ends of the conductor is : **[26 June 2022 (Evening)]**

(A) 5μV (B) 50μV
(C) 5mV (D) 50mV

57. A metallic rod of length 20 cm is placed in North-South direction and is moved at a constant speed of 20 m/s towards East. The horizontal component of the Earth's magnetic field at that place is 4×10^{-3} T and the angle of dip is 45°. The emf induced in the rod is ______ mV. **[27 June 2022 (Evening)]**

58. A coil is placed in a time varying magnetic field. If the number of turns in the coil were to be halved and the radius of wire doubled, the electrical power dissipated due to the current induced in the coil would be : **[28 June 2022 (Evening)]**
[Assume the coil to be short circuited]

(A) Halved (B) Quadrupled
(C) The same (D) Doubled

59. Magnetic flux (in weber) in a closed circuit of resistance 20 Ω varies with time t(s) as $\phi = 8t^2 - 9t + 5$. The magnitude of the induced current at $t = 0.25$ s will be ______ mA. **[25 July 2022 (Evening)]**

60. In a coil of resistance 8 Ω, the magnetic flux due to an external magnetic field varies with time as $\phi = \frac{2}{3}(9 - t^2)$. The value of total heat produced in the coil, till the flux becomes zero, will be ______ J. **[26 July 2022 (Evening)]**

61. A conducting circular loop is placed in $X - Y$ plane in presence of magnetic field $\vec{B} = (3t^3\hat{j} + 3t^2\hat{k})$ in SI unit. If the radius of the loop is 1 m, the induced emf in the loop, at time, $t = 2$s is $n\pi$ V. The value of n is ______. **[27 July 2022 (Evening)]**

62. A coil of inductance 1 H and resistance 100 Ω is connected to a battery of 6 V. Determine approximately :
(a) The time elapsed before the current acquires half of its steady-state value
(b) The energy stored in the magnetic field associated with the coil at an instant 15 ms after the circuit is switched on.
[Given : $\ln 2 = 0.693, e^{-3/2} = 0.25$] **[29 July 2022 (Morning)]**

(A) $t = 10$ ms; $U = 2$ mJ (B) $t = 10$ ms; $U = 1$ mJ
(C) $t = 7$ ms; $U = 1$ mJ (D) $t = 7$ ms; $U = 2$ mJ

63. A capacitor of capacitance 500 μF is charged completely using a dc supply of 100 V. It is now connected to an inductor of inductance 50 mH to form an LC circuit. The maximum current in LC circuit will be ______ A. **[29 July 2022 (Evening)]**

64. A conducting loop of radius $\frac{10}{\sqrt{\pi}}$ cm is placed perpendicular to a uniform magnetic field of $0.5T$. The magnetic field is decreased to zero in 0.5 s at a steady rate. The induced emf in the circular loop at 0.25 s is : **[24 Jan. 2023 (Morning)]**

(A) emf = 1 mV (B) emf = 10 mV
(C) emf = 100 mV (D) emf = 5 mV

65. A metallic rod of length L is rotated with an angular speed of ω normal to a uniform magnetic field B about an axis passing through one end of rod as shown in figure. The induced emf will be : **[24 Jan. 2023 (Evening)]**

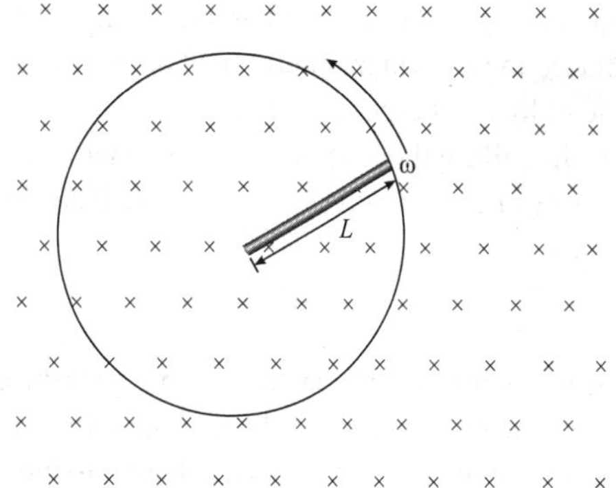

(A) $\frac{1}{4}B^2L\omega$ (B) $\frac{1}{4}BL^2\omega$
(C) $\frac{1}{2}BL^2\omega$ (D) $\frac{1}{2}B^2L^2\omega$

66. Match List-I with List-II : **[25 Jan. 2023 (Evening)]**

List-I	List-II
(a) Gauss's Law in Electrostatics	(i) $\oint \vec{E} \cdot d\vec{l} = -\frac{d\phi_B}{dt}$
(b) Faraday's Law	(ii) $\oint \vec{B} \cdot d\vec{A} = 0$
(c) Gauss's Law in Magnetism	(iii) $\oint \vec{B} \cdot d\vec{l} = \mu_0 i_C + \mu_0 \epsilon_0 \frac{d\phi_E}{dt}$
(d) Ampere-Maxwell Law	(iv) $\oint \vec{E} \cdot d\vec{s} = \frac{q}{\epsilon_0}$

Choose the correct answer from the options given below :
(A) (a)-(iv), (b)-(i), (c)-(ii), (d)-(iii)
(B) (a)-(i), (b)-(ii), (c)-(iii), (d)-(iv)
(C) (a)-(iii), (b)-(iv), (c)-(i), (d)-(ii)
(D) (a)-(ii), (b)-(iii), (c)-(iv), (d)-(i)

67. A wire of length 1 m moving with velocity 8 m/s at right angles to a magnetic field of $2T$. The magnitude of induced emf, between the ends of wire will be _____. **[25 Jan. 2023 (Evening)]**

(A) 20 V (B) 8 V
(C) 12 V (D) 16 V

68. A certain elastic conducting material is stretched into a circular loop. It is placed with its plane perpendicular to a uniform magnetic field $B = 0.8$ T. When released the radius of the loop starts shrinking at a constant rate of 2 cm^{-1}. The induced emf in the loop at an instant when the radius of the loop is 10 cm will be_____ mV. **[29 Jan. 2023 (Morning)]**

69. A square loop of area 25 cm^2 has a resistance of 10 Ω. The loop is placed in uniform magnetic field of magnitude 40.0 T. The plane of loop is perpendicular to the magnetic field. The work done in pulling the loop out of the magnetic field slowly and uniformly in 1.0 sec, will be : **[29 Jan. 2023 (Evening)]**

(A) 2.5×10^{-3} J (B) 1.0×10^{-3} J
(C) 1.0×10^{-4} J (D) 5×10^{-3} J

70. A square shaped coil of area 70 cm^2 having 600 turns rotates in a magnetic field of 0.4 wbm^{-2}, about an axis which is parallel to one of the side of the coil and perpendicular to the direction of field. If the coil completes 500 revolution in a minute, the instantaneous emf when the plane of the coil is inclined at 60° with the field, will be _____ V. $\left(\text{Take } \pi = \frac{22}{7}\right)$

[1 Feb. 2023 (Evening)]

71. The induced emf can be produced in a coil by

(a) Moving the coil with uniform speed inside magnetic field
(b) Moving the coil with non-uniform speed inside uniform magnetic field
(c) Rotating the coil inside the uniform magnetic field
(d) Changing the area of the coil inside the uniform magnetic field

Choose the correct answer from the options given below: **[6 April 2023 (Morning)]**

(A) (b) and (d) only (B) (b) and (c) only
(C) (a) and (c) only (D) (c) and (d) only

72. An emf of 0.08 V is induced in a metal rod of length 10 cm held normal to a uniform magnetic field of 0.4 T, when moves with a velocity of : **[8 April 2023 (Evening)]**

(A) 2 ms^{-1} (B) 3.2 ms^{-1}
(C) 0.5 ms^{-1} (D) 20 ms^{-1}

73. A 1 m long metal rod XY completes the circuit as shown in figure. The plane of the circuit is perpendicular to the magnetic field of flux density 0.15 T. If the resistance of the circuit is 5 Ω, the force needed to move the rod in direction, as indicated, with a constant speed of 4 m/s will be _____ 10^{-3} N. **[10 April 2023 (Morning)]**

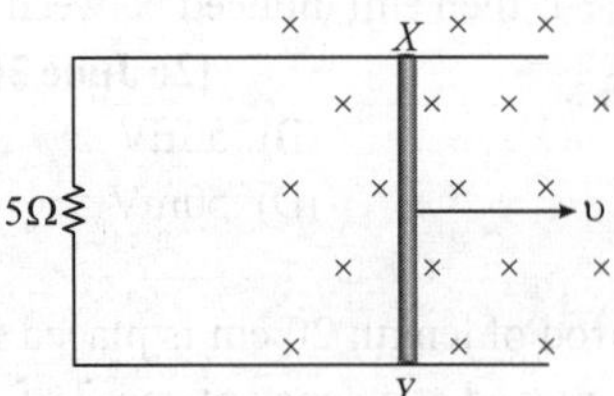

74. A square loop of side 2.0 cm is placed inside a long solenoid that has 50 turns per centimetre and carries a sinusoidally varying current of amplitude 2.5 A and angular frequency 700 rad s^{-1}. The central axes of the loop and solenoid coincide. The amplitude of the emf induced in the loop is $x \times 10^{-4}$ V. The value of x is_____. **[10 April 2023 (Evening)]**

75. The magnetic field B crossing normally a square metallic plate of area 4 m^2 is changing with time as shown in figure. The magnitude of induced emf in the plate during $t = 2$ s to $t = 4$ s, is_____ mV. **[11 April 2023 (Morning)]**

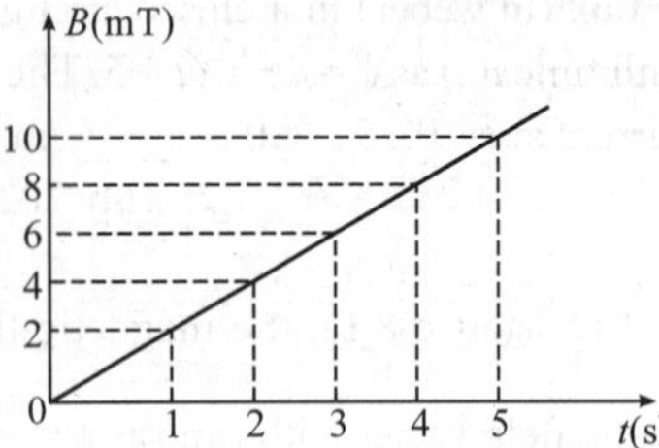

76. A metallic cube of side 15 cm moving along y-axis at a uniform velocity of 2 ms^{-1}. In a region of uniform magnetic field of magnitude 0.5 T directed along z-axis. In equilibrium the potential difference between the faces of higher and lower potential developed because of the motion through the field will be_____ mV. **[11 April 2023 (Evening)]**

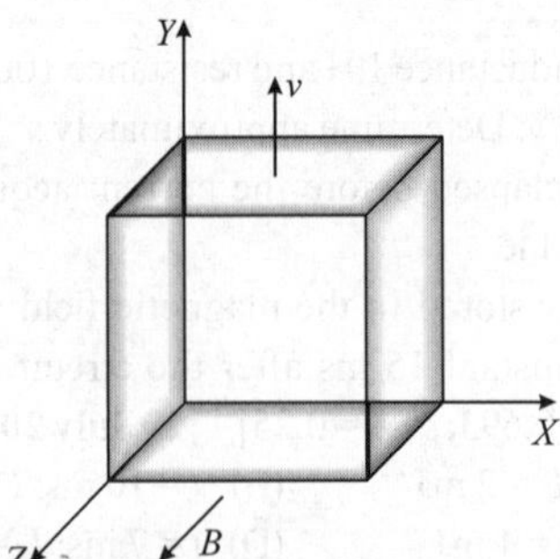

77. A conducting circular loop is placed in a uniform magnetic field of 0.4 T with its plane perpendicular to the field. Somehow, the radius of the loop starts expanding at a constant rate of

1 mm/s. The magnitude of induced emf in the loop at an instant when the radius of the loop is 2 cm will be _____ μV.

[12 April 2023 (Morning)]

78. The source of time varying magnetic field may be

(a) A permanent magnet

(b) An electric field changing linearly with time

(c) Direct current

(d) A decelerating charge particle

(e) An antenna fed with a digital signal

Choose the correct answer from the options given below :

[13 April 2023 (Morning)]

(A) (d) only
(B) (c) and (e) only
(C) (a) only
(D) (b) and (d) only

79. An insulated copper wire of 100 turns is wrapped around a wooden cylindrical core of the cross-sectional area 24 cm^2. The two ends of the wire are connected to a resistor. The total resistance in the circuit is 12 Ω. If an externally applied uniform magnetic field in the core along its axis changes from 1.5 T in one direction to 1.5 T in the opposite direction, the charge flowing through a point in the circuit during the change of magnetic field will be _____ mC. **[13 April 2023 (Evening)]**

80. A 12 V battery connected to a coil of resistance 6 Ω through a switch, drives a constant current in the circuit. The switch is opened in 1 ms. The emf induced across the coil is 20 V. The inductance of the coil is : **[15 April 2023 (Morning)]**

(A) 5 mH
(B) 12 mH
(C) 8 mH
(D) 10 mH

81. A 20 cm long metallic rod is rotated with 210 rpm about an axis normal to the rod passing through its one end. The order end of the rod is in contact with a circular metallic ring. A constant and uniform magnetic field 0.2 T parallel to the axis exists everywhere. The emf developed between the centre and the ring is _____ mV. [Take $\pi = \frac{22}{7}$] **[15 April 2023 (Morning)]**

* * * * *

CHAPTER 20

ELECTROMAGNETIC INDUCTION

20.2 SELF AND MUTUAL INDUCTION

1. The inductance between A and D is : **[2002]**

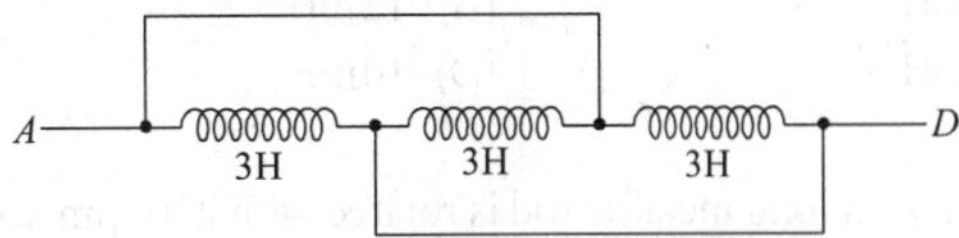

(A) 3.66 H (B) 9 H
(C) 0.66 H (D) 1 H

2. When the current changes from +2 A to – 2 A in 0.05 s, an emf of 8 V is induced in a coil. The coefficient of self-induction of the coil is : **[2003]**

(A) 0.2 H (B) 0.4 H
(C) 0.8 H (D) 0.1 H

3. Two coils are placed close to each other. The mutual inductance of the pair of coils depends upon : **[2003]**

(A) the rates at which currents are changing in the two coils
(B) relative position and orientation of the two coils
(C) the materials of the wires of the coils
(D) the currents in the two coils

4. A coil of inductance 300 mH and resistance 2 Ω is connected to a source of voltage 2 V. The current reaches half of its steady state value in : **[2005]**

(A) 0.05 s (B) 0.1 s
(C) 0.15 s (D) 0.3 s

5. The flux linked with a coil at any instant 't' is given by : $\phi = 10t^2 - 50t + 250$. The induced emf at $t = 3$s is : **[2006]**

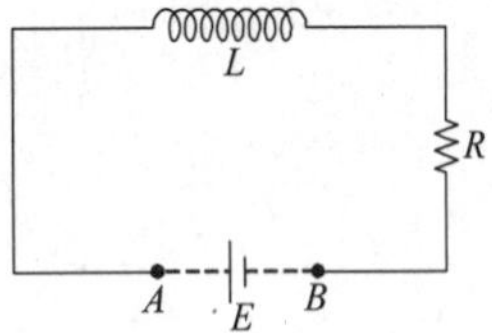

(A) – 190 V (B) – 10 V
(C) 10 V (D) 190 V

6. An inductor (L = 100 mH), a resistor (R = 100 Ω) and a battery (E = 100 V) are initially connected in series as shown in the figure. After a long time the battery is disconnected after short circuiting the points A and B. The current in the circuit 1 ms after the short circuit is : **[2006]**

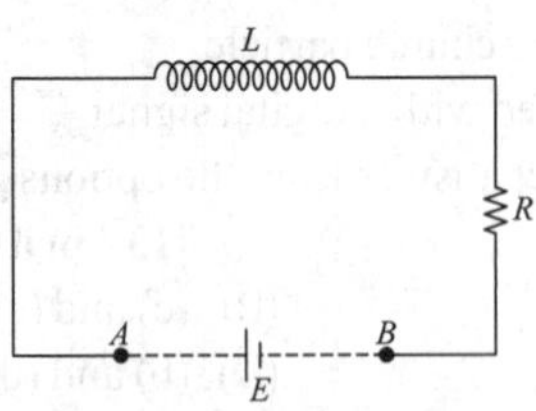

(A) $1/e$ A (B) e A
(C) 0.1 A (D) 1 A

7. An ideal coil of 10 H is connected in series with a resistance of 5 Ω and a battery of 5 V. 2 s after the connection is made, the current flowing (in ampere) in the circuit is : **[2007]**

(A) $(1 - e)$ (B) e
(C) e^{-1} (D) $(1 - e^{-1})$

8. Two coaxial solenoids are made by winding thin insulated wire over a pipe of cross-sectional area A = 10 cm^2 and length = 20 cm. If one of the solenoids has 300 turns and the other 400 turns, their mutual inductance is :
$[\mu_0 = 4\pi \times 10^{-7}\ \text{TmA}^{-1}]$ **[2008]**

(A) $2.4\,\pi \times 10^{-5}$ H (B) $4.8\,\pi \times 10^{-4}$ H
(C) $4.8\,\pi \times 10^{-5}$ H (D) $2.4\,\pi \times 10^{-4}$ H

9. An inductor of inductance L = 400 mH and resistors of resistances R_1 = 2 Ω and R_2 = 2 Ω are connected to a battery of emf 12 V as shown in the figure. The internal resistance of the battery is negligible. The switch S is closed at t = 0. The potential drop across L as a function of time is : **[2009]**

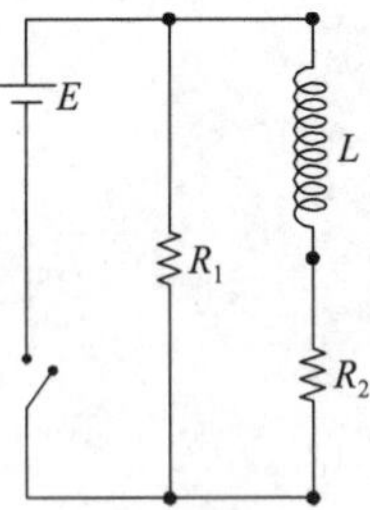

(A) $6(1 - e^{-t/0.2})$ V (B) $12e^{-5t}$ V
(C) $6e^{-5t}$ V (D) $\frac{12}{t}e^{-3t}$ V

10. In the circuit show below, the key K is closed at t = 0. The current through the battery is : **[2010]**

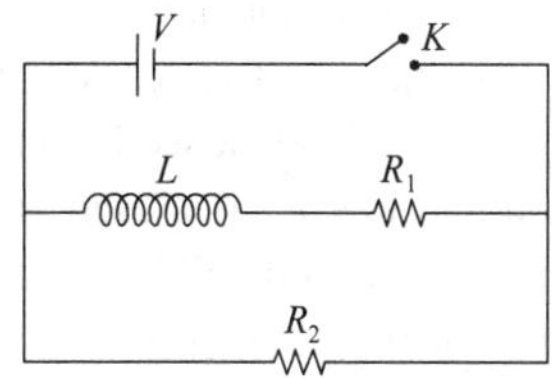

(A) $\dfrac{V(R_1+R_2)}{R_1R_2}$ at $t=0$ and $\dfrac{V}{R_2}$ at $t=\infty$

(B) $\dfrac{VR_1R_2}{\sqrt{R_1^2+R_2^2}}$ at $t=0$ and $\dfrac{V}{R_2}$ at $t=\infty$

(C) $\dfrac{V}{R_2}$ at $t=0$ and $\dfrac{V(R_1+R_2)}{R_1R_2}$ at $t=\infty$

(D) $\dfrac{V}{R_2}$ at $t=0$ and $\dfrac{VR_1R_2}{\sqrt{R_1^2+R_2^2}}$ at $t=\infty$

11. This question has Statement-1 and Statement-2. Of the four choices given after the Statements, choose the one that best describes the two Statements.

Statement-1 : Self inductance of a long solenoid of length L, total number of turns N and radius r is less than $\dfrac{\pi\mu_0 N^2 r^2}{L}$.

Statement-2 : The magnetic induction in the solenoid in Statement-1 carrying current I is $\dfrac{\mu_0 NI}{L}$ in the middle of the solenoid but becomes less as we move towards its ends : **[2012 (Online)]**

(A) Statement-1 is true, Statement-2 is false

(B) Statement-1 is true, Statement-2 is true, Statement-2 is the correct explanation of Statement-1

(C) Statement-1 is false, Statement-2 is true

(D) Statement-1 is true, Statement-2 is true, Statement-2 is not the correct explanation of Statement-1

12. A coil of self inductance L is connected at one end of two rails as shown in figure. A connector of length l, mass m can slide freely over the two parallel rails. The entire set up is placed in a magnetic field of induction B going into the page. At an instant $t=0$ an initial velocity v_0 is imparted to it and as a result of that it starts moving along x-axis. The displacement of the connector is represented by the figure : **[2012 (Online)]**

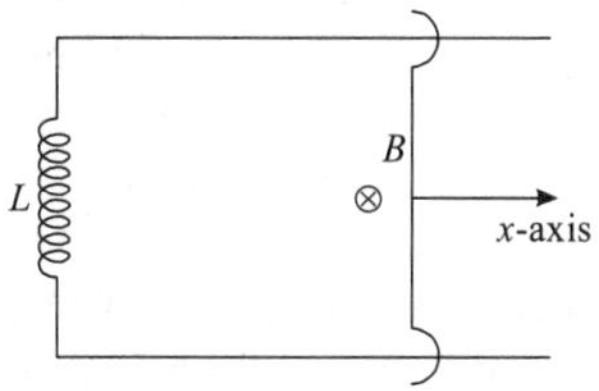

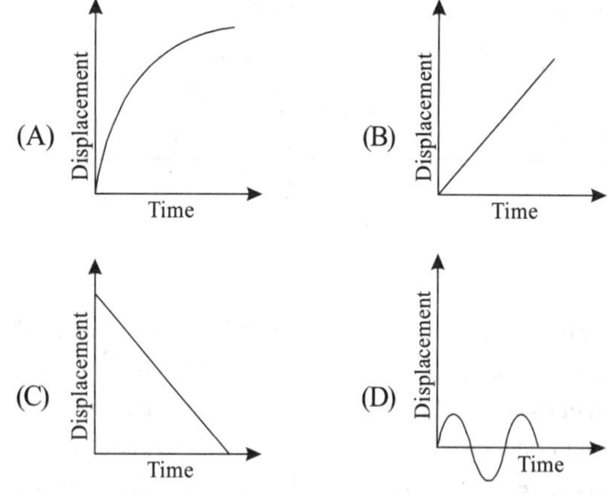

13. In an LCR circuit as shown below both switches are open initially. Now switch S_1 is closed, S_2 kept open. (q is charge on the capacitor and $\tau = RC$ is Capacitive time constant). Which of the following statement is correct? **[2013]**

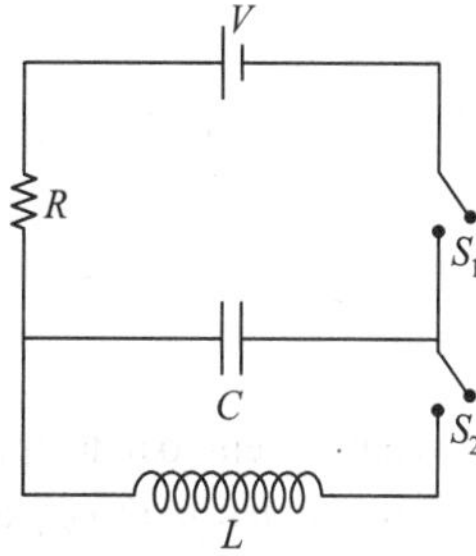

(A) Work done by the battery is half of the energy dissipated in the resistor

(B) At, $t=\tau$, $q = CV/2$

(C) At, $t=2\tau$, $q = CV(1-e^{-2})$

(D) At, $t=2\tau$, $q = CV(1-e^{-1})$

14. Two coils, X and Y, are kept in close vicinity of each other. When a varying current, $I(t)$, flows through coil X, the induced emf ($V(t)$) in coil Y, varies in the manner shown here. The variation of $I(t)$, with time, can then be represented by the graph labelled as graph : **[2013 (Online)]**

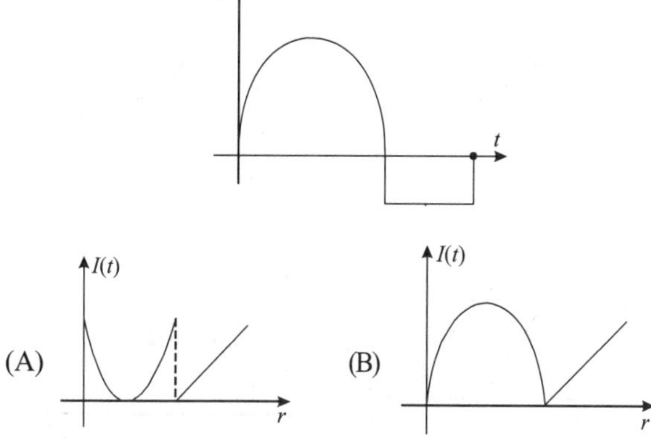

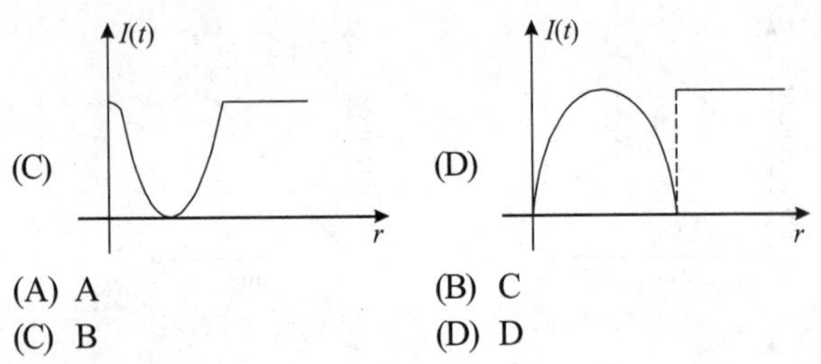

(A) A (B) C
(C) B (D) D

15. In the circuit shown here, the point 'C' is kept connected to point 'A' till the current flowing through the circuit becomes constant. Afterward, suddenly, point 'C' is disconnected from point 'A' and connected to point 'B' at time $t = 0$. Ratio of the voltage across resistance and the inductor at $t = L/R$ will be equal to : **[2014]**

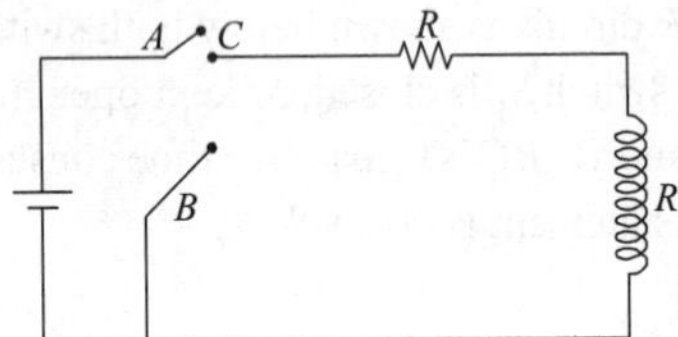

(A) $\frac{e}{1-e}$ (B) 1

(C) -1 (D) $\frac{1-e}{e}$

16. An LCR circuit is equivalent to a damped pendulum. In an LCR circuit the capacitor is charged to Q_0 and then connected to the L and R as shown below :

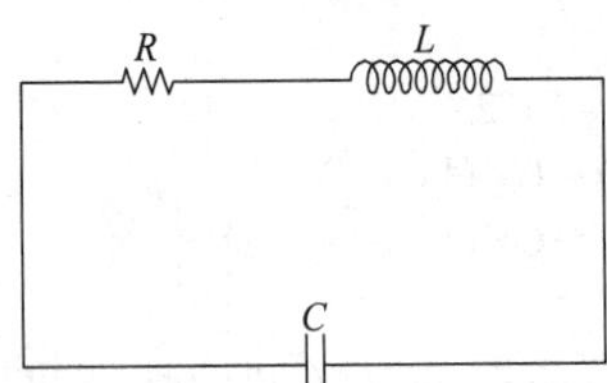

If a student plots graphs of the square of maximum charge (Q^2_{Max}) on the capacitor with time (t) for two different values L_1 and L_2 ($L_1 > L_2$) of L then which of the following represents this graph correctly? (plots are schematic and not drawn to scale) : **[2015]**

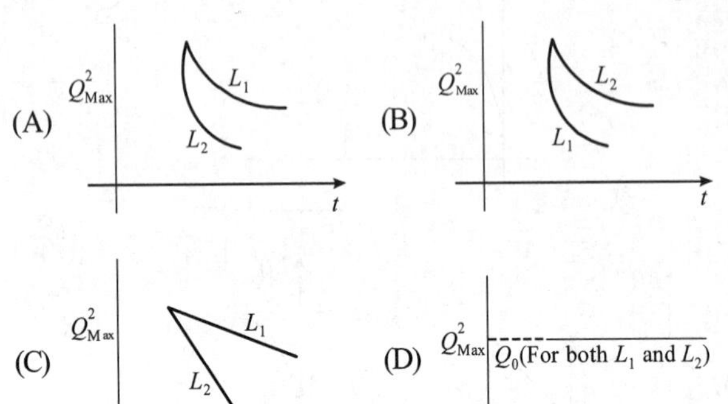

17. An inductor ($L = 0.03$ H) and a resistor ($R = 0.15$ kΩ) are connected in series to a battery of 15 V EMF in a circuit shown below. The key K_1 has been kept closed for a long time. Then at $t = 0$, K_1 is opened and key K_2 is closed simultaneously. At $t = 1$ ms, the current in the circuit will be: ($e^5 \cong 150$) : **[2015]**

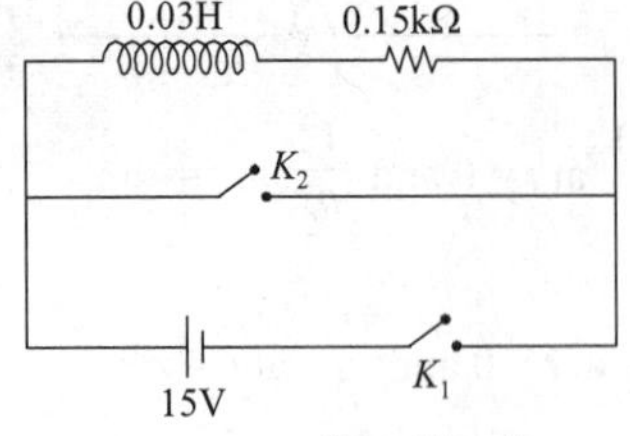

(A) 100 mA (B) 67 mA
(C) 6.7 mA (D) 0.67 mA

18. When current in a coil changes from 5 A to 2A in 0.1 s, average voltage of 50 V is produced. The self-inductance of the coil is : **[2015 (Online)]**

(A) 6 H (B) 0.67 H
(C) 3 H (D) 1.67 H

19. In the circuits (a) and (b) switches S_1 and S_2 are closed at $t = 0$ and are kept closed for a long time. The variation of current in the two circuits for $t \geq 0$ are roughly shown by figure (figures are schematic and not drawn to scale) : **[2015 (Online)]**

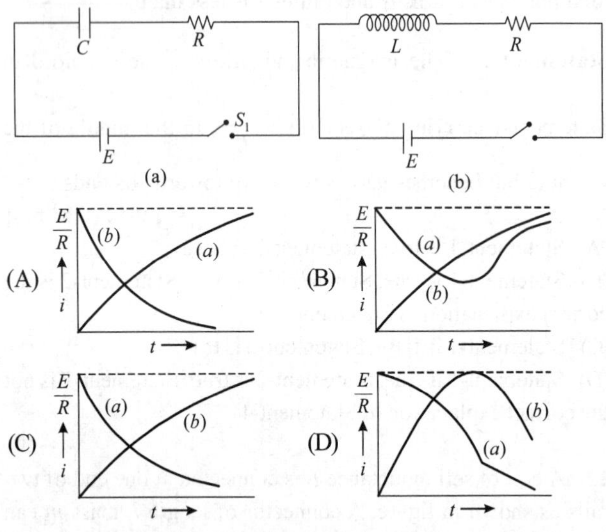

20. A power transmission line feeds input power at 2300 V to a step down transformer with its primary windings having 4000 turns. The output power is delivered at 230 V by the transformer. If the current in the primary of the transformer is 5 A and its efficiency is 90%, the output current would be : **[9 Jan. 2019 (Evening)]**

(A) 25 A (B) 50 A
(C) 35 A (D) 45 A

21. The self induced emf of a coil is 25 volt. When the current in it is changed at uniform rate from 10 A to 25 A in 1 s, the change in the energy of the inductance is :

[10 Jan. 2019 (Evening)]

(A) 437.5 J (B) 637.5 J

(C) 740 J (D) 540 J

22. In the circuit shown, the switch S_1 is closed at time $t = 0$ and the switch S_2 is kept open. At some later time (t_0), the switch S_1 is opened and S_2 is closed. The behaviour of the current I as a function of time 't' is given by :

[11 Jan. 2019 (Morning)]

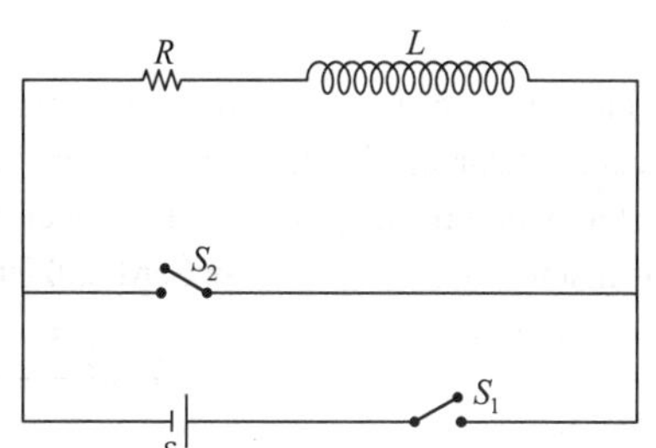

(A) 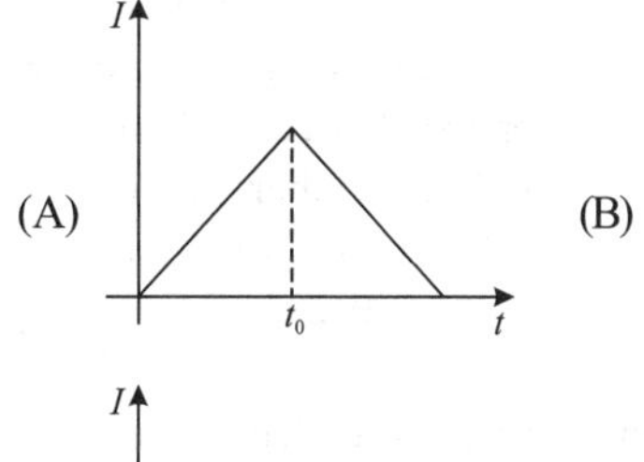

(B) 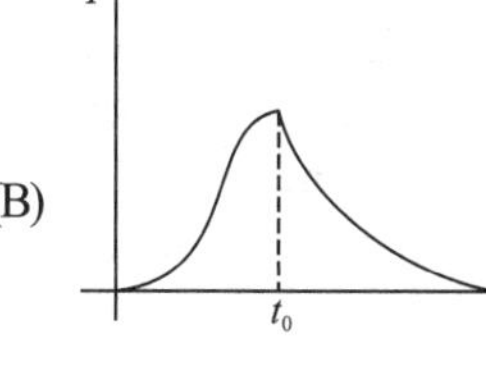

(C) 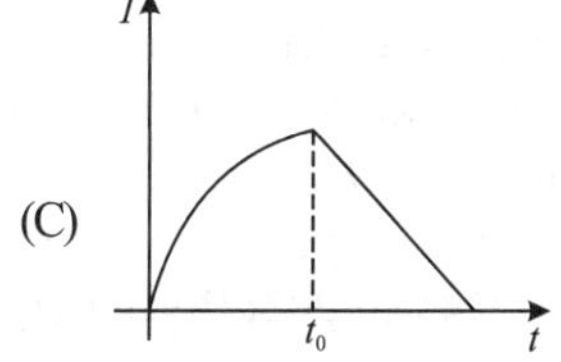

(D) 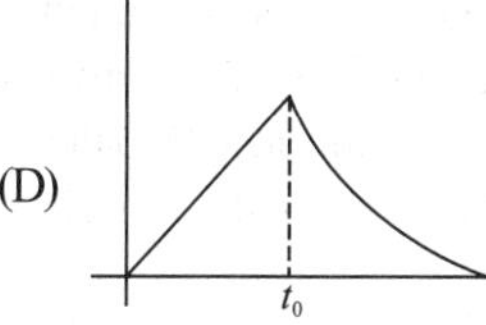

23. There are two long co-axial solenoids of same length l. The inner and outer coils have radii r_1 and r_2 and number of turns per unit length n_1 and n_2 respectively. The ratio of mutual inductance to the self-inductance of the inner-coil is :

[11 Jan. 2019 (Morning)]

(A) $\frac{n_2}{n_1} \cdot \frac{r_2^2}{r_1^2}$ (B) $\frac{n_2}{n_1} \cdot \frac{r_1}{r_2}$

(C) $\frac{n_1}{n_2}$ (D) $\frac{n_2}{n_1}$

24. A copper wire is wound on a wooden frame, whose shape is that of an equilateral triangle. If the linear dimension of each side of the frame is increased by a factor of 3, keeping the number of turns of the coil per unit length of the frame the same, then the self inductance of the coil : **[11 Jan. 2019 (Evening)]**

(A) Decreases by a factor of $9\sqrt{3}$

(B) Increases by a factor of 3

(C) Decreases by a factor of 9

(D) Increases by a factor of 27

25. In the figure shown, a circuit contains two identical resistors with resistance $R = 5\ \Omega$ and an inductance with $L = 2$ mH. An ideal battery of 15 V is connected in the circuit. What will be the current through the battery long after the switch is closed?

[12 Jan. 2019 (Morning)]

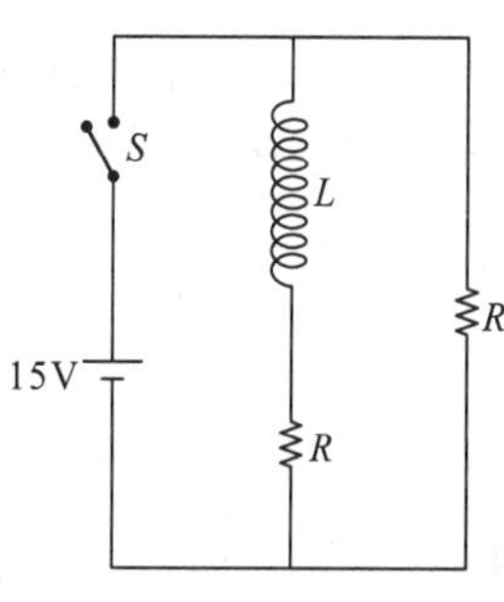

(A) 6 A (B) 7.5 A

(C) 5.5 A (D) 3 A

26. The total number of turns and cross-section area in a solenoid is fixed. However, its length L is varied by adjusting the separation between windings. The inductance of solenoid will be proportional to : **[9 April 2019 (Morning)]**

(A) $1/L^2$ (B) $1/L$

(C) L (D) L^2

27. Two coils 'P' and 'Q' are separated by some distance. When a current of 3 A flows through coil 'P', a magnetic flux of 10^{-3} Wb passes through 'Q'. No current is passed through 'Q'. When no current passes through 'P' and a current of 2A passes through 'Q', the flux through 'P' is : **[9 April 2019 (Evening)]**

(A) 6.67×10^{-3} Wb (B) 6.67×10^{-4} Wb

(C) 3.67×10^{-4} Wb (D) 3.67×10^{-3} Wb

28. A very long solenoid of radius R is carrying current $I(t) = kte^{-\alpha t}(k > 0)$, as a function of time ($t \geq 0$). Counter clockwise current is taken to be positive. A circular conducting coil of radius $2R$ is placed in the equatorial plane of the solenoid and concentric with the solenoid. The current induced in the outer coil is correctly depicted, as a function of time, by :

[9 April 2019 (Evening)]

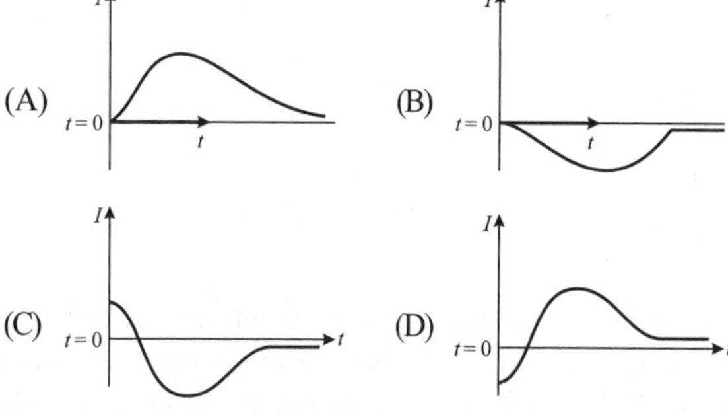

29. A coil of self inductance 10 mH and resistance 0.1 Ω is connected through a switch to a battery of internal resistance

0.9 Ω. After the switch is closed, the time taken for the current to attain 80% of the saturation value is : (Take ln 5 = 1.6)

[10 April 2019 (Evening)]

(A) 0.103 s (B) 0.016 s
(C) 0.002 s (D) 0.324 s

30. Consider the LR circuit shown in the figure. If the switch S is closed at $t = 0$ then the amount of charge that passes through the battery between $t = 0$ and $t = \frac{L}{R}$ is :

[12 April 2019 (Evening)]

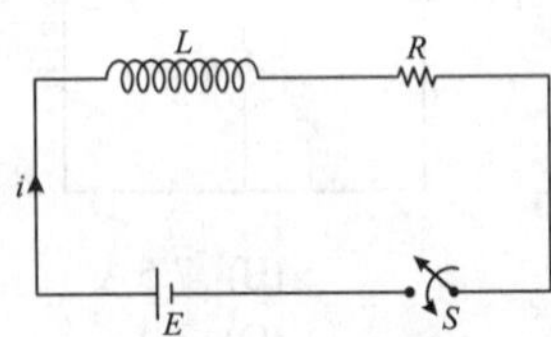

(A) $\frac{EL}{7.3R^2}$ (B) $\frac{EL}{2.7R^2}$

(C) $\frac{7.3EL}{R^2}$ (D) $\frac{2.7EL}{R^2}$

31. An emf of 20 V is applied at time $t = 0$ to a circuit containing in series 10 mH inductor and 5 Ω resistor. The ratio of the currents at time $t = \infty$ and at $t = 40$ s is close to : (Take $e^2 = 7.389$)

[7 Jan. 2020 (Evening)]

(A) 1.06 (B) 1.15
(C) 1.46 (D) 0.84

32. A shown in the figure, a battery of emf ε is connected to an inductor L and resistance R in series. The switch is closed at $t = 0$. The total charge that flows from the battery, between $t = 0$ and $t = t_c$ (t_c is the time constant of the circuit) is :

[8 Jan. 2020 (Evening)]

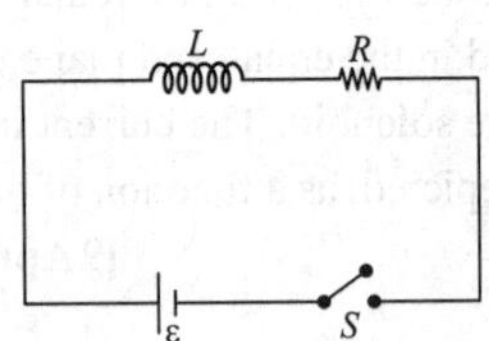

(A) $\frac{\varepsilon L}{R^2}\left(1-\frac{1}{e}\right)$ (B) $\frac{\varepsilon R}{eL^2}$

(C) $\frac{\varepsilon L}{R^2}$ (D) $\frac{\varepsilon L}{eR^2}$

33. In a fluorescent lamp choke (a small transformer) 100 V of reverse voltage is produced when the choke current changes uniformly from 0.25 A to 0 in a duration of 0.025 ms. The self-inductance of the choke (in mH) is estimated to be ______.

[9 Jan. 2020 (Morning)]

34. A series L-R circuit is connected to a battery of emf V. If the circuit is switched on at $t = 0$, then the time at which the energy stored in the inductor reaches $(1/n)$ times of its maximum value, is :

[4 Sept. 2020 (Evening)]

(A) $\frac{L}{R}\ln\left(\frac{\sqrt{n}-1}{\sqrt{n}}\right)$ (B) $\frac{L}{R}\ln\left(\frac{\sqrt{n}}{\sqrt{n}+1}\right)$

(C) $\frac{L}{R}\ln\left(\frac{\sqrt{n}}{\sqrt{n}-1}\right)$ (D) $\frac{L}{R}\ln\left(\frac{\sqrt{n}+1}{\sqrt{n}-1}\right)$

35. A part of a complete circuit is shown in the figure. At some instant, the value of current I is $1A$ and it is decreasing at a rate of $10^2 A\ s^{-1}$. The value of the potential difference $V_P - V_Q$, (in volts) at that instant, is :

[6 Sept. 2020 (Morning)]

36. A common transistor radio set requires 12 V (D.C.) for its operation. The D.C. source is constructed by using a transformer and a rectifier circuit, which are operated at 220 V (A.C.) on standard domestic A.C. supply. The number of turns of secondary coil are 24, then the number of turns of primary are ______.

[24 Feb. 2021 (Morning)]

37. Figure shows a circuit that contains four identical resistors with resistance $R = 2.0\ \Omega$, two identical inductors with inductance $L = 2.0$ mH and an ideal battery with emf $E = 9$ V. The current 'i' just after the switch 'S' is closed will be :

[24 Feb. 2021 (Evening)]

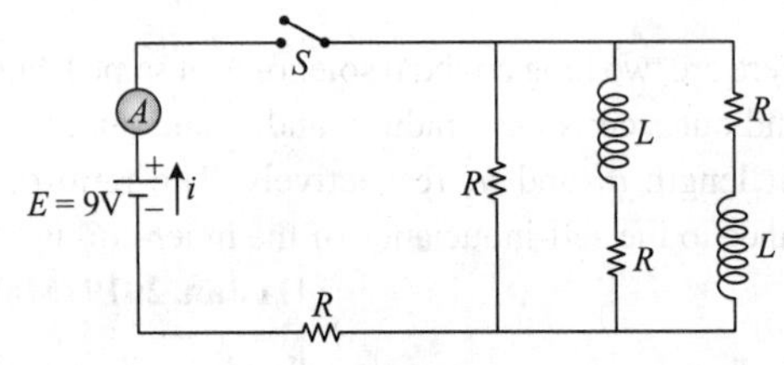

(A) 2.25 A (B) 3.0 A
(C) 3.37 A (D) 9 A

38. The current (i) at time $t = 0$ and $t = \infty$ respectively for the given circuit is :

[25 Feb. 2021 (Morning)]

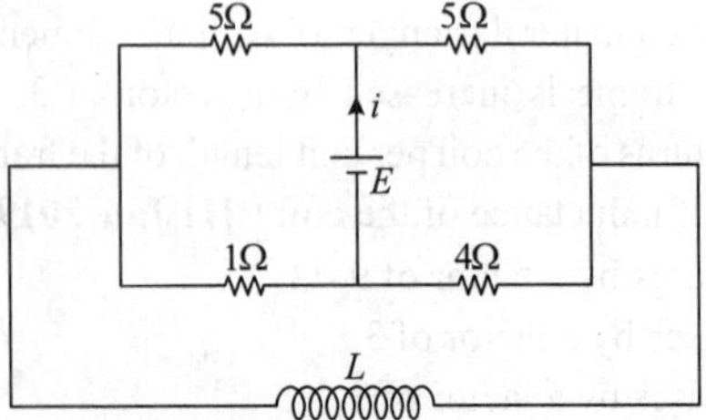

(A) $\frac{5E}{18}, \frac{10E}{33}$ (B) $\frac{10E}{33}, \frac{5E}{18}$

(C) $\frac{5E}{18}, \frac{18E}{55}$ (D) $\frac{18E}{55}, \frac{5E}{18}$

39. A coil of inductance 2 H having negligible resistance is connected to a source of supply whose voltage is given by $e = 3t$ volt. (Where t is in second). If the voltage is applied when $t = 0$, then the energy stored in the coil after 4 s is ______ J.
[25 Feb. 2021 (Morning)]

40. For the given circuit, comment on the type of transformer used : **[16 March, 2021 (Evening)]**

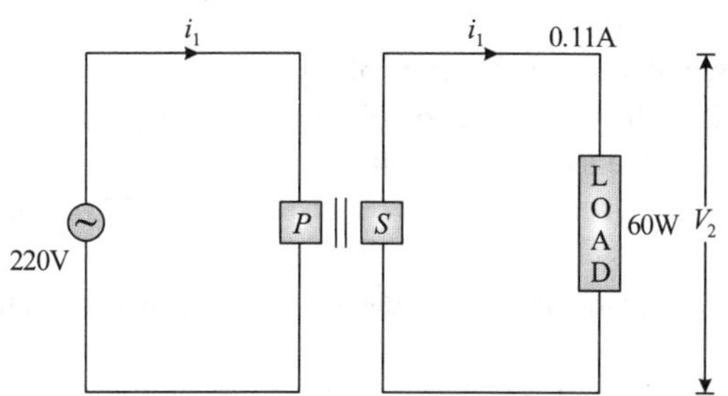

(A) Auxilliary transformer (B) Auto transformer
(C) Step-up transformer (D) Step down transformer

41. Four identical long solenoids A, B, C and D are connected to each other as shown in the figure. If the magnetic field at the center of A is 3 T, the field at the center of C would be : (Assume that the magnetic field is confined with in the volume of respective solenoid). **[18 March 2021 (Morning)]**

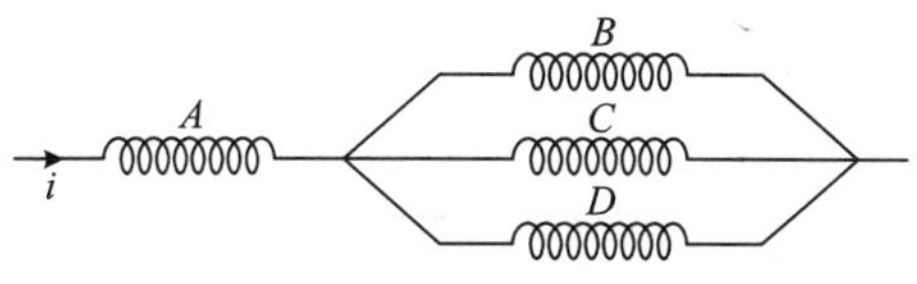

(A) 12 T (B) 6 T
(C) 9 T (D) 1 T

42. A small square loop of side a and one turn is placed inside a larger square loop of side b and one turn ($b >> a$). The two loops are coplanar with their centres coinciding. If a current I is passed in the square loop of side b, then the coefficient of mutual inductance between the two loops is:
[31 Aug. 2021 (Morning)]

(A) $\frac{\mu_0}{4\pi} 8\sqrt{2} \frac{b^2}{a}$ (B) $\frac{\mu_0}{4\pi} \frac{8\sqrt{2}}{b}$

(C) $\frac{\mu_0}{4\pi} \frac{8\sqrt{2}}{a}$ (D) $\frac{\mu_0}{4\pi} 8\sqrt{2} \frac{a^2}{a}$

43. The current in a coil of self inductance 2.0 H is increasing according to $I = 2 \sin(t^2)$ A. The amount of energy spent during the period when current changes from 0 to 2 A is _____ J.
[25 June 2022 (Morning)]

44. Two coils of self inductance L_1 and L_2 are connected in series combination having mutual inductance of the coils as M. The equivalent self inductance of the combination will be :
[26 June 2022 (Evening)]

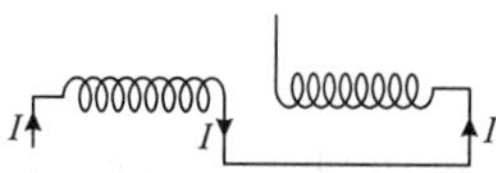

(A) $\frac{1}{L_1} + \frac{1}{L_2} + \frac{1}{M}$ (B) $L_1 + L_2 + M$

(C) $L_1 + L_2 + 2M$ (D) $L_1 + L_2 - 2M$

45. A small square loop of wire of side l is placed inside a large square loop of wire L ($L >> l$). Both loops are coplanar and their centres coincide at point O as shown in figure. The mutual inductance of the system is : **[25 July 2022 (Morning)]**

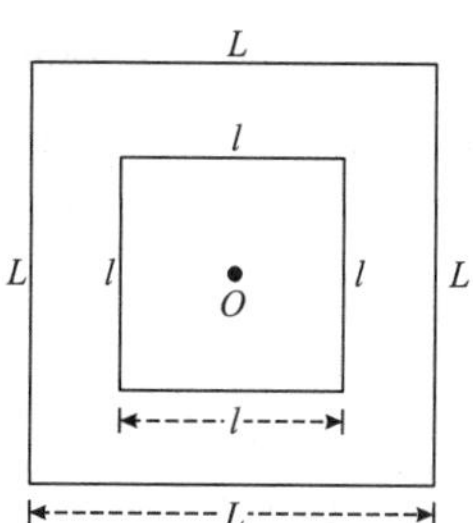

(A) $\frac{2\sqrt{2}\mu_0 L^2}{\pi l}$ (B) $\frac{\mu_0 l^2}{2\sqrt{2}\pi L}$

(C) $\frac{2\sqrt{2}\mu_0 l^2}{\pi L}$ (D) $\frac{\mu_0 L^2}{2\sqrt{2}\pi l}$

46. For the given circuit the current through battery of 6 V just after closing the switch S will be ______ A.
[28 July 2022 (Evening)]

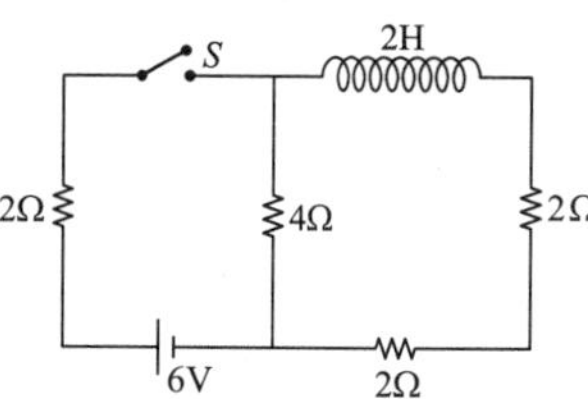

47. Three identical resistors with resistance $R = 12\ \Omega$ and two identical inductors with sell inductance $L = 5$ mH are connected to an ideal battery with emf of 12 V as shown in figure. The current through the battery long after the switch has been closed will be_____A. **[24 Jan. 2023 (Evening)]**

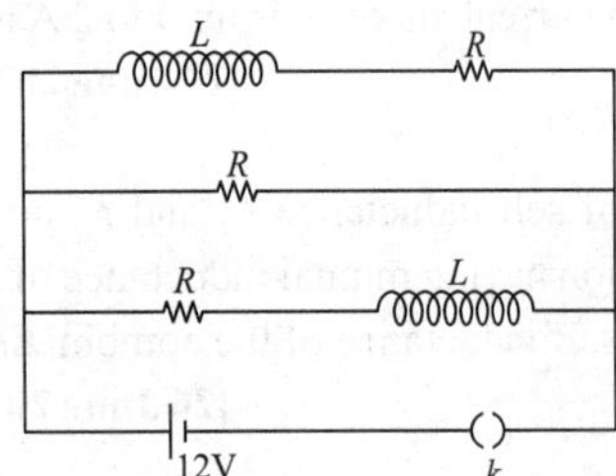

48. In an LC oscillator, if values of inductance and capacitance become twice and eight times, respectively, then the resonant frequency of oscillator becomes x times its initial resonant frequency ω_0. The value of x is : **[25 Jan. 2023 (Morning)]**

(A) 1/4 (B) 16

(C) 1/16 (D) 4

49. As per the given figure, if $\frac{dI}{dt} = -1$ A/s then the value of V_{AB} at this instant will be_____ V. **[30 Jan. 2023 (Morning)]**

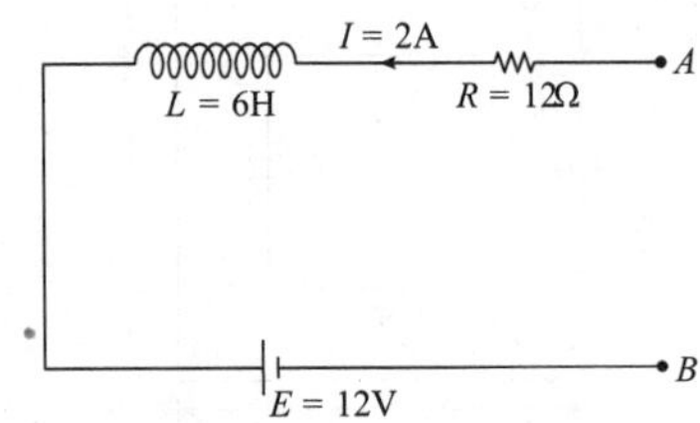

50. An ideal transformer with purely resistive load operates at 12 kV on the primary side. It supplies electrical energy to a number of nearby houses at 120 V. The average rate of energy consumption in the houses served by the transformer is 60 kW. The value of resistive load (Rs) required in the secondary circuit will be______ m. **[6 April 2023 (Morning)]**

51. Two concentric circular coils with radii 1 cm and 1000 cm, and number of turns 10 and 200 respectively are placed coaxially with centers coinciding. The mutual inductance of this arrangement will be_____ $\times 10^{-8}$ H. [Take, $\pi^2 = 10$] **[6 April 2023 (Evening)]**

52. An oscillating LC circuit consists of a 75 mH inductor and a 1.2 μF capacitor. If the maximum charge to the capacitor is 2.7 μC. The maximum current in the circuit will be_____ mA. **[8 April 2023 (Morning)]**

53. A series combination of resistor of resistance 100 Ω, inductor of inductance 1 H and capacitor of capacitance 6.25 μF is connected to an AC source. The quality factor of the circuit will be_____. **[8 April 2023 (Evening)]**

54. A coil has an inductance of 2 H and resistance of 4 Ω. A 10 V is applied across the coil. The energy stored in the magnetic field after the current has built up to its equilibrium value will be__0__ $\times 10^{-2}$ J. **[11 April 2023 (Evening)]**

55. In the given figure, an inductor and a resistor are connected in series with a battery of emf E volt. $\frac{E^a}{2b}$ J/s represents the maximum rate at which the energy is stored in the magnetic field (inductor). The numerical value of $\frac{b}{a}$ will be_____. **[13 April 2023 (Morning)]**

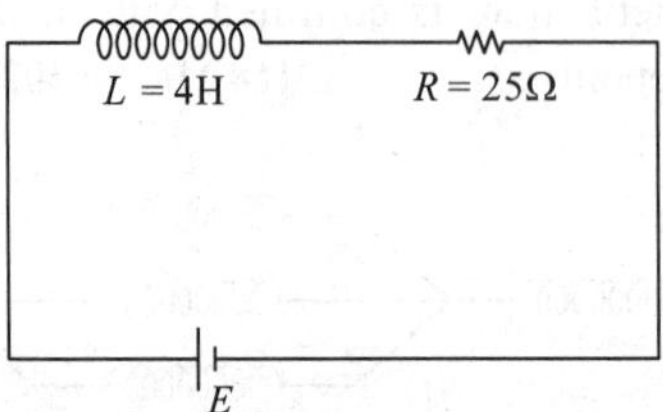

* * * * *

CHAPTER

21 ALTERNATING CURRENT

21.1 ALTERNATING CURRENT

1. In a transformer, number of turns in the primary are 140 and that in the secondary are 280. If current in primary is 4 A, then that in the secondary is : **[2002]**

(A) 4 A (B) 2 A

(C) 6 A (D) 10 A

2. The power factor of an *AC* circuit having resistance *R* and inductance *L* (connected in series) and an angular velocity ω is: **[2002]**

(A) $\frac{R}{\omega L}$ (B) $\frac{R}{(R^2+\omega^2 L^2)^{1/2}}$

(C) $\frac{\omega L}{R}$ (D) $\frac{R}{(R^2-\omega^2 L^2)^{1/2}}$

3. The core of any transformer is laminated so as to : **[2003]**

(A) reduce the energy loss due to eddy currents

(B) make it light weight

(C) make it robust and strong

(D) increase the secondary voltage

4. In an oscillating *LC* circuit the maximum charge on the capacitor is *Q*. The charge on the capacitor when the energy is stored equally between the electric and magnetic fields is : **[2003]**

(A) $Q/2$ (B) $Q/\sqrt{3}$

(C) $Q/\sqrt{2}$ (D) Q

5. A metal wire of linear mass density of 9.8 g/m is stretched with a tension of 10 kg-wt between two rigid supports 1 m apart. The wire passes at its middle point between the poles of a permanent magnet and it vibrates in resonance when carrying an alternating current of frequency *n*. The frequency *n* of the alternating source is : **[2003]**

(A) 50 Hz (B) 100 Hz

(C) 200 Hz (D) 25 Hz

6. In an *LCR* circuit, capacitance is changed from *C* to 2 C. For the resonant frequency to remain unchanged, the inductance should be changed from *L* to : **[2004]**

(A) $4L$ (B) $2L$

(C) $L/2$ (D) $L/4$

7. In an *LCR* series ac circuit, the voltage across each of the components. *L*, *C* and *R* is 50 V. The voltage across the *LC* combination will be : **[2004]**

(A) $50\ V$ (B) $50\sqrt{2}V$

(C) $100\ V$ (D) 0

8. Alternating current can not be measured by *DC* ammeter because : **[2004]**

(A) *AC* cannot pass through *DC* ammeter

(B) *AC* changes direction

(C) Average value of current for complete cycle is zero

(D) *DC* ammeter will get damaged

9. The phase difference between the alternating current and emf is π/2. which of the following cannot be the constituent of the circuit ? **[2005]**

(A) *C* alone (B) *R*, *L*

(C) *L*, *C* (D) *L* alone

10. A circuit has a resistance of 12 Ω and an impedance of 15 Ω. The power factor of circuit will be : **[2005]**

(A) 0.8 (B) 0.4

(C) 1.25 (D) 0.125

11. The self-inductance of the motor of an electric fan is 10 H. In order to impart maximum power at 50 Hz, it should be connected to a capacitance of : **[2005]**

(A) $4\ \mu F$ (B) $8\ \mu F$

(C) $1\ \mu F$ (D) $2\ \mu F$

12. In an *AC* generator, a coil with *N* turns, all of the same area *A* and total resistance *R*, rotates with frequency ω in a magnetic field *B*. The maximum value of emf generated in the coil is : **[2006]**

(A) *NABR*ω (B) *NAB*

(C) *NABR* (D) *NAB*ω

13. In a series resonant *LCR* circuit, the voltage across *R* is 100 volts and $R = 1\ k\Omega$ with $C = 2\ \mu F$. The resonant frequency ω is 200 rad/s. At resonance the voltage across *L* is : **[2006]**

(A) 2.5×10^{-2} V (B) 40 V

(C) 250 V (D) 4×10^{-3} V

14. In an *AC* circuit the voltage applied is $E = E_0 \sin\omega t$. The resulting current in the circuit is $I = I_0 \sin\left(\omega t - \frac{\pi}{2}\right)$. The power consumption in the circuit is given by : **[2007]**

(A) $P=\frac{E_0I_0}{\sqrt{2}}$ (B) $P=$ zero

(C) $P=\frac{E_0I_0}{2}$ (D) $P=\sqrt{2}\,E_0I_0$

15. In a series LCR circuit $R=200\ \Omega$ and the voltage and the frequency of the main supply is 220 V and 50 Hz respectively. On taking out the capacitance from the circuit the current lags behind the voltage by 30°. On taking out the inductor from the circuit the current leads the voltage by 30°. The power dissipated in the LCR circuit is : **[2010]**

(A) 242 W (B) 305 W
(C) 210 W (D) Zero

16. A fully charged capacitor C with initial charge q_0 is connected to a coil of self inductance L at $t=0$. The time at which the energy is stored equally between the electric and the magnetic fields is : **[2011]**

(A) $2\pi\sqrt{LC}$ (B) $\sqrt{LC}$

(C) $\pi\sqrt{LC}$ (D) $\frac{\pi}{4}\sqrt{LC}$

17. The capacitor of an oscillatory circuit is enclosed in a container. When the container is evacuated, the resonance frequency of the circuit is 10 kHz. When the container is filled with a gas, the resonance frequency changes by 50 Hz. The dielectric constant of the gas is : **[2012 (Online)]**

(A) 1.001 (B) 2.001
(C) 1.01 (D) 3.01

18. In an LCR circuit shown in the following figure, what will be the readings of the voltmeter across the resistor and ammeter if an a.c. source of 220 V and 100 Hz is connected to it as shown ? **[2012 (Online)]**

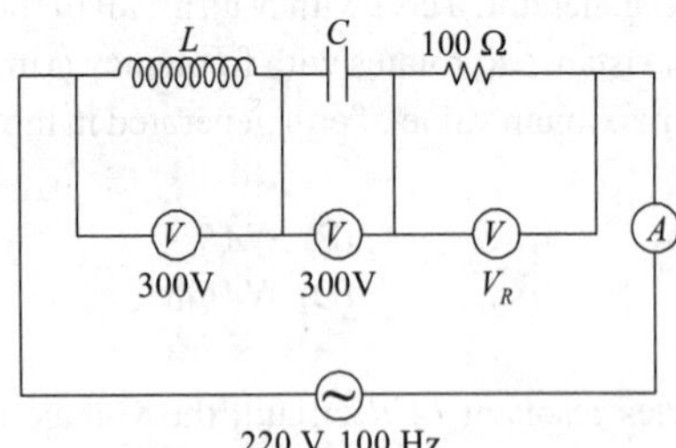

(A) 800 V, 8 A (B) 110 V, 1.1 A
(C) 300 V, 3 A (D) 220 V, 2.2 A

19. An LCR circuit as shown in the figure is connected to a voltage source V_{ac} whose frequency can be varied. The frequency, at which the voltage across the resistor is maximum, is : **[2013 (Online)]**

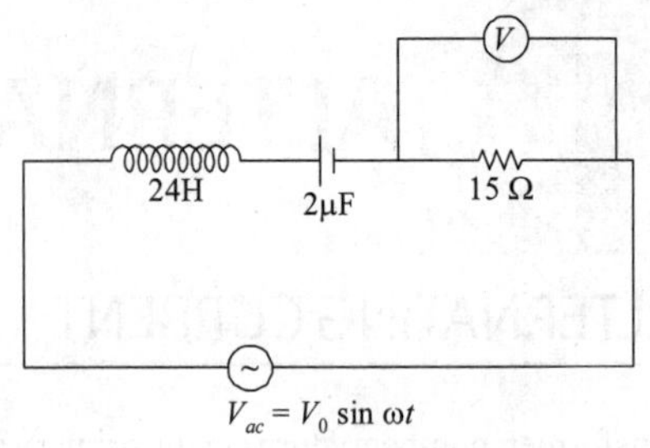

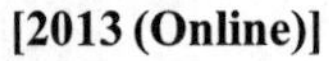

(A) 902 Hz (B) 143 Hz
(C) 23 Hz (D) 345 Hz

20. In a series LCR circuit, $C=10^{-11}$ Farad, $L=10^{-5}$ Henry and $R=100$ Ohm, when a constant D.C. voltage E is applied to the circuit, the capacitor acquires a charge 10^{-9} C. The D.C. source is replaced by a sinusoidal voltage source in which the peak voltage E_0 is equal to the constant D.C. voltage E. At resonance the peak value of the charge acquired by the capacitor will be : **[2013 (Online)]**

(A) 10^{-15} C (B) 10^{-6} C
(C) 10^{-10} C (D) 10^{-8} C

21. The plot given below is of the average power delivered to and LRC circuit versus frequency. The quality factor of the circuit is : **[2013 (Online)]**

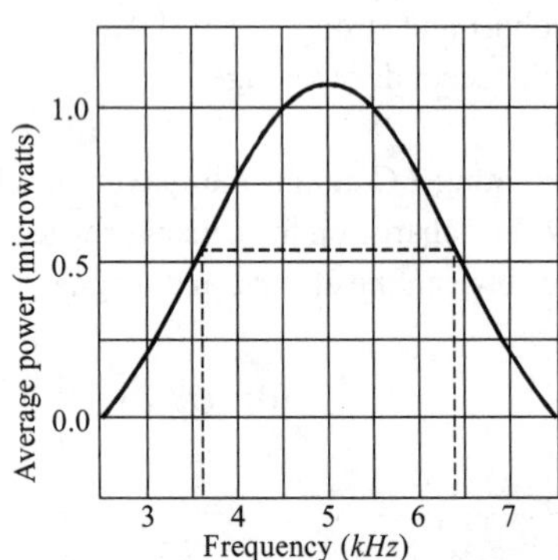

(A) 5.0 (B) 2.0
(C) 2.5 (D) 0.4

22. When resonance is produced in a series LCR circuit, then which of the following is not correct ? **[2013 (Online)]**

(A) Current in the circuit is in phase with the applied voltage
(B) Inductive and capacitive reactances are equal
(C) If R is reduced, the voltage across capacitor will increase
(D) Impedance of the circuit is maximum

23. A series LR circuit is connected to an ac source of frequency ω and the inductive reactance is equal to 2 R. A capacitance of capacitive reactance equal to R is added in series with L and R. The ratio of the new power factor to the old one is : **[2013 (Online)]**

(A) $\sqrt{\frac{2}{3}}$ (B) $\sqrt{\frac{2}{5}}$

(C) $\sqrt{\frac{3}{2}}$ (D) $\sqrt{\frac{5}{2}}$

24. When the rms voltages V_L, V_C and V_R are measured respectively across the inductor L, the capacitor C and the resistor R in a series LCR circuit connected to an AC source, it is found that the ratio $V_L : V_C : V_R = 1 : 2 : 3$. If the rms voltage of the AC sources is 100 V, the V_R is close to : **[2014 (Online)]**

(A) 50 V (B) 70 V

(C) 90 V (D) 100 V

25. A sinusoidal voltage $V(t) = 100 \sin (500t)$ is applied across a pure inductance of $L = 0.02$ H. The current through the coil is: **[2014 (Online)]**

(A) $10 \cos (500\,t)$ (B) $-10 \cos (500t)$

(C) $10 \sin (500t)$ (D) $-10 \sin (500t)$

26. The AC voltage across a resistance can be measured using a : **[2015 (Online)]**

(A) Hot wire voltmeter

(B) Moving coil galvanometer

(C) Potential coil galvanometer

(D) Moving magnet galvanometer

27. For the LCR circuit, shown here, the current is observed to lead the applied voltage. An additional capacitor C', when joined with the capacitor C present in the circuit, makes the power factor of the circuit unity. The capacitor C', must have been connected in : **[2015 (Online)]**

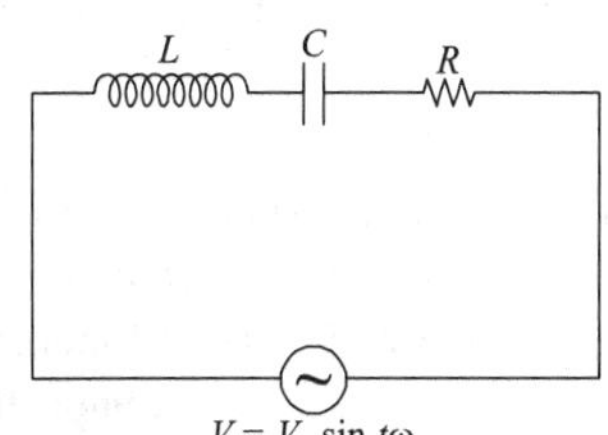

(A) Series with C and has a magnitude $\frac{C}{(\omega^2 LC - 1)}$

(B) Series with C and has a magnitude $\frac{1 - \omega^2 LC}{\omega^2 L}$

(C) Parallel with C and has a magnitude $\frac{1 - \omega^2 LC}{\omega^2 L}$

(D) Parallel with C and has a magnitude $\frac{C}{(\omega^2 LC - 1)}$

28. An arc lamp requires a direct current of 10 A at 80 V to function. If it is connected to a 220 V (rms), 50 Hz AC supply, the series inductor needed for it to work is close to : **[2016]**

(A) 80 H (B) 0.08 H

(C) 0.044 H (D) 0.065 H

29. A series LR circuit is connected to a voltage source with $V(t) = V_0 \sin \omega t$. After very large time, current (t) behaves as $\left(t_0 >> \frac{L}{R}\right)$: **[2016 (Online)]**

(A)

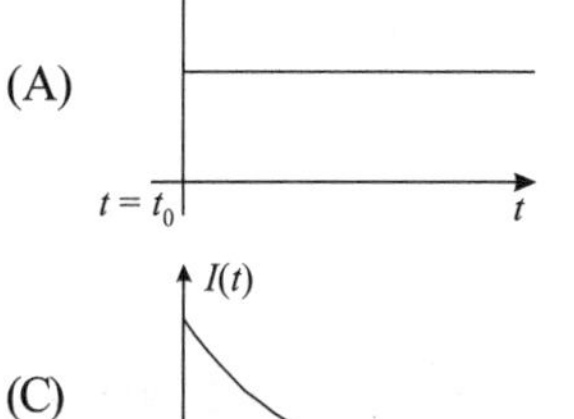

(B)

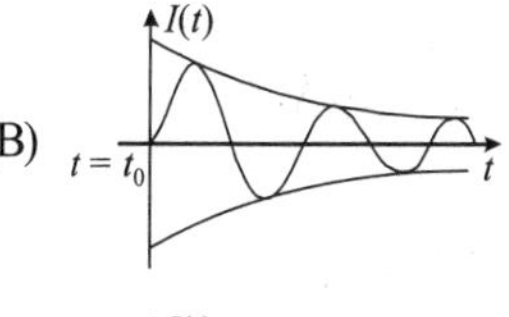

(C)

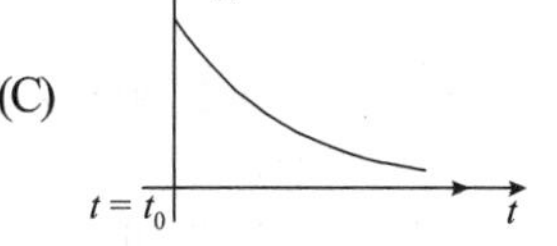

(D)

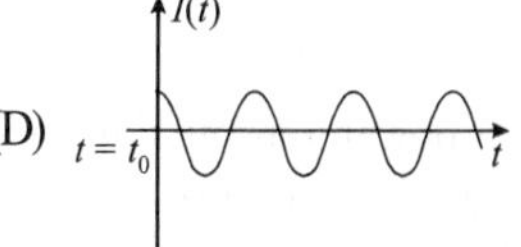

30. A sinusoidal voltage of peak value 283 V and angular frequency 320/s is applied to a series LCR circuit. Given that $R = 5\ \Omega$, $L = 25$ mH and $C = 1000\ \mu F$. The total impedance, and phase difference between the voltage across the source and the current will respectively be : **[2017 (Online)]**

(A) $10\ \Omega$ and $\tan^{-1}\left(\frac{5}{3}\right)$ (B) $7\ \Omega$ and $45°$

(C) $10\ \Omega$ and $\tan^{-1}\left(\frac{8}{3}\right)$ (D) $7\ \Omega$ and $\tan^{-1}\left(\frac{5}{3}\right)$

31. A small circular loop of wire of radius a is located at the centre of a much larger circular wire loop of radius b. The two loops are in the same plane. The outer loop of radius b carries an alternating current $I = I_0 \cos (\omega t)$. The emf induced in the smaller inner loop is nearly : **[2017 (Online)]**

(A) $\frac{\pi\mu_0 I_0}{2} \cdot \frac{a^2}{b}\ \omega \sin (\omega t)$ (B) $\frac{\pi\mu_0 I_0}{2} \cdot \frac{a^2}{b}\ \omega \cos (\omega t)$

(C) $\pi\mu_0 I_0 \frac{a^2}{b}\ \omega \sin (\omega t)$ (D) $\frac{\pi\mu_0 I_0 b^2}{a}\ \omega \cos (\omega t)$

32. For an RLC circuit driven with voltage of amplitude v_m and frequency $\omega_0 = \frac{1}{\sqrt{LC}}$ the current exhibits resonance. The quality factor, Q is given by : **[2018]**

(A) $\frac{\omega_0 L}{R}$ (B) $\frac{\omega_0 R}{L}$

(C) $\frac{R}{(\omega_0 C)}$ (D) $\frac{CR}{\omega_0}$

33. In an AC circuit, the instantaneous emf and current are given by

$e = 100 \sin 30t$

$i = 20 \sin\left(30t - \frac{\pi}{4}\right)$

In one cycle of AC, the average power consumed by the circuit and the wattless current are, respectively : **[2018]**

(A) 50 W, 10 A (B) $\frac{1000}{\sqrt{2}}$ W, 10 A

(C) $\frac{50}{\sqrt{2}}$ W, 0 (D) 50 W, 0

34. A power transmission line feeds input power at 2300 V to a step down transformer with its primary windings having 4000 turns, given the output power at 230 V. If the current in the primary of the transformer is 5 A, and its efficiency is 90%, the output current would be : **[2018 (Online)]**

(A) 20 A (B) 40 A

(C) 45 A (D) 25 A

35. A series *AC* circuit containing an inductor (20 mH), a capacitor (120 μF) and a resistor (60 Ω) is driven by an *AC* source of 24 V/50 Hz. The energy dissipated in the circuit in 60 s is : **[9 Jan. 2019 (Evening)]**

(A) 2.26×10^3 J (B) 3.39×10^3 J

(C) 5.65×10^2 J (D) 5.17×10^2 J

36.

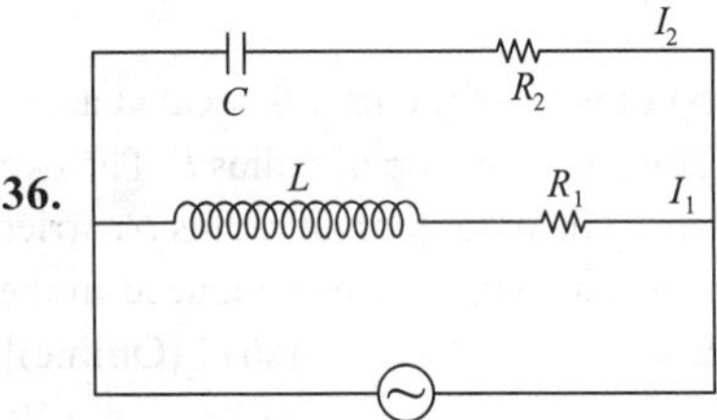

In the above circuit, $C = \frac{\sqrt{3}}{2} \mu F$, $R_2 = 20\ \Omega$, $L = \frac{\sqrt{3}}{10}$ H and $R_1 = 10\ \Omega$. Current in L-R_1 path is I_1 and in C-R_2 path it is I_2. The voltage of A.C source is given by $V = 200\sqrt{2}\sin(100t)$ volts. The phase difference between I_1 and I_2 is : **[12 Jan. 2019 (Evening)]**

(A) 30° (B) 0°

(C) 90° (D) 60°

37. An alternating voltage $v(t) = 220 \sin 100\pi t$ volt is applied to a purely resistance load of 50 Ω. The time taken for the current to rise from half of the peak value to the peak value is : **[8 April 2019 (Morning)]**

(A) 2.2 ms (B) 5 ms

(C) 3.3 ms (D) 7.2 ms

38. A circuit connected to an ac source of *emf* $e = e_0 \sin(100t)$ with t in seconds, gives a phase difference of $\pi/4$ between the emf e and current i. Which of the following circuits will exhibit this ? **[8 April 2019 (Evening)]**

(A) *RC* circuit with R = 1 kΩ and C = 1 μF

(B) *RL* circuit with R = 1 kΩ and L = 1 mH

(C) *RL* circuit with R = 1 kΩ and L = 10 mH

(D) *RC* circuit with R = 1 kΩ and C = 10 μF

39. A transformer consisting of 300 turns in the primary and 150 turns in the secondary gives output power of 2.2 kW. If the current in the secondary coil is 10 A, then the input voltage and current in the primary coil are : **[10 April 2019 (Morning)]**

(A) 220 V and 10 A (B) 440 V and 5 A

(C) 440 V and 20 A (D) 220 V and 20 A

40. A *LCR* circuit behaves like a damped harmonic oscillator. Comparing it with a physical spring-mass damped oscillator having damping constant 'b', the correct equivalence would be : **[7 Jan. 2020 (Morning)]**

(A) $L \leftrightarrow m, C \leftrightarrow \frac{1}{k}, R \leftrightarrow b$

(B) $L \leftrightarrow \frac{1}{b}, C \leftrightarrow \frac{1}{m}, R \leftrightarrow \frac{1}{k}$

(C) $L \leftrightarrow m, C \leftrightarrow k, R \leftrightarrow b$

(D) $L \leftrightarrow k, C \leftrightarrow b, R \leftrightarrow m$

41. In *LC* circuit the inductance L = 40 mH and capacitance C = 100 μF. If a voltage $V(t) = 10 \sin(314\,t)$ is applied to the circuit, the current in the circuit is given as : **[9 Jan. 2020 (Evening)]**

(A) $0.52 \cos 314\,t$ (B) $0.52 \sin 314\,t$

(C) $10 \cos 314\,t$ (D) $5.2 \cos 314\,t$

42. An inductance coil has a reactance of 100 W. When an *AC* signal of frequency 1000 Hz is applied to the coil, the applied voltage leads the current by 45°. The self-inductance of the coil is : **[2 Sept. 2020 (Evening)]**

(A) 1.1×10^{-2} H (B) 1.1×10^{-1} H

(C) 5.5×10^{-5} H (D) 6.7×10^{-7} H

43. A 750 Hz, 20 V (rms) source is connected to a resistance of 100 Ω, an inductance of 0.1803 H and a capacitance of 10 μF all in series. The time in which the resistance (heat capacity 2 J/°C) will get heated by 10°C. (assume no loss of heat to the surroundings) is close to : **[3 Sept. 2020 (Morning)]**

(A) 418 s (B) 245 s

(C) 348 s (D) 365 s

44. In a series *LR* circuit, power of 400 W is dissipated from a source of 250 V, 50 Hz. The power factor of the circuit is 0.8. In order to bring the power factor to unity, a capacitor of value C

is added in series to the L and R. Taking the value of C as $\left(\frac{n}{3\pi}\right)\mu F$, then value of n is : **[6 Sept. 2020 (Evening)]**

45. A resonance circuit having inductance and resistance 2×10^{-4} H and 6.28 Ω respectively oscillates at 10 MHz frequency. The value of quality factor of this resonator is ______. [$\pi = 3.14$] **[24 Feb. 2021 (Morning)]**

46. Match List-I with List-II : **[24 Feb. 2021 (Evening)]**

List-I	**List-II**
(a) Source of microwave frequency	(i) Radioactive decay of nucleus
(b) Source of infrared frequency	(ii) Magnetron
(c) Source of Gamma Rays	(iii) Inner shell electrons
(d) Source of X-rays	(iv) Vibration of atoms and molecules
	(v) LASER
	(vi) RC circuit

Choose the correct answer from the options given below :

(A) (a)-(vi), (b)-(v), (c)-(i), (d)-(iv)

(B) (a)-(ii), (b)-(iv), (c)-(i), (d)-(iii)

(C) (a)-(ii), (b)-(iv), (c)-(vi), (d)-(iii)

(D) (a)-(vi), (b)-(iv), (c)-(i), (d)-(v)

47. A series LCR circuit is designed to resonate at an angular frequency $\omega_0 = 10^5$ rad/s. The circuit draws 16 W power from 120 V source at resonance. The value of resistance 'R' in the circuit is ______ Ω. **[24 Feb. 2021 (Evening)]**

48. The angular frequency of alternating current in a LCR circuit is 100 rad/s. The components connected are shown in the figure. Find the value of inductance of the coil and capacity of condenser. **[25 Feb. 2021 (Morning)]**

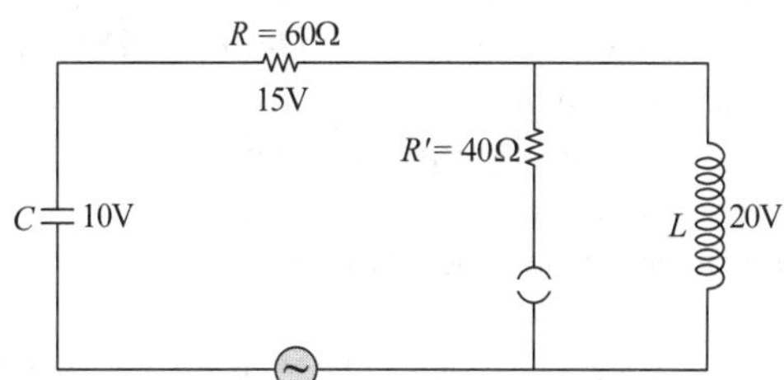

(A) 1.33 H and 250 μF (B) 1.33 H and 150 μF

(C) 0.8 H and 150 μF (D) 0.8 H and 250 μF

49. An LCR circuit contains resistance of 110 Ω and a supply of 220 V at 300 rad/s angular frequency. If only capacitance is removed from the circuit, current lags behind the voltage by 45°. If on the other hand, only inductor is removed the current leads by 45° with the applied voltage. The RMS current flowing in the circuit will be : **[25 Feb. 2021 (Evening)]**

(A) 2 A (B) 2.5 A

(C) 1.5 A (D) 1 A

50. Match List-I with List-II : **[25 Feb. 2021 (Evening)]**

List-I	**List-II**
(a) Rectifier	(i) Used either for stepping up or stepping down the a.c. voltage
(b) Stabilizer	(ii) Used to convert a.c. voltage into d.c. voltage
(c) Transformer	(iii) Used to remove any ripple in the rectified output voltage
(d) Filter	(iv) Used for constant output voltage even when the input voltage or load current change

(A) (a)-(ii), (b)-(iv), (c)-(i), (d)-(iii)

(B) (a)-(ii), (b)-(i), (c)-(iii), (d)-(iv)

(C) (a)-(ii), (b)-(i), (c)-(iv), (d)-(iii)

(D) (a)-(iii), (b)-(iv), (c)-(i), (d)-(ii)

51. An alternating current is given by the equation $i = i_1 \sin \omega t + i_2 \cos \omega t$. The rms current will be : **[26 Feb. 2021 (Morning)]**

(A) $\frac{1}{\sqrt{2}}(i_1 + i_2)^2$ (B) $\frac{1}{\sqrt{2}}(i_1 + i_2)$

(C) $\frac{1}{\sqrt{2}}(i_1^2 + i_2^2)^{1/2}$ (D) $\frac{1}{2}(i_1^2 + i_2^2)^{1/2}$

52. In a series LCR resonant circuit, the quality factor is measured as 100. If the inductance is increased by two fold and resistance is decreased by two fold, then the quality factor after this change will be ______. **[26 Feb. 2021 (Morning)]**

53. Find the peak current and resonant frequency of the following circuit (as shown in figure). **[26 Feb. 2021 (Evening)]**

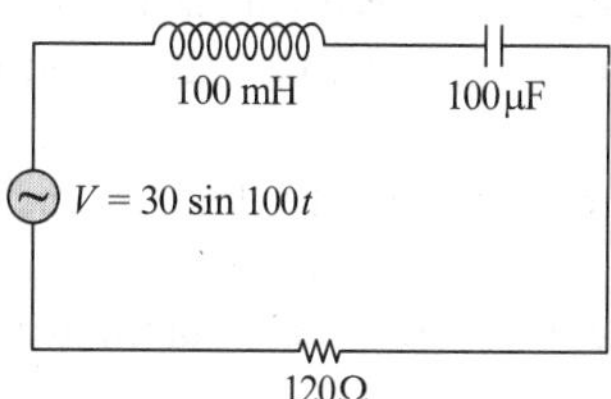

(A) 2 A and 50 Hz (B) 0.2 A and 100 Hz

(C) 2 A and 100 Hz (D) 0.2 A and 50 Hz

54. A sinusoidal voltage of peak value 250 V is applied to a series LCR circuit, in which $R = 8$ Ω, $L = 24$ mH and $C = 60$ μF. The value of power dissipated at resonant condition is 'x' kW. The value of x to the nearest integer is ______. **[16 March 2021 (Morning)]**

55. An AC current is given by $I = I_1 \sin\omega t + I_2 \cos\omega t$. A hot wire ammeter will give a reading : **[17 March 2021 (Morning)]**

(A) $\sqrt{\frac{I_1^2 - I_2^2}{2}}$ (B) $\sqrt{\frac{I_1^2 + I_2^2}{2}}$

(C) $\frac{I_1 + I_2}{\sqrt{2}}$ (D) $\frac{I_1 + I_2}{2\sqrt{2}}$

56. Match List-I with List-II

List-I	List-II
(a) Phase difference between current and voltage in a purely resistive AC circuit	(i) $\frac{\pi}{2}$; current leads voltage
(b) Phase difference between current and voltage in a pure inductive AC circuit	(ii) zero
(c) Phase difference between current and voltage in a pure capacitive AC circuit	(iii) $\frac{\pi}{2}$; current lags voltage
(d) Phase difference between current and voltage in an LCR series circuit	(iv) $\tan^{-1}\left(\frac{X_C - X_L}{R}\right)$

Choose the most appropriate answer from the options given below: **[17 March 2021 (Evening)]**

(A) (a)-(i), (b)-(iii), (c)-(iv), (d)-(ii)
(B) (a)-(ii), (b)-(iv), (c)-(iii), (d)-(i)
(C) (a)-(ii), (b)-(iii), (c)-(iv), (d)-(i)
(D) (a)-(ii), (b)-(iii), (c)-(i), (d)-(iv)

57. What happens to the inductive reactance and the current in a purely inductive circuit if the frequency is halved ? **[17 March 2021 (Evening)]**

(A) Both, inductive reactance and current will be halved.
(B) Inductive reactance will be halved and current will be doubled.
(C) Inductive reactance will be doubled and current will be halved.
(D) Both, inducting reactance and current will be doubled.

58. In a series LCR resonance circuit, if we change the resistance only, from a lower to higher value : **[18 March 2021 (Morning)]**

(A) The bandwidth of resonance circuit will increase
(B) The resonance frequency will increase
(C) The quality factor will increase
(D) The quality factor and the resonance frequency will remain constant

59. An AC source rated 220 V, 50 Hz is connected to a resistor. The time taken by the current to change from its maximum to the rms value is : **[18 March 2021 (Morning)]**

(A) 2.5 ms (B) 25 ms
(C) 2.5 s (D) 0.25 ms

60. In a series LCR circuit, the inductive reactance (X_L) is 10 Ω and the capacitive reactance (X_C) is 4 Ω. The resistance (R) in the circuit is 6 Ω. The power factor of the circuit is : **[18 March 2021 (Evening)]**

(A) $\frac{1}{2}$ (B) $\frac{1}{2\sqrt{2}}$

(C) $\frac{1}{\sqrt{2}}$ (D) $\frac{\sqrt{3}}{2}$

61. The time taken for the magnetic energy to reach 25% of its maximum value, when a solenoid of resistance R, inductance L is connected to a battery, is : **[18 March 2021 (Evening)]**

(A) $\frac{L}{R}\ln 5$ (B) Infinite

(C) $\frac{L}{R}\ln 2$ (D) $\frac{L}{R}\ln 10$

62. In an LCR series circuit, an inductor 30 mH and a resistor 1 Ω are connected to an AC source of angular frequency 300 rad/s. The value of capacitance for which, the current leads the voltage by 45° is $\frac{1}{x} \times 10^{-3}$ F. Then the value of x is ______.

[20 July 2021 (Morning)]

63. For a series LCR circuit with $R = 100\ \Omega$, $L = 0.5$ mH ad $C = 0.1$ pF connected across 220 V – 50 Hz AC supply, the phase angle between current and supplied voltage and the nature of the circuit is : **[20 July 2021 (Evening)]**

(A) 0°, resonance circuit
(B) ≈ 90°, predominantly capacitive circuit
(C) 0°, resistive circuit
(D) ≈ 90, predominantly inductive circuit

64. A series LCR circuit of $R = 5\ \Omega$, $L = 20$ mH and $C = 0.5\ \mu F$ is connected across an AC supply of 250 V, having variable frequency. The power dissipated at resonance condition is ______ $\times 10^2$ W. **[20 July 2021 (Evening)]**

65. In a circuit consisting of a capacitance and a generator with alternating emf $E_g = E_{g0} \sin \omega t$ V_C and I_C are the voltage and current. Correct phasor diagram for such circuit is : **[22 July 2021 (Evening)]**

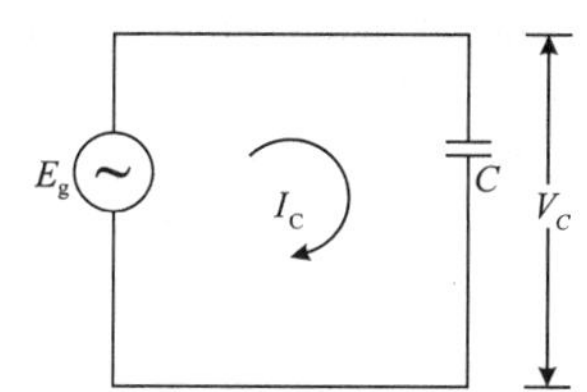

(A) 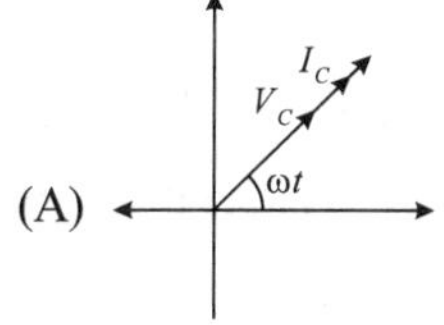

(B) 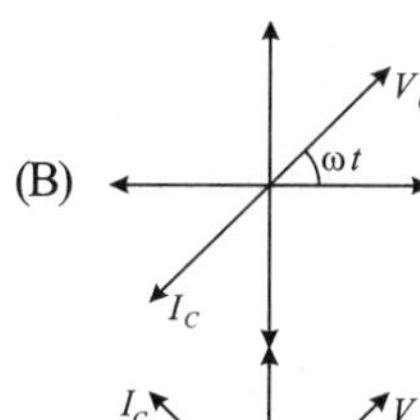

(C) 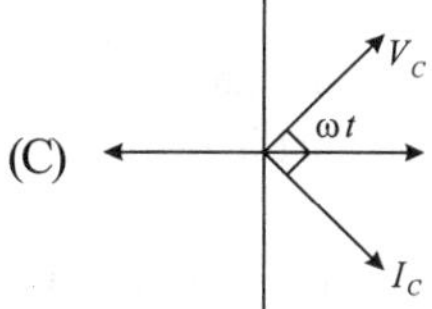

(D) 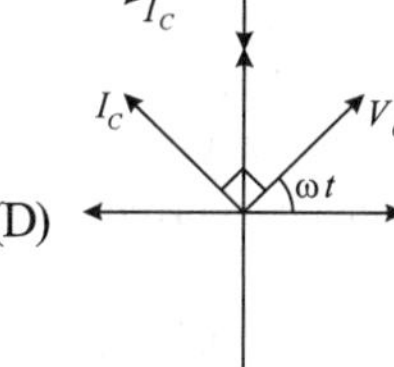

66. Match List-I with List-II : **[22 July 2021 (Evening)]**

List-I	**List-II**
(a) $\omega L > \frac{1}{\omega C}$	(i) Current is in phase with emf
(b) $\omega L = \frac{1}{\omega C}$	(ii) Current lags behind the applied emf
(c) $\omega L < \frac{1}{\omega C}$	(iii) Maximum current occurs
(d) Resonant frequency	(iv) Current leads the emf

Choose the correct answer from the option given below :

(A) (a)-(iv), (b)-(iii), (c)-(ii), (d)-(i)
(B) (a)-(ii), (b)-(i), (c)-(iv), (d)-(iii)
(C) (a)-(ii), (b)-(i), (c)-(iii), (d)-(iv)
(D) (a)-(iii), (b)-(i), (c)-(iv), (d)-(ii)

67. A 10 Ω resistance is connected across 220 V – 50 Hz AC supply. The time taken by the current to change from its maximum value to the rms value is : **[25 July 2021 (Evening)]**

(A) 3.0 ms (B) 1.5 ms
(C) 4.5 ms (D) 2.5 ms

68. Two circuits are shown in the figure (*a*) and (*b*). At a frequency of______ rad/s the average power dissipated in one cycle will be same in both the circuits :

[25 July 2021 (Evening)]

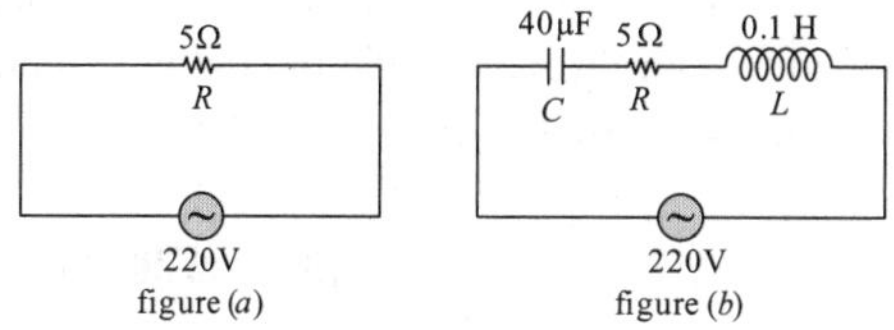

69. A 0.07 H inductor and 12 Ω resistor are connected in series to a 220 V, 50 Hz ac source. The approximate current in the circuit and phase angle between current and source voltage are respectively : [Take π as $\frac{22}{7}$] **[27 July 2021 (Morning)]**

(A) 8.8 A and $\tan^{-1}\left(\frac{6}{11}\right)$ (B) 0.88 A and $\tan^{-1}\left(\frac{11}{6}\right)$
(C) 8.8 A and $\tan^{-1}\left(\frac{11}{6}\right)$ (D) 88 A and $\tan^{-1}\left(\frac{11}{6}\right)$

70. A 100 Ω resistance, a 0.1 μF capacitor and an inductor are connected in series across a 250 V supply at variable frequency, Calculate the value of inductance of inductor at which resonance will occur, Given that the resonant frequency is 60 Hz :

[27 July 2021 (Evening)]

(A) 7.03×10^{-5} H (B) 0.70 H
(C) 70.3 mH (D) 70.3 H

71. An inductor coil stores 64 J of magnetic field energy and dissipates energy at the rate of 640 W when a current of 8 A is passed through it. If this coil is joined across an ideal battery, find the time constant of the circuit in seconds :

[26 Aug. 2021 (Morning)]

(A) 0.8 (B) 0.125
(C) 0.4 (D) 0.2

72. A series *LCR* circuit driven by 300 V at frequency of 50 Hz contains a resistance $R = 3$ kΩ, an inductor of inductive reactance $X_L = 250\,\pi\Omega$ and an unknown capacitor. The value of capacitance to maximize the average power should be : [Take $\pi^2 = 10$]

[26 Aug. 2021 (Morning)]

(A) 4 μF (B) 400 μF
(C) 40 μF (D) 25 μF

73. In the given circuit the *AC* source has $\omega = 100$ rad s^{-1}. Considering the inductor and capacitor to be ideal, what will be the current *I* flowing through the circuit ?

[26 Aug. 2021 (Evening)]

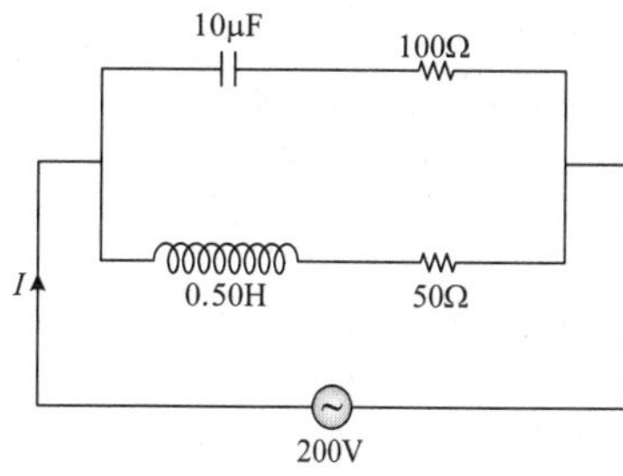

(A) 0.94 A (B) 5.9 A
(C) 4.24 A (D) 6 A

74. The alternating current is given by

$$i=\left\{\sqrt{42}\ \sin\left(\frac{2\pi}{T}t\right)+10\right\}A$$

The rms value of this current is ______ A. **[27 Aug. 2021 (Morning)]**

75. An AC circuit has an inductor and a resistor of resistance R in series, such that $X_L = 3R$. Now, a capacitor is added in series such that $X_C = 2R$. The ratio of new power factor with the old power factor of the circuit is $\sqrt{5}:x$. The value of x is______. **[27 Aug. 2021 (Evening)]**

76. In an ac circuit, an inductor, a capacitor and a resistor are connected in series with $X_L = R = X_C$. Impedance of this circuit is : **[31 Aug. 2021 (Morning)]**

(A) $R\sqrt{2}$ (B) R
(C) $2R^2$ (D) Zero

77. At very high frequencies, the effective impedance of the given circuit will be ______ Ω. **[31 Aug. 2021 (Evening)]**

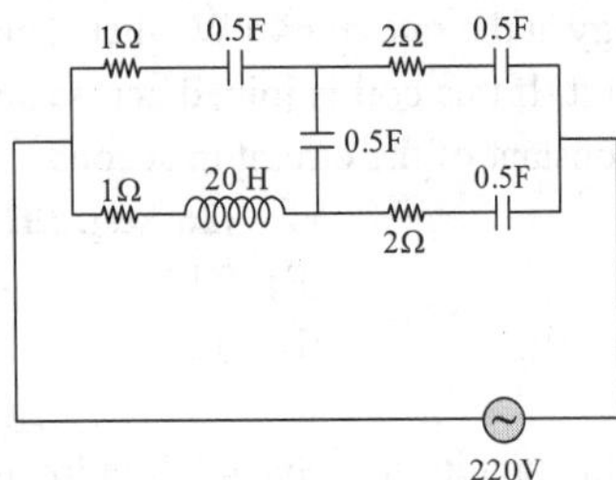

78. In the circuit shown in the figure, the ratio of the quality factor and the band width is _____ s. **[24 Jan. 2023 (Morning)]**

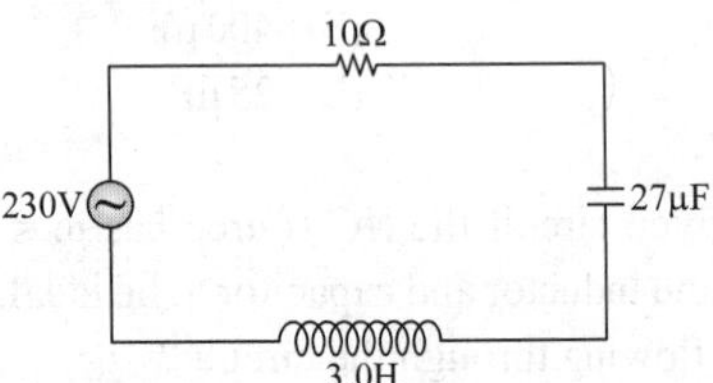

79. An LCR series circuit of capacitance 62.5 nF and resistance of 50 Ω. Is connected to an A.C. source of frequency 2.0 kHz. For maximum value of amplitude of current in circuit, the value of inductance is_____ mH. (Take $\pi^2 = 10$). **[25 Jan. 2023 (Morning)]**

80. A series LCR circuit is connected to an AC source of 220 V, 50 Hz. The circuit contains a resistance $R = 80\ \Omega$, an inductor of inductive reactance $X_L = 70\ \Omega$, and a capacitor of capacitive reactance $X_C = 130\ \Omega$. The power factor of circuit is $\frac{x}{10}$. The value of x is. **[25 Jan. 2023 (Evening)]**

81. For the given figures, choose the correct options : **[29 Jan. 2023 (Evening)]**

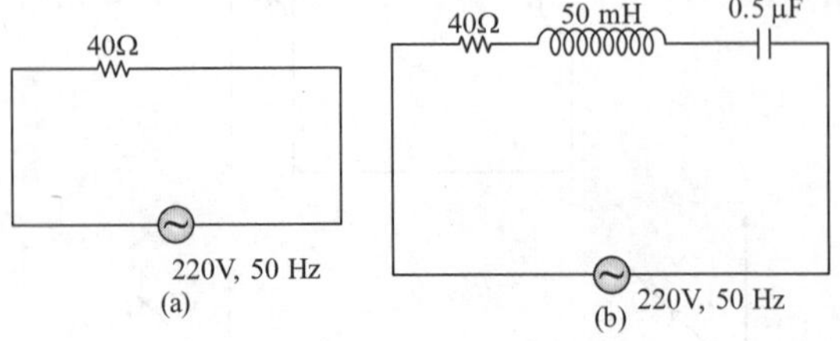

(A) The rms current in circuit (b) can never be larger than that in (a)
(B) The rms current in figure (a) is always equal to that in figure (b)
(C) The rms current in circuit (b) can be larger than that in (a)
(D) At resonance, current in (b) is less than that in (a)

82. An inductor of inductance 2 μH is connected in series with a resistance, a variable capacitor and an AC source of frequency 7 kHz. The value of capacitance for which maximum current is drawn into the circuit is $\frac{1}{x}F$, where the value of x is _____.

[Take $\pi = \frac{22}{7}$] **[29 Jan. 2023 (Evening)]**

83. In a series LR circuit with $X_L = R$. Power factor is P_1. If a capacitor of capacitance C with $X_C = X_L$ is added to the circuit the power factor becomes P_2. The ratio of P_1 to P_2 will be : **[30 Jan. 2023 (Morning)]**

(A) 1 : 3 (B) $1:\sqrt{2}$
(C) 1 : 1 (D) 1 : 2

84. In the given circuit, rms value of current (I_{rms}) through the resistor R is : **[30 Jan. 2023 (Evening)]**

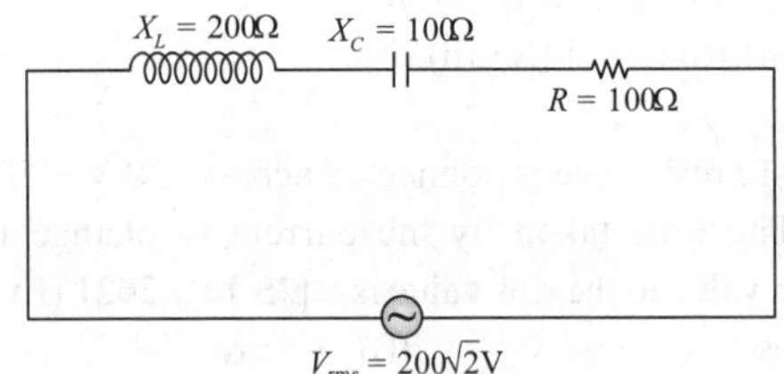

(A) 2 A (B) $\frac{1}{2}$ A
(C) 20 A (D) $2\sqrt{2}$ A

85. In an AC generator, a rectangular coil of 100 turns each having area $14 \times 10^{-2}\ m^2$ is rotated at 360 rev/min about an axis perpendicular to a uniform magnetic field of magnitude 3.0 T. The maximum value of the emf produced will be ______ V.

$\left(\text{Take } \pi = \frac{22}{7}\right)$ **[30 Jan. 2023 (Evening)]**

86. If R, X_L and X_C represent resistance, inductive reactance and capacitive reactance. Then which of the following is dimensionless : **[31 Jan. 2023 (Morning)]**

(A) RX_LX_C (B) $\frac{R}{\sqrt{X_LX_C}}$

(C) $\frac{R}{X_LX_C}$ (D) $R\frac{X_L}{X_C}$

87. An inductor of 0.5 mH, a capacitor of 20 μF and resistance of 20 Ω are connected in series with a 220 V ac source. If the current is in phase with the emf, the amplitude of current of the circuit is $\sqrt{x}A$. The value of x is : **[31 Jan. 2023 (Morning)]**

88. An alternating voltage source $V = 260 \sin(628t)$ is connected across a pure inductor of 5 mH. Inductive reactance in the circuit is : **[31 Jan. 2023 (Evening)]**

(A) 3.14 Ω (B) 6.28 Ω

(C) 0.5 Ω (D) 0.318 Ω

89. A series LCR circuit consists of $R = 80\ \Omega$. $X_L = 100\ \Omega$, and $X_C = 40\ \Omega$. The input voltage is $2500 \cos(100\pi t)$V. The amplitude of current, in the circuit, is_____ A. **[31 Jan. 2023 (Evening)]**

90. Match the List-I with List-II : **[1 Feb. 2023 (Morning)]**

List-I	**List-II**
(a) AC generator	(i) Presence of both L and C
(b) Transformer	(ii) Electromagnetic Induction
(c) Resonance phenomenon to occur	(iii) Quality factor
(d) Sharpness of resonance	(iv) Mutual Inductance

Choose the correct answer from the options given below :

(A) (a)-(iv), (b)-(ii), (c)-(i), (d)-(iii)
(B) (a)-(ii), (b)-(i), (c)-(iii), (d)-(iv)
(C) (a)-(ii), (b)-(iv), (c)-(i), (d)-(iii)
(D) (a)-(iv), (b)-(iii), (c)-(i), (d)-(ii)

91. A series LCR circuit is connected to an ac source of 220 V, 50 Hz. The circuit contain a resistance $R = 100\ \Omega$ and an inductor of inductive reactance $X_L = 79.6\ \Omega$. The capacitance of the capacitor needed to maximize the average rate at which energy is supplied will be _____ μF. **[1 Feb. 2023 (Morning)]**

92. Given below are two Statements:

Statement-I : Maximum power is dissipated in a circuit containing an inductor, a capacitor and a resistor connected in series with an AC source, when resonance occurs.

Statement-II : Maximum power is dissipated in a circuit containing pure resistor due to zero phase difference between current and voltage.

In the light of the above statements, choose the correct answer from the options given below: **[10 April 2023 (Morning)]**

(A) Statement-I is false but Statement-II is true
(B) Statement-I is true but Statement-II is false
(C) Both Statement-I and Statement-II are true
(D) Both Statement-I and Statement-II are false

93. As per the given graph choose the correct representation for curve A and curve B.

{Where X_C = reactance of pure capacitive circuit connected with A.C. source

X_L = reactance of pure inductive circuit connected with A.C. source

R = impedance of pure resistive circuit connected with A.C. source

Z = Impedance of the LCR series circuit}

[11 April 2023 (Morning)]

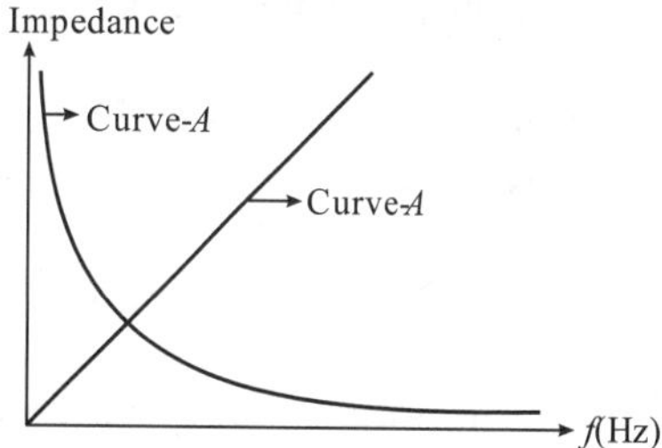

(A) $A = X_C, B = R$ (B) $A = X_L, B = Z$

(C) $A = X_C, B = X_L$ (D) $A = X_L, B = R$

94. Given below are two statements.

Statement-I: When the frequency of an a.c. source in a series LCR circuit increases, the current in the circuit first increases, attains a maximum value and then decreases.

Statement-II: In a series LCR circuit, the value of power factor at resonance is one.

In the light of given statements, choose the most appropriate answer from the options given below:

[12 April 2023 (Morning)]

(A) Statement-I is incorrect but Statement-II is true
(B) Both Statement-I and Statement-II are false
(C) Statement-I is correct but Statement-II is false
(D) Both Statement-I and Statement-II are true

95. Given below are two statements:

Statement-I : An AC circuit undergoes electrical resonance if it contains either a capacitor or an inductor.

Statement-II : An AC circuit containing a pure capacitor or a pure inductor consumes high power due to its non-zero power factor.

In the light of above statements, choose the correct answer from the options given below: **[13 April 2023 (Evening)]**

(A) Both Statement-I and Statement-II are false
(B) Statement-I is true but Statement-II is false
(C) Both Statement-I and Statement-II are true
(D) Statement-I is false but Statement-II is true

* * * * *

ELECTROMAGNETIC WAVES

22.1 ELECTROMAGNETIC WAVE

1. An electromagnetic wave of frequency $v = 3.0$ MHz passes from vacuum into a dielectric medium with permittivity $\in = 4.0$. Then : **[2004]**
(A) Wave length is halved and frequency remains unchanged
(B) Wave length is doubled and frequency becomes half
(C) Wave length is doubled and the frequency remains unchanged
(D) Wave length and frequency both remain unchanged

2. If a source of power 4 kW produces 10^{20} photons/second, the radiation belongs to apart of the spectrum called : **[2010]**
(A) γ-rays (B) X-rays
(C) ultraviolet rays (D) microwaves

3. An electromagnetic wave in vacuum has the electric and magnetic field $\vec{E}$ and $\vec{B}$, which are always perpendicular to each other. The direction of polarization is given by $\vec{X}$ and that of wave propagation by $\vec{k}$. Then : **[2012]**
(A) $\vec{X} \parallel \vec{B}$ and $\vec{k} \parallel \vec{B} \times \vec{E}$ (B) $\vec{X} \parallel \vec{E}$ and $\vec{k} \parallel \vec{E} \times \vec{B}$
(C) $\vec{X} \parallel \vec{B}$ and $\vec{k} \parallel \vec{E} \times \vec{B}$ (D) $\vec{X} \parallel \vec{E}$ and $\vec{k} \parallel \vec{B} \times \vec{E}$

4. The frequency of *X*-rays; γ-rays and ultraviolet rays are respectively *a*, *b* and *c* then : **[2012 (Online)]**
(A) $a < b; b > c$ (B) $a > b; b > c$
(C) $a < b < c$ (D) $a = b = c$

5. An electromagnetic wave with frequency ω and wavelength λ travels in the $+y$ direction. Its magnetic field is along $+x$-axis. The vector equation for the associated electric field (of amplitude E_0) is : **[2012 (Online)]**

(A) $\vec{E} = -E_0 \cos\left(\omega t + \frac{2\pi}{\lambda} y\right)\hat{x}$

(B) $\vec{E} = E_0 \cos\left(\omega t - \frac{2\pi}{\lambda} y\right)\hat{x}$

(C) $\vec{E} = E_0 \cos\left(\omega t - \frac{2\pi}{\lambda} y\right)\hat{z}$

(D) $\vec{E} = -E_0 \cos\left(\omega t + \frac{2\pi}{\lambda} y\right)\hat{z}$

6. The magnetic field in a travelling electromagnetic wave has a peak value of 20 nT. The peak value of electric field strength is : **[2013]**
(A) 3 V/m (B) 6 V/m
(C) 9 V/m (D) 12 V/m

7. Photons of an electromagnetic radiation has an energy 11 keV each. To which region of electromagnetic spectrum does it belong ? **[2013 (Online)]**
(A) X-rays region (B) Ultra violet region
(C) Infrared region (D) Visible region

8. Select the correct statement from the following : **[2013 (Online)]**
(A) Electromagnetic waves cannot travel in vacuum
(B) Electromagnetic waves are longitudinal waves
(C) Electromagnetic waves are produced by charges moving with uniform velocity
(D) Electromagnetic waves carry both energy and momentum as they propagate through space

9. A plane electromagnetic wave in a non-magnetic dielectric medium is given by $\vec{E} = \vec{E}_0 (4 \times 10^{-7} x - 50t)$ with distance being in meter and time in seconds. The dielectric constant of the medium is : **[2013 (Online)]**
(A) 2.4 (B) 5.8
(C) 8.2 (D) 4.8

10. During the propagation of electromagnetic waves in a medium: **[2014]**
(A) Electric energy density is double of the magnetic energy density
(B) Electric energy density is half of the magnetic energy density
(C) Electric energy density is equal to the magnetic energy density
(D) Both electric and magnetic energy densities are zero

11. If microwaves, X-rays, infrared, gamma rays, ultra-violet, radio waves and visible parts of the electromagnetic spectrum are denoted by *M, X, I, G, U, R* and *V* then which of the following is the arrangement in ascending order of wavelength ? **[2014 (Online)]**
(A) *R, M, I, V, U, X* and *G* (B) *M, R, V, X, U, G* and *I*
(C) *G, X, U, V, I, M* and *R* (D) *I, M, R, U, V, X* and *G*

12. An electromagnetic wave of frequency 1×10^{14} hertz is propagating along z-axis. The amplitude of electric field is 4 V/m. If $\varepsilon_0 = 8.8 \times 10^{-12}$ C²/N-m², then average energy density of electric field will be : **[2014 (Online)]**

(A) 35.2×10^{-10} J/m³ (B) 35.2×10^{-11} J/m³
(C) 35.2×10^{-12} J/m³ (D) 35.2×10^{-13} J/m³

13. A lamp emits monochromatic green light uniformly in all directions. The lamp is 3% efficient in converting electrical power to electromagnetic waves and consumes 100 W of power. The amplitude of the electric field associated with the electromagnetic radiation at a distance of 5 m from the lamp will be nearly : **[2014 (Online)]**

(A) 1.34 V/m (B) 2.68 V/m
(C) 4.02 V/m (D) 5.36 V/m

14. An electromagnetic wave travelling in the x-direction has frequency of 2×10^{14} Hz and electric field amplitude of 27 Vm⁻¹. From the options given below, which one describes the magnetic field for this wave ? **[2015 (Online)]**

(A) $\vec{B}(x,t) = (3\times10^{-8}\,\text{T})\hat{j}\,;\, \sin[2\pi(1.5\times10^{-8}x - 2\times10^{14}t)]$

(B) $\vec{B}(x,t) = (9\times10^{-8}\,\text{T})\hat{i}\,;\, \sin[2\pi(1.5\times10^{-8}x - 2\times10^{14}t)]$

(C) $\vec{B}(x,t) = (9\times10^{-8}\,\text{T})\hat{j}\,;\, \sin[1.5\times10^{-6}x - 2\times10^{14}t]$

(D) $\vec{B}(x,t) = (9\times10^{-8}\,\text{T})\hat{k}\,;\, \sin[2\pi(1.5\times10^{-6}x - 2\times10^{14}t)]$

15. For plane electromagnetic waves propagating in the z-direction, which one of the following combination gives the correct possible direction for $\vec{E}$ and $\vec{B}$ field respectively ? **[2015 (Online)]**

(A) $(2\hat{i}+3\hat{j})$ and $(\hat{i}+2\hat{j})$ (B) $(-2\hat{i}-3\hat{j})$ and $(3\hat{i}-2\hat{j})$

(C) $(3\hat{i}+4\hat{j})$ and $(4\hat{i}-3\hat{j})$ (D) $(\hat{i}+2\hat{j})$ and $(2\hat{i}-\hat{j})$

16. Arrange the following electromagnetic radiations per quantum in the order of increasing energy : **[2016]**

A : Blue B : Yellow light
C : X-ray D : Radiowave

(A) C, A, B, D (B) B, A, D, C
(C) D, B, A, C (D) A, B, D, C

17. Microwave oven acts on the principle of : **[2016 (Online)]**

(A) Giving rotational energy to water molecules
(B) Giving translational energy to water molecules
(C) Giving vibrational energy to water molecules
(D) Transferring electrons from lower to higher energy levels in water molecule

18. Consider an electromagnetic wave propagating in vacuum. Choose the correct statement : **[2016 (Online)]**

(A) For an electromagnetic wave propagating in $+y$ direction the electric field is $\vec{E} = \frac{1}{\sqrt{2}}E_{yz}(x,t)\hat{z}$ and the magnetic field is $\vec{B} = \frac{1}{\sqrt{2}}B_z(x,t)\hat{y}$

(B) For an electromagnetic wave propagating in $+y$ direction the electric field is $\vec{E} = \frac{1}{\sqrt{2}}E_{yz}(x,t)\hat{y}$ and the magnetic field is $\vec{B} = \frac{1}{\sqrt{2}}B_{yz}(x,t)\hat{z}$

(C) For an electromagnetic wave propagating in $+x$ direction the electric field is $\vec{E} = \frac{1}{\sqrt{2}}E_{yz}(y,z,t)\,(\hat{y}+\hat{z})$ and the magnetic field is $\vec{B} = \frac{1}{\sqrt{2}}B_{yz}(y,z,t)(\hat{y}+\hat{z})$

(D) For an electromagnetic wave propagating in $+x$ direction the electric field is $\vec{E} = \frac{1}{\sqrt{2}}E_{yz}(x,t)(\hat{y}-\hat{z})$ and the magnetic field is $\vec{B} = \frac{1}{\sqrt{2}}B_{yz}(x,t)(\hat{y}+\hat{z})$

19. Magnetic field in a plane electromagnetic wave is given by

$\vec{B} = B_0 \sin(kx + \omega t)\,\hat{j}T$

Expression for corresponding electric field will be :
Where c is speed of light. **[2017 (Online)]**

(A) $\vec{E} = B_0 c \sin(kx+\omega t)\,\hat{k}$ V/m

(B) $\vec{E} = \dfrac{B_0}{c}\sin(kx+\omega t)\,\hat{k}$ V/m

(C) $\vec{E} - B_0 c \sin(kx+\omega t)\,\hat{k}$ V/m

(D) $\vec{E} = B_0 c \sin(kx-\omega t)\,\hat{k}$ V/m

20. The electric field component of a monochromatic radiation is given by

$$\vec{E} = 2E_0\,\hat{i}\cos kz\cos\omega t$$

Its magnetic field $\vec{B}$ is then given by : **[2017 (Online)]**

(A) $\dfrac{2E_0}{c}\hat{j}\sin kz\cos\omega t$ (B) $\dfrac{2E_0}{c}\hat{j}\cos kz\sin\omega t$

(C) $\dfrac{2E_0}{c}\hat{j}\sin kz\sin\omega t$ (D) $\dfrac{2E_0}{c}\hat{j}\cos kz\cos\omega t$

21. An EM wave from air enters a medium. The electric fields are $\vec{E}_1 = E_{01}\,\hat{x}\cos\left[2\pi\nu\left(\frac{z}{c}-t\right)\right]$ in air and $\vec{E}_2 = E_{02}\,\hat{x}\cos[k(2z-ct)]$ in medium, where the wave number k and frequency

v refer to their values in air. The medium is nonmagnetic. If $\in_{r_1}$ and $\in_{r_2}$ refer to relative permittivities of air and medium respectively, which of the following options is correct ? **[2018]**

(A) $\frac{\in_{r_1}}{\in_{r_2}} = 4$ (B) $\frac{\in_{r_1}}{\in_{r_2}} = 2$

(C) $\frac{\in_{r_1}}{\in_{r_2}} = \frac{1}{4}$ (D) $\frac{\in_{r_1}}{\in_{r_2}} = \frac{1}{2}$

22. A monochromatic beam of light has a frequency $v = \frac{3}{2\pi} \times 10^{12}$ Hz and is propagating along the direction $\frac{\hat{i}+\hat{j}}{\sqrt{2}}$. It is polarized along the $\hat{k}$ direction. The acceptable form for the magnetic field is : **[2018 (Online)]**

(A) $k\frac{E_0}{C}\left(\frac{\hat{i}-\hat{j}}{\sqrt{2}}\right)\cos\left[10^4\left(\frac{\hat{i}-\hat{j}}{\sqrt{2}}\right)\cdot\vec{r}-(3\times10^{12})t\right]$

(B) $\frac{E_0}{C}\left(\frac{\hat{i}-\hat{j}}{\sqrt{2}}\right)\cos\left[10^4\left(\frac{\hat{i}+\hat{j}}{\sqrt{2}}\right)\cdot\vec{r}-(3\times10^{12})t\right]$

(C) $\frac{E_0}{C}\hat{k}\cos\left[10^4\left(\frac{\hat{i}+\hat{j}}{\sqrt{2}}\right)\cdot\vec{r}+(3\times10^{12})t\right]$

(D) $\frac{E_0}{C}\frac{(\hat{i}+\hat{j}+\hat{k})}{\sqrt{3}}\cos\left[10^4\left(\frac{\hat{i}+\hat{j}}{\sqrt{2}}\right)\cdot\vec{r}+(3\times10^{12})t\right]$

23. A plane electromagnetic wave of wavelength λ has an intensity I. It is propagating along the positive Y-direction. The allowed expressions for the electric and magnetic fields are given by : **[2018 (Online)]**

(A) $\vec{E} = \sqrt{\frac{I}{\in_0 c}}\cos\left[\frac{2\pi}{\lambda}(y-ct)\right]\hat{i}\ ;\ \vec{B} = \frac{1}{c}E\hat{k}$

(B) $\vec{E} = \sqrt{\frac{I}{\in_0 c}}\cos\left[\frac{2\pi}{\lambda}(y-ct)\right]\hat{k}\ ;\ \vec{B} = -\frac{1}{c}E\hat{k}$

(C) $\vec{E} = \sqrt{\frac{2I}{\in_0 c}}\cos\left[\frac{2\pi}{\lambda}(y-ct)\right]\hat{k}\ ;\ \vec{B} = +\frac{1}{c}E\hat{i}$

(D) $\vec{E} = \sqrt{\frac{2I}{\in_0 c}}\cos\left[\frac{2\pi}{\lambda}(y+ct)\right]\hat{k}\ ;\ \vec{B} = \frac{1}{c}E\hat{i}$

24. A plane polarized monochromatic *EM* wave is travelling a vacuum along z-direction such that at $t = t_1$ it is found that the electric field is zero at a spatial point z_1. The next zero that occurs in its neighbourhood is at z_2. The frequency of the electromagnetic wave is : **[2018 (Online)]**

(A) $\frac{3\times10^8}{|z_2-z_1|}$ (B) $\frac{6\times10^8}{|z_2-z_1|}$

(C) $\frac{1.5\times10^8}{|z_2-z_1|}$ (D) $\frac{1}{t_1+\frac{|z_2-z_1|}{3\times10^8}}$

25. A plane electromagnetic wave of frequency 50 MHz travels in free space along the positive x-direction. At a particular point in space and time, $\vec{E} = 6.3\hat{j}$ V/m. The corresponding magnetic field $\vec{B}$, at that point will be : **[9 Jan. 2019 (Morning)]**

(A) $18.9\times10^{-8}\ \hat{k}T$ (B) $6.6\times10^{-8}\ \hat{k}T$

(C) $2.1\times10^{-8}\ \hat{k}T$ (D) $18.9\times10^{8}\ \hat{k}T$

26. The energy associated with electric field is (U_E) and with magnetic field is (U_B) for an electromagnetic wave in free space. Then : **[9 Jan. 2019 (Evening)]**

(A) $U_E = \frac{U_B}{2}$ (B) $U_E < U_B$

(C) $U_E = U_B$ (D) $U_E > U_B$

27. If the magnetic field of a plane electromagnetic wave is given by (The speed of light = 3×10^8 /m/s)

$$B = 100\times10^{-6}\sin\left[2\pi\times2\times10^{15}\left(t-\frac{x}{c}\right)\right]$$

then the maximum electric field associated with it is : **[10 Jan. 2019 (Morning)]**

(A) 4×10^4 N/C (B) 4.5×10^4 N/C

(C) 6×10^4 N/C (D) 3×10^4 N/C

28. The electric field of a plane polarized electromagnetic wave in free space at time $t = 0$ is given by an expression $\vec{E}(x, y) = 10\hat{j}\cos[(6x+8z)]$.

The magnetic field $\vec{B}(x, z, t)$ is given by (c is the velocity of light) : **[10 Jan. 2019 (Evening)]**

(A) $\frac{1}{c}(6\hat{k}+8\hat{i})\cos[6x-8z+10ct]$

(B) $\frac{1}{c}(6\hat{k}-8\hat{i})\cos[6x+8z-10ct]$

(C) $\frac{1}{c}(6\hat{k}+8\hat{i})\cos[6x+8z-10ct]$

(D) $\frac{1}{c}(6\hat{k}-8\hat{i})\cos[6x+8z+10ct]$

29. An electromagnetic wave of intensity 50 Wm^{-2} enters in a medium of refractive index 'n' without any loss. The ratio of the magnitudes of electric fields, and the ratio of the magnitudes of magnetic fields of the wave before and after entering into the medium are respectively, given by : **[11 Jan. 2019 (Morning)]**

(A) $\left(\frac{1}{\sqrt{n}}, \frac{1}{\sqrt{n}}\right)$ (B) $\left(\sqrt{n}, \frac{1}{\sqrt{n}}\right)$

(C) $(\sqrt{n}, \sqrt{n})$ (D) $\left(\frac{1}{\sqrt{n}}, \sqrt{n}\right)$

30. A 27 mW laser beam has a cross-sectional area of 10 mm^2. The magnitude of the maximum electric field in this electromagnetic wave is given by [Given permittivity of space $\in_0 = 9 \times 10^{-12}$ SI units, Speed of light $c = 3 \times 10^8$ m/s]: **[11 Jan. 2019 (Evening)]**

(A) 1 kV/m (B) 2 kV/m

(C) 1.4 kV/m (D) 0.7 kV/m

31. A light wave is incident normally on a glass slab of refractive index 1.5. If 4% of light gets reflected and the amplitude of the electric field of the incident light is 30 V/m, then the amplitude of the electric field for the wave propagating in the glass medium will be : **[12 Jan. 2019 (Morning)]**

(A) 10 V/m (B) 24 V/m

(C) 30 V/m (D) 6 V/m

32. The mean intensity of radiation on the surface of the Sun is about 10^8 W/m^2. The rms value of the corresponding magnetic field is closest to : **[12 Jan. 2019 (Evening)]**

(A) 10^2 T (B) 10^{-4} T

(C) 1 T (D) 10^{-2} T

33. A plane electromagnetic wave travels in free space along the x-direction. The electric field component of the wave at a particular point of space and time is $E = 6$ V m^{-1} along y-direction. Its corresponding magnetic field component, B would be : **[8 April 2019 (Morning)]**

(A) 6×10^{-8} T along z-direction

(B) 6×10^{-8} T along x-direction

(C) 2×10^{-8} T along z-direction

(D) 2×10^{-8} T along y-direction

34. The magnetic field of an electromagnetic wave is given by.

$\vec{B} = 1.6 \times 10^{-6} \cos(2 \times 10^7 z + 6 \times 10^{15} t)(2\hat{i} + \hat{j})\frac{\text{Wb}}{\text{m}^2}$

The associated electric field will be : **[8 April 2019 (Evening)]**

(A) $\vec{E} = 4.8 \times 10^2 \cos(2 \times 10^7 z + 6 \times 10^{15} t)(\hat{i} - 2\hat{j})\frac{V}{m}$

(B) $\vec{E} = 4.8 \times 10^2 \cos(2 \times 10^7 z - 6 \times 10^{15} t)(2\hat{i} + \hat{j})\frac{V}{m}$

(C) $\vec{E} = 4.8 \times 10^2 \cos(2 \times 10^7 z - 6 \times 10^{15} t)(-2\hat{j} + \hat{i})\frac{V}{m}$

(D) $\vec{E} = 4.8 \times 10^2 \cos(2 \times 10^7 z + 6 \times 10^{15} t)(-\hat{i} + 2\hat{j})\frac{V}{m}$

35. The magnetic field of a plane electromagnetic wave is given by $\vec{B} = B_0 \hat{i}[\cos(kz - \omega t)] + B_1 \hat{j}\cos(kz + \omega t)$ where $B_0 = 3 \times 10^{-5}$ T and $B_1 = 2 \times 10^{-6}$ T. The rms value of the force experienced by a stationary charge $Q = 10^{-4}$ C at $z = 0$ is closest to : **[9 April 2019 (Morning)]**

(A) 0.9 N (B) 0.1 N

(C) 3×10^{-2} N (D) 0.6 N

36. The electric field of a plane electromagnetic wave is given by

$$\vec{E} = E_0 \hat{i}\cos(kz)\cos(\omega t)$$

The corresponding magnetic field $\vec{B}$ is then given by : **[10 April 2019 (Morning)]**

(A) $\vec{B} = \frac{E_0}{C}\hat{j}\sin(kz)\cos(\omega t)$ (B) $\vec{B} = \frac{E_0}{C}\hat{j}\sin(kz)\sin(\omega t)$

(C) $\vec{B} = \frac{E_0}{C}\hat{k}\sin(kz)\cos(\omega t)$ (D) $\vec{B} = \frac{E_0}{C}\hat{j}\cos(kz)\sin(\omega t)$

37. An electromagnetic wave is represented by the electric field $\vec{E} = E_0 \hat{n}\sin[\omega t + (6y - 8z)]$. Taking unit vectors in x, y and z directions to be $\hat{i}, \hat{j}, \hat{k}$, the direction of propagation $\hat{s}$, is: **[12 April 2019 (Morning)]**

(A) $\hat{s} = \frac{4\hat{j} - 3\hat{k}}{5}$ (B) $\hat{s} = \frac{3\hat{i} - 4\hat{j}}{5}$

(C) $\hat{s} = \left(\frac{-3\hat{j} + 4\hat{k}}{5}\right)$ (D) $\hat{s} = \frac{-4\hat{k} + 3\hat{j}}{5}$

38. A plane electromagnetic wave having a frequency $\nu = 23.9$ GHz propagates along the positive z-direction in free space. The peak value of the electric field is 60 V/m. Which among the following is the acceptable magnetic field component in the electromagnetic wave ? **[12 April 2019 (Evening)]**

(A) $\vec{B} = 2 \times 10^7 \sin(0.5 \times 10^3 z + 1.5 \times 10^{11} t)\hat{i}$

(B) $\vec{B} = 2 \times 10^{-7} \sin(1.5 \times 10^2 x + 0.5 \times 10^{11} t)\hat{j}$

(C) $\vec{B} = 2 \times 10^{-7} \sin(0.5 \times 10^3 z - 1.5 \times 10^{11} t)\hat{i}$

(D) $\vec{B} = 60 \sin(0.5 \times 10^3 x + 1.5 \times 10^{11} t)\hat{k}$

39. If the magnetic field in a plane electromagnetic wave is given by $\vec{B} = 3 \times 10^{-8} \sin(1.6 \times 10^3 x + 48 \times 10^{10} t)\hat{j}$ T, then what will be expression for electric field ? **[7 Jan. 2020 (Morning)]**

(A) $\vec{E} = (9 \sin(1.6 \times 10^3 x + 48 \times 10^{10} t)\hat{k}$ V/m)

(B) $\vec{E} = (3 \times 10^{-8} \sin(1.6 \times 10^3 x + 48 \times 10^{10} t)\hat{i}$ V/m)

(C) $\vec{E} = (60 \sin(1.6 \times 10^3 x + 48 \times 10^{10} t)\hat{k}$ V/m)

(D) $\vec{E} = (3 \times 10^{-8} \sin(1.6 \times 10^3 x + 48 \times 10^{10} t)\hat{j}$ V/m)

40. The electric field of a plane electromagnetic wave is given by

$$\vec{E} = E_0 \frac{\hat{i}+\hat{j}}{\sqrt{2}} \cos(kz + \omega t)$$

At $t = 0$, a positively charged particle is at the point $(x, y, z) = \left(0, 0, \frac{\pi}{k}\right)$. If its instantaneous velocity at $(t = 0)$ is $\upsilon_0 \hat{k}$, the force acting on it due to the wave is : **[7 Jan. 2020 (Evening)]**

(A) Zero (B) Parallel to $\frac{\hat{i}+\hat{j}}{\sqrt{2}}$

(C) Antiparallel to $\frac{\hat{i}+\hat{j}}{\sqrt{2}}$ (D) Parallel to $\hat{k}$

41. A plane electromagnetic wave of frequency 25 GHz is propagating in vacuum along the z-direction. At a particular point in space and time, the magnetic field is given by $\vec{B} = 5\times10^{-8}\hat{j}T$. The corresponding electric field $\vec{E}$ is (speed of light $c = 3 \times 10^8$ ms^{-1}) : **[8 Jan. 2020 (Evening)]**

(A) $1.66 \times 10^{-16}\hat{i}$ V/m (B) $15\hat{i}$ V/m

(C) $-1.66 \times 10^{-16}\hat{i}$ V/m (D) $-15\hat{i}$ V/m

42. The electric fields of two plane electromagnetic plane waves in vacuum are given by

$$\vec{E}_1 = E_0 \hat{j} \cos(\omega t - kx) \text{ and}$$

$$\vec{E}_2 = E_0 \hat{k} \cos(\omega t - ky)$$

At $t = 0$, a particle of charge q is at origin with a velocity $\vec{v} = 0.8c\hat{j}$ (c is the speed of light in vacuum). The instantaneous force experienced by the particle is : **[9 Jan. 2020 (Morning)]**

(A) $E_0 q(-0.8\hat{i} + \hat{j} + \hat{k})$ (B) $E_0 q(0.8\hat{i} - \hat{j} + 0.4\hat{k})$

(C) $E_0 q(0.8\hat{i} + \hat{j} + 0.2\hat{k})$ (D) $E_0 q(0.4\hat{i} - 3\hat{j} + 0.8\hat{k})$

43. A plane electromagnetic wave is propagating along the direction $\frac{\hat{i}+\hat{j}}{\sqrt{2}}$, with its polarization along the direction $\hat{k}$. The correct form of the magnetic field of the wave would be (here B_0 is an appropriate constant) : **[9 Jan. 2020 (Evening)]**

(A) $B_0 \frac{\hat{i}-\hat{j}}{\sqrt{2}} \cos\left(\omega t - k\frac{\hat{i}+\hat{j}}{\sqrt{2}}\right)$

(B) $B_0 \frac{\hat{i}+\hat{j}}{\sqrt{2}} \cos\left(\omega t - k\frac{\hat{i}+\hat{j}}{\sqrt{2}}\right)$

(C) $B_0 \hat{k} \cos\left(\omega t - k\frac{\hat{i}+\hat{j}}{\sqrt{2}}\right)$

(D) $B_0 \frac{\hat{j}-\hat{i}}{\sqrt{2}} \cos\left(\omega t + k\frac{\hat{i}+\hat{j}}{\sqrt{2}}\right)$

44. A plane electromagnetic wave, has frequency of 2.0×10^{10} Hz and its energy density is 1.02×10^{-8} J/m^3 in vacuum. The amplitude of the magnetic field of the wave is close to :

[Take: $\frac{1}{4\pi\varepsilon_0} = 9 \times 10^9 \frac{Nm^2}{C^2}$ and speed of light $= 3 \times 10^8$ ms^{-1}]

[2 Sept. 2020 (Morning)]

(A) 180 nT (B) 160 nT

(C) 150 nT (D) 190 nT

45. In a plane electromagnetic wave, the directions of electric field and magnetic field are represented by $\hat{k}$ and $2\hat{i} - 2\hat{j}$, respectively. What is the unit vector along direction of propagation of the wave : **[2 Sept. 2020 (Evening)]**

(A) $\frac{1}{\sqrt{2}}(\hat{i}+\hat{j})$ (B) $\frac{1}{\sqrt{5}}(\hat{i}+2\hat{j})$

(C) $\frac{1}{\sqrt{5}}(2\hat{i}+\hat{j})$ (D) $\frac{1}{\sqrt{2}}(\hat{j}+\hat{k})$

46. The magnetic field of a plane electromagnetic wave is given as $\vec{B} = 3 \times 10^{-8} \sin[200\pi(y + ct)]\hat{i}$ T. Where $c = 3 \times 10^8$ ms^{-1} is the speed of light. The corresponding electric field is : **[3 Sept. 2020 (Morning)]**

(A) $\vec{E} = 10^{-6} \sin[200\pi(y + ct)]\hat{k}$ V/m

(B) $\vec{E} = -9 \sin[200\pi(y + ct)]\hat{k}$ V/m

(C) $\vec{E} = 9 \sin[200\pi(y + ct)]\hat{k}$ V/m

(D) $\vec{E} = 3 \times 10^{-8} \sin[200\pi(y + ct)]\hat{k}$ V/m

47. The electric field of a plane electromagnetic wave propagating along the x direction in vacuum is $\vec{E} = E_0\hat{j}\cos(\omega t - kx)$. The magnetic field $\vec{B}$, at the moment $t = 0$ is : **[3 Sept. 2020 (Evening)]**

(A) $\vec{B} = E_0\sqrt{\mu_0 \epsilon_0}\cos(kx)\hat{j}$

(B) $\vec{B} = \frac{E_0}{\sqrt{\mu_0 \epsilon_0}}\cos(kx)\hat{k}$

(C) $\vec{B} = E_0\sqrt{\mu_0 \epsilon_0}\cos(kx)\hat{k}$

(D) $\vec{B} = \frac{E_0}{\sqrt{\mu_0 \epsilon_0}}\cos(kx)\hat{j}$

48. Choose the correct option relating wavelengths of different parts of electromagnetic wave spectrum : **[4 Sept. 2020 (Morning)]**

(A) $\lambda_{\text{x-rays}} < \lambda_{\text{micro waves}} < \lambda_{\text{radio waves}} < \lambda_{\text{visible}}$

(B) $\lambda_{\text{visible}} > \lambda_{\text{x-rays}} > \lambda_{\text{radio waves}} > \lambda_{\text{micro waves}}$

(C) $\lambda_{\text{radio waves}} > \lambda_{\text{micro waves}} > \lambda_{\text{visible}} > \lambda_{\text{x-rays}}$

(D) $\lambda_{\text{visible}} < \lambda_{\text{micro waves}} < \lambda_{\text{radio waves}} < \lambda_{\text{x-rays}}$

49. The electric field of a plane electromagnetic wave is given by $\vec{E} = E_0(\hat{x}+\hat{y})\sin(kz-\omega t)$. Its magnetic field will be given by : **[4 Sept. 2020 (Evening)]**

(A) $\frac{E_0}{c}(\hat{x}-\hat{y})\cos(kz-\omega t)$ (B) $\frac{E_0}{c}(-\hat{x}+\hat{y})\sin(kz-\omega t)$

(C) $\frac{E_0}{c}(\hat{x}-\hat{y})\sin(kz-\omega t)$ (D) $\frac{E_0}{c}(\hat{x}+\hat{y})\sin(kz-\omega t)$

50. An electron is constrained to move along the y-axis with a speed of 0.1 c (c is the speed of light) in the presence of electromagnetic wave, whose electric field is $\vec{E} = 30\hat{j}$ $\sin(1.5\times10^7 t - 5\times10^{-2}x)$ V/m The maximum magnetic force experienced by the electron will be :
(Given $c = 3\times10^8$ ms^{-1} and electron charge $= 1.6\times10^{-19}$ C)
[5 Sept. 2020 (Morning)]

(A) 1.6×10^{-19} N (B) 4.8×10^{-19} N
(C) 3.2×10^{-18} N (D) 2.4×10^{-18} N

51. The correct match between the entries in column-I and column-II are : **[5 Sept. 2020 (Evening)]**

Column-I	**Column-II**
Radiation	**Wavelength**
(a) Microwave	(i) 100 m
(b) Gamma rays	(ii) 10^{-15} m
(c) A.M. radio waves	(iii) 10^{-10} m
(d) X-rays	(iv) 10^{-3} m

(A) (a)-(ii), (b)-(i), (c)-(iv), (d)-(iii)
(B) (a)-(i), (b)-(iii), (c)-(iv), (d)-(ii)
(C) (a)-(iii), (b)-(ii), (c)-(i), (d)-(iv)
(D) (a)-(iv), (b)-(ii), (c)-(i), (d)-(iii)

52. An AC circuit has $R = 100\ \Omega$, $C = 2\ \mu$F and $L = 80$ mH, connected in series. The quality factor of the circuit is :
[6 Sept. 2020 (Morning)]

(A) 0.5 (B) 2
(C) 20 (D) 400

53. Suppose that intensity of a laser is $\left(\frac{315}{\pi}\right)$ W/m^2. The rms electric field, in units of V/m associated with this source is close to the nearest integer is ($\in_0 = 8.86\times10^{-12}$ C^2 Nm^{-2}; $c = 3\times10^8$ ms^{-1}) : **[6 Sept. 2020 (Morning)]**

54. For a plane electromagnetic wave, the magnetic field at a point x and time t is $\vec{B}(x,t) = [1.2\times10^{-7}\sin(0.5\times10^3x + 1.5\times10^{11}t)]T$. The instantaneous electric field $\vec{E}$ corresponding to $\vec{B}$ is : (speed of light $c = 3\times10^8$ ms^{-1})
[6 Sept. 2020 (Evening)]

(A) $\vec{E}(x,t) = [36\sin(0.5\times10^3x + 1.5\times10^{11}t)\hat{k}]\frac{v}{m}$

(B) $\vec{E}(x,t) = [36\sin(0.5\times10^3x + 1.5\times10^{11}t)\hat{j}]\frac{v}{m}$

(C) $\vec{E}(x,t) = [36\sin(1\times10^3x + 0.5\times10^{11}t)\hat{j}]\frac{v}{m}$

(D) $\vec{E}(x,t) = [36\sin(1\times10^3x + 1.5\times10^{11}t)\hat{j}]\frac{v}{m}$

55. An electromagnetic wave of frequency 5 GHz, is travelling in a medium whose relative electric permittivity and relative magnetic permeability both are 2. Its velocity in this medium is ______ $\times 10^7$ m/s. **[24 Feb. 2021 (Morning)]**

56. An electromagnetic wave of frequency 3 GHz enters a dielectric medium of relative electric permittivity 2.25 from vacuum. The wavelength of this wave in that medium will be ______ $\times 10^{-2}$ cm. **[24 Feb. 2021 (Evening)]**

57. The peak electric field produced by the radiation coming from the 8 W bulb at a distance of 10 m is $\frac{x}{10}\sqrt{\frac{\mu_0 c}{\pi}}\frac{V}{m}$. The bulb is a point source. The value of x is ______.
[25 Feb. 2021 (Evening)]

58. A radiation is emitted by 1000 W bulb and it generates an electric field and magnetic field at P, placed at a distance of 2 m. The efficiency of the bulb is 1.25%. The value of peak electric field at P is $x\times10^{-1}$ V/m. Value of x is ______. (Rounded-off to the nearest integer) **[26 Feb. 2021 (Morning)]**
[Take $\varepsilon_0 = 8.85\times10^{-12}$ C^{2}N^{-1} m^{-2}, $c = 3\times10^8$ ms^{-1}]

59. A plane electromagnetic wave of frequency 500 MHz is travelling in vacuum along y-direction. At a particular point in space and time, $\vec{B} = 8.0\times10^{-8}\ \hat{z}$ T. The value of electric field at this point is : (speed of light $= 3\times10^8$ ms^{-1})
$\hat{x}$, $\hat{y}$, $\hat{z}$ are unit vectors along x, y and z direction.
[16 March 2021 (Morning)]

(A) $-24\hat{x}$ V/m (B) $2.6\hat{x}$ V/m
(C) $24\hat{x}$ V/m (D) $-2.6\hat{y}$ V/m

60. For an electromagnetic wave travelling in free space, the relation between average energy densities due to electric (U_e) and magnetic (U_m) fields is : **[16 March 2021 (Morning)]**

(A) $U_e = U_m$ (B) $U_e > U_m$
(C) $U_e < U_m$ (D) $U_e \neq U_m$

61. The electric field intensity produced by the radiation coming from a 100 W bulb at a distance of 3m is E. The electric field intensity produced by the radiation coming from 60 W at the same distance is $\sqrt{\frac{x}{5}}E$. Where the value of $x =$ ______.

[17 March 2021 (Evening)]

62. Seawater at a frequency $f = 9 \times 10^2$ Hz, has permittivity $\varepsilon = 80\ \varepsilon_0$ and resistivity $\rho = 0.25\ \Omega$m. Imagine a parallel plate capacitor is immersed in seawater and is driven by an alternating voltage source $V(t) = V_0 \sin(2\pi ft)$. Then the conduction current density becomes 10^x times the displacement current density after time $t = \frac{1}{800}$ s. The value of x is ______.

(Given : $\frac{1}{4\pi\varepsilon_0} = 9 \times 10^9\ \text{Nm}^2\text{C}^{-2}$) **[17 March 2021 (Evening)]**

63. A plane electromagnetic wave of frequency 100 MHz is travelling in vacuum along the x-direction. At a particular point in space and time, $\vec{B} = 2.0 \times 10^{-8}\ \text{T}\hat{k}$. (Where, $\hat{k}$ is the unit vector along z-direction) What is $\vec{E}$ at this point ?

[18 March 2021 (Morning)]

(A) $0.6\hat{j}$ V/m (B) $6.0\hat{k}$ V/m

(C) $6.0\hat{j}$ V/m (D) $0.6\hat{k}$ V/m

64. A plane electromagnetic wave propagating along y-direction can have the following pair of electric field ($\vec{E}$) and magnetic field ($\vec{B}$) components. **[18 March 2021 (Evening)]**

(A) E_y, B_y or E_z, B_z (B) E_y, B_x or E_x, B_y

(C) E_x, B_z or E_z, B_x (D) E_x, B_y or E_y, B_x

65. AC voltage $V(t) = 20 \sin \omega t$ of frequency 50 Hz is applied to a parallel plate capacitor. The separation between the plates is 2 mm and the area is 1 m^2. The amplitude of the oscillating displacement current for the applied AC voltage is ______.

(Take $\varepsilon_0 = 8.85 \times 10^{-12}$ F/m) **[20 July 2021 (Morning)]**

(A) 55.58 μA (B) 21.14 μA

(C) 27.79 μA (D) 83.37 μA

66. In an electromagnetic wave the electric field vector and magnetic field vector are given as $\vec{E} = E_0\hat{i}$ and $\vec{B} = B_0\hat{k}$ respectively. The direction of propagation of electromagnetic wave is along : **[20 July 2021 (Evening)]**

(A) $(-\hat{j})$ (B) $\hat{j}$

(C) $(-\hat{k})$ (D) $\hat{k}$

67. Intensity of sunlight is observed as 0.092 Wm^{-2} at a point in free space. What will be the peak value of magnetic field at that point ? ($\varepsilon_0 = 8.85 \times 10^{-12}\ \text{C}^2\text{N}^{-1}\text{m}^{-2}$)

[22 July 2021 (Evening)]

(A) 1.96×10^{-8} T (B) 5.88 T

(C) 2.77×10^{-8} T (D) 8.31 T

68. A linearly polarized electromagnetic wave in vacuum is $E = 3.1 \cos[(1.8)z - (5.4 \times 10^6)t]\ \hat{i}$ N/C is incident normally on a perfectly reflecting wall at $z = a$. Choose the correct option :

[25 July 2021 (Morning)]

(A) The frequency of electromagnetic wave is 54×10^4 Hz

(B) The reflected wave will be

$3.1 \cos[(1.8)z + (5.4 \times 10^6)t]\ \hat{i}$ N/C

(C) The transmitted wave will be

$3.1 \cos[(1.8)z - (5.4 \times 10^6)t]\ \hat{i}$ N/C

(D) The wavelength is 5.4 m

69. The relative permittivity of distilled water is 81. The velocity of light in it will be : (Given $\mu_r = 1$) : **[27 July 2021 (Morning)]**

(A) 3.33×10^7 m/s (B) 2.33×10^7 m/s

(C) 5.33×10^7 m/s (D) 4.33×10^7 m/s

70. Light beam is described by $E = 800 \sin \omega\left(t - \frac{x}{c}\right)$. An electron is allowed to move normal to the propagation of light beam with a speed of 3×10^7. What is the maximum magnetic force exerted on the electron ? **[26 Aug. 2021 (Evening)]**

(A) 1.28×10^{-21} N (B) 1.28×10^{-18} N

(C) 12.8×10^{-18} N (D) 12.8×10^{-17} N

71. Electric field in a plane electromagnetic wave is given by $E = 50 \sin(500x - 10 \times 10^{10}\,t)$ V/m. The velocity of electromagnetic wave in this medium is : (Given C = speed of light in vacuum) :

[27 Aug. 2021 (Morning)]

(A) $\frac{2}{3}C$ (B) C

(C) $\frac{C}{2}$ (D) $\frac{3}{2}C$

72. A plane electromagnetic wave with frequency of 30 MHz travels in free space. At particular point in space and time, electric field is 6 V/m. The magnetic field at this point will be $x \times 10^{-8}$ T. The value of x is______. **[27 Aug. 2021 (Evening)]**

73. The electric field in an electromagnetic wave is given by $E = (50\ \text{NC}^{-1}) \sin \omega(t - x/c)$. The energy contained in a cylinder of volume V is 5.5×10^{-12} J. The value of V is ______cm^3. (Given $\epsilon_0 = 8.8 \times 10^{-12}\ \text{C}^2 - \text{N}^{-1}\text{m}^{-2}$) **[31 Aug. 2021 (Morning)]**

74. Electric of a plane electromagnetic wave propagating through a non magnetic medium is given by

$$E = 20\cos(2\times10^{10}t - 200x)\ \text{V/m}$$

The dielectric constant of the medium is equal to : (Take $\mu_r = 1$) **[01 Sep. 2021 (Evening)]**

(A) 3 (B) $\frac{1}{2}$

(C) 2 (D) 9

75. A plane electromagnetic wave travels in a medium of relative permeability 1.61 and relative permittivity 6.44. If magnitude of magnetic intensity is 4.5×10^{-2} Am^{-1} at a point, what will be the approximate magnitude of electric field intensity at that point ? (Given : permeability of free space $\mu_0 = 4\pi\times10^{-7}$ NA^{-2}, speed of light in vacuum $c = 3\times10^8$ ms^{-1}) **[24 June 2022 (Morning)]**

(A) 16.96 Vm^{-1} (B) 2.25×10^{-2} Vm^{-1}

(C) 5.26 Vm^{-1} (D) 6.75×10^{6} Vm^{-1}

76. An electric bulb is rated as 200 W. What will be the peak magnetic field at 4 m distance produced by the radiations coming from this bulb? Consider this bulb as a point source with 3.5% efficiency. **[24 June 2022 (Evening)]**

(A) 1.19×10^{-8} T (B) 1.71×10^{-8} T

(C) 0.84×10^{-8} T (D) 3.36×10^{-8} T

77. The electric field in an electromagnetic wave is given by $E = 56.5\sin\omega(t - x/c)$ NC^{-1}. Find the intensity of the wave if it is propagating along x-axis in the free space :
(Given $\varepsilon_0 = 8.85\times10^{-12}$ C^{2}N^{-1}m^{-2}) **[25 June 2022 (Morning)]**

(A) 5.65 Wm^{-2} (B) 4.24 Wm^{-2}

(C) 1.9×10^{-7} Wm^{-2} (D) 56.5 Wm^{-2}

78. The electromagnetic waves travel in a medium at a speed of 2.0×10^8 m/s. The relative permeability of the medium is 1.0. The relative permittivity of the medium will be : **[25 June 2022 (Evening)]**

(A) 2.25 (B) 4.25

(C) 6.25 (D) 8.25

79. If electric field intensity of a uniform plane electromagnetic wave is given as

$$\vec{E} = -301.6\sin(kz-\omega t)\,\hat{a}_x + 452.4\sin(kz-\omega t)\,\hat{a}_y\ \frac{V}{m}$$

Then, magnetic intensity H of this wave in Am^{-1} will be :
[Given: Speed of light in vacuum $c = 3\times10^8$ ms^{-1}, permeability of vacuum $\mu_0 = 4\pi\times10^{-7}$ NA^{-2}] **[26 June 2022 (Morning)]**

(A) $+0.8\sin(kz-\omega t)\,\hat{a}_y + 0.8\sin(kz-\omega t)\,\hat{a}_x$

(B) $+1.0\times10^{-6}\sin(kz-\omega t)\,\hat{a}_y + 1.5\times10^{-6}(kz-\omega t)\,\hat{a}_x$

(C) $-0.8\sin(kz-\omega t)\,\hat{a}_y - 1.2\sin(kz-\omega t)\,\hat{a}_x$

(D) $-1.0\times10^{-6}\sin(kz-\omega t)\,\hat{a}_y - 1.5\times10^{-6}\sin(kz-\omega t)\,\hat{a}_x$

80. Which is the correct ascending order of wavelengths? **[26 June 2022 (Evening)]**

(A) $\lambda_{\text{visible}} < \lambda_{X\text{-ray}} < \lambda_{\text{gamma-ray}} < \lambda_{\text{microwave}}$

(B) $\lambda_{\text{gamma-ray}} < \lambda_{X\text{-ray}} < \lambda_{\text{visible}} < \lambda_{\text{microwave}}$

(C) $\lambda_{X\text{-ray}} < \lambda_{\text{gamma-ray}} < \lambda_{\text{visible}} < \lambda_{\text{microwave}}$

(D) $\lambda_{\text{microwave}} < \lambda_{\text{visible}} < \lambda_{\text{gamma-ray}} < \lambda_{X\text{-ray}}$

81. Mach List-I with List-II : **[27 June 2022 (Morning)]**

List-I	List-II
(a) Ultraviolet rays	(i) Study crystal structure
(b) Microwaves	(ii) Greenhouse effect
(c) Infrared waves	(iii) Sterilizing surgical instrument
(d) X-rays	(iv) Radar system

Choose the correct answer from the options given below :

(A) (a)-(iii), (b)-(iv), (c)-(ii), (d)-(i)

(B) (a)-(iii), (b)-(i), (c)-(ii), (d)-(iv)

(C) (a)-(iv), (b)-(iii), (c)-(ii), (d)-(i)

(D) (a)-(iii), (b)-(iv), (c)-(i), (d)-(ii)

82. Given below are two statements :

Statement-I : A time varying electric field is a source of changing magnetic field and vice-versa. Thus a disturbance in electric or magnetic field creates *EM* waves.

Statement-II : In a material medium. The *EM* wave travels with speed $v = \frac{1}{\sqrt{\mu_0\varepsilon_0}}$

In the light of the above statements, choose the correct answer from the options given below : **[27 June 2022 (Evening)]**

(A) Both statement-I and statement-II are true

(B) Both statement-I and statement-II are false

(C) Statement-I is correct but statement-II is false

(D) Statement-I is incorrect but statement-II is true

83. An *EM* wave propagating in x-direction has a wavelength of 8 mm. The electric field vibrating y-direction has maximum magnitude of 60 Vm^{-1}. Choose the correct equations for electric and magnetic fields if the *EM* wave is propagating in vacuum : **[28 June 2022 (Evening)]**

(A) $E_y = 60\sin\left[\frac{\pi}{4}\times10^3(x - 3\times10^8 t)\right]\hat{j}\ \text{Vm}^{-1}$

$B_z = 2\sin\left[\frac{\pi}{4}\times10^3(x - 3\times10^8 t)\right]\hat{k}T$

(B) $E_y = 60\sin\left[\frac{\pi}{4}\times10^3(x - 3\times10^8 t)\right]\hat{j}\ \text{Vm}^{-1}$

$B_z = 2\times10^{-7}\sin\left[\frac{\pi}{4}\times10^3(x - 3\times10^8 t)\right]\hat{k}T$

(C) $E_y = 2\times10^{-7}\sin\left[\frac{\pi}{4}\times10^3(x - 3\times10^8 t)\right]\hat{j}\ \text{Vm}^{-1}$

$B_z = 60\sin\left[\frac{\pi}{4}\times10^3(x - 3\times10^8 t)\right]\hat{k}T$

(D) $E_y = 2 \times 10^{-7} \sin\left[\frac{\pi}{4} \times 10^4 (x - 4 \times 10^8 t)\right] \hat{j}$ Vm^{-1}

$$B_z = 60 \sin\left[\frac{\pi}{4} \times 10^4 (x - 4 \times 10^8 t)\right] \hat{k} T$$

84. The intensity of the light from a bulb incident on a surface is 0.22 W/m^2. The amplitude of the magnetic field in this light-wave is _____ $\times 10^{-9}$ T.
[Given : Permittivity of vacuum $\epsilon_0 = 8.85 \times 10^{-12}$ C^2 N^{-1} m^{-2}, speed of light in vacuum $c = 3 \times 10^8$ ms^{-1}]
[29 June 2022 (Morning)]

85. The electric field at the point associated with a light wave is given by $E = 200\,[\sin(6 \times 10^{15})\,t + \sin(9 \times 10^{15})\,t]$ Vm^{-1}. Given : $h = 4.14 \times 10^{-15}$ eVs. If this light falls on a metal surface having a work function of 2.50 eV, the maximum kinetic energy of the photoelectrons will be : **[29 June 2022 (Evening)]**

(A) 1.90 eV (B) 3.27 eV
(C) 3.60 eV (D) 3.42 eV

86. Light wave traveling in air along x-direction is given by $E_y = 540 \sin \pi \times 10^4 (x - ct)$ Vm^{-1}. Then, the peak value of magnetic field of wave will be : [Given : $c = 3 \times 10^8$ ms^{-1}]
[25 July 2022 (Evening)]

(A) 18×10^{-7} T (B) 54×10^{-7} T
(C) 54×10^{-8} T (D) 18×10^{-8} T

87. The magnetic field of a plane electromagnetic wave is given by $\vec{B} = 2 \times 10^{-8} \sin(0.5 \times 10^3 x + 1.5 \times 10^{11} t)\,\hat{j}\,T$. The amplitude of the electric field would be : **[26 July 2022 (Morning)]**

(A) 6 Vm^{-1} along x-axis (B) 3 Vm^{-1} along z-axis
(C) 6 Vm^{-1} along z-axis (D) 2×10^{-8} Vm^{-1} along z-axis

88. The oscillating magnetic field in a plane electromagnetic wave is given by $B_y = 5 \times 10^{-6} \sin 1000\,\pi\,(5x - 4 \times 10^8 t)$ T. The amplitude of electric field will be : **[26 July 2022 (Evening)]**

(A) 15×10^2 Vm^{-1} (B) 15×10^{-6} Vm^{-1}
(C) 16×10^{12} Vm^{-1} (D) 4×10^2 Vm^{-1}

89. A beam of light travelling along X-axis is described by the electric field $E_y = 900 \sin \omega(t - x/c)$. The ratio of electric force to magnetic force on a charge q moving along Y-axis with a speed of 3×10^7 ms^{-1} will be : [Given : speed of light $= 3 \times 10^8$ ms^{-1}]
[27 July 2022 (Morning)]

(A) 1 : 1 (B) 1 : 10
(C) 10 : 1 (D) 1 : 2

90. Identify the correct statements from the following descriptions of various properties of electromagnetic waves :
[27 July 2022 (Evening)]

(a) In a plane electromagnetic wave electric field and magnetic field must be perpendicular to each other and direction of propagation of wave should be along electric field or magnetic field
(b) The energy in electromagnetic wave is divided equally between electric and magnetic fields
(c) Both electric field and magnetic field are parallel to each other and perpendicular to the direction of propagation of wave
(d) The electric field, magnetic field and direction of propagation of wave must be perpendicular to each other
(e) The ratio of amplitude of magnetic field to the amplitude of electric field is equal to speed of light
Choose the most appropriate answer from the options given below :

(A) (e) only (B) (b) and (d) only
(C) (b), (c) and (e) only (D) (a), (b) and (e) only

91. Match List-I with List-II : **[29 July 2022 (Morning)]**

List-I	List-II
(a) UV-rays	(i) Diagnostic tool in medicine
(b) X-rays	(ii) Water purification
(c) Microwave	(iii) Communication, Radar
(d) Infrared wave	(iv) Improving visibility in foggy days

Choose the correct answer from the options given below :
(A) (a)-(iii), (b)-(ii), (c)-(i), (d)-(iv)
(B) (a)-(ii), (b)-(i), (c)-(iii), (d)-(iv)
(C) (a)-(ii), (b)-(iv), (c)-(iii), (d)-(i)
(D) (a)-(iii), (b)-(i), (c)-(ii), (d)-(iv)

92. If $\vec{E}$ and $\vec{K}$ represent electric field and propagation vectors of the electromagnetic waves in vacuum, then magnetic field vector is given by : (ω-angular frequency)
[24 Jan. 2023 (Morning)]

(A) $\frac{1}{\omega}(\vec{K} \times \vec{E})$ (B) $\omega(\vec{E} \times \vec{K})$
(C) $\omega(\vec{K} \times \vec{E})$ (D) $\vec{K} \times \vec{E}$

93. The electric field and magnetic field components of an electromagnetic wave going through vacuum is described by
$E_x = E_0 \sin(kz - \omega t)$
$B_y = B_0 \sin(kz - \omega t)$
Then the correct relation between E_0 and B_0 is given by :
[24 Jan. 2023 (Evening)]

(A) $kE_0 = \omega B_0$ (B) $E_0 B_0 = \omega k$
(C) $\omega E_0 = kB_0$ (D) $E_0 = kB_0$

94. All electromagnetic wave is transporting energy in the negative z direction. At a certain point and certain time the direction of electric field of the wave is along positive y direction. What will be the direction of the magnetic field of the wave at that point and instant ? **[25 Jan. 2023 (Morning)]**

(A) Positive direction of x (B) Positive direction of z
(C) Negative direction of x (D) Negative direction of y

95. Which of the following are true ?
(a) Speed of light in vacuum is dependent on the direction of propagation
(b) Speed of light in a medium in independent of the wavelength of light
(c) The speed of light is independent of the motion of the source
(d) The speed of light in a medium is independent of intensity
Choose the correct answer from the option given below :
[29 Jan. 2023 (Morning)]
(A) (a) and (c) only (B) (b) and (d) only
(C) (b) and (c) only (D) (c) and (d) only

96. Given below are two statements :
Statement-I : Electromagnetic waves are not deflected by electric and magnetic field.
Statement-II : The amplitude of electric field and the magnetic field in electromagnetic waves are related to each other as

$$E_0 = \sqrt{\frac{\mu_0}{\varepsilon_0}} B_0$$

In the light of the above statements, choose the correct answer from the options given below : **[29 Jan. 2023 (Evening)]**
(A) Statement-I is true but statement-II is false
(B) Both Statement-I and Statement-II are true
(C) Statement-I is false but statement-II is true
(D) Both Statement-I and Statement-II are false

97. A point source of 100 W emits light with 5% efficiency. At a distance of 5 m from the source, the intensity produced by the electric field component is : **[30 Jan. 2023 (Evening)]**

(A) $\frac{1}{2\pi}\frac{W}{m^2}$ (B) $\frac{1}{40\pi}\frac{W}{m^2}$

(C) $\frac{1}{10\pi}\frac{W}{m^2}$ (D) $\frac{1}{20\pi}\frac{W}{m^2}$

98. If a source of electromagnetic radiation having power 15 kW produces 10^{16} photons per second, the radiation belongs to a part of spectrum is. (Take Planck constant $h = 6 \times 10^{-34}$ Js) :
[31 Jan. 2023 (Morning)]
(A) Micro waves (B) Ultraviolet rays
(C) Gamma rays (D) Radio waves

99. In a medium the speed of light wave decreases to 0.2 times to its speed in free space. The ratio of relative permittivity to the refractive index of the medium is x : 1. The value of x is _____. (Given speed of light in free space = 3×10^8 ms^{-1} and for the given medium $\mu_r = 1$) **[31 Jan. 2023 (Morning)]**

100. Match List-I with List-II : **[31 Jan. 2023 (Evening)]**

List-I	List-II
(a) Microwaves	(i) Physiotherapy
(b) UV rays	(ii) Treatment of cancer
(c) Infra-red rays	(iii) Lasik eye surgery
(d) X-rays	(iv) Aircraft navigation

Choose the correct answer from the options given below :
(A) (a)-(ii), (b)-(iv), (c)-(iii), (d)-(i)
(B) (a)-(iv), (b)-(i), (c)-(ii), (d)-(iii)
(C) (a)-(iv), (b)-(iii), (c)-(i), (d)-(ii)
(D) (a)-(iii), (b)-(ii), (c)-(i), (d)-(iv)

101. The ratio of average electric energy density and total average energy density of electromagnetic wave is :
[1 Feb. 2023 (Evening)]
(A) 2 (B) 1
(C) 3 (D) $\frac{1}{2}$

102. For the plane electromagnetic wave given by $E = E_0 \sin(\omega t - kx)$ and $B = B_0 \sin(\omega t - kx)$, the ratio of average electric energy density to average magnetic energy density is:
[6 April 2023 (Morning)]
(A) 1 (B) 1 / 2
(C) 2 (D) 4

103. The energy density associated with electric field $\vec{E}$ and magnetic field $\vec{B}$ of an electromagnetic wave in free space is given by ($\in_0$ – permittivity of free space)
[6 April 2023 (Evening)]

(A) $U_E = \frac{E^2}{2\in_0}, U_B = \frac{B^2}{2\mu_0}$ (B) $U_E = \frac{E^2}{2\in_0}, U_B = \frac{\mu_0 B^2}{2}$

(C) $U_E = \frac{\in_0 E^2}{2}, U_B = \frac{\mu_0 B^2}{2}$ (D) $U_E = \frac{\in_0 E^2}{2}, U_B = \frac{B^2}{2\mu_0}$

104. The energy of an electromagnetic wave contained in a small volume oscillates with **[10 April 2023 (Morning)]**
(A) Zero frequency
(B) Half the frequency of the wave
(C) Double the frequency of the wave
(D) The frequency of the wave

105. The amplitude of magnetic field in an electromagnetic wave propagating along y-axis is 6.0×10^{-7} T. The maximum value of electric field in the electromagnetic wave is:
[10 April 2023 (Evening)]
(A) 5×10^{14} Vm^{-1} (B) 180 Vm^{-1}
(C) 2×10^{15} Vm^{-1} (D) 6.0×10^{-7} Vm^{-1}

106. The electric field in an electromagnetic wave is given as $\vec{E} = 20\sin\omega\left(t-\frac{x}{c}\right)\vec{j}\,\text{NC}^{-1}$ Where ω and c are angular frequency and velocity of electromagnetic wave respectively. The energy contained in a volume of $5 \times 10^{-4}\,\text{m}^3$ will be: (Given $\varepsilon_0 = 8.85 \times 10^{-12}\,\text{C}^2/\text{Nm}^2$) **[11 April 2023 (Morning)]**

(A) 28.5×10^{-13} J
(B) 17.7×10^{-13} J
(C) 8.85×10^{-13} J
(D) 88.5×10^{-13} J

107. A plane electromagnetic wave of frequency 20 MHz propagates in free space along x-direction. At a particular space and time, $\vec{E} = 6.6\,\hat{j}$ V/m. What is B at this point ? **[11 April 2023 (Evening)]**

(A) $-2.2 \times 10^{-8}\,\hat{i}$ T
(B) $2.2 \times 10^{-8}\,\hat{k}$ T
(C) $-2.2 \times 10^{-8}\,\hat{k}$ T
(D) $2.2 \times 10^{-8}\,\hat{i}$ T

108. Given below are two statement.

One is labelled as **Assertion-A** and the other is labelled as **Reason-R**.

Assertion-A: EM waves used for optical communication have longer wavelengths than that of microwave, employed in Radar technology.

Reason-R: Infrared EM waves are more energetic than microwaves. (used in Radar)

In the light of given statements, choose the correct answer from the options given below: **[12 April 2023 (Morning)]**

(A) **A** is false but **R** is true.
(B) **A** is true but **R** is false.
(C) Both **A** and **R** true but **R** is NOT the correct explanation of **A**.
(D) Both **A** and **R** true and r is the correct explanation of **A**.

109. Which of the following Maxwell's equations is valid for time varying conditions but not valid for static conditions : **[13 April 2023 (Morning)]**

(A) $\oint \vec{B}\cdot\vec{dl} = \mu_0 I$
(B) $\oint \vec{E}\cdot\vec{dl} = 0$
(C) $\oint \vec{E}\cdot\vec{dl} = -\frac{\partial\phi_B}{\partial t}$
(D) $\oint \vec{D}\cdot\vec{dA} = Q$

110. In an electromagnetic wave, at an instant and at a particular position, the electric field is along the negative z-axis and magnetic field is along the positive x-axis. Then the direction of propagation of electromagnetic wave is: **[13 April 2023 (Evening)]**

(A) At 45° angle from positive y-axis
(B) Negative y-axis
(C) Positive z-axis
(D) Positive y-axis

111. Match **List-I** with **List-II** of Electromagnetic waves with corresponding wavelength range : **[15 April 2023 (Morning)]**

List-I	List-II
(a) Microwave	(i) 400 nm to 1 nm
(b) Ultraviolet	(ii) 1 nm to 10^{-3} nm
(c) X-Ray	(iii) 1 mn to 700 nm
(d) Infra-red	(iv) 0.1 m to 1 mm

Choose the correct answer from the options given below :

(A) (a)-(i), (b)-(iv), (c)-(ii), (d)-(iii)
(B) (a)-(iv), (b)-(i), (c)-(ii), (d)-(iii)
(C) (a)-(iv), (b)-(ii), (c)-(i), (d)-(iii)
(D) (a)-(iv), (b)-(i), (c)-(iii), (d)-(ii)

* * * * *

CHAPTER 23

RAY OPTICS AND OPTICAL INSTRUMENTS

23.1 GEOMETRICAL OPTICS I - REFLECTION OF LIGHT

1. If two mirrors are kept at 60° to each other, then the number of images formed by them is : **[2002]**

(A) 5 (B) 6

(C) 7 (D) 8

2. To get three images of a single object, one should have two plane mirrors at an angle of : **[2003]**

(A) 60° (B) 90°

(C) 120° (D) 30°

3. A car is fitted with a convex side-view mirror of focal length 20 cm. A second car 2.8 m behind the first car is overtaking the first car at a relative speed of 15 m/s. The speed of the image of the second car as seen in the mirror of the first one is : **[2011]**

(A) 10 m/s (B) 15 m/s

(C) $\frac{1}{10}$ m/s (D) $\frac{1}{15}$ m/s

4. You are asked to design a shaving mirror assuming that a person keeps it 10 cm from his face and views the magnified image of the face at the closest comfortable distance of 25 cm. The radius of curvature of the mirror would then be : **[2015 (Online)]**

(A) 60 cm (B) –24 cm

(C) –60 cm (D) 24 cm

5. A particle is oscillating on the X-axis with an amplitude 2 cm about the point $x_0 = 10$ cm with a frequency ω. A concave mirror of focal length 5 cm is placed at the origin (see figure) Identify the correct statements : **[2018 (Online)]**

(a) The image executes periodic motion

(b) The image executes non-periodic motion

(c) The turning points of the image are asymmetric w.r.t. the image of the point at $x = 10$ cm

(d) The distance between the turning points of the oscillation of the image is $\frac{100}{21}$ cm.

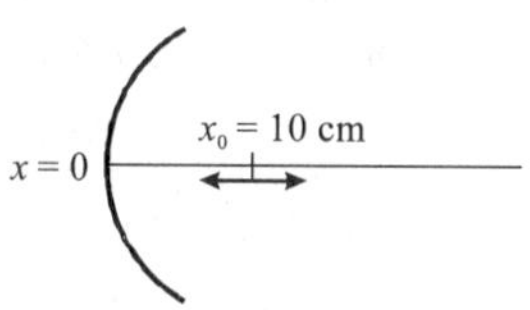

(A) (b), (d) (B) (b), (c)

(C) (a), (c), (d) (D) (a), (d)

6. Two plane mirrors are inclined to each other such that a ray of light incident on the first mirror (M_1) and parallel to the second mirror (M_2) is finally reflected from the second mirror (M_2) parallel to the first mirror (M_1). The angle between the two mirrors will be : **[9 Jan. 2019 (Evening)]**

(A) 90° (B) 45°

(C) 75° (D) 60°

7. A point source of light, S is placed at a distance L in front of the centre of plane mirror of width d which is hanging vertically on a wall. A man walks in front of the mirror along a line parallel to the mirror, at a distance $2L$ as shown below. The distance over which the man can see the image of the light source in the mirror is : **[12 Jan. 2019 (Morning)]**

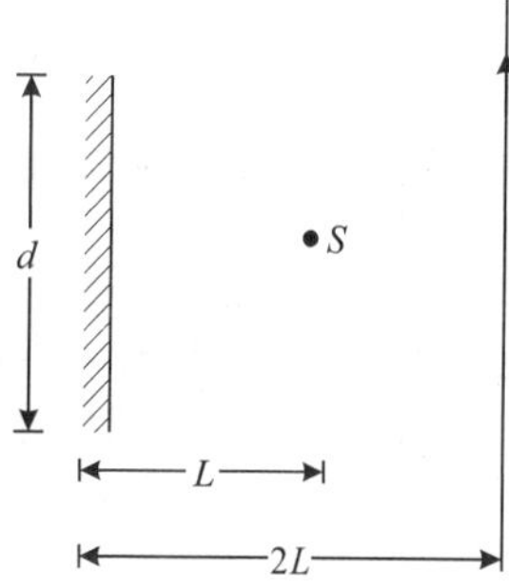

(A) $3d$ (B) $d/2$

(C) d (D) $2d$

8. A concave mirror for face viewing has focal length of 0.4 m. The distance at which you hold the mirror from your face in order to see your image upright with a magnification of 5 is : **[9 April 2019 (Morning)]**

(A) 1.60 m (B) 0.24 m

(C) 0.16 m (D) 0.32 m

9. A concave mirror has radius of curvature of 40 cm. It is at the bottom of a glass that has water filled up to 5 cm (see figure). If a small particle is floating on the surface of water, its image as seen, from directly above the glass, is at a distance d from the surface of water. The value of d is close to : (Refractive index of water = 1.33) **[12 April 2019 (Morning)]**

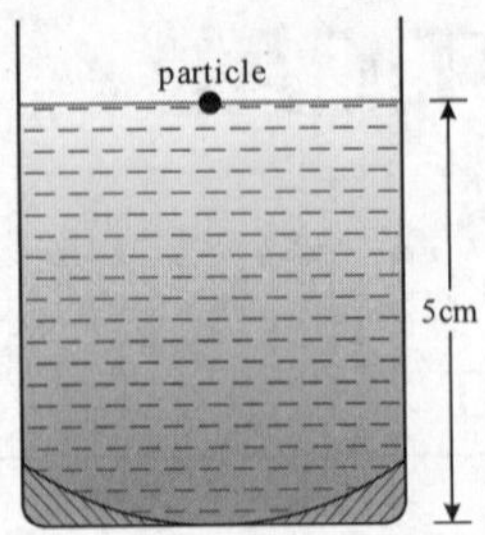

(A) 8.8 cm (B) 11.7 cm
(C) 6.7 cm (D) 13.4 cm

10. An object is gradually moving away from the focal point of a concave mirror along the axis of the mirror. The graphical representation of the magnitude of linear magnification (m) versus distance of the object from the mirror (x) is correctly given by : **[8 Jan. 2020 (Evening)]**
(Graphs are drawn schematically and are not to scale)

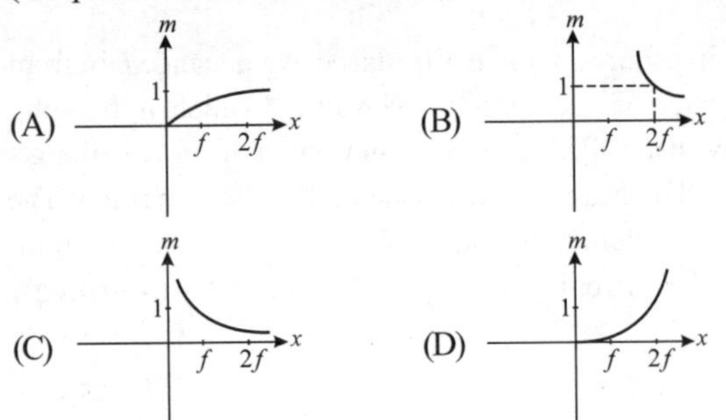

11. A spherical mirror is obtained as shown in the figure from a hollow glass sphere. If an object is positioned in front of the mirror, what will be the nature and magnification of the image of the object ? (Figure drawn as schematic and not to scale) **[2 Sept. 2020 (Morning)]**

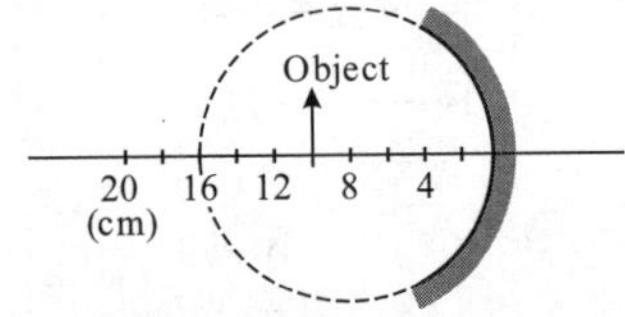

(A) Inverted, real and magnified
(B) Erect, virtual and magnified
(C) Erect, virtual and unmagnified
(D) Inverted, real and unmagnified

12. When an object is kept at a distance of 30 cm from a concave mirror, the image is formed at a distance of 10 cm from the mirror. If the object is moved with a speed of 9 cms^{-1}, the speed (in cms^{-1}) with which image moves at that instant is ______1. **[3 Sept. 2020 (Evening)]**

13. The focal length f is related to the radius of curvature r of the spherical convex mirror by : **[24 Feb. 2021 (Morning)]**

(A) $f = -r$ (B) $f = -\frac{1}{2}r$

(C) $f = +\frac{1}{2}r$ (D) $f = r$

14. A short straight object of height 100 cm lies before the central axis of a spherical mirror whose focal length has absolute value $|f| = 40$ cm. The image of object produced by the mirror is of height 25 cm and has the same orientation of the object. One may conclude from the information : **[26 Feb. 2021 (Morning)]**
(A) Image is virtual, opposite side of convex mirror
(B) Image is virtual, opposite side of concave mirror
(C) Image is real, same side of convex mirror
(D) Image is real, same side of concave mirror

15. The incident ray, reflected ray and the outward drawn normal are denoted by the unit vectors $\vec{a}$, $\vec{b}$ and $\vec{c}$ respectively. Then choose the correct relation for these vectors. **[26 Feb. 2021 (Evening)]**

(A) $\vec{b} = 2\vec{a} + \vec{c}$ (B) $\vec{b} = \vec{a} - 2(\vec{a}\cdot\vec{c})\vec{c}$

(C) $\vec{b} = \vec{a} - \vec{c}$ (D) $\vec{b} = \vec{a} + 2\vec{c}$

16. A point source of light S, placed at a distance 60 cm infront of the centre of a plane mirror of width 50 cm, hangs vertically on a wall. A man walks infront of the mirror along a line parallel to the mirror at a distance 1.2 cm from it (see in the figure). The distance between the extreme points where he can see the image of the light source in the mirror is ______ cm. **[26 Feb. 2021 (Evening)]**

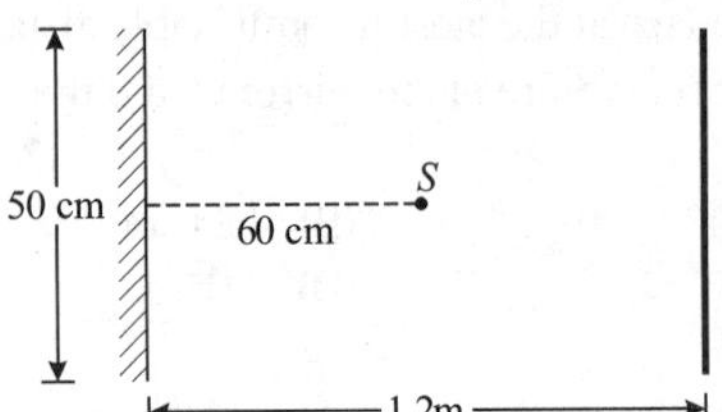

17. Car B overtakes another car A at relative speed of 40 ms^{-1}. How fast will the image of car B appear to move in the mirror of focal length 10 cm fitted in car A, when the car B is 1.9 m away from car A? **[26 Aug. 2021 (Morning)]**
(A) 0.2 ms^{-1} (B) 40 ms^{-1}
(C) 0.1 ms^{-1} (D) 4 ms^{-1}

18. An object is placed beyond the centre of curvature C of the given concave mirror. If the distance of the object is d_1 from C and the distance of the image formed is d_2 from C. The radius of curvature of this mirror is : **[27 Aug. 2021 (Morning)]**

(A) $\frac{d_1d_2}{d_1+d_2}$ (B) $\frac{2d_1d_2}{d_1+d_2}$

(C) $\frac{d_1d_2}{d_1-d_2}$ (D) $\frac{2d_1d_2}{d_1-d_2}$

19. Two plane mirror M_1 and M_2 are at right angle to each other shown. A point source P is placed at a and $2a$' meter away from M_1 and M_2 respectively. The shortest distance between the images thus formed is : (Take $\sqrt{5} = 2.3$)

[31 Aug. 2021 (Morning)]

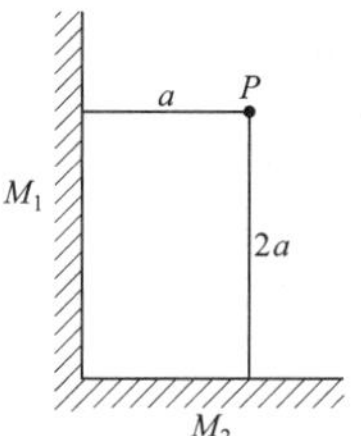

(A) $4.6\,a$ (B) $2\sqrt{10}\,a$
(C) $3a$ (D) $2.3\,a$

20. A light ray is incident, at an incident angle θ_1, on the system of two plane mirrors M_1 and M_2 having an inclination angle 75° between them (as shown in figure). After reflecting from mirror M_1 it gets reflected back by the mirror M_2 with an angle of reflection 30°. The total deviation of the ray will be _____ degree.

[26 June 2022 (Morning)]

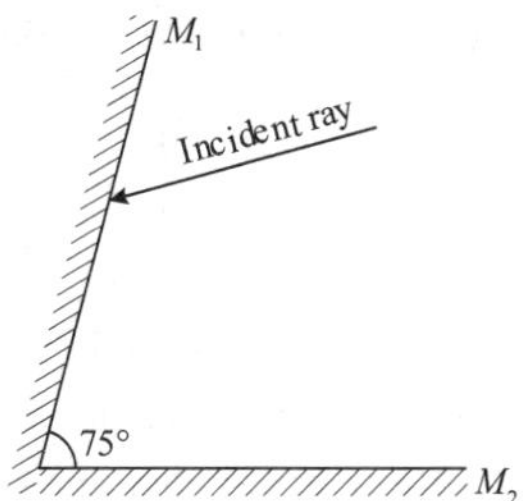

21. An object 'O' is placed at a distance of 100 cm in front of a concave mirror of radius of curvature 200 cm as shown in the figure. The object starts moving towards the mirror at a speed 2 cm/s. The position of the image from the mirror after 10 s will be at _____ cm.

[28 July 2022 (Evening)]

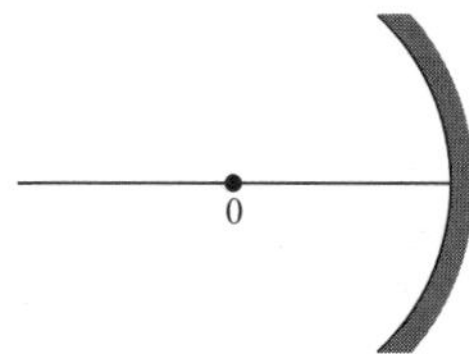

22. The light rays from an object have been reflected towards an observer from a standard flat mirror, the image observed by the observer are :

(a) Real
(b) Erect
(c) Smaller in size then object
(d) Laterally inverted

Choose the most appropriate answer from the options given below :

[25 Jan. 2023 (Evening)]

(A) (b) and (d) only (B) (b) and (c) only
(C) (a) and (d) only (D) (a), (c) and (d) only

23. In an experiment for estimating the value of focal length of converging mirror, image of an object placed at 40 cm from the pole of the mirror is formed at distance 120 cm from the pole of the mirror. These distances are measured with a modified scale in which there are 20 small divisions in 1 cm. The value of error in measurement of focal length of the mirror is 1/K cm. The value of K is _____.

[30 Jan. 2023 (Morning)]

24. A thin cylindrical rod of length 10 cm is placed horizontally on the principle axis of a concave mirror of focal length 20 cm. The rod is placed in a such a way that mid point of the rod is at 40 cm from the pole of mirror. The length of the image formed by the mirror will be $\frac{x}{3}$ cm. The value of x is _____.

[1 Feb. 2023 (Morning)]

25. Two objects A and B are placed at 15 cm and 25 cm from the pole in front of a concave mirror having radius of curvature 40 cm. The distance between images formed by the mirror is :

[1 Feb. 2023 (Evening)]

(A) 40 cm (B) 60 cm
(C) 160 cm (D) 100 cm

26. Two vertical parallel mirrors A and B are separated by 10 cm. A point object O is placed at a distance of 2 cm from mirror A. The distance of the second nearest image behind mirror A from the mirror A is _____ cm.

[8 April 2023 (Morning)]

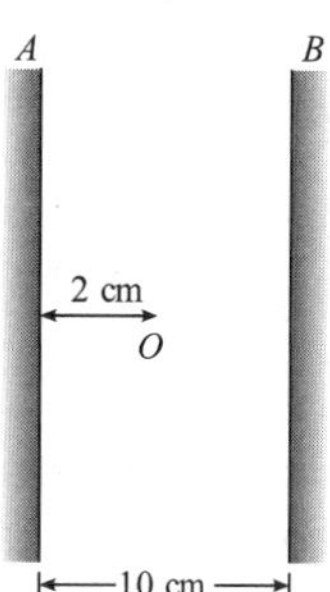

27. An object is placed at a distance of 12 cm in front of a plane mirror. The virtual and erect image is formed by the mirror. Now the mirror is moved by 4 cm towards the stationary object. The distance by which the position of image would be shifted, will be:

[10 April 2023 (Morning)]

(A) 4 cm towards mirror (B) 8 cm towards mirror
(C) 8 cm away from mirror (D) 2 cm towards mirror

28. When one light ray is reflected from a plane mirror with 30° angle of reflection, the angle of deviation of the ray after reflection is :

[11 April 2023 (Evening)]

(A) 140° (B) 120°
(C) 110° (D) 130°

* * * * *

CHAPTER 23

RAY OPTICS AND OPTICAL INSTRUMENTS

23.2 GEOMETRICAL OPTICS II - REFRACTION OF LIGHT

1. Which of the following is used in optical fibres ? **[2002]**

(A) Total internal reflection (B) Scattering

(C) Diffraction (D) Refraction

2. A light ray is incident perpendicular to one face of a 90° prism and is totally internally reflected at the glass-air interface. If the angle of reflection is 45°, we conclude that the refractive index n: **[2004]**

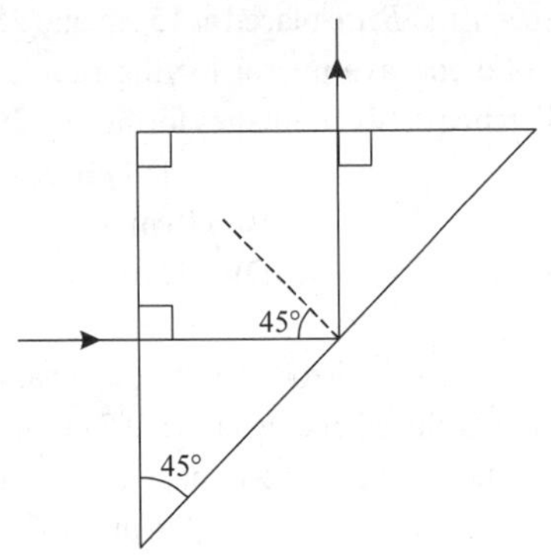

(A) $n < \frac{1}{\sqrt{2}}$ (B) $n > \sqrt{2}$

(C) $n > \frac{1}{\sqrt{2}}$ (D) $n < \sqrt{2}$

3. A fish looking up through the water sees the outside world, contained in a circular horizon. If the refractive index of water is 4/3 and the fish is 12 cm below the water surface, the radius of this circle in cm is : **[2005]**

(A) $36\sqrt{7}$ (B) $\frac{36}{\sqrt{7}}$

(C) $36\sqrt{5}$ (D) $4\sqrt{5}$

4. A transparent solid cylindrical rod has a refractive index of $\frac{2}{\sqrt{3}}$. It is surrounded by air. A light ray is incident at the mid-point of one end of the rod as shown in the figure :

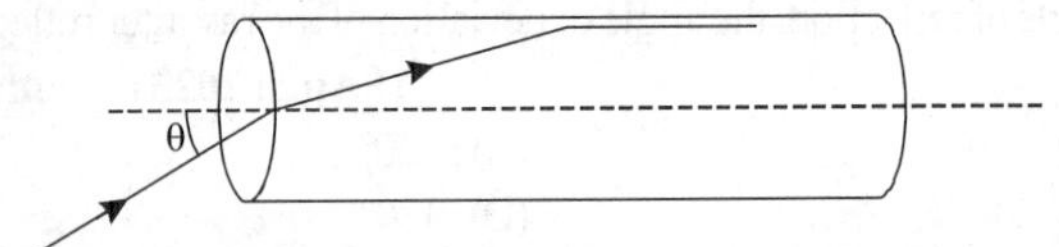

The incident angle θ for which the light ray grazes along the wall of the rod is : **[2009]**

(A) $\sin^{-1}\left(\frac{2}{\sqrt{3}}\right)$ (B) $\sin^{-1}\left(\frac{1}{\sqrt{3}}\right)$

(C) $\sin^{-1}\left(\frac{1}{2}\right)$ (D) $\sin^{-1}\left(\frac{\sqrt{3}}{2}\right)$

5. Let the x-y plane be the boundary between two transparent media. Medium 1 in $z \geq 0$ has a refractive index of $\sqrt{2}$ and medium 2 with $z < 0$ has a refractive index of $\sqrt{3}$. A ray of light in medium 1 given by the vector $\vec{A} = 6\sqrt{3}\hat{i} + 8\sqrt{3}\hat{j} - 10\hat{k}$ is incident on the plane of separation. The angle of refraction in medium 2 is : **[2011]**

(A) 60° (B) 75°

(C) 30° (D) 45°

6. A beaker contains water up to a height h_1 and kerosene of height h_2 above water so that the total height of (water + kerosene) is $(h_1 + h_2)$. Refractive index of water is μ_1 and that of kerosene is μ_2. The apparent shift in the position of the bottom of the beaker when viewed from above is : **[2011]**

(A) $\left(1-\frac{1}{\mu_1}\right)h_2+\left(1-\frac{1}{\mu_2}\right)h_1$ (B) $\left(1+\frac{1}{\mu_1}\right)h_1-\left(1+\frac{1}{\mu_2}\right)h_2$

(C) $\left(1-\frac{1}{\mu_1}\right)h_1+\left(1-\frac{1}{\mu_2}\right)h_2$ (D) $\left(1+\frac{1}{\mu_1}\right)h_2-\left(1+\frac{1}{\mu_2}\right)h_1$

7. A glass prism of refractive index 1.5 is immersed in water (refractive index 4/3) as shown in figure. A light beam incident normally on the face AB is totally reflected to reach the face BC, if: **[2012 (Online)]**

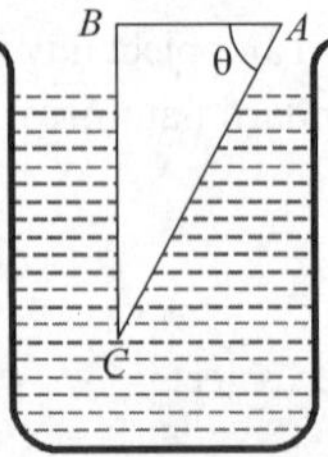

(A) $\sin\theta > \frac{5}{9}$ (B) $\sin\theta > \frac{2}{3}$

(C) $\sin\theta > \frac{8}{9}$ (D) $\sin\theta > \frac{1}{3}$

8. The graph between angle of deviation (δ) and angle of incidence (i) for a triangular prism is represented by : **[2013]**

(A) 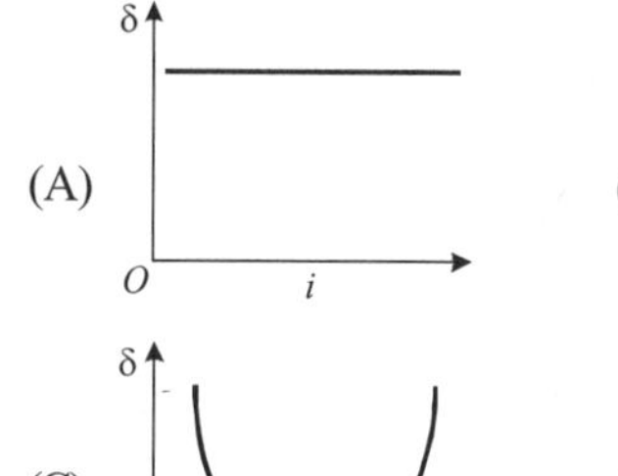

(B) 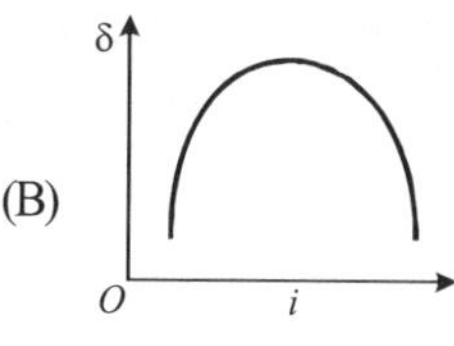

(C)

(D) 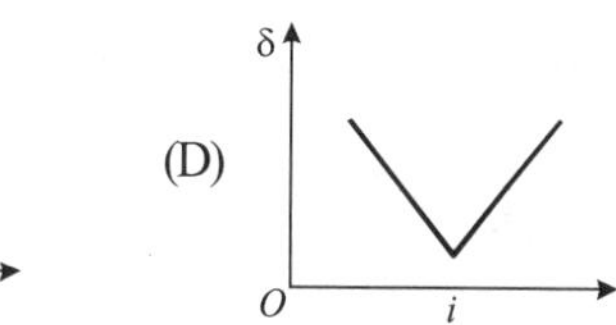

9. Light is incident from a medium into air at two possible angles of incidence (A) 20° and (B) 40°. In the medium light travels 3.0 cm in 0.2 ns. The ray will : **[2013 (Online)]**

(A) Suffer total internal reflection in both cases (A) and (B)
(B) Suffer total internal reflection in case (B) only
(C) Have partial reflection and partial transmission in case (B)
(D) Have 100% transmission in case (A)

10. A light ray falls on a square glass slab as shown in the diagram. The index of refraction of the glass, if total internal reflection is to occur at the vertical face, is equal to : **[2013 (Online)]**

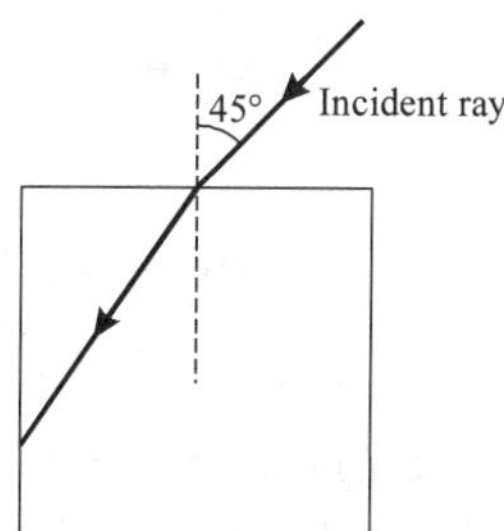

(A) $\frac{(\sqrt{2}+1)}{2}$ (B) $\sqrt{\frac{5}{2}}$

(C) $\frac{3}{2}$ (D) $\sqrt{\frac{3}{2}}$

11. A printed page is pressed by a glass of water. The refractive index of the glass and water is 1.5 and 1.33, respectively. If the thickness of the bottom of glass is 1 cm and depth of water is 5 cm, how much the page will appear to be shifted if viewed from the top ? **[2013 (Online)]**

(A) 1.033 cm (B) 3.581 cm
(C) 1.3533 cm (D) 1.90 cm

12. A diver looking up through the water sees the outside world contained in a circular horizontal. The refractive index of water is 4/3, and the diver's eyes are 15 cm below the surface of water. Then the radius of the circle is : **[2014 (Online)]**

(A) $15\times 3\times\sqrt{5}$ cm (B) $15\times 3\sqrt{5}$ cm

(C) $\frac{15\times\sqrt{7}}{3}$ cm (D) $\frac{15\times 3}{\sqrt{7}}$ cm

13. Monochromatic light is incident of a glass prism of angle A. If the refractive index of the material of the prism is μ, a ray, incident at an angle θ, on the face AB would get transmitted through the face AC of the prism provided : **[2015]**

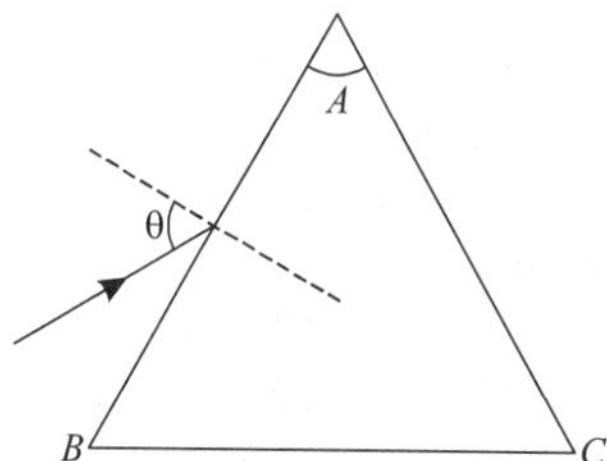

(A) $\theta > \sin^{-1}\left[\mu\sin\left(A-\sin^{-1}\left(\frac{1}{\mu}\right)\right)\right]$

(B) $\theta < \sin^{-1}\left[\mu\sin\left(A-\sin^{-1}\left(\frac{1}{\mu}\right)\right)\right]$

(C) $\theta > \cos^{-1}\left[\mu\sin\left(A+\sin^{-1}\left(\frac{1}{\mu}\right)\right)\right]$

(D) $\theta < \cos^{-1}\left[\mu\sin\left(A+\sin^{-1}\left(\frac{1}{\mu}\right)\right)\right]$

14. In an experiment for determination of refractive index of glass of a prism by $i-\delta$, plot, it was found that a ray incident at angle 35°, suffers a deviation of 40° and that it emerges at angle 79°. In that case which of the following is closest to the maximum possible value of the refractive index ? **[2016]**

(A) 1.5 (B) 1.6
(C) 1.7 (D) 1.8

15. A hemispherical glass body of radius 10 cm and refractive index 1.5 is silvered on its curved surface. A small air bubble is 6 cm below the flat surface inside it along the axis. The position of the image of the air bubble made by the mirror is seen : **[2016 (Online)]**

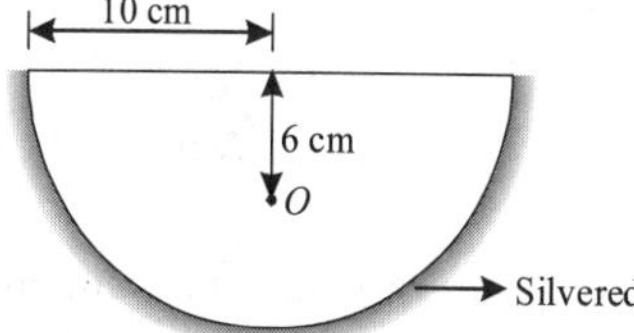

(A) 14 cm below flat surface (B) 20 cm below flat surface
(C) 16 cm below flat surface (D) 30 cm below flat surface

16. Let the refractive index of a denser medium with respect to a rarer medium be n_{12} and its critical angle be θ_C. At an angle of incidence A when light is travelling from denser medium to rarer medium, a part of the light is reflected and the rest is refracted and the angle between reflected and refracted rays is 90°. Angle A is given by : **[2017 (Online)]**

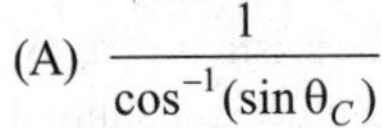

(A) $\frac{1}{\cos^{-1}(\sin\theta_C)}$ (B) $\frac{1}{\tan^{-1}(\sin\theta_C)}$

(C) $\cos^{-1}(\sin\theta_C)$ (D) $\tan^{-1}(\sin\theta_C)$

17. A ray of light is incident at an angle of 60° on one face of a prism of angle 30°. The emergent ray of light makes an angle of 30° with incident ray. The angle made by the emergent ray with second face of prism will be : **[2018 (Online)]**

(A) 30° (B) 90°

(C) 0° (D) 45°

18. The eye can be regarded as a single refracting surface. The radius of curvature of this surface is equal to that of cornea (7.8 mm). This surface separates two media of refractive indices 1 and 1.34. Calculate the distance from the refracting surface at which a parallel beam of light will come to focus : **[10 Jan. 2019 (Evening)]**

(A) 2 cm (B) 1 cm

(C) 3.1 cm (D) 4.0 cm

19. A monochromatic light is incident at a certain angle on an equilateral triangular prism and suffers minimum deviation. If the refractive index of the material of the prism is $\sqrt{3}$, then the angle of incidence is : **[11 Jan. 2019 (Evening)]**

(A) 30° (B) 45°

(C) 90° (D) 60°

20. In figure, the optical fiber is $l = 2$ m long and has a diameter of $d = 20$ μm. If a ray of light is incident on one end of the fiber at angle $\theta_1 = 40°$, the number of reflection it makes before emerging from the other end is close to :
(refractive index of fibre is 1.31 and sin 40° = 0.64) **[8 April 2019 (Morning)]**

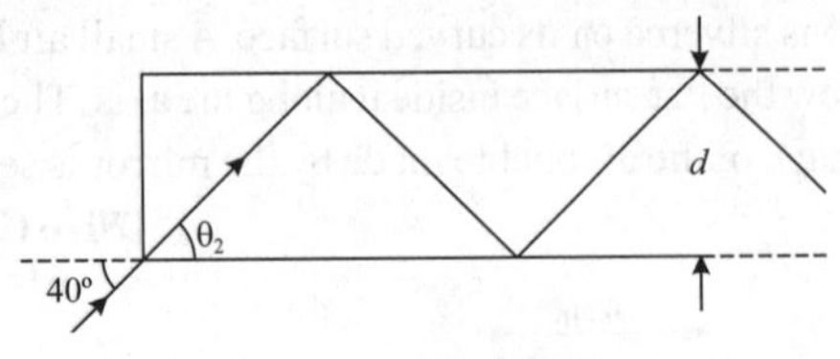

(A) 55000 (B) 57000

(C) 66000 (D) 45000

21. A ray of light AO in vacuum is incident on a glass slab at angle 60° and refracted at angle 30° along OB as shown in the figure. The optical path length of light ray from A to B is : **[10 April 2019 (Morning)]**

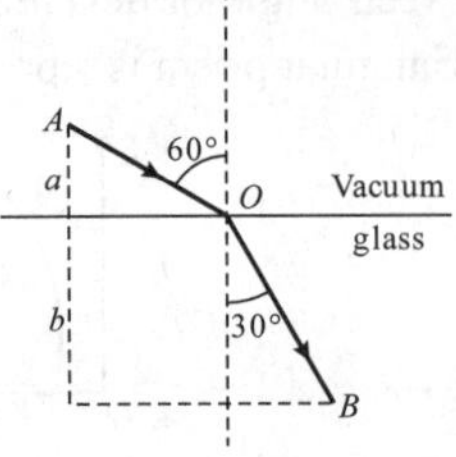

(A) $2a + 2b$ (B) $2a + \frac{2b}{3}$

(C) $\frac{2\sqrt{3}}{a} + 2b$ (D) $2a + \frac{2b}{\sqrt{3}}$

22. A transparent cube of side d, made of a material of refractive index μ_2, is immersed in a liquid of refractive index $\mu_1 (\mu_1 < \mu_2)$. A ray is incident on the face AB at an angle θ (shown in the figure). Total internal reflection takes place at point E on the face BC. The θ must satisfy : **[12 April 2019 (Evening)]**

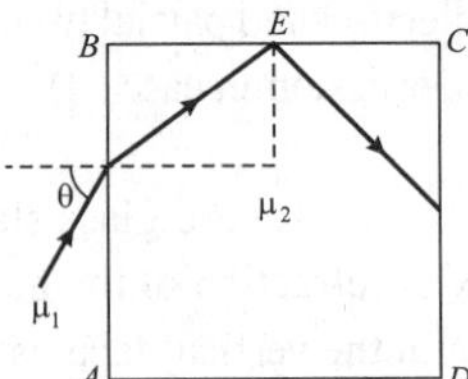

(A) $\theta < \sin^{-1}\frac{\mu_1}{\mu_2}$ (B) $\theta < \sin^{-1}\sqrt{\frac{\mu_2^2}{\mu_1^2} - 1}$

(C) $\theta > \sin^{-1}\frac{\mu_1}{\mu_2}$ (D) $\theta > \sin^{-1}\sqrt{\frac{\mu_2^2}{\mu_1^2} - 1}$

23. The critical angle of a medium for a specific wavelength, if the medium has relative permittivity 3 and relative permeability 4/3 for this wavelength, will be : **[8 Jan. 2020 (Morning)]**

(A) 60° (B) 15°

(C) 45° (D) 30°

24. A vessel of depth $2h$ is half filled with a liquid of refractive index $2\sqrt{2}$ and the upper half with another liquid of refractive index $\sqrt{2}$. The liquids are immiscible. The apparent depth of the inner surface of the bottom of vessel will be : **[9 Jan. 2020 (Morning)]**

(A) $\frac{h}{\sqrt{2}}$ (B) $\frac{3}{4}h\sqrt{2}$

(C) $\frac{h}{2(\sqrt{2}+1)}$ (D) $\frac{h}{3\sqrt{2}}$

25. There is a small source of light at some depth below the surface of water (refractive index = 4/3) in a tank of large cross sectional surface area. Neglecting any reflection from the bottom

and absorption by water, percentage of light that emerges out of surface is (nearly) :
[Use the fact that surface area of a spherical cap of height h and radius of curvature r is $2\pi rh$] **[9 Jan. 2020 (Evening)]**
(A) 17% (B) 21%
(C) 34% (D) 50%

26. A light ray enters a solid glass sphere of refractive index $\mu = \sqrt{3}$ at an angle of incidence 60°. The ray is both reflected and refracted at the farther surface of the sphere. The angle (in degrees) between the reflected and refracted rays at this surface is : **[2 Sept. 2020 (Evening)]**

27. An observer can see through a small hole on the side of a jar (radius 15 cm) at a point at height of 15 cm from the bottom (see figure). The hole is at a height of 45 cm. When the jar is filled with a liquid up to a height of 30 cm the same observer can see the edge at the bottom of the jar. If the refractive index of the liquid $N/100$, where N is an integer, the value of N is______. **[3 Sept. 2020 (Morning)]**

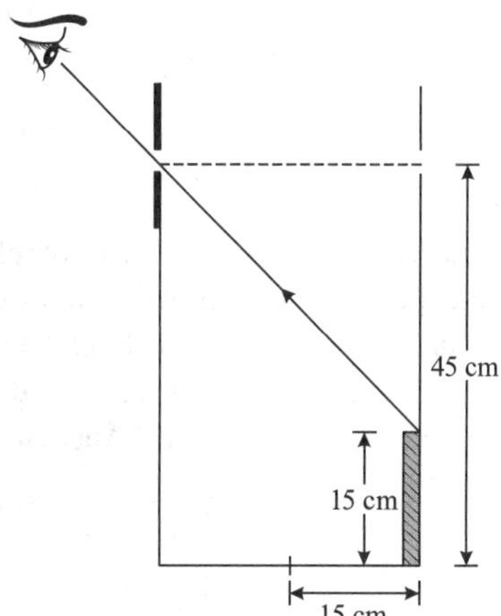

28. A prism of angle $A = 1°$ has a refractive index $\mu = 1.5$. A good estimate for the minimum angle of deviation (in degrees) is close to N/10. Value of N is ______. **[5 Sept. 2020 (Evening)]**

29. A double convex lens has power P and same radii of curvature R of both the surfaces. The radius of curvature of a surface of a plano-convex lens made of the same material with power $1.5\,P$ is : **[6 Sept. 2020 (Evening)]**
(A) $\frac{R}{2}$ (B) $2R$
(C) $\frac{3R}{2}$ (D) $\frac{R}{3}$

30. The refractive index of a converging lens is 1.4. What will be the focal length of this lens if it is placed in a medium of same refractive index ? (Assume the radii of curvature of the faces of lens are R_1 and R_2 respectively) **[16 March 2021 (Evening)]**
(A) 1 (B) Infinite
(C) $\frac{R_1 R_2}{R_1 - R_2}$ (D) Zero

31. The thickness at the centre of a plano convex lens is 3 mm and the diameter is 6 cm. If the speed of light in the material of the lens is $2 \times 10^8\ ms^{-1}$. The focal length of the lens is ______. **[17 March 2021 (Morning)]**
(A) 0.30 cm (B) 15 cm
(C) 1.5 cm (D) 30 cm

32. The image of an object placed in air formed by a convex refracting surface is at a distance of 10 m behind the surface. The image is real and is at $\left(\frac{2}{3}\right)^{rd}$ of the distance of the object from the surface. The wavelength of light inside the surface is 2/3 times the wavelength in air. The radius of the curved surface is $\frac{x}{13}$ m. The value of 'x' is______. **[17 March 2021 (Evening)]**

33. Three rays of light, namely red (R), green (G) and blue (B) are incident on the face PQ of a right angled prism PQR as shown in figure. The refractive indices of the material of the prism for red, green and blue wavelength are 1.27, 1.42 and 1.49 respectively. The colour of the ray(s) emerging out of the face PR is : **[18 March 2021 (Evening)]**

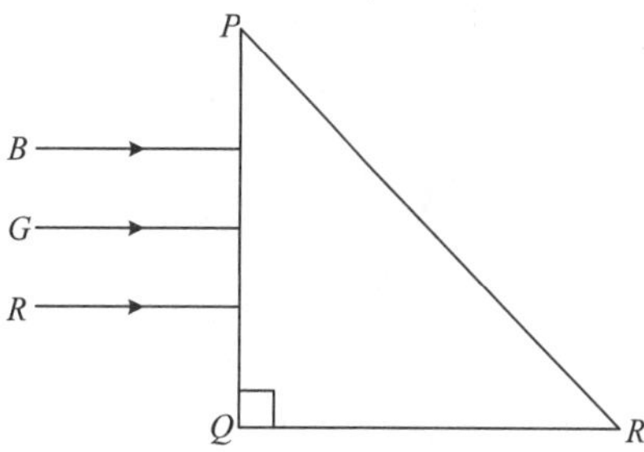

(A) Green (B) Red
(C) Blue and Green (D) Blue

34. Region I and II are separated by a spherical surface of radius 25 cm an object is kept in region I at a distance of 40 cm from the surface. The distance of the image from the surface : **[20 July 2021 (Morning)]**

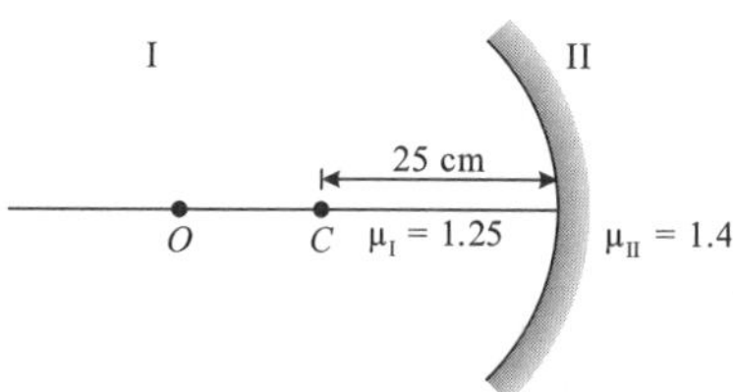

(A) 9.52 cm (B) 18.23 cm
(C) 55.44 cm (D) 37.58 cm

35. A ray of light passes from a denser medium to a rarer medium at an angle of incidence i. The reflected and refracted rays make an angle of 90° with each other. The angle of reflection and refraction are respectively r and r'. The critical angle is given by : **[22 July 2021 (Evening)]**

(A) $\sin^{-1}(\tan r)$ (B) $\sin^{-1}(\cot r)$
(C) $\tan^{-1}(\sin i)$ (D) $\sin^{-1}(\tan r')$

36. A ray of light passing through a prism ($\mu = \sqrt{3}$) suffers minimum deviation. It is found that the angle of incidence is double the angle of refraction within the prism. Then, the angle of prism is ______. (in degree) **[22 July 2021 (Evening)]**

37. A ray of laser of a wavelength 630 nm is incident at an angle of 30° at the diamond-air interface. It is going from diamond to air. The refractive index of diamond is 2.42 and that of air is 1. Choose the correct option : **[25 July 2021 (Morning)]**

(A) angle of refraction is 30° (B) angle of refraction is 24.41°
(C) refraction is not possible (D) angle of refraction is 53.4°

38. A ray of light entering from air into a denser medium of refractive index $\frac{4}{3}$, as shown in figure. The light ray suffers total internal reflection at the adjacent surface as shown. The maximum value of angle? Should be equal to : **[25 July 2021 (Evening)]**

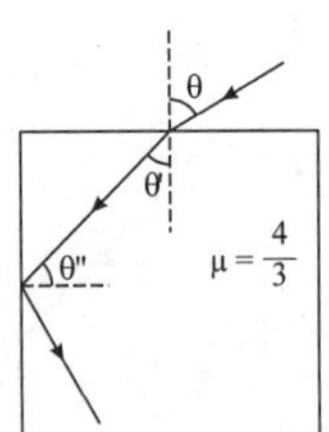

(A) $t^{1/2}$ (B) $t^{1/4}$
(C) $t^{3/2}$ (D) $t^{3/4}$

39. The expected graphical representation of the variation of angle of deviation 'δ' with angle of incidence 'i' in a prism is : **[27 July 2021 (Evening)]**

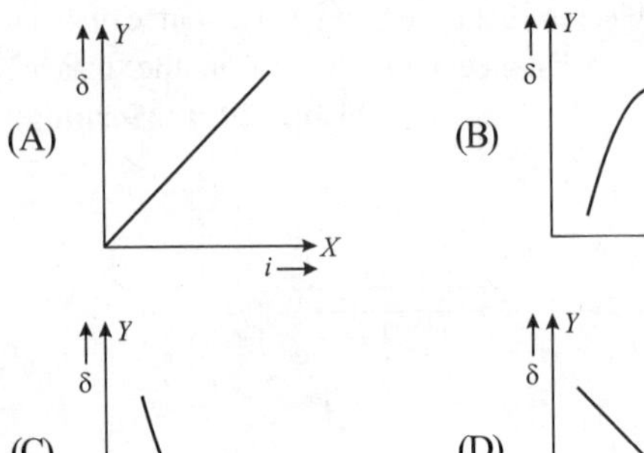
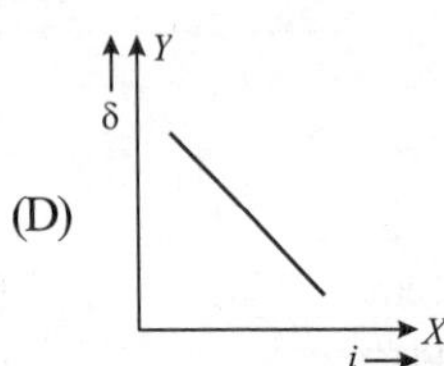

40. An object Is placed at a distance of 12 cm from a convex lens. A convex mirror of focal length 15 cm is placed on other side of lens at 8 cm as shown in the figure. Image of object coincides with the object. When the convex mirror is removed, a real and inverted image is formed at a position. The distance of the image from the object will be ______ (cm). **[26 Aug. 2021 (Evening)]**

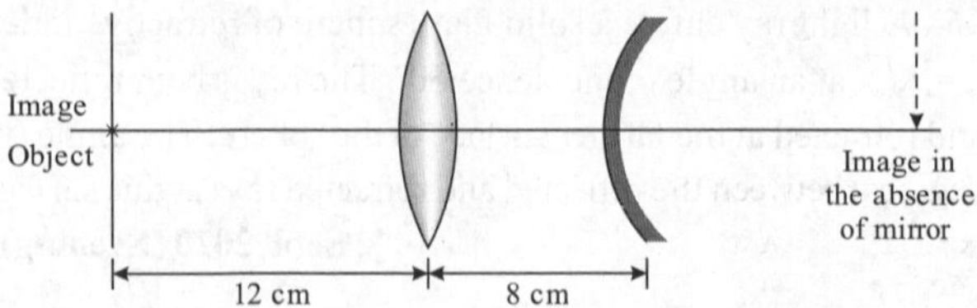

41. Find the distance of the image from object O, formed by the combination of lenses in the figure : **[27 Aug. 2021 (Morning)]**

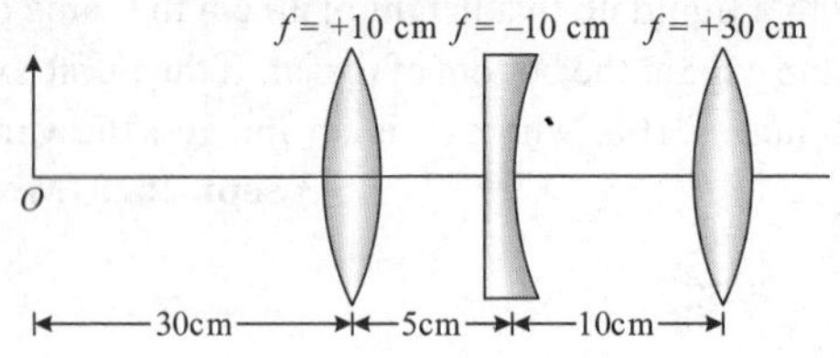

(A) 20 cm (B) 75 cm
(C) 10 cm (D) Infinity

42. Curved surfaces of a plano-convex lens of refractive index μ_1 and a plano-concave lens of refractive index μ_2 have equal radius of curvature as shown in figure. Find the ratio of radius of curvature to the focal length of the combined lenses : **[27 Aug. 2021 (Evening)]**

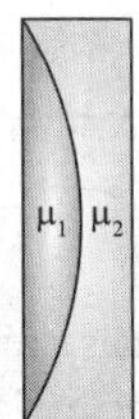

(A) $\frac{1}{\mu_2 - \mu_1}$ (B) $\frac{1}{\mu_1 - \mu_2}$
(C) $\mu_2 - \mu_1$ (D) $\mu_1 - \mu_2$

43. The temperature of an ideal gas in 3-dimensions is 300 K. The corresponding de-Broglie wavelength of the electron approximately at 300 K, is :
[m_e = mass of electron = 9×10^{-31} kg
h = Planck constant = 6.6×10^{-34} J s
k_B = Boltzmann constant = 1.38×10^{-23} JK^{-1}] **[01 Sep. 2021 (Evening)]**

(A) 6.26 nm (B) 3.25 nm
(C) 8.46 nm (D) 2.26 nm

44. A glass tumbler having inner depth of 17.5 cm is kept on a table. A student starts pouring water (μ = 4/3) into it while looking at the surface of water from the above. When he feels that the tumbler is half filled, he stops pouring water. Up to what height, the tumbler is actually filled ?

[01 Sep. 2021 (Evening)]

(A) 7.5 cm (B) 10 cm

(C) 8.75 cm (D) 11.7 cm

45. A ray of light is incident at an angle of incidence 60° on the glass slab of refractive index $\sqrt{3}$. After refraction, the light ray emerges out from other parallel faces and lateral shift between incident ray and emergent ray is $4\sqrt{3}$ cm. The thickness of the glass slab is ______ cm. **[24 June 2022 (Evening)]**

46. A light wave travelling linearly in a medium of dielectric constant 4, incident on the horizontal interface separating medium with air. The angle of incidence for which the total intensity of incident wave will be reflected back into the same medium will be : [Given : Relative permeability of medium $\mu_r = 1$]

[25 June 2022 (Morning)]

(A) 10° (B) 20°

(C) 30° (D) 60°

47. The difference of speed of light in the two media A and B $(v_A - v_B)$ is 2.6×10^7 m/s. If the refractive index of medium B is 1.47, then the ratio of refractive index of medium B to medium A is : [Given : Speed of light in vacuum $c = 3 \times 10^8$ ms^{-1}]

[25 June 2022 (Morning)]

(A) 1.303 (B) 1.318

(C) 1.13 (D) 0.12

48. A small bulb is placed at the bottom of a tank containing water to a depth of $\sqrt{7}m$. The refractive index of water is $\frac{4}{3}$. The area of the surface of water through which light from the bulb can emerge out is $x\pi$ m^2. The value of x is______.

[26 June 2022 (Evening)]

49. Consider a light ray travelling in air is incident into a medium of refractive index $\sqrt{2n}$. The incident angle is twice that of refracting angle. Then, the angle of incidence will be :

[27 June 2022 (Morning)]

(A) $\sin^{-1}(\sqrt{n})$ (B) $\cos^{-1}\left(\frac{\sqrt{n}}{2}\right)$

(C) $\sin^{-1}(\sqrt{2n})$ (D) $2\cos^{-1}\left(\sqrt{\frac{n}{2}}\right)$

50. If a wave gets refracted into a denser medium, then which of the following is true ? **[27 June 2022 (Evening)]**

(A) Wavelength speed and frequency decreases

(B) Wavelength increases, speed decreases and frequency remains constant

(C) Wavelength and speed decreases but frequency remains constant

(D) Wavelength, speed and frequency increases

51. The refracting angle of a prism is A and refractive index of the material of the prism is cot $(A/2)$. Then the angle of minimum deviation will be : **[28 June 2022 (Morning)]**

(A) $180 - 2A$ (B) $90 - A$

(C) $180 + 2A$ (D) $180 - 3A$

52. The speed of light in media A and B are 2.0×10^{10} cm/s and 1.5×10^{10} cm/s respectively. A ray of light enters from the medium B to A at an incident angle θ. If the ray suffers total internal reflection, then : **[29 June 2022 (Evening)]**

(A) $\theta = \sin^{-1}\left(\frac{3}{4}\right)$ (B) $\theta > \sin^{-1}\left(\frac{2}{3}\right)$

(C) $\theta = \sin^{-1}\left(\frac{3}{4}\right)$ (D) $\theta > \sin^{-1}\left(\frac{3}{4}\right)$

53. Time taken by light to travel in two different materials A and B of refractive indices μ_A and μ_B of same thickness is t_1 and t_2 respectively. If $t_2 - t_1 = 5 \times 10^{-10}$ s and the ratio of μ_A to μ_B is 1 : 2. Then the thickness of material, in meter is : (Given v_A and v_B are velocities of light in A and B materials respectively)

[25 July 2022 (Morning)]

(A) $5 \times 10^{-10} v_a$ m (B) 5×10^{-10} m

(C) 1.5×10^{-10} m (D) $5 \times 10^{-10} v_B$ m

54. As shown in the figure, after passing through the medium-1. The speed of light v_2 in medium 2 will be :
[Given : $c = 3 \times 10^8$ ms^{-1}] **[28 July 2022 (Morning)]**

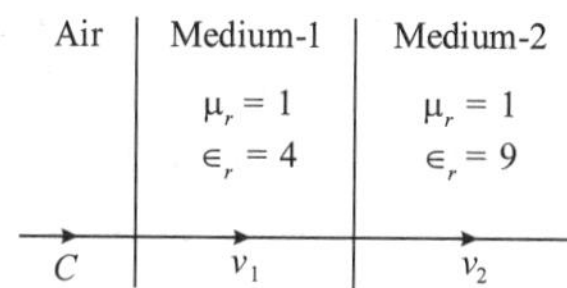

(A) 1.0×10^8 ms^{-1} (B) 0.5×10^8 ms^{-1}

(C) 1.5×10^8 ms^{-1} (D) 3.0×10^8 ms^{-1}

55. The X-Y plane be taken as the boundary between two transparent media M_1 and M_2. M_1 in $Z \geq 0$ has a refractive index of $\sqrt{2}$ and M_2 with $Z < 0$ has a refractive index of $\sqrt{3}$. A ray of light travelling in M_1 along the direction given by the vector $\vec{A} = 4\sqrt{3}\hat{i} - 3\sqrt{3}\hat{j} - 3\hat{k}$, is incident on the plane of separation. The value of difference between the angle of incident in M_1 and the angle of refraction in M_2 will be ______ degree.

[29 July 2022 (Morning)]

56. Light enters from air into a given medium at an angle of 45° with interface of the air-medium surface. After refraction, the light ray is deviated through an angle of 15° from its original direction. The refractive index of the medium is :

[29 July 2022 (Evening)]

(A) 1.732 (B) 1.333

(C) 1.414 (D) 2.732

57. A ray of light is incident from air on a glass plate having thickness $\sqrt{3}$ cm and refractive index $\sqrt{2}$. The angle of incidence of a ray is equal to the critical angle for glass-air interface. The lateral displacement of the ray when it passes through the plate is ______ $\times 10^{-2}$ cm. (Given $\sin 15° = 0.26$)

[25 Jan. 2023 (Morning)]

58. A thin prism P_1 with an angle 6° and made of glass of refractive index 1.54 is combined with another prism P_2 made from glass of refractive index 1.72 to produce dispersion without average deviation. The angle of prism P_2 is :

[30 Jan. 2023 (Evening)]

(A) 6° (B) 1.3°

(C) 7.8° (D) 4.5°

59. A microscope is focused on an object at the bottom of a bucket. If liquid with refractive index $\frac{5}{3}$ is poured inside the bucket, then microscope have to be raised by 30 cm to focus the object again. The height of the liquid in the bucket is :

[31 Jan. 2023 (Evening)]

(A) 75 cm (B) 50 cm

(C) 18 cm (D) 12 cm

60. A stone of mass 1 kg is tied to end of a massless string of length 1 m. If the breaking tension of the string is 400 N, then maximum linear velocity, the stone can have without breaking the string, while rotating in horizontal plane, is :

[31 Jan. 2023 (Evening)]

(A) 20 ms^{-1} (B) 40 ms^{-1}

(C) 400 ms^{-1} (D) 10 ms^{-1}

61. A monochromatic light wave with wavelength λ_1 and frequency v_1 in air enters another medium. If the angle of incidence and angle of refraction at the interface are 45° and 30° respectively, then the wavelength λ_2 and frequency v_2 of the refracted wave are :

[6 April 2023 (Morning)]

(A) $\lambda_2 = \lambda_1, v_2 = \sqrt{2}v_1$ (B) $\lambda_2 = \frac{1}{\sqrt{2}}\lambda_1, v_2 = v_1$

(C) $\lambda_2 = \sqrt{2}\lambda_1, v_2 = v_1$ (D) $\lambda_2 = \lambda_1, v_2 = \frac{1}{\sqrt{2}}v_1$

62. A pole is vertically submerged in swimming pool, such that it gives a length of shadow 2.15 m within water when sunlight is incident at an angle of 30° with the surface of water. If swimming pool is filled to a height of 1.5 m, then the height of the pole above the water surface in centimetres is ($n_w = 4/3$) ______.

[6 April 2023 (Morning)]

63. Two transparent media having refractive indices 1.0 and 1.5 are separated by a spherical refracting surface of radius of curvature 30 cm. The centre of curvature of surface is towards denser medium and a point object is placed on the principle axis in rarer medium at a distance of 15 cm from the pole of the surface. The distance of image from the pole of the surface is ______ cm.

[8 April 2023 (Evening)]

64. The critical angle for a denser-rarer interface is 45°. The speed of light in rarer medium is 3×10^8 ms. The speed of light in the denser medium is :

[11 April 2023 (Morning)]

(A) 5×10^7 m/s (B) 2.12×10^8 m/s

(C) 3.12×10^7 m/s (D) $\sqrt{2} \times 10^8$ m/s

65. The radius of curvature of each surface of a convex lens having refractive index 1.8 is 20 cm. The lens is now immersed in a liquid of refractive index 1.5. The ratio of power of lens in air to its power in the liquid will be $x : 1$. The value of x is______.

[11 April 2023 (Morning)]

66. As shown in the figure, a plane mirror is fixed at a height of 50 cm from the bottom of tank containing water $\left(\mu = \frac{4}{3}\right)$. The height of water in the tank is 8 cm. A small bulb is placed at the bottom of the water tank. The distance of image of the bulb formed by mirror from the bottom of the tank is______cm.

[11 April 2023 (Evening)]

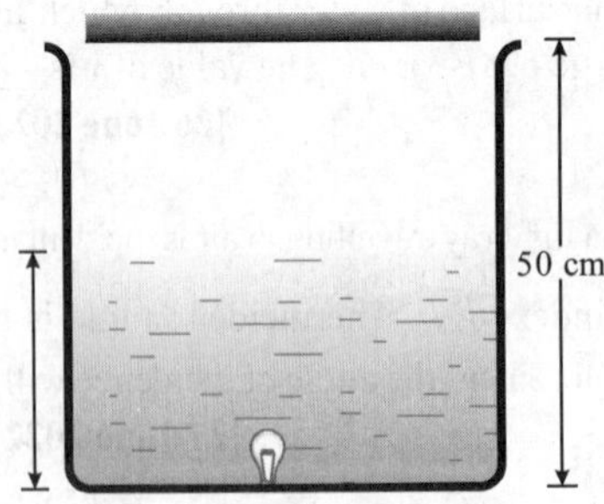

67. An ice cube has a bubble inside. When viewed from one side the apparent distance of the bubble is 12 cm. When viewed from the opposite side, the apparent distance of the bubble is observed as 4 cm. If the side of the ice cube is 24 cm, the refractive index of the ice cube is : **[12 April 2023 (Morning)]**

(A) $\frac{4}{3}$ (B) $\frac{3}{2}$

(C) $\frac{2}{3}$ (D) $\frac{6}{5}$

68. A vessel of depth d is half filled with oil of refractive index n_1 and the other half is filled with water of refractive index n_2. The apparent depth of this vessel when viewed from above will be : **[13 April 2023 (Morning)]**

(A) $\frac{d\, n_1 n_2}{(n_1 + n_2)}$ (B) $\frac{d\,(n_1 + n_2)}{2 n_1 n_2}$

(C) $\frac{d\, n_1 n_2}{2(n_1 + n_2)}$ (D) $\frac{2d\,(n_1 + n_2)}{n_1 n_2}$

69. A fish rising vertically upward with a uniform velocity of 8 ms^{-1}, observes that a bird is diving vertically downward towards the fish with the velocity of 12 ms^{-1}. If the refractive index of water is $\frac{4}{3}$, then the actual velocity of the diving bird to pick the fish, will be _____ ms^{-1}. **[13 April 2023 (Morning)]**

70. The refractive index of a transparent liquid filled in an equilateral hollow prism is $\sqrt{2}$. The angle of minimum deviation for the liquid will be _______°. **[15 April 2023 (Morning)]**

* * * * *

CHAPTER 23

RAY OPTICS AND OPTICAL INSTRUMENTS

23.3 GEOMETRICAL OPTICS III - THIN LENSES

1. A plano-convex lens of refractive index 1.5 and radius of curvature 30 cm is silvered at the curved surface. Now, this lens has been used to form the image of an object. At what distance from this lens, an object be placed in order to have a real image of the size of the object ? **[2004]**

(A) 20 cm (B) 30 cm
(C) 60 cm (D) 80 cm

2. A thin glass (refractive index 1.5) lens has optical power of – 5 D in air. Its optical power in a liquid medium with refractive index 1.6 will be : **[2005]**

(A) 1 D (B) – 1 D
(C) 25 D (D) – 25 D

3. Two lenses of power – 15 D and + 5 D are in contact with each other. The focal length of the combination is : **[2007]**

(A) – 20 cm (B) – 10 cm
(C) + 20 cm (D) + 10 cm

4. A student measures the focal length of a convex lens by putting an object pin at a distance 'u' from the lens and measuring the distance 'v' of the image pin. The graph between 'u' and 'v' plotted by the student should look like : **[2008]**

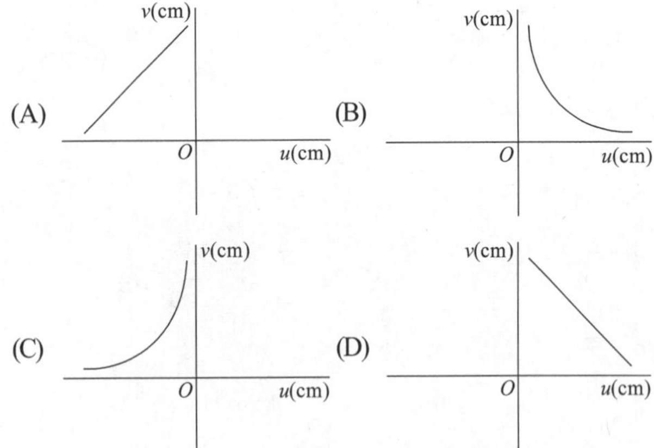

5. In an optics experiment, with the position of the object fixed, a student varies the position of a convex lens and for each position, the screen is adjusted to get a clear image of the object. A graph between the object distance u and the image distance v, from the lens, is plotted using the same scale for the two axes. A straight line passing through the origin and making an angle of 45° with the x-axis meets the experimental curve at P. The coordinates of P will be : **[2009]**

(A) $\left(\frac{f}{2}, \frac{f}{2}\right)$ (B) (f, f)
(C) $(4f, 4f)$ (D) $(2f, 2f)$

6. An object 2.4 m in front of a lens forms a sharp image on a film 12 cm behind the lens. A glass plate 1 cm thick, of refractive index 1.50 is interposed between lens and film with its plane faces parallel to film. At what distance (from lens) should object be shifted to be in sharp focus on film ? **[2012]**

(A) 5.6 m (B) 7.2 m
(C) 2.4 m (D) 3.2 m

7. Diameter of a plano-convex lens is 6 cm and thickness at the centre is 3 mm. If speed of light in material of lens is 2×10^8 m/s, the focal length of the lens is : **[2013]**

(A) 15 cm (B) 20 cm
(C) 30 cm (D) 10 cm

8. The image of an illuminated square is obtained on a screen with the help of a converging lens. The distance of the square from the lens is 40 cm. The area of the image is 9 times that of the square. The focal length of the lens is : **[2013 (Online)]**

(A) 36 cm (B) 27 cm
(C) 60 cm (D) 30 cm

9. A thin convex lens made from crown glass $\left(\mu = \frac{3}{2}\right)$ has focal length f. When it is measured in two different liquids having refractive indices 4/3 and 5/3, it has the focal lengths f_1 and f_2 respectively. The correct relation between the focal length is : **[2014]**

(A) $f_1 = f_2 < f$
(B) $f_1 > f$ and f_2 becomes negative
(C) $f_2 > f$ and f_1 becomes negative
(D) f_1 and f_2 both becomes negative

10. An object is located in a fixed position in front of a screen. Sharp image is obtained on the screen for two positions of a thin lens separated by 10 cm. The size of the images in two situations are in the ratio 3 : 2. What is the distance between the screen and the object ? **[2014 (Online)]**

(A) 124.5 cm (B) 144.5 cm
(C) 65.0 cm (D) 99.0 cm

11. The refractive index of the material of a concave lens is μ. It is immersed in a medium of refractive index μ_1. A parallel beam of light is incident on the lens. The path of the emergent rays when $\mu_1 > \mu$ is : **[2014 (Online)]**

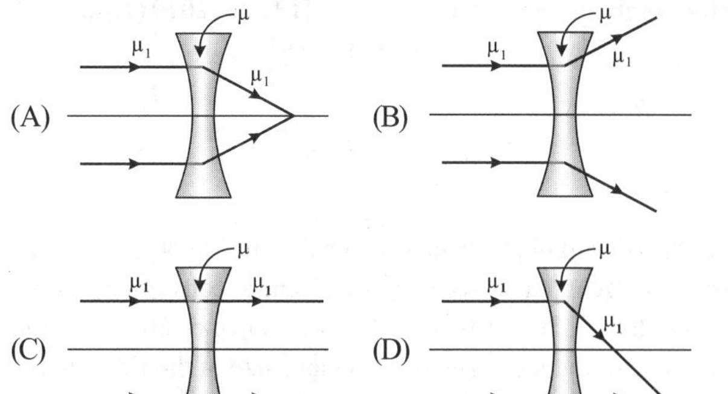

12. A thin convex lens of focal length 'f' is put on a plane mirror as shown in the figure. When an object is kept at a distance 'a' from the lens-mirror combination, its image is formed at a distance $a/3$ in front of the combination. The value of 'a' is : **[2015 (Online)]**

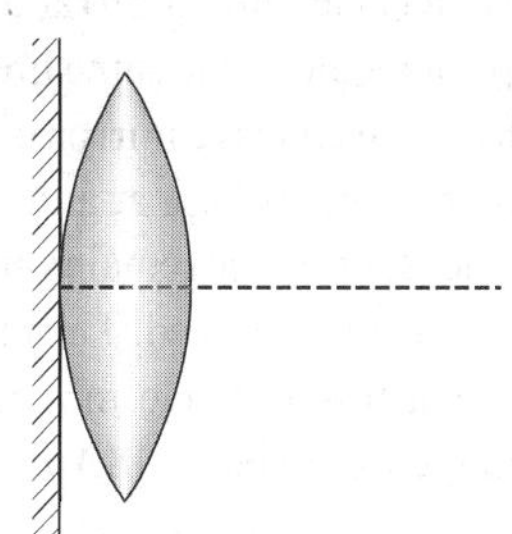

(A) $3f$ (B) $\frac{3}{2}f$
(C) f (D) $2f$

13. To find the focal length of a convex mirror, a student records the following data :

Object Pin	Convex Lens	Convex Mirror	Image Pin
22.2 cm	32.2 cm	45.8 cm	71.2 cm

The focal length of the convex lens is f_1 and that of mirror is f_2. Then taking index correction to be negligibly small, f_1 and f_2 are close to : **[2016 (Online)]**

(A) $f_1 = 7.8$ cm $f_2 = 12.7$ cm
(B) $f_1 = 12.7$ cm $f_2 = 7.8$ cm
(C) $f_1 = 15.6$ cm $f_2 = 25.4$ cm
(D) $f_1 = 7.8$ cm $f_2 = 25.4$ cm

14. A convex lens, of focal length 30 cm, a concave lens of focal length 120 cm, and a plane mirror are arranged as shown. For an object kept at a distance of 60 cm from the convex lens, the final image, formed by the combination, is a real image, at a distance of : **[2016 (Online)]**

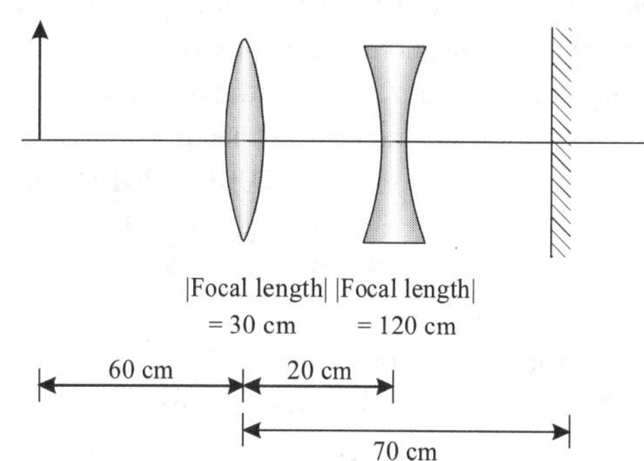

(A) 60 cm from the convex lens
(B) 60 cm from the concave lens
(C) 70 cm from the convex lens
(D) 70 cm from the concave lens

15. In an experiment a convex lens of focal length 15 cm is placed coaxially on an optical bench in front of a convex mirror at a distance of 5 cm from it. It is found that an object and its image coincide, if the object is placed at a distance of 20 cm from the lens. The focal length of the convex mirror is : **[2017 (Online)]**

(A) 27.5 cm (B) 20.0 cm
(C) 25.0 cm (D) 30.5 cm

16. A plano-convex lens becomes an optical system of 28 cm focal length when its plane surface is silvered and illuminated from left to right as shown in figure-A. If the same lens is instead silvered on the curved surface and illuminated from other side as in Figure-B, it acts like an optical system of focal length 10 cm. The refractive index of the material of lens if: **[2018 (Online)]**

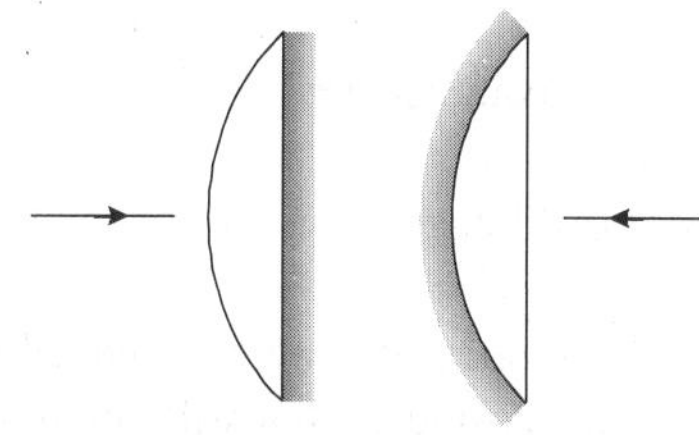

(A) 1.50 (B) 1.55
(C) 1.75 (D) 1.51

17. A convex lens is put 10 cm from a light source and it makes a sharp image on a screen, kept 10 cm from the lens. Now a glass block (refractive index 1.5) of 1.5 cm thickness is placed in contact with the light source. To get the sharp image again, the screen is shifted by a distance d. Then d is : **[9 Jan. 2019 (Morning)]**

(A) 0.55 cm away from the lens
(B) 1.1 cm away from the lens
(C) 0.55 cm towards the lens
(D) 0

18. A plano-convex lens of refractive index μ_1 and focal length f_1 is kept in contact with another plano-concave lens of refractive index μ_2 and focal length f_2. If the radius of curvature of their spherical faces is R each and $f_1 = 2f_2$, then μ_1 and μ_2 are related as : **[10 Jan. 2019 (Morning)]**

(A) $\mu_1 + \mu_2 = 3$ (B) $2\mu_1 - \mu_2 = 1$
(C) $2\mu_2 - \mu_1 = 1$ (D) $3\mu_2 - 2\mu_1 = 1$

19. An object is at a distance of 20 m from a convex lens of focal length 0.3 m. The lens forms an image of the object. If the object moves away from the lens at a speed of 5 m/s, the speed and direction of the image will be : **[11 Jan. 2019 (Morning)]**

(A) 0.92×10^{-3} m/s away from the lens
(B) 2.26×10^{-3} m/s away from the lens
(C) 1.16×10^{-3} m/s towards the lens
(D) 3.22×10^{-3} m/s towards the lens

20. What is the position and nature of image formed by lens combination shown in figure? (f_1, f_2 are focal lengths) : **[12 Jan. 2019 (Morning)]**

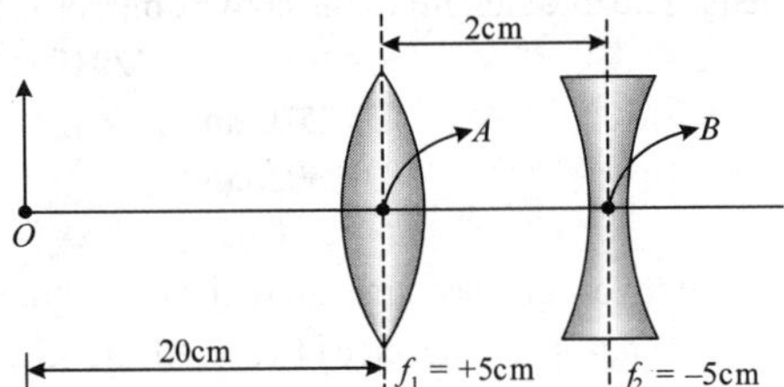

(A) 70 cm from point B at left, virtual
(B) 40 cm from point B at right, real
(C) $\frac{20}{3}$ cm from point B at right, real
(D) 70 cm from point B at right, real

21. Formation of real image using a biconvex lens is shown below.
If the whole set up is immersed in water without disturbing the object and the screen position, what will one observe on the screen ? **[12 Jan. 2019 (Evening)]**

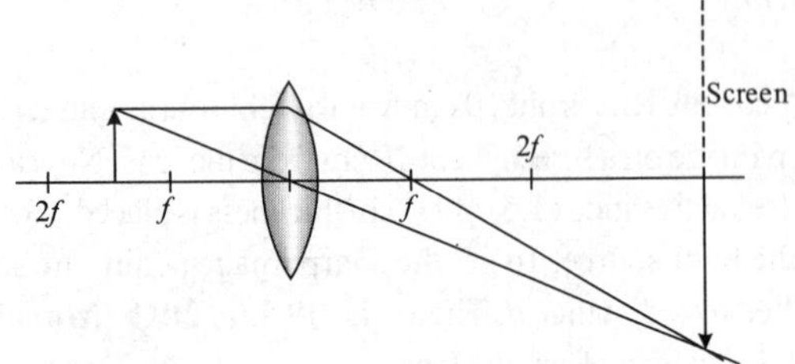

(A) Image disappears (B) No change
(C) Erect real image (D) Magnified image

22. A plano-convex lens (focal length f_2, refractive index μ_2, radius of curvature R) fits exactly into a plano-concave lens (focal length f_1, refractive index μ_1, radius of curvature R). Their plane surfaces are parallel to each other. Then, the focal length of the combination will be : **[12 Jan. 2019 (Evening)]**

(A) $f_1 - f_2$ (B) $f_1 + f_2$
(C) $\frac{R}{\mu_2 - \mu_1}$ (D) $\frac{2f_1 f_2}{f_1 + f_2}$

23. An upright object is placed at a distance of 40 cm in front of a convergent lens of focal length 20 cm. A convergent mirror of focal length 10 cm is placed at a distance of 60 cm on the other side of the lens. The position and size of the final image will be : **[8 April 2019 (Morning)]**

(A) 40 cm from the convergent mirror, same size as the object
(B) 20 cm from the convergent mirror, same size as the object
(C) 20 cm from the convergent mirror, twice the size of the object
(D) 40 cm from the convergent lens, twice the size of the object

24. A convex lens (of focal length 20 cm) and a concave mirror, having their principal axes along the same lines, are kept 80 cm apart from each other. The concave mirror is to the right of the convex lens. When an object is kept at a distance of 30 cm to the left of the convex lens, its image remains at the same position even if the concave mirror is removed. The maximum distance of the object for which this concave mirror, by itself would produce a virtual image would be : **[8 April 2019 (Evening)]**

(A) 20 cm (B) 10 cm
(C) 25 cm (D) 30 cm

25. A thin convex lens L (refractive index = 1.5) is placed on a plane mirror M. When a pin is placed at A, such that $OA = 18$ cm, its real inverted image is formed at A itself, as shown in figure. When a liquid of refractive index μ_1 is put between the lens and the mirror, The pin has to be moved to A', such that $OA' = 27$ cm, to get its inverted real image at A' itself. The value of μ_1 will be: **[9 April 2019 (Evening)]**

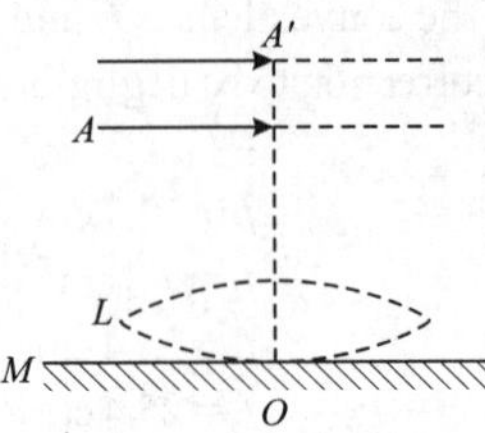

(A) $\sqrt{2}$ (B) $\frac{4}{3}$
(C) $\sqrt{3}$ (D) $\frac{3}{2}$

26. A convex lens of focal length 20 cm produces images of the same magnification 2 when an object is kept at two distances x_1 and $x_2 (x_1 > x_2)$ from the lens. The ratio of x_1 and x_2 is : **[9 April 2019 (Evening)]**

(A) 5 : 3 (B) 2 : 1
(C) 4 : 3 (D) 3 : 1

27. One plano-convex and one plano-concave lens of same radius of curvature 'R' but of different materials are joined side by side as shown in the figure. If the refractive index of the material of 1 is μ_1 and that of 2 is μ_2, Then the focal length of the combination is : **[10 April 2019 (Morning)]**

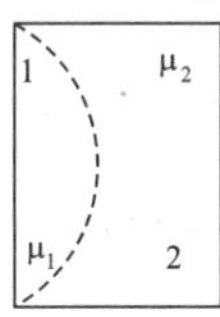

(A) $\frac{R}{2-(\mu_1-\mu_2)}$ (B) $\frac{2R}{\mu_1-\mu_2}$

(C) $\frac{R}{2(\mu_1-\mu_2)}$ (D) $\frac{R}{\mu_1-\mu_2}$

28. The graph shows how the magnification m produced by a thin lens varies with image distance v. What is the focal length of the lens used ? **[10 April 2019 (Evening)]**

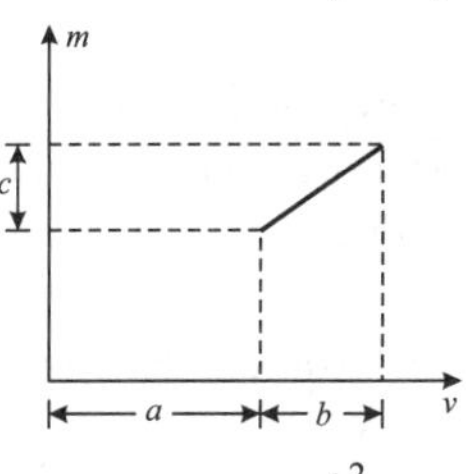

(A) $\frac{b^2c}{a}$ (B) $\frac{b^2}{ac}$

(C) $\frac{a}{c}$ (D) $\frac{b}{c}$

29. A thin lens made of glass (refractive index = 1.5) of focal length f = 16 cm is immersed in a liquid of refractive index 1.42. If its focal length in liquid is f_1, then the ratio f_1/f is closest to the integer : **[7 Jan. 2020 (Evening)]**

(A) 1 (B) 5
(C) 9 (D) 17

30. A point object in air is in front of the curved surface of a plano-convex lens. The radius of curvature of the curved surface is 30 cm and the refractive index of the lens material is 1.5, then the focal length of the lens (in cm) is ______. **[8 Jan. 2020 (Morning)]**

31. The distance between an object and a screen is 100 cm. A lens can produce real image of the object on the screen for two different positions between the screen and the object. The distance between these two positions is 40 cm. If the power of the lens is close to $\left(\frac{N}{100}\right)$D where N is an integer, the value of N is ______. **[4 Sept. 2020 (Evening)]**

32. For a concave lens of focal length f, the relation between object and image distance u and v, respectively, from its pole can best be represented by ($u = v$ is the reference line) : **[5 Sept. 2020 (Morning)]**

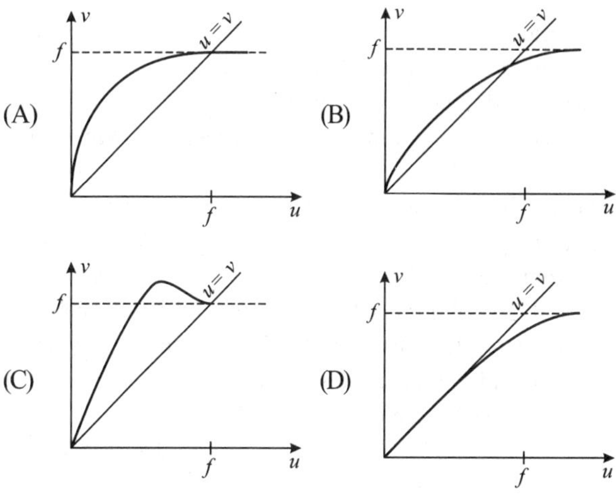

33. A point like object is placed at a distance of 1 m in front of a convex lens of focal length 0.5 m. A plane mirror is placed at a distance of 2 m behind the lens. The position and nature of the final image formed by the system is : **[6 Sept. 2020 (Morning)]**

(A) 1 m from the mirror, virtual
(B) 1 m from the mirror, real
(C) 2.6 m from the mirror, real
(D) 2.6 m from the mirror, virtual

34. An object is placed at the focus of concave lens having, focal length f. What is the magnification and distance of the image from the optical centre of the lens ? **[31 Aug. 2021 (Morning)]**

(A) 1/2, f/2 (B) 1/4, f/4
(C) 1, ∞ (D) very high, ∞

35. Two identical thin biconvex lenses of focal length 15 cm and refractive index 1.5 are in contact with each other. The space between the lenses is filled with a liquid of refractive index 1.25. The focal length of the combination is ______ cm. **[24 June 2022 (Morning)]**

36. A convex lens has power P. It is cut into two halves along its principal axis. Further one piece (out of the two halves) is cut into two halves perpendicular to the principal axis (as shown in figure). Choose the incorrect option for the reported pieces : **[27 June 2022 (Evening)]**

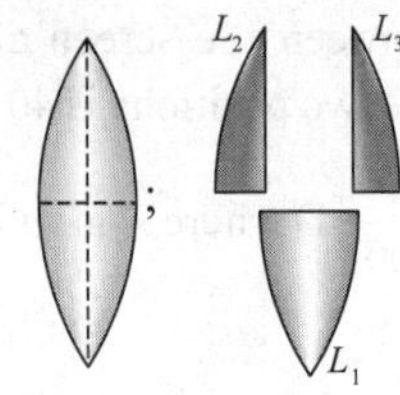

(A) Power of $L_1 = \frac{P}{2}$ (B) Power of $L_2 = \frac{P}{2}$

(C) Power of $L_3 = \frac{P}{2}$ (D) Power of $L_1 = P$

37. A parallel beam of light is allowed to fall on a transparent spherical globe of diameter 30 cm and refractive index 1.5. The distance from the centre of the globe at which the beam of light can converge is _____ mm. **[29 June 2022 (Morning)]**

38. For an object placed at a distance 2.4 m from a lens, a sharp focused image is observed on a screen placed at a distance 12 cm from the lens. A glass plate of refractive index 1.5 and thickness 1 cm is introduced between lens and screen such that the glass plate plane faces parallel to the screen. By what distance should the object be shifted so that a sharp focused image is observed again on the screen?

[25 July 2022 (Evening)]

(A) 0.8 m (B) 3.2 m
(C) 1.2 m (D) 5.6 m

39. A convex lens of focal length 20 cm is placed in front of convex mirror with principal axis coinciding each other. The distance between the lens and mirror is 10 cm. A point object is placed on principal axis at a distance of 60 cm from the convex lens. The image formed by combination coincides the object itself. The focal length of the convex mirror is _____ cm.

[25 July 2022 (Evening)]

40. The graph between $\frac{1}{u}$ and $\frac{1}{v}$ for a thin convex lens in order to determine its focal length is plotted as shown in the figure. The refractive index of length is 1.5 and its both the surfaces have same radius of curvatures R. The value of R will be _____ cm. (Where u = object distance, v = image distance)

[26 July 2022 (Morning)]

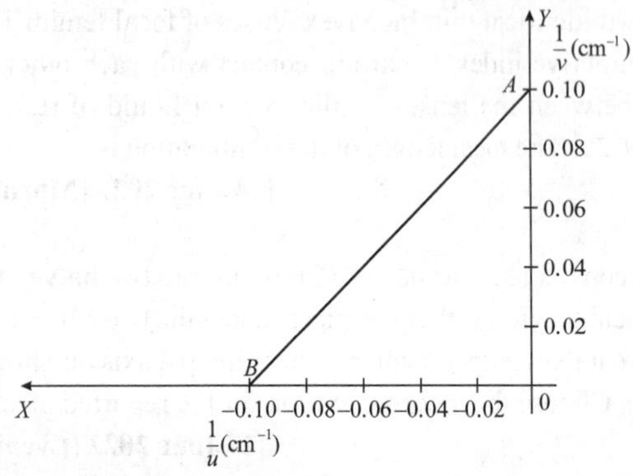

41. Light travels in two media M_1 and M_2 with speeds 1.5×10^8 ms^{-1} and 2.0×10^8 ms^{-1} respectively. The critical angle between them is : **[26 July 2022 (Evening)]**

(A) $\tan^{-1}\left(\frac{3}{\sqrt{7}}\right)$ (B) $\tan^{-1}\left(\frac{2}{3}\right)$

(C) $\cos^{-1}\left(\frac{3}{4}\right)$ (D) $\sin^{-1}\left(\frac{2}{3}\right)$

42. In the given figure, the face AC of the equilateral prism is immersed in a liquid of refractive index n. For incident angle 60° at the side AC, the refracted light beam just grazes along face AC. The refractive index of the liquid $n = \frac{\sqrt{x}}{4}$. The value of x is _____. [Given : Refractive index of glass = 1.5]

[26 July 2022 (Evening)]

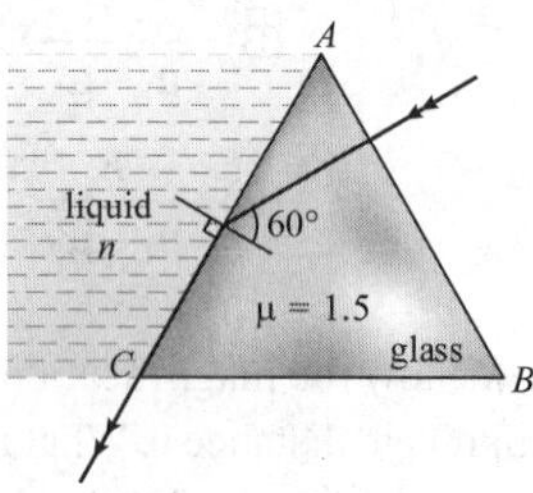

43. A thin prism of angle 6° and refractive index for yellow light (n_Y) 1.5 is combined with another prism of angle 5° and n_Y = 1.55. The combination produces no dispersion. The net average deviation (δ) produced by the combination is $\left(\frac{1}{x}\right)^\circ$. The value of x is _____. **[27 July 2022 (Evening)]**

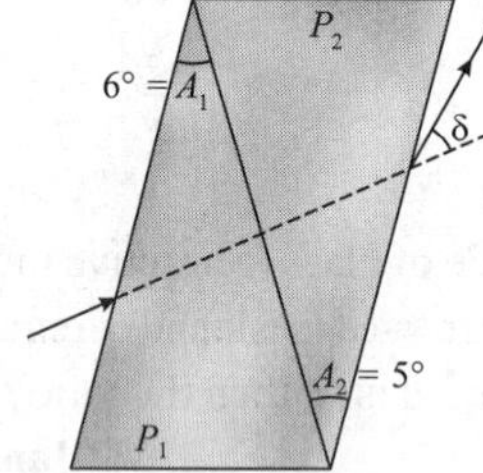

44. The power of a lens (biconvex) is 1.25 m^{-1} in particular medium. Refractive index of the lens is 1.5 and radii of curvature are 20 cm and 40 cm respectively. The refractive index of surrounding medium : **[28 July 2022 (Evening)]**

(A) 1.0 (B) $\frac{9}{7}$

(C) $\frac{3}{2}$ (D) $\frac{4}{3}$

45. In an experiment with a convex lens. The plot of the image distance (v') against the object distance (u') measured from the focus gives a curve $v'u' = 225$. If all the distances are measured in cm. The magnitude of the focal length of the lens is _____ cm. **[28 July 2022 (Evening)]**

46. As shown in the figure, a combination of a thin plano concave lens and a thin plano convex lens is used to image an object placed at infinity. The radius of curvature of both the lenses is 30 cm and refraction index of the material for both the lenses is 1.75. Both the lenses are placed at distance of 40 cm from each other. Due to the combination, the image of the object is formed at distance $x =$ _____ cm, from concave lens. **[24 Jan. 2023 (Morning)]**

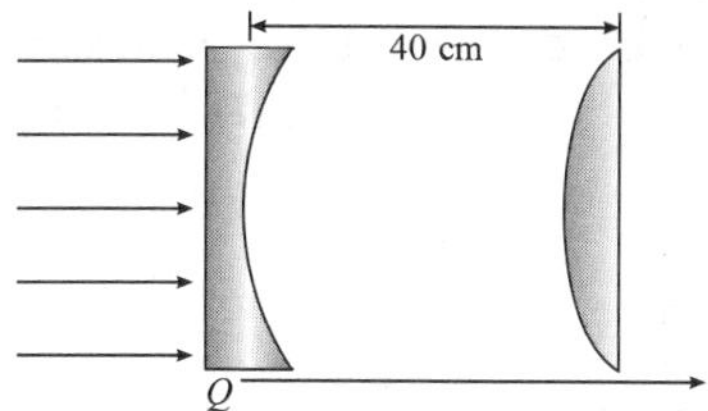

47. A convex lens of refractive index 1.5 and focal length 18 cm in air is immersed in water. The change in focal length of the lens will be _____ cm. (Given refractive index of water $= \frac{4}{3}$) **[24 Jan. 2023 (Evening)]**

48. An object is placed on the principal axis of convex lens of focal length 10 cm as shown. A plane mirror is placed on the other side of lens at a distance of 20 cm. The image produced by the plane mirror is 5 cm inside the mirror. The distance of the object from the lens is _____ cm. **[25 Jan. 2023 (Evening)]**

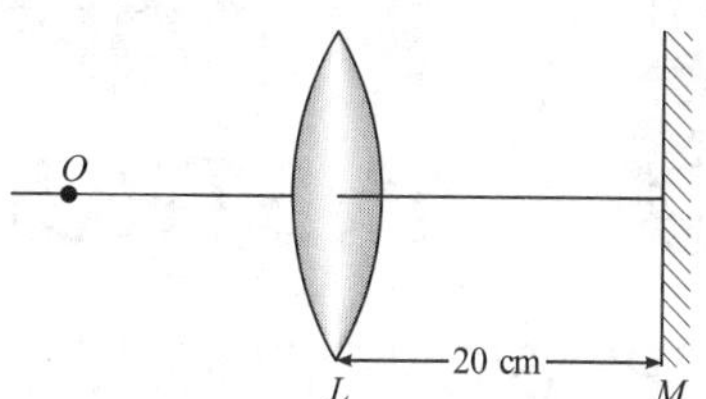

49. A point object, 'O' is placed in front of two thin symmetrical coaxial convex lenses L_1 and L_2 with focal length 24 cm and 9 cm respectively. The distance between two lenses is 10 cm and the object is placed 6 cm away from lens L_1 as shown in the figure. The distance between the object and the image formed by the system of two lenses is_____cm.

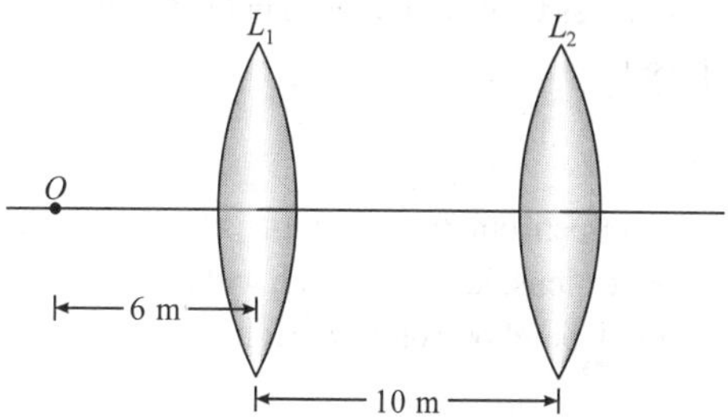

[10 April 2023 (Evening)]

50. Two convex lenses of focal length 20 cm each are placed coaxially with a separation of 60 cm between them. The image of the distant object formed by the combination is at _____ cm from the first lens. **[12 April 2023 (Morning)]**

51. A bi convex lens of focal length 10 cm is cut in two identical parts along a plane perpendicular to the principal axis. The power of each lens after cut is ______ D. **[13 April 2023 (Evening)]**

* * * * *

CHAPTER 23

RAY OPTICS AND OPTICAL INSTRUMENTS

23.4 DISPERSION OF LIGHT

1. The refractive index of glass is 1.520 for red light and 1.525 for blue light. Let D_1 and D_2 be angles of minimum deviation for red and blue light respectively in a prism of this glass. Then : **[2006]**

(A) $D_1 < D_2$
(B) $D_1 = D_2$
(C) D_1 can be less than or greater than D_2 depending upon the angle of prism
(D) $D_1 > D_2$

2. When monochromatic red light is used instead of blue light in a convex lens, its focal length will : **[2011]**

(A) Does not depend on colour of light
(B) Increase
(C) Decrease
(D) Remain same

3. Which of the following processes play a part in the formation of a rainbow ? **[2012 (Online)]**

(i) Refraction (ii) Total internal reflection
(iii) Dispersion (iv) Interference
(A) (i), (ii) and (iii) (B) (i) and (ii)
(C) (i), (ii) and (iv) (D) (iii) and (iv)

4. A beam of light consisting of red, green and blue colours is incident on a right-angled prism on face *AB*. The refractive indices of the material for the above red, green and blue colours are 1.39, 1.44 and 1.47 respectively. A person looking on surface *AC* of the prism will see : **[2012 (Online)]**

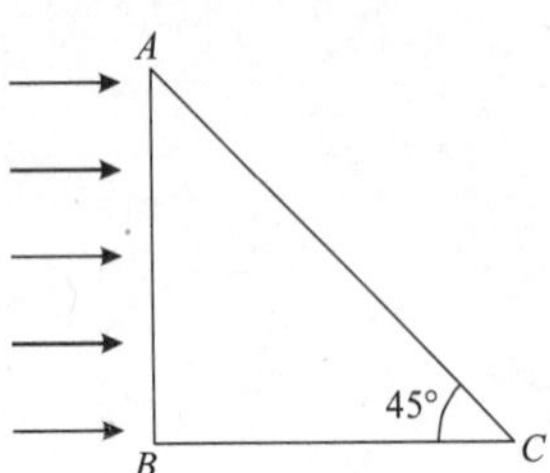

(A) No light (B) Green and blue colours
(C) Red and green colours (D) Red colour only

5. A convergent doublet of separated lenses, corrected for spherical aberration, has resultant focal length of 10 cm. The separation between the two lenses is 2 cm. The focal lengths of the component lenses : **[2018 (Online)]**

(A) 18 cm, 20 cm (B) 10 cm, 12 cm
(C) 12 cm, 14 cm (D) 16 cm, 18 cm

6. The variation of refractive index of a crown glass thin prism with wavelength of the incident light is shown. Which of the following graphs is the correct one, if D_m is the angle of minimum deviation? **[11 Jan. 2019 (Morning)]**

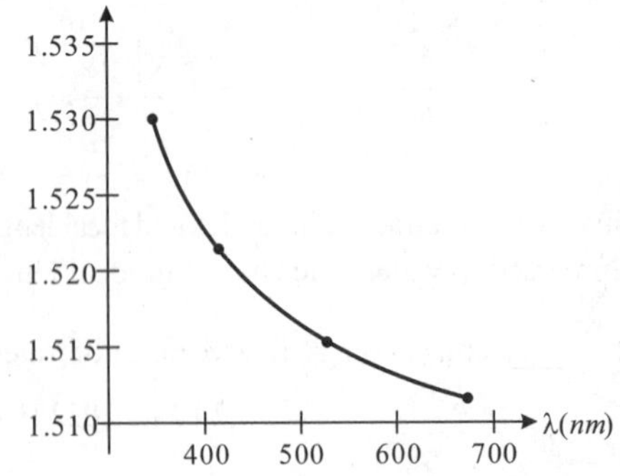

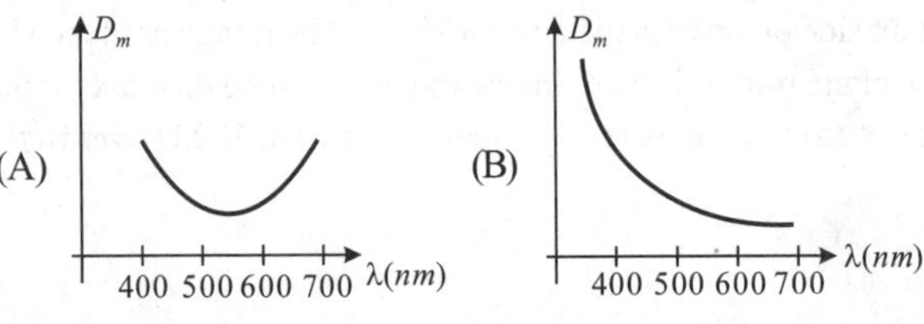

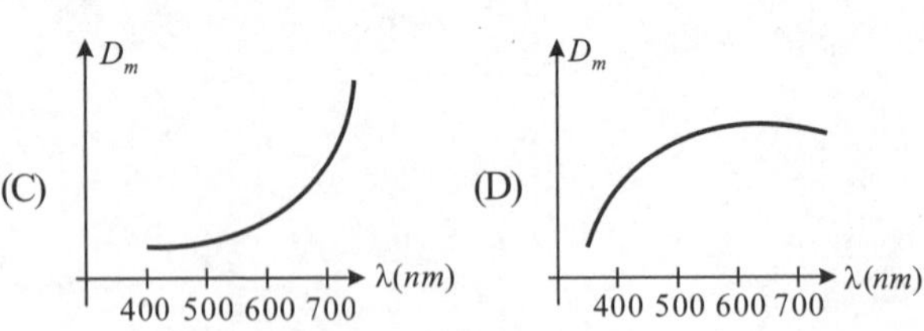

7. The angle of deviation through a prism is minimum when

(1) Incident ray and emergent ray are symmetric to the prism
(2) The refracted ray inside the prism becomes parallel to its base
(3) Angle of incidence is equal to that of the angle of emergence
(4) When angle of emergence is double the angle of incidence

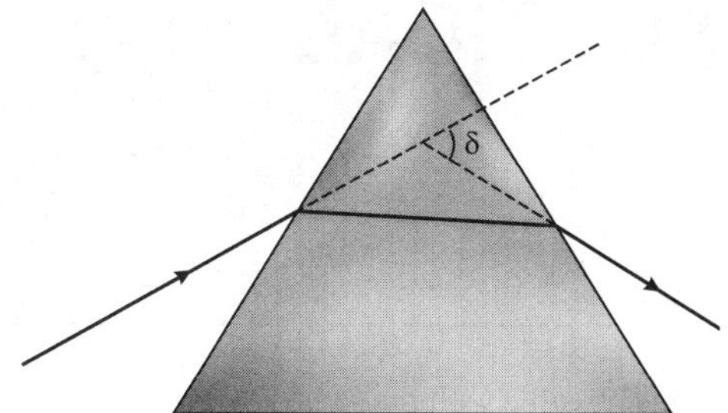

Choose the correct answer from the options given below :

[16 March 2021 (Morning)]

(A) Statements (1), (2) and (3) are true
(B) Only statement (4) is true
(C) Only statements (1) and (2) are true
(D) Statements (2) and (3) are true

8. A deviation of 2° is produced in the yellow ray when prism of crown and flint glass are achromatically combined. Taking dispersive powers of crown and flint glass are 0.02 and 0.03 respectively and refractive index for yellow light for these glasses are 1.5 and 1.6 respectively. The refracting angles for crown glass prism will be ______°. (in degree)
(Round off to the Nearest Integer) **[16 March 2021 (Evening)]**

9. A prism of refractive index μ and angle of prism A is placed in the position of minimum angle of deviation. If minimum angle of deviation is also A, then in terms of refractive index value of A is : **[25 July 2021 (Evening)]**

(A) $\sin^{-1}\left(\sqrt{\frac{\mu-1}{2}}\right)$ (B) $\sin^{-1}\left(\frac{\mu}{2}\right)$

(C) $\cos^{-1}\left(\frac{\mu}{2}\right)$ (D) $2\cos^{-1}\left(\frac{\mu}{2}\right)$

10. A prism of refractive index n_1 and another prism of refractive index n_2 are stuck together (as shown in the figure). n_1 and n_2 depend on λ, the wavelength of light, according to the relation

$$n_1 = 1.2 + \frac{10.8\times10^{-14}}{\lambda^2} \text{ and } n_2 = 1.45 + \frac{1.8\times10^{-14}}{\lambda^2}$$

The wavelength for which rays incident at any angle on the interface BC pass through without bending at that interface will be ______ nm. **[27 July 2021 (Morning)]**

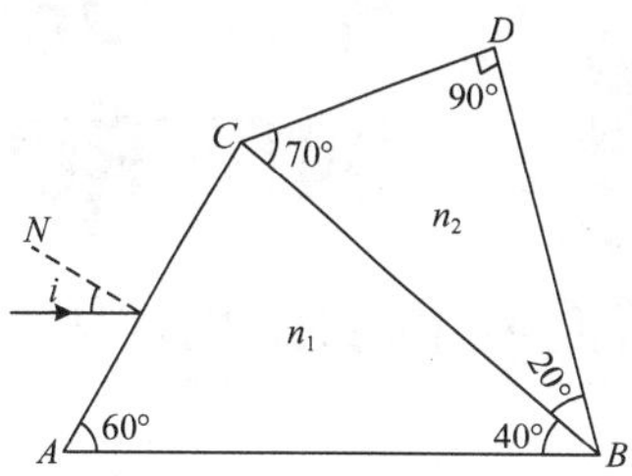

11. Cross-section view of a prism is the equilateral triangle ABC shown in the figure. The minimum deviation is observed using this prism when the angle of incidence is equal to the prism angle. The time taken by light to travel from P (mid point of BC) to A is ______ $\times 10^{-10}$ s. (Given, speed of light in vacuum $= 3 \times 10^8$ m/s and $\cos 30° = \frac{\sqrt{3}}{2}$) **[31 Aug. 2021 (Evening)]**

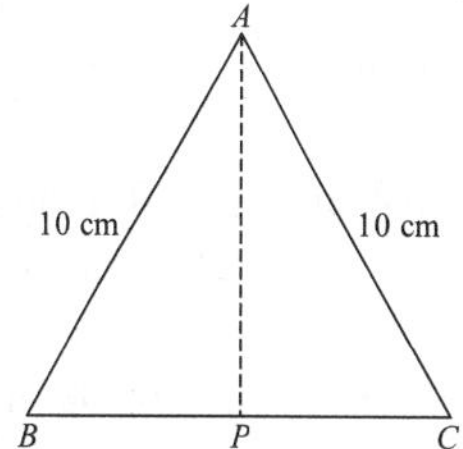

12. Which of the following statement is correct?

[25 July 2022 (Morning)]

(A) In primary rainbow, observer sees red colour on the top and violet on the bottom
(B) In primary rainbow, observer sees violet colour on the top and red on the bottom
(C) In primary rainbow, light wave suffers total internal reflection twice before coming out of water drops
(D) Primary rainbow is less bright than secondary rainbow

13. When a beam of white light is allowed to pass through convex lens parallel to principal axis, the different colours of light converge at different point on the principle axis after refraction. This is called : **[24 Jan. 2023 (Evening)]**

(A) Scattering (B) Chromatic aberration
(C) Spherical aberration (D) Polarisation

* * * * *

CHAPTER

23 RAY OPTICS AND OPTICAL INSTRUMENTS

23.5 OPTICAL INSTRUMENTS

1. An astronomical telescope has a large aperture to : **[2002]**
(A) Reduced spherical aberration
(B) Have high resolution
(C) Increase span of observation
(D) Have low dispersion

2. Wavelength of light used in a optical instrument are $\lambda_1 = 4000$ Å and $\lambda_2 = 5000$ Å, then ratio of their respective resolving powers (corresponding to λ_1 and λ_2) is : **[2002]**
(A) 16:25 (B) 9:1
(C) 4:5 (D) 5:4

3. The image formed by an objective of a compound microscope is : **[2003]**
(A) Virtual and diminished (B) Real and diminished
(C) Real and enlarged (D) Virtual and enlarged

4. Two point white dots are 1 mm apart on a black paper. They are viewed by eye of pupil diameter 3 mm. Approximately, what is the maximum distance at which these dots can be resolved by the eye ? [Take $\lambda = 500$ nm] : **[2005]**
(A) 5m (B) 1m
(C) 6m (D) 3m

5. An experiment is performed to find the refractive index of glass using a travelling microscope. In this experiment distances are measured by : **[2008]**
(A) A vernier scale provided on the microscope
(B) A standard laboratory scale
(C) A meter scale provided on the microscope
(D) A screws gauge provided on the microscope

6. We wish to make a microscope with the help of two positive lenses both with a focal length of 20 mm each and the object is positioned 25 mm from the objective lens. How far apart the lenses should be so that the final image is formed at infinity ? **[2012 (Online)]**
(A) 20mm (B) 100mm
(C) 120mm (D) 80mm

7. A telescope of aperture 3×10^{-2} m diameter is focused on a window at 80 m distance fitted with a wire mesh of spacing 2×10^{-3} m. Given : $\lambda = 5.5 \times 10^{-7}$ m, which of the following is true for observing the mesh through the telescope? **[2012 (Online)]**
(A) Yes, it is possible with the same aperture size
(B) Possible also with an aperture half the present diameter
(C) No, it is not possible
(D) Given data is not sufficient

8. The focal length of the objective and the eyepiece of a telescope are 50 cm and 5 cm respectively. If the telescope is focussed for distinct vision on a scale distant 2 m from its objective, then its magnifying power will be : **[2013 (Online)]**
(A) –4 (B) –8
(C) +8 (D) –2

9. This question has Statement-1 and Statement-2. Of the four choices given after the Statements, choose the one that best describes the two Statements. **[2013 (Online)]**

Statement-1 : Very large size telescopes are reflecting telescopes instead of refracting telescopes :

Statement-2 : It is easier to provide mechanical support to large size mirrors than large size lenses :
(A) Statement-1 is true and Statement-2 is false
(B) Statement-1 is false and Statement-2 is true
(C) Statement-1 and Statement-2 are true and Statement-2 is correct explanation for Statement-1
(D) Statement-1 and Statement-2 are true and Statement-2 is not the correct explanation for Statement-1

10. The focal lengths of objective lens and eye lens of a Galilean telescope are respectively 30 cm and 3.0 cm. Telescope produces virtual, erect image of an object situated far away from it at least distance of distinct vision from the eye lens. In this condition, the magnifying power of the Galilean telescope should be : **[2014 (Online)]**
(A) +11.2 (B) –11.2
(C) –8.8 (D) +8.8

11. In a compound microscope, the focal length of objective lens is 1.2 cm and focal length of eye piece is 3.0 cm. When object is kept at 1.25 cm in front of objective, final image is formed at infinity. Magnifying power of the compound microscope should be : **[2014 (Online)]**
(A) 200 (B) 100
(C) 400 (D) 150

12. Assuming human pupil to have a radius of 0.25 cm and a comfortable viewing distance of 25 cm, the minimum separation between two objects that human eye can resolve at 500 nm wavelength is : **[2015]**

(A) 1 μm (B) 30 μm
(C) 100 μm (D) 300 μm

13. A telescope has an objective lens of focal length 150 cm and an eyepiece of focal length 5 cm. If a 50 m tall tower at a distance of 1 km is observed through this telescope in normal setting, the angle formed by the image of the tower is θ, then θ is close to : **[2015 (Online)]**

(A) 30° (B) 15°
(C) 60° (D) 1°

14. An observer looks at a distant tree of height 10 m with a telescope of magnifying power of 20. To the observer the tree appears : **[2016]**

(A) 10 times taller (B) 10 times nearer
(C) 20 times taller (D) 20 times nearer

15. To determine refractive index of glass slab using a travelling microscope, minimum number of readings required are : **[2016 (Online)]**

(A) Two (B) Four
(C) Three (D) Five

16. Calculate the limit of resolution of a telescope objective having a diameter of 200 cm, if it has to detect light of wavelength 500 nm coming from a star : **[8 April 2019 (Evening)]**

(A) 305×10^{-9} radian (B) 152.5×10^{-9} radian
(C) 610×10^{-9} radian (D) 457.5×10^{-9} radian

17. Diameter of the objective lens of a telescope is 250 cm. For light of wavelength 600 nm coming from a distant object, the limit of resolution of the telescope is close to : **[9 April 2019 (Evening)]**

(A) 1.5×10^{-7} rad (B) 2.0×10^{-7} rad
(C) 3.0×10^{-7} rad (D) 4.5×10^{-7} rad

18. The value of numerical aperture of the objective lens of a microscope is 1.25. If light of wavelength 5000 Å is used, the minimum separation between two points, to be seen as distinct, will be : **[12 April 2019 (Morning)]**

(A) 0.24 μm (B) 0.48 μm
(C) 0.12 μm (D) 0.38 μm

19. If we need a magnification of 375 from a compound microscope of tube length 150 mm and an objective of focal length 5 mm, the focal length of the eye-piece, should be close to : **[7 Jan. 2020 (Morning)]**

(A) 22mm (B) 12mm
(C) 33mm (D) 2mm

20. The magnifying power of a telescope with tube 60 cm is 5. What is the focal length of its eye piece ? **[8 Jan. 2020 (Morning)]**

(A) 30 cm (B) 40 cm
(C) 20 cm (D) 10 cm

21. The aperture diameter of a telescope is 5 m. The separation between the moon and the earth is 4×10^5 km. With light of wavelength of 5500 Å, the minimum separation between objects on the surface of moon, so that they are just resolved, is close to : **[9 Jan. 2020 (Morning)]**

(A) 20 m (B) 600 m
(C) 60 m (D) 200 m

22. In a compound microscope, the magnified virtual image is formed at a distance of 25 cm from the eye-piece. The focal length of its objective lens is 1 cm. If the magnification is 100 and the tube length of the microscope is 20 cm, then the focal length of the eye-piece lens (in cm) is ______ . **[4 Sept. 2020 (Morning)]**

23. A compound microscope consists of an objective lens of focal length 1 cm and an eye piece of focal length 5 cm with a separation of 10 cm. The distance between an object and the objective lens, at which the strain on the eye is minimum is $\frac{n}{40}$ cm. The value of n is : **[5 Sept. 2020 (Morning)]**

24. Given below are two statements : one is labeled as **Assertion A** and the other is labeled as **Reason R**

Assertion-A : For a simple microscope, the angular size of the object equals the angular size of the image

Reason-R : Magnification is achieved as the small object can be kept much closer to the eye than 25 cm and hence it subtends a large angle

In the light of the above statements, choose the most appropriate answer from the options given below : **[26 Feb. 2021 (Evening)]**

(A) **A** is true but **R** is false
(B) Both **A** and **R** are true and **R** is the correct explanation of **A**
(C) Both **A** and **R** are true but **R** is NOT the correct explanation of **A**
(D) **A** is false but **R** is true

25. Your friend is having eye sight problem. She is not able lo see clearly a distant uniform window mesh and it appears to her as non-uniform and distorted. The doctor diagnosed the problem as : **[18 March 2021 (Morning)]**

(A) Astigmatism
(B) Myopia with Astigmatism
(C) Presbyopia with Astigmatism
(D) Myopia and hypermetropia

26. The speed of electrons in a scanning electron microscope is 1×10^7 ms^{-1}. If the protons having the same speed are used instead of electrons, then the resolving power of scanning proton microscope will be changed by a factor of : **[18 March 2021 (Evening)]**

(A) 1837 (B) $\frac{1}{1837}$

(C) $\sqrt{1837}$ (D) $\frac{1}{\sqrt{1837}}$

27. An object viewed from a near point distance of 25 cm, using a microscopic lens with magnification '6', gives an unresolved image. A resolved image is observed at infinite distance with a total magnification double the earlier using an eyepiece along with the given lens and a tube of length 0.6 m, if the focal length of the eyepiece is equal to ______ cm. **[20 July 2021 (Morning)]**

28. The aperture of the objective is 24.4 cm. The resolving power of this telescope. If a light of wavelength 2440 Å is used to see the object will be : **[28 June 2022 (Morning)]**

(A) 8.1×10^6 (B) 10.0×10^7

(C) 8.2×10^5 (D) 1.0×10^{-8}

29. A microscope was initially placed in air (refractive index 1). It is then immersed in oil (refractive index 2). For a light whose wavelength in air is λ, calculate the change of microscope's resolving power due to oil and choose the correct option : **[27 July 2022 (Morning)]**

(A) Resolving power will be $\frac{1}{4}$ in the oil than it was in the air

(B) Resolving power will be twice in the oil than it was in the air

(C) Resolving power will be four times in the oil than it was in the air

(D) Resolving power will be $\frac{1}{2}$ in the oil than it was in the air

30. In normal adjustment, for a refracting telescope, the distance between objective and eye piece is 30 cm. The focal length of the objective, when the angular magnification of the telescope is 2, will be : **[28 July 2022 (Morning)]**

(A) 20 cm (B) 30 cm

(C) 10 cm (D) 15 cm

31. A person has been using spectacles of power 1.0 diopter for distant vision and a separate reading glass of power 2.0 diopters. What is the least distance of distinct vision for this person : **[30 Jan. 2023 (Morning)]**

(A) 10 cm (B) 40 cm

(C) 30 cm (D) 50 cm

32. In a reflecting telescope, a secondary mirror is used to: **[8 April 2023 (Morning)]**

(A) Reduce the problem of mechanical support

(B) Remove spherical aberration

(C) Make chromatic aberration zero

(D) Move the eyepiece outside the telescopic tube

* * * * *

CHAPTER

24 WAVE OPTICS

24.1 INTERFERENCE OF LIGHT

1. To demonstrate the phenomenon of interference we require two sources which emit radiations of : **[2003]**
(A) nearly the same frequency
(B) the same frequency
(C) different wavelength
(D) the same frequency and having a definite phase relationship

2. The maximum number of possible interference maxima for slit-separation equal to twice the wavelength in Young's double-slit experiment, is : **[2004]**
(A) infinite
(B) five
(C) three
(D) zero

3. A Young's double slit experiment uses a monochromatic source. The shape of the interference fringes formed on a screen is : **[2005]**
(A) hyperbola
(B) circle
(C) straight line
(D) parabola

4. In a Young's double slit experiment the intensity at a point where the path difference is $\lambda/6$ (λ being the wavelength of the light used) is I. If I_0 denotes the maximum intensity, I/I_0 is equal to : **[2007]**

(A) $\frac{1}{\sqrt{2}}$
(B) $\frac{\sqrt{3}}{2}$
(C) $\frac{1}{2}$
(D) $\frac{3}{4}$

5. A mixture of light, consisting of wavelength 590 nm and an unknown wavelength, illuminates Young's double slit and gives rise to two overlapping interference patterns on the screen. The central maximum of both lights coincide. Further, it is observed that the third bright fringe of known light coincides with the 4th bright fringe of the unknown light. From this data, the wavelength of the unknown light is : **[2009]**
(A) 442.5 nm
(B) 776.8 nm
(C) 393.4 nm
(D) 885.0 nm

Direction : Below questions from 6 to 8 are based on the following paragraph.
An initially parallel cylindrical beam travels in a medium of refractive index $\mu(I) = \mu_0 + \mu_2 I$, where μ_0 and μ_2 are positive constants and I is the intensity of the light beam. The intensity of the beam is decreasing with increasing radius :

6. The initial shape of the wavefront of the beam is : **[2010]**
(A) planar
(B) convex
(C) concave
(D) convex near the axis and concave near the periphery

7. The speed of the light in the medium is : **[2010]**
(A) maximum on the axis of the beam
(B) minimum on the axis of the beam
(C) the same everywhere in the beam
(D) directly proportional to the intensity I

8. As the beam enters the medium, it will : **[2010]**
(A) travel as a cylindrical beam
(B) diverge
(C) converge
(D) diverge near the axis and converge near the periphery

9. Direction : The question has a paragraph followed by two statement, Statement-1 and statement-2. Of the given four alternatives after the statements, choose the one that describes the statements.
A thin air film is formed by putting the convex surface of a plane-convex lens over a plane glass plate. With monochromatic light, this film gives an interference pattern due to light reflected from the top (convex) surface and the bottom (glass plate) surface of the film
Statement-1 : When light reflects from the air-glass plate interface, the reflected wave suffers a phase change of π.
Statement-2 : The centre of the interference pattern is dark. **[2011]**
(A) Statement-1 is true, Statement-2 is true and Statement-2 is not the correct explanation of Statement-1
(B) Statement-1 is false, Statement-2 is true
(C) Statement-1 is true, Statement-2 is false
(D) Statement-1 is true, Statement-2 is true and Statement-2 is the correct explanation of Statement-1

10. In a Young's double slit experiment, the two slits act as coherent sources of waves of equal amplitude A and wavelength λ. In another experiment with the same arrangement the two slits are made to act as incoherent sources of waves of same amplitude and wavelength. If the intensity at the middle point of the screen in the first case is I_1 and in the second case I_2, then the ratio I_1/I_2 is : **[2011]**

(A) 4 (B) 2
(C) 1 (D) 0.5

11. At two points P and Q on screen in Young's double slit experiment, waves from slits S_1 and S_2 have a path difference of 0 and $\lambda/4$ respectively. The ratio of intensities at P and Q will be : **[2011]**

(A) 3 : 2 (B) 2 : 1
(C) $\sqrt{2}:1$ (D) 4 : 1

12. In Young's double slit experiment, one of the slit is wider than other, so that the amplitude of the light from one slit is double of that from other slit. If I_m be the maximum intensity, the resultant intensity I when they interfere at phase difference ϕ is given by : **[2012]**

(A) $\frac{I_m}{9}\left(1+8\cos^2\frac{\phi}{2}\right)$ (B) $\frac{I_m}{9}(4+5\cos\phi)$

(C) $\frac{I_m}{9}\left(1+2\cos^2\frac{\phi}{2}\right)$ (D) $\frac{I_m}{5}\left(1+4\cos^2\frac{\phi}{2}\right)$

13. In a Young's double slit experiment with light of wavelength λ, fringe pattern on the screen has fringe width β. When two thin transparent glass (refractive index μ) plates of thickness t_1 and t_2 $(t_1 > t_2)$ are placed in the path of the two beams respectively, the fringe pattern will shift by a distance : **[2012 (Online)]**

(A) $\frac{\beta(\mu-1)}{\lambda}\left(\frac{t_1}{t_2}\right)$ (B) $\frac{\mu\beta}{\lambda}\frac{t_1}{t_2}$

(C) $\frac{\beta(\mu-1)}{\lambda}(t_1-t_2)$ (D) $(\mu-1)\frac{\lambda}{\beta}(t_1+t_2)$

14. The maximum number of possible interference maxima for slit separation equal to 1.8 λ, where λ is the wavelength of light used, in a Young's double slit experiment is : **[2012 (Online)]**

(A) Zero (B) 3
(C) Infinite (D) 5

15. In Young's double slit interference experiment, the slit widths are in the ratio 1 : 25. Then the ratio of intensity at the maxima and minima in the interference pattern is : **[2012 (Online)]**

(A) 3 : 2 (B) 1 : 25
(C) 9 : 4 (D) 1 : 5

16. Two coherent plane light waves of equal amplitude makes a small angle $\alpha(<< 1)$ with each other. The fall almost normally on a screen. If λ is the wavelength of light waves, the fringe width Δx of interference patterns of the two sets of waves on the screen is : **[2012 (Online)]**

(A) $\frac{2\lambda}{\alpha}$ (B) $\frac{\lambda}{\alpha}$

(C) $\frac{\lambda}{(2\alpha)}$ (D) $\frac{\lambda}{\sqrt{\alpha}}$

17. Two coherent point sources S_1 and S_2 are separated by a small distance 'd' as shown. The fringes obtained on the screen will be : **[2013]**

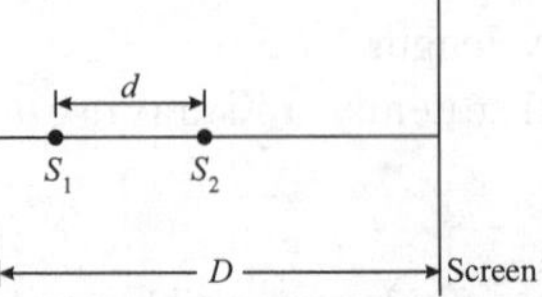

(A) points (B) straight lines
(C) semi-circles (D) concentric circles

18. This question has Statement-1 and Statement-2. Of the four choices given after the Statements, choose the one that best describes the two Statements.

Statement-1 : In young's double slit experiment, the number of fringes observed in the field of view is small with longer wavelength of light and is large with shorter wavelength of light.

Statement-2 : In the double slit experiment the fringe width depends directly on the wavelength of light. **[2013 (Online)]**

(A) Statement-1 is true, Statement-2 is true and the Statement-2 is correct explanation of the Statement-1
(B) Statement-1 is false and the Statement-2 is true
(C) Statement-1 is true Statement-2 is true and the Statement-2 is not correct explanation of the Statement-1
(D) Statement-1 is true and the Statement-2 is false

19. The source that illuminates the double-slit in 'double-slit interference experiment' emits two distinct monochromatic waves of wavelength 500 nm and 600 nm, each of them producing its own pattern on the screen. At the central point of the pattern when path difference is zero, maxima of both the patterns coincide and the resulting interference pattern is most distinct at the region of zero path difference. But as one moves out of this central region, the two fringe systems are gradually out of step such that maximum due to on wavelength coincides with the minimum due to the other and the combined fringe system becomes completely indistinct. This may happen when path difference in *nm* is : **[2013 (Online)]**

(A) 2000 (B) 3000
(C) 1000 (D) 1500

20. A ray of light of intensity I is incident on a parallel glass slab at point A as shown in diagram. It undergoes partial reflection and refraction. At each reflection, 25% of incident energy is reflected. The rays AB and $A'B'$ undergo interference. The ratio of I_{max} and I_{min} is : **[2013 (Online)]**

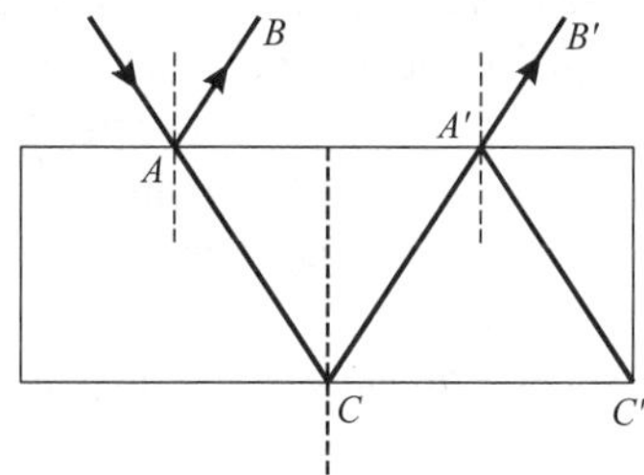

(A) 49 : 1 (B) 7 : 1
(C) 4 : 1 (D) 8 : 1

21. n identical waves each of intensity I_0 interfere with each other. The ratio of maximum intensities if the interference is (i) coherent and (ii) incoherent is : **[2013 (Online)]**

(A) n^2 (B) $\frac{1}{n}$

(C) $\frac{1}{n^2}$ (D) n

22. Two monochromatic light beams of intensity 16 and 9 units are interfering. The ratio of intensities of bright and dark parts of the resultant pattern is : **[2014 (Online)]**

(A) $\frac{16}{9}$ (B) $\frac{4}{3}$

(C) $\frac{7}{1}$ (D) $\frac{49}{1}$

23. On a hot summer night, the refractive index of air is smallest near the ground and increase with height from the ground. When a light beam is directed horizontally, the Huygen's principle leads us to conclude that as it travels, the light beam : **[2015]**

(A) becomes narrower
(B) goes horizontally without any deflection
(C) bends downwards
(D) bends upwards

24. In a Young's double slit experiment with light of wavelength λ the separation of slits is d and distance of screen is D such that $D >> d >> \lambda$. If the fringe width is β, the distance from point of maximum intensity to the point where intensity falls to half of maximum intensity on either side is : **[2015 (Online)]**

(A) $\frac{\beta}{6}$ (B) $\frac{\beta}{3}$

(C) $\frac{\beta}{4}$ (D) $\frac{\beta}{2}$

25. In a Young's double slit experiment, slits are separated by 0.5 mm, and the screen is placed 150 cm away. A beam of light consisting of two wavelengths, 650 nm and 520 nm, is used to obtain interference fringes on the screen. The least distance from the common central maximum to the point where the bright fringes due to both the wavelengths coincide is : **[2017]**

(A) 9.75 mm (B) 15.6 mm
(C) 1.56 mm (D) 7.8 mm

26. Two coherent sources produce waves of different intensities which interfere. After interference, the ratio of the maximum intensity to the minimum intensity is 16. The intensity of the waves are in the ratio : **[9 Jan. 2019 (Morning)]**

(A) 4 : 1 (B) 25 : 9
(C) 16 : 9 (D) 5 : 3

27. In a Young's double slit experiment, the slits are placed 0.320 mm apart. Light of wavelength $\lambda = 500$ nm is incident on the slits. The total number of bright fringes that are observed in the angular range $-30° \leq \theta \leq 30°$ is : **[9 Jan. 2019 (Evening)]**

(A) 320 (B) 641
(C) 321 (D) 640

28. In a Young's double slit experiment with slit separation 0.1 mm, one observes a bright fringe at angle 1/40 rad by using light of wavelength λ_1. When the light of wavelength λ_2 is used a bright fringe is seen at the same angle in the same set up. Given that λ_1 and λ_2 are in visible range (380 nm to 740 nm), their values are : **[10 Jan. 2019 (Morning)]**

(A) 380 nm, 500 nm (B) 625 nm, 500 nm
(C) 380 nm, 525 nm (D) 400 nm, 500 nm

29. Consider a Young's double slit experiment as shown in figure. What should be the slit separation d in terms of wavelength λ such that the first minima occurs directly in front of the slit (S_1) ? **[10 Jan. 2019 (Evening)]**

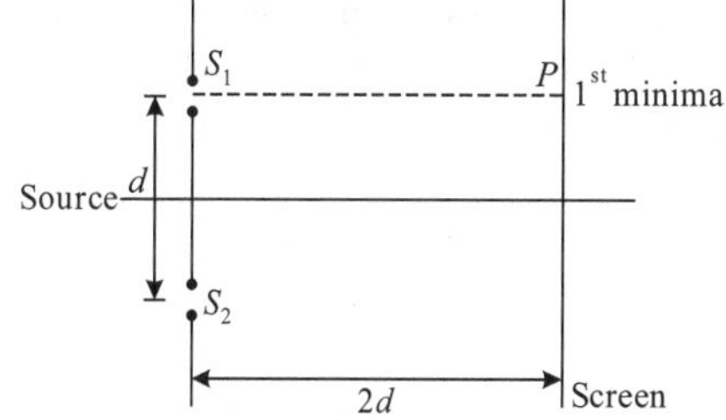

(A) $\frac{\lambda}{2(5-\sqrt{2})}$ (B) $\frac{\lambda}{(5-\sqrt{2})}$

(C) $\frac{\lambda}{(\sqrt{5}-2)}$ (D) $\frac{\lambda}{2(\sqrt{5}-2)}$

30. In a Young's double slit experiment, the path difference, at a certain point on the screen, between two interfering waves is 1/8th of wavelength. The ratio of the intensity at this point to that at the centre of a bright fringe is close to : **[11 Jan. 2019 (Morning)]**

(A) 0.94 (B) 0.74
(C) 0.85 (D) 0.80

31. In a double-slit experiment, green light (5303 Å) falls on a double slit having a separation of 19.44 μm and a width of 4.05 μm. The number of bright fringes between the first and the second diffraction minima is : **[11 Jan. 2019 (Evening)]**

(A) 09 (B) 10
(C) 04 (D) 05

32. In an interference experiment the ratio of amplitudes of coherent waves is $\frac{a_1}{a_2}=\frac{1}{3}$. The ratio of maximum and minimum intensities of fringes will be : **[8 April 2019 (Morning)]**

(A) 4 (B) 2
(C) 9 (D) 18

33. The figure shows a Young's double slit experimental setup. It is observed that when a thin transparent sheet of thickness t and refractive index μ is put in front of one of the slits, the central maximum gest shifted by a distance equal to n fringe widths. If the wavelength of light used is λ, t will be : **[9 April 2019 (Morning)]**

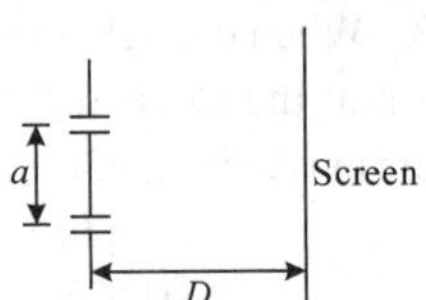

(A) $\frac{2D\lambda}{a(\mu-1)}$ (B) $\frac{D\lambda}{a(\mu-1)}$
(C) $\frac{2nD\lambda}{a(\mu-1)}$ (D) $\frac{nD\lambda}{a(\mu-1)}$

34. In a Young's double slit experiment, the ratio of the slit's width is 4 : 1. The ratio of the intensity of maxima to minima, close to the central fringe on the screen, will be : **[10 April 2019 (Evening)]**

(A) $(\sqrt{3}+1)^4:16$ (B) $9:1$
(C) $4:1$ (D) $25:9$

35. In a double slit experiment, when a thin film of thickness t having refractive index μ is introduced in front of one of the slits, the maximum at the centre of the fringe pattern shifts by one fringe width. The value of t is (λ is the wavelength of the light used) : **[12 April 2019 (Morning)]**

(A) $\frac{\lambda}{2(\mu-1)}$ (B) $\frac{\lambda}{(2\mu-1)}$
(C) $\frac{2\lambda}{(\mu-1)}$ (D) $\frac{\lambda}{(\mu-1)}$

36. In a Young's double slit experiment, the separation between the slits is 0.15 mm. In the experiment, a source of light of wavelength 589 nm is used and the interference pattern is observed on a screen kept 1.5 m away. The separation between the successive bright fringes on the screen is : **[7 Jan. 2020 (Evening)]**

(A) 6.9 mm (B) 5.9 mm
(C) 4.9 mm (D) 3.9 mm

37. In a double slit experiment, at a certain point on the screen the path difference between the two interfering waves is $\frac{1}{8}$ th of a wavelength. The ratio of the intensity of light at that point to that at the centre of a bright fringe is : **[8 Jan. 2020 (Evening)]**

(A) 0.568 (B) 0.672
(C) 0.760 (D) 0.853

38. Three harmonic waves having equal frequency ν and same intensity I_0, have phase angles 0, $\frac{\pi}{4}$ and $-\frac{\pi}{4}$ respectively. When they are superimposed the intensity of the resultant wave is close to : **[9 Jan. 2020 (Morning)]**

(A) $5.8\,I_0$ (B) $0.2\,I_0$
(C) I_0 (D) $3\,I_0$

39. In a Young's double slit experiment 15 fringes are observed on a small portion of the screen when light of wavelength 500 nm is used. Ten fringes are observed on the same section of the screen when another light source of wavelength λ is used. Then the value of λ is (in nm)______. **[9 Jan. 2020 (Evening)]**

40. Interference fringes are observed on a screen by illuminating two thin slits 1 mm apart with a light source (l = 632.8 nm). The distance between the screen and the slits is 100 cm. If a bright fringe is observed on a screen at a distance of 1.27 mm from the central bright fringe, then the path difference between the waves, which are reaching this point from the slits is close to : **[2 Sept. 2020 (Morning)]**

(A) 1.27 mm (B) 2 nm
(C) 2.87 nm (D) 2.05 mm

41. In a Young's double slit experiment, 16 fringes are observed in a certain segment of the screen when light of wavelength 700 nm is used. If the wavelength of light is changed to 400 nm, the number of fringes observed in the same segment of the screen would be : **[2 Sept. 2020 (Evening)]**

(A) 28 (B) 24
(C) 18 (D) 30

42. In a Young's double slit experiment, light of 500 nm is used to produce an interference pattern. When the distance between the slits is 0.05 mm, the angular width (in degree) of the fringes formed on the distance screen is close to :

[3 Sept. 2020 (Morning)]

(A) 0.07° (B) 0.17°
(C) 1.7° (D) 0.57°

43. Two light waves having the same wavelength λ in vacuum are in phase initially. Then the first wave travels a path L_1 through a medium of refractive index n_1 while the second wave travels a path of length L_2 through a medium of refractive index n_2. After this the phase difference between the two waves is : **[3 Sept. 2020 (Evening)]**

(A) $\frac{2\pi}{\lambda}(n_1L_1 - n_2L_2)$ (B) $\frac{2\pi}{\lambda}\left(\frac{L_2}{n_1} - \frac{L_1}{n_2}\right)$

(C) $\frac{2\pi}{\lambda}\left(\frac{L_1}{n_1} - \frac{L_2}{n_2}\right)$ (D) $\frac{2\pi}{\lambda}(n_2L_1 - n_1L_2)$

44. Two coherent sources of sound, S_1 and S_2, produce sound waves of the same wavelength, $\lambda = 1$ m, in phase. S_1 and S_2 are placed 1.5 m apart (see figure) A listener, located at L, directly in front of S_2 finds that the intensity is at a minimum when he is 2 m away from S_2. The listener moves away from S_1, keeping his distance from S_2 fixed. The adjacent maximum of intensity is observed when the listener is at a distance d from S_1. Then, d is : **[5 Sept. 2020 (Evening)]**

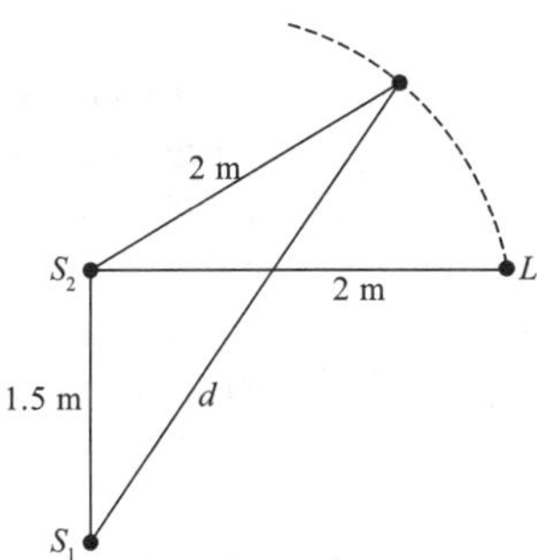

(A) 12m (B) 3m
(C) 5m (D) 2m

45. In the figure below, P and Q are two equally intense coherent sources emitting radiation of wavelength 20 m. The separation between P and Q is 5 m and the phase of P is ahead of that of Q by 90°. A, B and C are three distinct points of observation, each equidistant from the midpoint of PQ. The intensities of radiation at A, B, C will be in the ratio : **[6 Sept. 2020 (Morning)]**

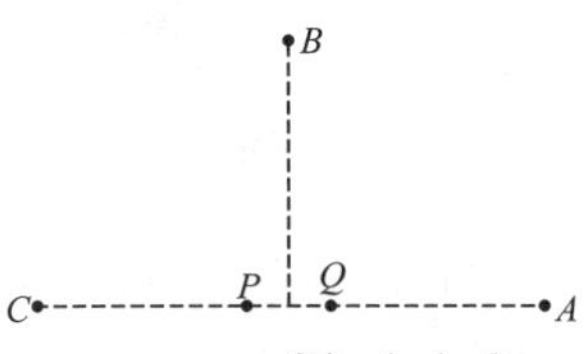

(A) 0 : 1 : 2 (B) 4 : 1 : 0
(C) 0 : 1 : 4 (D) 2 : 1 : 0

46. A Young's double-slit experiment is performed using monochromatic light of wavelength l. The intensity of light at a point on the screen, where the path difference is l, is K units. The intensity of light at a point where the path difference is $\frac{\lambda}{6}$ is given by $\frac{nK}{12}$, where n is an integer. The value of n is :

[6 Sept. 2020 (Evening)]

47. In a Young's double slit experiment, the width of the one of the slit is three times the other slit. The amplitude of the light coming from a slit is proportional to the slit-width. Find the ratio of the maximum to the minimum intensity in the interference pattern : **[24 Feb. 2021 (Morning)]**

(A) 2 : 1 (B) 4 : 1
(C) 3 : 1 (D) 1 : 4

48. If the source of light used in a Young's double slit experiment is changed from red to violet : **[24 Feb. 2021 (Evening)]**

(A) The intensity of minima will increase
(B) The fringes will become brighter
(C) The central bright fringe will become a dark fringe
(D) Consecutive fringe lines will come closer

49. Two coherent light sources having intensity in the ratio $2x$ produce an interference pattern. The ratio $\frac{I_{max} - I_{min}}{I_{max} + I_{min}}$ will be :

[25 Feb. 2021 (Morning)]

(A) $\frac{2\sqrt{2x}}{2x+1}$ (B) $\frac{\sqrt{2x}}{x+1}$

(C) $\frac{\sqrt{2x}}{2x+1}$ (D) $\frac{2\sqrt{2x}}{x+1}$

50. In a Young's double slit experiment two slits are separated by 2 mm and the screen is placed one meter away. When a light of wavelength 500 nm is used, the fringe separation will be :

[26 Feb. 2021 (Morning)]

(A) 0.75 mm (B) 0.25 mm
(C) 1 mm (D) 0.50 mm

51. A fringe width of 6 mm was produced for two slits separated by 1 mm apart. The screen is placed 10 m away. The wavelength of light used is 'x' nm. The value of 'x' to the nearest integer is ______. **[16 March 2021 (Morning)]**

52. In the Young's double slit experiment, the distance between the slits varies in time as $d(t) = d_0 + a_0 \sin \omega t$; where d_0, ω and a_0 are constants. The difference between the largest fringe width and the smallest fringe width obtained over time is given as :

[25 July 2021 (Morning)]

(A) $\frac{2\lambda D a_0}{(d_0^2 - a_0^2)}$ (B) $\frac{2\lambda D(d_0)}{(d_0^2 - a_0^2)}$

(C) $\frac{\lambda D}{d_0 - a_0}$ (D) $\frac{\lambda D}{d_0^2} a_0$

53. White light is passed through a double slit and interference is observed on a screen 1.5 m away. The separation between the slits is 0.3 mm. The first violet and red fringes are formed 2.0 mm and 3.5 mm away from the central white fringes. The difference in wavelength of red and violet light is______ nm : **[26 Aug. 2021 (Morning)]**

54. The light waves from two coherent sources have same intensity $I_1 = I_2 = I_0$. In interference pattern the intensity of light at minima is zero. What will be the intensity of light at maxima? **[27 Aug. 2021 (Evening)]**

(A) I_0 (B) $2I_0$
(C) $4I_0$ (D) $5I_0$

55. The width of one of the two slits in a Young's double slit experiment is three times the other slit. If the amplitude of the light coming from a slit is proportional to the slit-width, the ratio of maximum to minimum intensity in the interference pattern is $x : 4$ where x is______. **[01 Sep. 2021 (Evening)]**

56. Sodium light of wavelength 650 nm and 655 nm is used to study diffraction at a single slit of aperture 0.5 mm. The distance between the slit and the screen is 2.0 m. The separation between the positions of the first maximum of diffraction pattern obtained in the two cases is _____ $\times 10^{-5}$ m. **[24 June 2022 (Morning)]**

57. Two light beams of intensities in the ratio of 9 : 4 are allowed to interfere. The ratio of the intensity of maxima and minima will be : **[24 June 2022 (Evening)]**

(A) 2 : 3 (B) 16 : 81
(C) 25 : 169 (D) 25 : 1

58. The two light beams having intensities I and $9I$ interfere to produce a fringe pattern on a screen. The phase difference between the beams is $\frac{\pi}{2}$ at point P and π at point Q. Then the difference between the resultant intensities at P and Q will be : **[25 June 2022 (Morning)]**

(A) $2I$ (B) $6I$
(C) $5I$ (D) $7I$

59. The interference pattern is obtained with two coherent light sources of intensity ratio 4 : 1. And the ratio $\frac{I_{max} + I_{min}}{I_{max} - I_{min}}$ is $\frac{5}{x}$. Then, the value of x will be equal to : **[25 June 2022 (Evening)]**

(A) 3 (B) 4
(C) 2 (D) 1

60. In Young's double slit experiment the two slits are 0.6 mm distance apart. Interference pattern is observed on a screen at a distance 80 cm from the slits. The first dark fringe is observed on the screen directly opposite to one of the slits. The wavelength of light will be _____ nm. **[27 June 2022 (Morning)]**

61. In young's double slit experiment performed using a monochromatic light of wavelength λ, when a glass plate ($\mu = 1.5$) of thickness $x\lambda$ is introduced in the path of the one of the interfering beams, the intensity at the position where the central maximum occurred previously remains unchanged. The value of x will be : **[28 June 2022 (Evening)]**

(A) 3 (B) 2
(C) 1.5 (D) 0.5

62. In a Young's double slit experiment, an angular width of the fringe is 0.35° on a screen placed at 2 m away for particular wavelength of 450 nm. The angular width of the fringe, when whole system is immersed in a medium of refractive index 7/5, is $\frac{1}{\alpha}$. The value of α is_____. **[28 June 2022 (Evening)]**

63. Using Young's double slit experiment, a monochromatic light of wavelength 5000 Å produces fringes of fringe width 0.5 mm. If another monochromatic light of wavelength 6000 Å is used and the separation between the slits is doubled, then the new fringe width will be : **[29 June 2022 (Morning)]**

(A) 0.5 mm (B) 1.0 mm
(C) 0.6 mm (D) 0.3 mm

64. In a double slit experiment with monochromatic light, fringes are obtained on a screen placed at some distance from the plane of slits. If the screen is moved by 5×10^{-2} m towards the slits, the change in fringe width is 3×10^{-3} cm. If the distance between the slits is 1 mm, then the wavelength of the light will be _____ nm. **[29 June 2022 (Evening)]**

65. In young's double slit experiment, the fringe width is 12 mm. If the entire arrangement is placed in water of refractive index $\frac{4}{3}$, then the fringe width becomes : (in mm)

[26 July 2022 (Morning)]

(A) 16 (B) 9
(C) 48 (D) 12

66. Two beams of light having intensities I and $4I$ interfere to produce a fringe pattern on a screen. The phase difference between the two beams are $\pi/2$ and $\pi/3$ at points A and B

respectively. The difference between the resultant intensities at the two points is xI. The value of x will be _____.

[27 July 2022 (Morning)]

67. Two coherent sources of light interfere. The intensity ratio of two sources is 1 : 4. For this interference pattern if the value of $\frac{I_{max}+I_{min}}{I_{max}-I_{min}}$ is equal to $\frac{2\alpha+1}{\beta+3}$, then $\frac{\alpha}{\beta}$ will be :

[27 July 2022 (Evening)]

(A) 1.5 (B) 2
(C) 0.5 (D) 1

68. In a Young's double slit experiment, a laser light of 560 nm produces an interference pattern with consecutive bright fringes' separation of 7.2 mm. Now another light is used to produce an interference pattern with consecutive bright fringes' separation of 8.1 mm. The wavelength of second light is _____ nm.

[28 July 2022 (Morning)]

69. Two light beams of intensities $4I$ and $9I$ interfere on a screen. The phase difference between these beams on the screen at point A is zero and at point B is π. The difference of resultant intensities, at the point A and B, will be _____ I.

[29 July 2022 (Morning)]

70. Nearly 10% of the power of a 110 W light bulb is converted to visible radiation. The change in average intensities of visible radiation, at a distance of 1 m from the bulb to a distance of 5 m is $a \times 10^{-2}$ W/m^2. The value of a will be.

[29 July 2022 (Evening)]

71. In Young's double slits experiment, the position of 5th bright fringe from the central maximum is 5 cm. The distance between slits and screen is 1 m and wavelength of used monochromatic light is 600 nm. The separation between the slits is :

[25 Jan. 2023 (Morning)]

(A) 60 μm (B) 48 μm
(C) 12 μm (D) 36 μm

72. In a Young's double slit experiment, two slits are illuminated with a light of wavelength 800 nm. The line joining A_1P is perpendicular to A_1A_2 as shown in the figure. If the first minimum is detected at P, the value of slits separation a will be :

[29 Jan. 2023 (Morning)]

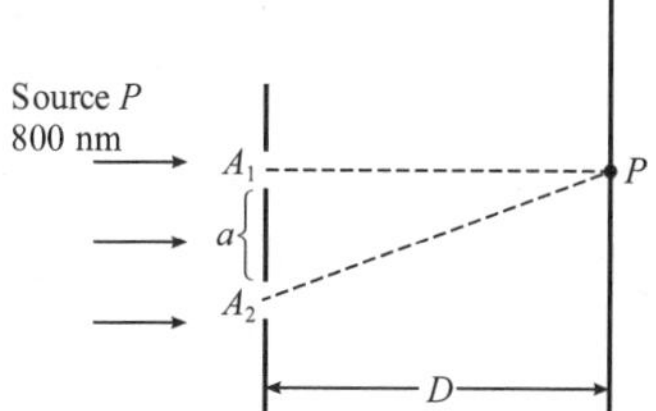

The distance of screen from slits $D = 5$ cm :

(A) 0.4 mm (B) 0.5 mm
(C) 0.2 mm (D) 0.1 mm

73. In Young's double slit experiment, two slits S_1 and S_2 are d distance apart and the separation from slits to screen is D (as shown in figure). Now if two transparent slabs of equal thickness 0.1 mm but refractive index 1.51 and 1.55 are introduced in the path of beam (λ = 4000 Å) from S_1 and S_2 respectively. The central bright fringe spot will shift by _____ number of fringes.

[30 Jan. 2023 (Morning)]

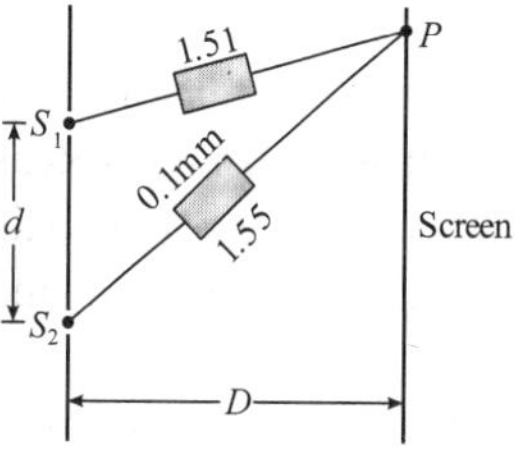

74. In a Young's double slit experiment, the intensities at two points, for the path difference $\frac{\lambda}{4}$ and $\frac{\lambda}{3}$ (λ being the wavelength of light used) are I_1 and I_2 respectively. If I_0 denotes the intensity produced by each one of the individual slits, then $\frac{I_1+I_2}{I_0}$ = _____.

[30 Jan. 2023 (Evening)]

75. Two light waves of wavelengths 800 and 600 nm are used in Young's double slit experiment to obtain interference fringes on a screen placed 7 m away from plane of slits. If the two slits are separated by 0.35 mm, then shortest distance from the central bright maximum to the point where the bright fringes of the two wavelength coincide will be _____ mm.

[31 Jan. 2023 (Evening)]

76. The displacement equations of two interfering waves are given by

$$y_1 = 10 \sin\left(\omega t + \frac{\pi}{3}\right) \text{ cm},$$

$$y_2 = 5[\sin(\omega t) + \sqrt{3}\cos\omega t] \text{ cm}$$

respectively. The amplitude of the resultant wave is _____ cm.

[31 Jan. 2023 (Evening)]

77. As shown in the figure, in Young's double slit experiment, a thin plate of thickness $t = 10$ μm and refractive index $\mu = 1.2$ is inserted infront of slit S_1. The experiment is conducted in air ($\mu = 1$) and uses a monochromatic light of wavelength $\lambda = 500$ nm. Due to the insertion of the plate, central maxima is shifted by a distance of $x\beta_0 \cdot \beta_0$ is the fringe-width before the insertion of the plate. The value of the x is_____. **[1 Feb. 2023 (Evening)]**

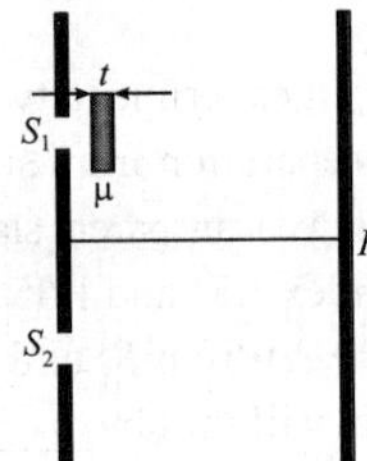

78. Given below are two statements: one is labelled as **Assertion-A** and the other is labelled as **Reason-R**.

Assertion-A: The phase difference of two light waves change if they travel through different media having same thickness, but different indices of refraction.

Reason-R: The wavelengths of waves are different in different media.

In the light of the above statements, choose the most appropriate answer from the options given below:

[6 April 2023 (Evening)]

(A) Both **A** and **R** are correct but **R** is **NOT** the correct explanation of **A**.

(B) **A** is correct but **R** is not correct.

(C) Both **A** and **R** are correct and **R** is the correct explanation of **A**.

(D) **A** is not correct but **R** is correct.

79. A beam of light consisting of two wavelengths 7000 Å and 5500 Å is used to obtain interference pattern in Young's double slit experiment. The distance between the slits is 2.5 mm and the distance between the plane of slits and the screen is 150 cm. The least distance from the central fringe, where the bright fringes due to both the wavelengths coincide, is $n \times 10^{-5}$ m. The value of n is_____. **[6 April 2023 (Evening)]**

80. The width of fringe is 2 mm on the screen in a double slits experiment for the light of wavelength of 400 nm. The width of the fringe for the light of wavelength 600 nm will be :

[8 April 2023 (Evening)]

(A) 4mm (B) 1.33 mm
(C) 3mm (D) 2mm

81. The ratio of intensities at two points P and Q on the screen in a Young's double slit experiment where phase difference between two wave of same amplitude are $\pi/3$ and $\pi/2$, respectively are: **[10 April 2023 (Evening)]**

(A) 1 : 3 (B) 3 : 1
(C) 3 : 2 (D) 2 : 3

82. If 917 Å be the lowest wavelength of Lyman series then the lowest wavelength of Balmer series will be______Å.

[10 April 2023 (Evening)]

83. In a Young's Double Slits Experiment, the ratio of amplitude of light coming from slits is 2 : 1. The ratio of the maximum to minimum intensity in the interference pattern is:

[13 April 2023 (Evening)]

(A) 9 : 4 (B) 9 : 1
(C) 2 : 1 (D) 25 : 9

84. A single slit of width a is illuminated by a monochromatic light of wavelength 600 nm. The value of a for which first minimum appears at $\theta = 30°$ on the screen will be :

[15 April 2023 (Morning)]

(A) 0.6 μm (B) 1.2 μm
(C) 1.8 μm (D) 3 μm

* * * * *

CHAPTER

24 WAVE OPTICS

24.2 DIFFRACTION OF LIGHT

1. If I_0 is the intensity of the principle maximum in the single slit diffraction pattern, then what will be its intensity when the slit width is doubled : **[2005]**

(A) $2I_0$ (B) $4I_0$
(C) I_0 (D) $I_0/2$

2. The first diffraction minimum due to the single slit diffraction is seen at $\theta = 30°$ for a light of wavelength 5000 Å falling perpendicularly on the slit. The width of the slit is : **[2012 (Online)]**

(A) 2.5×10^{-5} cm (B) 1.25×10^{-5} cm
(C) 10×10^{-5} cm (D) 5×10^{-5} cm

3. This question has Statement-1 and Statement-2. Of the four choices given after the Statements, choose the one that best describes the two Statements. **[2013 (Online)]**

Statement-1 : Out of radio waves and microwaves, the radio waves undergo more diffraction.

Statement-2 : Radio waves have greater frequency compared to microwaves.

(A) Statement-1 is true, Statement-2 is true and Statement-2 is the correct explanation of Statement-1
(B) Statement-1 is false, Statement-2 is true
(C) Statement-1 is true, Statement-2 is false
(D) Statement-1 is true, Statement-2 is true but Statement-2 is not the correct explanation of Statement-1

4. The diameter of the objective lens of microscope makes an angle β at the focus of the microscope. Further, the medium between the object and the lens is an oil of refractive index n. Then the resolving power of the microscope: **[2014 (Online)]**

(A) Increases with decreasing value of n
(B) Increases with decreasing value of β
(C) Increases with increasing value of $n \sin 2\beta$
(D) Increases with increasing value of $\frac{1}{n \sin 2\beta}$

5. In a Young's double slit experiment, the distance between the two identical slits is 6.1 times larger than the slit width. Then the number of intensity maxima observed within the central maximum of the single slit diffraction pattern is : **[2014 (Online)]**

(A) 3 (B) 6
(C) 12 (D) 24

6. In an experiment of single slit diffraction pattern, first minimum for red light coincides with first maximum of some other wavelength. If wavelength of red light is 6600 Å, then wavelength of first maximum will be : **[2014 (Online)]**

(A) 3300 Å (B) 4400 Å
(C) 5500 Å (D) 6600 Å

7. The box of a pin hole camera, of length L, has a hole of radius a. It is assumed that when the hole is illuminated by a parallel beam of light of wavelength λ the spread of the spot (obtained on the opposite wall of the camera) is the sum of its geometrical spread and the spread due to diffraction. The spot would then have its minimum size (say b_{min}) when : **[2016]**

(A) $a = \frac{\lambda^2}{L}$ and $b_{min} = \left(\frac{2\lambda^2}{L}\right)$

(B) $a = \sqrt{\lambda L}$ and $b_{min} = \left(\frac{2\lambda^2}{L}\right)$

(C) $a = \sqrt{\lambda L}$ and $b_{min} = \sqrt{4\lambda L}$

(D) $a = \frac{\lambda^2}{L}$ and $b_{min} = \sqrt{4\lambda L}$

8. In Young's double slit experiment, the distance between slits and the screen is 1.0 m and monochromatic light of 600 nm is being used. A person standing near the slits is looking at the fringe pattern. When the separation between the slits is varied, the interference pattern disappears for a particular distance d_0 between the slits. If the angular resolution of the eye is 1°/60, the value of d_0 is close to : **[2016 (Online)]**

(A) 1mm (B) 3mm
(C) 2mm (D) 4mm

9. Two stars are 10 light years away from the earth. They are seen through a telescope of objective diameter 30 cm. The wavelength of light is 600 nm. To see the stars just resolved by the telescope, the minimum distance between them should be (1 light year = 9.46×10^{15} m) of the order of: **[2016 (Online)]**

(A) 10^8 km (B) 10^{10} km
(C) 10^{11} km (D) 10^6 km

10. A single slit of width 0.1 mm is illuminated by a parallel beam of light of wavelength 6000 Å and diffraction bands are observed on a screen 0.5 m from the slit. The distance of the third dark band from the central bright band is : **[2017 (Online)]**

(A) 3mm (B) 9mm
(C) 4.5mm (D) 1.5mm

11. A single slit of width b is illuminated by a coherent monochromatic light of wavelength λ. If the second and fourth minima in the diffraction pattern at a distance 1 m from the slit are at 3 cm and 6 cm respectively from the central maximum, wheat is the width of the central maximum? (i.e. distance between first minimum on either side of the central maximum) : **[2017 (Online)]**

(A) 1.5 cm (B) 3.0 cm
(C) 4.5 cm (D) 6.0 cm

12. The angular width of the central maximum in a single slit diffraction pattern is 60°. The width of the slit is 1 μm. The slit is illuminated by monochromatic plane waves. If another slit of same width is made near it, Young's fringes can be observed on a screen placed at a distance 50 cm from the slits. If the observed fringe width is 1 cm, what is slit separation distance ? (i.e. distance between the centres of each slit) **[2018]**

(A) 25 μm (B) 50 μm
(C) 75 μm (D) 100 μm

13. Light of wavelength 550 nm falls normally on a slit of width 22.0×10^{-5} cm. The angular position of the second minima from the central maximum will be (in radians) : **[2018 (Online)]**

(A) $\frac{\pi}{8}$ (B) $\frac{\pi}{12}$

(C) $\frac{\pi}{4}$ (D) $\frac{\pi}{6}$

14. Visible light of wavelength 6000×10^{-8} cm falls normally on a single slit and produces a diffraction pattern. It is found that the second diffraction minimum is at 60° from the central maximum. If the first minimum is produced at θ_1, then θ_1 is close to : **[7 Jan. 2020 (Morning)]**

(A) 20° (B) 45°
(C) 30° (D) 25°

15. Orange light of wavelength 6000×10^{-10} m in illuminates a single slit of width 0.6×10^{-4} m. The maximum possible number of diffraction minima produced on both sides of the central maximum is ______. **[4 Sept. 2020 (Evening)]**

16. A beam of electrons of energy E scatters from a target having atomic spacing of 1 Å. The first maximum intensity occurs at $\theta = 60°$. Then E (in eV) is ______.
(Planck constant $h = 6.64 \times 10^{-34}$ Js, 1 eV = 1.6×10^{-19} J, electron mass $m = 9.1 \times 10^{-31}$ kg) : **[5 Sept. 2020 (Morning)]**

17. Consider the diffraction pattern obtained from the sunlight incident on a pinhole of diameter 0.1 μm. If the diameter of the pinhole is slightly increased, it will affect the diffraction pattern such that : **[25 Feb. 2021 (Evening)]**

(A) Its size decreases, but intensity increases
(B) Its size decreases, and intensity decreases
(C) Its size increases, and intensity increases
(D) Its size increases, but intensity decreases

18. In free space, an electromagnetic wave of 3 GHz frequency strikes over the edge of an object of size $\frac{\lambda}{100}$, where λ is the wavelength of the wave in free space. The phenomenon, which happens there will be : **[26 June 2022 (Morning)]**

(A) Reflection (B) Refraction
(C) Diffraction (D) Scattering

19. A scientist is observing a bacteria through a compound microscope. For better analysis and to improve its resolving power he should. (Select the best option) : **[29 Jan. 2023 (Evening)]**

(A) Increase the wave length of the light
(B) Increase the refractive index of the medium between the object and objective lens
(C) Decrease the focal length of the eye piece
(D) Decrease the diameter of the objective lens

* * * * *

CHAPTER

24 WAVE OPTICS

24.3 POLARIZATION OF LIGHT

1. Electromagnetic waves are transverse in nature is evident by : **[2002]**

(A) polarization (B) interference
(C) reflection (D) diffraction

2. The angle of incidence at which reflected light is totally polarized for reflection from air to glass (refractive index n), is : **[2004]**

(A) $\tan^{-1}(1/n)$ (B) $\sin^{-1}(1/n)$
(C) $\sin^{-1}(n)$ (D) $\tan^{-1}(n)$

3. When an unpolarized light of intensity I_0 is incident on a polarizing sheet, the intensity of the light which does not get transmitted is : **[2005]**

(A) $1/2I_0$ (B) $1/4I_0$
(C) Zero (D) I_0

4. Statement-1 : On viewing the clear blue portion of the sky through a Calcite Crystal, the intensity of transmitted light varies as the crystal is rotated. **[2011]**

Statement-2 : The light coming from the sky is polarized due to scattering of sun light by particles in the atmosphere. The scattering is largest for blue light.

(A) Statement-1 is false, statement-2 is true
(B) Statement-1 is true, statement-2 is false
(C) Statement-1 is true, statement-2 true; statement-2 is the correct explanation of statement-1
(D) Statement-1 is true, statement-2 is true; statement -2 is not correct explanation of statement-1

5. Two polaroids have their polarizing directions parallel so that the intensity of a transmitted light is maximum. The angle through which either polaroid must be turned if the intensity is to drop by one-half is : **[2012 (Online)]**

(A) 135° (B) 90°
(C) 120° (D) 180°

6. A beam of unpolarised light of intensity I_0 is passed through a polaroid A and then through another polaroid B which is oriented so that its principal plane makes an angle of 45° relative to that of A. The intensity of the emergent light is : **[2013]**

(A) I_0 (B) $I_0/2$
(C) $I_0/4$ (D) $I_0/8$

7. A person lives in a high-rise building on the bank of a river 50 m wide. Across the river is a well lit tower of height 40 m. When the person, who is at a height of 10 m, looks through a polarizer at an appropriate angle at light of the tower reflecting from the river surface, he notes that intensity of light coming from distance X from his building is the least and this corresponds to the light coming from light bulbs at height 'Y' on the tower. The values of X and Y are respectively close to (refractive index of water = 4/3) : **[2013 (Online)]**

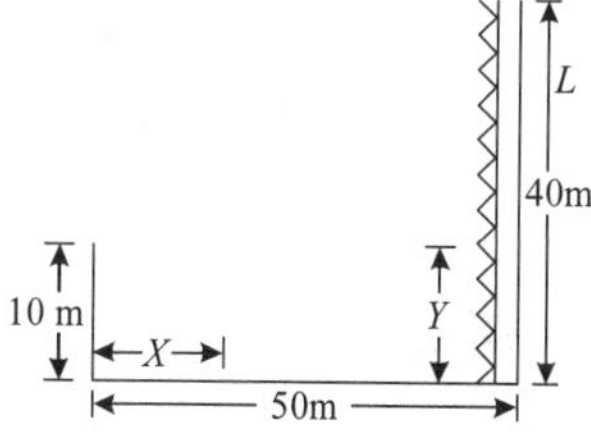

(A) 25 m, 10 m (B) 13 m, 27 m
(C) 22 m, 13 m (D) 17 m, 20 m

8. Two beams, A and B, of plane polarized light with mutually perpendicular planes of polarization are seen through a polaroid. From the position when the beam A has maximum intensity (and beam B has zero intensity), a rotation of polaroid through 30° makes the two beams appear equally bright. If the initial intensities of the two beams are I_A and I_B respectively, then I_A/I_B equals : **[2014]**

(A) 3 (B) 3/2
(C) 1 (D) 1/3

9. A ray of light is incident from a denser to a rarer medium. The critical angle for total internal reflection is θ_{iC} and Brewster's angle of incidence is θ_{iB}, such that $\sin\theta_{iC}/\sin\theta_{iB} = \eta = 1.28$. The relative refractive index of the two media is : **[2014 (Online)]**

(A) 0.2 (B) 0.4
(C) 0.8 (D) 0.9

10. Unpolarized light of intensity I_0 is incident on surface of a block of glass at Brewster's angle. In that case, which one of the following statements is true ? **[2015 (Online)]**

(A) Reflected light is completely polarized with intensity less than $I_0/2$
(B) Transmitted light is completely polarized with intensity less than $I_0/2$
(C) Transmitted light is partially polarized with intensity $I_0/2$
(D) Reflected light is partially polarized with intensity $I_0/2$

11. Unpolarized light of intensity I passes through an ideal polarizer A. Another identical polarizer B is placed behind A. The intensity of light beyond B is found to be $I/2$. Now another identical polarizer C is placed between A and B. The intensity beyond B is now found to be $I/8$. The angle between polarizer A and C is : **[2018]**

(A) 0° (B) 30°
(C) 45° (D) 60°

12. A plane polarized light is incident on a polarizer with its pass axis making angle θ with x-axis, as shown in the figure. At four different values of θ, θ = 8°, 38°, 188° and 218°, the observed intensities are same. What is the angle between the direction of polarization and x-axis : **[2018 (Online)]**

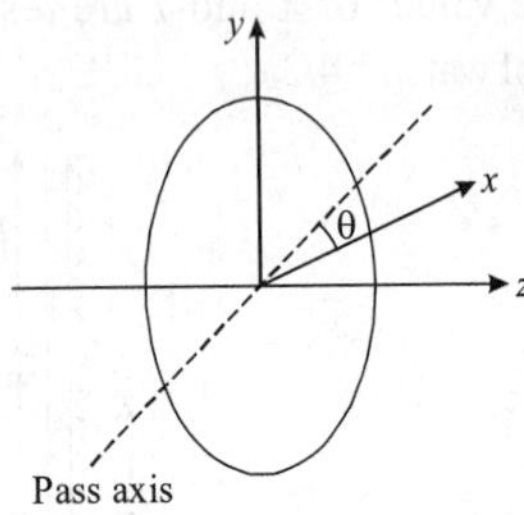

(A) 203° (B) 45°
(C) 98° (D) 128°

13. Unpolarized light of intensity I is incident on a system of two polarizers, A followed by B. The intensity of emergent light is $I/2$. If a third polarizer C is placed between A and B, the intensity of emergent light is reduced to $I/3$. The angle between the polarizers A and C is θ. Then : **[2018 (Online)]**

(A) $\cos\theta = \left(\frac{2}{3}\right)^{1/4}$ (B) $\cos\theta = \left(\frac{1}{3}\right)^{1/4}$

(C) $\cos\theta = \left(\frac{1}{3}\right)^{1/2}$ (D) $\cos\theta = \left(\frac{2}{3}\right)^{1/2}$

14. Consider a tank made of glass (refractive index 1.5) with a thick bottom. It is filled with a liquid of refractive index μ. A student finds that, irrespective of what the incident angle i (see figure) is for a beam of light entering the liquid, the light reflected from the liquid glass interface is never completely polarized. For this to happen, the minimum value of μ is : **[9 Jan. 2019 (Morning)]**

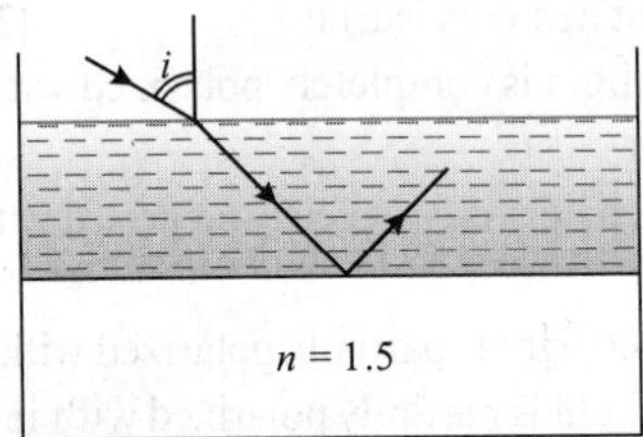

(A) $\frac{3}{\sqrt{5}}$ (B) $\frac{5}{\sqrt{3}}$

(C) $\sqrt{\frac{5}{3}}$ (D) $\frac{4}{3}$

15. A system of three polarizers P_1, P_2, P_3 is set up such that the pass axis of P_3 is crossed with respect to that of P_1. The pass axis of P_2 is inclined at 60° to the pass axis of P_3. When a beam of unpolarized light of intensity I_0 is incident on P_1, the intensity of light transmitted by the three polarizers is I. The ratio (I_0/I) equals (nearly) : **[12 April 2019 (Evening)]**

(A) 16.00 (B) 1.80
(C) 5.33 (D) 10.67

16. A polarizer-analyser set is adjusted such that the intensity of light coming out of the analyser is just 10% of the original intensity. Assuming that the polarizer-analyser set does not absorb any light, the angle by which the analyser need to be rotated further to reduce the output intensity to be zero, is : **[7 Jan. 2020 (Morning)]**

(A) 18.4° (B) 71.6°
(C) 90° (D) 45°

17. A beam of plane polarised light of large cross sectional area and uniform intensity of 3.3 Wm^{-2} falls normally on a polariser (cross sectional area $3 \times 10^{-4}\ m^2$) which rotates about its axis with an angular speed of 31.4 rad/s. The energy of light passing through the polariser per revolution, is close to : **[4 Sept. 2020 (Morning)]**

(A) 1.0×10^{-5} J (B) 5.0×10^{-4} J
(C) 1.0×10^{-4} J (D) 1.5×10^{-4} J

18. An unpolarized light beam is incident on the polarizer of a polarization experiment and the intensity of light beam emerging from the analyzer is measured as 100 Lumens. Now, if the analyzer is rotated around the horizontal axis (direction of light) by 30° in clockwise direction, the intensity of emerging light will be ______ Lumens. **[24 Feb. 2021 (Morning)]**

19. A source of light is placed in front of a screen. Intensity of light on the screen is I. Two Polaroids P_1 and P_2 are so placed in between the source of light and screen that the intensity of light on screen is $I/2$. P_2 should be rotated by an angle of______ (degrees) so that the intensity of light on the screen becomes $3I/8$. **[26 Aug. 2021 (Evening)]**

20. A light whose electric field vectors are completely removed by using a good Polaroid, allowed to incident on the surface of the prism at Brewster's angle. Choose the most suitable option for the phenomenon related to the prism : **[25 June 2022 (Evening)]**

(A) Reflected and refracted rays will be perpendicular to each other
(B) Wave will propagate along the surface of prism
(C) No refraction, and there will be total reflection of light
(D) No reflection and there will be total transmission of light

21. An unpolarised light beam of intensity $2I_0$ is passed through a polaroid P and then through another polaroid Q which is oriented in such a way that its passing axis makes an angle of 30° relative to that of P. The intensity of the emergent light is : **[29 July 2022 (Evening)]**

(A) $\frac{I_0}{4}$ (B) $\frac{I_0}{2}$
(C) $\frac{3I_0}{4}$ (D) $\frac{3I_0}{2}$

22. Given below are two statements :
Statement-I: If the Brewster's angle for the light propagating from air to glass is θ_B, then Brewster's angle for the light propagating from glass to air is $\frac{\pi}{2}-\theta_B$.
Statement-II: The Brewster's angle for the light propagating from glass to air is $\tan^{-1}(\mu_g)$ where μ_g is the refractive index of glass.
In the light of the above statements, choose the correct answer from the options given below : **[24 Jan. 2023 (Morning)]**
(A) Both Statements-I and Statement-II are true
(B) Statement-I is true but Statement-II is false
(C) Both Statement-I and Statement-II are false
(D) Statement-I is false but Statement-II is true

23. As shown in figures, three identical polaroids P_1, P_2 and P_3 are placed one after another. The pass axis of P_2 and P_3 are inclined at angle of 60° and 90° with respect to axis of P_1. The source S has an intensity of $256\frac{W}{m^2}$. The intensity of light at point O is _____ $\frac{W}{m^2}$. **[29 Jan. 2023 (Morning)]**

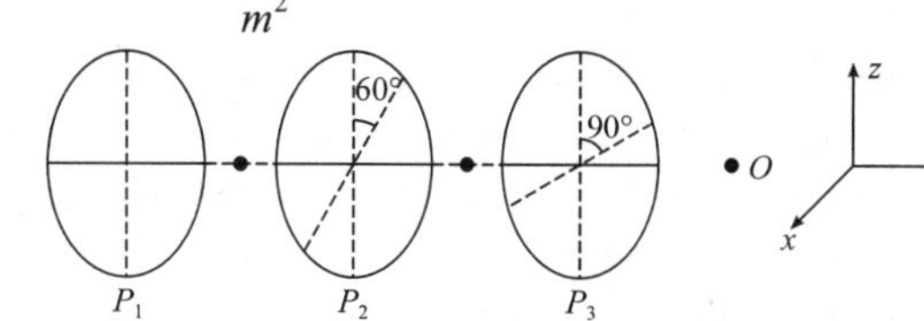

24. Unpolarised light is incident on the boundary between two dielectric media, whose dielectric constants are 2.8 (medium –1) and 6.8 (medium –2), respectively. To satisfy the condition, so that the reflected and refracted rays are perpendicular to each other, the angle of incidence should be $\tan^{-1}\left(1+\frac{10}{\theta}\right)^{1/2}$ the value of θ is _____. (Given for dielectric media, $\mu_r = 1$) **[29 Jan. 2023 (Evening)]**

25. Two polaroide A and B are placed in such a way that the pass-axis of polaroids are perpendicular to each other. Now, another polaroid C is placed between A and B bisecting angle between them. If intensity of unpolarised light is I_0 then intensity of transmitted light after passing through polaroid B will be : **[31 Jan. 2023 (Morning)]**

(A) $\frac{I_0}{4}$ (B) $\frac{I_0}{2}$
(C) $\frac{I_0}{8}$ (D) Zero

26. 'n' polarizing sheets are arranged such that each makes an angle 45° with the proceeding sheet. An unpolarized light of intensity I is incident into this arrangement. The output intensity is found to be $\frac{I}{6}$. The value of n will be : **[1 Feb. 2023 (Morning)]**
(A) 3 (B) 6
(C) 5 (D) 4

27. Unpolarised light of intensity 32 Wm^{-2} passes through the combination of three polaroids such that the pass axis of the last polaroid is perpendicular to that of the pass axis of first polaroid. If intensity of emerging light is 3 Wm^{-2}, then the angle between pass axis of first two polaroids is _____°. **[10 April 2023 (Morning)]**

* * * * *

CHAPTER 25

DUAL NATURE OF RADIATION, MATTER & X-RAYS

25.1 PHOTOELECTRIC EFFECT

1. Sodium and copper have work functions 2.3 eV and 4.5 eV respectively. Then the ratio of their threshold wavelengths is nearest to : **[2002]**

(A) 1 : 2 (B) 4 : 1
(C) 2 : 1 (D) 1 : 4

2. Two identical, photocathodes receive light of frequencies f_1 and f_2. If the velocities of the photoelectrons (of mass m) coming out are respectively v_1 and v_2, then : **[2003]**

(A) $v_1^2 - v_2^2 = \frac{2h}{m}(f_1 - f_2)$

(B) $v_1 + v_2 = \left[\frac{2h}{m}(f_1 + f_2)\right]^{1/2}$

(C) $v_1^2 + v_2^2 = \frac{2h}{m}(f_1 + f_2)$

(D) $v_1 + v_2 = \frac{2h}{m}\left[\frac{2h}{m}(f_1 - f_2)\right]^{1/2}$

3. The work function of a substance is 4.0 eV. The longest wavelength of light that can cause photoelectron emission from this substance is approximately : **[2004]**

(A) 540 nm (B) 400 nm
(C) 310 nm (D) 220 nm

4. According to Einstein's photoelectric equation, the plot of the kinetic energy of the emitted photoelectrons from a metal *Vs* the frequency, of the incident radiation gives a straight line whose slope : **[2004]**

(A) depends on the nature of the metal used
(B) depends on the intensity of the radiation
(C) depends both on the intensity of the radiation and the metal used
(D) is the same for all metals and independent of the intensity of the radiation

5. A photocell is illuminated by a small bright source placed 1 m away. When the same source of light is place 0.5 m away, the number of electrons emitted by photocathode would : **[2005]**

(A) decrease by a factor of 4 (B) increase by a factor of 4
(C) decrease by a factor of 2 (D) increase by a factor of 2

6. The anode voltage of a photocell is kept fixed. The wavelength λ of the light falling on the cathode is gradually changed. The plate current I of the photocell varies as follows : **[2006]**

(A)
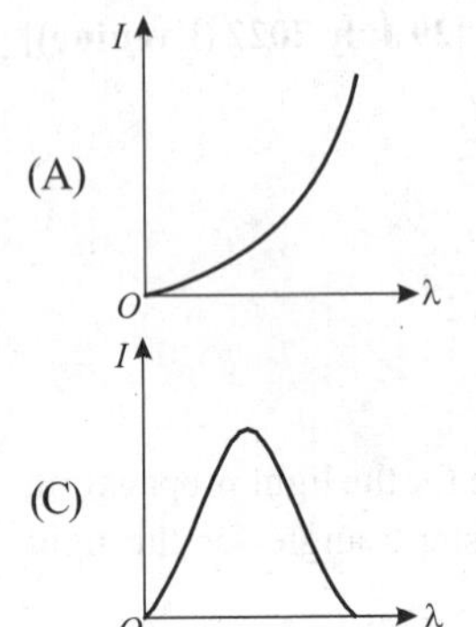

(B)
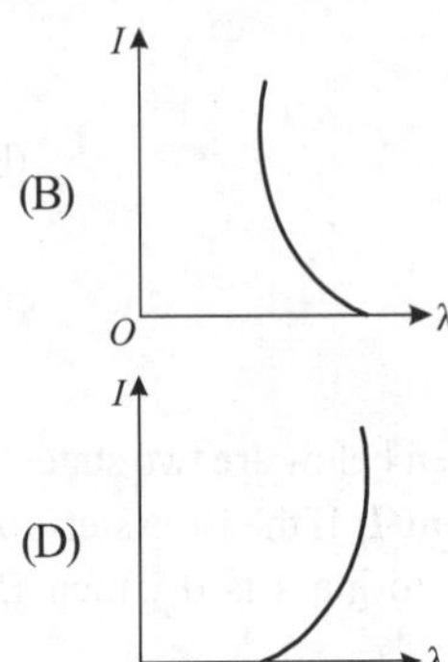

(C) I vs λ

(D) I vs λ

7. The threshold frequency for a metallic surface corresponds to an energy of 6.2 eV and the stopping potential for a radiation incident on this surface is 5 V. The incident radiation lies in : **[2006]**

(A) ultra-violet region (B) infra-red region
(C) visible region (D) X-ray region

8. The surface of a metal is illuminated with the light of 400 nm. The kinetic energy of the ejected photoelectrons was found to be 1.68 eV. The work function of the metal is : [$hc = 1240$ eV-nm] **[2009]**

(A) 1.51 eV (B) 1.68 eV
(C) 3.09 eV (D) 1.42 eV

9. Directions : Question contain Statement-1 and Statement-2. Of the four choices given after the statements, choose the one that best describes the two statements. **[2010]**

Statement-1 : When ultraviolet light is incident on a photocell, its stopping potential is V_0 and the maximum kinetic energy of the photoelectrons is K_{max}. When the ultraviolet light is replaced by X-rays, both V_0 and K_{max} increase.

Statement-2 : Photoelectrons are emitted with speeds ranging from zero to a maximum value because of the range of frequencies present in the incident light.

(A) Statement-1 is true, Statement-2 is false
(B) Statement-1 is true, Statement-2 is true; Statement-2 is the correct explanation of Statement-1
(C) Statement-1 is true, Statement-2 is true; Statement-2 is not the correct explanation of Statement-1
(D) Statement-1 is false, Statement-2 is true

10. This question has Statement-1 and Statement-2. Of the four choices given after the statements, choose the one that best describes the two statements. **[2011]**

Statement-1 : A metallic surface is irradiated by a monochromatic light of frequency $v > v_0$ (the threshold frequency). The maximum kinetic energy and the stopping potential are K_{max} and V_0 respectively. If the frequency incident on the surface is doubled, both the K_{max} and V_0 are also doubled.

Statement-2 : The maximum kinetic energy and the stopping potential of photoelectrons emitted from a surface are linearly dependent on the frequency of incident light.

(A) Statement-1 is true, Statement-2 is true, Statement-2 is not the correct explanation of Statement-1

(B) Statement-1 is false, Statement-2 is true

(C) Statement-1 is true, Statement-2 is false

(C) Statement-1 is true, Statement-2 is true, Statement-2 is the correct explanation of Statement-1

11. This question has Statement-1 and Statement-2. Of the four choices given after the Statements, choose the one that best describes the two Statements. **[2012 (Online)]**

Statement-1 : A metallic surface is irradiated by a monochromatic light of frequency $\upsilon > \upsilon_0$ (the threshold frequency). If the incident frequency is now doubled, the photocurrent and the maximum kinetic energy are also doubled.

Statement-2 : The maximum kinetic energy of photoelectrons emitted from a surface is linearly dependent on the frequency of the incident light. The photocurrent depends only on the intensity of the incident light :

(A) Statement-1 is true, Statement-2 is true, Statement-2 is the correct explanation of Statement-1

(B) Statement-1 is false, Statement-2 is true

(C) Statement-1 is true, Statement-2 is false

(D) Statement-1 is true, Statement-2 is true, Statement-2 is not the correct explanation of Statement-1

12. Photoelectrons are ejected from a metal when light of frequency ν falls on it. Pick out the wrong statement from the following : **[2012 (Online)]**

(A) No electrons are emitted if ν is less than W/h, where W is the work function of the metal

(B) The ejection of the photoelectrons is instantaneous

(C) The maximum energy of the photoelectrons is $h\nu$

(D) The maximum energy of the photoelectrons is independent of the intensity of the light

13. The anode voltage of photocell is kept fixed. The wavelength λ of the light falling on the cathode is gradually changed. The plate current I of the photocell varies as follows: **[2013]**

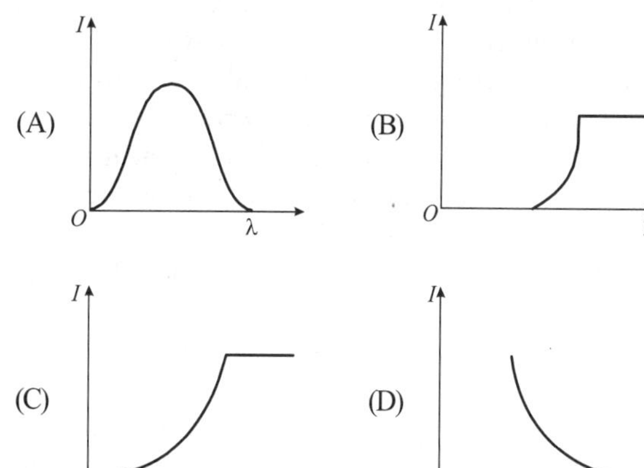

14. A copper ball of radius 1 cm and work function 4.47 eV is irradiated with ultraviolet radiation of wavelength 2500 Å. The effect of irradiation results in the emission of electrons from the ball. Further the ball will acquire charge and due to this there will be a finite value of the potential on the ball. The charge acquired by the ball is : **[2013 (Online)]**

(A) 5.5×10^{-13} C (B) 7.5×10^{-13} C

(C) 4.5×10^{-12} C (D) 2.5×10^{-11} C

15. In an experiment on photoelectric effect, a student plots stopping potential V_0 against reciprocal of the wavelength λ of the incident light for two different metals A and B. These are shown in the figure.

Looking at the graphs, you can most appropriately say that : **[2013 (Online)]**

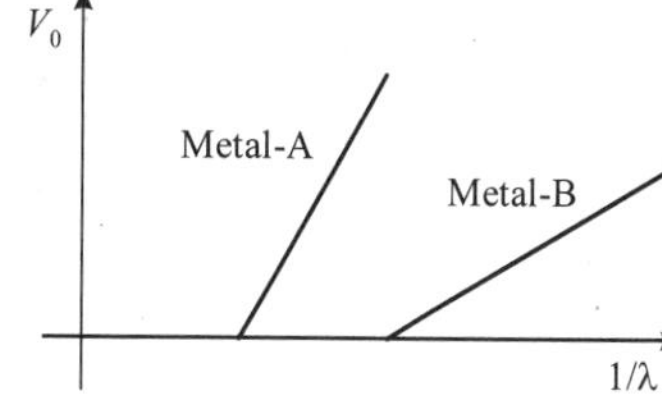

(A) Work function of metal B is greater than that of metal A

(B) For light of certain wavelength falling on both metal, maximum kinetic energy of electrons emitted from A will be greater than those emitted from B

(C) Work function of metal A is greater than that of metal B

(D) Students data is not correct

16. The radiation corresponding to 3 → 2 transition of hydrogen atom falls on a metal surface to produce photoelectrons. These electrons are made to enter a magnetic field of 3×10^{-4} T. If the radius of the largest circular path followed by these electrons is 10.0 mm, the work function of the metal is close to : **[2014]**

(A) 1.8 eV (B) 1.1 eV

(C) 0.8 eV (D) 1.6 eV

17. A beam of light has two wavelengths of 4972 Å and 6216 Å with a total intensity of 3.6×10^{-3} Wm^{-2} equally distributed among the two wavelengths. The beam falls normally on an area of 1 cm^2 of a clean metallic surface of work function 2.3 eV. Assume that there is no loss of light by reflection and that each capable photon ejects one electron. The number of photoelectrons liberated in 2 s is approximately : **[2014 (Online)]**

(A) 6×10^{11} (B) 9×10^{11}
(C) 11×10^{11} (D) 15×10^{11}

18. Radiation of wavelength λ, is incident on a photocell. The fastest emitted electron has speed v. If the wavelength is changed to $\frac{3\lambda}{4}$, the speed of the fastest emitted electron will be : **[2016]**

(A) $> v\left(\frac{4}{3}\right)^{\frac{1}{2}}$ (B) $< v\left(\frac{4}{3}\right)^{\frac{1}{2}}$

(C) $= v\left(\frac{4}{3}\right)^{\frac{1}{2}}$ (D) $= v\left(\frac{3}{4}\right)^{\frac{1}{2}}$

19. When photons of wavelength λ_1 are incident on an isolated sphere, the corresponding stopping potential is found to be V. When photons of wavelength λ_2 are used, the corresponding stopping potential was thrice that of the above value. If light of wavelength λ_3 is used then find the stopping potential for this case : **[2016 (Online)]**

(A) $\frac{hc}{e}\left[\frac{1}{\lambda_3}+\frac{1}{\lambda_2}-\frac{1}{\lambda_1}\right]$ (B) $\frac{hc}{e}\left[\frac{1}{\lambda_3}+\frac{1}{2\lambda_2}-\frac{1}{\lambda_1}\right]$

(C) $\frac{hc}{e}\left[\frac{1}{\lambda_3}-\frac{1}{\lambda_2}-\frac{1}{\lambda_1}\right]$ (D) $\frac{hc}{e}\left[\frac{1}{\lambda_3}+\frac{1}{2\lambda_2}-\frac{3}{2\lambda_1}\right]$

20. A photoelectric surface is illuminated successively by monochromatic light of wavelengths λ and $\frac{\lambda}{2}$. If the maximum kinetic energy of the emitted photoelectrons in the second case is 3 times that in the first case, the work function of the surface is : **[2016 (Online)]**

(A) $\frac{hc}{2\lambda}$ (B) $\frac{hc}{\lambda}$

(C) $\frac{hc}{3\lambda}$ (D) $\frac{3hc}{\lambda}$

21. The maximum velocity of the photoelectrons emitted from the surface is v when light of frequency n falls on a metal surface. If the incident frequency is increased to $3n$, the maximum velocity of the ejected photoelectrons will be : **[2017 (Online)]**

(A) Less than $\sqrt{3}v$ (B) v

(C) More than $\sqrt{3}v$ (D) Equal to $\sqrt{3}v$

22. A Laser light of wavelength 600 nm is used to weld retina detachment. If a Laser pulse of width 60 ms and power 0.5 kW is used the approximate number of photons in the pulse are : [Take Plank's constant $h = 6.62 \times 10^{-34}$ Js] **[2017 (Online)]**

(A) 10^{20} (B) 10^{18}
(C) 10^{22} (D) 10^{19}

23. Surface of certain metal is first illuminated with light of wavelength $\lambda_1 = 350$ nm and then, by light of wavelength $\lambda_2 = 540$ nm. It is found that the maximum speed of the photo electrons in the two cases differ by a factor of 2. The work function of the metal (in eV) is close to :

(Energy of photon $= \frac{1240}{\lambda(\text{in } nm)}$ eV) **[9 Jan. 2019 (Morning)]**

(A) 1.8 (B) 1.4
(C) 2.5 (D) 5.6

24. The magnetic field associated with a light wave is given, at the origin, by

$$B = B_0[\sin(3.14 \times 10^7)ct + \sin(6.28 \times 10^7)ct].$$

If this light falls on a silver plate having a work function of 4.7 eV, what will be the maximum kinetic energy of the photo electrons?

($c = 3 \times 10^8$ ms^{-1}, $h = 6.6 \times 10^{-34}$ J-s) : **[9 Jan. 2019 (Evening)]**

(A) 7.72 eV (B) 8.52 eV
(C) 12.5 eV (D) 6.82 eV

25. A metal plate of area 1×10^{-4} m^2 is illuminated by a radiation of intensity 16 mW/m^2. The work function of the metal is 5 eV. The energy of the incident photons is 10 eV and only 10% of it produces photo electrons. The number of emitted photo electrons per second and their maximum energy, respectively, will be. [1 eV $= 1.6 \times 10^{-19}$ J] **[10 Jan. 2019 (Evening)]**

(A) 10^{10} and 5 eV (B) 10^{14} and 10 eV
(C) 10^{12} and 5 eV (D) 10^{11} and 5 eV

26. In a photoelectric experiment, the wavelength of the light incident on a metal is changed from 300 nm to 400 nm. The decrease in the stopping potential is close to :

$\left(\frac{hc}{e} = 1240\, nm - V\right)$ **[11 Jan. 2019 (Evening)]**

(A) 0.5 V (B) 1.0 V
(C) 2.0 V (D) 1.5 V

27. When a certain photosensitive surface is illuminated with monochromatic light of frequency ν, the stopping potential for the photo current is $-V_0/2$. When the surface is illuminated by monochromatic light of frequency ν/2, the stopping potential is $-V_0$. The threshold frequency for photoelectric emission is: **[12 Jan. 2019 (Evening)]**

(A) $\frac{3\nu}{2}$ (B) 2ν

(C) $\frac{4}{3}\nu$ (D) $\frac{5\nu}{3}$

28. The electric field of light wave is given as

$$\vec{E} = 10^{-3}\cos\left(\frac{2\pi x}{5\times10^{-7}} - 2\pi\times6\times10^{14}t\right)\hat{x}\frac{N}{C}.$$

This light falls on a metal plate of work function 2 eV. The stopping potential of the photoelectrons is.

[Given, $E\,(\text{in } eV) = \frac{12375}{\lambda(\text{in Å})}$] **[9 April 2019 (Morning)]**

(A) 0.48 V (B) 2.0 V
(C) 2.48 V (D) 0.72 V

29. In a photoelectric effect experiment the threshold wavelength of the light is 380 nm. If the wavelength of incident light is 260 nm, the maximum kinetic energy of emitted electrons will be : Given $E\,(\text{in } eV) = \frac{1237}{\lambda(\text{in } nm)}$ **[10 April 2019 (Morning)]**

(A) 1.5 eV (B) 4.5 eV
(C) 15.1 eV (D) 3.0 eV

30. The stopping potential V_0 (in volt) as a function of frequency (ν) for a sodium emitter, is shown in the figure. The work function of sodium, from the data plotted in the figure, will be : (Given : Planck's constant $(h) = 6.63 \times 10^{-34}$ Js, electron charge $e = 1.6 \times 10^{-19}$ C) **[12 April 2019 (Morning)]**

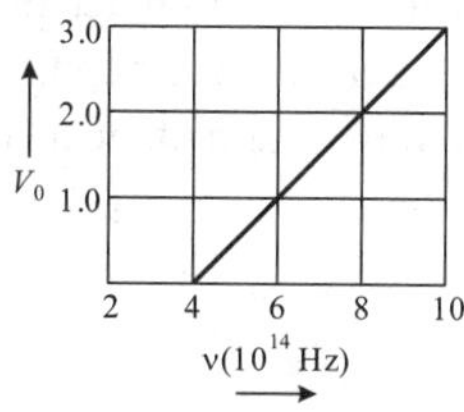

(A) 1.95 eV (B) 1.82 eV
(C) 1.66 eV (D) 2.12 eV

31. A beam of electromagnetic radiation of intensity 6.4×10^{-5} W/cm² is comprised of wavelength, λ = 310 nm. It falls normally on a metal (work function φ = 2 eV) of surface area of 1 cm². If one in 10^3 photons ejects an electron, total number of electrons ejected in 1 s is 10^x. (hc = 1240 $eVnm$, 1 eV = 1.6×10^{-19} J), then x is______. **[7 Jan. 2020 (Morning)]**

32. When photon of energy 4.0 eV strikes the surface of a metal A, the ejected photoelectrons have maximum kinetic energy $T_A eV$ end de-Broglie wavelength λ_A. The maximum kinetic energy of photoelectrons liberated from another metal B by photon of energy 4.50 eV is $T_B = (T_A - 1.5)$ eV. If the de-Broglie wavelength of these photoelectrons $\lambda_B = 2\lambda_A$, then the work function of metal B is : **[8 Jan. 2020 (Morning)]**

(A) 3 eV (B) 2 eV
(C) 4 eV (D) 1.5 eV

33. The dimension of stopping potential V_0 in photoelectric effect in units of Planck's constant 'h', speed of light 'c' and Gravitational constant 'G' and ampere A is : **[8 Jan. 2020 (Morning)]**

(A) $h^2\, G^{3/2}\, c^{1/3}\, A^{-1}$ (B) $h^{-2/3}\, c^{-1/3}\, G^{4/3}\, A^{-1}$
(C) $h^{1/3}\, G^{2/3}\, c^{1/3}\, A^{-1}$ (D) $h^{2/3}\, c^{5/3}\, G^{1/3}\, A^{-1}$

34. Radiation, with wavelength 6561 Å falls on a metal surface to produce photoelectrons. The electrons are made to enter a uniform magnetic field of 3×10^{-4} T. If the radius of the largest circular path followed by the electrons is 10 mm, the work function of the metal is close to : **[9 Jan. 2020 (Morning)]**

(A) 1.8 eV (B) 1.1 eV
(C) 0.8 eV (D) 1.6 eV

35. When radiation of wavelength λ is used to illuminate a metallic surface, the stopping potential is V. When the same surface is illuminated with radiation of wavelength 3λ, the stopping potential is $V/4$. If the threshold wavelength for the metallic surface is $n\lambda$ then value of n will be : **[2 Sept. 2020 (Morning)]**

36. When the wavelength of radiation falling on a metal is changed from 500 nm to 200 nm, the maximum kinetic energy of the photoelectrons becomes three times larger. The work function of the metal is close to : **[3 Sept. 2020 (Morning)]**

(A) 0.61 eV (B) 0.52 eV
(C) 0.81 eV (D) 1.02 eV

37. Given figure shows few data points in a photo electric effect experiment for a certain metal. The minimum energy for ejection of electron from its surface is : (Planck's constant $h = 6.62 \times 10^{-34}$ J.s) **[4 Sept. 2020 (Morning)]**

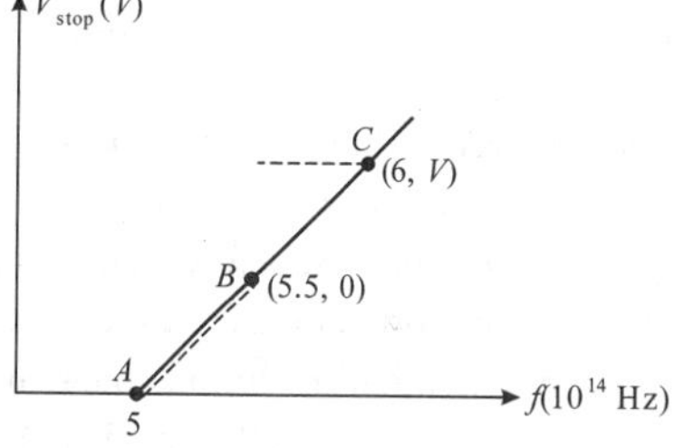

(A) 2.27 eV (B) 2.59 eV
(C) 1.93 eV (D) 2.10 eV

38. In a photoelectric effect experiment, the graph of stopping potential V versus reciprocal of wavelength obtained is shown in the figure. As the intensity of incident radiation is increased : **[4 Sept. 2020 (Evening)]**

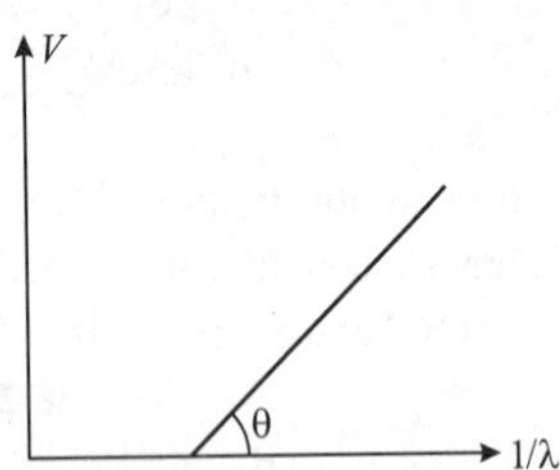

(A) Slope of the straight line get more steep
(B) Straight line shifts to left
(C) Graph does not change
(D) Straight line shifts to right

39. The surface of a metal is illuminated alternately with photons of energies $E_1 = 4$ eV and $E_2 = 2.5$ eV respectively. The ratio of maximum speeds of the photoelectrons emitted in the two cases is 2. The work function of the metal in (eV) is ______. **[5 Sept. 2020 (Evening)]**

40. The stopping potential for electrons emitted from a photosensitive surface illuminated by light of wavelength 491 nm is 0.710 V. When the incident wavelength is changed to a new value, the stopping potential is 1.43 V. The new wavelength is : (Consider $hc = 1240$ eV-nm) **[25 Feb. 2021 (Evening)]**
(A) 382 nm (B) 329 nm
(C) 309 nm (D) 400 nm

41. Two stream of photons, possessing energies equal to twice and ten times the work function of metal are incident on the metal surface successively. The value of ratio of maximum velocities of the photoelectrons emitted in the two respective cases is $x : y$. The value of x is ______. **[26 Feb. 2021 (Evening)]**

42. The stopping potential in the context of photoelectric effect depends on the following property of incident electromagnetic radiation : **[16 March 2021 (Morning)]**
(A) Phase (B) Intensity
(C) Amplitude (D) Frequency

43. If 2.5×10^{-6} N average force is exerted by a light wave on a non-reflecting surface of 30 cm^2 area during 40 minutes of time span, the energy flux of light just before it falls on the surface is ______ W/cm^2. (Round off to the Nearest Integer)
(Assume complete absorption and normal incidence conditions are there) **[17 March 2021 (Morning)]**

44. Two identical photocathodes receive the light of frequencies f_1 and f_2 respectively. If the velocities of the photo-electrons coming out are v_1 and v_2 respectively, then : **[17 March 2021 (Evening)]**

(A) $v_1^2 - v_2^2 = \frac{2h}{m}[f_1 - f_2]$

(B) $v_1^2 + v_2^2 = \frac{2h}{m}[f_1 + f_2]$

(C) $v_1 + v_2 = \left[\frac{2h}{m}(f_1 + f_2)\right]^{\frac{1}{2}}$

(D) $v_1 - v_2 = \left[\frac{2h}{m}(f_1 - f_2)\right]^{1/2}$

45. The radiation corresponding to $3 \rightarrow 2$ transition of a hydrogen atom falls on a gold surface to generate photoelectrons. These electrons are passed through a magnetic field of 5×10^{-4} T. Assume that the radius of the largest circular path followed by these electrons is 7 mm, the work function of the metal is : (Mass of electron $= 9.1 \times 10^{-31}$ kg) **[20 July 2021 (Morning)]**
(A) 0.82 eV (B) 0.16 eV
(C) 1.36 eV (D) 1.88 eV

46. A certain metallic surface is illuminated by monochromatic radiation of wavelength λ. The stopping potential for photoelectric current for this radiation is $3\ V_0$. If the same surface is illuminated with a radiation of wavelength 2λ, the stopping potential is V_0. The threshold wavelength of this surface for photoelectric effect is ______ λ. **[20 July 2021 (Evening)]**

47. When radiation of wavelength λ is incident on a metallic surface, the stopping potential of ejected photoelectrons is 4.8 V. If the same surface is illuminated by radiation of double the previous wavelength, then the stopping potential becomes 1.6 V. The threshold wavelength of the metal is : **[25 July 2021 (Evening)]**
(A) 8λ (B) 4λ
(C) 2λ (D) 6λ

48. A light beam of wavelength 500 nm is incident on a metal having work function of 1.25 eV, placed in a magnetic field of intensity B. The electrons emitted perpendicular to the magnetic field B, with maximum kinetic energy are bent into circular are of radius 30 cm. The value of B is ______ $\times 10^{-7}$ T.
(Given : $hc = 20 \times 10^{-26}$ J-m, mass of electron $= 9 \times 10^{-31}$ kg) **[25 July 2021 (Evening)]**

49. An electron and proton are separated by a large distance. The electron start approaching the proton with energy 3 eV.

The proton captures the electron and form a hydrogen atom in second excited state. The resulting photon is incident on a photosensitive metal of threshold wavelength 4000 Å. What is the maximum kinetic energy of the emitted photoelectron ?

[27 July 2021 (Evening)]

(A) No photoelectron would be emitted
(B) 1.41 eV
(C) 7.61 eV
(D) 3.3 eV

50. In a photoelectric experiment ultraviolet light of wavelength 280 nm used with lithium cathode having work function $\phi = 2.5$ eV. If the wavelength of incident light is switched to 400 nm, find out the change in the stopping potential. ($h = 6.63 \times 10^{-34}$ Js, c $= 3 \times 10^8$ ms^{-1}) **[26 Aug. 2021 (Morning)]**

(A) 0.6 V (B) 1.1 V
(C) 1.3 V (D) 1.9 V

51. The electric field in a plane electromagnetic wave is given by $\vec{E} = 200 \cos\left[\left(\frac{0.5\times10^3}{m}\right)x - \left(1.5\times10^{11}\frac{\text{rad}}{s}\times t\right)\right]\hat{j}$ V/m. If this wave falls normally on a perfectly reflecting surface having an area of 100 cm^2. If the radiation pressure exerted by the *E. M.* wave on the surface during a 10 minute exposure is $\frac{x}{10^9}\frac{N}{m^2}$. Find the value of x : **[26 Aug. 2021 (Morning)]**

52. In a photoelectric experiment, increasing the intensity of incident light : **[27 Aug. 2021 (Morning)]**

(A) Increases the number of photons incident and the K.E. of the ejected electrons remains unchanged.
(B) Increases the frequency of photons incident and the K.E. of the ejected electrons remains unchanged.
(C) Increases the number of photons incident and also increases the K.E. of the ejected electrons.
(D) Increases the frequency of photons incident and increases the K.E. of the ejected electrons

53. A monochromatic neon lamp with wavelength of 670.5 illuminates a photo-sensitive material which has a stopping voltage of 0.48 V. What will be the stopping voltage if the source light is changed with another source of wavelength of 476.6 nm ?

[27 Aug. 2021 (Evening)]

(A) 0.24 V (B) 1.5 V
(C) 1.25 V (D) 0.96 V

54. A free electron of 2.6 eV energy collides with a H^+ ion. This results in the formation of a hydrogen atoms in the first excites state and a photon is released. Find the frequency of the emitted photon. ($h = 6.6 \times 10^{-34}$ J S) **[31 Aug. 2021 (Evening)]**

(A) 1.45×1016 MHz (B) 1.45×10^9 MHz
(C) 9.0×10^{27} MHz (D) 0.19×10^{15} MHz

55. When light of frequency twice the threshold frequency is incident on the metal plate, the maximum velocity of emitted election is v_1. When the frequency of incident radiation is increased to five times the threshold value, the maximum velocity of emitted electron becomes v_2. If $v_2 = x\, v_1$, the value of x will be ______ . **[24 June 2022 (Morning)]**

56. The light of two different frequencies whose photons have energies 3.8 eV and 1.4 eV respectively, illuminate a metallic surface whose work function is 0.6 eV successively. The ratio of maximum speeds of emitted electrons for the two frequencies respectively will be : **[24 June 2022 (Evening)]**

(A) 1 : 1 (B) 2 : 1
(C) 4 : 1 (D) 1 : 4

57. The photodiode is used to detect the optical signals. These diodes are preferably operated in reverse biased mode because :

[25 June 2022 (Morning)]

(A) Fractional change in majority carriers produce higher forward bias current
(B) Fractional change in majority carriers produce higher reverse bias current
(C) Fractional change in minority carriers produce higher forward bias current
(D) Fractional change in minority carriers produce higher reverse bias current

58. A metal surface is illuminated by a radiation of wavelength 4500 Å. The ejected photo-electron enters a constant magnetic field of 2 mT making an angle of 90° with the magnetic field. If it starts revolving in a circular path of radius 2 mm, the work function of the metal is approximately :

[26 June 2022 (Evening)]

(A) 1.36 eV (B) 1.69 eV
(C) 2.78 eV (D) 2.23 eV

59. The stopping potential for photoelectrons emitted from a surface illuminated by light of wavelength 6630 Å is 0.42 V. If the threshold frequency is $x \times 10^{13}$/s, where x is _____ (nearest integer). [Given, speed light $= 3 \times 10^8$ m/s, Planck's constant $= 6.63 \times 10^{-34}$ Js] **[26 June 2022 (Evening)]**

60. Let K_1 and K_2 be the maximum kinetic energies of photo-electrons emitted when two monochromatic beams of wavelength λ_1 and λ_2, respectively are incident on a metallic surface. If $\lambda_1 = 3\lambda_3$ then : **[28 June 2022 (Evening)]**

(A) $K_1 > \frac{K_2}{3}$ (B) $K_1 < \frac{K_2}{3}$
(C) $K_1 = \frac{K_2}{3}$ (D) $K_2 = \frac{K_1}{3}$

61. Given below are two statements : One is labelled as Assertion-**A** and the other is labelled as Reason-**R**.

Assertion-A : The photoelectric effect does not takes place, if the energy of the incident radiation is less than the work function of a metal.

Reason-R : Kinetic energy of the photoelectrons is zero, if the energy of the incident radiation is equal to the work function of a metal

In the light of the above statements, choose the most appropriate answer from the options given below :

[29 June 2022 (Morning)]

(A) Both **A** and **R** are correct and **R** is the correct explanation of **A**.

(B) Both **A** and **R** are correct but **R** is not the correct explanation of **A**.

(C) **A** is correct but **R** is not correct.

(D) **A** is not correct but **R** is correct.

62. A metal exposed to light of wavelength 800 nm and emits photoelectrons with a certain kinetic energy. The maximum kinetic energy of photo-electron doubles when light of wavelength 500 nm is used. The work function of the metal is :

(Take $hc = 1230$ eV-nm) **[25 July 2022 (Morning)]**

(A) 1.537 eV (B) 2.46 eV

(C) 0.615 eV (D) 1.23 eV

63. With reference to the observations in photo-electric effect, identify the correct statements from below :

[27 July 2022 (Evening)]

(a) The square of maximum velocity of photoelectrons varies linearly with frequency of incident light

(b) The value of saturation current increases on moving the source of light away from the metal surface

(c) The maximum kinetic energy of photo-electrons decreases on decreasing the power of *LED* (light emitting diode) source of light

(d) The immediate emission of photo-electrons out of metal surface can not be explained by particle nature of light/ electromagnetic waves

(e) Existence of threshold wavelength can not be explained by wave nature of light/electromagnetic waves

Choose the correct answer from the options given below :

(A) (a) and (b) only (B) (a) and (e) only

(C) (c) and (e) only (D) (d) and (e) only

64. Two streams of photons, possessing energies to five and ten times the work function of metal are incident on the metal surface successively. The ratio of the maximum velocities of the photoelectron emitted, in the two cases respectively, will be :

[28 July 2022 (Evening)]

(A) 1 : 2 (B) 1 : 3

(C) 2 : 3 (D) 3 : 2

65. The kinetic energy of emitted electron is E when the light incident on the metal has wavelength λ. To double the kinetic energy, the incident light must have wavelength :

[29 July 2022 (Morning)]

(A) $\dfrac{hc}{E\lambda - hc}$ (B) $\dfrac{hc\lambda}{E\lambda + hc}$

(C) $\dfrac{h\lambda}{E\lambda + hc}$ (D) $\dfrac{hc\lambda}{E\lambda - hc}$

66. From the photoelectric effect experiment, following observations are made. Identify which of these are correct

[24 Jan. 2023 (Morning)]

(a) The stopping potential depends only on the work function of the metal

(b) The saturation current increases as the intensity of incident light increases

(c) The maximum kinetic energy of a photo electron depends on the intensity of the incident light

(d) Photoelectric effect can be explained using wave theory of light

Choose the correct answer from the options given below:

(A) (b), (c) only (B) (a), (c), (d) only

(C) (b) only (D) (a), (b), (d) only

67. Match List-I with List-II : **[24 Jan. 2023 (Morning)]**

List-I	**List-II**
(a) Planck's constant (h)	(i) $[M^1 L^2 T^{-2}]$
(b) Stopping potential (V_s)	(ii) $[M^1 L^1 T^{-1}]$
(c) Work function (ϕ)	(iii) $[M^1 L^2 T^{-1}]$
(d) Momentum (p)	(iv) $[M^1 L^2 T^{-3} A^{-1}]$

(A) (a)-(iii), (b)-(i), (c)-(ii), (d)-(iv)

(B) (a)-(iii), (b)-(iv), (c)-(i), (d)-(ii)

(C) (a)-(ii), (b)-(iv), (c)-(iii), (d)-(i)

(D) (a)-(i), (b)-(iii), (c)-(iv), (d)-(ii)

68. Given below are two statements :

Statement-I: Stopping potential in photoelectric effect does not depend on the power of the light source

Statement-II: For a given metal, the maximum kinetic energy of the photoelectron depends on the wavelength of the incident light.

In the light of above statements, choose the most appropriate answer from the options given below.

Options : **[25 Jan. 2023 (Evening)]**

(A) Statement-I is incorrect but statement-II is correct

(B) Both Statement-I and Statement-II are incorrect

(C) Statement-I is correct but statement-II is incorrect

(D) Both statement-I and statement-II are correct

69. The threshold wavelength for photoelectric emission from a material is 5500?. Photoelectrons will be emitted, when this material is illuminated with monochromatic radiation from a :

(a) 75 W infra-red lamp
(b) 10 W infra-red lamp
(d) 75 W ultra-violet lamp
(d) 10 W ultra-violet lamp

Choose the correct answer from the options given below :

[29 Jan. 2023 (Morning)]

(A) (b) and (c) only
(B) (a) and (d) only
(C) (c) only
(D) (c) and (d) only

70. If the two metals A and B are exposed to radiation of wavelength 350 nm. The work functions of metals A and B are 4.8 eV and 2.2 eV. Then choose the correct option :

[31 Jan. 2023 (Evening)]

(A) Metal B will not emit photoelectrons
(B) Both metals A and B will emit photoelectrons
(C) Both metals A and B will not emit photoelectrons
(D) Metal A will not emit photoelectrons

71. The threshold frequency of metal is f_0. When the light of frequency $2f_0$ is incident on the metal plate, the maximum velocity of photoelectron is v_1. When the frequency of incident radiation is increased to $5f_0$. The maximum velocity of photoelectrons emitted is v_2. The ratio of v_1 to v_2 is : **[1 Feb. 2023 (Evening)]**

(A) $\frac{v_1}{v_2} = \frac{1}{2}$
(B) $\frac{v_1}{v_2} = \frac{1}{8}$
(C) $\frac{v_1}{v_2} = \frac{1}{16}$
(D) $\frac{v_1}{v_2} = \frac{1}{4}$

72. The work functions of Aluminium and Gold are 4.1 eV and 5.1 eV respectively. The ratio of the slope of the stopping potential versus frequency plot for Gold to that of Aluminium is

[6 April 2023 (Evening)]

(A) 1.24
(B) 2
(C) 1
(D) 1.5

73. In photo electric effect :

(a) The photocurrent is proportional to the intensity of the incident radiation
(b) Maximum Kinetic energy with which photoelectrons are emitted depends on the intensity of incident light
(c) Max. K.E with which photoelectrons are emitted depends on the frequency of incident light
(d) The emission of photoelectrons require a minimum threshold intensity of incident radiation
(e) Max. K.E of the photoelectrons is independent of the frequency of the incident light

Choose the correct answer from the options given below :

[8 April 2023 (Evening)]

(A) (a) and (c) only
(B) (a) and (e) only
(C) (b) and (c) only
(D) (a) and (b) only

74. The variation of stopping potential (V_0) as a function of the frequency (ν) of the incident light for a metal is shown in figure. The work function of the surface is:

[10 April 2023 (Evening)]

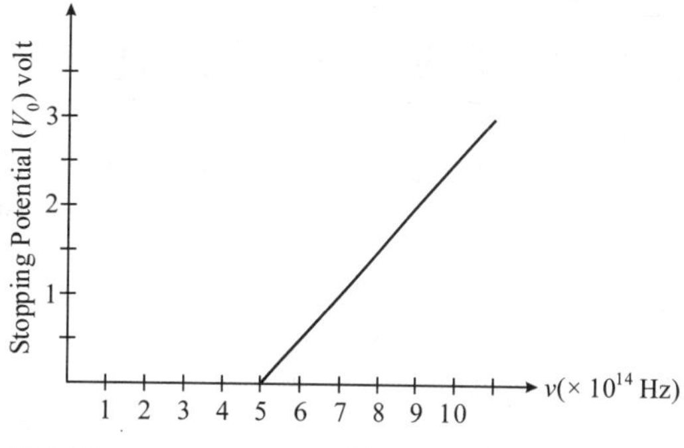

(A) 18.6 eV
(B) 2.98 eV
(C) 2.07 eV
(D) 1.36 eV

75. A metallic surface is illuminated with radiation of wavelength λ, the stopping potential is V_0. If the same surface is illuminated with radiation of wavelength 2λ, the stopping potential becomes $\frac{V_0}{4}$. The threshold wavelength for this metallic surface will be:

[11 April 2023 (Morning)]

(A) $\frac{\lambda}{4}$
(B) 4λ
(C) $\frac{3}{2}\lambda$
(D) 3λ

76. The difference between threshold wavelengths for two metal surfaces A and B having work function $\phi_A = 9$ eV and $\phi_B = 4.5$ eV in nm is : (Given, $hc = 1242$ eV nm)

[13 April 2023 (Morning)]

(A) 264
(B) 138
(C) 276
(D) 540

77. Given below are two statements:

Statement-I : Out of microwaves, infrared rays and ultraviolet rays, ultraviolet rays are the most effective for the emission of electrons from a metallic surface.

Statement-II : Above the threshold frequency, the maximum kinetic energy of photoelectrons is inversely proportional to the frequency of the incident light.

In the light of above statement, choose the correct answer from the options given below **[13 April 2023 (Evening)]**

(A) Statement-I is true but Statement-II is false
(B) Both Statement-I and Statement-II are true
(C) Statement-I is false but Statement-II is true
(D) Both Statement-I and Statement-II are false

* * * * *

CHAPTER 25

DUAL NATURE OF RADIATION, MATTER & X-RAYS

25.2 MATTER WAVES & X-RAYS

1. Formation of covalent bonds in compounds exhibits : **[2002]**

(A) wave nature of electron
(B) particle nature of electron
(C) both wave and particle nature of electron
(D) none of the above

2. A radiation of energy E falls normally on a perfectly reflecting surface. The momentum transferred to the surface is: **[2004]**

(A) E/c (B) $2E/c$
(C) Ec (D) E/c^2

3. If the kinetic energy of a free electron doubles, its de-Broglie wavelength changes by the factor : **[2005]**

(A) $\frac{1}{2}$ (B) 2
(C) $\frac{1}{\sqrt{2}}$ (D) $\sqrt{2}$

4. Photon of frequency ν has a momentum associated with it. If c is the velocity of light, the momentum is : **[2007]**

(A) ν/c (B) $h\nu c$
(C) $h\nu/c^2$ (D) $h\nu/c$

5. **Directions For question 5 and 6 :** These questions are based on the following paragraph.

Wave property of electrons implies that they will show diffraction effects. Davisson and Germer demonstrated this by diffracting electrons from crystals. The law governing the diffraction from a crystal is obtained by requiring that electron waves reflected from the planes of atoms in a crystal interfere constructively (see figure).

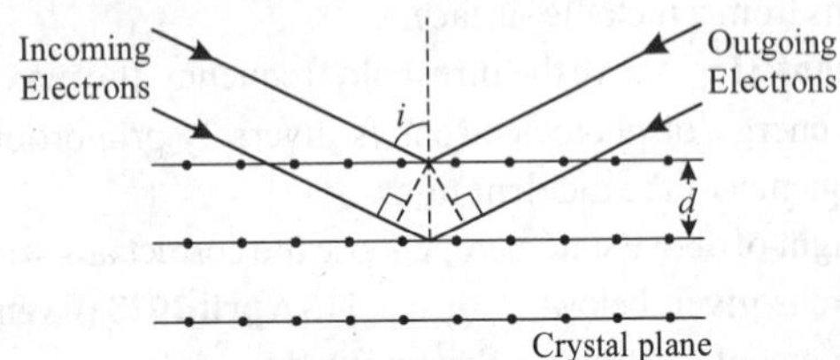

Electrons accelerated by potential V are diffracted from a crystal. If d = 1 Å and i = 30° V should be about ($h = 6.6 \times 10^{-34}$ Js, $m_e = 9.1 \times 10^{-31}$ kg, $e = 1.6 \times 10^{-19}$ C) : **[2008]**

(A) 2000 V (B) 50 V
(C) 500 V (D) 1000 V

6. If a strong diffraction peak is observed when electrons are incident at an angle 'i' from the normal to the crystal planes with distance 'd' between them (see figure), de-Broglie wavelength λ_{dB} of electrons can be calculated by the relationship (n is an integer) **[2008]**

(A) $d \sin i = n\lambda_{dB}$ (B) $2d \cos i = n\lambda_{dB}$
(C) $2d \sin i = n\lambda_{dB}$ (D) $d \cos i = n\lambda_{dB}$

7. This question has Statement-1 and Statement-2. Of the four choices given after the statements, choose the one that best describes the two statements. **[2012]**

Statement-1 : Davisson-Germer experiment established the wave nature of electrons.

Statement-2 : If electrons have wave nature, they can interfere and show diffraction.

(A) Statement-1 is true, Statement-2 is true and Statement-2 is not the correct explanation of Statement-1
(B) Statement-1 is false, Statement-2 is true
(C) Statement-1 is true, Statement-2 is false
(D) Statement-1 is true, Statement-2 is true and Statement-2 is the correct explanation of statement-1

8. Electrons are accelerated through a potential difference V and protons are accelerated through a potential difference 4 V. The de-Broglie wavelengths are λ_e and λ_p for electrons and protons respectively. The ratio of $\frac{\lambda_e}{\lambda_p}$ is given by: (given m_e is mass of electron and m_p is mass of proton) : **[2013 (Online)]**

(A) $\frac{\lambda_e}{\lambda_p} = \sqrt{\frac{m_p}{m_e}}$ (B) $\frac{\lambda_e}{\lambda_p} = \sqrt{\frac{m_e}{m_p}}$
(C) $\frac{\lambda_e}{\lambda_p} = \frac{1}{2}\sqrt{\frac{m_e}{m_p}}$ (D) $\frac{\lambda_e}{\lambda_p} = 2\sqrt{\frac{m_p}{m_e}}$

9. Match List-I (Experiment performed) with List-II (Phenomena discovered/associated) and select the correct option from the options given below the lists : **[2014 (Online)]**

List-I	List-II
(a) Davisson and germer experiment	(i) Wave nature of electrons
(b) Millikan's oil drop experiment	(ii) Charge of an electron
(c) Rutherford experiment	(iii) Quantisation of energy levels
(d) Franck-Hertz experiment	(iv) Existence of nucleus

(A) (a)-(i), (b)-(ii), (c)-(iii), (d)-(iv)
(B) (a)-(i), (b)-(ii), (c)-(iv), (d)-(iii)
(C) (a)-(iii), (b)-(iv), (c)-(i), (d)-(ii)
(D) (a)-(iv), (b)-(iii), (c)-(ii), (d)-(i)

10. A photon of wavelength λ is scattered from an electron, which was at rest. The wavelength shift $\Delta\lambda$ is three times of λ and the angle of scattering θ is 60°. The angle at which the electron recoiled is ϕ. The value of $\tan\phi$ is : (Electron speed is much smaller than the speed of light) **[2014 (Online)]**

(A) 0.16 (B) 0.22
(C) 0.25 (D) 0.28

11. For which of the following particles will it be most difficult to experimentally verify the de-Broglie relationship ? **[2014 (Online)]**

(A) An electron (B) A proton
(C) An α-particle (D) A dust particle

12. Match List-I (Fundamental Experiment) with List-II (its conclusion) and select the correct option from the choices given below the list : **[2015]**

List-I	List-II
(a) Franck-Hertz Experiment	(i) Particle nature of light
(b) Photo-electric experiment	(ii) Discrete energy levels of atom
(c) Davisson-Germer Experiment	(iii) Wave nature of electron
	(iv) Structure of atom

(A) (a)-(i); (b)-(iv); (c)-(iii)
(B) (a)-(ii); (b)-(iv); (c)-(iii)
(C) (a)-(ii); (b)-(i); (c)-(iii)
(D) (a)-(iv); (b)-(iii); (c)-(ii)

13. If one were to apply Bohr model to a particle of mass 'm' and charge 'q' moving in a plane under the influence of a magnetic field 'B', the energy of the charged particle in the nth level will be : **[2015 (Online)]**

(A) $n\left(\frac{hqB}{2\pi m}\right)$ (B) $n\left(\frac{hqB}{8\pi m}\right)$

(C) $n\left(\frac{hqB}{4\pi m}\right)$ (D) $n\left(\frac{hqB}{\pi m}\right)$

14. The de-Broglie wavelength associated with the electron in the $n = 4$ level is : **[2015 (Online)]**

(A) 1/4th of the de-Broglie wavelength of the electron in the ground state
(B) Four times the de-Broglie wavelength of the electron in the ground state
(C) Two times the de-Broglie wavelength of the electron in the ground state
(D) Half of the de-Broglie wavelength of the electron in the ground state

15. de-Broglie wavelength of an electron accelerated by a voltage of 50 V is close to ($|e| = 1.6 \times 10^{-19}$ C, $m_e = 9.1 \times 10^{31}$ kg, $h = 6.6 \times 10^{-34}$ Js) **[2015 (Online)]**

(A) 2.4 Å (B) 0.5 Å
(C) 1.7 Å (D) 1.2 Å

16. A parallel beam of electrons travelling in x-direction falls on a slit of width d (see figure). If after passing the slit, an electron acquires momentum P_y in the y-direction then for a majority of electrons passing through the slit : (h is Plank's constant) **[2015 (Online)]**

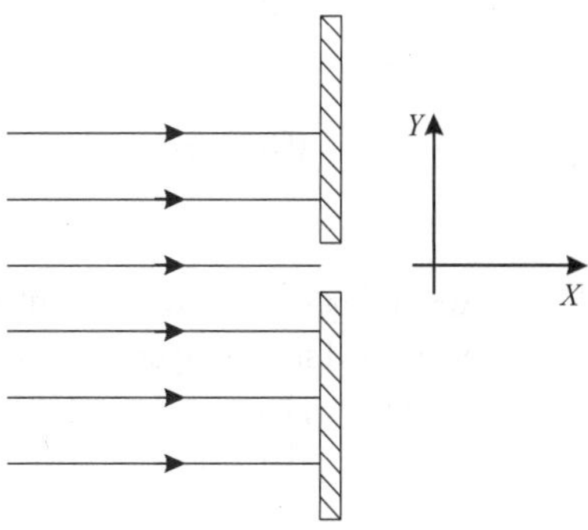

(A) $|P_y| d > h$ (B) $|P_y| d < h$
(C) $|P_y| d = h$ (D) $|P_y| d >> h$

17. A particle A of mass m and initial velocity v collides with a particle B of mass $m/2$ which is at rest. The collision is head on, and elastic. The ratio of the de-Broglie wavelengths λ_A to λ_B after the collision is : **[2017]**

(A) $\frac{\lambda_A}{\lambda_B} = \frac{2}{3}$ (B) $\frac{\lambda_A}{\lambda_B} = \frac{1}{2}$

(C) $\frac{\lambda_A}{\lambda_B} = \frac{1}{3}$ (D) $\frac{\lambda_A}{\lambda_B} = 2$

18. An electron beam is accelerated by a potential difference V to hit a metallic target to produce X-rays. It produces continuous as well as characteristic X-rays. If λ_{min} is the smallest possible wavelength of X-ray in the spectrum, the variation of $\log\lambda_{min}$ with $\log V$ is correctly represented in : **[2017]**

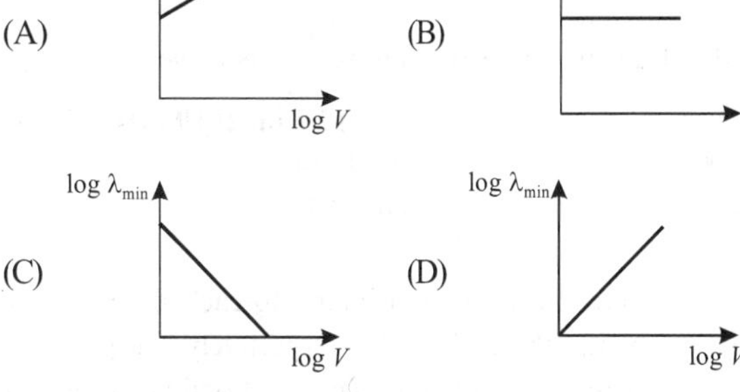

19. If the de Broglie wavelengths associated with a proton and an α-particle are equal, then the ratio of velocities of the proton and the α- particle will be : **[2018 (Online)]**

(A) 1 : 4 (B) 1 : 2
(C) 4 : 1 (D) 2 : 1

20. Two electrons are moving with non-relativistic speeds perpendicular to each other. If corresponding de Broglie wavelengths are λ_1 and λ_2, their de Broglie wavelength in the frame of reference attached to their centre of mass is : **[2018 (Online)]**

(A) $\lambda_{CM} = \lambda_1 = \lambda_2$ (B) $\frac{1}{\lambda_1} = \frac{1}{\lambda_1} + \frac{1}{\lambda_2}$

(C) $\lambda_{CM} = \frac{2\lambda_1\lambda_2}{\sqrt{\lambda_1^2 + \lambda_2^2}}$ (D) $\lambda_{CM} = \left(\frac{\lambda_1 + \lambda_2}{2}\right)$

21. The de-Broglie wavelength (λ_B) associated with the electron orbiting in the second excited state of hydrogen atom is related to that in the ground state (λ_G) by : **[2018 (Online)]**

(A) $\lambda_B = \lambda_G/3$ (B) $\lambda_B = \lambda_G/2$
(C) $\lambda_B = 2\lambda_G$ (D) $\lambda_B = 3\lambda_G$

22. In an electron microscope, the resolution that can be achieved is of the order of the wavelength of electrons used. To resolve a width of 7.5×10^{-12} m, the minimum electron energy required is close to : **[10 Jan. 2019 (Morning)]**

(A) 100 keV (B) 500 keV
(C) 25 keV (D) 1 keV

23. If the de-Broglie wavelength of an electron is equal to 10^{-3} times the wavelength of a photon of frequency 6×10^{14} Hz, then the speed of electron is equal to : **[11 Jan. 2019 (Morning)]**
(Speed of light = 3×10^8 m/s
Planck's constant = 6.63×10^{-34} J.s
Mass of electron = 9.1×10^{-31} kg)

(A) 1.45×10^6 m/s (B) 1.7×10^6 m/s
(C) 1.8×10^6 m/s (D) 1.1×10^6 m/s

24. A particle *A* of mass '*m*' and charge '*q*' is accelerated by a potential difference of 50 V. Another particle *B* of mass '4 m' and charge '*q*' is accelerated by a potential difference of 2500 *V*.

The ratio of de-Broglie wavelengths $\frac{\lambda_A}{\lambda_B}$ is close to : **[12 Jan. 2019 (Morning)]**

(A) 10.00 (B) 14.14
(C) 4.47 (D) 0.07

25. Two particles move at right angle to each other. Their de-Broglie wavelengths are λ_1 and λ_2 respectively. The particles suffer perfectly inelastic collision. The de-Broglie wavelength λ, of the final particle, is given by : **[8 April 2019 (Morning)]**

(A) $\lambda = \frac{\lambda_1 + \lambda_2}{2}$ (B) $\frac{2}{\lambda} = \frac{1}{\lambda_1} + \frac{1}{\lambda_2}$

(C) $\lambda = \sqrt{\lambda_1\lambda_2}$ (D) $\frac{1}{\lambda^2} = \frac{1}{\lambda_1^2} + \frac{1}{\lambda_2^2}$

26. A nucleus *A*, with a finite de-Broglie wavelength λ_A, undergoes spontaneous fission into two nuclei *B* and *C* of equal mass. *B* flies in the same direction as that of *A*, while *C* flies in the opposite direction with a velocity equal to half of that of *B*. The de-Broglie wavelengths λ_B and λ_C of *B* and *C* are respectively : **[8 April 2019 (Evening)]**

(A) $2\lambda_A, \lambda_A$ (B) $\lambda_A, 2\lambda_A$

(C) $\lambda_A, \frac{\lambda_A}{2}$ (D) $\frac{\lambda_A}{2}, \lambda_A$

27. A particle '*P*' is formed due to a completely inelastic collision of particles '*x*' and '*y*' having de-Broglie wavelengths 'λ_x' and 'λ_y' respectively. If *x* and *y* were moving in opposite directions, then the de-Broglie wavelength of '*P*' is : **[9 April 2019 (Evening)]**

(A) $\lambda_x + \lambda_y$ (B) $\frac{\lambda_x\lambda_y}{\lambda_x + \lambda_y}$

(C) $\frac{\lambda_x\lambda_y}{|\lambda_x - \lambda_y|}$ (D) $\lambda_x - \lambda_y$

28. 50 W/m^2 energy density of sunlight is normally incident on the surface of a solar panel. Some part of incident energy (25%) is reflected from the surface and the rest is absorbed. The force exerted on 1 m^2 surface area will be close to ($c = 3 \times 10^8$ m/s) : **[9 April 2019 (Evening)]**

(A) 15×10^{-8} N (B) 35×10^{-8} N
(C) 10×10^{-8} N (D) 20×10^{-8} N

29. Light is incident normally on a completely absorbing surface with an energy flux of 25 Wcm^{-2}. If the surface has an area of 25 cm^{-2}, the momentum transferred to the surface in 40 min time duration will be : **[10 April 2019 (Evening)]**

(A) 5.0×10^{-3} Ns (B) 3.5×10^{-6} Ns
(C) 1.4×10^{-6} Ns (D) 6.3×10^{-4} Ns

30. A 2 mW laser operates at a wavelength of 500 nm. The number of photons that will be emitted per second is :
[Given Planck's constant $h = 6.6 \times 10^{-34}$ Js, speed of light $c = 3.0 \times 10^8$ m/s] **[10 April 2019 (Evening)]**

(A) 2×10^{16} (B) 1.5×10^{16}
(C) 5×10^{15} (D) 1×10^{16}

31. An electron (of mass *m*) and a photon have the same energy *E* in the range of a few *eV*. The ratio of the de-Broglie wavelength associated with the electron and the wavelength of the photon is (c = speed of light in vacuum) **[7 Jan. 2020 (Evening)]**

(A) $\left(\frac{E}{2m}\right)^{1/2}$ (B) $\frac{1}{c}\left(\frac{E}{2m}\right)^{1/2}$

(C) $c(2mE)^{1/2}$ (D) $\frac{1}{c}\left(\frac{2E}{m}\right)^{1/2}$

32. An electron (mass m) with initial velocity $\vec{v} = v_0\hat{i} + v_0\hat{j}$ is in an electric field $\vec{E} = -E_0\hat{k}$. If λ_0 is initial de-Broglie wavelength of electron, its de-Broglie wave length at time t is given by : **[8 Jan. 2020 (Evening)]**

(A) $\frac{\lambda_0\sqrt{2}}{\sqrt{1+\frac{e^2E^2t^2}{m^2v_0^2}}}$ (B) $\frac{\lambda_0}{\sqrt{2+\frac{e^2E^2t^2}{m^2v_0^2}}}$

(C) $\frac{\lambda_0}{\sqrt{1+\frac{e^2E^2t^2}{2m^2v_0^2}}}$ (D) $\frac{\lambda_0}{\sqrt{1+\frac{e^2E_0^2t^2}{m^2v_0^2}}}$

33. A particle moving with kinetic energy E has de Broglie wavelength λ. If energy ΔE is added to its energy, the wavelength become $\lambda/2$. Value of ΔE, is : **[9 Jan. 2020 (Morning)]**

(A) $2E$ (B) E

(C) $3E$ (D) $4E$

34. An electron of mass m and magnitude of charge $|e|$ initially at rest gets accelerated by a constant electric field E. The rate of change of de-Broglie wavelength of this electron at time t ignoring relativistic effects is : **[9 Jan. 2020 (Evening)]**

(A) $\frac{-h}{|e|Et^2}$ (B) $\frac{|e|Et}{h}$

(C) $-\frac{h}{|e|E\sqrt{t}}$ (D) $-\frac{h}{|e|Et}$

35. A particle is moving 5 times as fast as an electron. The ratio of the de-Broglie wavelength of the particle to that of the electron is 1.878×10^{-4}. The mass of the particle is close to : **[2 Sept. 2020 (Evening)]**

(A) 4.8×10^{-27} kg (B) 1.2×10^{-28} kg

(C) 9.1×10^{-31} kg (D) 9.7×10^{-28} kg

36. Two sources of light emit X-rays of wavelength 1 nm and visible light of wavelength 500 nm, respectively. Both the sources emit light of the same power 200 W. The ratio of the number density of photons of X-rays to the number density of photons of the visible light of the given wavelengths is : **[3 Sept. 2020 (Evening)]**

(A) $\frac{1}{500}$ (B) 500

(C) 250 (D) $\frac{1}{250}$

37. Particle A of mass $m_A = \frac{m}{2}$ moving along the x-axis with velocity v_0 collides elastically with another particle B at rest having mass $m_B = \frac{m}{3}$. If both particles move along the x-axis after the collision, the change $\Delta\lambda$ in de-Broglie wavelength of particle A, in terms of its de-Broglie wavelength (λ_0) before collision is : **[4 Sept. 2020 (Morning)]**

(A) $\Delta\lambda = 4\lambda_0$ (B) $\Delta\lambda = \frac{5}{2}\lambda_0$

(C) $\Delta\lambda = 2\lambda_0$ (D) $\Delta\lambda = \frac{3}{2}\lambda_0$

38. An electron, a doubly ionized helium ion (He^{++}) and a proton are having the same kinetic energy. The relation between their respective de-Broglie wavelengths λ_e, $\lambda_{He^{++}}$ and λ_p is : **[6 Sept. 2020 (Morning)]**

(A) $\lambda_e < \lambda_p < \lambda_{He^{++}}$ (B) $\lambda_e < \lambda_{He^{++}} = \lambda_p$

(C) $\lambda_e > \lambda_{He^{++}} > \lambda_p$ (D) $\lambda_e > \lambda_p > \lambda_{He^{++}}$

39. Assuming the nitrogen molecule is moving with rms velocity at 400 K, the de-Broglie wavelength of nitrogen molecule is close to :

(Given : Nitrogen molecule weight 4.64×10^{-26} kg, Boltzmann constant 1.38×10^{-23} J/K, Planck constant 6.63×10^{-34} J.s) **[6 Sept. 2020 (Evening)]**

(A) 0.34 Å (B) 0.24 Å

(C) 0.20 Å (D) 0.44 Å

40. Given below are two statements :

Statement-I : Two photons having equal linear momenta have equal wavelengths

Statement-II : If the wavelength of photon is decreased, then the momentum and energy of a photon will also decrease

In the light of the above statements, choose the correct answer from the options given below. **[24 Feb. 2021 (Morning)]**

(A) Statement-I is false but Statement-II is true

(B) Both Statement-I and Statement-II are true

(C) Statement-I is true but Statement-II is false

(D) Both Statement-I and Statement-II are false

41. An X-ray tube is operated at 1.24 million volt. The shortest wavelength of the produced photon will be : **[24 Feb. 2021 (Evening)]**

(A) 10^{-4} nm (B) 10^{-2} nm

(C) 10^{-3} nm (D) 10^{-1} nm

42. The de-Broglie wavelength of a proton and α-particle are equal. The ratio of their velocities is : **[24 Feb. 2021 (Evening)]**

(A) 4 : 3 (B) 1 : 4
(C) 4 : 2 (D) 4 : 1

43. An α particle and a proton are accelerated from rest by a potential difference of 200 V. After this, their de-Broglie wavelengths are λ_α and λ_p respectively. $\frac{\lambda_p}{\lambda_\alpha}$ is : **[25 Feb. 2021 (Morning)]**

(A) 7.8 (B) 8
(C) 2.8 (D) 3.8

44. An electron of mass m_e and a proton of mass $m_p = 1836\, m_e$ are moving with the same speed. The ratio of their de-Broglie wavelength $\frac{\lambda_{\text{electron}}}{\lambda_{\text{proton}}}$ will be : **[25 Feb. 2021 (Evening)]**

(A) $\frac{1}{1836}$ (B) 918
(C) 1836 (D) 1

45. The wavelength of an X-ray beam is 10 Å. The mass of a fictitious particle having the same energy as that of the X-ray photons is $\frac{x}{3}h$ kg. The value of x is ______.
(h = Planck's constant) **[25 Feb. 2021 (Evening)]**

46. Given below are two statements : one is labelled as **Assertion A** and the other is labelled as **Reason R.**

Assertion-A : An electron microscope can achieve better resolving power than an optical microscope.

Reason-R : The de-Broglie's wavelength of the electrons emitted from an electron gun is much less than wavelength of visible light.

In the light of the above statements, choose the correct answer from the options given below : **[26 Feb. 2021 (Morning)]**

(A) **A** is true but **R** is false.
(B) **A** is false but **R** is true.
(C) Both **A** and **R** are true but **R** is NOT the correct explanation of **A**.
(D) Both **A** and **R** are true and **R** is the correct explanation of **A**.

47. The recoil speed of a hydrogen atom after it emits a photon in going from $n = 5$ state to $n = 1$ state will be : **[26 Feb. 2021 (Evening)]**

(A) 3.25 m/s (B) 4.34 m/s
(C) 4.17 m/s (D) 2.19 m/s

48. The de-Broglie wavelength associated with an electron and a proton were calculated by accelerating them through same potential of 100 V. What should nearly be the ratio of their wavelengths ? (m_p = 1.00727 u, m_e = 0.00055 u) **[16 March 2021 (Evening)]**

(A) 1860 : 1 (B) $(1860)^2 : 1$
(C) 41.4 : 1 (D) 43 : 1

49. Red light differs from blue light as they have : **[16 March 2021 (Evening)]**

(A) Different frequencies and different wavelengths
(B) Different frequencies and same wavelengths
(C) Same frequencies and same wavelengths
(D) Same frequencies and different wavelengths

50. An electron of mass m and a photon have same energy E. The ratio of wavelength of electron to that of photon is : (c being the velocity of light) **[17 March 2021 (Morning)]**

(A) $\frac{1}{c}\left(\frac{2m}{E}\right)^{1/2}$ (B) $\frac{1}{c}\left(\frac{E}{2m}\right)^{1/2}$
(C) $\left(\frac{E}{2m}\right)^{1/2}$ (D) $c\,(2mE)^{1/2}$

51. A particle is travelling 4 times as fast as an electron. Assuming the ratio of de-Broglie wavelength of a particle to that of electron is 2 : 1, the mass of the particle is : **[18 March 2021 (Morning)]**

(A) $\frac{1}{16}$ times the mass of e^- (B) 8 times the mass of e^-
(C) 16 times the mass of e^- (D) $\frac{1}{8}$ times the mass of e^-

52. An electron having de-Broglie wavelength λ is incident on a target in a X-ray tube. Cut-of wavelength of emitted X-ray is : **[20 July 2021 (Evening)]**

(A) $\frac{2m^2c^2\lambda^2}{h^2}$ (B) $\frac{hc}{mc}$
(C) 0 (D) $\frac{2mc\lambda^2}{h}$

53. An electron of mass m_e and a proton of mass m_p are accelerated through the same potential difference. The ratio of the de-Broglie wavelength associated with the electron to that with the proton is : **[22 July 2021 (Evening)]**

(A) $\frac{m_p}{m_e}$ (B) 1
(C) $\frac{m_e}{m_p}$ (D) $\sqrt{\frac{m_p}{m_e}}$

54. A particle of mass $4M$ at rest disintegrates into two particles of mass M and $3M$ respectively having non zero velocities. The ratio of de-Broglie wavelength of particle of mass M to that of mass $3M$ will be : **[25 July 2021 (Morning)]**

(A) $1:\sqrt{3}$ (B) $3:1$
(C) $1:1$ (D) $1:3$

55. What should be the order of arrangement of de-Broglie wavelength of electron (λ_e), an α-particle (λ_α) and proton (λ_p) given that all have the same kinetic energy? **[25 July 2021 (Morning)]**

(A) $\lambda_e=\lambda_p>\lambda_\alpha$ (B) $\lambda_e<\lambda_p<\lambda_\alpha$
(C) $\lambda_e=\lambda_p=\lambda_\alpha$ (D) $\lambda_e>\lambda_p>\lambda_\alpha$

56. An electron moving with speed v and a photon moving with speed c, have same de-Broglie wavelength. The ratio of kinetic energy of electron to that of photon is : **[25 July 2021 (Evening)]**

(A) $\frac{v}{3c}$ (B) $\frac{3c}{v}$
(C) $\frac{v}{2c}$ (D) $\frac{2c}{v}$

57. A particle of mass 9.1×10^{-31} kg travels in a medium with a speed of 10^6 m/s and a photon of a radiation of linear momentum 10^{-27} kg m/s travels in vacuum. The wavelength of photon is ______ times the wavelength of the particle. **[27 July 2021 (Morning)]**

58. The K_α X-ray of molybdenum has wavelength 0.071 nm. If the energy of a molybdenum atom with a K electron knocked out is 27.5 keV, the energy of this atom when an L electron is knocked out will be_____keV. (Round off to the nearest integer) ($h=4.14\times10^{-15}$ eVs, $c=3\times108$ ms^{-1}). **[27 July 2021 (Evening)]**

59. The de-Broglie wavelength of a particle having kinetic energy E is λ. How much extra energy must be given to this particle so that the de-Broglie wavelength reduces to 75% of the initial value? **[26 Aug. 2021 (Evening)]**

(A) E (B) $\frac{7}{9}E$
(C) $\frac{16}{9}E$ (D) $\frac{7}{9}E$

60. A monochromatic neon lamp with wavelength of 670.5 illuminates a photo-sensitive material which has a stopping voltage of 0.48 V. What will be the stopping voltage if the source light is changed with another source of wavelength of 476.6 nm? **[27 Aug. 2021 (Evening)]**

(A) 0.24 V (B) 1.5 V
(C) 1.25 V (D) 0.96 V

61. A moving proton and electron have the same de-Broglie wavelength. If K and P denote the kinetic energy and momentum respectively. Then choose the correct option : **[31 Aug. 2021 (Morning)]**

(A) $K_p<K_e$ and $P_p=P_e$ (B) $K_p>K_e$ and $P_p=P_e$
(C) $K_p=K_e$ and $P_p=P_e$ (D) $K_p<K_e$ and $P_p<P_e$

62. Given below are two statements : **[25 June 2022 (Morning)]**
Statement-I : Davisson-Germer experiment establishes the wave nature of electrons
Statement-II : If electrons have wave nature, they can interfere and show diffraction
In the light of the above statements choose the correct answer from the options given below :
(A) Both Statement-I and Statement-II are true
(B) Both Statement-I and Statement-II are false
(C) Statement-I is true but Statement-II is false
(D) Statement-I is false but Statement-II is true

63. A proton, a neutron, an electron and an α-particle have same energy. If λ_p, λ_n, λ_e and λ_α are the de-Broglie's wavelengths of proton, neutron, electron and α particle respectively, then choose the correct relation from the following : **[25 June 2022 (Evening)]**

(A) $\lambda_p=\lambda_n>\lambda_e>\lambda_\alpha$ (B) $\lambda_\alpha<\lambda_n<\lambda_e<\lambda_e$
(C) $\lambda_e<\lambda_p=\lambda_n>\lambda_\alpha$ (D) $\lambda_e=\lambda_p=\lambda_n=\lambda_\alpha$

64. An electron with speed v and a photon with speed c have the same de-Broglie wavelength. If the kinetic energy and momentum of electron are E_e and p_e and that of photon are E_{ph} and p_{ph} respectively. Which of the following is correct? **[26 June 2022 (Morning)]**

(A) $\frac{E_e}{E_{ph}}=\frac{2c}{v}$ (B) $\frac{E_e}{E_{ph}}=\frac{v}{2c}$

(C) $\frac{p_e}{p_{ph}}=\frac{2c}{v}$ (D) $\frac{p_e}{p_{ph}}=\frac{v}{2c}$

65. An α particle and a carbon 12 atom has same kinetic energy K. The ratio of their de-Broglie wavelength $f(\lambda_\alpha:\lambda_{C12})$ is : **[27 June 2022 (Morning)]**

(A) $1:\sqrt{3}$ (B) $\sqrt{3}:1$
(C) $3:1$ (D) $2:\sqrt{3}$

66. The de-Broglie wavelengths for an electron and a photon are λ_e and λ_p respectively. For the same kinetic energy of electron and photon. Which of the following presents the correct relation between the de-Brogue wavelengths of two? **[28 June 2022 (Morning)]**

(A) $\lambda_p\propto\lambda_e^2$ (B) $\lambda_p\propto\lambda_e$

(C) $\lambda_p\propto\sqrt{\lambda_e}$ (D) $\lambda_p\propto\sqrt{\frac{1}{\lambda_e}}$

67. The ratio of wavelengths of proton and deuteron accelerated by potential V_p and V_d is $1:\sqrt{2}$. Then, the ratio of V_p to V_d will be : **[25 July 2022 (Evening)]**

(A) 1 : 1 (B) $\sqrt{2}:1$
(C) 2 : 1 (D) 4 : 1

68. A parallel beam of light of wavelength 900 nm and intensity 100 Wm^{-2} is incident on a surface perpendicular to the beam. Tire number of photons crossing 1 cm^2 area perpendicular to the beam in one second is : **[26 July 2022 (Morning)]**

(A) 3×10^{16} (B) 4.5×10^{16}
(C) 4.5×10^{17} (D) 4.5×10^{20}

69. A nucleus of mass M at rest splits into two parts having masses $\frac{M'}{3}$ and $\frac{2M'}{3}$ $(M' < M)$. The ratio of de-Broglie wavelength of two parts will be : **[26 July 2022 (Evening)]**

(A) 1 : 2 (B) 2 : 1
(C) 1 : 1 (D) 2 : 3

70. An electron (mass m) with an initial velocity $\vec{v} = v_0\hat{i}$ $(v_0 > 0)$ is moving in an electric field $\vec{E} = -E_0\hat{i}$ $(E_0 > 0)$ where E_0 is constant. If at $t = 0$ de Broglie wavelength is $\lambda_0 = \frac{h}{mv_0}$, then its de Broglie wavelength after time t is given by : **[27 July 2022 (Morning)]**

(A) λ_0 (B) $\lambda_0\left(1+\frac{eE_0t}{mv_0}\right)$

(C) $\lambda_0 t$ (D) $\frac{\lambda_0}{\left(1+\frac{eE_0t}{mv_0}\right)}$

71. The equation $\lambda = \frac{1.227}{x}$ *nm* can be used to find the de-Broglie wavelength of an electron. In this equation x stands for : **[28 July 2022 (Morning)]**

Where,
m = mass of electron
P = momentum of electron
K = Kinetic energy of electron
V = Accelerating potential in volts for electron

(A) $\sqrt{mK}$ (B) $\sqrt{P}$
(C) $\sqrt{K}$ (D) $\sqrt{V}$

72. Sun light falls normally on a surface of area 36 cm^2 and exerts an average force of 7.2×10^{-9} N within a time period of 20 minutes. Considering a case of complete absorption, the energy flux of incident light is : **[28 July 2022 (Evening)]**

(A) 25.92×10^2 W/cm^2 (B) 8.64×10^{-6} W/cm^2
(C) 6.0 W/cm^2 (D) 0.06 W/cm^2

73. An α-particle, a proton and an electron have the same kinetic energy. Which one of the following is correct in case of their De-Broglie wavelength : **[24 Jan. 2023 (Evening)]**

(A) $\lambda_\alpha > \lambda_p > \lambda_e$ (B) $\lambda_\alpha < \lambda_p < \lambda_e$
(C) $\lambda_\alpha = \lambda_p = \lambda_e$ (D) $\lambda_\alpha > \lambda_p < \lambda_e$

74. Electron beam used in an electron microscope, when accelerated by a voltage of 20 kV, has a de-Broglie wavelength of λ_0. If the voltage is increased to 40 kV, then the de-Broglie wavelength associated with the electron beam would be : **[25 Jan. 2023 (Morning)]**

(A) $3\lambda_0$ (B) $9\lambda_0$

(C) $\frac{\lambda_0}{2}$ (D) $\frac{\lambda_0}{\sqrt{2}}$

75. The ratio of de-Broglie wavelength of an α-particle and a proton accelerated from rest by the same potential is $\frac{1}{\sqrt{m}}$, the value of m is : **[29 Jan. 2023 (Evening)]**

(A) 4 (B) 16
(C) 8 (D) 2

76. A small object at rest, absorbs a light pulse of power 20 mW and duration 300 ms. Assuming speed of light as 3×10^8 m/s. The momentum of the object becomes equal to : **[30 Jan. 2023 (Morning)]**

(A) 0.5×10^{-17} kg m/s (B) 2×10^{-17} kg m/s
(C) 3×10^{-17} kg m/s (D) 1×10^{-17} kg m/s

77. A point source of light is placed at the centre of curvature of a hemispherical surface. The source emits a power of 24 W. The radius of curvature of hemisphere is 10 cm and the inner surface is completely reflecting. The force on the hemisphere due to the light falling on it is _____ $\times 10^{-8}$ N. **[30 Jan. 2023 (Morning)]**

78. An electron accelerated through a potential difference V_1 has a de-Broglie wavelength of λ. When the potential is changed to V_2, its de-Broglie wavelength increases by 50%. The value of $\left(\frac{V_1}{V_2}\right)$ is equal to : **[30 Jan. 2023 (Evening)]**

(A) 3 (B) $\frac{9}{4}$

(C) $\frac{3}{2}$ (D) 4

79. Given below are two statements : One is labelled as **Assertion-A** and the other is labelled as **Reason-R**.

Assertion-A : The beam of electrons shows wave nature and exhibit interference and diffraction.

Reason-R : Davisson Germer Experimentally verified the wave nature of electrons.

In the light of the above statements. Choose the most appropriate answer from the options given below : **[31 Jan. 2023 (Morning)]**

(A) **A** is correct but **R** is not correct.

(B) **A** is not correct but **R** is correct.

(C) Both **A** and **R** are correct but **R** is Not the correct explanation of **A**.

(D) Both **A** and **R** are correct and **R** is the correct explanation of **A**.

80. A proton moving with one tenth of velocity of light has a certain de-Broglie wavelength of λ. An alpha particle having certain kinetic energy has the same de-Broglie wavelength λ. The ratio of kinetic energy of proton and that of alpha particle is : **[1 Feb. 2023 (Morning)]**

(A) 2 : 1 (B) 4 : 1
(C) 1 : 2 (D) 1 : 4

81. The kinetic energy of an electron, α-particle and a proton are given as 4 K, 2 K and K respectively. The de-Broglie wavelength associated with electron (λ_e) α – particle (λ_α) and the proton (λ_p) are as follows : **[06 April 2023 (Morning)]**

(A) $\lambda_\alpha = \lambda_p < \lambda_e$ (B) $\lambda_\alpha > \lambda_p > \lambda_e$
(C) $\lambda_\alpha < \lambda_p < \lambda_e$ (D) $\lambda_\alpha = \lambda_p > \lambda_e$

82. Proton (P) and electron (e) will have same de-Broglie wavelength when the ratio of their momentum is z: (Assume, $m_P = 1849\, m_e$) **[8 April 2023 (Morning)]**

(A) 1 : 43 (B) 43 : 1
(C) 1 : 1849 (D) 1 : 1

83. The waves emitted when a metal target is bombarded with high energy electrons are : **[8 April 2023 (Evening)]**

(A) Microwaves (B) X-rays
(C) Infrared rays (D) Radio Waves

84. The de Broglie wavelength of a molecule in a gas at room temperature (300 K) is λ_1. If the temperature of the gas is increased to 600 K, then the de Broglie wavelength of the same gas molecule becomes: **[10 April 2023 (Morning)]**

(A) $\frac{1}{\sqrt{2}}\lambda_1$ (B) $2\lambda_1$

(C) $\frac{1}{2}\lambda_1$ (D) $\sqrt{2}\lambda_1$

85. The ratio of the de-Broglie wavelengths of proton and electron having same kinetic energy : (Assume $m_p = m_e \times 1849$) **[11 April 2023 (Evening)]**

(A) 1 : 43 (B) 1 : 30
(C) 1 : 62 (D) 2 : 43

86. A proton and an α-particle are accelerated from rest by 2 V and 4 V potentials, respectively. The ratio of their de-Broglie wavelength is : **[12 April 2023 (Morning)]**

(A) 4 : 1 (B) 2 : 1
(C) 8 : 1 (D) 16 : 1

87. The de Broglie wavelength of an electron having kinetic energy E is λ. If the kinetic energy of electron becomes $\frac{E}{4}$, then its de-Broglie wavelength will be : **[15 April 2023 (Morning)]**

(A) $\frac{\lambda}{\sqrt{2}}$ (B) $\frac{\lambda}{2}$

(C) 2λ (D) $\sqrt{2}\lambda$

* * * * *

CHAPTER 26

ATOMS

26.1 ATOMIC STRUCTURE

1. If the binding energy of the electron in a hydrogen atom is 13.6 eV, the energy required to remove the electron from the first excited state of Li^{2+} is : **[2002]**

(A) 30.6 eV (B) 13.6 eV
(C) 3.4 eV (D) 122.4 eV

2. The wavelengths involved in the spectrum of deuterium ($^2_1 D$) are slightly different from that of hydrogen spectrum, because : **[2003]**

(A) sizes of the two nuclei are different
(B) nuclear forces are different in the two cases
(C) masses of the two nuclei are different
(D) attraction between the electron and the nucleus is different in the two cases

3. If 13.6 eV energy is required to ionize the hydrogen atom, then the energy required to remove an electron from $n = 2$ is : **[2003]**

(A) 10.2 eV (B) 0 eV
(C) 3.4 eV (D) 6.8 eV

4. The diagram shows the energy levels for an electron in a certain atom. Which transition shown represents the emission of a photon with the most energy? **[2005]**

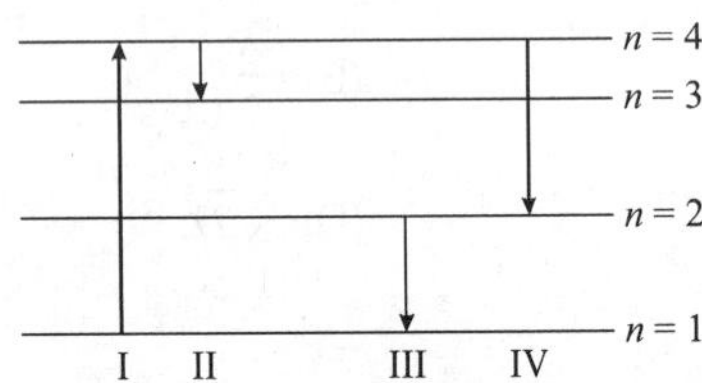

(A) III (B) IV
(C) I (D) II

5. Which of the following transitions in hydrogen atoms emit photons of highest frequency ? **[2007]**

(A) $n = 2$ to $n = 6$ (B) $n = 6$ to $n = 2$
(C) $n = 2$ to $n = 1$ (D) $n = 1$ to $n = 2$

6. Suppose an electron is attracted towards the origin by a force k/r where 'k' is a constant and 'r' is the distance of the electron from the origin. By applying Bohr model to this system, the radius of the nth orbital of the electron is found to be 'r_n' and the kinetic energy of the electron to be 'T_n'. Then which of the following is true? **[2008]**

(A) $T_n \propto \frac{1}{n^2}, r_n \propto n^2$

(B) T_n independent of n, $r_n \propto n$

(C) $T_n \propto \frac{1}{n}, r_n \propto n$

(D) $T_n \propto \frac{1}{n}, r_n \propto n^2$

7. The transition from the state $n = 4$ to $n = 3$ in a hydrogen like atom results in ultraviolet radiation. Infrared radiation will be obtained in the transition from : **[2009]**

(A) $4 \rightarrow 2$ (B) $5 \rightarrow 4$
(C) $2 \rightarrow 1$ (D) $3 \rightarrow 2$

8. Energy required for the electron excitation in Li^{++} from the first to the third Bohr orbit is : **[2011]**

(A) 108.8 eV (B) 122.4 eV
(C) 12.1 eV (D) 36.3 eV

9. A diatomic molecule is made of two masses m_1 and m_2 which are separated by a distance r. If we calculate its rotational energy by applying Bohr's rule of angular momentum quantization, its energy will be given by : (n is an integer) : **[2012]**

(A) $\dfrac{(m_1 + m_2)n^2h^2}{2m_1m_2r^2}$ (B) $\dfrac{(m_1 + m_2)^2n^2h^2}{2m_1^2m_2^2r^2}$

(C) $\dfrac{n^2h^2}{2(m_1 + m_2)r^2}$ (D) $\dfrac{2n^2h^2}{(m_1 + m_2)r^2}$

10. Hydrogen atom is excited from ground state to another state with principal quantum number equal to 4. Then the number of spectral lines in the emission spectra will be : **[2012]**

(A) 6 (B) 2
(C) 3 (D) 5

11. The electron of a hydrogen atom makes a transition from the $(n + 1)^{th}$ orbit to the nth orbit. For large n the wavelength of the emitted radiation is proportional to : **[2012 (Online)]**

(A) n (B) n^3
(C) n^4 (D) n^2

12. A hypothetical atom has only three energy levels. The ground level has energy, $E_1 = -8$ eV. The two excited states have energies, $E_2 = -6$ eV and $E_3 = -2$ eV. Then which of the following wavelengths will not be present in the emission spectrum of this atom? **[2012 (Online)]**

(A) 207 nm (B) 465 nm
(C) 310 nm (D) 620 nm

13. A doubly ionised Li atom is excited from its ground state ($n = 1$) to $n = 3$ state. The wavelengths of the spectral lines are given by λ_{32}, λ_{31} and λ_{21}. The ratio $\lambda_{32}/\lambda_{31}$ and $\lambda_{21}/\lambda_{31}$ are, respectively : **[2012 (Online)]**

(A) 8.1, 0.67 (B) 8.1, 1.2
(C) 6.4, 1.2 (D) 6.4, 0.67

14. Which of the plots shown in the figure represents speed (v_n) of the electron in a hydrogen atom as a function of the principal quantum number (n) ? **[2012 (Online)]**

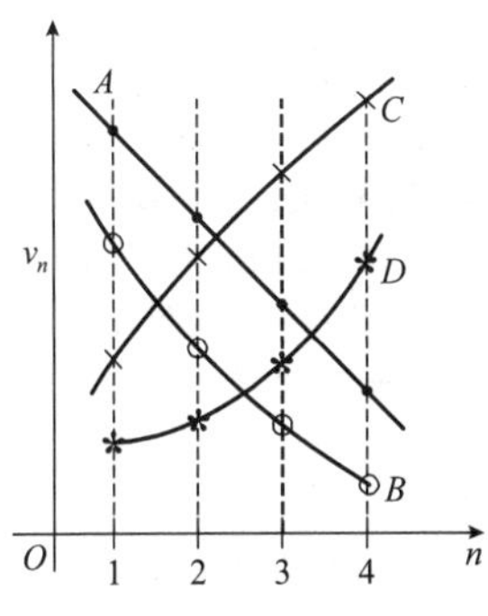

(A) B (B) D
(C) C (D) A

15. In the Rutherford experiment, α-particles are scattered from a nucleus as shown. Out of the four paths, which path is not possible ? **[2012 (Online)]**

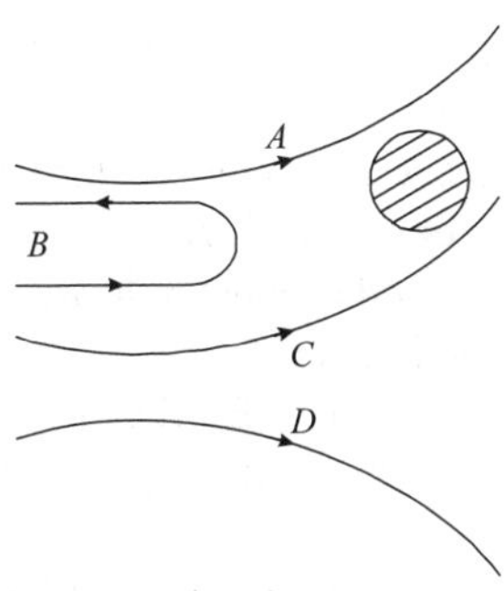

(A) D (B) B
(C) C (D) A

16. Ionisation energy of Li (Lithium) atom in ground state is 5.4 eV. Binding energy of an electron in Li^+ ion in ground state is 75.6 eV. Energy required to remove all three electrons of Lithium (Li) atom is : **[2012 (Online)]**

(A) 81.0 eV (B) 135.4 eV
(C) 203.4 eV (D) 156.6 eV

17. In a hydrogen like atom electron makes transition from an energy level with quantum number n to another with quantum number $(n - 1)$. If $n >> 1$, the frequency of radiation emitted is proportional to : **[2013]**

(A) $\frac{1}{n}$ (B) $\frac{1}{n^2}$
(C) $\frac{1}{n^{3/2}}$ (D) $\frac{1}{n^3}$

18. In the Bohr model an electron moves in a circular orbit around the proton. Considering the orbiting electron to be a circular current loop, the magnetic moment of the hydrogen atom, when the electron is in n^{th} excited state, is : **[2013 (Online)]**

(A) $\left(\frac{e}{2m}\frac{n^2h}{2\pi}\right)$ (B) $\left(\frac{e}{m}\right)\frac{nh}{2\pi}$
(C) $\left(\frac{e}{2m}\right)\frac{nh}{2\pi}$ (D) $\left(\frac{e}{m}\right)\frac{n^2h}{2\pi}$

19. Orbits of a particle moving in a circle are such that the perimeter of the orbit equals an integer number of de-Broglie wavelengths of the particle. For a charged particle moving in a plane perpendicular to a magnetic field, the radius of the nth orbital will therefore be proportional to : **[2013 (Online)]**

(A) n^2 (B) n
(C) $n^{1/2}$ (D) $n^{1/4}$

20. A 12.5 eV electron beam is used to bombard gaseous hydrogen at room temperature. It will emit : **[2013 (Online)]**

(A) 2 lines in the Lyman series and 1 line in the Balmer series
(B) 3 lines in the Lyman series
(C) 1 line in the Lyman series and 2 lines in the Balmer series
(D) 3 lines in the Balmer series

21. In the Bohr's model of hydrogen-like atom the force between the nucleus and the electron is modified as $F = \frac{e^2}{4\pi\varepsilon_0}\left(\frac{1}{r^2}+\frac{\beta}{r^3}\right)$, where β is a constant. For this atom, the radius of the nth orbit in terms of the Bohr radius $\left(a_0 = \frac{\varepsilon_0 h^2}{m\pi e^2}\right)$ is : **[2013 (Online)]**

(A) $r_n = a_0 n - \beta$ (B) $r_n = a_0 n^2 + \beta$
(C) $r_n = a_0 n^2 - \beta$ (D) $r_n = a_0 n + \beta$

22. Hydrogen ($_1H^1$), Deuterium ($_1H^2$), singly ionised Helium

$({}_2He^4)^+$ and doubly ionised lithium $({}_3Li^6)^{++}$ all have one electron around the nucleus. Consider an electron transition from $n=2$ to $n=1$. If the wave lengths of emitted radiation are $\lambda_1, \lambda_2, \lambda_3$ and λ_4 respectively then approximately which one of the following is correct? **[2014]**

(A) $4\lambda_1 = 2\lambda_2 = 2\lambda_3 = \lambda_4$ (B) $\lambda_1 = 2\lambda_2 = 2\lambda_3 = \lambda_4$
(C) $\lambda_1 = \lambda_2 = 4\lambda_3 = 9\lambda_4$ (D) $\lambda_1 = 2\lambda_2 = 3\lambda_3 = 4\lambda_4$

23. The binding energy of the electron in a hydrogen atom is 13.6 eV, the energy required to remove the electron from the first excited state of Li^{++} is: **[2014 (Online)]**

(A) 122.4 eV (B) 30.6 eV
(C) 13.6 eV (D) 3.4 eV

24. As an electron makes a transition from an excited state to the ground state of a hydrogen-like atom/ion: **[2015]**

(A) its kinetic energy increases but potential energy and total energy decrease
(B) kinetic energy, potential energy and total energy decrease
(C) kinetic energy decreases, potential energy increases but total energy remains same
(D) kinetic energy and total energy decrease but potential energy increases

25. A neutron moving with a speed 'v' makes a head on collision with a stationary hydrogen atom in ground state. The minimum kinetic energy of the neutron for which inelastic collision will take place is: **[2016 (Online)]**

(A) 20.4 eV (B) 10.2 eV
(C) 12.1 eV (D) 16.8 eV

26. A hydrogen atom makes a transition from $n=2$ to $n=1$ and emits a photon. This photon strikes a doubly ionized lithium atom ($Z=3$) in excited state and completely removes the orbiting electron. The least quantum number for the excited state of the ion for the process is: **[2016 (Online)]**

(A) 2 (B) 4
(C) 5 (D) 3

27. Some energy levels of a molecules are shown in the figure. The ratio of the wavelengths $r = \lambda_1/\lambda_2$, is given by: **[2017]**

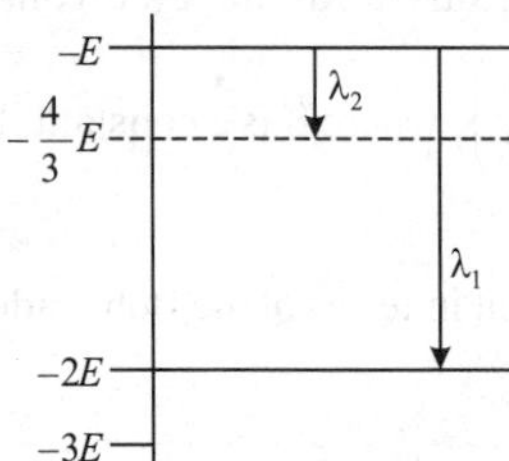

(A) $r = \frac{3}{4}$ (B) $r = \frac{1}{3}$
(C) $r = \frac{4}{3}$ (D) $r = \frac{2}{3}$

28. According to Bohr's theory, the time averaged magnetic field at the centre (i.e. nucleus) of a hydrogen atom due to the motion of electrons in the nth orbit is proportional to:
(n = principal quantum number) **[2017 (Online)]**

(A) n^{-4} (B) n^{-5}
(C) n^{-3} (D) n^{-2}

29. The acceleration of an electron in the first orbit of the hydrogen atom ($z=1$) is: **[2017 (Online)]**

(A) $\frac{h^2}{\pi^2 m^2 r^3}$ (B) $\frac{h^2}{8\pi^2 m^2 r^3}$
(C) $\frac{h^2}{4\pi^2 m^2 r^3}$ (D) $\frac{h^2}{4\pi m^2 r^3}$

30. If the series limit frequency of the Lyman series is v_L, then the series limit frequency of the P-fund series at: **[2018]**

(A) $25\,v_L$ (B) $16\,v_L$
(C) $v_L/16$ (D) $v_L/25$

31. An electron from various excited states of hydrogen atom emit radiation to come to the ground state. Let λ_n, λ_g be the de Broglie wavelength of the electron in the nth state and the ground state respectively. Let Λ_A be the wavelength of the emitted photon in the transition from the nth state to the ground state. For large n, (A, B are constants) **[2018]**

(A) $\Lambda_n \approx A + \frac{B}{\lambda_n^2}$ (B) $\Lambda_n \approx A + B\lambda_n$
(C) $\Lambda_n^2 \approx A + B\lambda_n^2$ (D) $\Lambda_n^2 \approx \lambda$

32. Muon (μ^{-1}) is negatively charged ($|q| = |e|$) with a mass $m_\mu = 200\, m_e$, where m_e is the mass of the electron and e is the electronic charge. If μ^{-1} is bound to a proton to form a hydrogen like atom, identify the correct statements: **[2018 (Online)]**

(a) Radius of the muonic orbit is 200 times smaller than that of the electron
(b) the speed of the μ^{-1} in the nth orbit is 1/200 times that of the electron in the nth orbit
(c) The ionization energy of muonic atom is 200 times more than that of an hydrogen atom
(d) The momentum of the muon in the nth orbit is 200 times more than that of the electron

(A) (a), (b), (d) (B) (b), (d)
(C) (c), (d) (D) (a), (c), (d)

33. The energy required to remove the electron from a singly ionized Helium atom is 2.2 times the energy required to remove an electron from Helium atom. The total energy required to ionize the Helium atom completely is: **[2018 (Online)]**

(A) 20 eV (B) 79 eV
(C) 109 eV (D) 34 eV

34. A hydrogen atom, initially in the ground state is excited by absorbing a photon of wavelength 980 Å. The radius of the atom in the excited state, in terms of Bohr radius a_0, will be : ($h_c = 12500$ eV – Å) **[11 Jan. 2019 (Morning)]**

(A) $9\ a_0$ (B) $25\ a_0$

(C) $4\ a_0$ (D) $16\ a_0$

35. In a hydrogen like atom, when an electron jumps from the *M*-shell to the *L*-shell, the wavelength of emitted radiation is λ. If an electron jumps from *N*-shell to the *L*-shell, the wavelength of emitted radiation will be : **[11 Jan. 2019 (Evening)]**

(A) $\frac{27}{20}\lambda$ (B) $\frac{16}{25}\lambda$

(C) $\frac{20}{27}\lambda$ (D) $\frac{25}{16}\lambda$

36. A particle of mass *m* moves in a circular orbit in a central potential field $U(r) = 1/2kr^2$. If Bohr's quantization conditions are applied, radii of possible orbitals and energy levels vary with quantum number *n* as : **[12 Jan. 2019 (Morning)]**

(A) $r_n \propto n^2, E_n \propto \frac{1}{n^2}$ (B) $r_n \propto \sqrt{n}, E_n \propto \frac{1}{n}$

(C) $r_n \propto n, E_n \propto n$ (D) $r_n \propto \sqrt{n}\ , E_n \propto n$

37. In a Frank-Hertz experiment, an electron of energy 5.6 eV passes through mercury vapour and emerges with an energy 0.7 eV. The minimum wavelength of photons emitted by mercury atoms is close to : **[12 Jan. 2019 (Evening)]**

(A) 2020 nm (B) 220 nm

(C) 250 nm (D) 1700 nm

38. Radiation coming from transitions $n = 2$ to $n = 1$ of hydrogen atoms fall on He^+ ions in $n = 1$ and $n = 2$ states. The possible transition of helium ions as they absorb energy from the radiation is : **[8 April 2019 (Morning)]**

(A) $n = 1 \rightarrow n = 4$ (B) $n = 2 \rightarrow n = 4$

(C) $n = 2 \rightarrow n = 5$ (D) $n = 2 \rightarrow n = 3$

39. Taking the wavelength of first Balmer line in hydrogen spectrum ($n = 3$ to $n = 2$) as 660 nm, the wavelength of the 2nd Balmer line ($n = 4$ to $n = 2$) will be : **[9 April 2019 (Morning)]**

(A) 889.2 nm (B) 642.7 nm

(C) 488.9 nm (D) 388.9 nm

40. A He^+ ion is in its first excited state. Its ionization energy is: **[9 April 2019 (Evening)]**

(A) 6.04 eV (B) 13.60 eV

(C) 54.40 eV (D) 48.36 eV

41. In Li^{++}, electron in first Bohr orbit is excited to a level by a radiation of wavelength λ. When the ion gets deexcited to the ground state in all possible ways (including intermediate emissions), a total of six spectral lines are observed. What is the value of λ? (Given : $h = 6.63 \times 10^{-34}$ Js ; $c = 3 \times 10^8$ ms^{-1}) **[10 April 2019 (Evening)]**

(A) 9.4 nm (B) 12.3 nm

(C) 10.8 nm (D) 11.4 nm

42. An excited He^+ ion emits two photons in succession, with wavelengths 108.5 nm and 30.4 nm, in making a transition to ground state. The quantum number *n*, corresponding to its initial excited state is : (For photon of wavelength λ, energy $E = \frac{1240\text{eV}}{\lambda(\text{in } nm)}$) **[12 April 2019 (Morning)]**

(A) $n = 5$ (B) $n = 4$

(C) $n = 6$ (D) $n = 7$

43. The electron in a hydrogen atom first jumps from the third excited state to the second excited state and subsequently to the first excited state. The ratio of the respective wavelengths, λ_1/λ_2, of the photons emitted in this process is : **[12 April 2019 (Evening)]**

(A) 9/7 (B) 7/5

(C) 27/5 (D) 20/7

44. Consider an electron in a hydrogen atom, revolving in its second excited state (having radius 4.65 Å). The de-Broglie wavelength of this electron is : **[12 April 2019 (Evening)]**

(A) 12.9 Å (B) 3.5 Å

(C) 9.7 Å (D) 6.6 Å

45. The time period of revolution of electron in its ground state orbit in a hydrogen atom is 1.6×10^{-16} s. The frequency of revolution of the electron in its first excited state (in s^{-1}) is : **[7 Jan. 2020 (Morning)]**

(A) 6.2×10^{15} (B) 5.6×10^{12}

(C) 7.8×10^{14} (D) 1.6×10^{14}

46. The graph which depicts the results of Rutherford gold foil experiment with α-particles is :

θ : Scattering angle

Y : Number of scattered α-particles detected (Plots are schematic and not to scale) : **[8 Jan. 2020 (Morning)]**

(A)
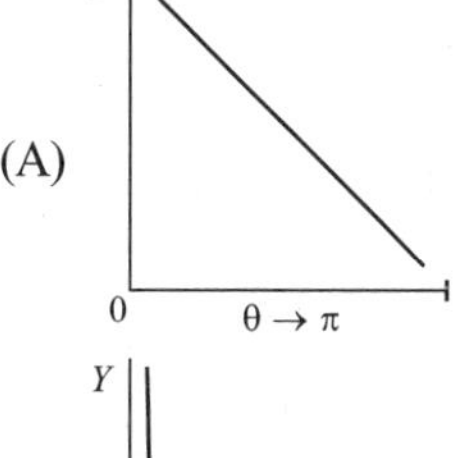

(B)
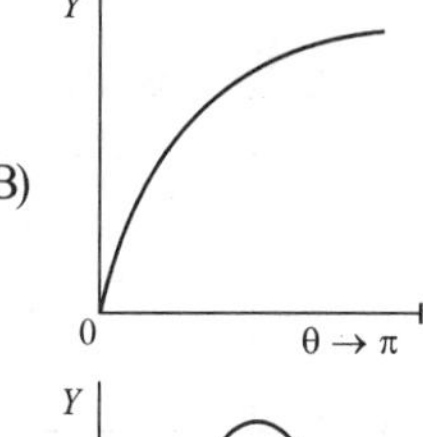

(C)
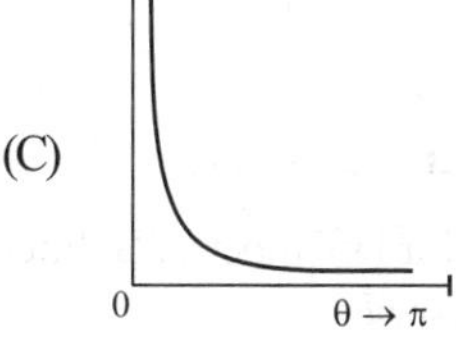

(D)
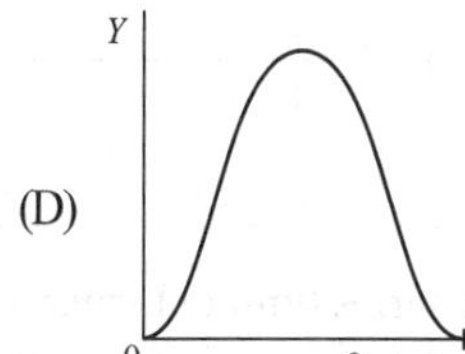

47. The first member of the Balmer series of hydrogen atom has a wavelength of 6561 Å. The wavelength of the second member of the Balmer series (in *nm*) is ______. **[8 Jan. 2020 (Evening)]**

48. The energy required to ionise a hydrogen like ion in its ground state is 9 Rydbergs. What is the wavelength of the radiation emitted when the electron in this ion jumps from the second excited state to the ground state ? **[9 Jan. 2020 (Evening)]**

(A) 35.8 nm (B) 24.2 nm
(C) 8.6 nm (D) 11.4 nm

49. In a hydrogen atom the electron makes a transition from $(n+1)^{th}$ level to the nth level. If $n >> l$, the frequency of radiation emitted is proportional to : **[2 Sept. 2020 (Evening)]**

(A) $\frac{1}{n^4}$ (B) $\frac{1}{n^3}$

(C) $\frac{1}{n^2}$ (D) $\frac{1}{n}$

50. In the line spectra of hydrogen atom, difference between the largest and the shortest wavelengths of the Lyman series is 304 Å. The corresponding difference for the Paschen series in Å is ______. **[4 Sept. 2020 (Morning)]**

51. A particle of mass 200 MeV/c^2 collides with a hydrogen atom at rest. Soon after the collision the particle comes to rest, and the atom recoils and goes to its first excited state. The initial kinetic energy of the particle (in *eV*) is $\frac{N}{4}$. The value of N is :
(Given the mass of the hydrogen atom to be 1 GeV/c^2) : **[5 Sept. 2020 (Morning)]**

52. In the given figure, the energy levels of hydrogen atom have been shown along with some transitions marked *A*, *B*, *C*, *D* and *E*. The transitions *A*, *B* and *C* respectively represent : **[24 Feb. 2021 (Morning)]**

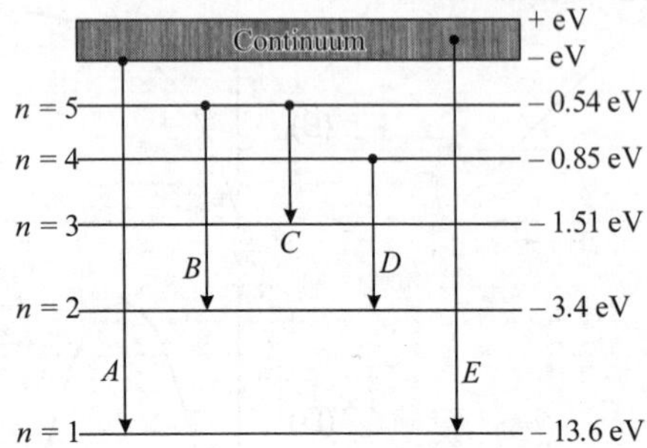

(A) The series limit of Lyman series, third member of Balmer series and second member of Paschen series

(B) The ionization potential of hydrogen, second member of Balmer series and third member of Paschen series

(C) The first member of the Lyman series, third member of Balmer series and second member of Paschen series

(D) The series limit of Lyman series, second member of Balmer series and second member of Paschen series

53. According to Bohr atom model, in which of the following transitions will the frequency be maximum ? **[24 Feb. 2021 (Evening)]**

(A) $n=2$ to $n=1$ (B) $n=4$ to $n=3$
(C) $n=3$ to $n=2$ (D) $n=5$ to $n=4$

54. The wavelength of the photon emitted by a hydrogen atom when an electron makes a transition from $n=2$ to $n=1$ state is : **[25 Feb. 2021 (Evening)]**

(A) 121.8 nm (B) 194.8 nm
(C) 913.3 nm (D) 490.7 nm

55. If λ_1 and λ_2 are the wavelength of the third member of Lyman and first member of the Paschen series respectively, then the value of $\lambda_1 : \lambda_2$ is : **[26 Feb. 2021 (Morning)]**

(A) 1 : 9 (B) 7 : 108
(C) 1 : 3 (D) 7 : 135

56. The first three spectral lines of H-atom in the Balmer series are given $\lambda_1, \lambda_2, \lambda_3$ considering the Bohr atomic model, the wave lengths of first and third spectral lines $\left(\frac{\lambda_1}{\lambda_3}\right)$ are related by a factor of approximately 'x' × 10^{-1}. The value of x, to the nearest integer, is ______. **[16 March 2021 (Morning)]**

57. If an electron is moving in the nth orbit of the hydrogen atom, then its velocity (v_n) for the nth orbit is given as : **[17 March 2021 (Morning)]**

(A) $v_n \propto n$ (B) $v_n \propto \frac{1}{n}$

(C) $v_n \propto n^2$ (D) $v_n \propto \frac{1}{n^2}$

58. Which level of the single ionized carbon has the same energy as the ground state energy of hydrogen atom? **[17 March 2021 (Morning)]**

(A) 1 (B) 6
(C) 4 (D) 8

59. The atomic hydrogen emits a line spectrum consisting of various series. Which series of hydrogen atomic spectra is lying in the visible region ? **[17 March 2021 (Evening)]**

(A) Brackett series (B) Paschen series
(C) Lyman series (D) Balmer series

60. Imagine that the electron in a hydrogen atom is replaced by a muon (μ). The mass of muon particle is 207 times that of an electron and charge is equal to the charge of an electron. The ionization potential of this hydrogen atom will be :

[18 March 2021 (Morning)]

(A) 13.6 eV (B) 2815.2 eV
(C) 331.2 eV (D) 27.2 eV

61. Consider a situation in which reverse biased current of a particular P-N junction increases when it is exposed to a light of wavelength $\leq$ 621 nm. During this process, enhancement in carrier concentration takes place due to generation of hole-electron pairs. The value of band gap is nearly :
(Take hc/e = 1224 nm) **[22 July 2021 (Evening)]**

(A) 4 eV (B) 1 eV
(C) 2 eV (D) 0.5 eV

62. In Bohr's atomic model, the electron is assumed to revolve in a circular orbit of radius 0.5 Å. If the speed of electron is 2.2×10^6 m/s, then the current associated with the electron will be ______ $\times 10^{-2}$ mA. [Take π as $\frac{22}{7}$] **[27 July 2021 (Morning)]**

63. Consider the following statements :

[27 July 2021 (Evening)]

(1) Atoms of each element emit characteristics spectrum
(2) According to Bohr's postulate, an electron in a hydrogen atom, revolves in a certain stationary orbit
(3) The density of nuclear matter depends on the size of the nucleus
(4) Neutrons always decay into protons
(5) Radioactivity is an indication of the instability of nuclei.
Choose the correct answer from the options given below :

(A) 1, 2 and 5 only (B) 1, 3 and 5 only
(C) 2 and 4 only (D) 1, 2, 3, 4 and 5

64. The difference in the number of waves when yellow light propagates through air and vacuum columns of the same thickness is one. The thickness of the column is ______ mm. [Refractive index of air = 1.0003, wavelength of yellow light in vaccum = 6000 Å] **[27 July 2021 (Evening)]**

65. A particular hydrogen like ion radiation of frequency 2.92×10^{15} Hz when it makes transition from $n = 3$ to $n = 1$. The frequency in Hz of radiation emitted in transition from $n = 2$ to $n = 1$ will be : **[26 Aug. 2021 (Morning)]**

(A) 6.57×10^{15} (B) 2.46×10^{15}
(C) 4.38×10^{15} (D) 0.44×10^{15}

66. X different wavelength may be observed in the spectrum from a hydrogen sample if the atoms are exited to states with principle quantum number $n = 6$? The value of X is ______.

[27 Aug. 2021 (Evening)]

67. Choose the correct option from the following options given below : **[24 June 2022 (Morning)]**

(A) In the ground state of Rutherford's model electrons are in stable equilibrium. While in Thomson's model electrons always experience a net-force.
(B) An atom has a nearly continuous mass distribution in a Rutherford's model but has a highly non-uniform mass distribution in Thomson's model.
(C) A classical atom based on Rutherford's model is doomed to collapse.
(D) The positively charged part of the atom possesses most of the mass in Rutherford's model but not in Thomson's model.

68. In Bohr's atomic model of hydrogen, let K, P and E are the kinetic energy, potential energy and total energy of the electron respectively. Choose the correct option when the electron undergoes transitions to a higher level :

[24 June 2022 (Evening)]

(A) All K, P and E increase
(B) K decreases P and E increase
(C) P decreases K and E increase
(D) K increases P and E decrease

69. The ratio for the speed of the electron in the 3rd orbit of He^+ to the speed of the electron in the 3rd orbit of hydrogen atom will be : **[25 June 2022 (Morning)]**

(A) 1 : 1 (B) 1 : 2
(C) 4 : 1 (D) 2 : 1

70. A hydrogen atom in is ground state absorbs 10.2 eV of energy. The angular momentum of electron of the hydrogen atom will increase by the value of :
[Given : Plank's constant = 6.6×10^{-34} Js]

[27 June 2022 (Morning)]

(A) 2.10×10^{-34} Js (B) 1.05×10^{-34} Js
(C) 3.15×10^{-34} Js (D) 4.2×10^{-34} Js

71. A beam of monochromatic light is used to excite the electron in Li^{++} from the first orbit to the third orbit. The wavelength of monochromatic light is found to be $x \times 10^{-10}$ m. The value of x is ______. [Given hc = 1242 eV nm] **[27 June 2022 (Morning)]**

72. Given below are two statements :
Statement-I : In hydrogen atom, the frequency of radiation emitted when an electron jumps from lower energy orbit (E_1) to higher energy orbit (E_2), is given as $hf = E_1 - E_2$
Statement-II : The jumping of electron from higher energy orbit (E_2) to lower energy orbit (E_1) is associated with frequency of radiation given as $f = (E_2 - E_1)/h$
This condition is Bohr's frequency condition. In the light of the above statements, choose the correct answer from the options given below : **[27 June 2022 (Evening)]**

(A) Both statement-I and statement-II are true
(B) Both statement-I and statement-II are false
(C) Statement-I is correct but statement-II is false
(D) Statement-I is incorrect but statement-II is true

73. $\sqrt{d_1}$ and $\sqrt{d_2}$ are the impact parameters corresponding to scattering angles 60° and 90° respectively, when an α particle is approaching a gold nucleus. For $d_1 = xd_2$, the value of x will be_____. **[29 June 2022 (Morning)]**

74. The momentum of an electron revolving in n^{th} orbit is given by : (Symbols have their usual meanings) **[25 July 2022 (Morning)]**

(A) $\frac{nh}{2\pi r}$ (B) $\frac{nh}{2r}$

(C) $\frac{nh}{2\pi}$ (D) $\frac{2\pi r}{nh}$

75. The magnetic moment of an electron (e) revolving in an orbit around nucleus with an orbital angular momentum is given by : **[25 July 2022 (Morning)]**

(A) $\vec{\mu}_L = \frac{e\vec{L}}{2m}$ (B) $\vec{\mu}_L = -\frac{e\vec{L}}{2m}$

(C) $\vec{\mu}_1 = -\frac{e\vec{L}}{m}$ (D) $\vec{\mu}_1 = \frac{2e\vec{L}}{m}$

76. An electron with energy 0.1 keV moves at right angle to the earth's magnetic field of 1×10^{-4} Wbm^{-2}. The frequency of revolution of the electron will be :
[Take : Mass of electron $= 9.0 \times 10^{-31}$ kg] **[25 July 2022 (Evening)]**

(A) 1.6×10^5 Hz (B) 5.6×10^5 Hz
(C) 2.8×10^6 Hz (D) 1.8×10^6 Hz

77. Hydrogen atom from excited state comes to the ground by emitting a photon of wavelength λ. The value of principal quantum number n of the excited state will be :
[R : Rydberg constant] **[25 July 2022 (Evening)]**

(A) $\sqrt{\frac{\lambda R}{\lambda - 1}}$ (B) $\sqrt{\frac{\lambda R}{\lambda R - 1}}$

(C) $\sqrt{\frac{\lambda}{\lambda R - 1}}$ (D) $\sqrt{\frac{\lambda R^2}{\lambda R - 1}}$

78. $\frac{x}{x+4}$ is the ratio of energies of photons produced due to transition of an electron of hydrogen atom from its.
(i) Third permitted energy level to the second level and
(ii) The highest permitted energy level to the second permitted level
The value of x will be. **[25 July 2022 (Evening)]**

79. In a hydrogen spectrum, λ be the wavelength of first transition line of Lyman series. The wavelength difference will be "$a\lambda$" between the wavelength of 3rd transition line of Paschen series and that of 2nd transition line of Balmer Series where $a =$ _____. **[26 July 2022 (Morning)]**

80. Find the ratio of energies of photons produced due to transition of an election of hydrogen atom from its (i) second permitted energy level to the first level, and (ii) the highest permitted energy level to the first permitted level : **[29 July 2022 (Morning)]**

(A) 3 : 4 (B) 4 : 3
(C) 1 : 4 (D) 4 : 1

81. A photon is emitted in transition from $n = 4$ to $n = 1$ level in hydrogen atom. The corresponding wavelength for this transition is : (Given, $h = 4 \times 10^{-15}$ eVs) **[24 Jan. 2023 (Evening)]**

(A) 94.1 nm (B) 941 nm
(C) 97.4 nm (D) 99.3 nm

82. The wavelength of the radiation emitted is λ_0 when an electron jumps from the second excited state to the first excited state of hydrogen atom. If the electron jumps from the third excited state to the second orbit of the hydrogen atom, the wavelength of the radiation emitted will be $\frac{20}{x}\lambda_0$. The value of x is _____. **[25 Jan. 2023 (Morning)]**

83. The energy levels of an atom is shown is figure. Which one of these transitions will result in the emission of a photon of wavelength 124.1 nm ? (Given: $h = 6.62 \times 10^{-34}$ Js) **[25 Jan. 2023 (Evening)]**

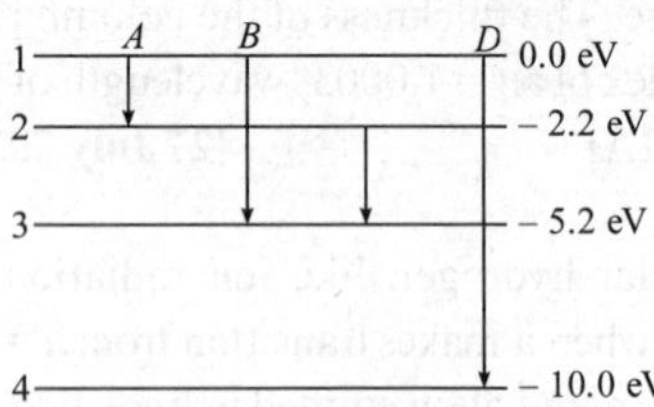

(A) B (B) A
(C) C (D) D

84. Speed of an electron in Bohr's 7th orbit for Hydrogen atom is 3.6×10^6 m/s. The corresponding speed of the electron in 3rd orbit, in m/s is : **[30 Jan. 2023 (Morning)]**

(A) (1.8×10^6) (B) (7.5×10^6)
(C) (3.6×10^6) (D) (8.4×10^6)

85. For hydrogen atom, λ_1 and λ_2 are the wavelengths corresponding to the transitions 1 and 2 respectively as shown in figure. The ratio of λ_1 and λ_2 is $\frac{x}{32}$. The value of x is_____.

[31 Jan. 2023 (Morning)]

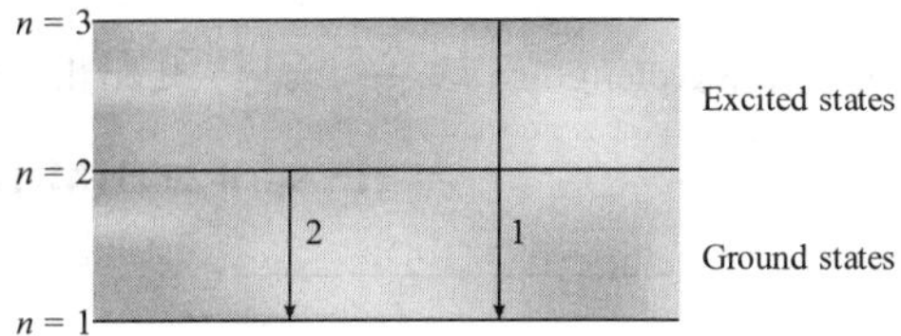

86. The radius of electron's second stationary orbit in Bohr's atom is R. The radius of 3rd orbit will be :

[31 Jan. 2023 (Evening)]

(A) $\frac{R}{3}$ (B) $2.25R$

(C) $3R$ (D) $9R$

87. If the binding energy of ground state electron in a hydrogen atom is 13.6 eV, then, the energy required to remove the electron from the second excited state of Li^{2+} will be : $x \times 10^{-1}$ eV. The value of x is _____. **[31 Jan. 2023 (Evening)]**

88. A light of energy 12.75 eV is incident on a hydrogen atomic its ground state. The atom absorbs the radiation and reaches to one of its excited states. The angular momentum of the atom in the excited state is $\frac{x}{\pi} \times 10^{-17}$ eVs. The value of x is _____ (Use $h = 4.14 \times 10^{-15}$ eVs, $c = 3 \times 10^8$ ms^{-1})

[1 Feb. 2023 (Morning)]

89. An electron of a hydrogen like atom, having $Z = 4$, jumps from 4th energy state to 2nd energy state. The energy released in this process, will be :
(Given $Rch = 13.6$ eV)
Where R = Rydberg constant
c = Speed of light in vacuum
h = Planck's constant : **[1 Feb. 2023 (Evening)]**

(A) 13.6 eV (B) 10.5 eV

(C) 3.4 eV (D) 40.8 eV

90. The energy levels of an hydrogen atom are shown below. The transition corresponding to emission of shortest wavelength is **[6 April 2023 (Morning)]**

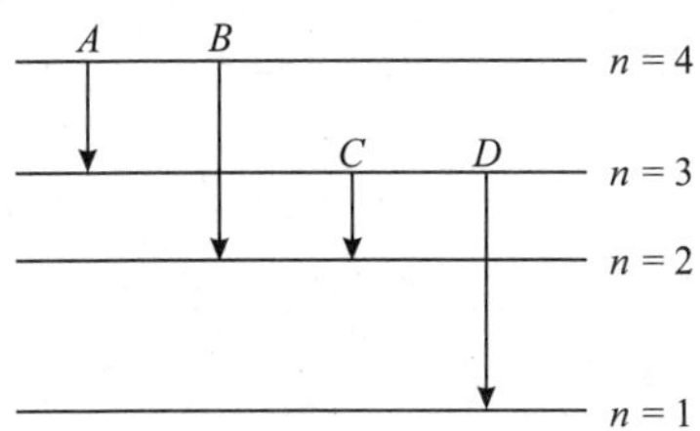

(A) C (B) D

(C) B (D) A

91. The radius of fifth orbit of the Li^{++} is _____ $\times 10^{-12}$ m.
[Take : radius of hydrogen atom = 0.51 Å]

[6 April 2023 (Morning)]

92. A small particle of mass m moves in such a way that its potential energy $U = \frac{1}{2} m\omega^2 r^2$ where ω is constant and r is the distance of the particle from origin. Assuming Bohr's quantization of momentum and circular orbit, the radius of nth orbit will be proportional to. **[6 April 2023 (Evening)]**

(A) $\sqrt{n}$ (B) n

(C) n^2 (D) $\frac{1}{n}$

93. Experimentally it is found that 12.8 eV energy is required to separate a hydrogen atom into a proton and an electron. So, the orbital radius of the electron in a hydrogen atom is $\frac{9}{x} \times 10^{-10}$ m. The value of the x is _____.

($1 \text{ eV} = 1.6 \times 10^{-19}$ J, $\frac{1}{4\pi \epsilon_0} = 9 \times 10^9$ Nm^2/C^2 and electronic charge $= 1.6 \times 10^{-19}$ J C) **[6 April 2023 (Evening)]**

94. The ratio of wavelength of spectral lines H_α and H_β in the Balmer series is $\frac{x}{20}$. The value of x is.

[8 April 2023 (Evening)]

95. The angular momentum for the electron in Bohr's orbit is L. If the electron is assumed to revolve in second orbit of hydrogen atom, then the change in angular momentum will be:

[10 April 2023 (Morning)]

(A) $\frac{L}{2}$ (B) Zero

(C) L (D) $2L$

96. A monochromatic light is incident on a hydrogen sample in ground state. Hydrogen atoms absorb a fraction of light and subsequently emit radiation of six different wavelengths. The frequency of incident light is $x \times 10^{15}$ Hz. The value of x is_____.
(Given $h = 4.25 \times 10^{-15}$ eVs) **[11 April 2023 (Morning)]**

97. The energy of He^+ ion in its first excited state is : (The ground state energy for the Hydrogen atom is –13.6 eV)

[11 April 2023 (Evening)]

(A) –3.4 eV (B) –54.4 eV

(C) –13.6 eV (D) –27.2 eV

98. A 12.5 eV electron beam is used to bombard gaseous hydrogen at room temperature. The number of spectral lines emitted will be: **[12 April 2023 (Morning)]**

(A) 2 (B) 1

(C) 3 (D) 4

99. The radius of 2nd orbit of He^+ of Bohr's model is r_1 and that of fourth orbit of Be^{3+} is represented as r_2. Now the ratio $\frac{r_2}{r_1}$ is $x : 1$. The value of x is _____. **[13 April 2023 (Morning)]**

100. An atom absorbs a photon of wavelength 500 nm and emits another photon of wavelength 600 nm. The net energy absorbed by the atom in this process is $n \times 10^{-4}$ eV. The value of n is _____. **[13 April 2023 (Evening)]**

101. As per given figure A, B and C are the first, second and third excited energy level of hydrogen atom respectively. If the ratio of the two wavelengths $\left(\text{i.e.} \frac{\lambda_1}{\lambda_2}\right)$ is $\frac{7}{4n}$, then the value of n will be _____. **[15 April 2023 (Morning)]**

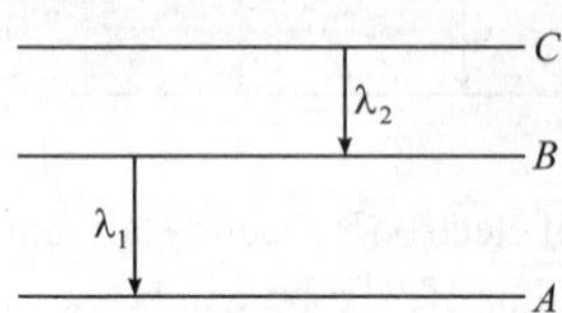

* * * * *

NUCLEI

27.1 RADIOACTIVITY

1. At a specific instant emission of radioactive compound is deflected in a magnetic field. The compound can emit : **[2002]**

(i) electrons (ii) protons

(iii) He^{2+} (iv) neutrons

The emission at the instant can be :

(A) i, ii, iii (B) i, ii, iii, iv

(C) iv (D) ii, iii

2. If N_0 is the original mass of the substance of half-life period $t_{1/2} = 5$ years, then the amount of substance left after 15 years is : **[2002]**

(A) $\frac{N_0}{8}$ (B) $\frac{N_0}{16}$

(C) $\frac{N_0}{2}$ (D) $\frac{N_0}{4}$

3. A radioactive sample at any instant has its disintegration rate 5000 disintegration per minute. After 5 minutes, the rate is 1250 disintegrations per minute. Then, the decay constant (per minute) is : **[2003]**

(A) 0.4 ln 2 (B) 0.2 ln 2

(C) 0.1 ln 2 (D) 0.8 ln 2

4. Which of the following radiations has the least wavelength ? **[2003]**

(A) γ-rays (B) β-rays

(C) α-rays (D) X-rays

5. A nucleus disintegrates into two nuclear parts which have their velocities in the ratio 2 : 1. The ratio of their nuclear sizes will be : **[2004]**

(A) $2^{1/3} : 1$ (B) $1 : 3^{1/2}$

(C) $3^{1/2} : 1$ (D) $1 : 2^{1/3}$

6. If radius of the $^{27}_{13}Al$ nucleus is estimated to be 3.6 fermi, then the radius of $^{125}_{52}Te$ nucleus be nearly : **[2005]**

(A) 6 fermi (B) 8 fermi

(C) 4 fermi (D) 5 fermi

7. Starting with a sample of pure ^{66}Cu, 7/8 of it decays into Zn in 15 min. The corresponding half-life is : **[2005]**

(A) 10 min (B) 15 min

(C) 5 min (D) $7\frac{1}{2}$ min

8. The half-life period of a radioactive element X is same as the mean life time of another radioactive element Y. Initially they have the same number of atoms. Then : **[2007]**

(A) X will decay faster than Y

(B) Y will decay faster than X

(C) Y and X have same decay rate initially

(D) X and Y decay at same rate always

9. The half life of a radioactive substance is 20 minutes. The approximate time interval $(t_2 - t_1)$ between the time t_2 when 2/3 of it has decayed and time t_1 when 1/3 of it had decayed is : **[2011]**

(A) 20 min (B) 28 min

(C) 7 min (D) 14 min

10. Half-lives of two radioactive elements A and B are 20 minutes and 40 minutes, respectively. Initially, the samples have equal number of nuclei. After 80 minutes, the ratio of decayed numbers of A and B nuclei will be : **[2012]**

(A) 1 : 16 (B) 4 : 1

(C) 1 : 4 (D) 5 : 4

11. The counting rate observed from a radioactive source at $t = 0$ was 1600 counts s^{-1}, and $t = 8$ s, it was 100 counts s^{-1}. The counting rate observed as counts s^{-1} at $t = 6$ s will be : **[2012 (Online)]**

(A) 250 (B) 400

(C) 300 (D) 200

12. A sample originally contained 10^{20} radioactive atoms, which emit α-particles. The ratio of α-particles emitted in the third year to that emitted during the second year is 0.3. How many α-particles were emitted in the first year ? **[2012 (Online)]**

(A) 3×10^{18} (B) 3×10^{19}

(C) 5×10^{18} (D) 7×10^{19}

13. Which of the following Statements is correct ? **[2012 (Online)]**

(A) The rate of radioactive decay cannot be controlled but that of nuclear fission can be controlled

(B) Nuclear forces are short range, attractive and charge dependent

(C) Nuclei of atoms having same number of neutrons are known as isobars

(D) Wavelength of matter waves is given by de-Broglie formula but that of photons is not given by the same formula

14. The decay constants of a radioactive substance for α and β emission are λ_α and λ_β respectively. If the substance emits α and β simultaneously, then the average half life of the material will be : **[2012 (Online)]**

(A) $\frac{2T_\alpha T_\beta}{T_\alpha + T_\beta}$ (B) $T_\alpha + T_\beta$

(C) $\frac{T_\alpha T_\beta}{T_\alpha + T_\beta}$ (D) $\frac{1}{2}(T_\alpha + T_\beta)$

15. The half-life of a radioactive element A is the same as the mean-life of another radioactive element B. Initially both substances have the same number of atoms, then : **[2013 (Online)]**

(A) A and B decay at the same rate always
(B) A and B decay at the same rate initially
(C) A will decay at a faster rate than B
(D) B will decay at a faster rate than A

16. A radioactive nuclei with decay constant 0.5/s is being produced at a constant rate of 100 nuclei/s. If at $t = 0$ there were no nuclei, the time when there are 50 nuclei is : **[2014 (Online)]**

(A) 1 s (B) $2\ln\left(\frac{4}{3}\right)$ s

(C) ln 2 s (D) $\ln\left(\frac{4}{3}\right)$ s

17. A piece of bone of an animal from a ruin is found to have ^{14}C activity of 12 disintegrations per minute per *gm* of its carbon content. The ^{14}C activity of a living animal is 16 disintegrations per minute per *gm*. How long ago nearly did the animal die? (Given half life of ^{14}C is $t_{1/2} = 5760$ years) : **[2014 (Online)]**

(A) 1672 years (B) 2391 years
(C) 3291 years (D) 4453 years

18. A piece of wood from a recently cut tree shows 20 decays per minute. A wooden piece of same size placed in a museum (obtained from a tree cut many years back) shows 2 decays per minute. If half life of C^{14} is 5730 years, then age of the wooden piece placed in the museum is approximately : **[2014 (Online)]**

(A) 10439 years (B) 13094 years
(C) 19039 years (D) 39049 years

19. Let N_β be the number of β particles emitted by 1 gram of Na^{24} radioactive nuclei (half life = 15 hrs) in 7.5 hours, N_β is close to (Avogadro number = 6.023×10^{23}/g. mole) : **[2015 (Online)]**

(A) 6.2×10^{21} (B) 7.5×10^{21}
(C) 1.25×10^{22} (D) 1.75×10^{22}

20. A radioactive nucleus A with a half life T, decays into a nucleus B. At $t = 0$, there is no nucleus B. At sometime t, the ratio of the number of B to that of A is 0.3. Then, t is given by : **[2017]**

(A) $t = T\log(1.3)$ (B) $t = \frac{T}{\log(1.3)}$

(C) $t = T\frac{\log(2)}{\log(1.3)}$ (D) $t = T\frac{\log(1.3)}{\log(2)}$

21. A solution containing active cobalt $^{60}_{27}Co$ having activity of 0.8 μCi and decay constant λ is injected in an animal's body. If 1 cm^3 of blood is drawn from the animal's body after 10 hrs of injection, the activity found was 300 decays per minute. What is the volume of blood that is flowing in the body ? (1Ci = 3.7×10^{10} decay per second and at $t = 10$ hrs $e^{-\lambda t} = 0.84$) **[2018 (Online)]**

(A) 6 litres (B) 7 litres
(C) 4 litres (D) 5 litres

22. At some instant, a radioactive sample S_1 having an activity 5 μCi has twice the number of nuclei as another sample S_2 which has an activity of 10 μCi. The half lives of S_1 and S_2 are : **[2018 (Online)]**

(A) 10 years and 20 years, respectively
(B) 5 years and 20 years, respectively
(C) 20 years and 10 years, respectively
(D) 20 years and 5 years, respectively

23. A sample of radioactive material A, that has an activity of 10 mCi (1 Ci = 3.7×10^{10} decays/s), has twice the number of nuclei as another sample of a different radioactive material B which has an activity of 20 mCi. The correct choices for half-lives of A and B would then be respectively : **[9 Jan. 2019 (Morning)]**

(A) 20 days and 5 days (B) 20 days and 10 days
(C) 5 days and 10 days (D) 10 days and 40 days

24. At a given instant, say $t = 0$, two radioactive substances A and B have equal activities. The ratio R_B/R_A of their activities after time t itself decays with time t as e^{-3t}. If the half-life of A is ln2, the half-life of B is : **[9 Jan. 2019 (Evening)]**

(A) $\frac{\ln 2}{2}$ (B) 2 ln2

(C) $\frac{\ln 2}{4}$ (D) 4 ln2

25. Using a nuclear counter the count rate of emitted particles from a radioactive source is measured. At $t = 0$ it was 1600 counts per second and $t = 8$ seconds it was 100 counts per second. The count rate observed, as counts per second, at $t = 6$ seconds is close to : **[10 Jan. 2019 (Morning)]**

(A) 150 (B) 360
(C) 200 (D) 400

26. In a radioactive decay chain, the initial nucleus is $^{232}_{90}$Th. At the end there are 6 α-particles and 4 β-particles which are emitted. If the end nucleus, If $^{A}_{Z}X$, A and Z are given by : **[12 Jan. 2019 (Evening)]**

(A) $A=208; Z=80$ (B) $A=202; Z=80$
(C) $A=200; Z=81$ (D) $A=208; Z=82$

27. Two radioactive materials A and B have decay constants 10λ and λ, respectively. It initially they have the same number of nuclei, then the ratio of the number of nuclei of A to that of B will be $1/e$ after a time : **[10 April 2019 (Morning)]**

(A) $\frac{11}{10\lambda}$ (B) $\frac{1}{9\lambda}$
(C) $\frac{1}{10\lambda}$ (D) $\frac{1}{11\lambda}$

28. Two radioactive substances A and B have decay constants 5λ and λ respectively. At $t=0$, a sample has the same number of the two nuclei. The time taken for the ratio of the number of nuclei to become $\left(\frac{1}{e}\right)^2$ will be : **[10 April 2019 (Evening)]**

(A) $1/4\lambda$ (B) $1/\lambda$
(C) $1/2\lambda$ (D) $2/\lambda$

29. Half lives of two radioactive nuclei A and B are 10 minutes and 20 minutes, respectively. If, initially a sample has equal number of nuclei, then after 60 minutes, the ratio of decayed numbers of nuclei A and B will be : **[12 April 2019 (Evening)]**

(A) 9 : 8 (B) 1 : 8
(C) 8 : 1 (D) 3 : 8

30. The activity of a radioactive sample falls from 700 s^{-1} to 500 s^{-1} in 30 minutes. Its half life is close to : **[7 Jan. 2020 (Evening)]**

(A) 66 min (B) 52 min
(C) 72 min (D) 62 min

31. In a radioactive material, fraction of active material remaining after time t is 9/16. The fraction that was remaining after $t/2$ is **[3 Sept. 2020 (Morning)]**

(A) $\frac{3}{4}$ (B) $\frac{7}{8}$
(C) $\frac{4}{5}$ (D) $\frac{3}{5}$

32. Activities of three radioactive substances A, B and C are represented by the curves A, B and C, in the figure. Then their half-lives $T_{1/2}(A) : T_{1/2}(B) : T_{1/2}(C)$ are in the ratio : **[5 Sept. 2020 (Morning)]**

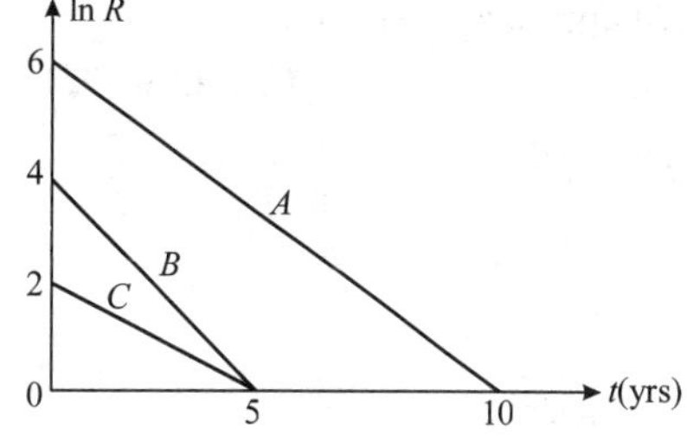

(A) 3 : 2 : 1 (B) 4 : 3 : 1
(C) 2 : 1 : 3 (D) 2 : 1 : 1

33. A radioactive nucleus decays by two different processes. The half life for the first process is 10 s and that for the second is 100 s. The effective half life of the nucleus is close to : **[5 Sept. 2020 (Evening)]**

(A) 9 s (B) 55 s
(C) 6 s (D) 12 s

34. Two radioactive substances X and Y originally have N_1 and N_2 nuclei respectively. Half life of X is half of the half life of Y. After three lives of Y, number of nuclei of both are equal. The ratio $\frac{N_1}{N_2}$ will be equal to : **[25 Feb. 2021 (Morning)]**

(A) $\frac{8}{1}$ (B) $\frac{3}{1}$
(C) $\frac{1}{3}$ (D) $\frac{1}{8}$

35. A radioactive sample is undergoing α decay. At any time t_1, its activity is A and another time t_2, the activity is $A/5$. What is the average life time for the sample ? **[26 Feb. 2021 (Evening)]**

(A) $\frac{t_2-t_1}{\ln 5}$ (B) $\frac{\ln(t_2+t_1)}{2}$
(C) $\frac{\ln 5}{t_2-t_1}$ (D) $\frac{t_1-t_2}{\ln 5}$

36. The half-life of Au^{198} is 2.7 days. The activity of 1.50 mg of Au^{198} if its atomic weight is 198 g mol^{-1} is, ($N_A = 6\times10^{23}$/mol) **[16 March 2021 (Evening)]**

(A) 240 Ci (B) 357 Ci
(C) 535 Ci (D) 252 Ci

37. Calculate the time interval between 33% decay and 67% decay if half-life of a substance is 20 minutes : **[16 March 2021 (Evening)]**

(A) 60 minutes (B) 20 minutes
(C) 40 minutes (D) 13 minutes

38. A radioactive sample disintegrates via two independent decay processes having half lives $T_{1/2}^{(1)}$ and $T_{1/2}^{(2)}$ respectively. The effective half-life $T_{1/2}$ of the nuclei is : **[18 March 2021 (Morning)]**

(A) $T_{1/2} = \frac{T_{1/2}^{(1)} + T_{1/2}^{(2)}}{T_{1/2}^{(1)} - T_{1/2}^{(2)}}$ (B) $T_{1/2} = T_{1/2}^{(1)} + T_{1/2}^{(2)}$

(C) $T_{1/2} = \frac{T_{1/2}^{(1)} T_{1/2}^{(2)}}{T_{1/2}^{(1)} + T_{1/2}^{(2)}}$ (D) None of the above

39. A radioactive material decays by simultaneous emissions of two particles with half lives of 1400 years and 700 years respectively. What will be the time after which one third of the material remains ? (Take ln 3 = 1.1) **[20 July 2021 (Morning)]**

(A) 700 years (B) 340 years
(C) 740 years (D) 1110 years

40. For a certain radioactive process the graph between ln$R$ (logarithm of activity) and t(in second) is obtained as shown in the figure. Then the value of half-life for the unknown radioactive material is approximately : **[20 July 2021 (Evening)]**

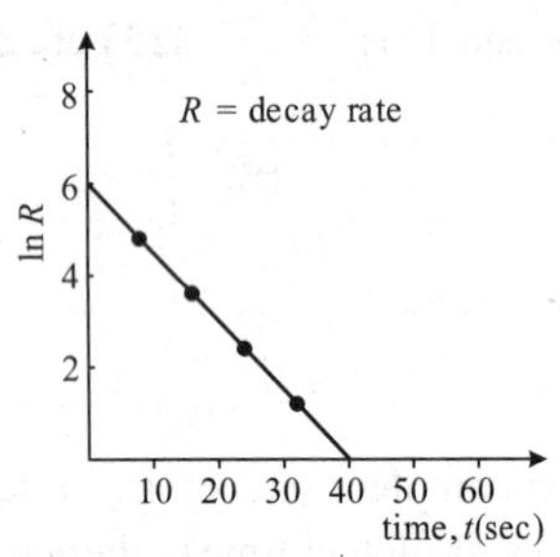

(A) 4.62 s (B) 2.62 s
(C) 9.15 s (D) 6.93 s

41. A radioactive substance decays to $\left(\frac{1}{16}\right)^{\text{th}}$ of its initial activity in 80 days. The half life of the radioactive substance expressed in days is ______. **[20 July 2021 (Evening)]**

42. Some nuclei of a radioactive material are undergoing radioactive decay. The time gap between the instances when a quarter of the nuclei have decayed and when half of the nuclei have decayed is given as: (where λ is the decay constant) **[25 July 2021 (Morning)]**

(A) $\frac{\ln\frac{3}{2}}{\lambda}$ (B) $\frac{1}{2}\frac{\ln 2}{\lambda}$

(C) $\frac{\ln 2}{\lambda}$ (D) $\frac{2\ln 2}{\lambda}$

43. The half-life of ^{198}Au is 3 days. If atomic weight of ^{198}Au is 198 g/mol then the activity of 2 mg ^{198}Au is [in dps] **[25 July 2021 (Morning)]**

(A) 2.67×10^{12} (B) 6.06×10^{18}
(C) 32.36×10^{12} (D) 16.18×10^{12}

44. The nuclear activity of a radioactive element becomes $\left(\frac{1}{8}\right)^{th}$ of its initial value in 30 years. The half-life of radioactive element is ______ years : **[25 July 2021 (Evening)]**

45. If f' denotes the ratio of the number of nuclei decayed (N_d) to the number of nuclei at $t = 0(N_0)$ then for a collection of radioactive nuclei, the rate of change of f with respect to time is given as : [x is the radioactive decay constant] : **[27 July 2021 (Morning)]**

(A) $\lambda e^{-\lambda t}$ (B) $\lambda(1-e^{-\lambda t})$
(C) $(-\lambda e^{-\lambda t})$ (D) $-\lambda(1-e^{-\lambda t})$

46. A radioactive sample has an average life of 30 ms and is decaying. A capacitor of capacitance 200 μF is first charged and later connected with resistor 'R'. If the ratio of charge on capacitor to the activity of radioactive sample is fixed with respect to time then the value of 'R' should be ______ Ω. **[27 July 2021 (Morning)]**

47. At time $t = 0$, material is composed of two radioactive atoms A and B, where $N_A(0) = 2N_B(0)$. The decay constant of both kind of radioactive atoms is λ. However, A disintegrates to B and B disintegrates to C. Which of the following figures represents the evolution of $N_B(t)/N_B(0)$ with respect to time t ?

$\left[\begin{array}{l} N_A(0) = \text{No. of } A \text{ atoms at } t = 0 \\ N_B(0) = \text{No. of } B \text{ atoms at } t = 0 \end{array}\right]$ **[26 Aug. 2021 (Evening)]**

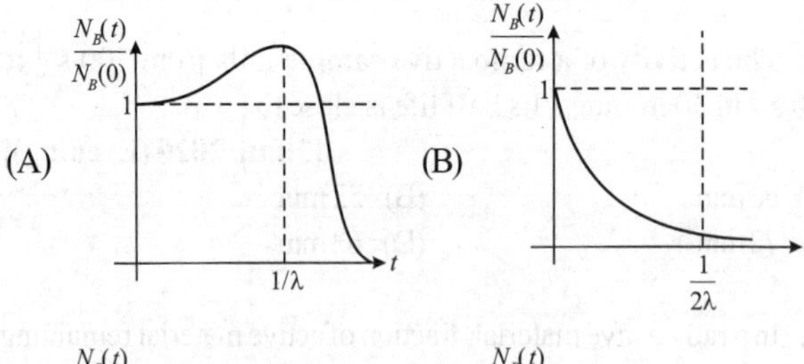

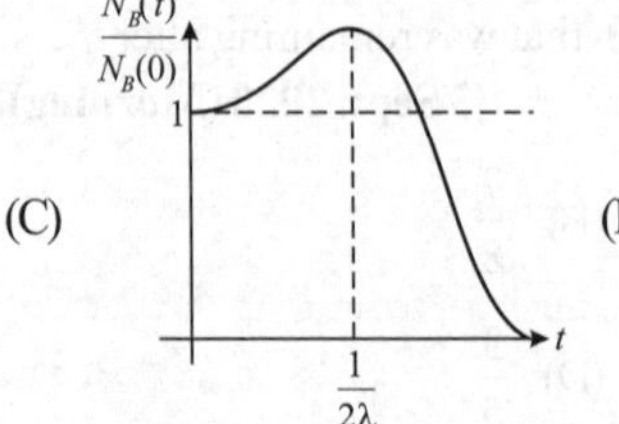

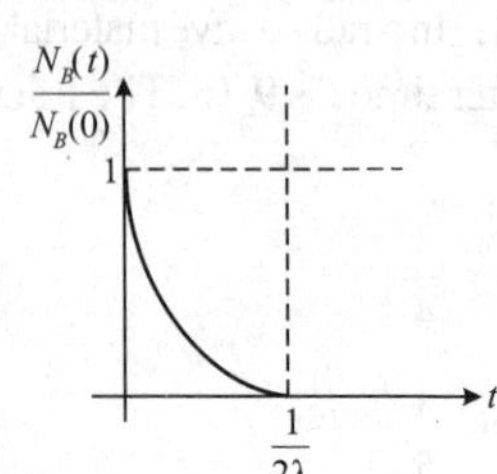

48. There are 10^{10} radioactive nuclei in a given radioactive element. Its half–life time is 1 minute. How many nuclei will remain after 30 seconds ? ($\sqrt{2} = 1.414$)

[27 Aug. 2021 (Morning)]

(A) 4×10^{10} (B) 7×10^{9}
(C) 10^{5} (D) 2×10^{10}

49. A sample of a radioactive nucleus A disintegrates to another radioactive nucleus B, which in turn disintegrates to some other stable nucleus C. Plot of a graph showing the variation of number of atoms of nucleus B vesus time is : (Assume that at $t = 0$, there are no B atoms in the sample)

[31 Aug. 2021 (Morning)]

(A) 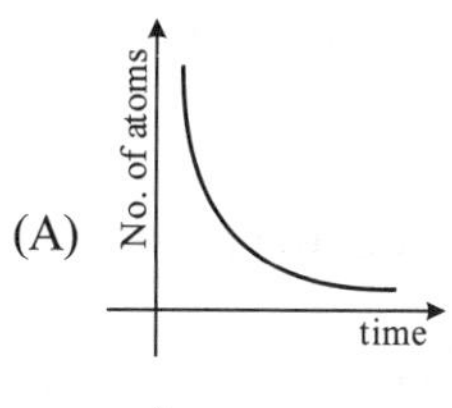

(B) 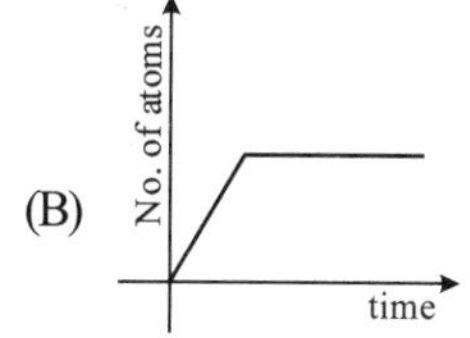

(C) 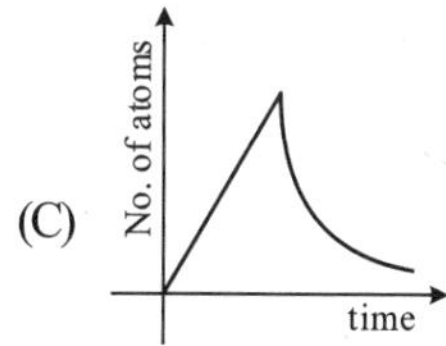

(D) 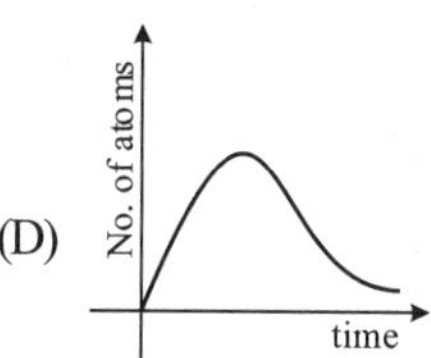

50. The half life period of a radioactive element x is same as the mean life time of another radioactive element y. Initially they have the same number of atoms. Then :

[1 Sep. 2021 (Evening)]

(A) x and y have same decay rate initially and later on different decay rate
(B) x and y decay at the same rate always
(C) x – will decay faster than y
(D) y – will decay faster than x

51. A sample contains 10^{-2} kg each of two substances A and B with half lives 4 s and 8 s respectively. The ratio of their atomic weights is 1 : 2. The ratio of the amounts of A and B after 16 s is $\frac{x}{100}$ the value of x is ______. **[24 June 2022 (Evening)]**

52. Which of the following figure represents the variation of $\left(\frac{R}{R_0}\right)$ with ln(A) : (If R = radius of a nucleus and A = its mass number) **[25 June 2022 (Evening)]**

(A) 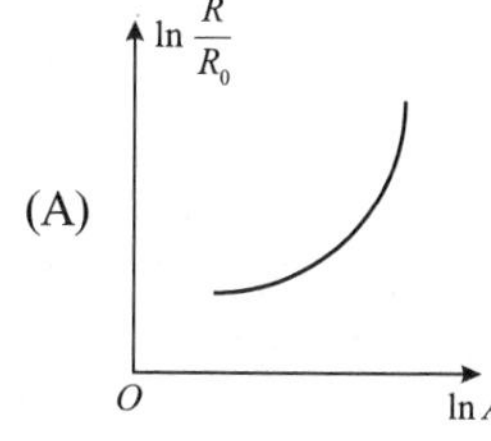

(B) 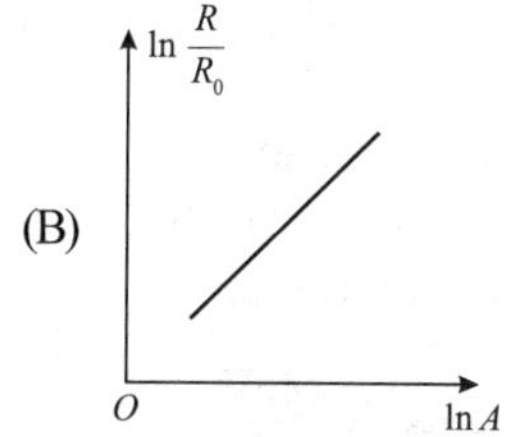

(C) 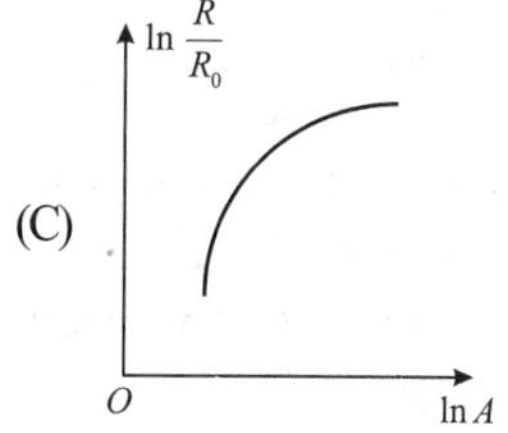

(D) 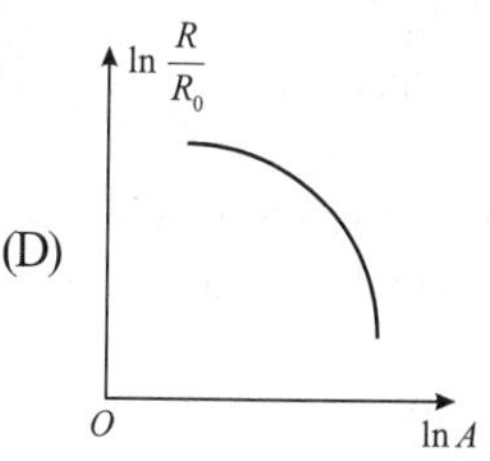

53. A radioactive nucleus can decay by two different processes. Half-life for the first process is 3.0 hours while it is 4.5 hours for the second process. The effective half-life of the nucleus will be : **[26 June 2022 (Evening)]**

(A) 3.75 hours (B) 0.56 hours
(C) 0.26 hours (D) 1.80 hours

54. Following statements related to radioactivity are given below : **[28 June 2022 (Evening)]**

(A) Radioactivity is a random and spontaneous process and is dependent on physical and chemical conditions
(B) The number of un-decayed nuclei in the radioactive sample decays exponentially with time
(C) Slope of the graph of $\log_e$ (no. of undecayed nuclei) vs time represents the reciprocal of mean life time (t)
(D) Product of decay constant (λ) and half-life time ($T_{1/2}$) is not constant

Choose the most appropriate answer from the options given below :

(A) (A) and (B) only (B) (B) and (D) only
(C) (B) and (C) only (D) (C) and (D) only

55. The activity of a radioactive material is 2.56×10^{-3} Ci. If the half life of the material is 5 days, after how many days the activity will become 2×10^{-5} Ci? **[29 June 2022 (Morning)]**

(A) 30 days (B) 35 days
(C) 40 days (D) 25 days

56. The half life of a radioactive substance is 5 years. After x years a given sample of the radioactive substance gets reduced to 6.25% of its initial value of x is ______.

[29 June 2022 (Evening)]

57. The disintegration rate of a certain radioactive sample at any instant is 4250 disintegrations per minute. 10 minutes later, the rate becomes 2250 disintegrations per minute. The approximate decay constant is : (Take $\log_{10} 1.88 = 0.274$)

[26 July 2022 (Morning)]

(A) $0.02\ \text{min}^{-1}$ (B) $2.7\ \text{min}^{-1}$
(C) $0.063\ \text{min}^{-1}$ (D) $6.3\ \text{min}^{-1}$

58. What is the half-life period of a radioactive material if its activity drops to 1/16th of its initial value of 30 years ?

[27 July 2022 (Morning)]

(A) 9.5 years (B) 8.5 years
(C) 7.5 years (D) 10.5 years

59. The activity of a radioactive material is 6.4×10^{-4} curie. Its half life is 5 days. The activity will become 5×10^{-6} curie after :
[27 July 2022 (Evening)]

(A) 7 days (B) 15 days
(C) 25 days (D) 35 days

60. The half life period of a radioactive substance is 60 days. The time taken for $\frac{7}{8}^{\text{th}}$ of its original mass to disintegrate will be : **[28 July 2022 (Morning)]**

(A) 120 days (B) 130 days
(C) 180 days (D) 20 days

61. A freshly prepared radioactive source of half life 2 hours 30 minutes emits radiation which is 64 times the permissible safe level. The minimum time, after which it would be possible to work safely with source, will be _____ hours.
[28 July 2022 (Morning)]

62. A radioactive sample decays $\frac{7}{8}^{\text{th}}$ times its original quantity in 15 minutes. The half-life of the sample is :
[28 July 2022 (Evening)]

(A) 5 min (B) 7.5 min
(C) 15 min (D) 30 min

63. Two radioactive materials A and B have decay constants 25λ and 16λ respectively. If initially they have the same number of nuclei, then the ratio of the number of nuclei of B to that of A will be "e" after a time $\frac{1}{a\lambda}$. The value of a is _____.

[29 July 2022 (Evening)]

64. If a radioactive element having half-life of 30 min is undergoing beta decay, the fraction of radioactive element remains undecayed after 90 min will be :
[29 Jan. 2023 (Morning)]

(A) $\frac{1}{8}$ (B) $\frac{1}{16}$

(C) $\frac{1}{4}$ (D) $\frac{1}{2}$

65. A radioactive element $^{242}_{92}X$ emits two α-particles, one electron and two positrons. The product nucleus is represented by $^{234}_{P}Y$. The value of P is _____. **[29 Jan. 2023 (Morning)]**

66. Substance A has atomic mass number 16 and half life of 1 day. Another substance B has atomic mass number 32 and half life of $\frac{1}{2}$ day. If both A and B simultaneously start undergo radio activity at the same time with initial mass $320g$ each, how many total atoms of A and B combined would be left after 2 days.
[29 Jan. 2023 (Evening)]

(A) 3.38×10^{24} (B) 6.76×10^{24}
(C) 6.76×10^{23} (D) 1.69×10^{24}

67. A radioactive nucleus decays by two different process. The half life of the first process is 5 minutes and that of the second process is 30 s. The effective half-life of the nucleus is calculated to be $\frac{\alpha}{11}$ s. The value of α is _____.

[30 Jan. 2023 (Evening)]

68. A radioactive material is reduced to 1/8 of its original amount in 3 days. If 8×10^{-3} kg of the material is left after 5 days. The initial amount of the material is :
[8 April 2023 (Evening)]

(A) 64 g (B) 40 g
(C) 32 g (D) 256 g

69. The decay constant for a radioactive nuclide is 1.5×10^{-5} s^{-1}. Atomic of the substance is 60 g mole^{-1}, ($N_A = 6 \times 10^{23}$). The activity of 1.0 μg of the substance is _____ 10^{10} Bq. **[10 April 2023 (Morning)]**

70. The half-life of a radioactive substance is T. The time taken, for disintegrating $\frac{7}{8}^{\text{th}}$ part of its original mass will be:

[10 April 2023 (Evening)]

(A) $3T$ (B) $8T$
(C) T (D) $2T$

71. Two radioactive elements A and B initially have same number of atoms. The half life of A is same as the average life of B. If λ_A and λ_B are decay constants of A and B respectively, then choose the correct relation from the given options :
[11 April 2023 (Morning)]

(A) $\lambda_A = \lambda_B$ (B) $\lambda_A = 2\lambda_B$
(C) $\lambda_A = \lambda_B \ln 2$ (D) $\lambda_A \ln 2 = \lambda_B$

72. A common example of alpha decay is
$$^{238}_{92}U \longrightarrow {}^{231}_{90}Th + {}_2He^4 + Q$$
Given :
$^{238}_{92}U = 238.05060\,u$,
$^{237}_{90}Th = 234.04360\,u$,
$^{4}_{2}He = 4.00260\,u$,

$1u = 931.5\,\frac{MeV}{c^2}$

The energy released (Q) during the alpha decay of $^{238}_{92}U$ is _____ MeV **[12 April 2023 (Morning)]**

73. The half-life of a radioactive nucleus is 5 years, The fraction of the original sample that would decay in 15 years is :
[15 April 2023 (Morning)]

(A) $\frac{1}{8}$ (B) $\frac{1}{4}$

(C) $\frac{7}{8}$ (D) $\frac{3}{4}$

* * * * *

CHAPTER 27

NUCLEI

27.2 NUCLEAR PHYSICS

1. In the nuclear fusion reaction, ${}_1^2\text{H} + {}_1^3\text{H} \rightarrow {}_2^4\text{He} + n$ given that the repulsive potential energy between the two nuclei is 7.7×10^{-14} J, the temperature at which the gases must be heated to initiate the reaction is nearly [Boltzmann's constant $k = 1.38 \times 10^{-23}$ J/K] **[2003]**

(A) 10^7 K (B) 10^5 K
(C) 10^3 K (D) 10^9 K

2. Which of the following cannot be emitted by radioactive substances during their decay? **[2003]**

(A) Protons (B) Neutrinos
(C) Helium nuclei (D) Electrons

3. A nucleus with $Z = 92$ emits the following in a sequence : α, α, β^-, β^-, α, α, α, α, β^-, β^-, α, β^+, β^+, α. The Z of the resulting nucleus is : **[2003]**

(A) 76 (B) 78
(C) 82 (D) 74

4. When U^{238} nucleus originally at rest, decays by emitting an alpha particle having a speed u, the recoil speed of the residual nucleus is : **[2003]**

(A) $\frac{4u}{238}$ (B) $-\frac{4u}{234}$
(C) $\frac{4u}{234}$ (D) $-\frac{4u}{238}$

5. An α-particle of energy 5 MeV is scattered through 180° by a fixed uranium nucleus. The distance of the closest approach is of the order of: **[2004]**

(A) 1 Å (B) 10^{-10} cm
(C) 10^{-12} cm (D) 10^{-15} cm

6. The binding energy per nucleon of deuteron $\left({}_1^2\text{H}\right)$ and helium nucleus $\left({}_2^4\text{He}\right)$ is 1.1 MeV and 7 MeV respectively. If two deuteron nuclei react to form a single helium nucleus, then the energy released is : **[2004]**

(A) 13.9 MeV (B) 26.9 MeV
(C) 23.6 MeV (D) 19.2 MeV

7. A nuclear transformation is denoted by $X(n, \alpha) \rightarrow {}_3^7\text{Li}$. Which of the following is the nucleus of element X? **[2005]**

(A) ${}_6^{12}\text{C}$ (B) ${}_5^{10}\text{B}$
(C) ${}_5^9\text{B}$ (D) ${}_4^{11}\text{Be}$

8. The intensity of gamma radiation from a given source is I. On passing through 36 mm of lead, it is reduced to $I/8$. The thickness of lead, which will reduce the intensity to $I/2$ will be : **[2005]**

(A) 6mm (B) 9mm
(C) 18mm (D) 12mm

9. The energy spectrum of β-particles [number $N(E)$ as a function of β-energy E] emitted from a radioactive source is: **[2006]**

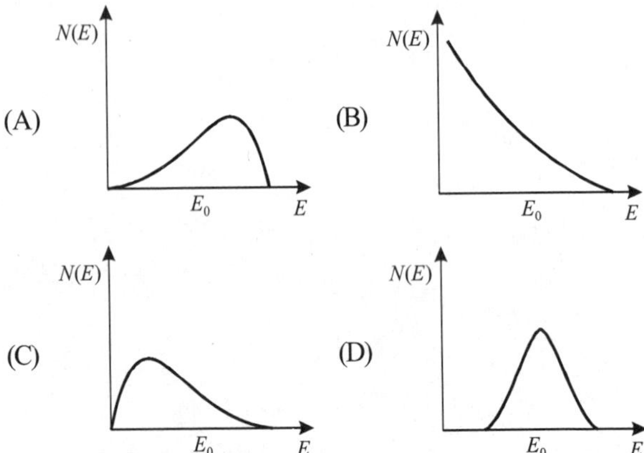

10. When 3Li^7 nuclei are bombarded by protons and the resultant nuclei are ${}_4\text{Be}^8$, the emitted particles will be : **[2006]**

(A) alpha particles (B) beta particles
(C) gamma photons (D) neutrons

11. An alpha nucleus of energy 1/2 mv^2 bombards a heavy nuclear target of charge Ze. Then the distance of closest approach for the alpha nucleus will be proportional to : **[2006]**

(A) v^2 (B) $1/m$
(C) $1/v^4$ (D) $1/Ze$

12. If the binding energy per nucleon in ${}_3^7\text{Li}$ and ${}_2^4\text{He}$ nuclei are 5.60 MeV and 7.06 MeV respectively, then in the reaction: $p + {}_3^7\text{Li} \rightarrow 2\,{}_2^4\text{He}$ energy of proton must be : **[2006]**

(A) 28.24 MeV (B) 17.28 MeV
(C) 1.46 MeV (D) 39.2 MeV

13. In gamma ray emission from a nucleus : **[2007]**
(A) both the neutron number and the proton number change
(B) there is no change in the proton number and the neutron number
(C) only the neutron number changes
(D) only the proton number changes

14. If M_0 is the mass of an oxygen isotope $_8O^{17}$, M_p and M_n are the masses of a proton and a neutron, respectively, the nuclear binding energy of the isotope is : **[2007]**
(A) $(M_o - 8M_p)c^2$ (B) $(M_o - 8M_p - 9M_n)c^2$
(C) M_oc^2 (D) $(M_o - 17M_n)c^2$

15. Which of the following are the constituents of the nucleus? **[2007]**
(A) Electrons and protons (B) Neutrons and protons
(C) Electrons and neutrons (D) Neutrons and positrons

16. This question contains Statement-1 and Statement-2. Out of the four choices given after the statements, choose the one that best describes the two statements. **[2008]**
Statement-1 : Energy is released when heavy nuclei undergo fission or light nuclei undergo fusion.
Statement-2 : For heavy nuclei, binding energy per nucleon increases with increasing Z while for light nuclei it decreases with increasing Z.
(A) Statement-1 is false, Statement-2 is true
(B) Statement-1 is true, Statement-2 is true; Statement-2 is a correct explanation for Statement-1
(C) Statement-1 is true, Statement-2 is true; Statement-2 is not a correct explanation for Statement-1
(D) Statement-1 is true, Statement-2 is false

17. The above is a plot of binding energy per nucleon E_b, against the nuclear mass M; A, B, C, D, E, F correspond to different nuclei. Consider four reactions : **[2009]**

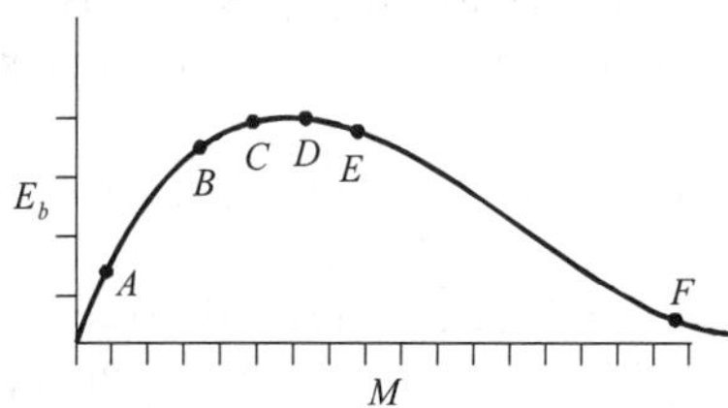

(i) $A+B \rightarrow C+\varepsilon$ (ii) $C \rightarrow A+B+\varepsilon$
(iii) $D+E \rightarrow F+\varepsilon$ (iv) $F \rightarrow D+E+\varepsilon$
where ε is the energy released ? In which reactions is ε positive ?
(A) (ii) and (iv) (B) (ii) and (iii)
(C) (i) and (iv) (D) (i) and (iii)

18. A radioactive nucleus (initial mass number A and atomic number Z) emits 3 α-particles and 2 positrons. The ratio of number of neutrons to that of protons in the final nucleus will be : **[2010]**
(A) $\frac{A-Z-4}{Z-2}$ (B) $\frac{A-Z-8}{Z-4}$
(C) $\frac{A-Z-4}{Z-8}$ (D) $\frac{A-Z-12}{Z-4}$

19. A nucleus of mass $M + \Delta m$ is at rest and decays into two daughter nuclei of equal mass $M/2$ each. Speed of light is c. The binding energy per nucleon for the parent nucleus is E_1 an that for the daughter nuclei is E_2. Then : **[2010]**
(A) $E_1 = 2E_2$ (B) $E_2 = 2E_1$
(C) $E_1 > E_2$ (D) $E_2 > E_1$

20. A nucleus of mass $M + \Delta m$ is at rest and decays into two daughter nuclei of equal mass $M/2$ each. Speed of light is c. The speed of daughter nuclei is : **[2010]**
(A) $c\sqrt{\frac{\Delta m}{M+\Delta m}}$ (B) $c\frac{\Delta m}{M+\Delta m}$
(C) $c\sqrt{\frac{2\Delta m}{M}}$ (D) $c\sqrt{\frac{\Delta m}{M}}$

21. Statement-1 : A nucleus having energy E_1 decays β^- emission to daughter nucleus having energy E_2, but the β^- rays are emitted with a continuous energy spectrum having end point energy $E_1 - E_2$. **[2011]**
Statement-2 : To conserve energy and momentum in β-decay at least three particles must take part in the transformation.
(A) Statement-1 is incorrect, statement-2 is correct
(B) Statement-1 is correct, statement-2 is incorrect
(C) Statement-1 is correct, statement-2 correct; statement-2 is the correct explanation of statement-1
(D) Statement-1 is correct, statement-2 is correct; statement-2 is not the correct explanation of statement-1

22. After absorbing a slowly moving neutron of mass m_N (momentum ~ 0) a nucleus of mass M breaks into two nuclei of masses m_1 and 5 m_1 ($6\ m_1 = M + m_N$), respectively. If the de Broglie wavelength of the nucleus with mass m_1 is λ, then de Broglie wavelength of the other nucleus will be : **[2011]**
(A) 25λ (B) 5λ
(C) $\frac{\lambda}{5}$ (D) λ

23. When Uranium is bombarded with neutrons, it undergoes fission. The fission reaction can be written as :

$$_{92}U^{235} + {}_0n^1 \rightarrow {}_{56}Ba^{141} + {}_{36}Kr^{92} + 3x + Q\,(\text{energy})$$

where three particles named x are produced and energy Q is released. What is the name of the particle x ? **[2013 (Online)]**
(A) Electron (B) α-particle
(C) Neutron (D) Neutrino

24. Two deuterons undergo nuclear fusion to form a Helium nucleus. Energy released in this process is : (given binding energy per nucleon for deuteron = 1.1 MeV and for helium = 7.0 MeV) **[2017 (Online)]**

(A) 30.2 MeV (B) 32.4 MeV
(C) 23.6 MeV (D) 25.8 MeV

25. Imagine that a reactor converts all given mass into energy and that it operates at a power level of 10^9 watt. The mass of the fuel consumed per hour in the reactor will be : (velocity of light, c is 3×10^8 m/s) **[2017 (Online)]**

(A) 0.96 gm (B) 0.8 gm
(C) 4×10^{-2} gm (D) 6.6×10^{-5} gm

26. Both the nucleus and the atom of some element are in their respectively first excited states. They get de-excited by emitting photons of wavelengths λ_N, λ_A respectively. The ratio $\frac{\lambda_N}{\lambda_A}$ is closet to : **[2018 (Online)]**

(A) 10^{-6} (B) 10
(C) 10^{-1} (D) 10^{-10}

27. An unstable heavy nucleus at rest breaks into two nuclei which move away with velocities in the ratio of 8 : 27. The ratio of the radii of the nuclei (Assumed to be spherical) is : **[2018 (Online)]**

(A) 8 : 27 (B) 2 : 3
(C) 3 : 2 (D) 4 : 9

28. Consider the nuclear fission

$$Ne^{20} \rightarrow 2He^4 + C^{12}$$

Given that the binding energy/nucleon of Ne^{20}, He^4 and C^{12} are, respectively, 8.03 MeV, 7.07 MeV and 7.86 MeV, identify the correct statement : **[10 Jan. 2019 (Evening)]**

(A) 8.3 MeV energy will be released
(B) Energy of 12.4 MeV will be supplied
(C) Energy of 11.9 MeV has to be supplied
(D) Energy of 3.6 MeV will be released

29. The ratio of mass densities of nuclei of ^{40}Ca and ^{16}O is close to : **[8 April 2019 (Evening)]**

(A) 1 (B) 2
(C) 0.1 (D) 5

30. In a reactor, 2 kg of $_{92}U^{235}$ fuel is fully used up in 30 days. The energy released per fission is 200 MeV. Given that the Avogadro number, $N = 6.023 \times 10^{26}$ per kilo mole and 1 eV = 1.6×10^{-19} J. The power output of the reactor is close to : **[2 Sept. 2020 (Morning)]**

(A) 125 MW (B) 60 MW
(C) 35 MW (D) 54 MW

31. The radius R of a nucleus of mass number A can be estimated by the formula $R = (1.3 \times 10^{-15})\, A^{1/3}$ m. It follows that the mass density of a nucleus is of the order of : ($M_{prot.} \cong M_{neut.} \simeq 1.6 \times 10^{-27}$ kg) **[3 Sept. 2020 (Evening)]**

(A) 10^{24} kg m^{-3} (B) 10^3 kg m^{-3}
(C) 10^{17} kg m^{-3} (D) 10^{10} kg m^{-3}

32. Find the binding energy per nucleon for $^{120}_{50}$Sn. Mass of proton $m_p = 1.00783$ U, mass of neutron $m_n = 1.00867$ U and mass of tin nucleus $m_{Sn} = 119.902199$ U. (Take 1 U = 931 MeV) **[4 Sept. 2020 (Evening)]**

(A) 8.5 MeV (B) 7.5 MeV
(C) 8.0 MeV (D) 9.0 MeV

33. You are given that Mass of $^7_3$Li = 7.0160 u, Mass of $^4_2$He = 4.0026 u and Mass of $^1_1$H = 1.0079 u. When 20 g of $^7_3$Li is converted into $^4_2$He by proton capture, the energy liberated, (in kWh), is : [Mass of nucleon = 1 GeV/c^2] **[6 Sept. 2020 (Morning)]**

(A) 8×10^6 (B) 1.33×10^6
(C) 6.82×10^5 (D) 4.5×10^5

34. Given the masses of various atomic particles $m_p = 1.0072$ u, $m_n = 1.0087$ u, $m_e = 0.000548$ u, $m_{\bar{\nu}} = 0$, $m_d = 2.0141$ u, where p is proton, n is neutron, e is electron, $\bar{\nu}$ is antineutrino and d is deuteron. Which of the following process is allowed by momentum and energy conservation : **[6 Sept. 2020 (Evening)]**

(A) $n + p \rightarrow d + g$ (B) $e^+ + e^- \rightarrow g$
(C) $n + n \rightarrow$ deuterium atom (electron bound to the nucleus)
(D) $p \rightarrow n + e^+ + \bar{\nu}$

35. The decay of a proton to neutron is : **[18 March 2021 (Evening)]**

(A) not possible as proton mass is less than the neutron mass
(B) possible only inside the nucleus
(C) not possible but neutron to proton conversion is possible
(D) always possible as it is associated only with β^+ decay

36. A nucleus of mass M emits γ-rays photon of frequency 'ν'. The loss of internal energy by the nucleus is : [Take 'c' as the speed of electromagnetic wave] **[20 July 2021 (Morning)]**

(A) 0 (B) $h\nu\left[1+\frac{h\nu}{2Mc^2}\right]$
(C) $h\nu$ (D) $h\nu\left[1-\frac{h\nu}{2Mc^2}\right]$

37. A nucleus with mass number 184 initially at rest emits an α-particle. If the Q value of the reaction is 5.5 MeV, calculate the kinetic energy of the α-particle : **[22 July 2021 (Evening)]**

(A) 5.5 MeV (B) 5.0 MeV
(C) 5.38 MeV (D) 0.12 MeV

38. From the given data, the amount of energy required to break the nucleus of aluminium $^{27}_{13}$Al is ______ $x \times 10^{-3}$ J.
Mass of neutron = 1.00866 u Mass of proton = 1.00726 u Mass of Aluminium nucleus = 27.18846 u (Assume 1u corresponds to x J of energy) (Round off to the nearest integer)
[25 July 2021 (Evening)]

39. Nucleus A is having mass number 220 and its binding energy per nucleon is 5.6 MeV. It splits in two fragments B and C of mass numbers 105 and 115. The binding energy of nucleons in B and C is 6.4 MeV per nucleon. The energy Q released per fission will be : **[24 June 2022 (Morning)]**
(A) 0.8 MeV (B) 275 MeV
(C) 220 MeV (D) 176 MeV

40. How many alpha and beta particles are emitted when Uranium $_{92}U^{238}$ decays to lead $_{82}Pb^{206}$?
[26 June 2022 (Morning)]
(A) 3 alpha particles and 5 beta particles
(B) 6 alpha particles and 4 beta particles
(C) 4 alpha particles and 5 beta particles
(D) 8 alpha particles and 6 beta particles

41. The Q-value of a nuclear reaction and kinetic energy of the projectile particle, K_p are related as : **[28 June 2022 (Morning)]**
(A) $Q = K_p$ (B) $(K_p + Q) < 0$
(C) $Q < K_p$ (D) $(K_p + Q) > 0$

42. In the following nuclear reaction,

$$D \xrightarrow{\alpha} D_1 \xrightarrow{\beta^-} D_2 \xrightarrow{\alpha} D_3 \xrightarrow{\gamma} D_4$$

Mass number of D is 182 and atomic number is 74. Mass number and atomic number of D_4 respectively will be _____.
[29 June 2022 (Evening)]
(A) 174 and 71 (B) 174 and 69
(C) 172 and 69 (D) 172 and 71

43. Mass numbers of two nuclei are in the ratio of 4 : 3. Their nuclear densities will be in the ratio of: **[26 July 2022 (Evening)]**
(A) 4 : 3 (B) $\left(\frac{3}{4}\right)^{1/3}$
(C) 1 : 1 (D) $\left(\frac{4}{3}\right)^{1/3}$

44. Two lighter nuclei combine to form a comparatively heavier nucleus by the relation given below :
$^2_1X + ^2_1X = ^4_2Y$
The binding energies per nucleon ^2_1X and ^4_2Y are 1.1 MeV and 7.6 MeV respectively. The energy released in this process is _____. MeV. **[26 July 2022 (Evening)]**

45. Read the following statements : **[29 July 2022 (Evening)]**
(A) Volume of the nucleus is directly proportional to the mass number
(B) Volume of the nucleus is independent of mass number
(C) Density of the nucleus is directly proportional to the mass number
(D) Density of the nucleus is directly proportional to the cube root of the mass number
(E) Density of the nucleus is independent of the mass number.
Choose the correct option from the following options.
(A) (A) and (D) only (B) (A) and (E) only
(C) (B) and (E) only (D) (A) and (C) only

46. Consider the following radioactive decay process

$$^{218}_{84}A \xrightarrow{\alpha} A_1 \xrightarrow{\beta^-} A_2 \xrightarrow{\gamma} A_3 \xrightarrow{\alpha} A_4 \xrightarrow{\beta^+} A_5 \xrightarrow{\gamma} A_6$$

The mass number and the atomic number A_6 are given by :
[24 Jan. 2023 (Morning)]
(A) 210 and 82 (B) 210 and 84
(C) 210 and 80 (D) 211 and 80

47. Assume that protons and neutrons have equal masses. Mass of a nucleon is 1.6×10^{-27} kg and radius of nucleus is $1.5 \times 10^{-15} A^{1/3}$ m. The approximate ratio of the nuclear density and water density is $n \times 10^{13}$. The value of n is _____.
[24 Jan. 2023 (Morning)]

48. The energy released per fission of nucleus of ^{240}X is 200 MeV. The energy released if all the atoms in 120g of pure ^{240}X undergo fission is _____ $\times 10^{25}$ MeV.
(Given $N_A = 6 \times 10^{23}$) **[24 Jan. 2023 (Evening)]**

49. The ratio of the density of oxygen nucleus ($^{16}_{8}$O) and helium nucleus ($^4_2$He) is : **[25 Jan. 2023 (Morning)]**
(A) 4 : 1 (B) 8 : 1
(C) 1 : 1 (D) 2 : 1

50. A nucleus disintegrates into two smaller parts, which have their velocities in the ratio 3 : 2. The ratio of their nuclear sizes will be $\left(\frac{x}{3}\right)^{1/3}$. The value of '$x$' is : **[25 Jan. 2023 (Evening)]**

51. Given below are two statements : One is labelled as **Assertion-A** and the other is labelled as **Reason-R**.
Assertion-A : The nuclear density of nuclides $^{10}_{5}$B, $^6_3$Li, $^{56}_{26}$Fe, $^{20}_{10}$Ne and $^{209}_{83}$Bi can be arranged as $\rho^N_{Bi} > \rho^N_{Fe} > \rho^N_{Ne} > \rho^N_{B}\ \rho^N_{Li}$.

Reason-R : The radius R of nucleus is related to its mass number A as $R = R_0 A^{1/3}$, where R_0 is a constant.
In the light of the above statement, choose the correct answer from the options given below : **[30 Jan. 2023 (Evening)]**
(A) Both **A** and **R** are true and **R** is the correct explanation of **A**.
(B) **A** is false but **R** is true.
(C) **A** is true but **R** is false.
(D) Both **A** and **R** are true but **R** is NOT the correct explanation of **A**.

52. A free neutron decays into a proton but a free proton does not decay into neutron. This is because : **[31 Jan. 2023 (Morning)]**
(A) Neutron is an uncharged particle
(B) Proton is a charged particle
(C) Neutron is a composite particle made of a proton and an electron
(D) Neutron has larger rest mass than proton

53. Match the List-I with List-II : **[1 Feb. 2023 (Morning)]**

List-I	**List-II**
(a) Microwaves	(i) Radio active decay of the nucleus
(b) Gamma rays	(ii) Rapid acceleration and deceleration of electron in aerials
(c) Radio waves	(iii) Inner shell electrons
(d) X-rays	(iv) Klystron valve

Choose the correct answer from the options given below :
(A) (a)-(i), (b)-(ii), (c)-(iii), (d)-(iv)
(B) (a)-(iv), (b)-(i), (c)-(ii), (d)-(iii)
(C) (a)-(i), (b)-(iii), (c)-(iv), (d)-(ii)
(D) (a)-(iv), (b)-(iii), (c)-(ii), (d)-(i)

54. The mass of proton, neutron and helium nucleus are respectively $1.0073u$, $1.0087u$ and $4.0015u$. The binding energy of helium nucleus is : **[1 Feb. 2023 (Morning)]**
(A) 14.2 MeV (B) 28.4 MeV
(C) 56.8 MeV (D) 7.1 MeV

55. Nucleus a having $Z = 17$ and equal number of protons and neutrons has 1.2 MeV binding energy per nucleon. Another nucleus B of $Z = 12$ has total 26 nucleons and 1.8 MeV binding energy per nucleons. The difference of binding energy of B and A will be _____ MeV. **[1 Feb. 2023 (Evening)]**

56. For a nucleus ${}^{A}_{Z}X$ having mass number A and atomic number Z:
(1) The surface energy per nucleon $(b_s) = a_1 A^{2/3}$
(2) The Coulomb contribution to the binding energy

$$b_c = -a_2 \frac{Z(Z-1)}{A^{4/3}}$$

(3) The volume energy $b_v = a_3 A$
(4) Decrease in the binding energy is proportional to surface area
(5) While estimating the surface energy, it is assumed that each nucleon interacts with 12 nucleons, (a_1, a_2 and a_3 are constants)
Choose the most appropriate answer from the options given below : **[8 April 2023 (Morning)]**
(A) (3), (4) only (B) (2), (3), (5) only
(C) (1), (2), (3), (4) only (D) (2), (3) only

57. A nucleus with mass number 242 and binding energy per nucleon as 7.6 MeV breaks into fragment each with mass number 121. If each fragment nucleus has binding energy per nucleon as 8.1 MeV, the total gain in binding energy is _____ MeV. **[8 April 2023 (Morning)]**

58. A nucleus disintegrates into two nuclear parts, in such a way that ratio of their nuclear sizes is $1 : 2^{1/3}$. Their respective speed have a ratio of $n : 1$. The value of n is______. **[11 April 2023 (Evening)]**

59. ${}^{238}_{92}A \rightarrow {}^{234}_{90}B + {}^{4}_{2}D + Q$
In the given nuclear reaction, the approximate amount of energy released will be :
[Given, mass of ${}^{238}_{92}A = 238.05079 \times 931.5$ MeV/c^2.
mass of ${}^{234}_{90}B = 234.04363 \times 931.5$ MeV/c^2,
mass of ${}^{4}_{2}D = 4.00260 \times 931.5$ MeV/c^2] **[13 April 2023 (Morning)]**
(A) 3.82 MeV (B) 5.9 MeV
(C) 2.12 MeV (D) 4.25 MeV

60. Given below are two statements : one is labelled as **Assertion-A** and the other is labelled as **Reason-R**.
Assertion-A : The binding energy per nucleon is practically independent of the atomic number for nuclei of mass number in the range 30 to 170.
Reason-R : Nuclear force is short ranged.
In the light of the above statements, choose the correct answer from the options given below. **[13 April 2023 (Evening)]**
(A) Both **A** and **R** are true but **R** is NOT the correct explanation of **A**.
(B) **A** is true but **R** is false.
(C) **A** is false but **R** is true.
(D) Both **A** and **R** are true and **R** is the correct explanation of **A**.

* * * * *

CHAPTER

28 SEMICONDUCTORS ELECTRONICS

28.1 SEMICONDUCTOR

1. The energy band gap is maximum in : **[2002]**
(A) Metals (B) Superconductors
(C) Insulators (D) Semiconductors

2. By increasing the temperature, the specific resistance of a conductor and a semiconductor : **[2002]**
(A) Increases for both (B) Decreases for both
(C) Increases, decreases (D) Decreases, increases

3. At absolute zero, *Si* acts as : **[2002]**
(A) Non-metal (B) Metal
(C) Insulator (D) None of these

4. A strip of copper and another of germanium are cooled from room temperature to 80 K. The resistance of: **[2003]**
(A) Each of these decreases
(B) Copper strip increases and that of germanium decreases
(C) Copper strip decreases and that of germanium increases
(D) Each of these increases

5. The difference in the variation of resistance with temperature in a metal and a semiconductor arises essentially due to the difference in the : **[2003]**
(A) Crystal structure
(B) Variation of the number of charge carriers with temperature
(C) Type of bonding
(D) Variation of scattering mechanism with temperature

6. When *p-n* junction diode is forward biased then : **[2004]**
(A) Both the depletion region and barrier height are reduced
(B) The depletion region is widened and barrier height is reduced
(C) The depletion region is reduced and barrier height is increased
(D) Both the depletion region and barrier height are increased

7. The electrical conductivity of a semiconductor increases when electromagnetic radiation of wavelength shorter than 2480 nm is incident on it. The band gap in (eV) for the semiconductor is : **[2005]**
(A) 2.5 eV (B) 1.1 eV
(C) 0.7 eV (D) 0.5 eV

8. If the ratio of the concentration of electrons to that of holes in a semiconductor is 7/5, and the ratio of currents is 7/4, then what is the ratio of their drift velocities ? **[2006]**
(A) $\frac{5}{8}$ (B) $\frac{4}{5}$
(C) $\frac{5}{4}$ (D) $\frac{4}{7}$

9. A solid which is not transparent to visible light and whose conductivity increases with temperature is formed by : **[2006]**
(A) Ionic bonding (B) Covalent bonding
(C) Vander Waals bonding (D) Metallic bonding

10. If the lattice constant of this semiconductor is decreased, then which of the following is correct ? **[2006]**

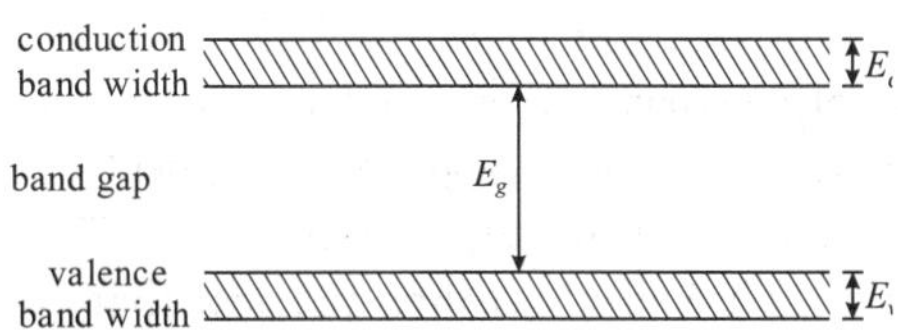

(A) All E_C, E_g, E_v increase
(B) E_c and E_v increase, but E_g decrease
(C) E_c and E_v decrease, but E_g increases
(D) All E_c, E_g, E_v decrease

11. Carbon, silicon and germanium have four valence electrons each. At room temperature which one of the following statements is most appropriate ? **[2007]**
(A) The number of free electrons for conduction is significant only in *Si* and *Ge* but small in *C*
(B) The number of free conduction electrons is significant in *C* but small in *Si* and *Ge*
(C) The number of free conduction electrons is negligibly small in all the three
(D) The number of free electrons for conduction is significant in all the three

12. This question has Statement-1 and Statement-2. Of the four choices given after the Statements, choose the one that best describes the two Statements. **[2012 (Online)]**
Statement-1 : A pure semiconductor has negative temperature coefficient of resistance.
Statement-2 : On raising the temperature, more charge carriers

are released into the conduction band

(A) Statement-1 is false, Statement-2 is true

(B) Statement-1 is true, Statement-2 is false

(C) Statement-1 is true, Statement-2 is true, Statement-2 is not a correct explanation of Statement-1

(D) Statement-1 is true, Statement-2 is true, Statement-2 is the correct explanation of Statement-1

13. The temperature dependence of resistances of Cu and undoped Si in the temperature range 300 – 400K, is best described by : **[2016]**

(A) Linear increase for Cu, exponential decrease of Si

(B) Linear decrease for Cu, linear decrease for Si

(C) Linear increase for Cu, linear increase for Si

(D) Linear increase for Cu, exponential increase for Si

14. What is the conductivity of a semiconductor sample having electron concentration of 5×10^{18} m^{-3}, hole concentration of 5×10^{19} m^{-3}, electron mobility of 2.0 m^2 V^{-1} s^{-1} and hole mobility of 0.01 m^2 V^{-1} s^{-1}? **[2017 (Online)]**

(Take charge of electron as 1.6×10^{-19} C)

(A) 1.68 $(\Omega - m)^{-1}$ (B) 1.83 $(\Omega - m)^{-1}$

(C) 0.59 $(\Omega - m)^{-1}$ (D) 1.20 $(\Omega - m)^{-1}$

15. Mobility of electrons in a semiconductor is defined as the ratio of their drift velocity to the applied electric field. If, for an *n*-type semiconductor, the density of electrons is 10^{19} m^{-3} and their mobility is 1.6 m^2/(V.s) then the resistivity of the semiconductor (since it is an *n*-type semiconductor contribution of holes is ignored) is close to : **[9 Jan. 2019 (Morning)]**

(A) 2 Ωm (B) 0.4 Ωm

(C) 4 Ωm (D) 0.2 Ωm

16. For extrinsic semiconductors; when doping level is increased; **[25 Feb. 2021 (Evening)]**

(A) Fermi-level of *p*-type semiconductors will go downward and Fermi-level of *n*-type semiconductor will go upward

(B) Fermi-level of *p* and *n*-type semiconductors will not be affected

(C) Fermi-level of *p*-type semiconductor will go upward and Fermi-level of *n*-type semiconductors will go downward

(D) Fermi-level of both *p*-type and *n*-type semiconductors will go upward for $T > T_F K$ and downward for $T < T_F K$, where T_F is Fermi temperature

17. In a semiconductor, the number density of intrinsic charge carries at 27°C is 1.5×10^{16}/m^3. If the semiconductor is doped with impurity atom, the hole density increases to 4.5×10^{22}/m^3. The electron density in the doped semiconductor is ______ 10^9/m^3. **[25 July 2021 (Evening)]**

18. A transistor is connected in common emitter circuit configuration, the collector supply voltage is 10 V and the voltage drop across a resistor of 1000 Ω in the collector circuit is 0.6 V. If the current gain factor (β) is 24, then the base current is ______ μ*A*. (Round off to the nearest integer) **[27 July 2021 (Morning)]**

19. Statement-I : By doping semiconductor with pentavalent material, the density increases.

Statement-II : The *n*-type semiconductor has net negative charge.

In the light of the above statements, choose the most appropriate answer from the options given below : **[26 Aug. 2021 (Morning)]**

(A) Both Statement I and Statement II are false

(B) Both Statement I and Statement II are true

(C) Statement I is false but Statement II is true

(D) Statement I is true but Statement II is false

20. The effect of increase in temperature on the number of electrons in conduction band (n_e) and resistance of a semiconductor will be as : **[31 Jan. 2023 (Morning)]**

(A) Both n_e and resistance decrease

(B) Both n_e and resistance increase

(C) n_e increases, resistance decreases

(D) n_e decreases, resistance increases

21. The resistivity (ρ) of semiconductor varies with temperature. Which of the following curve represents the correct behaviour **[6 April 2023 (Morning)]**

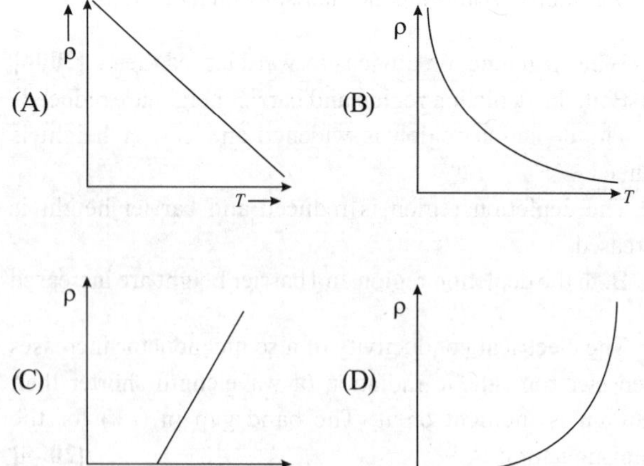

* * * * *

CHAPTER 28 SEMICONDUCTORS ELECTRONICS

28.2 PN JUNCTION AND DIODES

1. In the middle of the depletion layer of a reverse-biased *p-n* junction, the : **[2003]**

(A) Electric field is zero (B) Potential is maximum
(C) Electric field is maximum (D) Potential is zero

2. A strip of copper and another of germanium are cooled from room temperature to 80 K. The resistance of : **[2004]**

(A) Each of these decreases
(B) Copper strip increases and that of germanium decreases
(C) Copper strip decreases and that of germanium increases
(D) Each of these increases

3. The circuit has two oppositely connected ideal diodes in parallel. What is the current flowing in the circuit ? **[2006]**

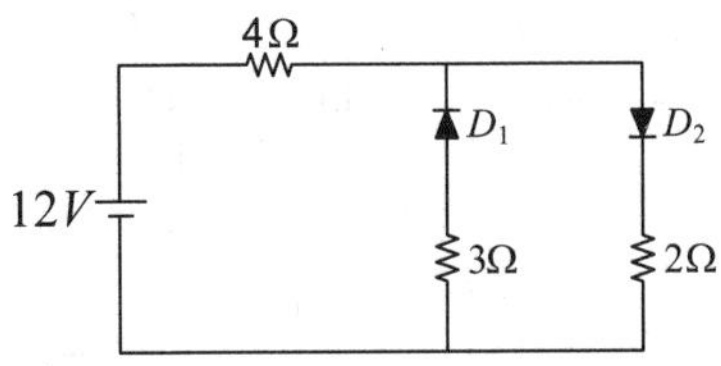

(A) 1.71 A (B) 2.00 A
(C) 2.31 A (D) 1.33 A

4. In the following, which one of the diodes reverse biased ? **[2006]**

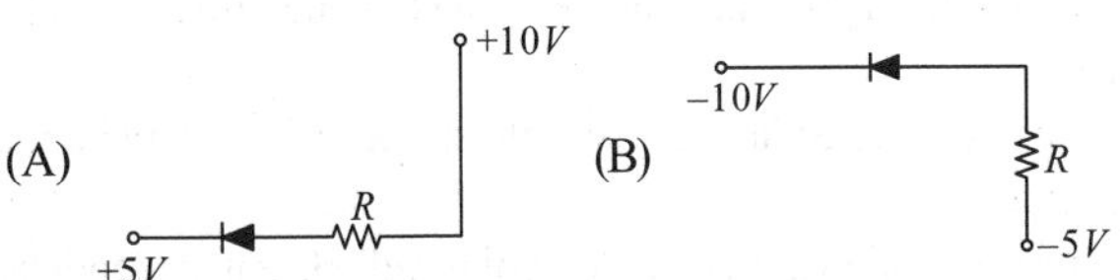

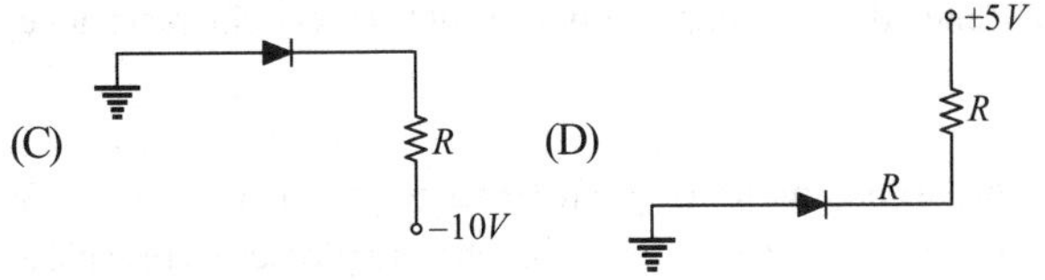

5. If in a *p-n* junction diode, a square input signal of 10 V is applied as shown **[2007]**

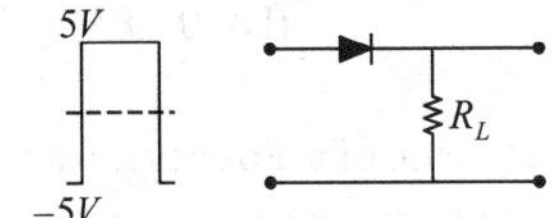

Then the output signal across R_L will be :

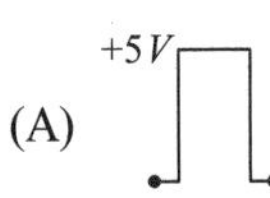

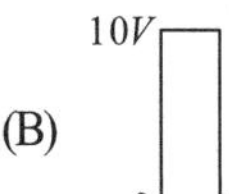

(C) −10V

(D) −5V

6. A p-n junction (D) shown in the figure can act as a rectifier. An alternating current source (V) is connected in the circuit. **[2009]**

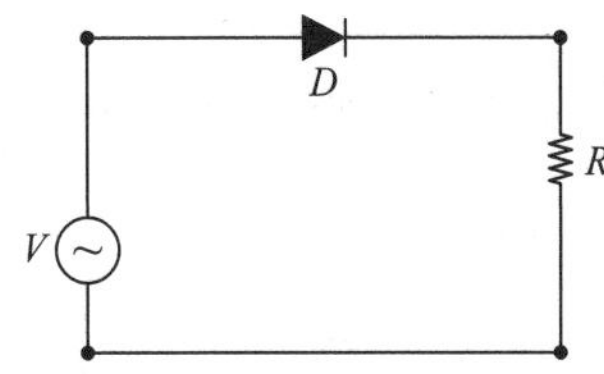

The current (*I*) in the resistor (*R*) can be shown by

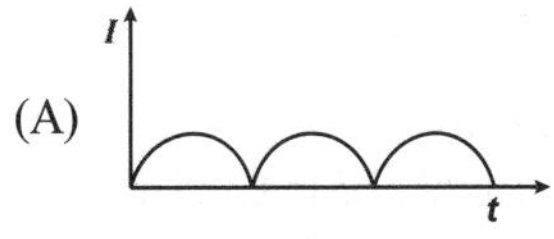

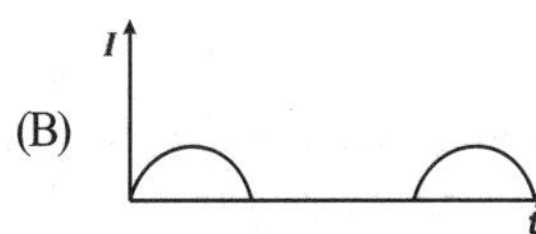

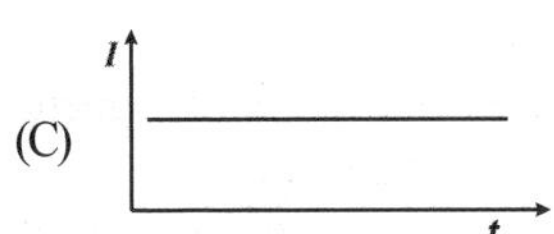

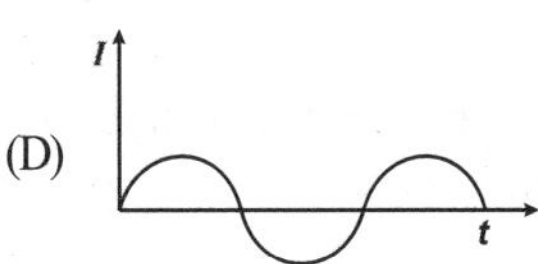

7. The I-V characteristic of an LED is : **[2013]**

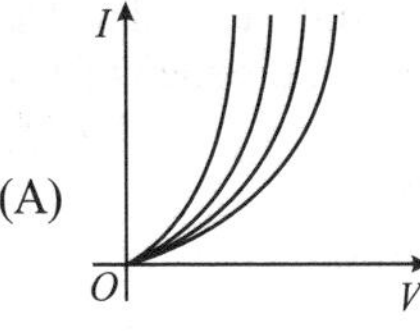

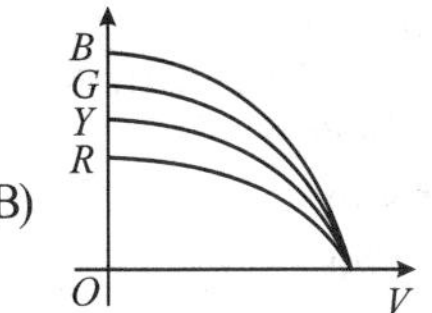

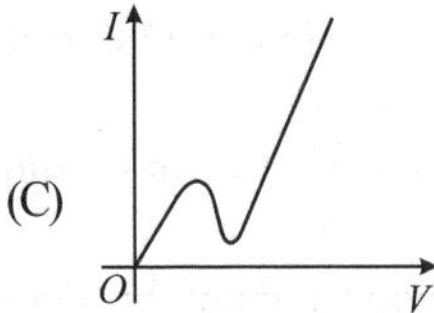

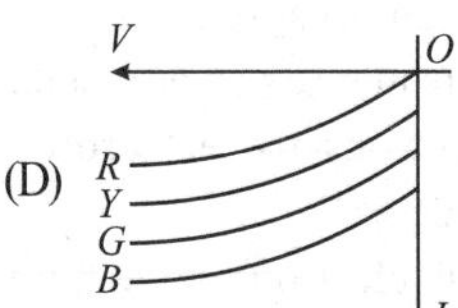

8. Figure shows a circuit in which three identical diodes are used. Each diode has forward resistance of 20 Ω and infinite backward resistance. Resistors $R_1 = R_2 = R_3 = 50\ \Omega$. Battery voltage is 6 V. The current through R_3 is : **[2013 (Online)]**

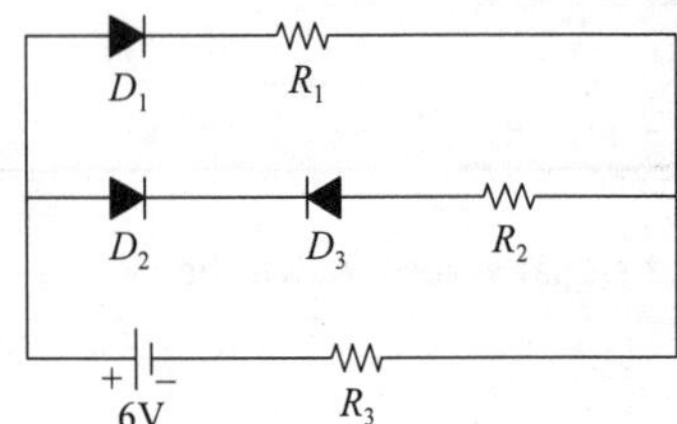

(A) 50 mA (B) 100 mA
(C) 60 mA (D) 25 mA

9. The forward biased diode connection is : **[2014]**

(A) $+2V$ ▷| ── ww $-2V$
(B) $-3V$ ▷| ── ww $-3V$
(C) $2V$ ▷| ── ww $4V$
(D) $-2V$ ▷| ── ww $+2V$

10. A Zener diode is connected to a battery and a load as shown below : **[2014 (Online)]**

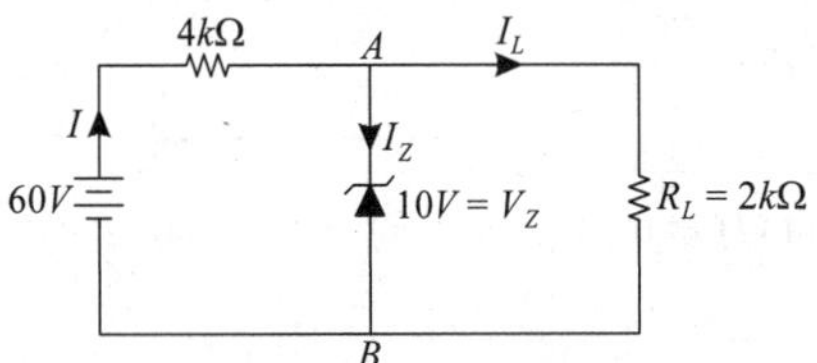

The currents, I, I_Z and I_L are respectively :
(A) 15 mA, 5 mA, 10 mA (B) 15 mA, 7.5 mA, 7.5 mA
(C) 12.5 mA, 5 mA, 7.5 mA (D) 12.5 mA, 7.5 mA, 5 mA

11. For LED's to emit light in visible region of electromagnetic light, it should have energy band gap in the range of : **[2014 (Online)]**
(A) 0.1 eV to 0.4 eV (B) 0.5 eV to 0.8 eV
(C) 0.9 eV to 1.6 eV (D) 1.7 eV to 3.0 eV

12. A red LED emits light at 0.1 watt uniformly around it. The amplitude of the electric field of the light at a distance of 1 m from the diode is : **[2015]**
(A) 5.48 V/m (B) 7.75 V/m
(C) 1.73 V/m (D) 2.45 V/m

13. In an unbiased *n-p* junction electrons diffuse from *n*-region to *p*-region because : **[2015 (Online)]**
(A) Hole in *p*-region attract them
(B) Electrons travel across the junction due to potential difference
(C) Only electrons move from *n* to *p* region and not the vice-versa
(D) Electron concentration in *n*-region is more compared to that in *p*-region

14. A 2 V battery is connected across *AB* as shown in the figure. The value of the current supplied by the battery when in one case battery's positive terminal is connected to *A* and in other case when positive terminal of battery is connected to *B* will respectively be : **[2015 (Online)]**

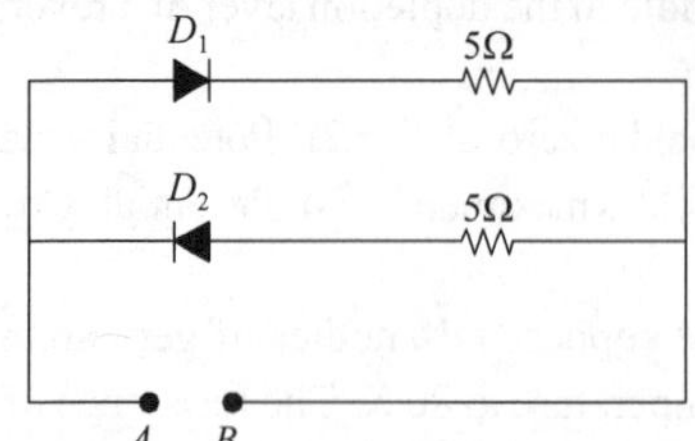

(A) 0.4 A and 0.2 A (B) 0.2 A and 0.4 A
(C) 0.1 A and 0.2 A (D) 0.2 A and 0.1 A

15. Identify the semiconductor devices whose characteristics are given below, in the order (i), (ii), (iii), (iv) : **[2016]**

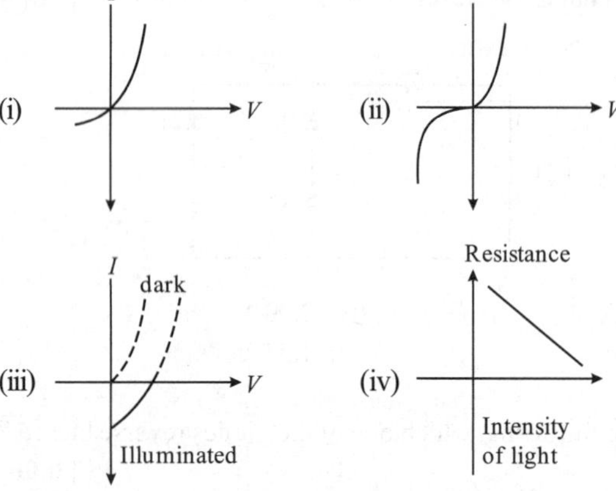

(A) Solar cell, Light dependent resistance, Zener diode, simple diode
(B) Zener diode, Solar cell, simple diode, Light dependent resistance
(C) Simple diode, Zener diode, Solar cell, Light dependent resistance
(D) Zener diode, Simple diode, Light dependent resistance, Solar cell

16. An experiment is performed to determine the 1-V characteristic of a Zener diode, which has a protective resistance of $R = 100\ \Omega$, and a maximum power of dissipation rating of 1 W. The minimum voltage range of the *DC* source in the circuit is : **[2016 (Online)]**
(A) 0 – 5 V (B) 0 – 24 V
(C) 0 – 12 V (D) 0 – 8 V

17. The V-I characteristic of a diode is shown in the figure. The ratio of forward to reverse bias resistance is : **[2017 (Online)]**

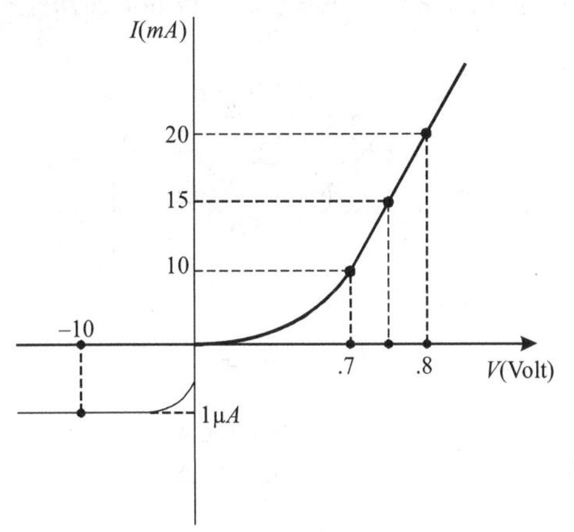

(A) 10 (B) 10^{-6}
(C) 10^{6} (D) 100

18. The reading of the ammeter for a silicon diode in the given circuit is : **[2018]**

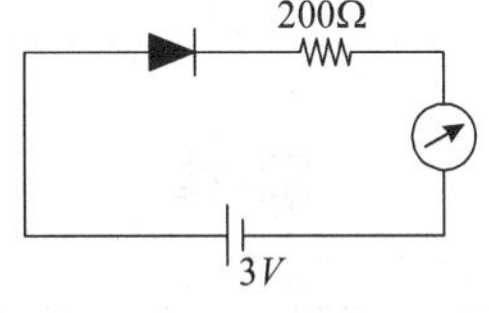

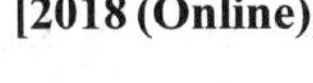

(A) 0 (B) 15 mA
(C) 11.5 mA (D) 13.5 mA

19. In the given circuit, the current through zener diode is : **[2018 (Online)]**

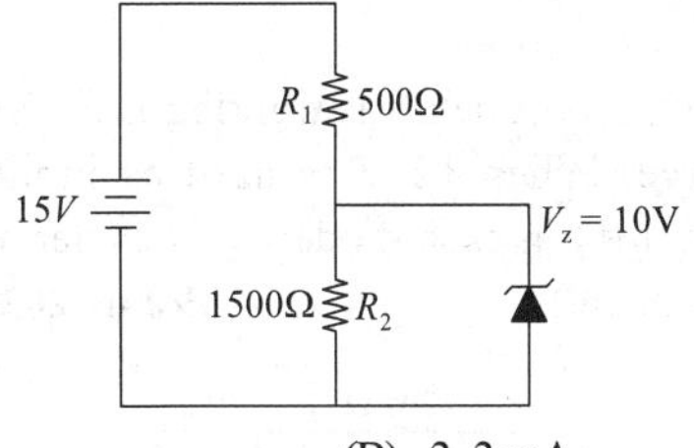

(A) 2.5 mA (B) 3.3 mA
(C) 5.5 mA (D) 6.7 mA

20. *Ge* and *Si* diodes start conducting at 0.3 V and 0.7 V respectively. In the following figure if *Ge* diode connection are reversed, the value of V_0 changes by : (Assume that the *Ge* diode has large breakdown voltage) **[9 Jan. 2019 (Evening)]**

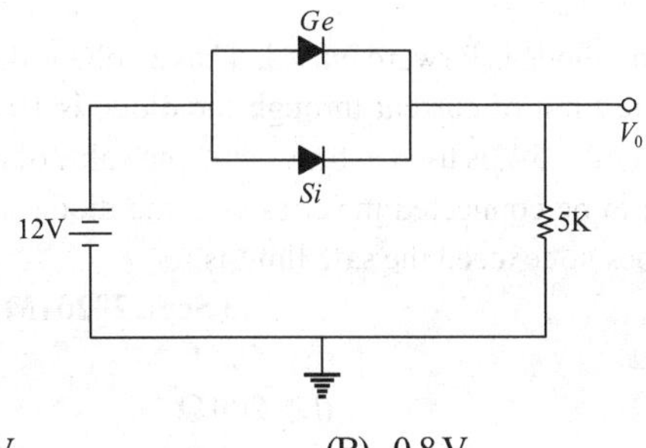

(A) 0.6 V (B) 0.8 V
(C) 0.4 V (D) 0.2 V

21. For the circuit shown below, the current through the Zener diode is : **[10 Jan. 2019 (Evening)]**

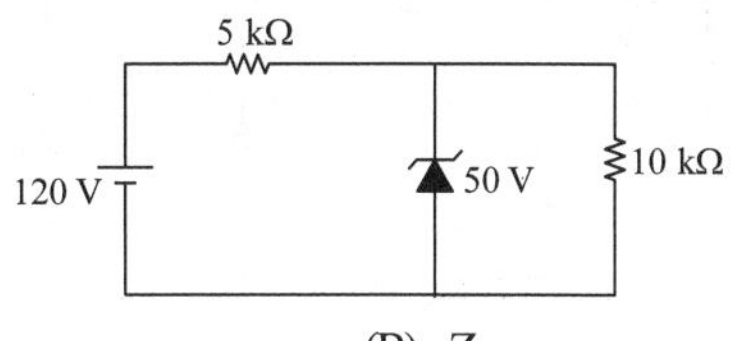

(A) 5 mA (B) Zero
(C) 14 mA (D) 9 mA

22. In the given circuit the current through Zener Diode is close to : **[11 Jan. 2019 (Morning)]**

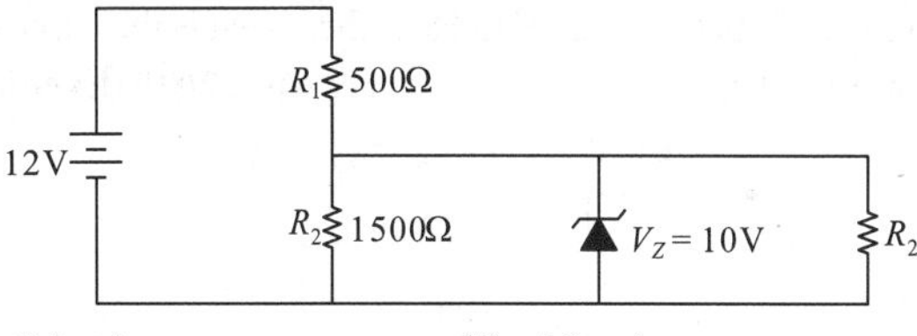

(A) 6.0 mA (B) 4.0 mA
(C) 6.7 mA (D) 0.0 mA

23. The circuit shown below contains two ideal diodes, each with a forward resistance of 50 Ω. If the battery voltage is 6 V, the current through the 100 Ω resistance (in Amperes) is : **[11 Jan. 2019 (Evening)]**

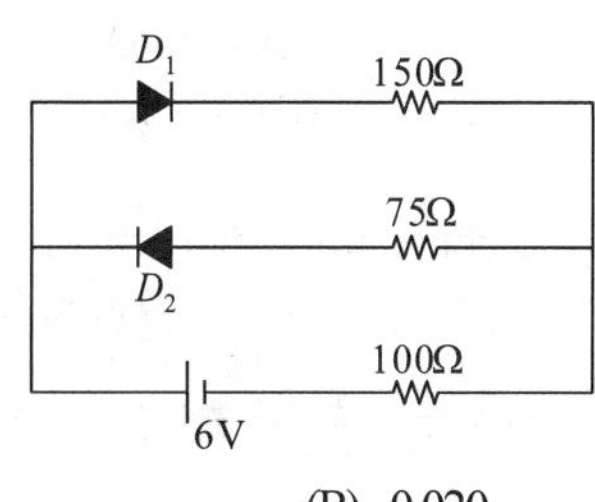

(A) 0.027 (B) 0.020
(C) 0.030 (D) 0.036

24. The reverse breakdown voltage of a Zener diode is 5.6 V in the given circuit. The current I_Z through the Zener is : **[8 April 2019 (Morning)]**

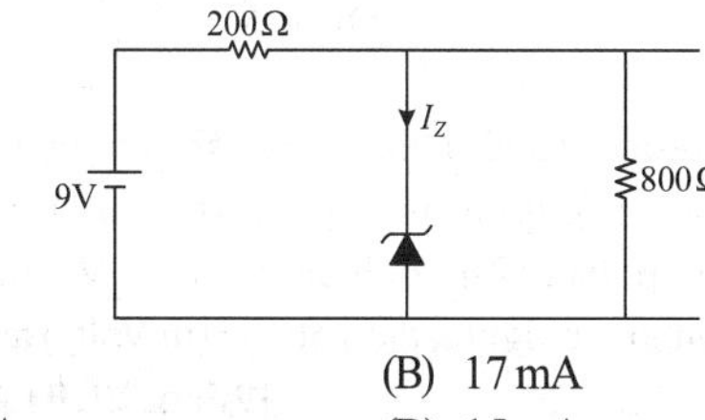

(A) 7 mA (B) 17 mA
(C) 10 mA (D) 15 mA

25. The figure represents a voltage regulator circuit using a Zener diode. The breakdown voltage of the Zener diode is 6 V and the load resistance is $R_L = 4$ kΩ. The series resistance of the circuit is $R_i = 1$ kΩ. If the battery voltage V_B varies from 8 V to

16 V, what are the minimum and maximum values of the current through Zener diode ? **[10 April 2019 (Evening)]**

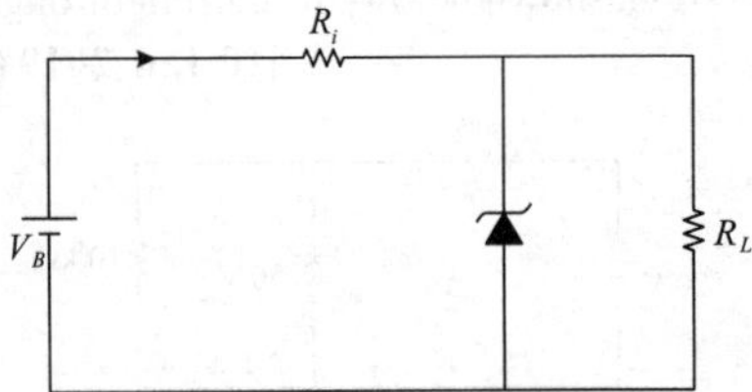

(A) 0.5 mA ; 6 mA (B) 0.5 mA ; 8.5 mA
(C) 1.5 mA ; 8.5 mA (D) 1 mA ; 8.5 mA

26. Figure shown a *DC* voltage regulator circuit, with a Zener diode of breakdown voltage = 6 V. If the unregulated input voltage varies between 10 V to 16 V, then what is the maximum Zener current ? **[12 April 2019 (Evening)]**

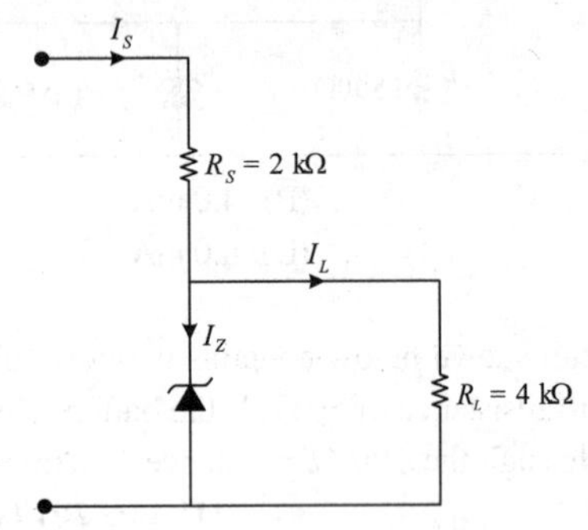

(A) 2.5 mA (B) 3.5 mA
(C) 7.5 mA (D) 1.5 mA

27. In the figure, potential difference between *A* and *B* is : **[7 Jan. 2020 (Evening)]**

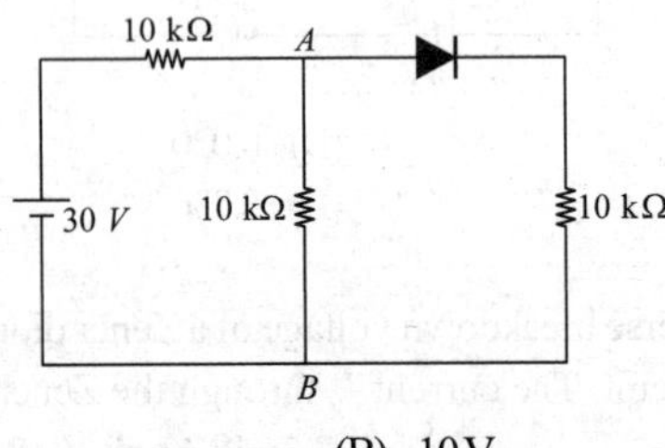

(A) 5V (B) 10V
(C) Zero (D) 15V

28. Both the diodes used in the circuit shown are assumed to be ideal and have negligible resistance when these are forward biased. Built in potential in each diode is 0.7 V. For the input voltages shown in the figure, the voltage (in Volts) at point *A* is ______. **[9 Jan. 2020 (Morning)]**

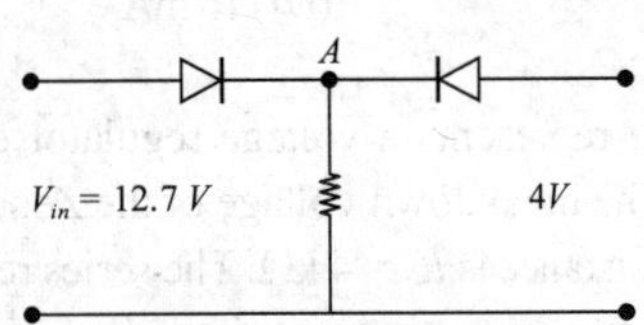

29. The current *i* in the network is : **[9 Jan. 2020 (Evening)]**

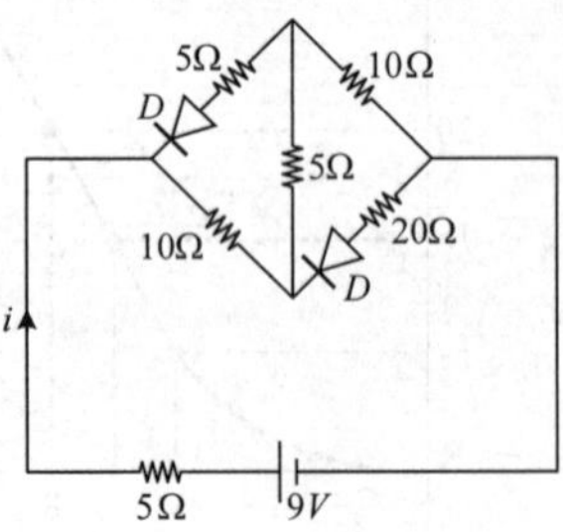

(A) 0 A (B) 0.6 A
(C) 0.3 A (D) 0.2 A

30. Two identical capacitors *A* and *B*, charged to the same potential $5V$ are connected in two different circuits as shown below at time $t = 0$. If the charge on capacitors *A* and *B* at time $t = CR$ is Q_A and Q_B respectively, then (Here *e* is the base of natural logarithm) **[9 Jan. 2020 (Evening)]**

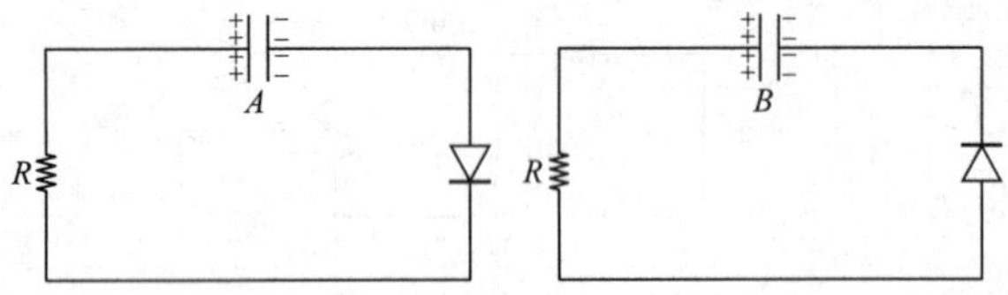

(A) $Q_A = VC, Q_B = \frac{VC}{e}$ (B) $Q_A = \frac{CV}{2}, Q_B = \frac{VC}{e}$

(C) $Q_A = VC, Q_B = CV$ (D) $Q_A = \frac{VC}{e}, Q_B = \frac{CV}{2}$

31. The circuit shown below is working as a 8 V dc regulated voltage source. When 12 V is used as input, the power dissipated (in *mW*) in each diode is; (considering both zener diodes are identical) ______. **[9 Jan. 2020 (Evening)]**

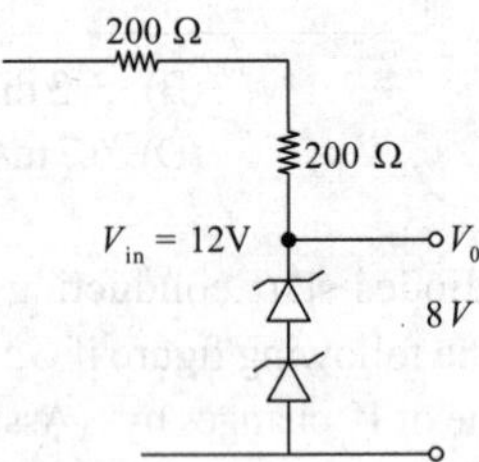

32. When a diode is forward biased, it has a voltage drop of 0.5 V. The safe limit of current through the diode is 10 mA. If a battery of emf 1.5 V is used in the circuit, the value of minimum resistance to be connected in series with the diode so that the current does not exceed the safe limit is :

[3 Sept. 2020 (Morning)]

(A) 100 Ω (B) 50 Ω
(C) 300 Ω (D) 200 Ω

33. If a semiconductor photodiode can detect a photon with a maximum wavelength of 400 nm, then its band gap energy is : **[3 Sept. 2020 (Evening)]**

Planck's constant $h = 6.63 \times 10^{-34}$ Js

Speed of light $c = 3 \times 10^{8}$ m/s

(A) 2.0 eV (B) 1.5 eV

(C) 3.1 eV (D) 1.1 eV

34. Take the breakdown voltage of the zener diode used in the given circuit as 6 V. For the input voltage shown in figure below, the time variation of the output voltage is : (Graphs drawn are schematic and not to scale) : **[4 Sept. 2020 (Morning)]**

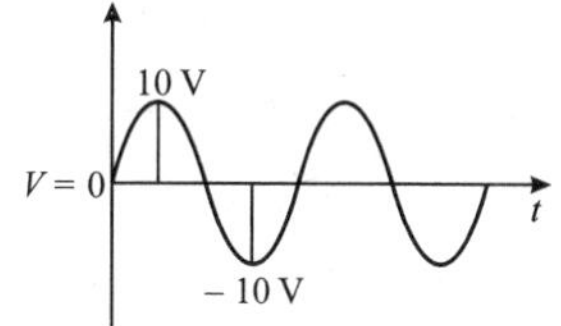

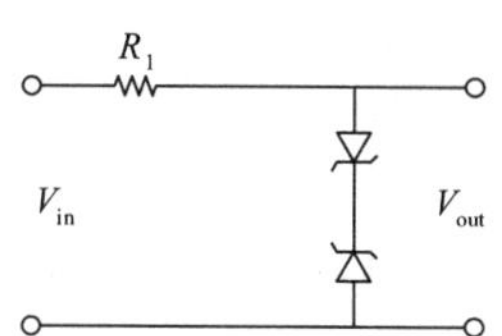

(A)

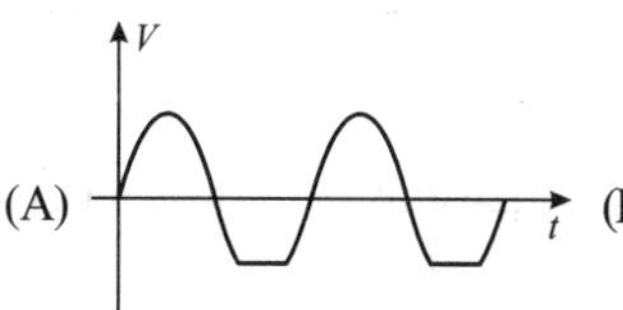

(B)

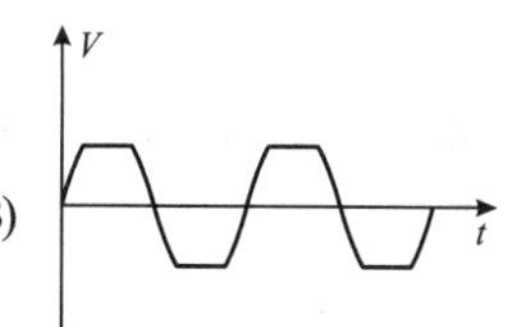

(C)

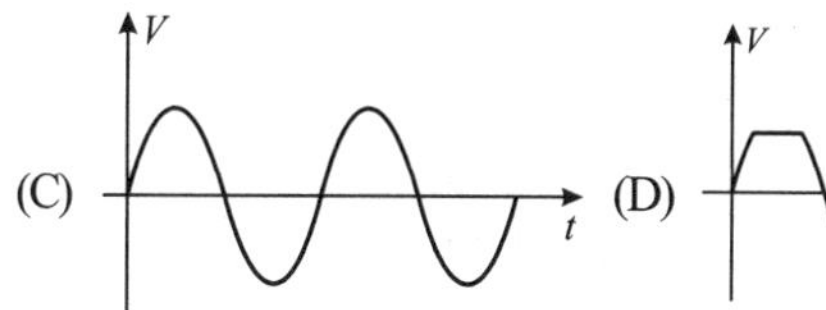

(D)

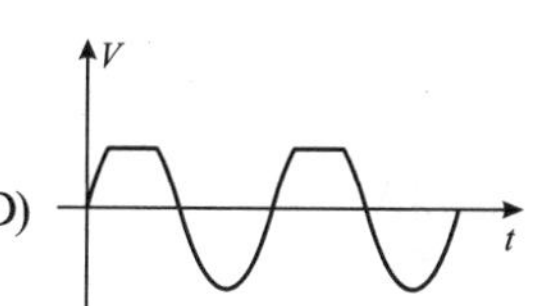

35. With increasing biasing voltage of a photodiode, the photocurrent magnitude : **[5 Sept. 2020 (Morning)]**

(A) Increases initially and saturates finally

(B) Increases initially and after attaining certain value, it decreases

(C) Increases linearly

(D) Remains constant

36. Two Zener diodes (*A* and *B*) having breakdown voltages of 6 V and 4 V respectively, are connected as shown in the circuit below. The output voltage V_O variation with input voltage linearly increasing with time, is given by : ($V_{input} = 0\ V$ at $t = 0$) (figures are qualitative) : **[5 Sept. 2020 (Evening)]**

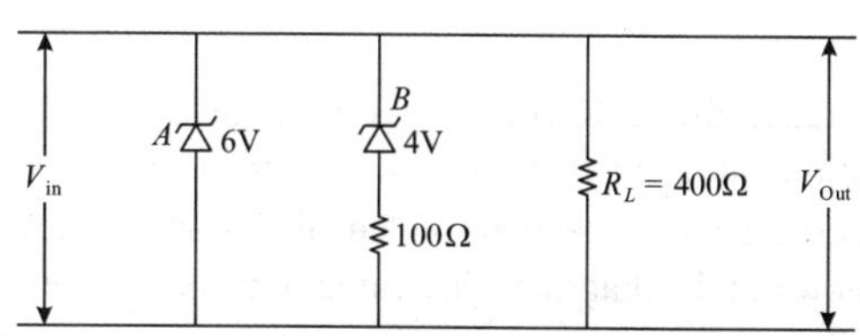

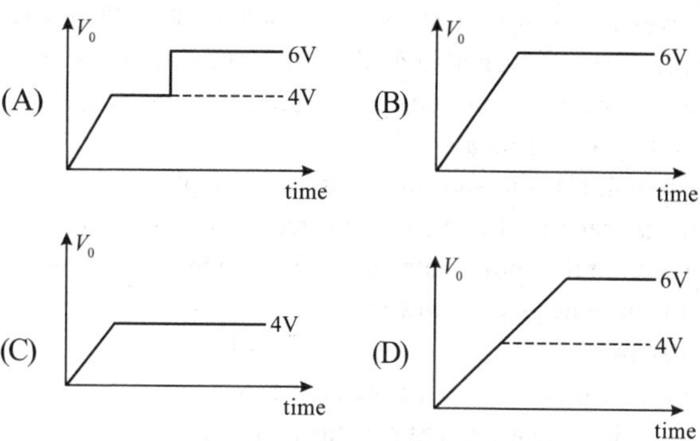

37. For the given input voltage waveform $V_{in}(t)$, the output voltage waveform $V_D(t)$, across the capacitor is correctly depicted by : **[6 Sept. 2020 (Morning)]**

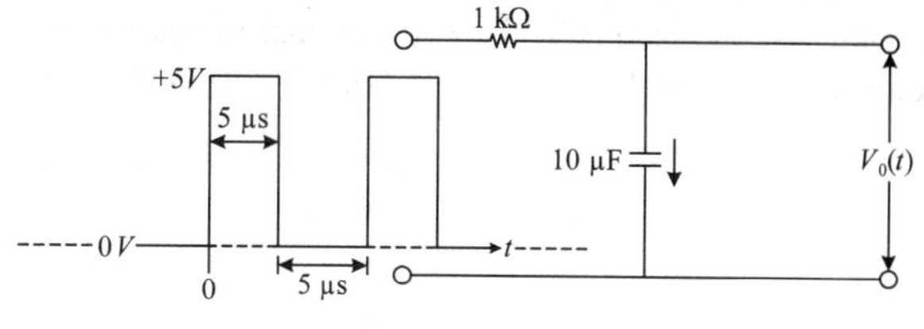

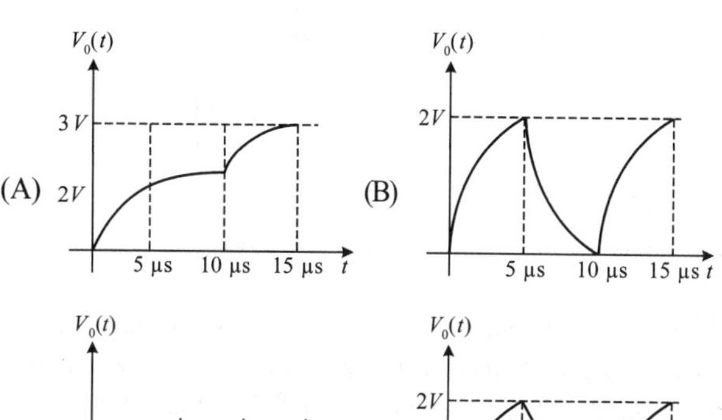

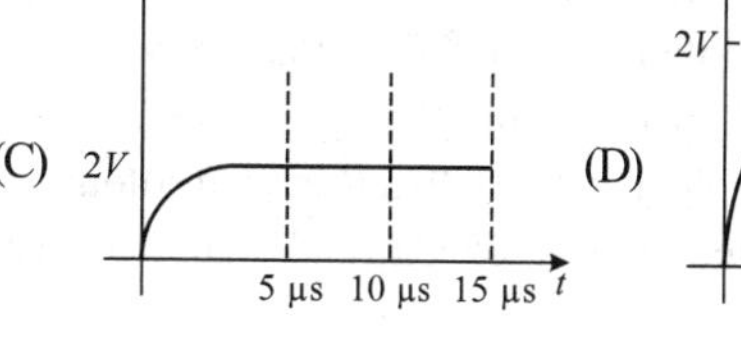

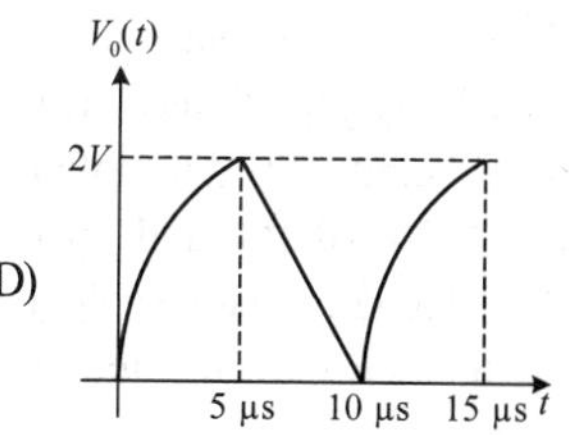

38. In connection with the circuit drawn below, the value of current flowing through 2 kΩ resistor is ______ $\times 10^{-4}$ A. **[24 Feb. 2021 (Morning)]**

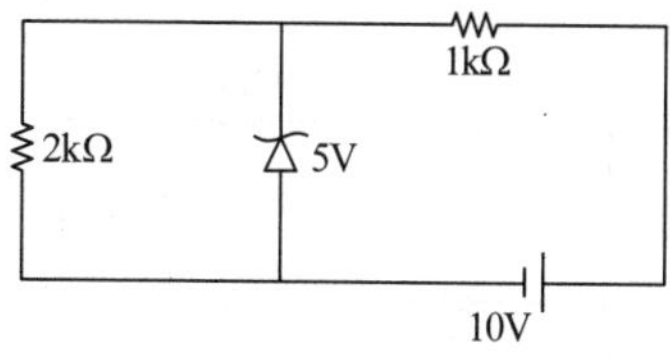

39. Zener breakdown occurs in a *p*-*n* junction having *p* and *n* both : **[24 Feb. 2021 (Evening)]**

(A) Heavily doped and have wide depletion layer

(B) Heavily doped and have narrow depletion layer

(C) Lightly doped and have wide depletion layer

(D) Lightly doped and have narrow depletion layer

40. Given below are two statements : **[24 Feb. 2021 (Evening)]**

Statement-I : PN-Junction diodes can be used to function as transistor, simply by connecting two diodes, back to back, which acts as the base terminal.

Statement-II : In the study of transistor, the amplification factor β indicates ratio of the collector current to the base current

In the light of the above statements, choose the correct answer from the options given below :

(A) Statement-I is true but Statement-II is false
(B) Statement-I is false but Statement-II is true
(C) Both Statement-I and Statement-II are true
(D) Both Statement-I and Statement-II are false

41. A 5 V battery is connected across the points X and Y. Assume D_1 and D_2 to be normal silicon diodes. Find the current supplied by the battery if the $+ve$ terminal of the battery is connected to point X : **[25 Feb. 2021 (Morning)]**

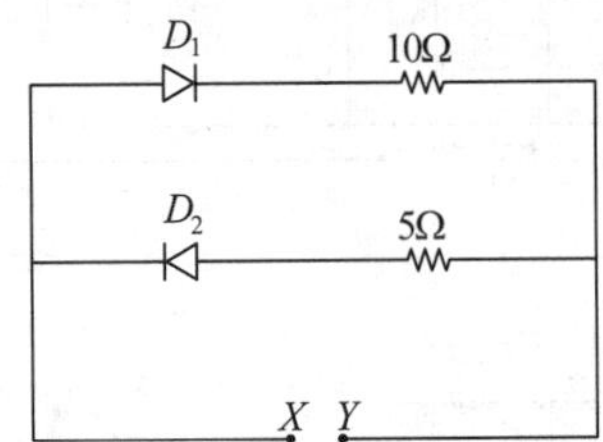

(A) ~0.86 A (B) ~0.5 A
(C) ~0.43 A (D) ~1.5 A

42. LED is constructed from *Ga-As-P* semiconducting material. The energy gap of this LED is 1.9 eV. Calculate the wavelength of light emitted and its colour. **[26 Feb. 2021 (Morning)]**
[$h = 6.63 \times 10^{-34}$ Js and $c = 3 \times 10^8$ ms^{-1}]

(A) 1046 nm and blue colour (B) 1046 nm and red colour
(C) 654 nm and red colour (D) 654 nm and orange colour

43. The circuit contains two diodes each with a forward resistance of 50 Ω and with infinite reverse resistance. If the battery voltage is 6 V, the current through the 120 Ω resistance is ______ *mA*. **[26 Feb. 2021 (Morning)]**

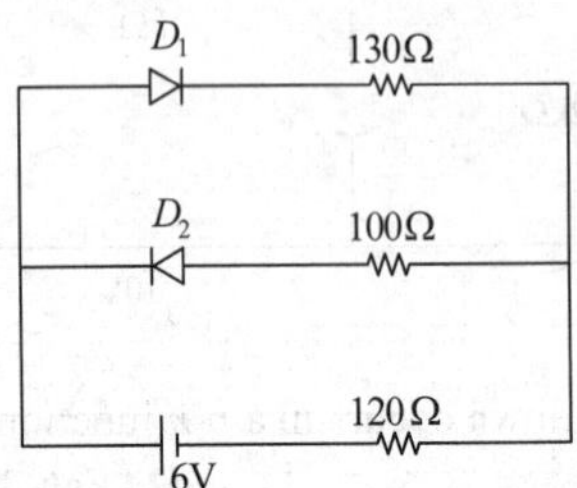

44. The zener diode has a V_z = 30 V. The current passing through the diode for the following circuit is ______ *mA*. **[26 Feb. 2021 (Evening)]**

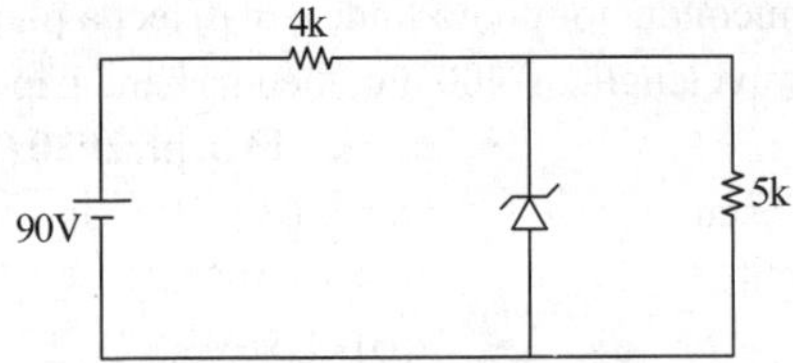

45. The value of power dissipated across the zener diode (V_z = 15 V) connected in the circuit as shown in the figure is $x \times 10^{-1}$ watt.

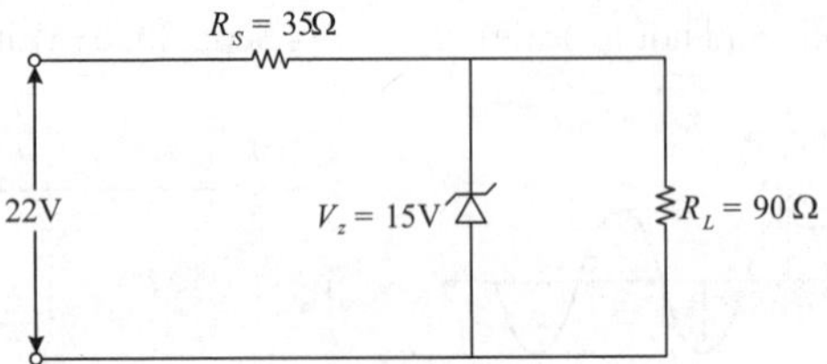

The value of x, to the nearest integer, is ______.
[16 March 2021 (Morning)]

46. For the circuit shown below, calculate the value of I_z : **[20 July 2021 (Morning)]**

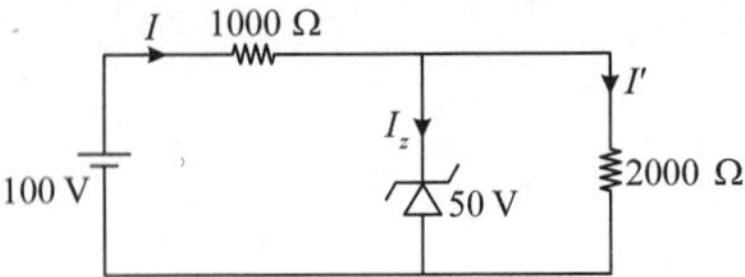

(A) 0.15 A (B) 25 mA
(C) 0.05 A (D) 0.1 A

47. For the forward biased diode characteristics shown in the figure, the dynamic resistance at I_D = 3 mÅ will be ______ Ω. **[20 July 2021 (Evening)]**

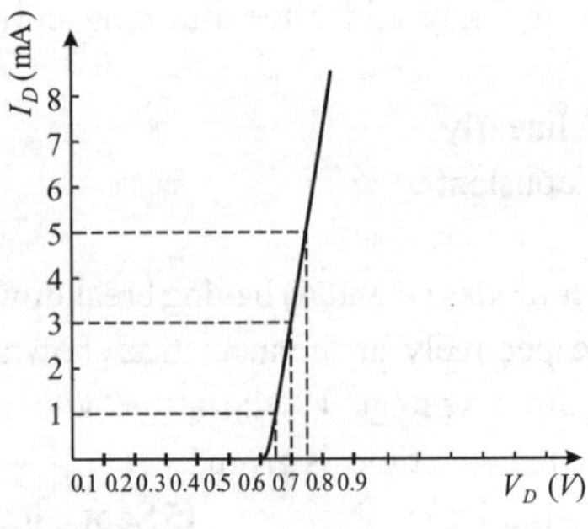

48. A zener diode having zener voltage 8 V and power dissipation rating of 0.5 W is connected across a potential divider arranged with maximum potential drop across zener diode is as shown in the diagram. The value of protective resistance

R_P is ______ Ω. **[20 July 2021 (Evening)]**

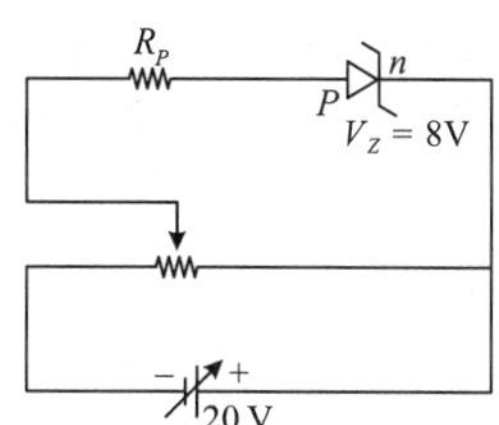

49. In a given circuit diagram, a 5 V zener diode along with a series resistance is connected across a 50 V power supply. The minimum value of the resistance required, if the maximum zener current is 90 mA will be ______ Ω. **[22 July 2021 (Evening)]**

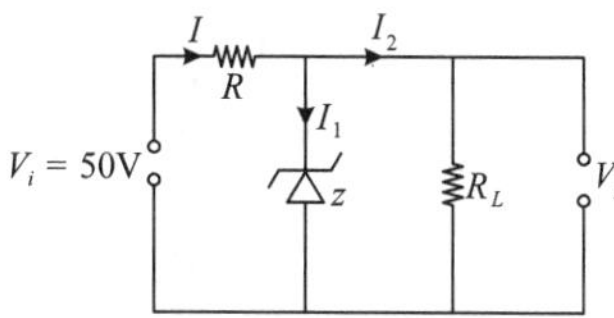

50. For the given circuit, the power across zener diode is ______ *mW*. **[26 Aug. 2021 (Evening)]**

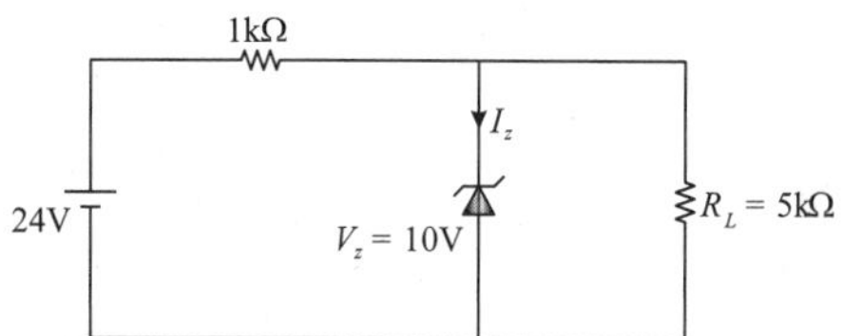

51. A zener diode of power rating 2 W is to be used as a voltage regulator. If the zener diode has a breakdown of 10 V and it has to regulate voltage fluctuated between 6 V and 14 V, the value of R_S for safe operation should be ______ Ω.
[27 Aug. 2021 (Evening)]

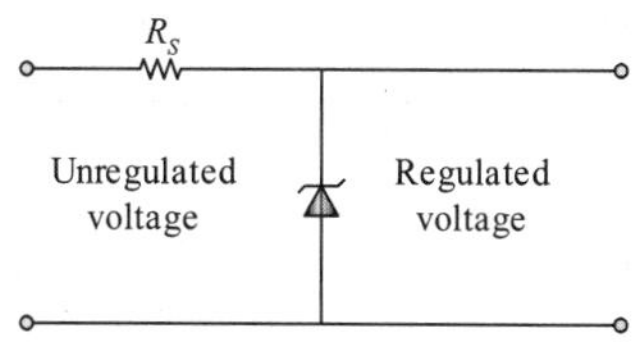

52. Choose the correct waveform that can represent the voltage across R of the following circuit, assuming the diode is ideal one : **[31 Aug. 2021 (Morning)]**

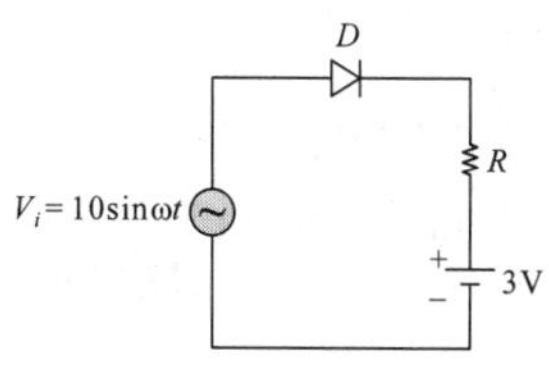

(A) 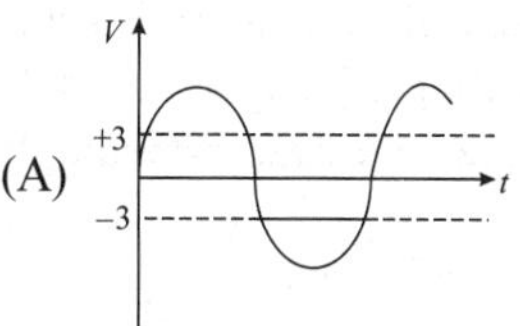

(B) 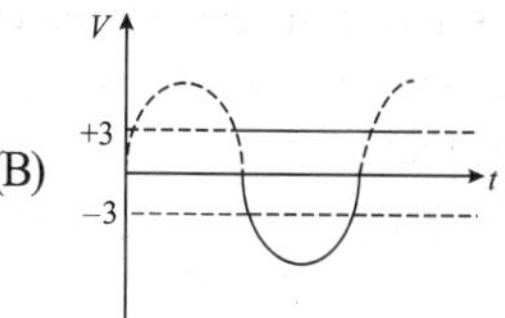

(C) 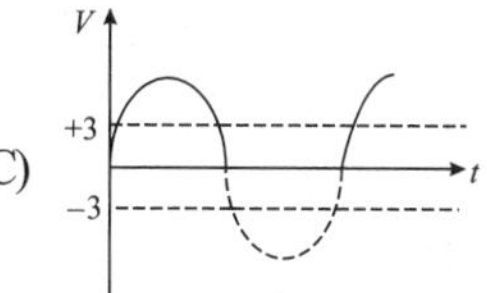

(D) 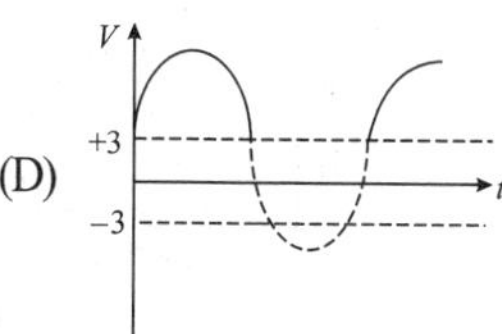

53. Statement-I : To get a steady dc output from the pulsating voltage received from a full wave rectifier we can connect a capacitor across the output parallel to the load R_L.
Statement-II : To get a steady dc output from the pulsating voltage received from a full wave rectifier we can connect an inductor in series with R_L.
In the light of the above statements, choose the most appropriate answer from the options given below :
[31 Aug. 2021 (Evening)]

(A) Both Statement-I and Statement-II are false
(B) Statement-I is false but Statement-II is true
(C) Both Statement-I and Statement-II are true
(D) Statement-I is true but Statement-II is false

54. In the given figure, each diode has a forward bias resistance of 30 Ω and infinite resistance in reverse bias. The current I_1 will be : **[01 Sep. 2021 (Evening)]**

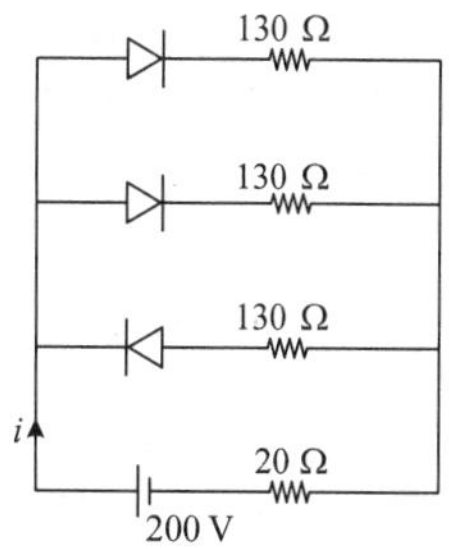

(A) 2 A (B) 3.75 A
(C) 2.73 A (D) 2.35 A

55. In the given circuit-the value of current I_L will be ______ *mA*. (When $R_L = 1$ kΩ) **[24 June 2022 (Evening)]**

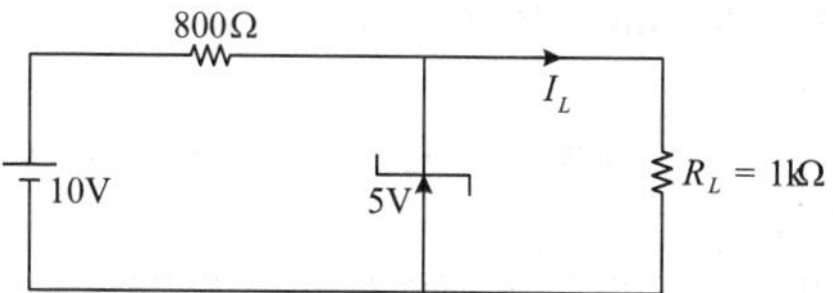

56. The I-V characteristics of a *p-n* junction diode in forward bias is shown in the figure. The ratio of dynamic resistance, corresponding to forward bias voltages of 2 V and 4 V respectively, is : **[26 June 2022 (Morning)]**

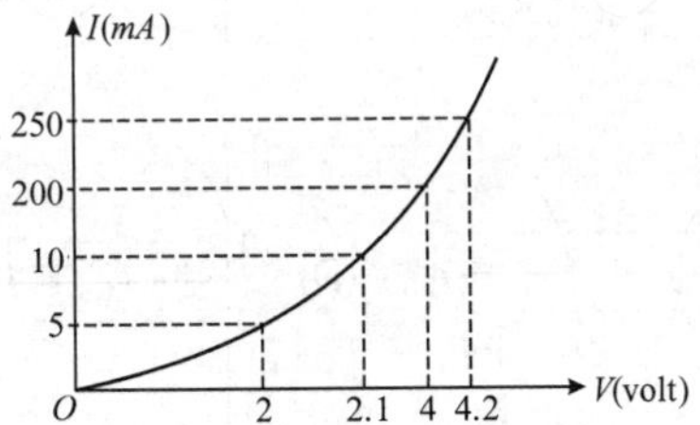

(A) 1:2 (B) 5:1
(C) 1:40 (D) 20:1

57. As per the given circuit, the value of current through the battery will be _____ A. **[26 June 2022 (Morning)]**

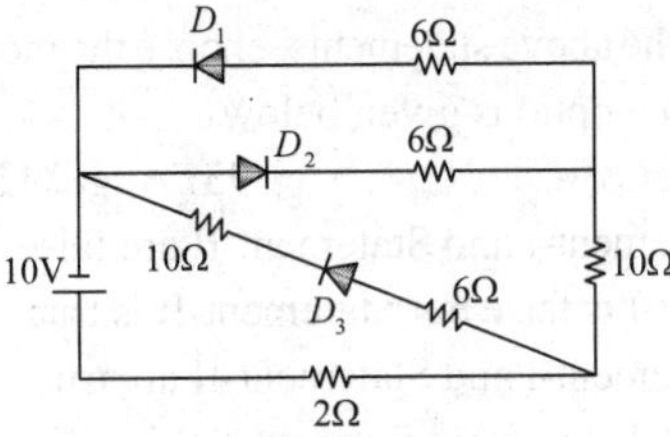

58. The cut-off voltage of the diodes (shown in figure) in forward bias is 0.6 V. The current through the resister of 40 Ω is _____ mA. **[27 June 2022 (Evening)]**

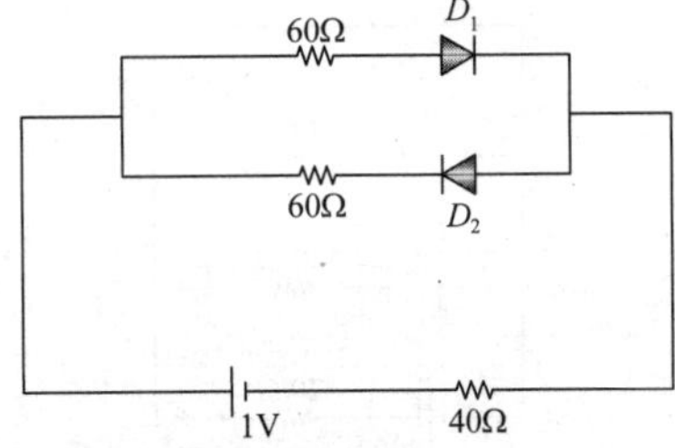

59. For using a multimeter to identify diode from electrical components, choose the correct statement out of the following about the diode : **[28 June 2022 (Morning)]**

(A) It is two terminal device which conducts current in both directions

(B) It is two terminal device which conducts current in one direction only

(C) It does not conduct current gives an initial deflection which decays to zero

(D) It is three terminal device which conducts current in ne direction only between central terminal and either of the remaining two terminals

60. In the given circuit the input voltage V_{in} is shown in figure. The cut-in voltage of *p-n* junction diode (D_1 or D_2) is 0.6 V. Which of the following output voltage (V_0) waveform across the diode is correct ? **[28 June 2022 (Evening)]**

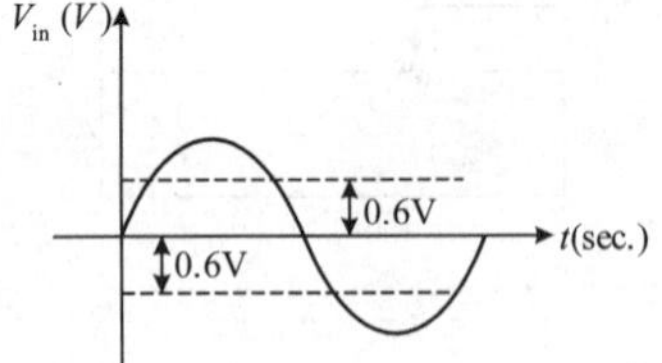

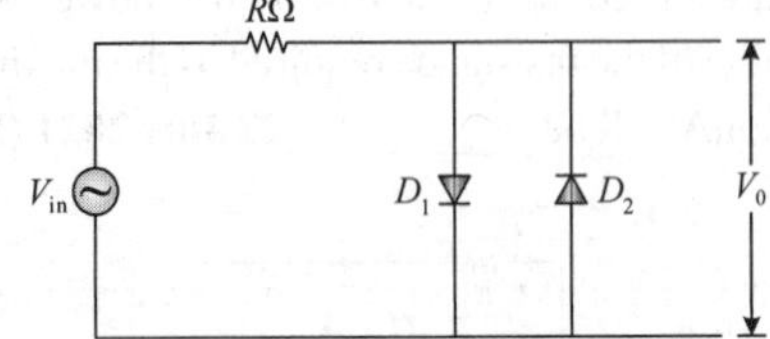

(A) 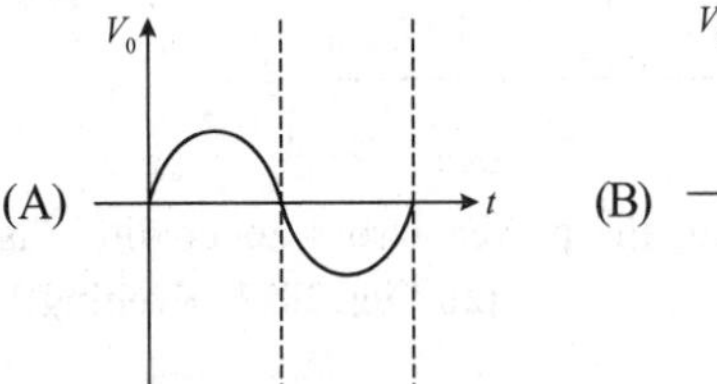

(B) 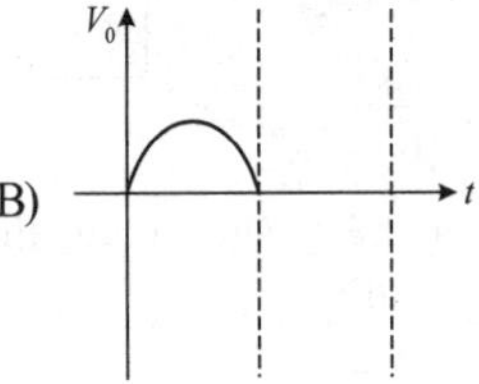

(C) 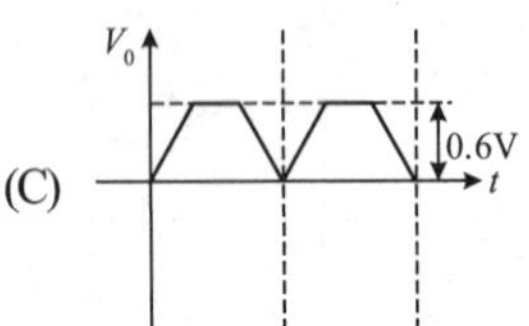

(D) 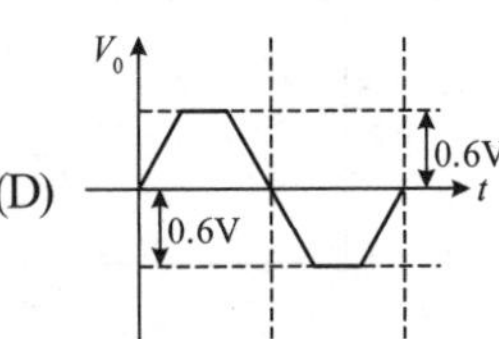

61. A Zener of breakdown voltage $V_Z = 8$V and maximum zener current, $I_{ZM} = 10$ mA is subjected to an input voltage $V_i = 10$ V with series resistance $R = 100\ \Omega$. In the given circuit R_L represents the variable load resistance. The ratio of maximum and minimum value of R_L is _____. **[28 June 2022 (Evening)]**

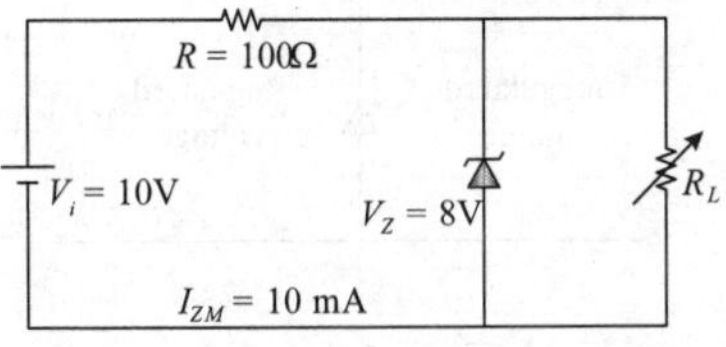

62. A potential barrier of 0.4 V exists across a *p-n* junction. An electron enters the junction from the *n*-side with a speed of 6.0×10^5 ms^{-1}. The speed with which electron enters the *p* side will be $\frac{x}{3} \times 10^5$ ms^{-1} the value of *x* is _____. (Given mass of electron $= 9 \times 10^{-31}$ kg, charge on electron $= 1.6 \times 10^{-19}$C) **[29 June 2022 (Evening)]**

63. The energy band gap of semiconducting material to produce violet light (wavelength = 4000 Å) *LED* is_____ eV. (Round off to the nearest integer) **[25 July 2022 (Morning)]**

64. Two ideal diodes are connected in the network as shown in figure. The equivalent resistance between *A* and *B* is_____ Ω. **[25 July 2022 (Evening)]**

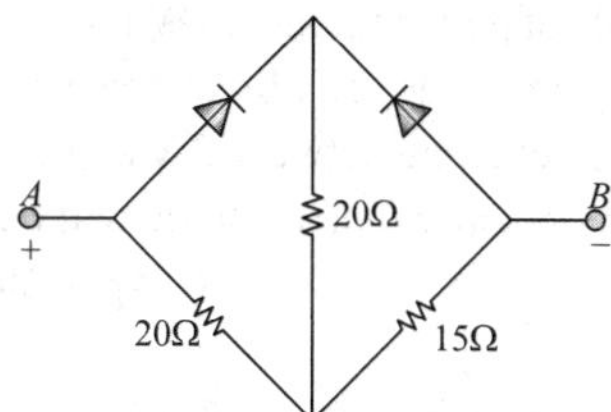

65. In the circuit shown below, maximum zener diode current will be_______ mA. **[26 July 2022 (Morning)]**

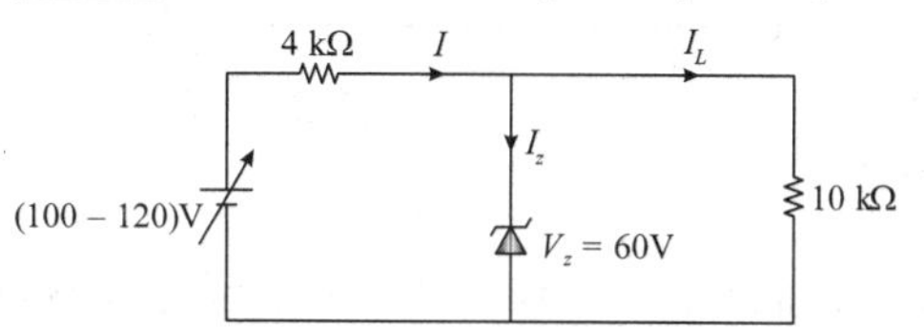

66. Identify the solar cell characteristics from the following options : **[28 July 2022 (Morning)]**

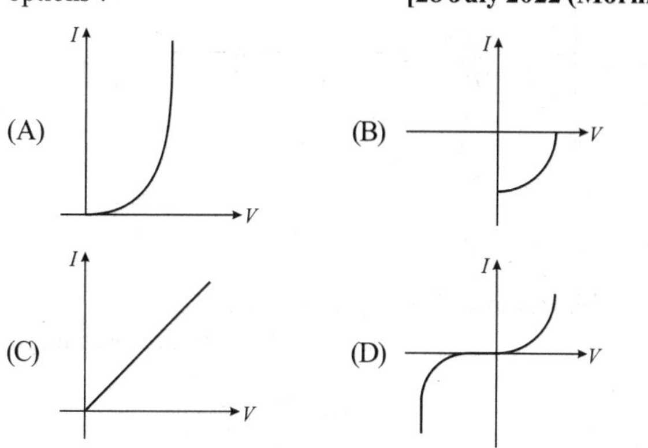

67. If the potential barrier across a *p-n* junction is 0.6V. Then the electric field intensity, in the depletion region having the width of 6×10^{-6} m, will be_____ $\times 10^5$ N/C. **[29 July 2022 (Morning)]**

68. A 8V Zener diode along with a series resistance *R* is connected across a 20 V supply (as shown in the figure). If the maximum Zener current is 25 mA, then the minimum value of *R* will be_____ Ω. **[29 July 2022 (Evening)]**

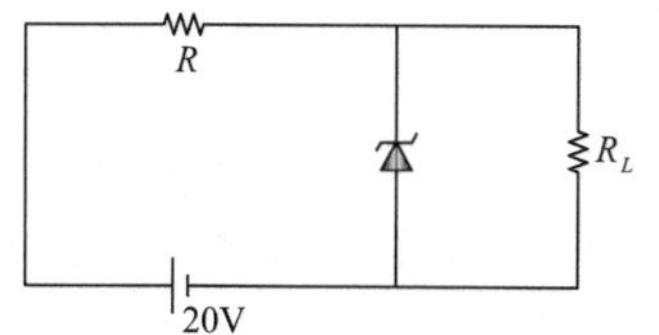

69. Given below are two statements:
One is labelled as **Assertion-A** and the other is labelled as **Reason-R**

Assertion-A : Photodiodes are preferably operated in reverse bias condition for light intensity measurement.

Reason-R : The current in the forward bias is more than the current in the reverse bias for a *p-n* junction diode

In the light of the above statement, choose the correct answer from the options given below : **[24 Jan. 2023 (Morning)]**

(A) **A** is false but **R** is true.
(B) Both **A** and **R** are true but **R** is NOT the correct explanation of **A**.
(C) **A** is true but **R** is false.
(D) Both **A** and **R** are true and **R** is the correct explanation of **A**.

70. Given below are two statements:
One is labeled as **Assertion-A** and the other is labeled as **Reason-R.**

Assertion-A: Photodiodes are used in forward bias usually for measuring the light intensity

Reason-R: For a *p-n* junction diode, at applied voltage *V* the current in the forward bias is more than the current in the reverse bias for $|V_Z| > \pm V \geq |V_0|$ where V_0 is the threshold voltage and V_z is the breakdown voltage. In the light of the above statements, choose the correct answer from the options given below: **[25 Jan. 2023 (Morning)]**

(A) Both **A** and **R** are true and **R** is correct explanation **A**.
(B) Both **A** and **R** are true but **R** is NOT the correct explanation of **A**.
(C) **A** is false but **R** is true.
(D) **A** is true but **R** is true.

71. Which of the following statement is not correct in the case of light emitting diodes ?
(a) It is a heavily doped *p-n* junction
(b) It emits light only when it is forward biased
(c) It emits light only when it is reverse biased
(d) The energy of the light emitted is equal to or slightly less than the energy gap of the semiconductor used
Choose the correct answer from the options given below : **[29 Jan. 2023 (Morning)]**

(A) (c) and (d) (B) (a)
(C) (c) (D) (b)

72. Statement-I: When a Si sample is doped with Boron, it becomes *P* type and when doped by Arsenic it becomes *N*-type semi conductor such that *P*-type has excess holes and *N*-type has excess electrons

Statement-II: When such *P*-type and *N*-type semi-conductors, are fused to make a junction, a current will automatically flow which can be detected with an externally connected ammeter. In the light of above statements, choose the most appropriate answer from the options given below.

Options : [25 Jan. 2023 (Evening)]
(A) Both Statement-I and statement-II are incorrect
(B) Statement-I is incorrect but statement-II is correct
(C) Both Statement-I and statement-II are correct
(D) Statement-I is correct but statement-II is incorrect

73. Match the List-I with List-II :

[1 Feb. 2023 (Morning)]

List-I	List-II
(a) Intrinsic Semiconductor	(i) Fermi-level near conduction band
(b) *n*-type semiconductor	(ii) Fermi-level at middle
(c) *p*-type semiconductor	(iii) Fermi-level near valance band
(d) Metals	(iv) Fermi-level inside conduction band

Choose the correct answer from the options given below :
(A) (a)-(i), (b)-(ii), (c)-(iii), (d)-(iv)
(B) (a)-(ii), (b)-(i), (c)-(iii), (d)-(iv)
(C) (a)-(ii), (b)-(iii), (c)-(i), (d)-(iv)
(D) (a)-(iii), (b)-(i), (c)-(ii), (d)-(iv)

74. Choose the correct statement about Zener diode :

[1 Feb. 2023 (Evening)]

(A) It works as a voltage regulator in reverse bias and behaves like simple pn junction diode in forward bias
(B) It works as a voltage regulator in both forward and reverse bias
(C) It works a voltage regulator only in forward bias
(D) It works as a voltage regulator in forward bias and behaves like simple *pn*-junction diode in reverse bias

75. Given below are two statements: one is labelled as **Assertion-A** and the other is labelled as **Reason-R**.
Assertion-A: Diffusion current in a *p-n* junction is greater than the drift current in magnitude if the junction is forward biased.
Reason-R: Diffusion current in a *p-n* junction is from the n side to the *p*-side if the junction is forward biased.
In the light of the above statements, choose the most appropriate answer from the options given below :

[6 April 2023 (Evening)]

(A) Both **A** and **R** are correct and **R** is the correct explanation of **A**.
(B) Both **A** and **R** are correct but **R** is NOT the correct explanation of **A**.
(C) **A** is correct but **R** is not correct.
(D) **A** is not correct but **R** is correct.

76. A zener diode of power rating 1.6 W is to be used as voltage regulator. If the zener diode has a breakdown of 8V and it has to regulate voltage fluctuating between 3 V and 10 V. The value of resistance R_s for safe operation of diode will be :

[10 April 2023 (Morning)]

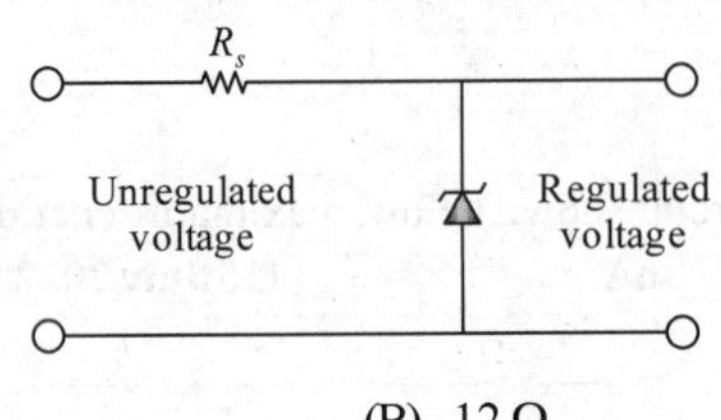

(A) 13.3 Ω (B) 12 Ω
(C) 10 Ω (D) 13 Ω

77. If each diode has a forward bias resistance of 25 Ω in the below circuit,

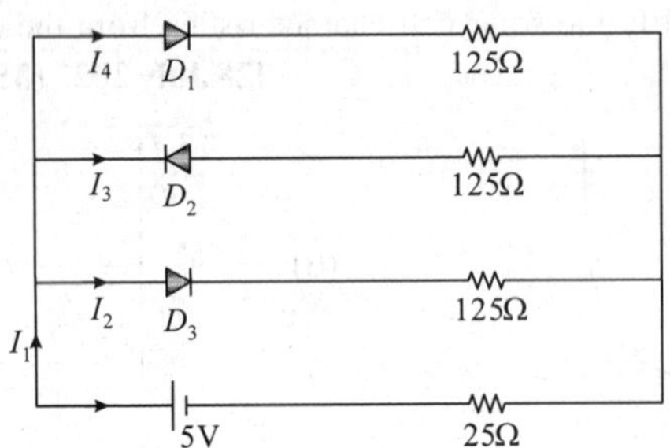

Which of the following options is correct:

[10 April 2023 (Evening)]

(A) $\frac{I_3}{I_4} = 1$ (B) $\frac{I_2}{I_3} = 1$

(C) $\frac{I_1}{I_2} = 1$ (D) $\frac{I_1}{I_2} = 2$

* * * * *

SEMICONDUCTORS ELECTRONICS

28.3 LOGIC GATES

1. In the circuit below, A and B represent two inputs and C represents the output. **[2008]**

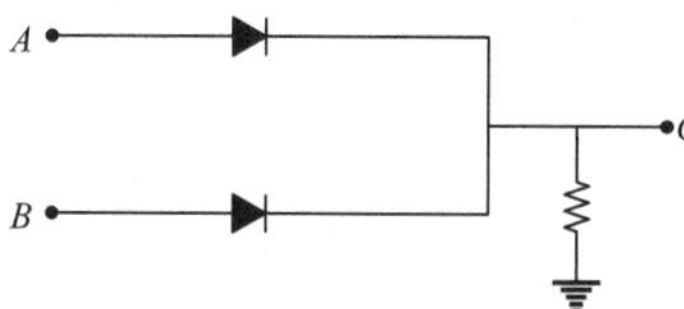

The circuit represents :

(A) NOR gate
(B) AND gate
(C) NAND gate
(D) OR gate

2. The logic circuit shown below has the input waveforms 'A' and 'B' as shown. Pick out the correct output waveform : **[2009]**

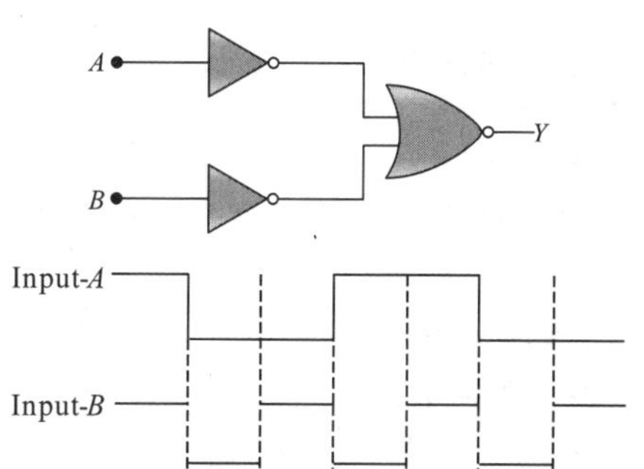

Output is :

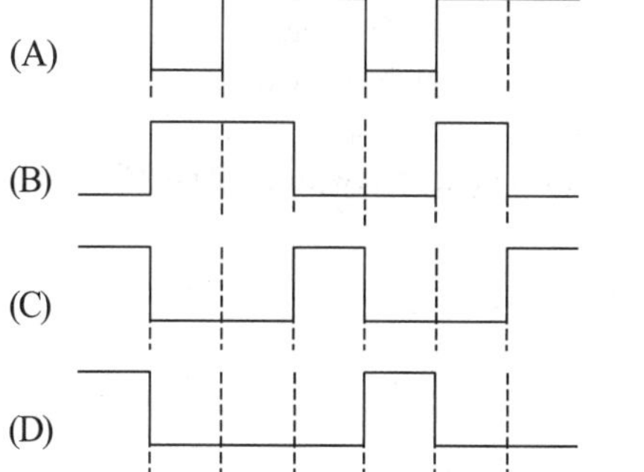

3. The combination of gates shown below yields : **[2010]**

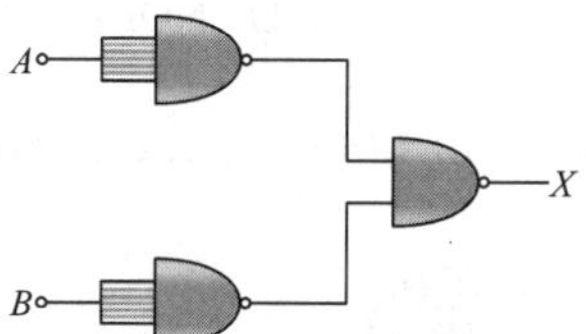

(A) OR gate
(B) NOT gate
(C) XOR gate
(D) NAND gate

4. The output of an OR gate is connected to both the inputs of a NAND gate. The combination will serve as : **[2011]**

(A) NOT gate
(B) NOR gate
(C) AND gate
(D) OR gate

5. Truth table for system of four NAND gates as shown in figure is : **[2012]**

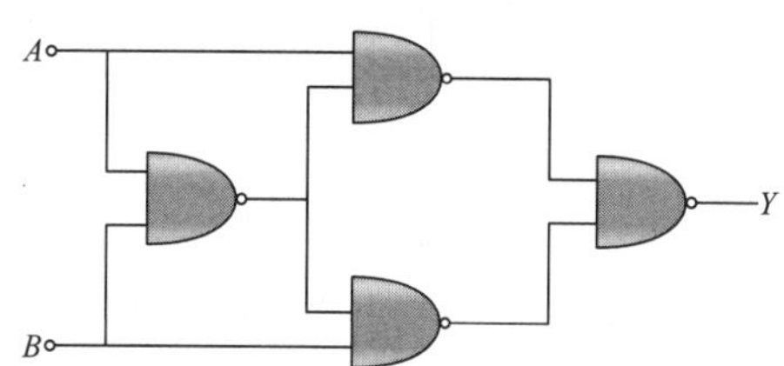

(A)

A	B	Y
0	0	0
0	1	1
1	0	1
1	1	0

(B)

A	B	Y
0	0	0
0	1	0
1	0	1
1	1	1

(C)

A	B	Y
0	0	1
0	1	1
1	0	0
1	1	0

(D)

A	B	Y
0	0	1
0	1	0
1	0	1
1	1	1

6. Which logic gate with inputs A and B performs the same operation as that performed by the following circuit ? **[2012 (Online)]**

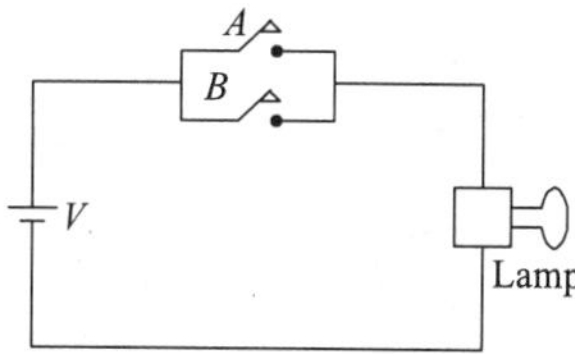

(A) NAND gate
(B) OR gate
(C) NOR gate
(D) AND gate

7. Which one of the following is the Boolean expression for NOR gate ? **[2012 (Online)]**

(A) $Y = \overline{A+B}$
(B) $Y = \overline{A \cdot B}$
(C) $Y = A \cdot B$
(D) $Y = \overline{A}$

8. The figure shows a combination of two NOT gates and a NOR gate. The combination is equivalent to a : **[2012 (Online)]**

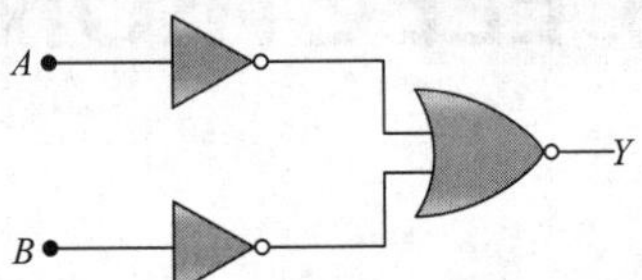

(A) NAND gate (B) NOR gate
(C) AND gate (D) OR gate

9. A system of four gates is set up as shown. The 'truth table' corresponding to this system is : **[2013 (Online)]**

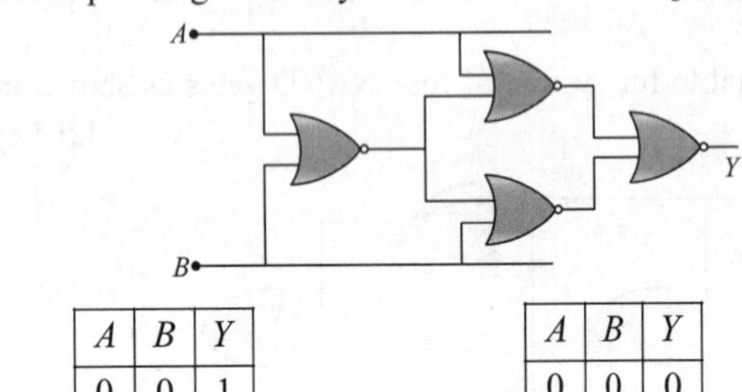

(A)

A	B	Y
0	0	1
0	1	0
1	0	0
1	1	1

(B)

A	B	Y
0	0	0
0	1	0
1	0	1
1	1	0

(C)

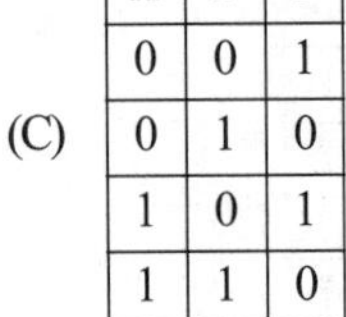

A	B	Y
0	0	1
0	1	0
1	0	1
1	1	0

(D)

A	B	Y
0	0	1
0	1	1
1	0	0
1	1	0

10. Which of the following circuits correctly represents the following truth table ? **[2013 (Online)]**

A	B	C
0	0	0
0	1	0
1	0	1
1	1	0

(A)
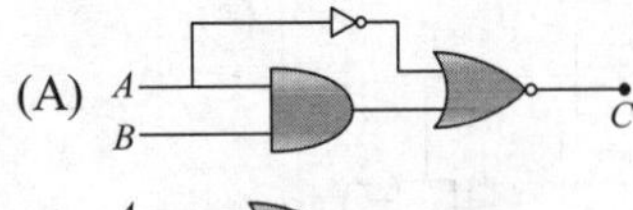

(B)
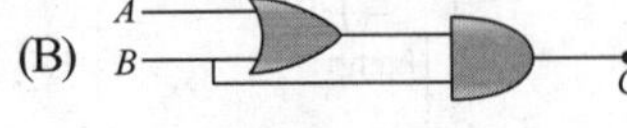

(C)
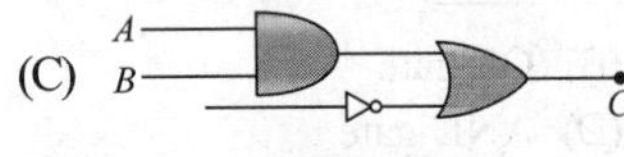

(D) A B C

11. Consider two *npn* transistors as shown in figure. If 0 Volts corresponds to false and 5 Volts correspond to true then the output at C corresponds to : **[2013 (Online)]**

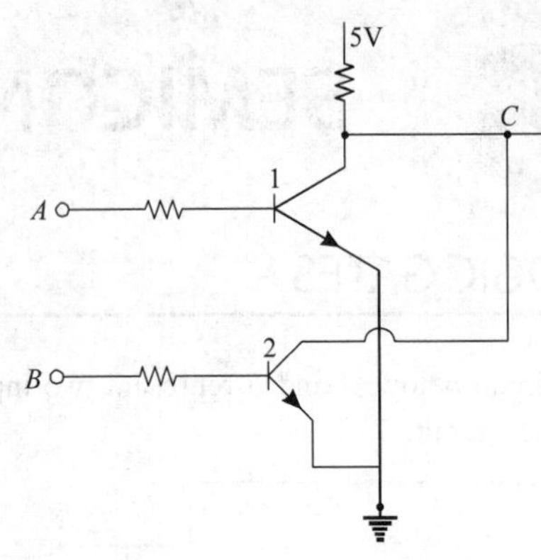

(A) A NAND B (B) A OR B
(C) A AND B (D) A NOR B

12. Identify the gate and match A, B, Y in bracket to check : **[2014 (Online)]**

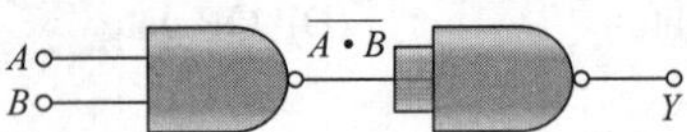

(A) AND ($A = 1, B = 1, Y = 1$) (B) OR ($A = 1, B = 1, Y = 0$)
(C) NOT ($A = 1, B = 1, Y = 1$) (D) XOR ($A = 0, B = 0, Y = 0$)

13. Given : A and B are input terminals
Logic 1 => 5 V
Logic 0 =< 1 V
Which logic gate operation, the above circuit does? **[2014 (Online)]**

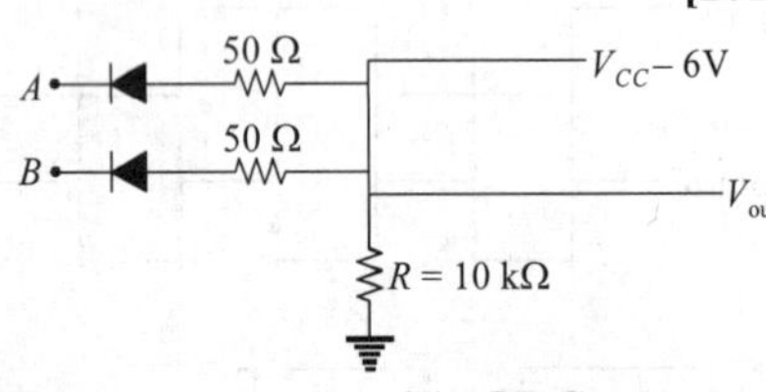

(A) AND Gate (B) OR Gate
(C) XOR Gate (D) NOR Gate

14. The truth table given in figure represents : **[2016 (Online)]**

A	B	Y
0	0	0
0	1	1
1	0	1
1	1	1

(A) OR – Gate (B) NAND – Gate
(C) AND – Gate (D) NOR – Gate

15. To get an output of 1 from the circuit shown in figure the input must be : **[2016 (Online)]**

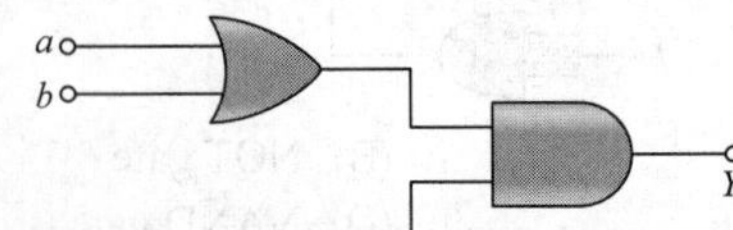

(A) $a=0, b=0, c=1$ (B) $a=1, b=0, c=0$
(C) $a=1, b=0, c=1$ (D) $a=0, b=1, c=0$

16. Truth table for the given circuit will be : **[2018 (Online)]**

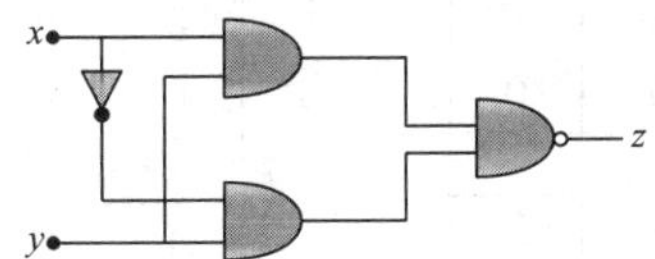

(A)

x	y	z
0	0	1
0	1	1
1	0	1
1	1	0

(B)

x	y	z
0	0	0
0	1	0
1	0	0
1	1	1

(C)

x	y	z
0	0	1
0	1	1
1	0	1
1	1	1

(D)

x	y	z
0	0	0
0	1	1
1	0	1
1	1	1

17. To get output '1' at R, for the given logic gate circuit the input values must be : **[10 Jan. 2019 (Morning)]**

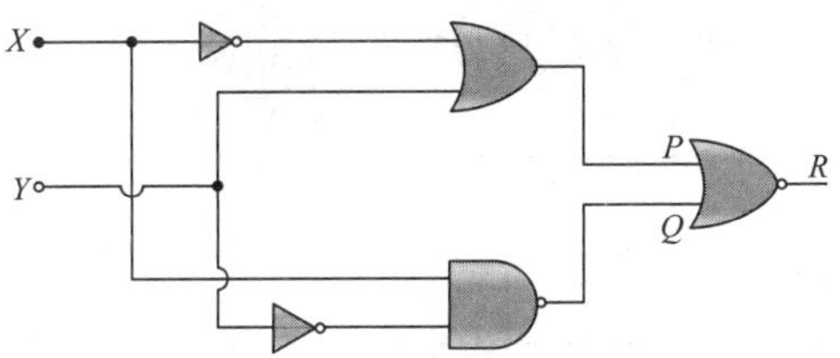

(A) $X=0, Y=1$ (B) $X=1, Y=1$
(C) $X=0, Y=0$ (D) $X=1, Y=0$

18. The output of the given logic circuit is : **[12 Jan. 2019 (Morning)]**

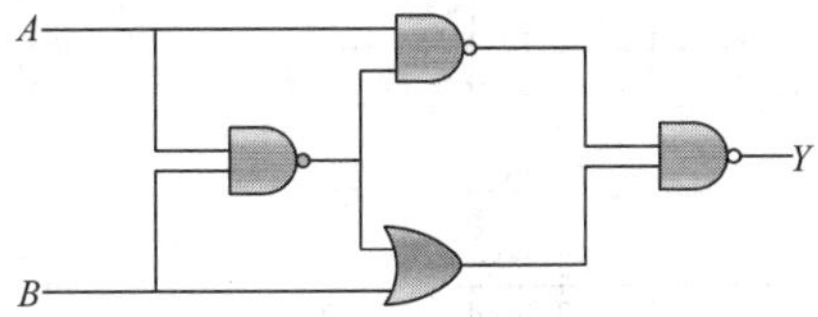

(A) $\overline{A}B$ (B) $A\overline{B}$
(C) $AB+\overline{AB}$ (D) $A\overline{B}+\overline{A}B$

19. The logic gate equivalent to the given logic circuit is : **[9 April 2019 (Evening)]**

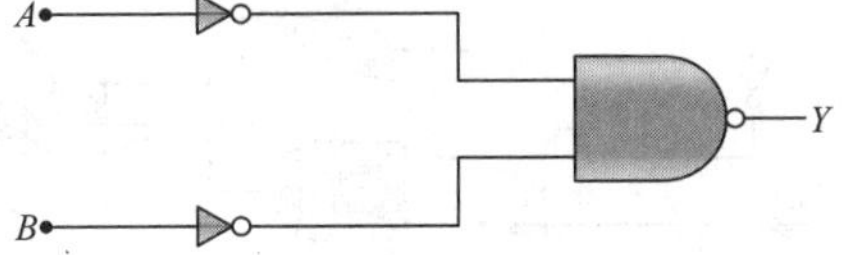

(A) OR (B) AND
(C) NOR (D) NAND

20. The truth table for the circuit given in the figure is: **[12 April 2019 (Morning)]**

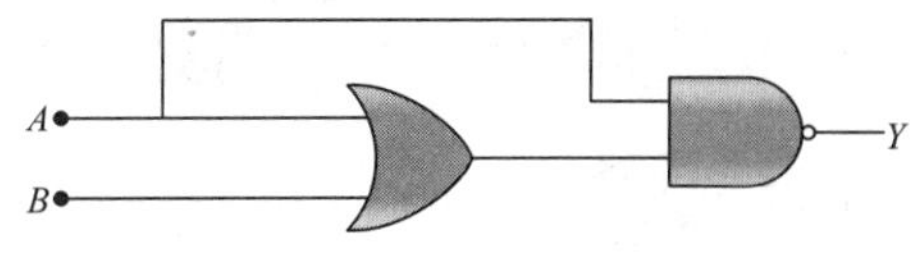

(A)

A	B	Y
0	0	1
0	1	1
1	0	0
1	1	0

(B)

A	B	Y
0	0	0
0	1	0
1	0	1
1	1	1

(C)

A	B	Y
0	0	1
0	1	0
1	0	0
1	1	0

(D)

A	B	Y
0	0	1
0	1	1
1	0	1
1	1	1

21. Which of the following gives a reversible operation ? **[7 Jan. 2020 (Morning)]**

(A)  (B) 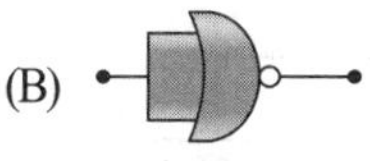

(C)  (D)

22. In the given circuit, value of Y is : **[8 Jan. 2020 (Evening)]**

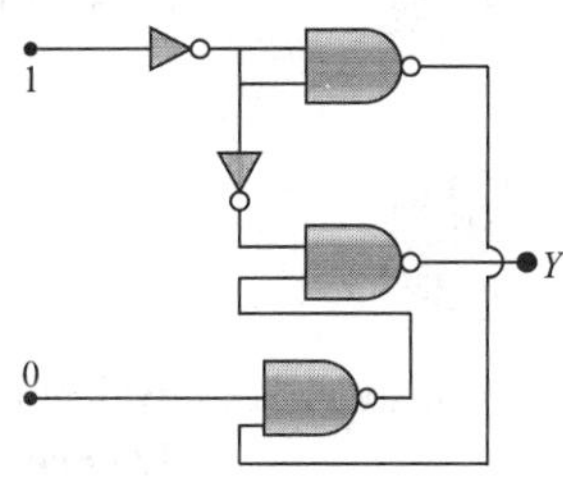

(A) Will not execute (B) 0
(C) Toggles between 0 and 1 (D) 1

23. In the following digital circuit, what will be the output at 'Z', when the input (A, B) are $(1, 0), (0, 0), (1, 1), (0, 1)$: **[2 Sept. 2020 (Evening)]**

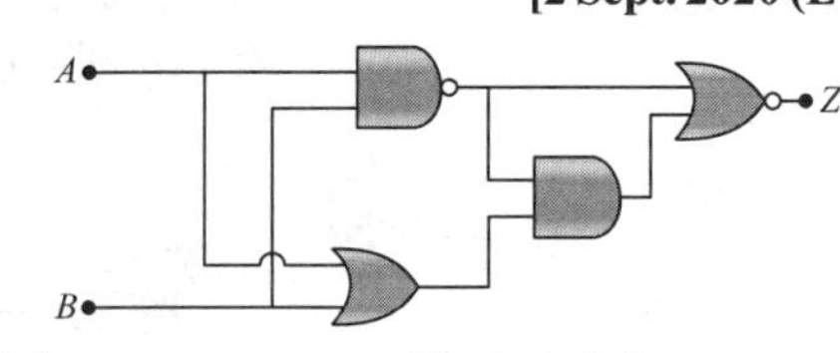

(A) 1, 0, 1, 1 (B) 0, 1, 0, 0
(C) 0, 0, 1, 0 (D) 1, 1, 0, 1

24. Identify the operation performed by the circuit given below : **[4 Sept. 2020 (Evening)]**

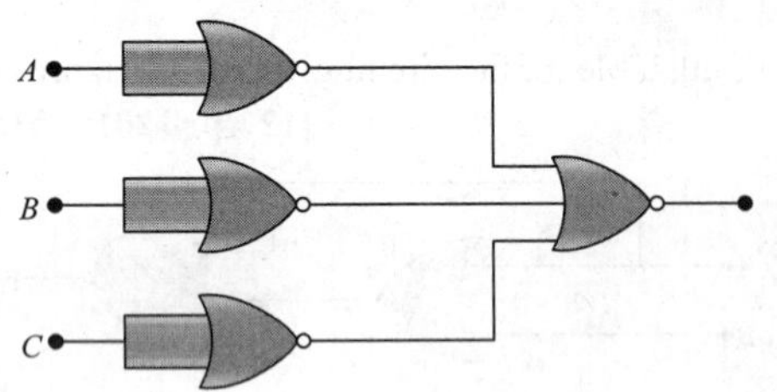

(A) AND (B) NAND
(C) OR (D) NOT

25. Identify the correct output signal *Y* in the given combination of gates (as shown) for the given inputs *A* and *B* : **[6 Sept. 2020 (Morning)]**

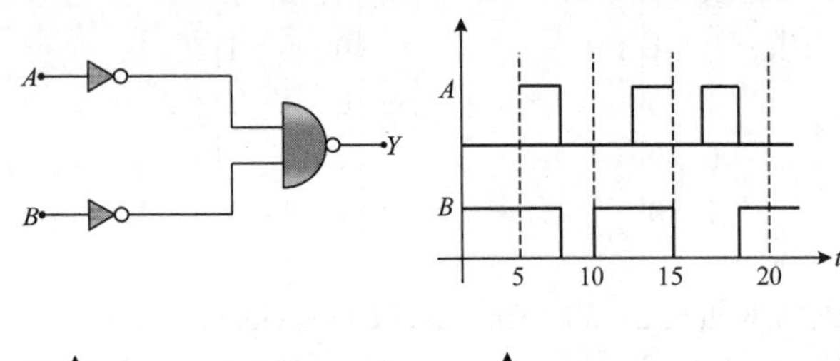

(A) (B)

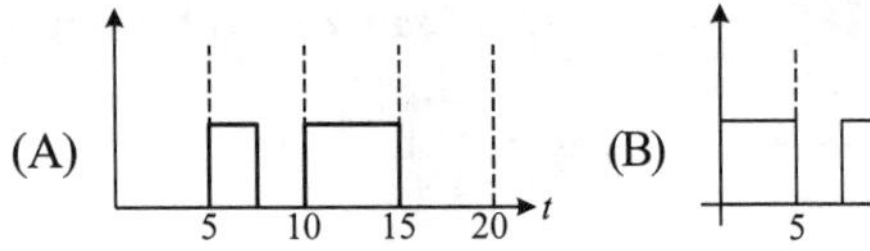

(C) (D)

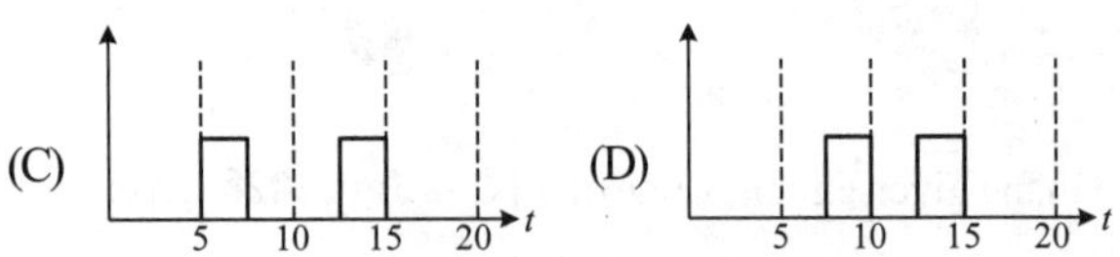

26.

The logic circuit shown above is equivalent to : **[24 Feb. 2021 (Evening)]**

(A) 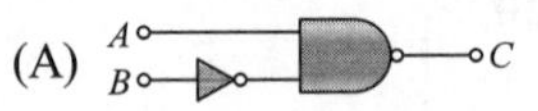 (B) 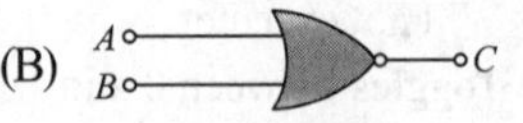

(C) 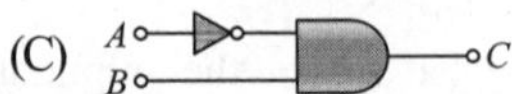 (D)

27. The truth table for the following logic circuit is : **[25 Feb. 2021 (Evening)]**

A
B
Y

(A)

A	*B*	*Y*
0	0	1
0	1	0
1	0	1
1	1	0

(B)

A	*B*	*Y*
0	0	0
0	1	1
1	0	0
1	1	1

(C)

A	*B*	*Y*
0	0	1
0	1	0
1	0	0
1	1	1

(D)

A	*B*	*Y*
0	0	0
0	1	1
1	0	1
1	1	0

28. Draw the output signal *Y* in the given combination of gates. **[26 Feb. 2021 (Evening)]**

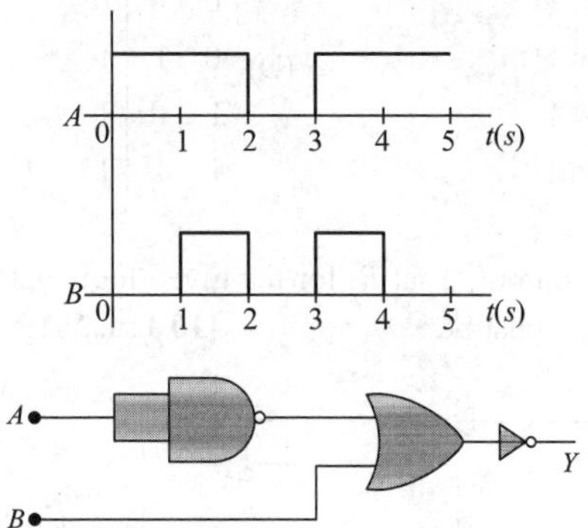

(A) 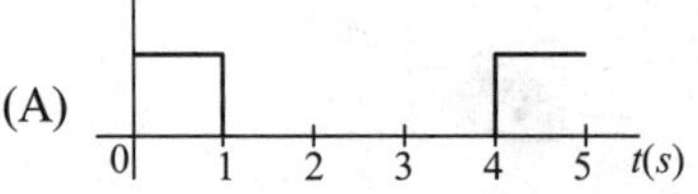

(B) 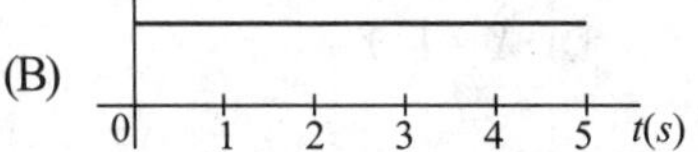

(C) 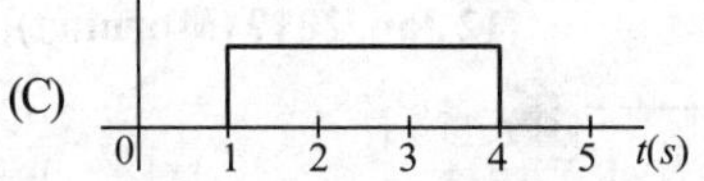

(D) 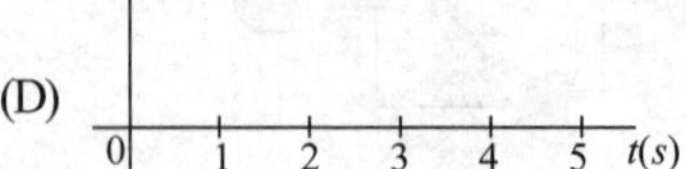

29. In the logic circuit shown in the figure, if input *A* and *B* are 0 to 1 respectively, the output at *Y* would be '*x*'. The value of *x* is _______. **[16 March 2021 (Morning)]**

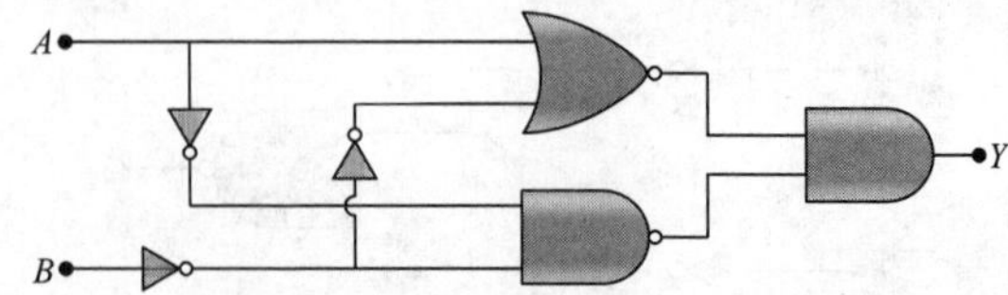

30. The following logic gate is equivalent to : **[16 March 2021 (Evening)]**

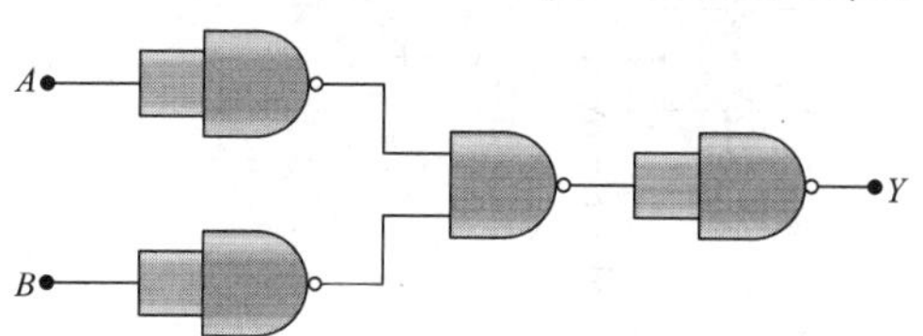

(A) NOR Gate (B) OR Gate
(C) AND Gate (D) NAND Gate

31. The output of the given combination gates represents : **[17 March 2021 (Morning)]**

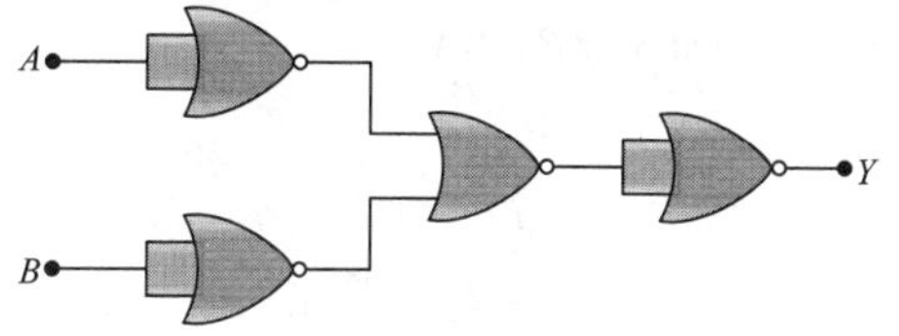

(A) XOR Gate (B) NAND Gate
(C) AND Gate (D) NOR Gate

32. Which one of the following will be the output of the given circuit ? **[17 March 2021 (Evening)]**

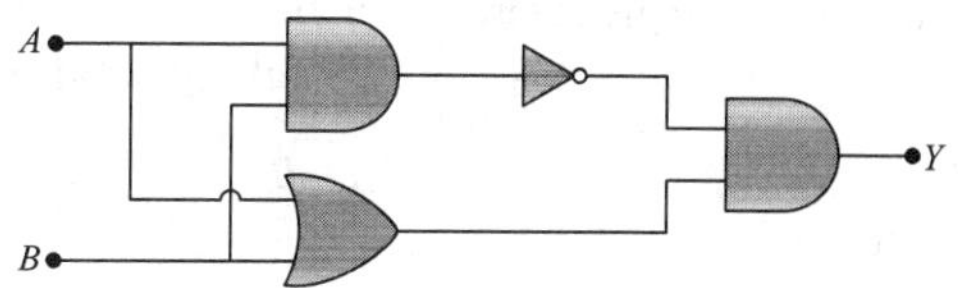

(A) NOR Gate (B) NAND Gate
(C) AND Gate (D) XOR Gate

33. Identify the logic operation carried out : **[25 July 2021 (Morning)]**

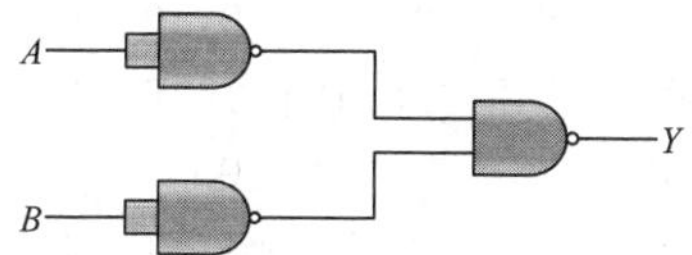

(A) *OR* (B) *NOR*
(C) *NAND* (D) *AND*

34. Find the truth table for the function *Y* for inputs *A* and *B* represented in the following figure : **[27 July 2021 (Evening)]**

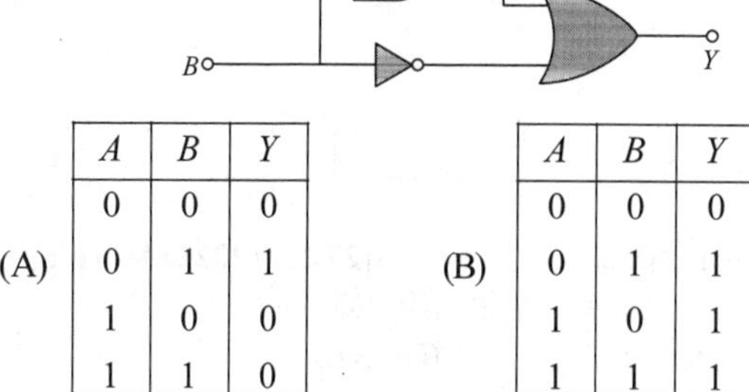

(A)

A	*B*	*Y*
0	0	0
0	1	1
1	0	0
1	1	0

(B)

A	*B*	*Y*
0	0	0
0	1	1
1	0	1
1	1	1

(C)

A	*B*	*Y*
0	0	1
0	1	0
1	0	1
1	1	1

(D)

A	*B*	*Y*
0	0	0
0	1	0
1	0	0
1	1	1

35. Identify the logic operation carried out by the given circuit : **[26 Aug. 2021 (Morning)]**

(A) *NOR* (B) *OR*
(C) *NAND* (D) *AND*

36. Four NOR gates are connected as shown in figure. The truth table for the given figure is : **[26 Aug. 2021 (Evening)]**

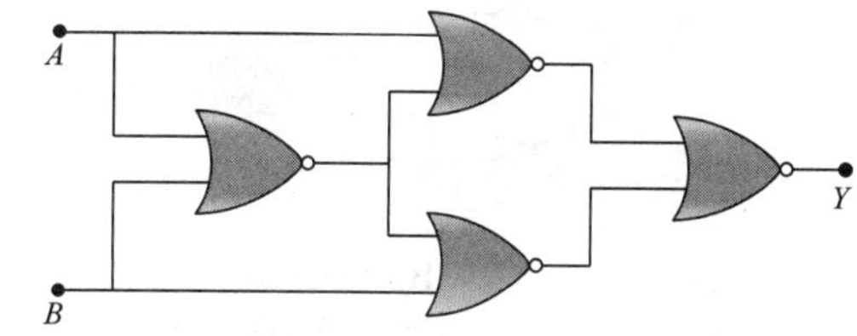

(A)

A	*B*	*Y*
0	0	1
0	1	0
1	0	0
1	1	1

(B)

A	*B*	*Y*
0	0	1
0	1	0
1	0	1
1	1	0

(C)

A	*B*	*Y*
0	0	0
0	1	1
1	0	0
1	1	1

(D)

A	*B*	*Y*
0	0	0
0	1	1
1	0	1
1	1	0

37. In the following logic circuit the sequence of the inputs *A*, *B* are (0, 0), (0, 1), (1, 0) and (1, 1). The output *Y* for this sequence will be : **[31 Aug. 2021 (Morning)]**

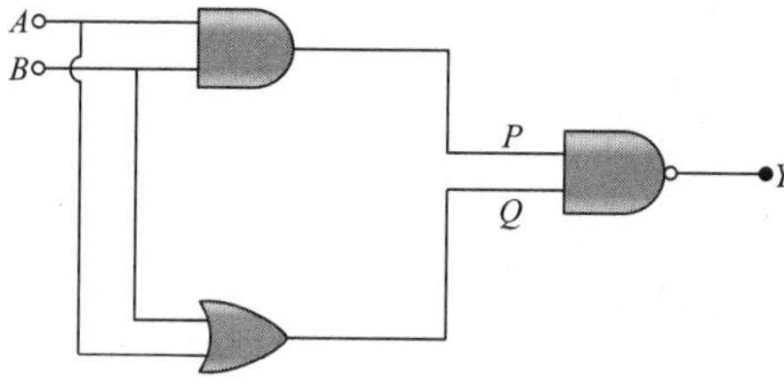

(A) 0, 0, 1, 1 (B) 1, 0, 1, 0
(C) 0, 1, 0, 1 (D) 1, 1, 1, 0

38. If V_A and V_B are the input voltages (either 5 V or 0 V) and V_0 is the output voltage then the two gates represented in the following circuits (*A*) and (*B*) are : **[31 Aug. 2021 (Evening)]**

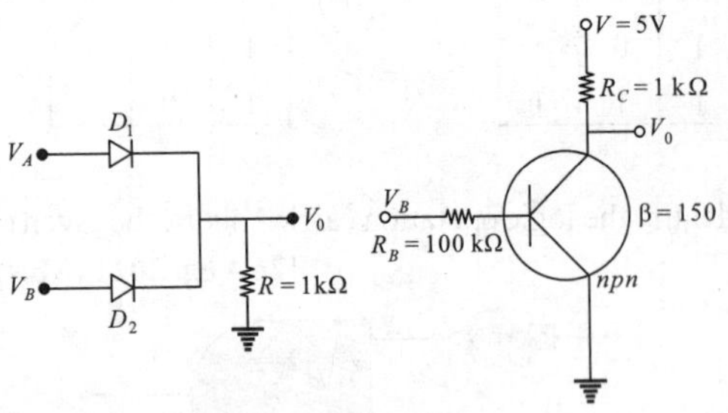

(A) AND and NOT Gate (B) OR and NOT Gate
(C) AND and OR Gate (D) NAND and NOR Gate

39. Identify the logic operation performed by the given circuit : **[25 June 2022 (Evening)]**

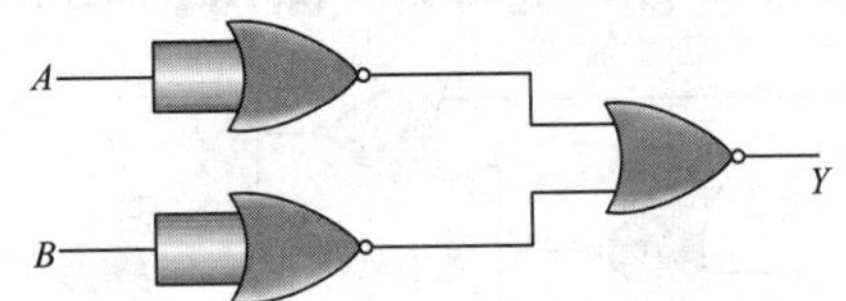

(A) AND gate (B) OR gate
(C) NOR gate (D) NAND gate

40. Identify the correct Logic Gate for the following output (*Y*) of two inputs *A* and *B* : **[27 June 2022 (Morning)]**

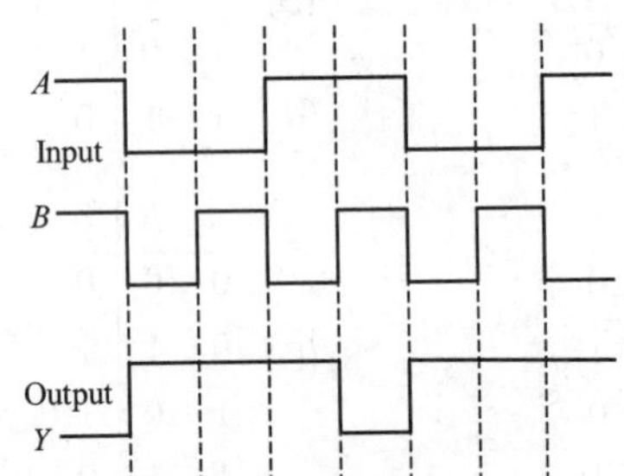

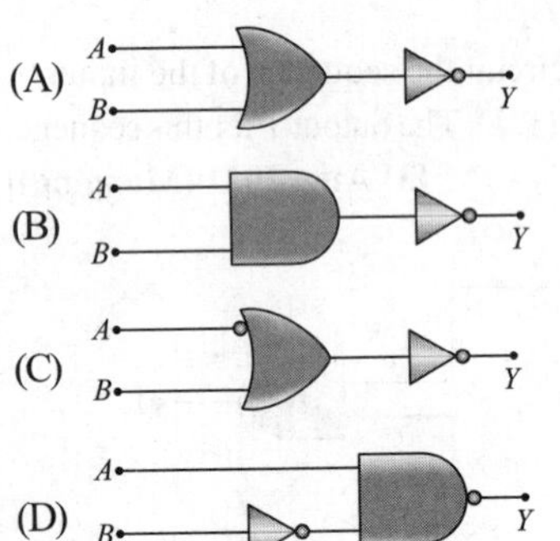

41. In the following circuit, the correct relation between output (*Y*) and inputs *A* and *B* will be : **[28 June 2022 (Morning)]**

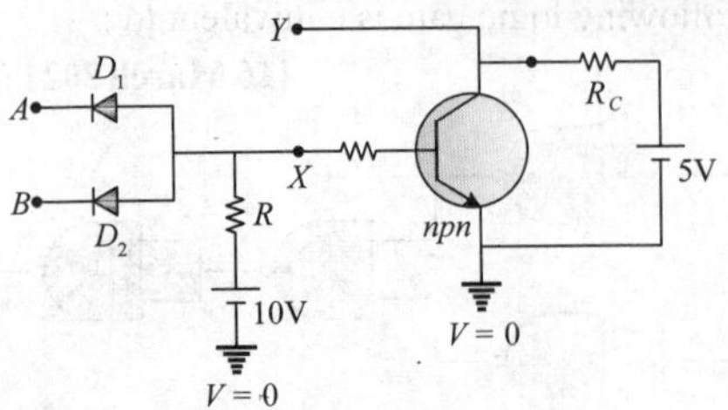

(A) $Y = AB$ (B) $Y = A + B$
(C) $Y = \overline{AB}$ (D) $Y = \overline{A + B}$

42. In the circuit, the logical value of $A = 1$ or $B = 1$ when potential at *A* or *B* is 5V and the logical value of $A = 0$ or $B = 0$ when potential at *A* or *B* is 0 V.

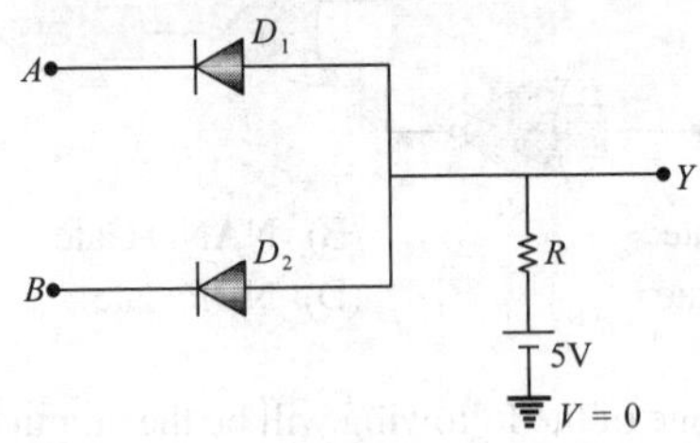

The truth table of the given circuit will be : **[25 July 2022 (Morning)]**

(A)

A	B	Y
0	0	0
1	0	0
0	1	0
1	1	1

(B)

A	B	Y
0	0	0
1	0	1
0	1	1
1	1	1

(C)

A	B	Y
0	0	0
1	0	0
0	1	0
1	1	0

(D)

A	B	Y
0	0	1
1	0	1
0	1	1
1	1	0

43. A logic gate circuit has two inputs *A* and *B* and output *Y*. The voltage waveforms of *A*, *B* and *Y* are shown below

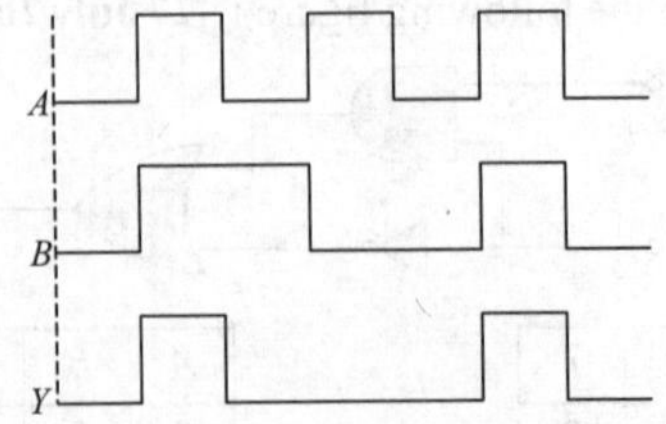

The logic gate circuit is : **[27 July 2022 (Morning)]**

(A) AND gate (B) OR gate
(C) NOR gate (D) NAND gate

44. The logic gate equivalent to the given circuit diagram is : **[24 Jan. 2023 (Evening)]**

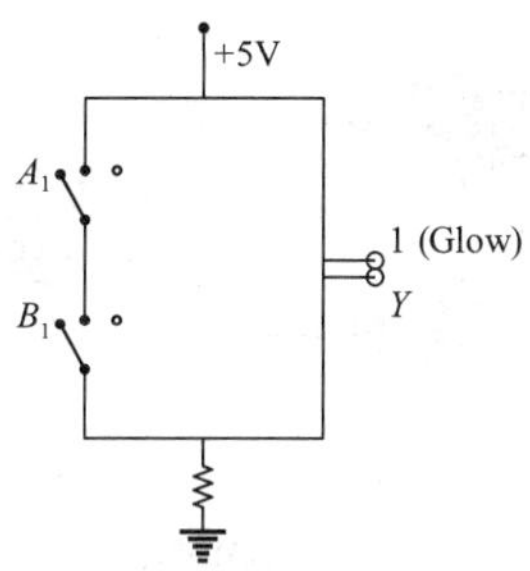

(A) OR (B) NAND
(C) NOR (D) AND

45. For the given logic gates combination, the correct truth table will be : **[29 Jan. 2023 (Evening)]**

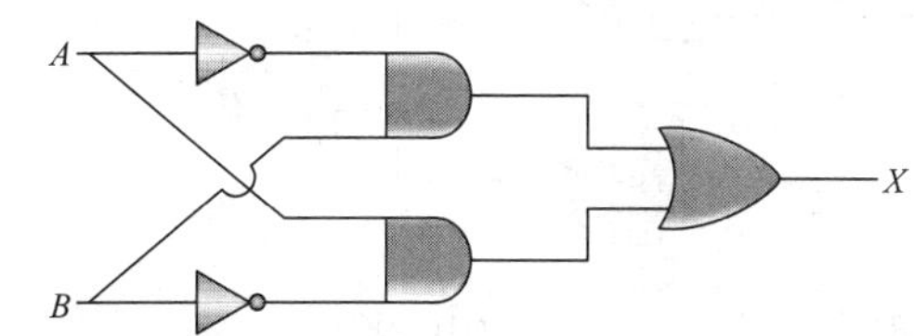

(A)

A	B	X
0	0	1
0	1	0
1	0	0
1	1	0

(B)

A	B	X
0	0	0
0	1	1
1	0	1
1	1	0

(C)

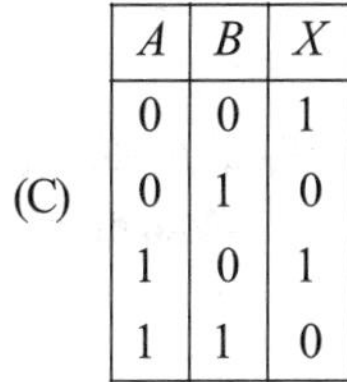

A	B	X
0	0	1
0	1	0
1	0	1
1	1	0

(D)

A	B	X
0	0	0
0	1	1
1	0	1
1	1	1

46. The output waveform of the given logical circuit for the following inputs A and B as shown below, is : **[30 Jan. 2023 (Morning)]**

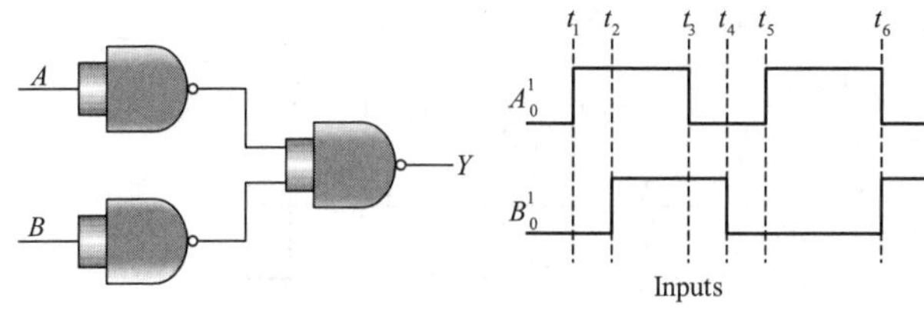

(A)

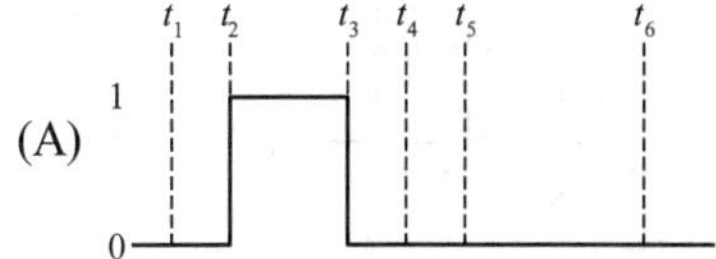

(B)

(C)

(D)

47. The output Y for the inputs A and B of circuit is given by

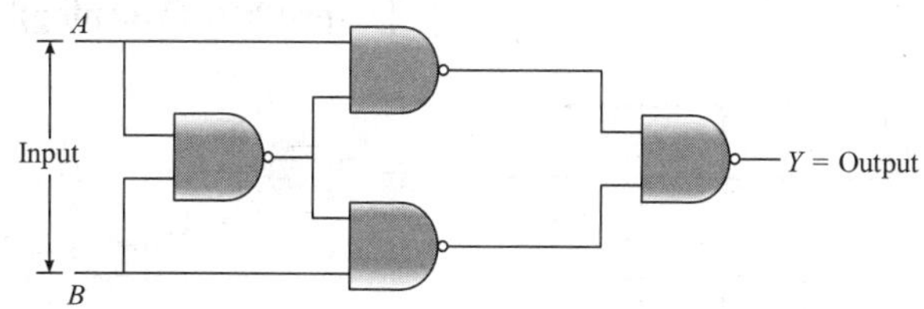

Truth table of the shown circuit is : **[30 Jan. 2023 (Evening)]**

(A)

A	B	Y
0	0	1
0	1	1
1	0	1
1	1	0

(B)

A	B	Y
0	0	1
0	1	0
1	0	0
1	1	1

(C)

A	B	Y
0	0	0
0	1	1
1	0	1
1	1	1

(D)

A	B	Y
0	0	0
0	1	1
1	0	1
1	1	0

48. Given below are two statements :

Statement-I : In a typical transistor, all three regions emitter, base and collector have same doping level

Statement-II : In a transistor, collector is the thickest and base is the thinnest segment

In the light of the above statements, choose the most appropriate answer from the options given below. **[31 Jan. 2023 (Evening)]**

(A) Both Statement-I and Statement-II are correct
(B) Both Statement-I and Statement-II are incorrect
(C) Statement-I is incorrect but Statement-II is correct
(D) Statement-I is correct but Statement-II is incorrect

49. Name the logic gate equivalent to the diagram attached

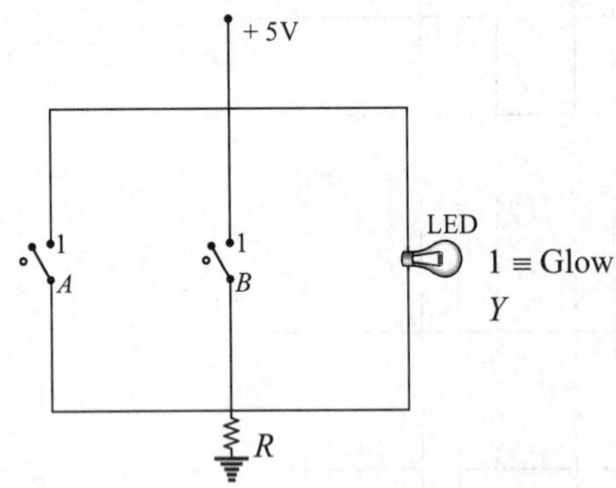

[6 April 2023 (Morning)]

(A) OR (B) NOR
(C) NAND (D) AND

50. For the logic circuit shown, the output waveform at Y is :
[8 April 2023 (Morning)]

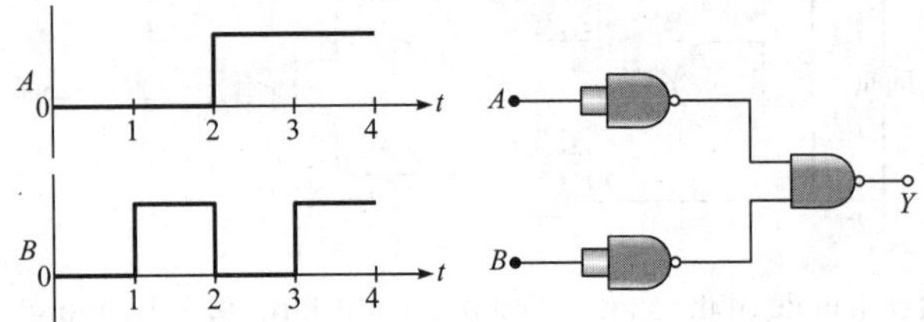

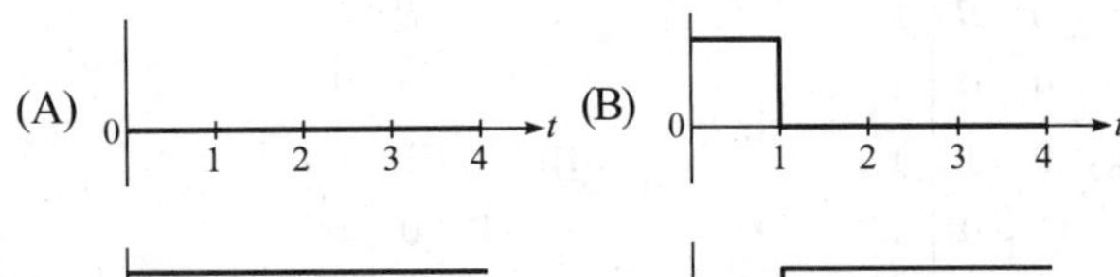

(A) (B) (C) (D)

51. For a given transistor amplifier circuit in CE configuration $V_{CC} = 1$ V, $R_c = 1\ k\Omega$, $R_b = 100\ k\Omega$ and $\beta = 100$. Value of base current I_b is : **[8 April 2023 (Evening)]**

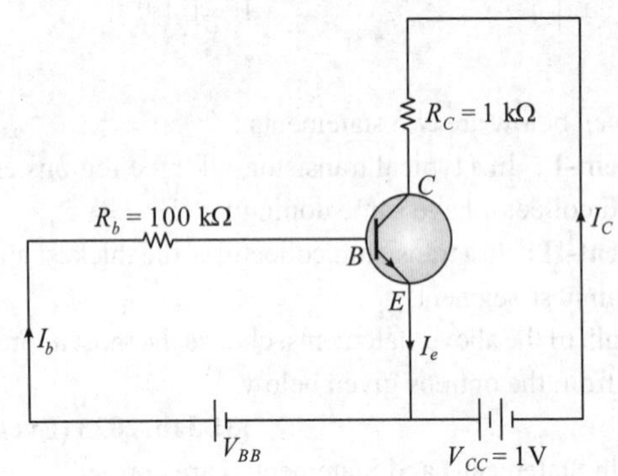

(A) $I_b = 1.0\ \mu A$ (B) $I_b = 1.10\ \mu A$
(C) $I_b = 100\ \mu A$ (D) $I_b = 10\ \mu A$

52. The logic performed by the circuit shown in figure is equivalent to : **[11 April 2023 (Morning)]**

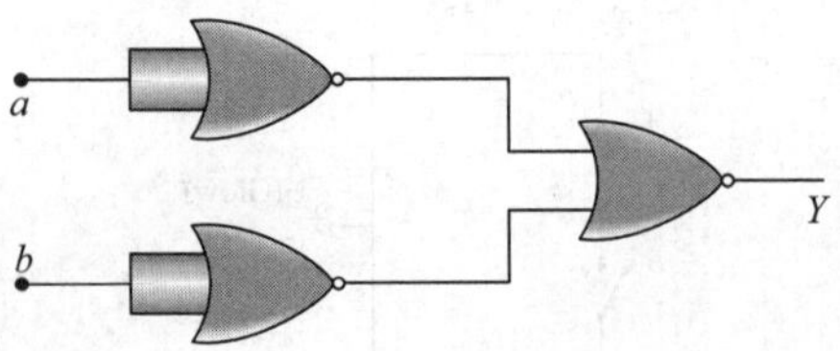

(A) AND (B) NAND
(C) OR (D) NOR

53. The logic operations performed by the given digital circuit is equivalent to : **[11 April 2023 (Evening)]**

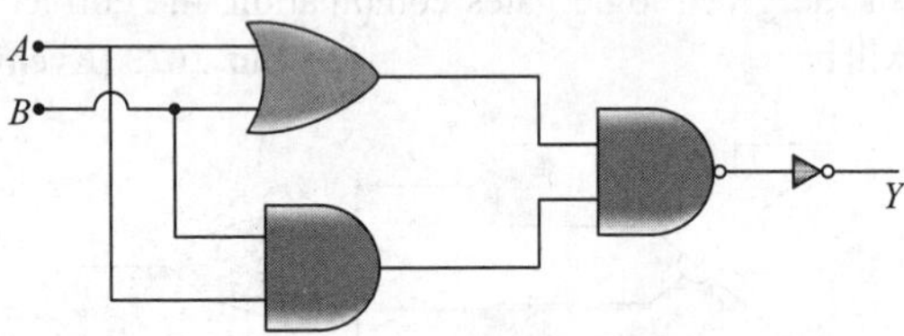

(A) AND (B) NOR
(C) OR (D) NAND

54. In an n-p-n common emitter (CE) transistor the collector current changes from 5 mA to 16 mA for the change in base current from 100 μA and 200 μA, respectively. The current gain of transistor is_____. **[12 April 2023 (Morning)]**

(A) 110 (B) 0.9
(C) 210 (D) 9

55. For the following circuit and given inputs A and B, choose the correct option for output Y : **[13 April 2023 (Morning)]**

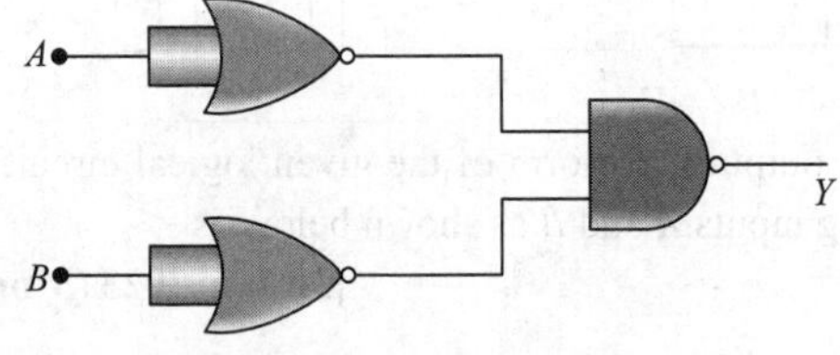

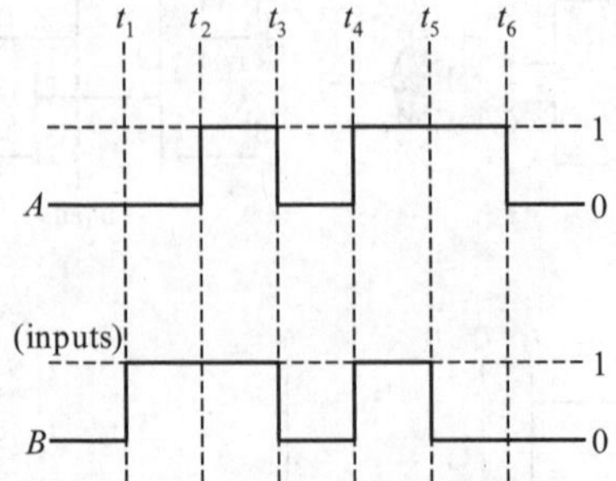

(A)

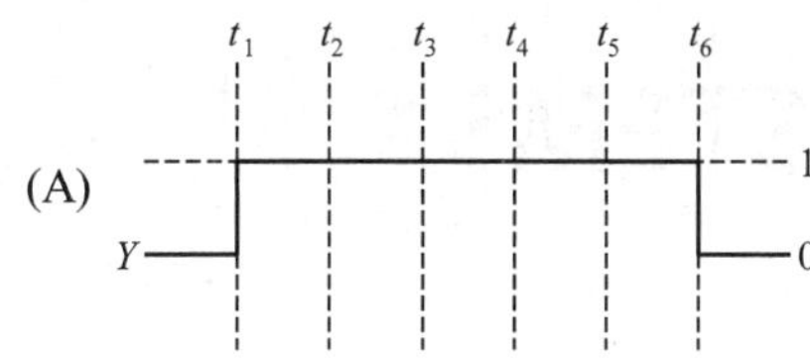

(B)

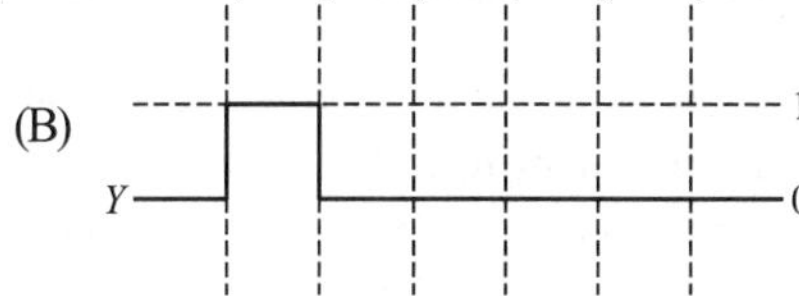

(C)

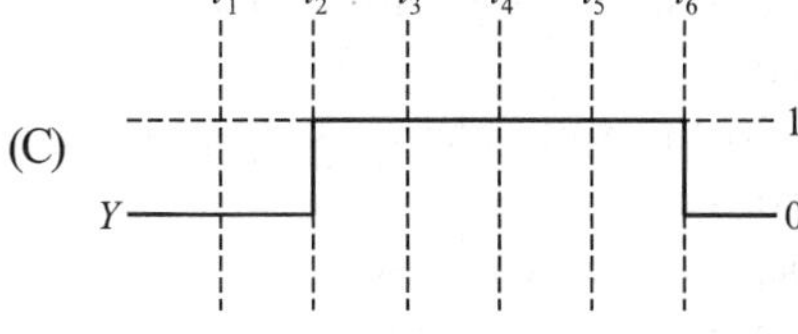

(D)

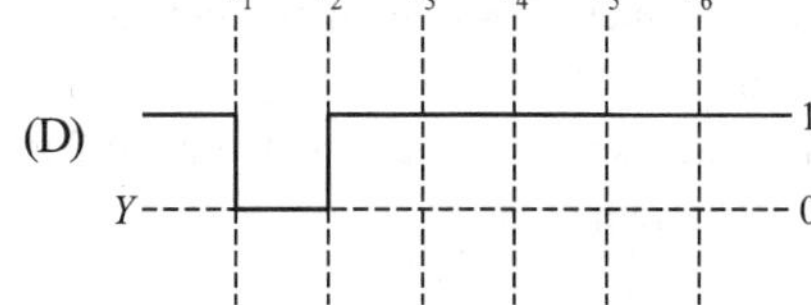

56. From the given transfer characteristic of a transistor in *CE* configuration, the value of power gain of this configuration is 10^x, for $R_B = 10\ k\Omega$, and $R_C = 1\ k\Omega$. The value of x is _____.

[13 April 2023 (Morning)]

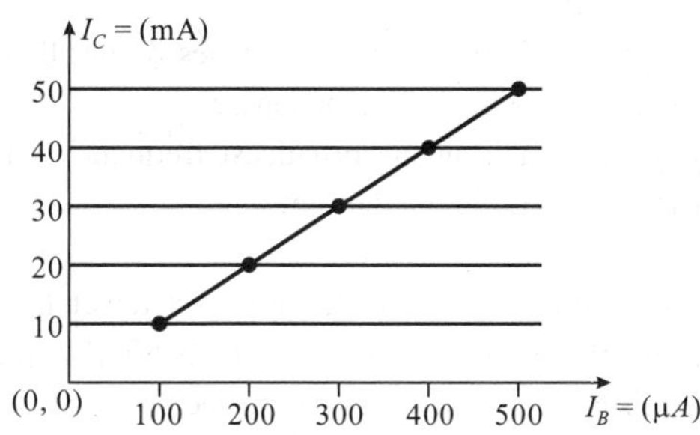

57. The output from a NAND gate having inputs *A* and *B* given below will be **[13 April 2023 (Evening)]**

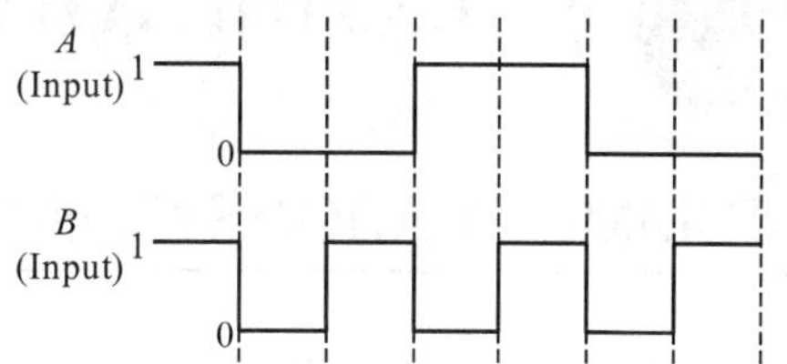

(A)

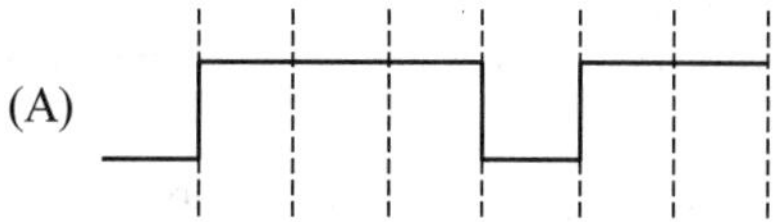

(B)

(C)

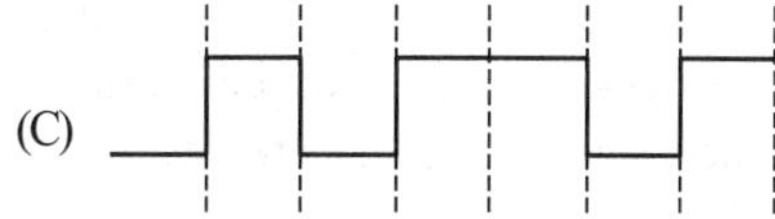

(D)

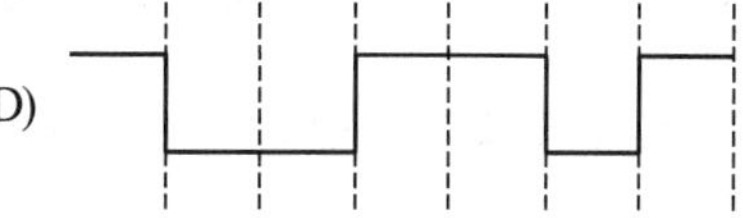

* * * * *

COMMUNICATION SYSTEMS

29.1 COMMUNICATION SYSTEMS

1. Consider telecommunication through optical fibres. Which of the following statements is not true? **[2003]**
(A) Optical fibres can be of graded refractive index
(B) Optical fibres are subject to electromagnetic interference from outside
(C) Optical fibres have extremely low transmission loss
(D) Optical fibres may have homogeneous core with a suitable cladding

2. This question has Statement-1 and Statement-2. Of the four choices given after the statements, choose the one that best describes the two statements. **[2011]**
Statement-1 : Sky wave signals are used for long distance radio communication. These signals are in general, less stable than ground wave signals.
Statement-2 : The state of ionosphere varies from hour to hour, day to day and season to season.
(A) Statement-1 is true, Statement-2 is true, Statement-2 is the correct explanation of Statement-1
(B) Statement-1 is true, Statement-2 is true, Statement-2 is not the correct explanation of Statement-1
(C) Statement-1 is false, Statement-2 is true
(D) Statement-1 is true, Statement-2 is false

3. A radar has a power of 1 kW and is operating at a frequency of 10 GHz. It is located on a mountain top of height 500 m. The maximum distance upto which it can detect object located on the surface of the earth **[2012]**
(Radius of earth = 6.4×10^6 m) is :
(A) 80 km (B) 16 km
(C) 40 km (D) 64 km

4. A 10 kW transmitter emits radio waves of wavelength 500 m. The number of photons emitted per second by the transmitter is of the order of : **[2012 (Online)]**
(A) 10^{37} (B) 10^{31}
(C) 10^{25} (D) 10^{43}

5. A radio transmitter transmits at 830 kHz. At a certain distance from the transmitter magnetic field has amplitude 4.82×10^{-11} T. The electric field and the wavelength are respectively : **[2012 (Online)]**
(A) 0.014 N/C, 36 m (B) 0.14 N/C, 36 m
(C) 0.14 N/C, 360 m (D) 0.014 N/C, 360 m

6. This question has Statement-1 and Statement-2. Of the four choices given after the Statements, choose the one that best describes the two Statements. **[2013]**
Statement-1 : Short wave transmission is achieved due to the total internal reflection of the *e-m* wave from an appropriate height in the ionosphere.
Statement-2 : Refractive index of a plasma is independent of the frequency of *e-m* waves.
(A) Statement-1 is true, Statement-2 is false
(B) Statement-1 is false, Statement-2 is true
(C) Statement-1 is true, Statement-2 is true but Statement-2 is not the correct explanation of Statement-1
(D) Statement-1 is true, Statement-2 is true and Statement-2 is the correct explanation of Statement-1

7. Which of the following statement is NOT correct ?
(A) Ground wave signals are more stable than the sky wave signals **[2013]**
(B) The critical frequency of an ionospheric layer is the highest frequency that will be reflected back by the layer when it is vertically incident
(C) Electromagnetic waves of frequencies higher than about 30 MHz cannot penetrate the ionosphere
(D) Sky wave signals in the broadcast frequency range are stronger at night than in the day time

8. A transmitting antenna at the top of a tower has height 32 m and height of the receiving antenna is 50 m. What is the maximum distance between them for satisfactory communication in line of sight (LOS) mode ? **[2014 (Online)]**
(A) 55.4 km (B) 45.5 km
(C) 54.5 km (D) 455 km

9. For sky wave propagation, the radio waves must have a frequency range in between : **[2014 (Online)]**
(A) 1 MHz to 2 MHz (B) 5 MHz to 25 MHz
(C) 35 MHz to 40 MHz (D) 45 MHz to 50 MHz

10. Long range radio transmission is possible when the radio waves are reflected from the ionosphere. For this to happen the frequency of the radio waves must be in the range : **[2014 (Online)]**
(A) 80 - 150 MHz (B) 8 - 25 MHz
(C) 1 - 3 MHz (D) 150 - 1500 kHz

11. A signal is to be transmitted through a wave of wavelength λ, using a linear antenna. The length l of the antenna and effective power radiated P_{eff} will be given respectively as : (K is a constant of proportionality)

[2017 (Online)]

(A) $\lambda, P_{eff} = K\left(\frac{1}{\lambda}\right)^2$ (B) $\frac{\lambda}{8}, P_{eff} = K\left(\frac{1}{\lambda}\right)$

(C) $\frac{\lambda}{16}, P_{eff} = K\left(\frac{1}{\lambda}\right)^3$ (D) $\frac{5}{\lambda}, P_{eff} = K\left(\frac{1}{\lambda}\right)^{\frac{1}{2}}$

12. A *TV* transmission tower has a height of 140 m and the height of the receiving antenna is 40 m. What is the maximum distance upto which signals can be broadcasted from this tower in *LOS* (Line of Sight) mode ? (Given : radius of earth = 6.4×10^6 m)

[10 Jan. 2019 (Morning)]

(A) 80 km (B) 48 km
(C) 40 km (D) 65 km

13. To double the covering range of a *TV* transmission tower, its height should be multiplied by : **[12 Jan. 2019 (Evening)]**

(A) $\frac{1}{\sqrt{2}}$ (B) 4
(C) $\sqrt{2}$ (D) 2

14. In a line of sight radio communication, a distance of about 50 km is kept between the transmitting and receiving antennas. If the height of the receiving antenna is 70 m, then the minimum height of the transmitting antenna should be.
(Radius of the Earth = 6.4×10^6 m) **[8 April 2019 (Evening)]**

(A) 40 m (B) 51 m
(C) 32 m (D) 20 m

15. Given below in the left column are different modes of communication using the kinds of waves given the right column:

[10 April 2019 (Morning)]

	Column-Left		Column-Right
(a)	Optical Fibre communication	(i)	Ultrasound
(b)	Radar	(ii)	Infrared Light
(c)	Sonar	(iii)	Microwaves
(d)	Mobile Phones	(iv)	Radio Waves

(A) (a)-(iv), (b)-(ii), (c)-(iii), (d)-(i)
(B) (a)-(iii), (b)-(i), (c)-(iv), (d)-(ii)
(C) (a)-(ii), (b)-(iv), (c)-(iii), (d)-(i)
(D) (a)-(ii), (b)-(iv), (c)-(i), (d)-(iii)

16. A signal of 0.1 kW is transmitted in a cable. The attenuation of cable is – 5 dB per km and cable length is 20 km. The power received at receiver is 10^{-x} W. The value of x is ______.

[Given in dB = $10\log_{10}\left(\frac{P_0}{P_i}\right)$] **[24 Feb. 2021 (Evening)]**

17. A transmitting station releases waves of wavelength 960 m. A capacitor of 2.56 μF is used in the resonant circuit. The self inductance of coil necessary for resonance is ______ $\times 10^{-8}$ H.

[25 Feb. 2021 (Morning)]

18. If the highest frequency modulating a carrier is 5 kHz, then the number of AM broadcast stations accommodated in a 90 kHz bandwidth are ______. **[26 Feb. 2021 (Evening)]**

19. A 25 m long antenna is mounted on an antenna tower. The height of the antenna tower is 75 m. The wavelength (in meter) of the signal transmitted by this antenna would be :

[16 March 2021 (Morning)]

(A) 300 (B) 400
(C) 200 (D) 100

20. Two identical antennas mounted on identical towers are separated from each other by a distance of 45 km. What should nearly be the minimum height of receiving antenna to receive the signals in line of sight ? (Assume radius of earth is 6400 km)

[16 March 2021 (Evening)]

(A) 19.77 m (B) 39.55 m
(C) 79.1 m (D) 158.2 m

21. For VHF signal broadcasting, ______ km^2 of maximum service area will be covered by an antenna tower of height 30 m, if the receiving antenna is placed at ground. Let radius of the earth be 6400 km. (Round off to the Nearest Integer)
(Take π as 3.14) **[17 March 2021 (Morning)]**

22. A carrier signal $C(t) = 25\sin(2.512 \times 10^{10} t)$ is amplitude modulated by a message signal $m(t) = 5\sin(1.57 \times 10^8 t)$ and transmitted through an antenna. What will be the bandwidth of the modulated signal ? **[17 March 2021 (Evening)]**

(A) 8 GHz (B) 2.01 GHz
(C) 1987.5 MHz (D) 50 MHz

23. Match List-I with List-II : **[18 March 2021 (Morning)]**

	List-I		List-II
(a)	10 km height over earth's surface	(i)	Thermosphere
(b)	70 km height over earth's surface	(ii)	Mesosphere
(c)	180 km height over earth's surface	(iii)	Stratosphere
(d)	270 km height over earth's surface	(iv)	Troposphere

(A) (a)-(iv), (b)-(iii), (c)-(ii), (d)-(i)
(B) (a)-(i), (b)-(iv), (c)-(iii), (d)-(ii)
(C) (a)-(iii), (b)-(ii), (c)-(i), (d)-(iv)
(D) (a)-(ii), (b)-(i), (c)-(iv), (d)-(iii)

24. A *TV* transmission tower antenna is at a height of 20 m. Suppose that the receiving antenna is at
(i) ground level
(ii) a height of 5 m
The increase in antenna range in case (ii) relative to case (i) is $n\%$. Calculate the value of n, to the nearest integer. **[18 March 2021 (Evening)]**

25. What should be the height of transmitting antenna and the population covered if the television telecast is to cover a radius of 150 km ? The average population density around the tower is 2000/ km^2 and the value of $R_e = 6.5 \times 10^6$ m : **[22 July 2021 (Evening)]**
(A) Height = 1731 m, Population Covered = 1413×10^5
(B) Height = 1600 m, Population Covered = 2×10^5
(C) Height = 1800 m, Population Covered = 1413×10^8
(D) Height = 1241 m, Population Covered = 7×10^5

26. In amplitude Modulation, the message signal is given as
$$V_m(t) = 10 \sin(2\pi \times 10^5 t) \text{ volts}$$
and Carrier signal
$$V_C(t) = 20 \sin(2\pi \times 10^7 t) \text{ volts and}$$
The modulated signal now contains the message signal with lower side band and upper side band frequency, therefore the bandwidth of modulated signal is α kHz. The value of is α is : **[25 July 2021 (Morning)]**
(A) 50 kHz (B) 200 kHz
(C) 100 kHz (D) 0

27. The amplitude of upper and lower side bands of A.M. wave where a carrier signal with frequency 11.21 MHz, peak voltage 15 V is amplitude modulated by a 7.7 kHz sine wave of 5 V amplitude are $\frac{a}{10}$ V and $\frac{b}{10}$ V respectively. Then the value of $\frac{a}{b}$ is ______. **[27 July 2021 (Morning)]**

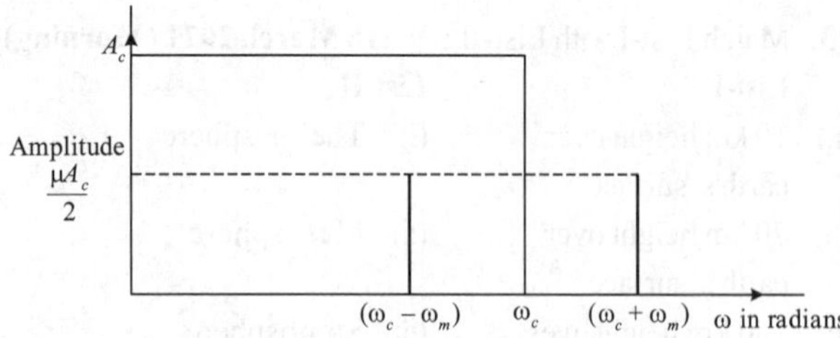

28. A transmitting antenna at top of a tower has a height of 50 m and the height of receiving antenna is 80 m. What is the range of communication for Line of Sight (LoS) mode ?
[use radius of earth = 6400 km] **[26 Aug. 2021 (Evening)]**
(A) 80.2 km (B) 57.28 km
(C) 144.1 km (D) 45.5 km

29. A transmitting antenna has a height of 320 m and that of receiving antenna is 2000 m. The maximum distance between them for satisfactory communication in line of sight mode is d. The value of d is______ km. **[27 Aug. 2021 (Morning)]**

30. An antenna is mounted on a 400 m tall building. What will be the wavelength of signal that can be radiated effectively by the transmission tower upto a range of 44 km ? **[27 Aug. 2021 (Evening)]**
(A) 605 m (B) 302 m
(C) 37.8 m (D) 75.6 m

31. If the sum of the heights of transmitting and receiving antennas in the line of sight of communication is fixed at 160 m, then the maximum range of *LOS* communication is _____ km (Take radius of Earth = 6400 km) **[31 Aug. 2021 (Morning)]**

32. A bandwidth of 6 MHz is available for *A.M.* transmission. If the maximum audio signal frequency used for modulating the carrier wave is not to exceed 6 kHz. The number of stations that can be broadcasted within this band simultaneously without interfering with each other will be ______. **[31 Aug. 2021 (Evening)]**

33. An antenna is placed in a dielectric medium of dielectric constant 6.25. If the maximum size of that antenna is 5.0 mm it can radiate a signal of minimum frequency of _____ GHz. (Given $\mu_r = 1$ for dielectric medium) **[24 June 2022 (Evening)]**

34. A signal of 100 THz frequency can be transmitted with maximum efficiency by : **[25 June 2022 (Morning)]**
(A) Coaxial cable
(B) Optical fibre
(C) Twisted pair of copper wires
(D) Water

35. Match List-I with List-II : **[25 June 2022 (Evening)]**

List-I	List-II
(a) Facsimile	(i) Static Document Image
(b) Guided media Channel	(ii) Local Broadcast Radio
(c) Frequency Modulation	(iii) Rectangular wave
(d) Digital Signal	(iv) Optical Fiber

Choose the correct answer from the following options :
(A) (a)-(iv), (b)-(iii), (c)-(ii), (d)-(i)
(B) (a)-(i), (b)-(iv), (c)-(ii), (d)-(iii)
(C) (a)-(iv), (b)-(ii), (c)-(iii), (d)-(i)
(D) (a)-(i), (b)-(ii), (c)-(iii), (d)-(iv)

36. The height of a transmitting antenna at the top of a tower is 25 m and that of receiving antenna is, 49 m. The maximum distance between them, for satisfactory communication in LOS (Line-Of-Sight) is $K\sqrt{5} \times 10^2$ m. The value of K is _____.

[Assume radius of Earth is $64 \times 10^{+5}$ m] (Calculate upto nearest integer value) **[27 June 2022 (Morning)]**

37. We do not transmit low frequency signal to long distance because : **[27 June 2022 (Evening)]**
(a) The size of the antenna should be comparable to signal wavelength which is unreal solution for a signal of longer wavelength
(b) Effective power radiated by a long wavelength baseband signal would be high
(c) We want to avoid mixing up signals transmitted by different transmitter simultaneously
(d) Low frequency signal can be sent to long distances by superimposing with a high frequency wave as well
Therefore, the most suitable options will be :
(A) All statements are true
(B) (a), (b) and (c) are true only
(C) (a), (c) and (d) are true only
(D) (b), (c) and (d) are true only

38. A radar sends an electromagnetic signal of electric field $(E_0) = 2.25$ V/m and magnetic field $(B_0) = 1.5 \times 10^{-8}$ T which strikes a target on line of sight at a distance of 3 km in a medium. After that, a pail of signal (echo) reflects back towards the radar with same velocity and by same path. If the signal was transmitted at time $t = 0$ from radar. Then after how much time echo will reach to the radar ? **[28 June 2022 (Morning)]**
(A) 2.0×10^{-5} s (B) 4.0×10^{-5} s
(C) 1.0×10^{-5} s (D) 8.0×10^{-5} s

39. Match List-I with List-II : **[28 June 2022 (Morning)]**

List-I		List-II
(a) Television signal	(i)	0.3 KHz
(b) Radio signal	(ii)	20 KHz
(c) High Quality Music	(iii)	0.2 MHz
(d) Human speed	(iv)	06 MHz

Choose the correct answer from the options given below :
(A) (a)-(i), (b)-(ii), (c)-(iii), (d)-(iv)
(B) (a)-(iv), (b)-(iii), (c)-(i), (d)-(ii)
(C) (a)-(iv), (b)-(iii), (c)-(ii), (d)-(i)
(D) (a)-(i), (b)-(ii), (c)-(iv), (d)-(iii)

40. A telegraph line of length 100 km has a capacity of 0.01 μF/km and it carries an alternating current at 0.5 kilo cycle per second. If minimum impedance is required, then the value of the inductance that needs to be introduced in series is _____ mH. (If $\pi = \sqrt{10}$) **[28 June 2022 (Morning)]**

41. Only 2% of the optical source frequency is the available channel bandwidth for an optical communicating system operating at 1000 nm. If an audio signal requires a bandwidth of 8 kHz, how many channels can be accommodated for transmission : **[29 June 2022 (Morning)]**
(A) 375×10^7 (B) 75×10^7
(C) 375×10^8 (D) 75×10^9

42. The *TV* transmission tower at a particular station has a height of 125 m. For doubling the coverage of its range, the height of the tower should be increased by : **[29 June 2022 (Evening)]**
(A) 125 m (B) 250 m
(C) 375 (D) 500 m

43. The required height of a *TV* tower which can cover the population of 6.03 lakh is h. If the average population density is 100 per square km and the radius of earth is 6400 km, then the value of h will be _____ m. **[25 July 2022 (Morning)]**

44. A radio can tune to any station in 6 MHz to 10 MHz band. The value of corresponding wavelength bandwidth will be : **[26 July 2022 (Morning)]**
(A) 4 m (B) 20 m
(C) 30 m (D) 50 m

45. At a particular station, the TV transmission tower has a height of 100 m. To triple its coverage range, height of the tower should be increased to : **[27 July 2022 (Morning)]**
(A) 200 m (B) 300 m
(C) 600 m (D) 900 m

46. Match List-I with List-II : **[25 Jan. 2023 (Evening)]**

List-I		List-II
(a) Troposphere	(i)	Approximate 65-75 km over Earth's surface
(b) *E*-Part of Stratosphere	(ii)	Approximate 300 km over Earth's surface
(c) F_2-Part of Thermosphere	(iii)	Approximate 10 km over Earth's surface
(d) *D*-Part of Stratosphere	(iv)	Approximate 100 km over Earth's surface

Choose the correct answer from the options given below :
(A) (a)-(iii), (b)-(iv), (c)-(ii), (d)-(i)
(B) (a)-(i), (b)-(ii), (c)-(iv), (d)-(iii)
(C) (a)-(i), (b)-(iv), (c)-(iii), (d)-(ii)
(D) (a)-(iii), (b)-(ii), (c)-(i), (d)-(iv)

47. If the height of transmitting and receiving antennas are 80 m each, the maximum line of sight distance will be :
(Given : Earth's radius = 6.4×10^6 m) **[29 Jan. 2023 (Morning)]**
(A) 32 km (B) 28 km
(C) 36 km (D) 64 km

48. Match List-I with List-II : **[30 Jan. 2023 (Evening)]**

List-I	List-II
(a) Attenuation	(i) Combination receiver and transmitter
(b) Transducer	(ii) Process of retrieval of information from the carrier wave at received
(c) Demodulation	(iii) Converts one form of energy into another
(d) Repeater	(iv) Loss of strength of a signal while propagating through a medium

Choose the correct answer from the options given below :

(A) (a)-(i), (b)-(ii), (c)-(iii), (d)-(iv)
(B) (a)-(ii), (b)-(iii), (c)-(iv), (d)-(i)
(C) (a)-(iv), (b)-(iii), (c)-(i), (d)-(ii)
(D) (a)-(iv), (b)-(iii), (c)-(ii), (d)-(i)

49. Given below are two statements

Statement-I : For transmitting a signal, size of antenna (l) should be comparable to wavelength of signal (at least $l = \frac{\lambda}{4}$ in dimension)

Statement-II : In amplitude modulation, amplitude of carrier wave remains constant (unchanged). In the light of the above statements, choose the most appropriate answer from the options given below. **[31 Jan. 2023 (Evening)]**

(A) Both Statement-I and Statement-II are correct
(B) Both Statement-I and Statement-II are incorrect
(C) Statement-I is incorrect but Statement-II is correct
(D) Statement-I is correct but Statement-II is incorrect

50. By what percentage will the transmission range of a TV tower be affected when the height of the tower is increased by 21%? **[6 April 2023 (Morning)]**

(A) 14% (B) 12%
(C) 10% (D) 15%

51. A TV transmitting antenna is 98 m high and the receiving antenna is at the ground level. If the radius of the earth is 6400 km, the surface area covered by the transmitting antenna is approximately : **[8 April 2023 (Morning)]**

(A) 1240 km^2 (B) 3942 km^2
(C) 4868 km^2 (D) 1549 km^2

52. The power radiated from a linear antenna of length l is proportional to : (Given, λ = Wavelength of wave) **[8 April 2023 (Evening)]**

(A) $\frac{l}{\lambda}$ (B) $\frac{l}{\lambda^2}$
(C) $\frac{l^2}{\lambda}$ (D) $\left(\frac{l}{\lambda}\right)^2$

53. A transmitting antenna is kept on the surface of the earth. The minimum height of receiving antenna required to receive the signal in line of sight at 4 km distance from it is $x \times 10^{-2}$ m. The value of x is (Let radius of earth $R = 6400$ km) **[11 April 2023 (Morning)]**

(A) 125 (B) 12.5
(C) 1.25 (D) 1250

54. In satellite communication, the uplink frequency band used is : **[11 April 2023 (Evening)]**

(A) 3.7–4.2 GHz (B) 5.925–6.422 MHz
(C) 76–8 GHz (D) 420–890 MHz

55. Match List-I with List-II

List-I (Layer of atmosphere)	List-II (Approximate height over earth' surface)
(a) F_1-Layer	(i) 10 km
(b) D-Layer	(ii) 170–190 km
(c) Troposphere	(iii) 100 km
(d) E-layer	(iv) 65–75 km

Choose the correct answer from the options given below: **[13 April 2023 (Morning)]**

(A) (a)-(iii), (b)-(iv), (c)-(i), (d)-(ii)
(B) (a)-(ii), (b)-(iv), (c)- (iii), (d)-(i)
(C) (a)-(ii), (b)-(i), (c)-(iv), (d)-(iii)
(D) (a)-(ii), (b)-(iv), (c)-(i), (d)-(iii)

56. To radiate EM signal of wavelength λ with high efficiency, the antennas should have a minimum size equal to: **[13 April 2023 (Evening)]**

(A) $\frac{\lambda}{2}$ (B) $\frac{\lambda}{4}$
(C) 2λ (D) λ

57. The height of transmitting antenna is 180 m and the height of the receiving antenna is 245 m. The maximum distance between them for satisfactory communication in line of sight will be : (Given $R = 6400$ km) **[15 April 2023 (Morning)]**

(A) 48 km (B) 56 km
(C) 96 km (D) 104 km

* * * * *

COMMUNICATION SYSTEMS

29.2 MODULATION AND DEMODULATION

1. Which of the following four alternatives is not correct ? We need modulation : **[2011]**

(A) To reduce the time lag between transmission and reception of the information signal

(B) To reduce the size of antenna

(C) To reduce the fractional band width, that is the ratio of the signal band width to the centre frequency

(D) To increase the selectivity

2. Given the electric field of a complete amplitude modulated wave as

$$\vec{E} = \hat{i}E_c\left(1+\frac{E_m}{E_c}\cos\omega_m t\right)\cos\omega_c t$$

Where the subscript stands for the carrier wave and m for the modulating signal. The frequencies present in the modulated wave are : **[2012 (Online)]**

(A) ω_c and $\sqrt{\omega_c^2+\omega_m^2}$ (B) ω_c, $\omega_c+\omega_m$ and $\omega_c-\omega_m$

(C) ω_c and ω_m (D) ω_c and $\sqrt{\omega_c\omega_m}$

3. Which of the following modulated signal has the best noise-tolerance ? **[2013]**

(A) Long-wave (B) Short-wave

(C) Medium-wave (D) Amplitude-modulated

4. A diode detector is used to detect an amplitude modulated wave of 60% modulation using a condenser of capacity 250 picofarad in parallel with a load resistance 100 kilo ohm. Find the maximum modulated frequency which could be detected by it : **[2013]**

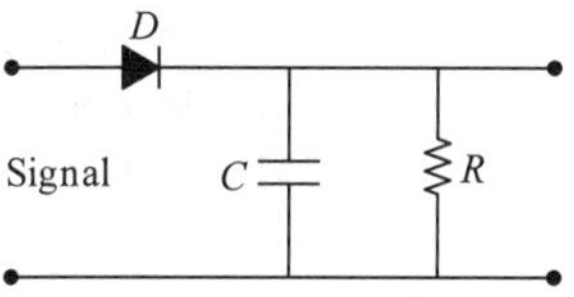

(A) 10.62 MHz (B) 10.62 kHz

(C) 5.31 MHz (D) 5.31 kHz

5. If a carrier wave $c(t) = A\sin\omega_c t$ is amplitude modulated by a modulator signal $m(t) = A\sin\omega_m t$ then the equation of modulated signal $[C_m(t)]$ and its modulation index are respectively : **[2013 (Online)]**

(A) $C_m(t) = A(1+\sin\omega_m t)\sin\omega_c t$ and 2

(B) $C_m(t) = A(1+\sin\omega_m t)\sin\omega_m t$ and 1

(C) $C_m(t) = A(1+\sin\omega_m t)\sin\omega_c t$ and 1

(D) $C_m(t) = A(1+\sin\omega_c t)\sin\omega_m t$ and 2

6. A signal of 5 kHz frequency is amplitude modulated on a carrier wave of frequency 2 MHz. The frequencies of the resultant signal is/are : **[2015]**

(A) 2005 kHz, 2000 kHz and 1995 kHz

(B) 2000 kHz and 1995 kHz

(C) 2 MHz only (D) 2005 kHz and 1995 kHz

7. Choose the correct statement : **[2016]**

(A) In frequency modulation the amplitude of the high frequency carrier wave is made to vary in proportion to the amplitude of the audio signal

(B) In frequency modulation the amplitude of the high frequency carrier wave is made to vary in proportional to the frequency of the audio signal

(C) In amplitude modulation the amplitude of the high frequency carrier wave is made to vary in proportion to the amplitude of the audio signal

(D) In amplitude modulation the frequency of the high frequency carrier wave is made to vary in proportion to the amplitude of the audio signal

8. A modulated signal $C_m(t)$ has the form $C_m(t) = 30\sin 300\pi t + 10(\cos 200\pi t - \cos 400\pi t)$. The carrier frequency f_c, the modulating frequency (message frequency) f_w and the modulation index μ are respectively given by : **[2016 (Online)]**

(A) $f_c = 200$ Hz; $f_w = 50$ Hz; $\mu = 1/2$

(B) $f_c = 150$ Hz; $f_w = 50$ Hz; $\mu = 2/3$

(C) $f_c = 150$ Hz; $f_w = 30$ Hz; $\mu = 1/3$

(D) $f_c = 200$ Hz; $f_w = 30$ Hz; $\mu = 1/2$

9. An audio signal consists of two distinct sounds : one a human speech signal in the frequency band of 200 Hz to 2700 Hz, while the other is a high frequency music signal in the frequency band of 10200 Hz to 15200 Hz. The ratio of the AM signal band width required to send both the signals together to the *AM* signal bandwidth required to send just the human speech is : **[2016 (Online)]**

(A) 2 (B) 5

(C) 6 (D) 3

10. In amplitude modulations, sinusoidal carrier frequency used is denoted by ω_c and the signal frequency is denoted by ω_m. The bandwidth ($\Delta\omega_m$) of the signal is such that $\Delta\omega_m < \omega_c$. Which of the following frequencies is not contained in the modulated wave ? **[2017]**

(A) $\omega_m + \omega_c$ (B) $\omega_c - \omega_m$
(C) ω_m (D) ω_c

11. A signal of frequency 20 kHz and peak voltage of 5 Volt is used to modulate a carrier wave of frequency 1.2 MHz and peak voltage 25 Volts. Choose the correct statement :

(A) Modulation index = 5, side frequency bands are at 1400 kHz and 1000 kHz
(B) Modulation index = 5, side frequency bands are at 21.2 kHz and 18.8 kHz
(C) Modulation index = 0.8, side frequency bands are at 1180 kHz and 1220 kHz
(D) Modulation index = 0.2, side frequency bands are at 1220 kHz and 1180 kHz **[2017 (Online)]**

12. A carrier wave of peak voltage 14 V is used for transmitting a message signal. The peak voltage of modulating signal given to achieve a modulation index of 80% will be : **[2018]**

(A) 11.2 V (B) 7 V
(C) 22.4 V (D) 28 V

13. A telephonic communication service is working at carrier frequency of 10 GHz. Only 10% of it is utilized for transmission. How many telephonic channels can be transmitted simultaneously if each channel requires a bandwidth of 5 kHz ? **[2018]**

(A) 2×10^3 (B) 2×10^4
(C) 2×10^5 (D) 2×10^6

14. The carrier frequency of a transmitter is provided by a tank circuit of a coil of inductance 49 μH and a capacitance of 2.5 nF. It is modulated by an audio signal of 12 kHz. The frequency range occupied by the side bands is : **[2018 (Online)]**

(A) 18 kHz – 30 kHz (B) 63 kHz – 75 kHz
(C) 442 kHz – 466 kHz (D) 13482 kHz – 13494 kHz

15. The number of amplitude modulated broadcast stations that can be accommodated in a 300 kHz band width for the highest modulating frequency 15 kHz will be : **[2018 (Online)]**

(A) 20 (B) 10
(C) 8 (D) 15

16. In a communication system operating at wavelength 800 nm, only one percent of source frequency is available as signal bandwidth. The number of channels accommodated for transmitting *TV* signals of band width 6 MHz are (Take velocity of light $c = 3 \times 10^8$ m/s, $h = 6.6 \times 10^{-34}$ J-s) : **[9 Jan. 2019 (Evening)]**

(A) 3.75×10^6 (B) 4.87×10^5
(C) 3.86×10^6 (D) 6.25×10^5

17. The modulation frequency of an *AM* radio station is 250 kHz, which is 10% of the carrier wave. If another *AM* station approaches you for license what broadcast frequency will you allot ? **[10 Jan. 2019 (Evening)]**

(A) 2750 kHz (B) 2000 kHz
(C) 2250 kHz (D) 2900 kHz

18. An amplitude modulated signal is given by $V(t) = 10[1 + 0.3 \cos(2.2 \times 10^4)] \sin(5.5 \times 10^5 t)$. Here t is in seconds. The sideband frequencies (in kHz) are, [Given $\pi = 22/7$] : **[11 Jan. 2019 (Morning)]**

(A) 1785 and 1715 (B) 892.5 and 857.5
(C) 89.25 and 85.75 (D) 178.5 smf 171.5

19. An amplitude modulated signal is plotted below : **[11 Jan. 2019 (Evening)]**

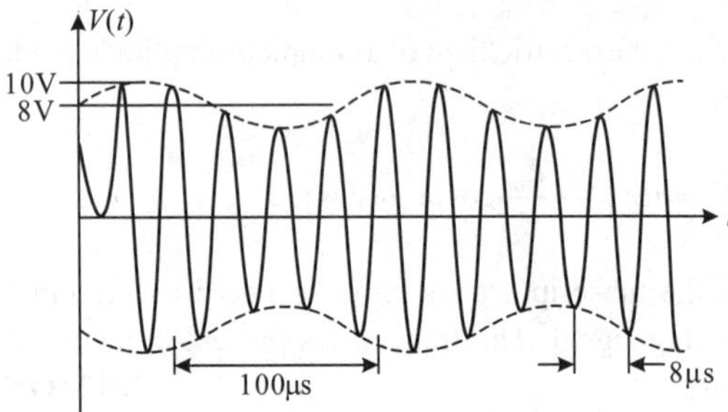

Which one of the following best describes the above signal ?

(A) $(9 + \sin(2.5\pi \times 10^5 t)) \sin(2\pi \times 10^4 t)V$
(B) $(9 + \sin(4\pi \times 10^4 t)) \sin(5\pi \times 10^5 t)V$
(C) $(1 + 9\sin(2\pi \times 10^4 t)) \sin(2.5\pi \times 10^5 t)V$
(D) $(9 + \sin(2\pi \times 10^4 t)) \sin(2.5\pi \times 10^5 t)V$

20. A 100 V carrier wave is made to vary between 160 V and 40 V by a modulating signal. What is the modulation index? **[12 Jan. 2019 (Morning)]**

(A) 0.6 (B) 0.5
(C) 0.3 (D) 0.4

21. The wavelength of the carrier waves in a modern optical fiber communication network is close to : **[8 April 2019 (Morning)]**

(A) 600 nm (B) 900 nm
(C) 2400 nm (D) 1500 nm

22. A signal $A \cos \omega t$ is transmitted using $v_0 \sin \omega_0 t$ as carrier wave. The correct amplitude modulated (*AM*) signal is : **[9 April 2019 (Morning)]**

(A) $v_0 \sin \omega_0 t + A \cos \omega t$
(B) $v_0 \sin \omega_0 t + \frac{A}{2}\sin(\omega_0 - \omega)t + \frac{A}{2}\sin(\omega_0 + \omega)t$
(C) $(v_0 + A) \cos \omega t \sin \omega_0 t$
(D) $v_0 \sin[\omega_0 (1 + 0.01 A \sin \omega t) t]$

23. The physical sizes of the transmitter and receiver antenna in a communication system are : **[9 April 2019 (Evening)]**
(A) Proportional to carrier frequency
(B) Inversely proportional to modulation frequency
(C) Inversely proportional to carrier frequency
(D) Independent of both carrier and modulation frequency

24. A message signal of frequency 100 MHz and peak voltage 100 V is used to execute amplitude modulation on a carrier wave of frequency 300 GHz and peak voltage 400 V. The modulation index and difference between the two side band frequencies are : **[10 April 2019 (Morning)]**
(A) 4; 1×10^8 Hz (B) 0.25; 1×10^8 Hz
(C) 4; 2×10^8 Hz (D) 0.25; 2×10^8 Hz

25. In an amplitude modulator circuit, the carrier wave is given by, $C(t) = 4 \sin(20000\,\pi t)$ while modulating signal is given by, $m(t) = 2 \sin(2000\,\pi t)$. The values of modulation index and lower side band frequency are : **[12 April 2019 (Evening)]**
(A) 0.5 and 9 kHz (B) 0.5 and 10 kHz
(C) 0.3 and 9 kHz (D) 0.4 and 10 kHz

26. An amplitude modulated wave is represented by the expression $v_m = 5(1 + 0.6 \cos 6280t) \sin(2\pi \times 10^4 t)$ volts. The minimum and maximum amplitudes of the amplitude modulated wave are, respectively : **[2 Sept. 2020 (Morning)]**
(A) 5 V, 8 V (B) $\frac{3}{2}$ V, 5 V
(C) $\frac{5}{2}$ V, 8 V (D) 3 V, 5 V

27. An audio signal $v_m = 20 \sin 2\pi(1500\,t)$ amplitude modulates a carrier $v_c = 80 \sin 2\pi(100{,}000t)$. The value of percent modulation is ______. **[24 Feb. 2021 (Morning)]**

28. Given below are two statements :
Statement-I : A speech signal of 2 kHz is used to modulate a carrier signal of 1 MHz. The bandwidth requirement for the signal is 4 kHz
Statement-II : The side band frequencies are 1002 kHz and 988 kHz
In the light of the above statements, choose the correct answer from the options given below : **[25 Feb. 2021 (Morning)]**
(A) Both Statement-I and Statement-II are true
(B) Statement-I is true but Statement-II is false
(C) Both Statement-I and Statement-II are false
(D) Statement-I is false but Statement-II is true

29. If a message signal of frequency 'f_m' is amplitude modulated with a carrier signal of frequency 'f_c' and radiated through an antenna, the wavelength of the corresponding signal in air is : **[25 Feb. 2021 (Evening)]**
(A) $\frac{c}{f_m}$ (B) $\frac{c}{f_c + f_m}$
(C) $\frac{c}{f_c - f_m}$ (D) $\frac{c}{f_c}$

30. The maximum and minimum amplitude of an amplitude modulated wave is 16 V and 8 V respectively. The modulation index for tis amplitude modulated wave is $x \times 10^{-2}$. The value of x is ______. **[26 Feb. 2021 (Morning)]**

31. A carrier wave $V_C(t) = 160 \sin(2\pi \times 10^6 t)$ volts is made to vary between $V_{max} = 200$ V and $V_{min} = 120$ V by a message signal $V_m(t) = A_m \sin(2\pi \times 10^3 t)$ volt. The peak voltage A_m of the modulating signal is ______. **[20 July 2021 (Morning)]**

32. A message signal of frequency 20 kHz and peak voltage of 20 volt is used to modulate a carrier wave of frequency 1 MHz and peak voltage of 20 volt. The modulation index will be ______ . **[25 July 2021 (Evening)]**

33. The maximum amplitude for an amplitude modulated wave is found to be 12 V while the minimum amplitude is found to be 3 V. The modulation index is $0.6x$ where x is ______. **[27 July 2021 (Evening)]**

34. An amplitude modulated wave is represented by $C_m(t) = 10(1 + 0.2 \cos 12560\,t) \sin(111 \times 10^4\,t)$ volts. The modulating frequency in kHz will be______. **[26 Aug. 2021 (Morning)]**

35. A carrier wave with amplitude of 250 V is amplitude modulated by a sinusoidal base band signal of amplitude 150 V. The ratio of minimum amplitude to maximum amplitude for the amplitude modulated wave is 50 : x, then value of x is______. **[01 Sep. 2021 (Evening)]**

36. A baseband signal of 3.5 MHz frequency is modulated with a carrier signal of 3.5 GHz frequency using amplitude modulation method. What should be the minimum size of antenna required to transmit the modulated signal ? **[24 June 2022 (Morning)]**
(A) 42.8 m (B) 42.8 mm
(C) 21.4 mm (D) 21.4 m

37. Choose the correct statement for amplitude modulation : **[26 June 2022 (Morning)]**
(A) Amplitude of modulating signal is varied in accordance with the information signal
(B) Amplitude of modulated signal is varied in accordance with the information signal
(C) Amplitude of carrier signal is varied in accordance with the information signal
(D) Amplitude of modulated signal is varied in accordance with the modulating signal

38. A sinusoidal wave $y(t) = 40 \sin(10 \times 10^6 \pi t)$ is amplitude modulated by another sinusoidal wave $x(t) = 20 \sin(1000 \pi t)$. The amplitude of minimum frequency component of modulated signal is : **[26 June 2022 (Evening)]**

(A) 0.5 (B) 0.25
(C) 20 (D) 10

39. Amplitude modulated wave is represented by
$V_{AM} = 10[1 + 0.4 \cos(2\pi \times 10^4 t)] \cos(2\pi \times 10^7 t)$
The total bandwidth of the amplitude modulated wave is : **[28 June 2022 (Evening)]**

(A) 10 kHz (B) 20 MHz
(C) 20 kHz (D) 10 MHz

40. In AM modulation, a signal is modulated on a carrier wave such that maximum and minimum amplitude are found to be 6 V and 2 V respectively. The modulation index is : **[25 July 2022 (Evening)]**

(A) 100% (B) 80%
(C) 60% (D) 50%

41. The maximum and minimum voltage of an amplitude modulated signal are 60 V and 20 V respectively. The percentage modulation index will be : **[26 July 2022 (Evening)]**

(A) 0.5% (B) 50%
(C) 2% (D) 30%

42. A square wave of the modulating signal is shown in the figure. The carrier wave is given by $C(t) = 5 \sin(8\pi t)$ Volt. The modulation index is : **[27 July 2022 (Evening)]**

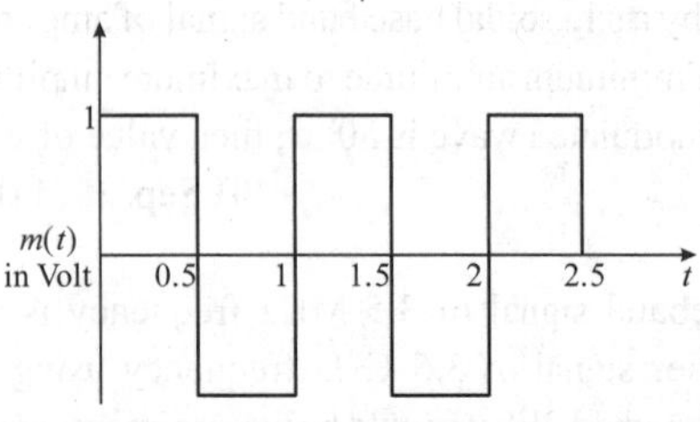

(A) 0.2 (B) 0.1
(C) 0.3 (D) 0.4

43. In the case of amplitude modulation to avoid distortion the modulation index (μ) should be : **[28 July 2022 (Morning)]**

(A) $\mu \leq 1$ (B) $\mu \geq 1$
(C) $\mu = 2$ (D) $\mu = 0$

44. A *FM* Broad cast transmitter, using modulating signal of frequency 20 kHz has a deviation ratio of 10. The Bandwidth required for transmission is : **[28 July 2022 (Evening)]**

(A) 220 kHz (B) 180 kHz
(C) 360 kHz (D) 440 kHz

45. Find the modulation index of an AM wave having 8 V variation where maximum amplitude of the AM wave is 9 V : **[29 July 2022 (Morning)]**

(A) 0.8 (B) 0.5
(C) 0.2 (D) 0.1

46. A modulating signal $2\sin(6.28 \times 10^6)t$ is added to the carrier signal $4\sin(12.56 \times 10^9)t$ for amplitude modulation. The combined signal is passed through a non-linear square law device. The output is then passed through a band pass filter. The bandwidth of the output signal of band pass filter will be_____ MHz. **[29 July 2022 (Evening)]**

47. A modulating signal is a square wave, as shown in the figure.

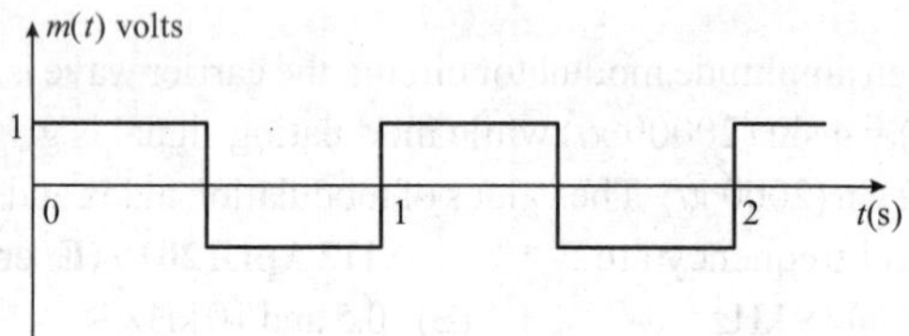

If the carrier wave is given as $c(t) = 2 \sin(8 \pi t)$ volts, the modulation index is : **[24 Jan. 2023 (Morning)]**

(A) 1/4 (B) 1
(C) 1/3 (D) 1/2

48. Match List-I with List-II **[24 Jan. 2023 (Evening)]**

List-I	List-II
(a) AM Broadcast	(i) 88 – 108 MHz
(b) FM Broadcast	(ii) 540 – 1600 kHz
(c) Television	(iii) 3.7 – 4.2 GHz
(d) Satellite Communication	(iv) 54 MH_z – 590 MHz

Choose the correct answer from the options given below :

(A) (a)-(ii), (b)-(i), (c)-(iv), (d)-(iii)
(B) (a)-(iv), (b)-(iii), (c)-(i), (d)-(ii)
(C) (a)-(ii), (b)-(iii), (c)-(i), (d)-(iv)
(D) (a)-(i), (b)-(iii), (c)-(ii), (d)-(iv)

49. A message signal of frequency 5 kHz is used to modulate a carrier signal of frequency 2 MHz. The bandwidth for amplitude modulation is : **[25 Jan. 2023 (Morning)]**

(A) 5 kHz (B) 20 kHz
(C) 10 kHz (D) 2.5 kHz

50. The modulation index for an A.M. wave having maximum and minimum peak to peak voltages of 14 mV and 6 mV respectively is : **[29 Jan. 2023 (Evening)]**

(A) 1.4 (B) 0.4
(C) 0.2 (D) 0.6

51. A sinusoidal carrier voltage is amplitude modulated. The resultant amplitude modulated wave has maximum and minimum amplitude of 120 V and 80 V respectively. The amplitude of each sideband is : **[30 Jan. 2023 (Morning)]**

(A) 15V (B) 10V
(C) 20V (D) 5V

52. The amplitude of 15 sin(1000 πt) is modulated by 10 sin(4 πt) signal. The amplitude modulated signal contains frequency(ies) of: **[31 Jan. 2023 (Morning)]**

(a) 500 Hz (b) 2 Hz
(c) 250 Hz (d) 498 Hz
(e) 502 Hz

Choose the correct answer from the options given below :

(A) (a) only (B) (a), (d) and (e) only
(C) (b) only (D) (a) and (b) only

53. Which of the following frequencies does not belong to *FM* broadcast. **[1 Feb. 2023 (Morning)]**

(A) 106 MHz (B) 64 MHz
(C) 99 MHz (D) 89 MHz

54. In an amplitude modulation, a modulating signal having amplitude of *XV* is superimposed with a carrier signal of amplitude *YV* in first case. Then, in second case, the same modulating signal is superimposed with different carrier signal of amplitude 2*YV*. The ratio of modulation index in the two case respectively will be : **[1 Feb. 2023 (Evening)]**

(A) 1 : 2 (B) 1 : 1
(C) 2 : 1 (D) 4 : 1

55. For an amplitude modulated wave the minimum amplitude is 3 V, while the modulation index is 60%. The maximum amplitude of the modulated wave is: **[6 April 2023 (Evening)]**

(A) 15V (B) 12V
(C) 10V (D) 5V

56. A carrier wave of amplitude 15 V is modulated by a sinusoidal base band signal of amplitude 3 V. The ratio of maximum amplitude to minimum amplitude in an amplitude modulated wave is: **[10 April 2023 (Morning)]**

(A) 2 (B) $\frac{3}{2}$
(C) 5 (D) 1

57. A message signal of frequency 3 kHz is used to modulate a carrier signal of frequency 1.5 MHz. The bandwidth of the amplitude modulated wave is: **[10 April 2023 (Evening)]**

(A) 3 kHz (B) 6 MHz
(C) 3 MHz (D) 6 kHz

58. The amplitude of 15 sin (1000 πt) is modulated by 10 sin (4πt) signal. The amplitude modulated signal contains frequencies of: **[12 April 2023 (Morning)]**

(1) 500 Hz (2) 2 Hz
(3) 250 Hz (4) 498 Hz
(5) 502 Hz

Choose the correct answer from the options given below:

(A) (1) and (3) only (B) (1) and (4) only
(C) (1) and (2) only (D) (1), (4) and (5) only

* * * * *

CHAPTER

30 ERRORS AND EXPERIMENTS

30.1 ERRORS AND EXPERIMENTS

1. Two full turns of the circular scale of a screw gauge cover a distance of 1 mm on its main scale. The total number of divisions on the circular scale is 50. Further, it is found that the screw gauge has a zero error of – 0.03 mm. While measuring the diameter of a thin wire, a student notes the main scale reading of 3 mm and the number of circular scale divisions in line with the main scale as 35. The diameter of the wire is : **[2008]**

(A) 3.32 mm (B) 3.73 mm
(C) 3.67 mm (D) 3.38 mm

2. A body of mass $m = 3.513$ kg is moving along the x-axis with a speed of 5.00 ms^{-1}. The magnitude of its momentum is recorded as : **[2008]**

(A) 17.6 kg ms^{-1} (B) 17.565 kg ms^{-1}
(C) 17.56 kg ms^{-1} (D) 17.57 kg ms^{-1}

3. In an experiment the angles are required to be measured using an instrument, 29 divisions of the main scale exactly coincide with the 30 divisions of the vernier scale. If the smallest division of the main scale is half a degree (= 0.5°), then the least count of the instrument is : **[2009]**

(A) half minute (B) one degree
(C) half degree (D) one minute

4. The respective number of significant figures for the numbers 23.023, 0.0003 and 2.1×10^{-3} are : **[2009]**

(A) 5, 1, 2 (B) 5, 1, 5
(C) 5, 5, 2 (D) 4, 4, 2

5. A screw gauge gives the following reading when used to measure the diameter of a wire.
Main scale reading : 0 mm
Circular scale reading : 52 divisions **[2011]**
Given that 1 mm on main scale corresponds to 100 divisions of the circular scale. The diameter of wire from the above data is :

(A) 0.052 cm (B) 0.026 cm
(C) 0.005 cm (D) 0.52 cm

6. A spectrometer gives the following reading when used to measure the angle of a prism.
Main scale reading : 58.5 degree
Vernier scale reading : 09 divisions
Given that 1 division on main scale corresponds to 0.5 degree. Total divisions on the Vernier scale is 30 and match with 29 divisions of the main scale. The angle of the prism from the above data is : **[2012]**

(A) 58.59 degree (B) 58.77 degree
(C) 58.65 degree (D) 59 degree

7. Resistance of a given wire is obtained by measuring the current flowing in it and the voltage difference applied across it. If the percentage errors in the measurement of the current and the voltage difference are 3% each, then error in the value of resistance of the wire is : **[2012]**

(A) 6% (B) Zero
(C) 1% (D) 3%

8. A student measured the diameter of a wire using a screw gauge with the least count 0.001 cm and listed the measurements. The measured value should be recorded as : **[2012 (Online)]**

(A) 5.3200 cm (B) 5.3 cm
(C) 5.32 cm (D) 5.320 cm

9. N divisions on the main scale of a vernier calliper coincide with $(N + 1)$ divisions of the vernier scale. If each division of main scale is 'a' units, then the least count of the instrument is : **[2012 (Online)]**

(A) a (B) $\frac{a}{N}$
(C) $\frac{N}{N+1} \times a$ (D) $\frac{a}{N+1}$

10. A student measured the length of a rod and wrote it as 3.50 cm. Which instrument did he use to measure it ? **[2014]**

(A) A meter scale
(B) A vernier calliper where the 10 divisions in vernier scale matches with 9 division in main scale and main scale has 10 divisions in 1 cm
(C) A screw gauge having 100 divisions in the circular scale and pitch as 1 mm
(D) A screw gauge having 50 divisions in the circular scale and pitch as 1 mm

11. The current voltage relation of a diode is given by $I = (e^{1000\,V/T} - 1)$ mA, where the applied voltage V is in volt and the temperature T is in degree kelvin. If a student makes an error measuring ± 0.01 V while measuring the current of 5 mA at 300 K, what will be the error in the value of current in mA ? **[2014]**

(A) 0.2 mA (B) 0.02 mA
(C) 0.5 mA (D) 0.05 mA

12. An experiment is performed to obtain the value of acceleration due to gravity g by using a simple pendulum of length L. In this experiment time for 100 oscillations is measured by using a watch of 1 second least count and the value is 90.0 seconds. The length L is measured by using a meter scale of least count 1 mm and the value is 20.0 cm. The error in the determination of g would be : **[2014 (Online)]**

(A) 1.7% (B) 2.7%
(C) 4.4% (D) 2.27%

13. In the experiment of calibration of voltmeter, a standard cell of e.m.f. 1.1 volt is balanced against 440 cm of potential wire. The potential difference across the ends of resistance is found to balance against 220 cm of the wire. The corresponding reading of voltmeter is 0.5 volt. The error in the reading of voltmeter will be : **[2014 (Online)]**

(A) – 0.15 volt (B) 0.15 volt
(C) 0.5 volt (D) – 0.05 volt

14. Match List-I (Event) with List-II (Order of the time interval for happening of the event) and select the correct option from the options given below the lists : **[2014 (Online)]**

List-I	List-II
(a) Rotation period of earth	(i) 10^5 s
(b) Revolution period of earth	(ii) 10^7 s
(c) Period of light wave	(iii) 10^{-15} s
(d) Period of sound wave	(iv) 10^{-3} s

(A) (a)-(i), (b)-(ii), (c)-(iii), (d)-(iv)
(B) (a)-(ii), (b)-(i), (c)-(iv), (d)-(iii)
(C) (a)-(i), (b)-(ii), (c)-(iv), (d)-(iii)
(D) (a)-(ii), (b)-(i), (c)-(iii), (d)-(iv)

15. The period of oscillation of a simple pendulum is $T = 2\pi\sqrt{\frac{L}{g}}$. Measured value of L is 20.0 cm know to 1 mm accuracy and time for 100 oscillations of the pendulum is found to be 90 s using a wrist watch of 1 s resolution. The accuracy in the determination of g is : **[2015]**

(A) 1% (B) 5%
(C) 2% (D) 3%

16. Diameter of a steel ball is measured using a Vernier callipers which has divisions of 0.1 cm on its main scale (MS) and 10 divisions of its vernier scale (VS) match 9 divisions on the main scale. Three such measurements for a ball are given as : **[2015 (Online)]**

S.No.	MS(cm)	VS divisions
1.	0.5	8
2.	0.5	4
3.	0.5	6

If the zero error is – 0.03 cm, then mean corrected diameter is :

(A) 0.52 cm (B) 0.59 cm
(C) 0.56 cm (D) 0.53 cm

17. A student measures the time period of 100 oscillations of a simple pendulum four times. The data set is 90 s, 91 s, 95 s, and 92 s. If the minimum division in the measuring clock is 1 s, then the reported mean time should be : **[2016]**

(A) 92 ± 1.8 s (B) 92 ± 3 s
(C) 92 ± 1.5 s (D) 92 ± 5.0 s

18. A screw gauge with a pitch of 0.5 mm and a circular scale with 50 divisions is used to measure the thickness of a thin sheet of Aluminium. Before starting the measurement, it is found that wen the two jaws of the screw gauge are brought in contact, the 45th division coincides with the main scale line and the zero of the main scale is barely visible. What is the thickness of the sheet if the main scale reading is 0.5 mm and the 25th division coincides with the main scale line ? **[2016]**

(A) 0.70 mm (B) 0.50 mm
(C) 0.75 mm (D) 0.80 mm

19. The following observations were taken for determining surface tension T of water by capillary method :
Diameter of capillary, $D = 1.25 \times 10^{-2}$ m, rise of water, $h = 1.45 \times 10^{-2}$ m. Using $g = 9.80$ m/s^2 and the simplified relation $T = \frac{rhg}{2} \times 10^3$ N/m, the possible error in surface tension is closest to : **[2017]**

(A) 2.4% (B) 10%
(C) 0.15% (D) 1.5%

20. A physical quantity P is described by the relation $P = a^{1/2} b^2 c^3 d^{-4}$. If the relative errors in the measurement of a, b, c, and d respectively, are 2%, 1%, 3% and 5%, then the relative error in P will be : **[2017 (Online)]**

(A) 8% (B) 12%
(C) 32% (D) 25%

21. The percentage errors in quantities P, Q, R and S are 0.5%, 1%, 3% and 1.5% respectively in the measurement of a physical quantity $A = \frac{P^3Q^2}{\sqrt{RS}}$.

The maximum percentage error in the value of A will be : **[2018 (Online)]**

(A) 8.5% (B) 6.0%
(C) 7.5% (D) 6.5%

22. In a screw gauge, 5 complete rotations of the screw cause it to move a linear distance of 0.25 cm. There are 100 circular scale divisions. The thickness of a wire measured by this screw gauge gives a reading of 4 main scale divisions and 30 circular scale divisions. Assuming negligible zero error, the thickness of the wire is : **[2018 (Online)]**

(A) 0.0430 cm (B) 0.3150 cm
(C) 0.4300 cm (D) 0.2150 cm

23. The relative error in the determination of the surface area of a sphere is α. Then the relative error in the determination of its volume is : **[2018 (Online)]**

(A) $\frac{2}{3}\alpha$ (B) $\frac{1}{3}\alpha$
(C) $\frac{3}{2}\alpha$ (D) α

24. The relative uncertainty in the period of a satellite orbiting around the earth is 10^{-2}. If the relative uncertainty in the radius of the orbit is negligible, the relative uncertainty in the mass of the earth is : **[2018 (Online)]**

(A) 3×10^{-2} (B) 10^{-2}
(C) 2×10^{-2} (D) 6×10^{-2}

25. The density of a material in the shape of a cube is determined by measuring three sides of the cube and its mass. If the relative errors in measuring the mass and length are respectively 1.5% and 1%, the maximum error in determining the density is : **[2018 (Online)]**

(A) 2.5% (B) 3.5%
(C) 4.5% (D) 6%

26. A copper wire is stretched to make it 0.5% longer. The percentage change in its electrical resistance, if its volume remains unchanged is : **[9 Jan. 2019 (Morning)]**

(A) 2.5% (B) 0.5%
(C) 1.0% (D) 2.0%

27. The pitch and the number of divisions, on the circular scale, for a given screw gauge are 0.5 mm and 100 respectively. When the screw gauge is fully tightened without any object, the zero of its circular scale lies 3 divisions below the reference line. The readings of the main scale and the circular scale, for a thin sheet, are 5.5 mm and 48 respectively, the thickness of this sheet is: **[9 Jan. 2019 (Evening)]**

(A) 5.755 m (B) 5.725 mm
(C) 5.740 m (D) 5.950 mm

28. The diameter and height of a cylinder are measured by a meter scale to be 12.6 ± 0.1 cm and 34.2 ± 0.1 cm, respectively. What will be the value of its volume in appropriate significant figures ? **[10 Jan. 2019 (Evening)]**

(A) 4260 ± 80 cm^3 (B) 4300 ± 80 cm^3
(C) 4264.4 ± 81.0 cm^3 (D) 4264 ± 81 cm^3

29. The least count of the main scale of a screw gauge is 1 mm. The minimum number of divisions on its circular scale required to measure 5 μm diameter of wire is : **[12 Jan. 2019 (Morning)]**

(A) 50 (B) 100
(C) 200 (D) 500

30. In a simple pendulum experiment for determination of acceleration due to gravity (g), time taken for 20 oscillations is measured by using a watch of 1 second least count. The mean value of time taken comes out to be 30 s. The length of pendulum is measured by using a meter scale of least count 1 mm and the value obtained is 55.0 cm. The percentage error in the determination of g is close to : **[8 April 2019 (Evening)]**

(A) 0.7% (B) 0.2%
(C) 3.5% (D) 6.8%

31. In the density measurement of a cube, the mass and edge length are measured as (10.00 ± 0.10) kg and (0.10 ± 0.01) m, respectively. The error in the measurement of density is : **[9 April 2019 (Morning)]**

(A) 0.10 kg/m^3 (B) 0.31 kg/m^3
(C) 0.07 kg/m^3 (D) 0.01 kg/m^3

32. The area of a square is 5.29 cm^2. The area of 7 such squares taking into account the significant figures is : **[9 April 2019 (Evening)]**

(A) 37 cm^2 (B) 37.0 cm^2
(C) 37.03 cm^2 (D) 37.030 cm^2

33. A simple pendulum is being used to determine the value of gravitational acceleration g at a certain place. The length of the pendulum is 25.0 cm and a stop watch with 1s resolution measures the time taken for 40 oscillations to be 50 s. The accuracy in g is : **[8 Jan. 2020 (Evening)]**

(A) 3.40% (B) 5.40%
(C) 4.40% (D) 2.40%

34. If the screw on a screw-gauge is given six rotations, it moves by 3 mm on the main scale. If there are 50 divisions on the circular scale the least count of the screw gauge is : **[9 Jan. 2020 (Morning)]**

(A) 0.001 mm (B) 0.001 cm
(C) 0.02 mm (D) 0.01 cm

35. For the four sets of three measured physical quantities as given below. Which of the following options is correct ?

(i) $A_1 = 24.36, B_1 = 0.0724, C_1 = 256.2$
(ii) $A_2 = 24.44, B_2 = 16.082, C_2 = 240.2$
(iii) $A_3 = 25.2, B_3 = 19.2812, C_3 = 236.183$
(iv) $A_4 = 25, B_4 = 236.191, C_4 = 19.5$ **[9 Jan. 2020 (Evening)]**

(A) $A_4 + B_4 + C_4 < A_1 + B_1 + C_1 < A_3 + B_3 + C_3 < A_2 + B_2 + C_2$

(B) $A_1+B_1+C_1<A_3+B_3+C_3<A_2+B_2+C_2<A_4+B_4+C_4$
(C) $A_1+B_1+C_1=A_2+B_2+C_2=A_3+B_3+C_3=A_4+B_4+C_4$
(D) $A_4+B_4+C_4<A_1+B_1+C_1=A_2+B_2+C_2=A_3+B_3+C_3$

36. The least count of the main scale of a vernier callipers is 1 mm. Its vernier scale is divided into 10 divisions and coincide with 9 divisions of the main scale. When jaws are touching each other, the 7th division of vernier scale coincides with a division of main scale and the zero of vernier scale is lying right side of the zero of main scale. When this vernier is used to measure length of a cylinder, the zero of the vernier scale is between 3.1 cm and 3.2 cm and 4th *VSD* coincides with a main scale division. The length of the cylinder is : (*VSD* is vernier scale division) **[2 Sept. 2020 (Morning)]**

(A) 3.21 cm (B) 2.99 cm
(C) 3.2 cm (D) 3.07 cm

37. Using screw gauge of pitch 0.1 cm and 50 divisions on its circular scale, the thickness of an object is measured. It should correctly be recorded as : **[3 Sept. 2020 (Morning)]**

(A) 2.123 cm (B) 2.125 cm
(C) 2.121 cm (D) 2.124 cm

38. A physical quantity z depends on four observables a, b, c and d, as $z=\dfrac{a^2 b^{\frac{2}{3}}}{\sqrt{c}\,d^3}$. The percentage of error in the measurement of a, b, c and d 2%, 1.5%, 4% and 2.5% respectively. The percentage of error in z is : **[5 Sept. 2020 (Morning)]**

(A) 12.25% (B) 14.5%
(C) 16.5% (D) 13.5%

39. A screw gauge has 50 divisions on its circular scale. The circular scale is 4 units ahead of the pitch scale marking, prior to use. Upon one complete rotation of the circular scale, a displacement of 0.5 mm is noticed on the pitch scale. The nature of zero error involved, and the least count of the screw gauge, are respectively : **[6 Sept. 2020 (Morning)]**

(A) Negative, 2 μm (B) Positive, 10 μm
(C) Positive, 0.1 μm (D) Positive, 0.1 mm

40. The density of a solid metal sphere is determined by measuring its mass and its diameter. The maximum error in the density of the sphere is $\left(\dfrac{x}{100}\right)$%. If the relative errors in measuring the mass and the diameter are 6.0% and 1.5% respectively, the value of x is. **[6 Sept. 2020 (Morning)]**

41. A student measuring the diameter of a pencil of circular cross-section with the help of a vernier scale records the following four readings 5.50 mm, 5.55 mm, 5.45 mm; 5.65 mm. The average of these four readings is 5.5375 mm and the standard deviation of the data is 0.07395 mm. The average diameter of the pencil should therefore be recorded as : **[6 Sept. 2020 (Evening)]**

(A) (5.5375 ± 0.0739) mm (B) (5.538 ± 0.074) mm
(C) (5.54 ± 0.07) mm (D) (5.5375 ± 0.0740) mm

42. The pitch of the screw gauge is 1 mm and there are 100 divisions on the circular scale. When nothing is put in between the jaws, the zero of the circular scale lies 8 division below the reference line. When a wire is placed between the jaws, the first linear scale division in clearly visible while 72nd division on circular scale coincides with the reference line. The radius of the wire is : **[25 Feb. 2021 (Morning)]**

(A) 1.64 mm (B) 0.90 mm
(C) 1.80 mm (D) 0.82 mm

43. A wire of 1 Ω has a length of 1 m. It is stretched till its length increases by 25%. The percentage change in resistance to the nearest integer is : **[26 Feb. 2021 (Evening)]**

(A) 12.5% (B) 25%
(C) 76% (D) 56%

44. The resistance $R=\dfrac{V}{I}$, where $V=(50\pm2)V$ and $I=(20\pm0.2)$A. The percentage error in R is 'x' %. The value of 'x' to the nearest integer is ______. **[16 March 2021 (Morning)]**

45. The vernier scale used for measurement has a positive zero error of 0.2 mm. If while taking a measurement it was noted that '0' on the vernier scale lies between 8.5 cm and 8.6 cm, vernier coincidence is 6, then the correct value of measurement is______ cm. (least count = 0.01 cm) **[17 March 2021 (Morning)]**

(A) 8.36 cm (B) 8.54 cm
(C) 8.58 cm (D) 8.56 cm

46. The time period of a simple pendulum is given by $T=2\pi\sqrt{\dfrac{l}{g}}$. The measured value of the length of pendulum is 10 cm known to a 1 mm accuracy. The time for 200 oscillations of the pendulum is found to be 100 second using a clock of 1s resolution. The percentage accuracy in the determination of 'g' using this pendulum is 'x'. The value of 'x' to the nearest integer is : **[18 March 2021 (Morning)]**

(A) 2% (B) 3%
(C) 5% (D) 4%

47. The radius of a sphere is measured to be (7.50 ± 0.85) cm. Suppose the percentage error in its volume is x. The value of x, to the nearest x, is______. **[18 March 2021 (Evening)]**

48. Three students S_1, S_2 and S_3 perform an experiment for determining the acceleration due to gravity (g) using a simple pendulum. They use different lengths of pendulum and record time for different number of oscillations. The observations are shown in the table. **[22 July 2021 (Evening)]**

Students No.	Length of Pendulum (cm)	No. of oscillations (n)	Total time for n oscillations	Time period (s)
1	64.0	8	128.0	16.0
2	64.0	4	64.0	16.0
3	20.0	4	36.0	9.0

(Least count of length = 0.1 cm least count for time = 0.1 s)

If E_1, E_2 and E_3 are the percentage errors in 'g' for students 1, 2 and 3 respectively, then the minimum percentage error is obtained by student no. ______ .

49. Student A and student B used two screw gauges of equal pitch and 100 equal circular divisions to measure the radius of a given wire. The actual value of the radius of the wire is 0.322 cm. The absolute value of the difference between the final circular scale readings observed by the students A and B is ______.
[Figures shows position of reference 'O' when jaws of screw gauge are closed] Given pitch = 0.1 cm : **[25 July 2021 (Morning)]**

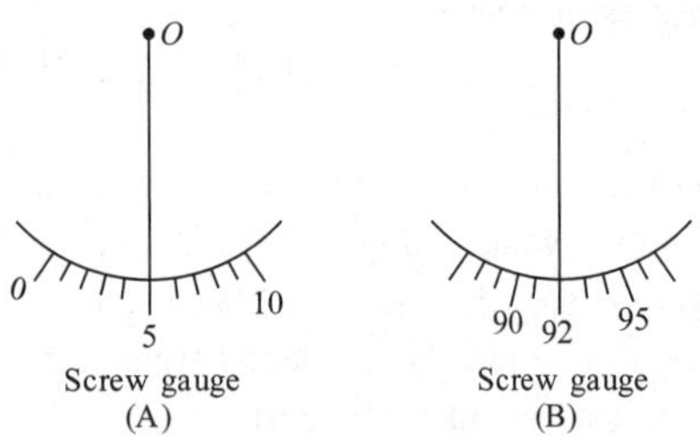

50. The value of aluminum susceptibility is 2.2×10^{-5}. The percentage increase in the magnetic field if space within a current carrying toroid is filled with aluminum is $\frac{x}{10^4}$. Then the value of x is ______. **[25 July 2021 (Morning)]**

51. Assertion-A : If in five complete rotations of the circular scale, the distance travelled on main scale of the screw gauge is 5 mm and there are 50 total divisions on circular scale, then least count is 0.001 cm

Reason-R : $\text{Least Count} = \dfrac{\text{Pitch}}{\text{Total division on circular scale}}$

In the light of the above statements, choose the most appropriate answer from the options given below : **[27 July 2021 (Morning)]**

(A) Both **A** and **R** are correct and R is the correct explanation of **A**.

(B) **A** is not correct but **R** is correct.

(C) **A** is correct but **R** is not correct.

(D) Both **A** and **R** are correct and **R** is not the correct explanation of **A**.

52. A physical quantity 'y' is represented by the formula $y = m^2 r^{-4} g^x l^{-3/2}$. If the percentage errors found in y, m, r, l and g are 18, 1, 0.5, 4 and p respectively, then find the value of x and p : **[27 July 2021 (Evening)]**

(A) 4 and ± 3 (B) 8 and ± 2

(C) $\frac{16}{3}$ and $\pm \frac{3}{2}$ (D) 5 and ± 2

53. The acceleration due to gravity is found upto an accuracy of 4% on a planet. The energy supplied to a simple pendulum of known mass m to undertake oscillations of time period T is being estimated. If time period is measured to an accuracy of 3%, the accuracy to which E is known as ______ %. **[26 Aug. 2021 (Evening)]**

54. If $Z = \dfrac{A^2 B^3}{C^4}$, then the relative error in Z will be : **[25 June 2022 (Morning)]**

(A) $\dfrac{\Delta A}{A} + \dfrac{\Delta B}{B} + \dfrac{\Delta C}{C}$ (B) $\dfrac{2\Delta A}{A} + \dfrac{3\Delta B}{B} - \dfrac{4\Delta C}{C}$

(C) $\dfrac{2\Delta A}{A} + \dfrac{3\Delta B}{B} + \dfrac{4\Delta C}{C}$ (D) $\dfrac{\Delta A}{A} + \dfrac{\Delta B}{B} - \dfrac{\Delta C}{C}$

55. For $z = a^2 x^3 y^{1/2}$, where 'a' is a constant. If percentage error in measurement of 'x' and 'y' are 4% and 12%, respectively, then the percentage error for 'z' will be %. **[25 June 2022 (Evening)]**

56. In a vernier callipers, each cm on the main scale is divided into 20 equal parts. If tenth vernier scale division coincides with ninth main scale division. Then the value of vernier constant will be ______ $\times 10^{-2}$ mm. **[26 June 2022 (Morning)]**

57. A travelling microscope is used to determine the refractive index of a glass slab. If 40 divisions are there in 1 cm on main scale and 50 Vernier scale divisions are equal to 49 main scale divisions, then least count of the travelling microscope is ______ $\times 10^{-6}$ m. **[26 June 2022 (Evening)]**

58. A sliver wire has mass $(0.6 \pm 0.006)\, g$, radius (0.5 ± 0.005) mm and length (4 ± 0.04) cm. The maximum percentage error in the measurement of its density will be : **[27 June 2022 (Morning)]**

(A) 4% (B) 3%

(C) 6% (D) 7%

59. A student in the laboratory measures thickness of a wire using screw gauge. The readings are 1.22 mm, 1.23 mm, 1.19 mm and 1.20 mm. The percentage error is $\frac{x}{121}\%$. The value of x is ______. **[28 June 2022 (Evening)]**

60. The vernier constant of Vernier callipers is 0.1 mm and it has zero error of (– 0.05) cm. While measuring diameter of a sphere, the main scale reading is 1.7 cm and coinciding vernier division is 5. The corrected diameter will be _____ $\times 10^{-2}$ cm. **[29 June 2022 (Evening)]**

61. The maximum error in the measurement of resistance, current and time for which current flows in an electrical circuit are 1%, 2% and 3% respectively. The maximum percentage error in the detection of the dissipated heat will be : **[25 July 2022 (Evening)]**

(A) 2 (B) 4
(C) 6 (D) 8

62. A screw gauge of pitch 0.5 mm is used to measure the diameter of uniform wire of length 6.8 cm, the main scale reading is 1.5 mm and circular scale reading is 7. The calculated curved surface area of wire to appropriate significant figures is :
[Screw gauge has 50 divisions on the circular scale] **[26 July 2022 (Morning)]**

(A) 6.8 cm^2 (B) 3.4 cm^2
(C) 3.9 cm^2 (D) 2.4 cm^2

63. In a Vernier Calipers. 10 divisions of Vernier scale is equal to the 9 divisions of main scale. When both jaws of Vernier calipers touch each other, the zero of the Vernier scale is shifted to the left of zero of the main scale and 4th Vernier scale division exactly coincides with the main scale reading. One main scale division is equal to 1 mm. While measuring diameter of a spherical body, the body is held between two jaws. It is now observed that zero of the Vernier scale lies between 30 and 31 divisions of main scale reading and 6th Vernier scale division exactly coincides with the main scale reading. The diameter of the spherical body will be : **[26 July 2022 (Evening)]**

(A) 3.02 cm (B) 3.06 cm
(C) 3.10 cm (D) 3.20 cm

64. A torque meter is calibrated to reference standards of mass, length and time each with 5% accuracy. After calibration, the measured torque with this torque meter will have net accuracy of: **[27 July 2022 (Morning)]**

(A) 15% (B) 25%
(C) 75% (D) 5%

65. The one division of main scale of vernier callipers reads 1 mm and 10 divisions of Vernier scale is equal to the 9 divisions on main scale. When the two jaws of the instrument touch each other the zero of the Vernier lies to the right of zero of the main scale and its fourth division coincides with a main scale division. When a spherical bob is tightly placed between the two jaws, the zero of the Vernier scale lies in between 4.1 cm and 4.2 cm and 6th Vernier division coincides with a main scale division. The diameter of the bob will be ______ 10^{-2} cm. **[27 July 2022 (Morning)]**

66. In an experiment to find acceleration due to gravity (g) using simple pendulum, time period of 0.5 s is measured from time of 100 oscillation with a watch of 1 s resolution. If measured value of length is 10 cm known to 1 mm accuracy. The accuracy in the determination of g is found to be x%. The value of x is. **[28 July 2022 (Evening)]**

67. A travelling microscope has 20 divisions per cm on the main scale while its Vernier scale has total 50 divisions and 25 Vernier scale divisions are equal to 24 main scale divisions, what is the least count of the travelling microscope ? **[29 July 2022 (Morning)]**

(A) 0.001 cm (B) 0.002 mm
(C) 0.002 cm (D) 0.005 cm

68. In an experiment to find out the diameter of wire using screw gauge, the following observation were noted : **[29 July 2022 (Morning)]**

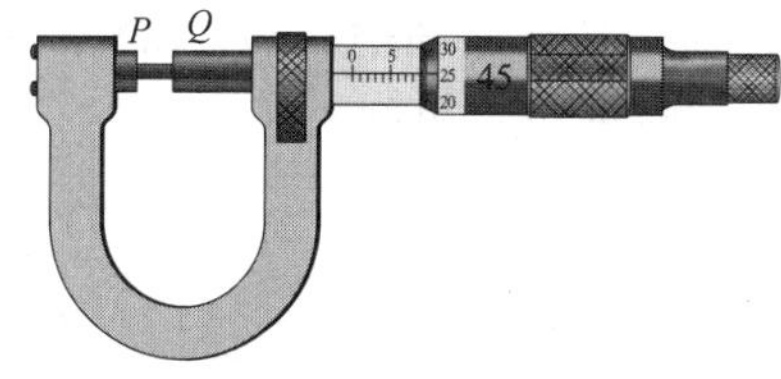

(a) Screw moves 0.5 mm on main scale in one complete rotation
(b) Total divisions on circular scale = 50
(c) Main scale reading is 2.5 mm
(d) 45th division of circular scale is in the pitch line
(e) Instrument has 0.03 mm negative error

Then the diameter of wire is :

(A) 2.92 mm (B) 2.54 mm
(C) 2.98 mm (D) 3.45 mm

69. In an experiment of measuring the refractive index of a glass slab using travelling microscope in physics lab, a student measures real thickness of the glass slab as 5.25 mm and apparent thickness of the glass slab at 5.00 mm. Travelling microscope has 20 divisions in one cm on main scale and 50 divisions on Vernier scale is equal to 49 divisions on main scale. The estimated uncertainty in the measurement of refractive index of the slab is $\frac{x}{10} \times 10^{-3}$, where x is ______.

[29 Jan. 2023 (Evening)]

70. In a screw gauge, there are 100 divisions on the circular scale and the main scale moves by 0.5 mm on a complete rotation of the circular scale. The zero of circular scale lies 6 divisions below the line of graduation when two studs are brought in contact with each other. When a wire is placed between the studs, 4 linear scale divisions are clearly visible while 46th division the circular scale coincide with the reference line. The diameter of the wire is _____ $\times 10^{-2}$ mm.

[30 Jan. 2023 (Morning)]

71. Two resistances are given as $R_1 = (10 \pm 0.5)\ \Omega$ and $R_2 = (15 \pm 0.5)\ \Omega$. The percentage error in the measurement of equivalent resistance when they are connected in parallel is :

[6 April 2023 (Morning)]

(A) 6.33 (B) 2.33

(C) 4.33 (D) 5.33

72. The length of a metallic wire is increased by 20% and its area of cross section is reduced by 4%. The percentage change in resistance of the metallic wire is _______.

[6 April 2023 (Morning)]

73. A 2 meter long scale with least count of 0.2 cm is used to measure the locations of objects on an optical bench. While measuring the focal length of a convex lens, the object pin and the convex lens are placed at 80 cm mark and 1 m mark, respectively. The image of the object pin on the other side of lens coincides with image pin that is kept at 180 cm mark. The % error in the estimation of focal length is:

[6 April 2023 (Evening)]

(A) 1.02 (B) 0.85

(C) 1.70 (D) 0.51

74. A cylindrical wire of mass $(0.4 \pm 0.01)g$ has length (8 ± 0.04) cm and radius (6 ± 0.03) mm. The maximum error in its density will be : **[8 April 2023 (Morning)]**

(A) 1% (B) 3.5%

(C) 4% (D) 5%

75. A physical quantity P is given as

$$P = \frac{a^2 b^3}{c\sqrt{d}}$$

The percentage error in the measurement of a, b, c and d are 1%, 2%, 3% and 4% respectively. The percentage error in the measurement of quantity P will be: **[10 April 2023 (Morning)]**

(A) 13% (B) 14%

(C) 12% (D) 16%

76. In an experiment with Vernier callipers of least count 0.1 mm, when two jaws are joined together the zero of Vernier scale lies right to the zero of the main scale and 6th division of. Vernier scale coincides with the main scale division. While measuring the diameter of a spherical bob, the zero of vernier scale lies in between 3.2 cm and 3.3 cm marks, and 4th division of vernier scale coincides with the main scale division. The diameter of bob is measured as: **[10 April 2023 (Evening)]**

(A) 3.18 cm (B) 3.25 cm

(C) 3.26 cm (D) 3.22 cm

* * * * *

ANSWERS KEY

1.1 Measurements and Dimensional Analysis

1. (A) 2. (B) 3. (C) 4. (A) 5. (D)
6. (B) 7. (C) 8. (A) 9. (B) 10. (D)
11. (C) 12. (C) 13. (B) 14. (D) 15. (C)
16. (B) 17. (A, D) 18. (A) 19. (B) 20. (D)
21. (D) 22. (B) 23. (B) 24. (C) 25. (B)
26. (C) 27. (C) 28. (A) 29. (B) 30. (D)
31. (C) 32. (B) 33. (B) 34. (A) 35. (C)
36. (D) 37. (D) 38. (D) 39. (25) 40. (D)
41. (A) 42. (A) 43. (B) 44. (D) 45. (D)
46. (B) 47. (A) 48. (C) 49. (C) 50. (A)
51. (D) 52. (B) 53. (D) 54. (52) 55. (D)
56. (A) 57. (D) 58. (C) 59. (C) 60. (A)
61. (D) 62. (A) 63. (D) 64. (C) 65. (D)
66. (D) 67. (B) 68. (A) 69. (B) 70. (C)
71. (B) 72. (B) 73. (D) 74. (B) 75. (D)
76. (C) 77. (C) 78. (B) 79. (B) 80. (A)
81. (A)

2.1 Motion in One Dimension

1. (A) 2. (B) 3. (D) 4. (C) 5. (C)
6. (D) 7. (C) 8. (C) 9. (A) 10. (C)
11. (C) 12. (B) 13. (B) 14. (D) 15. (A)
16. (B) 17. (C) 18. (C) 19. (C) 20. (A)
21. (C) 22. (B) 23. (B) 24. (C) 25. (A)
26. (D) 27. (B) 28. (A) 29. (D) 30. (C)
31. (C) 32. (A) 33. (D) 34. (B) 35. (8)
36. (3) 37. (20) 38. (A) 39. (C) 40. (D)
41. (C) 42. (A) 43. (B) 44. (B) 45. (A)
46. (12) 47. (C) 48. (B) 49. (B) 50. (C)
51. (A) 52. (C) 53. (B) 54. (A) 55. (C)
56. (50) 57. (12) 58. (D) 59. (1) 60. (3)
61. (C) 62. (B) 63. (A) 64. (5) 65. (6)
66. (B) 67. (A) 68. (18) 69. (D) 70. (D)
71. (C) 72. (3) 73. (100) 74. (A) 75. (A)
76. (B) 77. (392) 78. (B) 79. (B) 80. (D)
81. (D) 82. (D) 83. (D) 84. (A) 85. (A)
86. (50) 87. (D) 88. (C) 89. (200) 90. (D)
91. (A) 92. (A) 93. (C) 94. (A) 95. (C)
96. (C) 97. (B) 98. (B) 99. (C) 100. (D)
101. (D) 102. (C)

3.1 Motion in Two Dimensions

1. (C) 2. (D) 3. (B) 4. (A) 5. (C)
6. (A) 7. (C) 8. (A) 9. (D) 10. (A)
11. (D) 12. (A) 13. (B) 14. (B) 15. (C)
16. (C) 17. (A) 18. (A) 19. (B) 20. (C)
21. (D) 22. (B) 23. (D) 24. (C) 25. (C)
26. (D) 27. (B) 28. (A) 29. (C) 30. (B)
31. (580) 32. (A) 33. (C) 34. (C) 35. (B)
36. (D) 37. (180) 38. (C) 39. (82) 40. (None)
41. (5) 42. (5) 43. (B) 44. (C) 45. (C)
46. (3) 47. (B) 48. (B) 49. (A) 50. (B)
51. (30) 52. (B) 53. (C) 54. (B) 55. (A)
56. (A) 57. (D) 58. (30) 59. (B) 60. (B)
61. (B) 62. (D) 63. (20) 64. (60) 65. (B)
66. (C) 67. (B) 68. (D) 69. (C) 70. (5)
71. (C) 72. (2) 73. (1) 74. (5) 75. (C)
76. (15) 77. (C) 78. (1) 79. (D) 80. (4)
81. (D) 82. (D) 83. (D) 84. (80) 85. (B)
86. (A) 87. (B) 88. (D) 89. (A) 90. (C)
91. (80) 92. (A) 93. (Bonus) 94. (A) 95. (D)

3.2 Relative Motion

1. (B) 2. (D) 3. (C) 4. (C) 5. (D)
6. (D) 7. (B) 8. (D) 9. (120) 10. (B)
11. (C) 12. (C) 13. (B) 14. (3)

4.1 Newton's Laws of Motion

1. (A) 2. (B) 3. (C) 4. (D) 5. (C)
6. (B) 7. (B) 8. (A) 9. (A) 10. (B)
11. (D) 12. (D) 13. (A) 14. (D) 15. (D)
16. (C) 17. (D) 18. (C) 19. (A) 20. (D)
21. (A) 22. (D) 23. (D) 24. (A) 25. (D)
26. (A) 27. (B) 28. (B) 29. (A) 30. (A)
31. (90°) 32. (B) 33. (D) 34. (B) 35. (492)
36. (500) 37. (B) 38. (30) 39. (10) 40. (A)
41. (C) 42. (D) 43. (D) 44. (D) 45. (D)
46. (D) 47. (B) 48. (2) 49. (12) 50. (B)
51. (A) 52. (36) 53. (3) 54. (6) 55. (C)
56. (A) 57. (C) 58. (C) 59. (B) 60. (C)
61. (2) 62. (D) 63. (B) 64. (B) 65. (B)
66. (B) 67. (C) 68. (A) 69. (D) 70. (A)

4.2 Friction

1. (D) 2. (C) 3. (A) 4. (A) 5. (C)
6. (B) 7. (D) 8. (A) 9. (A) 10. (D)
11. (B) 12. (A) 13. (C) 14. (B) 15. (B)
16. (D) 17. (A) 18. (B) 19. (A) 20. (B)
21. (D) 22. (25) 23. (25) 24. (10) 25. (B)
26. (21) 27. (5) 28. (C) 29. (3) 30. (15)
31. (3) 32. (A) 33. (B) 34. (D) 35. (D)
36. (C) 37. (B)

5.1 Work Energy Power

1. (B) 2. (B) 3. (B) 4. (A) 5. (B)
6. (B) 7. (B) 8. (D) 9. (A) 10. (B)
11. (B) 12. (A) 13. (D) 14. (A) 15. (D)
16. (D) 17. (D) 18. (B) 19. (D) 20. (A)
21. (A) 22. (B) 23. (C) 24. (D) 25. (D)
26. (A) 27. (B) 28. (C) 29. (D) 30. (C)
31. (C) 32. (C) 33. (A) 34. (C) 35. (B)
36. (A) 37. (C) 38. (B) 39. (B) 40. (A)
41. (A) 42. (A) 43. (B) 44. (B) 45. (C)
46. (C) 47. (10) 48. (C) 49. (B) 20. (3)
51. (150) 52. (C) 53. (346) 54. (B) 55. (18)
56. (C) 57. (1) 58. (B) 59. (A) 60. (3)
61. (B) 62. (10) 63. (6) 64. (B) 65. (C)
66. (D) 67. (450) 68. (A) 69. (C) 70. (40)
71. (2) 72. (1) 73. (A) 74. (A) 75. (16)
76. (C) 77. (120) 78. (C) 79. (2) 80. (600)
81. (A) 82. (A) 83. (C) 84. (C) 85. (B)
86. (D) 87. (D) 88. (B) 89. (B) 90. (B)
91. (D) 92. (24) 93. (C) 94. (40) 95. (100)
96. (2) 97. (300) 98. (A) 99. (7) 100. (40)
101. (132) 102. (67) 103. (2) 104. (48) 105. (32)
106. (300) 107. (A) 108. (784) 109. (D) 110. (A)
111. (D) 112. (3) 113. (30)

5.2 Circular Motion

1. (B) 2. (C) 3. (B) 4. (C) 5. (D)
6. (C) 7. (A) 8. (C) 9. (C) 10. (B)
11. (B) 12. (A) 13. (D) 14. (D) 15. (B)
16. (D) 17. (D) 18. (B) 19. (D) 20. (B)
21. (A) 22. (A) 23. (5) 24. (A) 25. (D)
26. (D) 27. (728) 28. (D) 29. (200) 30. (2)
31. (02) 32. (D) 33. (B) 34. (C) 35. (B)
36. (B) 37. (C) 38. (24) 39. (C) 40. (B)
41. (C) 42. (A) 43. (5) 44. (B) 45. (A)
46. (C) 47. (C) 48. (C) 49. (B) 50. (40)
51. (4) 52. (125) 53. (D) 54. (A) 55. (B)
56. (D) 57. (D) 58. (D) 59. (A)

6.1 System of Particles

1. (C) 2. (C) 3. (C) 4. (B) 5. (C)
6. (C) 7. (B) 8. (C) 9. (C) 10. (A)
11. (A) 12. (B) 13. (B) 14. (D) 15. (A)
16. (C) 17. (D) 18. (C) 19. (D) 20. (C)
21. (D) 22. (D) 23. (D) 24. (B) 25. (C)
26. (B) 27. (B) 28. (D) 29. (B) 30. (B)
31. (C) 32. (A) 33. (A) 34. (B) 35. (A)
36. (A) 37. (D) 38. (A) 39. (C) 40. (C)
41. (A) 42. (A) 43. (A) 44. (D) 45. (C)
46. (B) 47. (B) 48. (C) 49. (C) 50. (D)
51. (B) 52. (A) 53. (C) 54. (B) 55. (D)
56. (1) 57. (C) 58. (D) 59. (B) 60. (D)
61. (C) 62. (D) 63. (10) 64. (23) 65. (A)
66. (C) 67. (D) 68. (120) 69. (D) 70. (3)
71. (1) 72. (A) 73. (2) 74. (1) 75. (B)
76. (C) 77. (30) 78. (C) 79. (D) 80. (4)
81. (20) 82. (A) 83. (C) 84. (2) 85. (A)
86. (400) 87. (D) 88. (6) 89. (12) 90. (C)
91. (C) 92. (10) 93. (B) 94. (6) 95. (B)
96. (C) 97. (B) 98. (B) 99. (2) 100. (B)
101. (D) 102. (C) 103. (B) 104. (B) 105. (A)
106. (8) 107. (A) 108. (4) 109. (120) 110. (C)
111. (D) 112. (B) 113. (5) 114. (B) 115. (375)
116. (125) 117. (D) 118. (30) 119. (C) 120. (–245)
121. (B) 122. (B)

6.2 Rigid Body Dynamics

1. (C) 2. (A) 3. (B) 4. (D) 5. (A)
6. (D) 7. (B) 8. (C) 9. (D) 10. (D)
11. (C) 12. (B) 13. (D) 14. (C) 15. (A)
16. (A) 17. (C) 18. (D) 19. (C) 20. (D)
21. (B) 22. (B) 23. (D) 24. (D) 25. (B)
26. (B) 27. (D) 28. (D) 29. (B) 30. (B)
31. (C) 32. (B) 33. (A) 34. (A) 35. (C)
36. (B) 37. (C) 38. (D) 39. (D) 40. (B)
41. (A) 42. (C) 43 (C) 44. (A) 45. (D)
46. (A, C) 47. (C) 48. (B) 49. (D) 50. (D)
51. (B) 52. (B) 53. (B) 54. (C) 55. (C)
56. (D) 57. (A) 58. (C) 59. (A) 60. (B)
61. (C) 62. (D) 63. (B) 64. (A) 65. (A)
66. (C) 67. (C) 68. (C) 69. (D) 70. (D)
71. (B) 72. (B) 73. (A) 74. (None) 75. (B)
76. (B) 77. (D) 78. (A) 79. (C) 80. (C)
81. (D) 82. (A) 83. (D) 84. (C) 85. (D)
86. (B) 87. (B) 88. (D) 89. (D) 90. (D)
91. (D) 92. (C) 93. (B) 94. (A) 95. (75)
96. (B) 97. (A) 98. (A) 99. (15) 100. (B)
101. (D) 102. (D) 103. (D) 104. (C) 105. (B)
106. (9) 107. (B) 108. (25) 109. (11) 110. (20)
111. (B) 112. (D) 113. (C) 114. (195) 115. (20)
116. (A) 117. (B) 118. (D) 119. (C) 120. (8)
121. (None) 122. (C) 123. (D) 124. (C) 125. (C)
126. (20) 127. (20) 128. (3) 129. (A) 130. (D)
131. (4) 132. (C) 133. (C) 134. (C) 135. (D)
136. (C) 137. (4) 138. (2) 139. (C) 140. (3)
141. (A) 142. (2) 143. (C) 144. (B) 145. (A)
146. (25) 147. (10) 148. (4) 149. (A) 150. (D)
151. (9) 152. (52) 153. (D) 154. (A) 155. (C)
156. (C) 157. (B) 158. (6) 159. (B) 160. (5)
161. (C) 162. (B) 163. (3) 164. (A) 165. (91)
166. (B) 167. (C) 168. (10) 169. (C) 170. (D)

171. (8) **172.** (D) **173.** (5) **174.** (5) **175.** (18)
176. (120) **177.** (3) **178.** (30) **179.** (C) **180.** (110)
181. (32) **182.** (C) **183.** (17) **184.** (5) **185.** (40)
186. (800) **187.** (3) **188.** (54) **189.** (C) **190.** (10)
191. (5) **192.** (2) **193.** (3) **194.** (176) **195.** (5)
196. (1) **197.** (75) **198.** (B) **199.** (35) **200.** (A)
201. (2) **202.** (A) **203.** (4) **204.** (15) **205.** (5)

7.1 Gravitational Field

1. (C) **2.** (B) **3.** (D) **4.** (C) **5.** (A)
6. (A) **7.** (B) **8.** (C) **9.** (B) **10.** (A)
11. (A) **12.** (B) **13.** (C) **14.** (A) **15.** (D)
16. (A) **17.** (D) **18.** (A) **19.** (B) **20.** (B)
21. (C) **22.** (C) **23.** (C) **24.** (B) **25.** (A)
26. (D) **27.** (B) **28.** (B) **29.** (C) **30.** (B)
31. (B) **32.** (D) **33.** (D) **34.** (D) **35.** (A)
36. (B) **37.** (A) **38.** (D) **39.** (D) **40.** (D)
41. (B) **42.** (C) **43.** (4) **44.** (3) **45.** (A)
46. (A) **47.** (D) **48.** (D) **49.** (4) **50.** (C)
51. (D) **52.** (C) **53.** (2) **54.** (C) **55.** (A)
56. (C) **57.** (B) **58.** (D) **59.** (A) **60.** (6)
61. (D) **62.** (A) **63.** (A) **64.** (C) **65.** (B)
66. (A) **67.** (A) **68.** (A) **69.** (D) **70.** (C)
71. (2) **72.** (A) **73.** (D) **74.** (C) **75.** (C)
76. (B) **77.** (C) **78.** (D) **79.** (B) **80.** (B)
81. (B) **82.** (B) **83** (C) **84.** (C) **85.** (B)
86. (D) **87.** (A) **88.** (A) **89.** (D) **90.** (16)
91. (B) **92.** (C) **93.** (C) **94.** (C) **95.** (A)

7.2 Satellite Motion

1. (B) **2.** (A) **3.** (A) **4.** (C) **5.** (C)
6. (A) **7.** (A) **8.** (D) **9.** (C) **10.** (B)
11. (C) **12.** (A) **13.** (D) **14.** (C) **15.** (C)
16. (C) **17.** (C) **18.** (C) **19.** (B) **20.** (D)
21. (A) **22.** (C) **23.** (C) **24.** (D) **25.** (B)
26. (B) **27.** (B) **28.** (16) **29.** (A) **30.** (A)
31. (C) **32.** (C) **33.** (C) **34.** (A) **35.** (B)
36. (B) **37.** (D) **38.** (C) **39.** (C) **40.** (C)
41. (10) **42.** (B) **43.** (A) **44.** (64) **45.** (C)
46. (B) **47.** (C) **48.** (B) **49.** (3) **50.** (D)
51. (2) **52.** (C) **53.** (C) **54.** (A) **55.** (B)
56. (B) **57.** (B) **58.** (C) **59.** (A) **60.** (D)
61. (A) **62.** (D) **63.** (D) **64.** (D) **65.** (B)
66. (D) **67.** (C) **68.** (C) **69.** (C) **70.** (D)
71. (B) **72.** (B) **73.** (B)

8.1 Elasticity

1. (D) **2.** (D) **3.** (B) **4.** (A) **5.** (B)
6. (A) **7.** (A) **8.** (C) **9.** (B) **10.** (C)
11. (C) **12.** (D) **13.** (C) **14.** (A) **15.** (C)
16. (C) **17.** (D) **18.** (B) **19.** (C) **20.** (D)
21. (B) **22.** (A) **23.** (None) **24.** (None) **25.** (B)
26. (4) **27.** (D) **28.** (35) **29.** (B) **30.** (A)
31. (A) **32.** (2) **33.** (A) **34.** (D) **35.** (D)
36. (D) **37.** (B) **38.** (32) **39.** (A) **40.** (C)
41. (C) **42.** (5) **43.** (D) **44.** (A) **45.** (20)
46. (40) **47.** (B) **48.** (500) **49.** (D) **50.** (5)
51. (B) **52.** (C) **53.** (25) **54.** (C) **55.** (C)
56. (C) **57.** (5) **58.** (2) **59.** (A) **60.** (25)
61. (48) **62.** (D) **63.** (2) **64.** (C) **65.** (A)
66. (50) **67.** (D) **68.** (C) **69.** (5) **70.** (A)
71. (D) **72.** (60) **73.** (D) **74.** (B) **75.** (C)
76. (1) **77.** (C) **78.** (25) **79.** (11) **80.** (A)
81. (4) **82.** (20) **83.** (B) **84.** (2) **85.** (B)
86. (40) **87.** (B)

9.1 Fluid Statics

1. (D) **2.** (C) **3.** (B) **4.** (C) **5.** (C)
6. (C) **7.** (B) **8.** (D) **9.** (A) **10.** (C)
11. (C) **12.** (B) **13.** (25600) **14.** (A) **15.** (A)
16. (B) **17.** (D) **18.** (B) **19.** (4) **20.** (C)
21. (C) **22.** (B)

9.2 Fluid Dynamics

1. (B) **2.** (B) **3.** (A) **4.** (A) **5.** (B)
6. (C) **7.** (B) **8.** (B) **9.** (B) **10.** (A)
11. (A) **12.** (B) **13.** (D) **14.** (B) **15.** (C)
16. (B) **17.** (D) **18.** (C) **19.** (B) **20.** (3)
21. (C) **22.** (6) **23.** (363) **24.** (300) **25.** (24)
26. (D) **27.** (C) **28.** (D) **29.** (2) **30.** (80)
31. (4) **32.** (A)

9.3 Viscosity

1. (B) **2.** (C) **3.** (A) **4.** (B) **5.** (A)
6. (C) **7.** (B) **8.** (C) **9.** (D) **10.** (C)
11. (D) **12.** (C) **13.** (D) **14.** (B) **15.** (D)
16. (C) **17.** (100) **18.** (B) **19.** (D) **20.** (20)
21. (11) **22.** (25) **23.** (7) **24.** (B) **25.** (25)
26. (A) **27.** (10) **28.** (B) **29.** (D)

9.4 Surface Tension

1. (C) **2.** (D) **3.** (C) **4.** (D) **5.** (B)
6. (A) **7.** (D) **8.** (C) **9.** (B) **10.** (C)
11. (A) **12.** (A) **13.** (B) **14.** (D) **15.** (D)
16. (C) **17.** (C) **18.** (B) **19.** (C) **20.** (A)
21. (101) **22.** (A) **23.** (B) **24.** (C) **25.** (2)
26. (A) **27.** (A) **28.** (C) **29.** (2) **30.** (A)
31. (1) **32.** (C) **33.** (C) **34.** (B) **35.** (A)
36. (264) **37.** (1150)

10.1 Thermal Expansion

1. (C)	**2.** (A)	**3.** (A)	**4.** (C)	**5.** (B)
6. (A)	**7.** (A)	**8.** (C)	**9.** (C)	**10.** (B)
11. (C)	**12.** (B)	**13.** (B)	**14.** (60)	**15.** (A)
16. (B)	**17.** (20)	**18.** (D)	**19.** (A)	**20.** (D)
21. (D)	**22.** (C)	**23.** (8)	**24.** (B)	**25.** (D)
26. (60)	**27.** (12)	**28.** (B)	**29.** (313)	**30.** (D)

10.2 Calorimetry

1. (B)	**2.** (C)	**3.** (C)	**4.** (B)	**5.** (A)
6. (B)	**7.** (D)	**8.** (D)	**9.** (D)	**10.** (B)
11. (A)	**12.** (B)	**13.** (C)	**14.** (C)	**15.** (40)
16. (50)	**17.** (B)	**18.** (A)	**19.** (B)	**20.** (C)
21. (C)	**22.** (C)	**23.** (D)	**24.** (C)	**25.** (31)
26. (C)	**27.** (42)	**28.** (90)	**29.** (A)	**30.** (300)

10.3 Heat Transfer

1. (A)	**2.** (B)	**3.** (A)	**4.** (D)	**5.** (D)
6. (D)	**7.** (D)	**8.** (C)	**9.** (B)	**10.** (C)
11. (D)	**12.** (B)	**13.** (A)	**14.** (A)	**15.** (C)
16. (D)	**17.** (C)	**18.** (C)	**19.** (B)	**20.** (B)
21. (D)	**22.** (A)	**23.** (D)	**24.** (A)	**25.** (B)
26. (A)	**27.** (D)	**28.** (A)	**29.** (A)	**30.** (A)
31. (57)	**32.** (C)	**33.** (2)	**34.** (D)	**35.** (B)
36. (16)	**37.** (21)	**38.** (B)	**39.** (C)	**40.** (A)
41. (B)	**42.** (28)	**43.** (C)	**44.** (C)	**45.** (40)

11.1 Laws of Thermodynamics

1. (A)	**2.** (C)	**3.** (B)	**4.** (D)	**5.** (C)
6. (A)	**7.** (B)	**8.** (A)	**9.** (C)	**10.** (D)
11. (A)	**12.** (A)	**13.** (A)	**14.** (A)	**15.** (D)
16. (D)	**17.** (C)	**18.** (D)	**19.** (A)	**20.** (C)
21. (C)	**22.** (B)	**23.** (D)	**24.** (B)	**25.** (D)
26. (B)	**27.** (D)	**28.** (D)	**29.** (D)	**30.** (C)
31. (D)	**32.** (C)	**33.** (B)	**34.** (C)	**35.** (C)
36. (D)	**37.** (A)	**38.** (B)	**39.** (C)	**40.** (A)
41. (B)	**42.** (C)	**43.** (A)	**44.** (D)	**45.** (B)
46. (A)	**47.** (B)	**48.** (D)	**49.** (A)	**50.** (D)
51. (B)	**52.** (D)	**53.** (A)	**54.** (C)	**55.** (A)
56. (B)	**57.** (B)	**58.** (A)	**59.** (C)	**60.** (D)
61. (B)	**62.** (A)	**63.** (C)	**64.** (A)	**65.** (C)
66. (D)	**67.** (B)	**68.** (A)	**69.** (A)	**70.** (B)
71. (600)	**72.** (B)	**73.** (C)	**74.** (A)	**75.** (D)
76. (C)	**77.** (B)	**78.** (B)	**79.** (D)	**80.** (1818 to 1819)
81. (B)	**82.** (4)	**83.** (C)	**84.** (B)	**85.** (D)
86. (8791)	**87.** (266)	**88.** (A)	**89.** (150)	**90.** (A)
91. (C)	**92.** (D)	**93.** (D)	**94.** (19)	**95.** (D)
96. (A)	**97.** (D)	**98.** (B)	**99.** (3600)	**100.** (50)
101. (B)	**102.** (208)	**103.** (26)	**104.** (B)	**105.** (25)
106. (60)	**107.** (A)	**108.** (–113)	**109.** (A)	**110.** (D)
111. (B)	**112.** (A)	**113.** (B)	**114.** (100)	**115.** (17258)
116. (C)	**117.** (D)	**118.** (A)	**119.** (D)	**120.** (C)
121. (D)	**122.** (C)	**123.** (A)	**124.** (C)	**125.** (500)
126. (C)	**127.** (480)	**128.** (25)	**129.** (B)	**130.** (C)
131. (2)	**132.** (B)	**133.** (A)	**134.** (540)	**135.** (16)
136. (1400)	**137.** (A)	**138.** (7479)	**139.** (B)	**140.** (102)
141. (D)	**142.** (B)	**143.** (B)	**144.** (C)	**145.** (A)
146. (C)	**147.** (C)	**148.** (750)	**149.** (4)	**150.** (3)
151. (B)	**152.** (A)	**153.** (C)	**154.** (D)	**155.** (B)
156. (B)	**157.** (C)	**158.** (D)	**159.** (D)	**160.** (D)
161. (B)	**162.** (A)	**163.** (D)	**164.** (D)	**165.** (D)
166. (D)	**167.** (C)	**168.** (D)	**169.** (D)	**170.** (A)
171. (B)	**172.** (C)	**173.** (B)	**174.** (B)	

12.1 Kinetic Theory of Gases

1. (C)	**2.** (D)	**3.** (C)	**4.** (C)	**5.** (A)
6. (B)	**7.** (C)	**8.** (A)	**9.** (A)	**10.** (D)
11. (D)	**12.** (A)	**13.** (B)	**14.** (C)	**15.** (A)
16. (A)	**17.** (B)	**18.** (A)	**19.** (D)	**20.** (D)
21. (C)	**22.** (A)	**23.** (B)	**24.** (D)	**25.** (A)
26. (C)	**27.** (D)	**28.** (D)	**29.** (C)	**30.** (D)
31. (B)	**32.** (A)	**33.** (None)	**34.** (C)	**35.** (C)
36. (B)	**37.** (D)	**38.** (C)	**39.** (C)	**40.** (C)
41. (D)	**42.** (C)	**43.** (D)	**44.** (C)	**45.** (A)
46. (41)	**47.** (5)	**48.** (C)	**49.** (D)	**50.** (400)
51. (C)	**52.** (C)	**53.** (D)	**54.** (B)	**55.** (B)
56. (A)	**57.** (D)	**58.** (C)	**59.** (D)	**60.** (D)
61. (B)	**62.** (B)	**63.** (A)	**64.** (25)	**65.** (D)
66. (A)	**67.** (C)	**68.** (D)	**69.** (C)	**70.** (B)
71. (B)	**72.** (A)	**73.** (500)	**74.** (12)	**75.** (B)
76. (250)	**77.** (B)	**78.** (D)	**79.** (B)	**80.** (A)
81. (B)	**82.** (B)	**83.** (C)	**84.** (C)	**85.** (C)
86. (A)	**87.** (B)	**88.** (B)	**89.** (C)	**90.** (B)
91. (D)	**92.** (C)	**93.** (C)	**94.** (B)	**95.** (C)
96. (D)	**97.** (C)	**98.** (B)	**99.** (D)	**100.** (B)
101. (B)	**102.** (A)	**103.** (A)	**104.** (B)	**105.** (A)
106. (D)	**107.** (A)	**108.** (C)	**109.** (C)	**110.** (D)
111. (B)	**112.** (A)	**113.** (C)	**114.** (A)	**115.** (B)
116. (A)				

13.1 Simple Harmonic Motion

1. (B)	**2.** (C)	**3.** (B)	**4.** (C)	**5.** (C)
6. (A)	**7.** (C)	**8.** (D)	**9.** (C)	**10.** (B)
11. (C)	**12.** (B)	**13.** (C)	**14.** (D)	**15.** (D)
16. (B)	**17.** (A)	**18.** (A)	**19.** (A)	**20.** (A)
21. (B)	**22.** (A)	**23.** (D)	**24.** (D)	**25.** (A)
26. (D)	**27.** (B)	**28.** (D)	**29.** (B)	**30.** (A)
31. (A)	**32.** (A)	**33.** (C)	**34.** (C)	**35.** (A)
36. (A)	**37.** (D)	**38.** (D)	**39.** (D)	**40.** (C)
41. (B)	**42.** (A)	**43.** (D)	**44.** (C)	**45.** (C)
46. (A)	**47.** (A)	**48.** (B)	**49.** (D)	**50.** (C)

51. (C) 52. (D) 53. (B) 54. (D) 55. (B)
56. (C) 57. (B) 58. (D) 59. (B) 60. (B)
61. (C) 62. (D) 63. (B) 64. (D) 65. (B)
66. (C) 67. (D) 68. (None) 69. (C) 70. (A)
71. (C) 72. (A) 73. (D) 74. (D) 75. (A)
76. (D) 77. (B) 78. (D) 79. (D) 80. (A)
81. (B) 82. (A) 83. (C) 84. (C) 85. (C)
86. (D) 87. (D) 88. (C) 89. (A) 90. (B)
91. (A) 92. (C) 93. (3) 94. (7) 95. (D)
96. (A) 97. (C) 98. (2) 99. (None) 100. (D)
101. (6) 102. (D) 103. (C, D) 104. (1) 105. (D)
106. (B) 107. (A) 108. (10) 109. (A) 110. (A)
111. (8) 112. (C) 113. (2) 114. (A) 115. (2)
116. (D) 117. (1) 118. (8) 119. (B) 120. (A)
121. (D) 122. (B) 123. (C) 124. (D) 125. (B)
126. (1) 127. (700) 128. (C) 129. (A) 130. (D)
131. 60) 132. (B) 133. (2) 134. (A) 135. (16)
136. (2) 137. (A) 138. (5) 139. (5) 140. (1)
141. (B) 142. (D) 143. (120) 144. (40) 145. (8)
146. (B) 147. (88) 148. (C) 149. (20) 150. (2)
151. (C) 152. (C) 153. (99) 154. (A) 155. (A)
156. (20) 157. (D) 158. (D) 159. (D) 160. (4)
161. (3) 162. (A)

14.1 Wave Motion

1. (A) 2. (D) 3. (A) 4. (A) 5. (A)
6. (B) 7. (D) 8. (C) 9. (B) 10. (B)
11. (D) 12. (A) 13. (C) 14. (A) 15. (B)
16. (A) 17. (C) 18. (C) 19. (C) 20. (C)
21. (B) 22. (B) 23. (D) 24. (B) 25. (D)
26. (2) 27. (1215) 28. (D) 29. (2) 30. (1)
31. (B) 32. (A) 33. (C) 34. (15) 35. (C)
36. (18) 37. (B) 38. (20) 39. (B) 40. (1152)
41. (80)

14.2 Stationary Waves and Beats

1. (C) 2. (B) 3. (B) 4. (B) 5. (C)
6. (C) 7. (A) 8. (B) 9. (B) 10. (A)
11. (B) 12. (D) 13. (B) 14. (D) 15. (B)
16. (B) 17. (D) 18. (C) 19. (B) 20. (D)
21. (A) 22. (B) 23. (A) 24. (D) 25. (A)
26. (D) 27. (A) 28. (D) 29. (B) 30. (A)
31. (A) 32. (D) 33. (B) 34. (D) 35. (D)
36. (C) 37. (106.06) 38. (D) 39. (C) 40. (C)
41. (B) 42. (B) 43. (4) 44. (7) 45. (34)
46. (10) 47. (A) 48. (5) 49. (80) 50. (152)
51. (D) 52. (50) 53. (104) 54. (3) 55. (900)
56. (60) 57. (1) 58. (500) 59. (90)

14.3 Doppler's Effect

1. (D) 2. (C) 3. (D) 4. (C) 5. (C)
6. (D) 7. (B) 8. (B) 9. (A) 10. (C)
11. (D) 12. (D) 13. (A) 14. (B) 15. (C)
16. (A) 17. (B) 18. (C) 19. (B) 20. (A)
21. (D) 22. (A) 23. (D) 24. (D) 25. (8)
26. (6) 27. (132) 28. (C) 29. (2025) 30. (1210)
31. (C) 32. (A) 33. (20) 34. (200) 35. (340)
36. (400) 37. (B) 38. (420) 39. (B) 40. (B)

15.1 Electric Force and Electric Field

1. (D) 2. (B) 3. (B) 4. (D) 5. (C)
6. (A) 7. (D) 8. (A) 9. (C) 10. (A)
11. (D) 12. (C) 13. (A) 14. (C) 15. (D)
16. (A) 17. (B) 18. (A) 19. (D) 20. (C)
21. (B) 22. (A) 23. (D) 24. (B) 25. (A)
26. (D) 27. (A) 28. (D) 29. (C) 30. (C)
31. (C) 32. (D) 33. (C) 34. (D) 35. (C)
36. (D) 37. (C) 38. (D) 39. (C) 40. (D)
41. (C) 42. (A) 43. (C) 44. (B) 45. (36)
46. (20) 47. (C) 48. (D) 49. (A) 50. (B)
51. (12) 52. (D) 53. (1) 54. (B) 55. (6000)
56. (B) 57. (C) 58. (D) 59. (A) 60. (D)
61. (B) 62. (A) 63. (A) 64. (C) 65. (B)
66. (D) 67. (A) 68. (B) 69. (D) 70. (A)
71. (17) 72. (1) 73. (B) 74. (B) 75. (C)
76. (B) 77. (A) 78. (2) 79. (A) 80. (24)
81. (B) 82. (2) 83. (8) 84. (300) 85. (6)
86. (B) 87. (5440)

15.2 Electric Flux and Gauss's Law

1. (B) 2. (A) 3. (B) 4. (B) 5. (C)
6. (D) 7. (A) 8. (A) 9. (C) 10. (B)
11. (D) 12. (−48.00) 13. (A) 14. (226) 15. (1)
16. (D) 17. (A) 18. (640) 19. (4) 20. (C)
21. (B) 22. (B) 23. (45) 24. (D) 25. (12)
26. (640) 27. (288)

16.1 Electric Potential and Dipole

1. (A) 2. (C) 3. (D) 4. (B) 5. (C)
6. (C) 7. (D) 8. (D) 9. (B) 10. (D)
11. (B) 12. (A) 13. (D) 14. (C) 15. (D)
16. (C) 17. (C) 18. (C) 19. (A) 20. (D)
21. (C) 22. (D, C) 23. (A) 24. (B) 25. (C)
26. (B) 27. (B) 28. (B) 29. (C) 30. (D)
31. (B) 32. (A) 33. (B) 34. (B) 35. (C)
36. (A) 37. (A) 38. (C) 39. (D) 40. (C)
41. (D) 42. (B) 43. (A) 44. (A) 45. (D)
46. (D) 47. (C) 48. (A) 49. (A) 50. (D)
51. (C) 52. (D) 53. (A) 54. (C) 55. (C)
56. (128) 57. (243) 58. (B) 59. (B) 60. (198)
70. (D) 72. (A) 73. (B) 74. (6) 75. (A)

76. (B) **77.** (B) **78.** (45) **79.** (C) **80.** (12)
81. (D) **82.** (C) **83.** (A) **84.** (A) **85.** (Bonus)
86. (18) **87.** (C) **88.** (5) **89.** (A) **90.** (A)
91. (160) **92.** (B) **93.** (C)

16.2 Capacitance

1. (A) **2.** (B) **3.** (D) **4.** (B) **5.** (A)
6. (D) **7.** (A) **8.** (D) **9.** (D) **10.** (C)
11. (C) **12.** (D) **13.** (D) **14.** (B) **15.** (C)
16. (C) **17.** (B) **18.** (A) **19.** (D) **20.** (D)
21. (C) **22.** (C) **23.** (B) **24.** (A) **25.** (C)
26. (D) **27.** (A) **28.** (B) **29.** (D) **30.** (C)
31. (A) **32.** (C) **33.** (D) **34.** (D) **35.** (C)
36. (None) **37.** (A) **38.** (C) **39.** (D) **40.** (D)
41. (A) **42.** (B) **43.** (C) **44.** (D) **45.** (D)
46. (D) **47.** (A) **48.** (C) **49.** (D) **50.** (A)
51. (A) **52.** (B) **53.** (A) **54.** (6) **55** (C)
56. (D) **57.** (–4.03) **58.** (B) **59.** (A) **60.** (A)
61. (D) **62.** (C) **63.** (B) **64.** (B) **65.** (None)
66. (C) **67.** (C) **68.** (2) **69.** (864) **70.** (16)
71. (161) **72.** (A) **73.** (D) **74.** (D) **75.** (B)
76. (B) **77.** (C) **78.** (C) **79.** (C) **80.** (D)
81. (4) **82.** (D) **83.** (A) **84.** (6) **85.** (A)
86. (A) **87.** (A) **88.** (C) **89.** (A) **90.** (125)
91. (23) **92.** (A) **93.** (100) **94.** (A) **95.** (8)
96. (A) **97.** (6) **98.** (A) **99.** (60) **100.** (C)
101. (240) **102.** (A) **103.** (A) **104.** (C) **105.** (105)
106. (C) **107.** (6) **108.** (225) **109.** (55) **110.** (D)
111. (A) **112.** (3) **113.** (A) **114.** (5) **115.** (6)
116. (C) **117.** (C) **118.** (3) **119.** (B) **120.** (C)
121. (4) **122.** (3) **123.** (5)

17.1 Current Electricity

1. (D) **2.** (C) **3.** (B) **4.** (A) **5.** (B)
6. (D) **7.** (D) **8.** (C) **9.** (A) **10.** (B)
11. (A) **12.** (B) **13.** (D) **14.** (C) **15.** (D)
16. (C) **17.** (A) **18.** (C) **19.** (C) **20.** (D)
21. (D) **22.** (B) **23.** (D) **24.** (C) **25.** (B)
26. (D) **27.** (D) **28.** (C) **29.** (B) **30.** (D)
31. (D) **32.** (5) **33.** (A) **34.** (3) **35.** (B)
36. (A) **37.** (B) **38.** (A) **39.** (01) **40.** (B)
41. (A) **42.** (300) **43.** (C) **44.** (48) **45.** (12)
46. (A) **47.** (144) **48.** (4) **49.** (A) **50.** (A)
51. (A) **52.** (136) **53.** (B) **54.** (10) **55.** (A)
56. (B) **57.** (A) **58.** (B) **59.** (2) **60.** (44)
61. (B) **62.** (C) **63.** (D) **64.** (A) **65.** (25)
66. (125) **67.** (100) **68.** (B) **69.** (3) **70.** (A)
71. (15) **72.** (B) **73.** (A)

17.2 Circuit Analysis

1. (B) **2.** (B) **3.** (B) **4.** (B) **5.** (D)
6. (C) **7.** (C) **8.** (D) **9.** (B) **10.** (C)
11. (C) **12.** (C) **13.** (C) **14.** (B) **15.** (A)
16. (B) **17.** (A) **18.** (B) **19.** (C) **20** (D)
21. (D) **22.** (A) **23.** (B) **24.** (C) **25.** (C)
26. (B) **27.** (30) **28.** (B) **29.** (D) **30.** (2)
31. (A) **32.** (C) **33.** (C) **34.** (2) **35.** (C)
36. (300) **37.** (4) **38.** (B) **39.** (C) **40.** (70)
41. (4) **42.** (B) **43.** (10) **44.** (A) **45.** (15)
46. (74) **47.** (45) **48.** (3) **49.** (D) **50.** (A)
51. (B) **52.** (20) **53.** (9) **54.** (1) **55.** (3)
56. (A) **57.** (D) **58.** (C) **59.** (B) **60.** (2)
61. (C) **62.** (3) **63.** (8) **64.** (10) **65.** (A)
66. (4) **67.** (D) **68.** (2) **69.** (D) **70.** (10)
71. (1) **72.** (2) **73.** (A) **74.** (2) **75.** (5)
76. (D) **77.** (A) **78.** (2) **79.** (A) **80.** (A)
81. (B) **82.** (1) **83.** (D) **84.** (5) **85.** (D)
86. (3)

17.3 Thermal Effects of Current & RC Circuits

1. (B) **2.** (B) **3.** (B) **4.** (A) **5.** (C)
6. (C) **7.** (D) **8.** (C) **9.** (A) **10.** (B)
11. (A) **12.** (B) **13.** (B) **14.** (D) **15.** (C)
16. (D) **17.** (D) **18.** (A) **19.** (D) **20.** (A)
21. (C) **22.** (C) **23.** (B) **24.** (A) **25.** (D)
26. (B) **27.** (C) **28.** (C) **29.** (B) **30.** (A)
31. (None) **32.** (A) **33.** (C) **34.** (A) **35.** (A)
36. (A) **37.** (B) **38.** (C) **39.** (C) **40.** (D)
41. (C) **42.** (A) **43.** (B) **44.** (D) **45.** (8)
46. (D) **47.** (D) **48.** (B) **49.** (C) **50.** (D)
51. (2500) **52.** (2) **53.** (50) **54.** (D) **55.** (B)
56. (100) **57.** (3840) **58.** (C) **59.** (C) **60.** (4)
61. (450) **62.** (15) **63.** (D) **64.** (C) **65.** (14)
66. (A) **67.** (B) **68.** (1) **69.** (B) **70.** (A)
71. (16) **72.** (A) **73.** (75)

17.4 Electrical Measuring Instruments

1. (C) **2.** (D) **3.** (D) **4.** (A) **5.** (D)
6. (B) **7.** (D) **8.** (B) **9.** (D) **10.** (C)
11. (C) **12.** (D) **13.** (A) **14.** (C) **15.** (D)
16. (D) **17.** (D) **18.** (C) **19.** (A) **20.** (A)
21. (D) **22.** (A) **23.** (D) **24.** (C) **25.** (D)
26. (D) **27.** (B) **28.** (C) **29.** (B) **30.** (B)
31. (D) **32.** (A) **33.** (D) **34.** (B) **35.** (B)
36. (C) **37.** (D) **38.** (A) **39.** (B) **40.** (D)
41. (A) **42.** (C) **43.** (C) **44.** (B) **45.** (A)
46. (A) **47.** (D) **48.** (D) **49.** (C) **50.** (12)
51. (D) **52.** (10) **53.** (A) **54.** (40) **55.** (A)
56. (A) **57.** (B) **58.** (B) **59.** (A) **60.** (A)
61. (1) **62.** (48) **63.** (C) **64.** (A) **65.** (B)
66. (B) **67.** (B) **68.** (25) **69.** (25) **70.** (B)
71. (D) **72.** (B) **73.** (8) **74.** (19) **75.** (10)

76. (8)	**77.** (18)	**78.** (20)	**79.** (20)	**80.** (2400)
81. (40)	**82.** (780)	**83.** (A)	**84.** (2)	**85.** (B)
86. (30)	**87.** (5)	**88.** (25)	**89.** (A)	**90.** (C)
91. (20)	**92.** (50)	**93.** (C)		

18.1 Magnetic Effect of Current

1. (A)	**2.** (A)	**3.** (B)	**4.** (B)	**5.** (C)
6. (A)	**7.** (B)	**8.** (D)	**9.** (D)	**10.** (C)
11. (D)	**12.** (D)	**13.** (A)	**14.** (B)	**15.** (B)
16. (B)	**17.** (A)	**18.** (A)	**19.** (A)	**20.** (D)
21. (A)	**22.** (B)	**23.** (A)	**24.** (C)	**25.** (D)
26. (B)	**27.** (A)	**28.** (A)	**29.** (C)	**30.** (B)
31. (D)	**32.** (A)	**33.** (C)	**34.** (A)	**35.** (C)
36. (C)	**37.** (A)	**38.** (C)	**39.** (A)	**40.** (B)
41. (B)	**42.** (A)	**43.** (B)	**44.** (D)	**45.** (B)
46. (D)	**47.** (D)	**48.** (C)	**49.** (B)	**50.** (A)
51. (5)	**52.** (D)	**53.** (B)	**54.** (B)	**55.** (B)
56. (C)	**57.** (3)	**58.** (C)	**59.** (B)	**60.** (C)
61. (B)	**62.** (68)	**63.** (C)	**64.** (D)	**65.** (C)
66. (D)	**67.** (C)	**68.** (C)	**69.** (C)	**70.** (D)
71. (B)	**72.** (C)	**73.** (628)	**74.** (2)	**75.** (8)
76. (2)	**77.** (40)			

18.2 Magnetic Force on Moving Charges

1. (C)	**2.** (A)	**3.** (B)	**4.** (A)	**5.** (A)
6. (A)	**7.** (B)	**8.** (C)	**9.** (D)	**10.** (A)
11. (B)	**12.** (B)	**13.** (A)	**14.** (A)	**15.** (D)
16. (B)	**17.** (C)	**18.** (A)	**19.** (A)	**20.** (C)
21. (A)	**22.** (B)	**23.** (D)	**24.** (C)	**25.** (C)
26. (B)	**27.** (B)	**28.** (A)	**29.** (C)	**30.** (C)
31. (B)	**32.** (A)	**33.** (C)	**34.** (A)	**35.** (D)
36. (C)	**37.** (D)	**38.** (B)	**39.** (C)	**40.** (B)
41. (A)	**42.** (C)	**43.** (A)	**44.** (A)	**45.** (A)
46. (None)	**47.** (D)	**48.** (A)	**49.** (A)	**50.** (B)
51. (B)	**52.** (B)	**53.** (A)	**54.** (C)	**55.** (B)
56. (None)	**57.** (B)	**58.** (A)	**59.** (C)	**60.** (D)
61. (B)	**62.** (B)	**63.** (B)	**64.** (B)	**65.** (A)
66. (C)	**67.** (A)	**68.** (C)	**69.** (20)	**70.** (A)
71. (A)	**72.** (D)	**73.** (B)	**74.** (A)	**75.** (C)
76. (D)	**77.** (A)	**78.** (B)	**79.** (D)	**80.** (B)
81. (C)	**82.** (543)	**83.** (C)	**84.** (D)	**85.** (A)
86. (D)	**87.** (C)	**88.** (C)	**89.** (2)	**90.** (10)
91. (C)	**92.** (C)	**93.** (C)	**94.** (B)	**95.** (B)
96. (B)	**97.** (A)	**98.** (B)	**99.** (A)	**100.** (C)
101. (11)	**102.** (A)	**103.** (A)	**104.** (9)	**105.** (B)
106. (C)	**107.** (C)	**108.** (D)	**109.** (B)	**110.** (A)
111. (D)	**112.** (144)	**113.** (40)	**114.** (C)	**115.** (D)
116. (A)	**117.** (B)	**118.** (A)	**119.** (2)	

19.1 Classical and Terrestrial Magnetism

1. (D)	**2.** (B)	**3.** (B)	**4.** (C)	**5.** (C)
6. (B)	**7.** (D)	**8.** (D)	**9.** (A)	**10.** (C)
11. (C)	**12.** (B)	**13.** (C)	**14.** (C)	**15.** (B)
16. (D)	**17.** (B)	**18.** (C)	**19.** (D)	**20.** (C)
21. (A)	**22.** (B)	**23.** (A)	**24.** (A)	**25.** (B)
26. (A)	**27.** (A)	**28.** (D)	**29.** (D)	**30.** (B)
31. (B)	**32.** (6)	**33.** (A)	**34.** (A)	**35.** (243)

19.2 Magnetic Properties of Materials

1. (A)	**2.** (B)	**3.** (B)	**4.** (B)	**5.** (C)
6. (B)	**7.** (B)	**8.** (D)	**9.** (B)	**10.** (B)
11. (C)	**12.** (C)	**13.** (D)	**14.** (B)	**15.** (D)
16. (A)	**17.** (B)	**18.** (B)	**19.** (D)	**20.** (A)
21. (D)	**22.** (D)	**23.** (1)	**24.** (3)	**25.** (3)
26. (250)	**27.** (D)	**28.** (C)	**29.** (A)	**30.** (C)
31. (A)	**32.** (B)	**33.** (B)	**34.** (C)	**35.** (A)

20.1 Electromagnetic Induction

1. (D)	**2.** (B)	**3.** (B)	**4.** (B)	**5.** (D)
6. (C)	**7.** (B)	**8.** (B)	**9.** (A)	**10.** (C)
11. (A)	**12.** (D)	**13.** (B)	**14.** (A)	**15.** (A)
16. (D)	**17.** (B)	**18.** (B)	**19.** (A)	**20.** (C)
21. (B)	**22.** (A)	**23.** (A)	**24.** (A)	**25.** (C)
26. (B)	**27.** (D)	**28.** (B)	**29.** (B)	**30.** (A)
31. (A)	**32.** (175)	**33.** (D)	**34.** (C)	**35.** (15)
36. (A)	**37.** (A)	**38.** (C)	**39.** (5)	**40.** (B)
41. (D)	**42.** (C)	**43.** (C)	**44.** (A)	**45.** (D)
46. (80)	**47.** (60)	**48.** (60)	**49.** (D)	**50.** (B)
51. (D)	**52.** (A)	**53.** (440)	**54.** (A)	**55.** (400)
56. (B)	**57.** (16)	**58.** (B)	**59.** (250)	**60.** (2)
61. (12)	**62.** (C)	**63.** (10)	**64.** (B)	**65.** (C)
66. (A)	**67.** (D)	**68.** (10)	**69.** (B)	**70.** (44)
71. (D)	**72.** (A)	**73.** (18)	**74.** (44)	**75.** (8)
76. (150)	**77.** (50)	**78.** (A)	**79.** (60)	**80.** (D)
81. (88)				

20.2 Self and Mutual Induction

1. (D)	**2.** (D)	**3.** (B)	**4.** (B)	**5.** (B)
6. (A)	**7.** (D)	**8.** (D)	**9.** (B)	**10.** (C)
11. (B)	**12.** (D)	**13.** (C)	**14.** (None)	**15.** (C)
16. (A)	**17.** (D)	**18.** (D)	**19.** (C)	**20.** (D)
21. (A)	**22.** (B)	**23.** (D)	**24.** (B)	**25.** (A)
26. (B)	**27.** (B)	**28.** (B)	**29.** (B)	**30.** (B)
31. (A)	**32.** (D)	**33.** (10)	**34.** (C)	**35.** (33)
36. (440)	**37.** (A)	**38.** (A)	**39.** (144)	**40.** (C)
41. (D)	**42.** (D)	**43.** (4)	**44.** (D)	**45.** (C)
46. (1)	**47.** (3)	**48.** (A)	**49.** (30)	**50.** (240)
51. (4)	**52.** (9)	**53.** (4)	**54.** (625)	**55.** (25)

21.1 Alternating Current

1. (B)	**2.** (B)	**3.** (A)	**4.** (C)	**5.** (A)

6. (C)	**7.** (D)	**8.** (C)	**9.** (B)	**10.** (A)
11. (C)	**12.** (D)	**13.** (C)	**14.** (B)	**15.** (A)
16. (D)	**17.** (C)	**18.** (D)	**19.** (B)	**20.** (D)
21. (B)	**22.** (D)	**23.** (D)	**24.** (C)	**25.** (B)
26. (A)	**27.** (C)	**28.** (D)	**29.** (D)	**30.** (B)
31. (A)	**32.** (A)	**33.** (B)	**34.** (C)	**35.** (D)
36. (None)	**37.** (C)	**38.** (D)	**39.** (B)	**40.** (A)
41. (A)	**42.** (A)	**43.** (C)	**44.** (400)	**45.** (2000)
46. (B)	**47.** (900)	**48.** (D)	**49.** (A)	**50.** (A)
51. (C)	**52.** (283)	**53.** (D)	**54.** (4)	**55.** (B)
56. (D)	**57.** (B)	**58.** (A)	**59.** (A)	**60.** (C)
61. (C)	**62.** (3)	**63.** (B)	**64.** (125)	**65.** (D)
66. (B)	**67.** (D)	**68.** (500)	**69.** (C)	**70.** (D)
71. (D)	**72.** (A)	**73.** (None)	**74.** (11)	**75.** (1)
76. (B)	**77.** (2)	**78.** (10)	**79.** (100)	**80.** (8)
81. (A)	**82.** (3872)	**83.** (B)	**84.** (A)	**85.** (1584)
86. (B)	**87.** (242)	**88.** (A)	**89.** (25)	**90.** (C)
91. (40)	**92.** (3)	**93.** (C)	**94.** (D)	**95.** (A)

22.1 Electromagnetic Wave

1. (A)	**2.** (B)	**3.** (B)	**4.** (A)	**5.** (C)
6. (B)	**7.** (A)	**8.** (D)	**9.** (B)	**10.** (C)
11. (C)	**12.** (C)	**13.** (B)	**14.** (D)	**15.** (B)
16. (C)	**17.** (C)	**18.** (D)	**19.** (A)	**20.** (C)
21. (C)	**22.** (B)	**23.** (C)	**24.** (C)	**25.** (C)
26. (C)	**27.** (D)	**28.** (B)	**29.** (B)	**30.** (C)
31. (B)	**32.** (B)	**33.** (C)	**34.** (D)	**35.** (D)
36. (B)	**37.** (C)	**38.** (C)	**39.** (A)	**40.** (C)
41. (B)	**42.** (C)	**43.** (A)	**44.** (B)	**45.** (A)
46. (B)	**47.** (C)	**48.** (C)	**49.** (B)	**50.** (B)
51. (D)	**52.** (B)	**53.** (194)	**54.** (B)	**55.** (15)
56. (667)	**57.** (2)	**58.** (137)	**59.** (A)	**60.** (A)
61. (3)	**62.** (6)	**63.** (C)	**64.** (C)	**65.** (C)
66. (A)	**67.** (C)	**68.** (B)	**69.** (A)	**70.** (C)
71. (A)	**72.** (2)	**73.** (500)	**74.** (D)	**75.** (C)
76. (B)	**77.** (B)	**78.** (A)	**79.** (C)	**80.** (B)
81. (A)	**82.** (C)	**83.** (B)	**84.** (43)	**85.** (D)
86. (A)	**87.** (C)	**88.** (D)	**89.** (C)	**90.** (B)
91. (B)	**92.** (B)	**93.** (A)	**94.** (A)	**95.** (D)
96. (A)	**97.** (B)	**98.** (C)	**99.** (5)	**100.** (C)
101. (D)	**102.** (A)	**103.** (D)	**104.** (C)	**105.** (B)
106. (C)	**107.** (B)	**108.** (A)	**109.** (C)	**110.** (B)
111. (B)				

23.1 Geometrical Optics I - Reflection of Light

1. (A)	**2.** (B)	**3.** (D)	**4.** (C)	**5.** (C)
6. (D)	**7.** (A)	**8.** (D)	**9.** (A)	**10.** (B)
11. (D)	**12.** (1)	**13.** (A)	**14.** (B)	**15.** (B)
16. (150)	**17.** (C)	**18.** (D)	**19.** (A)	**20.** (210)
21. (400)	**22.** (A)	**23.** (32)	**24.** (32)	**25.** (C)
26. (18)	**27.** (B)	**28.** (B)		

23.2 Geometrical Optics II - Refraction of Light

1. (A)	**2.** (B)	**3.** (B)	**4.** (B)	**5.** (D)
6. (C)	**7.** (C)	**8.** (C)	**9.** (B)	**10.** (D)
11. (None)	**12.** (D)	**13.** (A)	**14.** (A)	**15.** (B)
16. (D)	**17.** (C)	**18.** (C)	**19.** (D)	**20.** (B)
21. (A)	**22.** (B)	**23.** (D)	**24.** (B)	**25.** (A)
26. (90)	**27.** (158)	**28.** (5.00)	**29.** (D)	**30.** (B)
31. (D)	**32.** (30)	**33.** (B)	**34.** (D)	**35.** (A)
36. (60)	**37.** (C)	**38.** (C)	**39.** (C)	**40.** (50)
41. (B)	**42.** (D)	**43.** (A)	**44.** (B)	**45.** (12)
46. (D)	**47.** (C)	**48.** (9)	**49.** (D)	**50.** (C)
51. (A)	**52.** (D)	**53.** (A)	**54.** (A)	**55.** (15)
56. (C)	**57.** (52)	**58.** (D)	**59.** (A)	**60.** (A)
61. (B)	**62.** (50)	**63.** (30)	**64.** (B)	**65.** (4)
66. (98)	**67.** (B)	**68.** (B)	**69.** (3)	**70.** (30)

23.3 Geometrical Optics III - Thin Lenses

1. (A)	**2.** (A)	**3.** (B)	**4.** (C)	**5.** (D)
6. (A)	**7.** (C)	**8.** (D)	**9.** (B)	**10.** (D)
11. (A)	**12.** (D)	**13.** (A)	**14.** (A)	**15.** (A)
16. (B)	**17.** (A)	**18.** (B)	**19.** (C)	**20.** (D)
21. (A)	**22.** (C)	**23.** (B)	**24.** (B)	**25.** (B)
26. (D)	**27.** (D)	**28.** (D)	**29.** (C)	**30.** (60)
31. (476)	**32.** (D)	**33.** (C)	**34.** (A)	**35.** (10)
36. (A)	**37.** (225)	**38.** (B)	**39.** (10)	**40.** (10)
41. (A)	**42.** (27)	**43.** (4)	**44.** (B)	**45.** (15)
46. (120)	**47.** (54)	**48.** (30)	**49.** (34)	**50.** (100)
51. (5)				

23.4 Dispersion of Light

1. (A)	**2.** (B)	**3.** (A)	**4.** (D)	**5.** (A)
6. (B)	**7.** (A)	**8.** (12)	**9.** (D)	**10.** (600)
11. (5)	**12.** (A)	**13.** (B)		

23.5 Optical Instruments

1. (B)	**2.** (D)	**3.** (C)	**4.** (A)	**5.** (A)
6. (C)	**7.** (A)	**8.** (D)	**9.** (C)	**10.** (D)
11. (A)	**12.** (B)	**13.** (C)	**14.** (C)	**15.** (C)
16. (A)	**17.** (C)	**18.** (A)	**19.** (A)	**20.** (D)
21. (C)	**22.** (5)	**23.** (50)	**24.** (C)	**25.** (B)
26. (A)	**27.** (25)	**28.** (C)	**29.** (C)	**30.** (A)
31. (D)	**32.** (D)			

24.1 Interference of Light

1. (D)	**2.** (C)	**3.** (A)	**4.** (D)	**5.** (A)
6. (A)	**7.** (B)	**8.** (C)	**9.** (D)	**10.** (B)
11. (B)	**12.** (A)	**13.** (C)	**14.** (B)	**15.** (C)
16. (B)	**17.** (D)	**18.** (A)	**19.** (D)	**20.** (A)
21. (D)	**22.** (D)	**23.** (D)	**24.** (C)	**25.** (D)

26. (B) **27.** (B) **28.** (B) **29.** (D) **30.** (C)
31. (D) **32.** (A) **33.** (None) **34.** (B) **35.** (D)
36. (B) **37.** (D) **38.** (A) **39.** (750) **40.** (A)
41. (A) **42.** (D) **43.** (A) **44.** (B) **45.** (D)
46. (9) **47.** (B) **48.** (D) **49.** (A) **50.** (A)
51. (600) **52.** (A) **53.** (300) **54.** (C) **55.** (16)
56. (3) **57.** (D) **58.** (B) **59.** (D) **60.** (450)
61. (B) **62.** (4) **63.** (D) **64.** (600) **65.** (B)
66. (2) **67.** (B) **68.** (630) **69.** (24) **70.** (84)
71. (A) **72.** (C) **73.** (10) **74.** (3) **75.** (48)
76. (20) **77.** (4) **78.** (C) **79.** (462) **80.** (C)
81. (C) **82.** (3668) **83.** (D) **84.** (B)

24.2 Diffraction of Light

1. (B) **2.** (C) **3.** (C) **4.** (C) **5.** (C)
6. (B) **7.** (C) **8.** (C) **9.** (A) **10.** (B)
11. (B) **12.** (B) **13.** (D) **14.** (D) **15.** (198)
16. (50) **17.** (A) **18.** (D) **19.** (B)

24.3 Polarization of Light

1. (A) **2.** (D) **3.** (A) **4.** (C) **5.** (A)
6. (C) **7.** (B) **8.** (D) **9.** (C) **10.** (A)
11. (C) **12.** (A) **13.** (A) **14.** (A) **15.** (D)
16. (A) **17.** (C) **18.** (75) **19.** (30) **20.** (D)
21. (C) **22.** (B) **23.** (24) **24.** (7) **25.** (C)
26. (B) **27.** (30, 60)

25.1 Photoelectric Effect

1. (C) **2.** (A) **3.** (C) **4.** (D) **5.** (B)
6. (B) **7.** (A) **8.** (D) **9.** (A) **10.** (B)
11. (B) **12.** (C) **13.** (D) **14.** (A) **15.** (D)
16. (B) **17.** (B) **18.** (A) **19.** (None) **20.** (A)
21. (C) **22.** (A) **23.** (A) **24.** (A) **25.** (D)
26. (B) **27.** (None) **28.** (A) **29.** (A) **30.** (C)
31. (11) **32.** (C) **33.** (Bonus) **34.** (B) **35.** (9)
36. (A) **37.** (A) **38.** (C) **39.** (2) **40.** (A)
41. (1) **42.** (D) **43.** (25) **44.** (A) **45.** (A)
46. (4) **47.** (B) **48.** (125) **49.** (B) **50.** (C)
51. (1) **52.** (A) **53.** (C) **54.** (B) **55.** (2)
56. (B) **57.** (D) **58.** (A) **59.** (35) **60.** (B)
61. (B) **62.** (C) **63.** (B) **64.** (C) **65.** (B)
66. (C) **67.** (B) **68.** (D) **69.** (D) **70.** (D)
71. (A) **72.** (C) **73.** (A) **74.** (C) **75.** (D)
76. (B) **77.** (A)

25.2 Matter Waves and X-Rays

1. (A) **2.** (B) **3.** (C) **4.** (D) **5.** (B)
6. (B) **7.** (D) **8.** (D) **9.** (B) **10.** (C)
11. (D) **12.** (C) **13.** (C) **14.** (B) **15.** (C)
16. (A) **17.** (D) **18.** (C) **19.** (C) **20.** (C)
21. (D) **22.** (C) **23.** (A) **24.** (B) **25.** (D)
26. (D) **27.** (C) **28.** (D) **29.** (A) **30.** (B)
31. (B) **32.** (C) **33.** (C) **34.** (A) **35.** (D)
36. (A) **37.** (A) **38.** (D) **39.** (B) **40.** (C)
41. (C) **42.** (D) **43.** (C) **44.** (C) **45.** (10)
46. (D) **47.** (C) **48.** (D) **49.** (A) **50.** (B)
51. (D) **52.** (D) **53.** (D) **54.** (C) **55.** (D)
56. (C) **57.** (910) **58.** (10) **59.** (B) **60.** (C)
61. (A) **62.** (A) **63.** (B) **64.** (B) **65.** (B)
66. (A) **67.** (D) **68.** (B) **69.** (C) **70.** (D)
71. (D) **72.** (D) **73.** (B) **74.** (D) **75.** (C)
76. (B) **77.** (4) **78.** (B) **79.** (D) **80.** (B)
81. (C) **82.** (D) **83.** (B) **84.** (A) **85.** (A)
86. (A) **87.** (C)

26.1 Atomic Structure

1. (A) **2.** (C) **3.** (C) **4.** (A) **5.** (C)
6. (B) **7.** (B) **8.** (A) **9.** (A) **10.** (A)
11. (B) **12.** (B) **13.** (C) **14.** (A) **15.** (C)
16. (C) **17.** (D) **18.** (C) **19.** (C) **20.** (A)
21. (C) **22.** (C) **23.** (B) **24.** (A) **25.** (A)
26. (B) **27.** (B) **28.** (B) **29.** (C) **30.** (D)
31. (A) **32.** (D) **33.** (B) **34.** (D) **35.** (C)
36. (D) **37.** (C) **38.** (B) **39.** (C) **40.** (B)
41. (C) **42.** (A) **43.** (D) **44.** (C) **45.** (C)
46. (C) **47.** (486) **48.** (D) **49.** (B) **50.** (10553)
51. (51) **52.** (A) **53.** (A) **54.** (A) **55.** (D)
56. (15) **57.** (B) **58.** (B) **59.** (D) **60.** (B)
61. (C) **62.** (112) **63.** (A) **64.** (2) **65.** (B)
66. (15) **67.** (C) **68.** (B) **69.** (D) **70.** (B)
71. (114) **72.** (D) **73.** (3) **74.** (A) **75.** (B)
76. (C) **77.** (B) **78.** (5) **79.** (5) **80.** (A)
81. (A) **82.** (27) **83.** (D) **84.** (D) **85.** (27)
86. (B) **87.** (136) **88.** (828) **89.** (D) **90.** (B)
91. (425) **92.** (A) **93.** (16) **94.** (27) **95.** (C)
96. (3) **97.** (C) **98.** (C) **99.** (2) **100.** (4125)
101. (5)

27.1 Radioactivity

1. (A) **2.** (A) **3.** (A) **4.** (A) **5.** (D)
6. (A) **7.** (C) **8.** (B) **9.** (A) **10.** (D)
11. (D) **12.** (D) **13.** (A) **14.** (C) **15.** (D)
16. (B) **17.** (B) **18.** (C) **19.** (B) **20.** (D)
21. (D) **22.** (D) **23.** (A) **24.** (C) **25.** (C)
26. (D) **27.** (B) **28.** (C) **29.** (A) **30.** (D)
31. (A) **32.** (C) **33.** (A) **34.** (A) **35.** (A)
36. (B) **37.** (B) **38.** (C) **39.** (C) **40.** (A)
41. (3) **42.** (A) **43.** (D) **44.** (10) **45.** (A)
46. (150) **47.** (A) **48.** (B) **49.** (D) **50.** (D)
51. (50) **52.** (B) **53.** (D) **54.** (C) **55.** (B)
56. (20) **57.** (C) **58.** (C) **59.** (D) **60.** (C)
61. (15) **62.** (A) **63.** (9) **64.** (A) **65.** (87)

66. (A) 67. (300) 68. (D) 69. (15) 70. (A)
71. (C) 72. (4) 73. (C)

27.2 Nuclear Physics

1. (D) 2. (A) 3. (B) 4. (C) 5. (C)
6. (C) 7. (B) 8. (D) 9. (C) 10. (C)
11. (B) 12. (B) 13. (B) 14. (B) 15. (B)
16. (D) 17. (C) 18. (C) 19. (C) 20. (C)
21. (C) 22. (D) 23. (C) 24. (C) 25. (C)
26. (A) 27. (C) 28. (C) 29. (A) 30. (B)
31. (C) 32. (A) 33. (B) 34. (A) 35. (B)
36. (B) 37. (C) 38. (27) 39. (D) 40. (D)
41. (D) 42. (A) 43. (C) 44. (26) 45. (B)
46. (C) 47. (11) 48. (6) 49. (C) 50. (2)
51. (B) 52. (D) 53 (B) 54. (B) 55. (6)
56. (A) 57. (121) 58. (2) 59. (D) 60. (D)

28.1 Semiconductor

1. (C) 2. (C) 3. (C) 4. (C) 5. (B)
6. (A) 7. (D) 8. (C) 9. (B) 10. (C)
11. (A) 12. (D) 13. (A) 14. (A) 15. (B)
16. (A) 17. (5) 18. (25) 19. (D) 20. (C)
21. (B)

28.2 PN Junction and Diodes

1. (C) 2. (C) 3. (B) 4. (D) 5. (A)
6. (B) 7. (A) 8. (A) 9. (A) 10. (C)
11. (D) 12. (D) 13. (D) 14. (A) 15. (C)
16. (B) 17. (B) 18. (C) 19. (B) 20. (C)
21. (D) 22. (D) 23. (B) 24. (C) 25. (B)
26. (B) 27. (B) 28. (12) 29. (C) 30. (A)
31. (40) 32. (A) 33. (C) 34. (B) 35. (A)
36. (D) 37. (A) 38. (25) 39. (B) 40. (B)
41. (C) 42. (C) 43. (20) 44. (9) 45. (5)
46. (B) 47. (25) 48. (192) 49. (500) 50. (120)
51. (20) 52. (D) 53. (C) 54. (A) 55. (5)
56. (B) 57. (1) 58. (4) 59. (B) 60. (D)
61. (2) 62. (14) 63. (3) 64. (25) 65. (9)
66. (B) 67. (1) 68. (480) 69. (B) 70. (C)
71. (C) 72. (D) 73. (C) 74. (A) 75. (C)
76. (C) 77. (D)

28.3 Logic Gates

1. (D) 2. (D) 3. (A) 4. (B) 5. (A)
6. (B) 7. (A) 8. (C) 9. (A) 10. (A)
11. (A) 12. (A) 13. (None) 14. (A) 15. (C)
16. (C) 17. (D) 18. (B) 19. (A) 20. (A)
21. (B) 22. (B) 23. (C) 24. (A) 25. (None)
26. (D) 27. (C) 28. (A) 29. (0) 30. (A)
31. (B) 32. (D) 33. (A) 34. (C) 35. (A)
36. (A) 37. (D) 38. (B) 39. (A) 40. (B)
41. (C) 42. (A) 43. (A) 44. (B) 45. (B)
46. (B) 47. (D) 48. (C) 49. (B) 50. (D)
51. (D) 52. (A) 53. (A) 54. (A) 55. (A)
56. (3) 57. (A)

29.1 Communication Systems

1. (B) 2. (A) 3. (A) 4. (B) 5. (D)
6. (A) 7. (C) 8. (B) 9. (B) 10. (B)
11. (A) 12. (D) 13. (B) 14. (C) 15. (D)
16. (8) 17. (10) 18. (9) 19. (D) 20. (B)
21. (1206) 22. (D) 23. (A) 24. (50) 25. (A)
26. (B) 27. (1) 28. (B) 29. (224) 30. (A)
31. (64) 32. (500) 33. (6) 34. (B) 35. (B)
36. (192) 37. (C) 38. (B) 39. (C) 40. (100)
41. (B) 42. (C) 43. (150) 44. (B) 45. (D)
46. (A) 47. (D) 48. (D) 49. (D) 50. (C)
51. (B) 52. (D) 53. (A) 54. (B) 55. (D)
56. (B) 57. (D)

29.2 Modulation and Demodulation

1. (A) 2. (B) 3. (B) 4. (B) 5. (C)
6. (A) 7. (C) 8. (B) 9. (C) 10. (C)
11. (D) 12. (A) 13. (C) 14. (C) 15. (B)
16. (D) 17. (B) 18. (C) 19. (D) 20. (A)
21. (D) 22. (B) 23. (C) 24. (D) 25. (A)
26. (None) 27. (25) 28. (A) 29. (D) 30. (33)
31. (40) 32. (1) 33. (1) 34. (2) 35. (200)
36. (C) 37. (C) 38. (D) 39. (C) 40. (D)
41. (B) 42. (A) 43. (A) 44. (D) 45. (A)
46. (2) 47. (D) 48. (A) 49. (C) 50. (B)
51. (B) 52. (B) 53. (B) 54. (C) 55. (B)
56. (B) 57. (D) 58. (D)

30.1 Errors and Experiments

1. (D) 2. (A) 3. (D) 4. (A) 5. (A)
6. (C) 7. (A) 8. (D) 9. (D) 10. (B)
11. (A) 12. (B) 13. (D) 14. (A) 15. (D)
16. (B) 17. (C) 18. (D) 19. (D) 20. (C)
21. (D) 22. (D) 23. (C) 24. (C) 25. (C)
26. (C) 27. (B) 28. (A) 29. (C) 30. (D)
31. (None) 32. (C) 33. (C) 34. (B) 35. (Bonus)
36. (D) 37. (D) 38. (B) 39. (B) 40. (1050)
41. (C) 42. (D) 43. (D) 44. (5) 45. (B)
46. (B) 47. (34) 48. (1) 49. (13) 50. (22)
51. (B) 52. (C) 53. (14) 54. (C) 55. (18)
56. (5) 57. (5) 58. (A) 59. (150) 60. (180)
61. (D) 62. (B) 63. (C) 64. (B) 65. (412)
66. (5) 67. (C) 68. (C) 69. (41) 70. (220)
71. (C) 72. (25) 73. (C) 74. (C) 75. (A)
76. (A)

* * * * *

SOLUTIONS

1.1 Measurements and Dimensional Analysis

Sol. 1 (A) Torque $\tau = Fr \rightarrow [\text{MLT}^{-2}]\,[\text{L}] = [\text{ML}^2\text{T}^{-2}]$
Work $W = Fs \rightarrow [\text{MLT}^{-2}]\,[\text{L}] = [\text{ML}^2\text{T}^{-2}]$ } same

(B) Stress $= \dfrac{F}{A} \rightarrow \dfrac{[\text{MLT}^{-2}]}{[\text{L}^2]} = [\text{ML}^{-1}\text{T}^{-2}]$
Energy $\rightarrow$ [Work] $\rightarrow [\text{ML}^2\text{T}^{-2}]$ } different

(C) Force $\rightarrow [\text{MLT}^{-2}]$
Stress $\rightarrow [\text{ML}^{-1}\text{T}^{-2}]$ } different

(D) Force $\rightarrow [\text{MLT}^{-2}]$
Work $\rightarrow [\text{ML}^2\text{T}^{-2}]$ } different

Ans. (A)

Sol. 2 (A) Torque $\tau = Fr \rightarrow [\text{MLT}^{-2}][\text{L}] = [\text{ML}^2\text{T}^{-2}]$
Work $W = Fs \rightarrow [\text{ML}^2\text{T}^{-2}]$ } same

(B) Momentum $P = mv \rightarrow [\text{M}][\text{LT}^{-2}] = [\text{MLT}^{-1}]$
Plank's constant $h = \dfrac{E}{\nu} \rightarrow \dfrac{[\text{ML}^2\text{T}^{-2}]}{[\text{T}^{-1}]} = [\text{ML}^2\text{T}^{-1}]$ } different

(C) Stresss $= \dfrac{F}{A} \rightarrow \dfrac{[\text{MLT}^{-2}]}{[\text{L}^2]} = [\text{ML}^{-1}\text{T}^{-2}]$
Young's modulus $Y = \dfrac{\text{Stress}}{\text{strain}} = [\text{stress}] = \text{ML}^{-1}\text{T}^{-2}]$ } same

(D) speed (m/s) and $\dfrac{1}{\sqrt{\mu_0 \in_0}}$ = speed of light (m/s) } same

Ans. (B)

Sol. 3 Speed of light in free space is given as

$$c = \frac{1}{\sqrt{\mu_0 \in_0}}$$

$$\Rightarrow \quad \frac{1}{\mu_0 \in_0} = c^2$$

$$\Rightarrow \quad D\left(\frac{1}{\mu_0 \in_0}\right) = D(c^2) = [\text{LT}^{-1}]^2 = [\text{L}^2\text{T}^{-2}]$$

Ans. (C)

Sol. 4 From stokes rule we have viscous force on a sphere of radius r moving at a speed v in a medium of coefficient of viscosity η is given as

$$F = 6\pi\eta rv$$

$$\Rightarrow \quad \eta = \frac{F}{6\pi rv}$$

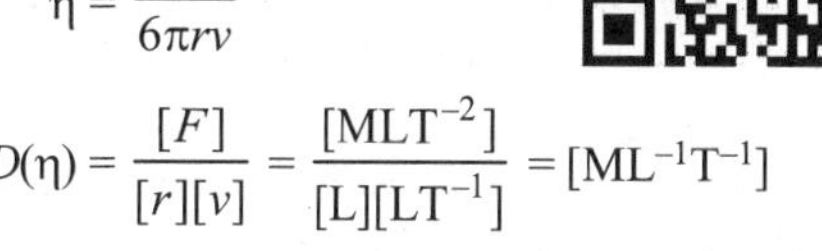

$$\Rightarrow \quad D(\eta) = \frac{[F]}{[r][v]} = \frac{[\text{MLT}^{-2}]}{[\text{L}][\text{LT}^{-1}]} = [\text{ML}^{-1}\text{T}^{-1}]$$

Ans. (A)

Sol. 5 (A) Impulse $J = Ft \rightarrow [\text{MLT}^{-2}][\text{T}] = [\text{MLT}^{-1}]$
Momentum $P = mv \rightarrow [\text{M}][\text{LT}^{-1}] = [\text{MLT}^{-1}]$ } same

(B) Angular momentum $L = mvr \rightarrow [\text{M}][\text{LT}^{-1}][\text{L}] = [\text{ML}^2\text{T}^{-1}]$
Planck's constant $h = \dfrac{E}{\nu} \rightarrow \dfrac{[\text{ML}^2\text{T}^{-2}]}{[\text{T}^{-1}]} = [\text{ML}^2\text{T}^{-1}]$ } same

(C) Work $W \rightarrow [\text{ML}^2\text{T}^{-2}]$
Torque $\tau = Fr \rightarrow [\text{MLT}^{-2}][\text{L}] = [\text{ML}^2\text{T}^{-2}]$ } same

(D) Moment of inertia $I = Mr^2 \rightarrow [\text{ML}^2]$
Moment of a force $\tau = Fr \rightarrow [\text{ML}^2\text{T}^{-2}]$ } Different

Ans. (D)

Sol. 6 Dimensions of $\dfrac{\text{ML}^2}{Q^2}$ is given as

$$D\left(\frac{\text{ML}^2}{Q^2}\right) = \frac{[\text{ML}^2]}{[\text{AT}]^2} = [\text{ML}^2\text{A}^{-2}\text{T}^{-2}]$$

(A) Wb/m^2 $\rightarrow$ unit of magnetic induction (B)

$$D(B) = [\text{MT}^{-2}\text{A}^{-1}]$$

(B) Henry $\rightarrow$ unit for self Induction (L)

We use energy in inductance $E = \dfrac{1}{2}LI^2 \Rightarrow L = \dfrac{2E}{I^2}$

$$\Rightarrow \quad D(L) = \frac{[\text{ML}^{-2}\text{T}^{-2}]}{[\text{A}^2]} = [\text{ML}^2\text{T}^{-2}\text{A}^{-2}]$$

(C) H/m^2 $\rightarrow D\left(\dfrac{\text{H}}{\text{m}^2}\right) = \dfrac{[\text{ML}^2\text{T}^{-2}\text{A}^{-2}]}{[\text{L}^2]} = [\text{MT}^2\text{A}^{-2}]$

(D) Wb $\rightarrow$ unit for magnetic flux ϕ

$$\Rightarrow \quad D(\phi) = D(\text{B.S}) = [\text{MT}^{-2}\text{A}^{-1}][\text{L}^2] = [\text{ML}^2\text{T}^{-2}\text{A}^{-1}]$$

Ans. (B)

Sol. 7 The force on moving charge in magnetic field is given as $F = qvB$

$$\Rightarrow \quad B = \frac{F}{qv}$$

$$\Rightarrow \quad D[B] = \frac{[F]}{[q][v]} = \frac{[MLT^{-2}]}{[C][LT^{-1}]} = [MT^{-1}C^{-1}]$$ **Ans. (C)**

Sol. 8 If surface tension S is expressed in terms of K, V and T with dimensions a, b and c then we use

$$[S] = [K]^a [V]^b [T]^c$$

$$\Rightarrow \quad [MT^{-2}] = [ML^2T^{-2}]^a [LT^{-1}]^b [T]^c$$

Comparing dimensions on LHS & RHS we have

$$\Rightarrow \quad a = 1$$

$$2a + b = 0 \Rightarrow b = -2$$

$$-2a - b + c = -2 \Rightarrow c = 2a + b - 2 \Rightarrow c = -2$$

$$\Rightarrow \quad [S] = [K^1 V^{-2} T^{-2}]$$ **Ans. (A)**

Sol. 9 By Coulomb's law we use the force between two point charges is given as

$$F = \frac{1}{4\pi \in_0} \frac{q_1 q_2}{r^2}$$

$$\Rightarrow \quad \in_0 = \frac{1}{4\pi} \frac{q_1 q_2}{Fr^2}$$

$$\Rightarrow \quad D(\in_0) = \frac{[\text{charge}]^2}{[F][L^2]} = \frac{[IT]^2}{[MLT^{-2}][L^2]}$$

$$\Rightarrow \quad D(\in_0) = [M^{-1}L^{-3}T^4I^2] = [M^{-1}L^{-3}T^4A^2]$$ **Ans. (B)**

Sol. 10 Angular momentum is expressed as

$$L = mvr$$

$$\Rightarrow \quad D[L] = [M][LT^{-1}][L] = [ML^2T^{-1}]$$

Latent heat of a substance is expressed as

$$L_f = \frac{Q}{m}$$

$$\Rightarrow \quad D(L_f) = \frac{[ML^2T^{-2}]}{[M]} = [L^2T^{-2}]$$

Capacitance is expressed as

$$C = \frac{q}{V}$$

Potential difference V is expressed as

$$V = \frac{W}{q}$$

$$\Rightarrow \quad D(V) = \frac{[ML^2T^{-2}]}{[AT]}$$

$$\Rightarrow \quad D(V) = [ML^2T^{-3}A^{-1}]$$

$$\Rightarrow \quad D(C) = \frac{D(q)}{D(V)} = \frac{[AT]}{[ML^2T^{-3}A^{-1}]}$$

$$= [M^{-1}L^{-2}T^4]$$ **Ans. (D)**

Sol. 11 Time period $t = r^b s^{c/2} d^{a/4}$ ...(1)

and it is also given that $t \propto \sqrt{\frac{d}{s}} = d^{1/2}s^{-1/2}$ ...(2)

$$t = kd^{1/2}s^{-1/2} \quad \text{...(3)}$$

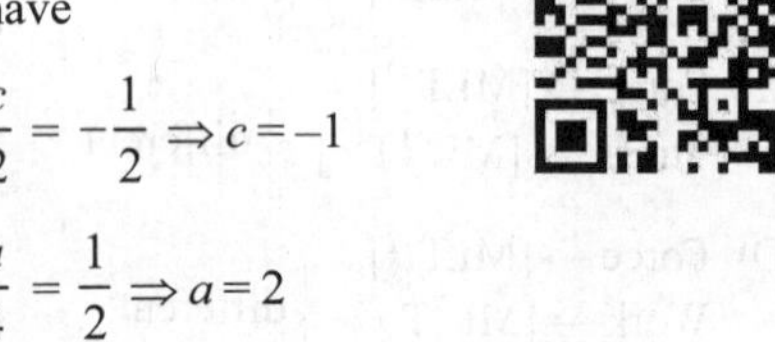

Here k involves the terms of r

From (1) and (3) we have

$$\frac{c}{2} = -\frac{1}{2} \Rightarrow c = -1$$

and $$\frac{a}{4} = \frac{1}{2} \Rightarrow a = 2$$

Writing dimensions of LHS & RHS in equation (1), gives

$$[T] = [L]^b [MT^{-2}]^{c/2} [ML^{-3}]^{a/4}$$

in above equation on RHS dimensions of L and M must be zero

$$\Rightarrow \quad b - \frac{3a}{4} = 0 \Rightarrow b = \frac{3a}{4} = \frac{3 \times 2}{4} = \frac{3}{2}$$ **Ans. (C)**

Sol. 12 Dimensions of R, h, $\in$ are

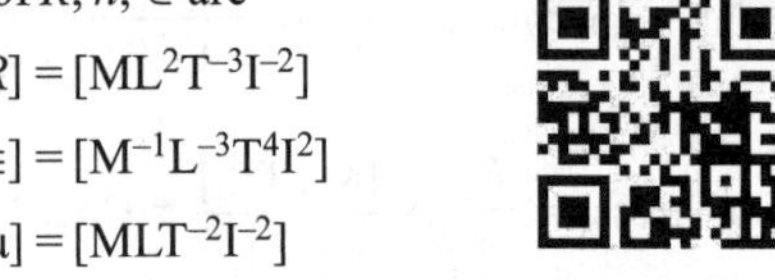

$$[R] = [ML^2T^{-3}I^{-2}]$$

$$[\in] = [M^{-1}L^{-3}T^4I^2]$$

$$[\mu] = [MLT^{-2}I^{-2}]$$

$$\Rightarrow \quad D\left(\frac{\mu}{\in}\right) = \frac{[MLT^{-2}I^{-2}]}{[M^{-1}L^{-3}T^4I^2]} = [M^2L^4T^{-6}I^{-4}]$$

$$= [R^2]$$ **Ans. (C)**

Sol. 13 Only a dimensionless constants (without units) have same values in different system of unit.

$$[e] = [IT]$$

$$[\in_0] = [M^{-1}L^{-3}T^4I^2]$$

$$[h] = [ML^2T^{-1}]$$

$$[c] = [LT^{-1}]$$

$$[G] = [M^{-1}L^3T^{-2}]$$

We can also use $$c = \frac{1}{\sqrt{\mu_0 \in_0}}$$

(A) $$D\left(\frac{ch}{2\pi \in_0^2}\right) = \frac{[LT^{-1}][ML^2T^{-1}]}{[M^{-1}L^{-3}T^4I^2]}$$

$\rightarrow$ Not dimensionless

(B) $$D\left(\frac{e^2}{2\pi \in_0 Gm_e^2}\right) = \frac{[IT]^2}{[M^{-1}L^{-3}T^4I^2][M^{-1}L^3T^{-2}][M]^2}$$

$\rightarrow$ Dimensionless

So this value does not change with system of units.

(C) $D\left(\frac{\mu_0 \in_0 G}{c^2 he^2}\right) = D\left(\frac{G}{c^4 he^2}\right)$

$$= \frac{[M^{-1}L^3T^{-2}]}{[LT^{-1}]^2[ML^2T^{-1}][IT]^2}$$

→ Not dimensionless

(D) $D\left(\frac{2\pi\sqrt{\mu_0 \in_0}}{ce^2}\frac{h}{G}\right) = D\left(\frac{h}{c^2e^2G}\right)$

$$= \frac{[M^1L^2T^{-1}]}{[LT^{-1}]^2[IT]^2[M^{-1}L^{-3}T^{-2}]}$$

→ Not dimensionless **Ans. (B)**

Sol. 14 By dimensional analysis we use

$$D(u) = [e]^a[a_0]^b[h]^c[c]^d$$

Comparing dimensions on LHS & RHS we have

$$\Rightarrow \quad c = -1$$

$$b + 2c + d = -2 \Rightarrow b = 1$$

$$a - c - d = 4 \Rightarrow d = -1$$

$$a = 2$$

$$\Rightarrow \quad u = e^2 a_0^1 h^{-1}c^{-1}$$

$$\Rightarrow \quad u = \frac{e^2 a_0}{hc}$$

Ans. (D)

Sol. 15 Here we use

$$\mu_0 = ke^a m^b c^c h^d$$

As speed of light $c = \frac{1}{\sqrt{\mu_0 \in_0}}$

$$\Rightarrow \quad D(\mu_0) = \frac{1}{D(c)^2 . D(\in_0)}$$

Then to find dimensions of e, m, c and h we compare the dimensions on LHS & RHS

$$\Rightarrow \quad D(\mu_0) = [IT]^a[M]^b[LT^{-1}]^c[ML^2T^{-1}]^d$$

$$\Rightarrow \quad [MLT^{-2}I^{-2}] = [IT]^a[M]^b[LT^{-1}]^c[ML^2T^{-1}]^d$$

$$\Rightarrow \quad b + d = 1 \Rightarrow b = 0$$

$$c + 2d = 1 \Rightarrow d = 1$$

$$a - c - d = -2 \Rightarrow c + d = 0 \Rightarrow c = -1$$

$$a = -2$$

$$\Rightarrow \quad \mu_0 = ke^{-2}m^0c^{-1}h^1$$

$$\Rightarrow \quad \mu_0 = k\left(\frac{h}{ce^2}\right)$$

Ans. (C)

Sol. 16 Resistance of a conductor is given as

$$R = \rho\frac{l}{A}$$

Conductivity $\sigma = \frac{1}{\rho} = \frac{l}{RA}$

by Ohm's law $V = IR \Rightarrow R = \frac{V}{I}$

$\Rightarrow \quad \sigma = \frac{lI}{VA}$ (for dimensions)

Electric potential $V = \frac{W}{q}$

$$\Rightarrow \quad D(V) = \frac{[ML^2T^{-2}]}{[I][T]} = [ML^2T^{-3}I^{-1}]$$

and $D(\sigma) = \frac{[L][I]}{[V][L^2]} = \frac{[L][I]}{[ML^2T^{-3}I^{-1}][L^2]}$

$$= [M^{-1}L^{-3}I^2T^3]$$

Ans. (B)

Sol. 17 From given equation $[BD]$ → dimensionless

and $[AD] \equiv [C]$

(A) $D\left[\frac{C}{BD}\right] \equiv D[C]$ and $D\left[\frac{AD^2}{C}\right] = D[D]$

$$\Rightarrow \quad \underbrace{\frac{C}{BD}}_{[C]} - \underbrace{\frac{AD^2}{C}}_{[D]}$$

⇒ Two physical quantities having different dimensions are subtracted here so this is not a meaningful quantity.

(B) $D(A^2) \equiv D\left(\frac{C^2}{D^2}\right)$ and $D(B^2C^2] \equiv D\left(\frac{C^2}{D^2}\right)$

$$\Rightarrow \quad \underbrace{A^2}_{\left[\frac{C^2}{D^2}\right]} - \underbrace{B^2C^2}_{\left[\frac{C^2}{D^2}\right]}$$

⇒ This equation is dimensionally homogeneous, so it is meaningful

(C) $D\left(\frac{A}{B}\right) \equiv D(AD)$ and $D(C) \equiv D(AD)$

$$\Rightarrow \quad \underbrace{\frac{A}{B}}_{[AD]} - \underbrace{C}_{[AD]}$$

⇒ This equation is dimensionally homogeneous.

This is a meaningful quantity.

(D) $D(A) \neq D(C)$

⇒ A-C will not give a meaningful quantity. **Ans. (A) & (D)**

Sol. 18 Dimensions of mass in terms of T, C and h are given as

$$D[\mathrm{M}] = D(T)^a.\, D(C)^b.\, D(h)^c$$
$$[\mathrm{M}] = [\mathrm{T}]^a[\mathrm{LT^{-1}}]^b[\mathrm{ML^2T^{-1}}]^c$$

Comparing LHS & RHS dimensions, we get

$$c = 1$$
$$a - b - c = 0 \Rightarrow a = -1$$
$$b + 2c = 0 \Rightarrow b = -2c = -2$$
$$\Rightarrow \quad D[\mathrm{M}] = [\mathrm{T^{-1}C^{-2}h^1}]$$

Ans. (A)

Sol. 19 Let the plank length (l_p) in terms of G, h and c is given as

$$l_p = KG^a h^b c^c \quad \ldots(1)$$
$$\Rightarrow \quad D(l_p) = D(G^a)\cdot D(h^b)\cdot D(c^c)$$
$$\Rightarrow \quad [\mathrm{L}] = [\mathrm{M^{-1}L^3T^{-2}}]^a[\mathrm{ML^2T^{-1}}]^b[\mathrm{LT^{-1}}]^c$$

Comparing dimensions of L, M & T on LHS & RHS gives

$$1 = 3a + 2b + c$$
$$0 = -a + b \Rightarrow a = b$$
$$0 = -2a - b - c \Rightarrow c = -3a$$
$$\Rightarrow \quad 1 = 5a - 3a \Rightarrow a = 1/2$$
$$\Rightarrow \quad b = 1/2$$
$$c = -3/2$$

From equation-(1) $l_p = KG^{1/2}h^{1/2}c^{-3/2} = K\sqrt{\dfrac{Gh}{c^3}}$

For all given options we take $K = 1$ **Ans. (B)**

Sol. 20 We use gravitational force as

$$F = \frac{GM^2}{R^2}$$
$$\Rightarrow \quad [G] = [\mathrm{M^{-1}L^3T^{-2}}]$$

Also as $E = h\nu$

$$\Rightarrow \quad [h] = [\mathrm{ML^2T^{-1}}]$$

and $[c] = [\mathrm{LT^{-1}}]$

Now we use

$$t \propto G^x\, h^y\, C^z$$
$$[\mathrm{T}] = [\mathrm{M^{-1}L^3T^{-2}}]^x[\mathrm{ML^2T^{-1}}]^y[\mathrm{LT^{-1}}]^z$$
$$[\mathrm{M^0L^0T^1}] = [\mathrm{M}^{-x+y}\,\mathrm{L}^{3x+2y+z}\,\mathrm{T}^{-2x-y-z}]$$

Comparing the powers of M, L, T in LHS & RHS, we have

$$-x + y = 0$$
$$\Rightarrow \quad x = y$$

and $3x + 2y + z = 0$

$$\Rightarrow \quad 5x + z = 0 \quad \ldots(1)$$

and $-2x - y - z = 1$

$$\Rightarrow \quad 3x + z = -1 \quad \ldots(2)$$

On solving (1) & (2) $x = y = \dfrac{1}{2}, z = -\dfrac{5}{2}$

$$\Rightarrow \quad t \propto \sqrt{\frac{Gh}{c^5}}$$

Ans. (D)

Sol. 21 Density of material is

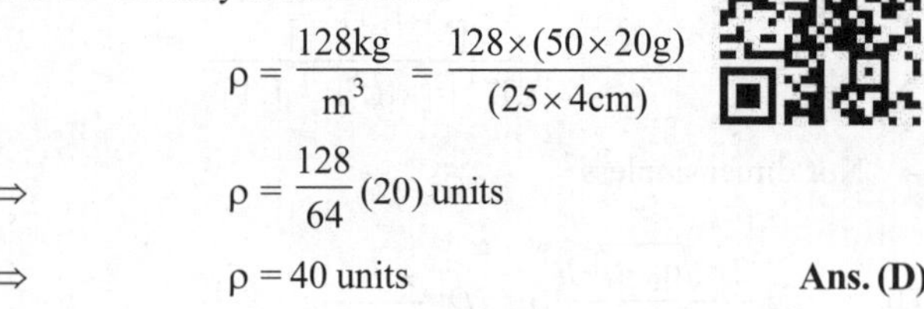

$$\rho = \frac{128\text{kg}}{\text{m}^3} = \frac{128\times(50\times 20\text{g})}{(25\times 4\text{cm})}$$
$$\Rightarrow \quad \rho = \frac{128}{64}(20)\text{ units}$$
$$\Rightarrow \quad \rho = 40\text{ units}$$

Ans. (D)

Sol. 22 Force is given as

$$F = \alpha\beta e^{\left(\frac{-x^2}{\alpha KT}\right)}$$

Here the exponential power must be dimensionless, so we have

$$D\left(\frac{x^2}{\alpha KT}\right) = \mathrm{M^0L^0T^0}$$
$$\Rightarrow \quad \frac{L^2}{[\alpha]ML^2T^{-2}} = \mathrm{M^0L^0T^0}$$
$$\Rightarrow \quad [\alpha] = \mathrm{M^{-1}T^2}$$

And we use

$$[F] = [\alpha]\,[\beta]$$
$$\mathrm{ML^1T^{-2}} = \mathrm{M^{-1}T^2}[\beta]$$
$$\Rightarrow \quad [\beta] = \mathrm{M^2L^1T^{-4}}$$

Ans. (B)

Sol. 23 We use young modulus $= \dfrac{\text{stress}}{\text{strain}}$

$$\Rightarrow \quad \frac{F}{S} = Y\cdot\frac{\Delta l}{l}$$
$$\Rightarrow \quad [\mathrm{Y}] = \frac{[\mathrm{F}]}{[\mathrm{S}]}$$

If dimensions of Y in unit of V, A and F are x, y, z then we use

$$[\mathrm{Y}] = \frac{[\mathrm{F}]}{[\mathrm{S}]}$$
$$\Rightarrow \quad [\mathrm{V}]^x[\mathrm{A}]^y[\mathrm{F}]^z = \frac{[\mathrm{MLT^{-2}}]}{[\mathrm{L^2}]}$$
$$[\mathrm{LT^{-1}}]^x[\mathrm{LT^{-2}}]^y[\mathrm{MLT^{-2}}]^z = [\mathrm{ML^{-1}\,T^{-2}}]$$

Comparing dimensions in LHS & RHS, we get

$$z = 1$$
$$x + y + z = -1$$
$$\Rightarrow \quad x + y = -2$$
$$-x - 2y - 2z = -2$$
$$\Rightarrow \quad x + 2y = 0$$
$$\Rightarrow \quad y = 2 \quad \text{and} \quad x = -4$$
$$\Rightarrow \quad [\mathrm{Y}] = \mathrm{V^{-4}A^2F^1}$$

Ans. (B)

Sol. 24 $\left[\frac{l}{r}\right] = T$

and $[CV] = AT$

$$\Rightarrow \quad \left[\frac{l}{rCV}\right] = \frac{T}{AT} = A^{-1}$$

Ans. (C)

Sol. 25 $p = kS^a I^b h^c$

Where k is dimensionless constant

$$MLT^{-1} = (MT^{-2})^a (ML^2)^b (ML^2T^{-1})^c$$

$$a + b + c = 1$$

$$2b + 2c = 1$$

$$-2a - c = -1$$

Solving, we get

$$a = \frac{1}{2}$$

$$b = \frac{1}{2}$$

$$c = 0$$

$$[P] = S^{1/2} I^{1/2} h^0$$

Ans. (B)

Sol. 26 Given that $X = 5YZ^2$

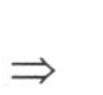

$$\Rightarrow \quad Y = \frac{X}{5Z^2}$$

Dimensionally we use

$$[Y] = \frac{[X]}{[Z^2]}$$

$$\Rightarrow \quad [Y] = \frac{A^2 \cdot M^{-1}L^{-2} \cdot T^4}{(MA^{-1}T^{-2})^2}$$

$$\Rightarrow \quad [Y] = M^{-3} \cdot L^{-2} \cdot T^8 \cdot A^4$$

Ans. (C)

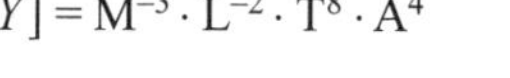

Sol. 27 Dimension of h, c and G are

$$[h] = M^1L^2T^{-1}$$

$$[c] = L^1T^{-1}$$

$$[G] = M^{-1}L^3T^{-2}$$

So dimensions of f can be calculated as

$$[f] = \sqrt{\frac{M^1L^2T^{-1} \times L^5T^{-5}}{M^{-1}L^3T^{-2}}} = M^1L^2T^{-2}$$

Ans. (C)

Sol. 28 Dimensions of Young's modulus are considered as $x\ y\ z$ then we have

$$[Y] = F^x A^y V^z$$

Comparing dimensions in term of MLT, we have

$$M^1L^{-1}T^{-2} = [MLT^{-2}]^x [L^2]^y [LT^{-1}]^z$$

$$M^1L^1T^{-2} = [M]^x [L]^{x+2y+z} [T]^{-2x-z}$$

Comparing LHS and RHS dimensions, we get

$$x = 1 \qquad \ldots(1)$$

$$x + 2y + z = -1 \qquad \ldots(2)$$

$$-2x - z = -2 \qquad \ldots(3)$$

Solving (1), (2) and (3), we get

$$x = 1$$

$$y = -1$$

$$z = 0$$

$$[Y] = FA^{-1}V^0$$

Ans. (A)

Sol. 29 If dimensions of fundamental quantities are taken as x, y and z, we have

$$D[E] = [P]^x [A]^y [T]^z$$

$$\Rightarrow \quad ML^2T^{-2} = [MLT^{-1}]^x [L^2]^y [T]^z$$

$$\Rightarrow \quad ML^2T^{-2} = M^x L^{x+2y} T^{-x+z}$$

Comparing dimensions of LHS and RHS, we have

$$x = 1$$

and $x + 2y = 2$

and $1 + 2y = 2$

$$\Rightarrow \quad y = \frac{1}{2}$$

$$\Rightarrow \quad -x + z = -2$$

$$\Rightarrow \quad -1 + z = -2$$

$$\Rightarrow \quad z = -1$$

$$D[E] = [PA^{1/2}\ T^{-1}]$$

Ans. (B)

Sol. 30 By steady state conduction, heat flow rate is given as

$$\frac{d\theta}{dt} = \frac{kA(T_2 - T_1)}{l}$$

$$\Rightarrow \quad D[k] = \frac{D[d\theta/dt][L]}{[L^2][\theta]} = \frac{[ML^2T^{-3}][L]}{[L^2][\theta]}$$

$$\Rightarrow \quad D[k] = [MLT^{-3}\ \theta^{-1}]$$

Ans. (D)

Sol. 31 Dimensions of x is given in terms of MLT as

$$[x] = \frac{[ML^2][MLT^{-2}][LT^{-1}]^2}{[ML^2T^{-2}][L^4]}$$

$$[x] = [ML^{-1}T^{-2}] = \frac{[ML^2T^{-2}]}{[L^3]} = \frac{\text{Energy}}{\text{volume}}$$

Thus dimensional formula of x matches with energy density.

Ans. (C)

Sol. 32 Dimensions of x, y and z are

$$x = \frac{1}{\sqrt{\mu_0 \varepsilon_0}} = \text{speed of light}$$

$$\Rightarrow \quad [x] = [LT^{-1}]$$

$$y = \frac{E}{B} = \text{speed of light}$$

$$\Rightarrow \quad [y] = [L^1T^{-1}]$$

$$z = \frac{l}{RC} = \frac{\text{length}}{\text{time}} = \text{speed}$$

$$\Rightarrow \quad [z] = [L^1T^{-1}]$$

So, x, y, z all have the same dimensions. **Ans. (B)**

Sol. 33 The dimensions of any value in exponent must be dimensionless so in the given expression $\frac{x^2}{\alpha kT}$ is dimensionless, this gives

$$\Rightarrow \quad D[\alpha] = \frac{[x^2]}{[kT]} = \frac{L^2}{ML^2T^{-2}} = M^{-1}T^2$$

and $\quad D[W] = [\alpha][\beta]^2$

$$\Rightarrow \quad D[\beta] = \sqrt{\frac{ML^2T^{-2}}{M^{-1}T^2}} = M^1L^1T^{-2}$$ **Ans. (B)**

Sol. 34 By direct dimensions of the given physical quantities, we can match the values in given columns. **Ans. (A)**

Sol. 35 By the generic expressions of force and energy, we have

$$F = \frac{1}{4\pi \epsilon_0}\frac{e^2}{r^2} \text{ and } E = \frac{hc}{\lambda}$$

$$\left[\frac{e^2}{4\pi\varepsilon_0} \times \frac{1}{hc}\right] = \frac{Fr^2}{E\lambda}$$

$$\Rightarrow \quad D\left(\frac{Fr^2}{E\lambda}\right) = [M^0 L^0 T^0]$$ **Ans. (C)**

Sol. 36 The dimensions of any value in exponent must be dimensionless so in the given expression kT which has dimensions of energy is such that $\frac{\beta x^2}{kT}$ is dimensionless, so we use

$$[\beta][L^2] = [ML^2T^{-2}]$$

$$\Rightarrow \quad D[\beta] = [MT^{-2}]$$

Here $\alpha^2\beta$ has dimensions of work, so we use

$$[\alpha^2][MT^{-2}] = [ML^2T^{-2}]$$

$$\Rightarrow \quad D[\alpha] = [M^0 L T^0]$$ **Ans. (D)**

Sol. 37 The given expression can be rewritten as

$$\lambda = \frac{C}{V} = \frac{Q/V}{V} = \frac{Q}{V^2}$$

Voltage, can be expressed as

$$V = \frac{\text{work}}{Q}$$

$$\Rightarrow \quad \lambda = \frac{Q^3}{(\text{work})^2} = \frac{(\text{It})^3}{(F.s)^2}$$

$$\Rightarrow \quad D[\lambda] = \frac{[I^3T^3]}{[ML^2T^{-2}]^2} = [M^{-2}L^{-4}I^3T^7]$$ **Ans. (D)**

Sol. 38 If one vernier scale division length is a' then we use

$$(n-1)a = n(a')$$

$$\Rightarrow \quad a' = \frac{(n-1)a}{n}$$

Least count of the vernier callipers is given as

$$LC = 1\text{ MSD} - 1\text{ VSD}$$

$$\Rightarrow \quad LC = (a - a')\text{ cm}$$

$$\Rightarrow \quad LC = a - \frac{(n-1)a}{n}$$

$$\Rightarrow \quad LC = \frac{na - na + a}{n} = \frac{a}{n}\text{ cm}$$

$$\Rightarrow \quad LC = \left(\frac{10a}{n}\right)\text{mm}$$ **Ans. (D)**

Sol. 39 Volume of 100 drops is equal to the volume of solution layer and if the acid layer of thickness t is there in the solution, we have

$$4t = 100 \times \frac{4}{3}\pi r^3 \times 0.01$$

$$\Rightarrow \quad 4t = 100 \times \frac{4\pi}{3} \times \frac{3}{40\pi} \times 10^{-9} \times 0.01$$

$$\Rightarrow \quad x = 25$$ **Ans. (25)**

Sol. 40 Dimensions of entropy, Boltzman's constant, gas constant, mechanical equivalent of heat are given as

$$[S] = \left[\frac{Q}{\Delta T}\right] = \frac{ML^2T^{-2}}{K}$$

$$[K] = [S] = \frac{ML^2T^{-2}}{K}$$

$$[R] = \left[\frac{\text{Energy}}{\text{nT}}\right] = \frac{ML^2T^{-2}}{\text{mol K}}$$

$$[J] = M^0L^0T^0$$
$$[\mu KR] = [J\beta^2]$$
$$[\beta] = ML^2T^{-2}K^{-1}$$
$$[\alpha^2] = \left[\frac{S}{\beta}\right] = \frac{ML^2T^{-2}}{K \times ML^2T^{-2}K^{-1}};$$
$$\alpha = M^0L^0T^0$$ **Ans. (D)**

Sol. 41 Mass can be written in terms of t, v and l with dimensions x, y and z respectively and given as

$$M \propto t^x v^y l^z$$
$$M^0L^0T^0 = t^x [LT^{-1}]^y [ML^2T^{-1}]^z$$
$$M^1L^0T^0 = t^{x-y-z} L^{y+2z} M^z$$

On comparing powers at LHS and RHS, we get

$$z = 1 \quad \ldots(1)$$
$$x - y - z = 0 \quad \ldots(2)$$
$$y + 2z = 0 \quad \ldots(3)$$
$$\Rightarrow \quad y + 2 \times 1 = 0$$
$$\Rightarrow \quad y = -2$$
$$\Rightarrow \quad x - (-2) - 1 = 0$$
$$\Rightarrow \quad x = -1$$
$$\Rightarrow \quad M \propto t^{-1} v^{-2} l^1$$
$$\Rightarrow \quad [M] \propto [t^{-1}v^{-2}l]$$ **Ans. (A)**

Sol. 42 By homogeneity of given equation, the dimensions of constants are taken as

$$A = MLT^{-2}, B = L^{-1}, D = T^{-1}$$

Thus dimension formula of $\frac{AD}{B}$ can be written as

$$\left[\frac{AD}{B}\right] = \frac{MLT^{-2}T^{-1}}{L^{-1}} = ML^2T^{-3}$$ **Ans. (A)**

Sol. 43 (a) Capacitance

$$C = \frac{q}{V} = \frac{q}{\left(\frac{W}{q}\right)} = \frac{q^2}{W} = \frac{(it)^2}{f \cdot x} = \frac{[AT]^2}{[M^1L^2T^{-2}]}$$
$$C = [M^{-1}L^{-2}T^4A^2]$$

(b) $$F = \frac{kq^2}{r^2} = \frac{1}{4\pi\varepsilon_0}\frac{q^2}{r^2}$$

$$\varepsilon_0 = \frac{q^2}{4\pi Fr^2} = \frac{[AT]^2}{[M^1L^1T^{-2}][L^2]}$$
$$\varepsilon_0 = [M^{-1}L^{-3}T^4A^2]$$

(c) $$C = \frac{1}{\sqrt{\mu_0\varepsilon_0}} \Rightarrow C^2 = \frac{1}{\mu_0\varepsilon_0}$$

$$\mu_0 = \frac{1}{[C^2\varepsilon_0]} = \frac{1}{[L^2T^{-2}][M^{-1}L^{-3}T^4A^2]}$$
$$\mu_0 = [M^1L^1T^{-2}A^{-2}]$$

(d) $$F = qE$$

$$E = \frac{F}{q} = \frac{M^1L^1T^{-2}}{A^1T^1}$$
$$E = [M^1L^1T^{-3}A^{-1}]$$

So correct option is (B) **Ans. (B)**

Sol. 44 Lease Count of screw Gauge is given as

$$LC = \frac{0.5}{50} = 0.01 \text{ mm}$$

Zero error of screw Gauge = $5 \times LC = 5 \times 0.01 = 0.05$ mm

Reading of screw gauge given as

R = main scale reading + circular scale reading × $L.C.$ – Zero error

$\Rightarrow \quad R = 5 + 20 \times 0.01 - 0.05 = 5.15$ mm **Ans. (D)**

Sol. 45 Dimensions of all physical quantities in the expression of P are substituted as

$$P = EL^2M^{-2}G^{-2}$$
$$\Rightarrow \quad P = [ML^2T^{-2}]\,[ML^2T^{-1}]2\,[M^{-5}]\,[M^{-1}L^3T^{-2}]^{-2}$$
$$\Rightarrow \quad P = [M^0L^0T^0]$$ **Ans. (D)**

Sol. 46 By dimensional analysis lists can be directly matched with appropriate options. **Ans. (B)**

Sol. 47 Permeability of free space has dimensions and all other option are dimensionless quantities. **Ans. (A)**

Sol. 48 The dimensional expression of electric field and magnetising field can be written as

$$D\left[\frac{E}{H}\right] = \frac{\text{newton/coulomb}}{\text{ampere/m}}$$
$$\Rightarrow \quad D\left[\frac{E}{H}\right] = \text{Joule/(second-Ampere}^2)$$ **Ans. (C)**

Sol. 49 By dimensional analysis list I can be matched directly with List II. **Ans. (C)**

Sol. 50 Dimension of density if taken as a, b and c then we have

$$d \propto F^a L^b T^c$$
$$\Rightarrow \quad [ML^{-3}] = [MLT^{-2}]^a [L^b] [T^c]$$
$$\Rightarrow \quad [ML^{-3}] = [M^a L^{a+b} T^{-2a+c}]$$

Comparing dimensions of LHS and RHS, we have

$$a = 1 \quad \ldots(1)$$
$$a + b = -3 \quad \ldots(2)$$
$$-2a + c = 0 \quad \ldots(3)$$

Solving equations (1), (2) and (3), we have

$$c = 2$$
$$b = -4$$
$$\Rightarrow \quad [d] = [FL^{-4}T^2]$$ **Ans. (A)**

Sol. 51 List-I can directly be matched with the dimensions of the physical quantities listed in List-II. **Ans. (D)**

Sol. 52 Option-(A) is the expression of displacement current. option-(C) is the expression of capillary height and option-(D) is the work calculation in rotational motion. Expression of option -(B) is that of volume flow rate whereas in its LHS volume is given so this is dimensionally wrong. **Ans. (B)**

Sol. 53 If the dimensions are taken as *a*, *b* and *c* respectively, we use

$$[M] = [V]^a[T]^b[F]^c$$
$$\Rightarrow \quad [M] = [LT^{-1}]^a[T]^b[MLT^{-2}]^c$$
$$\Rightarrow \quad [M] = [M]^c[L]^{a+c}[T]^{-a+b-2c}$$

Comparing dimensions at LHS and RHS gives

$$c = 1, \; a + c = 0 \text{ and } b - a - 2c = 0$$

Solving these equations, we get

$$a = -1 \text{ and } b = -1 + 2 = 1$$
$$\Rightarrow \quad [M] = [V]^{-1}[T]^1[F]^1$$ **Ans. (D)**

Sol. 54 For 9 main scale divisions are coinciding with 10 vernier scale divisions, we use

$$9\,\text{MSD} = 10\,\text{VSD}$$
$$\Rightarrow \quad 1\,\text{VSD} = 0.9\,\text{mm}$$

Least count of vernier callipers can be calculated as

$$LC = \frac{\text{MSD} - \text{VSD}}{\text{MSD}} = \frac{1 - 0.9}{1} = 0.1\,\text{mm}$$

Reading = Main scale reading + LC × vernier scale reading

$$\text{Reading} = 10 + 0.1 \times 8$$
$$\text{Reading} = 10.8$$

Considering zero error into account, final diameter of spherical bob is given as

$$d = 10.8\,\text{mm} - 0.04\,\text{cm}$$
$$\Rightarrow \quad d = 10.8\,\text{mm} - 0.4\,\text{mm}$$
$$\Rightarrow \quad d = 10.4\,\text{mm}$$

Eadius of the bob is calculated as

$$r = 5.2\,\text{mm}$$
$$\Rightarrow \quad r = 0.52\,\text{cm} = 52 \times 10^{-2}\,\text{cm}$$ **Ans. (52)**

Sol. 55 (A) Wave number is measured in units of m^{-1} and same is the unit for Rydberg's constant so both are same.

(B) Coefficient of elasticity is defined as ratio of stress to strain and strain is a dimensionless quantity hence both stress and coefficient of elasticity have same dimensions.

(C) Coercivity and Magnetisation both are measured in units of A/m so both have same dimensions.

(D) Specific heat capacity of a body is the amount of heat required to increase the temperature of unit mass of substance by unity so it is measured in units as given below

$$s = \left(\frac{Q}{m\Delta T}\right) \frac{J}{\text{Kg}^\circ C}$$

Latent heat of a body is the heat required to change the phase of unit mass of substance, measured in units as given below.

$$L = \left(\frac{Q}{m}\right) \frac{J}{\text{Kg}}$$

Here both are different. **Ans. (D)**

Sol. 56 (A) Velocity gradient is measured in s^{-1} unit and same is the unit of decay constant.

(B) Wien's constant has units m-K and Stefan's constant has units W/m^2K^4

(C) Angular frequency has units s^{-1} and that of angular momentum is J-s.

(D) Wave number is measured in units of s^{-1} whereas Avogadro number is measured in numbers and dimensionless. **Ans. (A)**

Sol. 57 After time *t* the velocities are given as

$$X_P(t) = \alpha t + \beta t^2 \qquad X_Q = ft - t^2$$
$$V_P(t) = \alpha + 2\beta t \qquad V_Q = f - 2t$$

When velocities become equal, we have

$$V_P = V_Q$$
$$\Rightarrow \quad \alpha + 2\beta t = f - 2t$$
$$\Rightarrow \quad t = \frac{f - \alpha}{2\beta + 2}$$ **Ans. (D)**

Sol. 58 Given that *P* is dimensionless means both α and β have same dimensions

$$P = \frac{\alpha}{\beta}\log_e\left(\frac{kt}{\beta x}\right)$$

Also the we have another dimensionless quantity

$$\left[\frac{kt}{\beta x}\right] = \text{dimensionless}$$
$$\Rightarrow \quad [\beta] = \left[\frac{kt}{x}\right] = \frac{ML^2T^{-2}}{L}$$

As *P* is dimensionless

$$\Rightarrow \quad [\alpha] = [\beta] = [MLT^{-2}]$$ **Ans. (C)**

Sol. 59 EMF induced in a pair of coil having mutual inductance is given as

$$e_2 = -M\frac{di_1}{dt}$$

e_2 : Induced emf in secondary coil
i_1 : Current in primary coil
M : Mutual inductance

$$\Rightarrow \quad M = -\frac{e_2}{\frac{di_1}{dt}}$$

$$\Rightarrow \quad [M] = \frac{[e_2]}{\left[\frac{di_1}{dt}\right]} = \frac{\left[\frac{W}{q}\right]}{\left[\frac{di_1}{dt}\right]} = \frac{\frac{[ML^2T^{-2}]}{[AT]}}{[AT^{-1}]}$$

$$\Rightarrow \quad [M] = [ML^2T^{-2}A^{-2}]$$ **Ans. (C)**

Sol. 60 Pascal second can be expressed in form of force, area and time as

$$\left[\frac{F}{A}t\right] = \frac{MLT^{-2}}{L^2}T = ML^{-1}T^{-1}$$ **Ans. (A)**

Sol. 61 $\left(\frac{L}{C}\right)$ does not have dimension of time, RC, $\frac{L}{R}$ are time constant while $\sqrt{LC}$ is reciprocal of angular frequency or having dimension of time. **Ans. (D)**

Sol. 62 Unit of viscosity is expressed as

U[Viscosity] = pascal–second

$$P^xA^yT^z = [M^1L^{-1}T^{-1}]$$

$$[M^1L^{+1}T^{-1}]^x[L^2]^y[T^1]^z = M^1L^{-1}T^{-1}$$

$$\Rightarrow \quad M^xL^{+x+2y}T^{-x+z} = M^1L^{-1}T^{-1}$$

$$\Rightarrow \quad x = 1 \quad x+2y = -1 \quad -x+z = -1$$

$$\Rightarrow \quad y = -1$$

$$\Rightarrow \quad z = 0$$

$$\Rightarrow \quad [\text{Viscosity}] = P^1A^{-1}T^0$$ **Ans. (A)**

Sol. 63 As the term inside log is dimensionless, we use

$$\frac{\alpha[L]}{[ML^2T^{-2}]} = [M^0L^0T^0]$$

$$\Rightarrow \quad [\alpha] = [ML^1T^{-2}]$$

$$\Rightarrow \quad \left[\frac{\alpha}{\beta}\right] = \frac{[ML^2T^{-2}]}{[L^3]}$$

$$\Rightarrow \quad [\beta] = \frac{[ML^1T^{-2}][L^3]}{ML^2T^{-2}}$$ **Ans. (D)**

Sol. 64 Expression of energy density is in magnetic field is given as

$$u = \frac{B^2}{2\mu_0}$$

$$\Rightarrow \quad \left[\frac{B^2}{\mu_0}\right] = [u] = \frac{[ML^2T^{-2}]}{[L^3]} = [ML^{-1}T^{-2}]$$ **Ans. (C)**

Sol. 65 As $\sin\theta$ is dimensionless, $\alpha\beta$ will also be dimensionless.

$$[\alpha\beta] = [\eta] = [\sin\theta] = \text{Dimensionless}$$

$$[\eta^{-1}\sin\theta] = [\alpha\beta] = \text{Dimensionless}$$ **Ans. (D)**

Sol. 66 In the expression of time period given in question

$$T = k\sqrt{\frac{\rho r^3}{s^{3/2}}}$$

The dimensions of RHS $= \frac{[M^{1/2}L^{-3/2}]}{[MT^{-2}]^{3/4}} = [M^{1/8}L^0T^{3/2}]$

Dimensions of LHS $\neq$ Dimensions of RHS
Thus option (D) is correct. **Ans. (D)**

Sol. 67 Units of given physical quantitites can be calculated by their fundamental expressions, given as

(A) Torque $= F \times r_\perp$ N–m

(B) Stress $= \frac{\text{Force}}{\text{Area}}$ N/m^2

(C) Latent heat $= \frac{\text{Energy}}{\text{Mass}}$ J Kg^{-1}

(D) Power $= \frac{\text{Work}}{\text{Time}}$ N ms^{-1}

$\Rightarrow$ A $\to$ III, B $\to$ IV, C $\to$ II, D $\to$ I **Ans. (B)**

Sol. 68 Expression of oscillation frequency can be expressed with dimensions a, b and c for radius, density and surface tension of the liquid. Using dimensions in terms of M, L and T, we have

$$[T^{-1}] = [L^1]^a[M^1L^{-3}]^b\left[\frac{MLT^{-2}}{L}\right]^c$$

$$\Rightarrow \quad T^{-1} = M^{b+c} \cdot L^{a-3b} \cdot T^{-2c}$$

Comparing dimensions at LHS and RHS gives

$$c = \frac{1}{2}, b = -\frac{1}{2} \text{ and } a - 3b = 0$$

$$\Rightarrow \quad a + \frac{3}{2} = 0$$

$$\Rightarrow \quad a = -\frac{3}{2}$$ **Ans. (A)**

Sol. 69 (a) Dimensions of Young's modulus is expressed as

$$[Y] = [\frac{\text{Strees}}{\text{Strain}}] = [\frac{F/A}{\Delta l/l}] = \frac{[MLT^{-2}]}{[L^2]} = [ML^{-1}T^{-2}]$$

(b) Dimensions of coefficient of viscosity is expressed by Stoke's Rule given as

$$F = 6\pi\eta rv$$

$$\Rightarrow \quad \eta = \frac{F}{6\pi rv}$$

$$\Rightarrow \quad [\eta] = \frac{[MLT^{-2}]}{[L][LT^{-1}]} = [ML^{-1}T^{-1}]$$

(c) Dimensions of Plank's constant can be expressed by photon energy given as

$$E = hv$$

$$\Rightarrow \quad [h] = [\frac{E}{v}] = \frac{[ML^2T^{-2}]}{[T^{-1}]} = [ML^2T^{-1}]$$

(d) Work function has same dimension as that of energy which can be given as

$$[\phi] = [ML^2T^{-2}]$$ **Ans. (B)**

Sol. 70 (a) Dimensions of pressure gradient is calculated as

$$\left[\frac{dp}{dx}\right] = \frac{[ML^{-1}T^{-2}]}{[L]} = [M^1L^{-2}T^{-2}]$$

(b) Dimensions of energy density is calculated as

$$\left[\frac{\text{energy}}{\text{volume}}\right] = \frac{[ML^2T^{-2}]}{[L^3]} = [M^1L^{-1}T^{-2}]$$

(c) Dimensions of Electric field is calculated as

$$\left[\frac{\text{Force}}{\text{charge}}\right] = \frac{[MLT^{-2}]}{[A.T]}$$

$$= [M^1L^1T^{-3}A^{-1}]$$

(d) Dimensions of latent heat is calculated as

$$\left[\frac{\text{heat}}{\text{mass}}\right] = \frac{[ML^2T^{-2}]}{[M]} = [M^0L^2T^{-2}]$$ **Ans. (C)**

Sol. 71 Using dimensional analysis and dimensional homogeneity on the given equation, we have

$$(x - At)^2 + \left(y - \frac{t}{B}\right)^2 = a^2$$

Dimensions of At as well as that of t/B should be that of length so we use

$$[At] = [A][T^{-1}] = [L]$$

$$\Rightarrow \quad [A] = [TL]$$

And in second term of equation dimensions of $\frac{1}{B}$ should be in meters, so we have

$$\Rightarrow \quad \frac{1}{T[B]} = L$$

$$\Rightarrow \quad [B] = T^{-1}L^{-1}$$ **Ans. (B)**

Sol. 72 Given electric field in the region is

$$\vec{E} = \frac{A}{x^2}\hat{i} + \frac{B}{y^3}\hat{j}$$

Using dimensions of every term, we have

$$\left[\frac{A}{x^2}\right] = NC^{-1}$$

$$\Rightarrow \quad [A] = Nm^2C^{-1}$$

and $$\left[\frac{B}{y^3}\right] = NC^{-1}$$

$$\Rightarrow \quad [B] = Nm^3C^{-1}$$ **Ans. (B)**

Sol. 73 Torque $\tau = \vec{r} \times \vec{F} = [M^1L^2T^{-2}]$

$$\text{Energy density} = \frac{\text{Energy}}{\text{Volume}} = [M^1L^{-1}T^{-2}],$$

$$\text{Pressure gradient} = \frac{\Delta p}{\Delta x} = [M^1L^{-2}T^{-2}]$$

$$\text{Impulse} = F\Delta t = [M^1L^1T^{-1}]$$ **Ans. (D)**

Sol. 74 Angular momentum is calculated as

$$\vec{L} = \vec{r} \times \vec{p}$$

$$\Rightarrow \quad [L] = [M^0L^1T^0][M^1L^1T^{-1}]$$

$$\Rightarrow \quad [L] = [M^1L^2T^{-1}]$$

Torque is calculated as

$$\vec{\tau} = \vec{r} \times \vec{F}$$

$$\Rightarrow \quad [\tau] = [L^1][MLT^{-2}]$$

$$\Rightarrow \quad [\tau] = [ML^2T^{-2}]$$

Stress has dimensions of pressure, calculated as

$$\text{Stress} = \frac{F}{A}$$

$$\Rightarrow \quad [\text{Stress}] = [ML^{-1}T^{-2}]$$

Pressure Gradient is calculated as $\frac{dP}{dx}$ so we use

$$\Rightarrow \quad [\text{Pressure Gradient}] = [ML^{-2}T^{-2}]$$ **Ans. (B)**

Sol. 75 If we consider dimensional formula of mass is $H^xC^yG^z$ then we use

$$M^1 = (ML^2T^{-1})^x(LT^{-1})(M^{-1}L^3T^{-2})^Z$$

$$\Rightarrow \quad M^1L^0T^0 = M^{x-z}L^{2x+y+3z}T^{-x-y-2z}$$

On comparing dimensions on both side, we have

$$x - z = 1$$
and $$2x + y + 3z = 0$$
and $$-x - y - 2z = 0$$
On solving above equations we get

$$x = \frac{1}{2}\ ;\ y = \frac{1}{2} \text{ and } z = \frac{-1}{2}$$ **Ans. (D)**

Sol. 76 Speed of electromagnetic waves is given as

$$c = \frac{1}{\sqrt{\mu_0 \varepsilon_0}}$$

Here, μ_0 is permeability and ε_0 is permittivity hence dimensionally we can write

$$\frac{1}{\mu_0 \epsilon_0} = c^2$$

$$\Rightarrow \left[\frac{1}{\mu_0 \epsilon_0}\right] = [c^2] = [L^2T^{-2}]$$ **Ans. (C)**

Sol. 77 (a) Torque is calculated as

$$\vec{\tau} = \vec{r} \times \vec{F}$$
$$\Rightarrow [\tau] = [L]\,[MLT^{-2}]$$
(b) Stress is calculated as

$$\text{Stress} = \frac{F}{A} \Rightarrow \frac{MLT^{-2}}{L^2}$$
$$\Rightarrow [\text{Stress}] = ML^{-1}T^{-2}$$
(c) Pressure gradient is calculated as

$$\text{Pressure gradient} = \frac{\Delta P}{\Delta X}$$
$$\Rightarrow \frac{[F/A]}{[L]} = \left[\frac{MLT^{-2}}{L^3}\right] = [ML^{-2}T^{-2}]$$
(d) Coefficient of viscosity is related to viscous force as
$$F = 6\pi\eta rv$$
$$\Rightarrow [MLT^{-2}] = [\eta]L^2T^{-1}$$
$$\Rightarrow [\eta] = ML^{-1}T^{-1}$$ **Ans. (C)**

Sol. 78 Unit weight of each given unit can be written in meter are listed below.
$$1\text{ AU} = 1.496 \times 10^{11}\text{ m}$$
$$1\text{ par sec} = 3.08 \times 10^{16}\text{ m}$$
$$1\text{ light year} = 9.46 \times 10^{15}\text{ m}$$
$$\Rightarrow \text{AU} < ly < \text{par sec}$$ **Ans. (B)**

Sol. 79 By using dimensional analysis, the density dimensions can be written as
$$[d] = [F]^a[V]^b[T]^c$$
$$[ML^{-3}] = [MLT^{-2}]^a[LT^{-1}]^b[T]^c$$
$$\Rightarrow [ML^{-3}] = [M^aL^{a+b}T^{-2a-b+c}]$$

By comparing equations, we have
$$a = 1$$
and $$a + b = -3,$$
and $$b = -4,$$
$$\Rightarrow -2a - b + c = 0$$
$$\Rightarrow c = -2$$ **Ans. (B)**

Sol. 80 From the given equation it is clear that X and $\frac{a}{Y^2}$ have same dimensions and Y and b also have same dimensions so we have
$$[a] = [ML^5T^{-2}]$$
and $$[b] = [L^3]$$
$$\Rightarrow \frac{[a]}{[b]} = [ML^2T^{-5}]$$

thus this ratio has dimensions of energy. **Ans. (A)**

Sol. 81 Using dimensions of wave speed in terms of given physical quantitites, we have
$$v = \lambda^a g^b \rho^c$$
Using dimension formula, we can write
$$\Rightarrow [M^0L^1T^{-1}] = [L^1]^a[L^1T^{-2}]^b[M^1L^{-3}]^c$$
$$\Rightarrow [M^0L^1T^{-1}] = [M^cL^{a+b-3c}T^{-2b}]$$
Comparing dimensions of LHS and RHS gives
$$c = 0,\ a + b - 3c = 1,\ -2b = -1$$
$$\Rightarrow b = \frac{1}{2}$$
and $$a + b - 3c = 1$$
$$\Rightarrow a + \frac{1}{2} - 0 = 1$$
$$\Rightarrow a = \frac{1}{2}$$
$$\Rightarrow a = \frac{1}{2}, b = \frac{1}{2}, c = 0$$ **Ans. (A)**

2.1 Motion in One Dimension

Sol. 1 Considering uniform retardation

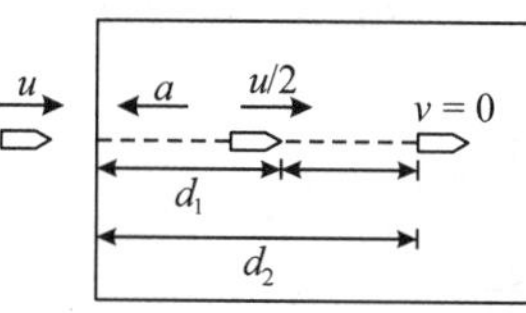

Wooden block

We use $$v^2 = u^2 - 2ad$$

Initially $$\left(\frac{u}{2}\right)^2 = u^2 - 2ad_1$$

$$\Rightarrow \frac{3}{4}u^2 = 2a(3)$$

$$\Rightarrow \quad a = \frac{u^2}{8}$$

Finally $\quad (0)^2 = u^2 - 2ad_2$

$$\Rightarrow \quad d_2 = 4\text{ cm}$$

$\Rightarrow$ more distance is $d_2 - d_1 = 4 - 3 = 1$ cm **Ans. (A)**

Sol. 2

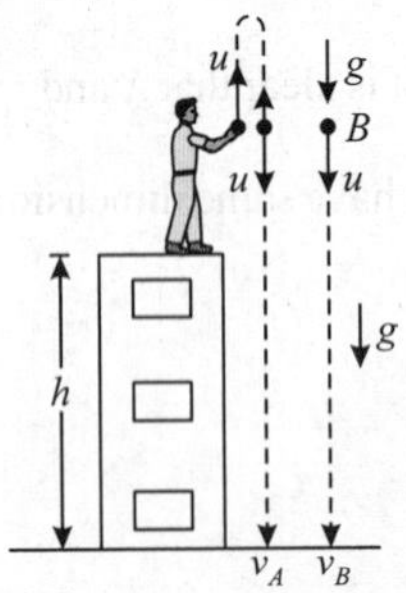

for both balls A & B, we use

$$v^2 = u^2 + 2gh$$

$$\Rightarrow \quad v_A = v_B$$ **Ans. (B)**

Sol. 3

$$s_1 = \frac{u^2}{2a}$$

for car having initial speed $4u$

$$s_2 = \frac{(4u)^2}{2a} = \frac{8u^2}{a}$$

$$\frac{s_1}{s_2} = \frac{1}{16}$$ **Ans. (D)**

Sol. 4 We use $\quad v^2 = u^2 - 2as$; $\quad a$ = constant

For stopping $\quad v = 0$

$\Rightarrow$ stopping distance

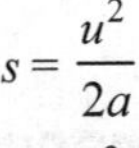

$$s = \frac{u^2}{2a}$$

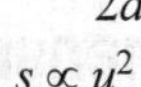

$$\Rightarrow \quad s \propto u^2$$

if u becomes doubled

$$\Rightarrow \quad s_f = 4s_i = 4 \times 6 = 24\text{ m}$$ **Ans. (C)**

Sol. 5

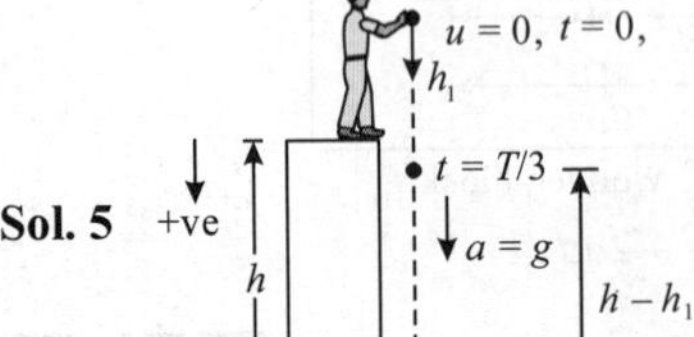

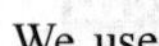

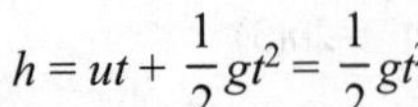

We use $\quad h = ut + \frac{1}{2}gt^2 = \frac{1}{2}gt^2$

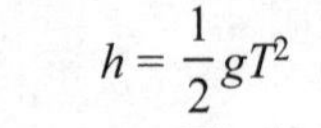

$$h = \frac{1}{2}gT^2$$

at $\quad t = T/3, h = h_1$

$$\Rightarrow \quad h_1 = \frac{1}{2}g\left(\frac{T}{3}\right)^2 = \frac{1}{2}g\left(\frac{T}{9}\right)^2 = \frac{h}{9}$$

distance from ground

$$h - h_1 = h - \frac{h}{9} = \frac{8h}{9}$$ **Ans. (C)**

Sol. 6 With reference to solution of question 4 came in 2003, we use

stopping distance $\quad s \propto u^2$

As u is getting doubled

$$\Rightarrow \quad s_f = 4s_i$$

$$s_f = 4 \times 20$$

$$= 80\text{ m}$$ **Ans. (D)**

Sol. 7 We are given $t = ax^2 + bx$

differentiate w.r. to time gives

$$1 = 2ax\left(\frac{dx}{dt}\right) + b\left(\frac{dx}{dt}\right) \quad \ldots(1)$$

$$\Rightarrow \quad x = \frac{1 - b\left(\frac{dx}{dt}\right)}{2a\left(\frac{dx}{dt}\right)}$$

again differentiating equation-(1) w.r. to time gives

$$0 = 2a\left[\left(\frac{dx}{dt}\right)^2 + x\left(\frac{d^2x}{dt^2}\right)\right] + b\left[\frac{d^2x}{dt^2}\right]$$

Acceleration is given as

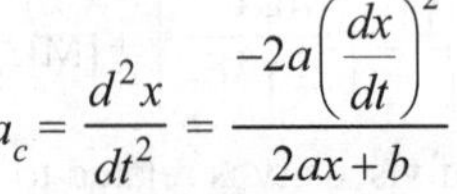

$$\Rightarrow \quad a_c = \frac{d^2x}{dt^2} = \frac{-2a\left(\frac{dx}{dt}\right)^2}{2ax + b}$$

[considering $v = dx/dt$]

$$a_c = \frac{-2av^2}{2a\left(\frac{1 - bv}{2av}\right) + b} = -2av^3$$ **Ans. (C)**

Sol. 8 We draw v-t curve for motion of car as shown in figure and for first part of journey, we use

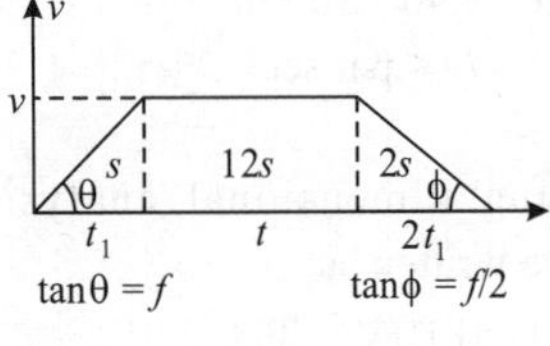

$$v = \sqrt{2fs}$$

$$\Rightarrow \quad 12s = vt$$

$$\Rightarrow \quad 12s = \sqrt{2fs}\,.\,t$$

$$\Rightarrow \quad 144s^2 = 2fs.\,t^2$$

$$\Rightarrow \quad s = \frac{1}{72}ft^2$$ **Ans. (C)**

Sol. 9 Average acceleration is given as

$$\bar{a} = \left|\frac{\vec{v}_f - \vec{v}_i}{\Delta t}\right| = \frac{|\Delta\vec{v}|}{\Delta t} = \frac{5\sqrt{2}}{10}\ \text{m/s}^2$$

$$\bar{a} = \frac{1}{\sqrt{2}}\ \text{m/s}^2$$

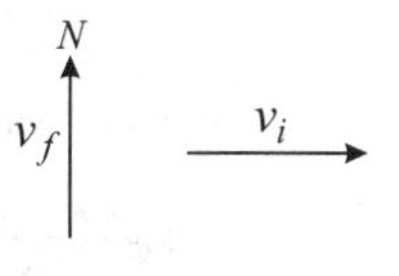

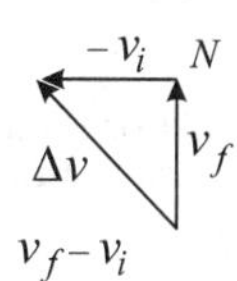

here $$|\overrightarrow{\Delta v}| = \sqrt{5^2+5^2} = 5\sqrt{2}\ \text{m/s}$$

along N-W direction. **Ans. (A)**

Sol. 10 We use

$$v^2 = u^2 - 2ah \quad \text{(for motion after parachute opens)}$$

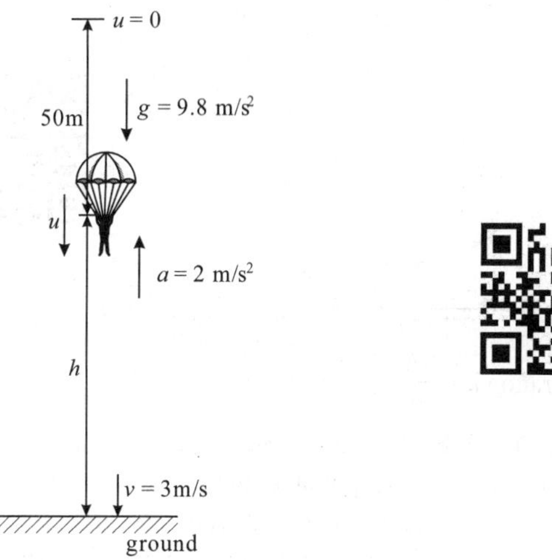

Here we use

$$u = \sqrt{2gs} = \sqrt{2\times 9.8\times 50} = \sqrt{980}\ \text{m/s}$$

$$\Rightarrow \quad (3)^2 = 980 - 2\times 2\times h$$

$$\Rightarrow \quad h = \frac{971}{4} = 242.7\ \text{m}$$

Total height

$$h_T = 242.7 + 50 = 292.7\ \text{m}$$ **Ans. (C)**

Sol. 11 We are given $v = \alpha\sqrt{x}$

$$\Rightarrow \quad \frac{dx}{dt} = \alpha\sqrt{x}$$

$$\Rightarrow \quad \frac{dx}{\sqrt{x}} = \alpha dt$$

Integrating above relation gives

$$\int_{x=0}^{x=x}\frac{dx}{\sqrt{x}} = \int_{t=0}^{t=t}\alpha dt$$

$$\Rightarrow \quad [2\sqrt{x}]_0^x = \alpha[t]_0^t$$

$$\Rightarrow \quad 2\sqrt{x} = \alpha t$$

$$\Rightarrow \quad x = \frac{1}{4}\alpha^2 t^2$$

$$\Rightarrow \quad x \propto t^2$$ **Ans. (C)**

Sol. 12 We are given $v = v_0 + gt + ft^2$

$$\Rightarrow \quad \frac{dx}{dt} = v_0 + gt + ft^2$$

$$\Rightarrow \quad dx = (v_0 + gt + ft^2)\,dt$$

integrating above relation gives

$$\int_{x=0}^{x=x} dx = \int_{t=0}^{t=1}(v_0 + gt + ft^2)dt$$

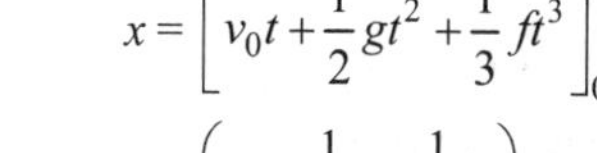

$$\Rightarrow \quad x = \left[v_0 t + \frac{1}{2}gt^2 + \frac{1}{3}ft^3\right]_0^1$$

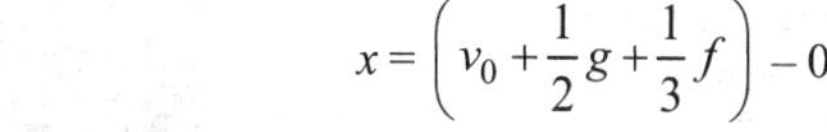

$$\Rightarrow \quad x = \left(v_0 + \frac{1}{2}g + \frac{1}{3}f\right) - 0$$ **Ans. (B)**

Sol. 13 Deceleration of particle is

$$a = \frac{dv}{dt} = -2.5\sqrt{v}$$

$$\Rightarrow \quad \frac{dv}{\sqrt{v}} = -2.5\,dt$$

Integrating above relation gives

$$\int_{6.25}^{0}\frac{dv}{\sqrt{v}} = \int_{t=0}^{t=t}-2.5dt$$

$$\Rightarrow \quad \left[2\sqrt{v}\right]_{6.25}^{0} = -2.5[t]_0^t$$

$$\Rightarrow \quad -2\times 2.5 = -2.5t \Rightarrow t = 2\text{s}$$ **Ans. (B)**

Sol. 14 As train is moving at constant acceleration, we use

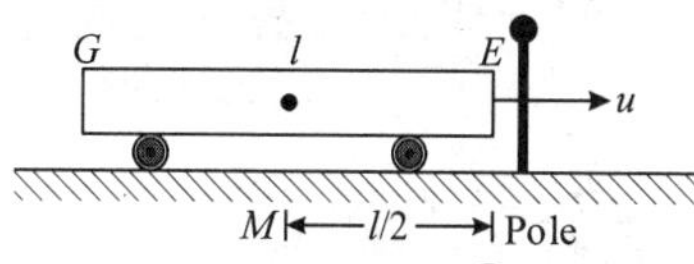

$$v^2 = u^2 + 2al \qquad \ldots(1)$$

and $$v_m^2 = u^2 + 2a(l/2) \qquad \ldots(2)$$

from (1) $$al = \frac{v^2 - u^2}{2}$$

from (2) $$v_m^2 = u^2 + \frac{v^2 - u^2}{2}$$

$\Rightarrow$ $$v_m = \sqrt{\frac{u^2 + v^2}{2}}$$ **Ans. (D)**

Sol. 15 As graph is a straight line the instantaneous velocity is equal to the average velocity.

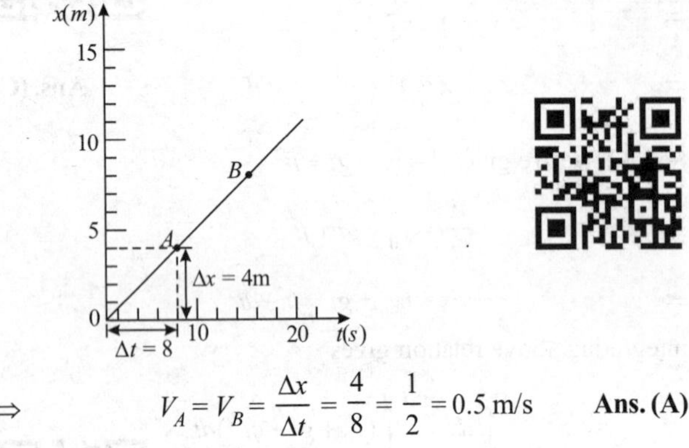

$\Rightarrow$ $$V_A = V_B = \frac{\Delta x}{\Delta t} = \frac{4}{8} = \frac{1}{2} = 0.5 \text{ m/s}$$ **Ans. (A)**

Sol. 16 We are given that

$$x = Ct^3$$

Velocity of particle is given as

$$v = \frac{dx}{dt} = 3Ct^2$$

$\Rightarrow$ Acceleration of particle

$$a = \frac{dv}{dt} = 6Ct \Rightarrow a \propto t$$ **Ans. (B)**

Sol. 17 Total time taken

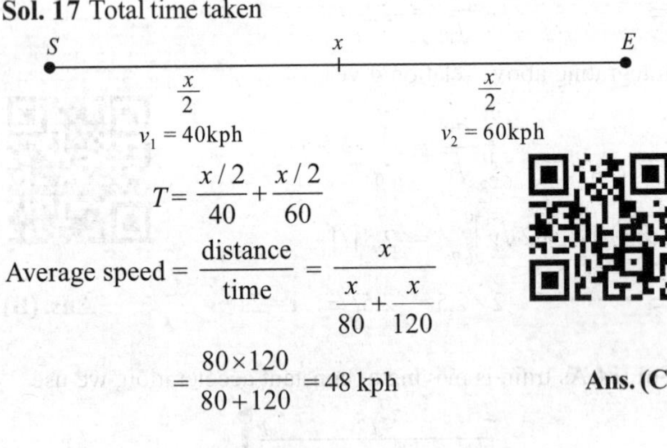

$$T = \frac{x/2}{40} + \frac{x/2}{60}$$

$$\text{Average speed} = \frac{\text{distance}}{\text{time}} = \frac{x}{\frac{x}{80} + \frac{x}{120}}$$

$$= \frac{80 \times 120}{80 + 120} = 48 \text{ kph}$$ **Ans. (C)**

Sol. 18

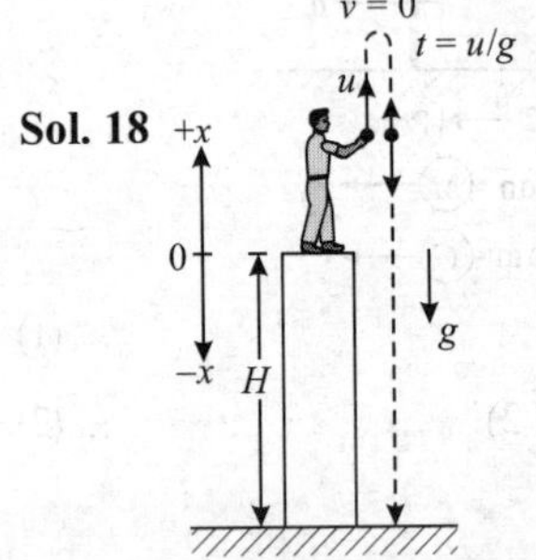

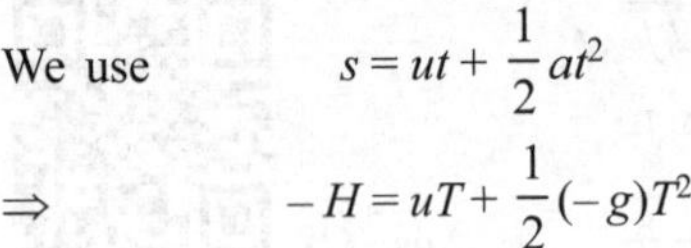

We use $$s = ut + \frac{1}{2}at^2$$

$\Rightarrow$ $$-H = uT + \frac{1}{2}(-g)T^2$$

Given that $$T = nt = \frac{nu}{g}$$

$\Rightarrow$ $$-H = u\left(\frac{nu}{g}\right) - \frac{1}{2}g\left(\frac{nu}{g}\right)^2$$

$\Rightarrow$ $$H = -\frac{nu^2}{g} + \frac{n^2u^2}{2g}$$

$\Rightarrow$ $$2gH = nu^2(n-2)$$ **Ans. (C)**

Sol. 19 If length of escalator is l and velocity of man's walks is v_1 and velocity escalator's motion is v_2.

Then we have $60 = \frac{l}{v_1} \Rightarrow v_1 = \frac{l}{60}$

and $40 = \frac{l}{v_2} \Rightarrow v_2 = \frac{l}{40}$

When man is walking up on moving escalator then time taken is

$$t = \frac{l}{v_1 + v_2} = \frac{l}{l/60 + l/40} = 24 \text{ s}$$ **Ans. (C)**

Sol. 20

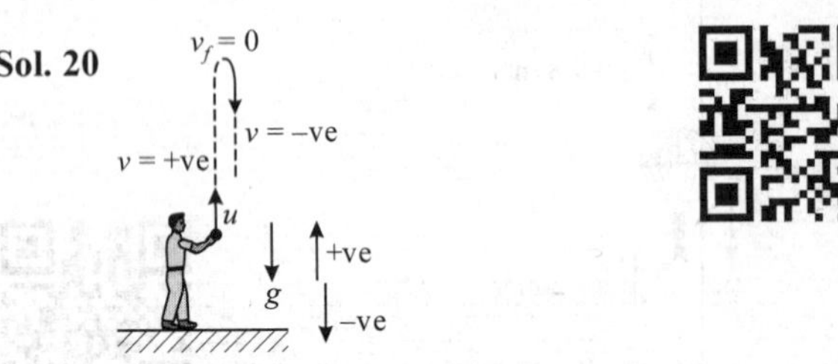

Acceleration $a = \frac{dv}{dt} = -g$ (constant)

Option (D) can be discarded as at one time instant velocity in two directions is NOT possible and options (B) can also be discarded as after topmost point direction (sign) of velocity is not changing in this case and option (C) can be discarded as velocity increases from beginning in this case. **Ans. (A)**

Sol. 21 v +ve, $v_f = 0$, x, a −ve, speed decreases

Acceleration of object is taken as

$$a = -v\frac{dv}{dx}$$

slope of v-x curve $\frac{dv}{dx} = -\frac{a}{v}$

with time initially v decreases so

$$\left|\frac{dv}{dx}\right| = \frac{a}{v} \text{ increases}$$

This is only shown in option (C). **Ans. (C)**

Sol. 22 (C)

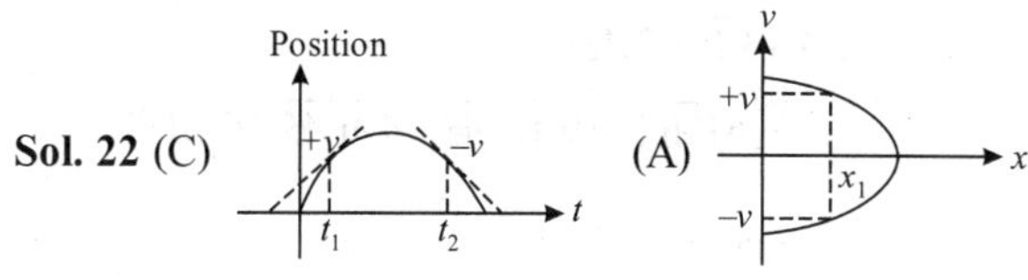

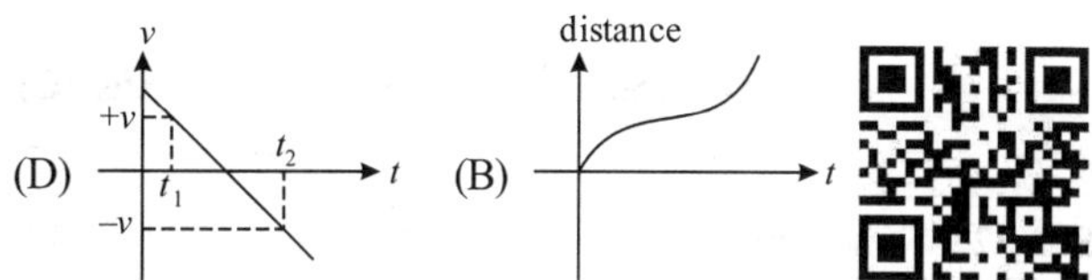

Actual distance-time curve for the given situation is shown above which speed is first increasing then decreasing as in the option (B) curve. **Ans. (B)**

Sol. 23 In first case $u_1 = 40$ kph

Stopping distance $s_1 = 40$ m

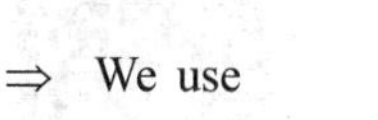

$\Rightarrow$ We use $s_1 = \dfrac{u_1^2}{2a}$

$$\Rightarrow \quad a = \frac{u_1^2}{2s_1} = \frac{1600}{2\times 40} = 20 \text{ units}$$

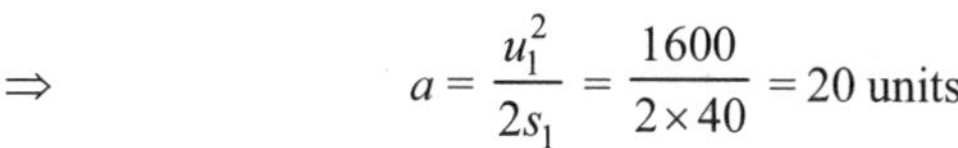

In second case $u_2 = 80$ km/h

Stopping distance $s_2 = \dfrac{u_2^2}{2a} = \dfrac{6400}{2\times 20} = 160$ m **Ans. (B)**

Sol. 24 Distance travelled by car in 15s is

$$S_{\text{car}} = \frac{1}{2}\left(\frac{45}{15}\right)(15)^2 = \frac{1}{2}\times 45\times 15 = \frac{675}{2}\text{ m}$$

Distance travelled by scooter in 15s is

$$S_{\text{scooter}} = vt = 30\times 15 = 450 \text{ m}$$

difference $S_{\text{scooter}} - S_{\text{car}} = 450 - \dfrac{675}{2} = \mathbf{112.5\ m}$

To verify: If car catches the scooter at time t then we use

$$S_{\text{car}\,t} = S_{\text{scooter}\,t} \qquad [t > 15\text{s}]$$

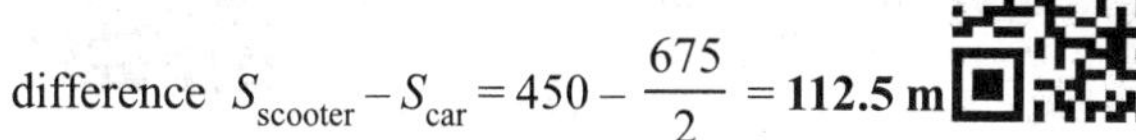

$$\frac{675}{2} + 45(t-15) = 30t$$

$$\Rightarrow \quad 15t = \frac{675}{2} \Rightarrow t = \mathbf{22.5\ s}$$ **Ans. (C)**

Sol. 25 The time taken by car to reach point P is

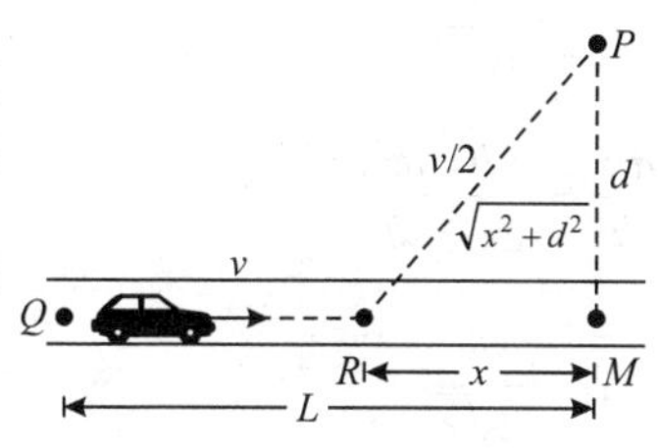

$$t = \frac{L-x}{v} + \frac{\sqrt{x^2+d^2}}{v/2}$$

t is minimum when $\dfrac{dt}{dx} = 0$

$$\Rightarrow \quad \frac{dt}{dx} = \frac{1}{v}\left[-1 + 2\frac{1}{2\sqrt{x^2+d^2}}\cdot 2x\right] = 0$$

$$\Rightarrow \quad x = \frac{d}{\sqrt{4-1}}$$

$$\Rightarrow \quad x = \frac{d}{\sqrt{3}}$$ **Ans. (A)**

Sol. 26 For A & B let time taken by A is t_0

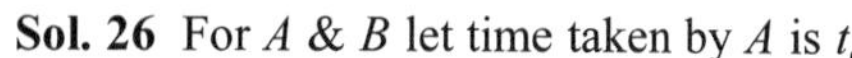

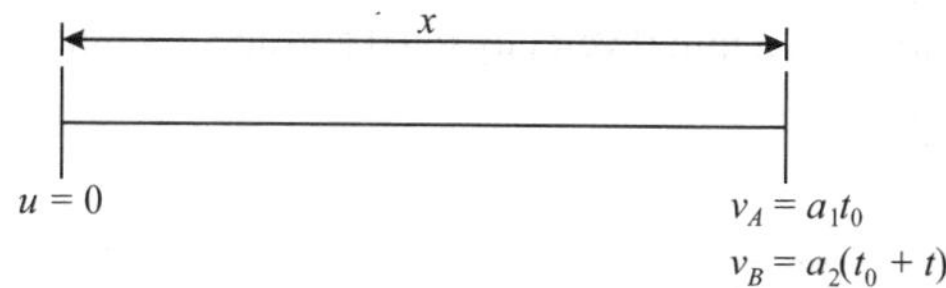

Given that

$$v_A - v_B = v = (a_1 - a_2)t_0 - a_2 t \qquad \ldots(1)$$

$$\Rightarrow \quad x_B = x_A = \frac{1}{2}a_1 t_0^2 = \frac{1}{2}a_2(t_0+t)^2$$

$$\Rightarrow \quad \sqrt{a_1}t_0 = \sqrt{a_2}(t_0+t)$$

$$\Rightarrow \quad (\sqrt{a_2} - \sqrt{a_2})t_0 = \sqrt{a_2}t \qquad \ldots(2)$$

Putting t_0 in equation, we have

$$\Rightarrow \quad v = (a_1 - a_2)\left(\frac{\sqrt{a_2}t}{\sqrt{a_1}-\sqrt{a_2}}\right) - a_2 t$$

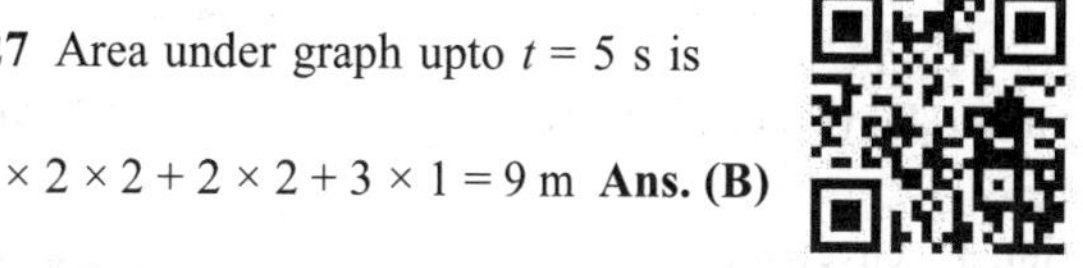

$$\Rightarrow \quad v = (\sqrt{a_1}+\sqrt{a_2})\sqrt{a_2}t - a_2 t$$

$$\Rightarrow \quad v = \sqrt{a_1 a_2}t + a_2 t - a_2 t$$

$$\Rightarrow \quad v = \sqrt{a_1 a_2}t$$ **Ans. (D)**

Sol. 27 Area under graph upto $t = 5$ s is

$s = \dfrac{1}{2}\times 2\times 2 + 2\times 2 + 3\times 1 = 9$ m **Ans. (B)**

Sol. 28 Displacement of particle is

$$\vec{s} = \vec{u}t + \frac{1}{2}at^2$$

$$\Rightarrow \quad \vec{s} = (5\hat{i}+4\hat{j})2 + \frac{1}{2}(4\hat{i}+4\hat{j})4$$

$$\Rightarrow \quad \vec{s} = 10\hat{i}+8\hat{j}+8\hat{i}+8\hat{j}$$

$$\Rightarrow \quad \vec{r}_f - \vec{r}_i = 18\hat{i}+16\hat{j}$$

$\Rightarrow \quad \vec{r}_f = 20\hat{i} + 20\hat{j}$ (As $\vec{r}_i = 2\hat{i} + 4\hat{j}$)

$\Rightarrow \quad |\vec{r}_f| = 20\sqrt{2}$ **Ans. (A)**

Sol. 29 When trains travel in same direction, time taken is

$$t_1 = \frac{180\times18}{(80-30)\times5} = \frac{180\times18}{50\times5}\text{ s} \quad \ldots(1)$$

When trains travel in opposite direction, time taken is

$$t_2 = \frac{180\times18}{(80+30)\times5} = \frac{180\times18}{110\times5}\text{ s} \quad \ldots(2)$$

From (1) and (2), we have

$$\frac{t_1}{t_2} = \frac{11}{5}$$ **Ans. (D)**

Sol. 30 Given initial velocity $u = 0$ and acceleration is constant

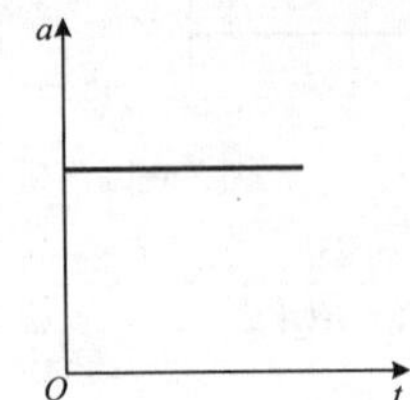

At time t, velocity is given as

$$v = 0 + at$$

$\Rightarrow \quad v = at$

and displacement is given as

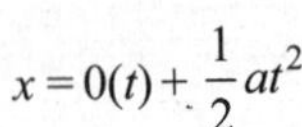

$$x = 0(t) + \frac{1}{2}at^2$$

$\Rightarrow \quad x = \frac{1}{2}at^2$

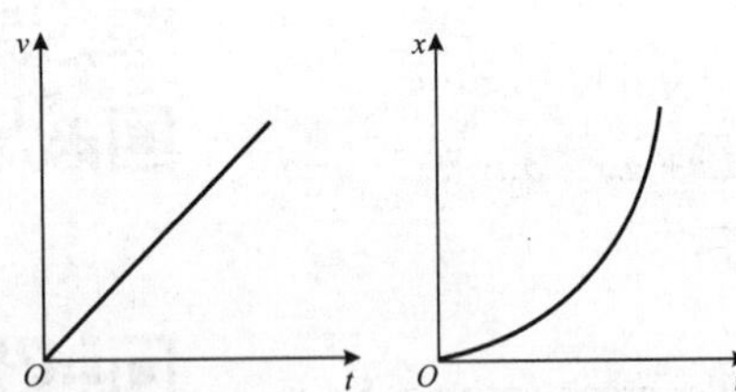

Graph (1) ; (2) and (4) are correct. **Ans. (C)**

Sol. 31 Given that

$$|\vec{A}_1| = 3$$

$$|\vec{A}_2| = 5$$

$$|\vec{A}_1 + \vec{A}_2| = 5$$

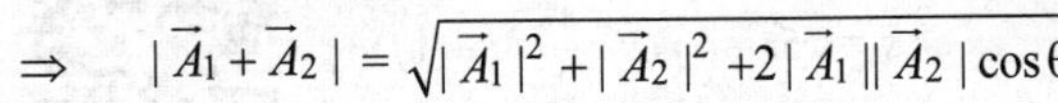

$\Rightarrow \quad |\vec{A}_1 + \vec{A}_2| = \sqrt{|\vec{A}_1|^2 + |\vec{A}_2|^2 + 2|\vec{A}_1||\vec{A}_2|\cos\theta}$

$\Rightarrow \quad 5 = \sqrt{9 + 25 + 2\times3\times5\cos\theta}$

$\Rightarrow \quad \cos\theta = -\frac{9}{2\times3\times5} = -\frac{3}{10}$

$\Rightarrow \quad (2\vec{A}_1 + 3\vec{A}_2)\cdot(3\vec{A}_1 - 2\vec{A}_2)$

$$= 6|\vec{A}_1|^2 + 9\vec{A}_1\cdot\vec{A}_2 - 4\vec{A}_1\vec{A}_2 - 6|\vec{A}_2|^2$$

$$= 54 + 5\times3\times5\left(-\frac{3}{10}\right) - 6\times25$$

$$= 54 - 150 - \frac{45}{2} = -118.5$$ **Ans. (C)**

Sol. 32 Momentum

$$p = mv \quad \ldots(1)$$

For motion under gravity, we use

$$h = \frac{u^2 - v^2}{2g} \quad \ldots(2)$$

$$h = \frac{u^2 - p^2/m^2}{2g}$$

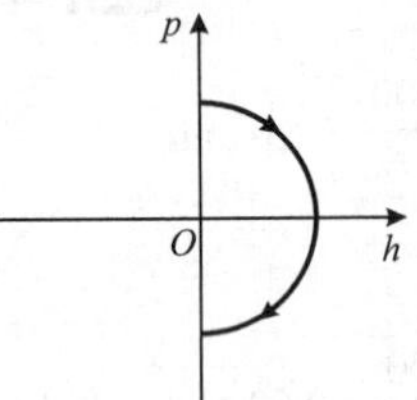

Ans. (A)

Sol. 33 Given that

$$x = at + bt^2 - ct^3$$

$\Rightarrow \quad v = \frac{dx}{dt} = a + 2bt - 3ct^2$

$\Rightarrow \quad a = \frac{dv}{dt} = 2b - 6ct = 0$

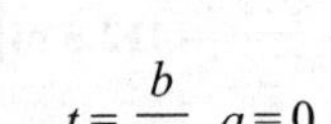

$\Rightarrow$ At $\quad t = \frac{b}{3c}, a = 0$

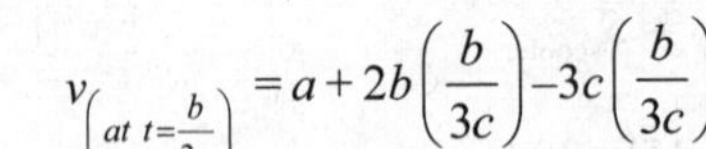

$\Rightarrow \quad v_{\left(at\ t=\frac{b}{3c}\right)} = a + 2b\left(\frac{b}{3c}\right) - 3c\left(\frac{b}{3c}\right)$

$$= a + \frac{b^2}{3c}$$ **Ans. (D)**

Sol. 34 Given that

$$v = b\sqrt{x}$$

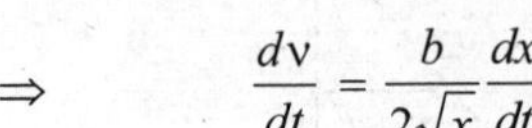

$\Rightarrow \quad \frac{dv}{dt} = \frac{b}{2\sqrt{x}}\frac{dx}{dt}$

$\Rightarrow \quad a = \frac{bv}{2\sqrt{x}}$

$$\Rightarrow \quad a = \frac{b(b\sqrt{x})}{2\sqrt{x}} = \frac{b^2}{2}$$

For constant acceleration velocity after time $t = \tau$ is given as

$$v = 0 + at$$

$$\Rightarrow \quad v = \frac{b^2}{2}\tau$$ **Ans. (B)**

Sol. 35 Total time taken by ball to travel 81 m is t s and for complete 100 m is $\left(t+\frac{1}{2}\right)$ s, by speed equation, we use

$$\Rightarrow \quad 81 = \frac{1}{2} \times g \times t^2$$

$$\Rightarrow \quad t = 9\sqrt{\frac{2}{g}}$$

$$\Rightarrow \quad 100 = \frac{1}{2} \times g \times \left(t+\frac{1}{2}\right)^2$$

$$\Rightarrow \quad t + \frac{1}{2} = 10\sqrt{\frac{2}{g}}$$

$$\Rightarrow \quad 9\sqrt{\frac{2}{g}} + \frac{1}{2} = 10\sqrt{\frac{2}{g}}$$

$$\Rightarrow \quad \frac{1}{2} = \sqrt{\frac{2}{g}}$$

$$\Rightarrow \quad a = 8 \text{ m/s}^2$$ **Ans. (8)**

Sol. 36 From the given expression, we have

$$x = \sqrt{at^2 + 2bt + c}$$

Velocity of particle is given as

$$v = \frac{dx}{dt} = \frac{1}{2\sqrt{at^2+2bt+c}} \times (2at + 2b)$$

$$\Rightarrow \quad v = \frac{at+b}{x}$$

$$\Rightarrow \quad vx = at + b$$

Differentiating above expression gives

$$vdx + xdv = adt$$

$$\Rightarrow \quad x\frac{dv}{dx} + v = a \times \frac{dt}{dx}$$

Multiplying both side by v, we get

$$\left(v\frac{dv}{dx}\right)x + v^2 = a$$

We use acceleration $f = v\frac{dv}{dx}$ so we have

$$fx = a - v^2$$

$$\Rightarrow \quad fx = a - \left(\frac{at+b}{x}\right)^2$$

$$\Rightarrow \quad fx = \frac{ax^2 - (at+b)^2}{x^2}$$

$$\Rightarrow \quad fx = \frac{a(at^2+2bt+c) - (at+b)^2}{x^2}$$

$$\Rightarrow \quad fx = \frac{ac - b^2}{x^2}$$

$$\Rightarrow \quad f = \frac{ac - b^2}{x^3}$$

$$\Rightarrow \quad f \propto \frac{1}{x^3}$$

$$\Rightarrow \quad n = 3$$ **Ans. (3.00)**

Sol. 37 Distance = Area under curve

$$s = \frac{1}{2} \times 5 \times 8 = 20 \text{ m}$$ **Ans. (20)**

Sol. 38

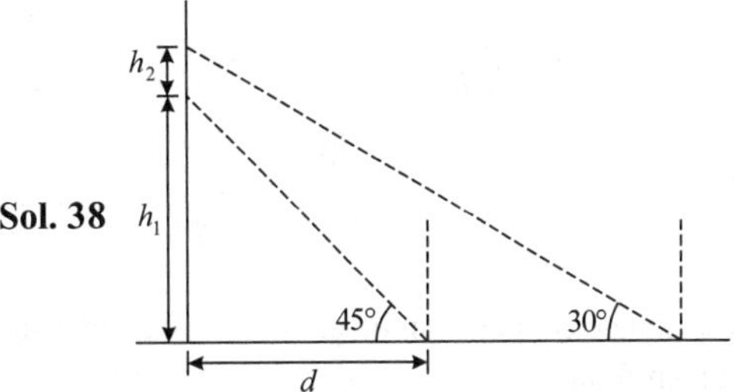

From above figure, we use

$$\frac{h_1}{d} = \tan 45°$$

$$\Rightarrow \quad h_1 = d \qquad \ldots(1)$$

and

$$\frac{h_1 + h_2}{d + 2.464d} = \tan 30°$$

$$\Rightarrow \quad (h_1 + h_2) \times \sqrt{3} = 3.46d$$

$$\Rightarrow \quad (h_1 + h_2) = \frac{3.46d}{\sqrt{3}}$$

$$\Rightarrow \quad d + h_2 = \frac{3.46d}{\sqrt{3}}$$

$$\Rightarrow \quad h_2 = d$$ **Ans. (A)**

Sol. 39

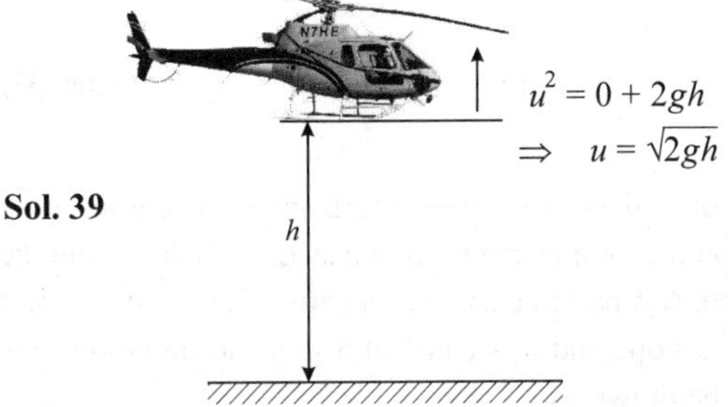

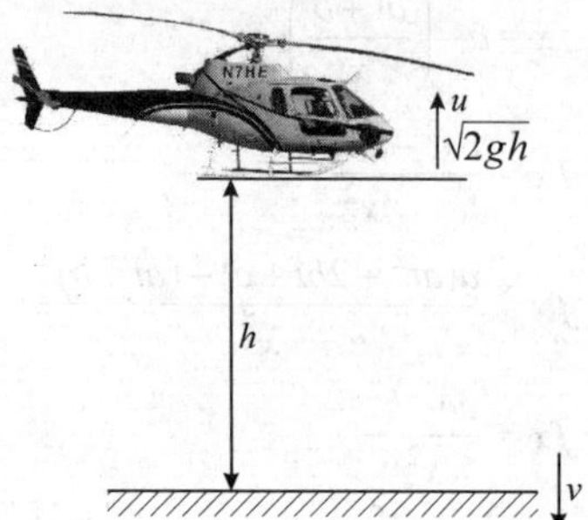

For food packet we use

$$-h = \sqrt{2gh}t - \frac{1}{2}gt^2$$

$$\Rightarrow \quad t^2 - 2\sqrt{\frac{2h}{g}}t - \frac{2h}{g} = 0$$

$$\Rightarrow \quad t = 2\sqrt{\frac{2h}{g}} \pm \sqrt{\frac{8h}{g} + \frac{2h}{g}}$$

$$\Rightarrow \quad t = \sqrt{\frac{2h}{g}} + 2\sqrt{\frac{h}{g}}$$

$$\Rightarrow \quad t = (2+\sqrt{2})\sqrt{\frac{h}{g}} = 3.4\sqrt{\frac{h}{g}}$$ **Ans. (C)**

Sol. 40 Total distance covered can be given by area under the *v-t* curve. In given figure

$$OS = 4 + \frac{1}{3} = \frac{13}{3}$$

$$SD = 2 - \frac{1}{3} = \frac{5}{3}$$

Area of *OABS* is

$$A_1 = \frac{1}{2}\left[\frac{13}{3} + 1\right] \times 4 = \frac{32}{3}\text{ m}$$

Area of *SCD* is

$$A_2 = \frac{1}{2} \times \frac{5}{3} \times 2 = \frac{5}{3}\text{ m}$$

Total distance covered

$$s = |A_1| + |A_2|$$

$$\Rightarrow \quad s = \frac{32}{3} + \frac{5}{3}\text{ m}$$

$$\Rightarrow \quad s = \frac{37}{3}\text{ m}$$ **Ans. (D)**

Sol. 41 Slope of velocity-time graph gives acceleration so acceleration time graph can be drawn as given below from the given graph. AM part of curve has negative slope and MB part has positive slope and it is constant also so acceleration time curve can be drawn as

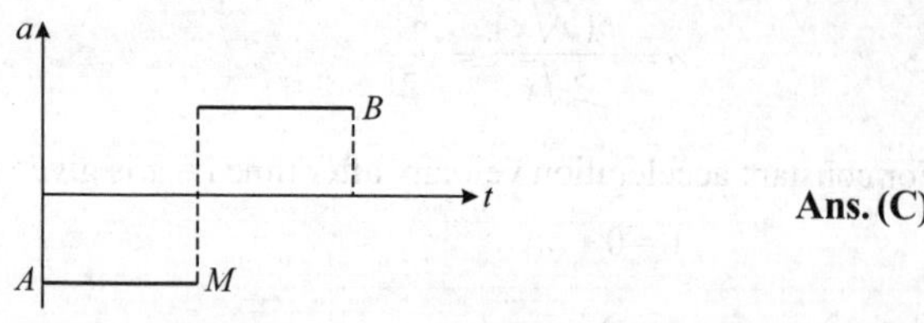

Ans. (C)

Sol. 42 Same as Question asked in 2012 Online. **Ans. (A)**

Sol. 43 Speed of first particle after falling 5m is given as

$$v = \sqrt{2gh} = 10\text{ m/s}$$

Time to fall 5m is

$$t = \sqrt{\frac{2h}{g}} = 1\text{ s}$$

Time required for first particle to meet second one can be given as

$$t' = \frac{20}{10} = 2\text{ s}$$

Time taken by I[st] particle to reach ground = 1 + 2 = 3 s

Thus height of building is given as the distance fallen by first stone in 3s, calculated as

$$H = \frac{1}{2}g(3)^2 = 45\text{ m}$$ **Ans. (B)**

Sol. 44 As in this situation scooter is accelerating and retarding in same journey so we should draw *v-t* curve of the situation as

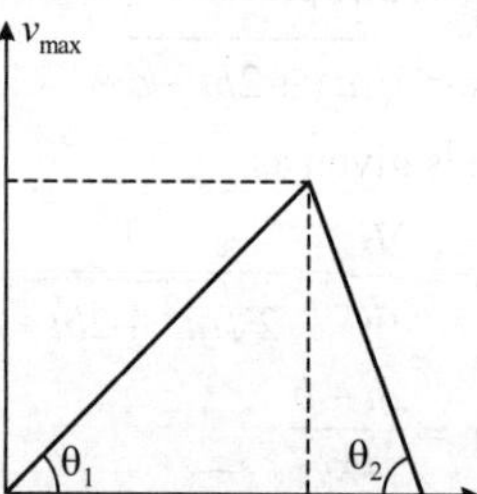

In above curve, we use

$$\tan\theta_1 = a_1 = \frac{v_{max}}{t_1}$$

and $$\tan\theta_2 = a_2 = \frac{v_{max}}{t_2}$$

From above two equations, we have

$$\frac{t_1}{t_2} = \frac{a_2}{a_1}$$ **Ans. (B)**

Sol. 45 For motion of bicycle from $0 \le x \le 200$, velocity is given as

$$v = \frac{1}{5}x + 10$$

Thus acceleration is calculated as

$$a = \frac{vdv}{dx} = \left(\frac{x}{5} + 10\right)\left(\frac{1}{5}\right)$$

$$\Rightarrow \quad a = \frac{x}{25} + 2$$

At $x = 0$, $a = 2$ m/s^2 and at $x = 200$, $a = 10$ m/s^2 so upto $x = 200$ m, acceleration vs displacement curve will be a straight line and then onward $a = 0$. Hence the curve should be drawan like the one shown in figure below.

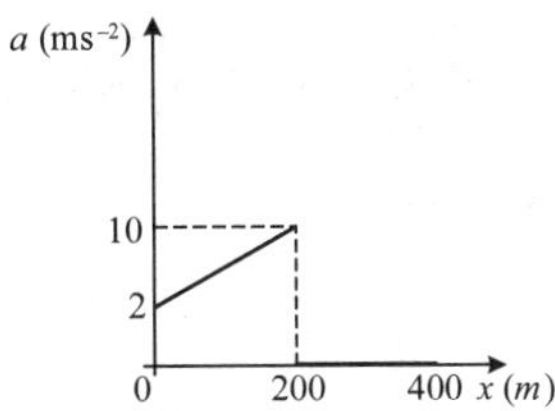

None of the given option matches with this. Most approriate shape of curve matches with option (A), or none would be correct. **Ans. (A)**

Sol. 46 Acceleration of the body is given as

$$\vec{a} = \frac{\vec{F}}{m} = \frac{2\hat{i} + 3\hat{j} + 5\hat{k}}{2} = \hat{i} + 1.5\hat{j} + 2.5\hat{k}$$

After time t, the displacement is given as

$$\vec{s} = \vec{u}t + \frac{1}{2}\vec{a}t^2$$

$$\Rightarrow \quad \vec{s} = 0 + \frac{1}{2}(\hat{i} + 1.5\hat{j} + 2.5\hat{k})(16)$$

$$\Rightarrow \quad \vec{s} = 8\hat{i} + 12\hat{j} + 20\hat{k}$$

$$\Rightarrow \quad b = 12$$ **Ans. (12)**

Sol. 47 If car accelerates for time t_1 and retards for time t_2 and the maximum speed of car in this journey is v_0, the velocity time curve of the car can be drawn as shown in figure below

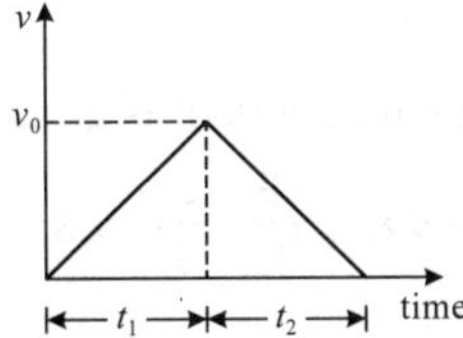

From graph we can use

$$v_0 = \alpha t_1 = \beta t_2$$

Total time elapsed is given as

$$t_1 + t_2 = t$$

$$\Rightarrow \quad v_0\left(\frac{1}{\alpha} + \frac{1}{\beta}\right) = t$$

$$\Rightarrow \quad v_0 = \frac{\alpha\beta t}{\alpha + \beta}$$

Total distance travelled can be calculated by the area under velocity time curve, given as

$$s = \frac{1}{2} \times t \times v_0 = \frac{1}{2} \times t \times \frac{\alpha\beta t}{\alpha + \beta} = \frac{\alpha\beta t^2}{2(\alpha + \beta)}$$ **Ans. (C)**

Sol. 48 The displacement of the particle can be calculated as

$$v = v_0 + gt + Ft^2$$

$$\Rightarrow \quad \frac{ds}{dt} = v_0 + gt + Ft^2$$

$$\Rightarrow \quad \int ds = \int_0^1 (v_0 + gt + Ft^2)\, dt$$

$$\Rightarrow \quad s = \left[v_0 t + \frac{gt^2}{2} + \frac{Ft^3}{3}\right]_0^1$$

$$\Rightarrow \quad s = v_0 + \frac{g}{2} + \frac{F}{3}$$ **Ans. (B)**

Sol. 49 For constant acceleration velocity time graph is straight line with positive slope and x-t graph should be an opening upward parabola so option (B) is appropriate among all given options. **Ans. (B)**

Sol. 50 From the given graph, velocity can be expressed as displacement function as

$$v = -\left(\frac{v_0}{x_0}\right)x + v_0$$

Acceleration of particle can be given as

$$a = \frac{vdv}{dx}$$

$$\Rightarrow \quad a = \left[-\left(\frac{v_0}{x_0}\right)x + v_0\right]\left[-\frac{v_0}{x_0}\right]$$

$$\Rightarrow \quad a = \left(\frac{v_0}{x_0}\right)^2 x - \frac{v_0^2}{x_0}$$

Thus appropriate graph matches with option (C). **Ans. (C)**

Sol. 51 If a drop comes from the tab after every t seconds then distance travelled by 1st drop in 4 s is given as

$$s_1 = \frac{1}{2}at_1^2 = 78.4 \text{ m}$$

Distance travelled by succeeding drop in $(4 - t)$ s

$$s_2 = \frac{1}{2}a(4 - t)^2$$

$$s_1 - s_2 = 34.3$$

$$78.4 - 4.9(4-t)^2 = 34.3$$
$$(4-t)^2 = 9$$
$$4 - t = 3$$
$$t = 1 \text{ s}$$ **Ans. (A)**

Sol. 52 Differentiating the given time equation, we have

$$1 = 2mxv + nv$$
$$1 = v(2mx + n)$$

Again differentiating we have

$$\frac{dv}{dt} \times (2mx+n) + 2mv^2 = 0$$
$$\Rightarrow \quad a = -2mv^3$$ **Ans. (C)**

Sol. 53 For the given expression of velocity, we use

$$dx = vdt$$
$$\Rightarrow \quad x = \int_1^2 \alpha t dt + \int_1^2 \beta t^2 dt$$
$$\Rightarrow \quad x = \left[\frac{\alpha t^2}{2}\right]_1^2 + \left[\frac{\beta t^3}{3}\right]_1^2$$
$$\Rightarrow \quad x = \frac{3}{2}\alpha + \frac{7}{3}\beta$$ **Ans. (B)**

Sol. 54 For stone moving upward at 10 m/s, we use

$$75 = -10t + \frac{1}{2}gt^2$$
$$\Rightarrow \quad 75 = -10t + 5t^2$$
$$\Rightarrow \quad t^2 = 2t - 15 = 0$$
$$\Rightarrow \quad t = 5 \text{ s}$$

Height of balloon at time t is given as

$$H = vt + 75$$
$$\Rightarrow \quad H = 10 \times 5 + 75 = 125 \text{ m}$$ **Ans. (A)**

Sol. 55 Throw velocity of the ball to reach a height h is given as

$$v = \sqrt{2gh}$$

If t is the time required to reach a height $h/3$ then we use

$$\frac{h}{3} = \sqrt{2gh}t - \frac{1}{2}gt^2$$
$$\Rightarrow \quad gt^2 - 2\sqrt{2gh}t + \frac{2h}{3} = 0$$
$$\Rightarrow \quad t = \frac{2\sqrt{2gh} \pm \sqrt{8gh - 4g \times \frac{2h}{3}}}{2g} = \frac{2\sqrt{2gh} \pm \sqrt{\frac{16gh}{3}}}{2g}$$
$$\Rightarrow \quad t = \frac{2\sqrt{2gh} \pm 4\sqrt{\frac{gh}{3}}}{2g}$$
$$\Rightarrow \quad \frac{t_1}{t_2} = \frac{2\sqrt{2gh} - 4\sqrt{\frac{gh}{3}}}{2\sqrt{2gh} + 4\sqrt{\frac{gh}{3}}}$$
$$\Rightarrow \quad \frac{t_1}{t_2} = \frac{2\sqrt{2} - \frac{4}{\sqrt{3}}}{2\sqrt{2} + \frac{4}{\sqrt{3}}} = \frac{\sqrt{3} - \sqrt{2}}{\sqrt{3} + \sqrt{2}}$$ **Ans. (C)**

Sol. 56 $S_1 = S_2$

$$35t + \frac{1}{2}(-g)t^2 = 35(t-3) + \frac{1}{2}(-g)(t-3)^2$$
$$35t - \frac{1}{2}gt^2 = 35t - 35 \times 3 - \frac{1}{2}g(t^2 - 6t + 9)$$
$$35 \times 3 + 45 = 30t$$
$$t = \frac{150}{30} = 5$$

height $h = 35 \times 5 - \frac{1}{2} \times 10 \times 5^2$

$$h = 175 - 125 = 50 \text{ m}$$ **Ans. (50)**

Sol. 57 Acceleration of body is given as

$$a = \frac{vdv}{dx}$$
$$\Rightarrow \quad v = \sqrt{5000 + 24x} \times \frac{1}{2\sqrt{5000 + 24x}} \times 24$$
$$\Rightarrow \quad a = 12 \text{ m/s}^2$$ **Ans. (12)**

Sol. 58 First drop strikes the floor at $t = 2$ s when third drop begins to fall. The distance from the nozzle is given as

$$h = \frac{1}{2}g(2)^2$$

2nd drop starts at $t = 1$ s so its depth below the nozzle is given as

$$h' = \frac{1}{2}g(1)^2$$
$$\Rightarrow \quad \frac{h}{h'} = \frac{4}{1}$$
$$\Rightarrow \quad h' = \frac{h}{4} = \frac{9.8}{4}$$

Height of 2nd drop above the floor is given as

$$H = h - h' = 9.8 - \frac{9.8}{4} = \frac{3}{4} \times 9.8 = 7.35 \text{ m}$$ **Ans. (D)**

Sol. 59 Acceleration of particle is given as

$$a = v\frac{dv}{dx}$$

From the graph we have the function of v^2 variation with x, given as

$$v^2 = 2x + 20$$

$$\Rightarrow \quad a = v\frac{dv}{dx} = 1 \text{ m/s}^2 \qquad \textbf{Ans. (1)}$$

Sol. 60 Let height of tower be h and speed of projection in first two cases be u.

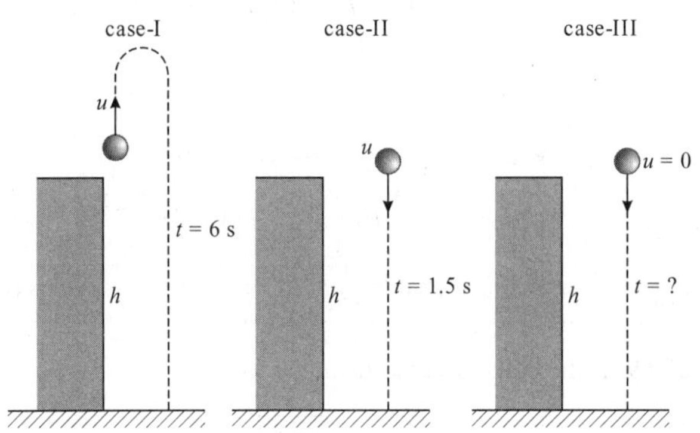

For **case-I :** Using speed equation for constant acceleration, we have

$$s = ut + \frac{1}{2}at^2$$

$$\Rightarrow \quad h = -u(6) + \frac{1}{2}g(6)^2$$

$$\Rightarrow \quad H = -6u + 18g \qquad \ldots(1)$$

For **case-II :** With same equation, we have

$$h = u(1.5) + \frac{1}{2}g(1.5)^2$$

$$\Rightarrow \quad h = 1.5u + \frac{2.25g}{2} \qquad \ldots(2)$$

Multiplying equation-(2) by 4 we get

$$4h = 6u + 4.5g \qquad \ldots(3)$$

Equation-(1) + (3) gives

$$5h = 22.5g$$

$$\Rightarrow \quad h = 4.5g \qquad \ldots(4)$$

For **case-III :** With same equation, we have

$$h = 0 + \frac{1}{2}gt^2 \qquad \ldots(5)$$

Using equation-(4) and (5), we have

$$4.5g = \frac{1}{2}gt^2$$

$$\Rightarrow \quad t^2 = 9 \Rightarrow t = 3 \text{ s} \qquad \textbf{Ans. (3)}$$

Sol. 61 As $|\vec{A}| \neq 0$, we have

$$\vec{A} \times \vec{A} = |\vec{A}||\vec{A}| \sin 0^\circ \, \hat{n} = 0 \qquad \textbf{Ans. (C)}$$

Sol. 62 Sum of two unit vectors is calculated as

$$|\hat{A} + \hat{B}| = \sqrt{|\hat{A}|^2 + |\hat{B}|^2 + 2|\hat{A}||\hat{B}|\cos\theta}$$

$$\Rightarrow \quad |\hat{A} + \hat{B}| = \sqrt{1 + 1 + 2\cos\theta}$$

$$\Rightarrow \quad |\hat{A} + \hat{B}| = \sqrt{2(1 + \cos\theta)}$$

$$\Rightarrow \quad |\hat{A} + \hat{B}| = \sqrt{2 \times 2\cos^2\frac{\theta}{2}}$$

$$\Rightarrow \quad |\hat{A} + \hat{B}| = 2\cos\frac{\theta}{2}$$

Difference of above two unit vectors is calculated as

$$|\hat{A} - \hat{B}| = \sqrt{|\hat{A}|^2 + |\hat{B}|^2 - 2|\hat{A}||\hat{B}|\cos\theta}$$

$$\Rightarrow \quad |\hat{A} - \hat{B}| = \sqrt{2 - 2\cos\theta}$$

$$\Rightarrow \quad |\hat{A} - \hat{B}| = 2\sin\frac{\theta}{2}$$

$$\Rightarrow \quad \frac{|\hat{A} + \hat{B}|}{|\hat{A} - \hat{B}|} = \cos\frac{\theta}{2} \qquad \textbf{Ans. (B)}$$

Sol. 63 For same range of projectiles as shown below, we use

$$\theta_1 + \theta_2 = 90^\circ$$

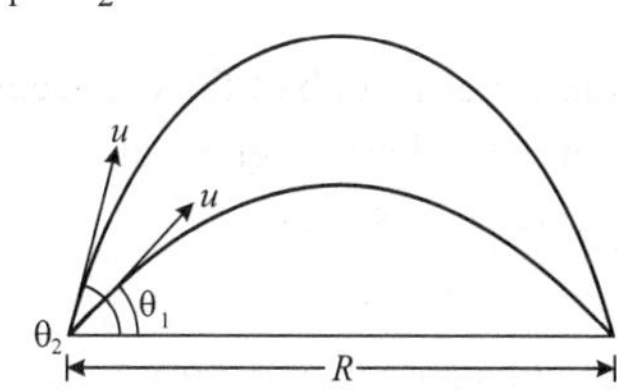

Here $\quad h_1 \quad \dfrac{u^2 \sin^2\theta_1}{2g}$

and $\quad h_2 = \dfrac{u^2 \sin^2\theta_2}{2g}$

$$\Rightarrow \quad h_1 h_2 = \frac{u^2 \sin^2\theta_1}{2g} \times \frac{u^2 \sin^2\theta_2}{2g}$$

As $\theta_2 = 90 - \theta_1$, we use

$$h_1 h_2 = \frac{u^2 \sin^2\theta_1}{2g} \cdot \frac{u^2 \sin^2\theta_2}{2g}$$

$$\Rightarrow \quad h_1 h_2 = \left[\frac{u^2 \sin\theta_1 \cos\theta_1}{2g}\right]^2$$

$$\Rightarrow \quad h_1 h_2 = \left[\frac{u^2 \sin\theta_1 \cos\theta_1}{2g} \times \frac{2}{2}\right]^2 = \frac{R^2}{16}$$

$$\Rightarrow \quad R = 4\sqrt{h_1 h_2}$$

So **R** is correct explanation of **A**. **Ans. (A)**

Sol. 64 Using speed equation for constant acceleration, we have

$$v^2 = u^2 + 2as$$

$$\Rightarrow \quad 100 = 0 + 2(10)s$$

$$\Rightarrow \quad s = 5\text{ m}$$

Height from ground = 10 – 5 = 5 m **Ans. (5)**

Sol. 65 Let they meet at $t = t$, so first ball gets t s & 2nd gets $(t-2)$ s until they will meet at same height, so we use

$$h_1 = 50t - \frac{1}{2}gt^2$$

$$\Rightarrow \quad h_2 = 50(t-2) - \frac{1}{2}g(t-2)^2$$

$$\Rightarrow \quad h_1 = h_2$$

$$\Rightarrow \quad 50t - \frac{1}{2}gt^2 = 50(t-2) - \frac{1}{2}g(t-2)^2$$

$$\Rightarrow \quad 100 = \frac{1}{2}g[t^2 - (t-2)^2]$$

$$\Rightarrow \quad 100 = \frac{10}{2}[4t-4]$$

$$\Rightarrow \quad 5 = t - 1$$

$$\Rightarrow \quad t = 6\text{ s}$$

Ans. (6)

Sol. 66 Above the water level ball falls at acceleration due to gravity so speed at water level is given as

$$v^2 = 2 \times 9.8 \times 4.9$$

$$v = 9.8\text{ m/s}$$

Now ball sinks at constant velocity so we use

Depth = distance travelled in 3 seconds

Depth = 9.8 × 3 = 29.4 m **Ans. (B)**

Sol. 67 Using velocity relation in the two systems, the distance and time dimensions are related as

$$\frac{L_2}{T_2} = \frac{n}{m^2}\frac{L_1}{T_1}$$

Using acceleration relation in the two systems, the distance and time dimensions are related as

$$\frac{L_2}{T_2^2} = \frac{L_1}{T_1^2 \times mn}$$

$$\Rightarrow \quad \frac{n}{m^2} \times \frac{T_2}{T_1} = \frac{T_2^2}{T_1^2 \times mn}$$

$$\Rightarrow \quad \frac{n^2}{m} = \frac{T_2}{T_1}$$

$$\Rightarrow \quad \frac{L_2}{L_1} = \frac{n^4}{m^2} \times \frac{1}{mn}$$

$$\Rightarrow \quad \frac{L_2}{L_1} = \frac{n^3}{m^3}$$

Ans. (A)

Sol. 68 For coverage of displacement l, the average velocity of car is given as

$$\langle \vec{v} \rangle = \frac{\text{Displacement}}{\text{time}}$$

$$\Rightarrow \quad \langle \vec{v} \rangle = \frac{l}{\left(\frac{l}{V_3} + \frac{l}{V_2} + \frac{l}{V_1}\right)\frac{1}{3}}$$

$$\Rightarrow \quad \langle \vec{v} \rangle = \frac{3}{\frac{1}{V_1} + \frac{1}{V_2} + \frac{1}{V_3}} = \frac{3}{\frac{1}{11} + \frac{1}{22} + \frac{1}{33}}$$

$$\Rightarrow \quad \langle \vec{v} \rangle = 18\text{ m/s}$$

Ans. (18)

Sol. 69 If the balls meet at time t, we use

$$t = \sqrt{\frac{2h}{g}} = \sqrt{\frac{2 \times 80}{10}}$$

$$\Rightarrow \quad t = 4\text{ s}$$

Time taken by ball B to meet A is 2 s, so using speed equation for constant acceleration, we have

$$s = ut + \frac{1}{2}at^2$$

$$\Rightarrow \quad -80 = -u \times 2 + \frac{1}{2}(-10)(2)^2$$

$$\Rightarrow \quad u = 30$$

Ans. (D)

Sol. 70 Distance covered by the charged particle before it comes to rest is given as

$$s = \frac{v^2}{2a} = \frac{v^2 m}{2qE}$$

$$\Rightarrow \quad s = \frac{(200)^2 \times 100 \times 10^{-6}}{2 \times 40 \times 10^{-6} \times 10^5} = 0.5\text{ m}$$

Ans. (D)

Sol. 71 As initial speed is zero so for constant acceleration motion of toy, we use for t seconds

$$10 = \frac{1}{2}at^2$$

For $2t$ seconds, we use, if it travels a further distance x, we have

$$10 + x = \frac{1}{2}a(2t)^2$$

$$\Rightarrow \quad \frac{10 + x}{10} = \frac{4}{1}$$

$$\Rightarrow \quad x = 30\text{ m}$$

Ans. (C)

Sol. 72 Stopping distance of car considering uniform acceleration is given as

$$d = \frac{v^2}{2a}$$

If speed is made one third, stopping distance becomes

$$d' = \frac{1}{9}d$$

$$\Rightarrow \quad d' = \frac{27}{9} = 3\text{ m}$$ **Ans. (3)**

Sol. 73 Rate of velocity increment per unit distance is given as

$$\frac{dv}{ds} = 5$$

$$\Rightarrow \quad a = v\frac{dv}{ds} = 20 \times 5 = 100\text{ m/s}^2$$ **Ans. (100)**

Sol. 74 Velocity of bullet as a function of time in downward direction is calculated by speed equation under gravity, which is given as

$$v = u - gt$$
$$v = -100 - 10t$$

Appropriate option for above linear variation of speed with time is represented by option (A). **Ans. (A)**

Sol. 75 Monkey drops a mango which falls on cadet after covering a distance *vt* as shown in figure below

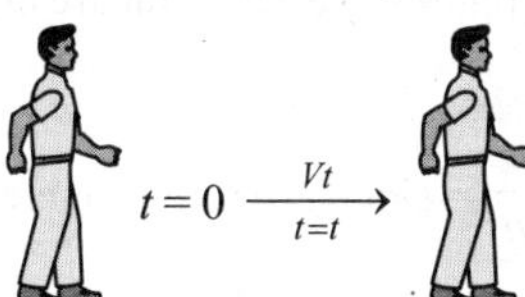

Time taken by mango to fall is given as

$$t = \sqrt{\frac{2n}{g}}$$

$$\Rightarrow \quad t = \sqrt{\frac{2 \times 19.6}{9.8}} = 2\text{ s}$$

Distance travelled by cadet is calculated as

$$s = vt$$

$$\Rightarrow \quad s = 9 \times \frac{5}{18} \times 2 = 5\text{ m}$$ **Ans. (A)**

Sol. 76 Position vector of the particle as a function of time is given as

$$\vec{r} = 3t\hat{i} + 5t^3\hat{j} + 7k$$

Acceleration of particle is given as

$$\frac{d^2\vec{r}}{dt^2} = 30t\,\hat{j}$$

At $t = 1$ s, we have

$$\frac{d^2\vec{r}}{dt^2} = 30\hat{j}$$ **Ans. (B)**

Sol. 77 Time taken by the ball to reach the top is given as

$$t_a = \frac{u}{g} = \frac{19.6}{9.8} = 2\text{s}$$

Time of downward journey of ball is calculated as

$$t_d = 6 - 2\text{s} = \sqrt{\frac{2h_{max}}{g}}$$

$$\Rightarrow \quad h_{max} = \frac{16 \times 9.8}{2} = \frac{392}{5}$$ **Ans. (392)**

Sol. 78 Figure below shows the situation explained in the question.

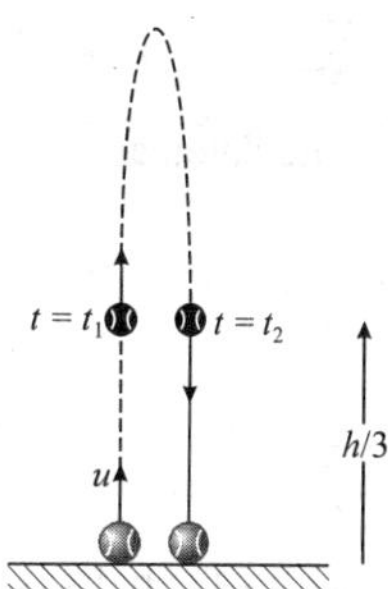

Maximum height attained by the ball is given as

$$h = \frac{u^2}{2g}$$

$$\Rightarrow \quad u = \sqrt{2gh}$$

Using speed equation under gravity for upward motion, we have

$$s = ut + \frac{1}{2}at^2$$

$$\Rightarrow \quad \frac{h}{3} = \sqrt{2gh}t + \frac{1}{2}(-g)t^2$$

$$\Rightarrow \quad \frac{gt^2}{2} - \sqrt{2gh}t + \frac{h}{3} = 0$$

Roots of above equation t_1 & t_2 are the times when ball is at a height $h/3$ so we use

$$\frac{t_2}{t_1} = \frac{\sqrt{2gh} + \sqrt{2gh - 4 \times \frac{g}{2} \times \frac{h}{3}}}{\sqrt{2gh} - \sqrt{2gh - 4 \times \frac{g}{2} \times \frac{h}{3}}}$$

$$\Rightarrow \quad \frac{t_2}{t_1} = \frac{\sqrt{2gh} + \sqrt{\frac{4gh}{3}}}{\sqrt{2gh} - \sqrt{\frac{4gh}{3}}} = \frac{\sqrt{3}+\sqrt{2}}{\sqrt{3}-\sqrt{2}}$$ **Ans. (B)**

Sol. 79 For the given expression, we use

$$t = \sqrt{x} + 4$$

$$\Rightarrow \quad x = (t-4)^2 = t^2 - 8t + 16$$

$$\Rightarrow \quad \frac{dx}{dt} = 2t - 8$$

$$\Rightarrow \quad \left.\frac{dx}{dt}\right|_{t=4} = 2 \times 4 - 8 = 0$$ **Ans. (B)**

Sol. 80 Time taken by ball to reach highest point under gravity is given as

$$t = \frac{u}{g}$$

Frequency of throw is calculated as

$$\frac{1}{t} = \frac{g}{u} = n$$

$$\Rightarrow \quad u = \frac{g}{n}$$

Maximum height attained by the balls is given as

$$H_{max} = \frac{u^2}{2g} = \frac{\left(\frac{g}{n}\right)^2}{2g} = \frac{g}{2n^2}$$ **Ans. (D)**

Sol. 81 For first half $\frac{h}{2}$, we use

$$\frac{h}{2} = \frac{1}{2}gt_1^2$$

For total height h, we use

$$h = \frac{1}{2}g(t_1 + t_2)^2$$

$$\Rightarrow \quad \frac{1}{\sqrt{2}} = \frac{t_1}{t_1 + t_2}$$

$$\Rightarrow \quad 1 + \frac{t_2}{t_1} = \sqrt{2}$$

$$\Rightarrow \quad \frac{t_1}{t_2} = \frac{1}{\sqrt{2}-1}$$

$$\Rightarrow \quad t_2 = (\sqrt{2}-1)t_1$$ **Ans. (D)**

Sol. 82 Displacement of body is calculated by area bounded by the graph with time axis including sign as displacement

$$= 16 - 8 + 16 - 8 = 16\text{ m}$$

Distance travelled by the body is calculated by area bounded by the graph with time axis considering all positive distance

$$= 48\text{ m}$$

$$\Rightarrow \quad \frac{\text{displacement}}{\text{distance}} = \frac{1}{3}$$ **Ans. (D)**

Sol. 83 For the two parts of journey of car as shown in figure below the average velocity is calculated as

x x
A v_1 B v_2 C

$$\text{Average velocity} = \frac{\text{Total displacement}}{\text{Total time}}$$

$$\langle v \rangle = \frac{x+x}{\frac{x}{v_1} + \frac{x}{v_2}} = \frac{2v_1v_2}{v_1 + v_2}$$ **Ans. (D)**

Sol. 84 The velocity of particle is calculated as

$$x = 4t^2$$

$$\Rightarrow \quad v = \frac{dx}{dt} = 8t$$

At $t = 5$ s we have

$$v = 8 \times 5 = 40\text{ m/s}$$ **Ans. (A)**

Sol. 85 From the figures, we can compare the slope of graphs, given as

$$\frac{dx}{dt} = \text{slope} \geq 0 \text{ always increasing}$$

$\Rightarrow$ (a) $\to$ (ii)

$$\frac{dx}{dt} < 0\text{; and at } t \to \infty \ \frac{dx}{dt} \to 0$$

$\Rightarrow$ (b) $\to$ (iv)

$$\frac{dx}{dt} > 0 \text{ for first half } \frac{dx}{dt} < 0 \text{ for second half}$$

$\Rightarrow$ (c) $\to$ (iii)

$$\frac{dx}{dt} = \text{constant}$$

$\Rightarrow$ (d) $\to$ (i) **Ans. (A)**

Sol. 86 If the rider travels the total distance $2x$ from A to C, we use

x mters x mters
A B C

$$t_{AB} = \frac{x}{5\text{m/s}}$$

In motion BC

$$x = d_1 + d_2$$

Here d_1 & d_2 are the distances travelled with speeds 10 m/s and 15 m/s respectively in equal time intervals $t/2$ so we have

$$d_1 = \frac{10t}{2}$$

and $$d_2 = \frac{15t}{2}$$

We have total distance in part BC given as

$$d_1 + d_2 = x = \frac{t}{2}(10+15) = \frac{25t}{2}$$

Thus total time taken in travelling the distance BC is given as

$$t = \frac{2x}{25}$$

$$\Rightarrow \quad <v> = \frac{2x}{\frac{x}{5}+\frac{2x}{25}} = \frac{2\times 25}{5+2} = \frac{50}{7} \text{ m/s}$$

$$\Rightarrow \quad <v> = 50 \text{ m/s}$$ **Ans. (50)**

Sol. 87 Average speed $= \dfrac{\text{Total distance}}{\text{Total time}}$

$$\Rightarrow \quad v = \frac{4+4}{\left(\frac{4}{3}+\frac{4}{5}\right)}$$

$$\Rightarrow \quad v = \frac{15}{4} = 3.75 \text{ km/hr}$$ **Ans. (D)**

Sol. 88 Let us consider the distances on given line as

$AB = x$

and $BC = x$

and $2x + CD = 3x$

and $CD = x$

Average speed of the object can be calculated as

$$<v> = \frac{3x}{\frac{x}{v_1}+\frac{x}{v_2}+\frac{x}{v_3}} = \frac{3v_1v_2v_3}{v_2v_3+v_1v_3+v_1v_2}$$ **Ans. (C)**

Sol. 89 By third equation of motion for the train engine, we have

$$u = 20 \text{ m/s}, s_1 = 500\, m, v = 0$$

$$0 = (20)^2 - 2a \cdot 500$$

$$\Rightarrow \quad a = \frac{4}{10} \text{ m/s}^2$$

If brakes were applied at half of the distance, we use

$$u = 20 \text{ m/s}, s_2 = 250 \text{ m}, v = ?$$

$$\Rightarrow \quad v^2 = (20)^2 - 2a.250$$

$$\Rightarrow \quad v = \sqrt{200} \text{ m/s} = \sqrt{x} \text{ m/s}$$

$$\Rightarrow \quad x = 200$$ **Ans. (200)**

Sol. 90 By first equation of uniformly accelerated motion

$$\Rightarrow \quad v = u + at$$

$$\Rightarrow \quad 60 = 10 + 2t$$

$$\Rightarrow \quad t = 25 \text{ s}$$ **Ans. (D)**

Sol. 91 Trajectory of particle projected from ground is written as

$$y = x - \frac{x^2}{20}$$

For maximum height of projectile, we use

$$\frac{dy}{dx} = 0$$

$$\Rightarrow \quad 1 - \frac{2x}{20} = 0$$

$$\Rightarrow \quad x = 10$$

$$\Rightarrow \quad y_{max} = 10 - \frac{100}{20} = 5 \text{ m}$$ **Ans. (A)**

Sol. 92 Area under velocity time graph gives displacement of body in given time. Area under acceleration time graph gives change in velocity in the given time. Thus statement-I false and statement-II True. **Ans. (A)**

Sol. 93 Range of projectile projected at angle of 15°

$$\Rightarrow \quad R = \frac{u^2 \sin 2\theta}{g}$$

$$\Rightarrow \quad R \propto \sin(2\theta)$$

Thus ratio of ranges at 15° and 45° is given as

$$\frac{R_1}{R_2} = \frac{\sin(2\theta_1)}{\sin(2\theta_2)} = \frac{\sin(2\times 15)}{\sin(2\times 45)} = \frac{\sin 30^\circ}{\sin 90^\circ}$$

$$\Rightarrow \quad \frac{50}{R_2} = \frac{1}{2}$$

$$\Rightarrow \quad R_2 = 100 \text{ m}$$ **Ans. (C)**

Sol. 94 As slope of B > Slope of A

Slope of (distance-time) graph gives velocity

$$\Rightarrow \quad V_B > V_A$$

Also, As velocity is more, time taken will be less.

$$\Rightarrow \quad t_B < t_A$$ **Ans. (A)**

Sol. 95 Average velocity of the person is calculated as

$$v = \frac{\text{Total displacement}}{\text{Total time}}$$

$$\Rightarrow \quad v = \frac{x+x}{\frac{x}{v_1}+\frac{x}{v_2}} = v$$

$$\Rightarrow \quad \frac{1}{v_1}+\frac{1}{v_2} = \frac{2}{v}$$ **Ans. (C)**

Sol. 96 Area under the graph from $t = 0$ to $t = 20$ s = 200 m
Area under the graph from $t = 20$ to $t = 25$ s = 50 m
Area under graph (speed-time) gives distance
So distance covered = (200 + 50) m = 250 m
Displacement = (200 – 50) m = 150 m
Thus ratio of distance to displacement is calculated as

$$r = \frac{250}{150} = \frac{5}{3}$$ **Ans. (C)**

Sol. 97 Velocity of body is given as

$$v = 10\sqrt{x}$$

$$\Rightarrow \quad v^2 = 100x$$

By differentiating above equestion gives

$$2v\frac{dv}{dx} = 100$$

$$\Rightarrow \quad a = 50 \text{ m/s}^2$$

$$\Rightarrow \quad F = 25 \text{ N}$$ **Ans. (B)**

Sol. 98 By first equation of uniformly accelerated motion, we have

$$\vec{v} = \vec{u} + \vec{a}t$$

$$v = 150 - 10t$$

Ratio of velocity after 3 s and 5 s can be calculated as

$$v(3) = 150 - 30 = 120$$

$$v(5) = 150 - 50 = 100$$

$$\Rightarrow \quad \frac{120}{100} = \frac{x+1}{x} = \frac{6}{5}$$

$$\Rightarrow \quad x = 5$$ **Ans. (B)**

Sol. 99 Velocity of A in m/s is calculated as

$$v_A = 30 \text{ m/s}$$

Velocity of B in m/s is calculated as

$$v_B = 20 \text{ m/s}$$

Time taken by$B$ to cross the tunnel is calculated as

$$t_A = \frac{64l}{20}$$

Time taken by$A$ to cross the tunnel is calculated as

$$t_B = \frac{61l}{30}$$

If train A takes 35 s less than train B, so we use

$$\frac{60l + 4l}{20} - \frac{61l}{30} = 35$$

$$\Rightarrow \quad l = \frac{1050}{35}$$

$$\Rightarrow \quad L = 60l = \frac{1050}{35} \times 60 = 1800 \text{ m}$$ **Ans. (C)**

Sol. 100 Velocity of train A is given as

$$V_A = 90\frac{\text{km}}{hr} = 90 \times \frac{5}{18} = 25 \text{ m/s}$$

Velocity of train B is given as

$$V_B = 54\frac{\text{km}}{hr} = 54 \times \frac{5}{18} = 15 \text{ m/s}$$

Velocity of train B w.r.t. train A is calculated as

$$V_{BA} = \vec{V}_B - \vec{V}_A$$

$$V_{BA} = 15 - (-25) \text{ m/s}$$

$$\Rightarrow \quad V_{BA} = 40 \text{ m/s}$$

$$\text{Time of crossing} = \frac{\text{length of train}}{\text{relative velocity}}$$

$$\Rightarrow \quad 8 = \frac{l}{40}$$

Thus the length of train B is given as

$$l = 8 \times 40 = 320 \text{ m}$$ **Ans. (D)**

Sol. 101 Given that distance travelled by a particle is given as

$$(s) = (2.5)t^2$$

By differentiating, we get speed of particle as

$$v = \frac{ds}{dt} = \frac{d}{dt}\{(2.5)t^2\} = 5t$$

$$\Rightarrow \quad v = 5t$$

At $t = 5$ s

$$v = 5 \times 5 = 25 \text{ m/s}$$ **Ans. (D)**

Sol. 102 Given that position of particle is

$$x = 5t^2 - 4t + 5$$

By differentiating above equation, we get

$$v = 10t - 4$$

Thus velocity at t = 2s is calculated as

$$v = 16 \text{ m/s}$$ **Ans. (C)**

3.1 Motion in Two Dimensions

Sol. 1

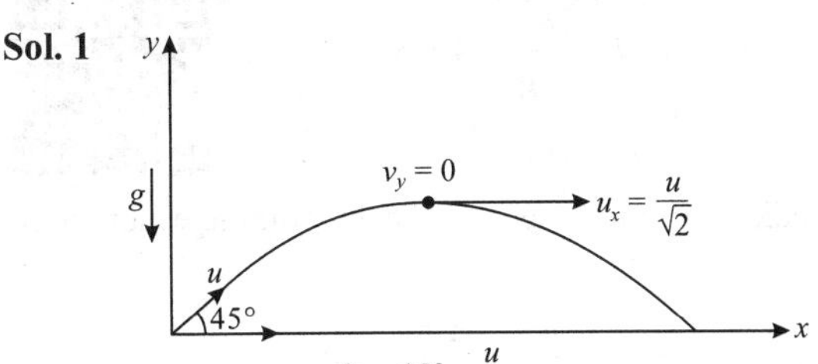

Initial KE of ball $E = \frac{1}{2}mu^2$

KE of ball at highest point of flight

$$E_H = \frac{1}{2}m\left(\frac{u}{\sqrt{2}}\right)^2 = \frac{1}{4}mu^2 = E/2$$ **Ans. (C)**

Sol. 2

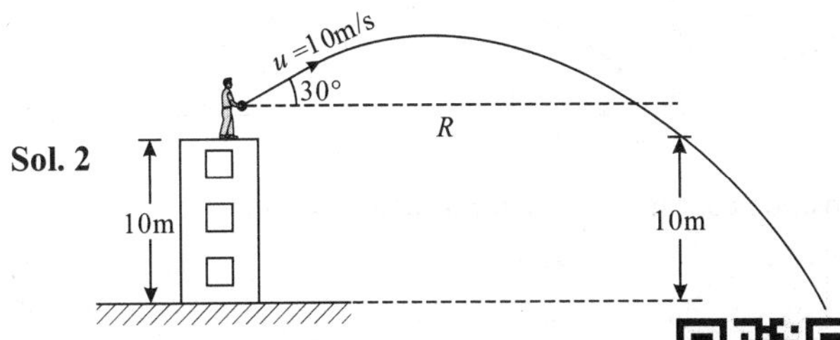

Horizontal Range of ball

$$R = \frac{u^2 \sin 2\theta}{g}$$

$$\Rightarrow \quad R = \frac{(10)^2 \sin(60°)}{10} = 10 \times \frac{\sqrt{3}}{2} = 5\sqrt{3}\text{ m}$$

$$\Rightarrow \quad R = 8.66\text{ m}$$ **Ans. (D)**

Sol. 3 As $x = \alpha t^3; \quad y = \beta t^3$

$$v_x = \frac{dx}{dt} = 3\alpha t^2; \quad v_y = \frac{dy}{dt} = 3\beta t^2$$

velocity of particle

$$v = \sqrt{v_x^2 + v_y^2}$$

$$v = \sqrt{(3\alpha t^2)^2 + (3\beta t^2)^2}$$

$$v = 3t^2\sqrt{\alpha^2 + \beta^2}$$ **Ans. (B)**

Sol. 4

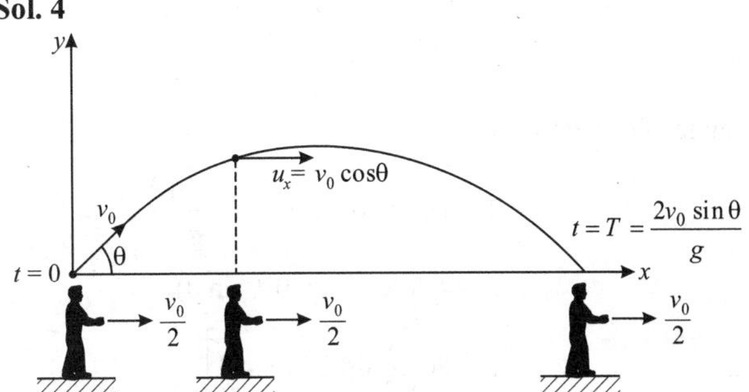

$$R = \frac{v_0}{2}T$$

$$\Rightarrow \quad \frac{v_0^2 \sin 2\theta}{g} = \frac{v_0}{2} \times \frac{2v_0 \sin\theta}{g}$$

$$2\sin\theta\cos\theta = \sin\theta \Rightarrow \cos\theta = 1/2$$

Person will be able to catch the ball if

$$v_0 \cos\theta = \frac{v_0}{2}$$

$$\Rightarrow \quad \cos\theta = \frac{1}{2}$$

$$\Rightarrow \quad \theta = 60°$$ **Ans. (A)**

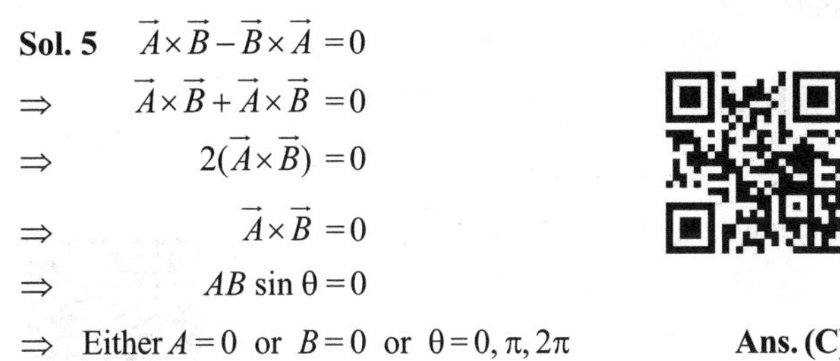

Sol. 5 $\vec{A}\times\vec{B} - \vec{B}\times\vec{A} = 0$

$\Rightarrow \quad \vec{A}\times\vec{B} + \vec{A}\times\vec{B} = 0$

$\Rightarrow \quad 2(\vec{A}\times\vec{B}) = 0$

$\Rightarrow \quad \vec{A}\times\vec{B} = 0$

$\Rightarrow \quad AB\sin\theta = 0$

$\Rightarrow$ Either $A = 0$ or $B = 0$ or $\theta = 0, \pi, 2\pi$ **Ans. (C)**

Sol. 6 Same range of projectile angles are complementary θ, 90 – θ so time of flight in the two cases are

$$T_1 = \frac{2u\sin\theta}{g}$$

and $$T_2 = \frac{2u\cos\theta}{g}$$

here $$T_1T_2 = \frac{4u^2\sin\theta\cos\theta}{g^2} = \frac{2u^2\sin 2\theta}{g^2} = \frac{2R}{g}$$

$$\Rightarrow \quad T_1T_2 \propto R$$ **Ans. (A)**

Sol. 7

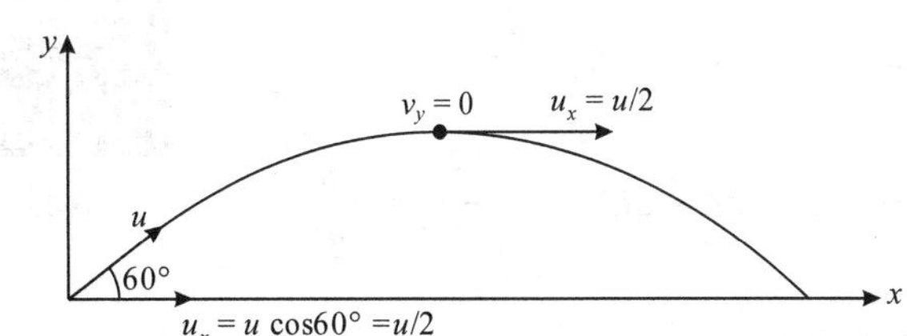

Initial kE $\quad K = \frac{1}{2}mu^2$

KE of particle at highest point

$$K_H = \frac{1}{2}mu_x^2 = \frac{1}{2}m\left(\frac{4}{2}\right)^2 = \frac{1}{8}mu^2$$

$$\Rightarrow \quad K_H = \frac{K}{4}$$ **Ans. (C)**

Sol. 8 We can use vector speed equation

$$\vec{v} = \vec{u} + \vec{a}\,t$$

After 10 s, find velocity is

$$\vec{v} = (3\hat{i} + 4\hat{j}) + (0.4\hat{i} + 0.3\hat{j}) \times 10$$

$$\vec{v} = 7\hat{i} + 7\hat{j}$$

$$|\vec{v}| = \sqrt{7^2 + 7^2}$$

$$= 7\sqrt{2} \text{ units}$$ **Ans. (A)**

Sol. 9 We are given velocity components of particle as

$$v_x = \frac{dx}{dt} = Ky \quad \ldots(1)$$

and $$v_y = \frac{dy}{dt} = Kx \quad \ldots(2)$$

By (1)/(2) $\Rightarrow$ $$\frac{dx}{dy} = \frac{y}{x}$$

$$x\,dx = y\,dy$$

integrating above relation gives

$$\int x\,dx = \int y\,dx$$

$$\Rightarrow \quad \frac{x^2}{2} = \frac{y^2}{2} + C$$

$$\Rightarrow \quad x^2 = y^2 + C$$

$$\Rightarrow \quad y^2 = x^2 + C_1$$ **Ans. (D)**

Sol. 10

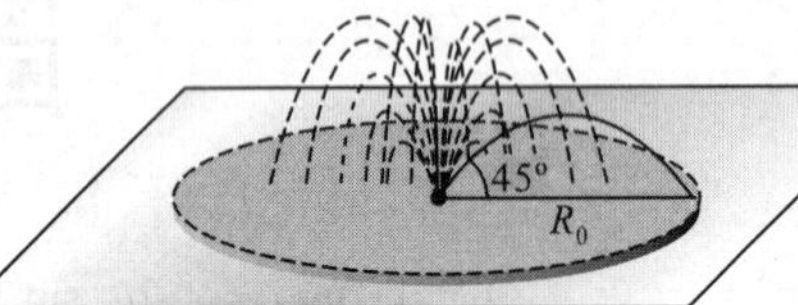

Maximum horizontal range

$$R = \frac{u^2 \sin 2\theta}{g}$$

at $\theta = 45° \rightarrow$ $R = R_0$ (max.)

$$\Rightarrow \quad R_0 = \frac{u^2}{g}$$

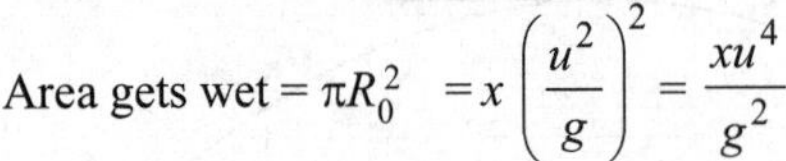

Area gets wet $= \pi R_0^2 = x\left(\frac{u^2}{g}\right)^2 = \frac{xu^4}{g^2}$ **Ans. (A)**

Sol. 11

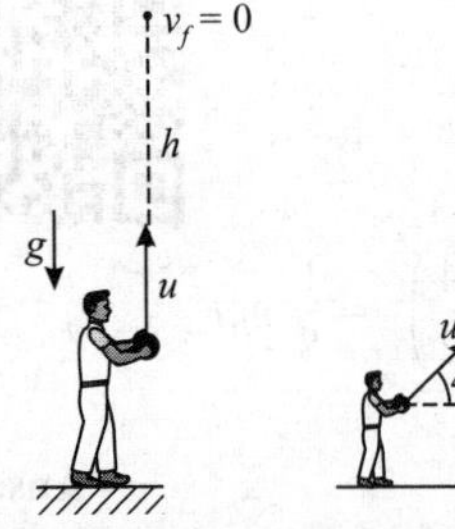

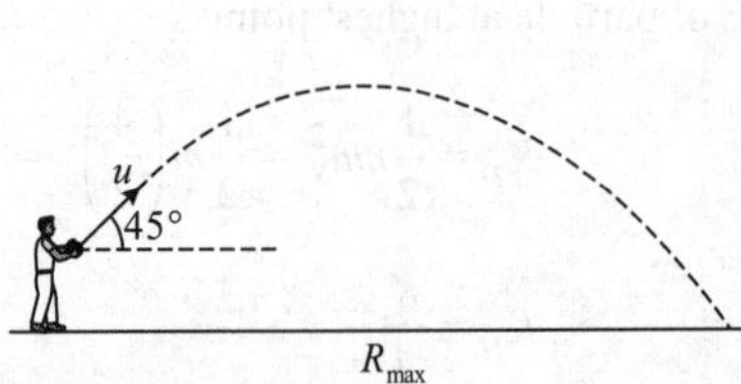

When stories thrown up, maximum height is given as

$$h = \frac{u^2}{2g} = 10 \text{ m}$$

$$\Rightarrow \quad u^2 = 200$$

For projectile maximum range is given for same projection speed, is

$$R_{max} = \frac{u^2}{g} = \frac{200}{10} = 20 \text{ m}$$ **Ans. (D)**

Sol. 12

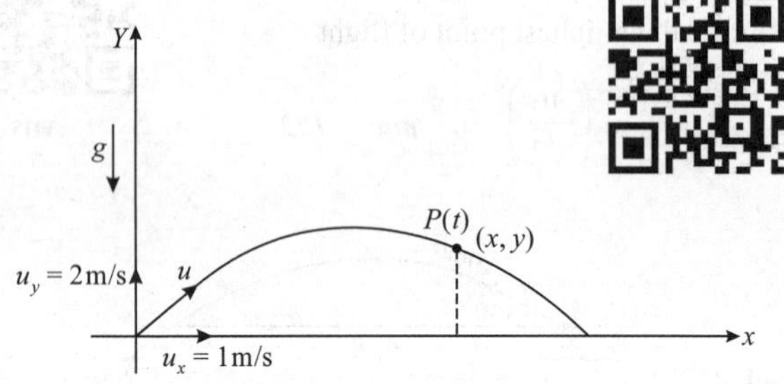

Position coordinates of particle at a time t are

$$x = u_x t = t \quad \ldots(1)$$

and $$y = u_y t - \frac{1}{2} gt^2 - 5t^2 \quad \ldots(2)$$

Eliminating t we get

$$\Rightarrow \quad y = 2x - 5x^2$$ **Ans. (A)**

Sol. 13 When car is at rest

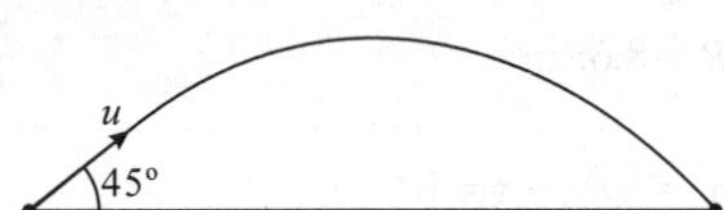

We use (max) $$R_{max} = \frac{u^2}{g} = 40$$

$$\Rightarrow \quad u = \sqrt{400} = 20 \text{ m/s}$$

When car is moving

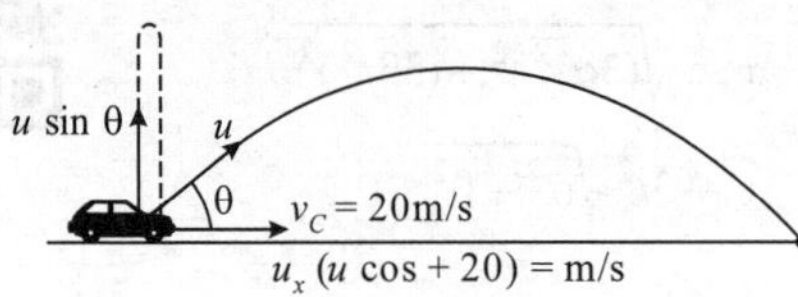

time of flight $$t = \frac{2u\sin\theta}{g}$$

Range of projectile $R_x = u_x t$

$$\Rightarrow \quad R_x = (u\cos\theta + 20) \times \frac{2u\sin\theta}{g}$$

$$= 80(\cos\theta\sin\theta + \sin\theta)$$

$$\Rightarrow \quad R_x = 80\left[\frac{1}{2}\sin 2\theta + \sin\theta\right]$$

R_x is max when $\frac{dR_x}{d\theta} = 0$

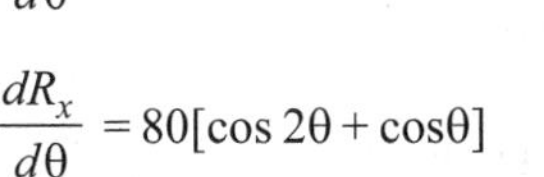

$$\frac{dR_x}{d\theta} = 80[\cos 2\theta + \cos\theta]$$

$\Rightarrow \quad \cos 2\theta = -\cos\theta$

$\Rightarrow \quad \cos 2\theta = \cos(180 - \theta) \Rightarrow 2\theta = 180 - \theta$

$\Rightarrow \quad \theta = 60°$ **Ans. (B)**

Sol. 14

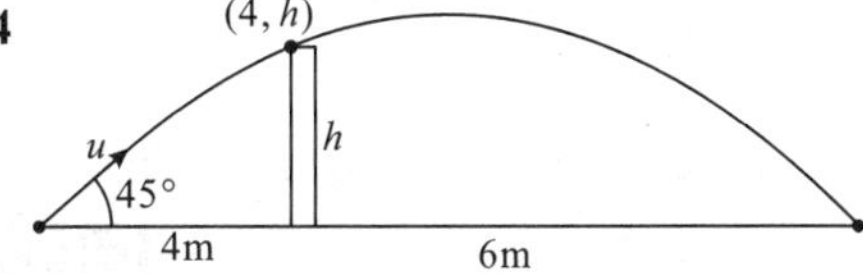

Maximum Range of projectile is

$$\frac{u^2}{g} = 10$$

$$\Rightarrow \quad u = \sqrt{100} = 10 \text{ m/s}$$

Equation of trajectory of projectile

$$y = x \tan\theta - \frac{gx^2}{2u^2\cos^2\theta}$$

$$\Rightarrow \quad y = x - \frac{10x^2}{2(10)^2\left(\frac{1}{\sqrt{2}}\right)^2}$$

$$\Rightarrow \quad y = x - \frac{x^2}{10}$$

at $x = 4$ m, $y = h$

$\Rightarrow \quad h = 4 - \frac{16}{10} = 2.4$ m **Ans. (B)**

Sol. 15

Rifle, $u = 630$m/s, θ, 700m, Target, A_{in}, h, g

As θ small the time taken by bullet to hit the target is

$$t = \frac{700}{4\cos\theta} = \frac{700}{630} = \frac{10}{9} \text{ s}$$

($\cos\theta \sim 1$ for small θ)

Distance fallen by bullet in time t is

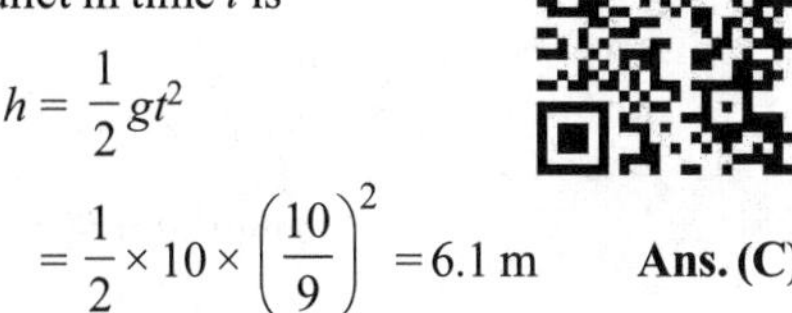

$$h = \frac{1}{2}gt^2$$

$$= \frac{1}{2} \times 10 \times \left(\frac{10}{9}\right)^2 = 6.1 \text{ m}$$ **Ans. (C)**

Alternative :

Considering projectile motion we use

$$\frac{u^2 \sin 2\theta}{g} = 700 \text{ m}$$

$$\frac{u^2(2\sin\theta\cos\theta)}{g} = 700$$

For small θ, we take $\cos\theta \sim 1$ and $\sin\theta \sim \tan\theta$

$$\Rightarrow \quad 2\sin\theta = \frac{700 \times 10}{(630)^2}$$

here $\quad \tan\theta = \frac{h}{700}$

$\Rightarrow \quad h = 700 \tan\theta$ m

$\Rightarrow \quad h \simeq 700 \sin\theta$

$\Rightarrow \quad h = 700 \times \frac{700 \times 10}{2 \times (630)^2} = 6.1$ m **Ans. (C)**

Sol. 16

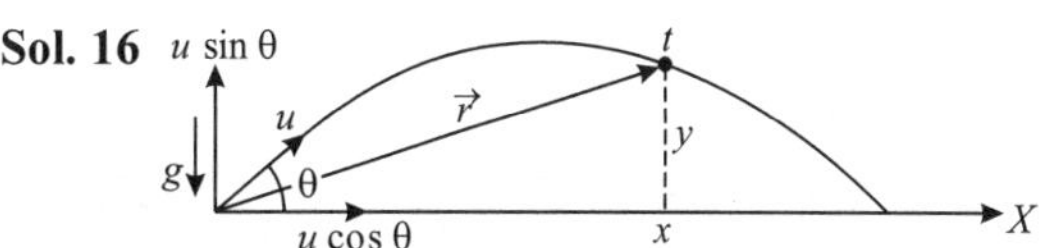

For a given projectile position vector at time t is given as

$$\vec{r} = (u\cos\theta.t)\,\hat{i} + \left(u\sin\theta \cdot t - \frac{1}{2}gt^2\right)\hat{i}$$

$$\vec{r} = 40\hat{i} + 50\hat{j} \text{ (given at } 2s)$$

By Comparing with given values

$u\cos\theta\,(2) = 40$

$\Rightarrow \quad u\cos\theta = 20 \quad \ldots(1)$

$u\sin\theta\,(2) - 20 = 50 \quad \ldots(2)$

By (2)/(1) $\Rightarrow \quad \tan\theta = \frac{35}{20} = \frac{7}{4}$ **Ans. (C)**

Sol. 17 $|\vec{A}| = |\vec{B}| = A$

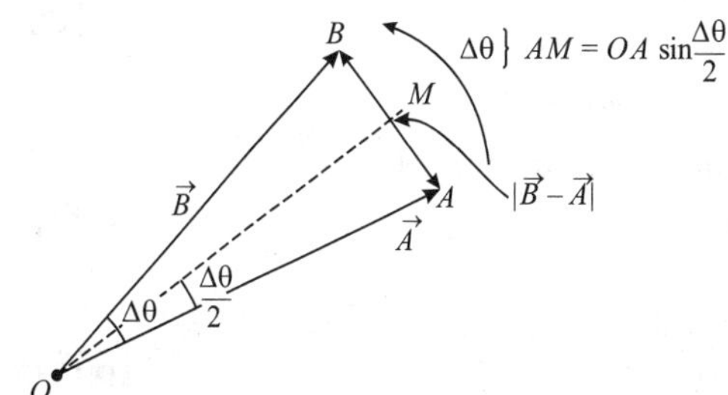

$$|\vec{A}| = |\vec{B}| = A$$

The magnitude of $\vec{B} - \vec{A}$ is given as

$$|\vec{B} - \vec{A}| = AB = 2A\sin\left(\frac{\Delta\theta}{2}\right)$$

$$\Rightarrow \quad |\vec{B}-\vec{A}| = 2A \times \frac{\Delta\theta}{2} \quad [\text{for } \theta << 1, \text{ we use } \sin\theta \sim \theta]$$

$$\Rightarrow \quad |\vec{B}-\vec{A}| = A\,\Delta\theta$$ **Ans. (A)**

Alternatively : From below figure it can be directly seen the magnitude of $|\vec{B}-\vec{A}| = A\,\Delta\theta$.

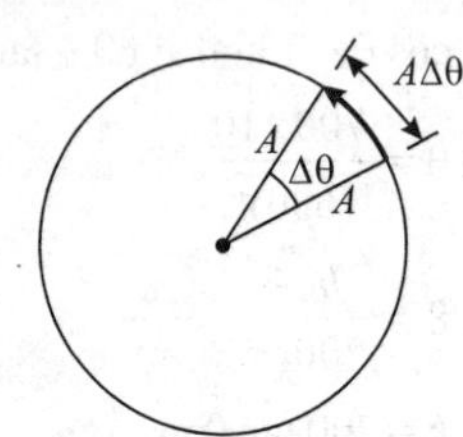

Sol. 18 If we consider

$$\vec{C} = a\hat{i} + b\hat{j}$$

$$\vec{A}.\vec{C} = a + b$$

$$\vec{B}.\vec{C} = 2a - b$$

$$\vec{A}.\vec{B} = 2 - 1 = 1$$

$$a + b = 1 \quad \ldots(1)$$

$$\Rightarrow \quad 2a - b = 1 \quad \ldots(2)$$

Solving-(1) and (2) gives

$$a = \frac{1}{3};\ b = \frac{2}{3}$$

$$\Rightarrow \quad |\vec{C}| = \sqrt{a^2 + b^2}$$

$$= \sqrt{\left(\frac{1}{3}\right)^2\left(\frac{2}{3}\right)^2} = \sqrt{\frac{5}{9}}$$ **Ans. (A)**

Sol. 19 From the given velocity function, we use

$$\frac{dx}{dt} = ky \quad \text{and} \quad \frac{dy}{dt} = kx$$

$$\Rightarrow \quad \frac{dy}{dx} = \frac{\frac{dy}{dt}}{\frac{dx}{dt}} = \frac{x}{y}$$

$$\Rightarrow \quad ydy = xdx$$

Integrating both side, we get

$$y^2 = x^2 + C$$ **Ans. (B)**

Sol. 20 Velocity components of particle are

$$v_x = -a\omega\sin\omega t$$

and $$v_y = a\omega\cos\omega t$$

and $$v_z = a\omega$$

$$\Rightarrow \quad v = \sqrt{v_x^2 + v_y^2 + v_z^2}$$

$$\Rightarrow \quad v = \sqrt{2}a\omega$$ **Ans. (C)**

Sol. 21 Range of projectile is

$$R = \frac{u^2 \sin 2\theta}{g}$$

Area covered on ground is

$$A = \pi R^2$$

$$\Rightarrow \quad A \propto R^2$$

$$\Rightarrow \quad A \propto u^4$$

$$\Rightarrow \quad \frac{A_1}{A_2} = \frac{u_1^4}{u_2^4} = \left[\frac{1}{2}\right]^4 = \frac{1}{16}$$ **Ans. (D)**

Sol. 22 We have position vectors of points G and H are

$$\vec{r}_g = \frac{a}{2}\hat{i} + \frac{a}{2}\hat{k}$$

$$\vec{r}_H = \frac{a}{2}\hat{j} + \frac{a}{2}\hat{k}$$

$$\vec{r}_H - \vec{r}_g = \frac{a}{2}(\hat{j} - \hat{i})$$ **Ans. (B)**

Sol. 23 Resultant of $2F$ and $3F$ is given as

$$4F^2 + 9F^2 + 12F^2\cos\theta = R^2 \quad \ldots(1)$$

When second force becomes $6F$ then resultant becomes $2R$, so we use

$$4F^2 + 36F^2 + 24F^2\cos\theta = 4R^2 \quad \ldots(2)$$

From equation-(1) and (2), we have

$$4F^2 + 36F^2 + 24F^2\cos\theta$$

$$= 4(13F^2 + 12F^2\cos\theta)$$

$$= 52F^2 + 48F^2\cos\theta$$

$$\Rightarrow \quad \cos\theta = -\frac{12F^2}{24F^2} = -\frac{1}{2}$$

$$\Rightarrow \quad \theta = 120°$$ **Ans. (D)**

Sol. 24 If θ is angle between $\vec{A}$ and $\vec{B}$, and $|\vec{A}| = |\vec{B}| = a$ then we have

$$|\vec{A}+\vec{B}| = 2a\cos\theta/2 \quad \ldots(1)$$

$$|\vec{A}-\vec{B}| = 2a\cos\frac{(\pi-\theta)}{2} = 2a\sin\theta/2 \quad \ldots(2)$$

$$\Rightarrow \quad n\left(2a\cos\frac{\theta}{2}\right) = 2a\frac{\sin\theta}{2}$$

$$\Rightarrow \quad \tan\frac{\theta}{2} = n$$

$$\Rightarrow \quad \theta = 2\tan^{-1} n = \cos^{-1}\left(\frac{n^2-1}{n^2+1}\right)$$ **Ans. (C)**

Sol. 25 Projection velocity components are

$$u_x = 10\cos 60° = 5 \text{ m/s}$$

$$u_y = 10\cos 30° = 5\sqrt{3} \text{ m/s}$$

velocity after $t = 1$s is given as

$$v_x = 5 \text{ m/s}$$

$$v_y = (5\sqrt{3} - 10) \text{ m/s} (-\text{ ve} \Rightarrow \text{downward motion})$$

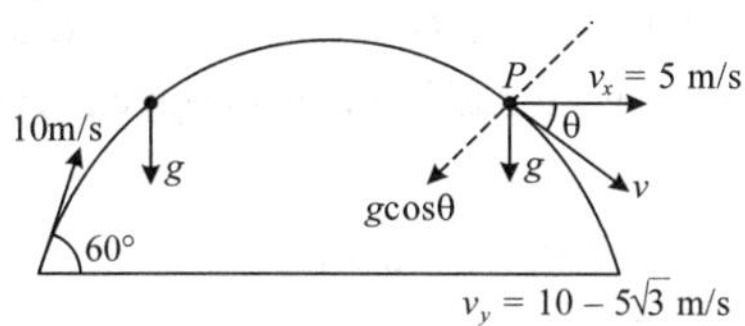

At point P, normal acceleration of body is

$$a_n = \frac{v^2}{R}$$

$$\Rightarrow \quad R = \frac{v_x^2 + v_y^2}{a_n} = \frac{25 + 100 + 75 - 100\sqrt{3}}{10\cos\theta}$$

Where $\tan\theta = \left|\frac{v_y}{v_x}\right| = \frac{10 - 5\sqrt{3}}{5} = 2 - \sqrt{3}$

$$\Rightarrow \quad \theta = 15°$$

$$\Rightarrow \quad R = \frac{100(2-\sqrt{3})}{10\cos 15°} = 2.8 \text{ m}$$

Ans. (C)

Sol. 26 Given $\vec{r} = 15t^2\hat{i} + (4 - 20t^2)\hat{j}$

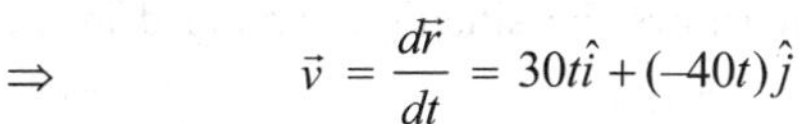

$$\Rightarrow \quad \vec{v} = \frac{d\vec{r}}{dt} = 30t\hat{i} + (-40t)\hat{j}$$

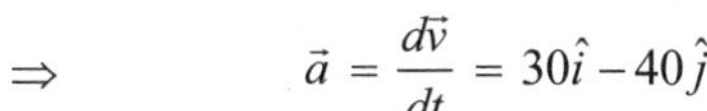

$$\Rightarrow \quad \vec{a} = \frac{d\vec{v}}{dt} = 30\hat{i} - 40\hat{j}$$

$$\Rightarrow \quad |\vec{a}| = 50 \text{ m/s}^2$$

Ans. (D)

Sol. 27 Time of flight is

$$t = \frac{2 \times 2 \times \sin 15°}{g\cos 30°}$$

Range on inclined plane is

$$R = 2\cos 15° \times t - \frac{1}{2} g \sin 30° t^2$$

Solving, we get

$$R \simeq 20 \text{ cm}$$

Ans. (B)

Sol. 28 Equation of trajectory is given as

$$y = 2x - 9x^2 \quad \ldots(1)$$

Comparing with general equation of projectile

$$y = x\tan\theta - \frac{g}{2u^2\cos^2\theta} \cdot x^2 \quad \ldots(2)$$

We get $\tan\theta = 2$

$$\Rightarrow \quad \cos\theta = \frac{1}{\sqrt{5}}$$

Also, $\frac{g}{2u^2\cos^2\theta} = 9$

$$\Rightarrow \quad \frac{10}{2 \times 9 \times \left(\frac{1}{\sqrt{5}}\right)^2} = u^2$$

$$\Rightarrow \quad u^2 = \frac{25}{9}$$

$$\Rightarrow \quad u = \frac{5}{3} \text{ m/s}$$

Ans. (A)

Sol. 29 Range will be same for time t_1 & t_2, so angles of projection will be 'θ' & '90° – θ'

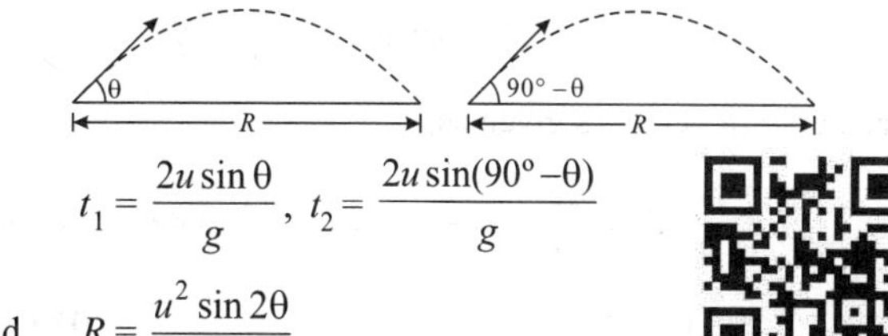

$$t_1 = \frac{2u\sin\theta}{g}, \quad t_2 = \frac{2u\sin(90° - \theta)}{g}$$

and $R = \frac{u^2 \sin 2\theta}{g}$

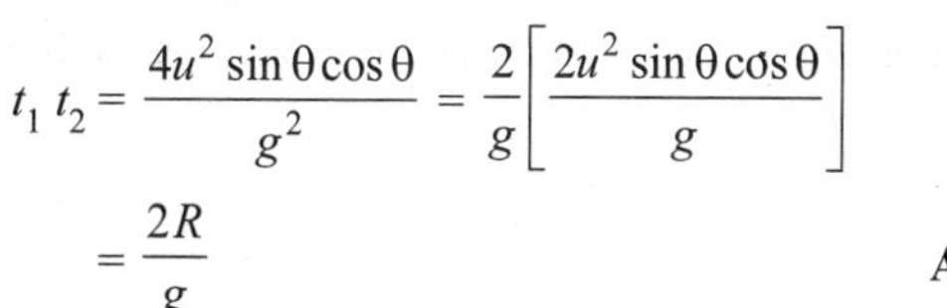

$$t_1 t_2 = \frac{4u^2 \sin\theta\cos\theta}{g^2} = \frac{2}{g}\left[\frac{2u^2\sin\theta\cos\theta}{g}\right]$$

$$= \frac{2R}{g}$$

Ans. (C)

Sol. 30

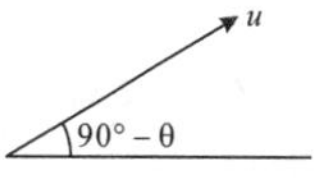

For same range angle of projection will be θ & 90 – θ, so we use

$$R = \frac{u^2 2\sin\theta\cos\theta}{g}$$

and time of flights are

$$h_1 = \frac{u^2\sin^2\theta}{g}$$

$$h_2 = \frac{u^2\sin^2(90 - \theta)}{g}$$

$$\Rightarrow \quad \frac{R^2}{h_1 h_2} = 16$$

Ans. (B)

Sol. 31 Position along x-direction is given as

$$x = 10 + 8t - 3t^2$$

$$\Rightarrow \quad v_{x_1} = 8 - 6t$$

$$\Rightarrow \quad (v_{x_1})_{t=1} = 2\hat{i}$$

Position along y-direction is given as

$$y = 5 - 8t^3$$

$$\Rightarrow \quad v_{y_2} = -24t^2$$

$$\Rightarrow \quad (v_{y_2})_{t=1} = -24\hat{j}$$

Now given that

$$\sqrt{v} = \sqrt{v_{x_1}^2 + v_{y_2}^2}$$

$$\sqrt{v} = \sqrt{(24)^2 + (2)^2} = \sqrt{580}$$

$$\Rightarrow \quad v = 580 \text{ m/s}$$ **Ans. (580.00)**

Sol. 32 Given that

$$\vec{r}(t) = \cos\omega t\,\hat{i} + \sin\omega t\,\hat{j}$$

Velocity vector is given as

$$\vec{v} = \frac{\vec{dr}}{dt} = -\omega\sin\omega t\hat{i} + \omega\cos\omega t\hat{j}$$

Acceleration vector is given as

$$\vec{a} = \frac{\vec{dv}}{dt} = -\omega^2\cos\omega t\,\hat{i} - \omega^2\sin\omega t\,\hat{j} = -\omega^2\,\vec{r}$$

$$\Rightarrow \quad \vec{v}\cdot\vec{r} = 0$$ **Ans. (A)**

Sol. 33 For uniform acceleration motion in both x and y direction, we use

$$x = u_x t + \frac{1}{2}a_x t^2 \quad \ldots(1)$$

and

$$y = u_y t + \frac{1}{2}a_y t^2 \quad \ldots(2)$$

$$\Rightarrow \quad 32 = 0 \times t + \frac{1}{2}(4)(t)^2 \quad [\text{Given that } u_y = 0]$$

$$\Rightarrow \quad t^2 = 16$$

$$\Rightarrow \quad t = 4 \text{ s}$$

Along x-direction, we use from equation-(1)

$$x = 3 \times 4 + \frac{1}{2} \times 6 \times 4^2$$

$$\Rightarrow \quad x = 12 + 48 = 60 \text{ m}$$ **Ans. (C)**

Sol. 34 Given that

$$\vec{u} = 5\hat{j} \text{ m/s}$$

$$\vec{a} = 10\hat{i} + 4\hat{j} \text{ m/s}$$

The coordinate $(20, y_0)$ after in time t are given as

Along x-direction

$$s_x = u_x t + \frac{1}{2}a_x t^2$$

$$\Rightarrow \quad 20 - 0 = 0 + \frac{1}{2} \times 10 \times t^2$$

$$\Rightarrow \quad t = 2 \text{ s}$$

Along y-direction

$$s_y = u_y \times t + \frac{1}{2}a_y t^2$$

$$\Rightarrow \quad y_0 = 5 \times 2 + \frac{1}{2}4 \times 2^2 = 18 \text{ m}$$ **Ans. (C)**

Sol. 35 As we know that for a closed polygon

$$\overrightarrow{OA} + \overrightarrow{OB} + \overrightarrow{OC} + \overrightarrow{OD} + \overrightarrow{OE} + \overrightarrow{OF} + \overrightarrow{OG} + \overrightarrow{OH} = \vec{0}$$

By triangle law of vector addition, in the given octagon, we can use

$$\overrightarrow{AB} = \overrightarrow{AO} + \overrightarrow{OB};$$

$$\overrightarrow{AC} = \overrightarrow{AO} + \overrightarrow{OC}$$

$$\overrightarrow{AD} = \overrightarrow{AO} + \overrightarrow{OD};$$

$$\overrightarrow{AE} = \overrightarrow{AO} + \overrightarrow{OE}$$

$$\overrightarrow{AF} = \overrightarrow{AO} + \overrightarrow{OF};$$

$$\overrightarrow{AG} = \overrightarrow{AO} + \overrightarrow{OG}$$

$$\overrightarrow{AH} = \overrightarrow{AO} + \overrightarrow{OH}$$

$$\Rightarrow \quad \overrightarrow{AB} + \overrightarrow{AC} + \overrightarrow{AD} + \overrightarrow{AE} + \overrightarrow{AF} + \overrightarrow{AG} + \overrightarrow{AH}$$

$$= (7\,\overrightarrow{AO}) + \overrightarrow{OB} + \overrightarrow{OC} + \overrightarrow{OD} + \overrightarrow{OE} + \overrightarrow{OF} + \overrightarrow{OG} + \overrightarrow{OH}$$

$$= (7\,\overrightarrow{AO}) + \overrightarrow{AO} = 8\overrightarrow{AO}$$

$$= 8(2\hat{i} + 3\hat{j} - 4\hat{k}) = 16\hat{i} + 24\hat{j} - 32\hat{k}$$ **Ans. (B)**

Sol. 36 Figure below shows the situation explained in the question

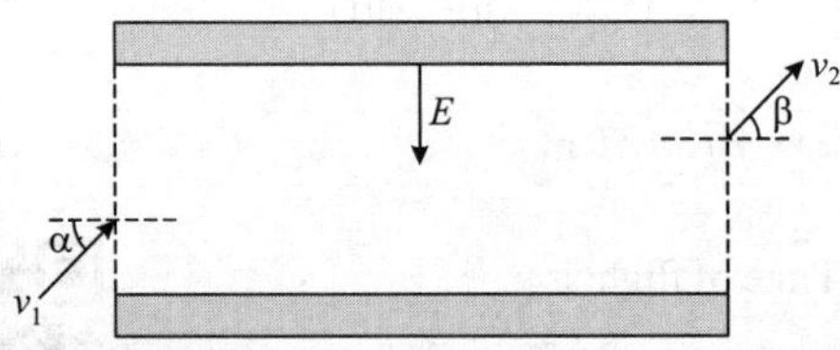

As acceleratoin is perpendicular to plates so velocity along the plate will remain constant

$$\Rightarrow \quad v_1\cos\alpha = v_2\cos\beta$$

$$\Rightarrow \quad \frac{K_1}{K_2} = \frac{v_1^2}{v_2^2} = \frac{\cos^2\beta}{\cos^2\alpha}$$ **Ans. (D)**

Sol. 37 For the given vector equation, we can write

$$-PQ\sin\theta = PQ\sin\theta$$

$$\Rightarrow \quad \theta = 180°$$ **Ans. (180)**

Sol. 38 From the given trajectory equation

$$y = \alpha x - \beta x^2$$

By comparing with the standard trajectory equation

$$y = x\tan\theta - \frac{1}{2}\frac{gx^2}{u^2\cos^2\theta}$$

We have

$$\tan\theta = \alpha$$

$$\Rightarrow \quad \theta = \tan^{-1}\alpha$$

and

$$\beta = \frac{1}{2}\frac{g}{u^2\cos^2\theta}$$

$$\Rightarrow \quad u^2 = \frac{g}{2\beta\cos^2\theta}$$

Thus maximum height during flight can be given as

$$H = \frac{u^2\sin^2\theta}{2g} = \frac{g}{2\beta\cos^2\theta}\frac{\sin^2\theta}{2g}$$

$$\Rightarrow \quad H = \frac{\tan^2\theta}{4\beta} = \frac{\alpha^2}{4\beta}$$ **Ans. (C)**

Sol. 39 Then angles between rods and vertical are shownin figure below.

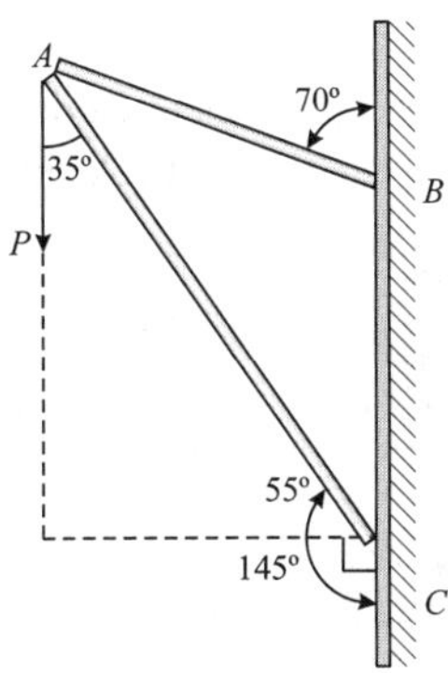

The component along AC of applied force is given as

$$P_c = 100\cos 35^\circ\,\text{N}$$

$$\Rightarrow \quad P_c = 100\times 0.819\,\text{N}$$

$$\Rightarrow \quad P_c = 81.9\,\text{N} \approx 82\,\text{N}$$ **Ans. (82)**

Sol. 40 Velocity of mosquito after 2 s is given as

$$\vec{v}_{t=2} = 2\hat{i} + 6\hat{j} + 9\hat{k}$$

Angle made by direction of motion of mosquito from the coordinate axes are given as

$$\cos^{-1}\frac{2}{11}\ (\text{from } x\text{-axis}) = \tan^{-1}\frac{\sqrt{117}}{2}$$

$$\cos^{-1}\frac{6}{11}\ (\text{from } y\text{-axis}) = \tan^{-1}\frac{\sqrt{85}}{6}$$

$$\cos^{-1}\frac{9}{11}\ (\text{from } z\text{-axis}) = \tan^{-1}\frac{\sqrt{40}}{9}$$

None of the given option is correct. **Ans. (None)**

Sol. 41 To reach a point directly opposite on other bank of river, swimmer has to swimg such that the resultant velocity of swimmer should be directly toward the point on other bank of river where he has to reach. Below diagram shows the resultant velocity.

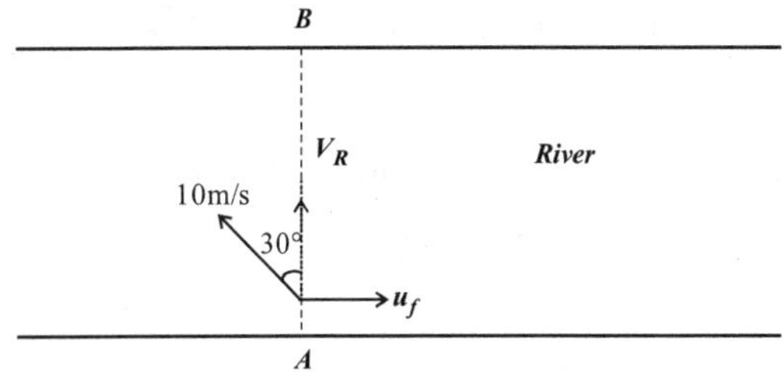

From above figure line AB is perpendicular to rivier flow when

$$10\sin 30^\circ = x$$

$$\Rightarrow \quad x = 5\text{ m/s}$$ **Ans. (5)**

Sol. 42 Velocity vectors of projectile at points A and B are given as

$$\vec{u} = u\cos 45\,\hat{i} + u\sin 45\,\hat{j}$$

$$\vec{v} = v\cos 45\,\hat{i} - v\sin 45\,\hat{j}$$

Change in momentum can be calculated as

$$|\overrightarrow{\Delta P}| = m(\vec{v} - \vec{u})$$

$$\Rightarrow \quad \Delta P = 2mu\sin 45^\circ$$

$$\Rightarrow \quad \Delta P = 2\times 5\times 10^{-3}\times 5\sqrt{2}\times\frac{1}{\sqrt{2}}\text{ N-s}$$

$$\Rightarrow \quad \Delta P = 50\times 10^{-3} = 5\times 10^{-2}\text{N-s}$$ **Ans. (5)**

Sol. 43 From the given relation in question, we use

$$\vec{A}\cdot\vec{B} = |\vec{A}\times\vec{B}|$$

$$\Rightarrow \quad AB\cos\theta = AB\sin\theta$$

$$\Rightarrow \quad \theta = 45^\circ$$

$$\Rightarrow \quad |\vec{A}-\vec{B}| = \sqrt{A^2 + B^2 - 2AB\cos 45^\circ}$$

$$\Rightarrow \quad |\vec{A}-\vec{B}| = \sqrt{A^2 + B^2 - \sqrt{2}AB}$$ **Ans. (B)**

Sol. 44 As per the given condition in question, we have

$$|\vec{P}+\vec{Q}| = n|\vec{P}-\vec{Q}|$$

$$\Rightarrow |\vec{P}|^2 + |\vec{Q}|^2 + 2|\vec{P}||\vec{Q}|\cos\theta = n^2(|\vec{P}|^2 + |\vec{Q}|^2 - 2|\vec{P}||\vec{Q}|\cos\theta)$$

$$\Rightarrow \quad 2 + 2\cos\theta = n^2(2 - 2\cos\theta)$$

$$\Rightarrow \quad \theta = \cos^{-1}\left(\frac{n^2-1}{n^2+1}\right)$$ **Ans. (C)**

Sol. 45 Projection of vector $\vec{A}$ on $\vec{B}$ is given as

$$p = |\vec{A}|\cos\theta\,\hat{B}$$

$$p = \frac{\vec{A}\cdot\vec{B}}{|\vec{B}|}\,\hat{B} = \hat{i} + \hat{j}$$ **Ans. (C)**

Sol. 46 Direction vector of motion of particle $\vec{P}$ can be given as

$$\vec{P} = K(\vec{A}\times\vec{B})$$

$$\begin{bmatrix} i & j & k \\ 1 & 1 & 0 \\ 0 & 1 & 1 \end{bmatrix} = K(\hat{i}-\hat{j}+\hat{k})$$

Direction vector of motion of particle $\vec{Q}$ can be given as

$$\vec{Q} = F(\vec{A}\times\vec{C})$$

$$\begin{bmatrix} \hat{i} & \hat{j} & \hat{k} \\ 1 & 1 & 0 \\ -1 & 1 & 0 \end{bmatrix} = 2F(\hat{k})$$

Angle between the direction of motion of P and Q can be calculated as

$$\cos\theta = \frac{\vec{P}\cdot\vec{Q}}{|\vec{P}||\vec{Q}|} = \frac{1}{\sqrt{3}}$$

Sol. 47 All the options in List-I can be directly related with the vector triangles given in List-II using triangle law of vector addition. **Ans. (B)**

Sol. 48 It is given that

$$|\vec{X} - \vec{Y}| = n|\vec{X} + \vec{Y}|$$

$$\Rightarrow |\vec{X}|^2+|\vec{Y}|^2-2|\vec{X}||\vec{Y}|\cos\theta = n^2[|\vec{X}|^2+|\vec{Y}|^2+2|\vec{X}||\vec{Y}|\cos\theta]$$

As magnitude of X is equal to that of Y, we use

$$2|\vec{X}|^2 - 2|\vec{X}|^2\cos\theta = 2n^2|\vec{X}|^2 + 2n^2|\vec{X}|^2\cos\theta$$

$$\Rightarrow \quad 1-\cos\theta = n^2 + n^2\cos\theta$$

$$\Rightarrow \quad \cos\theta = \frac{1-n^2}{1+n^2}$$

$$\Rightarrow \quad \theta = \cos^{-1}\frac{1-n^2}{1+n^2}$$ **Ans. (B)**

Sol. 49 Acceleration vector of body is given as

$$\vec{a} = 8\hat{i} + 2\hat{j}$$

As acceleration is constant, displacement is given as

$$\vec{s} = \vec{u}t + \frac{1}{2}\vec{a}t^2$$

$$\Rightarrow \quad \vec{s} = \frac{1}{2}(8\hat{i}+2\hat{j})\times 100$$

$$\Rightarrow \quad \vec{s} = 400\hat{i} + 100\hat{j}$$ **Ans. (A)**

Sol. 50 By law of vector addition we have

$$\overrightarrow{OA} + \overrightarrow{AB} = \overrightarrow{OB} \quad \ldots(1)$$

$$\overrightarrow{OB} + \overrightarrow{BC} = \overrightarrow{OC} \quad \ldots(2)$$

$$\overrightarrow{OC} + \overrightarrow{CD} = \overrightarrow{OD} \quad \ldots(3)$$

$$\Rightarrow \quad \overrightarrow{AB} = \overrightarrow{OB} - \overrightarrow{OA}$$

$$\Rightarrow \quad \overrightarrow{AC} = \overrightarrow{OC} - \overrightarrow{OA}$$

$$\Rightarrow \quad \overrightarrow{AD} = \overrightarrow{OD} - \overrightarrow{OA} = -\overrightarrow{OA} - \overrightarrow{OA}$$

$$\Rightarrow \quad \overrightarrow{AB} + \overrightarrow{AC} + \overrightarrow{AD} = \overrightarrow{OB} + \overrightarrow{OC} - 4\overrightarrow{OA}$$

$$\Rightarrow \quad \overrightarrow{AB} + \overrightarrow{AC} + \overrightarrow{AD} = \overrightarrow{OB} + \overrightarrow{OC} + 4\overrightarrow{AO} \quad \ldots(4)$$

Hence Assertion-A is correct. Now from equations-(1), (2) & (3), we get

$$\overrightarrow{AB} + \overrightarrow{BC} + \overrightarrow{CD} = \overrightarrow{OB} + \overrightarrow{OC} + \overrightarrow{OD} - \overrightarrow{OA} - \overrightarrow{OB} - \overrightarrow{OC}$$

$$\overrightarrow{AB} + \overrightarrow{BC} + \overrightarrow{CD} = -\overrightarrow{OA} - \overrightarrow{OA}$$

$$\overrightarrow{AB} + \overrightarrow{BC} + \overrightarrow{CD} = -2\overrightarrow{OA}$$

$$\overrightarrow{AB} + \overrightarrow{BC} + \overrightarrow{CD} = 2\overrightarrow{AO} \quad \ldots(5)$$

With above analysis, we can state that both A and R are correct but R is not the correct explanation of A. **Ans. (B)**

Sol. 51 Velocity of man with respect to ground can be given as

$$\vec{V}_M = \vec{V}_{MR} + \vec{V}_R$$

Vector addition of the velocities is shown below in which V_M should be along line AB. Given that velocity of man with respect to river and that of river is same so we use

$$V_{MR} = V_R$$

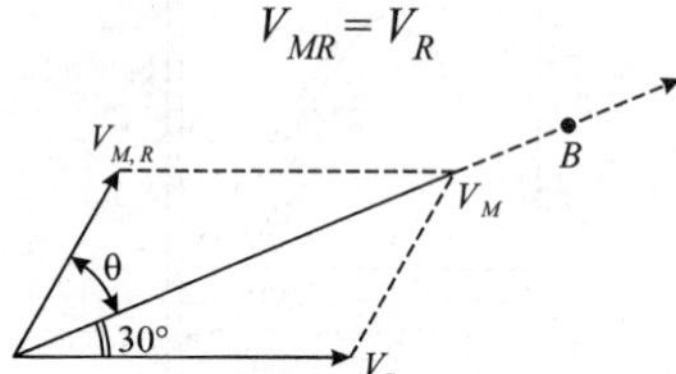

As the velocities of man and that of river are same the resultant velocity of man with respect to ground vector will be the angle bisector of the two vectors in the diagram shown so we have $\theta = 30°$. **Ans. (30)**

Sol. 52 From the given figure, the given vectors in terms of unit vectors are given as

$$\overrightarrow{OA} = R[\cos30°\,\hat{i} + \sin30°\,\hat{j}]$$

and $$\overrightarrow{OB} = R[\cos60°\,\hat{i} + (-\sin60°)\,\hat{j}]$$

and $$\overrightarrow{OC} = R[-\cos45°\,\hat{i} + \sin45°\,\hat{j}]$$

$$\Rightarrow \overrightarrow{OA} + \overrightarrow{OB} - \overrightarrow{OC} = R\left[\frac{1}{2}+\frac{\sqrt{3}}{2}+\frac{1}{\sqrt{2}}\right]\hat{i} + R\left[\frac{1}{2}-\frac{\sqrt{3}}{2}-\frac{1}{\sqrt{2}}\right]\hat{j}$$

Angle of above vector with x-axis is given as

$$\tan\alpha = \frac{\frac{1}{2}-\frac{\sqrt{3}}{2}-\frac{1}{\sqrt{2}}}{\frac{1}{2}+\frac{\sqrt{3}}{2}+\frac{1}{\sqrt{2}}} = \frac{1-\sqrt{3}-\sqrt{2}}{1+\sqrt{3}+\sqrt{2}}$$

$$\Rightarrow \qquad \alpha = \tan^{-1}\left[\frac{1-\sqrt{3}-\sqrt{2}}{1+\sqrt{3}+\sqrt{2}}\right] \qquad \textbf{Ans. (B)}$$

Sol. 53 Figure below shows the vector subtraction of the two given vectors

In above triangle, we have

$$\text{tna}\,\phi = \frac{B\sin 60^\circ}{A - B\cos 60^\circ}$$

$$\Rightarrow \qquad \phi = \tan^{-1}\left(\frac{B\sin 60^\circ}{A - B\cos 60^\circ}\right) = \tan^{-1}\left(\frac{\sqrt{3}B}{2A - B}\right) \qquad \textbf{Ans. (C)}$$

Sol. 54 Due to horizontally flying plane the horizontal component of velocity of bomb & fighter jet will be same so, bomb will remains just below the jet always during its projetile motion with respect to ground and with respect to an observer sitting in the plane the bomb appear to fall vertically in a straight line. **Ans. (B)**

Sol. 55 Resultant can be calculated by resolving all vectors into the components alsong x and y directions as

$$(\vec{R}) = \hat{i}\,(10\cos 30^\circ + 20\cos 60^\circ - 15\cos 30^\circ - 15\cos 45^\circ + 20\cos 45^\circ) + \hat{j}\,(10\sin 30^\circ + 20\sin 60^\circ + 15\sin 30^\circ - 15\sin 45^\circ - 20\cos 45^\circ)$$

$$(\vec{R}) = 9.25\hat{i} + 5\hat{j} \qquad \textbf{Ans. (A)}$$

Sol. 56 Maximum height of football during projectile is given as

$$H = \frac{u^2\sin^2\theta}{2g} = \frac{(25)^2 \times \left(\frac{1}{2}\right)}{2\times 10} = \frac{128}{8}m$$

Time to reach maximum height is given as

$$t = \frac{T}{2}$$

$$\Rightarrow \qquad t = \frac{u\sin\theta}{g} = \frac{25\left(\frac{1}{\sqrt{2}}\right)}{10}$$

$$\Rightarrow \qquad td = \frac{25}{10\sqrt{2}} = \frac{5}{2\sqrt{2}} \qquad \textbf{Ans. (A)}$$

Sol. 57 Situation described in question is shown in figure below

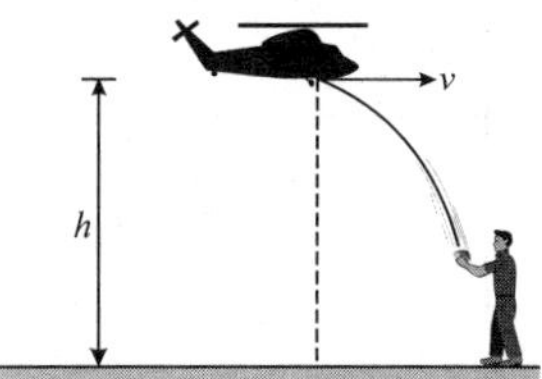

The food packet starts with the same speed of helicopter horizontally when dropped so horizontal range of the food packet is given as

$$R = v \times \sqrt{\frac{2h}{g}}$$

Horizontal distance, travelled by helicopter in this period will be same as that of R as horizontal speed of helicopter and food packet is same. Thus, vertical distance between helicopter and person is h. Thus from the point food packet is dropped, the distance between man and helicopter is given as

$$l = \sqrt{\frac{2v^2h}{g} + h^2} \qquad \textbf{Ans. (D)}$$

Sol. 58 Figure below shows the car ascending above the inclined plane and the pendulum inside it.

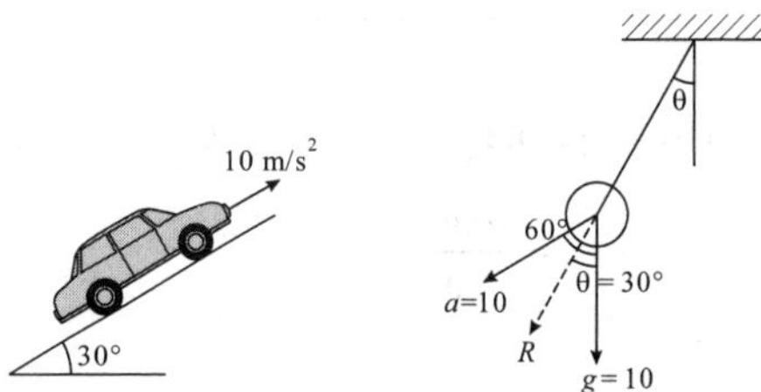

As shown in free body diagram of the pendulum, it experiences a pseudo force in direction opposite to the motion of car. We can clearly see that the resultant of this pseudo force and the weight of bob will be acting along the line at 30° from vertical so the string will be aligned along this line. **Ans. (30)**

Sol. 59 Three force vectors which represents the three sides of a triangle in same order then sum of these three must be zero thus statement-I is true. Statement-II is also stating the same thing in slightly changed language so this is also true.

Ans. (B)

Sol. 60 For forces, $\vec{F}_1$ and $\vec{F}_2$ at θ_1

$$F_{\text{net }1} = \sqrt{P^2 + Q^2 + P^2 + Q^2 + 2(P^2 + Q^2)\cos\theta_1}$$

$$F_{\text{net }2} = \sqrt{P^2 + Q^2 + P^2 + Q^2 + 2(P^2 + Q^2)\cos\theta_2}$$

If $\quad F_{\text{net }1} = \sqrt{3(P^2 + Q^2)} = \sqrt{2(P^2 + Q^2) + 2(P^2 + Q^2)\cos\theta_1}$

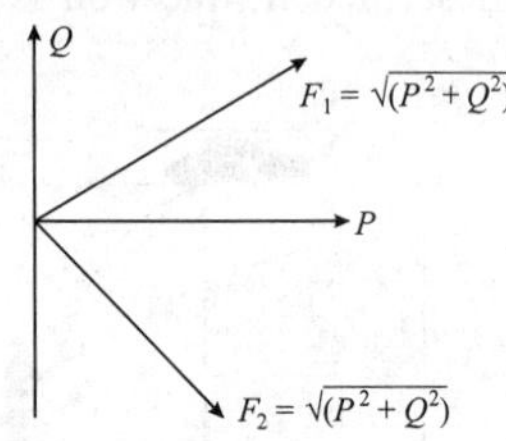

$$\Rightarrow \quad \cos\theta_1 = \frac{P^2+Q^2}{2(P^2+Q^2)}$$

$$\Rightarrow \quad \theta_1 = 60^\circ$$

$$\Rightarrow \quad F_{net\,2} = \sqrt{2(P^2+Q^2)}$$

$$\Rightarrow \quad F_{net\,2} = \sqrt{2(P^2+Q^2)+2(P^2+Q^2)\cos\theta_2}$$

$$\Rightarrow \quad \cos\theta_2 = 0$$

$$\Rightarrow \quad \theta_2 = 90^\circ$$

Ans. (B)

Sol. 61 As the given angles are complimentary angles, as

$$42^\circ + 48^\circ = 90^\circ$$

Thus range for the two projectiles will be same so $R_1 = R_2$ and as we have $\theta_1 < \theta_2$ we can conclude that $H_1 < H_2$ **Ans. (B)**

Sol. 62 Horizontal range of projectile is given as

$$R = \frac{2u^2 \sin\theta\cos\theta}{g}$$

Time taken by projectile to reach maximum height is given as

$$t = \frac{u\sin\theta}{g}$$

$$\Rightarrow \quad u = \frac{gt}{\sin\theta}$$

$$\Rightarrow \quad R = \frac{g^2t^2}{\sin^2\theta}\cdot\frac{2\sin\theta\cos\theta}{g}$$

$$\Rightarrow \quad \tan\theta = \frac{2gt^2}{R} = \frac{20t^2}{R}$$

$$\Rightarrow \quad \cot\theta = \frac{R}{20t^2}$$

Ans. (D)

Sol. 63 If u_x and u_y are the projection speeds at $t = 0$ then we have $u_x = u_y$ as projection angle is 45°. After $t = 2$ s velocity in vertical direction of projectile is given as

$$v_y = u_y - gt = u_x - 20$$

Velocity along horizontal remain same so after $t = 2$ s net velocity is given as

$$\sqrt{(u_x-20)^2+u_x^2} = 20$$

$$\Rightarrow \quad 2u_x^2 - 40u_x = 0$$

$$\Rightarrow \quad u_x = 20 \text{ m/s}$$

Thus maximum height reached by the body is given as

$$h = \frac{u_y^2}{2g} = \frac{(20)^2}{2(10)} = 20 \text{ m}$$

Ans. (20)

Sol. 64 To hit the jet, both jet and bullet will have same horizontal component of velocity, so we use

$$200 = 400\cos\theta$$

$$\Rightarrow \quad \theta = 60^\circ$$

Ans. (60)

Sol. 65 Figure below shows the situation described in the question.

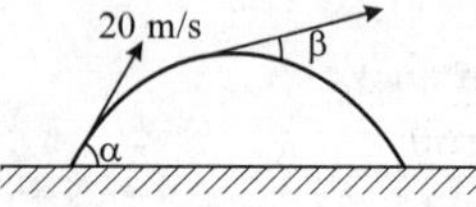

At launch point and after 10 s we use

$$v_x = u_x = 20\cos\alpha$$

and

$$v_y = 20\sin\alpha - 10\times 10$$

$$\Rightarrow \quad \tan\beta = \frac{v_y}{v_x} = \frac{20\sin\alpha - 100}{20\cos\alpha}$$

$$\Rightarrow \quad \tan\beta = \tan\alpha - 5\sec\alpha$$

Ans. (B)

Sol. 66 For the given condition between addition and subtraction of the two vectors we use

$$(a^2 + b^2 + 2ab\cos\theta) = 4(a^2 + b^2 - 2ab\cos\theta)$$

For $a = b$ we have

$$2a^2 + 2a^2\cos\theta = 8a^2 - 8a^2\cos\theta$$

$$\Rightarrow \quad \cos\theta = \frac{3}{5}$$

Ans. (C)

Sol. 67 For projectile throw of ball, maximum range is given as

$$R = \frac{u^2\sin 2\theta}{g}$$

$$\Rightarrow \quad R_{max} = \frac{u^2}{g} = 100 \text{ m}$$

With the same speed if the ball is thrown upward, maximum height attained by the ball is given as

$$H_{max} = \frac{u^2}{2g} = \frac{100}{2} = 50 \text{ m}$$

Ans. (B)

Sol. 68 Given that range and maximum height of projectile are equal so we use

$$R = H$$

$$\Rightarrow \quad \frac{2u^2 \sin\theta\cos\theta}{g} = \frac{u^2 \sin^2\theta}{2g}$$

$$\Rightarrow \quad u\cos\theta = \frac{u\sin\theta}{4}$$

$$\Rightarrow \quad \tan\theta = 4$$

Ans. (D)

Sol. 69 Time taken to reach maximum height for the projectile is given as

$$t = \frac{u\sin\theta}{g}$$

$$\Rightarrow \quad \frac{u_1\sin\theta_1}{g} = \frac{u_2\sin\theta_2}{g}$$

$$\Rightarrow \quad u_1\sin 30° = u_2\sin 45°$$

$$\Rightarrow \quad \frac{u_1}{u_2} = \frac{1/\sqrt{2}}{1/2} = \frac{\sqrt{2}}{1}$$

Ans. (C)

Sol. 70 Given that horizontal velocity of projectile at $t = 0$ is

$$u_x = 1 \text{ m/s}$$

Equation of trajectory is

$$y = 5x(1-x)$$

Differentiating with respect to time, we get

$$\frac{dy}{dt} = 5\frac{dx}{dt} - 10x\cdot\frac{dx}{dt}$$

For initial y-component of velocity, we have

$$\Rightarrow \quad u_y = 5(1) = 5 \text{ m/s}$$

Initial velocity vector in y direction is given as

$$\vec{u}_y = 5\hat{j}$$

Ans. (5)

Sol. 71 For the projection speed is u the range of projectiles for the two given angles are

$$R_1 = \frac{u^2\sin(90°)}{g}$$

and $$R_2 = \frac{u^2\sin(60°)}{g}$$

$$\Rightarrow \quad \frac{R_1}{R_2} = \frac{2}{\sqrt{3}}$$

Ans. (C)

Sol. 72 For the given vectors

$$\vec{A} = (2\hat{i} + 3\hat{j} - \hat{k})\text{m}$$

and $$\vec{B} = (\hat{i} + 2\hat{j} + 2\hat{k})\text{m}$$

Component of $\vec{A}$ along $\vec{B}$ is calculated as

$$\vec{B}_c = \vec{A}\cdot\hat{B}$$

$$\Rightarrow \quad \vec{B}_c = \frac{\vec{A}\cdot\vec{B}}{|\vec{B}|} = \frac{2+6-2}{\sqrt{1^2+2^2+2^2}}$$

$$\Rightarrow \quad \vec{B}_c = \frac{6}{3} = 2\text{ m}$$

Ans. (2)

Sol. 73 Time of flight for both the balls is same so the vertical component of velocity of both balls are same thus maximum heights attained by the balls in vertical direction will also be same. **Ans. (1)**

Sol. 74 If the projection of one vector on another is zero indicates that the vectors are perpendicular to each other hence we use

$$\vec{a}\cdot\vec{b} = 0$$

$$\Rightarrow \quad 2\times 1 + 4\times 2 - 2\times\alpha = 0$$

$$\Rightarrow \quad \alpha = 5$$

Ans. (5)

Sol. 75 The situation described in question is shown in figure below

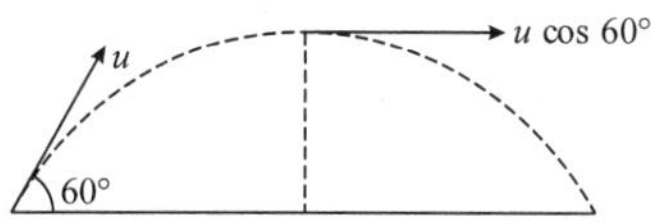

Initial kinetic energy of the ball is taken as

$$E = \frac{1}{2}mu^2$$

At highest point, velocity of projectile is taken as

$$v = u\cos 60° = \frac{u}{2}$$

Thus kinetic energy at the topmost point is calculated as

$$K = \frac{1}{2}m\left(\frac{u}{2}\right)^2 = \frac{E}{4}$$

Ans. (C)

Sol. 76 For first object the range is maximum, so we have

$$R_{max} = \frac{u^2\sin 2(45°)}{g} = \frac{u^2}{g}$$

For second object range is half, so we have

$$\frac{R}{2} = \frac{u^2}{2g} = \frac{u^2\sin 2\theta}{g}$$

$$\Rightarrow \quad \sin 2\theta = \frac{1}{2}$$

$$\Rightarrow \quad 2\theta = 30°, 150°$$

$$\Rightarrow \quad \theta = 15°, 75°$$

Ans. (15)

Sol. 77 Maximum vertical height to which a ball reaches if projected at speed v is given as

$$H_{max} = \frac{v^2}{2g} = 136\text{ m}$$

Maximum range of a ball in projectile motion is calculated as

$$R_{max} = \frac{v^2}{g} = 2H_{max}$$

$\Rightarrow$ $$R_{max} = 2(136) = 272\text{ m}$$ **Ans. (C)**

Sol. 78 For two perpendicular vectors, we use

$$(a\hat{i} + b\hat{j} + \hat{k}) \cdot (2\hat{i} - 3\hat{j} + 4\hat{k}) = 0$$

$\Rightarrow$ $$2a - 3b + 4 = 0$$

On solving, we get

$$2a - 3b = -4$$

Also it is given that $3a + 2b = 7$, solving this with above equation gives

$$a = 1, b = 2$$

$\Rightarrow$ $$\frac{a}{b} = \frac{x}{2}$$

$\Rightarrow$ $$x = 1$$ **Ans. (1)**

Sol. 79 As the two vectors are perpendicular to each other, we use

$$\vec{P} \cdot \vec{Q} = 0$$

$\Rightarrow$ $$(\hat{i} + 2m\hat{j} + m\hat{k}) \cdot (4\hat{i} - 2\hat{j} + m\hat{k}) = 0$$

$\Rightarrow$ $$4 - 4m + m^2 = 0$$

$\Rightarrow$ $$(m-2)^2 = 0$$

$\Rightarrow$ $$m = 2$$ **Ans. (D)**

Sol. 80 Cross product of the two given vectors is calculated as

$$\vec{P} \times \vec{Q} = \begin{vmatrix} \hat{i} & \hat{j} & k \\ 3 & \sqrt{3} & 2 \\ 4 & \sqrt{3} & 2.5 \end{vmatrix} = \sqrt{3}\frac{\hat{i}}{2} + \frac{\hat{j}}{2} - \sqrt{3}\hat{k}$$

Unit vector in the direction of above cross product is given as

$$\frac{\vec{P} \times \vec{Q}}{|\vec{P} \times \vec{Q}|} = \frac{1}{2}\left(\sqrt{3}\frac{\hat{i}}{2} + \frac{\hat{j}}{2} - \sqrt{3}\hat{k}\right)$$

$\Rightarrow$ $$\frac{\vec{P} \times \vec{Q}}{|\vec{P} \times \vec{Q}|} = \frac{1}{4}(\sqrt{3}\hat{i} + \hat{j} - 2\sqrt{3}\hat{k})$$

$\Rightarrow$ $$x = 4$$ **Ans. (4)**

Sol. 81 Range of projectile is given as

$$R = \frac{u^2 \sin 2\theta}{g}$$

Given that $\alpha + \beta = 90°$ and for these complementary angles we know that the range of projectile is same hence we use

$$\frac{R_1}{R_2} = \frac{1}{1}$$ **Ans. (D)**

Sol. 82 If initial speed of projection is u, the speed at the topmost point of projectile will be u cos 30° so the ratio of kinetic energies at these two points is given as

$$\frac{KE_i}{KE_{top}} = \frac{\frac{1}{2}M(u)^2}{\frac{1}{2}M(u\cos 30°)^2} = \frac{4}{3}$$ **Ans. (D)**

Sol. 83 Figure below shows the situation described in the question.

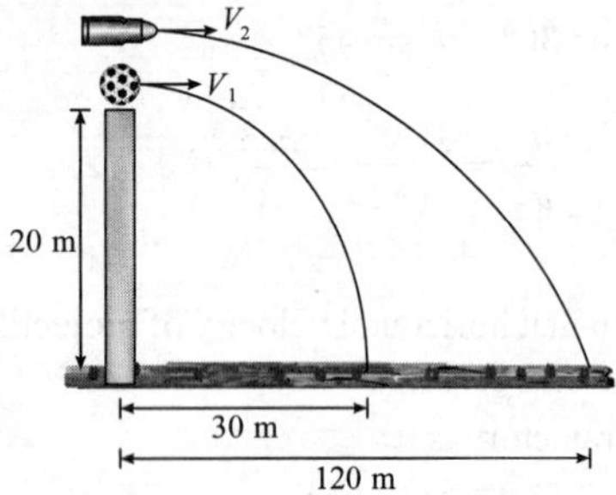

The velocity of bullet and ball after collision can be given as

$$v_1 = \frac{30}{\sqrt{\frac{2h}{g}}}$$

and $$v_2 = \frac{120}{\sqrt{\frac{2h}{g}}}$$

If the initial speed of bullet is u then by conservation of momentum, we have

$$(0.01)u = (0.2)\frac{30\sqrt{g}}{\sqrt{2h}} + (0.01)\frac{120\sqrt{g}}{\sqrt{2h}}$$

$\Rightarrow$ $$u = 300 + 60 = 360\text{ ms}^{-1}$$ **Ans. (D)**

Sol. 84 As range of the two projectiles is same, we use

$$\theta_1 + \theta_2 = 90°$$

$\Rightarrow$ $$\theta_2 = 30°$$

Thus sum of maximum heights attained by two projectiles is given as

$$(H_{max})_1 + (H_{max})_2 = \frac{U^2 \sin^2\theta_1}{2g} + \frac{U^2 \sin^2\theta_2}{2g}$$

$\Rightarrow$ $$(H_{max})_1 + (H_{max})_2 = \frac{40^2}{20}\left(\frac{1}{4} + \frac{3}{4}\right) = 80\text{ m}$$ **Ans. (80)**

Sol. 85 Picture in figure below shows the situation described in the question.

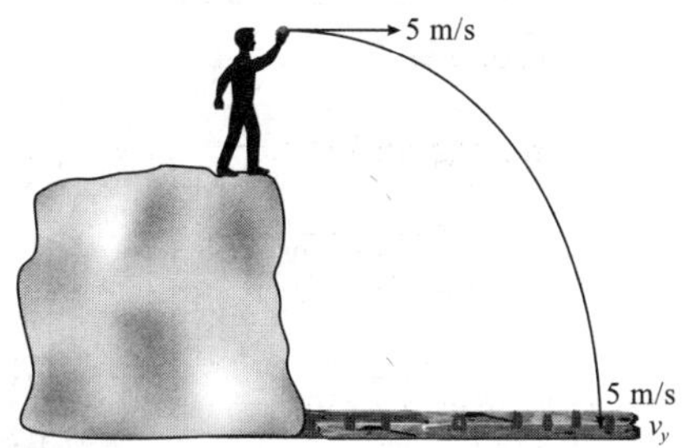

As the stone falls by a distance of 10 m, the velocity component gained in vertical direction by the stone is calculated as

$$v_y = \sqrt{2gh} = \sqrt{200}$$

Net velocity with which stone will hit the ground is given as

$$v_{net} = \sqrt{25+200} = 15 \text{ m/s} \qquad \textbf{Ans. (B)}$$

Sol. 86 At highest point of projectile, we use

$$V_y = 0$$

Horizontal component remain same so we use

$$V_x = u_x = u\cos\theta$$

Gravitational potential energy is given as

$$U_g = mgh$$

it is maximum at H_{max}. **Ans. (A)**

Sol. 87 Range of projectile is given as

$$R = \frac{u^2 \sin^2\theta}{g}$$

At 45°, $\sin^2\theta = 1$ so, range is maximum and this is for $2\theta = 90°$.

Ans. (B)

Sol. 88 Horizontal range of projectile when projected at angle θ_1 and θ_2.

$$R_1 = \frac{u_1^2 \sin 2\theta_1}{g};$$

$$R_2 = \frac{u_2^2 \sin 2\theta_2}{g}$$

So, ratio of ranges is calculated as

$$\frac{R_1}{R_2} = \frac{u_1^2 \sin 2\theta_1}{u_2^2 \sin 2\theta_2} = \frac{40^2 \sin(2\times 30°)}{60^2 \sin(2\times 60°)}$$

$$\Rightarrow \quad \frac{R_1}{R_2} = \frac{4}{9} \qquad \textbf{Ans. (D)}$$

Sol. 89 Given that velocity of particle at any instant is given as

$$\vec{v} = 2t\hat{i} + 3t^2\hat{j}$$

By differentiating, we have

$$\vec{a} = 2\hat{i} + 6t\hat{j}$$

at $t = 1$ s, $\quad \vec{a} = 2\hat{i} + 6\hat{j}$

Force $\quad \vec{F} = m\vec{a} = 0.5(2\hat{i} + 6\hat{j}) = \hat{i} + 3\hat{j}$

$$\Rightarrow \quad \vec{F} = \hat{i} + x\hat{j}$$

$$\Rightarrow \quad x = 3 \qquad \textbf{Ans. (A)}$$

Sol. 90 Maximum height attained by the projectile is

$$H_{max} = \frac{u^2 \sin^2\theta}{2g}$$

So when two projectiles are projected at complimentary angle 30° and 60°

$$\Rightarrow \quad \frac{H_1}{H_2} = \frac{\sin^2\theta_1}{\sin^2\theta_2} = \frac{1}{3} \qquad \textbf{Ans. (C)}$$

Sol. 91 Time of flight $t_1 + t_2 = 3 + 5 = 8$ s

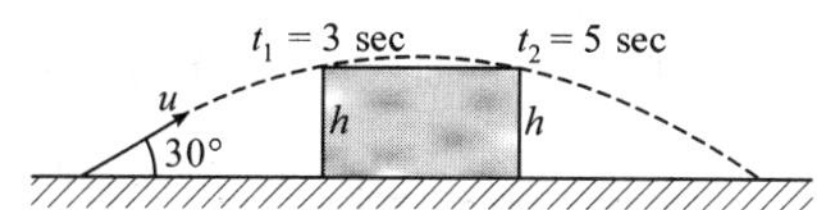

$$\Rightarrow \quad T = \frac{2u\sin 30°}{g}$$

$$\Rightarrow \quad 8 = \frac{2u\sin(30°)}{10}$$

$$\Rightarrow \quad u = 80 \text{ m/s} \qquad \textbf{Ans. (80)}$$

Sol. 92 At $t = 2$ s particle is at maximum height moving with velocity

$$v = u\cos\theta$$

Here u is projection velocity. As at height point only horizontal component exists.

$$v = 40\cos 30° = 20\sqrt{3} \text{ ms}^{-1} \qquad \textbf{Ans. (A)}$$

Sol. 93 By vector substraction, we have

$$\vec{B} - \vec{A} = 2\hat{j}$$

$$\Rightarrow \quad \vec{B} = 2\hat{i} + 5\hat{j} + 3\hat{k}$$

Thus magnitude of vector $\vec{B}$ is given as

$$|\vec{B}| = \sqrt{33} \qquad \textbf{Ans. (Bonus)}$$

Sol. 94 Position vector of particle with time t is given as

$$\vec{r} = 10t\hat{i} + 15t^2\hat{j} + 7\hat{k}$$

By differentiating above relateion, velocity is given as

$$\vec{v} = 10\hat{i} + 30t\hat{j}$$

By differentiating velocity, we get acceleration, given as

$$\vec{a} = 30t\hat{j}$$

Thus net force is along $+y$ direction. **Ans. (A)**

Sol. 95 y-component of given vector is written as

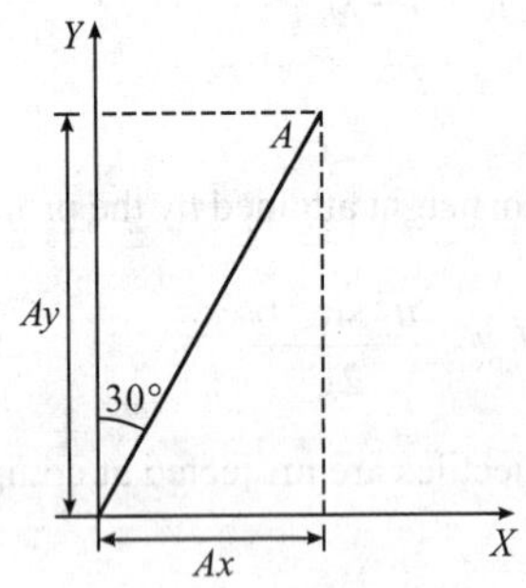

$$A_y = A\cos 30° = 2\sqrt{3}$$

$$\Rightarrow \quad A\frac{\sqrt{3}}{2} = 2\sqrt{3}$$

$$\Rightarrow \quad A = 4$$

Thus magnitude of x-component of velocity is calculated as

$$A_x = A\sin 30° = 4\times\frac{1}{2} = 2$$ **Ans. (D)**

3.2 Relative Motion

Sol. 1

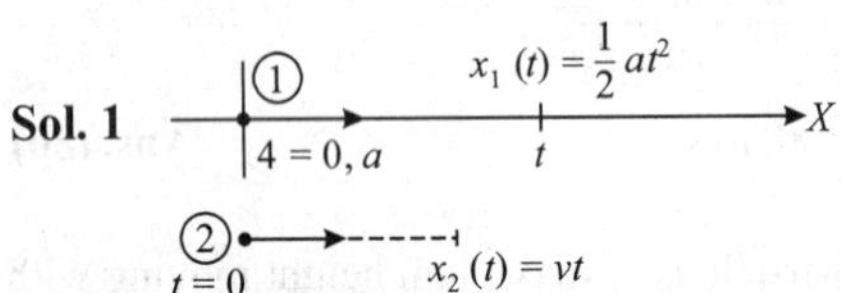

$t = 0 \quad x_2(t) = vt$

$$\Rightarrow \quad x_1 - x_2 = \frac{1}{2}at^2 - vt$$

at $t = 0+$ we have $x_1 - x_2 \rightarrow -\text{ve}$

Initially $x_1 - x_2$ is negative in option-(B) only. **Ans. (B)**

Sol. 2 Vertical acceleration of A is

$$a_{Ay} = g\sin 60° = \frac{\sqrt{3}g}{2}$$

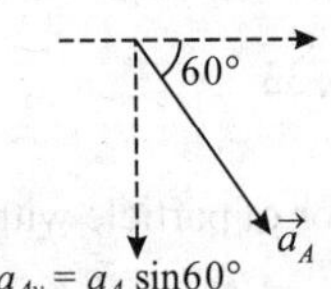

$$\Rightarrow \quad a_{Ay} = \frac{\sqrt{3}g}{2}\left(\frac{\sqrt{3}}{2}\right) = \frac{3g}{4}$$

Vertical acceleration of B is

$$a_{By} = g\sin 30° = \frac{g}{2}$$

$$\Rightarrow \quad a_{By} = \frac{g}{2}\left(\frac{1}{2}\right) = \frac{g}{4}$$

Acceleration of B w.r.t. A is vertical direction is

$$a_{yAB} = a_{Ay} - a_{By} = \frac{3g}{u} - \frac{g}{u} = \frac{g}{2}$$

$$= 4.9 \text{ m/s}^2$$ **Ans. (D)**

Sol. 3 For $u = 10$ m/s, $h_1 = \frac{u_1^2}{2g} = \frac{100}{2\times 10} = 5$ m

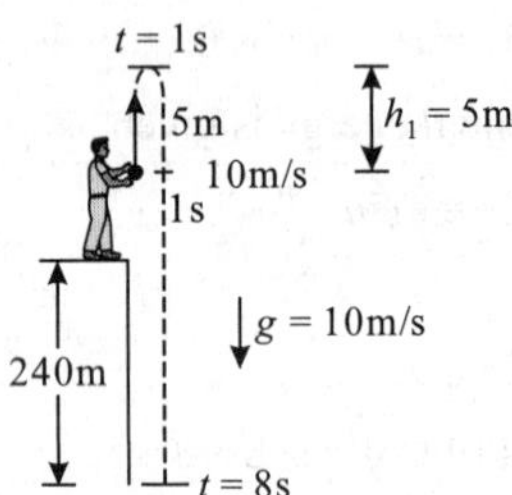

Time of downward journey

$$t_2 = \sqrt{\frac{2h}{g}}$$

$$t_2 = \sqrt{\frac{2\times 245}{10}} = 7 \text{ s}$$

For $u = 40$ m/s, $h_2 = \frac{u^2}{2g} = \frac{(40)^2}{2\times 10} = 80$ m

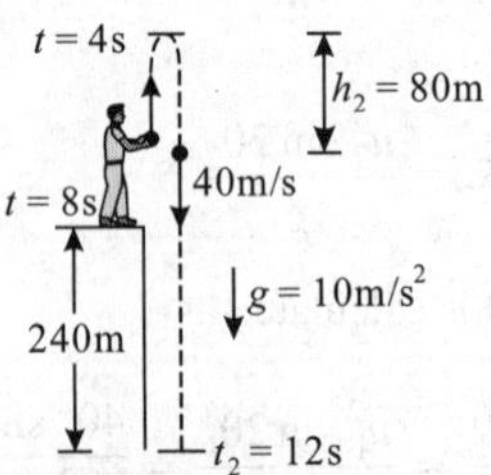

Time for downward journey of stone

$$t_2 = \sqrt{\frac{2h}{g}}$$

$$t_2 = \sqrt{\frac{2\times 320}{10}} = 8\text{s}$$

If y_1 and y_2 are height above ground for the two stones, we have

$$y_1 = 240 + (10t - 5t^2)$$
$$y_1 = 240 + 10t - 5t^2$$
$$y_2 = 240 + (40t - 5t^2)$$
$$y_2 = 240 + 40t - 5t^2$$

$\Rightarrow$ from $t = 0$ to $t = 8$ s when first stone reaches ground

$$y_2 - y_1 = 30t$$

at $t = 8$s $\quad y_2 - y_1 = 240$ m

after $t = 8$s, first stone is at rest $\Rightarrow y_1 = 0$

$\Rightarrow \quad \Delta y = y_2 - y_1 = 240 + 40t - 5t^2$

slope of graph

$$\frac{d(\Delta y)}{dt} = 40 - 10t$$

As for $t > 8$ s slope of graph is increasing **Ans. (C)**

Sol. 4

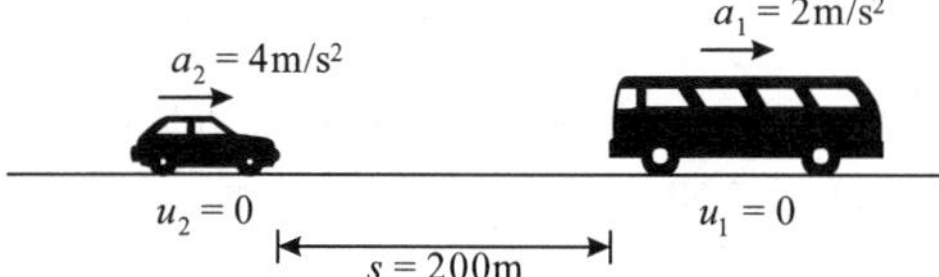

Considering bus to be at rest and relative to bus, speed and acceleration of car are

$$u_{CB} = 0 = u_2 - 4,$$
$$a_{CB} = a_2 - a_1 = 4 - 2 = 2\text{m/s}^2$$

time to bus by car (rel. to bus) a distance of 200 m is

$$t = \sqrt{\frac{25}{a}} = \sqrt{\frac{2 \times 200}{2}}$$
$$= 10\sqrt{2}\text{ s}$$ **Ans. (C)**

Sol. 5 If we take the position of ship 'A' as origin then positions and velocities of both ships can be given as :

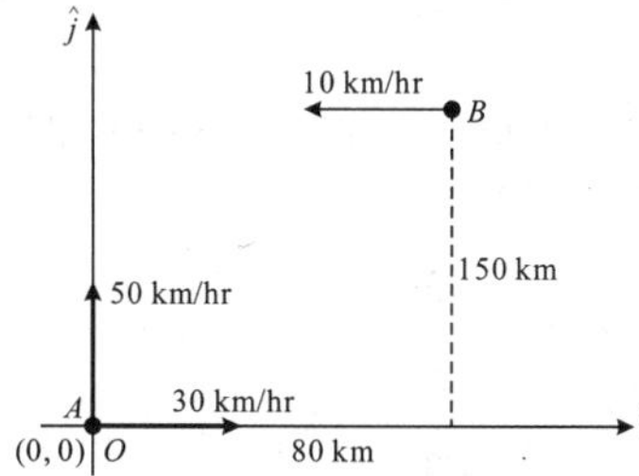

$$\vec{v}_A = (30\hat{i} + 50\hat{j})\text{ km/hr}$$
$$\vec{v}_B = -10\hat{i}\text{ km/hr}$$
$$\vec{r}_A = 0\hat{i} + 0\hat{j}$$
$$\vec{r}_B = (80\hat{i} + 150\hat{j})\text{ km}$$

Time after which distance between them will be minimum

$$t = -\frac{\vec{r}_{BA} \cdot \vec{v}_{BA}}{|\vec{v}_{BA}|^2};$$

where $\vec{r}_{BA} = (80\hat{i} + 150\hat{j})$ km

$$\vec{v}_{BA} = -10\hat{i} - (30\hat{i} + 50\hat{j})$$
$$(-40\hat{i} - 50\hat{j})\text{ km/hr}$$

$$\Rightarrow \quad t = -\frac{(80\hat{i} + 150\hat{j}) \cdot (-40\hat{i} - 50\hat{j})}{|-40\hat{i} - 50\hat{j}|^2}$$

$$= \frac{3200 + 7500}{4100}\text{ hr} = \frac{10700}{4100}\text{ hr} = 2.6\text{ hrs}$$ **Ans. (D)**

Sol. 6

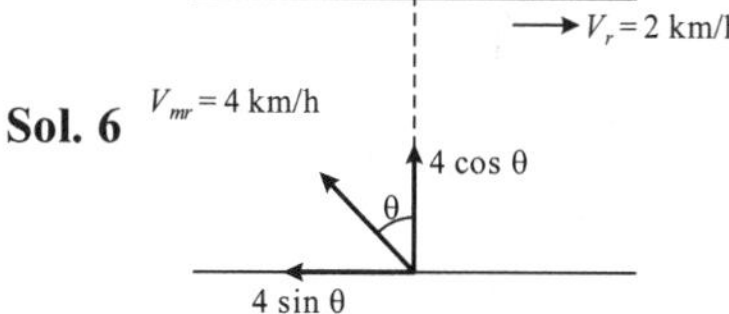

For swimmer to cross the river straight

$$\Rightarrow \quad 4\sin\theta = 2$$
$$\Rightarrow \quad \sin\theta = \frac{1}{2}$$
$$\Rightarrow \quad \theta = 30°$$

So, angle with direction of river flow

$$= 90° + \theta = 120°$$ **Ans. (D)**

Sol. 7 Figure below shows the given situation

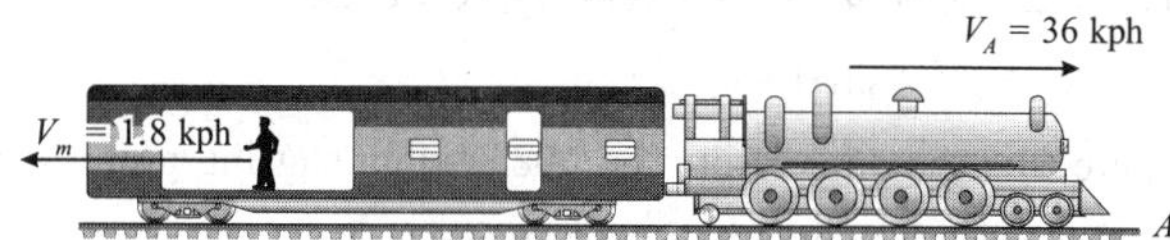

Speed of man w.r. to train B is

$$\vec{V}_{mB} = \vec{V}_m + \vec{V}_A - \vec{V}_B$$
$$= 1.8 - 36 - 72$$
$$= -106.2\text{ km/hr}$$
$$= 29.5\text{ m/s}$$ **Ans. (B)**

Sol. 8 When speed of car is v, we use

$$\overrightarrow{V_{RC}} = \overrightarrow{V_R} - \overrightarrow{V_C}$$

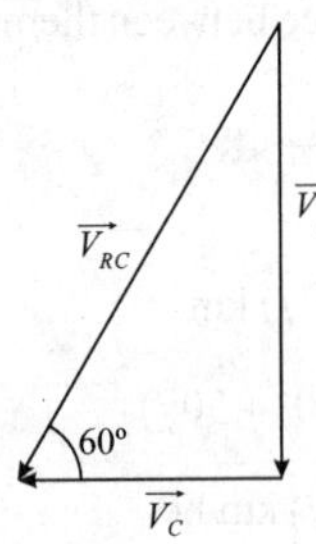

$$\tan 60° = \frac{V_R}{V} = \sqrt{3}$$

$$\Rightarrow \quad V_R = \sqrt{3}\,v \quad \ldots(1)$$

When speed of car changes to $V_C' = V(1+\beta)$, we use

$$\overrightarrow{V_{RC'}} = \overrightarrow{V_R} - \overrightarrow{V_{C'}}$$

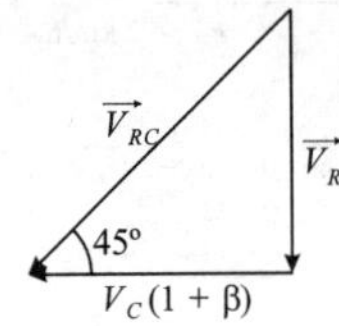

$$\Rightarrow \quad \tan 45° = \frac{V_R}{V(1+\beta)} = 1$$

$$\Rightarrow \quad V_R = v(1+\beta) \quad \ldots(2)$$

From (1) and (2)

$$1+\beta = \sqrt{3}$$

$$\Rightarrow \quad \beta = 1.732 - 1 = 0.732 \qquad \textbf{Ans. (D)}$$

Sol. 9 To reach just opposite point on the other bank from where he started, velocity component of swimmer along the river must be equal to rivir velocity, so we use

$$12 \sin\theta = v_r$$

where θ is the angle from the line joining starting point and opposite point on the other bank of river.

$$\Rightarrow \quad \sin\theta = \frac{1}{2}$$

$$\Rightarrow \quad \theta = 30°$$

Thus angle made by the swimmer velocity with the river will be given as

$$\alpha = 30° + 90° = 120° \qquad \textbf{Ans. (120)}$$

Sol. 10 If we consider $\hat{i}$ and $\hat{j}$ as unit vectors along north and east directions, the net displacement of butterfly can be given as

$$\vec{D} = vt$$

$$\Rightarrow \quad \vec{D} = \{(4\hat{i} + 4\hat{j} + (-\hat{j}))\} \times 3\text{s} = (4\hat{i} + 3\hat{j}) \times 3\text{s}$$

$$\Rightarrow \quad |\vec{D}| = 15\text{ m} \qquad \textbf{Ans. (B)}$$

Sol. 11 If the length of escalator is L, then from the situation and information given in the question we have

Speed of man relative to escalator is calculated as

$$v_m = \frac{L}{t_1}$$

Speed of escalator motion is calculated as

$$v_e = \frac{L}{t_2}$$

Speed of man relative to ground when escalator is moving can be given as

$$v_{mG} = \frac{L}{t_1} + \frac{L}{t_2}$$

Thus time taken by the man to walk on the moving escalator is calculated as

$$t = \frac{L}{\frac{L}{t_1} + \frac{L}{t_2}} = \frac{t_1 t_2}{t_1 + t_2} \qquad \textbf{Ans. (C)}$$

Sol. 12 Vector triangle below shows the relative velocity of rain drops with respect to moving girl.

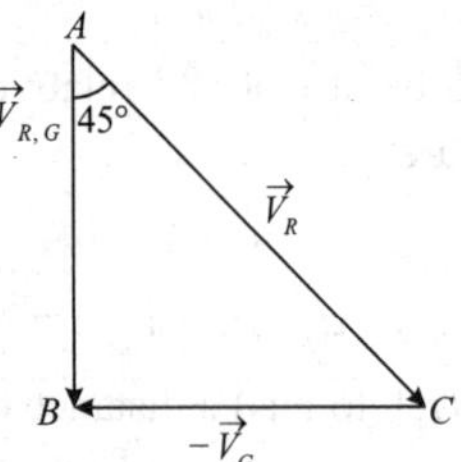

Here we use

$$\tan\theta = \frac{V_G}{V_{RG}}$$

$$\Rightarrow \quad 1 = \frac{V_G}{V_{RG}}$$

$$\Rightarrow \quad 15\sqrt{2} = V_{RG} \qquad \textbf{Ans. (C)}$$

Sol. 13 The horizontal component of the projectile at the projection point is given as

$$u\cos\theta = \frac{\sqrt{3}u}{2}$$

$$\Rightarrow \quad \cos\theta = \frac{\sqrt{3}}{2}$$

$$\Rightarrow \quad \theta = 30°$$

Thus time of flight of projectile is calculated as

$$T = \frac{2u\sin 30°}{g} = \frac{u}{g} \qquad \textbf{Ans. (B)}$$

Sol. 14 Figure below shows the path of swimmer affected by the river current.

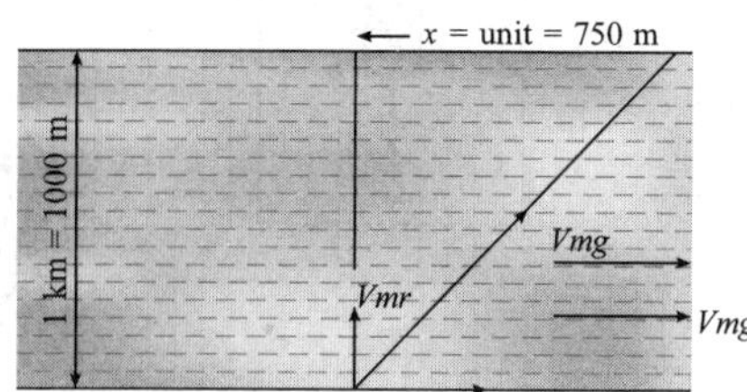

Time taken to cross the river width $w = 1000$ m is

$$t = \frac{1 \text{ km}}{4 \text{ km/h}}$$

Drift $\quad x = vm/g \times t$

Where vm/g is velocity of river w.r. to ground

$$x = v/mg \times \frac{1}{4} = 750 \text{ m} = \frac{3}{4} \text{ km}$$

$\Rightarrow \quad vm/g = 3\text{km/hr}$ **Ans. (3)**

4.1 Newton's Laws of Motion

Sol. 1 As particle remain stationary

$$\Rightarrow \quad \vec{F_1} + \vec{F_2} + \vec{F_3} = 0$$

$$\vec{F_1} + \vec{F_R} = 0$$

$$\vec{F_1} = -\vec{F_R}$$

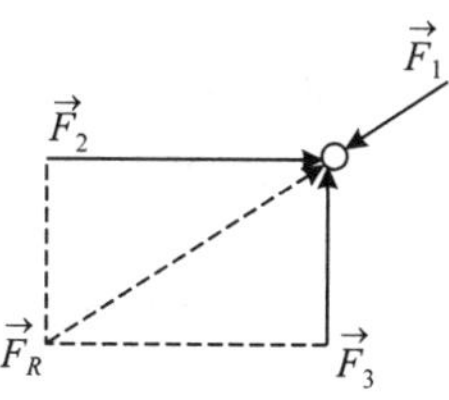

when $\vec{F_1}$ is removed, particle will move in under the influence of $\vec{F_R}$

$\Rightarrow$ acceleration

$$\vec{a} = \frac{\vec{F_R}}{m} \Rightarrow |\vec{a}| = \frac{F_1}{m}$$ **Ans. (A)**

Sol. 2

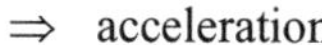

For block C, its equation of motion is

$$T_2 = m_C a$$
$$= 2 \times 0.6$$
$$= 1.2 \text{ N}$$ **Ans. (B)**

Sol. 3 When man is at rest

$$T = 60g = 600 \text{ N}$$

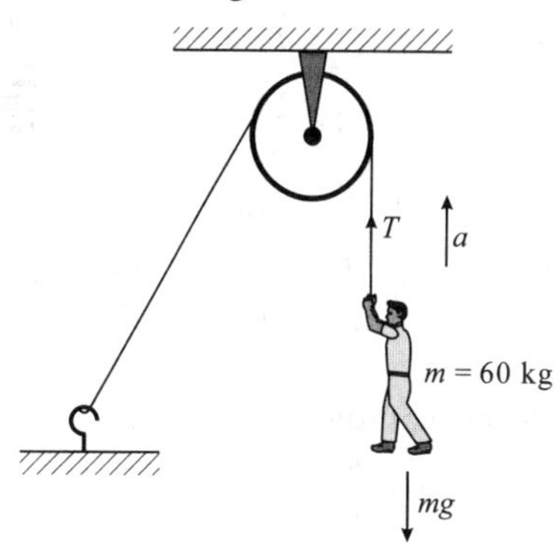

When man starts climbing up the rope at acceleration a, its equation of motion is

$$T - mg = ma$$

$$\Rightarrow \quad 840 - 600 = 60a$$

$$\Rightarrow \quad a = 4 \text{ m/s}^2$$ **Ans. (C)**

Sol. 4

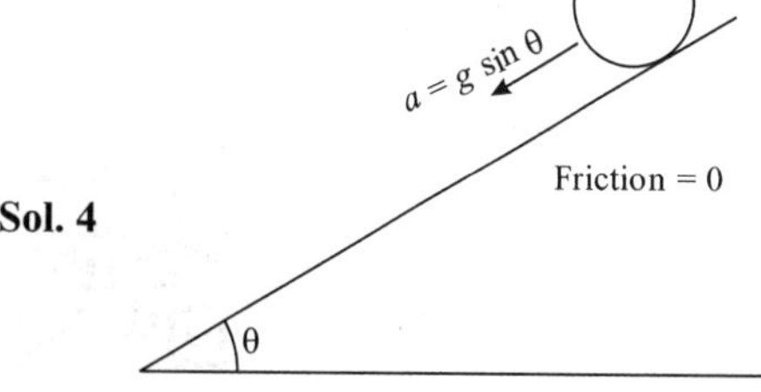

As no friction is there, all bodies slide with acceleration $g \sin \theta$.

Ans. (D)

Sol. 5 Acceleration of ball w.r.t. lift

$$\vec{a}_{BL} = \vec{a}_B - \vec{a}_L$$
$$a_{BL} = g - a$$

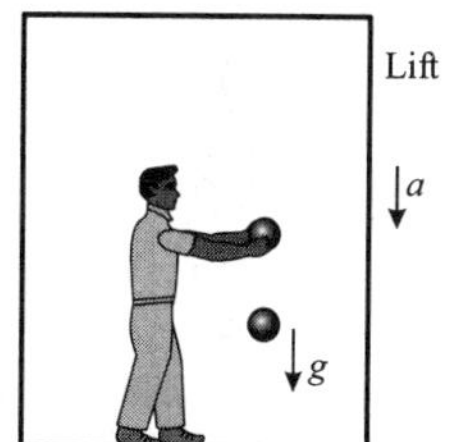

Acceleration of ball w.r.t. ground

$$a_B = g$$ **Ans. (C)**

Sol. 6 Given that

$$F_1 + F_2 = 18 \text{ N} \quad \dots(1)$$

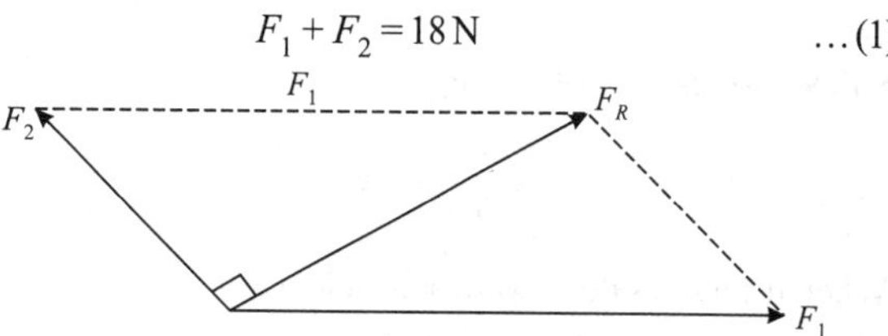

From the given vector triangle, we use

$$F_2^2 + F_R^2 = F_1^2$$

$$\Rightarrow \quad F_1^2 - F_2^2 = F_R^2 = 144$$

$$\Rightarrow \quad (F_1 + F_2)(F_1 - F_2) = 144$$

from (1) $\Rightarrow$ $$F_1 - F_2 = \frac{144}{18} = 8 \qquad \ldots(2)$$

from (1) and (2) $$2F_1 = 26 \Rightarrow F_1 = 13\text{ N}$$

$$2F_2 = 10 \Rightarrow F_2 = 5\text{ N}$$ **Ans. (B)**

Sol. 7 Acceleration of Atwood's machine is given as

$$a = \frac{m_2 - m_1}{m_2 + m_1} \cdot g$$

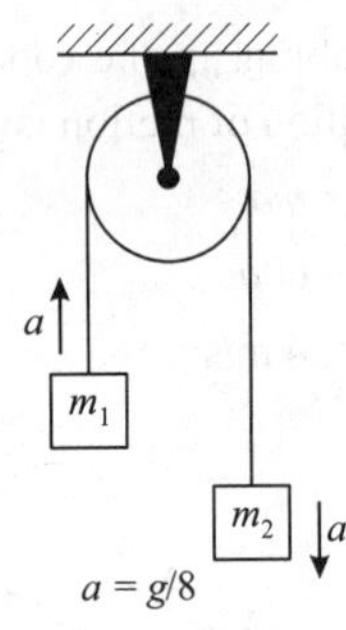

$$m_2 + m_1 = 8m_2 - 8m_1$$

$$9m_1 = 7m_2$$

$$\Rightarrow \quad \frac{m_2}{m_1} = \frac{9}{7}$$ **Ans. (B)**

Sol. 8 As spring balances are light, throughout the string length tension remain same.

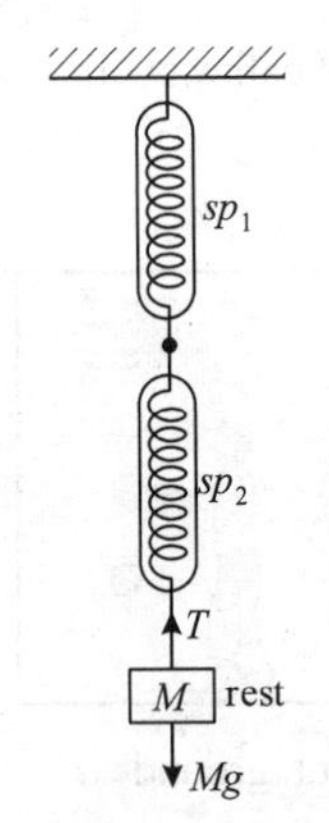

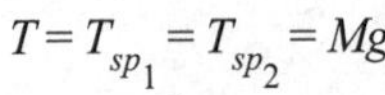

$$T = T_{sp_1} = T_{sp_2} = Mg$$

$\Rightarrow$ Readings of both sp_1 and sp_2 are same **Ans. (A)**

Sol. 9 At rest $T = 49\text{N} = mg$

$$\Rightarrow \quad m = \frac{49}{9.8} = 5\text{ kg}$$

When lift moves downward at acceleration a

Then its equation of motion is (for bag)

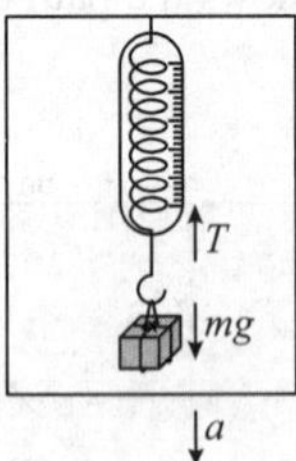

$$mg - T_1 = ma$$

$$\Rightarrow \quad T_1 = mg - ma = m(g - a)$$

$$= 5(9.8 - 5)$$

$$= 5 \times 4.8 = 24\text{ N}$$ **Ans. (A)**

Sol. 10

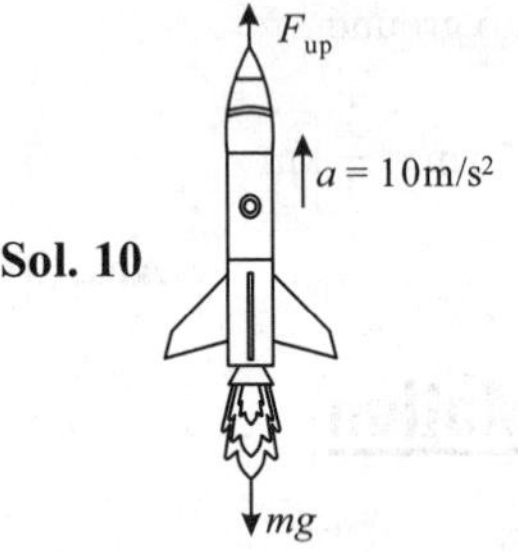

For upward acceleration of rocket

$$a = 10\text{ m/s}^2$$

We use

$$F_{up} - mg = ma$$

$$\Rightarrow \quad F_{up} = mg + ma$$

$$\Rightarrow \quad F_{up} = m(g + a) = 3.5 \times 10^4[10 + 10]$$

$$= 7 \times 10^5\text{ N}$$ **Ans. (B)**

Sol. 11 As vector sum of the three forces is equal to zero

$$\Rightarrow \quad \vec{a} = 0$$

$\Rightarrow$ $\vec{v} \rightarrow$ remain uncharged. **Ans. (D)**

Sol. 12 *FBD* of block is shown in figure

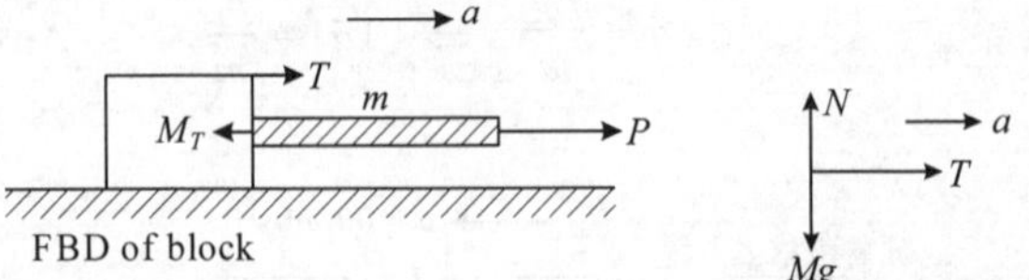

Acceleration of system (block + rope) is given as

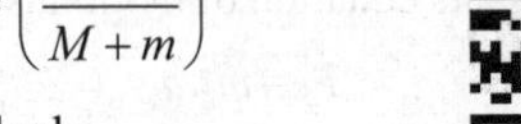

$$a = \left(\frac{P}{M + m}\right)$$

Equation of motion of block

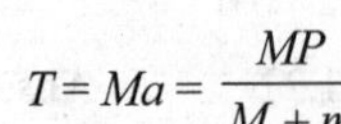

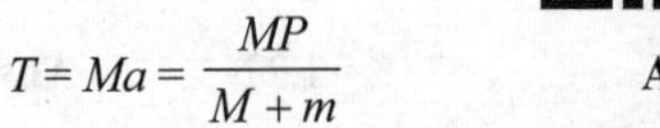

$$T = Ma = \frac{MP}{M + m}$$

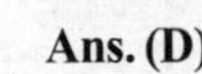

Ans. (D)

Sol. 13 Acceleration of masses in Atwood's machine is given as

$$a = \left(\frac{m_1 - m_2}{m_1 + m_2}\right)g$$

$$= \left(\frac{5-4.8}{5+4.8}\right) \times 9.8 \text{ m/s}^2$$

$$= 0.2 \text{ m/s}^2$$

Ans. (A)

Sol. 14

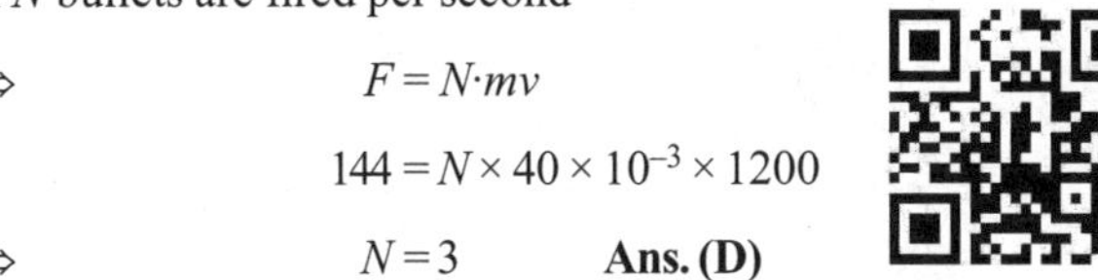

Man has to apply an equal force to keep the gun at rest

$$F = \text{rate of charge of linear momentum}$$

If N bullets are fired per second

$$\Rightarrow \quad F = N \cdot mv$$

$$144 = N \times 40 \times 10^{-3} \times 1200$$

$$\Rightarrow \quad N = 3$$ **Ans. (D)**

Sol. 15

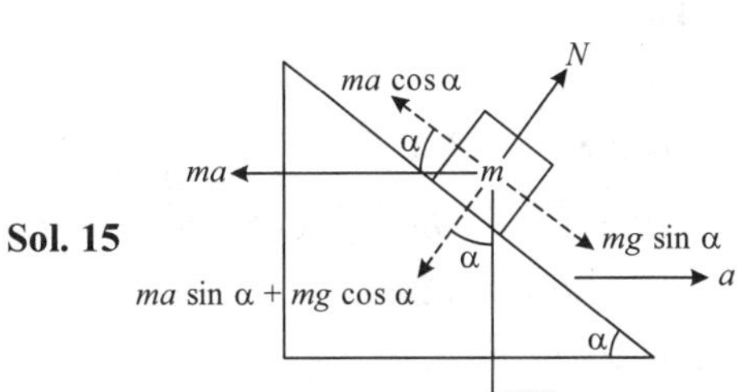

To keep the block at rest, we use

$$mg \sin\alpha = ma \cos\alpha$$

$$a = g \tan\alpha$$ **Ans. (D)**

Sol. 16

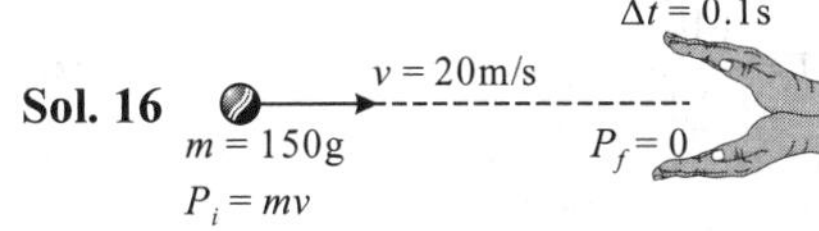

In the process of catching the ball, force on hand is

$$F = \frac{\Delta P}{\Delta t} = \frac{P_i - P_f}{\Delta t} = \frac{150 \times 10^{-3} \times 20}{0.1}$$

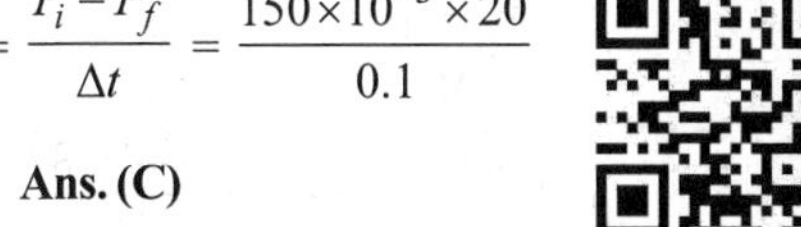

$$\Rightarrow \quad F = 30 \text{ N}$$ **Ans. (C)**

Sol. 17 In free fall motion of ball we use

$$v^2 = u^2 - 2gh$$

$$\Rightarrow \quad 0 = u^2 - 2 \times 10 \times 2$$

$$\Rightarrow \quad u = \sqrt{40} \text{ m/s}$$

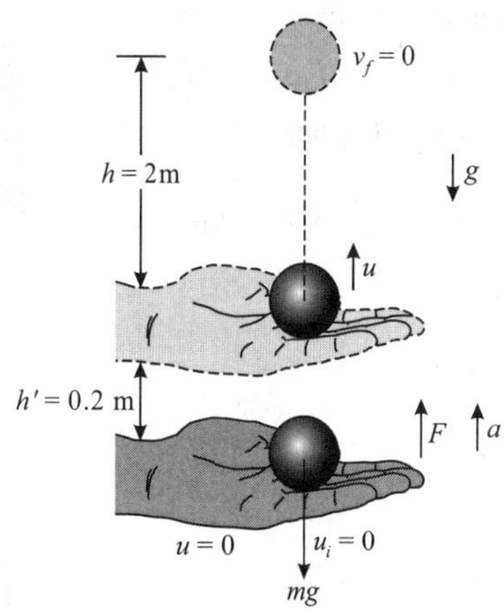

When ball was being pushed by hand, we use

$$u^2 = u_i^2 + 2ah'$$

$$40 = 0 + 2a(0.2)$$

$$\Rightarrow \quad a = 100 \text{ m/s}^2$$

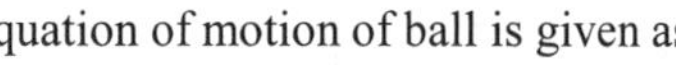

equation of motion of ball is given as

$$F - mg = ma$$

$$\Rightarrow \quad F = m(g + a) = 0.2(10 + 100)$$

$$\Rightarrow \quad F = 0.2 \times 110 = 22 \text{ N}$$ **Ans. (D)**

Sol. 18

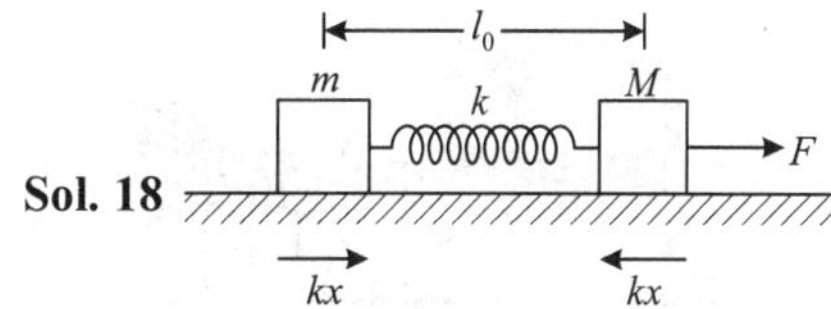

When both the blocks will start moving at same acceleration, say spring has an extension x_0

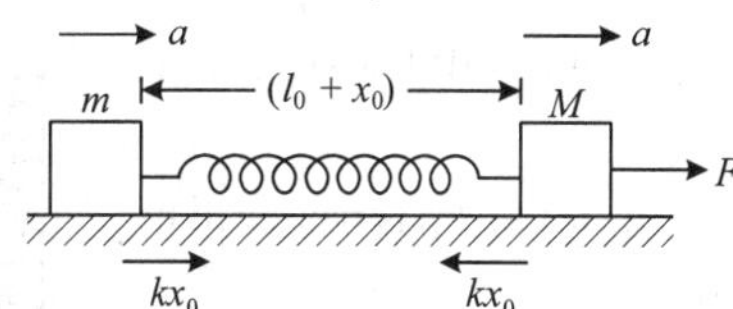

Common acceleration

$$a = \frac{F}{m + M}$$

equation of motion of blocks are

and $$F - kx_0 = Ma$$

$$kx_0 = ma$$

$$kx_0 = \frac{mF}{m + M}$$ **Ans. (C)**

Sol. 19 If a spring of length l is cut in different parts then force constant of a part of length l_1 is given as

$$k_1 = \frac{kl}{l_1}$$ [k = force constant of original spring]

For spring A and B we have

$$l_A = 2l_0;\ l_B = 3l_0;\ l = l_A + l_B = 5l_0$$

Thus force constant of spring A is

$$k_A = \frac{kl}{l_A} = k\left(\frac{5l_0}{2l_0}\right) = \frac{5}{2}k \quad \textbf{Ans. (A)}$$

Sol. 20 Acceleration of particle

$$a = \frac{F}{m} = \frac{F_0}{m}e^{-bt}$$

$$\Rightarrow \quad a = \frac{dv}{dt} = \frac{F_0}{m}e^{-bt}$$

$$\Rightarrow \quad dv = \frac{F_0}{m}e^{-bt}dt$$

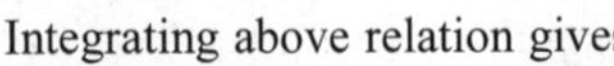

Integrating above relation gives

$$\int_{v=0}^{v=v} dv = \int_{t=0}^{t=t} \frac{F_0}{m}e^{-bt}dt$$

$$\Rightarrow \quad v = \frac{F_0}{m}\left[\frac{e^{-bt}}{-b}\right]_0^t$$

$$\Rightarrow \quad v_{(t)} = \frac{F_0}{mb}[1 - e^{-bt}] \quad \textbf{Ans. (D)}$$

Sol. 21 Statement-I :

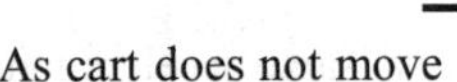

As cart does not move

$\Rightarrow$ $F_H = F_B$ (ignoring friction on cart)

Acceleration to Newton's III law statement-I is true

Statement-II : In cart as F_H is balanced by F_B it does not move

$\Rightarrow$ Statement-II is also true and it is explaining statement-I also. **Ans. (A)**

Sol. 22

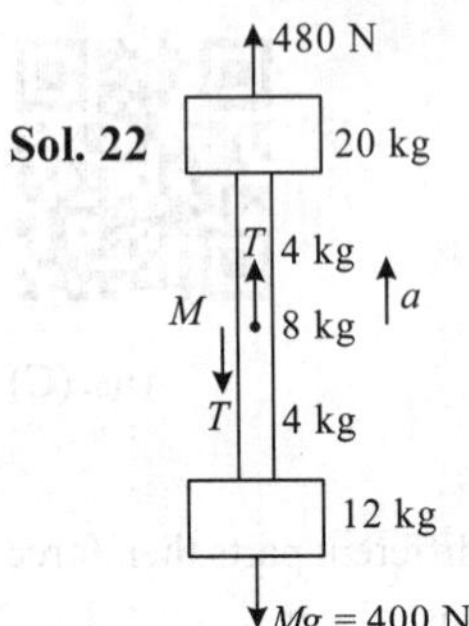

Equation of motion of system is

$$480 - 400 = 40a$$

$$\Rightarrow \quad a = \frac{80}{40} = 2\text{m/s}^2$$

for mass below point M, we use below FBD as shown

Equation of motion for above FBD is

$$T - 160 = 16 \times 2 = 32$$

$$\Rightarrow \quad T = 192\,\text{N} \quad \textbf{Ans. (D)}$$

Sol. 23

$$R = 5\text{ cm}$$

$$AC = 8 + 5 = 13\text{ cm}$$

and $$AB = \sqrt{AC^2 - BC^2} = \sqrt{13^2 - 5^2} = \sqrt{144} = 12\text{ cm}$$

For equilibrium of sphere, we use

$$T\cos\theta = W$$

$$T\sin\theta = N$$

$$\Rightarrow \quad T = \frac{W}{\cos\theta}$$

Here $$\cos\theta = \frac{AB}{AC} = \frac{12}{13}$$

$$\Rightarrow \quad T = \frac{13W}{12} \quad \textbf{Ans. (D)}$$

Sol. 24 As $\vec{F}$ = constant

$\Rightarrow$ $\vec{a}$ = constant

We can use $\vec{a} = \dfrac{\vec{v}_2 - \vec{v}_1}{\Delta t}$

$$\Rightarrow \quad \vec{a} = \frac{6\hat{i} - (6\hat{i} - 2\hat{j})}{10} = \frac{-6\hat{i} + 8\hat{j}}{10}$$

$$\Rightarrow \quad \vec{a} = \frac{-3\hat{i} + 4\hat{j}}{5}\text{ m/s}^2$$

Force on particle is

$$\vec{F} = m\vec{a} = 5\left(\frac{-3\hat{i} + 4\hat{j}}{5}\right) = -3\hat{i} + 4\hat{j}\text{ N} \quad \textbf{Ans. (A)}$$

Sol. 25

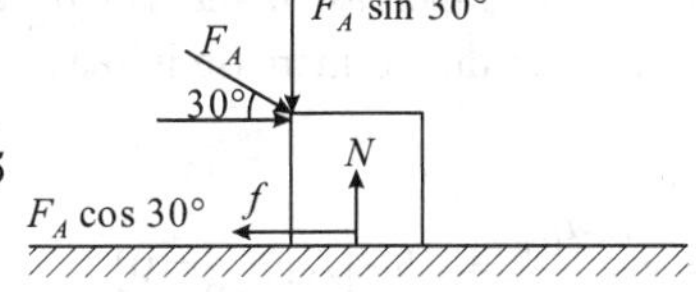

$$N = Mg + F_A \sin 30°$$

$$f = \mu N$$

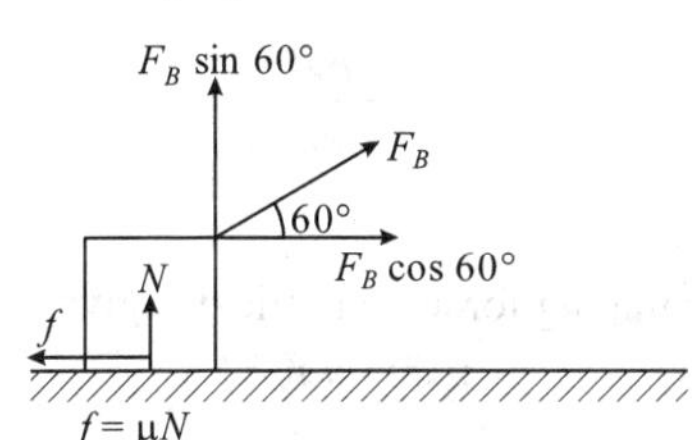

$$N = Mg - F_B \sin 60°$$

to drag the box by person A, we use

$$F_A \cos 30° \geq \mu(Mg + F_A \sin 30°)$$

$$F_A \geq \frac{2\mu Mg}{\sqrt{3} - \mu} \quad \ldots(1)$$

$$\Rightarrow \quad F_A\left(\frac{\sqrt{3}}{2} - \frac{\mu}{2}\right) \geq \mu\, Mg$$

to drag the box by person B, we use

$$F_B \cos 60° \geq \mu(Mg - F_B \sin 60°)$$

$$\Rightarrow \quad F_B \geq \frac{2\mu Mg}{1 + \mu\sqrt{3}} \quad \ldots(2)$$

$$\Rightarrow \quad F_B\left(\frac{1}{2} + \frac{\mu\sqrt{3}}{2}\right) \geq \mu Mg$$

The ratio $\frac{F_A}{F_B}$ can be obtained by equation-(1) & (2)

$$\Rightarrow \quad \frac{F_A}{F_B} = \frac{1 + \mu\sqrt{3}}{\sqrt{3} - \mu}$$

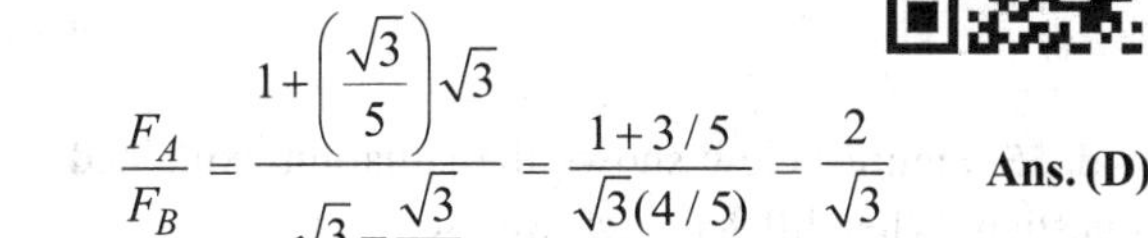

$$\Rightarrow \quad \frac{F_A}{F_B} = \frac{1 + \left(\frac{\sqrt{3}}{5}\right)\sqrt{3}}{\sqrt{3} - \frac{\sqrt{3}}{5}} = \frac{1 + 3/5}{\sqrt{3}(4/5)} = \frac{2}{\sqrt{3}}$$ **Ans. (D)**

Sol. 26 Given that

$$F = ma = \frac{R}{t^2} v$$

$$\Rightarrow \quad \frac{dv}{dt} = \frac{R}{mt^2} v$$

$$\Rightarrow \quad \int \frac{dv}{v} = \int \frac{R}{mt^2} dt$$

$$\Rightarrow \quad \ln(v) = -\frac{R}{mt} + C$$

$$\Rightarrow \quad \ln(v) = -\left(\frac{R}{m}\right)\frac{1}{t} + C$$

here above expression stations that graph of ln v and $\frac{1}{t}$ is a straight line with slope $-\frac{R}{m}$ **Ans. (A)**

Sol. 27

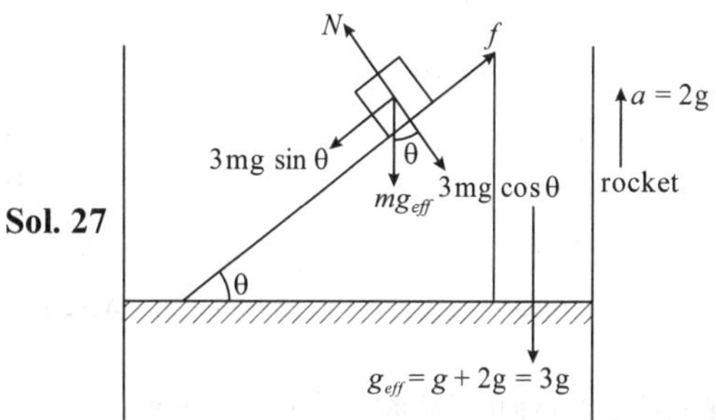

Limiting friction on the block is

$$f_L = \mu N = 3\mu\, mg \cos\theta$$

for block not to slide (f_L)

$$3\, mg \sin\theta < 3\text{m}\, mg \cos\theta$$

$$\mu > \tan\theta$$

$$\Rightarrow \quad \mu_{\text{min}} = \tan\theta$$ **Ans. (B)**

Sol. 28

45°

45°

F

100N

Equilibrium, we use

$$\tan 45° = \frac{100}{F}$$

$$\Rightarrow \quad F = 100\text{ N}$$ **Ans. (B)**

Sol. 29 By Newton's second law, we use

$$\frac{dp}{dt} = F = kt$$

$$\Rightarrow \quad \int_p^{3p} dp = \int_O^T kt\, dt$$

$$\Rightarrow \quad 2p = \frac{kT^2}{2}$$

$$\Rightarrow \quad T = 2\sqrt{\frac{p}{k}}$$ **Ans. (A)**

Sol. 30 FBD of the mass is shown in figure

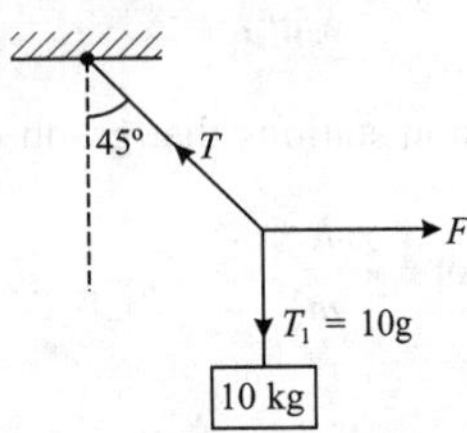

For equilibrium of system, we use

$$T\sin 45° = F \quad \text{...(1)}$$

and $$T\cos 45° = 10g \quad \text{...(2)}$$

$\Rightarrow$ $$F = 10g$$

$\Rightarrow$ $$F = 100\text{N}$$ **Ans. (A)**

Sol. 31 Figure-(1) shows the initial given condition

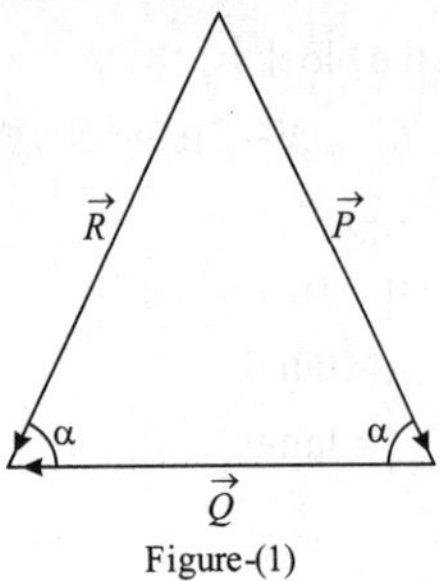

Figure-(1)

After $\vec{P}$ is made twice then the situation is shown in figure-(2) below

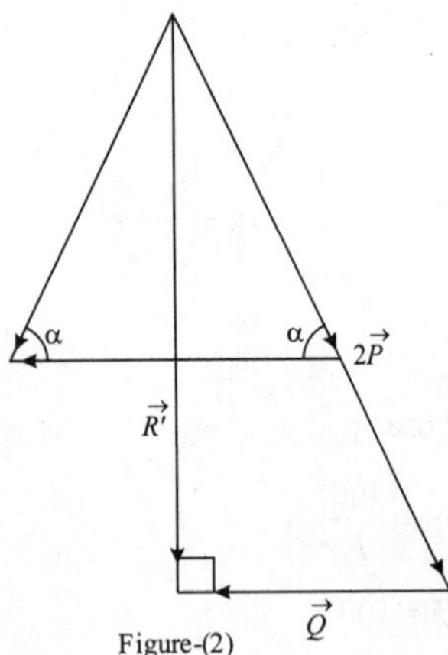

Figure-(2)

Thus angle here is 90°. **Ans. (90°)**

Sol. 32 White going up acceleration of ball is given as

$$\frac{vdv}{dy} = a = -(kv^2 + g)$$

$\Rightarrow$ $$\int_u^0 \frac{vdv}{kv^2+g} = \int_0^H -dy$$

$\Rightarrow$ $$\frac{1}{2k}\ln\left(\frac{ku^2+g}{g}\right) = H$$ **Ans. (B)**

Sol. 33 Force on planet due to increase in mass an be taken as Opposing force on spaceship due to increase in mass is given as

$$F = v\frac{dM_{(t)}}{dt} = -bv^3[\text{As } F_{ext} = 0]$$

Thus acceleration of spaceship is given as

$$a = \frac{F}{m_{(t)}} = -\frac{bv^3}{m_{(t)}}$$ **Ans. (D)**

Sol. 34 The damping force on particle is givn as

$$F = -\alpha x^2 = ma$$

$\Rightarrow$ $$a = \frac{-\alpha x^2}{m}$$

$\Rightarrow$ $$\frac{vdv}{dx} = -\frac{\alpha}{m}x^2$$

For the motion of particle from starting instant to coming to rest can be integrated as

$\Rightarrow$ $$\int_{v_0}^0 v\,dv = \int_0^x -\frac{\alpha}{m}x^2 dx$$

$\Rightarrow$ $$\left(\frac{v^2}{2}\right)_{v_0}^0 = -\frac{\alpha}{m}\left(\frac{x^3}{3}\right)_0^x$$

$\Rightarrow$ $$\frac{-v_0^2}{2} = -\frac{\alpha}{m}\frac{x^3}{3}$$

$\Rightarrow$ $$x = \left(\frac{3mv_0^2}{2\alpha}\right)^{1/3}$$

Option (B) out of all options given is most appropriate here.

Ans. (B)

Sol. 35 Figure below shows the situation explained in the question. When lift is at rest, we use

$$N = mg$$

$\Rightarrow$ $$60 \times 10 = 600\text{N}$$

When lift is moving with downward acceleration. In the frame of lift pseudo force will act on man in upward direction, so in frame of life, as man is at rest, we have

$$N' = M(g-a) = 60\,(10-1.8) = 492\text{ N}$$

Ans. (492)

Sol. 36 The applied force is given as

$$\vec{F} = 20\hat{i} + 10\hat{j}$$

Thus acceleration of box can be given as

$$\vec{a} = \frac{\vec{F}}{m} = \frac{20\hat{i}+10\hat{j}}{2} = 10\hat{i}+5\hat{j}$$

Displacement of box can now be given as

$$\vec{s} = \frac{1}{2}\vec{a}t^2 = \frac{1}{2}(10\hat{i}+5\hat{j}) \times (10)^2$$

$$= 50(10\hat{i}+5\hat{j})m$$

Displacement along x-axis here is 500 m **Ans. (500)**

Sol. 37 As friction provides the necessary centripetal force for circular turning, we have

$$\mu_s N = \frac{mv^2}{R}$$

If F_L is the negative lift acting on the car in downward direction, then for its equilibrium in vertical direction, we use

$$N = \frac{mv^2}{\mu_s R} = mg + F_L$$

$$\Rightarrow \quad F_L = \frac{mv^2}{\mu_s R} - mg$$ **Ans. (B)**

Sol. 38 The free body diagram of the man can be drawn as

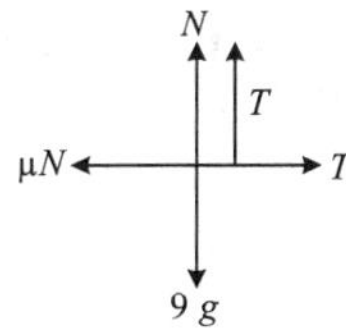

For equilibrium of man, we use

$$N + T = 90$$

$$\Rightarrow \quad T = \mu N = 0.5\,(90 - T)$$

$$\Rightarrow \quad 1.5\,T = 45$$

$$\Rightarrow \quad T = 30\,\text{N}$$

Tension in string is the force exerted by man on rope.

Ans. (30)

Sol. 39 The acceleration of bullet due to resistance force of wood can be calculated as

$$v^2 = u^2 + 2as$$

$$\Rightarrow \quad 0 = (10)^2 + 2\,(-a)\left(\frac{1}{2}\right)$$

$$\Rightarrow \quad a = 100\ \text{m/s}^2$$

Thus retarding force on bullet is given as

$$F = ma = (0.1)(100) = 10\ \text{N}$$ **Ans. (10)**

Sol. 40 As system is going down at acceleration 0.2 m/s^2, equation of motion of system can be written as

$$70\,g - N = 70 \times 0.2$$

$$\Rightarrow \quad N = 700 - 14 = 686\,\text{N}$$ **Ans. (A)**

Sol. 41 Acceleration of ball is given as

$$a = \frac{F}{m}$$

$$a = \frac{F_0}{m}\left[1 - \left(\frac{t-T}{T}\right)^2\right]$$

Velocity of ball as a function of time can be calculated as

$$\frac{dv}{dt} = \frac{F_0}{m}\left[1 - \left(\frac{t-T}{T}\right)^2\right]$$

$$\Rightarrow \quad \int_0^v dv = \frac{F_0}{m}\int_0^{2T}\left[1 - \left(\frac{t-T}{T}\right)^2\right]dt$$

$$\Rightarrow \quad v = \frac{F_0}{m}\left[t - \frac{1}{3T^2}(t-T)^3\right]_0^{2T}$$

$$\Rightarrow \quad v = \frac{F_0}{m}\left\{\left[2T - \frac{1}{3T^2}(2T-T)^3\right] - \left[0 - \frac{(-T)^3}{3T^2}\right]\right\}$$

$$\Rightarrow \quad v = \frac{4F_0 T}{3M}$$ **Ans. (C)**

Sol. 42 As rocket accelerates upward so thurst due to gases ejection on rocket is given as

$$f_T = mg + ma$$

$$\Rightarrow \quad \frac{dm}{dt}v_r = mg + ma$$

$$\Rightarrow \quad \frac{dm}{dt}(500) = 1000\,(10 + 20)$$

$$\Rightarrow \quad \frac{dm}{dt} = 60\ \text{kg/s}$$ **Ans. (D)**

Sol. 43 Tension in the string of an Atwood's machine is given as

$$T = \frac{2m_1 m_2}{(m_1 + m_2)}g$$

$$\Rightarrow \quad T = \frac{2\times 3\times 5\times 10}{8} = \frac{75}{2}$$

Stress in wire is given as

$$\text{Stress} = \frac{T}{A}$$

$$\Rightarrow \quad \frac{24}{\pi}\times 10^{-2} = \frac{75}{2(\pi r^2)}$$

$$\Rightarrow \quad r^2 = \frac{75}{48}\times 100$$

$$\Rightarrow \quad r^2 = \left(\frac{25}{2}\right)^2$$

$$\Rightarrow \quad r = 12.5\ \text{cm}$$ **Ans. (D)**

Sol. 44 Figure below shows the tension in strings and acceleration of blocks considered for equation of motion

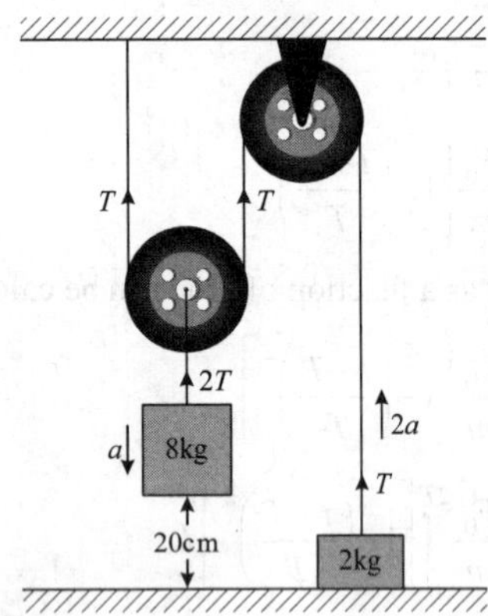

Equation of motion of the two blocks is written as

$$8g - 2T = 8a$$

and $$T - 2g = 4a$$

$\Rightarrow$ $$4g = 16a$$

$\Rightarrow$ $$a = \frac{5}{2}\text{ m/s}^2$$

By speed equation, displacement of 8 kg block is written as

$$s = ut + \frac{1}{2}at^2$$

$\Rightarrow$ $$0.2 = 0 + \frac{1}{2} \times \frac{5}{2} \times t^2$$

$\Rightarrow$ $$t = 0.4\text{ s}$$ **Ans. (D)**

Sol. 45 As the object is moving at constant velocity so net force on object must be zero. In this case while moving up the force would be positive and balances its weight component along the incline and while moving down force must be negative(acting upward) and balacing the weight component along the incline. Thus most appropriate choice here is option-(D) among all options. **Ans. (D)**

Sol. 46 Figure below shows the free body diagram representing forces acting on the wedge and block

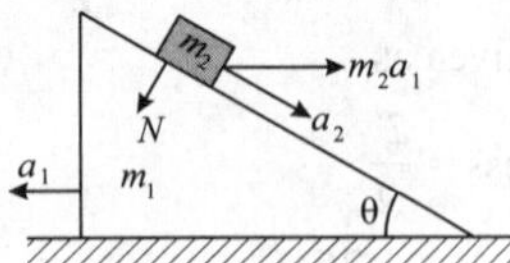

Equaton of motion of wedge is written as

$$N \cos 60° = Ma_1 = 16a_1$$

$\Rightarrow$ $$N = 32a_1$$

For block along the direction normal to incline, including pseudo force, we balance forces as

$$N = 8g \cos 30° - 8a_1 \sin 30°$$

$\Rightarrow$ $$32a_1 = 4\sqrt{3} - 4a_1$$

$\Rightarrow$ $$a_1 = \frac{\sqrt{3}}{9}g$$

Along the incline for the block we use

$$8g \sin 30° + 8a_1 \cos 30° = ma_2 = 8a_2$$

$\Rightarrow$ $$a_2 = \frac{g}{2} + \frac{\sqrt{3}}{9}g \cdot \frac{\sqrt{3}}{2} = \frac{2g}{3}$$ **Ans. (D)**

Sol. 47 About any point, torque is calculated as

$$\vec{\tau} = \vec{r} \times \vec{F}$$

$\Rightarrow$ $$\vec{\tau} = \begin{vmatrix} \hat{i} & \hat{j} & \hat{k} \\ 2 & 1 & 2 \\ 3 & 4 & -2 \end{vmatrix}$$

$\Rightarrow$ $$\vec{\tau} = \hat{i}\,(-2-8) - \hat{j}\,(-4-6) + \hat{k}\,(8-3)$$

$\Rightarrow$ $$\vec{\tau} = -10\hat{i} + 10\hat{j} + 5\hat{k}$$ **Ans. (B)**

Sol. 48 The force on object of mass 100 g is given as

$$\vec{F} = 10\hat{i} + 5\hat{j}$$

Acceleration of object is calculated as

$$\vec{a} = \frac{\vec{F}}{m} = 100\hat{i} + 50\hat{j}$$

As object starts from rest, its displacement after time t is calculated as

$$\vec{s} = \vec{u}t + \frac{1}{2}\vec{a}t^2 = \frac{1}{2}\vec{a}t^2$$

$\Rightarrow$ $$\vec{s} = \frac{1}{2}(100\hat{i} + 50\hat{j})2^2$$

$\Rightarrow$ $$\vec{s} = 200\hat{i} + 100\hat{j}$$

$\Rightarrow$ $$\vec{s} = a\hat{i} + b\hat{j}$$

$\Rightarrow$ $$a = 200$$

and $$b = 100$$

$\Rightarrow$ $$\frac{a}{b} = 2$$ **Ans. (2)**

Sol. 49 As the block is at rest, for its equilibrium the forces acting on the block are shown in figure below.

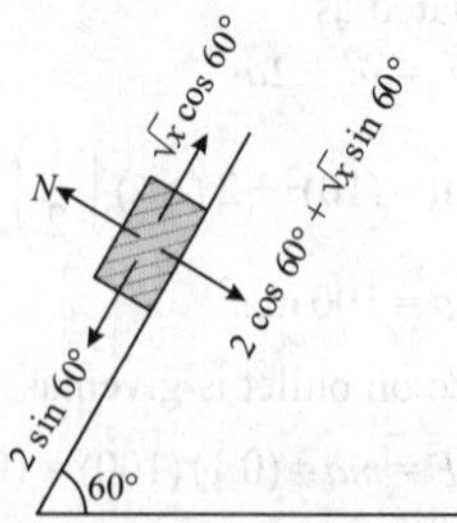

For its equilibrium, we use

$$\sqrt{x}\frac{1}{2} = \frac{2\sqrt{3}}{2}$$

$\Rightarrow$ $$x = 12$$ **Ans. (12)**

Sol. 50 For the elevator going down at acceleration a as shown in figure, we use

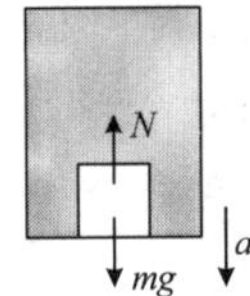

$$mg - N = ma$$

$$\Rightarrow \quad N = m(g - a) < mg$$

Person experiences weightloss, when acceleration of lift is downward. **Ans. (B)**

Sol. 51 In figure below we consider accelerations of the mass all in downward directions as shown.

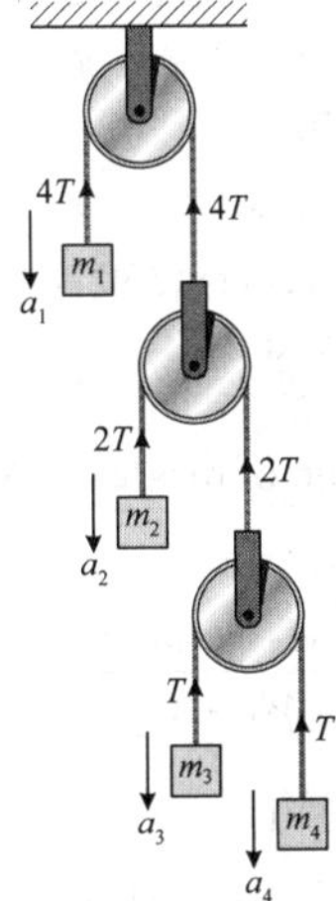

Using constrained relation for tension in ideal strings and acceleration of mass, we have

$$\sum \vec{T} \cdot \vec{a} = 0$$

$$\Rightarrow \quad -4Ta_1 - 2Ta_2 - Ta_3 - Ta_4 = 0$$

$$\Rightarrow \quad 4a_1 + 2a_2 + a_3 + a_4 = 0$$ **Ans. (A)**

Sol. 52 When sixth ball just leaves the table total weight acting on the string at the edge of table is 6 mg. Situation is as shown in figure below.

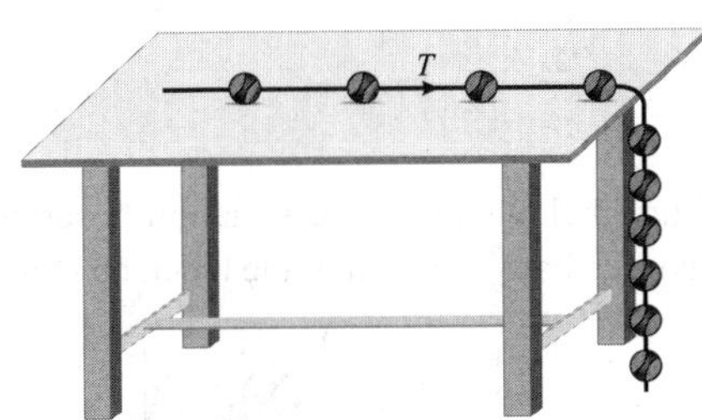

The acceleration of all the balls will be given as

$$a = \frac{6\text{ mg}}{10\text{ m}} = \frac{6g}{10} = \frac{3g}{5}$$

To calculate the tension in string between ball 7 and 8, we consider balls 8, 9 and 10 together accelerated by this tension as shown in figure below.

The equation of motion of these three balls is written as

$$T = 3\,ma$$

$$\Rightarrow \quad T = 3m \times \frac{3g}{5}$$

$$\Rightarrow \quad T = 36\text{ N}$$ **Ans. (36)**

Sol. 53 The situation described in question is shown in figure below.

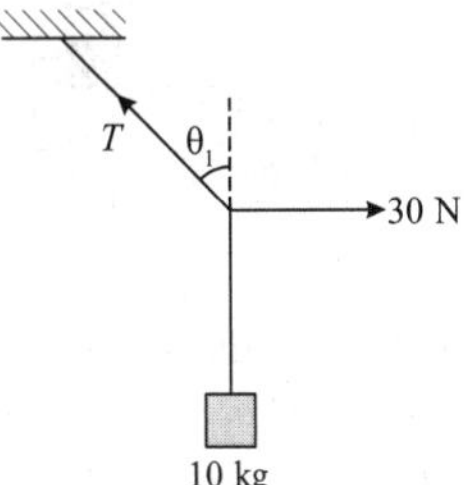

For equilibrium of mass, we use

$$T \sin\theta = 30$$

and $$T\cos\theta = 100$$

$$\Rightarrow \quad \tan\theta = 0.3$$ **Ans. (3)**

Sol. 54 For motion of blocks with acceleration a, we use

$$2\,Mg - T = 4\,Ma$$

and $$T - Mg = Ma$$

$$\Rightarrow \quad a = \frac{g}{5}$$

$$\Rightarrow \quad T = Mg + Ma = Mg + \frac{Mg}{5} = \frac{6}{5}Mg$$ **Ans. (6)**

Sol. 55 Figure below shows the mass m placed inside a box of which the floor is shown below which is exerting a normal reaction $mg/4$ on box.

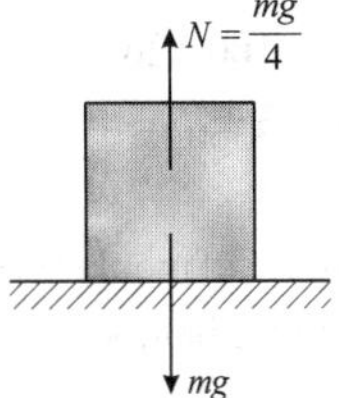

For upward acceleration of box we use

$$mg - N = ma$$

$$\Rightarrow \quad a = g - \frac{g}{4}$$

$$\Rightarrow \quad a = \frac{3g}{4}$$ **Ans. (C)**

Sol. 56 Let addition force required is $\vec{F}$ such that resultant of all forces will become zero so we use

$$\vec{F} + 5\hat{i} - 6\hat{i} + 7\hat{j} - 8\hat{j} = 0$$

$$\Rightarrow \quad \vec{F} = \hat{i} + \hat{j}$$

$$\Rightarrow \quad |\vec{F}| = \sqrt{2}$$

Angle made by the force vector with x-axis is 45° as coefficient of unit vectors along x and y axes is same. **Ans. (A)**

Sol. 57 Let acceleration of 100 kg block is a_1, the free body diagram of 100 kg block w.r.t. ground is shown in figure below.

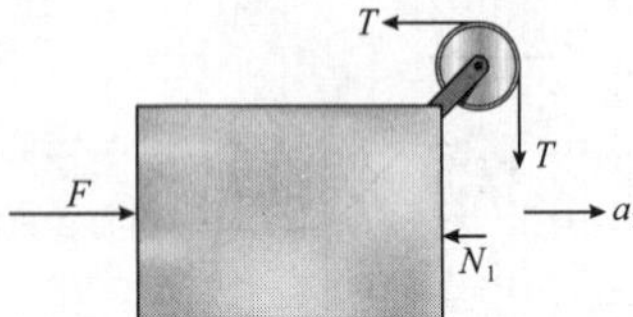

The equation of motion of body is written as

$$F - T - N_1 = 100\, a_1 \qquad \ldots(i)$$

Free body diagram of 20 block w.r.t. 100 kg block is drawn with pseudo force as shown in figure below.

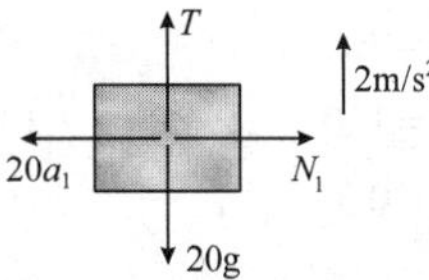

The equation of motion of 20 kg block is written as

$$T - 20\,g = 20(2)$$

$$\Rightarrow \quad T = 240 \qquad \ldots(ii)$$

$$\Rightarrow \quad N_1 = 20a_1 \qquad \ldots(iii)$$

Free body diagram of 10 kg block w.r.t. 100 kg is drawn with pseudo force as shown in figure below.

The equation of motion of 10 kg block is written as

$$10a_1 - 240 = 10(2)$$

$$\Rightarrow \quad a_1 = 26 \text{ m/s}^2$$

$$\Rightarrow \quad F - 240 - 20(26) = 100 \times 26$$

$$\Rightarrow \quad F = 3360 \text{ N}$$

Ans. (C)

Sol. 58 Free body diagram of monkey while moving downward is drawn as shown in figure below

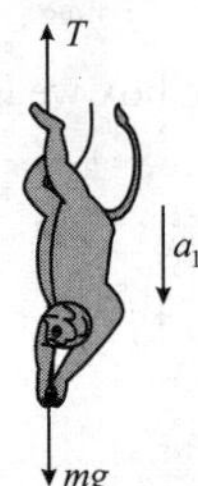

Equation of motion of monkey's motion is written as

$$mg - T = ma_1$$

$$\Rightarrow \quad 500 - T = 50 \times 4$$

$$\Rightarrow \quad T = 300 \text{ N}$$

Free body diagram of monkey while moving up is drawn as shown in figure below

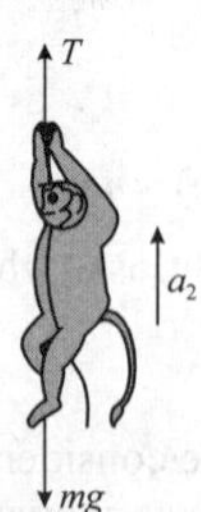

Equation of motion of monkey going up is written as

$$T - mg = ma_2$$

$$\Rightarrow \quad T - 500 = 50 \times 5$$

$$\Rightarrow \quad T = 750 \text{ N}$$

As breaking strength of string is 350 N so string will break while monkey is moving upward. **Ans. (C)**

Sol. 59 The acceleration of masses in Atwood's machine in is given as

$$a = \frac{M_2 g - M_1 g}{M_1 + M_2}$$

Case-I : $M_2 = 2M_1$

$$a_1 = \frac{2M_1 g - M_1 g}{3M_1}$$

$$\Rightarrow \quad a_1 = g/3$$

Case-II : $M_2 = 3M_1$

$$a_2 = \frac{3M_1 g - M_1 g}{4M_1}$$

$$\Rightarrow \quad a_2 = \frac{g}{2}$$

$$\Rightarrow \quad \frac{a_1}{a_2} = \frac{\frac{g}{3}}{\frac{g}{2}} = \frac{2}{3}$$

Ans. (B)

Sol. 60 Figure below shows the situation described in the question and free body diagram of the block on incline.

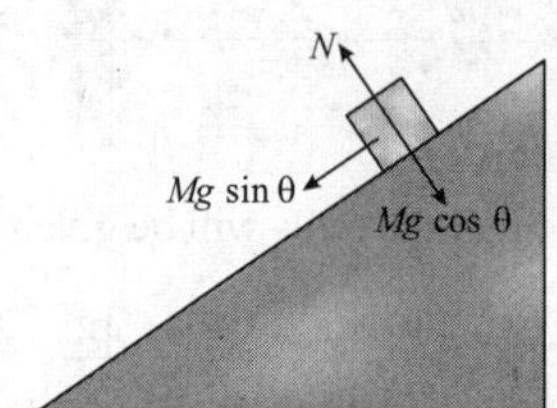

As lift is going up with uniform velocity, the acceleration of block is given as

$$a = g \sin \theta$$

When incline is at angle 30°, distance l covered by the block in 2 seconds so we use

$$l = \frac{1}{2} g \sin 30° (2)^2 \quad \ldots (1)$$

When angle of incline is changed to 45°, we use

$$l = \frac{1}{2} g \sin 45°\, t^2 \quad \ldots (2)$$

$$\Rightarrow \quad \left(\frac{1}{2}\right)(4) = \frac{1}{\sqrt{2}} t^2$$

$$\Rightarrow \quad t = \sqrt{2\sqrt{2}} = 1.68 \text{ s}$$

Ans. (C)

Sol. 61 When block reaches the top point B, its velocity becomes zero as shown in figure below.

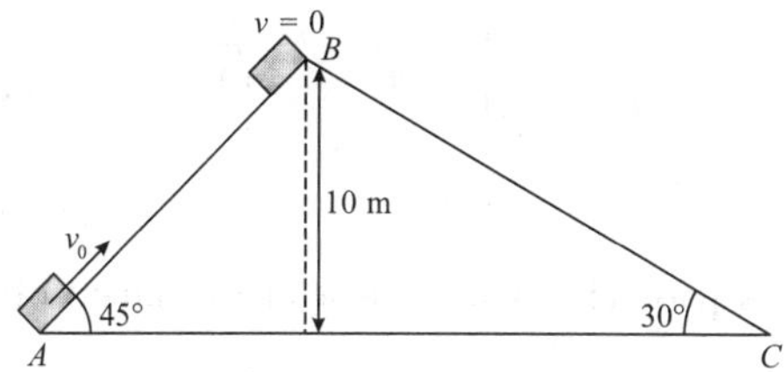

By energy conservation at points A and B, we use

$$\frac{1}{2} m v_0^2 = mgh$$

$$\Rightarrow \quad v_0 = 10\sqrt{2} \text{ m/s}$$

For motion of block from A to B, we use

$$a = -g \sin 45° = \frac{-10}{\sqrt{2}} \text{ m/s}^2$$

and $\quad v = u + at_1$

$$\Rightarrow \quad 0 = 10\sqrt{2} - \frac{10}{\sqrt{2}} t_1$$

$$\Rightarrow \quad t_1 = 2 \text{ s}$$

For motion of block from B to C, we use

$$s = ut_2 + \frac{1}{2} a t_2^2$$

$$\Rightarrow \quad \frac{10}{\sin 30°} = \frac{1}{2}(10 \sin 30°) t_2^2$$

$$\Rightarrow \quad t_2 = 2\sqrt{2} \text{ s}$$

Thus total time taken is given as

$$T = t_1 + t_2$$

$$\Rightarrow \quad T = 2\sqrt{2} + 2$$

$$\Rightarrow \quad T = 2(\sqrt{2} + 1) \text{ s}$$

Ans. (2)

Sol. 62

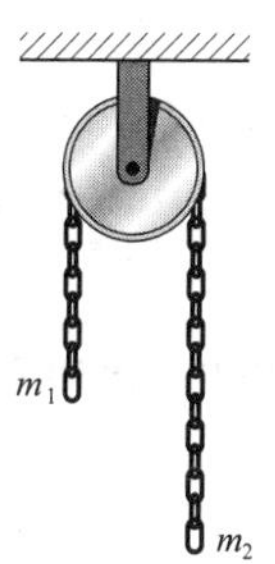

$$a = \frac{(m_2 - m_1)}{(m_2 + m_1)} g$$

$$\frac{g}{2} = \frac{(\lambda(L - l) - \lambda l) g}{\lambda L}$$

$$\Rightarrow \quad L = \frac{L}{4} = \frac{L}{x}$$

$$x = 4$$

Ans. (D)

Sol. 63 As system is at rest, for equilibrium of masses we use

$$m_2 g = m_1 g \sin \theta$$

$$\Rightarrow \quad \sin \theta = \frac{m_2}{m_1} = \frac{3}{5}$$

$$\Rightarrow \quad \cos \theta = \frac{4}{5}$$

Normal force on mass m_1 is calculated as

$$N = 5g \cos \theta$$

$$\Rightarrow \quad N = 5 \times 10 \times \frac{4}{5} = 40 \text{ N}$$

Ans. (B)

Sol. 64 Statement-I : When elevator is moving with uniform speed tension in cable will be equal to the weight of block hence statement-I is true.

Statement-II : When elevator is going down with increasing speed or downward acceleration the person experiences pseudo force in upward direction so the normal reaction on the foot will be decreased and it is less than the weight of person so statement-II is not true.

Ans. (B)

Sol. 65 The free body diagram of the two blocks are shown in figure below

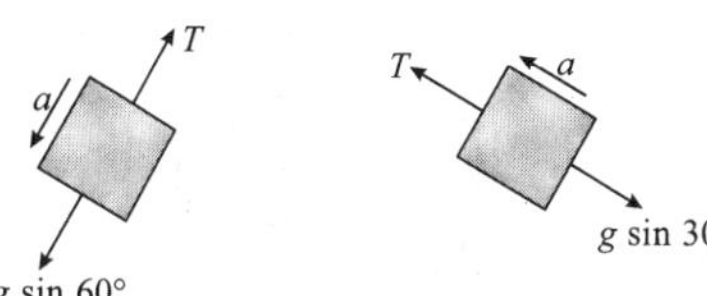

The equation of motion of the two blocks are written as

$$4g \sin 60° - T = 4a \quad \ldots (1)$$

$$T - g \sin 30° = a \quad \ldots (2)$$

Solving equations-(1) and (2) gives

$$20\sqrt{3} - T = 4T$$

$$\Rightarrow \quad T = 4(\sqrt{3}+1) \text{ N} \qquad \textbf{Ans. (B)}$$

Sol. 66 Figure below shows the situation described in the question.

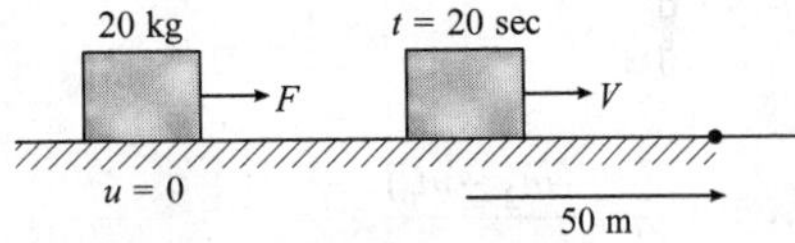

After force stops acting, the body moves with uniform speed so we use

$$50 = v \times 10$$

$$v = 5 \text{ m/s}$$

While force was acting, we use

$$v = 0 + a \times 20$$

$$\Rightarrow \quad 5 = a \times 20$$

$$\Rightarrow \quad a = \frac{1}{4} \text{ m/s}^2$$

Thus force is calculated as

$$F = ma = 20 \times \frac{1}{4} = 5 \text{ N} \qquad \textbf{Ans. (B)}$$

Sol. 67 Magnitude of force is given by the slope of the momentum time curve given as

$$\left|\frac{d\vec{p}}{dt}\right| = |\vec{F}|$$

$$\Rightarrow \quad \frac{d\vec{p}}{dt} = \text{Slope of curve}$$

From the graph we can see that maximum slope is in region (c) and minimum slope is in region (b). **Ans. (C)**

Sol. 68 Figure below shows the free body diagram of the force junction.

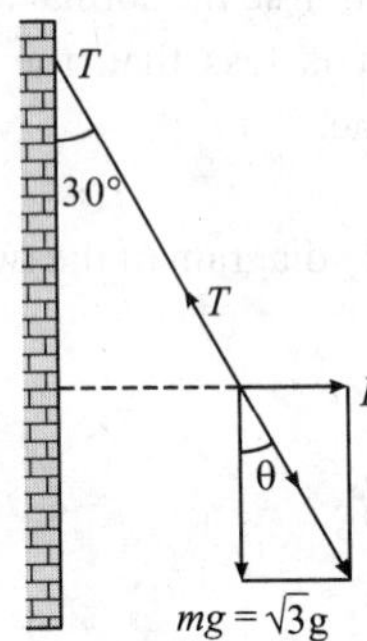

From force components, we have

$$\cos\theta = \frac{\sqrt{3}g}{T}$$

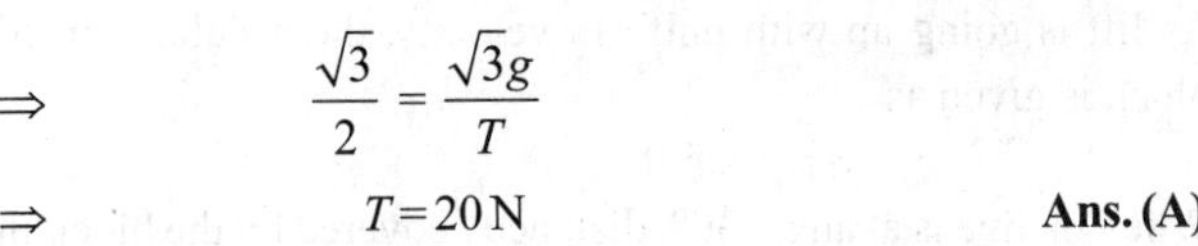

$$\Rightarrow \quad \frac{\sqrt{3}}{2} = \frac{\sqrt{3}g}{T}$$

$$\Rightarrow \quad T = 20 \text{ N} \qquad \textbf{Ans. (A)}$$

Sol. 69 Net force when two forces are mutually perpendicular is calculated as

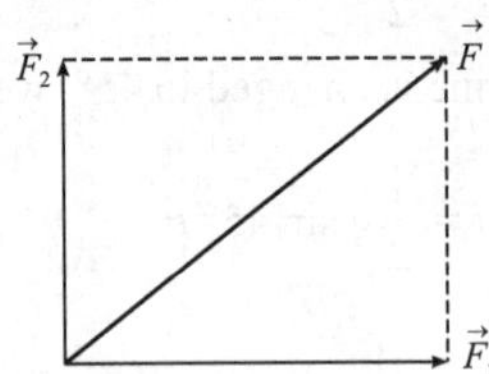

$$\vec{F} = (\vec{F}_1 + \vec{F}_2)$$

$$\Rightarrow \quad |\vec{F}| = \sqrt{F_1^2 + F_2^2 + 2F_1F_2\cos 90^\circ}$$

Given that $F_1 = A$; $F_2 = \frac{A}{2}$, it gives

$$\Rightarrow \quad |\vec{F}| = \sqrt{A^2 + \frac{A^2}{4}} = \frac{A\sqrt{5}}{2} \qquad \textbf{Ans. (D)}$$

Sol. 70 As particle remain at rest, the resultant of $\vec{F}_2$ and $\vec{F}_3$ should be opposite to $\vec{F}_1$. Thus from the free body diagram we have

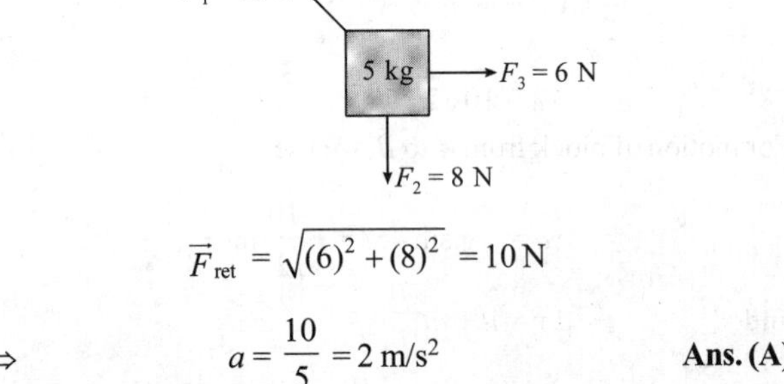

$$\vec{F}_{\text{ret}} = \sqrt{(6)^2 + (8)^2} = 10 \text{ N}$$

$$\Rightarrow \quad a = \frac{10}{5} = 2 \text{ m/s}^2 \qquad \textbf{Ans. (A)}$$

4.2 Friction

Sol. 1

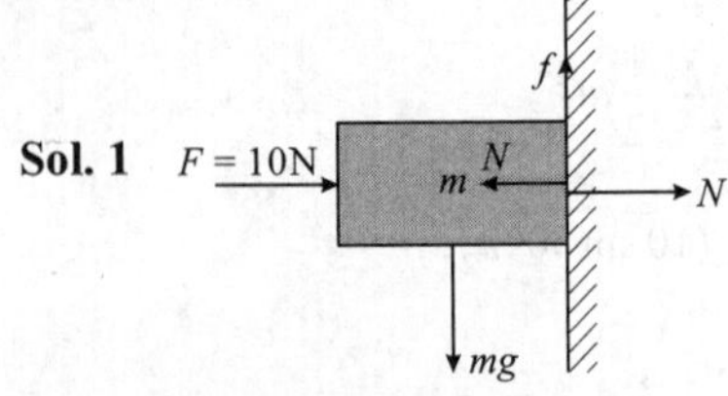

For equilibrium of block, we have

$$N = F = 10 \text{ N}$$

$$mg = f = \mu N$$

$$\Rightarrow \quad mg = 0.2 \times 10 = 2 \text{ N} \qquad \textbf{Ans. (D)}$$

Sol. 2

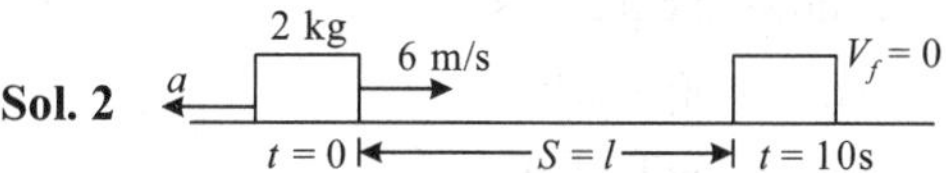

Retardation due to friction on block is

$$a = -\mu g$$

we use $\quad V = u + at$

$\Rightarrow \quad 0 = 6 - \mu(10)(10)$

$\Rightarrow \quad \mu = 0.06$ **Ans. (C)**

Sol. 3

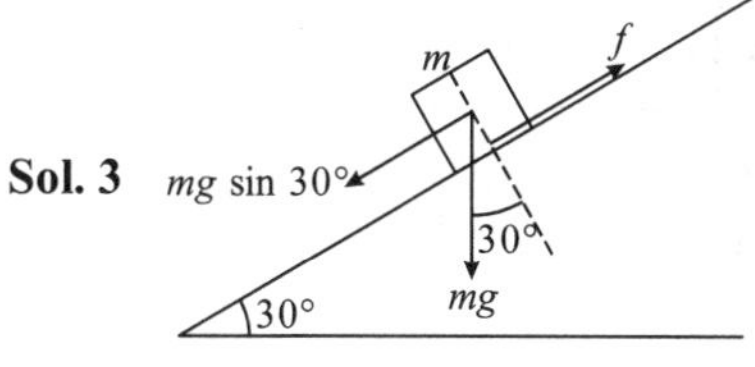

As block is rest on incline plane we use

$$f = mg \sin 30°$$

$\Rightarrow \quad 10 = m(10)\left(\frac{1}{2}\right)$

$\Rightarrow \quad m = 2 \text{ kg}$ **Ans. (A)**

Sol. 4

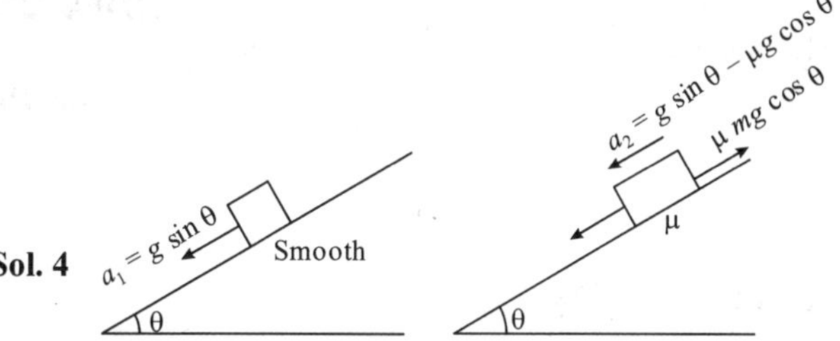

To slide a distance l along incline, time required for smooth incline

$$t_1 = \sqrt{\frac{2l}{a_1}} = \sqrt{\frac{2l}{g \sin \theta}}$$

For rough incline

$$t_2 = \sqrt{\frac{2l}{a_2}} = \sqrt{\frac{2l}{g \sin \theta - \mu g \cos \theta}}$$

Given that $\quad t_2 = nt_1$

$\Rightarrow \quad \sqrt{\frac{2l\sqrt{2}}{g(1-\mu)}} = n\sqrt{\frac{2l\sqrt{2}}{g}}$

$\Rightarrow \quad \sqrt{\frac{1}{1-\mu}} = n$

$\Rightarrow \quad 1 - \mu = \frac{1}{n^2}$

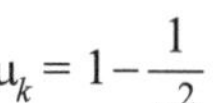

$\Rightarrow \quad \mu_k = 1 - \frac{1}{n^2}$ **Ans. (A)**

Sol. 5

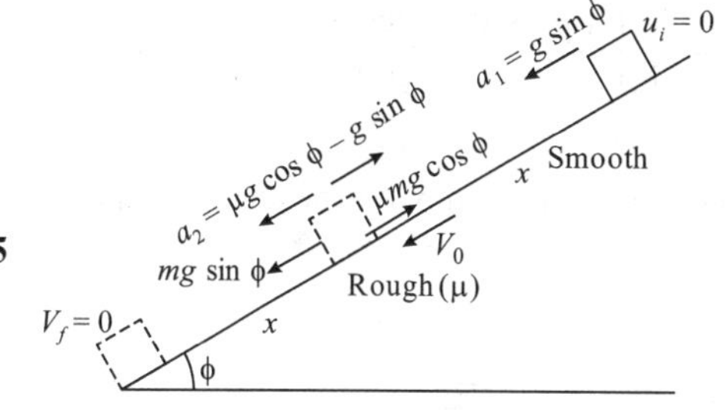

for upper half of motion, we use

$$V_0 = \sqrt{2a_1 x} = \sqrt{2g \sin \phi \cdot x} \quad \ldots(1)$$

for lower half of motion, we use

$$V_0 = \sqrt{2a_2 x} = \sqrt{2g(\mu \cos \phi - \sin \phi)x} \quad \ldots(2)$$

from (1) & (2), we have

$$2xg \sin \phi = 2xg(\mu \cos \phi - \sin \phi)$$

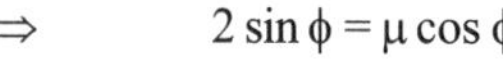

$\Rightarrow \quad 2 \sin \phi = \mu \cos \phi$

$\Rightarrow \quad \mu = 2 \tan \phi$ **Ans. (C)**

Sol. 6

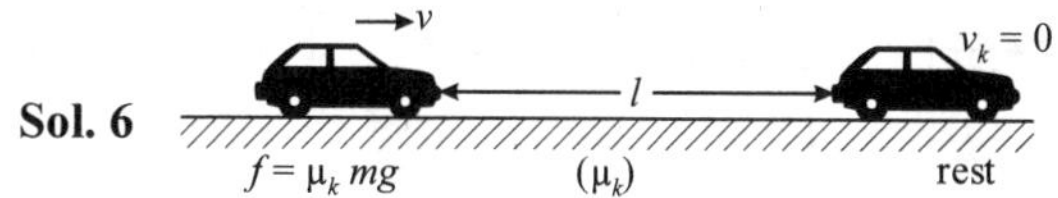

Due to friction maximum retardation in car is

$$a_{max} = \mu_x g$$

Using $\quad v_f^2 = v^2 - 2al$

$\Rightarrow \quad l = \frac{v^2}{2a} \quad$ [As $v_f = 0$]

$\Rightarrow \quad l = \frac{(100)^2}{2 \times 0.5 \times 10} = \frac{10000}{10}$

$\Rightarrow \quad l = 1000 \text{ m}$ **Ans. (B)**

Sol. 7

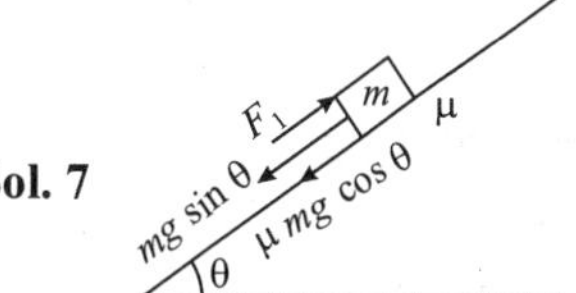

To push the block up on incline plane, minimum, force required is

$$F_1 = mg \sin \theta + \mu \, mg \cos \theta$$

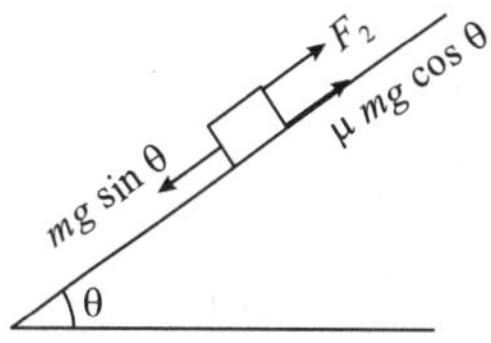

To prevent the body sliding down

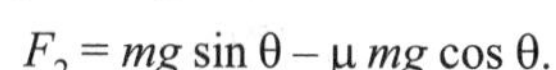

$$F_2 = mg \sin \theta - \mu \, mg \cos \theta.$$

$$\Rightarrow \quad \frac{F_1}{F_2}=\frac{\sin\theta+\mu\cos\theta}{\sin\theta-\mu\cos\theta}=\frac{\tan\theta+\mu}{\tan\theta-\mu}=\frac{3\mu}{\mu}=3$$ **Ans. (D)**

Sol. 8 For insect not to slip α is the angle of repose

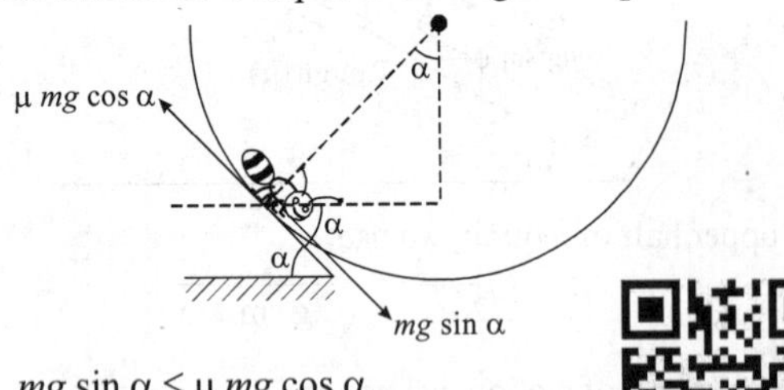

$$mg\sin\alpha \le \mu\, mg\cos\alpha$$

$$\Rightarrow \quad \mu \ge \tan\alpha \;\Rightarrow\; \tan\alpha=\frac{1}{3}$$

$$\Rightarrow \quad \cot\alpha = 3$$ **Ans. (A)**

Sol. 9

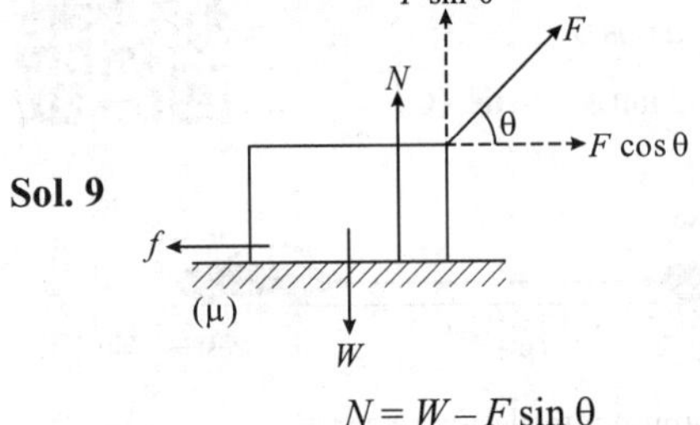

$$N = W - F\sin\theta$$

To slide the block we use

$$F\cos\theta = \mu(W - F\sin\theta)$$

$$\Rightarrow \quad F=\frac{\mu W}{\cos\theta+\mu\sin\theta} \quad \ldots(1)$$

F is minimum when

$$\frac{dF}{d\theta}=0$$

$$\Rightarrow \quad -\sin\theta+\mu\cos\theta=0$$

$$\Rightarrow \quad \mu=\tan\theta$$

$$\Rightarrow \quad \theta=\tan^{-1}(\mu)$$

$$\Rightarrow \quad \sin\theta=\frac{\mu}{\sqrt{1+\mu^2}};\ \cos\theta=\frac{1}{\sqrt{1+\mu^2}}$$

Minimum value of force can be given as

$$F_{\min}=\frac{\mu W}{\frac{1}{\sqrt{1+\mu^2}}+\frac{\mu^2}{\sqrt{1+\mu^2}}}=\frac{\mu W}{\sqrt{1+\mu^2}}$$ **Ans. (A)**

Sol. 10

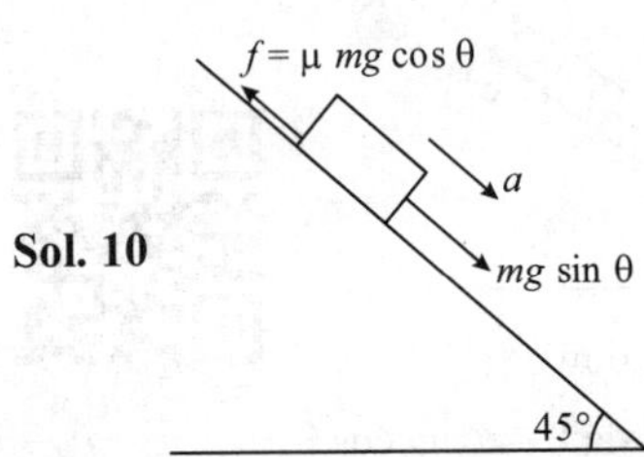

Acceleration body down the plane is

$$a = g\sin\theta - mg\cos\theta(\mu = 0.3x)$$

Speed of body increases until $a = 0$

After some time at a specific distance

$$a=0$$

$$\Rightarrow \quad g\left(\frac{1}{\sqrt{2}}\right)-0.3x\, g\cos 45°=0$$

$$\Rightarrow \quad x=\frac{1}{0.3}=3.33\text{ m}$$ **Ans. (D)**

Sol. 11

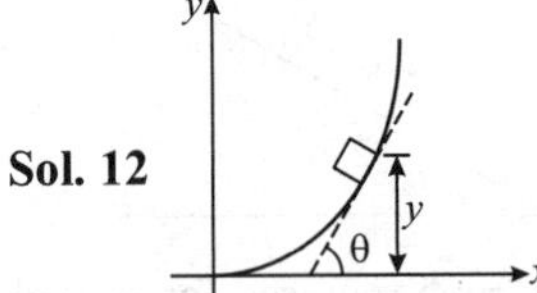

Limiting value of friction $= \mu mg$

for small values of F,

$$f = F$$

static friction and block static sliding at time $t = t_0$ when

$$kt_0 = \mu mg$$

and for $t > t_0$, equation of motion, we use

$$kt - \mu mg = ma$$

$$\Rightarrow \quad a=\frac{kt}{m}-mg$$

The graph will be a line with $+ve$ slope after

$$t = t_0$$ **Ans. (B)**

Sol. 12

Block will remain at rest at a height y where

if $\tan\theta = \mu$

$\theta \rightarrow$ Angle of Repose

here $$\tan\theta=\frac{dy}{dx}=\frac{d}{dx}\left(\frac{x^3}{6}\right)=\frac{x^2}{2}=0.5$$

$$\Rightarrow \quad x^2 = 0.5\times 2 = 1$$

$$\Rightarrow \quad x=\pm 1$$

$$\Rightarrow \quad y=\frac{1}{6}m$$ **Ans. (A)**

Sol. 13

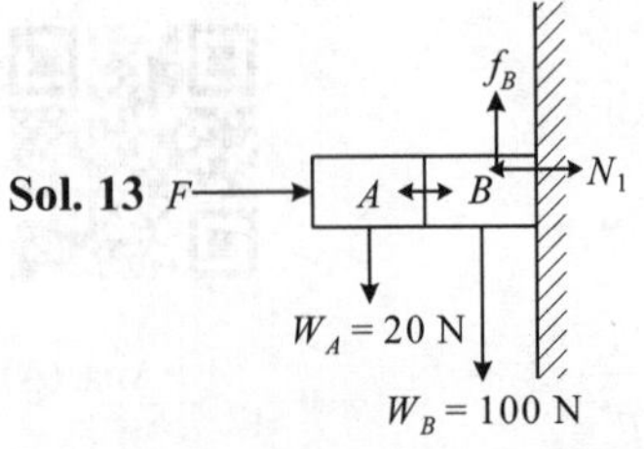

We consider blocks to be at rest and combined *FBD* of blocks with static friction is shown as

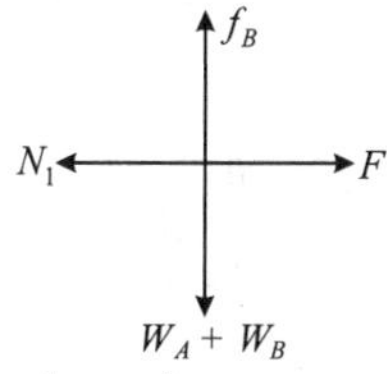

For equilibrium of blocks we have

$$f_B = W_A + W_B = 120\,\text{N}$$ **Ans. (C)**

Sol. 14

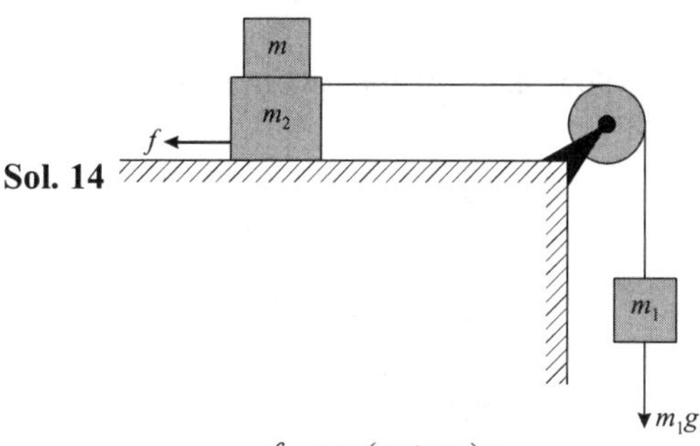

$$f_{max} = \mu(m + m_2)g$$

To stop the motion, we use

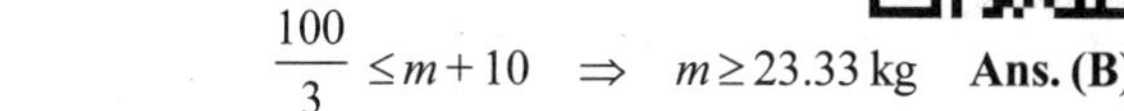

$$m_1g \le \mu(m + m_2)g$$

$$\Rightarrow \quad 5 \le 0.15\,(m + 10)$$

$$\Rightarrow \quad \frac{100}{3} \le m + 10 \quad \Rightarrow \quad m \ge 23.33\text{ kg}$$ **Ans. (B)**

Sol. 15

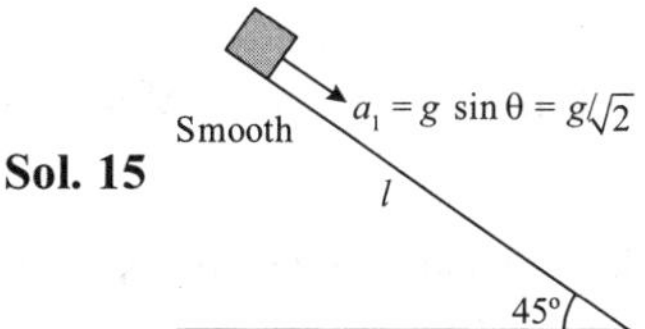

Time taken to slide is

$$t_1 = \sqrt{\frac{2l}{a_1}} = \sqrt{\frac{2\sqrt{2}l}{g}}$$

Given that $\quad t_2 = nt_1$

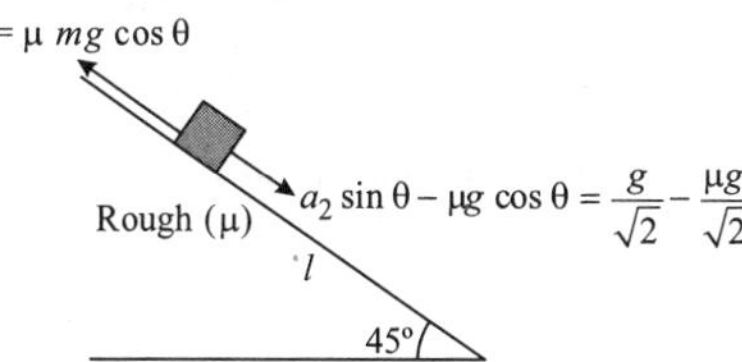

Time taken to slide is

$$t_2 = \sqrt{\frac{2l}{a_2}} = \sqrt{\frac{2\sqrt{2}l}{g(1-\mu)}}$$

$$\Rightarrow \quad \sqrt{\frac{2\sqrt{2}l}{g(1-\mu)}} = n\sqrt{\frac{2\sqrt{2}l}{g}}$$

$$\Rightarrow \quad \frac{1}{1-\mu} = n^2 \quad \Rightarrow \quad \mu = 1 - \frac{1}{n^2}$$ **Ans. (B)**

Sol. 16

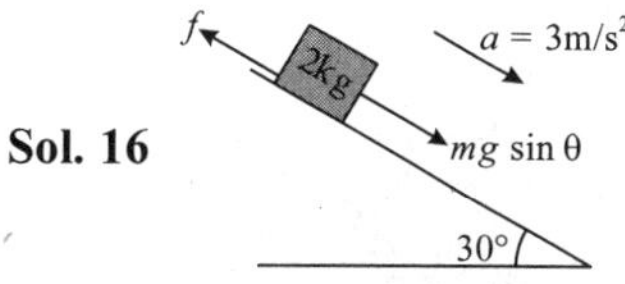

Equation motion of block

$$mg\sin\theta - f = ma$$

$$\Rightarrow \quad 2 \times 10 \times \frac{1}{2} - f = 2 \times 3$$

$$\Rightarrow \quad f = 10 - 6 = 4\,\text{N}$$

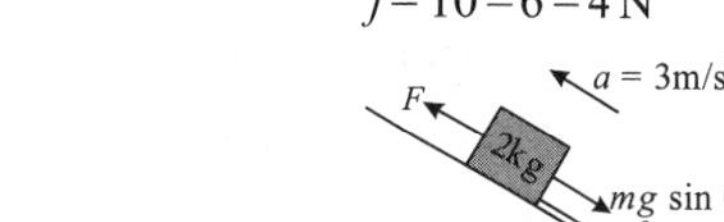

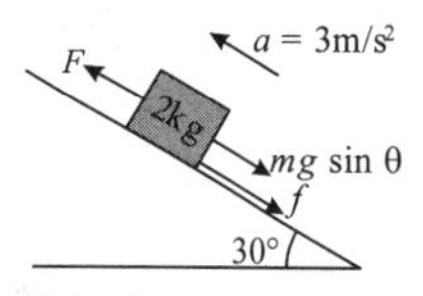

Equation of motion of block

$$F - mg\sin\theta - f = ma$$

$$\Rightarrow \quad F - 10 - 4 = 6$$

$$\Rightarrow \quad F = 20\,\text{N}$$ **Ans. (D)**

Sol. 17 To keep the block at rest *FBD* is shown in figure. At equilibrium of block we use

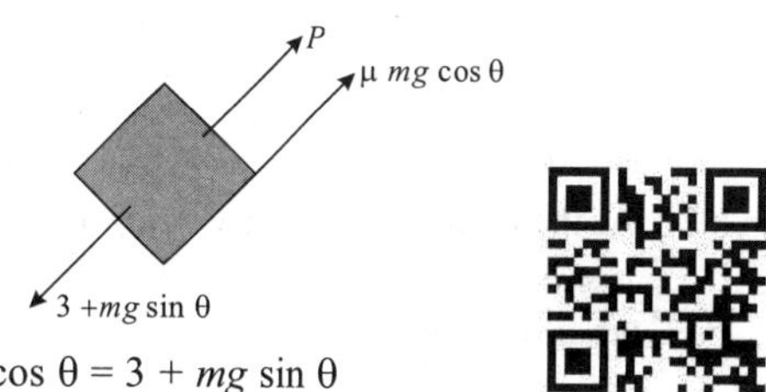

$$P + \mu mg\cos\theta = 3 + mg\sin\theta$$

$$\Rightarrow \quad P = 3 + 10 \times 10 \times \frac{1}{\sqrt{2}} - 0.6 \times 10 \times 10 \times \frac{1}{\sqrt{2}}$$

$$\Rightarrow \quad P = 3 + 20\sqrt{2} = 31.28\,\text{N}$$ **Ans. (A)**

Sol. 18 For limiting equilibrium down the plane, we use

$$2 + mg\sin 30 = \mu mg\cos 30° \quad \ldots(1)$$

for limiting equilibrium up the plane, we use

$$10 = mg\sin 30 + \mu mg\cos 30° \quad \ldots(2)$$

from (1) and (2)

$$\Rightarrow \quad 6 = \mu mg\cos 30°$$

and $\quad 4 = mg\sin 30°$

$$\Rightarrow \quad \frac{3}{2} = \mu \times \sqrt{3}$$

$$\Rightarrow \quad \mu = \frac{\sqrt{3}}{2}$$ **Ans. (B)**

Sol. 19

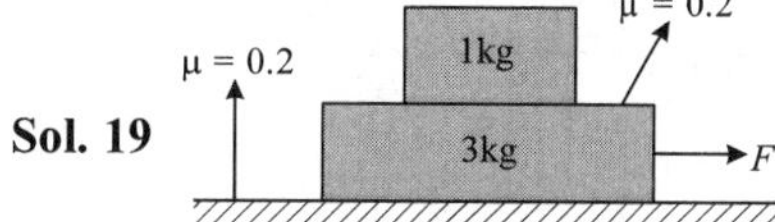

$$a_{Amax} = \mu g = 2\ \text{m/s}^2$$

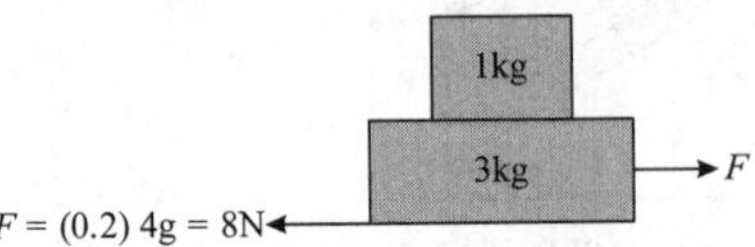

Equation of motion of blocks for combined motion is

$$F - 8 = 4 \times 2$$

$$F = 16\,\text{N}$$ **Ans. (A)**

Sol. 20 FBD of the two cases are shown below

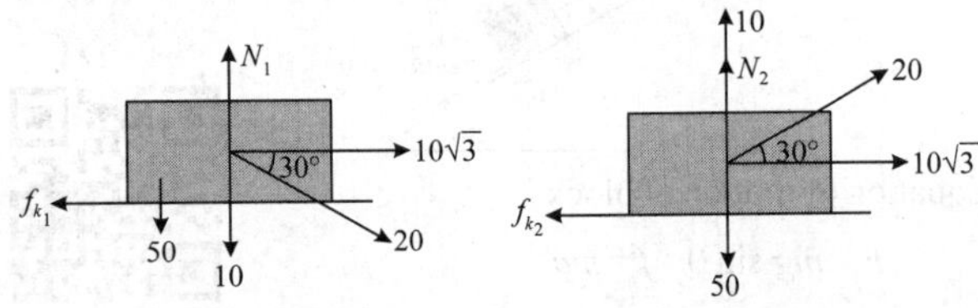

$$N_1 = 60 N_2 = 40$$

$$a_1 = \frac{10\sqrt{3} - 0.2 \times 60}{5}$$

$$a_2 = \frac{10\sqrt{3} - 0.2 \times 40}{5}$$

$$a_1 - a_2 = 0.8$$ **Ans. (B)**

Sol. 21 At the point of slipping as shown in figure, we use

$$\tan\theta = \mu = \frac{3}{4}$$

$$\Rightarrow \quad \cos\theta = \frac{4}{5}$$

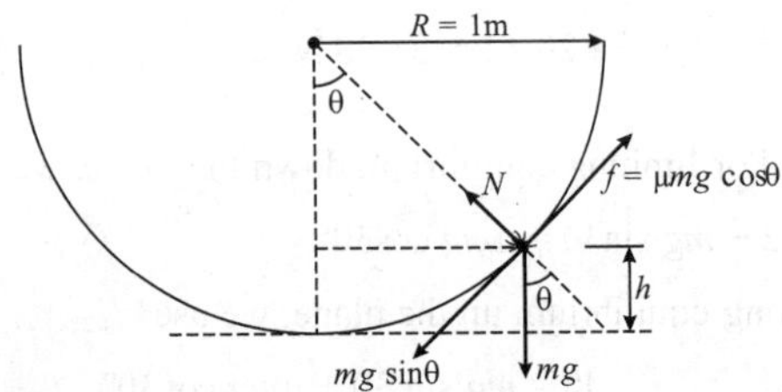

Value of h is given as

$$h = R(1 - \cos\theta) = R\left(1 - \frac{4}{5}\right) = \frac{R}{5}$$ **Ans. (D)**

Sol. 22 Figure below shows the free body diagram of the block as described in the question

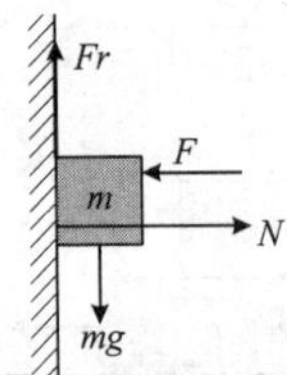

As block is at rest, in this case, we use

$$fr - mg = 0 \quad \ldots(1)$$

and $$F - N = 0 \quad \ldots(2)$$

$$\Rightarrow \quad fr = \mu N = \mu F \quad \ldots(3)$$

From equations-(1) and (3), we use

$$\mu F = mg$$

$$\Rightarrow \quad F = \frac{0.5 \times 10}{0.2} = 25\,\text{N}$$ **Ans. (25)**

Sol. 23 Figure below shows the situation explained in the question

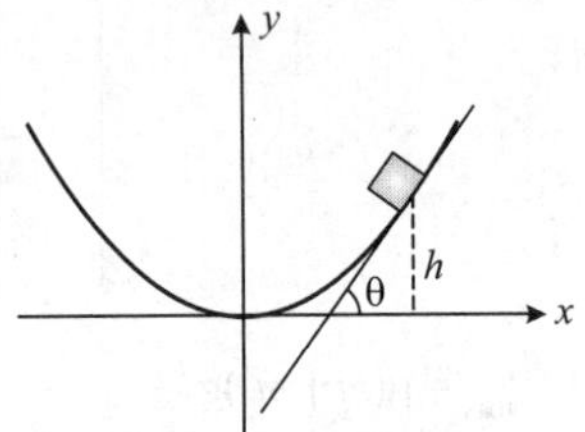

In this situation, at maximum height, block will experience the maximum friction. Thus at this height the slope angle of tangent must be angle of repose, so we use

$$\tan\theta = \frac{dy}{dx} = \frac{2x}{4} = \frac{x}{2} = 0.5$$

$$\Rightarrow \quad x = 1$$

$$\Rightarrow \quad y = \frac{x^2}{4} = 0.25\,\text{m} = 25\,\text{cm}$$ **Ans. (25)**

Sol. 24 Figure below shows the free body diagram of the block with force components shown on it

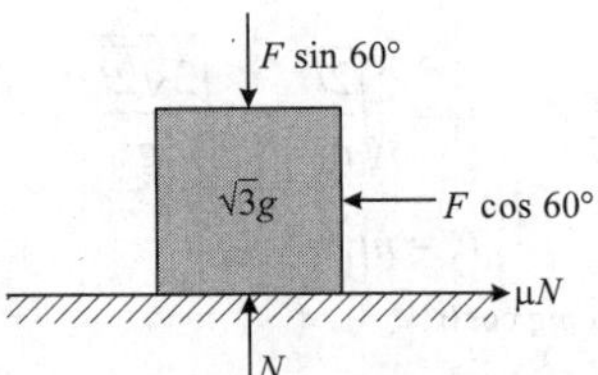

For limiting situation of block, we use

$$F\cos 60^\circ = \mu N$$

$$\Rightarrow \quad \frac{F}{2} = \frac{1}{3\sqrt{3}} N \quad \ldots(1)$$

and $$N = \sin 60^\circ + \sqrt{3}g \quad \ldots(2)$$

From equations-(1) & (2), we use

$$\frac{F}{2} = \frac{1}{3\sqrt{3}}\left(\frac{F\sqrt{3}}{2} + \sqrt{3}g\right)$$

$$\Rightarrow \quad F = g = 10\,\text{N} = x$$ **Ans. (10)**

Sol. 25

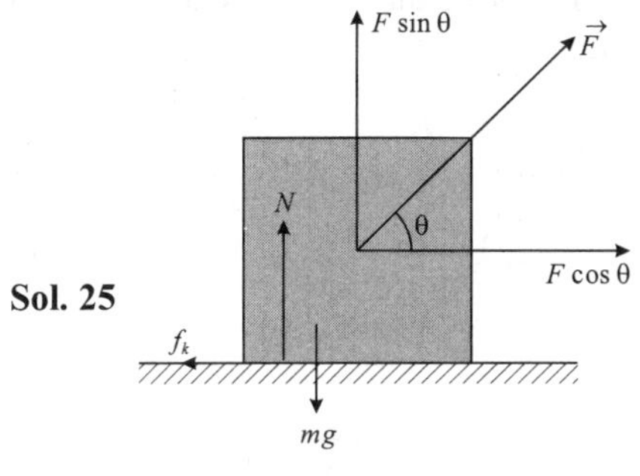

$$N = mg - F\sin\theta$$

$$\Rightarrow \quad F\cos\theta - \mu_k N = ma$$

$$\Rightarrow \quad F\cos\theta - \mu_k(mg - F\sin\theta) = ma$$

$$\Rightarrow \quad a = \frac{F}{m}\cos\theta - \mu_k\left(g - \frac{F}{m}\sin\theta\right)$$ **Ans. (B)**

Sol. 26 Maximum acceleration of upper block can be given as

$$a_{max} = \mu g = \frac{3}{7} \times 9.8 \text{ m/s}^2$$

The maximum force can be calculated as

$$F = (M+m)\,a_{max} = 5\,a_{max}$$

$$\Rightarrow \quad F = 21 \text{ Newton}$$ **Ans. (21)**

Sol. 27 The minimum force required to slide a body placed on a horizontal surface with friction coefficient μ, is given as

$$F_{min} = \frac{\mu mg}{\sqrt{1+\mu^2}} = \frac{\frac{1}{\sqrt{3}} \times 10}{\frac{2}{\sqrt{3}}} = 5 \text{ N}$$ **Ans. (5)**

Sol. 28 Figure below shows the cross section of the vehicle turing on a banked road.

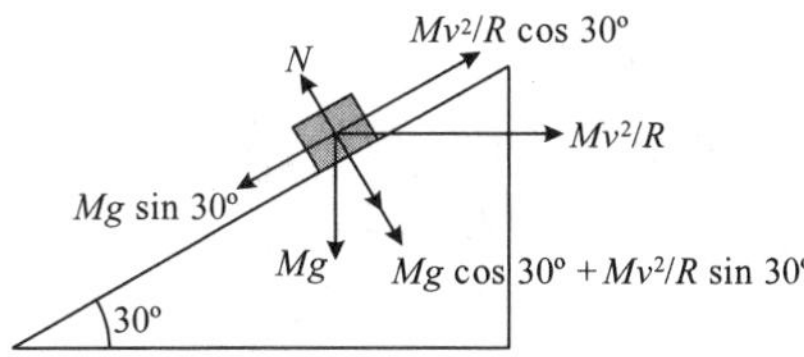

Along the direction normal to inclined plane, forces can be balanced, which gives

$$N = mg\cos 30^\circ + \frac{mv^2}{R}\sin 30^\circ$$

$$\Rightarrow \quad N - mg\cos 30^\circ = \frac{mv^2}{R}\sin 30^\circ \qquad \ldots(1)$$

Along inclined plane, force equation is written as

$$mg\sin 30^\circ + \mu_s N = \frac{mv^2}{R}\cos 30^\circ \qquad \ldots(2)$$

From equations-(1) & (2), we have

$$\frac{N - mg\cos 30^\circ}{mg\sin 30^\circ + \mu_s N} = \tan 30^\circ$$

$$\Rightarrow \quad N = 10.2 \times 10^3 \text{ kg m/s}^2$$ **Ans. (C)**

Sol. 29 If a_A and a_D are the accelerations during ascent and descent of the inclined plane then times of ascending and descending are related as

$$\frac{t_A^2}{t_D^2} = \frac{a_C}{a_A}$$

$$\Rightarrow \quad \frac{t_A^2}{t_D^2} = \frac{g\sin\theta - \mu g\cos\theta}{g\sin\theta + \mu g\cos\theta}$$

$$\Rightarrow \quad \frac{t_A}{t_D} = \sqrt{\frac{g\sin\theta - \mu g\cos\theta}{g\sin\theta + \mu g\cos\theta}}$$

$$\Rightarrow \quad \frac{1}{2} = \sqrt{\frac{1-\sqrt{3}\mu}{1+\sqrt{3}\mu}}$$

$$\Rightarrow \quad 1 + \sqrt{3}\mu = 4 - 4\sqrt{3}$$

$$\Rightarrow \quad 5\sqrt{3}\mu = 3$$

$$\Rightarrow \quad \mu = \frac{\sqrt{3}}{5}$$ **Ans. (3)**

Sol. 30 Maximum horizontal force F_{max} at which the two blocks will move together will be such that friction between the blocks f_s between 2kg & 1 kg will be maximum. Common acceleration of the blocks is given as

$$a_{max} = \frac{F}{3}$$

The acceleration of 1 kg block is only due to the static friction, which is given as

$$f_s = 0.5 \times 10 = 5 \text{ N}$$

Thus maximum acceleration of 1 kg block due to static friction is given as

$$a_{1\text{ kg}} = f_s / m = 5 \text{ m/s}^2$$

For the two blocks to move at common acceleration, we have

$$F = (1+2)\,a_{max} = 3 \times 5 = 15 \text{ N}$$ **Ans. (15)**

Sol. 31 Acceleration of body on a smooth incline is given as

$$a = g\sin 30^\circ$$

Displacement covered by the body is given by

$$s = ut + \frac{1}{2}at^2$$

$$\Rightarrow \quad s = \frac{1}{2}\frac{g}{2}T^2 = \frac{g}{4}T^2 \qquad \ldots(1)$$

On rough incline, acceleration is given as

$$a = g\sin 30^\circ - \mu g\cos 30^\circ$$

Displacement covered by the body is given by

$$s = ut + \frac{1}{2}at^2$$

$$\Rightarrow \quad s = \frac{1}{4}g(1 - \sqrt{3}\mu)(\alpha T)^2 \quad \ldots(2)$$

Using equations-(1) and (2), we have

$$\frac{1}{4}gT^2 = \frac{1}{4}g(1-\sqrt{3}\mu)\alpha^2T^2$$

$$\Rightarrow \quad 1 - \sqrt{3}g = \frac{1}{a^2}$$

$$\Rightarrow \quad g = \left(\frac{\alpha^2 - 1}{\alpha^2}\right)\frac{1}{\sqrt{3}}$$

$$\Rightarrow \quad x = 3$$ **Ans. (3)**

Sol. 32 Figure below shows the forces acting on the block when it is pushed up for sliding.

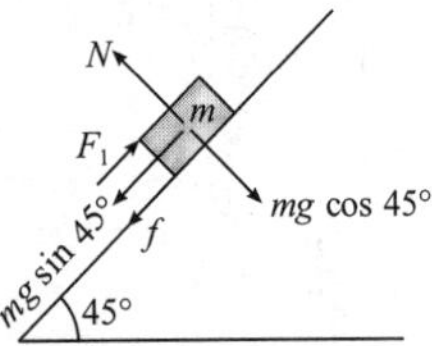

For sliding the upward required force is calculated as

$$F_1 = mg\sin 45° + f = mg\sin 45° + \mu N$$

$$\Rightarrow \quad F_1 = \frac{mg}{\sqrt{2}} + \mu mg\cos 45°$$

$$\Rightarrow \quad F_1 = \frac{mg}{\sqrt{2}}(1+\mu)$$

To prevent the block from sliding down the force required is calculated from the figure below when friction will act in upward direction.

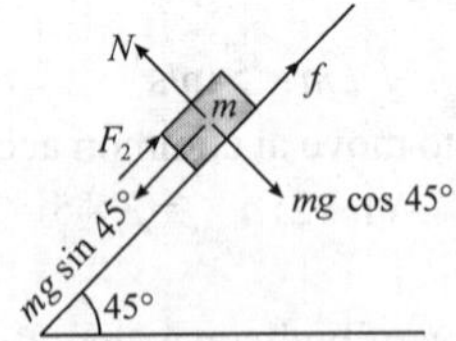

$$F_2 = mg\sin 45° - f = mg\sin 45° - \mu N$$

$$\Rightarrow \quad F_2 = \frac{mg}{\sqrt{2}}(1-\mu)$$

Given that $F_1 = 2F_2$ so we have

$$\frac{mg}{\sqrt{2}}(1+\mu) = 2\frac{mg}{\sqrt{2}}(1-\mu)$$

$$\Rightarrow \quad 1+\mu = 2 - 2\mu$$

$$\Rightarrow \quad \mu = 1/3 = 0.33$$ **Ans. (A)**

Sol. 33 The figure below shows all the forces acting on the block placed on incline.

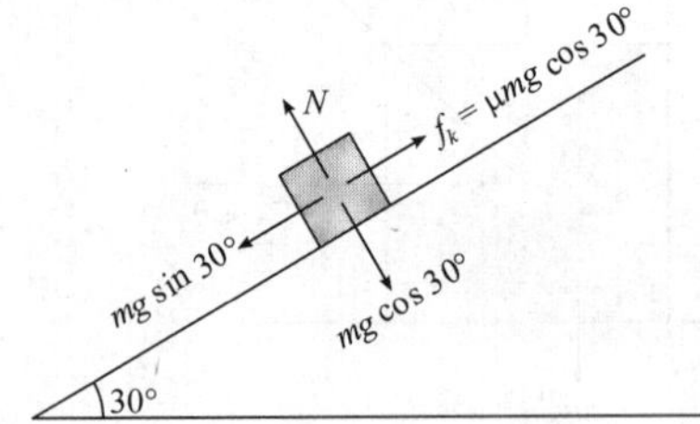

Equation of motion for sliding of the block down the plane is written as

$$Mg\sin 30° - \mu mg\cos 30° = ma$$

$$\Rightarrow \quad \frac{g}{2} - \frac{\sqrt{3}}{2}\cdot\mu g = \frac{g}{4}$$

$$\Rightarrow \quad \frac{\sqrt{3}}{2}\mu = \frac{1}{4}$$

$$\Rightarrow \quad \mu = \frac{1}{2\sqrt{3}}$$ **Ans. (B)**

Sol. 34 For smooth inclined plane acceleration of object is given as

$$a_1 = g\sin\theta = g/\sqrt{2}$$

For rough inclined plane if coefficient of friction is m, acceleration is given as

$$a_2 = g\sin\theta - \mu g\cos\theta = \frac{g}{\sqrt{2}} - \frac{\mu g}{\sqrt{2}}$$

Given that

$$t_2 = nt_1$$

As distance travelled by object in the two cases is same, we use

$$a_1t_1^2 = a_2t_2^2$$

$$\Rightarrow \quad \frac{g}{\sqrt{2}}t_1^2 = \left(\frac{g}{\sqrt{2}} - \frac{\mu g}{\sqrt{2}}\right)n^2t_1^2$$

$$\Rightarrow \quad \mu = 1 - \frac{1}{n^2}$$ **Ans. (D)**

Sol. 35 Acceleration due to friction is given as

$$a = -\mu g$$

By first equation of uniformly accelerates motion, we have

$$v = u + at$$

$$\Rightarrow \quad 0 = 20 + (-\mu \times 10) \times 5$$

$$\Rightarrow \quad 50\mu = 20$$

$$\Rightarrow \quad \mu = \frac{2}{5} = 0.4$$ **Ans. (D)**

Sol. 36 Distance travelled by the block in 10 s is calculated as

$$s = ut + \frac{1}{2}at^2$$

$$\Rightarrow \quad 50 = 0 + \frac{1}{2} \times a \times 100$$

$$\Rightarrow \quad a = 1 \text{ m/s}^2$$

Equation of motion of block is written as

$$F - \mu mg = ma$$

$$\Rightarrow \quad 30 - \mu \times 50 = 5 \times 1$$

$$\Rightarrow \quad 50\,\mu = 25$$

$$\Rightarrow \quad \mu = \frac{1}{2}$$

Ans. (C)

Sol. 37 Figure below shows the free body diagram of the block as given in question.

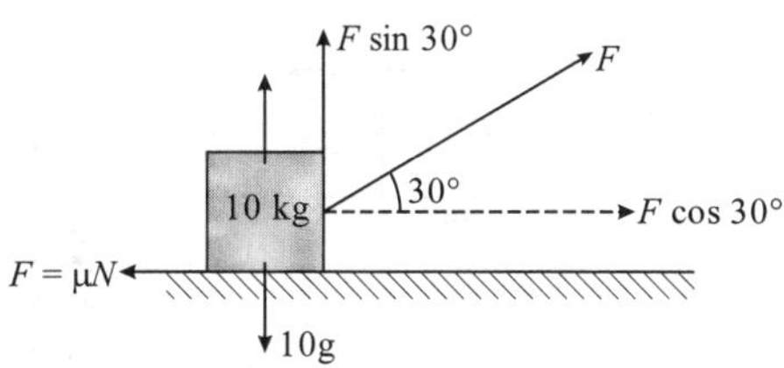

For vertical direction forces on block we have

$$N = Mg - F\sin 30°$$

$$\Rightarrow \quad = mg - \frac{F}{2} = 100 - \frac{F}{2} = \frac{200 - F}{2}$$

For horizontal motion of block we have

$$F\cos 30° = \mu N$$

$$\Rightarrow \quad \sqrt{3}\frac{F}{2} = 0.25 \times \left(\frac{200 - F}{2}\right)$$

$$\Rightarrow \quad 4\sqrt{3}F = 200 - F$$

$$\Rightarrow \quad F = \frac{200}{4\sqrt{3} + 1} = 25.22$$

Ans. (B)

5.1 Work Energy Power

Sol. 1 The energy stored in a spring having extension x is

$$U = \frac{1}{2}kx^2$$

Work done in extending the extension in spring from x_1 to x_2 is

$$W = U_f - U_i$$

$$\Rightarrow \quad W = \frac{1}{2}k\,(x_2^2 - x_1^2)$$

$$\Rightarrow \quad W = \frac{1}{2} \times 800\,[(0.15)^2 - (0.05)^2]$$

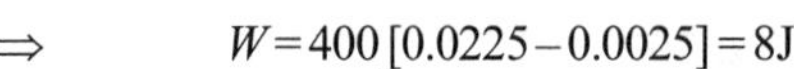

$$\Rightarrow \quad W = 400\,[0.0225 - 0.0025] = 8\text{J}$$

Ans. (B)

Sol. 2 Work done in extending spring extension from x_1 to x_2 is

$$W = U_f - U_i = \frac{1}{2}kx_2^2 - \frac{1}{2}kx_1^2$$

$$\Rightarrow \quad W = \frac{1}{2}(5 \times 10^3) \times [(0.1)^2 - (0.05)^2]$$

$$\Rightarrow \quad W = 18.75 \text{ J}$$

Ans. (B)

Sol. 3 Along a straight line motion, power is given as

$$P = F.v = mav$$

$$\Rightarrow \quad m\left(\frac{dv}{dt}\right)v = P \rightarrow \text{constant}$$

$$\Rightarrow \quad vdv = \frac{P}{m}dt$$

Integrating this relation gives

$$\int_0^v vdv = \frac{P}{m}\int_0^t dt$$

$$\Rightarrow \quad \frac{v^2}{2} = \frac{P}{m}t$$

$$\Rightarrow \quad v = \sqrt{\frac{2P}{m}t}$$

$$\Rightarrow \quad \frac{dx}{dt} = \sqrt{\frac{2P}{m}}\sqrt{t}$$

$$\Rightarrow \quad dx = \sqrt{\frac{2P}{m}}\sqrt{t}\,dt$$

$$\Rightarrow \quad \int_0^x dx = \sqrt{\frac{x}{m}}\int_0^t \sqrt{t}dt$$

$$\Rightarrow \quad x = \sqrt{\frac{x}{m}}\,\frac{t^{3/2}}{(3/2)}$$

$$\Rightarrow \quad x \propto t^{3/2}$$

Ans. (B)

Sol. 4 Given that

$$a \propto -x$$

$$\Rightarrow \quad v\frac{dv}{dx} = a = -Cx$$

$$\Rightarrow \quad \int_{v_i}^{v_f} vdv = -\int_{x=0}^{x} Cx\,dx$$

$$\Rightarrow \quad \left[\frac{v^2}{2}\right]_{v_i}^{v_f} = -\frac{Cx^2}{2}$$

$$\Rightarrow \quad \frac{v_i^2}{2} - \frac{v_f^2}{2} = -\frac{Cx^2}{2}$$

If equation is multiplied with mass of particle m

Here $v_i \rightarrow$ initial speed at $x = 0$

$v_f \rightarrow$ final speed after covering displacement x

$$\frac{1}{2}mv_i^2 - \frac{1}{2}mv_f^2 = \frac{1}{2}mCx^2$$

$$\Rightarrow \quad \Delta K = \frac{1}{2}mCx^2$$

$$\Rightarrow \quad \Delta K \propto x^2$$

Ans. (A)

Sol. 5 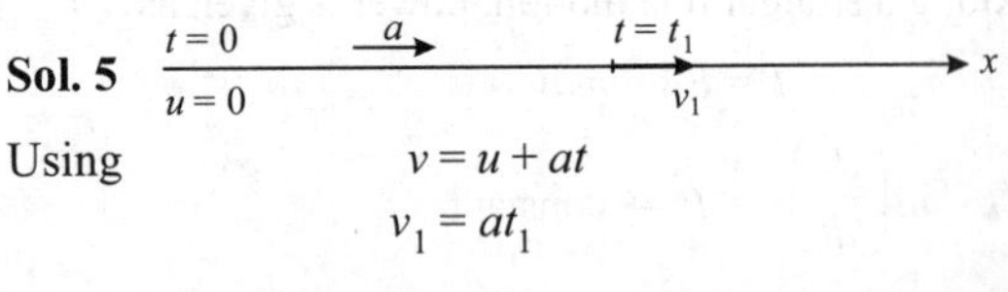

Using $\quad v = u + at$

$$v_1 = at_1$$

$$\Rightarrow \quad a = \frac{v_1}{t_1}$$

Instantaneous power

$$P = F.v = m\,a\,v$$

After time t_1 speed of particle is $v = at$

$$\Rightarrow \quad P = ma^2t = \frac{mv_1^2}{t_1^2}\cdot t$$

Ans. (B)

Sol. 6 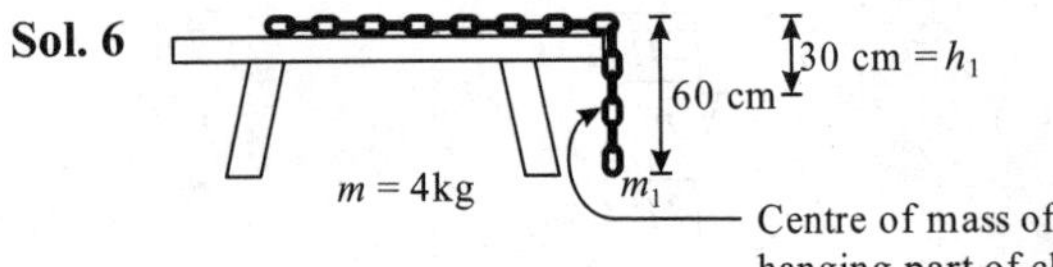

Mass of hanging part of chain is

$$m_1 = \frac{4}{2}\times 0.6 = 1.2\text{ kg}$$

Work done in lifting hanging part of chain is

$$W = m_1\,g\,h_1 = 1.2\times 10\times 0.3$$

$$= 3.6\text{ J}$$

Ans. (B)

Sol. 7 Work done in displacing a particle is given as

$$w = \vec{F}\cdot\vec{r}$$

$$\Rightarrow \quad w = (5\hat{i} + 3\hat{j} + 2\hat{k})\cdot(2\hat{i} - \hat{j})$$

$\Rightarrow \quad w = 10 - 3 = 7\text{ J}$ **Ans. (B)**

Sol. 8 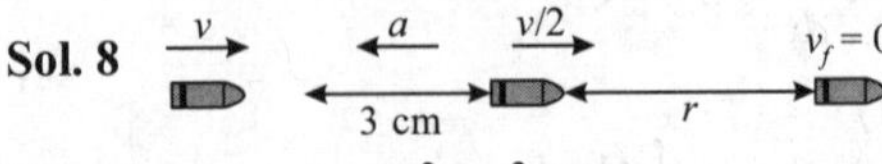

We use $\quad v^2 = u^2 + 2as$

For first part of motion

$$\left(\frac{v}{2}\right)^2 = (v)^2 - 2a(3)$$

$$\Rightarrow \quad \frac{3v^2}{4} = 6a$$

$$\Rightarrow \quad a = \frac{v^2}{8}$$

For second part of motion

$$(0)^2 = \left(\frac{v}{2}\right)^2 - 2as$$

$$\Rightarrow \quad s = \frac{v^2}{8a} = 1\text{ cm}$$

Ans. (D)

Sol. 9 By conservation of energy

$$\frac{1}{2}Mv^2 = \frac{1}{2}kL^2$$

$$\Rightarrow \quad v = \sqrt{\frac{k}{M}}L$$

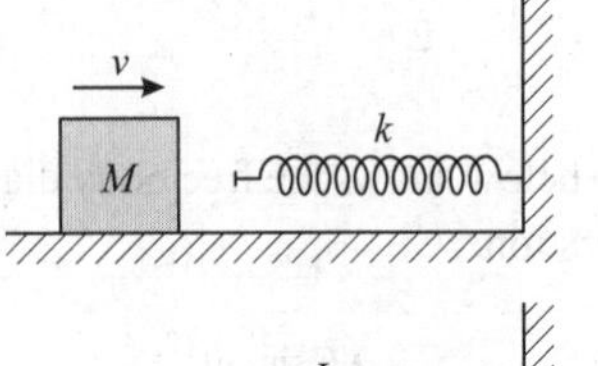

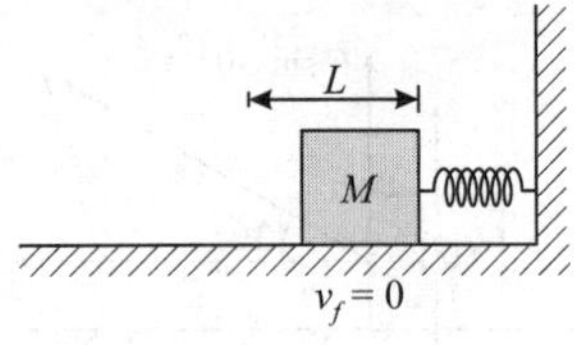

Maximum momentum of the block is

$$P_{max} = Mv = M\sqrt{\frac{k}{M}}\cdot L = \sqrt{kM}\cdot L$$

Ans. (A)

Sol. 10 All surface are smooth

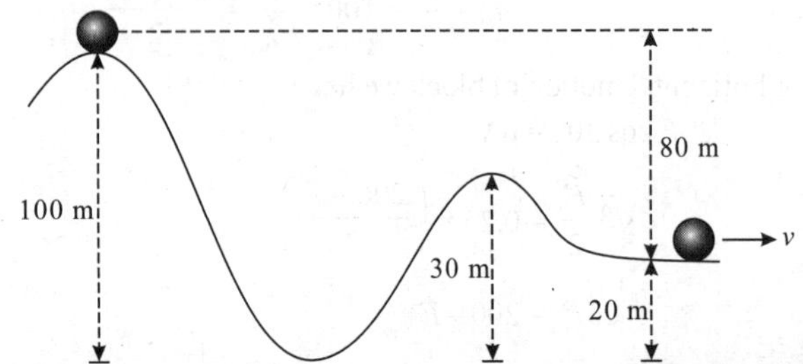

Using work energy theorem, we have

$$0 + mg(80) = \frac{1}{2}mv^2$$

$$\Rightarrow \quad v = \sqrt{1600} = 40\text{ m/s}$$

Ans. (B)

Sol. 11

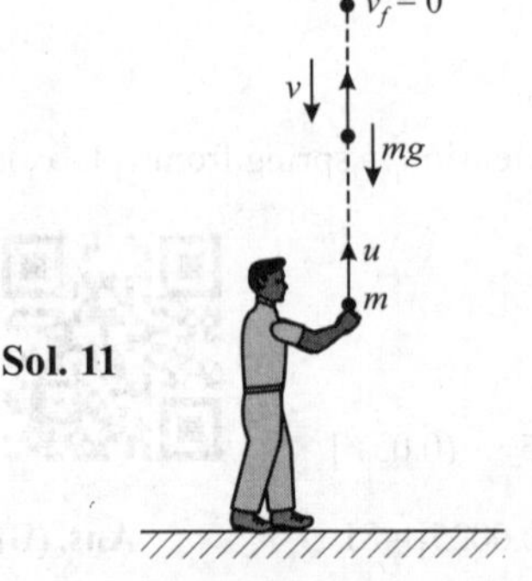

By applying work-energy theorem, we have

$$\frac{1}{2}mu^2 + W_g = 0$$

$$\Rightarrow \quad W_g = \frac{1}{2}mu^2 = -\frac{1}{2} \times 100 \times 10^{-3} \times (5)^2$$

$$\Rightarrow \quad W_g = -1.25\text{ J}$$

Ans. (B)

Sol. 12 At any instant total mechanical energy is given as

$$E = K + U$$

$$\Rightarrow \quad K = E - U = 2 - \left(\frac{x^4}{4} - \frac{x^2}{2}\right)$$

K is maximum when

$$\frac{dK}{dx} = 0 = -\frac{4x^3}{4} + \frac{2x}{2}$$

$$\Rightarrow \quad -x^3 + x = 0$$

$$\Rightarrow \quad x = 0, -1, +1$$

$$K_{x=0} = 2\text{J}$$

and $$K_{x=\pm 1} = 2 - \frac{1}{4} + \frac{1}{2} = \frac{9}{4}\text{ J (max)}$$

$$\Rightarrow \quad \frac{1}{2}mv^2_{max} = \frac{9}{4}$$

$$\Rightarrow \quad v_{max} = \sqrt{\frac{9}{2}} = \frac{3}{\sqrt{2}}\text{ m/s}$$

Ans. (A)

Sol. 13 As initial and final kinetic energy of mass is zero, using work-energy theorem, we have

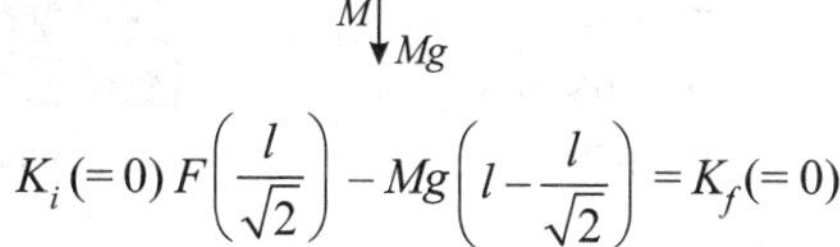

$$K_i(=0)\, F\left(\frac{l}{\sqrt{2}}\right) - Mg\left(l - \frac{l}{\sqrt{2}}\right) = K_f(=0)$$

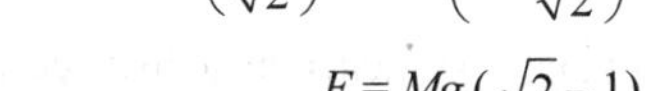

$$\Rightarrow \quad F = Mg(\sqrt{2} - 1)$$

Ans. (D)

Sol. 14 Using work-energy theorem for the compression of spring, we have

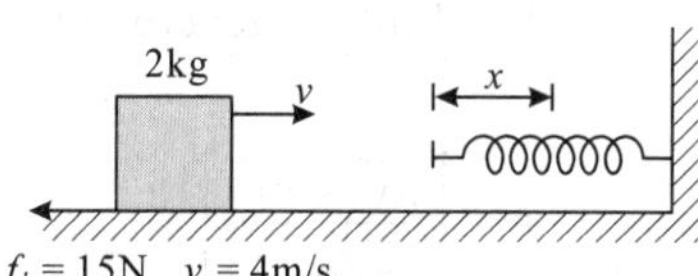

$$\Rightarrow \quad \frac{1}{2}mv^2 - f_k x - \frac{1}{2}kx^2 = 0$$

$$\Rightarrow \quad \frac{1}{2} \times 2 \times (4)^2 - 15x - \frac{1}{2} \times 10000\,x^2 = 0$$

$$\Rightarrow \quad 5000x^2 + 15x - 16 = 0$$

$$\Rightarrow \quad x = \frac{-15 \pm \sqrt{225 + 4 \times 5000 \times 16}}{10000} = 0.055\text{ m}$$

$$\Rightarrow \quad x = 5.5\text{ cm}$$

Ans. (A)

Sol. 15 Speed of athlete can be taken as

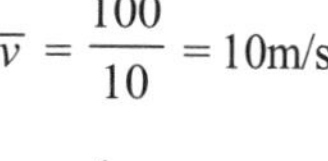

$$\bar{v} = \frac{100}{10} = 10\text{m/s}$$

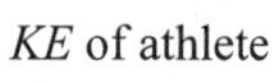

KE of athlete $$K = \frac{1}{2}m\bar{v}^2$$

$$= \frac{1}{2} \times 60 \times (10)^2 = 3000\text{ J}$$

mass of an athlete can be estimated ~ 60 kg

Ans. (D)

Sol. 16 At $x = \infty$, $U(x) = 0$

at equilibrium, force on atoms = 0

we use $$F = -\frac{dU}{dx} = \frac{12a}{x^{13}} - \frac{6b}{x^7} = 0$$

$$\Rightarrow \quad x^6 = \frac{2a}{b}$$

$$\Rightarrow \quad x = \left(\frac{2a}{b}\right)^{1/6}$$

PE at equilibrium is given as

$$U_{\text{at eqm}} = \frac{a}{\left(\frac{2a}{b}\right)^{12/6}} - \frac{b}{\left(\frac{2a}{b}\right)^{6/6}}$$

$$= \frac{a}{\left(\frac{4a^2}{b^2}\right)} - \frac{b}{\frac{2a}{b}}$$

$$= \frac{b^2}{4a} - \frac{b^2}{2a} = -\frac{b^2}{4a}$$

$$\Rightarrow \quad D = U_{\infty} - U_{\text{at eqm}} = 0 - \left(-\frac{b^2}{4a}\right) = \frac{b^2}{4a}$$

Ans. (D)

Sol. 17 We are given that

$$K = \frac{1}{2}mv^2 = Ct$$

$$\Rightarrow \quad v = \sqrt{\frac{2Ct}{m}}$$

$\Rightarrow$ Acceleration of particle

$$a = \frac{dv}{dx} = \sqrt{\frac{2C}{m}} \cdot \frac{1}{2\sqrt{t}}$$

$\Rightarrow$ Force on particle is given as

$$a = \frac{dv}{dx} = \sqrt{\frac{2C}{m}} \cdot \frac{1}{2\sqrt{t}} = \sqrt{\frac{Cm}{2t}}$$

$$\Rightarrow \quad F \propto \frac{1}{\sqrt{t}}$$

Ans. (D)

Sol. 18 We are given that applied force $F = k_1x_1 = k_2x_2$

and $W_{s_1} > W_{s_2}$

$$\frac{1}{2}k_1x_1^2 > \frac{1}{2}k_1x_2^2$$

$$\Rightarrow \quad \frac{1}{2}Fx_1 > \frac{1}{2}Fx_2$$

$$\Rightarrow \quad x_1 > x_2$$

$$\Rightarrow \quad k_1 < k_2$$

$\Rightarrow$ S_{II} is correct

If in the two springs extension is equal, x

$$\Rightarrow \quad W_{s_1} = \frac{1}{2}k_1x^2 \text{ and } W_{s_2} = \frac{1}{2}k_2x^2$$

$$\Rightarrow \quad W_{s_1} < W_{s_2}$$

$\Rightarrow$ S_I is false **Ans. (B)**

Sol. 19 Using work-energy theorem we have

$$\frac{1}{2}mv_i^2 + W_F = \frac{1}{2}mv_f^2$$

By area under F-x curve

$$W_{0.2s} = \frac{1}{2} \times 2 \times 2 = 2\text{ J}$$

$$W_{2.5s} = -\frac{1}{2} \times 4 \times 1 = -2\text{ J}$$

$$W_{5.8s} = \frac{1}{2} \times 3 \times 3 = 4.5\text{ J}$$

$$\Rightarrow \quad k_f = \frac{1}{2} \times 2 \times (5)^2 + [W_{0.2} + W_{2.5} + W_{5.8}]$$

$$\Rightarrow \quad k_f = 25 + [2 - 2 + 4.5] = 29.5\text{ J}$$

Ans. (D)

Sol. 20 By work-energy theorem

Gain (change) in KE = Work done on particle

$$\Delta k = \vec{F} \cdot \vec{\Delta r}$$

$$\Rightarrow \quad \Delta k = (7\hat{i} + 4\hat{j} + 3\hat{k}) \cdot (2\hat{i} + 3\hat{j} + 4\hat{k})$$

$$\Rightarrow \quad \Delta k = 14 + 12 + 12 = 38\text{ J}$$ **Ans. (A)**

Sol. 21 For springs in series, tension remain same

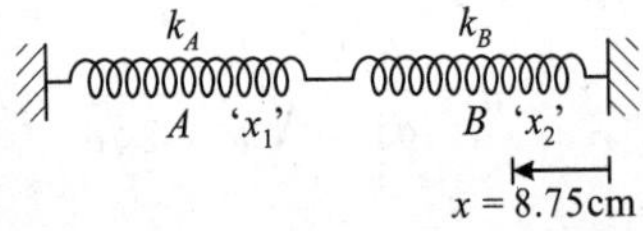

If compression in A and B are x_1 and x_2

$$\Rightarrow \quad k_A x_1 = k_B x_2$$

and $3x_1 = 4x_2$

$$x_1 + x_2 = 8.75$$

$$\Rightarrow \quad x_1 + \frac{3}{4}x_1 = 8.75$$

$$\Rightarrow \quad x_1 = \frac{8.75 \times 4}{7} = 5\text{ cm}$$

$$\Rightarrow \quad x_2 = 3.75\text{ cm}$$

Ratio of energy stored in A & B is

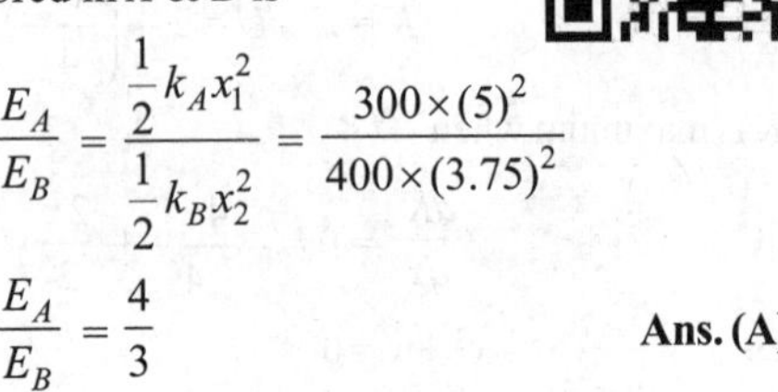

$$\frac{E_A}{E_B} = \frac{\frac{1}{2}k_Ax_1^2}{\frac{1}{2}k_Bx_2^2} = \frac{300 \times (5)^2}{400 \times (3.75)^2}$$

$$\Rightarrow \quad \frac{E_A}{E_B} = \frac{4}{3}$$

Ans. (A)

Sol. 22 If a is upward acceleration of man, we can use

$$v = \sqrt{2ah_1} = \sqrt{2gh_2}$$

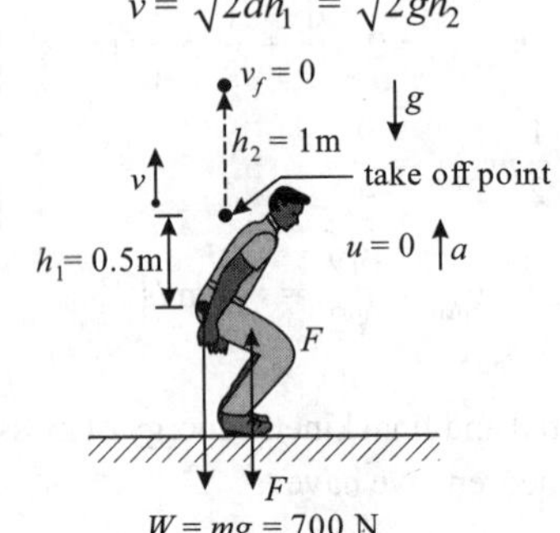

$$\Rightarrow \quad a = \frac{gh_2}{h_1} = \frac{10 \times 1}{0.5} = 20\text{ m/s}^2$$

$\Rightarrow$ Velocity at take off is

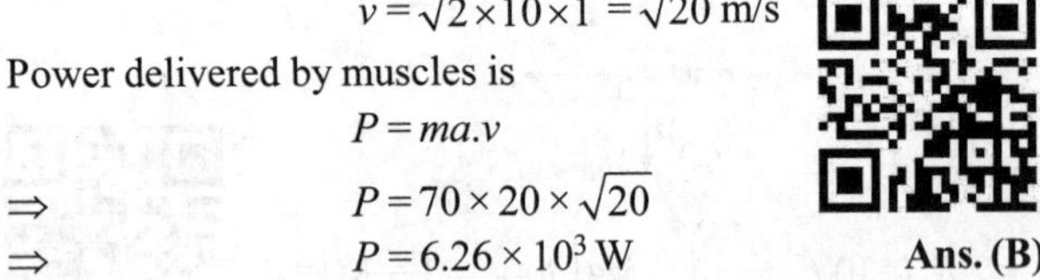

$$v = \sqrt{2 \times 10 \times 1} = \sqrt{20}\text{ m/s}$$

Power delivered by muscles is

$$P = ma.v$$

$$\Rightarrow \quad P = 70 \times 20 \times \sqrt{20}$$

$$\Rightarrow \quad P = 6.26 \times 10^3\text{ W}$$

Ans. (B)

Sol. 23 When the rubber band is stretched by a distance x than work done in further stretching it by dx is given as

$$dW = Fdx = (ax + bx^2)\,dx$$

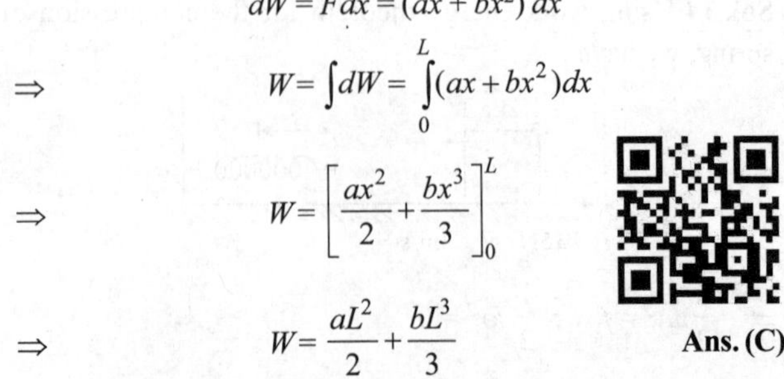

$$\Rightarrow \quad W = \int dW = \int_0^L (ax + bx^2)dx$$

$$\Rightarrow \quad W = \left[\frac{ax^2}{2} + \frac{bx^3}{3}\right]_0^L$$

$$\Rightarrow \quad W = \frac{aL^2}{2} + \frac{bL^3}{3}$$

Ans. (C)

Sol. 24 To find KE of spring, we consider an elemental part of spring as shown mass of element is $dm = \frac{m}{l}dx$

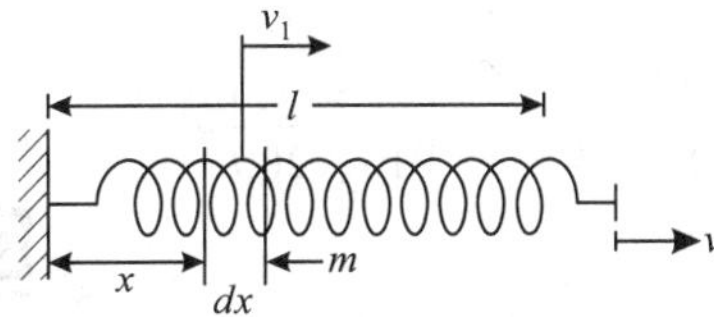

As spring is uniformly stretched, the speed of element is

$$v_1 = \frac{v}{l}\cdot x$$

KE of element is $\quad dK = \frac{1}{2}dmv_1^2$

$$\Rightarrow \quad dK = \frac{1}{2}\left(\frac{m}{l}dx\right)\left(\frac{v}{l}x\right)^2$$

$$\Rightarrow \quad \int dK = \frac{1}{2}\frac{mv^2}{l^3}\int_0^l x^2 dx$$

Total KE of spring is given as

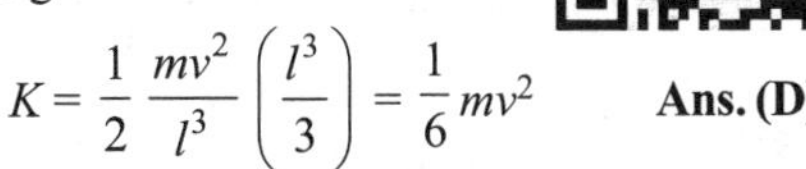

$$K = \frac{1}{2}\frac{mv^2}{l^3}\left(\frac{l^3}{3}\right) = \frac{1}{6}mv^2$$ **Ans. (D)**

Sol. 25 Wind volume flowing through the cross sectional area of blades (rotating) of wind mill per unit time is given as

$$V = Av$$

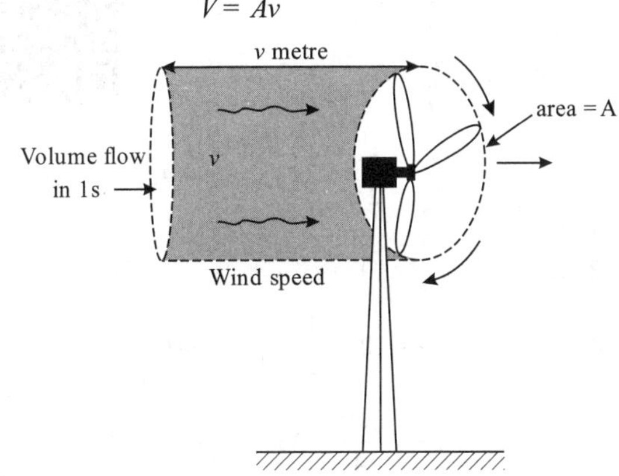

Mass of wind intercepting the blades per second is

$$m = \rho Av \quad [\rho \rightarrow \text{wind density of air}]$$

Kinetic energy of wind mass intercepting the section of blades per second is

$$\text{KE/sec} = \frac{1}{2}mv^2 = \frac{1}{2}\rho Av^3$$

$\Rightarrow$ Electrical power (J/s) $\propto v^3$ **Ans. (D)**

Sol. 26 If plank thickness is x, we have

$$u^2\left(1-\frac{1}{n}\right)^2 = u^2 - 2ax$$

$$\Rightarrow \quad 2ax = u^2\left[1-\left(\frac{n-1}{n}\right)^2\right]$$

$$\Rightarrow \quad 2ax = u^2\left[\frac{2n-1}{n^2}\right] \quad \ldots(1)$$

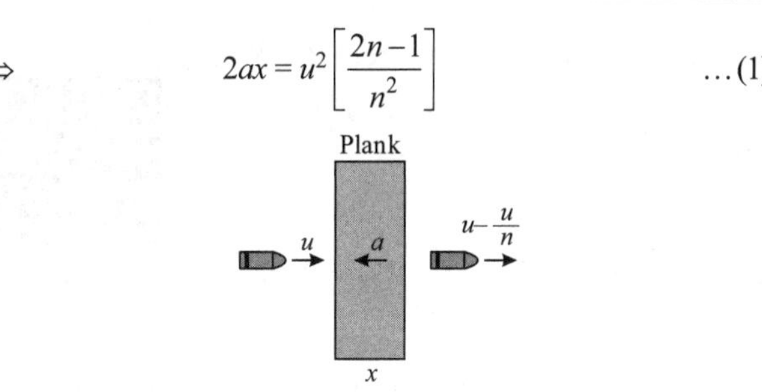

Using constant retardation, we have

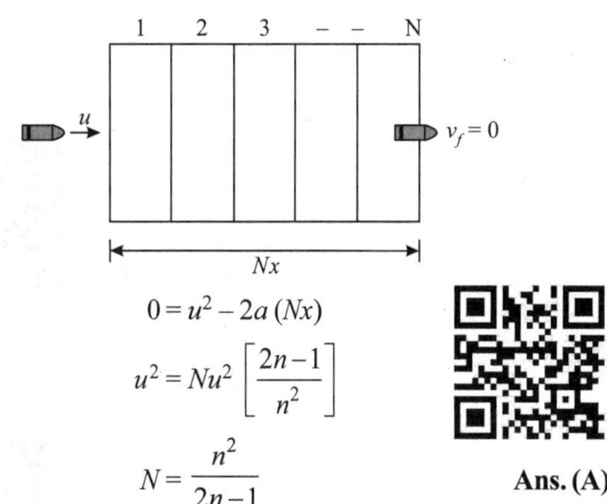

$$0 = u^2 - 2a(Nx)$$

$$\Rightarrow \quad u^2 = Nu^2\left[\frac{2n-1}{n^2}\right]$$

$$\Rightarrow \quad N = \frac{n^2}{2n-1}$$ **Ans. (A)**

Sol. 27 Using work energy theorem from point A to C, we have

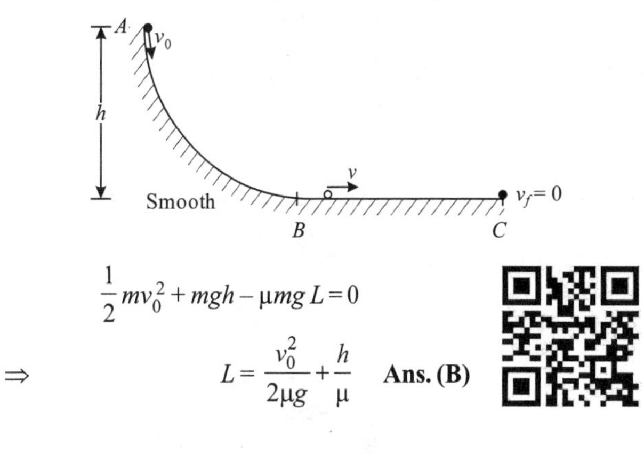

$$\frac{1}{2}mv_0^2 + mgh - \mu mg\,L = 0$$

$$\Rightarrow \quad L = \frac{v_0^2}{2\mu g} + \frac{h}{\mu}$$ **Ans. (B)**

Sol. 28

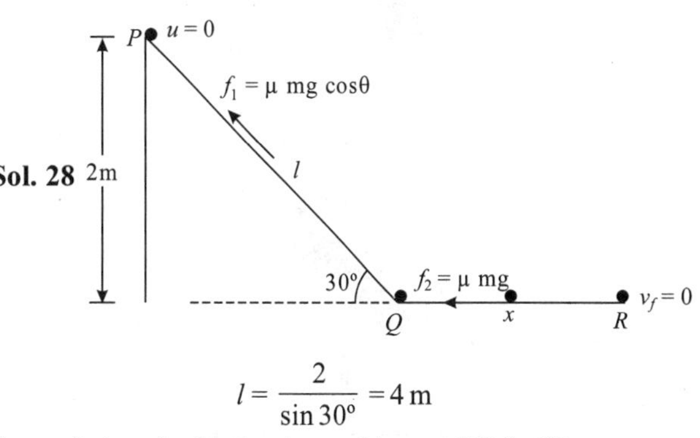

$$l = \frac{2}{\sin 30^\circ} = 4\text{ m}$$

As work done by friction is equal in part PQ & QR

$$\mu mg\cos 30^\circ(\mu) = \mu mg\,x$$

$$\Rightarrow \quad x = 2\sqrt{3}\text{ m} = 3.464\text{ m} \simeq 3.5\text{ m}$$

We use work energy theorem from P to R as

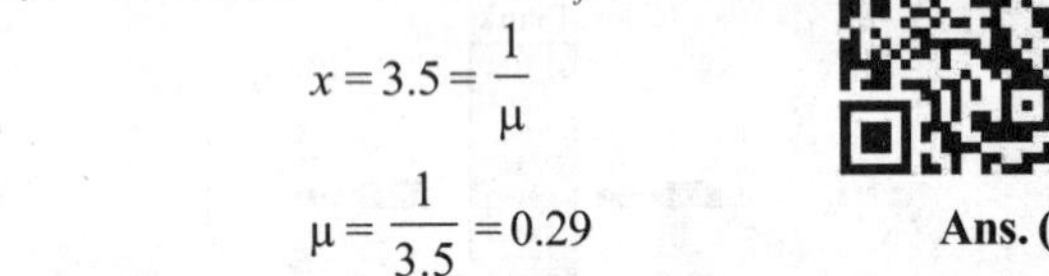

$$K_i(=0) + mg\,(2) - 2\mu\, mg\, x = K_f(=0)$$

$$\Rightarrow \quad x = 3.5 = \frac{1}{\mu}$$

$$\Rightarrow \quad \mu = \frac{1}{3.5} = 0.29$$

Ans. (C)

Sol. 29 Work done in lifting 10 kg mass by 1 m, 1000 times is

$$W = mgh \times 1000 = 10 \times 9.8 \times 1 \times 1000 \text{ J}$$

If m in kg mass of fat will be used in doing this work then energy released is

$$E = m \times 3.8 \times 10^7 \text{ J} \times \frac{20}{100}$$

By conservation of energy

$$m \times 3.8 \times 10^7 \times \frac{20}{100} = 10 \times 9.8 \times 1 \times 1000$$

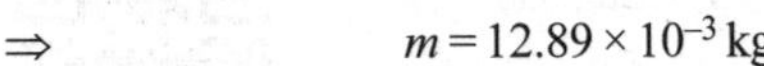

$$\Rightarrow \quad m = 12.89 \times 10^{-3} \text{ kg}$$

Ans. (D)

Sol. 30 Time function of body's speed

$$v = -5t + 50 \text{ m/s}$$

at $t = 0$; $v_1 = 50$ m/s

at $t = 2s$; $v_2 = -5 \times 2 + 50 = 40$ m/s

Work done on body is equal to gain in kinetic energy

$$\Rightarrow \quad W = \frac{1}{2} m\,(v_2^2 - v_1^2) = \frac{1}{2} \times 10 \times ((40)^2 - (50)^2) = -4500 \text{ J}$$

Ans. (C)

Sol. 31 To rise up at constant speed v_0, we have driving force

$$f_D = \left(\frac{W}{20} + \frac{W}{10}\right)$$

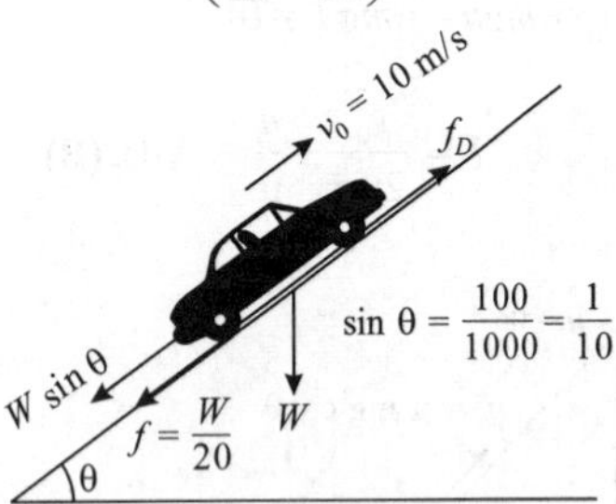

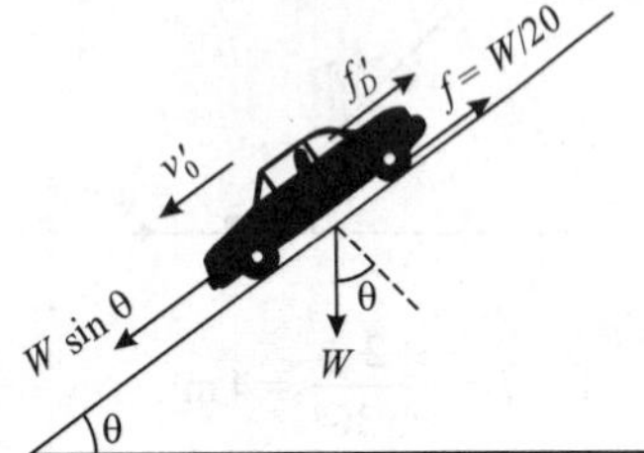

Given that power needed is

$$P = f_D v_0 = W\left(\frac{1}{20} + \frac{1}{10}\right) \times 10 \qquad \ldots(1)$$

To move down at constant speed v, we have

$$f'_D = \left(\frac{W}{10} - \frac{W}{20}\right) \text{ and power is } \frac{P}{2}$$

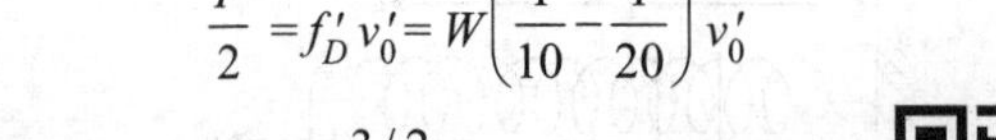

$$\Rightarrow \quad \frac{P}{2} = f'_D v'_0 = W\left(\frac{1}{10} - \frac{1}{20}\right) v'_0 \qquad \ldots(2)$$

$$\text{By (1)/(2)} \Rightarrow \quad 2 = \frac{3/2}{1/20v'_0}$$

$$\Rightarrow \quad v'_0 = \frac{30}{2} = 15 \text{ m/s}$$

Ans. (C)

Sol. 32 Work done by force = Gain in KE = $\frac{1}{2}mv^2$

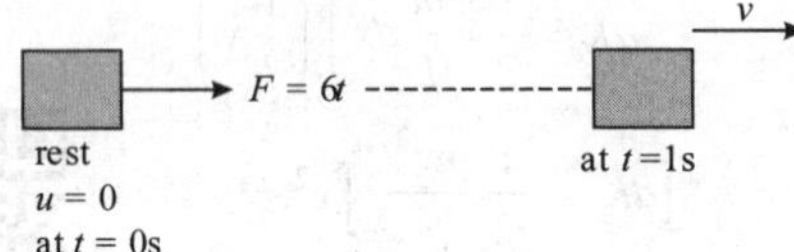

Acceleration of particle $a = \dfrac{F}{m} = \dfrac{6t}{1} = 6t$ m/s^2

$$\Rightarrow \quad \frac{dv}{dt} = 6t$$

$$\Rightarrow \quad \int_0^v dv = \int_0^1 6t\,dt$$

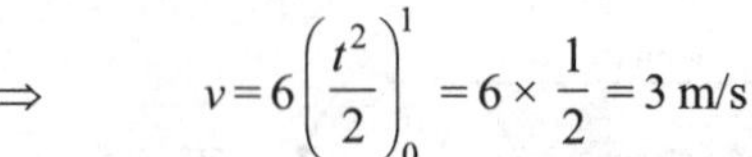

$$\Rightarrow \quad v = 6\left(\frac{t^2}{2}\right)_0^1 = 6 \times \frac{1}{2} = 3 \text{ m/s}$$

$$= \frac{1}{2} \times 1 \times (3)^2 = \frac{9}{2} = 4.5 \text{ J}$$

Ans. (C)

Sol. 33 At $t = 0$, $\quad v = 10$ m/s

and at $\quad t = 10\text{s}, \quad \frac{1}{2}mv_f^2 = \frac{1}{8}m\,(10)^2$

$$\Rightarrow \quad v_f^2 = \frac{100}{4} = 25 \Rightarrow v_f = 5 \text{ m/s}$$

$$F = -kv^2$$

m → v, F, t = 0, v = 10 m/s; t = 10s, v_f = 5 m/s

$$\Rightarrow \text{Acceleration } a = \frac{F}{m} = -\frac{k}{m}v^2$$

$$\Rightarrow \quad \frac{dv}{dt} = -\frac{k}{m}v^2$$

$$\Rightarrow \quad \int_{10}^{5} \frac{dv}{v^2} = \int_0^{10} \frac{k}{m}\,dt$$

$$\Rightarrow \quad \left[+\frac{1}{v}\right]_{10}^{5} = +\frac{k}{m}[t]_0^{10}$$

$$\Rightarrow \quad \frac{1}{5}-\frac{1}{10}=\frac{k}{m}\times 10$$

$$\Rightarrow \quad k=10^{-3}\left[\frac{1}{5}-\frac{1}{10}\right]=10^{-4}\text{ kg/m}$$

Ans. (A)

Sol. 34 Force on particle in the given potential field is given as

$$\vec{F}=-\frac{dU}{dr}\hat{r}=\frac{k}{r^3}\hat{r}$$

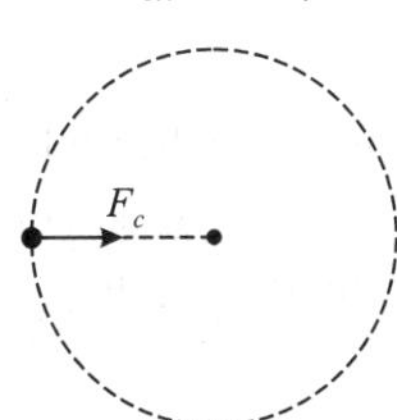

For stable circular motion we use

$$\frac{k}{r^3}=\frac{mv^2}{r}$$

$$\Rightarrow \quad mv^2=\frac{k}{r^2}$$

$\Rightarrow$ The kinetic energy of particle is

$$\text{KE}=\frac{1}{2}\text{mv}^2=\frac{k}{2r^2}$$

Total energy of particle TE = KE + PE

$$=\frac{k}{2r^2}-\frac{k}{2r^2}=0$$

Ans. (C)

Sol. 35 If particle moves at speed v in circular orbit then

$$\frac{16}{r}+r^3=\frac{mv^2}{r} \Rightarrow mv^2=\left(\frac{16}{r}+r^3\right)\times r=16+r^4$$

Kinetic energy of particle $\frac{1}{2}mv^2=\frac{1}{2}(16+r^4)$

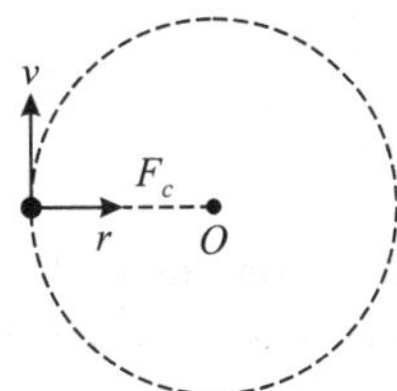

Ratio of kinetic energy of the two particles is given as

$$\frac{k_1}{k_2}=\frac{\frac{1}{2}(16+(1)^4)}{\frac{1}{2}(16+(4)^4)}=\frac{16+1}{16+256}=\frac{17}{272}=0.06$$

Ans. (B)

Sol. 36 Velocity of body is

$$v=a\sqrt{s}$$

$$\Rightarrow \quad \frac{ds}{dt}=a\sqrt{s}$$

$$\Rightarrow \quad \int_0^s\frac{ds}{\sqrt{s}}=\int_0^t a\,dt$$

$$\Rightarrow \quad 2\sqrt{s}=at$$

$$\Rightarrow \quad s=\frac{1}{4}a^2t^2$$

$$\Rightarrow \quad v=\frac{ds}{dt}=\frac{1}{2}a^2t$$

By work-energy theorem we have work done by all forces on body is

$$W=\Delta K=\frac{1}{2}mv^2-0$$

$$\Rightarrow \quad W=\frac{1}{2}m\left(\frac{1}{2}a^2t\right)^2=\frac{1}{8}ma^4t^2$$

Ans. (A)

Alternative:

$$W=F\cdot s \quad \left[\text{As } F=m\left(\frac{1}{2}a^2\right)\rightarrow\text{constant}\right]$$

$$\Rightarrow \quad W=\frac{1}{2}ma^2\left(\frac{1}{4}a^2t^2\right)=\frac{1}{8}ma^4t^2$$

Ans. (A)

Sol. 37 Maximum speed is at mean position (equilibrium) where

$$x=\frac{F}{k}$$

Using work energy theorem, we have

$$F(x)-\frac{1}{2}kx^2=\frac{1}{2}mv^2-0$$

$$\Rightarrow \quad F\left(\frac{F}{k}\right)-\frac{1}{2}k\left(\frac{F}{k}\right)^2=\frac{1}{2}mv^2$$

$$\Rightarrow \quad v_{max}=\frac{F}{\sqrt{mk}}$$

Ans. (C)

Sol. 38 Given that

$$x=3t^2+5$$

$$\Rightarrow \quad v=\frac{dx}{dt}=6t+0$$

at $\quad t=0\ ;\ v=0$

and at $\quad t=5\text{ sec}\ ;\ v=30\text{ m/s}$

By work energy theorem, we use

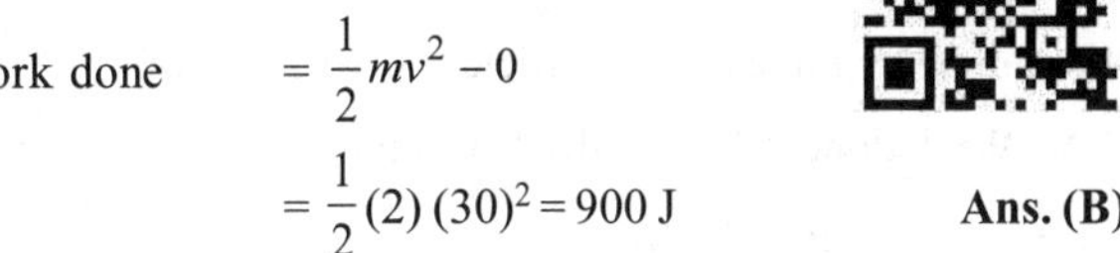

Work done $=\frac{1}{2}mv^2-0$

$$=\frac{1}{2}(2)(30)^2=900\text{ J}$$

Ans. (B)

Sol. 39 For block we use

$$N-mg=\frac{mg}{2}$$

$$\Rightarrow \quad N=\frac{3mg}{2}$$

Work done by normal reaction is

$$W = Ns = \left(\frac{3mg}{2}\right)\left[\frac{1}{2}\left(\frac{g}{2}\right)t^2\right]$$

$$\Rightarrow \quad W = \frac{3mg^2t^2}{8} \quad \textbf{Ans. (B)}$$

Sol. 40 Work done

$$W = \vec{F}\cdot\vec{d} = (3\hat{i} - 12\hat{j})\cdot(4\hat{i})$$

$$\Rightarrow \quad W = 12\text{ J}$$

By work energy theorem, we use

$$K_i + W_{net} = K_f$$

$$\Rightarrow \quad 3 + 12 = K_f$$

$$\Rightarrow \quad K_f = 15\text{ J}$$

Ans. (A)

Sol. 41 Velocity of 1 kg block just before it collides with 3kg block

$$= \sqrt{2gh} = \sqrt{2000}\text{ m/s}$$

Applying momentum conversation just before and just after collision

$$1 \times \sqrt{2000} = 4v$$

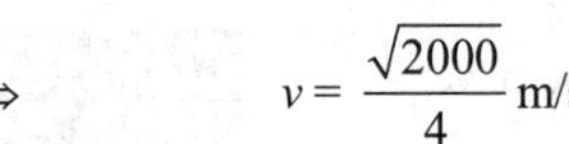

$$\Rightarrow \quad v = \frac{\sqrt{2000}}{4}\text{ m/s}$$

Initial compression of spring

$$1.25 \times 10^6 x_0 = 30$$

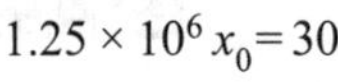

$$\Rightarrow \quad x_0 \simeq 0$$

Applying work energy theorem, we have

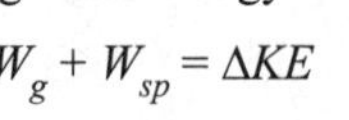

$$W_g + W_{sp} = \Delta KE$$

$$\frac{1}{2}(m_1 + m_2)v^2 + (m_1 + m_2)gx + \frac{1}{2}kx^2 = 0$$

$$\frac{1}{2} \times 4 \times v^2 + 40 \times x - \frac{1}{2} \times 1.25 \times 10^6 x^2 = 0$$

Solving, we get $x \simeq 4$ cm **Ans. (A)**

Sol. 42 According to work energy theorem

Work done by force on the particle = Change in KE

Work done = Area under F-x graph

$$\Rightarrow \quad W = \int F.dx$$

$$\Rightarrow \quad W = 2\times 2 + \frac{(2+3)\times 1}{2}$$

$$\Rightarrow \quad W = KE_{final} - KE_{initial} = 6.5$$

As $KE_{initial} = 0$

$$\Rightarrow \quad KE_{final} = 6.5\text{ J}$$

Ans. (A)

Sol. 43 Mass of the hanging part $= \dfrac{M}{n}$

$$h_{COM} = \frac{L}{2n}$$

Work done by gravity in lifting the cable is

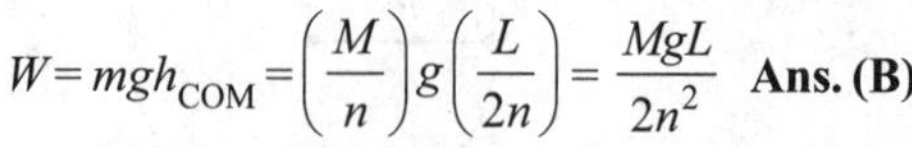

$$W = mgh_{COM} = \left(\frac{M}{n}\right)g\left(\frac{L}{2n}\right) = \frac{MgL}{2n^2} \quad \textbf{Ans. (B)}$$

Sol. 44 Considering gravity, acceleration of ball is

$$\frac{dv}{dt} = -(g + \gamma v^2)$$

$$\Rightarrow \quad \frac{g}{\gamma}\left(\frac{dv}{\frac{g}{\gamma} + v^2}\right) = -gdt$$

Integrating in limits from

$$t = 0 \text{ to } t = t$$

and

$$v = V_0 \text{ to } v = 0$$

$$\int_{V_0}^{0} \frac{g}{\gamma}\left(\frac{dv}{\frac{g}{\gamma} + v^2}\right) = -\int_0^t gdt$$

$$\Rightarrow \quad -gt = -\sqrt{\frac{g}{\gamma}}\tan^{-1}\left(\frac{V_0}{\sqrt{\frac{g}{\gamma}}}\right)$$

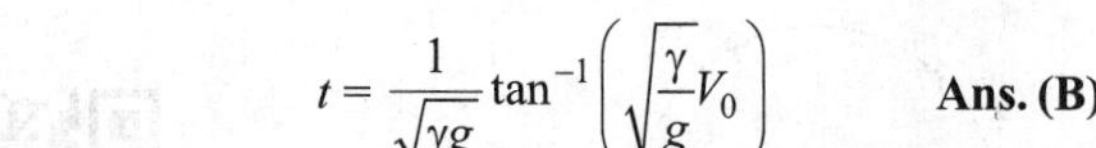

$$\Rightarrow \quad t = \frac{1}{\sqrt{\gamma g}}\tan^{-1}\left(\sqrt{\frac{\gamma}{g}}V_0\right) \quad \textbf{Ans. (B)}$$

Sol. 45 For a spring of length l and force constant k when cut in length l_1 and l_2 then force constants of these parts are given as

$$k_1 = \frac{kl}{l_1} \quad \ldots(1)$$

$$k_2 = \frac{kl}{l_2}$$

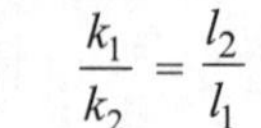

$$\Rightarrow \quad \frac{k_1}{k_2} = \frac{l_2}{l_1}$$

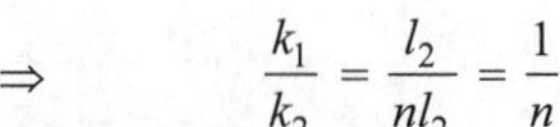

$$\Rightarrow \quad \frac{k_1}{k_2} = \frac{l_2}{nl_2} = \frac{1}{n}$$

Ans (C)

Sol. 46 Upward force by motor on elevator is

$$F = 2000\,g + f_r = 20000 + 4000$$

$$F = 24000\,\text{N}$$

At full load with speed v, power of motor is

$$\Rightarrow \quad 60 \times 746 = (24000)\,v$$

$$\Rightarrow \quad v = \frac{60 \times 746}{24000} = 1.865$$

$$\Rightarrow \quad v \approx 1.9\ \text{m/s}$$

Ans. (C)

Sol. 47 Using energy conservation between point A & P, we have

$$mgh_1 = mgh_2 + KE$$

$$mg \times 2 = mg \times 1 + KE$$

$$\Rightarrow \quad KE = mg \times 1 = 10\ \text{J}$$

Ans. (10)

Sol. 48 Free body diagram of elevator is

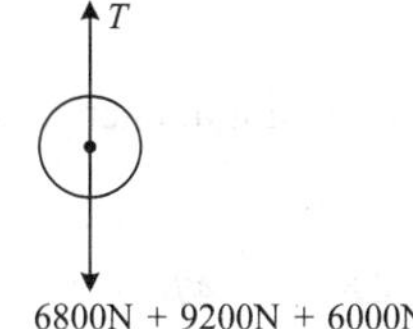

As elevator moving with constant speed, so we use

$$T = 6800 + 9200 + 6000$$

$$\Rightarrow \quad T = 22000\,\text{N}$$

Max power is given as

$$P_{max} = T.v = 22000 \times 3$$

$$\Rightarrow \quad P_{max} = 66000\ \text{W}$$

Ans. (C)

Sol. 49 Work done by force is given as

$$W = \int_{\vec{r}_i}^{\vec{r}_g} \vec{F} \cdot \vec{dr}$$

We use $\vec{F} = -x\hat{i} + y\hat{i}$

and $\vec{dr} = dx\hat{i} + dy\hat{j}$

$$\Rightarrow \quad W = \int_1^0 -x\,dx + \int_0^1 y\,dy$$

$$\Rightarrow \quad W = \left[\frac{-x^2}{2}\right]_1^0 + \left[\frac{y^2}{2}\right]_0^1$$

$$\Rightarrow \quad W = -\left(\frac{0^2}{2} - \frac{1^2}{2}\right) + \left(\frac{1^2}{2} - \frac{0^2}{2}\right)$$

$$\Rightarrow \quad W = 1\ \text{J}$$

Ans. (B)

Sol. 50 Using work-energy theorem, we use

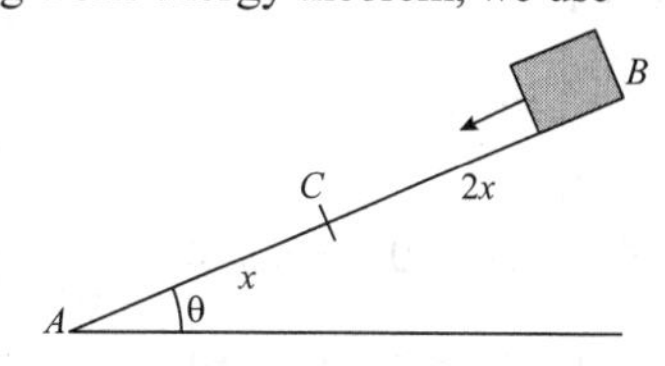

$$mg\sin\theta\,(x + 2x) - \mu mg\cos\theta\, x = 0$$

$$\Rightarrow \quad 3\sin\theta - \mu\cos\theta = 0$$

$$\Rightarrow \quad \mu = 3\tan\theta$$

$$\Rightarrow \quad k = 3$$

Ans. (3)

Sol. 51 By work energy theorem, we use

$$Fl = mgh$$

$$\Rightarrow \quad F \times 0.2 = 0.15 \times 10 \times 20$$

$$\Rightarrow \quad F = \frac{0.15 \times 200}{0.2} = 150\ \text{N}$$

Ans. (150)

Sol. 52 If power is constant, we use

$$K = Pt$$

$$\Rightarrow \quad \frac{1}{2}mv^2 = Pt$$

$$\Rightarrow \quad \frac{ds}{dt} = v = \sqrt{\frac{2Pt}{m}}$$

$$\Rightarrow \quad \int_0^s ds = \int_0^t \sqrt{\frac{2Pt}{m}}\,dt$$

$$\Rightarrow \quad s = \sqrt{\frac{2P}{m}}\left(\frac{2}{3}t^{3/2}\right)$$

$$\Rightarrow \quad s \propto t^{3/2}$$

Ans. (C)

Sol. 53 When block goes up its acceleration is

$$a_1 = -g(\sin\theta + \mu\cos\theta)$$

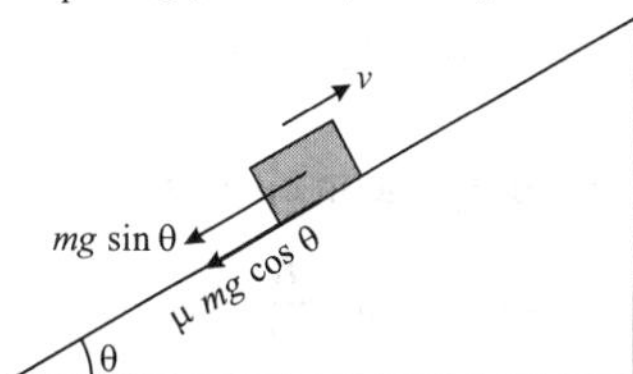

distance it goes up the plane is given as

$$l = \frac{v_0^2}{2a_1}$$

When it comes down its acceleration is given as

$$a_2 = g(\sin\theta - \mu\cos\theta)$$

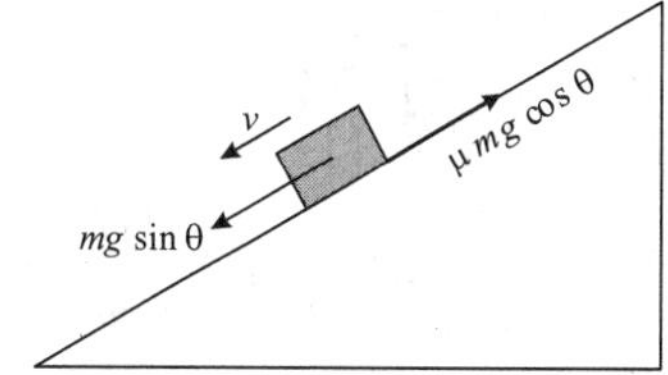

Velocity at bottom point is given as

$$\left(\frac{v_0}{2}\right)^2 = 2a_2 l$$

$$\Rightarrow \quad \frac{v_0^2}{4} = 2g(\sin\theta - \mu\cos\theta)\cdot\frac{v_0^2}{2g(\sin\theta+\mu\cos\theta)}$$

$\Rightarrow$ $\sin\theta + \mu\cos\theta = 4(\sin\theta - \mu\cos\theta)$

At $\theta = 30°$

$$\frac{1}{2}+\frac{\mu\sqrt{3}}{2} = 4\left(\frac{1}{2}-\frac{\mu\sqrt{3}}{2}\right)$$

$$\Rightarrow \quad 1+\mu\sqrt{3} = 4-4\sqrt{3}\mu$$

$$\Rightarrow \quad 5\sqrt{3}\mu = 3$$

$$\Rightarrow \quad \mu = \frac{\sqrt{3}}{5} = 0.3464 = \frac{I}{1000}$$

$$\Rightarrow \quad I = 346$$ **Ans. (346)**

Sol. 54 Given that applied force with position is

$$F = 200\text{ N for } x \le 15\text{ m}$$

$$\left.F = 200 - \frac{100}{15}(x-15)\right\}$$

for $\quad 15 \le x < 30$

$$\Rightarrow \quad F = 300 - \frac{100}{15}x$$

Total work done by force is

$$W = 200\times 15 + \int_{15}^{30}\left(30-\frac{100}{15}x\right)dx$$

$$\Rightarrow \quad W = 300 + 300\times 15 - \frac{100}{15}\cdot\left[\frac{(30)^2}{2}-\frac{(15)^2}{2}\right]$$

$$\Rightarrow \quad W = 7500 - \frac{100}{30}[900-225]$$

$$\Rightarrow \quad W = 7500 - 2250 = 5250\text{ J}$$ **Ans. (B)**

Sol. 55 As power of engine in constant we use

$$P = mva = mv\frac{dv}{dt}$$

$$\Rightarrow \quad \int_0^v v\,dv = \frac{P}{m}\int_0^t dt$$

$$\Rightarrow \quad v = \sqrt{\frac{2Pt}{m}} = \frac{dx}{dt}$$

$$\Rightarrow \quad \int_0^s dx = \sqrt{\frac{2P}{m}}\int_0^t t^{1/2}dt$$

$$\Rightarrow \quad s = \sqrt{\frac{2P}{m}}\cdot\frac{t^{3/2}}{(3/2)} = \frac{2}{3}\sqrt{\frac{2P}{m}}t^{3/2}$$

$$\Rightarrow \quad s = \frac{2}{3}\sqrt{\frac{2\times 1}{2}}\times(9)^{3/2}$$

$$\Rightarrow \quad s = \frac{2}{3}\times 27 = 18\text{ m}$$ **Ans. (18.00)**

Sol. 56 The field force between particles is given as

$$F = \left|\frac{dU}{dr}\right| = -(A(-6r^{-7})) + B(-12r^{-13})$$

$$\Rightarrow \quad F = \frac{6A}{r^7} - \frac{12B}{r^{13}}$$

At equilibrium, we have $F = 0$, so we use

$$0 = \frac{6A}{r^7} - \frac{12B}{r^{13}}$$

$$\Rightarrow \quad \frac{6A}{12B} = \frac{1}{r^6}$$

$$\Rightarrow \quad r = \left(\frac{2B}{A}\right)^{1/6}$$

At equilibrium position, potential energy is given as

$$U_{\left(r=\left(\frac{2B}{A}\right)^{1/6}\right)} = -\frac{A}{2B/A} + \frac{B}{4B^2/A^2}$$

$$\Rightarrow \quad U = \frac{-A^2}{2B} + \frac{A^2}{4B} = \frac{-A^2}{4B}$$ **Ans. (C)**

Sol. 57 For equilibrium of molecules, the net field force acting must be zero, so we use

$$\frac{dU}{dr} = 0$$

$$\Rightarrow \quad \frac{-10\alpha}{r^{11}} + \frac{5\beta}{r^6} = 0$$

$$\Rightarrow \quad \frac{5\beta}{r^6} = \frac{10\alpha}{r^{11}}$$

$$\Rightarrow \quad r = \left(\frac{2\alpha}{\beta}\right)^{\frac{1}{5}}$$

$$\Rightarrow \quad a = 1$$ **Ans. (1)**

Sol. 58 Initial kinetic energy of ball is given as

$$K_i = \frac{1}{2}mu^2 = \frac{1}{2}\times 0.5\times 20\times 20 = 100\text{ J}$$

After deflection it moves with 5% of K_i, thus final kinetic energy of ball is given as

$$K_f = \frac{5}{100}\times K_i$$

$$\Rightarrow \quad K_f = \frac{5}{100}\times 100$$

$\Rightarrow \quad K_f = 5\text{ J}$

Now, let the final speed be v m/s, then

$$K_f = 5 = \frac{1}{2}mv^2$$

$$\Rightarrow \quad v^2 = 20$$

$$\Rightarrow \quad v = \sqrt{20} = 4.47\text{ m/s}$$ **Ans. (B)**

Sol. 59 As C collides elastically, it comes to rest and A moves with speed v. Then spring compression will be maximum when A and B moves with equal speeds i.e. $v/2$. At this instant maximum compression can be given by conservation of energy, calculated as

$$\frac{1}{2}(2m)\left(\frac{v}{2}\right)^2 = \frac{1}{2}kx^2$$

$$\Rightarrow \quad x = \sqrt{\frac{m}{2k}}v$$ **Ans. (A)**

Sol. 60 Centripetal force provided by potential field is given as

$$F = \frac{-dU}{dr} = -4U_0r^3 = \frac{mv^2}{r}$$

$$\Rightarrow \quad mv^2 = 4U_0r^4$$

$$\Rightarrow \quad v \propto r^2$$

By using Bohr's quantization postulate, we have

$$mvr = \frac{nh}{2\pi}$$

$$\Rightarrow \quad r^3 \propto n$$

$$\Rightarrow \quad r \propto n^{1/3}$$

$$\Rightarrow \quad \alpha = 3$$ **Ans. (3)**

Sol. 61 For the machine delivering constant power to a box, the force exerted on the box and its speed are related as

$$Fv = C$$

$$\Rightarrow \quad m\left(\frac{dv}{dt}\right)v = C$$

$$\Rightarrow \quad \frac{v^2}{2} \propto t$$

$$\Rightarrow \quad v \propto t^{1/2}$$

$$\Rightarrow \quad \frac{dx}{dt} \propto t^{1/2}$$

$$\Rightarrow \quad x \propto t^{3/2}$$ **Ans. (B)**

Sol. 62 By work energy theorem from point A and B, we have

$$mgh = \frac{1}{2}mv^2$$

$$\Rightarrow \quad 10 \times 10 \times 5 = \frac{1}{2}(10)v^2$$

$$\Rightarrow \quad v = 10\text{ m/s}$$ **Ans. (10)**

Sol. 63 If the spring gets compressed by a distance y, by work energy theorem we use

$$\frac{1}{2}mv^2 = \frac{1}{2}ky^2$$

$$\Rightarrow \quad y = \sqrt{\frac{m}{k}} \cdot v = \sqrt{\frac{4}{100}} \times 10 = 2\text{m}$$

Thus final length of spring is given as

$$x = 8 - 2 = 6\text{ m}$$ **Ans. (6)**

Sol. 64 Kinetic energy of a moving body is related to its liniear momentum as

$$K = \frac{P^2}{2m}$$

$$\Rightarrow \quad P \propto \sqrt{K}$$

$$\Rightarrow \quad \frac{P_2}{P_1} = \sqrt{\frac{K_2}{K_1}} = \sqrt{\frac{4K}{K}} = 2$$

Percentage change in momentum is given as

$$\frac{P_2 - P_1}{P_1}\% = \left(\frac{P_2}{P_1} - 1\right) \times 100 = (2-1) \times 100 = 100$$

$$\Rightarrow \quad \frac{\Delta P}{P_1}\% = 100\%$$ **Ans. (B)**

Sol. 65 As power supplied is constant to the body, its energy after time t is given as Pt.

$$Pt = \frac{1}{2}mv^2$$

$$v \propto \sqrt{t}$$

$$\Rightarrow \quad \frac{ds}{dt} = C\sqrt{t}$$

Distance covered by the body in t seconds can be calculated as

$$\int_0^L ds = C\int_0^t t^{1/2}dt$$

$$\Rightarrow \quad L = \frac{2Ct^{3/2}}{3}$$

$$\Rightarrow \quad L \propto t^{3/2}$$ **Ans. (C)**

Sol. 66 The work done in lowering the suitcase by the porter can be calculated as

$$W_{\text{person}} = mgh\cos 180° = -784 \times 80 \times 10^{-2}$$

$$= -627.2\text{ J}$$ **Ans. (D)**

Sol. 67 Work done by the force is given as

$$W = \int F \cdot dy$$

$$\Rightarrow \quad W = \int_0^{10}(5y+20)dy = \left[\frac{5y^2}{2}+20y\right]_0^{10}$$

$$\Rightarrow \quad \frac{5\times100}{2}+200 = 450\text{ J}$$ **Ans. (450)**

Sol. 68 Energy supplied by engine in time t is given as

$$E = Pt = \frac{1}{2}mv^2$$

$$v = \sqrt{\frac{2pt}{m}}$$

$$\Rightarrow \quad \frac{ds}{dt} = \sqrt{\frac{2p}{m}}\sqrt{t}$$

$$\Rightarrow \quad \int_0^s ds = \sqrt{\frac{2p}{m}}\int_0^t t^{1/2}dt$$

$$\Rightarrow \quad s = \frac{2\sqrt{\frac{2p}{m}}t^{3/2}}{3}$$

$$\Rightarrow \quad t^{3/2} = \frac{3s}{2\sqrt{\frac{2p}{m}}}$$

$$\Rightarrow \quad s = \left(\frac{8P}{9m}\right)^{1/3} t^{3/2}$$ **Ans. (A)**

Sol. 69 By conservation of energy for the particle we use

$KE + U =$ Total energy = constant

At $x = x_2$ from graph we can see that its potential energy is zero thus its kinetic energy is maximum. **Ans. (C)**

Sol. 70 The initial and final situation of chain are shown in the diagram below

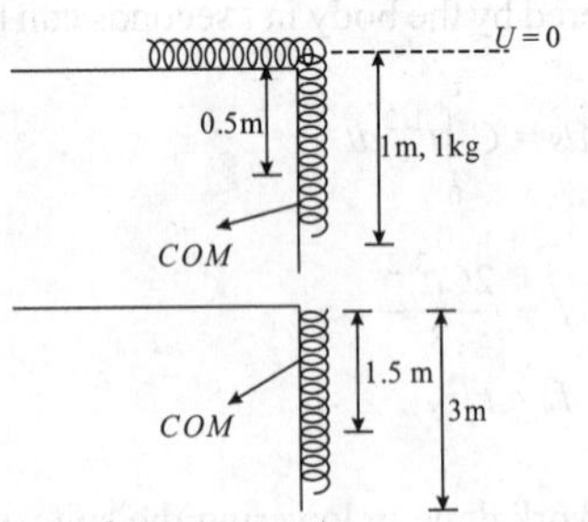

Using conservation of energy, we have

$$U_i + K_i = U_f + K_f$$

$$\Rightarrow \quad -1\times10\times\left(\frac{1}{2}\right)+0 = -3\times10\times1.5+K_f$$

$$\Rightarrow \quad K_f = 45-5$$

$$\Rightarrow \quad K_f = 40\text{ J}$$ **Ans. (40)**

Sol. 71 As work done given is same in the two cases, we use

$$W_1 = W_2$$

$$\Rightarrow \quad F_1 s\cos45° = F_2 s\cos60°$$

$$\Rightarrow \quad \frac{F_1}{F_2} = \frac{1}{\sqrt{2}}$$

$$\Rightarrow \quad x = 2$$ **Ans. (2)**

Sol. 72 Figure below shows the situation described in question for the splitting process

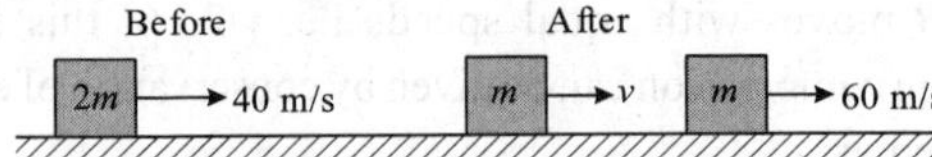

By conservation of linear momentum, we have

$$2m\times40 = 60\times m + mv$$

$$\Rightarrow \quad v = 20\text{ m/s}$$

Change in kinetic energy is given as

$$\Delta K = K_f - K_i = \frac{1}{2}m\times(20)^2+\frac{1}{2}m\times(60)^2-\frac{1}{2}2m\times(40)^2$$

$$\Rightarrow \quad \Delta K = \frac{1}{2}m(800)$$

$$K_i = \frac{1}{2}2m\times(40)^2 = \frac{1}{2}m\times3200$$

Fractional change in kinetic energy is given as

$$f = \frac{\Delta K}{K_i} = \frac{1}{4}$$ **Ans. (1)**

Sol. 73 Figure below describes the situation explained in the question.

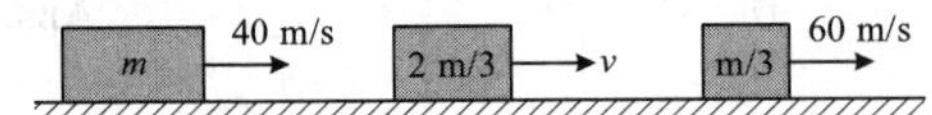

By conservation of linear momentum, we have

$$m\times40 = \frac{2}{3}mv+\frac{m}{3}\times60$$

$$\Rightarrow \quad 120 = 2v+60$$

$$\Rightarrow \quad v = 30\text{ m/s}$$

Now $$K_i = \frac{1}{2}\times m\times40^2 = 800\text{ m}$$

$$K_f = \frac{1}{2}\times\frac{2m}{3}(30)^2+\frac{1}{2}\times\frac{m}{3}\times(60)^2$$

$$\Rightarrow \quad K_T = \frac{1}{6}\{m(1800+3600)\} = \frac{m(5400)}{6} = 900\text{ m}$$

$$\Rightarrow \quad \frac{\Delta KE}{K_i} = \frac{900m-800}{800m} = \frac{1}{8}$$ **Ans. (A)**

Sol. 74 By work energy theorem, we use

$$0 + mgh + W_f = \frac{1}{2}mv^2$$

$$\Rightarrow \quad W_f = -mgh + \frac{1}{2}m(0.8)^2 gh$$

$$\Rightarrow \quad W_f = -mgh + 0.32\,mgh$$

$$\Rightarrow \quad W_f = -mgh(0.68) = -6.8\,mgh$$ **Ans. (A)**

Sol. 75 By work energy theorem, we can write

$$\frac{1}{2}mv^2 + W_{\text{friction}} + W_{\text{spring}} = 0$$

$$\Rightarrow \quad -\frac{90}{100}\left(\frac{1}{2}mv^2\right) + W_{\text{spring}} = -\frac{1}{2}mv^2$$

$$\Rightarrow \quad W_{\text{spring}} = -\frac{10}{100} \times \frac{1}{2}mv^2$$

$$\Rightarrow \quad -\frac{1}{2}kx^2 = -\frac{1}{20}mv^2$$

$$\Rightarrow \quad k = \frac{40000 \times (20)^2}{10 \times (1)^2} = 16 \times 10^5\,\text{N/m}$$ **Ans. (16)**

Sol. 76 Using work energy theorem we use work done equal change in kinetic energy, given as

$$W = \Delta KE$$

Using $F = 4x\hat{i} + 3y^2\hat{j}$, work done is calculated as

$$W = \int \vec{F} \cdot (dx\hat{i} + dy\hat{j})$$

$$\Rightarrow \quad W = \int_1^2 4x\,dx + \int_2^3 3y^2\,dx$$

$$\Rightarrow \quad W = [2x^2]_1^2 + [y^3]_2^3$$

$$\Rightarrow \quad W = (8-2) + (27-8)$$

$$\Rightarrow \quad W = 6 + 19 = 25\,\text{J}$$ **Ans. (C)**

Sol. 77 By conservation of energy, we use

$$\Delta PE = \Delta KE$$

$$\Rightarrow \quad mg\left(H + \frac{H}{2}\right) = \frac{1}{2}kx^2\left(x = \frac{H}{2}\right)$$

$$\Rightarrow \quad 0.100 \times 10 \times \frac{3}{2}(0.10) = \frac{1}{2}k(0.05 \times 0.05)$$

$$\Rightarrow \quad k = \frac{3 \times 0.10}{0.05 \times 0.05}$$

$$\Rightarrow \quad k = \frac{3 \times 1000}{25} = 120\,\text{N/m}$$ **Ans. (120)**

Sol. 78 The force on atoms is given as

$$F = -\frac{dU}{dr} = -\left(\frac{-10A}{r^{11}} + \frac{5B}{r^6}\right) = 0$$

$$\Rightarrow \quad r^5 = \frac{10A}{5B} = \frac{2A}{B}$$ **Ans. (C)**

Sol. 79 Mass per unit length is λ, the normal reaction on the table due to weight of chain resting on table is given as

$$N = mg = \lambda(L-x)g$$

Limiting friction to start chain sliding is given as

$$fs_{\max} = \mu_s N$$

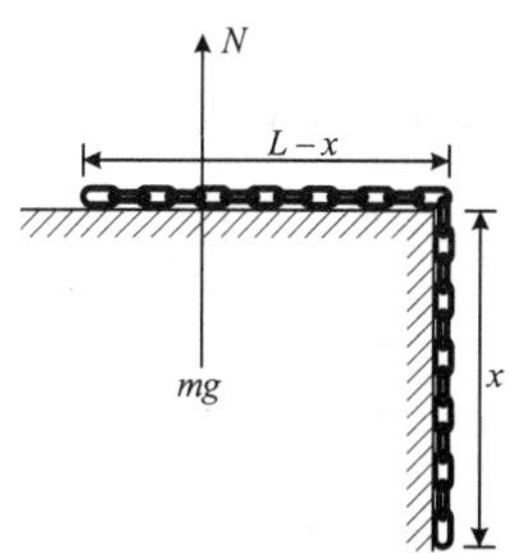

$$\Rightarrow \quad fs_{\max} = (0.5)(\lambda)(L-x)g$$

As chain starts sliding due to the weight of hanging part, we use

$$fs_{\max} = m_x g$$

$$\Rightarrow \quad 0.5\lambda(L-x)g = \lambda xg$$

$$\Rightarrow \quad \frac{L-x}{2} = x$$

$$\Rightarrow \quad \frac{L}{2} = \frac{3x}{2}$$

$$\Rightarrow \quad x = \frac{L}{3} = \frac{6}{3} = 2\,\text{m}$$ **Ans. (2)**

Sol. 80 Using energy conservation from initial to final state, we have

$$U_i + K_i = U_f + K_f$$

$$\Rightarrow \quad 0 + \frac{1}{2}m(12)^2 = \frac{1}{2}K(0.3)^2 + \frac{1}{2}m(6)^2$$

$$\Rightarrow \quad 0.5(12^2 - 6^2) = K(0.3)^2$$

$$K = 600\,\text{N/m}$$ **Ans. (600)**

Sol. 81 At maximum height, speed of object becomes zero hence momentum of object is zero. **Ans. (A)**

Sol. 82 Work done is equal to area under $F - x$ curve. Area below x-axis is taken negative & area above x-axis is taken positive. So for the given options, we have

$$W_3 > W_2 > W_1 > W_4$$ **Ans. (A)**

Sol. 83 When the object rotates, due to centrifugal force spring gets elongated, say by a distance Δx as shown in figure below. For equilibrium of object in rotating frame, we use

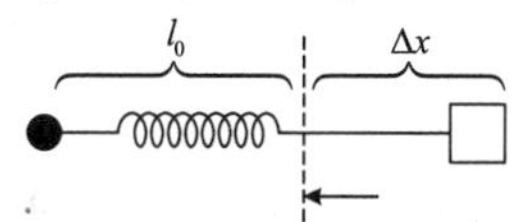

$$K\Delta x = m(l_0 + \Delta x)\omega^2$$

$$\Rightarrow \quad K\Delta x = ml_0\omega^2 + m\omega^2\Delta x$$

$$\Rightarrow \quad \Delta x = \frac{ml_0\omega^2}{k - m\omega^2}$$ **Ans. (C)**

Sol. 84 On rough surface acceleration of block is given as

$$a = \frac{-kx}{2} = \frac{-12x}{2} = -6x$$

$$\Rightarrow \quad \frac{vdv}{dx} = -6x$$

$$\Rightarrow \quad \int_4^v vdv = -\int_{1/2}^{3/2} 6xdx$$

$$\Rightarrow \quad \frac{v^2 - 4^2}{2} = -\frac{6}{2}\left[\left(\frac{3}{2}\right)^2 - \left(\frac{1}{2}\right)^2\right]$$

$$\Rightarrow \quad v^2 - 16 = -6\left(\frac{9}{4} - \frac{1}{4}\right)$$

$$\Rightarrow \quad v^2 = 16 - 6 \times 2 = 4$$

$$\Rightarrow \quad v = 2 \text{ m/s}$$ **Ans. (C)**

Sol. 85 Power produced by loss of gravitational energy of water is calculated as

$$P = \frac{mgh}{t} = \frac{9 \times 10^4 \times g \times 40}{3600} \times 0.5 = n \times 100$$

$$\Rightarrow \quad \frac{mgh}{t} \frac{10^4 \times 0.5}{100} = n$$

$$\Rightarrow \quad \frac{mgh}{t} \quad 100 \times 0.5 = n$$

$$n = 50$$ **Ans. (B)**

Sol. 86 Using work energy theorem gives the change in kinetic energy ΔKE is calculated by the work done as

$$W = \frac{1}{2}m\upsilon_f^2 - \frac{1}{2}m\upsilon_i^2$$

$$\Rightarrow \quad W = \frac{1}{2} \times 0.5 \times (0.25)^2 \times (4)^5$$

$$\Rightarrow \quad W = 16 \text{ J}$$ **Ans. (D)**

Sol. 87 Work done by gravity in this case is calculated as

$$W = K_B - K_A$$

$$\Rightarrow \quad mg\,y_0 = K_B - 0$$

$$\Rightarrow \quad K_B = mg\,y_0$$ **Ans. (D)**

Sol. 88 Body's velocity at the given points is calculated as

$$v_i = 3(0^2) + 4 = 4 \quad \text{at} \quad x = 0$$

and

$$v_F = 3(2)^2 + 4 = 16 \quad \text{at} \quad x = 2$$

Work done is calculated as

$$W = \Delta K = \frac{1}{2}m(16^2 - 4^2)$$

$$\Rightarrow \quad W = \frac{1}{2} \times \frac{1}{2}(256 - 16)$$

$$\Rightarrow \quad W = \frac{240}{4} = 60 \text{ J}$$ **Ans. (B)**

Sol. 89 If at the point of maximum elongation, the spring is stretched by a distance x, half will be travelled by both the blocks as shown in figure.

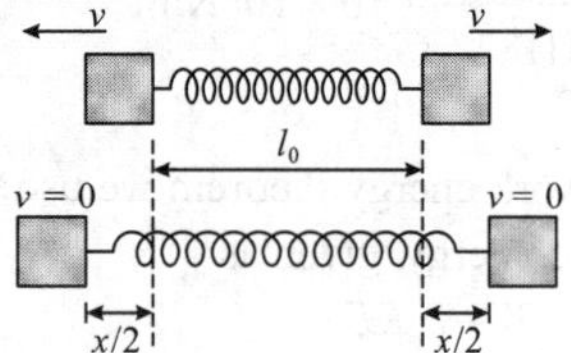

Using energy conservation, we have

$$\frac{1}{2}mv^2 \times 2 = \frac{1}{2}kx^2$$

$$\Rightarrow \quad \frac{1}{4}v^2 = \frac{1}{2} \times 2 \times x^2$$

$$\Rightarrow \quad x = \frac{v}{2}$$ **Ans. (B)**

Sol. 90 Kinetic energy of a particle with momentum p is calculated as

$$K = \frac{p^2}{2m}$$

As $K_1 = K_2$, we use

$$4p_2^2 = p_1^2$$

$$\Rightarrow \quad \frac{p_1}{p_2} = 2$$ **Ans. (B)**

Sol. 91 Force needed to keep the belt moving at uniform velocity is the change in momentum due to addition of mass, given as

$$F = \lambda v = 2.5 \text{ N}$$

Power of the applied force is given as

$$P = Fv = 12.5 \text{ W}$$ **Ans. (D)**

Sol. 92 Using work-energy theorem, we write

$$W_{\text{net}} = (K_f - K_i)$$

$$\Rightarrow \quad -\frac{1}{2}Kx^2 = \frac{1}{2}m\left(\frac{v}{2}\right)^2 - \frac{1}{2}mv^2 = \frac{E}{4} - E$$

$$\Rightarrow \quad \frac{1}{2}Kx^2 = \frac{3E}{4}$$

$$\Rightarrow \quad K = \frac{3E}{2x^2}$$

$$\Rightarrow \quad K = \frac{3E}{2\times\left(\frac{1}{4}\right)^2} = 24E$$

$$\Rightarrow \quad n = 24$$ **Ans. (24)**

Sol. 93 New momentum of the body is given as

$$P' = P + \frac{20}{100}P = 1.2P$$

Percentage increase in kinetic energy is given as

$$\Delta K = \frac{K' - K}{K} \times 100$$

$$\Delta K = \left(\frac{\frac{P'^2}{2m} - \frac{P^2}{2m}}{\frac{P^2}{2m}}\right) \times 100$$

$$\Delta K = [(1.2)^2 - 1] \times 100 = 44\%$$ **Ans. (C)**

Sol. 94 By work energy theorem, the kinetic energy of body is given as

$$\frac{1}{2} \times 2 \times v^2 = 10000$$

$$\Rightarrow \quad v^2 = 10000$$

$$\Rightarrow \quad v = 100 \text{ m/s}$$

If a be the acceleration of the body, we have

$$v = at = a \times 5 = 100$$

$$\Rightarrow \quad a = 20 \text{ m/s}^2$$

The force acting on body is calculated as

$$F = ma = 2 \times 20 = 40 \text{ N}$$ **Ans. (40)**

Sol. 95 Force is given as

$$\vec{F} = t\hat{i} + 3t^2\hat{j}$$

$$\Rightarrow \quad \frac{md\vec{v}}{dt} = t\hat{i} + 3t^2\hat{j}$$

$$\Rightarrow \quad \int_0^{\vec{v}} dv = \int_0^t tdt\hat{i} + \int_0^t 3t^2 dt\hat{j}$$

$$\Rightarrow \quad \vec{v} = \frac{t^2}{2}\hat{i} + t^3\hat{j}$$

Power developed by this force is calculated as

$$P = \vec{F}\cdot\vec{V} = \frac{t^3}{2} + 3t^5$$

At $t = 2$ s, we have

$$P = \frac{8}{2} + 3 \times 32 = 100 \text{ W}$$ **Ans. (100)**

Sol. 96 Forces acting on the block are shown in figure below

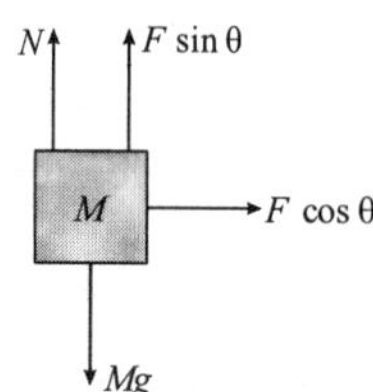

Equation of motion of block is written as

$$F\cos\theta = ma$$

$$\Rightarrow \quad 2\cos(kx) = \frac{mvdv}{dx}$$

$$\Rightarrow \quad m\int_0^v vdv = 2\int_0^x \cos(kx)dx$$

$$\Rightarrow \quad \frac{mv^2}{2} = \frac{2}{k}\sin kx$$

$$\Rightarrow \quad K = \frac{2}{k}\sin\theta$$

$$\Rightarrow \quad n = 2$$ **Ans. (2)**

Sol. 97 Displacement of mass in 8th second of motion is calculated as

$$S_8 = 0 + \frac{1}{2} \times 10 \times (2 \times 8 - 1)$$

$$\Rightarrow \quad S_8 = 5 \times 15 = 75 \text{ m}$$

Loss in potential energy in last second of motion is calculated as

$$\Delta U = 0.4 \times 10 \times 75 = 300 \text{ J}$$ **Ans. (300)**

Sol. 98 Whenever a force is supportive in motion of a body or it acts along the direction of motion of the body then work done by such a force is considered as positive and whenever a force opposes the motion of a body, its work is taken negative. Hence options (b) and (e) are correct. **Ans. (A)**

Sol. 99 Speed of lift after falling through a distance of 6 m is calculated as

$$v^2 = u^2 + 3as$$

$$\Rightarrow \quad v^2 = 2^2 + 2(2)(6)$$

$$\Rightarrow \quad v^2 = 4 + 24 = 28$$

Kinetic energy of the lift after falling through a distance of 6 m is given as

$$K = \frac{1}{2}mv^2$$

$$\Rightarrow \quad K = \frac{1}{2}(500)\,28$$

$$\Rightarrow \quad K = 7000\text{ J}$$

$$\Rightarrow \quad K = 7\text{ kJ}$$

Ans. (7)

Sol.100 Work done under the given force is calculated as

$$W = \vec{F} \cdot (\vec{r}_f - \vec{r}_i)$$

$$\Rightarrow \quad W = (5\hat{i} + 2\hat{j} + 7\hat{k}) \cdot ((5\hat{i} - 2\hat{j} + \hat{k}) - (2\hat{i} + 3\hat{j} - 4\hat{k}))$$

$$\Rightarrow \quad W = 40\text{ J}$$

Ans. (40)

Sol. 101 Given that force F acts on a particle in the y-direction

$$F = 5 + 3y^2$$

Work done by force in displacement from $y = 2$ m to $y = 5$ m is calculated as

$$W = \int_2^5 (5 + 3y^2)\,dy$$

$$\Rightarrow \quad W = \left[5y + \frac{3y^3}{3}\right]_2^5$$

$$\Rightarrow \quad W = 132\text{ J}$$

Ans. (132)

Sol. 102 Figure below shows the situation described in question.

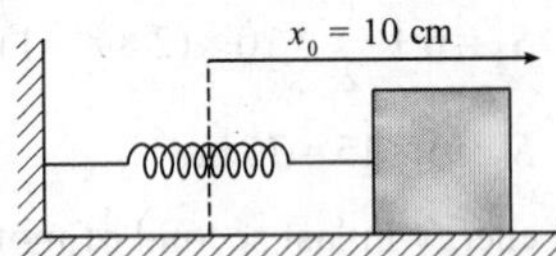

Initial energy when block is pulled to a distance $x = 10$ cm from equilibrium position

$$U_i = \frac{1}{2}kx_0^2$$

and $\quad K_i = 0$

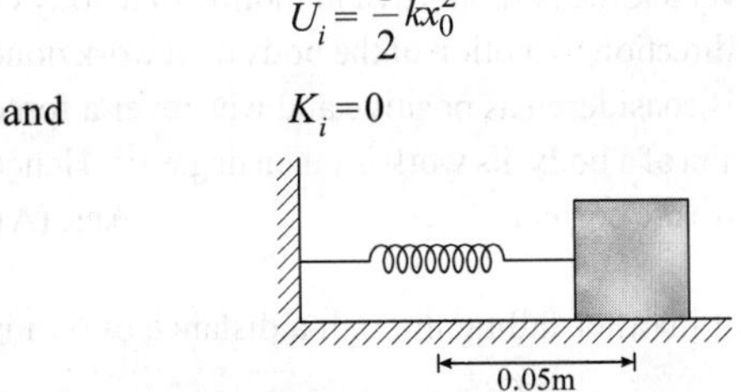

If block is released from $x = 10$ m to $x = 5$ m then energies are calculated as

$$U_f = \frac{1}{2}k\left(\frac{x_0}{2}\right)^2$$

and $\quad K_f = 0.25$ J

By energy conservation, we use

$$\frac{1}{2}kx_0^2 + 0 = \frac{1}{2}k\frac{x_0^2}{4} + 0.25$$

$$\Rightarrow \quad \frac{1}{2}kx_0^2\frac{3}{4} = \frac{1}{4}$$

$$\Rightarrow \quad \frac{1}{2}k\frac{3}{100} = 1$$

$$\Rightarrow \quad k = \frac{200}{3}\text{ N/m}$$

$$\Rightarrow \quad k = 67\text{ N/m}$$

Ans. (67)

Sol. 103 Loss of Kinetic Energy is equal to the work done against retarding force. Retardation of particle is given as $a = 2x$, so the work done is calculated as

$$W = \int_0^x ma\,dx = \int_0^x m2x\,dx = mx^2$$

$$\Rightarrow \quad W = (10^{-2}\text{ kg})\,x^2\text{ J} = \left(\frac{10}{x}\right)^{-2}\text{ J}$$

$$\Rightarrow \quad n = 2$$

Ans. (2)

Sol. 104 Maximum power used by the motor is

$$P_{max} = F_{max}v$$

Maximum force is calculated as

$$F_{max} = 1400\,g + \text{friction}$$

$$\Rightarrow \quad F_{max} = 14000 + 2000 = 16000\text{ N}$$

Thus maximum power is given as

$$P_{max} = 16000 \times 3 = 48000\text{ W}$$

$$\Rightarrow \quad P_{max} = 48\text{ KW}$$

Ans. (48)

Sol. 105 Work done by force during displacement from $x = 0$ to $x = 4$ m.

$$W = \int_0^4 (2 + 3x)\,dx$$

$$\Rightarrow \quad W = \left[2x + \frac{3x^2}{2}\right]_0^4$$

$$\Rightarrow \quad W = 8 + 3 \times 8 = 32\text{ J}$$

Ans. (32)

Sol. 106 Figure below shows the free body diagram of the block as described in the question.

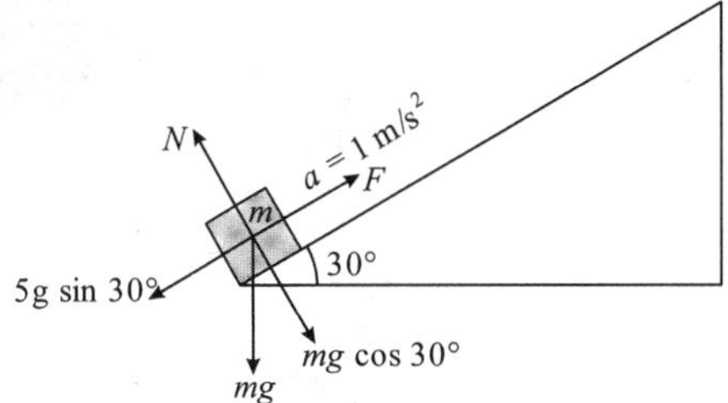

Equation of motion of block along the incline is written as

$$F - 5\,g\sin 30° = 5a$$

$$\Rightarrow \quad F = 5 + 25 = 30\text{ N}$$

By first equation of uniformly accelerated motion, we have

$$V_{10} = u + \text{at}$$

$$\Rightarrow \quad V_{10} = 0 + 1(10) = 10\text{ m/s}$$

Thus power delivered by pulling source at $t = 10$ s from start is given as

$$P = Fv = 300\text{ W}$$ **Ans. (300)**

Sol. 107 Work done is the change in kinetic energy so as both car and truck are brought to rest, we have

$$v = 0$$

and $$K_f = 0$$

Work done is calculated as

$$W = -Fs = 0 - K$$

$$\Rightarrow \quad s = \frac{K}{F}$$

Thus distance for both will be same as K is same for both hence statement-I is correct and statement-II is incorrect because direction of speed is changing hance acceleration is non-zero.

Ans. (A)

Sol. 108 For constant speed, workdone by engine + workdone by friction = 0 so we have

$$W_{\text{engine}} = -W_{\text{friction}} = -[-\mu mgx]$$

$$\Rightarrow \quad W_{\text{engine}} = 0.04 \times 500 \times 9.8 \times 4 \times 10^3$$

$$\Rightarrow \quad W_{\text{engine}} = 784\text{ KJ}$$ **Ans. (784)**

Sol. 109 Average Power is calculated as

$$P = \frac{\text{total work done}}{\text{total time}}$$

$$\Rightarrow \quad P = \frac{mgh}{t}$$

Ratio of power of two motors is calculated as

$$\Rightarrow \quad \frac{P_1}{P_2} = \frac{\dfrac{m_1 gh}{t_1}}{\dfrac{m_2 gh}{t_2}} = \frac{m_1}{t_1}\frac{t_2}{m_2}$$

$$\Rightarrow \quad \frac{P_1}{P_2} = \frac{300 \times 2}{5 \times 50} = \frac{12}{5} = \frac{3\sqrt{x}}{\sqrt{x}+1}$$

$$\Rightarrow \quad 12\sqrt{x} + 12 = 15\sqrt{x}$$

$$\Rightarrow \quad 3\sqrt{x} = 12$$

$$\Rightarrow \quad x = 16$$ **Ans. (D)**

Sol. 110 Kinetic energy of object i sgiven as

$$k = \frac{1}{2}mv^2$$

$$\Rightarrow \quad k = \frac{1}{2} \times 5 \times 400 = 5 \times 200 = 1000\text{ J}$$

Error in measurement of kinetic energy is calculated as

$$\frac{\Delta k}{2k} = \frac{\Delta m}{m} + \frac{2\Delta v}{v} = \frac{0.5}{4} + \frac{2 \times 0.4}{20}$$

$$\Rightarrow \quad \Delta k = 1000\left(\frac{1}{10} + \frac{4}{100}\right) = 1000\left(\frac{10+4}{100}\right) = 140\text{ J}$$

Ans. (A)

Sol. 111 By the momentum kinetic energy relation is written as

$$p^2 = 2mK$$

$$\Rightarrow \quad K = \frac{P^2}{2m}$$

$$\Rightarrow \quad \frac{K_1}{K_2} = \frac{p_1^2}{2m_1} \times \frac{2m_2}{p_2^2} = \frac{m_2}{m_1} = \frac{16}{9}$$

$$\Rightarrow \quad \frac{m_1}{m_2} = \frac{9}{16}$$ **Ans. (D)**

Sol. 112 Initial kinetic energy of car is given as

$$E_1 = \frac{1}{2}\,mu^2 - 0 = \frac{1}{2}mu^2 = E$$

Final kinetic energy when speed changes from u to $2u$ is calculated as

$$E_2 = \frac{1}{2}\text{m}\,(2u)^2 - \frac{1}{2}mu^2 = \frac{3}{2}\,mu^2$$

$$= 3E$$ **Ans. (3)**

Sol. 113 Work done by the force in moving block from $x = 2$ m to $x = 4$ m is calculated as

$$\text{Work done} = \int F dx$$

$$W = \int_2^4 5x dx = 5\left[\frac{x^2}{2}\right]_2^4$$

$$\Rightarrow \quad W = \frac{5}{2}[16-4]$$

$$\Rightarrow \quad W = 30\text{ J}$$

Ans. (30)

5.2 Circular Motion

Sol. 1 Friction provides the necessary centripetal force for circular motion so we use

$$f = \frac{mv^2}{R}$$

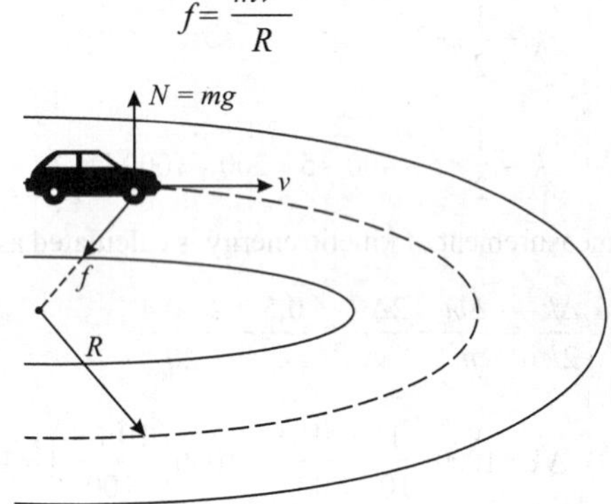

For skidding $\quad f = \dfrac{mv^2}{R} \geq \mu mg$

to avoid skidding

$$f = \frac{mv^2}{R} \leq \mu mg$$

$$\Rightarrow \quad v_{max} = \sqrt{\mu Rg} = \sqrt{0.6 \times 150 \times 10}$$

$$\Rightarrow \quad v_{max} = \sqrt{900} = 30\text{ m/s}$$

Ans. (B)

Sol. 2

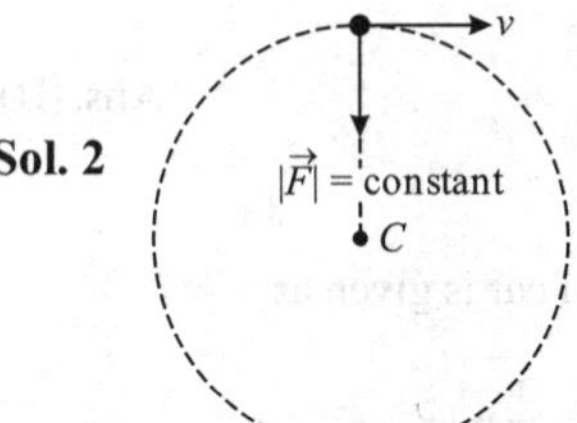

$$|\vec{F}| = \frac{mv^2}{R} \text{ (centripetal force)} = \text{constant}$$

$$\Rightarrow \quad |v| \to \text{constant}$$

$$\Rightarrow \quad \text{KE} \to \text{constant}$$

Ans. (C)

Sol. 3

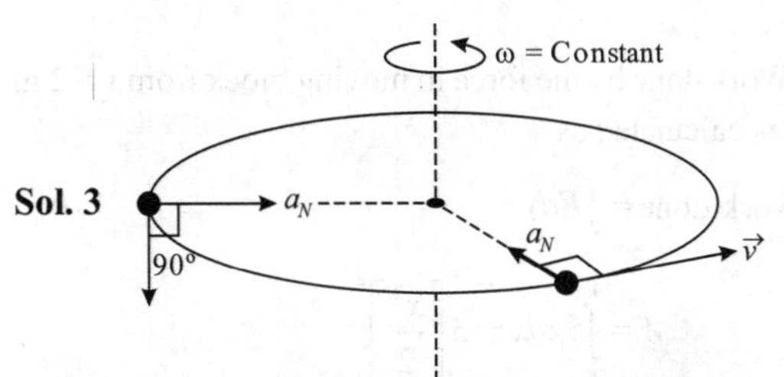

$$v = R\omega \text{ constant}$$

$\Rightarrow$ tangential acceleration = 0

Here normal acceleration $a_N = \dfrac{v^2}{R} = \omega^2 R$

(constant in magnitude) Option (B) is false here. **Ans. (B)**

Sol. 4

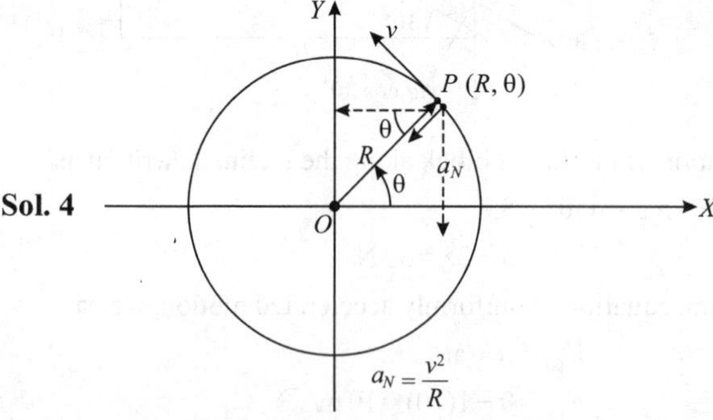

As in uniform circular motion $a_{tangential} = 0$ and only centripetal acceleration is there so its vector is given as

$$\vec{a} = -a_N \cos\theta\,\hat{i} - a_N \sin\theta\,\hat{j}$$

$$\vec{a} = -\frac{v^2}{R}\cos\theta\,\hat{i} - \frac{v^2}{R}\sin\theta\,\hat{j}$$

Ans. (C)

Sol. 5

Tangential acceleration of point P is given as

$$a_t = \frac{dv}{dt} = \frac{d^2v}{dt^2} = 6t$$

at $t = 2$s, $a_t = 6 \times 2 = 12\text{ m/s}^2$

$$s = t^3 + 5$$

$$\Rightarrow \quad v = \frac{ds}{dt} = 3t^2$$

at $\quad t = 2$s

$$v = 3 \times 4$$

$$\Rightarrow \quad v = 12\text{ m/s}$$

Normal acceleration of point P is given as

at $\quad t = 2\text{s}, \; a_N = \dfrac{v^2}{R} = \dfrac{144}{20} = 7.2\text{ m/s}^2$

Total acceleration of point P at $t = 2$s is

$$a_T = \sqrt{a_t^2 + a_N^2}$$

$$\Rightarrow \quad a_T = \sqrt{(12)^2 + (7.2)^2}$$

$$\Rightarrow \quad a_T = 13.99\text{ m/s}^2$$

Ans. (D)

Sol. 6 As cars complete the circles in same time $\Rightarrow$ Their angular speeds are equal

Centripetal acceleration of cars

$$a_{N_1} = \omega^2 r_1$$

$$a_{N_2} = \omega^2 r_2$$

$$\Rightarrow \quad \frac{a_{N_1}}{a_{N_2}} = \frac{r_1}{r_2}$$

Ans. (C)

Sol. 7

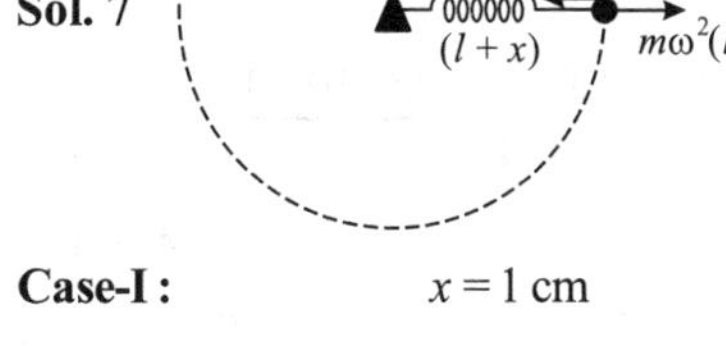

Case-I : $x = 1$ cm

If original length of spring is l, we use

For circular motion of body we have along radial direction

$$x = 1 \text{ cm}$$

$$k(1) = m\omega^2 (l+1) \quad \ldots(1)$$

Case-II : $x = 5$ cm

$$\omega \to 2\omega$$

$$k(5) = m(2\omega)^2 (l+5) \quad \ldots(2)$$

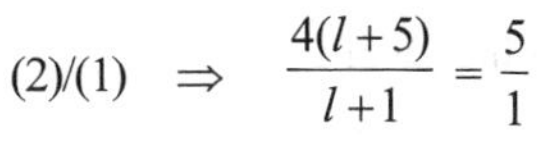

$$(2)/(1) \quad \Rightarrow \quad \frac{4(l+5)}{l+1} = \frac{5}{1}$$

$$\Rightarrow \quad 4l + 20 = 5l + 5$$

$$\Rightarrow \quad l = 15 \text{ cm}$$

Ans. (A)

Sol. 8

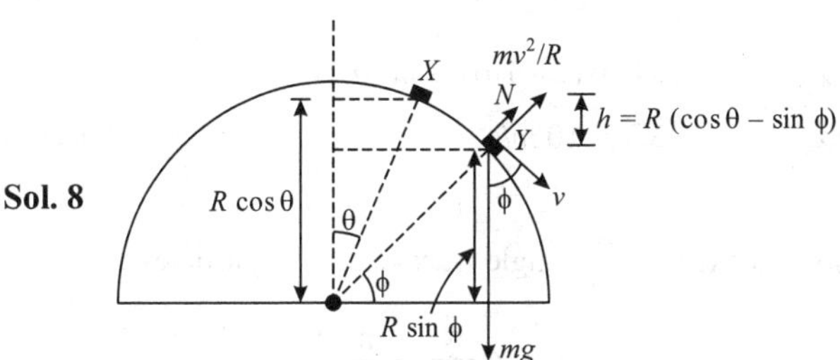

By work energy theorem, we use

$$mgh = \frac{1}{2} mv^2$$

$$\Rightarrow \quad v = \sqrt{2gR(\cos\theta - \sin\phi)}$$

At point Y we have along radial direction in circular motion (in frame of particle)

$$N + \frac{mv^2}{R} = mg\sin\phi$$

As $N = 0$

$$\Rightarrow \quad 2\,mg(\cos\theta - \sin\phi) = mg\sin\phi$$

$$\Rightarrow \quad 2\cos\theta = 3\sin\phi$$

Ans. (C)

Sol. 9

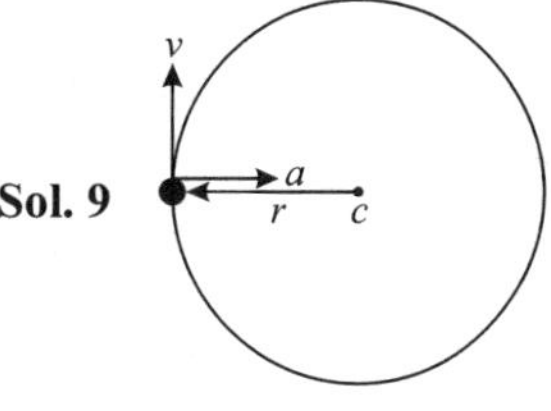

For uniform circular motion

$$a_{\tan} = 0$$

and $$a_N = \frac{v^2}{r} \text{ (Centripetal acceleration)}$$

$$\Rightarrow \quad a_{\text{total}} = a_N = \frac{v^2}{r}$$

$$a \propto \frac{1}{r}$$

Ans. (C)

Sol. 10

The centripetal force $F = \alpha r^2$

The PE of particle in the field of this force is given as

$$U = -\int_0^r \vec{F}\cdot\vec{dr} = \int_0^r F\,dr = \int_0^r \alpha r^2\,dr = \frac{\alpha r^3}{3} - 0$$

We also use

$$F = \frac{mv^2}{r} = \alpha r^2 \quad \Rightarrow \quad KE = \frac{1}{2}mv^2 = \frac{1}{2}\alpha r^3$$

Total mechanical energy $= \frac{1}{2}\alpha r^3 + \frac{1}{3}\alpha r^3 = \frac{5}{6}\alpha r^3$ **Ans. (B)**

Sol. 11

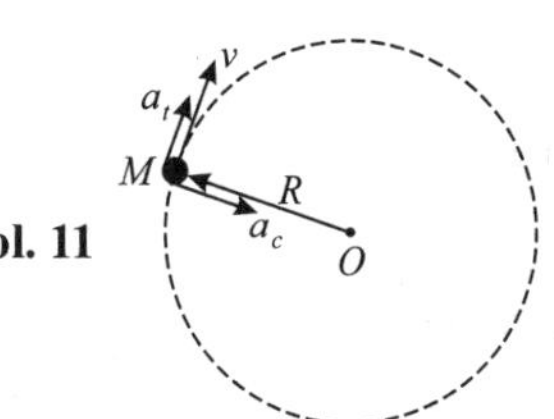

Centripetal acceleration of particle is given as

$$a_c = n^2 R t^2 = \frac{v^2}{R}$$

$$\Rightarrow \quad v^2 = n^2 R^2 t^2$$

$$\Rightarrow \quad v^2 = nRt \quad \Rightarrow \quad a_t = \frac{dv}{dt} = nR$$

As v increases, tangential acceleration is in the direction of velocity

Power delivered to the particle is given as

$$P = ma_t v = m(nR)(nRt)$$

$$\Rightarrow \quad P = mn^2R^2t$$

Ans. (B)

Sol. 12

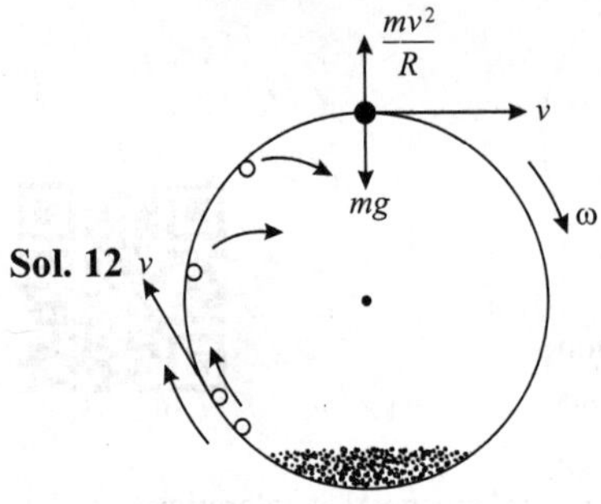

In the frame of rotating drum at the topmost point particle (mixture ingredients) will not loose contact from drum wall if

$$\frac{mv^2}{R} \geq mg$$

$$\Rightarrow \quad v \geq \sqrt{Rg}$$

If $v \geq \sqrt{Rg}$ than proper mixing will not take place

Thus for proper mixing

$$v < \sqrt{Rg}$$

$$\Rightarrow \quad \omega = \frac{v}{R} = \sqrt{\frac{g}{R}} = \sqrt{\frac{10}{1.25}} \text{ rad/s}$$

ω in rpm is

$$\omega_{RPM} = \frac{60}{2\pi}\sqrt{\frac{10}{1.25}} \gtrsim 27 \text{ rpm}$$

$\Rightarrow$ If $\omega \simeq 27$ rpm then before reaching top, ingredients fall and mix properly **Ans. (A)**

Sol. 13

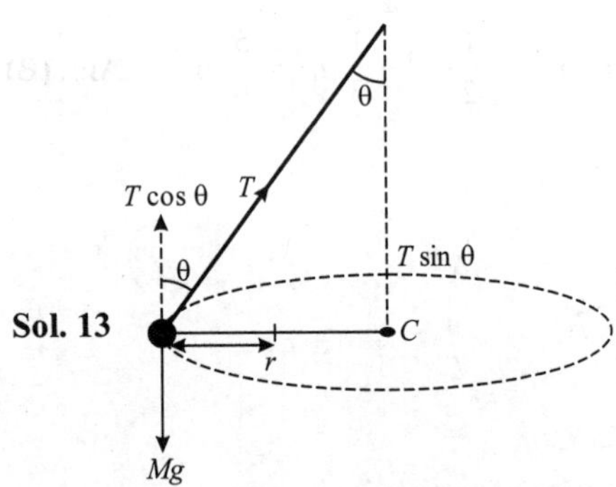

For stable circular motion, we use

$$T\sin\theta = \frac{mv^2}{r} \quad \ldots(1)$$

$$T\cos\theta = mg \quad \ldots(2)$$

$$(1)/(2) \Rightarrow \quad \tan\theta = \frac{v^2}{rg}$$

$$\Rightarrow \quad v = \sqrt{rg\tan\theta}$$

$$\Rightarrow \quad v = \sqrt{0.4 \times 10 \times 1}$$

$$= 2 \text{ m/s}$$ **Ans. (D)**

Sol. 14

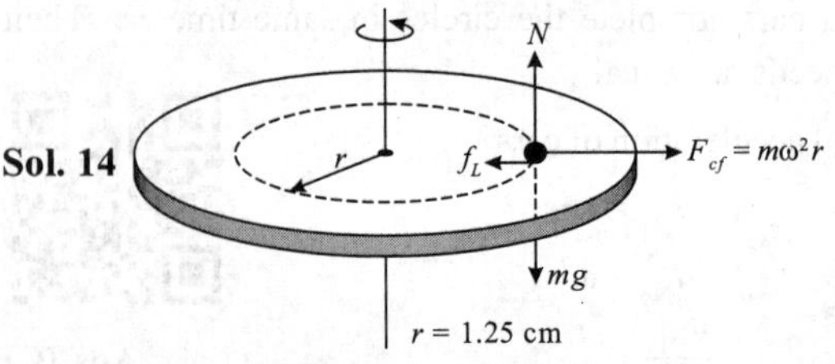

Angular speed $\omega = 2\pi n = 2\pi \times 3.5 = 7\pi$ rad/s

in frame of disc $\quad F_{cf} = m\omega^2 r$

If coin is at rest on disc

$$\Rightarrow \quad m\omega^2 r \leq f_L$$

$$\Rightarrow \quad m\omega^2 r \leq \mu g m$$

$$\Rightarrow \quad \mu \geq \frac{\omega^2 r}{g} = \frac{(7\times 3.14)^2 \times (1.25\times 10^{-2})}{10}$$

$$\Rightarrow \quad \mu = 0.6$$ **Ans. (D)**

Sol. 15

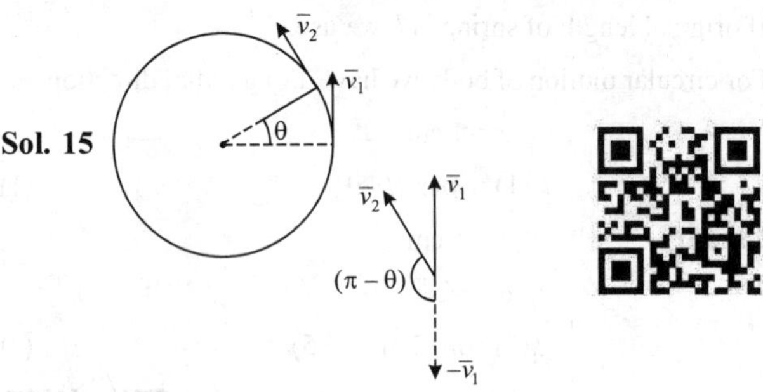

For above circular motion, change in velocity is

$$|\Delta\vec{v}| = \sqrt{v_1^2 + v_2^2 + 2v_1v_2\cos(\pi-\theta)}$$

$$\Rightarrow \quad |\Delta\vec{v}| = 2v\sin\frac{\theta}{2} \qquad \text{since } [|\vec{v}_1| = |\vec{v}_2|]$$

$$\Rightarrow \quad |\Delta\vec{v}| = (2\times 10)\times\sin(30°)$$

$$\Rightarrow \quad |\Delta\vec{v}| = 10 \text{ m/s}$$ **Ans. (B)**

Sol. 16 At $t = \frac{\pi}{2\omega}$ angle transversed by particles is

$$\theta = \omega t = \omega\frac{\pi}{2\omega} = \frac{\pi}{2}$$

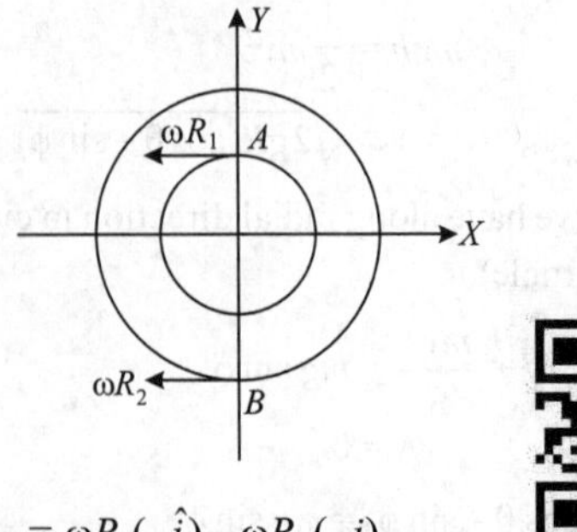

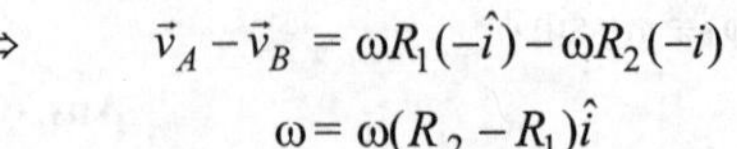

$$\Rightarrow \quad \vec{v}_A - \vec{v}_B = \omega R_1(-\hat{i}) - \omega R_2(-i)$$

$$\omega = \omega(R_2 - R_1)\hat{i}$$ **Ans. (D)**

Sol. 17 FBD of bead is shown in figure below

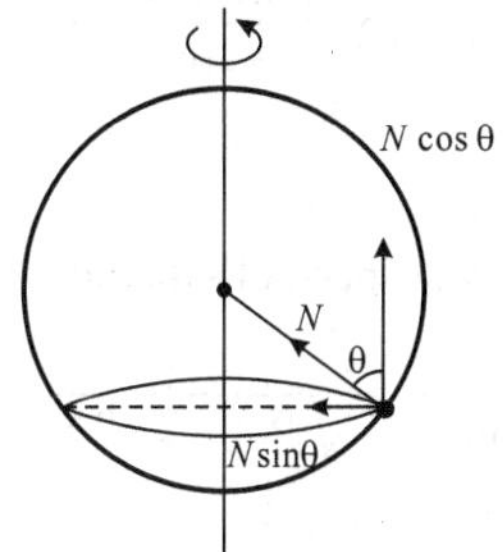

For equilibrium of bead in frame of wire, we have

$$N \sin \theta = m\frac{r}{2}\omega^2 \quad \dots(1)$$

$$N \cos \theta = mg \quad \dots(2)$$

$$\Rightarrow \quad \tan \theta = \frac{r\omega^2}{2g}$$

$$\Rightarrow \quad \frac{r}{2\frac{\sqrt{3}r}{2}} = \frac{r\omega^2}{2g}$$

$$\Rightarrow \quad \omega^2 = \frac{2g}{\sqrt{3}r}$$

Ans. (D)

Sol. 18 FBD of particle is shown in figure

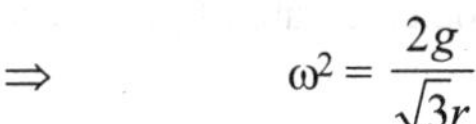

For uniform circular motion, we use

$$kx = ml\omega^2 (l + x)$$

$$\Rightarrow \quad x = \frac{ml\omega^2}{k - m\omega^2}$$

Ans. (B)

Sol. 19 Free body diagram of mass in frame of disc is shown in figure

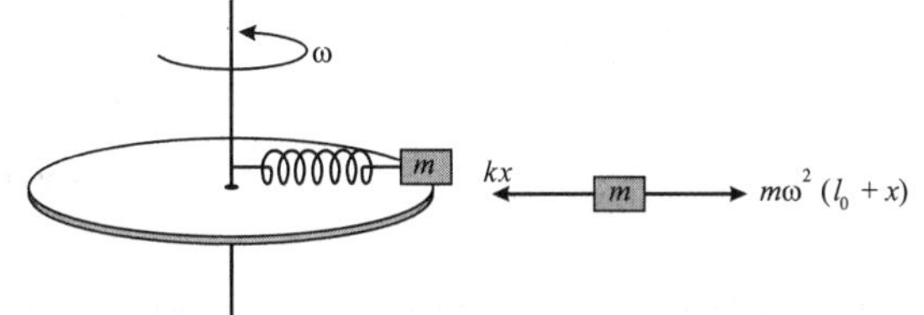

For equilibrium of mass m disc frame, we use

$$kx = m\omega^2(l_0 + x)$$

$$x = \frac{m\omega^2 l_0}{k - m\omega^2} \approx \frac{m\omega^2 l_0}{k}$$

Relative change in spring length is

$$\frac{x}{l_0} = \frac{m\omega^2}{k}$$

Ans. (D)

Sol. 20 Figure below shows the forces acting on bead

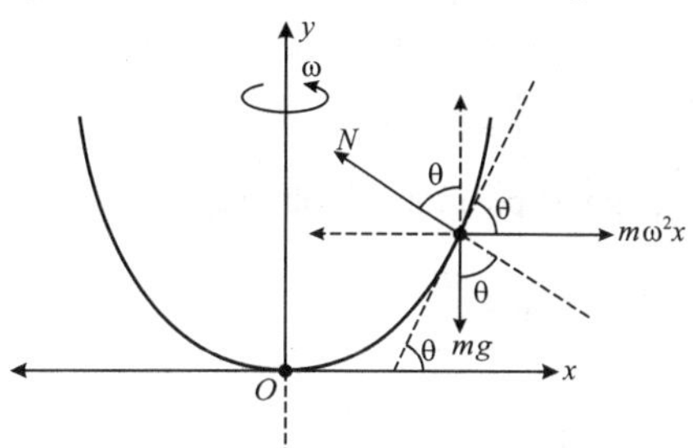

For equilibrium of bead in frame of wire, we use

$$m\omega^2 x \cos \theta = mg \sin \theta$$

$$\Rightarrow \quad \omega^2 x = g \tan \theta$$

Slope of parabola at point P is $\frac{dy}{dx} = \tan \theta$

$$\Rightarrow \quad \frac{dy}{dx} = \frac{\omega^2 x}{g} = 8Cx \qquad [\text{As } y = 4Cx^2]$$

$$\Rightarrow \quad \omega^2 = 8Cg$$

$$\Rightarrow \quad \omega = \sqrt{8Cg}$$

Ans. (B)

Sol. 21 As given is theory explanation shape of curved surface is given as

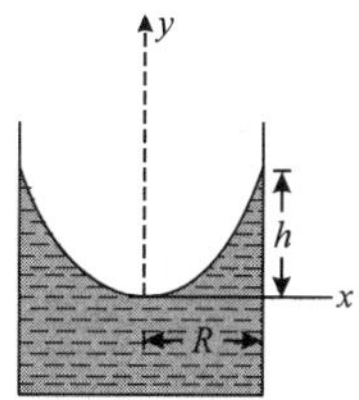

$$y = \frac{\omega^2 x^2}{2g}$$

$\Rightarrow$ at $x = R$, we have

$$h = \frac{\omega^2 R^2}{2g} = \frac{\omega^2 (5)^2}{2g} = \frac{25\omega^2}{2g}$$

Ans. (A)

Sol. 22 Angular speed of clock's second's hand is given as

$$\omega = \frac{2\pi}{T} = \frac{2\pi}{60} = 0.105 \text{ rad/sec}$$

Acceleration of tip of hand is

$$a = \omega^2 R$$

$$\Rightarrow \quad a = (0.105)^2 (0.1) \text{ m/s}^2$$

$$\Rightarrow \quad a = 0.0011 \text{ m/s}^2$$

$$\Rightarrow \quad a = 1.1 \times 10^{-3} \text{ m/s}^2$$

Ans. (A)

Sol. 23 In the vertical circular motion if the speed of bob at bottom most position be v_1 and at the topmost position be v_2 and during motion, we know that maximum tension occurs at bottom most position and minimum at the topmost position. By conservation of energy for bob, we have

$$\frac{1}{2}mv_1^2 = \frac{1}{2}mv_2^2 + mg(2l)$$

$$\Rightarrow \quad v_1^2 = v_2^2 + 4gl \quad \ldots(1)$$

At the bottommost position of bob, we have

$$T_{max} - mg = \frac{mv_1^2}{l}$$

$$\Rightarrow \quad T_{max} = mg + \frac{mv_1^2}{l} \quad \ldots(2)$$

At the topmost position of bob, we have

$$T_{min} + mg = \frac{mv_2^2}{l}$$

$$\Rightarrow \quad T_{min} = \frac{mv_2^2}{l} - mg \quad \ldots(3)$$

From equations-(2) and (3), we have

$$\frac{T_{max}}{T_{min}} = \frac{5}{1}$$

$$\Rightarrow \quad \frac{mg + \frac{mv_1^2}{l}}{\frac{mv_2^2}{l} - mg} = \frac{5}{1}$$

$$\Rightarrow \quad mg + \frac{m}{l}[v_2^2 + 4gl] = \frac{5mv_2^2}{l} - 5mg$$

$$\Rightarrow \quad 10\,mg = \frac{4mv_2^2}{l}$$

$$\Rightarrow \quad v_2^2 = \frac{10\times10\times1}{4} = 25$$

$$\Rightarrow \quad v_2 = 5 \text{ m/s}$$

Ans. (5)

Sol. 24 Given central force is expressed as

$$F \propto \frac{1}{R^3}$$

$$\Rightarrow \quad \frac{K}{R^3} = m\omega^2 R$$

$$\Rightarrow \quad \omega^2 = \frac{K}{m}\times\frac{1}{R^4}$$

If T is the time period of revolution, we use

$$\left(\frac{2\pi}{T}\right)^2 = \frac{K}{m}\times\frac{1}{R^4}$$

$$\Rightarrow \quad T^2 \propto R^4$$

$$\Rightarrow \quad T \propto R^2$$

Ans. (A)

Sol. 25 Normal force in this case provides the necessary centripetal force for uniform circular motion to block, so we use

$$N = m\omega^2 R$$

$$\Rightarrow \quad N = m\left[\frac{4\pi^2}{T^2}\right]R$$

$$\Rightarrow \quad N = 0.2\times\left[\frac{4(3.14)^2}{(40)^2}\right]\times 0.2 = 9.859\times10^{-4}\text{N}$$

Ans. (D)

Sol. 26 The maximum speed of safe turning without skidding is given as

$$v_{max} = \sqrt{\mu Rg} = \sqrt{(0.2)\times2\times9.8}$$

$$\Rightarrow \quad v_{max} = 1.97 \text{ m/s} = 7.1 \text{ km/hr}$$

The given peed is lower than v_{max} so it can take safe turn. For safe turning speed on banked road, we use

$$v_{max} = \sqrt{Rg\left[\frac{\tan\theta+\mu}{1-\mu\tan\theta}\right]}$$

$$\Rightarrow \quad v_{max} = \sqrt{2\times9.8\left[\frac{1+0.2}{1-0.2}\right]} = 5.42 \text{ m/s} = 19.5 \text{ km/hr}$$

The given speed is lower than v_{max} so it can take safe turn.

Ans. (D)

Sol. 27 The angle of revolution can be calculated by average angular speed and can be given as

$$\theta = \left(\frac{\omega_1+\omega_2}{2}\right)t$$

If total number of revolutions are N, we use

$$2\pi N = 2\pi\left(\frac{900+2460}{60\times2}\right)\times 26$$

$$\Rightarrow \quad N = 728$$

Ans. (728)

Sol. 28 In the given potential field, force on particle is calculated as

$$F = -\frac{dU}{dr} = -\frac{C}{r^2}$$

For circular motion of particle under this force, we have

$$|F| = \frac{mv^2}{r}$$

$$\Rightarrow \quad \frac{C}{r^2} = \frac{mv^2}{r}$$

$$\Rightarrow \quad v^2 \propto \frac{1}{r}$$

Ans. (D)

Sol. 29 Initial and final angular speeds in the process are given as

$$\omega_0 = 600 \text{ rpm} = 10 \text{ rps}$$

$$\omega = 1800 \text{ rpm} = 30 \text{ rps}$$

Angular acceleration of the body can be calculated as

$$\alpha = \frac{30-10}{10} = 2\text{rps}^2$$

Number of revolutions body turned in the given time duration can be calculated as

$$\theta = \omega_0 t + \frac{1}{2}\alpha t^2 = 10 \times 10 + \frac{1}{2} \times 2 \times (100) = 200 \text{ rev}$$

Ans. (200)

Sol. 30 By conservation of energy, we have

$$\frac{1}{2}mu^2 = \frac{1}{2}mv^2 + mgl(1 - \cos 60°)$$

$\Rightarrow$ $u^2 = v^2 + 2gl(1 - \cos 60°)$

$\Rightarrow$ $9 = v^2 + 20 \times 1/2 \times 1/2$

$\Rightarrow$ $9 = v^2 + 5$

$\Rightarrow$ $v = 2$ m/s **Ans. (2)**

Sol. 31 Situation shown in figure below describes the state of block when it leaves the surface

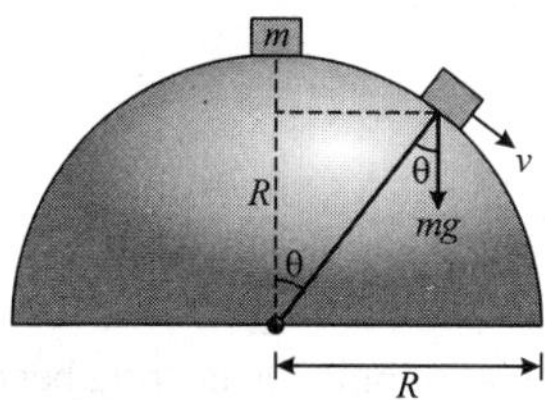

Using work energy theorem gives

$$Mg(R - R\cos\theta) = 1/2\, mv^2$$

$\Rightarrow$ $v = \sqrt{2gR(1 - \cos\theta)}$

To loose contact, normal reaction at that point should be zero, so we use

$$\frac{mv^2}{R} = mg\cos\theta$$

$\Rightarrow$ $2mg(1 - \cos\theta) = mg\cos\theta$

$\Rightarrow$ $2 - 2\cos\theta = \cos\theta$

$\Rightarrow$ $\cos\theta = 2/3 = \frac{h}{R} = \frac{h}{3}$

$\Rightarrow$ $h = 2$ m **Ans. (02.00)**

Sol. 32 Figure below shows the situation described in the question. For stable horizontal circular motion, forces on bob must be balanced in the rotating frame of reference. In this case from the given data we have $\theta = 45°$ so we use

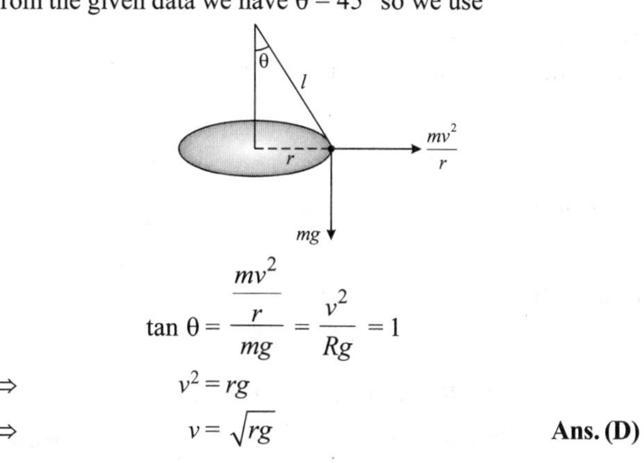

$$\tan\theta = \frac{\frac{mv^2}{r}}{mg} = \frac{v^2}{Rg} = 1$$

$\Rightarrow$ $v^2 = rg$

$\Rightarrow$ $v = \sqrt{rg}$ **Ans. (D)**

Sol. 33 Length of Arc in terms of radius is given as

$$l = R\theta$$

$$4.4 \times 9.46 \times 10^{15} = R\theta$$

Where $\theta = 4\text{ s} = 4 \times 4.843 \times 10^{-6} = 1.94 \times 10^5$ rad

$\Rightarrow$ $4.4 \times 9.46 \times 10^{15} = R \times 1.94 \times 10^{-5}$

$\Rightarrow$ $R = 2.1455 \times 10^{21}$ m

$\Rightarrow$ $v = 8\,AU = 8 \times 1.5 \times 10^{11}$ m/s

$\Rightarrow$ $v = 12 \times 10^{11}$ m/s

Distance covered in 4 revolution is given as

$$s = 4 \times 2\pi R \text{ metre}$$

Time taken in covering 4 revolution is given as

$$t = \frac{\text{distance}}{\text{speed}} = \frac{4 \times 2\pi R}{12 \times 10^{11}};$$

$\Rightarrow$ $t = \frac{8 \times 3.14 \times 2.1455 \times 10^{21}}{12 \times 10^{11}}$

$\Rightarrow$ $t = 4.5 \times 10^{10}$ sec **Ans. (B)**

Sol. 34 During revolution of stone, tension in string is given as

$$T = M\omega^2 R = 80 \text{ N}$$

$\Rightarrow$ $80 = 0.1\,\omega^2(2)$

$\Rightarrow$ $\omega^2 = 400$

$\Rightarrow$ $\omega = 20$ rad/s

$\Rightarrow$ $2\pi f = 20$

$\Rightarrow$ $f = \frac{10}{\pi}$ rev/s

$\Rightarrow$ $f = \frac{600}{\pi}$ rev/min **Ans. (C)**

Sol. 35 In vertical circular motion of a body tied with a string, tension is maximum at the lowest position and minimum at the highest position. **Ans. (B)**

Sol. 36 For beaker to move with disc the centripetal force on it should be less than limiting friction, given as

$$f_{max} \geq m\omega^2 R$$

$\Rightarrow$ $m\omega^2 R \leq \mu mg$

$\Rightarrow$ $R \leq \frac{\mu g}{\omega^2}$ **Ans. (B)**

Sol. 37 For uniform circular motion the acceleration of particle is given as

$$a = |\vec{a}| = \frac{V^2}{R}$$

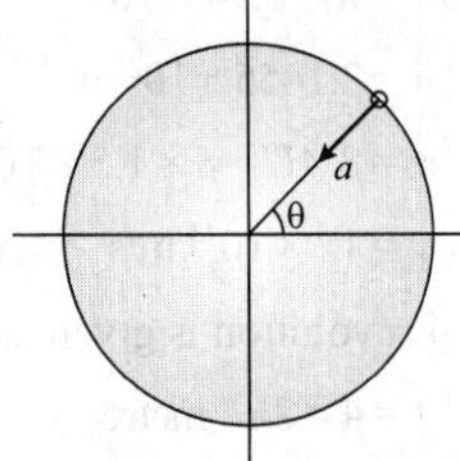

The direction of acceleration is always acting toward the centre of the circle so from above figure, acceleration vector is written as

$$\vec{a} = -a\cos\theta\,\hat{i} - a\sin\theta\,\hat{j}$$

$$\Rightarrow \quad \vec{a} = -\frac{V^2}{R}\cos\theta\hat{i} - \frac{V^2}{R}\sin\theta\hat{j}$$ **Ans. (C)**

Sol. 38 For turning of car without skidding the limiting friction is the maximum force which provides centripetal acceleration so we use

$$f_{s\max} = \frac{mv^2}{R}$$

$$\Rightarrow \quad \mu mg = \frac{mv^2}{R}$$

$$\Rightarrow \quad v = \sqrt{\mu Rg}$$

$$\Rightarrow \quad \frac{v_2}{v_1} = \sqrt{\frac{R_2}{R_1}}$$

$$\Rightarrow \quad \frac{v_2}{30} = \sqrt{\frac{48}{75}}$$

$$\Rightarrow \quad v_2 = 24 \text{ m/s}$$ **Ans. (24)**

Sol. 39 When ball starts from point P its kinetic energy at Q is calculated as

$$\frac{1}{2}mv^2 = mgh = mgR\sin\alpha$$

$$\Rightarrow \quad v = \sqrt{2gR\sin\alpha}$$

Centripetal force at this point is given as

$$\frac{mv^2}{R} = 2mg\sin\alpha$$

Free body diagram of particle at point Q is shown in figure below. The normal reaction at this point is calculated by the equation of motion along radial direction, written as

$$N - mg\sin\alpha = \frac{mv^2}{R} = 2mg\sin\alpha$$

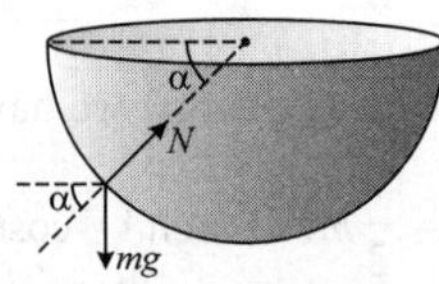

The ratio of normal reaction to centripetal force is given as

$$A = \frac{N}{2mg\sin\alpha} = \frac{1}{2} + 1 = \frac{3}{2}$$

$$\Rightarrow \quad A = \text{Constant}$$

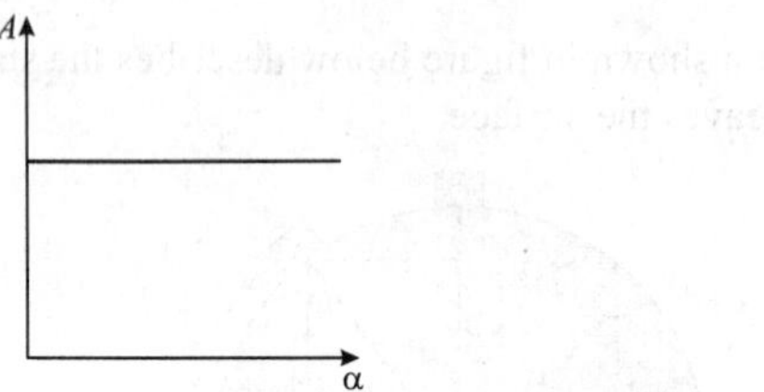

Ans. (C)

Sol. 40 The velocity of bob when string becomes horizontal is given as

$$v = \sqrt{u^2 - 2gL}$$

Above velocity is in vertical direction and velocity at the bottom point was horizontal so vector change in velocity is calculated as

$$\Delta v = \sqrt{u^2 + v^2}$$

$$\Rightarrow \quad \Delta v = \sqrt{u^2 + v^2 - 2gL}$$

$$\Rightarrow \quad \Delta v = \sqrt{2u^2 - 2gL}$$

$$\Rightarrow \quad \Delta v = \sqrt{2(u^2 - gL)}\ x = 2$$ **Ans. (B)**

Sol. 41 Centripetal acceleration of particle is written as

$$a = k^2rt^2 = \frac{v^2}{r}$$

$$\Rightarrow \quad v = krt$$

Tangential acceleration of particle is given as

$$a_t = \frac{dv}{dt} = kr$$

Thus tangential force on particle is calculated as

$$F_t = ma_t = mkr$$

Power of force is calculated as

$$P = \vec{F}\cdot\vec{v}$$

$$\Rightarrow \quad P = Fv\cos\theta = F_t v = mkr(krt)$$

$$\Rightarrow \quad P = mk^2r^2t$$ **Ans. (C)**

Sol. 42 Position coordinates of particle x and y are given as

$$x = 4\sin\left(\frac{\pi}{2} - \omega t\right) = 4\cos(\omega t)$$

$$y = 4\sin(\omega t)$$

Eliminating term of time t gives relation between x and y as

$$x^2 + y^2 = y^2\cos^2\omega t + y^2\sin^2\omega t = 4^2$$

$$\Rightarrow \quad x^2 + y^2 = 4^2 \qquad \textbf{Ans. (A)}$$

Sol. 43 After falling through a vertical height h, velocity attained by the bob is given as

$$v_{max} = \sqrt{2gh}$$

The speed will be highest at the lowest position, so we use

$$h = (l - l\cos 60°) = \frac{l}{2}$$

$$\Rightarrow \quad v_{max} = \sqrt{2 \times g \times \frac{l}{2}} = \sqrt{10 \times 2.5}$$

$$\Rightarrow \quad v_{max} = 5 \text{ m/s} \qquad \textbf{Ans. (5)}$$

Sol. 44 Arc length covered for angular displacement θ is given as

$$d = R\theta$$

$$\Rightarrow \quad 60 = R\left(\frac{3\pi}{4}\right)$$

$$\Rightarrow \quad R = \frac{60 \times 4}{3\pi} = \frac{80}{\pi} m$$

Displacement of particle from A to B is calculated as

$$d = \sqrt{R^2 + R^2 - 2R^2\cos 135}$$

$$\Rightarrow \quad d = \sqrt{2R^2 - 2R^2(-0.7)}$$

$$\Rightarrow \quad d = \sqrt{3.4R^2} = \sqrt{3.4\left(\frac{80}{\pi}\right)^2} = 47 \text{ m} \qquad \textbf{Ans. (B)}$$

Sol. 45 In the circular motion of block the normal reaction will provide the centripetal force so we use

$$N = \frac{mv^2}{r}$$

As normal reaction is proportional to v^2, the curve will be a parabola. **Ans. (A)**

Sol. 46 Figure below shows the situation described in the question.

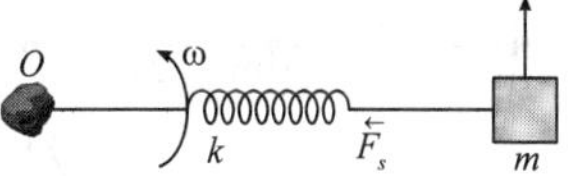

If extension in the spring is x, we use for uniform circular motion of body

$$kx = m(L_0 + x)\omega^2$$

$$\Rightarrow \quad 12.5x = \frac{1}{5}(L_0 + x)25$$

$$\Rightarrow \quad 1.5x = L_0$$

$$\Rightarrow \quad \frac{x}{L_0} = \frac{2}{3} \qquad \textbf{Ans. (C)}$$

Sol. 47 Figure below shows the free body diagram of the bob in circular motion

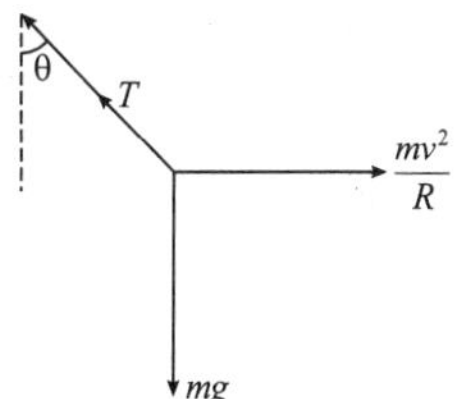

For equilibrium of bob in frame of car, we have

$$T\cos\theta = mg$$

and $$T\sin\theta = \frac{mv^2}{R}$$

$$\Rightarrow \quad \tan\theta = \frac{v^2}{Rg}$$

$$\Rightarrow \quad \tan\theta = \frac{20^2}{40 \times 10}$$

$$\Rightarrow \quad \tan\theta = 1$$

$$\Rightarrow \quad \theta = \frac{\pi}{4} \qquad \textbf{Ans. (C)}$$

Sol. 11 On curved road the friction will provide the necessary centripetal force for circular motion so we use

$$f_s = \frac{mv^2}{r}$$

For maximum speed in safe turning, we use

$$f_s = f_s \max = \mu mg$$

$$v_{max} = \sqrt{\mu rg}$$

$$v_{max} = \sqrt{0.34 \times 50 \times 10} \approx 13 \text{ m/s} \qquad \textbf{Ans. (C)}$$

Sol. 49 Figure below shows the situation as described in the question.

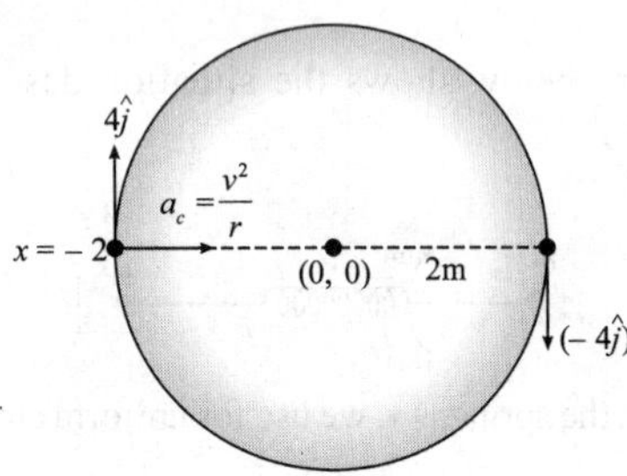

Centripetal acceleration of the object is given as

$$a_c = \frac{V^2}{r} = \frac{4^2}{2} = \frac{16}{2} = 8 \text{ m/s}^2$$

At $x = -2$ m, we have

$$\vec{V} = 4\hat{j}$$

$$\Rightarrow \quad \vec{a}_c = 8\hat{i} \qquad \textbf{Ans. (B)}$$

Sol. 50 Tangential acceleration of the car is given as

$$v\frac{dv}{dx} = \frac{v^2}{R}$$

$$\Rightarrow \quad \int_{15}^{v}\frac{dv}{v} = \frac{1}{R}\int_{0}^{x}dx$$

$$\Rightarrow \quad v = 15e^{x/R}$$

$$\Rightarrow \quad \frac{dx}{dt} = 15e^{x/R}$$

$$\Rightarrow \quad \int_{0}^{\frac{\pi R}{2}} e^{-x/R}dx = 15\int_{0}^{t_0}dt$$

$$\Rightarrow \quad t_0 = 40(1 - e^{-\pi/2}) \qquad \textbf{Ans. (40)}$$

Sol. 51 We use power energy relaiton as

Energy = Power × time

$$E = P \times t$$

$$\Rightarrow \quad \frac{1}{2}mv^2 = Pt$$

We use kinetic energy of body as

$$E = \frac{1}{2}mv^2$$

$$\Rightarrow \quad v = \sqrt{\frac{2Pt}{m}}$$

As $\quad v = \frac{dx}{dt}$

$$\Rightarrow \quad \frac{dx}{dt} = \sqrt{\frac{2Pt}{m}}$$

By integrating on both sides we have

$$x = \sqrt{\frac{2P}{m}}\frac{2}{3}[t^{3/2}]_0^4$$

$$\Rightarrow \quad x = \frac{16\sqrt{P}}{3} = \frac{1}{3} \times 16\sqrt{P}$$

$$\Rightarrow \quad \alpha = 4 \qquad \textbf{Ans. (4)}$$

Sol. 52 When stone is revolving in horizontal circle, a force is acting toward centre called centripetal force, so acceleration

$$a = \frac{v^2}{r} = \omega^2 r$$

$$\omega = \frac{28 \times 2\pi}{60}$$

$$\Rightarrow \quad a = \omega^2 R = \left(\frac{28 \times 2\pi}{60}\right)^2 \times 1.8$$

$$\Rightarrow \quad a = \left(\frac{56}{60} \times \frac{22}{7}\right)^2 \times 1.8$$

$$\Rightarrow \quad a = \frac{(44)^2}{225} \times 1.8$$

$$\Rightarrow \quad a = \frac{1936 \times 1.8}{225} = \frac{1936}{x}$$

$$\Rightarrow \quad x = 125 \qquad \textbf{Ans. (125)}$$

Sol. 53 $\quad r = 10$ m; $T = 4$ sec

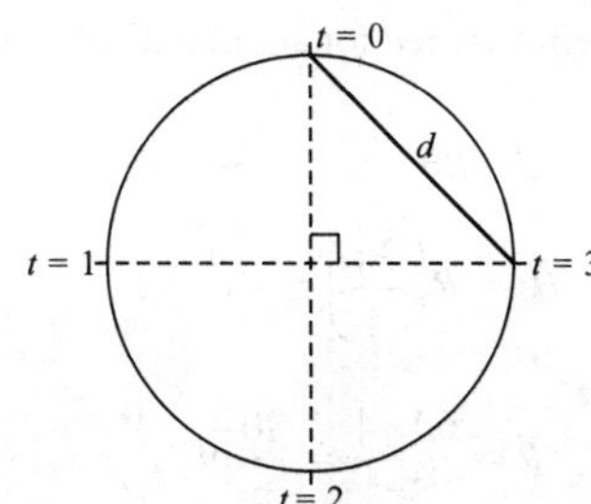

Displacement of body from its starting point is calculated as

$$d = \sqrt{2}(10)\text{ m} \qquad \textbf{Ans. (D)}$$

Sol. 54 Let instantaneous velocity of particle is v. The time taken by particle to travel from A to B is calculated as

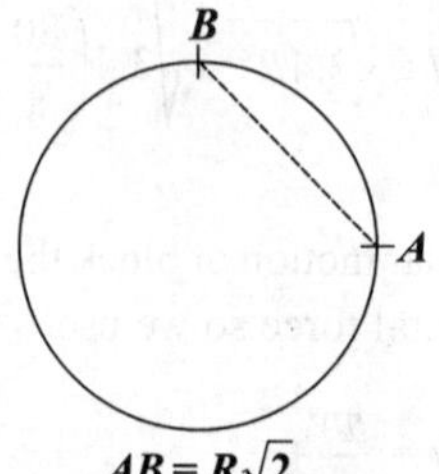

$$AB = R\sqrt{2}$$

$$t = \frac{Arc\ length}{v} = \frac{2\pi \frac{R}{4}}{v} = \frac{\pi R}{2v}$$

Average velocity in going from A to B is given as–

$$< v > = \frac{AB}{t} = \frac{R\sqrt{2}(2v)}{\pi r} = \frac{2\sqrt{2}v}{\pi}$$

Ratio of instantaneous velocity to average velocity is calculated as

$$\frac{v}{< v >} = \frac{\pi}{2\sqrt{2}} = \frac{\pi}{x\sqrt{2}}$$

$\Rightarrow$ $x = 2$ **Ans. (A)**

Sol. 55 For a block connected with spring which is free to move on circular path on smooth surface, if extension in length of spring is x, as shown in figure below, we have

← 0.2 + x →

kx ← → mω²r

For circular motion of block centripetal force is provided by spring force so we have

$$kx = m\omega^2 r$$

$$\Rightarrow \quad 7.5x = \left(\frac{1}{10}\right)(5^2)(0.2 + x)$$

$$\Rightarrow \quad \frac{15}{2}x = \frac{5}{2}\left(x + \frac{1}{5}\right)$$

$$\Rightarrow \quad x = \frac{1}{10}$$

Thus tension in spring is calculated as

$$F = kx = 7.5 \times \frac{1}{10} = 0.75\ \text{N}$$ **Ans. (B)**

Sol. 56 Average velocity is defined as total displacement travelled per unit time. Here displacement is from A to B and $\theta = 120°$ so we have

$$|\vec{r}_f - \vec{r}_i| = 2\cos\left(\frac{\pi - \theta}{2}\right)$$

Time taken from A to B is calculated as

$$t = \frac{\text{distance}}{\text{speed}} = \frac{2\pi R/3}{\pi} = \frac{2\pi R}{3\pi}$$

Thus average velocity is calculated as

$$|\langle \vec{v} \rangle| = \frac{|\vec{r}_f - \vec{r}_i|}{\Delta t}$$

$$\Rightarrow \quad |\langle \vec{v} \rangle| = \frac{2R\cos\left[\frac{\pi - \theta}{2}\right]}{\frac{2\pi R}{3\pi}} = 3\cos 30°$$

$$\Rightarrow \quad |\langle \vec{v} \rangle| = 1.5\sqrt{3}\ \text{m/s}$$ **Ans. (D)**

Sol. 57 Angular velocity of child on merry-go-round is given as

$$\omega = \frac{2\pi}{3.14} = 2\ \text{rad/s}$$

Here in circular motion centrifugal force is acting in non-inertial frame provided by angular velocity.

$$|\vec{f}_{centrifugal}| = |-m\vec{a}_{Ref.}|$$

$$\Rightarrow \quad |\vec{f}_{centrifugal}| = M\omega^2 R$$

$$\Rightarrow \quad |\vec{f}_{centrifugal}| = 40\ \text{N}$$ **Ans. (D)**

Sol. 58 Maximum friction on coin on rotating table is calculated as

$$f_{s\max} = \mu mg = m\omega^2 R$$

$$\Rightarrow \quad R = \frac{\mu g}{\omega^2}$$

Thus if ω becomes $\frac{\omega}{2}$, R will become $4R$ and the distance from the center will be 4 cm. **Ans. (D)**

Sol. 59 Centripetal force acting on the vehicle is given as

$$F_c = m\omega^2 r = 200 \times (0.2)^2 \times 70 = 560\ \text{N}$$ **Ans. (A)**

6.1 System of Particles

Sol. 1 m → 2V V ← m

– ve ←→ + ve

Velocity of centre of mass is given as

$$\vec{v}_{cm} = \frac{m_1\vec{V}_1 + m_2\vec{V}_2}{m_1 + m_2} = \frac{m(2V) + m(-v)}{2m} = \frac{v}{2}$$

Ans. (C)

Sol. 2 Linear Momentum → vector

Kinetic energy → scalar

If KE = 0 $\Rightarrow$ Speeds of all particles of system will be individually zero

$\Rightarrow$ linear momentum = 0

Ans. (C)

Sol. 3

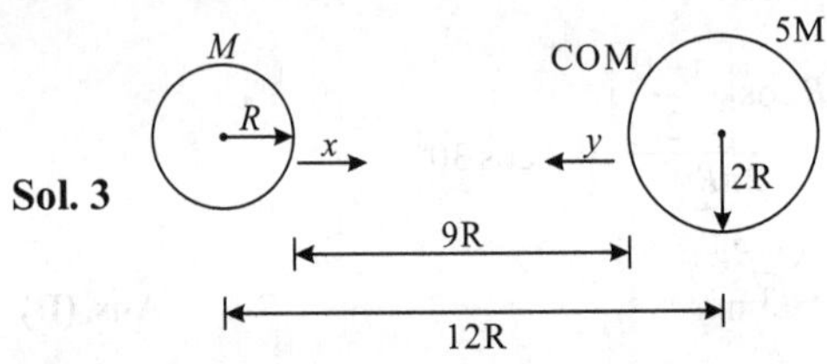

Due to internal gravitational force bodies move such that their centre of mass remain at rest

If M and 5 M travels distances x and y before collision

$\Rightarrow \quad x+y=9R$

In motion displacement of centre of mass is zero

$$\Rightarrow \quad \Delta x_{CM}=\frac{M\Delta x_1+5M\Delta x_2}{M+5M}=0$$

$$\Rightarrow \quad \Delta x_{CM}=Mx-5M(9R-x)=0$$

$$\Rightarrow \quad x=7.5R$$

Ans. (C)

Sol. 4

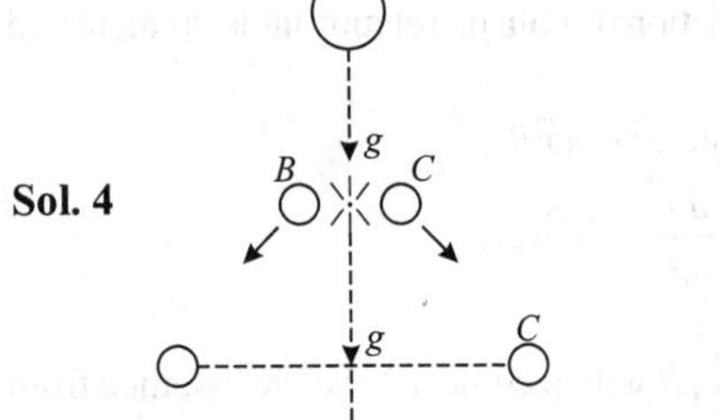

Breaking is due to internal forces which do not affect the motion of centre of mass

$\Rightarrow$ Centre of mass of bodies B & C continue to move as if body A is not broken. **Ans. (B)**

Sol. 5

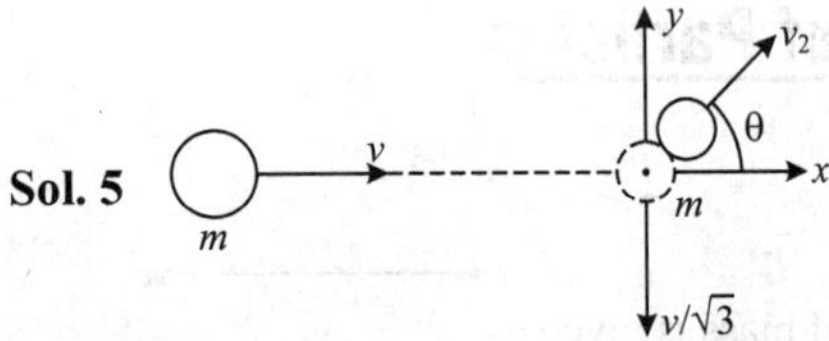

As no external force is present linear momentum of system will remain conserved so we use

in x-dir $\quad mv=mv_2\cos\theta$

in y-dir $\quad O=mv_2\sin\theta-\frac{mv}{\sqrt{3}}$

$\Rightarrow \quad v_2\cos\theta=v \quad \ldots(1)$

and $\quad v_2\sin\theta=\frac{v}{\sqrt{3}} \quad \ldots(2)$

Squaring and adding equation-(1) and (2) gives

$$v_2=\sqrt{v^2+\frac{v^2}{3}}=\sqrt{\frac{4v^2}{3}}=\frac{2v}{\sqrt{3}}$$ **Ans. (C)**

Sol. 6 For only translational motion force $\vec{F}$ must be applied at COM of system

Distance PC is given as

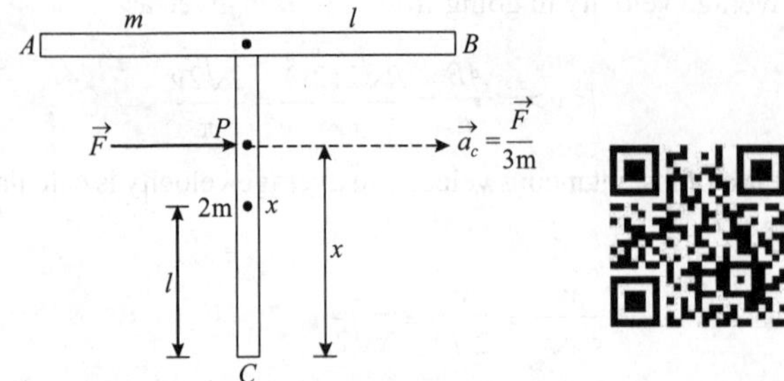

$$x=\frac{m_1x_1+m_2x_2}{M_1+m_2}=\frac{2m(l)+m(2l)}{2m+m}=\frac{4l}{3}$$

Ans. (C)

Sol. 7

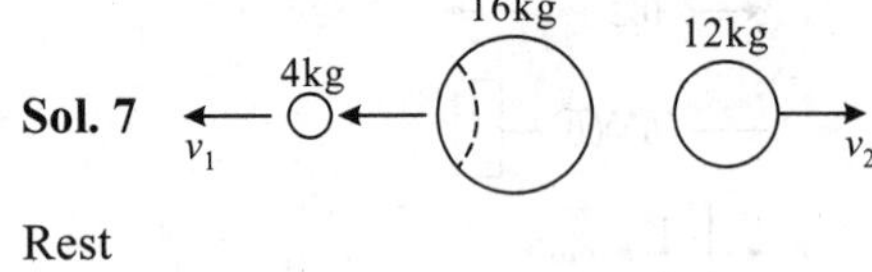

Rest

$$P_i=0=P_f$$

$$\Rightarrow \quad 0=12v_2-4v_1$$

$$\Rightarrow \quad 0=12\times4-4v_1$$

$$\Rightarrow \quad v_1=12\text{ m/s}$$

KE of 4 kg mass is

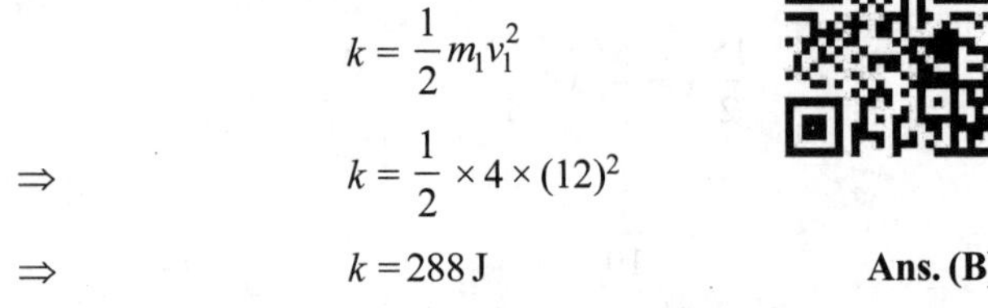

$$k=\frac{1}{2}m_1v_1^2$$

$$\Rightarrow \quad k=\frac{1}{2}\times4\times(12)^2$$

$$\Rightarrow \quad k=288\text{ J}$$

Ans. (B)

Sol. 8

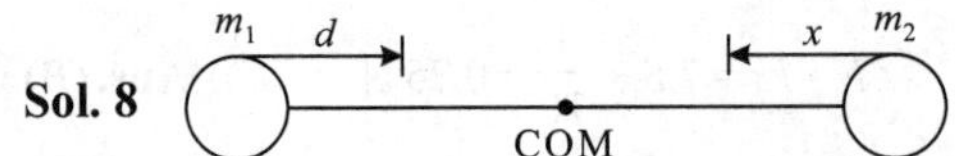

Displacement of COM

$$\Delta x_{COM}=0$$

We use

$$\vec{\Delta x}_{COM}=\frac{m_1\overrightarrow{\Delta x_1}+m_2\overrightarrow{\Delta x_2}}{m_1+m_2}$$

$$\Rightarrow \quad 0=\frac{m_1d+m_2x}{m_1+m_2}$$

$$\Rightarrow \quad x=-\frac{m_1}{m_2}d$$

[– *ve* sign indicates opposite direction]

Ans. (C)

Sol. 9

By conservation of linear momentum

$$m_1u_1 = (m_1 + m_2)V_f$$

$$\Rightarrow \quad V_f = \frac{m_1u_1}{m_1 + m_2}$$

loss in energy $\quad \Delta E_L = k_i - k_f$

$$\Rightarrow \quad \Delta E_L = \frac{1}{2}m_1u_1^2 - \frac{1}{2}(m_1 + m_2)v_f^2$$

$$\Rightarrow \quad \Delta E_L = \frac{1}{2}\left(\frac{m_1m_2}{m_1 + m_2}\right)u_1^2$$

$$\Rightarrow \quad \Delta E_L = \frac{1}{2} \times \frac{0.5 \times 1}{0.5 + 1} \times (2)^2 = \frac{2}{3}\text{J}$$

Ans. (C)

Sol. 10

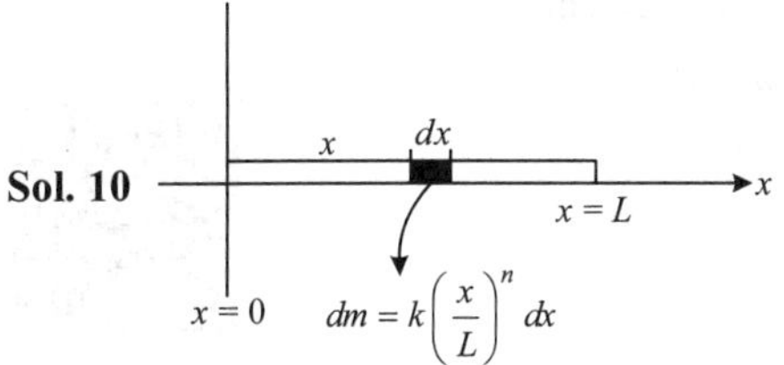

Mass of rod $\quad M = \int dm = \int_0^L \frac{k}{L^n} \cdot \left[\frac{x^{n+1}}{n+1}\right]_0^L = \left(\frac{k}{n+1}\right)L$

Centre of mass position of rod

$$x_{cm} = \frac{1}{M}\int dm x$$

$$\Rightarrow \quad x_{cm} = \frac{1}{\left(\frac{k}{n+1}\right)L} \cdot \int_0^L \frac{k}{L^n} \cdot x^{n+1}dx$$

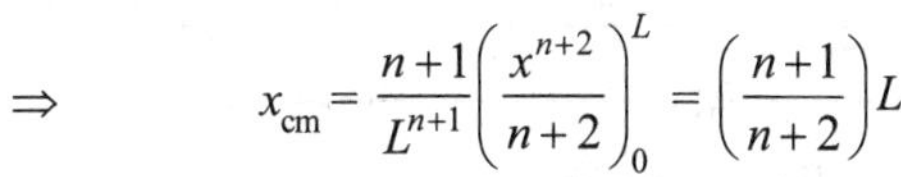

$$\Rightarrow \quad x_{cm} = \frac{n+1}{L^{n+1}}\left(\frac{x^{n+2}}{n+2}\right)_0^L = \left(\frac{n+1}{n+2}\right)L$$

at $n = 0 \rightarrow x_{cm} = \frac{L}{2}$ and at $n \rightarrow \infty$ we have $x_{cm} = L$ **Ans. (A)**

Sol. 11

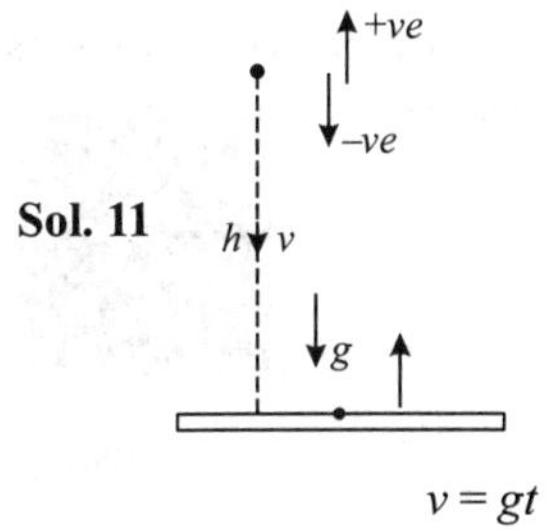

$$v = gt$$

Ans. (A)

Sol. 12 Slope of x-t curve from $t = 0$ to $t = 2$ s

$$v_1 = \frac{2}{2} = 1 \text{ m/s} \quad +ve$$

Slope of x-t curve from $t = 2$ s to $t = 4$ s

$$v_2 = \frac{2}{2} = -1 \text{ m/s}$$

Impulse

$$\Rightarrow \quad \vec{J} = \Delta\vec{P} = m\vec{v}_2 - m\vec{v}_1$$

$$\Rightarrow \quad |\vec{J}| = |0.4(-1-1)|$$

$$|\vec{J}| = 0.4 \times 2 = 0.8 \text{ Ns}$$

Ans. (B)

Sol. 13

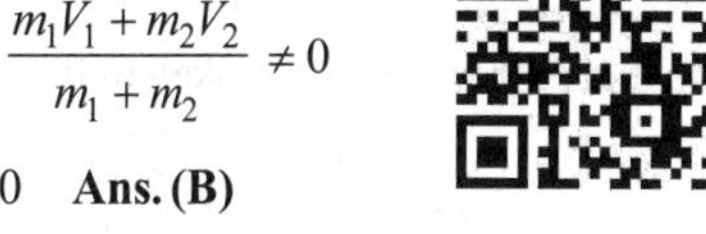

$$m_1v_1 + m_2v_2 = (m_1 + m_2)v_f$$

$$\Rightarrow \quad v_f = \frac{m_1V_1 + m_2V_2}{m_1 + m_2} \neq 0$$

$\Rightarrow \quad k_f \neq 0 \quad$ **Ans. (B)**

$\Rightarrow$ S-1 and S-2 are true and S-2 is explaining S-1

Sol. 14

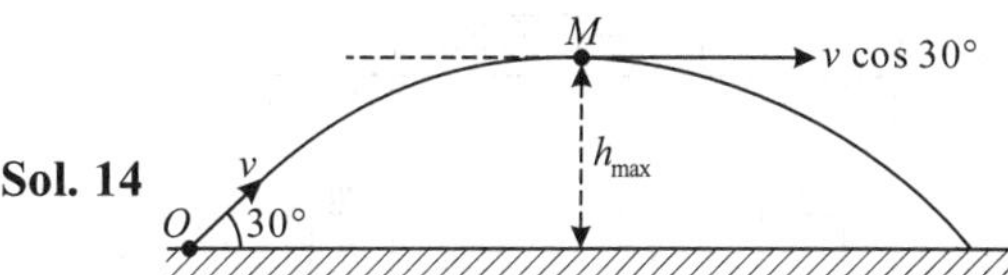

Maximum height of projectile

$$h_{max} = \frac{v^2 \sin^2(30°)}{2g}$$

Angular momentum of particle at M about O is given as

$$L_{MO} = m(V\cos 30°) \times h_{max}$$

$$\Rightarrow \quad L_{MO} = mv\left(\frac{\sqrt{3}}{2}\right) \times \frac{v^2}{2g}\left(\frac{1}{4}\right) = \frac{\sqrt{3}mv^3}{16g}$$

Ans. (D)

Sol. 15 As finally blocks come to rest

$$\Rightarrow \quad m_1V_1 = m_2V_2$$

$$\Rightarrow \quad V_{CM} = 0$$

Thus COM of system will remain at rest

$\Rightarrow$ displacement of $COM = 0$

Thus we use

$$\Delta x_{cm} = \frac{m_2x_2 - m_1x_1}{m_1 + m_2} = 0$$

$$\Rightarrow \quad m_1x_1 = m_2x_2$$

$$\Rightarrow \quad \frac{x_1}{x_2} = \frac{m_2}{m_1}$$

Ans. (A)

Sol. 16

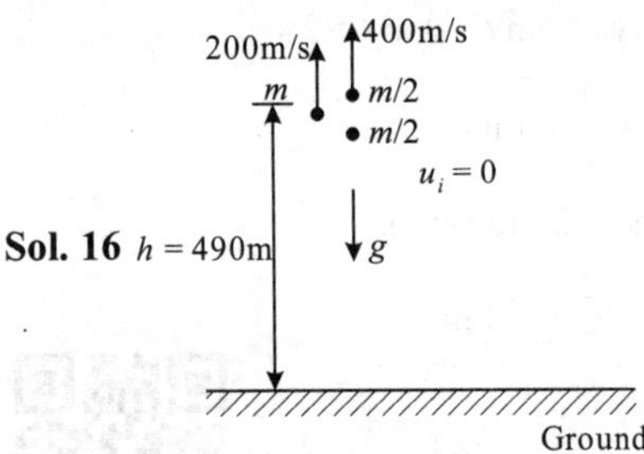

Initial upward momentum of moving body is

$$P_i = m(200)$$

Final upward momentum of $m/2$ (one part) is

$$P_f = \frac{m}{2}(400) = m(200)$$

As the moving part carries the same momentum as that of the initial body, No momentum is left in the other part

$$\Rightarrow \quad v_{\text{other part}} = 0$$

As it is in free fall, time taken to hit the ground is

$$t = \sqrt{\frac{2h}{g}} = \sqrt{\frac{2 \times 490}{9.8}} = \sqrt{100} = 10 \text{ s}$$ **Ans. (C)**

Sol. 17

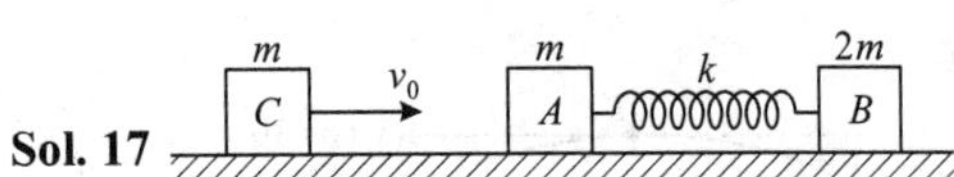

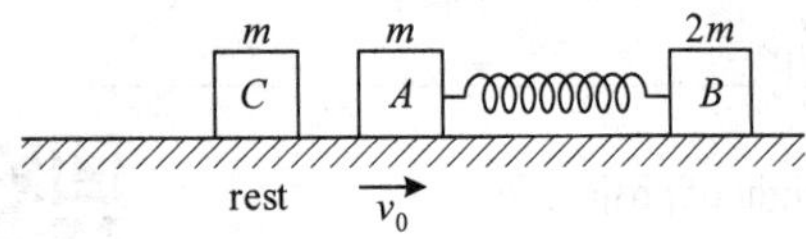

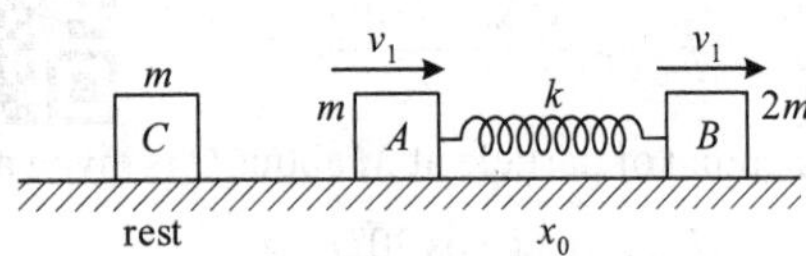

By conservation of linear momentum, we have

$$mv_0 = 3mv_1 \quad \Rightarrow \quad v_1 = \frac{v_0}{3}$$

By conservation of energy, we have

$$\frac{1}{2}mv_0^2 = \frac{1}{2}(3m)\left(\frac{v_0}{3}\right)^2 + \frac{1}{2}kx_0^2$$

$$\Rightarrow \quad \frac{1}{2}mv_0^2\left(1 - \frac{1}{3}\right) = \frac{1}{2}kx_0^2$$

$$\Rightarrow \quad k = \frac{2}{3}mv_0^2\left(\frac{1}{x_0^2}\right)$$ **Ans. (D)**

Sol. 18

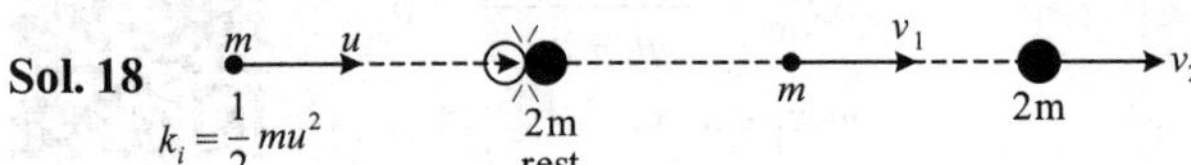

For elastic collision (Head-on), The final velocity of particle m is given as

$$v_1 = \left(\frac{m_1 - m_2}{m_1 + m_2}\right)u_1 + \left(\frac{2m_2}{m_1 + m_2}\right)u_2$$

$$\Rightarrow \quad v_1 = \left(\frac{m - 2m}{m + 2m}\right)u + 0 = -\frac{1}{3}u$$

Final KE of particle – 1 is

$$k_f = \frac{1}{2}m\left(\frac{u}{3}\right)^2 = \frac{1}{2}mu^2\left(\frac{1}{9}\right)$$

% loss in KE of first particle is % loss

$$\Delta k_L = \frac{k_i - k_f}{k_i} \times 100 = \left(1 - \frac{1}{9}\right) \times 100\ \%$$

$$\Rightarrow \quad \Delta k_L \simeq 89\%$$ **Ans. (C)**

Sol. 19 As maximum energy loss occurs in completely inelastic collision

$\Rightarrow$ Statement-2 is true

$m \xrightarrow{v}$ M $\longrightarrow$ $(m + M) \xrightarrow{} v_f$

by conservation of momentum

$$mv = (m + M)v_f$$

$$\Rightarrow \quad v_f = \frac{mv}{m + M}$$

Energy loss (maximum)

$$\Delta E_{\max} = \frac{1}{2}mv^2 - \frac{1}{2}(m + M)\left(\frac{mv}{m + M}\right)^2$$

$$\Delta E_{\max} = \frac{1}{2}\left(\frac{mM}{m + M}\right)v^2 = \underbrace{\frac{M}{m + M}}_{f}\left(\frac{1}{2}mv^2\right)$$ **Ans. (D)**

Sol. 20

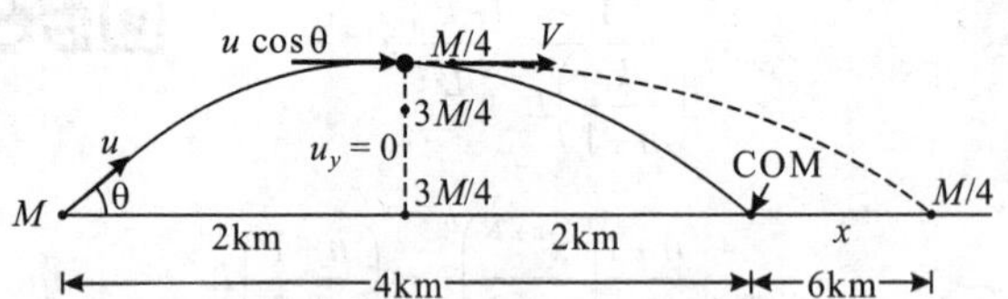

By Conservation of linear momentum

$$M(u \cos \theta) = \frac{M}{u} \cdot v$$

$$\Rightarrow \quad v = 4u \cos \theta$$

$3M/4$ and $M/4$ will fall on ground such that their centre of mass is located at the original path of projectile

$$\Rightarrow \quad \frac{3M}{4} \times 2 = \frac{M}{4} \times x$$

$$\Rightarrow \quad x = 6 \text{ km}$$

$$\text{Range} = 4 + 6 = 10 \text{ km}$$ **Ans. (C)**

Sol. 21

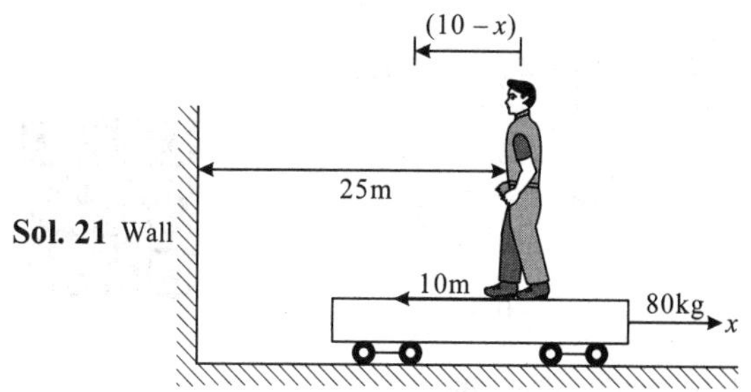

Due to motion of boy, if cart displaces toward right by x such that

$$\Delta x_{COM} = 0 = \frac{M_1x_1 + M_2x_2}{M_1 + M_2}$$

$$\Rightarrow \quad 0 = \frac{80x - 20(10 - x)}{100}$$

$$\Rightarrow \quad 80x = 20(10 - x)$$

$$\Rightarrow \quad 5x = 10$$

$$\Rightarrow \quad x = 2\text{ m}$$

Final distance of boy from wall is

$$d = 25 - (10 - 2) = 17\text{ m}$$ **Ans. (D)**

Sol. 22 After collision if the resulting mass $6m$ has velocity components v_x and v_y then by conservation of linear momentum, we have. Along x-direction

$$+m(3u) - 2m(2u)\cos 60° - 3mu\cos 60° = 6mv_x$$

$$\Rightarrow \quad 3u - 2u - \frac{3}{2}u = 6v_x$$

$$\Rightarrow \quad v_x = -\frac{u}{12}$$

Along y-direction

$$-(2m)(2u)\sin 60° + 3mu\sin 60° = 6mv_y$$

$$\Rightarrow \quad -4mu\left(\frac{\sqrt{3}}{2}\right) + 3mu\left(\frac{\sqrt{3}}{2}\right) = 6mv_y$$

$$\Rightarrow \quad -\frac{\sqrt{3}}{2}u = 6v_y$$

$$\Rightarrow \quad v_y = -\frac{\sqrt{3}}{12}u$$

$$\Rightarrow \quad \vec{v} = v_x\hat{i} + v_y\hat{i}$$

$$\Rightarrow \quad \vec{v}_{6m} = -\frac{u}{12}\hat{i} - \frac{\sqrt{3}}{12}u\hat{j}$$

$$= \frac{u}{12}(-i - \sqrt{3}\hat{j})$$ **Ans. (D)**

Sol. 23

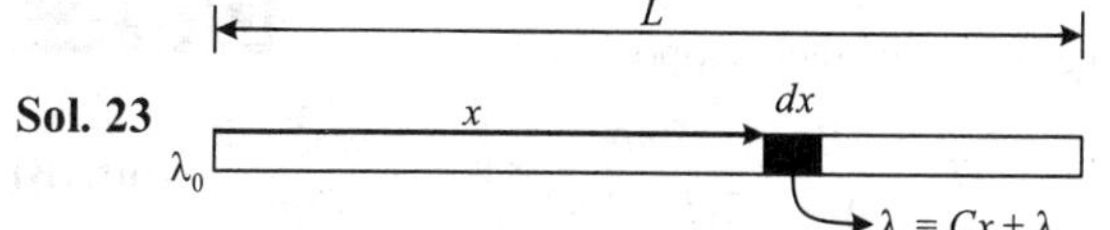

Mass of element is

$$M = \int dm = \int_0^L \lambda dx = \int_0^L (Cx + \lambda_0)dx$$

$$\Rightarrow \quad M = \int_0^L (Cx + \lambda_0)dx$$

$$\Rightarrow \quad M = \frac{CL^2}{2} + \lambda_0 L$$

$$\Rightarrow \quad C = \frac{2}{L^2}(M - \lambda_0 L)$$

Distance of COM from lighter end is

$$x_{COM} = \frac{1}{M}\int dmx = \frac{1}{M}\int_0^L (Cx^2 + \lambda_0 x)dx$$

$$\Rightarrow \quad x_{COM} = \frac{1}{M}\left(\frac{LC^3}{3} + \frac{\lambda_0 L^2}{2}\right) = \frac{L^2}{M}\left(\frac{CL}{3} + \frac{\lambda_0}{2}\right)$$

$$\Rightarrow \quad x_{COM} = \frac{L^2}{M}\left(\frac{2}{3L}(M - \lambda_0 L) + \frac{\lambda_0}{2}\right)$$

$$\Rightarrow \quad x_{COM} = \frac{2}{3}L - \frac{L^2}{M}\left(-\frac{2}{3}\lambda_0 + \frac{\lambda_0}{2}\right)$$

$$\Rightarrow \quad x_{COM} = \frac{2L}{3} - \frac{\lambda_0 L^2}{6M}$$ **Ans. (D)**

Sol. 24

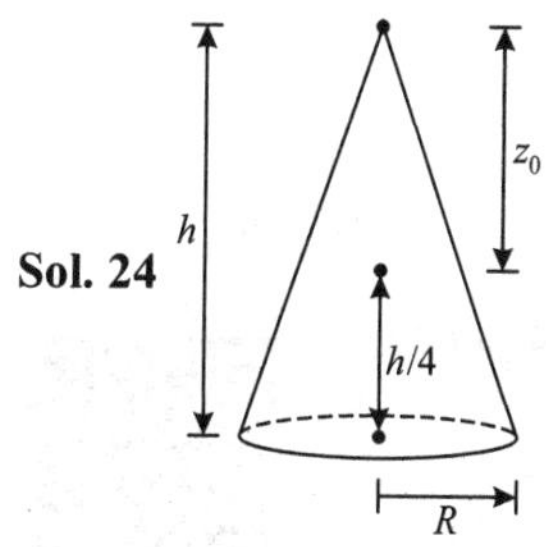

COM distance from vertex

$$z_0 = h - \frac{h}{4} = \frac{3h}{4}$$ **Ans. (B)**

Sol. 25

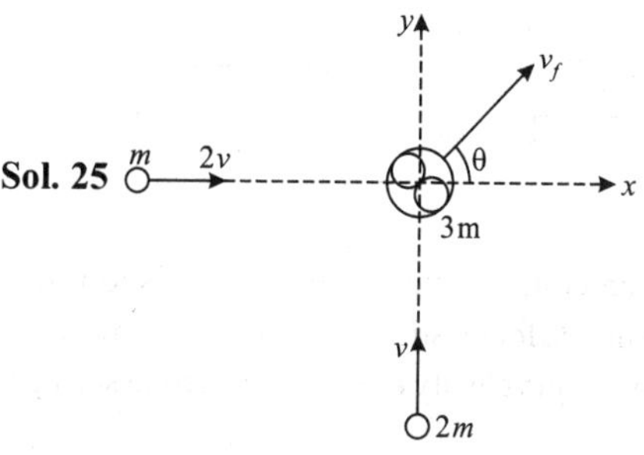

By conservation of linear momentum

in x-direction $\quad 2mv = 3mv_f\cos\theta$ ...(1)

in y-direction $\quad 2mv = 3mv_f\sin\theta$ ...(2)

Squaring and adding (1) and (2)

$$2v_f = \sqrt{8}v$$

$$\Rightarrow \quad v_f = \frac{2\sqrt{2}}{3}v$$

Initial KE of system is

$$k_i = \frac{1}{2}m(2V)^2 + \frac{1}{2}2m(V)^2 = 3mv^2$$

Final KE of system is

$$k_f = \frac{1}{2}3m\left(\frac{2\sqrt{2}}{3}v\right)^2 = \frac{4}{3}mv^2$$

Loss in energy during collision

$$\Delta E_L = k_i - k_y = \frac{5}{3}mv^2$$

$$\% \text{ loss} = \frac{k_i - k_f}{k_i} \times 100\%$$

$$\Rightarrow \quad \% \text{ loss} = \frac{500}{9} = 55.55\%$$

Ans. (C)

Sol. 26

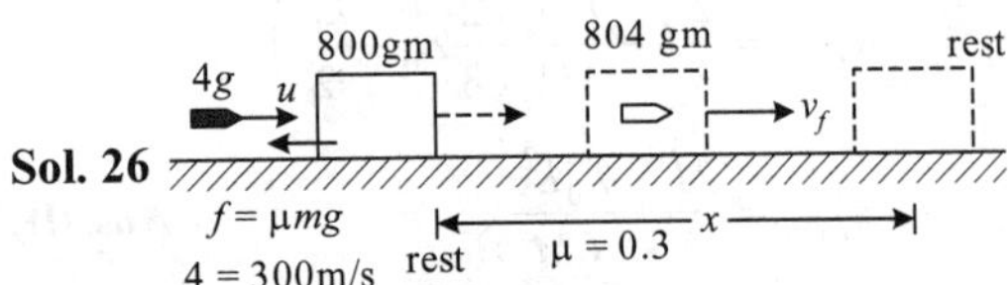

By conservation of momentum we use

$$4u + 0 = 804v_f$$

$$v_f = \frac{4 \times 300}{804} = \frac{1200}{804} = 1.49 \text{ m/s}$$

using work energy theorem for block sliding on rough surface, we have

$$\frac{1}{2}mv_f^2 - \mu mgx = 0$$

$$\Rightarrow \quad x = \frac{V_f^2}{2\mu g} = \frac{(1.49)^2}{2 \times 0.3 \times 10} = 0.379 \text{ m}$$

Ans. (B)

Sol. 27

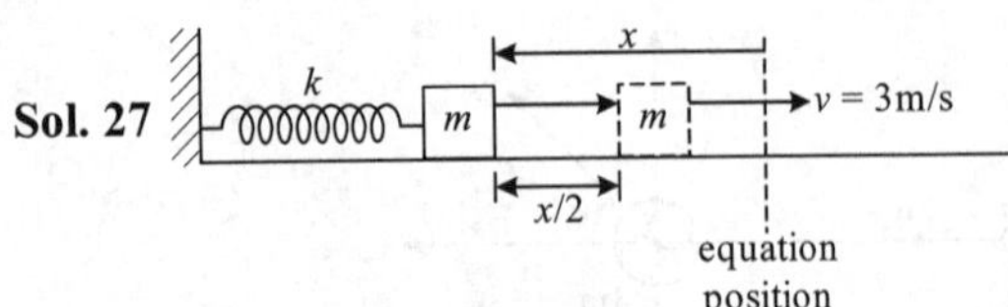

As it is given that on collision moving block comes to rest & other moves, this implies for elastic collision masses are equal. Thus just before collision velocity of mass attached to spring is 3 m/s.

By conservation of energy, we use

$$\frac{1}{2}kx^2 = \frac{1}{2}k\left(\frac{x}{2}\right)^2 + \frac{1}{2}mv^2$$

$$\Rightarrow \quad \frac{1}{2}kx^2\left(1 - \frac{1}{4}\right) = \frac{1}{2} \times 0.1 \times (3)^2$$

$$\Rightarrow \quad \frac{1}{2}kx^2 = \frac{0.9}{2} \times \frac{4}{3}$$

$$\Rightarrow \quad \frac{1}{2}kx^2 = 0.6 \text{ J}$$

initial stored energy in spring. **Ans. (B)**

Sol. 28

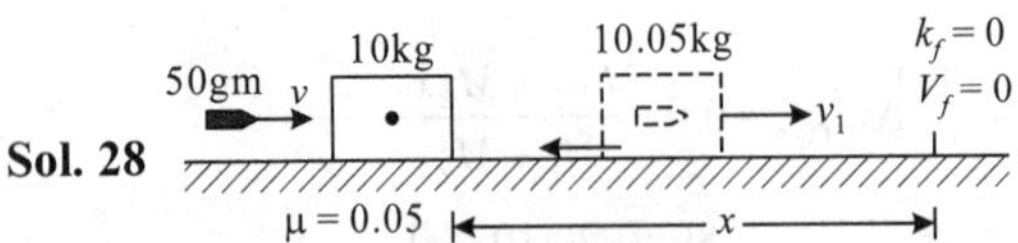

If after collision block moves at speed v_1 we have by momentum conservation

$$0.05V = 10.05v_1$$

$$\Rightarrow \quad v_1 = \frac{0.05v}{10.05}$$

Using work-energy theorem after collision we have

$$\frac{1}{2}(10.05)v_1^2 - \mu(m+M)gx = 0$$

$$\Rightarrow \quad \frac{1}{2}(10.05)\left(\frac{0.05v}{10.05}\right)^2 - 0.05 \times 10.05 \times 10 \times 2 = 0$$

$$\Rightarrow \quad v = 201\sqrt{2} \text{ m/s}$$

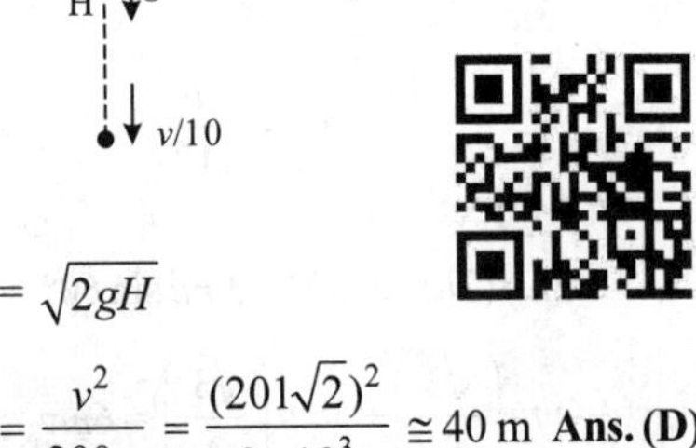

We have

$$\Rightarrow \quad \frac{v}{10} = \sqrt{2gH}$$

$$\Rightarrow \quad H = \frac{v^2}{200g} = \frac{(201\sqrt{2})^2}{2 \times 10^3} \cong 40 \text{ m}$$

Ans. (D)

Sol. 29 When a bead strikes with speed v and rebond the change in momentum is

$$\Delta P = mv - (-me) = 2mv$$

Between two supoort, the distance beads travel on rod is

$$r = L - n(2r) \text{ (diameter of each bead)}$$

Time between two successive collisions on same support is

$$\Delta t = \frac{2r}{v} = \frac{2[L - 2nr]}{v}$$

Average force on each support is

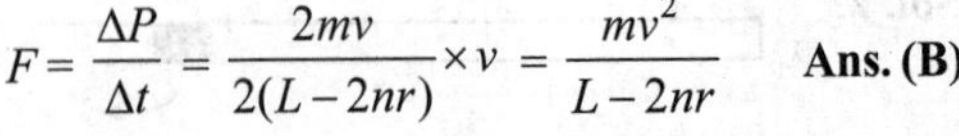

$$F = \frac{\Delta P}{\Delta t} = \frac{2mv}{2(L-2nr)} \times v = \frac{mv^2}{L-2nr}$$

Ans. (B)

Sol. 30

$$dm = \mu\, dx = \left(a + \frac{bx}{L}\right) dx$$

COM of rod is given as

$$x_{cm} = \frac{\int dm x}{\int dm} = \frac{\int_0^L \left(a + \frac{bx}{L}\right) x dx}{\int_0^L \left(a + \frac{bx}{L}\right) dx}$$

$$\Rightarrow \quad x_{cm} = \frac{\left(\frac{aL^2}{2} + \frac{bL^2}{3L}\right)}{\left(aL + \frac{bL^2}{2L}\right)} = \frac{7L}{12}$$

$$\Rightarrow \quad \frac{7L}{12} = \frac{\frac{a}{2} + \frac{b}{3}}{a + \frac{b}{2}}$$

$$\Rightarrow \quad 7a + \frac{7b}{2} = 6a + 4b$$

$$\Rightarrow \quad a = \frac{b}{2}$$

$$\Rightarrow \quad b = 2a$$ **Ans. (B)**

Sol. 31

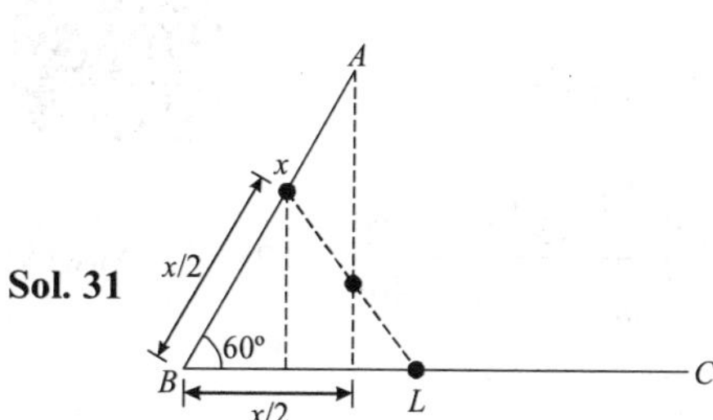

From point B, x coordinate of COM of ABC is located at a distance $x/2$ from B and for BC, COM is at a distance $L/2$ from B.

$$\Rightarrow \quad x_{cm} = \frac{m_1 x_1 + m_2 x_2}{m_1 + m_2}$$

$$\Rightarrow \quad \frac{x}{2} = \frac{\lambda x\left(\frac{x}{2}\cos 60°\right) + \lambda L\left(\frac{L}{2}\right)}{\lambda x + \lambda L}$$

If λ kg/m is linear mass density of wire

$\Rightarrow m_1 = \lambda x$ (AB) & $m_2 = \lambda L$ (BC)

$$\Rightarrow \quad \frac{x}{2}(x + L) = \frac{x^2}{4} + \frac{L^2}{2}$$

$$\Rightarrow \quad \frac{x^2}{2} + \frac{xL}{2} = \frac{x^2}{4} + \frac{L^2}{2}$$

$$\Rightarrow \quad \frac{x^2}{4} + \frac{xL}{2} = \frac{L^2}{2}$$

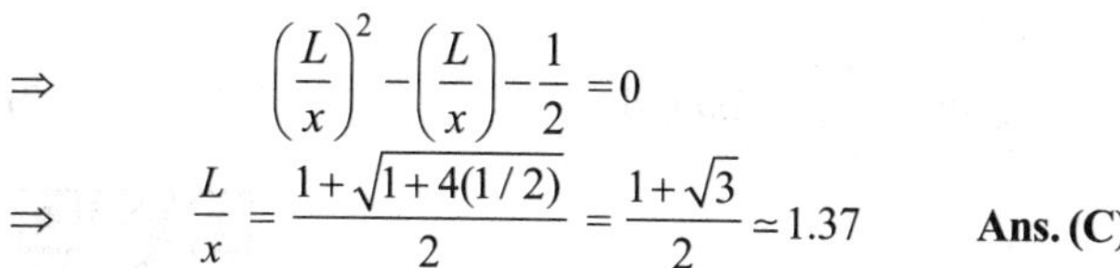

$$\Rightarrow \quad \left(\frac{L}{x}\right)^2 - \left(\frac{L}{x}\right) - \frac{1}{2} = 0$$

$$\Rightarrow \quad \frac{L}{x} = \frac{1 + \sqrt{1 + 4(1/2)}}{2} = \frac{1 + \sqrt{3}}{2} \simeq 1.37$$ **Ans. (C)**

Sol. 32

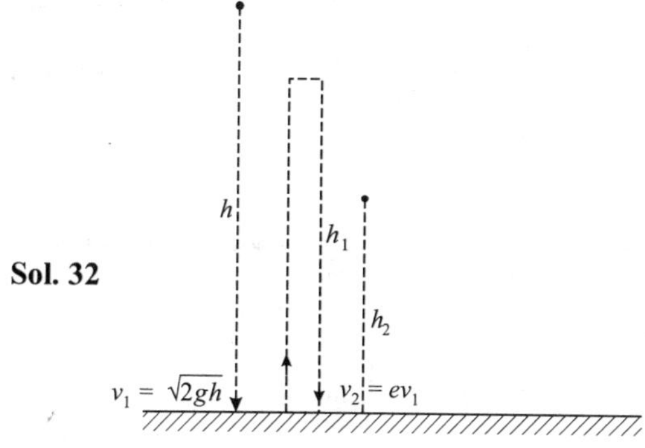

Given that

$$\frac{1}{2} m v_2^2 = \frac{1}{2}\left(\frac{1}{2} m v_1^2\right)$$

$$\Rightarrow \quad v_2 = \frac{1}{\sqrt{2}} v_1$$

We consider coefficient of restitution for this collision to be

$$e = \frac{1}{\sqrt{2}}$$

height attained by object after first impact

$$h_1 = \frac{v_2^2}{2g} = e^2\left(\frac{v_1^2}{2g}\right) = e^2 h$$

after second impact

$$h_2 = e^2 h_1 = e^4 h$$

at $t \to \infty$ total distance covered is

$$s = h + 2h_1 + 2h_2 + \ldots \infty$$

$$\Rightarrow \quad s = h(1 + 2e^2 + 2e^4 + \ldots \infty)$$

$$\Rightarrow \quad s = h\left[1 + 2e^2\left(\frac{1}{1 - e^2}\right)\right] = h\left(\frac{1 + e^2}{1 - e^2}\right)$$

$$\Rightarrow \quad s = h\left(\frac{1 + \frac{1}{2}}{1 - \frac{1}{2}}\right) = 3h$$ **Ans. (A)**

Sol. 33 By conservation of linear momentum, we have along x-direction $\to mv \sin 30° - mv \sin 45° = 2mv_1 \cos\theta$

$$\Rightarrow \quad \frac{mv}{2} - \frac{mv}{\sqrt{2}} = 2mv_1 \cos\theta \qquad \ldots(1)$$

along y-direction $\to$

$$mv \cos 30° + mv \cos 45° = 2mv_1 \sin\theta$$

$$\Rightarrow \quad \frac{\sqrt{3}}{2}mv + \frac{mv}{\sqrt{2}} = 2mv_1 \sin\theta \qquad \ldots(2)$$

Here $\frac{(2)}{(1)}$ gives

$$\Rightarrow \quad \tan\theta = \frac{\frac{\sqrt{3}}{2} + \frac{1}{\sqrt{2}}}{\frac{1}{2} - \frac{1}{\sqrt{2}}} = \frac{\sqrt{3}+\sqrt{2}}{1-\sqrt{2}}$$

Ans. (A)

Sol. 34 m (v_0) → m rest ⟶ m (v_1), m (v_2)

Given that

$$\frac{1}{2}mv_1^2 + \frac{1}{2}mv_2^2 = \frac{3}{2}\left(\frac{1}{2}mv_0^2\right)$$

$$\Rightarrow \quad v_1^2 + v_2^2 = \frac{3}{2}v_0^2 \qquad \ldots(1)$$

By conservation of linear momentum, we have

$$mv_0 = mv_1 + mv_2$$

$$\Rightarrow \quad v_1 + v_2 = b_0 \qquad \ldots(2)$$

$$(2)^2 \Rightarrow \quad v_1^2 + v_2^2 + 2v_1v_2 = v_0^2$$

$$(2)^2 - (1) \Rightarrow \quad 2v_1v_2 = -\frac{v_0^2}{2}$$

The relative velocity of separation is

$$(v_2 - v_1)^2 = v_1^2 + v_2^2 - 2v_1v_2$$

$$\Rightarrow \quad (v_2 - v_1)^2 = \frac{3}{2}v_0^2 + \frac{v_0^2}{2} = 2v_0^2$$

$$\Rightarrow \quad v_2 - v_1 = \sqrt{2}v_0$$

Ans. (B)

Sol. 35

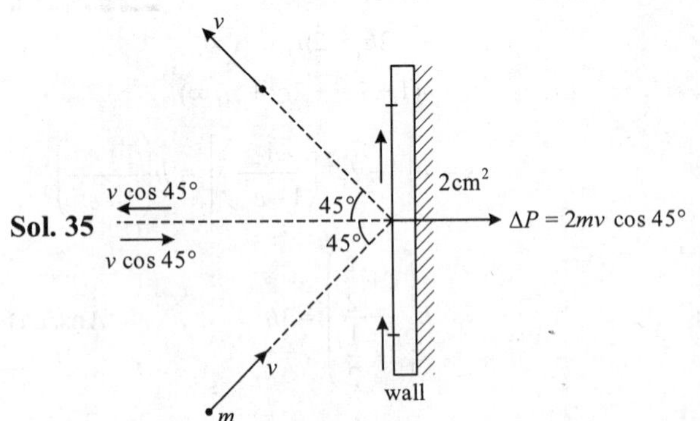

Change in momentum of H-molecule in one collision

Force on wall = change in momentum/sec due to collision on wall.

$$= 2mv\cos 45° \times 10^{23}\text{ N}$$

$$\text{Pressure on wall} = \frac{F}{A}$$

$$\Rightarrow \quad P = \frac{2mv\left(\frac{1}{\sqrt{2}}\right)\times 10^{23}}{2\times 10^{-4}}$$

$$\Rightarrow \quad P = \frac{2\times 3.32\times 10^{-27}\times 10^3 \times \frac{1}{\sqrt{3}}\times 10^{23}}{2\times 10^{-4}}$$

$$= 2.35 \times 10^3\text{ N/m}^2$$

Ans. (A)

Sol. 36 n (m) (u) → D/C $(2m)/(12m)$ ⟶ n (m) (v_1), D/C $(2m)/(12m)$ (v_2)

For elastic collision, final speed of neutron is given as

$$v_1 = \frac{m_1 - m_2}{m_1 + m_2}\cdot u_1 + \frac{2m_2}{m_1 + m_2}u_2$$

$$\Rightarrow \quad v_1 = -\frac{1}{3}u$$

Frictional loss in KE of neutron

$$P_d = \frac{\frac{1}{2}mu^2 - \frac{1}{2}m\left(\frac{u}{3}\right)^2}{\frac{1}{2}mu^2} = \frac{8}{9} = 0.89$$

For collision with C-nucleus we use $m_2 = 12m$ so final speed of neutron is given as

$$v_1' = -\frac{11}{13}u$$

fractional loss in KE of neutron

$$P_c = \frac{\frac{1}{2}mu^2 - \frac{1}{2}m\left(\frac{4}{13}u\right)^2}{\frac{1}{2}mu^2} = \frac{48}{469} = 0.28$$

Ans. (A)

Sol. 37

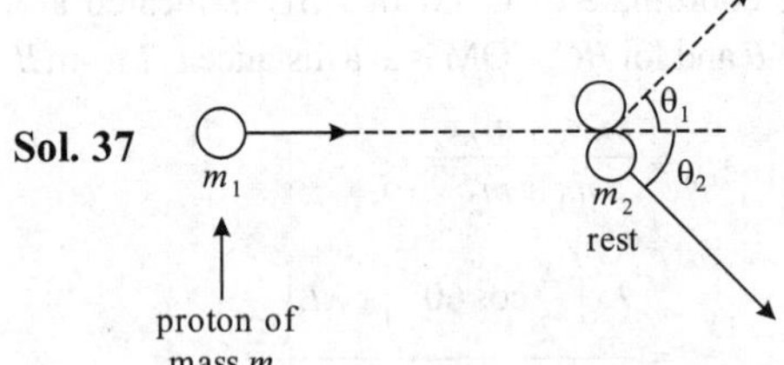

We have discussed that for an elastic collision if $m_1 = m_2$

Then divergence angle is

$$\theta_1 + \theta_2 = 90°$$

here $\quad m_2 = m$ for $\theta_1 + \theta_2 = 90°$

Ans. (D)

Sol. 38 $\quad k_i = \frac{1}{2}mv_0^2$

From linear momentum conservation

$$mv_0 = (2m + M)\,v_f$$

$$\Rightarrow \qquad v_f = \frac{mv_0}{2m+M}$$

Given that

$$\frac{k_i}{k_f} = 6$$

$$\Rightarrow \qquad \frac{\frac{1}{2}mv_0^2}{\frac{1}{2}(2m+M)\left(\frac{mv_0}{2m+M}\right)^2} = 6$$

$$\Rightarrow \qquad \frac{2m+M}{m} = 6$$

$$\Rightarrow \qquad \frac{M}{m} = 4$$

Ans. (A)

Sol. 39

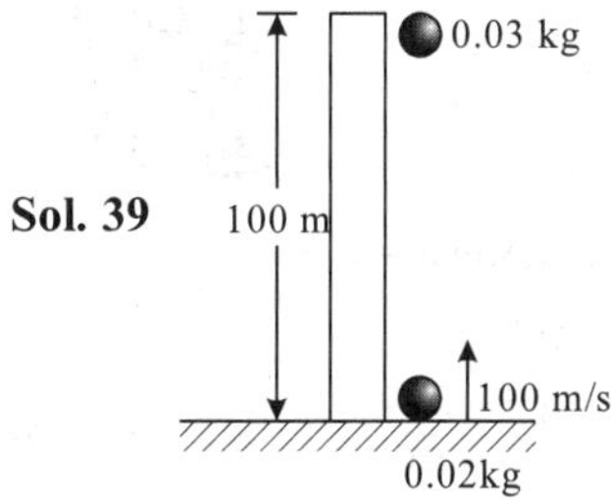

Time taken for the particles to collide,

$$t = \frac{d}{V_{rel}} = \frac{100}{100} = 1 \text{ sec}$$

In 1s distance fallen by wood is

$$h_1 = \frac{1}{2}gt^2 = \frac{1}{2} \times 10 \times (1)^2 = 5 \text{ m}$$

Speed of wood just before collision = gt = 10 m/s & speed of bullet just before collision v-gt

$$= 100 - 10 = 90 \text{ m/s}$$

Before | After
0.03kg ↓10 m/s
↑90 m/s 0.05 m/s
0.02kg
↑ v

Now, conservation of linear momentum just before and after the collision

$$-(0.03)(1v) + (0.02)(9v) = (0.05)v$$

$$\Rightarrow \qquad 150 = 5v$$

$$\Rightarrow \qquad v = 30 \text{ m/s}$$

Maximum height reached by body $h = \dfrac{v^2}{2g}$

$$\Rightarrow \qquad h = \frac{30 \times 30}{2 \times 10} = 45 \text{ m}$$

$\Rightarrow$ Height above tower = 45 – 5 = 40 m **Ans. (C)**

Sol. 40

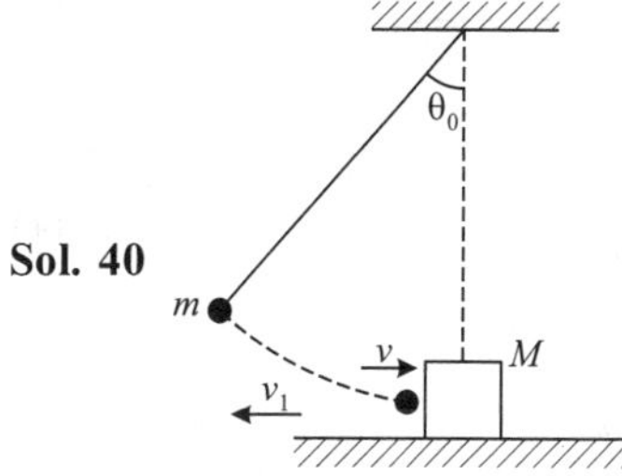

If m strikes M at speed v and rebound at v_1, we have

$$v = \sqrt{2gl(1-\cos\theta_0)}$$

and

$$v_1 = \sqrt{2gl(1-\cos\theta_1)}$$

By momentum conservation, we use

$$m\sqrt{2gl(1-\cos\theta_0)} = MV_m - m\sqrt{2gl(1-\cos\theta)}$$

$$\Rightarrow \quad m\sqrt{2gl}\{\sqrt{1-\cos\theta_0} + \sqrt{1-\cos\theta_1}\} = MV_m \qquad \ldots(1)$$

For elastic collision, we use

$$e = 1 = \frac{V_m + \sqrt{2gl(1-\cos\theta_1)}}{\sqrt{2gl(1-\cos\theta_0)}}$$

$$\Rightarrow \quad \sqrt{2gl}(\sqrt{1-\cos\theta_0} - \sqrt{1-\cos\theta_1}) = V_m \qquad \ldots(2)$$

(1)/(2) gives

$$\frac{(\sqrt{1-\cos\theta_0} + \sqrt{1-\cos\theta_1})}{(\sqrt{1-\cos\theta_0} - \sqrt{1-\cos\theta_1})} = \frac{M}{m}$$

$$\Rightarrow \qquad \frac{m-M}{m+M} = \frac{\sqrt{1-\cos\theta_1}}{\sqrt{1-\cos\theta_0}} = \frac{\sin\left(\frac{\theta_1}{2}\right)}{\sin\left(\frac{\theta_0}{2}\right)}$$

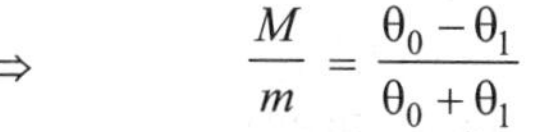

$$\Rightarrow \qquad \frac{M}{m} = \frac{\theta_0 - \theta_1}{\theta_0 + \theta_1}$$

$$\Rightarrow \qquad M = m\left(\frac{\theta_0 - \theta_1}{\theta_0 + \theta_1}\right)$$

Ans. (C)

Sol. 41

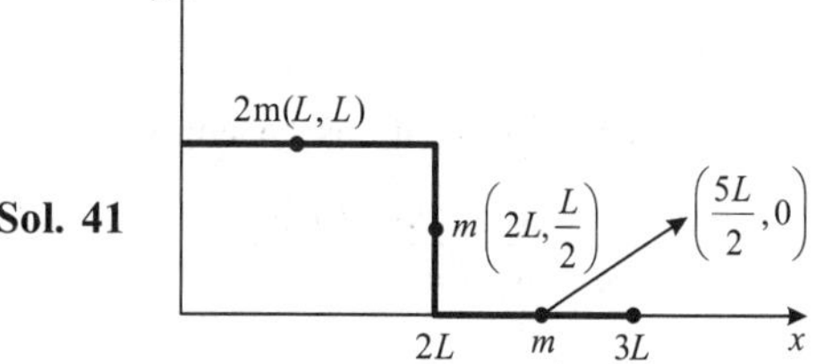

Centre of mass of three segments of bar are shown in figure. So centre of mass coordinates (X_{cm}, Y_{cm}) are given as

$$X_{cm} = \frac{2mL + 2mL + \frac{5mL}{2}}{4m} = \frac{13}{8}L$$

$$Y_{cm} = \frac{2m \times L + m \times \left(\frac{L}{2}\right) + m \times 0}{4m} = \frac{5L}{8}$$

Ans. (A)

Sol. 42 α N V_1 V_2

Using momentum conservation

$$mv_0 = Mv_2 - mv_1 \qquad \dots(1)$$

Given that

$$\frac{1}{2}mv_1^2 = 0.36\times\frac{1}{2}mv_0^2$$

$$\Rightarrow \quad v_1 = 0.6v_0$$

$$\Rightarrow \quad \frac{1}{2}Mv_2^2 = 0.64\times\frac{1}{2}mv_0^2$$

$$\Rightarrow \quad v_2 = \sqrt{\frac{m}{M}}\times 0.8v_0$$

From equation-(1)

$$mv_0 = \sqrt{mM}\times 0.8v_0 - m\times 0.6v_0$$

$$\Rightarrow \quad 1.6\,\text{m} = 0.8\sqrt{mM}$$

$$\Rightarrow \quad 4\,\text{m}^2 = mM$$

$$\Rightarrow \quad M = 4\,\text{m}$$

Ans. (A)

Sol. 43

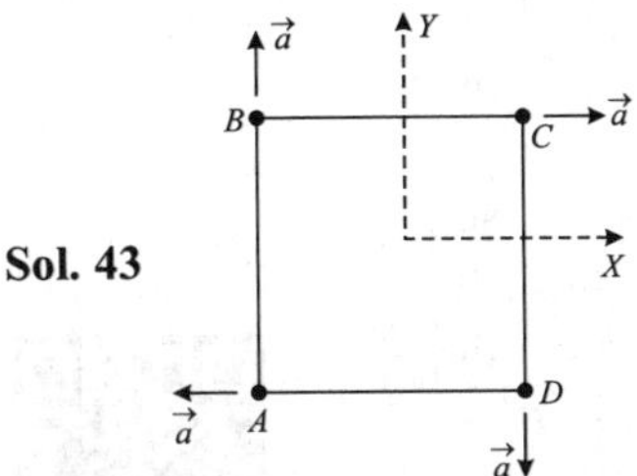

The acceleration vector of four particles are given as

$$\vec{a}_A = -a\hat{i}$$

$$\vec{a}_B = a\hat{j}$$

$$\vec{a}_C = a\hat{i}$$

$$\vec{a}_D = -a\hat{j}$$

Thus acceleration of centre of mass of this system is

$$\vec{a}_{cm} = \frac{m_a\vec{a}_a + m_b\vec{a}_b + m_c\vec{a}_c + m_d\vec{a}_d}{m_a + m_b + m_c + m_d}$$

$$\Rightarrow \quad \vec{a}_{cm} = \frac{-ma\hat{i} + 2ma\hat{j} + 3ma\hat{i} - 4ma\hat{j}}{10\text{m}}$$

$$= \frac{2ma\hat{i} - 2ma\hat{j}}{10\text{m}}$$

$$= \frac{a}{5}\hat{i} - \frac{a}{5}\hat{j}$$

$$= \frac{a}{5}(\hat{i} - \hat{j})$$

Ans. (A)

Sol. 44 In figure below O is COM of complete sheet and C is COM of the shaded portion which is cut off

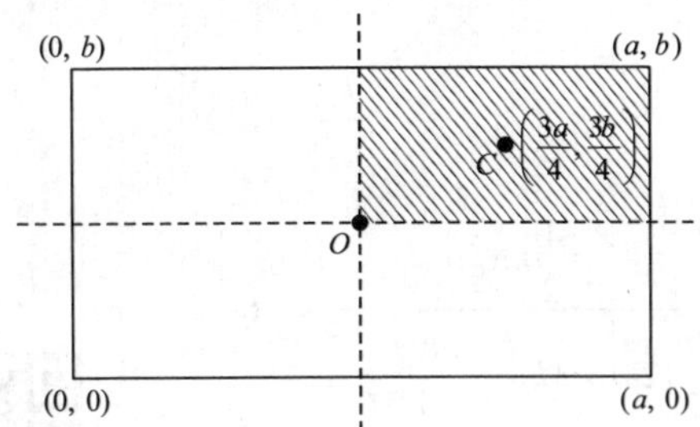

x-coordinate of COM of remaing sheet will be

$$x = \frac{M\frac{a}{2} - \frac{M}{4}\times\frac{3a}{4}}{M - \frac{M}{4}}$$

$$\Rightarrow \quad x = \frac{\frac{a}{2} - \frac{3a}{16}}{\frac{3}{4}} = \frac{\frac{5a}{16}}{\frac{3}{4}} = \frac{5a}{12}$$

y-coordinate of COM of remaining sheet will be

$$y = \frac{M\frac{b}{2} - \frac{M}{4}\times\frac{3b}{4}}{M - \frac{M}{4}} = \frac{5b}{12}$$

Ans. (D)

Sol. 45 Using linear momentum conservation, we have

$$m_1v_1\hat{i} + m_2v_2\hat{i} = m_1v_3\hat{i} + m_2v_4\hat{i}$$

$$m_1v_1 + 0.5m_1v_2 = m_1(0.5v_1) + 0.5m_1v_4$$

$$0.5m_1v_1 = 0.5m_1(v_4 - v_2)$$

$$v_1 = v_4 - v_2$$

Ans. (C)

Sol. 46

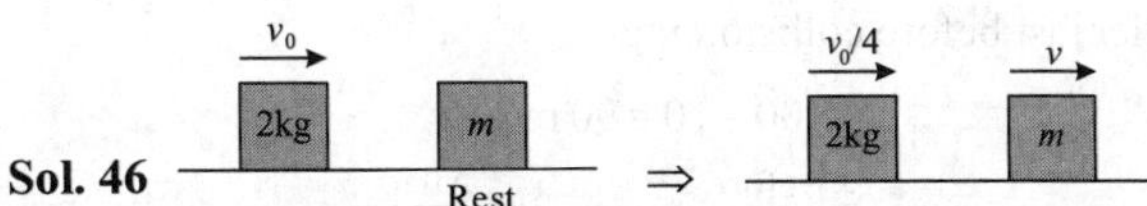

Using conservation of linear momentum, we have

$$2v_0 = 2\left(\frac{v_0}{4}\right) + mv$$

$$\Rightarrow \quad 2v_0 = \frac{v_0}{2} + mv$$

$$\Rightarrow \quad \frac{3v_0}{2} = mv \qquad \dots(1)$$

As collision is elastic, we use $e = 1$, so we have

$$v_{\text{separation}} = v_{\text{approch}}$$

$$\Rightarrow \quad v - \frac{v_0}{4} = v_0$$

$$\Rightarrow \quad \frac{5v_0}{4} = v \qquad \dots(2)$$

From (1) and (2), we use

$$\Rightarrow \quad \frac{3v_0}{2} = m\left(\frac{5v_0}{4}\right)$$

$$\Rightarrow \quad m = \frac{6}{5} = 1.2 \text{ kg}$$ **Ans. (B)**

Sol. 47

Before collision — After collision

Using linear momentum conservation, we have

$$m(2v) + 2m(v) = m \times 0 + m\left(\frac{v'}{\sqrt{2}} \times 2\right)$$

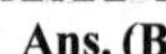

$$\Rightarrow \quad v' = 2\sqrt{2}\, v$$ **Ans. (B)**

Sol. 48

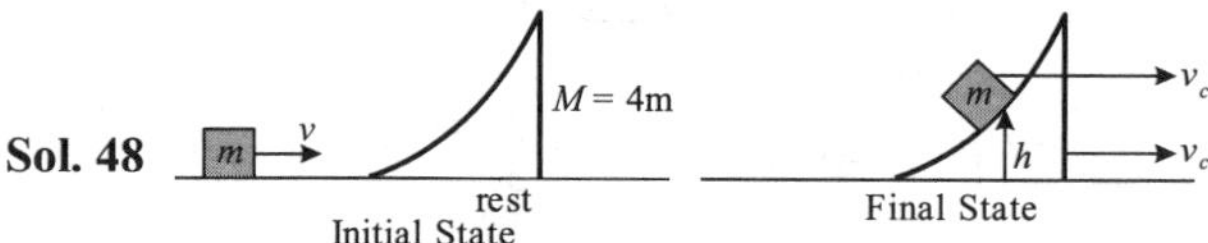

Using linear momentum conservation, we have

$$mv = (m+M)v_c$$

$$\Rightarrow \quad v_c = \frac{v}{5}$$

Using work energy theorem, we have

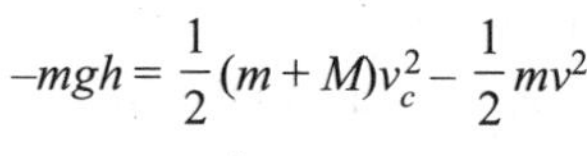

$$-mgh = \frac{1}{2}(m+M)v_c^2 - \frac{1}{2}mv^2$$

$$\Rightarrow \quad h = \frac{2v^2}{5g}$$ **Ans. (C)**

Sol. 49 By conservation of momentum, we use

in X-direction $M(10)\cos 30° + 2M(5)\cos 45° = 2Mv_1 \cos 30° + Mv_2 \cos 45°$

$$5\sqrt{3} + 5\sqrt{2} = \sqrt{3}\, v_1 = \frac{v_2}{\sqrt{2}} \qquad \ldots(1)$$

in Y-direction $10M\sin 30° - 2M(5)\sin 45° = Mv_2 \sin 45° - 2Mv_1 \sin 30°$

$$5 - 5\sqrt{2} = \frac{v_2}{\sqrt{2}} - v_1 \qquad \ldots(2)$$

(1)(2) $5 \times 1.732 + 10 \times 1.414 - 5 = (\sqrt{3}+1)v_1$

$$\Rightarrow \quad v_1 = \frac{8.66 + 14.14 - 5}{2.732}$$

$$\Rightarrow \quad v_1 = 6.51 \text{ m/s}$$

From (2) we use

$$v_2 = (5 - 7.07 + 6.51) \times \sqrt{2}$$

$$\Rightarrow \quad v_2 = 4.44 \times 1.414$$

$$\Rightarrow \quad v_2 = 6.28 \text{ m/s}$$ **Ans. (C)**

Sol. 50 $m = 20g$,

$u = 1$ m/s,

$v = ?$

$S = 20 \times 10^{-2}$ m

$$a = \frac{-2.5 \times 10^{-2}}{20 \times 10^{-3}} \text{ m/s}^2$$

$$v^2 = u^2 + 2as$$

$$\Rightarrow \quad v^2 = 1 - 2 \times \frac{2.5 \times 10^{-2}}{20 \times 10^{-3}} \times \frac{20}{100}$$

$$\Rightarrow \quad v = \frac{1}{\sqrt{2}} \simeq 0.7 \text{m/s}$$ **Ans. (D)**

Sol. 51 By conservation of angular momentum we use

$$Mv_0 L = Mv_1(L - l)$$

$$\Rightarrow \quad v_1 = v_0\left(\frac{L}{L-l}\right) \qquad \ldots(1)$$

Gain in KE of system in

$$\Delta k = \frac{1}{2}M(v_1^2 - v_0^2)$$

$$= \frac{1}{2}Mv_0^2\left[\left(\frac{L}{L-l}\right)^2 - 1\right]$$

$$= \frac{1}{2}Mv_0^2\left[\left(1 - \frac{l}{L}\right)^{-2} - 1\right]$$

$$= Mv_0^2 \cdot \frac{l}{L}$$

Total W/D by person is $W = Mgl + \dfrac{Mv_0^2 l}{L}$

$$= Mgl + \frac{Ml}{L} \cdot gL\theta_0^2$$

$$= mgl(1 + \theta_0^2)$$ **Ans. (B)**

Sol. 52

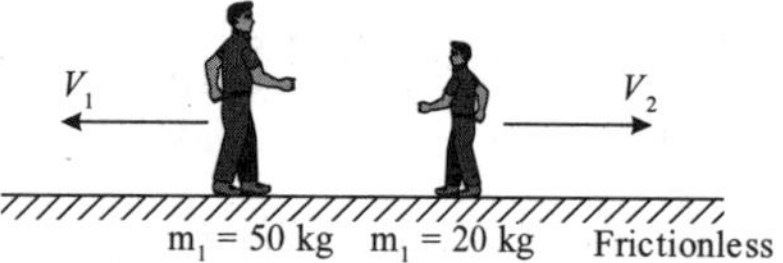

By conservation of momentum, we use

$$m_1 v_1 = m_2 v_2$$

$$50v_1 = 20v_2$$

and $\quad v_1 + v_2 = 0.7$

$$\Rightarrow \quad v_1 + 2.5v_1 = 0.7$$

$$\Rightarrow \quad v_1 = \frac{0.7}{3.5} = 0.2 \text{ m/a}$$ **Ans. (A)**

Sol. 53

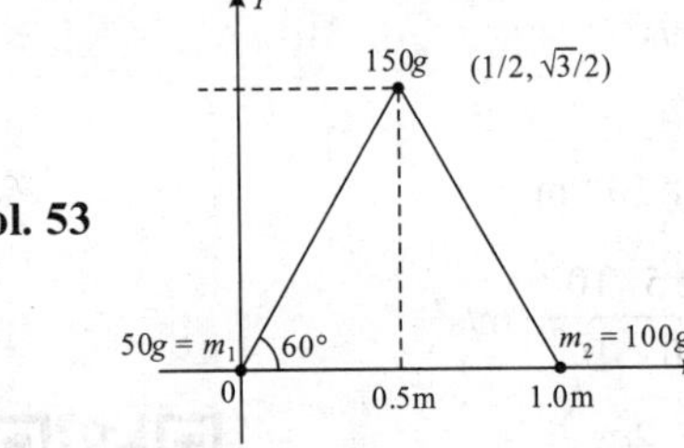

Co-ordinates of COM are

$$x_c = \frac{m_1x_1 + m_2x_2 + m_3x_3}{m_1 + m_2 + m_3} = \frac{50\times 0 + 100\times 1 + 150\times 0.5}{300}$$

$$= \frac{175}{300} = \frac{7}{12}\text{ m}$$

$$y_c = \frac{m_1y_1 + m_2y_2 + m_3y_3}{m_1 + m_2 + m_3} = \frac{50\times 0\times 100\times 0 + 150\times\sqrt{3}/2}{300}$$

$$= \frac{\sqrt{3}}{4}\text{ m}$$ **Ans. (C)**

Sol. 54 Using coordinate axis in given situation, considering 1 kg mass at origin we use

$$x_{cm} = \frac{1(0) + 1.5(3) + 2.5(0)}{5} = 0.9\text{ cm}$$

$$y_{cm} = \frac{1(0) + 1.5(0) + 2.5(4)}{5} = 2\text{ cm}$$ **Ans. (B)**

Sol. 55 In figure we consider masses m_1 and m_2 of two sections shown of 3 kg and 1 kg at centre coordinates (0.5, 1.5) and (1.5, 2.5)

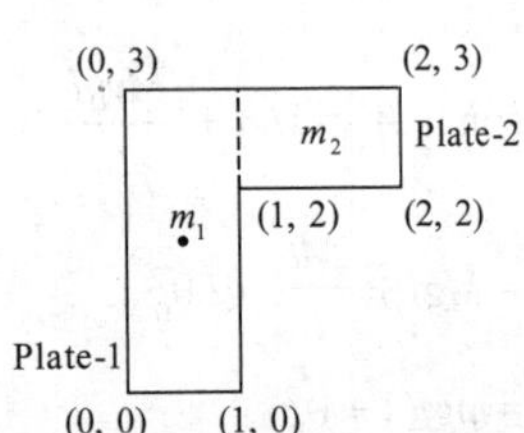

Thus coordinates of centre of mass are given as

$$x_{cm} = \frac{m_1x_1 + m_2x_2}{m_1 + m_2} = \frac{3\times 0.5 + 1\times 1.5}{4} = 0.75$$

$$y_{cm} = \frac{m_1y_1 + m_2y_2}{m_1 + m_2} = \frac{3\times 1.5 + 1\times 2.5}{4} = 1.75$$ **Ans. (D)**

Sol. 56 Using conservation of momentum, we have

$$(0.1)(3\hat{i}) + (0.1)(5\hat{j}) = (0.1)(4)(\hat{i} + \hat{j}) + (0.1)\vec{v}$$

$$\Rightarrow \quad \vec{v} = -\hat{i} + \hat{j}$$

$$\Rightarrow \quad |\vec{v}| = \sqrt{2}$$

Kinetic energy of B is

$$K = \frac{1}{2}mv^2 = \frac{1}{2}(0.1)(2) = \frac{1}{10}$$

$$\Rightarrow \quad x = 1$$ **Ans. (1)**

Sol. 57 Let $m_1 \to$ mass of body

$m_2 \to$ mass of cavity

Using mass moment concept of centre of mass, we have

$$m_1R_1 = m_2R_2$$

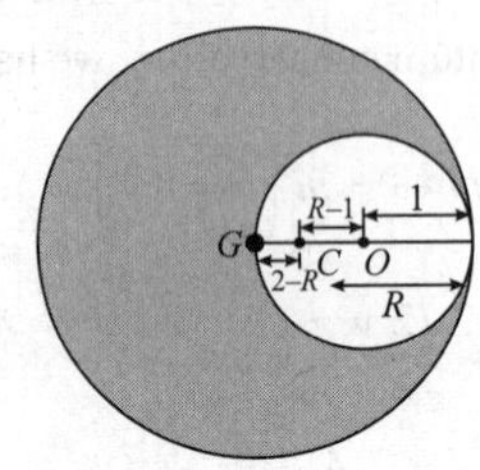

$$\Rightarrow \left(\frac{4}{3}\pi R^3\rho - \frac{4}{3}\pi(1)^3\rho\right)(2-R) = \frac{4}{3}\pi(1)^3\rho\times(R-1)$$

$$\Rightarrow (R^3 - 1)(2-R) = R - 1$$

$$\Rightarrow (R^2 + R + 1)(2-R) = 1$$ **Ans. (C)**

Sol. 58 Particles will collide in air after time given as

$$t_0 = \frac{h}{\sqrt{2gh}} = \sqrt{\frac{h}{2g}}$$

At the instant of collision velocities of the two particle are

$$v_A = gt_0 = \sqrt{\frac{gh}{2}}$$

$$v_B = u_B - gt_0 = \sqrt{\frac{gh}{2}}$$

$$\Rightarrow \quad v_A = v_B = \sqrt{\frac{gh}{2}}$$

Height at which collision takes place is

$$h_1 = h - \frac{1}{2}gt_0^2 = \frac{3h}{4}$$

As after collision, combined mass comes to rest, time taken to reach ground is given as

$$t_1 = \sqrt{\frac{2\times(3h/4)}{g}} = \sqrt{\frac{3h}{2g}}$$ **Ans. (D)**

Sol. 59 Using momentum conservation, we have

$$mu\,\hat{i} + mu\left(\frac{\hat{i}+\hat{j}}{2}\right) = (m+m)\vec{v}$$

$$\Rightarrow \quad \vec{v} = \frac{3}{4}u\hat{i} + \frac{u}{4}\hat{j}$$

$$\Rightarrow \quad v = \frac{u}{4}\sqrt{10}$$

After collision final kinetic energy is given as

$$K_i = \frac{1}{2}2\text{m}\left(\frac{u}{4}\sqrt{10}\right)^2 = \frac{5}{8}\text{mu}^2$$

Initial kinetic energy before collision is

$$K_i = \frac{1}{2}\text{mu}^2 + \frac{1}{2}\text{m}\left(\frac{u}{\sqrt{2}}\right)^2 = \frac{6}{8}\text{mu}^2$$

Loss in kinetic energy is given as

$$\Delta K = K_i - K_f = \frac{1}{8}\text{mu}^2$$ **Ans. (B)**

Sol. 60 Distance of centre of mass of rod from one end is given as

$$x_{\text{cm}} = \frac{\int x dm}{\int dm} = \frac{\int (\rho dx)x}{\int dm}$$

$$\Rightarrow \quad x_{\text{cm}} = \frac{\int_0^L \left(a + \frac{bx^2}{L^2}\right) x dx}{\int_0^L \left(a + \frac{bx^2}{L^2}\right) dx}$$

$$\Rightarrow \quad x_{\text{cm}} = \frac{\frac{aL^2}{2} + \frac{b}{L^2}\cdot\frac{L^4}{4}}{aL + \frac{b}{L^2}\cdot\frac{L^3}{3}}$$

$$\Rightarrow \quad x_{\text{cm}} = \frac{\left(\frac{4a+2b}{8}\right)L}{\frac{(3a+b)}{3}} = \frac{3}{4}\frac{(2a+b)L}{(3a+b)}$$ **Ans. (D)**

Sol. 61 Velocity of particle at highest point is given as

$$V_x = u\cos 60° = \frac{u}{2}$$

Using momentum conservation at the highest point, final velocity of combined mass is given as

$$\frac{mu}{2} + mu = 2mv'$$

$$\Rightarrow \quad v' = \frac{3v}{4}$$

Range of combined particle after collision is given as

$$R = \frac{3v}{4}\sqrt{\frac{2H}{g}}$$

$$\Rightarrow \quad R = \frac{3v}{4}\sqrt{\frac{2}{g}\left(\frac{u^2\sin^2 60°}{2g}\right)}$$

$$\Rightarrow \quad R = \frac{3}{4}\frac{\sqrt{3}}{2}\cdot\frac{u^2}{g}$$

$$\Rightarrow \quad R = \frac{3\sqrt{3}u^2}{8g}$$ **Ans. (C)**

Sol. 62 By momentum conservation principle, we have

$$m(u\hat{i}) + 3m(0) = mv\hat{j} + 3m\vec{v}_1$$

$$\Rightarrow \quad mu\hat{i} - mv\hat{j} = 3m\vec{v}_1$$

$$\Rightarrow \quad \vec{v}_1 = \frac{u\hat{i} - v\hat{j}}{3}$$

$$\Rightarrow \quad |\vec{v}_1| = \frac{\sqrt{u^2+v^2}}{3}$$

$$\Rightarrow \quad v_1^2 = \frac{u^2+v^2}{9} \qquad \ldots(1)$$

As collision is perfectly elastic, we have

$$KE_i = KE_j$$

$$\Rightarrow \quad \frac{1}{2}mu^2 + \frac{1}{2}3m(0)^2 = \frac{1}{2}mv^2 + \frac{1}{2}3mv_1^2$$

$$\Rightarrow \quad u^2 = v^2 + 3v_1^2$$

$$\Rightarrow \quad u^2 = v^2 + 3\frac{(u^2+v^2)}{9}$$

$$\Rightarrow \quad 3u^2 = 3v^2 + u^2 + v^2$$

$$\Rightarrow \quad 2u^2 = 4v^2$$

$$\Rightarrow \quad v = \frac{u}{\sqrt{2}}$$ **Ans. (D)**

Sol. 63

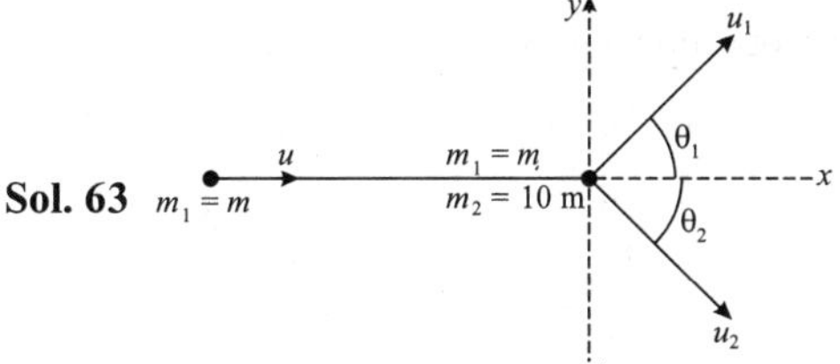

By momentum conservation along y direction, we have

$$m_1u_1\sin\theta_1 = m_2u_2\sin\theta_2$$

$$\Rightarrow \quad mu_1\sin\theta_1 = 10mu_2\sin\theta_2 \qquad \ldots(1)$$

Given that

$$kE_{m_{1(\text{final})}} = \frac{1}{2}kE_{m_{1(\text{initial})}}$$

$$\Rightarrow \quad \frac{1}{2}mu_1^2 = \frac{1}{2}\times\frac{1}{2}mu^2$$

$$\Rightarrow \quad u_1 = \frac{u}{\sqrt{2}} \qquad \ldots(2)$$

As collision is elastic, we use

$$\frac{1}{2}mu^2 = \frac{1}{2}mu_1^2 + \frac{1}{2}(10\,m)u_2^2$$

$$\Rightarrow \quad \frac{1}{2}mu^2 = \frac{1}{2} \times \frac{1}{2}mu^2 + \frac{1}{2} \times 10m \cdot u_2^2$$

$$\Rightarrow \quad \frac{1}{4}mu^2 = \frac{1}{2} \times 10 \times mu_2^2$$

$$\Rightarrow \quad u_2 = \frac{u}{\sqrt{20}} \qquad \dots(3)$$

From (1), (2) and (3), we have

$$\frac{u}{\sqrt{2}}\sin\theta_1 = 10 \cdot \frac{u}{\sqrt{20}}\sin\theta_2$$

$$\Rightarrow \quad \sin\theta_1 = \sqrt{10}\,\sin\theta_2$$

$$\Rightarrow \quad n = 10$$

Ans. (10)

Sol. 64

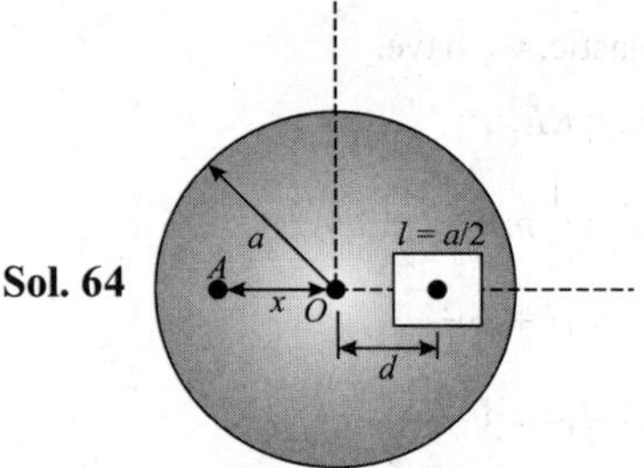

If centre of mass of remaining portion is at point A by mass moment concept, we have

$$m_1 x = m_2\left(\frac{a}{2}\right)$$

Hence m_1 is the mass of complete disc and m_2 is the removed mass from square hole.

As masses are proportional to surface are, we use

$$\left(\pi a^2 - \frac{a^2}{4}\right)x = \frac{a^2}{4}\left(\frac{a}{2}\right)$$

$$\Rightarrow \quad x = \frac{a}{2(4\pi - 1)}$$

$$\Rightarrow \quad X = 2(4\pi - 1)$$

$$\Rightarrow \quad X = 8\pi - 2 = 23.12$$

Thus nearest integral value of X is 23. **Ans. (23)**

Sol. 65

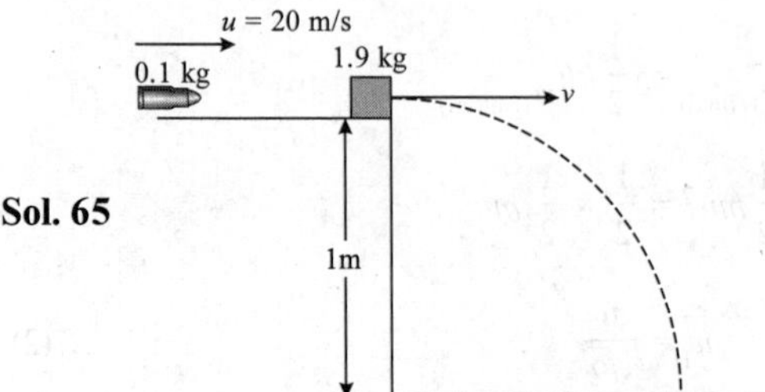

By linear momentum conservation we use

$$0.1 \times 20 = 2v$$

$$\Rightarrow \quad v = 1 \text{ m/s}$$

Final kinetic energy of combined system is given as

$$K_f = mgh + \frac{1}{2}mv^2$$

$$\Rightarrow \quad K_f = 2 \times 10 \times 1 + \frac{1}{2} \times 2\,(1)^2$$

$$\Rightarrow \quad K_f = 20 + 1 = 21 \text{ J}$$

Ans. (A)

Sol. 66 At a height h speed is given as

$$v = \sqrt{2g(H-h)}$$

The graph is parabolic and most appropriate option is (C) which can be analyzed by slope variation. **Ans. (C)**

Sol. 67 By conservation of momentum when last block starts moving

$$mv = 16\,mv'$$

$$\Rightarrow \quad v' = \frac{v}{16}$$

Initial energy $\quad E_1 = \frac{1}{2}mv^2$

Final energy $\quad E_2 = \frac{1}{2}(16m)\left(\frac{v}{16}\right)^2 = \frac{1}{32}mv^2$

Energy loss $\quad \Delta E = E_1 - E_2 = \frac{15}{32}mv^2$

Percentage energy loss is given as

$$\%\text{ loss} = \frac{\frac{15}{32}mv^2}{\frac{1}{2}mv^2} \times 100 = \frac{15}{16}\%$$

Ans. (D)

Sol. 68 Figure below shows the oblique collision in the given situation

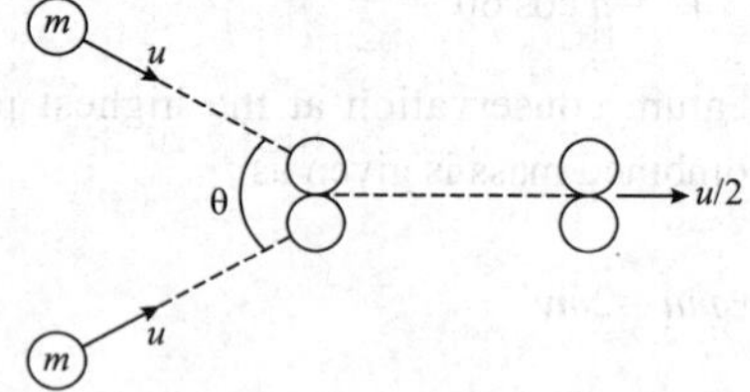

By conservation of momentum, we use

$$2mu\cos\frac{\theta}{2} = 2m\left(\frac{u}{2}\right)$$

$$\Rightarrow \quad \cos\frac{\theta}{2} = \frac{1}{2}$$

$$\Rightarrow \quad \frac{\theta}{2} = 60^\circ$$

$$\Rightarrow \quad \theta = 120^\circ$$

Ans. (120)

Sol. 69 Using conservation of linear momentum before and after collision, we have

$$m_1 \vec{u} = m_1 \vec{v}_1 + m_2 \vec{v}_2$$

$$\Rightarrow \quad 2m_2(\sqrt{3}\hat{i} + \hat{j}) + 0 = 2m_2(\hat{i} + \sqrt{3}\hat{j}) + m_2 \vec{v}_2$$

$$\Rightarrow \quad \vec{v}_2 = 2(\sqrt{3}\hat{i} + \hat{j}) - 2(\hat{i} + \sqrt{3}\hat{j})$$

$$\Rightarrow \quad \vec{v}_2 = 2(\sqrt{3}\hat{i} - \hat{j}) + 2(\hat{i} - \sqrt{3}\hat{j})$$

$$\Rightarrow \quad \vec{v}_2 = 2(\sqrt{3}\hat{i} - \hat{j})(\hat{i} - \hat{j})$$

Angle between $\vec{v}_1$ and $\vec{v}_2$ is given as

$$\cos\theta = \frac{\vec{v}_1 \cdot \vec{v}_2}{|\vec{v}||\vec{v}_2|} = \frac{2(\sqrt{3}-1)(1-\sqrt{3})}{2\times 2\sqrt{2}(\sqrt{3}-1)}$$

$$\Rightarrow \quad \cos\theta = \frac{1-\sqrt{3}}{2\sqrt{2}}$$

$$\Rightarrow \quad \theta = 105^\circ$$

Ans. (D)

Sol. 70 Centre of mass of hemisphere from its base centre is given as

$$x = \frac{3R}{8} = 3 \text{ cm}$$

$$x = 3$$

Ans. (3)

Sol. 71 The situation described in the question is shown in figure below

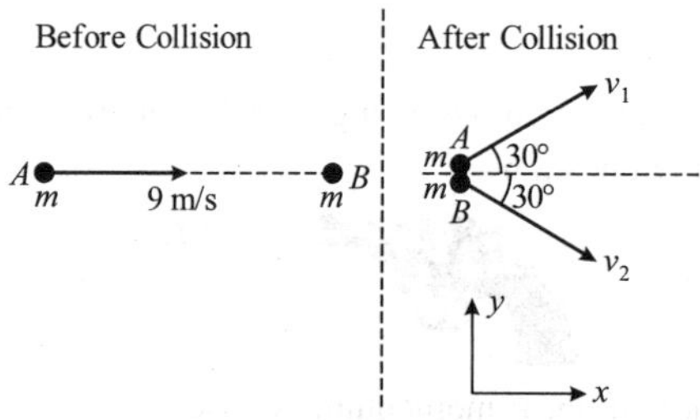

From conservation of momentum along the direction perpendicular to initial line of motion, we have

$$mv_2 \sin 30^\circ = mv_1 \sin 30^\circ$$

$$\Rightarrow \quad v_2 = v_1$$

$$\Rightarrow \quad \frac{v_1}{v_2} = 1$$

Ans. (1)

Sol. 72 If σ be the uniform mass density of disc then center of mass of the given object can be calculated as

$$x_{\text{COM}} = \frac{(\sigma\pi a^2)a - \sigma\pi\left(\frac{a^2}{4}\right)\times\frac{3a}{2}}{\sigma\pi a^2 - \frac{\sigma\pi a^2}{4}}$$

$$\Rightarrow \quad x_{\text{COM}} = \frac{a - \frac{3a}{8}}{1-\frac{1}{4}} = \frac{5a}{6}$$

Ans. (A)

Sol. 73 The kinetic energy of a particle is related to its momentum as

$$K = \frac{P^2}{2m}$$

As momentum are equal for both solids, so we have

$$K \propto \frac{1}{m}$$

$$\Rightarrow \quad \frac{K_A}{K_B} = \frac{m_B}{m_A} = \frac{2}{1}$$

Ans. (2)

Sol. 74 The kinetic energy of a particle is related to its momentum as

$$K = \frac{P^2}{2m}$$

As kinetic energies of both are same, we use

$$\frac{p_1^2}{2\times 4} = \frac{p_2^2}{2\times 16}$$

$$\Rightarrow \quad \frac{p_1}{p_2} = \frac{1}{2}$$

Ans. (1)

Sol. 75 For elastic collision, we use coefficient of restitution value $e = 1$ & second body was initially at rest, so final velocity can be given as

$$v_2 = \frac{2m_1u_1}{m_1+m_2} = \frac{2u(M)}{M+m}$$

As $M >> m$, we have

$$v_2 = 2u$$

Ans. (B)

Sol. 76 The acceleration of center of mass of the given system can be written as

$$a_{\text{cm}} = \frac{m_1a_1 + m_2a_2}{m_1+m_2} = \frac{F}{2M}$$

$$\Rightarrow \quad \frac{F}{2M} = \frac{Ma + Ma_B}{2M}$$

$$\Rightarrow \quad a_B = \frac{F - Ma}{M}$$

Ans. (C)

Sol. 77 From conservation of momentum along X-axis, we have

$$\vec{P}_i = \vec{P}_f$$

$$\Rightarrow \quad 10\times 10\sqrt{3} = 200\cos\theta$$

$$\Rightarrow \quad \cos\theta = \frac{\sqrt{3}}{2}$$

$$\Rightarrow \quad \theta = 30^\circ$$

Ans. (30)

Sol. 78 If v_1 is the speed of combined system after collision, then by energy conservation, after collision, we use

$$(M+m)gh = \frac{1}{2}(M+m)v_1^2$$

$$\Rightarrow \quad v_1 = \sqrt{2gh}$$

Using momentum conservation, just before and just after collision, we have

$$mv = (M+m)v_1$$

$$\Rightarrow \quad v = \left(\frac{M+m}{m}\right)$$

$$\Rightarrow \quad v_1 = \frac{6}{10\times10^{-3}} \times \sqrt{2\times9.8\times9.8\times10^{-2}}$$

$$\Rightarrow \quad v_1 \approx 831.55 \text{ m/s}$$ **Ans. (C)**

Sol. 79 Speed of ball after falling through a distance h is given as

$$v_0 = \sqrt{2gh}$$

Bouncing speed is calculated as

$$v = e\sqrt{2gh} = \sqrt{2gh'}$$

$$\Rightarrow \quad e = 0.9$$

Total distance travelled by ball until it comes to rest after infinte bounces is calculated as

$$s = h + 2e^2h + 2e^4h + \ldots$$

Total time of flight of ball after infinite bounces is calculated as

$$t = \sqrt{\frac{2h}{g}} + 2e\sqrt{\frac{2h}{g}} + 2e^2\sqrt{\frac{2h}{g}} + \ldots$$

Thus average speed of ball is given as

$$v_{av} = \frac{s}{t} = 2.5 \text{ m/s}$$ **Ans. (D)**

Sol. 80 The center of mass of the quarter disc is derived in Physics Galaxy Vol 1, chapter 4 so we use the direct result as $\left(\frac{4a}{3\pi}, \frac{4a}{3\pi}\right)$. Thus $x = 4$. **Ans. (4)**

Sol. 81 If the velocity of second piece is taken as v then by conservation of linear momentum, we have

$$10(10\sqrt{3})\hat{i} = (10)(10\hat{j}) + 10\vec{v}$$

$$\Rightarrow \quad \vec{v} = 10\sqrt{3}\hat{i} - 10\hat{j}$$

$$\Rightarrow \quad |\vec{v}| = \sqrt{300+100} = \sqrt{400} = 20 \text{ m/s}$$ **Ans. (20)**

Sol. 82 Figure below shows the situation described in the question

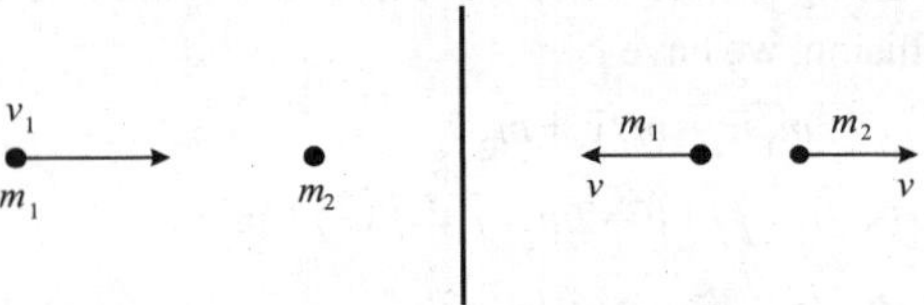

By conservation of linear momentum, we have

$$m_1v_1 = -m_1v + m_2v$$

$$\Rightarrow \quad v_1 = -v + \frac{m_2}{m_1}v$$

$$\Rightarrow \quad \frac{(v_1+v)}{v} = \frac{m_2}{m_1} \quad \ldots(1)$$

For elastic collision, we use coefficient of restitution as

$$e = \frac{2v}{v_1} = 1$$

$$\Rightarrow \quad v = \frac{v_1}{2}$$

From equation-(1), we have

$$\frac{v_1 + v_1/2}{v_1/2} = \frac{m_2}{m_1}$$

$$\Rightarrow \quad 3 = \frac{m_2}{m_1}$$ **Ans. (A)**

Sol. 83 Figure below shows firing of a bullet from gun as described in the question

By conservation of linear momentum, we use

$$0 = m_Gv_G + M_Bv_b$$

$$\Rightarrow \quad 0 = 4v_G + \frac{4}{1000}v_b$$

$$\Rightarrow \quad v_b = -1000\,v_G \quad \ldots(1)$$

$$v_{bG} = v_b - v_G$$

$$\Rightarrow \quad 50 = v_B - v_G \quad \ldots(2)$$

$$\Rightarrow \quad 50 = -1001\,v_G$$

$$v_G \approx 0.05 \text{ m/s}$$

Impulse $= m_Gv_G = 4 \times 0.05 = 0.2$ kg m/s **Ans. (C)**

Sol. 84 The center of mass of a half ring is located at the bisector of the half ring at a distance y_{cm} from its center, given as

$$y_{cm} = \frac{2R}{\pi}$$

$$\Rightarrow \quad x = 2$$ **Ans. (2)**

Sol. 85 By conservation of linear momentum, we have

$$9m = mv_A + 2mv_B$$

$$\Rightarrow \quad 9 = v_A + 2v_B \quad \ldots(1)$$

$$\Rightarrow \quad 9 = v_B - v_A \quad \ldots(2)$$

Solving equations-(1) and (2) gives

$$v_B = 6 \text{ m/s}$$

For collision between B & C, we have

$$2mv_B = 4mv_C$$

$$\Rightarrow \quad v_C = 3 \text{ m/s} \qquad \textbf{Ans. (A)}$$

Sol. 86 For pendulum to complete the vertical circle, final velocity of system (bullet + mass) is given as

$$v_f = \sqrt{5gR}$$

Using conservation of linear momentum before and after collision is given as

$$M\sqrt{5gR} = m(v+100)$$

$$\Rightarrow \quad v = \frac{M}{m}\sqrt{5gr} - 100$$

$$\Rightarrow \quad v = 500 - 100 = 400 \text{ m/s} = 400 \text{ m/s} \qquad \textbf{Ans. (400)}$$

Sol. 87 Figure below shows the situation described in the question.

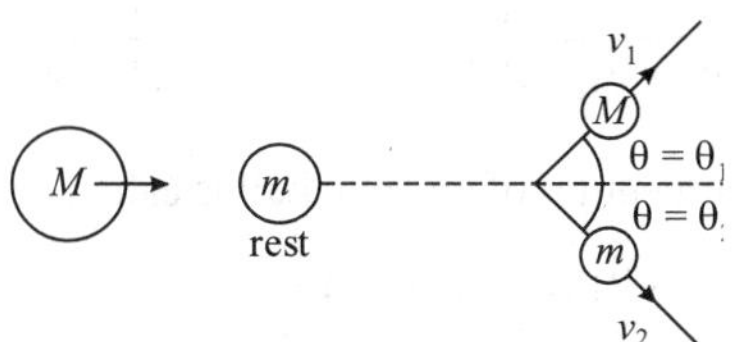

Using linear momentum conservationalong y-direction, we have

$$mv_2 \sin\theta = Mv_1 \sin\theta \quad \ldots(1)$$

$$\Rightarrow \quad mv_2 = Mv_1$$

Along x-direction, we have

$$Mv_0 = mv_2 \cos\theta + Mv_1 \cos\theta \quad \ldots(2)$$

From equations-(1) and (2), we have

$$Mv_0 = mv_2 \cos\theta + mv_2 \cos\theta$$

$$\Delta\Rightarrow \quad mv_0 = 2mv_2 \cos\theta \quad \ldots(3)$$

As collision is elastic, we use $e = 1$

$$\frac{v_2 - v_1 \cos 2\theta}{v_0 \cos\theta} = 1$$

$$\Rightarrow \quad v_2 - v_1 \cos 2\theta = v_0 \quad \ldots(4)$$

$$\Rightarrow \quad v_2 - \frac{mv_2}{M}\cos 2\theta = v_0 \cos\theta$$

From equation-(3), we have

$$v_2\left[1 - \frac{m}{M}\cos 2\theta\right] = v_0 \cos\theta$$

$$\Rightarrow \quad \frac{mv_0}{2m\cos\theta}\left[1 - \frac{m}{M}\cos 2\theta\right] = v_0 \cos\theta$$

$$\Rightarrow \quad M - m\cos 2\theta = 2m\cos 2\theta$$

$$\Rightarrow \quad M = 2m\cos^2\theta + m(2\cos^2\theta - 1)$$

$$\Rightarrow \quad \frac{M}{m} = 4\cos^2\theta - 1$$

$$\Rightarrow \quad \left(\frac{M}{m}\right)_{max} = 4 - 1 = 3 \qquad \textbf{Ans. (D)}$$

Sol. 88 Figure below shows the situation described in the question. If mass of meter scale is m, its weight act at centre of scale as shown.

40 cm
10 cm 30 cm 10 cm
20 gm
0.02 kg
0.02 g
knife
mg

As meter scale is balanced, torque about knife edge for all forces is equal to zero so we have

$$(0.02\text{ g}) \times (30 \times 10^{-2}) = mg \times (10 \times 10^{-2})$$

$$\Rightarrow \quad m = 0.06 \text{ kg} = 6 \times 10^{-2} \text{ kg} \qquad \textbf{Ans. (6)}$$

Sol. 89 Impulse imparted is equal to the change in momentum of the ball, given as

$$J = m[v - (-v)] = 2\,mv$$

$$\Rightarrow \quad J = 2 \times 0.4 \times 15 = 12 \text{ Ns} \qquad \textbf{Ans. (12)}$$

Sol. 90 The displacement of centre of mass is given as

$$\Delta x_G = \frac{m_1 \Delta x_1 + m_2 \Delta x_2}{m_1 + m_2}$$

As centre of mass of system remain unchanged for the given displacement of the masses, we use

$$\Rightarrow \quad 0 = \frac{10 \times 6 + 30(\Delta x_2)}{40}$$

$$\Rightarrow \quad \Delta x_2 = -2 \text{ cm}$$

Block of mass 30 kg will to move towards 10 kg. **Ans. (C)**

Sol. 91 Velocity of second particle after collision is calculated as

$$v_2 = \frac{(m_2 - m_1)u_2 + 2m_1 u_1}{m_1 + m_2}$$

$$\Rightarrow \quad v_2 = \frac{(5m - m)0 + 2m \cdot u_0}{m + 5m} = \frac{u_0}{3}$$

Percentage change in kinetic energy of particle is calculated as

$$\Delta K = \frac{\frac{1}{2}5m\left(\frac{u_0}{3}\right)^2 - 0}{\frac{1}{2}mu_0^2} \times 100$$

$$\Rightarrow \quad \Delta K = \frac{5u_0^2}{9u_0^2} \times 100 = \frac{500}{9} = 55.6\% \quad \textbf{Ans. (C)}$$

Sol. 92 Figure below shows the situation described in the question. For the collision of bullet with bob we consider only horizontal direction motion and time of contact is considered to be very small.

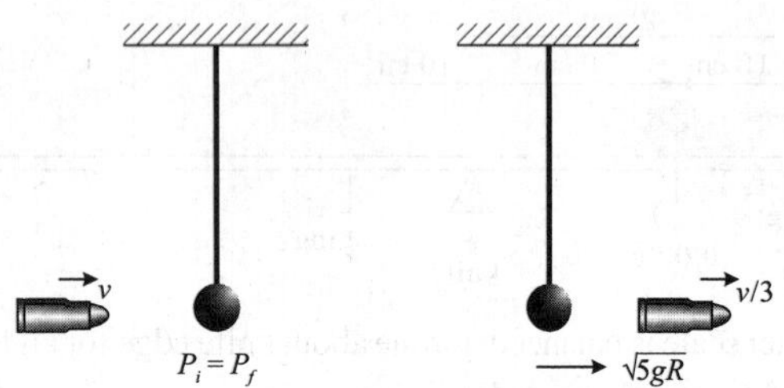

To complete vertical circular motion bob should attain a speed $\sqrt{5gR}$ so by conservation of linear momentum, we use

$$(75v) + 0 = 50(\sqrt{5gR}) + 75\frac{v}{3}$$

$$\Rightarrow \quad 75\left(v - \frac{v}{3}\right) = 50\sqrt{100}$$

$$\Rightarrow \quad v = 10 \text{ m/s} \quad \textbf{Ans. (10)}$$

Sol. 93 Energy supplied by the bullet kinetic energy is used in first raising the temperature of bullet to its melting point then it is used in melting the bullet, so we use

$$m \times 125 \times 200 + m \times 2.5 \times 10^4 = \frac{1}{2}mv^2 \times \frac{40}{100}$$

$$\Rightarrow \quad v = 500 \text{ m/s} \quad \textbf{Ans. (B)}$$

Sol. 94 Taking the system as man and trolley and using conservation of linear momentum, we have

$$60 \times v = (60 + 120) \times 2$$

$$\Rightarrow \quad v = 6 \text{ m/s} \quad \textbf{Ans. (6)}$$

Sol. 95 Mass of pieces after explosion are taken in the given ratio as

$$m_1 = \frac{M}{4}, m_2 = \frac{M}{4}, m_3 = \frac{M}{2}$$

Using conserving of momentum as the body was initially at rest, after explosion we use

$$\vec{P}_1 + \vec{P}_2 + \vec{P}_3 = 0$$

$$\Rightarrow \quad \vec{P}_3 = -(\vec{P}_1 + \vec{P}_2)$$

As $\vec{P}_1$ & $\vec{P}_2$ are perpendicular to each other in magnitude, we use

$$P_3 = \sqrt{P_1^2 + P_2^2}$$

$$\Rightarrow \quad P_3 = (50)\frac{M}{4} = \frac{M}{2}v$$

$$\Rightarrow \quad v = 25 \text{ m/s} \quad \textbf{Ans. (B)}$$

Sol. 96 Force acting on block due to water jet is given as

$$F = \frac{dp}{dt} = v\frac{dm}{dt}$$

$$\Rightarrow \quad Ma = 10 \times 1$$

$$\Rightarrow \quad 2a = 10$$

$$\Rightarrow \quad a = 5 \text{ m/s}^2 \quad \textbf{Ans. (C)}$$

Sol. 97 Figure below shows the situation described in the question

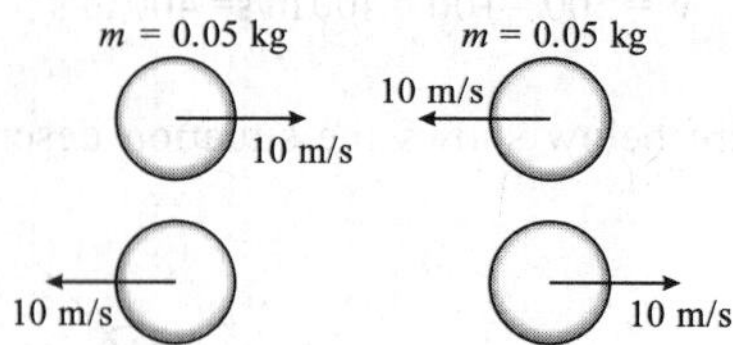

Change in momentum of any one ball is calculated as

$$|\Delta\vec{P}| = 2 \times 0.05 \times 10$$

$$\Rightarrow \quad |\Delta\vec{P}| = 1$$

Applied force on ball is given as

$$|\vec{F}_{av}| = \frac{|\Delta\vec{P}|}{\Delta t}$$

$$\Rightarrow \quad F_{av} = 200 \text{ N} \quad \textbf{Ans. (B)}$$

Sol. 98 By conservation of momentum, we use

$$0.2 \times 10 = 10 \times v$$

$$\Rightarrow \quad v = 0.2 \text{ m/s}$$

Loss in kinetic energy is calculated as

$$\Delta K = \frac{1}{2} \times (0.2) \times 10^2 - \frac{1}{2} \times 10(0.2)^2$$

$$\Rightarrow \quad \Delta K = \frac{1}{2} \times 10 \times (0.2)[10 - 0.2]$$

$$\Rightarrow \quad \Delta K = 9.8 \text{ J} \quad \textbf{Ans. (B)}$$

Sol. 99 Figure below shows the situation described in the question

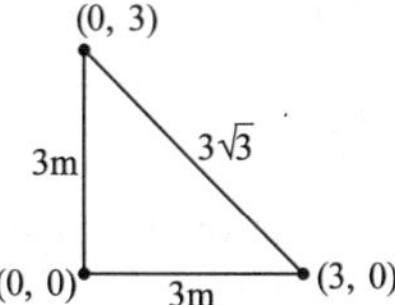

Position vector of centre of mass is given as

$$\vec{r}_{com} = \frac{M(0\hat{i}+0\hat{j})+M(3\hat{i})+M(3\hat{j})}{3M}$$

$$\Rightarrow \quad \vec{r}_{com} = \hat{i}+\hat{j}$$

$$\Rightarrow \quad |\vec{r}_{com}| = \sqrt{2} = \sqrt{x}$$

$$\Rightarrow \quad x = 2$$

Ans. (2)

Sol. 100 Initial momentum of the ball before hitting the ball is given as

$$\vec{P}_i = 0.15 \times 12(\hat{i})$$

Final momentum of ball after hitting the wall is given as

$$\vec{P}_f = 0.15 \times 12(-\hat{i})$$

Change in momentum of ball is calculated as

$$|\overrightarrow{\Delta P}| = 3.6 \text{ kg} - \text{m/s}$$

$$\Rightarrow \quad 3.6 = F\Delta t$$

$$\Rightarrow \quad 3.6 = 100\,\Delta t$$

$$\Rightarrow \quad \Delta t = 0.036 \text{ s}$$

Ans. (B)

Sol. 101 Equation of trajectory of the projected body is written as

$$y = x - \frac{10x^2}{2u^2\left(\frac{1}{2}\right)}$$

$$\Rightarrow \quad 10 = 20 - \frac{(10)(100)}{u^2}$$

$$\Rightarrow \quad u = 20$$

Time of flight of body is calculated as

$$T = \frac{(2)(20)}{\sqrt{2}(10)} = 2\sqrt{2} \text{ s}$$

Velocity vector of body at $t = \frac{T}{\sqrt{2}} = 2$ s is calculated as

$$\vec{v} = 10\sqrt{2}\hat{i} + (10\sqrt{2} - 10(2)]\hat{j}$$

Momentum of body is now written as

$$\vec{p} = M\vec{v} = 100\sqrt{2}\hat{i} + (100\sqrt{2} - 200)\hat{j}$$

Ans. (D)

Sol. 102 Figure below shows the situation described in question.

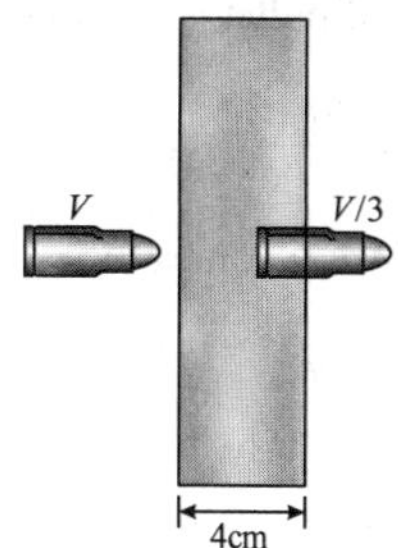

For first case, we use

$$\left(\frac{v}{3}\right)^2 = v^2 - 2a(4)$$

$$\Rightarrow \quad a = \frac{8v^2}{9(8)} = \frac{v^2}{9}$$

For second case, we use

$$0 = v^2 - 2a(4+x)$$

$$\Rightarrow \quad v^2 = 2\left(\frac{v^2}{9}\right)(4+x)$$

$$\Rightarrow \quad 4.5 = 4 + x$$

$$\Rightarrow \quad x = 0.5$$

Ans. (C)

Sol. 103 Impulse is the change in momentum so we use

$$J = \Delta P$$

Average force acting on object is calculated as

$$F_{avg} = \frac{\Delta P}{\Delta t}$$

Change in momentum is same so impulse will also be same but time is different so F_{avg} in case (i) is more than (ii). **Ans. (B)**

Sol. 104 Average force on ballon can be given by rate of change of momentum of escaping air, given as

$$F = \frac{dm}{dt}v$$

$$\Rightarrow \quad F = \frac{10g}{5s}\left(4.5\frac{\text{cm}}{\text{s}}\right) = 9\frac{g\,\text{cm}}{s^2}$$

$$\Rightarrow \quad F = 9 \text{ dyne}$$

Ans. (B)

Sol. 105 Momentum of bullet is calculated as

$$mv = \sqrt{2mk}$$

Initial speed of bullet is given as

$$u = \frac{1}{0.2}\sqrt{2 \times 0.2 \times 90} = 30 \text{ m/s}$$

After 1 s speed of bullet becomes

$$v = \frac{1}{0.2}\sqrt{2 \times 0.2 \times 40} = 20 \text{ m/s}$$

Average acceleration of bullet is calculated as

$$a=\frac{20-30}{1}=-10\text{ m/s}^2$$

Distance covered before it comes to rest is calculated as

$$s=\frac{-u^2}{2a}=45\text{ m} \qquad \textbf{Ans. (A)}$$

Sol. 106 Figure below shows the rod mentioned in question. From its left end we consider an element of width dx as shown.

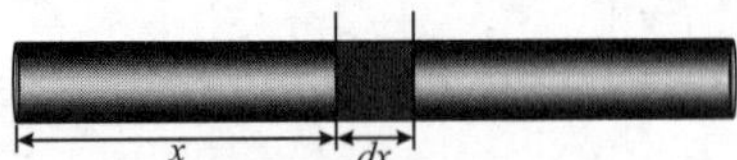

Mass of the element dm is given as

$$dm=\rho dx=\rho_0\left(1-\frac{x^2}{l^2}\right)$$

Position coordinate of centre of mass of rod is given as

$$x_{cm}=\frac{\int x dm}{\int dm_l}$$

$$\Rightarrow \quad x_{cm}=\frac{\rho_0\int_0^l\left(x-\frac{x^3}{l^2}\right)dx}{\int_0^l\rho_0\left(1-\frac{x^2}{l^2}\right)dx}=\frac{\frac{l^2}{2}-\frac{l^4}{4l^2}}{l-\frac{l^3}{3l^2}}$$

$$\Rightarrow \quad x_{cm}=\frac{3l}{8} \qquad \textbf{Ans. (8)}$$

Sol. 107 The position vector of centre of mass is given as

$$\vec{r}_{com}=\frac{m_1\vec{r}_1+m_2\vec{r}_2}{m_1+m_2}$$

$$\Rightarrow \quad \vec{r}_{com}=\frac{1(\hat{i}+2\hat{j}+\hat{k})+3(-3\hat{i}-2\hat{j}+\hat{k})}{1+3}$$

$$\Rightarrow \quad \vec{r}_{com}=-2\hat{i}-\hat{j}+\hat{k}$$

$$\Rightarrow \quad |\vec{r}_{com}|=\sqrt{(2)^2+(1)^2+(1)^2}=\sqrt{6} \qquad \textbf{Ans. (A)}$$

Sol. 108 Figure below shows the collision as described in question.

$u = 0$

u_1

1 kg 3 kg

After collision situation is shown below

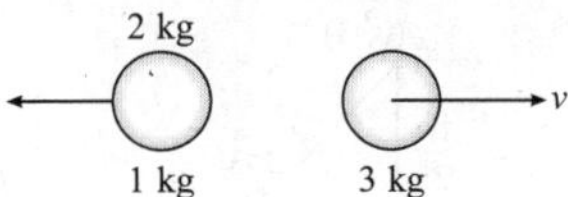

Using momentum conservation we have

$$1\times u_1=-2+3v$$

$$\Rightarrow \quad 1=\frac{v+2}{u_1}$$

$$\Rightarrow \quad u_1=-2+3v \qquad \ldots(1)$$

For elastic collision we can use same approach and separation velocity of bodies so we have

$$v+2=u_1 \qquad \ldots(2)$$

Solving equations-(1) and (2) gives

$$u_1=4\text{ m/s} \qquad \textbf{Ans. (4)}$$

Sol. 109 Velocity of ball after falling through a height h_i is calculated as

$$v_i=\sqrt{2gh_i}$$

$$\Rightarrow \quad v_i=\sqrt{2\times10\times9.8}=14\text{ m/s}$$

After rebound, it starts with a velocity in upward direction, given as

$$v_f=\sqrt{2gh_f}=\sqrt{2\times10\times5}=10\text{ m/s}$$

Average acceleration of ball is calculated as

$$|\vec{a}_{avg}|=\left|\frac{\Delta\vec{v}}{\Delta t}\right|=\frac{24}{0.2}=120\text{ m/s}^2 \qquad \textbf{Ans. (120)}$$

Sol. 110 The velocities will be swapped after collision and the speed of P just before collision is given as

$$v=\sqrt{2gh}$$

$$\Rightarrow \quad v=\sqrt{2\times10\times0.2}=2\text{ m/s} \qquad \textbf{Ans. (C)}$$

Sol. 111 By momentum conservation, we have

$$20\times10^{-3}\times\frac{180}{60}\times100=10v$$

$$\Rightarrow \quad v=0.6\text{ m/s} \qquad \textbf{Ans. (D)}$$

Sol. 112 Momentum in N balls travelling at speed v is given as

$$P_i=Nmv\hat{i}$$

After reflection from the wall, momentum is given as

$$\vec{P}_f=-Nmv\hat{i}$$

Change in momentum due to reflection of balls is given as

$$\Delta\vec{P}=\vec{P}_f-\vec{P}_i=-2Nmv\hat{i}$$

$$\Rightarrow \quad \Delta\vec{P}=-200\,Nmv\hat{i}$$

Force on wall is given as

$$\vec{F}_{\text{Total}} = \frac{\Delta\vec{P}}{\Delta t} = -\frac{200mv}{t}$$

$$\Rightarrow \quad |\vec{F}| = \frac{200mv}{t}$$ **Ans. (B)**

Sol. 113 We know by collision that

$$h' = e^2 h$$

$$\Rightarrow \quad h' = (0.25)^2 \times 20\,\text{m} = 5\,\text{m}$$ **Ans. (5)**

Sol. 114 Impulse = Area under $F - t$ curve which is calculated for all options as given below.

(a) $\frac{1}{2} \times 1 \times 0.5 = \frac{1}{4}$ N.s

(b) $0.5 \times 2 = 1$ N.s (maximum)

(c) $\frac{1}{2} \times 1 \times 0.75 = \frac{3}{8}$ N.s

(d) $\frac{1}{2} \times 2 \times 0.5 = \frac{1}{2}$ N.s **Ans. (B)**

Sol. 115 Figure below shows the situation described in the question. Let v_1 and v_2 are velocity just before and just after hitting the floor.

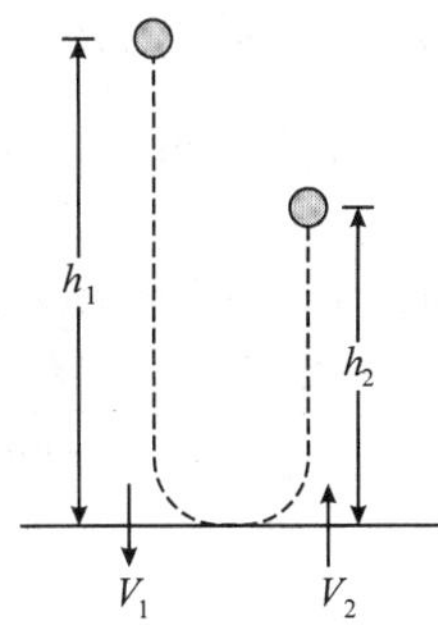

$$\frac{v_1}{v_2} = 4$$

$$\Rightarrow \quad v_1 = 4v_2$$

Kinetic energy before collision is given as

$$K_{before} = \frac{1}{2}mv_1^2$$

Kinetic energy after collision is given as

$$K_{after} = \frac{1}{2}mv_2^2 = \frac{1}{2}\frac{mv_1^2}{16}$$

Change in kinetic energy is given as

$$\Delta K = \frac{1}{2}mv_1^2\left(\frac{1}{16} - 1\right) = \frac{-15}{32}mv_1^2$$

Percentage change in kinetic energy is calculated as

$$p = \frac{\Delta K}{K_{before}} \times 100\,\%$$

$$\Rightarrow \quad p = \frac{-15}{16} \times 100 = \frac{-375}{4}\,\%$$ **Ans. (375)**

Sol. 116 Kinetic energy of body is given as

$$K = \frac{p^2}{2m}$$

When momentum is increased by 50% the final momentum is given as

$$p_f = p_i + \frac{50}{100}p_i$$

Thus final kinetic energy is calculated as

$$K = \frac{p_f^2}{2m} = \frac{(1.5p_i)^2}{2m}$$

$$\Rightarrow \quad K = \frac{2.25p_i^2}{2m}$$

Percentage increase in kinetic energy is calculated as

$$K = \frac{2.25\frac{p_i^2}{2m} - \frac{p_i^2}{2m}}{\frac{p_i^2}{2m}} \times 100$$

$$\Rightarrow \quad K = 125\%$$ **Ans. (125)**

Sol. 117 Figure below shows the situation described in the question.

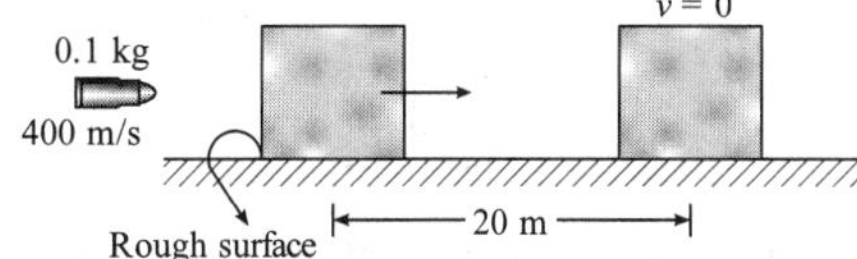

By conservation of momentum during collision we have

$$p_i = p_f$$

$$\Rightarrow \quad (0.1)(400) = (0.1 + 3.9)v$$

$$\Rightarrow \quad v = \frac{0.1 \times 400}{4} = 10\,\text{m/s}$$

Acceleration of block due to friction when bullet gets embedded into block and exert force is given as

$$a = \frac{\mu mg}{m} = \mu g$$

Using third equation of motion, we have

$$v^2 = u^2 + 2as$$

$$\Rightarrow \quad 0 = (10)^2 - 2\mu g \times 20$$

$\Rightarrow \quad 40\mu g = 100$

$\Rightarrow \quad \mu = \dfrac{100}{2\times 10\times 20} = \dfrac{1}{4}$ **Ans. (D)**

Sol. 118 Body mass is given as

$$M = 5\ kg$$

Initial Momentum of body is given as

$$P_i = 10\ \text{kg m/s}$$

Impulse momentum equation is written as

$$F\Delta t = \Delta P = P_f - P_i$$

$\Rightarrow \quad 2\times 5 = P_f - 10$

$\Rightarrow \quad P_f = 20\ \text{kg m/s}$

Increase in kinetic energy is calculated as

$$\Delta K = K_f - K_i$$

$\Rightarrow \quad \Delta K = \dfrac{P_f^2}{2m} - \dfrac{P_i^2}{2m}$

$\Rightarrow \quad \Delta K = \dfrac{400}{2\times 5} - \dfrac{100}{2\times 5}$

$\Rightarrow \quad \Delta K = 40 - 10 = 30\ \text{J}$ **Ans. (30)**

Sol. 119 Using conservation of linear momentum

$$\vec{P}_i = \vec{P}_f$$

$\Rightarrow \quad mv + 2m\times 0 = (3m)v'$

$\Rightarrow \quad mv = 3mv'$

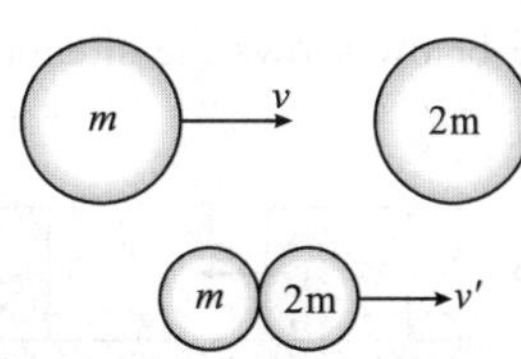

After collision velocity of combination is given as

$$v' = \frac{v}{3}$$ **Ans. (C)**

Sol. 120 Average radius of circular tube is given as

$$r_{avg} = 15\ \text{cm}$$

By work-energy theorem, we use the sum of work done by gravity and that by friction on the block is the total change in its kinetic energy so we have

$$W_f + W_g = \Delta K$$

$\Rightarrow \quad W_f + 10\times 0.3 = -\dfrac{1}{2}\times 484$

$\Rightarrow \quad w_f = -245\ \text{J}$ **Ans. (– 245)**

Sol. 121 Force exerted by gun when n bullets fired per second

$$F = nmv$$

where n = number of bullets fired per second then

$$n = \frac{f}{mv} = \frac{125}{10\times 10^{-3}\times 250} = 25$$ **Ans. (B)**

Sol. 122 By momentum conservation, we have

$$0 = 3(-v) + 0.01(600 - v)$$

$\Rightarrow \quad v = 2\ \text{m/s}$

$$\text{Impulse} = F\Delta t = m\Delta v$$

Impulse on gun is calculated as

$$J = 3\times 2 = 6\ \text{Ns}$$ **Ans. (B)**

6.2 Rigid Body Dynamics

Sol. 1

Initial MI of disc

$$I_1 = \frac{1}{2}MR^2$$

Initial MI of disc with small spheres

$$I_2 = \frac{1}{2}MR^2 + 2mR^2$$

As no external torque is applied on disc, its angular momentum remain conserved.

$\Rightarrow \quad I_1\omega_1 = I_2\omega_2$ (where ω_2 is final angular speed)

$\Rightarrow \quad \omega_2 = \dfrac{I_1\omega_1}{I_2} = \left(\dfrac{M}{M+4m}\right)\omega_1$ **Ans. (C)**

Sol. 2

MI of a circular ring about its diametrical axis is given as

$$I = \frac{1}{2}MR^2$$ **Ans. (A)**

Sol. 3

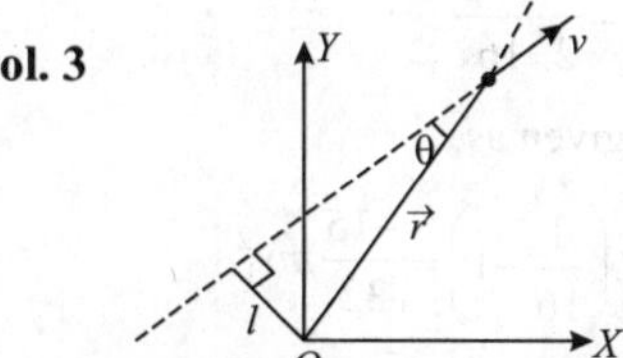

Angular momentum

$$\vec{L} = m(\vec{r} \times \vec{v}) = mvr\sin\theta = mvl \quad \textbf{Ans. (B)}$$

Sol. 4

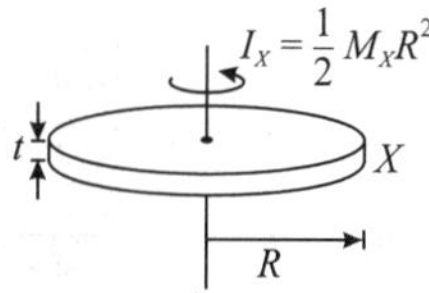

Mass of disc X is

$$M_x = \rho(\pi R^2 t)$$

$$\Rightarrow \quad I_x = \frac{1}{2}\rho\pi R^4 t$$

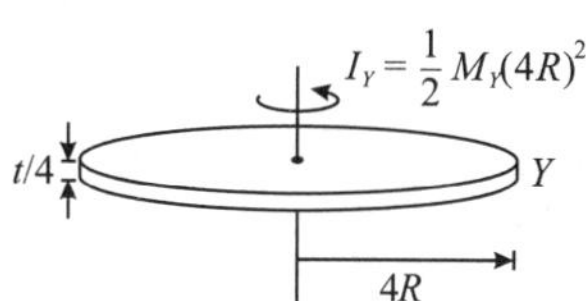

Mass of disc Y is

$$M_y = \rho(\pi(4R)^2(t/4)) = 4\rho\pi R^2 t$$

$$\Rightarrow \quad I_Y = \frac{1}{2}(64\rho\pi R^4 t)$$

$$\Rightarrow \quad I_y = 64I_X \quad \textbf{Ans. (D)}$$

Sol. 5 For a particle in circular motion, we have

Angular momentum $L = I\omega \Rightarrow I = \dfrac{L}{\omega}$

Kinetic Energy $\quad K = \dfrac{1}{2}I\omega^2 = \dfrac{1}{2}\left(\dfrac{L}{\omega}\right)\omega^2$

$$\Rightarrow \quad L = \frac{2K}{\omega}$$

If $\quad K \to K/2$ and $\omega \to 2\omega$

Then final angular momentum is

$$L_f = \frac{2(K/2)}{2\omega} = \frac{K}{2\omega} = \frac{L}{4} \quad \textbf{Ans. (A)}$$

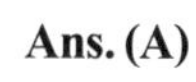

Sol. 6 For the given particle, we know

$$\vec{\tau} = \vec{r} \times \vec{F}$$

$\Rightarrow \quad \vec{\tau}$ is $\perp$ to $\vec{r}$ and also $\vec{\tau}$ is $\perp$ to $\vec{F}$

$$\Rightarrow \quad \vec{r}\cdot\vec{\tau} = 0 \text{ and } \vec{F}\cdot\vec{\tau} = 0 \quad \textbf{Ans. (D)}$$

Sol. 7

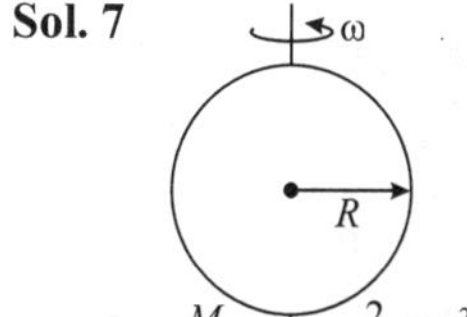

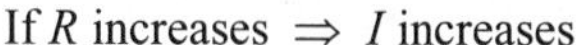

If R increases $\Rightarrow$ I increases

As no external torque is applied

$\Rightarrow$ Angular momentum remain conserved

So as I increases, ω decreases

Rotational KE $\quad K = \dfrac{1}{2}I\omega^2 = \dfrac{1}{2}L\omega \quad$ **Ans. (B)**

Sol. 8

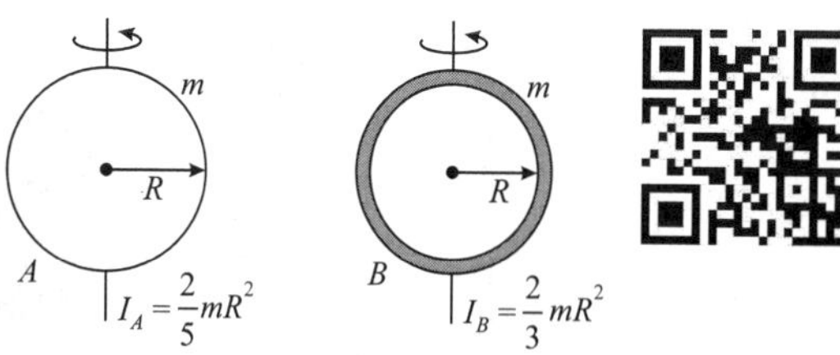

$\Rightarrow \quad I_B > I_A \quad$ **Ans. (C)**

Sol. 9

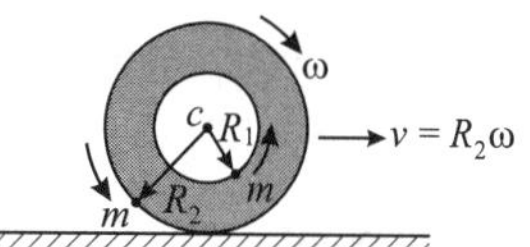

With respect to ring centre the two particles on inner and outer parts of ring are in uniform circular motion at speed ω.

$\Rightarrow$ Forces on the two particles are

$$F_1 = m\omega^2 R_1 \text{ and } F_2 = m\omega^2 R_2$$

$$\Rightarrow \quad \frac{F_1}{F_2} = \frac{R_1}{R_2} \quad \textbf{Ans. (D)}$$

Sol. 10

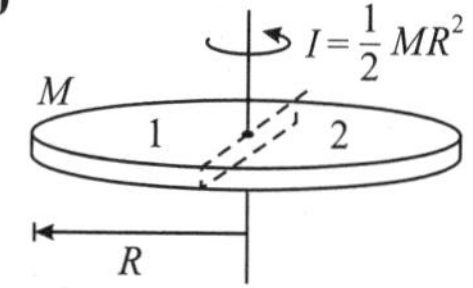

By property of mass distribution for all identical parts of a body, expression of moment of inertia remain same.

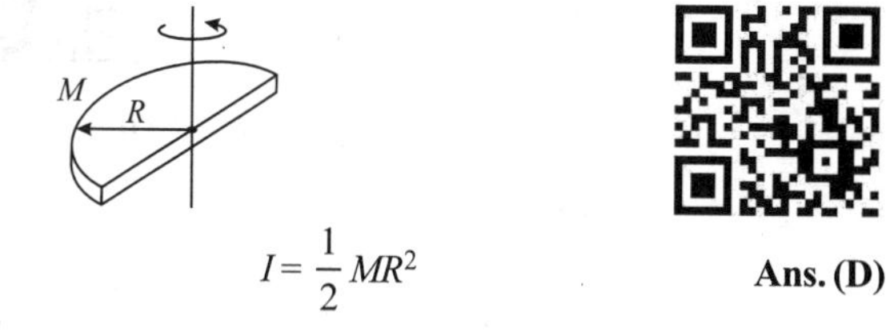

$$\Rightarrow \quad I = \frac{1}{2}MR^2 \quad \textbf{Ans. (D)}$$

Sol. 11

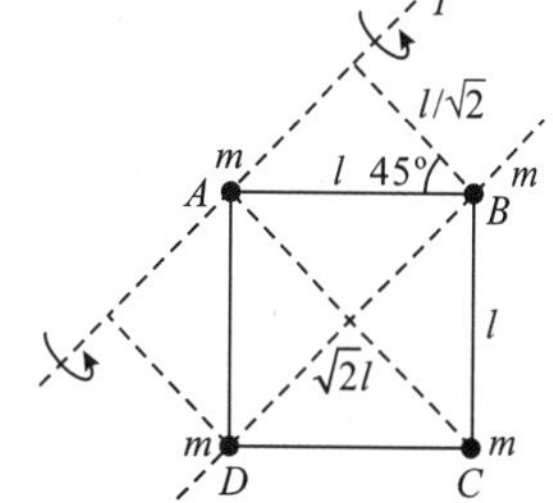

MI of all the masses about given AOR is

$$I = I_A + I_B + I_C + I_D$$

$$\Rightarrow \quad I = 0 + m\left(\frac{l}{\sqrt{2}}\right)^2 + m(\sqrt{2}l)^2 + m\left(\frac{l}{\sqrt{2}}\right)^2$$

$$\Rightarrow \quad I = \frac{ml^2}{2} + 2ml^2 + \frac{ml^2}{2} = 3ml^2$$

Ans. (C)

Sol. 12

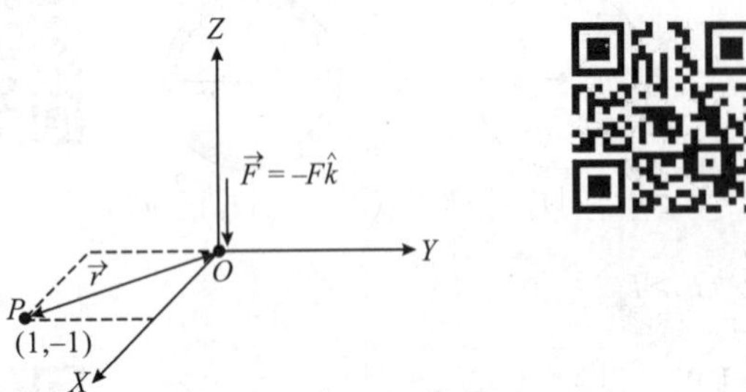

Position vector of origin about the given point is

$$\vec{r} = -\hat{i} + \hat{j}$$

Torque of the given force about point P is

$$\vec{\tau} = \vec{r} \times \vec{F} = (-\hat{i} + \hat{j}) \times (-F\hat{k})$$

$$\Rightarrow \quad \vec{\tau} = F(-\hat{j} - \hat{i}) = -F(\hat{i} + \hat{j})$$

Ans. (B)

Sol. 13

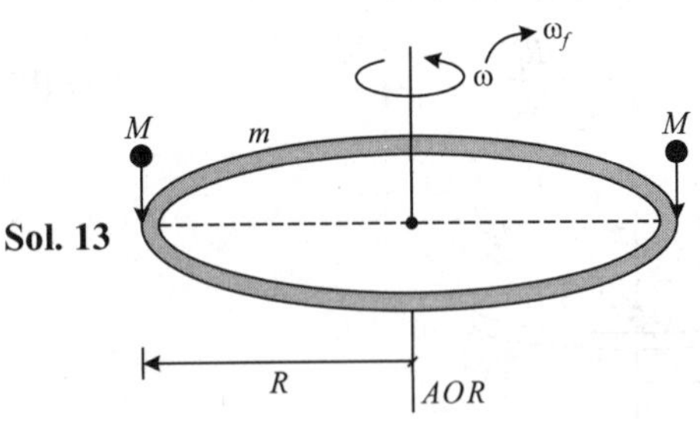

As no external torque is applied on ring, its angular momentum remain conserved

$$L = I\omega = I_f\omega_f$$

here $\quad I = mR^2$ and $I_f = mR^2 + 2MR^2$

$$\Rightarrow \quad mR^2\omega = (m + 2M)R^2\omega_f$$

$$\Rightarrow \quad \omega_f = \frac{m\omega}{m + 2M}$$

Ans. (D)

Sol. 14

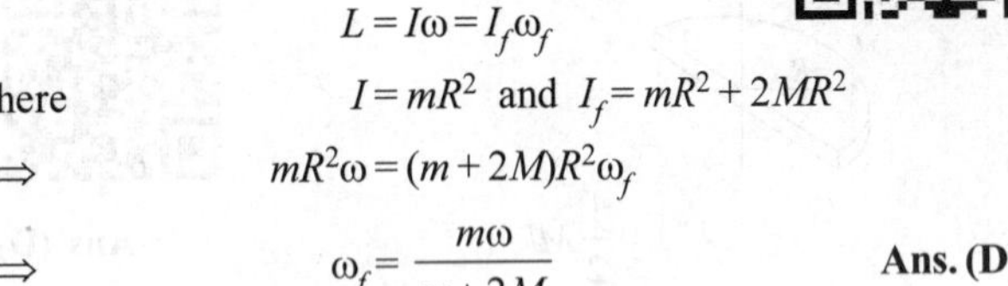
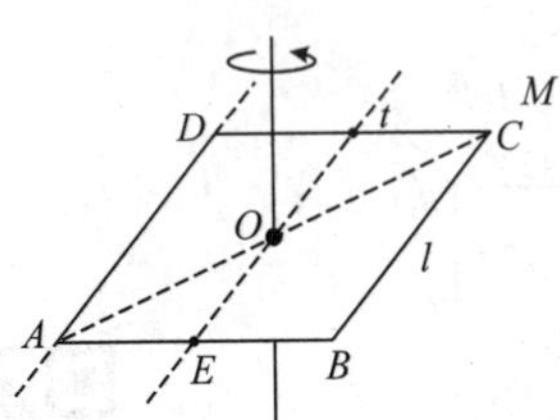

MI of a square plate about AOR passing through its center and $\perp$ to its plane is given as

$$I_0 = \frac{1}{6}Ml^2$$

For any AOR in the plane of lamina passing through its center MI can be given as

$$2I_P = \frac{1}{6}Ml^2$$

$$\Rightarrow \quad I_P = \frac{1}{12}Ml^2 = I_{EF} = I_{AC}$$

$$\Rightarrow \quad I_{AD} = I_{EF} + \frac{Ml^2}{4} = \frac{Ml^2}{3}$$

Ans. (C)

Sol. 15

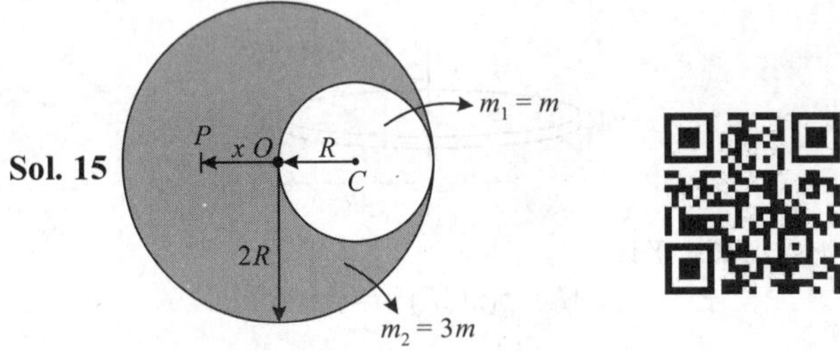

As disc is uniform, its mass is proportional to its surface area

$$\Rightarrow \quad m_1 = C\cdot\pi R^2 = m$$

and $\quad m_2 = C[\pi(2R)^2 - \pi R^2] = C(3\pi R^2) = 3m$

As point O is the COM of bigger disc of radius $2R$, we use

$$3mx = mR$$

$$\Rightarrow \quad x = \frac{R}{3} = \alpha R$$

$$\Rightarrow \quad \alpha = \frac{1}{3}$$

Ans. (A)

Sol. 16

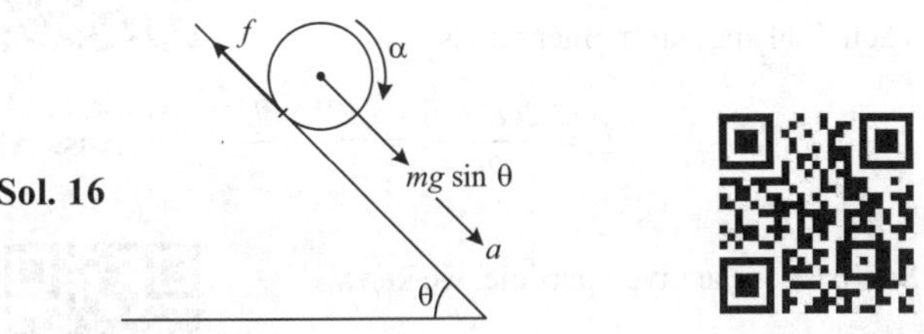

Equation of motion of body

$$mg\sin\theta - f = ma \qquad \ldots(1)$$

$$\Rightarrow \quad fR = I\left(\frac{a}{R}\right) \qquad \ldots(2)$$

[For pure rolling $a = R\alpha$]

$$(1) + (2) \Rightarrow \quad mg\sin\theta = a\left(M + \frac{I}{R^2}\right)$$

$$\Rightarrow \quad a = \frac{g\sin\theta}{\left(1 + \frac{I}{MR^2}\right)}$$

Ans. (A)

Sol. 17

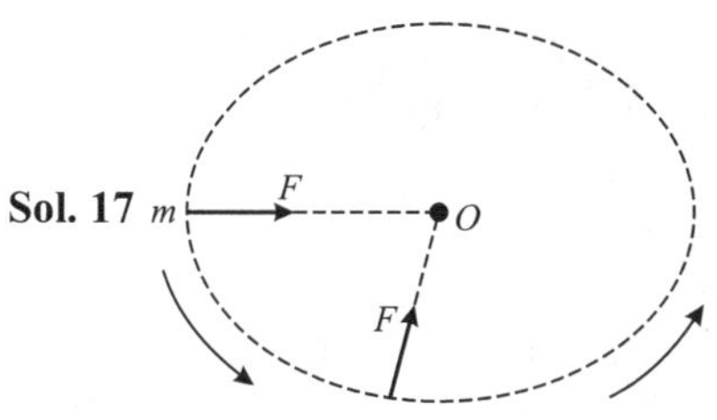

Due to central force which always passes through centre so the torque due to this force about O is always zero during motion.

As $\qquad \vec{\tau}_0 = 0$

$\Rightarrow$ Angular momentum about O remain conserved. **Ans. (C)**

Sol. 18

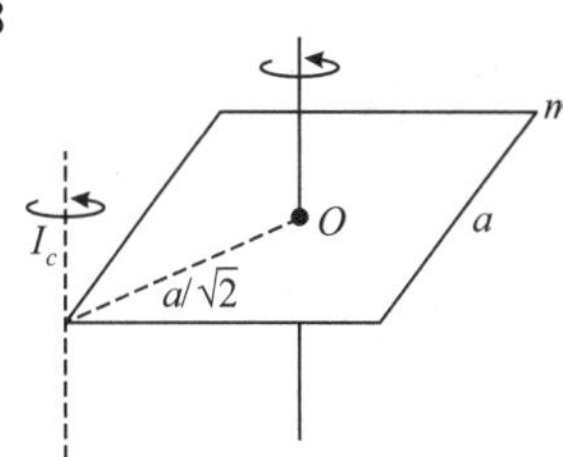

MI of square plate about its central AOR is

$$I_0 = \frac{1}{6}Ma^2$$

From parallel AOR, we have

$$I_c = I_0 + Md^2$$

$$\Rightarrow \quad I_c = \frac{1}{6}Ma^2 + \frac{1}{2}Ma^2 = \frac{2}{3}Ma^2$$ **Ans. (D)**

Sol. 19 By conservation of energy, we have

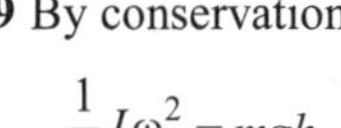

$$\frac{1}{2}I\omega^2 = mgh$$

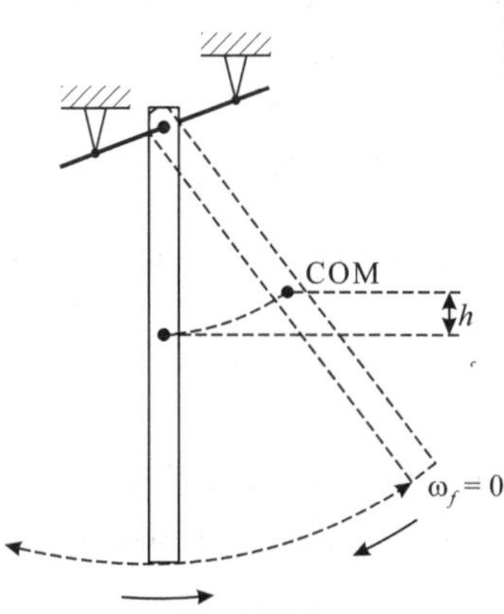

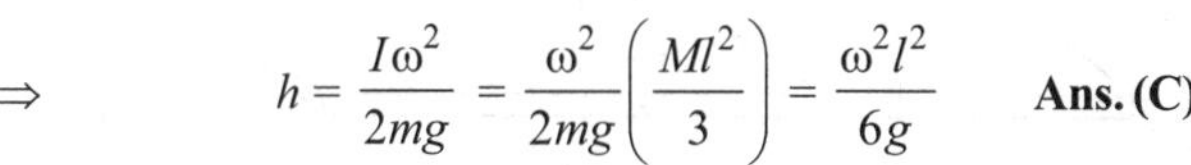

$$\Rightarrow \quad h = \frac{I\omega^2}{2mg} = \frac{\omega^2}{2mg}\left(\frac{Ml^2}{3}\right) = \frac{\omega^2 l^2}{6g}$$ **Ans. (C)**

Sol. 20 Angular momentum of particle is given as

$$\vec{L} = m(\vec{r} \times \vec{v})$$

where position vector of particle at time t is given as

$$\vec{r} = (v_0 \cos\theta t)\,\hat{i} + (v_0 \sin\theta t - \frac{1}{2}gt^2)\,\hat{j}$$

velocity vector of particle at time t is given as

$$\vec{v} = (v_0 \cos\theta)\,\hat{i} + (v_0 \sin\theta - gt)\,\hat{j}$$

$$\Rightarrow \quad \vec{L} = -\frac{1}{2}mgv_0t^2\cos\theta\,\hat{k}$$ **Ans. (D)**

Sol. 21 Equation of motion for mass m

$$mg - T = ma \qquad \ldots(1)$$

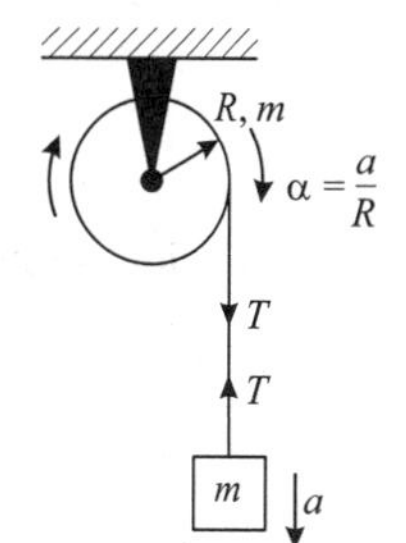

For pulley

$$T.R = \left(\frac{1}{2}mR^2\right)\left(\frac{a}{R}\right) \qquad \ldots(2)$$

$$(1) + (2) \quad \Rightarrow \quad mg = \frac{3}{2}ma$$

$$\Rightarrow \quad a = \frac{2g}{3}$$ **Ans. (B)**

Sol. 22

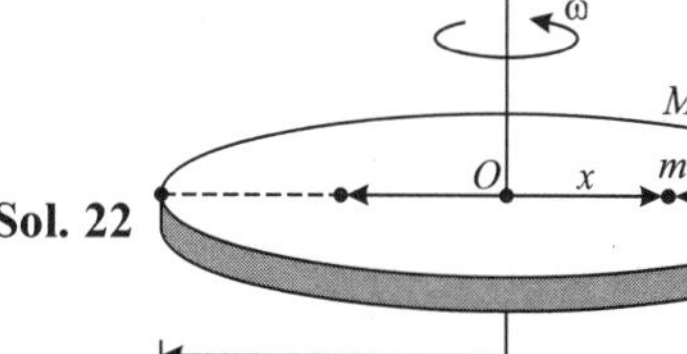

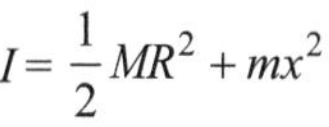

$$I = \frac{1}{2}MR^2 + mx^2$$

An no external torque is acting on disc, its angular momentum remain conserved

$$\Rightarrow \quad L = I\omega = \text{constant}$$

$$\Rightarrow \quad \left(\frac{1}{2}MR^2 + mx^2\right)\omega = \text{constant}$$

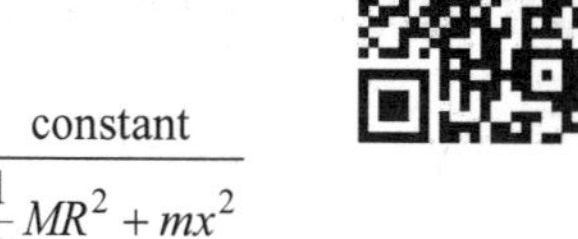

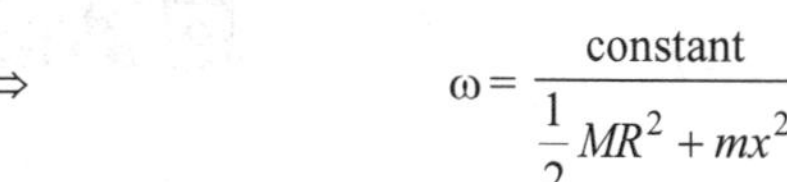

$$\Rightarrow \quad \omega = \frac{\text{constant}}{\frac{1}{2}MR^2 + mx^2}$$

As $x \downarrow \Rightarrow \omega\uparrow$ and after passing centre O $x \uparrow \Rightarrow \omega\downarrow$ **Ans (B)**

Sol. 23 Torque on pulley

$$\tau = rF = 40t - 10t^2 = I\alpha$$

Angular acceleration of pulley

$$\alpha = 4t - t^2$$

$$\Rightarrow \quad \frac{d\omega}{dt} = 4t - t^2$$

$$\Rightarrow \quad \int_0^{\omega} d\omega = \int_0^{t}(4t - t^2)dt$$

$$\Rightarrow \quad \omega = 2t^2 - \frac{t^3}{3}$$

here ω first ↑ then ↓ and it becomes zero when

$$2t^2 - \frac{t^3}{3} = 0$$

$$\Rightarrow \quad t = 6\text{s}$$

at $t = 6$s direction of pulley rotation is reversed

$$\frac{d\theta}{dt} = 2t^2 - \frac{t^3}{3}$$

$$\int_0^{\theta} d\theta = \int_0^{6}\left(2t^2 - \frac{t^3}{3}\right)dt$$

$$\Rightarrow \quad \theta = \left[\frac{2t^3}{3} - \frac{t^4}{12}\right]_0^6 = \frac{2}{3}\times 216 - \frac{(6)^4}{12} - 0$$

$$\Rightarrow \quad \theta = 36 \text{ rad}$$

$$\text{No of turns} = \frac{\theta}{2\pi} = \frac{36}{2\pi} = 5.73$$ **Ans. (D)**

Sol. 24

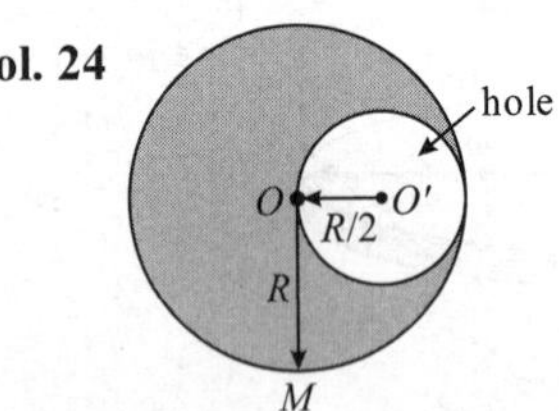

MI of whole disc is

$$I_0 = \frac{1}{2}MR^2$$

Mass of small (hole) disc is

$$m = \frac{M}{\pi R^2}\times\frac{\pi R^2}{4} = \frac{M}{4}$$

MI of small disc about axis at O is

$$I_{hole} = I_{O'} + \left(\frac{M}{4}\right)\left(\frac{R}{2}\right)^2$$

$$\Rightarrow \quad I_{hole} = \frac{1}{2}\left(\frac{M}{4}\right)\left(\frac{R}{2}\right)^2 + \left(\frac{M}{4}\right)\left(\frac{R}{2}\right)^2$$

$$\Rightarrow \quad I_{hole} = \frac{3}{2}\times\frac{M}{4}\times\frac{R^2}{4} = \frac{3}{32}MR^2$$

MI of remaining disc is

$$I = I_0 - I_{hole}$$

$$\Rightarrow \quad I = \frac{1}{2}MR^2 - \frac{3}{32}MR^2 = \frac{13}{32}MR^2$$ **Ans. (D)**

Sol. 25 Translational $KE_T = \frac{1}{2}mv^2$ ...(1)

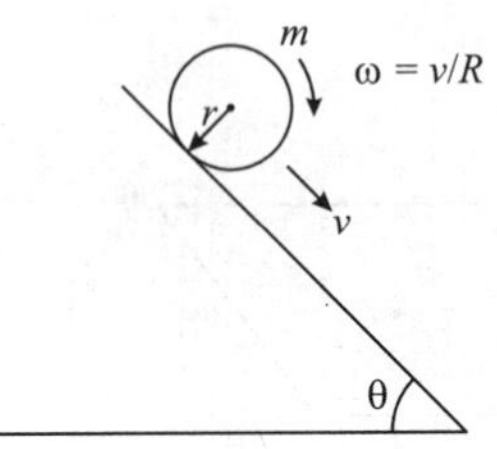

Rotational $KE_R = \frac{1}{2}I\omega^2 = \frac{1}{2}\left(\frac{2}{5}mR^2\right)\left(\frac{v}{R}\right)^2$

$$KE_R = \frac{1}{5}mv^2$$ ...(2)

By (2)/(1) ⇒ $\frac{KE_R}{KE_T} = \frac{2}{5}$

$\Rightarrow \quad KE_R = \frac{2}{7}$ times of TE

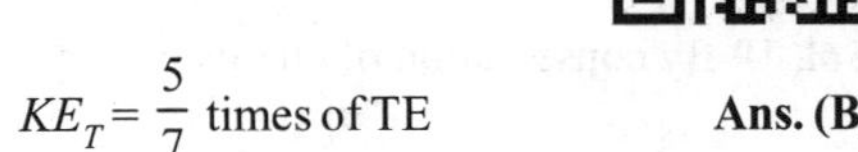

$KE_T = \frac{5}{7}$ times of TE **Ans. (B)**

Sol. 26 If there is No torque ⇒ $L \to$ constant

we use $L = I\omega$

and $K = \frac{1}{2}I\omega^2 = \frac{L^2}{2I}$

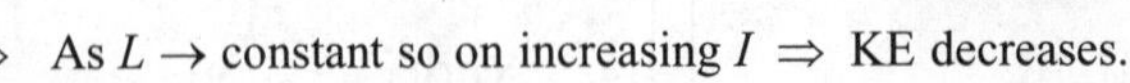

⇒ As $L \to$ constant so on increasing I ⇒ KE decreases.

S-1 is false.

S-2 is true. **Ans. (B)**

Sol. 27 Angular momentum of revolving stone is

$$L = mvr = C \text{ (constant)}$$

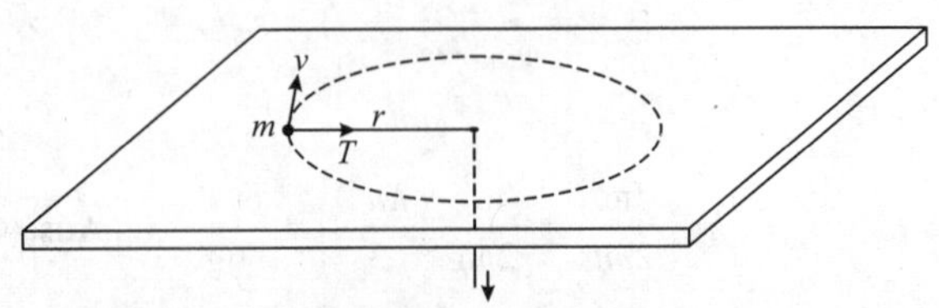

Considering stable AB (r is gradually decreased) circular motion we use

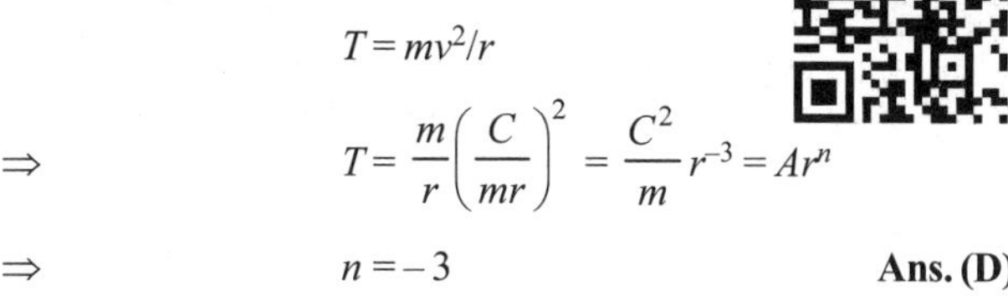

$$T = mv^2/r$$

$$\Rightarrow \quad T = \frac{m}{r}\left(\frac{C}{mr}\right)^2 = \frac{C^2}{m}r^{-3} = Ar^n$$

$$\Rightarrow \quad n = -3$$

Ans. (D)

Sol. 28 Initial speed of sphere should be sufficient to take sphere to top when its $v_f = 0$.

So using work-energy theorem we have

$$\frac{1}{2}mv^2 + \frac{1}{2}\left(\frac{2}{5}mr^2\right)\left(\frac{v}{r}\right)^2 - mgh = 0$$

$$\left(\frac{1}{2}+\frac{1}{5}\right)mv^2 = mgh$$

$$\Rightarrow \quad v = \sqrt{\frac{10}{7}gh}$$

Ans. (D)

Sol. 29 In case of pure rolling, the speed attained by the body at bottom of incline is

$$v_0 = \sqrt{\frac{2gh}{1+\frac{k^2}{R_0^2}}} \quad \ldots(1)$$

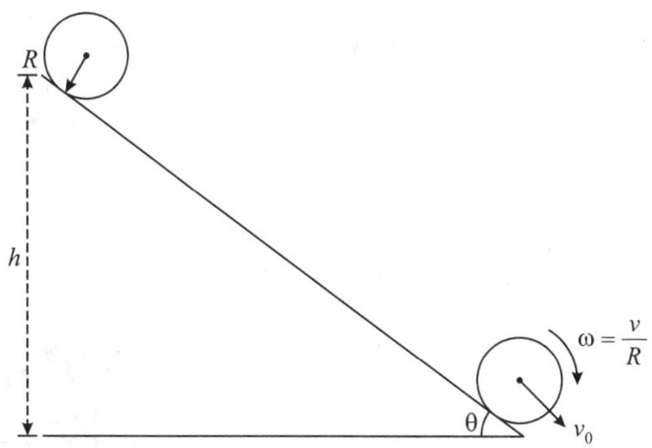

In case of smooth incline no rotational KE is gained by body so speed attained is

$$v = \sqrt{2gh} = \frac{5v_0}{4} \quad \ldots(2)$$

Using (2)/(1) gives $\frac{5}{4} = \sqrt{1+\frac{k^2}{R_0^2}}$

$$\Rightarrow \quad 1+\frac{k^2}{R_0^2} = \frac{25}{16}$$

$$\Rightarrow \quad \frac{k^2}{R_0^2} = \frac{9}{16} \Rightarrow k = \frac{3}{4}R_0$$

Ans. (B)

Sol. 30

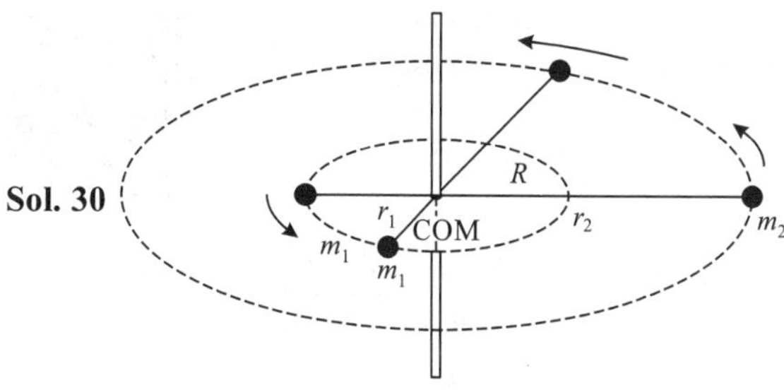

As no external force is acting on the system of these two particles

$\Rightarrow$ Their COM will remain at rest

$\Rightarrow$ During revolution the two masses will always remain opposite to each other as shown on the two sides of COM.

$\Rightarrow$ Their time period will be same for them to be on opposite sides of COM

$\Rightarrow$ $\omega_1 = \omega_2$ (independent of m_1 & m_2) **Ans. (B)**

Sol. 31 At pure rotting we use

$$v_f = r\omega_f$$

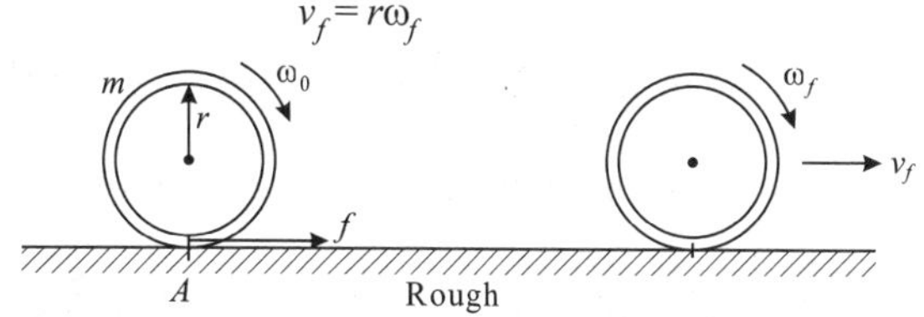

By conversation of angular momentum about point A, we use

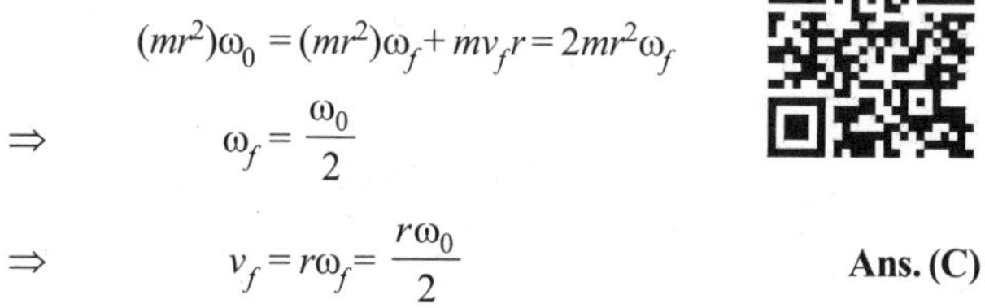

$$(mr^2)\omega_0 = (mr^2)\omega_f + mv_f r = 2mr^2\omega_f$$

$$\Rightarrow \quad \omega_f = \frac{\omega_0}{2}$$

$$\Rightarrow \quad v_f = r\omega_f = \frac{r\omega_0}{2}$$

Ans. (C)

Sol. 32 In this process of collision the angular momentum of door + bullet system remain conserved.

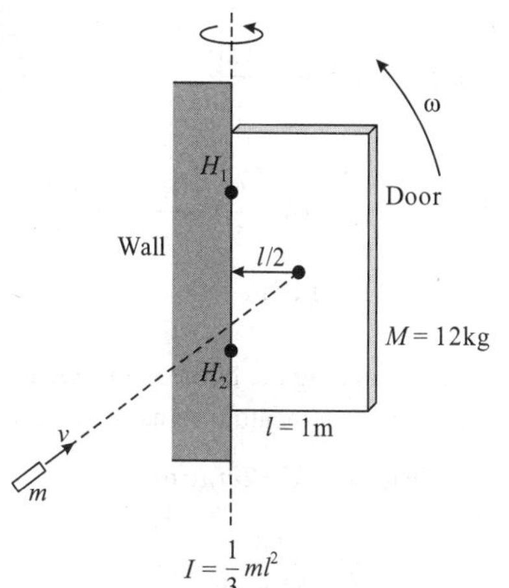

As angular momentum of door + bullet system remain conserved, we use

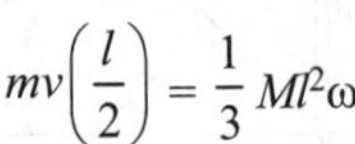

$$mv\left(\frac{l}{2}\right) = \frac{1}{3}Ml^2\omega$$

$$\Rightarrow \quad \omega = \frac{3mvl}{2Ml^2} = \frac{3\times 0.01\times 500}{2\times 12\times 1} = \frac{15}{24}$$

$$\Rightarrow \quad \omega = 0.625 \text{ rad/s}$$

Ans. (B)

Sol. 33 Angular momentum

$$\vec{L} = m(\vec{r}\times\vec{v})$$

Where $\vec{r} = 5\hat{i} - 2t^2\hat{j}$

and $\vec{v} = \frac{d\vec{r}}{dt} = -4t\,\hat{j}$

$$\Rightarrow \quad \vec{L} = m[(5\hat{i} - 2t^2\hat{j})\times(-4t\hat{j})]$$

$$\Rightarrow \quad \vec{L}_{(t=2s)} = 2[(5\hat{i} - 8\hat{j})\times(-8\hat{j})]$$

$$\vec{L}_{(t=2s)} = 2\times[-40\hat{k}] = -80\hat{k}$$

Ans. (A)

Sol. 34 By work-energy theorem from 0 to A, we have

$$0 + mg(1.8) = \frac{1}{2}mv^2 + \frac{1}{2}\left(\frac{2}{3}mR^2\right)\left(\frac{v}{R}\right)^2 = \frac{5}{6}mv^2$$

$$\Rightarrow \quad v = \sqrt{\frac{6}{5}\times 10\times 1.8}$$

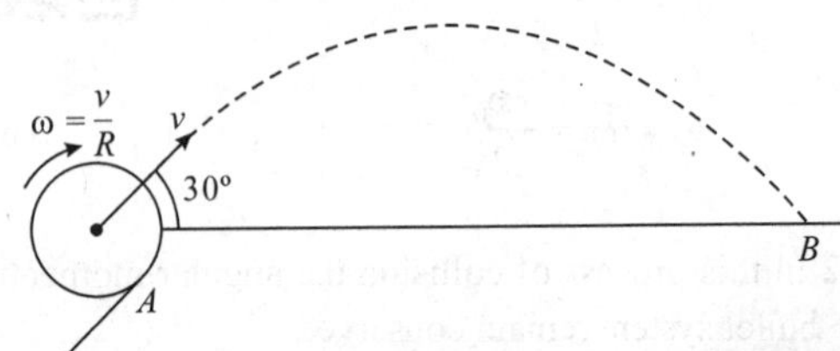

Horizontal Range $R = \dfrac{v^2\sin 2\theta}{g}$

$$\Rightarrow \quad R = AB = \frac{6}{5}\times\frac{10\times 1.8}{10}\times\frac{\sqrt{3}}{2}$$

$$\Rightarrow \quad R = 1.87\text{ m}$$

Ans. (A)

Sol. 35 In gently attaching the bodies, no external torque acts on system, so angular momentum remain conserved.

$$MR^2\omega = (M + 2m)R^2\omega_f$$

$$\Rightarrow \quad \omega_f = \frac{m\omega}{M + 2m}$$

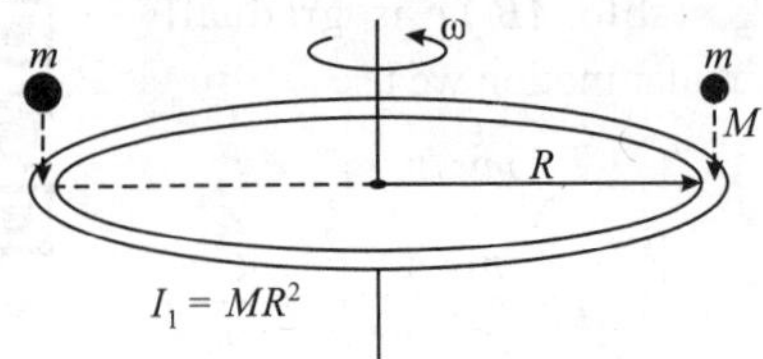

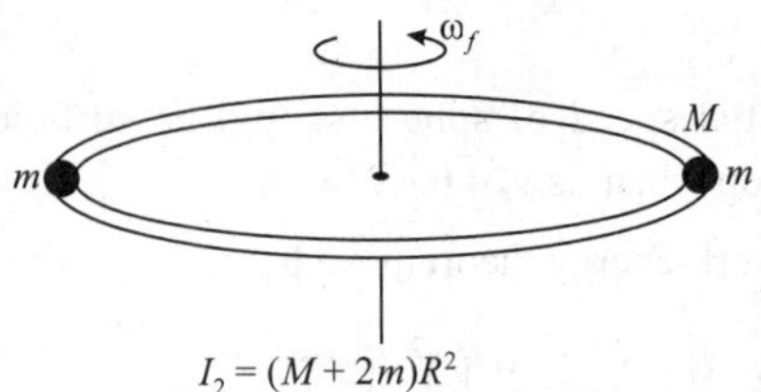

Loss in kinetic energy is given as

$$\Delta K_{\text{Loss}} = K_i - K_f$$

$$\Rightarrow \quad \Delta K_{\text{Loss}} = \frac{1}{2}I_1\omega^2 - \frac{1}{2}I_1\omega_f^2$$

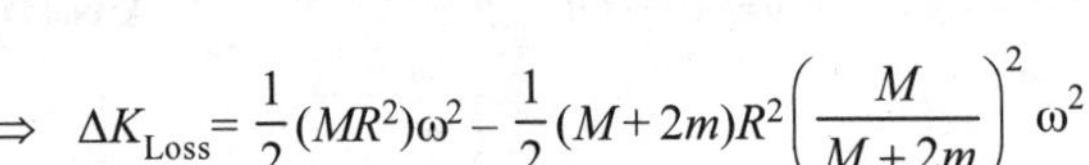

$$\Rightarrow \quad \Delta K_{\text{Loss}} = \frac{1}{2}(MR^2)\omega^2 - \frac{1}{2}(M+2m)R^2\left(\frac{M}{M+2m}\right)^2\omega^2$$

$$\Rightarrow \quad \Delta K_{\text{Loss}} = \frac{1}{2}MR^2\omega^2 - \frac{1}{2}\frac{M^2R^2}{(M+2m)}\omega^2$$

$$\Rightarrow \quad \Delta K_{\text{Loss}} = \frac{1}{2}R^2\omega^2\left[M - \frac{M^2}{M+2m}\right]$$

$$\Rightarrow \quad \Delta K_{\text{Loss}} = \frac{1}{2}R^2\omega^2\left[\frac{2mM}{M+2m}\right] = \frac{mMR^2\omega^2}{(M+2m)}$$

Ans. (C)

Sol. 36

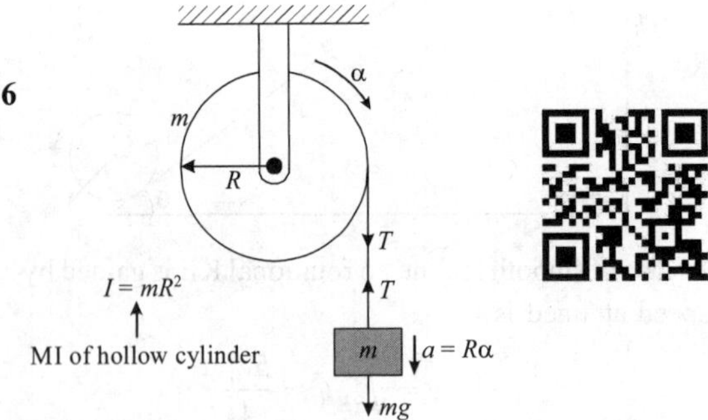

Equation of motion for falling mass

$$mg - T = ma \quad \ldots(1)$$

For cylinder $\quad TR = (mR^2)\left(\dfrac{a}{R}\right) \quad \ldots(2)$

$(1) + (2) \Rightarrow \quad mg = 2ma$

$$\Rightarrow \quad a = g/2$$

Ans. (B)

Sol. 37

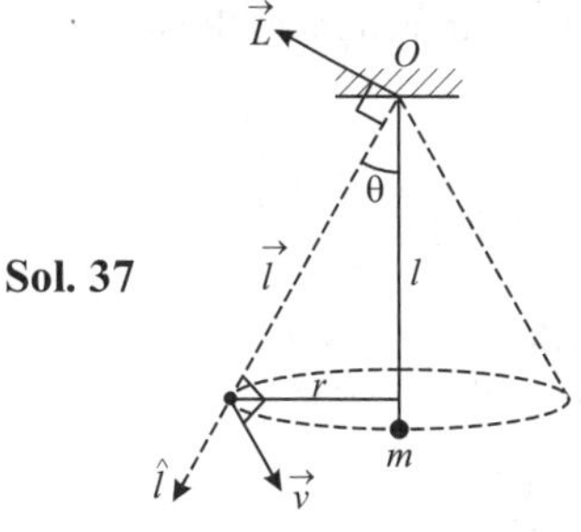

$$|\vec{v}| = \text{constant}$$

Angular momentum of bob about point O is

$$\vec{L} = m(\vec{l} \times \vec{v})$$

Direction of $\vec{L}$ is $\perp$ to both $\vec{l}$ and $\vec{v}$ as shown in figure and its magnitude is

$$L = mlv \rightarrow \text{constant}$$ **Ans. (C)**

Sol. 38 relative to rug we consider cylinder is having acceleration a, for which equations of motion of cylinder are

$$ma - f = Ma_1 \quad \ldots(1)$$

$$fR = \left(\frac{1}{2}MR^2\right)\left(\frac{a_1}{R}\right) \quad \ldots(2)$$

From (1) and (2), we have

$$Ma - f = 2f$$

$\Rightarrow$ $$3f = Ma$$

$\Rightarrow$ $$f = \frac{1}{3}Ma$$ **Ans. (D)**

Sol. 39 For pure rolling of a body on incline plane, acceleration is given as

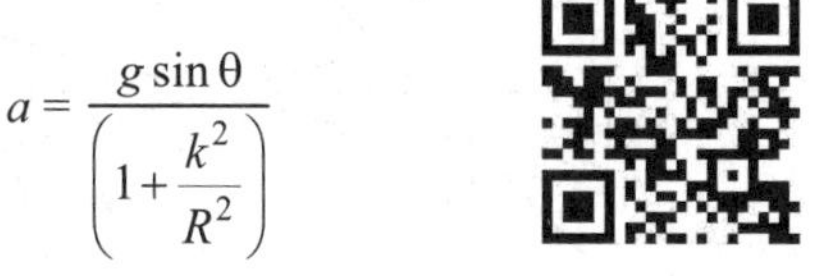

$$a = \frac{g\sin\theta}{\left(1+\frac{k^2}{R^2}\right)}$$

for cylinder $$k_c = \frac{R}{\sqrt{2}} \Rightarrow a_c = \frac{g\sin\theta}{1+\frac{1}{2}} = \frac{2}{3}g\sin\theta_c$$

for sphere $$k_s = \sqrt{\frac{2}{5}}R \Rightarrow a_s = \frac{g\sin\theta_s}{1+\frac{2}{5}} = \frac{5}{7}g\sin\theta_s$$

$$a_c = a_s \Rightarrow \frac{2}{3}g\sin\theta_c = \frac{5}{7}g\sin\theta_s$$

$\Rightarrow$ $$\frac{\sin\theta_c}{\sin\theta_s} = \frac{1}{2/3\big/5/7} = \frac{15}{14}$$ **Ans. (D)**

Sol. 40 Angular momentum of particle at highest point of trajectory about O is given as

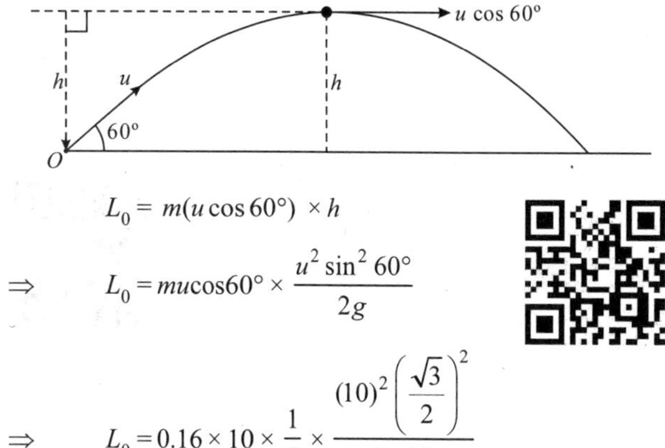

$$L_0 = m(u\cos 60°) \times h$$

$\Rightarrow$ $$L_0 = mu\cos 60° \times \frac{u^2\sin^2 60°}{2g}$$

$\Rightarrow$ $$L_0 = 0.16 \times 10 \times \frac{1}{2} \times \frac{(10)^2\left(\frac{\sqrt{3}}{2}\right)^2}{2\times 10}$$

$\Rightarrow$ $$L_0 = 3 \text{ kg m}^2\text{/s}$$ **Ans. (B)**

Sol. 41 Angle α is given as in ΔPCA

$$\alpha = 180° - (2\theta + 90 - \theta)$$

$\Rightarrow$ $$\alpha = 90° - \theta$$

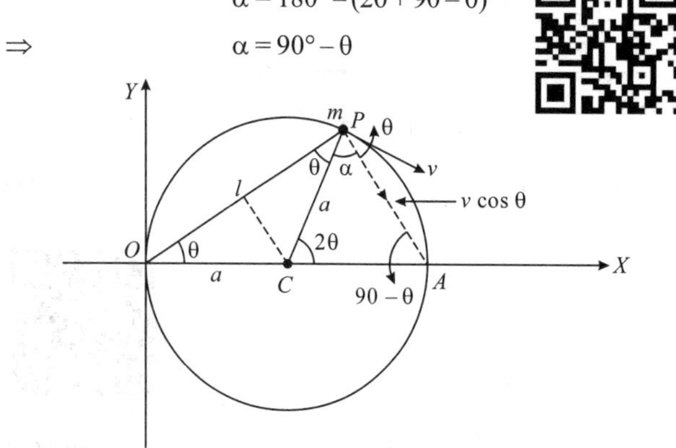

Angular momentum of mass m about O is given as

$$L_0 = m(v\cos\theta)l$$

Here $$l = 2a\cos\theta$$

$\Rightarrow$ $$L_0 = mv\cos\theta \times 2a\cos\theta$$

$\Rightarrow$ $$L_0 = 2mva\cos^2\theta$$

$\Rightarrow$ $$L_0 = mva(1 + \cos 2\theta)$$ **Ans. (A)**

Sol. 42 For largest cube we use

$$2R = \sqrt{3}l$$

$\Rightarrow$ $$l = \frac{2R}{\sqrt{3}}$$

2D view   3D view

Mass of cube $\quad m = \dfrac{M}{\dfrac{4}{3}\pi R^3} \times \left(\dfrac{2R}{\sqrt{3}}\right)^3$

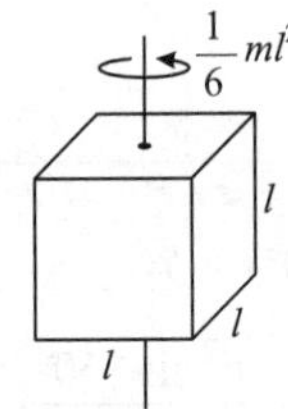

MI of cube about its central axis is

$$I = \frac{1}{6}\left(\frac{3M}{4\pi R^3}\,\frac{8R^3}{9\sqrt{3}}\right)\left(\frac{2R}{\sqrt{3}}\right)^2$$

$\Rightarrow \quad I = \dfrac{4MR^2}{9\sqrt{3}\pi}$ **Ans. (C)**

Sol. 43 Equations of motion of cylinder are

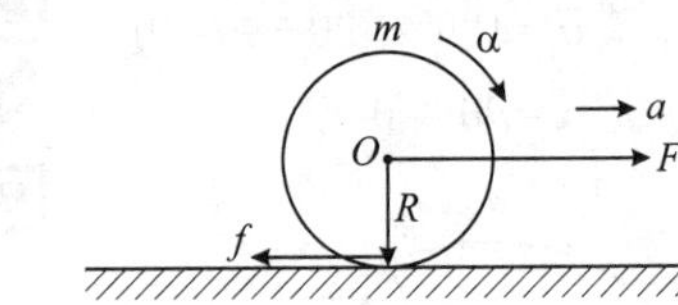

$$F - f = ma \qquad \dots(1)$$

$$fR = \left(\frac{1}{2}mR^2\right)\left(\frac{a}{R}\right)$$

$$f = \frac{1}{2}ma \qquad \dots(2)$$

equation (1) + (2)

$F = \dfrac{3}{2}ma$ **Ans. (C)**

Sol. 44 $\quad l = \sqrt{r^2 + h^2}$

$$l = \sqrt{(0.6)^2 + (0.8)^2} = 1\text{ m}$$

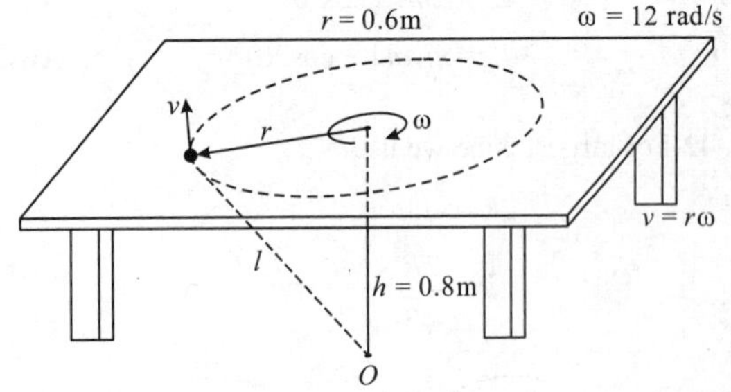

In figure angular momentum of particle about O is given as

$$L = mvl$$

$\Rightarrow \quad L = 2 \times (0.6 \times 12) \times 1$

$\Rightarrow \quad L = 14.4\text{ kg m}^2\text{/s}$ **Ans. (A)**

Sol. 45 MI of a square lamina about its central AOR is given as

$$I_0 = \frac{Ma^2}{6}$$

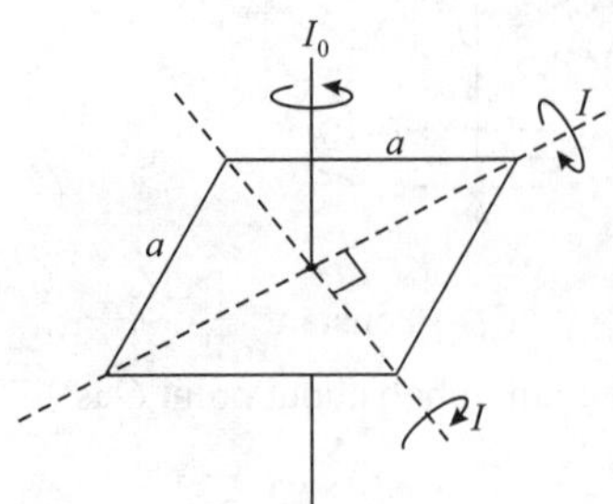

By perpendicular axes theorem we have

$$I + I = I_0$$

$\Rightarrow \quad I = \dfrac{1}{2}I_0 = \dfrac{ma^2}{12}$ **Ans. (D)**

Sol. 46 Angular momentum of particle about origin for its motion

along AB $\quad \vec{L} = mv\left(\dfrac{R}{\sqrt{2}}\right)(-\hat{k})$

along BC $\quad \vec{L} = mv\left(a + \dfrac{R}{\sqrt{2}}\right)\hat{k}$

along CD $\quad \vec{L} = mv\left(a + \dfrac{R}{\sqrt{2}}\right)\hat{k}$

along DA $\quad \vec{L} = mv\left(\dfrac{R}{\sqrt{2}}\right)(-\hat{k})$ **Ans. (A, C)**

Sol. 47 Here we analyze friction on the bottom contact points of cones and rails

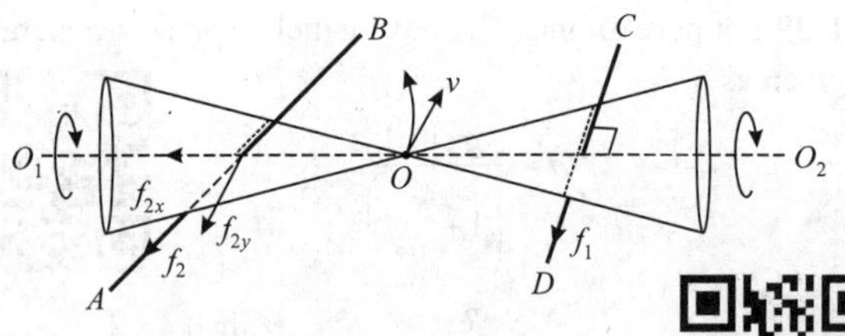

Due to a component of friction f_2 on roller at rail AB it will turn toward left. **Ans. (C)**

Sol. 48 As no torque is acting about O at the instant of impact so angular momentum of cube remain conserved about O.

$$mv\frac{a}{2} = I\omega$$

$\Rightarrow \quad \omega = \dfrac{mva}{2I}$

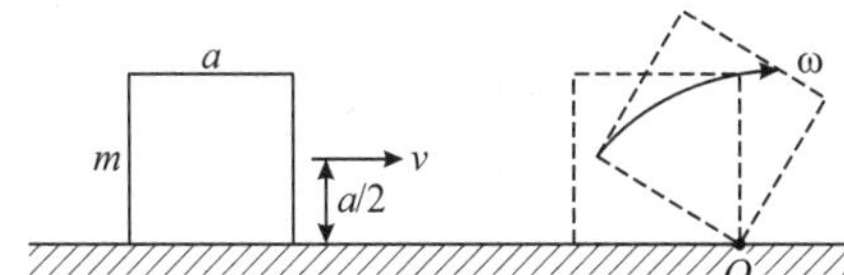

MI of cube about its corner

$$I = I_c + M\left(\frac{a}{\sqrt{2}}\right)^2 = \frac{Ma^2}{6} + \frac{Ma^2}{2}$$

$$\Rightarrow \quad I = \frac{2}{3}Ma^2$$

$$\Rightarrow \quad \omega = \frac{mva}{2\left(\frac{2}{3}Ma^2\right)} = \frac{3v}{4a}$$

$$\Rightarrow \quad \omega = \frac{3\times 2}{4\times 0.3} = 5 \text{ rad/s}$$

Ans. (B)

Sol. 49 MI of the given cylinder about AOR shown is

$$I = \frac{1}{12}Ml^2 + \frac{1}{4}MR^2$$

$$\Rightarrow \quad I = \frac{1}{12}Ml^2 + \frac{1}{4}M\left(\frac{M}{\rho\pi l}\right)$$

for I to be minimum. We use $\frac{dI}{dl} = 0$

$$\Rightarrow \quad \frac{dI}{dl} = \frac{1}{6}Ml + \frac{M^2}{4\rho\pi}\left(-\frac{1}{l^2}\right) = 0$$

$$\Rightarrow \quad l^3 = \frac{6M}{4\rho\pi} = \frac{3}{2}\frac{\rho\pi R^2 l}{\rho\pi}$$

$$\Rightarrow \quad \frac{l^2}{R^2} = \frac{3}{2} \Rightarrow \frac{l}{R} = \sqrt{\frac{3}{2}}$$

Ans. (D)

Sol. 50 Torque of rod's weight about point O is

$$\tau = Mg \times \frac{l}{2}\sin\theta$$

If α be the angular acceleration of rod, we use

$$Mg\frac{l}{2}\sin\theta = \left(\frac{Ml^2}{3}\right)\alpha$$

$$\Rightarrow \quad \alpha = \frac{3g}{2l}\sin\theta$$

Ans. (D)

Sol. 51 MI of whole disc is

$$I_0 = \frac{1}{2}MR^2$$

Mass of small disc is

$$m = \frac{M}{\pi R^2}\pi\left(\frac{R}{4}\right)^2 = \frac{M}{16}$$

MI of small disc about O is

$$I_{\text{hole}} = I_c + md^2$$

$$I_{\text{hole}} = \frac{1}{2}m\left(\frac{R}{4}\right)^2 + m\left(\frac{3R}{4}\right)^2$$

$$\Rightarrow \quad I_{\text{hole}} = \frac{1}{2}\left(\frac{M}{16}\right)\left(\frac{R^2}{16}\right) + \left(\frac{M}{16}\right)\left(\frac{9R^2}{16}\right)$$

$$\Rightarrow \quad I_{\text{hole}} = \frac{19}{512}MR^2$$

MI of given remaining portion is

$$I = I_0 - I_{\text{hole}}$$

$$\Rightarrow \quad I = \left(\frac{1}{2} - \frac{19}{512}\right)MR^2 = \frac{237}{512}MR^2$$

Ans. (B)

Sol. 52 By mass distribution property for same shape of objects of different dimensions MI expression remain same.

For lamina ABC $\quad I_0 = \gamma ML^2$

$$\Rightarrow \quad I_{DEF} = \gamma ml^2; \text{ where } m_{DEF} = \frac{1}{4}M \text{ and } l = \frac{l}{2}$$

$$\Rightarrow \quad I_{DEF} = \gamma\left(\frac{M}{4}\right)\left(\frac{L}{2}\right)^2 = \frac{\gamma}{16}ML^2 = \frac{I_0}{16}$$

MI of remaining part of lamina after removing DEF is

$$I = I_0 - I_{DEF} = \frac{15I_0}{16}$$

Ans. (B)

Sol. 53 In the given situation we have

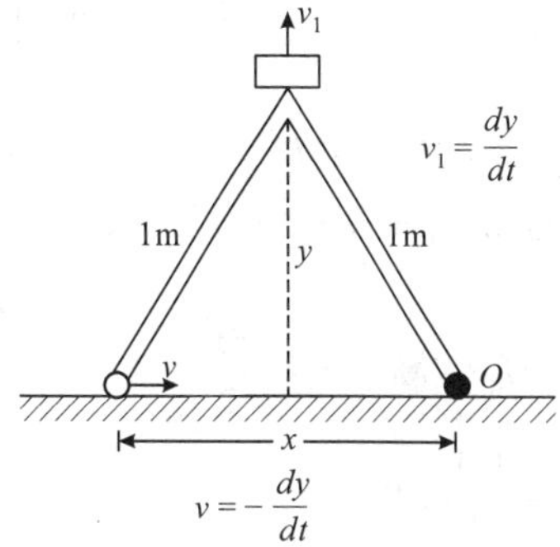

$$\left(\frac{x}{2}\right)^2 + y^2 = 1$$

Differentiate w.r.t. to time

$$\frac{x}{2}\frac{dx}{dt} + 2y\frac{dy}{dt} = 0$$

$$\Rightarrow \quad -\frac{x}{2}v + 2yv_1 = 0 \Rightarrow v_1 = \frac{x}{4y}v$$

[v = constant]

As with time $x \downarrow$ and $y \uparrow \Rightarrow v_1$ decreases **Ans. (B)**

Sol. 54 For equilibrium of beam balance we use

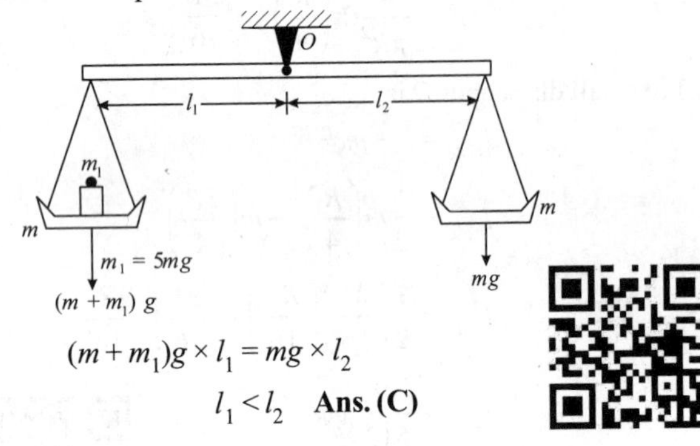

$$(m + m_1)g \times l_1 = mg \times l_2$$

$$\Rightarrow \quad l_1 < l_2 \quad \textbf{Ans. (C)}$$

Sol. 55 The central force on particle is given as

$$F = \frac{k}{R^n} = m\omega^2 R \quad \text{(for circular motion)}$$

$$\Rightarrow \quad \omega = \sqrt{\frac{k}{mR^{n+1}}}$$

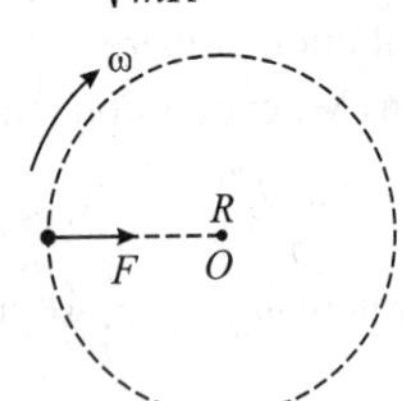

Time period of revolution is

$$T = \frac{2\pi}{\omega} = 2\pi\sqrt{\frac{mR^{n+1}}{k}}$$

$$\Rightarrow \quad T \propto R^{\frac{n+1}{2}}$$ **Ans. (C)**

Sol. 56 MI of this system about centre O of central disc is

$$I_0 = \frac{1}{2}MR^2 + 6\left(\frac{1}{2}MR^2 + M(2R)^2\right)$$

$$I_0 = \frac{7}{2}MR^2 + 24MR^2 = \frac{55}{2}MR^2$$

Now using parallel axes theorem for the system, its MI about P is given

$$I_P = I_0 + 7M(3R)^2 = \left(\frac{55}{2} + 63\right)MR^2$$

$$\Rightarrow \quad I_P = \frac{181}{2}MR^2$$ **Ans. (D)**

Sol. 57 Mass per unit area of disc $= \dfrac{9M}{\pi R^2}$

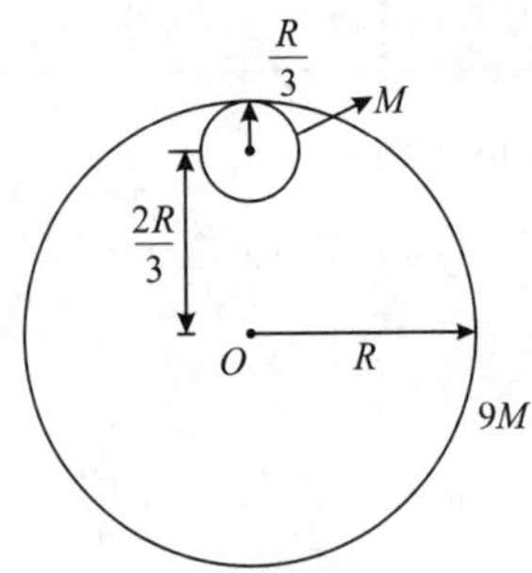

Mass of small disc is

$$m = \frac{9M}{\pi R^2} \times \pi\left(\frac{R}{3}\right)^2 = M$$

MI of remaining disc is

$$I_0 = (I_{9M} - I_M)_{\text{about } O}$$

$$I_0 = \frac{1}{2}(9M)R^2 - \left[\frac{1}{2}M\left(\frac{R}{3}\right)^2 + M\left(\frac{2R}{3}\right)^2\right]$$

$$\Rightarrow \quad I_0 = \frac{9}{2}MR^2 - \frac{1}{2}MR^2 = 4MR^2$$ **Ans. (A)**

Sol. 58 Moment of force (Torque) about point A will be maximum when line AB is $\perp$ direction of $\vec{F}$.

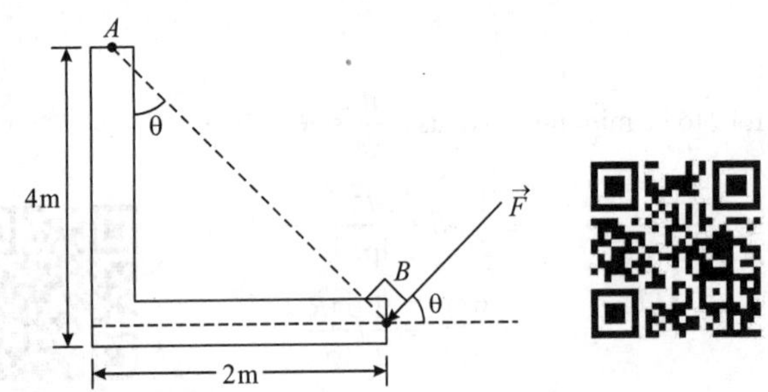

For this we use $\tan\theta = \dfrac{2}{4} = \dfrac{1}{2}$ **Ans. (C)**

Sol. 59 To keep the rod horizontal net force and torque on it must be zero.

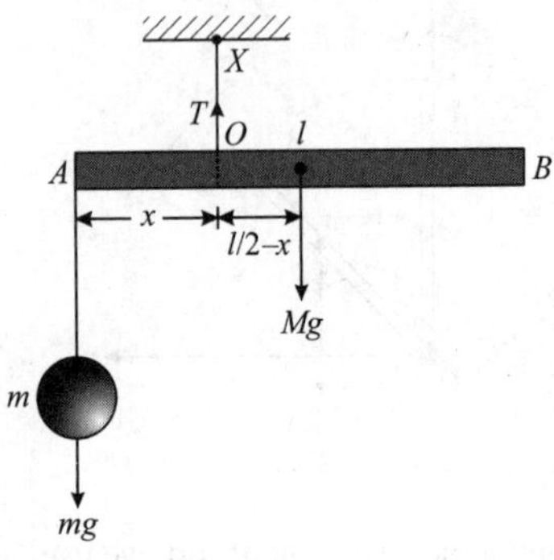

$$mg\,x = Mg\left(\frac{l}{2} - x\right)$$

Where $mgx \rightarrow$ Anticlockwise torque about O

and $Mg\left(\frac{l}{2} - x\right) \rightarrow$ Clockwise torque about O

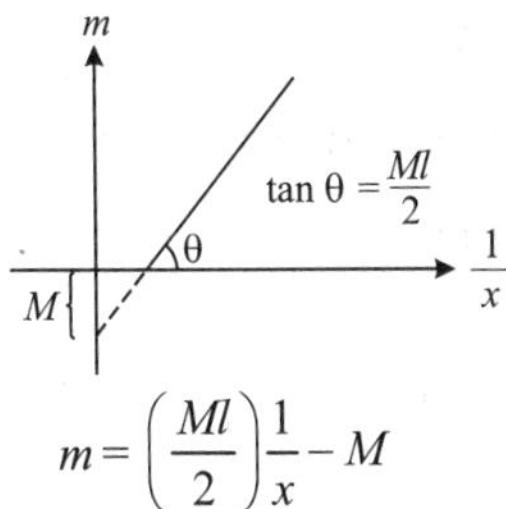

$$m = \left(\frac{Ml}{2}\right)\frac{1}{x} - M$$

where $\frac{Ml}{2} \rightarrow$ Slope

and $M \rightarrow$ Intercept on Y axis **Ans. (A)**

Sol. 60 After masses stick to rod, final COM of rod remains at mid point of rod as mass moments of m and $2m$ about O are equal & on opposite sides.

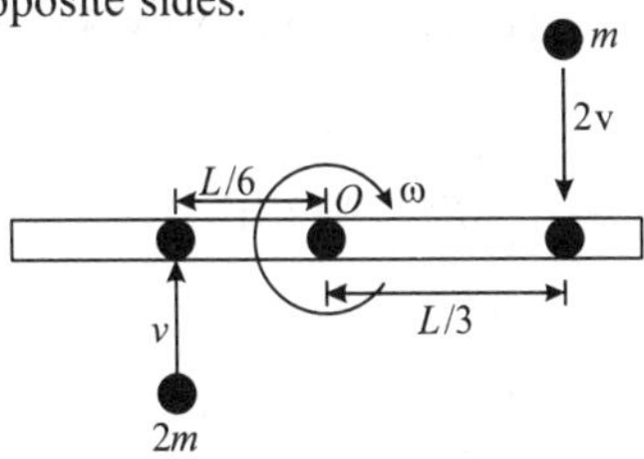

As no external torque is acting on system.

$\Rightarrow$ Total angular momentum will remain conserved about O.

$$L_i = L_f$$

$$2m(v)\left(\frac{L}{6}\right) + m(2v)\left(\frac{L}{3}\right) = I\omega$$

$$\Rightarrow \quad \frac{mvL}{3} + \frac{2mvL}{3} = \left[\frac{1}{12}(8m)L^2 + m\left(\frac{L}{3}\right)^2 + 2m\left(\frac{L}{6}\right)^2\right]\omega$$

$$mvL = mL^2\omega\left[\frac{2}{3} + \frac{1}{9} + \frac{1}{18}\right] = \frac{15}{18}mL^2\omega$$

$$\Rightarrow \quad mvL = \frac{5}{6}mL^2\omega$$

$$\Rightarrow \quad \omega = \frac{6v}{5L}$$ **Ans. (B)**

Sol. 61 Ratio of I_z and $I_{z'}$ is

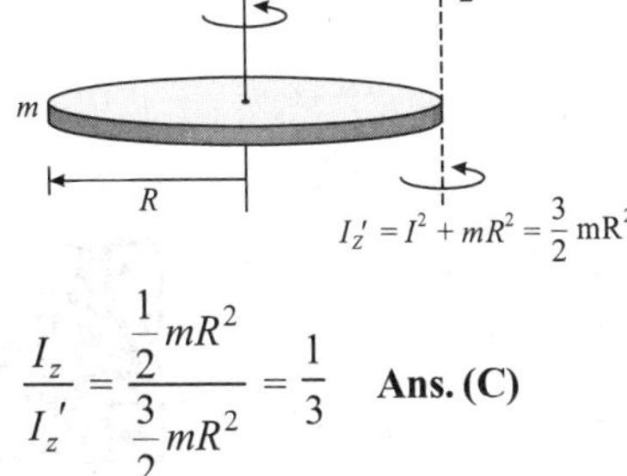

$$\frac{I_z}{I_z'} = \frac{\frac{1}{2}mR^2}{\frac{3}{2}mR^2} = \frac{1}{3}$$ **Ans. (C)**

Sol. 62 The COM of rod has fallen by

$$h = \frac{l}{2}\sin\alpha$$

By work-energy theorem, we have

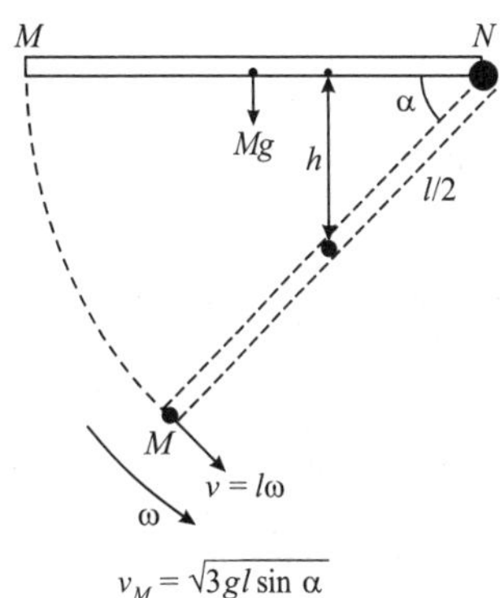

$$mg\left(\frac{l}{2}\sin\alpha\right) = \frac{1}{2}I\omega^2$$

$$\Rightarrow \quad mg\frac{l}{2}\sin\alpha = \frac{1}{2}\left(\frac{ml^2}{3}\right)\omega^2$$

$$\Rightarrow \quad \omega = \sqrt{\frac{3g\sin\alpha}{l}}$$ **Ans. (D)**

Sol. 63 Centre of mass must be vertically below the suspension point.

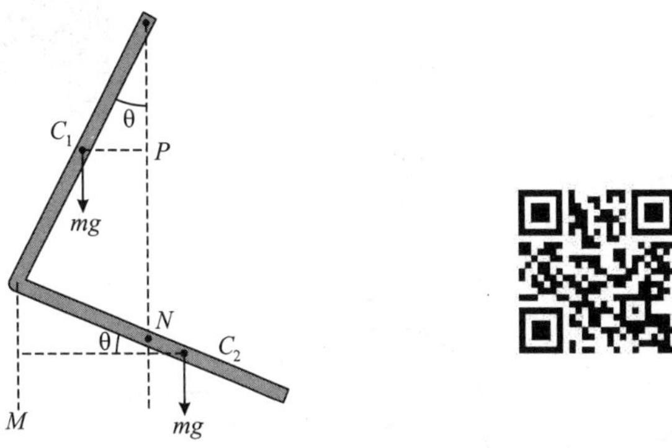

Balancing torque about suspension point, we have

$$mg\left(\frac{L}{2}\sin\theta\right) = mg\left(\frac{L}{2}\cos\theta - L\sin\theta\right)$$

$$\Rightarrow \quad \frac{3}{2}mgL\sin\theta = \frac{mgL}{2}\cos\theta$$

$$\Rightarrow \quad \tan\theta = \frac{1}{3}$$ **Ans. (B)**

Sol. 64

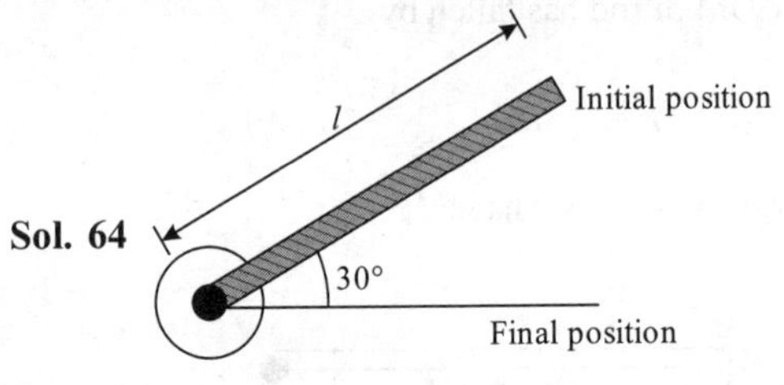

$$mg\left(\frac{l}{2}\right)\sin 30° = \frac{1}{2}I\omega^2$$

$$\Rightarrow \quad \frac{1}{2}\left(\frac{ml^2}{3}\right)\omega^2 = \frac{mgl}{4}$$

$$\Rightarrow \quad \omega = \sqrt{\frac{3g}{2l}} = \sqrt{\frac{3\times10}{2\times0.5}}$$

$$\Rightarrow \quad \omega = \sqrt{30}\text{ rad/s}$$

Ans. (A)

Sol. 65

Consider a strip of radius x & thickness dx,

Torque due to friction on this strip

$$\Rightarrow \quad \tau = \int d\tau = \int_0^R \frac{\mu F\cdot 2\pi x dx}{\pi R^2}\times x$$

$$\Rightarrow \quad \tau = \frac{2\mu F}{R^2}\cdot\frac{R^3}{3}$$

$$\Rightarrow \quad \tau = \frac{2\mu FR}{3}$$

Ans. (A)

Sol. 66

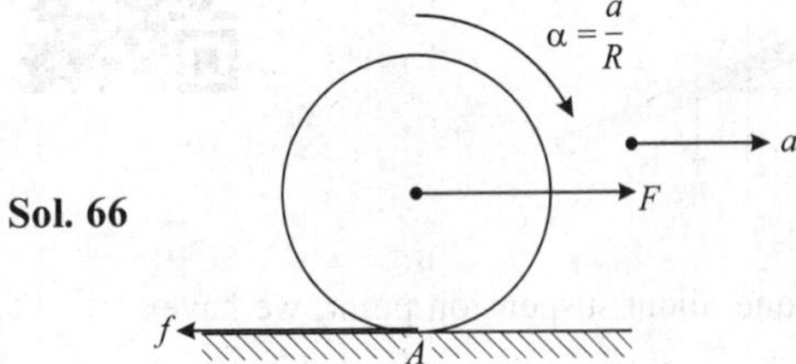

Equation of motion of cylinder about bottom point of contact A is written as

$$FR = \frac{3}{2}MR^2\alpha$$

$$\Rightarrow \quad \alpha = \frac{2F}{3MR}$$

Ans. (C)

Sol. 67

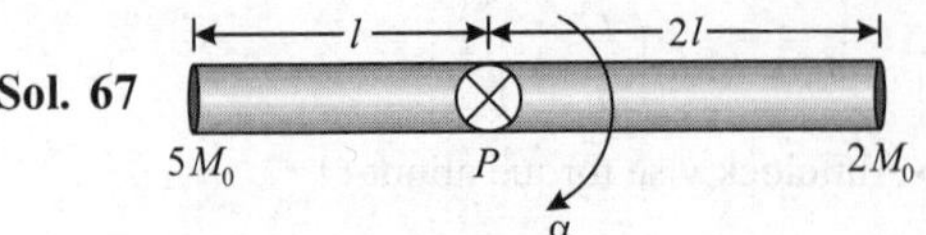

Applying torque equation about point P, we have

$$2M_0g(2l) - 5M_0\,gl = I\alpha$$

Where $\quad I = 2M_0(2l)^2 + 5M_0$

$$l^2 = 13M_0l^2d$$

$$\Rightarrow \quad \alpha = -\frac{M_0gl}{13M_0l^2}$$

$$\Rightarrow \quad \alpha = -\frac{g}{13l}$$

$$\Rightarrow \quad \alpha = \frac{g}{13l}\text{ anticlockwise}$$

Ans. (C)

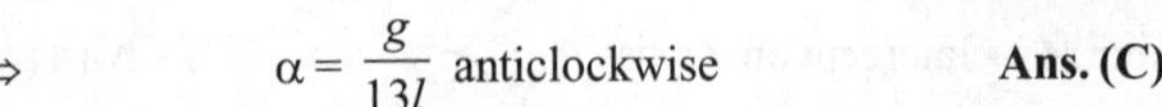

Sol. 68 For Balls (considered solid spheres)

Using parallel axis theorem, the moment of inertia of each balls is

$$I_{ball} = \frac{2}{5}MR^2 + M(2R)^2$$

$$\Rightarrow \quad I_{ball} = \frac{22}{5}MR^2$$

For two balls it is $\frac{44}{5}MR^2$

Moment of inertia of rod is

$$I_{rod} = \frac{ML^2}{12} = \frac{M(2R)^2}{12} = \frac{MR^2}{3}$$

$$I_{system} = I_{Ball} + I_{rod}$$

$$\Rightarrow \quad I_{system} = \frac{44}{5}MR^2 + \frac{MR^2}{3}$$

$$\Rightarrow \quad I_{system} = \frac{137}{15}MR^2$$

Ans. (C)

Sol. 69 Suppose M is mass and a is side of larger triangle, then $\frac{M}{4}$ and $\frac{a}{2}$ will be mass and side length of smaller triangle. By symmetry expression of moment of inertia of a triangle remain same as

$$I = \gamma ML^2$$

So we use

$$\frac{I_{removed}}{I_{original}} = \frac{\frac{M}{4}\cdot\left(\frac{a}{2}\right)^2}{M\ (a)^2}$$

$$I_{removed} = \frac{I_0}{16}$$

Thus moment of inertia of remaining triangle is

$$I = I_0 - \frac{I_0}{16} = \frac{15I_0}{16}$$ **Ans. (D)**

Sol. 70 Torque for F_2 force

$$\overrightarrow{F_2} = \frac{F}{2}(-\hat{i}) + \frac{F\sqrt{3}}{2}(-\hat{j})$$

$$\vec{r}_2 = 0\hat{i} + 6\hat{j}$$

$$\vec{\tau}_{F_2} = \vec{r}_2 \times \overrightarrow{F_2} = 3F\hat{k}$$

Torque for F_1 force

$$\overrightarrow{F_2} = F\hat{k}$$

$$\vec{r}_1 = 2\hat{i} + 3\hat{j}$$

$$\vec{\tau}_{F_1} = \vec{r}_1 \times \overrightarrow{F_1} = 3F\hat{i} + 2F(-\hat{j})$$

$$\vec{\tau}_{net} = \vec{\tau}_{F_1} + \vec{\tau}_{F_2}$$

$\Rightarrow$ $$\vec{\tau}_{net} = 3F\hat{i} + 2F(-\hat{j}) + 3F(\hat{k})$$ **Ans. (D)**

Sol. 71

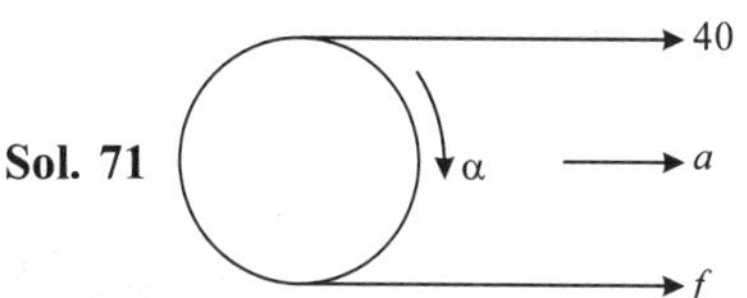

For pure rolling

$$\alpha = \frac{a}{R}$$

Equation of motion of cylinder is

$$40 + f = m(R\alpha) \quad \ldots(1)$$

$$40 \times R - f \times R = mR^2\alpha$$

$\Rightarrow$ $$40 - f = mR\alpha \quad \ldots(2)$$

From (1) and (2)

$\Rightarrow$ $$\alpha = \frac{40}{mR} = \frac{40}{5 \times 0.5}$$

$$= 16 \text{ rad/s}^2$$ **Ans. (B)**

Sol. 72 Torque on particle is given as

$$\tau = Fr \sin\theta$$

$\Rightarrow$ $$2.5 = 1 \times 5 \sin\theta$$

$\Rightarrow$ $$\sin\theta = 0.5 = \frac{1}{2}$$

$\Rightarrow$ $$\theta = \frac{\pi}{6}$$ **Ans. (B)**

Sol. 73 Moment of inertia of system is given as

$$I = \frac{MR^2}{2} + 2\left(\frac{MR^2}{4} + MR^2\right)$$

$\Rightarrow$ $$I = \frac{MR^2}{2} + \frac{MR^2}{2} + 2MR^2$$

$\Rightarrow$ $$I = 3MR^2$$ **Ans. (A)**

Sol. 74 Moment of inertia of hollow cylinder is

$$I = \frac{1}{2}M(R_1^2 + R_2^2) = \frac{1}{2}M(10^2 + 20^2) = 250\,M$$

Moment of inertia of a solid cylinder is

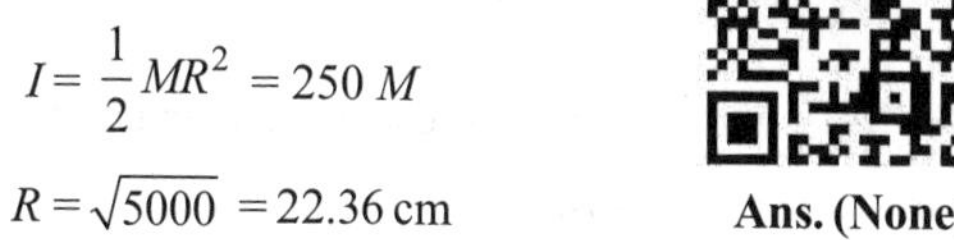

$$I = \frac{1}{2}MR^2 = 250\,M$$

$\Rightarrow$ $$R = \sqrt{5000} = 22.36 \text{ cm}$$ **Ans. (None)**

Sol. 75 Using Work Energy Theorem from A to B, we have

$$\frac{1}{2}mv_A^2 + mgh = \frac{1}{2}mv_B^2$$

$\Rightarrow$ $$2gh = v_B^2 - v_A^2$$

$\Rightarrow$ $$2 \times 10 \times 10 = v_B^2 - 5^2$$

$\Rightarrow$ $$v_B = 15 \text{ m/s}$$

Angular momentum about O is given as

$$L_0 = mvr$$

$\Rightarrow$ $$L_0 = 20 \times 10^{-3} \times 20$$

$\Rightarrow$ $$L_0 = 6 \text{ kg.m}^2\text{/s}$$ **Ans. (B)**

Sol. 76 By parallel axes theorem, we use

$$I = \frac{2}{5}mR^2 + mx^2$$ **Ans. (B)**

Sol. 77 By parallel axes theorem, we have

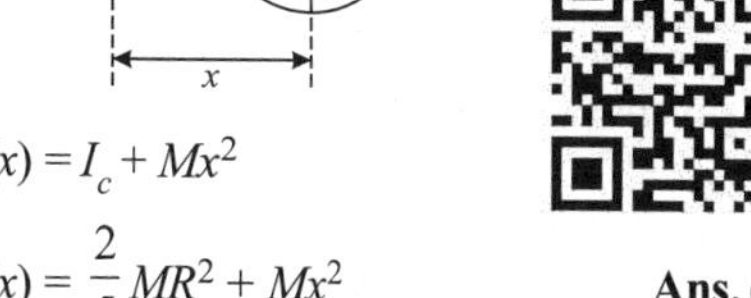

$$I(x) = I_c + Mx^2$$

$\Rightarrow$ $$I(x) = \frac{2}{5}MR^2 + Mx^2$$ **Ans. (D)**

Sol. 78 As the box is in contact with the edge for $\Delta t = 0.01$ s. So angular impulse on box is

$$\Delta L = \tau \cdot \Delta t = mg \cdot \frac{l}{2} \times \Delta t = I\omega$$

Gain in angular speed of box

$$\omega = \frac{mgl\Delta t}{2I} = \frac{mgl\Delta t}{2\left(\frac{ml^2}{3}\right)} = \frac{3g\Delta t}{2l}$$

$$\Rightarrow \quad \omega = \frac{3\times 10\times 0.01}{2\times 0.3} = 0.5 \text{ rad/s}$$

Time taken in reaching ground is

$$t = \sqrt{\frac{2h}{g}} = \sqrt{\frac{2\times 5}{10}} = 1\text{s}$$

In 1s angle rotated by box is

$$\theta = \omega t = 0.5\times 1 = 0.5 \text{ rad}$$

Ans. (A)

Sol. 79 Angular impulse = change in angular momentum

$$\tau\Delta t = \Delta L$$

$$\Rightarrow \quad mg\frac{l}{2}\times .01 = \frac{ml^2}{3}\omega$$

$$\Rightarrow \quad \omega = \frac{3g\times 0.01}{2l}$$

$$\omega = \frac{3\times 10\times .01}{2\times 0.3}$$

$$\omega = \frac{1}{2} = 0.5 \text{ rad/s}$$

time taken by rod to hit the ground

$$t = \sqrt{\frac{2h}{g}} = \sqrt{\frac{2\times 5}{10}} = 1 \text{ s}$$

in this time angle rotate by rod

$$\theta = \omega t = 0.5\times 1 = 0.5 \text{ rad}$$

Ans. (C)

Sol. 80 If moment of inertia of rolling object is taken as I and speed of centre of mass at bottom of incline is v, we use

$$\frac{1}{2}\left(m + \frac{I}{R^2}\right)v^2 = mgh$$

If radius of gyration is k, then, we use $I = mk^2$, so we have

$$h = \frac{\left(1+\frac{k^2}{R^2}\right)v^2}{2g} \quad \ldots(1)$$

For given object we use

$$\frac{k_{\text{ring}}}{R_{\text{ring}}} = 1$$

$$\frac{k_{\text{solid cylinder}}}{R_{\text{solid cylinder}}} = \frac{1}{\sqrt{2}}$$

$$\Rightarrow \quad \frac{k_{\text{solid sphere}}}{R_{\text{solid sphere}}} = \sqrt{\frac{2}{5}}$$

From equation-(1) we have

$$h_1 : h_2 : h_3 :: (1+1) : \left(1+\frac{1}{2}\right) : \left(1+\frac{2}{5}\right) :: 20 : 15 : 14$$

Therefor **most appropriate option is (B)** which in not in proper sequence.

Ans. (C)

Sol. 81 Kinetic energy of disc is given as

$$KE = \frac{1}{2}I\omega^2 = k\theta^2$$

$$\Rightarrow \quad \omega^2 = \frac{2k\theta^2}{I}$$

$$\Rightarrow \quad \omega = \sqrt{\frac{2k}{I}}\theta \quad \ldots(1)$$

Differentiating-(1) wrt time, we get angular acceleration

$$\alpha = \frac{d\omega}{dt} = \sqrt{\frac{2k}{I}}\left(\frac{d\theta}{dt}\right)$$

$$\Rightarrow \quad \alpha = \sqrt{\frac{2k}{I}}\cdot\sqrt{\frac{2k}{I}}\theta$$

$$\Rightarrow \quad \alpha = \frac{2k}{I}\theta$$

Ans. (D)

Sol. 82 Kinetic energy of rotating body is

$$KE = \frac{1}{2}I\omega^2$$

$$\Rightarrow \quad 1200 = \frac{1}{2}1.5\times\omega^2$$

$$\Rightarrow \quad \omega^2 = \frac{1200\times 2}{1.5} = 1600$$

$$\Rightarrow \quad \omega = 40 \text{ rad/s}^2$$

If body rotates for time t at angular acceleration $\alpha = 20$ rad/s^2, we use

$$\omega = \omega_0 + \alpha t$$

$$40 = 0 + 20t$$

$$t = 2 \text{ s}$$

Ans. (A)

Sol. 83 Applying angular momentum conservation, about axis of rotation, we have

$$L_i = L_f$$

$$\frac{ML^2}{12}\omega_0 = \left(\frac{ML^2}{12} + m\left(\frac{L}{2}\right)^2\times 2\right)\omega$$

$$\Rightarrow \quad \omega = \frac{M\omega_0}{M+6m}$$

Ans. (D)

Sol. 84 We consider an elemental strip of radius r and width dr as shown in figure mass of elemental strip is

$$dm = \sigma(r)\cdot 2\pi r dr$$

Moment of inertia of elemental strip is

$$dI = dmr^2$$

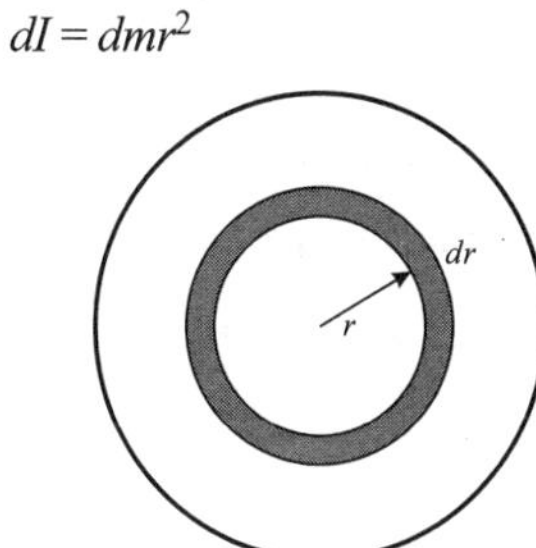

$$\Rightarrow \quad I_{\text{Disc}} = \int_0^R (dm)r^2$$

$$\Rightarrow \quad I_{\text{Disc}} = \int_0^R (\sigma 2\pi r dr)r^2$$

$$\Rightarrow \quad I_{\text{Disc}} = \int_0^R (kr^2 2\pi r dr)r^2$$

$$\Rightarrow \quad I_{\text{Disc}} = 2\pi k\int_0^R r^5 dr = \frac{2\pi kR^6}{6} = \frac{\pi kR^6}{3} \qquad \ldots(1)$$

Total mass of disc is

$$M = \int dm = \int_0^R 2\pi r dr\,(kr^2)$$

$$\Rightarrow \quad M = 2\pi k\int_0^R r^3 dr$$

$$\Rightarrow \quad M = 2\pi k\left.\frac{r^4}{4}\right|_0^R = \frac{\pi kR^4}{2} \qquad \ldots(2)$$

From equation-(1) and (2), we have

$$I_{\text{Disc}} = \frac{\pi kR^6}{3} = \left(\frac{\pi kR^4}{2}\right)\times\frac{2}{3}R^2$$

$$\Rightarrow \quad I_{\text{Disc}} = M\times\frac{2}{3}R^2$$

$$\Rightarrow \quad I_{\text{Disc}} = \frac{2}{3}MR^2$$

Ans. (C)

Sol. 85 Initial *KE* of system is

$$E_i = \frac{1}{2}I_1\times\omega_1^2 + \frac{1}{2}\left(\frac{I_1}{2}\right)\left(\frac{\omega_1^2}{4}\right)$$

$$\Rightarrow \quad E_i = \frac{I_1\omega_1^2}{2}\left(\frac{9}{8}\right) = \frac{9}{16}I_1\omega_1^2$$

By conservation of angular momentum, we have

$$I_1\omega_1 + \frac{I_1\omega_1}{4} = \frac{3I_1}{2}\omega$$

$$\Rightarrow \quad \frac{5}{4}I_1\omega_1 = \frac{3I_1}{2}\omega$$

$$\Rightarrow \quad \omega = \frac{5}{6}\omega_1$$

Final KE of system is

$$E_f = \frac{1}{2}I_f\,\omega^2 = \frac{1}{2}\left(\frac{3}{2}I_1\right)\left(\frac{5}{6}\omega_1\right)^2$$

$$\Rightarrow \quad E_f = \frac{25}{48}I_1\omega_1^2$$

Difference in kinetic energies $(E_f - E_i)$ is given as

$$E_f - E_i = I_1\omega_1^2\left(\frac{25}{48}-\frac{9}{16}\right) = \frac{-2}{48}I_1\omega_1^2$$

$$\Rightarrow \quad E_f - E_i = \frac{-I_1\omega_1^2}{24}$$

Ans. (D)

Sol. 86 Acceleration of particle

$$\vec{a_x} = \frac{d^2x}{dt^2} = -a\omega_1^2\cos\omega_1 t\,\hat{i}$$

$$\vec{a_y} = \frac{d^2y}{dt^2} = -b\omega_2^2\sin\omega_2 t\,\hat{i}$$

$$\vec{a}_{t=0} = -a\omega_1^2\,\hat{i}$$

Position vector of particle at $t = 0 \rightarrow \vec{r} = (x_0 + a)\,\hat{i} + y_0\,\hat{j}$

Torque on particle is $\vec{\tau} = \vec{r}\times\vec{F} = \vec{r}\times m\,\vec{a}_{t=0}$

$$= [(x_0 + a)\,\hat{i} + y_0\hat{j}]\times m\,[-a\omega_1^2\,\hat{i}\,]$$

$$= my_0\,a\omega_1^2\,\hat{k}$$

Ans. (B)

Sol. 87 Angular acceleration of coin is

$$\alpha = \frac{\Delta\omega}{\Delta t} = \frac{25\times 2\pi}{5} = 10\pi \text{ rad/s}^2$$

Torque on coin is

$$\tau = \left(\frac{5}{4}MR^2\right)\alpha$$

$$\Rightarrow \quad \tau = \frac{5}{4}\times 5\times 10^{-3}\times(10^{-2})^2\times 10\pi$$

$$\Rightarrow \quad \tau = 1.9625\times 10^{-5}\,\text{Nm} \simeq 2.0\times 10^{-5}\,\text{Nm}$$ **Ans. (B)**

Sol. 88 Velocity at $t = 2$s is

$$\vec{v}_{(t)} = \frac{\vec{dr}}{dt} = 2\,\hat{i} - 6t\,\hat{j}\,\big|_{t=2\text{s}}\,(2\,\hat{i} - 12\,\hat{j}\,)$$

Position vector at $t = 2$s is

$$\vec{r}_{(t)} = (4\,\hat{i} - 12\,\hat{j}\,)$$

Angular momentum at $t = 2$s is

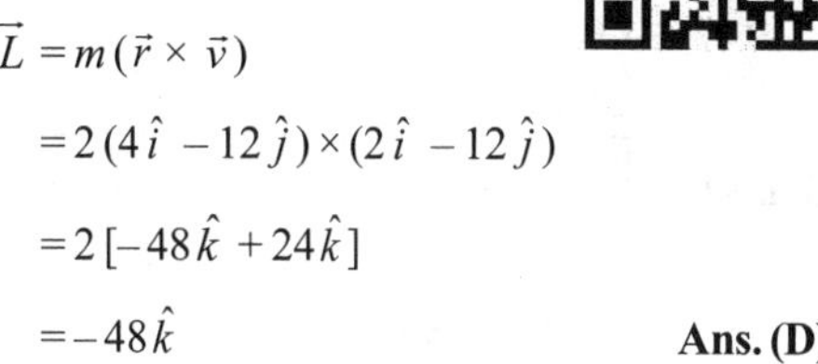

$$\vec{L} = m(\vec{r}\times\vec{v})$$

$$= 2(4\,\hat{i} - 12\,\hat{j}\,)\times(2\,\hat{i} - 12\,\hat{j}\,)$$

$$= 2[-48\hat{k} + 24\hat{k}]$$

$$= -48\hat{k}$$

Ans. (D)

Sol. 89 For disc $m_1 = \frac{7M}{8}$, $R_1 = 2R$

$$MI \text{ of disc } I_1 = \frac{1}{2}m_1R_1^2 = \frac{1}{2}\left(\frac{7M}{8}\right)(2R)^2 = \frac{7}{4}MR^2 \qquad \ldots(1)$$

For new solid sphere $m_2 = \frac{M}{8}$,

$$R_2 = \frac{R}{2} \text{ [we use } \frac{1}{8}\left(\frac{4}{3}\pi R^3\right) = \frac{4}{3}\pi R_2^3 \Rightarrow R_2 = \frac{R}{2}\text{]}$$

$$MI \text{ of new sphere } I_2 = \frac{2}{5}m_2R_2^2 = \frac{2}{5}\left(\frac{M}{8}\right)\left(\frac{R}{2}\right)^2$$

$$= \frac{1}{80}MR^2 \quad \ldots(2)$$

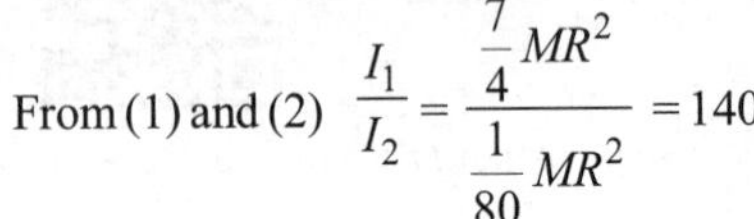

From (1) and (2) $\frac{I_1}{I_2} = \frac{\frac{7}{4}MR^2}{\frac{1}{80}MR^2} = 140$ **Ans. (D)**

Sol. 90 Area of elemental strip $= ds = 2\pi x\, dx$

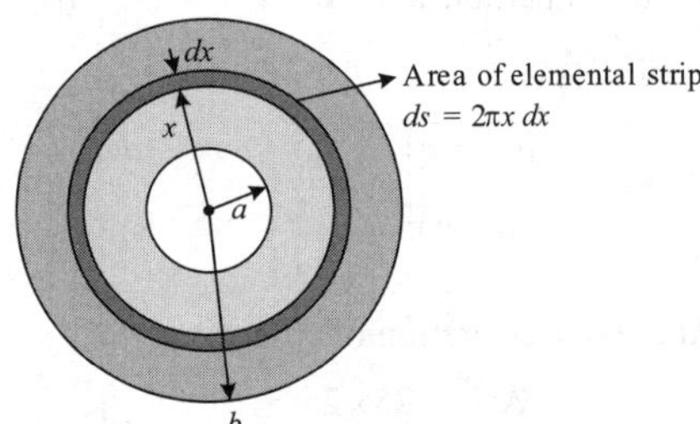

$$\text{Mass of elemental strip } dm = \frac{\sigma_0 ds}{x} = \frac{\sigma_0}{x}\cdot 2\pi x\, dx = 2\pi\sigma_0\, dx$$

$$\text{Total mass of disc } m = \int dm = 2\pi\sigma_0 = \int_a^b dx = 2\pi\sigma_0(b-a)$$

MI of elemental strip $dI = dmx^2$

$\Rightarrow$ $dI = 2\pi\sigma_0 x^2 dx$

$$\Rightarrow \quad I = \int dI = \int_a^b 2\pi\sigma_0 x^2 dx = \frac{2\pi\sigma_0}{3}\cdot(b^3 - a^3)$$

Radius of gyration of disc is

$$k = \sqrt{\frac{I}{m}} = \sqrt{\frac{(b^3 - a^3)}{3(b-a)}} = \sqrt{\frac{a^2 + b^2 + ab}{3}}$$ **Ans. (D)**

Sol. 91

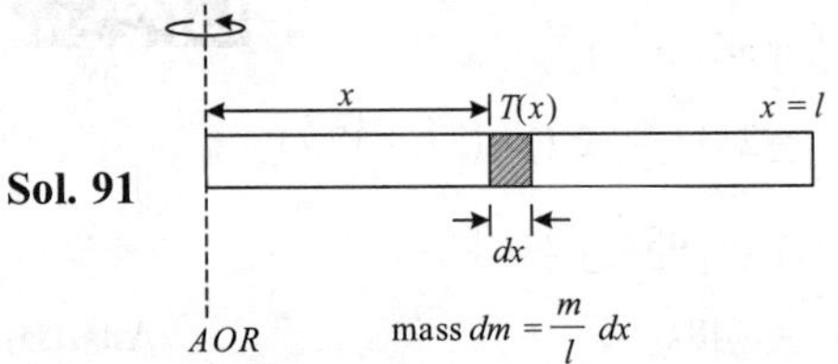

At a distance x from AOR, tension in rod will be due to the centrifugal force on part of rod outside it in frame of rod. Centrifugal force on dm is

$$dF = dm\omega^2 x$$

Total centrifugal force on outer part of rod i.e. from x to l is

$$F = T_{(x)} = \int dF = \int_x^l \frac{m}{l}\omega^2 x dx = \frac{m\omega^2}{2l}(l^2 - x^2)$$

This is a parabolic function **Ans. (D)**

Sol. 92 Using work energy theorem, we use

$$mgh = \frac{1}{2}mv^2 + \frac{1}{2}\times\left(\frac{1}{2}mr^2\right)\times\frac{v^2}{r^2} = \frac{3}{4}mv^2$$

$$\Rightarrow \quad v = \sqrt{\frac{4}{3}gh}$$

Angular speed is given as

$$\omega = \frac{v}{r} = \frac{1}{r}\sqrt{\frac{4gh}{3}}$$ **Ans. (C)**

Sol. 93 Moment of inertia of given system is taken as

$$m\frac{l^2}{12} + m\frac{l^2}{16} = mk^2$$

$$\Rightarrow \quad \frac{7l^2}{48} = k^2$$

$$\Rightarrow \quad k = \sqrt{\frac{7}{48}}\cdot l$$ **Ans. (B)**

Sol. 94 We consider an elemental ring in disc as shown

Moment of inertia of this ring is given as

$$dI = dmr^2$$

$$\Rightarrow \quad dI = \sigma 2\pi r\, dr\, r^2$$

$$\Rightarrow \quad dI = 2\pi(A + Br)\cdot r^3\cdot dr$$

Total moment of inertia of disc is

$$I = \int dI = 2\pi\int_0^a (Ar^3 + Br^4)dr$$

$$\Rightarrow \quad I = 2\pi a^4\left(\frac{A}{4} + \frac{aB}{5}\right)$$ **Ans. (A)**

Sol. 95 Forces are shown in figure below

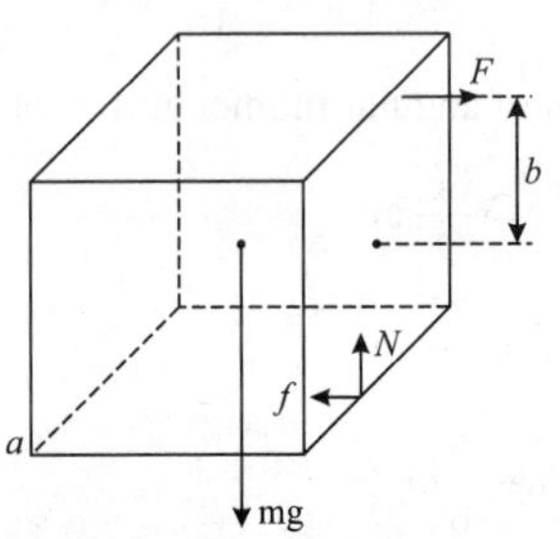

To move the block

$$F = \mu mg \quad \ldots(1)$$

For block not to topple, we use

$$F\left(b+\frac{a}{2}\right) = mg\,\frac{a}{2} \quad \ldots(2)$$

$$\Rightarrow \quad \mu mg\left(b+\frac{a}{2}\right) = mg \times \frac{a}{2}$$

$$\Rightarrow \quad \left(b+\frac{a}{2}\right)\mu = \frac{a}{2}$$

$$\Rightarrow \quad 0.8b + 0.4a = a$$

$$\Rightarrow \quad 0.8b = 0.6a$$

$$\Rightarrow \quad \frac{b}{a} = \frac{3}{4}$$

Ans. (75)

Sol. 96 Figure below shows the given situation

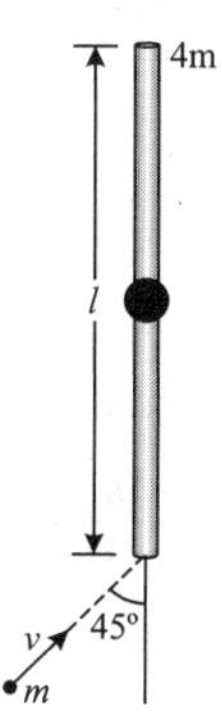

Using by conservation of angular momentum about the hinge, we have

$$m\left(\frac{v}{\sqrt{2}}\right)\left(\frac{l}{2}\right) = \left[\frac{4ml^2}{12} + \frac{ml^2}{4}\right]\omega$$

$$\Rightarrow \quad \omega = \frac{3\sqrt{2}}{7}\left(\frac{v}{l}\right)$$

Ans. (B)

Sol. 97 Kinetic energy of rolling sphere is given as

$$KE = \frac{1}{2}mv^2 + \frac{1}{2}I\omega^2$$

$$\Rightarrow \quad KE = \frac{1}{2}mv^2\left(1+\frac{K^2}{R^2}\right)$$

$$\Rightarrow \quad KE = 8.75 \times 10^{-4}\,\text{J}$$

Ans. (A)

Sol. 98 Using parallel axis theorem, moment of inertia of system is given as

$$I_0 = 3 \times \left[\frac{2}{5}M\left(\frac{d}{2}\right)^2 + M\left(\frac{d}{\sqrt{3}}\right)^2\right] = \frac{13}{10}\text{Md}^2 \quad \ldots(1)$$

Moment of inertia about axis passing through A is given as

$$I_A = I_0 + 3M\left(\frac{d}{\sqrt{3}}\right)^2$$

$$\Rightarrow \quad I_A = \frac{13}{10}Md^2 + Md^2$$

$$\Rightarrow \quad I_A = \frac{23}{10}Md^2 \quad \ldots(2)$$

From (1) and (2), we have

$$\frac{I_0}{I_A} = \frac{13}{23}$$

Ans. (A)

Sol. 99 Using work energy theorem, we have

$$\Rightarrow \quad mg\left(\frac{l}{2}\sin 30^\circ\right) = \frac{1}{2}I\omega^2$$

$$\Rightarrow \quad mg \times \frac{1}{2} \times \frac{1}{2} = \frac{1}{2}\left(\frac{m(1)^2}{3}\right)\omega^2$$

$$\Rightarrow \quad \omega^2 = \frac{3g}{2}$$

$$\Rightarrow \quad \omega = \sqrt{15}\text{ rad/s}$$

$$\Rightarrow \quad n = 15$$

Ans. (15.00)

Sol. 100 Using work energy theorem, we have

$$(m_1 - m_2)\,gh = \frac{1}{2}(m_1 + m_2)\,v^2 + \frac{1}{2}I\omega^2$$

$$\Rightarrow \quad (m_1 - m_2)\,gh = \frac{1}{2}(m_1 + m_2)\,(\omega R)^2 + \frac{1}{2}I\omega^2$$

$$\Rightarrow \quad (m_1 - m_2)\,gh = \frac{\omega^2}{2}[(m_1 + m_2)\,R^2 + I]$$

$$\Rightarrow \quad \omega = \sqrt{\frac{2(m_1 - m_2)gh}{(m_1 + m_2)R^2 + I}}$$

Ans. (B)

Sol. 101 Figure below shows the forces acting on rod

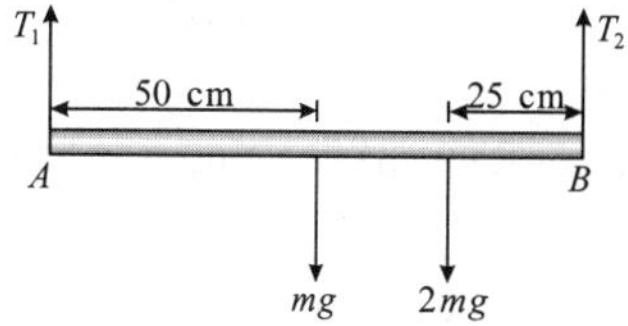

For equilibrium of rod, we use torque about point B

$$\tau_b = 0$$

$$2mg \times 25 + mg \times 50 - T_1 \times 100 = 0$$

$$\Rightarrow \quad 100\,T_1 = 100mg$$

$$\Rightarrow \quad T_1 = mg$$

Ans. (D)

Sol. 102 For maximum torque F must be applied normal to radius as shown in figure

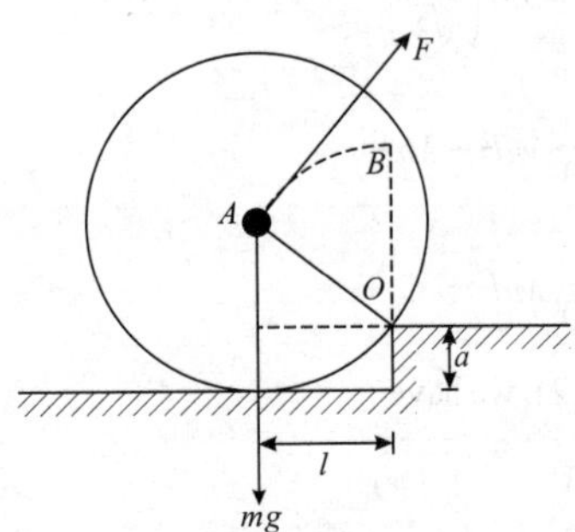

To pull over the edge force should be able to lift cylinder centre from A to B so at point A torque of F must be more than that of gravity about O, so we use

$$FR \geq Mgl$$

$$\Rightarrow \quad F \geq \frac{Mg}{R}\sqrt{R^2-(R-a)^2}$$

$$\Rightarrow \quad F \geq Mg\sqrt{1-\left(\frac{R-a}{R}\right)^2} \qquad \textbf{Ans. (D)}$$

Sol. 103 By angular momentum conservation, we use

$$L_1 + L_2 = L_{\text{final}}$$

$$\Rightarrow \quad I_1\omega_1 + I_2\omega_2 = (I_1+I_2)\omega_f$$

$$\Rightarrow \quad 0.1\times 10 + 0.2\times 5 = (0.1+0.2)\times\omega_f$$

$$\Rightarrow \quad \omega_f = \frac{20}{3}$$

Kinetic energy of combined disc system can be given as

$$\Rightarrow \quad K = \frac{1}{2}(I_1+I_2)\,\omega_f^2$$

$$\Rightarrow \quad K = \frac{1}{2}(0.1+0.2).\left(\frac{20}{3}\right)^2$$

$$\Rightarrow \quad K = \frac{0.3}{2}\times\frac{400}{9} = \frac{120}{18} = \frac{20}{3}\text{J} \qquad \textbf{Ans. (D)}$$

Sol. 104 For constant mass, volume is also constant, so we use

$$\pi R^2 L = \text{Constant}$$

$$\Rightarrow \quad R^2L = C \qquad \ldots(1)$$

Moment of inertia as given as

$$I = M\left(\frac{R^2}{4}+\frac{L^2}{12}\right)$$

$$\Rightarrow \quad I = M\left(\frac{C}{4L}+\frac{L^2}{12}\right)$$

For I to be minimum, we use

$$\frac{dI}{dL} = M\left[-\frac{C}{4L^2}+\frac{L}{6}\right] = 0$$

$$\Rightarrow \quad \frac{C}{4L^2} = \frac{L}{6}$$

$$\Rightarrow \quad L = \left[\frac{3C}{2}\right]^{1/3}$$

From equation-(1), we use

$$R^2 = \frac{C}{L} = \frac{C}{\left[\frac{3}{2}C\right]^{1/3}} = \frac{C^{2/3}}{\left(\frac{3}{2}\right)^{1/3}}$$

$$\Rightarrow \quad R = \frac{C^{1/3}}{\left(\frac{3}{2}\right)^{1/6}}$$

$$\Rightarrow \quad \frac{L}{R} = \frac{\left(\frac{3}{2}\right)^{1/3}\cdot(C)^{1/3}}{C^{1/3}}\times\left(\frac{3}{2}\right)^{1/6}$$

$$\Rightarrow \quad \frac{L}{R} = \left(\frac{3}{2}\right)^{1/2} = \sqrt{\frac{3}{2}} \qquad \textbf{Ans. (C)}$$

Sol. 105 By angular momentum conservation, we have

$$mvl = \frac{Ml^2}{3}\,\omega\times ml^2\omega$$

$$\Rightarrow \quad \omega = \frac{mvl}{\left(ml^2+\frac{Ml^2}{3}\right)}$$

$$\Rightarrow \quad \omega = \frac{1\times 6\times 1}{\frac{2}{3}+1} = \frac{18}{5}\ \text{rad/s}$$

By energy conservation, we have

$$\frac{1}{2}\left(M\frac{l^2}{3}\right)\omega^2 + \frac{1}{2}(ml^2)\omega^2 = mgl(1-\cos\theta) + \frac{Mgl}{2}(1-\cos\theta)$$

$$\Rightarrow \quad \frac{1}{2}\left(\frac{2\times(1)^2}{3}\right)\left(\frac{18}{5}\right)^2 + \frac{1}{2}(1\times(1)^2)\left(\frac{18}{5}\right)^2 = (1+1)\times 10\times 1(1-\cos\theta)$$

$$\Rightarrow \quad \frac{5}{6}\times\left(\frac{18}{5}\right)^2 = 20(1-\cos\theta)$$

$$\Rightarrow \quad 1-\cos\theta = \frac{18}{5}\times\frac{3}{20}$$

$$\Rightarrow \quad \cos\theta = 1-\frac{27}{50}$$

$$\Rightarrow \quad \theta \simeq 63° \qquad \textbf{Ans. (B)}$$

Sol. 106 By conservation of angular momentum, we use

$$\left(80R^2+\frac{200R^2}{2}\right)\omega_i=\left(0+\frac{200R^2}{2}\right)\omega_f$$

$$\Rightarrow \qquad 180\,\omega_i=100\,\omega_f$$

$$\Rightarrow \qquad \omega_f=1.8\,\omega_i=1.8\times 5=9 \text{ rev/min}$$

Ans. (9)

Sol. 107 Horizontal and vertical forces given in system are shown in below figure are

$$F_H=m\omega^2\frac{l}{2}\sin\theta$$

$$F_V=mg$$

As stated in question, we use

$$mg\times\frac{l}{2}\sin\theta-m\omega^2\frac{l}{2}\sin\theta\cdot\frac{l}{2}\cos\theta=\frac{ml^2}{12}\omega^2\sin\theta\cos\theta$$

$$\Rightarrow \qquad \cos\theta=\frac{3g}{2\omega^2 l}$$

Ans. (B)

Sol. 108 Moment of inertia about axis *EX* is given as

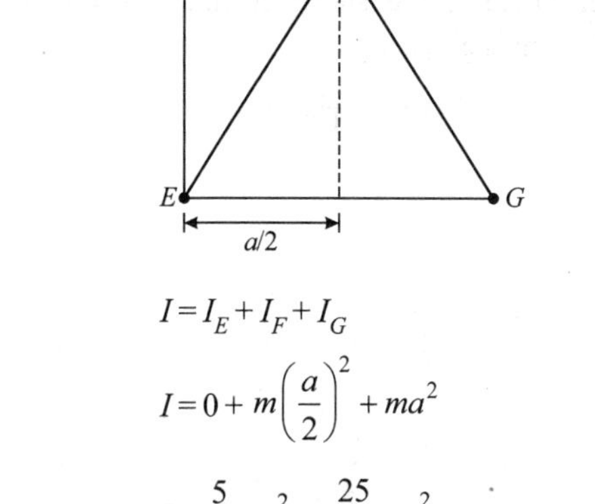

$$I=I_E+I_F+I_G$$

$$\Rightarrow \qquad I=0+m\left(\frac{a}{2}\right)^2+ma^2$$

$$\Rightarrow \qquad I=\frac{5}{4}ma^2=\frac{25}{20}ma^2$$

$$\Rightarrow \qquad N=25$$

Ans. (25)

Sol. 109 Moment of inertia of an equilateral triangle about its centroid is

$$I_0=\frac{ma^2}{12}$$

ΔADE is also an equilateral triangle of side $\frac{a}{2}$ so its moment of inertia about its centroid is given as

$$I_{ADE}=\frac{m_1\left(\frac{a}{2}\right)^2}{12}=\frac{m_1a^2}{48}=\frac{ma^2}{192}\qquad\left(m_1=\frac{m}{4}\right)$$

MI of system about *G* after removing ΔADE is given as

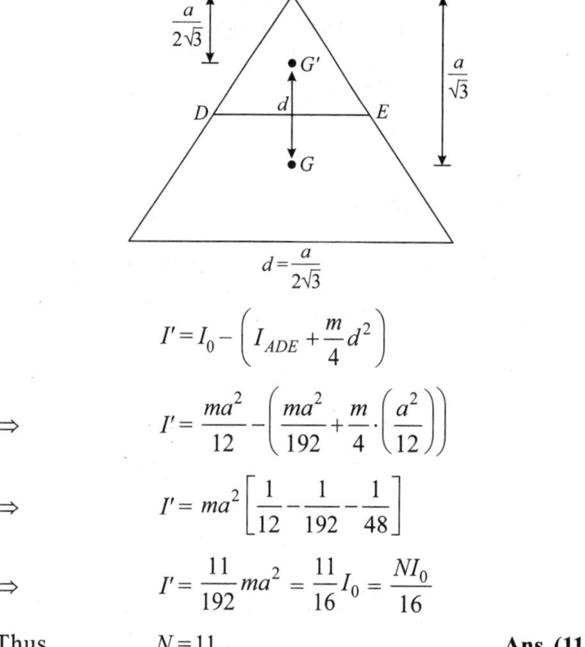

$$I'=I_0-\left(I_{ADE}+\frac{m}{4}d^2\right)$$

$$\Rightarrow \qquad I'=\frac{ma^2}{12}-\left(\frac{ma^2}{192}+\frac{m}{4}\cdot\left(\frac{a^2}{12}\right)\right)$$

$$\Rightarrow \qquad I'=ma^2\left[\frac{1}{12}-\frac{1}{192}-\frac{1}{48}\right]$$

$$\Rightarrow \qquad I'=\frac{11}{192}ma^2=\frac{11}{16}I_0=\frac{NI_0}{16}$$

Thus $\qquad N=11$

Ans. (11)

Sol. 110 By angular momentum conservation, we use

$$\frac{1}{2}MR^2\omega_1=\left[\frac{1}{2}MR^2+\frac{1}{2}M\left(\frac{R}{2}\right)^2\right]\omega_2$$

$$\Rightarrow \qquad \omega_2=\frac{4\omega_1}{5}$$

Initial kinetic energy

$$K_1=\frac{1}{2}\left(\frac{1}{2}MR^2\right)\omega_1^2$$

Final kinetic energy

$$K_2=\frac{1}{2}\left(\frac{5}{8}MR^2\right)\left(\frac{4\omega_1}{5}\right)^2$$

$$\Rightarrow \qquad K_2=\frac{1}{2}\left(\frac{2}{5}MR^2\right)\omega_1^2$$

Kinetic energy loss

$$\Delta K=K_1-K_2=\frac{1}{2}\left(\frac{1}{10}MR^2\right)\omega^2$$

Percentage loss P is given as

$$P = \frac{\Delta K}{K_1} \times 100 = \frac{\frac{1}{10}MR^2}{\frac{1}{2}MR^2} \times 100$$

$\Rightarrow$ $P = 20\%$ **Ans. (20)**

Sol. 111 Moment of inertia of the two discs are given as

$$I_1 = \frac{1}{2}MR^2 = \frac{1}{2}(\rho\pi R^2)R^2$$

and $$I_2 = \frac{1}{2}M'(\alpha R)^2 = \frac{1}{2}(\rho\pi\alpha^2R^2)\alpha^2R^2$$

Where ρ is surface mass density of discs

Given that $$\frac{I_1}{I_2} = \frac{1}{\alpha^4} = \frac{1}{16}$$

$\Rightarrow$ $\alpha = 2$ **Ans. (B)**

Sol. 112 Moment of inertia about axis passing through O is

$$I_0 = \frac{M}{12}(l^2 + b^2) = \frac{M}{12}((80)^2 + (60)^2)$$

MI about axis passing through O' is

$$I'_0 = I_0 + Md^2$$

$\Rightarrow$ $$I_{0'} = I_0 + M(50)^2$$

$\Rightarrow$ $$I_{0'} = \frac{M}{12}((80)^2 + (60)^2) + M(50)^2$$

$\Rightarrow$ $$\frac{I_0}{I_{0'}} = \frac{\frac{M}{12}((80)^2 + (60)^2)}{\frac{M}{12}((80)^2 + (60)^2) + M(50)^2}$$

$$= \frac{1}{4}$$ **Ans. (D)**

Sol. 113 On coupling of another wheel we use conservation of angular momentum as

$$I\omega = (I + 3I)\omega_1$$

$\Rightarrow$ $$\omega_1 = \frac{\omega}{4}$$

Initial kinetic energy of wheel

$$K_1 = \frac{1}{2}I\omega^2$$

Final kinetic energy of coupled wheels

$$K_2 = \frac{1}{2}(4I)\left(\frac{\omega}{4}\right)^2 = \frac{1}{8}I\omega^2$$

Fractional loss in kinetic energy

$$f = \frac{K_1 - K_2}{K_1} = \frac{\frac{1}{2} - \frac{1}{8}}{\frac{1}{2}} = \frac{3}{4}$$ **Ans. (C)**

Sol. 114 Torque of force is given as

$$\vec{\tau} = \vec{r} \times \vec{F}$$

Here $$\vec{r} = \vec{r}_2 - \vec{r}_1 = (3\hat{i} + \hat{j} - 2\hat{k})$$

$\Rightarrow$ $$\vec{\tau} = (3\hat{i} + \hat{j} - 2\hat{k}) \times (\hat{i} + 2\hat{j} + 3\hat{k})$$

$\Rightarrow$ $$\vec{\tau} = (7\hat{i} - 11\hat{j} + 5\hat{k})$$

$\Rightarrow$ $$|\vec{\tau}| = \sqrt{49 + 121 + 25} = \sqrt{195} \text{ N-m}$$ **Ans. (195)**

Sol. 115 By conservation of angular momentum, we use

$$L_i = L_f$$

$$mvl = \left(ml^2 + \frac{ML^2}{3}\right)\omega$$

$\Rightarrow$ $$0.1 \times 80 \times 1 = \left(0.1 \times (1)^2 + \frac{0.9 \times (1)^2}{3}\right)\omega^2$$

$\Rightarrow$ $$\omega^2 = \frac{8}{0.4} = 20 \text{ rad/s}$$ **Ans. (20)**

Sol. 116 Moment of inertia of a hollow cone is $\frac{1}{2}MR^2$ as already derived in theory. **Ans. (A)**

Sol. 117 Moment of inertia of all point masses fixed on square about given axis of rotation is given as

$$I = 0 + 2\left(m\left(\frac{l}{\sqrt{2}}\right)^2\right) + m(\sqrt{2}\,l)^2$$

$\Rightarrow$ $$I = ml^2 + 2ml^2 = 3ml^2$$

Angular momentum of system is given as

$$L = I\omega = 3ml^2\omega$$ **Ans. (B)**

Sol. 118

We consider an element as shown in figure. The moment of inertia of element is given as

$$I = \int dm x^2 = \int \lambda dx \cdot x^2$$

$\Rightarrow$ $$I = \int_0^L x^2\lambda_0\left(1 + \frac{x}{L}\right)dx$$

$\Rightarrow$ $$I = \lambda_0\int_0^L\left(x^2 + \frac{x^3}{L}\right)dx$$

$\Rightarrow$ $$I = \lambda\left[\frac{L^3}{3} + \frac{L^3}{4}\right]$$

$\Rightarrow$ $$I = \frac{7L^3\lambda_0}{12} \quad \ldots(1)$$

Mass of rod is given as

$$M = \int_0^L \lambda dx = \int_0^L \lambda_0 \left(1 + \frac{x}{L}\right) dx$$

$$\Rightarrow \quad M = \lambda_0 \left(L + \frac{L}{2}\right) = \frac{3\lambda_0 L}{2} \quad \ldots(2)$$

From (1) & (2), we use

$$I = \frac{7}{12} L^3 \left(\frac{2M}{3L}\right)$$

$$\Rightarrow \quad I = \frac{7}{18} ML^2$$

Ans. (D)

Sol. 119 The given moment of inertias are standard values written below

For a ring about diameter $I_1 = \dfrac{MR^2}{2}$

For a Disc about central axis $I_2 = \dfrac{MR^2}{2}$

For a Solid cylinder about central axis $I_3 = \dfrac{MR^2}{2}$

For a Solid sphere about central axis $I_4 = \dfrac{2}{5} MR^2$

$$\Rightarrow \quad I_1 = I_2 = I_3 > I_4$$

Ans. (C)

Sol. 120 The situation described in question is shown in figure below

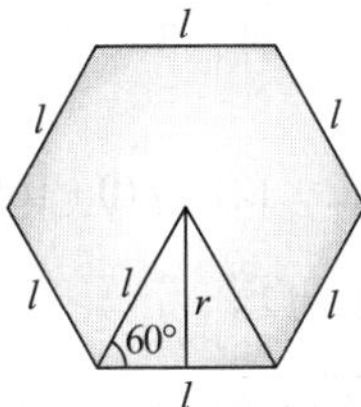

If m is the mass of one side of hexagon and r is the distance of side center from the center of hexagon, we have

$$6l = 2.4\,\text{m}$$

$$\Rightarrow \quad l = 0.4\,\text{m}$$

and $\quad \sin 60° = \dfrac{r}{l}$

$$\Rightarrow \quad r = l \sin 60° = \frac{l\sqrt{3}}{2}$$

Moment of inertia of the system is given as

$$I = \left[\frac{ml^2}{12} + mr^2\right] 6$$

$$\Rightarrow \quad I = \left[\frac{ml^2}{12} + m\left(\frac{l\sqrt{3}}{2}\right)^2\right] 6$$

$$\Rightarrow \quad I = 5\,ml^2 = 5 \times 1 \times 0.16 = 0.8\,\text{kg-m}^2$$

Ans. (8)

Sol. 121 When the sphere starts up the plane, its angular speed changes which can be calculated by using angular momentum conservation as shown in figure below

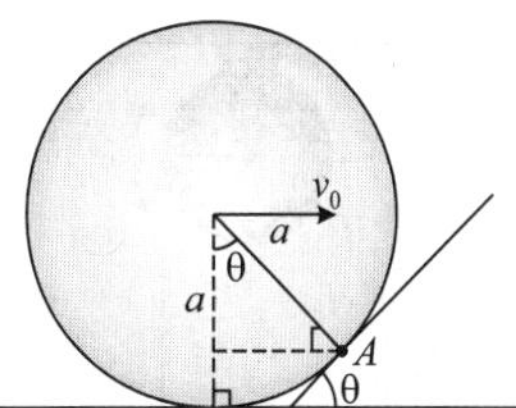

As initial speed of sphere on horizontal plane is v_0 and sphere starts moving up the incline plane with speed v, we use angular momentum conservation about point A, gives

$$mv_0\, a \cos\theta + \frac{2}{5} ma^2 \omega = mva + \frac{2}{5} ma^2 \omega'$$

$$\Rightarrow \quad mv_0 a \left[\frac{2}{5} + \cos\theta\right] = \frac{7}{5} mva$$

$$\Rightarrow \quad v = \frac{5}{7} = v_0 \left[\frac{2}{5} + \cos\theta\right]$$

If sphere ascends up the plan to a height h, by conservation of energy, we use

$$\frac{1}{2} mv^2 + \frac{1}{2} I\omega^2 = \frac{7}{10} mv^2 = mgh$$

$$\Rightarrow \quad v = \sqrt{\frac{10gh}{7}} = v_0 \left[\frac{2}{5} + \cos\theta\right]$$

$$\Rightarrow \quad h = \frac{7v_0^2 (2 + 5\cos\theta)^2}{250g} = l \sin\theta$$

Thus none of the given options are correct. **Ans. (None)**

Sol. 122 The situation described in question is shown in figure below

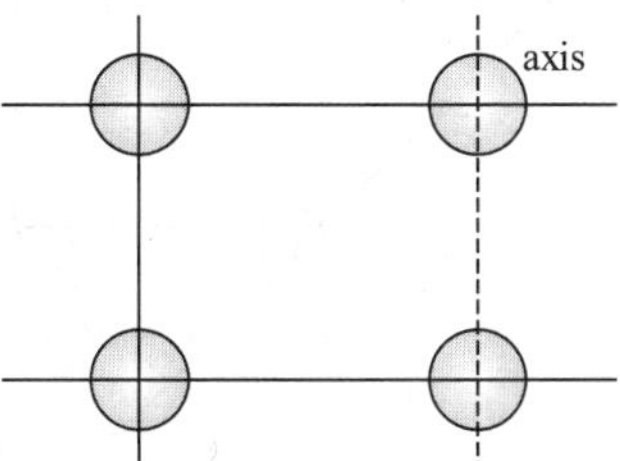

The moment of inertia of the system is given by using parallel axes theorem as

$$I = 2\times\left(\frac{2}{5}ma^2\right) + 2\times\left(\frac{2}{5}ma^2 + mb^2\right)$$

$$\Rightarrow \quad I = \frac{8}{5}ma^2 + 2mb^2 \qquad \textbf{Ans. (C)}$$

Sol. 123 Figure below shows the situation described in the question

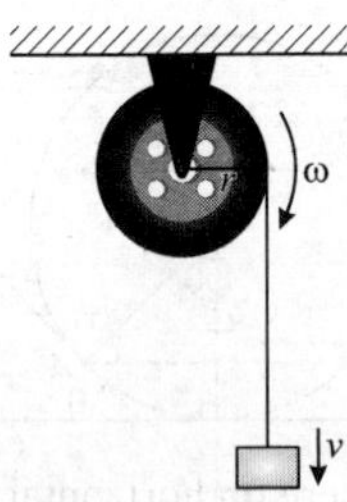

By conservation of energy after falling through a distance h, we have

$$mgh = \frac{1}{2}I\omega^2 + \frac{1}{2}mv^2$$

As $v = \omega r$, we have

$$mgh = \frac{1}{2}I\omega^2 + \frac{1}{2}m\omega^2 r^2$$

$$\Rightarrow \quad \frac{2mgh}{(I+mr^2)} = \omega^2 \qquad \textbf{Ans. (D)}$$

Sol. 124 By angular momentum conservation in elliptical orbit of comet, we use

$$mv_1r_1 = mv_2r_2$$

$$\Rightarrow \quad v_1 = \frac{48\times10^{14}}{1.6\times10^{12}} = 3000 \text{ m/sec}$$

$$\Rightarrow \quad v_1 = 3\times10^3 \text{ m/sec} \qquad \textbf{Ans. (C)}$$

Sol. 125 Moment of inertia of the masses about the axis of rotation AA' as shown in figure is calculated as

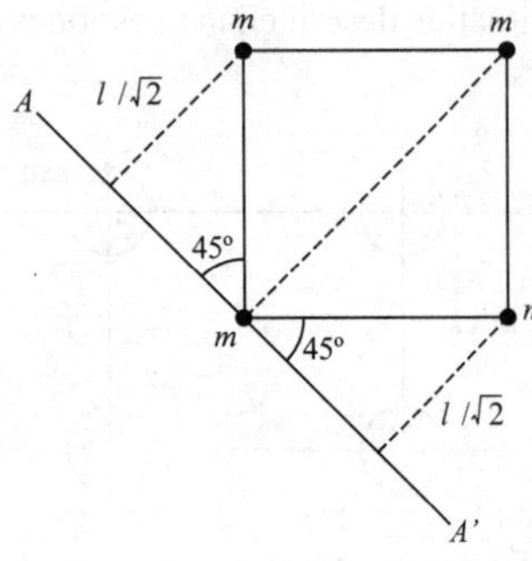

Moment of Inertia of the masses about axis AA' is given as

$$I = m(0)^2 + m(l\sqrt{2})^2 + m\left(\frac{l}{\sqrt{2}}\right)^2 + m\left(\frac{l}{\sqrt{2}}\right)^2 = 3\,ml^2$$

Ans. (C)

Sol. 126 The angular acceleration of the disc can be calculated as

$$\alpha = \frac{\tau}{I} = \frac{F.R.}{mR^2/2} = \frac{2F}{mR}$$

$$\Rightarrow \quad \alpha = \frac{2\times200}{20\times(0.2)} = 10 \text{ rad/s}^2$$

Angular speed after revolution by an angle $\Delta\theta$ is given as

$$\omega^2 = \omega_0^2 + 2\alpha\Delta\theta$$

$$\Rightarrow \quad (50)^2 = 0^2 + 2(10)\,\Delta\theta$$

$$\Rightarrow \quad \Delta\theta = \frac{2500}{20}$$

$$\Rightarrow \quad \Delta\theta = 125 \text{ rad}$$

$$\text{No. of revolution} = \frac{125}{2\pi} = \frac{125}{6.28} \approx 20 \text{ revolution} \qquad \textbf{Ans. (20)}$$

Sol. 127 Torque is calculated as

$$\vec{\tau} = \vec{r}\times\vec{F}$$

Position vector of the point of appliation of force is given as

$$\vec{r} = (2\hat{i}) - (2\hat{i} + 3\hat{j} + 4\hat{k}) = -3\hat{j} - 4\hat{k}$$

Torque calculation is done now as

$$\vec{\tau} = \vec{r}\times\vec{F} = \begin{vmatrix} \hat{i} & \hat{j} & \hat{k} \\ 0 & -3 & -4 \\ 4 & 3 & 4 \end{vmatrix}$$

$$\Rightarrow \quad \vec{\tau} = \hat{i}\,(-12+12) - \hat{j}\,(0+16) + \hat{k}\,(0+12)$$

$$\Rightarrow \quad \vec{\tau} = -16\hat{i} + 12\hat{k}$$

$$\Rightarrow \quad |\vec{\tau}| = \sqrt{16^2+12^2} = 20 \qquad \textbf{Ans. (20)}$$

Sol. 128 Acceleration of disc in pure rolling on incline plance can be given as

$$a = \frac{g\sin\theta}{1+\dfrac{I}{mR^2}} = \frac{g\sin\theta}{1+\dfrac{1}{2}} = \frac{2}{3}g\sin\theta$$

$$\Rightarrow \quad b = 3 \qquad \textbf{Ans. (3)}$$

Sol. 129 Position vectors of point P with respect to O is given as

$$\vec{r}_1 = 5\hat{i} + 5\sqrt{3}\hat{j}$$

And position vector of point P with respect to Q is given as

$$\vec{r}_2 = -5\hat{i} + 5\sqrt{3}\hat{j}$$

Torque about O is calculated as

$$\vec{\tau}_O = \vec{r}_1 \times \vec{F} = (-15 - 20\sqrt{3})\hat{k} = (15 + 20\sqrt{3})(-\hat{k})$$

Torque about Q is calculated as

$$\vec{\tau}_Q = \vec{r}_2 \times \vec{F} = (-15 + 20\sqrt{3})\hat{k} = (15 - 20\sqrt{3})(-\hat{k}) \quad \textbf{Ans. (A)}$$

Sol. 130 The angular momentum of a moving particle is given as

$$\vec{L} = m(\vec{r} \times \vec{v})$$

Now for point A, always the direction of angular momentum will be along the axis of circle by right hand thumb rule and it also remain constant in magnitude as mvr. With respect to point B, the magnitude remain constant but direction will continuously changing as it is always in normal direction to line joining the mass M and point B. **Ans. (D)**

Sol. 131 The acceleration of the rolling body on the inclined plane without slipping is given as

$$a = \frac{g\sin\theta}{1 + \frac{k^2}{R^2}}$$

Thus time of motion can be calculated as

$$t = \sqrt{\frac{2s}{a}} = \sqrt{\frac{2s}{g\sin\theta}\left(1 + \frac{k^2}{R^2}\right)}$$

For least time, radius of gyration k should be least and out of given options for solid sphere it is minimum. **Ans. (4)**

Sol. 132 Acceleration of a body in pure rolling on inclined plane is given as

$$a = \frac{g\sin\theta}{1 + \frac{I}{mR^2}} = \frac{5}{7} \times \frac{10}{2} = \frac{25}{7}$$

Time taken by sphere to come back to starting point can be calculated as

$$t = \frac{2v_0}{a} = \frac{2 \times 1 \times 7}{25} = 0.56 \text{ s} \quad \textbf{Ans. (C)}$$

Sol. 133 Using conservation of angular momentum, we have

$$(Mr^2)\omega = (Mr^2 + 2mr^2)\omega'$$

$$\Rightarrow \quad \omega' = \frac{M\omega}{M + 2m} \quad \textbf{Ans. (C)}$$

Sol. 134 As the wire is bent into a semicircle of radius r, we have

$$\pi r = L$$

$$\Rightarrow \quad r = \frac{L}{\pi}$$

The moment of inertia of the wire can be given as

$$I = Mr^2 = \frac{ML^2}{\pi^2} \quad \textbf{Ans. (C)}$$

Sol. 135 In Physics Galaxy vol. 1 chapter 6, we have derived that the areal velocity of a planet around the Sun is given as

$$\frac{dA}{dt} = \frac{L}{2m} \quad \textbf{Ans. (D)}$$

Sol. 136 As per situation described in question the cylinder is placed on the rough inclined plane. It is not given that it is sliding or moving so first we consider it in equilibrium. Forces acting on it are shown in figure below.

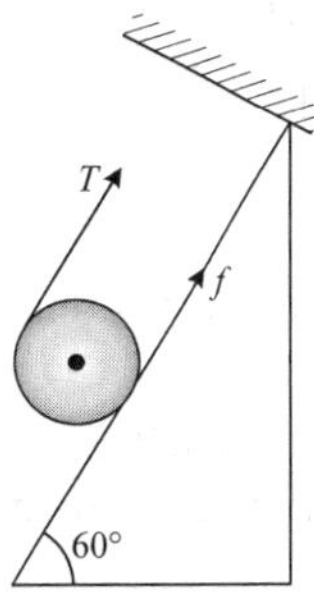

$$T + f = mg\sin\theta \quad \ldots(1)$$

$$TR - fR = 0 \quad \ldots(2)$$

Solving we get

$$T = f = \frac{mg\sin\theta}{2}$$

At $\theta = 60°$ and $\mu = 0.4$, we have

$$\mu mg\cos 60° < \frac{mg\sin 60°}{2}$$

Thus cylinder will not remain in equilibrium and will start moving down and friction will be maintained at kinetic friction. As separately kinetic friction coefficient is not specified, the friction value is given as

$$f = \mu N$$

$$\Rightarrow \quad f = \mu mg\cos 60° = \frac{mg}{5} \quad \textbf{Ans. (C)}$$

Sol. 137 Figure below shows the situation described in the question for collision of a particle with the rod.

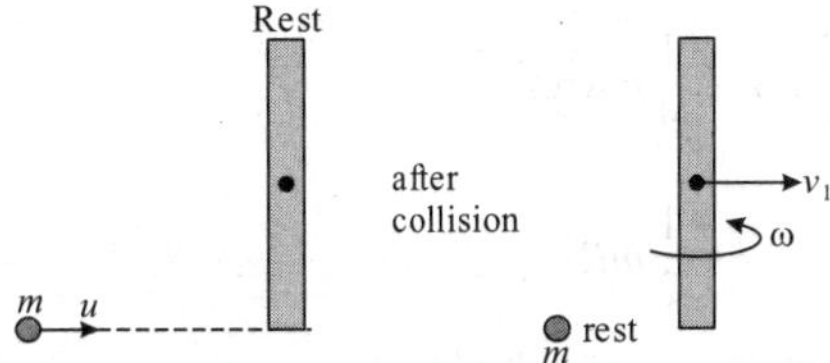

Conservation of angular momentum about centre of mass of rod

$$mu\left(\frac{L}{2}\right) = \frac{ML^2}{12}(\omega) \qquad \dots(1)$$

By conservation of linear momentum, we have

$$mu = Mv_1 \qquad \dots(2)$$

For elastic collision, we use coefficient of restitution as

$$1 = \frac{v_1 + \omega\frac{L}{2}}{u} \qquad \dots(3)$$

From equations-(1), (2) and (3), we have

$$u = \frac{m}{M}u + \frac{6mu}{2M}$$

$$1 = \frac{4m}{M}$$

$$\Rightarrow \qquad \frac{m}{M} = \frac{1}{4} \qquad \textbf{Ans. (4)}$$

Sol. 138 When the disc slides along the inclined plane, its acceleration is given as

$$a_1 = g \sin\theta$$

Thus displacement of disc can be written as

$$L = ut_1 + \frac{1}{2}a_1t_1^2 = \frac{1}{2}g \sin\theta \cdot t_1^2 \qquad \dots(1)$$

When disc will be in pure rolling and coming down the inclined plane, its acceleration can be given as

$$a_2 = \frac{g\sin\theta}{1+k^2/R^2} = \frac{g\sin\theta}{1+1/2} = \frac{2}{3}g\sin\theta$$

Now displacement of disc can be written as

$$L = ut_2 + \frac{1}{2}a_2t_2^2 = \frac{1}{2}\cdot\frac{2}{3}g\sin\theta\cdot t_2^2 \qquad \dots(2)$$

From equations-(1) & (2), we have

$$\frac{t_2}{t_1} = \sqrt{\frac{3}{2}} \qquad \textbf{Ans. (2)}$$

Sol. 139 In the situation described in question, we have

$$\frac{1}{2}I\omega^2 = \frac{1}{2}\times\frac{1}{2}\text{mv}^2$$

For pure rolling without slipping, we use

$$v = R\omega$$

$$\Rightarrow \qquad \frac{1}{2}I\omega^2 = \frac{1}{4}mR^2\omega^2$$

$$\Rightarrow \qquad I = \frac{1}{2}mR^2$$

So the body is a solid cylinder. **Ans. (C)**

Sol. 140 Considering bodies of same mass, given that

$$\text{K.E}_{\text{ring}} = \text{K.E}_{\text{solid cylinder}}$$

$$\frac{1}{2}M_Rv_R^2 + \frac{1}{2}I_R\omega_R^2 = \frac{1}{2}M_Cv_C^2 + \frac{1}{2}I_C\omega_C^2$$

Moment of inertia of ring and that of solid cylinder are given as

$$I_R = M_RR^2 \text{ and } I_C = 1/2\, M_CR^2$$

Considering same mass, we have

$$2M_RV_R^2 = \frac{3}{2}M_CV_C^2$$

$$\Rightarrow \qquad \frac{V_R}{V_C} = \frac{\sqrt{3}}{2} \qquad \textbf{Ans. (3)}$$

Sol. 141 By conservation of energy, we have

$$mgh = \frac{1}{2}I_{\text{cm}}\omega^2 + \frac{1}{2}mv^2$$

$$\Rightarrow \qquad v = \sqrt{\frac{2gh}{1+\frac{I_{\text{cm}}}{mR^2}}}$$

From above expression, it can be seen with increase in moment of inertia, speed decreases. **Ans. (A)**

Sol. 142 Situation described in question is shown in figure below. The speed of the particle on rim at the level of center is calculated by the vector sum of the two velocities in the particle is calculated below

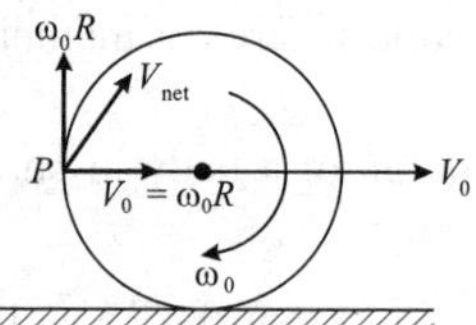

$$(V_P)_{\text{net}} = \sqrt{V_0^2 + (\omega_0R)^2} = \sqrt{V_0^2 + V_0^2} = \sqrt{2}V_0 \qquad \textbf{Ans. (2)}$$

Sol. 143 Statement in Assertion ***A*** is false as moment of inertia about all given axes are different and are related by perpendicular axes theorem. Statement in Reason ***R*** is true as rotational motion for a rigid body is defined for a body which has fixed mass and shape and is called a rigid body. **Ans. (C)**

Sol. 144 Figure below shows the situation described in question.

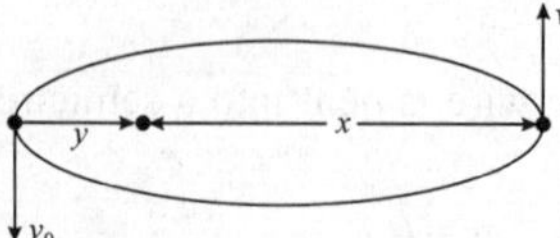

About the Sunc, by angular momentum conservation for motion of the planet, we use

$$mv_0x_2 = mvx_1$$

$$\Rightarrow \quad v = \frac{v_0x_2}{x_1} \qquad \textbf{Ans. (B)}$$

Sol. 145 Ratio of impulse

$$\frac{J_1}{J_2} = \frac{\Delta P_1}{\Delta P_2} = \frac{m(u+u)}{m\left(\frac{u}{\sqrt{2}}+\frac{u}{\sqrt{2}}\right)} = \sqrt{2} : 1 \qquad \textbf{Ans. (A)}$$

Sol. 146 From linear momentum conservation for the collision explained in the question, we have

$$2 \times 4 + 0 = 2 \times 1 + m_2v_2$$

As collision is elastic ($e = 1$), we use

$$v_2 - v_1 = e(u_1 - u_2)$$

$$\Rightarrow \quad v_2 - 1 = 1(4-0)$$

$$\Rightarrow \quad v_2 = 5$$

$$\Rightarrow \quad 8 = 2 + m_2 + 5$$

$$\Rightarrow \quad m_2 = 6/5$$

The speed of centre of mass of the two bodies is calculated as

$$v_{cm} = \frac{m_1v_1 + m_2v_2}{m_1 + v_2} = \frac{2\times4+0}{2+\frac{6}{5}} = 2.5 \text{ m/s}$$

$$\Rightarrow \quad x = 25 \qquad \textbf{Ans. (25)}$$

Sol. 147 The velocity vector of the particle at an instant t is given as

$$\vec{v} = \frac{d\vec{r}}{dt} = 20\alpha t\hat{i} + 5\beta\hat{j}$$

The angular momentum of the particle about origin is given as

$$\vec{L} = m(\vec{r}\times\vec{v})$$

$$\Rightarrow \quad \vec{L} = m(10\alpha t^2\hat{i} + 5\beta(t-5)\hat{j})\times(20\alpha t\hat{i} + 5\beta\hat{j})$$

From above expression, at $t = 0$, $\vec{L} = 0$, thus at any time t, $L = 0$ occurs when

$$\vec{L} = m(50\alpha\beta t\hat{k} - 100\,\alpha\beta(t-5))\hat{k}$$

$$\Rightarrow \quad 0 = 50\,\text{m}\,\alpha\beta\,[t - 2\,(t-5)]\,\hat{k}$$

$$\Rightarrow \quad t - 2t + 10 = 0$$

$$\Rightarrow \quad t = 10 \text{ s} \qquad \textbf{Ans. (10)}$$

Sol. 148 Angular speed at 600 rpm is calculated as

$$\omega = \frac{600\times2\pi}{60} = 20\pi \text{ rad/s}$$

For constant acceleration, we use

$$\omega_t = \omega_i + \alpha t$$

$$\Rightarrow \quad 0 = 20\pi - \alpha(10)$$

$$\Rightarrow \quad \alpha = 2\pi \text{ rad/s}^2$$

Torque required at this angular accelration is given as

$$\tau = I\times\alpha$$

$$\Rightarrow \quad \tau = \frac{mR^2}{2}\times2\pi$$

$$\Rightarrow \quad \tau = \frac{10\times0.04}{2}\times2\pi = 4\times10^{-4}\pi$$

$$\Rightarrow \quad x = 4 \qquad \textbf{Ans. (4)}$$

Sol. 149 If surface mass density of discs is considered to be σ, then their masses can be written as

$$M_1 = \pi R^2\sigma,$$

$$M_2 = \pi r^2\sigma$$

Moment of inertia of the two discs is given as

$$I_1 = \frac{1}{2}M_1R^2 = \frac{\pi\sigma R^4}{2}$$

and

$$I_2 = \frac{1}{4}M_2r^2 = \frac{\pi\sigma r^4}{4}$$

$$\Rightarrow \quad \frac{I_1}{I_2} = \frac{2R^4}{r^4} \qquad \textbf{Ans. (A)}$$

Sol. 150 Moment of inertia of the rod in respective options of List-I are given below -

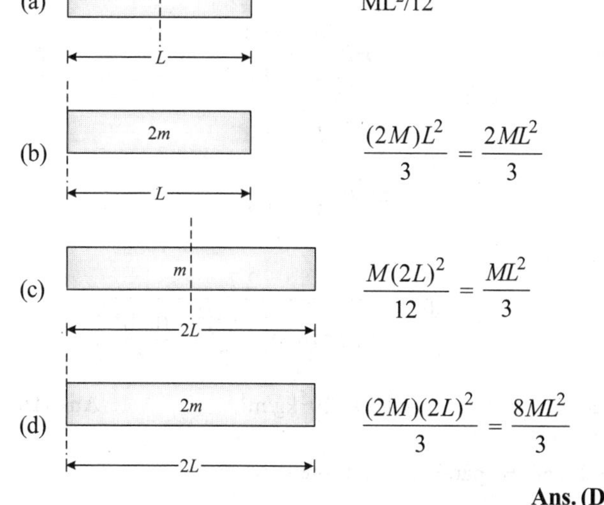

Ans. (D)

Sol. 151 As tangential velocities of the two wheels are same due to chain, we use

$$3R\times\omega_P = R\times\omega_Q$$

$$\Rightarrow \quad \frac{\omega_Q}{\omega_P} = 3$$

As kinetic energy of the two are same, we use

$$\frac{1}{2}I_P\omega_P^2 = \frac{1}{2}I_Q\omega_Q^2;$$

$$\Rightarrow \quad \frac{I_P}{I_Q} = \left(\frac{\omega_Q}{\omega_P}\right)^2 = 9 \qquad \textbf{Ans. (9)}$$

Sol. 152 Moment of inertia of racket about the given axis of rotation

$$I_A = \left[\frac{Mr^2}{2} + M\left(7r - \frac{r}{2}\right)^2\right]$$

$$+ \left[\frac{M(6r)^2}{12} + \left(M\left(3r - \frac{r}{2}\right)^2\right)\right]$$

$$\Rightarrow I_A = \left\{\frac{Mr^2}{2} + M\frac{(169)r^2}{4}\right\} + \left\{M3r^2 + \frac{M25r^2}{4}\right\}$$

$$\Rightarrow I_A = \frac{171Mr^2}{4} + \frac{37Mr^2}{4} = \frac{208}{4}Mr^2 = 52Mr^2 \qquad \textbf{Ans. (52)}$$

Sol. 153 Moment of inertia of the cylinder can be calculated by using parallel axes theorem as

$$I = \frac{Mr^2}{2} + Ml^2 = \frac{\rho(\pi r^2)lr^2}{2} + (\rho\pi r^2 l)\frac{l^2}{4}$$

$$\Rightarrow \quad I = \rho\pi r^2 l\left(\frac{r^2}{2} + \frac{l^2}{4}\right)$$

$$\Rightarrow \quad \rho = \frac{I}{\pi r^2 l\left(\frac{r^2}{2} + \frac{l^2}{4}\right)}$$

$$\Rightarrow \quad \rho = \frac{2.7}{\pi(0.2)^2(0.8)\left(\frac{(0.2)^2}{2} + \frac{(0.8)^2}{4}\right)}$$

$$\Rightarrow \quad \rho = 1.49 \times 10^2 \text{ kg/m}^3 \qquad \textbf{Ans. (D)}$$

Sol. 154 By parallel axes theorem, we have

$$I_Z = I_{cm} + M\left(\frac{l}{\sqrt{2}}\right)^2$$

$$I_Z = \frac{Ml^2}{6} + \frac{Ml^2}{2} = \frac{4Ml^2}{6} = \frac{2Ml^2}{3} \qquad \textbf{Ans. (A)}$$

Sol. 155 Using angular momentum conservation before and after contact of the discs, we use

$$I_1\omega_1 - I_2\omega_2 = (I_1 + I_2)\omega$$

$$\Rightarrow \quad \omega = \frac{I_1\omega_1 - I_2\omega_2}{I_1 + I_2}$$

Loss of energy can be calculated as

$$\Delta E = \frac{1}{2}I_1\omega_1^2 + \frac{1}{2}I_2\omega_2^2 - \frac{1}{2}(I_1 + I_2)\omega^2$$

$$\Rightarrow \quad \Delta E = \frac{1}{2}I_1\omega_1^2 + \frac{1}{2}I_2\omega_2^2 - \frac{1}{2}(I_1 + I_2)\left(\frac{I_1\omega_1 - I_2\omega_2}{I_1 + I_2}\right)^2$$

$$\Rightarrow \quad \Delta E = \frac{1}{2}\frac{I_1 I_2}{(I_1 + I_2)}(\omega_1 - \omega_2)^2$$

$$\Rightarrow \quad \Delta E = \frac{I_1 I_2(\omega_1 - \omega_2)^2}{2(I_1 + I_2)} \qquad \textbf{Ans. (C)}$$

Sol. 156 As it is uniform circular motion, angular momentum is given as

$$\vec{L} = \vec{r} \times \vec{p} = \vec{r} \times m\vec{v} = mvr$$

Thus it will remain same in magnitude and direction about its center. **Ans. (C)**

Sol. 157 Figure below shows the situation described in the question.

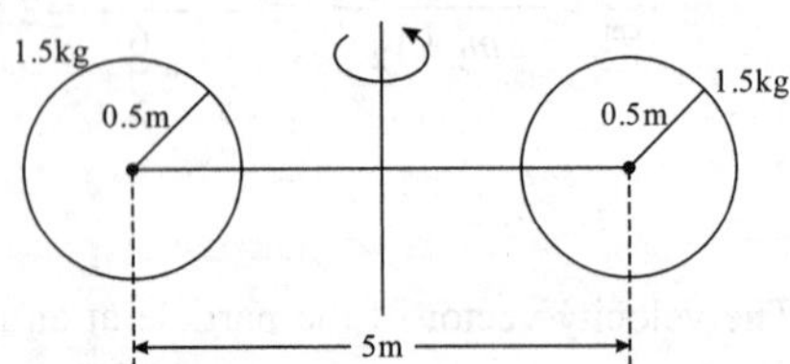

The moment of inertia of the above system can be calculated by parallel axes theorem as

$$I = \left(\frac{2}{5}MR^2 + Mx^2\right) \times 2$$

$$\Rightarrow \quad I = \left[\frac{2}{5} \times 1.5 \times (0.5)^2 + 1.5 \times (2.5)^2\right] \times 2$$

$$\Rightarrow \quad I = [0.15 + 9.375] \times 2 = 19.05 \text{ kgm}^2 \qquad \textbf{Ans. (B)}$$

Sol. 158 By work energy theorem, we have

$$mgl = \frac{1}{2}\frac{ml^2}{3}\omega^2$$

$$\Rightarrow \quad \omega = \sqrt{\frac{6g}{l}}$$

$$\Rightarrow \quad v = l\omega = \sqrt{3.6g} = 6 \text{ m/s} \qquad \textbf{Ans. (6)}$$

Sol. 159 In one second flywheel rotates by 5 rad so we use angular speed equation for uniform angular acceleration, given as

$$\theta = \frac{1}{2}\alpha t^2$$

$$\Rightarrow \quad 5 = \frac{1}{2}\alpha(1)^2$$

$$\Rightarrow \quad \alpha = 10 \text{ rad/s}^2$$

Angle rotated in two seconds is calculated as

$$\theta_1 = \frac{1}{2}\alpha(2)^2$$

$$\theta_1 = 0.5 \times 10 \times 4 = 20 \text{ rad}$$

Angle of rotation in 2nd second is calculated as

$$\theta_2 = 20 - 5 = 15 \text{ rad}$$ **Ans. (B)**

Sol. 160 For the given objects in question, standard moment of inertia are given as

$$I_1 = \frac{2}{5}M(2R)^2 = \frac{8}{5}MR^2$$

$$I_2 = \frac{1}{2}M(2R)^2 = 2MR^2$$

$$I_3 = \frac{1}{4}M(2R)^2 = MR^2$$

$$I_4 = \frac{M(2R)^2}{2} = 2MR^2$$

Substituting the values in given equation gives

$$2(I_2 + I_3) + I_4 = xI_1$$

$$\Rightarrow \quad 8MR^2 = x\frac{8}{5}MR^2$$

$$\Rightarrow \quad x = 5$$ **Ans. (5)**

Sol. 161 Using conservation of angular momentum before and after adding the masses, gives

$$MR^2\omega = (MR^2 + 2mR^2)\omega'$$

$$\Rightarrow \quad \omega' = \frac{2M}{M+2m}$$ **Ans. (C)**

Sol. 162 Total kinetic energy of the rolling ball is written as

$$K_{\text{total}} = K_{\text{rotational}} + K_{\text{Translational}}$$

$$K_{\text{total}} = \frac{1}{2}I_{\text{cm}}\omega^2 + \frac{1}{2}mV_{\text{cm}}^2$$

For pure rolling, we use $v_{\text{cm}} = R\omega$ and moment of inertia of ball is given as

$$I_{\text{cm}} = \frac{2}{5}mR^2$$

Rotational kinetic energy of ball is calculated as

$$K_{\text{Rot}} = \frac{1}{2}I_{\text{cm}}\omega^2 = \frac{1}{2}\times\frac{2}{5}mR^2 \times \frac{v_{\text{cm}}^2}{R^2} = \frac{1}{5}mv_{\text{cm}}^2$$

$$\Rightarrow \quad K_{\text{Total}} = \frac{1}{5}mv_{\text{cm}}^2 + \frac{1}{2}mv_{\text{cm}}^2 = \frac{7}{10}mv_{\text{cm}}^2$$

$$\Rightarrow \quad \frac{K_{\text{Rot}}\,\frac{1}{5}mv_{\text{cm}}^2}{K_{\text{Total}}\,\frac{7}{10}mv_{\text{cm}}^2} = \frac{2}{7}$$ **Ans. (B)**

Sol. 163 For motion of bodies as stated in question, using energy conservation gives

$$12gh - 3gh = \frac{1}{2}3v^2 + \frac{1}{2}12v^2 + \frac{1}{2}[12r^2]\left(\frac{v}{r}\right)^2$$

$$\Rightarrow \quad 9gh = \frac{1}{2}[3 + 12 + 12]v^2$$

$$\Rightarrow \quad v^2 = \frac{2gh}{3}$$

$$\Rightarrow \quad v = \frac{1}{2}\sqrt{\frac{8}{3}gh}$$

$$\Rightarrow \quad x = \frac{8}{3} = 3$$ **Ans. (3)**

Sol. 164 For solid sphere about tangential axis, we use

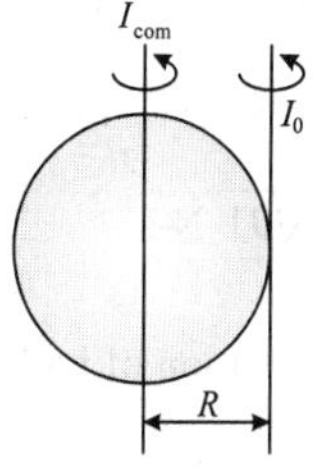

$$I_0 = I_{\text{com}} + MR^2 = \frac{2}{5}MR^2 + MR^2$$

$$\Rightarrow \quad I_0 = \frac{7}{5}MR^2$$

For hollow sphere about tangential axis, we use

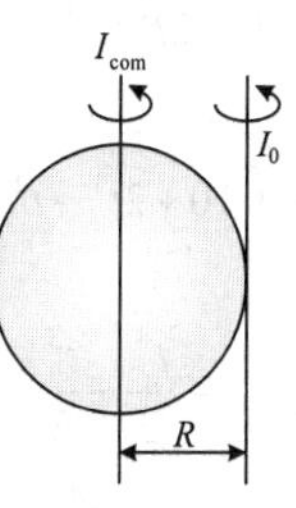

$$I_0 = I_{\text{com}} + MR^2$$

$$\Rightarrow \quad I_0 = \frac{2}{3}MR^2 + MR^2 = \frac{5}{3}MR^2$$

For a ring about diametrical axis, we use

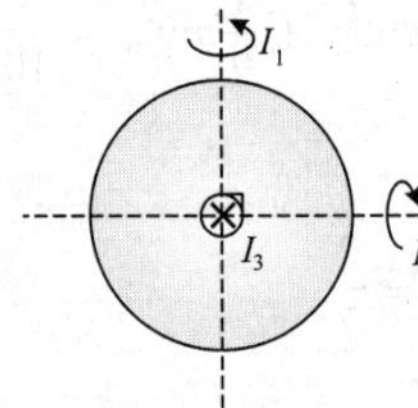

By perpendicular axes theorem, we have

$$I_1 + I_2 = I_3$$

By symmetry moment of inertia about axis 1 and 2 are same so we have

$$I_1 = I_2$$

$$\Rightarrow \quad 2I_1 = I_3 = MR^2$$

$$I_1 = \frac{MR^2}{2}$$

For a disc about diametrical axis, we use

$$I_1 = \frac{MR^2}{4}$$ **Ans. (A)**

Sol. 165 Angular momentum is given as

$$\vec{L} = \vec{r} \times \vec{p} = \vec{r} \times m\vec{v}$$

$$\Rightarrow \quad \vec{L} = (3\hat{i} - \hat{j}) \times (3\hat{i} + \hat{k})$$

$$\Rightarrow \quad \vec{L} = (9\hat{k} - 3\hat{j} - \hat{i}) \text{ N-s}$$

$$\Rightarrow \quad |\vec{L}| = \sqrt{91} \text{ N-s}$$ **Ans. (91)**

Sol. 166 Angular acceleration of ball is given as

$$\frac{d\omega}{dt} = 6t^2 - 2t$$

$$\Rightarrow \quad \int_{10}^{\omega} d\omega = 2t^3 - t^2$$

$$\Rightarrow \quad \omega = 10 + 2t^3 - t^2$$

Now to calculate angular position of ball we use

$$\frac{d\theta}{dt} = 10 + 2t^3 - t^2$$

$$\Rightarrow \quad \int_4^{\theta} d\theta = \int_0^t (10 + 2t^3 - t^2)\,dt$$

$$\Rightarrow \quad \theta - 4 = 10t + \frac{t^4}{2} - \frac{t^3}{3}$$

$$\Rightarrow \quad \theta = 4 + 10t + \frac{t^4}{2} - \frac{t^3}{3}$$ **Ans. (B)**

Sol. 167 Figure below shows the forces acting on the ladder.

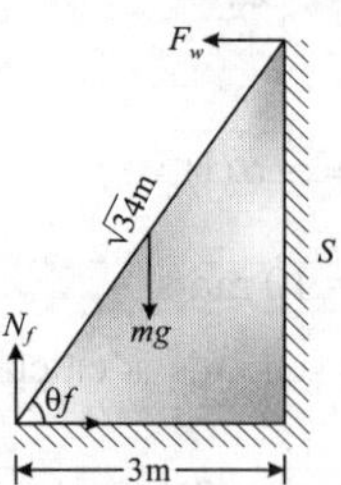

For equilibrium of ladder we use

$$f = F_w$$

$$N_f = mg$$

Net reaction at the contact point of floor on ladder is given as

$$F_f = \sqrt{N_f^2 + f^2}$$

Torque on all forces on ladder must be zero about any point. Here we consider torque about the bottom contact of ladder, given as

$$F_w \times l \sin\theta = mg\frac{l}{2}\cos\theta$$

$$\Rightarrow \quad F_w = \frac{mg}{2}\cot\theta$$

$$\Rightarrow \quad \frac{F_w}{F_f} = \frac{\frac{mg}{2}\cot\theta}{\sqrt{(mg)^2 + \left(\frac{mg}{2}\cot\theta\right)^2}}$$

$$\Rightarrow \quad \frac{F_w}{F_f} = \frac{1}{\sqrt{1 + \frac{4}{\cot^2\theta}}}$$

$$\Rightarrow \quad \frac{F_w}{F_f} = \frac{3}{\sqrt{109}}$$ **Ans. (C)**

Sol. 168 When mass goes down with acceleration a, disc rotates with angular acceleration a/R.

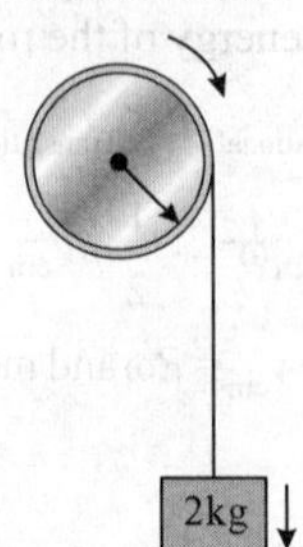

Equation of motion of block and disc are written as

$$2g - T = 2a \quad \ldots(1)$$

$$TR = \frac{MR^3}{2}\alpha \qquad \ldots(2)$$

$\Rightarrow \quad T = 2a$

$\Rightarrow \quad 2g - T = 2a$

$\Rightarrow \quad T = g = 10\,\text{N}$ **Ans. (10)**

Sol. 169 Angular momentum of shell about O is calculated as

$$L_0 = mV_{cm}R + I\omega$$

$$\Rightarrow \quad L_0 = 1 \times \omega R \times R + \frac{2}{3}R^2\omega$$

$$\Rightarrow \quad L_0 = \frac{5}{3}R^2\omega$$

$\Rightarrow \quad a = 5$ **Ans. (C)**

Sol. 170 If string tension is taken as T, for 4 kg block, equation of motion is written as

$$4g - T = 4a$$

For 40 kg block, equation of motion is written as

$$T - 40g \times 0.02 = 40a$$

Adding both equation gives

$$40 - 8 = 44a$$

$$\Rightarrow \quad a = \frac{32}{44} = \frac{8}{11}\ \text{m/s}^2$$ **Ans. (D)**

Sol. 171 For the rod of length l the moment of inertia about its side axis is given as

$$I_1 = \frac{ml^2}{3}$$

When the rod is bent to form a ring of radius r, we use

$$l = 2\pi r$$

$$\Rightarrow \quad \frac{l}{r} = 2\pi$$

Moment of inertia of ring about diametrical axis is given as

$$I_1 = \frac{mr^2}{2}$$

$$\Rightarrow \quad \frac{I_1}{I_2} = \frac{2}{3}\left(\frac{l}{r}\right)^2$$

$$\Rightarrow \quad \frac{I_1}{I_2} = \frac{2}{3} \times 4\pi^2 = \frac{8\pi^2}{3}$$

$\Rightarrow \quad x = 8$ **Ans. (8)**

Sol. 172 Velocity attained by the rolling body at the bottom of incline of height H is given as

$$v = \sqrt{\frac{2gH}{1 + k^2/R^2}}$$

$$\Rightarrow \quad \frac{v_{\text{cylinder}}}{v_{\text{sphere}}} = \sqrt{\frac{(1 + k^2/R^2)_{\text{sphere}}}{(1 + k^2/R^2)_{\text{cylinder}}}}$$

$$\Rightarrow \quad \frac{v_{\text{cylinder}}}{v_{\text{sphere}}} = \sqrt{\frac{1 + 2/5}{1 + 1/2}} = \sqrt{\frac{7}{5} \times \frac{2}{3}} = \sqrt{\frac{14}{15}}$$ **Ans. (D)**

Sol. 173 Figure below shows the initial and final positions of the point mass attached to the disc

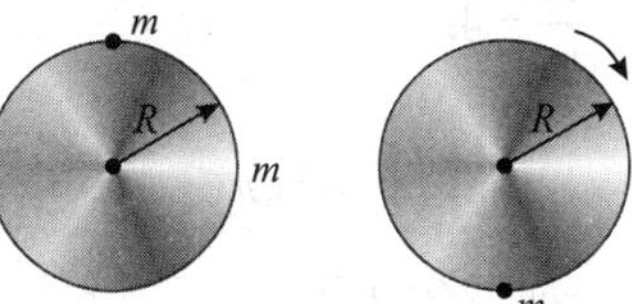

Using conservation of energy, we have

$$mg(2R) = \frac{1}{2}I_{\text{dics}}\,\omega^2 + \frac{1}{2}I_{\text{particle}}\omega^2$$

$$mg(2R) = \frac{\omega^2}{2}\left[\frac{mR^2}{2} + mR^2\right]$$

$$mg(2R) = \frac{\omega^2}{2}\frac{3}{2}mR^2$$

$$\Rightarrow \quad \frac{3}{4}\omega^2 = \frac{2g}{R}$$

$$\Rightarrow \quad \omega^2 = \frac{8g}{3R}$$

$$\Rightarrow \quad \omega = \sqrt{\frac{80}{3R}} = 4\sqrt{\frac{x}{3R}}$$

$$\Rightarrow \quad 16\frac{x}{3R} = \frac{80}{3R}$$

$\Rightarrow \quad x = 5$ **Ans. (5)**

Sol. 174 For a uniform cylinderical rod rotating about the central axis, the moment of inertia is given as

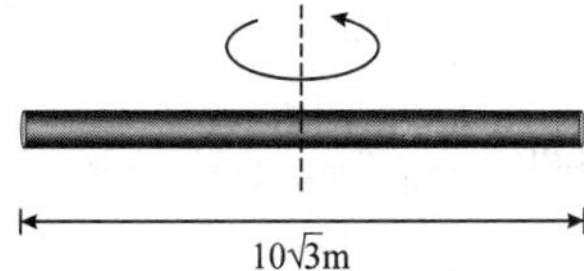

$$I = \frac{ml^2}{12} = mk^2$$

$$\Rightarrow \quad k^2 = \frac{l^2}{12}$$

$$\Rightarrow \quad k = \frac{l}{\sqrt{12}} = \frac{l}{2\sqrt{3}} = \frac{10\sqrt{3}}{2\sqrt{3}} = 5$$ **Ans. (5)**

Sol. 175 Using Newton's second law for rotational motion, we have

$$\tau = I\alpha$$

$$\Rightarrow \quad (12t - 3t^2)1.5 = 4.5\alpha$$

$$\Rightarrow \quad \alpha = 4t - t^2$$

$$\Rightarrow \quad \frac{d\omega}{dt} = 4t - t^2$$

$$\Rightarrow \quad \omega = \int_0^t (4t - t^2)dt$$

$$\Rightarrow \quad \omega = 2t^2 - t^3/3$$

With time angular speed of pulley will first increase then becomes zero and then it get reversed so we use

$$0 = 2t^2 - \frac{t^3}{3}$$

$$\Rightarrow \quad t^2\left(2 - \frac{t}{3}\right) = 0$$

$$\Rightarrow \quad t = 0, 6$$

Thus pulley will reverse its direction of rotation at $t = 6$ s. As it is time function so we use

$$\frac{d\theta}{dt} = 2t^2 - \frac{t^3}{3}$$

$$\Rightarrow \quad \theta = \int_0^6 \left(2t^2 - \frac{t^3}{3}\right)dt$$

$$\Rightarrow \quad \theta = \left[\frac{2t^3}{3} - \frac{t^4}{12}\right]_0^6$$

$$\Rightarrow \quad \theta = 6^3\left(\frac{2}{3} - \frac{6}{12}\right) = 6^3\left(\frac{8-6}{12}\right)$$

$$\Rightarrow \quad \theta = \frac{6^3}{6} = 36$$

$$\text{No. of revolutions} = \frac{36}{2\pi} = \frac{18}{\pi}$$

$$\Rightarrow \quad K = 18$$ **Ans. (18)**

Sol. 176 From energy conservation for falling cylinder, we use

$$mgh = \frac{1}{2}mv^2 + \frac{1}{2}I\omega^2$$

$$\Rightarrow \quad mgh = \frac{1}{2}mv^2 + \frac{1}{2}\frac{mR^2}{2}\omega^2$$

$$\Rightarrow \quad 10h = \frac{16}{2} + \frac{16}{4}$$

$$\Rightarrow \quad h = 1.2\text{ m} = 120\text{ cm}$$ **Ans. (120)**

Sol. 177 Moment of inertia of individual discs about the axis OO' are calculated as

$$I_1 = I_3 = \frac{MR^2}{4}$$

and $$I_2 = I_4 = \frac{MR^2}{4} + MR^2 = \frac{5}{4}MR^2$$

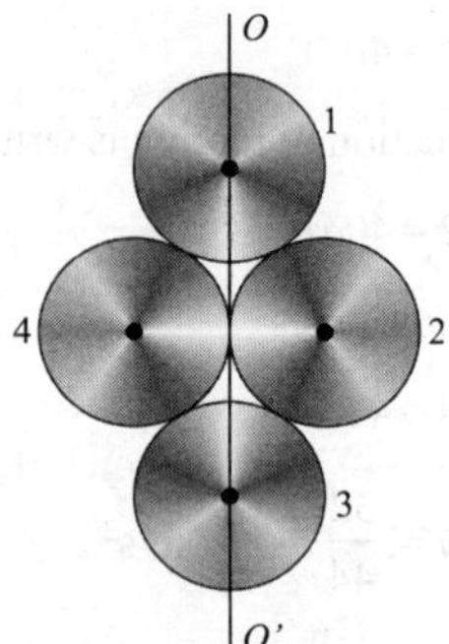

Total moment of inertia is calculated as

$$I = I_1 + I_2 + I_3 + I_4$$

$$\Rightarrow \quad I = \frac{MR^2}{2} + \frac{5}{2}MR^2$$

$$\Rightarrow \quad I = 3MR^2,\ \text{Putting } R = \frac{a}{2}$$

$$\Rightarrow \quad I = \frac{3Ma^2}{4}$$

$$\Rightarrow \quad x = 3$$ **Ans. (3)**

Sol. 178 Figure below shows the situation described in question.

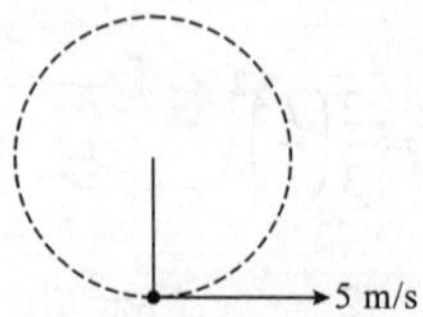

Strain in the string is given as

Strain $= F/AY$

$$\Rightarrow \quad \textbf{Strain} = \frac{mg + \frac{mv^2}{R}}{AY}$$

$$\Rightarrow \quad \textbf{Strain} = \frac{20 + \frac{2(5)^2}{0.5}}{3\times10^{-6}\times10^{11}} = 30 \times 10^{-5} \quad \textbf{Ans. (30)}$$

Sol. 179 The torque of force is calculated about a given point as

$$\vec{\tau} = \vec{r}\times\vec{F} = \begin{vmatrix} i & j & k \\ 2 & 2 & 1 \\ 5 & 3 & -7 \end{vmatrix}$$

$$\vec{\tau} = i(-14-3) - j(-14-5) + \hat{k}(6-10)$$

$$\vec{\tau} = -17\hat{i} + 19\hat{j} - 4\hat{k} \quad \textbf{Ans. (C)}$$

Sol. 180 The moment of inertia of a solid sphere about its central axis is given as

$$I_{cm} = \frac{2}{5}MR^2$$

The moment of inertia of the given sphere about the given axis PQ is calculated by parallel axes theorem as

$$I_{PQ} = I_{cm} + md^2$$

$$\Rightarrow \quad I_{PQ} = \frac{2}{5}mR^2 + m(10)^2$$

Radius of gyration about axis PQ is calculated as

$$I_{PQ} = mK^2$$

$$\Rightarrow \quad K^2 = \frac{2}{5}R^2 + (10)^2$$

$$\Rightarrow \quad K^2 = 10 + 100 = 110$$

$$\Rightarrow \quad K = \sqrt{110} \text{ cm}$$

$$\Rightarrow \quad x = 110 \quad \textbf{Ans. (110)}$$

Sol. 181 The two cases of solid cylinders are shown below in the given diagram.

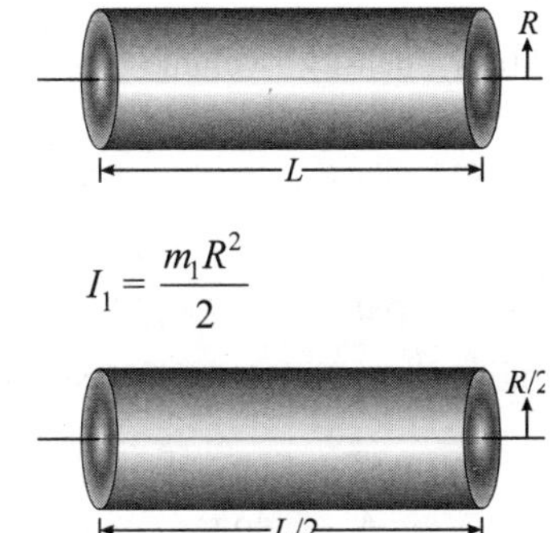

$$I_1 = \frac{m_1R^2}{2}$$

$$I_2 = \frac{m_2(R/2)^2}{2}$$

$$\Rightarrow \quad \frac{I_1}{I_2} = \frac{4m_1}{m_2} = \frac{4\cdot\rho\pi R^2 l}{\rho\cdot\frac{\pi R^2}{4}\times\frac{l}{2}}$$

$$\Rightarrow \quad \frac{I_1}{I_2} = 32 \quad \textbf{Ans. (32)}$$

Sol. 182 Figure below shows the forces acting on the rod

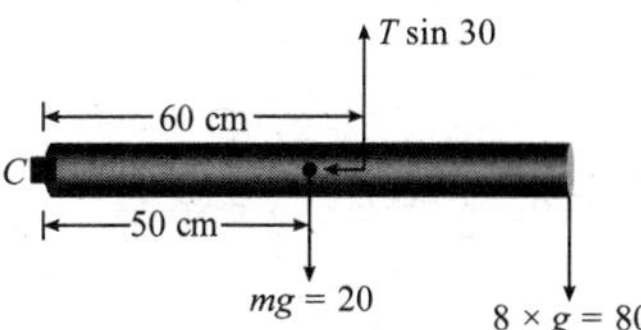

As the rod is in equilibrium, balancing torque about point C, we have

$$\frac{T}{2}\times 60 = 20\times 50 + 80\times 100$$

$$\Rightarrow \quad 3T = 100 + 800$$

$$\Rightarrow \quad T = 300\text{ N} \quad \textbf{Ans. (C)}$$

Sol. 183 Moment of inertia of the circular disc about its central axis is given as

$$I_{cm} = \frac{mR^2}{2}$$

Moment of inertia about the given axis AB is calculated by parallel axes theorem, given as

$$I_{AB} = \frac{mR^2}{2} + m\left(\frac{2R}{3}\right)^2 = \frac{17}{18}mR^2$$

$$\Rightarrow \quad \frac{I_{AB}}{I_{cm}} = \frac{17}{9}$$

$$\Rightarrow \quad x = 17 \quad \textbf{Ans. (17)}$$

Sol. 184 The moment of inertia about tangential axis of rotation for sphere is given by parallel axes theorem, given as

$$I_1 = \frac{2}{5}m_1R^2 + m_1R^2$$

$$\Rightarrow \quad I_1 = m_1R^2\left(\frac{7}{5}\right) = 7R^2$$

For a disc the moment of inertia about tangential axis is given by

$$I_2 = \frac{m_2R^2}{4} + m_2R^2$$

$$\Rightarrow \quad I_2 = \frac{5}{4} m_2 R^2 = 5R^2$$

$$\Rightarrow \quad \frac{I_2}{I_1} = \frac{5}{7}$$

$$\Rightarrow \quad x = 5$$ **Ans. (5)**

Sol. 185 Kinetic energy of the sphere rolling on the surface is given as

$$K = \frac{1}{2}mv^2 + \frac{1}{2}I\omega^2$$

$$\Rightarrow \quad 2240 = \frac{1}{2}2(v)^2 + \frac{1}{2}\frac{2}{5}(2)R^2 \cdot \left(\frac{v}{R}\right)^2$$

$$\Rightarrow \quad 2240 = v^2 + \frac{2}{5}v^2$$

$$\Rightarrow \quad v = 40 \text{ m/s}$$ **Ans. (40)**

Sol. 186 For the given question, the total angular momentum is given as

$$L_0 = \int_0^t \tau dt$$

$$\Rightarrow \quad L_0 = \int_0^2 mg(v_x t)dt$$

$$\Rightarrow \quad L_0 = mg\, v_x \frac{t^2}{2} = (0.1)(10)\,(10\sqrt{2})\,\frac{2^2}{2}$$

$$\Rightarrow \quad L_0 = 20\sqrt{2}$$

$$\Rightarrow \quad L_0 = \sqrt{800} \text{ kg m}^2\text{/s}$$ **Ans. (800)**

Sol. 187 Kinetic Energy of rod is calculated as

$$(KE)_{\text{Rotational}} = \frac{1}{2}I\omega^2$$

$$\Rightarrow \quad E = \frac{1}{2}\frac{ml^2}{12}\omega^2$$

$$\Rightarrow \quad E = \frac{1}{2}\frac{dAl^3}{12}\omega^2$$

$$\Rightarrow \quad E = \frac{dA(2)^3}{24}\omega^2$$

$$\Rightarrow \quad \sqrt{\frac{3E}{dA}} = \omega$$

$$\alpha = 3$$ **Ans. (3)**

Sol. 188 Acceleration due to friction

$$a = -\mu_k g = -3$$

Velocity after 2 second is given as

$$v = 18 - 3 \times 2$$

$$\Rightarrow \quad v = 12 \text{ m/s}$$

As the ball is having translational and rotational kinetic energy then

$$K = \frac{1}{2}mv^2 + \frac{1}{2}\frac{mr^2}{2}\frac{v^2}{r^2}$$

$$\Rightarrow \quad K = \frac{3}{4}mv^2$$

$$\Rightarrow \quad K = 3 \times 18 = 54 \text{ J}$$ **Ans. (54)**

Sol. 189 In the free body diagram shown in figure below for equilibrium of roller in vertical direction, we have

$$N = mg + F\sin 30°$$

$$\Rightarrow \quad N = 700 + 200 \times \frac{1}{2} = 800 \text{ N}$$

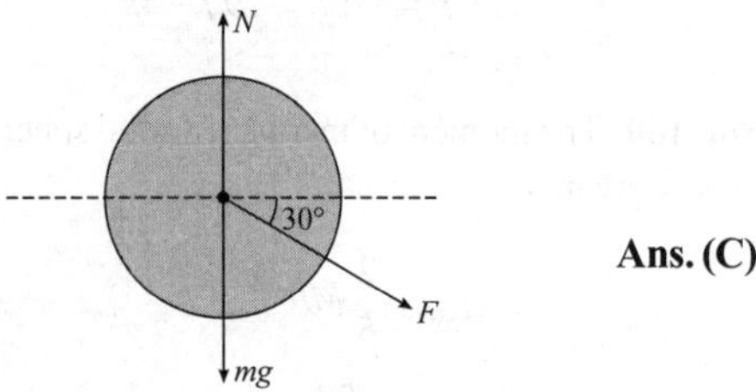

Ans. (C)

Sol. 190 Total kinetic energy of the solid sphere is given as

$$\frac{1}{2}mv^2 + \frac{1}{2}I\omega^2 = 7 \times 10^{-3}$$

$$\Rightarrow \quad \frac{1}{2}mv^2 + \frac{1}{2}\left(\frac{2}{5}mR^2\right)\left(\frac{v}{R}\right)^2 = 7 \times 10^{-3}$$

$$\Rightarrow \quad \frac{1}{2}Mv^2\left[1 + \frac{2}{5}\right] = 7 \times 10^{-3}$$

$$\Rightarrow \quad \frac{1}{2}(1)(v^2)\left(\frac{7}{5}\right) = 7 \times 10^{-3}$$

$$\Rightarrow \quad v^2 = 10^{-2}$$

$$\Rightarrow \quad v = 10^{-1} = 0.1 \text{ m/s}$$

$$\Rightarrow \quad v = 10 \text{ cm/s}$$ **Ans. (10)**

Sol. 191 Mass = density × volume

$$m = \rho \times \pi R^2 t$$

$$\Rightarrow \quad m = \rho\pi R^2 t$$

$$\Rightarrow \quad R^2 = \frac{m}{\rho\pi t}$$

Thus moment of inertia of disc about diameter is given as

$$I = \frac{mR^2}{4} = \frac{m^2}{4\rho\pi t}$$

$$\Rightarrow \quad \frac{l_1}{l_2} = \frac{\rho_2 t_2}{\rho_1 t_1} = \frac{5}{3} \times \frac{0.5}{1} = \frac{5}{6}$$

$$\Rightarrow \quad x = 5$$ **Ans. (5)**

Sol. 192 Speed at the bottom of inclined plane is calculated as

$$v = \sqrt{\frac{2gh}{1+\frac{k^2}{R^2}}}$$

Where $h = 60 \sin 30° = 30$ cm

and $k^2 = \frac{R^2}{2}$

Solving above expression gives

$$v = 2 \text{ ms}^{-1}$$ **Ans. (2)**

Sol. 193 Moment of inertia of disc about an axis normal to disc and passing through a point on its edge as shown in figure below is calculated by parallel axes theorem as

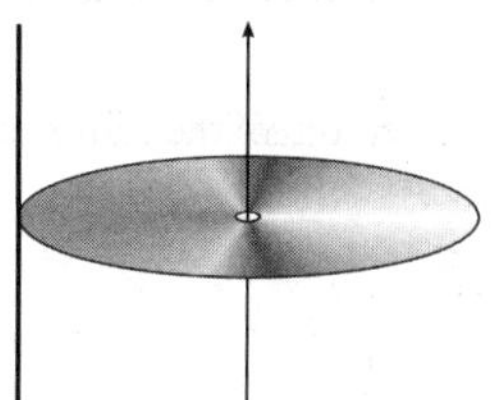

$$I = I_{cm} + Md^2$$

$$\Rightarrow \quad I = \frac{MR^2}{2} + MR^2$$

$$\Rightarrow \quad I = \frac{3}{2}MR^2 = \frac{x}{2}MR^2$$

$$\Rightarrow \quad x = 3$$ **Ans. (3)**

Sol. 194 If I_{cm} is the moment of inertia about centre of mass of solid sphere then the moment of inertia of system about an axis perpendicular to rod passing through centre by parallel axis theorem.

$$I = 2(I_{cm} + md^2)$$

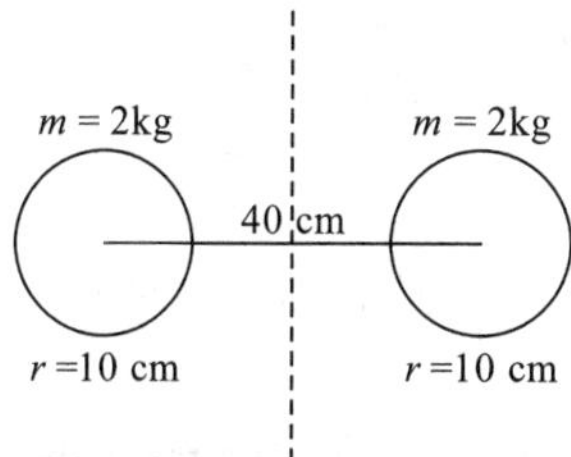

As we know the moment of inertia of solid sphere about centre of mass is given as

$$I_{cm} = \frac{2}{5}mr^2$$

$$\Rightarrow \quad I = 2\left(\frac{2}{5}mr^2 + md^2\right)$$

$$\Rightarrow \quad I = \frac{4}{5} \times 2 \times (0.1)^2 + 2(2)(0.20)^2$$

$$\Rightarrow \quad I = \frac{8}{5} \times 10^{-2} + 2(2)(0.20)^2$$

$$\Rightarrow \quad I = (1.6 + 16) \times 10^{-2}$$

$$\Rightarrow \quad I = 17.6 \times 10^{-2}$$

$$\Rightarrow \quad I = 176 \times 10^{-3} \text{ kg m}^2$$ **Ans. (176)**

Sol. 195 R is radius of ring and K is radius of gyration so we have

$$I = mR_1^2 = mK_1^2$$

$$\Rightarrow \quad K_1 = R_1 \quad \ldots(1)$$

For a solid sphere, we have

$$I' = \frac{2}{5}m'R_2^2 = m'K_2^2$$

$$\Rightarrow \quad K_2 = \sqrt{\frac{2}{5}}R_2 \quad \ldots(2)$$

From equations-(1) and (2), we have

$$K_1 = K_2$$

$$\Rightarrow \quad R_2 = \sqrt{\frac{2}{5}}R_2$$

$$\Rightarrow \quad \frac{R_1}{R_2} = \sqrt{\frac{2}{5}} = \sqrt{\frac{2}{x}}$$

$$\Rightarrow \quad x = 5$$ **Ans. (5)**

Sol. 196 The moment of inertia of semicircular ring about axis passing through centre of ring and perpendicular to plane of ring is MR^2 so the value of x is 1. **Ans. (1)**

Sol. 197 At maximum height velocity of ball becomes zero so by work energy theorem, we use

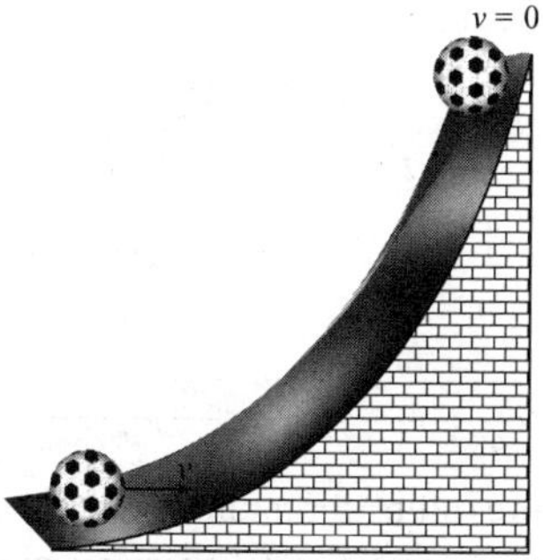

$$\Rightarrow \quad mgh = \frac{1}{2}mv^2 + \frac{1}{2}I\omega^2$$

In case of pure rolling, we use

$$v = R\omega$$

$$\Rightarrow \quad mgh = \frac{1}{2}mv^2 + \frac{1}{2}\times\frac{2}{3}mR^2\times\frac{v^2}{R^2}$$

$$\Rightarrow \quad mgh = \frac{5}{6}mv^2$$

$$\Rightarrow \quad h = \frac{5}{6\times 10}\times 9\,\text{m} = \frac{15}{20}$$

$$\Rightarrow \quad h = 75\,\text{cm}$$

Ans. (75)

Sol. 198 Due to inertia an object maintain its state so option *B* is correct. **Ans. (B)**

Sol. 199 Moment of inertia of sphere about tangent I_T is given as

$$I_t = x\times 10^{-2}L = \frac{2}{5}mR^2 + mR^2$$

$$\Rightarrow \quad \frac{7}{5}mR^2 = x\times 10^{-2}\left(\frac{2}{5}mR^2\omega\right)$$

$$\Rightarrow \quad \frac{7}{2\omega} = x\times 10^{-2} = \frac{7}{2\times 10}$$

$$\Rightarrow \quad x = 35$$

Ans. (35)

Sol. 200 Spring Constant dimensions are written as

$$[K] = \frac{[F]}{[x]} = \frac{MLT^{-2}}{L} = MT^{-2}$$

By definition of angular, velocity its dimensions can be written as

$$[\omega] = \frac{[\theta]}{[t]} = \frac{1}{T} = T^{-1}$$

Ans. (A)

Sol. 201 Figure below shows the rolling spherical shell as in described in question.

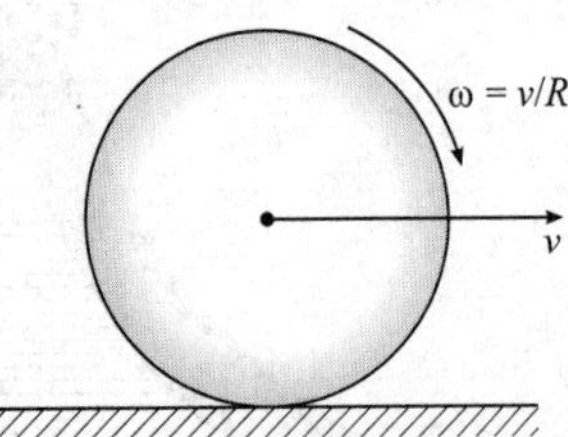

Considering pure rolling for rolling spherical shell, the ratio of rotational kinetic energy to total energy is given as

As $\quad I = \frac{2}{3}mR^2$

$$\frac{K_{rot}}{K_{Total}} = \frac{\frac{1}{2}\left(\frac{2}{3}mR^2\right)\left(\frac{V}{R}\right)^2}{\frac{1}{2}mv^2 + \frac{1}{2}\left(\frac{2}{3}mR^2\right)\left(\frac{V}{R}\right)^2}$$

$$\Rightarrow \quad \frac{x}{5} = \frac{2}{5}$$

$$\Rightarrow \quad x = 2$$

Ans. (2)

Sol. 202 When disc completes half of its rotation, top most point at *A* reaches to bottom so

$$\text{Displacement} = \sqrt{(2R)^2 + (\pi R)^2} = R\sqrt{4+\pi^2}$$

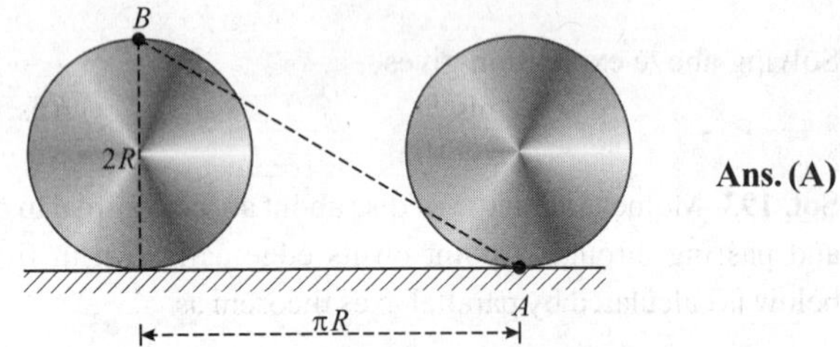

Ans. (A)

Sol. 203 Angular momentum of the rolling sphere is calculated as

$$L = I_{com}\omega$$

$$\Rightarrow \quad L = \frac{2}{5}MR^2\frac{V_{com}}{R}$$

$$\Rightarrow \quad L = \frac{2MRV_{com}}{5}$$

Rotational kinetic energy of the sphere is given as

$$K = \frac{1}{2}(I_{com})(\omega^2) + \frac{1}{2}MV_{com}^2$$

$$\Rightarrow \quad K = \frac{1}{2}\left(\frac{2}{5}MR^2\right)\frac{V_{com}^2}{R^2} + \frac{1}{2}MV_{com}^2$$

$$\Rightarrow \quad K = \frac{7}{10}MV_{com}^2$$

Ratio of angular momentum to kinetic energy is calculated as

$$\frac{L}{K} = \frac{4}{7}\frac{R}{V_{com}} = \frac{\pi}{22}$$

$$\Rightarrow \quad \omega = \frac{4}{7}\times\frac{22}{22}\times 7 = 4\,\text{rad/s}$$

Ans. (4)

Sol. 204 When rope is pulled with force, then hollow cylinder rotates and its angular acceleration is related to the applied torque as

$$\tau = I\alpha$$

$$\Rightarrow \quad FR = mR^2\alpha$$

Thus angular acceleration of hollow cylinder is calculated as

$$\alpha = \frac{F}{mR} = \frac{52.5}{5\times0.7} = 15 \text{ rad s}^{-2} \qquad \textbf{Ans. (15)}$$

Sol. 205 For solid sphere, we use

$$\frac{2}{5}mR^2 = mk_{\text{sph}}^2$$

$$k_{\text{sph}} = \sqrt{\frac{2}{5}}R$$

For solid cylinder, we use

$$\frac{mR^2}{2} = mk_{\text{cyl}}^2$$

$$\Rightarrow \quad k_{\text{cyl}} = \frac{R}{\sqrt{2}}$$

Ratio radius of gyration for sphere and cylinder

$$\frac{k_{\text{sph}}}{k_{\text{cyl}}} = \frac{\sqrt{\frac{2}{5}}}{\frac{1}{\sqrt{2}}} = \frac{2}{\sqrt{5}} = \frac{2}{\sqrt{x}}$$

$$\Rightarrow \quad x = 5 \qquad \textbf{Ans. (5)}$$

7.1 Gravitational Field

Sol. 1

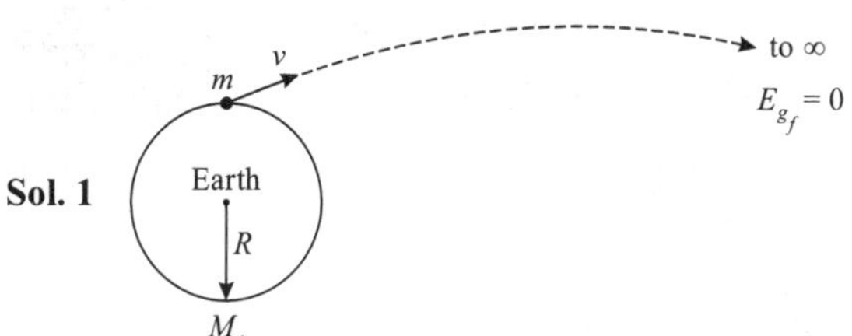

On surface of earth for the body mechanical energy is

$$E_{gi} = -\frac{GM_em}{R} + \frac{1}{2}mv^2 = E_{gf} = 0$$

$$\Rightarrow \quad KE = \frac{GM_em}{R} = (gR)\,\text{m} \quad [\text{As } g = \frac{GM_e}{R^2}]$$

$$\Rightarrow \quad KE = mgR \qquad \textbf{Ans. (C)}$$

Sol. 2

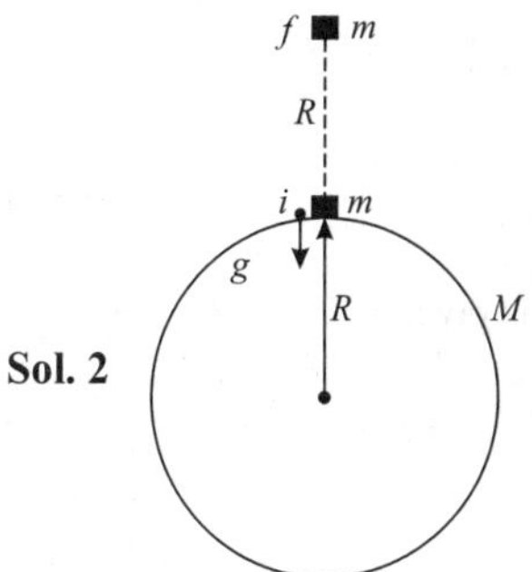

Acceleration due to gravity on

At earth surface

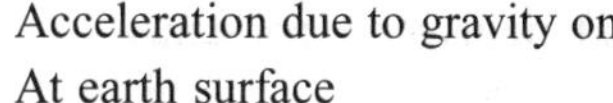

$$g = \frac{GM}{R^2}$$

$$\Rightarrow \quad \frac{GM}{R} = gR$$

Gain in PE when body of mass m is raised from position i to f is

$$\Delta U = U_f - U_i$$

$$\Rightarrow \quad \Delta U = \left(-\frac{GMm}{2R}\right) - \left(-\frac{GMm}{R}\right)$$

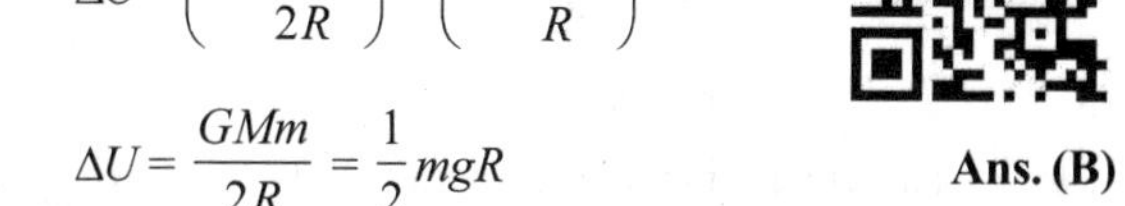

$$\Rightarrow \quad \Delta U = \frac{GMm}{2R} = \frac{1}{2}mgR \qquad \textbf{Ans. (B)}$$

Sol. 3

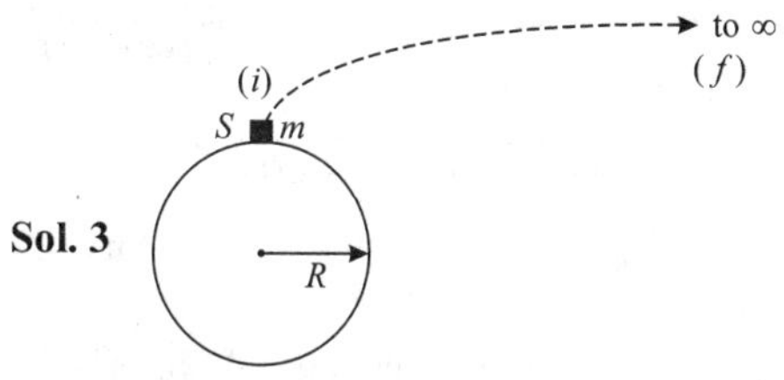

Work done in shifting mass from point S to ∞, is given as

$$W = m(V_f - V_i)$$

$$\Rightarrow \quad W = m\left(0 - \left(-\frac{GM}{R}\right)\right)$$

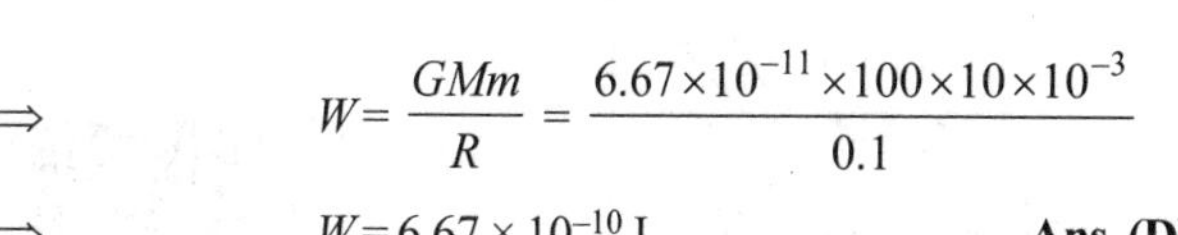

$$\Rightarrow \quad W = \frac{GMm}{R} = \frac{6.67\times10^{-11}\times100\times10\times10^{-3}}{0.1}$$

$$\Rightarrow \quad W = 6.67\times10^{-10}\text{ J} \qquad \textbf{Ans. (D)}$$

Sol. 4 At a small height h above earth surface, value of g is given as

$$g_h = g_s\left(1 - \frac{2h}{R_e}\right) \qquad \ldots(1)$$

At a depth d below earth surface, value of g is given as

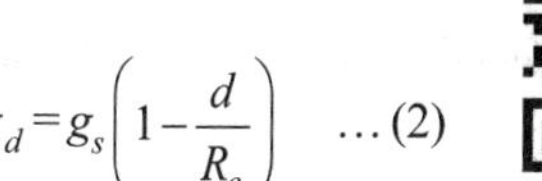

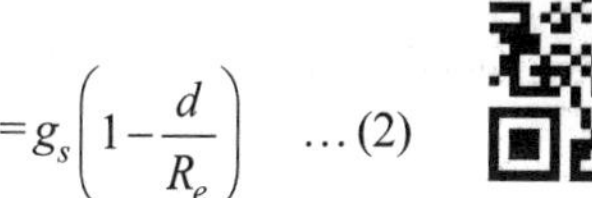

$$g_d = g_s\left(1 - \frac{d}{R_e}\right) \qquad \ldots(2)$$

Given that $\quad g_h = g_d$

or $\quad g_s - g_h = g_s - g_d$

$$\Rightarrow \quad \frac{2h}{R_e} = \frac{d}{R_e}$$

$$\rightarrow \quad 2h = d \qquad \textbf{Ans. (C)}$$

Sol. 5 For earth, its mass M and radius R are constants

$\Rightarrow$ Average density $\rho = \dfrac{M}{\frac{4}{3}\pi R^3} = \dfrac{3M}{4\pi R^3} \rightarrow$ constant

If for a planet its mass M is constant and its radius R is changed

Acceleration due to gravity $g = \dfrac{GM}{R^2} \Rightarrow R = \sqrt{\dfrac{GM}{g}}$

Average density $\rho = \dfrac{3M}{4\pi\left(\dfrac{GM}{g}\right)^{3/2}} = \dfrac{3g^{3/2}}{4\pi G^{3/2} M^{1/2}}$

$\Rightarrow \quad \rho \propto g^{3/2}$

If for a planet its radius R is constant and its mass M is changed

$$M = \frac{gR^2}{G}$$

Average density $\quad \rho = \dfrac{3}{4\pi R^3} \times \dfrac{gR^2}{G} = \dfrac{3}{4\pi RG} \cdot g$

$\Rightarrow \quad \rho \propto g$ **Ans. (C)**

Sol. 6 Electronic charge is independent of g.

$\Rightarrow \quad \dfrac{q_m}{q_e} = 1$ **Ans. (A)**

Sol. 7 By Gauss's law for gravitational field we use

$$\phi_m = \int \vec{g} \cdot \vec{ds} = 4\pi GM$$

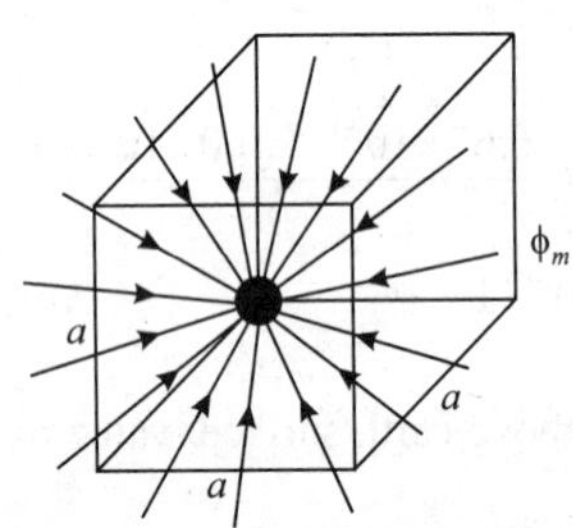

S-1 is True

By Gauss's law the gravitational field due to a point m is radially inward (radial) and for the surface enclosing the mass $\phi = 4\pi$GM.

Ans. (B)

Sol. 8 At a height h above the earth surface the value of g is given as

$$g_h = \frac{g_s}{\left(1+\dfrac{h}{R_e}\right)^2} = \frac{g_s}{9}$$

$\Rightarrow \quad \left(1+\dfrac{h}{g}\right)^2 = 9$

$\Rightarrow \quad 1+\dfrac{h}{R} = 3 \Rightarrow h = 2R$ **Ans. (C)**

Sol. 9

Gravitational field strength due to m and $4m$ at point P are given as

$$g_1 = \frac{Gm}{x^2} = \frac{4Gm}{(r-x)^2}$$

$\Rightarrow \quad 2x = r - x$

$\Rightarrow \quad x = \dfrac{r}{3}$

Gravitational potential at point P is given as

$$V_P = -\frac{Gm}{x} - \frac{G(4m)}{r-x}$$

$\Rightarrow \quad V_P = -\dfrac{3Gm}{r} - \dfrac{3(4Gm)}{2r} = -\dfrac{9Gm}{r}$ **Ans. (B)**

Sol. 10

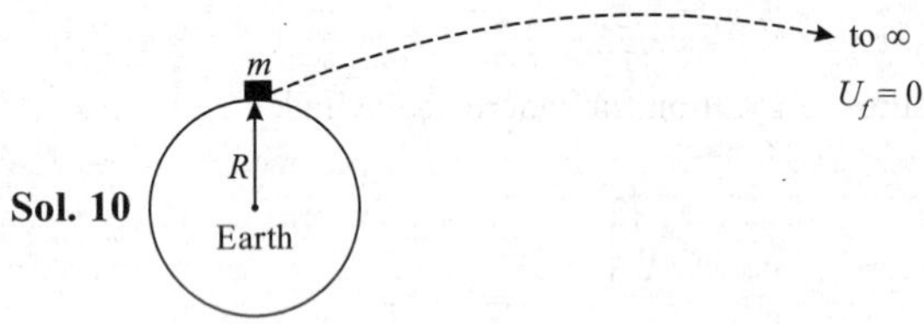

Initial gravitational PE of space ship on surface of earth is

$U_i = -\dfrac{GM_e m}{R_e} = -(gR)m$ [As $g = \dfrac{GM_e}{R_e^2}$]

Work done required to launch space ship to ∞ (free space) is

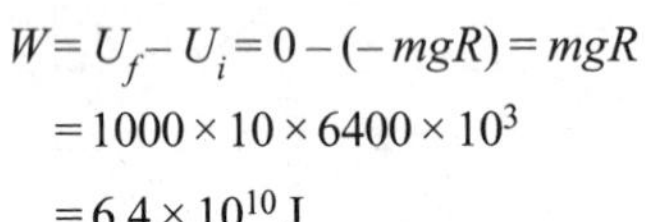

$$W = U_f - U_i = 0 - (-mgR) = mgR$$

$$= 1000 \times 10 \times 6400 \times 10^3$$

$$= 6.4 \times 10^{10} \text{ J}$$

Ans. (A)

Sol. 11

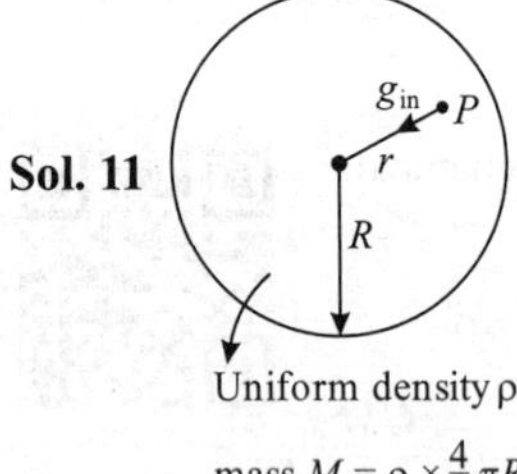

For $r < R$ acceleration due to gravity inside a sphere is given as

$$g_{in} = \frac{GMr}{R^3} = \frac{G}{R^3}\left(\rho \cdot \frac{4}{3}\pi R^3\right) \cdot r$$

$\Rightarrow \quad g_{in} = \dfrac{4}{3}\pi G\rho r$

$\Rightarrow \quad g_{in} \propto r$ **Ans. (A)**

Sol. 12 Inside earth, value of g is given as

For $x < R_e \quad g_{in} = \dfrac{GM_e x}{R_e^3}$

For $x \geq R_e \quad g_{s/out} = \dfrac{GM_e}{x^2}$

(Where x is distance from earth centre)

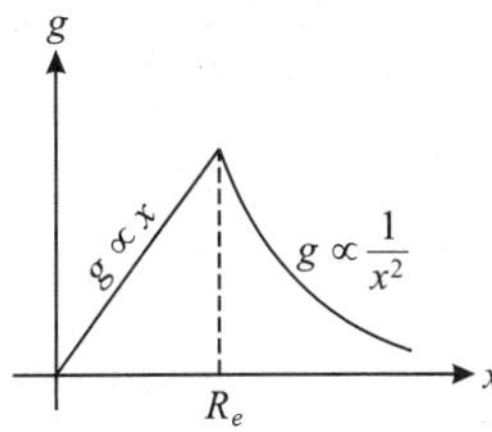

The closest/best possible option given is (B) **Ans. (B)**

Sol. 13 The mass of removed part of sphere is given as

$$m=\left(\frac{M}{\frac{4}{3}\pi R^3}\right)\times\frac{4}{3}\pi\left(\frac{R}{2}\right)^3=\frac{M}{8}$$

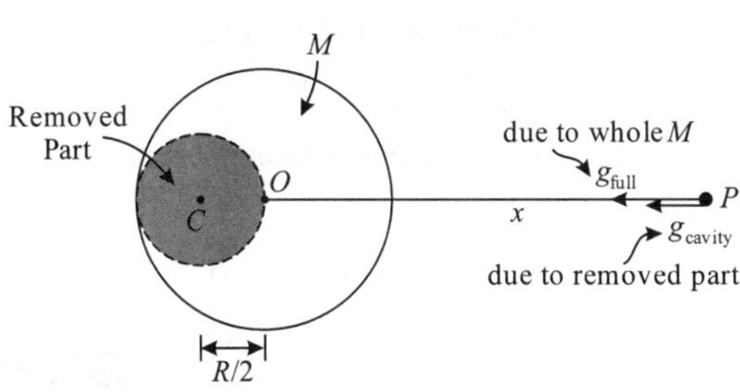

Net gravitational field at point P due to remaining mass is

$$g_P = g_{full} - g_{cavity}$$

$$\Rightarrow \quad g_P=\frac{GM}{x^2}-\frac{G\left(\frac{M}{8}\right)}{\left(x+\frac{R}{2}\right)^2}$$

$$\Rightarrow \quad g_P=\frac{GM}{x^2}-\frac{GM}{8x^2}$$

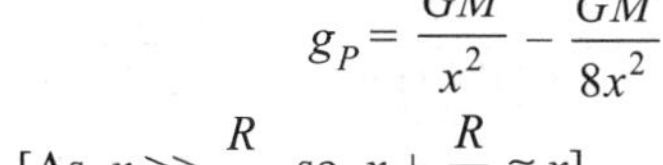

$$\left[\text{As } x>>\frac{R}{2} \text{ so } x+\frac{R}{2}\simeq x\right]$$

$$\Rightarrow \quad g_P=\frac{7}{8}\frac{GM}{x^2}$$ **Ans. (C)**

Sol. 14

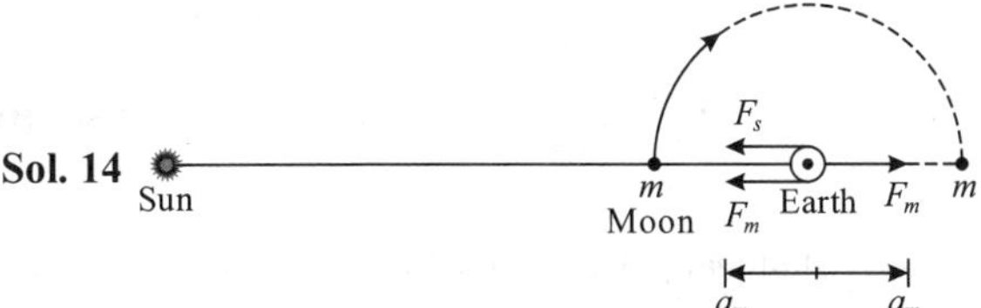

Change in acceleration of earth toward Sun due to change in position of moon is

$$\Delta a = a_m - (-a_m)$$

$$\Rightarrow \quad \Delta a = 2a_m$$

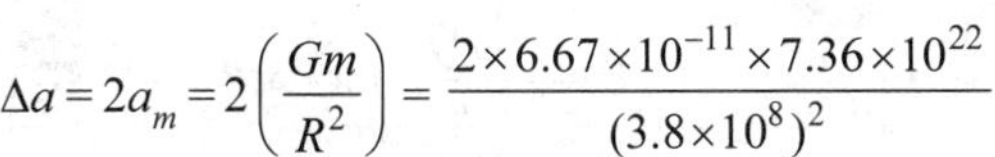

$$\Delta a = 2a_m = 2\left(\frac{Gm}{R^2}\right)=\frac{2\times 6.67\times 10^{-11}\times 7.36\times 10^{22}}{(3.8\times 10^8)^2}$$

$$\Rightarrow \quad \Delta a = 6.75\times 10^{-5}\ \text{m/s}^2$$ **Ans. (A)**

Sol. 15

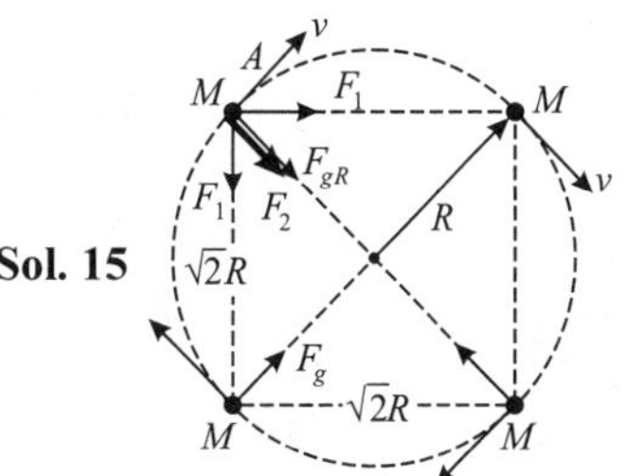

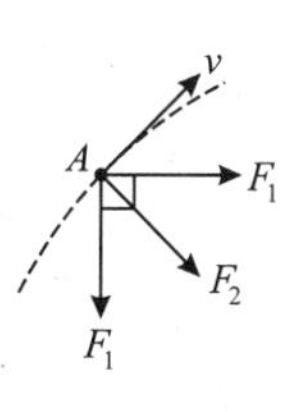

Gravitational forces acting on particle A are

$$F_1=\frac{GM^2}{(\sqrt{2}R)^2};\ F_2=\frac{GM^2}{4R^2}$$

Net gravitational force on particle A, is given as

$$F_{gA}=F_2+\sqrt{2}F_1$$

$$\Rightarrow \quad F_{gA}=\frac{GM^2}{4R^2}+\sqrt{2}\left(\frac{GM^2}{2R^2}\right)=\frac{Mv^2}{R}$$

$$\Rightarrow \quad \frac{GM^2}{2R^2}\left(\frac{1}{2}+\sqrt{2}\right)=\frac{Mv^2}{R}$$

$$\Rightarrow \quad v=\sqrt{\frac{GM}{4R}(2\sqrt{2}+1)}$$ **Ans. (D)**

Sol. 16 Capillary height above the water level in tube is given as

$$h=\frac{2T\cos\theta}{r\rho g}$$

at earth surface $\quad x=\dfrac{2T\cos\theta}{r\rho g}\quad$ ...(1)

at a depth d below the surface of earth

$$y=\frac{2T\cos\theta}{r\rho\left[g\left(1-\frac{d}{R}\right)\right]} \quad ...(2)$$

(1)/(2) $\Rightarrow \quad \dfrac{x}{y}=1-\dfrac{d}{R}$ **Ans. (A)**

Sol. 17

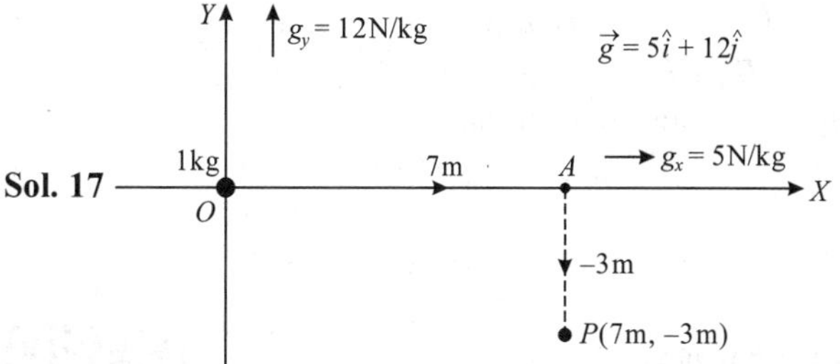

In the given situation work done in taking particle from $O\rightarrow A$ and $A\rightarrow P$ are given as

$$V_0 - V_A = 5\times 7 = 35\ \text{J}$$

$$V_A - V_P = 12\times(-3) = -36\ \text{J}$$

Work done or change in PE of 1 kg mass in

shifting from O to P is

$$V_P - V_0 = 1\text{ J}$$ **Ans. (D)**

Sol. 18

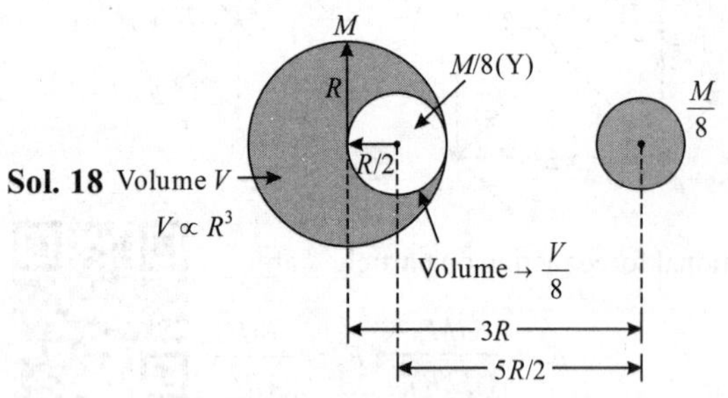

If force F_1 acts between M and $\frac{M}{8}$ than it is given as

$$F_1 = \frac{GM\left(\frac{M}{8}\right)}{(3R)^2} = \frac{GM^2}{72R^2} \quad \ldots(1)$$

If F_2 is the force between M/8 (X) and M/8 (Y) than it is given as

$$F_2 = \frac{G\left(\frac{M}{8}\right)\left(\frac{M}{8}\right)}{\left(5\frac{R}{2}\right)^2} = \frac{GM^2}{64 \times \frac{25}{4}R^2}$$

$$\Rightarrow \quad F_2 = \frac{GM^2}{400R^2} \quad \ldots(2)$$

The net force between the sphere (with cavity) and other part (x) is given as

$$F_{net} = F_1 - F_2$$

$$\Rightarrow \quad F_{net} = \frac{GM^2}{72R^2} - \frac{GM^2}{400R^2}$$

$$\Rightarrow \quad F_{net} = \frac{41GM^2}{3600R^2}$$ **Ans. (A)**

Sol. 19

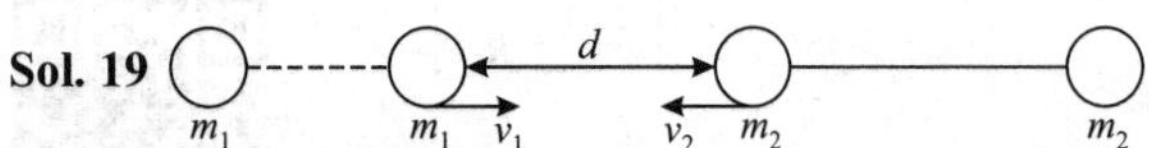

By conservation of energy

$$\Rightarrow \quad E_i = E_f$$

$$\underset{PE}{O} + \underset{KE}{O} = -\frac{Gm_1m_2}{d} + \frac{1}{2}m_1v_1^2 + \frac{1}{2}m_2v_2^2 \quad \ldots(1)$$

By conservation of momentum

$$\Rightarrow \quad m_1v_1 = m_2v_1 \quad \ldots(2)$$

$$\Rightarrow \quad v_2 = \left(\frac{m_1}{m_2}\right)v_1$$

We use v_2 in equation-(1) as

$$\Rightarrow \quad \frac{Gm_1m_2}{d} = \frac{1}{2}m_1v_1^2\left[1 + \frac{m_1}{m_2}\right]$$

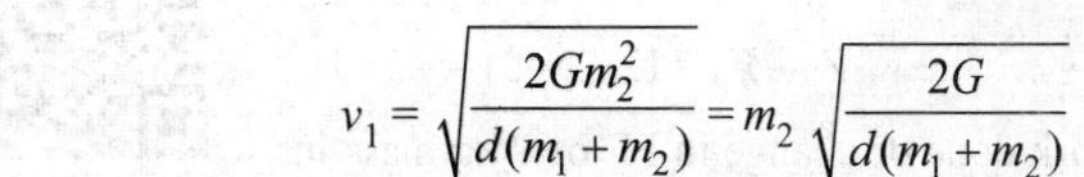

$$\Rightarrow \quad v_1 = \sqrt{\frac{2Gm_2^2}{d(m_1+m_2)}} = m_2\sqrt{\frac{2G}{d(m_1+m_2)}}$$

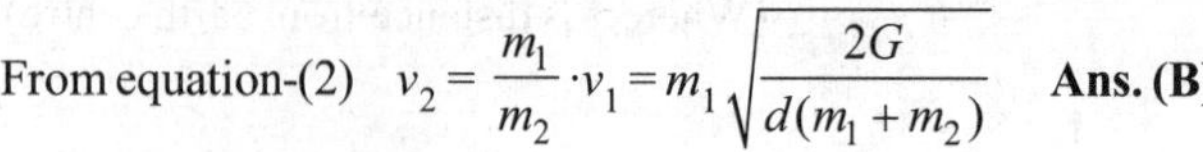

From equation-(2) $v_2 = \frac{m_1}{m_2}\cdot v_1 = m_1\sqrt{\frac{2G}{d(m_1+m_2)}}$ **Ans. (B)**

Sol. 20

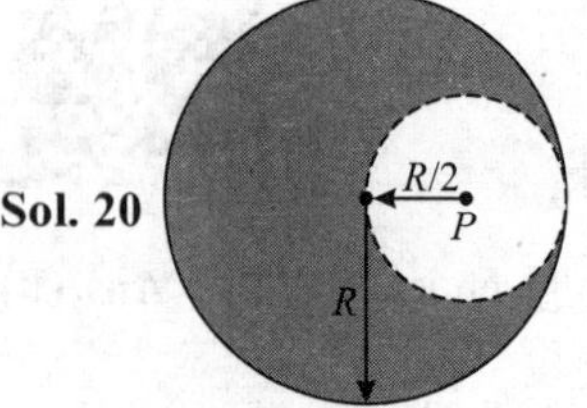

Removed mass

$$m = \frac{M}{\frac{4}{3}\pi R^3} \times \frac{4}{3}\pi\left(\frac{R}{2}\right)^3 = \frac{M}{8}$$

Gravitational potential at centre of cavity is

$$V_P = V_{\text{sphere }P} - V_{mP} \quad \ldots(1)$$

Due to a solid sphere gravitational potential at its interior point is given as

$$V_{in} = -\frac{GM}{2R^3}(3R^2 - x^2)$$

$\Rightarrow$ at $P\left(x = \frac{R}{2}\right)$ so we have

$$V_{in} = -\frac{GM}{2R^3}\left(3R^2 - \frac{R^2}{4}\right)$$

$$\Rightarrow \quad V_{in} = -\frac{11}{8}\frac{GM}{R}$$

Gravitational potential due to removed mass m at centre of cavity is

$$V_{Pm} = -\frac{3}{2}\left(\frac{GM}{(R/2)}\right) = -\frac{3}{2}\frac{G(M/8)}{(R/2)}$$

$$V_{Pm} = -\frac{3}{8}\frac{GM}{R}$$

From equation-(1)

$$V_P = \left(-\frac{11}{8}\frac{GM}{R}\right) - \left(-\frac{3}{8}\frac{GM}{R}\right)$$

$$\Rightarrow \quad V_P = -\frac{GM}{R}$$ **Ans. (B)**

Sol. 21 Gravitational potential due to uniform solid spherical planet is given as

$$V_{in} = -\frac{GM}{2R^3}(3R^2 - r^2) \text{ for } r < R$$

at $r = 0$ $\quad V_c = -\frac{3}{2}\frac{GM}{R}$

$$V_s = -\frac{GM}{R} \quad \text{for } r = R$$

$$V_{out} = -\frac{GM}{r} \quad \text{for } r > R$$ **Ans. (C)**

Sol. 22

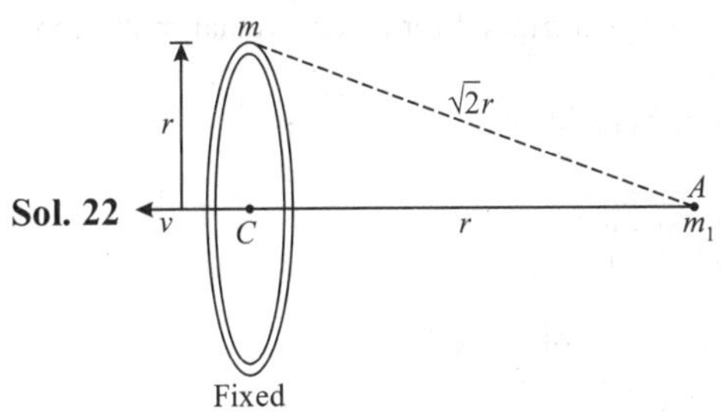

By using energy conservation, we have

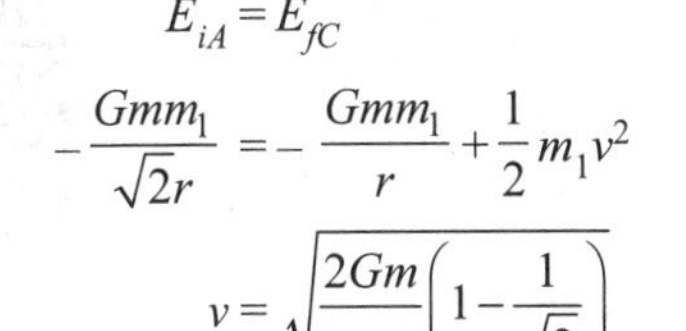

$$E_{iA} = E_{fC}$$

$$-\frac{Gmm_1}{\sqrt{2}r} = -\frac{Gmm_1}{r} + \frac{1}{2}m_1 v^2$$

$$\Rightarrow \qquad v = \sqrt{\frac{2Gm}{r}\left(1 - \frac{1}{\sqrt{2}}\right)}$$ **Ans. (C)**

Sol. 23 Weight of a person of mass m on earth surface is given as

$$W = mg_s$$

If earth is rotating at angular speed ω, value of g on earth's equator is

$$g_{eq} = g_s - \omega^2 R_e$$

For weight to be $\frac{3}{4}W$, we use

$$g_{eq} = \frac{3}{4}g_s = g_s - \omega^2 R_e$$

$$\Rightarrow \qquad \omega^2 R_e = \frac{g_s}{4}$$

$$\Rightarrow \qquad \omega = \sqrt{\frac{g_s}{4R_e}} = \sqrt{\frac{10}{4 \times 6400 \times 1000}}$$

$$\Rightarrow \qquad \omega = \frac{1}{16 \times 10^2} = 0.625 \times 10^{-3} \text{ rad/s}$$ **Ans. (C)**

Sol. 24

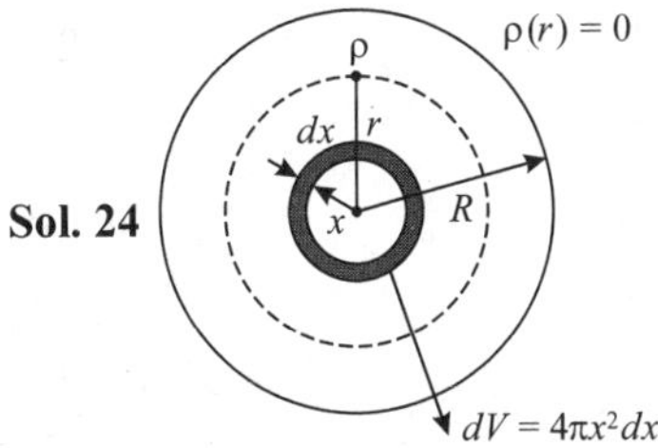

The mass of elemental shell of radius x and thickness dx is given as

$$dm = \rho(x) \times 4\rho x^2 dx$$

Total mass contained within the inner sphere of radius r is given as

$$m = \int dm = \int_0^r \frac{k}{x} \cdot 4\pi x^2 dx$$

$$\Rightarrow \qquad m = 4\pi k \int_0^r x\, dx = 2\pi k r^2$$

acceleration of a test particle at point P is given as

$$a = -\frac{Gm}{r^2} = -\frac{G}{r^2} \cdot 2\pi k r^2 = -2\pi k G$$

– ve sign only denotes inward direction.

$|a| \to$ constant

independent of r for $r < R$ **Ans. (B)**

Sol. 25

Similarly for Sun and earth this difference of force is given as

Force between earth and moon is

$$F_1 = \frac{GM_e M_m}{r_1^2}$$

$$\Rightarrow \qquad \Delta F_1 = -\frac{2GM_e M_m}{r_1^3} \cdot \Delta r_1 \qquad \ldots(1)$$

$$\text{and} \qquad \Delta F_2 = -\frac{2GM_e M_S}{r_2^3} \cdot \Delta r_2 \qquad \ldots(2)$$

From (a) (2) $\Rightarrow$

$$\frac{\Delta F_1}{\Delta F_2} = \frac{M_m}{M_S} \times \left(\frac{r_2}{r_1}\right)^3$$

$$\Rightarrow \qquad \frac{\Delta F_1}{\Delta F_2} = \left(\frac{8 \times 10^{22}}{2 \times 10^{30}}\right) \times \left(\frac{150 \times 10^6}{0.4 \times 10^6}\right)^3 = 2.1$$

Ans. (A)

Sol. 26

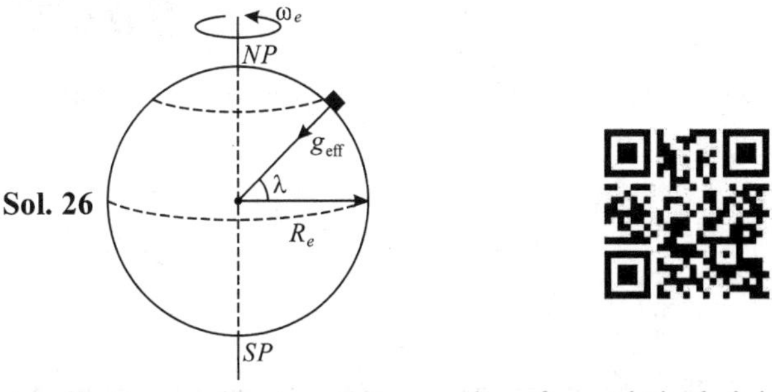

As effective gravity at a point on earth surface at latitude λ is given as

$$g_{eff} = g_s - \omega_e^2 R_e \cos^2\lambda \qquad \ldots(1)$$

at poles $\qquad \lambda = \pm 90°$

$$\Rightarrow \qquad g_{eff} = g_s$$

at all points on earth other than poles

$$g_{eff} < g_s$$

If ω_e increases $\Rightarrow$ g_{eff} everywhere decreases except poles

$$\text{weight} = mg_{eff}$$ **Ans. (D)**

Sol. 27 Acceleration due to gravity is

$$g = \frac{GM}{R^2}$$

$$\frac{g_p}{g_e} = \frac{M_e}{M_e}\left(\frac{R_e}{R_p}\right)^2 = 3\left(\frac{1}{3}\right)^2 = \frac{1}{3}$$

For a simple pendulum, we have

$$T \propto \frac{1}{\sqrt{g}}$$

$$\Rightarrow \quad \frac{T_p}{T_e} = \sqrt{\frac{g_e}{g_p}} = \sqrt{3}$$

$$\Rightarrow \quad T_p = 2\sqrt{3}\text{s}$$

Ans. (B)

Sol. 28

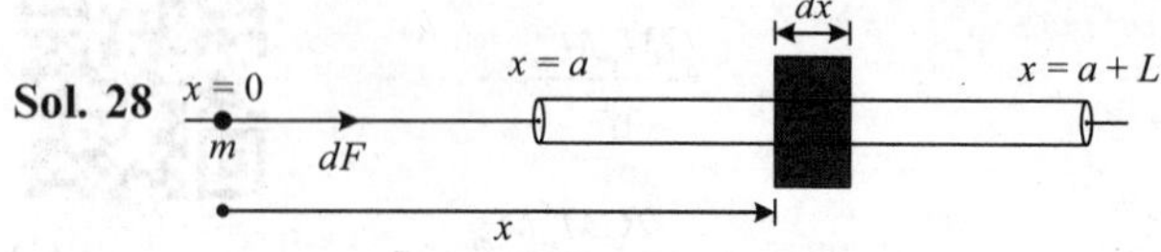

Mass of element of width dx considered is

$$dm = (A + Bx^2)dx$$

Force on this element due to m is given as

$$dF = \frac{Gm\,dm}{x^2} = \frac{Gm}{x^2}(A + Bx^2)dx$$

Net force on rod is

$$F = \int_a^{a+L} \frac{Gm}{x^2}(A + Bx^2)dx$$

$$\Rightarrow \quad F = Gm\left[-\frac{A}{x} + Bx\right]_a^{a+L}$$

$$\Rightarrow \quad F = Gm\left[A\left(\frac{1}{a} - \frac{1}{a+L}\right) + BL\right]$$

Ans. (B)

Sol. 29

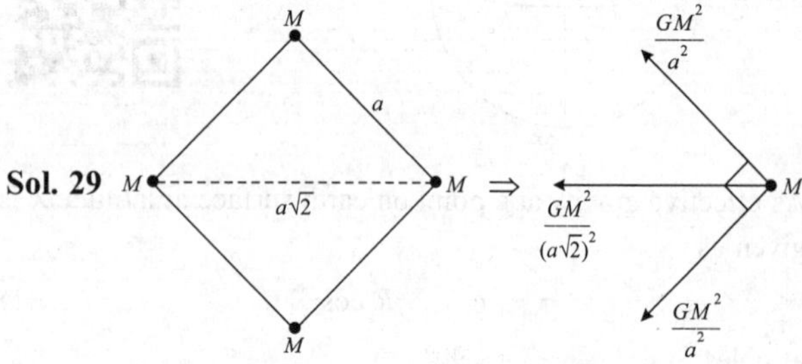

Net force on particle towards centre of circle is

$$F_C = \frac{GM^2}{2a^2} + \frac{GM^2}{a^2}\sqrt{2}$$

$$\Rightarrow \quad F_C = \frac{GM^2}{a^2}\left(\frac{1}{2} + \sqrt{2}\right)$$

This force will act as centripetal force for circular motion in a circle of radius $\frac{a}{\sqrt{2}}$, given as

$$F_C = \frac{mv^2}{r}$$

$$\Rightarrow \quad \frac{mv^2}{\frac{a}{\sqrt{2}}} = \frac{GM^2}{a^2}\left(\frac{1}{2} + \sqrt{2}\right)$$

$$\Rightarrow \quad v^2 = \frac{GM}{a}\left(\frac{1}{2\sqrt{2}} + 1\right)$$

$$\Rightarrow \quad v^2 = \frac{GM}{a}(1.35)$$

$$\Rightarrow \quad v = 1.16\sqrt{\frac{GM}{a}}$$

Ans. (C)

Sol. 30

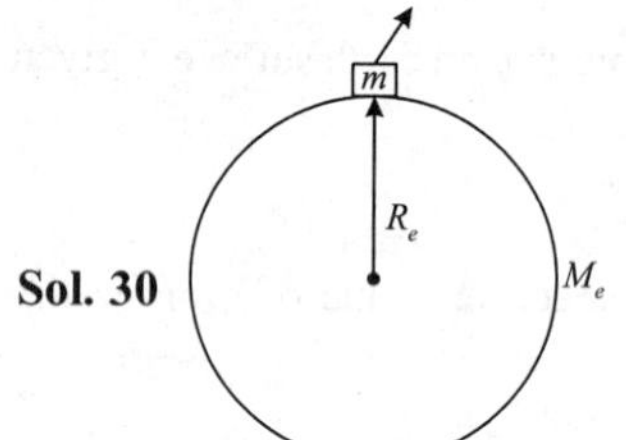

Energy required (minimum) for rocket to escape is

$$E = \frac{GM_e m}{R_e}$$

For moon, we have

$$R_m = \frac{R_e}{4} \Rightarrow V_m = \frac{V_e}{64}$$

$$\Rightarrow \quad M_m = \frac{M_e}{64}$$

Minimum energy required for rocket to escape from moon is

$$E_{\text{moon}} = \frac{GM_m m}{R_m} \qquad \left(\because M_m = \frac{M_e}{64}, R_m = \frac{R_e}{4}\right)$$

$$E_{\text{moon}} = \frac{E}{16}$$

Ans. (B)

Sol. 31

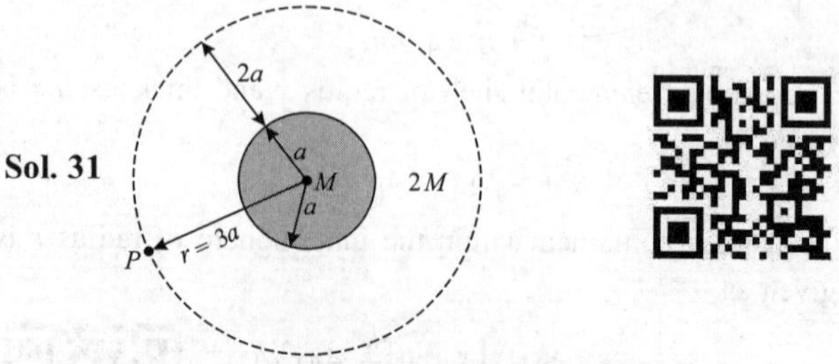

At point P gravitational field is due to the whole spherically symmetric mass $3M$ which will act as concentrated at centre for this point, so

$$g_P = \frac{G(3M)}{r^2} = \frac{G(3M)}{(3a)^2} = \frac{GM}{3a^2}$$ **Ans. (B)**

Sol. 32 Acceleration due to gravity at height h above earth surface is given as

$$g_h = \frac{GM_e}{(R_e + h)^2}$$

Given that $\frac{GM_e}{(R_e + h)^2} = \frac{1}{2}\frac{GM_e}{R_e^2}$

$$\Rightarrow \quad \sqrt{2}\,R_e = R_e + h$$

$$\Rightarrow \quad h = (\sqrt{2} - 1)\,R_e$$

$$= 0.414 \times 6.4 \times 10^6\,\text{m}$$

$$= 2.6 \times 10^6\,\text{m}$$ **Ans. (D)**

Sol. 33 Weight at equator is given as

$$W_{eq} = W_{pole} - m\omega^2 R_e$$

$$\Rightarrow \quad W_{eq} = 196 - \frac{196}{10} \times \left(\frac{2\pi}{86400}\right)^2 \times 6.4 \times 10^5$$

$$= 195.34\,\text{N}$$ **Ans. (D)**

Sol. 34 Gravitational field on the surface of a solid sphere is given as

$$g_s = \frac{GM}{R^2}$$

From the graphs, we have

$$\frac{GM_1}{(1)^2} = 2 \text{ and } \frac{GM_2}{(2)^2} = 3$$

$$\Rightarrow \quad \frac{M_1}{M_2} = \frac{1}{6}$$ **Ans. (D)**

Sol. 35 Value of g at a height h above earth surface is

$$g_{out} = \frac{GM}{(R + h)^2}$$

Value of g at a depth h below the earth surface is

$$g_{in} = \frac{GM(R - h)}{R^3}$$

Given that

$$g_{out} = g_{in}$$

$$\Rightarrow \quad \frac{GM}{(R + h)^2} = \frac{GM(R - h)}{R^3}$$

$$\Rightarrow \quad (R - h)(R + h)^2 = R^3$$

Simplifying above equation gives

$$h^2 + Rh - R^2 = 0$$

$$\Rightarrow \quad h = \frac{-R + \sqrt{R^2 + 4R^2}}{2}$$

$$\Rightarrow \quad h = \frac{(\sqrt{5} - 1)R}{2}$$ **Ans. (A)**

Sol. 36 Mass contained in a region of radius r is

$$m = \int_0^r \rho(r) \cdot 4\pi r^2 dr$$

$$m = \int_0^r \rho_0 \left(1 - \frac{r^2}{R^2}\right) 4\pi r^2 dr$$

$$\Rightarrow \quad m = 4\pi\rho_0 \left[\frac{r^3}{3} - \frac{r^5}{5R^2}\right]$$

Gravitational field due to this expression of EF, we use

$$g = \frac{Gm}{r^2}$$

$$\Rightarrow \quad g = \frac{G}{r^2}\left(4\pi\rho_0\left(\frac{r^3}{3} - \frac{r^5}{5R^2}\right)\right)$$

$$\Rightarrow \quad g = 4\pi G\rho_0\left(\frac{r}{3} - \frac{r^3}{5R^2}\right)$$

g is maximum when

$$\frac{dg}{dr} = 0$$

$$\Rightarrow \quad \frac{dg}{dr} = 4\pi G\rho_0\left(\frac{1}{3} - \frac{3r^2}{5R^2}\right) = 0$$

$$\Rightarrow \quad r = \sqrt{\frac{5}{9}}R$$ **Ans. (B)**

Sol. 37 Potential on X-axis is given as

$$V = -\int_\infty^x E_g \cdot dx = -\int_\infty^x \frac{Ax}{(x^2 + a^2)^{3/2}} dx$$

$$\Rightarrow \quad V = \left[\frac{A}{(x^2 + a^2)^{1/2}}\right]_\infty^x = \frac{A}{\sqrt{x^2 + a^2}}$$ **Ans. (A)**

Sol. 38 Gravity at a height $h = \frac{R}{2}$ above earth surface is given as

$$g_{h\,up} = \frac{GM}{\left(R + \frac{R}{2}\right)^2}$$

gravity at a depth d below earth surface is given as

$$g_{d\,in} = \frac{GM(R - d)}{R^3}$$

Given that $g_{h\,up} = g_{d\,in}$

$$\Rightarrow \quad \frac{1}{\left(\frac{3R}{2}\right)^2} = \frac{R-d}{R^3}$$

$$\Rightarrow \quad \frac{4R}{9} = R - d$$

$$\Rightarrow \quad \frac{d}{R} = \frac{5}{9}$$ **Ans. (D)**

Sol. 39 Gravity on equator of earth is given as

$$g_e = g_s - \omega^2 R$$

Gravity at a height h above earth surface is

$$g_h = g_s\left(1 - \frac{2h}{R}\right) \qquad [\text{For } h << R]$$

Given that $g_e = g_h$

$$\Rightarrow \quad g_s - \omega^2 R = g_s - \frac{2g_s h}{R}$$

$$\Rightarrow \quad h = \frac{\omega^2 R^2}{2g}$$ **Ans. (D)**

Sol. 40 Weight of body on pole is given to be

$$mg = 49\text{ N}$$

At equator due to rotation, the effective value of g is given as

$$g_e = g - R\omega^2$$

Thus weight at equator is given as

$$W_e = mg_e = m(g - R\omega^2)$$

$$W_P > W_e$$

As $W_P = 49$ N so the only possible option is $W_e = 48.83$ N

Ans. (D)

Sol. 41 If mass of complete sphere is m', the mass of removed part will be $m'/8$. The force on A due to complete sphere is calculated as

$$F_1 = mg_O = \frac{Gmm'}{9R^2}$$

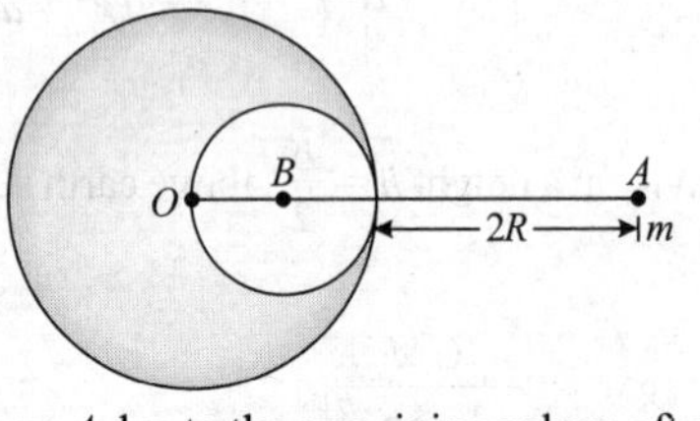

The net force on A due to the remaining sphere after removal of the small sphere from it is given as

$$F_2 = m\left[\frac{G.m'}{(3R)^2} - \frac{G.m'/8}{(5R/2)^2}\right]$$

$$\Rightarrow \quad F_2 = \frac{Gm'}{9R^2} - \frac{Gm' \times 4}{8 \times 25} = \left(\frac{1}{9} - \frac{1}{50}\right)\frac{Gm'}{R^2}$$

$$\Rightarrow \quad F_2 = \frac{41}{50 \times 9} \cdot \frac{Gm'}{R^2}$$

$$\Rightarrow \quad \frac{F_1}{F_2} = \frac{1}{9} \times \frac{50 \times 9}{41} = \frac{50}{41}$$ **Ans. (B)**

Sol. 42 Gravitational field of ring at the center of sphere is given as

$$g_Y = -\frac{Gmx}{(R^2 + x^2)^{3/2}}$$

Force between sphere & ring can be given as

$$F = \frac{GmM(\sqrt{8}R)}{(R^2 + 8R^2)^{3/2}} = \frac{GmM}{R^2} \times \frac{\sqrt{8}}{27}$$ **Ans. (C)**

Sol. 43 Gravitational field at point A is given as

$$g_A = \frac{GM(r)}{R^3}$$

Gravitational field at point C is given as

$$g_C = \frac{GM}{\left(R + \frac{R}{2}\right)^2}$$

It is given that field at A is equal to that at point C, so we use

$$g_A = g_C$$

$$\Rightarrow \quad \frac{r}{R^3} = \frac{1}{\frac{9}{4}R^2}$$

$$\Rightarrow \quad OA = r = \frac{4R}{9}$$

The distance AB is given as

$$AB = R - r = \frac{5R}{9}$$

$$\Rightarrow \quad OA : AB = 4 : 5$$ **Ans. (4)**

Sol. 44 As already derived in Physics Galaxy book volume III, Self energy of a solid sphere of mass M and radius R which is the amount of energy required to break it up into its constituents and separate upto infinity is given as

$$U = \frac{3}{5}\frac{GM^2}{R}$$

$$\Rightarrow \quad x = 3$$ **Ans. (3)**

Sol. 45 At any point between Earth and Mars, gravitational field can be given as

$$\vec{g}\text{ (at any point)} = \vec{g}_{\text{Earth}} + \vec{g}_{\text{mars}}.$$

As we move away from earth, at some point between value of g decrease to zero where $\vec{g}_E + \vec{g}_M = 0$ then again increases upto the surface of Mars. **Ans. (A)**

Sol. 46 In the given system of binary star both the planets will move by the mutual gravitational interaction, so they will have same angular velocity and hence same time period. **Ans. (A)**

Sol. 47 By conservation of energy, the speed of body at a height y is given by

$$\frac{1}{2}mv^2 - \frac{GMm}{R+y} = 0$$

$$\Rightarrow \quad v = \sqrt{\frac{2GM}{R+y}}$$

$$\Rightarrow \quad \frac{dy}{dt} = \sqrt{\frac{2Gm}{R+y}}$$

If t is the time take for body to heach a height h above the surface of earth, we have

$$\Rightarrow \quad \int_0^h \sqrt{R+y}\,dy = \sqrt{2Gm}\int_0^t dt$$

$$\Rightarrow \quad \left[\frac{2(R+y)^{3/2}}{3}\right]_0^h = \sqrt{2Gm}\,t$$

$$\Rightarrow \quad \frac{2}{3}[(R+h)^{3/2} - R^{3/2}] = \sqrt{2Gm}\,t$$

$$\Rightarrow \quad t = \frac{\frac{2}{3}[(R+h)^{3/2} - R^{3/2}]}{\sqrt{2Gm}}$$

$$\Rightarrow \quad \frac{1}{3}\sqrt{\frac{2R_e}{g}}\left[\left(1+\frac{h}{R_e}\right)^{3/2} - 1\right]$$ **Ans. (D)**

Sol. 48 Given that mass of planet is twice that of Earth so we use

$$2M_E = M_P$$

If ρ is the density of planet and the Earth, we have

$$2\rho \times \frac{4}{3}R_E^3 = \rho \times \frac{4}{3}\pi R_P^3$$

$$\Rightarrow \quad R_P = 2^{1/3}R_E$$

Acceleration due to gravity on surface of planet is given as

$$g_P = \frac{GM_P}{R_P^2}$$

$$\Rightarrow \quad g_P = \frac{G2M_E}{(2^{1/3}R_E)^2} = \frac{G2M_E}{2^{2/3}R_E^2}$$

$$\Rightarrow \quad g_P = 2^{1/3}g_e$$

Object weight on planet = $2^{1/3}$ of the object weight on earth

$$\Rightarrow \quad W_P = 2^{1/3}\,W$$ **Ans. (D)**

Sol. 49 Let at a distance x from planet A, the net gravitational field becomes zero then x is given as

$$\frac{GM}{x^2} = \frac{G\times 9M}{(8R-x)^2}$$

$$(3x)^2 = (8R-x)^2$$

$$x = 2R$$

Thus, particle should be projected so that it cross a minimum distance of $2R$ as after this point other planet gravity will pull it, so we have

$$\frac{1}{2}mv^2 - \frac{GMm}{R} - \frac{G(9M)m}{7R} = -\frac{GMm}{2R} - \frac{G(9M)m}{6R}$$

$$\Rightarrow \quad \frac{1}{2}v^2 = \frac{2}{7}\frac{GM}{R}$$

$$\Rightarrow \quad v = \sqrt{\frac{4}{7}\frac{GM}{R}}$$ **Ans. (4)**

Sol. 50 Resulting gravitational attraction provides the necessary centripetal force for uniform circular motion, so we use

$$\frac{Gm^2}{4R^2} = m\omega^2 R$$

$$\Rightarrow \quad \omega = \sqrt{\frac{Gm}{4R^3}}$$

$$\Rightarrow \quad \omega = \frac{1}{2}\sqrt{\frac{G}{R^3}}$$ **Ans. (C)**

Sol. 51 For revolution of moon in circular motion, we use

$$m\omega^2 R = \frac{GmM}{R^2}$$

$$\Rightarrow \quad \omega^2 = \frac{GM}{R^3}$$

$$\Rightarrow \quad T = \frac{2\pi}{\omega} = \frac{2\pi}{\sqrt{\frac{GM}{R^3}}}$$

$$\Rightarrow \quad \frac{GM}{R^3} = \frac{4\pi^2}{T^2}$$

$$\Rightarrow \quad M = \frac{4\pi^2 R^3}{GT^2} = \frac{6\times 10^{11}\times(9\times 10^6)^3}{(7\times 3600 + 30\times 60)^2}$$

$$\Rightarrow \quad M = 6\times 10^{23}\,\text{kg}$$ **Ans. (D)**

Sol. 52 Gravitational field inside a spherical shell is zero everywhere inside so option (A) and (C) are correct. Gravitational potential due to uniform spherical shell is same everywhere inside so option (D) is also correct. **Ans. (C)**

Sol. 53 Gravitational potential energy of the system is given as

$$U = \frac{Gm^2}{a} + \frac{G(M-m)^2}{a} + \frac{2Gm(M-m)}{a} + \frac{2Gm(M-m)}{\sqrt{2}a}$$

The energy becomes maximum when $\frac{dU}{dm} = 0$

$$\Rightarrow \frac{dU}{dm} = \frac{G}{a}(2m - 2(M-m) + (M-2m)2 + \sqrt{2}(M-2m))$$

$$(4m - 4m - 2\sqrt{2}m) + (-2M + 2M + \sqrt{2}M) = 0$$

$$\Rightarrow \sqrt{2}M = 2\sqrt{2}m$$

$$\Rightarrow \frac{M}{m} = 2$$ **Ans. (2)**

Sol. 54 Inside the given spherical shell, the gravitational potential can be written as

$$V_P = -\frac{GM}{R} - \frac{Gm}{r}$$

$$\Rightarrow V_P = -\frac{G(100)}{50} - \frac{G(50)}{25}$$

$$\Rightarrow V_P = -4G$$ **Ans. (C)**

Sol. 55 Value of g at a height r above the Earth's surface and at the same depth below the surface is given in the ratio calculated below as

$$\frac{g_h}{g_d} = \frac{\frac{GM}{(R_E + r)^2}}{\frac{GM}{R_E^3}(R_E - r)} = \frac{R_E^3}{(R_E - r)^2 (R_E - r)}$$

$$\Rightarrow \frac{g_d}{g_h} = \frac{(R_E + r)^2 (R_E - r)}{R_E^3}$$

$$\Rightarrow \frac{g_d}{g_h} = \frac{R_E^3 - R_E r^2 + r R_E^2 - r^3}{R_E^3}$$

$$\Rightarrow \frac{g_d}{g_h} = 1 - \frac{r^2}{R_E^2} + \frac{r}{R_E} - \left(\frac{r}{R_E}\right)^3$$ **Ans. (A)**

Sol. 56 Figure below shows the forces acting on each particle due to others. Net force on one particle is given as

$$F_{net} = F_1 + 2F_2 \cos 45°$$

$$\Rightarrow F_{net} = \frac{GM^2}{(2R)^2} + 2 \times \frac{GM^2}{(\sqrt{2}R)^2} \times \frac{1}{\sqrt{2}}$$

$$\Rightarrow F_{net} = \frac{GM^2}{4R^2} + \frac{GM^2}{\sqrt{2}R^2} = \frac{(4+\sqrt{2})GM^2}{4\sqrt{2}R^2}$$

M
45°
R
M M
R R
M

For ciruclar motion this force will provide the necessary centripetal force so we have

$$F_C = F_{net}$$

$$\frac{Mv^2}{R} = \frac{(4+\sqrt{2})GM^2}{4\sqrt{2}R^2}$$

$$\Rightarrow v = \sqrt{\frac{(4+\sqrt{2})}{4\sqrt{2}} \frac{GM}{R}}$$ **Ans. (C)**

Sol. 57 If g' is the effective gravity at the point where weight becomes one third then we have

$$Mg' = \frac{M}{3} g$$

$$\Rightarrow g' = \frac{g}{3}$$

At a height h, gravity is calculated as

$$g' = g\left(\frac{R}{R+h}\right)^2 = \frac{g}{3}$$

$$\Rightarrow \frac{R}{R+h} = \frac{1}{\sqrt{3}}$$

$$\Rightarrow h = (\sqrt{3} - 1)R$$

$$\Rightarrow h = (1.732 - 1)\,6400 \text{ km}$$

$$\Rightarrow h = 4685 \text{ km}$$ **Ans. (B)**

Sol. 58 Gravitational acceleration at a distance r from the centre of earth is given as

$$g = \frac{GM}{r^2}$$

At a distance $3r$ from the centre of earth, acceleration due to gravity is calculated as

$$g' = \frac{GM}{(3r)^2} = \frac{GM}{9r^2}$$

$$\Rightarrow g' = \frac{g}{9}$$ **Ans. (D)**

Sol. 59 Acceleration due to gravity due to earth inside and outside the earth surface is calculated by the expression given as

$$g = \begin{cases} \frac{GMr}{R^3}, & r \le R \\ \frac{GM}{r^2}, & r \ge R \end{cases}$$

Hence option (A) is correct. **Ans. (A)**

Sol. 60 Elongation of a wire is proportional to the weight of body suspended from the wire so we use

$$\Delta l \propto g$$

$$\Rightarrow \quad \frac{\Delta l_{\text{earth}}}{\Delta l_{\text{planet}}} = \frac{g_{\text{earth}}}{g_{\text{planet}}} = \frac{10^{-4}}{6\times10^{-5}}$$

$$\Rightarrow \quad g_{\text{planet}} = 6 \text{ m/s}^2$$

Ans. (6)

Sol. 61 Effective acceleration due to gravity is the resultant of g & $r\omega^2$ which is not directed toward earth's centre as well as its magnitude is also not constant. Hence assertion is false. At equator, effective acceleration is radially inward so reason is true. **Ans. (D)**

Sol. 62 Since law of gravitation is a universal law so it holds good for any pair of bodies hence statement I is true. The value of g at centre of earth is zero, so weight of any body at earth's centre will also be zero. Thus statement-I and Statement-II are true. **Ans. (A)**

Sol. 63 The total gravitational potential energy is calculated by potential energy of all the pair of masses in the given configuration of masses

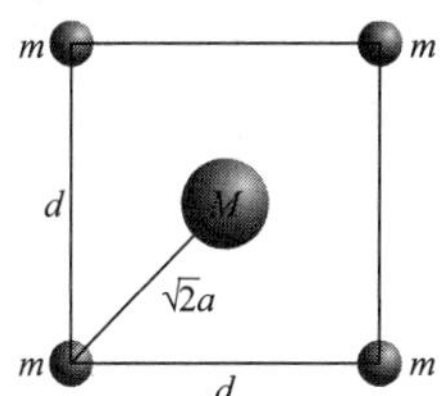

$$U = -\frac{Gm^2}{d}\times 4 - \frac{Gm^2}{\sqrt{2}d}\times 2 - \frac{GMm}{d}\times 4\sqrt{2}$$

$$\Rightarrow \quad U = -\frac{Gm}{d}[(4+\sqrt{2})m + 4\sqrt{2}M]$$

Ans. (A)

Sol. 64 Initially the force between masses is given as

$$F = \frac{Gm^2}{r^2}$$

Now if one third of mass is shifted from one object to another, new force between them is given as

$$F' = \frac{G\left(\frac{4m}{3}\right)\times\left(\frac{2m}{3}\right)}{r^2}$$

$$\Rightarrow \quad F' = \frac{8}{9}F$$

Ans. (C)

Sol. 65 Figure below shows the situation described in question.

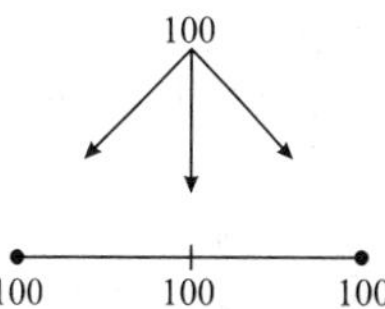

The force acting on the fourth particle due to the particles located at A, B and C is given as

$$F = \frac{GMM}{r^2} + \sqrt{2}\frac{GMM}{(\sqrt{2}r)^2}$$

$$\Rightarrow \quad F = \frac{GMM}{r^2}\left(1+\frac{1}{\sqrt{2}}\right)$$

$$\Rightarrow \quad F = \frac{G\times10^4}{13^2}\left(1+\frac{1}{\sqrt{2}}\right)$$

$$\Rightarrow \quad F \approx 100\,\text{G}$$

Ans. (B)

Sol. 66 Figure below shows the situation described in question where object is taken to a position above earth surface at a distance $5R/4$ from the centre of earth.

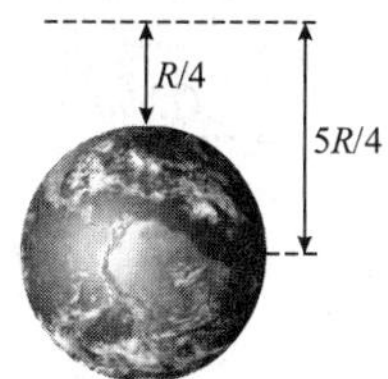

Acceleration due to gravity at this point is calculated as

$$g_h = \frac{g}{\left(1+\frac{h}{R}\right)^2}$$

$$\Rightarrow \quad g_h = \frac{g}{\left(1+\frac{1}{4}\right)^2} = \frac{16g}{25}$$

Percentage change in weight of object is calculated by percentage change in value of g, given as

$$\frac{\Delta g}{g}\times 100 = \frac{|g_h - g|}{g}\times 100 = \frac{\left|\frac{16}{25}-1\right|}{1}\times 100$$

$$\Rightarrow \quad \frac{\Delta g}{g}\times 100 = \frac{9}{25}\times 100 = 36\%$$

Ans. (A)

Sol. 67 Acceleration due to gravity at a height $h << R$ is

$$g' = g\left(1-\frac{2h}{R}\right)$$

$$\Rightarrow \quad \frac{\Delta g}{g} = \frac{2h}{R}$$

$$\Rightarrow \quad \frac{\Delta g}{g} \times 100 = \frac{2h}{R} \times 100$$

$$\Rightarrow \quad \frac{\Delta g}{g} \times 100 = 2 \times \frac{32}{6400} \times 100 = 1\% \qquad \textbf{Ans. (A)}$$

Sol. 68 When a body is thrown upward from earth surface, it reaches to a maximum height where its final velocity becomes zero

Escape velocity from earth's surface is given as

$$v_e = \sqrt{\frac{2Gm}{R}}$$

By conservation of energy, if maximum height attained is h then we use

$$\frac{-GMm}{R} + \frac{1}{2}m\left(\frac{v_e}{3}\right)^2 = -\frac{GMm}{R+h}$$

$$\Rightarrow \quad \frac{GM}{R+h} = \frac{GM}{R} - \frac{v_e^2}{18}$$

$$\Rightarrow \quad \frac{GM}{R+h} = \frac{GM}{R} - \frac{GM}{9R}$$

$$\Rightarrow \quad \frac{GM}{R+h} = \frac{8GM}{9R}$$

$$\Rightarrow \quad \frac{1}{R+h} = \frac{8}{9R}$$

$$\Rightarrow \quad 9R = 8R + 8h$$

$$\Rightarrow \quad h = \frac{R}{8} = \frac{6400}{8} = 800\,\text{km} \qquad \textbf{Ans. (A)}$$

Sol. 69 Acceleration due to gravity on earth surface is given as

$$g = \frac{GM}{R^2}$$

As mass remain constant, variation of gravity with radius of earch is calculated as

$$\frac{\Delta g}{g} \times 100 = 2\frac{\Delta R}{R} \times 100$$

$$\Rightarrow \quad \frac{\Delta g}{g} \times 100 = 2(2)$$

$$\Rightarrow \quad \frac{\Delta g}{g} \times 100 = 4\% \qquad \textbf{Ans. (D)}$$

Sol. 70 For a simple pendulum clock, time period is inversely proportional to acceleration due to gravity, given as

$$t \propto \frac{1}{\sqrt{g}}$$

In this case, value of g at a height h above the earth surface is given as

$$g \propto \frac{1}{(R+h)^2}$$

$$\Rightarrow \quad \frac{t_1}{t_2} = \sqrt{\frac{g'}{g}} = \sqrt{\frac{R^2}{(R+h)^2}}$$

$$\Rightarrow \quad \frac{t_1}{t_2} = \frac{4}{6} = \frac{R}{(R+h)}$$

$$\Rightarrow \quad h = 3200\,\text{km} \qquad \textbf{Ans. (C)}$$

NOTE: Here as per the language of question we will solve it for value of g at a height h above the earth surface but practically in a space station which is a kind of satellite, effective gravity will be zero.

Sol. 71 As per given data in question gravity at a height h is same at a depht αh so we use

$$g\left(1 - \frac{2h}{R}\right) = g\left(1 - \frac{d}{R}\right)$$

$$\Rightarrow \quad \frac{2h}{R} = \frac{\alpha h}{R}$$

$$\Rightarrow \quad \alpha = 2 \qquad \textbf{Ans. (2)}$$

Sol. 72 Initial and final potential energies of the object are given as

$$U_i = \frac{-GMm}{R}$$

$$U_f = -\frac{GMm}{4R}$$

Gain in potential energy is calculated as

$$\Delta U = U_f - U_i = \frac{3GMm}{4R}$$

$$\Rightarrow \quad \Delta U = \frac{3}{4}mgR$$

$$\Rightarrow \quad \Delta U = \frac{3}{4} \times 1 \times 10 \times 64 \times 10^5 = 48\,\text{MJ} \qquad \textbf{Ans. (A)}$$

Sol. 73 Acceleration due to gravity at height h is given as

$$g' = \frac{g}{\left[1 + \frac{h}{R}\right]^2}$$

Thus weight at the given height is expressed as

$$mg' = \frac{mg}{\left[1+\frac{h}{R}\right]^2} = \frac{18}{\left[1+\frac{1}{2}\right]^2} = 8\text{ N} \qquad \textbf{Ans. (D)}$$

Sol. 74 The acceleration due to earth's gravity at a height and at a depth is given by the expressions given below.

$$g' = \frac{g}{\left(1+\frac{h}{R}\right)^2}$$

$$g' = g\left\{1-\frac{d}{R}\right\}$$

Both are less than g hence Statement-I is correct & we can see from above expressions that for same height and depth values of g are not equal hence Statement-II is incorrect. **Ans. (C)**

Sol. 75 Oscillation period of a pendulum clock is given as

$$T \propto \frac{1}{\sqrt{g}}$$

Hence at Mount Everest as g decreases, time period increases and clock slows down hence Assertion-A is wrong and Reason-R is correct. **Ans. (C)**

Sol. 76 Figure below shows the SHM of the given particle inside the tunnel along earth's diameter.

When particle is at a distance x from centre of Earth, then at that point, gravitational acceleration at particle's position is given as

$$g = \frac{GM}{R^3}x$$

Acceleration of particle is inward so it is written as

$$\vec{a} = -\frac{GM}{R^3}\vec{x}$$

Comparing with acceleration of particle in SHM $a = -\omega^2 x$ it gives

$$\omega = \sqrt{\frac{GM}{R^3}} = \sqrt{\frac{g}{R}}$$

Thus the time period of oscillations is calculated as

$$T = \frac{2\pi}{\omega} = 2\pi\sqrt{\frac{R}{g}}$$

$$\Rightarrow \quad T = 2\times 3.14\times\sqrt{\frac{6400\times 10^3}{10}}$$

$$\Rightarrow \quad T = 2\times 3.14\times 800\text{ sec}$$

$$\Rightarrow \quad T \approx 1\text{ hour }24\text{ minutes} \qquad \textbf{Ans. (B)}$$

Sol. 77 Potential energy of a body at a distance r from the centre of earth is given as

$$U = \frac{-GM_e m}{r}$$

Initially when the body is on earth surface, its potential energy is given as

$$U_i = \frac{-GM_e m}{R_e}$$

When body is taken to a height h above the earth surface, its potential energy is given as

$$U_f = \frac{-GM_e m}{(R_e+h)} = \frac{-GM_e m}{R_e+2R_e} = \frac{-GM_e m}{3R_e}$$

Increase in potential energy is calculated as

$$\Delta U = U_f - U_i$$

$$\Rightarrow \quad \Delta U = \frac{2}{3}\frac{GM_e m}{R_e} = \frac{2}{3}\frac{GM_e}{R_e^2}mR_e$$

$$\Rightarrow \quad \Delta U = \frac{2}{3}mgR_e \qquad \textbf{Ans. (C)}$$

Sol. 78 For circular motion gravitational force will provide the centripetal force, so we use

$$\frac{Gm^2}{4r^2} = \frac{mv^2}{r}$$

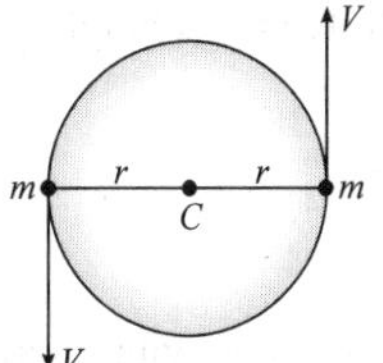

$$\Rightarrow \quad v = \sqrt{\frac{Gm}{4r}} \qquad \textbf{Ans. (D)}$$

Sol. 79 Gravitational potential in space is given as

$$-\frac{dV}{dr} = -\frac{k}{r^2}$$

$$\Rightarrow \quad \int_{10}^{V} dV = \int_{2}^{3}\frac{k}{r^2}dr$$

$$\Rightarrow \quad V - 10 = k\left[\frac{1}{2}-\frac{1}{3}\right]$$

$$\Rightarrow \quad V-10=\frac{k}{6}$$

$$\Rightarrow \quad V=11\text{ J/kg}$$ **Ans. (B)**

Sol. 80 During fall of the object, work done by gravity is equal to the gain in kinetic energy so when object is allowed full from height R at rest to surface of earth then

$$\left(-\frac{GMm}{2R}\right)-\left(-\frac{GMm}{R}\right)=\frac{1}{2}mv^2$$

$$\Rightarrow \quad v^2=\frac{GM}{R}=gR$$

$$\Rightarrow \quad v=\sqrt{gR}$$ **Ans. (B)**

Sol. 81 As per given condition in the question we can use

$$g_d=g_h$$

$$\Rightarrow \quad \frac{GM}{R^2}\left[1-\frac{d}{R}\right]=\frac{4\times GM}{(4R)^2}$$

$$\Rightarrow \quad 1-\frac{d}{R}=\frac{1}{4}$$

$$\Rightarrow \quad \frac{d}{R}=\frac{3}{4}$$

$$\Rightarrow \quad d=\frac{3}{4}R=4800\text{ km}$$ **Ans. (D)**

Sol. 82 Acceleration due to gravity at the height of $10R$ is calculated as

$$h=R+9R=10R$$

$$g'=\frac{GM}{(10R)^2}=\left(\frac{g}{100}\right)$$

Thus weight of the body at this height is calculated as

$$W'=\left(\frac{W}{100}\right)$$ **Ans. (B)**

Sol. 83 Due to variation in gravitational acceleration on earth surface, effective gravity at a co-latitude angle θ is given as

$$g_{eff}=g-\omega^2R_e\sin^2\theta$$

Acceleration due to gravity at a depth below the earth surface is given as

$$g_{eff}=g\left(1-\frac{d}{R_e}\right)$$ **Ans. (C)**

Sol. 84 As earth is spherical, its volume is calculated as

$$V=\frac{4}{3}\pi R^3$$

mass of earth is taken as

$$m=\rho\times\frac{4}{3}\pi R^3$$

Considering ρ = constant, we use

$$R\propto m^{\frac{1}{3}}$$

As weight of a body is mg and acceleration due to gravity on the surface of earth is given as $g=\frac{GM}{R^2}$, body weiht is written as

$$W\propto g\propto\frac{Gm}{R^2}$$

$$\Rightarrow \quad W\propto\frac{m}{m^{2/3}}\propto m^{1/3}$$

Thus weight of body on the planet having double the mass of earth is given as

$$W_p=(2)^{1/3}W$$ **Ans. (C)**

Sol. 85 Acceleration due to gravity at height h is given as

$$\Rightarrow \quad g'=\frac{gR^2}{r^2}=\frac{gR^2}{\left(R+\frac{R}{4}\right)^2}=\frac{16g}{25}$$

$$\Rightarrow \quad \text{Weight}=\frac{16}{25}\times100=64\text{ N}$$ **Ans. (B)**

Sol. 86 Given the weight of body is

$$W=mg=400\text{ N}$$

At depth d, gravity is calculated as

$$g'=g\left(1-\frac{d}{R}\right)$$

At $d=\frac{R}{2}$, we have

$$g'=g\left(1-\frac{R}{2R}\right)=\frac{g}{2}$$

Thus apparent weight of body at $\frac{R}{2}$ depth is given as

$$W'=mg'=\frac{mg}{2}=200\text{ N}$$ **Ans. (D)**

Sol. 87 Electric field of solid sphere (uniformly charged) is given as

$$E(r)=\begin{cases}\dfrac{Q}{4\pi\epsilon_0 r^2} & r\geq R\text{; for outside solid sphere}\\ \dfrac{Qr}{4\pi\epsilon_0 R^3} & r\leq R\text{; for inside solid sphere}\end{cases}$$

Graphically it can be plotted as shown in figure below

$$E(r) \propto r \text{ for } r \leq R$$

$$E(r) \propto \frac{1}{r^2} \text{ for } r \geq R$$

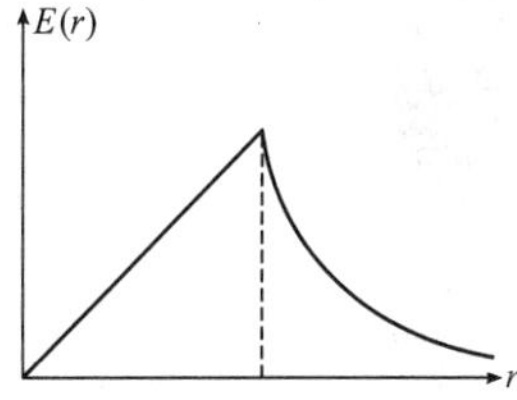

Ans. (A)

Sol. 88 For points outside the surface of earth, gravitational acceleration at a distance r is given as

$$g = \frac{GM}{r^2}$$

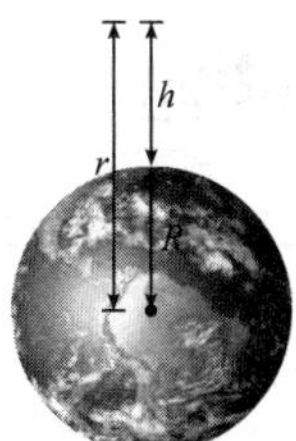

Acceleration due to gravity at h height above surface of earth is given as

$$g(h) = \frac{GM}{(R+h)^2}$$

$$\Rightarrow \quad g(h) = \frac{GM}{R^2\left(1+\frac{h}{R}\right)^2}$$

$$\Rightarrow \quad g(h) = \frac{GM}{R^2}\left(1+\frac{h}{R}\right)^{-2}$$

If $h << R$, as height is very less compared to radius of earth, using binomial expansion and approximation, we have

$$\left(1+\frac{h}{R}\right)^{-2} \approx 1 - \frac{2h}{R}$$

$$\Rightarrow \quad g(h) = \frac{GM}{R^2}\left(1-\frac{2h}{R}\right)$$

$$\Rightarrow \quad g(h) = g_{\text{surface}}\left(1-\frac{2h}{R}\right)$$

Ans. (A)

Sol. 89 Mass of the object is given as

$$\Rightarrow \quad M = \frac{W}{g} = \frac{200}{10} = 20 \text{ kg}$$

Acceleration due to gravity at a depth d is calculated as

$$g' = g\left(1 - \frac{d}{R}\right)$$

At $d = \frac{R}{2}$, we use

$$\Rightarrow \quad g' = g\left(1 - \frac{R/2}{2}\right) = \frac{g}{2} = 5 \text{ m/s}^2$$

Weight of the body at the given depth is calculated as

$$W = mg'$$

$$\Rightarrow \quad W = 20 \times 5 = 100 \text{ N}$$

Ans. (D)

Sol. 90 From conservation of angular momentum, we have

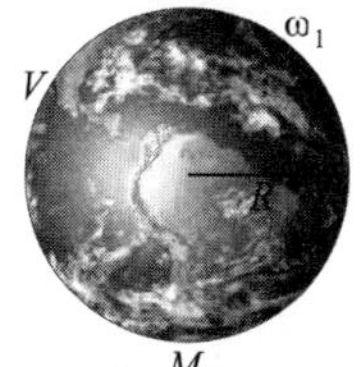

$$I_1\omega_1 = I_2\omega_2 \quad \text{...(1)}$$

When volume shrinks to $\frac{1}{64}$ of original volume, we have

$$\frac{1}{64}\left(\frac{4}{3}\pi R^3\right) = \left(\frac{4}{3}\pi r^3\right)$$

$$\Rightarrow \quad r = \left(\frac{R}{4}\right)$$

From equation-(1), we have

$$\frac{2}{5}MR^2\omega_1 = \frac{2}{5}M\left(\frac{R}{4}\right)^2\omega_2$$

$$\Rightarrow \quad MR^2\omega_1 = \frac{MR^2}{16}\omega_2$$

$$\Rightarrow \quad \frac{\omega_1}{\omega_2} = \frac{1}{16}$$

$$\Rightarrow \quad \frac{T_2}{T_2} = \frac{1}{16}$$

$$\Rightarrow \quad \frac{T_1}{T_2} = \frac{16}{1} = \frac{t_1}{x}$$

$$\Rightarrow \quad t_1 = 24 \text{ s}$$

$$\Rightarrow \quad \frac{1}{16} = \frac{24}{t_2}$$

$$\Rightarrow \quad x = 16 \text{ s}$$

Ans. (16)

Sol. 91 Statement-I is true due to centrifugal force. Statement-**II** is incorrect.
At pole $g = g_s$ as no effect of rotation of earth.
At equator value of g is given as

$$g = g_s - r\omega^2\cos^2\lambda = g_s - r\omega^2$$

Thus $\cos^2\lambda_{maximum}$ at $\lambda = 0°$ i.e. at equator so effect is maximum at equator. **Ans. (B)**

Sol. 92 Acceleration due to gravity on the surface of earth is given as

$$g = \frac{GM}{R^2} = \frac{G}{R^2} \times \rho \times \frac{4\pi}{3}R^3 = \left(\frac{4\pi}{3}G\right)\rho R$$

Thus ratio of acceleration due to gravity on the surface of planets whose densities are ρ and $\frac{\rho}{3}$

$$\Rightarrow \quad \frac{g_A}{g_B} = \frac{R \times \rho}{4R \times \frac{\rho}{3}} = \frac{3}{4}$$ **Ans. (C)**

Sol. 93 Acceleration due to gravity on the surface of planets

$$g = \frac{GM}{R^2} = \frac{4}{3}\pi G\rho R$$

Ratio of acceleration due to gravity at the surface of B to A is

$$\Rightarrow \quad \frac{g_2}{g_1} = \frac{\rho_2}{\rho_1} \times \frac{R_2}{R_1} = \frac{1}{2} \times 1.5 = \frac{3}{4}$$ **Ans. (C)**

Sol. 94 Figure below shows the situation described in question.

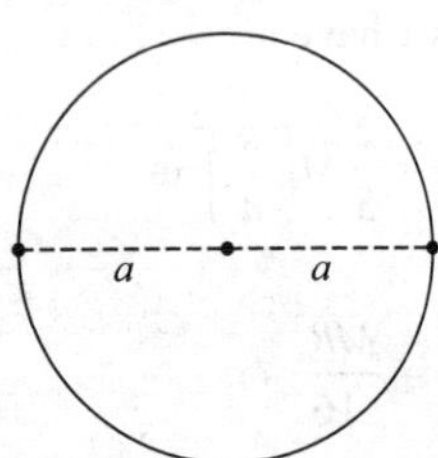

Force on circular path under mutual gravitation of masses will provide the centripetal acceleration

$$F = m\omega^2 r$$

$$\Rightarrow \quad \frac{Gmm}{(2a^2)} = m\omega^2 a$$

So angular speed of particles is given as

$$\Rightarrow \quad \omega = \sqrt{\frac{Gm}{4a^3}}$$ **Ans. (C)**

Sol. 95 By conservation of mechanical energy from the initial point to the point on surface of earth where body strikes is written as

$$U_i + K_i = U_f + K_i$$

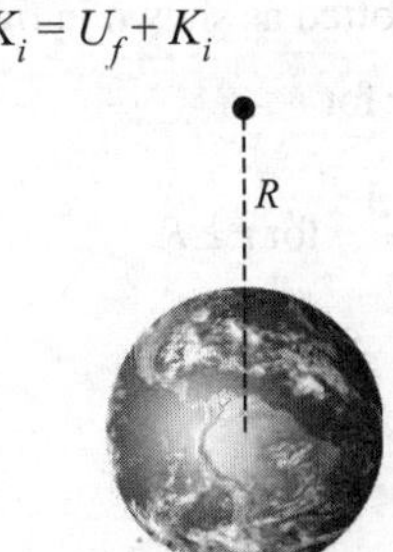

$$\Rightarrow \quad -\frac{GMm}{2R} + 0 = -\frac{GMm}{R} + \frac{1}{2}mv^2$$

$$\Rightarrow \quad \frac{GMm}{2R} = \frac{1}{2}mv^2$$

Velocity of body when it strikes the surface is calculated as

$$\Rightarrow \quad v = \sqrt{\frac{GM}{R}} = \sqrt{gR}$$ **Ans. (A)**

7.2 Satellite Motion

Sol. 1

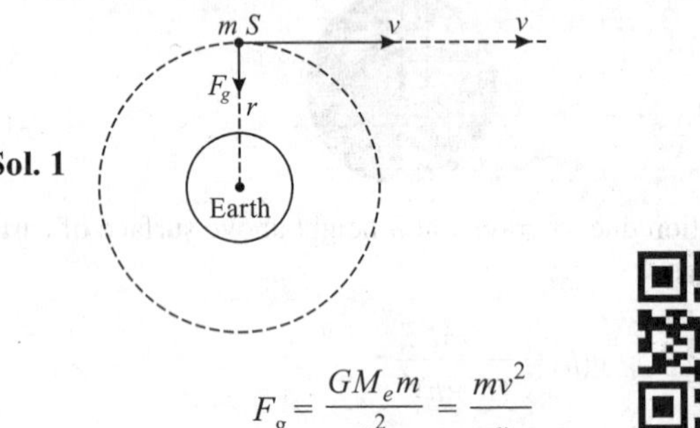

$$F_g = \frac{GM_e m}{r^2} = \frac{mv^2}{r}$$

If gravitational force between earth and satellite becomes zero than there will be no centripetal force. So no circular motion will be continued and particle will move with $\vec{v}$ due to its inertia tangentially. **Ans. (B)**

Sol. 2

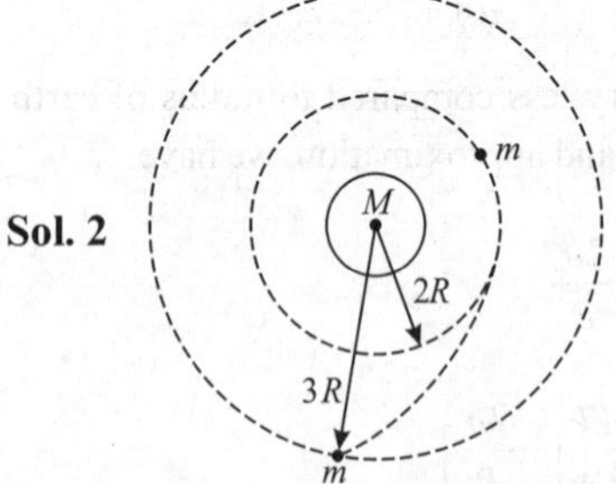

For a satellite of mass m, total energy in an orbit of radius r is given as

$$E = -\frac{GMm}{2r}$$

Initial energy of satellite ($r = 2R$) $E_i = -\frac{GMm}{2(2R)} = -\frac{GMm}{4R}$

Final energy of satellite ($r = 3R$) $E_f = -\frac{GMm}{2(3r)} = -\frac{GMm}{6R}$

Energy required $\Delta E = E_f - E_i = \left(-\dfrac{GMm}{6R}\right) - \left(-\dfrac{GMm}{4R}\right)$

$\Rightarrow \qquad \Delta E = \dfrac{GMm}{2R}\left(\dfrac{1}{2} - \dfrac{1}{3}\right)$

$\Rightarrow \qquad \Delta E = \dfrac{GMm}{12R}$ **Ans. (A)**

Sol. 3 v_{es} is such that its KE is numerically equal to the magnitude of gravitational PE so that $TE = 0$

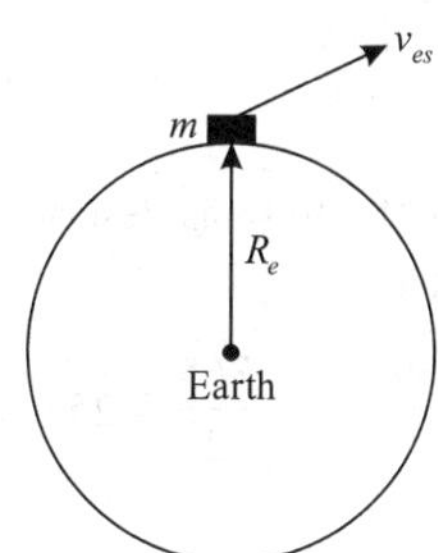

$$\frac{1}{2}mv_{es}^2 + \left(-\frac{GM_e m}{R_e}\right) = 0$$

$\Rightarrow \qquad v_{es} = \sqrt{\dfrac{2GM_e}{R_e}} = \sqrt{2gR_e}$

This is independent of m **Ans. (A)**

Sol. 4 By Kepler's III law we have

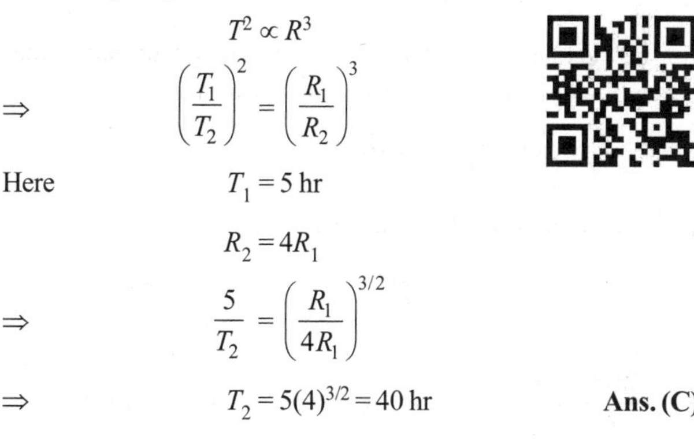

$$T^2 \propto R^3$$

$\Rightarrow \qquad \left(\dfrac{T_1}{T_2}\right)^2 = \left(\dfrac{R_1}{R_2}\right)^3$

Here $\qquad T_1 = 5$ hr

$\qquad R_2 = 4R_1$

$\Rightarrow \qquad \dfrac{5}{T_2} = \left(\dfrac{R_1}{4R_1}\right)^{3/2}$

$\Rightarrow \qquad T_2 = 5(4)^{3/2} = 40$ hr **Ans. (C)**

Sol. 5 Escape velocity is required to overcome the $-ve$ PE in body due to Earth's gravity

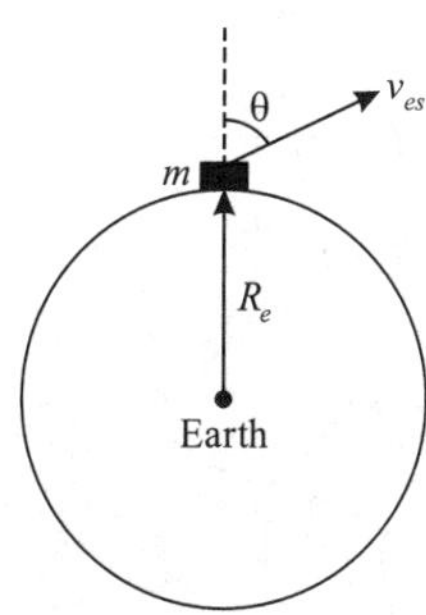

For escape velocity

$$\frac{1}{2}mv_{es}^2 + \left(-\frac{GM_e m}{R_e}\right) = 0$$

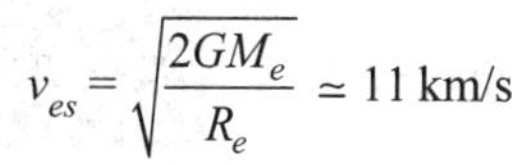

$$v_{es} = \sqrt{\frac{2GM_e}{R_e}} \simeq 11 \text{ km/s}$$

This is independent of θ **Ans. (C)**

Sol. 6

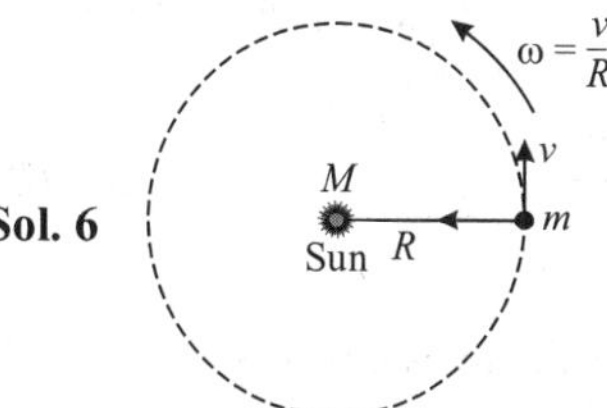

Here gravitational force provides the centripetal force to planet for its circular motion, so we have

$$F_g = \frac{GMm}{R^n} = m\omega^2 R$$

$\Rightarrow \qquad \omega = \sqrt{\dfrac{GM}{R^{n+1}}}$

time period of orbital motion is

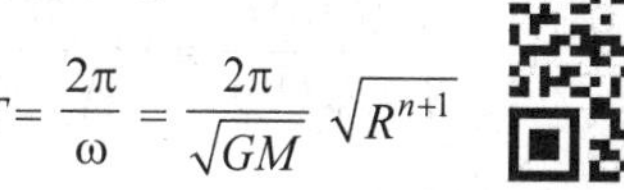

$$T = \frac{2\pi}{\omega} = \frac{2\pi}{\sqrt{GM}}\sqrt{R^{n+1}}$$

$\Rightarrow \qquad T \propto R^{\left(\frac{n+1}{2}\right)}$ **Ans. (A)**

Sol. 7

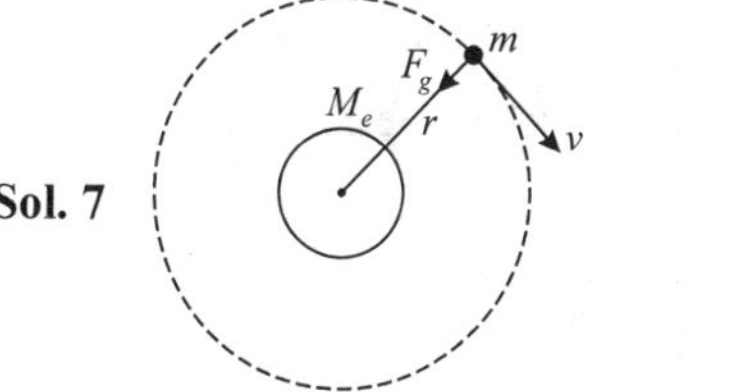

Orbital speed of a satellite in an orbit radius r is given as

$$v = \sqrt{\frac{GM_e}{r}}$$

time period of satellite is

$$T = \frac{2\pi r}{v} = \frac{2\pi}{\sqrt{GM_e}} \cdot r^{3/2}$$

T is independent of mass of satellite **Ans. (A)**

Sol. 8

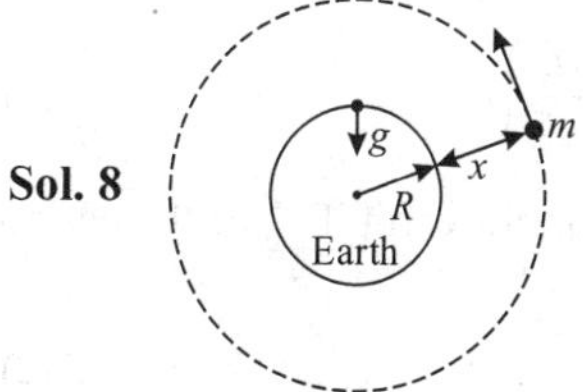

If g is acceleration due to gravity on earth surface then we have

$$g = \frac{GM}{R^2}$$

orbital speed of satellite in the orbit of radius $r = R + x$ is given as

$$v = \sqrt{\frac{GM}{r}} = \sqrt{\frac{GM}{R+x}} = \sqrt{\frac{GM}{R^2} \cdot \frac{R^2}{R+x}}$$

$$= \sqrt{\frac{gR^2}{R+x}}$$ **Ans. (D)**

Sol. 9 From a planet escape velocity is given as

$$v_e = \sqrt{\frac{2GM}{R}} \quad \text{for earth } v_{e_1} = \sqrt{\frac{2GM_e}{R_e}} = 11 \text{ km/s}$$

$$\Rightarrow \quad v_e \propto \sqrt{\frac{M}{R}}$$

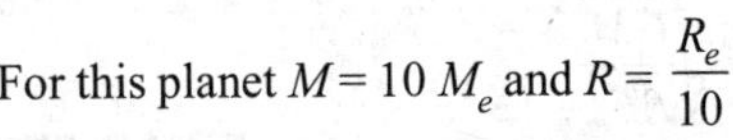

For this planet $M = 10\, M_e$ and $R = \frac{R_e}{10}$

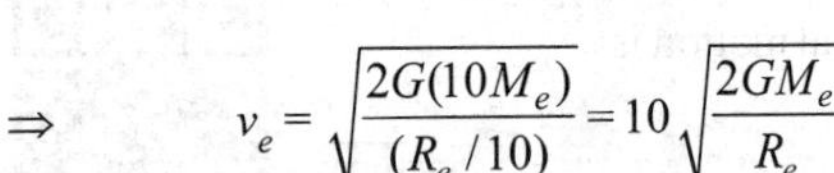

$$\Rightarrow \quad v_e = \sqrt{\frac{2G(10M_e)}{(R_e/10)}} = 10\sqrt{\frac{2GM_e}{R_e}}$$

$$\Rightarrow \quad v_e = 110 \text{ km/s}$$ **Ans. (C)**

Sol. 10

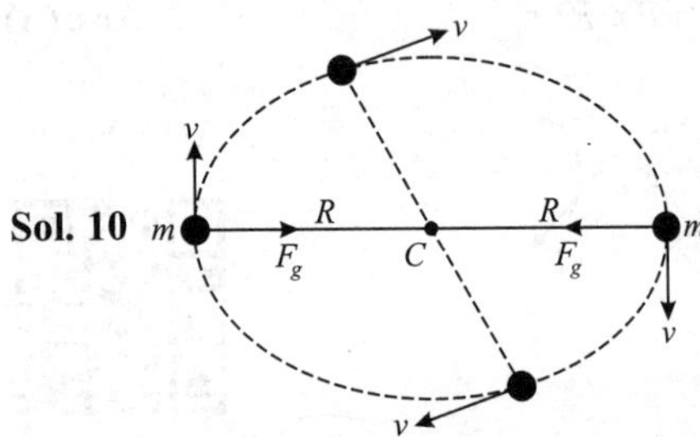

In the given double star system the gravitational force will provide the centripetal force for this circular motion, so we use

$$\frac{Gm^2}{(2R)^2} = \frac{mv^2}{R}$$

$$\Rightarrow \quad v = \sqrt{\frac{Gm}{4R}}$$ **Ans. (B)**

Sol. 11 For EOM elliptical orbit if time period of motion in this orbit is T then we use

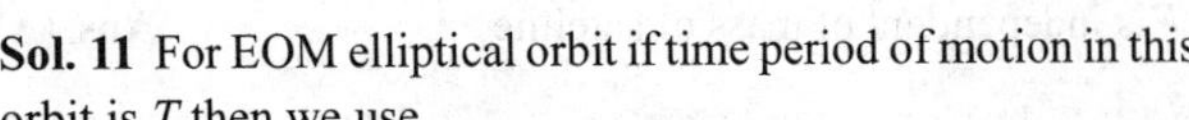

$$T^2 \propto \left(\frac{a_m + a_e}{2}\right)^3 \quad \ldots(1)$$

Semi major axis of elliptical orbit for earth if T_e be the period around sun then we use

$$T_e^2 \propto a_e^3 \quad \ldots(2)$$

Here (1)/(2) gives

$$\frac{T^2}{T_e^2} = \left[\frac{(a_m + a_e)/2}{a_e}\right]^3$$

$$T = \left[\frac{(a_m + a_e)/2}{a_e}\right]^{3/2} T_e$$

$$T = \left[\frac{(1.5 + 2.28)/2}{1.5}\right]^{3/2} \times 365 \text{ days}$$

$$T = \left(\frac{1.89}{1.5}\right)^{3/2} \times 365 \text{ days}$$

Time of motion in half ellipse as described (path EOM) is given as

$$t = \frac{T}{2} = \left[\frac{1.89}{1.5}\right]^{3/2} \times \frac{365}{2} \cong 258 \text{ days}$$ **Ans. (C)**

Sol. 12

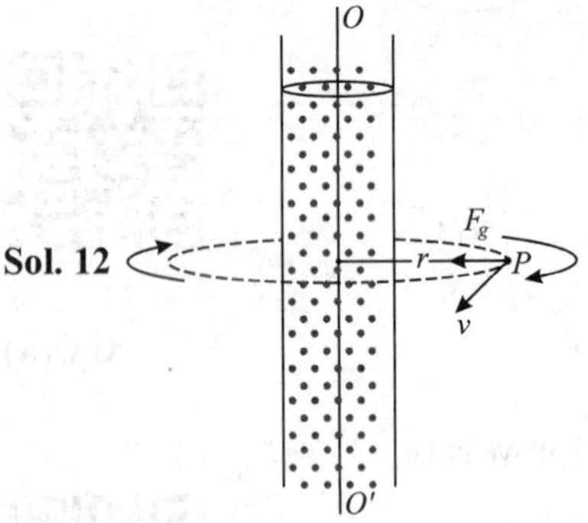

If λ be the linear mass density along the length of galaxy then gravitational field at the location of star is given as

$$g_P = \frac{2G\lambda}{r}$$

For circular motion of star, we use

$$mg_P = \frac{mv^2}{r}$$

$$\Rightarrow \quad \frac{2G\lambda}{r} = \frac{v^2}{r} \quad \Rightarrow \quad v = \sqrt{2G\lambda} \quad \text{orbital speed of star}$$

Period of revolution of star in circular orbit is

$$T = \frac{2\pi r}{v}$$

$$\Rightarrow \quad T = \frac{2\pi r}{\sqrt{2G\lambda}}$$

$$\Rightarrow \quad T \propto r$$ **Ans. (A)**

Sol. 13

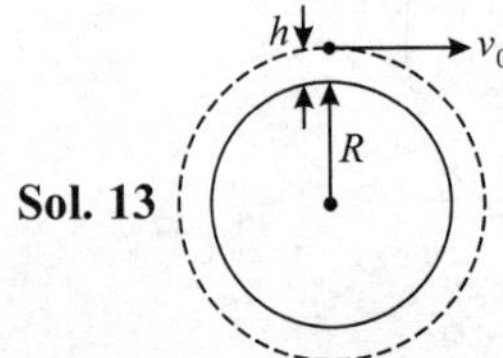

Orbital speed velocity close to earth surface is given as

$$v_0 = \sqrt{gR}$$

Escape velocity very close to earth surface is given as

$$v_e = \sqrt{2gR}$$

Additional velocity required for satellite to escape is

$$\Delta v = v_e - v_0 = \sqrt{gR}\ (\sqrt{2} - 1)$$ **Ans. (D)**

Sol. 14

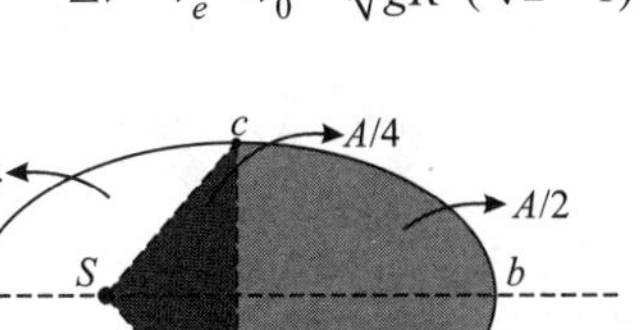

Total area = A

According Kepler's law of areas, rate of sweeping out area by position vector of planet w.r. to sun remain constant.

As t ∝ area swept, we have

$$\Rightarrow \quad t_1 \propto \frac{3A}{4} \quad \ldots(1)$$

and $$t_2 \propto \frac{A}{4} \quad \ldots(2)$$

$$(1)/(2) \Rightarrow \quad \frac{t_1}{t_2} = 3 \Rightarrow t_1 = 3t_2$$ **Ans. (C)**

Sol. 15

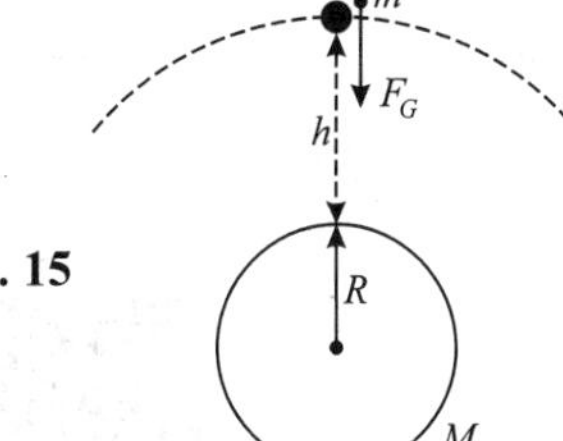

Along with satellite we consider the astronaut is also revolving around earth in orbit of radius $(R + h)$ so gravitational force on astronaut is given as

$$F_G = \frac{GMm}{(R+h)^2}$$

Here we neglect the gravitational force between astronaut and satellite. **Ans. (C)**

Sol. 16 Initial State

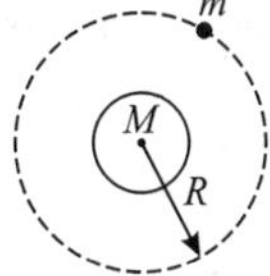

Total energy of body is

$$E_i = -\sqrt{\frac{GMm}{2R}}$$

Final State

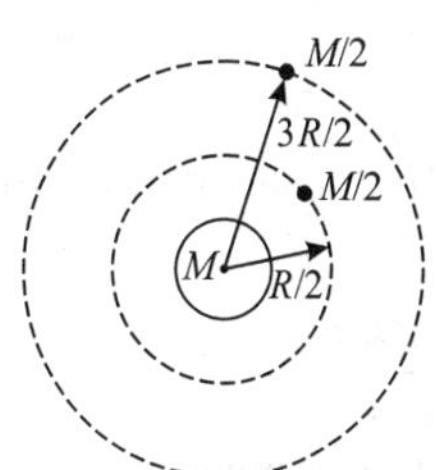

$$E_f = -\frac{GM\left(\frac{m}{2}\right)}{2\left(\frac{R}{2}\right)} - \frac{GM\left(\frac{m}{2}\right)}{2\left(3\frac{R}{2}\right)}$$

$$\Rightarrow \quad E_f = -\frac{2}{3}\frac{GMm}{R}$$

Difference in final and initial energies is

$$\Delta E = E_f - E_i$$

$$\Rightarrow \quad \Delta E = -\frac{GMm}{R}\left(\frac{2}{3} - \frac{1}{2}\right)$$

$$\Rightarrow \quad \Delta E = -\frac{GMm}{6R}$$ **Ans. (C)**

Sol. 17 By Kepler's II law areal velocity of a planet is constant even in elliptical path and given as

$$\frac{dA}{dt} = \frac{L}{2m}$$ **Ans. (C)**

Sol. 18 $U_{surface} + E_1 = U_h$

KE of satellite is zero at earth surface & at height *h*, we use

$$-\frac{GM_e m}{R_e} + E_1 = -\frac{GM_e m}{(R_e + h)}$$

$$E_1 = GM_e m\left(\frac{1}{R_e} - \frac{1}{R_e + h}\right)$$

$$E_1 = \frac{GM_e m}{(R_e + h)} \times \frac{h}{R_e}$$

Gravitational attraction at height *h* will be

$$F_G = mg = \frac{mv^2}{(R_e + h)}$$

$$\Rightarrow \quad \frac{GM_e m}{(R_e + h)^2} = \frac{mv^2}{(R_e + h)}$$

$$\Rightarrow \quad mv^2 = \frac{GM_e m}{(R_e + h)}$$

$$E_2 = \frac{1}{2}mv^2 = \frac{GM_e m}{2(R_e + h)}$$

Given that $E_1 = E_2$

$$\Rightarrow \quad \frac{h}{R_e} = \frac{1}{2}$$

$$\Rightarrow \quad h = \frac{R_e}{2} = 3200 \text{ km} \qquad \textbf{Ans. (C)}$$

Sol. 19 At height r from center of earth orbital velocity is given as

$$v = \sqrt{\frac{GM}{r}}$$

By energy conservation, we have

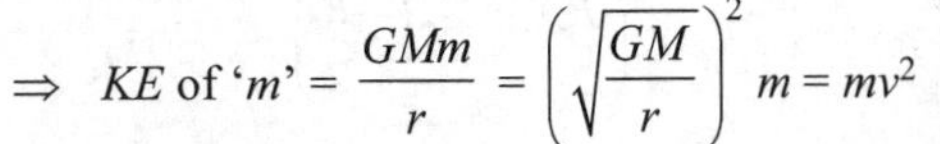

$$KE \text{ of 'm'} + \left(-\frac{GMm}{r}\right) = 0 + 0$$

$$\Rightarrow \quad KE \text{ of 'm'} = \frac{GMm}{r} = \left(\sqrt{\frac{GM}{r}}\right)^2 m = mv^2 \qquad \textbf{Ans. (B)}$$

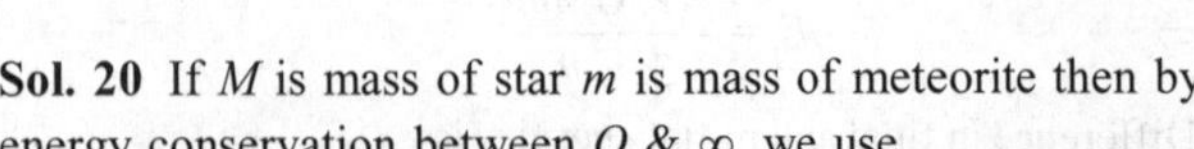

Sol. 20 If M is mass of star m is mass of meteorite then by energy conservation between O & ∞, we use

$$-\frac{GMm}{r} + \frac{-GMm}{r} + \frac{1}{2}mv^2 = 0 + 0$$

$$\Rightarrow \quad v = \sqrt{\frac{4GM}{r}} = 2.8 \times 10^5 \text{ m/s} \qquad \textbf{Ans. (D)}$$

Sol. 21 Orbital speed of satellite is

$$v_0 = \sqrt{g(R+h)} \approx \sqrt{gR}$$

Escape velocity is

$$v_e = \sqrt{2g(R+h)} \approx \sqrt{2gR}$$

Increment in speed required is

$$\Delta v = v_e - v_0 = (\sqrt{2} - 1)\sqrt{gR} \qquad \textbf{Ans. (A)}$$

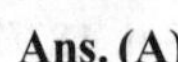

Sol. 22 As after impact the resulting velocity of the combined body is at some angle to the radial direction so resulting motion will be an elliptical orbit. **Ans. (C)**

Sol. 23 Orbital speed of satellite in orbit of radius P_c is given

$$v = \sqrt{\frac{GM}{r}}$$

Kinetic energies of the two satellites are

$$T_A = \frac{1}{2}m_A v_A^2$$

and $$T_B = \frac{1}{2}m_B v_B^2$$

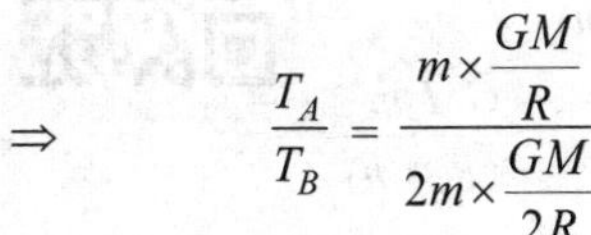

$$\Rightarrow \quad \frac{T_A}{T_B} = \frac{m \times \frac{GM}{R}}{2m \times \frac{GM}{2R}}$$

$$\Rightarrow \quad \frac{T_A}{T_B} = 1 \qquad \textbf{Ans (C)}$$

Sol. 24 Mass contained in the elemental shell is

$$dm = \rho \cdot 4\pi x^2 dx$$

$$M = \int dm = \int_0^R \frac{k}{x^2} \cdot 4\pi x^2 \, dx = 4\pi kR$$

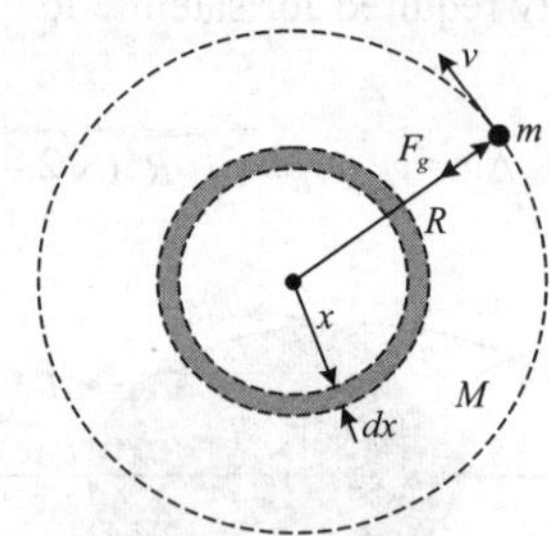

Gravitational force on mass in orbit is

$$F_g = \frac{GMm}{R^2} = \frac{G(4\pi kR)m}{R^2}$$

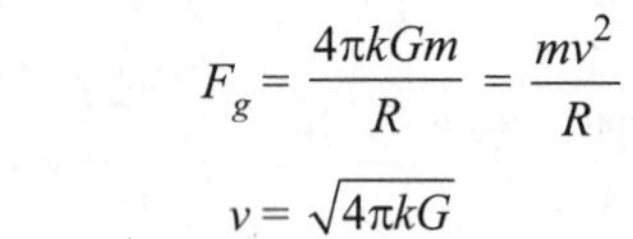

$$\Rightarrow \quad F_g = \frac{4\pi kGm}{R}$$

For circular motion, we use

$$F_g = \frac{4\pi kGm}{R} = \frac{mv^2}{R}$$

$$\Rightarrow \quad v = \sqrt{4\pi kG}$$

Time period $$T = \frac{2\pi R}{\sqrt{4\pi kG}}$$

$$\Rightarrow \quad \frac{T}{R} = \text{constant} \qquad \textbf{Ans. (D)}$$

Sol. 25 Orbital speed of spaceship is

$$v = \sqrt{\frac{Gm}{r}}$$

Where $r = R_P + h = 2.02 \times 10^6$ m

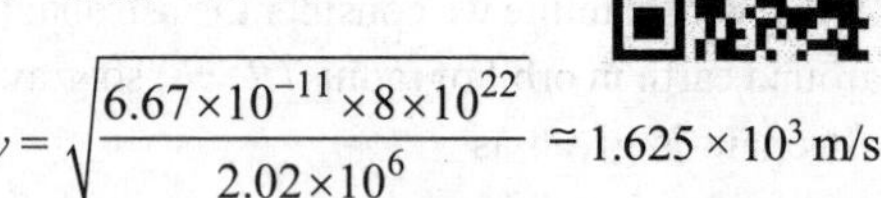

$$v = \sqrt{\frac{6.67 \times 10^{-11} \times 8 \times 10^{22}}{2.02 \times 10^6}} \simeq 1.625 \times 10^3 \text{ m/s}$$

Time period of revolutions of spaceship is

$$T = \frac{2\pi r}{v}$$

Number of revolutions in 24 hrs are

$$n = \frac{24 \times 3600}{T} = \frac{24 \times 3600 \times 1.625 \times 10^3}{2 \times 3.14 \times 2.02 \times 10^6}$$

$$\simeq 11 \qquad \textbf{Ans. (B)}$$

Sol. 26 Gravity on surface of a planet is given as

$$g_P = \frac{GM_P}{R_P^2}$$

$$\frac{g_e}{g_P} = \frac{9}{4}\frac{M_e/R_e^2}{M_P/R_P^2} \quad \left(\text{here } M_P = \frac{1}{9}M_e\right)$$

$$\Rightarrow \quad \frac{9}{4} = \frac{9R_P^2}{R_e^2} \Rightarrow R_P = \frac{R_e}{2} = \frac{R}{2}$$ **Ans. (B)**

Sol. 27 Using energy conservation, we have

$$K_i + U_i = K_f + U_f$$

$$\frac{1}{2}mu^2 + \left(-\frac{GMm}{R}\right) = \frac{1}{2}mv^2 - \frac{GMm}{2R}$$

$$v = \sqrt{u^2 - \frac{GM}{R}} \quad \ldots(1)$$

At height R rocket must carry all the momentum radial direction and along tangent to orbit it should carry the momentum such that satellite has left only with orbital speed $\sqrt{\frac{GM}{2R}}$ for circular motion. Thus by momentum conservation, we use

Along orbital direction (along tangent)

$$\frac{m}{10}v_{R0} = \frac{9m}{10}\sqrt{\frac{GM}{2R}}$$

$$v_{R0} = 9\sqrt{\frac{GM}{2R}} \quad \ldots(2)$$

Along radial direction

$$\frac{m}{10}v_{RR} = mv$$

$$\Rightarrow \quad \frac{m}{10}v_{RR} = m\sqrt{u^2 - \frac{GM}{R}} \quad \ldots(3)$$

$$\Rightarrow \quad v_{RR} = 10\sqrt{u^2 - \frac{GM}{R}}$$

Now kinetic energy of rocket is given as

$$K = \frac{1}{2}m(v_{R0}^2 + v_{RR}^2)$$

$$\Rightarrow \quad K = \frac{m}{20}\left(81\frac{GM}{2R} + 100u^2 - 100\frac{GM}{R}\right)$$

$$\Rightarrow \quad K = \frac{m}{20}\left(100u^2 - \frac{119GM}{2R}\right)$$

$$\Rightarrow \quad K = 5m\left(u^2 - \frac{119GM}{200R}\right)$$ **Ans. (B)**

Sol. 28 Using conservation of energy, we have

$$P_1 + K_1 = P_2 + K_2$$

$$\Rightarrow \quad -\frac{GM_em}{10R} + \frac{1}{2}mv_0^2 = -\frac{GM_em}{R} + \frac{1}{2}mv^2$$

$$\Rightarrow \quad +\frac{9}{10}\times\frac{GM_em}{R} + \frac{1}{2}mv_0^2 = \frac{1}{2}mv^2$$

$$\Rightarrow \quad \frac{9}{10}\times\frac{1}{2}M \times v_e^2 + \frac{1}{2}mv_0^2 = \frac{1}{2}mv^2$$

$$\Rightarrow \quad v^2 = \frac{9}{10}v_e^2 + v_0^2$$

$$\Rightarrow \quad v^2 = \frac{9}{10}\times(11.2)^2 + (12)^2$$

$$\Rightarrow \quad v^2 = 112.896 + 144$$

$$\Rightarrow \quad v = 16.027 \text{ km/s}$$

$$\Rightarrow \quad v = 16 \text{ km/s}$$ **Ans. (16)**

Sol. 29 The orbital speed of body is given as

$$v_0 = \sqrt{\frac{GM}{R}}$$

After collision, if combined mass moves with speed v_1, we use by conservation of momentum

$$mv_0 + \frac{m}{2}\frac{v_0}{2} = \left(\frac{3m}{2}\right)v_1$$

$$\Rightarrow \quad v_1 = \frac{5v_0}{6}$$

As this speed after collision is less than escape speed $\sqrt{2}v_0$, so the combined body cannot from escape gravitational field. So its motion will be elliptical around the planet. **Ans. (A)**

Sol. 30 Escape velocity from a planet is given as

$$v_e = \sqrt{\frac{2GM}{R}}$$

For planets A and B, we use

$$v_A = \sqrt{\frac{2GM}{R}}$$

and $$v_B = \sqrt{\frac{2G(M/2)}{R/2}} = \sqrt{\frac{2GM}{R}}$$

$$\Rightarrow \quad \frac{v_A}{v_B} = 1 = \frac{n}{4} \Rightarrow n = 4$$ **Ans. (A)**

Sol. 31 Mass of galaxy inside radius R is given as

$$M = \int_0^R \rho_{(r)} \cdot 4\pi r^2 dr$$

$$\Rightarrow \quad M = \int_0^R \frac{K}{r}\cdot 4\pi r^2 dr$$

$$\Rightarrow \quad M = 4\pi K\left[\frac{R^2}{2}\right] = 2\pi K R^2$$

For orbital motion of star of mass m, we use

$$\frac{GMm}{R^2} = \frac{mv^2}{R}$$

$$\Rightarrow \qquad v = \sqrt{\frac{GM}{R}} = \sqrt{2\pi KGR}$$

Time period of revolution is given as

$$T = \frac{2\pi R}{v} = \frac{2\pi R}{\sqrt{2\pi KGR}}$$

$$\Rightarrow \qquad T = \sqrt{\frac{2\pi R}{KG}}$$

$$\Rightarrow \qquad T \propto \sqrt{R}$$

$$\Rightarrow \qquad T^2 \propto R$$

Ans. (C)

Sol. 32 Figure below shows the trajectory of satellite after firing rockets

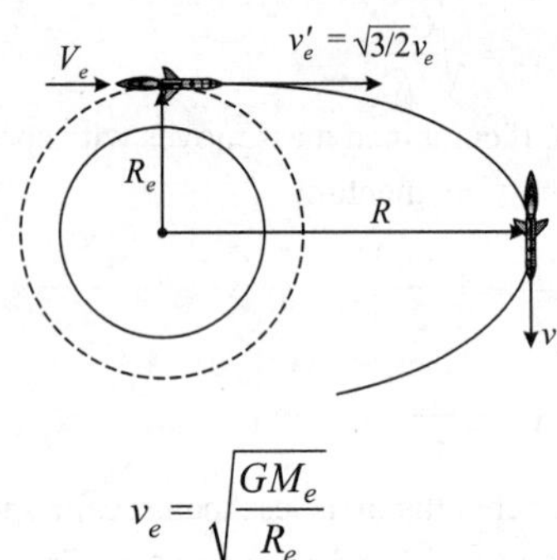

$$v_e = \sqrt{\frac{GM_e}{R_e}}$$

By conservation of angular momentum, we have

$$v_e' \cdot R_e = v_1 R$$

$$\Rightarrow \qquad \sqrt{\frac{3}{2}\frac{GM_e}{R_e}} \cdot R_e = v_1 R$$

$$\Rightarrow \qquad v_1 = \sqrt{\frac{3}{2}\frac{GM_e m}{R_e}} \cdot \frac{R_e}{R}$$

By conservation of energy we use

$$-\frac{GM_e m}{R_e} + \frac{3}{4}\frac{GM_e m}{R_e} = -\frac{GM_e m}{R} + \frac{1}{2}mv_1^2$$

$$\Rightarrow \qquad -\frac{GM_e m}{4R_e} = -\frac{GM_e m}{R} + \frac{1}{2}m\left(\frac{3}{2}\frac{GM_e m}{R_e}\right) \cdot \frac{R_e^2}{R^2}$$

$$\Rightarrow \qquad \frac{3}{4}\frac{R_e}{R^2} - \frac{1}{R} + \frac{1}{4R_e} = 0$$

$$\Rightarrow \qquad \frac{1}{R} = \frac{1 \pm \sqrt{1 - 4\left(\frac{3}{4}\right)\left(\frac{1}{4}\right)}}{2 \times \frac{3}{4}R_e}$$

$$\Rightarrow \qquad \frac{1}{R} = \frac{1 \pm \frac{1}{2}}{\frac{3}{2}R_e}$$

$$\Rightarrow \qquad R = 3R_e$$

Here R_e is not possible as $R > R_e$ **Ans. (B)**

Sol. 33 Orbital speed very close to planet surface is given as

$$v_0 = \sqrt{\frac{GM}{R}}$$

escape velocity from the planet surface is given as

$$v_e = \sqrt{\frac{2GM}{R}}$$

$$\Rightarrow \qquad \frac{v_0}{v_e} = \frac{1}{\sqrt{2}}$$

Ans. (C)

Sol. 34 By angular momentum conservation, we use

$$v_{max} r_{min} = v_{min} r_{max}$$

$$\Rightarrow \qquad \frac{r_{min}}{r_{max}} = \frac{v_{min}}{v_{max}} = \frac{1}{6}$$

Ans. (A)

Sol. 35

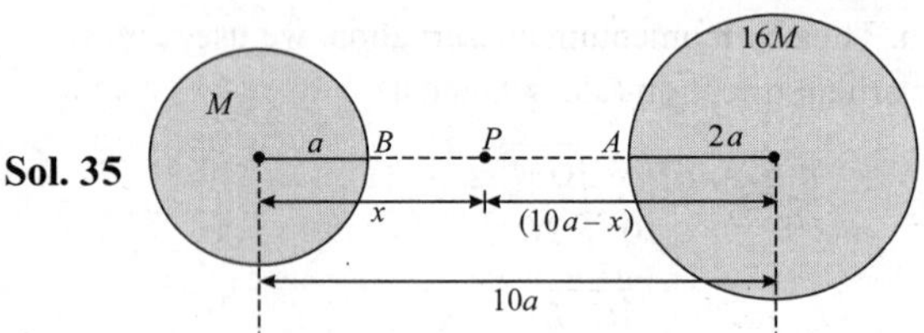

Null point P can be obtained as

$$\frac{GM}{x^2} = \frac{G(16M)}{(10a - x)^2}$$

$$\Rightarrow \qquad \frac{1}{x} = \frac{4}{(10a - x)}$$

$$\Rightarrow \qquad 4x = 10a - x$$

$$\Rightarrow \qquad x = 2a \qquad \ldots(1)$$

For body fired from point A at speed v, it should cross null point P in order to reach point B so by energy conservation, we use

$$-\frac{GMm}{8a} - \frac{G(16M)m}{2a} + \frac{1}{2}mv^2 = -\frac{GMm}{2a} - \frac{G(16M)m}{8a}$$

$$\Rightarrow \qquad v^2 = 2GM\left[\frac{1}{8a} + \frac{16}{2a} - \frac{1}{2a} - \frac{16}{8a}\right]$$

$$\Rightarrow \qquad v^2 = 2GM\left[\frac{1 + 64 - 4 - 16}{8a}\right]$$

$$\Rightarrow \qquad v^2 = 2GM\left[\frac{45}{8a}\right]$$

$$\Rightarrow \qquad v = \sqrt{\frac{90GM}{8a}}$$

$$\Rightarrow \qquad v = \frac{3}{2}\sqrt{\frac{5GM}{a}} \qquad \textbf{Ans. (B)}$$

Sol. 36 The situation described in question is shown in figure below

d
c.o.m
m 2m
2d/3 d/3

The force acting between the masses will provide the necessary centripetal force for circular motion which is given as

$$F = \frac{G(2m)m}{d^2} = (2m)\omega^2 (d/3)$$

$$\Rightarrow \qquad \frac{Gm}{d^2} = \omega^2 \frac{d}{3}$$

$$\Rightarrow \qquad \omega = \sqrt{\frac{3Gm}{d^3}}$$

Thus time period of revolution is given as

$$T = \frac{2\pi}{\omega} = 2\pi\sqrt{\frac{d^3}{3Gm}} \qquad \textbf{Ans. (B)}$$

Sol. 37 The ratio of time period of revolution is taken as

$$\frac{T_1}{T_2} = \frac{1}{8}$$

$$\Rightarrow \qquad \frac{2\pi/\omega_1}{2\pi/\omega_2} = \frac{1}{8}$$

$$\Rightarrow \qquad \frac{\omega_1}{\omega_2} = \frac{8}{1} \qquad \textbf{Ans. (D)}$$

Sol. 38 Figure below shows the situation described in the question. The net gravitational force on one particle provides the necessary centripetal force for circular motion

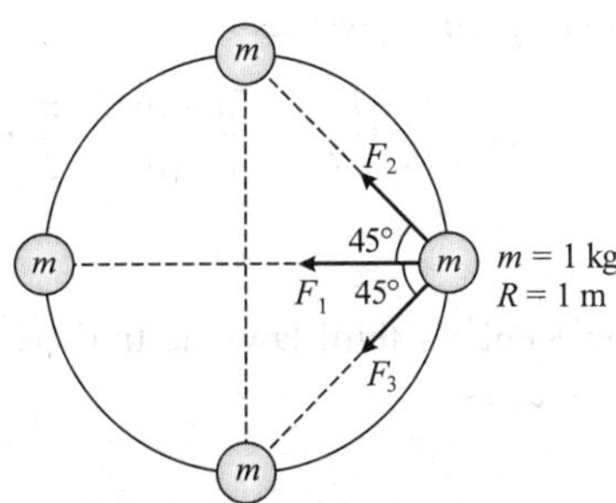

The forces shown in figure are calculated as

$$F_1 = \frac{Gmm}{(2R)^2} = \frac{Gm^2}{4R^2}$$

$$F_2 = \frac{Gmm}{(\sqrt{2}R)^2} = \frac{Gm^2}{2R^2}$$

$$F_3 = \frac{Gmm}{(\sqrt{2}R)^2} = \frac{Gm^2}{2R^2}$$

Resultant of these forces on mass can be given as

$$F_{net} = F_1 + F_2 \cos 45° + F_3 \cos 45°$$

$$\Rightarrow \qquad F_{net} = \frac{Gm^2}{4R^2} + \frac{Gm^2}{2R^2}\frac{1}{\sqrt{2}} + \frac{Gm^2}{2R^2}\frac{1}{\sqrt{2}}$$

$$\Rightarrow \qquad F_{net} = \frac{Gm^2}{R^2}\left(\frac{1}{4} + \frac{1}{2\sqrt{2}} + \frac{1}{2\sqrt{2}}\right)$$

$$\Rightarrow \qquad F_{net} = \frac{Gm^2}{R^2}\left(\frac{1}{4} + \frac{1}{\sqrt{2}}\right) = \frac{Gm^2}{4R^2}(1 + 2\sqrt{2})$$

$$\Rightarrow \qquad F_{net} = \frac{Gm^2}{4R^2}(1 + 2\sqrt{2}) = \frac{mv^2}{R}$$

$$\Rightarrow \qquad v = \frac{\sqrt{G(1+2\sqrt{2})}}{2} \qquad \textbf{Ans. (C)}$$

Sol. 39 Escape velocity on the surface of a planet is given as

$$v_e = \sqrt{\frac{2GM}{R}}$$

For the given two planets, as escape velocities are equal, we use

$$\frac{M_1}{R_1} = \frac{M_2}{R_2}$$

$$\Rightarrow \qquad M_1 R_2 = M_2 R_1$$

Hence reason **R** is not correct. **Ans. (C)**

Sol. 40 The time period of revolution of satelllite is given as

$$T = 2\pi\sqrt{\frac{r^3}{GM}}$$

For planets A and B, we use

$$T_A = 2\pi\sqrt{\frac{(6400+600)\times 10^3}{GM}}$$

$$= 2\pi \times 10^9\sqrt{\frac{7^3}{GM}}$$

and $$T_B = 2\pi \times 10^9\sqrt{\frac{8^3}{GM}}$$

$$\Rightarrow \qquad T_B - T_A = \frac{2\pi 10^9}{\sqrt{GM}}[8\sqrt{8} - 7\sqrt{7}]$$

$$= 314 \times 4.107 = 1.289 \times 10^3 \text{ s} \qquad \textbf{Ans. (C)}$$

Sol. 41 By conservation of energy from surface of earth to the point of maximum height is given as

$$\frac{-GMm}{11R} = \frac{-GMm}{R} + \frac{1}{2}mv^2$$

$$\Rightarrow \qquad v = \sqrt{\frac{20GM}{11R}} \qquad \textbf{Ans. (10)}$$

Sol. 42 According to Keplar's second law, areal velocity of the position vector of planet remain constant. **Ans. (B)**

Sol. 43 Situation described in question is shown in figure below

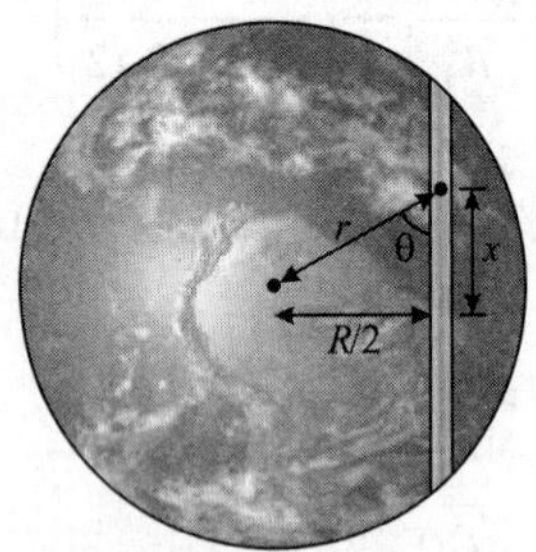

When particle is at a distance x from the center of tunnel, the force acting along the tunnel is given as

$$F = -\left(\frac{GMmr}{R^3}\right)\cos\theta = -\frac{gm}{R}x$$

$$\Rightarrow \qquad a = -\frac{g}{R}x$$

Comparing above acceleration with the acceleration of SHM, it gives

$$\omega^2 = \frac{g}{R}$$

$$\Rightarrow \qquad T = 2\pi\sqrt{\frac{R}{g}} \qquad \textbf{Ans. (A)}$$

Sol. 44 Escape velocity is calculated as

$$v_e = \sqrt{\frac{2Gm}{R}} \qquad \dots(1)$$

If escape velocity is to be increased to 10 times, we use

$$10\, v_e = \sqrt{\frac{2GM}{R'}} \qquad \dots(2)$$

From equations-(1) and (2), we have

$$10 = \sqrt{\frac{R}{R'}}$$

$$\Rightarrow \qquad R' = \frac{R}{100} = \frac{6400}{100} = 64 \text{ km} \qquad \textbf{Ans. (64)}$$

Sol. 45 By Kepler's third law, we have

$$T \propto R^{3/2}$$

$$\Rightarrow \qquad \frac{24}{T} = \left(\frac{12R}{3R}\right)^{3/2}$$

$$\Rightarrow \qquad T = 3 \text{ hr} \qquad \textbf{Ans. (C)}$$

Sol. 46 From Keplar's third law we have

$$T^2 \propto R^3$$

$$\Rightarrow \qquad \left(\frac{T'}{T}\right)^2 = \left(\frac{9R}{R}\right)^3$$

$$\Rightarrow \qquad T'^2 = T^2 \times 9^3$$

$$\Rightarrow \qquad T' = 27\,\text{T} \qquad \textbf{Ans. (B)}$$

Sol. 47 Time period of a satellite in circular orbit around earth is related to its orbit radius as

$$T \propto R^{3/2}$$

$$\Rightarrow \qquad \frac{\Delta T}{T} = \frac{3}{2} \times \frac{\Delta R}{R} = \frac{3}{2} \times 0.02 \times 100 = 3\% \qquad \textbf{Ans. (C)}$$

Sol. 48 Situation described in question is shown in figure below

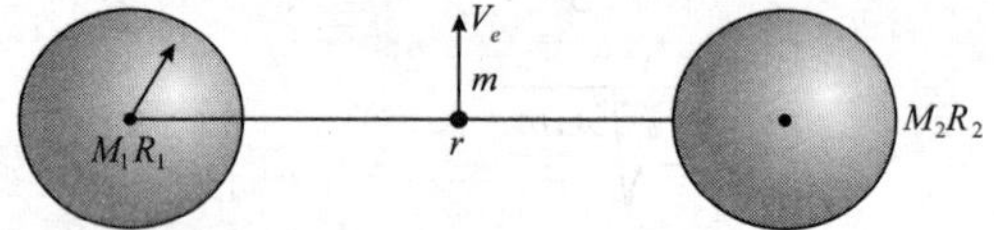

For mass m to escape final interaction energy of this mass to become zero, so we use

$$-\left(\frac{GM}{r/2} + \frac{GM_2}{r/2}\right)m + \frac{1}{2}mv_e^2 = 0$$

$$\Rightarrow \qquad v_e = \sqrt{\frac{4G(M_1 + M_2)}{r}} \qquad \textbf{Ans. (B)}$$

Sol. 49 By Keplar's third law, we use

$$T^2 \propto R^3$$

$$\Rightarrow \qquad \left(\frac{R_2}{R_1}\right)^3 = \left(\frac{T_2}{T_1}\right)^2$$

$$\Rightarrow \qquad \frac{R_2}{R_1} = \left(\frac{1}{8}\right)^{2/3} = 4$$

$$\Rightarrow \qquad R = 8 \times 10^3 \text{ km}$$

$$v_1 = \omega_1 R_1 = 4\pi \times 10^3 \text{ km/h}$$

$$v_2 = \omega_2 R_2 = 2\pi \times 10^3 \text{ km/h}$$

Relative angular speed is given as

$$\omega = \frac{V_1 - V_2}{R_2 - R_1} = \frac{2\pi \times 10^3}{6 \times 10^3} = \frac{\pi}{3} \text{ rad/hour}$$

$$\Rightarrow \qquad x = 3 \qquad \textbf{Ans. (3)}$$

Sol. 50 Using Keplar's third law, the time period of earth revolution is given as

$$T' = T\left(\frac{3R}{R}\right)^{3/2} = 3\sqrt{3}T \qquad \textbf{Ans. (D)}$$

Sol. 51 Orbital speed of satellite in an orbit radius r is given as

$$v = \frac{GM}{r}$$

$$\Rightarrow \quad \frac{v_1}{v_2} = \sqrt{\frac{800}{3200}} = \frac{1}{2}$$ **Ans. (2)**

Sol. 52 Figure below shows the angular view of Sun from earth.

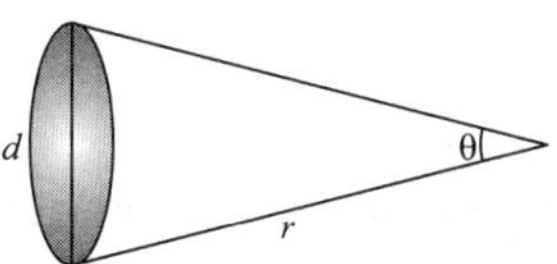

Here the angular diameter is calculated as

$$\theta = \frac{d}{r}$$

$$\Rightarrow \quad \frac{2000}{60\times60}\times\frac{\pi}{180} = \frac{d}{1.5\times10''}$$

$$\Rightarrow \quad d = \frac{2000}{60\times60}\times\frac{\pi}{180}\times1.5\times10''$$

$$\Rightarrow \quad d = \frac{\pi\times1.5}{3\times6\times18}\times10''$$

$$= 1.45\times10^9$$ **Ans. (C)**

Sol. 53 Using Keplar's law for the revolving planets, we have

$$T = \frac{2\pi}{\sqrt{Gm_a}}r^{3/2}$$

$$T^2 \propto r^3$$

$$\Rightarrow \quad \left(\frac{T_A}{T_B}\right)^2 = \left(\frac{r_A}{r_B}\right)^3$$

$$\Rightarrow \quad \left(\frac{2}{1}\right)^2 = \left(\frac{r_A}{r_B}\right)^3$$

$$\Rightarrow \quad r_A^3 = 4r_B^3$$ **Ans. (C)**

Sol. 54 Escape velocity of a planet is given as

$$v_{\text{escape}} = \sqrt{\frac{2GM}{R}}$$

$$\Rightarrow \quad v_{\text{escape}} = \sqrt{\frac{2G\rho\times\frac{4}{3}\pi R^3}{R}}$$

$$v_{\text{escape}} \propto \sqrt{\rho R^2}$$

If ρ is 4 times and Radius is halved, escape velocity will remain same. **Ans. (A)**

Sol. 55 By Keplar's third law the time period of revolving satellite is given as

$$T = \frac{2\pi}{\sqrt{GM}}r^{3/2}$$

$$\Rightarrow \quad \frac{T_1}{T_2} = \left(\frac{r_1}{r_2}\right)^{3/2} = \left(\frac{1}{3}\right)^{3/2}$$

$$\Rightarrow \quad T_2 = T_1 3\sqrt{3} = 21\sqrt{3} \text{ hours}$$

$$\Rightarrow \quad T_2 \approx 36 \text{ hours}$$ **Ans. (B)**

Sol. 56 Mass ratio and orbit radii ratio of satellites are given as

$$\frac{m_1}{m_2} = \frac{4}{3} \text{ and } \frac{r_1}{r_2} = \frac{3}{4}$$

Total energy of a revolving satellite around a planet of mass M is given as

$$E = \frac{1}{2}mv^2 + \left(\frac{-GMm}{r}\right)$$

$$\Rightarrow \quad E = -\frac{GMm}{2r}$$

$$\Rightarrow \quad \frac{E_1}{E_2} = \frac{m_1}{m_2}\cdot\frac{r_2}{r_1} = \frac{4}{3}\times\frac{4}{3} = \frac{16}{9}$$ **Ans. (B)**

Sol. 57 If body can go upto a maximum height h from the centre of earth where its final speed becomes zero, we use by conservation of energy

$$-\frac{GMm}{R} + \frac{1}{2}m\lambda^2 v_e^2 = -\frac{GMm}{h}$$

$$\Rightarrow \quad -\frac{GMm}{R} + \frac{1}{2}\lambda^2\frac{2GMm}{R} = -\frac{GMm}{h}$$

$$\Rightarrow \quad \frac{\lambda^2}{R} - \frac{1}{R} = \frac{-1}{h}$$

$$\Rightarrow \quad \frac{1}{h} = \frac{1-\lambda^2}{R}$$

$$\Rightarrow \quad h = \frac{R}{1-\lambda^2}$$ **Ans. (B)**

Sol. 58 By Keplar's third law, we have

$$T^2 \propto R^3$$

$$\Rightarrow \quad \left(\frac{T_1}{T_2}\right)^2 = \left(\frac{R_1}{R_2}\right)^3$$

$$\Rightarrow \quad \left(\frac{1}{2.83}\right)^2 = \left(\frac{1.5\times10^6}{R_2}\right)^3$$

$$\Rightarrow \quad R_2 = [(2.83)^2\times(1.5\times10^6)^3]^{1/3}$$

$$\Rightarrow \quad R_2 = 8^{1/3}\times1.5\times10^6 = 3\times10^6 \text{ km}$$ **Ans. (C)**

Sol. 59 Gravitational force on planets by Sun in its orbit when planet is at a distance r is given as

$$F=\frac{Gm_1m_2}{r^2}$$

$$\Rightarrow \qquad F\propto\frac{1}{r^2}$$

and $\qquad F\propto m_1m_2$

Thus option (a) is correct and (b) is NOT correct. This force provides centripetal force and it acts towards sun so option (c) is not correct. According to Kepler's third law we have

$$T^2\propto a^3$$

Hence option (d) is correct. **Ans. (A)**

Sol. 60 By Keplar's third law, we have

$$T^2\propto R^3$$

$$\Rightarrow \qquad \frac{T_1^2}{T_2^2}=\frac{R_1^3}{R_2^3}$$

$$\Rightarrow \qquad \left(\frac{T_1}{T_2}\right)^2=\left(\frac{R}{\frac{R}{4}}\right)^3$$

$$\Rightarrow \qquad \frac{T_1^2}{T_2^2}=64$$

$$\Rightarrow \qquad T_2^2=\frac{T_1^2}{64}$$

$$\Rightarrow \qquad T_2=\frac{24}{8}=3$$

Ans. (D)

Sol. 61 Escape velocity on the surface of a planet is given as

$$v_{\text{(escape) planet}}=\sqrt{\frac{2GM_P}{R_P}}$$

$$\Rightarrow \qquad v_{\text{(escape) planet}}=\sqrt{\frac{2G\left(\frac{M_e}{9}\right)}{\left(\frac{R_e}{2}\right)}}=\frac{v_e\sqrt{2}}{3}$$

$$\Rightarrow \qquad x=2$$

Ans. (A)

Sol. 62 Escape velocity on surface of a planet is given as

$$V_e=\sqrt{\frac{2GM}{R}}=\sqrt{\frac{2G\rho\frac{4}{3}\pi R^3}{R}}=C\sqrt{\rho}\cdot R$$

Ratio of escape velocities of 2 planets is written as

$$\frac{V_{e_1}}{V_{e_2}}=\frac{R_1}{R_2}\sqrt{\frac{\rho_1}{\rho_2}}=\frac{1}{2}$$

$$\Rightarrow \qquad \frac{R_1^2}{R_2^2}\times\frac{\rho_1}{\rho_2}=\frac{1}{4}$$

$$\Rightarrow \qquad \frac{R_1}{R_2}=\frac{1}{3}$$

Acceleration due to gravity on the surface of planet

$$g=\frac{GM}{R^2}=\frac{G\frac{4}{3}\pi R^3\times\rho}{R^2}C\cdot\rho R$$

So ratio of acceleration due to gravity of planet A and planet B

$$\frac{g_1}{g_2}=\frac{\rho_1R_1}{\rho_2R_2}=\frac{1}{4}\frac{R_2^2}{R_1^2}\times\frac{R_1}{R_2}$$

$$\Rightarrow \qquad =\frac{1}{4}\times\frac{R_2}{R_1}=\frac{3}{4}$$

Ans. (D)

Sol. 63 At Moon, due to low escape velocity, the rms velocity of molecules is greater than escape velocity.

As Escape velocity is given as

$$v_e=\sqrt{\frac{2GM}{R}}=\sqrt{2gR}$$

As $g_m<g_e$ escape velocity on moon less than escape velocity on earth.Hence molecules escape and there is no atmosphere at Moon. **Ans. (D)**

Sol. 64 Planets revolve in elliptical paths around sun. Thus their linear speed is not constant as distance of planet from Sun continuously changes and so gravitational forces changes. **Ans. (D)**

Sol. 65 Energy of satellite in its orbit is given as

$$E=\frac{-GMm}{2R}$$

Potential Energy of satellite in orbit is given as

$$U=\frac{-GMm}{R}$$

$$\Rightarrow \qquad U=2E$$

Kinetic Energy of satellite in orbit is given as

Kinetic Energy, $\quad K=E-U$

$$\Rightarrow \qquad K=\frac{GMm}{2R}\ (-E)$$

Thus both Statement-I and II are incorrect. **Ans. (B)**

Sol. 66 Orbital angular momentum of satellite is given as

$$L = mvr$$

Figure below shows the change in orbit of the satellite.

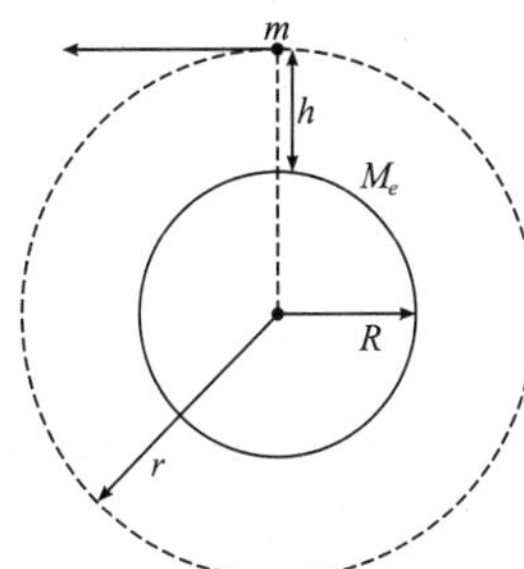

Orbital velocity of satellite is given as

$$v = \sqrt{\frac{GM_e}{r}}$$

Thus angular momentum is given as

$$L = m\sqrt{\frac{GM_e}{r}} \cdot r$$

$$\Rightarrow \quad L \propto r^{1/2}$$

As distance from centre is increased by 8 times so new distance from centre becomes $r + 8r = 9r$ hence we have new angular momentum given as

$$L' \propto (9r)^{1/2}$$

$$\Rightarrow \quad \frac{L}{L'} = \frac{r^{1/2}}{(9r)^{1/2}} = \frac{1}{3}$$

$$\Rightarrow \quad L' = 3L$$ **Ans. (D)**

Sol. 67 Orbital speed of satellite is given as

$$v = \sqrt{\frac{GM}{r}}$$

$$\Rightarrow \quad v \propto \frac{1}{\sqrt{r}}$$

So ratio of orbital speeds of satellites

$$\Rightarrow \quad \frac{v_1}{v_2} = \sqrt{\frac{r_2}{r_1}} = \sqrt{\frac{3r}{r}}$$

$$\Rightarrow \quad \frac{v_1}{v_2} = \sqrt{3} : 1$$ **Ans. (C)**

Sol. 68 The orbital velocity of satellite is given as

$$v = \sqrt{\frac{GM}{2R}} = \sqrt{\frac{Rg}{2}}$$

Thus time period of satellite is given as

$$T = \frac{2\pi(2R)}{v} = \frac{4\pi R\sqrt{2}}{\sqrt{Rg}} = \sqrt{32R}$$ **Ans. (C)**

Sol. 69 Orbital velocity of spaceship is given as

$$v_{\text{orbit}} = \sqrt{\frac{GM}{R}} = \sqrt{gR}$$

Velocity required to overcome gravitational pull is the escape velocity given as

$$v_{\text{escape}} = \sqrt{\frac{2GM}{R}} = \sqrt{2gR}$$

Thus additional velocity required in orbit is calculated as

$$\Delta v = (\sqrt{2} - 1)\sqrt{gR} = 8(\sqrt{2} - 1) \text{ km/s}$$ **Ans. (C)**

Sol. 70 Given that for the two satellites, we have

$$M_A = 2M_B$$

Potential energy of a satellite is given as

$$\text{P} = -\frac{GM_P M_A}{R}$$

Kinetic energy of a satellite is given as

$$\text{K} = +\frac{GM_P M_A}{2R}$$

Total energy of a satellite is given as

$$\text{T} = -\frac{GM_P M_A}{2R}$$

Orbital speed of a satellite is given as

$$v = \sqrt{\frac{GM_P}{R}}$$

From above relations we can see that speed of satellite is independent from mass of satellite. **Ans. (D)**

Sol. 71 Escape velocity from the surface of planet is given as

$$v_e = \sqrt{\frac{2GM}{R}}$$

$$\Rightarrow \quad v_{eP} = \sqrt{\frac{2G(16M)}{(4R)}} = 2\sqrt{\frac{2GM_E}{R_E}}$$

$$\Rightarrow \quad v_{eP} = 2v_{eP}$$ **Ans. (B)**

Sol. 72 Escape velocity at planet and earth surface is written as

$$v_P = \sqrt{\frac{2GM_p}{R_p}}$$

and $$v_E = \sqrt{\frac{2GM_E}{R_E}}$$

Ratio of escape velocities is caluclated as

$$\frac{v_P}{v_E} = \frac{\sqrt{\frac{2GM_P}{R_P}}}{\sqrt{\frac{2GM_E}{R_E}}} = \sqrt{\frac{R_E}{R_P} \times \frac{M_P}{M_E}}$$

$$\Rightarrow \quad v_P = \sqrt{\frac{1}{4} \times 9} \times V_E = \frac{3}{2} V_E$$

$$\Rightarrow \quad v_P = \frac{3}{2} \times 11.2 \text{ km/s}$$

$$\Rightarrow \quad v_P = 16.8 \text{ km/s}$$ **Ans. (B)**

Sol. 73 Escape velocity for surface a planet is given as

$$v_e = \sqrt{\frac{2GM}{R}}$$

$$\Rightarrow \quad v_e \propto \sqrt{\frac{M}{R}}$$

As $\frac{M}{R}$ increases, v_e increases hence statement-I is correct.

Also from above relation as we have

$$v_e \propto \frac{1}{\sqrt{R}}$$

As v_e depends upon R so statement-II is incorrect, hence option (B) is correct. **Ans. (B)**

8.1 Elasticity

Sol. 1  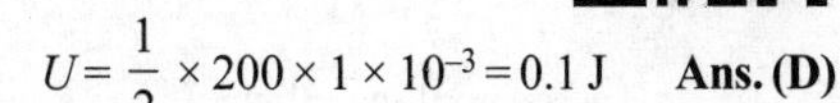

Work done by external force is stretching the wire, so $W = fx$

Half of above work is stored in elastic PE of wire, so we use

and $$U = \frac{1}{2} fx$$

$$\Rightarrow \quad U = \frac{1}{2} \times 200 \times 1 \times 10^{-3} = 0.1 \text{ J}$$ **Ans. (D)**

Sol. 2

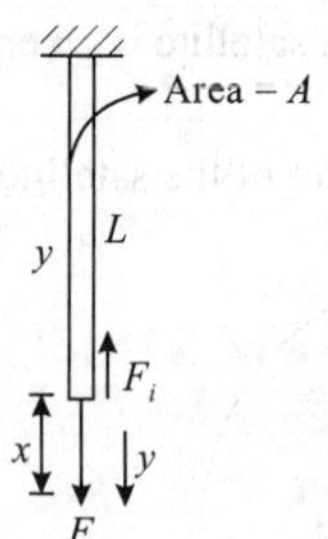

Work done by external force $W = Fx$

Due to stretching by internal force because of elasticity it develops stress in wire, using Young's modulus we have

$$Y = \frac{\text{stress}}{\text{stress}} = \left(\frac{f_i / A}{y / L}\right) \Rightarrow F_i = \frac{YA}{L} \cdot y$$

y increases from 0 to x.

$y \rightarrow$ instantaneous extension is wire

Work done by internal force

$$W_i = \int_0^x F_i day = -\frac{yA}{L} \int_0^x y\, dy$$

$$\Rightarrow \quad W_i = \frac{1}{-2} \frac{yA}{L} x^2 \quad \ldots (1)$$

At final extension x, we have for equilibrium of wire

$$F = \frac{YA}{L} \cdot x \Rightarrow \frac{YA}{L} = \frac{F}{x} \quad \ldots (2)$$

from (1) and (2)

$$W_i = -\frac{1}{2} \frac{YA}{L} x^2 = -\frac{1}{2}\left(\frac{F}{x}\right) x^2 = -\frac{1}{2} Fx$$

– ve sign denotes absorption of energy

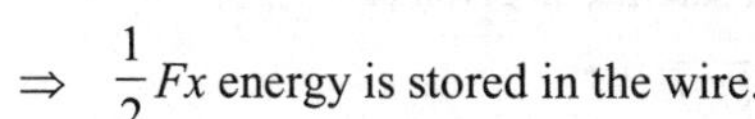

$\Rightarrow \quad \frac{1}{2} Fx$ energy is stored in the wire.

Work done in stretching by external force = Fl

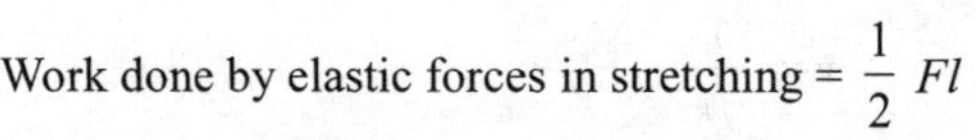

Work done by elastic forces in stretching = $\frac{1}{2} Fl$

Heat dissipation = $Fl - \frac{1}{2} Fl = \frac{1}{2} Fl$ **Ans. (D)**

Sol. 3 If a wire is stretched by an external force then elastic PE density (J/m^3) in the wire is given as

$$u_E = \frac{1}{2} \times \text{stress} \times \text{strain}$$

By definition of Young's modulus, we use

$$y = \frac{\text{stress}}{\text{strain}} = \frac{S}{\text{strain}}$$

$$\Rightarrow \quad \text{Strain} = \frac{S}{Y}$$

$$u_E = \frac{1}{2} \times S \times \frac{S}{Y} = \frac{S^2}{2Y}$$ **Ans. (B)**

Sol. 4

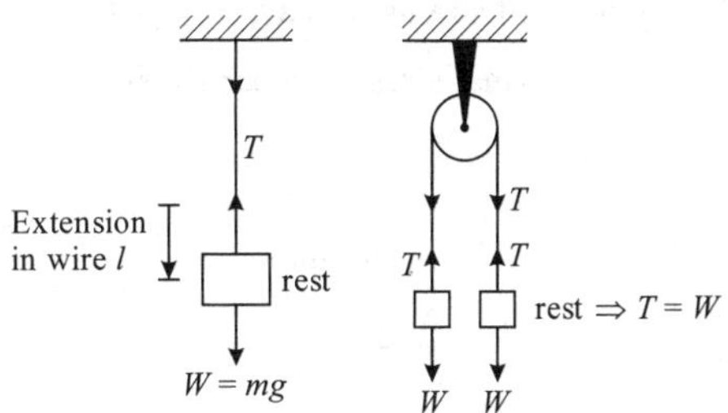

For equilibrium of load, tension in wire $T = W$

As tension in wire in the two cases is same

$\Rightarrow$ extension in these cases must be same.

Ans. (A)

Sol. 5

Volume of wire $V = Al$

By definition of young's modulus, we have

$$Y = \frac{\text{Stress}}{\text{strain}} = \frac{(F/A)}{\Delta l/l}$$

$$\Rightarrow \quad F = \frac{YA}{l}\Delta l \times \frac{A}{A}$$

$$\Rightarrow \quad F = \frac{YA^2\Delta l}{V}$$

For same Δl and V we have

$$F \propto A^2$$

$$\Rightarrow \quad \frac{F_1}{F_2} = \left(\frac{A_1}{A_2}\right)^2$$

If $\quad A_2 = 3A \ \& \ A_1 = A$

$$\Rightarrow \quad F_2 = F_1(3)^2 = 9F_1 = 9F$$

Ans. (B)

Sol. 6

For a given value of load (external force)

$$x_A > x_B > x_C > x_D$$

As x_A is maximum

$\Rightarrow$ Strain is maximum in wire OA

$\Rightarrow$ Stress is maximum in wire OA

$$\text{stress} = \frac{F_{ext}}{A}$$

$\Rightarrow$ A is minimum for wire OA **Ans. (A)**

Sol. 7 Young's modulus $Y = \frac{\text{Stress}}{\text{Strain}}$

$$\Rightarrow \quad \text{Strain} = \frac{\text{Stress}}{Y} = \frac{F/A}{Y} = \frac{F}{\pi r^2 Y}$$

$$\Rightarrow \quad \text{Strain} = \frac{10^5}{3.14 \times (10 \times 10^{-3})^2 \times 2 \times 10^{11}}$$

$$\Rightarrow \quad \text{Strain} = \frac{1}{628}$$

$$\% \text{ strain in wire} = \frac{1}{628} \times 100\% = 0.16\%$$

Ans. (A)

Sol. 8 Breaking force of a wire is the force at which stress in wire becomes equal to rupture strength of the wire material and it is independent of length of wire. **Ans. (C)**

Sol. 9 Stress produced in wire is given as

$$\text{Stress} = \frac{T}{A}$$

In the simple Atwood's machine shown in figure tension in wire is given as

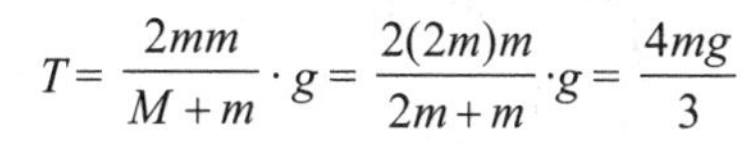

$$T = \frac{2mm}{M+m} \cdot g = \frac{2(2m)m}{2m+m} \cdot g = \frac{4mg}{3}$$

$$\Rightarrow \quad \text{Stress} = \frac{4mg}{3A}$$

Ans. (B)

Sol. 10 Given that $\frac{l_{st}}{l_{br}} = a$; $\frac{r_{st}}{r_{br}} = b$ and $\frac{1/_{st}}{1/_{br}} = C$

We use $\quad Y = \frac{\text{Stress}}{\text{Strain}} = \frac{(T/\pi r^2)}{\Delta l/l}$

$$\Rightarrow \quad \Delta l = \frac{Tl}{\pi r^2 Y}$$

For steel wire $\quad \Delta l_{st} = \frac{3mg.l_{st}}{\pi r_{st}^2 Y_{st}}$

For brass wire $\quad \Delta l_{br} = \frac{2mg\, l_{br}}{\pi r_{br}^2 Y_{br}}$

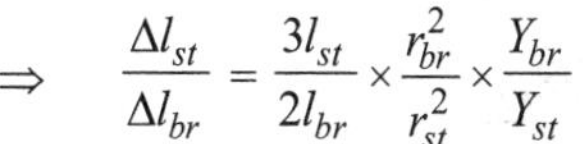

$$\Rightarrow \quad \frac{\Delta l_{st}}{\Delta l_{br}} = \frac{3l_{st}}{2l_{br}} \times \frac{r_{br}^2}{r_{st}^2} \times \frac{Y_{br}}{Y_{st}}$$

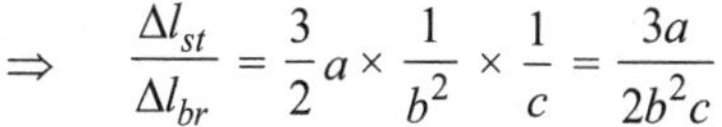

$$\Rightarrow \quad \frac{\Delta l_{st}}{\Delta l_{br}} = \frac{3}{2}a \times \frac{1}{b^2} \times \frac{1}{c} = \frac{3a}{2b^2 c}$$

Ans. (C)

Sol. 11 Young's modulus of wire

$$Y = \frac{\text{Stress}}{\text{Strain}} = \frac{\text{Stress}}{\Delta l/l}$$

[$l \to$ wire length & $\Delta l \to$ extension]

$$\Rightarrow \quad \frac{\Delta l}{l} = \frac{\text{Stress}}{Y} = \frac{5\times10^7}{2\times10^{11}} = 2.5\times10^{-4}$$

If wire is having a cross-sectional radius r than its value is

$$V = \pi r^2 l$$

$$\Rightarrow \quad \Delta V = \pi r^2 \Delta l + 2\pi r\,\Delta r l$$

$$\Rightarrow \quad \frac{\Delta V}{V} = \frac{\Delta r^2 \Delta l + 2\pi r \Delta r l}{\pi r^2 l} = \frac{0.02}{100}$$

$$\Rightarrow \quad \frac{\Delta l}{l} + \frac{2\Delta r}{r} = \frac{0.02}{100} = 2\times10^{-4}$$

$$\Rightarrow \quad \frac{2\Delta r}{r} = (2-2.5)\times10^{-4} = -0.5\times10^{-4}$$

$$\Rightarrow \quad \frac{\Delta r}{r} = -0.25\times10^{-4}$$ **Ans. (C)**

Sol. 12 As cross-sectional areas of the two wires are equal

$$\Rightarrow \quad (\text{Stress})_{\text{Cu}} = (\text{Stress})_{\text{steel}}$$

$$Y_{\text{cm}} \times \frac{\Delta l_{Cu}}{l_{Cu}} = Y_s \times \frac{\Delta l_{st}}{l_{st}}$$

$$1.0\times10^{11} \times \frac{1\times10^{-3}}{1} = 2.0\times10^{11} \times \frac{\Delta l_{st}}{0.5}$$

$$\Rightarrow \quad \Delta l_s = \frac{1}{4}\times10^{-3}\,\text{m}$$

$$= 0.25\,\text{mm}$$

Cu 1m
Steel 0.5m
mg

Total extension is composite wire

$$\Delta l = \Delta l_{\text{cu}} + \Delta l_{\text{st}}$$

$$= 1 + 0.25$$

$$\Rightarrow \quad \Delta l = 1.25\,\text{mm}$$ **Ans. (D)**

Sol. 13

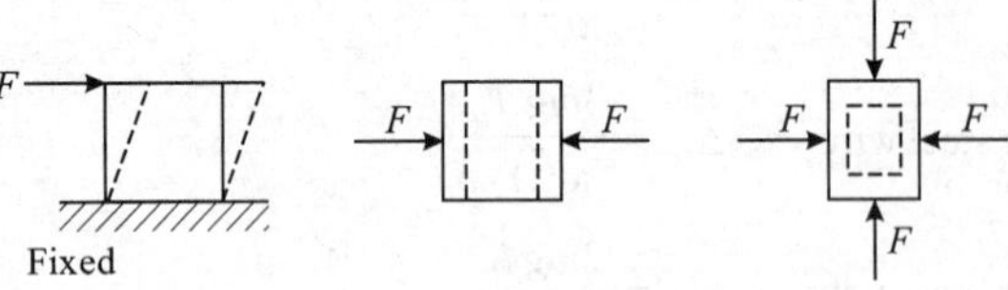

Modulus of elasticity

Shear modulus

$$\eta = \frac{\text{shear strain}}{s.\text{strain}}$$

Young's modulus

$$Y = \frac{\text{long stress}}{\text{long strain}}$$

Bulk modulus

$$B = \frac{\text{Bulk stress}}{\text{Volume strain}}$$

For equal magnitude of strains we have

Bulk stress > longitudinal stress > shear stress

$\Rightarrow \; B > Y > \eta$ **Ans. (C)**

Sol. 14 By definition of Bulk modulus

$$B = \left|\frac{\Delta P_{ex}}{\Delta V / V}\right|$$

Here $\frac{\Delta V}{V}$ is fractional compression in volume due to excess pressure ΔP_{ex}

$\Rightarrow$ For ΔP_{ex} = constant for all liquids

We have $\frac{\Delta V}{V} \propto \frac{1}{B}$

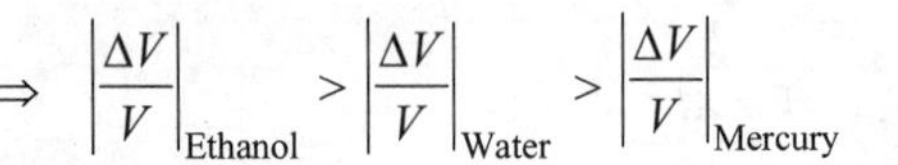

$$\Rightarrow \quad \left|\frac{\Delta V}{V}\right|_{\text{Ethanol}} > \left|\frac{\Delta V}{V}\right|_{\text{Water}} > \left|\frac{\Delta V}{V}\right|_{\text{Mercury}}$$ **Ans. (A)**

Sol. 15

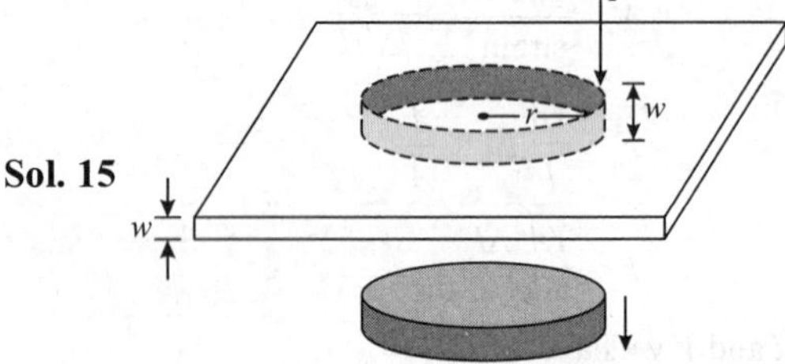

Shear stress by the applied force on the part of steel sheet (disc shaped) which is to be cut/removed is given as

Shear stress $\quad \sigma = \frac{F}{2\pi r w}$

When the shear stress becomes equal to rapture strength of steel than hole is cut

$\Rightarrow$ required force is given as

$$F = 2\pi r w \times \sigma_{\max}$$

$$\Rightarrow \quad F = 2\times3.14\times0.5\times10^{-2}\times0.3\times10^{-2}\times3.5\times10^8$$

$$\Rightarrow \quad F = 3.29\times10^4\,\text{N}$$ **Ans. (C)**

Sol. 16

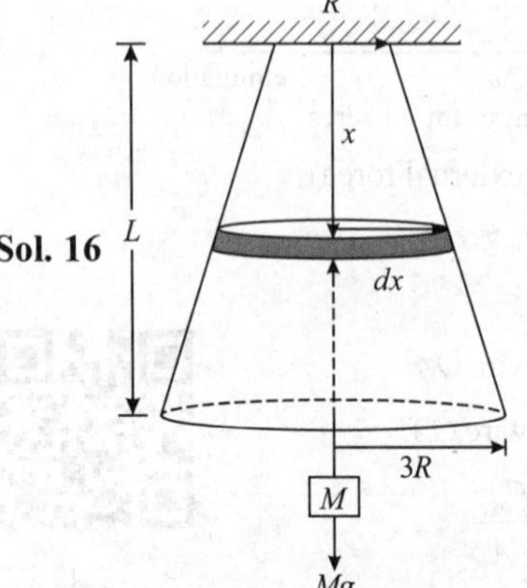

Radius of elemental disc from top end at a distance x is

$$R=\left(\frac{3R-R}{L}\right)x+R=\frac{2R}{L}x+R$$

For whole wire we can use on all such elements

$$Y=\frac{\text{Stress}}{\text{Strain}}$$

If dy is the elongation in width dx

$$\Rightarrow \quad Y=\frac{Mg/\pi r^2}{dy/dx}$$

$$\Rightarrow \quad dy=\frac{Mg\,dx}{Y\pi\left(\frac{2R}{L}x+R\right)^2}$$

Total extension in wire is

$$y=\int dy=\frac{Mg}{Y\pi}\int_0^L\frac{dx}{\left(\frac{2R}{L}x+R\right)^2}$$

$$\Rightarrow \quad y=\frac{Mg}{Y\pi}\left[\frac{L}{2R}\left(-\frac{1}{\left(\frac{2R}{L}x+R\right)}\right)\right]_0^L$$

$$\Rightarrow \quad y=\frac{Mg\,L}{2Y\pi R}\left[-\frac{1}{3R}-\left(-\frac{1}{R}\right)\right]=\frac{Mg\,L}{3\pi YR^2}$$

Total extended length of wire

$$L_f=L+y=L\left(1+\frac{Mg}{3\pi YR^2}\right)$$ **Ans. (C)**

Sol. 17

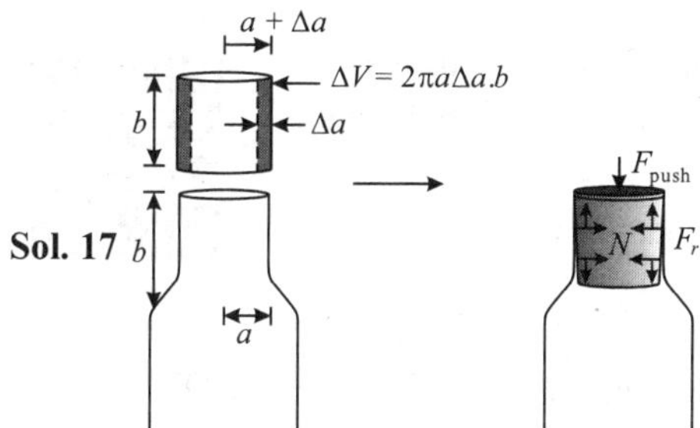

To fit the cork in the bottle opening a contact force compressive in nature is there and stress due to this, contact force on cork volume is

excess pressure $\Delta P=\frac{N}{3}=\frac{N}{2\pi ab}$

Volume compression in cork

$$\Delta V=(2\pi a\Delta a).b$$

Volume of cork

$$V=\pi a^2 b$$

Volume strain in cork

$$\Rightarrow \quad \frac{\Delta V}{V}=\frac{2\pi a\Delta a.b}{\pi a^2 b}$$

Bulk modulus of cork is given as

$$B=\frac{\Delta P_{ex}}{\left|\frac{\Delta V}{V}\right|}$$

$$\Rightarrow \quad B=\frac{N/2\pi ab}{(2\pi a\Delta ab)/\pi a^2 b}=\frac{N.a^2b}{2\pi\Delta ab.2\pi ab}$$

$$\Rightarrow \quad N=4\pi\Delta abB$$

to push the cork inside the bottle the external force has to overcome the friction on cork

$$\Rightarrow \quad F_{\text{push}}\geq\mu N=4\pi\mu\Delta abB$$ **Ans. (D)**

Sol. 18

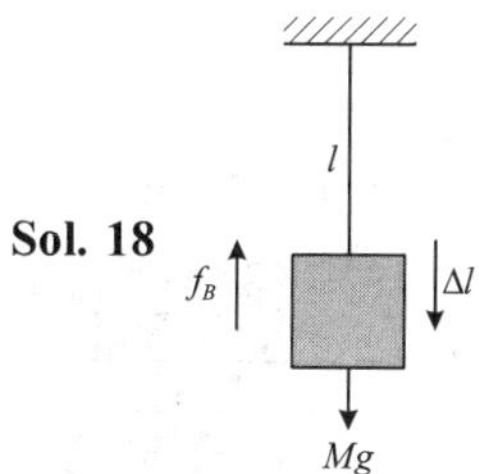

Young's modulus of wire is given as

$$Y=\frac{\text{Stress}}{\text{Strain}}=\frac{Mg/s}{\Delta l/l}$$

$$\Rightarrow \quad \Delta l\ \alpha\ MG \qquad \ldots(1)$$

If load is immersed in liquid then the new extension in wire can be given as

$$\Delta l'\ \alpha\ Mg_{\text{eff}}=Mg\left(1-\frac{\rho_l}{\rho_s}\right)=Mg\left(1-\frac{2}{8}\right)=\frac{3}{4}Mg \quad \ldots(2)$$

$$(1)/(2) \Rightarrow \quad \frac{\Delta l}{\Delta l'}=\frac{4}{3}$$

$$\Rightarrow \quad \Delta l'=\frac{3}{4}\times\Delta l=\frac{3}{4}\times 4\text{ mm}=3\text{ mm}$$ **Ans (B)**

Sol. 19 Due to stretch cord stores elastic PE and on releasing, this PE is transformed into KE of stone.

Energy density of cord

$$=\frac{1}{2}\times\text{stress}\times\text{strain}$$

$$=\frac{1}{2}\times Y\times(\text{strain})^2=\frac{1}{2}Y\left(\frac{\Delta l}{l}\right)^2$$

PE in cord $\quad E_C=\frac{1}{2}Y\left(\frac{\Delta l}{l}\right)^2\times A\times l$

By energy conservation

$$E_C=\frac{1}{2}mv^2$$

$$\Rightarrow \quad \frac{1}{2}Y\frac{(\Delta l)^2}{l}A=\frac{1}{2}mv^2$$

$$\Rightarrow \quad Y = \frac{mv^2 l}{(\Delta l^2)A} = \frac{0.02\times(20)^2\times 42\times 10^{-2}}{(20\times 10^{-2})^2\times\pi\times(3\times 10^{-3})^2}$$

$$\Rightarrow \quad Y \cong 3\times 10^6\ \text{N/m}^2$$ **Ans. (C)**

Sol. 20 Tensile stresss in wire will be

$$\text{Stress} = \frac{\text{Tensile force}}{\text{Cross section Area}}$$

$$= \frac{mg}{\pi R^2} = \frac{4\times 3.1\pi}{\pi\times 4\times 10^{-6}}\text{Nm}^{-2}$$

$$= 3.1\times 10^6\ \text{Nm}^{-2}$$ **Ans. (D)**

Sol. 21 Given that

$$\frac{Y_A}{Y_B} = \frac{7}{4};\ L_A = 2\text{ m};\ L_B = 1.5\text{ m};\ r_A = R;\ r_B = 2\text{ mm}$$

Using $$Y = \frac{\text{Sress}}{\text{Strain}} = \frac{F/\pi r^2}{x/L}$$

$$\Rightarrow \quad \frac{F}{x} = \frac{Y\pi r^2}{L}$$ Same for A and B

$$\Rightarrow \quad \frac{Y_A\pi r_A^2}{L_A} = \frac{Y_B\pi r_B^2}{L_B}$$

$$\Rightarrow \quad \frac{7}{4}\times\frac{R^2}{2} = \frac{(2\times 10^{-3})^2}{1.5}$$

$$\Rightarrow \quad R \simeq 1.74\times 10^{-3}\text{ m} \simeq 1.7\text{ mm}$$ **Ans. (B)**

Sol. 22 Stress in rod is

$$\text{Stress} = \frac{F}{A} = \frac{400\times 4}{\pi d^2} = 379\times 10^6$$

$$\Rightarrow \quad d^2 = \frac{1600}{\pi\times 379\times 10^6} = 1.34\times 10^{-6}$$

$$\Rightarrow \quad d = \sqrt{1.34}\times 10^{-3} = 1.15\times 10^{-3}\text{m}$$ **Ans. (A)**

Sol. 23

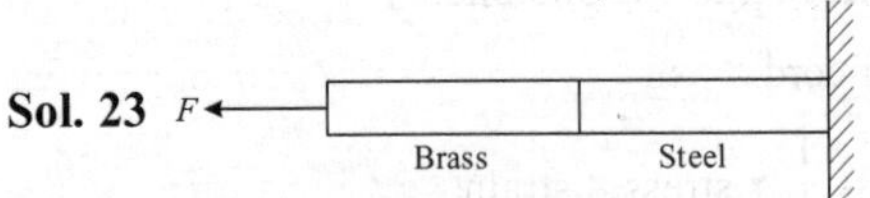

Equivalent force constaints for brass and steel wires are

$$k_1 = \frac{Y_1 A_1}{l_1} = \frac{120\times 10^9\times A}{1}$$

$$k_2 = \frac{Y_2 A_2}{l_2} = \frac{60\times 10^9\times A}{1}$$

For series combination, combined force constent is

$$k_{eq} = \frac{k_1 k_2}{k_1\times k_2} = \frac{120\times 60}{180}\times 10^9\times A$$

$$\Rightarrow \quad k_{eq} = 40\times 10^9\times \text{A}$$

For applied force F if net elongation is x, we have

$$F = k_{eq}(x)$$

$$\Rightarrow \quad F = (40\times 10^9)A\cdot(0.2\times 10^{-3})$$

$$\Rightarrow \quad \text{Stress} = \frac{F}{A} = 8\times 10^6\ \text{N/m}^2$$

None of the given option is correct **Ans. (None)**

Sol. 24 Thermal contraction is given as

$$\Delta l = l\alpha\Delta T$$

If Δl is elongation in wire due to hanging mass, we use Young's Modulus

$$Y = \frac{\text{Sress}}{\text{Strain}} = \frac{Mg/A}{\Delta l/l} = \frac{Mg}{A\alpha\Delta T}$$

$$\Rightarrow \quad M = \frac{YA\alpha\Delta T}{g}$$

$$= \frac{10^4\times\pi(10^{-3})^2\times 10^{-5}\times 20}{10} = 2\pi = 6.28\text{ kg}$$

None of the given options is correct. **Ans. (None)**

Sol. 25 The expension in rod due to heating is

$$\Delta L = \alpha LT \qquad \text{(Coefficient of linear expension)}$$

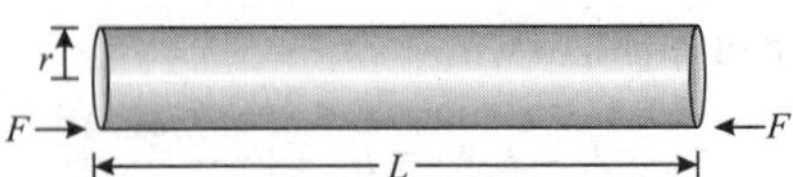

Compressional stress in rod is $= \dfrac{F}{\pi r^2}$

The compression due to F, exactly balances the thermal expension so its length remain unchanged.

Using young's modulus, we have

$$Y = \frac{\text{Sress}}{\text{Strain}} = \frac{F/\pi r^2}{\Delta L/L} = \frac{F}{\pi r^2 dT} \Rightarrow \alpha = \frac{F}{\pi r^2 YT}$$

Coefficient of volume expension

$$\gamma = 3\alpha = \frac{3F}{\pi r^2 YT}$$ **Ans. (B)**

Sol. 26 Maximum tension in wire is given as

$$T = m\omega^2 l$$

Breaking stress is wire is given as

$$S = \frac{T}{A} = \frac{m\omega^2 l}{A}$$

$$\Rightarrow \quad \omega^2 = \frac{4.8\times 10^7\times(10^{-2}\times 10^{-4})}{10\times 0.3} = 16$$

$$\Rightarrow \quad \omega = 4$$ **Ans. (4)**

Sol. 27 Elastic energy stored per unit volume is given as

$$u = \frac{1}{2} \times \text{stress} \times \text{strain}$$

$$\Rightarrow \qquad u = \frac{(\text{stress})^2}{2Y}$$

Given that

$$\frac{u_1}{u_2} = \frac{1}{4}$$

$$\Rightarrow \qquad 4u_1 = u_2$$

$$4\left(\frac{1}{Y}\right)\left[\frac{mg}{\pi d_1^2/4}\right]^2 = \left(\frac{1}{2Y}\right)\left[\frac{mg}{\pi d_2^2/4}\right]^2$$

$$\Rightarrow \qquad 4 = \left(\frac{d_1}{d_2}\right)^4$$

$$\Rightarrow \qquad \frac{d_1}{d_2} = \sqrt{2} : 1$$

Ans. (D)

Sol. 28 Strain in wire is 4.9×10^{-4}

Stress in wire is given as

$$\text{Stress} = Y \times \text{Strain}$$
$$= 9 \times 10^{10} \times 4.9 \times 10^{-4}\ \text{N/m}^2$$

Lowest vibration frequency of clamped wire is

$$n_f = \frac{1}{2l}\sqrt{\frac{T}{\mu}} = \frac{1}{2l}\sqrt{\frac{T}{\rho A}} = \frac{1}{2l}\sqrt{\frac{\text{Stress}}{\rho}}$$

$$\Rightarrow \qquad n_f = \frac{1}{2\times 1} \times \sqrt{\frac{9\times 10^{10} \times 4.9\times 10^{-4}}{9\times 10^3}}$$

$$\Rightarrow \qquad n_f = \frac{1}{2} \times \sqrt{4.9\times 10^3}$$

$$\Rightarrow \qquad n_f = \frac{7}{2}\times 10 = 35\ \text{Hz}$$

Ans. (35)

Sol. 29 Bulk modulus of material is given as

$$B = \frac{\Delta P}{\left|\frac{\Delta V}{V}\right|}$$

$$\Rightarrow \qquad \left|\frac{\Delta V}{V}\right| = \frac{\Delta P}{B} = \frac{4\times 10^9}{8\times 10^{10}} = \frac{1}{20}$$

$$\Rightarrow \qquad \frac{\Delta l}{l} = \frac{1}{3}\left|\frac{DV}{V}\right| = \frac{1}{3}\times\frac{1}{20} = \frac{1}{60}$$

Percentage change in length of cube is given as

$$\frac{\Delta l}{l} \times 100 = \frac{1}{60} \times 100 = \frac{5}{3} = 1.67\%$$

Ans. (B)

Sol. 30 An elastic wire can be considered as a spring of force constant k given as

$$k = \frac{YA}{l}$$

Thus time period of oscillations of object can be given as

$$T = 2\pi\sqrt{\frac{m}{k}}$$

So frequency of oscillations is given as

$$f = \frac{1}{2\pi}\sqrt{\frac{k}{m}} = \frac{1}{2\pi}\sqrt{\frac{YA}{ml}}$$

Ans. (A)

Sol. 31 Modulus of elasticity are related as

$$Y = 3K(1-2\sigma)$$

$$\Rightarrow \qquad \sigma = \frac{1}{2}\left(1 - \frac{Y}{3K}\right) \qquad \ldots(2)$$

$$\text{and} \qquad Y = 2\eta(1+\sigma)$$

$$\Rightarrow \qquad \sigma = \frac{Y}{2\eta} - 1 \qquad \ldots(2)$$

From equation-(1) and (2), we have

$$\frac{1}{2}\left(1 - \frac{Y}{3k}\right) = \frac{Y}{2\eta} - 1$$

$$\Rightarrow \qquad \frac{Y}{3K} = \frac{3\eta - Y}{\eta}$$

$$\Rightarrow \qquad K = \frac{\eta Y}{9\eta - 3Y}$$

Ans. (A)

Sol. 32 The force on the wire can be related to the elongation in wire as

$$F = Y.A.\frac{\Delta l}{l}$$

$$\Rightarrow \qquad \Delta l = \frac{F}{Y.A.}\cdot l = \frac{F.l}{Y.\pi r^2}$$

$$\Rightarrow \qquad \Delta l \propto \frac{l}{r^2}$$

$$\Rightarrow \qquad \frac{\Delta l_2}{\Delta l_1} = \left(\frac{l_2}{l_1}\right)\left(\frac{r_1}{r_2}\right)^2 = (2)\left(\frac{1}{2}\right)^2 = \frac{1}{2}$$

$$\Rightarrow \qquad \Delta l_2 = \frac{\Delta l_1}{2} = \frac{0.04}{2} = 0.02\ \text{m}$$

$$\Rightarrow \qquad \Delta l_2 = 2\ \text{cm}$$

Ans. (2)

Sol. 33 Density of the material can be written as

$$\rho = \frac{M}{V}$$

$$\Rightarrow \qquad \frac{d\rho}{\rho} = -\frac{dV}{V} \qquad \text{(As mass is constant)}$$

Bulk modulus of material is given as

$$K = -\frac{P}{\frac{dV}{V}}$$

$$\Rightarrow \quad -\frac{dV}{V} = \frac{P}{K}$$

$$\Rightarrow \quad \frac{d\rho}{\rho} = \frac{P}{k}$$

$$\Rightarrow \quad d\rho = \frac{\rho P}{K}$$

Ans. (A)

Sol. 34 For the elastic wire we can use tension in wire is proportional to its elongation, so we use

$$T = k(\Delta l)$$

If l_0 is the natural original length of wire then we use

$$T_1 = k(l_1 - l_0)$$

and

$$T_2 = k(l_2 - l_0)$$

$$\Rightarrow \quad \frac{T_1}{T_2} = \frac{l_1 - l_0}{l_2 - l_0}$$

$$\Rightarrow \quad l_0 = \frac{T_2 l_1 - T_1 l_2}{T_2 - T_1}$$

Ans. (D)

Sol. 35 Young's Modulus of wire can be given as

$$Y = \frac{\text{Stress}}{\text{Strain}} = \frac{FL}{Al} = \frac{mg \cdot L}{\pi R^2 \cdot l}$$

Fractional error in calculation of Young's modulus is given as

$$\frac{\Delta Y}{Y} = \frac{\Delta m}{m} + \frac{\Delta L}{L} + 2 \cdot \frac{\Delta R}{R} + \frac{\Delta l}{l}$$

Percentage error is calculated as

$$\frac{\Delta Y}{Y} \times 100 = 100 \left[\frac{1}{1000} + \frac{1}{1000} + 2\left(\frac{0.001}{0.2}\right) + \frac{0.001}{0.5} \right]$$

$$\Rightarrow \quad \frac{\Delta Y}{Y} \times 100 = \frac{1}{10} + \frac{1}{10} + 1 + \frac{1}{5} = \frac{14}{10} = 1.4\%$$

Ans. (D)

Sol. 36 Pressure at a depth h is given as

$$P = h\rho g$$

Bulk modulus of the medium which is the requierd ratio asked in question, is given as

$$\beta = \frac{p}{\frac{\Delta V}{V}} = \frac{2 \times 10^3 \times 10^3 \times 9.8}{1.36 \times 10^{-2}}$$

$$\Rightarrow \quad \beta = 1.44 \times 10^9 \text{ N/m}^2$$

Ans. (D)

Sol. 37 Fringe width in YDSE setup is given as

$$\beta = \frac{\lambda D}{d} = \frac{5890 \times 10^{-10} \times 0.5}{0.5 \times 10^{-3}} = 589 \times 10^{-6} \text{ m}$$

Distance between first and third bright fringe can be calculated as

$$2\beta = 2 \times 589 \times 10^{-6} \text{m} = 1178 \times 10^{-6} \text{ m}$$

Ans. (B)

Sol. 38 For wire A stress can be given as

$$\frac{F}{\pi r^2} = Y\left(\frac{2mm}{aL}\right) \quad \ldots(1)$$

For wire B stress is given as

$$\frac{E}{\pi.16r^2} = Y\left(\frac{4mm}{bL}\right) \quad \ldots(2)$$

From equations-(1) & (2), we get

$$16 = \frac{2b}{4a}$$

$$\Rightarrow \quad \frac{a}{b} = \frac{1}{32}$$

Ans (32)

Sol. 39 For adiabatic process, we use

$$PV^\gamma = \text{constant}$$

Differentiating the above equation, gives

$$\frac{dP}{dV} = -\frac{\gamma P}{V}$$

$$\Rightarrow \quad \frac{dP}{P} = -\frac{\gamma dV}{V}$$

Ans. (A)

Sol. 40 Let initial length of wire be L_0 and cross sectional area A, then stress in wire can be given as

$$\frac{T}{A} = Y\frac{\Delta l}{l}$$

So for the two values, we use

$$\frac{T_1}{A} = \frac{Y(L_1 - L_0)}{L_0}$$

and

$$\frac{T_2}{A} = \frac{Y(L_2 - L_0)}{L_0}$$

$$\Rightarrow \quad \frac{T_1}{T_2} = \frac{L_1 - L_0}{L_2 - L_0}; \; T_1 L_2 - T_1 L_0 = T_2 L_0;$$

$$L_0 = \frac{L_2 T_1 - L_1 T_2}{T_1 - T_2}$$

Ans. (C)

Sol. 41 Let initial length of rod be l_0 and area A, the stress in wire at the two states can be given as

$$\frac{T_1}{A} = \frac{Y(l_1 - l_0)}{l_0}$$

and $$\frac{T_2}{A} = \frac{Y(l_2 - l_0)}{l_0}$$

$$\Rightarrow \quad \frac{T_1}{T_2} = \frac{l_1 - l_0}{l_2 - l_0}$$

$$\Rightarrow \quad T_1 l_2 - T_1 l_0 = T_2 l_0 - T_2 l_0$$

$$\Rightarrow \quad l_0 = \frac{l_1 T_2 - l_2 T_1}{T_2 - T_1}$$ **Ans. (C)**

Sol. 42 Elastic energy stored in the track is calculated as

$$U = \frac{1}{2} \times \text{stress} \times \text{strain} \times \text{volume}$$

$$U = \frac{1}{2} \times \text{stress} \times \text{strain} \times \text{volume} \times Al$$

Energy stored per unit length in track is given as

$$\Rightarrow \quad \frac{U}{l} = \frac{1}{2} \times \text{stress} \times \text{strain} \times \text{volume} \times A$$

$$\Rightarrow \quad \frac{U}{l} = \frac{1}{2} \times Y \times (\text{strain})^2 \times A$$

$$\Rightarrow \quad \frac{U}{l} = \frac{1}{2} \times Y\left(\frac{\Delta l}{l}\right)^2 \times A = \frac{1}{2} Y \left(\frac{l\alpha\Delta t}{l}\right)^2 \times A = \frac{1}{2} Y\alpha^2 \Delta t^2 A$$

$$\Rightarrow \quad \frac{U}{l} = \frac{1}{2} \times 10^{11} \times 10^{-10} \times 10 \times 10 \times 10^{-2} = 5 \text{ J/m}$$ **Ans. (5)**

Sol. 43 As the two wires are joined end to end (series combination) so for series combination of two wires or springs, equivalent spring constant is given as

$$K_{eff} = \frac{K_1 K_2}{K_1 + K_2}$$

$$\Rightarrow \quad \frac{YA}{2l} = \frac{\frac{Y_1 A}{l} \times \frac{Y_2 A}{l}}{\frac{Y_1 A}{l} + \frac{Y_2 A}{l}}$$

$$\Rightarrow \quad Y = \frac{2Y_1 Y_2}{Y_1 + Y_2}$$ **Ans. (D)**

Sol. 44 Fringe width in YDSE is given as

$$\beta = \frac{\lambda D}{d}$$

As $\lambda_{blue} < \lambda_{orange} \Rightarrow \beta_{blue} < \beta_{orange}$ **Ans. (A)**

Sol. 45 When the stone is projected, it gains kinetic energy by elastic energy of catapult, so we use

$$\frac{1}{2}kx^2 = \frac{1}{2}mv^2$$

$$\Rightarrow \quad \left(\frac{YA}{l}\right)x^2 = mv^2$$

$$\Rightarrow \quad \frac{0.5 \times 10^9 \times 10^{-6}}{0.1} \times 0.04 \times 0.04 = 20 \times 10^{-3} v^2$$

$$\Rightarrow \quad v = 20 \text{ m/s}$$ **Ans. (20)**

Sol. 46 Maximum tension supported by the two wires can be written as

$$F_{1max} = \text{stress} \times \text{area} = 1.25 \times 8 \times 10^2 = 1000 \text{ N}$$

$$F_{2max} = 1.25 \times 8 \times 10^2 = 500 \text{ N}$$

To calcluate the maximum weight which can be placed in pan, it should increase the tension in wire W_2 to the maximum value so we use

$$F_{2max} = (10 + M)g = 500 \text{ N}$$

$$\Rightarrow \quad M = 40 \text{ kg}$$ **Ans. (40)**

Sol. 47 Elongation in rod due to its self weight can be given as

$$\Delta L = \frac{Wl}{2YA}$$

$$\Rightarrow \quad \Delta L = \frac{10 \times 0.2}{2 \times 2 \times 10^{11} \times 100 \times 10^{-4}} = 5 \times 10^{-10} \text{ m}$$ **Ans. (B)**

Sol. 48 Volume strain in rubber ball is calculated as

$$\frac{-dV}{V} = \frac{0.5}{100} = 0.005 = 0.005$$

Bulk modulus of the ball can be written as

$$B = \frac{-dP}{\frac{dV}{V}} = \frac{\rho g h}{\frac{-dV}{V}}$$

$$\Rightarrow \quad 9.8 \times 10^8 = \frac{10^3 \times 9.8 \times h}{5 \times 10^{-3}}$$

$$\Rightarrow \quad h = 500 \text{ m}$$ **Ans. (500)**

Sol. 49 In the situation described in question, we can consider total area under stress is four times that of each columns so longitudinal strain can be given as

$$\frac{\Delta L}{L} = \frac{mg}{4AY}$$

$$\Rightarrow \quad \frac{\Delta L}{L} = \frac{50 \times 10^3 \times 9.8}{4\pi(100^2 - 50^2) \times 10^{-4} \times 2 \times 10^{11}}$$

$$\Rightarrow \quad \frac{\Delta L}{L} = \frac{5 \times 10^5}{4 \times \pi \times 150 \times 50 \times 10^{-4} \times 2 \times 10^{11}}$$

$$\Rightarrow \quad \frac{\Delta L}{L} = \frac{1}{12\pi} \times 10^{-5} = 2.60 \times 10^{-7}$$ **Ans. (D)**

Sol. 50 The distance between nth bright fridge on both side of central maxima is 2.4 cm. So this would be the fringe width on interference pattern. Distance between fourth bright fringe on the two sides of central bright fringe will be 8 time sthe fringe width, given as

$$\Delta y = \frac{D\lambda}{d} \times 8$$

The wavelength of light can be obtained as

$$\Rightarrow \quad \lambda = \frac{d\Delta y}{8D} = \frac{0.3\times10^{-3}\times2.4\times10^{-2}}{8\times1.5}$$

$$\Rightarrow \quad \lambda = \frac{0.12\times10^{-5}}{2} = 0.06\times10^{-5}$$

$$\Rightarrow \quad \lambda = 6\times10^{-7}\,\text{m}$$

Thus frequency of light is given as

$$f = \frac{c}{6\times10^{-7}} = 0.5\times10^{+15}$$

$$\Rightarrow \quad f = 5\times10^{14}\,\text{Hz}$$

Ans. (5)

Sol. 51 From the given expression of Young's modulus, we calculate the fractional error in measurement of Y as

$$Y = \frac{MgL^3}{4bd^3\delta}$$

$$\Rightarrow \quad \frac{dY}{Y} = \frac{dM}{M} + 3\frac{dL}{L} + 3\frac{dd}{d} + \frac{db}{b} + \frac{d\delta}{\delta}$$

$$\Rightarrow \quad \frac{dY}{Y} = 0.0005 + 3\times0.001 + 3\times0.0025 + 0.0025 + 0.002$$

$$\Rightarrow \quad \frac{dY}{Y} = 0.0155$$

Ans. (B)

Sol. 52 Bulk modulus of liquid is given as

$$B = 3\times10^{10}$$

Volume strain in liquid is given as

$$-\frac{\Delta V}{V} = 0.02$$

We use definition of bulk modulus as

$$B = \frac{\Delta P}{-\frac{\Delta V}{v}}$$

$$\Rightarrow \quad \Delta P = -B\left(\frac{\Delta V}{V}\right)$$

$$\Rightarrow \quad \Delta P = (3\times10^{10})(0.02)$$

$$\Rightarrow \quad \Delta P = 6\times10^{8}\,\text{N/m}^2$$

Ans. (C)

Sol. 53 Young Modulus of material from graph is calculated as

$$Y = \frac{\text{stress}}{\text{strain}} = 2.0\times10^{10}$$

Energy density due to linear strain in material is given as

$$\text{Energy density} = \frac{1}{2}\text{stress}\times\text{strain}$$

$$\Rightarrow \quad u_e = \frac{1}{2}(\text{strain})^2 y = \frac{1}{2}(5\times10^{-4})^2\times20\times10^{10}$$

$$\Rightarrow \quad u_e = 25\times10^2\times10 = 25\frac{\text{kJ}}{\text{m}^3}$$

Ans. (25)

Sol. 54 Reynold's number for a flowing viscous fluid by definition is given as $\frac{\rho v d}{\eta}$. **Ans. (C)**

Sol. 55 Units of Pressure and time are written as

$$u[p] = \text{N/m}^2$$

and

$$u[t] = \text{s}$$

Unit for measurement of coefficient of viscosity is

$$u[\eta] = \text{N-s/m}^2$$

Comparing above units, Assertion-A is true.

The expression of viscous force in a flowing fluid is given as

$$F = -\eta A\frac{dv}{dx}$$

From this expression coefficient of viscosity is can be written as

$$\text{Coefficient of Viscosity} = \frac{\text{Force}}{\text{Velocity Gradient}\times\text{Area}}$$

Hence Reason-R is false. **Ans. (C)**

Sol. 56 By Hooke's Law, the stress causing force is directly proportional to the extension in wire so we have

$$F \,\alpha\, \Delta L$$

$$\Rightarrow \quad \frac{F_1}{F_2} = \frac{\Delta L_1}{\Delta L_2}$$

$$\Rightarrow \quad \frac{10}{20} = \frac{(L_1 - L)}{(L_2 - L)}$$

$$\Rightarrow \quad L = 2L_1 - L_2$$

Ans. (C)

Sol. 57 The extension in a wire due to applied force is calculated as

$$\Delta l_1 = \frac{Fl}{AY} = \frac{Fl}{\pi r^2 Y} = 5\,\text{cm}$$

and

$$\Delta l_2 = \frac{4F4l}{\pi16r^2Y} = \frac{Fl}{\pi r^2 Y} = 5\,\text{cm}$$

Ans. (5)

Sol. 58 Using Hooke's law for extension of a wire by applied force, we have

$$\frac{F}{A} = Y\frac{\Delta L}{L}$$

$$\Rightarrow \quad Y = \frac{FL}{A\Delta L} = \frac{(mg)\cdot(1)}{\left(\frac{\pi d^2}{4}\right)0.4\times10^{-3}}$$

$$\Rightarrow \quad Y = \frac{10\times4}{\pi(0.4\times10^{-3})^2\times0.4\times10^{-3}}$$

$$\Rightarrow \quad Y = \frac{40}{\pi(0.4\times10^{-3})^3}$$

$$\Rightarrow \quad Y = \frac{40\times7}{22\times64\times10^{-3}\times10^{-9}}$$

$$\Rightarrow \quad Y = 0.199\times10^{-12}\ \text{N/m}^2$$

Calculating fractional error in measurement of Young's Modulus, we have

$$\frac{\Delta Y}{Y} = \frac{\Delta F}{F} + \frac{\Delta L}{L} + \frac{\Delta A}{A} + \frac{\Delta(\Delta L)}{(\Delta L)}$$

$$\Rightarrow \quad \frac{\Delta Y}{Y} = \frac{0.02}{0.4} + 2\frac{\Delta d}{d} = \frac{0.2}{4} + 2\times\frac{0.01}{0.4}$$

$$\Rightarrow \quad \frac{\Delta Y}{Y} = \frac{0.1}{2} + \frac{0.1}{2} = 0.1$$

$$\Rightarrow \quad \Delta Y = 0.1\times Y$$

$$\Rightarrow \quad \Delta Y = 0.199\times10^{11} = 1.99\times10^{10}$$ **Ans. (2)**

Sol. 59 Since breaking stress (Maximum lifting capacity) is the property of material so it will remain same which is calculated as

$$\text{breaking stress} = \frac{\text{Maximum lifting cpacity}}{\text{Area of cross section of rope}}$$

$$\Rightarrow \quad \frac{10}{2.5\times10^{-4}} = \frac{25}{A}$$

$$\Rightarrow \quad A = 625\times10^{-6}$$

$$\Rightarrow \quad A = 6.25\times10^{-4}\ \text{m}^2$$ **Ans. (A)**

Sol. 60 Figure below shows the hanging rod. To find the elongation due to its own weight, we consider an element in rod at a distance x from its bottom end of width dx. If elongation in this element is dy due to the tension in rod at this point, we use by definition of Young's Modulus

$$Y = \frac{\text{stress}}{\text{strain}}$$

$$\Rightarrow \quad Y = \frac{\frac{T}{A}}{\frac{dy}{dx}} = \frac{T}{A}\cdot\frac{dx}{dy}$$

$$\Rightarrow \quad dy = \frac{Tdx}{AY}$$

Tension at a distance x from lower end is given as

$$T = \frac{mg}{l}x$$

Total elongation in rod can be calculated as

$$\int_0^{\Delta l} dy = \int_0^l \frac{mg}{l}x\frac{dx}{AY}$$

$$\Rightarrow \quad \Delta l = \frac{mg}{lAY}\left[\frac{x^2}{2}\right]_0^l$$

$$\Rightarrow \quad \Delta l = \frac{mgl}{2AY}$$

$$\Rightarrow \quad \Delta l = \frac{20\times10\times20}{2\times0.4\times2\times10^{11}}$$

$$\Rightarrow \quad \Delta l = 2500\times10^{-11}$$

$$\Rightarrow \quad \Delta l = 25\times10^{-9} = x\times10^{-9}$$

$$\Rightarrow \quad x = 25$$ **Ans. (25)**

Sol. 61 The shear stress is related to shear strain as

$$\frac{F}{A} = \eta\frac{x}{l}$$

$$\Rightarrow \quad \frac{Fl}{A\eta} = x$$

$$\Rightarrow \quad x = \frac{18\times10^4\times60\times10^{-2}}{60\times10^{-2}\times15\times10^{-2}\times25\times10^9}$$

$$\Rightarrow \quad x = 48\times10^{-6}\ \text{m} = 48\ \mu\text{m}$$ **Ans. (48)**

Sol. 62 Figure below shows the situation described in question.

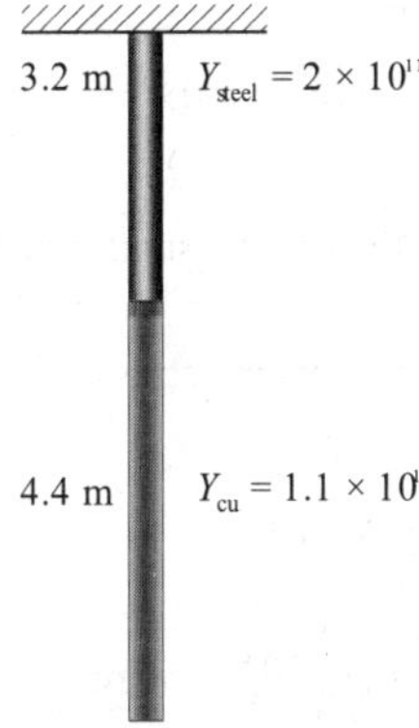

Due to the applied load at the bottom of composite wire, net elongation in wire is sum of the elongation in the two wires under same tension, so we use

$$\Delta l_1 + \Delta l_2 = \Delta l$$

$$\Rightarrow \quad \frac{Fl_1}{A_1Y_1}+\frac{Fl_2}{A_2Y_2}=\Delta l$$

$$\Rightarrow \quad F=\frac{\Delta l}{\frac{l_1}{A_1y_1}+\frac{l_2}{A_2y_2}}=1.54\times 10^2$$

$$\Rightarrow \quad F=154\,\text{N}$$ **Ans. (D)**

Sol. 63 From the given graph, we can calculate the slope of graph as

$$\text{Slope}=\frac{\Delta l/w}{L}=\frac{\Delta l/L}{w}=\frac{1}{YA}$$

$$\Rightarrow \quad Y=\frac{1}{(\text{slope})A}$$

$$\Rightarrow \quad Y=\frac{1}{2\times10^{-6}(0.25\times10^{-5})}$$

$$\Rightarrow \quad Y=2\times 10^{11}\,\text{N/m}^2$$ **Ans. (2)**

Sol. 64 By definition of Young's modulus, the applied force can be given as

$$F=YA\frac{\Delta l}{l}$$

$$\Rightarrow \quad F=2\times 10^{11}\times 10^{-4}\left(\frac{2l-l}{l}\right)$$

$$\Rightarrow \quad F=2\times 10^{7}\,\text{N}$$ **Ans. (C)**

Sol. 65 Young's modulus of a material only depends on material of wire so it will remain same. **Ans. (A)**

Sol. 66 Due to revolution in uniform circular motion, tension developed in wire is given as

$$T=\frac{mv^2}{l}=\frac{10\times v^2}{0.5}=20v^2$$

For maximum speed the tension should be limited by breaking stress, given as

$$T_{max}=\text{Breaking stress}\times\text{Area}$$

$$\Rightarrow \quad T_{max}=5\times 10^8\times 10^{-4}=5\times 10^4$$

$$\Rightarrow \quad 20v^2=5\times 10^4$$

$$\Rightarrow \quad v=\sqrt{\frac{1}{4}10^4}=50\,\text{m/s}$$ **Ans. (50)**

Sol. 67 Elongation in wire due to external force F is given as

$$\delta=\frac{Fl}{AY}$$

$$\Rightarrow \quad \delta=\frac{250\times100}{6.25\times10^{-4}\times10^{10}}=4\times 10^{-3}\,\text{m}$$ **Ans. (D)**

Sol. 68 By definition both (A) and (R) are true and (R) is also explaining the assertion-(A). **Ans. (C)**

Sol. 69 From graph as slope is unity we use

$$F=\Delta L$$

Young's modulus of the material is given as

$$Y=\frac{FL}{A\Delta L}$$

$$\Rightarrow \quad Y=\frac{L}{A}$$

$$\Rightarrow \quad Y=\frac{62.8\times10^{-2}}{\pi(2\times10^{-3})^2}=5\times 10^4\,\text{N/m}^2$$ **Ans. (5)**

Sol. 70 The relation between different modulus of elasticity are given as

$$Y=3\eta(1+\sigma)$$

$$\Rightarrow \quad Y=3K(1-\sigma)$$

$$\Rightarrow \quad 2\eta(1+\sigma)=3K(1-2\sigma)$$

$$\Rightarrow \quad \sigma=\left(\frac{3K-2\eta}{6K+2\eta}\right)$$ **Ans. (A)**

Sol. 71 By definition of young's modulus, we know

$$Y=\frac{\text{Stress}}{\text{Strain}}$$

$$\Rightarrow \quad Y=\frac{F/A}{\frac{\Delta l}{l}}$$

$$\Rightarrow \quad F=\frac{YA}{l}\Delta l$$

As force applied on the two wires is same, we use

$$\left(\frac{A\Delta l}{l}\right)_1=\left(\frac{A\Delta l}{l}\right)_2$$

$$\Rightarrow \quad \frac{\Delta l_2}{\Delta l_1}=\frac{A_1}{A_2}\times\frac{l_2}{l_1}$$

$$\Rightarrow \quad \frac{\Delta l_2}{0.2}=\frac{1}{2.4\times2.4}\times\frac{2}{1}$$

$$\Rightarrow \quad \Delta l_2=6.9\times 10^{-2}\,\text{mm}$$ **Ans. (D)**

Sol. 72 If Δl is decease in length of rod due to decease in temperature, this can be calculated as

$$\Delta l=l\alpha\Delta T$$

$$\Rightarrow \quad \Delta l=1\times 2\times 10^{-5}\times 50=10^{-3}\,\text{m}$$

Young Modulus of the material of rod is expressed as

$$Y = \frac{F/A}{\Delta l/l}$$

$$\Rightarrow \quad 2\times10^{11} = \frac{Mg/3\times10^{-6}}{10^{-3}/1}$$

$$\Rightarrow \quad Mg = 2\times10^{11}\times3\times10^{-9} = 6\times10^{-2}$$

$$\Rightarrow \quad M = 60\text{ kg}$$

Ans. (60)

Sol. 73 Velocity of longitudinal wave in the rod is given as

$$v = \sqrt{\frac{Y}{\rho}} = \sqrt{\frac{3.2\times10^{11}}{8\times10^{3}}}$$

$$\Rightarrow \quad v = 2\times10^{3}\sqrt{10}$$

$$\Rightarrow \quad v = 6.32\times10^{3}\text{ m/s}$$

Ans. (D)

Sol. 74 Young's modulus is given as

$$\Delta l = \frac{Fl}{SY}$$

F is same for both wire and Δl is also same so we have

$$\frac{\Delta l}{F} = \frac{l}{SY}$$

$$\Rightarrow \quad \frac{l_A}{S_AY_A} = \frac{l_B}{S_BY_B}$$

$$\Rightarrow \quad \frac{5}{2.5\times Y_A} = \frac{6}{3\times Y_B}$$

$$\Rightarrow \quad \frac{Y_A}{Y_B} = 1$$

Ans. (B)

Sol. 75

$$[b] = [V]$$

$$\left[\frac{a}{b^2}\right] = [P]$$

$$\Rightarrow \quad \left[\frac{b^2}{a}\right] = \frac{1}{[P]} = \frac{1}{[B]} = [K]$$

Ans. (C)

Sol. 76 Bulk modulus of water is calculated as

$$B_{\text{water}} = \frac{-\Delta P}{\left(\frac{\Delta V}{V}\right)} = \frac{-\Delta P}{\frac{0.01}{100}}$$

Bulk modulus of liquid is calculated as

$$B_{\text{liquid}} = \frac{-\Delta P}{\frac{0.03}{100}}$$

$$\Rightarrow \quad \frac{B_{\text{water}}}{B_{\text{liquid}}} = 3$$

$$\Rightarrow \quad x = 1$$

Ans. (1)

Sol. 77 Tension $(F) = mg$

$$\Rightarrow \quad F = 4\times\frac{10}{4} = 10\text{ N}$$

By Young's modulus, change in length

$$\Delta L = \frac{FL}{AY}$$

$$\Rightarrow \quad \Delta L = \frac{10\times6}{3\times10^{-6}\times2\times10^{11}}$$

$$\Rightarrow \quad \Delta L = 10^{-4}\text{ m} = 0.1\text{ mm}$$

Ans. (C)

Sol. 78 By the definition of Young's modulus

$$\text{Strain} = \frac{\text{stress}}{Y}$$

$$\Rightarrow \quad \text{Strain} = \frac{\frac{62.8\times10^{3}}{\pi\times(0.02)^2}}{2\times10^{11}}$$

Stress in wire is calculated as

$$\text{Stress} = \frac{F}{A} = \frac{62.8\times10^{3}}{\pi\times(2\times10^{-2})^2}$$

$$\Rightarrow \quad \text{Strain} = \frac{62.8\times10^{3}}{3.14\times4\times10^{-4}\times2\times10^{11}}$$

$$\Rightarrow \quad \text{Strain} = 2.5\times10^{-4} = 25\times10^{-5}$$

Ans. (25)

Sol. 79 Tensile stress, $\sigma = \frac{F}{A} = \frac{4mg}{\pi D^2}$

As $r = \frac{D}{2}$, cross sectional area of wire is $A = \pi r^2 = \frac{\pi D^2}{4}$

$$\Rightarrow \quad m = \frac{\pi D^2\sigma}{4g}$$

$$\Rightarrow \quad m = \frac{22}{7}\times\frac{(14\times10^{-3})^2\times7\times10^{5}}{4\times9.8}$$

$$\Rightarrow \quad m = 11\text{ kg}$$

Ans. (11)

Sol. 80 Young's modulus

$$Y = 7\times10^{10}\text{ N/m}^2$$

$$\text{Strain} = \frac{0.04}{100}$$

$$\text{Energy} = \frac{1}{2}\left(\frac{YA}{l}\right)\Delta x^2$$

$$\Rightarrow \quad \text{Energy} = \frac{1}{2}YA\left(\frac{\Delta x}{l}\right)^2 \times l$$

Energy per unit volume i.e. energy density is given as

$$u_e = \frac{E}{V} = \frac{1}{2} \times Y \times \text{strain}^2$$

$$\Rightarrow \quad u_e = \frac{1}{2} \times 7 \times 10^{10} \times \frac{0.04 \times 0.04}{10^4}$$

$$\Rightarrow \quad u_e = 56 \times 10^2 \text{ J/m}^3$$ **Ans. (A)**

Sol. 81 By the definition of Young's modulus, we have

$$Y = \frac{\text{Stress}}{\text{Strain}}$$

$$\Rightarrow \quad \text{Stress} = Y \times \text{strain}$$

As strain in the rod is given as $\frac{\Delta l}{l}$, we have

$$\text{Stress} = Y \times \frac{\Delta l}{l}$$

$$\Rightarrow \quad \text{Stress} = Y \times \frac{l\alpha\Delta T}{l} = Y\alpha\Delta T$$

Compressive Tension in the rod is calculated as

$$T = \text{Stress} \times \text{Area of cross section}$$

$$\Rightarrow \quad T = YA\alpha\Delta T = 4 \times 10^4 \text{N}$$ **Ans. (4)**

Sol. 82 Tension in steel wire is calculated as

$$T_2 = 2g + T_1$$

$$T_2 = 20 + 11.4 = 31.4 \text{ N}$$

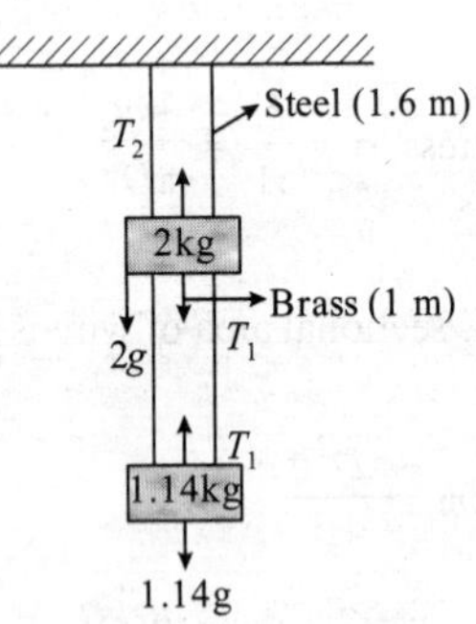

Elongation in steel wire is given as

$$\Delta L = \frac{T_2 L}{Ay}$$

$$\Rightarrow \quad \Delta L = \frac{31.4 \times 1.6}{\pi(0.2 \times 10^{-2})^2 \times 2 \times 10^{11}}$$

$$\Rightarrow \quad \Delta L = \frac{16}{2 \times 4 \times 10^{-6} \times 10^{11}}$$

$$\Rightarrow \quad \Delta L = 2 \times 10^{-5} \text{ m}$$

$$\Rightarrow \quad \Delta L = 20 \times 10^{-6} \text{ m}$$ **Ans. (20)**

Sol. 83 By the definition of young's modulus, we have

$$Y = \frac{\text{Stress}}{\text{Strain}} = \frac{Fl}{A\Delta l}$$

$$\Rightarrow \quad \Delta L = \frac{FL}{AY}$$

Thus ratio of elongation in wire A and B is

$$\Rightarrow \quad \frac{\Delta L_A}{\Delta L_B} = \frac{A_B}{A_A}\frac{Y_B}{Y_A} = 12$$ **Ans. (B)**

Sol. 84 Let the original length of wire is l_0 so when tension in wire $T_1 = 100$ N, extension in wire will be $l_1 - l_0$. Similarly when tension is $T_2 = 120$ N, extension will be $l_2 - l_0$.

$$\Rightarrow \quad 100 = K(l_2 - l_0) \quad \ldots(1)$$

$$\Rightarrow \quad 120 = K(l_2 - l_0) \quad \ldots(2)$$

Ratio of equation-(1) and (2) gives

$$\frac{5}{6} = \frac{l_1 - l_0}{l_2 - l_0}$$

$$\Rightarrow \quad 5l_2 - 5l_0 = 6l_1 = 6l_0$$

$$\Rightarrow \quad l_0 = 6l_1 - 5l_2$$

$$\Rightarrow \quad l_0 = 6l_1 - 5\left(\frac{11l_1}{10}\right)$$

$$\Rightarrow \quad l_0 = 6l_1 - \frac{11l_1}{2}$$

$$\Rightarrow \quad l_0 = \frac{l_1}{2} = \frac{1}{x}l_1$$

$$\Rightarrow \quad x = 2$$ **Ans. (2)**

Sol. 85 Bulk modulus at constant temperature

$$B = -\frac{dP}{dV/V}$$

Given that at isothermal condition

$$PV^3 = a$$

Differentiating w.r.t to pressure

$$V^2 + P3V^2\frac{dv}{dP} = 0$$

$$\Rightarrow \quad V = -3\frac{PdV}{dP} = 0$$

$$\Rightarrow \quad V = -3\frac{PdV}{dP}$$

$$\Rightarrow \quad \frac{dPV}{dV} = -3P$$

Thus bulk modulus is given as

$$B = -\left(\frac{dPv}{dv}\right) = -(-3P) = 3P$$ **Ans. (B)**

Sol. 86 Elastic energy per unit volume is given as

$$u_e = \frac{1}{2}\text{stress} \times \text{strain}$$

Total elastic energy stored in wire is calculated as

$$E = \frac{1}{2}\text{stress} \times \text{strain} \times \text{volume}$$

$$\Rightarrow \quad 80 = \frac{1}{2} \times Y \times \text{strain}^2\, A \times l$$

$$\Rightarrow \quad 80 = \frac{1}{2} \times 2 \times 10^{11} \times \frac{(2\times10^{-2})^2}{400} \times A \times 20$$

$$\Rightarrow \quad 20 = \frac{10^{+7}}{20} \times A$$

$$\Rightarrow \quad 40 \times 10^{-6}\,\text{m}^2 = A$$

$$\Rightarrow \quad A = 40\,\text{mm}^2$$

Ans. (40)

Sol. 87 Figure below shows the wire being pulled by the force f as described in the question.

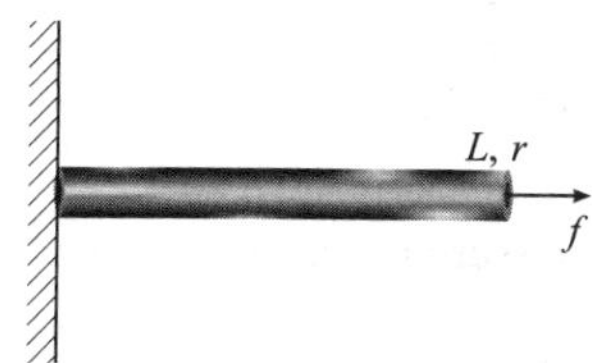

By the definition of Young's modulus, we have

$$Y = \frac{F/A}{l/L} = \frac{fL}{A\cdot l}$$

$$\Rightarrow \quad \frac{f}{\pi r^2} = Y\frac{l}{L} \qquad \ldots(1)$$

For another wire of same material but length $2L$ and radius $2r$ which is pulled with the force $2f$, we use

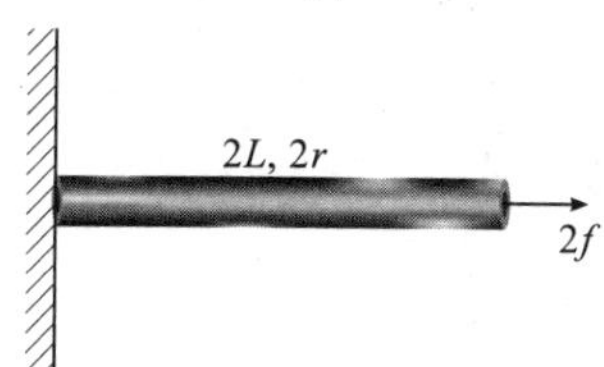

$$\frac{2f}{\pi(2r)^2} = Y\frac{l'}{2L} \qquad \ldots(2)$$

From equations-(1) and (2), we have

$$\frac{2}{1} = \frac{2l'}{l}$$

$$\Rightarrow \quad l' = l$$

Ans. (B)

9.1 Fluid Statics

Sol. 1 As ball sinks in liquid-1

$$\Rightarrow \quad \rho_3 > \rho_1 \qquad \ldots(1)$$

and ball is not sinking in liquid-2

$$\Rightarrow \quad \rho_3 < \rho_2 \qquad \ldots(2)$$

Here in jar liquid-1 is above liquid-2

$$\Rightarrow \quad \rho_1 < \rho_2 \qquad \ldots(3)$$

From-(1), (2) and 3

$$\rho_1 < \rho_3 < \rho_2$$

Ans. (D)

Sol. 2 As $\rho_{oil} < \rho_{water}$

$\Rightarrow$ Oil will be above water in container

$$\rho_{oil} < \rho$$

$\Rightarrow$ Ball will sink in oil

$$\rho < \rho_{water}$$

$\Rightarrow$ Ball will float in water

Ans. (C)

Sol. 3

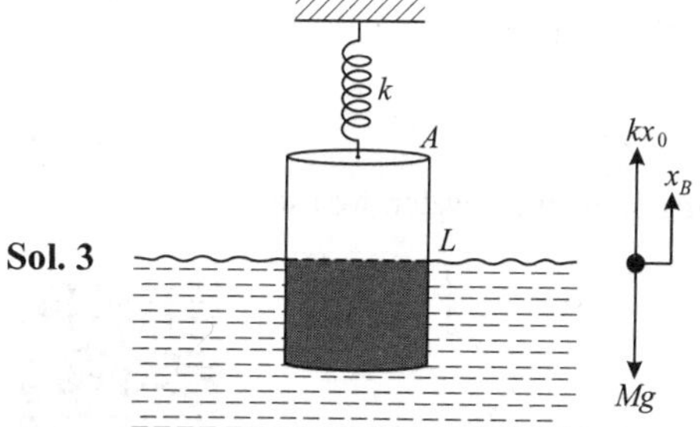

Buoyant force on cylinder due to liquid displaced by submerged part is

$$F_B = V_{\sigma g} = \left(\frac{L}{2}A\right)\sigma g$$

For equilibrium, we use

$$kx_0 + F_B = Mg$$

$$\Rightarrow \quad kx_0 = Mg - \frac{LA\sigma g}{2}$$

$$\Rightarrow \quad x_0 = \frac{1}{k}\left(Mg - \frac{LA\sigma g}{2}\right)$$

$$\Rightarrow \quad x_0 = \frac{Mg}{k}\left(1 - \frac{\sigma LA}{2M}\right)$$

Ans. (B)

Sol. 4 For equilibrium of liquids pressure at any point of a liquid is same at one horizontal level.

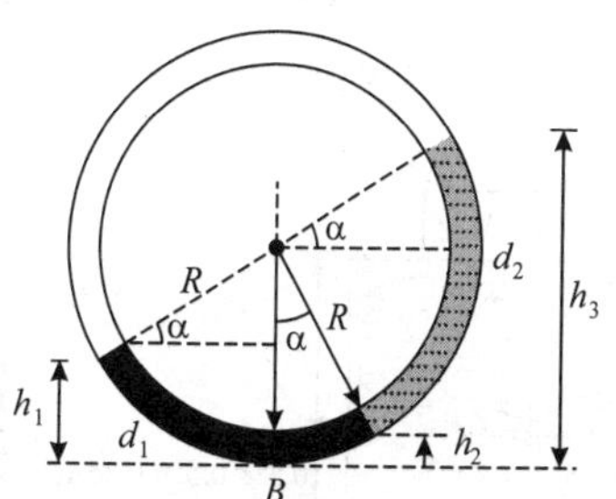

In figure if R is the radius of tube, we have

$$h_1 = R - R\sin\alpha$$
$$h_2 = R - R\cos\alpha$$
$$h_3 = R - R\sin\alpha$$

Pressure due to liquid at point B is given as

$$P_B = h_1 d_1 g$$

and $$P_B = (h_3 - h_2)d_2 g + h_2 d_1 g$$

$$\Rightarrow \quad h_1 d_1 g = (h_3 - h_2)\, d_2 g + h_2 d_1 g$$

$$\Rightarrow \quad (h_1 - h_2)d_1 = (h_3 - h_2)d_2$$

$$\Rightarrow \quad \frac{d_1}{d_2} = \frac{h_3 - h_2}{h_1 - h_2} = \frac{\sin\alpha + \cos\alpha}{\cos\alpha - \sin\alpha}$$

$$\Rightarrow \quad \frac{d_1}{d_2} = \frac{1 + \tan\alpha}{1 - \tan\alpha}$$ **Ans. (C)**

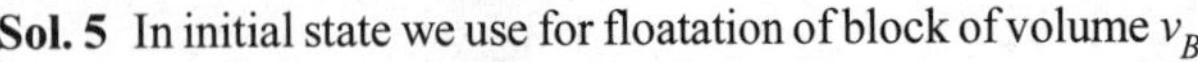

Sol. 5 In initial state we use for floatation of block of volume v_B

$$V_B \rho_B g = \frac{4}{5} V_B \rho_w g$$

$$\Rightarrow \quad \rho_B = \frac{4}{5}\rho_w \quad \ldots(1)$$

Later after pouring oil over the water, we use

$$V_B \rho_B g = \frac{V_B}{2}\rho_w g + \frac{V_B}{2}\rho_0 g$$

$$\Rightarrow \quad \frac{4}{5}\rho_w = \frac{1}{2}\rho_w + \frac{1}{2}\rho_0$$

$$\Rightarrow \quad \frac{3}{10}\rho_w = \frac{1}{2}\rho_0$$

$$\Rightarrow \quad \frac{P_0}{\rho_w} = \frac{6}{10} = 0.6$$ **Ans. (C)**

Sol. 6 At a depth d_1, we have

$$P_{atm} + d_1 \rho g = P_1 = 5.05 \times 10^6 \quad \ldots(1)$$

At a depth d_2, we have

$$P_{atm} + d_2 \rho g = P_2 = 8.08 \times 10^6 \quad \ldots(2)$$

$$(2) - (1) \Rightarrow (d_2 - d_1)\,\rho g = 3.03 \times 10^6$$

$$\Rightarrow \quad d_2 - d_1 = \frac{3.03 \times 10^6}{10^3 \times 10}$$
$$= 3.03 \times 10^2$$
$$= 303 \text{ m}$$ **Ans. (C)**

Sol. 7

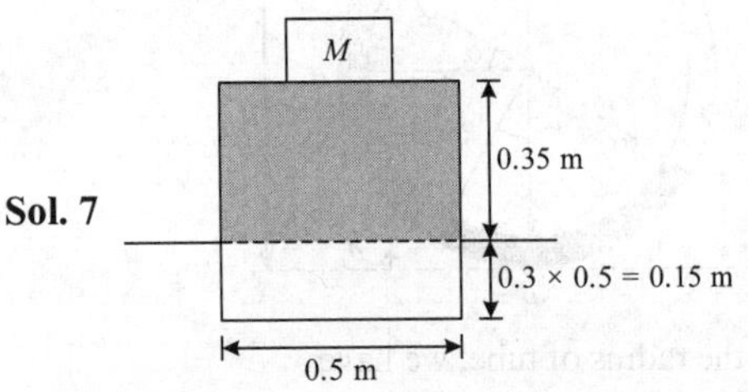

To fully submerge first submerge the complete block weight of mass M is to be balanced by buoyancy due to the 70% part of cubical block outside initially

$$\Rightarrow \quad Mg = (0.5 \times 0.5 \times 0.35) \times \rho_w \cdot g$$

$$\Rightarrow \quad M = (0.25 \times 0.35) \times 10^3 = 87.5 \text{ kg}$$ **Ans. (B)**

Sol. 8 For minimum density of liquid, sphere will be floating while completely submerged so the buoyant force must be balancing the weight. Mass of sphere is given as

$$m = \int_0^R \rho(4\pi r^2\, dr)$$

$$\Rightarrow \quad m = \rho_0 \int_0^R \left(1 - \frac{r^2}{R^2}\right) 4\pi r^2\, dr$$

$$\Rightarrow \quad m = 4\pi\rho_0 \left[\frac{R^3}{3} - \frac{R^3}{5}\right] = \frac{8}{15}\pi\rho_0 R^3$$

For floating, we use

$$mg = V\rho_l g$$

$$\Rightarrow \quad \frac{8}{15} 4\pi\rho_0 R^3 = \frac{4}{3}\pi R^3 \rho_l g$$

$$\Rightarrow \quad \rho_l = \frac{2}{5}\rho_0$$ **Ans. (D)**

Sol. 9 Average pressure on upper part side wall of liquid is $\frac{\rho g h}{2}$, so force on wall MN is given as

$$F_1 = \frac{\rho g h}{2} \times A$$

Average pressure on lower part side wall of liquid is $\left(\rho g h + \frac{2\rho g h}{2}\right)$, so force on side wall ON is given as

$$F_2 = \left(\rho g h + \frac{2\rho g h}{2}\right) A$$

$$\Rightarrow \quad \frac{F_1}{F_2} = \frac{1}{4}$$ **Ans. (A)**

Sol. 10 For bubble, we use

$$F_B - mg = ma$$

$$\Rightarrow \quad V\rho_\omega g - mg = ma$$

$$\Rightarrow \quad m = \frac{V\rho_\omega g}{g + a} = \frac{\frac{4}{3} \times 3.14 \times (1)^3 \times 1 \times 980}{980 + 9.8}$$

$$\Rightarrow \quad m \simeq 4.15 \text{ gm}$$ **Ans. (C)**

Sol. 11 Initial energy of liquid in the two vessels is

$$U_1 = (\rho S x_1) g \left(\frac{x_1}{2}\right) + (\rho S x_2) g \left(\frac{x_2}{2}\right)$$

If after equilibrium, final level becomes x_0, we use

$$U_2 = 2(\rho S x_0) g \left(\frac{x_0}{2}\right)$$

As volume of liquid is constant we use

$$x_0 = \frac{x_1 + x_2}{2}$$

Change in energy is given as

$$\Delta U = U_1 - U_2 = \rho S g \left[\left(\frac{x_1 + x_2}{4}\right) - x_0^2\right]$$

$$\Rightarrow \quad \Delta U = \rho S g \left[\frac{x_1^2}{2} + \frac{x_2^2}{2} - \left(\frac{x_1 + x_2}{2}\right)^2\right]$$

$$\Rightarrow \quad \Delta U = \frac{1}{4} \rho S g (x_1 - x_2)^2$$ **Ans. (C)**

Sol. 12 For floating of shell, we use

$$\frac{4}{3}\pi (R^3 - r^3)\, \rho_s g = \frac{4}{3}\pi R^3 \rho_w g$$

$$\Rightarrow \quad 1 - \left(\frac{r}{R}\right)^3 = \frac{8}{27}$$

$$\Rightarrow \quad \frac{r}{R} = \left(\frac{19}{27}\right)^{1/3} = \frac{19^{1/3}}{3}$$

$$\Rightarrow \quad r = 0.88 \simeq \frac{8}{9} R$$ **Ans. (B)**

Sol. 13 For a hydraulic press in the given situation, we can use pressure on both sides are equal so we have

$$\frac{100 \times g}{A_2} = \frac{mg}{A_1} \quad \ldots(1)$$

Let m mass can lift M_0 in second case then, we use

$$\frac{M_0 g}{16 A_2} = \frac{mg}{A_1/16} \quad \ldots(2)$$

From equation (1) and (2) we get

$$\frac{M_0}{16 \times 100} = 16$$

$$\Rightarrow \quad M_0 = 25600 \text{ kg}$$ **Ans. (25600)**

Sol. 14 At a depth d, pressure on submarine is given as

$$P_1 = \rho g d + P_0 = 3 \times 10^5 \text{ Pa}$$

$$\Rightarrow \quad \rho g d = 2 \times 10^5 \text{ Pa}$$

When the depth is doubled, pressure becomes

$$P_2 = 2\rho g d + P_0$$

$$\Rightarrow \quad P_2 = 4 \times 10^5 + 10^5$$

$$\Rightarrow \quad P_2 = 5 \times 10^5 \text{ Pa}$$

Percentage increase in pressure is given as

$$\%\Delta = \frac{P_2 - P_1}{P_1} \times 100$$

$$\Rightarrow \quad \%\Delta = \frac{5 \times 10^5 - 3 \times 10^5}{3 \times 10^5} \times 100 = \frac{200}{3}\%$$ **Ans. (A)**

Sol. 15 Terminal speed of raindrop is given as

$$v_T = \frac{2R^2 (f_w - f_a) g}{9\eta}$$

$$\Rightarrow \quad v_T = \frac{2 \times (0.002)^2 \times (1000 - 1.2) \times 10}{9 \times 1.8 \times 10^{-5}}$$

$$\Rightarrow \quad v_T \approx 4.94$$ **Ans. (A)**

Sol. 16 Figure below shows the situation described in the quesion.

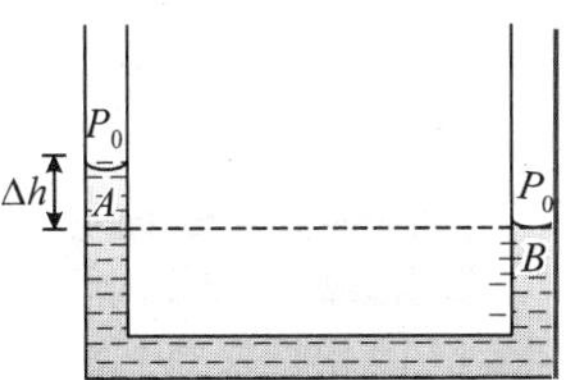

The pressure at a point just below the meniscus is $2T/R$ lower than that of atmospheric pressure so at the bottom of the tube, from left and right arm of tube, pressure is calculated from the two arms as

$$P_0 - \frac{2T}{R_1} + h_1 \rho g = P_0 - \frac{2T}{R_2} + h_2 \rho g$$

$$\Rightarrow \quad (h_1 - h_2)\rho g = 2T \left[\frac{1}{\frac{5}{2} \times 10^{-3}} - \frac{1}{\frac{8}{2} \times 10^{-3}}\right];$$

$$\Rightarrow \quad h_1 - h_2 = \frac{2 \times 7.3 \times 10^{-2}}{1000 \times 10 \times 10^{-3}} \times \frac{3}{20}$$

$$= 2.19 \text{ mm}$$ **Ans. (B)**

Sol. 17 For an oil drop in experiment no electric force acts on uncharged drop so viscous force balances its weight, given as

Viscous force = weight

$$\Rightarrow \quad \text{Viscous force} = mg = \rho \times \left(\frac{4}{3}\pi r^3\right) \times g$$

$$\Rightarrow \quad \text{Viscous force} = 3.9 \times 10^{-10} \text{ N}$$ **Ans. (D)**

Sol. 18 Figure below shows the process described in question. When water level equalizes, the final height of water in both vessels becomes

$$h = \frac{h_1 + h_2}{2}$$

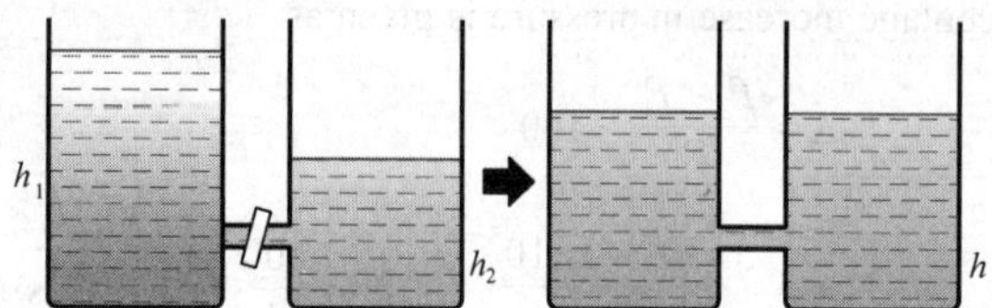

The work done by force of gravity in this process is equal to change in gravitational potential energy of system, calculated as

$$W = U_i - U_f$$

$$\Rightarrow \quad W = (\rho A h_1) g \frac{h_1}{2} + (\rho A h_2) g \frac{h_2}{2} - \rho A (h_1 + h_2) g \left(\frac{h_1 + h_2}{4} \right)$$

$$\Rightarrow \quad W = \frac{\rho A g}{2} \left[h_1^2 + h_2^2 - \frac{(h_1 + h_2)^2}{2} \right]$$

$$\Rightarrow \quad W = 1\text{ J}$$ **Ans. (B)**

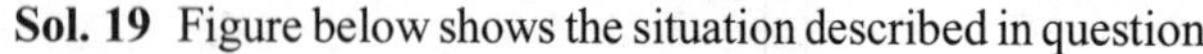

Sol. 19 Figure below shows the situation described in question.

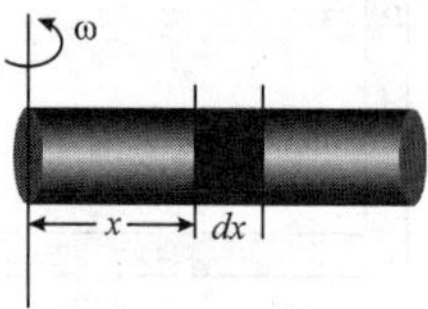

The force on elemental length of fluid dx as shown in figure due to rotation is given as

$$F = \int (dm) \omega^2 x$$

$$\Rightarrow \quad F = \int_0^L \left(\frac{m}{L} dx \right) \omega^2 x$$

$$\Rightarrow \quad F = \frac{m}{L} \omega^2 \frac{L^2}{2} = \frac{m\omega^2 L}{2}$$

$$\Rightarrow \quad \omega = \sqrt{\frac{2}{mL}} \sqrt{F}$$

$$\Rightarrow \quad \omega = \sqrt{\frac{2}{0.25 \times 0.5}} \sqrt{F}$$

$$\Rightarrow \quad \omega = \sqrt{16}\sqrt{F} = 4\sqrt{F}$$ **Ans. (4)**

Sol. 20 (R) is the statement of Pascal's principle & which explains the assertion (A). According to Pascal's law, when there is change in pressure inside fluid at rest then this change in pressure is equally transmitted in all directions at same level. **Ans. (C)**

Sol. 21 Weight capacity of hydraulic lift

$$\text{Force} = mg = 5000\text{ g}$$

Area of cross section

$$A = 250\text{ cm}^2 = 250 \times 10^{-4}\text{ m}^2$$

Maximum pressure can be calculated as

$$P = \frac{5000g}{250 \times 10^{-4}}$$

$$\Rightarrow \quad P = \frac{20 \times g}{10^{-4}}$$

$$\Rightarrow \quad P = 2 \times 10^6\text{ Pa}$$ **Ans. (C)**

Sol. 22 Pressure in a static liquid will be same at each point on same horizontal level.

$$P = P_{atm} + \rho g h$$

As per Pascal law, same pressure applied to enclosed water is transmitted in all directions equally. **Ans. (B)**

9.2 Fluid Dynamics

Sol. 1

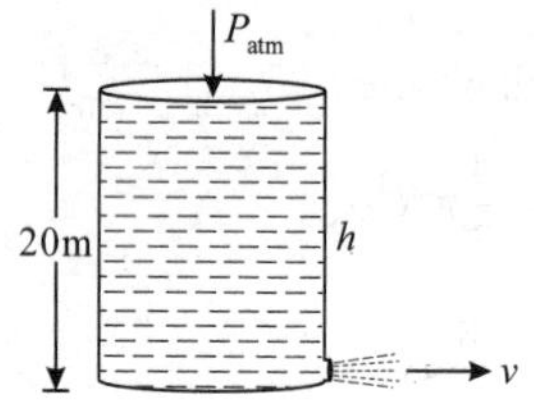

Efflux velocity $\quad v = \sqrt{2gh}$

$$\Rightarrow \quad v = \sqrt{2 \times 10 \times 20} = 20\text{ m/s}$$ **Ans. (B)**

Sol. 2

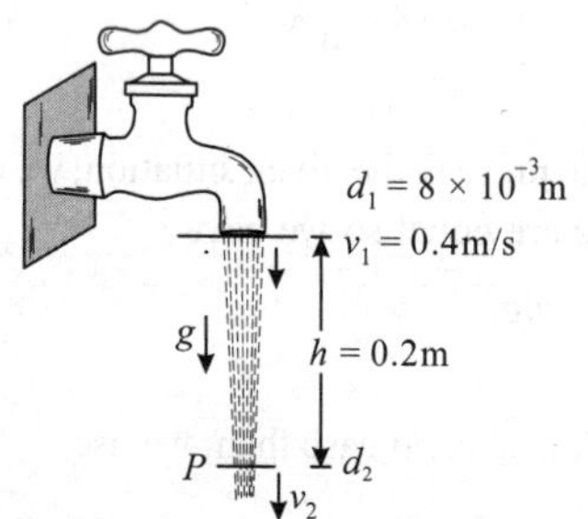

By speed equation for free fall motion of water stream we have

$$v_2^2 = v_1^2 + 2gh$$

$$\Rightarrow \quad v_2^2 = (0.4)^2 + 2 \times 10 \times 0.2 = 4.16$$

$$\Rightarrow \quad v_2 = \sqrt{4.16}\text{ m/s}$$

By continuity equation is stream of water, we use

$$A_1 v_1 = A_2 v_2$$

$$\left(\frac{\pi d_1^2}{4} \right) v_1 = \left(\frac{\pi d_2^2}{u} \right) v_2$$

$$d_2 = d_1 \sqrt{\frac{v_1}{v_2}} = 8 \times 10^{-3} \times \sqrt{\frac{0.4}{4.16}}$$

$$\Rightarrow \quad d_2 \cong 3.6 \times 10^{-3}$$ **Ans. (B)**

Sol. 3 Efflux velocity of water coming out from holes

$$v_1 = \sqrt{2gh}$$

and $$v_2 = \sqrt{2g(4h)} = 2\sqrt{2gh}$$

As flows rate from holes is same, we use

$$A_1v_1 = A_2v_2$$

$$\Rightarrow \quad l^2\sqrt{2gh} = \pi r^2 \times 2\sqrt{2gh}$$

$$\Rightarrow \quad r^2 = \frac{l^2}{2\pi}$$

$$\Rightarrow \quad r = \frac{l}{\sqrt{2\pi}}$$ **Ans. (A)**

Sol. 4

$$P - P' = 3 \times 10^5$$

Using Bernoulli's equation at the ends of tube we have

$$P + \frac{1}{2}\rho v^2 = P' + \frac{1}{2}\rho V'^2$$

$$\Rightarrow \quad P - P' = \frac{1}{2}\rho(v'^2 - v^2) \quad \ldots(1)$$

By continuity equation, we use

$$Av = A'v'$$

$$\Rightarrow \quad v' = \frac{A}{A'}v = 5v$$

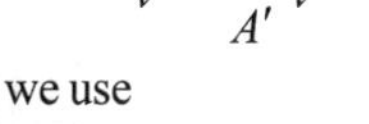

From (1), we use

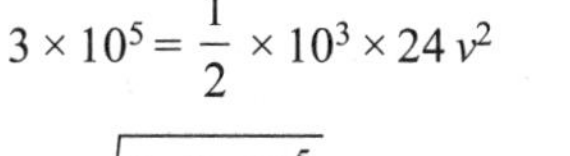

$$3 \times 10^5 = \frac{1}{2} \times 10^3 \times 24\,v^2$$

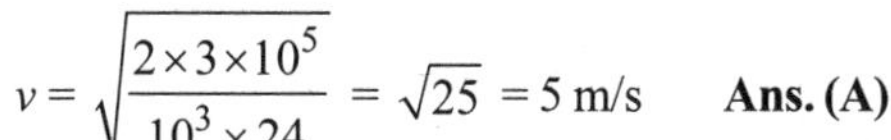

$$\Rightarrow \quad v = \sqrt{\frac{2 \times 3 \times 10^5}{10^3 \times 24}} = \sqrt{25} = 5 \text{ m/s}$$ **Ans. (A)**

Sol. 5 Efflux velocity of water from hole A is

$$v_1 = \sqrt{2gh}$$

Range of water from hole A is

$$R_A = v_1 \times \sqrt{\frac{2(H - h_1)}{g}}$$

$$R_A = 2\sqrt{h_1(H - h_1)} \quad \ldots(1)$$

Efflux velocity of water from hole B is

$$v_2 = \sqrt{2g(H - h_2)}$$

Range of water from hole B is

$$R_B = v_2 \times \sqrt{\frac{2h_2}{g}} = 2\sqrt{h_2(H - h_2)} \quad \ldots(2)$$

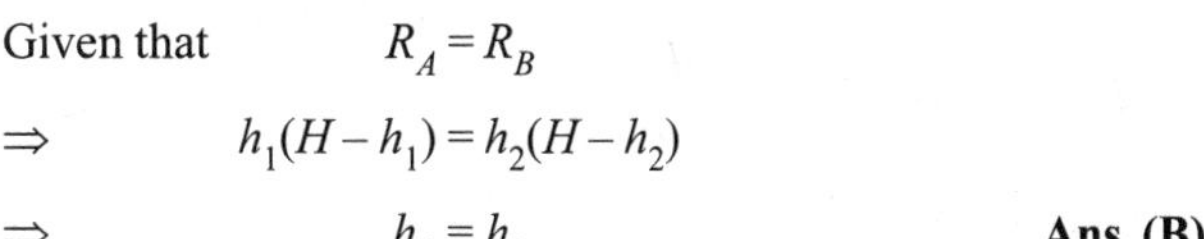

Given that $R_A = R_B$

$$\Rightarrow \quad h_1(H - h_1) = h_2(H - h_2)$$

$$\Rightarrow \quad h_1 = h_2$$ **Ans. (B)**

Sol. 6

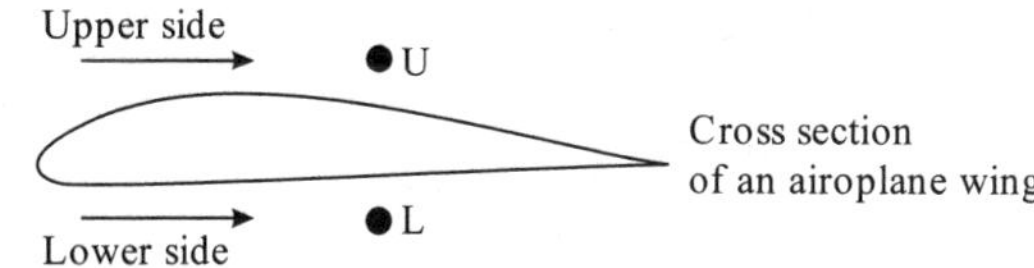

Wind velocity

$$v_U = 150 \text{ m/s}$$

$$v_L = 100 \text{ m/s}$$

If P_U and P_L are the air pressures at upper and lower sides of wings then we use Barnoullie's theorem, we have

$$P_U + \frac{1}{2}\rho v_U^2 = P_L + \rho v_L^2$$

$$\Rightarrow \quad P_L - P_U = \frac{1}{2}\rho(v_U^2 - v_L^2)$$

$$\Rightarrow \quad P_L - P_U = \frac{1}{2} \times 1.2 \times ((150)^2 - (100)^2)$$

$$\Rightarrow \quad P_L - P_U = 0.6 \times 10^4 \times (2.25 - 1) = 6 \times 1.25 \times 103$$

$$\Rightarrow \quad P_L - P_U = 7500 \text{ N/m}^2$$ **Ans. (C)**

Sol. 7

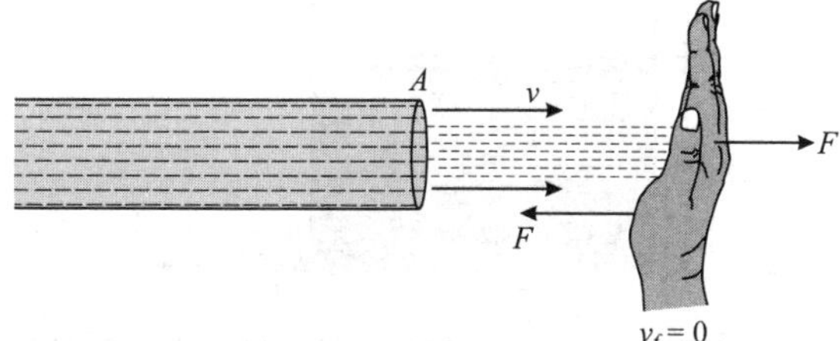

Water stops when hit the palm and it transfers momentum to the hand.

Force exerted = rate of change of momentum
= momentum in hitting water/sec

$$F = \rho(Av) \times V = \rho Av^2$$

$$= 10^3 \times 10^{-2} \times (1.5)^2$$

$$\Rightarrow \quad F = 22.5 \text{ N}$$ **Ans (B)**

$\rho \rightarrow$ density; $Av \rightarrow$ volume flow rate

Sol. 8

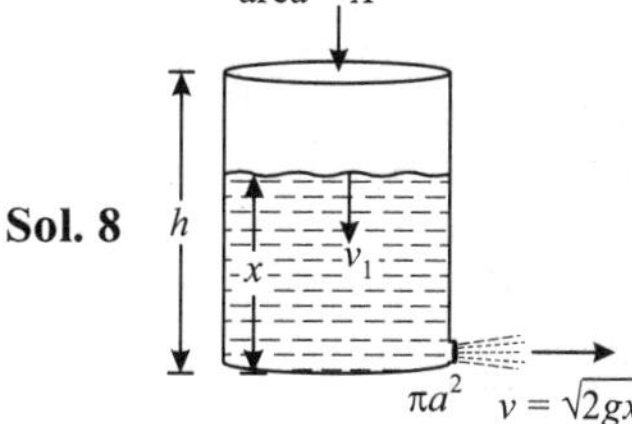

Here we can use $v_1 = -\dfrac{dx}{dt}$

By equation of contains we use

$$V_1 A = \sqrt{2gx}\,.(\pi a^2)$$

$$\Rightarrow \quad -A\frac{dx}{dt} = \pi a^2\sqrt{2g}.\sqrt{x}$$

$$\Rightarrow \quad \frac{dx}{\sqrt{x}} = -\frac{\pi a^2}{A}\sqrt{2g}\,dt$$

with time x is decreasing and if in time T if the vessel is emptied than we use

$$\int_{x=h}^{x=0}\frac{dx}{\sqrt{x}} = -\frac{\pi a^2}{A}\sqrt{2g}\int_{t=0}^{t=T} dt$$

$$\Rightarrow \quad \left[2\sqrt{x}\right]_h^0 = -\frac{\pi a^2}{A}\sqrt{2g}[t]_0^T$$

$$\Rightarrow \quad 2\sqrt{h} = \frac{\pi a^2}{A}\sqrt{2g}\,.\,T$$

$$\Rightarrow \quad T = \frac{A}{\pi a^2}\sqrt{\frac{2h}{g}} \qquad \textbf{Ans. (B)}$$

Sol. 9

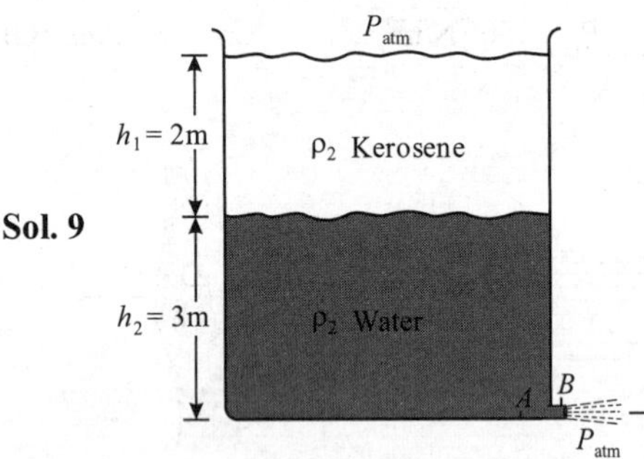

Applying Beronoulli's theorem at points A and B, we have

$$P_{atm} + h_1\rho_1 g + h_2\rho_2 g = P_{atm} + \frac{1}{2}\rho_2 v^2$$

$$\Rightarrow \quad v = \sqrt{\frac{2h_1\rho_1 g}{\rho_2} + 2h_2 g}$$

$$\Rightarrow \quad v = \sqrt{\frac{2\times 2\times 800\times 10}{1000} + 2\times 3\times 10}$$

$$\Rightarrow \quad v = \sqrt{32+60} = \sqrt{92} = 9.59 \text{ m/s} \qquad \textbf{Ans. (B)}$$

Sol. 10 We apply Beronoulli's equation at A and B

$$P_A + \frac{1}{2}\rho v_A^2 = P_B + \frac{1}{2}\rho v_B^2$$

$$\Rightarrow \quad P_B - P_A = \rho\left(v_A^2 - v_B^2\right)$$

$$\Rightarrow \quad h\rho g = \frac{1}{2}\rho v_A^2\left(-\frac{A_A^2}{A_B^2}\right)$$

$$\Rightarrow \quad v_A = \sqrt{\frac{2gh}{\left(-\frac{A_A^2}{A_B^2}\right)}} = \sqrt{\frac{2\times 10\times 0.05}{1-(3/5)^2}}$$

$$\Rightarrow \quad v_A = \sqrt{\frac{1}{16/25}} = \frac{5}{4}\text{ m/s}$$

By continuity equation

$$A_A v_B = A_B v_B \qquad \ldots(1)$$

We have $\quad P_B - P_A = h\rho g$

From (1) $\quad v_B = \dfrac{A_A v_A}{A_B}$

Fluid flow rate through tube is

$$\frac{dV}{dt} = A_A v_A = 6\times 10^{-6}\times 1.25$$

$$\Rightarrow \quad \frac{dV}{dt} = 7.5\times 10^{-6}\text{ m}^3\text{/s} \qquad \textbf{Ans. (A)}$$

Sol. 11 As $r << R$, the efflux velocity of water coming out from hole is given as

$$u = \sqrt{2gh}$$

If v is the speed at which it hits ground

$$\Rightarrow \quad v^2 = u^2 + 2gh = 2g(H+h)$$

$$\Rightarrow \quad v = \sqrt{2g(H+h)}$$

Now we use equation of continuity at hole and where water stream strikes the ground

$$\Rightarrow \quad A_1 v_1 = A_2 v_2$$

$$\Rightarrow \quad \pi r^2\sqrt{2gH} = \pi x^2\sqrt{2g(H+h)}$$

$$\Rightarrow \quad x = r\left(\frac{H}{H+h}\right)^{1/4} \qquad \textbf{Ans. (A)}$$

Sol. 12 By equation of continuity, we use

In flow volume = Out flow volume

$$\Rightarrow \quad \frac{0.74}{60} = (\pi\times 4\times 10^{-4})\times\sqrt{2gh}$$

$$\Rightarrow \quad \sqrt{2gh} = \frac{74\times 100}{240\pi}$$

$$\Rightarrow \quad 2gh = \frac{740\times 740}{24\times 24\times 10}$$

$$\Rightarrow \quad h = \frac{74\times 74}{2\times 24\times 24}$$

$$\Rightarrow \quad h \approx 4.8\text{ m} \qquad \textbf{Ans. (B)}$$

Sol. 13

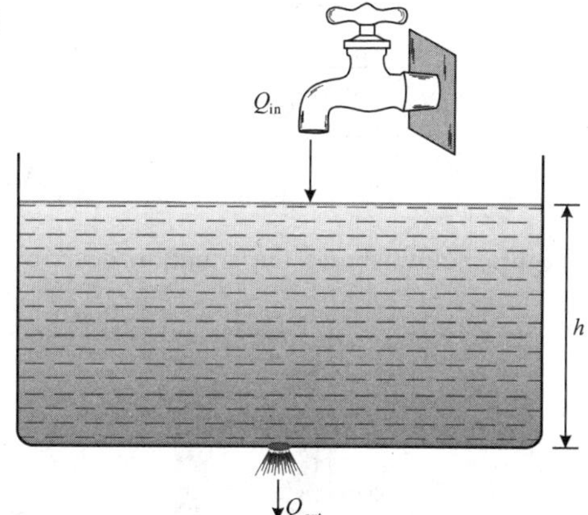

As height of water column is constant, we use water inflow rate (Q_{in}) = water outflow rate (Q_{out})

$\Rightarrow \quad Q_{in} = 10^{-4}\ \text{m}^3/\text{s}$

$\Rightarrow \quad Q_{out} = Av = 10^{-4} \times \sqrt{2gh}$

$\Rightarrow \quad 10^{-4} = 10^{-4}\sqrt{20 \times h}$

$\Rightarrow \quad h = \frac{1}{20}\ \text{m} = 0.05\ \text{m} = 5\ \text{cm}$ **Ans. (D)**

Sol. 14 Momentum per second carried by liquid per second is $\rho A v^2$ where area of hose pipe is $A = \pi a^2$

Net force due to reflected liquid $= 2\times\left[\frac{1}{4}\rho A v^2\right]$

Net force due to stopped liquid $= \frac{1}{4}\rho A v^2$

Total force

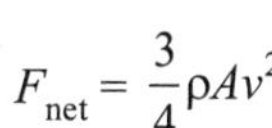

$$F_{net} = \frac{3}{4}\rho A v^2$$

Net pressure $= p = \frac{F_{net}}{A} = \frac{3}{4}\rho v^2$ **Ans. (B)**

Sol. 15

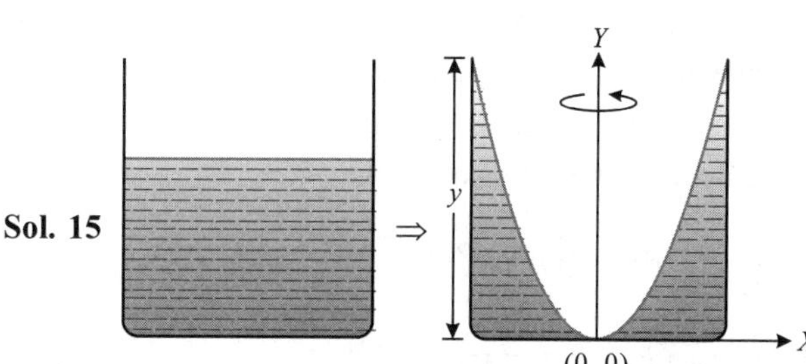

For the figure shown above equation of the liquid surface profile in a rotating container is given as

$$y = \frac{\omega^2 x^2}{2g} = \frac{(2\times 2\pi)^2 \times (0.05)^2}{20} \simeq 2\ \text{cm}$$ **Ans. (C)**

Sol. 16

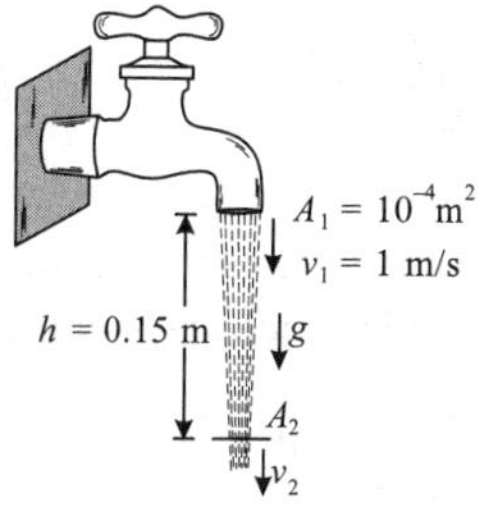

By equation of continuity we use

$$A_1 v_1 = A_2 v_2 \quad \ldots(1)$$

By acceleration due to gravity, we have

$$v_2^2 = v_1^2 + 2gh$$

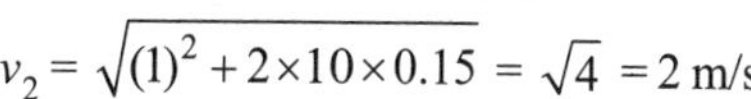

$$v_2 = \sqrt{(1)^2 + 2\times 10\times 0.15} = \sqrt{4} = 2\ \text{m/s}$$

From equation-(1)

$$A_2 = \frac{A_1 v_1}{v_2} = \frac{10^{-4}\times 1}{2} = 5\times 10^{-5}\ \text{m}^2$$ **Ans. (B)**

Sol. 17 By continuity equation, we use

$$A_1 v_1 = A_2 v_2$$

$\Rightarrow \quad \frac{v_{min}}{v_{max}} = \frac{A_{min}}{A_{max}}$

$\Rightarrow \quad \frac{v_{min}}{v_{max}} = \left(\frac{4.8}{6.4}\right)^2$

$\Rightarrow \quad \frac{v_{min}}{v_{max}} = \frac{9}{16}$ **Ans. (D)**

Sol. 18 Rate of flow of water through the tube is given as

$$Q = A_A v_A = A_B v_B$$

$\Rightarrow \quad (40)v_A = (20)v_B$

$\Rightarrow \quad v_B = 2v_A \quad \ldots(1)$

Applying Bernoulli's theorem at A and B, we have

$$P_A + \frac{1}{2}\rho v_A^2 = P_B + \frac{1}{2}\rho v_B^2$$

$\Rightarrow \quad P_A - P_B = \frac{1}{2}\rho\,(v_B^2 - v_A^2)$

$\Rightarrow \quad 700 = \frac{1}{2}\times 1000\,(4v_A^2 - v_A^2)$

$\Rightarrow \quad v_A = 0.68\ \text{m/s} = 68\ \text{cm/s}$

$\Rightarrow \quad Q = A_A v_A$

$\Rightarrow \quad Q = (40)(68) = 2720\ \text{cm}^3/\text{s}$ **Ans. (C)**

Sol. 19 Using Bernoulli's Equation at the two points, we have

$$P + \frac{1}{2}\rho v^2 = \frac{P}{2} + \frac{1}{2}\rho V^2$$

$\Rightarrow \quad \frac{2P}{2\rho} + \frac{1}{2}\frac{\rho v^2}{\rho} \times 2 = V^2$

$$\Rightarrow \quad V=\sqrt{\frac{P}{\rho}+v^2} \qquad \textbf{Ans. (B)}$$

Sol. 20 In the given situation described in question, we use Bernoulli's equation just inside and outside the opening, it gives

$$\left(P_0+\frac{mg}{A}\right)+\rho gH+\frac{1}{2}\rho v_1^2=P_0+0+\frac{1}{2}\rho v^2 \qquad \ldots(1)$$

As cross sectional area of vessel is very large, we can neglect v_1 which gives

$$v=\sqrt{2gH+\frac{2mg}{A\rho}}$$

$$\Rightarrow \quad v=\sqrt{8+1.2}=3.033\text{ m/s}\sim 3\text{m/s} \qquad \textbf{Ans. (3)}$$

Sol. 21 Force exerted by ejection of water on the container can be given by rate of momentum ejection per unit time, calculated as

$$F=\left(\frac{dm}{dt}\right)v=(\rho av)v$$

$$\Rightarrow \quad F=\rho av^2$$

The velocity of efflux of water is given as

$$v=\sqrt{2gh}$$

For vessel not to slide, we use

$$\mu mg\geq \rho a(2gh)$$

$$\Rightarrow \quad \mu\rho(Ah)g\geq \rho a2gh$$

$$\Rightarrow \quad \mu_{min}=\frac{2a}{A} \qquad \textbf{Ans. (C)}$$

Sol. 22 R is maximum $dR/dh=0$, which gives

$$h'=h=H/2=12/2=6\text{ m} \qquad \textbf{Ans. (6)}$$

Sol. 23 From continuity equation, we have

$$av_1=\frac{a}{2}v_2$$

$$\Rightarrow \quad v_2=2v_1$$

From Bernoulli's theorem, we have

$$P_1+\rho gh_1+\frac{1}{2}\rho v_1^2=P_2+\rho gh_2+\frac{1}{2}\rho v_2^2$$

$$\Rightarrow \quad P_1-P_2=\rho\left[\left(\frac{v_2^2-v_1^2}{2}\right)+g(h_2-h_1)\right]$$

$$\Rightarrow \quad 4100=800\left[\left(\frac{v_1^2-v_1^2}{2}\right)+10\times(0-1)\right]$$

$$\Rightarrow \quad \frac{41}{8}+10=\frac{3v_1^2}{2}$$

$$\Rightarrow \quad \frac{121}{8}\times\frac{2}{3}=v_1^2$$

$$\Rightarrow \quad v_1=\sqrt{\frac{121}{4\times3}\times\frac{3}{3}}$$

$$\Rightarrow \quad v_1=\frac{\sqrt{363}}{6}\text{ m/s}$$

$$\Rightarrow \quad X=363 \qquad \textbf{Ans. (363)}$$

Sol. 24 Figure below shows the situation described in the question.

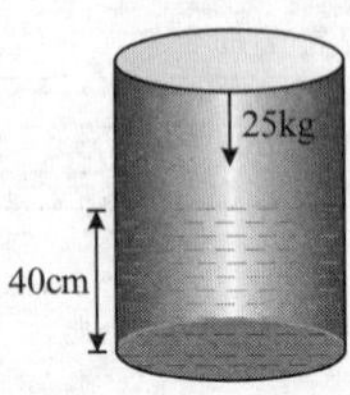

Using Bernoulli's equation, we have

$$P_0+\frac{250}{0.5}+\rho g(40\times10^{-2})=P_0+\frac{1}{2}\rho v^2$$

$$\Rightarrow \quad 500+\frac{1000\times10\times40}{100}=\frac{1}{2}\times1000\times v^2$$

$$\Rightarrow \quad v=3\text{ m/s}$$

$$\Rightarrow \quad v=300\text{ cm/s} \qquad \textbf{Ans. (300)}$$

Sol. 25 Figure below shows the situation described in the question.

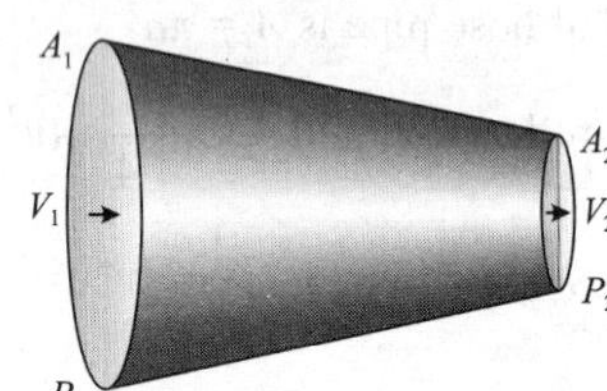

As $\quad A_2=\frac{A_1}{2}$

Using continuity equation, we have

$$A_1v_1=A_2v_2$$

$$\Rightarrow \quad v_2=2v_1 \qquad \ldots(1)$$

Pressure difference across pipe is given as

$$P_1-P_2=4500\text{ Pa}$$

Using Bernoulli's equation, we have

$$P_1+\frac{1}{2}\rho v_1^2+\rho gh=P_2+\frac{1}{2}\rho v_2^2+\rho gh$$

$$\Rightarrow \quad P_1-P_2=\frac{1}{2}\rho(v_2^2-v_1^2) \qquad \ldots(2)$$

$$\Rightarrow \quad 4500=\frac{1}{2}\times750\times3v_1^2$$

$$\Rightarrow \quad v_1=2\text{ m/s}$$

Volume flow rate of liquid is given as

$$Q=A_1v_1=24\times10^{-3}\text{ m}^3\text{s}^{-1} \qquad \textbf{Ans. (24)}$$

Sol. 26 Force exerted on the wall due to momentum in water is given as

$$F = \rho a v^2 = 10^3 \times 10 \times 10^{-4} \times 20 \times 20$$

$$\Rightarrow \quad F = 400 \qquad \textbf{Ans. (D)}$$

Sol. 27 Using Bernoulli's theorem at points just below the piston and just outside the hole gives

$$P_A + \rho g h = P_0 + \frac{1}{2}\rho v_e^2$$

$$\Rightarrow \quad \frac{5\times10^5}{\pi} + 10^3 \times 10 \times 10 = 1.01 \times 10^5 + \frac{1}{2} \times 10^3 \times v_e^2$$

$$\Rightarrow \quad v_e = 17.8 \text{ m/s} \qquad \textbf{Ans. (C)}$$

Sol. 28 The weight of aircraft is balanced by the upward force due to pressure difference on wings, given as

$$P_2 A - P_1 A = 5.4 \times 10^5 \times g$$

$$\Rightarrow \quad P_2 - P_1 = \frac{5.4\times10^6}{500} = 5.4 \times 2 \times 10^2 \times 10$$

$$\Rightarrow \quad P_2 - P_1 = 10.8 \times 10^3 \text{ Pa}$$

Using Bernoulli's theorem above and below point of wings is written as

$$P_2 + 0 + \frac{1}{2}\rho v_2^2 = P_1 + 0 + \frac{1}{2}\rho v_1^2$$

$$\Rightarrow \quad P_2 - P_1 = \frac{1}{2}\rho(v_1^2 - v_2^2) = \frac{1}{2}\rho(v_1 - v_2)(v_1 + v_2)$$

$$\Rightarrow \quad 10.8 \times 10^3 = \frac{1}{2} \times 1.2\,(v_1 - v_2) \times 2 \times 3 \times 10^2$$

$$\Rightarrow \quad 10.8 \times 10 = 3.6(v_1 - v_2)$$

$$\Rightarrow \quad v_1 - v_2 = 30$$

$$\Rightarrow \quad \left(\frac{v_1 - v_2}{v}\right) \times 100 = \frac{30}{300} \times 100 = 10\% \qquad \textbf{Ans. (D)}$$

Sol. 29 By continuity equation, we have

$$A_1 v_1 = A_2 v_2$$

$$\Rightarrow \quad 750 \times 10^{-4} v_1 = 500 \times 10^{-6} \times 0.3$$

$$\Rightarrow \quad v_1 = \frac{500\times3\times10^{-3}}{750} \text{ m/s}$$

$$\Rightarrow \quad v_1 = 2 \times 10^{-3} \text{ m/s}$$

$$v_1 = \frac{dh}{dt} = -2 \times 10^{-3} \text{ m/s} \qquad \textbf{Ans. (2)}$$

Sol. 30 By equation of continuity, we have

$$A_1 V_1 = A_2 V_2$$

$$\Rightarrow \quad V_2 = \frac{A_1 V_1}{A_2}$$

$$\Rightarrow \quad V_2 = \frac{2\times4}{10\times10^{-2}} = 80 \text{ cm/s} \qquad \textbf{Ans. (80)}$$

Sol. 31 Figure below shows the flow of glycerine as described in the question.

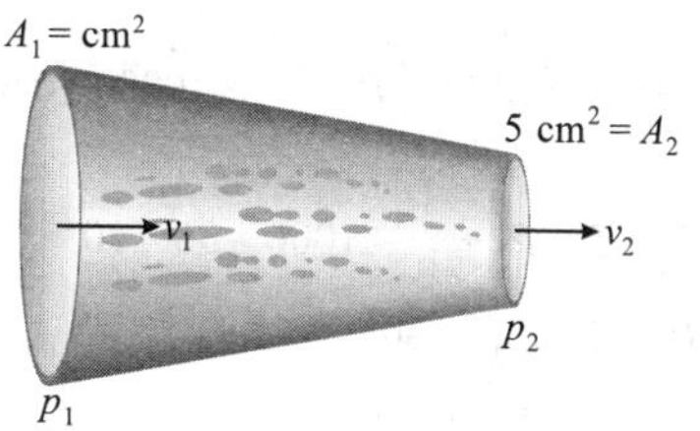

Pressure drop across the length of pipe is given as

$$\Delta P = P_1 - P_2 = 3 \text{ N/m}^2 \text{ (given)}$$

By continuity equation, we have

$$A_1 v_1 = A_2 v_2$$

$$\Rightarrow \quad v_1 = \frac{A_2}{A_1} v_2 \qquad \ldots(1)$$

By Bernoulli's equation, we have

$$P_1 + \frac{1}{2}\rho v_1^2 = P_1 + \frac{1}{2}\rho v_1^2$$

$$\Rightarrow \quad P_1 - P_2 = \frac{1}{2}\rho(v_2^2 - v_1^2)$$

$$\Rightarrow \quad \Delta P = \frac{1}{2}\rho\left(v_2^2 - \frac{A_2^2}{A_1^2}v_2^2\right)$$

$$\Rightarrow \quad \Delta P = \frac{1}{2}\rho\left[1 - \left(\frac{A_2}{A_1}\right)^2\right]v_2^2$$

$$\Rightarrow \quad 3 = \frac{1}{2} \times 1.25 \times 10^3 \left[1 - \left(\frac{5}{10}\right)^2\right]v_2^2$$

$$\Rightarrow \quad 3 = \frac{1}{2} \times 1.25 \times 10^3 \left[1 - \frac{1}{4}\right]v_2^2$$

$$\Rightarrow \quad 3 = \frac{1}{2} \times 1.25 \times 10^3 \times \frac{3}{4}v_2^2$$

$$\Rightarrow \quad v_2 = 8 \times 10^{-2} \text{ m/s}$$

Discharge rate of glycerine is calculated as

$$Q = A_2 v_2$$

$$\Rightarrow \quad Q = 5 \times 10^{-4} \times 8 \times 10^{-2}$$

$$\Rightarrow \quad Q = 4 \times 10^{-5} \text{ m}^3\text{/s}$$

$$\Rightarrow \quad x = 4 \qquad \textbf{Ans. (4)}$$

Sol. 32 From continuity equation, we have

$$A_1 V_1 = A_2 V_2$$

$$\Rightarrow \quad 1.5 \times V_1 = 25 \times 10^{-2} \times 60$$

$$\Rightarrow \quad V_1 = \frac{25\times60\times10^{-2}\times10}{1.5}$$

$$\Rightarrow \quad V_1 = 10 \text{ cm/s}$$

By Bernoulli's theorem

$$P_1 + \frac{1}{2}\delta_1 V_1^2 = P_2 + \frac{1}{2}\delta_2 V_2^2$$

$$P_1 + \frac{1}{2} \times 1000 \times (0.1)^2 = P_2 + \frac{1}{2} \times 1000 \times (0.6)^2$$

$$\Rightarrow \quad P_1 + 5 = P_2 + \frac{1}{2} \times 1000 \times 36 \times 10^{-2}$$

$$\Rightarrow \quad P_1 + 5 = P_2 + 180$$

$$\Rightarrow \quad P_1 - P_2 = 175 \text{ Pa}$$

Ans. (A)

9.3 Viscosity

Sol. 1

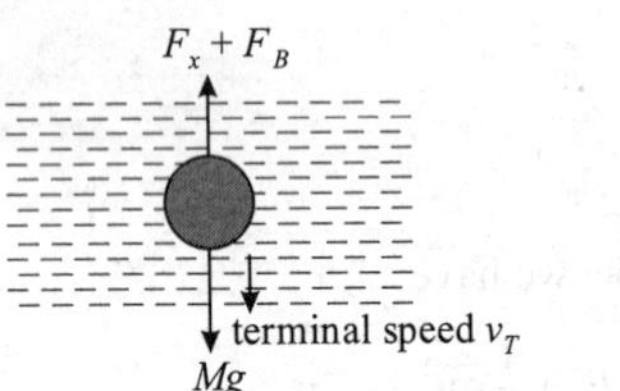

Using stoke's Rule we know before terminal speed is obtained, ball will have an acceleration which decreases because viscous force on ball is given as

$$F_v = 6\pi\eta R v$$

$$\Rightarrow \quad F_v \propto Rv$$

Ans. (B)

Sol. 2 In a liquid, terminal speed of sphere is given as

$$v_T = \frac{2}{9}\frac{gR^2}{\eta}(\rho_s - \rho_l)$$

For gold sphere, we have

$$v_1 = \frac{2}{9}\frac{9R^2}{\eta}(19.5 - 1.5) \quad \ldots(1)$$

For silver sphere, we have

$$v_2 = \frac{2}{9}\frac{gR^2}{\eta}(10.5 - 1.5) \quad \ldots(2)$$

Using (1)/(2) gives

$$\frac{v_1}{v_2} = \frac{18}{9} = 2$$

$$\Rightarrow \quad v_2 = \frac{v_1}{2} = \frac{0.2}{2} = 0.1 \text{ m/s}$$

Ans. (C)

Sol. 3

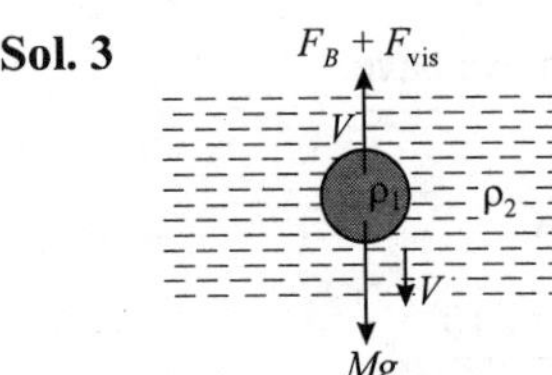

Terminal speed is attained by ball when its acceleration $a = 0$

$$\Rightarrow \quad F_B + F_{vis} = mg$$

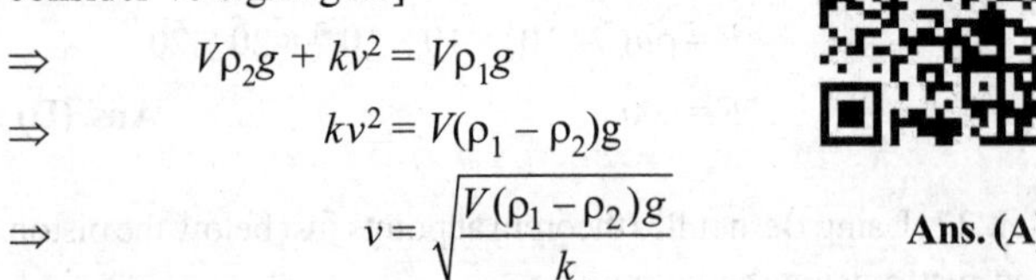

[F_{vis} is in opposite direction of v so need to consider-ve sign again]

$$\Rightarrow \quad V\rho_2 g + kv^2 = V\rho_1 g$$

$$\Rightarrow \quad kv^2 = V(\rho_1 - \rho_2)g$$

$$\Rightarrow \quad v = \sqrt{\frac{V(\rho_1 - \rho_2)g}{k}}$$

Ans. (A)

Sol. 4 Terminal velocity of a ball in a viscous liquid is given as

$$v = \frac{2}{9}\frac{gR^2}{\eta}(\rho_S - \rho_L)$$

For ball falling in water

$$10 = v_w = \frac{2}{9}\frac{gR^2}{8.5 \times 10^{-4}} \times (7.8 - 1) \quad \ldots(1)$$

For ball falling in glycerine

$$v_g = \frac{2}{9}\frac{gR^2}{13.2} \times (7.8 - 1.2) \quad \ldots(2)$$

From (2)/(1) we have

$$\frac{v_g}{10} = \frac{8.5 \times 10^{-4}}{13.2} \times \left(\frac{7.8 - 1.2}{7.8 - 1}\right)$$

$$\Rightarrow \quad v_g = 6.25 \times 10^{-4} \text{ cm/s}$$

Ans. (B)

Sol. 5

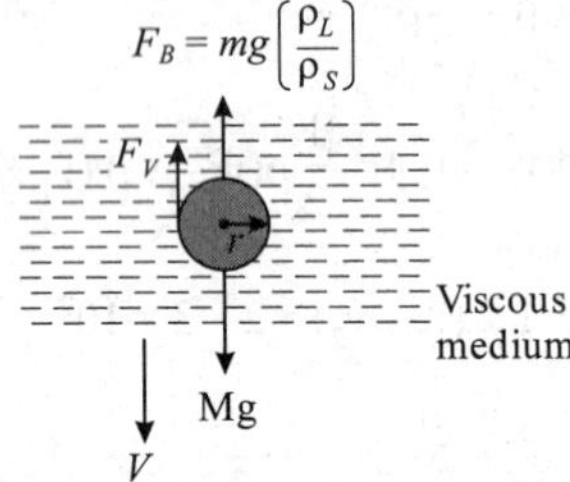

Upward viscous force on ball is given by stoke's rule as

$$F_v = 6\pi\eta r v$$

At terminal speed, all forces on sphere are balanced, so we have

$$mg = mg\left(\frac{\rho_L}{\rho_S}\right) + 6\pi\eta r v$$

$$\Rightarrow \quad \rho_3\left(\frac{4}{3}\pi r^3\right)g\left(1 - \frac{\rho_L}{\rho_S}\right) = 6\pi\eta r v$$

$$\Rightarrow \quad v = \frac{2}{9}\frac{gr^2}{\eta}(\rho_S - \rho_L)$$

$$\Rightarrow \quad v \propto r^2$$

Ans. (A)

Sol. 6 Case-1

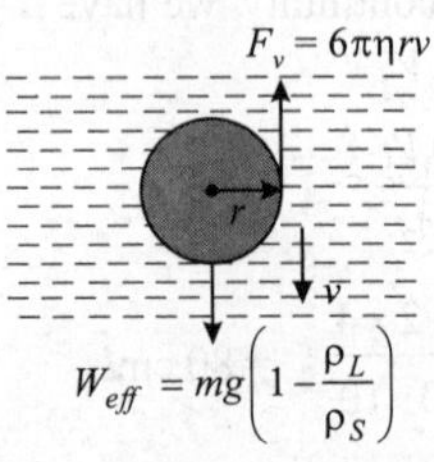

At terminal speed of ball

$$W_{eff} = 6\pi\eta r v \quad \ldots(1)$$

When ball is pulled upward at $F_{ext} = 2W_{eff}$ then at terminal speed v', we have

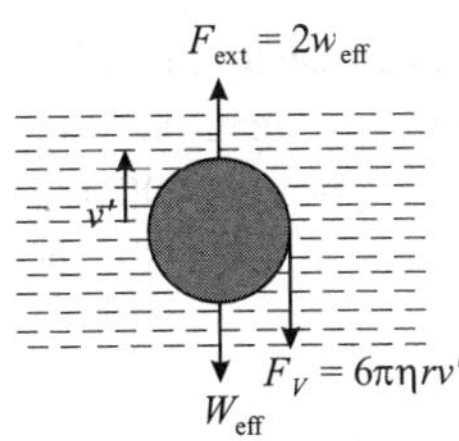

$$2W_{eff} = W_{eff} + 6\pi\eta r v'$$

$$\Rightarrow \quad W_{eff} = 6\pi\eta r v' \quad \ldots(2)$$

From (1) and (2)

$$\Rightarrow \quad v' = v = 10 \text{ cm/s}$$ **Ans. (C)**

Sol. 7

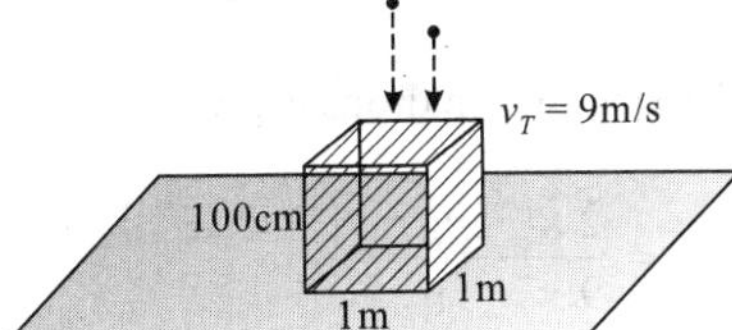

Total calculate volume of rain falling on the place per square meter in one year is

$$V = 1 \times 1 \times 100 \text{ cm} = 1 \text{ m}^3$$

KE of the rain volume which becomes zero on striking at ground is

$$\text{KE} = \frac{1}{2}(PV).\, V^2 = \frac{1}{2} \times 10^3 \times (9)^2$$

$$\Rightarrow \quad \text{KE} = \frac{81}{2} \times 1000$$

$$\Rightarrow \quad \text{KE} = 4.05 \times 10^4 \text{J}$$ **Ans. (B)**

Sol. 8

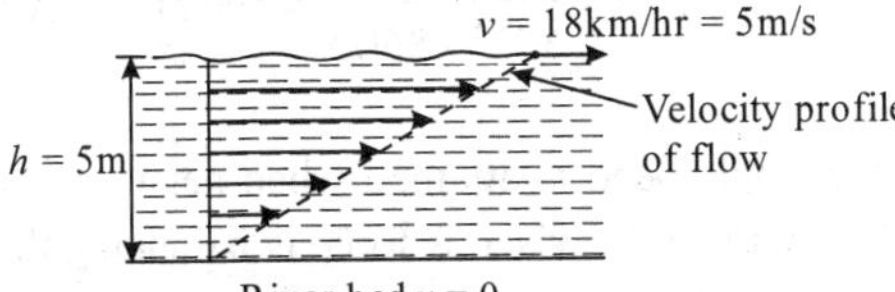

Viscous force between layers of river water is given as

$$F_v = \eta A \frac{dv}{d\pi} = \eta A\left(\frac{\Delta v}{h}\right)$$

(we assume velocity profile to be linear as shown in figure)

Shearing stress between layers of flowing water is given as

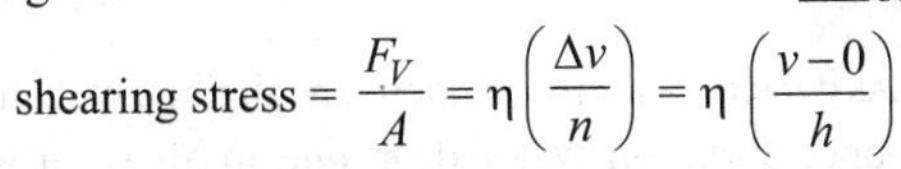

$$\text{shearing stress} = \frac{F_V}{A} = \eta\left(\frac{\Delta v}{n}\right) = \eta\left(\frac{v-0}{h}\right)$$

$$\Rightarrow \quad \sigma_s = 10^{-3} \times \frac{5}{5} = 10^{-3} \text{N/m}^2$$

Using 1 poise $= \frac{1}{10}$ poiseuille $= \frac{1}{10}$ N·s/m^2 **Ans. (C)**

Sol. 9 For a liquid flow in a pipeline, Reynolds number is given as

$$R = \frac{PvD}{\eta}$$

$v \rightarrow$ flow velocity

$P \rightarrow$ density

$D \rightarrow$ internal daintier of pipeline

The volume flow rate of water through tap is given as

$$Av = \frac{V}{t} = \frac{15 \times 10^3}{5 \times 60} \text{cm}^3\text{/s}$$

$$A = \pi r^2$$

Av = Cross-section area of tap

$$\Rightarrow \quad v = \frac{15 \times 10^3}{300 \times \pi \times \left(\frac{1}{\sqrt{\pi}}\right)^2} = 50 \text{ cm/s} = 0.5 \text{ m/s}$$

$\Rightarrow$ Reynolds number is given as

$$R = \frac{pvd}{\eta} = \frac{10^3 \times 0.5 \times \frac{2}{\sqrt{3.14}} \times 10^{-2}}{10^{-3}}$$

$$R = \frac{1}{\sqrt{3.14}} \times 10^4 = 5643$$ **Ans. (D)**

Sol. 10

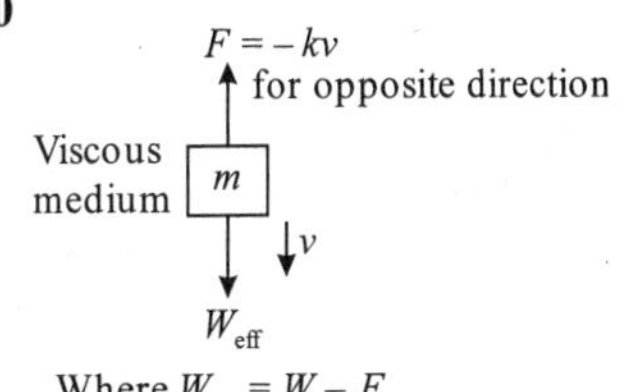

If a be the acceleration of mass, we use

$$W_{eff} - kv = ma$$

$$\Rightarrow \quad a = \frac{W_{eff} - kv}{m}$$

[If we put $a = \frac{dv}{dt}$ and integrative then it given $v(t)$ as exponent]

Here we can see that as $v\uparrow \Rightarrow a\downarrow$ and after some (long time)

time when $v = \frac{W_{eff}}{k}$ (constant) $\Rightarrow a = 0$ **Ans. (C)**

Sol. 11

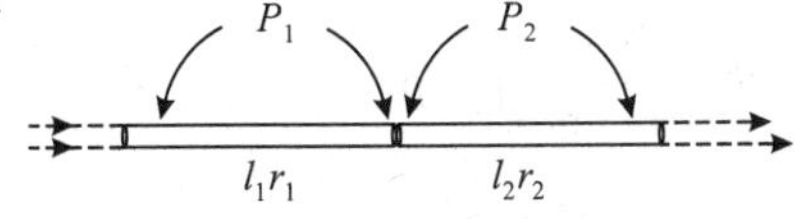

For viscous fluid flowing in a capillary tube we use

$$\text{flow rate} \quad \frac{dQ}{dt} = \frac{\pi r^4}{8\eta l}\Delta P$$

As tubes in this case are in series, so we use

$$\left.\frac{dQ}{dt}\right|_1 = \left.\frac{dQ}{dt}\right|_2$$

$$\frac{\pi r_1^4}{8\eta l_1} P_1 = \frac{\pi r_2^4}{8\eta l_2} . P_2$$

$$\Rightarrow \quad \frac{r_1^4}{l_1} P_1 = \frac{r_2^4}{l_1 / 4}(4P_1)$$

$$\Rightarrow \quad r_1^4 = 16 r_2^4$$

$$\Rightarrow \quad r_2 = \left(\frac{r_1^4}{16}\right)^{1/4} = \frac{r_1}{2}$$

Ans. (D)

Sol. 12 Volume flow rate of water is

$$Q = vA = v(\pi r^2)$$

$$\Rightarrow \quad v = \frac{Q}{\pi r} = \frac{100\times10^{-3}}{60} \times \frac{1}{\pi\times25\times10^{-4}}$$

$$\Rightarrow \quad v = \frac{2}{3\pi}\text{ m/s}$$

Reynolds Number is given as

$$R = \frac{\rho v d}{\eta} = \frac{10^3\times2\times10\times10^{-2}}{10^{-3}\times3\pi}$$

$$\simeq 2\times10^4$$

Thus order R is 10^4 **Ans. (C)**

Sol. 13 We have

$$V_T = \frac{2}{9}\left(\frac{r^2 g}{\eta}\right)(\rho_0 - \rho_l)$$

$$\Rightarrow \quad v_T \propto r^2$$

As mass of sphere will remain same before and after breaking, we have

$$\Rightarrow \quad \rho\left(\frac{4}{3}\pi R^3\right) = \rho\left(\frac{4}{3}\pi R^2\right)\times 27$$

$$\Rightarrow \quad r = \frac{R}{3}$$

$$\Rightarrow \quad \frac{v_1}{v_2} = \frac{R^2}{r^2} = 9$$

An.s (D)

Sol. 14 After falling through a height h as the velocity of ball is equal to terminal velocity, we use

$$\sqrt{2gh} = \frac{2}{9}\frac{r^2 g}{\eta}(\rho_l - \rho)$$

$$\Rightarrow \quad h = \frac{2}{81}\frac{r^4 g(\rho_l - \rho)^2}{\eta^2}$$

$$\Rightarrow \quad h \propto r^4$$

Ans. (B)

Sol. 15 The nature of flow whether streamline or unsteady or turbulent is determined by Reynolds Number, calculated as

$$R_e = \frac{\rho v D}{\eta}$$

As a standard data we can use for the value of Raynolds number less than 1000 flow remains steady and above 1000 upto 2000 it remain unsteady and above 2000 it becomes turbulent. Calculating the Raynolds number for the given situtions, we have

$$R_i = 10^3 \times \frac{0.18\times10^{-3}}{\pi\times(0.5\times10^{-2})^2\times60} \times \frac{1\times10^{-2}}{10^{-3}} = 382.16$$

$$R_f = 10^3 \times \frac{0.48\times10^{-3}}{\pi\times(0.5\times10^{-2})^2\times60} \times \frac{1\times10^{-2}}{10^{-3}} = 1019.09$$

Thus flow changes from steady to unsteady. **Ans. (D)**

Sol. 16 Terminal speed of rain drops is given as

$$v_t = \frac{2}{9}\frac{gr^2(\rho_p - \rho_1)}{\eta}$$

$$\Rightarrow \quad v_t \propto r^2$$

Ans. (C)

Sol. 17 Viscous force between layers of a flowing viscous fluid is given as

$$F = \eta A \frac{\Delta v_x}{\Delta y}$$

$$\Rightarrow \quad \frac{F}{A} = \eta\frac{\Delta v_x}{\Delta y}$$

$$\Rightarrow \quad 10^{-3} = 10^{-2}\times\frac{36\times1000}{h\times3600}$$

$$\Rightarrow \quad h = 10^{-2}\times\frac{36\times1000}{10^{-3}\times3600} = 100\text{ m}$$

Ans. (100)

Sol. 18 Terminal speed is attained when net force on ball is balanced so in this state viscous force is balancing the weight and buoyancy force, so we use

$$F_v = mg - F_B$$

$$\Rightarrow \quad F_v = mg - \left(\frac{m}{d_1}\times d_2\right)g$$

$$\Rightarrow \quad F_v = mg\left(1 - \frac{d_2}{d_1}\right)$$

Ans. (B)

Sol. 19 As density of air is considered negligible, we can neglect the buoyance force on water drop due to air so at terminal speed viscous force and weight of water drop gets balanced, given as

$$6\pi\eta r v_t = \frac{4}{3}\pi r^3 \rho g$$

$F_v = 6\pi\eta r v_t$

$\Rightarrow \quad v_t = \frac{4}{3} \times \frac{\pi r^3 \rho g}{6\pi\eta r}$

$\Rightarrow \quad v_t = \frac{4}{3} \times \frac{\pi r^3 \rho g}{6\pi\eta r} = \frac{2\times 10^{-12}\times 10^3 \times 10}{9\times 1.8\times 10^{-5}}$

$\Rightarrow \quad v_t = 123.4 \times 10^{-6}$ m/s **Ans. (D)**

Sol. 20 Speed after falling by a height *h,* it remain constant inside water so it is equal to the terminal speed of ball, given as

$$\sqrt{2gh} = \frac{2}{9}\frac{r^2(d-\rho)g}{\eta}$$

$\Rightarrow \quad \sqrt{2gh} = \frac{2}{9}\frac{10^{-8}(10000-1000)\times 10}{10^{-5}}$

$\Rightarrow \quad \sqrt{2gh} = \frac{2}{9} \times 10^{-8}\frac{9\times 10^4}{10^{-5}} = 20$

$\Rightarrow \quad 2 \times 10 \times h = 400$

$\Rightarrow \quad h = 20$ m **Ans. (20)**

Sol. 21 As the bubble starts rising uniformly, the net force acting on it will be zero, so we use

$$F_B = F_v$$

$\Rightarrow \quad \frac{4\pi}{3} R^3 \rho g = 6\pi\eta R v$

$\Rightarrow \quad \eta = \frac{10}{9} \approx 1.11$ SI unit

$= 11$ poise **Ans. (11)**

B, $\vec{v}$, mg, $F = 6\pi\eta R v$

Sol. 22 When the velocity of ball becomes constant, we use net force on ball becomes zero, which gives

$$F_V + F_B = mg$$

$\Rightarrow \quad F_V = mg - F_B$

$\Rightarrow \quad F_V = \rho_B V g - \rho_L V g$

$\Rightarrow \quad F_V = (\rho_B - \rho_L) V g$

$\Rightarrow \quad F_V = (8-1.3)\times 10^{+3} \times \frac{0.3\times 10^{-3}}{8\times 10^3} \times 10$

$\Rightarrow \quad F_V = \frac{6.7\times 0.3}{8} \times 10^{-2}$

$\Rightarrow \quad F_V = \frac{67\times 3}{8} \times 10^{-4} = 25.125 \times 10^{-4}$

$\Rightarrow \quad F_V = 25.125$ N **Ans. (25)**

Sol. 23 When the ball attain terminal velocity, net force on it will be zero so we use

$$F_v = (mg - F_B)$$

$\Rightarrow \quad F_v = V\sigma_b g - V\rho_l g$

$\Rightarrow \quad F_v = Vg(\sigma_b - \rho_l)$

$\Rightarrow \quad F_v = \frac{4}{3}\pi(10^{-3})^3 \times 9.8\,(10.5-1.5)\times 10^3$

$\Rightarrow \quad F_v = 3696 \times 10^{-7}$ N

$\Rightarrow \quad x = 7$ **Ans. (7)**

Sol. 24 (a) Dimensions of Surface Tension is given as

$$[S] = \left[\frac{F}{l}\right] = \frac{\text{MLT}^{-2}}{\text{L}} = \text{ML}^{-1}\text{T}^{-2}$$

Thus unit of Surface Tension is written as Kgs^{-2}.

(b) Dimensions of Pressure is given as

$$[p] = \left[\frac{F}{A}\right] = \frac{\text{MLT}^{-2}}{\text{L}^2}$$

Thus unit of pressure is written as kg m^{-1} s^{-2}.

(c) Dimensions of Viscosity is given as

$$[\eta] = \left[\frac{F}{A\left(\frac{dV}{dz}\right)}\right] = \frac{\text{MLT}^{-2}}{L^2\left(\frac{\text{LT}^{-1}}{\text{L}}\right)} = \text{ML}^{-1}\text{T}^{-1}$$

Thus unit of viscosity is written as kg m^{-1} s^{-1}.

(d) Dimensions of Impulse is given as

$$[J] = [Ft] = \text{MLT}^{-2} \times \text{T} = \text{MLT}^{-1}$$

Thus unit of Impulse is written as Kg ms^{-1}. **Ans. (B)**

Sol. 25 Viscous force on the block is equal to the drag force required at constant speed which is given as

$$|F| = \eta A \frac{\Delta v}{\Delta h} = 0.1$$

$\Rightarrow \quad 0.1 = 5 \times 10^{-3} \times 0.2 \times \frac{v}{.25\times 10^{-3}}$

$\Rightarrow \quad v = 0.025$ m/s

$\Rightarrow \quad v = 25 \times 10^{-3}$ m/s **Ans. (25)**

Sol. 26 When ball is dropped in a viscous fluid, viscous forces and Buoyant forces act on the ball as shown in figure below.

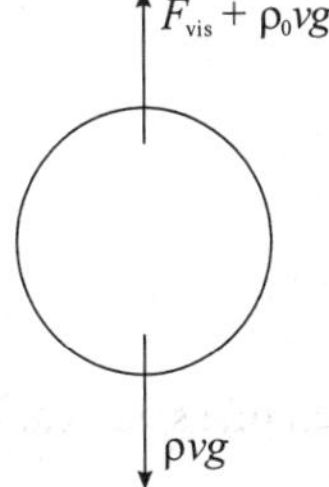

For constant velocity $F_{net} = 0$ as ball attains terminal velocity during motion so we use

$$F_{vis} + \rho_0 vg = \rho vg$$

$\Rightarrow \quad F_{vis} = (\rho - \rho_0)\, vg$

$$\Rightarrow \qquad F_{vis} = \rho v g\left(1 - \frac{\rho_0}{\rho}\right)$$

$$\Rightarrow \qquad F_{vis} = Mg\left(1 - \frac{\rho_0}{\rho}\right)$$ **Ans. (A)**

Sol. 27 As the bubble is moving at constant speed after attaining terminal velocity, the force acting on it is zero. Thus the buoyant force and viscous forces are balanced so we have

$$B = F_V$$

B is Buoyant force given as $B = V\rho g$ and F_V is viscous force, given by Stoke's rule as

$$F_V = 6\pi\eta R v$$

$$\Rightarrow \qquad \frac{4}{3}\pi R^3 \rho g = 6\pi\eta R v$$

Coefficient of Viscosity is calculated as

$$\eta = \frac{2R^2\rho g}{9v} = \frac{2\times(3\times10^{-3})^2\times1750\times10}{9\times0.35\times10^{-2}}$$

$$\Rightarrow \qquad \eta = 10 \text{ Pa.s}$$ **Ans. (10)**

Sol. 28 Terminal velocity of drops is given as

$$v_T = \frac{2}{9}\frac{(\rho_{water} - \rho_{air})r^2 g}{\eta}$$

$$\Rightarrow \qquad v_T \propto r^2$$

If drops coalesce then new velocity is given as

$$\Rightarrow \qquad \frac{v_1}{v_2} = \left(\frac{r}{R}\right)^2$$

Volume of 8 small drops of radius r is equal to one big drop of radius R

$$8\cdot\frac{4}{3}\pi r^3 = \frac{4}{3}\pi r^3$$

$$\Rightarrow \qquad R = 2r$$

$$\Rightarrow \qquad \frac{10}{v_2} = \left(\frac{1}{2}\right)^2$$

$$\Rightarrow \qquad v_2 = 40 \text{ cm/s}$$ **Ans. (B)**

Sol. 29 Terminal velocit of a spherical body in liquid is given as

$$V_t \propto r^2$$

For percentage error calculation we use

$$\frac{\Delta V_t}{V_t} = 2\cdot\frac{\Delta r}{r}$$

$$\Rightarrow \qquad \frac{\Delta V_t}{V_t}\times100\% = 2\frac{(0.1)}{5}\times100 = 4\%$$

Hence Assertion-A is correct and we also have

$$V_t \propto r^2$$

Thus **Reason-R** is false so option (D) is correct. **Ans. (D)**

9.4 Surface Tension

Sol. 1

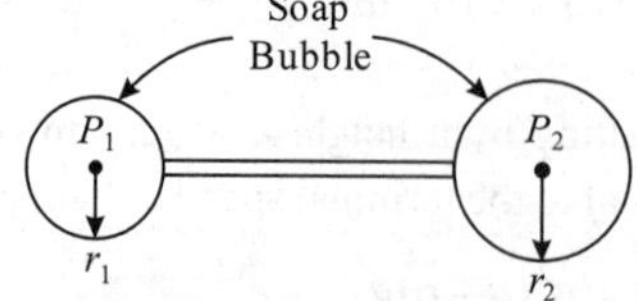

Excess pressure inside a soap bubble of radius r is given as-

$$P_{ex} = \frac{4T}{r}$$

Where $T \rightarrow$ surface tension of soap solution.

Excess pressure inside the two given soap bubble are

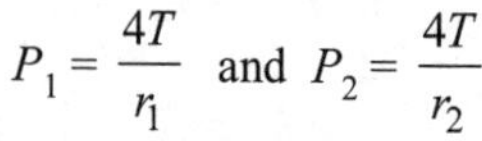

$$P_1 = \frac{4T}{r_1} \text{ and } P_2 = \frac{4T}{r_2}$$

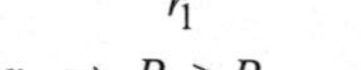

As $r_1 < r_2 \Rightarrow P_1 > P_2$

$\Rightarrow$ Air flows from smaller bubble to bigger bubble **Ans. (C)**

Sol. 2

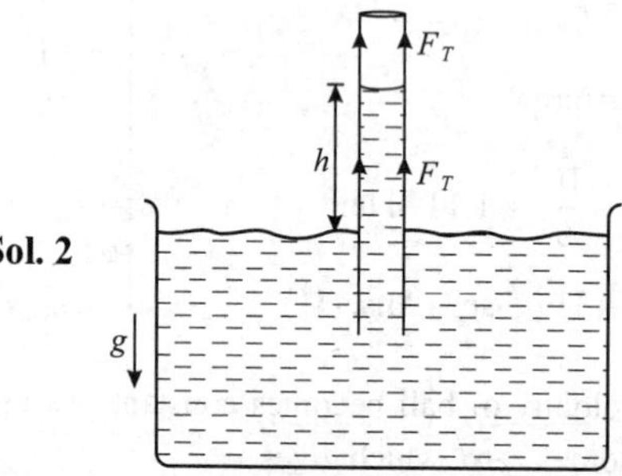

Inside capillary liquid is raised upto the level where force of surface tension balances the weight of liquid raised.

In a free falling state $g_{eff} = 0$

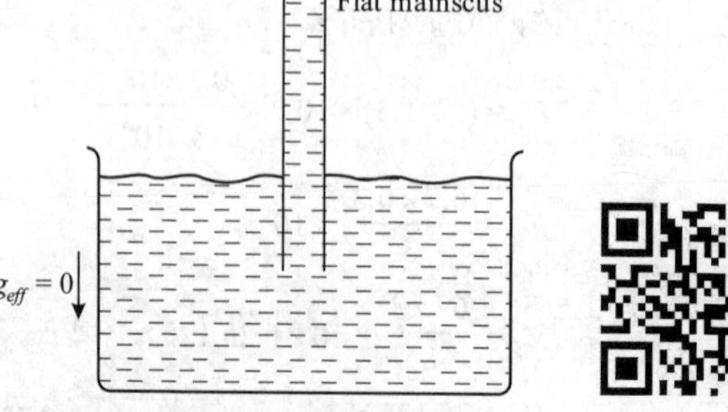

$\Rightarrow$ Effective weight of liquid raised = 0

$\Rightarrow F_T(\text{upward}) = 0 \Rightarrow$ liquid will fill the complete height of capillary. **Ans. (D)**

Sol. 3 As $T_{\text{soap sol.}} < T_{\text{water}}$

$\Rightarrow$ Capillary height $h = \dfrac{2T\cos\theta}{r\rho g}$

$\Rightarrow \quad h_B < h_A$ **Ans. (C)**

Sol. 4

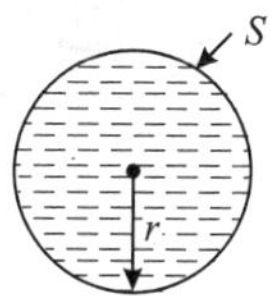

A liquid drop has only one surface of area outside

$$S = 4\pi r^2$$

When two mercury drops merge to form a bigger drop radius R then we have

$$2 \times \frac{4}{3}\pi r^3 = \frac{4}{3}\pi R^3$$

$$\Rightarrow \quad R = 2^{1/3} r$$

Surface energy of bigger drop

$$E = T \times S$$

$$\Rightarrow \quad E = T(4\pi R^2)$$

$$\Rightarrow \quad E = 4\pi T(2^{1/3} r)^2$$

$$\Rightarrow \quad E = 2^{8/3}\pi r^2 T$$

Ans. (D)

Sol. 5

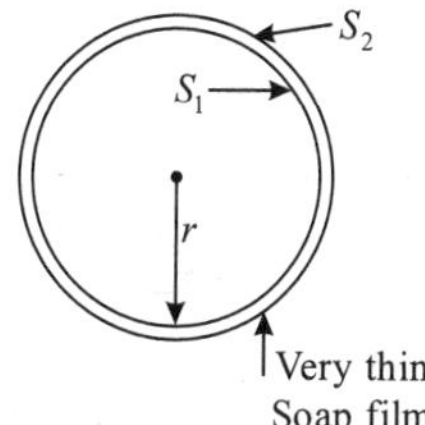

Soap bubble has the surface S_1 and S_2 which carries surface energy, given as

$$E = T \times 2(4\pi r^2) = 8\pi T r^2$$

If radius of bubble is increased from r_1 to r_2 than work done in this process is

$$W = E_f - E_i = 8\pi T(r_2^2 - r_1^2)$$

$$\Rightarrow \quad W = 8\pi \times 0.03\,[(0.05)^2 - (0.03)^2]$$

$$\Rightarrow \quad W = 0.384\,\pi\,\text{mJ} \sim 0.4\,\pi\,\text{mJ}$$

Ans. (B)

Sol. 6

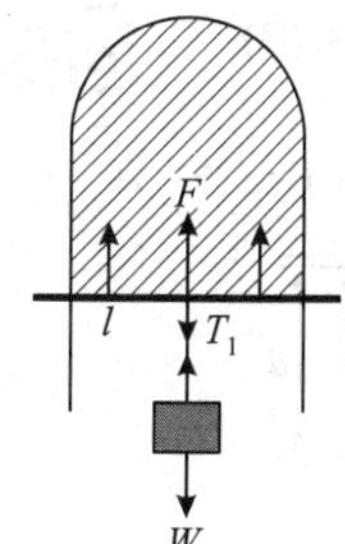

If surface tension of liquid is T than total upward force on slider due to two surfaces of the film is

$$F = 2Tl = w$$

$$\Rightarrow \quad T = \frac{w}{2l} = \frac{1.5 \times 10^{-2}}{2 \times 0.3} = 0.025\ \text{N/m}$$

Ans. (A)

Sol. 7 If n small drops coalesce to from the big drop of radius R, we have

$$n \times \frac{4}{3}\pi r^3 = \frac{4}{3}\pi R^3 \Rightarrow n = \frac{R^3}{r^3}$$

Amount of surface energy released in process

$$\Delta E = n(4\pi r^2 T) - 4\pi R^2 T$$

$$\Delta E = 4\pi T\left[r^2 \cdot \frac{R^3}{r^2} - R^2\right] = 4\pi T R^3\left[\frac{1}{r} - \frac{1}{R}\right]$$

If v is the speed attained by big drop then we have

$$\Delta E = \frac{1}{2}mv^2$$

$$\frac{1}{2}\left(\rho \times \frac{4}{3}\pi R^3\right)v^2 = 4\pi T R^3\left[\frac{1}{r} - \frac{1}{R}\right]$$

$$\Rightarrow \quad v = \sqrt{\frac{6T}{\rho}\left[\frac{1}{r} - \frac{1}{R}\right]}$$

Ans. (D)

Sol. 8 Due to surface energy the outer layer (very thin) of a liquid drop if evaporates by absorption of liquid heat, whole drop can get evaporated at same temperature.

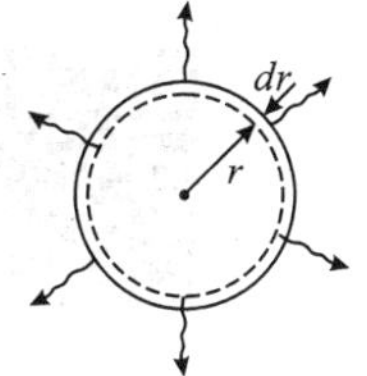

Due to evaporation of thin layer of liquid at the surface of drop the change in surface energy is

$$\Delta E = T(4\pi(r + dr)^2) - T(4\pi r^2)$$

$$\Rightarrow \quad \Delta E = T(8\pi r\,dr)$$

for evaporation of the thin elemental layer of liquid of width dr and out of heat to be absorbed by it is

$$dQ = dm\,L \qquad dm \rightarrow \text{mass of layer}$$

$$\Rightarrow \quad dQ = (\rho dv)L \qquad dv \rightarrow \text{volume of layer}$$

$$\Rightarrow \quad dQ = \rho L(4\pi r^2\,dr) \qquad [dv = 4\pi r^2\,dr]$$

The drop gets evaporated only by its surface energy if

$$\Delta E = dQ$$

$$T(8\pi r\,dr) = \rho L(4\pi r^2 dr)$$

$$\Rightarrow \quad r = \frac{2T}{\rho L}$$

Ans. (C)

Sol. 9 Capillary height in tube is given as

$$h = \frac{2T\cos\theta}{r\rho g}$$

As with temperature surface temperature describes

$\Rightarrow$ as temperature interface, h decreases

S-1 Is false

S-2 Is true as it is a fact studied in theory **Ans. (B)**

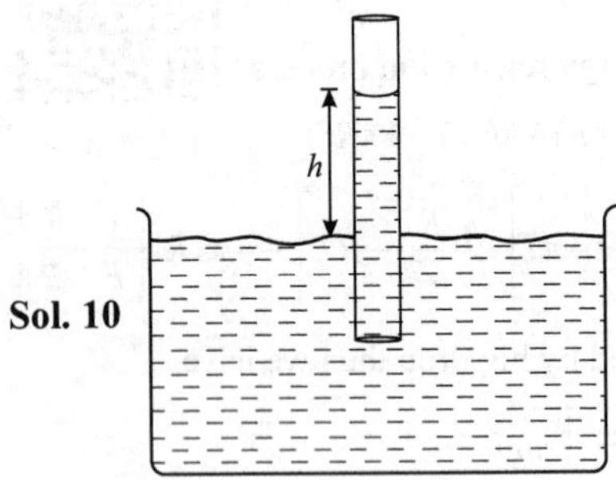

Sol. 10

Capillary height is given as

$$h = \frac{2T\cos\theta}{r\rho g}$$

$\theta < 90°$ for glass- water interface as water wets the glass surface. When capillary is coated with wax which is not wet by water so angle of contact for wax - water surface is $\theta > 90°$

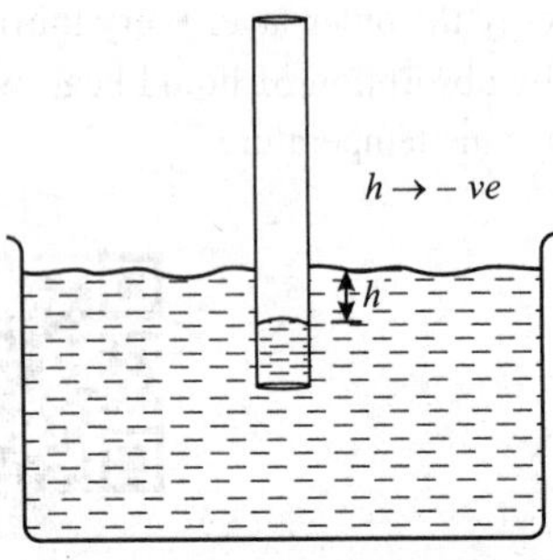

$\Rightarrow$ $\cos\theta = -ve$

$\Rightarrow$ $h = -ve$ **Ans. (C)**

Sol. 11 The excess pressure inside soap bubble is

$$P_{ex} = \frac{2T}{R}$$

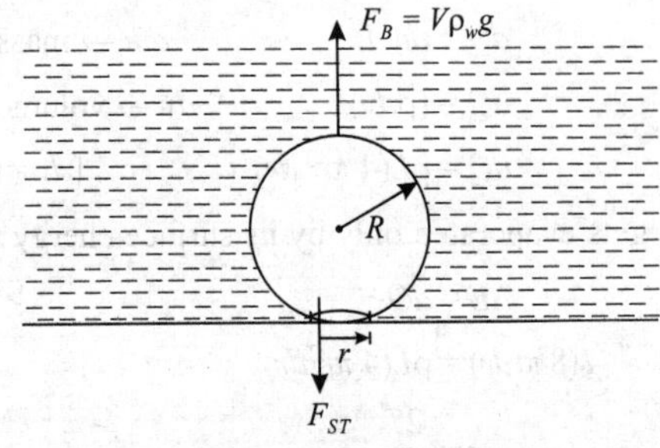

Force of surface tension at bottom on bubble is

$$F_{ST} = P_{ex} \times \pi r^2$$

At the moment spherical bubble detach from base, we use

$$F_B = F_{ST}$$

$$\Rightarrow \quad \frac{4}{3}\pi R^3 \rho_\omega g = \frac{2T}{R} \times \pi r^2$$

$$\Rightarrow \quad r = R^2\sqrt{\frac{2\rho_\omega g}{3T}}$$

Ans. (A)

Sol. 12

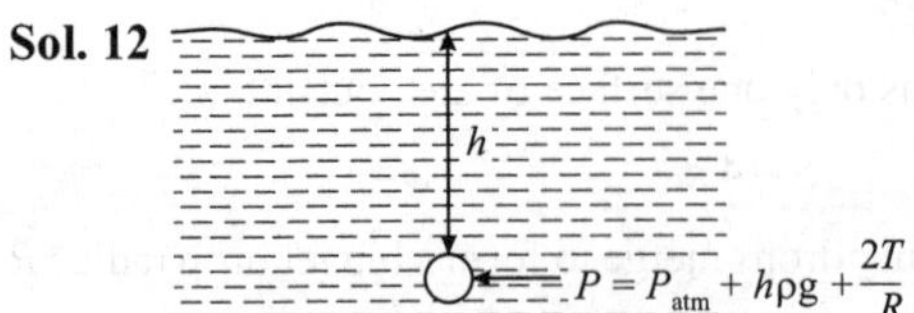

Inside the bubble excess pressure is given as

$$P_{ex} = P - P_{atm} = h\rho g + \frac{2T}{R} = 1100$$

$$\Rightarrow \quad h \times 1000 \times 9.8 + \frac{2 \times 0.06}{10^{-3}} = 1100$$

$$\Rightarrow \quad 9800h + 120 = 1100$$

$$\Rightarrow \quad h = \frac{980}{9800} = 0.1 \text{ m}$$

Ans. (A)

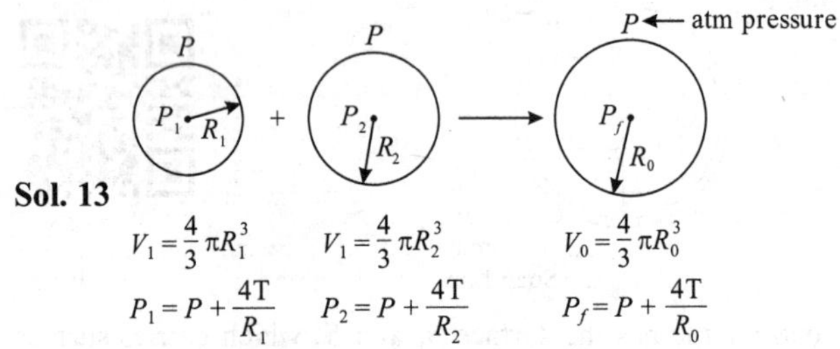

Sol. 13

$V_1 = \frac{4}{3}\pi R_1^3$ $V_1 = \frac{4}{3}\pi R_2^3$ $V_0 = \frac{4}{3}\pi R_0^3$

$P_1 = P + \frac{4T}{R_1}$ $P_2 = P + \frac{4T}{R_2}$ $P_f = P + \frac{4T}{R_0}$

If temperature is contained constant then we use

$$P_1V_1 + P_2V_2 = P_fV_0$$

$$\Rightarrow \quad \left(P + \frac{4T}{R_1}\right)\left(\frac{4}{3}\pi R_1^3\right) + \left(P + \frac{4T}{R_2}\right)\left(\frac{4}{3}\pi R_2^3\right) = \left(P + \frac{4T}{R_o}\right)\left(\frac{4}{3}\pi R_o^3\right)$$

$$\Rightarrow P\left[\frac{4}{3}\pi R_1^3 + \frac{4}{3}\pi R_2^3 - \frac{4}{3}\pi R_o^3\right] + \frac{4T}{3}(4\pi R_1^2) + \frac{4T}{3}(4\pi R_2^2) - \frac{4T}{3}(4\pi R_o^2) = 0$$

$$\Rightarrow \quad PV + \frac{4T}{3}S = 0$$

$$\Rightarrow \quad 3PV + 4TS = 0$$ **Ans. (B)**

Sol. 14

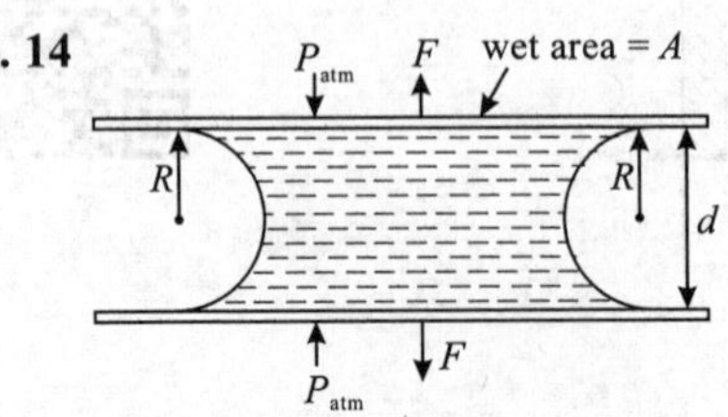

To pull apart the plates, the required force is given as

$$F=\frac{2TA}{d}=\frac{2TA}{2R}=\frac{TA}{R}$$

The pressure in water will be lower by

$$\Delta P_L=\frac{F}{A}=\frac{T}{R}\quad \textbf{Ans. (D)}$$

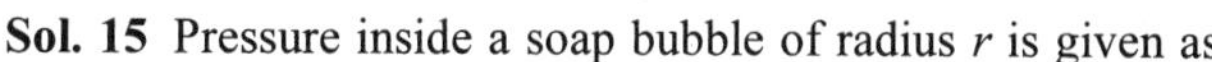

Sol. 15 Pressure inside a soap bubble of radius r is given as

$$P_{in}=P_{atm}+\frac{4T}{r}\quad \ldots(1)$$

If volume of soap bubble increases at constant rate, we have

$$V\propto t$$

$$\Rightarrow\quad r\propto t^{1/3}$$

$$\Rightarrow\quad r=kt^{1/3}$$

From equation-(1), we have

$$P_{in}=P_{atm}+\frac{4T}{kt^{1/3}}$$

Thus graph between P and $\frac{1}{t^{1/3}}$ is a straight line with positive slope and positive intercept. **Ans. (D)**

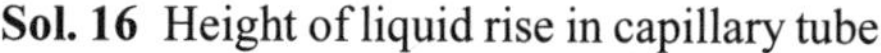

Sol. 16 Height of liquid rise in capillary tube

$$h=\frac{2T\cos\theta_C}{\rho rg}$$

$$\Rightarrow\quad h\propto\frac{1}{r}$$

When radius becomes double height become half

$$\Rightarrow\quad h'=\frac{h}{2}$$

Now, $\quad M=\pi r^2h\times\rho$

and $\quad M'=\pi(2r)^2(h/2)\times\rho=2M$ **Ans. (C)**

Sol. 17 Capillary height is given as

$$h=\frac{2T\cos\theta}{r\rho g}$$

For H_g, we use $h=\frac{2T_{Hg}\cos\theta_{Hg}}{r_1\rho_{Hg}g}$

and for water, we use

$$h=\frac{2T_w\cos\theta_w}{r_2\rho_w g}$$

$$\Rightarrow\quad \frac{T_{Hg}\cos\theta_{Hg}}{r_1\rho_{Hg}}=\frac{T_w\cos\theta_w}{r_2\rho_w}$$

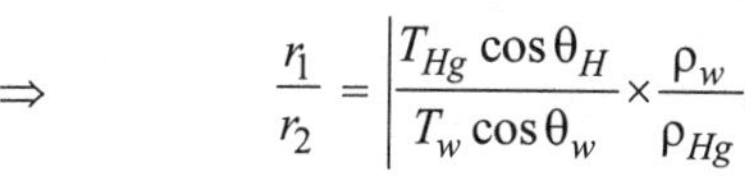

$$\Rightarrow\quad \frac{r_1}{r_2}=\left|\frac{T_{Hg}\cos\theta_H}{T_w\cos\theta_w}\times\frac{\rho_w}{\rho_{Hg}}\right|$$

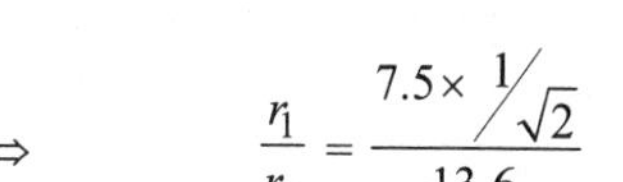

$$\Rightarrow\quad \frac{r_1}{r_2}=\frac{7.5\times{}^{1}\!/\!\sqrt{2}}{13.6}$$

$$\Rightarrow\quad \frac{r_1}{r_2}=\frac{5}{9\sqrt{2}}=\frac{5}{12.72}\approx\frac{2}{5}\quad \textbf{Ans. (C)}$$

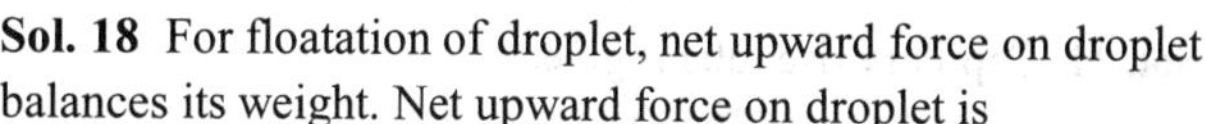

Sol. 18 For floatation of droplet, net upward force on droplet balances its weight. Net upward force on droplet is

$$F_{up}=F_{\text{Buoyancy}}+F_{\text{surface tension}}$$

$$\Rightarrow\quad F_{up}=\left(\frac{2}{3}\pi R^3\right)\rho g+T(2\pi R)$$

For equilibrium of droplet, we use

$$\left(\frac{2}{3}\pi R^3\right)\rho g+T(2\pi R)=d\left(\frac{4}{3}\pi R^3\right)g$$

$$\Rightarrow\quad T(2\pi R)=\left(\frac{2}{3}\pi R^3\right)g\,[2d-\rho]$$

$$\Rightarrow\quad R=\sqrt{\frac{3T}{(2d-\rho)g}}\quad \textbf{Ans. (B)}$$

Force of surface tension can be upward (as given in question) so in this case liquid does not wet the droplet.

Sol. 19 Figure below shows that angle of contact is $\theta=30°$

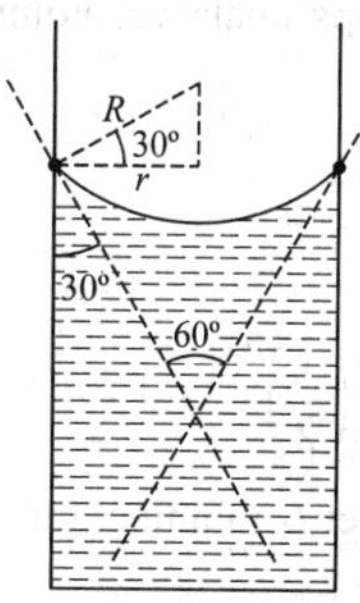

Capillary height of liquid in tube is given as

$$h=\frac{2T}{\rho gr\cos\theta}$$

$$\Rightarrow\quad h=\frac{2\times0.05}{667\times10\times\left(\frac{0.15\times10^{-3}\times2}{\sqrt{3}}\right)}$$

$$\Rightarrow\quad h=0.087\,\text{m}\quad \textbf{Ans. (C)}$$

Sol. 20 By excess pressure formula, we use

$$0.01=\frac{4T}{R_1}$$

and $0.02 = \frac{4T}{R_2}$

$\Rightarrow \frac{R_1}{R_2} = 2$

$\Rightarrow R_1 = 2R_2$

Ratio of volumes is given as

$$\frac{V_1}{V_2} = \frac{R_1^3}{R_2^3} = 8$$ **Ans. (A)**

Sol. 21 Capillary rise tube is given as

$$h = \frac{2T\cos\theta}{r\rho g}$$

$\Rightarrow T = \frac{\rho grh}{2\cos\theta}$

$\Rightarrow T = \frac{(900)(10)(15\times10^{-5})(15\times10^{-2})}{2}$

$\Rightarrow = 101.25\times10^{-3}$ N/m

$\Rightarrow = 101.25$ *m*N/m **Ans. (101)**

Sol. 22 Excess pressure at common surface of the two bubbles is given as

$$P_{ex} = 4T\left(\frac{1}{a}-\frac{1}{b}\right) = \frac{4T}{r}$$

$\Rightarrow \frac{1}{r} = \frac{1}{a} - \frac{1}{b}$

$\Rightarrow r = \frac{ab}{b-a}$ **Ans. (A)**

Sol. 23 When the drops coalesce, volume of liquid remain conserved, given as

$$2\times\frac{4}{3}\pi R^3 = \frac{4}{3}\pi r^3$$

$\Rightarrow \frac{R}{r} = \left(\frac{1}{2}\right)^{1/3}$ ...(1)

Ratio of before and after combining, ratio of surface energy of the liquid is given as

$$\frac{U_{before}}{U_{after}} = \frac{2\times S\times4\pi R^2}{S\times4\pi r^2} = 2\times\left(\frac{R}{r}\right)^2 = \frac{2^{1/3}}{1}$$

Ans. (B)

Sol. 24 When the two bubbles combine, the total surface energy remain conserved as temperature of system is not increasing, thus for surface tension σ we use

$$\sigma A_1 + \sigma A_2 = \sigma A$$

$\Rightarrow \sigma[2\times4\pi r_1^2] + \sigma[2\times4\pi r_2^2] = \sigma[2\times4\pi r^2]$

$\Rightarrow r_1^2 + r_2^2 = r^2$

$\Rightarrow r = \sqrt{r_1^2 + r_2^2}$ **Ans. (C)**

Sol. 25 If R is the radius of equivalent soap bubble then pressure inside the 3 cm bubble is given as

$$P_0 + \frac{4T}{6} + \frac{4T}{3} = P_0 + \frac{4T}{R}$$

$\Rightarrow \frac{1}{6} + \frac{1}{3} = \frac{1}{R}$

$\Rightarrow R = 2$ cm **Ans. (2)**

Sol. 26 In the process of breaking of water drop, surface area increases so gain in surface energy is calculated as

$$\Delta E = T\Delta A$$

$\Rightarrow \Delta E = 0.075\,(A_f - A_1)$

Initial and final surface area of drops is given as

$$A_i = 4\pi r^2$$

and $A_f = 4\pi r_0^2\times64$

Radius of drops after splitting is calculated by volume conservation as

$$\frac{4}{3}\pi r^3 = 64\cdot\frac{4}{3}\pi r_0^3$$

$\Rightarrow r_0 = \frac{r}{4}$

$\Rightarrow A_f = 4\pi\left(\frac{r}{4}\right)^2\cdot64 = 16\pi r^2$

Thus surface energy is calculated as

$$\Delta E = 0.075\,(16\pi r^2 - 4\pi r^2)$$

$\Rightarrow \Delta E = 0.075\,(12\pi(0.01)^2)$

$\Rightarrow \Delta E = 2.8\times10^{-4}$ J **Ans. (A)**

Sol. 27 Figure below shows the situation described in question.

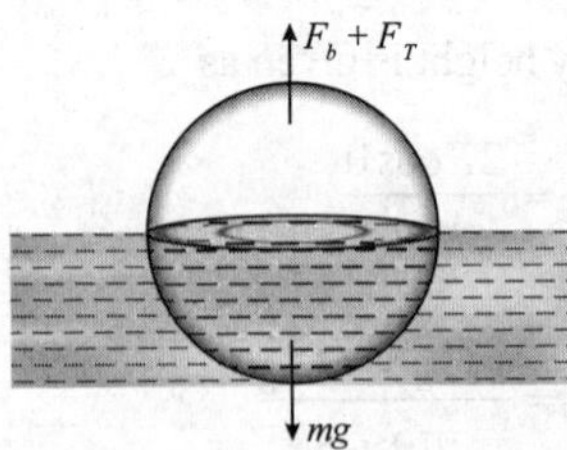

For equilibrium of drop, we use

Buoyant force + surface tension = mg

$$\sigma\frac{V}{2}g + 2\pi RT = \rho Vg$$

$\Rightarrow 2\pi RT = \frac{(2\rho-\sigma)}{2}\cdot\frac{4}{3}\pi R^3 g$

$$\Rightarrow \quad R^3 = \frac{3T}{(2\rho-\sigma)g}$$

$$\Rightarrow \quad R = \sqrt{\frac{3\times 7.5\times 10^{-2} N-m^{-1}}{(2\rho-\sigma)\times 10}}$$

$$\Rightarrow \quad R = \frac{3}{20\sqrt{(2\rho-\sigma)}} m = \frac{15}{\sqrt{2\rho-\sigma}} \text{ cm}$$

Ans. (A)

Sol. 28 Initial surface energy of the water drop is given as

$$U_i = TA$$

Where T is surface tension and A is surface area

$$U_i = \left(\frac{75\times 10^{-5}}{10^{-2}}\frac{N}{m}\right) \times [4\pi(1\times 10^{-2})^2]$$

$$\Rightarrow \quad U_i = 75\times 10^{-3}\times 4\pi\times 10^{-4}$$

$$\Rightarrow \quad U_i = 942\times 10^{-7}\text{ J}$$

Final radius of drops after splitting is calculated by volume conservation as

$$\frac{4}{3}\pi R^3 = 729\left(\frac{4}{3}\pi r^3\right)$$

$$\Rightarrow \quad r = \frac{R}{(729)^{1/3}} = \frac{R}{9} = \frac{1}{9}\text{ cm}$$

Final surface energy of all drops after splitting is calculated as

$$U_f = 729[TA]$$

$$\Rightarrow \quad U_f = 729\left[\frac{75\times 10^{-5}}{10^{-2}}\frac{N}{m}\right]\times\left[4\pi\left(\frac{1}{9}\times 10^{-2}\right)^2\right]$$

$$\Rightarrow \quad U_f = 729\left[75\times 10^{-3}\times\frac{4\pi\times 10^{-4}}{81}\right]$$

$$\Rightarrow \quad U_f = 9(942\times 10^{-7})\text{ J}$$

Gain in surface energy is calculated as

$$\Delta U = 9\times 942\times 10^{-7} - 942\times 10^{-7}$$

$$\Rightarrow \quad \Delta U = 7.5\times 10^{-4}\text{ J}$$

Ans. (C)

Sol. 29 Figure below shows the two bubbles one inside another as described in question. Here we consider pressures of air inside the bubbles as indicated in figure.

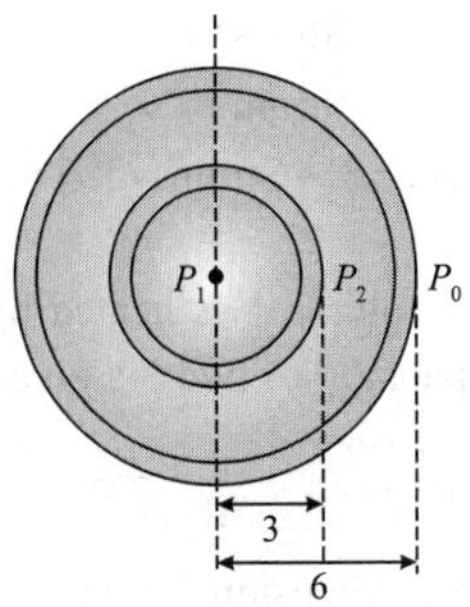

By using the expression of excess pressure inside a soap bubble, we have

$$P_2 - P_0 = \frac{4T}{6} \quad ...(1)$$

and $$P_1 - P_2 = \frac{4T}{3} \quad ...(2)$$

If an independent soap bubble of radius r has same pressure as that of inner bubble then we use

$$P_1 - P_0 = \frac{4T}{r} \quad ...(3)$$

Adding equations-(1) and (2) we get

$$P_1 - P_0 = \frac{4T}{2} \quad ...(4)$$

From equations-(3) and (4) we get

$$r = 2\text{ cm}$$

Ans. (2)

Sol. 30 **Assertion-A** is true as the oil or grease stains do not have affinity with water molecules or adhesive forces with water are very less so thes stains cannot be cleaned with water. **Reason-R** is explaining the same hence option (A) is correct.

Ans. (A)

Sol. 31 If we consider R as the radius of bigger drop and r is the radius of smaller drop. During splitting of bigger drop to 1000 smaller drops, volume remain same so we use

$$\frac{4}{3}\pi R^3 = 1000\times\frac{4}{3}\pi r^3$$

$$\Rightarrow \quad R = 10r$$

Initial total surface energy of the drop is calculated as

$$u_i = T\cdot 4\pi R^2$$

After splitting the total surface energy of all the drops are calculated as

$$u_f = T\cdot 4\pi r^2\times 1000$$

$$\Rightarrow \quad \frac{u_f}{u_i} = \frac{1000r^2}{R^2}$$

$$\Rightarrow \quad \frac{u_f}{u_i} = \frac{10}{1}$$

$$\Rightarrow \quad x = 1$$

Ans. (1)

Sol. 32 Work done in the required case is equal to the increase in surface energy of the soap bubble given as

$$W = T_S\times 8\pi\times(R_2^2 - R_1^2)$$

$$\Rightarrow \quad W = 2\times 10^{-2}\times 8\times\frac{22}{7}\times 49\times\frac{3}{4}\times 10^{-4}$$

$$\Rightarrow \quad W = 18.48\times 10^{-4}\text{ J}$$

Ans. (C)

Sol. 33 Capillary height of the liquid is given as

$$h = \frac{2S\cos\theta}{r\rho g}$$

As angle of contact is not given for the two liquids, we consider it to be approximately same to proceed, so we have

$$\frac{h_1}{h_2} = \frac{S_1}{S_2}\frac{\rho_2}{\rho_1}$$

$$\Rightarrow \quad \frac{5}{h_2} = \left[\frac{1}{2}\right]\left[\frac{2}{1}\right]$$

$$\Rightarrow \quad h_2 = 5\text{ cm} = 0.05\text{ m}$$ **Ans. (C)**

Sol. 34 When 1000 droplets combine to form a single drop of radius R, by conserving volume we use

$$1000 \times \frac{4\pi}{3}(1)^3 = \frac{4\pi}{3}R^3$$

$$\Rightarrow \quad R = 10\text{ mm}$$

Total change in surface energy due to reduction in area is calculated as

$$\Delta E = T \times 1000 \times 4\pi(10^{-3})^2 - T \times 4\pi(10 \times 10^{-3})^2$$

$$\Rightarrow \Delta E = 4 \times \pi \times 7 \times 10^{-2}(1000 - 100) \times 10^{-6}$$

$$\Rightarrow \Delta E = 7.92 \times 10^{-4}\text{ J}$$ **Ans. (B)**

Sol. 35 Initial surface energy

$$\Delta E = 0.45 \times 4\pi(10^{-3})^2$$

Using volume of liquid remain conserved when big drop is broken into 125 small drops so we use

$$\frac{4}{3}\pi(10^{-3})^3 = 125 \times \frac{4\pi}{3}R_{new}^3$$

$$\Rightarrow \quad 10^{-3} = 5R_{new}$$

$$\Rightarrow \quad R_{new} = \frac{10^{-3}}{5}m$$

So, final surface energy

$$= 0.45 \times 125 \times 4\pi\left(\frac{10^{-3}}{5}\right)^2$$

Increase in energy is calculated as

$$= 0.45 \times 4\pi \times (10^{-3})^2\left[\frac{125}{25} - 1\right]$$

$$= 4 \times 0.45 \times 4\pi \times 10^{-6}$$

$$\Delta E = 2.26 \times 10^{-5}\text{ J}$$ **Ans. (A)**

Sol. 36 Amount of work done required to increase the radius of soap.

$$W = T\cdot(\Delta A)$$

In the soap bubble, change in surface area of both surfaces is given as

$$\Delta A = 2[4\pi(r_2^2 - r_1^2)]$$

Thus work done is calculated as

$$W = T(8\pi(r_2^2 - r_1^2))$$

$$\Rightarrow \quad W = 264 \times 10^{-4}\text{ J}$$ **Ans. (264)**

Sol. 37 Figure below shows the situation described in the question.

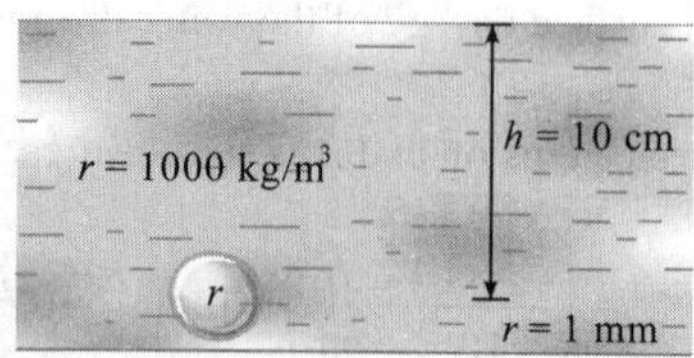

Pressure inside the bubble is given as

$$P = P_0 + h\rho g + \frac{2T}{r}$$

$$\Rightarrow \quad P - P_0 = h\rho g + \frac{2T}{r}$$

$$\Rightarrow \quad P - P_0 = 0.1 \times 1000 \times 10 + \frac{2 \times .075}{10^{-3}}$$

$$\Rightarrow \quad P - P_0 = 1000 + (0.15)(1000)$$

$$\Rightarrow \quad P - P_0 = 1150\text{ Pa}$$ **Ans. (1150)**

10.1 Thermal Expansion

Sol. 1

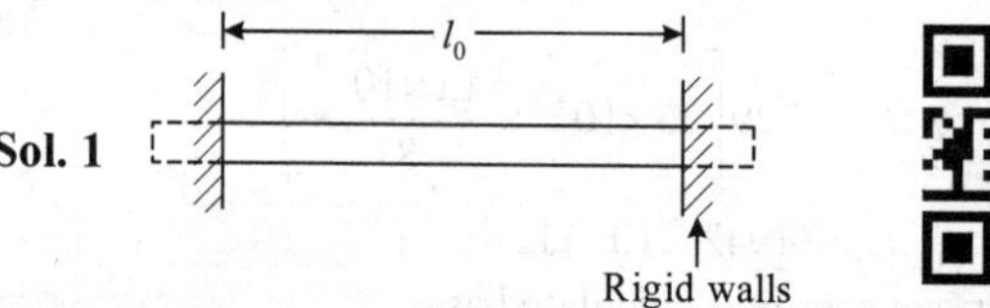

If on rising temperature by t°C rod expands by Δl then due to rigid walls, it is compressed by the same length elastically strain developed in rod due to this is

$$\text{strain} = \frac{\Delta l}{l} = \frac{\alpha l \Delta t}{l} = \alpha\Delta t = \alpha t$$

$\Rightarrow$ By definition of Young's modulus

$$Y = \frac{\text{Stress}}{\text{Strain}}$$

$$\Rightarrow \quad \text{linear stress} = Y \times \text{Strain} = Y\alpha t$$ **Ans. (C)**

Sol. 2 If a body of volume V is heated by ΔT then volume expansion of body is given as

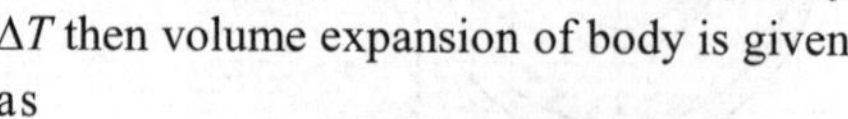

$$\Delta V = \gamma V \Delta T$$

Coefficient of volume expansion $\gamma = 3\alpha$

$$\Rightarrow \quad \Delta V = 3\alpha V \Delta T = 3\alpha\left(\frac{4}{3}\pi r^3\right)\Delta T$$

$$= 3 \times 23 \times 10^{-6} \times \left(\frac{4}{3} \times 3.14 \times (10)^3\right) \times 100$$

$$= 28.88 \text{ cm}^3 \sim 28.9 \text{ cc}$$ **Ans. (A)**

Sol. 3 Here we consider wood is having negligible expansion coefficient compared to metal so on rising temperature by ΔT, strain due to thermal expansion which is compensated by elastic compression on cooling down is given as

$$\text{Strain} = \frac{\Delta l}{l} = \alpha\Delta T$$

$$\Rightarrow \quad \text{Stress} = \frac{F}{S} = Y \times \text{Strain} = Y\alpha\Delta T$$

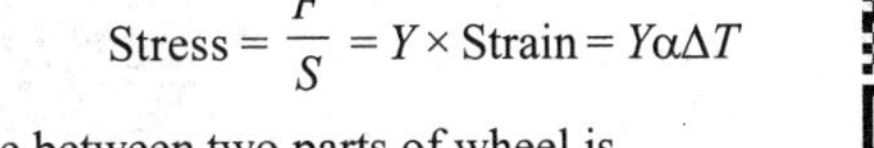

force between two parts of wheel is

$$2F = 2YS\alpha\Delta T$$

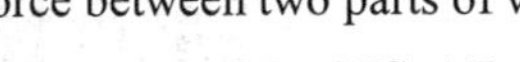

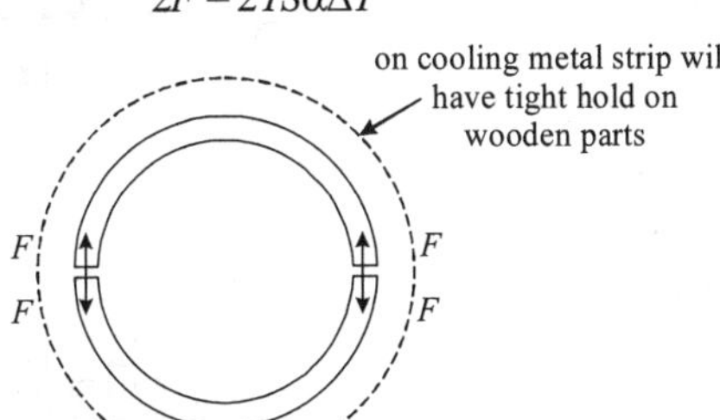

Ans. (A)

Sol. 4 For all linear scales, we use

$$\frac{\text{Reading} - LFP}{UFP - LFP} = \text{Constant}$$

For Kelvin scale & Y scale we use

$$\frac{Y-(-160)}{-50-(-160)} = \frac{340-273}{373-273} = \frac{67}{100}$$

$$Y + 160 = \frac{67}{100} \times 110 = 73.7$$

$$\Rightarrow \quad Y = 73.7 - 160 = -86.3°Y$$ **Ans. (C)**

Sol. 5 Given that

$$\frac{\gamma_G}{\gamma_L} = \frac{1}{4}$$

Remain constant with temperature

V_G

V_L

$\Rightarrow$ For any rise in temperature (say ΔT) the expansion in container is equal to expansion in liquid volume.

$$\Rightarrow \quad \Delta V_G = \Delta V_L$$

$$\Rightarrow \quad \gamma_G V_G \Delta T = \gamma_L V_L \Delta T$$

$$\Rightarrow \quad \frac{V_L}{V_G} = \frac{\gamma_G}{\gamma_L} = \frac{1}{4}$$ **Ans. (B)**

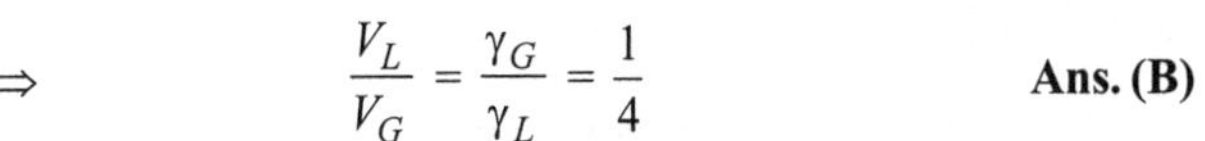

Sol. 6 On the ends of steel wire if F is the force needed to keep its length constant then stress in wire is

$$\text{Stress} = \frac{F}{A} = Y \times \text{Strain}$$

$$\text{Strain} = \frac{\Delta l}{l} = \alpha\Delta T$$

$$\Rightarrow \quad \text{Stress} = \frac{F}{A} = Y\alpha\Delta T$$

pressure required

$$p = 2 \times 10^{11} \times 1.1 \times 10^{-5} \times 100$$

$$\Rightarrow \quad p = 2.2 \times 10^8 \text{ Pa}$$ **Ans. (A)**

Sol. 7. Fractional time lost/second by a pendulum clock is given as

$$\frac{\delta t}{t} = \frac{1}{2}\alpha\Delta T$$

Case-1 : When $T = 40°C$ and we consider T_g is graduation temperature of clock when it reads correct time then we have

$$\delta t = 12 \text{ s}$$

$$t = 1 \text{ day} \equiv 86400 \text{ s}$$

$$\Rightarrow \quad \frac{12}{86400} = \frac{1}{2}\alpha(40 - T_g) \quad \ldots(1)$$

Case-2 :

$$\delta t = 4\text{s}$$

$$t = 1 \text{ day}$$

$$\Rightarrow \quad \frac{4}{86400} = \frac{1}{2}\alpha(T_g - 20) \quad \ldots(2)$$

using $\frac{(1)}{(2)}$ gives

$$\frac{12}{4} = \frac{40 - T_g}{T_g - 20}$$

$$\Rightarrow \quad 3T_g - 60 = 40 - T_g$$

$$\Rightarrow \quad 4T_g = 100 \Rightarrow T_g = 25°C$$

From equation-(1)

$$\frac{12}{86400} = \frac{1}{2}\alpha(40 - 25)$$

$$\Rightarrow \quad \alpha = \frac{12 \times 2}{86400 \times 15} = 1.85 \times 10^{-5} \,°C^{-1}$$ **Ans. (A)**

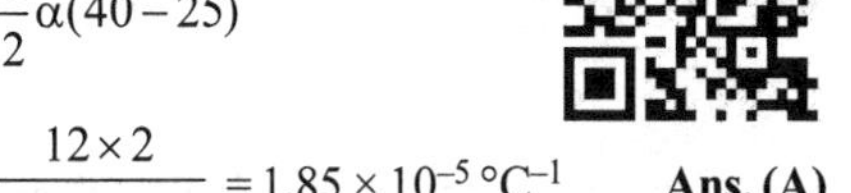

Sol. 8 For a cube of volume V, if on heating its volume increases by ΔV then we use for ΔT rise in temperature

$$\Delta V = V\gamma\Delta T = V(3\alpha)\Delta T \quad \ldots(1)$$

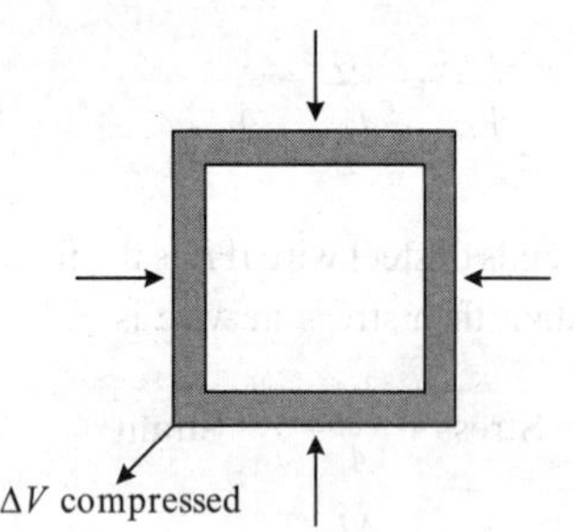

On applying pressure to cube if its volume is compressed by ΔV then we have

Bulk modulus

$$K = \frac{P}{\left|\frac{\Delta V}{V}\right|} = \frac{P}{3\alpha\Delta T}$$

$$\Rightarrow \quad \Delta T = \frac{P}{3\alpha K}$$

for same size ΔV increased (due to thermal expansion).

Ans. (C)

Sol. 9 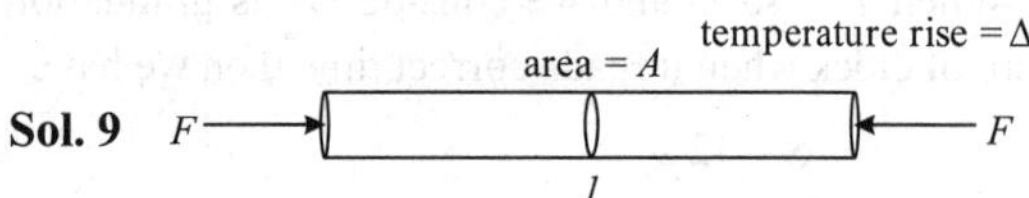

As length of rod $\rightarrow$ constant

$\Rightarrow$ Thermal expansion due to rise in temperature is compensated by elastic compression due to compressive force F.

Here we use

$$Y = \frac{\text{Stress}}{\text{Strain}} = \frac{F/A}{\Delta l/l}$$

Thermal expansion

$$\Delta l = \alpha l \Delta T$$

$$\Rightarrow \quad Y = \frac{F/A}{\alpha\Delta T}$$

$$\Rightarrow \quad F = YA\alpha\Delta T$$

Ans. (C)

Sol. 10 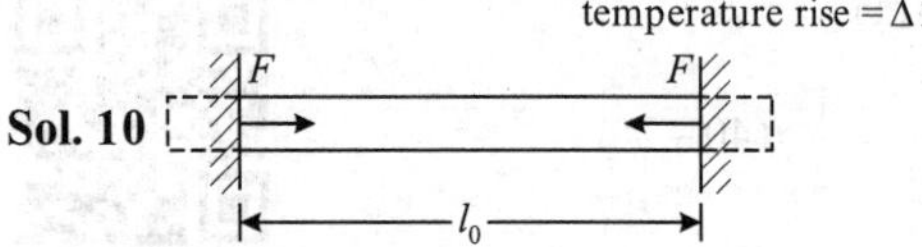

Due to the external force at ends, then thermal expansion in rail due to rise in temperature in compensated by elastic compression due to external force.

$$\Rightarrow \quad Y = \frac{\text{Stress}}{\text{Strain}} = \frac{F/A}{\Delta l/l_0}$$

Thermal expansion

$$\Delta l = \alpha l_0 \Delta T$$

$$\Rightarrow \quad Y = \frac{F/A}{\alpha\Delta T}$$

$$\Rightarrow \quad F = YA\alpha\Delta T$$

$$\Rightarrow \quad F = 2 \times 10^{11} \times 40 \times 10^{-4} \times 1.2 \times 10^{-5} \times 10$$

$$\Rightarrow \quad F = 9.6 \times 10^4 \text{ N}$$

$$\Rightarrow \quad F \simeq 1 \times 10^5 \text{ N}$$

Ans. (B)

Sol. 11 Young's modulus

$$Y = \frac{\text{Stress}}{\text{Strain}}$$

$$\Rightarrow \quad Y = \frac{F/A}{(\Delta l/l)}$$

$$\Rightarrow \quad Y = \frac{F}{A(\alpha\Delta T)}$$

Ans. (C)

Sol. 12 Given that

$$\Delta l_1 = \Delta l_2$$

$$\Rightarrow \quad l\alpha_1\Delta T_1 = l\alpha_2\Delta T_2$$

$$\Rightarrow \quad \frac{\alpha_1}{\alpha_2} = \frac{\Delta T_1}{\Delta T_2}$$

$$\Rightarrow \quad \frac{4}{3} = \frac{T-30}{180-30}$$

$$\Rightarrow \quad T = 230°\text{C}$$

Ans. (B)

Sol. 13 For the given temperature and scale, we use

$$\frac{\frac{x_0}{2} - \frac{x_0}{3}}{x_0 - \frac{x_0}{3}} = \frac{R-0}{100-0}$$

$$\Rightarrow \quad R = 100 \times \left(\frac{x_0/6}{2x_0/3}\right)$$

$$\Rightarrow \quad R = 25°\text{C}$$

Ans. (B)

Sol. 14 Coefficient of volume expansion is given as

$$\gamma = \alpha_x + \alpha_y + \alpha_z$$

$$\Rightarrow \quad \gamma = 5 \times 10^{-5} + 5 \times 10^{-6} + 5 \times 10^{-6}$$

$$\Rightarrow \quad \gamma = (50 + 5 + 5) \times 10^{-6}$$

$$\Rightarrow \quad \gamma = 60 \times 10^{-6}$$

$$\Rightarrow \quad C = 60$$

Ans. (60)

Sol. 15 Initially we use

$$\rho_0 S(80)\, g = mg \quad \ldots(1)$$

later $\quad \rho S(79)\, g = mg \quad \ldots(2)$

$$\Rightarrow \quad \frac{\rho}{\rho_0} = \frac{80}{79} = 1.012$$

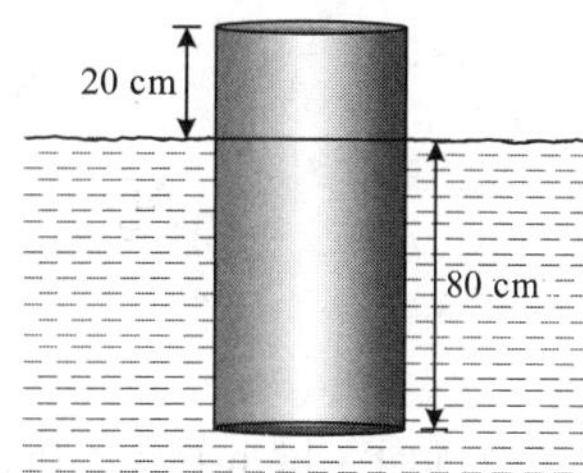

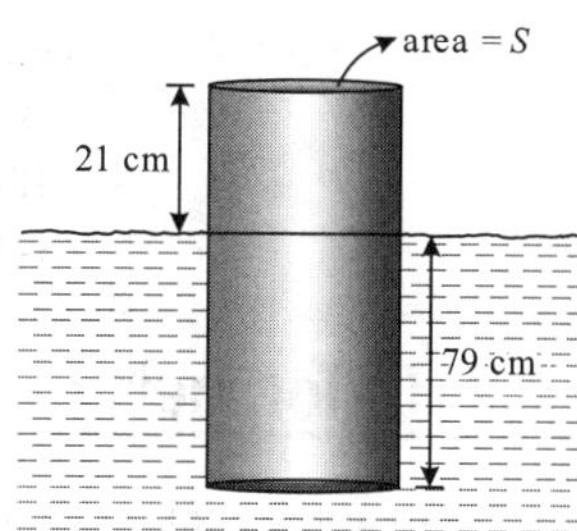

Ans. (A)

Sol. 16 It is given that by thermal expansion

$$\frac{\Delta l}{l} = \frac{0.02}{100} = \propto \Delta T$$

$$\Rightarrow \quad \propto = \frac{2\times10^{-4}}{\Delta T}$$

Thus volume expansion coefficient is

$$\gamma = 3\alpha = \frac{6\times10^{-4}}{\Delta T}$$

Density variation is given as

$$\Delta\rho = -\rho\gamma\Delta T$$

$$\Rightarrow \quad \frac{\Delta\rho}{\rho} = -\gamma\Delta T = -6\times10^{-4} = -0.06\%$$ **Ans. (B)**

Sol. 17 The unfitted volume of beaker is given as

$$\Delta V = (V_0 - V_m)$$

After increasing temperature by ΔT, we use

$$\Delta V' = (V_0' - V_m')$$

As $\Delta V' = \Delta V$, we use

$$V_0 - V_m = V_0(1+\gamma_b\Delta T) - V_m(1+\gamma_M\Delta T)$$

$$\Rightarrow \quad V_0\gamma_b = V_m\gamma_m$$

$$\Rightarrow \quad V_m = \frac{V_0\gamma_b}{\gamma_m} = \frac{(500)(6\times10^{-6})}{(1.5\times10^{-4})}$$

$$\Rightarrow \quad V_m = 20\text{ cm}^3$$ **Ans. (20)**

Sol. 18 At T°C, Total length of combination s

$$L = L_1 + L_2$$

L_1, α_1 $\quad$ L_2, α_2

At $T + \Delta T$, Total length becomes

$$L'_{eq} = L'_1 + L'_2$$

$(L_1 + L_2), \alpha_{avg}$

There we us $\quad L'_1 = L_1(1+\alpha_1\Delta T)$

and $\quad L'_2 = L_2(1+\alpha_2\Delta T)$

$$\Rightarrow \quad L'_{eq} = (L_1+L_2)(1+\alpha_{avg}\Delta T)$$

$$\Rightarrow \quad (L_1+L_2)(1+\alpha_{avg}\Delta T) = L_1+L_2+L_1\alpha_1\Delta T+L_2\alpha_2\Delta T$$

$$\Rightarrow \quad (L_1+L_2)\alpha_{avg} = L_1\alpha_1 + L_2\alpha_2$$

$$\Rightarrow \quad \alpha_{avg} = \frac{L_1\alpha_1 + L_2\alpha_2}{L_1 + L_2}$$ **Ans. (D)**

Sol. 19 Volume increment can be directly given as

$$\Delta V = V\gamma\Delta T$$

$$\Rightarrow \quad \Delta V = 3a^3\alpha\,\Delta T$$ **Ans. (A)**

Sol. 20 As rod is lying freely on smooth surface, it expands as no force is developed in rod so no stress will be there and in above case R is not the correct explanation of A. **Ans. (D)**

Sol. 21 As $\alpha_A > \alpha_B$ so on placing in the cold bath lenth of A will decrease more than length of B due to which system will bend toward left. **Ans. (D)**

Sol. 22 Resistance as a function of temperature is given as

$$R' = R(1+\alpha\Delta t)$$

$$\Rightarrow \quad 20 = 16(1+\alpha.85)$$

$$\Rightarrow \quad \alpha = \frac{\frac{20}{16}-1}{85} = \frac{1}{4\times85} = 0.003°\text{C}^{-1}$$ **Ans. (C)**

Sol. 23 Thermal force acting on the rod can be written as

$$F = Ay\alpha\Delta T$$

$$\Rightarrow \quad F = (10\times10^{-4})\times(2\times10^{11})(10^{-5})(400) = 8\times10^{-5}\text{ N}$$

$$\Rightarrow \quad x = 8$$ **Ans. (8)**

Sol. 24 Increase in volume of the cube is given as

$$\Delta V = \gamma V_0\Delta T$$

$$\Rightarrow \quad \Delta V = (3\alpha)V_0\Delta T$$

Total surface area of cube is $6a^2$, where a is its edge length, so we use

$$24 = 6a^2$$

$$\Rightarrow \quad a = 2\text{ m}$$

Total volume of cube is calculated as

$$V_0 = (2)^3 = 8\text{ m}^3$$

$$\Rightarrow \quad \Delta V = (3\times5\times10^{-4})(8)\times10$$

$$\Rightarrow \quad \Delta V = 1.2\times10^5\text{ cm}^3$$ **Ans. (B)**

Sol. 25 The difference in the two given diameter are

$$\Delta l = 6.241 - 6.230 = 0.011\text{ cm}$$

After heating the gold diameter should fit on wooden bangle so we use

$$\Delta l = l\alpha\Delta\theta$$

$\Rightarrow$ $$0.011 = 6.230 \times 1.4 \times 10^{-5}(\theta - 27)$$

$\Rightarrow$ $$\theta - 27 = \frac{0.011 \times 10^5}{6.230 \times 1.4}$$

$\Rightarrow$ $$\theta \approx 153.11 \text{ (Nearest value is 152.7°C)}$$

Ans. (D)

Sol. 26 Difference between the lengths is given as

$$l_B(1 + \alpha_B\Delta T) - l_i(1 + \alpha_i\Delta T) = l_B - l_i$$

As it should remain constant, we use

$$\alpha_B l_B = l_i \alpha_i$$

$\Rightarrow$ $$1.8 \times 10^{-5} \times 40 = l_i \times 1.2 \times 10^{-5}$$

$\Rightarrow$ $$l_i = \frac{1.8 \times 10^{-5} \times 40}{1.2 \times 10^{-5}} = \frac{3 \times 40}{2} = 60 \text{ cm}$$

Ans. (60)

Sol. 27 If we consider diameter of hole is d_0 at 27°C & d_1 at 177°C, we use

$$d_1 = d_0(1 + \alpha\Delta T)$$

$\Rightarrow$ $$d_1 - d_0 = 5 \times 1.6 \times 10^{-5} \times 150 \text{ cm}$$

$\Rightarrow$ $$d_1 - d_0 = 12 \times 10^{-3} \text{ cm}$$

Ans. (12)

Sol. 28 For any given temperature scale we use below relation as a constant for all scales.

$$\frac{\text{reading on scale-Lower fixed point}}{\text{upper fixed point-lower fixed point}} = \text{constant}$$

$\Rightarrow$ $$\frac{t_P - 30}{180 - 30} = \frac{t_Q - 0}{100 - 0}$$

$\Rightarrow$ $$\frac{t_P - 30}{150} = \frac{t_Q}{100}$$

Ans. (B)

Sol. 29 By thermometric relation for two temperature scales, we use

$$\frac{T - LEF}{UFP - LFP} = \text{constant}$$

$\Rightarrow$ $$\frac{41° - 5°}{95° - 5°} = \frac{C - 0°}{100° - 0°}$$

$\Rightarrow$ $$C = \frac{36}{90} \times 100 = 40°\text{C} = 313 \text{ K}$$

Ans. (313)

Sol. 30 By thermometric relation of temperature scales, we use

$$\frac{X - X_{\text{freez}}}{X_{\text{boil}} - X_{\text{freez}}} = \frac{t - 32}{212 - 32}$$

$\Rightarrow$ $$\frac{-95 - (-15)}{65 - (-15)} = \frac{t - 32}{180}$$

$\Rightarrow$ $$\frac{-80}{80} = \frac{t - 32}{180}$$

$\Rightarrow$ $$t = -180 + 32$$

$\Rightarrow$ $$t = -148°\text{F}$$

Ans. (D)

10.2 Calorimetry

Sol. 1 Specific heat $\rightarrow$ amount of heat which raises the temperature of unit mass of a substance by 1°C

Total heat required to raise the temperature of mass m by temperature t°C is

$$Q = mst \text{ (specific heat)}$$

If $t = 1$°C

$\Rightarrow$ $$Q = ms \rightarrow \text{Thermal capacity}$$

amount of heat to raise a body temperature by 1°C **Ans. (B)**

Sol. 2 By mass energy equivalence we have

$$E = mc^2$$

In case of water forming ice, it releases latent heat to surrounding

As heat is released from substance $\Rightarrow$ some mass is converted to heat (energy)

$\Rightarrow$ mass of water should decrease. **Ans. (C)**

Sol. 3 Amount of heat required to change the phase of 1g water from liquid to vapour is

$$\Delta Q = mL_v = 1 \times 2256 \text{ J/g}$$

$\Rightarrow$ $$\Delta Q = 2256 \text{ J}$$

volume change due to vaporization is

$$\Delta V = 1671 - 1 = 1670 \text{ cm}^3$$

As expansion take place at constant atmospheric pressure (10^5 Pa), the work done by steam is

$$W = P\Delta V = 10^5 \times 1670 \times 10^{-6} = 167 \text{ J}$$

By first law of thermodynamics we use

$$\Delta Q = \Delta U + W$$

$\Rightarrow$ $$\Delta U = \Delta Q - W$$

$$= 2256 - 167 = 2089 \text{ J}$$

Ans. (C)

Sol. 4 Mixture = 500 gm water at 0°C + 100 gm ice at 0°C + 40 gm water (calorimeter) at 0°C

+ 10 gm steam at 100°C

heat required to melt 100 gm ice is

$$Q_{\text{ice}} = 100 \times 80 = 8000 \text{ cal}$$

Max heat which can be supplied by steam is

$$Q_{\text{steam}} = 10 \times 540 + 10 \times 1 \times 100$$

$$= 6400 \text{ cal}$$

As $\quad Q_{\text{steam}} < Q_{\text{ice}}$

$\Rightarrow$ Whole ice will not melt and find temperature mixture is 0°C

$\Rightarrow$ Amount of ice melted is given as

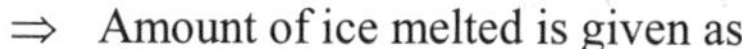

$$m = \frac{Q_{\text{steam}}}{L_{\text{ice}}} = \frac{6400}{80} = 80 \text{ gm}$$

$\Rightarrow$ Mixture= 500 gm water at 0°C + 80 gm water at 0°C + 20 gm ice at 0°C + Calorimeter (40 gm water) + 10 gm water at 0°C

In mixture finally there is 590 gm water at 0°C. **Ans. (B)**

Sol. 5 For heating 2 L water (2 kg) by 50°C (from 27°C to 77°C) amount of heat required is

$$Q = ms\Delta T = 2 \times 4200 \times 50 \text{ J}$$

Net heat supplied (rate) to water is

$$r = 1000 - 160 = 840 \text{ W}$$

(supply by coil is 1000W, loss is 160 J/s)

If in time t it happens then

$$840\,t = 2 \times 4200 \times 50$$

$\Rightarrow \quad t = 500 \text{ s} = 8 \text{ m } 20 \text{ s}$ **Ans. (A)**

Sol. 6 Let mass of water is m in gm and the uniform rate at which heat is being supplied is r cal/s, then we use

for first 10 min. $\rightarrow$ Total heat supplied

$$r \times 10 \times 60 = m \times 1 \times 100 \quad \ldots(1)$$

for next 55 min $\rightarrow$ Total heat supplied

$$r \times 55 \times 60 = m \times L_v \quad \ldots(2)$$

$$\frac{(2)}{(1)} \Rightarrow \frac{L_v}{100} = \frac{55}{10}$$

$\Rightarrow \quad L_v = 550 \text{ cal/gm}$ **Ans. (B)**

Sol. 7

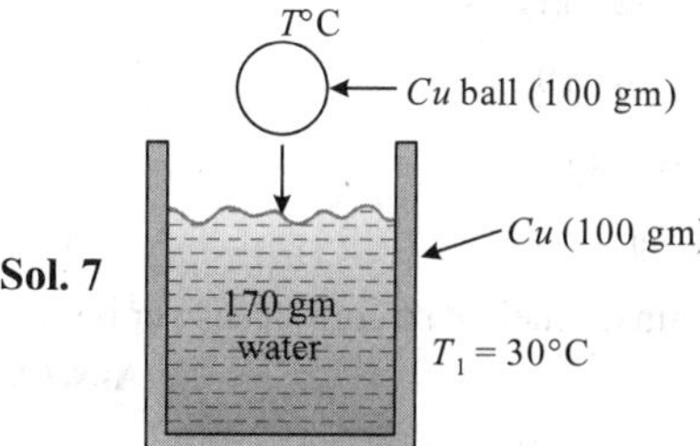

As $\quad T_f = 75°C$

So rise in temperature of calorimeter + water

$$\Delta T = 75° - 30°$$

$\Rightarrow \quad \Delta T = 45°C$

As final temperature of mixture is 75°C

$\Rightarrow$ Heat lost by Cu-ball = Heat gained by Cu-calorimeter + water

$\Rightarrow \quad 100 \times 0.1 \times (T - 75) = 100 \times 0.1 \times 45 + 170 \times 1 \times 45$

$\Rightarrow \quad 10\,T - 750 = 450 + 7650$

$\Rightarrow \quad 10\,T = 8100 + 750 = 8850$

$\Rightarrow \quad T = \dfrac{8850}{10} = 885°C$ **Ans. (D)**

Sol. 8

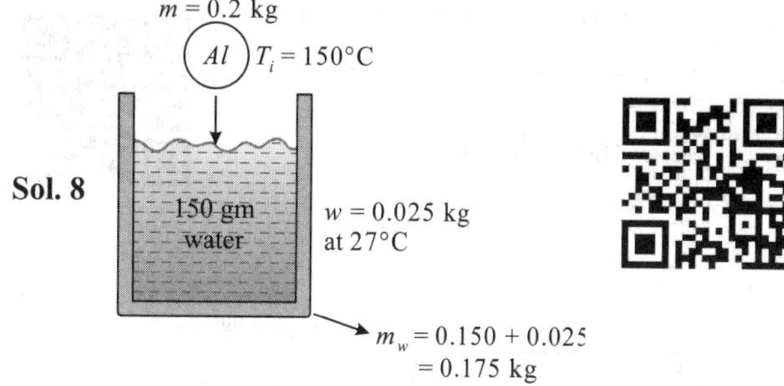

Equilibrium temperature attained = 40°C

Heat lost by Al - sphere = heat gained by calorimeter + water

$$0.2 \times s_M \times 110 = 0.175 \times 4200 \times 13$$

$$\Rightarrow \quad s_{Al} = \frac{0.175 \times 4200 \times 13}{0.2 \times 110} = 434.31 \text{ J/kg°C}$$

Ans. (D)

Sol. 9 By energy conservation, we use

$$192 \times s \times (100 - 21.5)$$

$$= 128 \times 394 \times (21.5 - 8.4)$$

$$+ 240 \times 4200 \times (21.5 - 8.4)$$

$\Rightarrow \quad s = 916 \text{ J/kg-K}$ **Ans. (D)**

Sol. 10 Let amount of ice is m gm. According to principal of calorimeter heat taken by ice = heat given by water

$$\Rightarrow \quad 20 \times 2.1 \times m + (m - 20) \times 334$$

$$= 50 \times 4.2 \times 40$$

$$376\text{m} = 8400 + 6680$$

$m = 40.1\text{g}$ **Ans. (B)**

Sol. 11 By conservation of energy, we use

$$100 \times s_A \times [100 - 90] = 50 \times s_B \times (90 - 75)$$

$$\Rightarrow \quad 2s_A = 1.5s_B$$

$$\Rightarrow \quad s_A = \frac{3}{4}s_B$$

And $\quad 100 \times s_A \times [100 - T] = 50 \times s_B\,(T - 50)$

$$\Rightarrow \quad 2\times\left(\frac{3}{4}\right)(100-T)=(T-50)$$

$$\Rightarrow \quad 300-3T=2T-100$$

$$400=5T$$

$$T=80°\text{C}$$ **Ans. (A)**

Sol. 12 By conservation of energy, we use

$$0.1\times 400\times(500-T)=0.5\times 4200\times(T-30)+800(T-30)$$

$$\Rightarrow \quad 40(500-T)=(T-30)(2100+800)$$

$$\Rightarrow \quad 20000-40T=2900\,T-30\times 2900$$

$$\Rightarrow \quad 20000+30\times 2900=T(2940)$$

$$\Rightarrow \quad T=36.4°\text{C}$$

$$\Rightarrow \quad \frac{\Delta T}{T}\times 100=\frac{6.4}{30}\times 100$$

$$=21.3\%$$ **Ans. (B)**

Sol. 13

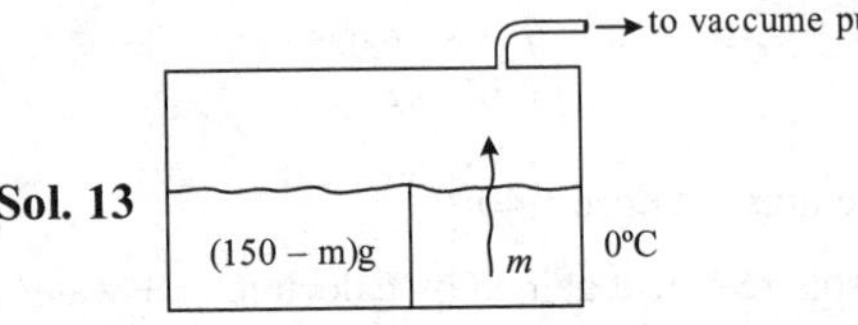

If mass m gets evaporated

$\Rightarrow$ heat required is

$$\Delta Q_{evap}=mL_v$$

This heat is supplied by remaining water in vessel which will transform into ice

$\Rightarrow$ heat supplied is $\Delta Q_{freeze}=(150-m)L_f$

$$\Rightarrow \quad \Delta Q_{evap}=\Delta Q_{freeze}$$

$$mL_v=(150-m)L_f$$

$$m=\frac{150L_f}{L_v+L_f}=\frac{150\times 3.36\times 10^5}{(21+3.36)\times 10^5}$$

$$m\simeq 20\text{g}$$ **Ans. (C)**

Sol. 14 Heat lost by water = Heat gained by ice

$$\Rightarrow \quad M_2S_w\Delta T_w=M_1S_i\Delta T_i+M_1L_f$$

$$\Rightarrow \quad M_2\times 1\times 50=M_1\times 0.5\times 10+M_1L_f$$

$$\Rightarrow \quad 50M_2=5M_1+M_1L_f$$

$$\Rightarrow \quad L_f=\frac{50M_2}{M_1}-5$$ **Ans. (C)**

Sol. 15 For exchange of heat in mixture, we use

Heat Supplied = Heat Gained

$$M\times 540+M\times 1\times 60=200\times 80+200\times 1\times 4$$

$$\Rightarrow \quad M=40\text{ gm}$$ **Ans. (40)**

Sol. 16 According to table using heat exchange between water amounts, we use

Case-I: $1(T_1-60)+2(T_2-60)=0$

$$\Rightarrow \quad 1T_1+2T_2=(1+2)60°$$

$$T_1+2T_2=180 \qquad \ldots(1)$$

Case-II: $2(T_1-60)+1(T_2-30)=0$

$$\Rightarrow \quad T_2+2T_3=(1+2)30°$$

$$T_2+2T_3=90 \qquad \ldots(2)$$

Case-III: $2(T_1-60)+1(T_3-60)=0$

$$\Rightarrow \quad 2T_1+1T_3=(1+2)60°$$

$$2T_1+T_3=180 \qquad \ldots(3)$$

Adding (1) + (2) + (3) gives

$$3(T_1+T_2+T_3)=450$$

$$\Rightarrow \quad T_1+T_2+T_3=150° \qquad \ldots(4)$$

Case-IV: $(T_1-\theta)+(T_2-\theta)+(T_3-\theta)=0$

$$\Rightarrow \quad T_1+T_2+T_3=(1+1+1)\theta$$

$$\Rightarrow \quad 150°=3\theta$$

$$\Rightarrow \quad \theta=50°\text{C}$$ **Ans. (50)**

Sol. 17 Given mixture can be taken as

Mixture = 200 g water at 25°C + m gram steam at 100°C

Heat supplied by steam = $m\times 540+m\times 1\times 69=609m$

Heat absorbed by water = $200\times 1\times 6=1200$ cal

by conservation of energy, we use

$$1200=609\,m$$

$$\Rightarrow \quad m=\frac{1200}{609}\simeq 2\text{ g}$$ **Ans. (B)**

Sol. 18 Here the water will supply heat to ice for melting

$$m_w s_w\Delta\theta=m_{ice}L_{ice}$$

$$m_{ice}=\frac{0.2\times 4200\times 25}{3.4\times 10^5}$$

$$\Rightarrow \quad m_{ice}=0.0617\text{ kg}$$

$$\Rightarrow \quad m_{ice}=61.7\text{ gm}$$

Remaining ice will remain un-melted so correct answer is (A). **Ans. (A)**

Sol. 19 Given that half of KE is converted to heat, we use

$$\frac{1}{2}\times\frac{1}{2}mv^2=ms\Delta T$$

$$\Rightarrow \quad \Delta T=\frac{(210)^2}{4\times 0.03\times 4200}\simeq 87.5°\text{C}$$ **Ans. (B)**

Sol. 20 If n drops combine to form a bigger drop, the radius of bigger drop is given as

$$n\times\frac{4}{3}\pi r^3 = \frac{4}{3}\pi R^3$$

$$\Rightarrow \quad n^{1/3}r = R$$

Total change in surface energy of the liquid is given by product of surface tension and change in area, given as

$$U = (n(4\pi r^2) - 4\pi R^2)T$$

$$\Rightarrow \quad U = 4\pi T(nr^2 - R^2)$$

Heat energy released is given as

$$U = \frac{4\pi T(nr^2 - R^2)}{J\times\frac{4}{3}\pi R^3} = \frac{3T}{J}\left(\frac{nr^2}{R^3} - \frac{1}{R}\right)$$

$$\Rightarrow \quad U = \frac{3T}{J}\left(\frac{1}{r} - \frac{1}{R}\right)$$ **Ans. (C)**

Sol. 21 When A & B are mixed, by conservation of energy we consider heat supplied by one liquid is equal to the heat gained by other liquid

$$m_1 S_1 (16-10) = m_2 S_2 (20-16)$$

$$\Rightarrow \quad 6m_1 S_1 = 4m_2 S_2 \quad \ldots(1)$$

When B & C are mixed, we use

$$m_2 S_2 (26-20) = m_3 S_3 (30-26)$$

$$\Rightarrow \quad 6m_2 S_2 = 4m_3 S_3 \quad \ldots(2)$$

From equation (1) and (2), we have

$$9m_1 S_1 = 4m_3 S_3$$

When A & C are mixed, we use

$$m_1 S_1 (T-10) = m_3 S_3 (30-T)$$

$$\Rightarrow \quad T = \frac{30m_3S_3 + 10m_1S_1}{m_1S_1 + m_3S_3}$$

$$\Rightarrow \quad T = \frac{10(3m_3S_3 + m_1S_1)}{m_3S_3 + m_1S_1}$$

$$\Rightarrow \quad T = \frac{10\left(3m_3S_3 + \frac{4}{9}m_3S_3\right)}{m_3S_3 + \frac{4}{9}m_3S_3}$$

$$\Rightarrow \quad T = \frac{10\times 31}{13} = 23.84°C$$ **Ans. (C)**

Sol. 22 By conservation of energy we have

$$mgh = ms\Delta\theta$$

$$\Rightarrow \quad g\times 63 = \Delta\theta\times 4.2\times 10^3$$

$$\Rightarrow \quad \Delta\theta = 0.147°C$$ **Ans. (C)**

Sol. 23 Heat supplied from resistance can be written as

$$i^2Rt = ms\Delta T + mL$$

$$\Rightarrow \quad (0.5)^2\times 4\times 10^3\times t = (10^3\times 1\times 10^{-4})\times 2400\times 10 + (10^3\times 1\times 10^{-4})\times 330\times 10^3$$

$$\Rightarrow \quad t = 24\times 10^3\times 10^{-4} + 33\times 10^4\times 10^{-4} = 2.4 + 33 = 35.4 \text{ s}$$

Ans. (D)

Sol. 24 Given that one fourth of energy of hammer goes into heating the nail, we have

$$\frac{1}{2}\times 1.5\times 10^2\times\frac{1}{4} = 0.1\times 420\times\Delta T$$

$$\Rightarrow \quad \Delta T = 16.07°C$$ **Ans. (C)**

Sol. 25 Heat rejected $= mL_f + mS\Delta T$

$$\Delta Q = (50\times 540) + 50(1)(100-20)$$

$$\Rightarrow \quad \Delta Q = 31000 \text{ Cal}$$

$$\Rightarrow \quad \Delta Q = 31\times 10^3 \text{ Cal}$$ **Ans. (31)**

Sol. 26 Heat given by block to get 0°C temperature

$$\Delta Q_1 = 5\times(0.39\times 10^3)\times(500-0)$$

$$\Rightarrow \quad \Delta Q_1 = 975\times 10^3 \text{ J}$$

Heat absorbed by ice to melt m mass

$$\Delta Q_2 = m\times(335\times 10^3)\text{ J}$$

As $\Delta Q_1 = \Delta Q_2$ we use

$$m\times(335\times 10^3) = 975\times 10^3$$

$$\Rightarrow \quad m = \frac{975}{335} = 2.910 \text{ kg}$$ **Ans. (C)**

Sol. 27 Water flow rate is 2000 gm/min so heat required by water per minute is calculated as

$$Q = mS\Delta T$$

$$\Rightarrow \quad Q = (2000)\times 4.2\times 40 \text{ J/min}$$

$$\Rightarrow \quad Q = 336000 \text{ J/min}$$

The rate of combustion is given as

$$Q = \left(\frac{dm}{dt}L\right) = 336000 \text{ J/min}$$

$$\Rightarrow \quad \frac{dm}{dt} = \frac{336000}{8\times 10^3} \text{ g/min}$$

$$\Rightarrow \quad \frac{dm}{dt} = 42 \text{ gm/min}$$ **Ans. (42)**

Sol. 28 Energy released by water

$$= 0.3\times 25\times 4200 = 31500 \text{ J}$$

Let m kg ice melts

$$m\times 3.5\times 10^5 = 31500$$

$$m = \frac{31500\times 10^{-5}}{3.5} = 9000\times 10^{-5}$$

$$m = 0.09 \text{ kg} = 90 \text{ gm}$$

$$x = 90$$ **Ans. (90)**

Sol. 29 Heat supplied to ice is given as

$$\Delta Q = 184\times 10^3 \text{ J}$$

Heat required to raise the temperature of ice to zero degree is

$$Q_1 = 0.600\times 2222.3\times 12 = 16000.56 \text{ J}$$

Remaining available heat is calculated as

$$\Delta Q_1 - 184000 - 16000.56$$

$$\Rightarrow \qquad \Delta Q_1 = 167999.44\ \text{J}$$

For melting of ice at 0°C, heat required is calculated as

$$\Delta Q_2 = 0.600 \times 336000 = 201600\ \text{J}$$

As this much heat is not available so complete ice is not melted in this case and amount of ice melted with the available heat is calculated as

$$167999.44 = m \times 336000$$

$$m = 0.5\ \text{kg}$$

Thus mass of ice left is calculated as

$$m_{ice} = 0.6 - 0.5 = 0.1\ \text{kg}$$

Hence options (a) and (d) are correct. **Ans. (A)**

Sol. 30 70% heat of heater is used to increase the temperature of water

$$\eta \times P \times \Delta t = M \times s \times \Delta T$$

Thus time required to heat 2 kg water is calculated as

$$\Delta t = \frac{2 \times 4200 \times (60 - 10)}{0.7 \times 2000}$$

$$\Rightarrow \qquad \Delta t = 300\ \text{s}$$

Ans. (300)

10.3 Heat Transfer

Sol. 1 By Stefan's-Boltzmann law, the heat energy radiated/sec (Thermal radiation Power) by body of surface area A is given as

$$P_{th} = \frac{dQ}{dt} = \sigma\varepsilon A T^4$$

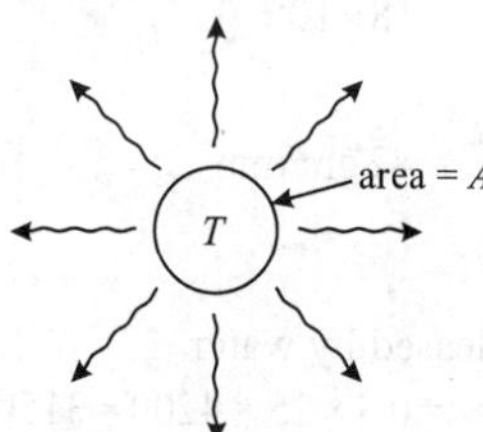

$e \rightarrow$ emissivity of body surface

$e < 1$ for general bodies

$e = 1$ for black body

For the given spheres we can use

$$P_1 = e\sigma(4\pi r_1^2)T_1^4$$

and

$$P_2 = e\sigma(4\pi r_2^2)T_2^4$$

$$\Rightarrow \qquad \frac{P_1}{P_2} = \frac{r_1^2 T_1^4}{r_2^2 T_2^4} = \frac{(1)^2(4000)^4}{(4)^2(2000)^4} = \frac{256}{256} = \frac{1}{1}$$

Ans. (A)

Sol. 2 (A) Spectrometer $\rightarrow$ used to analyse *EM* radiation spectrum

(B) Pyrometer $\rightarrow$ used to measure thermal radiations & radiations from a hot body

(C) Nanometer $\rightarrow$ used to measure small dimensions

(D) Photometer $\rightarrow$ used to measure light intensity **Ans. (B)**

Sol. 3 Black Body $\rightarrow$ which absorbs all the radiation incident on it and emits maximum radiation from it at a specific temperature In above options, options (A) is most suitable for choosing it close to a Black body. **Ans. (A)**

Sol. 4

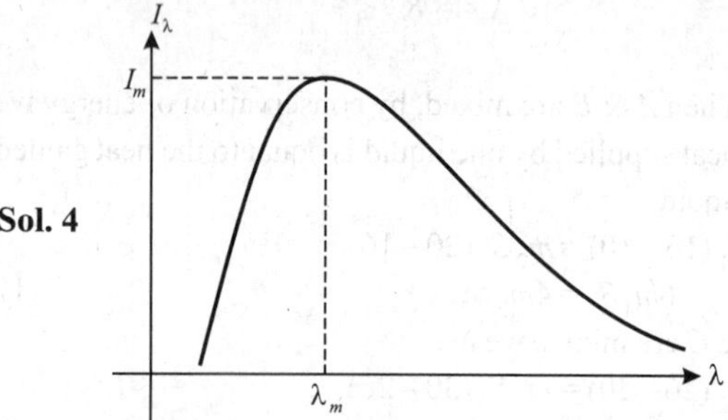

For any Black Body The *EM* radiation spectrum is given by Wien's Displacement law & Wien's fifth power law

$$\lambda_m \propto \frac{1}{T}$$

$$I_m \propto T^5$$

Ans. (D)

Sol. 5 By Newton's law of cooling the rate of cooling of a body is given as

$$\frac{\Delta\theta}{\Delta t} \propto \Delta\theta$$

($\Delta\theta$ = temperature difference of body with surrounding)

$$\Rightarrow \qquad n = 1$$

Ans. (D)

Sol. 6 We consider sum as a black body and by Stefan's law, radiant power emitted by sun is given as

$$P_{Rad} = \sigma A T^4 = \sigma(4\pi R^2)T^4$$

($R \rightarrow$ radius of Sun)

$$\Rightarrow \qquad P_{rad} \propto R^2 T^4$$

If $T \rightarrow 2T$ and $R \rightarrow 2R$

Then we have

$$\frac{P_f}{P_i} = \frac{(2R)^2(2T)^4}{R^2T^4}$$

$$= (2)^2(2)^4 = 64$$

$$\Rightarrow \qquad P_f = 64\,P_i$$

Ans. (D)

Sol. 7

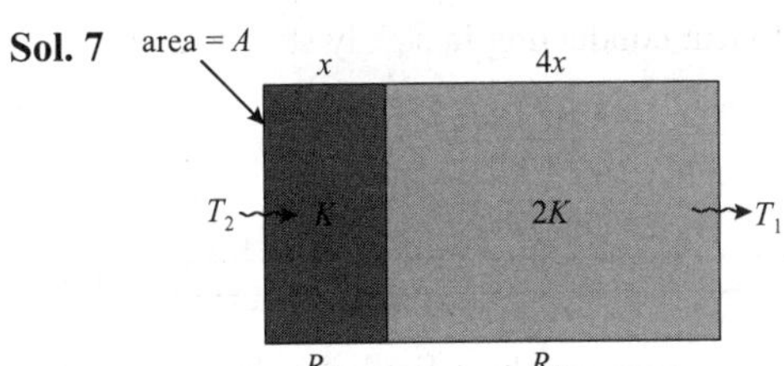

Where $$R_{th_1} = \frac{1}{k}\cdot\frac{x}{A}$$

and $$R_{th_2} = \frac{1}{2k}\cdot\frac{4x}{A}$$

Heat conduction through composite slab in steady state is given as

$$\frac{dQ}{dt} = \frac{T_2 - T_1}{(R_{th_1} + R_{th_2})}$$

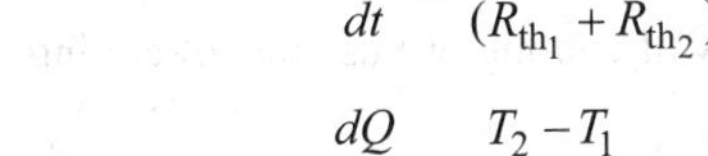

$$\Rightarrow \quad \frac{dQ}{dt} = \frac{T_2 - T_1}{\frac{x}{kA} + \frac{2x}{kA}} = \frac{kA(T_2 - T_1)}{3x}$$

$\Rightarrow \quad f = 1/3$ **Ans. (D)**

Sol. 8

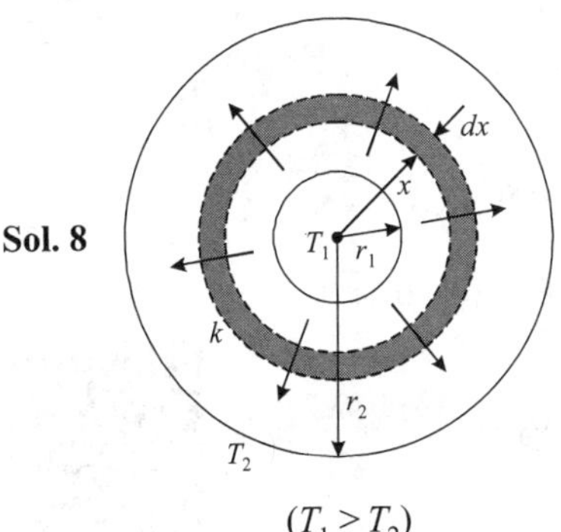

$(T_1 > T_2)$

Rate of heat flow between shells is given as

$$\frac{dQ}{dt} = \frac{T_1 - T_2}{R_{th}} \quad \ldots(1)$$

Where R_{th} is the thermal resistance for radial heat flow.

The thermal resistance of elemental shell for radial heat flow is given as

$$dR = \frac{1}{k}\cdot\frac{dx}{4\pi x^2}$$

Total thermal resistance between inner and outer shells is

$$R_{th} = \int dR = \frac{1}{4\pi k}\int_{r_1}^{r_2}\frac{dx}{x^2} = \frac{1}{4\pi k}\left[-\frac{1}{x}\right]_{r_1}^{r_2}$$

$$= \frac{1}{4\pi k}\left[\frac{1}{r_1} - \frac{1}{r_2}\right]$$

$$\Rightarrow \quad R_{th} = \frac{(r_2 - r_1)}{4\pi k r_1 r_2}$$

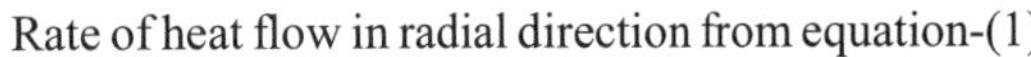

Rate of heat flow in radial direction from equation-(1)

$$\frac{dQ}{dt} = \frac{(T_1 - T_2)}{(r_2 - r_1)} \times 4\pi k r_1 r_2$$

$$\Rightarrow \quad \frac{dQ}{dt} \propto \frac{r_1 r_2}{r_2 - r_1}$$ **Ans. (C)**

Sol. 9

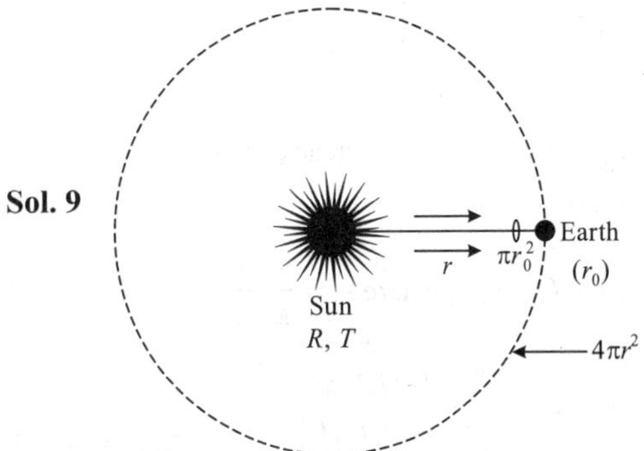

Surface area of spherical shell in which whole power of Sun will be distributed

Radiant power of Sun is

$$P_{rad} = \sigma(4\pi R^2)T^4$$

Radiant power of Sun falling per unit area of the earth surface (Normally) is

$$S = \frac{P_{rad}}{4\pi r^2} = \frac{\sigma(4\pi R^2)T^4}{4\pi r^2}$$

Solar constant $$S = \frac{\sigma R^2 T^4}{r^2}$$

Energy incident on Earth is

$$E_e = S.(\pi r_0^2)$$

$$\Rightarrow \quad E_e = \frac{\pi\sigma R^2 r_0^2 T^4}{r^2}$$ **Ans. (B)**

Sol. 10

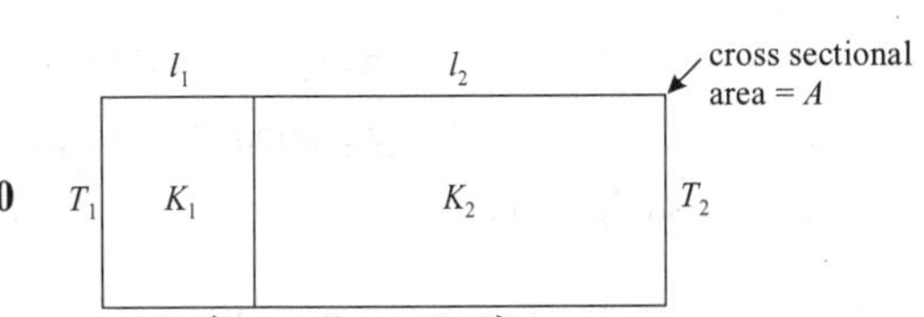

In steady state throughout the rod heat current is constant as heat absorption is zero, so we have

$$\left.\frac{dQ}{dt}\right|_1 = \left.\frac{dQ}{dt}\right|_2$$

$$\frac{T_i - x}{\left(\frac{1}{K_1}\cdot\frac{l_1}{A}\right)} = \frac{x - T_2}{\left(\frac{1}{K_2}\cdot\frac{l_2}{A}\right)}$$

$$\Rightarrow \quad x = \frac{K_1 l_2 T_1 + K_2 l_1 T_2}{K_1 l_2 + K_2 l_1}$$ **Ans. (C)**

Sol. 11

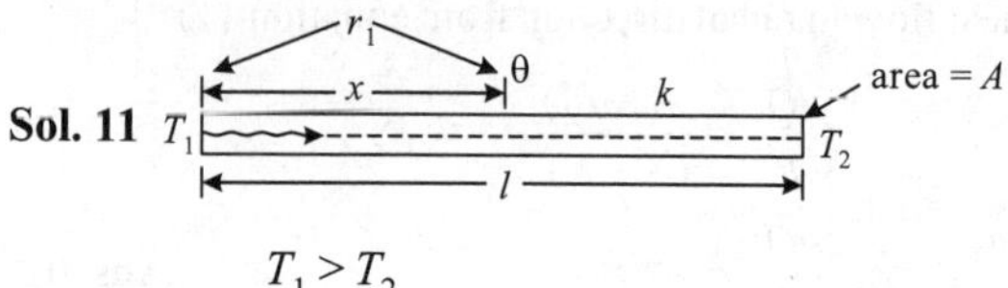

$T_1 > T_2$

Thermal current in rod is given as

$$I_{th} = \frac{dQ}{dt} = \frac{T_1 - T_2}{\left(\frac{1}{k}\cdot\frac{l}{A}\right)}$$

$\Rightarrow$ I_{th} = constant under steady state

Using Ohm's law we have

$$\theta = T_1 - I_{th} r_1 \text{ where } r_1 = \frac{1}{k}\cdot\frac{x}{A}$$

$$\Rightarrow \quad \theta = T_1 - \frac{(T_1 - T_2)}{\left(\frac{1}{k}\frac{l}{A}\right)}\left(\frac{1}{k}\frac{x}{A}\right)$$

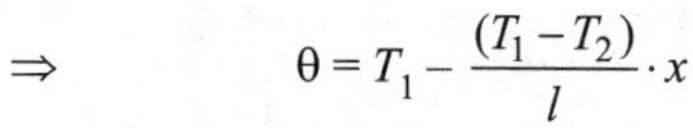

$$\Rightarrow \quad \theta = T_1 - \frac{(T_1 - T_2)}{l}\cdot x$$

Above equation has negative slope and positive intercept on y-axis. **Ans. (D)**

Sol. 12 By Newton's law of cooling, the rate of colling is given as

$$\frac{d\theta}{dt} = -k(\theta - \theta_0) \text{ °C/s}$$

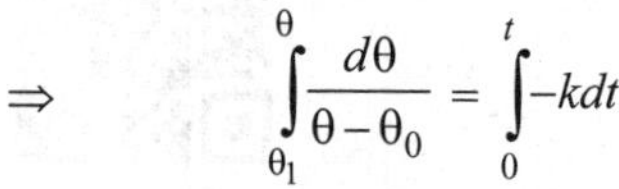

$$\Rightarrow \quad \int_{\theta_1}^{\theta} \frac{d\theta}{\theta - \theta_0} = \int_0^t -kdt$$

θ_1 is temperature of liquid at $t = 0$ $(\theta_1 > \theta_0)$

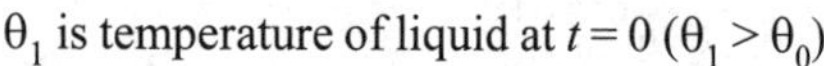

$$[\log_e(\theta - \theta_0)]_{\theta_1}^{\theta} = -kt$$

$$\Rightarrow \quad \log_e(\theta - \theta_0) = -kt + \log_e(\theta_1 - \theta_0) \text{ (constant)}$$

$y = mx + C$ [equation of a straight line]

$\Rightarrow$ – ve slope straight line **Ans. (B)**

Sol. 13 At such a high temperature (3000 K) we can consider the furnace like a black body so heat radiated per unit area/sec is given by Stefan's law as

$$E = \sigma T^4 \text{ J/m}^2\text{-s}$$

$\Rightarrow$ in 1 Hr. total heat/m² is given as

$$\Rightarrow \quad Q = Et = \sigma T^4.t$$

$$= 5.7 \times 10^{-8} \times (3000)^4 \times 3600$$

$$\Rightarrow \quad Q = 1.66 \times 10^{10} \text{ J/m}^2$$

$$Q \simeq 1.7 \times 10^{10} \text{ J/m}^2$$

Ans. (A)

Sol. 14 Rate of Heat conducting in steady state is given as

$$\frac{dQ}{dt} = \frac{kA(T_1 - T_2)}{l}$$

100°C	$C_u(9k)$, $L/2$	(T)	Steel (k), $L/2$	0°C

In steady state of conduction heat flow rate through the two rods will be same

$$\Rightarrow \quad \frac{(9k)A(100 - T)}{L/2} = \frac{kA(T - 0)}{L/2}$$

$$\Rightarrow \quad 9(100 - T) = T$$

$$\Rightarrow \quad 900 = T + 9T$$

$$\Rightarrow \quad T = \frac{900}{10} = 90\text{°C}$$

Ans. (A)

Sol. 15 By Newton's law of cooling, we use rate of colling of metal is given as

$$\frac{dT}{dt} = -k(T - \theta_0)$$

$$\Rightarrow \quad \int_{\theta}^{T} \frac{dT}{T - \theta_0} = \int_0^t -kdt$$

$$\Rightarrow \quad [\ln(T - \theta_0)]_{\theta}^{T} = -kt$$

$$\ln\left(\frac{T - \theta_0}{\theta - \theta_0}\right) = -kt$$

$$\Rightarrow \quad T = \theta_0 + (\theta - \theta_0)e^{-kt}$$

at $t = 0$; $T = \theta$

at $t \to \infty$ (long time), $T = \theta_0$

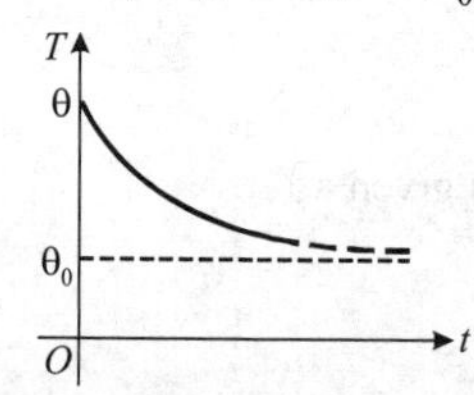

Ans. (C)

Sol. 16 By average from of Newton's law of cooling, we use

$$\frac{\Delta T}{t} = \frac{k}{(ms + \omega)}\left[\frac{T_1 + T_2}{2} - T_s\right]$$

where $\Delta T = T_2 - T_1$

Initially with water, we use

$$\frac{30 - 25}{2} = \frac{k}{50 \times 1 + 30}\left[\frac{30 + 25}{2} - T_s\right] \quad \ldots(1)$$

Then with given liquid, we use

$$\frac{30 - 25}{2} = \frac{k}{100s + 30}\left[\frac{30 + 25}{2} - T_s\right] \quad \ldots(2)$$

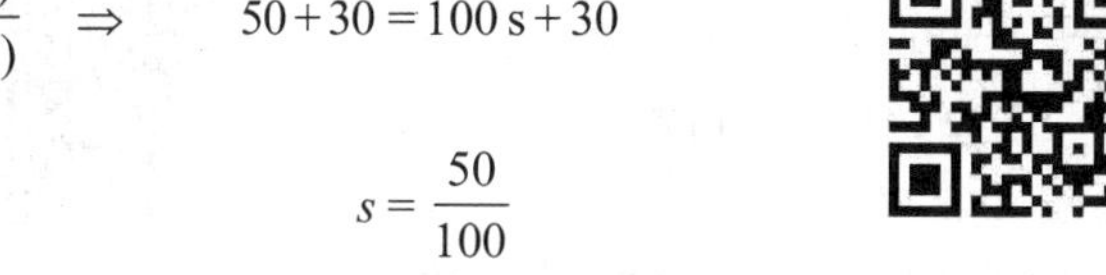

$\frac{(1)}{(2)} \Rightarrow \quad 50+30 = 100\,s + 30$

$\Rightarrow \quad s = \frac{50}{100}$

$= 0.5$ cal/gm $= 0.5$ kcal/kg **Ans. (D)**

Sol. 17

100°C 0°C C_n l_2 l_1 Br T l_3 st 0°C

$$A = 4\text{ cm}^2$$

$$l_1 = 46\text{ cm};\ k_1 = 0.92\text{ CGS units}$$

$$l_2 = 13\text{ cm};\ k_2 = 0.26\text{ CGS units}$$

$$l_3 = 12\text{ cm};\ k_3 = 0.12\text{ CGS units}$$

Using Junction rule at temperature T, we have

$$I_{Cu} = I_{Br} + I_{st} \quad \ldots(1)$$

Thermal resistances of rods are

$$R_{Cn} = \frac{1}{k_1}\cdot\frac{l_1}{A} = \frac{1}{0.92}\times\frac{46}{4} = \frac{100}{8}\,CGS\text{ Units}$$

$$R_{Br} = \frac{1}{k_2}\frac{l_2}{A} = \frac{1}{0.26}\times\frac{13}{4} = \frac{100}{8}\,CGS\text{ Units}$$

$$R_{st} = \frac{1}{k_3}\frac{l_3}{A} = \frac{1}{0.12}\times\frac{12}{4} = \frac{100}{4}\,CGS\text{ Units}$$

$$I_{Cu} = \frac{100-T}{R_{Cu}};\ I_{Br} = \frac{T-0}{R_{Br}};\ I_{st} = \frac{T-0}{R_{st}}$$

From equation (1) we have

$$\frac{100-T}{(100/8)} = \frac{T-0}{(100/8)} + \frac{T-0}{(100/4)}$$

$$\Rightarrow \quad 100 - T = T + \frac{T}{2} = \frac{3T}{2}$$

$$\Rightarrow \quad 200 = 5T$$

$$\Rightarrow \quad T = 40°\text{C}$$

Heat current through Cu-rod is

$$\left.\frac{dQ}{dt}\right|_{Cu} = \frac{100-T}{100/8}$$

$$= \frac{100-40}{100/8} = \frac{8\times 60}{100} = 4.8\text{ cal/s}$$

Ans. (C)

Sol. 18 Specific heat of sphere

$$s = \alpha T^3$$

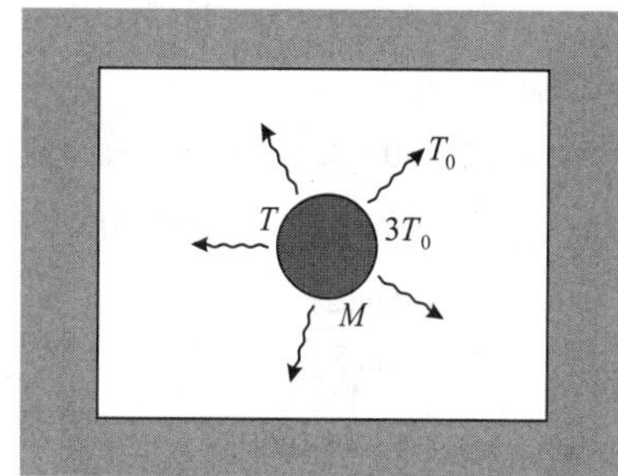

Due to radiation rate of heat loss by sphere is given as, when sphere temperature is T

$$2T_0 < T < 3T_0$$

$$\frac{dQ}{dt} = \sigma(4\pi R^2)(T^4 - T_0^4)$$

$$\Rightarrow \quad -\frac{MsdT}{dt} = 4\pi\sigma R^2(T^4 - T_0^4)$$

$$\Rightarrow \quad -\frac{M\alpha T^3 dT}{dt} = 4\pi\sigma R^2(T^4 - T_0^4)$$

$$\Rightarrow \quad \frac{T^3 dT}{T^4 - T_0^4} = -\frac{4\pi\sigma R^2}{M\alpha}dt$$

Integrating above expression, gives

$$\int_{T=3T_0}^{T=2T_0} \frac{T^3 dT}{T^4 - T_0^4} = -\frac{4\pi\sigma R^2}{M\alpha}\int_{t=0}^{t=t} dt$$

Let $\quad x = T^4 - T_0^4$

$\Rightarrow \quad dx = 4T^3 dT$

at $\quad T = 3T_0 \rightarrow x_1 = 80T_0^4$

and at $\quad T = 2T_0 \rightarrow x_2 = 15T_0^4$

$$\Rightarrow \quad \frac{1}{4}\int_{x_1}^{x_2}\frac{dx}{x} = -\frac{4\pi\sigma R^2}{M\alpha}\cdot t$$

$$\Rightarrow \quad \frac{1}{4}[\ln x]_{80T_0^4}^{15T_0^4} = -\frac{4\pi\sigma R^2}{M\alpha}t$$

$$\Rightarrow \quad [\ln(80T_0^4) - \ln(15T_0^4)]\cdot\frac{M\alpha}{16\pi\sigma R^2} = t$$

$$\Rightarrow \quad t = \frac{M\alpha}{16\pi\sigma R^2}\ln\left(\frac{16}{3}\right)$$

Ans. (C)

Sol. 19 By Newton's law of cooling for a body colling down, its temperature as a function of time is given as

$$T_f = T_0 + (T_H - T_0)e^{-kt}$$

Here $T_H \rightarrow$ initial high temperature

$T_0 \rightarrow$ surrounding temperature

For case-1, we use

$$40 = 30 + 50e^{-k(5)}$$

$$\Rightarrow \quad e^{5k} = 5 \quad \Rightarrow \quad 5k = \ln(5) \qquad \ldots(1)$$

For case-2, we use

$$32 = 30 + 32e^{-kt}$$

$$\Rightarrow \quad e^{kt} = 16 \quad \Rightarrow \quad kt = \ln(16) \qquad \ldots(2)$$

$$\frac{(2)}{(1)} \Rightarrow \quad \frac{t}{5} = \frac{4\ln(2)}{\ln 5}$$

$$\Rightarrow \quad t = \frac{5 \times 4 \times 0.693}{1.609} = 8.61 \text{ min}$$ **Ans. (B)**

Sol. 20 Here we use average form of Newton's law of colling as

$$\frac{T_2 - T_1}{t} = k\left(\frac{T_1 + T_2}{2} - T_s\right)$$

Case-I : First 10 minimum

$$\frac{60 - 50}{10} = k(55 - T_s) \qquad \ldots(1)$$

Case-II : Next 10 minimum

$$\frac{50 - 42}{10} = k(46 - T_s) \qquad \ldots(2)$$

$$\frac{(1)}{(2)} \Rightarrow \quad \frac{10}{8} = \frac{55 - T_s}{46 - T_s}$$

$$\Rightarrow \quad 230\sqrt{T_s} = 220 - 4T_s$$

$$\Rightarrow \quad T_s = 230 - 220 = 10°\text{C}$$ **Ans. (B)**

Sol. 21 Equivalent circuit of the given system of rods is shown in figure below.

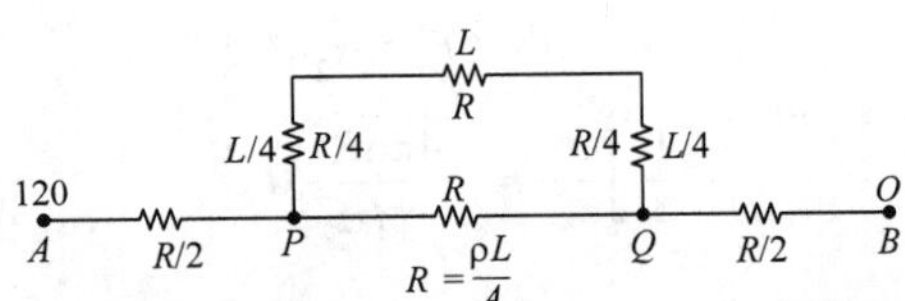

Thermal current is given as

$$I_{\text{th}} = \frac{\Delta T}{R_{eq.}} = \frac{(120)5}{8R} = \frac{120 \times 5}{8R}$$

Temperature difference of P and Q is given as

$$\Delta T_{PQ} = I_{\text{th}} R_{PQ} = \left(\frac{120 \times 5}{8R}\right) \times \frac{3}{5} R = \frac{360}{8} = 45°\text{C}$$ **Ans. (D)**

Sol. 22

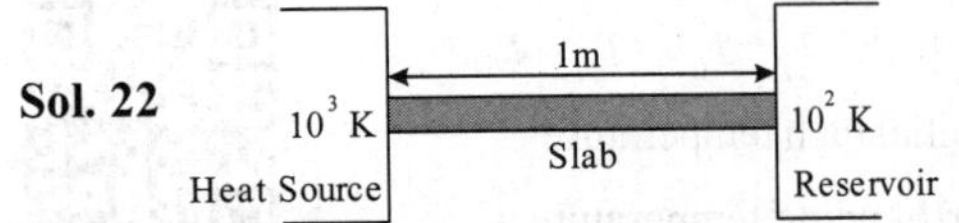

Rate of heat conduction through slab in steady state is given as

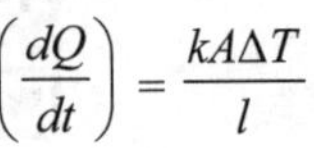

$$\left(\frac{dQ}{dt}\right) = \frac{kA\Delta T}{l}$$

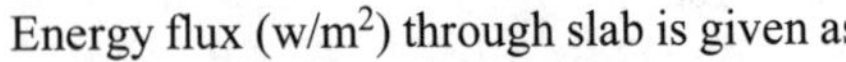

Energy flux (w/m^2) through slab is given as

$$\frac{1}{A}\left(\frac{dQ}{dt}\right) = \frac{(0.1)(900)}{1} = 90 \text{W/m}^2$$ **Ans. (A)**

Sol. 23

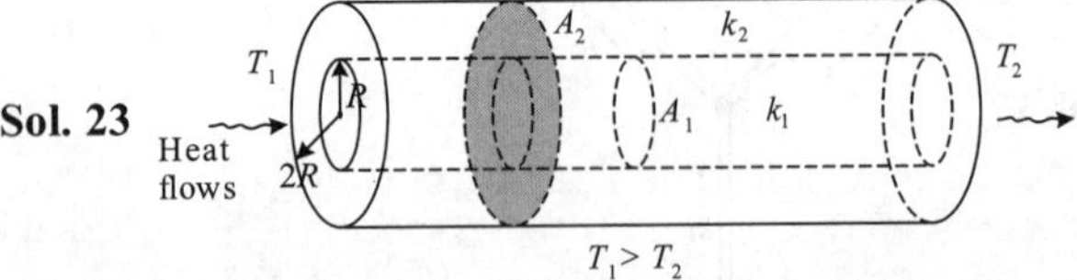

In parallel combination, equivalent thermal conductivity is given as

$$K_{\text{eq}} = \frac{k_1 A_1 + k_2 A_2}{A_1 + A_2}$$

$$K_{\text{eq}} = \frac{k_1(\pi R^2) + k_2(\pi(2R)^2 - \pi R^2)}{4\pi R^2}$$

$$K_{\text{eq}} = \frac{k_1 + 3k_2}{4}$$ **Ans. (D)**

Sol. 24 By Newton's law of cooling

$$\frac{dT}{dt} = -\frac{4eA\sigma T_0^3}{ms} \cdot (\Delta T)$$

$$\Rightarrow \quad \left|\frac{dT}{dt}\right| \propto \frac{1}{ms}$$

$$\Rightarrow \quad \left|\frac{dT}{dt}\right| \propto \frac{1}{\rho s}$$ (As for equal volume $m \propto \rho$)

Here $(\rho s)_A = 8 \times 10^2 \times 2000 = 16 \times 15^5$ SI units

$(\rho s)_B = 10^3 \times 4000 = 4 \times 10^6$ SI units

$$(\rho s)_B > (\rho s)_A \Rightarrow \left|\frac{dT}{dt}\right|_B < \left|\frac{dT}{dt}\right|_A$$

Here T vs t curve of A is steeper than that of B. **Ans. (A)**

Sol. 25 If interface temperature is T, we use in steady state

$$\frac{3kA(\theta_2 - T)}{d} = \frac{kA(T - \theta_1)}{3d}$$

$$\Rightarrow \quad 3\theta_2 - 3T = \frac{T - \theta_1}{3}$$

$$\Rightarrow \quad 10T = 9\theta_2 + \theta_1$$

$$\Rightarrow \quad T = \frac{9\theta_2 + \theta_1}{10}$$ **Ans. (B)**

Sol. 26 Using average form of Newton's law of cooling it gives

$$\frac{50-40}{300} = k\left(\frac{50+40}{2}-20\right) \quad \ldots(1)$$

and $$\frac{40-T}{300} = k\left(\frac{40+T}{2}-20\right) \quad \ldots(2)$$

Dividing (1) and (2), it gives

$$\frac{10}{40-T} = \frac{25}{T/2}$$

$$\Rightarrow \quad 5T = 100-25T$$

$$\Rightarrow \quad 30T = 1000$$

$$\Rightarrow \quad T = \frac{100}{3}\,^\circ\text{C}$$ **Ans. (A)**

Sol. 27 Solar constant has units of Joule/s-m^2 thus its dimensional formula is given as

$$\left[\frac{\text{Joule}}{\text{sec}-m^2}\right] \equiv \frac{[ML^2T^{-2}]}{[T][L^2]} = MT^{-3}$$ **Ans. (D)**

Sol. 28 100°C K_1 K_2 K_3 0°C

70°C 20°C

We consider rods have same length l and area of cross-section A so for steady state conductor, we use

$$\left(\frac{\Delta Q}{\Delta t}\right)_{AB} = \left(\frac{\Delta Q}{\Delta t}\right)_{BC} = \left(\frac{\Delta Q}{\Delta t}\right)_{CD}$$

$$\frac{(100-70)K_1A}{l} = \frac{(70-20)K_2A}{l} = \frac{(20-0)K_3A}{l}$$

$$\Rightarrow \quad 30K_1 = 50K_2 = 20K_3$$

$$\Rightarrow \quad 3K_1 = 2K_3$$

$$\Rightarrow \quad \frac{K_1}{K_3} = \frac{2}{3}.$$

and $$\frac{K_2}{K_3} = \frac{2}{5} = 2:5$$ **Ans. (A)**

Sol. 29 Heat flow rate through the two slabswill be same through both

$$\Rightarrow \quad \frac{\theta_1-\theta}{R_1} = \frac{\theta-\theta_2}{R_2}$$

$$R_2\theta_1 - R_2\theta = R_1\theta - R_1\theta_2$$

$$\theta = \frac{R_2\theta_1 + R_1\theta_2}{R_1+R_2}$$ **Ans. (A)**

Sol. 30 For the two wires connected in series, thermal resistance can be calculated by series combination as

$$R_{\text{eff}} = \frac{l}{K_1A} + \frac{l}{K_2A} = \frac{2l}{K_{eq}A}$$

$$\Rightarrow \quad K_{\text{eq}} = \frac{2K_1K_2}{K_1+K_2}$$ **Ans. (A)**

Sol. 31 Average rate of cooling of body is written as

$$\frac{\Delta T}{t} = K\left(\frac{T_1+T_2}{2}-T_0\right)$$

$$\Rightarrow \quad \frac{75-65}{5} = K\left(\frac{75+65}{2}-25\right) \quad \ldots(1)$$

$$\Rightarrow \quad \frac{65-T}{5} = K\left(\frac{T+65}{2}-25\right) \quad \ldots(2)$$

From equations-(1) and (2), we have

$$\frac{65-T}{75-65} = \frac{\frac{T+65}{2}-25}{\frac{75+65}{2}-25}$$

$$\Rightarrow \quad \frac{65-T}{10} = \frac{T+15}{90}$$

$$\Rightarrow \quad 90\times 65 - 90T = 10T + 10\times 15$$

$$\Rightarrow \quad 100T = 90\times 65 - 15\times 10$$

$$\Rightarrow \quad T = 57^\circ\text{C}$$ **Ans. (57)**

Sol. 32 By average form of Newton's law of cooling, we have

$$\frac{61-59}{4} = K\left[\frac{61+59}{2}-30\right] \quad \ldots(1)$$

$$\frac{51-49}{t} = K\left[\frac{51+49}{2}-30\right] \quad \ldots(2)$$

Dividing equation-(1) and (2), wehave

$$\frac{t}{4} = \frac{30}{20}$$

$$\Rightarrow \quad t = \frac{4\times 30}{20} = 6\text{ min}$$ **Ans. (C)**

Sol. 23 Replacing the rods with thermal resistance, equivalent circuit can be shown in figure below

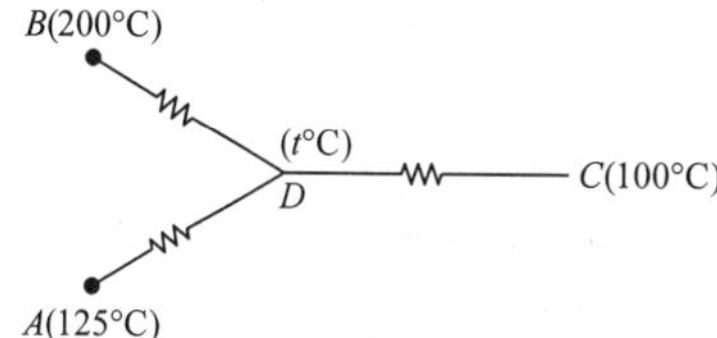

By Kirchoff's law total heat current at a junction must be zero, so we use

$$Q_{BD}+Q_{AD}=Q_{DC}$$

$$\Rightarrow \frac{KA(200-t)}{\frac{l}{2}}+\frac{KA(125-t)}{l}=\frac{(t-100)KA}{\frac{l}{2}}$$

$$\Rightarrow 400-2t+125-t=2t-200$$

$$\Rightarrow 725=5t$$

$$\Rightarrow t=145^{\circ}$$

$$\Rightarrow Q_{AD}=\frac{KA(125-145)}{l}=\left|-\frac{20}{\frac{l}{KA}}\right|=\left|-\frac{20}{10}W\right|=2W$$

Ans. (2)

Sol. 34 Figure below shows the situation described in the question.

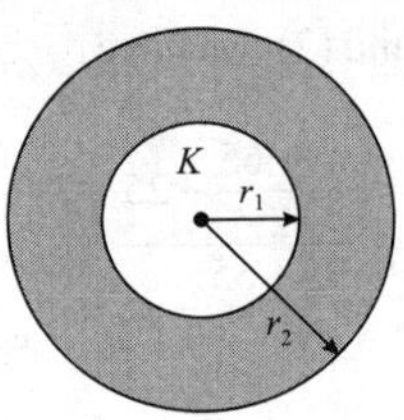

For this situation the equivalent thermal resistance of the shell is given as

Equivalent resistance $R_{th}=\frac{r_2-r_1}{K4\pi r_2 r_1}$

So heat current through the shell material in radial direction is calculated as

$$Q=\frac{\theta_2-\theta_1}{R_{th}}=\frac{4\pi K r_1 r_2(\theta_2-\theta_1)}{(r_2-r_1)}$$ **Ans. (D)**

Sol. 35 Considering steady state of thermal conduction, we use

$$\Delta T \;\alpha\; \frac{l}{k}$$

$$\Rightarrow \frac{\Delta T_1}{\Delta T_2}=\frac{l_1}{k_1}\times\frac{k_2}{l_2}=\frac{16}{k_1}\times\frac{k}{8}$$

$$\Rightarrow \frac{20}{80}=\frac{16}{k_1}\times\frac{k}{8}$$

$$\Rightarrow k_1=8k$$ **Ans. (B)**

Sol. 36 According to Newton's law of colling the temperature of water as a function of time is given as

$$T=T_0(T_i-T_0)\,e^{\frac{Bt}{ms}}$$

$$\Rightarrow 6\lambda=\ln 1.5$$

$$\Rightarrow 40=60\,e^{-\lambda(6)}$$

$$\Rightarrow 6\lambda=\ln 1.5$$

$$\Rightarrow 20=60\,e^{-\lambda t_2}$$

$$\Rightarrow t_2\lambda=\ln 3$$

$$\Rightarrow \frac{t_2}{6}=\frac{\ln 3}{\ln 1.5}=\frac{\log 3}{\log 3-\log 2}$$

$$\Rightarrow t_2=16.25\text{ min}$$

$$\Rightarrow t_2\approx 16$$ **Ans. (16)**

Sol. 37 For the given composit plate the equivalent series combinaiton of thermal resistance is calculated as

$$R_{eq}=R_1+R_2$$

$$\Rightarrow R_{eq}=\frac{L_1}{K(120)}+\frac{L_2}{(2K)(120)}$$

4.0cm	2.5cm
K	2K
(1)	(2)

R_{eq} of this series combination can be calculated by considering total length

$$L_{eq}=4+2.5=6.5\text{ cm}$$

This is given as

$$\frac{L_{eq}}{K_{eq}(A)}=\frac{4}{K(120)}+\frac{4}{\frac{2}{K(120)}}$$

$$\Rightarrow \frac{6.5}{K_{eq}(120)}=\frac{4}{K(120)}+\frac{5}{4K(120)}$$

$$\Rightarrow \frac{6.5}{K_{eq}}=\frac{21}{4K}$$

$$\Rightarrow K_{eq}=\frac{26}{21}K=\left(1+\frac{5}{21}\right)K$$

$$\Rightarrow \alpha=21$$ **Ans. (21)**

Sol. 38 For the ice cube, rate of heat condution through the walls of the insulation box is calculated as

$$\frac{dQ}{dt}=\frac{KA\Delta T}{l}$$

Wall area can be calculated as

$$A=2(0.6\times 0.5+0.5\times 0.2+0.2\times 0.6)$$

$$\Rightarrow A=2(0.3+0.1+0.12)$$

$$\Rightarrow A=2(0.4+0.12)=1.04\text{ m}^2$$

Thermal resistance of the insulation is calculated as

$$R_{th}=\frac{l}{KA}=\frac{1\times 10^{-2}}{0.05\times 1.04}=\frac{10^{-2}}{0.052}$$

$$\Rightarrow \frac{dQ}{dt}=\frac{\Delta T}{R_{th}}$$

$$\Rightarrow \frac{dQ}{dt}=\frac{40\times 0.052}{10^{-2}}=2.08\times 10^2\text{ J/s}$$

If ice melts at a rate of m kg per second, we use

$$2.08 \times 10^2 = m \times 3.4 \times 10^5$$

$$\Rightarrow \quad m = \frac{2.08}{3.4 \times 10^3} \Rightarrow 0.61 \times 10^{-3} \text{ kg/s}$$

$$\Rightarrow \quad m = 61 \times 10^{-5} \text{ Kg/s}$$ **Ans. (B)**

Sol. 39 In steady state rate of heat conduction in both the rods will be same so we use

$T_1 = 450°C$ Steel Copper $T_2 = 0°C$; A_1 L_1 L_2 A_2

$$\frac{dQ}{dt} = \frac{K_1A_1}{L_1}(T_1 - T) = \frac{K_2A_2}{L_2}(T - T_2)$$

$$\Rightarrow \quad \frac{450 - T}{T - 0} = \frac{K_2A_2L_1}{K_1A_1L_2} = 9 \times \frac{1}{2} \times 2$$

$$\Rightarrow \quad 450 - T = 9T$$

$$\Rightarrow \quad T = 45°C$$ **Ans. (C)**

Sol. 40 Heat Transfer

(a) According to Newton's law of colling, the rate of cooling is directly proportional to the temperature difference of body with surrounding hence this statement is true.

(b) By Stefan's radiation law the rate of loss of heat from a body is given as

$$H = \frac{dQ}{dt} = \sigma e A T^4$$

$$\Rightarrow \quad \frac{H_P}{H_Q} = \left(\frac{T_P}{T_Q}\right)^4 = \left(\frac{283}{293}\right)^4$$

$$\Rightarrow \quad H_P : H_Q = 1 : (1.03)^4 = 1 : 1.15$$

Hence this statement is true.

(c) Effciency of Carnot engine is given as

$$\eta = 1 - \frac{100}{400} = \frac{3}{4} = 75\%$$

Hence this statement is true.

(d) This is not true as according to Newton's law of colling, the rate of cooling is directly proportional to the temperature difference of body with surrounding. **Ans. (A)**

Sol. 41 Using average form of Newton's law of cooling, we have

$$\frac{\Delta T}{\Delta t} = \frac{T_2 - T_1}{t} = -\frac{K}{ms}\left(\frac{T_2 + T_1}{2} - T_s\right)$$

In first condition we use

$$\frac{12}{2} = -\frac{K}{ms}\left(\frac{98 + 86}{2} - 22\right)$$

$$\Rightarrow \quad 6 = -\frac{K}{\text{ms}}\left[\frac{98 + 86}{2} - 22\right]$$

$$\Rightarrow \quad 6 = -\frac{K}{\text{ms}}[70] \qquad \ldots(1)$$

In second condition we use

$$\frac{6}{\Delta t} = -\frac{K}{ms}\left(\frac{75 + 69}{2} - 22\right)$$

$$\frac{6}{\Delta t} = -\frac{K}{ms}(50) \qquad \ldots(2)$$

From equations-(1) and (2), we have

$$\frac{6}{\Delta t(6)} = \frac{50}{70}$$

$$\Rightarrow \quad \Delta t = \frac{7}{5} = 1.4 \text{ min}$$ **Ans. (B)**

Sol. 42 Using average form of Newton's law of cooling in first case we have

$$\frac{20}{6} = k(50 - 10) \qquad \ldots(1)$$

In second case, we have

$$\frac{40 - T}{6} = K\left(\frac{40 + T}{2} - 10\right) \qquad \ldots(2)$$

From equation (1) and (2), we use

$$\frac{20}{40 - T} = \frac{40}{10 + T/2}$$

$$\Rightarrow \quad 10 + \frac{T}{2} = 80 - 2T$$

$$\Rightarrow \quad \frac{5T}{2} = 70$$

$$\Rightarrow \quad T = 28°C$$ **Ans. (28)**

Sol. 43 Using average rate of Newton's law of cooling

$$\frac{T_1 - T_2}{t} = K\left(\frac{T_1 + T_2}{2} - T_S\right)$$

Here T_1 and T_2 are initial and final temperature and T_s is temperature of surrounding. With initial conditions we have

$$\frac{60 - 40}{7} = K(50 - 10) \qquad \ldots(1)$$

Final temperature after next 7 minutes can be calculated as

$$\frac{40 - T}{7} = K\left(\frac{40 + T}{2} - 10\right) \qquad \ldots(2)$$

From (1) and (2), we have

$$T = 28°C$$ **Ans. (C)**

Sol. 44 As rate of cooling is directly proportional to the temperature difference with the surrounding for a body by average form of Newton's law of cooling when temperature decreases from 80°C to 60°C is 5 minutes, we have

$$\frac{80-60}{5} = k\{70-20\} \quad \ldots(1)$$

When temperature decreases to 40°C in time t then we have

$$\frac{60-40}{t} = k\{50-20\} \quad \ldots(2)$$

$$\Rightarrow \quad \frac{4t}{20} = \frac{50}{30}$$

$$\Rightarrow \quad t = \frac{25}{3}\text{ min} = 500\text{ s}$$

$$\Rightarrow \quad t = 500\text{ s}$$

Ans. (C)

Sol. 45 Figure below shows the plates in contact as described in question.

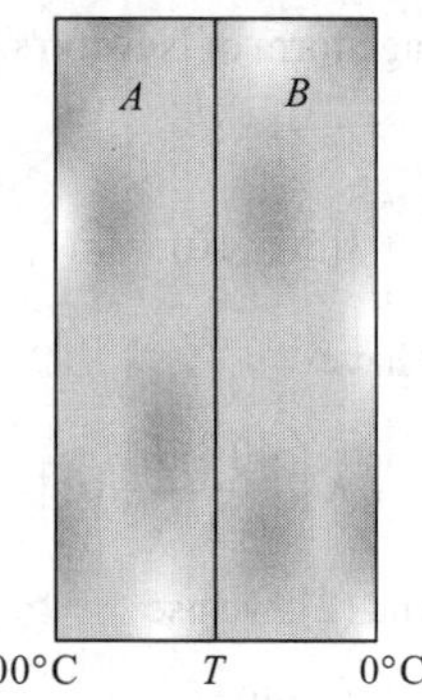

If the temperature of contact surface is T, then in steady state the rate of heat flow remain constant in both plates so we use

$$H_A = H_B$$

$$\Rightarrow \quad \frac{K_A A(T_A - T)}{L} = \frac{K_B A(T - T_B)}{L}$$

$$\Rightarrow \quad 84(100 - T) = 3T$$

$$\Rightarrow \quad 200 - 2T = 3T$$

$$\Rightarrow \quad T = 40^\circ\text{C}$$

Ans. (40)

11.1 Laws of Thermodynamics

Sol. 1 (A) False for reversible cycles efficiency

$$\eta = \frac{\text{work done}}{\text{heat input}} \times 100\%$$

for different thermodynamic processes w/o by gas in different for same heat input.

(B, C, D) True. **Ans. (A)**

Sol. 2 Efficiency of a carnot engine is given as

$$\eta = \left(1 - \frac{T_2}{T_1}\right) \times 100\%$$

Where $T_2 \rightarrow$ temperature sink

$T_1 \rightarrow$ temperature of source

To achieve $\quad \eta = 100\%$

either $\quad T_2 = 0 \rightarrow$ Not attainable

or $\quad T_1 = \infty$ [Impossible] **Ans. (C)**

Sol. 3 Efficiency of carnot engine is given as

$$\eta = 1 - \frac{T_{\text{Sink}}}{T_{\text{Source}}}$$

here $\quad T_{\text{Sources}} = 627 + 273 = 900\text{ K}$

and $\quad T_{\text{Sink}} = 27 + 273 = 300\text{ K}$

$$\Rightarrow \quad \eta = 1 - \frac{300}{900} = \frac{2}{3}$$

For engine efficiency relates work done and heat input as

$$\eta = \frac{\text{work done}}{\text{Heat input}}$$

$\Rightarrow$ Work done = η × heat supplied

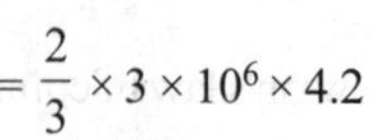

$$= \frac{2}{3} \times 3 \times 10^6 \times 4.2$$

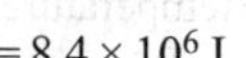

$$= 8.4 \times 10^6\text{ J}$$

Ans. (B)

Sol. 4 For an adiabatic process we use

$$PV^\gamma = \text{constant} \quad \ldots(1)$$

We are given that $\quad P \propto T^3$

$$\Rightarrow \quad T = kP^{1/3}$$

By Gas law we have

$$PV = nRT$$

$$\Rightarrow \quad PV = nRkP^{1/3}$$

$$\Rightarrow \quad P^{1-1/3}V = \text{Cost}$$

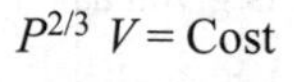

$$\Rightarrow \quad P^{2/3}V = \text{Cost}$$

$$\Rightarrow \quad PV^{3/2} = \text{Cost} \quad \ldots(2)$$

Comparing equation-(1) and (2) we have $\gamma = 3/2$ **Ans. (D)**

Sol. 5 State of a gas is indicate by thermodynamic coordinates (P, V, T) which helps in denoting gas state on indicator diagram and also we can plot process equation of *ID* curve. **Ans. (C)**

Sol. 6 Above statement is a consequence of "Second law of thermodynamics"

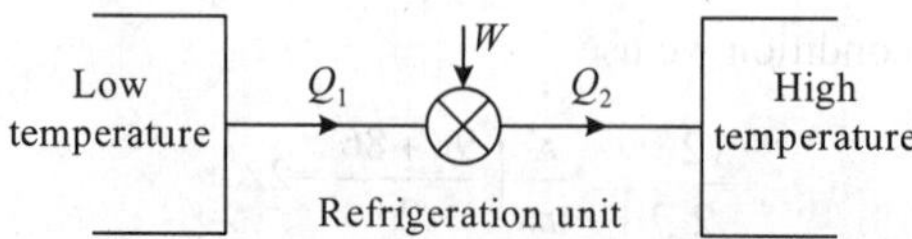

$$W + Q_1 = Q_2$$

Without external work it is impossible to transfer heat from low temperature zone to high temperature zone. **Ans. (A)**

Sol. 7. (A) As internal energy is a state function

$$U = \frac{f}{2}nRT \quad \text{'False'}$$

This is for an isothermal process as

$$\Delta T = 0\,;\, \Delta U = 0$$

(B) 'True' Both U & S are state functions

(C) As entropy is a state function so in cyclic process or whenever a system returns to its initial state $\Delta S = 0 \rightarrow$ 'False'

(D) For adiabatic process $\Delta Q = 0 \Rightarrow \Delta W = -\Delta U \rightarrow$ 'False'

Ans. (B)

Sol. 8 As we know internal energy of a gas is a 'State function'

$$U = \frac{f}{2}nRT$$

At state A and B if internal energy of system are U_A and U_B respectively

$$\Rightarrow \quad \Delta U_1 = \Delta U_2 = U_B - U_A$$

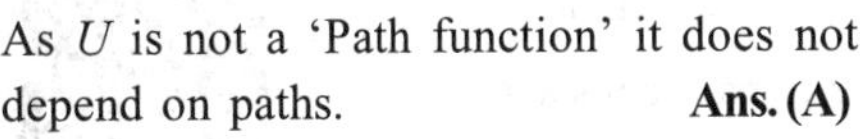

As U is not a 'Path function' it does not depend on paths. **Ans. (A)**

Sol. 9 Efficiency of a heat engine is given as

$$\eta = \frac{\text{Work done}}{\text{Heat input}}$$

Heat $\rightarrow$ Area under T-S diagram

heat supplied in cycle

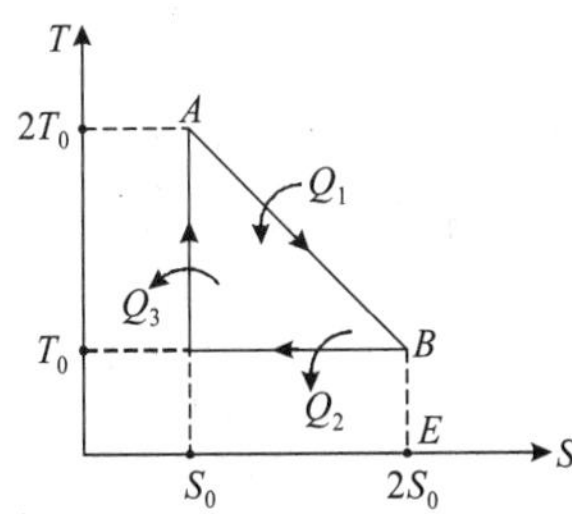

$$Q_1 = \text{Area } ABEDA$$

$$= T_0(2S_0 - S_0) + \frac{1}{2}T_0S_0 = \frac{3}{2}T_0S_0$$

Work done in cycle W = Area ABC

$$= \frac{1}{2}T_0S_0$$

$\Rightarrow$ Efficiency of cycle

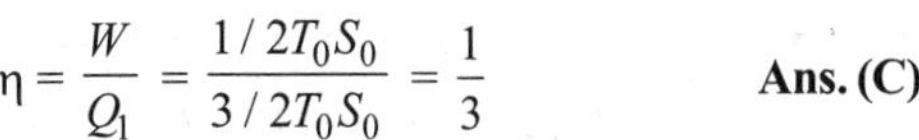

$$\eta = \frac{W}{Q_1} = \frac{1/2T_0S_0}{3/2T_0S_0} = \frac{1}{3}$$ **Ans. (C)**

Sol. 10 (A) **True :** for all process (cyclic or not)

$$\Delta Q = \Delta U + \Delta W$$

or for elemental duration FLT is valid as

$$dQ = dU + dW$$

(B) **True :** FLT is defined by using energy conservation

(C) **True :** For FLT we use internal energy of gas/substance along with heat supplied & work done.

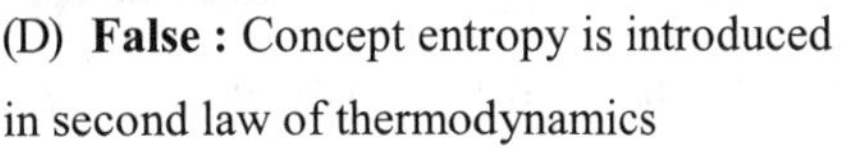

(D) **False :** Concept entropy is introduced in second law of thermodynamics

Sol. 11 For an adiabatic process work done is given as

$$W = \frac{nR(T_2 - T_1)}{\gamma - 1}$$

$$\Rightarrow \quad \gamma - 1 = \frac{nR(T_2 - T_1)}{W}$$

$$\Rightarrow \quad \gamma = 1 + \frac{10^3 \times 8.3 \times 7}{146 \times 10^3} = 1 + 0.4 = 1.4$$

$\gamma = 1.4$ for a diatomic gas

$$\left(f = 5 \Rightarrow \gamma = \frac{f+2}{f} = \frac{7}{5} = 1.4\right)$$ **Ans. (A)**

Sol. 12 As we know for same initial and final states of all the thermodynamic process we have

$$\overset{iaf}{\Delta U_1} = \overset{ibf}{\Delta U_2} \quad (\text{as } U \rightarrow \text{state function})$$

$$\Rightarrow \quad Q_1 - W_1 = Q_2 - W_L$$

$$\Rightarrow \quad 50 - 20 = 36 - W_2$$

$$\Rightarrow \quad W_2 = 36 - 30 = 6 \text{ cal}$$ **Ans. (A)**

Sol. 13 If same engine is used as refrigerator then relation in its efficiency (η) and coefficient of performance (k) is given as

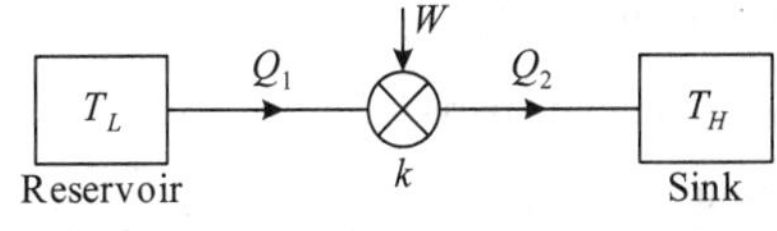

$$k = \frac{1}{\eta} - 1 = \frac{1}{1/10} - 1 = 9$$

COP of a refrigerator is given as

$$k = \frac{Q_1}{W} = \frac{\text{Heat extracted}}{\text{Work input}}$$

$$\Rightarrow \quad 9 = \frac{Q_1}{W} \Rightarrow Q_1 = 9 \times 10 = 90 \text{ J}$$ **Ans. (B)**

Sol. 14 Work done by gas in Isobaric Process AB is

$$W_{AB} = P(V_2 - V_1) = nR(T_2 - T_1)$$

$$W_{AB} = 2R(500 - 300) = 400R$$

Work done on the gas

$$= -W_{AB} = -400R \quad \textbf{Ans. (A)}$$

Sol. 15 Work done by gas in Isothermal process DA is

$$W_{DA} = nRT\ln\left(\frac{P_1}{P_2}\right)$$

$$\Rightarrow \quad W_{DA} = 2.303\, nR\,(300)\,\log_{10}\left(\frac{1}{2}\right)$$

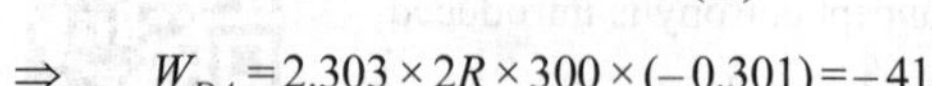

$$\Rightarrow \quad W_{DA} = 2.303 \times 2R \times 300 \times (-0.301) = -414R$$

Work done on the gas is

$$= -W_{DA} = 414R \quad \textbf{Ans. (D)}$$

Sol. 16 Total work done in complete cycle is given as

$$W_{\text{cycle}} = W_{AB} + W_{BC} + W_{CD} + W_{DA}$$

$$\Rightarrow \quad W_{\text{cycle}} = nR(T_B - T_A) + nRT_B \ln\frac{P_B}{P_C} + nR(T_D - T_C) + nRT_D \ln\frac{P_D}{P_A}$$

$$\Rightarrow \quad W_{\text{cycle}} = 2R(200) + 2.303 \times 2R \times 500 \log_{10}(2) + 2R(-200) + 2.303 \times 2R \times 300 \log_{10}(1/2)$$

$$\Rightarrow \quad W_{\text{cycle}} = 690\,R - 414\,R = 276\,R$$

Work done on the gas $= -W_{\text{cycle}} = -276\,R$ **Ans. (D)**

Sol. 17

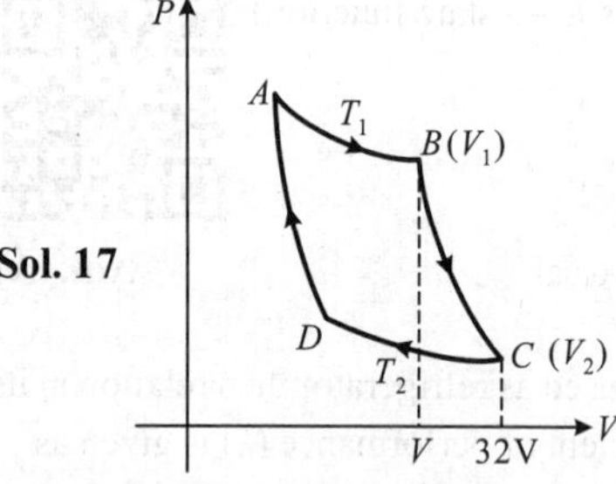

AB and $CD \rightarrow$ isothermal parts

BC and $DA \rightarrow$ adiabatic parts

$T_1 \rightarrow$ temperature of source

$T_2 \rightarrow$ temperature of sink

For adiabatic process BC, we use process equation

$$TV^{\gamma-1} = \text{constant} \quad (\text{for diatomic gases } \gamma = 7/5)$$

$$\Rightarrow \quad T_1V_1^{\gamma-1} = T_2V_2^{\gamma-1}$$

$$\Rightarrow \quad \frac{T_2}{T_1} = \left(\frac{V_1}{V_2}\right)^{\gamma-1} = \left(\frac{V_1}{32V}\right)^{7/5-1}$$

$$= \left(\frac{1}{2^5}\right)^{2/5} = \frac{1}{2^2} = \frac{1}{4}$$

Engine efficiency is given as

$$\eta = 1 - \frac{T_2}{T_1} = 1 - \frac{1}{4} = \frac{3}{4} = 0.75$$

in % $\quad \eta = \frac{3}{4} \times 100\% = 75\%$ **Ans. (C)**

Sol. 18 First law of thermodynamics increase in internal energy of gas = Amount of heat supplied

$$\Delta U = Q = mS\Delta T$$

$$\Rightarrow \quad \Delta U = 0.1 \times 4184 \times 20$$

$$\Rightarrow \quad \Delta U \cong 8.4\ \text{kJ} \quad \textbf{Ans. (D)}$$

Sol. 19

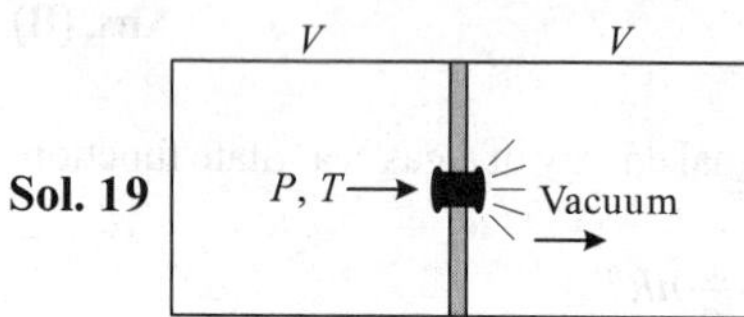

In case of 'Free expansion' (no opposition) no work is done by gas so as $\Delta Q = 0$ (insulating walls)

or $\quad U =$ constant

$\Rightarrow \quad T =$ constant

As gas temperature remain constant

we use $\quad P_1V_1 = P_2V_2$

$$PV = P_2(2V) \Rightarrow P_2 = P/2 \quad \textbf{Ans. (A)}$$

Sol. 20 The amount of heat to be last by metal in cooling from 20 K to 4 K is

$$Q = \int dQ = -\int mSdT = -\int_{20}^{4} m\left(32\left[\frac{T}{400}\right]^3\right)dT$$

heat lost

$$\Rightarrow \quad Q = -\frac{0.1 \times 32}{(400)^3} \cdot \left[\frac{T^4}{4}\right]_{20}^{4} = \frac{0.1 \times 32}{(400)^3 \times 4}[(20)^4 - (4)^2]$$

$$Q = 0.002\ \text{kJ}$$

Refrigeration

Reservation T_L — Q_1 → W → Q_2 — Sink T_H

$W = Q_1 - Q_2$

Coefficient of performance

$$K = \frac{Q_1}{W}$$

$$K = \frac{Q_1}{Q_2 - Q_1} = \frac{T_1}{T_2 - T_1}$$

At $\quad T_1 = 20\ \text{K}\ \&\ T_S = 27 + 273 = 300\ \text{K}$

The amount of work done in refrigeration unit is

$$\frac{Q_1}{W_1} = \frac{T_1}{T_2 - T_1} = \frac{20}{300-20} = \frac{20}{280}$$

$$\Rightarrow \quad W_1 = \frac{0.002\text{kJ}}{20/280} = 0.028\text{ kJ}$$

As metal is cooled to 4 K, so at T_L = 4 K, work done by refrigeration unit is

$$\frac{Q_1}{W_2} = \frac{T_1}{T_2 - T_1} = \frac{4}{300-4} = \frac{4}{296}$$

$$\Rightarrow \quad W_2 = \frac{0.002\text{kJ}}{(4/296)} = 0.148\text{ kJ}$$

As metal is cooling from 20 K to 4 K so any temperature of metal during cooling is between 20 K and 4 K so work required by refrigeration unit in between W_1 and W_2. **Ans. (C)**

Sol. 21 If $T_1 \rightarrow$ Source temperature

& $T_2 \rightarrow$ Sink temperature

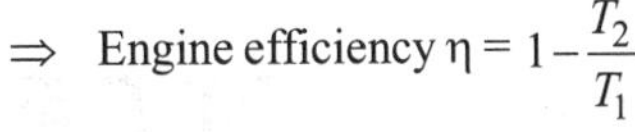

$\Rightarrow$ Engine efficiency $\eta = 1 - \frac{T_2}{T_1}$

Initially $\quad 1 - \frac{T_2}{T_1} = \frac{1}{6} \quad \Rightarrow \quad \frac{T_2}{T_1} = \frac{5}{6}$ ...(1)

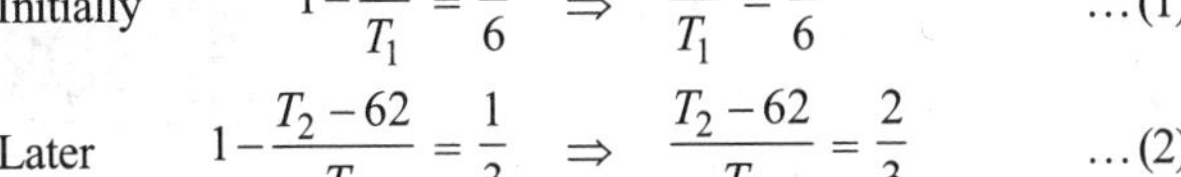

Later $\quad 1 - \frac{T_2 - 62}{T_1} = \frac{1}{3} \quad \Rightarrow \quad \frac{T_2 - 62}{T_1} = \frac{2}{3}$ ...(2)

Solving equation-(1) and (2) we get

$$T_1 = 372\text{ K and } T_2 = 310\text{ K}$$ **Ans. (C)**

Sol. 22

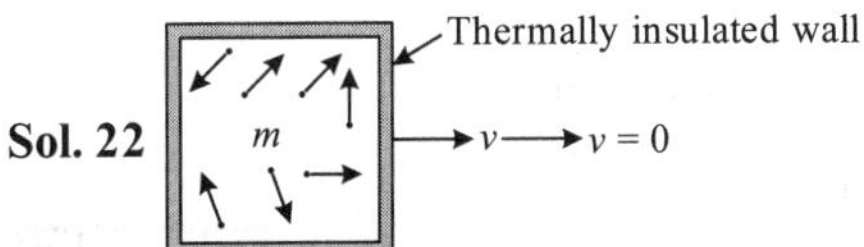

KE of gas particles due to the motion of vessel

$$K = \frac{1}{2}mv^2$$

When suddenly the vessel in brought to rest then above *KE* will increase the energy of random motion (thermal energy) of molecules as no energy can transmit through container walls

$$\Rightarrow \quad \frac{1}{2}mv^2 = \Delta U = n\left(\frac{R}{T-1}\right)\Delta T$$

$$\Rightarrow \quad \frac{1}{2}mv^2 = \frac{m}{M}\left(\frac{R}{\gamma - 1}\right)\Delta T$$

$$\Rightarrow \quad \Delta T = \frac{Mv^2}{2R}(\gamma - 1)$$ **Ans. (B)**

Sol. 23 Efficiency of carnot engine is given as

$$\eta = 1 - \frac{T_{\text{sink}}}{T_{\text{Source}}}$$

for first engine $T_{\text{Source}} = 500\text{ K}$ & $\eta = 0.4$

$$\Rightarrow \quad 0.4 = 1 - \frac{T_{\text{Sink}}}{500}$$

$$\Rightarrow \quad T_{\text{Sink}} = 500 \times 0.6$$

$$\Rightarrow \quad T_{\text{Sink}} = 300\text{ K}$$

For another engine with

$$\eta = 0.6 \;\&\; T_{\text{Sink}} = 300\text{ K}$$

$$\Rightarrow \quad 0.6 = 1 - \frac{300}{T_{\text{Source}}}$$

$$\Rightarrow \quad T_{\text{Source}} = \frac{300}{0.4} = 750\text{ K}$$ **Ans. (D)**

Sol. 24 Cycle efficiency is given as

$$\eta = \frac{\text{Work Done}}{\text{Heat input}} \times 100\%$$

Work done by gas in cycle W = Area enclosed by $ABCD = P_0V_0$

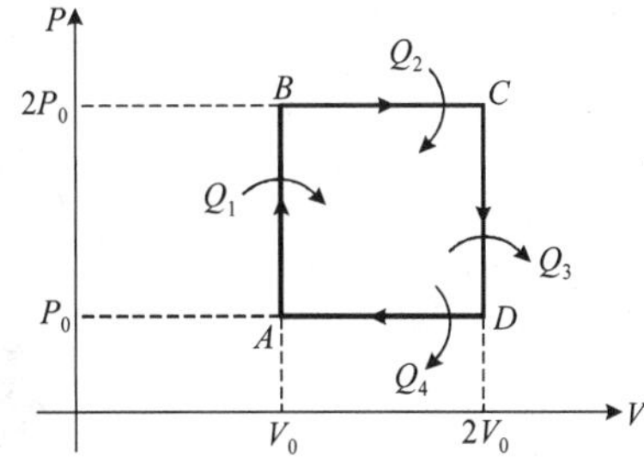

$$W = (Q_1 + Q_2) - (Q_3 - Q_4)$$

Heat input in cycle $Q_{\text{in}} = Q_1 + Q_2$

$$\Rightarrow \quad Q_{\text{in}} = nC_v\Delta T_{AB} + nC_p\Delta T_{BC}$$

$$\Rightarrow \quad Q_{\text{in}} = \frac{3}{2}nR\Delta T_{AB} + \frac{5}{2}nR\Delta T_{BC}$$

$$\Rightarrow \quad Q_{\text{in}} = \frac{3}{2}(P_0V_0) + \frac{5}{2}(2P_0V_0) = \frac{13}{2}P_0V_0$$

Cycle efficiency

$$\eta = \frac{W}{Q_{\text{in}}} \times 100\%$$

$$\Rightarrow \quad \eta = \frac{P_0V_0}{13/2P_0V_0} \times 100$$

$$\Rightarrow \quad \eta = \frac{200}{13}\% \sim 15.4\%$$ **Ans. (B)**

Sol. 25 Statement-2 : It is true

So carnot efficiency for source and sink at temperature 373 K and is given as

$$\eta = \left(1 - \frac{T_2}{T_1}\right) \times 100 = \left(1 - \frac{273}{373}\right) \times 100$$

$$\Rightarrow \quad \eta = (1 - 0.732) \times 100$$

$$= 26.8\% \text{ max possible}$$

Statement-1 : Is true & it is explained by Statement-2. **Ans. (D)**

Sol. 26 As in an adiabatic process there is no heat exchange

$$\Rightarrow \qquad \Delta Q = 0$$

So from first law of thermodynamics, we have

$$\Delta Q = \Delta U + W$$

$$\Rightarrow \qquad \Delta U = -W$$

$\Rightarrow$ Statement-1 is true

As $T =$ constant in isothermal process $\Rightarrow$ Statement-2 is false

In adiabatic process we use $TV^{\gamma-1} =$ constant **Ans. (B)**

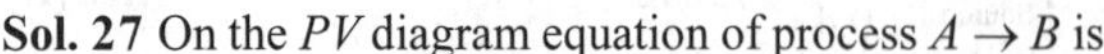

Sol. 27 On the PV diagram equation of process $A \to B$ is

$$P = mV + C \text{ where}$$

$$m = -\frac{P_0}{V_0} \text{ and } C = 3P_0$$

$$\Rightarrow \qquad P = -\frac{P_0}{V_0}\cdot V + 3P_0$$

By gas law, we use

$$T = \frac{PV}{nR} = -\frac{P_0}{nRV_0}\cdot V^2 + \frac{3P_0}{nR}\cdot V \qquad \dots(1)$$

T is max when $\dfrac{dT}{dV} = 0$

$$\Rightarrow \qquad -\frac{P_0}{nRV_0}(2V) + \frac{3P_0}{nR} = 0$$

$$\Rightarrow \qquad \frac{2V}{V_0} = 3 \;\Rightarrow\; V = \frac{3V_0}{2}$$

at maximum temperature

$\Rightarrow$ Maximum temperature gas in given by equation-(1) as

$$T_m = -\frac{P_0}{nRV_0}\left(\frac{3V_0}{2}\right)^2 + \frac{3P_0}{nR}\left(\frac{3V_0}{2}\right)$$

$$\Rightarrow \qquad T_m = -\frac{P_0}{nRV_0}\times\frac{9V_0^2}{4} + \frac{9}{2}\frac{P_0V_0}{nR}$$

$$= \frac{P_0V_0}{nR}\left(\frac{9}{2} - \frac{9}{4}\right) = \frac{9}{4}\frac{P_0V_0}{nR}$$ **Ans. (D)**

Sol. 28 Gas expands from volume $V \to mV$

Work done by gas in the process is given as

$$W = \int_V^{mV} PdV = \int_V^{mV} \alpha V dV$$

$$\Rightarrow \qquad W = \alpha\left[\frac{V^2}{2}\right]_V^{mV}$$

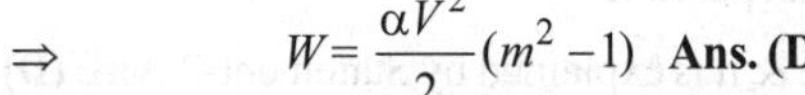

$$\Rightarrow \qquad W = \frac{\alpha V^2}{2}(m^2 - 1)$$ **Ans. (D)**

Sol. 29 Initial state of gas for ideal monatomic gas

$$\gamma = 5/3$$

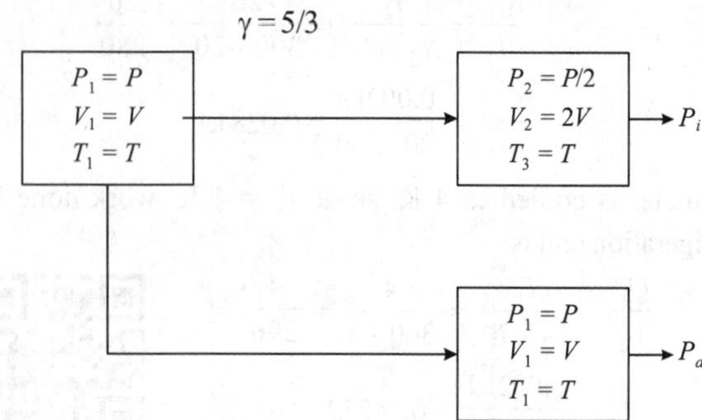

Isothermal expanded

$$T = \text{constant}$$

$$PV = \text{constant}$$

$$P.V = P_2(2V)$$

$$\Rightarrow \qquad P_2 = P/2$$

Adiabatic expanded

$$\Delta Q = 0$$

$$PV^{5/3} = \text{constant}$$

$$PV^{5/3} = P'_2(2V)^{5/3}$$

$$\Rightarrow \qquad P'_2 = \frac{P^{5/3}}{2}$$

Ratio $\qquad \dfrac{P_a}{P_i} = \dfrac{P/2^{5/3}}{P/2} = 2^{1-5/3} = 2^{-2/3}$ **Ans. (D)**

Sol. 30 For an adiabatic process state parameters V and T are related as

$$TV^{\gamma-1} = \text{constant}$$

$$\Rightarrow \qquad T_1V_1^{\gamma-1} = T_2V_2^{\gamma-1}$$

As $\qquad V_2 = \dfrac{V_1}{4}$

$$\Rightarrow \qquad T_1V_1^{0.5} = T_2\left(\frac{V_1}{4}\right)^{0.5}$$

$$\Rightarrow \qquad T_2 = \sqrt{4}T_1 = 2T_1 = 2T$$ **Ans. (C)**

Sol. 31 Refrigerator

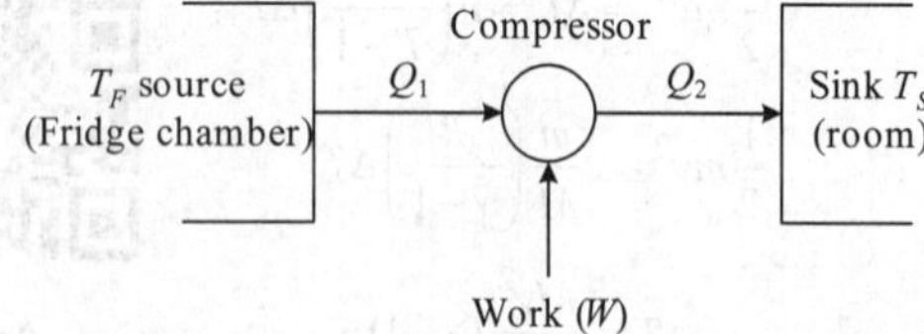

By conservation of energy

$$Q_2 = Q_1 + W$$

If door of refrigerator is kept open

$\Rightarrow$ Source and sink are same.

$\Rightarrow$ Net heat W will be given to room

$\Rightarrow$ Temperature of room increases **Ans. (D)**

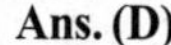

Sol. 32 In the given cycle, heat extracted from the source is given as

$$Q_{out} = Q_3 + Q_4$$

$$\Rightarrow \quad Q_{out} = nC_V\Delta T_{BC} + nC_P\Delta T_{CD}$$

$$\Rightarrow \quad Q_{out} = \frac{3}{2}nR\Delta T_{BC} + \frac{5}{2}nR\Delta T_{CD}$$

$$\Rightarrow \quad Q_{out} = \frac{3}{2}\times 2P_0V_0 + \frac{5}{2}P_0V_0 = \frac{11}{2}P_0V_0$$

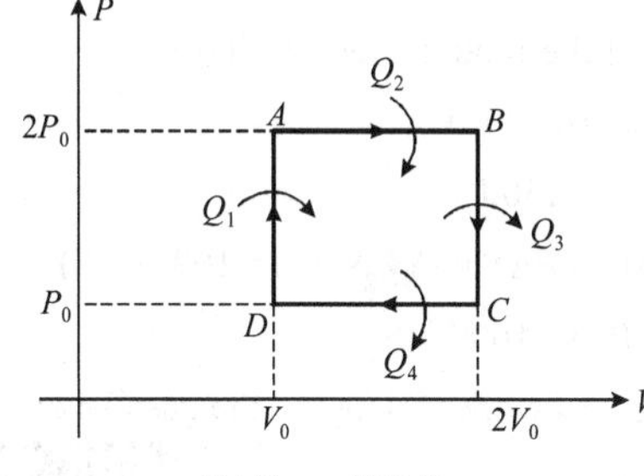

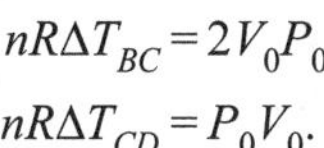

$$nR\Delta T_{BC} = 2V_0P_0$$

$$nR\Delta T_{CD} = P_0V_0.$$

Ans. (C)

Sol. 33

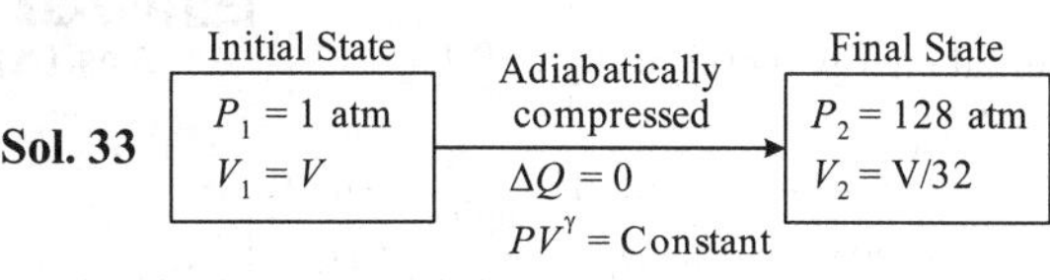

As density becomes 32 times

$$\Rightarrow \quad V \to V/32$$

$$\Rightarrow \quad P_1V_1^\gamma = P_2V_2^\gamma$$

$$\Rightarrow \quad (1)(V)^\gamma = 128\left(\frac{V}{32}\right)^\gamma$$

$$\Rightarrow \quad 32^\gamma = 128$$

$$\Rightarrow \quad 2^{5\gamma} = 2^7$$

$$\Rightarrow \quad 5^\gamma = 7$$

$$\Rightarrow \quad \gamma = \frac{7}{5} = 1.4$$

Ans. (B)

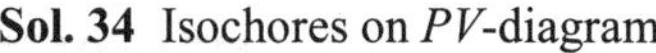

Sol. 34 Isochores on PV-diagram

P = constant

Isocharms on PV-diagram

V = constant

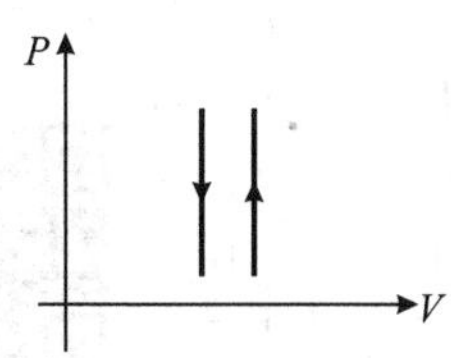

Isotherms on PV-diagram

T = constant

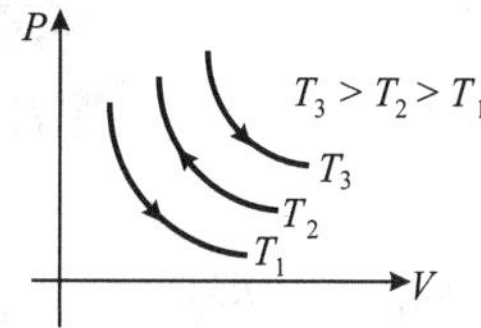

The given process has

$$\begin{bmatrix} \text{2 Isobars} \to AD \text{ and } BC \\ \text{1 Isotherm} \to CD \\ \text{1 Isochore} \to AB \end{bmatrix}$$

Ans. (C)

Sol. 35 Work done by a gas in isothermal expansion to given as

$$W = nRT\ln\left(\frac{V_2}{V_1}\right) = \frac{10\times 10^{-3}}{M}RT\ln(2)$$

$$\Rightarrow \quad 575 = \frac{10^{-2}}{M}\cdot RT\times 0.693$$

$$\Rightarrow \quad \frac{RT}{M} = \frac{57500}{0.693}$$

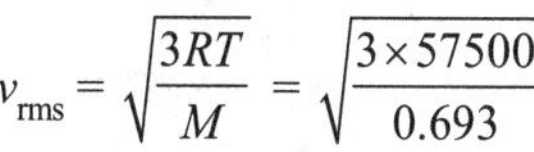

$$v_{rms} = \sqrt{\frac{3RT}{M}} = \sqrt{\frac{3\times 57500}{0.693}}$$

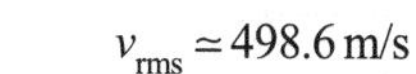

$$\Rightarrow \quad v_{rms} \simeq 498.6 \text{ m/s}$$

Ans. (C)

Sol. 36 Change in internal energy is given as

$$\Delta U = \frac{nR\Delta T}{\gamma - 1} = \frac{nR\Delta T}{0.4}$$

(for diatomic gas $\gamma = 1.4$)

$$\Rightarrow \quad \Delta U_{CA} = \frac{R}{0.4}(-200) = -500\text{ R}$$

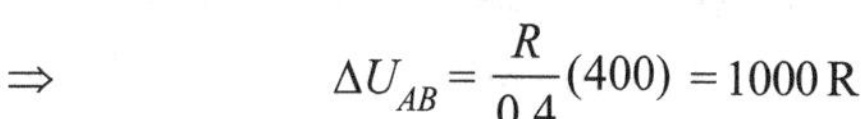

$$\Rightarrow \quad \Delta U_{AB} = \frac{R}{0.4}(400) = 1000\text{ R}$$

$$\Rightarrow \quad \Delta U_{BC} = \frac{R}{0.4}(-200) = -500\text{ R}$$

Ans. (D)

Sol. 37 Work done in an isobaric process is given as

$$W = P\Delta V = P(V_f - V_i)$$

as initial process is P_0

$$\Rightarrow \quad W = P_0(V_f - V_i) \qquad \ldots(1)$$

From the given equation, we have

$$PV = nRT + \alpha V$$

For $n = 1$ mole $\quad V = \dfrac{RT}{P - \alpha}$

at $T = T_0 \rightarrow V_i = \dfrac{RT_0}{P_0 - \alpha}$

at $T = 2T_0 \rightarrow V_f = \dfrac{2RT_0}{P_0 - \alpha}$

From (1)

$$W = P_0\left(\frac{2RT_0}{P_0-\alpha} - \frac{RT_0}{P_0-\alpha}\right) = \frac{RP_0T_0}{P_0-\alpha}$$ **Ans. (A)**

Sol. 38 Initial state of gas

$$P_1 = 10^5\ \text{N/m}^2$$
$$V_1 = 2.4 \times 10^{-3}\ \text{m}^3$$
$$T_1 = 300\ \text{K}$$

After expansion, gas process is

$$P_f = P_{atm} + \frac{kx}{A}$$

$$\Rightarrow \quad P_f = 10^5 + \frac{8000 \times 0.1}{8.0 \times 10^{-3}}$$

$$\Rightarrow \quad P_f = 10^5 + 10^5 = 2 \times 10^5\ \text{N/m}^2$$

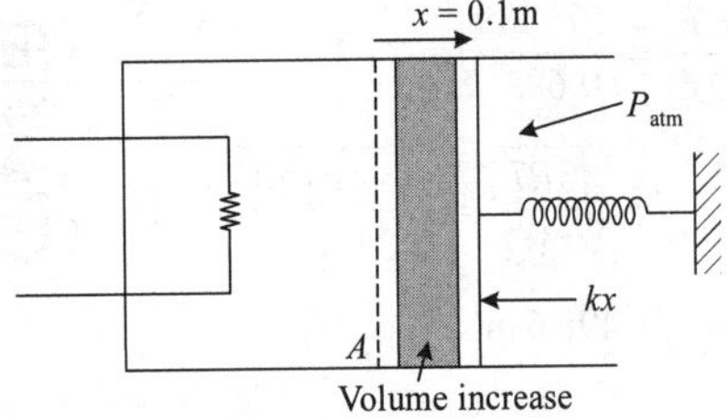

$$\Delta V = Ax$$
$$\Delta V = 8.0 \times 10^{-3} \times 0.1$$
$$\Delta V = 0.8 \times 10^{-3}\ \text{m}^3$$

Find state of gas

$$P_f = 2 \times 10^5\ \text{N/m}^2$$
$$V_f = V_i + \Delta V = 2.4 \times 10^{-3}\text{m}^3 + 0.8 \times 10^{-3}\text{m}^3$$
$$= 3.2 \times 10^{-3}\ \text{m}^2$$
$$T_f = \rho$$

As no of moles of gas are constant. So we use

$$\frac{P_1V_1}{T_1} = \frac{P_fV_f}{T_f}$$

$$\Rightarrow \quad \frac{10^5 \times 2.4 \times 10^{-3}}{300} = \frac{2 \times 10^5 \times 3.2 \times 10^{-3}}{T_F}$$

$$\Rightarrow \quad T_f = \frac{2 \times 3.2}{2.4} \times 300 = 800\ \text{K}$$ **Ans. (B)**

Sol. 39 In an adiabatic process, work done by gas is given as

$$W = -\Delta U = -\frac{nR\Delta T}{\gamma - 1}$$

given that $W = -830$ J (work done on gas) for $n = 2$ mole and $\gamma - 1.4$ for a diatomic ideal gas

$$\Rightarrow \quad W = -830 = -\frac{2 \times 8.3 \times \Delta T}{1.4 - 1}$$

$$\Rightarrow \quad \Delta T = \frac{830 \times 0.4}{2 \times 8.3} = \frac{40}{2} = 20\ \text{K}$$

$\Rightarrow$ Temperature gas increases by 20 K **Ans. (C)**

Sol. 40 By first law of thermodynamics we have

$$\Delta Q = \Delta U + \Delta W \quad \ldots(1)$$

In whole process $\Delta Q = 150$ J

and $\Delta W = P\Delta V = P(V_2 - V_1) = 100\,(1 - 2)$

$\Rightarrow$ $\Delta W = -100$ J

As gas is compressed $\Rightarrow$ work is done on the gas (– *ve*)

So from equation-(1), we use

$\Rightarrow$ $150 = \Delta U - 100$

$\Rightarrow$ $\Delta U = 250$ J

$\Rightarrow$ Internal energy increases by 250 J **Ans. (A)**

Sol. 41

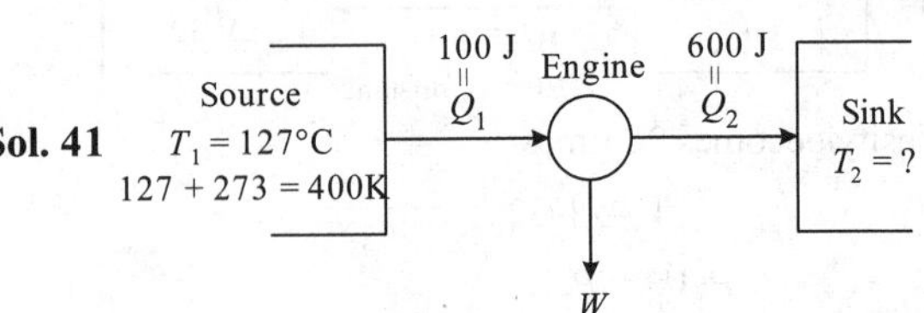

Work done per cycle by engine

$$W = Q_1 - Q_2$$
$$= 1000 - 600 = 400\ \text{J}$$

Efficiency of engine

$$\eta = \frac{W}{Q_1} \times 100 = \frac{400}{1000} \times 100 = 40\%$$

In carnot engine Q_1 and Q_2 are proportional to T_1 and T_2

$$\Rightarrow \quad \frac{Q_1}{Q_2} = \frac{T_1}{T_2}$$

$$\Rightarrow \quad \frac{1000}{600} = \frac{400}{T_2}$$

$$\Rightarrow \quad T_2 = 240\ \text{K} = -33°\text{C}$$ **Ans. (B)**

Sol. 42 Average time between collisions of gas molecules

$$t \propto \frac{V}{\sqrt{T}}$$

In Adiabatic process we use $TV^{\gamma-1}$ = constant

$$\Rightarrow \quad T = CV^{1-\gamma}$$

$$\Rightarrow \quad t \propto \frac{V}{\sqrt{CV^{1-\gamma}}} = C_1V^{1-\frac{1-\gamma}{2}} = C_1V^{\frac{1+\gamma}{2}}$$

$$\Rightarrow \quad t \propto V^{\frac{1+\gamma}{2}} \quad \Rightarrow \quad q = \frac{1+\gamma}{2}$$ **Ans. (C)**

Sol. 43 Given that

$$\frac{U}{V} = kT^4$$

$$\Rightarrow \text{ Gas pressure } \quad P = \frac{1}{3}kT^4$$

By Gas law $\quad PV = nR_gT$

$R_g \to$ Gas constant

$$\Rightarrow \quad \frac{1}{3}kT^4 \cdot \left(\frac{4}{3}\pi R^3\right) = nR_gT$$

$$\Rightarrow \quad T^3R^3 = \text{constant}$$

$$\Rightarrow \quad T \propto \frac{1}{R}$$ **Ans. (A)**

Sol. 44 Change in Entropy of a solid body is given as

$$\Delta S = \int \frac{dQ}{T}$$

for a solid body we use

$$dQ = CdT$$

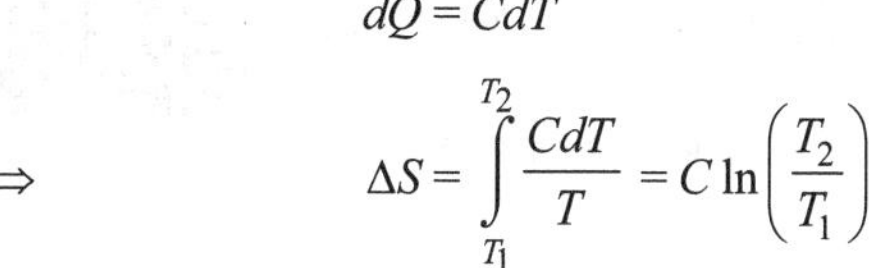

$$\Rightarrow \quad \Delta S = \int_{T_1}^{T_2} \frac{CdT}{T} = C\ln\left(\frac{T_2}{T_1}\right)$$

$$\Rightarrow \quad \Delta S = 1\ln\left(\frac{200}{100}\right) = \ln(2)$$

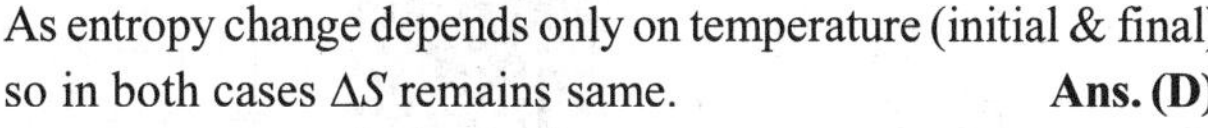

As entropy change depends only on temperature (initial & final) so in both cases ΔS remains same. **Ans. (D)**

Sol. 45 In *V-T* diagram '*ab*' and '*cd*' are straight line

$\Rightarrow$ These are isobars on *PV* diagram

Isobars on *VT* curve pass through *O* (assuming this by looking at given options)

$$\Rightarrow \quad P = \text{constant}$$

from '*bc*' and '*da*' which are adiabatic

Option (B) is the correct Ans.

(B) $b \to c\ (\to)$ volume is increasing

$d \to a\ (\to)$ volume is decreasing

Corresponding to given *VT* curve.

Also slope of curve in $-ve$ and decreasing

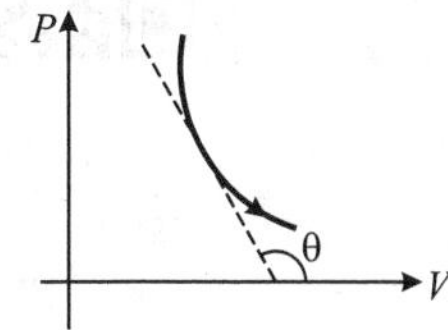

$$m = \tan\theta \to -ve$$

m is decreasing in magnitude with volume. **Ans. (B)**

Sol. 46 Equation of Straight line shown in indicator diagram is

$$P = -\frac{P_0}{V_0}\cdot V + 3P_0$$

$$\text{slope} = -\frac{P_0}{V_0}$$

$$\text{intercept} = 3P_0$$

$$\text{equation } y = mx + c$$

Temperature of gas during the process is

$$T = \frac{PV}{nR} = -\frac{P_0}{nRV_0}\cdot V^2 + \frac{3P_0}{nR}\cdot V$$

For maximum temperature we use

$$\frac{dT}{dV} = 0 = -\frac{2P_0}{nRV_0}\cdot V + \frac{3P_0}{nR}$$

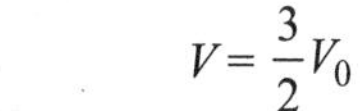

$$\Rightarrow \quad V = \frac{3}{2}V_0$$

The maximum temperature of gas is given as

$$T_{max} = -\frac{P_0}{nRV_0}\left(\frac{3V_0}{2}\right)^2 + \frac{3P_0}{nR}\left(\frac{3V_0}{2}\right)$$

$$\Rightarrow \quad T_{max} = -\frac{9}{4}\frac{P_0V_0}{nR} + \frac{9}{2}\frac{P_0V_0}{nR} = \frac{9}{4}\frac{P_0V_0}{nR}$$ **Ans. (A)**

Sol. 47 For a polytrophic process here $n \to$ Polytropic constant

$$PV^n = \text{Constant}$$

& for this process the molar specific heat of gas is given as

$$C = C_V + \frac{R}{1-n}$$

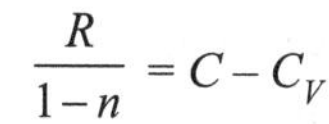

$$\Rightarrow \quad \frac{R}{1-n} = C - C_V$$

$$\Rightarrow \quad 1 - n = \frac{R}{C - C_V} = \frac{C_P - C_v}{C - C_V}$$

$$\Rightarrow \quad n = 1 - \frac{C_P - C_V}{C - C_V} = \frac{C - C_V - C_P + C_V}{C - C_V}$$

$$\Rightarrow \quad n = \frac{C - C_P}{C - C_V}$$ **Ans. (B)**

Sol. 48 Refrigerator

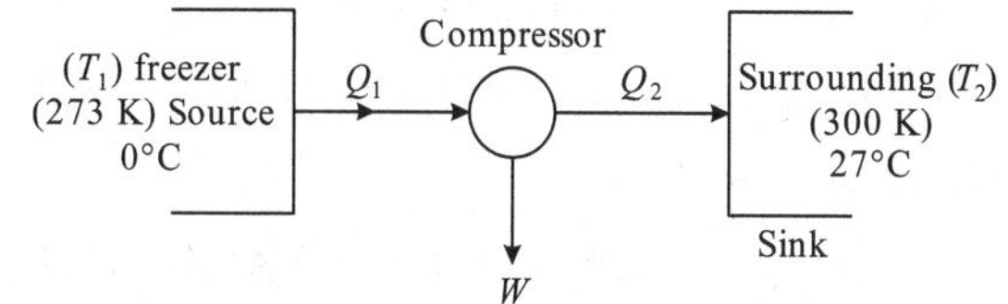

by conservation of energy

$$Q_1 + W_{ext} = Q_2$$

$$\Rightarrow \quad W_{ext} = Q_2 - Q_1$$

For a carnot freezer we use

$$\frac{Q_1}{Q_2} = \frac{T_1}{T_2} \quad \ldots(1)$$

here $Q_1 \rightarrow$ heat released by freezer in freezing ice

Here $\quad Q_1 = mL = 5 \times 336 \times 10^3\ (\text{J/kg}) \times \text{kg}$

$$= 5 \times 336 \times 10^3\ \text{J}$$

We use by equation-(1)

$$Q_2 = \left(\frac{T_2}{T_1}\right) \times Q_1 = \frac{300}{273} \times 5 \times 336 \times 10^3$$

Energy consumed by freezer in

$$W = Q_2 - Q_1$$

$$\Rightarrow \quad W = 5 \times 336 \times 10^3 \left(\frac{300}{273} - 1\right)$$

$$\Rightarrow \quad W = 1.67 \times 10^5\ \text{J}$$

Ans. (D)

Sol. 49 In an isobaric process (P = constant)

Heat supplied $\quad \Delta Q = nC_P \Delta T$

[for a monoatomic gas $C_P = \frac{J}{2} R$]

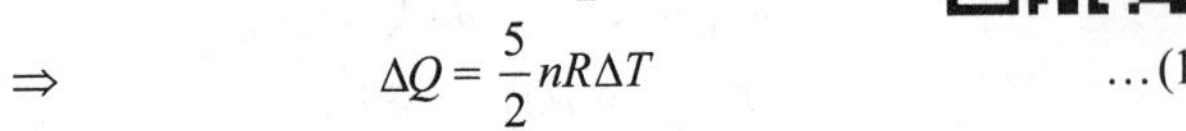

$$\Rightarrow \quad \Delta Q = \frac{5}{2} nR\Delta T \quad \ldots(1)$$

Work done $\quad W = P\Delta V = nRT \quad \ldots(2)$

$$\frac{(2)}{(1)} \Rightarrow \frac{W}{\Delta Q} = \frac{nR\Delta T}{5/2nR\Delta T} = \frac{2}{5}$$

Ans. (A)

Sol. 50 By first law of thermodynamics

$$\Delta Q = \Delta U + \Delta W$$

If expansion of water is neglected

$$\Rightarrow \quad \Delta W = 0\ \text{(heat supplied)}$$

$$\Delta U = mS\Delta T$$

$$\Rightarrow \quad \Delta U = 0.2 \times 4184 \times 20 = 16.7\ \text{kJ}$$

Ans. (D)

Sol. 51 Work done by the gas in the cycle is

$$W = P_0 V_0\ \text{(area enclosed)}$$

Heat supplied in cycle can be given as

$$\Delta Q_{supply} = Q_{AB} + Q_{BC} = nC_V(T_B - T_A) + nC_P(T_C - T_B)$$

Here we use $C_V = 1.5R$ and $C_P = C_V + R = 2.5R$

$$\Rightarrow \quad \Delta Q_{supply} = 1.5(nRT_B - nRT_A) + 2.5\,(nRT_C - nRT_B)$$

By Gas law, we use

$$\Delta Q_{supply} = 1.5(2P_0V_0 - P_0V_0) + 2.5\,(4P_0V_0 - 2P_0V_0)$$

$$\Rightarrow \quad \Delta Q_{supply} = 1.5P_0V_0 + 5P_0V_0$$

$$\Rightarrow \quad \Delta Q_{supply} = 6.5P_0V_0$$

Engine efficiency

$$\eta = \frac{W}{\Delta Q_{supply}}$$

$$= \frac{P_0V_0}{(13/2)P_0V_0} = \frac{2}{13} = 0.15$$

Ans. (B)

Sol. 52 By First law of thermodynamics, we have for ACB

$$\Delta Q_{ACB} = \Delta U_{ACB} + \Delta W_{ACB}$$

$$\Rightarrow \quad 60 = \Delta U_{ACB} + 30\ \text{J}$$

$$\Rightarrow \quad \Delta U_{ACB} = 30\ \text{J} = \Delta U_{ADB}$$

For process ADB

$$\Delta Q_{ADB} = \Delta U_{ADB} + \Delta W_{ADB}$$

$$\Delta Q_{ADB} = 30\ \text{J} + 10\ \text{J} = 40\ \text{J}$$

Ans. (D)

Sol. 53 For closed vessel, heat supplied is given as

$$\Delta Q = nC_v \Delta T$$

$$\Rightarrow \quad \Delta Q = \frac{15}{28} \times \frac{5R}{2} \times (4T - T) \quad [T = 300\ \text{K}]$$

$$\Rightarrow \quad \Delta Q = 10021\ \text{J} \cong 10\ \text{kJ}$$

Ans. (A)

Sol. 54 Given system of carnot engines can be drawn as

As we use $\quad W_1 = W_2$

$$\Rightarrow \quad \Delta Q_1 = \Delta Q_2$$

$$\Rightarrow \quad T_2 - T_1 = T_3 - T_2$$

$$\Rightarrow \quad T_2 = \frac{T_1 + T_3}{2} + \frac{600 + 400}{2}$$

$$= 500\ K$$

Ans. (C)

Sol. 55 As engines are equally efficient, we use

$$\eta = 1 - \frac{T_2}{T_1} = 1 - \frac{T_3}{T_2} = 1 - \frac{T_4}{T_3}$$

$$\Rightarrow \quad \frac{T_2}{T_1} = \frac{T_3}{T_4} = \frac{T_4}{T_3}$$

$$\Rightarrow \quad T_2 = \sqrt{T_1T_3}$$

and $\quad T_3 = \sqrt{T_2T_4}$

$$\Rightarrow \quad T_2^{3/4} = \sqrt{T_1^{1/2}T_4^{1/4}}$$

$$\Rightarrow \quad T_2 = T_1^{2/3}T_4^{1/3}$$

and $\quad T_3 = T_1^{1/3}T_4^{2/3}$

Ans. (A)

Sol. 56 Work done by gas is

$$W = P\Delta V = nR\Delta T$$

$$\Rightarrow \quad W = \frac{1}{2} \times 8.31 \times 70$$

$$\Rightarrow \quad W = 291 \text{ J}$$

Ans. (B)

Sol. 57 For adiabatic process we use $TV^{\gamma-1}$ = constant and for a diatomic gas we use

$$\gamma - 1 = \frac{7}{5} - 1 = \frac{2}{5}$$

$$\Rightarrow \quad x = \frac{2}{5}$$ **Ans. (B)**

Sol. 58 Given that $VT = K$

$$\Rightarrow \quad V\left(\frac{PV}{nR}\right) = K$$

$$\Rightarrow \quad PV^2 = K$$

For a polytropic process, molar specific heat of gas is given as

$$C = \frac{R}{1-n} + C_V \qquad (n = 2)$$

$$\Rightarrow \quad C = \frac{R}{1-2} + \frac{3R}{2} = \frac{R}{2}$$

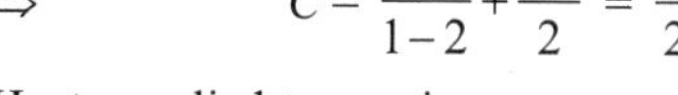

Heat supplied to gas is

$$\Delta Q = nC\,\Delta T$$

$$\Rightarrow \quad \Delta Q = 1 \times \frac{R}{2} \times \Delta T = \frac{R\Delta T}{2}$$

Ans. (A)

Sol. 59 Since P–V indicator diagram is given, so work done by gas is area enclosed by the cyclic diagram

$$\Rightarrow \quad \text{Work done by gas} = \frac{1}{2} \times 4 \times 5 \text{ J}$$

$$\Rightarrow \quad W = 10 \text{ J}$$ **Ans. (C)**

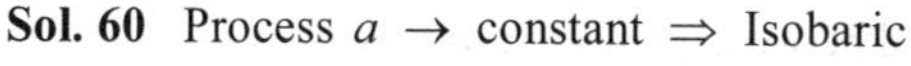

Sol. 60 Process $a \rightarrow$ constant $\Rightarrow$ Isobaric
Process $b \rightarrow$ Isothermal
Process $c \rightarrow$ Adiabatic
Process $d \rightarrow$ Isochoric

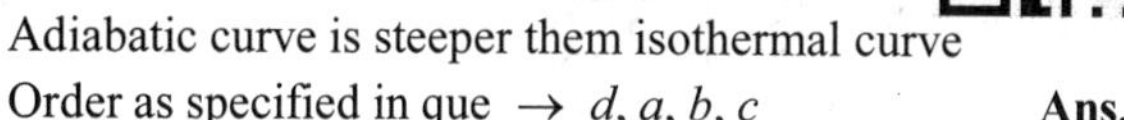

Adiabatic curve is steeper them isothermal curve
Order as specified in que $\rightarrow d, a, b, c$ **Ans. (D)**

Sol. 61

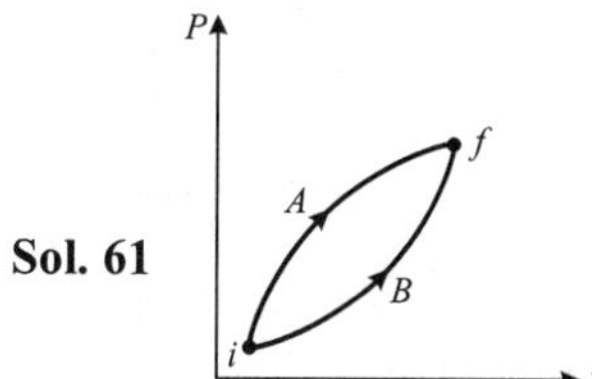

Initial and final states for both the processes are same so we use

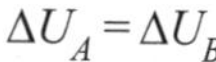

$$\Delta U_A = \Delta U_B$$

Work done during process A is greater than in process B so using First Law of thermodynamics, we have

$$\Delta Q = \Delta U + W$$

$$\Rightarrow \quad \Delta Q_A > \Delta Q_B$$ **Ans. (B)**

Sol. 62 For constant volume heating, we use

$$\Delta Q = nC_V\Delta T = n\frac{3}{2}R\Delta T$$

$$= \left(\frac{67.2}{22.4}\right)\left(\frac{3}{2}\times 8.31\right)(20)$$

$$\approx 748 \text{ J}$$

Ans. (A)

Sol. 63 For isobaric process work done by gas is given as

Heat supplied is $\quad W = nR\Delta T$

$$\Delta Q = (C_v + nR)\Delta T$$

$$\Rightarrow \quad \frac{W}{\Delta Q} = \frac{nR}{C_v + nR}$$ **Ans. (C)**

Sol. 64 Heat required for constant volume and constant pressure to raise gas temperature by ΔT is

$$Q = nC_v\Delta T$$

$$Q' = nC_p\Delta T$$

$$\Rightarrow \quad \frac{Q'}{Q} = \frac{C_p}{C_v}$$

For diatomic gas we have

$$\frac{C_p}{C_v} = \gamma = \frac{7}{5}$$

$$\Rightarrow \quad Q' = \frac{7}{5}Q$$

Ans. (A)

Sol. 65 Temperature of gas ($n = 1$) is given as

$$PV = RT \Rightarrow T = \frac{PV}{R} = \frac{P_0V}{R}\left(1 - \frac{1}{2}\left(\frac{V_0}{V}\right)^2\right)$$

At $V = V_0$, temperature is

$$T_1 = \frac{P_0V_0}{R}\left[1 - \frac{1}{2}\right] = \frac{P_0V_0}{2R}$$

At $V = 2V_0$, temperature is

$$T_2 = \frac{2P_0V_0}{R}\left[1 - \frac{1}{2}\left(\frac{1}{4}\right)\right] = \frac{7P_0V_0}{4R}$$

Temperature change

$$\Delta T = T_2 - T_1 = \frac{7P_0V_0}{4R} - \frac{P_0V_0}{2R} = \frac{5P_0V_0}{4R}$$ **Ans. (C)**

Sol. 66 Work done along path *abc* is

$$W_{abc} = W_{ab}$$

As $W_{bc} = 0$ (being an isochoric process)

By First law of thermodynamics, we use

$$W_{ab} = \Delta Q_{ab} - \Delta U_{ab}$$

In a cyclic process net

$$\Delta U_{cycle} = 0$$

$$\Delta U_{ab} + \Delta U_{bc} + \Delta U_{ca} = 0$$

$$\Rightarrow \quad \Delta U_{ab} + 60 - 180 = 0 \qquad [\Delta U_{bc} = \Delta Q_{bc}]$$

$$\Rightarrow \quad \Delta U_{ab} = 120\text{ J}$$

$$= 250 - 120 = 130\text{ J}$$

Ans. (D)

Sol. 67 Efficiency of Carnot engine

$$\eta = 1 - \frac{T_{sink}}{T_{source}}$$

$$\Rightarrow \quad \frac{1}{6} = 1 - \frac{T_{sink}}{T_{source}}$$

$$\Rightarrow \quad \frac{T_{sink}}{T_{source}} = \frac{5}{6} \qquad \ldots(1)$$

Efficiency gets doubled when temperature of sink decreases by 62°C so we use

$$\frac{2}{6} = 1 - \frac{T_{sink} - 62}{T_{source}}$$

$$\Rightarrow \quad \frac{62}{T_{source}} = \frac{1}{6} \qquad \ldots(2)$$

$$\Rightarrow \quad T_{source} = 372\text{ K} = 99°\text{C}$$

From equation-(1), we have

$$T_{sink} = \frac{5}{6} \times 372 = 310\text{ K} = 37°\text{C}$$

Ans. (B)

Sol. 68 For a diatomic gas, we have

$$C_p = \frac{7}{2}R$$

As gas undergoes isobaric process, heat absorbed by gas is

$$\Rightarrow \quad \Delta Q = nC_P\Delta T$$

In isobaric process, work done is

Also, $\Delta W = nR\Delta T = 10\text{J}$ (given)

$$\Rightarrow \quad \Delta Q = n\left(\frac{7}{2}R\Delta T\right) = \frac{7}{2}(nR\Delta T)$$

$$= \frac{7}{2} \times 10 = 35\text{ J}$$

Ans. (A)

Sol. 69 Work done in adiabatic process is given as

$$W = \frac{nR(T_1 - T_2)}{\gamma - 1} = \frac{P_1V_1 - P_2V_2}{0.4}$$

$$\Rightarrow \quad W = \frac{100 \times 1 - \frac{100}{4.6555} \times 3}{0.4} = 88.90$$

Thus closest answer is option (A) **Ans. (A)**

Sol. 70 For equivalent *r* for mixture we use

$$\frac{n_1 + n_2}{r_{eq} - 1} = \frac{n_1}{r_1 - 1} + \frac{n_2}{r_2 - 1}$$

$$\Rightarrow \quad \frac{5}{r_{eq} - 1} = \frac{2}{2/3} + \frac{3}{1/3}$$

$$\Rightarrow \quad r_{eq} = \frac{17}{12} = 1.416$$

$$\simeq 1.42$$

Ans. (B)

Sol. 71 For Carnot Cycle, we use

$$\frac{Q_1}{Q_2} = \frac{T_1}{T_2}$$

$$\Rightarrow \quad \frac{Q_2 + 1200}{Q_2} = \frac{900}{300} \qquad [\text{As } Q_1 = Q_2 + W]$$

$$\Rightarrow \quad Q_2 + 1200 = 3Q$$

$$\Rightarrow \quad Q_2 = 600\text{ J}$$

Ans. (600)

Sol. 72 Electric field due to each charged sheet is given as

$$E = \frac{\sigma}{2\varepsilon_0}$$

Which in direction normal to planes as shown in figure

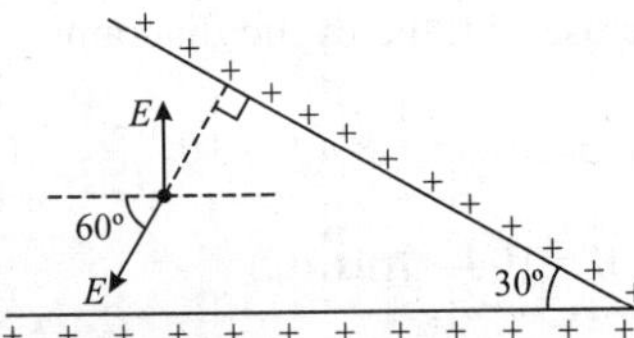

Net electric field between plates is given as

$$\vec{E}_R = E\cos 60°(-\hat{x}) + (E - E\sin 60°)(\hat{y})$$

$$\Rightarrow \quad \vec{E}_R = \frac{\sigma}{2\varepsilon_0}\left[-\frac{\hat{x}}{2} + \left(1 - \frac{\sqrt{3}}{2}\right)\hat{y}\right]$$

Ans. (B)

Sol. 73 For the engines, we use

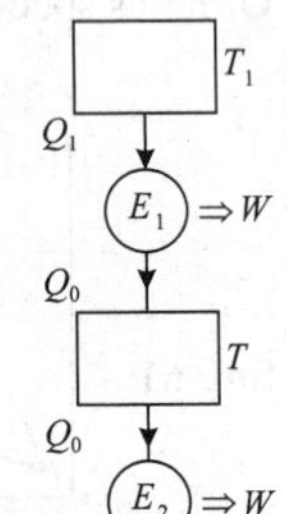

$$\frac{Q}{Q_0} = \frac{T_1}{T}$$

and

$$\frac{Q_0}{Q_2} = \frac{T_1}{T}$$

As work is equal by both engines, we use

$$\frac{Q_2 + W}{Q_2} = \frac{T_1}{T}$$

and $$\frac{Q_2}{Q_2 - W} = \frac{T}{T_2}$$

Solving we get

$$T = \frac{T_1 + T_2}{2}$$ **Ans. (C)**

Sol. 74 Mean collision time is taken as

$$t \propto \frac{V}{\sqrt{T}} \quad \ldots(1)$$

And for adiabatic process, we use

$$TV^{\gamma-1} = \text{constant} \quad \ldots(2)$$

$$\Rightarrow \quad t \propto V^{\frac{\gamma+1}{2}}$$ **Ans. (A)**

Sol. 75 Given processes are

$x \to y$ Isobaric, $y \to z \Rightarrow$ Isochoric, $z \to x \Rightarrow$ Isothermal

So by slope analysis correct option is (D) **Ans. (D)**

Sol. 76 Efficiency for refrigerator cycle is

$$\eta = \frac{W}{Q_1} = \frac{W}{W + Q_2}$$

$$\frac{1}{10} = \frac{10}{10 + Q_2}$$

$$\Rightarrow \quad Q_2 = 90\text{ J}$$

Heat absorbed from the reservoir at lower temperature is 90 J

Ans. (C)

Sol. 77 C_P/C_V for mixture is given as

$$\gamma_{\text{mixture}} = \frac{n_1 C_{P_1} + n_2 C_{P_2}}{n_1 C_{V_1} + n_2 C_{V_2}}$$

$$\Rightarrow \quad \gamma_{\text{mixture}} = \frac{n \times \left(\frac{5R}{2}\right) + 2n\left(\frac{7R}{2}\right)}{n \times \frac{3R}{2} + 2n\left(\frac{5R}{2}\right)}$$

$$\Rightarrow \quad \gamma_{\text{mixture}} = \frac{19}{13}$$ **Ans. (B)**

Sol. 78 For a diatomic gas molecule there are 5 degrees of freedom if vibration mode is absent and if vibration mode is there then degrees of freedom will be 7 so specific heats in the two cases are $C_V^A = \frac{5}{2}R$ and $C_V^B = \frac{7}{2}R \Rightarrow \frac{C_V^A}{C_V^B} = \frac{5}{7}$.

Ans. (B)

Sol. 79 In the given PV diagram process 2 to 3 is isobaric & process 3 to 1 is isochoric. This is shown only in option (D).

Ans. (D)

Sol. 80 For an adiabatic process, we use

$$PV^\gamma = \text{constant}$$

and $$TV^{\gamma-1} = C$$

For the given situation, we use

$$300 \times V^{\frac{7}{5}-1} = T_2\left(\frac{V}{16}\right)^{\frac{7}{5}-1}$$

$$\Rightarrow \quad T_2 = 300 \times 2^{4\times\frac{2}{5}}$$

For isobaric process, we use

$$\frac{V_2}{T_2} = \frac{V_3}{T_3}$$

Using $V_3 = 2V_2$, we get

$$\frac{1}{2} = \frac{T_2}{T_3} \Rightarrow T_3 = 2T_2$$

$$\Rightarrow \quad T_3 = 2 \times 300 \times 2^{\frac{8}{5}} = 1818.85$$

Ans. (1818 to 1819)

Sol. 81 Total internal energy of mixture is given as

$$U = \frac{f_1}{2} n_1 RT + \frac{f_2}{2} n_2 RT$$

$$\Rightarrow \quad U = \frac{5}{2} \times 3RT + \frac{3}{2} \times 5RT = 15RT$$ **Ans. (B)**

Sol. 82 As volume compresses to $V/10$, we use

$$TV^{\gamma-1} = \text{Constant} \quad [r = 7/5 \text{ for air}]$$

$$\Rightarrow \quad T_1 V_1^{2/5} = T_2\left(\frac{V}{10}\right)^{2/5}$$

$$\Rightarrow \quad T_2 = T(10)^{2/5}$$

Change in internal energy in adiabatic process is

$$\Delta U = \frac{R(T_2 - T_1)}{\gamma - 1}$$

$$\Rightarrow \quad \Delta U = \frac{5 \times 8.31 \times 293 \times (10^{2/5} - 1)}{\frac{7}{5} - 1}$$

$$\cong 4 \times 10^3\text{ J}$$

$$\Rightarrow \quad x = 4$$ **Ans. (4)**

Sol. 83 Efficiency of heat engine is given as

$$\eta = \frac{\text{Work done}}{\text{Heat supplied}}$$

$$\Rightarrow \quad \eta = \frac{1915 - 40 + 125 - Q}{1915 + 125} = 0.5$$

$$\Rightarrow \quad \frac{2000 - Q}{2040} = 0.5$$

$$\Rightarrow \quad 2000 - Q = 1020$$

$$\Rightarrow \quad Q = 980\text{ J}$$ **Ans. (C)**

Sol. 84 As all sudden process are considered irreversible and in these processes no time is available for heat exchange so considered adiabatic. **Ans. (B)**

Sol. 85 For triangular molecule degrees of freedom are given translational $dof = 3$

Rotational, $dof = 3$

Thus internal energy of gas is given as

$$U = \frac{f}{2}nRT = 3RT$$ **Ans. (D)**

Sol. 86 For minimum work, we use Carnot cycle so, coefficient of performance is given as

$$COP = \frac{Q_{in}}{W} = \frac{Q_{in}}{Q_{out} - Q_{in}} = \frac{273}{300 - 273} = \frac{273}{27}$$

$$\Rightarrow \quad W = \frac{27}{273} \times Q_{in} = \frac{27}{273} \times 100 \times 80 \text{ cal}$$

Heat rejected by refrigerator is given as

$$Q_{out} = W + Q_1$$

$$\Rightarrow \quad Q_{out} = \frac{27}{273} \times 100 \times 80 + 100 \times 80 \text{ cal}$$

$$\Rightarrow \quad Q_{out} = 8791.2 \text{ cal}$$ **Ans. (8791)**

Sol. 87 For mixture of gases, equilibrium temperature is given as

$$T_{eq} = \frac{n_1 T_1 + n_2 T_2}{n_1 + n_2} = \frac{0.1 \times 200 + 0.05 \times 400}{0.1 + 0.05}$$

$$\Rightarrow \quad T_{eq} = \frac{20 + 20}{0.15} = \frac{40}{0.15} = \frac{800}{3}$$

$$\Rightarrow \quad T_{eq} = 266.66 \text{ K}$$ **Ans. (266)**

Sol. 88 (I) In Adiabatic process no heat exchange take place so option (B) is correct match

(II) In Isothermal process no change internal energy so option (D) is correct match

(III) In Isochoric process V = constant. So $\Delta\omega = 0$

(IV) In isobaric process as P = constant option (C) is correct match. **Ans. (A)**

Sol. 89 By gas law we use

$$PV = nRT$$

At constant pressure

$$P\Delta V = nR\Delta T$$

$$\Rightarrow \quad \Delta T = \left(\frac{P}{nR}\right)\Delta V$$

At constant temperature

$$P\Delta V + V\Delta P = 0$$

$$\Rightarrow \quad \Delta P = -\frac{P\Delta V}{V}$$

Value of C is given as

$$C = \left|\frac{\Delta T}{\Delta P}\right| = \frac{V}{nR} = \frac{T}{P} = \frac{300}{2}$$

$$\Rightarrow \quad C = 150 \text{ K/atm}$$ **Ans. (150)**

Sol. 90 From the given indicter diagram, we can see for all processes as ΔT is same

$$\Delta U = nC_v \Delta T = \text{same}$$

$AB \rightarrow$ volume is increasing $\Rightarrow W > 0$

$AD \rightarrow$ volume is decreasing $\Rightarrow W < 0$

$AC \rightarrow$ volume is constant $\Rightarrow W = 0$ **Ans. (A)**

Sol. 91 Using gas law, we use

$$n = \frac{PV}{RT}$$

Mean kinetic energy of a molecule is given as

$$\frac{3}{2}kT = 4 \times 10^{-14}$$

$$\Rightarrow \quad RT = \frac{8}{3} \times 10^{-14} \times N_{AV} \qquad \left[\text{As } k = \frac{R}{N_{AV}}\right]$$

Total molecules in gas are given as

$$N = \frac{PV}{RT} \times N_{AV}$$

$$\Rightarrow \quad N = \frac{2 \times 13.6 \times 980 \times 4}{\frac{8}{3} \times 10^{-14}} = 3.99 \times 10^{18}$$ **Ans. (C)**

Sol. 92 In adiabatic process we use

$$PV^{\gamma} = \text{constant}$$

$$\Rightarrow \quad P\left(\frac{m}{\rho}\right)^{\gamma} = \text{constant}$$

$$\Rightarrow \quad P \propto \rho^{\gamma}$$

$$\Rightarrow \quad \frac{P_f}{P_i} = \left(\frac{\rho_f}{\rho_i}\right)^{\gamma} = (32)^{7/5} = 2^7 = 128$$ **Ans. (D)**

Sol. 93 Internal energy of gas for f degrees of freedom is given as

$$U = \frac{f}{2}nRT = \frac{5}{2}RT$$

Ratio of specific heats is given as

$$\gamma = \frac{C_P}{C_V}$$

$$\Rightarrow \qquad \gamma = \frac{f+2}{f} = \frac{5+2}{5} = \frac{7}{5}$$ **Ans. (D)**

Sol. 94 Work done in cycle $ABCDA$ is given as

$$W_{ABCDA} = 2P_0V_0$$

Total heat input in cycle is given as

$$Q_{in} = Q_{AB} + Q_{BC}$$

Here, we use in process AB

$$Q_{AB} = nC(T_B - T_A)$$

$$\Rightarrow \qquad Q_{AB} = \frac{n3R}{2}(T_B - T_A)$$

$$\Rightarrow \qquad Q_{AB} = \frac{3}{2}(P_BV_B - P_AV_A)$$

$$\Rightarrow \qquad Q_{AB} = \frac{3}{2}(3P_BV_0 = P_0V_0) = 3P_0V_0$$

For process BC, we use

$$Q_{BC} = nC_P(T_C - T_B)$$

$$\Rightarrow \qquad Q_{BC} = \frac{n5R}{2}(T_C - T_B)$$

$$\Rightarrow \qquad Q_{BC} = \frac{5}{2}(P_CV_C - P_BV_B)$$

$$\Rightarrow \qquad Q_{BC} = \frac{5}{2}(6P_0V_0 - 3P_0V_0) = \frac{15}{2}P_0V_0$$

Efficiency of cycle is given as

$$\eta = \frac{W}{Q_{in}} \times 100$$

$$\Rightarrow \qquad \eta = \frac{2P_0V_0}{3P_0V_0 + \frac{15}{2}P_0V_0} \times 100$$

$$\Rightarrow \qquad \eta = \frac{400}{21} = 19.04 \approx 19$$ **Ans. (19)**

Sol. 95 All the given processes in list-I are by definition related to the parameters given in list II. **Ans. (D)**

Sol. 96 Work done in the process AB is given as

$$W_{AB} = nRT\ln\left(\frac{v_2}{v_1}\right) = nRT\ln\left(\frac{2V}{V}\right) = nRT\ln 2$$

Work done in process BC and CA are given as

$$W_{BC} = P\Delta V = nR\Delta T = nR\left(\frac{T}{2} - T\right) = -nR\frac{T}{2}$$

$$W_{CA} = 0$$

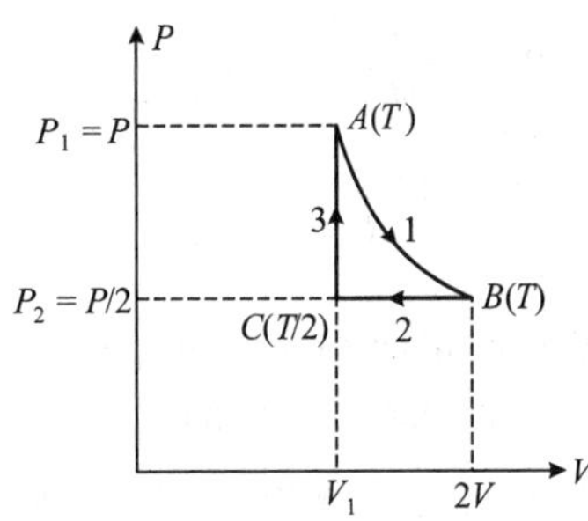

Net work done in the cycle is given by sum of all

$$W_{net} = nRT\left(\ln 2 - \frac{1}{2}\right)$$ **Ans. (A)**

Sol. 97 Work done in process AB is given as

$$W_{AB} = P_1V_1 \ln\left[\frac{2V_1}{V_1}\right] = P_1V_1\ln(2)$$

Work done in process BC is given as

$$W_{BC} = 0$$

Work done in process CA is given as

$$W_{CA} = \frac{P_1V_1 - \frac{P_1}{4} \times 2V_1}{1-\gamma}$$

$$\Rightarrow \qquad W_{CA} = \frac{P_1V_1\left[1 - \frac{1}{2}\right]}{1-\gamma} = \frac{P_1V_1}{2(1-\gamma)}$$

Net work done is calculated as

$$W_{net} = W_{AB} + W_{BC} + W_{CA}$$

$$\Rightarrow \qquad W_{net} = P_1V_1\ln(2) + 0 + \frac{P_1V_1}{2(1-\gamma)}$$

$$\Rightarrow \qquad W_{net} = RT\left[\ln(2) - \frac{1}{2(\gamma-1)}\right]$$ **Ans. (D)**

Sol. 98 For an isobaric process, we use

$$dU = nC_v dT$$

$$dQ = nC_p dT$$

$$dW = PdV = nRdT$$

$$\Rightarrow \qquad dU : dQ : dW = C_v : C_p : R$$

$$\Rightarrow \qquad dU : dQ : dW = \frac{5R}{2} : \frac{7R}{2} : R = 5:7:2$$ **Ans. (B)**

Sol. 99 When container is suddendly stopped and due to this the change in temperature is ΔT, we have

$$\frac{1}{2}mv^2 = nC_v\Delta T$$

$$\Rightarrow \quad \frac{1}{2}m\times(30)^2 = \frac{m}{M}\times\frac{3R}{2}\times\Delta T$$

$$\Rightarrow \quad \Delta T = \frac{3600}{3R}$$ **Ans. (3600)**

Sol. 100 For the given thermodynamic process, we have

$$\Delta T = 300 - 100 = 200°C$$

As $P = kV^3$ we use by gas law $PV = nRT$ we have

$$kV^4 = nRT$$

$$\Rightarrow \quad 4kV^3\, dV = nRdT$$

$$\Rightarrow \quad PdV = nRdT/4$$

Work done in the process is given as

$$W = \int PdV = \int\frac{nRdT}{4} = \frac{nR}{4}\Delta T$$

$$\Rightarrow \quad W = \frac{200}{4}\times nR = 50nR$$ **Ans. (50)**

Sol. 101 For the thermodynamic process $PV^{½} = Constant$, we use by gas law

$$\frac{nRT}{V}V^{½} = C$$

$$\Rightarrow \quad T = C_1 V^{½}$$

$$\Rightarrow \quad \frac{T_2}{T_1} = \left(\frac{V_2}{V_1}\right)^{1/2} = \left(\frac{2V_1}{V_1}\right)^{1/2}$$

$$\Rightarrow \quad \frac{T_2}{T_1} = \sqrt{2}$$ **Ans. (B)**

Sol. 102 Efficiency of given heat engine is

$$\eta = \frac{1}{4} = 1 - \frac{T_2}{T_1}$$

$$\Rightarrow \quad \frac{T_2}{T_1} = \frac{3}{4}$$

When temperature of sink is reduced by 52 K, efficiency is doubled so we have

$$\frac{T_2 - 52}{T_1} = \frac{1}{2}$$

$$\Rightarrow \quad \frac{0.75T_1 - 52}{T_1} = \frac{1}{2}$$

$$\Rightarrow \quad T_1 = 208\text{ K}$$ **Ans. (208)**

Sol. 103 If equilibrium pressure of mixture is P atm. then internal energy of mixture is given as

$$U_1 + U_2 = U_{\text{mixutre}}$$

$$\Rightarrow \quad \frac{f}{2}P_1V_1 + \frac{f}{2}P_2V_2 = \frac{f}{2}P(V_1 + V_2)$$

$$\Rightarrow \quad \frac{f}{2}(2)(4.5) + \frac{f}{2}(3)(5.5) = \frac{f}{2}P(4.5 + 5.5)$$

$$\Rightarrow \quad P = 2.55 = x\times 10^{-1}\text{ atm}$$

$$\Rightarrow \quad x = 25.5 \approx 26 \text{ (Nearest integer)}$$

Ans. (26)

Sol. 104 The thermodynamics process is given as

$$U = 3PV + 4$$

If the gas has f degrees of freedom then we can write

$$\frac{nf}{2}RT = 3PV + 4$$

$$\Rightarrow \quad \frac{f}{2}PV = 3PV + 4$$

$$\Rightarrow \quad f = 6 + \frac{8}{PV}$$

Since degree of freedom is more than 6 therefore gas is polyatomic. **Ans. (B)**

Sol. 105 By first law of thermodynamics, we have

$$Q = \Delta U + W$$

$$\Rightarrow \quad Q = \Delta U + \frac{Q}{5}$$

$$\Rightarrow \quad \Delta U = \frac{4Q}{5}$$

$$\Rightarrow \quad nC_v\Delta T = \frac{4}{5}nC\Delta T$$

$$\Rightarrow \quad \frac{5}{4}C_V = C$$

$$\Rightarrow \quad C = \frac{5}{4}\left(\frac{f}{2}\right)R = \frac{5}{4}\left(\frac{5}{2}\right)R$$

$$\Rightarrow \quad C = \frac{25}{8}R$$

$$\Rightarrow \quad x = 25$$ **Ans. (25)**

Sol. 106 Work done in the process is given as

$$W = \int PdV \quad \ldots(1)$$

$$\Rightarrow \quad W = \int\frac{nRT}{V}dV$$

$$\Rightarrow \quad W = \int\frac{nRT}{KT^{2/3}}\cdot dV$$

From the process equation, we have

$$dV = \frac{2}{3}KT^{-1/3}dT$$

$$\Rightarrow \quad W = \int_{T_1}^{T_2}\frac{nRT}{KT^{2/3}}\frac{2}{3}K\frac{1}{T^{1/3}}dT$$

$$\Rightarrow \quad W = \frac{2}{3}nR \times (T_2 - T_1)$$

Given that $T_2 - T_1 = 90$ K, so we have

$$\Rightarrow \quad W = \frac{2}{3}nR \times 90 = 60\,nR$$

Considering 1 mole of gas, $W = 60\,R$. **Ans. (60)**

Sol. 107 Heat and work are considered as path functions for thermodynamic processes and related by first law of thermodynamics as

$$\Delta Q = \Delta U + \Delta W$$

Since work done by gas depends on type of process which is determined by path and ΔU depends just on initial and final states, so ΔQ, the heat supplied , also depend on path of process. **Ans. (A)**

Sol. 108 Efficiency of heat engine is given as

$$\eta = 0.60 = 1 = \frac{T_L}{T_H}$$

$$\Rightarrow \quad \frac{T_L}{T_H} = 0.4$$

$$\Rightarrow \quad T_L = 0.4 \times 400 = 160\text{ K} = -113^\circ\text{C}$$ **Ans. (–113)**

Sol. 109 As each vibrational mode has 2 degrees of freedom so number of vibrational degrees of freedom will be 48 and total degrees of freedom are given as

$$f = 3 + 3 + 48 = 54$$

Thus ratio of specific heats is given as

$$\gamma = 1 + \frac{2}{f} = \frac{28}{27} = 1.03$$ **Ans. (A)**

Sol. 110 The efficiency of carnot engine is given as

$$\eta = \frac{T_2}{T_1} = \frac{Q_2}{Q_1} = \frac{Q_1 - W}{Q_1}$$

$$\Rightarrow \quad \frac{400}{800} = 1 - \frac{W}{Q_1}$$

$$\Rightarrow \quad \frac{W}{Q_1} = 1 - \frac{1}{2} = \frac{1}{2}$$

$$\Rightarrow \quad Q_1 = 2W = 2400\text{ J}$$ **Ans. (D)**

Sol. 111 Option (a) is wrong as as in adiabatic process $V \neq$ constant and 0ption (b) is wrong as in isothermal process $P \neq$ constant.

In option (c) the slope of adiabatic curve on V–T diagram is negative and increasing which matches with its process equation hence correct and similarly in option (d) the slope of adiabatic curve on P–T diagram is positive and increasing which also matches with its process equation hence correct. **Ans. (B)**

Sol. 112 The adiabatic process is from C to D can be calculated as

$$W_{CD} = \frac{P_2V_2 - P_1V_1}{1-\gamma} = \frac{P_DV_D - P_CV_C}{1-\gamma}$$

$$\Rightarrow \quad W_{CD} = \frac{200(3) - (100)(4)}{1 - 1.4} = -500\text{ J}$$ **Ans. (A)**

Sol. 113 As work is zero, for isochoric process, we use

$$Q = \Delta U$$

$$\Rightarrow \quad Q = nC_v\Delta T = 4 \times \frac{5R}{2} \times 50 = 500\,R$$ **Ans. (B)**

Sol. 114 In the given cyclic process the heat absorbed by the system can be calculated by area enclosed by the cyclic curve, given as

$$\Delta Q = \pi ab$$

$$\Rightarrow \quad \Delta Q = \left[\frac{40-20}{2} \times 10^3\right] \times \left[\frac{40-20}{2} \times 10^{-3}\right]$$

$$\Rightarrow \quad \Delta Q = 100\,\pi\text{ Joule}$$ **Ans. (100)**

Sol. 115 Work in the isothermal process AB shown in the indicator diagram is given as

$$W = nRT\ln\frac{V_2}{V_1}$$

$$\Rightarrow \quad W = 1 \times 8.3 \times 300\ln 2 = 1725.8\text{ J}$$ **Ans. (17258)**

Sol. 116 For adiabatic process, the process equation can be written as

$$\frac{T_1}{T_2} = \left(\frac{V_2}{V_1}\right)^{\gamma-1}$$

$$\Rightarrow \quad \frac{T_1}{T_2} = \left(\frac{V_2}{V_1}\right)^{5/3-1} = \left(\frac{Al_2}{Al_1}\right)^{2/3} = \left(\frac{l_2}{l_1}\right)^{2/3}$$ **Ans. (C)**

Sol. 117 Heat supplied can be written as

$$Q = nC\Delta T$$

Rate of heat supplied can be written as

$$\frac{dQ}{dt} = nC\frac{dT}{dt}$$

Rate of heat supply is same for both gases, so we use

$$n_1C_1\left(\frac{dT}{dt}\right)_1 = n_2C_2\left(\frac{dT}{dt}\right)_2$$

$$\Rightarrow \quad \frac{C_1}{C_2} = \frac{\left(\frac{dT}{dt}\right)_2}{\left(\frac{dT}{dt}\right)_1} = \frac{\frac{90}{6}}{\frac{120}{3}} = \frac{90 \times 3}{120 \times 6} = \frac{3}{8}$$ **Ans. (D)**

Sol. 118 Considering Carnot cycle in the given heat engine as temperature based efficiency calculation is to be done. Thus efficiency of a carnot engine is given as

$$\eta = \left(1 - \frac{T_2}{T_1}\right)$$

$$\Rightarrow \quad \frac{T_2}{T_1} = 1 - \eta = 1 - \frac{1}{6} \quad \ldots(1)$$

$$\Rightarrow \quad \frac{T_2 - 62}{T_1} = 1 - \frac{1}{3} \quad \ldots(2)$$

From equations-(1) & (2), we use

$$\frac{T_2}{T_2 - 62} = \frac{5}{6} \times \frac{3}{2} = \frac{5}{4}$$

$$\Rightarrow \quad T_2 = 5 \times 62$$

From equation-(1), we have

$$T_1 = \frac{T_2}{1-\eta} = \frac{5 \times 62}{1 - \frac{1}{6}} = 5 \times 62 \times \frac{6}{5} = 372\text{K}$$

$$\Rightarrow \quad T_1 = 372 - 273 = 99°\text{C}$$

Ans. (A)

Sol. 119 From the given curve in question, we have

$$T_A = T_B$$

and $$T_C = T_D$$

Work done in process BC is given as

$$W_{BC} = \frac{nR(T_B - T_C)}{\gamma - 1}$$

Work done in process AD is given as

$$W_{AD} = \frac{nR(T_A - T_D)}{\gamma - 1}$$

$$\Rightarrow \quad W_{AD} = W_{BC}$$

Ans. (D)

Sol. 120 For four vibrational modes total 8 vibrational degrees of freedom are added so total degress of freedom of gas molecules are

$$f = 6 + 8 = 14$$

Work done in adiabatic process is given as

$$W = \frac{P_1V_1 - P_2V_2}{\gamma - 1} = \frac{nR(T_1 - T_2)}{\gamma - 1}$$

$$\Rightarrow \quad W = \frac{1 \times 8.314 \times (-10)}{\left(\frac{8}{7} - 1\right)} \quad \left(as\ \gamma = 1 + \frac{2}{f} = \frac{8}{7}\right)$$

$$\Rightarrow \quad W = -582\text{ J}$$

Ans. (C)

Sol. 121 Figure below shows the two carnot engines working in series as described in question

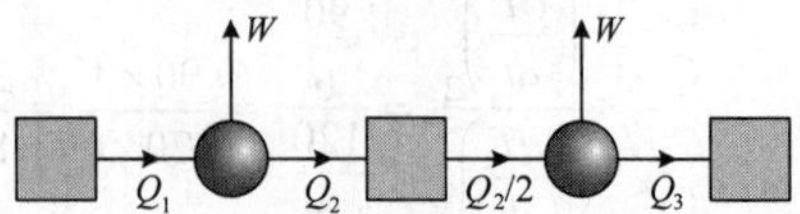

Given that work done in the two engines are equal so we use

$$W = Q_1 - Q_2$$

$$W = \frac{Q_2}{2} - Q_3$$

$$\Rightarrow \quad Q_1 - Q_2 = \frac{Q_2}{2} - Q_3$$

$$\Rightarrow \quad Q_1 + Q_3 = \frac{3}{2}Q_2$$

$$\Rightarrow \quad T_1 + \frac{T_3}{2} = \frac{3}{2}T$$

$$\Rightarrow \quad T = \frac{2}{3}T_1 + \frac{1}{3}T_3$$

Ans. (D)

Sol. 122 Rate of energy supply is given as

$$\frac{6000}{60}\text{ J/s} = 100\text{ J/s}$$

By first law of thermodynamics, for a time t, we have

$$\Delta U = Q - W$$

$$2500 = 100t - 90t$$

$$\Rightarrow \quad t = 250\text{ s}$$

Ans. (C)

Sol. 123 For a refrigerator if Q_1 is the amount of heat rejected per second to surrounding and Q_2 is the heat extracted per second from the cooling chamber and W is the work done per second by compressor then we use

$$W + Q_2 = Q_1$$

$$\Rightarrow \quad W = Q_1 - Q_2$$

$$\Rightarrow \quad \frac{Q_2}{W} = \frac{Q_2}{Q_1 - Q_2}$$

$$\Rightarrow \quad \frac{Q_2}{W} = \frac{T_2}{T_1 - T_2}$$

$$\Rightarrow \quad \frac{Q_2}{35} = \frac{263}{298 - 263}$$

$$\Rightarrow \quad Q_2 = 263\text{ J/s}$$

Ans. (A)

Sol. 124 Coefficient of volume expansion is given as

$$\gamma = \frac{1}{V}\frac{dV}{dt}$$

$$\Rightarrow \quad PT^3 = \text{constant}$$

$$\Rightarrow \quad \frac{1}{V}\frac{dV}{dt} = \text{Constant}$$

$$\Rightarrow \quad T^4V^{-1} = C$$

$$\Rightarrow \quad V^{-1}(4T^3dT) + T^4\left(\frac{-1}{V^2}dV\right) = 0$$

$$\Rightarrow \quad \frac{dV}{dT} = \frac{4T^3/V}{T^4/V^2} = \frac{4V}{T}$$

$$\Rightarrow \quad \gamma = \frac{4}{T}$$ **Ans. (C)**

Sol. 125 Efficiency of a heat engine is given as

$$\eta = \frac{W}{Q_H} = 1 - \frac{T_L}{T_H}$$

$$\Rightarrow \quad \frac{60}{300} = 1 - \frac{400}{T_H}$$

$$\Rightarrow \quad T_H = 500\,\text{K}$$ **Ans. (500)**

Sol. 126 Efficiency of the heat engine is given as

$$\eta = 1 - \frac{T_L}{T_H}$$

$$\Rightarrow \quad \frac{1}{4} = 1 - \frac{T_L}{T_H}$$

$$\Rightarrow \quad \frac{T_L}{T_H} = \frac{3}{4}$$

When temperature of sink is recuced by 58°C, its efficiency becomes double so we have

$$2\eta = \frac{1}{2} = 1 - \frac{T_L - 58}{T_H}$$

$$\Rightarrow \quad \frac{T_L - 58}{T_H} = \frac{1}{2}$$

$$\Rightarrow \quad \frac{T_L}{T_L - 58} = \frac{3/4}{1/2}$$

$$\Rightarrow \quad 2T_L = 3T_L - 174$$

$$\Rightarrow \quad T_L = 174\,°\text{C}$$ **Ans. (C)**

Sol. 127 As volume is compressed adiabatically, we use

$$P_i V_i^{\gamma} = P_f V_f^{\gamma}$$

$$\Rightarrow \quad 200 \times 10^3 \left(\frac{1200 \times 10^{-6}}{300 \times 10^{-6}}\right)^{\gamma} = P_f$$

$$\Rightarrow \quad P_f = 200 \times 10^3 \times 4^{3/2}$$

$$\Rightarrow \quad P_f = 200 \times 10^3 \times 8 = 1600 \times 10^3 = 16 \times 10^5\,Pa$$

Work done in adiabatic process is calculated as

$$W = \frac{P_i V_i - P_f V_f}{\gamma - 1}$$

$$\Rightarrow \quad W = \frac{2 \times 10^5 \times 1200 \times 10^{-6} - 16 \times 10^5 \times 300 \times 10^{-6}}{1.5 - 1}$$

$$\Rightarrow \quad W = \frac{240 - 480}{0.5} = -240 \times 2 = -480\,\text{J}$$ **Ans. (480)**

Sol. 128 Ratio of change in internal energy to the work done can be given as

$$\frac{\Delta U}{W} = \frac{\frac{5}{2} nR\Delta T}{nR\Delta T} = \frac{5}{2} = \frac{25}{10}$$

$$\Rightarrow \quad x = 25$$ **Ans. (25)**

Sol. 129 Carnot engine efficiency is given with sink temperature 300 K is given as

$$1 - \frac{300}{T} = 0.25$$

$$\Rightarrow \quad \frac{300}{T} = 0.75$$

$$\Rightarrow \quad T = 400\,\text{K}$$

If efficiency increased by 100% then new efficiency

$$n' = 50\%$$

$$\Rightarrow \quad 1 - \frac{300}{T'} = 0.5$$

$$\Rightarrow \quad T' = 600\,\text{K}$$

Increase in temperature is calculated as

$$\Delta T = 600 - 400$$

$$\Rightarrow \quad \Delta T = 200\,\text{K or } 200°\text{C}$$ **Ans. (B)**

Sol. 130 Efficiency of Carnot engine is given as

$$\eta = \frac{W}{Q_H}$$

$$\Rightarrow \quad W = Q_H\left(1 - \frac{T_L}{T_H}\right)$$

$$\Rightarrow \quad W = 5 \times 10^3\left(1 - \frac{400}{1000}\right)$$

$$\Rightarrow \quad W = 3000\,\text{kcal}$$ **Ans. (C)**

Sol. 131 Using first law of thermodynamics, we have

$$\Delta Q = \Delta U + W$$

$$\Rightarrow \quad Q = \Delta U + \frac{Q}{4}$$

$$\Rightarrow \quad \Delta U = n\frac{3R}{2}\Delta T = \frac{3Q}{4}$$

$$\Rightarrow \quad n\Delta T = \frac{Q}{2R}$$

$$\Rightarrow \quad C = 2R$$ **Ans. (2)**

Sol. 132 Molar heat capacity at constant volume for a gas having f degrees of freedom is given as

$$C_V = \frac{fR}{2}$$

Molar heat capacity at constant pressure for the gas can be written as

$$C_P = R + C_V = R + \frac{fR}{2} = \left(1 + \frac{f}{2}\right)R$$

$$\Rightarrow \quad \frac{C_P}{C_v} = 1 + \frac{2}{f} \qquad \textbf{Ans. (B)}$$

Sol. 133 The efficiency of Carnot engine is given as

$$\eta = \left[1 - \frac{T_L}{T_H}\right] \times 100\%$$

Steam point and ice point are taken as

$$T_L = 273\text{ K and } T_H = 373\text{ K}$$

$$\Rightarrow \quad \eta = 26.809\% \qquad \textbf{Ans. (A)}$$

Sol. 134 For the heat engine, heat supplied by hot reservoir and to cold reservoir are given as

$$Q_H = 300\text{ J}$$

$$Q_L = 180\text{ J}$$

Using Carnot cycle process, we have

$$\frac{Q_L}{Q_H} = \frac{T_L}{T_H}$$

$$\Rightarrow \quad T_H = \frac{Q_H}{Q_L} \times T_L = \frac{300}{180} \times 324$$

$$\Rightarrow \quad T_H = 540\text{ K} \qquad \textbf{Ans. (540)}$$

Sol. 135 Efficiency of Carnot engine is given as

$$\eta = 1 - \frac{T_2}{T_1} = \frac{W}{Q_1} = 1 - \frac{200}{800} = \frac{200\times10^6}{Q_1}$$

$$\Rightarrow \quad \frac{3}{4} = \frac{12\times10^6}{Q_1} = Q_1 = 16\times10^6\text{ J} \qquad \textbf{Ans. (16)}$$

Sol. 136 For isobaric process heat supplied to gas is given as

$$Q = nC_p\Delta T = n\left(\frac{\gamma R}{\gamma - 1}\right)\Delta T$$

$$\Rightarrow \quad Q = \left(\frac{\gamma}{\gamma - 1}\right)W = \frac{1.4}{0.4} \times 400 = 1400\text{ J} \qquad \textbf{Ans. (1400)}$$

Sol. 137 Work done in adiabatic process on μ moles of gas is written as

$$W_{\text{adiabatic}} = \frac{\mu R(T_f - T_i)}{1 - \gamma}$$

Using first law of thermodynamics, we have

$$Q = W + \Delta U$$

For adiabatic process $Q = 0$, so we use

$$0 = W + \Delta U$$

$$\Rightarrow \quad \Delta U = -W$$

If work is done on the gas, or when work is negative we have ΔU positive hence temperature will increase. **Ans. (A)**

Sol. 138 Internal energy of an ideal gas is given as

$$U = nC_vT$$

$$\Rightarrow \quad U = 2 \times \frac{3}{4}R \times 300$$

$$\Rightarrow \quad U = 900R = 900 \times 8{\cdot}31 = 7479\text{ J} \qquad \textbf{Ans. (7479)}$$

Sol. 139 Work and heat involved in different process of cycle are shown in the diagram below.

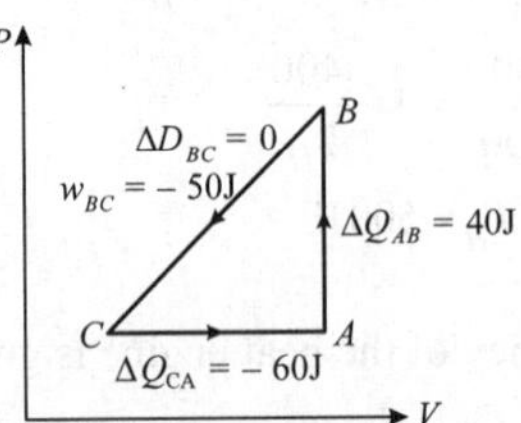

Total heat associated with the cycle is given as

$$\Delta Q_{\text{cycle}} = 40 - 60 = \Delta W$$

$$\Rightarrow \quad \Delta W = -20\text{ J} = W_{BC} + W_{CA}$$

$$\Rightarrow \quad W_{CA} = -20\text{ J} - W_{BC}$$

$$\Rightarrow \quad W_{CA} = -20 - (-50)$$

$$\Rightarrow \quad W_{CA} = 30\text{ J} \qquad \textbf{Ans. (B)}$$

Sol. 140 Efficiency of a Carnot engine is given as

$$1 - \frac{Q_2}{Q_1} = 1 - \frac{T_2}{T_1}$$

$$\Rightarrow \quad \frac{Q_2}{Q_1} = \frac{T_2}{T_1}$$

$$\Rightarrow \quad \frac{225}{300} = \frac{T_2}{500}$$

$$\Rightarrow \quad T_2 = \frac{500\times225}{300}$$

$$\Rightarrow \quad T_2 = 375\text{ K}$$

$$\Rightarrow \quad T_2 = 102°\text{C} \qquad \textbf{Ans. (102)}$$

Sol. 141 Figure below shows the indicator diagram for the three processes.

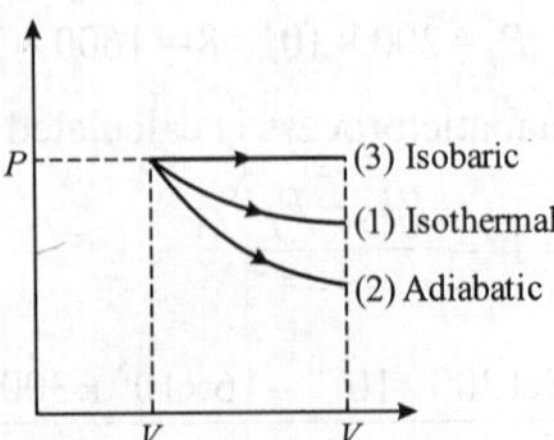

Area under curve is the work done by the gas so we have

$$W_2 < W_1 < W_3 \qquad \textbf{Ans. (D)}$$

Sol. 142 For isothermal expansion of gas, we use

$$P_1V_1 = P_2V_2$$

As $V_2 = 2V_1$, it gives

$$\Rightarrow \quad P_2 = P_1/2$$

$$\Rightarrow \quad P_2 = 1 \times 10^7 \text{ Pa}$$

For adiabatic explansion, we use

$$P_2(V_2)^\gamma = P_3(2V_2)^\gamma$$

$$\Rightarrow \quad P_3 = \frac{1\times10^7}{2^{1.5}} = 3.536 \times 10^6$$ **Ans. (B)**

Sol. 143 Efficiency of Carnot engine is given as

$$\eta = 1 - \frac{T_L}{T_H}$$

For the two engines mentioned in the question, the efficiencies are calculated as

$$\eta_1 = 1 - \frac{147+273}{447+273} = 1 - \frac{420}{720}$$

and

$$\eta_2 = 1 - \frac{47+273}{947+273} = 1 - \frac{320}{1220}$$

$$\Rightarrow \quad \eta_2 = \frac{900}{1220}$$

$$\Rightarrow \quad \frac{\eta_1}{\eta_2} = \frac{300}{720} \times \frac{1220}{900} = \frac{122}{72\times3} = 0.56$$ **Ans. (B)**

Sol. 144 Constant entropy is for an adiabatic process, so we use

$$PV^\gamma = \text{constant}$$

For $V_2 = \frac{V_1}{8}$ we use

$$P_1V_1^\gamma = P_2V_2^\gamma$$

$$\Rightarrow \quad P_1V_1^\gamma = P_2\left(\frac{V_1}{8}\right)^{5/3}$$

$$\Rightarrow \quad P_1V_1^{5/3} = \frac{P_2V_1^{5/3}}{32}$$

$$\Rightarrow \quad P_2 = 32P_1$$ **Ans. (C)**

Sol. 145 For n degrees of freedome the molar specific heats of a gas are given as

$$C_v = \frac{nR}{2}$$

and

$$C_p = \frac{(n+2)R}{2}$$

$$\Rightarrow \quad \frac{C_v}{C_p} = \frac{n}{n+2}$$ **Ans. (A)**

Sol. 146 For First case, the efficiency of Carnot engines is calculated as

$$\eta_I = 1 - \frac{100}{300} = \frac{2}{3}$$

In Second case total efficiency is calculated as

$$\eta_{II} = \eta_1 + \eta_2 - \eta_1\eta_2$$

Where

$$\eta_1 = 1 - \frac{200}{300} = \frac{1}{3}$$

and

$$\eta_2 = 1 - \frac{100}{200} = \frac{1}{2}$$

$$\Rightarrow \quad \eta_{net} = \frac{1}{3} + \frac{1}{2} - \frac{1}{6} = \frac{2}{3}$$

$$\Rightarrow \quad \eta_I = \eta_{II}$$ **Ans. (C)**

Sol. 147 Efficiency of Carnot engine is written as

$$\eta = 1 - \frac{T_L}{T_H}$$

$$\Rightarrow \quad \frac{1}{2} = 1 - \frac{T_L}{T_H}$$

$$\Rightarrow \quad \frac{1}{2}(1.3) = 1 - \left(\frac{T_L - 40}{T_H}\right)$$

$$\Rightarrow \quad \frac{1}{2}(1.3) = \frac{1}{2} + \frac{40}{T_H}$$

$$\Rightarrow \quad T_H = 266.7 \text{ K}$$ **Ans. (C)**

Sol. 148 For isobaric expansion of gas, work done is given as

$$W = nR\Delta T = 150 \text{ J}$$

Heat supplied in the process is calculated as

$$Q = \left(\frac{f}{2}+1\right)nR\Delta T = \left(\frac{8}{2}+1\right)150$$

$$\Rightarrow \quad Q = 750 \text{ J}$$ **Ans. (750)**

Sol. 149 For adiabatic process, we use

$$PV^\gamma = \text{constant}$$

Gas density is is cacluated as

$$d = \frac{m}{v}$$

$$\Rightarrow \quad p\left(\frac{m}{d}\right)^\gamma = \text{constant}$$

Given that $\frac{d_2}{d_1} = 32$, so we use

$$\frac{p_1}{p_2} = \left(\frac{d_1}{d_2}\right)^\gamma = \left(\frac{1}{32}\right)^{7/5} = \frac{1}{128}$$

$$\Rightarrow \quad \frac{T_1}{T_2} = \frac{P_1V_1}{P_2V_2} = \frac{1}{128} \times 32 = \frac{1}{4}$$ **Ans. (4)**

Sol. 150 Molar specific heat for the mixture of gases is given as

$$C_{v(\text{Mix})} = \frac{n_1 Cv_1 + n_2 Cv_2}{n_1 + n_2}$$

$$\Rightarrow \quad C_{v(\text{Mix})} = \frac{1 \cdot \frac{3R}{2} + 3 \cdot \frac{5R}{2}}{1+3}$$

$$\Rightarrow \quad C_{v(\text{Mix})} = \frac{9R}{4} = \frac{\alpha^2}{4} R$$

$$\Rightarrow \quad \alpha = 3$$

Ans. (3)

Sol. 151 Work done by gas in process DE is calculated as

$$W_{DE} = \frac{1}{2}(600+300)\,3\text{ J}$$

$$\Rightarrow \quad W_{DE} = 1350\text{ J}$$

Work done in process EF is calculated as

$$\Rightarrow \quad W_{EF} = -300 \times 3 = -900\text{ J}$$

$$\Rightarrow \quad W_{DEF} = 450\text{ J}$$

Ans. (B)

Sol. 152 Work done in the process at given pressure is calculated as

$$W = P\Delta V$$

$$W = 3 \times 10^5 \times 1600 \times 10^{-6} = 480\text{ J}$$

As only 10% of heat is used in work done so total heat supplied is given as

$$\Delta Q = 4800\text{ J}$$

Rest goes in internal energy, which is 90% of heat supplied and is equal to 4320 J. **Ans. (A)**

Sol. 153 For monoatomic gas, we use

$$\gamma_1 = \frac{5}{3}$$

For diatomic gas considering as a rigid rotator, we ignore vibrations and use

$$\gamma_2 = \frac{7}{5}$$

$$\Rightarrow \quad \frac{\gamma_1}{\gamma_2} = \frac{\frac{5}{3}}{\frac{7}{5}} = \frac{25}{21}$$

Ans. (C)

Sol. 154 For isothermal process P-V graph is rectangular hyperbola

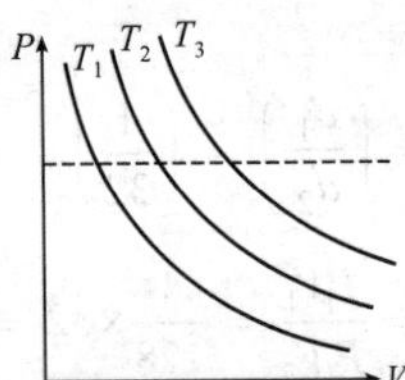

For the given graph $T_3 > T_2 > T_1$ as volume is increasing with rise in temperature. **Ans. (D)**

Sol. 155 Figure below shows the diagram of Carnot engine operating between temperature of source 600K and sink.

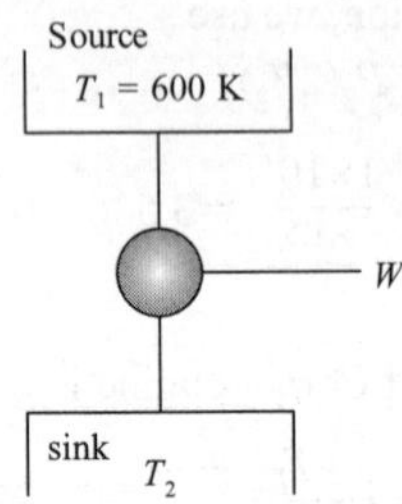

Initially the given efficiency of engine is

$$\eta = \frac{1}{2}$$

$$\Rightarrow \quad \eta = 1 - \frac{T_2}{T_1}$$

$$\Rightarrow \quad \frac{1}{2} = 1 - \frac{T_2}{600}$$

$$\Rightarrow \quad \frac{T_2}{600} = \frac{1}{2}$$

$$\Rightarrow \quad T_2 = 300\text{ K}$$

Now efficiency is increased to 70% and sink temperature will remain same so we keep $T_2 = 300$K and temperature of source will now become T, so we have

$$0.7 = 1 - \frac{300}{T}$$

$$\Rightarrow \quad \frac{300}{T} = 1 - \frac{7}{10}$$

$$\Rightarrow \quad \frac{300}{T} = \frac{3}{10}$$

$$\Rightarrow \quad T = 1000\text{ K}$$

Ans. (B)

Sol. 156 (a) In isothermal process the temperature of system remain constant so no change in internal energy take place hence option (ii) matches with this.

(b) In adiabatic process no heat is involved so from first law of thermodynamics, we have

$$\Delta Q = 0$$

$$\Rightarrow \quad \Delta Q = \Delta U + \Delta W$$

$$\Rightarrow \quad \Delta U = -\Delta W$$

Thus work done by gas is by its internal energy so it decreases hence option (i) matches with this.

(c) In Isochoric process as volume remain constant no work is done by the gas hence option (iv) matches with this.

(d) In Isobaric process some part of heat supplied is used in raising temperature and some is used for work done hence option (iii) is matching with this. **Ans. (B)**

Sol. 157 First law of thermodynamics is based on law of conservation of energy and it can be written as

$$dQ = dU - dW$$

where dW is work done on the system. **Ans. (C)**

Sol. 158 By First law of thermodynamics, we have

$$dQ = dU + dW$$

$$dU = nC_v\, dT$$

For isothermal process we have $dU = 0$

$\Rightarrow$ $U =$ constant

Also for heat is supplied we have

$$dQ > 0$$

Hence $dW > 0$ **Ans. (D)**

Sol. 159 Both **A** and **R** are true and **R** is the correct explanation of **A**, as efficiency of carnot engine depends on temperature of hot and cold reservoir. **Ans. (D)**

Sol. 160 As it is given that

$$\Delta Q = 0$$

$\Rightarrow$ $\Delta U = -W$

Work done by gas is calculated as

$$W = \int PdV = \text{Area under the } PV \text{ curve}$$

$\Rightarrow$ $\Delta U = -W = 30 \times 10^3 \times 150 \times 10^{-6}$ J

$\Rightarrow$ $\Delta U = 4500 \times 10^{-3}$ J

$\Rightarrow$ $\Delta U = 4.5$ J **Ans. (D)**

Sol. 161 For any gaseous system or a gas γ is independent of temperature. **Ans. (B)**

Sol. 162 Let γ be the ratio of $\frac{C_p}{C_v}$ so for adiabatic process we use

$$PV^{\gamma} = \text{Constant}$$

$\Rightarrow$ $$\frac{P_i}{P_f} = \left(\frac{V_f}{V_i}\right)^{\gamma}$$

$\Rightarrow$ $$\frac{81}{16} = \left(\frac{27}{8}\right)^{\lambda}$$

$\Rightarrow$ $$\gamma = \frac{4}{3}$$ **Ans. (A)**

Sol. 163 For adiabatic expansion, we use

$$T_1V_1^{\gamma-1} = T_2V_2^{\gamma-1}$$

$\Rightarrow$ $$TV^{1/2} = T_2(2V)^{1/2}$$

$\Rightarrow$ $$T_2 = \frac{T}{\sqrt{2}}$$

Work done in adiabatic expansion is calculated as

$$W = \frac{R(T_1 - T_2)}{\gamma - 1} = \frac{R\left(T - \frac{T}{\sqrt{2}}\right)}{\frac{1}{2}}$$

$\Rightarrow$ $$W = RT(2 - \sqrt{2})$$ **Ans. (D)**

Sol. 164 Temperature of hot reservoir is given as

$$T_H = 99°C = 99 + 273 = 372 \text{ K}$$

Efficiency of carnot engine is calculated as

$$1 - \frac{T_C}{T_H} = \frac{1}{3}$$

$\Rightarrow$ $$\frac{T_C}{T_H} = \frac{2}{3} \quad \ldots(1)$$

$\Rightarrow$ $$T_C = \frac{2}{3} \times 372 = 2 \times 124 = 248 \text{ K}$$

When temperature of cold reservoir is increases by x then we use new temperature to be $T_C + x$. The efficiency now becomes

$$1 - \frac{T_C + x}{T_H} = \frac{1}{6}$$

$\Rightarrow$ $$\frac{5}{6} = \frac{T_C + x}{T_H}$$

$\Rightarrow$ $$\frac{5}{6} = \frac{248 + x}{372}$$

$\Rightarrow$ $$248 + x = 5 \times 62$$

$\Rightarrow$ $$x = 310 - 248 = 62 \text{ K}$$ **Ans. (D)**

Sol. 165 By first law of thermodynamics

$$dQ = dU + dW$$

$\Rightarrow$ $$dU = dQ - dW$$

By differentiating with respect to time, we have

$$\frac{dU}{dt} = \frac{dQ}{dt} - \frac{dW}{dt}$$

Rate of heat supplied to system is $\frac{dQ}{dt} = 1000$ W

Rate of work done is $\frac{dW}{dt} = 200$ W

$\Rightarrow$ $$\frac{dU}{dt} = 1000 - 200 = 800 \text{ W}$$ **Ans. (D)**

Sol. 166 Statement-I: $\Delta Q > 0$

According to 1st law of thermodynamics

$$\Delta Q = \Delta U + W$$

If $\Delta Q > 0$, $\Delta U < 0$ then $W > 0$ is also possible.

Hence $\Delta T < 0$, so T decreases hence statement-I is false

Statement-II: $W > 0$

$\Rightarrow$ $$\int Pdv > 0$$

Therefore volume of system must increase during positive work done by the system hence statement-II is true. **Ans. (D)**

Sol. 167 Figure below shows the block diagram of the heat engine as described in the question.

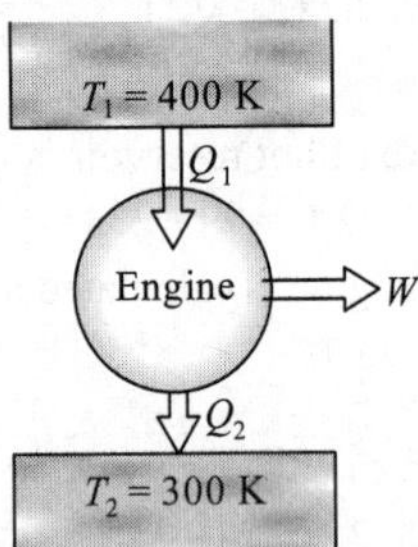

Efficiency of carnot engine when T_1 is temperature of source and T_2 is temperature of sink, is given as

$$\eta = 1 - \frac{T_2}{T_1} = \frac{W}{Q_1}$$

Here W is the workdone by engine and Q_1 is the heat supplied to engine.

$$\Rightarrow \quad \frac{W}{Q_1} = 1 - \frac{300}{400} = \frac{1}{4}$$

$$\Rightarrow \quad \frac{2\text{kJ}}{Q_1} = \frac{1}{4}$$

$$\Rightarrow \quad Q_1 = 8\,\text{kJ}$$ **Ans. (C)**

Sol. 168 For an isothermal process, T = constant so by ideal gas equation we have

$$PV = nRT = \text{constant}$$

$$\Rightarrow \quad P_1V_1 = P_2V_2$$

Gas B is compressed adiabatically to $\frac{1}{8}^{\text{th}}$ of its initial volume so

we use

$$PV = P_A(V/8)$$

$$\Rightarrow \quad P_A = 8P$$

For an adiabatic process we use PV^γ = constant and γ for monoatomic gas is $\frac{5}{3}$ so we use

$$P_1V_1^\gamma = P_2V_2^\gamma$$

$$\Rightarrow \quad \frac{P_B}{P} = \left(\frac{V_1}{V_2}\right)^\gamma = \left(\frac{V}{V/8}\right)^{\frac{5}{3}}$$

$$P_B = 32P$$

$$\Rightarrow \quad \frac{P_B}{P_A} = \frac{32P}{8P} = 4$$ **Ans. (D)**

Sol. 169 In adiabatic compression, we use

$$\Delta Q = 0$$

By I law of thermodynamics, we have

$$\Delta Q = \Delta U + \Delta W$$

$$\Rightarrow \quad \Delta U = -\Delta W$$

In adiabatic compression volume decreases so we have

$$\Delta W = -\text{ve}$$

$$\Rightarrow \quad \Delta U = +\text{ve}$$

$$\Rightarrow \quad \Delta U \neq 0$$

So there will be change in internal energy. **Ans. (D)**

Sol. 170 By I[st] law of thermodynamics we have

$$Q = mL_V$$

where L_V is the Latent heat of vapourization. Here work is calculated as

$$W = P\Delta V$$

and $\Delta Q = \Delta U + \Delta W$

Thus change in internal energy is given as

$$\Delta U = \Delta Q - \Delta W$$

$$\Delta U = mL_V - P\Delta V$$

$$\Rightarrow \quad \Delta U = (1\text{ kg})(2257 \times 10^3 \text{ J/kg}) - (1 \times 10^5 \text{ Pa})(1.671 \text{ m}^3 - 1 \times 10^{-3} \text{ m}^3)$$

$$\Rightarrow \quad \Delta U = 2275 \times 10^3 \text{ J} - 167 \times 10^3 \text{ J}$$

$$\Rightarrow \quad \Delta U = 2090 \text{ kJ}$$ **Ans. (A)**

Sol. Internal energy remain constant in isothermal process. **Ans. (B)**

Sol. 172 Efficiency of engine is given as

$$\eta = 1 - \frac{T_2}{T_1}$$

Using $T_1 = 373$ K, $T_2 = 273$ K, we have

$$\eta = \left(1 - \frac{273}{373}\right) \times 100 = 26.8\%$$ **Ans. (C)**

Sol. 173 As gas is suddenly compressed, the processes is adiabatic.

Equation of gas for adiabatic process is PV^γ = constant so we have

$$P_1V_1^\gamma = P_2V_2^\gamma$$

$$\Rightarrow \quad P_0V_0^\gamma = P_2\left(\frac{V_0}{4}\right)^\gamma$$

$$\Rightarrow \quad P_2 = P_0(4)^\gamma$$ **Ans. (B)**

$\Rightarrow$ **Sol. 174** Work done is calculated by the area of the loop, given as

$$\Rightarrow \quad W = +\frac{1}{2}(2) \times 300$$

$$\Rightarrow \quad W = 300 \text{ J}$$ **Ans. (B)**

12.1 Kinetic Theory of Gases

Sol. 1

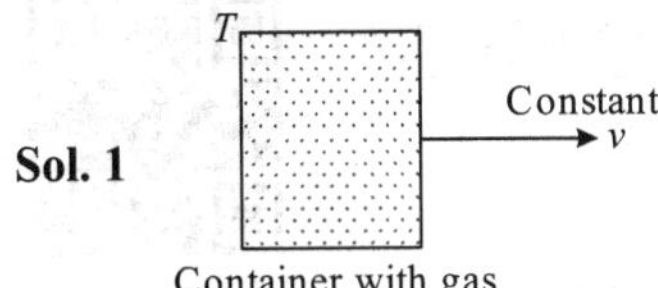

Temperature of gas is a measure of *KE* of random motion of gas molecules

$$E=\frac{f}{2}kT$$

where $f\rightarrow$ degrees of freedom

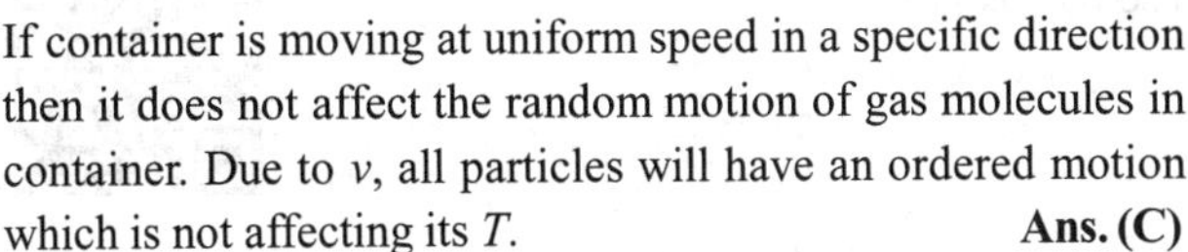

If container is moving at uniform speed in a specific direction then it does not affect the random motion of gas molecules in container. Due to v, all particles will have an ordered motion which is not affecting its T. **Ans. (C)**

Sol. 2 $T_{O_2}=273+47=320\text{ K}$

RMS velocity of molecules is given as

$$v=\sqrt{\frac{3RT}{M}}$$

If at T_{H_2} the velocity (*RMS*) of H_2 molecule is equal to that of O_2 molecule at 320 K

$$\Rightarrow \quad v_{H_2}=v_{O_2}$$

$$\Rightarrow \quad \sqrt{\frac{3RT_{H_2}}{2\times10^{-3}}}=\sqrt{\frac{3RT_{O_2}}{32\times10^{-3}}}$$

$$\Rightarrow \quad T_{H_2}=\frac{2}{32}\times320\text{K}=20\text{ K}$$ **Ans. (D)**

Sol. 3 The equivalent value of γ for a mixture of gas is given as

$$\frac{n_1+n_2}{\gamma_{eq}-1}=\frac{n_1}{\gamma_1-1}+\frac{n_2}{\gamma_2-1}$$

$$\Rightarrow \quad \frac{1+1}{\gamma_{eq}-1}=\frac{1}{\frac{7}{5}-1}+\frac{1}{\frac{5}{3}-1}$$

$$\Rightarrow \quad \frac{2}{\gamma_{eq}-1}=\frac{5}{2}+\frac{3}{2}=4$$

$$\Rightarrow \quad \gamma_{eq}-1=\frac{1}{2}$$

$$\Rightarrow \quad \gamma_{eq}=\frac{3}{2}=\frac{24}{16}$$ **Ans. (C)**

Sol. 4

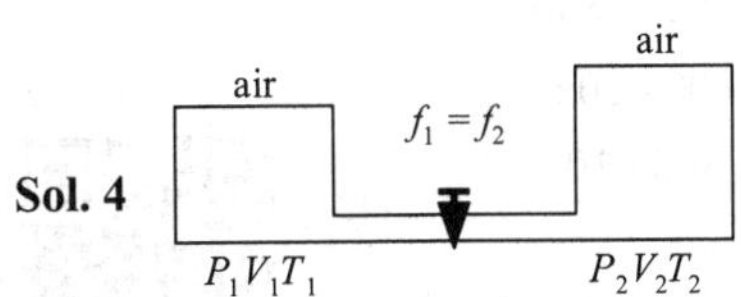

No of moles of gas (air) in the two vessels are

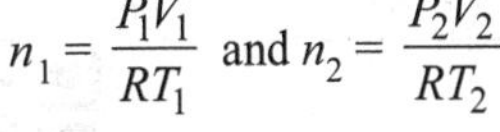

$$n_1=\frac{P_1V_1}{RT_1} \text{ and } n_2=\frac{P_2V_2}{RT_2}$$

By conservation of internal energy final equilibrium temperature of mixture is given as

$$T_f=\frac{f_1n_1T_1+f_2n_2T_2}{f_1n_1+f_2n_2}=\frac{n_1T_1+n_2T_2}{n_1+n_2}$$

$$\Rightarrow \quad T_f=\frac{P_1V_1+P_2V_2}{\frac{P_1V_1}{T_1}+\frac{P_2V_2}{T_2}}=\frac{T_1T_2(P_1V_1+P_2V_2)}{P_1V_1T_2+P_2V_2T_1}$$ **Ans. (C)**

Sol. 5 For equivalent γ of mixture of gases we use

$$\frac{n_1+n_2}{\gamma_{eq}-1}=\frac{n_1}{\gamma_1-1}+\frac{n_2}{\gamma_2-1}$$

$$\Rightarrow \quad \frac{1+1}{\gamma_{eq}-1}=\frac{1}{\frac{5}{3}-1}+\frac{1}{\frac{7}{5}-1}$$

$$\Rightarrow \quad \frac{2}{\gamma_{eq}-1}=\frac{3}{2}+\frac{5}{2}=4$$

$$\Rightarrow \quad \gamma_{eq}-1=\frac{1}{2} \Rightarrow \gamma_{eq}=\frac{3}{2}$$ **Ans. (A)**

Sol. 6 Equivalent γ for a gaseous mixture is given as

$$\frac{n_1+n_2}{\gamma_{eq}-1}=\frac{n_1}{\gamma_1-1}+\frac{n_2}{\gamma_2-1}$$

for H_e gas $\quad n_1=\frac{m_1}{M_1}=\frac{16}{4}=4\text{ mole}$

for O_2 gas $\quad n_2=\frac{m_2}{M_2}=\frac{16}{32}=0.5\text{ mole}$

$$\Rightarrow \quad \frac{4.5}{\gamma_{eq}-1}=\frac{4}{\frac{5}{3}-1}+\frac{0.5}{\frac{7}{5}-1}=6+\frac{2.5}{2}=7.25$$

$$\Rightarrow \quad \gamma_{eq}-1=\frac{4.5}{7.25} \Rightarrow \gamma_{eq}=1.62$$ **Ans. (B)**

Sol. 7 When heat flows between boxes, total internal energy of the two gases will remain constant so for this cases the final equilibrium temperature of gases is given as

$$T_{eq}=\frac{f_1n_1T_1+f_2n_2T_2}{f_1n_1+f_2n_2}$$

for $N_2\rightarrow \begin{bmatrix} n_1=1 \\ f_1=5 \\ T_1=T_0 \end{bmatrix}$ and for $He\rightarrow \begin{bmatrix} n_2=1 \\ f_2=3 \\ T_2=\frac{7}{3}T_0 \end{bmatrix}$

$$\Rightarrow \quad T_{eq} = \frac{5\times1\times T_0 + 3\times1\times(7/3)T_0}{5\times1+3\times1}$$

$$\Rightarrow \quad T_{eq} = \frac{12T_0}{8} = \frac{3}{2}T_0 \quad \textbf{Ans. (C)}$$

Sol. 8 C'_P and C'_V are the specific heats for an ideal gas per unit mole then we use the relation

$$C'_P - C'_V = R \qquad \ldots(1)$$

C_P and C_V per unit mass for N_2 gas can be given as

$$M = 28\text{gm}$$

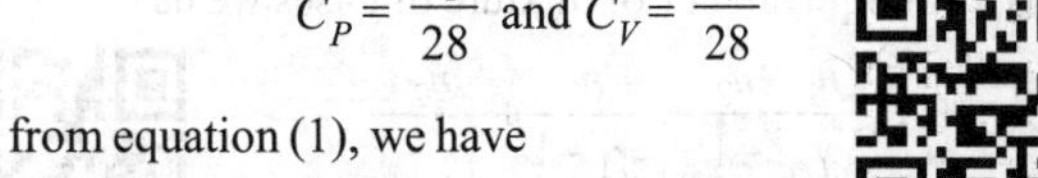

$$\Rightarrow \quad C_P = \frac{C'_P}{28} \text{ and } C_V = \frac{C'_V}{28}$$

$\Rightarrow$ from equation (1), we have

$$28C_P - 28C_V = R$$

$$\Rightarrow \quad C_P - C_V = \frac{R}{28} \qquad \textbf{Ans. (A)}$$

Sol. 9

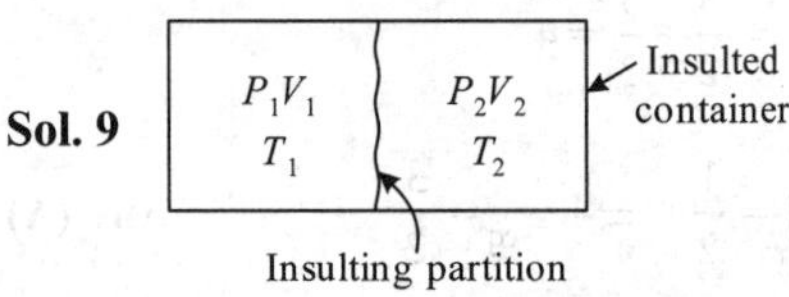

$$f_1 = f_2$$

$$n_1 = \frac{P_1V_1}{RT_1} \quad \text{and} \quad n_2 = \frac{P_2V_2}{RT_2}$$

When gases are mixed after removing partition then by internal energy conservation final equation temperature mixture is given as

$$T_f = \frac{f_1n_1T_1 + f_2n_2T_2}{f_1n_1 + f_2n_2} = \frac{n_1T_1 + n_2T_2}{n_1 + n_2}$$

$$\Rightarrow \quad T_f = \frac{T_1T_2(P_1V_1 + P_2V_2)}{P_1V_1T_2 + P_2V_2T_1} \qquad \textbf{Ans. (A)}$$

Sol. 10 Energy due to thermal motion $\rightarrow$ internal energy of gas (energy of rendom motion)

For a given sample of gas its internal energy is given as

$$U = \frac{f}{2}nRT$$

For a diatomic gas $f = 5$ and here gas volume

$$V = \frac{m}{\rho} = \frac{1}{4}m^3$$

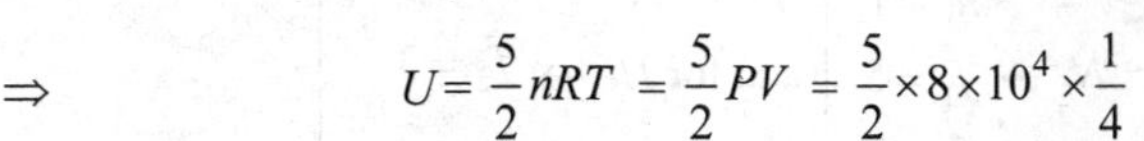

$$\Rightarrow \quad U = \frac{5}{2}nRT = \frac{5}{2}PV = \frac{5}{2}\times 8\times10^4\times\frac{1}{4}$$

$$\Rightarrow \quad U = 5\times10^4 \text{ J} \qquad \textbf{Ans. (D)}$$

Sol. 11 For gaseous mixture, final equilibrium temperature is given as

$$T_{eq} = \frac{f_1n_1T_1 + f_2n_2T_2 + f_3n_3T_3}{f_1n_1 + f_2n_2 + f_3n_3}$$

We consider $f_1 = f_2 = f_3$

As no of moles α no of molecules

$$\Rightarrow \quad T_{eq} = \frac{n_1T_1 + n_2T_2 + n_3T_3}{n_1 + n_2 + n_3} \qquad \textbf{Ans. (D)}$$

Sol. 12 As gas pressure is constant so we use by gas law

$$\frac{V_1}{T_1} = \frac{V_2}{T_2}$$

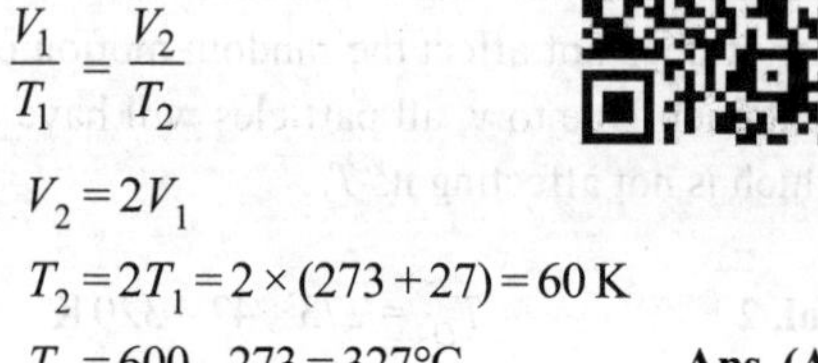

here $\quad V_2 = 2V_1$

$$\Rightarrow \quad T_2 = 2T_1 = 2\times(273+27) = 60\text{ K}$$

$$\Rightarrow \quad T_2 = 600 - 273 = 327°\text{C} \qquad \textbf{Ans. (A)}$$

Sol. 13 Initial situation

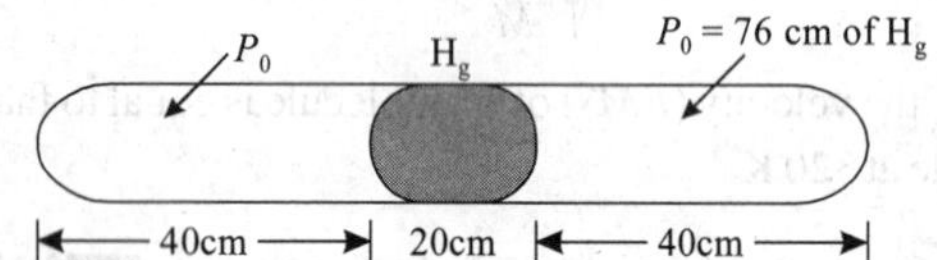

Final situation

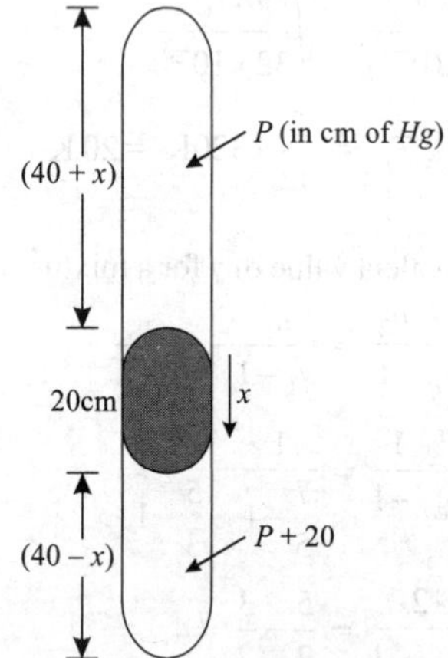

we consider constant temperature, so we have

upper section $\rightarrow P_0\times40 = P\times(40+x) \qquad \ldots(1)$

lower section $\rightarrow P_0\times40 = (P+20)\times(40-x) \qquad \ldots(2)$

from (1) and (2)

$$40P + P_x = 40P + 800 - P_x = 20x$$

$$\Rightarrow \quad P_x = 400 - 10x$$

$$\Rightarrow \quad P = \frac{400-10x}{x} \qquad \ldots(3)$$

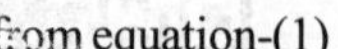

from equation-(1)

$$76 \times 40 = \frac{(400-10x)}{x}(40+x)$$

$\Rightarrow$ $$3040x = 1600 - 400x + 400x - 10x^2$$

$\Rightarrow$ $x^2 + 304x - 1600 = 0$

$\Rightarrow$ $$x = \frac{-304+\sqrt{(304)^2 - 6400}}{2} \simeq 5.17 \text{ cm.}$$

Ans. (B)

Sol. 14 For an isobaric process heat supplied is related to the rise in temperature as

$$Q = nC_P\Delta T$$

Slope of $Q - \Delta T$ curve $= \dfrac{Q}{\Delta T} = nC_P$

Here all gases are having same initial states $(P_1, V_1, T_1) \rightarrow$ same

$\Rightarrow$ Same have equal no of moles.

For a gas $C_P = \dfrac{f+2}{2}$

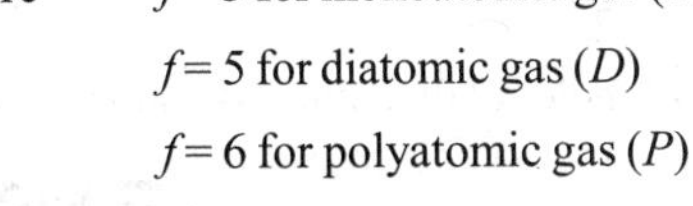

Where $f = 3$ for monoatomic gas (M)

$f = 5$ for diatomic gas (D)

$f = 6$ for polyatomic gas (P)

Maximum linear molecule

As $C_P(P) > C_P(D) > C_P(M)$ **Ans. (C)**

Sol. 15 Statement-1 : It is false because in free expansion temperature do not change as $W = 0$.

Statement-2 : True as for ideal gas we neglect PE as particles have no interaction but in real gas PE is there **Ans. (A)**

Sol. 16

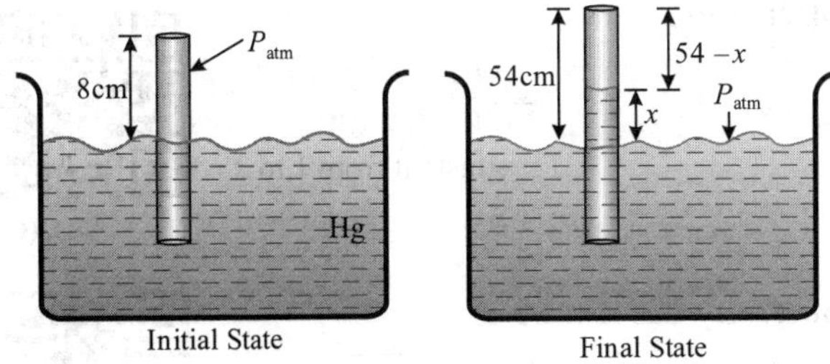

Initial State Final State

As total no of moles of air trapped in tube is constant we use

$$P_1V_1 = P_2V_2$$

$\Rightarrow$ $$76 \times 8 = (76 - x) \times (54 - x)$$

$\Rightarrow$ $$76 \times 8 = 76 \times 54 - 130x + x^2$$

$\Rightarrow$ $x^2 - 130x + 76 \times 46 = 0$

$\Rightarrow$ $$x = \frac{130 \pm \sqrt{(130)^2 - 4 \times 76 \times 46}}{2}$$

$\Rightarrow$ $$x = \frac{130-54}{2} = \frac{76}{2} = 38 \text{ cm}$$

Length of air column in final state

$$l = 54 - x = 54 - 38 = 16 \text{ cm}$$ **Ans. (A)**

Sol. 17 If there are N molecules in a vessel of volume V then mean distance between molecules in the vessel is given as

$$d_m = \left(\frac{V}{N}\right)^{1/3} \quad \ldots(1)$$

Using gas law, we have

$$PV = nRT$$

$\Rightarrow$ $$PV = \frac{N}{N_{AV}} \cdot RT$$

$\Rightarrow$ $$\frac{V}{N} = \frac{RT}{PN_{AV}} = \frac{8.0 \times 300}{4 \times 10^{-10} \times 6 \times 10^{23}}$$

$$= \frac{100}{10^{12}} = 10^{-11} \text{ m}^3\text{/molecule}$$

$\Rightarrow$ from equation-(1)

$$d_m = (10^{-11})^{1/2} = (10 \times 10^{-12})^{1/3}$$

$\Rightarrow$ $$d_m = 2.1 \times 10^{-4} \text{ m} = 0.21 \text{ mm.}$$ **Ans. (B)**

Sol. 18 *RMS* speed of a gas molecule is given as

$$v_{rms} = \sqrt{\frac{3RT}{M}}$$

If we consider room temperature to be at $\sim 27°C = 300$ K.

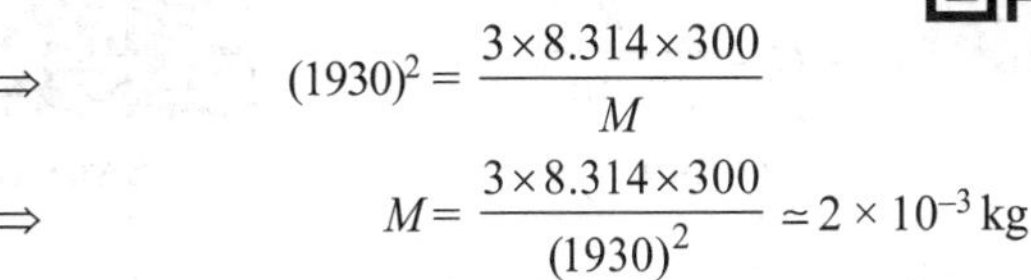

$\Rightarrow$ $$(1930)^2 = \frac{3 \times 8.314 \times 300}{M}$$

$\Rightarrow$ $$M = \frac{3 \times 8.314 \times 300}{(1930)^2} \simeq 2 \times 10^{-3} \text{ kg}$$

Thus gas is H_2 **Ans. (A)**

Sol. 19 The (Translational) *KE* of a gas molecule at temperature T is given as

$$KE = \frac{3}{2}k_BT$$

at $$T = 0°C = 273 \text{ K}$$

$\Rightarrow$ $$KE = \frac{3}{2}k_B \times 273 = \frac{819}{2}k_B$$

If it rises to a height h above the earth surface

$\Rightarrow$ $$KE = Mgh = \frac{819}{2}k_B$$

[As $h << R_e$ so we consider g = constant]

$\Rightarrow$ $$h = \frac{819k_B}{2Mg}$$ **Ans. (D)**

Sol. 20 For atoms in a solid lattice we take 6 vibrational degrees of freedom ($KE + PE$)

Then molar specific heat of Al at constant volume is given as

$$C_v = \frac{f}{2}R = \frac{6}{2}R = 3\text{R J/mol.K}$$

The specific heat in J/kg-K can be given as

$$s = \frac{C_v}{M} = \frac{3R}{M} = \frac{3 \times 8.314}{27 \times 10^{-3}} \text{ J/kg-K}$$

$$\Rightarrow \quad s \simeq 925 \text{ J/kg-K}$$ **Ans. (D)**

Sol. 21 The pressure exerted by gas molecules on container walls is

$$P = \frac{1}{3}\rho v_{rms}^2 \qquad \left[\text{We use } v_{\text{rms}} = \sqrt{\frac{3RT}{M}}\right]$$

$$\Rightarrow \quad P = \frac{\rho RT}{M} \Rightarrow P \,\alpha\, T$$

[for a closed container $\rho \to$ constant]

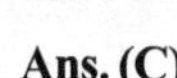

$\Rightarrow$ Force on walls α pressure by gas

Force $\alpha\, T^{(1)}$

$$\Rightarrow \quad q = 1$$ **Ans. (C)**

Sol. 22 If C_p and C_v are molar specific heats (J/mol-k) then we use for all gases

$$C_p - C_v = R = 8.314 \text{ J/mol-k}$$

for C_p and C_v taken as specific heats (J/kg-k) we use

for H_2 gas $\quad C_p - C_v = \frac{R}{M_{H_2}} = \frac{R}{2 \times 10^{-3}} = a$

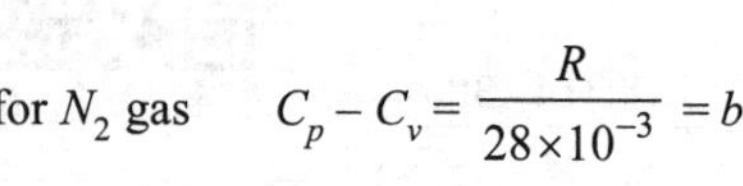

for N_2 gas $\quad C_p - C_v = \frac{R}{28 \times 10^{-3}} = b$

$$\Rightarrow \quad a = 14b$$ **Ans. (A)**

Sol. 23 From Gas law we use no of molecules in room are given as

$$N = n.A_v = \left(\frac{PV}{RT}\right) \cdot Av$$

Initial molecules in room

$$n_i = \frac{PVA_v}{RT_i} \quad (\text{where } T_i = 273 + 17 = 290 \text{ k})$$

Final molecules in room

$$n_f = \frac{PVA_v}{RT_f} \quad (\text{where } T_f = 273 + 27 = 300 \text{ k})$$

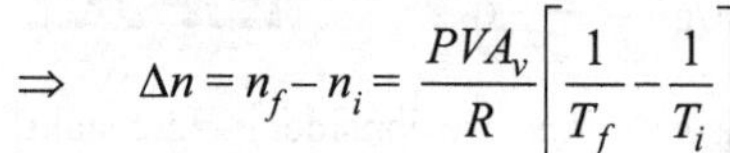

$$\Rightarrow \quad \Delta n = n_f - n_i = \frac{PVA_v}{R}\left[\frac{1}{T_f} - \frac{1}{T_i}\right]$$

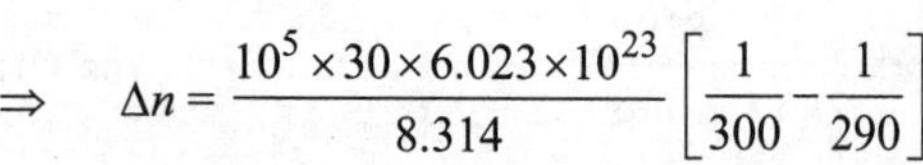

$$\Rightarrow \quad \Delta n = \frac{10^5 \times 30 \times 6.023 \times 10^{23}}{8.314}\left[\frac{1}{300} - \frac{1}{290}\right]$$

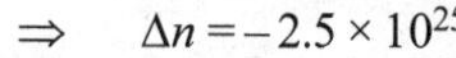

$$\Rightarrow \quad \Delta n = -2.5 \times 10^{25}$$ **Ans. (D)**

Sol. 24 For a gas molecule having f degrees of freedom, we use

$$C_v = \frac{f}{2}R \quad \text{and} \quad C_p = \frac{f+2}{2}R$$

for $f = 5$ we have

$$C_v = \frac{5}{2}R \text{ and } C_p = \frac{7}{2}R$$

$$\Rightarrow \quad \gamma = \frac{C_p}{C_v} = \frac{7}{5}$$ **Ans. (D)**

Sol. 25 The total KE of a gas having n moles is given as

$$U = \frac{f}{2}nRT$$

[where $f \to$ degrees of freedom]

for diatomic gas $f = 5$

for monoatomic gas $f = 3$

Initially N moles of diatomic gas

$$\Rightarrow \quad U_i = \frac{f}{2}NRT = \frac{5}{2}NRT \quad \ldots(1)$$

Initially there are $(N - n)$ moles of diatomic gas and $2n$ moles of monoatomic gas

$$\Rightarrow \quad U_f = \frac{5}{2}(N\text{-}n)RT + \frac{3}{2}(2n)RT$$

$$\Rightarrow \quad U_f = \frac{5}{2}NRT + 3nRT - \frac{5}{2}nRT$$

$$\Rightarrow \quad U_f = \frac{5}{2}NRT + \frac{1}{2}nRT \quad \ldots(2)$$

from (1) and (2) change in total KE of gas

$$\Delta U = U_f - U_i = \frac{1}{2}nRT$$ **Ans. (A)**

Sol. 26 By Gas law

$$P = \frac{nRT}{V}$$

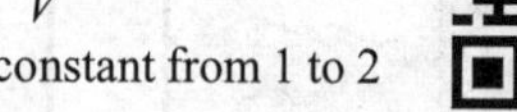

$\Rightarrow \quad T =$ constant from 1 to 2

$$P_1 > P_2$$ **Ans. (C)**

Sol. 27 By Gas law, we use

$$\frac{V_{\text{rms}}(He)}{V_{\text{rms}}(Ar)} = \sqrt{\frac{M_{Ar}}{M_{He}}}$$

$$= \sqrt{\frac{40}{4}} \cong 3.16$$ **Ans. (D)**

Sol. 28 For a monoatomic gas, energy of thermal motion

$$U = \frac{3}{2}kT \times N$$

$$\Rightarrow \quad U = \frac{3}{2}nRT = \frac{3}{2}PV \quad \left[\text{Volume } V = \frac{2}{8} = \frac{1}{4}\text{m}^3\right]$$

$$\Rightarrow \quad U = \frac{3}{2} \times 4 \times 10^4 \times \frac{1}{4}$$

$$\Rightarrow \quad U = 1.5 \times 10^4 \text{ J}$$

Order is $= 10^4$ J **Ans. (D)**

Sol. 29 Total internal energy of a gaseous mixture is given as

$$U = \frac{f_1}{2}n_1RT + \frac{f_2}{2}n_2RT$$

$$\Rightarrow \quad U = \frac{5}{2}(3RT) + \frac{3}{2} \times (5RT)$$

$$\Rightarrow \quad U = 15RT$$

Ans. (C)

Sol. 30 Energy $= \frac{f}{2}nRT = \frac{f}{2}PV$

$$\Rightarrow \quad E = \frac{f}{2}(3 \times 10^6)(2)$$

$$\Rightarrow \quad E = f \times 3 \times 10^6$$

Considering gas is monatomic for ideal case, we use

$$\Rightarrow \quad f = 3$$

$$\Rightarrow \quad E = 9 \times 10^6 \text{ J}$$

Ans. (D)

Sol. 31

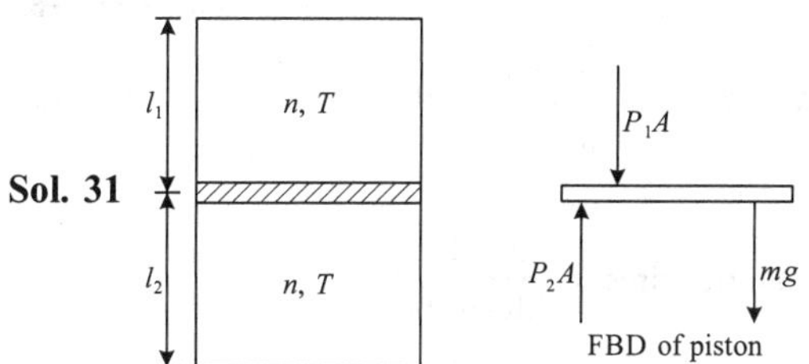

At equilibrium, we use

$$P_2A = P_1A + mg$$

$$\Rightarrow \quad \frac{nRT.A}{Al_2} = \frac{nRT.A}{Al_1} + mg$$

$$\Rightarrow \quad nRT\left(\frac{1}{l_2} - \frac{1}{l_1}\right) = mg$$

$$\Rightarrow \quad m = \frac{nRT}{g}\left(\frac{l_1 - l_2}{l_1 \cdot l_2}\right)$$

Ans (B)

Sol. 32 Mean time between two successive collisions is given as

$$t_m = \frac{\text{average separation}}{\text{mean velocity}}$$

Average separation between molecules is

$$\lambda_m = \frac{(V)^{1/3}}{N}$$

and

$$v_m = \sqrt{\frac{8RT}{\pi M}}$$

$$\Rightarrow \quad t_m = \frac{(V)^{1/3}}{N}\sqrt{\frac{\pi M}{8RT}}$$

$$\Rightarrow \quad t_m \propto \frac{T^{1/3}}{P^{1/3}\sqrt{T}}$$

$$\Rightarrow \quad t_m \propto \frac{1}{P^{1/3}}T^{1/6}$$

$$\Rightarrow \quad \frac{t_1}{t_2} = \left(\frac{P_2}{P_1}\right)^{1/3}\left(\frac{T_2}{T_1}\right)^{1/6}$$

$$\Rightarrow \quad t_2 = 6 \times 10^{-8} \times \left(\frac{1}{2}\right)^{1/3}\left(\frac{3}{5}\right)^{1/6}$$

$$\Rightarrow \quad t_2 = 6 \times 10^{-8}/1.26 \times 1.09 = 4.37 \times 10^{-8} \text{ s}$$

Ans. (A)

Sol. 33

Magnitude of change in momentum per collision $= 2mv$

$$\text{Pressure} = \frac{\text{Force}}{\text{Area}} = \frac{N(2mv)}{1}$$

$$p = \frac{10^{22} \times 2 \times 10^{-26} \times 10^4}{1}$$

$$\Rightarrow \quad p = 2 \text{ N/m}^2$$

Ans. (None)

Sol. 34 RMS velocity of hydrogen molecules is given as

$$v_{\text{rms}} = \sqrt{\frac{3RT}{M}}$$

If RMS velocity is equal to escape velocity then we use

$$\sqrt{\frac{3RT}{M}} = \sqrt{2gRe}$$

$$\Rightarrow \quad T = \frac{2gR_eM_e}{3R} = \frac{2 \times 10 \times 6.4 \times 10^6 \times 2 \times 10^{-3}}{3 \times 8.314}$$

$$\Rightarrow \quad T \simeq 10 \times 10^3 = 10^4 \; 6.4 \times 10^6$$

$$\Rightarrow \quad T = \frac{4 \times 10 \times 6.4 \times 10^6}{3 \times 1.38 \times 6.02 \times 10^3}$$

$$\Rightarrow \quad T = 10 \times 10^3 = 10^4 \text{ K}$$

Ans. (C)

Sol. 35 According to law of equipartition of energy, we use

$$\frac{1}{2}m(v_{rms}^2) = 3 \times \frac{1}{2}K_bT$$

$$\Rightarrow \quad T = \frac{m\bar{v}_{rms}^2}{3k}$$

Ans. (C)

Sol. 36 RMS speed is given as

$$v_{\text{rms}} = \sqrt{\frac{3RT}{M}}$$

$$\Rightarrow \quad v_{\text{rms}} \propto \sqrt{T}$$

For the given data, we use

$$\frac{v}{200} = \sqrt{\frac{500}{400}}$$

$$\Rightarrow \quad \frac{v}{200} = \frac{\sqrt{5}}{2}$$

$$\Rightarrow \quad v = 100\sqrt{5} \text{ m/s}$$

Ans. (B)

Sol. 37 For A, we use

$$R = C_p - C_v = 7$$

$$C_v = \frac{fR}{2} = 22$$

$$\Rightarrow \quad f = \frac{44}{7} = 6.3$$

$f \simeq 6$ → 5(Rotation + Translational), 1(Vibration)

For B, we use

$$R = C_p - C_v = 9$$

$$C_v = \frac{fR}{2} = 21$$

$$\Rightarrow \quad f = \frac{42}{9}$$

$f \simeq 5$ → 5(Rotation + Translational), 0(Vibration)

Ans. (D)

Sol. 38 Mean free path of O_2 gas

$$\lambda = \frac{RT}{\sqrt{2}\pi d^2 N_A P} = \frac{V}{\sqrt{2}\pi d^2 N_A}$$

Average collision rate is given as

$$r_c = \frac{v_{av}}{\lambda} = \frac{\sqrt{8/3\pi} \times v_{ms}}{\lambda}$$

$$\Rightarrow \quad r_c = \frac{\sqrt{8/3\pi} \times 200 \text{ m/s}}{25 \times 10^{-3}} \times \sqrt{2}\pi (2 \times 0.3 \times 10^{-9}) \times 6.023 \times 10^{23}$$

$$\Rightarrow \quad r_c = 0.18 \times 10^{10} \text{ sec}^{-1} \sim 10^{10} \text{ sec}^{-1}$$

Ans. (C)

Sol. 39 C_V of a mixture of gases is given as

$$C_{V_{eq}} = \frac{n_1 C_{V_1} + n_2 C_{V_2}}{n_1 + n_2}$$

For He, $C_{V_1} = \frac{3R}{2}$ and For H_2, $C_{V_2} = \frac{5R}{2}$

$$\Rightarrow \quad C_{V_{eq}} = \frac{2 \times 3R/2 + 3 \times 5R/2}{2 + 3}$$

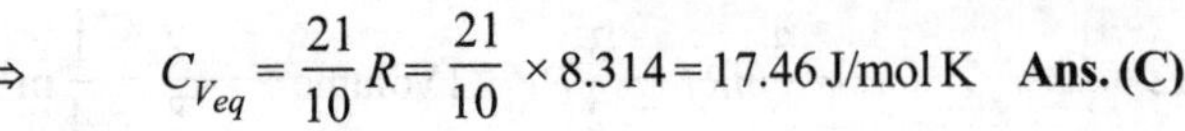

$$\Rightarrow \quad C_{V_{eq}} = \frac{21}{10}R = \frac{21}{10} \times 8.314 = 17.46 \text{ J/mol K}$$

Ans. (C)

Sol. 40

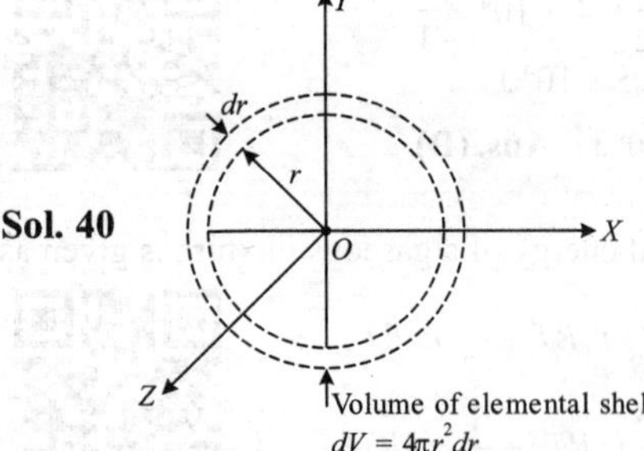

Number of molecule in the elemental shell is

$$dN = n(r) \cdot dV$$

$$\Rightarrow \quad dN = n_0 e^{-\alpha r^4} \cdot 4\pi r^2 dr$$

Total molecules

$$N = \int dN = 4\pi n_0 \int_0^\infty e^{-\alpha r^4} \cdot r^2 dr$$

Let $t = \alpha r^4 \Rightarrow dt = 4\pi r^3 dr$

$$\Rightarrow \quad r^2 dr = \frac{dt}{4\alpha r} \Rightarrow r = \frac{t^{1/4}}{\alpha^{1/4}}$$

$$\Rightarrow \quad N = 4\pi n_0 \int_0^\infty e^{-t} \cdot \frac{dt\,\alpha^{1/4}}{4\alpha t^{1/4}}$$

$$\Rightarrow \quad N = \pi n_0 \alpha^{-3/4} \int_0^\infty t^{-1/4} \cdot e^{-t} dt$$

Ans. (C)

Sol. 41 Mean free time $\propto \frac{1}{\text{Mean speed}}$

$$\Rightarrow \quad t \propto \frac{1}{\sqrt{T}}$$

Ans. (D)

Sol. 42 Mean free path is given as

$$\lambda = \frac{1}{\sqrt{2}\pi n_v d^2}$$

Mean time between successive collision is given as

$$\tau = \frac{\lambda}{v_m} = \frac{1}{\sqrt{2}\pi n_v d^2 v_m} = \frac{1}{\sqrt{2}\pi n_v d^2}\sqrt{\frac{\pi M}{8RT}}$$

For two gases, we use

$$\frac{\tau_1}{\tau_2} = \sqrt{\frac{M_1}{M_2}} \frac{d_2^2}{d_1^2}$$

$$\Rightarrow \quad \frac{\tau_1}{\tau_2} = \sqrt{\frac{40}{140}} \frac{(0.1)^2}{(0.07)^2} = 1.09$$

Thus most appropriate answer among given option is (C).

Ans. (C)

Sol. 43 The mean free path of molecules of an ideal gas is given as

$$\lambda = \frac{V}{\sqrt{2}\pi d^2 N}$$

Due to rise in temperature speed of molecules increases so mean collision time between molecules decreases and mean free path is independent of temperature. **Ans. (D)**

Sol. 44 Given that at constant pressure

$$\Delta Q = nC_P\Delta T$$

$$\Rightarrow \quad 160 = n \times C_P(50) \quad \ldots(1)$$

At constant volume, we use

$$240 = nC_V(100) \quad \ldots(2)$$

from (1) & (2)

$$\gamma = \frac{C_P}{C_V} = \frac{160/50}{240/100} = \frac{4}{3}$$

As gas has $\gamma = \frac{4}{3}$

$$\Rightarrow \quad \gamma = \frac{f+2}{f}$$

$$\Rightarrow \quad f = 6$$ **Ans. (C)**

Sol. 45 For monoatomic molecule $\to f = 3$

$$\Rightarrow \quad \gamma = \frac{f+2}{f} = \frac{5}{3}$$

For Diatomic rigid molecules $\to f = 5$

$$\Rightarrow \quad \gamma = \frac{7}{5}$$

For Diatomic non rigid molecules $\to f = 7$

$$\Rightarrow \quad \gamma = \frac{9}{7}$$

For Triatomic rigid molecules $\to f = 6$

$$\Rightarrow \quad \gamma = \frac{8}{6} = \frac{4}{3}$$ **Ans. (A)**

Sol. 46 RMS speed of gas molecule is given as

$$V_{\text{rms}} = \sqrt{\frac{3RT}{M}}$$

Given that

$$V_{N_2} = V_{H_2}$$

$$\Rightarrow \quad \sqrt{\frac{3RT_{N_2}}{M_{N_2}}} = \sqrt{\frac{3RT_{H_2}}{M_{H_2}}}$$

$$\Rightarrow \quad \frac{573}{28} = \frac{T_{H_2}}{2}$$

$$\Rightarrow \quad T_{H_2} = 40.928\,\text{K}$$ **Ans. (41)**

Sol. 47 By gas law we use

$$P_1V_1 = nR(250) \quad \ldots(1)$$

If $\frac{n}{4}$ mole molecules gets dissociated then final number of moles will become $n + \frac{n}{4} = \frac{5n}{4}$ so we use again by gas law

$$P_2(2V_1) = \frac{5n}{4}R(2000) \quad \ldots(2)$$

From (1) and (2), we have

$$\frac{P_1}{2P_2} = \frac{4\times 250}{5\times 2000}$$

$$\Rightarrow \quad \frac{P_1}{P_2} = \frac{1}{5} \text{ or } \frac{P_2}{P_1} = 5$$ **Ans. (5)**

Sol. 48 As mean speed of molecules is $v_m \propto \sqrt{T}$ mean time between collisions is given as

$$t_m = \frac{\lambda}{v_m}$$

$$\Rightarrow \quad t_m \propto \frac{1}{\sqrt{T}}$$ **Ans. (C)**

Sol. 49 From the assumption of Kinetic Theory, the molecules of gas collide with the container walls and continuously suffer momentum change which causes the force on the wall and hence pressure. **Ans. (D)**

Sol. 50 Root mean square speed of molecules of the gas is given as

$$v_{\text{rms}} = \sqrt{\frac{3RT}{M}}$$

$$\Rightarrow \quad v_{\text{rms}} \propto \sqrt{T}$$

$$\Rightarrow \quad \frac{(v_{rms})_2}{(v_{rms})_1} = \sqrt{\frac{T_2}{T_1}} = \sqrt{\frac{400}{300}} = \frac{2}{\sqrt{3}}$$

$$\Rightarrow \quad (v_{\text{rms}})_2 = \frac{2}{\sqrt{3}}(v_{\text{rms}})_1 = \frac{2}{\sqrt{3}}\times 200$$

$$\Rightarrow \quad (v_{\text{rms}})_2 = \frac{400}{\sqrt{3}}\,\text{m/s}$$ **Ans. (400)**

Sol. 51 In a diatomic molecule, translational degree of freedom are 3 and rotational degree of freedom are 2 so statement II is false. Statement I is by definition true. **Ans. (C)**

Sol. 52 Gas law for the mixture of gases is written as

$$PV = (n_1 + n_2 + n_3)RT$$

$$\Rightarrow \quad PV = \left[\frac{16}{32} + \frac{28}{28} + \frac{44}{44}\right]RT$$

$$\Rightarrow \qquad PV=\left[\frac{1}{2}+1+1\right]RT$$

$$\Rightarrow \qquad P=\frac{5}{2}\frac{RT}{V}$$ **Ans. (C)**

Sol. 53 Mean free path at temperature T for a gas is given as

$$\lambda=\frac{RT}{\sqrt{2}\pi d^2 N_A P}$$

$$\Rightarrow \qquad \lambda=102\text{ nm}$$ **Ans. (D)**

Sol. 54 If the final temperature of the gaseous mixture is T, by conservation of internal energy we use

$$\frac{f_1}{2}n_1R\,\Delta T+\frac{f_2}{2}n_2R\Delta T=0$$

$$\Rightarrow \quad \frac{f_1}{2}n_1R(T_1-T)+\frac{f_2}{2}n_2R(T_2-T)=0$$

$$\Rightarrow \quad T=\frac{f_1n_1RT_1+f_2n_2RT_2}{f_1n_1R+f_2n_2R}=\frac{f_1n_1T_1+f_2n_2T_2}{f_1n_1+f_2n_2}$$ **Ans. (B)**

Sol. 55 For two vibrational modes of gas molecule it has 4 vibrational degress of freedom and assuming non linear molecule, total degress of freedom can be given as

$$f=4+3+3=10$$

$$\Rightarrow \qquad \beta=\frac{C_p}{C_v}=1+\frac{2}{f}=\frac{12}{10}=1.2$$ **Ans. (B)**

Sol. 56 According to law of equipartition of energy, the energy associated with each degree of freedom per molecule is $\frac{1}{2}k_BT$.

Ans. (A)

Sol. 57 Conceptually it can be stated that at molecular level both the gas molecules (identical) get mixed and total number of possible states available for molecular motion will increase so final entropy of the system will be S_1+S_2. **Ans. (D)**

Sol. 58 Root mean square velocity of gas molecules is given as

$$v_{\text{rms}}=\sqrt{\frac{3RT}{M}}$$

Average speed of gas molecules is given as

$$v_{\text{avg}}=\sqrt{\frac{8}{\pi}\frac{RT}{M}}$$

$$\Rightarrow \qquad \frac{v_{\text{rms}}}{v_{\text{avg}}}=\sqrt{\frac{3\pi}{8}}$$ **Ans. (C)**

Sol. 59 Root mean square speed of gas molecules is given as

$$v_{\text{rms}}=\sqrt{\frac{3RT}{m}}$$

as it is inversely proportional to the square root of mass of molecules, option (D) is correct. **Ans. (D)**

Sol. 60 For a gas, its specific heats at constant volume and at constant pressure are given in terms of degrees of freedom as

$$C_V=\frac{fR}{2}$$

and $$C_P=\left(\frac{f}{2}+1\right)R$$

$$\Rightarrow \qquad \gamma=\frac{C_P}{C_V}=1+\frac{2}{f}$$

$$\Rightarrow \qquad f=\frac{2}{\gamma-1}$$ **Ans. (D)**

Sol. 61 For an ideal gas, by gas law, we have

$$PV=nRT$$

$$\Rightarrow \qquad PV\propto T$$ **Ans. (B)**

Sol. 62 By law of equipartition of energy the average kinetic energy for a molecule having 3 degrees of freedom (monoatomic gas) is given as

$$E=\frac{f}{2}k_BT=\frac{3}{2}k_BT$$ **Ans. (B)**

Sol. 63 For an ideal gas the difference of specific heats is R and as gas behaviour deviates from ideal the difference also deviates away from R. At very high temperature real gases also behave like ideal gases hence in this case state P can be considered at higher temperature. **Ans. (A)**

Sol. 64 Mean free path $=\frac{1}{\sqrt{2}d^2n}$

$$\frac{x_1}{x_2}=\left(\frac{d_2}{d_1}\right)^2$$

$$\left(\frac{1}{2}\right)^2=0.25=25\times10^{-2}$$ **Ans. (25)**

Sol. 65 Kinetic energy of a gas molecule at absolute temperature T is given as

$$K=\frac{3}{2}kT$$

By gas law, we have

$$PV=\frac{NRT}{N_A}$$

$$\Rightarrow \quad N = \frac{PV}{\left(\frac{2}{3}\right)k} = \frac{2\times10^{-3}}{2/3\times10^{-9}} = 1.5\times10^{11}$$ **Ans. (D)**

Sol. 66 RMS speed of gas molecules is given as

$$v_{rms} = \sqrt{\frac{3RT}{M}} \propto \frac{1}{\sqrt{M}}$$

As $M_{H_2} < M_{O_2} < M_{CO_2}$, we have

$$V_{H_2} > V_{O_2} > V_{CO_2}$$ **Ans. (A)**

Sol. 67 For gas A using gas law, we have

$$P_A V = n_A RT$$

$$\Rightarrow \quad P_A = \frac{(1)RT}{V}$$

For gas B, using gas law, we have

$$P_B V = n_B RT$$

$$\Rightarrow \quad P_B = \frac{(2)RT}{V}$$

The pressure in mixture of gases is given as

$$P = P_A + P_B$$

$$\Rightarrow \quad P = \frac{3RT}{V} = \frac{3\times8.314(400)}{4\times10^{-3}}$$

$$\Rightarrow \quad P = 24.9\times10^5\ \text{Pa}$$ **Ans. (C)**

Sol. 68 In floating condition of balloon, we use

$$V\rho_{air}g = mg$$

$$V\rho_{air} = 185$$

When pressure and temperature changes, if new load which the balloon can carry is m' then we use

$$V\rho_{air}g = m'g$$

$$\Rightarrow \quad \rho_{air}\left(\frac{P'}{P}\times\frac{T}{T'}\right)V = m'$$

$$\Rightarrow \quad 185\left[\frac{45}{76}\times\frac{300}{266}\right] = m'$$

$$\Rightarrow \quad m' = 123.54\ \text{kg}$$ **Ans. (D)**

Sol. 69 Root mean square speed of molecules are related to molecular moss as

$$V_{rms} \propto \frac{1}{\sqrt{M}}$$

$$\Rightarrow \quad \frac{(V_{rms})O_2}{(V_{rms})H_2} = \sqrt{\frac{m_{H_2}}{m_{O_2}}} = \sqrt{\frac{2}{32}} = \frac{1}{4}$$

$$(V_{RMS})_{H_2} = 4(V_{RMS})_{O_2} = 4\times160 = 640\ \text{m/s}$$ **Ans. (C)**

Sol. 70 Pressure variation with volume is given as

$$\frac{dp}{dV} = -aP$$

$$\Rightarrow \quad \int_{P_0}^{P}\frac{dP}{P} = -\int_0^V a\,dV$$

$$\Rightarrow \quad ln\frac{P}{P_0} = -aV$$

$$\Rightarrow \quad P = P_0e^{-aV}$$

By gas law for one mole of gas, we use

$$PV = RT$$

$$\Rightarrow \quad P + V\frac{dP}{dV} = RT$$

$$\Rightarrow \quad P + V\frac{dP}{dV} = R\frac{dT}{dV}$$

For maximum Temperature, we use

$$\frac{dT}{dV} = 0$$

$$\Rightarrow \quad P - (aP)V = 0$$

$$\Rightarrow \quad V = \frac{1}{a}$$

$$\Rightarrow \quad T = \frac{PV}{R} = \frac{P_0e^{-1}}{Ra} = \frac{P_0}{eRa}$$ **Ans. (B)**

Sol. 71 If m_1 and m_2 are the masses of oxygen and hydrogen in the misture then number of moles of O_2 are given as

$$n_1 = \frac{m_1}{32}$$

Number of moles of H_2 in mixture are given as

$$n_2 = \frac{m_2}{2}$$

By ideal gas equation, we have

$$PV = \left[\frac{m_1}{32} + \frac{m_2}{2}\right]\frac{25}{3}\times300$$

$$\Rightarrow \quad 400\times10^3\times500\times10^{-6} = \left[\frac{m_1}{32} + \frac{m_2}{2}\right]25\times100$$

$$\Rightarrow \quad 2\times10^2 = \left[\frac{m_1}{m_2} + \frac{1}{32} + \frac{1}{2}\right]m_2\times25\times100$$

$$\Rightarrow \quad \frac{2}{25} = \left[\frac{m_1}{m_2}\times\frac{1}{32} + \frac{1}{2}\right]m_2 \quad \ldots(1)$$

As the mass of mixture is 0.76 g, we use

$$m_1 + m_2 = 0.76$$

$$\Rightarrow \quad m_2\left[1 + \frac{m_1}{m_2}\right] = \frac{76}{100} \quad \ldots(2)$$

From equations-(1) and (2), we have

$$\frac{\left[\frac{m_1}{m_2}\times\frac{1}{32}+\frac{1}{2}\right]}{\left[\frac{m_1}{m_2}+1\right]}=\frac{2}{25}\times\frac{100}{76}$$

Here if we consider $\frac{m_1}{m_2}=x$, we have

$$\frac{x+16}{32(x+1)}=\frac{8}{76}$$

$\Rightarrow \quad 76x+76\times16=8\times32x+8\times32$

$\Rightarrow \quad 76x+1216=256x+256$

$\Rightarrow \quad 960=180x$

$\Rightarrow \quad x=\frac{96}{18}=\frac{16}{3}$ **Ans. (B)**

Sol. 72 Kinetic energy of each particle of either of electron or protons gases is given as

$$K=\frac{3}{2}kT$$

$\Rightarrow \quad \frac{P^2}{2m}=\frac{3}{2}kT$

$\Rightarrow \quad P=\sqrt{3kTm}$

Using Heisenberg uncertainty principle, we have

$$\Delta x.\,\Delta P=\frac{h}{4\pi}$$

$\Rightarrow \quad \Delta x=\frac{h}{4\pi\Delta P}=\frac{h}{4\pi\sqrt{3kTm}}$

$\Rightarrow \quad \Delta x\propto\frac{1}{\sqrt{m}}$

$\Rightarrow \quad \frac{\Delta x_e}{\Delta x_p}=\sqrt{\frac{m_p}{m_e}}$ **Ans. (A)**

Sol. 73 As stated in question that translation kinetic energy of N_2 molecule is equal to the kinetic energy of electron, we use

$$\frac{3}{2}kT=\text{eV}$$

$\Rightarrow \quad \frac{3}{2}\times1.38\times10^{-3}\,\text{T}=1.6\times10^{-19}\times0.1$

$\Rightarrow \quad T=773\text{ K}$

$\Rightarrow \quad T=773-273=500^\circ\text{C}$ **Ans. (500)**

Sol. 74 To increase the speed of molecule to double the initial value temperature of gas is to be increased to 4 times so from 400 K, it will be increased to 1600 K. Heat required for this purpose at constant volume is given as

$$Q=nC_v\Delta T$$

$\Rightarrow \quad Q=\frac{f}{2}nR\Delta T$

For a diatomic gas we use $f=5$, which gives

$$Q=12\text{ k cal}$$ **Ans. (12)**

Sol. 75 Root mean square speed and most probable speed of oxygen molecules are given as

$$v_{\text{rms}}=\sqrt{\frac{3RT}{M}}$$

and $\quad v_{\text{mp}}=\sqrt{\frac{2RT}{M}}$

$\Rightarrow \quad v_{\text{rms}}=\sqrt{\frac{3}{2}}v_{\text{mp}}$ **Ans. (B)**

Sol. 76 By ideal gas law, we have

$$pV=nRT$$

$\Rightarrow \quad \Delta PV=nR\Delta T$

$\Rightarrow \quad \frac{\Delta P}{P}=\frac{\Delta T}{T}=\frac{0.4}{100}$

$\Rightarrow \quad T=\frac{100\times1}{0.4}=250\text{ K}$ **Ans. (250)**

Sol. 77 For ratio of specific heats 1.4, degrees of freedom of gas are 5. As vessel is suddenly brought to rest, kinetic energy of gas molecules will be converted to its internal energy, which is given as

$$\frac{f}{2}nR\Delta T=\frac{1}{2}mv^2$$

$$\Delta T=\frac{mv^2}{fR}=\frac{Mv^2}{5R}$$ **Ans. (B)**

Sol. 78 Average kinetic energy per molecule of gas is given by law of equipartition of energy, written as

$$K=\frac{f}{2}kT$$

$\Rightarrow \quad \frac{K_{\text{Ar}}}{K_{O_2}}=\frac{\frac{3}{2}kT}{\frac{5}{2}kT}=\frac{3}{5}$ **Ans. (D)**

Sol. 79 From ideal gas law for the mixture we have

$$PV=nRT$$

Total number of moles in mixture are

$$n=\frac{100\times10^3\times2000\times10^{-6}}{\frac{25}{3}\times300}$$

$\Rightarrow \quad n=80\times10^{-3}$

$$\Rightarrow \quad n_1 + n_2 = 0.08$$

Total mass of mixture is 0.76 g, so we use

$$n_1 \times 2 + n_2 \times 32 = 0.76$$

$$\Rightarrow \quad (0.08 - n_2)2 + n_2(32) = 0.76$$

$$\Rightarrow \quad n_2 = 0.02$$

and $\quad n_1 = 0.06$

$$\Rightarrow \quad \frac{n_1}{n_2} = \frac{3}{1}$$ **Ans. (B)**

Sol. 80 From ideal gas law, we have

$$PV = nRT$$

$$\Rightarrow \quad \frac{V}{T} = \frac{nR}{P}$$

For the two gases, we have

$$\frac{nR}{P_1} < \frac{nR}{P_2}$$

$$\Rightarrow \quad P_2 < P_1$$ **Ans. (A)**

Sol. 81 (a) NOT correct as motion freezes at 0 K temperature.

(b) Mean free path of gas molecules is given as

$$\lambda = \frac{kT}{\sqrt{2}\pi d^2 P}$$

Hence this statement is correct.

(c) From above expression of mean free path this statement is also correct.

(d) From law of equipartition of energy average kinetic energy per molecule per degree of freedom is 0.5 kT so this statement is NOT correct. **Ans. (B)**

Sol. 82 Root mean square velocity of oxygen gas molecules is given as

$$v_{rms} = \sqrt{\frac{3RT}{M}}$$

When temperature is doubled and molar mass is reduced to half due to dissociation of oxygen molecules we use

$$v_{rms} \propto \sqrt{\frac{T}{M}}$$

$$\Rightarrow \quad (V_{rms})_O = (V_{rms})_{O_2} \times \sqrt{\frac{2}{1/2}}$$

$$\Rightarrow \quad (V_{rms})_O = 2(V_{rms})_{O_2}$$ **Ans. (B)**

Sol. 83 No of moles helium gas in cylinder are

$$n = \frac{44.8}{22.4} = 2$$

Gas is monoatomic so we use

$$C_V = \frac{3}{2}R$$

Heat needed to raise the temperature of gas by ΔT is given as

$$\Delta Q = nC_V \Delta T$$

$$\Rightarrow \quad Q = 2 \times \frac{3}{2}R(20)$$

$$\Rightarrow \quad Q = 60\,R = 60 \times 8.3 = 498\text{ J}$$ **Ans. (C)**

Sol. 84 By principle of homogeneity of equation, we use

$$[P] = \left[\frac{a}{V^2}\right]$$

and $\quad [b] = [V]$

$$\Rightarrow \quad \left[\frac{a}{b}\right] = \left[\frac{PV^2}{V}\right] = [PV]$$ **Ans. (C)**

Sol. 85 No of moles of H_2 in vessel are

$$n_1 = 8\text{ moles}$$

No of moles of O_2 in vessel are

$$n_2 = 4\text{ moles}$$

Total moles in vessel are

$$n = 12\text{ moles}$$

At STP 1 mole occupies 22.4 litre or 22.4×10^3 cm^3 so12 moles will occupy a volume

$$V = 12 \times 22.4 \times 10^3\text{ cm}^3 \approx 26.8 \times 10^4\text{ cm}^3$$

Ans. (C)

Sol. 86 The average kinetic energy of a gas molecule is given as

$$K_{avg} = \frac{3}{2}KT$$

Gas pressure is given as

$$P = \frac{1}{3}\rho v_{rms}^2$$

Note : Statement (4) is correct if we consider the situation at constant volume and not constant pressure. With the language of question it is clear that a printing error is there in statement (4). In this question bonus marks should be given but most appropriate answer is option (A) if printing error is corrected.

Ans. (A)

Sol. 87 Sound speed in a gaseous mixture is given as

$$v_s = \sqrt{\frac{\gamma RT}{M}}$$

Root mean squre speed of gas molecules in the mixture are given as

$$v_{rms} = \sqrt{\frac{3RT}{M}}$$

Given that rms speed is $\sqrt{2}$ times the speed of sound, we use

$$\frac{v_s}{v_{\text{rms}}} = \sqrt{\frac{\gamma}{3}} = \frac{1}{\sqrt{2}}$$

$$\Rightarrow \quad \frac{\gamma}{3} = \frac{1}{2}$$

$$\Rightarrow \quad \gamma = \frac{3}{2}$$

As $\gamma = 1 + \frac{2}{f_{\text{mix}}}$, we use

$$f_{\text{mix}} = \frac{2\times 3 + n\times 5}{n+2} = \frac{6+n\times 5}{(n+2)}$$

$$\Rightarrow \quad \gamma = 1 + \frac{2(n+2)}{6+n\times 5} = \frac{6+5n+2n+4}{6+5n}$$

$$\Rightarrow \quad \gamma = \frac{7n+10}{6+5n} = \frac{3}{2}$$

$$\Rightarrow \quad 14n+20 = 18+15n$$

$$\Rightarrow \quad n = 2$$

Ans. (B)

Sol. 88 For a quasi-static process the change in internal energy of an ideal gas is

$$\Delta U = nC_V\Delta T$$

$$\Rightarrow \quad \Delta U = n \times \frac{3R}{2} \times \Delta T$$

Molar heat capacity at constant volume for mono atomic gasis given as $\frac{3R}{2}$, so we use

$$\Delta U = 7 \times \frac{3}{2} \times 8.3 \times 40 = 3486 \text{ J}$$

Ans. (B)

Sol. 89 (a) RMS velocity of gas molecules is given as

$$v_{\text{rms}} = \sqrt{\frac{3RT}{M_W}}$$

Thus v_{rms} is same in the two vessels.

(b) The pressure ratio in the two vessels at same volume and same temperature is given as

$$\frac{P_1}{P_2} = \frac{N_1}{N_2}$$

Hence this statement is correct.

(c) As explained above this statement is NOT correct.

(d) As explained above this statement is NOT correct.

Ans. (C)

Sol. 90 Statement (b) is correct by definition of law of equipartition of energy and Methane molecule is tetrahedron so it has 3 translational and 3 rotational degrees of freedome hence statements (b) and (d) are correct. **Ans. (B)**

Sol. 91 Momentum is a vector quantitiy thus average momentum in a gas sample due to random motion of molecules is always zero. Hence statment I is false.

RMS speed of gas molecules at a given temperature is given as

$$v_{\text{rms}} = \sqrt{\frac{3RT}{M}}$$

As temperature of gas is doubled and molar mass is reduced to half due to dissociation of molecules, we use

$$T_{\text{new}} = 2T$$

$$M_{\text{new}} = \frac{M}{2}$$

$$\Rightarrow \quad \frac{v_{\text{new}}}{v} = \frac{\sqrt{\frac{2T}{M/2}}}{\sqrt{\frac{T}{M}}}$$

$$\Rightarrow \quad v_{\text{new}} = 2v$$

Ans. (D)

Sol. 92 To double the RMS speed of gas molecules, temperature of gas is to be made 4 times so in this case temperature is to be increased from 300 K to 1200 K for which heat required at constant volume in a vessel is calculated as

$$Q = nC_v\Delta T = \frac{1}{2} \times \frac{5R}{2} \times 900$$

$$\Rightarrow \quad Q = 9360 \text{ J}$$

Ans. (C)

Sol. 93 Root mean square speed of smoke particles is given as

$$v_{\text{rms}} = \sqrt{\frac{3kT}{m}} = \sqrt{\frac{3\times 1.38\times 10^{-23}\times 293}{5\times 10^{-17}}}$$

$$\Rightarrow \quad v_{\text{rms}} \approx 15 \text{ mm/s}$$

Ans. (C)

Sol. 94 Statement-I: The given temperatures are

$$T_1 = -73°\text{C} = 200 \text{ K}$$

$$T_2 = 527°\text{C} = 800 \text{ K}$$

The ratio of RMS speed of molecules are written as

$$\frac{v_1}{v_2} = \frac{\sqrt{\frac{3RT_1}{M}}}{\sqrt{\frac{3RT_2}{M}}} = \sqrt{\frac{T_1}{T_2}} = \sqrt{\frac{200}{800}} = \frac{1}{2}$$

$$\Rightarrow \quad v_2 = 2v_1$$

Hence statement-I is true.

Statement-II: Using ideal gas law we have

$$PV = nRT$$

Translational kinetic energy of gas molecules is given as

$$K = \frac{3}{2}nRT$$

Hence statement-II is false. **Ans. (B)**

Sol. 95 The rms speed of a gas molecule is given as

$$v_{rms} = \sqrt{\frac{3RT}{M}}$$

$$\Rightarrow \quad v_{rms} \propto \sqrt{T}$$ **Ans. (C)**

Sol. 96 A diatomic gas molecule has three translational degree of freedom and two rotational degree of freedom & corresponding to one vibrational mode there will be two additional degrees of freedom thus it has total seven degrees of freedom. So its molar specific heat at constant volume is given as

$$C_V = \frac{fR}{2} = \frac{7R}{2}$$ **Ans. (D)**

Sol. 97 At constant volume we use

$$\frac{P_1}{T_1} = \frac{P_2}{T_2}$$

$$\Rightarrow \quad P_2 = \frac{P_1}{T_1} \times T_2 = \frac{270 \times (309)}{300}$$

$$\Rightarrow \quad P_2 = 278 \text{ kPa}$$ **Ans. (C)**

Sol. 98 RMS speed of gas molecules and average speed are related as

$$\sqrt{\frac{3RT}{M}} = \sqrt{\frac{\alpha+5}{\alpha}} \sqrt{\frac{8}{\pi}\frac{RT}{M}}$$

$$\Rightarrow \quad 3 = \frac{\alpha+5}{\alpha}\frac{8}{\pi}$$

$$\Rightarrow \quad \alpha = 28$$ **Ans. (B)**

Sol. 99 The process equation of gas is given as

$$PT^2 = \text{constant},$$

Using $\quad PV = nRT$

$$P = \frac{nRT}{V}$$

$$\Rightarrow \quad PT^2 = \frac{nRT}{V} \times T^2 = \text{constant}$$

$$\Rightarrow \quad T^3 = KV$$

Differentiating the equation gives

$$\frac{d}{dT}(KV) = 3T^2$$

$$\Rightarrow \quad \frac{KdV}{dT} = 3T^2$$

$$\Rightarrow \quad dV = \frac{3T^2}{K}dT$$

Coefficient of expansion γ is related with volume expansion as

$$dV = V\gamma\, dT$$

$$\Rightarrow \quad \gamma V = \frac{3T^2}{K}$$

$$\Rightarrow \quad \gamma = \frac{3T^2}{KV} = \frac{3T^2}{T^3} = \frac{3}{T}$$ **Ans. (D)**

Sol. 100 Average kinetic energy per molecule is given as

$$K_{av} = \frac{5}{2}kT$$

So the ratio of average kinetic energy per molecule of hydrogen and oxygen atoms can be written as 1 : 1 **Ans. (B)**

Sol. 101 Heat supplied at constant pressure is calculated as

$$\Delta Q = nC_P\Delta T = 735 \text{ J}$$

$$\Rightarrow \quad \frac{5nR\Delta T}{2} = 735 \text{ J}$$

Increase in the internal energy of gas is calculated as

$$\Delta U = nC_V\Delta T = \frac{3}{2}(nR\Delta T)$$

$$\Rightarrow \quad \Delta U = \frac{3}{2} \times \frac{2}{5} \times 735 = 441 \text{ J}$$ **Ans. (B)**

Sol. 102 According to fundamentals of kinetic theory of gases, translational *K.E* on average of a molecule is $\frac{3}{2}kT$. This is independent of nature of molecule, pressure and volume. **Ans. (A)**

Sol. 103 For isochoric process, we use

$$\frac{P}{T} = n\frac{R}{V} = \text{constant}$$

$$\Rightarrow \quad P = \frac{nR}{V}(t+273)$$

At $P = 0$, we have

$$t = -273°\text{C}$$ **Ans. (A)**

Sol. 104 Collision frequency of air molecules in contianer is given as

$$f = \frac{V}{\lambda} = \frac{V}{\left(\frac{1}{\sqrt{2}\pi d^2 n_v}\right)} = \sqrt{2}\pi d^2 v n_v$$

$$\Rightarrow \quad f \propto n_v$$

Where n_v is number density. If frequency of air molecule before and after collision are f_1 and f_2 then we have

$$\frac{f_1}{f_2} = \frac{nv_1}{nv_2} = \frac{3\times10^{19}}{12\times10^{-19}} = 0.25$$ **Ans. (B)**

Sol. 105 Speed of sound in gas at T temperature

$$c = \sqrt{\frac{\gamma RT}{M}}$$

$$\Rightarrow \quad c \propto \frac{1}{\sqrt{M}}$$

Thus ratio of speed of sound in H_2 and O_2 is given as

$$\frac{C_{H_2}}{C_{O_2}} = \sqrt{\frac{32}{2}} = 4:1$$

Correct option (A). **Ans. (A)**

Sol. 106 RMS speed of gas at temperature T is given as

$$V_{\text{rms}} = \sqrt{\frac{3RT}{M}}$$

$$\Rightarrow V_{\text{rms}} \propto \sqrt{T}$$

So, by increasing temperature 4 times, rms speed gets doubled. **Ans. (D)**

Sol. 107 Pressure at the depth of 40 m is calculated as

$$P = P_0 + \rho g h = 10^5\ \text{Pa} + 10^3 \times 10 \times 40$$

$$\Rightarrow \quad P = 5 \times 10^5\ \text{Pa}$$

As T is constant, by gas equation, we use

$$PV = P_0 V_0$$

$$\Rightarrow \quad 5 \times 10^5\ \text{Pa} \times 1\ \text{cm}^3 = 10^5\ \text{Pa} \times V_0$$

$$\Rightarrow \quad V_0 = 5\ \text{cm}^3$$

Ans. (A)

Sol. 108 Kinetic energy of a molecule is given as $\frac{f}{2}kT$ where T is absolute temperature of gas. If K_1 is kinetic energy at 27°C K_2 is kinetic energy at new temperature T, we use

$$\frac{K_1}{K_2} = \frac{T_1}{T_2}$$

$$\Rightarrow \quad \frac{1}{2} = \frac{300}{T}$$

$$\Rightarrow \quad T = 600\ \text{K}$$

$$\Rightarrow \quad T = 327°\text{C}$$

Ans. (C)

Sol. 109 For different types of gases number of degrees of freedom are given in the table below.

Type of gases	No. of degrees of freedom
Monoatomic gas	$3T$
Diatomic + rigid	$3T + 2R$
Diatomic + non-rigid	$3T + 2R + 1V$
Polyatomic	$3T + 3R +$ More then 1V

Here

T = Translational degree of freedom

R = Rotational degree of freedom

V = Vibrational degree of freedom **Ans. (C)**

Sol. 110 Specific heat of mixture at constant volume is given as

$$(C_v)_{\text{mix}} = \frac{n_1 Cv_1 + n_2 Cv_2}{n_1 + n_2}$$

$$Cv_1\,(\text{di atomic}) = \frac{5}{2}R;\ Cv_2\,(\text{monnoatomic}) = \frac{3}{2}R$$

$$\Rightarrow \quad (C_v)_{\text{mix}} = \frac{2 \times \frac{5}{2}R + 4 \times \frac{3}{2}R}{2+4} = \frac{11R}{6}$$

$$\Rightarrow \quad \Delta U = n(C_v)_{\text{mix}} T$$

Total internal energy of system is given as

$$\Delta U = 6\frac{11R}{6} \times RT = 11R$$

Ans. (D)

Sol. 111 RMS speed of molecules in monoatomic gas (neon) is given as

$$v_{\text{rms}}(\text{mono}) = \sqrt{\frac{3RT}{4 \times 10^{-3}}}$$

RMS speed of molecules in diatomic gas (chlorine) is given as

$$v_{\text{rms}}(\text{dia}) = \sqrt{\frac{3RT}{71 \times 10^{-3}}}$$

RMS speed in molecules in polyatomic gas (uranium hexafluoride) is given as

$$v_{\text{rms}}(\text{poly}) = \sqrt{\frac{3RT}{146 \times 10^{-3}}}$$

Thus relation among these is given as

$$v_{\text{rms}}(\text{mono}) > v_{\text{rms}}(\text{dia}) > v_{\text{rms}}(\text{poly})$$

Ans. (B)

Sol. 112 RMS speed of N_2 gas at T temperature is given as

$$V_{\text{rms}} = \sqrt{\frac{3k_B T}{m}}$$

$$\Rightarrow \quad V_{\text{rms}} = \sqrt{\frac{3 \times 1.4 \times 10^{-23} \times 300}{4.6 \times 10^{-26}}}$$

$$\Rightarrow \quad V_{\text{rms}} = 523\ \text{m/s}$$

Ans. (A)

Sol. 113 RMS speed for chlorine gas at constant temperature is given as

$$v_{\text{rms}} = \sqrt{\frac{3RT}{M}}$$

v_{Ar} is rms speed of Argon molecule, which is given as

$$\Rightarrow \quad \frac{v_{Ar}}{v_{Cl}} = \sqrt{\frac{M_{\text{Cl}}}{M_{\text{Ar}}}}$$

$$\Rightarrow \quad v_{\text{Ar}} = 1.33 \times 490 = 651.7\ \text{m/s}$$

Ans. (C)

Sol. 114 Given that RMS speed of oxygen molecule is given as

$$v_{rms} = \sqrt{\frac{3RT}{M}}$$

Average speed of molecules is given as

$$v = \sqrt{\frac{8RT}{\pi M}}$$

Given that at a particular temperature, we have

$$\sqrt{\frac{3RT}{M}} = \left(1+\frac{5}{x}\right)^{\frac{1}{2}} \sqrt{\frac{8RT}{\pi M}}$$

$$\Rightarrow \quad \frac{3\times 22}{7\times 8} = 1 + \frac{5}{x}$$

$$\Rightarrow \quad x = 28$$

Ans. (A)

Sol. 115 Mean free path of molecules of a gas is given as

$$\lambda = \frac{RT}{\sqrt{2}\pi d^2 N_A P}$$

$$\Rightarrow \quad \lambda \propto T$$

$$\Rightarrow \quad \frac{1500d}{\lambda} = \frac{273}{373}$$

$$\Rightarrow \quad \lambda = 2049d$$

Ans. (B)

Sol. 116 Average kinetic energy per molecule is given as

$$E = \frac{3}{2}kT$$

As temperature is constant so ratio of average kinetic energy per molecule of two gases is calculated as

$$\frac{K_{Ar}}{K_H} = \frac{1}{1}$$

Ans. (A)

13.1 Simple Harmonic Motion

Sol. 1 k — n parts — $k' = nk$

Each part has force constant n times of the original spring if it is divided in n equal parts.

For a force constant k, the time period of oscillations is

$$T = 2\pi\sqrt{\frac{m}{k}}$$

If $\quad k \to k' = nk$

New time period will be

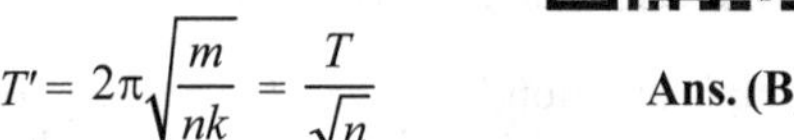

$$T' = 2\pi\sqrt{\frac{m}{nk}} = \frac{T}{\sqrt{n}}$$

Ans. (B)

Sol. 2 In SHM, KE and PE with position are given as

$$K_y = \frac{1}{2}m\omega^2(A^2 - y^2)$$

and $$P_y = \frac{1}{2}m\omega^2 y^2$$

At mean position $\quad y = 0$

$$K_m = \frac{1}{2}m\omega^2 A^2 \quad \text{(max)}$$

$$P_m = 0 \quad \text{(min)}$$

At extreme position $\; y = \pm A$

$$K_e = 0 \quad \text{(min)}$$

$$P_e = \frac{1}{2}m\omega^2 A^2 \quad \text{(max)}$$

Ans. (C)

Sol. 3 Time period of a swing (similar to simple pendulum) is given as $\quad T = 2\pi\sqrt{\frac{l}{g}}$

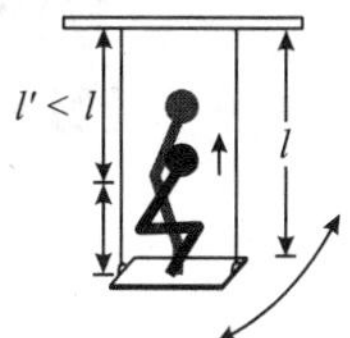

As boy stands up, length of pendulum decreases

$\Rightarrow$ new time period $T' = 2\pi\sqrt{\frac{l'}{g}} \Rightarrow T' < T$

Ans. (B)

Sol. 4

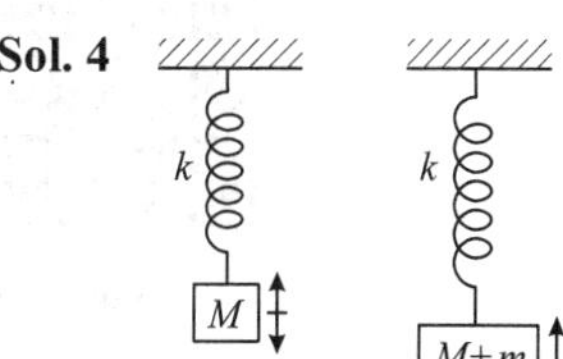

Time period of SHM in first condition

$$T = 2\pi\sqrt{\frac{M}{k}} \quad \ldots(1)$$

Time period of SHM in second condition

$$\frac{5T}{3} = T' = 2\pi\sqrt{\frac{M+m}{k}} \quad \ldots(2)$$

By (2)/(1) $\Rightarrow \quad \frac{5}{3} = \sqrt{\frac{M+m}{M}}$

$$\Rightarrow \quad \frac{25}{9} = \frac{M+m}{M}$$

$$\Rightarrow \quad \frac{m}{M} = \frac{16}{9}$$

Ans. (C)

Sol. 5 SHM equation is given as

$$x = 4\cos\pi t + 4\sin\pi t = A\sin(\omega t + \theta)$$

Where A is SHM amplitude

here we use $\quad A\cos\theta = 4$

and $\quad A\sin\theta = 4$

square above equations and add

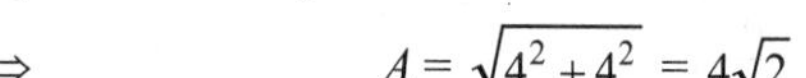

$$\Rightarrow \quad A = \sqrt{4^2 + 4^2} = 4\sqrt{2}$$

Ans. (C)

Sol. 6 Energies as a function of position x in SHM are given as

$$TE = \frac{1}{2}m\omega^2A^2 \rightarrow \text{constant}$$

$$K_x = \frac{1}{2}m\omega^2(A^2 - x^2)$$

$$P_x = \frac{1}{2}m\omega^2x^2$$

At mean position $x = 0$

$$K_x = \frac{1}{2}m\omega^2A^2 \text{ (max)}$$

$$P_x = 0 \text{ (min)}$$

At extreme position

$$x = \pm A$$

$$K_x = 0$$

$$P_x = \frac{1}{2}m\omega^2A^2 \text{ (max)}$$

Ans. (A)

Sol. 7 If amplitudes of A and B are A_1 and A_2 respectively then we use $v_{\max A} = v_{\max B}$

For SHM $v_{\max} = \omega A$

$$\omega_A A_1 = \omega_B A_2$$

$$\omega = \sqrt{\frac{k}{m}}$$

$$\Rightarrow \quad \frac{A_1}{A_2} = \sqrt{\frac{k_2}{k_1}}$$

Ans. (C)

Sol. 8 Time period of a simple pendulum is given as

$$T = 2\pi\sqrt{\frac{l}{g}}$$

If $l \rightarrow 1.21l \Rightarrow$ new time period will be

$$T' = 2\pi\sqrt{\frac{1.21l}{g}} = 1.1 \times 2\pi\sqrt{\frac{l}{g}}$$

$$\Rightarrow \quad T' = 1.1T$$

% increase in time period is

$$= \frac{T' - T}{T} \times 100$$

$$= \frac{(1.1 - 1)}{1} \times 100 = 10\%$$

Ans. (D)

Sol. 9 Time period of a simple pendulum is

$$T = 2\pi\sqrt{\frac{l}{g}} \text{ in air}$$

$$T' = 2\pi\sqrt{\frac{l}{g_{\text{eff}}}} \text{ in water}$$

Where $g_{\text{eff}} = g\left(1 - \frac{\rho_l}{\rho_s}\right)$

$$T' = 2\pi\sqrt{\frac{l}{g\left(\frac{\rho_s - \rho_l}{\rho_s}\right)}} = T\sqrt{\frac{\rho_s}{\rho_s - \rho_l}}$$

$$\Rightarrow \quad T' = t_0\sqrt{\frac{\frac{4}{3} \times 10^3}{\left(\frac{4}{3} - 1\right) \times 10^3}} = t_0\sqrt{\frac{4}{3} \times \frac{3}{1}} = 2t_0$$

Ans. (C)

Sol. 10 For first spring (force constant k_1), time period

$$t_1 = 2\pi\sqrt{\frac{m}{k_1}}$$

For second spring (force constant k_2), time period

$$t_2 = 2\pi\sqrt{\frac{m}{k_2}}$$

In series combination of springs, effective force constant k_{eq} is given by

$$\frac{1}{k_{eq}} = \frac{1}{k_1} + \frac{1}{k_2}$$

Time period of SHM using series combination is given as

$$T_{eq} = T = 2\pi\sqrt{\frac{m}{k_{eq}}} = 2\pi\sqrt{\frac{m}{k_1} + \frac{m}{k_2}}$$

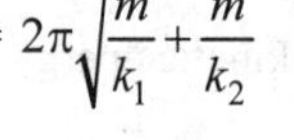

Squaring gives

$$T^2 = 4\pi^2\frac{m}{k_1} + 4\pi^2\frac{m}{k_2}$$

$$\Rightarrow \quad T^2 = t_1^2 + t_2^2$$

Ans. (B)

Sol. 11 Total energy in SHM is given as

$$TE = \frac{1}{2}m\omega^2A^2 + U_{\min}$$

Constant minimum PE does not change with position.

with position KE & PE oscillates and

$$K_x + P_x = \text{TE} \rightarrow \text{constant}$$

Ans. (C)

Sol. 12 For a forced oscillation in a damping medium if external force is $F(t) = F_0\cos\omega t$

then the oscillation amplitude is given as

$$A = \frac{F_0}{\sqrt{m^2(\omega_0^2 - \omega^2)^2 + b^2\omega^2}}$$

(where $b \rightarrow$ damping coefficient)

in this question $b = 0 \Rightarrow A = \dfrac{F_0}{m(\omega_0^2 - \omega^2)}$

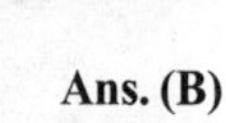

Ans. (B)

Sol. 13 Energy of oscillations $\propto$ (amplitude)^2

$\Rightarrow$ When amplitude is maximum, energy will also be maximum and this happens at same frequency of force

$$\Rightarrow \quad \omega_1 = \omega_2$$

Ans. (C)

Sol. 14 Acceleration of particle is given as

$$a = \frac{F}{m} = -\frac{kx}{m}$$

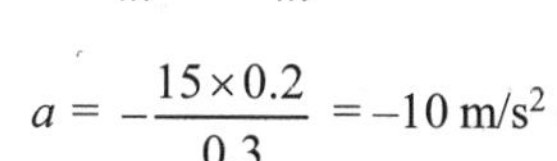

$$\Rightarrow \quad a = -\frac{15 \times 0.2}{0.3} = -10 \text{ m/s}^2$$

Negative sign shows opposite direction. **Ans. (D)**

Sol. 15 The position given is

$$y = \sin^2 \omega t = \frac{1}{2}(1 - \cos 2\omega t)$$

$$\Rightarrow \quad y = \frac{1}{2} - \frac{1}{2}\cos 2\omega t$$

$$\Rightarrow \quad y = y_0 + A\cos(2\omega t + \pi)$$

This represents a function of SHM with mean position at $y = y_0$, Amplitude A and angular frequency 2ω.

Where $A = 1/2$ and $y_0 = 1/2$

$$\Rightarrow \quad \text{Time period} = \frac{2\pi}{2\omega} = \frac{\pi}{\omega}$$ **Ans. (D)**

Sol. 16 The basic differential equation of SHM is given as

$$\frac{d^2x}{dt^2} + \omega^2 x = 0$$

where $\omega \rightarrow$ Angular frequency of SHM.

comparing both the above equation, we have

$$\omega^2 = \alpha \text{ or } \omega = \sqrt{\alpha}$$

Time period of SHM $T = \frac{2\pi}{\omega} = \frac{2\pi}{\sqrt{\alpha}}$ **Ans. (B)**

Sol. 17

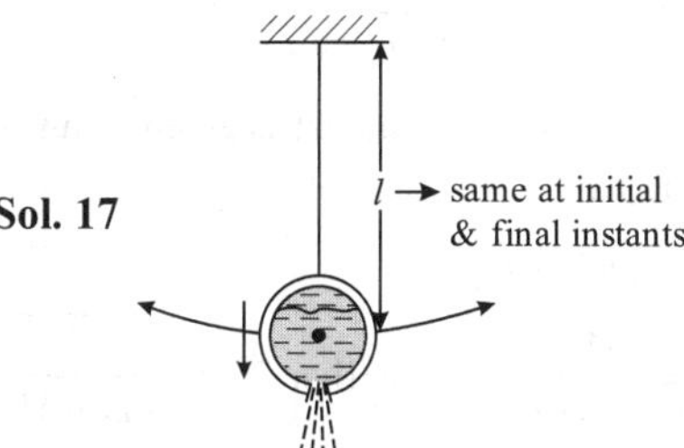

As water flows out, inside the shell water level comes down. Time period of oscillation of pendulum

$$T = 2\pi\sqrt{\frac{l}{g}}$$

As water level goes down

$\Rightarrow$ l increases as centre of mass is going down.

When shell becomes empty then again COM of shell is at its centre $\Rightarrow$ after some time l decreases.

$\Rightarrow$ T first increases then decreases to same value. **Ans. (A)**

Sol. 18 Velocity of particle-1 is

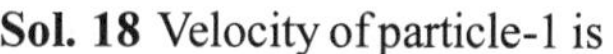

$$v_1 = \frac{dy_1}{dt} = 0.1 \times 100\pi \cos\left(100\pi t + \frac{\pi}{3}\right)$$

$$\Rightarrow \quad v_1 = 10\pi \cos\left(100\pi t + \frac{\pi}{3}\right) \quad \ldots(1)$$

Velocity of particle-2 is

$$v_2 = \frac{dy_2}{dt} = -0.1 \times 100\pi \sin(100\pi t)$$

$$\Rightarrow \quad v_2 = 10\pi \cos\left(100\pi t + \frac{\pi}{2}\right) \quad \ldots(2)$$

Phase difference between v_1 and v_2 is

$$\alpha = \left(100\pi t + \frac{\pi}{3}\right) - \left(100\pi t + \frac{\pi}{2}\right)$$

$$\Rightarrow \quad \alpha = -\frac{\pi}{6}$$ **Ans. (A)**

Sol. 19 In SHM maximum velocity of particle is given as

$$v_{max} = A\omega = A\left(\frac{2\pi}{T}\right)$$

$$\Rightarrow \quad T = \frac{2\pi A}{v_{max}} = \frac{2 \times 3.14 \times 7 \times 10^{-3}}{4.4}$$

$$\Rightarrow \quad T = 0.01\text{s}$$ **Ans. (A)**

Sol. 20 $-x \leftarrow\!\!-A \quad 0 \quad +A\!\!\rightarrow x$, $x = A\sin\omega t$

For the particle in SHM, its KE is given as

$$K = \frac{1}{2}m\omega^2(A^2 - x^2) \quad \ldots(1)$$

Given that at time t $K = \frac{3}{4}E = \frac{3}{4}\left(\frac{1}{2}m\omega^2 A^2\right)$

From (1) $\frac{1}{2}m\omega^2(A^2 - x^2) = \frac{3}{4}\left(\frac{1}{2}m\omega^2 A^2\right)$

$$\Rightarrow \quad A^2 - x^2 = \frac{3}{4}A^2$$

$$\Rightarrow \quad x = \pm A/2$$

As $T = 2$s

$$\Rightarrow \quad \omega = \frac{2\pi}{2} = \pi \text{ rad/s}$$

$$\Rightarrow \quad x = A\sin\pi t$$

at $\quad x = A/2 = A\sin\pi t$

$$\Rightarrow \quad \sin\pi t = \frac{1}{2} \Rightarrow \pi t = \frac{\pi}{6} \Rightarrow t = \frac{1}{6}\text{s}$$ **Ans. (A)**

Sol. 21 In SHM

$$a_{max} = \omega^2 A$$

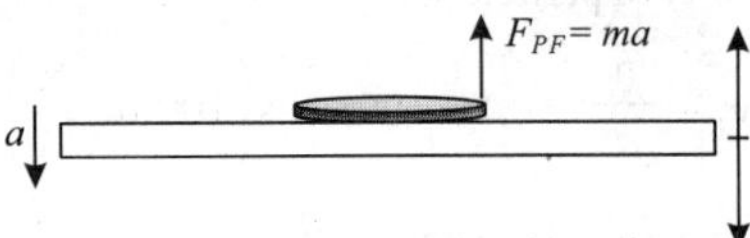

During SHM when platform goes down at acceleration a, coin experiences a pseudo force ma in upwards direction, it will leave contact when

$$ma = mg$$

$$\Rightarrow \quad \omega^2 A = g$$

$$\Rightarrow \quad A = \frac{g^2}{\omega^2} \quad \textbf{Ans. (B)}$$

Sol. 22 The SHM equation given is of the form

$$x = A\cos\omega t \quad [A = 2\times10^{-2}\,\text{m}, \omega = \pi\ \text{rad/s}]$$

Here particle starts SHM from external position as $t = 0$, $x = A$. Particle speed is maximum in SHM when it crosses the mean position which happens at

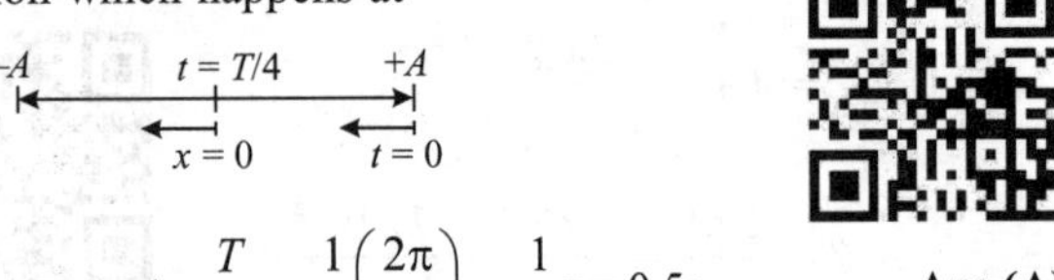

$$t = \frac{T}{4} = \frac{1}{4}\left(\frac{2\pi}{\omega}\right) = \frac{1}{2}\text{s} = 0.5\text{s} \quad \textbf{Ans.(A)}$$

Sol. 23 SHM equation as

$$x = x_0\cos(\omega t - \pi/4)$$

velocity in SHM is given as

$$v = \frac{dx}{dt} = -\omega x_0\sin(\omega t - \pi/4)$$

Acceleration in SHM is given as

$$a = \frac{dv}{dt} = -\omega^2 x_0\cos(\omega t - \pi/4)$$

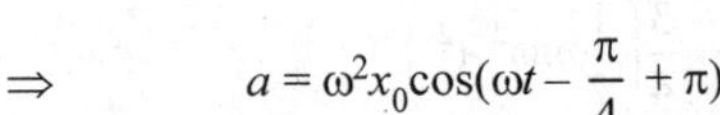

$$\Rightarrow \quad a = \omega^2 x_0\cos(\omega t - \frac{\pi}{4} + \pi)$$

$$\Rightarrow \quad a = \omega^2 x_0\cos(\omega t + \frac{3\pi}{4}) \quad \ldots(1)$$

Comparing equation-(1) with the given acceleration we get

$$A = \omega^2 x_0$$

and $\delta = \frac{3\pi}{4}$ **Ans. (D)**

Sol. 24 The frequency of oscillation of block in this case is given as

$$f = \frac{1}{2\pi}\sqrt{\frac{k_1 + k_2}{m}}$$

if $k_1 \leftarrow 4k_1$

and $k_2 \leftarrow 4k_2$

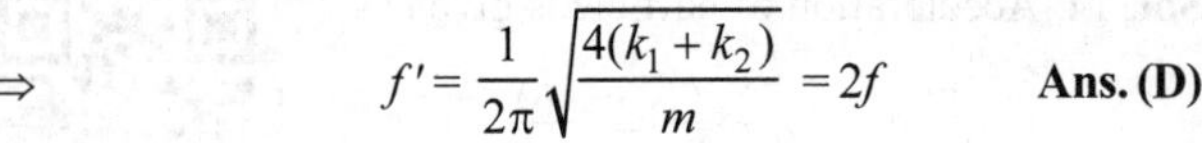

$$\Rightarrow \quad f' = \frac{1}{2\pi}\sqrt{\frac{4(k_1 + k_2)}{m}} = 2f \quad \textbf{Ans. (D)}$$

Sol. 25 Mean —— Extreme

Average KE of a particle in SHM for one period (or $T/4$ period from mean to extreme position) is given as

$$(\text{KE})_{\text{avg}} = \frac{1}{2}\text{TE} = \frac{1}{2}\left(\frac{1}{2}m\omega^2 A^2\right)$$

$$\Rightarrow \quad (\text{KE})_{\text{avg}} = \frac{1}{4}\text{m}(2\pi\nu)^2 a^2$$

$$\Rightarrow \quad (\text{KE})_{\text{avg}} = m\pi^2\nu^2 a^2 \quad \textbf{Ans. (A)}$$

Sol. 26 In SHM $\quad a = -\omega^2 x$ and $v = \omega\sqrt{A^2 - x^2}$

(A) $-\omega^2 xT + 2\pi\omega\sqrt{A^2 - x^2}$

(B) $\frac{aT}{v} = -\frac{\omega^2 xT}{\omega\sqrt{A^2 - x^2}}$

(C) $a^2T^2 + 4\pi^2v^2 = \omega^2x^2T^2 + 4\pi^2\omega^2(A^2 - x^2)$

above options are dependent on x which varies with time.

(D) $\frac{aT}{x} = -\omega^2 T$ independent of x. **Ans. (D)**

Sol. 27 [Figure: spring attached to wall with block M, block m on top; v_2, v_1; Mean position]

$$v_1 = A_1\omega_1 = A_1\sqrt{\frac{k}{M}}$$

Due to addition of m, the new speed of blocks is given by linear momentum conservation as

$$Mv_1 = (m + M)v_2$$

$$v_2 = \frac{Mv_1}{M + m} = A_2\omega_2 \text{ [here } \omega_2 = \sqrt{\frac{k}{m + M}}\text{]}$$

$$\Rightarrow \quad A_2 = \frac{Mv_1}{(M + m)}\cdot\sqrt{\frac{m + M}{k}}$$

$$\Rightarrow \quad A_2 = \frac{M}{\sqrt{M + m}}\cdot\frac{1}{\sqrt{k}}\cdot A_1\sqrt{\frac{k}{m}}$$

$$\Rightarrow \quad A_2 = A_1\sqrt{\frac{M}{M + m}}$$

$$\Rightarrow \quad \frac{A_1}{A_2} = \sqrt{\frac{M + m}{M}} \quad \textbf{Ans. (B)}$$

Sol. 28 SHM equation of the two particles are given as

$$x_1 = A\sin\omega t$$
$$x_2 = X_0 + A\sin(\omega t + \phi)$$

separation between the particles is

$$x_2 - x_1 = X_0 + A\sin(\omega t + \phi) - A\sin\omega t$$

given that

$$(x_2 - x_1)_{max} = X_0 + A$$
$$(A\sin(\omega t + \phi) - A\sin\omega t)_{max} = A$$

$$\Rightarrow \left[2A\cos\left(\omega t + \frac{\phi}{2}\right)\sin\frac{\phi}{2}\right]_{max} = A$$

$$\Rightarrow \left[\left(2A\sin\frac{\phi}{2}\right)\cos\left(\omega t + \frac{\phi}{2}\right)\right]_{max} = A$$

Here $\cos\left(\omega t + \frac{\phi}{2}\right) = (+1)_{max}$

$$\Rightarrow 2A\sin\frac{\phi}{2} = A$$
$$\Rightarrow \phi = \pi/3$$

Ans. (D)

Sol. 29 Under the state of equilibrium we have

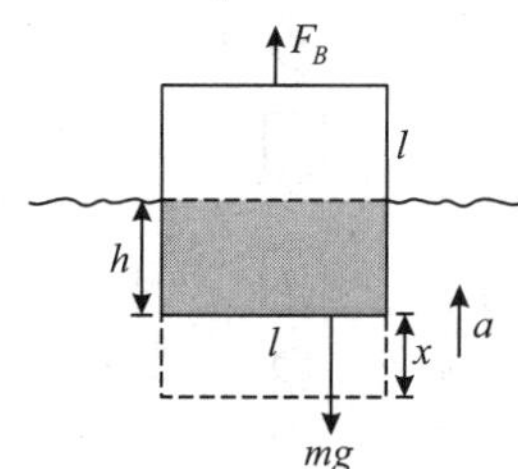

$$F_B = mg$$
$$\Rightarrow l^2 h\rho g = l^3 dg$$
$$\Rightarrow h\rho = ld \qquad \ldots(1)$$

After pushing the cube by x, restoring force on it upward will be $F_B' - mg = ma$

$$\Rightarrow (h + x)l^2\rho g - mg = ma$$
$$hl^2\rho g + xl^2\rho g - l^3 dg = l^3 da$$
$$\Rightarrow a = -\frac{\rho g}{ld}x$$

–ve sign denotes opposite direction

Comparing with $a = -\omega^2 x$

we get $\omega = \sqrt{\frac{\rho g}{ld}}$

$\Rightarrow$ Time period $T = \frac{2\pi}{\omega}$

$$\Rightarrow T = 2\pi\sqrt{\frac{ld}{\rho g}}$$

Ans. (B)

Sol. 30 For a simple pendulum having damped oscillations its equation of SHM in differential form is given as

$$\frac{d^2x}{dt^2} + bv + \omega^2 x = 0$$

Solution to above equation is

$$x = A_0 e^{-bt/2}\sin(\omega t + \phi)$$

Where $A_0 e^{-bt/2}$ = Amplitude envelope.

Given that at $t = \tau$, $A = \frac{A_0}{e}$

we use $A = A_0 e^{-bt/2}$

$$\Rightarrow \frac{A_0}{e} = A_0 e^{-b\tau/2}$$
$$\Rightarrow \frac{b\tau}{2} = 1$$
$$\Rightarrow \tau = \frac{2}{b}$$

Ans. (A)

Sol. 31 This ring oscillates as a compound pendulum for which oscillation period is given as

$$T = 2\pi\sqrt{\frac{I}{mgl}}$$

here $I = mR^2 + mR^2 = 2mR^2$; $l = R$

$$\Rightarrow 1 = 2\pi\sqrt{\frac{2mR^2}{mgR}} = 2\pi\sqrt{\frac{2R}{g}}$$

squaring, we get

$$\Rightarrow 1 = 4\pi^2 \times \frac{2R}{g} \Rightarrow R = \frac{1}{8}\text{m} = 0.125\text{m}$$

Ans. (A)

Sol. 32 General SHM equation

$$y(t) = A\sin\omega t$$

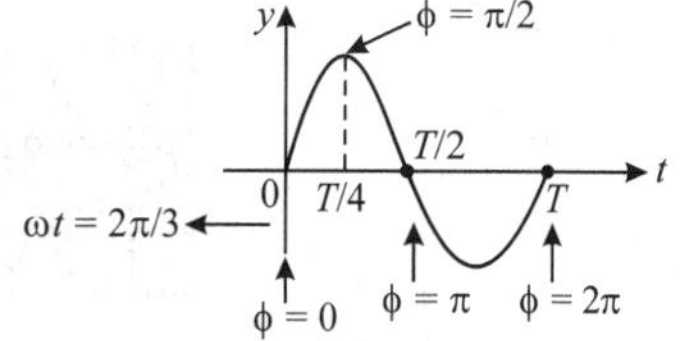

$$y = A\sin(\omega t + \phi) = A\sin\left(\omega t + \frac{2\pi}{3}\right)$$

at $\omega t = -\frac{2\pi}{3}$ the function should start from $y = 0$

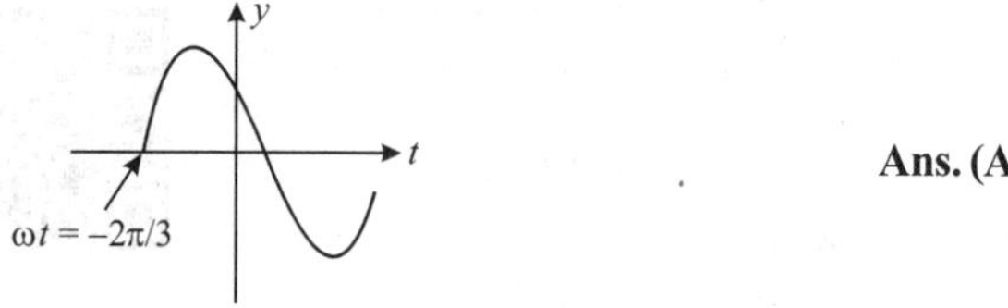

Ans. (A)

Sol. 33

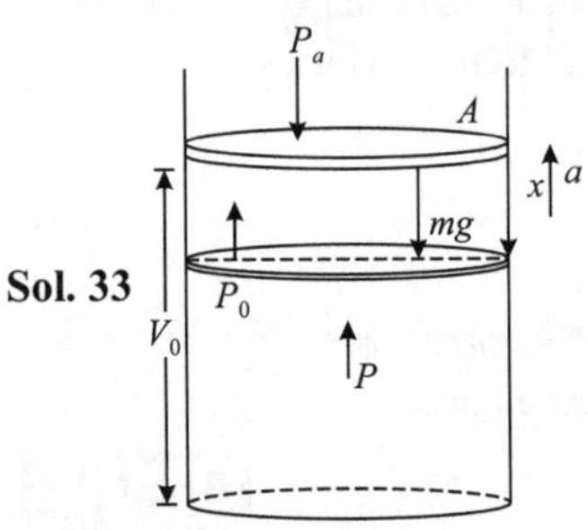

At equilibrium position, we use

$$P_0A = mg + P_aA \qquad \ldots(1)$$

If after displacing piston downward gas pressure becomes P then equation of motion of piston is given as

$$PA - mg - P_aA = ma \qquad \ldots(2)$$

After pushing the piston down, we use

$$P_0V_0^r = P(V_0 - Ax)^\gamma$$

$$\Rightarrow \quad P_0V_0^r = PV_0^r\left(1 - \frac{Ax}{V_0}\right)^\gamma$$

$$\Rightarrow \quad P = P_0\left(1 - \frac{Ax}{V_0}\right)^{-\gamma}$$

Using binomial expansion as $1 >> Ax/V_0$ we have

$$P = P_0\left(1 + \frac{\gamma Ax}{V_0}\right)$$

From equation-(2), we have

$$\Rightarrow \quad P_0A + \frac{P_0\gamma A^2x}{V_0} - mg - P_aA = ma$$

$$\Rightarrow \quad a = -\frac{P_0\gamma A^2}{mV_0} \qquad \ldots(3)$$

–ve sign denotes opposite direction of a compared to x.
Comparing equation-(3) with $a = -\omega^2x$

$$\Rightarrow \quad \omega = \sqrt{\frac{P_0 rA^2}{mV_0}}$$

SHM frequency of piston

$$f = \frac{\omega}{2\pi} = \frac{1}{2\pi}\sqrt{\frac{P_0\gamma A^2}{mV_0}}$$

Ans. (C)

Sol. 34 For a damped oscillator amplitude is given

$$A = A_0e^{-bt/2m}$$

Given that at $t = 5$s we use $A = 0.9A_0$

$$\Rightarrow \quad 0.9A_0 = A_0e^{-5b/2m}$$

$$\Rightarrow \quad e^{-5b/2m} = 0.9$$

At $t = 15$s we use

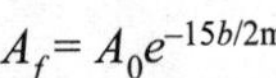

$$A_f = A_0e^{-15b/2m}$$

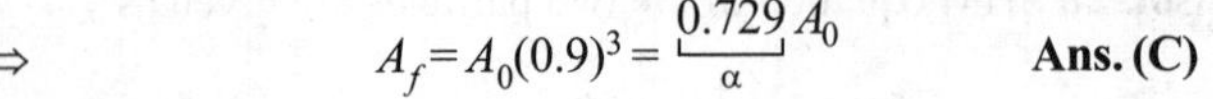

$$\Rightarrow \quad A_f = A_0(0.9)^3 = 0.729\,A_0$$

Ans. (C)

Sol. 35

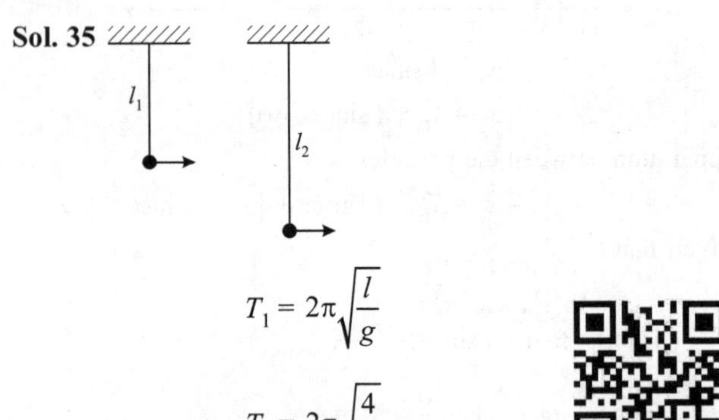

$$T_1 = 2\pi\sqrt{\frac{l}{g}}$$

$$T_2 = 2\pi\sqrt{\frac{4}{g}}$$

As $T_1 < T_2$, shorter pendulum will exceed in phase compared to longer pendulum & these will be in phase again when shorter pendulum will oscillate 1 oscillation more than longer pendulum. If by this time shorter pendulum completes N oscillations.

$$\Rightarrow \quad N.2\pi\sqrt{\frac{1}{g}} = (N-1)\,2\pi\sqrt{\frac{4}{g}}$$

$$\Rightarrow \quad N = 2N - 2 \Rightarrow N = 2$$

Ans. (A)

Sol. 36 Angular frequency of vertical SHM of mass is

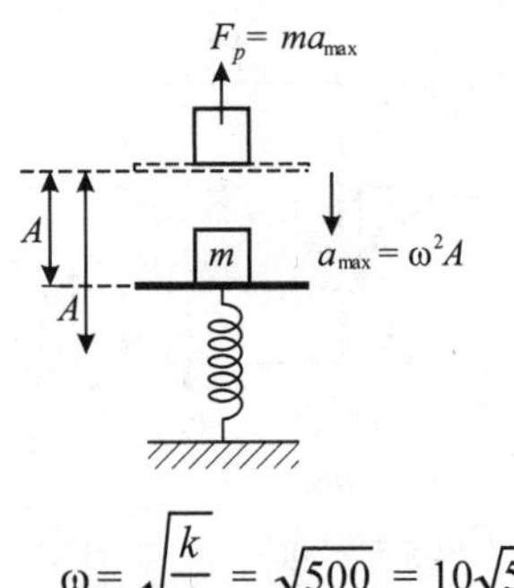

$$\omega = \sqrt{\frac{k}{m}} = \sqrt{500} = 10\sqrt{5} \text{ rad/s}$$

At upper extreme position block will experience a Pseudo force upward w.r. to pan, given as

$$F_P = ma_{max} = m\omega^2A$$

If $F_P > mg$ then block will have tendency to get detached from pan $\quad m\omega^2A > mg$

$$\Rightarrow \quad A > \frac{g}{\omega^2}$$

$$\Rightarrow \quad A > \frac{10}{500}$$

$$\Rightarrow \quad A > 0.02 \text{ m} = 2 \text{ cm}$$

for $A > 2$cm somewhere at position above mean position mass can get detached from pan. **Ans. (A)**

Sol. 37 If at equilibrium spring extension is h then we have

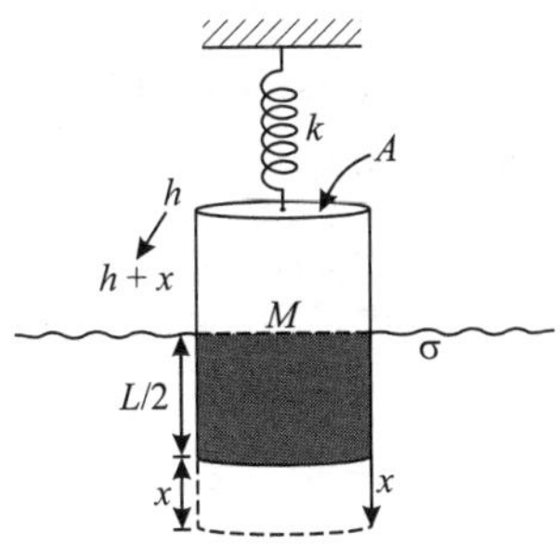

FBD of cylinder

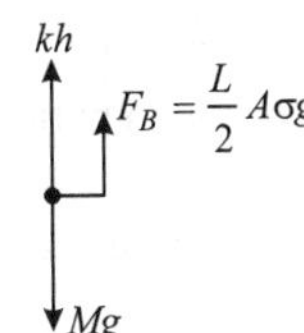

For equation of cylinder, we have

$$kh + \frac{L}{2}A\sigma g = Mg \qquad \ldots(1)$$

After pushing down the cylinder by x, its FBD is

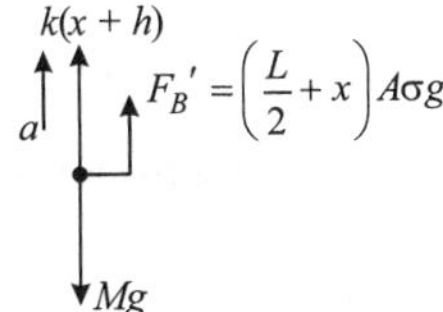

Equation of motion of cylinder

$$k(x+h) + \left(\frac{L}{2}+x\right)A\sigma g - Mg = Ma$$

$$\Rightarrow \qquad a = \left(\frac{k + A\sigma g}{M}\right)x$$

As a is in opposite direction to x, we use

$$a = -\left(\frac{k + A\sigma g}{M}\right)x$$

Comparing with SHM equation $a = -\omega^2 x$, we have frequency of its SHM is

$$\Rightarrow \qquad \omega = \sqrt{\frac{k + A\sigma g}{M}}$$

Time period of oscillation $T = \frac{2\pi}{\omega}$

$$\Rightarrow \qquad T = 2\pi\sqrt{\frac{M}{k + A\sigma g}}$$

As nothing is specified in question about damping. **Ans. (D)**

Sol. 38 Time period of oscillation for a simple pendulum

$$T = 2\pi\sqrt{\frac{l}{g}}$$

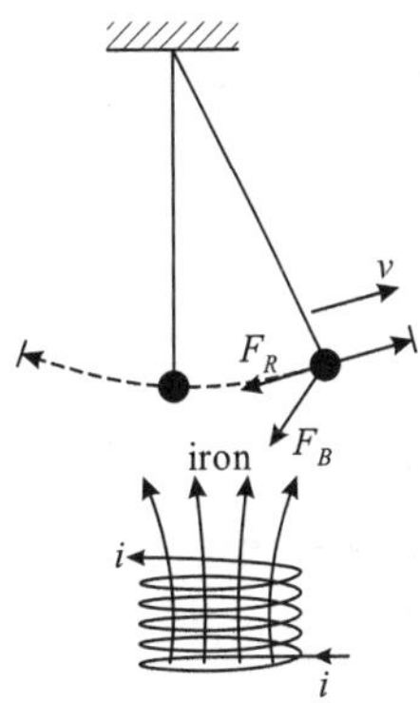

Due to coil the net restoring force on bob during oscillation increases (as shown) so angular frequency of oscillation increases $\Rightarrow$ T decreases.

$$\Rightarrow \qquad T < 2\pi\sqrt{\frac{l}{g}}$$

Coil behaves as a magnet and damping is more than in air alone. **Ans. (D)**

Sol. 39

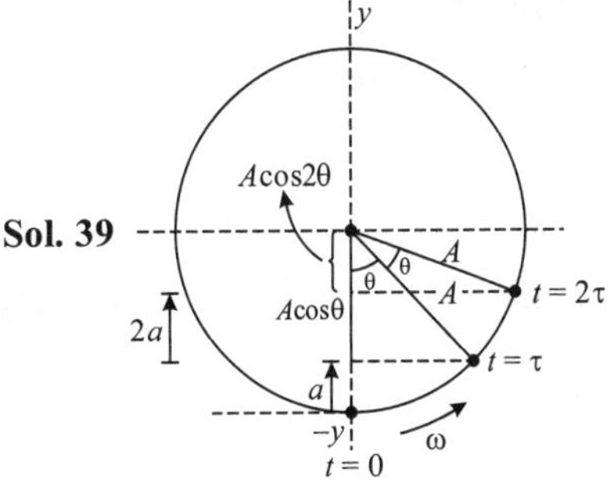

From $t = 0$ to $t = \tau$, displacement of particle in SHM is

$$a = A - A\cos(\omega\tau) \qquad \ldots(1)$$

from $t = \tau$ to $t = 2\tau$, displacement of particle is $2a$

$$\Rightarrow \qquad 3a = A - A\cos(2\omega\tau) \qquad \ldots(2)$$

from equation (1) $\cos(\omega\tau) = \left(\frac{A-a}{A}\right)$

from equation (2) $2\cos^2(\omega\tau) - 1 = \frac{A-3a}{A}$

$$\Rightarrow \qquad 2\left(\frac{A-a}{A}\right)^2 - 1 = \frac{A-3a}{A}$$

$$\Rightarrow \qquad \frac{2A^2 + 2a^2 - 4aA}{A^2} = \frac{2A-3a}{A}$$

$$\Rightarrow \qquad 2A^2 + 2a^2 - 4aA = 2A^2 - 3aA$$

$$\Rightarrow \qquad 2a^2 = aA$$

$$\Rightarrow \qquad A = 2a$$

From equation-(1) we have

$$\cos(\omega\tau) = \left(\frac{A-a}{A}\right) = \frac{1}{2}$$

$\Rightarrow \quad \omega\tau = \frac{\pi}{3}$

$\Rightarrow \quad \left(\frac{2\pi}{T}\right)\tau = \frac{\pi}{3}$

$\Rightarrow \quad T = 6\tau$ **Ans. (D)**

Sol. 40 Stokes's Law

$$F_v = -6\pi\eta r v$$

if $\quad F_v = -bv$

$\Rightarrow \quad b \propto \eta$

$\Rightarrow \quad \frac{b_{air}}{b_{CO_2}} = \frac{\eta_{air}}{\eta_{CO_2}} = 1.3$

For damped oscillation in air, we use

$$A = A_0 e^{-bt/2m}$$

at $\quad t = 0, A_0 = 10$ cm

and $t = 40$s, $A = 8$ cm

$\Rightarrow \quad 8 = 10e^{\frac{-b_{air}\times 40}{2m}}$ ...(1)

For damped oscillations in CO_2, we use

at $\quad t = 0, A_0 = 10$ cm

and $t = t_1, A = 5$ cm

$\Rightarrow \quad 5 = 10e^{\frac{-b_{CO_2}\times t_1}{2m}}$ ...(2)

from equation-(1) $\frac{40 b_{air}}{2m} = \ln\left(\frac{5}{4}\right)$ ...(3)

from equation-(2) $\frac{t_1 b_{CO_2}}{2m} = \ln(2)$ ...(4)

using (4)/(3) we have

$$\frac{t_1}{40(b_{air}/b_{CO_2})} = \frac{\ln(2)}{\ln(5) - \ln(4)}$$

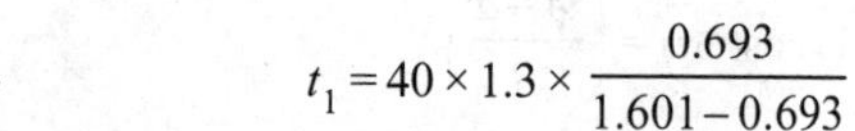

$\Rightarrow \quad t_1 = 40 \times 1.3 \times \frac{0.693}{1.601 - 0.693}$

$\Rightarrow \quad t_1 \approx 167$s **Ans. (C)**

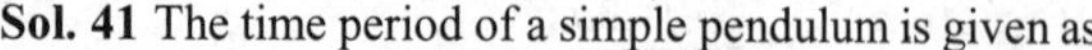

Sol. 41 The time period of a simple pendulum is given as

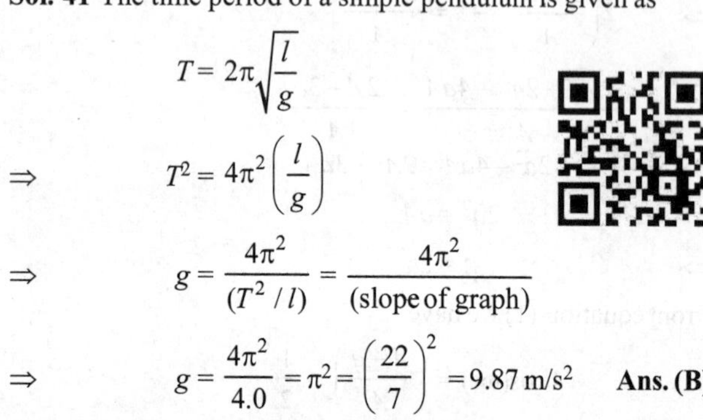

$$T = 2\pi\sqrt{\frac{l}{g}}$$

$\Rightarrow \quad T^2 = 4\pi^2\left(\frac{l}{g}\right)$

$\Rightarrow \quad g = \frac{4\pi^2}{(T^2/l)} = \frac{4\pi^2}{(\text{slope of graph})}$

$\Rightarrow \quad g = \frac{4\pi^2}{4.0} = \pi^2 = \left(\frac{22}{7}\right)^2 = 9.87 \text{ m/s}^2$ **Ans. (B)**

Sol. 42 As $\quad x = a_1\cos\omega t$

$$y = a_2\cos 2\omega t = a_2(2\cos^2\omega t - 1)$$

$\Rightarrow$ at $t = 0$, $x = a$ and $y = a_2$

$\Rightarrow \quad y = a_2\left(\frac{2x^2}{a_1^2} - 1\right)$

$\Rightarrow \quad x^2 = \frac{a_1^2}{2a_2}y + \frac{a_1^2}{2}$

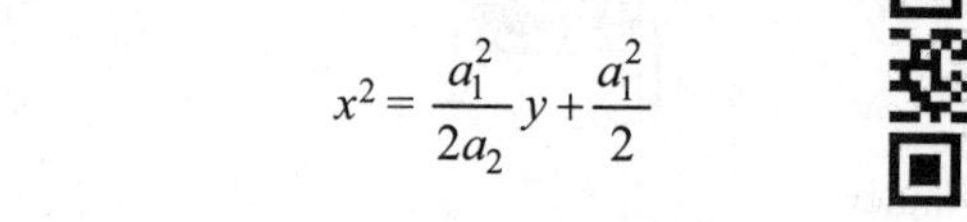

At $y = 0$, $\quad x = \pm a_1/2$

$\Rightarrow$ Trajectory is an upward open parabola. **Ans. (A)**

Sol. 43 In SHM the restoring force $\propto$ displacement from mean position $x = 0$.

If $x \neq 0$ at mean position than force is a linear function of x and direction of force always toward mean position.

In option (A), (B) and (C) function depends upon x^2

$\Rightarrow$ it can not be SHM

In option (D) it is $(-bx)$

$\Rightarrow \quad F = -bx$ (it is SHM) **Ans. (D)**

Sol. 44 $\quad A = 1$ cm

$$\omega = \frac{2\pi}{T} = \frac{2\pi}{0.5} = 4\pi \text{ rad/s}$$

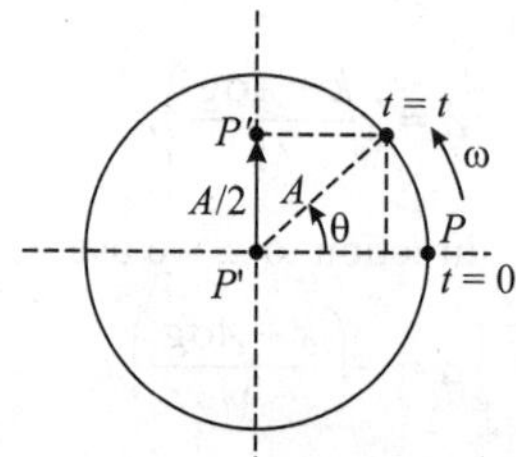

To cover distance from mean position to $A/2$.

Phase change is $\quad \theta = \sin^{-1}\frac{A/2}{A} = \sin^{-1}\frac{1}{2} = \frac{\pi}{6}$

time taken by SHM in going from mean position to $A/2$ is

$$t = \frac{\theta}{\omega} = \frac{\pi/6}{4\pi} = \frac{1}{24}\text{ s}$$

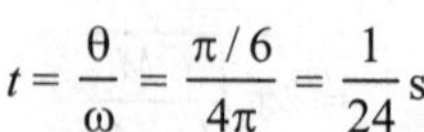

Average velocity in moving from $x = 0$ to $x = A/2$ is

$$v_{avg} = \frac{\Delta x}{\Delta t} = \frac{A/2}{1/24}$$

$= 12\text{A} = 12$ cm/s **Ans. (C)**

Sol. 45

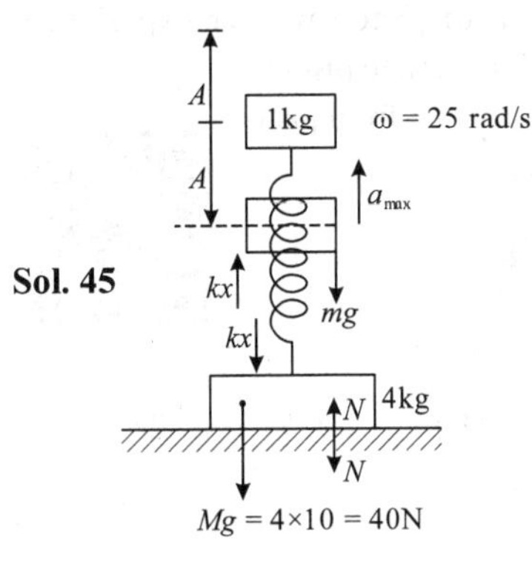

For 4 kg block, force on ground

$$N = Mg + kx|_{max} \quad \ldots(1)$$

Maximum force will be exerted on floor when 1 kg mass is at its lowest extreme so that spring compression is maximum.

Acceleration of 1 kg block at this position is

$$a_{max} = \omega^2 A = (25)^2 \times 0.016 = 10 \text{ m/s}^2$$

Equation of motion of 1 kg block is

$$kx - mg = ma_{max}$$

$$kx = mg + ma_{max} = 1 \times 10 + 1 \times 10$$

$$= 20 \text{ N (max)}$$

From equation-(1)

$$N_{max} = Mg + kx|_{max}$$

$$= 40 \times 10 + 20 = 60 \text{ N}$$

Ans. (C)

Sol. 46 Angular frequency of undamped oscillator

$$\omega_0 = \sqrt{\frac{k}{m}}$$

Change in angular frequency $\Delta\omega = \omega_0 - \omega$

$$\Rightarrow \quad \Delta\omega = \sqrt{\frac{k}{m}} - \sqrt{\frac{k}{m} - \frac{r^2}{4m^2}}$$

$$\Rightarrow \quad \Delta\omega = \sqrt{\frac{k}{m}}\left(1 - \sqrt{1 - \frac{r^2}{4mk}}\right)$$

$$\Rightarrow \quad \frac{\Delta\omega}{\omega_0} = \left(1 - \sqrt{1 - \frac{0.08}{4}}\right)$$

$$\Rightarrow \quad \frac{\Delta\omega}{\omega_0} = 1 - \sqrt{1 - 0.02} = 1 - (1 - 0.01) = 0.01$$

As $\quad T = \dfrac{2\pi}{\omega}$

$$\Rightarrow \quad \frac{\Delta T}{T_0} = -\frac{\Delta\omega}{\omega_0}$$

$\Rightarrow$ As ω decreases by 1 % $\Rightarrow$ T increases by 1% **Ans. (A)**

Sol. 47 If for length l of pendulum time period is T then

$$T = 2\pi\sqrt{\frac{l}{g}} \Rightarrow l = \frac{T^2 g}{4\pi^2} \quad \ldots(1)$$

after adding a mass M, if length becomes $l + \Delta l$

then we have $\quad T_M = 2\pi\sqrt{\dfrac{l + \Delta l}{g}} \Rightarrow l + \Delta l = \dfrac{T_M^2 g}{4\pi^2} \quad \ldots(2)$

If Y is Young's modulus of wire than we use

$$Y = \frac{\text{Stress}}{\text{Strain}} = \frac{Mg / A}{\Delta l / l}$$

$$\Rightarrow \quad \frac{\Delta l}{l} = \frac{Mg}{YA}$$

We use (2)/(1)

$$\Rightarrow \quad 1 + \frac{\Delta l}{l} = \left(\frac{T_M}{T}\right)^2$$

$$\Rightarrow \quad 1 + \frac{Mg}{YA} = \left(\frac{T_M}{T}\right)^2$$

$$\Rightarrow \quad \frac{1}{Y} = \left[\left(\frac{T_M}{T}\right)^2 - 1\right]\frac{A}{Mg}$$

Ans. (A)

Sol. 48 For SHM, KE and PE as a function of displacement are given as

$$K_x = \frac{1}{2}m\omega^2(A^2 - x^2)$$

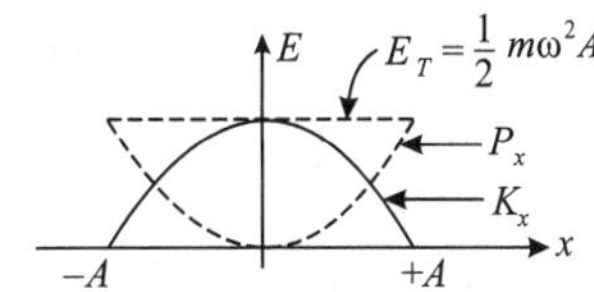

and $\quad P_x = \dfrac{1}{2}m\omega^2 x^2$ **Ans. (B)**

Sol. 49 In damped oscillations, mechanical energy as a function of time is given as

$$E = E_0 e^{-bt/m} \quad \ldots(1)$$

at $\quad t = 0, E_0 = 45$ J

and at $t = 15$s, $E = 15$ J

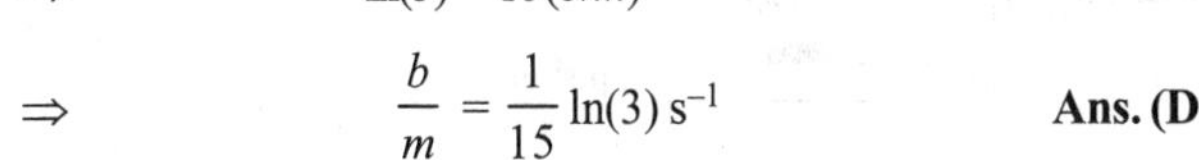

$$\Rightarrow \quad 15 = 45\, e^{-\left(b/m\right)15}$$

$$\Rightarrow \quad \frac{1}{3} = e^{-15\left(b/m\right)}$$

$$\Rightarrow \quad \ln(3) = 15(b/m)$$

$$\Rightarrow \quad \frac{b}{m} = \frac{1}{15}\ln(3) \text{ s}^{-1}$$

Ans. (D)

Sol. 50 As $\quad x = a\sin\omega t \quad$ at $t = 0, x = 0$

and $\quad y = a\sin 2\omega t \quad y = 0$

$\Rightarrow$ Answer could be option (A) or (C), so we check both here

(A) In one half oscillation of y, x completes its one oscillation

$\Rightarrow$ frequency of x is twice that of y.

(C) In one half oscillation of x, y completes its one oscillation

$\Rightarrow$ frequency of y is twice that of x, so this is correct answer for the given equation.

Ans. (C)

Sol. 51 Given that at $t = 0$, $x = 0$ and oscillator starts from rest $v = 0$. Checking given option for that

(A) $x = C(\sin t + \frac{1}{2}\cos t)|_{t=0} = C(0 + \frac{1}{2}) \neq 0 \quad \times$

(B) $x = C(\cot t - \frac{1}{2}\cos t)|_{t=0} \to \infty \quad \times$

(C) $x = C(\sin t + \frac{1}{2}\cos t)|_{t=0} = 0 \quad ✓$

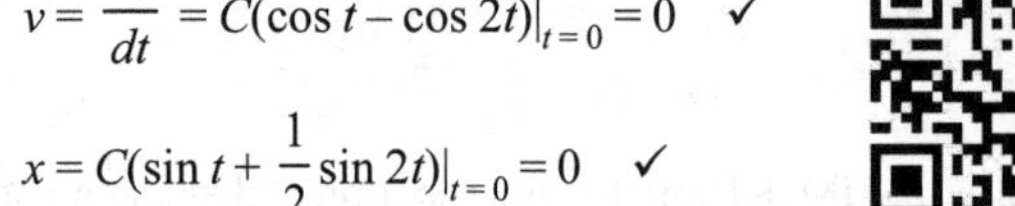

$$v = \frac{dx}{dt} = C(\cos t - \cos 2t)|_{t=0} = 0 \quad ✓$$

(D) $x = C(\sin t + \frac{1}{2}\sin 2t)|_{t=0} = 0 \quad ✓$

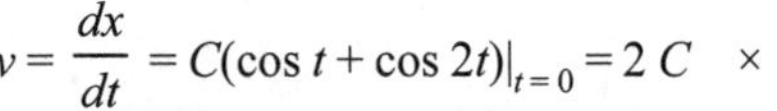

$$v = \frac{dx}{dt} = C(\cos t + \cos 2t)|_{t=0} = 2C \quad \times$$

Ans. (C)

Sol. 52 Speed of particle when it is at a distance $\frac{2A}{3}$ from mean position is

$$v = \omega\sqrt{A^2 - \left(\frac{2A}{3}\right)^2} = \frac{\sqrt{5}}{3}A\omega$$

After increasing the speed 3 times if new amplitude becomes A' then we use

$$3v = \omega\sqrt{(A')^2 - \left(\frac{2A}{3}\right)^2} = \sqrt{5}\,Aw$$

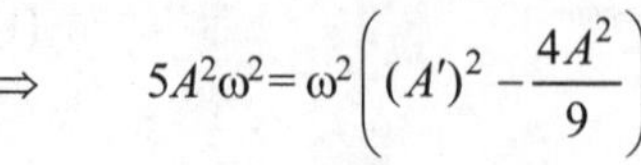

$$\Rightarrow \quad 5A^2\omega^2 = \omega^2\left((A')^2 - \frac{4A^2}{9}\right)$$

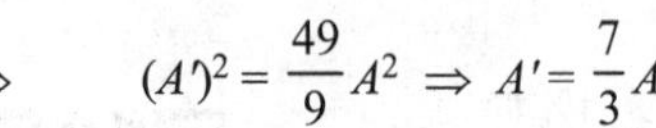

$$\Rightarrow \quad (A')^2 = \frac{49}{9}A^2 \Rightarrow A' = \frac{7}{3}A$$

Ans. (D)

Sol. 53

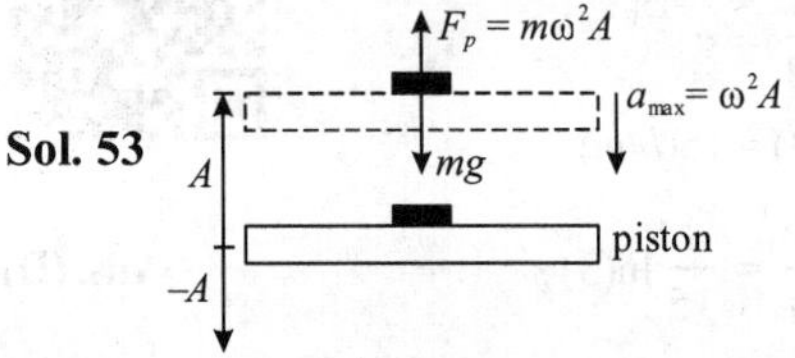

At upper extreme position of piston washer experiences maximum upward Pseudo force w.r. to piston.

If $\quad F_p \geq mg$ then it looses contact from piston

$$\Rightarrow \quad m\omega^2 A = mg$$

$$\Rightarrow \quad \omega = \sqrt{\frac{g}{A}}$$

$$\Rightarrow \quad f = \frac{1}{2\pi}\sqrt{\frac{10}{0.07}} = \frac{1}{2\pi}\sqrt{\frac{100}{7}} = \frac{1}{2}\sqrt{14.2}$$

$$\Rightarrow \quad f \approx \frac{3.8}{2} = 1.9\text{ Hz}$$

Ans. (B)

Sol. 54

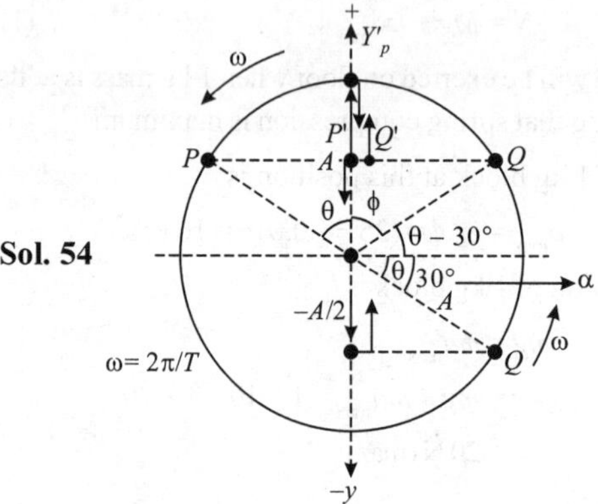

$$A\sin\alpha = A/2$$

$$\Rightarrow \quad \sin\alpha = 1/2$$

$$\Rightarrow \quad \alpha = 30°$$

P' and Q' are projections of P and Q in uniform circular motion along YY' which are in SHM.

In figure shown we have

$$\phi = 90° - (\theta - 30°) = 120° - \theta$$

and we have $\theta = \phi$

$$\Rightarrow \quad \theta = 120 - \theta \Rightarrow \theta = 60° = \pi/3$$

Time taken by P and Q to traverse angle $\pi/3$ is

$$t = \frac{\theta}{\omega} = \frac{\pi/3}{2\pi/T} = \frac{T}{6}$$

Ans. (D)

Sol. 55 For a particle starting SHM from mean position, equation of motion is given as

$$y = A\sin\omega t$$

$$\Rightarrow \text{ particle's velocity } v = \frac{dy}{dt} = A\omega\cos\omega t$$

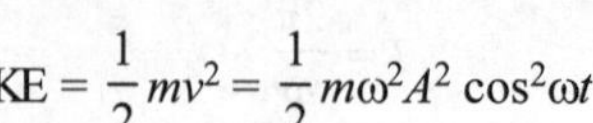

$$\Rightarrow \quad \text{KE} = \frac{1}{2}mv^2 = \frac{1}{2}m\omega^2 A^2\cos^2\omega t$$

KE_{max} at $t = 0$

In option (B), KE oscillates at twice the frequency of SHM.

Ans. (B)

Sol. 56 Case-1:

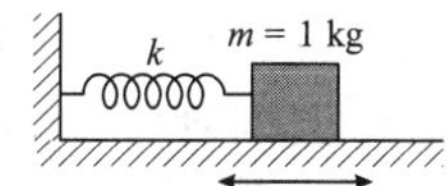

Oscillation frequency of block

$$f = \frac{1}{2\pi}\sqrt{\frac{k}{m}} = 1\text{ Hz}$$

$$\Rightarrow \quad k = 4\pi^2\text{ N/m}$$

Case-2:

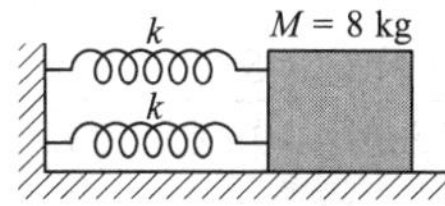

In parallel effective force constant of spring

$$k_{eff} = k_1 + k_2 = 2k = 8\pi^2\text{ N/m}$$

Oscillation frequency $f' = \frac{1}{2\pi}\sqrt{\frac{2k}{M}}$

$$= \frac{1}{2\pi}\sqrt{\frac{8\pi^2}{8}} = \frac{1}{2}\text{Hz}$$

Ans. (C)

Sol. 57 Time period of oscillation of a pendulum is given as

$$T = 2\pi\sqrt{\frac{l}{g}}$$

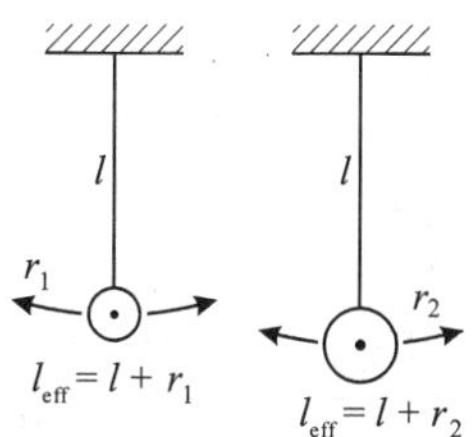

For uniform mass distribution COM is taken at centre of bob.

$$\ln T = \ln\frac{2\pi}{\sqrt{g}} + \ln\sqrt{l}$$

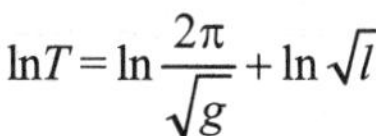

Relative change in period

$$\frac{dT}{T} = \frac{1}{2}\frac{dl}{l} \qquad \text{here } dl = |r_2 - r_1|$$

$$\Rightarrow \quad 5 \times 10^{-4} = \frac{1}{2}\frac{|r_2 - r_1|}{1\text{m}}$$

$$\Rightarrow \quad |r_2 - r_1| = 2 \times 5 \times 10^{-4} = 10^{-3}\text{ m} = 1\text{ mm}$$ **Ans.(B)**

Sol. 58 For damped oscillations, the oscillation amplitude is given as

$$A = A_0 e^{\frac{-bt}{2m}} \qquad \ldots(1)$$

$b \to$ damping coefficient

$m \to$ mass of block

As we know mechanical energy of oscillation $E \propto A^2$

E will be half when $A \to \frac{A}{\sqrt{2}}$

from equation-(1), if at $t = t$, $A_0 \to \frac{A}{\sqrt{2}}$

$$\Rightarrow \quad \frac{A_0}{\sqrt{2}} = A_0 e^{\frac{-bt}{2m}}$$

$$\Rightarrow \quad \frac{bt_1}{2m} = \ln\sqrt{2} = \frac{1}{2}\ln 2$$

$$\Rightarrow \quad t_1 = \frac{m}{b}\ln(2) = \frac{0.1}{0.01} \times 0.693$$

$$\Rightarrow \quad t_1 = 6.93\text{ s}$$

Ans. (D)

Sol. 59 If in SHM amplitude is A and angular frequency is ω, then we have

max. velocity $v_{max} = A\omega$

max. acceleration $a_{max} = \omega^2 A$

Given that $\frac{a_{max}}{v_{max}} = \omega = 10\text{ s}^{-1}$

General equation of SHM can be taken as

$$y = A\sin(\omega t + \alpha)$$

at $t = 0, y = 5\text{m}, \alpha = \frac{\pi}{4}$

$$5 = A\sin\left(\frac{\pi}{4}\right)$$

$$\Rightarrow \quad 5 = A\left(\frac{1}{\sqrt{2}}\right)$$

$$\Rightarrow \quad A = 5\sqrt{2}\text{ m}$$

$$\Rightarrow \quad a_{max} = \omega^2 A = (10)^2 \times 5\sqrt{2}$$

$$= 500\sqrt{2}\text{ m/s}^2$$

Ans. (B)

Sol. 60 Mass of one atom of silver is

$$m = \frac{M}{Av} = \frac{108 \times 10^{-3}}{6.02 \times 10^{23}}$$

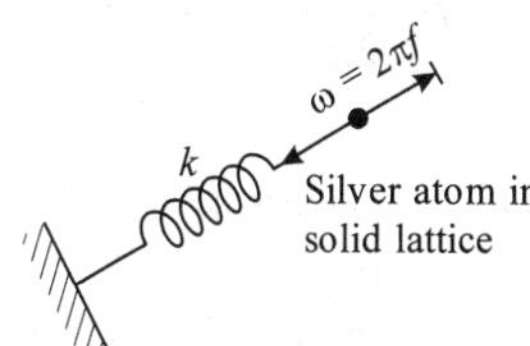

Spring represent the bonds connecting the silver atom with other atoms.

For SHM angular frequency of atom is given as

$$\omega=\sqrt{\frac{k}{m}}$$

$\Rightarrow$ Oscillation frequency $f=\frac{\omega}{2\pi}$

$$\Rightarrow \quad 10^{12}=\frac{1}{2\times 3.14}\sqrt{\frac{k\times 6.02\times 10^{23}}{108\times 10^{-3}}}$$

$$\Rightarrow \quad k=7.1\text{ N/m}$$ **Ans. (B)**

Sol. 61 For $a=b$, general equation of Lissajous's figure is given as

$$\frac{x^2}{A^2}+\frac{y^2}{B^2}-\frac{2xy}{AB}\cos\delta=\sin^2\delta$$

(B) $\frac{x^2}{A^2}+\frac{y^2}{B^2}-\frac{2xy}{A^2}(0)=1 \Rightarrow x^2+y^2=A^2 \rightarrow$ circle (✗)

(C) $\frac{x^2}{A^2}+\frac{y^2}{B^2}-\frac{2xy}{AB}(0)=1 \Rightarrow \frac{x^2}{A^2}+\frac{y^2}{B^2}=1 \rightarrow$ ellipse (✓)

(D) $\frac{x^2}{A^2}+\frac{y^2}{B^2}-\frac{2xy}{AB}(1)=0 \rightarrow$ Not a parabola. (✗) **Ans. (C)**

Sol. 62 If particle starts SHM from its mean position (say) then equation of SHM is

$$y=A\sin\omega t$$

With given conditions, we have

$$a=A\sin\omega t_0 \qquad \ldots(1)$$

$$b=A\sin 2\omega t_0 \qquad \ldots(2)$$

$$c=A\sin 3\omega t_0 \qquad \ldots(3)$$

$$(1)+(3) \Rightarrow \quad a+c=2A\sin 2\omega t_0\cos\omega t_0$$

$$\Rightarrow \quad \cos\omega t_0=\frac{a+c}{2b}$$

$$\Rightarrow \quad \omega t_0=\cos^{-1}\left(\frac{a+c}{2b}\right)$$

$$\omega=\frac{1}{t_0}\cos^{-1}\left(\frac{a+c}{2b}\right)$$

$$f=\frac{\omega}{2\pi}=\frac{1}{2\pi t_0}\cos^{-1}\left(\frac{a+c}{2b}\right)$$ **Ans. (D)**

Sol. 63 Restoring force on mass M is

$$F=\left|\frac{dV}{dx}\right|=k(x-X)=0 \text{ (a mean position)}$$

$$\Rightarrow \quad x=X$$

I collision: rest, M, u, m, $x=X$, $P_i=mu$

II collision: F, M, u, m, rest, $P_f=mu$

III collision: $P=mu$, $M+3m$

IV collision: rest, mu, $M+4m$

13th collision: v, $P=mu$, $M+13m$

$mu=(M+13m)v \Rightarrow v=\frac{mu}{M+13m}$

After 13th collision velocity of combined mass $M+13m$ is given as

$$v=\frac{mu}{M+13m}=\frac{5\times 1}{10+65}=\frac{5}{75}=\frac{1}{15}$$

at mean position

$v_f=0$, v, A, $(M+13m)$, $x=X-A$, $x=X$

When block reaches extreme position, its KE at mean position is transformed into PE at extreme position.

$\Rightarrow$ By conservation of energy

$$\frac{1}{2}(M+13m)v^2=\frac{1}{2}k(X-A-x)^2$$

$$\Rightarrow \quad \frac{1}{2}(10+65)\left(\frac{1}{15}\right)^2=\frac{1}{2}\times 1\times A^2$$

$$\Rightarrow \quad A^2=\frac{75}{15\times 15}=\frac{1}{3} \Rightarrow A=\frac{1}{\sqrt{3}}$$ **Ans. (B)**

Sol. 64 Angular frequency of SHM of rod is

$$\omega=\sqrt{\frac{k}{I}}$$

$$\Rightarrow \quad \omega=\sqrt{\frac{3k}{ml^2}} \qquad \left(I=\frac{ml^2}{9}+\frac{m}{2}\cdot\frac{4l^2}{9}=\frac{ml^2}{3}\right)$$

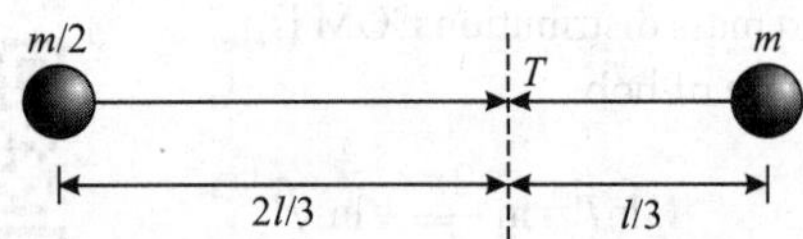

$$\dot{\theta}_{max}=\omega\theta_0 \qquad \text{[maximum angular speed]}$$

tension in rod is given as

$$\Rightarrow \quad T=m\dot{\theta}^2 r_1$$

$$\Rightarrow \quad T=m\dot{\theta}^2\frac{l}{3}$$

$$\Rightarrow \quad T=m\omega^2\theta_0^2\frac{l}{3}$$

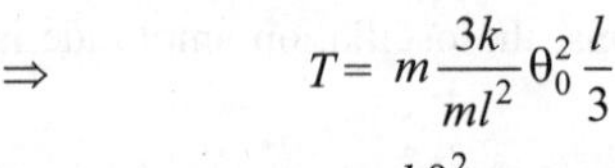

$$\Rightarrow \quad T=m\frac{3k}{ml^2}\theta_0^2\frac{l}{3}$$

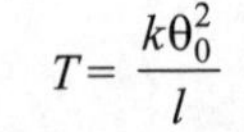

$$\Rightarrow \quad T=\frac{k\theta_0^2}{l}$$ **Ans. (D)**

Sol. 65 Angular frequency of torsional oscillations is given by

$$\omega = \sqrt{\frac{k}{I}}$$

$$\omega_1 = \sqrt{\frac{k}{\frac{M(2L)^2}{12}}} \qquad \text{[Oscillations of rod]}$$

$$\omega_2 = \sqrt{\frac{k}{\frac{M(2L)^2}{12} + 2m\left(\frac{L}{2}\right)^2}}$$

[Oscillations of rod with masses]

$$\omega_2 = 0.8\omega_1$$

$$\Rightarrow \quad \frac{m}{M} = 0.375$$

Ans. (B)

Sol. 66 Potential energy $(U) = \frac{1}{2}kx^2$

Kinetic energy $(K) = \frac{1}{2}kA^2 - \frac{1}{2}kx^2$

Given that, $U = k$

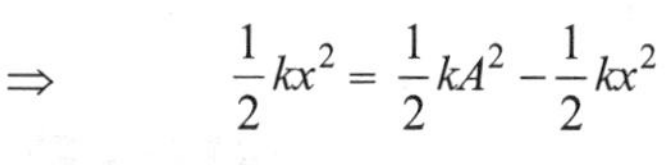

$$\Rightarrow \quad \frac{1}{2}kx^2 = \frac{1}{2}kA^2 - \frac{1}{2}kx^2$$

$$\Rightarrow \quad x = \pm\frac{A}{\sqrt{2}}$$

Ans. (C)

Sol. 67 Velocity and acceleration of particle in SHM's

$$v = \omega\sqrt{A^2 - x^2} \qquad \ldots(1)$$

$$a = -\omega^2 x \qquad \ldots(2)$$

Given that

$$|v| = |a| \qquad \ldots(3)$$

$$\omega\sqrt{A^2 - x^2} = \omega^2 x$$

$$\Rightarrow \quad A^2 - x^2 = \omega^2 x^2$$

$$\Rightarrow \quad 5^2 - 4^2 = \omega^2(4^2)$$

$$\Rightarrow \quad 3 = \omega \times 4$$

$$\Rightarrow \quad \omega = 3/4$$

$$T = 2\pi/\omega = \frac{8\pi}{3}$$

Ans. (D)

Sol. 68

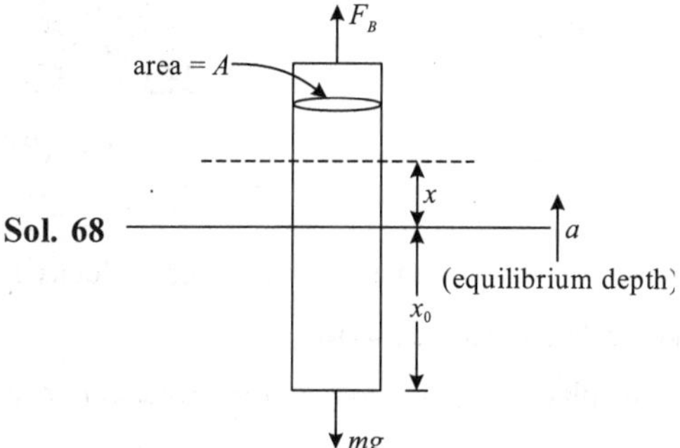

At equilibrium

$$mg = \rho_\omega A x_0 g$$

When further pushed down by x, we use

$$\rho_\omega A(x + x_0)g - mg = ma$$

$$\Rightarrow \quad \rho_\omega Axg = ma$$

$$\Rightarrow \quad a = \frac{\rho_\omega Ag}{m}x$$

Comparing with

$$a = \omega^2 x \text{ we get}$$

$$\omega = \sqrt{\frac{\rho_\omega Ag}{m}}$$

$$\Rightarrow \quad \omega = \sqrt{\frac{10^3 \times \pi(2.5)^2 \times 10^{-4} \times 10}{310 \times 10^{-6} \times 10^3}}$$

$$\Rightarrow \quad \omega = \sqrt{63.30} = 7.95 \text{ rad/s}$$

Ans. (None)

Sol. 69 Kinetic energy of particle in SHM is

$$K = \frac{1}{2}m\omega^2 A^2 \cos^2\omega t \qquad \ldots(1)$$

Potential energy of particle in SHM is

$$U = \frac{1}{2}m\omega^2 A^2 \sin^2\omega t \qquad \ldots(2)$$

$$\Rightarrow \quad \frac{K}{U} = \cot^2\omega t = \cot^2\frac{\pi}{90}(210) = \frac{1}{3}$$

Ans. (C)

Sol. 70 Angular frequency of simple pendulum is given as

$$\omega = \sqrt{\frac{g_{eff}}{l}}$$

$$\Rightarrow \quad \frac{\Delta\omega}{\omega} = \frac{1}{2}\frac{\Delta g_{eff}}{g_{eff}}$$

$$\Rightarrow \quad \Delta\omega = \frac{1}{2}\frac{\Delta g_{eff}}{g_{eff}} \times \omega \qquad \ldots(1)$$

Due to vertical oscillations of support change in g_{eff} is given as

$$\Delta g_{eff} = 2\omega^2 A$$

So from equation-(1), we have

$$\Delta\omega = \frac{1}{2} \times \frac{2A\omega_s^2}{10} \times 10$$

$$\Rightarrow \quad \Delta\omega = 10^{-3} \text{ rad/sec}$$

Ans. (A)

Sol. 71 Maximum kinetic energy at lowest point B is given as

$$K = mgl(1 - \cos\theta) \qquad \ldots(1)$$

Where θ = angular amplitude of pendulum

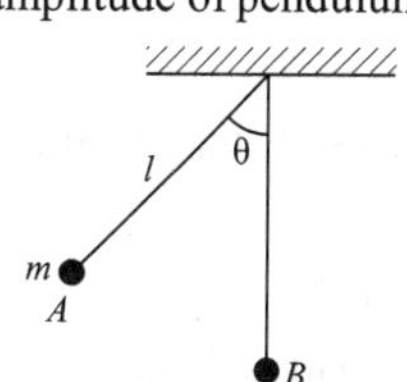

For the given conditions, we use from equation-(1)

$$K_1 = mgl(1 - \cos\theta)$$

and $$K_2 = mgl(2l)(1 - \cos\theta)$$

$\Rightarrow$ $K_2 = 2K_1$ **Ans. (C)**

Sol. 72

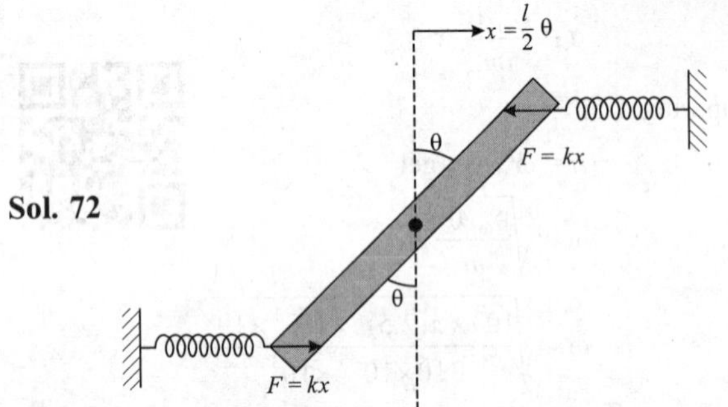

Restoring torque on rod due to its angular displacement is

$$\tau = -2kx + \frac{l}{2}$$

$$\Rightarrow \quad \tau = -\left(\frac{kl^2}{2}\right)\theta$$

Angular acceleration of rod is

$$\alpha = \frac{\tau}{I} = -\frac{\frac{kl^2}{2}\theta}{(Ml^2/12)}$$

$$\Rightarrow \quad \alpha = -\frac{6k}{M}\cdot\theta$$

Comparing with

$\alpha = -\omega^2\theta$, we get

$$\omega = \sqrt{\frac{6k}{M}}$$

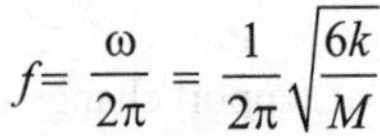

$$\Rightarrow \quad f = \frac{\omega}{2\pi} = \frac{1}{2\pi}\sqrt{\frac{6k}{M}}$$ **Ans. (A)**

Sol. 73 SHM equation given as

$$y = 5\left[\sin(3\pi t) + \sqrt{3}\cos(3\pi t)\right]$$

Above equation cab be rewritten as

$$y = 10\sin\left(3\pi t + \frac{\pi}{3}\right)$$

$\Rightarrow$ Amplitude = 10 cm

$$\Rightarrow \quad T = \frac{2\pi}{\omega} = \frac{2\pi}{3\pi} = \frac{2}{3}\text{ s}$$ **Ans. (D)**

Sol. 74 For a damped oscillator with time amplitude decreases exponentially

$$\Rightarrow \quad A_t = A_0 e^{-\gamma t} \qquad \ldots(1)$$

Given that after 10 oscilations (2 sec)

$$A = \frac{A_0}{2}$$

$$\Rightarrow \text{ From (1) } \frac{A_0}{2} = A_0 e^{-2\gamma}$$

$$\Rightarrow \ln(2) = 2\gamma \Rightarrow \gamma = \frac{\ln(2)}{2}$$

$$\frac{A_0}{1000} = A_0\, e^{-\gamma t}$$

$$\ln(1000) = \frac{\ln 2}{2}\cdot 2$$

$$\Rightarrow \quad t = \frac{6\ln(10)}{\ln 2} = 19.93\text{ s} \simeq 20\text{ s}$$ **Ans. (B)**

Sol. 75 For a simple pendulum

$$T = 2\pi\sqrt{\frac{L}{g_{eff}}}$$

Situation 1 : When pendulum is in air, we use $g_{\text{eff}} = g$

Situation 2 : When pendulum is in liquid, we use

$$g_{\text{eff}} = g\left(1 - \frac{\rho_{\text{liquid}}}{\rho_{\text{body}}}\right) = g\left(1 - \frac{1}{16}\right) = \frac{15g}{16}$$

$$\Rightarrow \quad \frac{T'}{T} = \frac{2\pi\sqrt{\frac{L}{15g/16}}}{2\pi\sqrt{\frac{L}{g}}}$$

$$\Rightarrow \quad T' = \frac{4T}{\sqrt{15}}$$ **Ans. (A)**

Sol. 76 By law of conservation of energy, we use

$$\frac{1}{2}kx^2 = (m_1s_1 + m_2s_2)\Delta T$$

$$\Rightarrow \quad \Delta T = \frac{16\times10^{-2}}{4384} = 3.65\times10^{-5}$$ **Ans. (D)**

Sol. 77 For damped oscillation, amplitude is given as

$$A = A_0 e^{-0.1t} = \frac{A_0}{2}$$

$$\Rightarrow \quad \ln 2 = 0.1t$$

$$\Rightarrow \quad t = 10\ln 2 = 6.93$$

$$\Rightarrow \quad \approx 7\text{ sec}$$ **Ans. (B)**

Sol. 78 **(i)** At $t = \frac{3T}{4}$ displacement is zero where velocity is maximum and acceleration force is zero.

(ii) At $t = T$, displacement is maximum so acceleration is maximum.

(iii) At $t = \frac{T}{4}$, displacement is zero so velocity is maximum.

(iv) at $t = \frac{T}{2}$ it shows displacement is maximum where kinetic energy = 0 and potential energy is maximum.

Hence (i), (ii) and (iii) are correct. **Ans. (D)**

Sol. 79 Angular frequency of SHM of block is given as

Initial $$\omega_1 = \sqrt{\frac{K}{m}}$$

Initial $$\omega_2 = \sqrt{\frac{K}{m/2}}$$

At mean position mass breaks off at maximum speed so we use

$$\omega_1 A_1 = \omega_2 A_2$$

$$\Rightarrow \quad A_2 = \frac{\omega_1 A_1}{\omega_2} = \frac{A_1}{\sqrt{2}}$$ **Ans. (D)**

Sol. 80 The ring acts like a compound pendulum. In its oscillations in plane, time period is given as

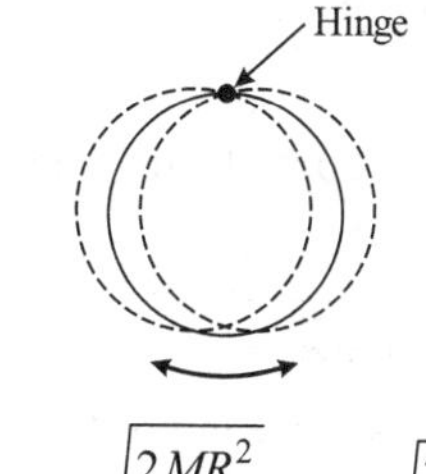

$$T_1 = 2\pi\sqrt{\frac{2MR^2}{MgR}} = 2\pi\sqrt{\frac{2R}{g}}$$

In its oscillations back and forth. Perpendicular to vertical plane, its time period is given as

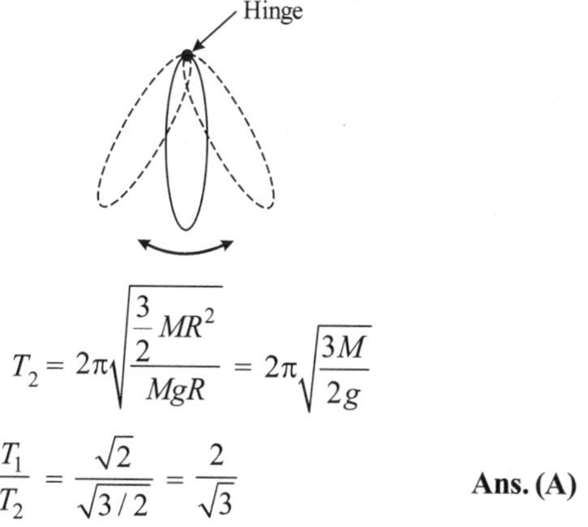

$$T_2 = 2\pi\sqrt{\frac{\frac{3}{2}MR^2}{MgR}} = 2\pi\sqrt{\frac{3M}{2g}}$$

$$\Rightarrow \quad \frac{T_1}{T_2} = \frac{\sqrt{2}}{\sqrt{3/2}} = \frac{2}{\sqrt{3}}$$ **Ans. (A)**

Sol. 81 Given motion equation is

$$y = y_0 \sin^2 \omega t = \frac{y_0}{2}(1 - \cos 2\omega t)$$

$$\Rightarrow \quad y - \frac{y_0}{2} = -\frac{y_0}{2}\cos 2\omega t$$

As y is measured from lower end of unstretched spring here $\frac{y_0}{2}$ is the amplitude of SHM. Thus we use

$$\frac{y_0}{2} = \frac{mg}{k}$$

SHM frequency is given as

$$2\omega = \sqrt{\frac{k}{m}} = \sqrt{\frac{2g}{y_0}}$$

$$\Rightarrow \quad \omega = \sqrt{\frac{g}{2y_0}}$$ **Ans. (B)**

Sol. 82 When mass m is placed on M, momentum of system remains conserved and after placing the mass m angular frequency of oscillations and amplitude changes. Here we use

$$MA\omega = (m + M)A'\omega'$$

$$\Rightarrow \quad MA\sqrt{\frac{k}{M}} = (m+M)A'\sqrt{\frac{k}{m+M}}$$

$$\Rightarrow \quad A' = A\sqrt{\frac{M}{M+m}}$$ **Ans. (A)**

Sol. 83 For a particle executing SHM, the equation of motion and velocity of particle is given as

$$x = A\sin(\omega t + \phi)$$

$$v = \omega A\cos(\omega t + \phi)$$

$$\Rightarrow \quad \frac{v^2}{\omega^2 A^2} + \frac{x^2}{A^2} = 1$$

Above is the equation of ellipse between v and x plotted on a coordinate system. **Ans. (C)**

Sol. 84 In the given situation when block is displaced, the two springs will deform by same distance and exert restoring force on block in same direction thus we can consider the two springs to be connected in parallel so we use

$$K_{eq} = K_1 + K_2 = K + K = 2K$$

Thus time period of oscillations of body is given as

$$T = 2\pi\sqrt{\frac{M}{K_{eq}}} = 2\pi\sqrt{\frac{M}{2K}}$$

$$f = \frac{1}{T} = \frac{1}{2\pi}\sqrt{\frac{2K}{M}}$$ **Ans. (C)**

Sol. 85 The time period of small oscillations of a simple pendulum is given as

$$T = 2\pi\sqrt{\frac{L}{g}}$$

$$\Rightarrow \quad g = \frac{4\pi^2 l}{T^2}$$

$$\Rightarrow \quad \frac{\Delta g}{g} = \frac{\Delta l}{l} + \frac{2\Delta T}{T}$$

$$\Rightarrow \quad \frac{\Delta g}{g} = \frac{1\times 10^{-3}}{1} + 2 \times \frac{0.01}{1.95} = 0.0113 \text{ or } 1.13\%$$

Ans. (C)

Sol. 86 For the simple pendulum, we have

$$T = 2\pi\sqrt{\frac{l}{g}}$$

$$\Rightarrow \quad 2 = 2\pi\sqrt{\frac{2}{g}}$$

$$\Rightarrow \quad g = 2\pi^2$$

Ans. (D)

Sol. 87 At $t=0$ if the initial displacement of particle is $A/2$, and it is moving along negative x-direction, the initial phase can be calculated by projection of a particle in uniform circular motion on its diameter as explained in the chapter of SHM in Physics Galaxy Book Vol 2. This is given as

$$\frac{\pi}{2} + \frac{\pi}{3} = \frac{5\pi}{6}$$

Ans. (D)

Sol. 88 In the given situation when block is displaced, the two springs will deform by same distance and exert restoring force on block in same direction thus we can consider the two springs to be connected in parallel so we use

$$k_{eq} = k_1 + k_2$$

$$\Rightarrow \quad k_{eq} = 4k$$

$$\Rightarrow \quad T = 2\pi\sqrt{\frac{m}{k_{eq}}}$$

Ans. (C)

Sol. 89 At M the acceleration of particle will be maximum as it is one extreme position. The time period of oscillation in this SHM is given as

$$T = \frac{0.1}{30°} \times 360° = 1.2 \text{ s}$$

Angular frequency of SHM in this case can be given as

$$\omega = \frac{2\pi}{T} = \frac{5\pi}{3} \text{ rad/s}$$

At M, the restroing force on particle per unit mass is given as

$$F = m\omega^2 A$$

$$\Rightarrow \quad \frac{F}{m} = \omega^2 A$$

Ans. (A)

Sol. 90 For series combination of springs, we use

$$\frac{1}{k_{eq}} = \frac{1}{k_1} + \frac{1}{k_2} = \frac{1}{k} + \frac{1}{k}$$

$$\Rightarrow \quad k_{eq} = \frac{k}{2}$$

Thus time period of oscillation can be given as

$$T = 2\pi\sqrt{\frac{M}{k_{eq}}}$$

$$\Rightarrow \quad T' = 2\pi\sqrt{\frac{M}{k}} \times \sqrt{2}$$

$$\Rightarrow \quad T' = \sqrt{2}T$$

Ans. (B)

Sol. 91 This question is same as the one asked on 24 Feb, 2021 evening shift. Refer solution of this question. **Ans. (A)**

Sol. 92 Second pendulum has a time period of 2 s thusstatement-I is false but with this it is clear that from one extreme to other it takes only half the time period i.e. 1 s thus statement-II is true. **Ans. (C)**

Sol. 93 The speed of particle in SHM as a function of displacement is given as

$$v = \omega\sqrt{A^2 - y^2}$$

Maximum speed occurs at mean position i.e. at $y=0$, thus we have

$$v_{max} = A\omega$$

So half of maximum speed occurs at a position, given as

$$\frac{A\omega}{2} = \omega\sqrt{A^2 - y^2}$$

$$\Rightarrow \quad \frac{A^2}{4} = A^2 - y^2$$

$$\Rightarrow \quad y^2 = \frac{3A^2}{4}$$

$$\Rightarrow \quad y = \frac{\sqrt{3}}{2}A$$

Ans. (3)

Sol. 94 In the given oscillation $\frac{5}{8}^{th}$ of oscillation can also be written as $\left(\frac{1}{2} + \frac{1}{8}\right)^{th}$ of the oscillation, which can be shown on projection diagram of uniform circular motion as given below.

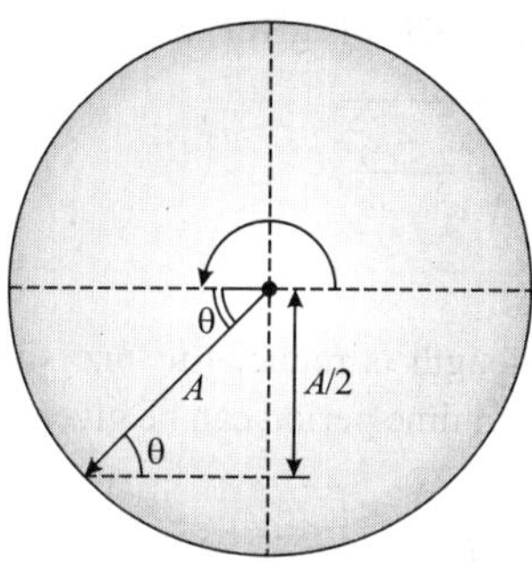

From this diagram if t is the time take for this part of oscillation, we use

$$\pi + \theta = \omega t$$

$$\Rightarrow \quad \frac{7\pi}{6} = \left(\frac{2\pi}{T}\right)t$$

$$\Rightarrow \quad t = \frac{7T}{12}$$

Ans. (7)

Sol. 95 When lift is stationary, we use

$$T = 2\pi\sqrt{\frac{L}{g}}$$

When lift is moving upwards, effective gravity is taken as

$$g_{\text{eff}} = g + \frac{g}{2} = \frac{3g}{2}$$

New time period now is given as

$$T' = 2\pi\sqrt{\frac{L}{g_{eff}}} = 2\pi\sqrt{\frac{2L}{3g}}$$

$$\Rightarrow \quad T' = \sqrt{\frac{2}{3}}\,T$$

Ans. (D)

Sol. 96 By radioactive decay law we have

$$A = A_0 e^{-\gamma t} = A_0 e^{-bt/2\text{m}}$$

$$\Rightarrow \quad \frac{A_0}{2} = A_0 e^{-bt/2\text{m}}$$

$$\Rightarrow \quad \frac{bt}{2m} = \ln 2$$

$$\Rightarrow \quad t = \frac{2m}{b}\ln 2 = \frac{2\times 500\times 0.693}{20} = 34.65 \text{ s}$$

Ans. (A)

Sol. 97 For SHM, at a point where kinetic and potential energy are equal, we have

$$KE = PE$$

$$\Rightarrow \quad \frac{1}{2}m\omega^2(A^2 - x^2) = \frac{1}{2}m\omega^2 x^2$$

$$\Rightarrow \quad A^2 - x^2 = x^2$$

$$\Rightarrow \quad 2x^2 = A^2$$

$$\Rightarrow \quad x = \pm\frac{A}{\sqrt{2}}$$

Ans. (C)

Sol. 98 Time period of oscillations of the two SHM cases in figure-1 and figure-2 are given as

$$T_a = 2\pi\sqrt{\frac{M}{K}}$$

and

$$T_b = 2\pi\sqrt{\frac{M}{K/2}}$$

$$\Rightarrow \quad \frac{T_b}{T_a} = \sqrt{2} = \sqrt{x}$$

$$\Rightarrow \quad x = 2$$

Ans. (2)

Sol. 99 Amplitude of damped oscillations is given as a function of time as

$$A = A_0 e^{-bt/2m}$$

$$\Rightarrow \quad \ln 2 = \frac{b}{2m}\times 120$$

$$\Rightarrow \quad \frac{0.693\times 2\times 1}{120} = b = 1.16\times 10^{-2}\text{ kg/sec}$$

Ans. (None)

Sol. 100 As the maximum velocities are equal, we use

$$A_1\omega_1 = A_2\omega_2$$

$$\Rightarrow \quad A_1\sqrt{\frac{k_1}{m}} = A_2\sqrt{\frac{k_2}{m}}$$

$$\Rightarrow \quad \frac{A_1}{A_2} = \sqrt{\frac{k_2}{k_1}}$$

Ans. (D)

Sol. 101 Half amplitude displacement corresponds to a phase change of $\pi/6$ rad which is one twelfth of the complete oscillation so the time taken can be given as

$$t = \frac{T}{12} = \frac{2}{12} = \frac{1}{6}$$

$$\Rightarrow \quad a = 6$$

Ans. (6)

Sol. 102 As objects on earth surface starts to float, we use

$$mg = m\omega^2 R$$

$$\Rightarrow \quad \omega = \sqrt{\frac{g}{R}} \quad \ldots(1)$$

If duration of the day (time period of rotation of earth) is T, we use

$$T = \frac{2\pi}{\omega} \quad \ldots(2)$$

$$\Rightarrow \quad T = 2\pi\sqrt{\frac{R}{g}} = 2\pi\sqrt{\frac{6400\times 10^3}{10}}$$

$$\Rightarrow \quad \frac{T}{60} = 83.775 \text{ min} \sim 84 \text{ min}$$

Ans. (D)

Sol. 103 The angular frequency of SHM having period π/ω is given as

$$\omega' = \frac{2\pi}{T} = 2\omega$$

The functions respresenting oscillations with angular frequency 2ω are given in option (C) and (D). Option (D) is having its mean position at origin whereas in option (C) it is displaced fro origin but both are the correct answer to this question. **Ans. (C, D)**

Sol. 104 Figure below shows the situation described in the question.

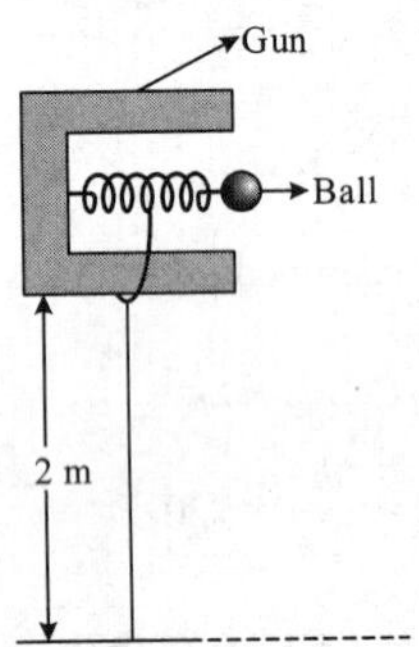

By conservation of energy, we can calculate the kinetic energy of the projected ball as

$$\frac{1}{2}kx^2 = \frac{1}{2}mv^2$$

$$\Rightarrow \quad v = x\sqrt{\frac{x}{m}}$$

$$\Rightarrow \quad v = 0.05 \times \sqrt{\frac{100}{0.1}} = 0.5\sqrt{10} \text{ m/s}$$

Time of flight of ball can be given as

$$T = \sqrt{\frac{2H}{g}} = \sqrt{\frac{2\times 2}{10}} = \frac{2}{\sqrt{10}} \text{ s}$$

Range of ball on ground can be calculated as

$$d = ut$$

$$\Rightarrow \quad d = 0.5\sqrt{10} \times \left(\frac{2}{\sqrt{10}}\right) = 1 \text{ m} \qquad \textbf{Ans. (1)}$$

Sol. 105 Velocity of particle during SHM at a displacement x from its equilibrium position is given as

$$v = \omega\sqrt{A^2 - x^2}$$

$$\Rightarrow \quad v_1 = \omega\sqrt{A^2 - x_1^2}$$

$$\Rightarrow \quad v_2 = \omega\sqrt{A^2 - x_2^2}$$

$$\Rightarrow \quad \left(\frac{v_1}{\omega}\right)^2 - \left(\frac{v_2}{\omega}\right)^2 = x_2^{\,2} - x_1^{\,2}$$

$$\Rightarrow \quad \omega^2 = \frac{v_1^2 - v_2^2}{x_2^2 - x_1^2}$$

$$\Rightarrow \quad \omega = \sqrt{\frac{v_1^2 - v_2^2}{x_2^2 - x_1^2}}$$

Thus time period of SHM is given as

$$T = \frac{2\pi}{\omega} = 2\pi\sqrt{\frac{x_2^2 - x_1^2}{v_1^2 - v_2^2}} \qquad \textbf{Ans. (D)}$$

Sol. 106 When length is reduced to one sixteenth of initial value, the modified time period can be given as

$$T = 2\pi\sqrt{\frac{l}{g}}$$

$$\Rightarrow \quad T' = 2\pi\sqrt{\frac{l}{16g}} = \frac{T}{4} \qquad \textbf{Ans. (B)}$$

Sol. 107 From the given equation of motion, we use

$$x(t) = A\sin\omega t + B\cos\omega t \qquad \ldots(1)$$

and $$v(t) = A\omega\cos\omega t - B\omega\sin\omega t \qquad \ldots(2)$$

$$\Rightarrow \quad x(0) = B \text{ and } v(0) = A\omega$$

In Equation-(1) we can substitute

$$A = C\sin\phi$$

and $$B = C\cos\phi$$

After substitution, it gives

$$x(t) = C\cos(\omega t - \phi)$$

Where $$C = \sqrt{A^2 + B^2} \Rightarrow C = \sqrt{\frac{v(0)^2}{\omega^2} + x(0)^2}$$

and $$\phi = \tan^{-1}\left(\frac{A}{B}\right) \Rightarrow \phi = \tan^{-1}\left(\frac{v(0)}{x(0)\omega}\right) \qquad \textbf{Ans. (A)}$$

Sol. 108 For free oscillations of the blocks placed on smooth surface, we analyse the motion by considering reduced mass of one body with respect to another, given as

$$\mu = \frac{m_1 m_2}{m_1 + m_2}$$

$$\Rightarrow \quad \mu = \frac{200\times 800}{200 + 800} = 160 \text{ g} = 0.16 \text{ kg}$$

As springs are connected in series, effective force constant is given as

$$K_{eq} = \frac{k_1 k_2}{k_1 + k_2} = \frac{4k \times k}{4k + k} = \frac{4}{5}k = \frac{4}{5}\times 20 = 16 \text{ N/m}$$

Angular frequency of oscillations can now be given as

$$\omega = \sqrt{\frac{K_{eq}}{\mu}} = \sqrt{\frac{16}{0.16}} = 10 \text{ rad/s} \qquad \textbf{Ans. (10)}$$

Sol. 109 At midway point, the speed of particle in SHM is given as

$$v_{A/2} = \omega\sqrt{A^2 - x^2}$$

$$\Rightarrow \quad v_{A/2} = \omega\sqrt{A^2 - \left(\frac{A}{2}\right)^2} = \omega\left(\frac{\sqrt{3}}{2}A\right) = \frac{\sqrt{3}}{2}V_{max}$$

$$\Rightarrow \quad KE = \frac{1}{2}m\left(\frac{\sqrt{3}}{2}V_{max}\right)^2$$

$$\Rightarrow \quad TE = \frac{1}{2}m(V_{max})^2$$

$$\Rightarrow \quad \frac{KE}{TE} = \frac{3}{4}$$ **Ans. (A)**

Sol. 110 Speed of particle at a displacement x is given as

$$v = \omega\sqrt{a^2 - x^2}$$

Given that

$$\frac{3}{4}KE_{max} = KE_x$$

$$\Rightarrow \quad \frac{3}{4}\left(\frac{1}{2}m\omega^2A^2\right) = \frac{1}{2}m\omega^2(a^2 - x^2)$$

$$\Rightarrow \quad \frac{3}{4}a^2 = a^2 - x^2$$

$$\Rightarrow \quad x^2 = \frac{a^2}{4}$$

$$\Rightarrow \quad x = \pm\frac{a}{2}$$ **Ans. (A)**

Sol. 111 Time period of oscillations of the magnetic needle in magnetic field is given as

$$T = 2\pi\sqrt{\frac{I}{MB}}$$

$$\Rightarrow \quad \frac{1}{2} = 2\pi\sqrt{\frac{5\times10^{-6}}{9.85\times10^{-2}\times B}}$$

$$\Rightarrow \quad \frac{1}{2} = 2\pi\sqrt{\frac{5\times10^{-6}}{\pi^2\times B\times10^{-2}}}$$

$$\Rightarrow \quad B = 8\times10^{-3}\,\text{T} = 8\,mT$$ **Ans. (8)**

Sol. 112 At $t = \frac{T}{4}$ particle is at extreme position where potential energy of SHM is maximum, given as

$$U_{max} = \frac{1}{2}m\omega^2A^2,$$

$$\Rightarrow \quad \frac{1}{2}\times0.5\times\left(\frac{2\pi}{0.2}\right)^2\times(0.05)^2 = 0.6186 \approx 0.62$$ **Ans. (C)**

Sol. 113 From the given SHM equation, we use

$$2 = A\sin(0 + \phi)$$

$$\Rightarrow \quad \sin\phi = \frac{2}{A}$$

$$\Rightarrow \quad \cos\phi\sqrt{1 - \frac{4}{A^2}} = \sqrt{\frac{A^2 - 4}{A^2}} \quad \ldots(1)$$

Velocity of particle in SHM for given equation is written as

$$v = A\omega\cos(\omega t + \phi)$$

$$\Rightarrow \quad 2\omega = A\omega\cos(0 + \phi)$$

$$\Rightarrow \quad 2 = A\cos\phi$$

$$\Rightarrow \quad A = \frac{2}{\cos\phi} \quad \ldots(2)$$

From equations-(1) & (2), we have

$$A = \frac{2}{\sqrt{\frac{A^2 - 4}{A^2}}}$$

$$\Rightarrow \quad A = \frac{2A}{\sqrt{A^2 - 4}}$$

$$\Rightarrow \quad \sqrt{A^2 - 4} = 2$$

$$\Rightarrow \quad A = \sqrt{8} = \pm 2\sqrt{2}$$ **Ans. (2)**

Sol. 114 Time period of simple pendulum is given as

$$T = 2\pi\sqrt{\frac{l}{g}}$$

$$\Rightarrow \quad T \propto l^{1/2}$$

$$\Rightarrow \quad \frac{\Delta T}{T} = \frac{1}{2}\frac{\Delta l}{l}$$

$$\Rightarrow \quad \frac{\Delta T}{24\times3600} = \frac{1}{2}\times\frac{0.1}{100}$$

$$\Rightarrow \quad \Delta T = \frac{1}{2}\times\frac{0.1}{100}\times24\times3600$$

$$\Rightarrow \quad \Delta T = 43.2\text{ s}$$ **Ans. (A)**

Sol. 115 Amplitude of firs SHM is given as

$$A_1 = 5$$

Second SHM equation can be re-written as

$$x_2 = 5\sqrt{2}\,[\sin 2\pi t + \cos 2\pi t]$$

$$\Rightarrow \quad x_2 = 10\left[\sin\left(2\pi T + \frac{\pi}{4}\right)\right]$$

Thus amplitude of second SHM is given as

$$A_2 = 10$$

$$\Rightarrow \quad A_2 = 2A_1$$ **Ans. (2)**

Sol. 116 For the given SHM of the particle, its equation is written as

$$x = x_0 \sin \omega t$$

Potential energy of particle in SHM is expressed as

$$U(x) = \frac{1}{2}kx^2 = \frac{1}{2}kx_0^2 \sin^2 \omega t$$

$$\Rightarrow \quad U(x) = C \sin^2 \omega t$$ **Ans. (D)**

Sol. 117 Amplitude of first wave is 10 and for second wave, amplitude can be calculated by rewriting the equation as

$$y_2 = 5[\sin 3\pi t + \sqrt{3} \cos \pi t]$$

$$\Rightarrow \quad y_2 = 10 \sin\left(3\pi t + \frac{\pi}{3}\right)$$

$$\Rightarrow \quad \frac{A_2}{A_1} = \frac{10}{10} = 1$$ **Ans. (1)**

Sol. 118 When kinetic and potential energies are equal in SHM at a position y, we have

$$K = U$$

$$\Rightarrow \quad \frac{1}{2}k(A^2 - y^2) = \frac{1}{2}ky^2$$

$$\Rightarrow \quad y^2 = \frac{A^2}{2}$$

$$\Rightarrow \quad y = \pm \frac{A}{\sqrt{2}}$$

SHM equation is taken as

$$y = A\sin\omega t$$

$$\Rightarrow \quad \frac{A}{\sqrt{2}} = A \sin\omega t$$

$$\Rightarrow \quad \sin\frac{\pi}{4} = \sin\omega t$$

At $\omega t = \frac{\pi}{4}$ we have $t = \frac{T}{8}$ which gives $x = 8$. **Ans. (8)**

Sol. 119 Time period of a simple pendulum is given as

$$T = 2\pi\sqrt{\frac{l}{g}}$$

Inside liquid, due to buoyancy force the effective gravity decreases, which is given as

$$g_{\text{eff}} = g - \frac{g}{4} = \frac{3g}{4}$$

Thus new time period is given as

$$T' = 2\pi\sqrt{\frac{4l/3}{3g/4}} = 2\pi\sqrt{\frac{16l}{9g}}$$

$$\Rightarrow \quad T' = \frac{4}{3}T$$ **Ans. (B)**

Sol. 120 In a body exeucting SHM, kinetic and potential energy of particles continuously changes such that total energy remain constant so option (a) and (b) are not correct and option (c0 and (d) are correct. **Ans. (A)**

Sol. 121 From figure, amplitude of oscillation is 2 m. The energy of oscillations can be given as

$$\frac{1}{2}m\omega^2A^2 = 10$$

$$\Rightarrow \quad \frac{1}{2}m\omega^2(2)^2 = 10$$

$$\Rightarrow \quad m\omega^2 = 5$$

$$\Rightarrow \quad 5(\omega^2) = 5$$

$$\Rightarrow \quad \omega^2 = 1$$

$$\Rightarrow \quad \omega = 1$$

Time period of oscillation can be calculated as

$$T = \frac{2\pi}{\omega} = 2\pi$$

$$\Rightarrow \quad 2\pi = \sqrt{\frac{l}{g_p}} = 2\pi$$

$$\Rightarrow \quad 2\pi = \sqrt{\frac{4}{g_p}} = 2\pi$$

$$\Rightarrow \quad g_p = 4 \text{ m/s}^2$$ **Ans. (D)**

Sol. 122 Maximum velocity of mass in SHM is given as

$$v_{\max} = \omega A$$

Where angular frequency of SHM for spring mass system is given as

$$\omega = \sqrt{\frac{k}{m}}$$

For the two SHMs, as maximum velocities are equal, we use

$$\frac{A_1}{A_2} = \frac{\omega_1}{\omega_2} = \sqrt{\frac{9}{2} \times \frac{1}{2}} = \frac{3}{2}$$ **Ans. (B)**

Sol. 123 Time period of a pendulum in a lift is given as

$$T = 2\pi\sqrt{\frac{l}{g_{eff}}}$$

When acceleration of lift $a = 0$, we use

$$T = 2\pi\sqrt{\frac{l}{g}}$$

When acceleration of lift is $a = \frac{g}{6}$ upward, we use

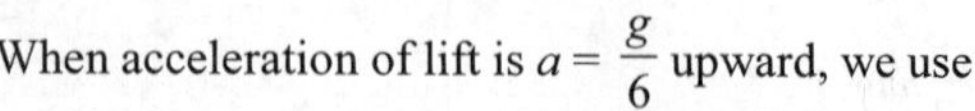

$$T' = 2\pi\sqrt{\frac{l}{g + \frac{g}{6}}}$$

$$\Rightarrow \quad T' = \sqrt{\frac{6}{7}}T$$ **Ans. (C)**

Sol. 124 As oscillator starts oscillations from mean position, its SHM equation is given as

$$x = A\sin\omega t$$

At $t = 3$ s, $x = A/3$, we use

$$\frac{A}{2} = A\sin 3\omega$$

$$\Rightarrow \quad \sin 3\omega = \frac{1}{2}$$

$$\Rightarrow \quad 3\omega = \frac{\pi}{6}$$

$$\Rightarrow \quad \omega = \frac{\pi}{18} = \frac{2\pi}{T}$$

$$\Rightarrow \quad T = 36\text{ s}$$

Ans. (D)

Sol. 125 The given SHM equation is

$$x = \sin\pi\left(t + \frac{1}{3}\right)$$

$$\Rightarrow \quad x = \sin\left(\pi t + \frac{\pi}{3}\right)$$

Speed of particle is given as

$$v = \frac{dx}{dt} = \cos\left(\pi t + \frac{\pi}{3}\right)\pi$$

$$\Rightarrow \quad v = -\pi \times \frac{1}{2} = 157\text{ cm/s}$$

Ans. (B)

Sol. 126 If a particle takes time t in displacement from two positions at a phase difference of $\Delta\phi$ then we use

$$t = \frac{\Delta\phi}{\omega}$$

From position of maximum displacement to the position of half of amplitude, phase difference is taken as $\Delta\phi = \pi/2 - \pi/6$, which gives

$$t = \frac{\pi/2 - \pi/6}{2\pi/6} = \frac{\pi/3}{\pi/3} = 1\text{ s}$$

Ans. (1)

Sol. 127 Total energy of body in SHM is given as

$$E = \frac{1}{2}KA^2$$

By energy conservation we can find speed v at $x = 5$ m as

$$\frac{1}{2}K(10)^2 = \frac{1}{2}K(5)^2 + \frac{1}{2}mv^2$$

$$\Rightarrow \quad v = \sqrt{\frac{75K}{m}}$$

Now, velocity is tripled through by some external means so the amplitude of SHM will change and also the total energy, but potential energy at this moment will be the same as potential energy depends upon position only. If new amplitude is A_1 so by energy conservation we use

$$\frac{1}{2}K(5)^2 + \frac{1}{2}m\left(3\sqrt{\frac{75K}{m}}\right)^2 = \frac{1}{2}KA_1^2$$

$$\Rightarrow \quad 25K + 675K = KA_1^2$$

$$\Rightarrow \quad A_1 = \sqrt{700}$$

$$\Rightarrow \quad x = 700$$

Ans. (700)

Sol. 128 For the given SHM, angular frequency is given as

$$\omega = \sqrt{\frac{g}{l}} = \pi$$

$$\Rightarrow \quad \frac{g}{l} = \pi^2$$

$$\Rightarrow \quad l = \frac{g}{\pi^2}$$

$$\Rightarrow \quad l = \frac{980}{\pi^2} \approx 99.4\text{ cm}$$

Ans. (C)

Sol. 129 In first case (A) the equivalent force constant of the oscillation will be $k + k = 2k$ so for oscillations of mass $3m$, we use

$$T_1 = 2\pi\sqrt{\frac{3m}{2k}}$$

For second case (B), the equivalent force constant of the oscillation will be $k + 2k = 3k$ so for oscillations of mass m, we use

$$T_2 = 2\pi\sqrt{\frac{m}{3k}}$$

$$\Rightarrow \quad \frac{T_1}{T_2} = \frac{2\pi\sqrt{\frac{3m}{2k}}}{2\pi\sqrt{\frac{m}{3k}}} = \frac{3}{\sqrt{2}}$$

Ans. (A)

Sol. 130 For a simple pendulum, the time period of oscillations is given as

$$T = 2\pi\sqrt{\frac{L}{g}}$$

At a height $h = 2R$, the acceleration due to gravity is given as

$$g' = \frac{GM}{9R^2} = \frac{g}{9} = \frac{\pi^2}{9}$$

Time period of seconds pendulum is 2 seconds so we use

$$2 = 2\pi\sqrt{\frac{L}{\pi^2} \times 9}$$

$$\Rightarrow \quad 1 = \pi\sqrt{L} \times \frac{3}{\pi}$$

$$\Rightarrow \quad L = \frac{1}{9}m$$

Ans. (D)

Sol. 131 Two SHMs of same frequency when superpose at a phase difference of ϕ, the resulting amplitude of oscillations is given as

$$A_R = \sqrt{A_1^2 + A_2^2 + 2A_1A_2\cos\phi}$$

$$\Rightarrow \quad \sqrt{3}A = \sqrt{A^2 + A^2 + 2A^2\cos\phi}$$

$$\Rightarrow \quad 3A^2 = 2A^2 + 2A^2\cos\phi$$

$$\Rightarrow \quad \cos\phi = \frac{1}{2}$$

$$\Rightarrow \quad \phi = 60°$$

Ans. (60)

Sol. 132 For a particle in SHM, its speed as a function of position is given as

$$v = \omega\sqrt{A^2 - x^2}$$

Here ω is the angular frequency of SHM and A is its oscillation amplitude

$$\Rightarrow \quad v^2 = \omega^2A^2 - \omega^2x^2$$

$$\Rightarrow \quad \frac{v^2}{(\omega A)^2} + \frac{x^2}{(A)^2} = 1$$

This is an equation of ellipse. **Ans. (B)**

Sol. 133 For figure-(a), equivalent force constant of springs in series is given as

$$K_{eq} = \frac{K \times 2K}{K + 2K} = \frac{2K}{3}$$

Thus time period of this spring mass system is given as

$$T = 2\pi\sqrt{\frac{m}{K_{eq}}} = 2\pi\sqrt{\frac{m}{2K/3}} = 2\pi\sqrt{\frac{3m}{2K}}$$

For figure-(b), equivalent force constant of springs in parallel is given as

$$K_{eq} = K + 2K = 3K$$

Thus time period of this spring mass system is given as

$$T' = 2\pi\sqrt{\frac{m}{3K}}$$

$$\Rightarrow \quad \frac{T'}{T} = \sqrt{\frac{m \times 2K}{3K \times 3m}} = \frac{\sqrt{2}}{3}$$

$$\Rightarrow \quad T' = \sqrt{2}$$

$$\Rightarrow \quad x = 2$$

Ans. (2)

Sol. 134 Figure below shows the situation described in the question. At a small displacement x of the point charge from mid point, the resulting force on this charge is calculated as

$$F = \frac{KQq_0}{(a-x)^2} - \frac{KQq_0}{(a+x)^2}$$

$$F = \frac{KQq_0[2a][2x]}{(a^2 - x^2)^2}$$

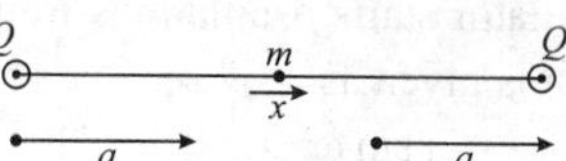

Acceleration of charge is calculated as

$$F = mf$$

$$\Rightarrow \quad f \approx \left(\frac{4KQq_0}{ma^3}\right)x \quad \ldots(1)$$

For oscillation to be simple harmonic, acceleration is given as

$$f = -\omega^2x \quad \ldots(2)$$

Comparing equations-(1) and (2), we get

$$\omega = \sqrt{\frac{Qq_0}{\pi\varepsilon_0 ma^3}}$$

Thus time period of SHM is given as

$$T = \frac{2\pi}{\omega} = 2\pi\sqrt{\frac{\pi\varepsilon_0 ma^3}{Qq_0}}$$

$$\Rightarrow \quad T = \sqrt{\frac{4\pi^3\varepsilon_0 ma^3}{Qq_0}}$$

Ans. (A)

Sol. 135 At mean position total energy of spring mass SHM will be kinetic so its momentum will remain conserved while adding a mass, so we use before and after adding the mass

$$\frac{1}{2}kA_1^2 = \frac{p^2}{2m_1}$$

and

$$\frac{1}{2}kA_2^2 = \frac{p^2}{2m_2}$$

$$\Rightarrow \quad \left(\frac{A_1}{A_2}\right)^2 = \frac{m_2}{m_1} = \frac{1024}{900}$$

$$\Rightarrow \quad \frac{A_1}{A_2} = \frac{32}{30} = \frac{16}{15} = \frac{16}{16-1}$$

$$\Rightarrow \quad \alpha = 16$$

Ans. (16)

Sol. 136 Using potential energy of particle during oscillation, we can calculate the restoring force on it, given as

$$U = 4(1 - \cos 4x)$$

$$F = -\frac{dU}{dx} = -4(+\sin 4x)4$$

$$\Rightarrow \quad F = -16\sin(4x)$$

For small θ, we have

$$\sin\theta \approx \theta$$

$$\Rightarrow \quad F = -64x$$

$$\Rightarrow \quad a = -64\,x/m = -16x$$

For SHM, we use

$$a = -\omega^2x$$

$$\Rightarrow \quad \omega^2 = 16$$

$$\Rightarrow \quad \omega = 4 \text{ rad/s}$$

Thus time period of SHM is given as

$$T = \frac{2\pi}{\omega} = \frac{\pi}{2} \quad \textbf{Ans. (2)}$$

Sol. 137 As the vehile is sliding without friction, its acceleration will be g sinα so effective acceleration due to gravity inside the vehicle will be considered as

$$g_{\text{eff}} = g \cos \alpha$$

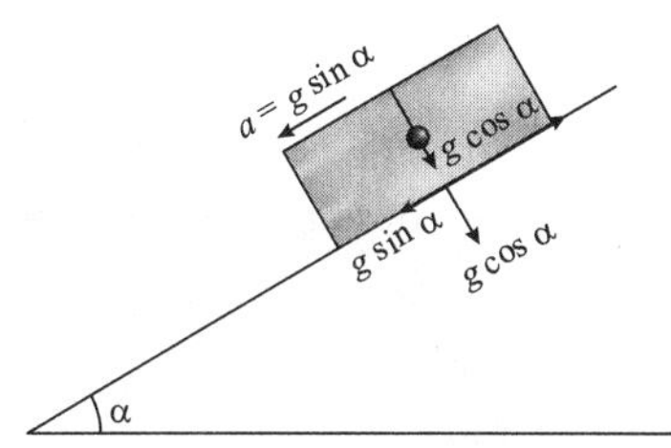

Thus time period of simple pendulm in the vehicle is given as

$$T = 2\pi\sqrt{\frac{L}{g\cos\alpha}} \quad \textbf{Ans. (A)}$$

Sol. 138 Inside water effective weight of bob is calculated as

$$mg' = mg - F_B$$

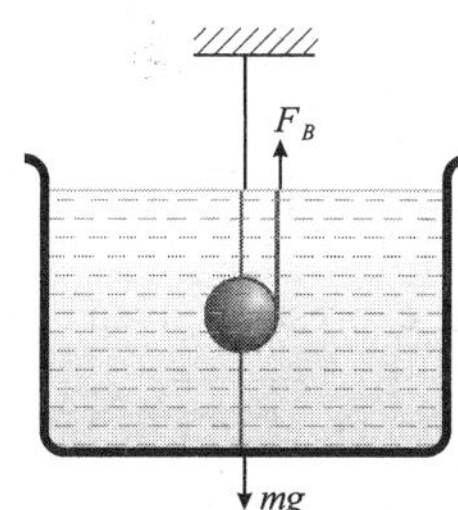

$$\Rightarrow \quad g' = \frac{mg - F_B}{m}$$

$$\Rightarrow \quad g' = \frac{\rho_B Vg - \rho_w Vg}{\rho_B V}$$

$$\Rightarrow \quad g' = \left(\frac{\rho_B - \rho_w}{\rho_B}\right) g$$

$$\Rightarrow \quad g' = \frac{5-1}{5} \times g = \frac{4}{5}g$$

Time period of oscillation of simple pendulum inside water is given as

$$T' = 2\pi\sqrt{\frac{l}{g'}}$$

If initial time period is T, we use

$$\frac{T'}{T} = \sqrt{\frac{g}{g'}} = \sqrt{\frac{g}{\frac{4}{5}g}} = \sqrt{\frac{5}{4}}$$

$$\Rightarrow \quad T' = T\sqrt{\frac{5}{4}} = \frac{10}{2}\sqrt{5}$$

$$\Rightarrow \quad T' = 5\sqrt{5} \text{ s} \quad \textbf{Ans. (5)}$$

Sol. 139 In the given situation the net force on block due to displacement x in both springs is in same direction, given as

$$F = -2kx$$

Acceleration of block is given as

$$a = -\frac{2kx}{m}$$

Comparing with the acceleration of block in SHM $a = -\omega^2 x$ where negative sign gives the restoring nature, we get

$$\omega = \sqrt{\frac{2k}{m}} = \sqrt{\frac{2\times 20}{2}} = \sqrt{20} \text{ rad/s}$$

The time period of oscillations is calculated as

$$T = \frac{2\pi}{\omega} = \frac{2\pi}{\sqrt{20}} = \frac{\pi}{\sqrt{5}} \text{ s}$$

$$\Rightarrow \quad x = 5 \quad \textbf{Ans. (5)}$$

Sol. 140 Time period of oscillations of spring block system is given as

$$T = 2\pi\sqrt{\frac{m}{k}} = 1$$

After adding 3 kg mass, the time period becomes 2 seconds so we have

$$T' = 2\pi\sqrt{\frac{m+m}{k}} = 2$$

$$\Rightarrow \quad \frac{T}{T'} = \sqrt{\frac{m}{m+3}} = \frac{1}{2}$$

$$\Rightarrow \quad \frac{m}{m+3} = \frac{1}{4}$$

$$\Rightarrow \quad m = 1 \quad \textbf{Ans. (1)}$$

Sol. 141 At surface of earth time period of pendulum is given as

$$T = 2\pi\sqrt{\frac{l}{g}}$$

At height $h = R$, we have acceleration due to gravity is given as

$$g_h = \frac{g}{\left(1+\frac{h}{R}\right)^2} = \frac{g}{4}$$

Time period of pendulum at $h = R$ is given as

$$xT = 2\pi\sqrt{\frac{l}{(g/4)}}$$

$$\Rightarrow \quad xT = 2\times 2\pi\sqrt{\frac{l}{g}}$$

$$\Rightarrow \quad xT = 2T$$

$$\Rightarrow \quad x = 2 \quad \textbf{Ans. (B)}$$

Sol. 142 Figure below shows the circular motion analysis of the given SHM along the diametrical axis AA'.

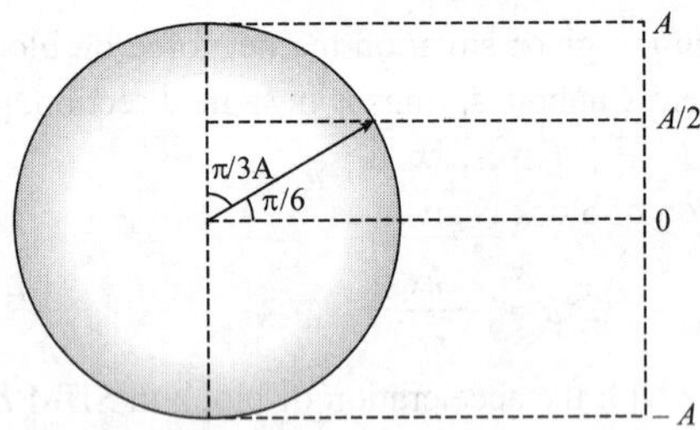

If time taken by particle in going from 0 to $A/2$ is t_1 & from $A/2$ to A is t_2 we have

$$\omega t_1 = \pi/6$$

and $$\omega t_2 = \pi/3$$

$$\Rightarrow \quad \frac{t_1}{t_2} = \frac{1}{2}$$

$$\Rightarrow \quad t_2 = 2t_1 = 2 \times 2 = 4 \text{ s}$$ **Ans. (D)**

Sol. 143 The resultant amplitude after superposition of waves is given as

$$R = 2A\cos\left(\frac{\Delta\phi}{2}\right) = A$$

$$\Rightarrow \quad \cos\left(\frac{\Delta\phi}{2}\right) = \frac{1}{2}$$

$$\Rightarrow \quad \frac{\Delta\phi}{2} = 60°$$

$$\Rightarrow \quad \Delta\phi = 120°$$ **Ans. (120)**

Sol. 144 Force on particle is related to its acceleration given as

$$\frac{1}{4}a = -25x$$

$$\Rightarrow \quad a = -100x$$

Comparing with $a = -\omega^2 x$, we have

$$\omega^2 = 100$$

$$\Rightarrow \quad \omega = 10 \text{ rad/s}$$

Maximum speed of particle is given as

$$\omega A = 4$$

$$\Rightarrow \quad A = \frac{4}{10} = 0.4 \text{ m}$$

$$\Rightarrow \quad A = 40 \text{ cm}$$ **Ans. (40)**

Sol. 145 Displacement of harmonic oscillator is given as

$$x = A\sin(\omega t)$$

The potential energy of the oscillator is given as

$$U_{(x)} = \frac{1}{2}kx^2$$

Differentiating this expression with respect to time gives

$$\frac{dU}{dt} = \frac{1}{2}k\left(2x\frac{dx}{dt}\right)$$

$$\Rightarrow \quad \frac{dU}{dt} = kA^2\,\omega\sin\omega t\cos\omega t$$

$$\Rightarrow \quad \left(\frac{dU}{dt}\right)_{max} = \frac{kA^2\omega}{2}(\sin 2\omega t)_{max}$$

$$\Rightarrow \quad 2\omega t = \frac{\pi}{2}$$

$$\Rightarrow \quad t = \frac{\pi}{4}\omega = \frac{T}{8}$$

$$\Rightarrow \quad \beta = 8$$ **Ans. (8)**

Sol. 146 Angular frequency of simple harmonic motion of block is given as

$$\omega = \sqrt{\frac{k}{m}}$$

Ratio of angular frequencies when masses of blocks are 1 kg and 2 kg are calculated as

$$\frac{\omega_2}{\omega_1} = \sqrt{\frac{m_1}{m_2}} = \sqrt{\frac{1}{2}}$$ **Ans. (B)**

Sol. 147 Velocity of particle executing SHM varies with displacement as

$$4v^2 = 50 - x^2$$

$$\Rightarrow \quad v = \frac{1}{2}\sqrt{50 - x^2}$$

Comparing v with the expression given below, we have

$$v = \pm\,\omega\sqrt{A^2 - x^2}$$

$$\omega = \frac{1}{2} \text{ rad/s}$$

Time period of oscillations is calculated as

$$T = \frac{2\pi}{\omega} = 4\pi = \frac{88}{7} = \frac{x}{7}$$

$$\Rightarrow \quad x = 88$$ **Ans. (88)**

Sol. 148 The maximum potential energy of oscillator is given as

$$U_{max} = \frac{1}{2}m\omega^2 A^2 = 25 \text{ J}$$

Kinetic Energy at $\frac{A}{2}$ position is given as

$$K = \frac{1}{2}mv_1^2 = \frac{1}{2}m\omega^2\left(A^2 - \frac{A^2}{4}\right)$$

$$\Rightarrow \quad K = \frac{1}{2}m\omega^2\frac{3A^2}{4} = \frac{3}{4}\left(\frac{1}{2}m\omega^2 A^2\right)$$

$$\Rightarrow \quad K = \frac{3}{4} \times 25 = 18.75 \text{ J}$$ **Ans. (C)**

Sol. 149 In the situation shown in figure when the block is displaced by some distance, both springs get deformed by same distance in such a way that their restoring forces will act in same direction on block so the two springs can be considered in parallel.

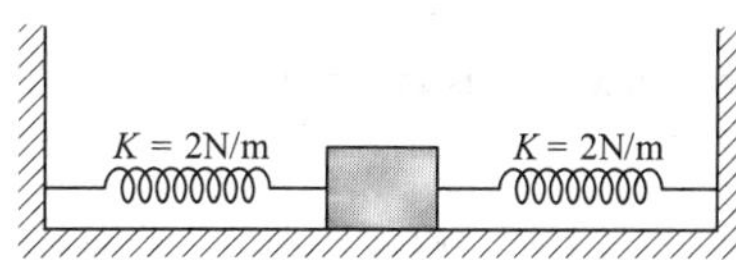

As both springs are in parallel, effective force constant of the springs can be written as

$$K_{eff} = K + K = 2K = 2 \times 2 = 4 \text{ N/m}$$

Time period of oscillations of the block is given as

$$T = 2\pi\sqrt{\frac{m}{K_{eff}}} = 2\pi\sqrt{\frac{0.49\text{ kg}}{4}}$$

$$\Rightarrow \quad T = 2\pi\sqrt{\frac{49}{400}} = 2\pi\frac{7}{20} = \frac{7\pi}{10}$$

Number of oscillation in the 14π s are calculated as

$$N = \frac{t}{T} = \frac{14\pi}{7\pi/10} = 20 \qquad \textbf{Ans. (20)}$$

Sol. 150 A position at which kinetic energy will be 1.25 times that of the potential energy can be calculated as

$$K_x = 1.25 P_x$$

$$\Rightarrow \quad \frac{1}{2}m\omega^2(A^2 - x^2) = \frac{5}{4} \times \frac{1}{2}m\omega^2 x^2$$

$$\Rightarrow \quad A^2 - x^2 = \frac{5}{4}x^2$$

$$\Rightarrow \quad \frac{9x^2}{4} = A^2$$

$$\Rightarrow \quad x = \frac{2}{3}A$$

$$\Rightarrow \quad x = \frac{2}{3} \times 3 \text{ cm}$$

$$\Rightarrow \quad x = 2 \text{ cm} \qquad \textbf{Ans. (2)}$$

Sol. 151 Time period of simple pendulum

$$T = 2\pi\sqrt{\frac{l}{g}}$$

$$\Rightarrow \quad T^2 = \frac{4\pi^2}{g} \times l$$

$$\Rightarrow \quad T^2 \alpha\, l$$

Thus graph will be linear. **Ans. (C)**

Sol. 152 In the given case the deformation in springs will always be equal and will exert restrong force on blocks in same direction, so these can be considered in parallel.

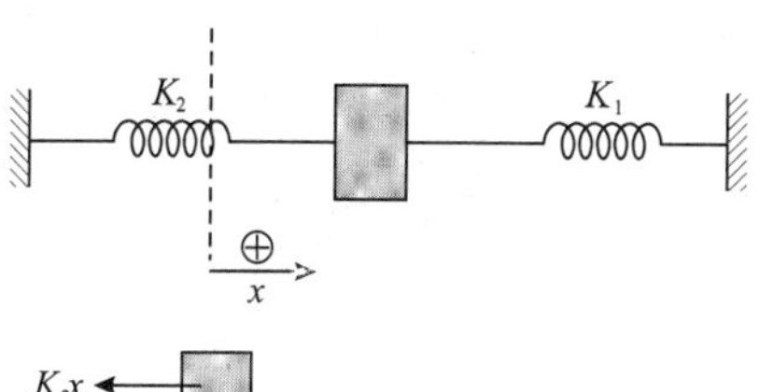

$$F = -(k_1x + k_2x) = -(k_1 + k_2)x$$

Acceleration of mass m is calculated as

$$a = \frac{F}{m} = -\left(\frac{k_1 + k_2}{m}\right)x = -\omega^2 x$$

As the block is oscillating, Angular frequency is given as

$$\omega = \sqrt{\frac{k_1 + k_2}{m}}$$

Thus, time period of oscillation is calculated as

$$T = \frac{2\pi}{\omega} = 2\pi\sqrt{\frac{m}{k_1 + k_2}} \qquad \textbf{Ans. (C)}$$

Sol. 153 When bob is displaced with amplitude of 10cm, we use

$$\sin\theta_0 = \frac{A}{l} = \frac{10}{100} = \frac{1}{10}$$

By using conservation of energy, we have

$$\frac{1}{2}mv^2 = mgl(1 - \cos\theta)$$

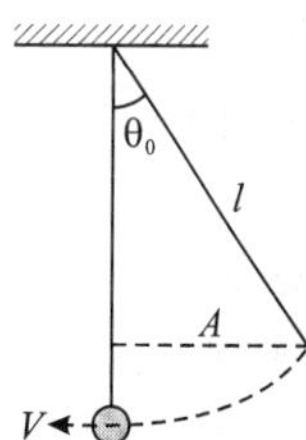

Maximum tension in string will be at mean position, given as

$$T - mg = \frac{mv^2}{l}$$

$$\Rightarrow \quad T = mg + \frac{mv^2}{l}$$

$$\Rightarrow \quad T = mg + 2mg(1 - \cos\theta)$$

$$\Rightarrow \quad T = mg\left[1 + 2\left(1 - \sqrt{1 - \sin^2\theta}\right)\right]$$

$$\Rightarrow \quad T = mg\left[3 - 2\sqrt{1 - \frac{1}{100}}\right]$$

$$\Rightarrow \quad T = \frac{250}{1000} \times 9.8\left[3 - 2\left(1 - \frac{1}{200}\right)\right]$$

$$\Rightarrow \quad T = \frac{99}{40} = \frac{x}{40}\text{N}$$

$$\Rightarrow \quad x = 99 \qquad \textbf{Ans. (99)}$$

Sol. 154 Figure below shows the phase angles of the particle.

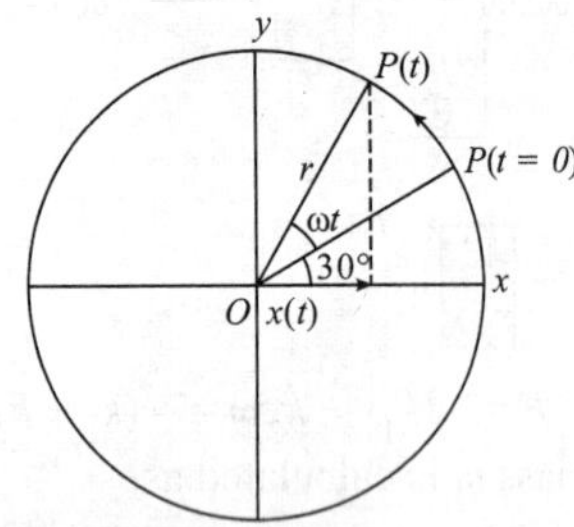

From the diagram shown, initial phase is $\pi / 6$

$$\Rightarrow \quad x(t) = r\cos(\omega t + 30^\circ)$$

$$\Rightarrow \quad x(t) = r\cos(\omega t + \pi/6) \qquad \textbf{Ans. (A)}$$

Sol. 155 Position of particle executing SHM is given as

$$x = A\sin(\omega t + \delta)$$

Velocity of particle in SHM is given as

$$v = A\omega\cos(\omega t + d)$$

At $t = 0$, $s = \frac{A}{2}$, we have

$$\frac{A}{2} = A\sin(\omega t + \delta)$$

velocity is positive so in this case phase angle has to lie in first or fourth quadrent and it is given as

$$\sin\delta = \frac{1}{2}$$

$$\Rightarrow \quad \delta = \frac{\pi}{6}, \frac{5\pi}{6}$$

For positive velocity the solution is $\delta = \frac{\pi}{6}$. **Ans. (A)**

Sol. 156 Given that time period of SHM is

$$T = 3.14\text{ s}$$

$$\Rightarrow \quad T = \pi = \frac{2\pi}{\omega}$$

$$\Rightarrow \quad \omega = 2\text{ rad/s}$$

For simple harmonic maximum force exerted by spring on block is given as

$$F_{max} = m\,a_{max}$$

$$\Rightarrow \quad F_{max} = m(A\omega^2)$$

$$\Rightarrow \quad F_{max} = mA(2)^2$$

$$\Rightarrow \quad F_{max} = 5 \times 1 \times 4$$

$$\Rightarrow \quad F_{max} = 20\text{ N} \qquad \textbf{Ans. (20)}$$

Sol. 157 For a particle executing SHM, its kinetic energy is given as

$$K = \frac{1}{2}m\omega^2(A^2 - x^2)$$

When $x = 0$, K is maximum & when $x = A$, K is zero thus we can state that K *vs* x graph is a parabola shown in option (D). **Ans. (D)**

Sol. 158 When displacement is half of its amplitude, then potential energy at $x = \frac{A}{2}$ is given as

$$P = \frac{1}{2}kx^2$$

The kinetic energy at a position x is given as

$$K = \frac{1}{2}k(A^2 - x^2)$$

$$K = \frac{1}{2}kA^2 - \frac{1}{2}kx^2$$

$$\Rightarrow \quad \frac{P}{K} = \frac{x^2}{A^2 - x^2} = \frac{A^2}{4\left(\frac{3A^2}{4}\right)} = \frac{1}{3} \qquad \textbf{Ans. (D)}$$

Sol. 159 The kinetic energy of the particle executing SHM with its distance from mean position is given as

$$K_x = \frac{1}{2}m\omega^2(A^2 - x^2)$$

This is the equation of downward parabola. **Ans. (D)**

Sol. 160 General equation for displacement of SHM is given as

$$x = A\sin(\omega t + \phi)$$

At a specific time the phase of particle is given as

$$\Rightarrow \quad \omega t + \phi = 30^\circ$$

$$\Rightarrow \quad x = 40 \times \frac{\sqrt{3}}{2}$$

$$\Rightarrow \quad x = 20\sqrt{3}\text{ cm}$$

$$\Rightarrow \quad A = 40\text{ cm}$$

Kinetic energy of the particle is given as

$$\Rightarrow \quad K = \frac{1}{2}k(A^2 - x^2) = 200$$

$$\Rightarrow \quad 200 = \frac{1}{2}k\left(\frac{1600 - 1200}{100 \times 100}\right)$$

$$\Rightarrow \quad 400 \times 100 \times 100 = k \times 400$$

$$\Rightarrow \quad k = 10^4$$

$$\Rightarrow \quad x = 4 \qquad \textbf{Ans. (4)}$$

Sol. 161 Let at x distance from mean position kinetic energy is equal to potential energy of particle so we use

$$K_x = P_x$$

$$\Rightarrow \quad \frac{1}{2}M\omega^2(A^2 - x^2) = \frac{1}{2}M\omega^2x^2$$

$$\Rightarrow \quad A^2 - x^2 = x^2$$

$\Rightarrow \qquad A^2 = 2 \times 2$

$\Rightarrow \qquad x = \pm \dfrac{A}{\sqrt{2}}$ **Ans. (3)**

Sol. 162 For simple harmonic motion, we have

$$F = -kx$$

$$a = -\omega^2 x$$

Hence (a) and (b) are true. The velocity is maximum at mean position so (c) is true and acceleration is maximum at extreme points hence (d) is not true. **Ans. (A)**

14.1 Wave Motion

Sol. 1 The given wave equation is

$$y = 10^{-4} \sin\left(600t - 2x + \frac{\pi}{3}\right) \qquad \ldots(1)$$

The general wave equation is expressed as

$$y = A \sin(\omega t - kx + \alpha) \qquad \ldots(2)$$

Comparing equation-(1) & (2), we have

$$\omega = 600 \text{ rad/s}$$

$$k = 2 \text{ m}^{-1}$$

$\Rightarrow$ Wave speed $\quad v = \dfrac{\omega}{k}$

$$= \frac{600}{2} = 300 \text{ m/s}$$ **Ans. (A)**

Sol. 2 If we use sound level of I, w.r.t. to I_{01}

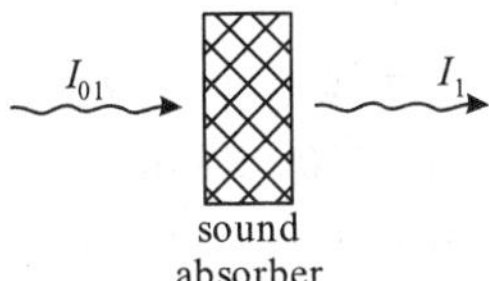

We have

$$L_{I_{01}} - L_{I_1} = 20 \text{ dB}$$

$\Rightarrow \qquad 10\log_{10}\left(\dfrac{I_{01}}{I_0}\right) - 10\log_{10}\left(\dfrac{I_1}{I_0}\right) = 20$

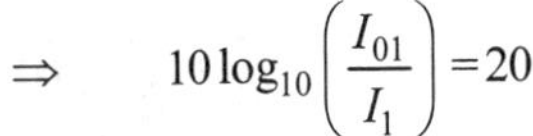

$\Rightarrow \qquad 10\log_{10}\left(\dfrac{I_{01}}{I_1}\right) = 20$

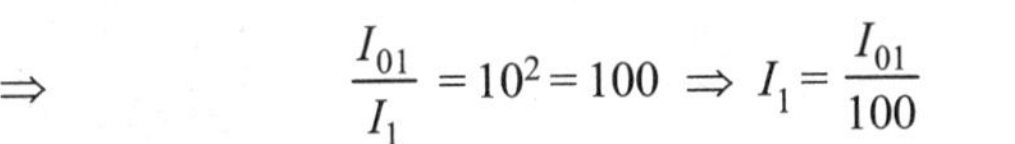

$\Rightarrow \qquad \dfrac{I_{01}}{I_1} = 10^2 = 100 \Rightarrow I_1 = \dfrac{I_{01}}{100}$ **Ans. (D)**

Sol. 3 The given wave equation is

$$y = 0.005 \cos(\alpha x - \beta t) \qquad \ldots(1)$$

General expression for a wave equation is

$$y = A \cos(kx - \omega t) \qquad \ldots(2)$$

$$A \cos(kx - \omega t) = A \sin(\omega t - kx + \phi)$$

Comparing (1) and (2) we have

$$\alpha = k = \frac{2\pi}{\lambda} = \frac{2\pi}{0.08} = 25\pi$$

and $\qquad \beta = \omega = \dfrac{2\pi}{T} = \dfrac{2\pi}{2} = \pi$ **Ans. (A)**

Sol. 4 In gaseous medium speed of sound is given as

$$v = \sqrt{\frac{\gamma RT}{M}}$$

$\Rightarrow \qquad \dfrac{v_{O_2}}{v_{He}} = \sqrt{\dfrac{\gamma_{O_2}}{M_{O_2}} \times \dfrac{M_{He}}{\gamma_{He}}}$

$\Rightarrow \qquad v_{He} = v_{O_2}\sqrt{\dfrac{\gamma_{He} M_{O_2}}{\gamma_{O_2} M_{He}}} = 460\sqrt{\dfrac{\frac{5}{3} \times 32}{\frac{7}{5} \times 4}}$

$\Rightarrow \qquad v_{He} = 460\sqrt{\dfrac{200}{21}}$ m/s **Ans. (A)**

Sol. 5 Given wave equation is

$$y = 0.02 \sin\left[2\pi\left(\frac{t}{0.04} - \frac{x}{0.05}\right)\right] \qquad \ldots(1)$$

General equation of a wave equation is of the form

$$y = A \sin\left[2\pi\left(\frac{t}{T_0} - \frac{x}{A}\right)\right] \qquad \ldots(2)$$

Comparing (1) and (2) gives

$$T_0 = 0.04 \text{ s}$$

$$\lambda = 0.5 \text{ m}$$

$\Rightarrow$ Wave speed $\quad v = \dfrac{\lambda}{T_0} = \dfrac{0.5}{0.04} = \dfrac{50}{4}$ m/s

Wave speed on a string is given as

$$v = \sqrt{\frac{T}{4}}$$

$\Rightarrow \qquad T = \mu v^2 = 0.04 \times \dfrac{2500}{16}$

$\Rightarrow \qquad T = \dfrac{25}{4} = 6.25 \text{ N}$ **Ans. (A)**

Sol. 6 As sound wave intensities are given as

$$I = 2\pi^2 n^2 A^2 \rho v$$

For a given medium ρ, $v \rightarrow$ constant $\Rightarrow I \propto x^2 A^2$

in S-I for first wave $n_1 = \dfrac{\omega}{2\pi}$; $A_1 = 2a$

For second wave $\quad n_2 = \dfrac{2\omega}{2\pi}$; $A_2 = a$

$\Rightarrow \quad I_1 = I_2$

S-I is true

As Intensity $\propto n^2 A^2$

$\Rightarrow$ S-II is false **Ans. (B)**

Sol. 7 The given expression is

$$y = e^{-(ax^2+bt^2+2\sqrt{ab}xt)}$$

$$y = e^{-[\sqrt{a}x+\sqrt{b}t]^2}$$

It is a function of $(c_1x + c_2t)$ thus it represents a propagating wave. +ve sign shows that wave is travelling in $-x$ direction. having $\omega = \sqrt{b}$ and $k = \sqrt{a}$

$\Rightarrow$ Wave speed $\quad v = \dfrac{\omega}{k} = \sqrt{\dfrac{b}{a}}$ **Ans. (D)**

Sol. 8 If a wave displacement function is given as

$$y = f(x) \text{ at } t = 0$$

For a wave travelling in $+x$ direction at speed v, then at time $t = t$, the displacement function is given as

$$y = f(x - vt)$$

Given at $t = 0$

$$y = \frac{1}{1+x^2}$$

So at time t the function becomes

$$y = \frac{1}{1+(x-vt)^2} \quad \ldots(1)$$

In question at $t = 2$s displacement function is given as

$$y = \frac{1}{1+(x-1)^2} \quad \ldots(2)$$

Comparing equation-(2) with equation-(1), we get

$$vt = 1\big|_{t=2s}$$

$\Rightarrow$ $\quad v = \dfrac{1}{2} = 0.5$ m/s **Ans. (C)**

Sol. 9 Standard wave equation in a medium can be given as

$$y = A\cos(kx - \omega t)$$

Comparing y_1 with above standard equation we get

$$A = 0.05 \text{ m}$$

and $\quad k_1 = \dfrac{2\pi}{\lambda_1} = 0.5\pi$

$\Rightarrow$ $\quad \lambda_1 = 4$ m

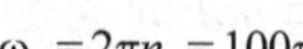

$$\omega_1 = 2\pi n_1 = 100\pi$$

$\Rightarrow$ $\quad n_1 = 50$ Hz

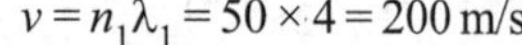

Wave velocity $\quad v = n_1\lambda_1 = 50 \times 4 = 200$ m/s **Ans. (B)**

Sol. 10

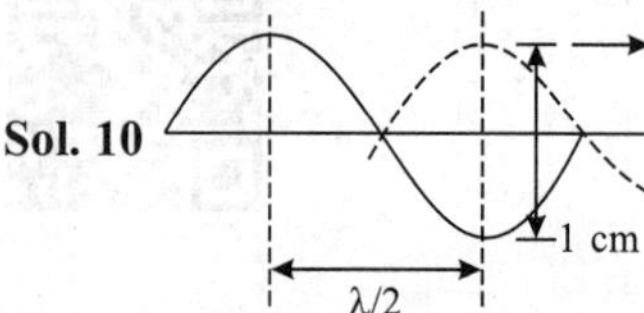

Distance between a crest and neighbouring trough is

$$\frac{\lambda}{2} = 0.4 \text{ cm}$$

$\Rightarrow$ $\quad \lambda = 0.8$ cm

It is given that $2A = 1$ cm $\Rightarrow A = 0.5$ cm

As distance between two neighbouring crests is λ

$\Rightarrow$ $\quad t = \dfrac{\lambda}{v} = 0.4$ s

$\Rightarrow$ $\quad v = \dfrac{\lambda}{0.4} = \dfrac{0.8}{0.4} = 2$ cm/s

$\Rightarrow$ Maximum speed of vibrating particles

$$v_{max} = A\omega = A\left(\frac{2\pi v}{\lambda}\right) = \frac{0.5\times 2\pi \times 2}{0.8}$$

$\Rightarrow$ $\quad v_{max} = \dfrac{5\pi}{2}$ cm/s **Ans. (B)**

Sol. 11

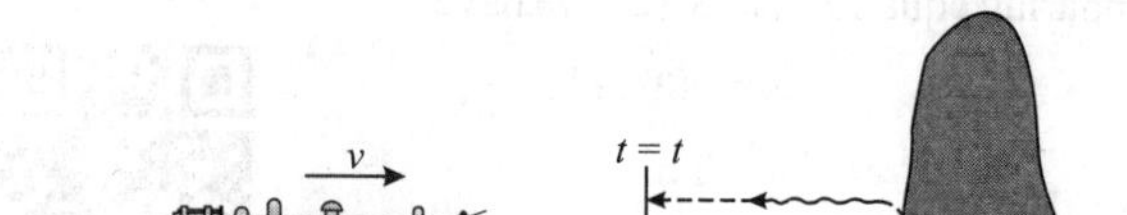

In time t when reflected sound is heard by the driver then distance travelled by sound is

$$900 + (900 - vt) = v_s.t$$

at $\quad t = 5$s

$\Rightarrow$ $\quad 900 + 900 - 5v = 330 \times 5 = 1650$

$\Rightarrow$ $\quad 5v = 1800 - 1650 = 150$

$\Rightarrow$ $\quad v = \dfrac{150}{5} = 30$ m/s **Ans. (D)**

Sol. 12 Given equation of SHW is

$$y = \frac{10}{\pi}\sin\left(\frac{2\pi}{T}t - \frac{2\pi}{\lambda}x\right)$$

Comparing with $\quad y = A\sin(\omega t - kx)$

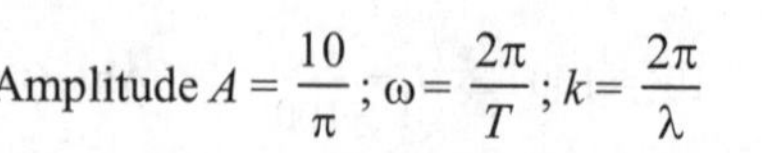

Amplitude $A = \dfrac{10}{\pi}$; $\omega = \dfrac{2\pi}{T}$; $k = \dfrac{2\pi}{\lambda}$

As wave velocity $\quad = 2 \times$ maximum particle velocity

$$\frac{\omega}{k} = 2 \times A\omega$$

$\Rightarrow$ $\quad \dfrac{\lambda}{2\pi} = 2 \times \dfrac{10}{\pi}$

$\Rightarrow$ $\quad \lambda = 40$ cm **Ans. (A)**

Sol. 13

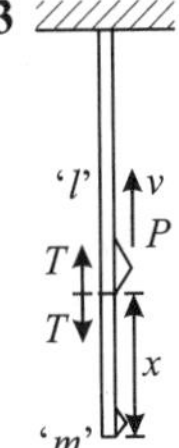

Tension at point P in string is

$$T=\left(\frac{m}{l}x\right)g$$

Wave speed in string at point P is

$$v=\sqrt{\frac{T}{\mu}}=\sqrt{\frac{\left(\frac{m}{l}\times g\right)}{\frac{m}{l}}}=\sqrt{xg}$$

$$\Rightarrow \quad \frac{dx}{dt}=\sqrt{xg}$$

$$\Rightarrow \quad \int_0^l \frac{dx}{\sqrt{x}}=\int_0^t \sqrt{g}\,dt$$

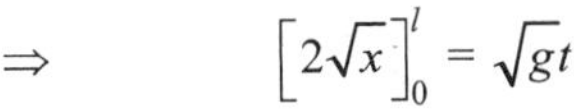

$$\Rightarrow \quad \left[2\sqrt{x}\right]_0^l=\sqrt{g}t$$

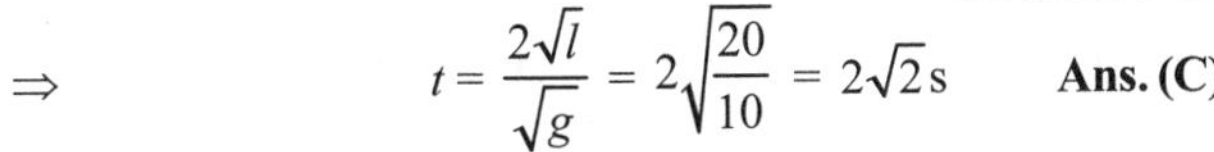

$$\Rightarrow \quad t=\frac{2\sqrt{l}}{\sqrt{g}}=2\sqrt{\frac{20}{10}}=2\sqrt{2}\,\text{s}$$ **Ans. (C)**

Sol. 14 When car is at rest, wave speed is given as

$$60=\sqrt{\frac{Mg}{\mu}}$$

When car is accelerating, we use

$$60.5=\sqrt{\frac{M(g^2+a^2)^{1/2}}{\mu}}$$

$$\Rightarrow \quad \frac{60.5}{60}=\sqrt{\sqrt{\frac{g^2+a^2}{g^2}}}$$

$$\Rightarrow \quad \left(1+\frac{0.5}{60}\right)^4=\frac{g^2+a^2}{g^2}=1+\frac{2}{60}$$

$$\Rightarrow \quad g^2+a^2=g^2+g^2\times\frac{2}{60}$$

$$\Rightarrow \quad a=g\sqrt{\frac{2}{60}}=\frac{g}{\sqrt{30}}=\frac{g}{5.47}$$

$$\Rightarrow \quad a\simeq\frac{g}{5}$$ **Ans. (A)**

Sol. 15 Given equation is

$$y=0.03\sin(450t-9x)$$

Comparing with

$$y=A\sin(\omega t-kx), \text{ we get}$$

$$v=\frac{\omega}{k}=\frac{450}{9}=50\text{ m/s}$$

Velocity of transverse waves on a string gives

$$v=\sqrt{\frac{T}{\mu}}$$

$$\Rightarrow \quad \frac{T}{\mu}=2500$$

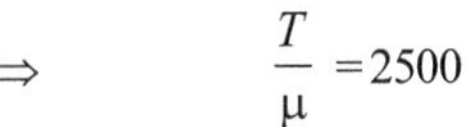

$$\Rightarrow \quad T=2500\times 5\times 10^{-3}$$

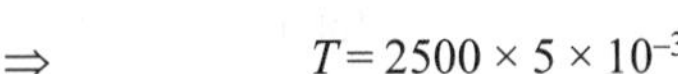

$$\Rightarrow \quad T=12.5\text{ N}$$ **Ans. (B)**

Sol. 16 Above wave is moving along $-ve$ x-axis with speed given as

$$v=\frac{\omega}{K}$$

$$\Rightarrow \quad v=\frac{50}{2}=25\text{ m/s}$$ **Ans. (A)**

Sol. 17 Speed of wave from wave equation is written as

$$v=-\frac{(\text{coeffecient of }t)}{(\text{coeffecient of }x)}$$

$$v=-\frac{1000}{(-3)}=\frac{1000}{3}\text{ m/s}$$

As speed of wave $v\propto\sqrt{T}$, we have

$$=\frac{1000/3}{336}=\sqrt{\frac{273}{T}}$$

$$\Rightarrow \quad T=277.41\text{ K}$$

$$\Rightarrow \quad T=4.41°\text{C}$$ **Ans. (C)**

Sol. 18

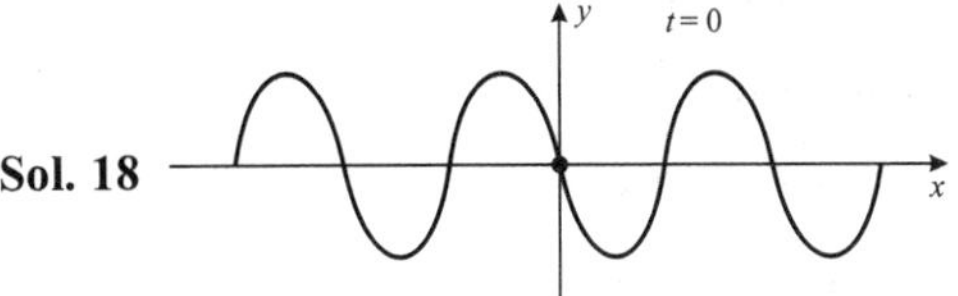

For the given wave equation

$$y=A\sin(kx-wt+\phi)$$

at $x=0, t=0, y=0$ and slope is negative

$$\Rightarrow \quad \frac{\partial y}{\partial x}=Ak\cos(kx-wt+\phi)=-\text{ve}$$

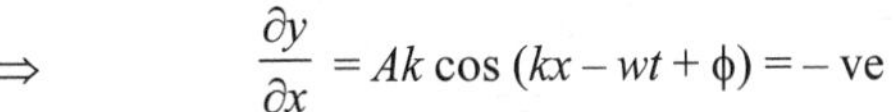

$$\Rightarrow \quad \phi=x \text{ (From the given options)}$$ **Ans. (C)**

Sol. 19 Loudness of sound is given by

$$L(dB) = 10 \log \frac{I}{I_0}$$

$$\Rightarrow \quad 120 = 10 \log \left(\frac{I}{I_0}\right)$$

$$\Rightarrow \quad I = 1 \text{W/m}^2$$

For isotropic point sources, intensity at a distance r is given as

$$I = \frac{P}{4\pi r^2} = \frac{2}{4\pi r^2}$$

$$\Rightarrow \quad r = \sqrt{\frac{2}{4\pi}} = \sqrt{\frac{1}{2\pi}} \, m \qquad (\text{Taking } I = 1 \text{ W/m}^2)$$

$$\Rightarrow \quad r = 0.399 \text{ m} \approx 40 \text{ cm}$$ **Ans. (C)**

Sol. 20 Velocity of waves on wire is given as

$$v = \sqrt{\frac{T}{\mu}}$$

$$\Rightarrow \quad 90 = \sqrt{\frac{\frac{YA}{l}\Delta l}{\frac{m}{l}}} = \sqrt{\frac{16\times10^{11}\times10^{-6}\times\Delta l}{6\times10^{-3}}}$$

$$\Rightarrow \quad \Delta l = \frac{8100\times3}{8}\times10^{-8} \cong 3\times10^{-5} \text{ m}$$

$$= 0.03 \text{ mm}$$ **Ans. (C)**

Sol. 21 Velocity of transverse wave in a string is $v \propto \sqrt{T}$. When velocity is changed to $v/2$ then tension changes to $T/4$, so new tension is

$$T' = \frac{2.06\times10^4}{4}$$

$$\Rightarrow \quad T' = 5.15\times10^3 \text{ N}$$ **Ans. (B)**

Sol. 22 Tension in repeat bottom is

$$T_1 = 2g = 20 \text{ N}$$

tension in rope at top is

$$T_2 = 8g = 80 \text{ N}$$

Wave speed in rope is given as

$$v = \sqrt{\frac{T}{\mu}}$$

As frequency of pulse is same at bottom and top, we use

$$\frac{v_1}{\lambda_1} = \frac{v_2}{\lambda_2}$$

$$\Rightarrow \quad \lambda_2 = \frac{v_2}{v_1}\times\lambda_1 = \sqrt{\frac{T_2}{T_1}}\times\lambda_1$$

$$\Rightarrow \quad \lambda_2 = \sqrt{\frac{80}{20}}\times6 = 12 \text{ cm}$$ **Ans. (B)**

Sol. 23 Given that

$$5\text{m} = N_2\lambda$$

and $$1.5\text{m} = (2N_1+1)\lambda/2$$

$$\Rightarrow \quad \frac{5}{1.5} = \frac{2N_2}{2N_1+1}$$

$$\Rightarrow \quad 10N_1 + 5 = 3N_2$$

Thus for $N_1 = 1 \Rightarrow N_2 = 5 \Rightarrow \lambda = 1$

and $N_1 = 4 \Rightarrow N_2 = 15 \Rightarrow \lambda = 1/3$

and $N_1 = 7 \Rightarrow N_2 = 25 \Rightarrow \lambda = 1/5$ **Ans. (D)**

Sol. 24 Pressure amplitude is given as

$$\Delta P_0 = \frac{2\pi}{\lambda} BA$$

$$\Rightarrow \quad \Delta P_0 = \frac{2\pi}{\lambda}\times\rho v^2\times A = \frac{w}{v}\times\rho v^2\times A$$

$$\Rightarrow \quad A = \frac{\Delta P_0}{\rho v\omega} = \frac{10}{1\times300\times2\times3.14\times1000}$$

$$\Rightarrow \quad A = \frac{10}{300\times6.28}\times10^{-3} \text{ m}$$

Among all given options, (B) is correct. **Ans. (B)**

Sol. 25 A travelling wave is represented by a function given as

$$y = F(x, t)$$

For travelling wave y should be a function of x and t and the function must be in form of $(x \pm vt)$ where $\pm$ represents the direction of propagation of wave. Out of all given option, option (D) is appropriate here. **Ans. (D)**

Sol. 26 The speed of transverse waves on a stretched string is given as

$$V = \sqrt{\frac{T}{\mu}}$$

$$\Rightarrow \quad \frac{\Delta V}{V} = \frac{1}{2}\frac{\Delta T}{T}$$ **Ans. (2)**

Sol. 27 From the given wave equation

$$y = -0.21 \sin(x + 30t)$$

The wave velocity is given as

$$v = \frac{\omega}{k} = \frac{30}{1} = 30 \text{ m/s}$$

The speed of transverse waves on a stretched string is given as

$$v = \sqrt{\frac{T}{\mu}}$$

$\Rightarrow \quad T = v^2\mu = (30)^2(0.0135) = 12.15$

$\Rightarrow \quad T = x \times 10^{-2}$ N

$\Rightarrow \quad x = 1215$ **Ans. (1215)**

Sol. 28 The angular frequency of the wave is given as

$$\omega = 2\pi f = 1.5 \times 10^3 \text{ rad/s}$$

Amplitude of the wave is calculated as

$$A = \frac{6}{2} = 3 \text{ cm} = 0.03 \text{ m}$$

This matches with option (D) only. **Ans. (D)**

Sol. 29 The general equation of a wave disturbance propagating with a speed v is given as

$$y = \frac{1}{1+(x-vt)^2}$$

By comparing it with given equation at $t = 1$s, we have

$v = 2$ m/s **Ans. (2)**

Sol. 30 At node of standing wave, displacement of particle remain zero always so we have

$$1\cos(1.57x)\sin(178.5)t = 0$$

$\Rightarrow \quad \cos(1.57x) = 0$

$\Rightarrow \quad 1.57 \times = \frac{\pi}{2}$

$\Rightarrow \quad x = 1$ cm **Ans. (1)**

Sol. 31 Given that maximum particle velocity in oscillation of medium particles is four times the wave velocity so we use

$$v_{P\max} = 4v_{\text{wave}}$$

$\Rightarrow \quad \omega A = 4\left(\frac{\omega}{k}\right)$

$\Rightarrow \quad A = \frac{4\lambda}{2\pi}$

$\Rightarrow \quad \lambda = \frac{2\pi A}{4} = \frac{20\pi}{4} = 5\pi$ **Ans. (B)**

Sol. 32 Given that maximum particle velocity of medium particles is equal to the wave velocity, so we use

$$\frac{\omega}{k} = A\omega$$

$\Rightarrow \quad k = \frac{1}{A} = \frac{2\pi}{\lambda}$

$\Rightarrow \quad \lambda = 2\pi A = 4\pi$ cm **Ans. (A)**

Sol. 33 For the given wave equation

$$y = 0.5\sin\left(\frac{2\pi}{\lambda}400t - \frac{2\pi}{\lambda}x\right)$$

Angular frequency of oscillation of medium particles is given as

$$\omega = \frac{2\pi}{\lambda}400$$

Angular wave number is given as

$$k = \frac{2\pi}{\lambda}$$

Wave speed is given as

$v = \frac{\omega}{k} = 400$ m/s **Ans. (C)**

Sol. 34 Speed of transverse wave on string is given as

$$v = \sqrt{\frac{T}{\mu}}$$

$\Rightarrow \quad 60 = \sqrt{\frac{T}{10\times10^{-3}} \times 0.5}$

$\Rightarrow \quad T = \frac{(60)^2 \times 10^{-2}}{0.5} = 72$ N

Extension in wire over its natural length is calculated as

$$\Delta l = \frac{Fl}{AY} = \frac{72\times0.5}{2\times10^{-6}\times1.2\times10^{11}}$$

$\Rightarrow \quad \Delta l = \frac{72\times5}{24} \times 10^{-5} = 15 \times 10^{-5}$ **Ans. (15)**

Sol. 35 From the given equation the angular wave number and angular frequency is given as

$$k = 8 \text{ m}^{-1} \text{ and } \omega = 4 \text{ rad/s}$$

Wave velocity is calculated as

$$v = \frac{\omega}{k}$$

$\Rightarrow \quad v = \frac{4}{8} = 0.5$ m/s **Ans. (C)**

Sol. 36 The phase difference between two points separated by a path of Δx is given as

$$\Delta\phi = \frac{2\pi}{\lambda}\Delta x$$

$\Rightarrow \quad \frac{\pi}{3} = \frac{2\pi}{\lambda}(6 \text{ m})$

$\Rightarrow \quad \lambda = 36$ m

$V = f\lambda = (500 \text{ Hz})(36 \text{ m})$

$= 18000$ m/s $= 18$ km/s **Ans. (18)**

Sol. 37 Speed of transverse waves in a wire is given as

$$v=\sqrt{\frac{T}{\mu}}=\sqrt{\frac{70}{70\times10^{-3}}}=100\text{ m/s} \qquad \textbf{Ans. (B)}$$

Sol. 38 Equation of transverse harmonic wave on a string is given as

$$y(x,t)=5\sin(6t+0.003x)$$

By comparing with

$$y=A\sin(wt+kx)$$

$$k=0.003\text{ cm}^{-1},\ \omega=6\text{ rad/s}$$

Wave velocity is given as

$$v=\frac{\omega}{k}$$

$$\Rightarrow \quad v=\frac{6}{0.003\times10^{2}}=20\text{ ms}^{-1} \qquad \textbf{Ans. (20)}$$

Sol. 39 Equation of periodic motion is given as

$$y=\sin\omega t+\cos\omega t$$

$$\Rightarrow \quad y=\sin\omega t+\sin\left(\omega t+\frac{\pi}{2}\right)$$

Phase difference between the components of motion is

$$\Delta\phi=\frac{\pi}{2}$$

Thus resulting amplitude of motion is calculated as

$$A_R=\sqrt{1^2+1^2+2\times1\times1\times\cos(\Delta\phi)}$$

$$\Rightarrow \quad A_R=\sqrt{2} \qquad \textbf{Ans. (B)}$$

Sol. 40 Wave velocity is given as

$$v=\frac{\omega}{k}=\frac{2\pi\times60}{2\pi\times0.5}=\frac{160}{0.5}\text{ m/s}$$

$$\Rightarrow \quad v=\frac{160}{0.5}\times\frac{18}{5}\text{ km/h}$$

$$\Rightarrow \quad v=1152\text{ km/h} \qquad \textbf{Ans. (1152)}$$

Sol. 41 Fundamental frequency of vibration in wire

$$f=\frac{1}{2L}\sqrt{\frac{T}{\mu}}=\frac{1}{2L}\sqrt{\frac{YA\Delta L}{\rho AL}}$$

By definition of Young's Modulus, we have

$$Y=\frac{(T/A)}{(\Delta l/l)}$$

$$\Rightarrow \quad T=YA\frac{\Delta l}{l}$$

$$\Rightarrow \quad f=80\text{ Hz} \qquad \textbf{Ans. (80)}$$

14.2 Stationary Waves and Beats

Sol. 1 Fundamental frequencies of A and B are

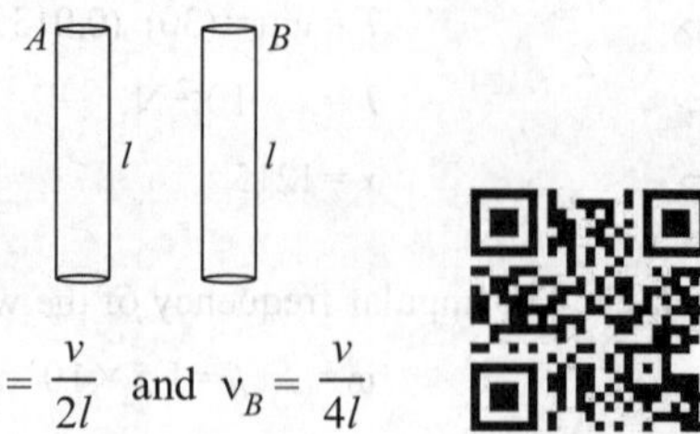

$$\nu_A=\frac{v}{2l} \text{ and } \nu_B=\frac{v}{4l}$$

Here we ignore end correction

$$\Rightarrow \quad \frac{\nu_A}{\nu_B}=\frac{2}{1} \qquad \textbf{Ans. (C)}$$

Sol. 2 If the frequency of unknown fork is ν_0 then we use Initially

$$|\nu_0-288|=4$$

$$\Rightarrow \quad \nu_0=292\text{ cps or }284\text{ cps}$$

On loading ν_0 with some wax its frequency decreases & later it produces 2 beats/s with 288 cps fork.

After loading with wax $\nu_0\to\nu_0'=292\downarrow$ or $284\downarrow$

for 292 → $f_B=290 \Rightarrow$ 2 beats/sec

for 284 → $f_B>4$ beats/s

⇒ 2 beats/s is not possible with 284 Hz. **Ans. (B)**

Sol. 3

AN
N N
40 cm

Maximum wavelength of stationary wave in a clamped string is for fundamental frequency given as

$$\nu_0=\frac{v}{2l}$$

⇒ wavelength $\lambda=2l=2\times40=80$ cm **Ans. (B)**

Sol. 4 General equation of a stationary wave can be

$$y=2A\sin kx\sin(\omega t+\alpha)$$

→ at $x=0$; $R=2A\sin kx=0$

⇒ Node

or $y=2A\cos kx\sin(\omega t+\alpha)$

→ at $x=0$; $R=2A\cos kx=2A$

⇒ Antinode

As one wave is $y_1=a\sin(\omega t-kx)$

with option (A) → $y_R=a\sin(\omega t-kx)+a\sin(\omega t+kx)$

$$=2A\sin\omega t\cos kx$$

with option (B) → $y_R = a\sin(\omega t - kx) + a\sin(\omega t + kx)$

$= -2A\cos\omega t\sin kx$ **Ans. (B)**

Sol. 5 If initial frequency of piano string is ν_0.

then we have $|\nu_0 - 256| = 5 \Rightarrow \nu_0 = 261$ Hz or 251 Hz

On increasing tension in string, its oscillation frequency increases.

$\Rightarrow \nu_0 \to \nu_0' = 261$ Hz (↑) $\to f_B' > 5$ Hz

or 251 Hz (↑) $\to f_B' < 5$ Hz

Giving that $f_B' = 2$ Hz

final beat frequency $\Rightarrow \nu_0 = 251$ Hz **Ans. (C)**

Sol. 6 Initially we have $\nu_1 = 200$ Hz and $\nu_2 = ?$

$$f_B = |200 - \nu_2| = 4 \text{ Hz}$$

$$\Rightarrow \nu_2 = 196 \text{ Hz or } 204 \text{Hz}$$

On loading some tape on fork-2, its frequency decreases

$$\nu_2 \to \nu_2'$$

we use $\nu_2' = 196\downarrow$ or $204\downarrow$

when 196 $\Rightarrow f_B' > 4$ Hz and for 204 $\Rightarrow f_B' < 4$ Hz

later it is given that

$$f_B' = |200 - \nu_2'| = 6 \text{ Hz}$$

$$\Rightarrow \nu_2 = 196 \text{ Hz}$$ **Ans. (C)**

Sol. 7 For a clamped string oscillations all the horizontal frequencies of fundamental frequency are available for resonance so for two successive resonant frequency, we use

$$n_2 - n_1 = n_0 \ (n_0 = \text{fundamental frequency})$$

$$\Rightarrow n_0 = 420 - 315 = 105 \text{ Hz}$$ **Ans. (A)**

Sol. 8 For resonance column experiment we know that second resonance occurs at 3 times the length of first resonance.

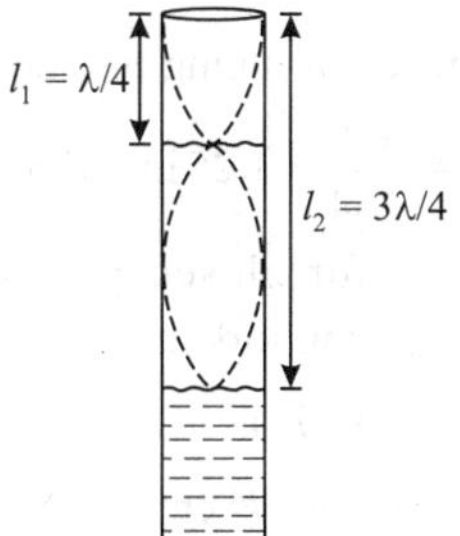

In winter $l_1 = 18$ cm $\Rightarrow l_2 = 54$ cm

As speed of sound $v \propto \sqrt{T}$ in summer speed of sound increases so frequencies of oscillating column are given as (frequency of tuning fork)

$$\nu_1 = \frac{v}{4l_1} = \frac{3v}{4l_2} \to \text{constant}$$

In summer as v increases $\Rightarrow$ for same experiment l_2 also increases. **Ans. (B)**

Sol. 9 Considering two waves at a time beat frequency can be given as

$$\left.\begin{matrix}\nu_1 = \nu - 1 \\ \nu_2 = \nu\end{matrix}\right] \to f_B = 1\text{Hz}$$ and

time of maxima $t = 0, 1, 2, 3, \ldots\ldots$ s

and $$\left.\begin{matrix}\nu_2 = \nu \\ \nu_2 = \nu + 1\end{matrix}\right] \to f_B' = 1\text{Hz}$$

$\Rightarrow t_m = 0, 1, 2, 3, \ldots\ldots$ s

and $$\left.\begin{matrix}\nu_2 = \nu - 1 \\ \nu_3 = \nu + 1\end{matrix}\right] \to f_B'' = 2\text{Hz}$$

$\Rightarrow t_m = 0, 0.5, 1.5, 2, \ldots$ s

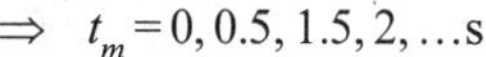

So 0.5, 1.5 .... Very low intensity maxima

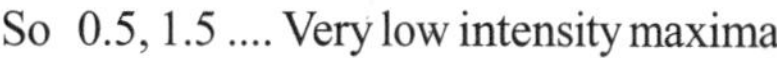

and 0, 1, 2, ... Higher maximas are produced with high intensity.

$\Rightarrow$ Time difference = 1s

$$\Rightarrow f_{Bf} = \frac{1}{T} = 1 \text{ Hz}$$ **Ans. (B)**

Sol. 10 Superposition of the two waves is giving resulting displacement as $y = y_1 + y_2$

$\Rightarrow y = A\sin(\omega t - kx) + A\sin(\omega t + kx)$

$\Rightarrow y = 2A\sin\omega t\cos kx$

$\Rightarrow y = R\sin\omega t$

Where $R = 2A\cos kx$ is amplitude of different medium particles.

For nodes $R = 2A\cos kx = 0$

$\Rightarrow \cos kx = 0$

$\Rightarrow kx = (2N+1)\frac{\pi}{2}; N = 0, 1, 2, \ldots\ldots$

$\Rightarrow \frac{2\pi}{\lambda}x = (2N+1)\frac{\pi}{2}$

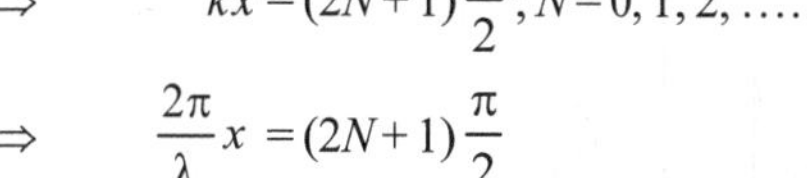

$\Rightarrow x = (2N+1)\frac{\lambda}{4}$ (position of Nodes) **Ans. (A)**

Sol. 11 Fundamental oscillations of an open pipe ⇒ Antinodes at open ends and 1 node at middle

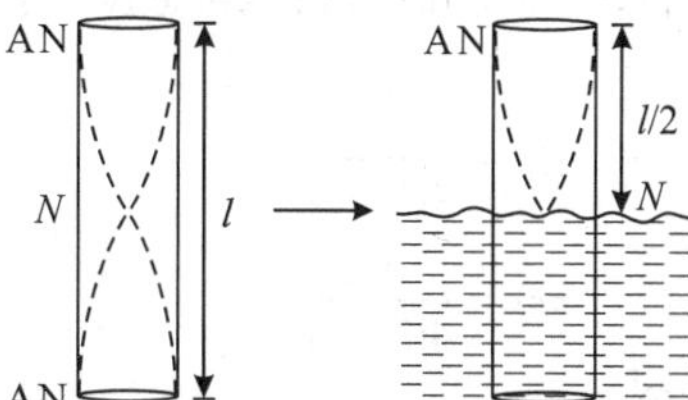

For closed pipe → Antinode at open end and node at closed end.

By looking at figures, the wavelength of sound is same in the two cases $\Rightarrow$ frequency is also same

To verify: $f_{open} = \frac{v}{2l} = f$

$$f_{closed} = \frac{v}{4l'} = \frac{v}{4(l/2)}$$

$$= \frac{v}{2l} = f$$

Ans. (B)

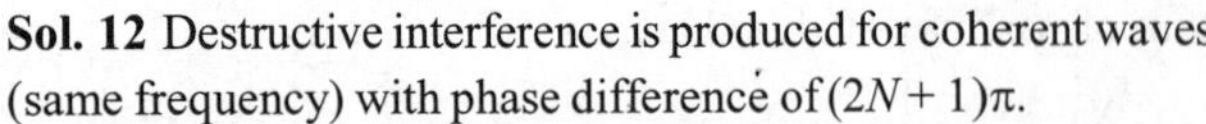

Sol. 12 Destructive interference is produced for coherent waves (same frequency) with phase difference of $(2N+1)\pi$.

In y_1 & y_2 frequency in same & phase difference is π so there SHW will produce Destructive interference.

Stationary waves are produced by superposition of two coherent waves travelling in opposite direction, it happens in y_3 & y_4. **Ans. (D)**

Sol. 13 A general equation of stationary wave can be given in two ways

$y = A_0 \sin kx \sin(\omega t + \phi)$ for a node at $x = 0$

$y = A_0 \cos kx \sin(\omega t + \phi)$ for an Antinode at $x = 0$

As one wave is

$$y_1 = a\cos(kx - \omega t)$$

if $y_2 = -a\cos(kx + \omega t) = a\cos(kx + \omega t + \pi)$

by superposition

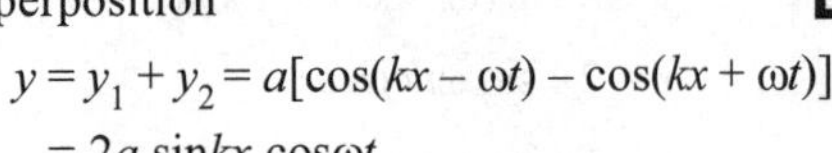

$$y = y_1 + y_2 = a[\cos(kx - \omega t) - \cos(kx + \omega t)]$$

$$= 2a \sin kx \cos\omega t$$

at $x = 0$ there will be a Node **Ans. (B)**

Sol. 14

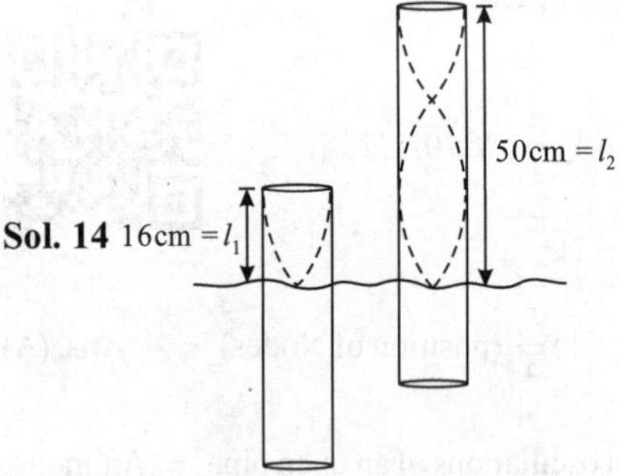

As l_2 is approx 3 times l_1 (but not exact) so we need to account end correction (say 'e'). If v is the velocity of sound than we use

$$500 = \frac{v}{4(l_1 + e)} \quad \ldots(1)$$

and

$$500 = \frac{3v}{4(l_1 + e)} \quad \ldots(2)$$

From (1) and (2)

$$\Rightarrow \frac{v}{4(0.16 + e)} = \frac{3v}{4(0.5 + e)}$$

$$\Rightarrow 0.5 + e = 0.48 + 3e$$

$$\Rightarrow 2e = 0.02$$

$$\Rightarrow e = 0.01 \text{ cm}$$

From equation-(1)

$$500 = \frac{v}{4(0.16 + 0.01)} = \frac{v}{4 \times 0.17}$$

$$\Rightarrow v = 500 \times 4 \times 0.17 = 20 \times 17 = 340 \text{ m/s}$$

When tube is pulled out of water, it behaves like an open tube

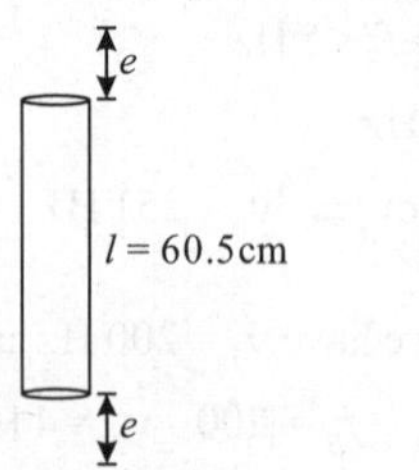

fundamental frequency of the open tube is

$$\nu_0 = \frac{v}{2(l + 2e)} = \frac{340}{2(0.605 + 0.02)}$$

$$\Rightarrow \nu_0 = \frac{340}{2 \times 0.625} = \frac{340}{1.25} = 272 \text{ Hz}$$

An open pipe has all restoring frequency which are integral multiple of its fundamental frequency. So lowest two frequencies are ν_0 and $2\nu_0 \Rightarrow 272$ Hz & 544 Hz **Ans. (D)**

Sol. 15 For a closed pipe, it resonates with a given frequency if its length is odd multiple of $\lambda/4$.

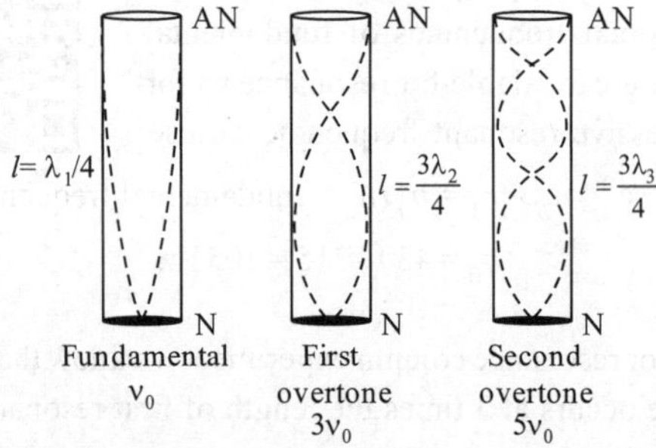

For the given frequency of tuning fork sound wavelength is

$$\lambda = \frac{v}{\nu} = \frac{330}{264} = 1.25 \text{ m} = 125 \text{ cm}$$

The possible length for closed pipe to resonate with above tuning fork are

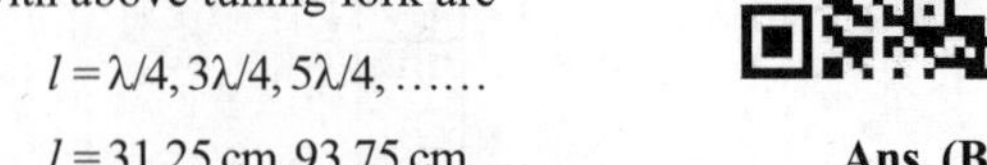

$$l = \lambda/4, 3\lambda/4, 5\lambda/4, \ldots\ldots$$

$$\Rightarrow l = 31.25 \text{ cm}, 93.75 \text{ cm}, \ldots$$

Ans. (B)

Sol. 16 For a stretched string if elastic deformation is there with tension T in it than we have

$$Y = \frac{\text{Stress}}{\text{Strain}} = \frac{T/A}{\text{Strain}}$$

$\Rightarrow$ String tension $T = YA \times$ Strain

Fundamental frequency of a clamped wire is given as

$$\nu_0 = \frac{1}{2l}\sqrt{\frac{T}{\mu}} = \frac{1}{2l}\sqrt{\frac{YA(\text{Strain})}{\rho A}}$$

$$\Rightarrow \quad \nu_0 = \frac{1}{2l}\sqrt{\frac{Y(\text{Strain})}{\rho}}$$

$$\Rightarrow \quad \nu_0 = \frac{1}{2\times 1.5}\sqrt{\frac{2.2\times 10^{11}\times 0.01}{7.7\times 10^{3}}}$$

$$\Rightarrow \quad \nu_0 = 178.2\ \text{Hz}$$

Ans. (B)

Sol. 17 In a damped sonometer wire if l is the length between bridges then

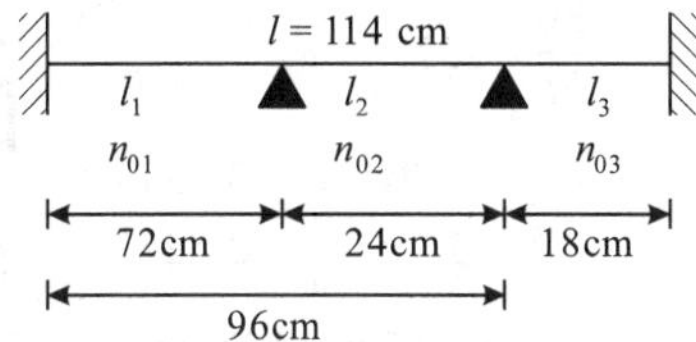

Its fundamental frequency is given as

$$\nu_0 = \frac{v}{2l} = \frac{1}{2l}\sqrt{\frac{T}{\mu}}$$

$$\text{As } n \propto \frac{1}{l} \Rightarrow \frac{1}{l_1} : \frac{1}{l_2} : \frac{1}{l_3} = 1:3:4$$

$$\Rightarrow \quad l_1 : l_2 : l_3 = 12:4:3$$

$$\Rightarrow \quad l_1 = \frac{12}{19}\times 114 = 72\ \text{cm}$$

$$l_2 = \frac{4}{19}\times 114 = 24\ \text{cm}$$

$$l_3 = \frac{3}{19}\times 114 = 18\ \text{cm}$$

Ans. (D)

Sol. 18 For a closed pipe, fundamental frequency is given as

$$n_0 = \frac{v}{4L} = \frac{340}{4\times 0.85} = \frac{85}{0.85} = 100\ \text{Hz}$$

For a closed pipe only odd harmonic frequency resonate

$\Rightarrow$ below 1250 Hz the resonating frequency are

100 Hz, 300 Hz, 500 Hz, 700 Hz, 900 Hz, & 1100 Hz

1 2 3 4 5 6 **Ans. (C)**

Sol. 19 The lengths of parts in which 110 cm wire is divided is

$l_1 = 60$ cm; $l_2 = 30$ cm; $l_3 = 20$ cm

Fundamental frequencies of the 3 parts of wire are

$$\nu_{1p} = \frac{1}{2l_1}\sqrt{\frac{T}{\mu}} = \frac{1}{2\times 0.6}\times\sqrt{\frac{400}{0.01}} = \frac{1}{2\times 0.6}\times\frac{20}{0.1}$$

$$= \frac{1000}{6}\ \text{Hz}$$

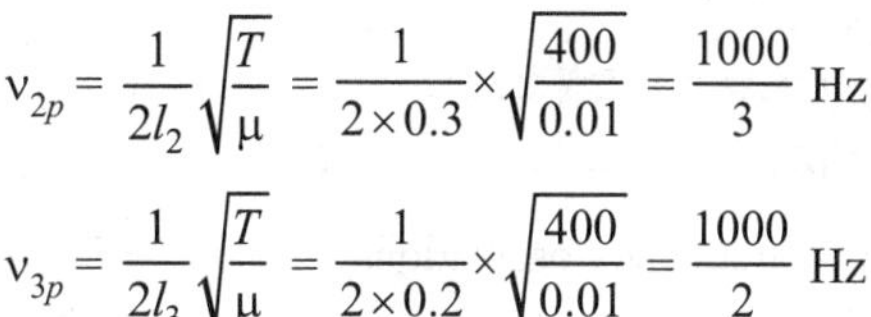

$$\nu_{2p} = \frac{1}{2l_2}\sqrt{\frac{T}{\mu}} = \frac{1}{2\times 0.3}\times\sqrt{\frac{400}{0.01}} = \frac{1000}{3}\ \text{Hz}$$

$$\nu_{3p} = \frac{1}{2l_3}\sqrt{\frac{T}{\mu}} = \frac{1}{2\times 0.2}\times\sqrt{\frac{400}{0.01}} = \frac{1000}{2}\ \text{Hz}$$

As there wires can oscillate at any multiple of their fundamental frequency.

$\Rightarrow$ The lowest common frequency here is

$$\nu_0 = 1000\ \text{Hz}$$

Ans. (B)

Sol. 20 Fundamental frequency of open & closed pipes are shown below

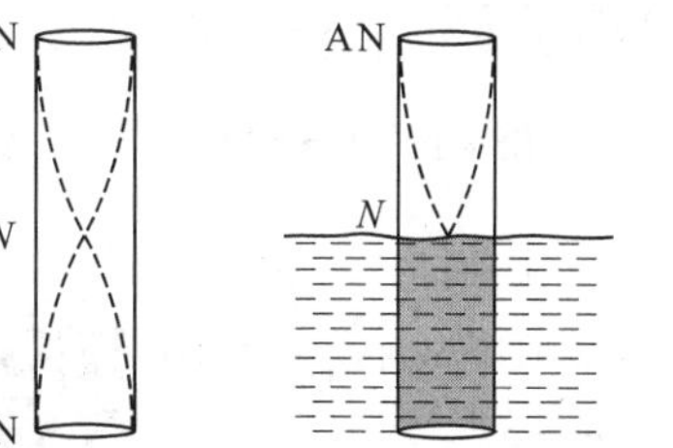

As wavelength in the above two cases is same for fundamental frequencies.

$\Rightarrow$ Their fundamental frequency must be same. **Ans. (D)**

Sol. 21 General equation for stationary wave is given as-

$$y(x, t) = 2A \sin kx \cos \omega t \qquad \ldots(1)$$

Comparing equation-(1) with the given equation we have

$$k = \frac{5\pi}{4} \text{ and } \omega = 200\,\pi$$

$\Rightarrow$ Wave speed is given as

$$v = \frac{\omega}{k} = \frac{200\pi}{5\pi/4} = \frac{200\times 4}{5}$$

$$\Rightarrow \quad v = 160\ \text{m/s}$$

Ans. (A)

Sol. 22 If wire W_1 oscillates in p loops then its frequency is

$$\nu_1 = \frac{p}{2l}\sqrt{\frac{T}{\mu_1}} = \frac{p}{2l}\sqrt{\frac{T}{\rho_1 A}}$$

If wire W_2 oscillates in q loops then its frequency is

$$\nu_2 = \frac{q}{2l}\sqrt{\frac{T}{\mu_2}} = \frac{q}{2l}\sqrt{\frac{T}{\rho_2 A}}$$

p and q are no. of Antinodes in wires W_1 and W_2.

For stationary wave in composite wire, we have

$$\nu_1 = \nu_2$$

$$\Rightarrow \quad \frac{p}{2l}\sqrt{\frac{T}{\rho_1 A}} = \frac{q}{2l}\sqrt{\frac{T}{\rho_2 A}}$$

$$\Rightarrow \quad \frac{p}{q} = \frac{1}{2}$$

Ans. (B)

Sol. 23

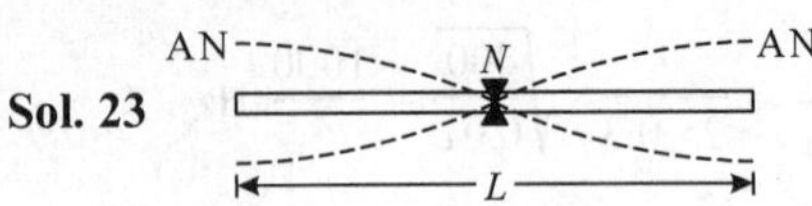

Fundamental frequency oscillations

here $L = \lambda/2$

$\Rightarrow \lambda = 2L$

Frequency of oscillation

$$\nu_0 = \frac{v}{\lambda} = \frac{v}{2L} = \frac{1}{2L}\sqrt{\frac{Y}{\rho}}$$

$$\Rightarrow \nu_0 = \frac{1}{2\times 0.6}\times\sqrt{\frac{9.27\times 10^{10}}{2.7\times 10^3}}$$

$$\Rightarrow \nu_0 = 4.88\times 10^3\text{ Hz} \approx 5\text{ kHz}$$ **Ans. (A)**

Sol. 24 Given that $|\nu_A - \nu_B| = 5$

either $\nu_B = 420$ Hz or 430 Hz As $\nu_A = 425$ Hz

As tension in B is increased

$\Rightarrow$ ν_B increases and new beat frequency with A becomes 3 Hz.

If $\nu_B = 420$ Hz then on increasing it can be 422 Hz & $f_B = 3$ Hz with ν_A.

If $\nu_B = 430$ Hz then on increasing always $f_B > 5$ Hz with ν_A.

Ans. (D)

Sol. 25 For a sonometer wire oscillation frequency is given as

$$n_s = \frac{v}{2l}$$

at $l = 0.95$ m, $n_{s_1} = \dfrac{v}{2\times 0.95}$

and at $l = 1$ m, $n_{s_2} = \dfrac{v}{2\times 1}$

$\Rightarrow n_{s_1} > n_{s_2}$

As tuning fork produces 5 beats/s with both n_{s_1} and n_{s_2}

$\Rightarrow n_{s_1} > n_f > n_{s_2}$

$$\Rightarrow n_f - 5 = n_{s_2} = \frac{v}{1.9} \quad \ldots(1)$$

and $$n_f - 5 = n_{s_1} = \frac{v}{2} \quad \ldots(2)$$

(2)/(1) gives $\dfrac{n_f + 5}{n_f - 5} = \dfrac{20}{19}$

$\Rightarrow 19n_f + 95 = 20n_f - 100$

$\Rightarrow n_f = 195$ Hz **Ans. (A)**

Sol. 26 Frequency of 3rd Harmonic of open pipe is

$$n_3 = \frac{3v}{2l}$$

As beat frequency is 1 beat/s

$$\Rightarrow \frac{3v}{2l} = 256 \pm 1 = 255\text{ Hz or } 257\text{ Hz}$$

Case-I: $\dfrac{3v}{2l} = 255$

$$\Rightarrow l = \frac{3v}{2\times 255} = \frac{3\times 340}{2\times 255}$$

$= 2$ m $= 200$ cm

Case-II: $\dfrac{3v}{2l} = 257$

$$\Rightarrow l = \frac{3v}{2\times 257} = \frac{3\times 340}{2\times 257}$$

$\Rightarrow l \simeq 1.97$ m $= 197$ cm **Ans. (D)**

Sol. 27 First resonance length is 10 cm

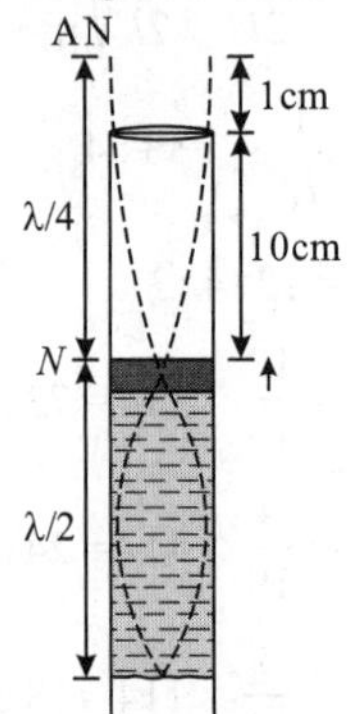

$$\Rightarrow \frac{\lambda}{4} = 10 + 1 = 11\text{ cm}$$

Second resonance occurs when length of column is approx $3\lambda/4$.

$$\frac{3\lambda}{4} = 3\times 11 = 33\text{ cm}$$

Resonance column length now is

$$\frac{3\lambda}{4} - e = 33 - 1 = 32\text{ cm}$$ **Ans. (A)**

Sol. 28 Frequency of the sound produced by flute is given as

$$f = 2\left(\frac{v}{2l}\right) = \frac{2\times 330}{2\times 0.5} = 660\text{Hz}$$

Velocity of observer,

$$v_0 = 10\times\frac{5}{18} = \frac{25}{9}\text{ m/s}$$

Frequency detected by observer is

$$f_{ap} = \left[\frac{v + v_0}{v}\right] f$$

$$\Rightarrow \qquad f' = \left[\frac{\frac{25}{9}+330}{330}\right]660$$

$$\Rightarrow \qquad f' = 335.56 \times 2 = 671.12\text{ Hz}$$ **Ans. (D)**

Sol. 29 Velocity of transverse wave on string

$$v = \sqrt{\frac{T}{\mu}} = \sqrt{\frac{8}{5}\times 1000} = 40\text{ m/s}$$

Now, wavelength of wave

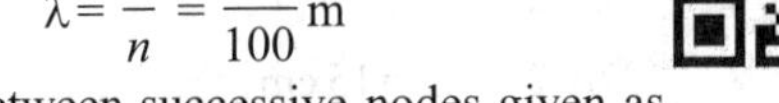

$$\lambda = \frac{v}{n} = \frac{40}{100}\text{ m}$$

Separation between successive nodes given as

$$\frac{\lambda}{2} = \frac{20}{100}\text{ m} = 20\text{ cm}$$ **Ans. (B)**

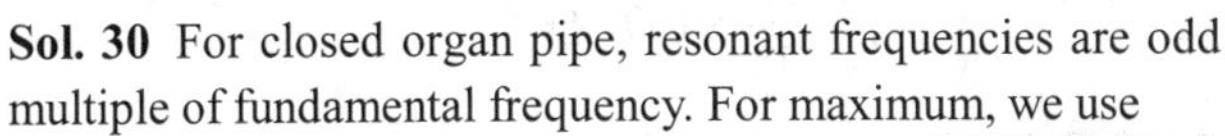

Sol. 30 For closed organ pipe, resonant frequencies are odd multiple of fundamental frequency. For maximum, we use

$(2n+1)f_0 \leq 20{,}000$

(f_0 is fundamental frequency = 1.5 kHz)

$$\Rightarrow \qquad n = 6$$

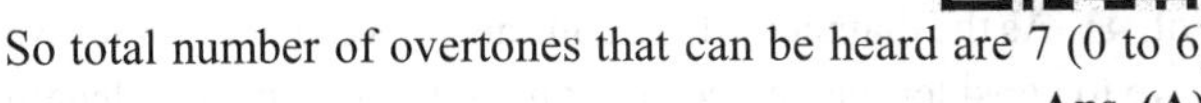

So total number of overtones that can be heard are 7 (0 to 6) **Ans. (A)**

Sol. 31 If antinode is produced at a distance x above the reference mark, with frequency 512 Hz, we have

$$512 = \frac{v}{4(11+x)} \qquad \ldots(1)$$

with 256 Hz frequency we use

$$256 = \frac{v}{4(27+x)} \qquad \ldots(2)$$

$$\frac{(1)}{(2)} \Rightarrow \qquad 2 = \frac{27+x}{11+x}$$

$$\Rightarrow \qquad 22 + 2x = 27 + x$$

$$\Rightarrow \qquad x = 5\text{ cm}$$

Thus sound speed in air is

$$v = 512 \times 4(11+x)$$

$$\Rightarrow \qquad v = 512 \times 4 \times 0.16$$

$$\Rightarrow \qquad v = 327.68\text{ m/s}$$ **Ans. (A)**

Sol. 32

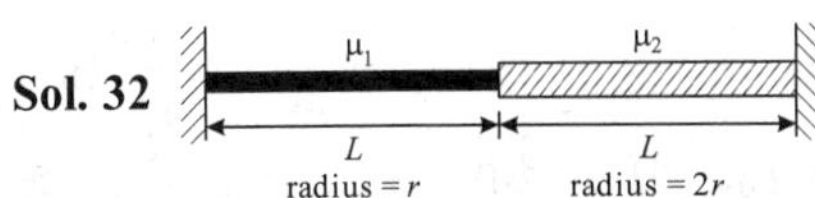

Let mass per unit length of wires are μ_1 and μ_2 respectively As materials are same, so density ρ is same, we use

$$\mu_1 = \frac{\rho\pi r^2 L}{L} = \mu$$

and $$\mu_2 = \frac{\rho 4\pi r^2 L}{L} = 4\mu$$

Tension in both are same, taken as T, let speed of wave in wires are v_1 and v_2, we have

$$v_1 = \sqrt{\frac{T}{\mu}} = v;$$

$$v_2 = \sqrt{\frac{T}{4\mu}} = \frac{v}{2}$$

So fundamental frequencies in both wires are given as

$$f_{01} = \frac{v_1}{2L} = \frac{v}{2L}$$

and $$f_{02} = \frac{v_2}{2L} = \frac{v}{4L}$$

Minimum frequency at which both wires resonate is L.C.M of both of these frequencies, given as

$$f_0 = \frac{v}{2L}$$

Hence number of loops in wires are 1 and 2 respectively as shown below

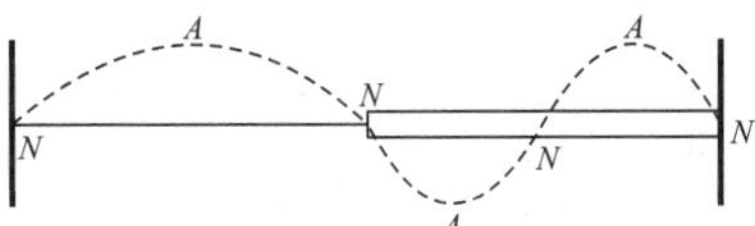

So, ratio of no. of antinodes is 1 : 2 **Ans. (D)**

Sol. 33 4th harmonic oscillations in clamped wire are shown below

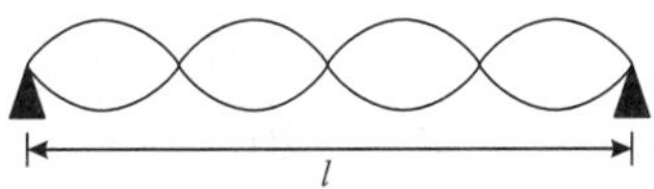

$$\Rightarrow \qquad 4\frac{\lambda}{2} = l$$

$$\Rightarrow \qquad 2\lambda = l \Rightarrow \lambda = \frac{l}{2}$$

From given equation, we use

$$\frac{2\pi}{\lambda} = 0.157$$

$$\lambda = 40$$

$$l = 2\lambda$$

$$\Rightarrow \qquad l = 80\text{ m}$$ **Ans. (B)**

Sol. 34 As string vibrates in 3rd harmonic, we use

$$3\left(\frac{v}{2l}\right) = 240$$

$$\Rightarrow \qquad 3\frac{v}{2\times 2} = 240$$

$$\Rightarrow \qquad v = 320\text{ m/s}$$

Fundamental frequency of string is given as

$$n_0 = \frac{v}{2l} = \frac{320}{2\times 2} = 80\text{ Hz}$$ **Ans. (D)**

Sol. 35 Beat frequency of the given waves is

$$f_{beat} = 11 - 9 = 2\text{ Hz}$$

Thus in one second there are two maxima and minima occur. This is corresponding to option (D). **Ans. (D)**

Sol. 36 In resonance tube experiment, wave speed is given as

$$v = 2f(l_2 - l_1)$$

$$\Rightarrow \quad v = 2 \times 480 \times (70 - 30) \times 10^{-2}$$

$$\Rightarrow \quad v = 960 \times 40 \times 10^{-2}$$

$$\Rightarrow \quad v = 38400 \times 10^{-2}\text{ m/s}$$

$$\Rightarrow \quad v = 384\text{ m/s}$$

Ans. (C)

Sol. 37 Speed of sound in air is given as

$$v_s = \sqrt{\frac{\gamma P}{\rho}}$$

$$\Rightarrow \quad \frac{v_{gas}}{v_{air}} = \sqrt{\frac{\rho_{air}}{\rho_{gas}}}$$

$$\Rightarrow \quad \frac{v_{gas}}{300} = \frac{1}{\sqrt{2}}$$

$$\Rightarrow \quad v_{gas} = \frac{300}{\sqrt{2}}\text{ m/s}$$

$$\Rightarrow \quad v_{gas} = 150\sqrt{2}\text{ m/s}$$

For open organ pipe, we use

$$n_2 - n_1 = \frac{v_{gas}}{2l} = \frac{150\sqrt{2}}{2(1)} = 75\sqrt{2}\text{ Hz}$$

$$\Rightarrow \quad \Delta n = 106.06\text{ Hz}$$

Ans. (106.06)

Sol. 38 For n^{th} harmonic, frequency is given as

$$\frac{nv}{2l} = 420$$

For next $(n+1)^{th}$ harmonic, we use

$$\frac{(n+1)v}{2l} = 490$$

$$\Rightarrow \quad \frac{v}{2l} = 70$$

$$\Rightarrow \quad l = \frac{v}{140} = \frac{1}{140}\sqrt{\frac{540}{6\times10^{-3}}} = \frac{1}{140}\sqrt{90\times10^3}$$

$$\Rightarrow \quad l = \frac{300}{140} = 2.142\text{ m}$$

Ans. (D)

Sol. 39 Fundamental frequency of oscillation for a clamped string is given as

$$f = \frac{l}{2l}\sqrt{\frac{T}{\mu}}$$

For identical string l and m will be same

$$\Rightarrow \quad f \propto \sqrt{T}$$

$$\Rightarrow \quad \frac{450}{300} = \sqrt{\frac{T_x}{T_y}}$$

$$\Rightarrow \quad \frac{T_x}{T_y} = \frac{9}{4} = 2.25$$

Ans. (C)

Sol. 40 For two successive resonance lengths in tube, we use

$$\Rightarrow \quad \lambda = 2(l_2 - l_1)$$

$$\Rightarrow \quad \lambda = 2 \times (24.5 - 17)$$

$$\Rightarrow \quad \lambda = 2 \times 7.5 = 15\text{ cm}$$

As $v = f\lambda$, we use

$$\Rightarrow \quad 330 = f \times 15 \times 10^{-2}$$

$$\Rightarrow \quad f = \frac{330}{15} \times 100\text{ Hz}$$

$$\Rightarrow \quad f = \frac{110 \times 100}{5}\text{ Hz}$$

$$\Rightarrow \quad f = 2200\text{ Hz}$$

Ans. (C)

Sol. 41 As the diameter of the column is given so it implies we have to consider the end correction $e = 0.3d$ in the tube length so, first resonance length can be given as

$$l + e = \frac{\lambda}{4} = \frac{V}{4f}$$

$$\Rightarrow \quad l = 16.66 - 0.3 \times 6 = 14.866\text{ cm}$$

$$\Rightarrow \quad l = 14.8\text{ cm}$$

Ans. (B)

Sol. 42 Initially beat frequency is given as 5 Hz so initial frequency of for A is either 345 Hz or 335 Hz and after filing the frequency of fork A increases. It is also given that beat frequency decreases so frequency of A must be 335 Hz and as its final frequency is increasing due to which beat frequency also decrease to 2 beats/s. **Ans. (B)**

Sol. 43 As both pipes are oscillating at same frequency, if length of open pipe of L', we use

$$f_c = f_0$$

$$\Rightarrow \quad \frac{3V_C}{4L} = \frac{2V_0}{2L'}$$

$$\Rightarrow \quad \frac{3V_C}{4L} = \frac{V_0}{L'}$$

$$\Rightarrow \quad L' = \frac{4L}{3}\frac{V_0}{V_C} = \frac{4L}{3}\sqrt{\frac{B.\rho_1}{\rho_2.B}} \quad (B\text{ is bulk modulus})$$

$$\Rightarrow \quad L' = \frac{4L}{3}\sqrt{\frac{\rho_1}{\rho_2}}$$

$$\Rightarrow \quad x = 4$$

Ans. (4)

Sol. 44 The phase difference between the two waves at the is given as

$$\Delta\phi = kx_0 = \frac{2\pi}{\lambda}x_0 = 2\pi \times 3.5 = 7\pi \text{ rad}$$

Amplitude of resulting wave can be given as

$$\Rightarrow \quad A_R = \sqrt{A_1^2 + A_2^2 + 2A_1A_2\cos(\Delta\phi)}$$

$$\Rightarrow \quad A_R = \sqrt{12^2 + 5^2 + 2\times 12\times 5\times \cos(7\pi)}$$

$$\Rightarrow \quad A_R = \sqrt{(12-5)^2} = 7 \qquad \textbf{Ans. (7)}$$

Sol. 45 Closed organ pipe resonate at frequencies given as

$$f = (2n+1)\frac{v}{4l}$$

Minimum length, at which resonance occurs can be calculated for

$$n = 0$$

$$\Rightarrow \quad f_0 = \frac{v}{4l}$$

$$\Rightarrow \quad l = \frac{v}{4f_0}$$

$$\Rightarrow \quad l = \frac{340}{4\times 250} = 34 \text{ cm} \qquad \textbf{Ans. (34)}$$

Sol. 46 Resonating frequency of the wire stretched between the rigid supports is given as

$$f_n = \frac{nv}{2l} = 500$$

Next higher frequency is given as

$$f_{n+1} = \frac{(n+1)v}{2l} = 500$$

$$\Rightarrow \quad \frac{n+1}{n} = \frac{11}{10}$$

$$\Rightarrow \quad n = 10$$

$$\Rightarrow \quad l = \frac{nv}{2f_n}\left(v = \sqrt{\frac{T}{\mu}}\right)$$

$$\Rightarrow \quad l = \frac{10}{2\times 500}\times\sqrt{\frac{900}{9\times 10^{-4}}} = 10 \text{ m} \qquad \textbf{Ans. (10)}$$

Sol. 47 Phase difference between the waves at the point of superposition is given as

$$\Delta\phi = 2\pi(1.5) = 3\pi$$

Resulting amplitude at the point of superposition is given as

$$A_R = \sqrt{A_1^2 + A_2^2 + 2A_1A_2\cos(3\pi)}$$

$$\Rightarrow \quad A_R = |A_1 - A_2| = 2 \text{ cm} \qquad \textbf{Ans. (A)}$$

Sol. 48 The equation of stationary wave gives the amplitude of the particle as

$$R = 10\cos\left(\frac{4\pi}{3}\right) = 5 \text{ cm} \qquad \textbf{Ans. (5)}$$

Sol. 49 First overtone frequency of an open organ pipe is given as

$$f_1 = \frac{2v}{2l_1}$$

Fundamental frequency of a closed organ pipe is given as

$$f_2 = \frac{v}{4l_2}$$

As stated in the question, we use

$$f_1 = f_2$$

$$\Rightarrow \quad \frac{2v}{2l_1} = \frac{v}{4l_2}$$

$$\Rightarrow \quad l_1 = 4l_2 = 80 \text{ cm} \qquad \textbf{Ans. (80)}$$

Sol. 50 The frequencies of the tuning forks are considered as

$f_1 = f$

and $f_2 = f + 4$

and $f_3 = f + 2\times 4$

and $f_4 = f + 3\times 4$

and $f_{20} = f + 19\times 4$

It is also specified that the last fork has frequency double of the first one, so we have

$$f + (19\times 4) = 2\times f$$

$$\Rightarrow \quad f = 76 \text{ Hz}$$

Thus frequency of last tuning forks is given as

$$2f = 152 \text{ Hz} \qquad \textbf{Ans. (152)}$$

Sol. 51 Beat frequency of two waves is given as

$$f_b = f_1 - f_2$$

$$\frac{v}{4.08} - \frac{v}{4.16} = \frac{40}{12}$$

$$\Rightarrow \quad v = 707.2 \text{ m/s} \qquad \textbf{Ans. (D)}$$

Sol. 52 As per the language of question if initially tuning fork is resonating in fundamental mode then it can never resonate at air column of lower length so after pouring water it can never achieve state of resonance. So here we will be solving the question without considering initial fundamental mode. The wavelength of wave is calculated as

$$\lambda = \frac{v}{f} = \frac{340}{340} = 1 \text{ m}$$

First resonating length of an air column is one quarter of the wavelength given as

$$l_1 = \frac{\lambda}{4} = 25 \text{ cm}$$

Second resonating length is three quarter of the wavelength, given as

$$l_2 = \frac{3\lambda}{4} = 75 \text{ cm}$$

At the surface of water level a node is obtained so water level height required is the difference between the distances of successive nodes, which is half wavelength. So height of water required to achieve resonance again is

$$\Delta l = 75 - 25 = 50 \text{ cm}$$ **Ans. (50)**

Sol. 53 Wavelength of wave is given as

$$\lambda = \frac{336}{400} \times 100 \text{ cm} = 84 \text{ cm}$$

For first resonance the column length is one quarter of wavelength, calculated as

$$\frac{\lambda}{4} = 21 \text{ cm}$$

But in question it is given as 20 cm that indicates 1 cm is the end correction involved in this case hence we use

$$l_1 + e = \frac{\lambda}{4}$$

For third resonance in air column we use

$$l_3 + e = \frac{5\lambda}{4} = 105 \text{ cm}$$

$$\Rightarrow \quad l_3 = 104 \text{ cm}$$ **Ans. (104)**

Sol. 54 For n^{th} harmonic of stretched string, the frequency is given as

$$\frac{nv}{2l} = \frac{nv}{0.6} = 400$$

For $(n+1)^{\text{th}}$ harmonic, it is given as

$$\frac{(n+1)v}{0.6} = 450$$

$$\Rightarrow \quad \left[\frac{0.6 \times 400}{v} + 1\right]\frac{v}{0.6} = 450$$

$$\Rightarrow \quad v = 30 \text{ m/s}$$

Wave velocity in a stretched string is given as

$$\sqrt{\frac{T}{\mu}} = 30$$

$$\Rightarrow \quad \frac{2700}{\mu} = 900$$

$$\Rightarrow \quad \mu = 3 \text{ kg/m}$$ **Ans. (3)**

Sol. 55 Figure below shows the state of second harmonic oscillations in the open organ pipe.

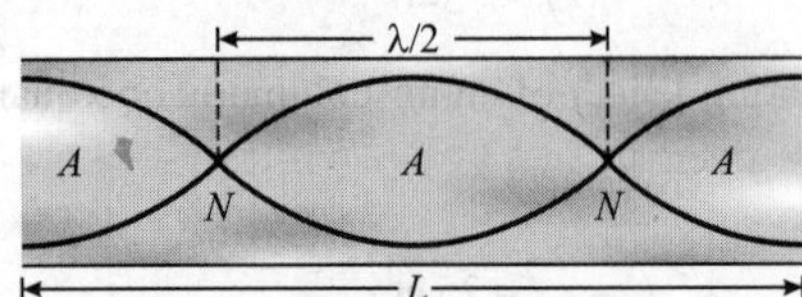

Given that length of organ pipe is 40 cm. For second harmonic of open organ pipe, we use

$$L = \lambda$$

So frequency of vibration is $f = \frac{v}{\lambda}$

$$\Rightarrow \quad f = \frac{v}{\lambda} = \frac{v}{L} = \frac{360}{\frac{40}{100}} = 900 \text{ Hz}$$ **Ans. (900)**

Sol. 56 Frequency in open organ pipe

$$f = \frac{nv}{2l}$$

For fundamental mode $n = 1$, so we have

$$f = \frac{v}{2l}$$

$$\Rightarrow \quad f \propto \frac{1}{l}$$

$$\Rightarrow \quad \frac{f_1}{f_2} = \frac{l_2}{l_1}$$

As guitar string of 90 cm vibrate with 120 Hz then for 180 Hz length will be

$$\Rightarrow \quad \frac{120}{180} = \frac{l_2}{90}$$

$$\Rightarrow \quad l_2 = 60 \text{ cm}$$ **Ans. (60)**

Sol. 57 Resonating frequency in closed organ pipe is given as

$$f = \frac{nv}{4l}$$

Where n is an odd integer and for 5th harmonic in closed organ pipe, the frequency is given as

$$f_5 = \frac{5v}{4l}$$

$$\Rightarrow \quad 405 = \frac{5 \times 324}{4l}$$

$$\Rightarrow \quad l = 1 \text{ m}$$ **Ans. (1)**

Sol. 58 Frequency of vibration in string is given as

$$f = \frac{1}{2l}\sqrt{\frac{T}{\mu}}$$

When in sonometer string a mass m is attached instead of 180g, the new frequency f_2 can be calculated as

$$\frac{f_2}{f_1} = \sqrt{\frac{T_2}{T_1}}$$

$$\Rightarrow \quad \left(\frac{50}{30}\right)^2 = \frac{mg}{180g}$$

$$\Rightarrow \quad m = \frac{25}{9} \times 180 = 500 \text{ g}$$ **Ans. (500)**

Sol. 59 Figure below shows the fundamental oscillations of a stretched string between two rigid supports.

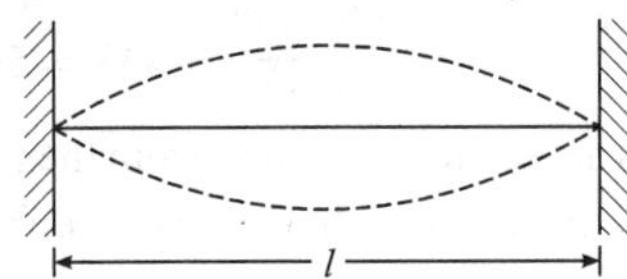

Fundamental frequency given as

$$n_0 = 50\text{ Hz}$$

Length of stringis given as

$$l = \frac{18}{20}\text{m} = \frac{9}{10}\text{m}$$

Distance between two nodes is given as

$$\frac{\lambda}{2} = l$$

$$\Rightarrow \quad \lambda = 2l = \frac{9}{5}\text{m}$$

Wave speed is calculated as

$$v = f\lambda = 50 \times \frac{9}{5} = 90\text{ m/s}$$ **Ans. (90)**

14.3 Doppler's Effect

Sol. 1

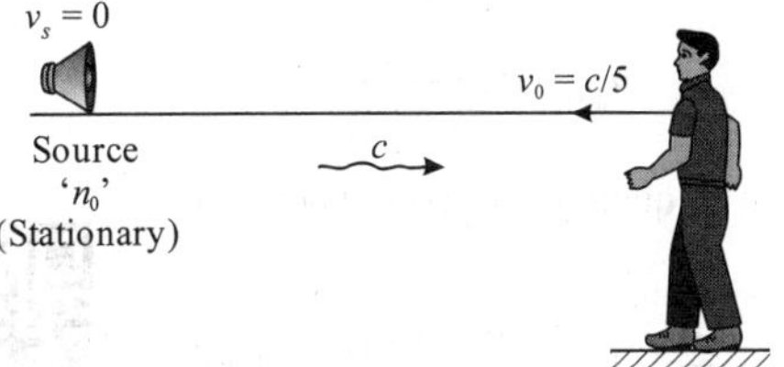

Apparent frequency is

$$\nu_{ap} = \nu_0\left(\frac{c - v_0}{c - v_s}\right) = \nu_0\left(\frac{c-(-c/5)}{c}\right)$$

$$\Rightarrow \quad \nu_{ap} = \nu_0\left(\frac{6}{5}\right)$$

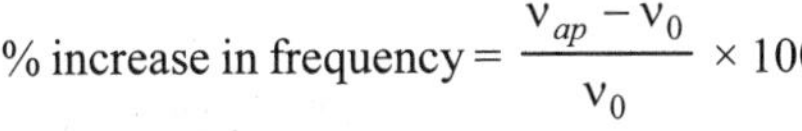

$$\%\text{ increase in frequency} = \frac{\nu_{ap} - \nu_0}{\nu_0} \times 100$$

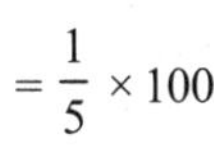

$$= \frac{1}{5} \times 100$$

$$= 20\,\%$$ **Ans. (D)**

Sol. 2

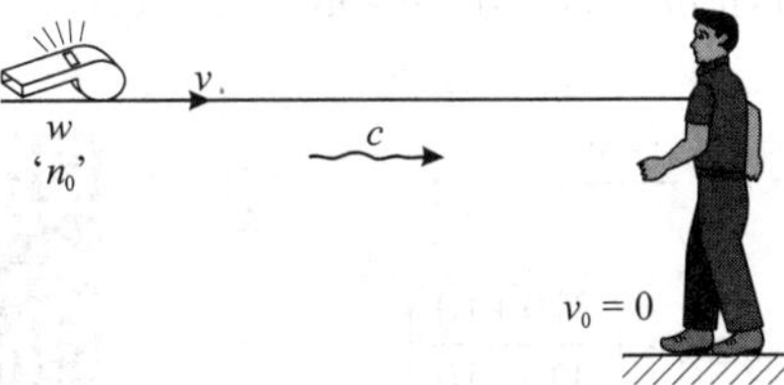

Apparent frequency heard by observer is

$$n_{ap} = n_0\left(\frac{c - v_0}{c - v_s}\right) = n_0\left(\frac{c}{c - v_s}\right)$$

As $n_{ap} = 10000\text{ Hz}$ and $n_0 = 9500\text{ Hz}$

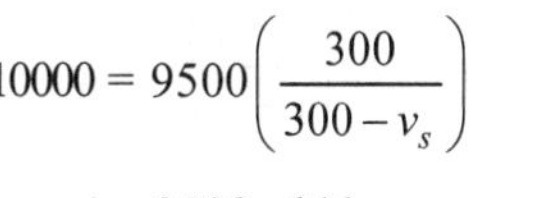

$$\Rightarrow \quad 10000 = 9500\left(\frac{300}{300 - v_s}\right)$$

$$\Rightarrow \quad 300 - v_s = \frac{9500 \times 300}{10000} = 285$$

$$\Rightarrow \quad v_s = 300 - 285 = 15\text{ m/s}$$ **Ans. (C)**

Sol. 3 S n_0 $u = 0$ (rest) $v_s = 0$, a, s, c, $v_M = \sqrt{2as}$

Apparent frequency is $n_{app} = n_0\left(\dfrac{c - v_M}{c}\right)$

$$\Rightarrow \quad \frac{94}{100}n_0 = n_0\left(\frac{c - \sqrt{2as}}{c}\right)$$

$$\Rightarrow \quad \frac{94 \times 330}{100} = 330 - \sqrt{2 \times 2 \times s}$$

$$\Rightarrow \quad s = 98\text{ m}$$ **Ans. (D)**

Sol. 4 **S-1: Echolocation :** A phenomenon by which bats detect the objects in their surrounding.

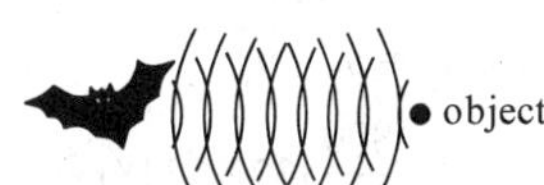

Waves reflected from object through which bats estimate the location of object.

S-1 is true. By Definition of Doppler's effects.

In general **S-2** is True.

As **S-2** is not linked with **S-1** directly. **Ans. (C)**

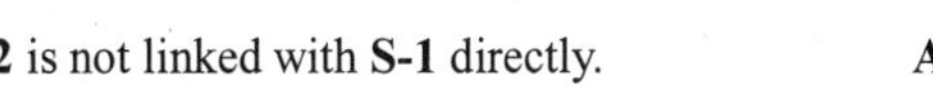

Sol. 5 **Case-I:**

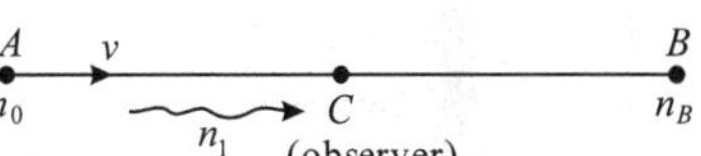

Apparent frequency of A heard by C is

$$n_1 = n_0\left(\frac{c}{c - v}\right)$$

$$\Rightarrow \quad n_1 = 500 \times \frac{340}{336} = 506\text{ Hz}$$

$$f_B = |n_1 - n_B| = 6$$

$$n_B = 500\text{ Hz or } 512\text{ Hz}$$

Case-2:

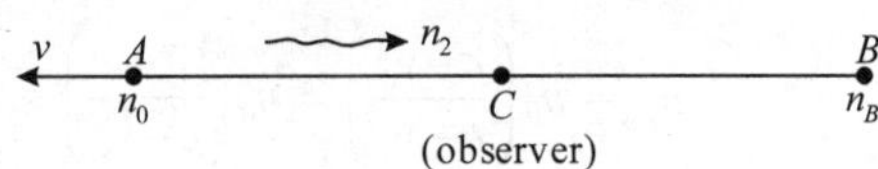

Apparent frequency of A heard by C is

$$n_2 = n_0\left(\frac{c}{c+v}\right)$$

$$\Rightarrow \quad n_2 = 500 \times \frac{340}{344} = 494\text{ Hz}$$

$$f_B' = |n_2 - n_B| = 18$$

$$\Rightarrow \quad n_B = 512\text{ Hz}$$

Ans. (C)

Sol. 6

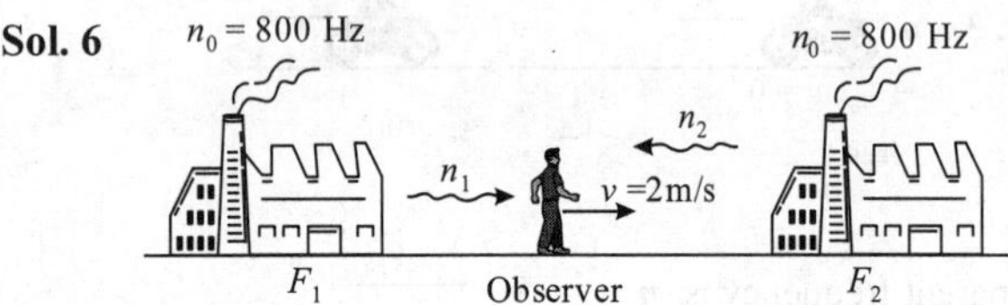

Apparent frequency of sound heard by observer are

from factory-1 it is $n_1 = n_0\left(\frac{c-v}{c}\right) = 800 \times \left(\frac{320-2}{320}\right)$

$$\Rightarrow \quad n_1 = \frac{800 \times 318}{320}$$

from factory-2 it is $n_2 = n_0\left(\frac{c+v}{c}\right) = 800 \times \left(\frac{320+2}{320}\right)$

$$\Rightarrow \quad n_2 = \frac{800 \times 322}{320}$$

Beat frequency heard $f_B = n_2 - n_1$

$$= \frac{800}{320}(322-318)$$

$$\Rightarrow \quad f_B = \frac{800 \times 4}{320} = 10\text{ Hz}$$

Ans. (D)

Sol. 7

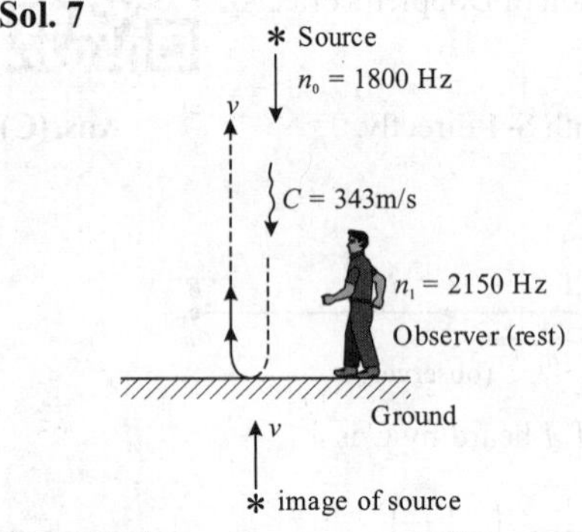

By Doppler's effect apparent frequency of sound received by observer is

$$n_1 = n_0\left(\frac{c}{c-v_s}\right)$$

$$2150 = 1800\left(\frac{343}{343-v}\right)$$

$$\Rightarrow \quad 343 - v = \frac{1800 \times 340}{2150}$$

$$\Rightarrow \quad v = -\frac{180 \times 343}{215} + 343 \simeq 56\text{ m/s}$$

Reflected waves appear to be coming from the image of source in the ground so the apparent frequency received by source is reflected waves is given as

$$n_{ap} = n_0\left(\frac{c+v}{c-v}\right) = 1800 \times \left(\frac{343+56}{343-56}\right)$$

$$\Rightarrow \quad n_{ap} = 1800 \times \frac{399}{287} = 2502\text{ Hz}$$

Ans. (B)

Sol. 8

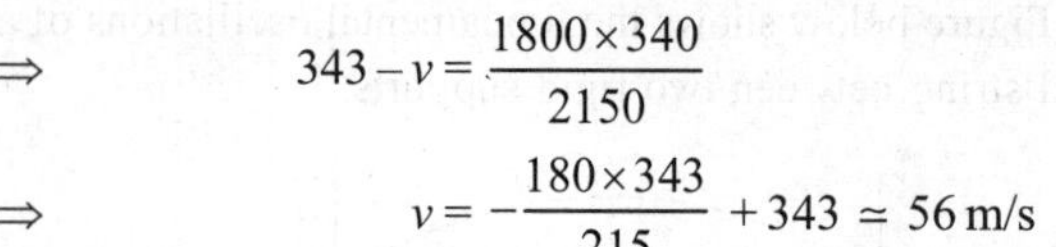

If n_{ap_2} is the frequency observed when train is approaching the person then we have

$$n_{ap_1} = n_0\left(\frac{c}{c-v_s}\right) \quad \ldots(1)$$

If v_{app_2} is frequency observed when train reaches away from person then we have

$$n_{ap_2} = n_0\left(\frac{c}{c+v_S}\right) \quad \ldots(2)$$

% change in frequency observed when train passes

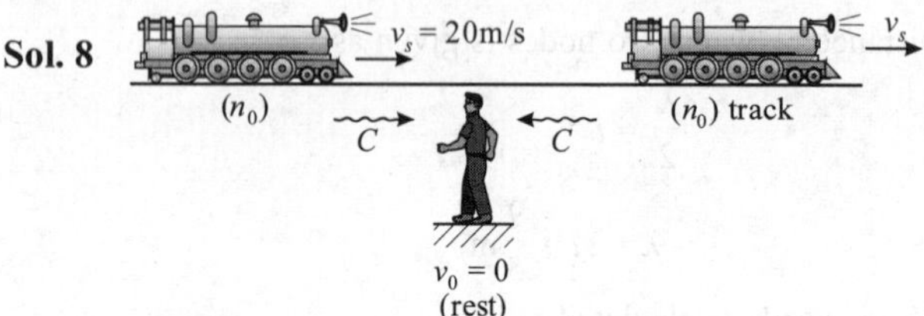

$$\frac{\Delta n}{n_0} \times 100 = \frac{n_{ap_1} - n_{ap_2}}{n_0} \times 100\%$$

$$\%\text{ change} = \left[\frac{c}{c-v_S} - \frac{c}{c+v_S}\right] \times 100$$

$$\Rightarrow \quad \%\text{ change} = \left[\frac{320}{300} - \frac{320}{340}\right] \times 100$$

$$\Rightarrow \quad \%\text{ change} = 106.66 - 94.11 \simeq 12\%$$

Ans. (B)

Sol. 9

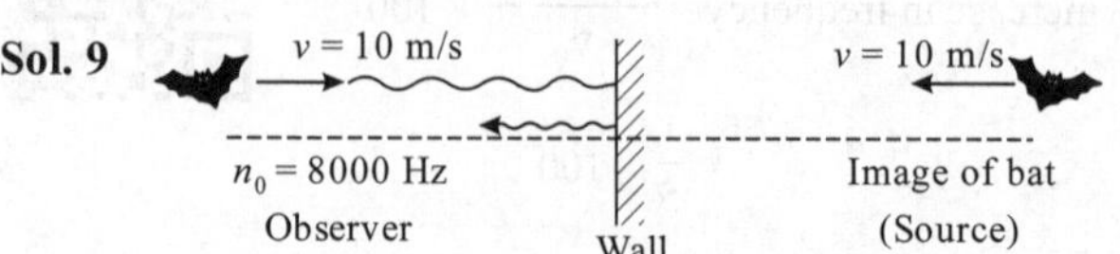

Frequency of reflected sound appear to be coming from image of bat as shown in figure so the apparent frequency heard is given as

$$f_{ap} = n_0\left(\frac{c+v}{c-v}\right)$$

$$= 8000 \times \left(\frac{320+10}{320-10}\right)$$

$$\Rightarrow \qquad f_{ap} = 8000 \times \frac{330}{310} \cong 8516\,\text{Hz} \qquad \textbf{Ans. (A)}$$

Sol. 10 By Doppler's effect apparent frequency of sound heard by observer is given as

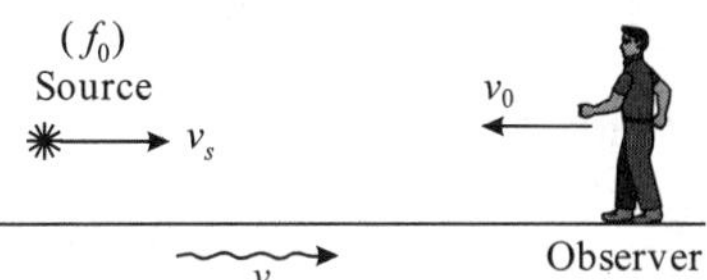

$$\text{Apparent frequency} \quad f = f_0\left(\frac{v+v_0}{v-v_s}\right)$$

$$\Rightarrow \qquad f = \left(\frac{f_0}{v-v_s}\right)v_0 + \left(\frac{f_0 v}{v-v_s}\right)$$

This is of form $\quad y = mx + C$

slope of graph is $\quad m = \dfrac{f_0}{v-v_s}$ **Ans. (C)**

Sol. 11

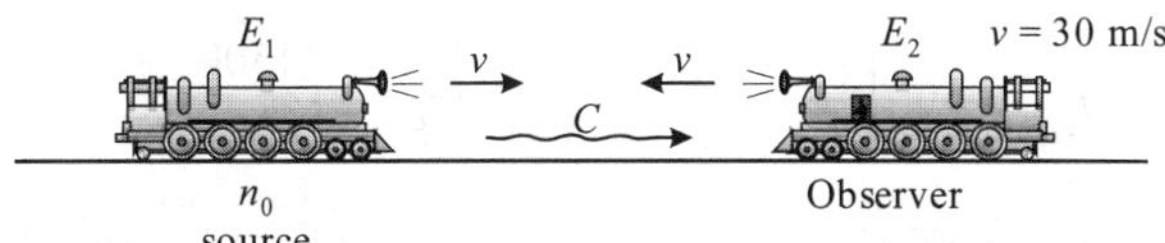

Apparent frequency heard by driver of second engine E_2 is

$$n_{ap} = n_0\left(\frac{c+v}{c-v}\right) = 540 \times \left(\frac{330+30}{330-30}\right)$$

$$\Rightarrow \quad n_{ap} = 540 \times \frac{360}{300} = 54 \times 12 = 648\,\text{Hz} \qquad \textbf{Ans. (D)}$$

Sol. 12

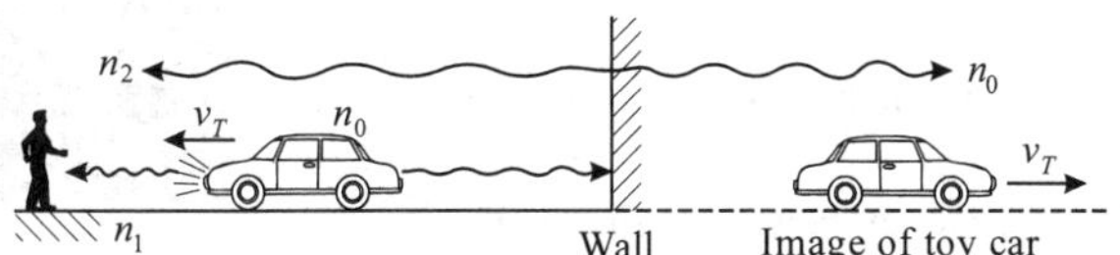

$n_1 \rightarrow$ Apparent frequency of horn from approaching toy car

$$\Rightarrow \qquad n_1 = n_0\left(\frac{c}{c-v_T}\right) = n_0\left(\frac{340}{340-5}\right)$$

$$\Rightarrow \qquad n_1 = \frac{340}{335}n_0$$

$n_2 \rightarrow$ Apparent frequency of horn from the receding image of toy car due to reflected wave from wall.

$$\Rightarrow \qquad n_2 = n_0\left(\frac{c}{c+v_T}\right) = n_0\left(\frac{340}{340+5}\right)$$

$$\Rightarrow \qquad n_2 = \frac{340}{345}n_0$$

As beat frequency heard by observer is 5 beats/s

$$\Rightarrow \qquad n_1 - n_2 = 5$$

$$\Rightarrow \qquad \left(\frac{340}{335} - \frac{340}{345}\right)n_0 = 5$$

$$\Rightarrow \qquad n_0 \simeq 170\,\text{Hz} \qquad \textbf{Ans. (D)}$$

Sol. 13

The apparent frequency observed by observer is given by relativistic relation as

$$n_{ap} = n_0\sqrt{\frac{C-V_0}{C+V_0}} = n_0\sqrt{\frac{C+C/2}{C-C/2}} \quad [\because v_0 = -C/2]$$

$$\Rightarrow \quad n_{ap} = n_0\left[\frac{3C/2}{C/2}\right]^{1/2} = 10 \times 10^9 \times \sqrt{3}$$

$$\Rightarrow \quad n_{ap} = 17.32\,\text{GHz} \qquad \textbf{Ans. (A)}$$

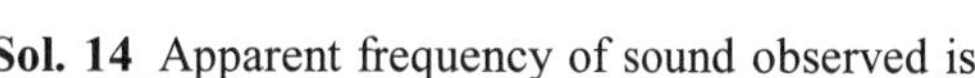

Sol. 14 Apparent frequency of sound observed is

$$f_{ap} = f_0\left[\frac{c_2 \pm v_0}{c_2 \mp v_s}\right]$$

when $\qquad v_s = 34$ m/s

$$\Rightarrow \qquad f_1 = f_0\left[\frac{340}{340-34}\right]$$

when $\qquad v_s = 17$ m/s

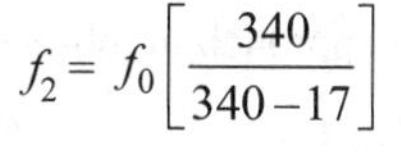

$$f_2 = f_0\left[\frac{340}{340-17}\right]$$

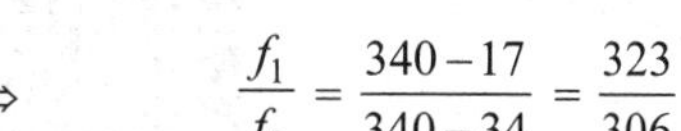

$$\Rightarrow \qquad \frac{f_1}{f_2} = \frac{340-17}{340-34} = \frac{323}{306}$$

$$\Rightarrow \qquad \frac{f_1}{f_2} = \frac{19}{18} \qquad \textbf{Ans. (B)}$$

Sol. 15 Sound appears to be coming from location B and plane in at A, so we use

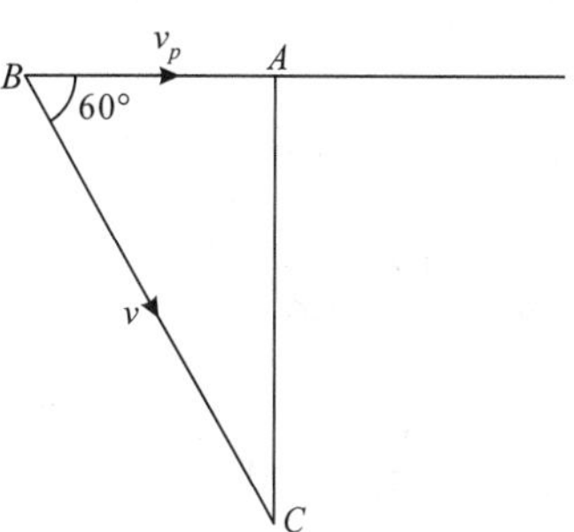

$AB = v_p \times t \qquad (v_p \rightarrow \text{velocity of plane})$

$BC = vt$

From figure

$$\cos 60° = \frac{AB}{BC}$$

$$\Rightarrow \quad \frac{1}{2} = \frac{v_p \times t}{vt}$$

$$\Rightarrow \quad v_p = \frac{v}{2}$$ **Ans. (C)**

Sol. 16 20m/s x — A(observer) x_0 20m/s — B(source)

Applying Doppler effect for sound

$$x = \frac{c - v_0}{c + v_s} n_0$$

(v_0 & v_s is taken ⊕ when approaching each other)

$$\Rightarrow \quad 2000 = \frac{340 - 20}{340 + 20} f_0$$

$$\Rightarrow \quad f_0 = 2250\,\text{Hz}$$ **Ans. (A)**

Sol. 17 Apparent frequency for the recording observer is given as

$$f_1 = 480 = \frac{300 - v_1}{300} \times 500$$

$$\Rightarrow \quad \frac{1440}{5} = 300 - v_1$$

$$\Rightarrow \quad v_1 = \frac{60}{5} = 12\,\text{m/s}$$

Apparent frequency for the approaching observer is given as

$$530 = \frac{300 + v_2}{300} \times 500$$

$$\Rightarrow \quad 1590 = 1500 + 5v_2$$

$$\Rightarrow \quad 5v_2 = 90$$

$$\Rightarrow \quad v_2 = 18\,\text{m/s}$$ **Ans. (B)**

Sol. 18 When source is approaching as shown below, we use

O ← 50 m/s S

$$f_{\text{app}} = \left(\frac{c - 0}{c - 50}\right) f_{\text{source}} \quad (c \to \text{speed of second})$$

$$\Rightarrow \quad 1000 = \left(\frac{350}{300}\right) f_{\text{source}}$$

$$\Rightarrow \quad f_{\text{source}} = \frac{1000 \times 300}{350}\,\text{Hz}$$

When source is recording as shown below, we use

← 50 m/s S O

$$f'_{\text{app}} = \left(\frac{c}{c + 50}\right) \cdot f_{\text{source}}$$

$$\Rightarrow \quad f'_{\text{app}} = \frac{350}{400} \times 1000 \times \frac{300}{350}$$

$$\Rightarrow \quad f'_{\text{app}} = 750\,\text{Hz}$$ **Ans. (C)**

Sol. 19 B 27 km/hr 7.5 m/s, $n_0 = 500$ Hz, C_w; A 18 km/hr 5 m/s, n_{ap_1}

Frequency of signal detected/received by A

$$n_{ap_1} = n_0\left(\frac{C_w - v_{OA}}{C_w - v_{SB}}\right) = 500\left(\frac{1500 - 5}{1500 - 7.5}\right)$$

$$\Rightarrow \quad n_{ap_1} = \frac{500 \times 1495}{1492.5}\,\text{Hz}$$

Now n_{ap_1} will be reflected by A as source and frequency detected by B for this reflected sound is

$$n_{ap_2} = n_{ap_1}\left(\frac{C_w + v_{OB}}{C_w + v_{SA}}\right) = \frac{500 \times 1495}{1492.5} \times \left(\frac{1500 + 7.5}{1500 + 5}\right)$$

$$\Rightarrow \quad n_{ap_2} \simeq 502\,\text{Hz}$$ **Ans. (B)**

Sol. 20 $n_0 = 660$ Hz, S_1, n_{ap_1}, O observer u, n_{ap_2}, S_2, $n_0 = 660$ Hz

Apparent frequency received from S_2 is

$$n_{ap_2} = n_0\left[\frac{C + 4}{C}\right]$$

Apparent frequency received from S_1 is

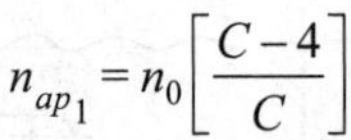

$$n_{ap_1} = n_0\left[\frac{C - 4}{C}\right]$$

Beat frequency heard by observer

$$f_B = n_{ap_2} - n_{ap_1} = n_0\left[\frac{C + 4}{C}\right] - n_0\left[\frac{C - 4}{C}\right] = n_0\left(\frac{24}{C}\right)$$

$$\Rightarrow \quad 10 = 660 \times \frac{24}{330} \Rightarrow 4 = \frac{10}{4} = 2.5\,\text{m/s}$$ **Ans. (A)**

Sol. 21

Frequency from fork A

$$v_A = \left(\frac{c}{c - v}\right) v_0$$

Frequency from fork B

$$\nu_B = \left(\frac{c}{c+v}\right)\nu_0$$

Beat frequency

$$\Rightarrow \quad f_B = \nu_1 - \nu_2$$

$$\Rightarrow \quad f_B = c\nu_0\left(\frac{1}{c-v} - \frac{1}{c+v}\right)$$

$$\Rightarrow \quad f_B = c\nu_0\left(\frac{c+v-c+v}{c^2-v^2}\right) = \frac{2c\nu_0^2 v}{c^2-v^2}$$

$$\Rightarrow \quad f_B \approx \frac{2c\nu_0 v}{c^2}$$

$$\Rightarrow \quad f_B = \frac{2\nu_0 v}{c}$$

$$\Rightarrow \quad 2 = \frac{2\times 1400\times v}{350}$$

$$\Rightarrow \quad v = \frac{1}{4}\text{ m/s}$$

Ans. (D)

Sol. 22 Apparent frequency of bus as received by wall is

$$f_w = 420\times\left(\frac{330}{330-v_B}\right)$$

Apparent frequency as received by driver is

$$f_D = 420\times\left(\frac{330+v_0}{330}\right)\left(\frac{330}{330-v_B}\right)$$

$$\Rightarrow \quad 490 = 420\times\left(\frac{330+v_B}{330-v_B}\right)$$

$$\Rightarrow \quad \frac{7}{6} = \frac{330+v_B}{330-v_B}$$

$$\Rightarrow \quad v_B = \frac{330}{13}\text{ m/s}$$

$$\Rightarrow \quad v_B = \frac{330}{13}\times\frac{18}{5}$$

$$\approx 91\text{ km/hr}$$

Ans. (A)

Sol. 23 Frequency received by wall is

$$h_\omega = h_0\left[\frac{C}{C-V_C}\right]$$

reflected frequency received by driver is

$$h_d = h_\omega\left[\frac{C+V_C}{C}\right] = n_0\left(\frac{C+V_C}{C-V_C}\right)$$

$$\Rightarrow \quad 480 = 440\frac{(C+V_C)}{(C-V_C)}$$

$$\Rightarrow \quad 12C - 12V_C = 11C + 11V_C$$

$$\Rightarrow \quad 23V_C = C$$

$$\Rightarrow \quad V_C = \frac{C}{23} = \frac{345}{23} = 15\text{ m/s}$$

$$\Rightarrow \quad V_C = 54\text{ kph}$$

Ans. (D)

Sol. 24 Observer is located at finite distance from track as shown in figure.

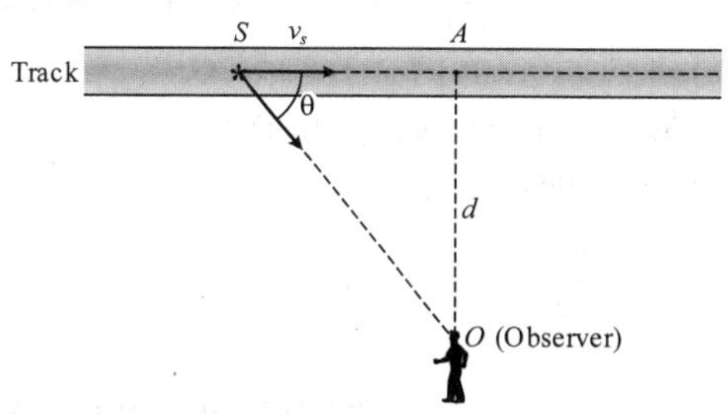

When source crosses point A the apparent frequency which was higher then ν_0 will drop below ν_0 as it started receding the observer. There correct option can be (B) or (D) but (B) is possible only when observer is located on track so in this case (D) is correct answer. This can also be mathematically verified with expression of apparent frequency, given as

$$f_{app} = \nu_0\left[\frac{v_{sound}}{v_{sound} - v_s\cos\theta}\right]$$

Ans. (D)

Sol. 25 Frequency of sound received by first car, which comes from second car is given as

$$f_1 = f_0\left(\frac{340+2}{340-2}\right) = 676\times\left(\frac{342}{338}\right) = 684\text{ Hz}$$

Frequency of sound from its own horn remains the same at 676 Hz so beat frequency is given as

$$f_2 = f_0 = 676\text{ Hz}$$

$$\Rightarrow \quad f_1 - f_2 = 684 - 676 = 8\text{ beats/s}$$

Ans. (8)

Sol. 26 By Doppler's effect in light we use

$$\frac{\Delta\lambda}{\lambda}c = v$$

$$\Rightarrow \quad \Delta\lambda = \frac{v}{c}\times\lambda$$

$$\Rightarrow \quad \Delta\lambda = \frac{286}{3\times 10^5}\times 630\times 10^{-9} = 6\times 10^{-10}$$

$$\Rightarrow \quad x = 6$$

Ans. (6)

Sol. 27 Frequency received by wall as observer is given as

$$f' = \left(\frac{v_s}{v_s - v}\right)f_0$$

Reflected frequency from wall as received by man is given as

$$f'' = \left(\frac{v_s+v}{v_s}\right)f'$$

$$\Rightarrow \quad f'' = \left(\frac{v_s+v}{v_s}\right)\left(\frac{v_s}{v_s-v}\right)f_0$$

$$\Rightarrow \quad f'' = \left(\frac{v_s + v}{v_s - v}\right) f_0$$

$$\Rightarrow \quad 500 = \left(\frac{330 + v}{330 - v}\right) 400$$

$$\Rightarrow \quad v = \frac{330}{9} \times \frac{18}{5} = 132 \text{ km/hr}$$ **Ans. (132)**

Sol. 28 The wavelength change $\Delta\lambda$ is related to the receding velocity of the galaxy as

$$\frac{v_{rel}}{c} = \frac{\Delta\lambda}{\lambda}$$

$$\Rightarrow \quad V_{rel} = \frac{6}{5890} \times 3 \times 10^8 \approx 306 \text{ km/s}$$ **Ans. (C)**

Sol. 29 The apparent frequency observed by detector in the situation described in question is expressed as

$$f' = f\left(\frac{v - v_0}{v + v_s}\right)$$

$$\Rightarrow \quad 1800 = f\left[\frac{340 - 20}{324 + 20}\right]$$

$$\Rightarrow \quad f = \frac{1800 \times 360}{320}$$

$$\Rightarrow \quad f = 2050 \text{ Hz}$$ **Ans. (2025)**

Sol. 30 The apparent frequency received by Y is given as

$$f_{app} = f_0\left(\frac{v + v_0}{v - v_s}\right)$$

$$1320 = f_0\left(\frac{340 + 20}{340 - 10}\right)$$

$$f_0 = 1320 \times \frac{33}{36} = 1210 \text{ Hz}$$ **Ans. (1210)**

Sol. 31 Actual wavelength of light emitted from galaxy is given as

$$\lambda_{emitted} = 670 \text{ nm}$$

Observed wavelength is given as

$$\lambda_{obs} = 670.7 \text{ nm}$$

The relative speed of galaxy v ($v << c$) receding away is calculated as

$$\frac{\lambda_{obs} - \lambda_{emitted}}{\lambda_{emitted}} = \frac{v}{c}$$

$$\Rightarrow \quad \frac{670.7 - 670}{670} = \frac{v}{c}$$

$$\Rightarrow \quad v = 3.13 \times 10^5 \text{ m/s}$$ **Ans. (C)**

Sol. 32 Apparent frequency of sound is calculated as

$$f_0 = \left(\frac{v + v_0}{v}\right) f_s$$

$$\Rightarrow \quad f_0 = \left(\frac{v + \frac{v}{5}}{v}\right) f_s$$

$$\Rightarrow \quad f_0 = \frac{5}{6} f_s$$

Percentage change in frequency is calculated as

$$\Delta f = \frac{f_0 - f_s}{f_s} \times 100$$

$$\Rightarrow \quad \Delta f = \frac{1}{5} \times 100 = 20\%$$ **Ans. (A)**

Sol. 33

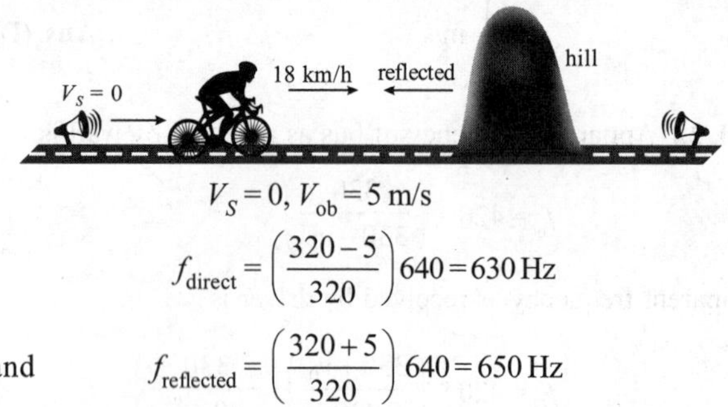

$$V_S = 0, V_{ob} = 5 \text{ m/s}$$

$$f_{direct} = \left(\frac{320 - 5}{320}\right) 640 = 630 \text{ Hz}$$

and $$f_{reflected} = \left(\frac{320 + 5}{320}\right) 640 = 650 \text{ Hz}$$

Beat frequency is calculated as

$$f_{beat} = 650 - 360 = 20 \text{ Hz}$$ **Ans. (20)**

Sol. 34 Apparent frequency of horn when car is approaching is given as

$$f_1 = 100 = f_0\left(\frac{c}{c - v_s}\right) \quad \ldots(1)$$

Apparent frequency of horn when car is receding the observer is given as

$$f_2 = 50 = f_0\left(\frac{c}{c + v_s}\right) \quad \ldots(2)$$

$$\Rightarrow \quad \frac{f_1}{f_2} = 2 = \frac{c + v_S}{c - v_S}$$

$$\Rightarrow \quad 2c - 2v_S = c + v_S$$

$$\Rightarrow \quad 3v_S = c$$

$$\Rightarrow \quad v_S = \frac{c}{3}$$

From equation-(1), we have

$$100 = f_0 \frac{c}{\frac{2c}{3}} = \frac{3}{2} f_0$$

$$\Rightarrow \quad f_0 = \frac{200}{3}$$ **Ans. (200)**

Sol. 35 Figure below shows the situation described in question.

If f_1 is the frequency of the whistle as received by hill, then it is given as

$$f_1 = \left(\frac{v}{v - v_s}\right) f = \left(\frac{330}{320}\right) \times 320 = 330 \text{ Hz}$$

If f_2 is the frequency of the sound reflected by hill then this echo frequency is calculated as

$$f_2 = \left(\frac{v + v_0}{v}\right) f_1 = \frac{(330+10)}{330} \times 330 = 340 \text{ Hz} \quad \textbf{Ans. (340)}$$

Sol. 36 Figure below shows the situation as described in question.

The apparent frequency of the whistle as received by observer is calculated as

$$f_{\text{app}} = f\left(\frac{v}{v - v_s}\right)$$

$$\Rightarrow \quad f_{\text{app}} = 320\left(\frac{330}{330-66}\right)$$

$$\Rightarrow \quad f_{\text{app}} = 400 \text{ Hz} \quad \textbf{Ans. (400)}$$

Sol. 37 By using Doppler's effect, the apparent frequency heard by person is given as

$$f_1 = 300\left(\frac{330-0}{330-(-30)}\right) = 275$$

$$\Rightarrow \quad f_2 = 300\left(\frac{330-0}{330-(-30)}\right) = 330$$

$$\Rightarrow \quad \Delta f = 330 - 275 = 55 \text{ Hz} \quad \textbf{Ans. (B)}$$

Sol. 38 By Doppler's effect, the frequency of reflected sound is given as

$$f_{ap} = \left(\frac{v + v_c}{v - v_c}\right) f_0$$

$$\Rightarrow \quad f_{ap} = \left(\frac{330+15}{330-15}\right) \times f_0$$

$$\Rightarrow \quad f_{ap} = \frac{345}{315} f_0$$

Change in frequency of Car's horn after reflection from wall is calculated as

$$\Delta f = \frac{345}{315} f_0 - f_0 = 40 \text{ Hz}$$

$$\Rightarrow \quad \frac{30}{315} f_0 = 40$$

Thus, frequency of horn is calculated as

$$f_0 = \frac{4 \times 315}{3} = 420 \text{ Hz} \quad \textbf{Ans. (420)}$$

Sol. 39 As passenger is inside train and train is moving with constant speed. The relative velocity of a passenger with source of sound (engine) is 0. So there will be no doppler's effect. So frequency heard is 400 Hz. **Ans. (B)**

Sol. 40 By Doppler's effect apparent frequency when observer is moving towards source and source is moving away from observer.

$$\Rightarrow \quad f = f_0\left(\frac{c + v_0}{c + v_s}\right)$$

$$\Rightarrow \quad f = 400\left(\frac{360+40}{360+20}\right)$$

$$\Rightarrow \quad f = 421 \text{ Hz} \quad \textbf{Ans. (B)}$$

15.1 Electric Force and Electric Field

Sol. 1 $\overset{F_{1Q}}{\longleftarrow} \underset{Q}{\bullet} \overset{F_{2q}}{\longrightarrow} \quad x \quad \underset{q}{\bullet} \quad x \quad \underset{Q}{\bullet}$

[For equation θ must be of opposite sign to Q]

For equilibrium of Q, we use

$$F_{1Q} = F_{2q}$$

$$\frac{KQ^2}{(2x)^2} = \frac{KQq}{x^2}$$

$$\Rightarrow \quad |q| = \frac{Q}{4} \text{ or } q = -\frac{Q}{4} \quad \textbf{Ans. (D)}$$

Sol. 2 By Colomb's law

$$F_{12} = \frac{Kq_1q_2}{b^2}; F_{13} = \frac{Kq_1q_3}{a^2}$$

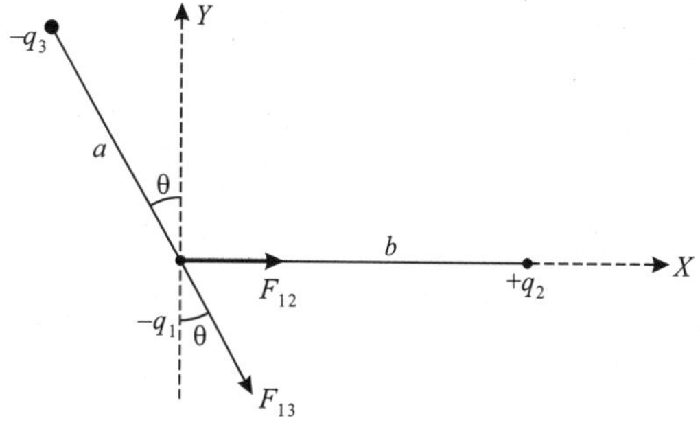

Net force on $-q_1$ along x direction is given as

$$F_x = F_{12} + F_{13}\sin\theta$$

$$\Rightarrow \quad F_x = Kq_1\left[\frac{q_2}{b^2} + \frac{q_3}{a^2}\sin\theta\right] \quad \textbf{Ans. (B)}$$

Sol. 3

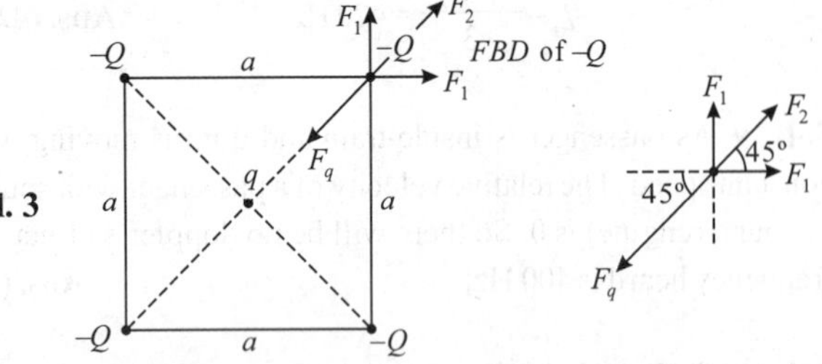

To keep net force on each of $-Q$, q has to be 'Positive'. So that for equilibrium of $-Q$ charge we have

Hare $\quad F_1 = \frac{KQ^2}{a^2}; F_2 = \frac{KQ^2}{2a^2}$

$$F_q = F_2 + \sqrt{2}\,F_1$$

and $\quad F_q = \frac{KqQ}{(a/\sqrt{2})^2}$

$$\Rightarrow \quad \frac{KqQ}{a^2 2} = \frac{KQ^2}{2a^2} + \frac{\sqrt{2}KQ^2}{a^2}$$

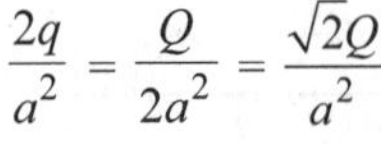

$$\Rightarrow \quad \frac{2q}{a^2} = \frac{Q}{2a^2} = \frac{\sqrt{2}Q}{a^2}$$

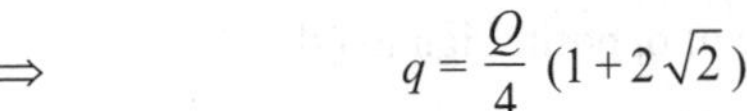

$$\Rightarrow \quad q = \frac{Q}{4}(1+2\sqrt{2}) \quad \textbf{Ans. (B)}$$

Sol. 4 Here to solve the case, we ignore induction effects.

$$\Rightarrow \quad r >> R$$

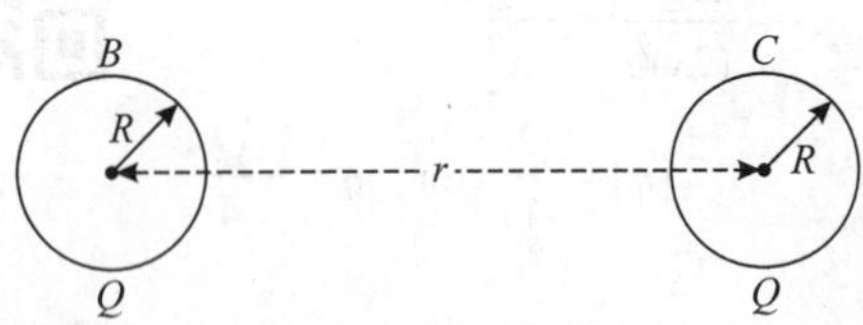

Initially the force between spheres is given as

$$F = \frac{KQ^2}{r^2}$$

In second case when the third identical sphere is located to B

Step-I

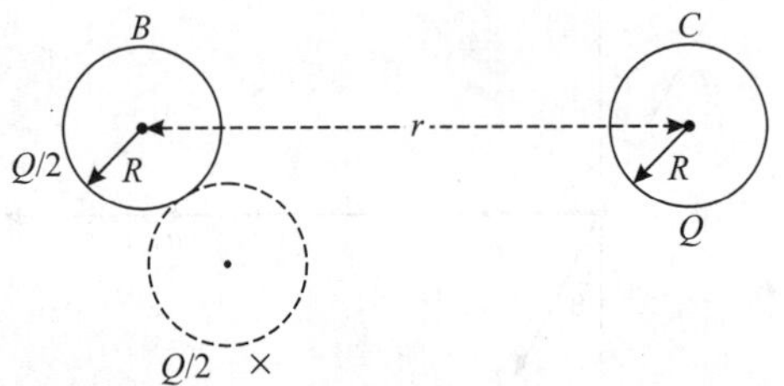

Step-II

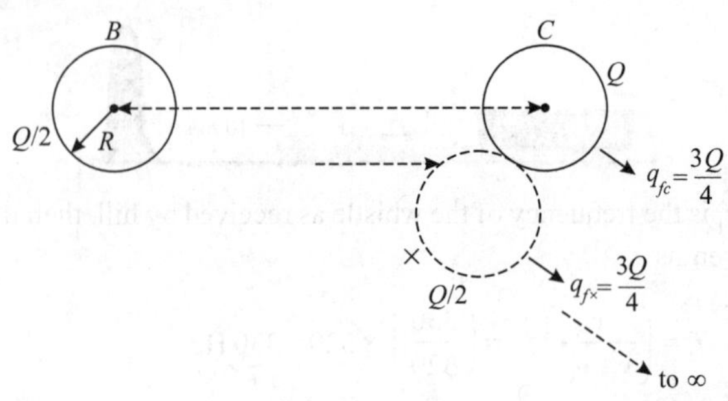

When sphere x is connected to sphere C then again their total charge will be equally distributed in than as these are identical.

$\Rightarrow$ Final charge on B and C are $Q/2$ and $3Q/4$

$\Rightarrow$ Final force between B and C is

$$F' = \frac{K(Q/2)(3Q/4)}{r^2}$$

$$= \frac{3}{8}\frac{KQ^2}{r^2} \quad \textbf{Ans. (D)}$$

Sol. 5 For equilibrium of drop, we have

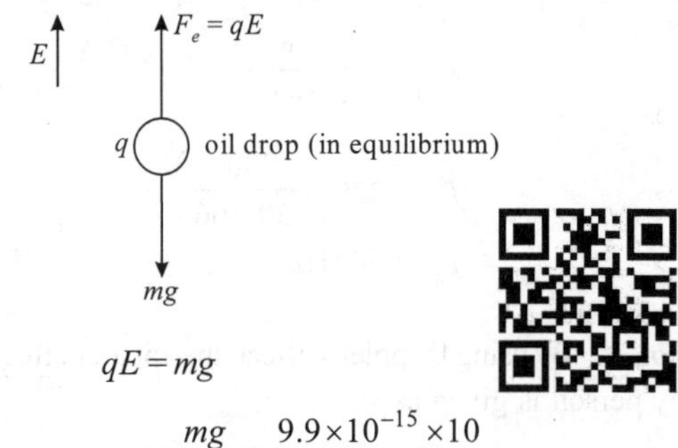

$$qE = mg$$

$$\Rightarrow \quad q = \frac{mg}{E} = \frac{9.9\times10^{-15}\times10}{3\times10^4}$$

$$\Rightarrow \quad q = 3.3\times10^{-18}\,\text{C} \quad \textbf{Ans. (C)}$$

Sol. 6

Y, $+8q$, $x = 0$, E_2, E_1, $-2q$, $x = L$, x, X, Null point, $x - L$

$E_1 \longrightarrow EF$ at x due to $8q$

$E_2 \longrightarrow EF$ at x due to $-2q$

If at x, $EF_{net} = 0$

$$\Rightarrow \quad E_1 = E_2$$

$$\frac{K(8q)}{x^2} = \frac{K(2q)}{(x-L)^2}$$

$$\Rightarrow \quad 8(x-L)^2 = 2x^2$$

$$\Rightarrow \quad x = 2L \quad \textbf{Ans. (A)}$$

Sol. 7

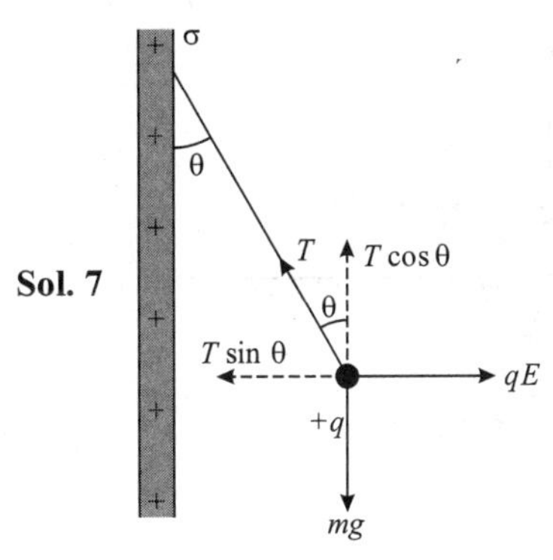

Electric field due to a large charged sheet

$$E = \frac{\sigma}{2\epsilon_0}$$

As ball is in equilibrium, we use

$$T\sin\theta = q\left(\frac{\sigma}{2\epsilon_0}\right) \quad \ldots(1)$$

$$T\cos\theta = mg \quad \ldots(2)$$

(1) ÷ (2) gives

$$\Rightarrow \quad \tan\theta = \frac{q\sigma}{2\epsilon_0\, mg}$$

$$\Rightarrow \quad \sigma \propto \tan\theta$$

Ans. (D)

Sol. 8

$$E = \frac{KQ}{r^2} \text{ (For } r > R)$$

$$E \propto \frac{1}{r^2}$$

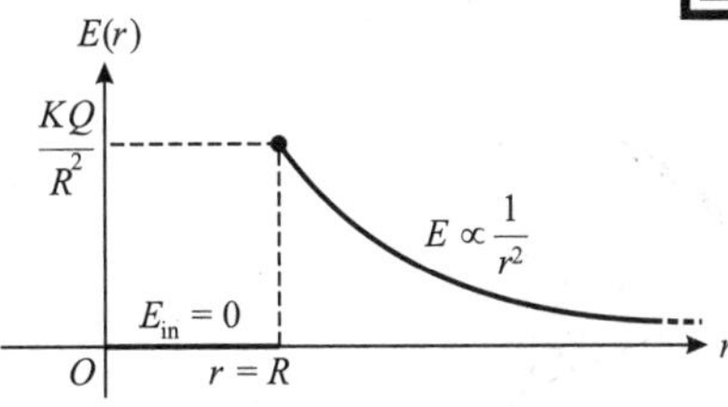

Ans. (A)

Sol. 9

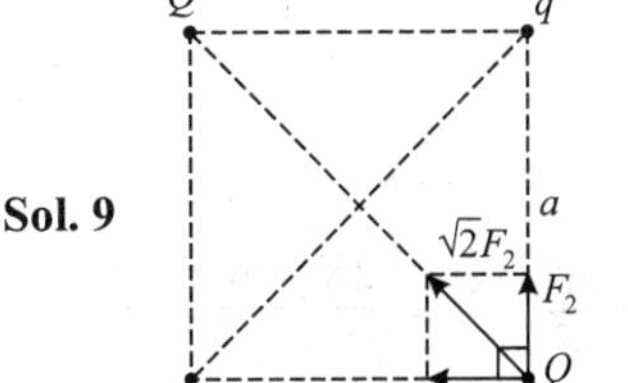

Force on Q due to other Q is $F_1 = \frac{KQ^2}{(\sqrt{2}a)^2}$

If not force on Q is to be zero

$\Rightarrow$ q has to be –ve so that resultant of two F_2 on Q balances F_1.

Force on Q due to each of q is $F_2 = \frac{KqQ}{a^2}$

As not force on Q is zero

$$\Rightarrow \quad \sqrt{2}F_2 = F_1$$

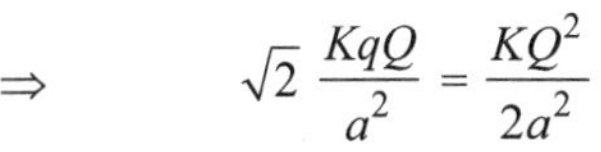

$$\Rightarrow \quad \sqrt{2}\,\frac{KqQ}{a^2} = \frac{KQ^2}{2a^2}$$

$$\Rightarrow \quad Q = 2\sqrt{2}\,q \text{ in magnitude}$$

$$\Rightarrow \quad Q = -2\sqrt{2}\,q$$

Ans. (C)

Sol. 10

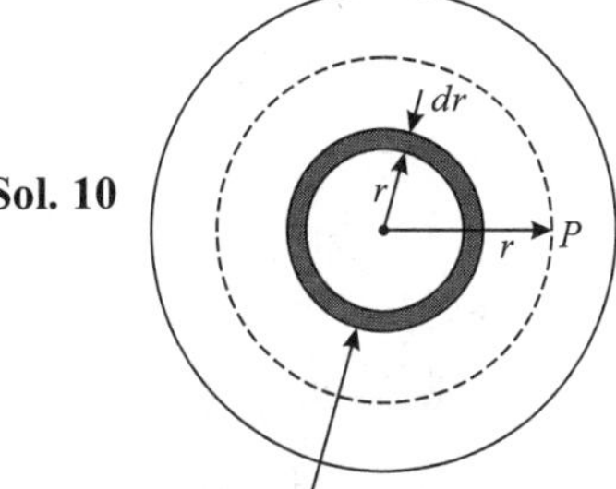

The charge contained inside the thin layer of elemental shell is

$$dq = \rho(r).dV$$

$$\Rightarrow \quad dq = \frac{Q}{\pi R^4}\cdot r\cdot(4\pi r^2 dr)$$

$$\Rightarrow \quad dq = \frac{4Q}{R^4}\cdot r^3 dr$$

Total charge inside a sphere of radius r_1 is

$$q = \int dq = \int_0^{r_1}\frac{4Q}{R^4}\cdot r^3 dr$$

$$\Rightarrow \quad q = \frac{4Q}{R^4}\int_0^{r_1} r^3 dr$$

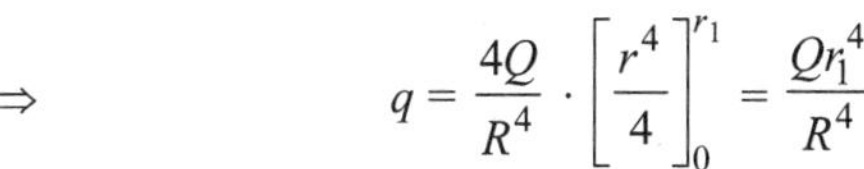

$$\Rightarrow \quad q = \frac{4Q}{R^4}\cdot\left[\frac{r^4}{4}\right]_0^{r_1} = \frac{Qr_1^4}{R^4}$$

Electric field at point 'P' due to the inside charge is given as

$$E = \frac{Kq}{r_1^2} = \frac{1}{4\pi\epsilon_0}\cdot\frac{(Qr_1^4/R^4)}{r_1^2}$$

$$\Rightarrow \quad E = \frac{1}{4\pi\epsilon_0}\,\frac{Qr_1^2}{R^4}$$

Ans. (A)

Sol. 11 In Air

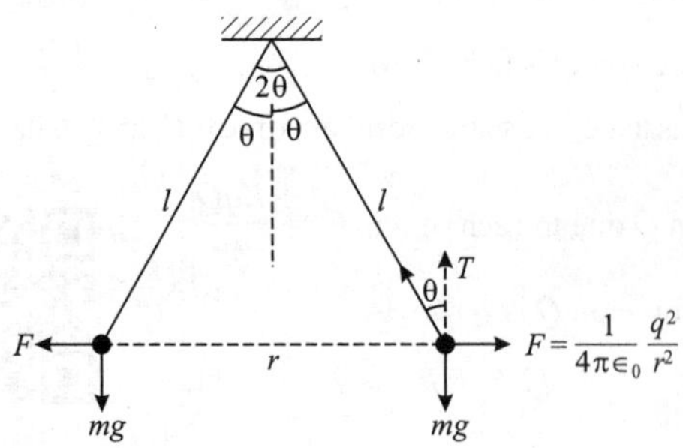

For equilibrium of bob

$$T\sin\theta = \frac{1}{4\pi\epsilon_0}\frac{q^2}{r^2} \quad \ldots(1)$$

$$T\cos\theta = mg \quad \ldots(2)$$

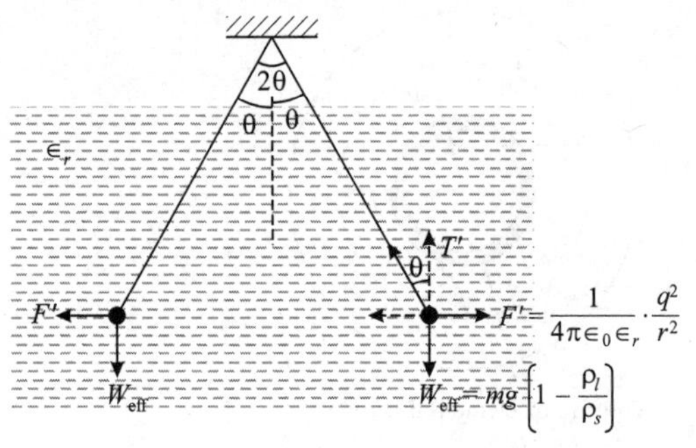

As $\qquad F_B = mg\cdot\frac{\rho_l}{\rho_s}$

Force of Buoyancy

$$T'\sin\theta = \frac{1}{4\pi\epsilon_0\epsilon_r}\cdot\frac{q^2}{r^2} \quad \ldots(3)$$

$$T'\cos\theta = mg\left(1-\frac{\rho_l}{\rho_s}\right) \quad \ldots(4)$$

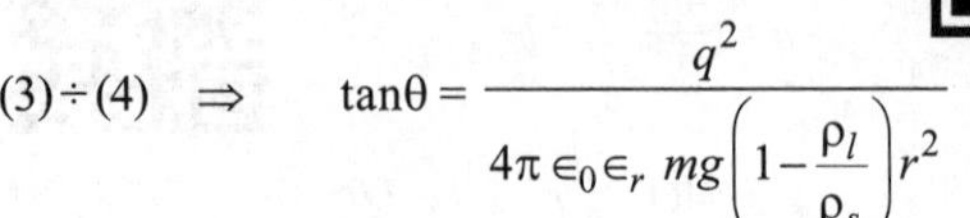

$$(1)\div(2) \Rightarrow \quad \tan\theta = \frac{q^2}{4\pi\epsilon_0 mgr^2}$$

$$(3)\div(4) \Rightarrow \quad \tan\theta = \frac{q^2}{4\pi\epsilon_0\epsilon_r mg\left(1-\frac{\rho_l}{\rho_s}\right)r^2}$$

Here we use

$$\frac{q^2}{4\pi\epsilon_0 mgr^2} = \frac{q^2}{4\pi\epsilon_0\epsilon_r mg\left(1-\frac{\rho_l}{\rho_s}\right)r^2}$$

$$\Rightarrow \quad \epsilon_r = \frac{1}{1-\frac{\rho_l}{\rho_s}} = \frac{1}{1-\frac{0.8}{1.6}} = 2$$ **Ans. (D)**

Sol. 12

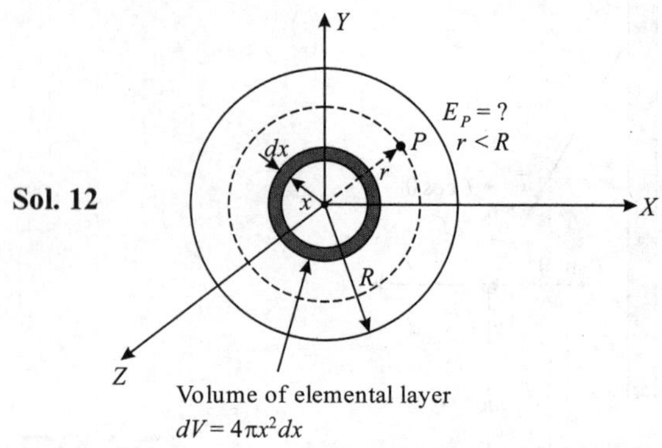

Charge contained in the elemental shell is

$$dq = \rho(r)\cdot dV|_{r=x}$$

$$\Rightarrow \quad dq = \rho_0\left(\frac{5}{4}-\frac{x}{R}\right)\cdot(4\pi x^2 dx)$$

Total charge contained in spherical region of radius r is given as

$$q = \int dq = \int_0^r \rho_0\left(\frac{5}{4}-\frac{x}{R}\right)\cdot 4\pi x^2 dx$$

$$\Rightarrow \quad q = 4\pi\rho_0\int_0^r\left(\frac{5}{4}x^2-\frac{x^3}{R}\right)dx$$

$$\Rightarrow \quad q = 4\pi\rho_0\left[\frac{5x^3}{12}-\frac{x^4}{4R}\right]_0^r$$

$$\Rightarrow \quad q = 4\pi\rho_0\left[\frac{5r^3}{12}-\frac{r^4}{4R}\right]_0^r$$

Electric field at the surface of inside sphere of radius r is

$$E = \frac{Kq}{r^2} = \frac{1}{4\pi\epsilon_0}\cdot\frac{4\pi\rho_0}{r^2}\left[\frac{5r^3}{12}-\frac{r^4}{4R}\right]$$

$$\Rightarrow \quad E = \frac{\rho_0 r}{4\epsilon_0}\left[\frac{5}{3}-\frac{r}{R}\right]$$ **Ans. (C)**

Sol. 13

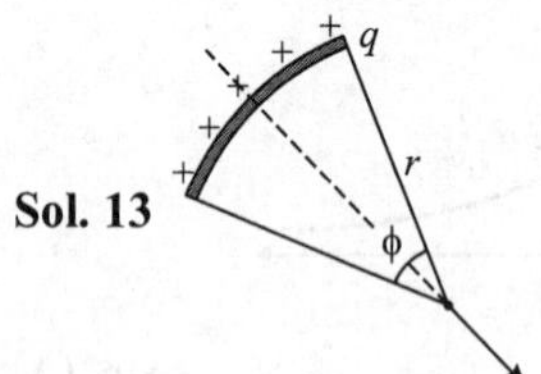

For a half ring

$\phi = \pi$

$$E = \frac{2Kq\sin(\phi/2)}{\phi r^2} = \frac{2Kq\sin(\pi/2)}{\pi r^2}$$

$$\Rightarrow \quad E = \frac{2q}{4\pi^2\epsilon_0 r^2} = \frac{q}{2\pi^2\epsilon_0 r^2}$$

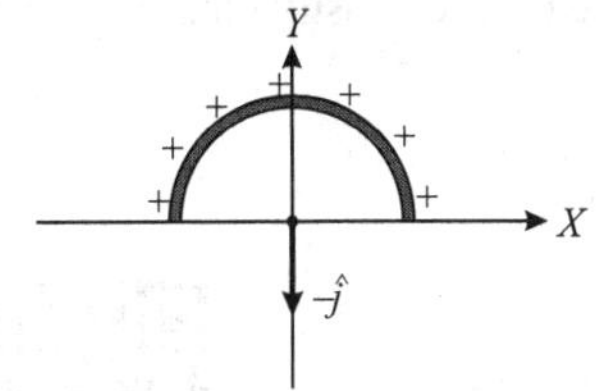

$$\Rightarrow \qquad \vec{E} = -\frac{q}{2\pi^2 \in_0 r^2}\hat{j}$$ **Ans. (D)**

Sol. 14

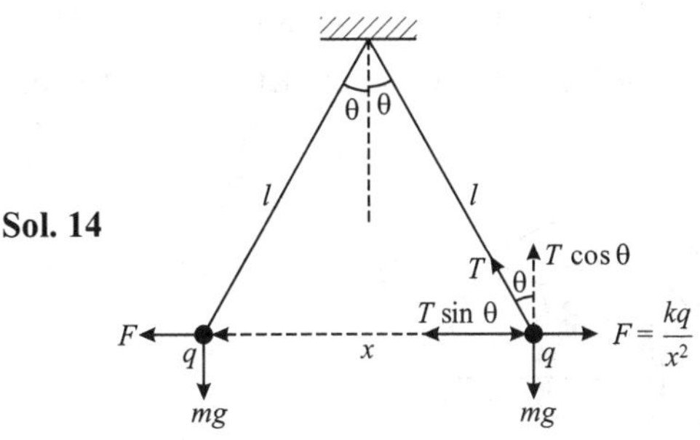

For equilibrium of spheres, we use

$$T\sin\theta = \frac{Kq^2}{x^2} \qquad \ldots(1)$$

$$T\cos\theta = mg \qquad \ldots(2)$$

(1) ÷ (2)

$$\Rightarrow \qquad \tan\theta = \frac{Kq^2}{mgx^2} = \frac{x}{2l}$$

$\Rightarrow$ as θ is small

$$\Rightarrow \qquad \tan\theta \simeq \theta = \frac{x}{2l}$$

$$\Rightarrow \qquad q = \sqrt{\frac{mg}{2Kl}} \cdot x^{3/2}$$

If q starts leaking (decreasing) than x will decreases and as θ is very small, we can consider very slow variation in x. So we have rate of charge leakage (constant)

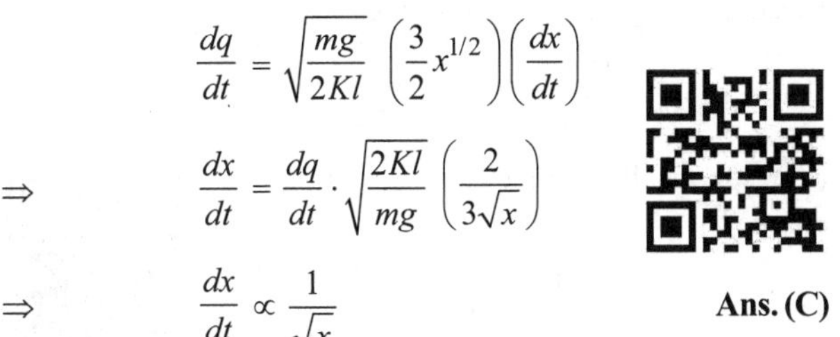

$$\frac{dq}{dt} = \sqrt{\frac{mg}{2Kl}}\left(\frac{3}{2}x^{1/2}\right)\left(\frac{dx}{dt}\right)$$

$$\Rightarrow \qquad \frac{dx}{dt} = \frac{dq}{dt} \cdot \sqrt{\frac{2Kl}{mg}}\left(\frac{2}{3\sqrt{x}}\right)$$

$$\Rightarrow \qquad \frac{dx}{dt} \propto \frac{1}{\sqrt{x}}$$ **Ans. (C)**

Sol. 15 In a uniformly charged sphere, electric field at a distance r from centre is given as

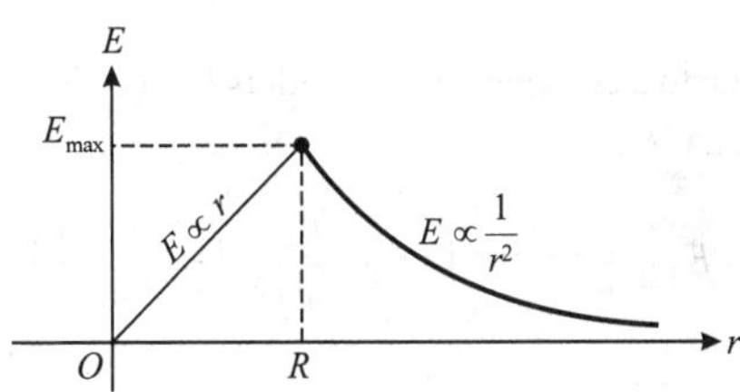

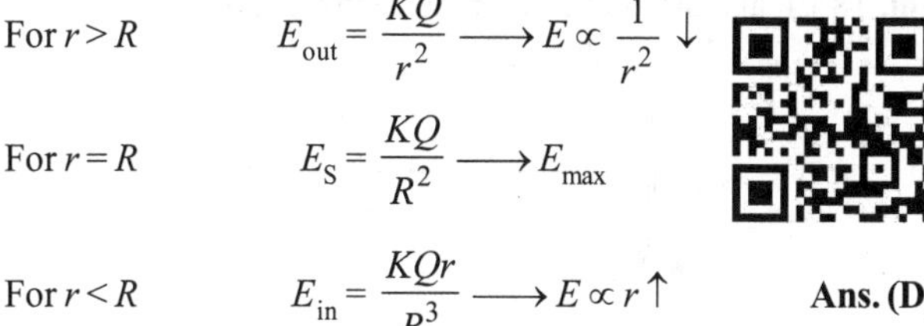

For $r > R$ $\qquad E_{out} = \frac{KQ}{r^2} \longrightarrow E \propto \frac{1}{r^2} \downarrow$

For $r = R$ $\qquad E_S = \frac{KQ}{R^2} \longrightarrow E_{max}$

For $r < R$ $\qquad E_{in} = \frac{KQr}{R^3} \longrightarrow E \propto r \uparrow$ **Ans. (D)**

Sol. 16

From Figure not force on q_0 due to charges on x-axis (considered fixed) is

$$F_{net} = 2F\sin\theta$$

$F \longrightarrow$ Coulombic Force

$$F = \frac{Kqq_0}{r^2} = \frac{Kqq_0}{(a^2+y^2)}$$

$$\Rightarrow \qquad F_{net} = \frac{2Kqq_0}{(a^2+y^2)} \cdot \frac{y}{\sqrt{a^2+y^2}}$$

As $y << a$

$$\Rightarrow \qquad a^2 + y^2 \approx a^2$$

$$\Rightarrow \qquad F_{net} = \frac{2Kqq_0}{(a^2+y^2)^{3/2}} \cdot y \approx \frac{2Kqq_0 y}{a^3}$$ **Ans. (A)**

Sol. 17 In space electric force must be in downward direction so that upward electric force on $-6e$ charge balances the weight of drop.

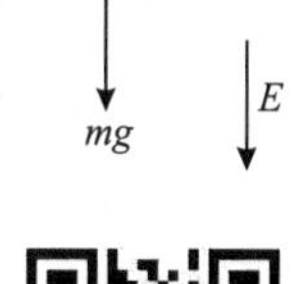

$\Rightarrow$ At equilibrium of drop

$$qE = mg$$

Mass of liquid drop

$$m = \frac{4}{3}\pi R^3 \cdot \rho$$

$$\Rightarrow \qquad 6eE = \frac{4}{3}\pi R^3 \rho g$$

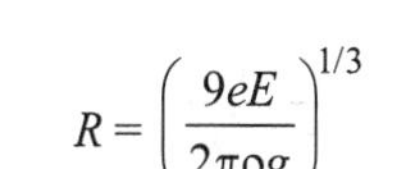

$$\Rightarrow \qquad R = \left(\frac{9eE}{2\pi\rho g}\right)^{1/3}$$

$$\Rightarrow \qquad R = \left(\frac{9 \times 1.6 \times 10^{-19} \times 25.5 \times 10^3}{6.28 \times 1.26 \times 10^3 \times 10}\right)^{1/3}$$

$$\Rightarrow \qquad R = 7.74 \times 10^{-7}\,\text{m}$$ **Ans. (B)**

Sol. 18 EF at a distance x from the centre of disc is

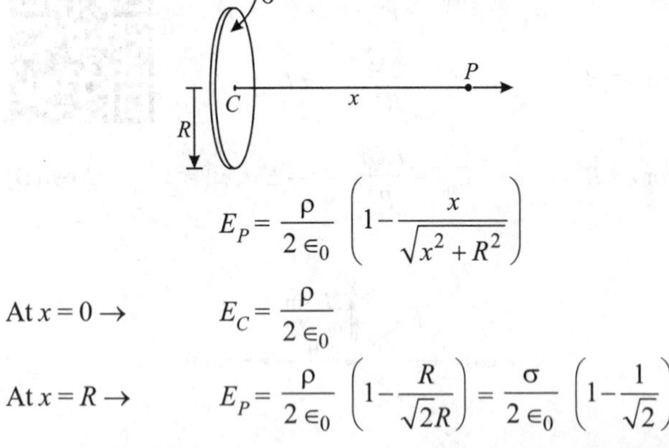

$$E_P = \frac{\rho}{2\in_0}\left(1 - \frac{x}{\sqrt{x^2+R^2}}\right)$$

At $x = 0 \rightarrow$ $$E_C = \frac{\rho}{2\in_0}$$

At $x = R \rightarrow$ $$E_P = \frac{\rho}{2\in_0}\left(1-\frac{R}{\sqrt{2}R}\right) = \frac{\sigma}{2\in_0}\left(1-\frac{1}{\sqrt{2}}\right)$$

% reduction in electric force at P w.r.t C is

$$= \frac{E_C - E_P}{E_C} = 100\%$$

$$= \frac{\frac{\sigma}{2\in_0}\cdot\frac{1}{\sqrt{2}}}{\sigma/2\in_0} \times 100\% = 70.7\%$$ **Ans. (A)**

Sol. 19

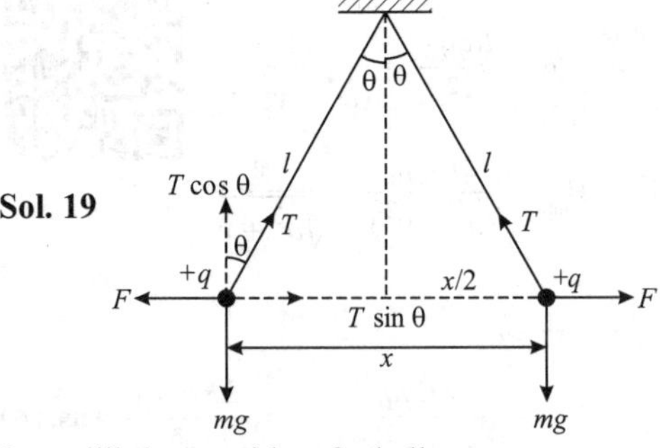

In equilibrium position, for balls, we use

$$T\sin\theta = F = \frac{Kq^2}{x^2} = \frac{1}{4\pi\in_0}\cdot\frac{q^2}{x^2} \quad \ldots(1)$$

and $$T\cos\theta = mg \quad \ldots(2)$$

(1) ÷ (2)

$$\Rightarrow \quad \tan\theta = \frac{F}{mg} = \frac{q^2}{4\pi\in_0 mgx^2}$$

For small θ, we use

$$\tan\theta \simeq \sin\theta = \frac{x}{2l}$$

$$\Rightarrow \quad \frac{x}{2l} = \frac{q^2}{4\pi\in_0 mgx^2}$$

$$\Rightarrow \quad x = \left[\frac{q^2 l}{2\pi\in_0 mg}\right]^{1/3}$$ **Ans. (D)**

Sol. 20

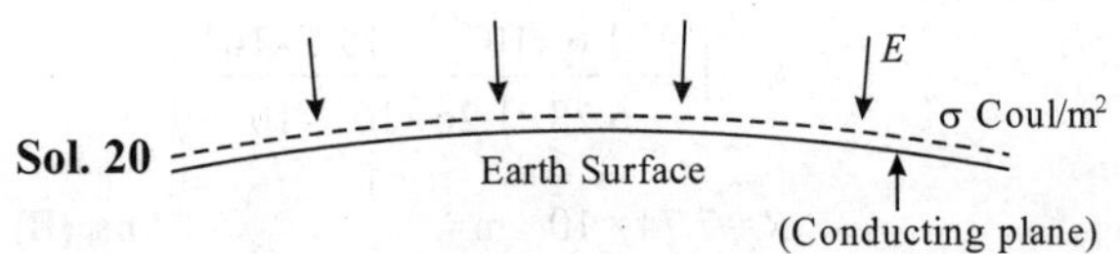

If σ C/m² (–ve) is the surface charge density on Earth surface then we use

$$E = \frac{\sigma}{\in_0}$$

$$\Rightarrow \quad \sigma = \in_0 E$$

The total surface area of earth = $4\pi R_e^2$

⇒ Total charge on earth surface is

$$q = -\sigma \times 4\pi R_e^2 = \in_0 E \cdot 4\pi R_e^2$$

$$\Rightarrow \quad q = -8.85\times10^{-12}\times150\times4\times3.14\times(6.37\times10^6)^2$$

$$\Rightarrow \quad q = -676.55\times10^3\text{ C} \simeq -680\text{ KC}$$ **Ans. (C)**

Sol. 21

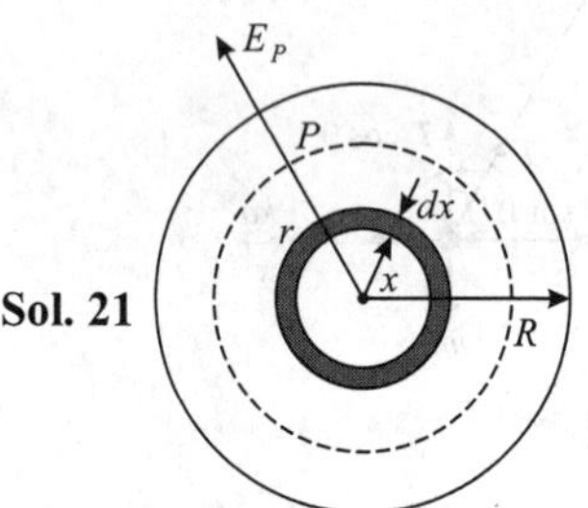

The charge contained in the elemental layer of radius x and width dx is

$$dq = \rho(x)\cdot dV = \rho(x)\cdot 4\pi x^2 dx$$

Total charge in spherical region from $x = 0$ to $x = r$ $(r < R)$ is

$$q = \int dq = \int_0^r \rho_0\left(1-\frac{x}{R}\right)\cdot 4\pi x^2 dx$$

$$\Rightarrow \quad q = 4\pi\rho_0\int_0^r\left(x^2 - \frac{x^3}{R}\right)dx$$

$$\Rightarrow \quad q = 4\pi\rho_0\left[\frac{x^3}{3} - \frac{x^4}{4R}\right]_0^r dx = 4\pi\rho_0 r^3\left[\frac{1}{3}-\frac{r}{4R}\right]$$

Electric force strength at point P $(x = r)$ is given as

$$E_P = \frac{Kq}{r^2}$$

(For spherically symmetric distribution of charge)

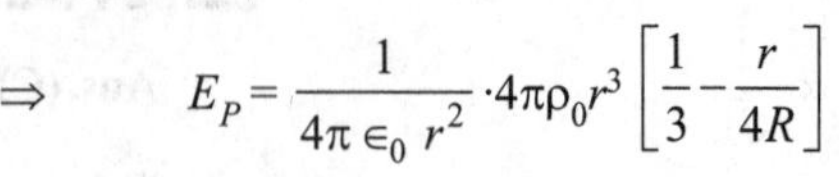
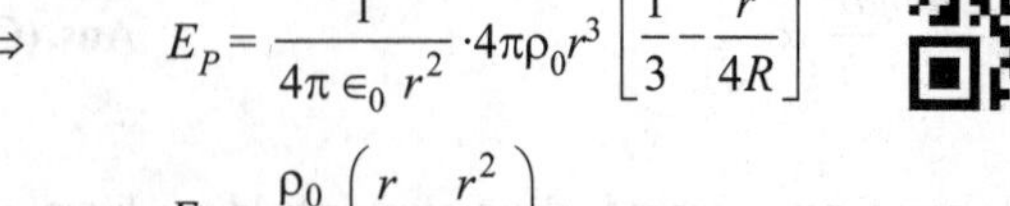

$$\Rightarrow \quad E_P = \frac{1}{4\pi\in_0 r^2}\cdot 4\pi\rho_0 r^3\left[\frac{1}{3}-\frac{r}{4R}\right]$$

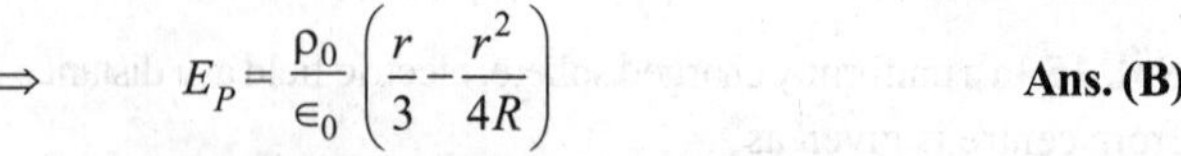

$$\Rightarrow \quad E_P = \frac{\rho_0}{\in_0}\left(\frac{r}{3}-\frac{r^2}{4R}\right)$$ **Ans. (B)**

Sol. 22 Due to a complete disc of radius b, electric field at point P is given as

$$E_{disc} = \frac{\sigma}{2\in_0}\left(1-\frac{h}{\sqrt{b^2+h^2}}\right) = \frac{\sigma}{2\in_0}\left(1-\frac{h}{b}\right) \ldots(1)$$

[As $h << b$]

Electric field at point P only due to the circular disc of radius 'a' is given as

$$E_{\text{disc}} = \frac{\sigma}{2\epsilon_0}\left(1-\frac{h}{\sqrt{a^2+h^2}}\right) = \frac{\sigma}{2\epsilon_0}\left(1-\frac{h}{b}\right) \quad \ldots(2)$$

[As $h << a$]

Electric force P due to disc with the hole of radius 'a' is given as

$$E_{\text{given object}} = E_{\text{disc}} - E_{\text{hole}}$$

$$= \frac{\sigma}{2\epsilon_0}\left[\left(1-\frac{h}{b}\right)-\left(1-\frac{h}{a}\right)\right]$$

$$= \frac{\sigma}{2\epsilon_0}\left[\frac{h}{a}-\frac{h}{2a}\right] \qquad (\text{As } b = 2a)$$

$$= \frac{\sigma h}{4\epsilon_0 a} = Ch \Rightarrow C = \frac{\sigma}{4\epsilon_0 a}$$ **Ans. (A)**

Sol. 23

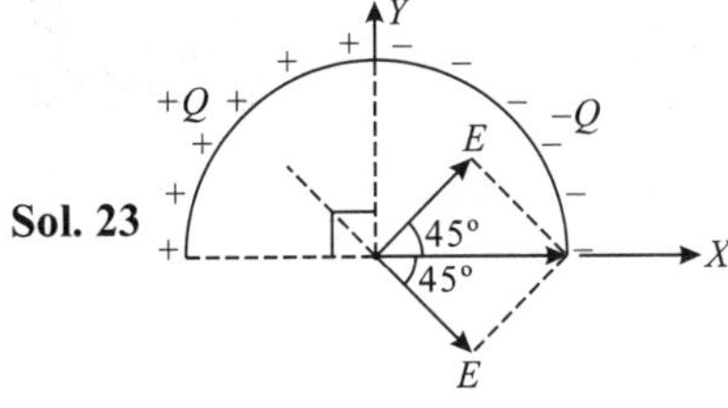

We use $\pi R = L$

$$\Rightarrow \quad R = \frac{20}{\pi}\text{ cm}$$

where $$E = \frac{2kQ\sin\frac{\phi}{2}}{\phi R^2} = \frac{2kQ\sin\left(\frac{\pi}{4}\right)}{\left(\frac{\pi}{2}\right)R^2} \quad \ldots(1)$$

$$E_{\text{net at } O} = \sqrt{2}E \quad \ldots(2)$$

From equation-(1) $$E = \frac{4}{\pi R^2}\cdot\frac{1}{4\pi\epsilon_0}\cdot Q\cdot\frac{1}{\sqrt{2}}$$

$$= \frac{10\epsilon_0}{\pi^2\epsilon_0\left(\frac{20}{\pi}\times 10^{-2}\right)^2\sqrt{2}}$$

$$\Rightarrow \quad E = \frac{1000}{\sqrt{2}\times 400\times 10^{-4}} = \frac{2.5}{\sqrt{2}}\times 10^{+4}\text{ N/C}$$

From equation-(2), net electric force at O is given as

$$E_{\text{net at } O} = \sqrt{2}E = 2.5\times 10^4 = 25\times 10^3\text{ N/C}$$

as $$\vec{E}_0 = 25\times 10^3\,\hat{i}\text{ N/C}$$ **Ans. (D)**

Sol. 24 As net charge enclosed in cavity is $+Q - Q = 0$ so net induced charge on inner surface of shell $Q_1 = 0$.

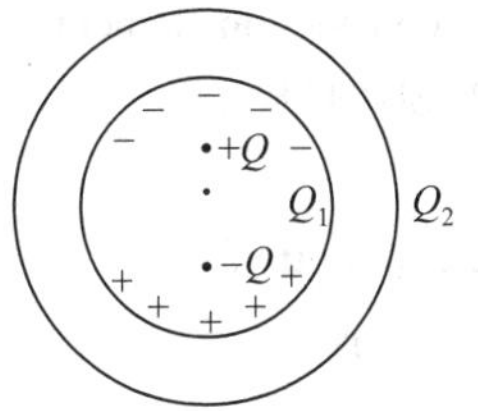

As shown we use $\sigma_1 \neq 0$ and varying with position inner surface of shell = 0

$\Rightarrow$ charge on outer surface

$$Q_2 = 0 \text{ and } \sigma_2 = 0$$ **Ans. (B)**

Sol. 25

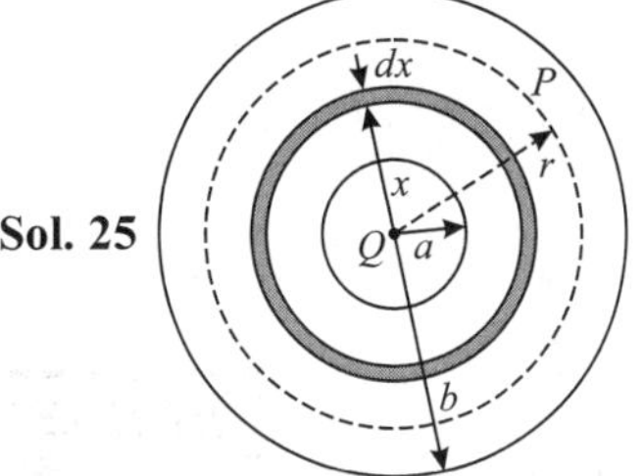

Electric field at point P is given as

$$E_P = \frac{K(Q+q_{en})}{r^2} \quad \ldots(1)$$

Where q_{en} is the charge between charge contained within the elemental layer of radius x and width dx is

$$dq = \rho\cdot dv$$

$$dq = \left(\frac{A}{x}\right)\cdot 4\pi x^2 dx$$

Total charge in region from $x = a$ to $x = r$ is

$$q_{en} = \int dq = 4\pi A\int_{x=a}^{x=r} x\,dx$$

$$\Rightarrow \quad q_{en} = 4\pi A\left[\frac{x^2}{2}\right]_a^r = 2\pi Ar^2 - 2\pi Aa^2$$

From equation-(1) $$E_P = \frac{K}{r^2}(Q + 2\pi Ar^2 - 2\pi Aa^2) \quad \ldots(2)$$

Given that E_p in equation-(2) is a constant (independent of r) if equation-(2)

$$Q = 2\pi Aa^2 \Rightarrow E = K(2\pi A) \rightarrow \text{constant}$$

$$\Rightarrow \quad A = \frac{Q}{2\pi a^2}$$ **Ans. (A)**

Sol. 26

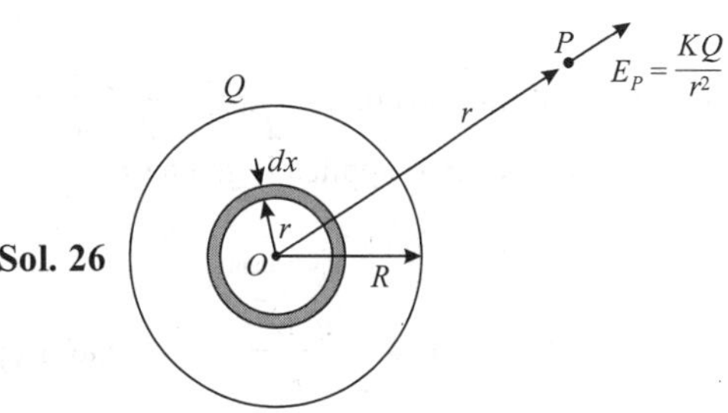

To calculate total charge on ball we consider on elemental shell inside the ball as shown of volume given as

$$dV = 4\pi r^2 dr$$

(For spherically symmetric charge distribution)

charge in the elemental shell thickness is

$$dq = \rho dV = \rho_0 = \left(1 - \frac{r}{R}\right) \cdot 4\pi r^2 dr$$

Total ball's charge is given as

$$Q = \int dq = \int_0^R 4\pi\rho_0 \left(1 - \frac{r}{R}\right) \cdot r^2 dr$$

$$\Rightarrow \quad Q = 4\pi\rho_0 \int_0^R \left(r^2 - \frac{r^3}{R}\right) dr$$

$$\Rightarrow \quad Q = 4\pi\rho_0 \left[\frac{r^3}{3} - \frac{r^4}{4R}\right]_0^R$$

$$\Rightarrow \quad Q = 4\pi\rho_0 \left[\frac{R^3}{3} - \frac{R^3}{4}\right]$$

$$\Rightarrow \quad Q = 4\pi\rho_0 \left(\frac{R^3}{12}\right) = \frac{1}{3}\pi\rho_0 R^3$$

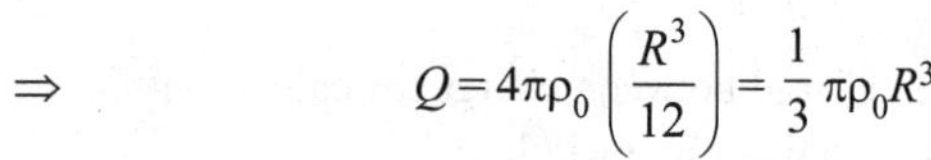

Electric field at an outer point of ball is given as

$$E_P = \frac{KQ}{r^2} = \frac{1}{4\pi\epsilon_0} \cdot \frac{\left(\frac{1}{3}\pi\rho_0 R^3\right)}{r^2} = \frac{\rho_0 R^3}{12\epsilon_0 r^2} \qquad \textbf{Ans. (D)}$$

Sol. 27

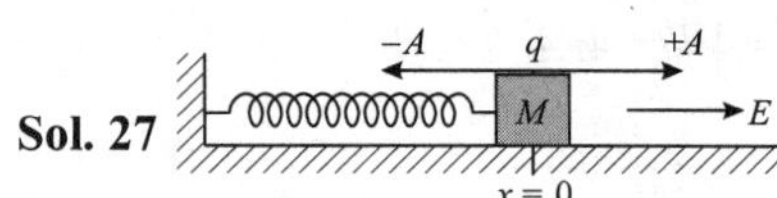

Oscillation energy of system $E_T = \frac{1}{2} m\omega^2 A^2$

PE of oscillations at equilibrium position = 0

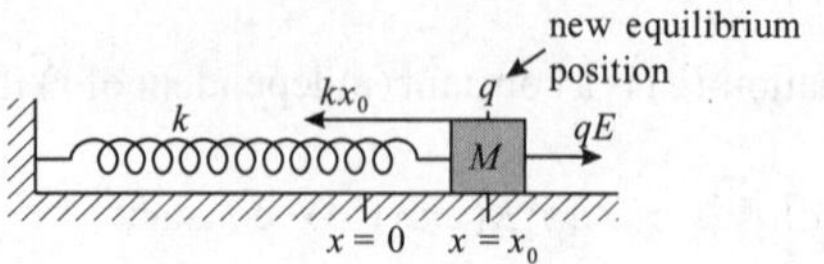

At equilibrium position $kx_0 = qE$

$$\Rightarrow \quad x_0 = \frac{qE}{k}$$

PE of oscillation at equilibrium position $= \frac{1}{2}kx_0^2 = \frac{1}{2}\frac{q^2E^2}{k}$

Total oscillation energy after EF is applied is given as

$$E_T = \frac{1}{2} m\omega^2 A^2 + U_{\text{equilibrium}}$$

$$E_T = \frac{1}{2} m\omega^2 A^2 + \frac{1}{2}\frac{q^2E^2}{k} \qquad \textbf{Ans. (A)}$$

Sol. 28

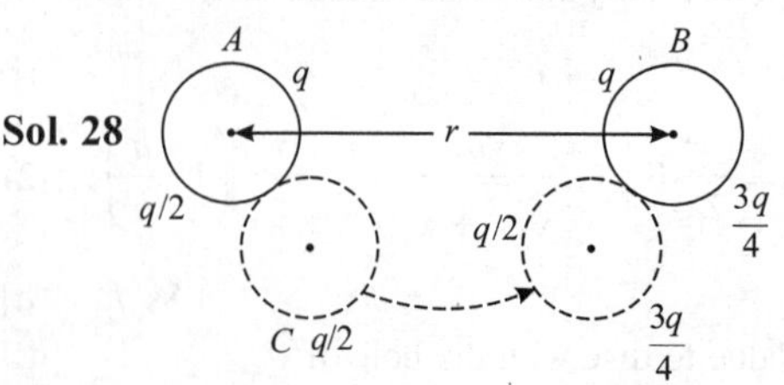

When C is touched to A, charge q is divided equally among A and C

$\Rightarrow$ final charge of A is $q_A = \frac{q}{2}$

Now C is touched to B, charge $\left(q + \frac{q}{2} = \frac{3q}{2}\right)$ is divided equally among B and C

$\Rightarrow$ Final charge on B is $q_B = \frac{3q}{4}$

So final force between A and B is

$$F' = \frac{K\left(\frac{a}{2}\right)\left(\frac{3q}{4}\right)}{r^2} = \frac{3}{8}\frac{Kq^2}{r^2} = \frac{3}{8}F \qquad \textbf{Ans. (D)}$$

Sol. 29

d; F_Q, F_q; $+Q$ — $d/2$ — q — $d/2$ — $+Q$

For equilibrium of charge Q, we use

$$\overrightarrow{F_Q} + \overrightarrow{F_q} = 0$$

$$\overrightarrow{F_Q} = -\overrightarrow{F_q}$$

$$\frac{kQQ}{d^2} = -\frac{kQq}{(d/2)^2}$$

$$\Rightarrow \quad q = -\frac{Q}{4} \qquad \textbf{Ans. (C)}$$

Sol. 30 Electric field on axis of ring at a distance h is given as

$$E = \frac{kQx}{(h^2 + R^2)^{3/2}}$$

For maximum electric field, we use

$$\frac{dE}{dx} = 0$$

Solving we get

$$x = \frac{R}{\sqrt{2}} \qquad \textbf{Ans. (C)}$$

Sol. 31 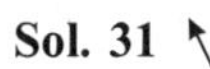

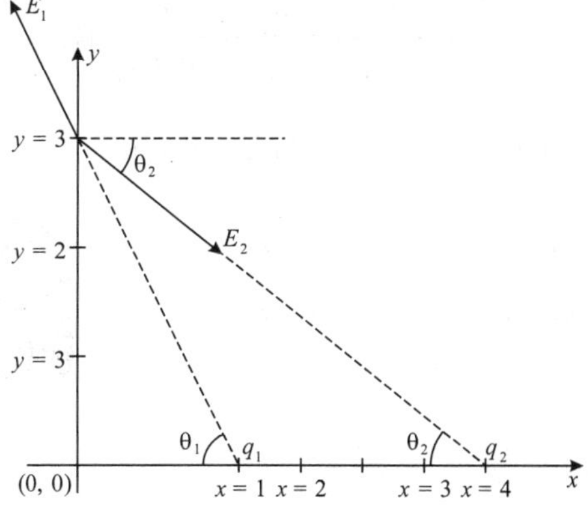

Let $\vec{E}_1$ & $\vec{E}_2$ are the values of electric field due to q_1 & q_2 respectively are magnitude of

$$E_2 = \frac{1}{4\pi \in_0}\frac{q_2}{r^2}$$

$$E_2 = \frac{9\times10^9\times(25)\times10^{-6}}{(4^2+3^2)}\text{ V/m}$$

$$E_2 = 9\times10^3\text{ V/m}$$

$$\Rightarrow \quad \vec{E}_2 = 9\times10^3(\cos\theta_2\hat{i}-\sin\theta_2\hat{j})\text{ V/m}$$

Where $\tan\theta_2 = \frac{3}{4}$

$$\Rightarrow \quad \vec{E}_2 = 9\times10^3\left(\frac{4}{5}\hat{i}-\frac{3}{5}\hat{j}\right) = (72\hat{i}-54\hat{j})\times10^2\text{ V/m}$$

Magnitude of

$$E_1 = \frac{1}{4\pi \in_0}\frac{\sqrt{10}\times10^{-6}}{(1^2+3^2)}$$

$$\Rightarrow \quad E_1 = (9\times10^9)\times\sqrt{10}\times10^{-7}\text{ V/m}$$

$$\Rightarrow \quad E_1 = 9\sqrt{10}\times10^2\text{ V/m}$$

$$\Rightarrow \quad \vec{E}_1 = 9\sqrt{10}\times10^2\left[\cos\theta_1(-\hat{i})+\sin\theta_1\hat{j}\right]\text{V/m}$$

Where $\tan\theta_1 = 3$

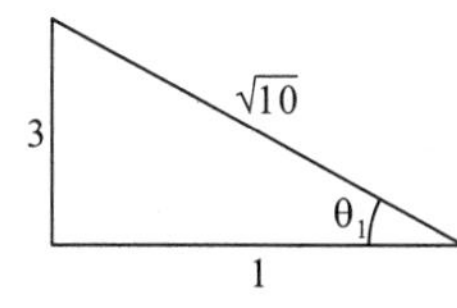

$$\Rightarrow \quad E_1 = 9\times\sqrt{10}\times10^2\left[\frac{1}{\sqrt{10}}(-\hat{i})+\frac{3}{\sqrt{10}}\hat{j}\right]\text{V/m}$$

$$\Rightarrow \quad E_1 = 9\times10^2\left[-\hat{i}+3\hat{j}\right] = \left[-9\hat{i}+27\hat{j}\right]10^2\text{ V/m}$$

$$\Rightarrow \quad \vec{E}_{net} = \vec{E}_1+\vec{E}_2 = (63\hat{i}-27\hat{j})\times10^2\text{V/m}$$

Ans. (C)

Sol. 32 Total charge in spherical volume is given as

$$Q = \int\rho dV$$

Where dV is the volume of elemental shall shown in figure

$$\Rightarrow \quad Q = \int_0^R \frac{A}{r^2}e^{-2r/a}(4\pi r^2 dr)$$

$$\Rightarrow \quad Q = 4\pi A\int_0^R e^{-2r/a}dr$$

$$\Rightarrow \quad Q = 4\pi A\left(\frac{e^{-2r/a}}{-\frac{2}{a}}\right)_0^R$$

$$\Rightarrow \quad Q = 4\pi A\left(-\frac{a}{2}\right)(e^{-2R/a}-1)$$

$$\Rightarrow \quad Q = 2\pi aA(1-e^{-2R/a})$$

$$\Rightarrow \quad R = \frac{a}{2}\log\left(\frac{1}{1-\frac{Q}{2\pi aA}}\right)$$

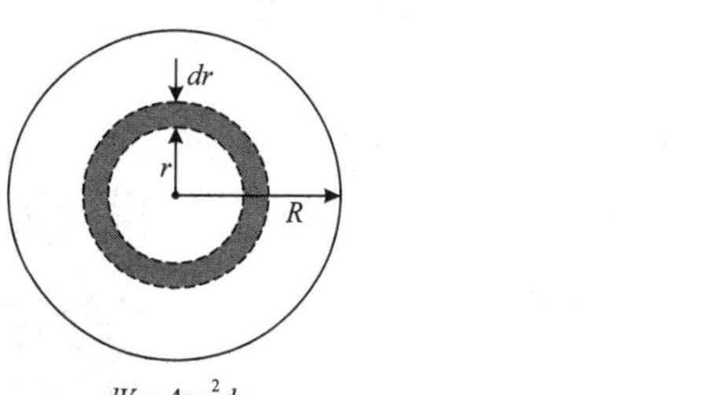

$dV = 4\pi r^2 dr$

Ans. (D)

Sol. 33

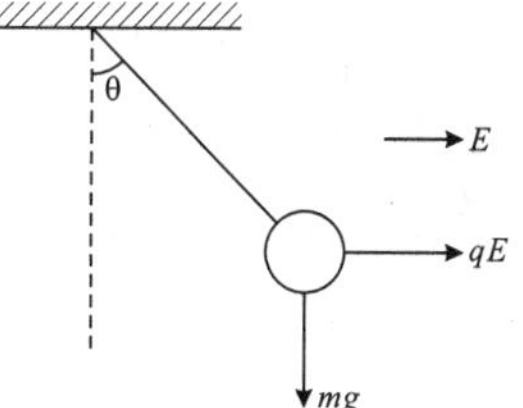

For equilibrium of bob, we use

$$\tan\theta = \frac{qE}{mg} = \frac{5\times10^{-6}\times2000}{2\times10^{-3}\times10}$$

$$\Rightarrow \quad \tan\theta = \frac{1}{2}$$

$$\Rightarrow \quad \theta = \tan^{-1}(0.5)$$

Ans. (C)

Sol. 34

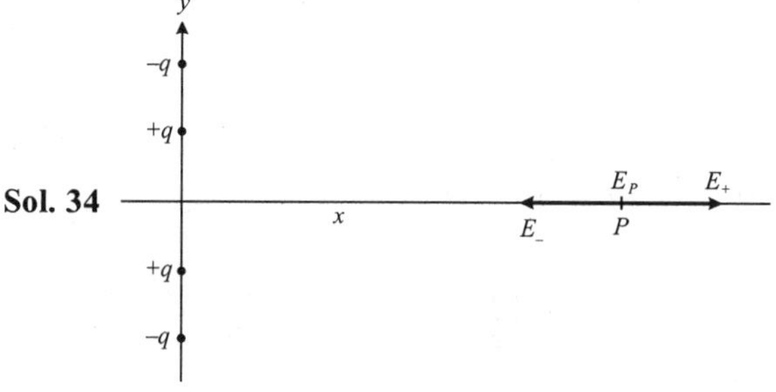

Electric field due to $+q$ charges at point P is

$$E_+ = \frac{K(2q)x}{(x^2+d^2)^{3/2}} = \frac{2kqD}{(D^2+d^2)^{3/2}}$$

Electric field due to $-q$ charges at point P is

$$E_- = \frac{K(2q)x}{(x^2+4d^2)^{3/2}} = \frac{2kqD}{(D^2+4d^2)^{3/2}}$$

Net Electric field at point P is given as

$$E_P = E_+ - E_- = 2kqD\left[\frac{1}{(D^2+d^2)^{3/2}} - \frac{1}{(D^2+4d^2)^{3/2}}\right]$$

For $D >> d$ we use $\Rightarrow E_P = E_P = \frac{2kqD}{D^3}$

$$\left[\left(1+\frac{d^2}{D^2}\right)^{-3/2} - \left(1+\frac{4d^2}{D^2}\right)^{-3/2}\right]$$

$$\Rightarrow \quad E_P = \frac{2kq}{D^2}\left[\left(1-\frac{3d^2}{2D^2}\right)-\left(1-\frac{3}{2}\cdot\frac{4d^2}{D^2}\right)\right]$$

$$\Rightarrow \quad E_P = \frac{2kq}{D^2}\left[1-\frac{3d^2}{2D^2}-1+\frac{6d^2}{D^2}\right] = \frac{9kqd^2}{D^4}$$

$$\Rightarrow \quad E_P \propto \frac{1}{D^4}$$ **Ans. (D)**

Sol. 35

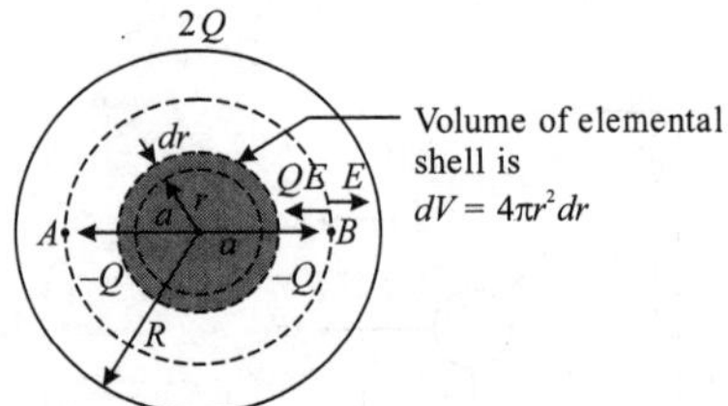

Charge in shell

$$dq = \rho(r)\cdot 4\pi r^2\, dr$$

$$\Rightarrow \quad 2Q = \int dq = \int_0^R kr\cdot 4\pi r^2 dr$$

$$\Rightarrow \quad 2Q = 4\pi\frac{R^4}{4} = \pi k R^4$$

$$\Rightarrow \quad k = \frac{2Q}{\pi R^4} \qquad \ldots(1)$$

Electric field at a distance a from centre is

$$E_a = \left(\frac{1}{4\pi\epsilon_0}\right)\frac{q_a}{a^2}$$

Where $\quad q_a = \int_0^a kr\cdot 4\pi r^2 dr$

$$\Rightarrow \quad q_a = 4\pi\frac{kq^4}{4}$$

$$\Rightarrow \quad E_a = \frac{1}{4\pi\epsilon_0}\cdot\frac{\pi k a^4}{a^2} = \frac{1}{4\pi\epsilon_0}\cdot \pi k a^2$$

Repulsive force between A and B is given as

$$F = \frac{1}{4\pi\epsilon_0}\cdot\frac{Q^2}{(2a)^2}$$

As net force on A and B is zero, we use

$$\Rightarrow \quad F = QE$$

$$\frac{1}{4\pi\epsilon_0}\cdot\frac{Q^2}{4a^2} = Q\cdot\frac{1}{4\pi\epsilon_0}\cdot\pi k a^2$$

$$\Rightarrow \quad \frac{Q}{4a^2} = \pi\left(\frac{2Q}{\pi R^4}\right)\cdot a^2 \Rightarrow a^4 = \frac{1}{8}R^4$$

$$\Rightarrow \quad a = 8^{-1/4}\cdot R$$ **Ans. (C)**

Sol. 36 Summing up all components of electric field along X-direction, we get

$$E_x = \frac{K(4q)}{R^2}\cos 30° + \frac{K(2q)}{R^2}\cos 30° + \frac{K(2q)}{R^2}\cos 30°$$

$$\Rightarrow \quad E_x = (2\sqrt{3}+\sqrt{3}+\sqrt{3})\,\text{kg/R}^2 = \frac{\sqrt{3}q}{\pi\varepsilon_0 d^2}$$ **Ans. (D)**

Sol. 37 Speed after travelling a distance x is given as

$$v^2 = u^2 + 2ax$$

$$\Rightarrow \quad v^2 = 0 + 2\left(\frac{qE}{m}\right)x$$

$$\Rightarrow \quad v^2 = \frac{2qE}{m}x$$

$$\Rightarrow \quad v \propto \sqrt{x}$$

Thus option (C) is correct. **Ans. (C)**

Sol. 38 If we fill the empty space of cavity with $+\rho$ and $-\rho$ charge density in this region, the electric fields at points A and B are

$$E_A = 0 + \frac{k\rho.\frac{4}{3}\pi\left(\frac{R}{2}\right)^3}{\left(\frac{R}{2}\right)^2} = k\rho\,\frac{4}{3}\pi\left(\frac{R}{2}\right) \qquad \ldots(1)$$

and $$E_B = \frac{k\rho\cdot\frac{4}{3}\pi R^3}{R^2} - \frac{k\rho\cdot\frac{4}{3}\pi\left(\frac{R}{2}\right)^3}{\left(\frac{3R}{2}\right)^2}$$

$$\Rightarrow \quad E_B = k\rho\frac{4}{3}\pi R - k\rho\frac{4}{3}\pi\frac{R}{18} = k\rho\cdot\frac{4}{3}\pi\left(\frac{17R}{18}\right) \qquad \ldots(2)$$

Using equation-(1) and (2), we have

$$\frac{E_A}{E_B} = \frac{9}{17} = \frac{18}{34}$$ **Ans. (D)**

Sol. 39 Let the charges on inner and outer spheres are q_1 and q_2.

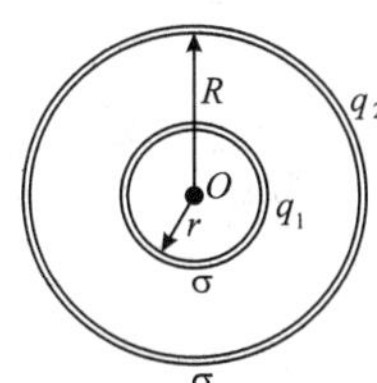

As charge density is same on both shells, we use

$$\sigma = \frac{q_1}{4\pi r^2} = \frac{q_2}{4\pi R^2}$$

$$\Rightarrow \quad \frac{q_1}{q_2} = \frac{r^2}{R^2}$$

$$q_1 + q_2 = Q$$

$$\Rightarrow \quad \frac{q_2 r^2}{R^2} + q_2 = Q$$

$$\Rightarrow \quad q_2 = \frac{qR^2}{(r^2 + R^2)}$$

$$\Rightarrow \quad q_1 = \frac{r^2}{R^2} \cdot \frac{qR^2}{(R^2 + r^2)} = \frac{qr^2}{(R^2 + r^2)}$$

Potential at common centre is given as

$$V_C = \frac{kQ_1}{r} + \frac{kQ_2}{R}$$

$$\Rightarrow \quad V_C = K\left[\frac{Qr^2}{r(R^2 + r^2)} + \frac{QR^2}{R(R^2 + r^2)}\right]$$

$$\Rightarrow \quad V_C = \frac{KQ(r + R)}{(R^2 + r^2)}$$

$$\Rightarrow \quad V_C = \frac{1}{4\pi \epsilon_0} \frac{(R + r)}{(R^2 + r^2)} Q$$ **Ans. (C)**

Sol. 40 As initial speed of particle is zero, particle starts moving in the direction of acceleration which is given as

$$a = \sqrt{g^2 + \left(\frac{qE}{m}\right)^2}$$

at angle $\theta = \tan^{-1}\left(\frac{qE}{mg}\right)$ from Y direction.

Thus particle follows a straight line path and acceleration along the direction of acceleration. **Ans. (D)**

Sol. 41 Electric field at O due to Q_1 and Q_2 are

$$\vec{E}_1 = -\frac{kQ_1}{x_1^2}\hat{i}$$

and $$\vec{E}_2 = -\frac{kQ_2}{x_2^2}\hat{i}$$

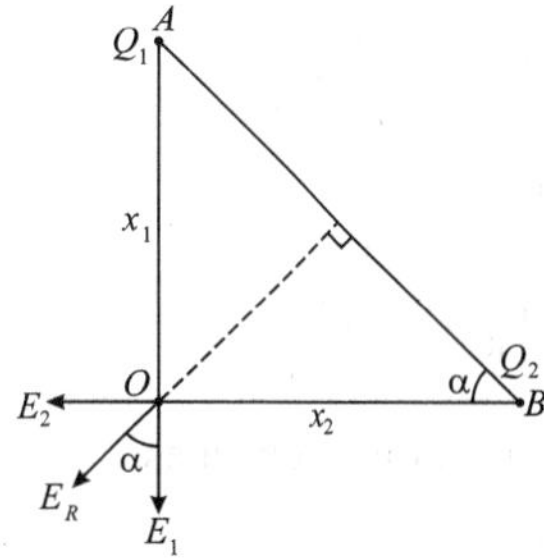

For $\overrightarrow{E_R}$ to be normal to line AB, we use

$$\tan\alpha = \frac{E_2}{E_1} = \frac{x_1}{x_2}$$

$$\Rightarrow \quad \frac{Q_2}{Q_1} = \frac{x_2}{x_1}$$ **Ans. (C)**

Sol. 42 For a uniformly charged body electric field is always in radial direction for all outside points so force on charge q (outside) is given as

$$F = \frac{1}{4\pi \epsilon_0} \cdot \frac{qQ}{r^2}$$

and at all interior points as

$$E = 0, \ F = 0$$ **Ans. (A)**

Sol. 43 If we replace $-Q$ charge at origin by a combination of one $+Q$ and one $-2Q$ charge then the situation remains the same. Now due to $+Q$ charge at every corner of cube the electric field at center of cube is zero and the net electric field at center is because of only $-2Q$ charge kept at origin, which is given as

$$\vec{E} = \frac{kq\vec{r}}{r^3} = \frac{1(-2Q)\frac{a}{2}(\hat{x} + \hat{y} + \hat{z})}{4\pi\varepsilon_0\left(\frac{a}{2}\sqrt{3}\right)^3}$$

$$\Rightarrow \quad \vec{E} = \frac{-2Q(\hat{x} + \hat{y} + \hat{z})}{3\sqrt{3}\pi a^2 \varepsilon_0}$$ **Ans. (C)**

Sol. 44 The given situation described in question is shown in figure below

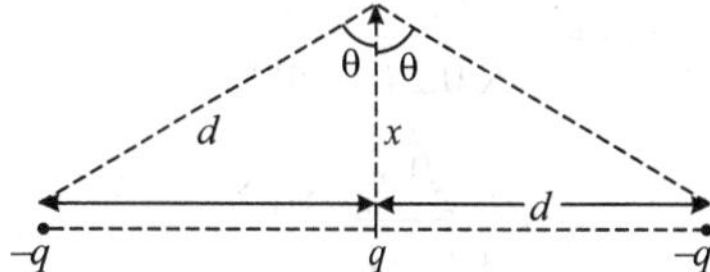

For the small displacement of charge at center, restoring force on it is given as

$$F_R = -[2F\cos\theta]$$

$$\Rightarrow \quad F_R = -2\cdot\frac{1}{4\pi\varepsilon_0}\cdot\frac{q^2}{(\sqrt{d^2+x^2})^2}\cdot\frac{x}{\sqrt{d^2+x^2}}$$

$$\Rightarrow \quad F_R = -\frac{q^2}{2\pi\varepsilon_0}\frac{x}{(d^2+x^2)^{3/2}}$$

For $x << d$, we have

$$F_R = -\frac{q^2}{2\pi\varepsilon_0 d^3}x$$

The acceleration of proton is given as

$$\Rightarrow \quad a = -\frac{q^2}{2\pi\varepsilon_0 .md^3}x$$

Comparing with equation of SHM ($a = -\omega^2 x$)

$$\Rightarrow \quad \omega = \sqrt{\frac{q^2}{2\pi\varepsilon_0 md^3}}$$ **Ans. (B)**

Sol. 45 The final charges after conduction on each sphere will be given as

$$q = \frac{(2.1-0.1)}{2}nC = 1nC$$

The force between the spheres can now be given as

$$F = \frac{9\times10^9\times10^{-18}}{(0.5)^2} = 36\times10^{-9}$$ **Ans. (36)**

Sol. 46 The situation described in question is shown in figure below

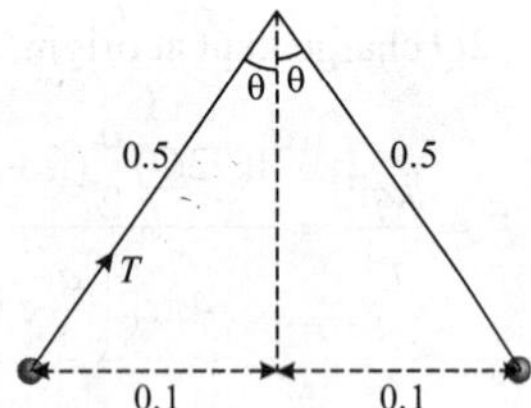

For equilibrium of sphere, we use

$$T\cos\theta = mg = 10\times10^{-6}\times10 = 10^{-4}$$

$$T\sin\theta = \frac{9\times10^9\times q^2}{0.04} = F$$

$$\Rightarrow \quad \tan\theta = \frac{0.1}{\sqrt{0.24}} = \frac{F}{mg}$$

$$\Rightarrow \quad q = \frac{2\sqrt{10}}{3\sqrt{\sqrt{24}}}\times10^{-8}$$

$$\Rightarrow \quad 0.95\times10^{-8} = \frac{a}{21}\times10^{-8}$$

$$\Rightarrow \quad a = 20$$ **Ans. (20)**

Sol. 47 The electric field due to the rod at point P is given as

$$E = \frac{k\lambda}{a}(\sin\theta_1 + \sin\theta_2)$$

$$\Rightarrow \quad E = \frac{1}{4\pi\varepsilon_0}\times\frac{Q}{L}\times\frac{1}{\left(\frac{\sqrt{3}L}{2}\right)}\times(2\sin\theta)$$

Where angle θ is given as

$$\tan\theta = \frac{L/2}{\frac{\sqrt{3}L}{2}} = \frac{1}{\sqrt{3}}$$

$$\Rightarrow \quad \sin\theta = \frac{1}{2}$$

$$\Rightarrow \quad E = \frac{1}{4\pi\varepsilon_0}\times\frac{2Q}{\sqrt{3}L^2}\times\left(2\times\frac{1}{2}\right) = \frac{Q}{2\sqrt{3}\pi\varepsilon_0 L^2}$$

Ans. (C)

Sol. 48 Free body diagram of the body is shown in figure below

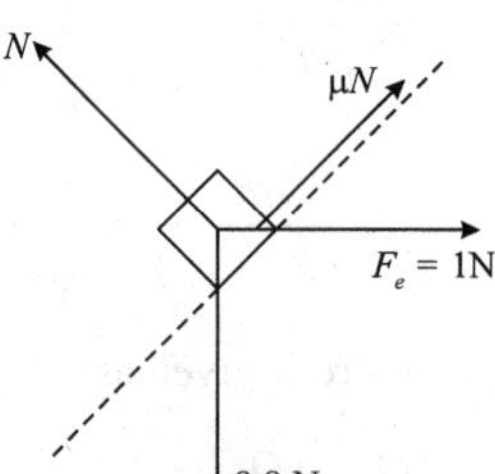

The normal reaction on body is given as

$$N = 9.8\cos30° + 1\sin30° \approx 9\text{ N}$$

Acceleration of body is given as

$$a = \frac{9.8\sin30° - 1\cos30° - \mu N}{1} = 2.233\text{ m/s}^2$$

Displacement of body is given as

$$s = ut + \frac{1}{2}at^2 = \frac{1}{2}(2.233)t^2\sin30°$$

$$\Rightarrow \quad t \approx 1.3\text{ sec}$$ **Ans. (D)**

Sol. 49 Inside the plane as electric field should be zero so for σ induced surface charge on plane, we have

$$\frac{2K\lambda}{r} = \frac{\sigma}{\varepsilon_0}$$

$$\Rightarrow \quad \sigma = \frac{2\times9\times10^9\times8\times10^{-9}\times8.85\times10^{-12}}{3}$$

$$\Rightarrow \quad \sigma = 0.424\times10^{-9}\frac{C}{m^2}$$ **Ans. (A)**

Sol. 50 For equilibrium of oil drop we have

$$qE = Mg$$

$$\Rightarrow \quad neE = \rho\left(\frac{4}{3}\pi r^3\right)\times g$$

$$\Rightarrow \quad n \times 1.6 \times 10^{-19} \times 3.55 \times 10^5$$

$$= 3 \times 10^3 \times \frac{4}{3} \times \pi \times (2 \times 10^{-3})^3 \times 9.81$$

$$\Rightarrow \quad n = 173 \times 10^{(3-9-5+19)}$$

$$\Rightarrow \quad n = 1.73 \times 10^{10}$$

Ans. (B)

Sol. 51 Force on 1 C charge can be calculated by Coulomb's law as

$$F = k(1\text{C})(1\mu\text{C})\left[1+\frac{1}{2^2}+\frac{1}{4^2}+\frac{1}{8^2}+...\right]$$

$$\Rightarrow \quad F = 9 \times 10^3\left[\frac{1}{1-\frac{1}{4}}\right] = 12 \times 10^3 \text{N}$$

Ans. (12)

Sol. 52 The force between the charges is given as

$$F = \frac{Kq_1q_2}{r^2} = \frac{K(q)(Q-q)}{r^2}$$

The force will be maximum when

$$\frac{dF}{dq} = 0$$

$$\Rightarrow \quad Q - 2q = 0$$

$$\Rightarrow \quad Q = 2q$$

Ans. (D)

Sol. 53 Acceleration of the body due to electric field is given as

$$a = \frac{qE}{m} = \frac{8\times10^{-6}}{10^{-3}} \times 100 = 0.8 \text{ m/s}^2$$

As electric field is switched on, ball first strikes to wall and returns back thus time of one oscillation can be calculated as

$$s = ut_1 + \frac{1}{2}at_1^2$$

$$0.1 = \frac{1}{2} \times 0.8t_1^2;\ t_1 = \frac{1}{2}\text{ s}$$

Thus time period $T = 2 \times \frac{1}{2} = 1$ s **Ans. (1)**

Sol. 54 Situation described in question is shown in figure below

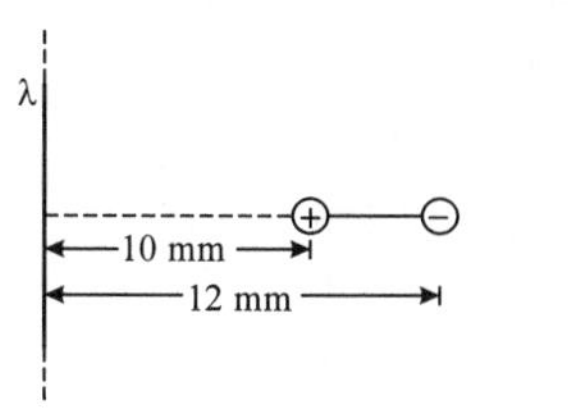

Let each charge in dipole is of magnitude Q, net force on dipole can be given as

$$F_{\text{net}} = \frac{2k\lambda Q}{r_1} + \frac{2k\lambda(-Q)}{r_2}$$

$$\Rightarrow \quad 4N = 2k\lambda Q\left[\frac{1}{r_2} - \frac{1}{r_2}\right]$$

$$\Rightarrow \quad 4N = 2 \times 9 \times 10^9 \times 3 \times 10^{-6} Q\left[\frac{1}{10} - \frac{1}{12}\right] \times 10^3$$

$$\Rightarrow \quad 4N = 54 \times 10^6\, Q \times \frac{1}{60}$$

$$\Rightarrow \quad Q = 4.44 \times 10^{-6}\text{C} = 4.44\ \mu\text{C}$$

Ans. (B)

Sol. 55 The situation described in question is shown in figure

q ●——————● q x ●——————● q (2m)

The restoring force on particle executing SHM is given as

$$F_{\text{net}} = \frac{Kq^2}{(1-x)^2} - \frac{Kq^2}{(1+x)^2} = Kq^2\left[\frac{4x}{(1^2-x^2)^2}\right]$$

$$\Rightarrow \quad a = \frac{4Kq^2x}{m}$$

$$\Rightarrow \quad \omega = 2\sqrt{\frac{Kq^2}{m}} = 2 \times \sqrt{\frac{9\times10^9\times10}{10^{-6}}} = 6000$$

Ans. (6000)

Sol. 56 Situation described in question is shown in figure below

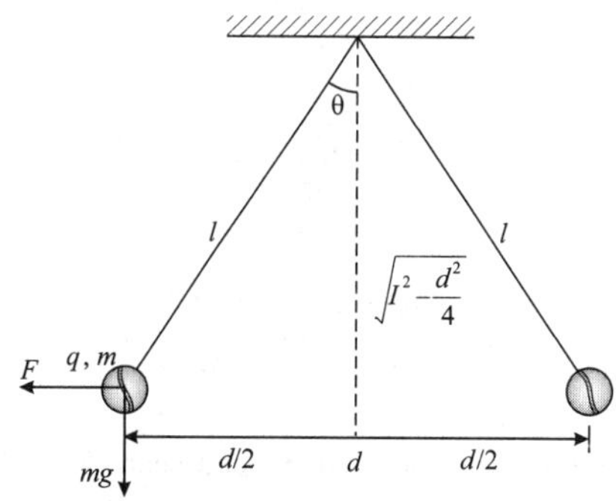

Force acting on charges is given as

$$F = \frac{kq^2}{d^2}$$

At equilibrium, we use

$$\tan\theta = \frac{\frac{kq^2}{d^2}}{mg} \approx \sin\theta$$

$$\Rightarrow \quad \frac{kq^2}{mgd^2} = \frac{d}{2l}$$

$$\Rightarrow \quad d^3 = \frac{2kq^2l}{mg}$$

$$\Rightarrow \quad d = \left(\frac{2kq^2l}{mg}\right)^{1/3}$$

$$\Rightarrow \quad d = \left(\frac{2q^2l}{4\pi\varepsilon_0 mg}\right)^{1/3} = \left(\frac{q^2l}{2\pi\varepsilon_0 mg}\right)^{1/3} \quad \textbf{Ans. (B)}$$

Sol. 57 Inside the metal shell electric field is zero and as charge is at the central sphere, electric lines of forces are radial and electric field will be inversely proportional to the square of distance from center of sphere so proper variation of electric field as a function of r is shown in figure below.

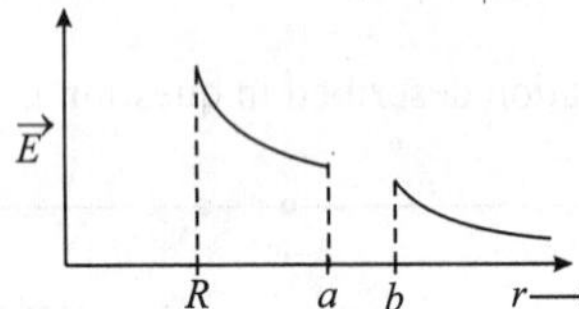

Out of given options, none of the option is correct but option (C) is most appropriate here. **Ans. (C)**

Sol. 58 This expression of electric field due to a uniformly surface charged disc is derived and explained in Physics Galaxy Vol 3A in detail. Final expression is written as

$$E = \frac{\sigma}{2\varepsilon_0}\left[1 - \frac{Z}{\sqrt{R^2+Z^2}}\right] \quad \textbf{Ans. (D)}$$

Sol. 59 Electric field due to a uniformly charged circular arc at its centre of curvature is given as

$$E = \frac{2k\lambda}{R}\sin\frac{\theta}{2}\hat{i}$$

$$\Rightarrow \quad E = \frac{2k}{R}\left(\frac{Q}{2R\frac{\pi}{3}}\right)\sin 60°\hat{i}$$

$$\Rightarrow \quad E = \frac{3kQ}{\pi R^2}\frac{\sqrt{3}}{2}\hat{i} = \frac{3\sqrt{3}\,kQ}{2\pi\ R^2}\hat{i} \quad \textbf{Ans. (A)}$$

Sol. 60 Third charge should be placed near $-5\mu C$ charge at x distance as shown in figure below for which it does not experience any electric force.

+20μC —— 5 cm —— −5μC —— x —— Q

The electric force on this charge due to the two charges are balanced, so we use

$$\frac{K(20)Q}{(5+x)^2} = \frac{K(5)Q}{X^2}$$

$$\Rightarrow \quad 4x^2 = (5+x)^2$$

$$\Rightarrow \quad 4x^2 = 25 + x^2 + 10x$$

$$\Rightarrow \quad 3x^2 - 10x - 25 = 0$$

$$\Rightarrow \quad x = 5 \quad \textbf{Ans. (D)}$$

Sol. 61 As water droplet is in equilibrium, we have

$$Mg = qE$$

$$\Rightarrow \quad (0.1\times10^{-3})(9.8) = 4.9\times10^5 q$$

$$\Rightarrow \quad \frac{2\times10^{-4}}{10^5} = q$$

$$\Rightarrow \quad q = 2\times10^{-9}\,\text{C} \quad \textbf{Ans. (B)}$$

Sol. 62 As the particles are in limiting equilibrium, the electrostatic force is exactly balancing the limiting friction, which is given as

$$\frac{kq^2}{L^2} = \mu mg$$

$$\Rightarrow \quad L = \sqrt{\frac{k}{\mu mg}}q = \sqrt{\frac{9\times10^9\times4\times10^{-14}}{0.25\times10\times10^{-3}\times10}} \quad \textbf{Ans. (A)}$$

Sol. 63 On the surface of a cylindrical volume, electric field at a distance R from the axis is given as

$$E = \frac{\rho R}{2\varepsilon_0}$$

Electric force on the charge q will provide the centipetal force for revolution, so we use

$$qE = \frac{mv^2}{r}$$

$$\Rightarrow \quad mv^2 = \frac{q\rho R^2}{2\varepsilon_0}$$

Kinetic energy of particle is given as

$$\Rightarrow \quad K = \frac{1}{2}mv^2 = \frac{q\rho R^2}{2\varepsilon_0} \quad \textbf{Ans. (A)}$$

Sol. 64 As the sheet is very large $\vec{E}$ is independent of distance from it and it is given as

$$E_1 = E_2 = \frac{\sigma}{2\varepsilon_0} \quad \textbf{Ans. (C)}$$

Sol. 65 If R is the radius of combined drop and r is the radius of smaller drop, we use by conserving volume of drops

$$\frac{4}{3}\pi R^3 = 64\times\frac{4}{3}\pi r^3$$

$$R = 4r$$

Charge on bigger drop is $Q = 64q$

The ratio of charge densities on bigger and smaller drops can be calculated as

$$\frac{\sigma_{\text{bigger}}}{\sigma_{\text{smaller}}} = \frac{\frac{Q}{4\pi R^2}}{\frac{q}{4\pi r^2}} = \frac{Q}{q}\cdot\frac{r^2}{R^2}$$

$$\Rightarrow \quad \frac{\sigma_{bigger}}{\sigma_{smaller}} = 64\frac{r^2}{16r^2} = 4$$

$$\Rightarrow \quad \frac{\sigma_{bigger}}{\sigma_{smaller}} = \frac{4}{1}$$ **Ans. (B)**

Sol. 66 Figure below shows the force acting on a suspended charged ball due to other two balls, force due to one ball is calculated as

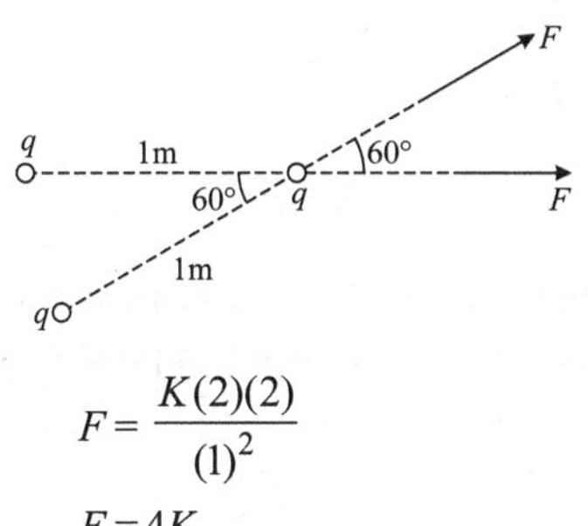

$$F = \frac{K(2)(2)}{(1)^2}$$

$$\Rightarrow \quad F = 4K$$

Resultant force on a ball due to both the forces is given as

$$F_{net} = 2F\cos 30° = 2F\left(\frac{\sqrt{3}}{2}\right) = \sqrt{3}F$$

$$\Rightarrow \quad \frac{F_{net}}{F} = \frac{\sqrt{3}F}{F} = \sqrt{3}$$

As the suspended balls are in equilibrium, net force on each ball is zero but here we are solving the question for electrostatic forces as no option is given for this case. **Ans. (D)**

Sol. 67 Figure below shows all the electric field produced at point D due to the charges at vertices A, B and C

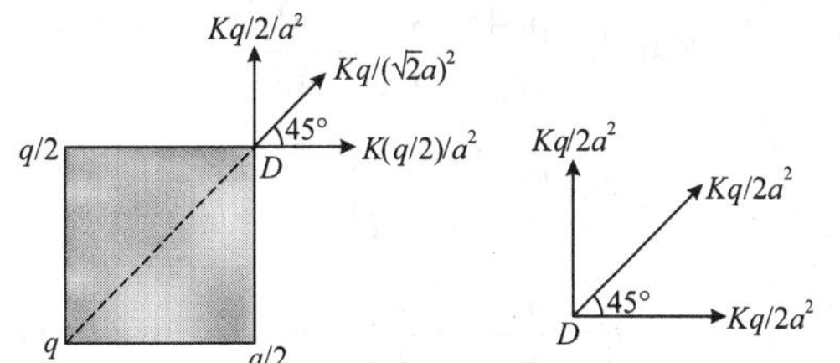

Net electric field at point D is calculated as

$$(E_{net})_D = \frac{kq}{2a^2} + \frac{\sqrt{2}kq}{2a^2}$$

$$\Rightarrow \quad (E_{net})_D = \frac{kq}{a^2}\left(\frac{1}{2} + \frac{1}{\sqrt{2}}\right)$$

$$\Rightarrow \quad (E_{net})_D = \frac{q}{4\pi \epsilon_0 a^2}\left(\frac{1}{2} + \frac{1}{\sqrt{2}}\right)$$ **Ans. (A)**

Sol. 68 Figure below shows the situation described in the question

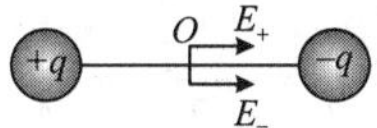

The electric field due to the two charges at mid point O is calculated as

$$E_0 = 2 \times \frac{Kq}{(d/2)^2}$$

$$\Rightarrow \quad E_0 = 8\frac{Kq}{d^2}$$

$$\Rightarrow \quad d^2 = \frac{8 \times 9 \times 10^9 \times 8 \times 10^{-6}}{6.4 \times 10^4}$$

$$\Rightarrow \quad d = 3\,\text{m}$$ **Ans. (B)**

Sol. 69 Figure below shows the forces acting on the point charge q place don the perpendicular bisector of the line joining the two charges Q. Each of the force on q by Q is given as

$$F = \frac{KQq}{\left(x^2 + \frac{d^2}{4}\right)}$$

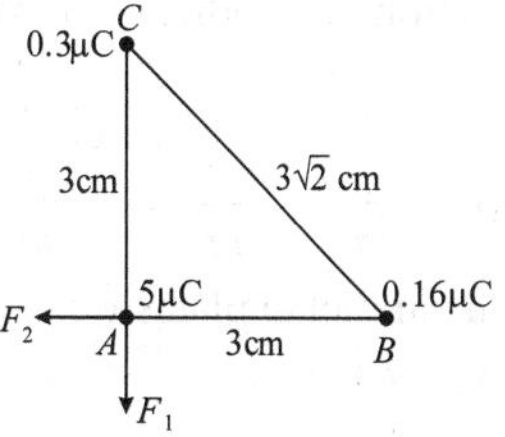

Net force on q is calculated as

$$F_{net} = 2F\cos\theta$$

$$\Rightarrow \quad F_{net} = \frac{2KQqx}{\left(x^2 + \frac{d^2}{4}\right)^{3/2}}$$

For F_{net} to be maximum, we use

$$\frac{dF_{net}}{dx} = 0$$

Solving this equation after differentiation, we get

$$\Rightarrow \quad x = \frac{d}{2\sqrt{2}}$$ **Ans. (D)**

Sol. 70 Electric displacement vector is written as

$$\vec{D} = \epsilon_0 \vec{E}$$

$$\Rightarrow \quad [D] = [\epsilon_0 E] = \left[\epsilon_0 \frac{\sigma}{\epsilon_0}\right]$$

$$\Rightarrow \quad [D] = [\sigma]$$

Thus electric displacement vector has same dimensions as that of surface charge density. **Ans. (A)**

Sol. 71 Figure below shows the situation described in the question

Force on charge located at point A are calculated as

$$F_1 = \frac{K\times 5\times 0.3\times 10^{-12}}{9\times 10^{-4}}$$

$$\Rightarrow \quad F_1 = \frac{9\times 10^9\times 5\times 0.3\times 10^{-12}}{9\times 10^{-4}}$$

$$\Rightarrow \quad F_1 = 1.5\times 10 = 15\,\text{N}$$

and $$F_2 = \frac{9\times 10^9\times 5\times 0.16\times 10^{-12}}{9\times 10^{-4}} = 8\,\text{N}$$

Net force experienced by charge at A is given as

$$F_R = \sqrt{F_1^2+F_2^2}$$

$$\Rightarrow \quad F_R = \sqrt{15^2+8^2}$$

$$\Rightarrow \quad F_R = \sqrt{289} = 17\,\text{N}$$ **Ans. (17)**

Sol. 72 Inside a long cylindrical uniformly charged volume of charge density ρ, the electric field strength is given as

$$E = \frac{\rho x}{2\varepsilon_0}$$

$$\Rightarrow \quad E = \frac{\rho}{2\varepsilon_0}\times\frac{2\varepsilon_0}{\rho} = 1$$ **Ans. (1)**

Sol. 73 Force between the given charges kept at a separation d is given as

q ——— d ——— $4-q$

$$F = \frac{Kq(4-q)}{d^2}$$

For the force to be maximum, we use

$$\frac{dF}{dq} = \frac{K}{d^2}[4-2q] = 0$$

$$\Rightarrow \quad q = 2\,\mu\text{C}$$ **Ans. (B)**

Sol. 74 During motion horizontal speed of electron remains constant as acceleration due to electric field is perpendicular to plates

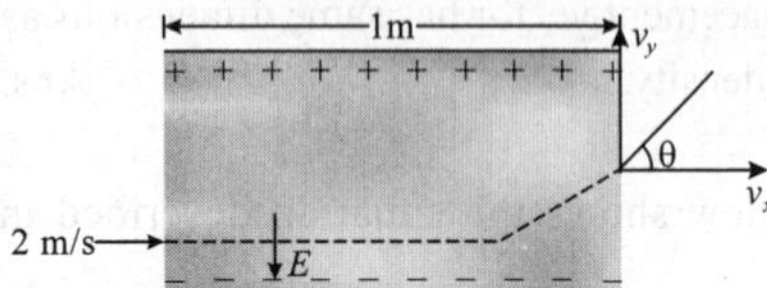

Acceleration of electron perpendicular to plates is given as

$$a_y = \frac{F_y}{m} = \frac{e(E)}{m} = \frac{e\left(\frac{8m}{e}\right)}{m} = 8\,\text{m/s}^2$$

Along the direction parallel to plates we use

$$s_x = u_x t$$

$$\Rightarrow \quad 1 = 2\times t$$

Time take by electron to come out of the plates is given as

$$t = \frac{1}{2}\,\text{s}$$

Perpendicular to plates we use speed equation for electron as

$$v_y = u_y + a_y t$$

$$\Rightarrow \quad v_y = 0 + 8\times\frac{1}{2}$$

$$\Rightarrow \quad v_y = 4\,\text{m/s}$$

Angle of deviation when it comes out is calculated as

$$\tan\theta = \frac{v_y}{v_x} = \frac{4}{2} = 2$$

$$\Rightarrow \quad \theta = \tan^{-1}(2)$$ **Ans. (B)**

Sol. 75 As point P is inside the spherical charge, we use a Gaussian surface of radius r such that point P is at its surface as shown in figure below

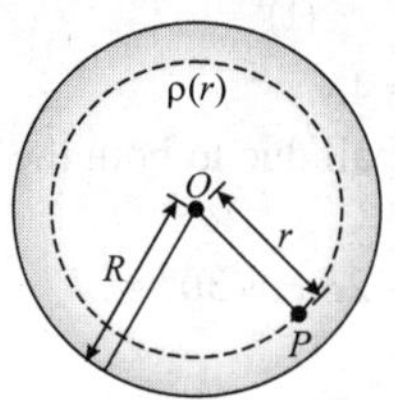

Using Gauss's law, we have

$$\oint \vec{E}\cdot d\vec{s} = \frac{Q_{in}}{\varepsilon_0}$$

$$\Rightarrow \quad E.4\pi r^2 = \frac{\int_0^r \rho_0\left(\frac{3}{4}-\frac{r}{R}\right)4\pi r^2 dr}{\varepsilon_0}$$

$$\Rightarrow \quad E.4\pi r^2 = \frac{\rho_0 4\pi}{\varepsilon_0}\left(\frac{3}{4}\frac{r^3}{3}-\frac{r^4}{4R}\right)$$

$$\Rightarrow \quad E.r^2 = \frac{\rho_0 r^3}{4\varepsilon_0}\left\{1-\frac{r}{R}\right\}$$

$$\Rightarrow \quad E = \frac{\rho_0 r}{4\varepsilon_0}\left\{1-\frac{r}{R}\right\}$$ **Ans. (C)**

Sol. 76 As the spheres are identical let their charges be

$$q_A = q_B = q$$

Force between them is calculated as

$$F = \frac{Kq^2}{r^2}$$

When C is placed in contact with A, charge on A & C will equally distributed and will become $q/2$. Now C is placed in contact with B, again charges will be equally distributed between B & C and their final charges will be

$$q_C = q_B = \frac{q+\frac{q}{2}}{2} = \frac{3q}{4}$$

Now, with final charges the three spheres are placed as shown in figure below

$$\frac{q}{2}\;(A)\text{-----}\overset{F_1}{\leftarrow}\overset{\frac{3q}{4}}{(C)}\overset{F_2}{\rightarrow}\text{-----}(B)\;\frac{3q}{2}$$
$$\frac{r}{2} \qquad \frac{r}{2}$$

Net force experienced by the sphere C will be calculated as

$$F' = F_2 - F_1 = K\left(\frac{3q}{4}\right)\frac{\left(\frac{3q}{4}-\frac{q}{2}\right)}{\left(\frac{r}{2}\right)^2}$$

$$\Rightarrow \qquad F' = \frac{3Kq^2}{4r^2} = \frac{3F}{4}$$ **Ans. (B)**

Sol. 77 Electrostatic force between two charges in a medium is given as

$$F = \frac{1}{(4\pi\varepsilon_0)}\frac{q_1q_2}{kd^2}$$

In air when the charges are separated by a distance d', the force is given as

$$F_{\text{Air}} = \frac{1}{4\pi\varepsilon_0}\frac{q_1q_2}{d'^2}$$

If the forces are equal then we use

$$F = F_{\text{Air}}$$

$$\Rightarrow \qquad \frac{q_1q_2}{4\pi\varepsilon_0 kd^2} = \frac{q_1q_2}{4\pi\varepsilon_0 d'^2}$$

$$\Rightarrow \qquad d' = d\sqrt{k}$$ **Ans. (A)**

Sol. 78 Figure below shows the region of electric field in 10 cm region where deflection of charge is happening

$V_0 = 3 \times 10^7$ N/m $\quad E = 1.8 \times 10^3$ N/m $\quad l = 10$ cm

Acceleration of charge in this region is given as

$$a = \frac{F}{m} = \frac{qE}{m} = (2\times 10^{11})(1.8\times 10^3)$$

$$\Rightarrow \qquad a = 3.6\times 10^{14}\ \text{m/s}^2$$

Time taken by the charge to cross the region is calculated as

$$t = \frac{d}{v} = \frac{0.10}{3\times 10^7}$$

The deflection of charge in y direction is calculated as

$$y = \frac{1}{2}at^2 = \frac{1}{2}(3.6\times 10^{14})\left(\frac{0.01}{9\times 10^{14}}\right)$$

$$\Rightarrow \qquad y = 0.2\times 0.01 = 0.002\ \text{m} = 2\ \text{mm}$$ **Ans. (2)**

Sol. 79 Figure below shows the charges placed on X-axis as described in question

10μc at 0; 40μc at x_0; point at 2 cm; 2cm, $(x_0 - 2)$ cm

Net electric field at $x = 2$ cm is calculated as

$$E_P = \frac{K\times 10}{2^2} - \frac{K\times 40}{(x_0-2)^2} = 0$$

$$\Rightarrow \qquad \frac{1}{2} = \frac{2}{x_0-2}$$

$$\Rightarrow \qquad x_0 - 2 = 4$$

$$\Rightarrow \qquad x_0 = 6\ \text{cm}$$ **Ans. (A)**

Sol. 80 Figure below shows the situation described in question.

$4q_0$ —12— $-q_0$ —x— q_0

For null point at the location of q_0, we use

$$\frac{q_0}{x^2} = \frac{4q_0}{(x+12)^2}$$

$$\Rightarrow \qquad x + 12 = 2x$$

$$\Rightarrow \qquad x = 12$$

Distance from origin $= x + 12 = 24$ cm. **Ans. (24)**

Sol. 81 Electric field inside material of conductor is zero, as all the charge reside only on the surface of conductor. **Ans. (B)**

Sol. 82 Figure below shows the situation of charges as described in the question. The net force on the charge q_0 is calculated as

$$F = \frac{2Kqq_0x}{(x^2+a^2)^{3/2}} \qquad \ldots(1)$$

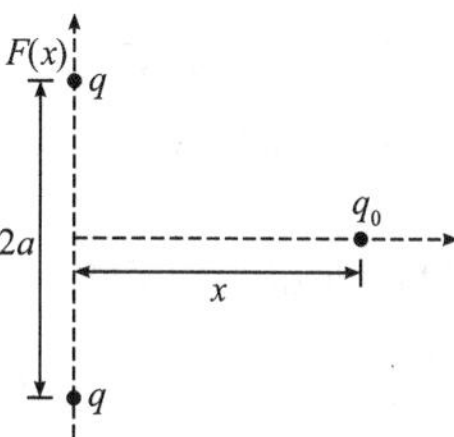

For F to be maximum, we use

$$\frac{dF}{dx} = 0$$

Differentiating equation-(1) and solving gives

$$x = \frac{a}{\sqrt{2}}$$ **Ans. (2)**

Sol. 83 Figure below shows the coulombic force acting on electron in circular motion which acts as centripetal force on it, this gives

$$eE = \frac{mv^2}{r}$$

Electric field due to cylindrical wire is

$$E = \frac{2k\lambda}{r}$$

$$\Rightarrow \quad e\cdot\frac{2K\lambda}{r} = \frac{mv^2}{r}$$

Thus velocity of electron is calculated as

$$v = \sqrt{\frac{e\cdot 2k\lambda}{m}}$$

$$\Rightarrow \quad v = \sqrt{\frac{1.6\times10^{-19}\times2\times9\times10^{9}\times2\times10^{-8}}{9\times10^{-31}}}$$

$$\Rightarrow \quad v = 8\times10^{6}\text{ m/s}$$ **Ans. (8)**

Sol. 84 For equilibrium along the plane we have

$$mg\sin\theta = \frac{1}{4\pi\epsilon_0}\times\frac{q_0^2}{(h\text{ cosec }30°)^2}$$

$$\Rightarrow \quad h^2 = \frac{1}{4\pi\epsilon_0}\times\frac{q_0^2}{mg\text{ cosec }30°}$$

$$\Rightarrow \quad h^2 = 9\times10^9\times\frac{(2\times10^{-6})}{0.02\times10\times2}$$

$$\Rightarrow \quad h = 3\times10^4\times\frac{2\times10^{-6}}{0.2}$$

$$\Rightarrow \quad h = 0.3\text{ m} = 300\text{ mm}$$ **Ans. (300)**

Sol. 85 Electric field intensity at point P

5m

At a point P with $\left\{r_P = \frac{3}{\pi}\right\}$ we have

$$E_P = \frac{\lambda}{2\pi\varepsilon_0 r_A} - \frac{\sigma}{2\varepsilon_0}$$

$$\Rightarrow \quad E_p = \frac{1}{2\varepsilon_0}\left[\frac{\lambda}{3} - \sigma\right]$$

Electric field intensity at Q with $\left\{r_Q = \frac{4}{\pi}\right\}$ we have

$$E_Q = \frac{\lambda}{2\pi\varepsilon_0 r_A} - \frac{\sigma}{2\varepsilon_0}$$

$$\Rightarrow \quad E_Q = \frac{1}{2\varepsilon_0}\left[\frac{\lambda}{4} - \sigma\right]$$

Ratio of electric field intensities is given as

$$\frac{E_P}{E_Q} = \frac{4}{3}\left(\frac{\lambda-3\sigma}{\lambda-4\sigma}\right)$$

$$\Rightarrow \quad \frac{E_P}{E_Q} = \frac{4}{3}\left[\frac{2\sigma-3\sigma}{2\sigma-4\sigma}\right]$$

$$\Rightarrow \quad \frac{E_P}{E_Q} = \frac{4}{3}\left[\frac{-\sigma}{-2\sigma}\right] = \frac{4}{6}$$ **Ans. (6)**

Sol. 86 Divide $q = 10\ \mu C$ into two parts x & $q - x$.

$$F = \frac{Fx(q-x)}{r^2}$$

For F to be maximum. By differentiating with respect to x.

$$\frac{dF}{dx} = \frac{K}{r^2}(q-2x) = 0$$

$$\Rightarrow \quad x = \frac{q}{2}$$ **Ans. (B)**

Sol. 87 Figure below shows the forces acting on the charge – $2q$.

q A — F_{BA} $-2q$ B F_{BC} — $2q$ C

By Coulomb's law, force between charge at A and B is calculated as

$$F_{BA} = \frac{Kq(2q)}{\left(\frac{3}{4}R\right)^2} = \frac{32Kq^2}{9R^2}$$

Force between charge at B and C is calculated as

$$F_{BC} = \frac{K(2q)(2q)}{\left(\frac{R}{4}\right)^2} = \frac{64Kq^2}{R^2}$$

So net force at B is given as

$$F_B = F_{BC} - F_{BA} = \frac{544Kq^2}{9R^2}$$

$$\Rightarrow \qquad F_B = \frac{544\times 9\times 10^9 \times (2\times 10^{-6})^2}{9\times (2\times 10^{-2})^2} = 5440\text{ N}$$

Ans. (5440)

15.2 Electric Flux and Gauss's Law

Sol. 1

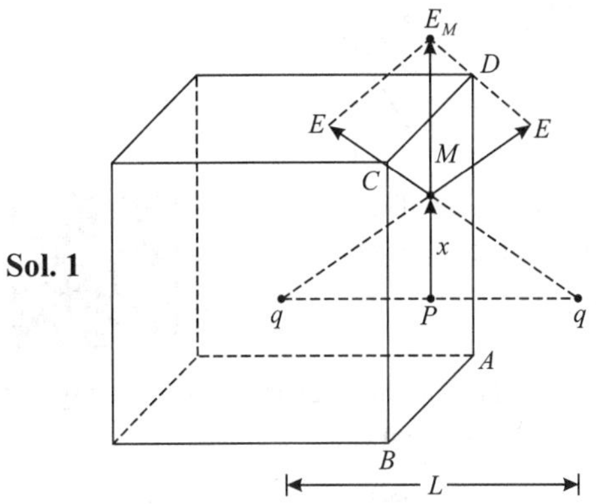

Here $P \rightarrow$ centre of square $ABCD$

$M \rightarrow$ a point in plane of square $ABCD$ at a distance x from P.

Due to the two charges, electric field resultant at any point M in plane of square $ABCD$ is parallel (tangential) to $ABCD$. So there is no electric field normal to surface $ABCD$ hence electric flux through surface $ABCD$ is zero. **Ans. (B)**

Sol. 2 Using Gauss's law, we use

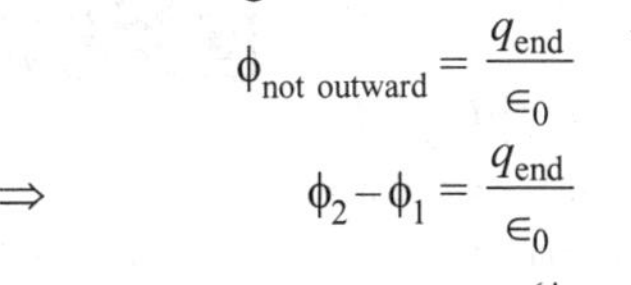

$$\phi_{\text{not outward}} = \frac{q_{\text{end}}}{\epsilon_0}$$

$$\Rightarrow \qquad \phi_2 - \phi_1 = \frac{q_{\text{end}}}{\epsilon_0}$$

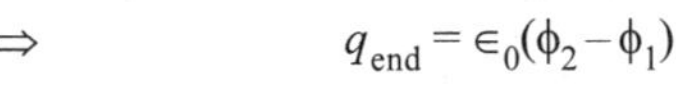

$$\Rightarrow \qquad q_{\text{end}} = \epsilon_0(\phi_2 - \phi_1)$$

Ans. (A)

Sol. 3

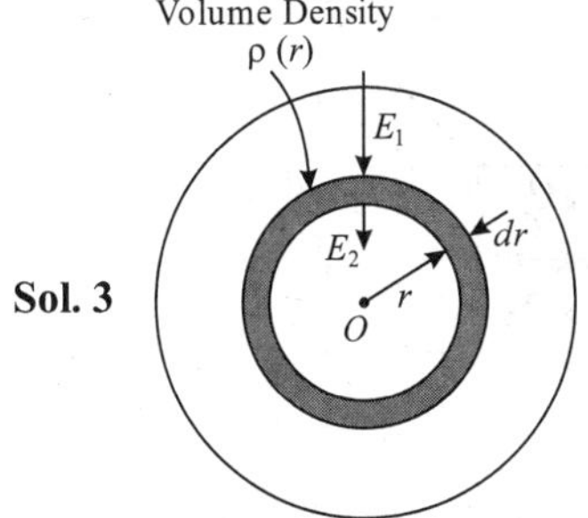

Potential at a distance r from centre is

$$\phi = ar^2 + b$$

$\Rightarrow$ EF at a distance r is given as

$$\vec{E} = -\frac{dv}{dr}\hat{r}$$

$$\Rightarrow \qquad \vec{E} = -2ar\hat{r} \qquad \ldots(1)$$

–ve sign implies field is in radially inward direction

$\Rightarrow$ charge on ball is –ve

From equation-(1) we can see that $E \propto r$ so this happens inside a uniformly charged ball and if its volume charge density is ρ then inside EF of this ball is given as

$$E_{\text{in}} = \frac{\rho r}{3\in_0} \qquad \ldots(2)$$

from (1) and (2), we have

$$\frac{\rho r}{3\in_0} = 2ar \quad \Rightarrow \quad \rho = 6a\in_0$$

Ans. (B)

Alternative : If E_1 and E_2 are EF at outer surfaces of the elemental shell as shown then by Gauss's law, we have

$$\phi_{\text{out}} - \phi_{\text{in}} = E_2(4\pi r^2) - E_1(4\pi(r+dr)^2) = \frac{q_{\text{encl}}}{\epsilon_0}$$

$$8\pi ar^3 - 8\pi ar^3\left(1+\frac{dr}{r}\right)^3 = \frac{\rho(r)[4\pi r^2 dr]}{\epsilon_0}$$

$$8\pi ar^3 - 8\pi ar^3\left(1+\frac{3dr}{r}\right) = \frac{\rho(r)[4\pi r^2 dr]}{\epsilon_0}$$

$$\Rightarrow \qquad \frac{\rho(r)dr}{\epsilon_0} = -6adr$$

$$\Rightarrow \qquad \rho(r) = -6a\epsilon_0$$

Ans. (B)

Sol. 4

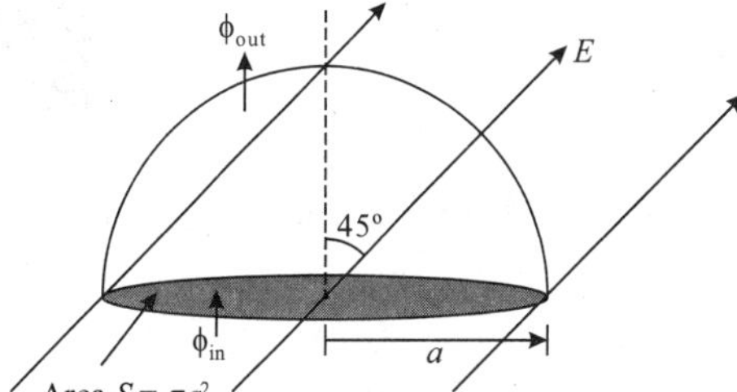

As no charge is enclosed in hemisphere

$$\Rightarrow \qquad \phi_{\text{in}} = \phi_{\text{out}}$$

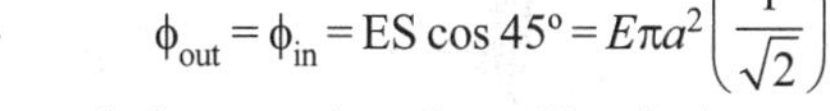

$$\Rightarrow \qquad \phi_{\text{out}} = \phi_{\text{in}} = \text{ES}\cos 45^\circ = E\pi a^2\left(\frac{1}{\sqrt{2}}\right)$$

Through the curved surface of hemisphere **Ans. (B)**

Sol. 5

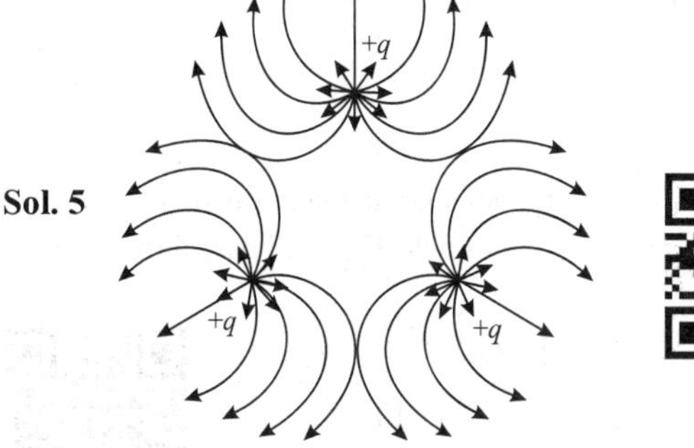

As electric lines originated from a +ve charge are repelled by other neighbouring +ve charges **Ans. (C)**

Sol. 6

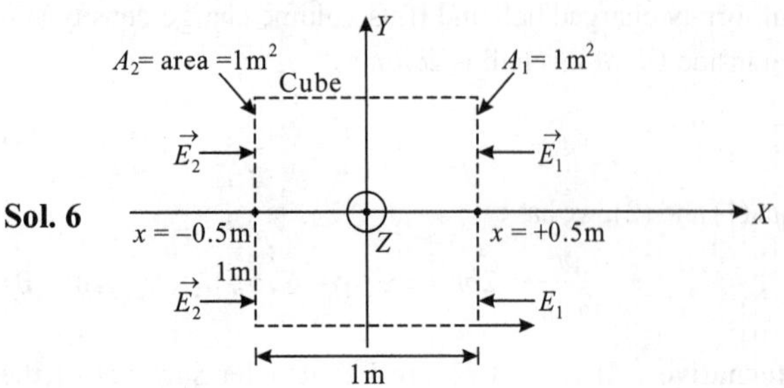

EF in the region is given as

$$\vec{E} = -\frac{dV}{dx}\hat{i}$$

at $\quad x = 0.5\text{ m} \rightarrow \vec{E_1} = -4\hat{i}$

at $\quad x = -0.5\text{ m} \rightarrow \vec{E_2} = 4\hat{i}$

Total flux going into the cubical surface is given as

$$\phi = \vec{E_1}\cdot\vec{A_1} + \vec{E_2}\cdot\vec{A}$$

$$\Rightarrow \quad \phi = 4.1 + 4.1 = 8 \text{ units}$$

By Gauss's Law $\quad \phi_{out} = \frac{q_{encl}}{\epsilon_0}$

$$\Rightarrow \quad -8 = \frac{q_{encl}}{\epsilon_0} \Rightarrow q_{encl} = -8\epsilon_0$$ **Ans. (D)**

Sol. 7

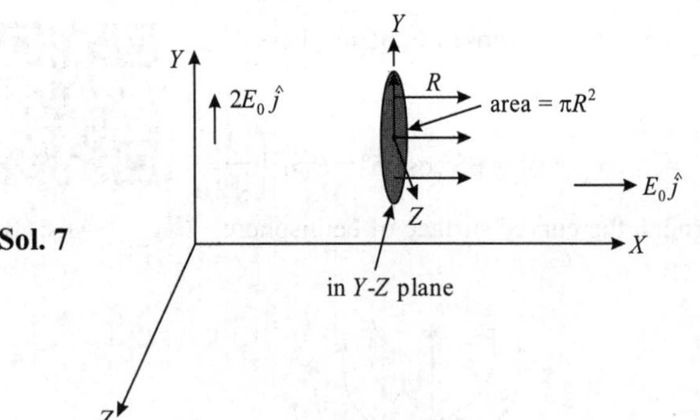

As circular area is perpendicular to X-axis (as in YZ plane) so only the electric flux due to field $E_0\hat{i}$ will pass through so we have

$$\phi = E_0 \times A = E_0 \cdot \pi r^2$$

$$= 100 \times 3.14 \times 4 \times 10^{-4}$$

$$= 12.56 \times 10^{-2}$$

$$= 0.1256 \text{ Nm}^2/\text{C}$$ **Ans. (A)**

Sol. 8

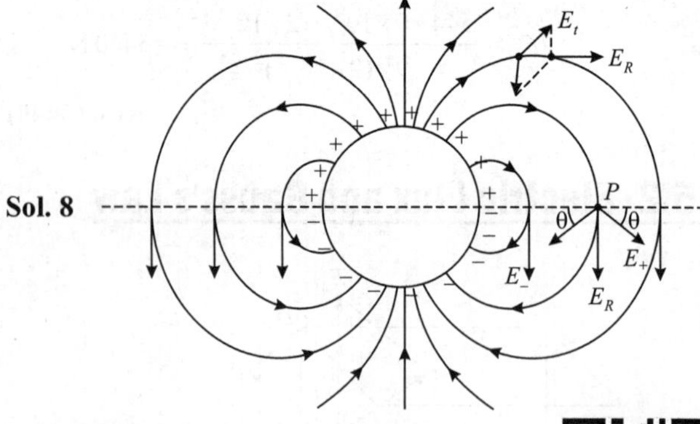

The preference of termination of electric lines is always for a –ve charge (nearly) rather at ∞. **Ans. (A)**

Sol. 9 Using Gauss's law we have electric flux coming out from any closed surface is given as

$$\phi_s = \frac{q_{\text{enclosed by S}}}{\epsilon_0}$$

flux through surfaces S_1, S_2, S_3 and S_4 are

$$\phi_1 = \frac{2q}{\epsilon_0};$$

$$\phi_2 = \frac{3q - q}{\epsilon_0} = \frac{2q}{\epsilon_0};$$

$$\phi_3 = \frac{2q}{\epsilon_0};$$

$$\phi_4 = \frac{8q - 2q - 4q}{\epsilon_0} = \frac{2q}{\epsilon_0}$$

$$\Rightarrow \quad \phi_1 = \phi_2 = \phi_3 = \phi_4$$ **Ans. (C)**

Sol. 10

As all six surfaces (squares of side a) of imaginary cube considered are symmetrically located w.r. to charge Q at centre. Total electric flux coming out from cubical surface is

$$\phi_0 = \frac{Q}{\epsilon_0}$$

By symmetry, flux through the given square is

$$\phi = \frac{\phi_0}{6} = \frac{Q}{6\epsilon_0}$$ **Ans. (B)**

Sol. 11 By Gauss's law we use

$$\int \vec{E}\cdot\vec{dA} = \frac{q_{enc}}{\varepsilon_0}$$

If $\vec{E}$ is normal to surface and uniform over the surface then we can use

$$\int \vec{E}\cdot\vec{dA} = EA$$

$$\Rightarrow \quad E = \frac{q_{enc}}{\varepsilon_0 A}$$

Thus option-(D) is correct. **Ans. (D)**

Sol. 12 Electric flux through face $ABCD$ is

$$\phi_{\text{I}} = 0 \quad \text{as} \quad E_z = 0$$

Electric flux through face $BCGF$ is

$$\phi_{\text{II}} = E_x \cdot S = (4\times 3)\times(2\times 2) = 48\ \text{Nm}^2/\text{C}$$

$$\Rightarrow \quad \phi_{\text{I}} - \phi_{\text{II}} = 0 - 48 = -48\ \text{Nm}^2/\text{C}$$ **Ans. (– 48)**

Sol. 13

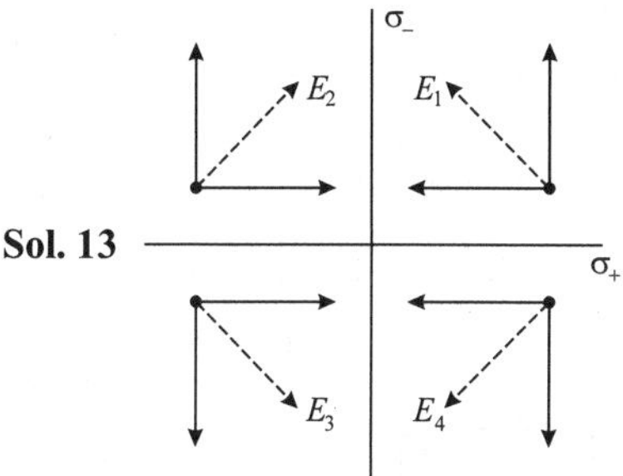

In figure above E_1, E_2, E_3 and E_4 represent the resultants electric field in region due to the two sheets. As $|\sigma_+| > |\sigma_-|$ option (A) is more appropriate. **Ans. (A)**

Sol. 14 By symmetry, the magnitude of electric flux through the square can be given as one sixth of the total flux originating from the charge as the square can be considered as one of the six symmetric faces of a cube at the center of which the charge is considered to be kept.

$$\phi = \frac{1}{6}\left(\frac{q}{\varepsilon_0}\right) = \frac{12\times 10^{-6}}{6\times 8.85\times 10^{-12}}$$

$$= 225.98\times 10^3\frac{Nm^2}{s} = 226\times 10^3\frac{Nm^2}{C}$$ **Ans. (226)**

Sol. 15 The electric field vector is given as

$$\vec{E} = \left(\frac{3E_0}{5}\hat{i} + \frac{4E_0}{5}\hat{j}\right)\frac{N}{C}$$

The area vectors of the given surfaces can be written as

$$\vec{A}_1 = 0.2\text{m}^2\,\hat{i} \quad \text{(parallel to } y\text{-}z \text{ plane)}$$

and

$$\vec{A}_2 = 0.3\text{m}^2\,\hat{j} \quad \text{(parallel to } x\text{-}z \text{ plane)}$$

The electric flux through these areas can be given as

$$\phi_a = \left[\frac{3E_0}{5}\hat{i} + \frac{4E_0}{5}\hat{j}\right]\cdot[0.2\hat{i}] = \frac{3\times 0.2}{5}E_0$$

and

$$\phi_b = \left[\frac{3E_0}{5}\hat{i} + \frac{4E_0}{5}\hat{j}\right]\cdot[0.3\hat{j}] = \frac{4\times 0.3}{5}E_0$$

The ratio as asked is given as

$$\frac{\phi_a}{\phi_b} = \frac{0.6}{1.2} = \frac{1}{2} = \frac{a}{b}$$

$$\Rightarrow \quad a : b = 1 : 2$$

$$\Rightarrow \quad a = 1$$ **Ans. (1)**

Sol. 16 Total flux through the cube can be given by symmetry of charge placed at one vertex, which is given as

$$\phi = \frac{q}{8\epsilon_0}$$

The flux through the cube side surfaces $ABEH$, $ADGH$, $ABCD$ will be zero and for the remaining three surfaces flux will be equal which can be given as

$$\phi(EFGH) = \phi(DCFG) = \phi(EBCF) = \frac{1}{3}\left(\frac{q}{8\epsilon_0}\right) = \frac{q}{24\epsilon_0}$$

Ans. (D)

Sol. 17 Figure below shows the situation described in statement-I, in which it can be observed that total charge enclosed in the hollow sphere is zero but due to electric dipole the electric field is not zero anywhere in surrounding of dipole thus statement-I is true.

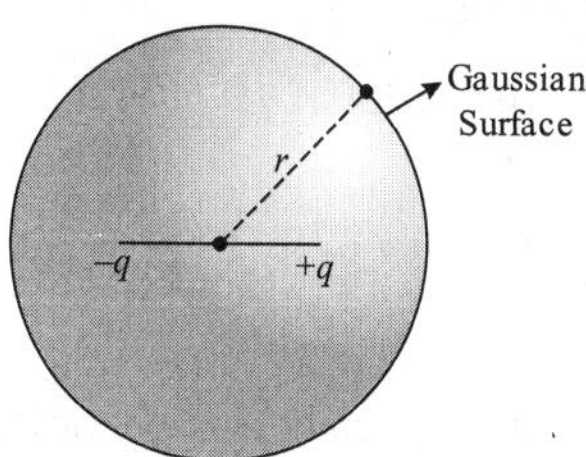

As per situation given in Statement-II, in case of a metal sphere always electric field inside is zero so electric flux through any surface considered inside the sphere is always zero thus statement-II is false. **Ans. (A)**

Sol. 18 Flux through the surface is calculated by using only x component of electric field as area is parallel to YZ plane. Thus required flux is given as

$$\phi = E_x A$$

$$\Rightarrow \quad \frac{2}{5}\times 4\times 10^3\times 0.4 = 640$$ **Ans. (640)**

Sol. 19 Using vector form of Gauss's law, we have

$$\vec{\nabla}\cdot\vec{E} = \frac{\rho}{\varepsilon_0}$$

$$\Rightarrow \quad \frac{\delta E}{\delta x}+\frac{\delta E}{\delta y}+\frac{\delta E}{\delta z}=\frac{\rho}{\varepsilon_0}$$

$$\Rightarrow \quad \rho=2\varepsilon_0$$

$$\Rightarrow \quad Q=2\times 10^{-9}\times \rho=4\varepsilon_0 \text{ nC}$$ **Ans. (4)**

Sol. 20 For a given surface ingoing flux is taken negative because field and surface area vectors are in opposite direction hence option (a) is correct. Due to symmetry of cube, flux through all faces due to a point charge is same hence option (b) is correct. In case of uniform field, number of lines entering into a closed surface are equal to the number of lines coming out hence option (c) is correct. When electric field is parallel to a surface, flux through the surface must be zero hence option (d) is incorrect. **Ans. (C)**

Sol. 21 Using gauss law, we have

$$\oint \vec{E}\cdot \vec{ds}=\frac{q_{in}}{\varepsilon_0}$$

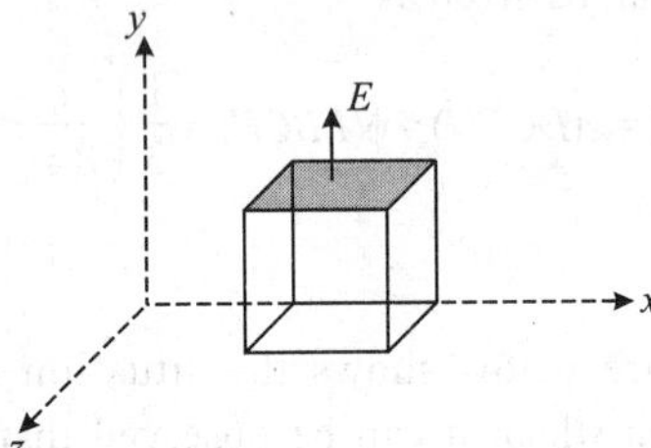

The electric flux coming out of the cube is from the top face as shown in above figure, which can be calculated as

$$\varepsilon_0(150)(0.5)^2\times(0.5)^2=q_{in}$$

$$\Rightarrow \quad q_{in}=\frac{8.85\times10^{-12}\times150}{16}$$

$$\Rightarrow \quad q_{in}=8.3\times10^{-11}\text{ C}$$ **Ans. (B)**

Sol. 22 Total flux through complete spherical surface is $\frac{q}{\varepsilon_0}$ as whole flux of charge q will be coming out from the spherical surface. From the hemispherical part electric flux will be half as the charge is placed at centre and electric field lines are radially outward. If we consider a lower hemisphere then same flux will be passing through that hemisphere also which eventually first cross the flat part shown in figure then passes through the lower hemisphere so flux through the flat surface is considered half of total flux. **Ans. (B)**

Sol. 23 Number of electric field lines per unit area or flux density is considerd equal to the magnitude of electric field. From the surface of a spherical uniform charge electric field is given as

$$E=\frac{\rho R}{3\in_0}=\frac{2\times6}{3\times8.85\times10^{-12}}$$

$$\Rightarrow \quad E=0.45\times10^{12}\text{ NC}^{-1}$$

$$\Rightarrow \quad E=45\times10^{10}\text{ N/C}$$ **Ans. (45)**

Sol. 24 Figure below shows the given cuboid and at centre of square face having area $4L^2$, a charge q is placed. To analyse the flux through its opposite face, we consider an identical cuboid below this one as shown in figure so that including this a cube of edge length $2L$ is formed as shown.

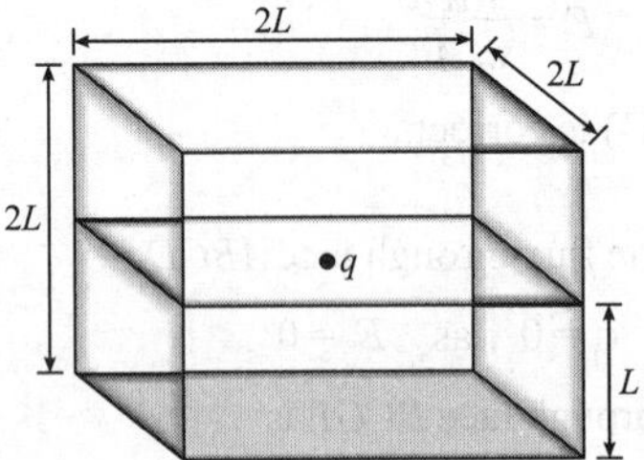

As the charge q is now at the centre of this cube thus formed so electric flux passing through shaded face in above figure is given as

$$\phi=\frac{q}{6\varepsilon_0}$$ **Ans. (D)**

Sol. 25 Given that electric field in a region is

$$\vec{E}=2x^2\hat{i}-4y\hat{j}+6\hat{k}$$

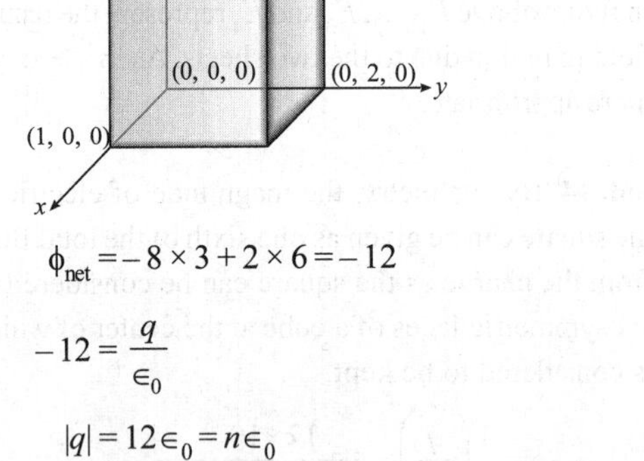

$$\phi_{net}=-8\times3+2\times6=-12$$

$$\Rightarrow \quad -12=\frac{q}{\epsilon_0}$$

$$\Rightarrow \quad |q|=12\epsilon_0=n\epsilon_0$$

So $n=12$ **Ans. (12)**

Sol. 26 In the given condition, electric flux will only come out of the right face of the cube which is given as

$$\phi=\vec{E}\cdot\vec{A}$$

$$\Rightarrow \quad \phi=4000(0\cdot2)^2\frac{V}{m}\cdot(0\cdot2)^2 m^2$$

$$\Rightarrow \quad \phi=4000\times16\times10^{-4}\text{ Vm}$$

$$\Rightarrow \quad \phi=640\text{ V cm}$$ **Ans. (640)**

Sol. 27 Figure below shows the cubical surface as described in the question

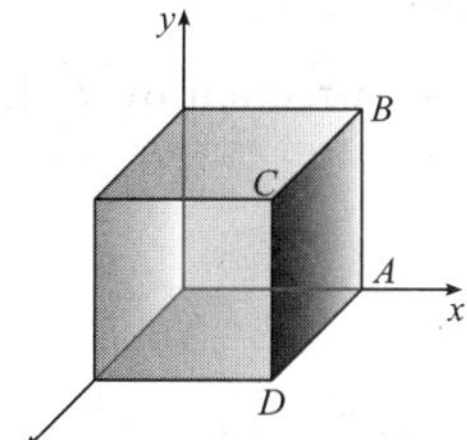

Electric field in the region is given as

$$\vec{E} = E_0 x\hat{i}$$

Net flux through the surface $ABCD$ is the total flux coming out from this cubical surface, which is calculated as

$$\phi_{\text{net}} = \phi_{ABCD} = E_0 a \cdot a^2$$

$$\Rightarrow \quad \phi_{\text{net}} = \frac{q_{en}}{\epsilon_0} = E_0 a^3$$

Thus charge enclosed by this region is calculated as

$$q_{en} = E_0 \epsilon_0 a^3$$

$$\Rightarrow \quad q_{en} = 4 \times 10^4 \times 9 \times 10^{-12} \times 8 \times 10^{-6}$$

$$\Rightarrow \quad q_{en} = 288 \times 10^{-14}\,C = Q \times 10^{-14}\,\text{C}$$

$$\Rightarrow \quad Q = 288$$ **Ans. (288)**

16.1 Electric Potential and Dipole

Sol. 1 Work done in moving a charge from one point to another (say A to B) is given as

$$W_{AB} = q \times (V_B - V_A) = \text{charge} \times \text{potential difference}$$

$\Rightarrow$ Potential difference

$$V_B - V_A = \frac{W_{AB}}{q} = \frac{2}{20} = 0.1\,\text{V}$$ **Ans. (A)**

Sol. 2

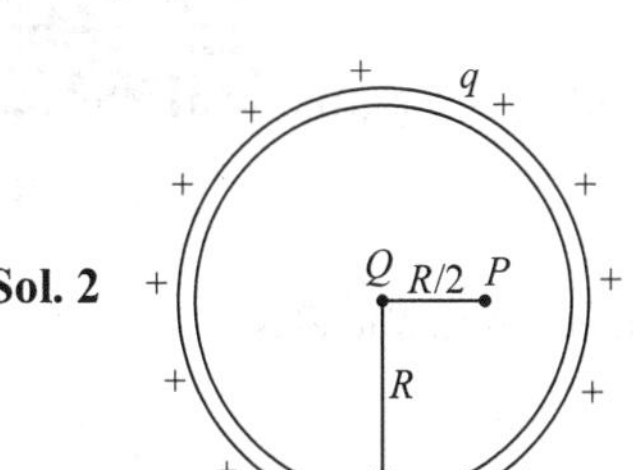

Electric potential at point P is given as

$$V_P = V_{QP} + V_{qP}$$

$$\Rightarrow \quad V_P = \frac{KQ}{\left(\frac{R}{2}\right)} + \frac{Kq}{R}$$

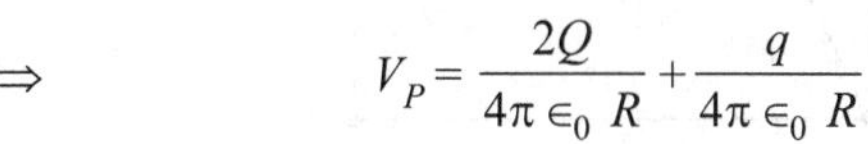

$$\Rightarrow \quad V_P = \frac{2Q}{4\pi \epsilon_0 R} + \frac{q}{4\pi \epsilon_0 R}$$ **Ans. (C)**

Sol. 3 At the point of closest approach, the KE of particle transforms into the PE of system. Thus by conservation of energy, we have

$$\Rightarrow \quad \frac{1}{2}mv^2 + O = O + \frac{KqQ}{r} \Rightarrow r = \frac{2KqQ}{mv^2}$$

$$\Rightarrow \quad r \propto \frac{1}{v^2}$$

If q was given a speed $2v$ then now closest approach distance will be

$$\frac{r_1}{r_2} = \left(\frac{v_2}{v_1}\right)^2 \Rightarrow r_2 = \left(\frac{v_1}{v_2}\right)^2 r_1 = \left(\frac{v_1}{2v}\right)^2 r = \frac{r}{4}$$ **Ans. (D)**

Sol. 4 Electric potential at point C_2 is

$$V_{C_2} = \left(+\frac{Kq}{\sqrt{d^2+R^2}}\right) + \left(-\frac{Kq}{R}\right)$$

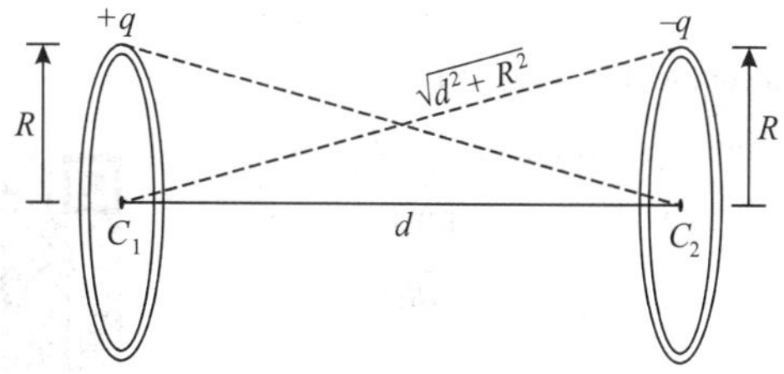

Electric potential at point C_1 is

$$V_{C_1} = V_{+q} + V_{-q}$$

$$\Rightarrow \quad V_{C_1} = \frac{Kq}{R} + \left(-\frac{Kq}{\sqrt{d^2+R^2}}\right)$$

Potential difference between centres of rings is

$$\Delta V = V_{C_1} - V_{C_2} = \frac{q}{2\pi \epsilon_0}\left(\frac{1}{R} - \frac{1}{\sqrt{d^2+R^2}}\right)$$ **Ans. (B)**

Sol. 5

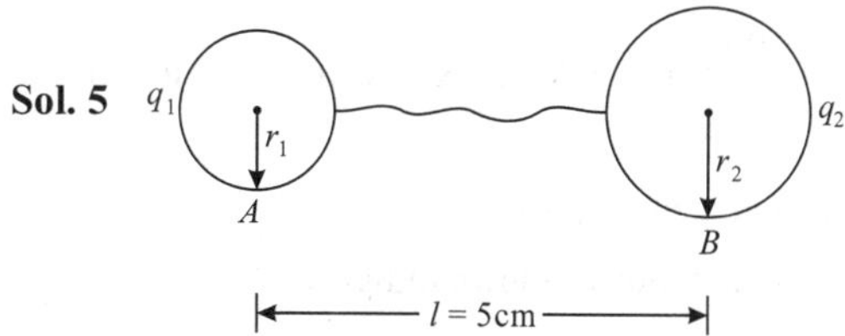

We can ignore induction effects on spheres due to charge on other sphere as separation

$$l >> r_1, r_2$$

If q_1 and q_2 are charges on A and B after connection, we have

$$V_A = V_B$$

$$\frac{Kq_1}{r_1} = \frac{Kq_2}{r_2}$$

Electric field magnitude on surfaces of A and B are

$$E_A = \frac{Kq_1}{r_1^2}$$

and $$E_B = \frac{Kq_2}{r_2}$$

$$\Rightarrow \quad \frac{E_A}{E_B} = \frac{r_2}{r_1} = \frac{2}{1} \quad \textbf{Ans. (C)}$$

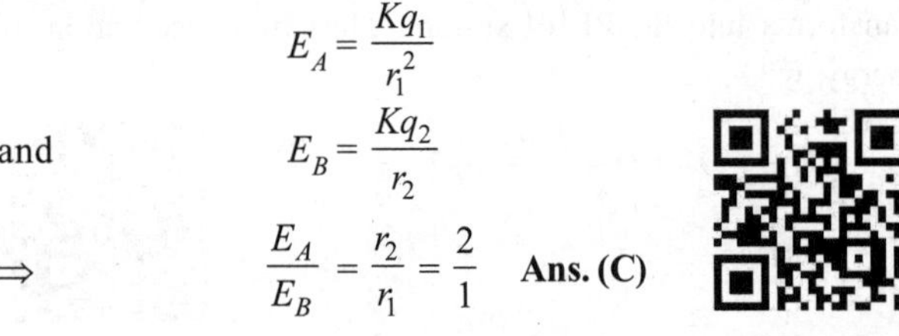

Sol. 6 Non-uniform EF

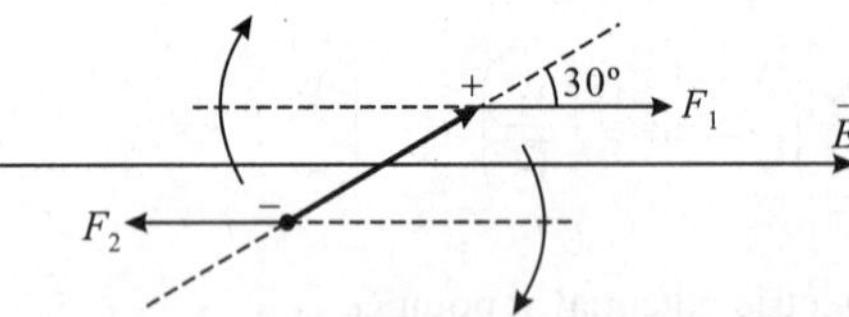

The electric force on a dipole placed in non-uniform EF is given as

$$F = \vec{P}\cdot\frac{\overrightarrow{dE}}{dx} = P\frac{dE}{dx}\cos 30°$$

Force along the direction of EF

As EF is non uniform here $F_1 \neq F_2$

so dipole will experience a translational force

$$|F_1 - F_2| = \left|\vec{P}\cdot\frac{\overrightarrow{dE}}{dx}\right|$$

As F_1 and F_2 are not acting along same lines of action so there will be a torque due to these forces present on dipole. **Ans. (C)**

Sol. 7 We know that EF in space is given as

$$\vec{E} = -\frac{dV}{dx}\hat{i} \quad \text{(for unidirectional EF)}$$

$$\Rightarrow \quad \vec{E} = -\left[20\left(-\frac{1}{(x^2-4)}\times 2x\right)\right]\hat{i}$$

$$\Rightarrow \quad \vec{E} = \frac{40x}{(x^3-4)}\hat{i}\Bigg|_{x=4\mu m} = \frac{40\times 4}{(16-4)^2} = \frac{160}{144}\hat{i}\ \text{V/}\mu\text{m}$$

$$\Rightarrow \quad \vec{E} = \frac{10}{9}\hat{i}\ \text{V/}\mu\text{m (along + X-direction)} \quad \textbf{Ans. (D)}$$

Sol. 8 Due to symmetry $V_0 = 0$

At O net EF is in direction as shown in figure-1

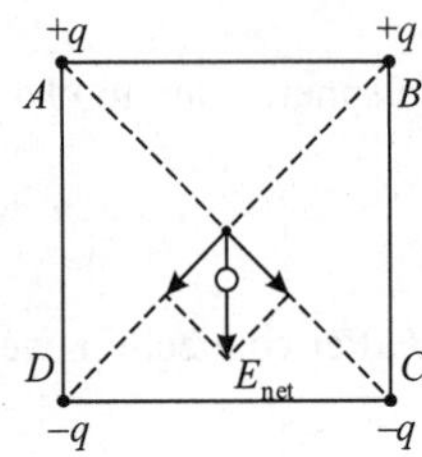

Figure-1

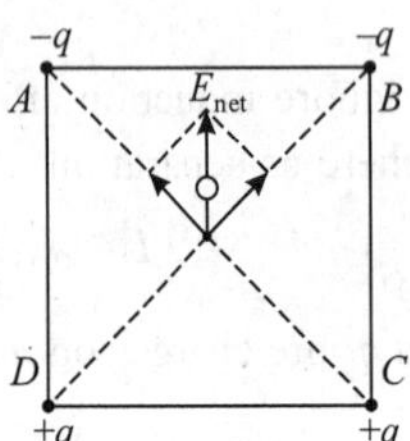

Figure-2

After interchanging charges as given in que

Still $V_0 = 0$ (remain unchanged)

As shown in figure-2 direction of $\vec{E}$ is reversed. **Ans. (D)**

Sol. 9 $r_1 = \sqrt{2^2 + 0^2} = 2$

$r_2 = \sqrt{(\sqrt{2})^2 + (\sqrt{2})^2} = 2$

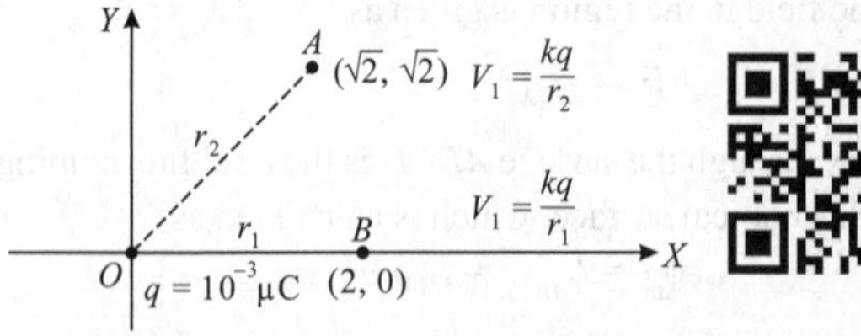

As $r_1 = r_2$ so electric potentials at points A and B are equal

$\Rightarrow V_A = V_B$ **Ans. (B)**

Sol. 10 Work done in moving charge q from point P to Q is

$$W = q(V_Q - V_P)$$

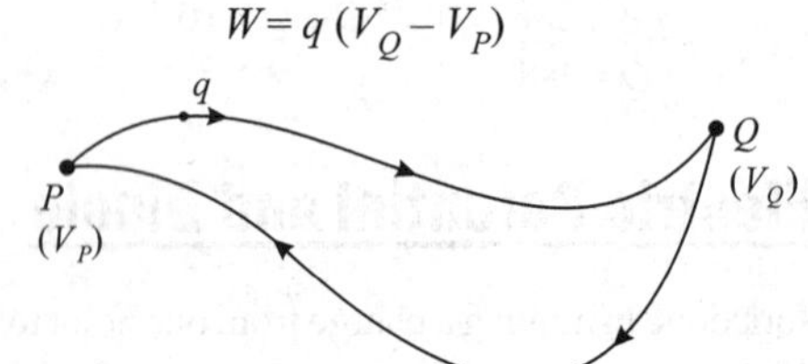

Being EF a conservative field work done does not depend on the path chosen.

Work done in moving 'q' from Q to P is

$$W' = q(V_P - V_Q)$$

$$W_{closed} = W + W' = 0 \quad \textbf{Ans. (D)}$$

Sol. 11 Work done in moving a charge q from point P to Q is given as

$$W = q(V_Q - V_P)$$

$$\Rightarrow \quad W = [100\times(-1.6\times10^{-19})]\times[-4-(10)]$$

$$\Rightarrow \quad W = 2.24\times10^{-16}\ \text{J} \quad \textbf{Ans. (B)}$$

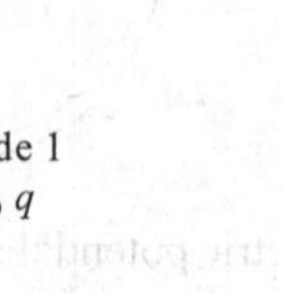

Sol. 12 KE_{gain} of Q when it moves from X to O is

$$KE_{gain} = Q(V_X - V_O)$$

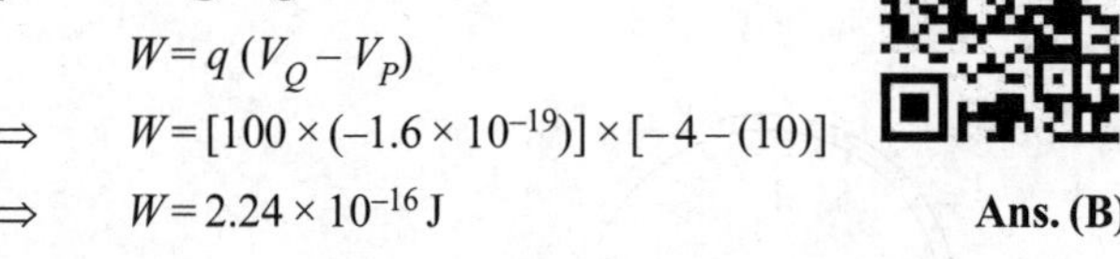

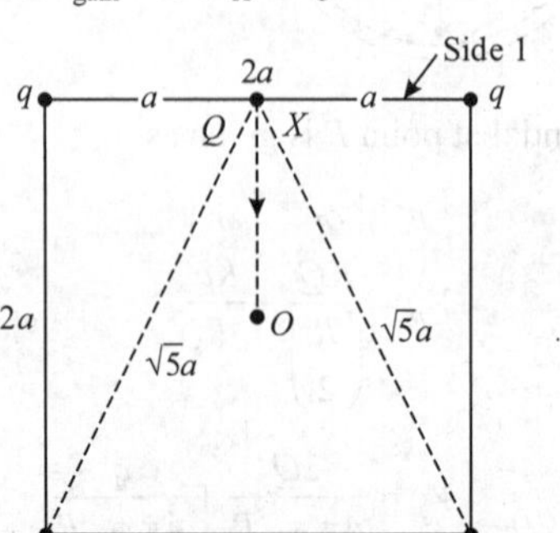

Potential at point X is

$$V_X = \frac{2Kq}{a} - \frac{2Kq}{\sqrt{5}a} = \frac{2Kq}{a}\left(1-\frac{1}{\sqrt{5}}\right)$$

Potential at point centre of square O is

$$V_O = O$$

$$\Rightarrow \quad KE_{gain} = Q(V_n - O) = QV_X$$

$$\Rightarrow \quad KE_{gain} = \frac{2KqQ}{a}\left(1-\frac{1}{\sqrt{5}}\right)$$

$$\Rightarrow \quad KE_{gain} = \frac{1}{4\pi\epsilon_0}\cdot\frac{2qQ}{a}\left(1-\frac{1}{\sqrt{5}}\right)$$

Ans. (A)

Sol. 13 Inside a uniformly charged sphere EF is given as

$$E_{in} = \frac{\rho r}{3\epsilon_0}$$

$\Rightarrow$ S-2 is true

Work done is moving 'q' from centre to surface of sphere is

$$W = \Delta U = q(V_S - V_C) = q\left[-\int_0^r E dr\right]$$

$$\Rightarrow \quad W = -q\int_0^r \frac{\rho r}{3\epsilon_0}\cdot dr$$

$$\Rightarrow \quad W = -\frac{q\rho}{3\epsilon_0}\frac{r^2}{2}$$

$\Rightarrow$ S-1 is false

Alternative : We can see that $\frac{q\rho}{3\epsilon_0}$ is not having dimensions of energy

$\Rightarrow$ S-1 is false **Ans. (D)**

Sol. 14 As the surface charge densities of the two shells are equal

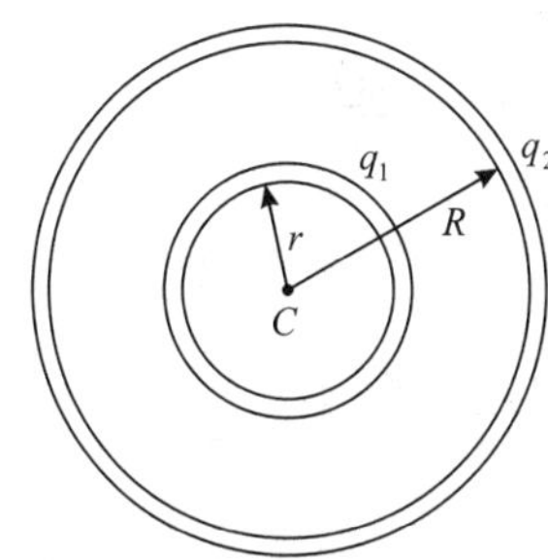

$$\Rightarrow \quad \frac{q_1}{4\pi r^2} = \frac{q_2}{4\pi R^2} = \frac{Q}{4\pi(r^2+R^2)}$$

Electric potential at common centre is given as

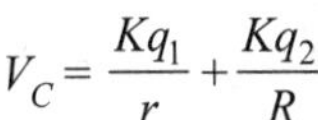

$$V_C = \frac{Kq_1}{r} + \frac{Kq_2}{R}$$

$$\Rightarrow \quad V_C = \frac{1}{4\pi\epsilon_0}\left[\frac{Qr^2}{r(r^2+R^2)} + \frac{QR^2}{R(r^2+R^2)}\right]$$

$$\Rightarrow \quad V_C = \frac{(r+R)Q}{4\pi\epsilon_0(r^2+R^2)}$$

Ans. (C)

Sol. 15

Charge on element

$$dq = \frac{Q}{L}dx$$

Potential at O due to dq is given as

$$dV = \frac{Kdq}{x} = \frac{KQ}{Lx}dx$$

Potential at O due to whole rod AB is

$$V = \int dV = \int_{x=L}^{x=2L}\frac{KQ}{Lx}dx$$

$$\Rightarrow \quad V = \frac{KQ}{L}[\ln x]_L^{2L} = \frac{KQ}{L}[\ln 2L \ln L]$$

$$\Rightarrow \quad V = \frac{KQ}{L}\ln(2) = \frac{Q}{4\pi\epsilon_0 L}\ln(2)$$

Ans. (D)

Sol. 16 As the force between dipoles placed 1 to X-axis will be zero so we consider these along X-axis as shown.

EF due to $\vec{P_1}$ at the location of $\vec{P_2}$ is given as

$$E_1 = \frac{2kP_1}{x^3}$$

In non uniform EF the force on a dipole placed along direction of EF is given as

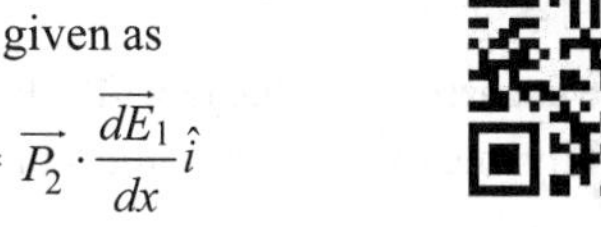

$$\vec{F} = \vec{P_2}\cdot\frac{\vec{dE_1}}{dx}\hat{i}$$

$$\Rightarrow \quad \vec{F} = P_2\cdot\frac{d}{dx}\left(\frac{2Kp_1}{x^3}\right)\hat{i} = 2Kp_1p_2\left(-\frac{3}{x^4}\right)\hat{i}$$

$$\Rightarrow \quad \vec{F} = -\frac{6Kp_1p_2}{x^4}\cdot\hat{i}$$

$$\Rightarrow \quad \vec{F} = -\frac{1}{4\pi\epsilon_0}\cdot\frac{6p_1p_2}{x^4}\hat{i}$$

Ans. (C)

Sol. 17 S-1 is true

Work done in moving a test charge between any two point is

$$W = q_0(V_A - V_B) = O$$

If $V_A = V_B$ (us A and B are on an equipotential surface)

S-2 is false

As at every point of equipotential surface, electric lines are $\perp$ to this surface not mutually $\perp$ to each other. **Ans. (C)**

Sol. 18

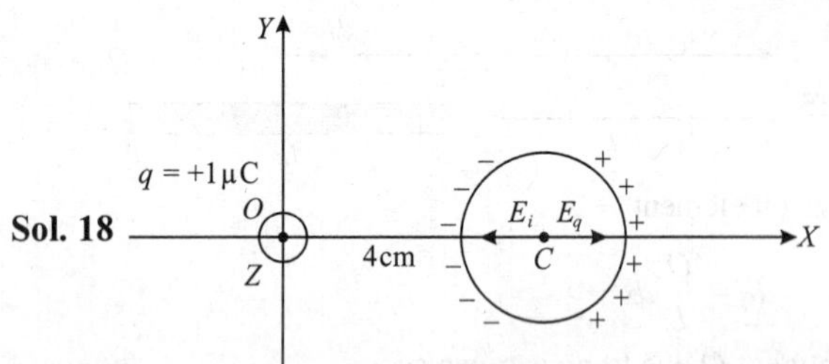

As net EF at C is zero (inside a conductor) we use

$$E_i = E_q = \frac{Kq}{x^2}$$

$$\Rightarrow \quad E_i = \frac{9\times10^9\times10^{-6}}{(4\times10^{-2})^2} = 5.625\times10^6 \text{ V/m}$$

Potential at C is only due to q (As $V_i = 0$) and it is given as

$$V = \frac{Kq}{x} = \frac{9\times10^9\times10^{-6}}{4\times10^{-2}} = 2.25\times10^5 \text{ V} \quad \textbf{Ans. (C)}$$

Sol. 19 As in the direction of EF always potential decreases

$\Rightarrow$ we can use $V_X > V_Y > V_\infty$

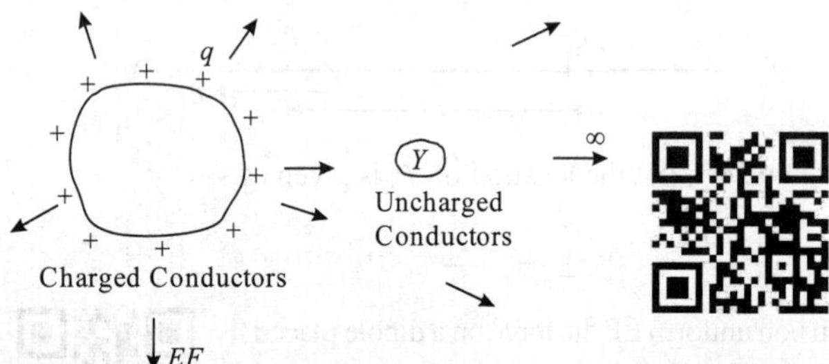

Because uncharged conductor will only affect its neighbourhood EF slightly due to induced charges on it **Ans. (A)**

Sol. 20 At equilibrium of charges, we have

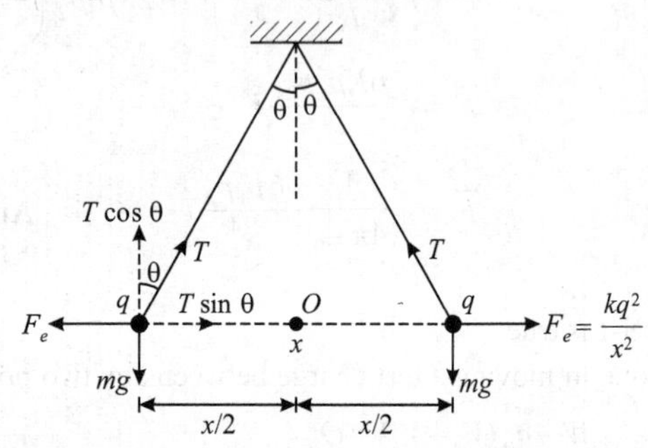

$$T \sin\theta = \frac{kq^2}{x^2} \quad \ldots(1)$$

$$T \cos\theta = mg \quad \ldots(2)$$

$$(1)(2) \Rightarrow \quad \tan\theta = \frac{kq^2}{mgx^2}$$

$$\Rightarrow \quad \frac{q}{x} = \sqrt{\frac{mg\tan\theta}{k}}$$

Electric potential at point O is $V_0 = \dfrac{kq}{x/2} \times 2 = \dfrac{4kq}{x}$

$$= 4k\sqrt{\frac{mg\tan\theta}{k}} = 4\sqrt{mgk\tan\theta} \quad \textbf{Ans. (D)}$$

Sol. 21 We use

$$V_A - V_O = -\int_0^2 E dx$$

here $V_A < V_O$ So $V_A - V_O \to$ –ve

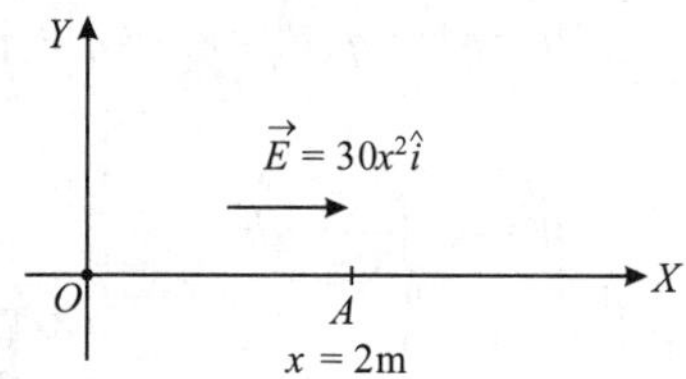

As along direction of EF potential decreases

$$\Rightarrow \quad V_A - V_0 = -\int_0^2 30x^2 dx$$

$$\Rightarrow \quad V_A - V_0 = -30\left[\frac{x^3}{3}\right]_0^2 = -10[8-0]$$

$$\Rightarrow \quad V_A - V_0 = -80 \text{ V} = -80 \text{ J/C} \quad \textbf{Ans. (C)}$$

Sol. 22 Potential on surface of a solid uniformly charged sphere is

$$V_O = \frac{KQ}{R}$$

If at $r = R_1$, $V = \dfrac{3V_O}{2}$ $(> V_0 \Rightarrow r < R)$

$$\Rightarrow \quad \frac{5V_O}{4} = \frac{V_O}{2}\left(3 - \frac{r^2}{R^2}\right)$$

$$\Rightarrow \quad 5 = 6 - \frac{2r^2}{R^2} \Rightarrow r = \frac{R}{\sqrt{2}} = R_2$$

If at $r = R_3$, $V = \dfrac{3V_O}{4}$ $(< V_O \Rightarrow r > R)$

$$\Rightarrow \quad \frac{V_O}{4} = \frac{KQ}{r} \Rightarrow \frac{KQ}{4R} = \frac{KQ}{r} \Rightarrow r = 4R = R_4 \quad \textbf{Ans. (D)}$$

Here $R_4 - R_3 = 4R - \frac{4R}{3} = \frac{8R}{3} > R_2$ (As $R_2 = \frac{R}{\sqrt{2}}$) **Ans. (C)**

Both are correct.

Sol. 23 Along the direction of EF, potential decreases and in uniform EF, potential difference between two points is given as

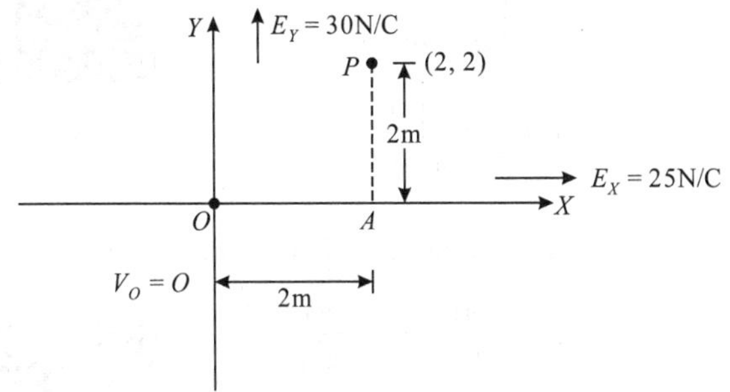

$$V_X - V_Y = Ed$$

In moving from O to P we can write

$$V_O - V_A = 25 \times 2$$

$$V_A - V_P = 30 \times 2$$

$\Rightarrow$ $$V_O - V_P = 110$$

$\Rightarrow$ $$V_P = -110\text{ V or } -110\text{ J/C}$$ **Ans. (A)**

Sol. 24 For $|Z| \leq |m$ $\quad \vec{E} = -\frac{dV}{dZ}\hat{k} = 10|Z|\hat{k}$

$-1\text{m} < Z < 1\text{m}$

and for $|Z| \geq 1\text{m}$ $\quad \vec{E} = -\frac{dV}{dZ}\hat{k} = \pm 10\hat{k}$

$Z > 1$m and $Z < -1$m constant EF

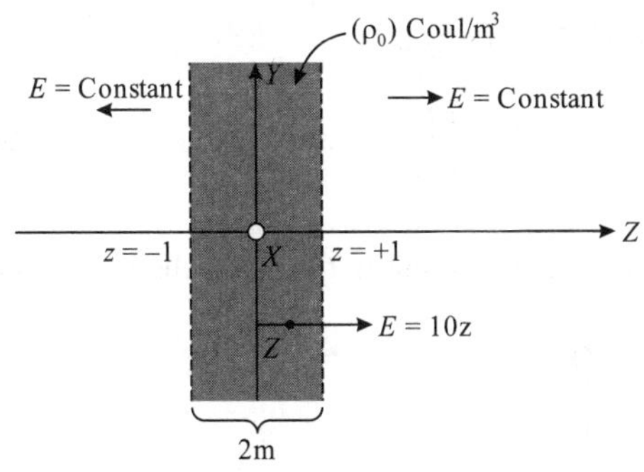

By analyzing the EF we can state that it is produced by a uniform charge distribution in a thick sheet of width 2m as shown.

Due to a thick volume charged sheet, inside at a distance Z from central plane the EF is given as

$$E_{in} = \frac{\rho z}{\epsilon_O} = 10z$$

$\Rightarrow$ $\rho_0 = 10\,\epsilon_0$ for $|z| \leq 1$m

$\Rightarrow$ $\rho_0 = 0$ for $|z| \geq 1$m **Ans. (B)**

Sol. 25 $r_2 - r_1 = r_1 - r_0 = r_N - r_{N-1}$

$\Delta V \quad \Delta V \quad \Delta V$ Potential differences

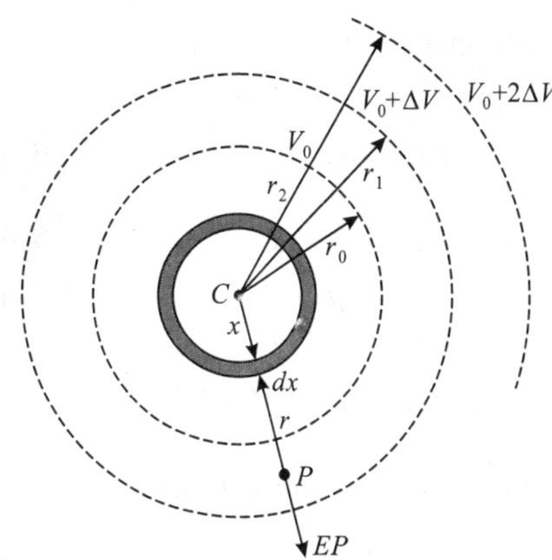

Only in uniform EF the separation between equipotential surface having same ΔV is constant.

To calculate EF at a distance r from centre of charge distribution then it is given by

$$E_P = \frac{kq}{r^2}$$

Where $q \to$ charge enclosed within sphere of rad $r = 0$ to $r = r$.

Charge in sphere of radius r is given as

$$q = \int dq = \int_0^r \rho(x) 4\pi x^2 dx$$

$\Rightarrow$ EF at point P is

$$E_P = \frac{K\int_0^r \rho(x)\cdot 4\pi x^2 dx}{r^2} = C \text{ (constant)}$$

Differentiating we get

$$\frac{d}{dr}\left[k\int_0^r \rho(x)\cdot 4\pi x^2 dx\right] = 2\text{Cr}$$

$$\frac{1}{4\pi\,\epsilon_0}\cdot[\rho(r)\,4\pi r^2 - 0] = 2\text{ Cr}$$

$\Rightarrow$ $\rho(r) = \frac{2C\,\epsilon_0}{r}$ $\Rightarrow$ $\rho(r) \propto \frac{1}{r}$ **Ans. (C)**

Sol. 26 Here $V_P > V_M > V_B$

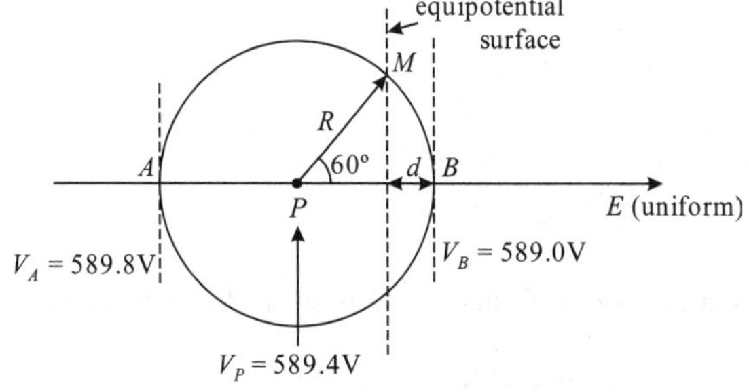

There exist only option (B) which is 589.2 V satisfying the given condition. **Ans. (B)**

Alternative : In given figure shown we use here

$$d = R(1 - \cos 60°) = \frac{R}{2}$$

We use from the given values, we have

$$V_A - V_B = E(2R) = 0.8$$

$$\Rightarrow \quad ER = 0.4$$

$$V_M - V_B = Ed = \frac{ER}{2}$$

$$\Rightarrow \quad V_M = \frac{ER}{2} + V_B$$

$$\Rightarrow \quad V_M = 0.2 + 589.0 = 589.2\ \text{V}$$

Ans. (B)

Sol. 27 Charges on shells A, B and C are

$$q_A = +\sigma \times 4\pi a^2$$

$$q_B = -\sigma \times 4\pi b^2$$

$$q_C = -\sigma \times 4\pi e^2$$

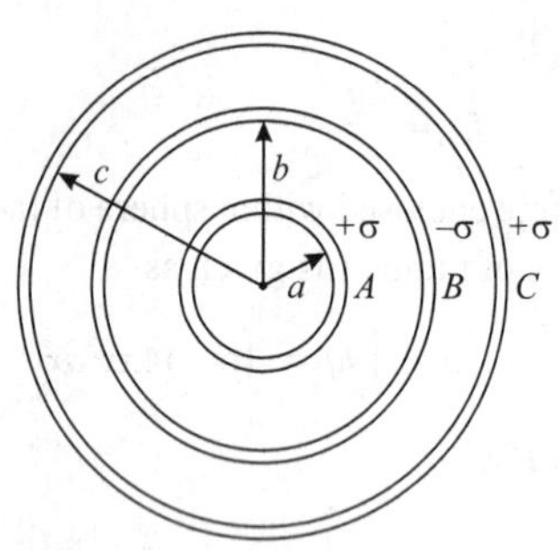

Potential of shell B due to all the shell charges is given as

$$V_B = V_{q_A} + V_{q_B} + V_{q_C}$$

$$\Rightarrow \quad V_B = \frac{Kq_A}{b} + \frac{Kq_B}{b} + \frac{Kq_C}{C}$$

$$\Rightarrow \quad V_B = \frac{1}{4\pi \epsilon_0}\left(\frac{\sigma 4\pi a^2}{b} - \frac{\sigma 4\pi b^2}{b} + \frac{\sigma 4\pi c^2}{c}\right)$$

$$\Rightarrow \quad V_B = \frac{\sigma}{\epsilon_0}\left(\frac{a^2}{b} - b + c\right) = \frac{\sigma}{\epsilon_0}\left(\frac{a^2 - b^2}{b} + c\right)$$

Ans. (B)

Sol. 28

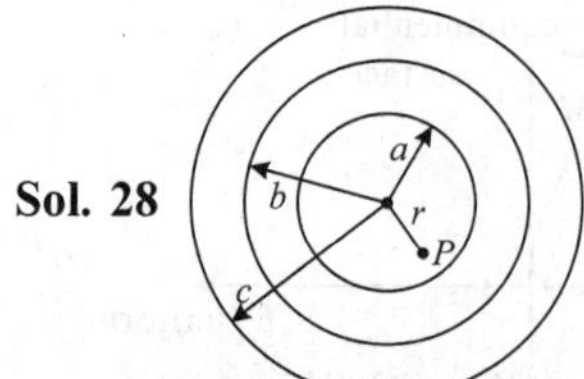

Potential at point P, due to the three shells is given as

$$V = \frac{kQ_a}{a} + \frac{kQ_b}{b} + \frac{kQ_c}{c}$$

As $Q_a : Q_b : Q_c :: a^2 : b^2 : c^2$ [since $\sigma_a = \sigma_b = \sigma_c$]

$$\Rightarrow \quad Q_a = \left[\frac{a^2}{a^2 + b^2 + c^2}\right]Q$$

$$\Rightarrow \quad Q_b = \left[\frac{b^2}{a^2 + b^2 + c^2}\right]Q$$

$$\Rightarrow \quad Q_c = \left[\frac{c^2}{a^2 + b^2 + c^2}\right]Q$$

$$\Rightarrow \quad V = \frac{Q}{4\pi \epsilon_0}\left[\frac{(a+b+c)}{a^2 + b^2 + c^2}\right]$$

Ans. (B)

Sol. 29 At a distance x on right of dipole B, we use

$$V = \frac{4qa}{(R+x)^2} = \frac{2qa}{(x^2)}$$

$$\Rightarrow \quad \sqrt{2}x = R + x$$

$$\Rightarrow \quad x = \frac{R}{\sqrt{2} - 1}$$

Distance of this point from A is

$$x_A = \frac{R}{\sqrt{2} - 1} + R = \frac{\sqrt{2}R}{\sqrt{2} - 1}$$

Ans. (C)

Sol. 30 Electric field at equatorial points of dipole is given as

$$E = \frac{Kp}{r^3}$$

At point P, force on Q is

$$F = \frac{Kp}{r^3}Q \quad \ldots (1)$$

At point P', force on Q is

$$F' = \frac{KpQ}{(r/3)^3} = 27F$$

Ans. (D)

Sol. 31 Figure below shows the given charge configuration

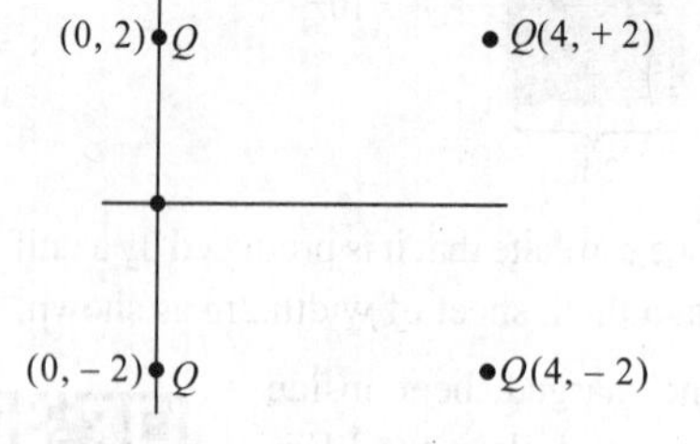

Potential at origin is given as

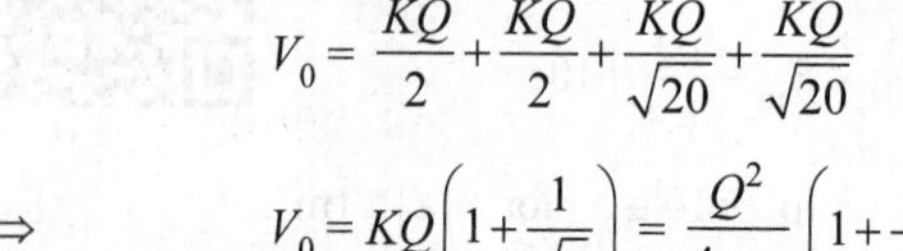

$$V_0 = \frac{KQ}{2} + \frac{KQ}{2} + \frac{KQ}{\sqrt{20}} + \frac{KQ}{\sqrt{20}}$$

$$\Rightarrow \quad V_0 = KQ\left(1 + \frac{1}{\sqrt{5}}\right) = \frac{Q^2}{4\pi\varepsilon_0}\left(1 + \frac{1}{\sqrt{5}}\right)$$

Work required to put a fifth charge Q at origin is given as

$$W = QV = \frac{Q^2}{4\pi\varepsilon_0}\left(1+\frac{1}{\sqrt{5}}\right) \quad \textbf{Ans. (B)}$$

Sol. 32 Total electrostatic energy of system is given as

$$U = \left[\frac{q^2}{a}+\frac{Qq}{a}+\frac{Qq}{a\sqrt{2}}\right] = 0$$

$$\Rightarrow \quad q = -Q\left[1+\frac{1}{\sqrt{2}}\right]$$

$$\Rightarrow \quad Q = \frac{-q\sqrt{2}}{\sqrt{2}+1}$$

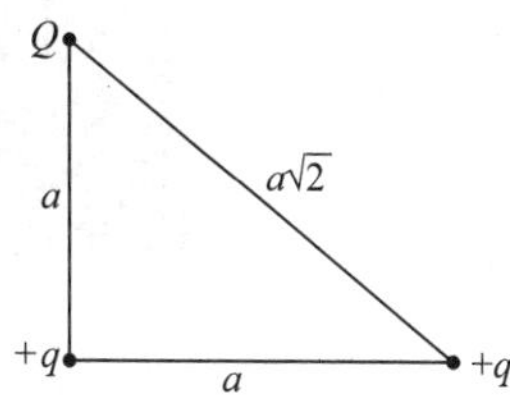

Ans. (A)

Sol. 33 Graph in figure shows that from centre to surface of sphere the physical quantity is constant than at outer points it decreases with distance so it must be electric potential due to a uniformly charged conducting shell. **Ans. (B)**

Sol. 34 Potential energy of a dipole in electric field given as

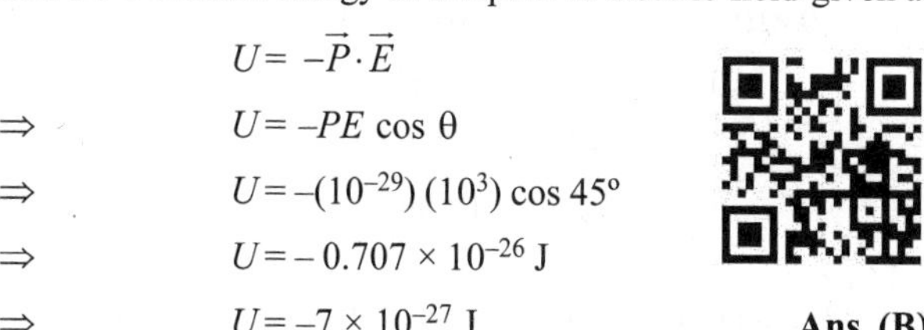

$$U = -\vec{P}\cdot\vec{E}$$

$$\Rightarrow \quad U = -PE\cos\theta$$

$$\Rightarrow \quad U = -(10^{-29})(10^3)\cos 45^\circ$$

$$\Rightarrow \quad U = -0.707\times 10^{-26}\text{ J}$$

$$\Rightarrow \quad U = -7\times 10^{-27}\text{ J}$$

Ans. (B)

Sol. 35

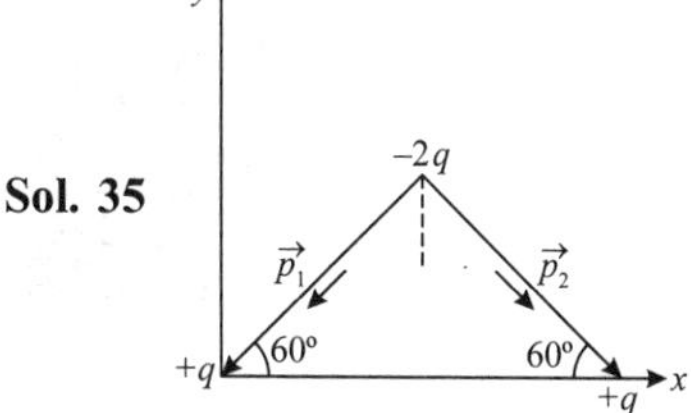

Given system is equivalent to two dipoles as shown above where dipole moments are given as

$$\vec{p}_1 = ql(-\cos 60^\circ\hat{i} - \sin 60^\circ\hat{j}) = \frac{ql}{2}(\hat{i}+\sqrt{3}\hat{j})$$

$$\vec{p}_2 = ql(\cos 60^\circ\hat{i} - \sin 60^\circ\hat{j}) = \frac{ql}{2}(\hat{i}-\sqrt{3}\hat{j})$$

Net dipole moment of system is

$$\vec{p}_{net} = \vec{p}_1+\vec{p}_2 = -ql\sqrt{3}\hat{j} \quad \textbf{Ans. (C)}$$

Sol. 36 At any instant 't' total energy of charge distribution is constant thus by energy conservation, we use

$$\frac{1}{2}mv^2 + \frac{KQ^2}{2R} = 0 + \frac{KQ^2}{2R_0}$$

$$\Rightarrow \quad \frac{1}{2}mV^2 = \frac{KQ^2}{2R_0} - \frac{KQ^2}{2R}$$

$$\Rightarrow \quad v = \sqrt{\frac{2}{m}\frac{KQ^2}{2}\cdot\left(\frac{1}{R_0}-\frac{1}{R}\right)}$$

$$\Rightarrow \quad v = \sqrt{\frac{KQ^2}{m}\left(\frac{1}{R_0}-\frac{1}{R}\right)}$$

From above expression we can see that with increase in R, v increases and $\frac{dv}{dR}$ decreases. **Ans. (A)**

Sol. 37 As given in the first case, outer shall is given charge $+Q$

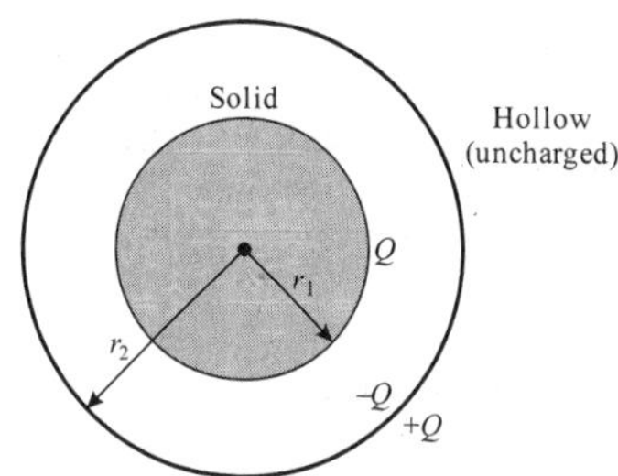

The potential difference of inner and outer spheres is given as

$$V_{in} - V_{out} = \left(\frac{kQ}{r_1}\right) - \left(\frac{kQ}{r_2}\right)$$

$$\Rightarrow \quad V_{in} - V_{out} = kQ\left(\frac{1}{r_1}-\frac{1}{r_2}\right) = V$$

In the second case, outer shall is given charge $-4Q$

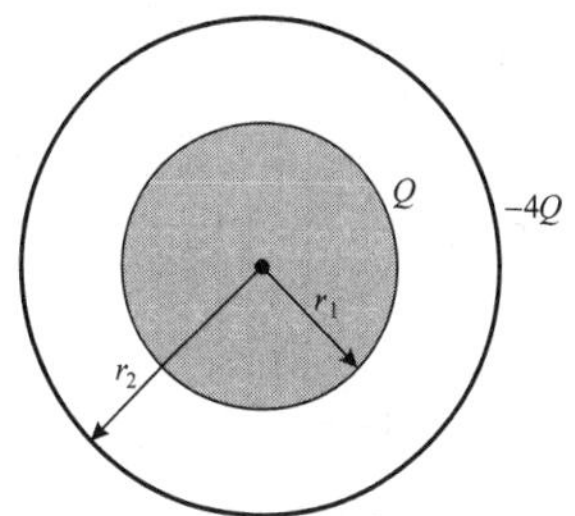

New potential difference is given as

$$V_{in} - V_{out} = \left(\frac{kQ}{r_1} - \frac{4kQ}{r_2}\right) - \left(\frac{kQ}{r_2} - \frac{4kQ}{r_2}\right)$$

$$= \frac{kQ}{r_1} - \frac{kQ}{r_2}$$

$$= kQ\left(\frac{1}{r_1}-\frac{1}{r_2}\right) = V \quad \textbf{Ans. (A)}$$

Sol. 38

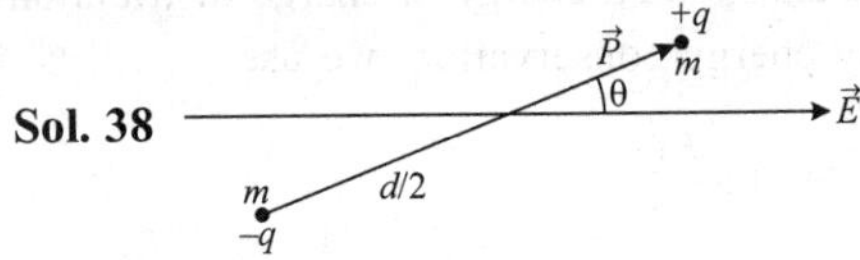

Moment of inertia of charges

$$I = 2\times m\left(\frac{d}{2}\right)^2 = \frac{md^2}{2}$$

Torque on dipole is

$$\tau = PE\sin\theta = (qd)E\sin\theta = I\alpha$$

$\Rightarrow$ $\alpha = -\frac{qEd}{I}\cdot\theta$ [As for small θ, sin θ ≃ θ]

– sign denotes θ to be in opposite direction to α

$\Rightarrow$ $\alpha = -\frac{2qE}{md}\cdot\theta$

Comparing with $\alpha = -\omega^2\theta$ we get $\omega = \sqrt{\frac{2qE}{md}}$ **Ans. (C)**

Sol. 39

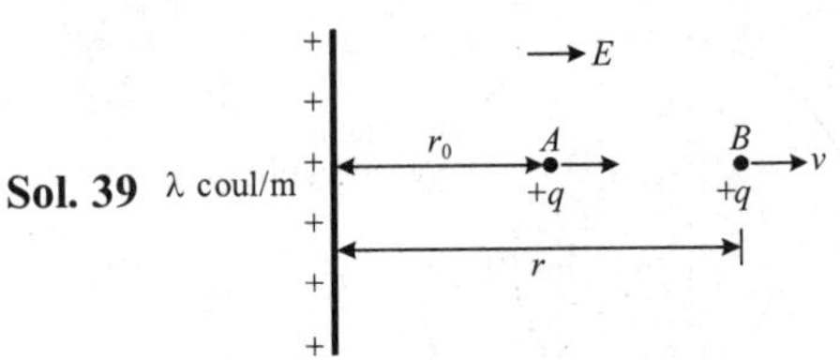

By work energy theorem, we use

$$\frac{1}{2}mv^2 = -q(V_A - V_B) \quad \ldots(1)$$

$\Rightarrow$ $V_A - V_B = \int_{r_0}^{r} E\cdot dr = \int_{r_0}^{r}\frac{\lambda}{2\pi\epsilon_0 r}dr = \frac{\lambda}{2\pi\epsilon_0}[\ln r]_{r_0}^{r}$

$\Rightarrow$ $V_A - V_B = \frac{\lambda}{2\pi\epsilon_0}\ln\left(\frac{r}{r_0}\right)$

From equation-(1), we use

$$\frac{1}{2}mv^2 = q\left[\frac{\lambda}{2\pi\epsilon_0}\ln\left(\frac{r}{r_0}\right)\right]$$

$\Rightarrow$ $v \propto \sqrt{\ln\left(\frac{r}{r_0}\right)}$ **Ans. (D)**

Sol. 40 Given electric field is

$$\vec{E} = (20x+10)\hat{i}$$

Potential difference is given as

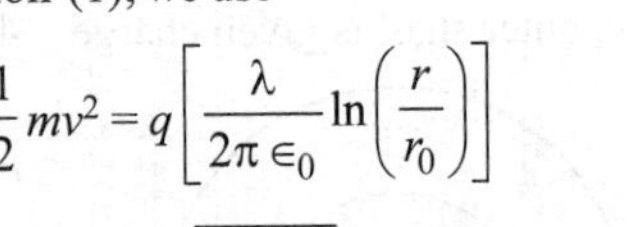

$$V_1 - V_2 = -\int_{-5}^{1}(20x+10)dx$$

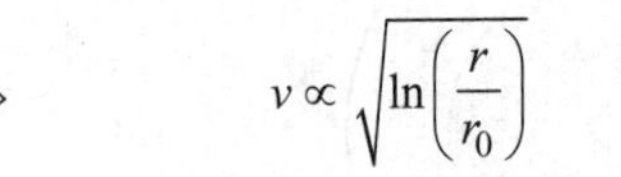

$\Rightarrow$ $V_1 - V_2 = -(10x^2+10x)_{-5}^{1}$

$\Rightarrow$ $V_1 - V_2 = 10(25-5-1-1)$

$\Rightarrow$ $V_1 - V_2 = 180\text{ V}$ **Ans. (C)**

Sol. 41

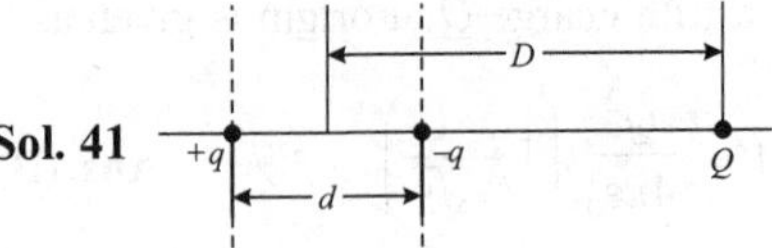

Total potential energy of this system is given as

$$U_{\text{total}} = U_{\text{self of dipole}} + U_{\text{interaction}}$$

$\Rightarrow$ $U_{\text{total}} = -\frac{kq^2}{d} - \left(\frac{kQ}{D^2}\right)qd$

$\Rightarrow$ $U_{\text{total}} = -k\left[\frac{q^2}{d} + \frac{qQd}{D^2}\right]$ **Ans. (D)**

Sol. 42 By conservation of energy, we have

$$U_i + K_i = U_f + K_f$$

$\Rightarrow$ $\frac{Kq^2}{\sqrt{16a^2+9a^2}} + \frac{1}{2}mv^2 = \frac{Kq^2}{3a}$

$\Rightarrow$ $\frac{1}{2}mv^2 = \frac{Kq^2}{a}\left(\frac{1}{3}-\frac{1}{5}\right) = \frac{2Kq^2}{15a}$

$\Rightarrow$ $v = \sqrt{\frac{4Kq^2}{15ma}}$ **Ans. (B)**

Sol. 43 By energy conservation, we have

$$\frac{Kq_1q_2}{r_1} - \frac{Kq_1q_2}{r_2} = \frac{1}{2}mv^2$$

$\Rightarrow$ $\frac{(9\times10^9)\times10^{-12}}{10^{-3}} - \frac{(9\times10^9)\times10^{-12}}{9\times10^{-3}} = \frac{1}{2}\times(4\times10^{-6})v^2$

$\Rightarrow$ $v^2 = 4\times10^6$

$\Rightarrow$ $v = 2\times10^3$ m/s **Ans. (A)**

Sol. 44 Due to inside dipole charge induced on inner surface will be non uniform as shown below

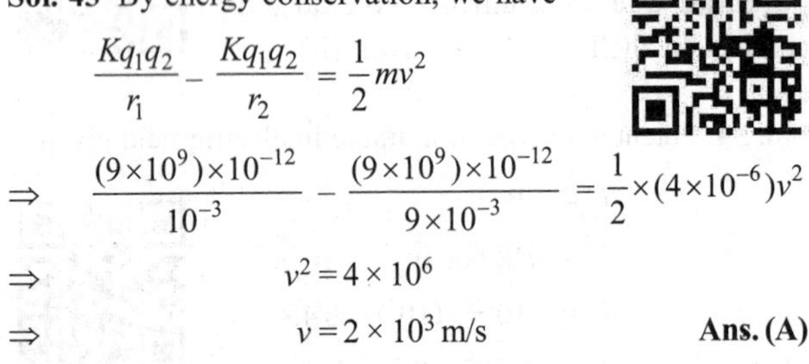

In this case as total charge of dipole is 0, so charge induced on outside surface is 0 due to it. Due to non uniform electric field of dipole, the charge induced on inner surface is non zero and non uniform so, for any observer outside the shell, the resultant electric field is due to Q which is uniformly distributed on outer surface only and it is given as

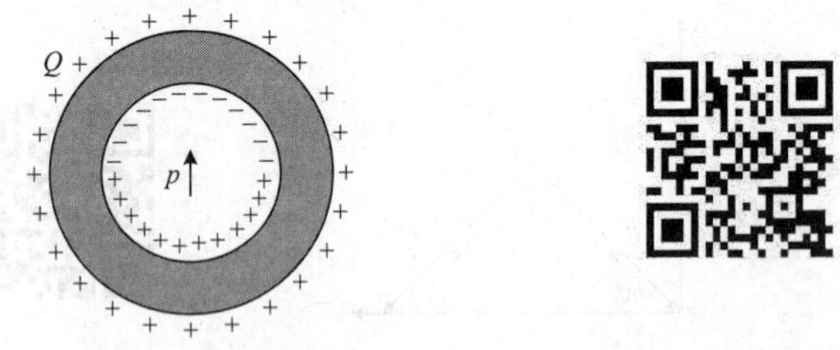

$$E = \frac{KQ}{r^2} = \frac{1}{4\pi\epsilon_0}\cdot\frac{Q}{r^2}$$ **Ans. (A)**

Sol. 45 At point M in figure as both charges of dipole are at same distance potential at M is

$$V = 0$$

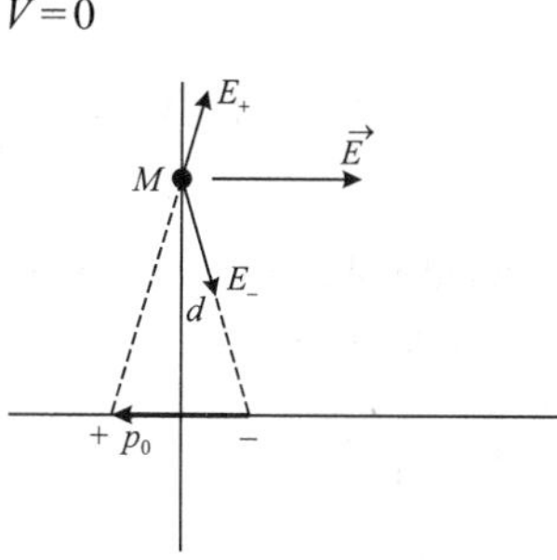

As M is an equatorial point of dipole electric field at M is given as

$$E = \frac{Kp}{d^3}$$

Vectorially, it is written as

$$\vec{E} = -\frac{1}{4\pi \in_0} \frac{p_0}{d^3}$$ **Ans. (D)**

Sol. 46 Electric fields on surfaces of spheres are

$$E_1 = \frac{KQ_1}{R_1^2}$$

$$E_2 = \frac{KQ_2}{R_2^2}$$

Given that ratio of electric fields is

$$\frac{E_1}{E_2} = \frac{R_1}{R_2}$$

$$\Rightarrow \quad \frac{\frac{KQ_1}{R_1^2}}{\frac{KQ_2}{R_2^2}} = \frac{R_1}{R_2}$$

$$\Rightarrow \quad \frac{Q_1}{Q_2} = \frac{R_1^3}{R_2^3}$$

$$\Rightarrow \quad \frac{V_1}{V_2} = \frac{KQ_1/R_1}{KQ_2/R_2} = \frac{R_1^2}{R_2^2}$$ **Ans. (D)**

Sol. 47 As $\vec{r}$ and $\vec{p}$ are perpendicular to each other so the point is on the equatorial plane and electric field at this point will be in direction opposite to the dipole moment so we use

$$\hat{E} = -\hat{p}$$

Thus $\vec{E}$ is parallel to $-\vec{p}$ **Ans. (C)**

Sol. 48 As total charge remain conserved, if final charges are q_1' and q_2', we use

$$q_1 + q_2 = q_1' + q_2' = 12\,\mu C - 3\,\mu C = 9\,\mu C$$

As spheres are connected their potential must be same

$$\Rightarrow \quad V_1 = V_2$$

$$\Rightarrow \quad \frac{Kq_1'}{\left(\frac{2R}{3}\right)} = \frac{Kq_2'}{\left(\frac{R}{3}\right)}$$

$$\Rightarrow \quad q_1' = 2q_2'$$

$$\Rightarrow \quad 2q_2' + q_2' = 9\,\mu C$$

$$\Rightarrow \quad q_2' = 3\,\mu C$$

and $$q_1' = 6\,\mu C$$ **Ans. (A)**

Sol. 49 Potential of inner sphere is given as

$$V_i = \frac{KQ_1}{R} + \frac{KQ_2}{4R} \quad \ldots(1)$$

and that of outer sphere is given as

$$V_O = \frac{KQ_1}{4R} + \frac{KQ_2}{4R} \quad \ldots(2)$$

As surface density of charges are equal, we use

$$\frac{Q_1}{4\pi R_1^2} = \frac{Q_2}{4\pi R_2^2}$$

$$\Rightarrow \quad \frac{Q_1}{R} = \frac{Q_2}{4R}$$

$$\Rightarrow \quad Q_2 = 4Q_1$$

From (1) & (2) we use

$$V_i - V_O = \frac{3}{4}\frac{KQ_1}{R} = \frac{3Q_1}{16\pi \in_0 R}$$ **Ans. (A)**

Sol. 50 For a charge, kinetic energy attained due to electric potential difference is given as

$$q\Delta V = \frac{1}{2}mv^2$$

$$\Rightarrow \quad v = \sqrt{\frac{2q\Delta V}{m}}$$

$$\Rightarrow \quad \frac{v_1}{v_2} = \sqrt{\frac{q \times m_0}{q \times m_1}} = 2$$ **Ans. (D)**

Sol. 51 Due to $-\frac{q}{2}$ there is no charge in energy of q so energy change is only due to $4q$, which is given as

$$\Delta U = 4kq^2\left[\frac{1}{\frac{3d}{2}} - \frac{1}{\frac{d}{2}}\right]$$

$$\Rightarrow \quad \Delta U = \frac{8kq^2}{d}\left[\frac{1}{3} - 1\right] = -\frac{16kq^2}{3d}$$

$$\Rightarrow \quad \Delta U = -\frac{4q^2}{3\pi \in_0 d} \text{ (Decreased)}$$ **Ans. (C)**

Sol. 52 Particle given kinetic energy due to work done by electric field so kinetic energy becomes zero again when

$$W=\int_0^{x_1} qEdx = \int_0^{x_1} qE_0(1-ax^2)dx = 0$$

$$\Rightarrow \quad qE_0\left[x_1 - \frac{ax_1^3}{3}\right] = 0$$

$$\Rightarrow \quad x_1 = \sqrt{\frac{3}{a}}$$ **Ans. (D)**

Sol. 53 By work energy theorem for mass m, we use

$$\frac{kQq}{R} + mgy = \frac{kQq}{R+y} + \frac{1}{2}mv^2$$

$$\Rightarrow \quad v^2 = 2gy + \frac{2kQqy}{mR(R+y)}$$ **Ans. (A)**

Sol. 54 By symmetry of charges $E_C = 0$ and sum of total charges is zero, so at centre $V_C = 0$. **Ans. (C)**

Sol. 55 Initial interaction energy of dipoles is given as

$$U = -\vec{p_1}\cdot\vec{E_2}$$

$$\Rightarrow \quad U = -p_1\left(\frac{2Kp_2}{a^3}\right)\cos 180^\circ$$

$$= \frac{2Kp_1p_2}{a^3} = \frac{1}{2\pi\epsilon_0}\cdot\frac{p^2}{a^3}$$

At infinite separation interaction energy becomes zero so by conservation of energy, we have

$$\frac{p^2}{2\pi\epsilon_0 a^3} = 2\times\frac{1}{2}\text{mv}^2$$

$$\Rightarrow \quad v = \sqrt{\frac{p^2}{2\pi\epsilon_0 ma^3}} = \frac{p}{a}\sqrt{\frac{1}{2\pi\epsilon_0 ma}}$$ **Ans. (C)**

Sol. 56 As 512 drops are joined to form a single drop, the final charge on this bigger drop is given as

$$Q = 512q$$

As total liquid volume is same and if R is the final radius of the bigger drop, we have

$$512\times\frac{4}{3}\pi r^3 = \frac{4}{3}\pi R^3$$

$$\Rightarrow \quad 2^9 r^3 = R^3$$

$$\Rightarrow \quad R = 8r$$

The final potential of the bigger drop thus formed is given as

$$V = \frac{KQ}{R} = \frac{K(512q)}{8r}$$

Given that potential of each small drop is 2 V so we have

$$2 = \frac{Kq}{R}$$

$$\Rightarrow \quad V = \frac{512}{8}\times 2 = 128\text{ V}$$ **Ans. (128)**

Sol. 57 As total liquid volume is same and if R is the final radius of the bigger drop, we have

$$27\times\frac{4}{3}\pi r^3 = \frac{4}{3}\pi R^3$$

$$\Rightarrow \quad 3^3 r^3 = R^3$$

$$\Rightarrow \quad R = 3r$$

Potential energy of smaller drop is given as

$$U_1 = \frac{Kq^2}{2r}$$

Potential energy of bigger drop is given as

$$U = \frac{KQ^2}{2R}$$

$$U = \frac{K(27q)^2}{2R} = \frac{K(27)(27)q^2}{2(3r)}$$

$$U = \frac{(27)(27)}{3}\left(\frac{Kq^2}{2r}\right) = 243\,U_1$$ **Ans. (243)**

Sol. 58 The electric field due to $\vec{P_1}$ at an axial point, is given as

$$E_a = \frac{2KP_1}{r^3}$$

$$\Rightarrow \quad E\cos 37^\circ = \frac{2KP_1}{r^3} \quad \ldots(1)$$

Electric field due to $\vec{P_2}$ at equatorial point which is the axis point of first dipole is given as

$$E_e = \frac{KP_2}{r^3}$$

$$\Rightarrow \quad E\sin 37^\circ = \frac{KP_2}{r^3} \quad \ldots(3)$$

$$\Rightarrow \quad \frac{\frac{2KP_1}{r^3}}{\frac{KP_2}{r^3}} = \frac{E\cos 37^\circ}{E\sin 37^\circ}$$

$$\Rightarrow \quad \frac{2P_1}{P_2} = \frac{4}{3}$$

$$\Rightarrow \quad \frac{P_1}{P_2} = \frac{2}{3}$$ **Ans. (B)**

Sol. 59 Potentials at the center of the two rings

$$V_1 = \frac{KQ}{a} + \frac{-KQ}{\sqrt{s^2+a^2}}$$

and

$$V_2 = \frac{KQ}{\sqrt{s^2+a^2}} + \frac{-KQ}{a}$$

$$\Rightarrow \quad V_1 - V_2 = \frac{Q}{2\pi\varepsilon_0}\left[\frac{1}{a} - \frac{1}{\sqrt{s^2+a^2}}\right]$$

Ans. (B)

Sol. 60 Charge of bigger drop will be $27q$ if q is the charge of one drop. If R is the radius of bigger drop, we use

$$n\frac{4}{3}\pi r^3 = \frac{4}{3}\pi R^3$$

$$\Rightarrow \quad R = n^{1/3} r = 3r$$

If potential of each drop is V then we have

$$V = \frac{Kq}{r}$$

Potential of bigger drop is calculated as

$$V' = \frac{K(nq)}{R} = \frac{K(nq)}{n^{1/3} r} = n^{2/3}\text{V}$$

$$\Rightarrow \quad V' = 9\,\text{V} = 9 \times 22 = 198\,\text{V}$$

Ans. (198)

Sol. 70 A : In nonpolar molecules, centre of positive charge coincides with centre of negative charge, thus net dipole moment of molecule is zero hence (A) is correct.

R : When non polar material is placed in external field, centre of charges of negative and positive poles shift and gives non zero dipole moment in presence of field hence (R) is NOT correct.

Ans. (C)

Sol. 71 Electric field in space is given as

$$E_x = -\frac{\partial V}{\partial x} = -6x$$

At point (1, 0, 3), it is calculated as

$$\vec{E} = -6\,\text{V/m}\,\hat{i}$$

Ans. (D)

Sol. 72 Capacitance of and isolated Conducting sphere is given as

$$C = 4\pi\varepsilon_0 R_1$$

By enclosing it inside another sphere of radius R_2 and outer shell is connected to earth, new capacitance becomes

$$C' = \frac{4\pi\varepsilon_0 R_1 R_2}{(R_2 - R_1)}$$

Given that

$$C' = nC$$

$$\Rightarrow \quad \frac{4\pi\varepsilon_0 R_1 R_2}{(R_2 - R_1)} = n \times 4\pi\varepsilon_0 R_1$$

$$\Rightarrow \quad \frac{R_2}{(R_2 - R_1)} = n$$

$$\Rightarrow \quad \frac{\frac{R_2}{R_1}}{\left(\frac{R_2}{R_1} - 1\right)} = n$$

$$\Rightarrow \quad \frac{R_2}{R_1} = n\frac{R_2}{R_1} - n$$

$$\Rightarrow \quad \frac{R_2}{R_1} = \frac{n}{(n-1)}$$

Ans. (A)

Sol. 73 As the spheres are connected with a conducting wire, their potentials will be same so we use

$$V_A = V_B$$

$$\Rightarrow \quad \frac{KQ_A}{R_A} = \frac{KQ_B}{R_B}$$

$$\Rightarrow \quad \frac{Q_A}{Q_B} = \frac{R_A}{R_B} = \frac{1}{2}$$

Electric fields at the surfaces of spheres are given as

$$E_A = \frac{KQ_A}{R_A}$$

and

$$E_B = \frac{KQ_B}{R_B^2}$$

$$\Rightarrow \quad \frac{E_A}{E_B} = \frac{Q_A}{Q_B} \times \frac{R_B^2}{R_A^2} = \frac{R_B}{R_A} = \frac{2}{1}$$

Ans. (B)

Sol. 74 Maximum torque on dipole is calculated as

$$|\tau|_{\max} = PE$$

$$\Rightarrow \quad \frac{\tau_1}{\tau_2} = \frac{P_1 E_1}{P_2 E_2} = \frac{1\cdot 2 \times 10^{-30} \times 5 \times 10^4}{2\cdot 4 \times 10^{-30} \times 15 \times 10^4} = \frac{1}{6}$$

$$\Rightarrow \quad x = 6$$

Ans. (6)

Sol. 75 Statement-I, As inside the body of conductor electric field due to static charges is zero so it acts as equipotential surface hence this statement is true.

Statement-II : As conductor is equipotential the tangential component of electric field on its surface will be zero. Thus electric field will be perpendicular to its surface hence this statement is true.

Ans. (A)

Sol. 76 When particle is accelerated at a potential difference, it gains kinetic energy qV so its momentum is given as

$$p = \sqrt{2mE} = \sqrt{2mqV}$$

$$\Rightarrow \quad \frac{p_\alpha}{p_p} = \sqrt{\frac{m_\alpha q_\alpha}{m_p q_p}} = \sqrt{\frac{4}{1} \times \frac{2}{1}}$$

$$\Rightarrow \quad \frac{p_\alpha}{p_p} = \frac{2\sqrt{2}}{1}$$

Ans. (B)

Sol. 77 Electric potential at centre is calculated as

$$V=\frac{(\lambda\cdot\pi R_2)}{4\pi\varepsilon_0 R_2}+\frac{(\lambda\cdot\pi R_1)}{4\pi\varepsilon_0 R_1}$$

$$\Rightarrow \quad V=\frac{\lambda}{2\varepsilon_0} \qquad \textbf{Ans. (B)}$$

Sol. 78 $$0.5e=\frac{1}{2}mv_x^2$$

$$\Rightarrow \quad v_x=\sqrt{\frac{e}{m}}$$

Along $x\ L=v_x t=\sqrt{\frac{e}{m}}t$

Along $y\ v_y=\frac{eE}{m}t$

dividing $\frac{v_y}{L}=E\sqrt{\frac{e}{m}}=Ev_x$

$$\Rightarrow \quad \tan\theta=\frac{v_y}{v_x}=E\times L=10\times 0.1=1$$

$$\theta=45° \qquad \textbf{Ans. (45)}$$

Sol. 79 Figure below shows the situation described in the question.

$S(0, 0, 0)$ $P(1, 2, 0)$ $E=30\hat{i}$

Thus work done is calculated as

$$W_E=q\vec{E}\cdot\vec{S}$$

$$\Rightarrow \quad W_E=2\times 10^{-2}[30\hat{i}\cdot(-\hat{i})]$$

$$\Rightarrow \quad W_E=2\times 10^{-2}(-30)$$

$$\Rightarrow \quad W_E=-60\times 10^{-2}$$

$$\Rightarrow \quad W_E=-\frac{60}{100}=-0.6\text{ J}=-600\text{ mJ} \qquad \textbf{Ans. (C)}$$

Sol. 80 Electric field is calculated as

$$E=-\frac{dV}{dr}=-4ar$$

As here field is directly proportional to r it indicates that the charge density is uniform inside spherical region for which field can be given as

$$E=\frac{\rho r}{3\varepsilon_0}=-4ar$$

Thus inside uniformly charged solid sphere the charge density is given as

$$\rho=-12a\varepsilon_0$$

$$\Rightarrow \quad \lambda=12 \qquad \textbf{Ans. (12)}$$

Sol. 81 Figure below shows the situation described in question for the two spheres.

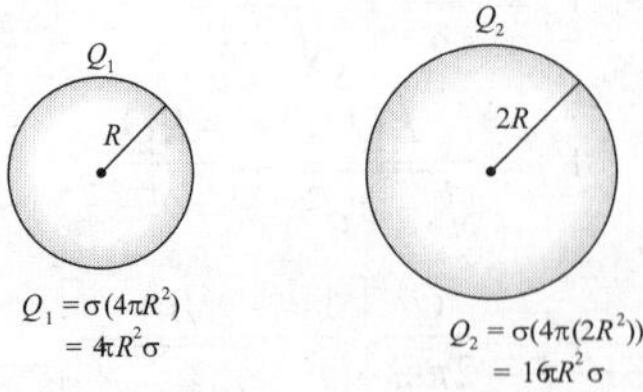

When the two spheres are connect via a conducting wire, the charge flows between them until their potential becomes equal so we have

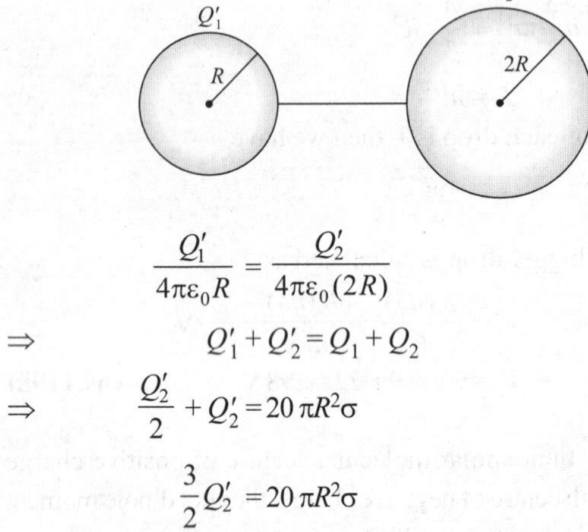

$$\frac{Q_1'}{4\pi\varepsilon_0 R}=\frac{Q_2'}{4\pi\varepsilon_0(2R)}$$

$$\Rightarrow \quad Q_1'+Q_2'=Q_1+Q_2$$

$$\Rightarrow \quad \frac{Q_2'}{2}+Q_2'=20\pi R^2\sigma$$

$$\frac{3}{2}Q_2'=20\pi R^2\sigma$$

New charge density on the bigger sphere after connection is given as

$$\sigma'=\frac{Q_2'}{4\pi(2R)^2}=\frac{2}{3}\cdot\frac{20\pi R^2\sigma}{16\pi R^2}$$

$$\Rightarrow \quad \frac{\sigma'}{\sigma}=\frac{5}{6} \qquad \textbf{Ans. (D)}$$

Sol. 82 Electric potential inside a charged spherical conductor remain constant and outside it decreases with inverse square of the distance from the center hence option (C) is correct.

Ans. (C)

Sol. 83 Net potential due to group of charges is calculated as

$$V=\frac{\Sigma KQ_i}{r_i}$$

Here, Q_i & r_i are positive so we have

$$V>0 \qquad \textbf{Ans. (A)}$$

Sol. 84 Electric potential inside uniformly charged thin spherical shell is constant and it is equal to that on its surface, which is given as

$$V_{\text{inside}}=\frac{KQ}{R}$$

Electric potential outside uniformly charged spherical shell varies inversely with the distance, which is given as

$$V_{\text{outside}} = \frac{KQ}{r}$$

Hence option (A) is correct. **Ans. (A)**

Sol. 85 If the dipole is released at some angle in electric field, it will oscillate about centre of mass. For small angle θ, the torque on dipole is given as

$$\tau = PE \cdot \theta$$

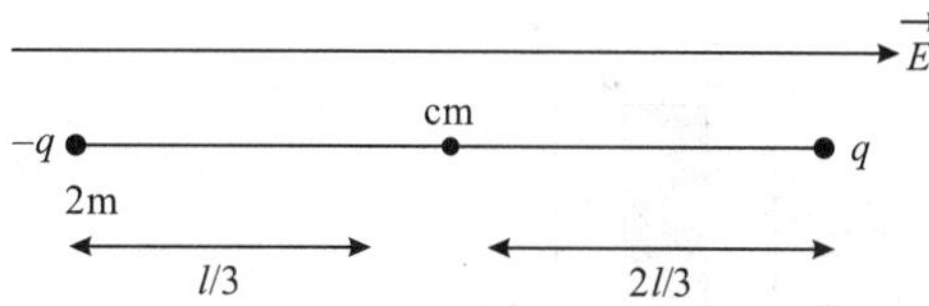

If angular acceleration of dipole is α, we have $\tau = I\alpha$

$$\Rightarrow \qquad \tau = \left[2m\frac{l^2}{9} + m\frac{4l^2}{9}\right]\alpha = -qlE \cdot \theta$$

$$\Rightarrow \qquad \frac{2ml^2}{3}\alpha = -qlE \cdot \theta$$

$$\Rightarrow \qquad \alpha = -\frac{3qE}{2ml}\theta$$

As $\alpha = -\omega^2\theta$ the angular frequency of oscillations of dipole is given as

$$\omega = \sqrt{\frac{3qE}{2ml}}$$ **Ans. (BONUS)**

Sol. 86 The work done in rotating the electric dipole is calculated as

$$W = \Delta U = U_f - U_i$$

Potential Energy of dipole in electric field is given as

$$U = -PE\cos\theta$$

When dipole is rotated by 180°, we use

$$\theta_f = 180^\circ\,;\, \theta_i = 0^\circ$$

$$\Rightarrow \qquad W = (-pE\cos(180^\circ)) - (-pE\cos(0^\circ))$$

$$\Rightarrow \qquad W = pE + pE = 2pE$$

$$\Rightarrow \qquad W = 2 \times 6 \times 10^{-6} \times 1.5 \times 10^3$$

$$\Rightarrow \qquad W = 18\text{ mJ}$$ **Ans. (18)**

Sol. 87 Potential at point P due to point charge is given as

$$V_P = \frac{KQ}{r}$$

$$\Rightarrow \qquad 50 = \frac{9\times10^9 \times 5\times10^{-9}}{r}$$

$$\Rightarrow \qquad r = \frac{45}{50} = \frac{9}{10} = 0.9\text{ m} = 90\text{ cm}$$ **Ans. (C)**

Sol. 88

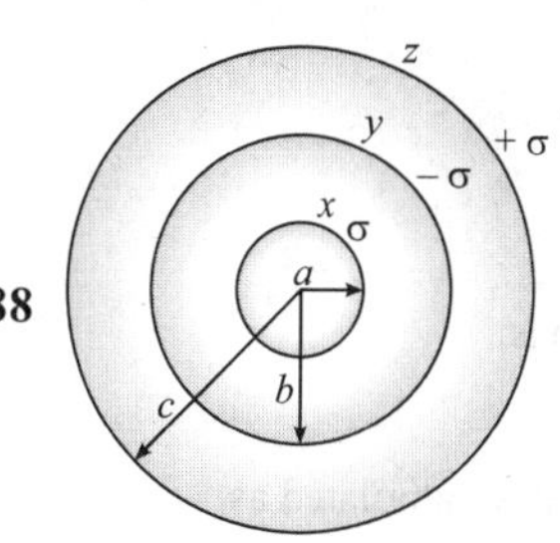

Surface charge density

$$\sigma = \frac{Q}{A}$$

So charges at shell surfaces x, y, and z are given as

$$q_x = \sigma(4\pi a^2)$$

$$\Rightarrow \qquad q_y = -\sigma(4\pi b^2)$$

$$\Rightarrow \qquad q_z = \sigma(4\pi c^2)$$

Given that $V_x = V_z$ which gives

$$\frac{q_x}{4\pi\varepsilon_0 a} + \frac{q_y}{4\pi\varepsilon_0 b} + \frac{q_z}{4\pi\varepsilon_0 c} = \frac{q_x}{4\pi\varepsilon_0 c} + \frac{q_y}{4\pi\varepsilon_0 c} + \frac{q_z}{4\pi\varepsilon_0 c}$$

$$\Rightarrow \qquad \frac{\sigma 4\pi a^2}{a} - \frac{\sigma 4\pi b^2}{b} + \frac{\sigma 4\pi c^2}{c} = \frac{4\pi\sigma[a^2 - b^2 + c^2]}{c}$$

$$\Rightarrow \qquad c(a - b + c) = a^2 - b^2 + c^2$$

$$\Rightarrow \qquad c(a - b) = a^2 - b^2$$

$$\Rightarrow \qquad c = a + b$$

$$\Rightarrow \qquad c = 5\text{ cm}$$ **Ans. (5)**

Sol. 89 Gravitation Potential inside surface of Earth

$$V = \frac{GM}{2R^3}(3R^2 - r^2) \text{ at } r = R$$

$$\Rightarrow \qquad V = \left(\frac{GM}{R}\right)$$

At $r = 0$, we have

$$\Rightarrow \qquad V_0 = \frac{3GM}{2R}$$

$$\Rightarrow \qquad V_0 = \left(\frac{3V}{2}\right)$$ **Ans. (A)**

Sol. 90 For a dipole as net charge is zero because it is made up of two equal and opposite charges, flux through the surface enclosing a dipole will be zero hence both statement-1 and Statement-2 are correct. **Ans. (A)**

Sol. 91 Potential of a drop having charge q is given as

$$V = \frac{Kq}{r} \qquad \ldots(1)$$

Now for combination of 64 drop the new drop radius is calculated as

$$64 \times \frac{4}{3}\pi r^3 = \frac{4}{3}\pi R^3$$

$$\Rightarrow \qquad R = 4r$$

Final charge on bigger drop is

$$Q = 64\,q$$

Potential of bigger drop is now calculated as

$$V_{new} = \frac{KQ}{R} = \frac{K64q}{4r} = 16\frac{Kq}{r}$$

$$\Rightarrow \qquad V_{new} = 16 \times 10\,\text{mV} = 160\,\text{mV}$$ **Ans. (160)**

Sol. 92 Dipole moment of electric dipole is given as

$$|\vec{P}| = qd$$

$$\Rightarrow \qquad |\vec{P}| = 0.01 \times 0.4 \times 10^{-3}$$

$$\Rightarrow \qquad |\vec{P}| = 4 \times 10^{-6}\,\text{Cm}$$

When dipole is placed in electric field at angle of 30° with $\vec{E}$, then torque on dipole is calculated as

$$|\vec{\tau}| = PE\sin\theta$$

$$\Rightarrow \qquad |\vec{\tau}| = 4 \times 10^{-6} \times 10 \times 10^{-5} \times 30$$

$$\Rightarrow \qquad |\vec{\tau}| = 4 \times 10^{-6-5+1} \times \frac{1}{2}$$

$$\Rightarrow \qquad |\vec{\tau}| = 2 \times 10^{-10}\,\text{Nm}$$ **Ans. (B)**

Sol. 93 Electric field due to a dipole at point on its axial point is given as

$$E = \frac{2kp}{r^3}$$

$$\Rightarrow \qquad E \propto \frac{1}{r^3}$$ **Ans. (C)**

16.2 Capacitance

Sol. 1 Capacitance of a spherical conductor is

$$C = 4\pi\epsilon_0 R$$

$$\Rightarrow \qquad C = \frac{1}{9\times10^9} \times 1$$

$$= 1.1 \times 10^{-10}\,\text{F}$$ **Ans. (A)**

Sol. 2 For n identical capacitors (C) parallel combination equivalent capacitance is

$$C_{eq} = C + C + C + \ldots n \text{ times} = nC$$

Across a voltage V the energy stored is

$$U = \frac{1}{2}C_{eq}V^2 = \frac{1}{2}nCV^2$$ **Ans. (B)**

Sol. 3 Work done in charging a capacitor of capacitance C with a charge q is

$$W = \frac{q^2}{2C}$$

$$= \frac{1}{2}\left(\frac{8\times10^{-18}}{100\times10^{-6}}\right)^2$$

$$= 32 \times 10^{-32}\,\text{J}$$

Electrostatic energy stored in capacitor. **Ans. (D)**

Sol. 4

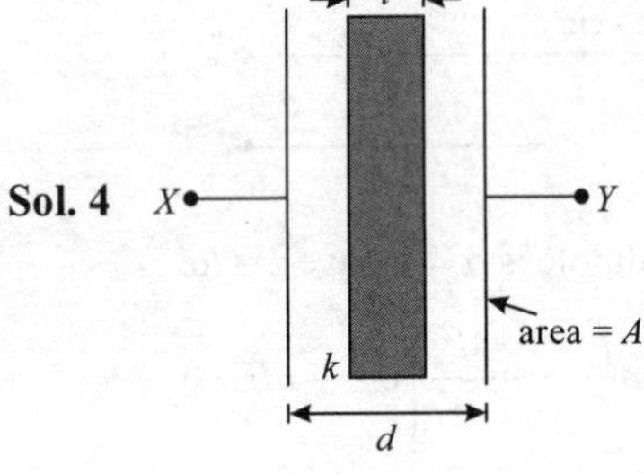

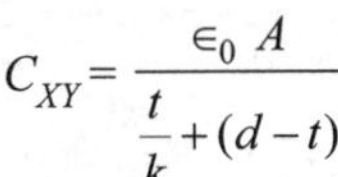

$$C_{XY} = \frac{\epsilon_0 A}{\frac{t}{k} + (d-t)}$$

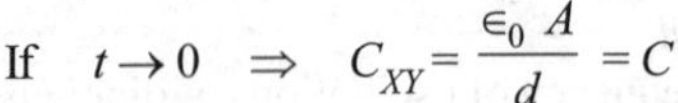

If $t \to 0 \Rightarrow C_{XY} = \frac{\epsilon_0 A}{d} = C$ **Ans. (B)**

Sol. 5

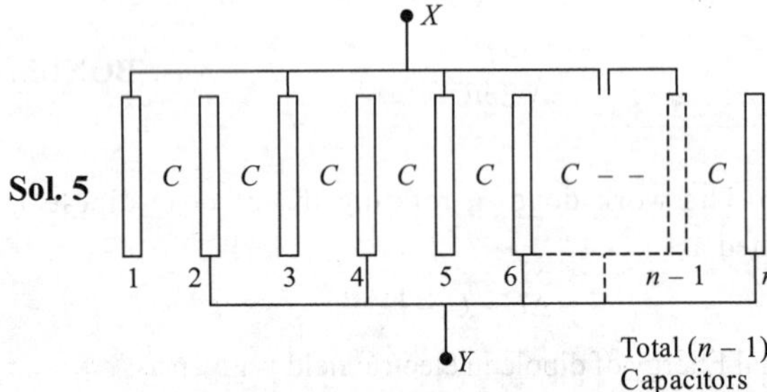

In above system of plates one plate of each capacitor is connected to Y and its other plate is connected to X. Thus all $(n-1)$ capacitors formed are in parallel combination.

$$\Rightarrow \qquad C_{XY} = C + C + \ldots (n-1) \text{ times}$$

$$= (n-1)C$$

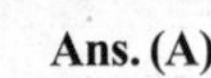

Ans. (A)

Sol. 6

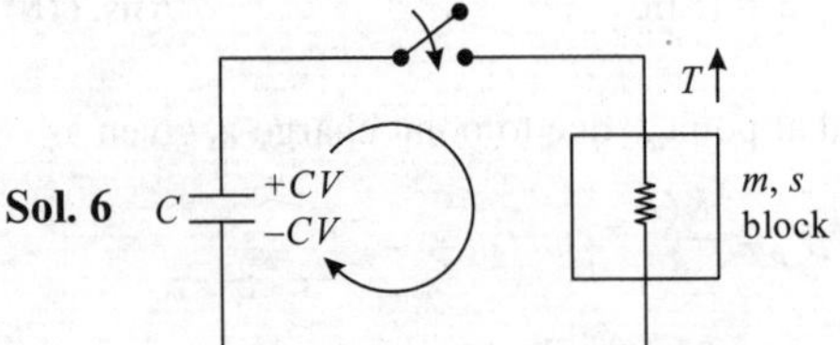

On discharging the capacitor, heat supplied to block is

$$H = U_{cap} = \frac{1}{2}CV^2$$

If temperature of block is raised by ΔT

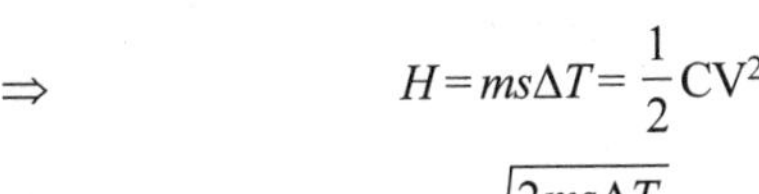

$$\Rightarrow \quad H = ms\Delta T = \frac{1}{2}CV^2$$

$$\Rightarrow \quad V = \sqrt{\frac{2ms\Delta T}{C}}$$

Ans. (D)

Sol. 7

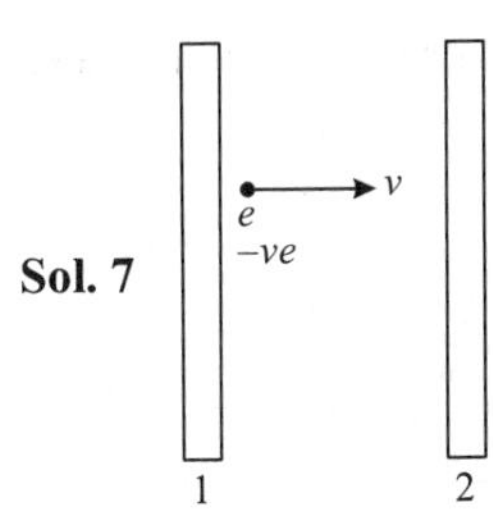

Work done in travelling from plate 1 to 2 on electron by the EF of plates is given as

$$W = e \times (V_2 - V_2) = \Delta K = \frac{1}{2}mv^2$$

$$\Rightarrow \quad W = (-1.6 \times 10^{-19}) \times (-20)$$

$$\Rightarrow \quad W = \frac{1}{2} \times 9.1 \times 10^{-31} \times v^2$$

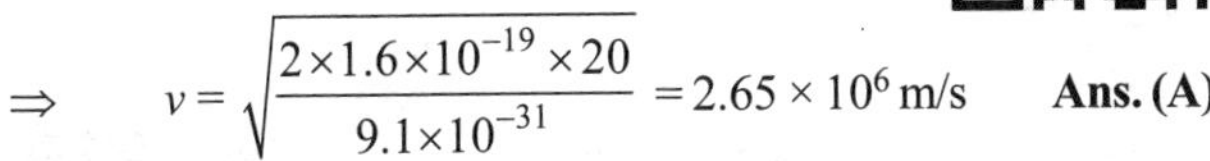

$$\Rightarrow \quad v = \sqrt{\frac{2 \times 1.6 \times 10^{-19} \times 20}{9.1 \times 10^{-31}}} = 2.65 \times 10^6 \text{ m/s}$$

Ans. (A)

Sol. 8

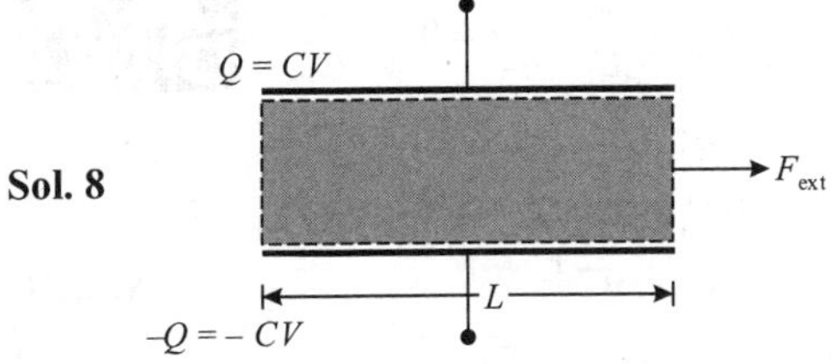

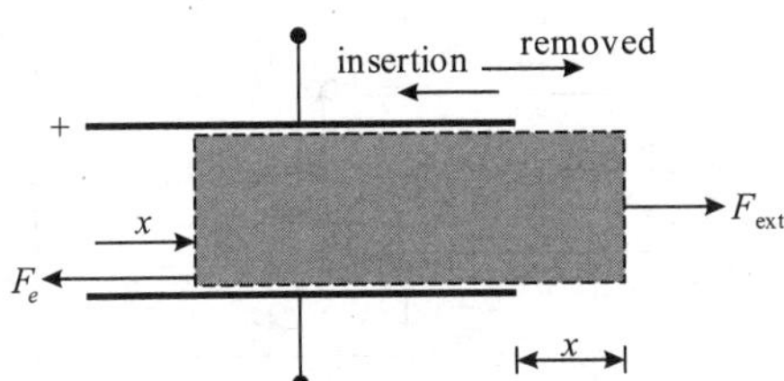

Work done in removing the dielectric slab slowly from capacitor is

$$\text{removal} \rightarrow W_1 = \int_0^l F_{ext} \cdot dx$$

$$\text{insertion} \rightarrow W_2 = \int_0^l F_{ext} \cdot dx$$

For slow insertion or removal of slab $F_{ext} = F_e$

Net work done is $W_1 + W_2 = 0$ **Ans. (D)**

Sol. 9

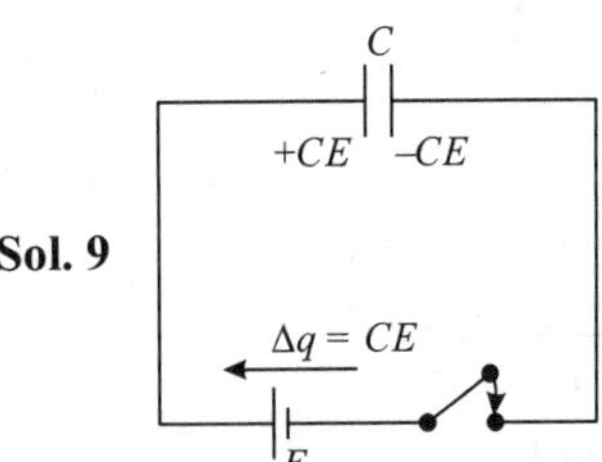

After capacitor is fully charged (when potential difference across plates is equal to that of battery EMF) charge on its plates is equal to

$$q_f = CE \text{ and } q_i = 0$$

Energy stored in capacitor is

$$U = \frac{1}{2}CE^2$$

Work done by battery $W = \Delta qE = CE^2$

$$\Rightarrow \quad \frac{U}{W} = \frac{\frac{1}{2}CE^2}{CE^2} = \frac{1}{2}$$

Ans. (D)

Sol. 10

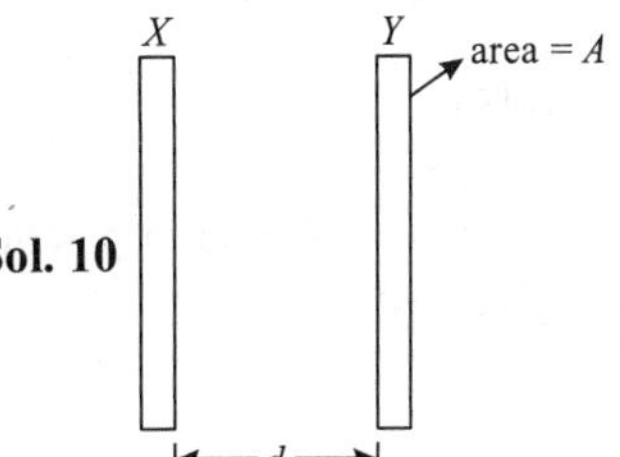

$$C = \frac{\epsilon_0 A}{d} = 9pF \Rightarrow \epsilon_0 A = 9d$$

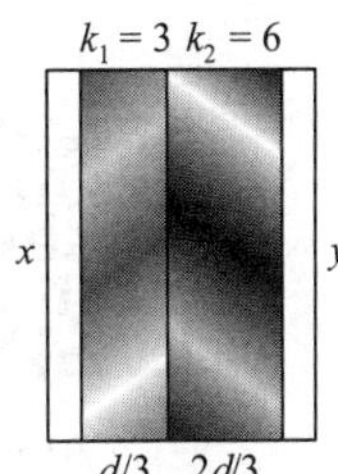

Capacitance of capacitor with multiple dielectric field slabs parallel to plates is given as

$$C' = \frac{\epsilon_0 A}{\frac{t_1}{k_1} + \frac{t_2}{k_2}}$$

$$\Rightarrow \quad C' = \frac{9d}{\frac{d/3}{3} + \frac{2d/3}{6}}$$

$$\Rightarrow \quad C' = \frac{9}{\frac{1}{9} + \frac{1}{9}} = \frac{81}{2} = 40.5 \text{ pF}$$

Ans. (C)

Sol. 11 (a) $V_L = \frac{2Q}{3C}$ $\quad V_R = \frac{Q}{C}$

As $V_R > V_L$

then on closing S flows from R to L

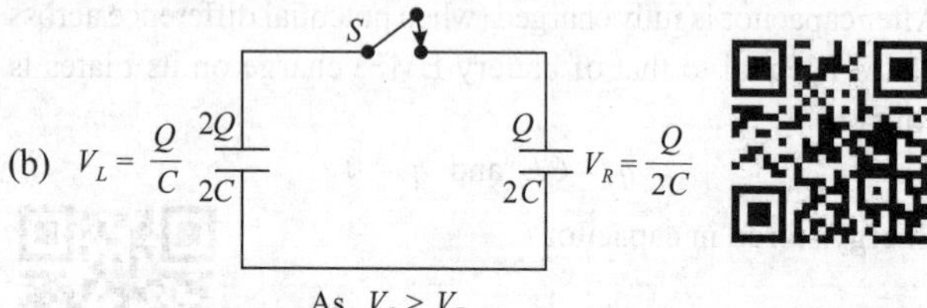

(b) $V_L = \frac{Q}{C}$ $\quad V_R = \frac{Q}{2C}$

As $V_L > V_R$

then on closing S charge flows from L to R. **Ans. (C)**

Sol. 12 Case-I : n_1 capacitors, each C_1 when connected in series then its equivalent capacitance is given as

$$C_{eq_1} = \frac{C_1}{n_1}$$

$\Rightarrow$ Energy stored with potential difference = 4V is given as

$$U_1 = \frac{1}{2}\left(\frac{C_1}{n_1}\right)(4V)^2 \quad \ldots(1)$$

Case-II : n_1 capacitors, each C_2 when connected in parallel then its equation capacitance is given as

$$C_{eq_2} = n_2 C_2$$

$\Rightarrow$ Energy stored with p.d. = V is given as

$$U_2 = \frac{1}{2}(n_2 C_2)V^2 \quad \ldots(2)$$

Given that $U_1 = U_2$

$$\Rightarrow \quad \frac{1}{2}\left(\frac{C_1}{n_1}\right)(4V)^2 = \frac{1}{2}(n_2 C_2)V^2$$

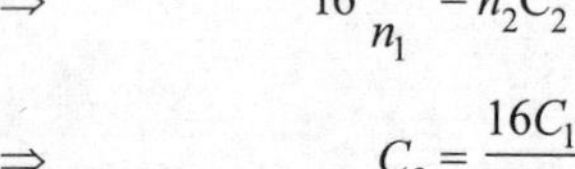

$$\Rightarrow \quad 16\frac{C_1}{n_1} = n_2 C_2$$

$$\Rightarrow \quad C_2 = \frac{16C_1}{n_1 n_2}$$

Ans. (D)

Sol. 13 Statement-1 : It is true because the radius of a sphere required for making 1F cap is gives as

$$C = 4\pi\epsilon_0 R$$

$$1 = 4\pi\epsilon_0 R$$

$$\Rightarrow \quad R = \frac{1}{4\pi\epsilon_0} = 9 \times 10^9 \text{ m}$$

Statement-2 : As $6.4 \times 10^6 \text{ m} < 9 \times 10^9 \text{ m}$

$\Rightarrow$ It is not possible with earth

$$C_{earth} < 1\text{F}$$

Ans. (D)

Sol. 14

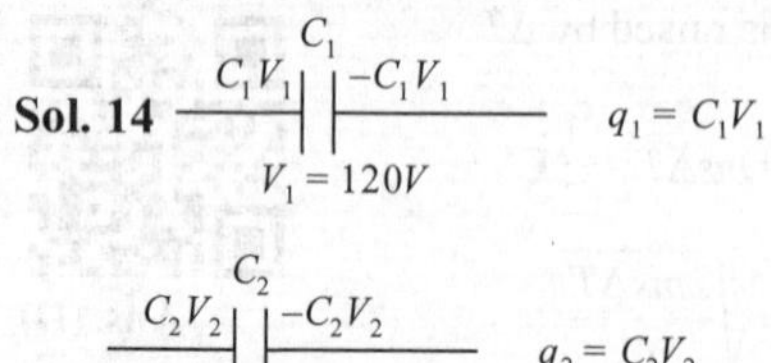

If the capacitors are joined in parallel with same polarity plates together then

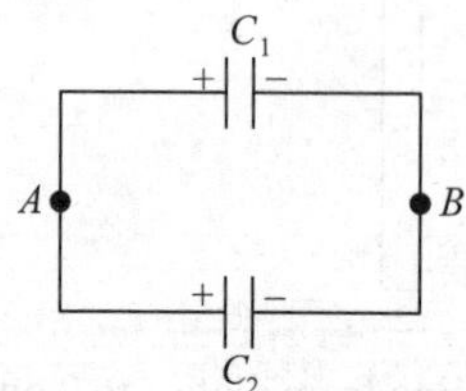

$$V_{AB} = \frac{q_1 + q_2}{C_1 + C_2} \neq 0$$

If the capacitors are joined in parallel with opposite polarity plates together then

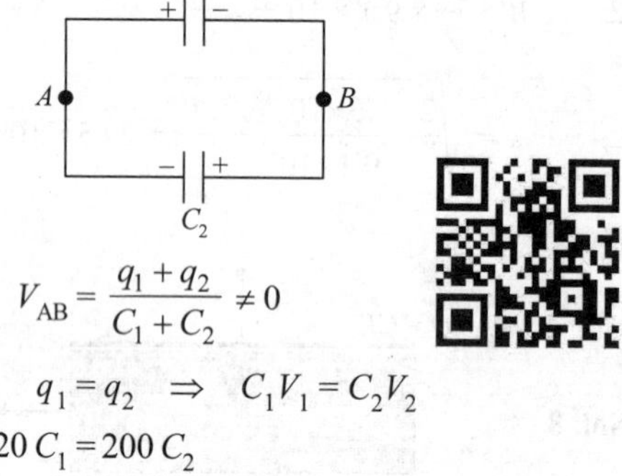

$$V_{AB} = \frac{q_1 + q_2}{C_1 + C_2} \neq 0$$

$$\Rightarrow \quad q_1 = q_2 \Rightarrow C_1V_1 = C_2V_2$$

$$\Rightarrow \quad 120\,C_1 = 200\,C_2$$

$$\Rightarrow \quad 3C_1 = 5C_2$$

Ans. (B)

Sol. 15

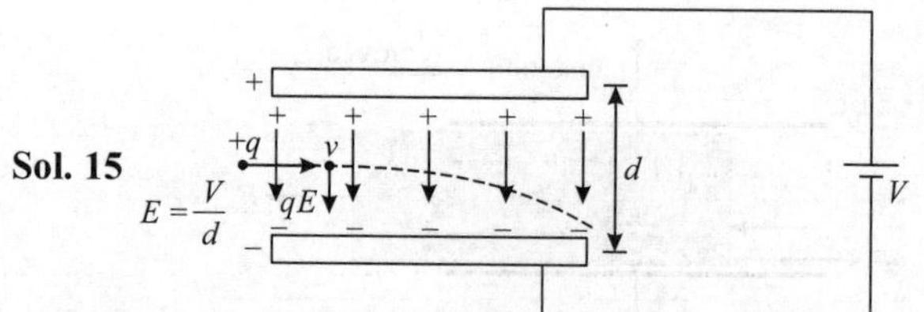

Acceleration of charged particle due to electric force is

$$a = \frac{F}{m} = \frac{qE}{m} \rightarrow \text{constant}$$

at constant acceleration, particle follows projectile motion with

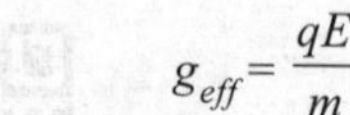

$$g_{eff} = \frac{qE}{m}$$

$\Rightarrow$ path $\rightarrow$ parabolic trajectory **Ans. (C)**

Sol. 16 When a capacitor is charged to a potential difference V volt then we have

$$q = CV$$

If V starts varying then q changes and current in connecting wires is given as

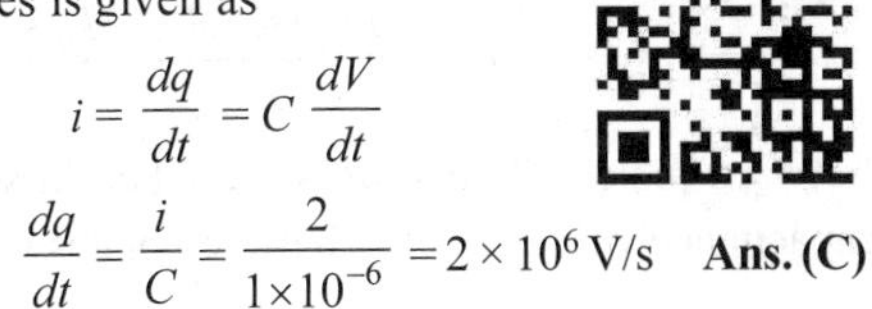

$$i = \frac{dq}{dt} = C\frac{dV}{dt}$$

$$\Rightarrow \qquad \frac{dq}{dt} = \frac{i}{C} = \frac{2}{1\times10^{-6}} = 2\times10^{6}\ \text{V/s} \qquad \textbf{Ans. (C)}$$

Sol. 17

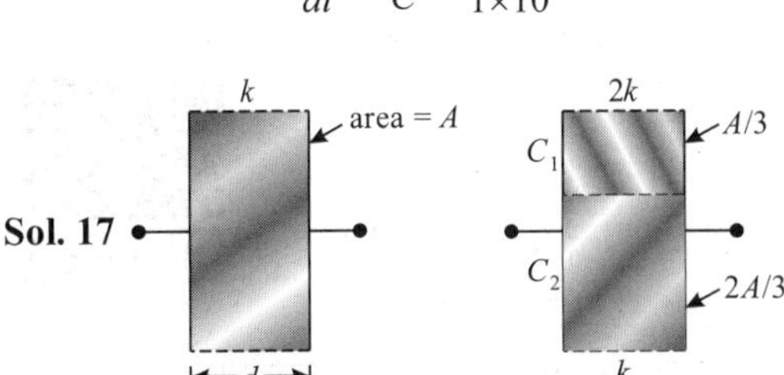

$$C_0 = \frac{K\in_0 A}{d} \qquad \ldots(1)$$

$$C = C_1 + C_2$$

$$\Rightarrow \qquad C = \frac{(2k)\in_0 A/3}{d} + \frac{k\in_0 (2A/3)}{d}$$

$$= \frac{4k\in_0 A}{3d} \qquad \ldots(2)$$

From (1) and (2) $\Rightarrow$ $\dfrac{C}{C_0} = \dfrac{4}{3}$ **Ans. (B)**

Sol. 18

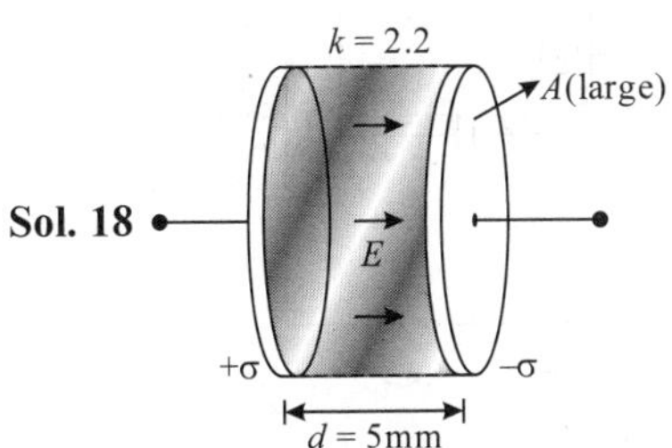

Electric field between plates of a capacitor inside dielectric is given as

$$E = \frac{\sigma}{k\in_0} = 3\times10^4\ \text{V/m}$$

$$\Rightarrow \qquad \sigma = 3\times10^4\times k\in_0$$

$$\Rightarrow \qquad \sigma = 3\times10^4\times2.2\times8.89\times10^{-12}$$

$$\Rightarrow \qquad \sigma = 5.87\times10^{-7}\ \text{C/m}^2 \qquad \textbf{Ans. (A)}$$

Sol. 19

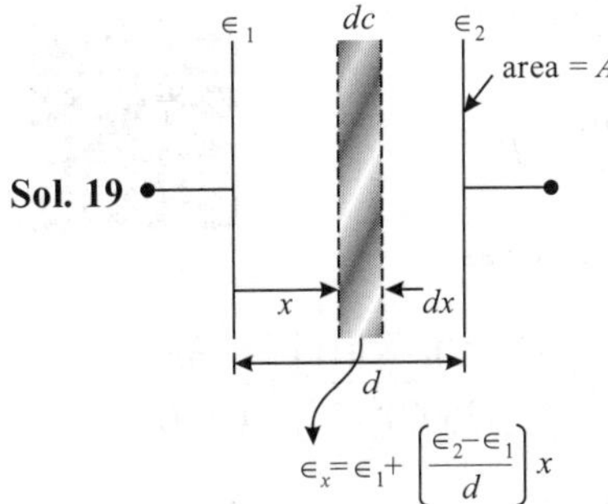

Using $\in_x = mx + C$

at $x = 0, \in = \in_1$

at $x = d, \in = \in_2$

The capacitance of elemental capacitor as shown in

$$dC = \frac{\in_x \in_0 A}{dx}$$

All such dC's are in series combination so we use

$$\frac{1}{C} = \int \frac{1}{dC}$$

$$\Rightarrow \qquad \frac{1}{C} = \int_0^d \frac{dx}{\in_0 A\left(\in_1 + \frac{\in_2 - \in_1}{d}\cdot x\right)}$$

$$\Rightarrow \quad \frac{1}{C} = \frac{1}{\in_0 A}\frac{d}{(\in_2 - \in_1)}\left(\ln\left[\in_1 + \frac{\in_2 - \in_1}{d}\cdot x\right]\right)_0^d$$

$$\Rightarrow \quad \frac{1}{C} = \frac{d}{\in_0 A(\in_2 - \in_1)}[\ln \in_2 - \ln \in_1]$$

$$\Rightarrow \quad \frac{1}{C} = \frac{d\ln\left(\frac{\in_2}{\in_1}\right)}{\in_0(\in_2 - \in_1)A}$$

$$\Rightarrow \quad C = \frac{\in_0(\in_2 - \in_1)A}{d\ln\left(\frac{\in_2}{\in_1}\right)} \qquad \textbf{Ans. (D)}$$

Sol. 20 Case-I : A —3μF—3μF—3μF— B

$$\frac{1}{C_{AB}} = \frac{1}{3} + \frac{1}{3} + \frac{1}{3} = 1\mu\text{F} \quad \Rightarrow \quad 1\mu\text{F}$$

Case-II : A — three 3 μF capacitors in parallel — B

$$C_{AB} = 3 + 3 + 3 = 9\mu\text{F}$$

Case-III : A — (3μF ∥ 3μF = 6μF) — 3μF — B

$$C_{AB} = \frac{6\times3}{6+3} = \frac{18}{9} = 2\mu\text{F}$$

Case-IV :

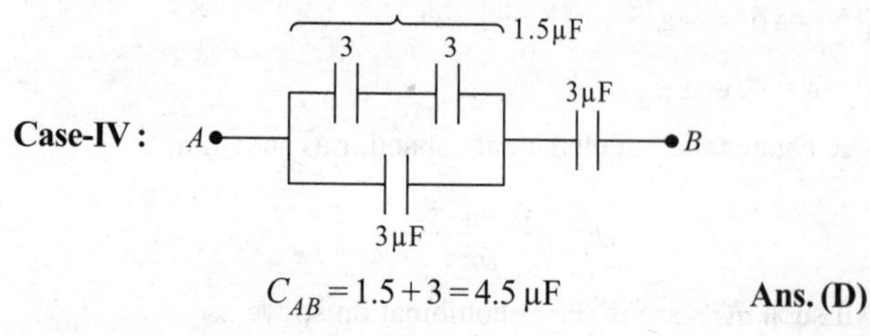

$$C_{AB} = 1.5 + 3 = 4.5\ \mu\text{F}$$ **Ans. (D)**

Sol. 21

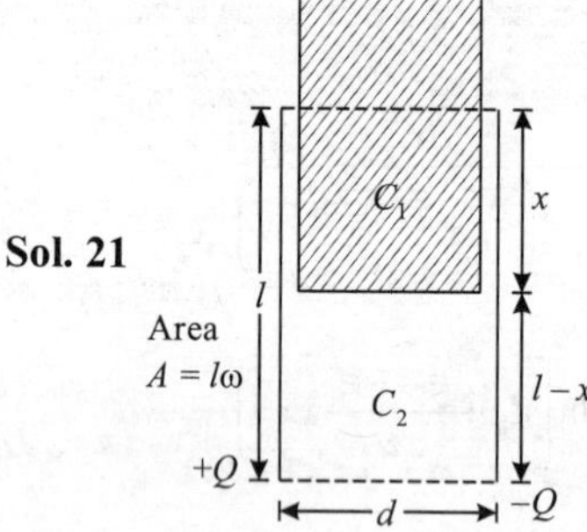

As C_1 and C_2 capacitors are in parallel with values

$$C_1 = \frac{K \in_0 (xw)}{d} \text{ and } C_2 = \frac{\in_0 (l-x)w}{d}$$

$$C_{eq} = C_1 + C_2 = \frac{\in_0 w}{d}\ [l + x(k-1)]$$

Energy in capacitor with dielectric filled upto a distance x is

$$U = \frac{Q^2}{2C} = \frac{Q^2}{2\frac{\in_0 w}{d}[l + x(k-1)]}$$

⇒ force on dielectric is given as

$$F = -\frac{\partial U}{\partial x}$$

$$\Rightarrow \quad F = \frac{Q^2 d}{2\in_0 w}\left[\frac{1}{[l + x(k-1)]^2} \times (k-1)\right]$$

at $x = 0$ or when dielectric is at the edge, we have

$$F_{x=0} = \frac{Q^2 d(k-1)}{2\in_0 wl^2}$$ **Ans. (C)**

Sol. 22

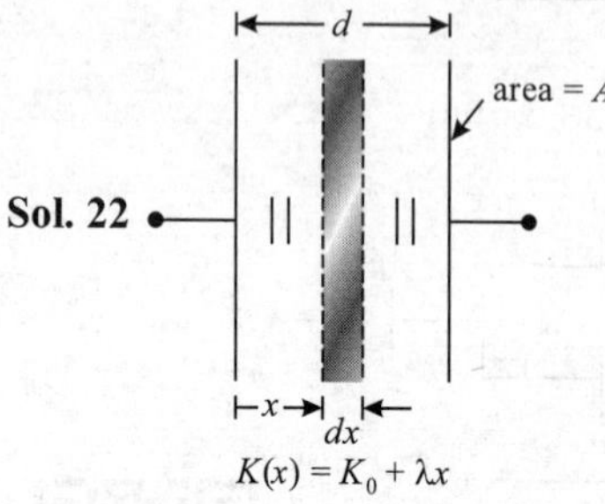

Capacitance with vacuum between plates is given as

$$C_0 = \frac{\in_0 A}{d} \quad \ldots(1)$$

Capacitance of elemental capacitor as shown in figure is

$$dC = \frac{k(x)\in_0 A}{dx} = \frac{(k_0 + \lambda x)\in_0 A}{dx}$$

Between plates all such dC's are taken in series so equation capacitance with dielectric between plates is given as

$$\frac{1}{C} = \int \frac{1}{dC}$$

$$\Rightarrow \quad \frac{1}{C} = \int_0^d \frac{1}{\in_0 A} \cdot \frac{dx}{k_0 + \lambda x}$$

$$\Rightarrow \quad \frac{1}{C} = \frac{1}{\lambda \in_0 A} \cdot [\ln (k_0 + \lambda d) - \ln k_0]$$

$$\Rightarrow \quad \frac{1}{C} = \frac{\lambda \in_0 A}{\ln\left(1 + \frac{\lambda d}{k_0}\right)} = \frac{\lambda C_0 d}{\ln\left(1 + \frac{\lambda d}{k_0}\right)}$$ **Ans. (C)**

Sol. 23 Using nodal analysis we distribute potential in given circuit as shown in figure.

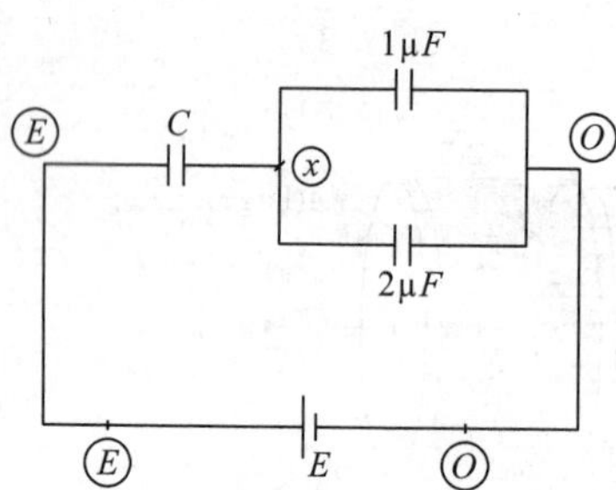

For unknown x, potential x, nodal equation is

$$1(x) + 2(x) + C(x - E) = 0$$

$$\Rightarrow \quad x = \frac{CE}{3 + C}$$

charge on $2\mu F$ capacitor is

$$q_{2\mu F} = Q_2 = 2x = \frac{2CE}{3 + C} \quad \text{(Not linear relation)}$$

When $C = 1\mu\text{F} \rightarrow Q_2 = \frac{2E}{4} = \frac{E}{2}$

When $C = 3\mu\text{F} \rightarrow Q_2 = \frac{6E}{6} = E$

Slope of Q_2 *vs* C graph is given as

$$\frac{dQ_2}{dC} = 2E\left[\frac{(3 + C)1 - C(1)}{(3 + C)^2}\right] = \frac{6E}{(3 + C)^2}$$

As C increase ⇒ $\frac{dQ_2}{dC}$ decreases **Ans. (B)**

Sol. 24

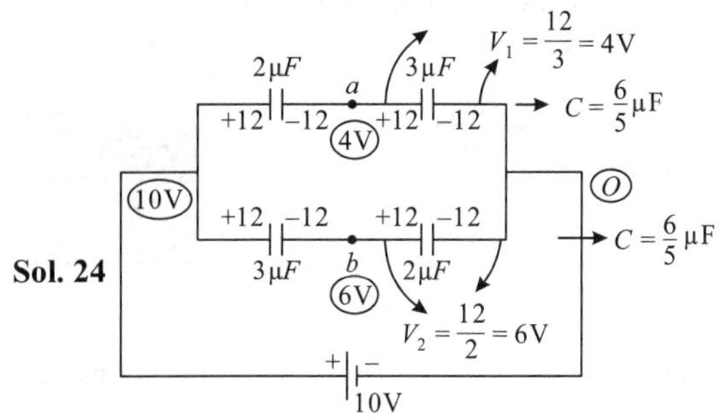

When S is open, charges on each capacitor in circuit is

$$q = CV$$

$$q = \frac{6}{5} \times 10 = 12\mu C$$

When S is closed

Charges flows from b to a

As $V_B > V_A$

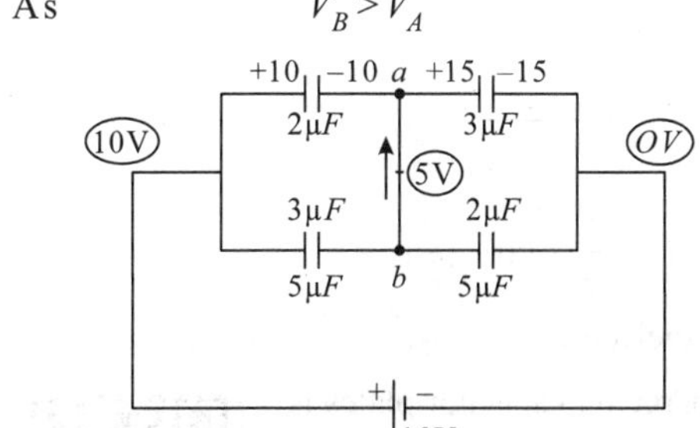

Charge connected to the two plates at point a is

$$q_a = -10 + 15 = +5\mu C$$

Initial charge on these plates

$$q_{a_i} = -12 + 12 = 0$$

On closing the switch 5μC charge flows from b to a. **Ans. (A)**

Sol. 25

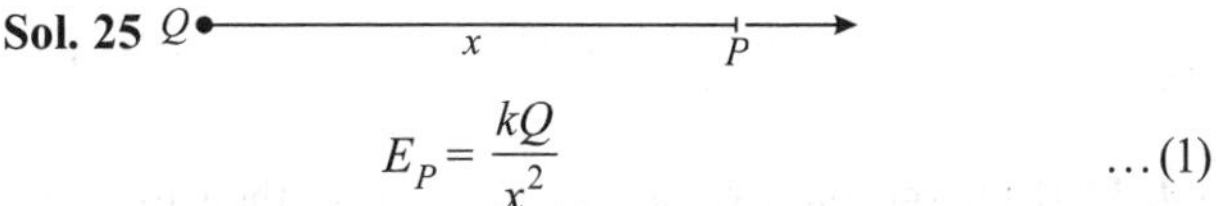

$$E_P = \frac{kQ}{x^2} \quad \dots(1)$$

Charges on capacitors in circuit can be calculated by using nodal analysis quickly

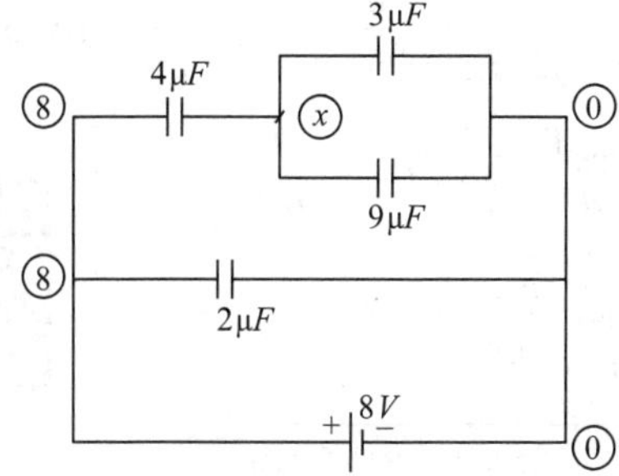

Nodal equation for potential x is given as

$$3x + 9x + 4(x - 8) = 0$$

$$\Rightarrow \quad 12x + 4(x - 8) = 0$$

$$\Rightarrow \quad 4x + 8 \Rightarrow x = 2V$$

$$q_{4\mu F} = 4 \times 6 = 24\mu C$$

and $$q_{9\mu F} = 9 \times 2 = 18\mu C$$

$$\Rightarrow \quad Q = 24 \times 18 = 42\mu C$$

From (1) $$E_P = \frac{9 \times 10^9 \times 42 \times 10^{-6}}{(30)^2}$$

$$\Rightarrow \quad E_P = 420 \text{ N/C}$$ **Ans. (C)**

Sol. 26 4μF 4μF 4μF

$$C_{eq} = \frac{4}{2} = 2\mu F$$

To get 6μF equivalent

Capacitance circuit combination is as shown in figure

Two in series and one in parallel

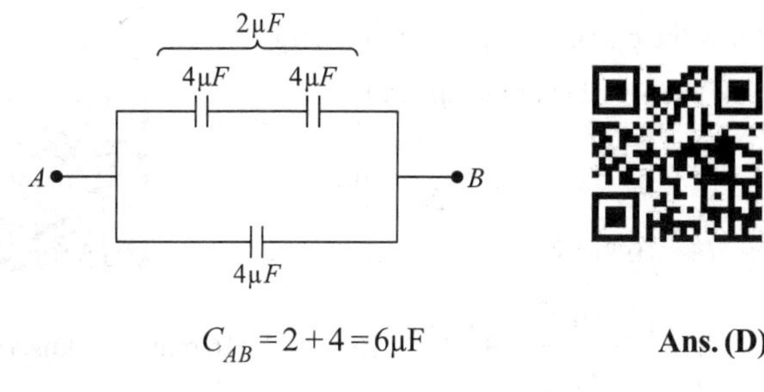

$$C_{AB} = 2 + 4 = 6\mu F$$ **Ans. (D)**

Sol. 27

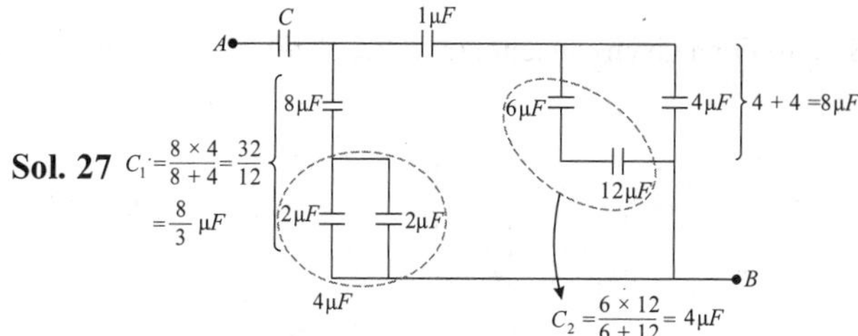

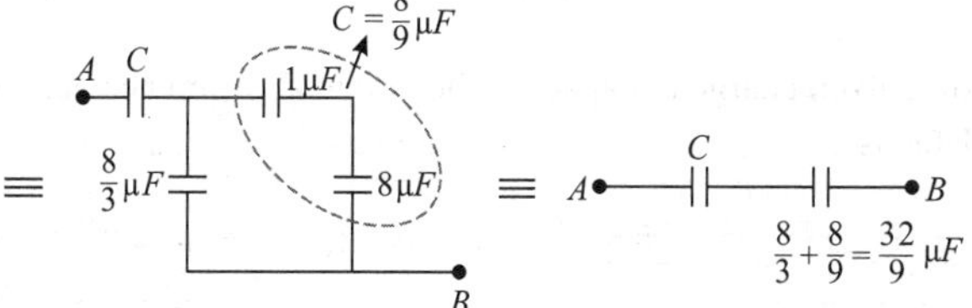

As capacitance across A and B is 1μF

$$\Rightarrow \quad \frac{1}{1} = \frac{1}{C} + \frac{1}{\frac{32}{9}}$$

$$\Rightarrow \quad \frac{1}{C} = 1 - \frac{9}{32} = \frac{23}{32} \quad \Rightarrow \quad C = \frac{32}{23}\mu F$$ **Ans. (A)**

Sol. 28 Supply voltage = 1000 V

Potential is equally divided in identical capacitors in series so for max 300V p.d. we need 4 capacitors (each 1μF) in series across supply.

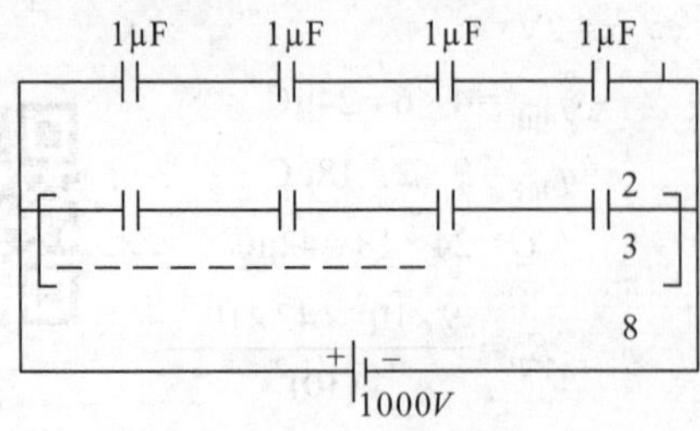

$$C_{eq} = \frac{C}{4} = 0.25\ \mu F$$

To make 2μF capacitance we have to connect 8 such branches (4 in series) in parallel

$\Rightarrow \quad C_F = 0.25 \times 8 = 2\mu F$

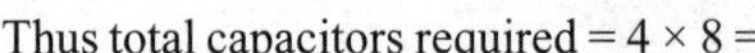

Thus total capacitors required = 4 × 8 = 32 **Ans. (B)**

Sol. 29 The capacitance of an isolated metal sphere is

$$C = 4\pi\epsilon_0 R$$

If q is the charge on sphere than energy stored in its EF is (Self energy of charged sphere)

$$U = \frac{q^2}{2C} = 4.5$$

$\Rightarrow \quad (4 \times 10^{-6}) = 4.5 \times 2 \times 4\pi\epsilon_0 R$

$\Rightarrow \quad R = \frac{16\times10^{-12}}{9} \times 9 \times 10^9 = 16 \times 10^{-3} = 16\ \text{mm}$ **Ans. (D)**

Sol. 30 For a given capacitor (parallel plate)

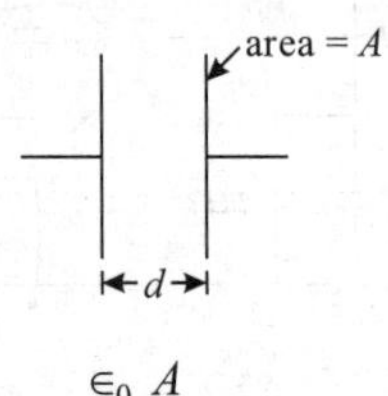

$$C = \frac{\epsilon_0 A}{d}$$

For a given charge to capacitor the potential difference across plates is

$$V = \frac{q}{C} \quad \Rightarrow \quad \text{As } C\uparrow \Rightarrow V\downarrow$$

To maintain the potential difference across capacitors, C must remain constant if d is increased to $d + 2.4$ ⇒ now capacitance with slab is given as

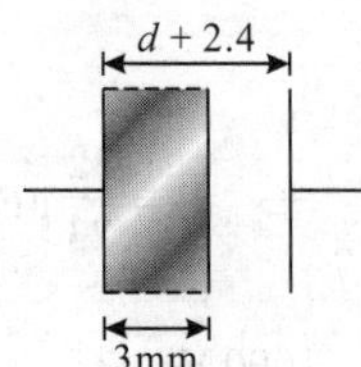

$$C' = \frac{\epsilon_0 A}{d' - t + \frac{t}{k}} \qquad t \to \text{thickness of slab}$$

To maintain the potential difference we use

$$C' = C$$

$$\frac{\epsilon_0 A}{d' - t + \frac{t}{k}} = \frac{\epsilon_0 A}{d}$$

$\Rightarrow \quad d + 2.4 - 3 + \frac{3}{k} = d$

$\Rightarrow \quad \frac{3}{k} = 0.6 \quad \Rightarrow \quad k = 5$ **Ans. (C)**

Sol. 31

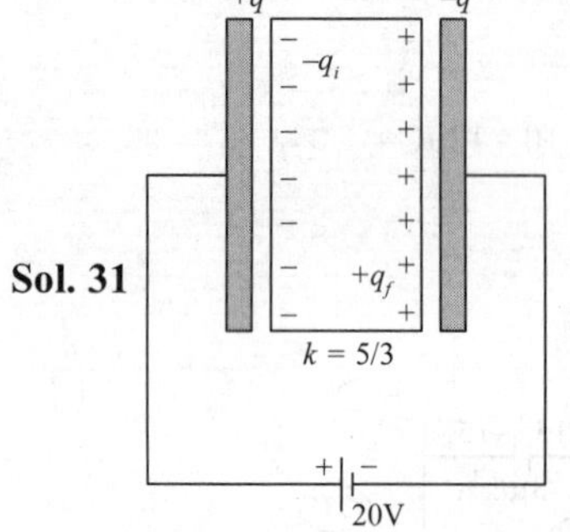

$$C = 90\text{pF} \to C' = \frac{5}{3} \times 90 = 150\ \text{pF}$$

Capacitance with dielectric

In steady state charge on capacitor plates is

$$q = C'V$$

$$q = 3000\ \text{pC}$$

Free charge on plates

Induced charges on the surface of dielectric is given as

$$q_i = q\left(1 - \frac{1}{k}\right) = 3000\left(1 - \frac{1}{5/3}\right)$$

$\Rightarrow \quad q_i = 3000 \times \frac{2}{5} = 1200\ \text{pC} = 1.2\ \text{nC}$ **Ans. (A)**

Sol. 32 If charge on a capacitor plates is q then force of attraction between plates is given as

$$F = \frac{q^2}{2\epsilon_0 A} = \frac{C^2V^2}{2\epsilon_0 A} \qquad (\text{As } q = CV)$$

and $\quad C = \frac{\epsilon_0 A}{d} \quad \Rightarrow \quad F = \frac{\epsilon_0 AV^2}{2d^2}$

$\Rightarrow \quad V = \sqrt{\frac{2Fd^2}{\epsilon_0 A}}$

$$= \sqrt{\frac{2\times 25\times10^{-6}\times(1.5\times10^{-2})^2}{8.85\times10^{-12}\times200\times10^{-4}}} \simeq 250\ \text{V}$$ **Ans. (C)**

Sol. 33 Simplifying the above circuit, it can be redrawn as

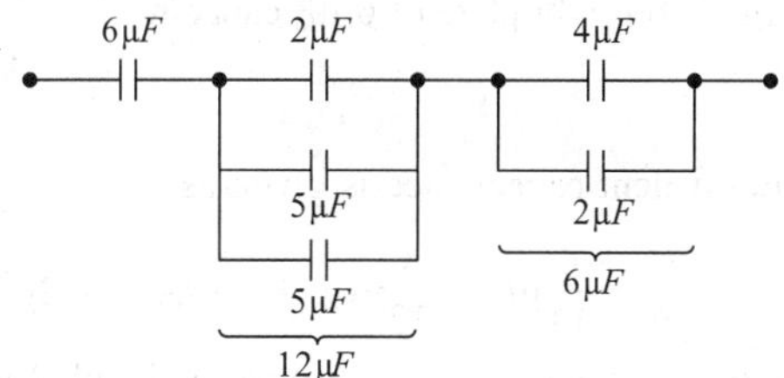

Here equivalent capacitance across AB is given as

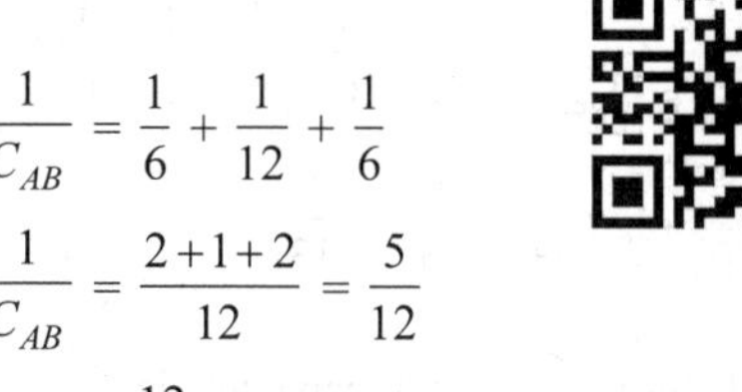

$$\frac{1}{C_{AB}}=\frac{1}{6}+\frac{1}{12}+\frac{1}{6}$$

$$\frac{1}{C_{AB}}=\frac{2+1+2}{12}=\frac{5}{12}$$

$$\Rightarrow \quad C_{AB}=\frac{12}{5}=2.4\,\mu\text{F} \qquad \textbf{Ans. (D)}$$

Sol. 34 When C_1 is charged by battery then its charge will be

$$q_{C_1}=C_1V=1\times 60=60\,\mu\text{C}$$

Combination of C_2 and C_3 in series is

$$C'=\frac{C_2C_3}{C_2+C_3}=\frac{3\times 6}{3+6}=2\mu\text{F}$$

After switch is shifted to position (2), the circuit will be

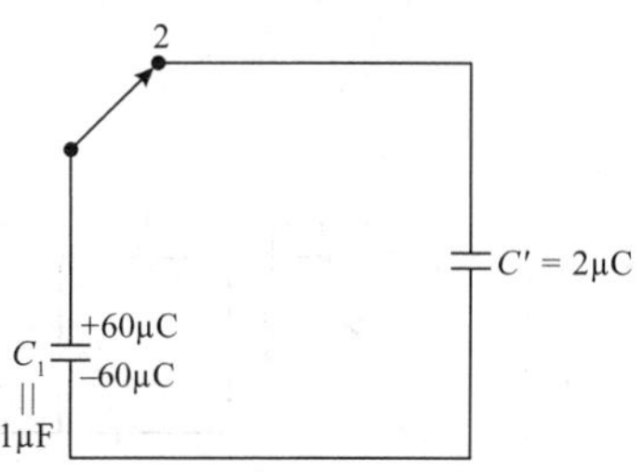

Final voltage across C_1 and C' is

$$V_f=\frac{q}{C_1+C'}=\frac{60}{1+2}=20\text{V}$$

$\Rightarrow$ Final charge on C' is

$$q_C=C'V_f=2\times 20=40\,\mu\text{C} \qquad \textbf{Ans. (D)}$$

Sol. 35

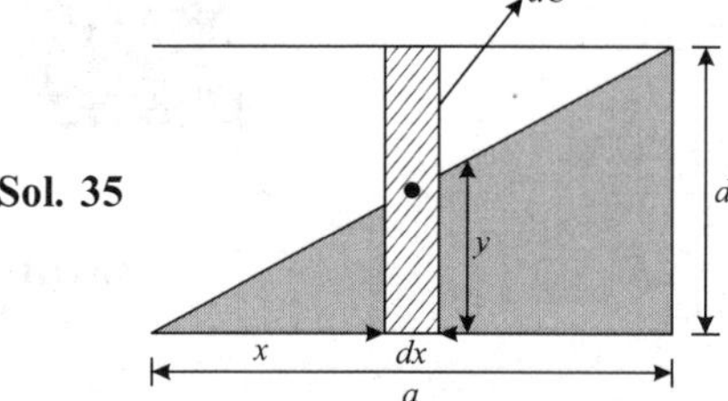

Here we consider elemental capacitor dC as shown in figure for which, we use

$$\frac{y}{x}=\frac{d}{a}$$

$$y=\frac{d}{a}x$$

$$dy=\frac{d}{a}(dx)$$

Capacitance of elemental capacitor is taken as

$$\frac{1}{dC}=\frac{y}{k\in_0\cdot adx}+\frac{(d-y)}{\in_0 adx}$$

$$\Rightarrow \quad \frac{1}{dC}=\frac{1}{\in_0 adx}\left(\frac{y}{k}+d-y\right)$$

All elemental capacitors are considered in parallel so equivalent capacitance is

$$C_{eq}=\int dC=\int\frac{\in_0 adx}{\frac{y}{k}+d-y}$$

$$\Rightarrow \quad C_{eq}=\in_0 a\cdot\frac{a}{d}\int_0^d\frac{dy}{d+y\left(\frac{1}{k}-1\right)}$$

$$\Rightarrow \quad C_{eq}=\frac{\in_0 a^2}{\left(\frac{1}{k}-1\right)d}\left[\ln\left(d+y\left(\frac{1}{k}-1\right)\right)\right]_0^d$$

$$\Rightarrow \quad C_{eq}=\frac{k\in_0 a^2}{(1-k)d}\ln\left(\frac{d+d\left(\frac{1}{k}-1\right)}{d}\right)$$

$$\Rightarrow \quad C_{eq}=\frac{k\in_0 a^2}{(1-k)d}\ln\left(\frac{1}{k}\right)=\frac{k\in_0 a^2\ln k}{(k-1)d} \qquad \textbf{Ans. (C)}$$

Sol. 36 Equivalent circuit of given system is shown below

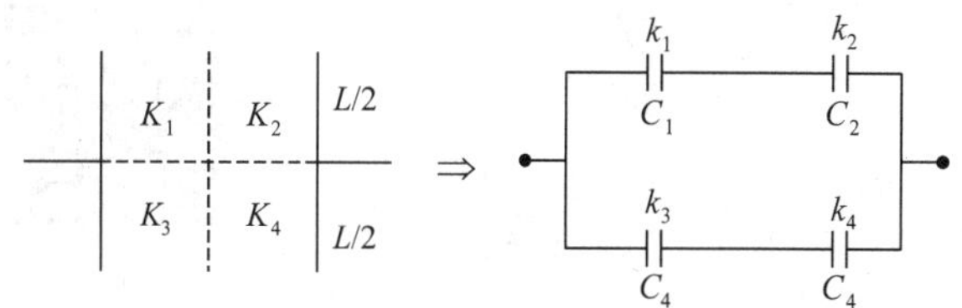

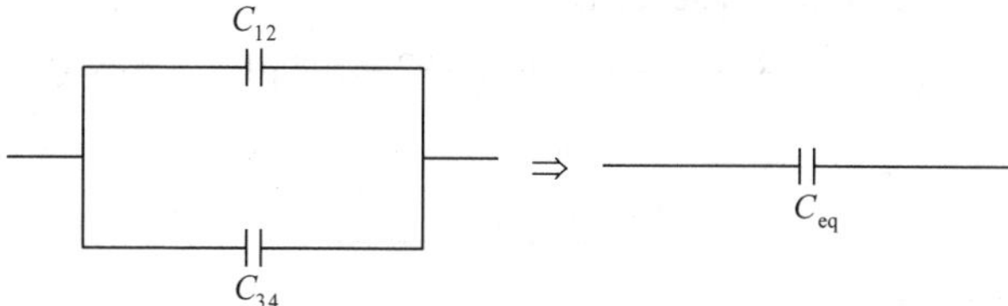

Thus equivalent capacitance is given as

$$C_{12}=\frac{C_1C_2}{C_1+C_2}=\frac{\frac{k_1\in_0\frac{L}{2}\times L}{d/2}\cdot\frac{k_2\left[\in_0\frac{L}{2}\times L\right]}{d/2}}{(k_1+k_2)\left[\frac{\in_0\cdot\frac{L}{2}\times L}{d/2}\right]}$$

$$C_{12} = \frac{k_1 k_2}{k_1 + k_2} \frac{\epsilon_0 L^2}{d}$$

In the same way we get,

$$C_{34} = \frac{k_3 k_4}{k_3 + k_4} \frac{\epsilon_0 L^2}{d}$$

$$\Rightarrow \quad C_{eq} = C_{12} + C_{34} = \left[\frac{k_1 k_2}{k_1 + k_2} + \frac{k_3 k_4}{k_3 + k_4}\right] \frac{\epsilon_0 L^2}{d} \quad \ldots (1)$$

Comparing with

$$C_{eq} = \frac{k \epsilon_0 L^2}{d} \quad \ldots (2)$$

$$\Rightarrow \quad k_{eq} = \frac{k_1 k_2 (k_3 + k_4) + k_3 k_4 (k_1 + k_2)}{(k_1 + k_2)(k_3 + k_4)}$$

This does not match with any of the options, so none of the given options is correct **Ans. (None)**

Sol. 37 Let dielectric constant of material used be K

$$\Rightarrow \frac{10 \epsilon_0 A/3}{d} + \frac{12 \epsilon_0 A/3}{d} + \frac{14 \epsilon_0 A/3}{d} = \frac{K \epsilon_0 A}{d}$$

$$\Rightarrow K = 12$$

Ans. (A)

Sol. 38 Initial energy of capacitor

$$U_i = \frac{1}{2} C V^2 = \frac{1}{2} \times 12 \times 10^{-12} \times (10)^2$$

$$\Rightarrow \quad U_i = 600 \text{ pJ}$$

Since battery is disconnected so charge remain same at 120pc. Final energy of capacitor will be

$$U_f = \frac{q^2}{2C}$$

$$\Rightarrow \quad U_f = \frac{1}{2} \times \frac{120 \times 120}{12 \times 6.5} = 92 \text{ pJ}$$

Work done by capacitor is

$$W = U_i - U_f = 600 - 92 = 508 \text{ pJ}$$

Ans. (C)

Sol. 39

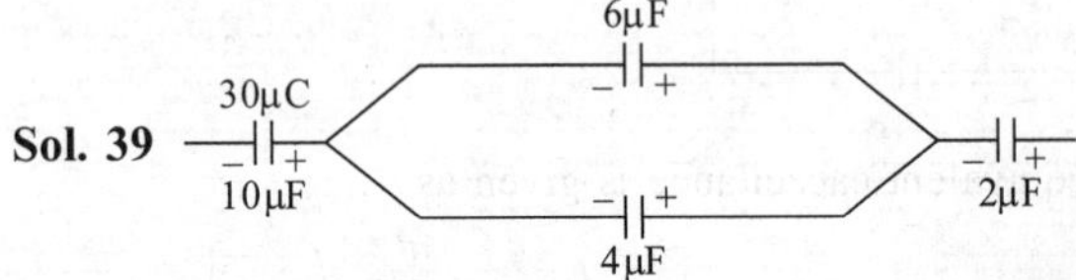

6μF & 4μF are in parallel & total charge on this combination is 30 μC

⇒ Charge on 6μF capacitor

$$= \frac{6}{6+4} \times 30 = 18 \, \mu\text{C}$$

Thus charge is on right plate of 6 μF capacitor is +18 μC.

Ans. (D)

Sol. 40 Equivalent capacitance is given as

$$C_{eq} = \frac{6}{13} \mu F = \frac{3C}{13} \quad [\text{As } C = 2\mu F \text{ each}]$$

Therefore three capacitors must be in parallel with 4 in series so, we have

$$\frac{1}{C_{eq}} = \frac{1}{3C} + \frac{1}{C} + \frac{1}{C} + \frac{1}{C} + \frac{1}{C}$$

$$C_{eq} = \frac{3C}{13} = \frac{6}{13} \mu F$$

Ans. (D)

Sol. 41 Initial energy in capacitor is

$$U_i = \frac{Q^2}{2C}$$

After turning the switch to position B, find energy in circuit is

$$U_f = \frac{Q^2}{8C}$$

Heat dissipated is given as

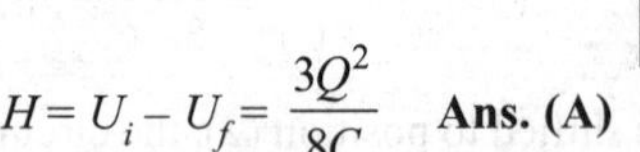

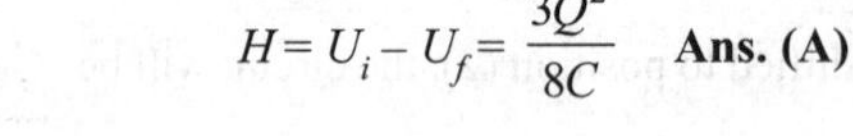

$$H = U_i - U_f = \frac{3Q^2}{8C}$$ **Ans. (A)**

Sol. 42

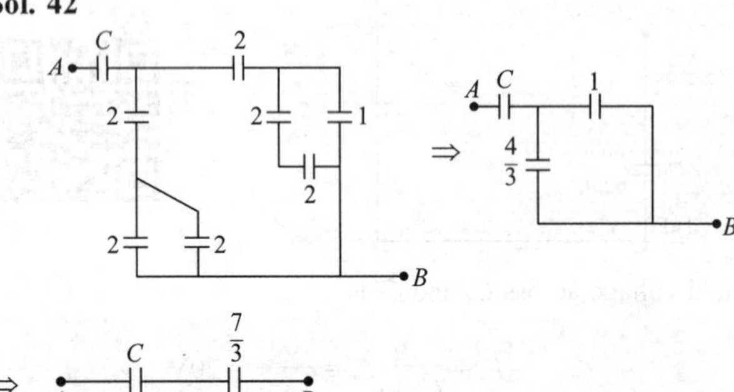

Equivalent capacitance across A and B is given as

$$\Rightarrow \quad \frac{\frac{7C}{3}}{\frac{7}{3} + C} = \frac{1}{2}$$

$$\Rightarrow \quad 14C = 7 + 3C$$

$$\Rightarrow \quad C = \frac{7}{11} \mu F$$

Ans (B)

Sol. 43 At $t = 4$s, charge on capacitor plates is constant so current in circuit will be zero.

Ans. (C)

Sol. 44 Electric field between plates of a capacitor is given as

$$E = \frac{\sigma}{\epsilon_0} = \frac{Q}{A\epsilon_0}$$

$\Rightarrow$ $Q = AE\epsilon_0$

$\Rightarrow$ $Q = (1)(100)(8.85 \times 10^{-12})$

$\Rightarrow$ $Q = 8.85 \times 10^{-10}\,\text{C}$ **Ans. (D)**

Sol. 45 Maximum electric field between plates

$$E_m = 10^6\,\text{V/m}$$

If d is the plate separation, we use

$$E_m = \frac{V_m}{d}$$

$\Rightarrow$ $d = \frac{V_m}{E_m} = \frac{500}{10^6} = 5 \times 10^{-4}\,\text{m}$

Capacitance $C = \frac{k\epsilon_0 A}{d}$

$\Rightarrow$ $k = \frac{Cd}{\epsilon_0 A} = \frac{15\times10^{-12}\times5\times10^{-4}}{8.85\times10^{-12}\times10^{-4}}$

$= \frac{15\times5}{8.85} = 8.465 \sim 8.5$ **Ans. (D)**

Sol. 46

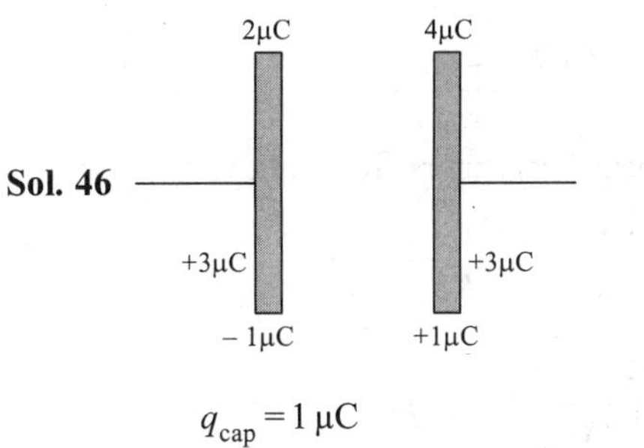

$q_{cap} = 1\,\mu\text{C}$

$\Rightarrow$ Potential difference across plates

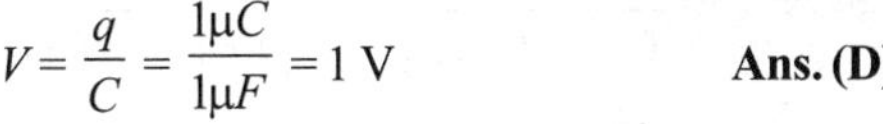

$$V = \frac{q}{C} = \frac{1\mu C}{1\mu F} = 1\,\text{V}$$ **Ans. (D)**

Sol. 47 Work done in pulling plates apart is given as

$W = U_f - U_i$

$\Rightarrow$ $W = \frac{q^2}{2C_f} - \frac{q^2}{2C_i}$

$\Rightarrow$ $W = \frac{(5\times10^{-6})^2}{2}\left(\frac{1}{2\times10^{-6}} - \frac{1}{5\times10^{-6}}\right)$

$\Rightarrow$ $W = \frac{15}{4} \times 10^{-6}$

$\Rightarrow$ $W = 3.75 \times 10^{-6}\,\text{J}$ **Ans. (A)**

Sol. 48

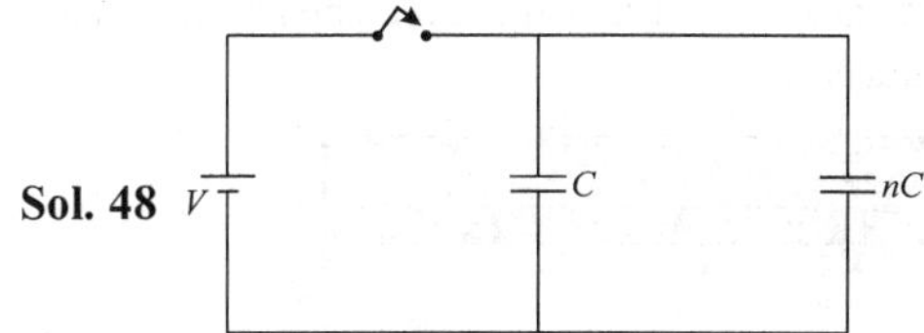

After fully charging, battery is disconnected from circuit so charge on system remain constant. after disconnecting battery system now is shown below

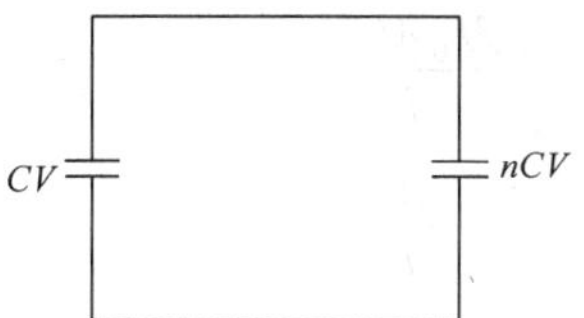

Total charge of the system

$= CV + nCV$

$= (n+1)CV$

After the insertion of dielectric of constant K, if final potential difference becomes V_f, we use

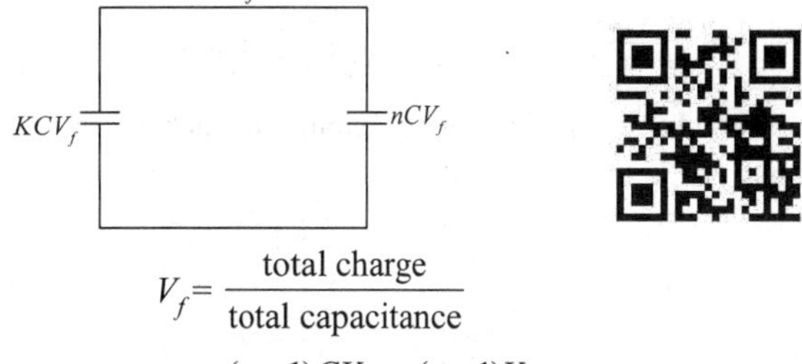

$$V_f = \frac{\text{total charge}}{\text{total capacitance}}$$

$\Rightarrow$ $V_f = \frac{(n+1)CV}{KC+nC} = \frac{(n+1)V}{K+n}$ **Ans. (C)**

Sol. 49 As $q = CV$ so the slope of graph will give capacitance. Slope will be more in parallel combination. Hence capacitance in parallel should be 50 μF & in series combination should be 8 μF. From the given option only in option 40 μF & 10 μF it is obtained as

$C_{parallel} = 40 + 10 = 50\mu\text{F}$

$C_{series} = \frac{40\times10}{40+10} = 8\,\mu\text{F}$ **Ans. (D)**

Sol. 50 In electric field effective gravity for pendulum is given as

$$g_{eff} = \sqrt{g^2 + \left(\frac{qE}{m}\right)^2}$$

Time period of pendulum is taken as

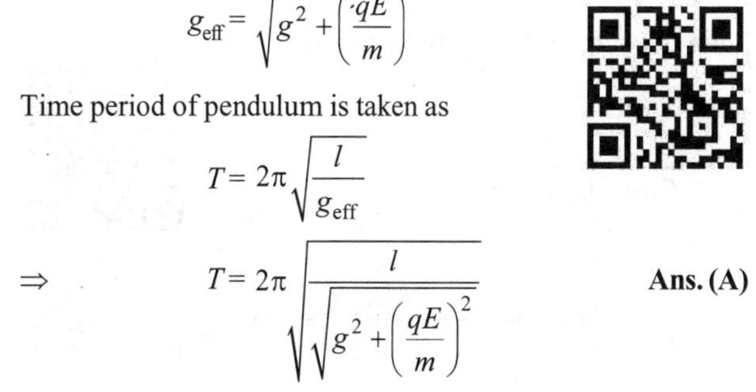

$$T = 2\pi\sqrt{\frac{l}{g_{eff}}}$$

$\Rightarrow$ $T = 2\pi\sqrt{\frac{l}{\sqrt{g^2 + \left(\frac{qE}{m}\right)^2}}}$ **Ans. (A)**

Sol. 51 In first case we consider the three capacitors of each dielectric are in series.

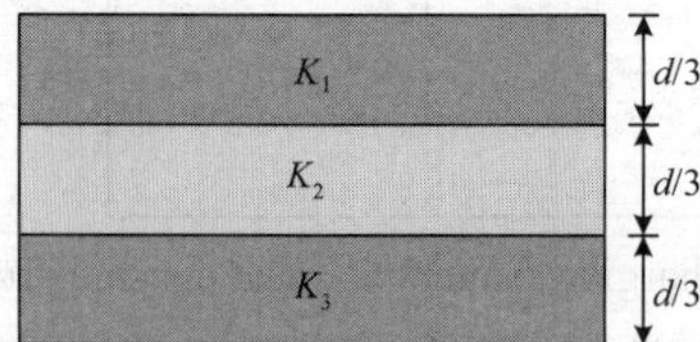

Capacitance for each dielectric is given as

$$C_1 = \frac{3\varepsilon_0 AK_1}{d}$$

$$C_2 = \frac{3\varepsilon_0 AK_2}{d}$$

$$C_3 = \frac{3\varepsilon_0 AK_3}{d}$$

Combination capacitance is taken as

$$\frac{1}{C_{eq}} = \frac{1}{C_1} + \frac{1}{C_2} + \frac{1}{C_3}$$

$$\Rightarrow \quad C_{eq} = \frac{3\varepsilon_0 AK_1K_2K_3}{d(K_1K_2 + K_2K_3 + K_3K_1)} \quad \ldots(1)$$

In second case we consider capacitance of each dielectric are in parallel

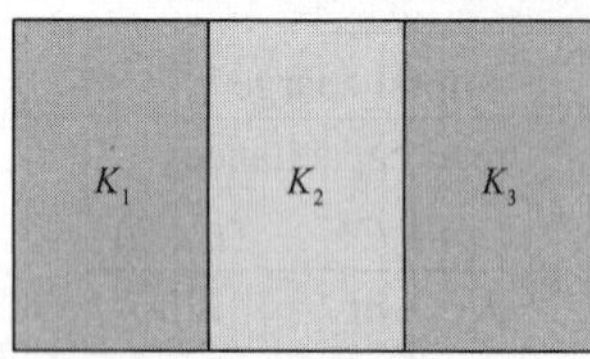

Capacitance of each dielectric is given as

$$C_1 = \frac{\varepsilon_0 K_1 A}{3d}$$

$$C_2 = \frac{\varepsilon_0 K_2 A}{3d}$$

$$C_3 = \frac{\varepsilon_0 K_3 A}{3d}$$

Combination capacitance is taken as

$$C'_{eq} = C_1 + C_2 + C_3$$

$$C'_{eq} = \frac{\varepsilon_0 A}{3d}(K_1 + K_2 + K_3) \quad \ldots(2)$$

Ratio of energy stored in two cases is

$$\frac{E_1}{E_2} = \frac{\frac{1}{2}C_{eq}.V^2}{\frac{1}{2}C'_{eq}V^2}$$

$$\Rightarrow \quad \frac{E_1}{E_2} = \frac{9K_1K_2K_3}{(K_1 + K_2 + K_3)(K_1K_2 + K_2K_3 + K_3K_1)}$$ **Ans. (A)**

Sol. 52

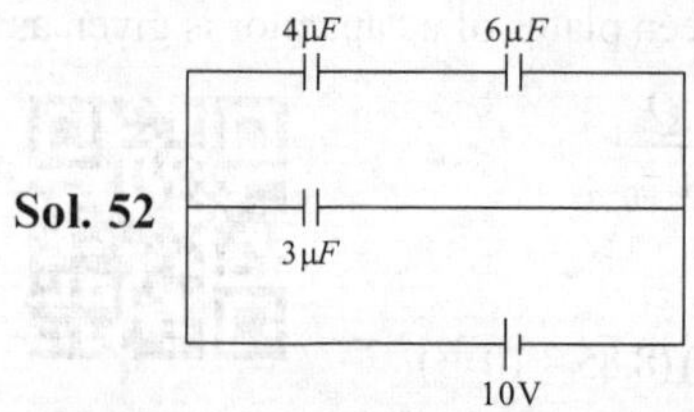

In above circuit we consider 4μF and 6 μF capacitors in series

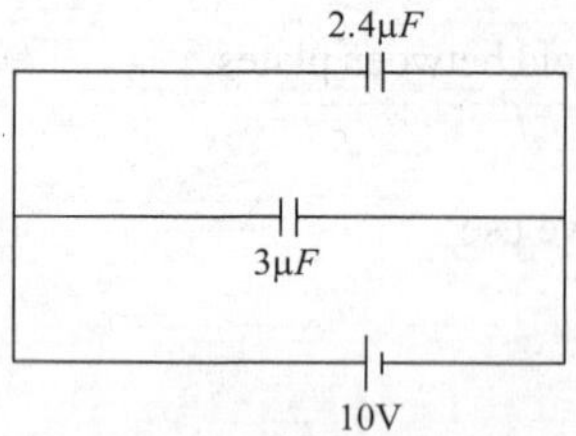

Charge supplied by battery to 2.4 μF capacitor is

$$q = 2.4 \times 10 = 24\ \mu C$$

In series both capacitors will have same charge of 24 μC.

Ans. (B)

Sol. 53 Here we use,

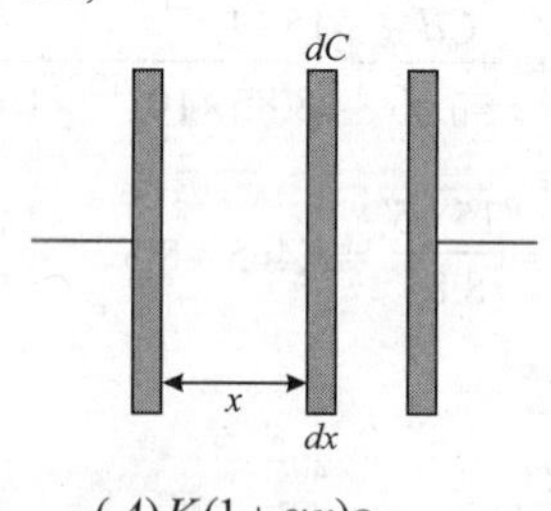

$$dC = \frac{(A)K(1+\alpha x)\varepsilon_0}{dx}$$

Now all such elemental capacitors are is series so we use,

$$\frac{1}{C} = \int \frac{1}{dC} = \int_0^d \frac{dx}{AK\varepsilon_0(1+\alpha x)}$$

$$\Rightarrow \quad \frac{1}{C} = \int_0^d \frac{1}{AK\varepsilon_0} \cdot (1-\alpha x)dx$$

$$\Rightarrow \quad \frac{1}{C} = \frac{1}{AK\varepsilon_0}\left[d - \frac{\alpha d^2}{2}\right] = \frac{d}{AK\varepsilon_0}\left(1 - \frac{\alpha d}{2}\right)$$

$$\Rightarrow \quad \frac{1}{C} = \frac{d}{AK\varepsilon_0}\left(1 - \frac{\alpha d}{2}\right)$$

$$\Rightarrow \quad C = \frac{AK\varepsilon_0}{d}\left(1 + \frac{\alpha d}{2}\right)$$ **Ans. (A)**

Sol. 54 After second connection charge is equally distributed in both capacitors

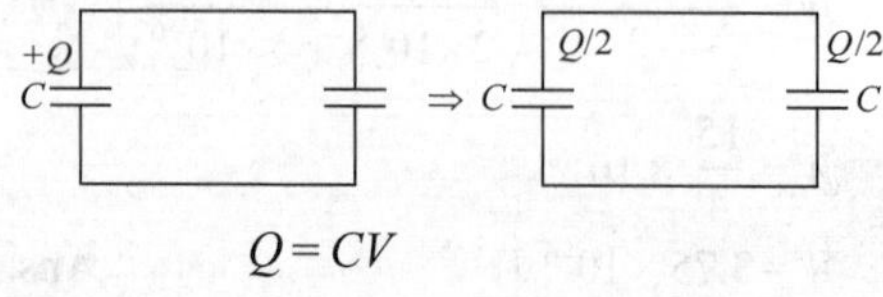

$$Q = CV$$

Energy loss is given as

$$\Delta Q_L = \frac{Q^2}{2C} - \left[\frac{(Q/2)^2}{2C}\times 2\right] = \frac{Q^2}{4C}$$

$$\Rightarrow \quad \Delta Q_L = \frac{1}{4}CV^2$$

$$\Rightarrow \quad \Delta Q_L = \frac{1}{4}\times 60\times 10^{-12}\times 4\times 10^2$$

$$\Rightarrow \quad \Delta Q_L = 6\text{ nJ}$$

Ans. (6)

Sol. 55 In parallel combination, we use

$$C_1 + C_2 = 10 \qquad \dots(1)$$

For energy stored, it is given that

$$\frac{1}{2}C_2V^2 = 4\times\frac{1}{2}C_1V^2$$

$$\Rightarrow \quad C_2 = 4C_1 \qquad \dots(2)$$

$$\Rightarrow \quad C_1 = 2$$

$$\& \quad C_2 = 8$$

For series combination, we use

$$C_{eq} = \frac{C_1C_2}{C_1+C_2} = 1.6\ \mu\text{F}$$

Ans. (C)

Sol. 56 Assume small element dx at a distance x from left end

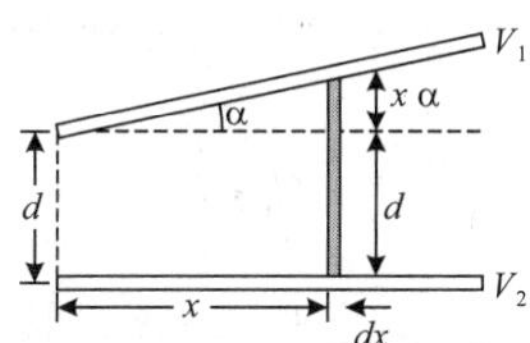

Capacitance of this element is given as

$$dC = \frac{\varepsilon_0(adx)}{d+x\alpha}$$

Thus total capacitance is given as

$$C = \int_0^a \frac{\varepsilon_0 adx}{d+x\alpha}$$

$$\Rightarrow \quad C = \frac{\varepsilon_0 a}{\alpha}\ln\left(\frac{1+\alpha a}{d}\right)\Bigg|_0^a \quad \left(\text{We use } \ln(1+x)\approx x - \frac{x^2}{2}\right)$$

$$\Rightarrow \quad C = \frac{\varepsilon_0 a^2}{d}\left(1-\frac{\alpha a}{2d}\right)$$

Ans. (D)

Sol. 57 Initial energy stored in capacitor is

$$U_i = \frac{1}{2}\times 5\times 10^{-6}(220)^2$$

Final common potential of capacitors is given as

$$V = \frac{220\times 5 + 0\times 2.5}{5+2.5} = 220\times\frac{2}{3}$$

Final energy stored in capacitors is

$$U_f = \frac{1}{2}(5+2.5)\times 10^{-6}\left(220\times\frac{2}{3}\right)^2$$

$$\Delta U = U_f - U_i$$

$$\Rightarrow \quad \Delta U = -403.33\times 10^{-4}$$

$$\Rightarrow \quad -403.33\times 10^{-4} = \frac{X}{100}$$

$$X = -4.03$$

Ans. (–4.03)

Sol. 58 Initial charges on capacitors are

+500μC –500μC 10μF C

After connection

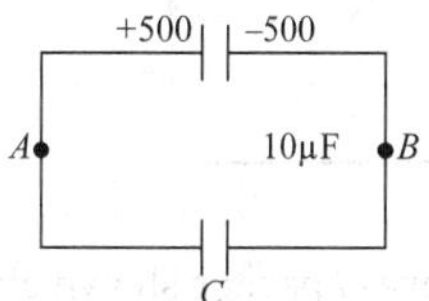

Potential difference across A and B is

$$V_{AB} = \frac{q_{\text{total}}}{q_{\text{total}}} = \frac{500}{C+10} = 20$$

$$\Rightarrow \quad 20C + 200 = 500$$

$$\Rightarrow \quad 20C = 300$$

$$\Rightarrow \quad C = 15\ \mu\text{F}$$

Ans. (B)

Sol. 59 If total charge is 750 μC then charge on C_2 is given as

$$q_{C_2} = 750 - q_{C_3} = 750 - 8\times 20 = 590\ \mu\text{C}$$

Ans. (A)

Sol. 60 Initial energy in capacitor is

$$U_1 = \frac{1}{2}CV_0^2$$

After connection with $\frac{C}{2}$ final potential of system is given as

$$V_f = \frac{CV_0}{C+\frac{C}{2}} = \frac{2V_0}{3}$$

Final energy in system is

$$U_2 = \frac{1}{2}\left(\frac{3C}{2}\right)\left(\frac{2V_0}{3}\right)^2 = \frac{1}{2}CV_0^2$$

Energy loss is given as

$$\Delta U_L = U_1 - U_2 = \left(\frac{1}{2}-\frac{1}{3}\right)CV_0^2 = \frac{1}{6}CV_0^2$$

Ans. (A)

Sol. 61 Initial charges on capacitors are given as

$$q_1 = CV$$

$$q_2 = 2C\times 2V = 4CV$$

When capacitors are connected with opposite polarity, find change on combination is

$$q_f = 4CV - CV = 3CV$$

Thus final energy in combination is

$$U_f = \frac{q_f^2}{2(3C)} = \frac{(3CV)^2}{6C} = \frac{3}{2}CV^2 \qquad \textbf{Ans. (D)}$$

Sol. 62

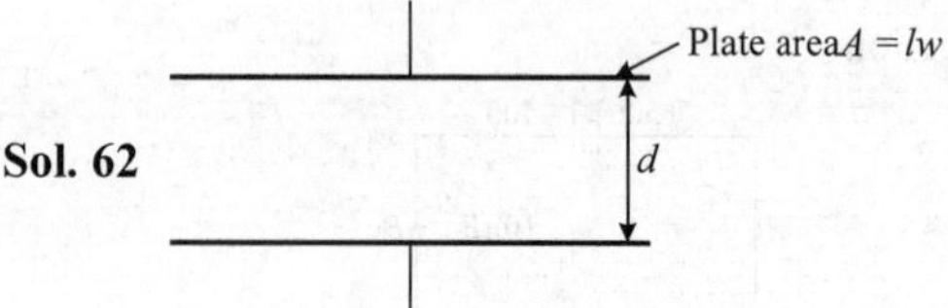

Capacitance of empty capacitor shown above is given as

$$C = \frac{\epsilon_0\, lw}{d}$$

Capacitance of capacitor with dielectric shown above is given as

$$C_f = \frac{\epsilon_0\,(4)xw}{d} + \frac{\epsilon_0\,(l-x)w}{d}$$

For given applied potential difference final energy be two times if capacitance is two times, so we use

$$C_f = 2C$$

$$\frac{4\epsilon_0\, xw}{d} + \frac{3\epsilon_0\, lw}{d} - \frac{\epsilon_0\, xw}{d} = 2\frac{\epsilon_0\, lw}{d}$$

$$\Rightarrow \qquad \frac{3\epsilon_0\, xw}{d} = \frac{\epsilon_0\, lw}{d}$$

$$\Rightarrow \qquad x = \frac{l}{3} \qquad \textbf{Ans. (C)}$$

Sol. 63 Figure below shows the potential distribution on different points of circuit by

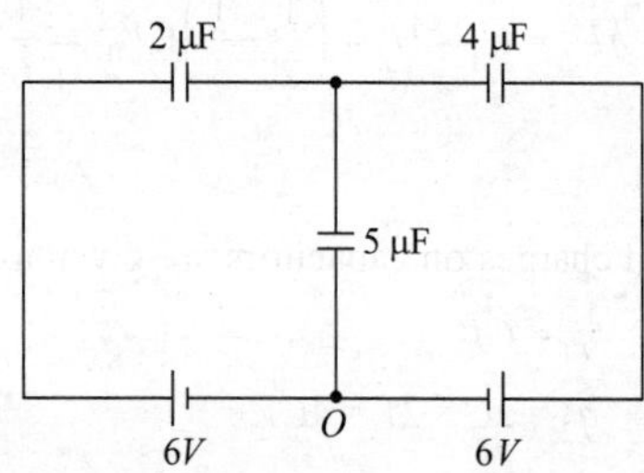

Taking potential of point O at 0

Writing equation for x, we have

$$q_1 + q_2 + q_3 = 0$$

$$2(x-6) + 4(x-6) + 5(x) = 0$$

$$\Rightarrow \qquad x = \frac{36}{11}$$

$$\Rightarrow \qquad q_3 = \frac{36(5)}{11} = \frac{180}{11}$$

$$q_3 = 16.36\ \mu C \qquad \textbf{Ans. (B)}$$

Sol. 64 For series combination of capacitances, we have

$$\frac{1}{C_{eq_1}} = \frac{1}{C} + \frac{1}{C}$$

$$\Rightarrow \qquad C_{eq_1} = \frac{C}{2}$$

For parallel combination, we have

$$C_{eq_2} = C + C$$

$$\Rightarrow \qquad C_{eq_2} = 2C$$

$$\Rightarrow \qquad \frac{C_{eq_1}}{C_{eq_2}} = \frac{(C/2)}{2C} = \frac{1}{4} = 1:4 \qquad \textbf{Ans. (B)}$$

Sol. 65 When capacitors are connected in parallel, we have

$$C_{eq} = C_1 + C_2$$

When capacitors are connected in series, we have

$$C'_{eq} = \frac{C_1 C_2}{C_1 + C_2}$$

Given that $\qquad C_1 + C_2 = \frac{15}{4}\left(\frac{C_1 C_2}{C_1 + C_2}\right)$

$$\Rightarrow \qquad 4(C_1 + C_2)^2 = 15 C_1 C_2$$

$$\Rightarrow \qquad 4C_1^2 + 4C_2^2 - 7C_1C_2 = 0$$

Dividing all terms of the equation by C_1^2, we have

$$4\left(\frac{C_2}{C_1}\right)^2 - \frac{7C_2}{C_1} + 4 = 0$$

Considering $\frac{C_2}{C_1} = x$, we get

$$4x^2 - 7x + 4 = 0$$

For this quadratic equation, $b^2 - 4ac = 49 - 64 < 0$ thus no solution exits to this equation. **Ans. (None)**

Sol. 66 The capacitance before insertion of slab is given as

$$C_0 = \frac{\epsilon_0\, A}{d}$$

After insertion of slab capacitor looks like shown in figure below which can be considered as combination of two capacitors in series, one with dielectric and one with air

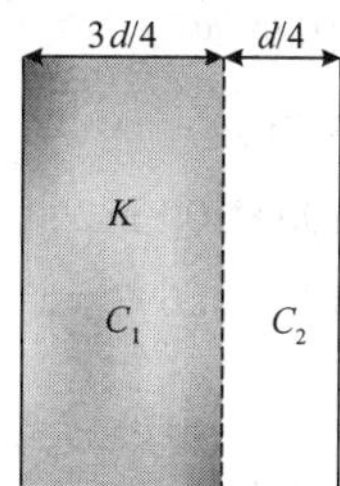

The equivalent capacitance of this system C' is given as

$$\frac{1}{C'} = \frac{1}{C_1} + \frac{1}{C_2}$$

$$\Rightarrow \quad \frac{1}{C'} = \frac{(3d/4)}{\epsilon_0 KA} + \frac{d/4}{\epsilon_0 A}$$

$$\Rightarrow \quad \frac{1}{C'} = \frac{d}{4\epsilon_0 A}\left(\frac{3+K}{K}\right)$$

$$\Rightarrow \quad C' = \frac{4KC_0}{(3+K)} \qquad \textbf{Ans. (C)}$$

Sol. 67 Capacitance of given capacitor partially filled with dielectric is given as

$$C = \frac{\varepsilon_0 A}{\frac{d}{2K} + \frac{d}{2}} = \frac{2\varepsilon_0 A}{\frac{d}{K} + d}$$

$$\Rightarrow \quad C = \frac{2 \times 2\varepsilon_0}{\frac{1}{3.2} + 1} = \frac{4 \times 3.2}{4.2}\varepsilon_0 = 3.04\,\varepsilon_0 \qquad \textbf{Ans. (C)}$$

Sol. 68 The equivalent combination of capacitor from the plates setup can be drawn as shown in figure below

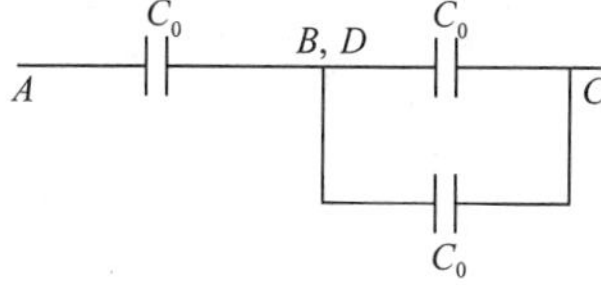

Thus equivalent capacitance is calculated as

$$C_{eq} = \frac{2C_0}{3} = \frac{2}{3}\frac{\epsilon_0 A}{d}$$

$$\Rightarrow \quad C_{eq} = \frac{2\epsilon_0}{3d} \times \left(2 \times \frac{3}{2}\right) = 2 \qquad \textbf{Ans. (2)}$$

Sol. 69 The initial energy stored in the capacitor is given as

$$U_i = \frac{1}{2} \times 14 \times 12 \times 12 \text{ pJ} = 1008 \text{ pJ}$$

Final energy stored in capacitor after insertion of plate is given as

$$U_f = \frac{1008}{7} \text{ pJ} = 144 \text{ pJ}$$

The mechanical energy of oscillation can be considered as the difference in energies as it remains there in the plate as kinetic energy at the mean position and as potential energy at extreme position, calculated as

$$\Delta U = 1008 - 144 = 864 \text{ pJ} \qquad \textbf{Ans. (864)}$$

Sol. 70 When first capacitor is charged to 10 V then it will be charged to $C_1 V = 20\ \mu C$ charge which will remain constant once battery is disconnected. When it is connected to other capacitor then same charge is distributed between the two and if final potential difference becomes V', we have

$$20 = (C_1 + C_2)\,V$$

$$\Rightarrow \quad V = 2 \text{ V}$$

Thus final charge on second capacitor can be calculated as

$$Q_2 = C_2 V = 16\ \mu C \qquad \textbf{Ans. (16)}$$

Sol. 71 The capacitance of the dielectric filled part of capacitor is calculated as

$$C_1 = \frac{k\epsilon_0 A}{d}$$

$$\Rightarrow \quad C_1 = \frac{10\epsilon_0 (100)}{5} = 200\,\epsilon_0$$

For the air filled part of capacitor, capacitance is given as

$$C_2 = \frac{\epsilon_0 (100)}{5} = 20\,\epsilon_0$$

As C_1 & C_2 are in series so, for equivalent capacitance, we use

$$C_{eq} = \frac{C_1 C_2}{C_1 + C_2} = \frac{4000\,\epsilon_0}{220}$$

$$\Rightarrow \quad C_{eq} = 160.9 \times 10^{-12} = 161 \text{ pF} \qquad \textbf{Ans. (161)}$$

Sol. 72 To calculate the capacitance, we consider an elemental capacitor of plate separation dx as shown in figure. All such elemental capacitors will be considered in series so the capacitance of given system is calculated as

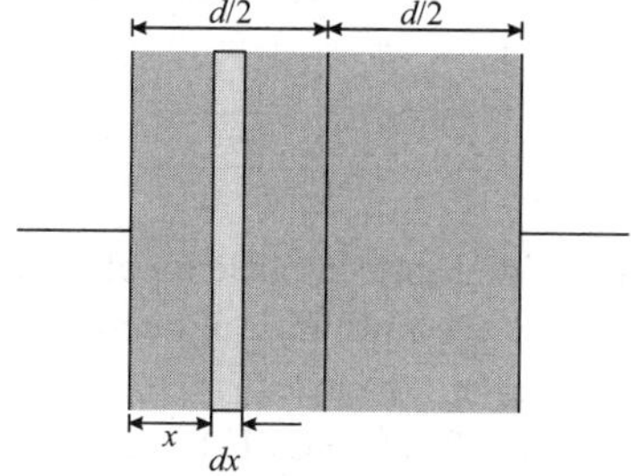

$$\frac{1}{C_{eq}} = \int \frac{1}{dC} = \int \frac{dx}{A\varepsilon}$$

$$\Rightarrow \quad \frac{1}{C_{eq}} = \int_0^{d/2} \frac{dx}{A(\varepsilon_0 + kx)} + \int_{d/2}^{d} \frac{dx}{A[\varepsilon_0 + k(d-x)]}$$

$$\Rightarrow \quad \frac{1}{C_{eq}} = \frac{1}{A}\left\{\frac{1}{k}\left[\ln(\varepsilon_0 + kx)\right]_0^{d/2} + \frac{1}{2}\left[\ln(\varepsilon_0 + k(d-x))\right]_{d/2}^{d}\right\}$$

$$\Rightarrow \quad \frac{1}{C_{eq}} = \frac{1}{Ak} \times 2 \ln \frac{\varepsilon_0 + \frac{kd}{2}}{\varepsilon_0}$$

$$\Rightarrow C_{eq} = \frac{kA}{2ln\left(\frac{2\varepsilon_0 + kx}{2\varepsilon_0}\right)}$$ **Ans. (A)**

Sol. 73 Bound charge on capacitor plate can be directly given by the expression

$$q_b = q_f\left(1 - \frac{1}{K}\right)$$ **Ans. (D)**

Sol. 74 The given figure can be considered as three capacitors in series combination as shown in figure below

The capacitances of the three capacitors can be given as

$$C_1 = \frac{A\varepsilon_0 K}{d}; C_2 = \frac{3KA\varepsilon_0}{2d}; C_3 = \frac{5KA\varepsilon_0}{3d}$$

Equivalent capacitance of the given combination is calculated as

$$\frac{1}{C_{eq}} = \frac{1}{C_1} + \frac{1}{C_2} + \frac{1}{C_3}$$

$$\Rightarrow \quad \frac{1}{C_{eq}} = \frac{d}{A\varepsilon_0 K}\left(1 + \frac{2}{3} + \frac{3}{5}\right) = \frac{34}{15}\frac{d}{A\varepsilon_0 K}$$

$$\Rightarrow \quad C_{eq} = \frac{15}{34}\frac{A\varepsilon_0 K}{d}$$ **Ans. (D)**

Sol. 75 After removing the battery, the combination is shown as

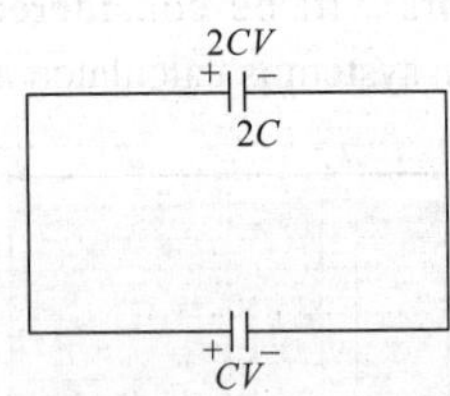

If after insertion of dielectric in capacitor of capacitance C, the voltage across the combination becomes V_1 then by conservation of charge, we have

$$2CV + CV = 2C(V_1 + Ck\,V_1)$$

$$\Rightarrow \quad V_1 = \frac{3CV}{2C + CK}$$

$$\Rightarrow \quad V_1 = \frac{3V}{2 + K}$$ **Ans. (B)**

Sol. 76 During charging capacitor charge is given as

$$V = V_0 e^{-\frac{t}{RC}}$$

$$\Rightarrow \quad 50 = 100e^{-\frac{t}{RC}}$$

$$\Rightarrow \quad t = RC \ln 2$$

$$\Rightarrow \quad t = 100 \times 10^{-6} \ln 2$$

$$\Rightarrow \quad t = 10^{-4} \times 0.693 \text{ sec} = 69.3\ \mu\text{sec}$$ **Ans. (B)**

Sol. 77 Considering one part of capacitor with dielectric as C_2 and remaining with pendulum as C_1, the equivalent circuit is drawn as shown below

The potential difference across the capacitors is given as

$$V = V_2 - (-V_1) = V_1 + V_2$$

Charge on each capacitor can be given as

$$q = \frac{C_1 C_2}{C_1 + C_2}(V_1 + V_2)$$

Potential difference across C_1 can be calculated as

$$V_1 = \frac{Q}{C_1} = \frac{C_2(V_1 + V_2)}{(C_1 + C_2)}$$

The electric field between plates of C_1 is given as

$$E = \frac{V_1}{(d - t)}$$

$$\Rightarrow \quad E = \frac{C_2(V_1 + V_2)}{(C_1 + C_2)(d - t)}$$

For equilibrium of pendulum bob, its free body diagram is shown in figure below

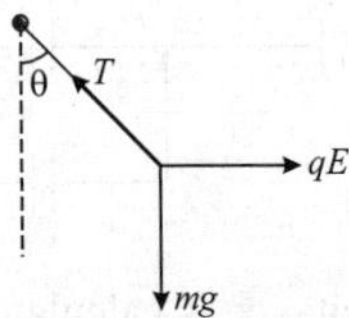

Force on pendulum bob due to electric field is given as

$$F = Eq = \frac{qC_2}{(C_1 + C_2)}\frac{(V_1 + V_2)}{(d - t)}$$

The equilibrium angle of string is given as

$$\theta = \tan^{-1}\left(\frac{qE}{mg}\right)$$

$$\Rightarrow \quad \theta = \tan^{-1}\left[\frac{q}{mg} \times \frac{C_2(V_1 + V_2)}{(C_1 + C_2)(d - t)}\right]$$ **Ans. (C)**

Sol. 78 For a capacitor with leaky dielectric we use

$$RC = \rho K \epsilon_0$$

Resistance of the leaky dielectric is given as

$$R=\frac{\rho \in_0 K}{C}$$

Leakage current in circuit is calculated as

$$I=\frac{V}{R}=\frac{V}{\rho}\cdot\frac{C}{\in_0 K}$$

$$\Rightarrow \quad I=\frac{40}{200}\times\frac{2\times10^{-12}}{8.85\times10^{-12}\times50}=0.9\text{ mA} \quad \textbf{Ans. (C)}$$

Sol. 79 The given capacitor can be drawn as equivalent circuit shown below considering it as a combination of three capacitors.

$C_1=\frac{(A/2)\varepsilon_0}{d}$ $\quad C_2=\frac{K_1(A/2)\varepsilon_0}{d/2}=\frac{k_1A\varepsilon_0}{d}$ $\quad C_3=\frac{K_2(A/2)\varepsilon_0}{d/2}=\frac{k_2A\varepsilon_0}{d}$

The equivalent capacitance of this system is given as

$$C_{eq}=C_1+\frac{C_3C_3}{C_2+C_3}$$

$$\Rightarrow \quad C_{eq}=\frac{A\varepsilon_0}{2d}+\frac{\frac{K_1A\varepsilon_0}{d}\times\frac{K_2A\varepsilon_0}{d}}{\frac{K_1A\varepsilon_0}{d}+\frac{K_2A\varepsilon_0}{d}}$$

$$\Rightarrow \quad C_{eq}=\frac{A\varepsilon_0}{2d}+\frac{A\varepsilon_0}{d}\left(\frac{K_1K_2}{K_1+K_2}\right)$$

$$\Rightarrow \quad C_{eq}=\frac{A\varepsilon_0}{d}\left(\frac{1}{2}+\frac{K_1K_2}{K_1+K_2}\right) \quad \textbf{Ans. (C)}$$

Sol. 80 Equivalent capacitance of the given system is calculated as

$$C_{eq}=C_1+\frac{C_2C_3}{C_2+C_3}$$

$$\Rightarrow \quad C_{eq}=2+\frac{6\times12}{6+12}=6\ \mu\text{F}$$

As C_2 and C_3 are in series, they have same charges and the ratio of charges on C_1 and C_2 are given as

$$\frac{Q_{C_1}}{Q_{C_2}}=\frac{C_1V}{\frac{C_2C_3}{C_2+C_3}V}=\frac{2\times(6+12)}{6\times12}=\frac{1}{2} \quad \textbf{Ans. (D)}$$

Sol. 81 Initially energy stored in capacitor is given as

$$E_i=\frac{1}{2}CV^2=\frac{1}{2}\times200\times20^{-6}\times(200)^2\text{ J}$$

After dielectric insertion, capacitance becomes

$$C'=kC=400\ \mu\text{F}$$

Final energy stored in capacitor is given as

$$E_f=\frac{1}{2}C'V^2=\frac{1}{2}\times400\times10^{-6}\times(200)^2\text{ J}$$

Change in energy stored in capacitor is given as

$$\Delta E=\frac{1}{2}\times(400-200)\times10^{-6}\times4\times10^4=4\text{J}$$

Ans. (4)

Sol. 82 Dielectric strength is the maximum electric field upto which it can withstand. Between plates of capacitor electric field is given as

$$E=\frac{\sigma}{k\in_0}=\frac{q}{Ak\in_0}$$

$$\Rightarrow \quad k=\frac{q}{A\in_0 E}=\frac{7\times10^{-6}}{30\pi\times10^{-4}\times\frac{1}{4\pi\times9\times10^9}\times3.6\times10^7}$$

$$\Rightarrow \quad k=\frac{36\times7}{30\times3.6}=2.33 \quad \textbf{Ans. (D)}$$

Sol. 83 Energy stored in a capacitor is given as $q^2/2C$, thus we have

$$U\propto q^2$$

The charge on capacitor is increased by 2 C due to which energy is increased by 44% or it becomes 1.44 times hence charge will become 1.2 times so we use

$$q_f-q=2$$

$$\Rightarrow \quad 1.2q-q=2$$

$$\Rightarrow \quad q=10\text{ C} \quad \textbf{Ans. (A)}$$

Sol. 84 Two capacitors in the given circuit are short circuited so we can remove these. Effective remaining circuit will be shown in figure below

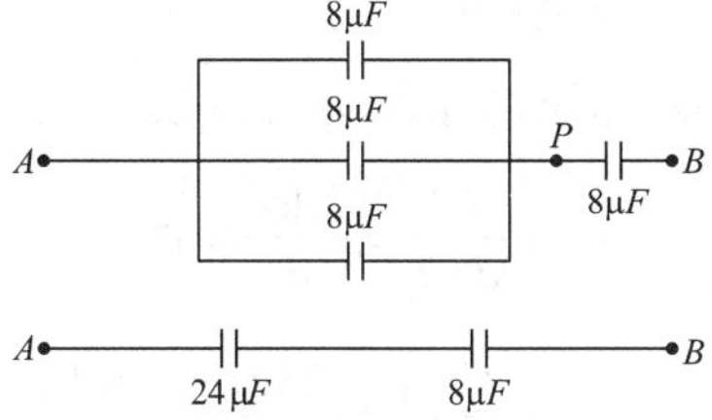

Finally equivalent capacitance across A and B is given as

$$C_{AB}=\frac{24\times8}{24+8}=\frac{24\times8}{32}=6\mu F \quad \textbf{Ans. (6)}$$

Sol. 85 Initial capacitance of capacitor is given as

$$C_1=\frac{\in_0 A}{d}$$

Final capcitance is given as

$$C_2 = \frac{\epsilon_0 A}{\frac{d}{2} + \frac{d/2}{\infty}} = \frac{2\epsilon_0 A}{d}$$

$$\Rightarrow \quad \frac{C_2}{C_1} = \frac{2}{1} \qquad \textbf{Ans. (A)}$$

Sol. 86 Charge on capacitor after charging is $C_1 V$ and now it is connected to the capacitor C_2. After distribution of charges, final charge on C_2 is given as

$$C_2 = \frac{C_2 \times Q_{\text{total}}}{C_{\text{total}}} = \frac{C_2[C_1 V]}{C_1 + C_2} = \frac{C_1 C_2 V}{C_1 + C_2} \qquad \textbf{Ans. (A)}$$

Sol. 87 As all capacitors are in series, charges will be same on all. The equivalent capacitance of this system is given as

$$\frac{1}{C_{eq}} = \frac{1}{10} + \frac{1}{15} + \frac{1}{20} = \frac{12+8+6}{120} = \frac{26}{120}$$

$$\Rightarrow \quad C_{eq} = \frac{60}{13}\,\mu F$$

Charge on capacitors is calculated as

$$Q = \frac{13 \times 60}{13} = 60\,\mu C \qquad \textbf{Ans. (A)}$$

Sol. 88 Original capacitance of the capacitor is given as

$$C_{\text{original}} = \frac{A\varepsilon_0}{d}$$

After dielectric is inserted, it can be considered two capacitors in series combination as shown in figure below

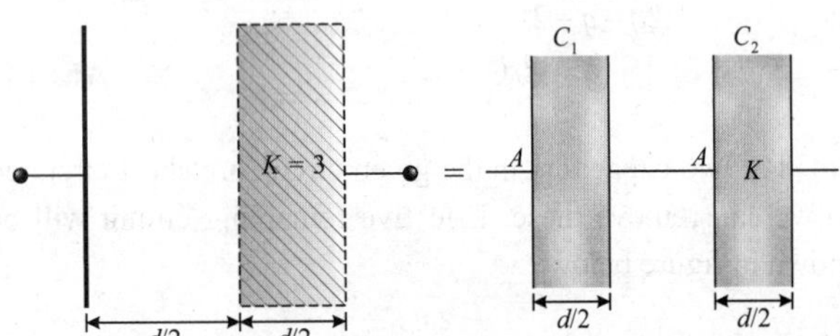

The capcitances of these individual capacitances are given as

$$C_1 = \frac{A\varepsilon_0}{d/2} = \frac{2A\varepsilon_0}{d} = C$$

$$\Rightarrow \quad C_2 = \frac{KA\varepsilon_0}{d/2} = \frac{2KA\varepsilon_0}{d} = \frac{6A\varepsilon_0}{d} = 3C$$

As C_1 & C_2 are in series, the equivalent capcitance of the combination is given as

$$C_{\text{new}} = \frac{C_1 C_2}{C_1 + C_2} = \frac{C \times 3C}{C + 3C} = \frac{3C}{4}$$

$$\Rightarrow \quad C_{\text{new}} = \frac{3}{4} \times \frac{2A\varepsilon_0}{d} = \frac{3}{2} \times \frac{A\varepsilon_0}{d}$$

$$\Rightarrow \quad C_{\text{new}} = \frac{3}{2} C_{\text{original}}$$

$$\Rightarrow \quad C_{\text{new}} = \frac{3}{2} \times 4 = 6\,\mu F \qquad \textbf{Ans. (C)}$$

Sol. 89 Figure below shows the situation described in the question

The force on charge particle is given as

$$F = qE = q\left(\frac{Q}{A\epsilon_0}\right) = \frac{qQ}{A\epsilon_0} = 10\text{ N}$$

When one plate is removed, electric field is reduced to half and given as

$$E' = \frac{Q}{2A\epsilon_0}$$

New force on charge particle will become

$$F = qE' = \frac{Qq}{2A\epsilon_0} = 5\text{ N} \qquad \textbf{Ans. (A)}$$

Sol. 90 Energy loss due to parallel combination with an identical capacitor is given as

$$\Delta E = \frac{1}{2}\frac{C_1 C_2}{C_1 + C_2}(V_1 - V_2)^2$$

$$\Rightarrow \quad \Delta E = \frac{1}{2}\frac{50 \times 50 \times 10^{-12} \times 10^{-12}}{(50+50)10^{-12}}(100-0)^2$$

$$\Rightarrow \quad \Delta E = 125\text{ nJ} \qquad \textbf{Ans. (125)}$$

Sol. 91 All the capacitances due to facing plates will be in parallel combination so effective capacitance of the given system will be calculated as

$$C_{eq} = \varepsilon_0 A\left[\frac{1}{5b} + \frac{1}{3b} + \frac{1}{b}\right] = \frac{23}{15}\frac{\varepsilon_0 A}{b} \qquad \textbf{Ans. (23)}$$

Sol. 92 Statement-I : When a charge is placed in external electric field then in direction parallel to the field the point ahead and behind of charge electric field is added and subtracted so at one point field is increased hence this statement is true.

Statement-II : If the dipole is kept at the point where the electric field is non uniform, at both the charges of dipole field can be different so force on it will be non zero hence this statement is also true. **Ans. (A)**

Sol. 93 Before closing the switch, charge on capacitor is

$$Q = C_1 V_0 = 5 \times 30 = 150\,\mu C$$

After closing the switch, common potential difference across the capacitor combination is given as

$$V = \frac{Q}{C_1 + C_2} = \frac{150}{10+5} = 10\text{ V}$$

Final charge on second capacitor is calculated as

$$Q_2 = C_2 V = 10 \times 10 = 100\ \mu\text{C}$$ **Ans. (100)**

Sol. 94 The energy stored in capacitor after charging is given as

$$E = \frac{1}{2}(KC)V^2$$

As battery remains connected, potential difference doesn't change only dielectric constant changes so percentage change in energy of capacitor will be given as

$$\frac{\Delta E}{E} \times 100 = \frac{\frac{1}{2}K_2CV^2 - \frac{1}{2}K_1CV^2}{\frac{1}{2}K_1CV^2} \times 100$$

$$\Rightarrow \quad \frac{\Delta E}{E} \times 100 = \frac{K_2 - K_1}{K_1} \times 100$$

$$\Rightarrow \quad \frac{\Delta E}{E} \times 100 = \frac{15-10}{10} \times 100 = 50\%$$ **Ans. (A)**

Sol. 95 For a capacitor with varying applied voltage across plates, displacement current is equal to conduction Current

$$I_c = \frac{dq}{dt}$$

$$I_d = \frac{\epsilon_0 A}{d}\frac{dV}{dt}$$

$$\Rightarrow \quad d = \frac{8.85 \times 10^{-12} \times 4 \times 10^{-3} \times 10^{6}}{4.425 \times 10^{-6}}$$

$$\Rightarrow \quad d = 8\text{ mm}$$

$$\Rightarrow \quad x = 8$$ **Ans. (8)**

Sol. 96 Charge on capacitor is given as

$$Q = CV$$

$$\Rightarrow \quad V = \frac{1}{C}Q$$

This shows that graph is a straight line with slope $\frac{1}{C}$, calculated as

$$\frac{1}{C} = \frac{1}{2 \times 10^{-6}} = 5 \times 10^5$$

This matches with option (A). **Ans. (A)**

Sol. 97 Figure below shows the situation described in the question

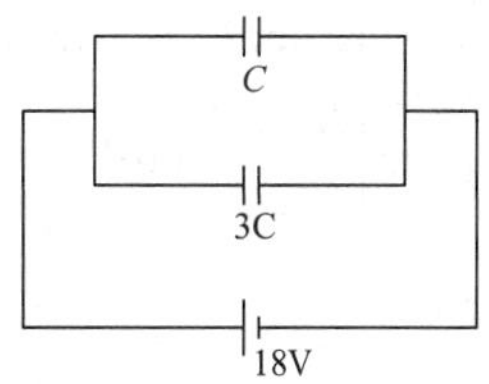

Initial charge on capacitor C is $18C$ and that on capacitor $3C$ is $54C$. Now battery is disconnected and dielectric is filled so capacitance C changes to $9C$. If common potential difference after filling dielectric becomes V', we use conservation of charge as

$$9CV' + 3CV' = 18C + 54C$$

$$\Rightarrow \quad 12CV' = 72C$$

$$\Rightarrow \quad V' = 6\text{ V}$$ **Ans. (6)**

Sol. 98 Figure below shows the circuit as described in question.

$q_1C_1 = 1\ \mu$F
$q_2C_2 = 2\ \mu$F
$q_3C_3 = 4\ \mu$F
$q_4C_4 = 3\ \mu$F
20 V

Total charge on system is given as

$$q_T = q_1 + q_2 + q_2 + q_4$$

$$\Rightarrow \quad q_T = 1 \times 20 + 2 \times 20 + 4 \times 20 + 3 \times 20$$

$$\Rightarrow \quad q_T = 200\ \mu\text{C}$$ **Ans. (A)**

Sol. 99 In the circuit shown below if low potential side of battery is considered at zero potential the potential difference across lower dielectric will be the potential of foil.

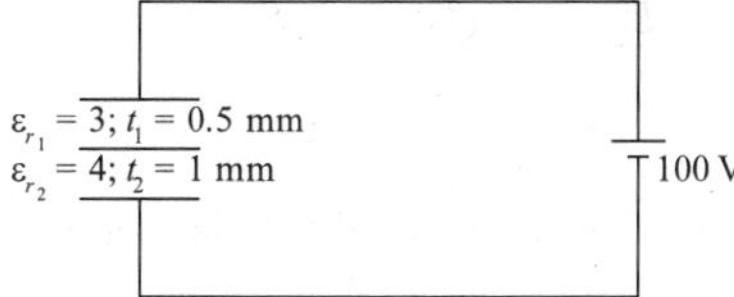

Capacitance of the two capacitor with different dielectrics are calculated as

$$C_1 = \frac{\epsilon_0 (3)A}{0.5} = 6A\epsilon_0$$

and $$C_2 = \epsilon_0(4)A = 4A\epsilon_0$$

Effective capacitance of the combined capacitor is calculated as

$$C_{eq} = \frac{C_1C_2}{C_1 + C_2} \Rightarrow \frac{24}{10}A\epsilon_0$$

Charge on the two capacitors in series is calculated as

$$q_{net} = C_{eq}(\Delta V) = 240A\epsilon_0$$

Potential drop across capacitor C_2 is calculated as

$$\Delta V_2 = \frac{240 A \epsilon_0}{4 A \epsilon_0} = 60\text{V}$$

$$\Rightarrow \quad V_{\text{foil}} = 60\text{V} \qquad \textbf{Ans. (60)}$$

Sol. 100 When K is closed the two capacitors in the circuit will be in parallel combination so effective capacitance is given as

$$C_{\text{eq}} = 2\text{C}$$

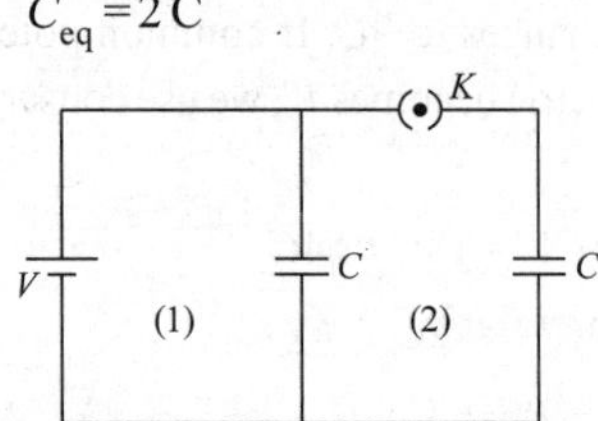

Energy stored in combination is calculated as

$$E_1 = \frac{1}{2} C_{\text{eq}} V^2$$

$$\Rightarrow \quad E_1 = \frac{1}{2} 2C \times V^2 = CV^2$$

When switch is opened charge on right capacitor remain constant at CV whereas potential difference on left capacitor remain same at V. When dielectric with $k = 5$ is inserted in both the capacitors, their capacitance becomes $kC = 5C$. Thus final energy stored in the two capacitors is now given as

$$E_2 = \frac{1}{2}(5C)V^2 + \frac{(CV)^2}{2(5C)}$$

$$\Rightarrow \quad E_2 = \frac{5CV^2}{2} + \frac{CV^2}{10} = \frac{13CV^2}{5}$$

Ratio of the energies before and after opening the key is given as

$$\frac{E_1}{E_2} = \frac{CV^2}{\frac{13CV^2}{5}} = \frac{5}{13} \qquad \textbf{Ans. (C)}$$

Sol. 101 Figure below shows the capacitor desribed in the question.

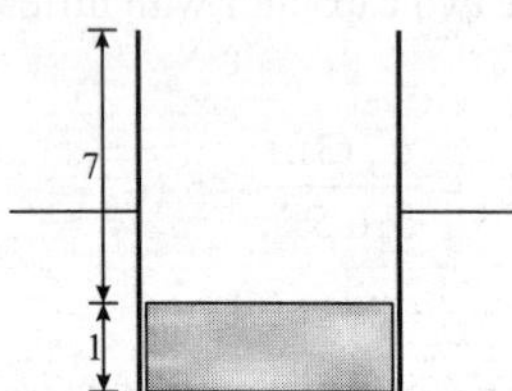

Capacitance of this partly dielectric filled capacitor is calculated as

$$C = \left[\frac{\varepsilon_0(7\times 4)}{4/10} + \frac{5\varepsilon_0(1\times 4)}{4/10}\right] \times 10^{-2}$$

$$\Rightarrow \quad C = 1.2\,\varepsilon_0$$

Energy stored in this capacitor is given as

$$E = \frac{1}{2} C_{\text{eff}} V^2$$

$$\Rightarrow \quad E = \frac{1}{2}(1.2)\varepsilon_0(20)(20) = 240\,\varepsilon_0 \qquad \textbf{Ans. (240)}$$

Sol. 102 Figure below shows the capacitances as described in question

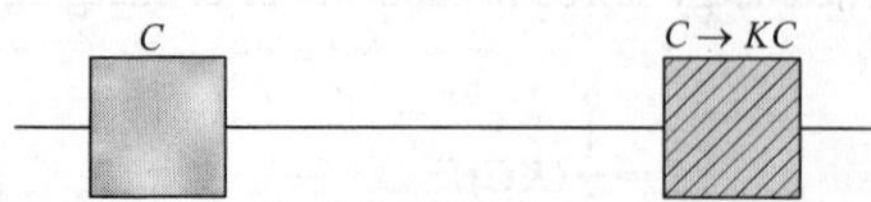

The equivalent capacitance of the series combination of these two capacitors will be given as

$$C_{\text{eq}} = \frac{C(kC)}{C + kC} = \frac{kC}{k+1}$$

$$\Rightarrow \quad 24 = \frac{k(40)}{k+1}$$

$$\Rightarrow \quad k = 1.5 \qquad \textbf{Ans. (A)}$$

Sol. 103 Figure below shows the capacitor descirbed in question.

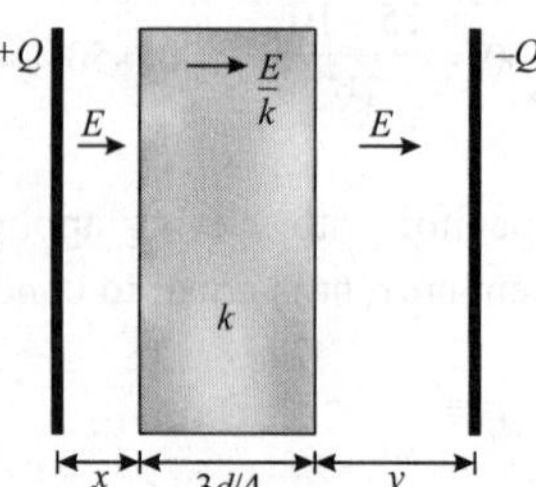

Capacitance of partly dielectric filled parallel plate capacitor is given as

$$C = \frac{\epsilon_0 A}{\frac{3d/4}{k} + \frac{d/4}{1}}$$

$$\Rightarrow \quad C = \frac{\epsilon_0 A}{d}\left(\frac{4k}{3+k}\right) = \frac{4kC_0}{3+k} \qquad \textbf{Ans. (A)}$$

Sol. 104 Charge on the facing surfaces of metal plates will be equal and opposite and is considered as charge on the formed capacitor C it is given as

$$q = \frac{q_1 - q_2}{2}$$

The electric field between the plates is calculated as

$$E = \frac{\sigma}{\epsilon_0} = \frac{q_1 - q_2}{2A\epsilon_0}$$

Potential difference between the plates is given as

$$V = Ed = \frac{q_1 - q_2}{2A\epsilon_0} d$$

$$\Rightarrow \quad V = \frac{q_1 - q_2}{2C} \qquad \textbf{Ans. (C)}$$

Sol. 105 Capacitance of air capacitor is given as

$$C_0 = \frac{\epsilon_0 A}{d} = 15 \text{ pF}$$

When the separation is doubled and it is filled with a medium of dielectric constant 3.5, the new capacitance is calculated as

$$C = \frac{K \epsilon_0 A}{2d} = \frac{3.5}{2} \times 15 \text{ pF}$$

$$\Rightarrow \quad C = \frac{105}{4} \text{ pF} \qquad \textbf{Ans. (105)}$$

Sol. 106 Figure below shows the capacitor as described in the question.

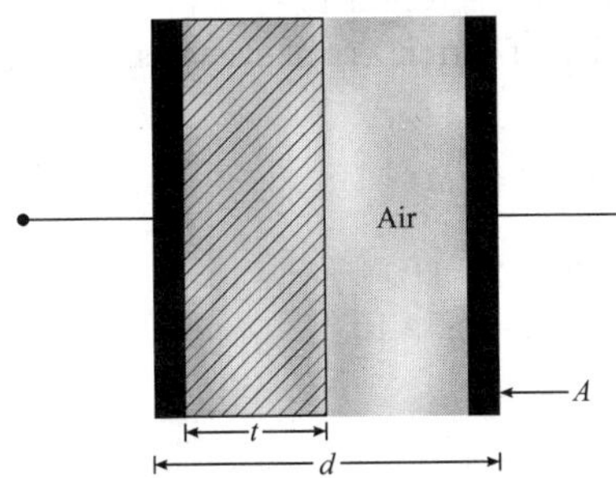

This can be considered as two capacitors in series combination for which equivalent capacitance is given as

$$\frac{1}{C_{eq}} = \frac{1}{C_1} + \frac{1}{C_2}$$

$$\Rightarrow \quad \frac{1}{C_{eq}} = \frac{1}{\frac{K \epsilon_0 A}{t}} + \frac{1}{\frac{\epsilon_0 A}{d-t}}$$

$$\Rightarrow \quad \frac{1}{C_{eq}} = \frac{t}{K \epsilon_0 A} + \frac{d-t}{\epsilon_0 A}$$

$$\Rightarrow \quad \frac{1}{C_{eq}} = \frac{1 \times 10^{-3}}{5 \epsilon_0 \times 40 \times 10^{-4}} + \frac{1 \times 10^{-3}}{\epsilon_0 \times 40 \times 10^{-4}}$$

$$\Rightarrow \quad \frac{1}{C_{eq}} = \frac{1}{20 \epsilon_0} + \frac{1}{4 \epsilon_0}$$

$$\Rightarrow \quad C_{eq} = \frac{24 \times 4 \epsilon_0}{24} = \frac{10 \epsilon_0}{3} \text{ F} \qquad \textbf{Ans. (C)}$$

Sol. 107 Figure below shows the capacitance as described in question.

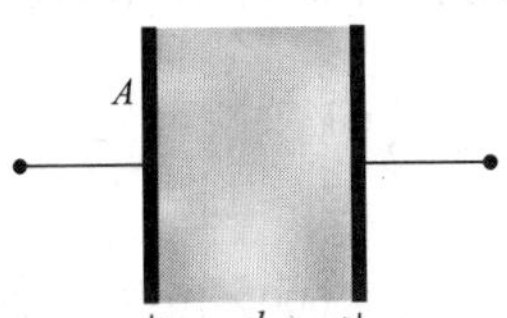

Capacitance of this capacitor is given as

$$\frac{\epsilon_0 A}{d} = 5 \text{ μF}$$

Now a dielectric slab is inserted in half the space as shown in figure.

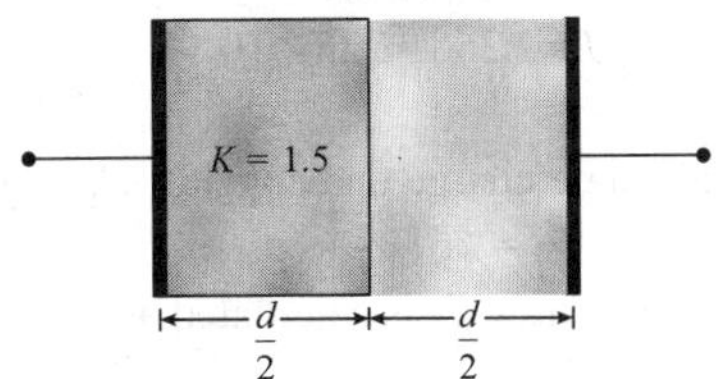

New capacitance of capacitor is calculated as

$$C_{\text{new}} = \frac{\epsilon_0 A}{\frac{\left(\frac{d}{2}\right)}{1.5} + \frac{\left(\frac{d}{2}\right)}{1}}$$

$$\Rightarrow \quad C_{\text{new}} = \frac{\epsilon_0 A}{\left(\frac{d}{3} + \frac{d}{2}\right)} = \frac{6 \epsilon_0 A}{5d}$$

$$\Rightarrow \quad C_{\text{new}} = \frac{6}{5} \times 5 \text{ μF} = 6 \text{ μF} \qquad \textbf{Ans. (6)}$$

Sol. 108 Figure below shows the situation described in the question when capacitor is connected to a battery.

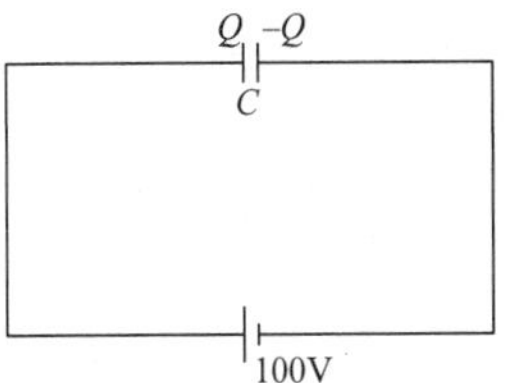

Steady state charge on capacitor is given as

$$Q = CV = 900 \times 10^{-6} \times 100 = 9 \times 10^{-2}$$

$$\Rightarrow \quad Q = 90 \text{ mC}$$

Now the charged capacitor is connected to another similar uncharged capacitor as shown below.

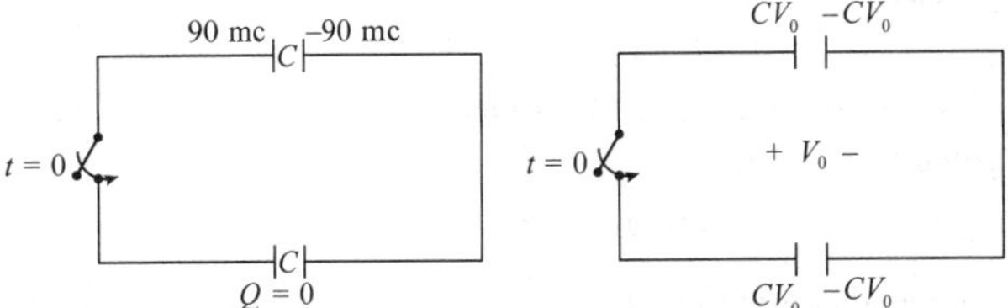

After connections, if the common potential developed across both capacitors is V_0 we use

Total charge on left plates of capacitors should be conserved.

$$2CV_0 = 90 \text{ mC}$$

$$\Rightarrow \quad CV_0 = 45 \text{ mC}$$

Heat dissipated in circuit is calculated as

$$H = U_i - U_f$$

$$\Rightarrow \quad H = \frac{1}{2} \frac{(90 \times 10^{-3})^2}{900 \text{μF}} - 2 \times \frac{1}{2} \frac{(45 \times 10^{-3})^2}{900 \text{μF}}$$

$$\Rightarrow \quad H = \frac{1}{2\times 900\times 10^{-6}}(8100-4050)\times 10^{-6}$$

$$\Rightarrow \quad H = 2.25\text{ J} = 225\times 10^{-2}\text{ J}$$ **Ans. (225)**

Sol. 109 Steady state charge on capacitor C_1 is given as

$$q_1 = kCE$$

And charge on capacitor C_2 is calculated as

$$q_2 = CE$$

When they are connected in parallel charge will be equally divided so charge on one capacitor is given as

$$q = \frac{k+1}{2}CV$$

Thus potential difference across first capacitor is calculated as

$$V = \frac{q}{kC} = \frac{k+1}{2k} = 55\text{ V}$$ **Ans. (55)**

Sol. 110 In the given situation shown in figure for right direction the unit vector is considered to be $\hat{n}$ so for the three regions, the electric fields are calculated as

$$\vec{E}_{\text{I}} = \frac{\sigma}{2\epsilon_0}(-\hat{n}) + \frac{\sigma}{2\epsilon_0}(-\hat{n}) = \frac{\sigma}{\epsilon_0}\hat{n}$$

and $\vec{E}_{\text{II}} = 0$,

and $$\vec{E}_{\text{III}} = \frac{\sigma}{2\epsilon_0}(\hat{n}) + \frac{\sigma}{2\epsilon_0}(\hat{n}) = \frac{\sigma}{\epsilon_0}(\hat{n})$$ **Ans. (D)**

Sol. 111 For solid as well as hollow sphere, the potential of a conducting sphere is given as

$$V = \frac{KQ}{R}$$

For $V_1 = V_2$

and $R_1 = R_2$

we have

$$Q_1 = Q_2$$

Thus assertion is false and the capacitance of metallic sphere is given as

$$C = 4\pi\varepsilon_0 R,$$

It depends on radius of sphere so reason is true. **Ans. (A)**

Sol. 112 For $x = \frac{d}{3}$ the capacitance of system is given as

$$C_1 = \frac{\epsilon_0 A}{\left(\frac{d/3}{k} + \frac{2d}{3}\right)} = \frac{\epsilon_0 A}{\frac{d}{12} + \frac{2d}{3}}$$

$$\Rightarrow \quad C_1 = \frac{\epsilon_0 A}{d}\times\left(\frac{12}{9}\right)$$

$$\Rightarrow \quad C_1 = \frac{4}{3}\frac{\epsilon_0 A}{d} = 2\ \mu\text{F}$$

For $x = \frac{2d}{3}$ the capacitance of system is given as

$$C_2 = \frac{\epsilon_0 A}{\left(\frac{2d/3}{k} + \frac{d}{3}\right)} = \frac{\epsilon_0 A}{d}\times 2$$

$$\Rightarrow \quad C_2 = \frac{6}{4}\times 2 = 3\ \mu\text{F}$$ **Ans. (3)**

Sol. 113 In a pure capacitive circuit, reactance is given as

$$\text{Reactance } X_C = \frac{1}{\omega C}$$

Peak value of EMF is given as $E_0 = 36$ V

Thus peak value of current is calculated as

$$I_0 = \frac{E_0}{x_c} = \frac{E_0}{\frac{1}{\omega_C}} = E_0\omega_C$$

$\Rightarrow \quad I_0 = 36\times 120\,\pi\times 150\times 10^{-6}$

$\Rightarrow \quad I_0 = 2.03$

$\Rightarrow \quad I_0 = 2$ A **Ans. (A)**

Sol. 114 Capacitance of capacitance when $C = 1$ mm

$$C = \frac{\varepsilon_0 A}{(d-c)}$$

$$\Rightarrow \quad C = \frac{\varepsilon_0\times 200\times 10^{-4}}{4\times 10^{-3}} = \varepsilon_0\times 5 = x\,\varepsilon_0\text{ F}$$

$$\Rightarrow \quad x = 5$$

This situation is equivalent to a conducting slab placed between the plates. **Ans. (5)**

Sol. 115 Figure below shows the circuit as described in the question.

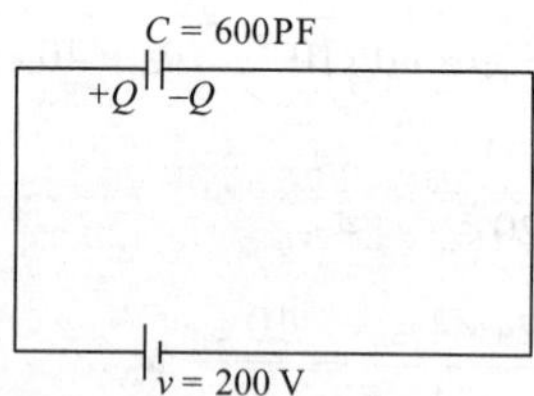

Charge stored in capacitor is given as

$$Q = CV = 600\times 10^{-12}\times 200 = 12\times 10^{-8}\text{ C}$$

Initial energy stored in capacitor is given as

$$U_i = \frac{1}{2}CV^2$$

$$\Rightarrow \quad U_i = \frac{1}{2}\times 600\times 10^{-12}\times (200)^2 = 12\ \mu\text{J}$$

When connected to another uncharged capacitor charge will be equally distributed on identical capacitor as shown in figure below

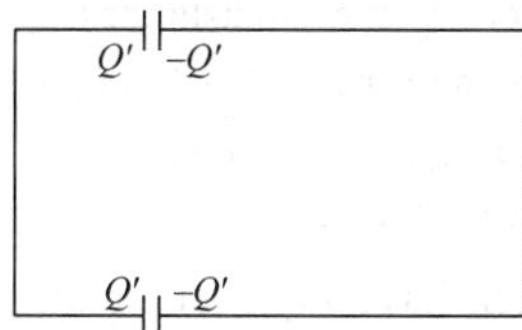

Thus final charges on the two capacitors will be

$$Q' = \frac{Q}{2} = 6 \times 10^{-8}$$

Final energy stored in the capacitors is calculated as

$$U_f = 2 \times \frac{Q'^2}{2C} = \frac{Q'^2}{C}$$

$$\Rightarrow \quad U_f = \frac{(6\times10^{-8})^2}{600\times10^{-12}} = 6\ \mu\text{J}$$

Energy lost is calculated as

$$\Delta E = (12-6)\ \mu\text{J} = 6\ \mu\text{J}$$ **Ans. (6)**

Sol. 116 In the circuit two capacitances are short circuited so the circuit can be simplified and redrawan as shown in figure below.

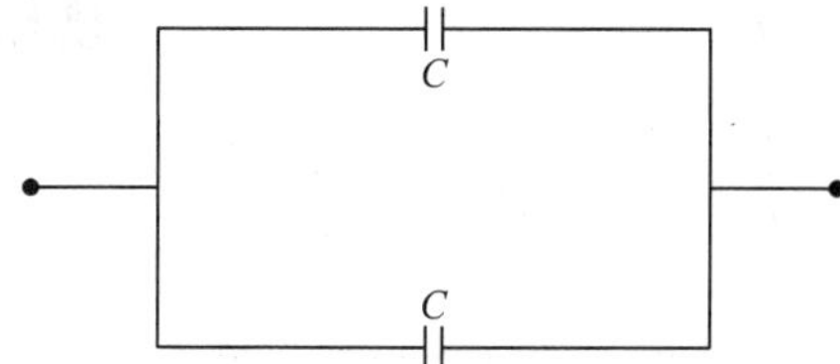

Thus in Parallel combination equivalent capacitance is

$$C_{eq} = C + C = 2C$$ **Ans. (C)**

Sol. 117 For metal dielectric constant is taken infinite so we have

$$K_{\text{metal sheet}} = \infty$$

Thickness of metal sheet

$$t = \frac{2d}{3}$$

When air is between the plates then capacitance is given as

$$C_1 = \frac{\epsilon_0 A}{d}$$

When metal sheet is introduced between plates, capacitance is given as

$$C_2 = \frac{\epsilon_0 A}{d - t + \frac{t}{k}} = \frac{\epsilon_0 A}{d - \frac{2d}{3} + 0} = 3C_1$$

Thus ratio of capacitances is calculated as

$$\Rightarrow \quad \frac{C_2}{C_1} = 3:1$$ **Ans. (C)**

Sol. 118 Torque about a given point is calculated as

$$\tau = \vec{r} \times \vec{F}$$

Where $\vec{r} = -2\hat{i} + 3\hat{j}$

Thus torque about (2, – 3) is

$$\tau = (-2\hat{i} + 3\hat{j}) \times (-P\hat{k})$$

$$\Rightarrow \quad \tau = P(-2\hat{j} - 3\hat{i}) = P(-3\hat{i} - 2\hat{j})$$

Thus $a = -3$ and $b = -2$ so we have

$$\frac{a}{b} = \frac{3}{2} = \frac{x}{2}$$

$$\Rightarrow \quad x = 3$$ **Ans. (3)**

Sol. 119 Initially charge on capacitor

$$Q_1 = CV = (2)V$$

Energy stored in capacitor is calculated as

$$E_1 = 1/2\, CV^2 = 1/2\,(2)\, V^2 = V^2$$

When charged capacitor connected to uncharged capcitor, charge on each capacitor, will become

$$Q_2 = \frac{Q_1}{2} = \frac{2V}{2} = V$$

Now energy stored will become

$$E_2 = 2\left(\frac{1}{2}\frac{Q_2^2}{C}\right) = \frac{V^2}{2}$$

$$\Rightarrow \quad \frac{E_2}{E_1} = \frac{1}{2}$$ **Ans. (B)**

Sol. 120 Flux through Gaussian surface enclosing positive charge plate of capacitor is given as

$$\phi = \frac{q_{in}}{\epsilon_0}$$

Charge stored on capacitor plate of capacitance C is

$$Q = CV$$

$$\Rightarrow \quad \phi = \frac{CV}{\epsilon_0}$$ **Ans. (C)**

Sol. 121 In the circuit shown below the equivalent capacitance is calculated as

$$C_{eq} = 0.5\ \mu\text{F}$$

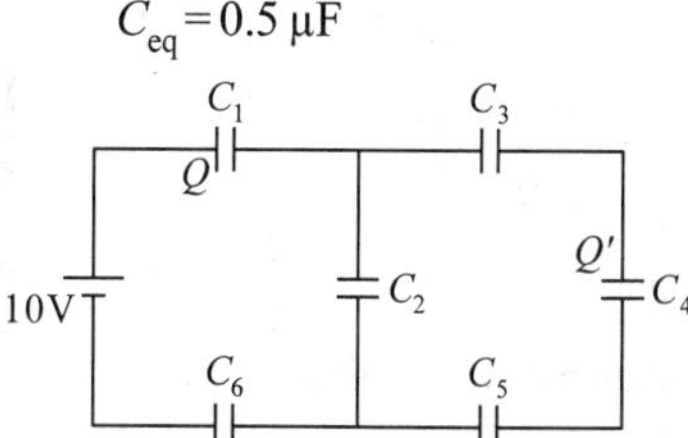

When potential difference of 10 V is applied then total charge supplied by the battery is

$$Q = 0.5 \times 10 = 5\ \mu\text{C}$$

Charge stored on capacitor C_4 can be calculated as

$$Q' = \frac{5\mu C \times 0.8}{0.8+0.2} = 4\,\mu\text{C}$$ **Ans. (4)**

Sol. 122 Rotational Kinetic Energy is given as

$$K = \frac{1}{2}I\omega^2$$

$$\Rightarrow \quad K = \frac{1}{2}\frac{(I\omega)^2}{I}$$

$$\Rightarrow \quad K = \frac{L^2}{2I}$$

Thus ratio of Kinetic Energies is given as

$$\frac{K_{\text{final}}}{K_{\text{initial}}} = \frac{I_{\text{initial}}}{I_{\text{final}}}$$

$$\Rightarrow \quad \frac{K_{\text{final}}}{E} = \frac{1}{3}$$

$$\Rightarrow \quad K_{\text{final}} = \frac{E}{3}$$ **Ans. (3)**

Sol. 123 In parallel potential remain constant and charge on capacitor is

$$Q = CV$$

Charge on C_1 is $Q_1 = 2 \times 10 = 20\,\mu\text{C}$ ...(1)

Charge on C_2 is $Q_2 = x \times 10 = 10x\,\mu\text{C}$ ...(2)

Charge on C_3 is $Q_3 = 3 \times 10 = 30\,\mu\text{C}$ ...(3)

Total charge $20 + 10x + 30 = 100$

$\Rightarrow x = 5$ **Ans. (5)**

17.1 Current Electricity

Sol. 1 If a resistance wire is of length l and cross sectional area A then its resistance is given as

$$R = \rho\frac{l}{A} \quad [\rho \to \text{wire resistivity}]$$

volume of wire $\quad V = RA$

$[r \to \text{wire resistivity}]$

If length of wire is stretched by 100%

$\Rightarrow \quad l_f = 2l$

As $V \to$ constant $\Rightarrow A_f = \frac{A}{2}$

$\Rightarrow$ Final resistance of wire will become

$$R_f = \rho\frac{l_f}{A_f} = \rho\left(\frac{2l}{A/2}\right) = 4\rho\frac{l}{A} = 4R$$

Change $= R_f - R = 4R - R = 3R$ **Ans. (D)**

Sol. 2 Thermistors are such resistances of which resistance changes rapidly with small change in this temperature. These are used mainly as temperature sensing devices.

Metal oxides which have high temperature coefficients of resistivity are most suitable for making thermistors. **Ans. (C)**

Sol. 3 $\frac{l_1}{l_2} = \frac{4}{3}$ and $\frac{r_1}{r_2} = \frac{2}{3} \Rightarrow \frac{A_1}{A_2} = \left(\frac{2}{3}\right)^2 = \frac{4}{9}$

Resistances of the given wires

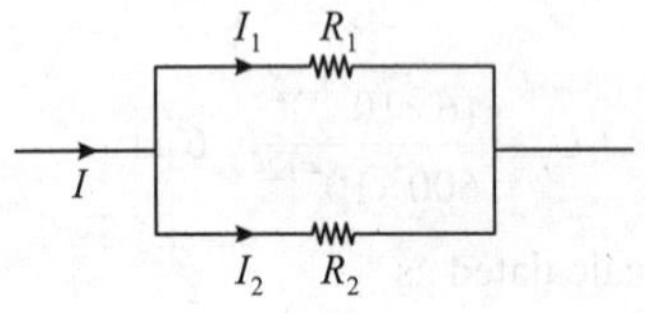

$$R_1 = \rho\frac{l_1}{A_1} \text{ and } R_2 = \rho\frac{l_2}{A_2}$$

$$\Rightarrow \quad \frac{R_1}{R_2} = \frac{l_1}{A_1} \times \frac{A_2}{l_2} = \frac{4}{3} \times \frac{9}{4} = \frac{3}{1}$$

$$\Rightarrow \quad I_1R_1 = I_2R_2$$

$$\Rightarrow \quad \frac{I_1}{I_2} = \frac{R_2}{R_1} = \frac{1}{3}$$ **Ans. (B)**

Sol. 4 If the given resistances are R_1 and R_2

$\Rightarrow S = R_1 + R_2$ and $P = \frac{R_1R_2}{R_1 + R_2}$

We use $\quad S = nP$

$$R_1 + R_2 = n\left(\frac{R_1R_2}{R_1 + R_2}\right)$$

$$\Rightarrow \quad n = \frac{(R_1 + R_2)^2}{R_1R_2} = \frac{R_1^2 + R_2^2 + 2R_1R_2}{R_1R_2}$$

$$\Rightarrow \quad n = \frac{R_1}{R_2} + \frac{R_2}{R_1} + 2$$

$$\Rightarrow \quad n = \left(\sqrt{\frac{R_1}{R_2}} - \sqrt{\frac{R_2}{R_1}}\right)^2 + 4$$

here n is minimum when $R_1 = R_2 \Rightarrow n_{\text{min}} = 4$ **Ans. (A)**

Sol. 5 We use $\quad R_T = R_0(1 + \alpha T)$

where $\quad R_T \to$ resistance at T°C

$R_0 \to$ resistance at 0°C

Initially at 100°C, $R_T = 100\Omega$

$\Rightarrow \quad 100 = R_0(1 + 100\alpha)$ ...(1)

Later at T°C, $\quad R_T = 200\Omega$

$\Rightarrow \quad 200 = R_0(1 + \alpha T)$ ...(2)

$$(2)\div(1) \Rightarrow \quad \frac{200}{100}=\frac{1+\alpha T}{1+100\alpha}$$

$$\Rightarrow \quad 2+200\,\alpha=1+\alpha T$$

$$\Rightarrow \quad T-200=\frac{1}{\alpha}=\frac{1000}{5}=200$$

$$\Rightarrow \quad T=200+200+400°\text{C}$$

Ans. (B)

Sol. 6 $\rho_B=2\rho_A$

and $r_B=2r_A \Rightarrow A_B=4A_A$

As resistances of the two wires are equal

$$\Rightarrow \quad R_A=R_B$$

$$\Rightarrow \quad \rho_A\cdot\frac{l_A}{A_A}=\rho_B\cdot\frac{l_B}{A_B}$$

$$\Rightarrow \quad \frac{l_B}{l_A}=\frac{\rho_A}{\rho_B}\cdot\frac{A_B}{A_A}=\frac{1}{2}\times\frac{4}{1}=2$$

Ans. (D)

Sol. 7 If resistance at 0°C is considered to be R_0, we use at t°C

$$R_t=R_0(1+\alpha t)$$

at 50°C, we have

$$5=R_0(1+50\alpha)$$

$$\Rightarrow \quad 5-R_0=50R_0\alpha \quad \ldots(1)$$

at 100°C, we have

$$6=R_0(1+100\alpha)$$

$$\Rightarrow \quad 6-R_0=100R_0\alpha \quad \ldots(2)$$

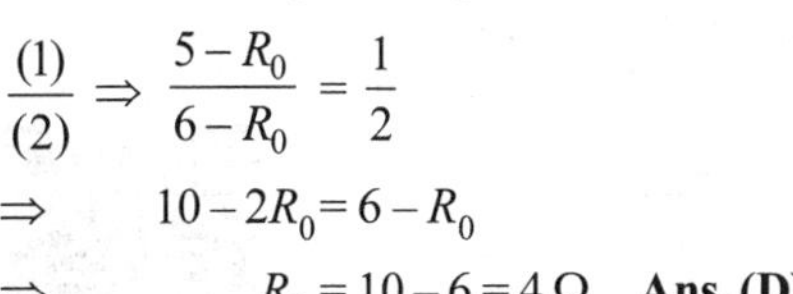

$$\frac{(1)}{(2)} \Rightarrow \frac{5-R_0}{6-R_0}=\frac{1}{2}$$

$$\Rightarrow \quad 10-2R_0=6-R_0$$

$$\Rightarrow \quad R_0=10-6=4\,\Omega$$

Ans. (D)

Sol. 8 Current density at a distance r from point A is given as

$$J=\frac{I}{2\pi r^2}$$

Electric field at a distance r from A is

$$E(r)=\rho J=\frac{\rho I}{2\pi r^2}$$

Ans. (C)

Sol. 9 Here potential difference between points B and C can be given as

$$\Delta V_{BC}=\int_B^C E_{(r)}\cdot dr$$

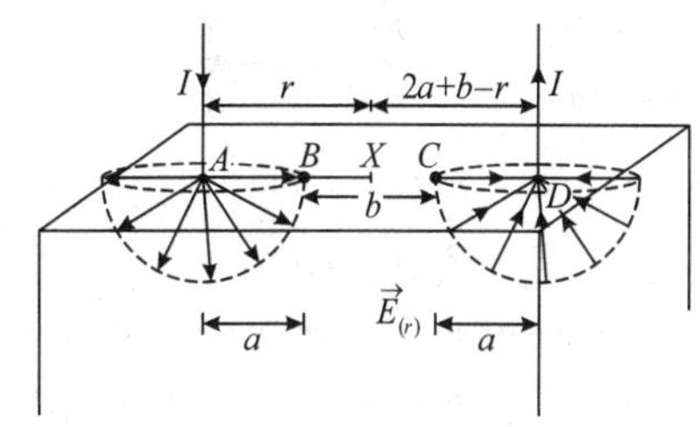

At point X (at a distance r from A) the net EF is given as

$$E_{(r)}=E_{\text{due to }A}+E_{\text{dut to }D}$$

$$E_{(r)}=\frac{\rho I}{2\pi r^2}+\frac{\rho I}{2\pi(2a+b-r)^2}$$

potential difference between points B and C is given as

$$\Delta V_{BC}=\int_a^{a+b}\frac{\rho I}{2\pi}\left(\frac{1}{r^2}+\frac{1}{(2a+b-r)^2}\right)\cdot dr$$

$$\Rightarrow \quad \Delta V_{BC}=\frac{\rho I}{2\pi}\left[-\frac{1}{r}+\frac{1}{2a+b-r}\right]_a^{a+b}$$

$$\Rightarrow \quad \Delta V_{BC}=\frac{\rho I}{2\pi}\left[\left(-\frac{1}{a+b}+\frac{1}{a}\right)-\left(-\frac{1}{a}+\frac{1}{a+b}\right)\right]$$

$$\Rightarrow \quad \Delta V_{BC}=\frac{\rho I}{2\pi}\left[\frac{2}{a}-\frac{2}{a+b}\right]$$

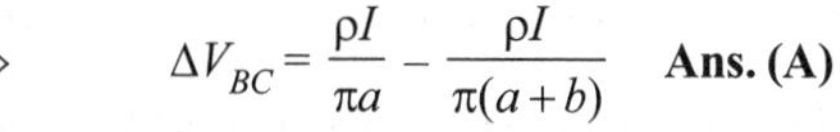

$$\Rightarrow \quad \Delta V_{BC}=\frac{\rho I}{\pi a}-\frac{\rho I}{\pi(a+b)}$$

Ans. (A)

Sol. 10 Using $R=R_0(1+\alpha\Delta T)$

$$\Rightarrow \quad 100=R_0(1+\alpha(27)) \quad \ldots(1)$$

$$\Rightarrow \quad 150=R_0(1+\alpha(227)) \quad \ldots(2)$$

$$(2)\div(1) \Rightarrow \quad \frac{150}{100}=\frac{1+\alpha(227)}{1+\alpha(27)}$$

$$\Rightarrow \quad \frac{3}{2}=(1+227\,\alpha)(1+27\,\alpha)^{-1}$$

For v small $\alpha \Rightarrow \frac{3}{2}=1+200\,\alpha$

$$\Rightarrow \quad \alpha=\frac{1}{400}=2.5\times10^{-3}\,°\text{C}^{-1}$$

As per given values for small variations in resistance and small range of temperature variation S-1 is false.

But 27°C to 227°C is large variation as α is considered constant.

$\Rightarrow$ for given values S-1 is true if α is considered constant.

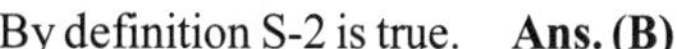

By definition S-2 is true. **Ans. (B)**

Sol. 11 For series combination at a temperature T we use

$$R_{eq}=2R_0(1+\alpha_S T) \quad \ldots(1)$$

(equivalent coefficient for series combination)

at 0°C R_0 R_0
at T°C R_1 R_2

Here we have $R_1 = R_0(1+\alpha_1 T)$

and $R_2 = R_0(1+\alpha_2 T)$

Equivalent resistance of series combination as (at T°C)

$$R_{eq} = R_1 + R_2 = R_0(1+\alpha_1 T) + R_0(1+\alpha_2 T)$$

$$\Rightarrow \quad R_{eq} = R_0(2+(\alpha_1+\alpha_2)T)$$

$$\Rightarrow \quad R_{eq} = 2R_0\left(1+\left(\frac{\alpha_1+\alpha_2}{2}\right)T\right) \quad \ldots(2)$$

Comparing (1) and (2) gives

$$\alpha_s = \frac{\alpha_1+\alpha_2}{2}$$

$R_0 \to R_1$

$R_0 \to R_2$

at 0°C at T°C

For parallel combination

At 0°C we have $R'_{eq} = \dfrac{R_0}{2}$

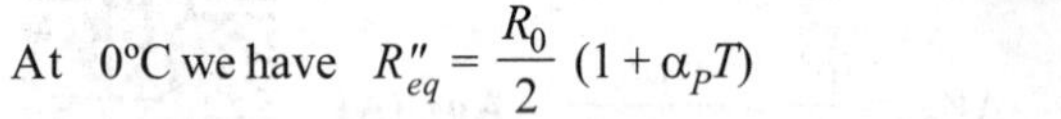

At 0°C we have $R''_{eq} = \dfrac{R_0}{2}(1+\alpha_p T)$ $\ldots(3)$

[$\alpha_p T$ = equal coefficient for parallel combination]

Equal resistance at T°C is given as

$$R''_{eq} = \frac{R_1R_2}{R_1+R_2} = \frac{R_0^2(1+\alpha_1 T)(1+\alpha_2 T)}{R_0(2+(\alpha_1+\alpha_2)T)}$$

[terms of consider $\alpha_1\,\alpha_2 << 1$]

$$\Rightarrow \quad R''_{eq} = \frac{R_0(1+(\alpha_1+\alpha_2)T)}{2\left(1+\left(\frac{\alpha_1+\alpha_2}{2}\right)T\right)} \quad \text{[term of consider } \alpha_1\,\alpha_2 << 1]$$

$$\Rightarrow \quad R''_{eq} = \frac{R_0}{2}(1+(\alpha_1+\alpha_2)T)\left(1+\left(\frac{\alpha_1+\alpha_2}{2}\right)T\right)^{-1}$$

$$\Rightarrow \quad R''_{eq} = \frac{R_0}{2}(1+(\alpha_1+\alpha_2)T)\left(1-\left(\frac{\alpha_1+\alpha_2}{2}\right)T\right)$$

$$\Rightarrow \quad R''_{eq} = \frac{R_0}{2}\left(1+\left(\frac{\alpha_1+\alpha_2}{2}\right)T\right) \quad \ldots(4)$$

[As $(\alpha_1+\alpha_2)^2 << 1$]

Comparing (3) and (4) we have

$$\alpha_P = \frac{\alpha_1+\alpha_2}{2}$$

Ans. (A)

Sol. 12 Tolerance of a resistance R is given as

$$\frac{dR}{R}\times 100 = 5\% \text{ for } R = 100\,\Omega$$

If we use

$$R_e = R+R+R+R = 4R$$

where $R_e = 400\,\Omega$ and $R = 100\,\Omega$

Tolerance of equivalent resistance

$$\text{Tolerance} = \frac{dR_e}{R_e}\times 100 = \frac{4dR}{4R}\times 100$$

$$= \frac{dR}{R}\times 100 = 5\%$$

Ans. (B)

Sol. 13 Resistance of a conductor is given as

$$R = \frac{\rho l}{A}$$

l $A \to$ area

When a conductor is stretched its volume remain constant

$$V = lA = \text{constant}$$

$$A = \frac{V}{l}$$

$$\Rightarrow \quad R = \left(\frac{\rho}{V}\right)l^2$$

$$\Rightarrow \quad R \propto l^2$$

Taking log on both sides

$$\ln R = \ln\left(\frac{\rho}{V}\right) + 2\ln(\rho)$$

differentiating $\dfrac{dR}{R} = \dfrac{2dl}{l}$

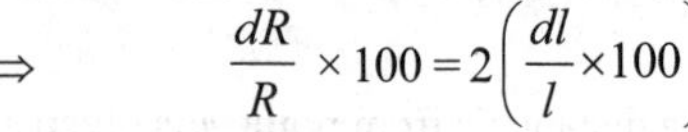

$$\Rightarrow \quad \frac{dR}{R}\times 100 = 2\left(\frac{dl}{l}\times 100\right)$$

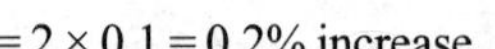

$$= 2\times 0.1 = 0.2\% \text{ increase}$$

Ans. (D)

Sol. 14 In final state the wire AOB behaves like two resistances $\dfrac{R}{2}$ in parallel combination so new resistance of shorter wire is given as

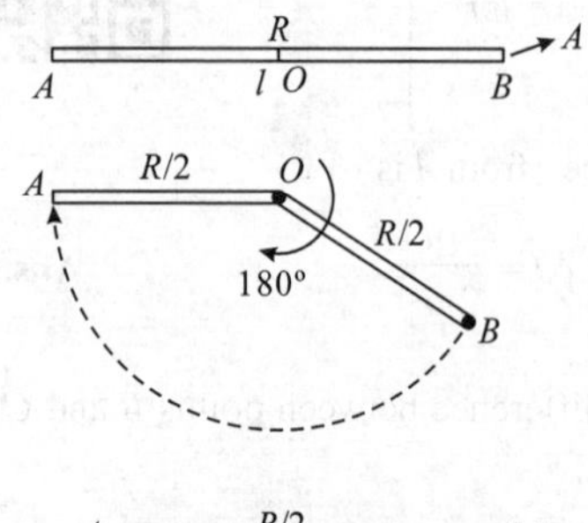

A R/2 2A
B R/2 O
l/2

$$R_{new} = \frac{\frac{R}{2} \times \frac{R}{2}}{\frac{R}{2} + \frac{R}{2}} = \frac{R}{4}$$ **Ans. (C)**

Sol. 15 Using ohm's law for a conductor, we have

$$V = IR \quad \text{where } I = neAv_d \quad \text{and } R = \rho\frac{l}{A}$$

$\Rightarrow$ $$V = neAv_d \times \frac{\rho l}{A}$$

$\Rightarrow$ $$\rho = \frac{V}{nev_d l}$$

$\Rightarrow$ $$\rho = \frac{5}{8\times10^{28}\times1.6\times10^{-19}\times2.3\times10^{-4}\times0.1}$$

$\Rightarrow$ $$\rho = \frac{2}{12.8} \times 10^{-4} = 1.6 \times 10^{-5}\ \Omega\text{-m}$$ **Ans. (D)**

Sol. 16 In a material current density is related to drift speed as

$$J = ne\, v_d$$

If $$v_d = C\sqrt{E}$$

$\Rightarrow$ $$\frac{I}{A} = ne\, C\sqrt{E}$$

As $$E \,\alpha\, V$$

$\Rightarrow$ $$I = IneC\sqrt{C_1 V}$$

$\Rightarrow$ $$I \,\alpha\, \sqrt{V} \quad \text{or} \quad I^2 \,\alpha\, V$$ **Ans. (C)**

Sol. 17

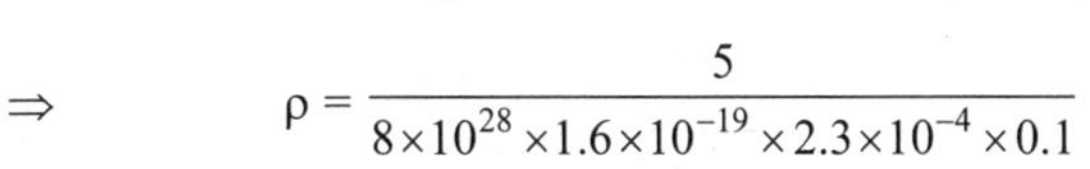

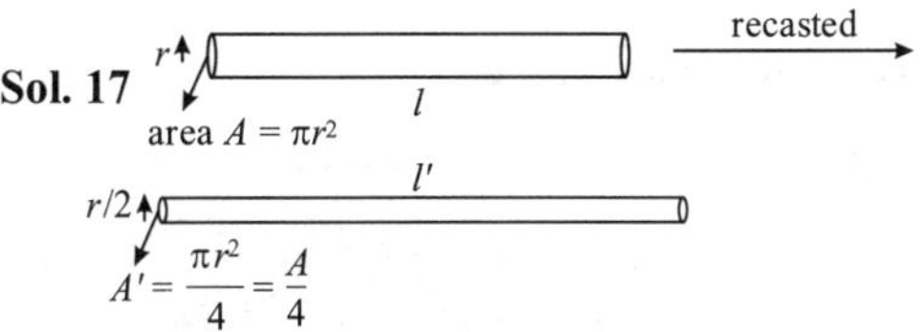

$$R = \rho\frac{l}{A} = 100 \qquad \ldots(1)$$

As volume of material remain same

$\Rightarrow$ $$Al = A'l'$$

$\Rightarrow$ $$l' = 4l$$

New resistance of wire will be

$$R' = \rho\frac{l'}{A'} = 16\rho\frac{l}{A}$$

$\Rightarrow$ $$R' = 1600\ \Omega$$ **Ans. (A)**

Sol. 18

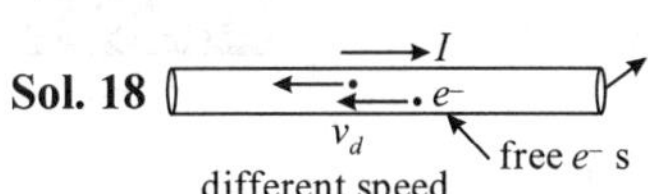

The drift speed of free e^-s (charge carries) is related to current density as

$$v_d = \frac{J}{ne} = \frac{J}{\rho}$$

(As in this case volume charge density (C/m³) is $\rho = ne$)

$\Rightarrow$ $$v_d = \frac{I}{\rho A} \quad [\text{as } J = \frac{I}{A}]$$

Time the charges take in traveling a distance d is

$$t = \frac{d}{v_d} = \frac{\rho A d}{I}$$ **Ans. (C)**

Sol. 19 We use $R_T = R_0(1 + \alpha T)$

at room temperature

$$R_T = 100\ \Omega$$

$\Rightarrow$ $$100 = R_0(1 + \alpha T_R) \qquad \ldots(1)$$

at temperature $T_R + 500°\text{C}$, $R'_T = \frac{220}{2} = 110\ \Omega$

$\Rightarrow$ $$110 = R_0(1 + \alpha(T_R + 500)) \qquad \ldots(2)$$

$(2) \div (1) \Rightarrow$ $$\frac{110}{100} = \frac{1 + \alpha(T_R + 500)}{1 + \alpha T_R}$$

$$= [1 + \alpha(T_R + 500)](1 + \alpha T_R)^{-1}$$

As α is very small we neglect terms of α^2 which gives

$$\frac{110}{100} = 1 + 500\,\alpha \qquad [RT_1 = RT_0(1 + \alpha\Delta T)]$$

$\Rightarrow$ $$\alpha = \frac{10}{500 \times 100}$$

$$= 2 \times 10^{-4}\ °\text{C}^{-1}$$ **Ans. (C)**

Sol. 20

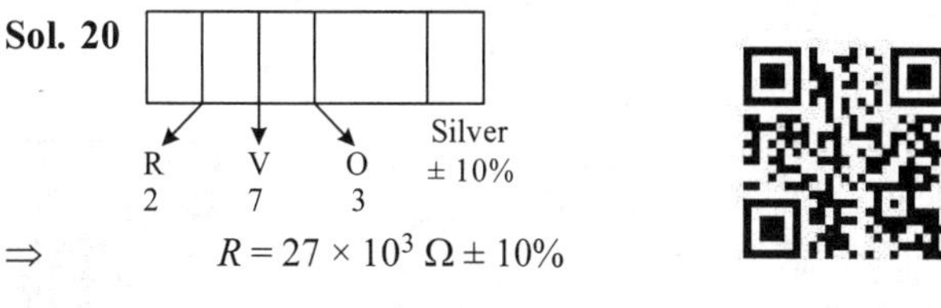

$\Rightarrow$ $$R = 27 \times 10^3\ \Omega \pm 10\%$$

$\Rightarrow$ $$R = 27\ \text{K}\Omega \pm 10\%$$ **Ans. (D)**

Sol. 21 Current in wire is given as

$$I = neAv_d$$

$\Rightarrow$ $$v_d = \frac{I}{neA} = \frac{1.5}{9\times10^{28}\times1.6\times10^{-19}\times5\times10^{-6}}$$

$\Rightarrow$ $$v_d = 0.02 \times 10^{-3}\ \text{m/s}$$ **Ans. (D)**

Sol. 22

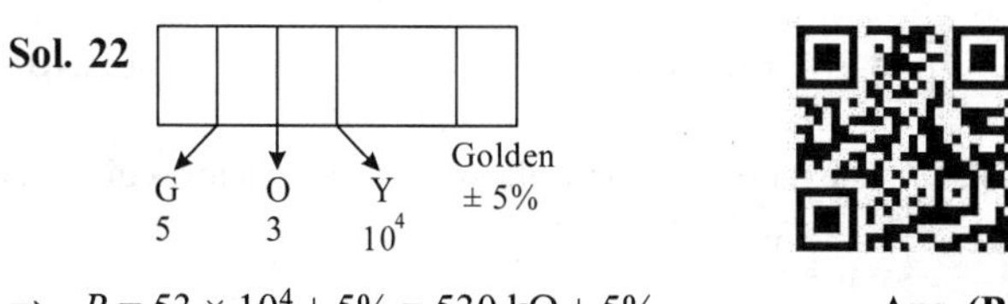

$\Rightarrow$ $R = 53 \times 10^4 \pm 5\% = 530\ \text{k}\Omega \pm 5\%$ **Ans. (B)**

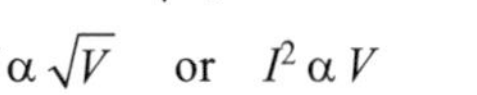
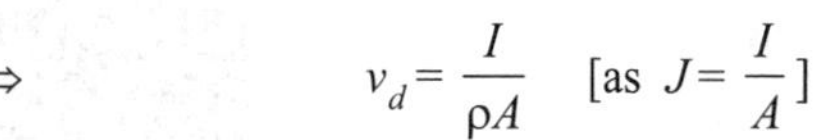

Sol. 23 $P = i^2R$

For i_{max}, R must be minimum

From color coding

$$R = 50 \times 10^2\,\Omega$$

Thus we use $2 = i^2_{max} \times 50 \times 10^2$

$$\Rightarrow \quad i^2_{max} = \frac{2}{5000} = 4 \times 10^{-4}\text{ A}$$

$$\Rightarrow \quad i_{max} = 20\text{ mA}$$ **Ans. (D)**

Sol. 24 When red is replaced with green 1st digit changes to 5 so new resistance will be 500 Ω **Ans. (C)**

Sol. 25 $R = 3\Omega = \dfrac{\rho l}{A}$

If $l' \to 2l \Rightarrow A' = \dfrac{A}{2}$ (As volume is constant)

New resistance of wire will be

$$R' = \frac{\rho l'}{A'} = \frac{\rho(2l)}{A/4} = 4\frac{\rho l}{A} = 4R = 4 \times 3 = 12\Omega$$

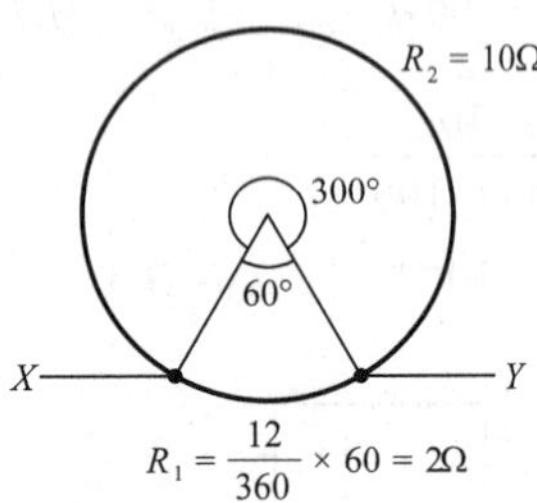

$$R_1 = \frac{12}{360} \times 60 = 2\Omega$$

Here $R_{XY} = \dfrac{R_1R_2}{R_1+R_2} = \dfrac{10\times 2}{10+2}$

$$\Rightarrow \quad R_{XY} = \frac{20}{12}\Omega = \frac{5}{3}\Omega$$ **Ans. (B)**

Sol. 26 Resistivity of a conductor is given as

$$\rho = \frac{m}{ne^2\tau}$$

$$= \frac{9.1\times10^{-31}}{8.5\times10^{28}\times(1.6\times10^{-19})\times25\times10^{-15}}$$

$\Rightarrow \quad \rho \cong 1.67 \times 10^{-8}\,\Omega\text{-m}$ **Ans. (D)**

Sol. 27 The mobility of free electrons in conductor is given as

$$\mu = \frac{v_d}{E}$$

and $E = \rho J$

$$\Rightarrow \quad \mu = \frac{1.1\times10^{-3}}{1.7\times10^{-8}\times\dfrac{5}{\pi\times25\times10^{-6}}}$$

$$\Rightarrow \quad \mu = \frac{1.1\times10^{-3}\times\pi\times25\times10^{-6}}{1.7\times10^{-8}\times5} \approx 1.01\text{ m}^2\text{/Vs}$$ **Ans. (D)**

Sol. 28

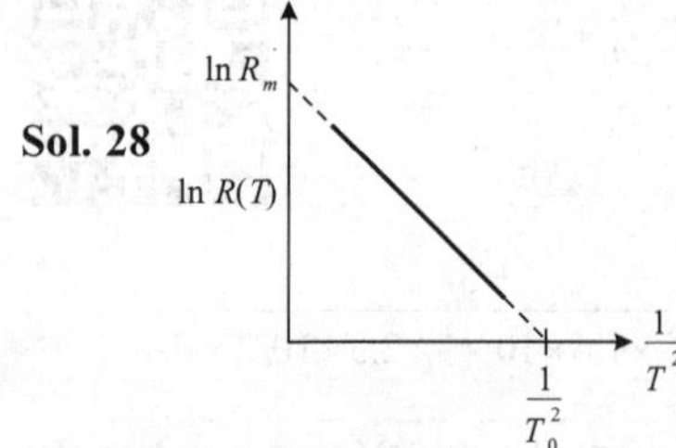

Equation of straight line

$$\ln R(\text{T}) = -\frac{\ln R_m}{(1/T_1^2)} \times \frac{1}{T^2} + \ln R_m$$

$$\Rightarrow \quad \ln R(T) = \ln R_m\left(1 - \frac{T_1^2}{T^2}\right)$$

$$\Rightarrow \quad \ln\left(\frac{R(T)}{R_m}\right) = -\ln R_m\left(\frac{T_1^2}{T^2}\right)$$

$$\Rightarrow \quad R(T) = R_m \cdot e^{-T_1^2/T^2(\ln R_m)} = R_0\, e^{-T_0^2/T^2}$$ **Ans. (C)**

Sol. 29 Elemental resistance of shell is

$$dR = \rho.\frac{dx}{4\pi x^2}$$

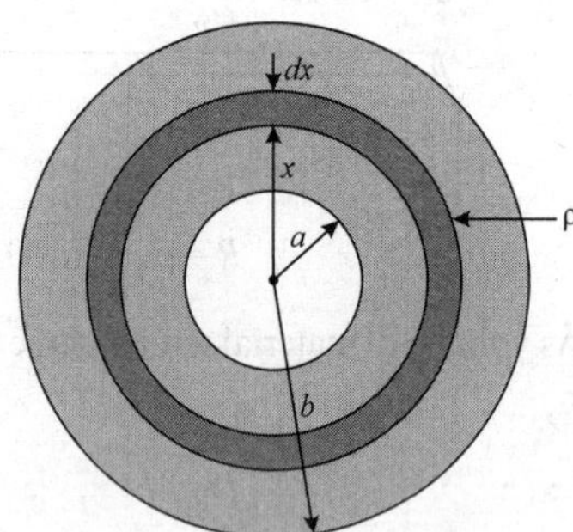

$$dR = \rho\cdot\frac{dx}{4\pi x^2}$$

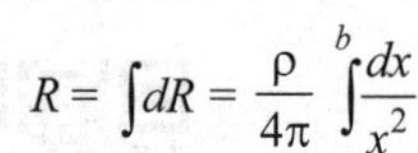

$$R = \int dR = \frac{\rho}{4\pi}\int_a^b \frac{dx}{x^2}$$

$$\Rightarrow \quad R = \frac{\rho}{4\pi}\left[-\frac{1}{x}\right]_a^b$$

$$= \frac{\rho}{4\pi}\left[\frac{1}{a} - \frac{1}{b}\right]$$

$$= \frac{\rho(b-a)}{4\pi ab}$$ **Ans. (B)**

Sol. 30 Resistivity of different commonly used metals are given in increasing order below

Silver
Copper
Gold
Aluminium
Magnetism
Tungsten
Zinc
Nickel
Iron
Lead
Mercury
Germanium
Silicon

$\Rightarrow \quad \rho_M > \rho_A > \rho_C$

Ans. (D)

Sol. 31 For the given expression of current, we use

$$i = 20t + 8t^2 = \frac{dq}{dt}$$

Total charge crossed through the wire can be calculated as

$$q = \int i dt$$

$$\Rightarrow \quad q = \int_0^{15} (20t + 8t^2) dt$$

$$\Rightarrow \quad q = \left(\frac{20t^2}{2} + \frac{8t^3}{3}\right)_0^{15}$$

$$\Rightarrow \quad q = 10 \times (15)^2 + \frac{8(15)^3}{3}$$

$$\Rightarrow \quad q = 2250 + 9000 = 11250 \text{ C}$$

Ans. (D)

Sol. 32 The electric field in the wire is given as

$$E = 10 \times 10^{-3} \frac{V}{m}$$

Current density in wire can be calculated as

$$J = \sigma E = 10 \times 10^{-3} \times 5 \times 10^7 \text{ A/m}^2$$

$$J = 5 \times 10^5 \text{ A/m}^2$$

If current in wire is i, it can be used as

$$\frac{i}{A} = 5 \times 10^5$$

$$\Rightarrow \quad i = 5 \times 10^5 \times \pi r^2 = 5 \times 10^5 \times \pi \times (5 \times 10^{-4})^2$$

$$\Rightarrow \quad i = 125\pi \times 10^{-3} \text{ A} = 125\,\pi\, mA$$

$$\Rightarrow \quad x = 5$$

Ans. (5)

Sol. 33 As described in the question, when length of wire is doubled and cross section area is halved, we have new resistance given as

$$R' = \frac{\rho(2l)}{(A/2)} = \frac{4\rho l}{A}$$

Circuit current is given as

$$i = \frac{V}{R} = \frac{VA}{4\rho l}$$

Ans. (A)

Sol. 34 The effective circuit is shown below

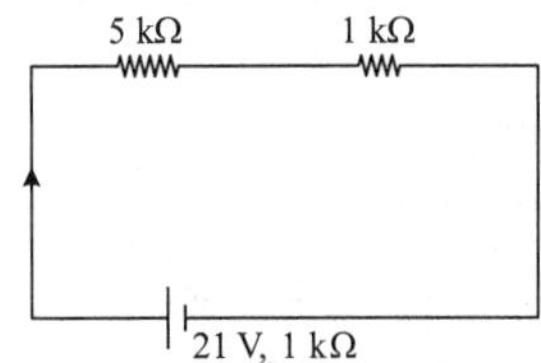

Current in circuit is given as

$$I = \frac{21}{5+1+1} = 3 \text{ mA}$$

Ans. (3)

Sol. 35 Current in wire is given as

$$i = neAv_d$$

$$\Rightarrow \quad 10 = n \times 1.6 \times 10^{-19} \times 5 \times 10^{-6} \times 2 \times 10^{-3}$$

$$\Rightarrow \quad n = 0.625 \times 10^{28} = 625 \times 10^{25}$$

Ans. (B)

Sol. 36 Resistance of conductor is given as

$$R = \frac{\rho l}{A} = \frac{V}{I}$$

$$\Rightarrow \quad \rho = \frac{AV}{Il} = \frac{\pi d^2 V}{4Il} \qquad \left(A = \frac{\pi d^2}{4}\right)$$

Maximum fractional error in resistivity calculation is given as

$$\frac{\Delta\rho}{\rho} = \frac{2\Delta d}{d} + \frac{\Delta V}{V} + \frac{\Delta I}{I} + \frac{\Delta l}{l}$$

$$\Rightarrow \quad \frac{\Delta\rho}{\rho} = 2\left(\frac{0.01}{5.00}\right) + \frac{0.1}{5.0} + \frac{0.01}{2.00} + \frac{0.1}{10.0}$$

$$\Rightarrow \quad \frac{\Delta\rho}{\rho} = 0.004 + 0.02 + 0.005 + 0.01$$

$$\Rightarrow \quad \frac{\Delta\rho}{\rho} = 0.039$$

Maximum percentage error can be given as

$$\%\text{ error} = \frac{\Delta\rho}{\rho} \times 100 = 0.039 \times 100 = 3.90\%$$

Ans. (A)

Sol. 37 Current passing through the wire can be given as

$$i = JA\cos\theta = \frac{JA}{2}$$

$$\Rightarrow \quad J = \frac{2i}{A}$$

If electric field in wire is E, we use

$$E = \rho J = \frac{2\rho i}{A} = \frac{2 \times 44 \times 10^{-8} \times 5}{4 \times 10^{-2}} = 11 \times 10^{-5} \text{ V/m}$$

Ans. (B)

Sol. 38 Current in the conductor is given as

$$i = nAev_d$$

$\Rightarrow$ v_d increases as area A decrease **Ans. (A)**

Sol. 39 From the given figure voltage at t = 3.2 s can be calculated by linear function as 6 V. So we use, from the below graph

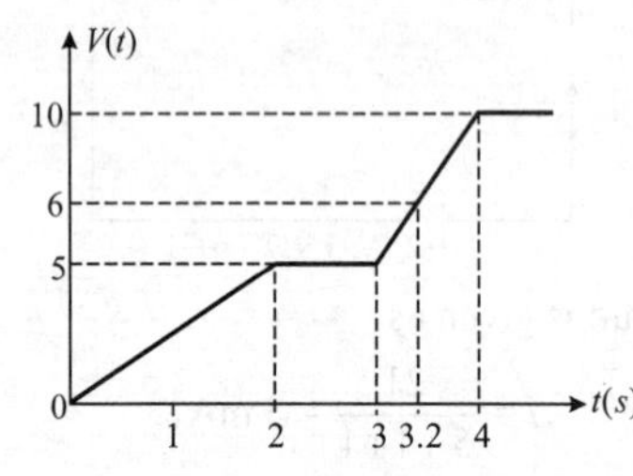

$$\frac{\Delta V}{(3.2-3)} = \frac{5}{1}$$

$$\Rightarrow \quad \Delta V = 1\text{ V}$$

Thus at $t = 3.2$ s, $V(t) = 5 + 1 = 6$ V

Thus circuit current at this instant is given as

$$I = \frac{6-5}{R} = 1\text{ A}$$ **Ans. (01)**

Sol. 40 For parallel combination of wires, we have

$$\frac{R_1R_2}{R_1+R_2} = 3\Omega$$

$$\Rightarrow \quad \frac{\rho_1\rho_2}{\rho_1+\rho_2}\times\frac{l}{A} = 3\Omega$$

$$\Rightarrow \quad l = 3\times A\times\frac{(\rho_1+\rho_2)}{\rho_1\rho_2}\times\frac{3\times\pi(10^{-3})^2\times 63\times10^{-8}}{12\times51\times10^{-16}};$$

$$\Rightarrow \quad l = 97\text{ m}$$ **Ans. (B)**

Sol. 41 From the colour coading of bands on resistor, we can get the below value of color coded resistor given in question.

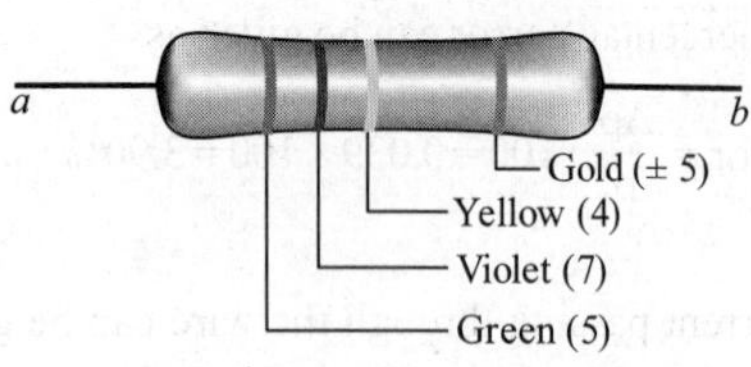

$$R = (7500 \pm 375)\Omega$$ **Ans. (A)**

Sol. 42 As the wire volume remain constant so due to increase in its length, new cross sectional area of wire is given as

$$l'A' = lA$$

$$\Rightarrow \quad 2lA' = lA$$

$$\Rightarrow \quad A' = \frac{A}{2}$$

Resistance of wire initially is calculated as

$$R = \rho\frac{l}{A} \quad \ldots(\text{i})$$

Now length and cross sectional area is changed to

$$l' = 2l \text{ and } A' = \frac{A}{2}$$

Thus new resistance becomes

$$R' = \frac{\rho l'}{A'} = \frac{\rho 2l}{\frac{A}{2}}$$

$$\Rightarrow \quad R' = \frac{4\rho l}{A} = 4R$$

Percentage increase in resistance is calculated as

$$p = \frac{R'-R}{R}\times100 = \frac{4R-R}{R}\times100$$

$$\Rightarrow \quad p = 300\%$$ **Ans. (300)**

Sol. 43 Due to increase in length from l to $1.004l$ the cross sectional area of wire is changed from A to $A/1.4$ thus resistance of the wire is calculated as

Initial resistance $R = \dfrac{\rho l}{A}$

Final resistance $R' = \dfrac{\rho(1.004l)}{(A/1.004)} = 1.008\,R$

Percentage change in resistance is calculated as

$$\Delta p = \frac{\Delta R}{R}\times100 = \frac{R'-R}{R}\times100$$

$$\Delta p = 0.008\times100 = 0.8\%$$ **Ans. (C)**

Sol. 44 Current through the cross sectional area A is calculated as

$$I = JA$$

$$\Rightarrow \quad I = 4\times10^6\times\left[\pi R^2 - \pi\left(\frac{R}{2}\right)^2\right]$$

$$\Rightarrow \quad I = 4\times10^6\times\pi R^2\times\frac{3}{4}$$

$$\Rightarrow \quad I = 4\times10^6\times\pi\times(4\times10^{-3})^2\times\frac{3}{4} = 48\,\pi\text{A}$$

Ans. (48)

Sol. 45 Figure below shows the outer portion as described in the question

The current through this outer portion between radius $r/2$ to r is calculated as

$$I = JA$$

$$\Rightarrow \quad I = 1 \times 10^6 \times \left[\pi r^2 - \pi\left(\frac{r}{2}\right)^2\right]$$

$$\Rightarrow \quad I = \pi \times 10^6 \left[r^2 - \frac{r^2}{4}\right] = 12\pi$$

$$\Rightarrow \quad x = 12$$

Ans. (12)

Sol. 46 The resistance at different temperatures are related as

$$R = R_0(1 + \alpha\Delta T)$$

$$\Rightarrow \quad 3 = R_0(1 + \alpha(30 - 0))$$

$$\Rightarrow \quad 2 = R_0(1 + \alpha(10 - 0))$$

$$\Rightarrow \quad \frac{3}{2} = \frac{1 + 30\alpha}{1 + 10\alpha}$$

$$\Rightarrow \quad \alpha = \frac{1}{30} = 0.033\ °C^{-1}$$

Ans. (A)

Sol. 47 The slope of graph is tan 45° = 1 which is resistance of wire, given as

$$R = \rho\frac{l}{A} = 1$$

$$\Rightarrow \quad 1 = \frac{\rho \times 31.4}{\frac{\pi(2.4)^2}{4}}$$

$$\Rightarrow \quad \frac{\pi(2.4)^2}{4} = \rho \times 314$$

$$\Rightarrow \quad \rho \times 10 = \frac{2.4 \times 2.4}{4}$$

$$\Rightarrow \quad \rho = \frac{0.6 \times 2.4}{10}$$

$$\Rightarrow \quad \rho = 0.144\ \Omega\,\text{cm}$$

$$\Rightarrow \quad \rho = 144 \times 10^{-3}\ \Omega\,\text{cm}$$

Ans. (144)

Sol. 48 Each wire has resistance given as

$$r = \rho\frac{4l}{\pi d^2}$$

Eight wire are in parallel, so equivalent resistance of combination is

$$\frac{r}{8} = \frac{\rho l}{2\pi d^2}$$

Single copper wire of length $2l$ has resistance

$$R = \rho\frac{2l \times 4}{\pi d_1^2} = \frac{\rho l}{2\pi d^2}$$

$$\Rightarrow \quad d_1 = 4d$$

Ans. (4)

Sol. 49 Given circuit is balanced wheat stone bridge so middle 2 Ω resistance can be removed

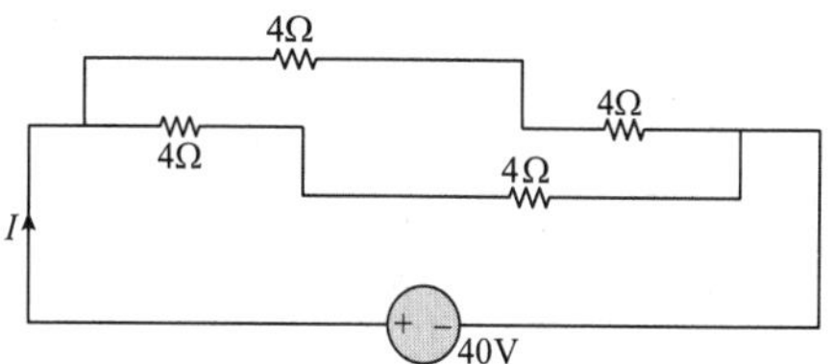

Thus equivalent resistance of circuit is given as

$$R_{net} = 4\ \Omega$$

Current supplied by source is given as

$$I = \frac{40}{4}$$

$$\Rightarrow \quad I = 10\ \text{A}$$

Ans. (A)

Sol. 50 Given circuit is a balanced wheat stone bridge so middle 5 Ω resistor can be removed from the circuit as shown in figure below.

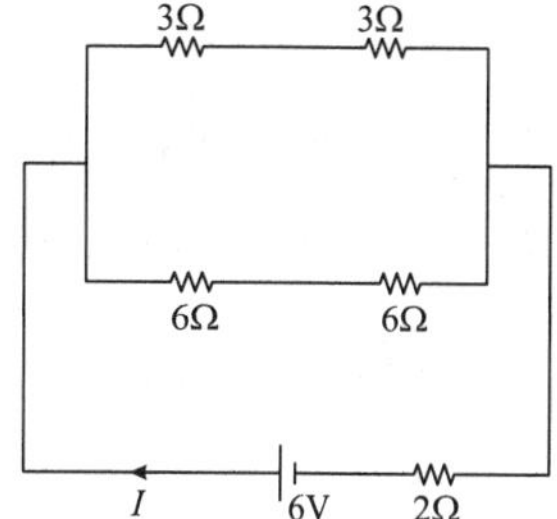

Equivalent resistance of this circuit is calculated as

$$R_{eq} = \frac{6 \times 12}{6 + 12} + 2 = \frac{6 \times 12}{18} + 2$$

$$R_{eq} = 6\ \Omega$$

Current supplied by the battery is given as

$$I = \frac{V}{R_{eq}} = \frac{6}{6} = 1\ \text{A}$$

Ans. (A)

Sol. 51 Figure below shows the circuit as described in the questoin.

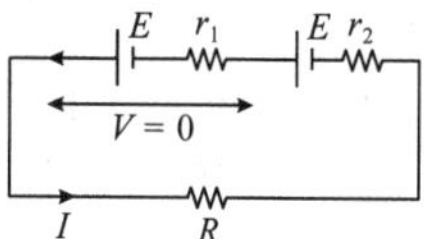

Current flowing in circuit is calculated as

$$I = \frac{2E}{r_1 + r_2 + R}$$

As potential drop across the first source is zero, by Kirchoff's Voltage law equation for the loop, we have

$$IR = E - Ir_2$$

$$\Rightarrow \quad I = \frac{E}{R + r_2}$$

$$\Rightarrow \quad \frac{2E}{r_1 + r_2 + R} = \frac{E}{R + r_2}$$

$$\Rightarrow \quad 2R + 2r_2 = r_1 + r_2 + R$$

$$\Rightarrow \quad R = r_1 - r_2$$

Ans. (A)

Sol. 52 Resistance of the given copper wire is given as

$$R = \frac{\rho l}{A} = \frac{1.7 \times 10^{-8} \times 1}{2 \times 10^{-5}} = \frac{1.7}{2} \times 10^{-2}$$

Potential difference across the copper wire is given as

$$V = IR = \frac{1.7}{2} \times 10^{-2}\,\text{V}$$

Force acting on moving electron in wire is given as

$$F = qE = q\left(\frac{V}{l}\right)$$

$$\Rightarrow \quad F = 1.6 \times 10^{-19} \times \frac{1.7}{2} \times 10^{-2}$$

$$\Rightarrow \quad F = 1.36 \times 10^{-21} = 136 \times 10^{-23}$$

Ans. (136)

Sol. 53 Drift velocity of electrons in a conductor are given as

$$v_d = \left(\frac{e\tau}{m}\right)E$$

$$\Rightarrow \quad v_d = \left(\frac{e\tau}{m}\right)\left(\frac{\Delta V}{l}\right)$$

As temperature increases, relaxation time decreases, hence v_d decreases hece option statement (a) is correct. As per above expression, we have

$$v_d \propto \frac{1}{l}$$

Hence option statment (d) is correct. **Ans. (B)**

Sol. 54 For the given circuit, steady state charge on capacitor is given as

$$q = CV_{100\Omega} = (1.1 \times 10^{-6})\left(\frac{10}{R + r}R\right)$$

$$\Rightarrow \quad q = 1.1 \times 10^{-6}\left(\frac{10}{110} \times 100\right)$$

$$\Rightarrow \quad q = 10\,\mu\text{C}$$

Ans. (10)

Sol. 55 Resistance R_1 is given as

$$R_1 = \rho\frac{L_1}{A_1}$$

As length of wire is increased to three times, its cross sectional area will decrease by three times, thus new resistance is given as

$$R_2 = \rho\left(\frac{3L_1}{A_1/3}\right) = 9\rho\frac{L_1}{A_1}$$

$$\Rightarrow \quad \frac{R_2}{R_1} = 9$$

Ans. (A)

Sol. 56 Figure below shows the composite wire and the equivalent wire of effective conductivity σ_{eq}.

Here we consider each wire of length l and cross sectional area A. To find effective conductivity we consider another wire of total length $2l$ and cross sectional area A. Equivalent resistance of equivalent wire is calculated as

$$R_{eq} = R_1 + R_2$$

$$\Rightarrow \quad \frac{2l}{\sigma_{eq}A} = \frac{l}{\sigma_1 A} + \frac{l}{\sigma_1 A}$$

$$\Rightarrow \quad \frac{2l}{\sigma_{eq}} = \frac{l}{\sigma_1} + \frac{l}{\sigma_2}$$

$$\Rightarrow \quad \sigma_{eq} = \frac{2\sigma_1\sigma_2}{\sigma_1 + \sigma_2}$$

Ans. (B)

Sol. 57 Constantan and manganin doesn't show much variation with temperature thats why they are used in making standard resistances and also used in making resistance boxes and other calibration based laboratory devices. Here Reason-(R) is correct as well as explaining Assertion-(A). **Ans. (A)**

Sol. 58 Figure below shows the parts X and Y of the long wire and the wire W which is made by stretching the wire X.

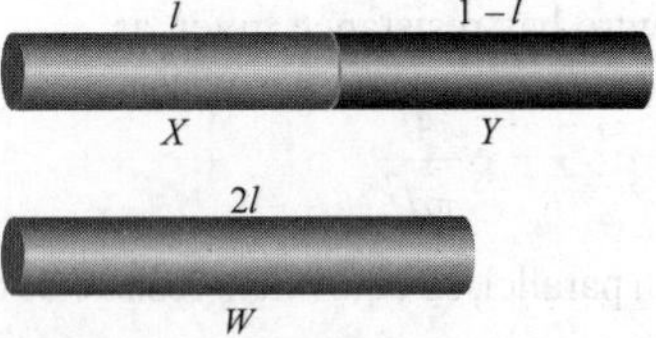

As wire W is made by stretching the wire X by making its length double so its cross sectional area will reduce to half so its resistnace will become four times that of X, so we use

$$R_W = 4R_X = 2R_Y$$

$$\Rightarrow \quad \frac{R_X}{R_Y} = \frac{1}{2}$$

As resistances of wires X and Y are proportional to their lengths, we use

$$\frac{l_X}{l_Y} = \frac{1}{2}$$

Ans. (B)

Sol. 59 The resistance of conductor is given as

$$R = \rho\frac{l}{A}$$

The cross-sectional area of conductor is taken as $\pi(b^2 - a^2)$

$$\Rightarrow \quad R = \rho\frac{l}{\pi(b^2 - a^2)} = \frac{2.4\times10^{-8}\times3.14}{3.14\times(4^2 - 2^2)\times10^{-6}}$$

$$\Rightarrow \quad R = 2\times10^{-3}\,\Omega$$

$$\Rightarrow \quad n = 2$$

Ans. (2)

Sol. 60 As volume is constant, the cross sectional area decreases when wire is stretched and its resistance can be written as

$$R \propto l^2$$

When length of wire is increased by 20%, its new resistance becomes

$$R' = 1.44R$$

Thus it increases by 44%. **Ans. (44)**

Sol. 61 Using Ohm's Law for the metallic wire, its resistance is given as

$$R = \frac{3.4}{2} = 1.7\,\Omega$$

$$\Rightarrow \quad 1.7 = \frac{\rho L}{A}$$

$$\Rightarrow \quad L = \frac{1.7(A)}{\rho}$$

If V is the volume of wire, we use $V = AL$ and volume is also given as $V = m/\rho$, which is calculated as

$$\text{Volume} = \frac{8.92\times10^{-3}}{8.92\times10^{3}} = 10^{-6}$$

$$\Rightarrow \quad L^2 = \frac{1.7}{\rho}(10^{-6}) = \frac{1.7}{1.7}\times10^2$$

$$\Rightarrow \quad L = 10\,\text{m}$$

Ans. (B)

Sol. 62 Initial resistance of wire is given as

$$R_{\text{initial}} = \frac{\rho l}{A} = 5\Omega$$

As volume of wire remain constant in stretching, we use

$$V_i = V_f$$

$$\Rightarrow \quad A_i l_i = A_f l_f$$

$$\Rightarrow \quad Al = A'(5l)$$

$$\Rightarrow \quad A' = \frac{A}{5}$$

Final resistance of wire is calculated as

$$R_f = \frac{\rho l_f}{A_f} = \frac{\rho(5l)}{\left(\frac{A}{5}\right)}$$

$$\Rightarrow \quad R_f = 25\left(\frac{\rho l}{A}\right)$$

$$\Rightarrow \quad R_f = 25\times5 = 125\,\Omega$$

Ans. (C)

Sol. 63 Current is calculated from charge as

$$Q = (\alpha t - \beta t^2 + \gamma t^3)$$

$$\Rightarrow \quad i = \frac{dQ}{dt} = (\alpha - 2\beta t + 3\gamma t^2)$$

Current will be minimum when

$$\frac{di}{dt} = (3\gamma t - 2\beta) = 0$$

$$\Rightarrow \quad t = \frac{\beta}{3\gamma}$$

Thus minimum current is calculated as

$$i = (\alpha - 2\beta t + 3\gamma t^2) = \left(\alpha - \frac{\beta^2}{3\gamma}\right)$$

Ans. (D)

Sol. 64 The drift velocity of free electrons in a conductor is given as

$$V_d = \frac{eE}{m}\tau$$

This expression is independent of area of cross section hence the drift velocity will remain the same. **Ans. (A)**

Sol. 65 Relation between current and drift velocity

$$I = neAv_d$$

$$\Rightarrow \quad v_d = \frac{I}{neA}$$

$$\Rightarrow \quad v_d = \frac{2}{2\times10^{28}\times1.6\times10^{-19}\times25\times10^{-6}}$$

$$\Rightarrow \quad v_d = 25\times10^{-6}\,\text{ms}^{-1}$$

Ans. (25)

Sol. 66 Number density of free electrons in copper is given as

$$n = 8\times10^{28}\,\text{m}^{-3}$$

Cross-section area of the copper wire is given as

$$A = 2\times10^{-6}\,\text{m}^2$$

Current in wire is given as

$$I = 3.2\,\text{A}$$

Current is related to drift speed of electrons as

$$I = neAv_d$$

$$\Rightarrow \quad v_d = \frac{I}{neA} = 125\times10^{-6}\,\text{m/s}$$

Ans. (125)

Sol. 67 Maximum resistance occurs when all the resisters are connected in series combination so it is given as

$$R_{max} = 10R$$

Where $R = 10\,\Omega$

Minimum resistance occurs when all the resistance are connected in parallel combination which is given as

$$R_{min} = \frac{R}{10}$$

$$\Rightarrow \quad \frac{R_{max}}{R_{min}} = 100 \qquad \textbf{Ans. (100)}$$

Sol. 68 Move in curve path

$$i = neAV_d$$

As in presence of electric field inside conductor electrons are accelerated. **Ans. (B)**

Sol. 69 Figure below shows the rectangular parallelopiped as described in the question.

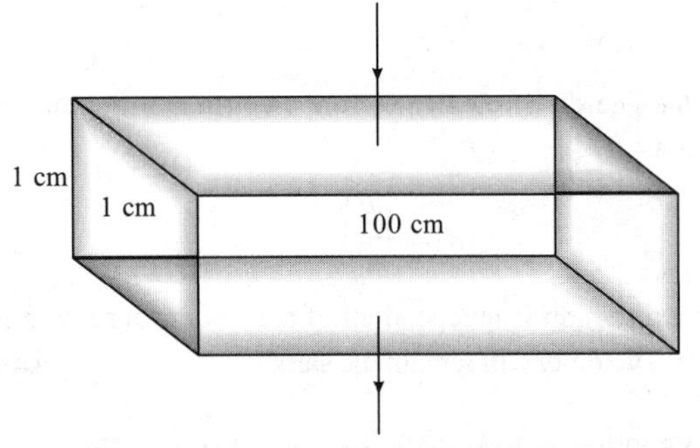

Resistance of rectangular parallelopiped across the terminals shown in figure is calculated as

$$R = \rho\frac{l}{A} = \frac{3\times10^{-7}\times(1\times10^{-2})}{100\times1\times10^{-4}}$$

$$\Rightarrow \quad R = 3\times10^{-7}\,\Omega \qquad \textbf{Ans. (3)}$$

Sol. 70 As the volume of wire is constant, we use

$$A_1L_1 = A_2L_2$$

$$\Rightarrow \quad A_1L = A_2\frac{L}{4}$$

$$\Rightarrow \quad 4A_1 = A_2$$

Let resistance of first wire be R_1 we have

$$R_1 = \frac{\rho L_1}{A_1}$$

The resistance of second wire is R_2 which is given as

$$R_2 = \frac{\rho L_2}{A_2}$$

Ratio of resistances is calculated as

$$\frac{R_2}{R_1} = \frac{L_2A_1}{A_2L_1} = \frac{L}{4}\frac{A_1}{4A_1L}$$

$$\Rightarrow \quad R_2 = \frac{1}{16}R_1 = 10\,\Omega \qquad \textbf{Ans. (A)}$$

Sol. 71 As the source potential difference is constant, we use

$$i_0R_0 = i_{100}R_{100}$$

As resistance depends on temperature as

$$R = R_0(1+\alpha\Delta T)$$

$$\Rightarrow \quad 2R_0 = 1.2\,R_0[1+100\alpha] \qquad \ldots(1)$$

$$\Rightarrow \quad 1+100\alpha = \frac{5}{3}$$

$$\Rightarrow \quad 100\alpha = \frac{2}{3}$$

$$\Rightarrow \quad 50\alpha = \frac{1}{3}$$

So current at 50°C is calculated as

$$i_{50}R_{50} = i_0R_0$$

$$\Rightarrow \quad i_{50} = \frac{i_0R_0}{R_{50}} = 2\frac{2\times R_o}{R_o(1+50\alpha)} = \frac{2}{1+\frac{1}{3}} = 1.5\,\text{A}$$

$$\Rightarrow \quad i_{50} = 15\times10^2\,\text{mA} \qquad \textbf{Ans. (15)}$$

Sol. 72 Equivalent resistance in series is more than maximum resistance

$$R_{eq} = R_1 + R_2 + R_3$$

Resistivity depends on temperature as

$$R = R_0(1+\alpha\Delta T)$$

Thus statement I and II are false. **Ans. (B)**

Sol. 73 In the circuit diodes D_1 and D_3 are forward biased and D_2 is reverse biased. So considering these states of diodes, simplified circuit is redrawn as shown in figure below.

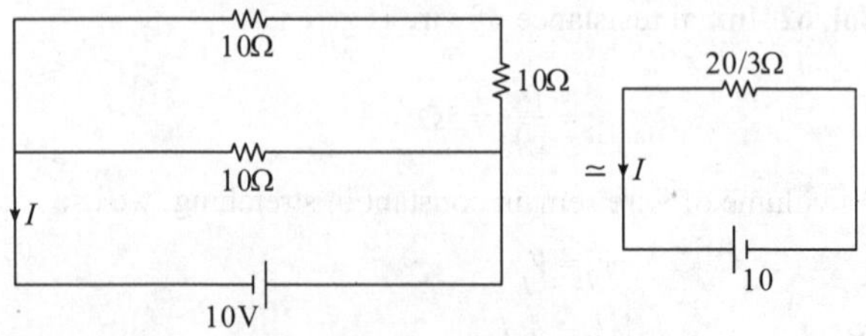

Thus current in the circuit is calculated as

$$I = \frac{10}{20/3} = \frac{3}{2}\text{A} = 1.5\,\text{A} \qquad \textbf{Ans. (A)}$$

17.2 Circuit Analysis

Sol. 1 $$R_{eq} = \frac{6\times 3}{6+3} = 2\,\Omega$$

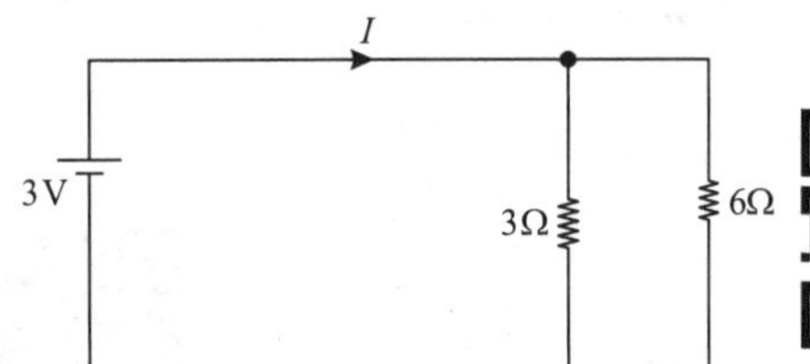

By Ohm's law, circuit current is

$$I = \frac{3}{2} = 1.5\,\text{A}$$ **Ans. (B)**

Sol. 2

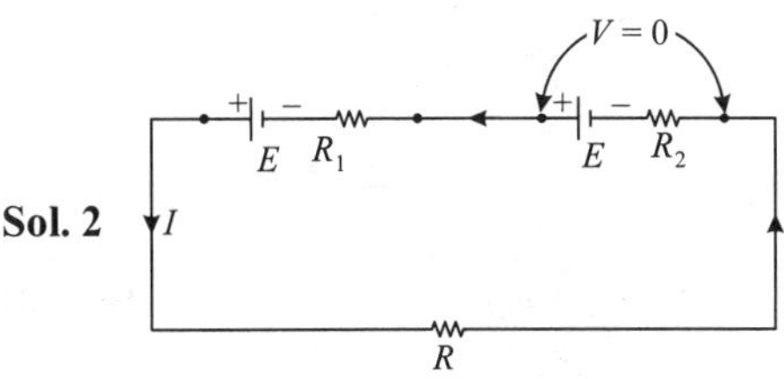

Current in circuit

$$I = \frac{2E}{R_1 + R_2 + R}$$

Potential difference across second source

$$V = E - IR_2 = 0$$

$\Rightarrow$ $$E = IR_2$$

$\Rightarrow$ $$E = \frac{2ER_2}{R_1 + R_2 + R}$$

$\Rightarrow$ $$R_1 + R_2 + R = 2R_2$$

$\Rightarrow$ $$R = R_2 - R_1$$ **Ans. (B)**

Sol. 3 As the circuit shown in figure is a balanced wheatstone bridge we can remove the middle resistor

$\Rightarrow$

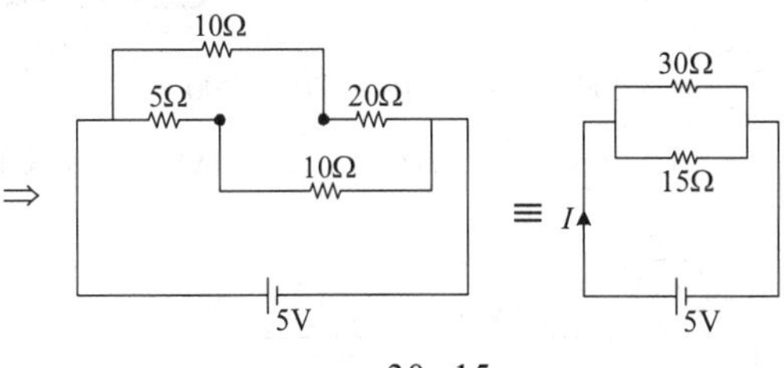

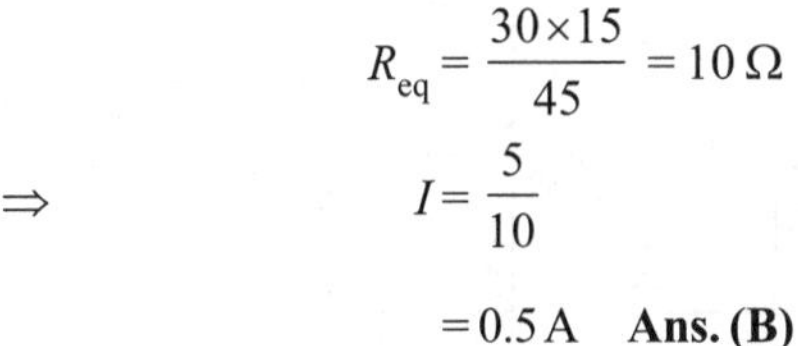

$$R_{eq} = \frac{30\times 15}{45} = 10\,\Omega$$

$\Rightarrow$ $$I = \frac{5}{10}$$

$$= 0.5\,\text{A}$$ **Ans. (B)**

Sol. 4

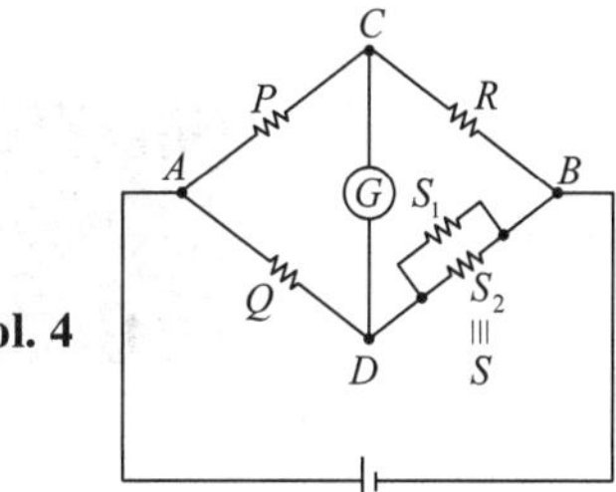

Balancing condition for the given wheatstone bridge is

$$\frac{P}{Q} = \frac{R}{S}$$

here $$S = \frac{S_1 S_2}{S_1 + S_2}$$

$\Rightarrow$ $$\frac{P}{Q} = \frac{R(S_1 + S_2)}{S_1 S_2}$$ **Ans. (B)**

Sol. 5 (1) Kirchhoff's first law (KCL) gives

Σi at a function = 0

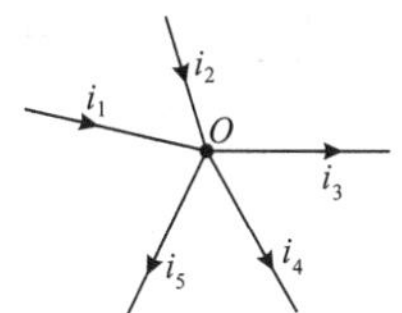

$$i_1 + i_2 = i_3 + i_4 + i_5$$

As $i \to$ charge flow/s

$\Rightarrow$ KCL is based on

"Conservation of charge".

(2) Kirchhoff's Second law (KVL)

$$\Sigma iR = \Sigma E$$

(potential drop across all resistances in a loop)

= (*EMF* in that loop)

As *EF* is conservative

$\Rightarrow$ Work done in a closed path = 0 (any loop of circuit)

$\Rightarrow$ Thus KVL is based on "Energy Conservation" **Ans. (D)**

Sol. 6

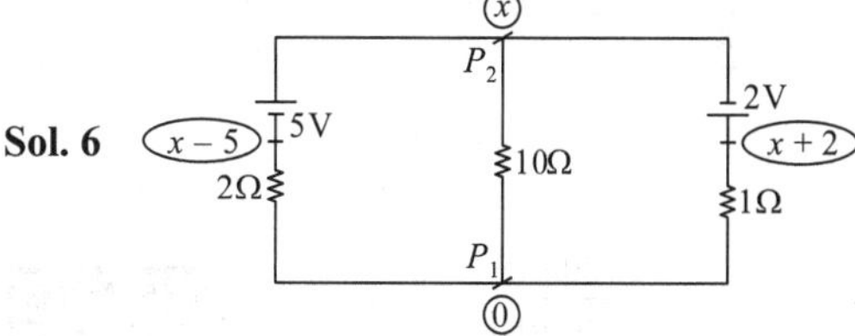

Distributing potentials in circuit and writing KCL equation for x we have

$$\frac{x-0}{10} + \frac{x+2}{1} + \frac{x-5}{2} = 0$$

$$\Rightarrow \quad x+10x+20+5x-25=0$$

$$\Rightarrow \quad 16x=5$$

$$\Rightarrow \quad x=\frac{5}{16}\simeq 0.3\text{V}$$

$$I_{10\Omega}=\frac{x-0}{10}=\frac{0.3}{10}=0.03\text{A}$$

As $x>0 \Rightarrow$ current flows from P_2 to P_1 **Ans. (C)**

Sol. 7 If E_2 is removed and potential difference across A and B is equal to $E_1=90$V then on placing E_2 back in circuit no current will flow through E_2

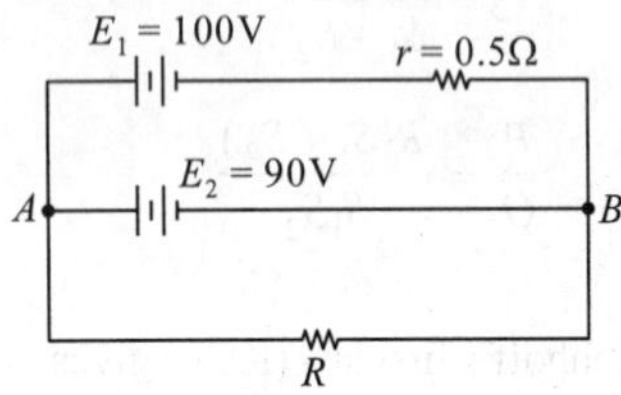

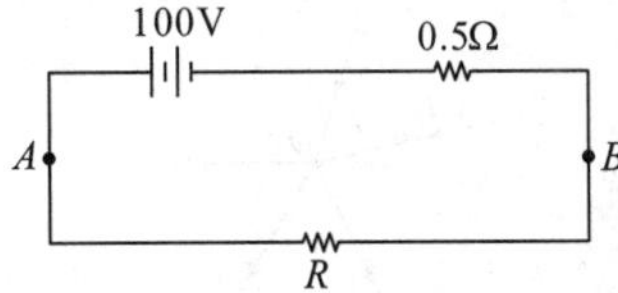

When E_2 is removed the potential difference across A and B is given as

$$V_A-V_B=\frac{100}{R+0.5}\times R=90$$

$$\Rightarrow \quad 100R=90R+45$$

$$\Rightarrow \quad 10R=45$$

$$\Rightarrow \quad R=4.5\,\Omega$$ **Ans. (C)**

Sol. 8

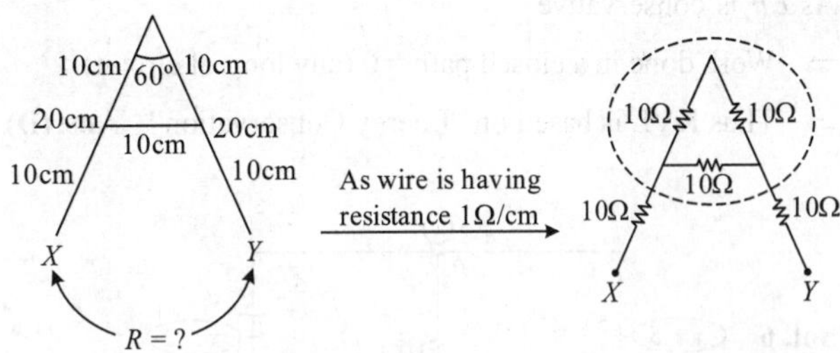

$$r_i=\frac{20\times 10}{20+10}=\frac{20}{3}\Omega$$

Here $$R_{xy}=10+\frac{20}{3}+10$$

$$\Rightarrow \quad R_{xy}=20+\frac{20}{3}=\frac{80}{3}\Omega=26.7\Omega$$ **Ans. (D)**

Sol. 9

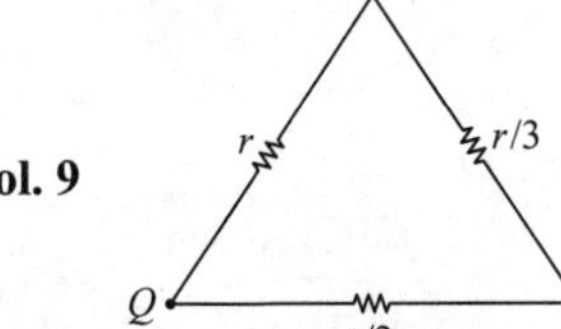

$$r_{PQ}=r\,\|\left(\frac{1}{3}+\frac{r}{2}\right)$$

$$\Rightarrow \quad r_{PQ}=\frac{r\times\frac{5r}{6}}{r+\frac{5r}{6}}=\frac{5r}{11}$$

$$r_{QR}=\frac{r}{2}\,\|\,\frac{4r}{3}=\frac{\frac{r}{2}\times\frac{4r}{3}}{\frac{r}{2}+\frac{4R}{3}}=\frac{4r}{11}$$

and $$r_{PR}=\frac{r}{3}\,\|\,\frac{3r}{2}=\frac{\frac{r}{3}\times\frac{3r}{2}}{\frac{r}{3}+\frac{3r}{2}}=\frac{3r}{11}$$

Highest

$$\Rightarrow \quad r_{PQ}>r_{QR}>r_{PR}$$ **Ans. (B)**

Sol. 10

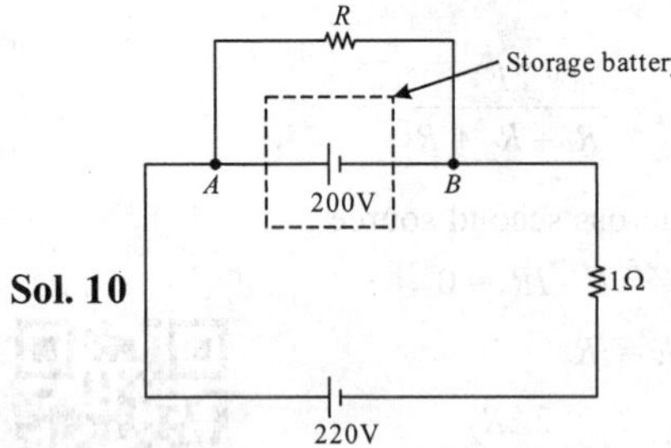

If storage battery is removed and if $V_A - V_B > 200$ V then a current passes through this battery to charge it

$\Rightarrow$ in open circuit condition (without 200 V battery)

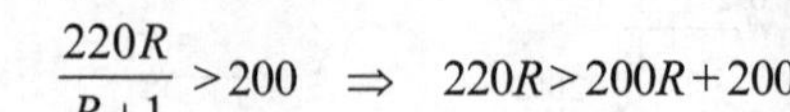

$$V_A-V_B=\frac{220}{R+1}\times R$$

For battery to be charge

$$V_A-V_B>200\text{V}$$

$$\Rightarrow \quad \frac{220R}{R+1}>200 \Rightarrow 220R>200R+200$$

$$\Rightarrow \quad R>10\,\Omega \Rightarrow \text{here } 11\,\Omega>10\,\Omega$$ **Ans. (C)**

Sol. 11

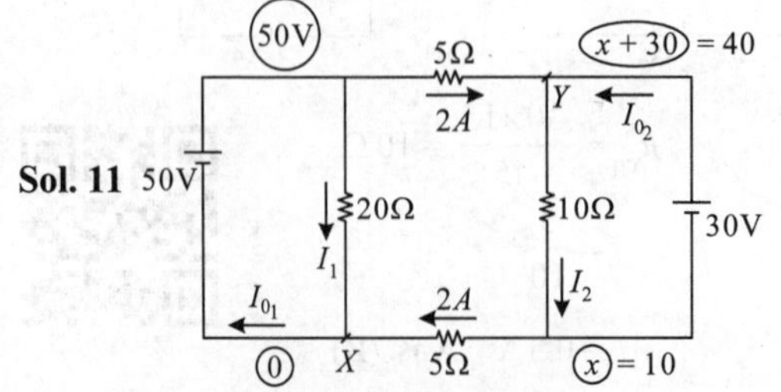

Currents is 20 Ω and 10 Ω resistances are

$$I_1 = \frac{50}{20} = 2.5\text{ A}$$

$$I_2 = \frac{30}{10} = 3\text{ A}$$

KCL equation for (x)

$$\frac{x+30-50}{5} = 0$$

$\Rightarrow$ $2x = 20\text{ V}$

$\Rightarrow$ $x = 10\text{ V}$

At junction (x) $I_{0_1} = 2 + 2.5 = 4.5\text{ A}$

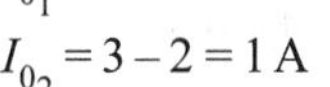

at junction (y) $I_{0_2} = 3 - 2 = 1\text{ A}$ **Ans. (C)**

Sol. 12 Modified circuit in

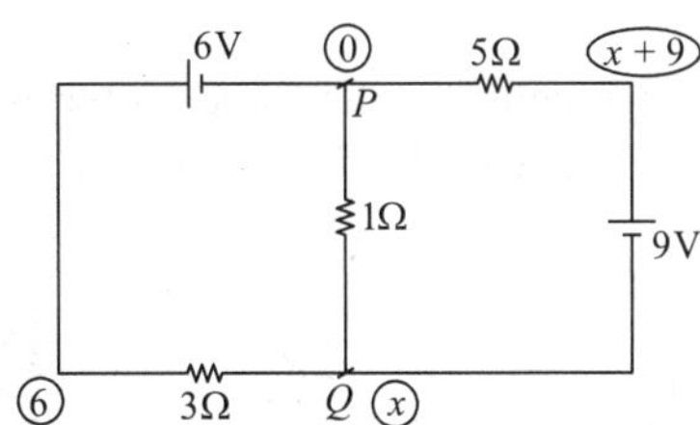

Distributing potentials in the circuit and writing KCL equation for (x), we have

$$\frac{x-6}{3} + \frac{x-0}{1} + \frac{x+9-0}{5} = 0$$

$\Rightarrow$ $5x - 30 + 15x + 3x + 27 = 0$

$\Rightarrow$ $23x = 3$

$\Rightarrow$ $x = \frac{3}{23}V$

(As $x > 0$ current flows from Q to P)

Current in 1Ω resistance is

$$I = \frac{x}{1} = \frac{3}{23} \simeq 0.13\text{ A} \quad \textbf{Ans. (C)}$$

Sol. 13 We can write equation of potential drop from point D to A along path $DEBA$

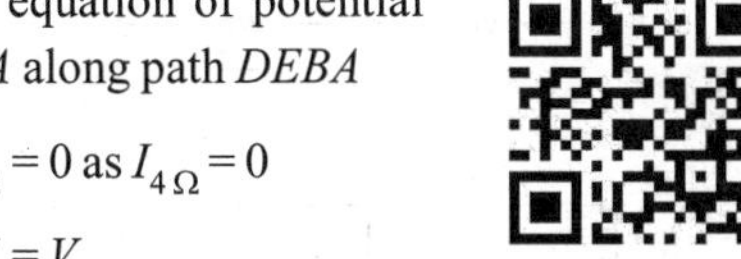

$$V_{4\Omega} = 0 \text{ as } I_{4\Omega} = 0$$

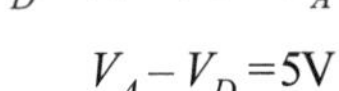

$$V_D - 4\text{V} + 9\text{V} = V_A$$

$\Rightarrow$ $V_A - V_D = 5\text{V}$ **Ans. (C)**

Sol. 14

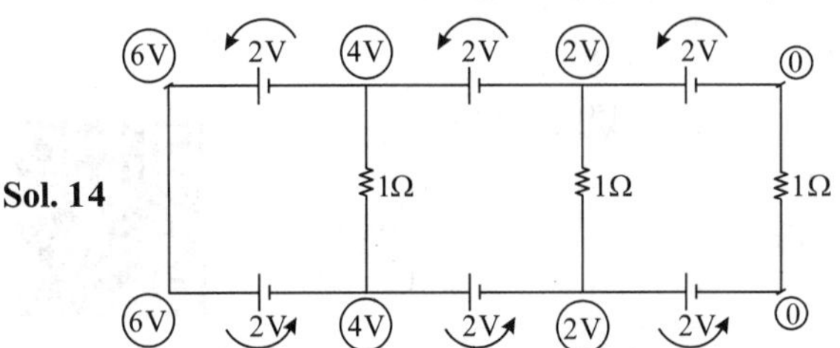

As shown in figure, after distribution of potentials in circuit, potential difference across each resistance is equal to zero.

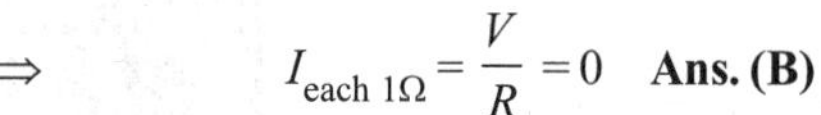

$\Rightarrow$ $I_{\text{each } 1\Omega} = \frac{V}{R} = 0$ **Ans. (B)**

Sol. 15

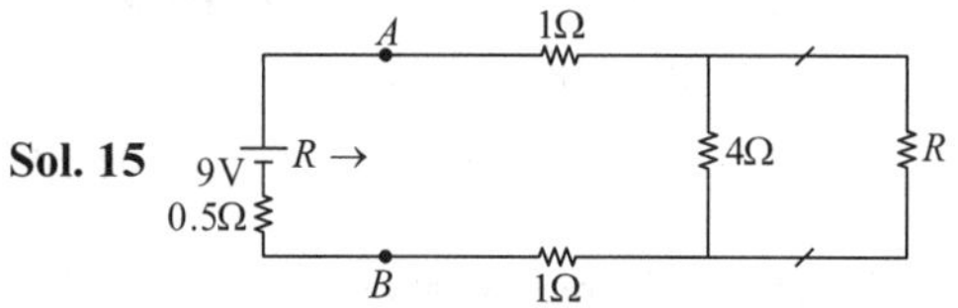

Here R is the equivalent Resistance across terminals A and B

For ∞ grid we use

Across $AB \rightarrow R = 2 + \frac{4R}{4+R}$

$\Rightarrow$ $4R + R^2 = 8 + 6R$

$\Rightarrow$ $R^2 - 2R - 8 = 0$

$\Rightarrow$ $R^2 - 2R - 8 = 0$

$\Rightarrow$ $R = \frac{2 \pm \sqrt{4+32}}{2} = \frac{2 \pm 6}{2}$

$\Rightarrow$ $R = \frac{8}{2} = 4\,\Omega$

$\Rightarrow$

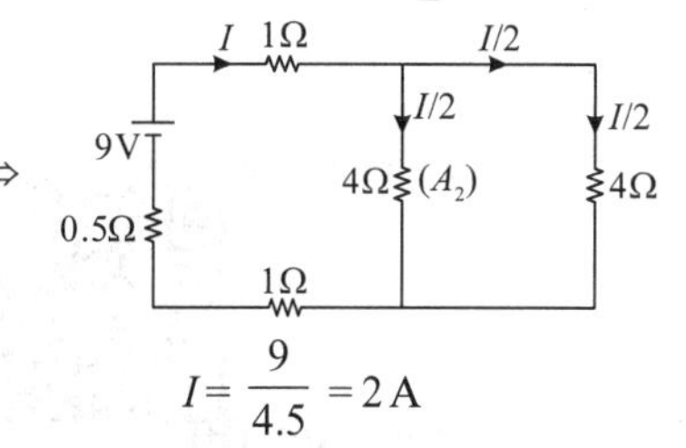

$$I = \frac{9}{4.5} = 2\text{ A}$$

Reading of (A_2) is

$$I_2 = \frac{2}{2} = 1\text{ A}$$

In further part of grid current will divide half in 4 Ω and half to next section

$\Rightarrow$ Reading of (A_3) is

$$\frac{I}{4} = \frac{2}{4} = 0.5\text{ A}$$

Reading of voltmeter (V) is

$$= IR = 2 \times 4 = 8\text{ V}$$

Here current supplied by battery is

$$I = \frac{9}{0.5+4} = \frac{9}{4.5} = 2\text{ A (Reading of } A_1) \quad \textbf{Ans. (A)}$$

Sol. 16

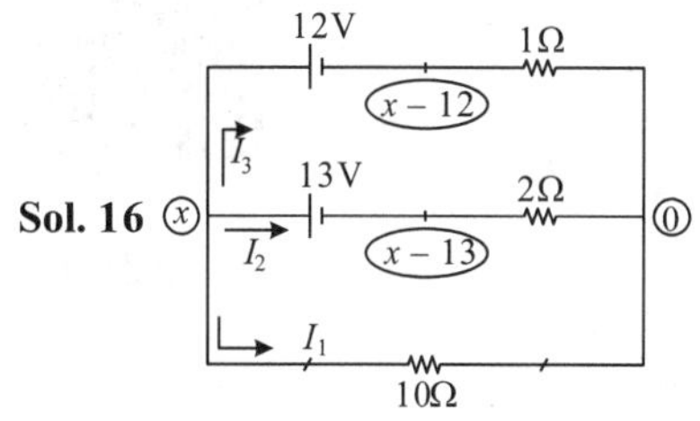

Distributing potentials at junctions of circuit as shown and using KCL for junction with potential x, it gives

$$I_1 + I_2 + I_3 = 0$$

$$\Rightarrow \quad \frac{x-0}{10} + \frac{x-13}{2} + \frac{x-12}{1} = 0$$

$$\Rightarrow \quad x + 5x - 65 + 10x - 120 = 0$$

$$\Rightarrow \quad 16x = 185$$

$$\Rightarrow \quad x = 11.56\,\text{V}$$

Ans. (B)

Sol. 17 Equivalent circuit of above situation is

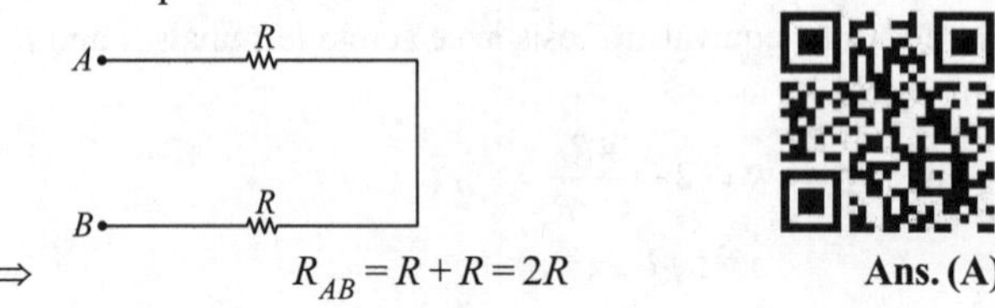

$$\Rightarrow \quad R_{AB} = R + R = 2R$$

Ans. (A)

Sol. 18

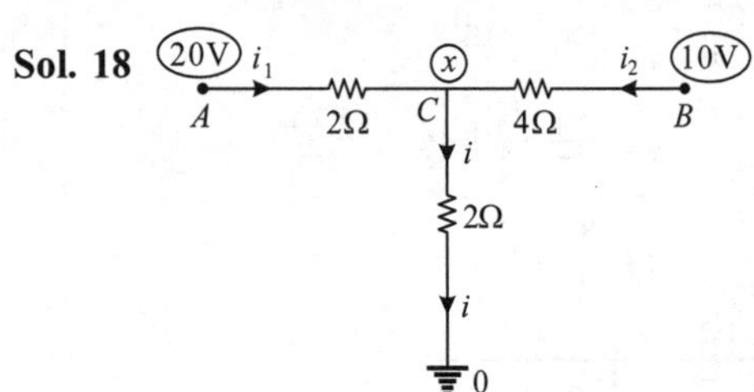

Using KCL equation at C in above circuit, we have

$$\frac{20-x}{2} + \frac{10-x}{4} = \frac{x-0}{2}$$

$$\Rightarrow \quad x = 10$$

$$\Rightarrow \quad i = \frac{x}{2} = \frac{10}{2} = 5\,\text{A}$$

Ans. (B)

Sol. 19

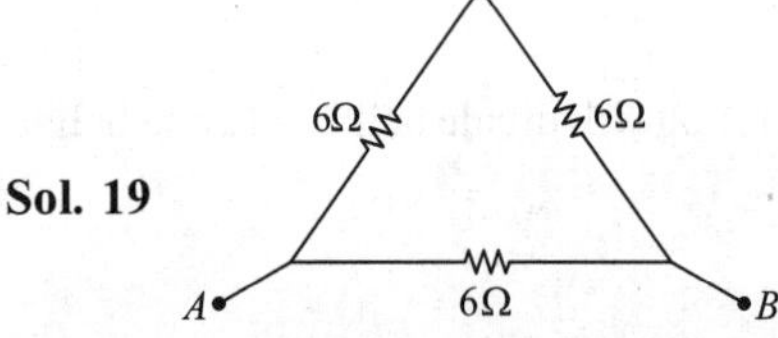

R_{eq} between any two vertex will be

$$\frac{1}{R_{eq}} = \frac{1}{12} + \frac{1}{6}$$

$$\Rightarrow \quad R_{eq} = 4\,\Omega$$

Ans. (C)

Sol. 20 From the given circuit, we can write

$$i_1 = \frac{10}{20} = 0.5\,\text{A}$$

$$i_2 = \frac{0}{20} = 0\,\text{A}$$

Ans. (D)

Sol. 21 Potential difference across AB will be equal to battery equivalent across CD, given as

$$V_{AB} = V_{CD} = \frac{\frac{E_1}{r_1} + \frac{E_2}{r_2} + \frac{E_3}{r_3}}{\frac{1}{r_1} + \frac{1}{r_2} + \frac{1}{r_3}} = \frac{\frac{1}{1} + \frac{2}{1} + \frac{3}{1}}{\frac{1}{1} + \frac{1}{1} + \frac{1}{1}}$$

$$V_{AB} = \frac{6}{3} = 2\,\text{V}$$

Ans. (D)

Sol. 22 Using KCL, we can write

$$I_3 = 0.8 - 0.4 = 0.4\,\text{A}$$

$$I_2 = 0.4 + 0.4 + 0.3$$

$$\Rightarrow \quad I_2 = 1.1\,\text{A}$$

$$I_6 = 0.4\,\text{A}$$

Ans. (A)

Sol. 23 Here we consider the three batteries in parallel combination so the potential difference across a and b is given as

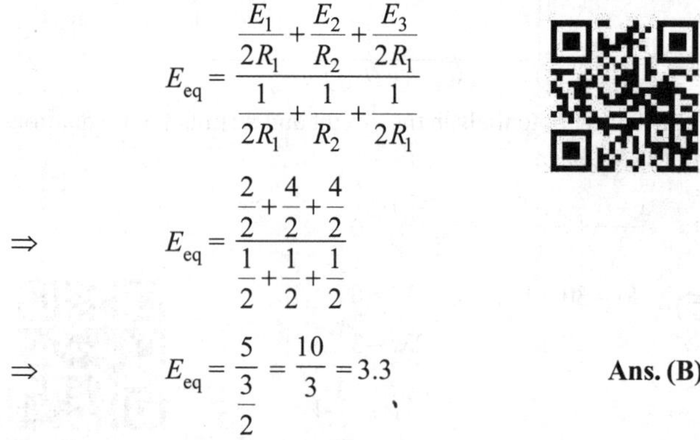

$$E_{eq} = \frac{\frac{E_1}{2R_1} + \frac{E_2}{R_2} + \frac{E_3}{2R_1}}{\frac{1}{2R_1} + \frac{1}{R_2} + \frac{1}{2R_1}}$$

$$\Rightarrow \quad E_{eq} = \frac{\frac{2}{2} + \frac{4}{2} + \frac{4}{2}}{\frac{1}{2} + \frac{1}{2} + \frac{1}{2}}$$

$$\Rightarrow \quad E_{eq} = \frac{5}{\frac{3}{2}} = \frac{10}{3} = 3.3$$

Ans. (B)

Sol. 24 The circuit given is reduced as shown below

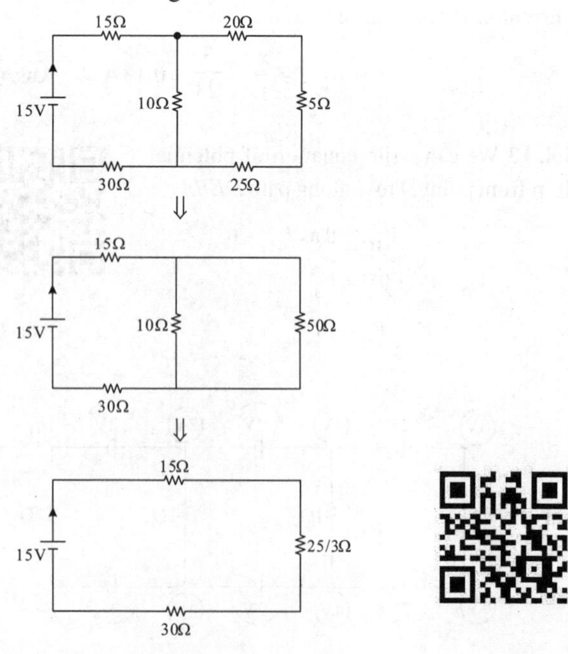

Equivalent resistance across battery is given as

$$R_{eq} = 15 + \frac{25}{3} + 30 = \frac{45+25+90}{3} = \frac{160}{3}\,\Omega$$

Thus current supplied by battery is given as

$$I = \frac{E}{R_{eq}} = \frac{15\times 3}{160} = \frac{9}{32}\,\text{A} \qquad \textbf{Ans. (C)}$$

Sol. 25 The equivalent circuit of the given square is redrawn as shown below

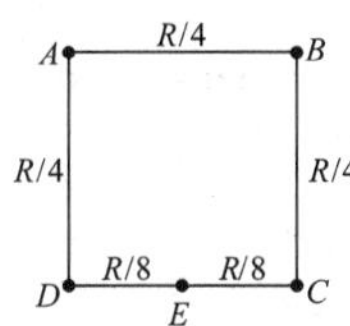

The equivalent resistance across E and C is given as

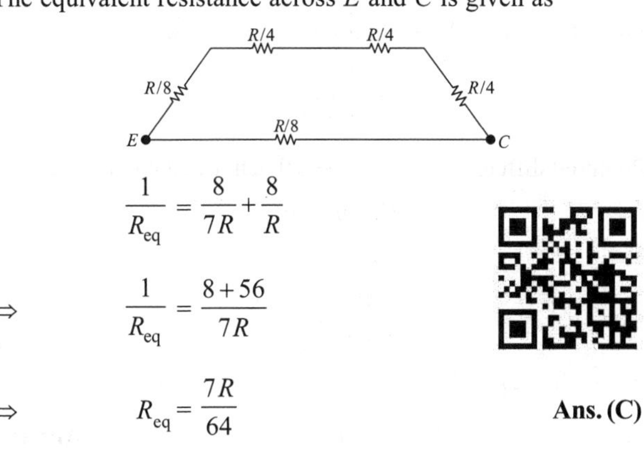

$$\frac{1}{R_{eq}} = \frac{8}{7R} + \frac{8}{R}$$

$$\Rightarrow \quad \frac{1}{R_{eq}} = \frac{8+56}{7R}$$

$$\Rightarrow \quad R_{eq} = \frac{7R}{64} \qquad \textbf{Ans. (C)}$$

Sol. 26 Equivalent resistance of upper branch of circuit is 2.5 Ω so total current in upper branch is 1/2.5 = 0.4A which divides equally in two 1Ω resistances so through each current is 0.2 A.

Ans. (B)

Sol. 27 Figure below shows the circuit for the described situation

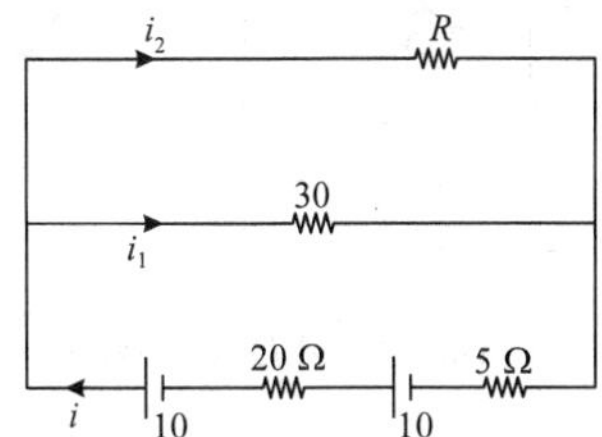

In above circuit, we use potential difference across batteries

$$E_1 = E - ir$$

$$\Rightarrow \quad E_1 = E - ir$$

$$\Rightarrow \quad i = 0.5\,\text{A}$$

$$\text{and} \quad E_2 = E - ir$$

$$\Rightarrow \quad E_2 = 10 - 0.5\times 5$$

$$\Rightarrow \quad E_2 = 7.5\,\text{V}$$

Using KCL at right most junction, we have

$$0.5 = \frac{7.5}{R} + \frac{7.5}{30}$$

$$R = 30\,\Omega \qquad \textbf{Ans. (30)}$$

Sol. 28 Considering potentials as shown in circuit below we write *KLL* equation for *x*

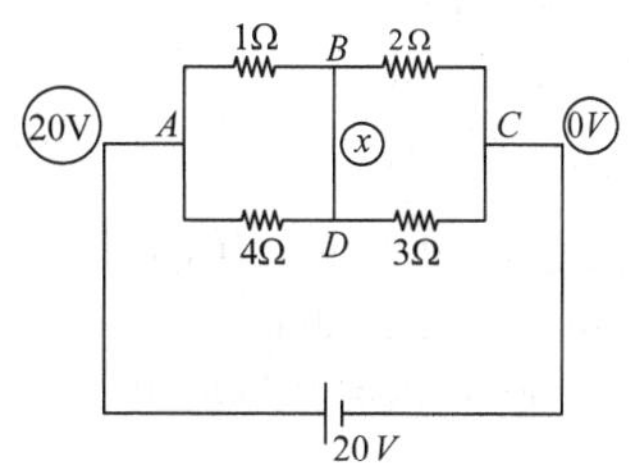

$$\frac{x}{2} + \frac{x}{3} + \frac{x-20}{1} + \frac{x-20}{4} = 0$$

$$\Rightarrow \quad 6x + 4x + 12x - 240 + 3x - 60 = 0$$

$$\Rightarrow \quad 25x = 300$$

$$\Rightarrow \quad x = 12\text{V}$$

Thus current in 1Ω and 2Ω resistance are

$$I_{1\Omega} = \frac{20-12}{1} = 8A$$

$$\text{and} \quad I_{2\Omega} = \frac{12}{2} = 6A$$

Current in branch *BD* is

$$I_{BD} = I_{1\Omega} - I_{2\Omega} = 2A \qquad \textbf{Ans. (B)}$$

Sol. 29 Current in branch of 8Ω resistance is given as

$$i = \frac{V}{R} = \frac{8}{8} = 1\,\text{A} \qquad \textbf{Ans. (D)}$$

Sol. 30 Current in upper & power branches are given as

$$i_1 = \frac{40}{100} = 0.4\,\text{A}$$

$$\text{and} \quad i_2 = \frac{40}{200} = 0.2\,\text{A}$$

We write equation of potential drop from *B* to *D* as

$$V_B - 60\times 0.4 + 110\times 0.2 = V_D$$

$$\Rightarrow \quad V_B - V_D = 24 - 22 = 2\,\text{V} \qquad \textbf{Ans. (2)}$$

Sol. 31 Writing equation of potential drop *A* to *B*, we have,

$$V_A + 1 + 1\times 2 - 2 = V_B$$

(Current in 2 Ω resistance is 1 *A* downward)

$$\Rightarrow \quad V_B - V_A = 1\,\text{V} \qquad \textbf{Ans. (A)}$$

Sol. 32 By principle of superposition current in 4Ω resistance is given as

$$I_{10V} = I_{4\Omega} = I_{20V} - I_{10V}$$

$$\Rightarrow \quad I_{10V} = \frac{20}{7+\frac{20}{7}} \times \frac{10}{14} - \frac{10}{4+\frac{70}{17}}$$

$$\Rightarrow \quad I_{10V} = \frac{140}{69} \times \frac{10}{14} - \frac{170}{138}$$

$$\Rightarrow \quad I_{10V} = \frac{100}{69} - \frac{170}{138} = \frac{30}{138} \simeq 0.22\,\text{A} \quad \textbf{Ans. (C)}$$

NOTE : This question can also be solved using KCL or KVL or by Thevenin's analysis. Students are advised to solve this question by all these methods for conceptual clarity.

Sol. 33 The current in circuit is given as

$$I = \frac{6-4}{10} = \frac{1}{5}A$$

By KVL equation in given circuit, we have

$$V_x + 4 + 8 \times \frac{1}{5} - V_y = 0$$

$$\Rightarrow \quad V_y - V_x = 5.6 \quad \textbf{Ans. (C)}$$

Sol. 34 For the given parallel combination of resistances, current is divided in inverse ratio of resistance, so we have

$$i_{PQ} = \frac{2}{6} \times 6\,\text{A} \quad \textbf{Ans. (2)}$$

Sol. 35 This circuit diagram given in question can also be drawn as shown in figure below as a wheat-stone bridge

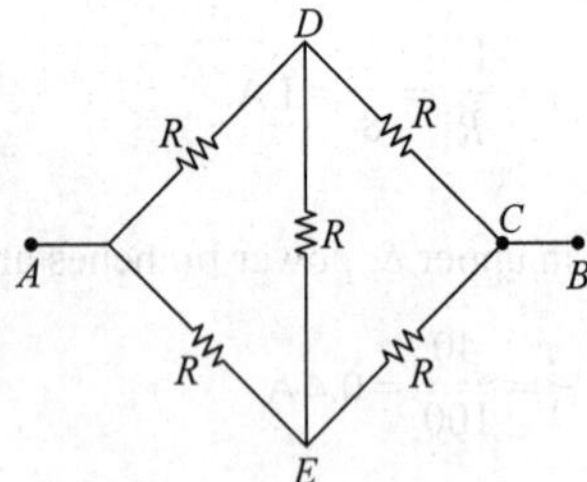

The equivalent resistance of this balanced wheat-stone bridge can be directly given as

$$R_{eq} = \frac{2R \times 2R}{2R + 2R} = R \quad \textbf{Ans. (C)}$$

Sol. 36 Work done by battery in passing a given amount of charge is given as

$$W = Q(\Delta V) = 20 \times 15 = 300\text{J} \quad \textbf{Ans. (300)}$$

Sol. 37 The equivelent resistance of the combinations as described in question are taken as

$$R_1 + R_2 = s \quad \ldots(1)$$

$$\frac{R_1R_2}{R_1 + R_2} = p \quad \ldots(2)$$

$$\Rightarrow \quad R_1R_2 = sp$$

$$\Rightarrow \quad R_1R_2 = np^2$$

$$\Rightarrow \quad R_1 + R_2 = \frac{nR_1R_2}{(R_1 + R_2)}$$

$$\Rightarrow \quad \frac{(R_1 + R_2)^2}{R_1R_2} = n$$

For minimum value of n happens when

$$R_1 = R_2 = R$$

$$\Rightarrow \quad n = \frac{(2R)^2}{R^2} = 4 \quad \textbf{Ans. (4)}$$

Sol. 38 The current in circuit is calculated as

$$i = \frac{3E}{R + r_1 + r_2}$$

Potential difference across first cell can be calculated as

$$V = 2E - ir_1 = 0$$

$$\Rightarrow \quad 2E = ir_1$$

$$\Rightarrow \quad 2E = \frac{3E \times r_1}{R + r_1 + r_2}$$

$$\Rightarrow \quad 2R + 2r_1 + 2r_2 = 3r_1$$

$$\Rightarrow \quad R = \frac{r_1}{2} - r_2 \quad \textbf{Ans. (B)}$$

Sol. 39 To calculate the current in circuit, we distribute the potentials at different junctions of circuit as shown in below figure.

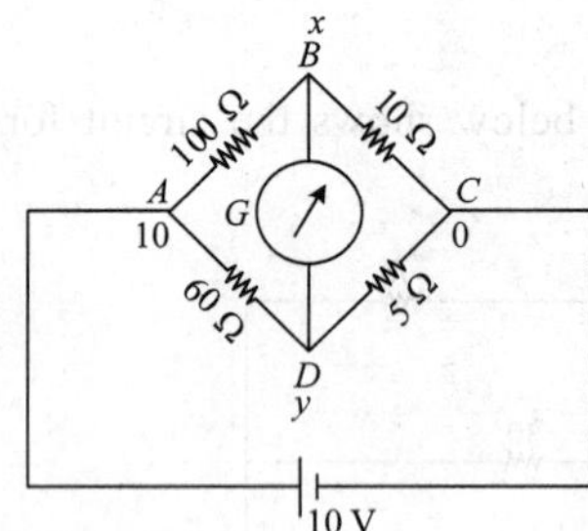

Writing KCL equations for junctions B and D, we have

$$\frac{x-10}{100} + \frac{x-y}{15} + \frac{x-0}{10} = 0$$

$$53x - 20y = 30 \quad \ldots(1)$$

$$\frac{y-10}{60} + \frac{y-x}{15} + \frac{y-0}{5} = 0$$

$$17y - 4x = 10 \quad \ldots(2)$$

Solving equations-(1) & (2), we get

$$x = 0.865\,\text{V}$$

$$y = 0.792\text{ V}$$
$$\Delta V = 0.073\ R = 15\Omega$$

Current through galvanometer is calculated as

$$i = \frac{V}{R} = \frac{x-y}{R} = \frac{0.073}{15} = 4.87\text{ mA}$$

Ans. (C)

Sol. 40 Equivalent resistance of the above circuit can be given as

$$R_{eq_1} = \frac{50\times 20}{70} + 10 = \frac{170}{7}\Omega$$

Voltage across 10Ω resistance is given as

$$V_1 = \left[\frac{170}{\frac{170}{7}}\right]\times 10 = 70\text{ V}$$

Ans. (70.00)

Sol. 41 In parallel combination, we use

$$R_{net} = \frac{R_1R_2}{R_1+R_2}$$

$$\Rightarrow \quad \frac{\rho l}{2A} = \frac{R_1R_2}{R_1+R_2}$$

$$\Rightarrow \quad \frac{\rho}{2} = \frac{6\times 3}{6+3} = 2$$

$$\Rightarrow \quad \rho = 4$$

Ans. (4)

Sol. 42 Distributing potentials in the circuit shown in figure below, gives

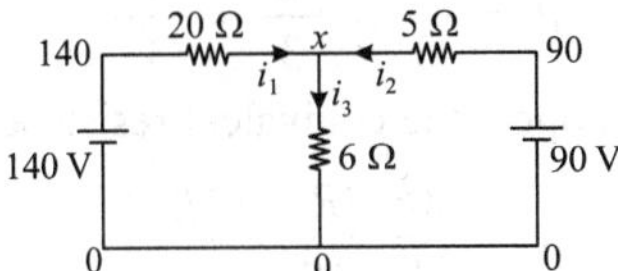

Writing KCL equation for unknown potential x, gives

$$\frac{140-x}{20} + \frac{90-x}{5} + \frac{0-x}{6} = 0$$

$$\Rightarrow \quad x = 60\text{ V}$$

Thus current in 6Ω resistance is given as

$$i_3 = \frac{x}{6} = \frac{60}{6} = 10\text{ A}$$

Ans. (B)

Sol. 43 When switches are closed the effective circuit across ab can be redrawn as shown in figure below

a •—4 Ω—•—2 Ω—•—4 Ω—•

For this, effective resistance across ab is given as

$$R_{ab} = 10\ \Omega$$

Ans. (10)

Sol. 44 The two rods can be considered to be in parallel combination so the effective resistance is given as

$$R_{eq} = \frac{R_1R_2}{R_1+R_2} = \frac{\rho_1\frac{l}{A}\rho_2\left(\frac{l}{A}\right)}{\rho_1\left(\frac{l}{A}\right)+\rho_2\left(\frac{l}{A}\right)}$$

$$\Rightarrow \quad R_{eq} = \frac{l}{A}\left(\frac{\rho_1\rho_2}{\rho_1+\rho_2}\right) = \frac{0.25}{3\times 10^{-6}}\left(\frac{1.7\times 10^{-8}\times 2.6\times 10^{-8}}{1.7\times 10^{-8}+2.6\times 10^{-8}}\right)$$

$$\Rightarrow \quad R_{eq} = 0.085\times 10^{-2} = 0.858\text{ m}\Omega$$

Ans. (A)

Sol. 45 Figure below shows the circuits as explained in the question.

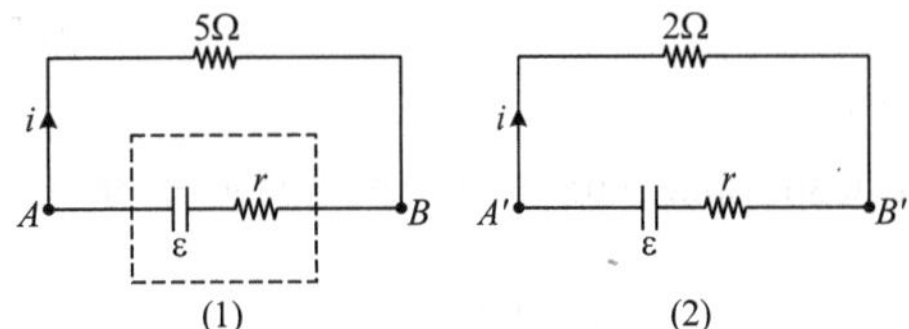

In case-(1), the current in circuit is given as

$$I = \frac{1.25}{5} = 0.25\text{ A}$$

The potential drop across the load resistance is given as

$$V_{AB} = 5V = \varepsilon - ir$$

$$\Rightarrow \quad 1.25 = \varepsilon - 0.25r \qquad \ldots(1)$$

In case-(2), the current in circuit

$$I = \frac{1}{2} = 0.5\text{ A}$$

$$\Rightarrow \quad V_{AB} = 1 = \varepsilon - 0.5r \qquad \ldots(2)$$

Solving equations-(1) & (2), we get

$$\varepsilon = 15\text{ V}$$

Ans. (15)

Sol. 46 The maximum current in circuit is given as

$$i_0 = \frac{20}{10} = 2\text{A}$$

After the current is switched off (battery is removed), current starts decaying through the given inductor so now current as a function of time is given as

$$i = i_0e^{-Rt/L}$$

At $t = 1$ ms = 0.001 s, we have

$$i = 2\times e^{-\frac{10\times 0.001}{10\times 10^{-3}}} = \frac{2}{2.718} = 0.74\text{ A}$$

$$\Rightarrow \quad x = 74$$

Ans. (74)

Sol. 47 Figure below shows the effective circuit formed for the situation described in the question

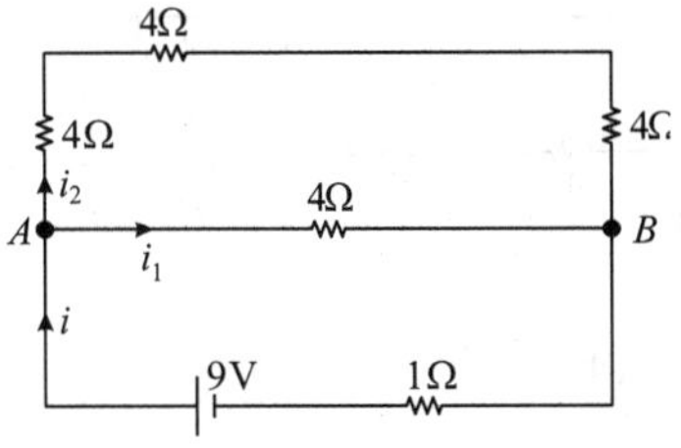

Resistance across battery is given as

$$R_{AB} = \frac{12\times 4}{12+4} = 3\,\Omega$$

Current through battery is given as

$$i = \frac{9}{3+1} = \frac{9}{4}\text{A}$$

Current i_2 is given as

$$i_2 = \frac{r_2}{r_2+r_1}i = \frac{1}{4}\times\frac{9}{4} = \frac{9}{16};$$

$$\Rightarrow \quad V_{AB} = \frac{9}{2} = 4.5\text{ V}$$ **Ans. (45)**

Sol. 48 In steady state, current in circuit is given as

$$I = \frac{6V}{6\Omega} = 1\text{ A}$$

Just after T_1 is connected to T_3, current will remain same due to inductor so potential drop across 3 Ω resistor is given as

$$V_{3\Omega} = 3\times 1 = 3\text{ V}$$ **Ans. (3)**

Sol. 49 Below figures show the calculation of equivalent resistance of the given circuit.

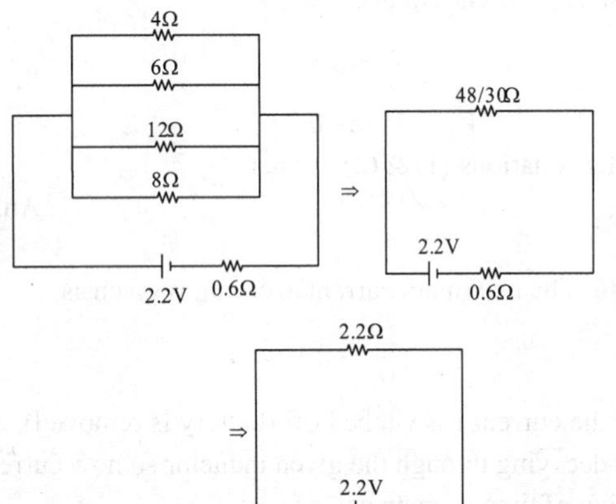

Thus power dissipated in the circuit can be calculated as

$$P = \frac{V^2}{R}, P = \frac{(2.2)^2}{2.2} = 2.2\text{ W}$$ **Ans. (D)**

Sol. 50 For the required equivalent resistance the desired combination is shown in figure below

2Ω
4Ω
8Ω
6Ω
$\frac{4}{3}\Omega$
14Ω

$$\Rightarrow R_{eq} = \frac{4}{3} + 14 = \frac{46}{3}\Omega$$ **Ans. (A)**

Sol. 51 For series combination of batteries current through effective battery is given as

$$i_1 = \frac{5\varepsilon}{R+5r}$$

For parallel combination of batteries current through effective battery is given as

$$i_2 = \frac{\varepsilon}{R+\frac{r}{5}}$$

As current in both cases is same, we use

$$i_1 = i_2$$

$$\Rightarrow \quad \frac{5\varepsilon}{R+5r} = \frac{\varepsilon}{R+\frac{r}{5}}$$

$$\Rightarrow \quad R = r = 1\,\Omega$$ **Ans. (B)**

Sol. 52 According to the given conditions, we have

$$i_p = 20 i_a$$

$$\Rightarrow \quad \frac{20}{\left(\frac{10}{n}+10\right)} = 20\left(\frac{20}{10n+10}\right)$$

$$\Rightarrow \quad \frac{20n}{10+10n} = 20\left(\frac{20}{10n+10}\right)$$

$$n = 20$$ **Ans. (20)**

Sol. 53 After closing the switch, equivalent resistance is given as

$$R_1 = \frac{2R}{3} + \frac{2R}{3} = \frac{4R}{3}$$

When switch is open, the equivalent resistance is given as

$$R_2 = \frac{3R\cdot 3R}{3R+3R} = \frac{3R}{2}$$

$$\Rightarrow \quad \frac{R_1}{R_2} = \frac{4R}{3}\times\frac{2}{3R} = \frac{8}{9} = \frac{8}{x}$$ **Ans. (9)**

Sol. 54 The equivalent circuit of the given circuit is drawn below

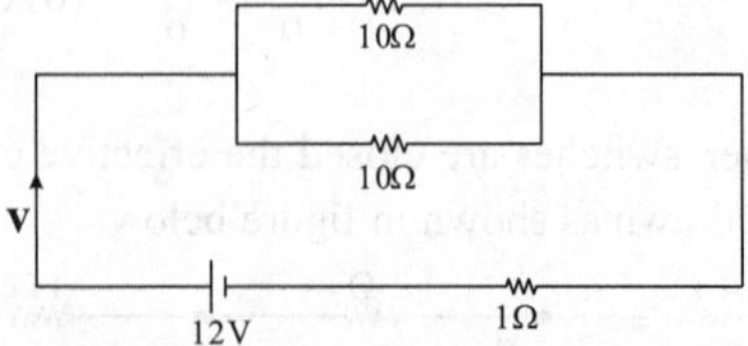

Current in circuit through the battery is given as

$$i = \frac{12}{6} = 2\text{A}$$

Currents in the top and lower branch will be same as both have equal 10 Ω resistance

$$i_1 = i_2 = i/2 = 1\text{ A}$$

Voltage drop across 15 Ω resistance can be calculated by ohms law as

$$V_{15\Omega} = i_1 \times \frac{10\times 15}{10+15} = 1\times 6 = 6\text{ V}$$ **Ans. (1)**

Sol. 55 Figure below shows the resistances described in the question. When the wire is bent to form a circle then across the diametrically opposite ends there will be two equal resistances of 6 Ω each considered to be connected in parallel as shown. From the equivalent resistance required is equal to 3 Ω.

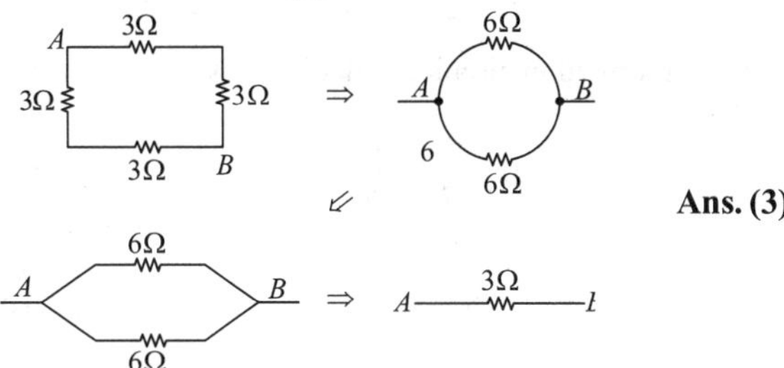

Ans. (3)

Sol. 56 In the given circuit leftmost resistance is short circuited and for right most resistances of 3 Ω in parallel and rest of resistances, we cam find equivalent by using series and parallel analysis so we have

$$R_{AB} = \frac{3\times\frac{3}{2}}{3+\frac{3}{2}} = 1\,\Omega$$ **Ans. (A)**

Sol. 57 The equivalent resistance of the two resistors can be given as

$$\frac{1}{R} = \frac{1}{R_1} + \frac{1}{R_2}$$

For error calculation, we differntiate the above expression as

$$\frac{\Delta R}{R^2} = \frac{\Delta R_1}{R_1^2} + \frac{\Delta R_2}{R_2^2}$$

$$\Rightarrow \quad \Delta R = (R)^2\left[\frac{\Delta R_1}{R_1^2} + \frac{\Delta R_2}{R_2^2}\right]$$

$$\Rightarrow \quad \Delta R = (2)^2\left[\frac{0.8}{4^2} + \frac{0.4}{4^2}\right]$$

$$\Rightarrow \quad \Delta R = [0.2 + 0.1] = 0.3$$

Thus equivalent resistance with error is given as

$$R_{net} = R + \Delta R = 2 \pm 0.3$$ **Ans. (D)**

Sol. 58 Figure below shows the circuit as described in question and its effective reduced circuit by taking equivalent battery combination and equivalent resistance combination.

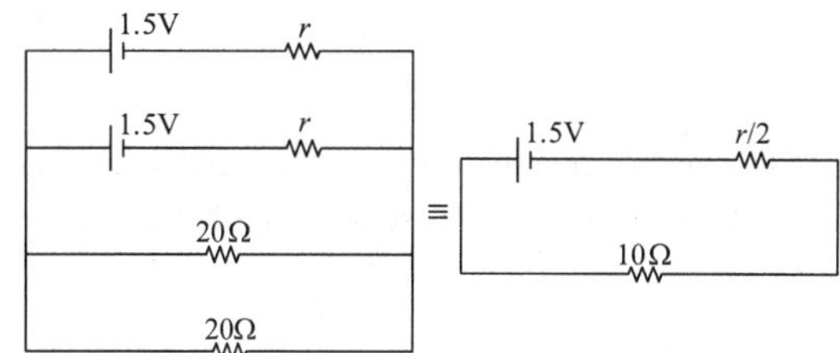

The potential difference across each cell is given as

$$V = E - ir/2$$

$$\Rightarrow \quad 1.2 = 1.5 - i\left(\frac{r}{2}\right)$$

$$\Rightarrow \quad i\frac{r}{2} = 0.3 \qquad \ldots(1)$$

Circuit current is given as

$$i = \frac{1.5}{10 + \frac{r}{2}}$$

$$\Rightarrow \quad 10i + \frac{ir}{2} = 1.5$$

$$\Rightarrow \quad 10i = 1.5 - 0.3$$

$$\Rightarrow \quad i = 0.12\text{ A}$$

From equation-(1), we have

$$r = \frac{0.6}{0.12} = 5\,\Omega$$ **Ans. (C)**

Sol. 59 From the options given, we can analyze that parallel combination of 2 Ω and 4 Ω in series with 6 Ω gives the desired equivalent resistance, calculated as

$$R_{eq} = \frac{4}{3} + 6 = \frac{22}{3}$$ **Ans. (B)**

Sol. 60 The equivalent circuit of the given circuit is drawn as

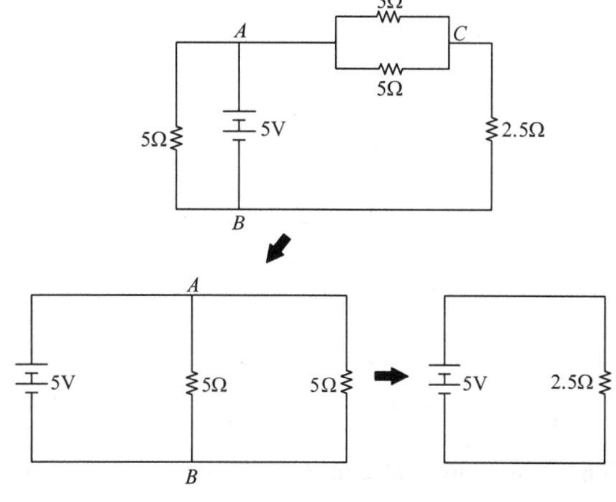

Current supplied by 5 V battery is calculated as

$$I=\frac{5}{2.5}=2\,\text{A} \qquad \textbf{Ans. (2)}$$

Sol. 61 The given circuit in question is reduced by using series and parallel combination of resistances, as shown below.

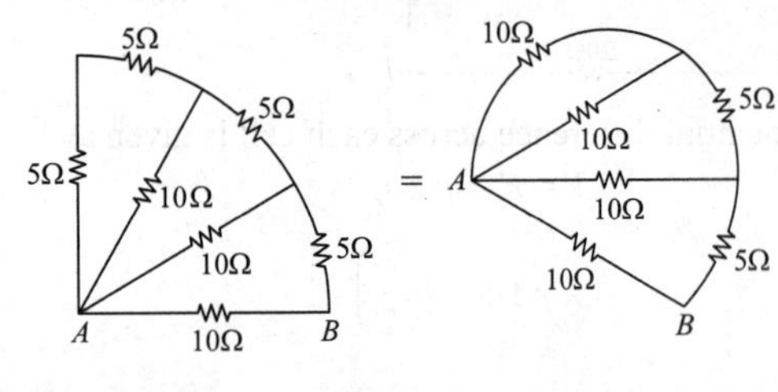

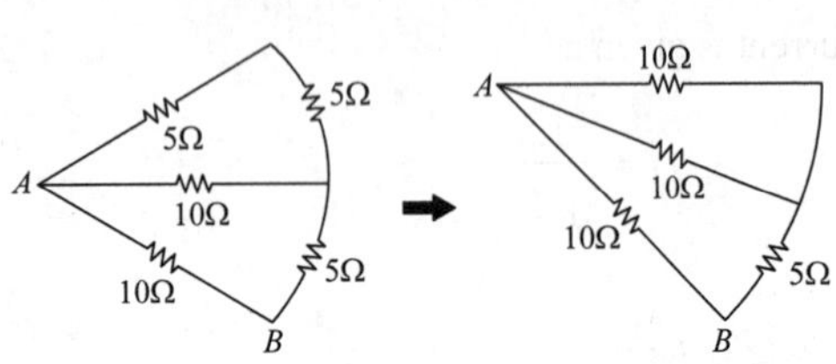

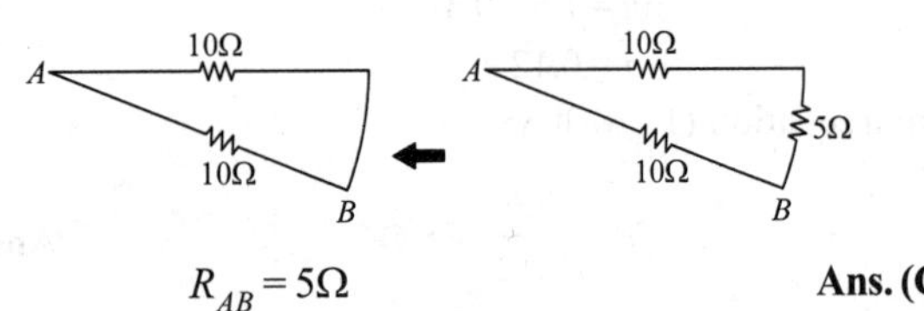

$$R_{AB}=5\Omega \qquad \textbf{Ans. (C)}$$

Sol. 62 For the given circuit, equivalent resistance is calculated as

$$R=\left(\frac{ma}{3}\right)+\left(\frac{a}{2m}\right)$$

R to be minimum, we use

$$\frac{dR}{dm}=\frac{a}{3}-\frac{a}{2m^2}=0$$

$$\Rightarrow \qquad \frac{a}{3}=\frac{a}{2m^2}$$

$$\Rightarrow \qquad m^2=\frac{3}{2}$$

$$\Rightarrow \qquad m=\sqrt{\frac{3}{2}}$$

$$\Rightarrow \qquad x=3 \qquad \textbf{Ans. (3)}$$

Sol. 63 The equivalent circuit of the given combination of resistance is redrawn as shown in figure below.

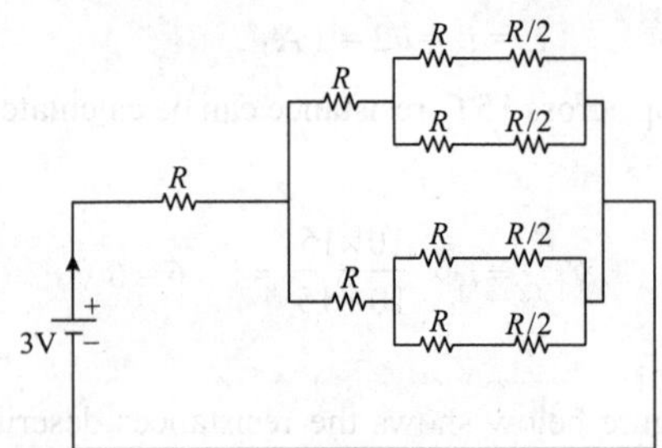

Here equivalent combination of all resistances of 1 Ω each is calculated as

$$R_{\text{eq}}=\frac{15R}{8}=\frac{15}{8}\Omega$$

Current supplied in circuit is calculated as

$$I=\frac{3}{\frac{15}{8}}=\frac{8}{5}\,\text{A}$$

$$\Rightarrow \qquad a=8 \qquad \textbf{Ans. (8)}$$

Sol. 64 The current in given circuit is calculated as

$$i=\frac{15}{3}=5\,\text{A}$$

The potential difference across terminals A and B is given as

$$V_A-V_B=iR=5\times 2=10\,\text{V} \qquad \textbf{Ans. (10)}$$

Sol. 65 Figure (a) and (b) below shows the circuit as described in question.

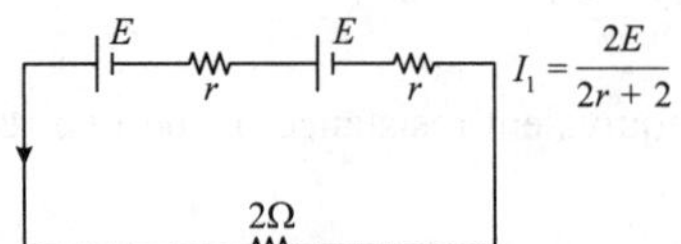

Figure-(a)

Current supplied in above circuit is given as

$$I_1=\frac{2E}{2r+2} \qquad \ldots(1)$$

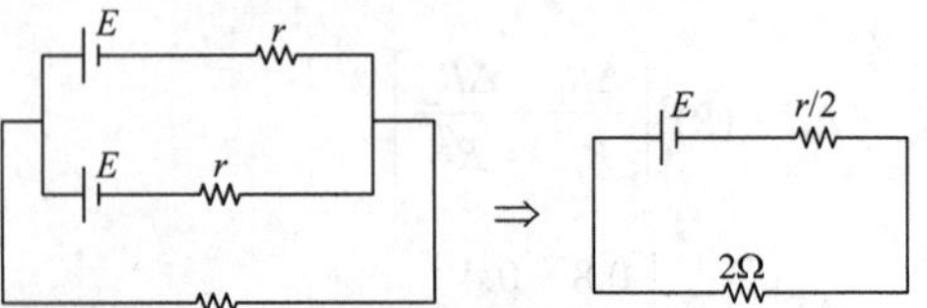

Figure-(b)

Current supplied in the above cases is given as

$$I_2=\frac{E}{\frac{r}{2}+2}=\frac{2E}{r+4} \qquad \ldots(2)$$

As it is given that the current in both of above cases is same so we use

$$I_1 = I_2$$

$\Rightarrow$ $2r+2 = r+4$

$\Rightarrow$ $2r - r = 2\,\Omega$

$\Rightarrow$ $r = 2\,\Omega$ **Ans. (A)**

Sol. 66 In the given circuit using Kirchoff's Current Law at junction where voltage is V_0, the equation is written as

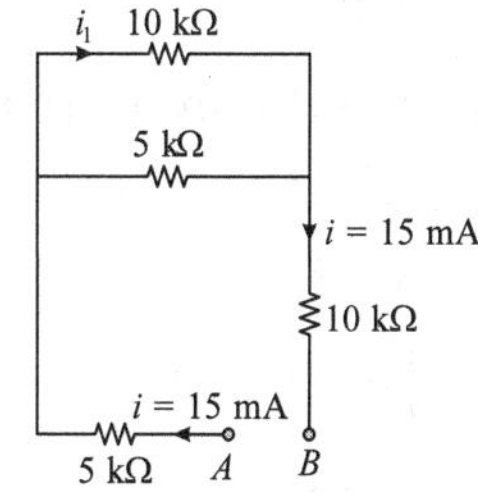

$$\frac{V_0 - 2}{1k\Omega} + \frac{V_0 - 4}{1k\Omega} + \frac{V_0 - 6}{1k\Omega} = 0$$

$\Rightarrow$ $3V_0 - 12 = 0$

$\Rightarrow$ $V_0 = 4$ **Ans. (4)**

Sol. 67 In the figure shown, the current i_1 is calculated as

$$i_1 = \frac{5}{10+5} \times 15\text{ mA} = 5\text{ mA}$$

Using Kirchoff's Voltage Law in the outer loop, we have

$$V_A - 5i - 10i_1 - 10i = V_B$$

$\Rightarrow$ $V_A - V_B = 75 + 50 + 150 = 275\text{ V}$ **Ans. (D)**

Sol. 68 Equivalent circuit of the given circuit can be drawn as all three resistances in parallel so equivalent resistance is 3 Ω

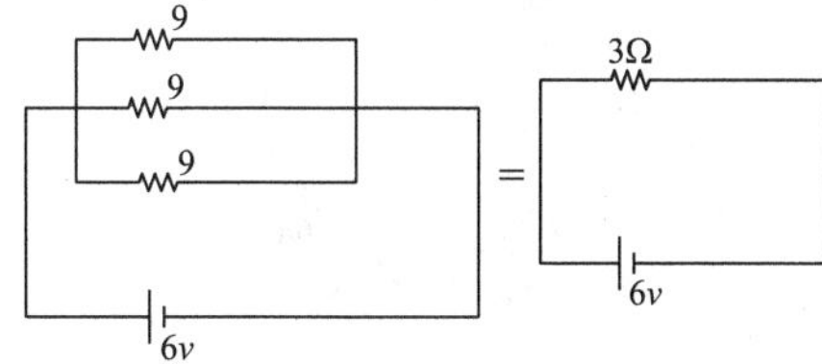

Current in circuit is calculated as

$$I = \frac{6}{3} = 2\text{ A}$$ **Ans. (2)**

Sol. 69 Equivalent resistance of given circuit is calculated as

$$R_{eq} = 3 + 1 + 2 + 4 + 2 = 12\,\Omega$$

Current through battery is calculated as

$$i = \frac{24}{12} = 2\text{ A}$$

Currents I_4 and I_5 as shown in the given circuit is calculated by current distribution as

$$I_4 = \frac{R_5}{R_4 + R_5} \times 2 = \frac{5}{20+5} \times 2 = \frac{2}{5}\text{A}$$

and $$I_5 = 2 - \frac{2}{5} = \frac{8}{5}\text{A}$$ **Ans. (D)**

Sol. 70 Due to straight wire the two 4 Ω resistance gets short circuited so we can remove these resistors and the equivalent circuit will be as shown below

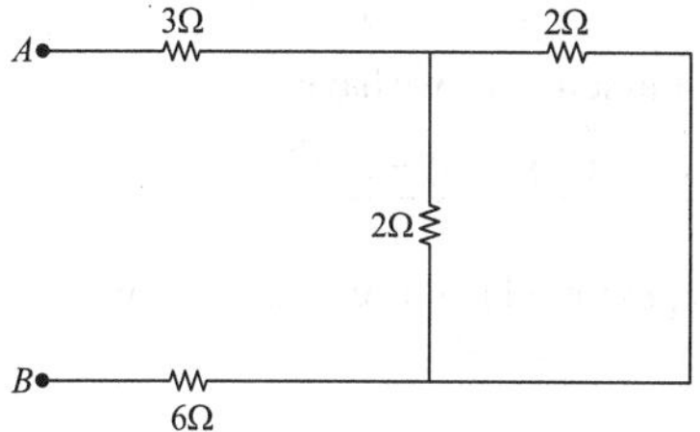

The equivalent resistance across A and B is calculated as

$$R_{eq} = 3 + (2 \,\|\, 2) + 6$$

$$R_{eq} = 3 + 1 + 6 = 10\,\Omega$$ **Ans. (10)**

Sol. 71 Figure below shows the circuit with external resistance as mentioned in the question

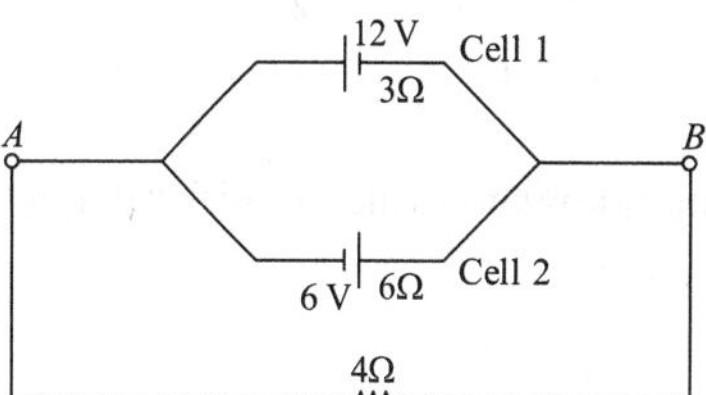

Equivalent emf of the parallel combination of cells is given as

$$E_{eq} = \frac{\frac{12}{3} - \frac{6}{6}}{\frac{1}{3} + \frac{1}{6}} = 6\text{ V}$$

Equivalent internal resistance of the cell combination is taken as r_{eq} = 2 Ω. The equivalent circuit after replacing the cell combination with equivalent cell is shown in figure below

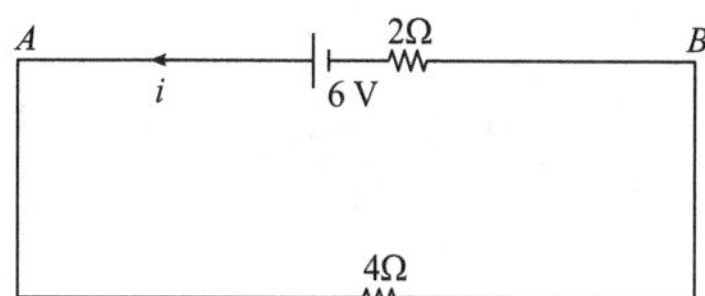

The current in external resistance is given as

$$i = \frac{6}{2+4} = 1\text{ A}$$ **Ans. (1)**

Sol. 72 In the circuit shown below, we distribute the potentials at different points as indicated

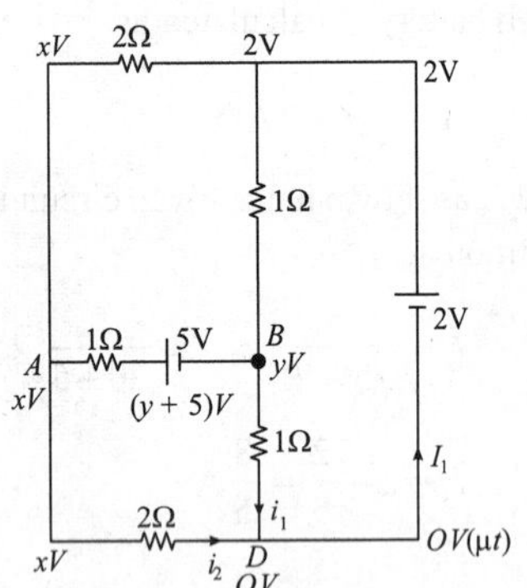

Using KCL at junction A, we have

$$\frac{x-(y+5)}{1}+\frac{x-2}{2}+\frac{x-0}{2}=0 \quad \ldots(1)$$

Using KCL at junction B, we have

$$\frac{y+5-x}{1}+\frac{y-0}{1}+\frac{y-2}{1}=0 \quad \ldots(2)$$

On solving equation-(1) and equation-(2), we get

$$x=3\text{ V}$$

&

$$y=0\text{ V}$$

At junction D, we use

$$I_1=i_1+i_2$$

$$\Rightarrow \quad I_1=\frac{y-0}{1}+\frac{x-0}{2}$$

$$\Rightarrow \quad I_1=\frac{0-0}{1}+\frac{3-0}{2}$$

$$\Rightarrow \quad I_1=1.5\text{ A}$$

Ans. (2)

Sol. 73 Figure below shows the simplified diagram of the given circuit in question

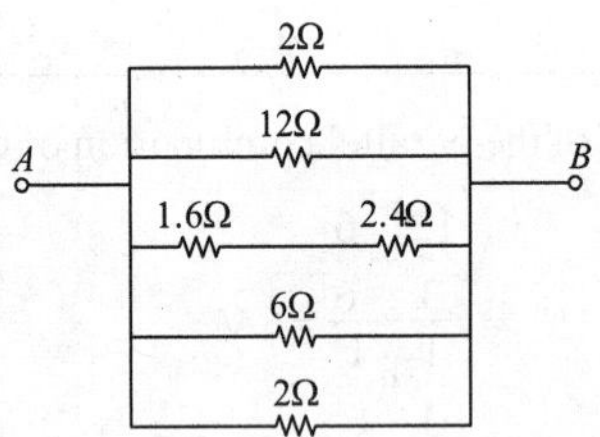

Equivalent resistance of given circuit can be calculated as

$$\frac{1}{R_{eq}}=\frac{1}{2}+\frac{1}{12}+\frac{1}{4}+\frac{1}{6}+\frac{1}{2}$$

$$\Rightarrow \quad \frac{1}{R_{eq}}=\frac{6+1+3+2+6}{12}=\frac{18}{12}=\frac{3}{2}$$

$$\Rightarrow \quad R_{eq}=\frac{2}{3}\Omega$$

Ans. (A)

Sol. 74 By the condition of balanced wheatstone bridge

$$\frac{2}{3}=\frac{\frac{x}{x+1}}{x}$$

$$\Rightarrow \quad \frac{2}{3}=\frac{1}{x+1}$$

$$\Rightarrow \quad x=0.5=\frac{1}{2}=\frac{1}{n}\Omega$$

$$\Rightarrow \quad n=2$$

Ans. (2)

Sol. 75 Series and parallel combination of the two cells are shown in figures below.

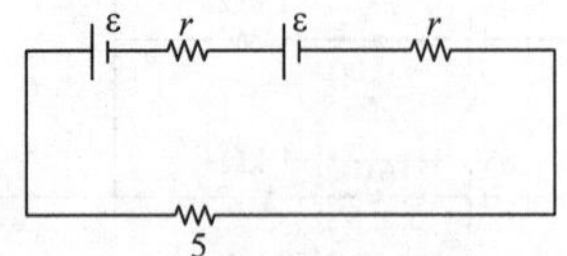

When the cells are in series, current is given as

$$i=\frac{2\varepsilon}{5+2r} \quad \ldots(1)$$

When the cells are in parallel, current is given as

$$i=\frac{\varepsilon}{\frac{r}{2}+5} \quad \ldots(2)$$

From equations (1) and (2), we have

$$\frac{2\varepsilon}{5+2r}=\frac{\varepsilon}{\frac{r}{2}+5}$$

$$\Rightarrow \quad r+10=5+2r$$

$$\Rightarrow \quad r=5\,\Omega$$

Ans. (5)

Sol. 76 The given wheat stone bridge is in balanced condition so across terminals A and B the equivalent resistance is calculated as

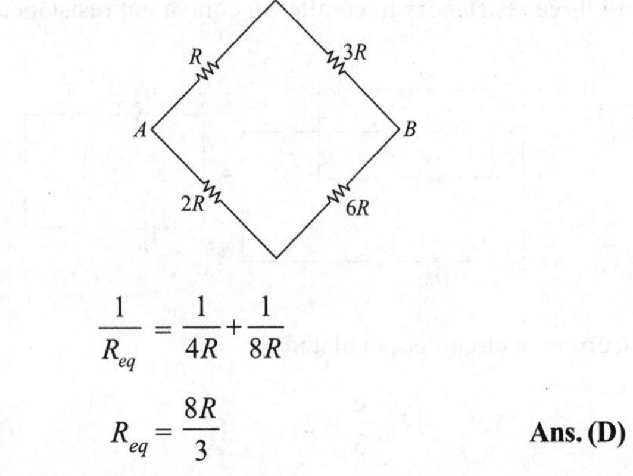

$$\frac{1}{R_{eq}}=\frac{1}{4R}+\frac{1}{8R}$$

$$\Rightarrow \quad R_{eq}=\frac{8R}{3}$$

Ans. (D)

Sol. 77 Figure below shows the polygon as described in the question.

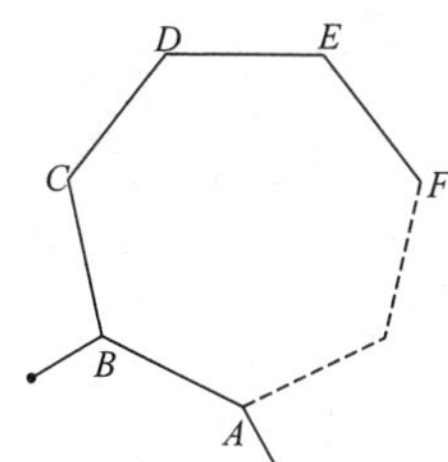

Suppose resistance of each arm of this polygon is r, then for n sides, we use $r = R/n$. The equivalent resistance across vertices A and B is given as

$$R_{eq\,(AB)} = \frac{R_1 R_2}{R_1 + R_2}$$

$$\Rightarrow \quad R_{eq\,(AB)} = \frac{r(n-1)r}{r+(n-1)r}$$

$$\Rightarrow \quad R_{eq\,(AB)} = \frac{r(n-1)r}{nr}$$

$$\Rightarrow \quad R_{eq\,(AB)} = \frac{n-1}{n}r$$

$$\Rightarrow \quad R_{eq\,(AB)} = \frac{(n-1)R}{n^2}$$

Ans. (A)

Sol. 78 In the circuit shown below, considering zero potential at the top right corner of the circuit, potentials are distributed on the circuit as shown. The curents in circuit are calculated as

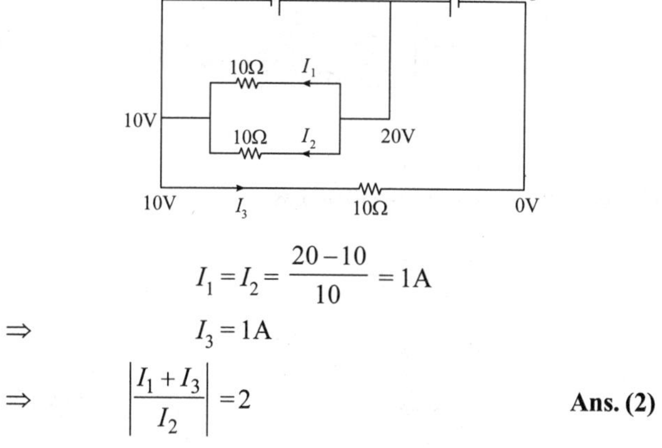

$$I_1 = I_2 = \frac{20-10}{10} = 1\text{A}$$

$$\Rightarrow \quad I_3 = 1\text{A}$$

$$\Rightarrow \quad \left|\frac{I_1 + I_3}{I_2}\right| = 2$$

Ans. (2)

Sol. 79 Sum of current at junction point will be zero by Kirchoff's current law.

Let Potential at junction is x, using KCL at the junction gives

$$\frac{x-30}{10} + \frac{x-12}{20} + \frac{x-2}{30} = 0$$

$$\Rightarrow \quad x\left(\frac{1}{10} + \frac{1}{20} + \frac{1}{30}\right) = \frac{30}{10} + \frac{12}{20} + \frac{2}{30}$$

$$\Rightarrow \quad x\left(\frac{6+3+2}{60}\right) = \frac{180+36+4}{60}$$

$$\Rightarrow \quad x = \frac{220}{11} = 20\text{ V}$$

Thus current through 20Ω resistance is calculated as

$$i = \frac{x-12}{20}$$

As $x = 20$ V, we have

$$\Rightarrow \quad i = \frac{20-12}{20} = \frac{2}{5} = 0.4\text{A}$$

Ans. (A)

Sol. 80 As electrical circuit is short-circuited between A and D and C and B.

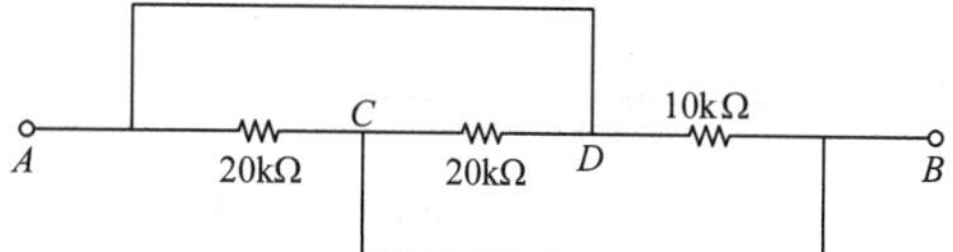

Thus potentials $V_A = V_D$ and $V_C = V_B$ thus circuit can be simplified and redrawn as shown in figure below.

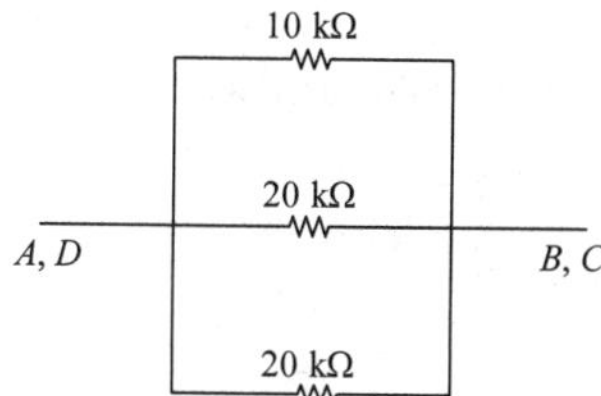

Thus all resistors are in parallel for which equivalent resistance is calculated as

$$\frac{1}{R_{eq}} = \frac{1}{10} + \frac{1}{20} + \frac{1}{20}$$

$$\Rightarrow \quad R_{eq} = 5\text{ k}\Omega$$

Ans. (A)

Sol. 81 The circuit can be simplified and redrawn as shown in figure below.

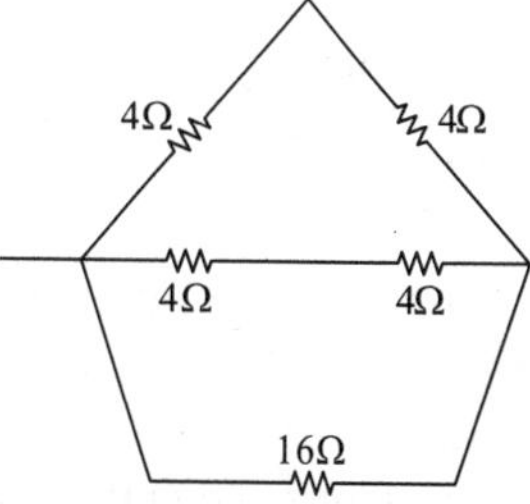

The equivalent resistance of this combination can be calculated as

$$R_{eq.} = \frac{16\times 4}{16+4} = \frac{16}{5}\,\Omega$$

$$\Rightarrow \quad R_{eq.} = 3.2\,\Omega \qquad \textbf{Ans. (B)}$$

Sol. 82 In the circuit shown in figure given in question, it can be simplified as given below by considering equivalent battery and equivalent resistances in circuit.

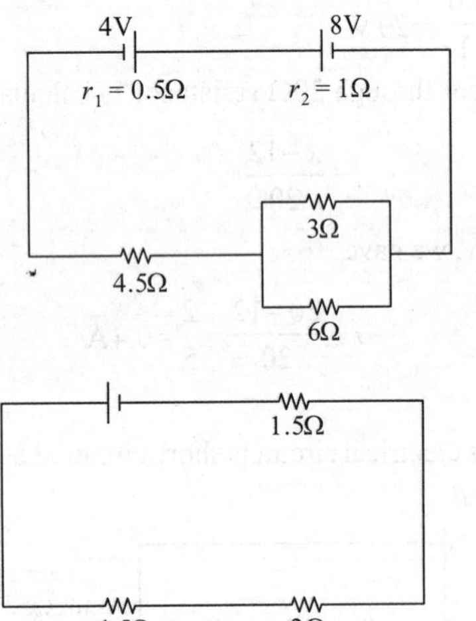

Equivalent resistance of the combination is given as

$$\frac{1}{3}+\frac{1}{6} = \frac{1}{2} = \frac{1}{R}$$

$$\Rightarrow \quad R = 2\,\Omega$$

Current in the circuit is calculated as

$$I = \frac{4}{8} = 0.5\,\text{A}$$

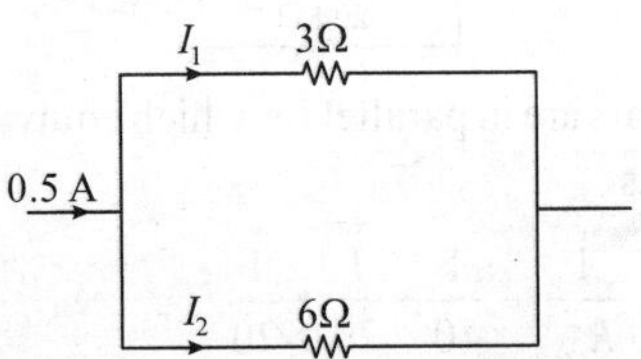

Current through 3 Ω resistance is calculated as

$$I_1 = \left(\frac{6}{3+6}\right)\times 0.5$$

$$\Rightarrow \quad I_1 = \frac{2}{3}\times 0.5 = \frac{1}{3}\,\text{A}$$

$$\Rightarrow \quad I_1 = \frac{x}{3} = \frac{1}{3}$$

$$\Rightarrow \quad x = 1 \qquad \textbf{Ans. (1)}$$

Sol. 83 Figure below shows the current distribution in all branches of the circuit.

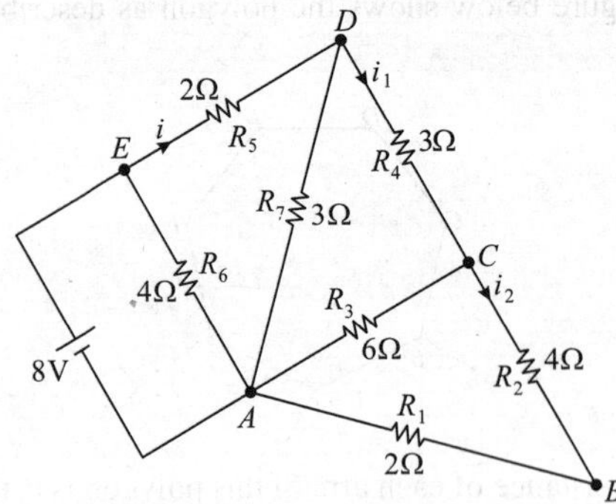

Equivalent resistance across the battery is given as

$$R_{eq} = 4\,\Omega$$

Thus net current through battery is given as

$$i = \frac{8}{4} = 2\,\text{A}$$

$$\Rightarrow \quad i_1 = \frac{2\times 3}{3+6} = \frac{2}{3}\,\text{A}$$

$$\Rightarrow \quad i_2 = \frac{2/3}{2} = \frac{1}{3}\,\text{A} \qquad \textbf{Ans. (D)}$$

Sol. 84 Potential across 10 Ω resistnace is given as

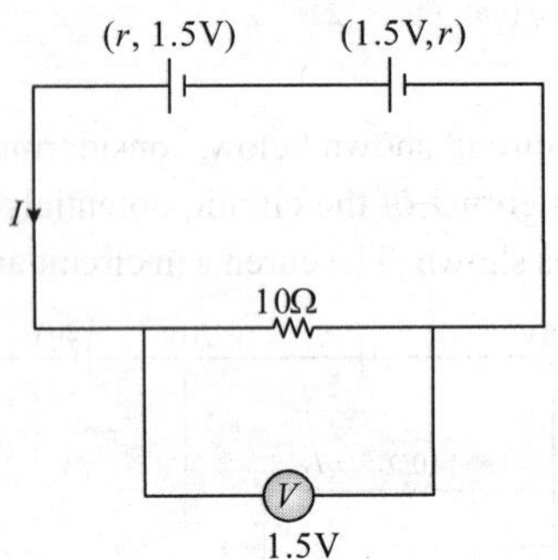

$$V = I\times 10$$

Current in the circuit is calculated as

$$I = \frac{E}{R+2V}$$

$$\Rightarrow \quad 1.5 = \left(\frac{3}{10+2r}\right)\times 10$$

$$\Rightarrow \quad r = 5\,\Omega \qquad \textbf{Ans. (5)}$$

Sol. 85 Power dissipation across a resistance is given as

$$P = I^2R$$

Equivalent resistance in combination *a*, *b*, *c* and *d* are calculated as

$$R_1 = \frac{3R}{2},\ R_2 = \frac{2R}{3},\ R_3 = \frac{R}{3},\ R_4 = 3R$$

Since *i* is same, hence $P \propto R$ so options (D) is correct.

Ans. (D)

Sol. 86 Figure below shows the currents flowing in the circuit as described in the question.

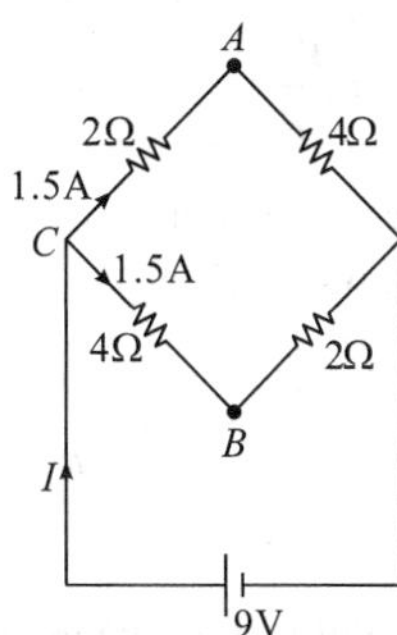

Equivalent resistance of the circuit across battery is 3 Ω. The current is divided equally as resistance of both branches are same.

In the circuit $I = \frac{9}{3} = 3\text{ A}$

$\Rightarrow \quad V_C - V_A = 2 \times 1.5 = 3 \quad \ldots(1)$

$\Rightarrow \quad V_C - V_B = 4 \times 1.5 = 6 \quad \ldots(2)$

Fom Equation-(1) & (2), we have

$V_A - V_B = 6 - 3 = 3\text{ V}$ **Ans. (3)**

17.3 Thermal Effects of Current & RC Circuits

Sol. 1 By Faraday's first law of electrolysis

$$m = Zq = ZIt$$

$m \propto It$ **Ans. (B)**

Sol. 2 Initial power dissipation in wire

$$P_1 = \frac{(220)^2}{R} \qquad [R \to \text{resistance of wire}]$$

When wire cut in two equal pieces, each will have resistance $R/2$ and these are connected in parallel

$$\Rightarrow \quad R_{eq} = \frac{R}{4}$$

New power dissipation will be

$$P_2 = \frac{(220)^2}{R/4} = 4\left[\frac{(220)^2}{R}\right] = 4P_1$$

$\Rightarrow \quad P_2/P_1 = 4$ **Ans. (B)**

Sol. 3 Here $R \| 2 = \frac{2R}{2+R}$ resistance is across 15V battery so power dissipation is given as

$$P = \frac{V^2}{R} = \frac{(15)^2}{2R} \times (2+R) = 150$$

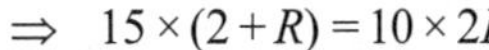

$\Rightarrow \quad 15 \times (2+R) = 10 \times 2R$

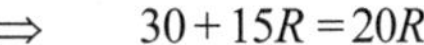

$\Rightarrow \quad 30 + 15R = 20R$

$\Rightarrow \quad R = \frac{30}{5} = 6\,\Omega$ **Ans. (B)**

Sol. 4 As minimum current that can be detected by galvanometer is

$$I_{min} = 10^{-5}\text{A}$$

If minimum temperature difference which can be measured is $\Delta\theta$ then we use

$$I_{min} = \frac{25 \times 10^{-6} \times \Delta\theta}{R}$$

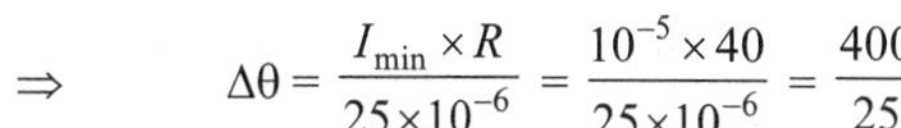

$$\Rightarrow \quad \Delta\theta = \frac{I_{min} \times R}{25 \times 10^{-6}} = \frac{10^{-5} \times 40}{25 \times 10^{-6}} = \frac{400}{25}$$

$\Rightarrow \quad \Delta\theta = 16\,°\text{C}$ **Ans. (A)**

Sol. 5 Mass of metal at negative pole deposited or at positive pole used is given as

$$m = Zq$$

for Zn & Cu electrodes, we use

$$\frac{m_1}{m_2} = \frac{Z_1}{Z_2}$$

$$\Rightarrow \quad m_2 = m_1 \times \frac{Z_2}{Z_1} = 0.13\text{g} \times \frac{31.5}{32.5}$$

$\Rightarrow \quad m_2 \simeq 0.126\text{g}$ **Ans. (C)**

Sol. 6 The resistance of bulb is

$$R = \frac{V^2}{P} = \frac{(220)^2}{1000}\Omega$$

Power consumed across 110V supply is given as

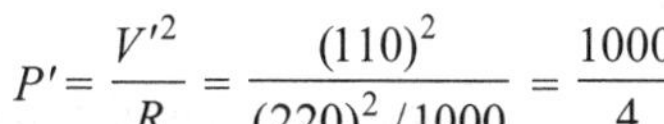

$$P' = \frac{V'^2}{R} = \frac{(110)^2}{(220)^2/1000} = \frac{1000}{4}$$

$\Rightarrow \quad P' = 250\text{ W}$ **Ans. (C)**

Sol. 7 The neutral temperature is the temperature of hot junction when thermo EMF is maximum and cold junction is kept at 0°C.

Here in this case

$$E = a\theta + b\theta^2$$

E is maximum when

$$\frac{dE}{d\theta} = a + 2b\theta = 0$$

$$\Rightarrow \quad \theta = -\frac{a}{2b} = -350°\text{C}$$

We check $\frac{d^2E}{d\theta^2} = 2b\,(+ve)$ value

$\Rightarrow$ E will never attain a maxima.

So no neutral temperature is possible is this case. **Ans. (D)**

Sol. 8 Mass of metal deposited is given as

$$m = Zq$$

charge flown $\quad q = It = 3 \times 2 = 6\text{C}$

mass $\quad m = 3.35 \times 10^{-7} \times 6$

$$= 20.10 \times 10^{-7}\ \text{kg} \quad \textbf{Ans. (C)}$$

Sol. 9 The heat required to raise the temperature of 1kg water from 10°C to 40°C is

$$\Delta Q = mS_{\omega}\Delta t$$

$\Rightarrow \quad \Delta Q = 1 \times 4180 \times 30\ \text{J} \qquad [s_{\omega} = 4180\ \text{J/°C kg}]$

It t is the time required then heat supplied by heater is

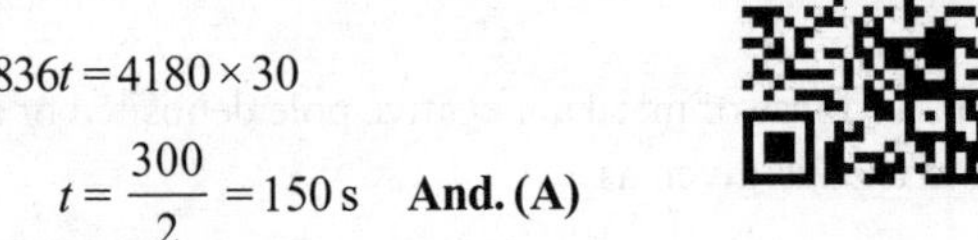

$$836t = 4180 \times 30$$

$$t = \frac{300}{2} = 150\ \text{s} \quad \textbf{And. (A)}$$

Sol. 10 Constant current source

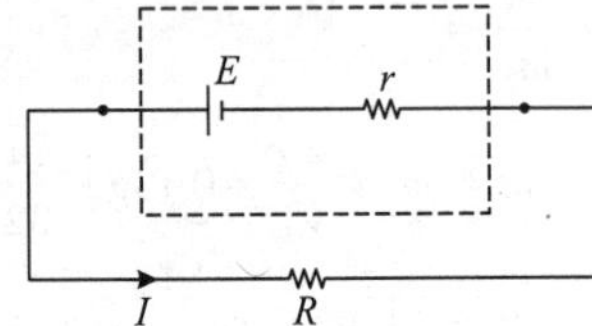

Current supplied

$$I = \frac{E}{r + R}$$

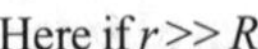

Here if $r >> R$

$\Rightarrow \quad I = \frac{E}{r} \rightarrow$ constant for small variations in R **Ans. (B)**

Sol. 11 If q_1 and q_2 are charges flown through copper & silver voltameter then we use

$$q = q_1 + q_2 \qquad \ldots(1)$$

[As the two voltmeter are in parallel]

mass of metal deposited is given as

$$m = Zq$$

As for both, mass deposited are same

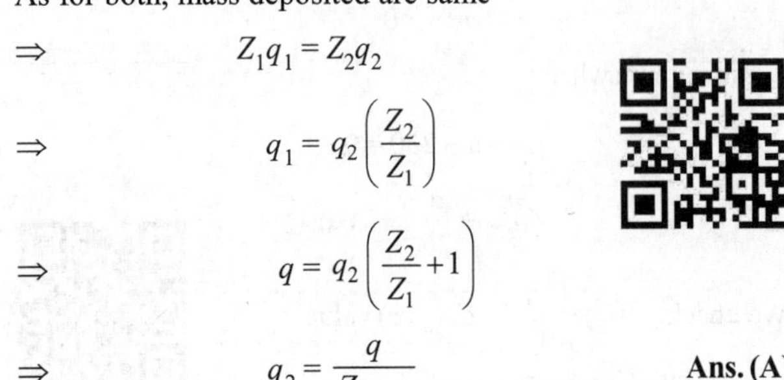

$\Rightarrow \quad Z_1q_1 = Z_2q_2$

$\Rightarrow \quad q_1 = q_2\left(\frac{Z_2}{Z_1}\right)$

$\Rightarrow \quad q = q_2\left(\frac{Z_2}{Z_1} + 1\right)$

$\Rightarrow \quad q_2 = \dfrac{q}{\frac{Z_2}{Z_1} + 1}$ **Ans. (A)**

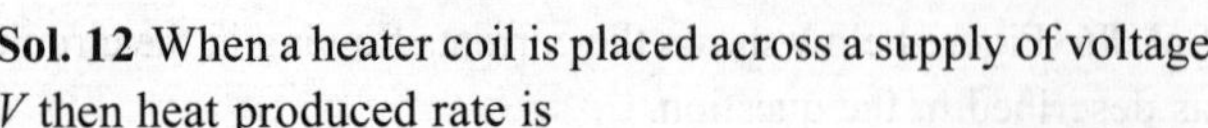

Sol. 12 When a heater coil is placed across a supply of voltage V then heat produced rate is

$$H_1 = \frac{V^2}{R}$$

Each part of coil after cut has a resistance $R/2$ so now heat generation rate is

$$H_2 = \frac{V^2}{R/2} = 2\left(\frac{V^2}{R}\right) = 2H_1 \qquad \textbf{Ans. (B)}$$

Sol. 13 For the given temperature its hot resistance is given as

$$R_{hot} = \frac{V^2}{P}$$

$$= \frac{(200)^2}{100} = 400\ \Omega$$

Given that $\quad R_{cold} = \frac{1}{10} \cdot R_{hot}$

$$= \frac{1}{10} \times 400 = 40\ \Omega \qquad \textbf{Ans. (B)}$$

Sol. 14

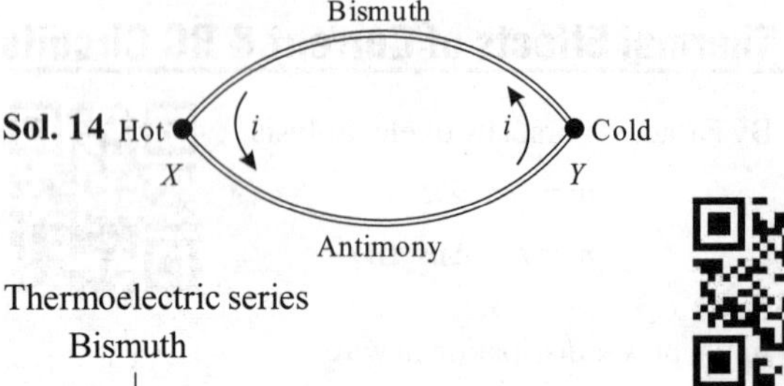

Thermoelectric series

Bismuth

↓

Antimony **Ans. (D)**

Sol. 15 The resistance of bulb is

$$R = \frac{V^2}{P} = \frac{(220)^2}{100}$$

Power consumed when operated at 110V is given as

$$P_C = \frac{V_1^2}{R} = \frac{(110)^2}{(220)^2/100} = \frac{100}{4}$$

$\Rightarrow \quad P_C = 25\ \text{W}$ **Ans. (C)**

Sol. 16 When a capacitor discharges through a resistor, its charge as time function is given as

$$q = Qe^{-t/RC}$$

Where Q is charge at $t = 0$

Energy in capacitor

$$U = \frac{q^2}{2C} = \left(\frac{Q}{2C}\right)^2 e^{-2t/RC} = U_0e^{-2t/RC}$$

Given that at $t = t_1$, $U = \frac{U_0}{2}$

$$\Rightarrow \quad \frac{U_0}{2} = U_0 e^{-2t_1/RC}$$

$$\Rightarrow \quad \frac{2t_1}{RC} = \ln(2)$$

$$\Rightarrow \quad t_1 = \frac{RC}{2}\ln(2) \qquad \ldots(1)$$

At $t = t_2$, $\quad q = \frac{Q}{4}$

$$\Rightarrow \quad \frac{Q}{4} = Qe^{-t_2/RC}$$

$$\Rightarrow \quad \frac{t_2}{RC} = \ln(4) = 2\ln(2)$$

$$\Rightarrow \quad t_2 = 2RC\ln(2) \qquad \ldots(2)$$

$$\frac{(1)}{(2)} \Rightarrow \quad \frac{t_1}{t_2} = \frac{\frac{RC}{2}\ln(2)}{2RC\ln(2)} = \frac{1}{4}$$ **Ans. (D)**

Sol. 17 For discharging of capacitors in RC circuit, voltage across capacitor combination is given as

$$V = V_0 e^{-t/RC_{eq}}$$

For parallel combination $C_{eq} = 2C$ and at $t = 10$s, $V = \frac{V_0}{2}$

$$\Rightarrow \quad \frac{V_0}{2} = V_0 e^{-10/2RC} \qquad \ldots(1)$$

for series combination

$$C_{eq} = C/2 \text{ and at } t = t_1,$$

$$V = \frac{V_0}{2}$$

$$\Rightarrow \quad \frac{V_0}{2} = V_0 e^{-2t_1/RC} \qquad \ldots(2)$$

Using (1) and (2) we have

$$\frac{2t_1}{RC} = \frac{10}{2RC}$$

$$\Rightarrow \quad t_1 = \frac{10}{4} = 2.5\text{s}$$ **Ans. (D)**

Sol. 18

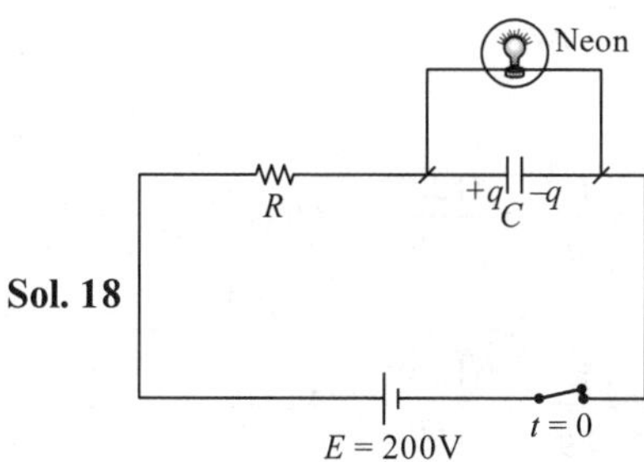

In RC circuit charge on capacitor as a function of time is given as

$$q = CE(1 - e^{-t/RC})$$

p.d. across capacitor as a function of time is given as

$$V = \frac{q}{C} = E(1 - e^{-t/RC})$$

at $t = 5$, $\quad V = 120$V

$$120 = 200(1 - e^{-5/RC})$$

$$\Rightarrow \quad e^{-5/RC} = 1 - \frac{120}{200} = \frac{2}{5}$$

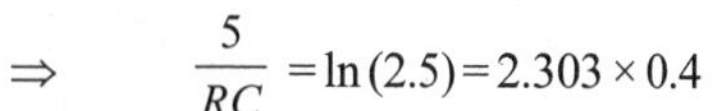

$$\Rightarrow \quad \frac{5}{RC} = \ln(2.5) = 2.303 \times 0.4$$

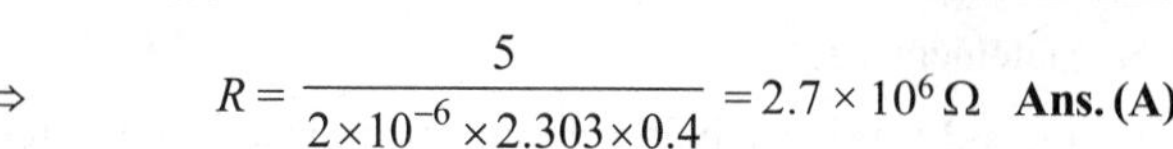

$$\Rightarrow \quad R = \frac{5}{2\times10^{-6}\times2.303\times0.4} = 2.7\times10^6\,\Omega$$ **Ans. (A)**

Sol. 19

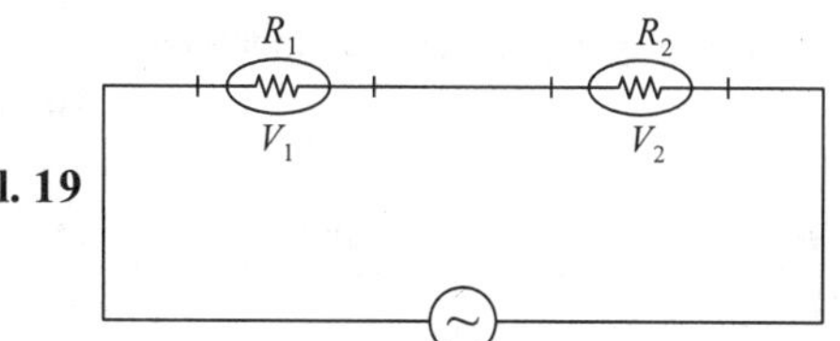

Resistances of bulbs are given as

$$R_1 = \frac{V_1^2}{P_1} = \frac{(220)^2}{25}$$

and $\quad R_2 = \frac{V_2^2}{P_2} = \frac{(220)^2}{100} \Rightarrow R_1 > R_2$

$$V_1 = \frac{440R_1}{R_1 + R_2} \quad \text{and} \quad V_2 = \frac{440R_2}{R_1 + R_2}$$

As $\quad V_1 + V_2 = 440$

$$\Rightarrow \quad V_1 > V_2$$

$$\Rightarrow \quad V_1 > 220\text{V} \quad \text{and} \quad V_2 < 220\text{V}$$

$\Rightarrow$ 25 W bulb will fuse **Ans. (D)**

Sol. 20 In RC discharging circuit if at $t = 0$, voltage across capacitor is V_0 then after time $t = t$, it is given as

$$V = V_0 e^{-t/RC}$$ [from graph at $t = 0$, $V_0 = 25$ V]

where time constant $\tau = RC$

$$\Rightarrow \quad V = V_0 e^{-t/\tau} \qquad \ldots(1)$$

at $t = 200$s from graph we can see that $V = 5$ V

from equation-(1), we use

$$5 = 25\, e^{-200/\tau}$$

$$\Rightarrow \quad \ln 5 = \frac{200}{\tau}$$

$$\Rightarrow \quad \tau = \frac{200}{\ln(5)} = \frac{200}{2.303\log(5)} = \frac{200}{2.303\times0.7}$$

[As $\log 5 = 0.698$]

$$\Rightarrow \quad \tau \cong \frac{200}{1.6} \simeq 125 \text{ s}$$ **Ans. (A)**

Sol. 21 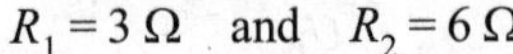 it burns when its temperature exceeds the melting point of resistance material.
Heat production rate in fuse

$$H = \frac{V^2}{R}$$

When the current is switched on, R is low as it is cold (at room temperature) and with flow of current R increases and H decreases.
(due to rise in temperature)

$\Rightarrow$ Statement-1 is true.

Statement-2 : When a bulb is lit up, due to heat its resistance increases

$\Rightarrow$ Statement-2 is true & explaining Statement-1 also

Ans. (C)

Sol. 22

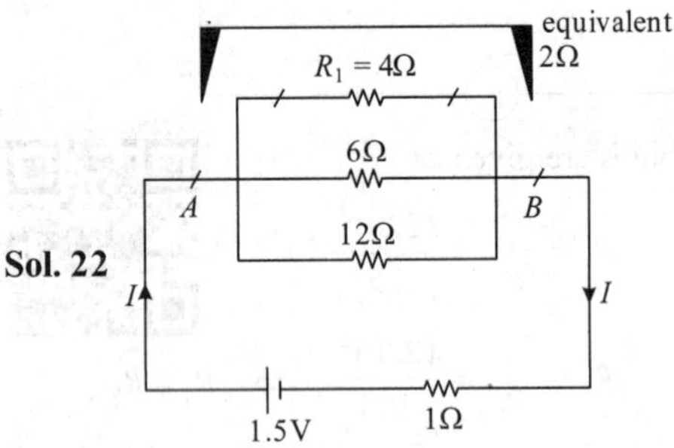

Equivalent resistance of parallel combination is

$$\frac{1}{R} = \frac{1}{4} + \frac{1}{6} + \frac{1}{12} = \frac{3+2+1}{12} = \frac{6}{12} = 0.5$$

$\Rightarrow \quad R = 2\,\Omega$ (Across A and B)

Circuit current

$$I = \frac{1.5}{1+2} = \frac{1.5}{3} = \frac{1}{2} = 0.5\text{ A}$$

Potential difference across A and B is

$$V_{AB} = IR = 0.5 \times 2 = 1\text{ V}$$

Rate of Joule heating in 4Ω resistance is

$$P_{th} = \frac{V_{AB}^2}{R_1} = \frac{(1)^2}{4} = 0.25\text{ W}$$ **Ans. (C)**

Sol. 23 Bulb brightness α thermal power produced in it

$$\Rightarrow \text{Brightness} \propto \frac{V^2}{R}$$

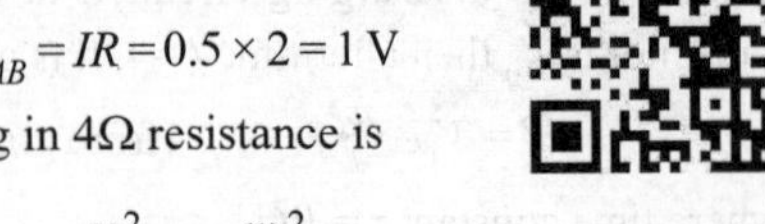

$$\Rightarrow \text{Brightness} \propto \frac{1}{R}$$ [As V is same for both bulbs]

$$R_1 = 3\,\Omega \quad \text{and} \quad R_2 = 6\,\Omega$$

Bulb-1 will glow brighter **Ans. (B)**

Sol. 24

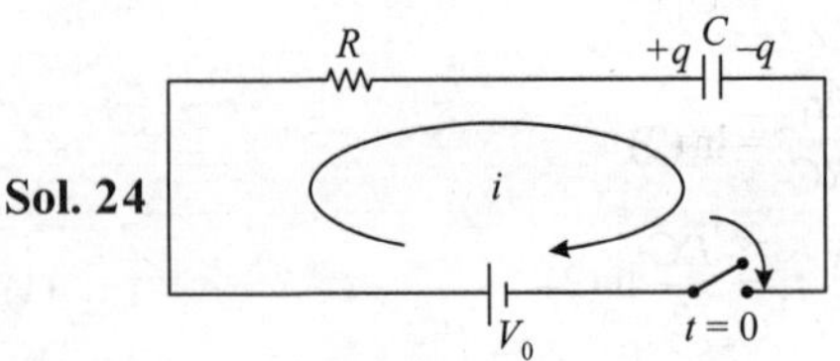

The charge on capacitor as a function of time is

$$q = CV_0(1 - e^{-t/RC})$$

Charge current in circuit is

$$i = \frac{dq}{dt} = \frac{V_0}{R}e^{-t/RC}$$

Voltage across capacitor at time t is

$$V_C = \frac{q}{C} = V_0(1 - e^{-t/RC}) \quad \ldots(1)$$

and voltage across resistance is

$$V_R = iR = V_0 e^{-t/RC} \quad \ldots(2)$$

At time t is given that

$$V_C = 7V_R$$

So from (1) and (2)

$$\%(1 - e^{-t/RC}) = 7\% e^{-t/RC}$$

$$\Rightarrow \quad 1 = 8e^{-t/RC}$$

$$\Rightarrow \quad \frac{t}{RC} = \ln(8)$$

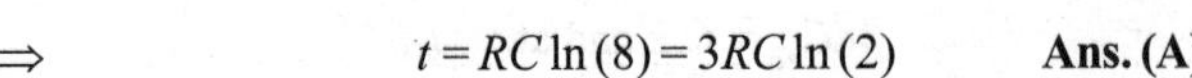

$$\Rightarrow \quad t = RC\ln(8) = 3RC\ln(2)$$ **Ans. (A)**

Sol. 25

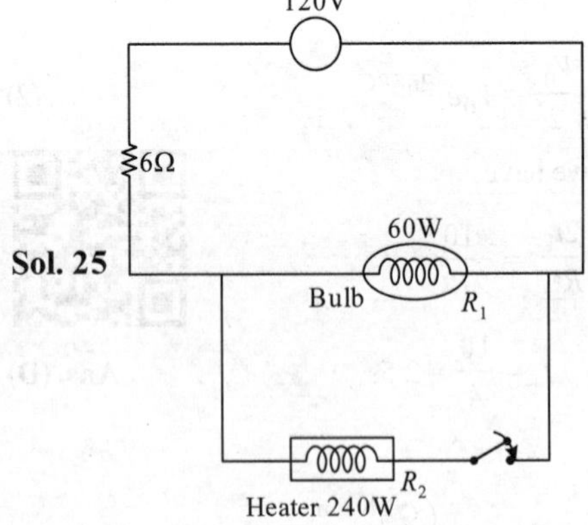

Resistance of heater

$$R_2 = \frac{V^2}{P} = \frac{(120)^2}{240} = 60\,\Omega$$

Resistance of bulb

$$R = \frac{V^2}{P} = \frac{(120)^2}{60} = 240\,\Omega$$

Initially voltage across bulb is

$$V_1 = \frac{120}{246} \times 240$$

$$\simeq 117.07\,\text{V}$$

When heater is switched on the equation equivalent of bulb and heater is

$$R_{eq} = \frac{240 \times 60}{240 + 60} = 48\,\Omega$$

Potential difference across bulb & heater combination is

$$V_2 = \frac{120}{54} \times 48 \simeq 106.6\,\text{V}$$

Change in potential difference across bulb is

$$\Delta V = V_1 - V_2 \simeq 10.04\,\text{V}$$ **Ans. (D)**

Sol. 26

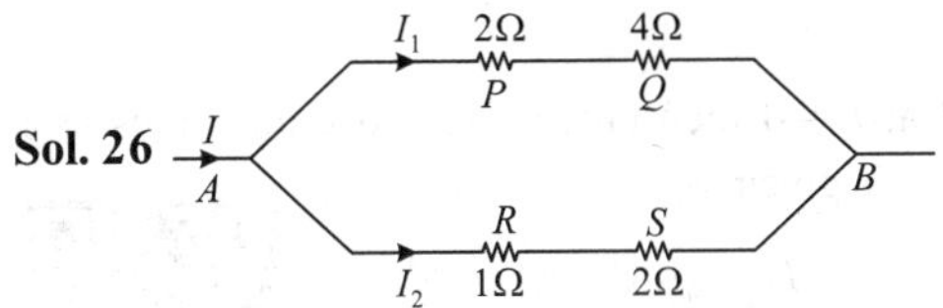

Here I_1 and I_2 are given as

$$I_1 = \frac{I}{3} \text{ and } I_2 = \frac{2I}{3}$$

Thermal power dissipated in resistances are

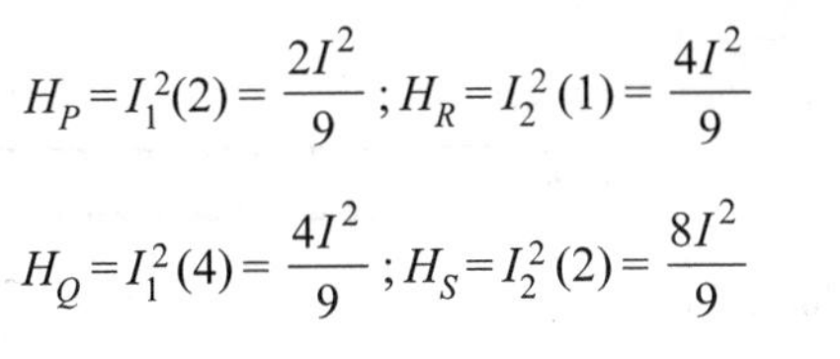

$$H_P = I_1^2(2) = \frac{2I^2}{9}\,;\, H_R = I_2^2(1) = \frac{4I^2}{9}$$

$$H_Q = I_1^2(4) = \frac{4I^2}{9}\,;\, H_S = I_2^2(2) = \frac{8I^2}{9}$$

$\Rightarrow$ $$H_S > (H_R = H_Q) > H_P$$ **Ans. (B)**

Sol. 27 Total power consumption of building at full load is

$$P = 15 \times 40 + 5 \times 100 + 5 \times 80 + 1000$$

$\Rightarrow$ $$P = 600 + 500 + 400 + 1000 = 1400\text{W}$$

The current drawn from mains supply at full load is

$$I = \frac{P}{V} = \frac{2500}{220} = \frac{125}{11} = 11.36\,\text{A}$$

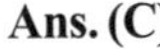

Maximum current supplied at full load

As fuse rating must be kept such that maximum current at full load can sustain. **Ans. (C)**

Sol. 28

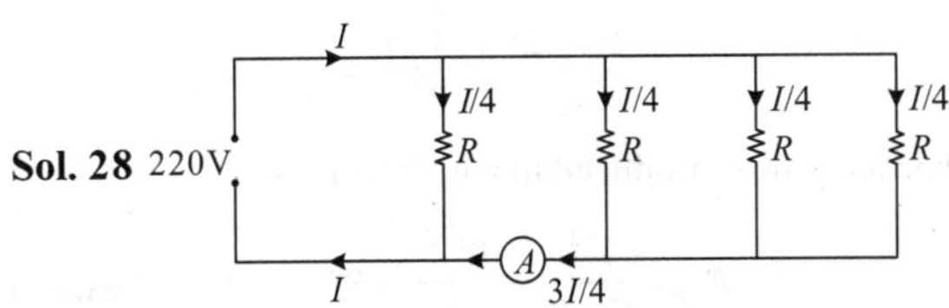

As each bulb is of 100W consumption

$\Rightarrow$ $$\frac{I}{4} = \frac{P}{V} = \frac{100}{220} = \frac{5}{11} = 0.45\,\text{A}$$

The current through ammeter is

$$I_A = \frac{3I}{4} = 3 \times 0.45 = 1.35\,\text{A}$$ **Ans. (C)**

Sol. 29 In steady state no current flows through capacitive branches of circuit

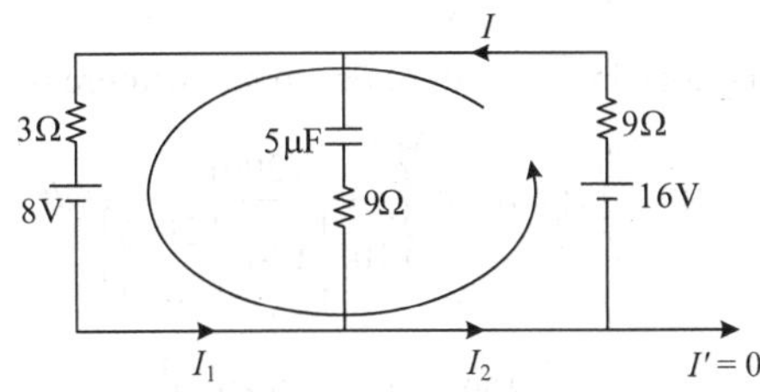

As current flows only in outer loop

$\Rightarrow$ $$I_1 = I_2 = I$$

Current in outer loop is

$$I = \frac{8}{9+3}$$

$$= \frac{8}{12} = \frac{2}{3} A = 0.67\,\text{A}$$ **Ans. (B)**

Sol. 30

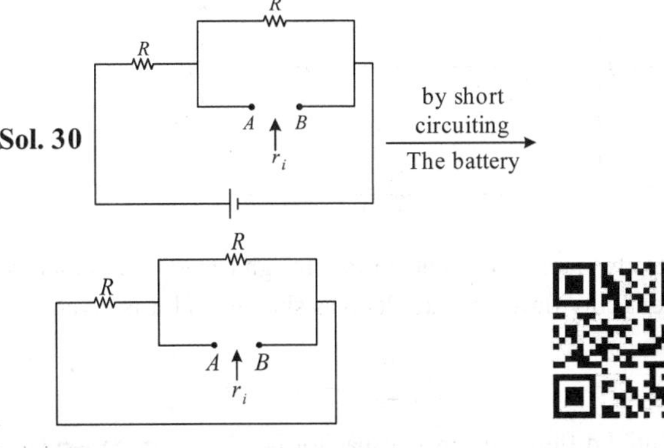

$\Rightarrow$ $$r_i = R \parallel R = R/2$$

According to maximum power transfer theorem, heat generation in r is maximum when

$$r = r_i = R/2 = fR$$

$\Rightarrow$ $$f = \frac{1}{2}$$ **Ans. (A)**

Sol. 31 At 300 K, $R = 100\Omega$

and at 500 K, $R = 120\Omega$

$$120 = 100\,[1 + \alpha(200)]$$

$\Rightarrow$ $$20 = \alpha(20000) \Rightarrow \alpha = 10^{-3}\,°\text{C}^{-1}$$

When toaster is connected across $V = 200$ volt then thermal power produced is

$$P_{th} = \frac{V^2}{R} = \frac{(200)^2}{100[1+\alpha(T-300)]}$$

at temperature T

$$300\text{K} < T < 500\text{ K}$$

As T is raised at a constant rate (linear variation) from 300K to 500 K, so time function of temperature is given as

$$T = 300 + \frac{500-300}{30} \cdot t = 300 + \frac{20}{3}t$$

Total work done in these 30s of rise in temperature is

$$W = \int_0^{30} P_{th} \cdot dt = \int_0^{30} \frac{(200)^2}{100\left[1+\alpha\left(\frac{20t}{3}\right)\right]} \cdot dt$$

$$\Rightarrow \quad W = 400 \times \frac{300}{20\alpha}\left[\ln\left(1+\frac{200\alpha t}{3}\right)\right]_0^{30}$$

$$\Rightarrow \quad W = \frac{60}{\alpha}[\ln(1+200\alpha) - \ln(1)]$$

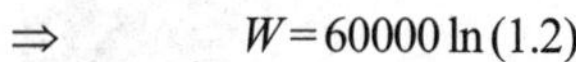

$$\Rightarrow \quad W = 60000 \ln(1.2)$$ **Ans. (None)**

None of the given option is correct.

Sol. 32

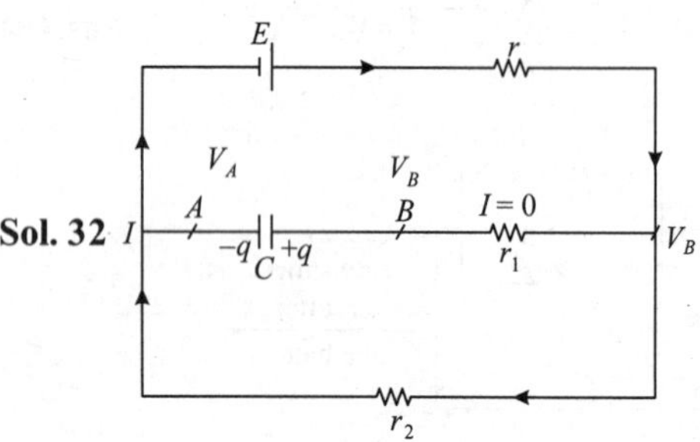

In steady state no current flows through capacitive branch so current only flows in outer loop as shown, and it is given as

$$I = \frac{E}{r+r_2}$$

Potential difference across capacitor is

$$V_B - V_A = Ir_2 = \frac{Er_2}{r+r_2}$$

$\Rightarrow$ Charge on capacitor is

$$q = C(V_B - V_A) = \frac{CEr_2}{r+r_2}$$ **Ans. (A)**

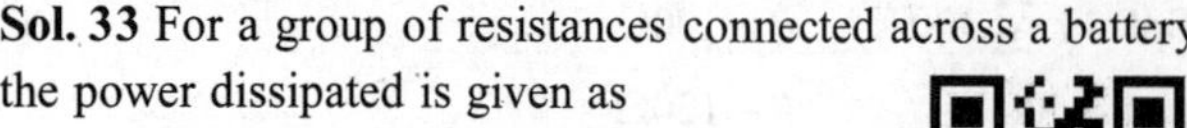

Sol. 33 For a group of resistances connected across a battery the power dissipated is given as

$$P = \frac{V^2}{R_{eq}}$$

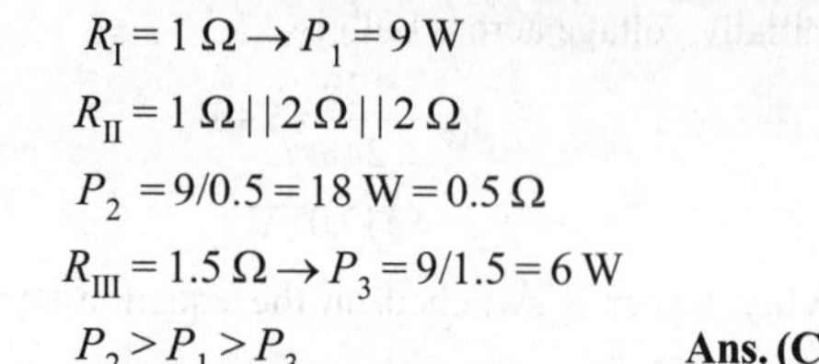

$$R_I = 1\,\Omega \rightarrow P_1 = 9\text{ W}$$

$$R_{II} = 1\,\Omega\,||\,2\,\Omega\,||\,2\,\Omega$$

$$P_2 = 9/0.5 = 18\text{ W} = 0.5\,\Omega$$

$$R_{III} = 1.5\,\Omega \rightarrow P_3 = 9/1.5 = 6\text{ W}$$

$$\Rightarrow \quad P_2 > P_1 > P_3$$ **Ans. (C)**

Sol. 34

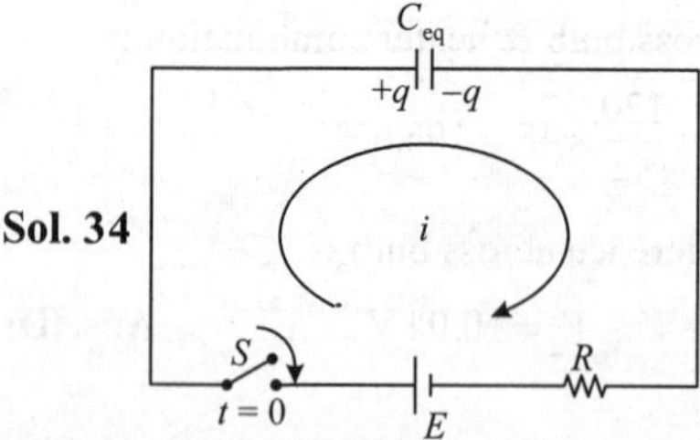

If S is closed at $t = 0$ then charge on capacitor plates as a function of time is given as

$$q = C_q E\,(1 - e^{-t/RC_{eq}})$$

As C_1 and C_2 are in series, charge on their plates must be equal to q only.

C_1 C_2 $+q$ $-q$ $+q$ $-q$

Ans. (A)

Sol. 35

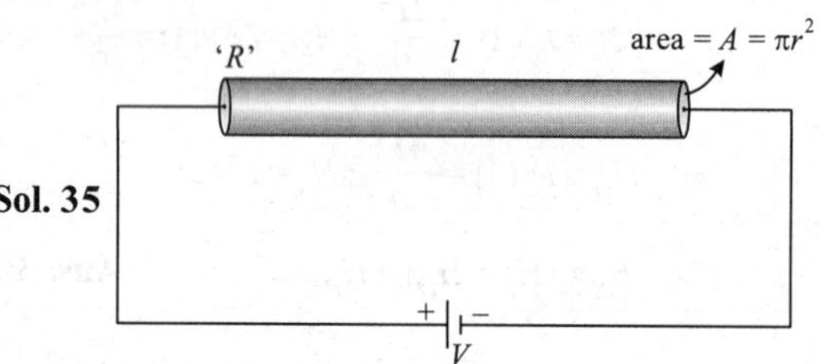

Resistance of wire is

$$R = \frac{\rho l}{A}$$

Thermal power produced in wire

$$P_H = \frac{V^2}{R}$$

when $l \rightarrow l/2$

and $r \rightarrow 2r$

$\Rightarrow \quad A \rightarrow 4A$

$\Rightarrow$ New resistance of wire becomes

$$R' = \frac{\rho(l/2)}{4A} = \frac{1}{8}\frac{\rho l}{A} = \frac{R}{8}$$

Now thermal power produced in wire becomes

$$P'_H = \frac{V^2}{R/8} = \frac{8V^2}{R} = 8P_H$$ **Ans. (A)**

Sol. 36 Power dissipation in resistance is

$$P = I^2R$$

$$\Rightarrow \quad 4.4 = 4 \times 10^{-6} R$$

$$\Rightarrow \quad R = 1.1 \times 10^6 \,\Omega$$

When 11 V power supply is used then

$$P' = \frac{11^2}{R} = \frac{11^2}{1.1} \times 10^{-6} = 11 \times 10^{-5}\text{ W}$$ **Ans. (A)**

Sol. 37 In series combination, equivalent resistance is 2R thus power consumed is $60\text{W} = \frac{\varepsilon^2}{2\text{R}}$

In parallel combination, equivalent resistance is $R/2$ thus new power is consumed is

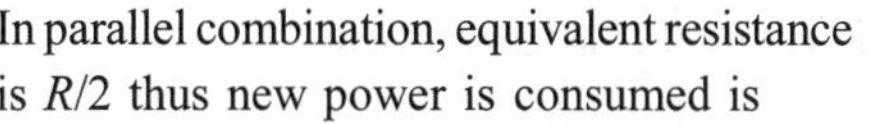

$$P' = \frac{\varepsilon^2}{(R/2)} = 4P = 4 \times 60 = 240 \text{ W}$$ **Ans. (B)**

Sol. 38 Resistances of the two bulbs are given as

$$R_1 = \frac{220^2}{25}$$

and $$R_2 = \frac{220^2}{100}$$

When bulbs are connected in series, we use

$$I = \frac{220}{R_1 + R_2}$$

Power consumed by the bulbs are

$$P_1 = i^2 R_1$$

$$\Rightarrow \quad P_1 = \frac{220^2}{\left(\frac{220^2}{25} + \frac{220^2}{100}\right)} \times \frac{220^2}{25}$$

$$\Rightarrow \quad P_1 = \frac{400}{25} = 16 \text{ W}$$

and $$P_2 = i^2 R_2 = \frac{220^2}{\left(\frac{220^2}{25} + \frac{220^2}{100}\right)} \times \frac{220^2}{25} = 4 \text{ W}$$

Ans. (C)

Sol. 39 Rate of magnetic energy storage in inductor is

$$\frac{d}{dt}\left(\frac{1}{2}LI^2\right) = LIdI$$

$$\Rightarrow \quad LIdI = I^2R$$

$$LdI = IR$$

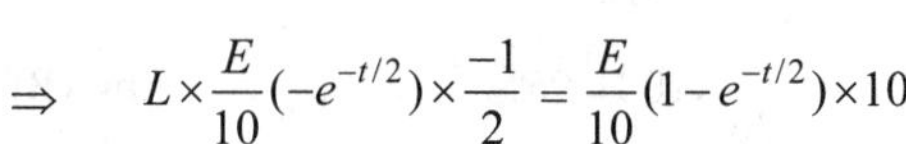

We use current as $I = \frac{E}{R}(1 - e^{-Rt/2})$

$$\Rightarrow \quad L \times \frac{E}{10}(-e^{-t/2}) \times \frac{-1}{2} = \frac{E}{10}(1 - e^{-t/2}) \times 10$$

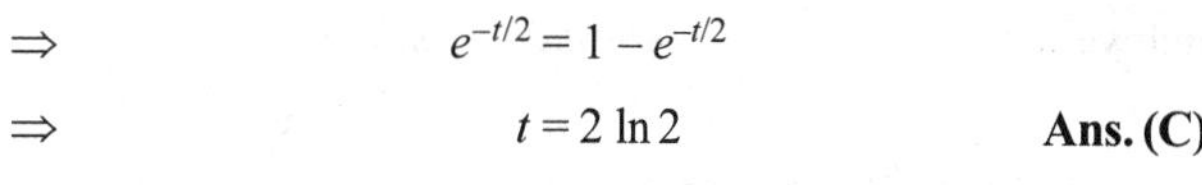

$$\Rightarrow \quad e^{-t/2} = 1 - e^{-t/2}$$

$$\Rightarrow \quad t = 2 \ln 2$$ **Ans. (C)**

Sol. 40

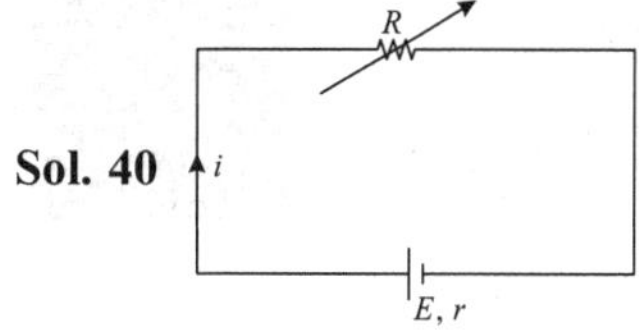

Current in external resistance is given as

$$i = \frac{E}{r + R}$$

Power generated in R is given as

$$P = i^2 R$$

$$P = \frac{E^2 R}{(r + R)^2}$$

For maximum power, we use

$$\frac{dP}{dR} = 0$$

$$\Rightarrow \quad E^2\left[\frac{(r + R)^2 \times 1 - R \times 2(r + R)}{(r + R)^4}\right] = 0$$

$$\Rightarrow \quad r = R$$ **Ans. (D)**

Sol. 41 Applying point potential method of potential distribution in given circuit as shown below

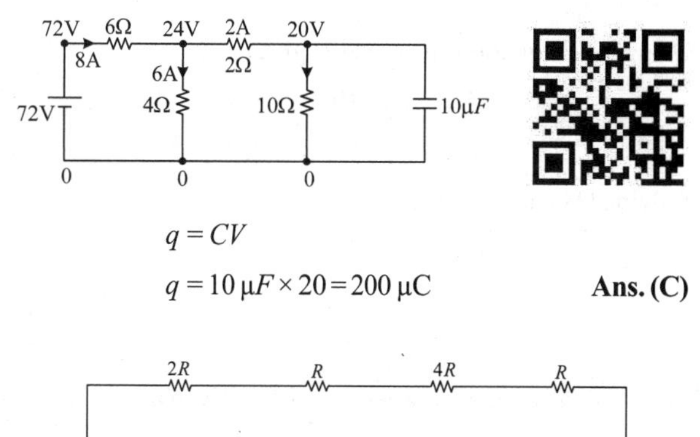

$$q = CV$$

$$q = 10\,\mu F \times 20 = 200\,\mu\text{C}$$ **Ans. (C)**

Sol. 42

2R R 4R R
16V

Power consumed by equivalent resistance $8R$ is circuit is

$$P = \frac{16^2}{8R} = 4$$

$$\Rightarrow \quad R = 8\,\Omega$$ **Ans. (A)**

Sol. 43 Total power supplied in time t is

$$Q = P \times t$$

and we use $$Q = mc\Delta T + mL$$

Power supplied by mains is $P = \frac{V_{rms}^2}{R}$

Substituting all values, we have

$$4200 \times 80 + 2260 \times 10^3 = \frac{(200)^2}{20} \times t$$

$\Rightarrow$ $$t = 1298 \text{ s}$$

$\Rightarrow$ $$t \simeq 22 \text{ min}$$ **Ans. (B)**

Sol. 44 Total power consumption with current I at full load of building is

$$220I = P = 15 \times 45 + 15 \times 100 + 15 \times 10 + 2 \times 10^3$$

$\Rightarrow$ $$I = \frac{4325}{220} = 19.66$$

$\Rightarrow$ $$I \simeq 20 \text{ A}$$ **Ans. (D)**

Sol. 45 Considering loop currents i_1 and i_2 in circuit as shown in figure

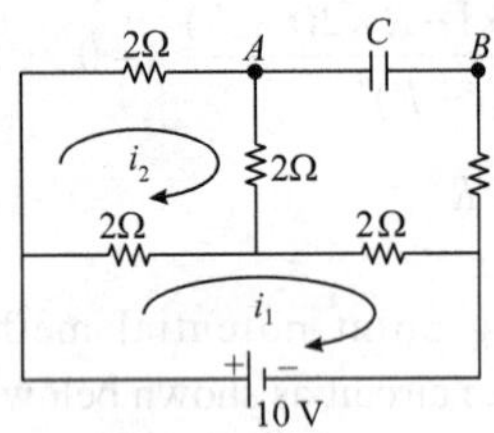

Using KVL equation for i_1 and i_2 gives

$$4i_1 - 2i_2 = 10 \quad \ldots(1)$$

$$6i_2 - 2i_1 = 0 \quad \ldots(2)$$

Solving (1) and (2) gives

$$12i_2 - 2i_2 = 10$$

$\Rightarrow$ $$i_2 = 1\text{A}$$

and $$i_1 = 3\text{A}$$

Potential difference between A and B is given as

$$V_A - 2 \times 1 - 2 \times 3 = V_B$$

$\Rightarrow$ $$V_A - V_B = 8V$$ **Ans. (8)**

Sol. 46 Resistance of an elemental shell of radius x and width dx inside the battery for radial flow of current as shown in figure is given as

$$dr = \frac{\rho dx}{2\pi x l}$$

$\Rightarrow$ $$r = \int dr = \int_a^b \frac{\rho}{2\pi l}\frac{dx}{x} = \frac{\rho}{2\pi l}\ln\left(\frac{b}{a}\right)$$

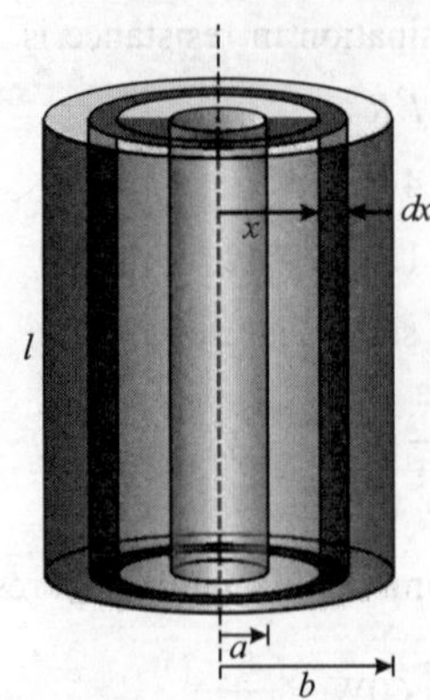

When battery is connected to external resistance R, maximum heat is produced in external resistance when

$$R = r = \frac{\rho}{2\pi l}\ln\left(\frac{b}{a}\right)$$ **Ans. (D)**

Sol. 47 If circuit current is i, we use

$$i^2R = 0.5$$

and $$iR = 2.5$$

Terminal voltage of battery is given as

$$V = E - ir$$

$\Rightarrow$ $$2.5 = 3 - ir$$

$\Rightarrow$ $$ir = 0.5$$

$\Rightarrow$ $$\frac{R}{r} = 5$$

Power dissipated in r is given as

$$P_r = i^2r$$

Power dissipated in external R is given as

$$P_R = i^2R$$

$\Rightarrow$ $$\frac{P_r}{P_R} = \frac{r}{R} = \frac{1}{5}$$

$\Rightarrow$ $$P_r = \frac{1}{5} \times 0.5 = 0.1 \text{ W}$$ **Ans. (D)**

Sol. 48 Power delivered by transmission line is

$$P = vi = 1000 \text{ W}$$

$\Rightarrow$ $$i = \frac{1000}{220} = \frac{50}{11}\text{A}$$

Power loss in line is

$$P_L = i^2R = \left(\frac{50}{11}\right)^2 \times 2 = \frac{5000}{121}\text{ W}$$

line efficiency is given as

$$\eta = \frac{1000}{1000 + \frac{5000}{121}} \times 100$$

$\Rightarrow$ $$\eta = \frac{121}{126} \times 100 \simeq 96\%$$ **Ans. (B)**

Sol. 49 When constant voltage is applied across the RC circuit, capacitor charges exponentially and when the voltage drops to zero, it discharges exponentially so most appropriate wave pattern across capacitor matches with option (C) when square wave EMF is applied across RC circuit. **Ans. (C)**

Sol. 50 Thermal energy can be expressed as

$$H = i^2 Rt$$

$$\Rightarrow \quad 500 = (1.5)^2 \times R \times 20$$

$$\Rightarrow \quad E = (3)^2 \times R \times 20 = 2000 \text{ J}$$ **Ans. (D)**

Sol. 51 Energy dissipated in resistor is given as

$$H = i^2 RT$$

$$\Rightarrow \quad R = \frac{Q}{i^2 t} = \frac{10\times10^{-3}}{4\times10^{-6}\times1} = 2500\,\Omega$$ **Ans. (2500)**

Sol. 52 Just after closing the switch, capacitor can be considered as short circuit so current in circuit can be given as

$$i_0 = \frac{V}{R} = \frac{30/3}{5\times10^6} = 2\times10^{-6}\text{A}$$ **Ans. (2)**

Sol. 53 Power of bulb is given as

$$P = \frac{V^2}{R_b}$$

$$\Rightarrow \quad 200 = \frac{(100)^2}{R_b}$$

Resistance of bulb is given as

$$R_b = 50\,\Omega$$

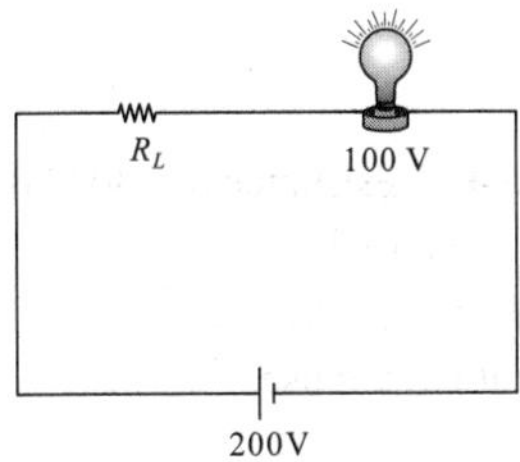

To produce same power in bulb when connected across 200 V supply as shown in figure above as explained in question, the current through bulb should remain same and potential difference across bulb must remain 10 V. Thus current in circuit must be

$$i = \frac{100}{50} = 2\text{ A}$$

The voltage across the series resistance now should be $200 - 100 = 100$ V thus resistance R can be given as

$$R_L = \frac{100}{2} = 50\,\Omega$$ **Ans. (50)**

Sol. 54 Resistance of bulb can be calculated as

$$P = \frac{V^2}{R}$$

$$R = \frac{(100)^2}{500} = \frac{100\times100}{500} = 20\,\Omega$$

Current through bulb for 500 W power production is calculated as

$$I = \frac{P}{V} = \frac{500}{100} = 5\text{ A}$$

For a supply of 200 V, current 5 A wll flow when total resistance of circuit will be

$$R_T = \frac{200}{5} = 40\,\Omega$$

Thus extra resistance required to be kept in series of bulb is 20 Ω. **Ans. (D)**

Sol. 55 The effective circuit is shown in figure below

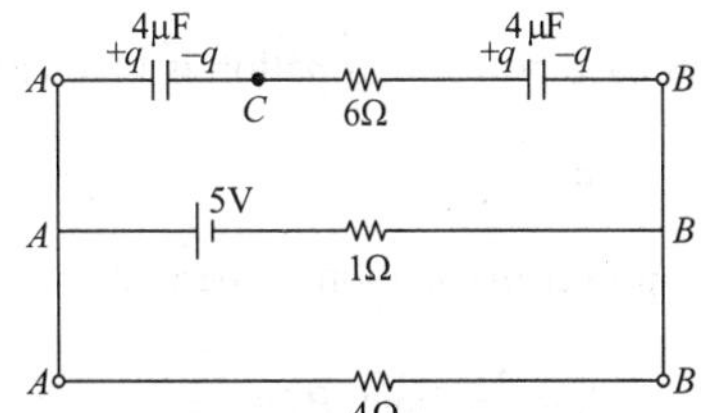

The current in lower loop of circuit is given as

$$i = \frac{5}{4+1} = 1\text{ A}$$

Potential difference across AB is given as

$$V_{AB} = i \times 4 = 4\text{ V}$$

$$\Rightarrow \quad V_{AC} = V_{CB} = 2\text{ V}$$

$$\Rightarrow \quad q \text{ on } 4\mu\text{F} = CV_{AC} = 8\mu\text{C}$$ **Ans. (B)**

Sol. 56 In steady state, the current through capacitor is zero. The potential difference across capacitor is 2 V across the resistance in parallel to which capacitor is connected. So charge on capacitor is given as

$$q = 50 \times 2 = 100\,\mu\text{C}$$ **Ans. (100)**

Sol. 57 Power dissipated in time t is given as

$$P = i^2 Rt$$

$$\Rightarrow \quad 192 = 4 \times 4 \times R \times 1$$

$$\Rightarrow \quad R = 12\,\Omega$$

If the current is doubled, amount of energy dissipated is calculated as

$$P' = i'^2 Rt'$$

$$\Rightarrow \quad P' = 8 \times 8 \times 12 \times 5 = 3840\text{ J}$$ **Ans. (3840)**

Sol. 58 The potential difference across capacitor as a function of time is given as

$$V = V(1 - e^{-t/RC})$$

$$\Rightarrow \quad 2 = 20\,(1 - e^{-1\mu S/10C})$$

$$\Rightarrow \quad \frac{1}{10} = (1 - e^{-1\times10^{-6}/10C})$$

$$\Rightarrow \quad e^{-1\times10^{-6}/10C} = \frac{9}{10}$$

$$\Rightarrow \quad \frac{1\times10^{-6}}{10C} = ln\frac{9}{10}$$

$$\Rightarrow \; C = \frac{1\times10^{-6}}{10\times\ln\left(\frac{9}{10}\right)} = \frac{10^{-7}}{1.05} = \frac{1}{1.05}\,\mu\text{F} = \frac{100}{105}\,\mu\text{F} = 0.95\,\mu\text{F}$$

Ans. (C)

Sol. 59 When steady state is reached, inductors in circuit behave like short circuit. So all 3 Ω resistance will act in parallel and current supplied by the battery can be given as

$$I = \frac{30}{2+1} = \frac{30}{3} = 10\,\text{A}$$ **Ans. (C)**

Sol. 60 In first case power dissipation is given as

$$P = \frac{V^2}{R}$$

In second case power dissipation in the two parts are given as

$$P_1 = \frac{V^2}{\frac{R}{2}} \text{ and } P_2 = \frac{V^2}{\frac{R}{2}}$$

Total power dissipation in second case is given as

$$P_1 + P_2 = \frac{4V^2}{R}$$

Ratio of final to initial power dissipation is given as

$$\frac{P_1+P_2}{P} = \frac{\frac{4V^2}{R}}{\frac{V^2}{R}} = 4$$ **Ans. (4)**

Sol. 61 Thermal energy in time t os calculated as

$$H = i^2Rt$$

$$\Rightarrow \quad 300 = 2^2\times R\times 15$$

$$\Rightarrow \quad R = \frac{300}{60} = 5\,\Omega$$

Total thermal energy when current is increased to 3 A, is calculated as

$$H = 3^2\times5\times10 = 450\,\text{J}$$ **Ans. (450)**

Sol. 62 Heat energy produced in coil in time t is given as

$$\Delta Q = \left(\frac{V^2}{R}\right)t$$

Given that for the two coils heat energy produced is same in 20 minutes and 60 minutes respectively, so we use

$$Q_0 = \frac{V^2}{R_1}\times20 = \frac{V^2}{R_1}\times60 \qquad \ldots(1)$$

$$\Rightarrow \quad R_2 = 3R_1 \qquad \ldots(2)$$

Now if they are connected in parallel then equivalent resistance is given as

$$R_{eq} = \frac{R_2R_1}{R_1+R_2} = \frac{3R_1\cdot R_1}{3R_1+R_1} = \left(\frac{3R_1}{4}\right)$$

To produce same heat, using equation-(1), we have

$$Q_0 = \frac{V^2}{R_1}\times20 = \frac{V^2}{\left(\frac{3R_1}{4}\right)}\times t$$

$$\Rightarrow \quad t = \frac{3\times20}{4} = 15\,\text{min}$$ **Ans. (15)**

Sol. 63 In t_1 time energy becomes half so charge will become $\frac{1}{\sqrt{2}}$ times the initial value so while discharging we use

$$q = Q_0e^{-\frac{t_1}{RC}} = \frac{Q_0}{\sqrt{2}} \qquad \ldots(1)$$

and

$$q = Q_0e^{-\frac{t_2}{RC}} = \frac{Q_0}{8} = \frac{Q_0}{\left(\sqrt{2}\right)^6} \qquad \ldots(2)$$

From equation-(1) and (2) we have

$$t_1 = \frac{1}{2}RC\ln2$$

and

$$t_2 = 3\,RC\ln2$$

$$\Rightarrow \quad t_2 = 6t_1$$

$$\Rightarrow \quad \frac{t_1}{t_2} = \frac{1}{6}$$ **Ans. (D)**

Sol. 64 Statement-I : Resistance $R = 80\,\Omega$ is cut into four equal parts so each will have resistance

$$R_1 = R_2 = R_3 = R_4 = 20\,\Omega$$

In parallel combination, we use

$$R_{eq} = \frac{20}{4} = 5\,\Omega$$

Hence statement I is correct.

Statement-II : Figure below shows the circuit described in statement II.

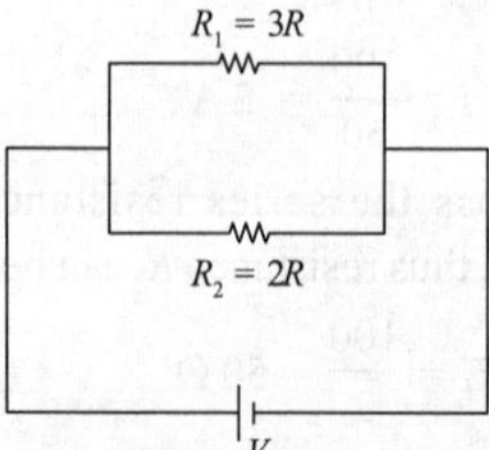

Thermal power in parallel resistance is given as

$$P_{th} = \frac{V^2}{R}$$

$$\Rightarrow \quad \frac{P_1}{P_2} = \left(\frac{R_2}{R_1}\right) = \frac{2}{3} \qquad \textbf{Ans. (C)}$$

Sol. 65 Resistance of the first and second bulb are given as

$$R_1 = \frac{V^2}{P} = \frac{220^2}{100} = 484\Omega$$

and $$R_2 = \frac{V^2}{P} = \frac{220^2}{60} = 484\left(\frac{10}{6}\right)\Omega$$

Current in circuit is given as

$$I = \frac{220}{484 + 484 \times \frac{10}{6}} = 0.17\,\text{A}$$

Thermal power consumed by 100 W bulb is given as

$$P_1 = I^2R_1 = (0.17)^2 \times 484 = 13.98\text{ W} \quad \textbf{Ans. (14)}$$

Sol. 66 For parallel combination voltage across resistors remain same so thermal energy released can be calculated as

$$H = \frac{V^2}{R} \times t$$

$$\Rightarrow \quad \frac{H_1}{H_2} = \frac{\frac{V^2t}{R}}{\frac{V^2t}{3R}} = 3:1 \qquad \textbf{Ans. (A)}$$

Sol. 67 By Joule's law thermal energy developed by the resistor is $H \propto i^2$ so for t = constant, we use

$$\Rightarrow \quad \frac{H}{H'} = \left(\frac{4}{16}\right)^2$$

$$\Rightarrow \quad H' = 16\text{H} \qquad \textbf{Ans. (B)}$$

Sol. 68 In steady state, capacitor behaves as an open circuit so the simplified form of given circuit can be drawn as shown in figure below.

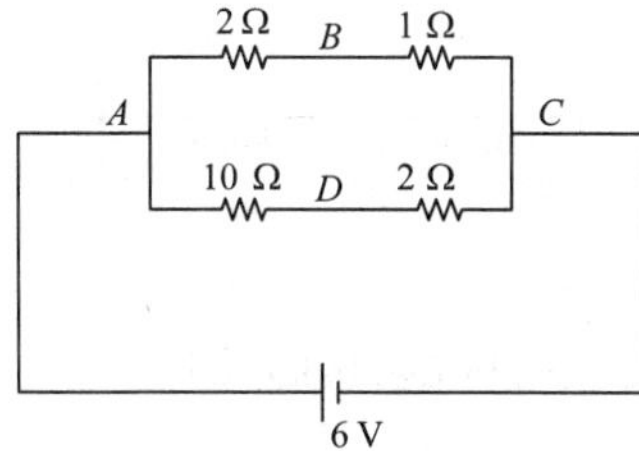

Current in branch AB and AD is calculated as

$$i_{AB} = \frac{6}{3} = 2\,\text{A}$$

& $$i_{AD} = \frac{6}{12} = 0.5\,\text{A}$$

$$\Rightarrow \quad V_B + 2 \times 2 - 10 \times 0.5 = V_D$$

$$\Rightarrow \quad V_B - V_D = 1\,\text{V} \qquad \textbf{Ans. (1)}$$

Sol. 69 At steady state, current in the circuit is given as

$$i = \frac{4}{6+2+8} = \frac{1}{4}\text{A}$$

Voltage across capacitor C_1 is given as

$$V_1 = V_{AC} = i(6+2) = \frac{1}{4} \times 8 = 2\,\text{V}$$

Voltage across capacitor C_2 is given as

$$V_2 = V_{BD} = i(2\Omega + 8\Omega) = \frac{1}{4} \times 10 = 2.5\,\text{V}$$

$$\Rightarrow \quad \frac{V_1}{V_2} = \frac{2}{2.5} = \frac{4}{5} \qquad \textbf{Ans. (B)}$$

Sol. 70 For parallel combination of the two heater filament, heat produced is calculated as

$$H_p = \left[\frac{V^2}{\left(\frac{R}{2}\right)}\right]t = \frac{2V^2t}{R}$$

For series combination of heater filaments, heat produced is calculated as

$$H_s = \left(\frac{V^2}{2R}\right)t$$

Thus ratio of heat produced in parallel to series combination is given as

$$\frac{H_p}{H_s} = 4 \qquad \textbf{Ans. (A)}$$

Sol. 71 Power dissipation across resistance is calculated as

$$P = VI = I^2R = \frac{V^2}{R}$$

$$\Rightarrow \quad R = \frac{\rho l}{A}$$

If wire is cut in two equal halves, we use

$$R' = \frac{R}{2}$$

Initial power dissipation is given as

$$P_1 = \frac{V_0^2}{R}$$

Final power dissipation is given as

$$P_2 = \frac{V_0^2}{R'} \times 2$$

$$\Rightarrow \quad P_2 = \frac{V_0^2}{R} \times 4$$

$$\Rightarrow \quad \frac{P_2}{P_1} = 4 = \frac{\sqrt{x}}{1}$$

$$\Rightarrow \quad x = 16 \qquad \textbf{Ans. (16)}$$

Sol. 72 Figure below shows the flow of current in the circuit.

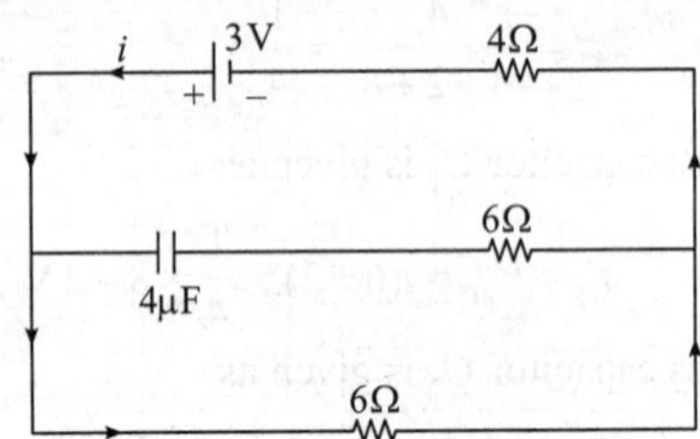

No current will flow in capacitor in steady state, current flowing in the circuit in steady state

$$I = \frac{3}{6+4} = \frac{3}{10}$$

Potential difference on 6Ω resistance

$$V = 6 \times \frac{3}{10} = 1.8\text{ V}$$

Capacitor will have same potential so charge,

$$q = CV = (4\mu F)\cdot(1.8\text{ volt}) = 7.2\mu C.$$ **Ans. (A)**

Sol. 73 Figure below shows the distribution of currents flowing in the circuit given in the question

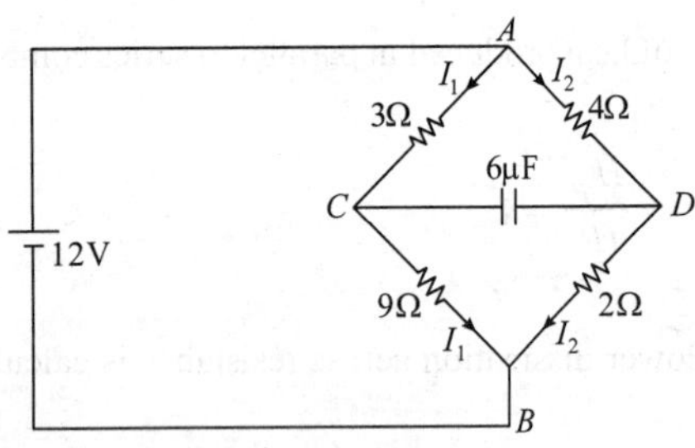

As potential difference between A and B is 12 V, we use

$$I_1 = \frac{12}{3+9} = 1\text{ A}$$

and $$I_2 = \frac{12}{4+2} = 2\text{ A}$$

The potential difference between A and C is calculated as

$$V_A - V_C = 3I_1 = 3\text{ V} \quad \ldots(1)$$

and potential difference between A and D is calculated as

$$V_A - V_D = 2 \times 4 = 8\text{ V} \quad \ldots(2)$$

Subtracting equation-(1) from (2) gives

$$V_C - V_D = 5\text{V}$$

$$\Rightarrow \quad V = 5\text{V}$$

Thus energy stored in capacitor is calculated as

$$U = \frac{1}{2}CV^2 = \frac{1}{2} \times 6 \times 5^2 = 75\ \mu\text{J}$$ **Ans. (75)**

17.4 Electrical Measuring Instruments

Sol. 1 Ammeter → its internal resistance is very low

Voltmeter → its internal resistance is very high

When we wish to convert an ammeter into a voltmeter then we connect a high resistance in series. **Ans. (C)**

Sol. 2 To convert a galvanometer/ammeter into a higher range ammeter we need to connect a shunt resistance to bypass the excess current.

The required shunt resistance can be given as

$$(I_R - I_m)R_S = I_m Rg$$

$$\Rightarrow \quad R_S = \frac{I_m R_g}{I_R - I_m} = \frac{1 \times 0.81}{10-1} = \frac{0.81}{9}$$

$$\Rightarrow \quad R_S = 0.09\ \Omega$$ **Ans. (D)**

Sol. 3 Across potentiometer wire, potential gradient is

$$\lambda = \frac{E}{L} = \frac{E}{100\text{ cm}}\frac{L}{}$$

For the given battery if unknown emf is E_{uk} and this balances at l length on wire so we use

$$E_{uk} = \lambda l = \frac{E}{100} \times 30 = \frac{30E}{100}$$ **Ans. (D)**

Sol. 4 Case-I

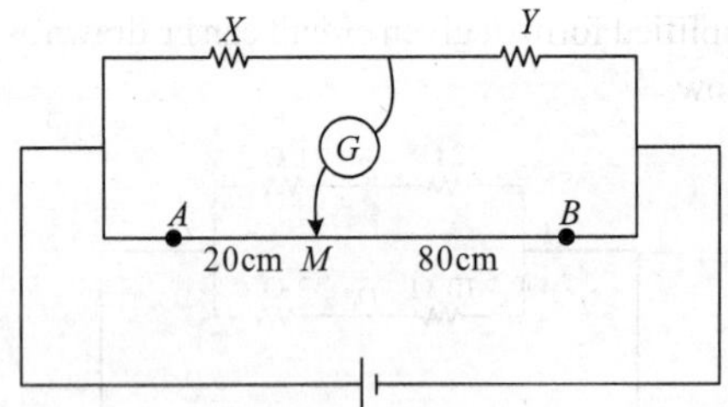

At null point we have bridge balanced

$$\Rightarrow \quad \frac{X}{20} = \frac{Y}{80} \quad \Rightarrow \quad y = 4x$$

At null point as bridge is balanced

$$\Rightarrow \quad \frac{4X}{l} = \frac{Y}{100-l}$$

$$\Rightarrow \quad 2l = 100$$

$$\Rightarrow \quad l = 50 \text{ cm}$$ **Ans. (A)**

Sol. 5 If λ be the potential gradient on potentiometer wire then

Case-I : emf of cell is

$$E = \lambda(240) \quad \ldots(1)$$

Case-II : potential difference across cells is

$$V_{AB} = \frac{2E}{2+r} = \lambda(120) \quad \ldots(2)$$

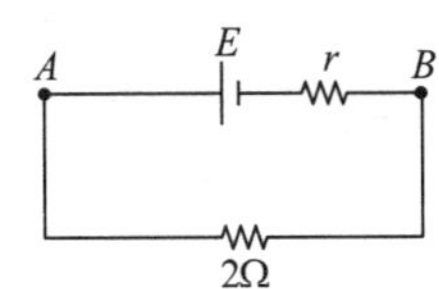

$$V_{AB} = \left(\frac{E}{r+2}\right) \cdot 2$$

$$\frac{(1)}{(2)} \Rightarrow \quad \frac{2+r}{2} = \frac{240}{120} = 2$$

$$\Rightarrow \quad 2 + r = 4$$

$$\Rightarrow \quad r = 2\,\Omega$$ **Ans. (D)**

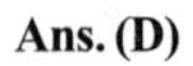

Alternative : The internal resistance of a cell by potentiometer experiment is given as

$$r = \left(\frac{l_1}{l_2} - 1\right)R$$

$$\Rightarrow \quad r = \left(\frac{240}{120} - 1\right) \times 2 = 2\,\Omega$$ **Ans. (D)**

Sol. 6

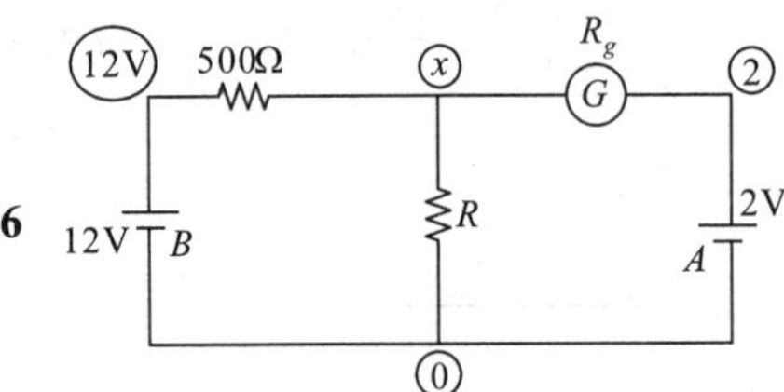

Distributing potentials at different junctions of circuit by considering zero volt reference at bottom junctions.

As current through 'G' is zero

$$\Rightarrow \quad x = 2\text{ V}$$

Writing KCL equation at junction x, we have

$$\frac{x-12}{500} + \frac{x-0}{R} + \frac{x-2}{R_g} = 0$$

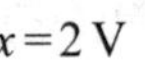

$x = 2\text{ V}$

$$\Rightarrow \quad -\frac{10}{500} + \frac{2}{R} = 0 \quad \Rightarrow \quad R = 100\,\Omega$$ **Ans. (B)**

Sol. 7 Full deflection current of galvanometer

$$I_{max} = \frac{150}{10} = 15 \text{ mA}$$

Full deflection voltage of galvanometer

$$V_{max} = \frac{150}{10} = 75 \text{ mV}$$

at a potential difference of 75mV across galvanometer, a current of 15mA flows through it. So coil resistance is given as

$$R_C = \frac{V_{max}}{I_{max}} = \frac{75}{15} = 5\,\Omega$$

New range of galvanometer to be used as voltmeter is

$$V_R = 150 \times 1 = 150\text{ V}$$

So if R is external resistance connected in series, we use

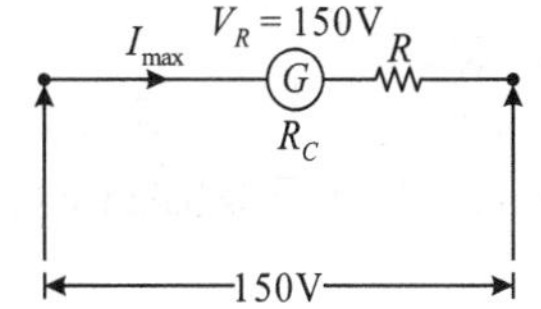

$$\text{So use} \quad \frac{150}{R + R_C} = I_{max} = 15 \text{ mA}$$

$$\Rightarrow \quad R + R_C = \frac{150}{15 \times 10^{-3}} = 10^4\,\Omega$$

$$\Rightarrow \quad R = 10^4\,\Omega - 5\,\Omega = 9995\,\Omega$$ **Ans. (D)**

Sol. 8 For bridge to be balanced we use

$$\frac{55}{20} = \frac{R}{100-20}$$

$$\Rightarrow \quad R = \frac{55}{20} \times 80 = 55 \times 4$$

$$= 220\,\Omega$$ **Ans. (B)**

Sol. 9

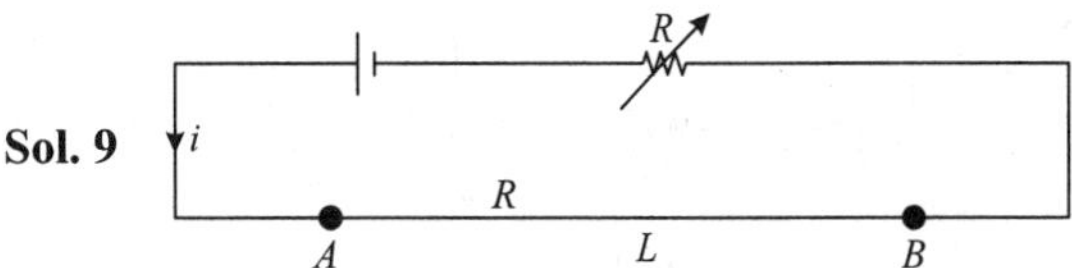

Resistance of potentiometer wire is given as

$$R = \frac{\rho L}{A}$$

Potential gradient on wire AB is given as

$$\lambda = \frac{V_{AB}}{L} = \frac{iR}{L} = \frac{i\rho}{A} = \frac{0.2 \times 4 \times 10^{-7}}{8 \times 10^{-1}}$$

$$\Rightarrow \quad \lambda = 0.1 \text{ V/m}$$ **Ans. (D)**

Sol. 10 (i) $\rightarrow$ As no current flows through cell, its terminal p.d.

is equal to its emf so (i) is correct.

(ii) $\rightarrow$ As potential gradient on potentiometer wire is given as

$$\lambda = \frac{V_{AB}}{L}$$

Measured emf $\quad E = \lambda\, l_{bal} \quad$ [If λ is small then l_{bal} is large]

$\Rightarrow$ Precision will be more closer to actual value

So (ii) is correct

(iii) & (iv) are NOT correct **Ans. (C)**

Sol. 11

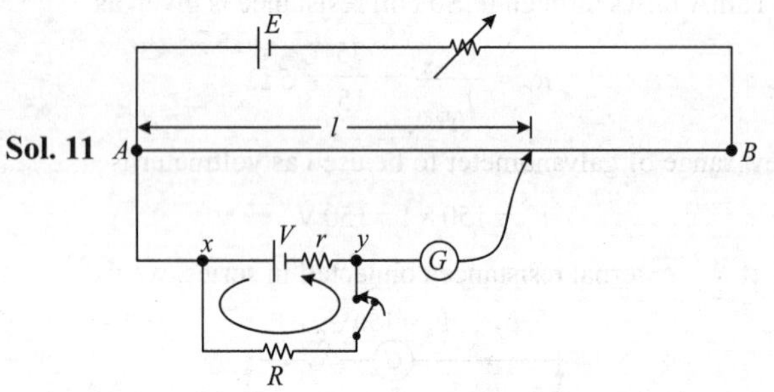

Initially when the cell of EMF V is balanced at length l then we use

$$V = \lambda l \qquad \ldots(1)$$

$$r = R$$

After connecting external resistance, we use

$$V_{xy} = \frac{V}{2R}\cdot R = \frac{V}{2}$$

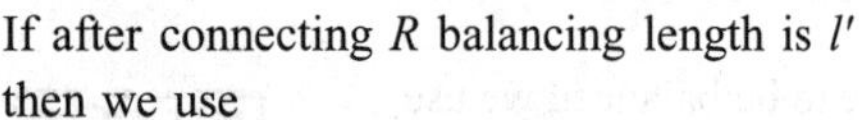

If after connecting R balancing length is l' then we use

$$\frac{V}{2} = \lambda l' \qquad \ldots(2)$$

from (1) and (2) $\Rightarrow \quad l' = l/2$ **Ans. (C)**

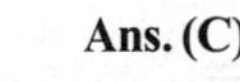

Sol. 12

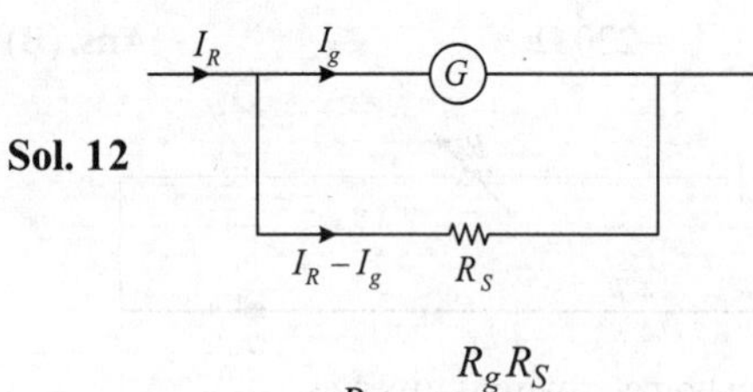

$$R_A = \frac{R_g R_S}{R_g + R_S}$$

Internal resistance of Ammeter.

If I_R increases then more current has to be bypassed through shunt resistance so R_S should be decreased or to decrease it an additional shunt is required across then ammeter.

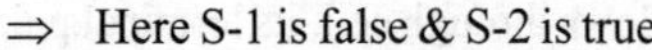

$\Rightarrow$ Here S-1 is false & S-2 is true. **Ans. (D)**

Sol. 13 To verify Ohm's law we use

$$V = IR$$

By varying R, we measure V across R and I through R & plot the graph in V and I

The graph obtained in s straight line which verifies Ohm's law.

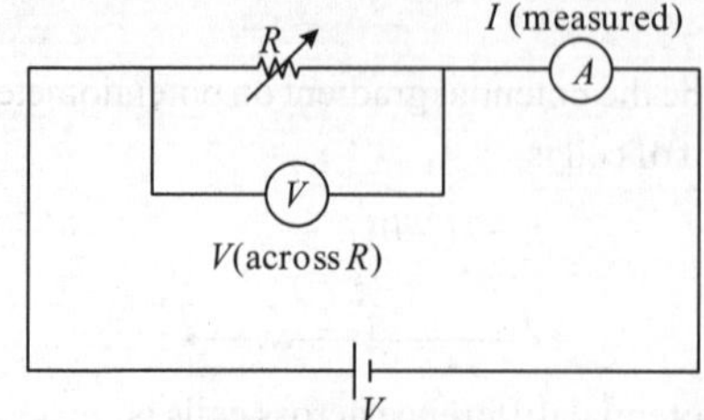

Only option (A) closely matches with above circuit used for experimental verification of Ohm's law.

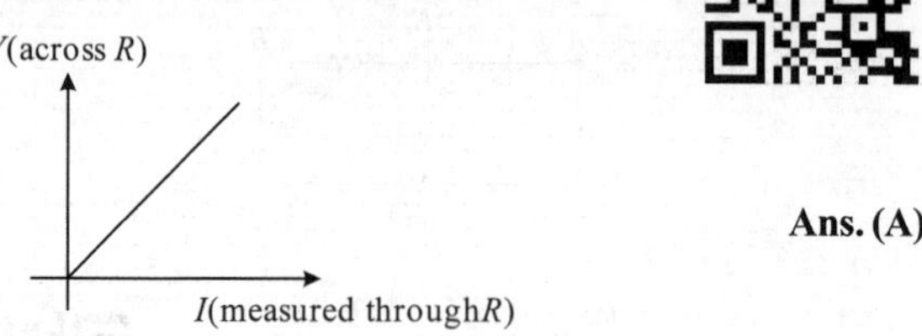

Ans. (A)

Sol. 14 Initially when balancing length is 40 cm, the balancing condition is

$$\frac{X}{40} = \frac{Y}{60} \qquad \ldots(1)$$

later when $3x$ is balanced against y, if balancing length is l, then we use

$$\frac{3X}{l} = \frac{Y}{100 - l} \qquad \ldots(2)$$

$$\frac{(1)}{(2)} \Rightarrow \quad \frac{l}{40\times 3} = \frac{100 - l}{60}$$

$$\Rightarrow \quad 6l = 1200 - 12l$$

$$\Rightarrow \quad l = \frac{1200}{18} = \frac{200}{3} = 66.66\text{ cm} \sim 67\text{ cm}$$

Ans. (C)

Sol. 15

$S = 1\Omega$

$I - I_g$

$I = 5.5\text{A}$ $\quad I_g$ $\quad$ G $\quad R_g = 120\Omega$

The supplied current of 5.5A will divide in inverse ratio of S and R_g and through "G" its full deflection current I_g flows so, we use

$$I_g = I \times \frac{S}{S + R_g} = \frac{5.5\times 1}{1 + 120}$$

$$= \frac{5.5}{1210} = \frac{55}{1210}$$

$\Rightarrow \quad I_g \simeq 0.045\,\text{A}$ **Ans. (D)**

Sol. 16 If resistance of galvanometer is G then we use

Case-1 : When $R_3 = 0$, current through galvanometer is

$$I = \frac{E}{R_1 + \frac{R_2 G}{R_2 + G}} \times \frac{R_2}{R_2 + G} \quad \ldots(1)$$

Produces deflection d

Case-2 : $R_3 = 107\ \Omega$, current through galvanometer is $[G' = G + R_3^*]$

$$I' = \frac{E}{\left(R_1 + \frac{R_2 G'}{R_2 + G'}\right)} \times \frac{R_2}{R_2 + G'} \quad \ldots(2)$$

Produces deflection $d/2$ for half deflection in galvanometer, we use

$$I' = \frac{I}{2}$$

$$\frac{E}{R_1 + \frac{R_2 G'}{R_2 + G'}} \times \frac{R_2}{R_2 + G'} = \frac{1}{2}\left(\frac{E}{R_1 + \frac{R_2 G}{R_2 + G}}\right) \times \frac{R_2}{R_2 + G}$$

$\Rightarrow \quad 2R_1R_2 + 2R_1G + 2R_2G = R_1R_2 + R_1G' + R_2G'$ [here $G' = G + 107$]

$\Rightarrow \quad 9970 \times 30 + 2 \times 9970 \times G + 60\,G$

$= 9970\,G + 9970 \times 107 + 30\,G + 30 \times 107$

$\Rightarrow \quad 10000\,G = 10000 \times 107 - 9970 \times 30$

$\Rightarrow \quad 10000\,G = (10700 - 2991) \times 100 = 7709 \times 100$

$\Rightarrow \quad G = \frac{7709 \times 100}{10000} = 77.09\,\Omega$ **Ans. (D)**

Sol. 17 Among given circuits D is most appropriate. When K_1 is closed and K_2 is open then current in galvanometer is

$$I = \frac{E}{R + G} \quad \ldots(1)$$

When both K_1 and K_2 are closed current supplied by battery is

$$I' = \frac{E}{R + \frac{GS}{G + S}} \quad \ldots(2)$$

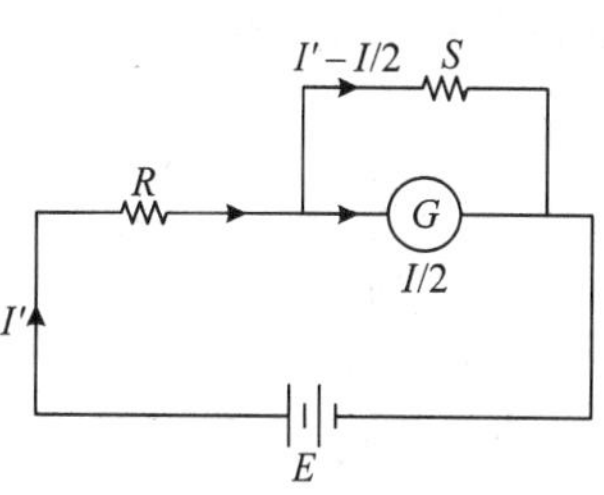

Here I' is divided in inverse ratio of S and G so we use

$$I' \times \frac{S}{S + G} = \frac{I}{2}$$

$$\frac{E.S}{RG + RS + GS} = \frac{E}{2(R + G)}$$

$$2RS + 2SG = RG + RS + GS$$

$\Rightarrow \quad G = \frac{RS}{R - S}$ **Ans. (D)**

Sol. 18 When two batteries are connected in parallel then the terminal potential difference of this combination is given as

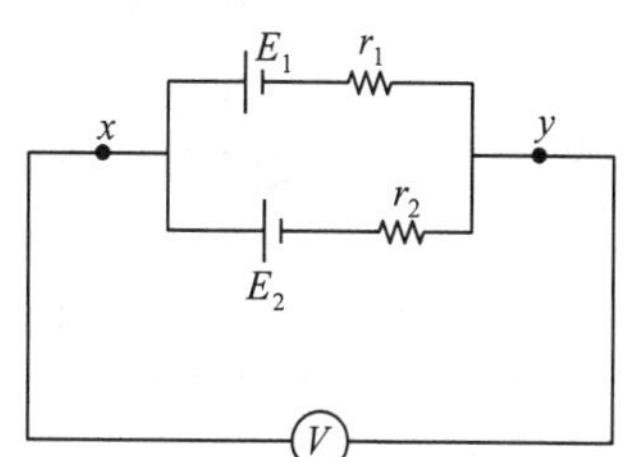

$$V_{xy} = \frac{\frac{E_1}{r_1} + \frac{E_2}{r_2}}{\frac{1}{r_1} + \frac{1}{r_2}}$$

[Already discussed in theory]

$\Rightarrow \quad V_{xy} = \frac{\frac{10}{1} + \frac{15}{0.6}}{\frac{1}{1} + \frac{1}{0.6}} = \frac{10 + 25}{1 + \frac{5}{3}} = \frac{35}{8/3} = \frac{105}{8}$

$\Rightarrow \quad V_{xy} = 13.1\,\text{V}$ **Ans. (C)**

Sol. 19

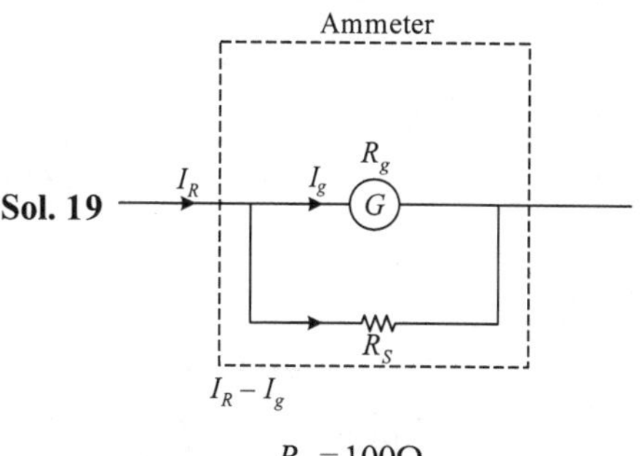

$R_g = 100\Omega$

$I_g = 10^{-3}\,\text{A}$

$I_R = 10\,\text{A}$

As R_g and R_S are in parallel we use

$$I_g R_g = (I_R - I_g)R_S$$

$$\Rightarrow \quad R_S = \frac{I_g R_g}{I_R - I_g}$$

$$\text{As } I_g << I_R \quad \Rightarrow \quad R_S = \frac{I_g R_g}{I_R} = \frac{10^{-3}\times 100}{10} = 10^{-2}\,\Omega$$

Ans. (A)

Sol. 20 Step-1 :

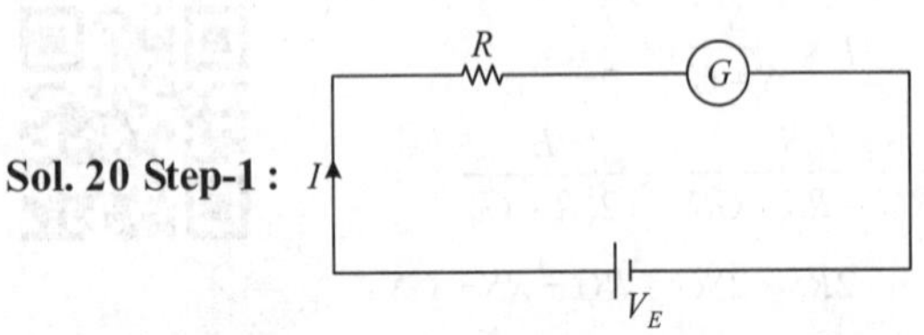

Current through galvanometer is given as

$$I = \frac{V_E}{(R+G)} \qquad \ldots(1)$$

Step-2 :

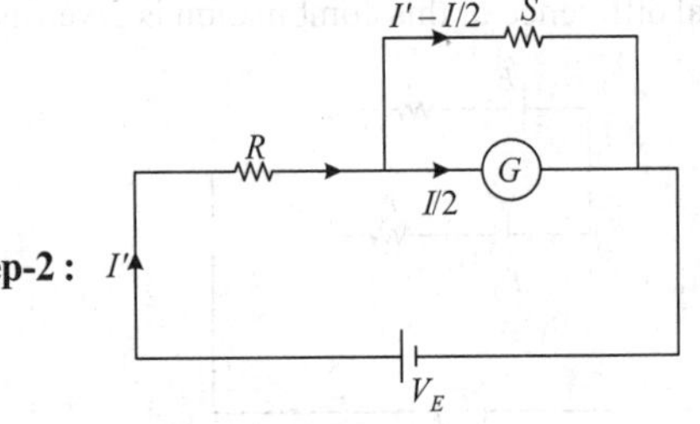

Current supplied by battery

$$I' = \frac{V_E}{R + \frac{SG}{S+G}} \qquad \ldots(2)$$

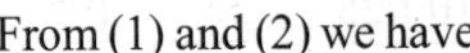

Here I' is distributed in inverse ratio of S and G so we use

$$I/2 = \frac{I'S}{S+G}$$

From (1) and (2) we have

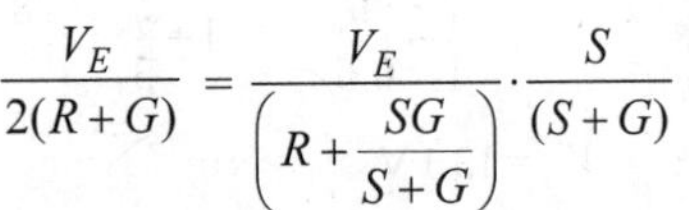

$$\frac{V_E}{2(R+G)} = \frac{V_E}{\left(R+\frac{SG}{S+G}\right)} \cdot \frac{S}{(S+G)}$$

$$\Rightarrow \quad \frac{1}{2(R+G)} = \frac{S(S+G)}{[R(S+G)SG](S+G)}$$

$$\Rightarrow \quad RS + RG + SG = 2RS + 2SG$$

$$\Rightarrow \quad RG = S(R+G) \qquad \Rightarrow \quad \textbf{Ans. (A)}$$

Sol. 21

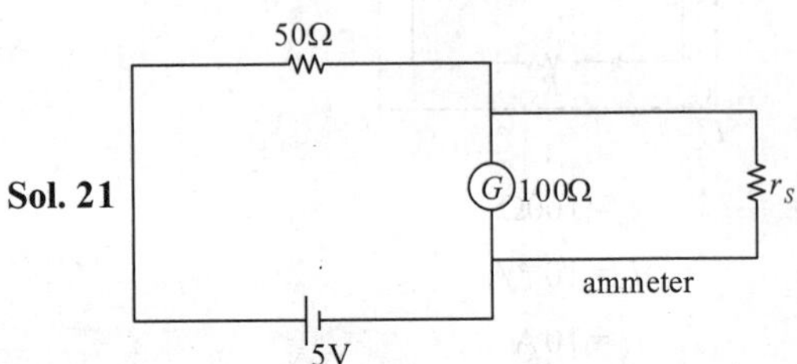

Current in circuit without the ammeter

$$I = \frac{5}{50} = 0.1\text{ A}$$

When ammeter is there is circuit, the new current will be

$$I' = \frac{5}{50 + \frac{100 r_S}{100 + r_S}}$$

As $I' < I \simeq I' = 0.991$ (within 1% initial value I)

$$\Rightarrow \quad \frac{5}{50 + \frac{100 r_S}{100 + r_S}} = 0.99 \times 0.1\text{ A}$$

$$\Rightarrow \quad \frac{5}{50 + \frac{100 r_S}{100 + r_S}} = 0.099\text{ A}$$

$$\Rightarrow \quad 50 + \frac{100 r_S}{100 + r_S} = \frac{5}{0.099} = 50.50\,\Omega$$

$$\Rightarrow \quad \frac{100 r_S}{100 + r_S} = 0.50\,\Omega$$

$$\Rightarrow \quad 100\, r_S = 50 + 0.50\, r_S$$

$$\Rightarrow \quad r_S = \frac{50}{99.50} = 0.502\Omega$$

(to be connected in parallel to G)

Ans. (D)

Sol. 22 If resistance of galvanometer is R_g, we have

Case-1 : At 40 div with $R = 2400\,\Omega$ if current sensitivity of "G" is i_d A/div then we use

$$40 i_d = \frac{2}{2400 + R_g} \qquad \ldots(1)$$

Case-2 : At 20 div with $R = 4900\,\Omega$, we use

$$20 i_d = \frac{2}{4900 + R_g} \qquad \ldots(2)$$

$$\frac{(1)}{(2)} \quad \Rightarrow \quad 2 = \frac{4900 + R_g}{2400 + R_g}$$

$$\Rightarrow \quad 4800 + 2R_g = 4900 + R_g$$

$$\Rightarrow \quad R_g = 100\,\Omega$$

from equation-(1)

$$40 i_d = \frac{2}{2400 + 100}$$

$$\Rightarrow \quad i_d = \frac{2}{40 \times 2500} = \frac{1}{5} \times 10^{-4}$$

$$= 20 \times 10^{-6}\, A/\text{div}$$

$$= 20\,\mu\text{A/div}$$

Case-3 : from 10 div if $R = R$

$$\Rightarrow \quad 10 i_d = \frac{2}{R+10}$$

$$\Rightarrow \quad 10 \times 20 \times 10^{-6} = \frac{2}{R+100}$$

$$\Rightarrow \quad R + 100 = 10000$$

$$\Rightarrow \quad R = 9900\,\Omega$$

As there are total 50 div in "G"

$\Rightarrow$ Range of "G" is

$$I_{max} = 50 \times 20 \times 10^{-6}$$

$$= 10^{-3} A = 1\text{ mA}$$

Ans. (A)

Sol. 23 All options (A), (B), (C) and (D) are covered in the theory option (D) is false.

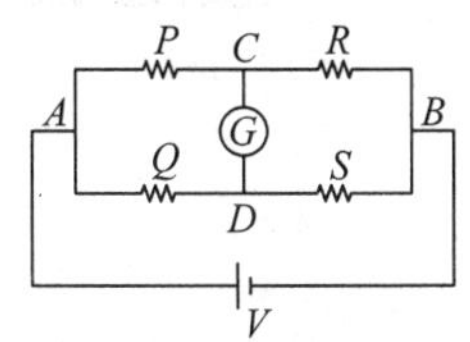

G and V are interchanged

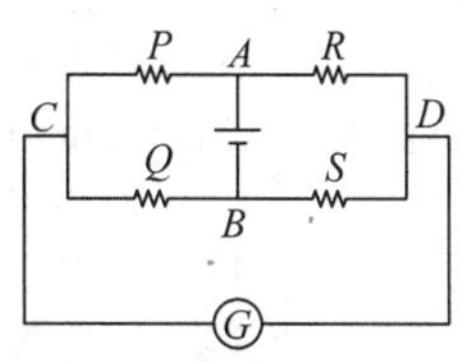

Balancing condition

$$\rightarrow \frac{P}{Q} = \frac{R}{S}$$

Balancing condition

$$\rightarrow \frac{Q}{S} = \frac{P}{R}$$

Option (D) is false. **Ans. (D)**

Sol. 24

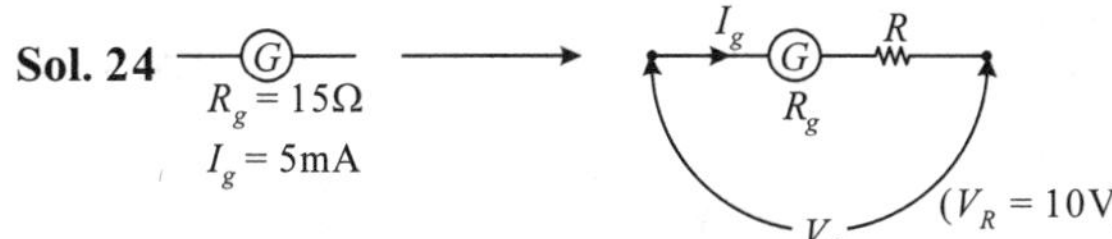

So for full deflection at voltage V_R, we use

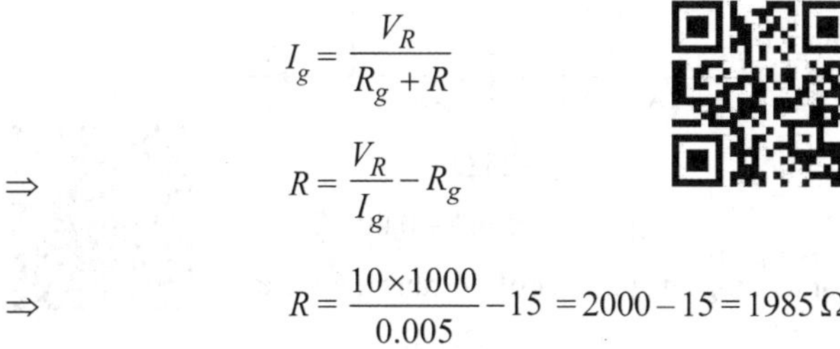

$$I_g = \frac{V_R}{R_g + R}$$

$$\Rightarrow \quad R = \frac{V_R}{I_g} - R_g$$

$$\Rightarrow \quad R = \frac{10 \times 1000}{0.005} - 15 = 2000 - 15 = 1985\,\Omega$$

Ans. (C)

Sol. 25 In first case we use

$$IR_1 = \lambda l, \qquad \ldots(1)$$

where $I \rightarrow$ current flowing through R_1 and R_2 in series

$\lambda \rightarrow$ potential gradient on wire PQ

In second case when K_3 is plugged into points 1 and 3 then R_1 and R_2 in series are balanced across points P and M where

$$PM = l_2$$

$$\Rightarrow \quad I(R_1 + R_2) = \lambda l_2 \qquad \ldots(2)$$

$$\frac{(2)}{(1)} \Rightarrow \quad \frac{R_1 + R_2}{R_1} = \frac{l_2}{l_1}$$

$$\Rightarrow \quad 1 + \frac{R_2}{R_1} = \frac{l_2}{l_1}$$

$$\Rightarrow \quad \frac{R_2}{R_1} = \frac{l_2 - l_1}{l_1}$$

$$\Rightarrow \quad \frac{R_1}{R_2} = \frac{l_1}{l_2 - l_1}$$

Ans. (D)

Sol. 26 For meter bridge initially we use

$$AN = 60\text{ cm}$$

$$NB = 40\text{ cm}$$

$$\Rightarrow \quad \frac{P}{60} = \frac{Q}{40}$$

$$\Rightarrow \quad \frac{4}{60} = \frac{Q}{40} \quad Q = \frac{8}{3}\,\Omega$$

In next state $P \rightarrow 4 + R$ and

$$AN = 80\text{ cm}$$

$$NB = 20\text{ cm}$$

$$\Rightarrow \quad \frac{4+R}{80} = \frac{8/3}{20}$$

$$\Rightarrow \quad 12 + 3R = 32$$

$$\Rightarrow \quad 3R = 20 \Rightarrow R = 20/3\,\Omega$$

Ans. (D)

Sol. 27 The internal resistance of cell using potentiometer experiment is given as

$$r = R\left(\frac{l_1}{l_2} - 1\right) = 5 \times \left(\frac{52}{40} - 1\right)$$

$$= 5 \times \frac{12}{40} = \frac{12}{8} = 1.5\,\Omega$$

Alternative : If cell of EMF E is balanced at l_1 length of wire

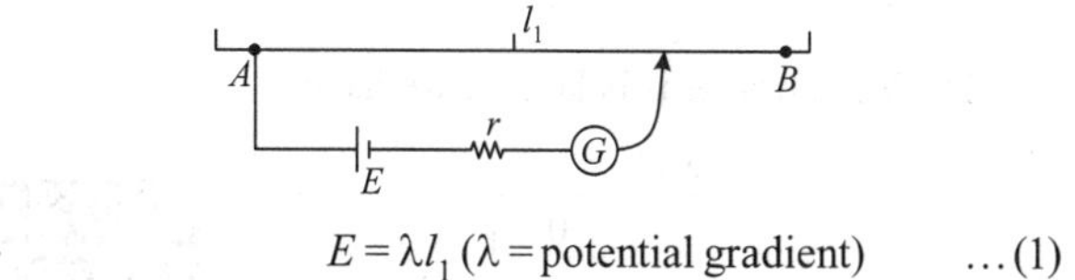

$$\Rightarrow \quad E = \lambda l_1 \ (\lambda = \text{potential gradient}) \qquad \ldots(1)$$

When call is shunted with a resistance R

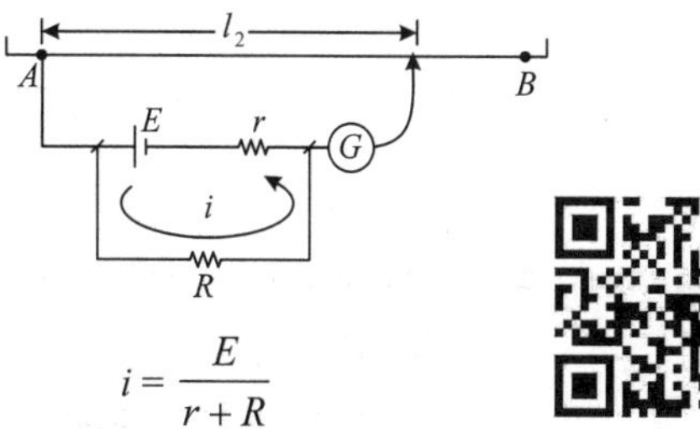

$$i = \frac{E}{r+R}$$

The potential difference across cell terminals after shunting is

$$V_E = iR = \frac{ER}{r+R} = \lambda l_2 \qquad \dots(2)$$

$$\frac{(2)}{(1)} \Rightarrow \frac{R}{r+R} = \frac{l_2}{l_1}$$

$$\Rightarrow \quad r = R\left(\frac{l_1}{l_2} - 1\right) = 1.5\,\Omega \qquad \textbf{Ans. (B)}$$

Sol. 28

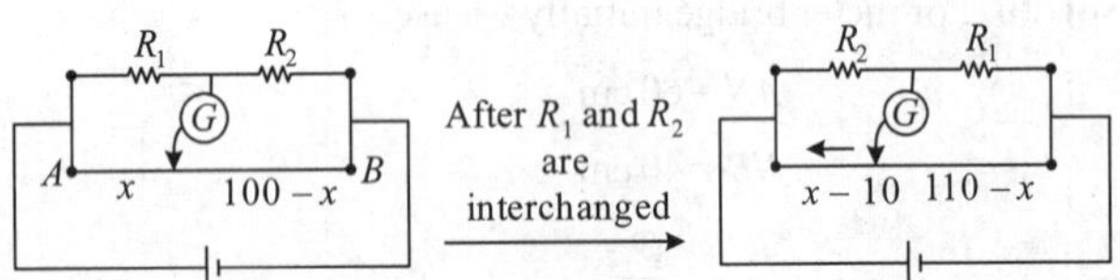

For balanced meter bridge, we use

$$\frac{R_1}{x} = \frac{R_2}{100-x} \qquad \dots(1)$$

$$\frac{R_2}{x-10} = \frac{R_1}{110-x} \qquad \dots(2)$$

$$\frac{(1)}{(2)} \Rightarrow \frac{110-x}{x} = \frac{x-10}{100-x}$$

$$\Rightarrow (110-x)(100-x) = x(x-10)$$

$$\Rightarrow 11000 + x^2 - 210x = x^2 - 10x$$

$$\Rightarrow 200x = 11000$$

$$\Rightarrow x = \frac{110}{2} = 55 \text{ cm}$$

From equation-(1) we have

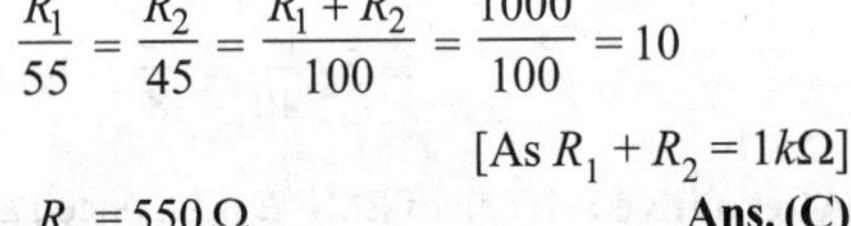

$$\frac{R_1}{55} = \frac{R_2}{45} = \frac{R_1+R_2}{100} = \frac{1000}{100} = 10$$

$$[\text{As } R_1 + R_2 = 1k\Omega]$$

$$\Rightarrow R_1 = 550\,\Omega \qquad \textbf{Ans. (C)}$$

Sol. 29 When bridge is balanced, we have

$$\frac{x}{l_1} = \frac{y}{100-l_1}$$

$$\Rightarrow \frac{x}{39.5} = \frac{y}{60.5}$$

$$\Rightarrow x = y \times \frac{39.5}{60.5} = 12.5 \times \frac{39.5}{60.5}$$

$$= 8.16\,\Omega$$

When x and y are interchanged then balancing length will be

$$l_2 = 100 - l_1$$

$$\Rightarrow l_2 = 100 - 39.5 = 60.5 \text{ cm} \qquad \textbf{Ans. (B)}$$

Sol. 30 Case-I : Without using a shunt resistance

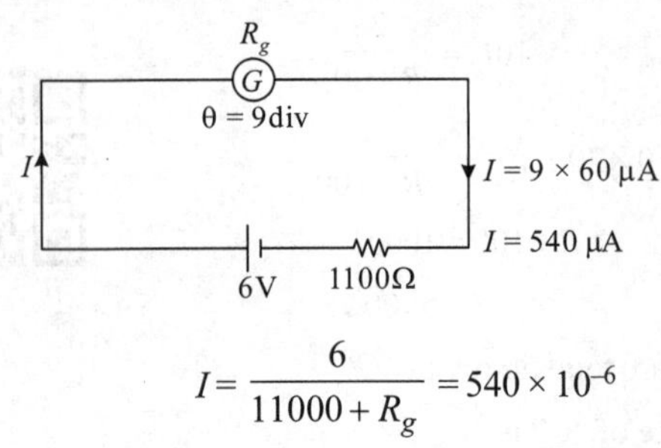

$$\Rightarrow \quad I = \frac{6}{11000 + R_g} = 540 \times 10^{-6}$$

$$\Rightarrow \quad 11000 + R_g = \frac{10^5}{9}$$

$$= 11111.11\,\Omega$$

$$\Rightarrow \quad R_g = 11111.11 - 1000$$

$$= 111.11\,\Omega$$

Case-II : with a shunt resistance

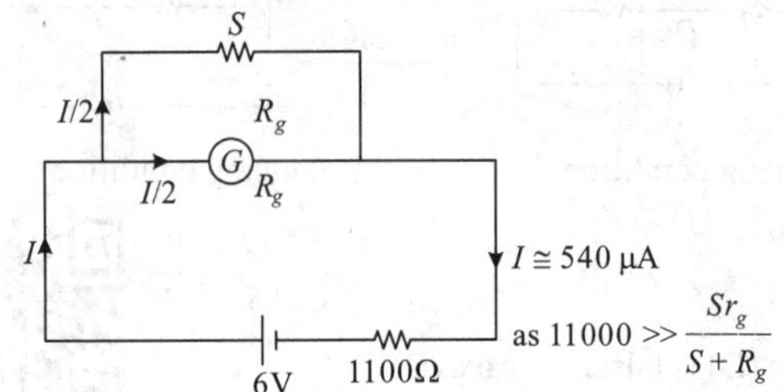

For half deflection in galvanometer we should use

$$S \simeq R_g = 111.11\,\Omega \qquad \textbf{Ans. (B)}$$

Sol. 31

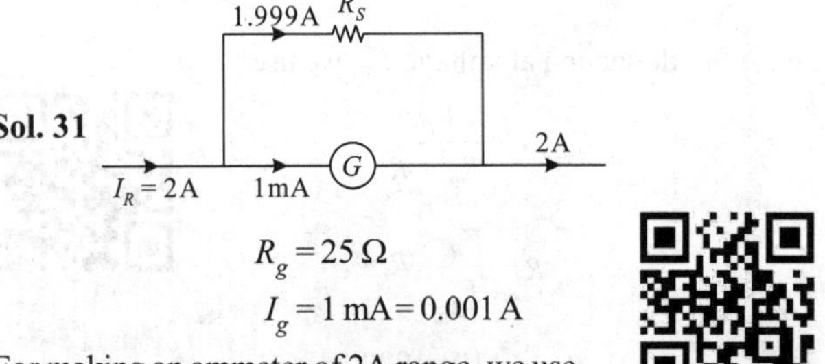

$$R_g = 25\,\Omega$$

$$I_g = 1 \text{ mA} = 0.001 \text{ A}$$

For making an ammeter of 2A range, we use

$$I_g R_g = (I_R - I_g) R_S$$

$$0.001 \times 25 = 1.999 \times R_S$$

$$\Rightarrow \quad R_S = \frac{25}{1999} \simeq 1.25 \times 10^{-2}\,\Omega \qquad \textbf{Ans. (D)}$$

Sol. 32

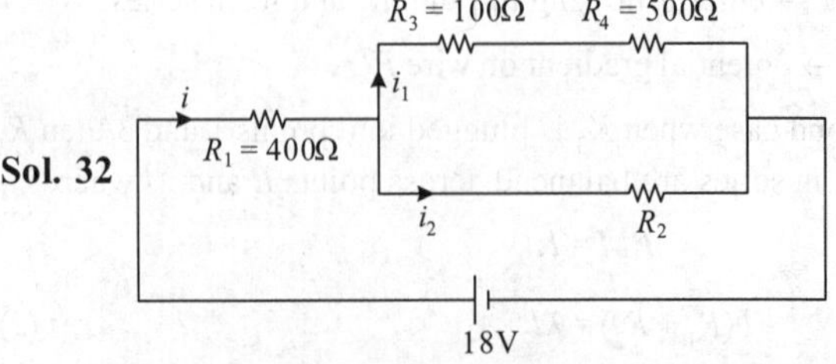

Given that

$$V_4 = 5\text{ V}$$

$$\Rightarrow \quad i_1 = \frac{V_4}{R_4} = 0.01\text{ A}$$

$$\Rightarrow \quad V_3 = i_1R_3 = 1\text{ V}$$

$$\Rightarrow \quad V_3 + V_4 = 6\text{ V} = V_2$$

$$\Rightarrow \quad V_1 + V_3 + V_4 = 18\text{ V}$$

$$\Rightarrow \quad V_1 = 12\text{ V}$$

$$\Rightarrow \quad i = \frac{V_1}{R_1} = 0.03\text{ A}$$

$$\Rightarrow \quad i - i_1 = 0.02\text{ A}$$

$$\Rightarrow \quad R_2 = \frac{V_2}{i_2} = \frac{6}{0.02} = 300\ \Omega$$ **Ans. (A)**

Sol. 33 Current in primary circuit is

$$i = \frac{\varepsilon}{13r}$$

At no deflection

$$i \times \left(\frac{12r}{L}\times x\right) = \frac{\varepsilon}{2}$$

$$\Rightarrow \quad \frac{\varepsilon}{13r}\left(\frac{12r}{L}\times x\right) = \frac{\varepsilon}{2}$$

$$\Rightarrow \quad x = \frac{13L}{24}$$ **Ans. (D)**

Sol. 34 Using V and I readings measured resistance will be parallel combination of R_V and R, so we use

$$0.95\,R = \frac{RR_V}{R+R_V}$$

$$\Rightarrow \quad 0.95\,R^2 + 0.95\,RR_V = RR_V$$

$$\Rightarrow \quad 0.95\,R = 0.05\,R_V$$

$$\Rightarrow \quad R_V = 19 \times 30 = 570\ \Omega$$ **Ans. (B)**

Sol. 35 $R_1 = 32 \times 10 = 320\ \Omega$

For wheat stone bridge, balancing condition is

$$\Rightarrow \quad \frac{R_1}{R_3} = \frac{R_2}{R_4}$$

$$\Rightarrow \quad \frac{320}{R_3} = \frac{80}{40}$$

$$\Rightarrow \quad R_3 = 160\ \Omega$$

Colour code of 160Ω resistance is

(Brown, Blue, Brown) **Ans. (B)**

Sol. 36 Potential gradient with $R_h = 2\Omega$ is

$$\lambda = \left(\frac{6}{2+4}\right)\times\frac{4}{L} = \frac{4}{100}\frac{\text{V}}{\text{cm}}$$

For emf ε_1 let null point be at l cm at J, we use

$$\varepsilon_1 = 0.5\text{ V} = 0.04l \quad \ldots(1)$$

Now with $R_h = 6\Omega$ new potential gradient is

$$\lambda' = \left(\frac{6}{4+6}\right)\times\frac{4}{L} = \frac{2.4}{100} = 0.024\frac{\text{V}}{\text{cm}}$$

For emf ε_2 null deflection length is same, so we use

$$\varepsilon_2 = 0.024l \quad \ldots(2)$$

Dividing equation-(1) by (2) we get

$$\frac{0.5}{\varepsilon_2} = \frac{0.04}{0.024}$$

$$\Rightarrow \quad \varepsilon_2 = 0.3\text{ V}$$ **Ans. (C)**

Sol. 37 At $R = 400\ \Omega$ bridge is balanced so we use

$$\frac{P}{400} = \frac{Q}{X} \quad \ldots(1)$$

On interchanging P and Q for balance $R = 405\ \Omega$. So we use

$$\frac{Q}{405} = \frac{P}{X} \quad \ldots(2)$$

(1) × (2) gives

$$X = \sqrt{400\times405} = 402.5\ \Omega$$ **Ans. (D)**

Sol. 38 Given that

$$R_g = 20\ \Omega$$

$$N_{\text{Left}} = N_{\text{Right}} = 30$$

Figure of merit of galvonometer is

$$\text{FOM} = \frac{I}{\phi} = 0.005\text{ A/Div}$$

$$\text{Current sensitivity} = \left(\frac{1}{0.005}\right)\text{Div/A}$$

$$Ig_{\max} = 0.005 \times 30$$

$$= 15 \times 10^{-2} = 0.15\text{ A}$$

For this galvanometer to be used as 15 V range voltmeter, we use

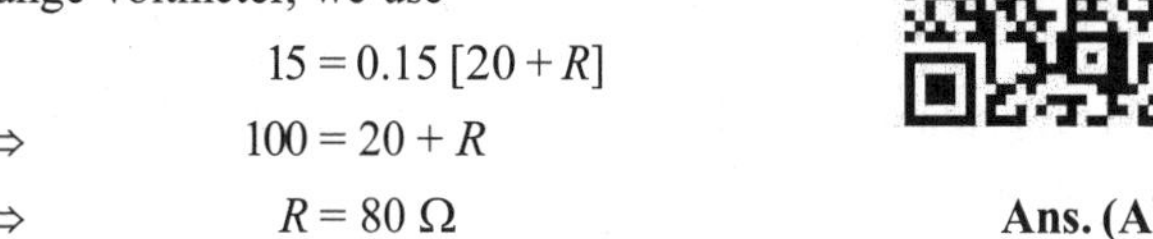

$$15 = 0.15\,[20 + R]$$

$$\Rightarrow \quad 100 = 20 + R$$

$$\Rightarrow \quad R = 80\ \Omega$$ **Ans. (A)**

Sol. 39 Initial balancing condition of bridge is

$$\frac{R_1}{R_2} = \frac{2}{3} \quad \ldots(1)$$

When 10 Ω is connected in series with R_1, we use

$$\frac{R_1+10}{R_2} = 1$$

$$\Rightarrow \quad R_1 + 10 = R_2 \quad \ldots(2)$$

From (1) and (2)

$$\frac{2R_2}{3} + 10 = R_2$$

$$\Rightarrow \quad 10 = \frac{R_2}{3}$$

$$\Rightarrow \quad R_2 = 30\ \Omega$$

and $\quad R_1 = 20\ \Omega$

If a resistance R is connected in parallel with $R_1 + 10$ then we use

$$\frac{\frac{30 \times R}{30 + R}}{30} = \frac{2}{3}$$

$$\Rightarrow \quad R = 60\ \Omega \quad \textbf{Ans. (B)}$$

Sol. 40 In first case galvanometer current is

$$i_g = \frac{E}{220 + R_g} = C\theta_0 \qquad \ldots(1)$$

In second case galvanometer current is

$$i_g = \left(\frac{E}{220 + \frac{5R_g}{5 + R_g}}\right) \times \frac{5}{(R_g + 5)} = \frac{C\theta_0}{5}$$

$$\Rightarrow \quad \frac{5E}{225R_g + 1100} = \frac{C\theta_0}{5} \qquad \ldots(2)$$

(1)/(2) gives

$$\Rightarrow \quad \frac{225R_g + 1100}{1100 + 5R_g} = 5$$

$$\Rightarrow \quad 5500 + 25R_g = 225R_g + 1100$$

$$\Rightarrow \quad 200R_g = 4400$$

$$\Rightarrow \quad R_g = 22\ \Omega$$

Ans. (D)

Sol. 41 If resistance of part AP is R_1 and PB is R_2 so for null deflection, we use

$$\frac{R'}{R'} = \frac{R_1}{R_2}$$

$$\Rightarrow \quad R_1 = R_2$$

On the given wire AB the resistance of an elemental length dl is given as

$$dR = \frac{C}{\sqrt{l}}dl \qquad \text{(where } C \rightarrow \text{constant)}$$

Thus resistance of length AB is given as

$$R_1 = C\int_0^l l^{-1/2}dl = C(2, \sqrt{l})$$

Similarly of resistance of length PB is given as

$$R_2 = C\int_l^1 l^{-1/2}dl = C(2 - 2\sqrt{l})$$

Using $\quad R_1 = R_2$, we have

$$C_2\sqrt{l} = C(2 - 2\sqrt{l})$$

$$\Rightarrow \quad 2\sqrt{l} = 1$$

$$\Rightarrow \quad \sqrt{l} = \frac{1}{2}$$

$$\Rightarrow \quad l = \frac{1}{4}\text{m} = 0.25\text{ m}$$

Ans. (A)

Sol. 42

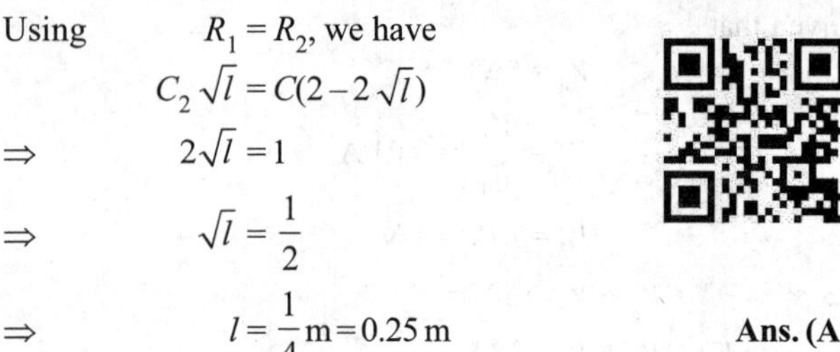

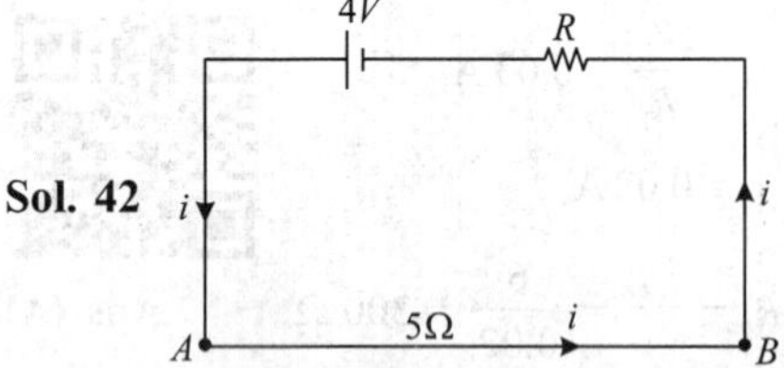

If current flowing in the wire is i, then we use

$$i = \left(\frac{4}{R + 5}\right)\text{A}$$

Resistance of 10 cm length of wire is

$$\frac{5}{10} = 0.5\ \Omega$$

Potential difference on 10 cm of wire is

$$5 \times 10^{-3} = \left(\frac{4}{R + 5}\right) \times 0.5$$

$$\Rightarrow \quad \frac{4}{R + 5} = 10^{-2}$$

$$\Rightarrow \quad R + 5 = 400\ \Omega$$

$$\Rightarrow \quad R = 395\ \Omega$$

Ans. (C)

Sol. 43 Full deflection current of galvanometer is

$$I_g = 4 \times 10^{-4} \times 25 = 10^{-2}\text{ A}$$

To convert the galvanometer into a voltmeter, we connect a resistance R in series, such that

$$2.5 = (50 + R)\ 10^{-2}$$

$$\Rightarrow \quad R = 200\ \Omega \quad \textbf{Ans. (C)}$$

Sol. 44

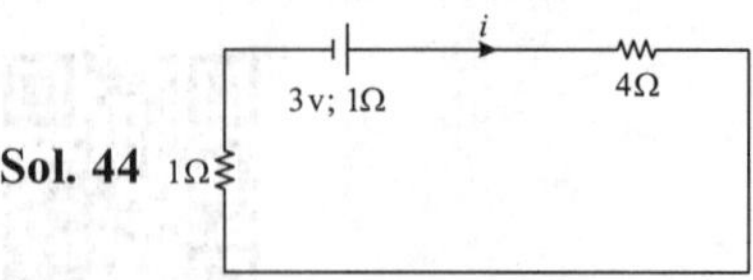

Resistance of wire AB is given as

$$R_{AB} = 400 \times 0.01 = 4\ \Omega$$

Circuit current is

$$i = \frac{1.5 + 1.5}{1 + 1 + 4} = \frac{3}{6} = 0.5\text{A}$$

Voltmeter reading is

$$V_R = i \times R$$

$\Rightarrow \quad V_R = 0.5 \times 0.01 \times 50$

$\Rightarrow \quad V_R = 0.25\,\text{V}$ **Ans. (B)**

Sol. 45

Shunt resistance is calculated as

$$R_S(0.5-0.002) = 50 \times 0.002$$

$$R_S = \frac{50 \times 0.002}{(0.5-0.002)} = \frac{0.1}{0.498} = 0.2$$ **Ans. (A)**

Sol. 46 For balanced state of bridge we use

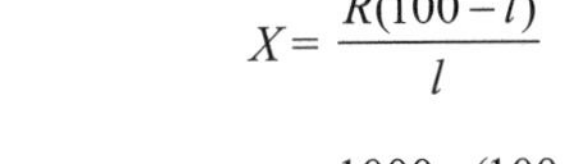

$$X = \frac{R(100-l)}{l}$$

for (1) $X = \frac{1000 \times (100-60)}{60} \approx 666.67\,\Omega$

for (2) $X = \frac{100 \times (100-13)}{13} \approx 669.23\,\Omega$

for (3) $X = \frac{10 \times (100-1.5)}{1.5} \approx 656.67\,\Omega$

for (4) $X = \frac{1 \times (100-1)}{1} \approx 99\,\Omega$

So option (A) is completely different so inconsistent with given data. **Ans. (A)**

Sol. 47 Equivalent resistance of circuit across battery is

$$R_{eq} = \frac{15 \times 10}{25} + 2 + 2r$$

$$= 8 + 2r$$

Current in circuit is

$$i = \frac{3}{8+2r}$$

Potential difference across 10Ω resistance is

$$2 = iR_{eq} = \frac{3}{8+2r} \times 6$$

$\Rightarrow \quad 16 + 4r = 18$

$\Rightarrow \quad r = 0.5\,\Omega$ **Ans. (D)**

Sol. 48 Full deflection current

$$I_g = 20 \times 10^{-6} \times 50 = 10^{-3}\,\text{A} = 1\,\text{mA}$$

For 2 V range of voltmeter, we use

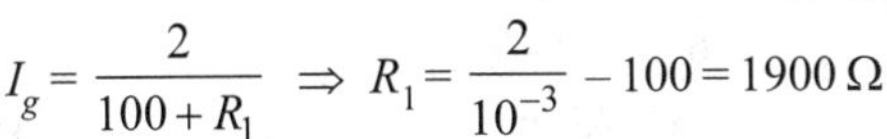

$$I_g = \frac{2}{100+R_1} \Rightarrow R_1 = \frac{2}{10^{-3}} - 100 = 1900\,\Omega$$

For 10 V range of voltmeter, we use

$$I_g = \frac{10}{2000+R_2} \Rightarrow R_2 = \frac{2}{10^{-3}} - 2000 = 8000\,\Omega$$

For 20 V range of voltmeter, we use

$$I_g = \frac{20}{10000+R_3} \Rightarrow R_3 = \frac{2}{10^{-3}} - 10000 = 10000\,\Omega$$

Ans. (D)

Sol. 49 Reading of voltmeter V and current I in circuit (reading of ammeter) can be related as

$$V = E - Ir$$

at $\quad I = 1A,\ V = V_0 \simeq 0$

$\Rightarrow \quad 0 = E - r$

$\Rightarrow \quad E = r$

at $\quad V = 1.5\,\text{V}, I = 0$

$\Rightarrow \quad V = E = 1.5\,\text{V}$

$\Rightarrow \quad r = 1.5\,\Omega$ **Ans. (C)**

Sol. 50 Internal resistance of cell by potentiometer is given as

$$r = R\left(\frac{l_1 - l_2}{l_2}\right)$$

$\Rightarrow \quad r = 10 \times \frac{60}{500} r$

$\Rightarrow \quad r = 1.2\,\Omega = \frac{12}{10}\Omega$ **Ans. (12)**

Sol. 51 If λ is potential gradient of potentiometer wire & L is total length of wire, then we use

$$5 = \lambda l$$

$\Rightarrow \quad 5 = \frac{\Delta V}{L} l$

$\Rightarrow \quad \Delta V = \frac{5 \times L}{l} = 5 \times \frac{12}{10} = 6\text{V}$

By Ohm's Law, we use

$$\Delta V = IR$$

$\Rightarrow \quad R = \frac{\Delta V}{I} = \frac{6}{60 \times 10^{-3}} = 100\,\Omega$ **Ans. (D)**

Sol. 52 For balancing the wheatstone network as shown in figure, we use

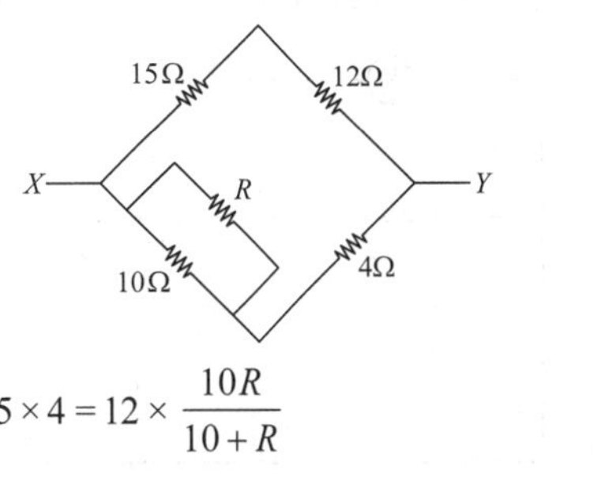

$$15 \times 4 = 12 \times \frac{10R}{10+R}$$

$\Rightarrow \quad R = 10\,\Omega$ **Ans. (10)**

Sol. 53 For the given galvanometer

$$I_g = 1\text{ mA},$$
$$R_g = 100\,\Omega$$

To convert it into a voltmeter, we use shunt resistance R, so we have

$$V = I_g(R + R_g)$$
$$\Rightarrow \quad 10 = 1 \times 10^{-3}(R + 100)$$
$$\Rightarrow \quad R = 9.9\text{ k}\Omega$$

Ans. (A)

Sol. 54 In balancing condition we use

$$\frac{R}{S} = \frac{25}{75}$$

If R is replaced by a wire half length and one fourth cross-section then resistance becomes $2R$, so we use

$$\frac{2R}{S} = \frac{l'}{100 - l'}$$
$$\Rightarrow \quad 2 \times \frac{1}{3} = \frac{l'}{100 - l'} = 3l' = 200 - 2l'$$
$$\Rightarrow \quad 5l' = 200$$
$$\Rightarrow \quad l' = 40$$

Ans. (40.00)

Sol. 55 If λ be the potential gradient then for balancing length $l = 100 - 49 = 51$ cm, we use

$$E_2 = \lambda l$$
$$\Rightarrow \quad 1.05 = \lambda \times 51$$
$$\Rightarrow \quad \lambda = \frac{1.02}{51}$$
$$\Rightarrow \quad \lambda = 0.02\text{ V/cm}$$

Ans. (A)

Sol. 56 (A) Is incorrect as there is no connectivity between plates of capacitor so infinite resistance

(B) Is correct because *LED* works only in forward based mode

(C) For metal wire if considered as a perfect conductor, this is correct

(D) For a resistance both sides resistance value is equal so this is correct.

Ans. (A)

Sol. 57 Effective circuit is shown below

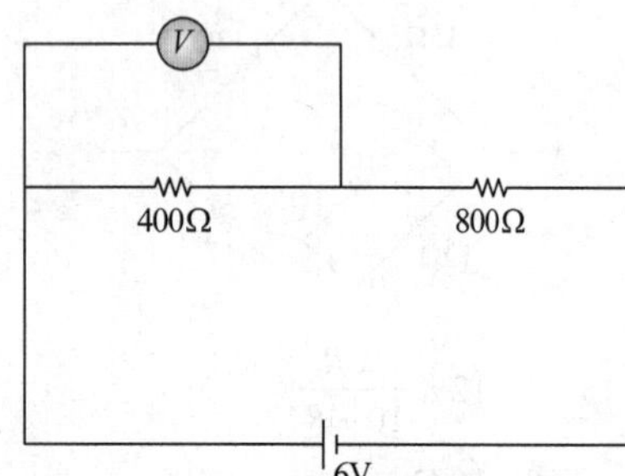

Equivalent resistance across battery is

$$R_{eq} = \frac{400 \times 10000}{10400} + 800 = 1185\,\Omega$$

Potential difference measured is

$$V_{400\Omega} = \frac{6}{1185} \times 400 \cong 1.95\text{ V}$$

Ans. (B)

Sol. 58 In first case, we use

$$1 = I_g(G + R_1) \qquad \ldots(1)$$

In second case, we use

$$2 = I_g(G + R_1 + R_2) \qquad \ldots(2)$$

From (1) and (2), we use

$$\frac{G + R_1 + R_2}{G + R_1} = 1$$
$$\Rightarrow \quad 1 + \frac{R_2}{G + R_1} = 2$$
$$\Rightarrow \quad R_2 = G + R_1$$

Ans. (B)

Sol. 59 Figure of Merit of galvanometer is given as

$$C = \frac{i}{\theta}$$
$$\Rightarrow \quad C = \frac{6 \times 10^{-3}}{2} = 3 \times 10^{-3}\text{ Am}^2$$

Ans. (A)

Sol. 60 For normal operation of circuit ammeter is to be connected in series because it has low resistance and voltmeter is to be connected in parallel as it has high resistance.

Ans. (A)

Sol. 61 Length of potentiometer wire is given as

$$AB = 10\text{ m}$$

For battery E_1, balancing length is $l_1 = 380$ cm

For battery E_2, balancing length is $l_2 = 760$ cm

For the given value of potential gradient on wire, we can use

$$\frac{E_1}{E_2} = \frac{l_1}{l_2}$$
$$\Rightarrow \quad \frac{E_1}{E_2} = \frac{380}{760} = \frac{1}{2} = \frac{a}{b}$$
$$a = 1 \quad \& \quad b = 2$$
$$a = 1$$

Ans. (1)

Sol. 62 Under balanced state, by Wheatstone bridge condition, we have

$$\frac{12}{6} = \frac{x}{72 - x}$$
$$\Rightarrow \quad x = 48\text{ cm}$$

Ans. (48)

Sol. 63 At null deflection in galvanometer, the maximum value of unknown EMF that can be measured is V_{AB} which is given as

$$V_{AB}=\frac{6}{100+20}\times 100=5\text{ V} \qquad \textbf{Ans. (C)}$$

Sol. 64 When the sliding contact is at the middle of the wire, the equivalent circuit of the given potentiometer is given as

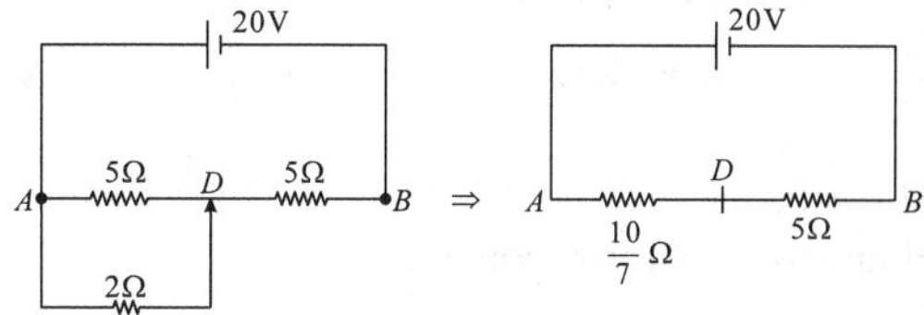

The potential drop across 2 Ω resistance can now be given as

$$V_{2\Omega}=\frac{20}{\frac{10}{7}+5}\times\frac{10}{7}$$

$$\Rightarrow \qquad V_{2\Omega}=\frac{40}{9}\text{V} \qquad \textbf{Ans. (A)}$$

Sol. 65 The ratio of secondary circuit emf at the two balancing lengths can be given as

$$\frac{\varepsilon_1}{\varepsilon_1+\varepsilon_2}=\frac{\lambda\times 250}{\lambda\times 400}$$

$$\Rightarrow \qquad 8\varepsilon_1=5\varepsilon_1+5\varepsilon_2$$

$$\Rightarrow \qquad 3\varepsilon_1=5\varepsilon_2$$

$$\Rightarrow \qquad \frac{\varepsilon_1}{\varepsilon_2}=\frac{5}{3} \qquad \textbf{Ans. (B)}$$

Sol. 66 Current sensitivity of the galvanometer is given as

$$C=\frac{\theta}{I_g}$$

$$\Rightarrow \qquad I_g=\frac{\theta}{C}=\frac{25\text{ div}}{2\text{ div/mA}}=25\text{ mA}$$

Potential difference required for full deflection is 50 mV at which 25 mA current flows so resistance of galvanometer is given as

$$R=\frac{V}{I}=\frac{50\text{mV}}{25\text{mA}}=2\,\Omega \qquad \textbf{Ans. (B)}$$

Sol. 67 Figure below shows the case described in the question

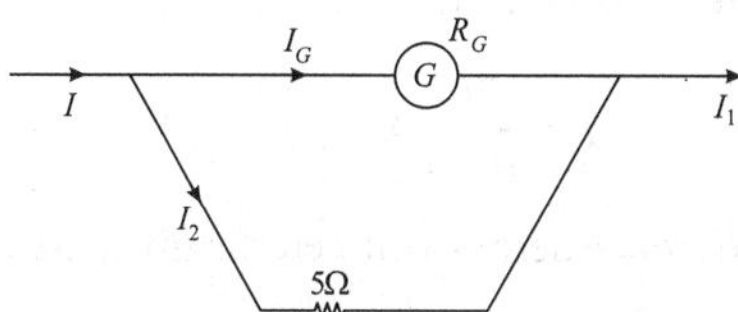

In above circuit galvanometer current which is 2% of the total current, can be given as

$$I_G=\frac{2}{100}\times I=\frac{5}{R_G+5}\times I$$

$$\Rightarrow \qquad \frac{5}{R_G+5}=\frac{1}{50}$$

$$\Rightarrow \qquad 250=R_G+5$$

$$\Rightarrow \qquad R_G=245\,\Omega \qquad \textbf{Ans. (B)}$$

Sol. 68 The ratio of two emf's is given by ratio of their balancing lengths so we use

$$\frac{\varepsilon_1}{\varepsilon_2}=\frac{l_1}{l_2}$$

$$\Rightarrow \qquad \frac{3}{2}=\frac{75\text{cm}}{l_2}$$

$$\Rightarrow \qquad l_2=50\text{ cm}$$

$$\Rightarrow \qquad l_1-l_2=75-50$$

$$\Rightarrow \qquad l_1-l_2=25\text{ cm} \qquad \textbf{Ans. (25)}$$

Sol. 69 Figure below shows the potentiometer circuit as explained in question

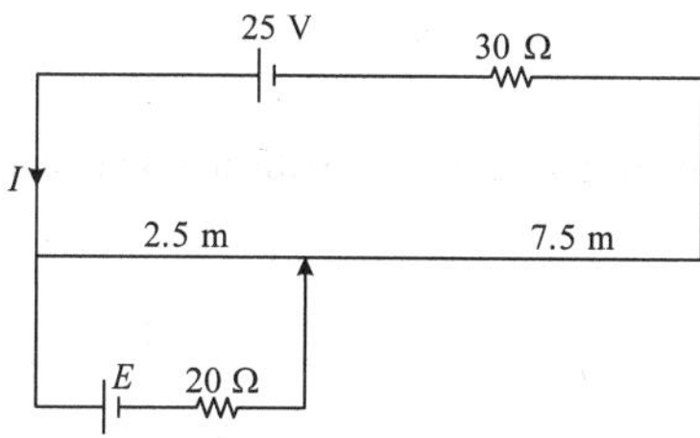

Current in primary circuit is calculated as

$$I=\frac{25}{50}=0.5\text{ A}$$

Potential difference across potentiometer wire is calculated as

$$\Delta V=IR=0.5\times 20=10\text{ V}$$

Potential gradient on potentiometer wire is calculated as

$$\lambda=\frac{\Delta V}{L}=\frac{10}{10}=1\text{ V/m}$$

As cell of emf E is balanced at 2.5 m of wire, its emf is given as

$$E=\lambda x=1\times 2.5=2.5\text{ V} \qquad \textbf{Ans. (25)}$$

Sol. 70 In galvanometer experiment, circuit is shown in figure below. If shunt resistance is S, we use

$$(I-I_g)S=I_gG$$

$$\Rightarrow \qquad \frac{I_g}{I}=\frac{S}{S+G}$$

$$\Rightarrow \quad \frac{1}{3}=\frac{S}{S+G}$$

$$\Rightarrow \quad S+G=3S$$

$$\Rightarrow \quad G=2S$$ **Ans. (B)**

Sol. 71 Figure below shows the conversion of galvanometer by connecting the shunt resistance

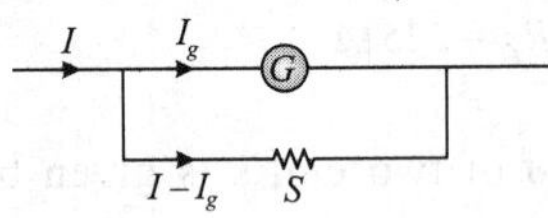

In above circuit we use

$$I_gG=(I-I_g)S \quad \ldots(1)$$

Figure of merit of galvanometer is defined as

$$K=\frac{I_g}{\theta}$$

$$\Rightarrow \quad I_g=Kn$$

From equation-(1), we have

$$I=\frac{I_g}{S}(G+S)$$

$$\Rightarrow \quad I=\frac{nK}{S}(G+S)$$ **Ans. (D)**

Sol. 72 For a galvanometer with a shunt resistance, we can write

$$S=\frac{R_G}{\frac{I}{I_g}-1}$$

$$\Rightarrow \quad 8=\frac{72}{\frac{I}{I_g}-1}$$

$$\Rightarrow \quad \frac{I}{I_g}-1=9$$

$$\Rightarrow \quad \frac{I}{I_g}=10$$

$$\Rightarrow \quad \frac{I_g}{I}=\frac{1}{10}$$

Percentage of current which passes through galvanometer is calculated as

$$p=\frac{I_g}{I}\times 100=10\%$$ **Ans. (B)**

Sol. 73 For the two conditions of cell with a shunt, we use

$$\frac{V_1}{V_2}=\frac{3}{2}=\frac{E-i_1r}{E-i_2r}$$

$$\Rightarrow \quad \frac{3}{2}=\frac{E-\frac{E}{8+r}\times r}{E-\frac{E}{4+r}\times r}$$

$$\Rightarrow \quad \frac{3}{2}=\frac{8(4+r)}{4(8+r)}$$

$$\Rightarrow \quad 24+3r=16+4r$$

$$\Rightarrow \quad r=8\,\Omega$$ **Ans. (8)**

Sol. 74 Using the conditions of a balanced wheat stone bridge and adding the end correction, we have

$$\frac{15}{(43+2)}=\frac{R}{(102-45)}$$

$$\Rightarrow \quad R=\frac{57}{45}\times 15$$

$$\Rightarrow \quad R=19\,\Omega$$ **Ans. (19)**

Sol. 75 Using the condition of a balanced wheat stone bridge,

$$\Rightarrow \quad \frac{R}{3}=\frac{4}{6}$$

$$\Rightarrow \quad R=2\,\Omega$$

So the effective resistance of the circuit is given as

$$R_{eq}=\frac{6\times 9}{6+9}=\frac{18}{5}\Omega$$

Circuit current is calculated as

$$i=\frac{36}{R_{eq}}=10\,\text{A}$$ **Ans. (10)**

Sol. 76 Figure below shows the circuit when voltmeter is connected across 500 Ω resistance. The equivalent resistance by parallel combination of 500 W and 2000 W is given as

$$R_e=\frac{2000\times 500}{2000+500}=400\,\Omega$$

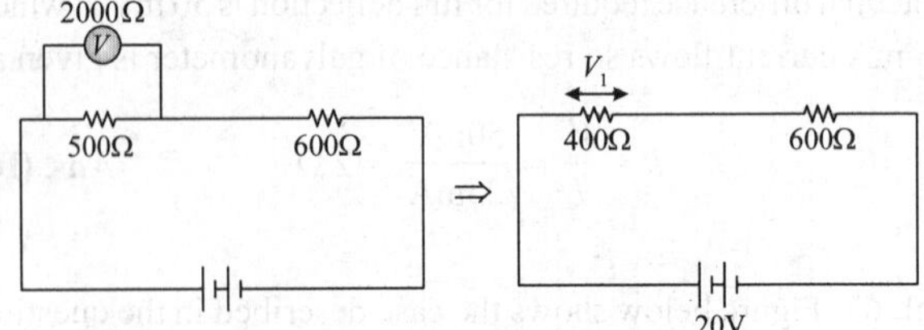

Current in circuit is given as

$$I=\frac{20}{1000}\text{A}$$

Potential difference across voltmeter is given as

$$V_1=I\times 400=\frac{20}{1000}\times 400=8\text{ V}$$ **Ans. (8)**

Sol. 77 For the two cases of cells we use

$$1.2 = \lambda \times 36$$
$$1.8 = \lambda \times x$$

Dividing above two equations, we have

$$\frac{2}{3} = \frac{36}{x}$$

$$\Rightarrow \quad x = 18 \times 3 = 54 \text{ cm}$$

Thus difference in balancing lengths is given as

$$\Delta x = 54 - 36 = 18 \text{ cm}$$

Ans. (18)

Sol. 78 Initially for balancing of bridge we use

$$\frac{P}{Q} = \frac{40}{60} = \frac{2}{3} \quad \ldots(1)$$

Later we use

$$\frac{P+x}{Q} = \frac{80}{20} = \frac{4}{1} \quad \ldots(2)$$

Dividing equation-(2) by (1) gives

$$\frac{P+x}{P} = 4 \times \frac{3}{2} = 6$$

$$\Rightarrow \quad 1 + \frac{x}{P} = 6$$

$$\Rightarrow \quad \frac{x}{P} = 5$$

$$\Rightarrow \quad x = 5P = 5 \times 4 = 20\,\Omega$$

Ans. (20)

Sol. 79 Figure below shows the potentiometer circuit as described in the question.

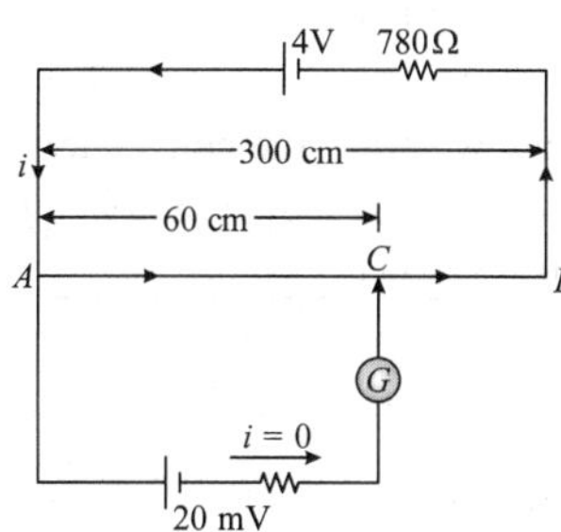

If resistance of potentiometers wire is taken as R, current through primary circuit is given as

$$i = \frac{4}{R + 780}$$

Potential difference across AB is calculated as

$$V_{AB} = \frac{4R}{R + 780}$$

Potential difference across balancing length AC is calculated as

$$V_{AC} = i\left(\frac{60}{300}\right) = \frac{4R \times 60}{(R+780) \times 300} = \frac{4R}{5(R+780)}$$

This is equal to 20 mV so we have

$$\frac{4R}{5(R+780)} = 20 \times 10^{-3} = 2 \times 10^{-2}$$

$$\Rightarrow \quad 4R = 10^{-1}(R + 780)$$

$$\Rightarrow \quad 4R = \frac{R}{10} + 78$$

$$\Rightarrow \quad 4R - \frac{R}{10} = 78$$

$$\Rightarrow \quad \frac{39R}{10} = 78$$

$$\Rightarrow \quad R = 20\,\Omega$$

Ans. (20)

Sol. 80 For balanced bridge condition, we have

$$\frac{S}{30} = \frac{5.6 \times 10^3}{70}$$

$$\Rightarrow \quad S = \frac{3}{7} \times 5.6 \times 10^3 = 2400\,\Omega$$

Ans. (2400)

Sol. 81 Balancing length is independent of cross sectional area of the potentiometer wire so it will remain same. **Ans. (40)**

Sol. 82 The emf of secondary circuit cell is given as

$$E = \frac{AC}{AB}(V_A - V_B)$$

$$\Rightarrow \quad 20 \times 10^{-3} = \frac{60}{300} \times \frac{4 \times 20}{R + 20}$$

$$\Rightarrow \quad R = 780\,\Omega$$

Ans. (780)

Sol. 83 Figure of circuit shown below is what is explained in question

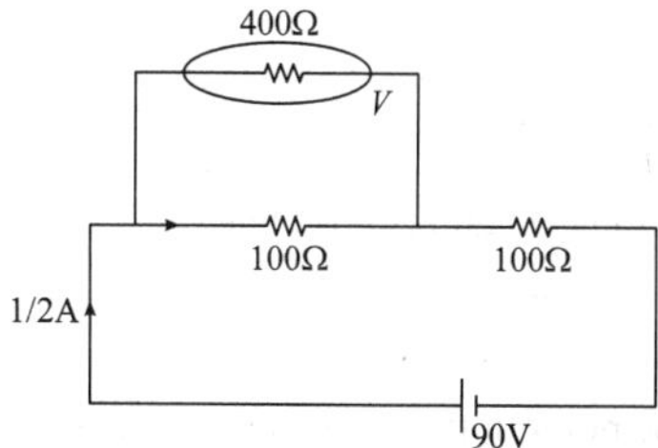

Equivalent resistance across the battery is calculated as

$$R_{eq} = \frac{400 \times 100}{500} + 100$$

$$R_{eq} = 180\,\Omega$$

Current through battery is calculated as

$$i = \frac{90}{180} = \frac{1}{2}\text{A}$$

Reading of voltmeter is calculated as

$$V = \frac{1}{2} \times \frac{400}{500} \times 100$$

$$\Rightarrow \quad V = 40\text{ V}$$

Ans. (A)

Sol. 84 For the meter bridge experiment, at balance point, we use

$$\frac{2}{\left(\frac{3x}{3+x}\right)}=\frac{40+22.5}{60-22.5}=\frac{62.5}{37.5}=\frac{5}{3}$$

$$\Rightarrow \quad \frac{6}{5}=\frac{3x}{3+x}$$

$$\Rightarrow \quad 6+2x=5x$$

$$\Rightarrow \quad x=2$$

Ans. (2)

Sol. 85 The sensitivity of potentiometer wire is inversely proportional to potential gradient and directly proportional to the length of the potentiometer wire hence option (B) is correct. **Ans. (B)**

Sol. 86 For balanced bridge, we have

$$\frac{R_1+R_2}{10}=\frac{60}{40}=\frac{3}{2}$$

$$\Rightarrow R_1+R_2=15\,\Omega$$

Now after resistances are connected in parallel, we use

$$\frac{R_1R_2}{(R_1+R_2)\times 3}=\frac{40}{60}=\frac{2}{3}$$

$$\Rightarrow \quad R_1R_2=30\,\Omega^2$$

Ans. (30)

Sol. 87 For internal resistance r the null point obtained for shunt resistance 5 Ω, is 200 cm so we use

$$\frac{\varepsilon}{r+5}\times 5=200x \qquad \ldots(1)$$

When shunt resistance of 15 Ω is used, null point is at 300 cm so we use

$$\frac{\varepsilon\times 15}{r+15}=300x \qquad \ldots(2)$$

Solving equations (1) and (2), we get

$$r=5\,\Omega$$

Ans. (5)

Sol. 88 For the potentiometer the ratio of the two emf is written as

$$\frac{E_1}{E_2}=\frac{l_1}{l_2}$$

$$\Rightarrow \quad \frac{1.5}{E_2}=\frac{60}{60+40}=\frac{6}{10}=\frac{3}{5}$$

$$\Rightarrow \quad E_2=\frac{5}{2}=\frac{x}{10}$$

$$\Rightarrow \quad x=25$$

Ans. (25)

Sol. 89 Error of voltmeter decreases with increase in its resistance. For ideal voltmeter resistance is infinite. **Ans. (A)**

Sol. 90 To convert galvanometer into ammeter low resistances should be added into parallel and for voltmeter conversion, a very high resistance should be added in series. By connecting low resistance parallel to galvanometer equivalent resistance decreases. **Ans. (C)**

Sol. 91 As voltmeter reads 2 V. So, potential across 5 Ω is 2 V and current through 5 Ω resistance is calculated as

$$i_1=\frac{2V}{5\Omega}=\frac{2}{5}\text{A}$$

Thus voltage across 2 Ω is 1 V and current i flowing through 2 Ω is calculated as

$$i=\frac{1V}{2\Omega}=\frac{1}{2}\text{A}$$

Current through voltmeter is calculated as

$$i-i_1=\frac{1}{2}-\frac{2}{5}=\frac{5-4}{10}=\frac{1}{10}\text{A}$$

For voltmeter, we use $V=IR$, which gives

$$2=\left(\frac{1}{10}\right)R$$

$$\Rightarrow \quad R=20\,\Omega$$

Ans. (20)

Sol. 92 Figure below shows the conversion of galvanometer to an ammeter with a shunt as specified in question.

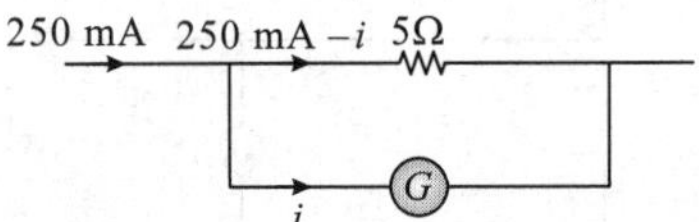

In this ammeter, we use

$$I_gG=(I_R-I_g)R$$

$$\Rightarrow \quad I_g(G+R)=I_RR$$

$$\Rightarrow \quad I_g=\frac{I_RR}{G+R}$$

Full scale deflection current is given as

$$I=250\text{ mA}$$

$$\Rightarrow \quad \frac{250mA\times 5}{5+R_G}=i \qquad \ldots(1)$$

For conversion into voltmeter, we have

$$I(G+R)=V$$

$$\Rightarrow \quad I=\frac{V}{G+R}$$

$$\Rightarrow \quad i=\frac{25}{1050+R_G} \qquad \ldots(2)$$

From equations-(1) and (2), we have

$$\frac{25}{1050+R_G}=\frac{5}{4(5+R_G)}$$

$$\Rightarrow \quad 100(5+R_G)=1050\times 5+R_G\times 5$$

$$\Rightarrow \quad 95R_G=4750$$

$$\Rightarrow \quad R_G=50\,\Omega$$

Ans. (50)

Sol. 93 For voltmeter, we use high resistance in series such that

$$R=\frac{V}{I_g}-G$$

$$\Rightarrow \quad R=\frac{50}{10^{-3}}-54\approx 50\,\text{k}\Omega\,(A)$$

For Ammeter, we use shunt resistance such that

$$S=\frac{I_gG}{I-I_g}=\frac{10^{-3}\times 54}{(10-1)\times 10^{-3}}$$

$$\Rightarrow \quad S=6\,\Omega\,(C)$$

Ans. (C)

18.1 Magnetic Effects of Current

Sol. 1

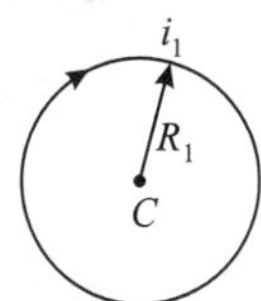

Magnetic induction at C is

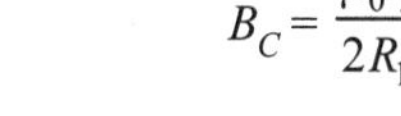

$$B_C=\frac{\mu_0 i_1}{2R_1}$$

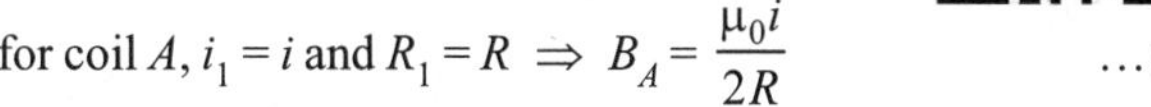

for coil A, $i_1=i$ and $R_1=R \Rightarrow B_A=\dfrac{\mu_0 i}{2R}$ $\ldots(1)$

for coil B, $i_2=2i$ and $R_2=2R \Rightarrow B_B=\dfrac{\mu_0(2i)}{2(2R)}=\dfrac{\mu_0 i}{2R}$ $\ldots(2)$

$$\Rightarrow \quad (1)/(2)\Rightarrow \quad \frac{B_A}{B_B}=1$$

Ans. (A)

Sol. 2

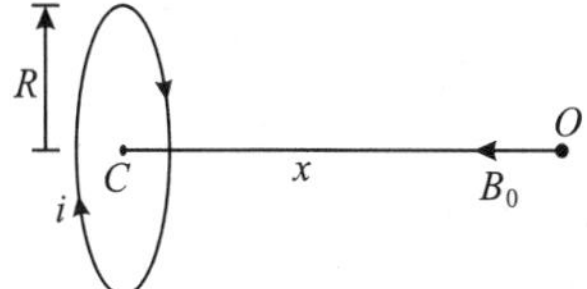

Magnetic field induction at point O is given as

$$B_0=\frac{\mu_0 iR^2}{2(x^2+R^2)^{3/2}}=54\,\mu\text{T} \qquad \ldots(1)$$

Magnetic field induction at point C is given as

$$B_C=\frac{\mu_0 i}{2R} \qquad \ldots(2)$$

$$\Rightarrow \quad (2)/(1)\Rightarrow \frac{B_C}{54}=\frac{(x^2+R^2)^{3/2}}{R^3}=\frac{(4^2+3^2)}{3^3}=\frac{125}{27}$$

$$\Rightarrow \quad B_C=54\times\frac{125}{27}=250\,\mu\text{T}$$

Ans. (A)

Sol. 3

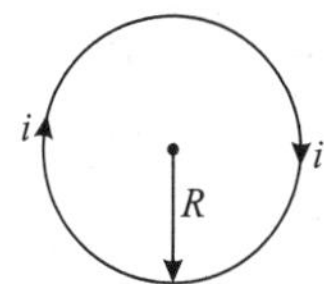

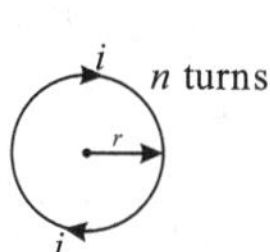

Magnetic induction at centre of one turn loop is given as

$$B=\frac{\mu_0 i}{2R} \qquad \ldots(1)$$

As length of wire is constant, we use

$$2\pi R=2\pi r\times n \Rightarrow r=R/n$$

Magnetic induction at centre of n turn loop is

$$B'=\frac{\mu_0 i}{2r}\times n=\frac{\mu_0 in^2}{2R}=n^2B$$

Ans. (B)

Sol. 4 Using Ampere's law on path M, we get

$$\int_M \vec{B}\cdot\vec{dl}=\mu_0 i_{\text{end}}$$

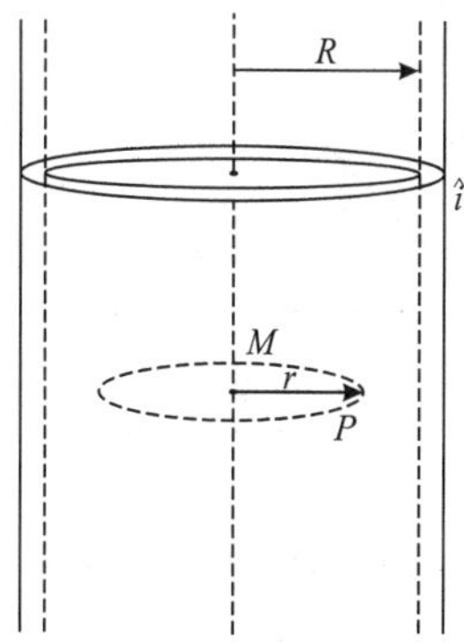

As the conductor is hollow

$$\Rightarrow \quad i_{\text{enclosed } M}=0$$

$$\Rightarrow \quad \int_M \vec{B}\cdot\vec{dl}=0$$

$$\Rightarrow \quad B_{\text{in } M}=0$$

Ans. (B)

Sol. 5

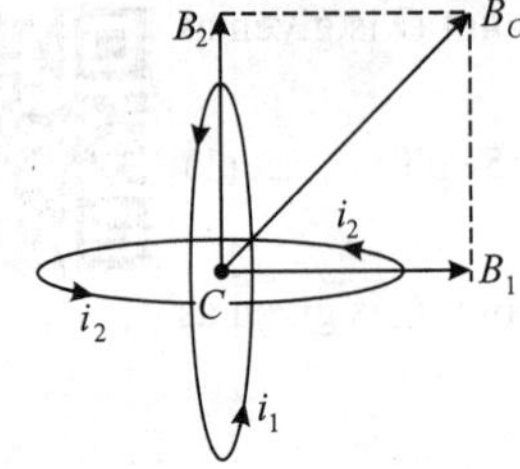

If B_1 and B_2 are magnetic induction at centre of the two coils, we use

$$B_1 = \frac{\mu_0 i_1}{2r_1} \text{ and } B_2 = \frac{\mu_0 i_2}{2r_2}$$

$$r_1 = r_2 = r = 2\pi \text{ cm}$$

Net magnetic induction at common centre of coils is given as

$$B_C = \sqrt{B_1^2 + B_2^2} = \frac{\mu_0}{2r}\sqrt{i_1^2 + i_2^2}$$

$$\Rightarrow \quad B_C = \frac{4\pi\times10^{-7}}{2\times2\pi\times10^{-2}}\sqrt{(3)^2+(4)^2}$$

$$\Rightarrow \quad B_C = 5\times10^{-5}\ \text{Wb/m}^2$$

Ans. (C)

Sol. 6 For a long solenoid having turns per unit length n and current i the magnetic induction as

$$B = \mu_0 ni$$

For case-I: $n = 200$ turns/cm

$$\Rightarrow \quad 6.28\times10^{-2} = \mu_0\times200\times10^2\times i \qquad \ldots(1)$$

For case-II: $n = 100$ turns/cm; $i = i/3$

$$\Rightarrow \quad B = \mu_0\times100\times10^2\times i/3 \qquad \ldots(2)$$

$$\text{By (2)/(1)} \Rightarrow \quad \frac{B}{6.28\times10^{-2}} = \frac{1}{3\times2}$$

$$\Rightarrow \quad B = \frac{6.28\times10^{-2}}{6}$$

$$\Rightarrow \quad B = 1.05\times10^{-2}\ \text{T}$$

Ans. (A)

Sol. 7

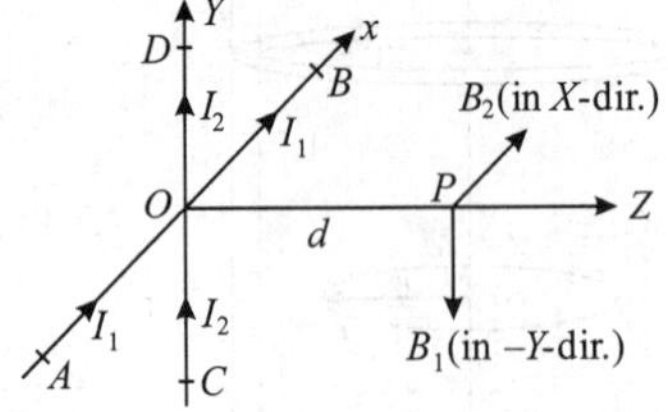

The magnetic inductions at point P due to currents I_1 and I_2 are

$$B_1 = \frac{\mu_0 I_1}{2\pi d} \text{ and } B_2 = \frac{\mu_0 I_2}{2\pi d}$$

As B_1 and B_2 are mutually perpendicular resulting magnetic induction at point P is given as

$$B_P = \sqrt{B_1^2 + B_2^2} = \frac{\mu_0}{2\pi d}\sqrt{I_1^2 + I_2^2}$$

Ans. (B)

Sol. 8 Magnetic field at interior point P of wire is given as

$$B_P = \frac{\mu_0 ix}{2\pi a^2} \qquad \ldots(1)$$

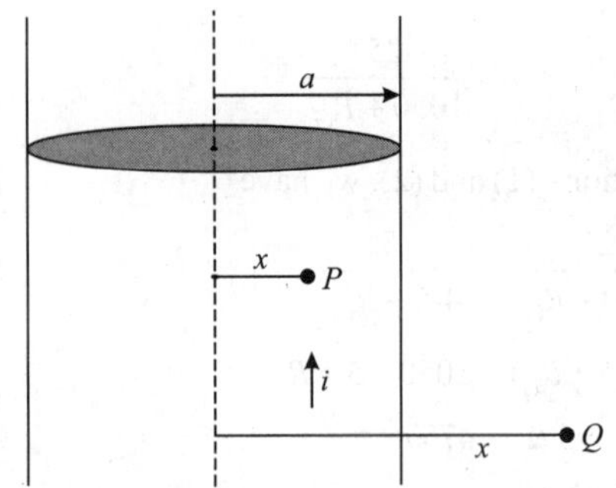

Magnetic field at outer point Q is given as

$$B_Q = \frac{\mu_0 i}{2\pi x} \qquad \ldots(2)$$

at $x = \frac{a}{2}$

$$B_P = \frac{\mu_0 i\left(\frac{a}{2}\right)}{2\pi a^2}$$

$$\Rightarrow \quad B_P = \frac{\mu_0 i}{4\pi a}$$

at $x = 2a$

$$B_Q = \frac{\mu_0 i}{2\pi(2a)}$$

$$\Rightarrow \quad B_Q = \frac{\mu_0 i}{4\pi a}$$

$$\Rightarrow \quad \frac{B_P}{B_Q} = 1$$

Ans. (D)

Sol. 9 If we consider a path M at interior points of pipe of radius x and use Ampere's law, then

$$\int_M \vec{B}\cdot\vec{dl} = \mu_0 . i_{\text{end}}$$

Being a hollow pipe, here

$$i_{\text{enclosed}} = 0$$

$$\Rightarrow \quad \int_M \vec{B}\cdot\vec{dl} = 0$$

$$\Rightarrow \quad B_{\text{in } M} = 0$$ **Ans. (D)**

Sol. 10

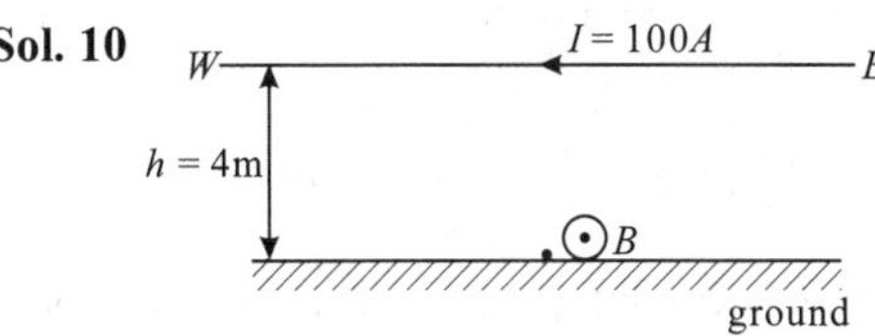

N E W S → (horizontal direction)

Outward → South

Inward → North

Magnetic induction on ground is

$$B = \frac{\mu_0 I}{2\pi h} = \frac{4\pi\times10^{-7}\times100}{2\pi\times4}$$

$$\Rightarrow \quad B = 5\times10^{-6}\text{ T in south}$$ **Ans. (C)**

Sol. 11 The magnetic induction due to a circular arc carrying a current I and subtending angle θ at centre is given as

$$B = \frac{\mu_0 I\theta}{4\pi R} \qquad [R \to \text{radius of arc}]$$

At point O, $B_{AB} = 0$

$$B_{BC} = \frac{\mu_0 I(\pi/6)}{4\pi b} = \frac{\mu_0 I}{24b} \otimes \text{ (inward)}$$

$$B_{CD} = 0$$

$$B_{DA} = \frac{\mu_0 I(\pi/6)}{4\pi a} = \frac{\mu_0 I}{24a} \odot \text{ (outward)}$$

Net magnetic induction at O is given as

$$B_0 = B_{DA} - B_{BC}$$

$$\Rightarrow \quad B_0 = \frac{\mu_0 I}{24a} - \frac{\mu_0 I}{24b} = \frac{\mu_0 I}{24}\left[\frac{1}{a} - \frac{1}{b}\right]$$

$$= \frac{\mu_0 I(b-a)}{24ab}$$ **Ans. (D)**

Sol. 12

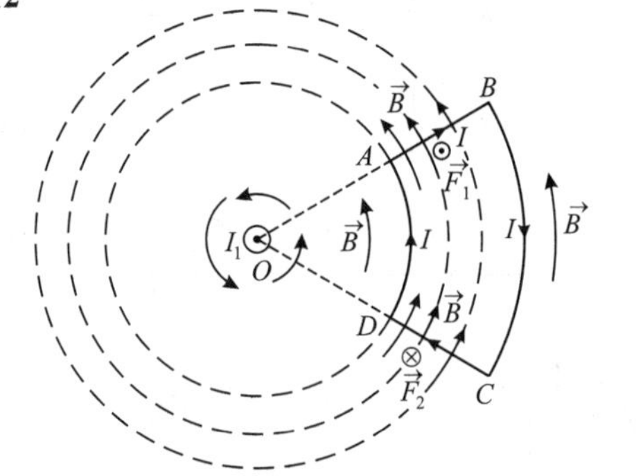

Force on wire AB due to I_1 is given in outward direction → $\vec{F_1}$

and Force on wire CD due to I_1 is in inward direction → $\vec{F_2}$

by symmetry we have $|\vec{F_1}| = |\vec{F_2}|$

Force on DA and BC are zero as field due to I_1 is parallel to all segments of these arc. **Ans. (D)**

Sol. 13 Due to a long current carrying wire magnetic induction in its surrounding is as shown below for given situation.

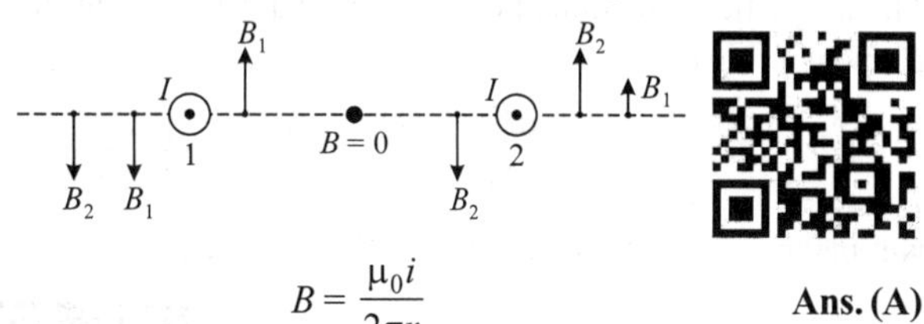

$$B = \frac{\mu_0 i}{2\pi r}$$ **Ans. (A)**

Sol. 14

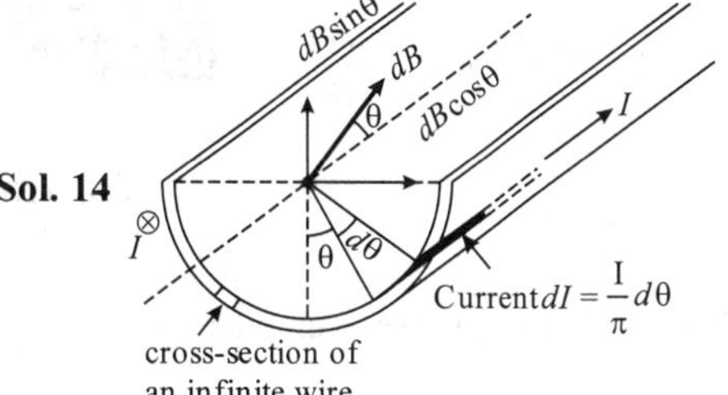

Current flowing in the elemental strip considered on surface of the half cylindrical infinite wire of width '$Rd\theta$'.

The magnetic induction due to this elemental current dI at any point on axis is given as

$$dB = \frac{\mu_0 dI}{2\pi R}$$

Net magnetic induction along axis can be given as $B = \int dB\cos\theta$

$$\Rightarrow \quad B = \int_{-\frac{\pi}{2}}^{+\frac{\pi}{2}} \frac{\mu_0\left(\frac{I}{\pi}d\theta\right)}{2\pi R}\cos\theta$$

$$\Rightarrow \quad B = \frac{\mu_0 I}{2\pi^2 R}\int_{-\frac{\pi}{2}}^{+\frac{\pi}{2}} \cos d\theta$$

$$\Rightarrow \quad B = \frac{\mu_0 I}{2\pi^2 R}[\sin\theta]_{-\pi/2}^{+\pi/2} = \frac{\mu_0 I}{2\pi^2 R}[1-(-1)]$$

$$\Rightarrow \quad B = \frac{\mu_0 I}{\pi^2 R}$$ **Ans. (C)**

Sol. 15

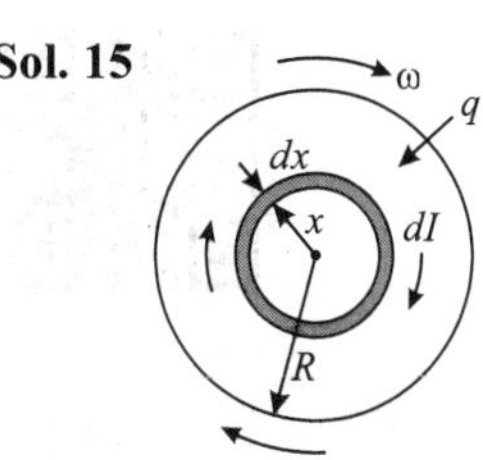

Charge on elemental ring (strip) of radius x and width dx is given as

$$dq = \frac{q}{\pi R^2} 2\pi x dx = \frac{2qxdx}{R^2}$$

Current in this elemental ring is

$$dI = \frac{dq\omega}{2\pi} = \frac{qxdx\omega}{\pi R^2}$$

The magnetic induction at centre of disc due to current in the elemental ring is

$$dB = \frac{\mu_0 dI}{2x} = \frac{\mu_0}{2x} \times \frac{q\omega dx}{\pi R^2} = \frac{\mu_0 q\omega}{2\pi R^2} dx$$

Net magnetic induction at centre of disc is given as

$$B = \int dB = \frac{\mu_0 q\omega}{2\pi R^2} \int_0^R dx$$

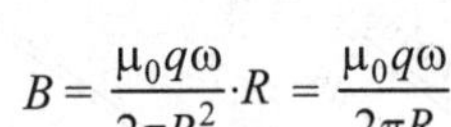

$$\Rightarrow \quad B = \frac{\mu_0 q\omega}{2\pi R^2} \cdot R = \frac{\mu_0 q\omega}{2\pi R}$$

$$\Rightarrow \quad B \propto \frac{1}{R}$$

Ans. (B)

Sol. 16 Magnetic induction at centre of a circular coil

$$B_1 = \frac{\mu_0 I}{2R} \quad \ldots(1)$$

MI at an axial point at a distance x from coil centre is

$$B_2 = \frac{\mu_0 I R^2}{2(R^2 + x^2)^{3/2}}$$

at $x = 2\sqrt{2}\,R$, $$B_2 = \frac{\mu_0 I R^2}{2(R^2 + 8R^2)^{3/2}} = \frac{\mu_0 I}{54R} \quad \ldots(2)$$

from (1) and (2)

$$\Rightarrow \quad \frac{B_1}{B_2} = \frac{\mu_0 I / 2R}{\mu_0 I / 54R} = 27$$

Ans. (B)

Sol. 17

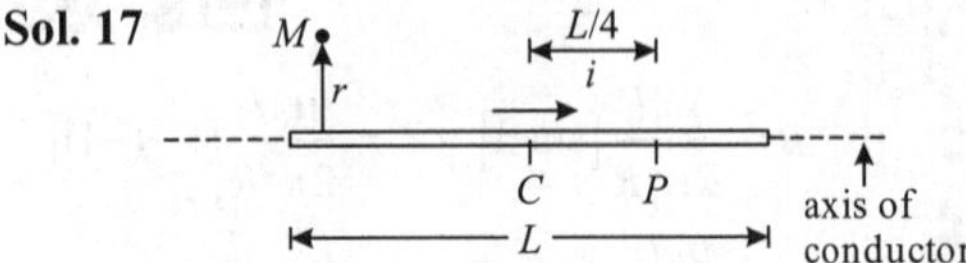

Along the axis of a current carrying conductor magnetic induction is always zero.

$$B_M = \frac{\mu_0 i}{2\pi r}$$

and $$B_{\text{inside}} = \frac{\mu_0 i r}{2\pi r^2} = \frac{1}{2}\mu_0 J r$$

at P as $r = 0$ (inside) $\Rightarrow B_P = 0$ **Ans. (A)**

Sol. 18

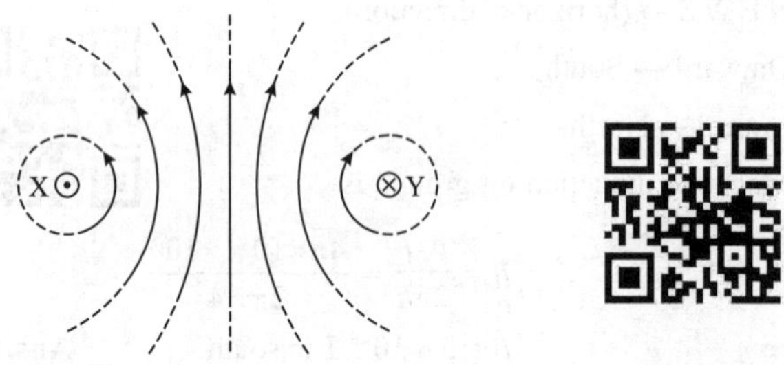

3D-view of magnetic lines configuration due to a circular loop.

Cross-sectional view **Ans. (A)**

Sol. 19 If length of wires is l then

Case I : $l = 2\pi R_1 \Rightarrow 2R_1 = l/\pi$

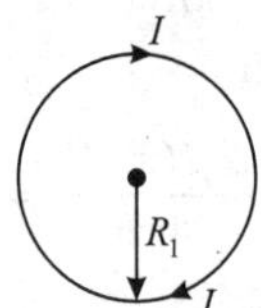

Magnetic Induction at Centre of loop is

$$B_1 = \frac{\mu_0 I}{2R_1} = \frac{\mu_0 I\pi}{l} \quad \ldots(1)$$

Case II : $l = 3 \times 2\pi R_2 \Rightarrow 2R_2 = l/3\pi$

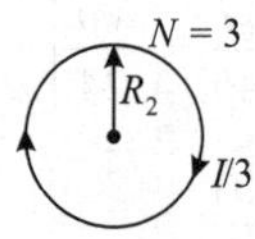

Magnetic Induction at Centre of loop is

$$B_2 = \frac{\mu_0 (I/3)}{2R_2} \times 3 = \frac{3\mu_0 I\pi}{l} \quad \ldots(2)$$

By (1)/(2) $\Rightarrow$ $$\frac{B_1}{B_2} = \frac{1}{3}$$ **Ans. (B)**

Sol. 20 l, I

Case-A:

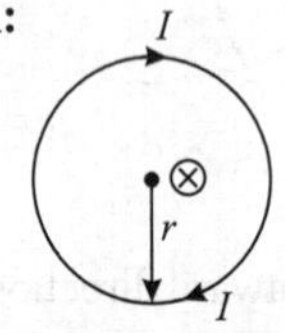

Magnetic induction at the centre of coil is

$$B_A = \frac{\mu_0 I}{2r} = \frac{\mu_0 I\pi}{l} \qquad [l = 2\pi r]$$

Case-B:

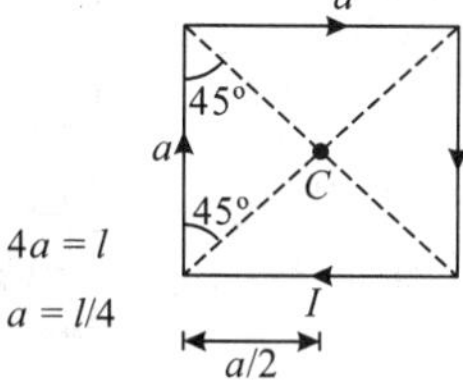

Magnetic induction due to square coil at its centre is given as

$$B_B = 4\left[\frac{\mu_0 I}{4\pi(a/2)}[\cos 45° + \cos 45°]\right]$$

$$\Rightarrow \quad B_B = \frac{2\mu_0 I}{\pi a}\sqrt{2}$$

$$\Rightarrow \quad B_B = \frac{2\sqrt{2}\mu_0 I}{\pi\left(\frac{l}{4}\right)} = \frac{8\sqrt{2}\mu_0 I}{\pi l} \qquad \ldots(2)$$

By (1)/(2) $\Rightarrow \quad \frac{B_A}{B_B} = \frac{\pi^2}{8\sqrt{2}}$ **Ans. (D)**

Sol. 21

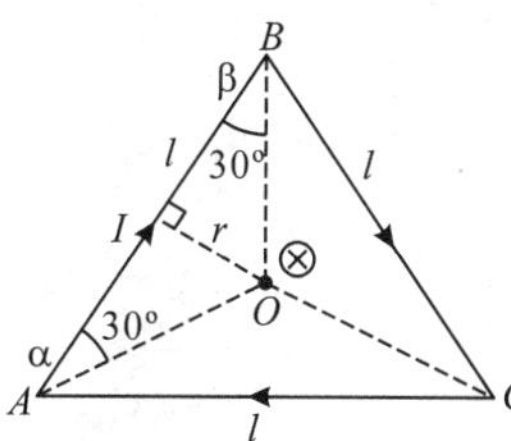

Magnetic induction due to a finite length wire is given as

$$B_0 = \frac{\mu_0 I}{4\pi r}(\cos\alpha + \cos\beta)$$

$$\Rightarrow \quad B_0 = \frac{\mu_0 I}{4\pi\left(\frac{l}{2\sqrt{3}}\right)}(2\cos 30°) \qquad \left[\because r = \frac{l}{2\sqrt{3}}\right]$$

$$\Rightarrow \quad B_0 = \frac{\sqrt{3}\times 4\pi\times 10^{-7}\times 1}{2\pi\times 4.5\times 10^{-2}}\times 2\times\frac{\sqrt{3}}{2}$$

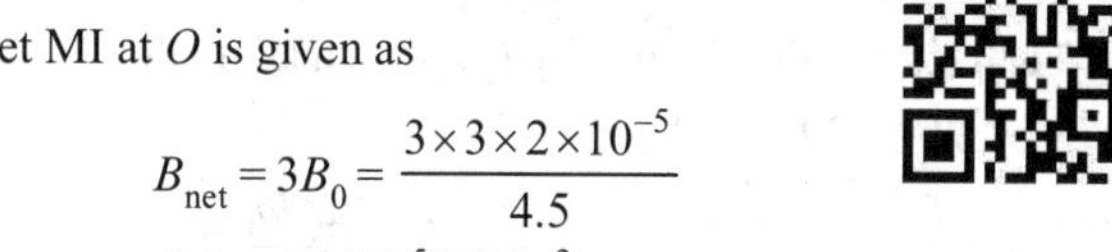

Net MI at O is given as

$$B_{net} = 3B_0 = \frac{3\times 3\times 2\times 10^{-5}}{4.5}$$

$$\Rightarrow \quad B_{net} = 4\times 10^{-5}\ \text{Wb/m}^2$$ **Ans. (A)**

Sol. 22 Magnetic induction due to the two coils at P are in same direction so we have.

$$B = \frac{\mu_0 IR^2}{2(x^2+R^2)^{3/2}}\times 2\Bigg|_{x=R/2}\times N$$

$$\Rightarrow \quad B = \frac{\mu_0 IR^2 N}{\left(\frac{R^2}{4}+R^2\right)^{3/2}} = \frac{\mu_0 IR^2 N}{\left(\frac{5}{4}\right)^{3/2} R^3}$$

$$\Rightarrow \quad B = \frac{8\mu_0 IN}{(5)^{3/2} R}$$ **Ans. (B)**

Sol. 23 Magnetic field at O is given as

$$\vec{B} = \frac{\mu_0 i}{4\pi}\theta\left[\frac{1}{r_1} - \frac{1}{r_2}\right]$$

$$\Rightarrow \quad \vec{B} = \frac{4\pi\times 10^{-7}}{16}\times 10\times\left[\frac{1}{3\times 10^{-2}} - \frac{1}{5\times 10^{-2}}\right]$$

$$\Rightarrow \quad \vec{B} = \frac{\pi}{3}\times 10^{-5}\ \text{T}$$ **Ans. (A)**

Sol. 24

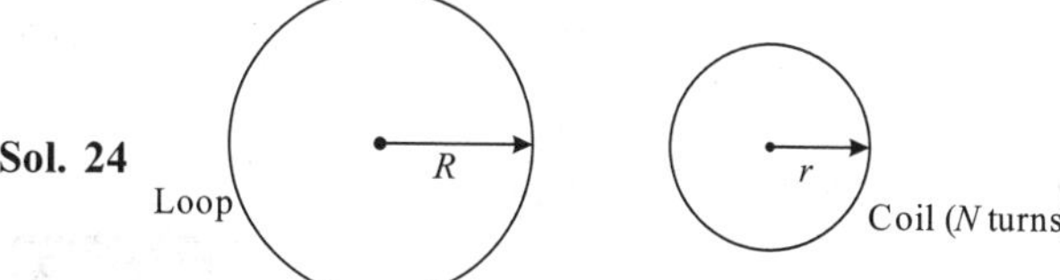

For loop: $L = 2\pi R$

For coil: $L = N\times 2\pi r$

$\Rightarrow \quad R = Nr$

Magnetic field at the centres are given as

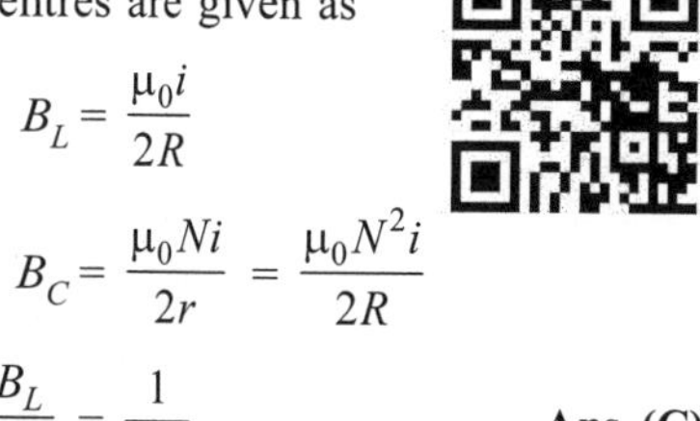

$$B_L = \frac{\mu_0 i}{2R}$$

and $$B_C = \frac{\mu_0 Ni}{2r} = \frac{\mu_0 N^2 i}{2R}$$

$$\Rightarrow \quad \frac{B_L}{B_C} = \frac{1}{N^2}$$ **Ans. (C)**

Sol. 25 Magnetic field at 'O' will be only due to 'PS' and 'QN' only, so we have

$$B_0 = B_{PS} + B_{QN}$$

Let current in each wire is i so magnetic field at O is given as

$$B_0 = \frac{\mu_0 i}{4\pi d} + \frac{\mu_0 i}{4\pi d}$$

$$\Rightarrow \quad 10^{-4} = \frac{\mu_0 i}{2\pi d} = \frac{2\times 10^{-7}\times i}{4\times 10^{-2}}$$

$$\Rightarrow \quad i = 20\ \text{A (into the page)}$$ **Ans. (D)**

Sol. 26 Dimension of $\sqrt{\frac{\varepsilon_0}{\mu_0}}$

$$[\varepsilon_0] = [M^{-1}L^{-3}T^4A^2]$$

$$[\mu_0] = [MLT^{-2}A^{-2}]$$

Dimensions of

$$\sqrt{\frac{\varepsilon_0}{\mu_0}} = \left[\frac{M^{-1}L^{-3}T^4A^2}{MLT^{-2}A^{-2}}\right]^{\frac{1}{2}}$$

$$= [M^{-2}L^{-4}T^6A^4]^{1/2}$$

$$= [M^{-1}L^{-2}T^3A^2]$$ **Ans. (B)**

Sol. 27

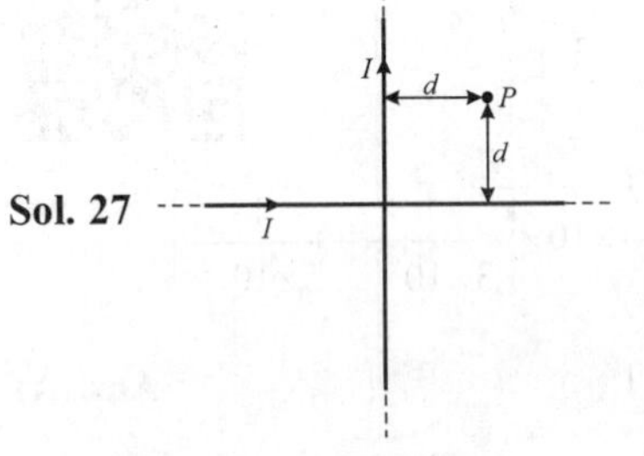

Magnetic field at point P is given as

$$\vec{B}_{net} = \frac{\mu_0 i}{2\pi d}(-\hat{k}) + \frac{\mu_0 i}{2\pi d}(\hat{k}) = 0$$ **Ans. (A)**

Sol. 28 Magnetic field at centroid O is given as

$$B = 3\left[\frac{\mu_0 i}{4\pi r}(\sin 60° + \sin 60°)\right]$$

Here r is given as

$$r = \frac{a}{2\sqrt{3}} = \frac{1}{2\sqrt{3}}$$

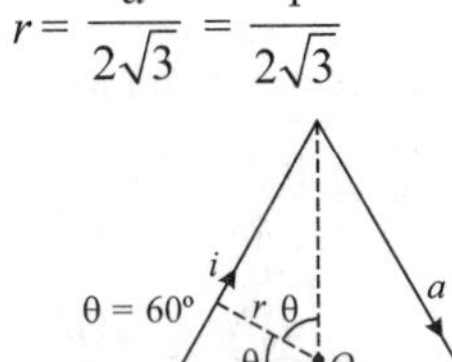

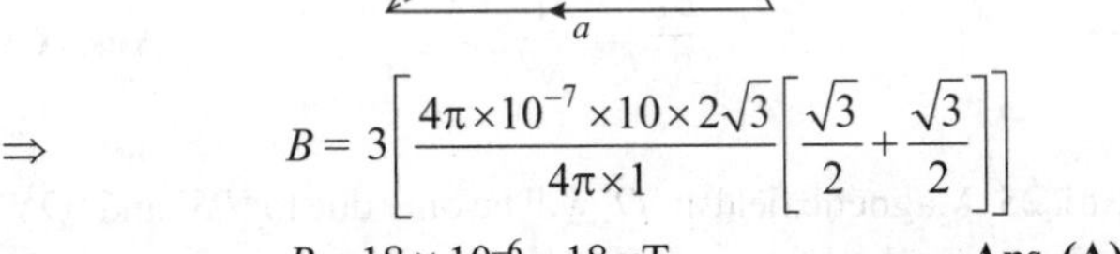

$$\Rightarrow \quad B = 3\left[\frac{4\pi\times10^{-7}\times10\times2\sqrt{3}}{4\pi\times1}\left[\frac{\sqrt{3}}{2}+\frac{\sqrt{3}}{2}\right]\right]$$

$$\Rightarrow \quad B = 18\times10^{-6} = 18\,\mu T$$ **Ans. (A)**

Sol. 29 Dimensions of given physical quantities are

$$[\varepsilon_0] = M^{-1}\,L^{-3}\,T^4\,A^2$$

$$[\mu_0] = M\,L\,T^{-2}\,A^{-2}$$

$$[R] = M\,L^2\,T^{-3}\,A^{-2}$$

$$\Rightarrow \quad [R] = \left[\sqrt{\frac{\mu_0}{\varepsilon_0}}\right]$$ **Ans. (C)**

Sol. 30 Magnetic field due to a rotating ring is given as

$$B = \frac{\mu_0 i}{2R} = \frac{\mu_0 q\omega}{2R2\pi}$$

$$\Rightarrow \quad q = 3\times10^{-5}\,C$$ **Ans. (B)**

Sol. 31

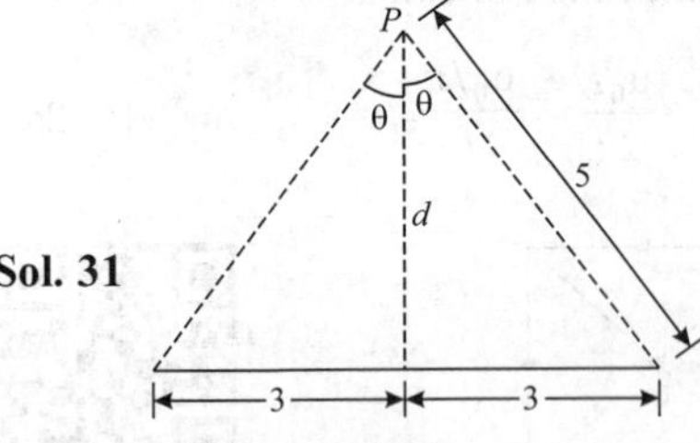

Magnetic field at point P is given as

$$B = \frac{\mu_0 I}{4\pi d}(2\sin\theta)$$

Here $\quad d = 4$ cm and $\sin\theta = \frac{3}{5}$

$$\Rightarrow \quad B = \frac{10^{-7}\times5}{0.04}\times2\times\frac{3}{5} = 1.5\times10^{-5}\,T$$ **Ans. (D)**

Sol. 32 The given expression is of magnetic energy stored per unit volume so its dimensions are $ML^{-1}\,T^{-2}$. **Ans. (A)**

Sol. 33 The given wire loop can be considered in three parts A, B and C as shown

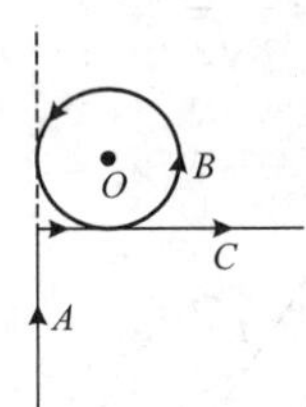

Magnetic field at point O is given as

$$B_0 = B_A + B_B + B_C$$

$$\Rightarrow \quad B_0 = \frac{\mu_0 I}{4\pi R}(\sin 90° - \sin 45°)\otimes + \frac{\mu_0 I}{2R}\odot$$

$$+ \frac{\mu_0 I}{4\pi R}(\sin 45° + \sin 90°)\odot$$

$$\Rightarrow \quad B_0 = \frac{\mu_0 I}{2\pi R}(\sin 45° + \pi)$$

$$\Rightarrow \quad B_0 = \frac{\mu_0 I}{2\pi R}\left(\pi + \frac{1}{\sqrt{2}}\right)$$ **Ans. (C)**

Sol. 34 If current density in wire is J, then magnitude field due to wire is outside is given as

$$B_1 = \frac{\mu_0 I}{2\pi r} = \frac{\mu_0(J\pi a^2)}{2\pi(2a)} = \frac{\mu_0 Ja}{4}$$

Due to wire inside field is given as

$$B_2 = \frac{1}{2}\mu_0 Jr = \frac{1}{2}\mu_0 J\left(\frac{a}{3}\right) = \frac{\mu_0 Ja}{6}$$

$$\Rightarrow \quad \frac{B_2}{B_1} = \frac{\mu_0 Ja\times4}{\mu_0 J6a} = \frac{2}{3}$$ **Ans. (A)**

Sol. 35 In figure shown below, magnetic field due to coil at centre is given as

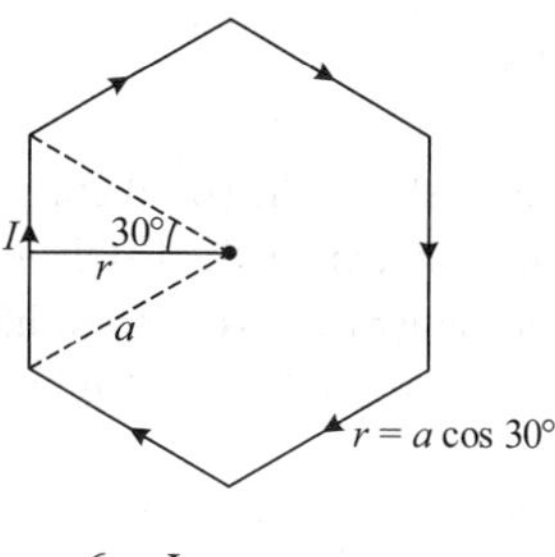

$$B = \frac{6\mu_0 I}{4\pi r}(2\sin 30^\circ) \times 50$$

$$\Rightarrow \quad B = \frac{6\mu_0 I}{4\pi a \cos 30^\circ} \times 2\sin 30^\circ \times 50$$

$$\Rightarrow \quad B = \frac{\mu_0 I}{\pi}\frac{150}{\sqrt{3}a} = \frac{50\sqrt{3}}{0.1}\frac{\mu_0 I}{\pi}$$

$$\Rightarrow \quad B = 500\sqrt{3}\frac{\mu_0 I}{\pi}$$ **Ans. (C)**

Sol. 36 In given figure magnetic fields due to wires A and B are

$$B_A = \frac{\mu_0 I\left(\frac{3\pi}{2}\right)}{4\pi R_A}$$

and $$B_B = \frac{\mu_0 I\left(\frac{5\pi}{3}\right)}{4\pi R_B}$$

$$\Rightarrow \quad \frac{B_A}{B_B} = \frac{I_A(3/2)}{R_A} \times \frac{R_B}{I_B(5/6)}$$

$$\Rightarrow \quad \frac{B_A}{B_B} = \frac{2(3/2)\times 4}{2\times 3\times(5/3)} = \frac{12}{10} = \frac{6}{5}$$ **Ans. (C)**

Sol. 37 The magnetic field on the axis of a current carrying circular coil is is given as

$$B = \frac{\mu_0}{4\pi}\frac{2NIA}{(R^2+x^2)^{3/2}}$$

$$\Rightarrow \quad \frac{B_1}{B_2} = \frac{8}{1} = \left[\frac{R^2+(0.2)^2}{R^2+(0.05)^2}\right]^{3/2}$$

$$\Rightarrow \quad 4[R^2+(0.05)^2] = [R^2+(0.2)^2]$$

$$\Rightarrow \quad 4R^2 - R^2 = (0.2)^2 - 4\times(0.05)^2$$

$$\Rightarrow \quad 3R^2 = 0.3 \times 0.1$$

$$\Rightarrow \quad R^2 = (0.1)^2$$

$$\Rightarrow \quad R = 0.1$$ **Ans. (A)**

Sol. 38 As force on a charge by magnetic field is always perpendicular to $\vec{v}$, the work by magnetic force on the point charge is zero. **Ans. (C)**

Sol. 39 The magnetic flux density (or magnetic induction) due to a long solenoid can be given as

$$B = \mu nI = \mu_0\mu_r nI$$

$$\Rightarrow \quad B = 4\pi \times 10^{-7} \times 500 \times 1000 \times 5 = \pi \text{ Tesla}$$ **Ans. (A)**

Sol. 40 Net magnetic field at point P can be given as

$$B = 2 \times B_{\text{wires}} + B_{\text{loop}}$$

$$\Rightarrow \quad B = 2 \times \frac{\mu_0 i}{4\pi r} + \frac{\mu_0 i}{2r}\left(\frac{\pi}{2\pi}\right)$$

$$\Rightarrow \quad B = \frac{\mu_0 i}{4\pi r}(2+\pi)$$ **Ans. (B)**

Sol. 41 Magnetic induction B due to a current carrying straight wire of circular cross section inside the body of wire is given as

$$B = \frac{\mu_0 Jr}{2}$$

Where current density is given as

$$J = \frac{i}{\pi R^2}$$

Thus in the given curve, slope of straight line is proportional to the current density J and outside the wire magnetic induction varies with inverse relation to distance r. The maximum magnetic induction on surface of wire is given as

$$B_{\max} = \frac{\mu_0 JR}{2} = \frac{\mu_0 iR}{2\pi R^2} = \frac{\mu_0 i}{2\pi R}$$

As $b > a$ so slope of line for second wire is less than that of first wire and maximum magnetic induction will also be more for the wire with smaller radius hence option (B) is most appropriate. **Ans. (B)**

Sol. 42 Figure below shows the vectors of eletric field due to the charges specified in question.

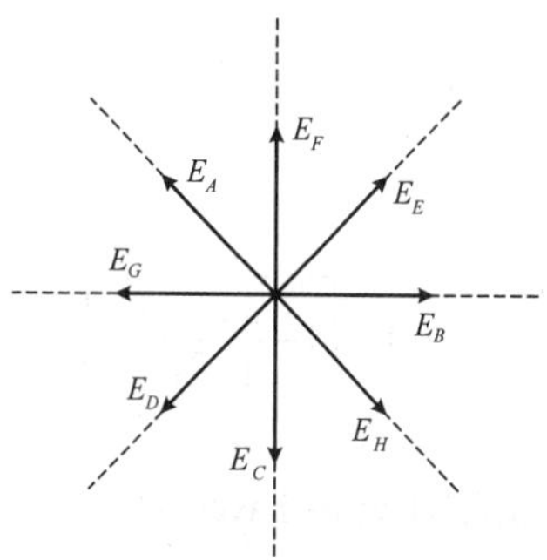

The magnitude of electric fields due to the charges are given as

$$E_A = \frac{kq}{(\sqrt{2}l)^2} = \frac{1}{2}\frac{kq}{l^2}$$

$$E_B = \frac{kq}{(l)^2} = \frac{kq}{l^2}$$

$$E_C = \frac{2kq}{(l)^2} = 2\frac{kq}{l^2}$$

$$E_D = \frac{kq}{(\sqrt{2}l)^2} = \frac{1}{2}\frac{kq}{l^2}$$

$$E_E = \frac{2kq}{(\sqrt{2}l)^2} = \frac{kq}{l^2}$$

$$E_F = \frac{kq}{(l)^2} = \frac{kq}{l^2}$$

$$E_G = \frac{2kq}{(l)^2} = 2\frac{kq}{l^2}$$

$$E_H = \frac{kq}{(\sqrt{2}l)^2} = \frac{1}{2}\frac{kq}{l^2}$$

Calculating the resultant of all fields gives

$$E_{\text{net}} = \sqrt{2}\frac{kq}{l^2} - \frac{kq}{2l^2} = \frac{1}{4\pi\varepsilon_0}\frac{q}{(2l^2)}(2\sqrt{2}-1)$$

Ans. (A)

Sol. 43 Due to a current carrying coil magnetic field at a distance x along its axis is given as

$$B_{\text{axis}} = \frac{\mu_0 ia^2}{2(a^2+r^2)^{3/2}}$$

Magnetic field at centre of the coil is given as

$$B_{\text{centre}} = \frac{\mu_0 i}{2R}$$

Fractional change in magnetic field is given as

$$f = \frac{\dfrac{\mu_0 i}{2a} - \dfrac{\mu_0 ia^2}{2(a^2+r^2)^{3/2}}}{\dfrac{\mu_0 i}{2a}}$$

$$\Rightarrow \quad f = 1 - \frac{1}{\left[1+\left(\dfrac{r^2}{a^2}\right)\right]^{3/2}} \quad (\text{As } r << a)$$

$$\Rightarrow \quad f \approx 1 - \left[1 - \frac{3}{2}\frac{r^2}{a^2}\right] = \frac{3}{2}\frac{r^2}{a^2}$$

Ans. (B)

Sol. 44 By Ampere's law, we have

$$B_1 \times 2\pi(x) = \frac{\mu_0 I_0 (x)^2}{a^2}$$

and $\quad B_2 \times 2\pi(x) = \mu_0 I_0$

$$\Rightarrow \quad \frac{B_1}{B_2} = \frac{x^2}{a^2}$$

Ans. (D)

Sol. 45 Figure below shows the spiral coil as described in the question. To calculate magnetic field at the center we consider an elemental circular coil of radius x and width dx in it as shown. The number of turns in this elemental coil is given as

$$dN = \frac{N}{b-a}dx$$

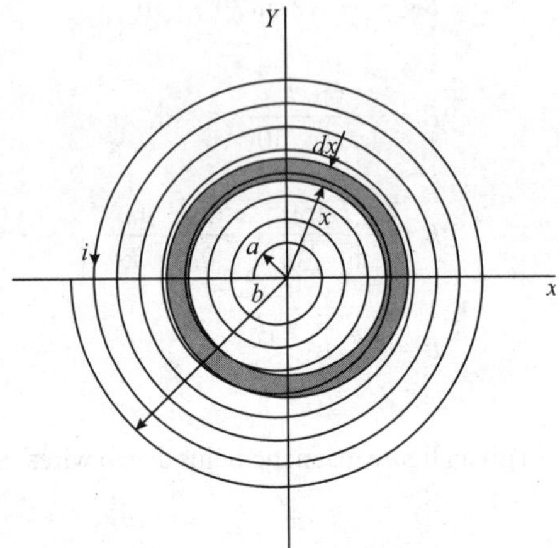

The magnetic induction due to the elemental coil is given as

$$dB = \frac{\mu_0 i}{2x} \times dN$$

Total magnetic induction due to whole coil at center is given as

$$B = \int \frac{\mu_0 dNi}{2x} = \int \frac{\mu_0\left(\dfrac{N}{b-a}dx\right)i}{2x}$$

$$\Rightarrow \quad B = \frac{\mu_0 Ni}{2(b-a)} ln\frac{b}{a}$$

Ans. (B)

Sol. 46 Figure below shows the triangular coil as described in the question

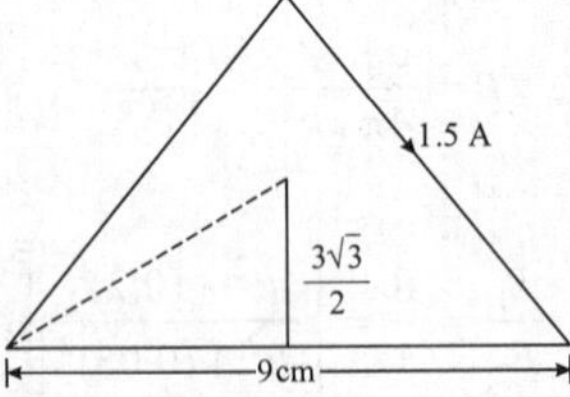

The magnetic field at the centroid of the triangle due to one side of triangle is given as

$$B_1 = \frac{\mu_0 I}{4\pi R}(2\sin 60°) = \frac{4\pi\times10^{-7}\times1.5}{4\pi.\left(\dfrac{3\sqrt{3}}{2}\right)\times10^{-2}}\times2\times\frac{\sqrt{3}}{2} = 1\times10^{-5}$$

Magnetic field due to whole triangle is given as

$$B_{net} = 3B_1 = 3 \times 10^{-5}\,T$$

As the current is in clockwise manner, direction of field will be into the plane of triangle. **Ans. (D)**

Sol. 47 The magnetic field due to the wire at point P is given as

$$B = \frac{\mu_0 I}{4\pi y}(1+\sin\theta_1) + \frac{\mu_0 I}{4\pi x}(1+\sin\theta_2)$$

$$\Rightarrow \quad B = \frac{\mu_0 I}{4\pi y}\left(1+\frac{x}{\sqrt{x^2+y^2}}\right) + \frac{\mu_0 I}{4\pi x}\left(1+\frac{y}{\sqrt{x^2+y^2}}\right)$$

$$\Rightarrow \quad B = \frac{\mu_0 I}{4\pi x} + \frac{\mu_0 I}{4\pi y} + \frac{\mu_0 I}{4\pi}\left[\frac{x}{y\sqrt{x^2+y^2}} + \frac{y}{x\sqrt{x^2+y^2}}\right]$$

$$\Rightarrow \quad B = \frac{\mu_0 I}{4\pi}\left[\frac{1}{x}+\frac{1}{y}\right] + \frac{\mu_0 I}{4\pi}\frac{\sqrt{x^2+y^2}}{xy}$$

$$\Rightarrow \quad B = \frac{\mu_0 I}{4\pi}\left[\frac{x+y+\sqrt{x^2+y^2}}{xy}\right]$$ **Ans. (D)**

Sol. 48 Magnetic field at the centre of a circular coil of radius r is given as

$$B = \frac{\mu_0 I}{2r}$$

At an axial point at a distance x from centre of coil, magnetic field is given as

$$B_a = \frac{\mu_0 I r^2}{2(x^2+r^2)^{3/2}}$$

At $x = \frac{r}{2}$, we have

$$B_a = \frac{\mu_0 I r^2}{2\left(\frac{r^2}{4}+r^2\right)^{3/2}}$$

$$\Rightarrow \quad B_a = \frac{\mu_0 I r^2}{2\left(\frac{5}{4}r^2\right)^{3/2}} \frac{\mu_0 I}{2r}\left(\frac{4}{5}\right)^{3/2}$$

$$\Rightarrow \quad B_a = B\left(\frac{2}{\sqrt{5}}\right)^3$$ **Ans. (C)**

Sol. 49 Magnetic field inside the wire at a dsitance r is considered as B and current is uniformly distributed over its cross section then curren in the inner cylinder as shown in figure below is given as

$$i = \frac{I}{\pi R^2}\cdot \pi r^2$$

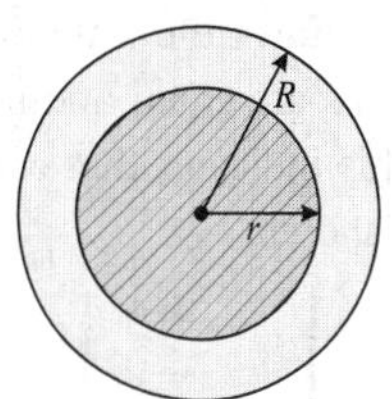

Using Ampere's law, we have

$$B.2\pi r = \mu_0 \cdot \frac{I}{\pi R^2}\cdot \pi r^2$$

$$\Rightarrow \quad B \propto r$$ **Ans. (B)**

Sol. 50 Magnetic field inside a long solenoid is given as

$$B_1 = \mu_0 n I$$

If number of turns per unitlength is halved and current is doubled, new magnetic field is given as

$$B_2 = \mu_0\left(\frac{n}{2}\right)(2I)$$

$$\Rightarrow \quad B_1 = B_2$$ **Ans. (A)**

Sol. 51 The force experienced by a length L of the wire is given as

$$F = \frac{\mu_0 I_1 I_2 L}{2\pi d}$$

$$\Rightarrow \quad 10^{-5} = \frac{2\times10^{-7}\times5\times5}{d}\times 10\times10^{-2}$$

$$\Rightarrow \quad d = 50\times10^{-3}\,m = 5\,cm$$ **Ans. (5)**

Sol. 52 For the hollow cylindrical wire as shown in figure below, its inside mangetic field is zero as no current is enclosed in a symmetrically chosen Ampere's path inside it, so we use

For $r < R$

$$B_p = 0$$

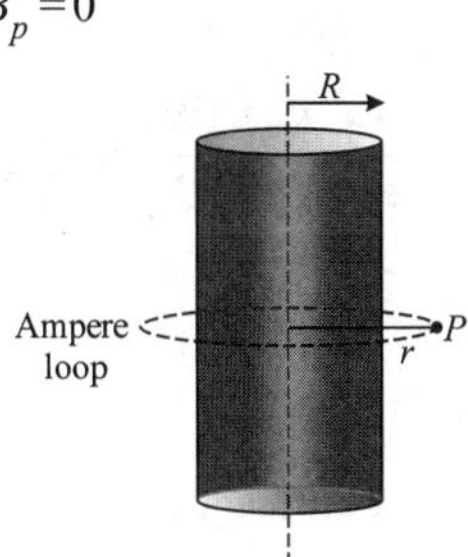

For $r \geq R$, using a coaxial Ampere's loop outside the wire as shown in figure, we get

$$B_p = \frac{\mu_0 1}{2\pi r}$$

$$\Rightarrow \quad B_p \propto \frac{1}{r}$$

Thus option (D) is most appropriate one. **Ans. (D)**

Sol. 53 Figure below shows the situation described in the question. As the currents are equal and magnetic field at mid point is non-zero so this indicates that currents are flowing in opposite direction as shown

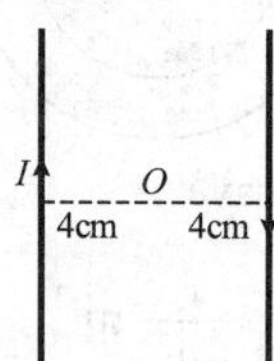

B at O is calculated by sum of the magnetic fields produced by the two wires as opposite currents will produce magnetic fields at O in same direction, given as

$$B_o = 2\frac{\mu_0 I}{2\pi r}$$

$$\Rightarrow \quad \frac{2\times 4\pi\times 10^{-7} I}{2\pi(4\times 10^{-2})} = 3\times 10^{-4}\,\text{T}$$

$$\Rightarrow \quad I = 30\,\text{A}$$

Ans. (B)

Sol. 54 Magnetic field at centre of a current carrying coil having N turns is given as

$$B = \frac{N\mu_0 i}{2R}$$

$$\Rightarrow \quad B_1 = \frac{N_1\mu_0 i}{2R_1}$$

For $N_2 = 5$, the radious of coil will be changed for same length of wire, given as

$$2\pi R_2 N_2 = 2\pi R_1 N_1$$

$$\Rightarrow \quad R_2 = \frac{N_1 \times R_1}{N_2}$$

New magnetic field at the centre of coil is given as

$$B_2 = \frac{N_2\mu_0 i}{R_2}$$

$$\Rightarrow \quad \frac{B_2}{B_1} = \frac{N_2}{N_1}\frac{R_1}{R_2} = \frac{N_2}{N_1}\times\frac{N_2}{N_1};$$

$$\Rightarrow \quad \frac{B_2}{B_1} = \frac{25}{4}$$

Ans. (B)

Sol. 55 At centre of a coil magnetic field due to N turns is given as

$$B = N\left(\frac{\mu_0 i}{2R}\right)$$

$$\Rightarrow \quad B_X = 200\left(\frac{\mu_0 i}{2\times 20\,\text{cm}}\right)$$

and $$B_Y = 400\left(\frac{\mu_0 i}{2\times 20\,\text{cm}}\right)$$

NOTE : As ratio of B_X and B_Y is to be caluclated, radius is taken in same units

$$\Rightarrow \quad \frac{B_x}{B_y} = \frac{1}{2}$$

Ans. (B)

Sol. 56 Figure below shows the situation described in question for a circular coil and magnetic fields at centre and the given axial point are given as

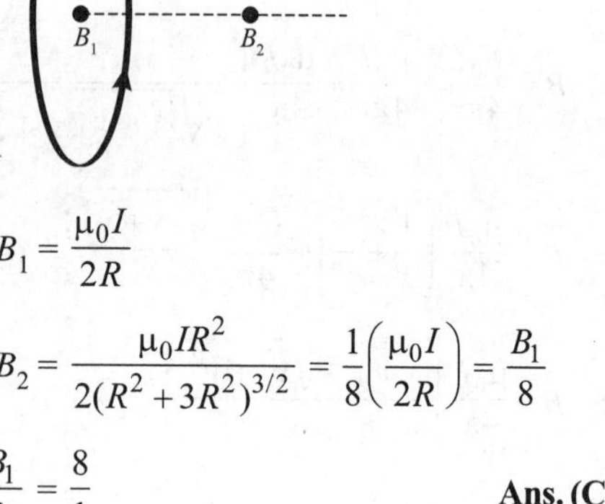

$$B_1 = \frac{\mu_0 I}{2R}$$

and $$B_2 = \frac{\mu_0 I R^2}{2(R^2+3R^2)^{3/2}} = \frac{1}{8}\left(\frac{\mu_0 I}{2R}\right) = \frac{B_1}{8}$$

$$\Rightarrow \quad \frac{B_1}{B_2} = \frac{8}{1}$$

Ans. (C)

Sol. 57 Magnetic field at the centre of a coil having N turns is given as

$$B_{\text{centre}} = \frac{N\mu_0 1}{2R}$$

$$\Rightarrow \quad 37.68\times 10^{-4} = \frac{100\times 4\pi\times 10^{-7}\times I}{2\times 5\times 10^{-2}}$$

$$\Rightarrow \quad I = 3\text{A}$$

Ans. (3)

Sol. 58 Magnetic field due to current carrying circular loop at a point on its axis is given as

$$B = \frac{\mu_0 i r^2}{2(r^2+x^2)^{3/2}}$$

At centre of loop we use $x = 0$, so magnetic field is given as

$$B_1 = \frac{\mu_0 i}{2r}$$

At a point on axis at $x = r$, magnetic field is given as

$$B_2 = \frac{\mu_0 i}{2\times 2\sqrt{2}r}$$

$$\Rightarrow \quad \frac{B_1}{B_2} = 2\sqrt{2}$$

Ans. (C)

Sol. 59 Magnetic field due to a long solenoid inside is given as

$$B = \mu_0 n I$$

$$\Rightarrow \quad B = 4\pi \times 10^{-7} \times 70 \times 10^2 \times 2$$

$$\Rightarrow \quad B = 56\pi \times 10^{-4}\,\text{T} = 176 \times 10^{-4}\,\text{T}$$ **Ans. (B)**

Sol. 60 (a) The magnetic induction due to the circular loop and the straight wire segments at point O is given as

$$B_{ab} = \frac{\mu_0}{4\pi}\frac{I}{r} \text{ (out of the plane)}$$

$$B_{bcd} = \frac{\mu_0 I}{2r} \text{ (in the plane)}$$

$$B_{de} = \frac{\mu_0}{4\pi}\frac{I}{r} \text{ (out of the plane)}$$

Hence magnetic field at O is calculated as

$$B_0 = -\frac{\mu_0}{4\pi}\frac{I}{r} + \frac{\mu_0 I}{2r} - \frac{\mu_0}{4\pi}\frac{I}{r}$$

$$B_0 = \frac{\mu_0}{2\pi}\frac{I}{r}(\pi - 1) \quad \ldots\text{(iii)}$$

(b) The magnetic induction due to the semicircular loop and the straight wire segments at point O is given as

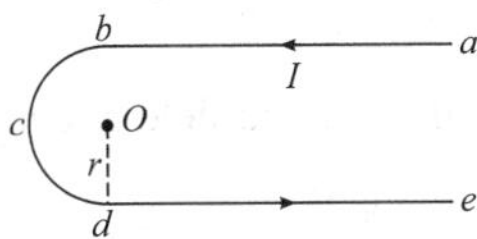

$$B_{ab} = \frac{\mu_0}{4\pi}\frac{I}{r} \text{ (out of the plane)}$$

$$B_{bcd} = \frac{\mu_0 I}{4r} \text{ (out of the plane)}$$

$$B_{de} = \frac{\mu_0}{4\pi}\frac{I}{r} \text{ (out of the plane)}$$

Hence magnetic field at O is calculated as

$$B_0 = \frac{\mu_0}{4\pi}\frac{I}{r} + \frac{\mu_0 I}{4r} + \frac{\mu_0}{4\pi}\frac{I}{r}$$

$$B_0 = \frac{\mu_0}{4\pi}\frac{I}{r}(\pi + 2) \quad \ldots\text{(i)}$$

(c) The magnetic induction due to the semicircular loop and the straight wire segments at point O is given as

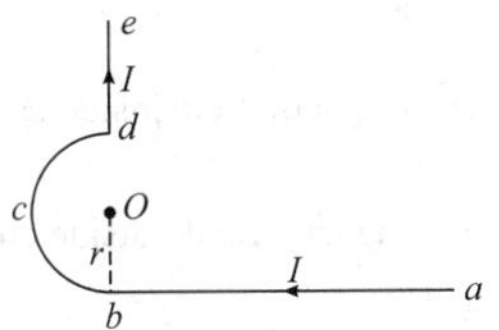

$$B_{ab} = \frac{\mu_0}{4\pi}\frac{I}{r} \text{ (into the plane)}$$

$$B_{bcd} = \frac{\mu_0 I}{4r} \text{ (into the plane)}$$

$$B_{de} = 0$$

Hence magnetic field at O is calculated as

$$B_0 = \frac{\mu_0}{4\pi}\frac{I}{r}(1 + \pi) \quad \ldots\text{(iv)}$$

(d) The magnetic induction due to the semicircular loop and the straight wire segments at point O is given as

$$B_{ab} = 0$$

$$B_{bcd} = \frac{\mu_0 I}{4r} \text{ (out of the plane)}$$

$$B_{de} = 0$$

Hence magnetic field at O is calculated as

$$B_0 = \frac{\mu_0 I}{4r} \quad \ldots\text{(ii)}$$

Ans. (C)

Sol. 61 Magnetic field at centre inside the solenoid is given by

$$B = \mu_0 nI$$

Thus magnetic field intensity at centre is calculated as

$$H = \frac{B}{\mu_0} = nI = \left(\frac{1200}{2}\right)(2)$$

$$\Rightarrow \quad H = 1.2 \times 10^3\,\text{Am}^{-1}$$ **Ans. (B)**

Sol. 62 Figure below is showing the two wires as described in the question

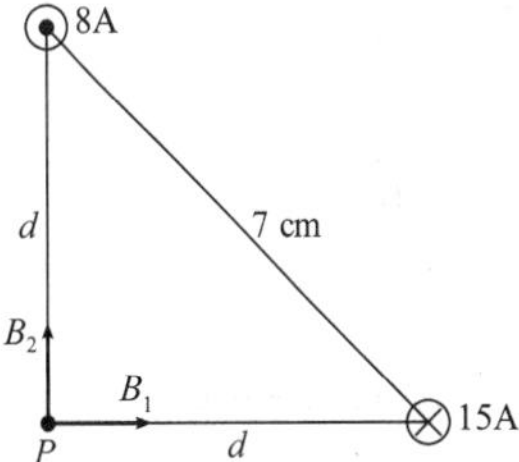

Magnetic fields due to both wires at point P will be perpendicular to each other which are given as

$$B_1 = \frac{\mu_0 i_1}{2\pi d}$$

and $$B_2 = \frac{\mu_0 i_2}{2\pi d}$$

Resulting magnetic field at point P is given as

$$B_{net} = \sqrt{B_1^2 + B_2^2}$$

$$\Rightarrow \quad B_{net} = \frac{\mu_0}{2\pi d}\sqrt{i_1^2 + i_2^2}$$

$$\Rightarrow \quad B_{net} = \frac{4\pi \times 10^{-7}}{2\pi \times (7/\sqrt{2}) \times 10^{-2}} \times \sqrt{8^2 + 15^2}$$

$$\Rightarrow \quad B_{net} = 68 \times 10^{-6}\text{ T}$$

Ans. (68)

Sol. 63 Figure below shows a magnetic line passing through the plane of coil a distance a from the central axis of coil.

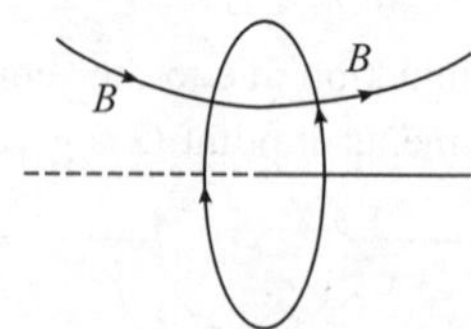

We can see that magnetic field component B_y in the plane of coil is zero and B_y is in opposite direction of each other in $-z$ and $+z$ position on the magnetic line as shown in figure above.

Ans. (C)

Sol. 64 Magnetic field due to current in wire segments BC and ET are outward at point O and of magnitude given as

$$B_0 = \frac{\mu_0 i}{4\pi r} + \frac{\mu_0 i}{4\pi r} = \frac{\mu_0 i}{2\pi r} = \frac{\mu_0 i}{\pi a}$$

Ans. (D)

Sol. 65 Magnetic induction due to the circular coil having four turns is given as

$$B = \frac{\mu_0 i}{2R} \times 4 = 32$$

For the single turn coil made by same wire now radius will become $R' = 4R$, the magnetic induction at centre is given as

$$B' = \frac{\mu_0 i}{2R'}$$

$$\Rightarrow \quad B' = \frac{\mu_0 i}{8R}$$

$$\Rightarrow \quad \frac{B'}{B} = \frac{1}{16}$$

$$\Rightarrow \quad B' = 2\text{T}$$

Ans. (C)

Sol. 66 In the given figure, we use

$$d \tan 60° = 2\sqrt{3}$$

$$\Rightarrow \quad d = 2\text{ cm}$$

Magnetic field at the centroid O of triangle is calculated as

$$B = 3 \times \frac{\mu_0 i}{2\pi d} \sin 60°$$

$$\Rightarrow \quad B = 3 \times \frac{2 \times 10^{-7} \times 2}{2 \times 10^{-2}} \times \frac{\sqrt{3}}{2}$$

$$\Rightarrow \quad B = 3\sqrt{3} \times 10^{-5}\text{T}$$

Ans. (D)

Sol. 67 Flux linked with the windings is calculated as

$$\phi = \mu_r \mu_0 \frac{N}{l} IA$$

$$\Rightarrow \quad \mu_r = 125$$

Ans. (C)

Sol. 68 Long wire of length l is bent into a coil of N turns and of radius r then we have

$$l = (2\pi r)n$$

$$\Rightarrow \quad r \propto \left(\frac{l}{n}\right)$$

Magnetic field at the centre of coils is given as

$$B = n\left(\frac{\mu_0 i}{2r}\right) \propto \left(\frac{\mu_0 i}{2l}\right) n^2$$

So ratio of magnetic fields is calculated as

$$\Rightarrow \quad \frac{B_1}{B_2} = \left(\frac{N^2}{n^2}\right)$$

Ans. (C)

Sol. 69 The magnetic induction at the point P is given by the sum of magnetic fields due to the three wires, two straight and one semicircular. In these also the magnetic field due to vertical wire is zero at point P as it lies on the line of extension of this wire. For rest two wires magnetic field is given as

$$B_P = \left(\frac{\mu_0 i}{4r} + \frac{\mu_0 i}{4\pi r}\right) = \frac{\mu_0 i}{2r}\left(\frac{I}{2} + \frac{I}{2\pi}\right)$$

Ans. (C)

Sol. 70 Magnitude of magnetic field at centre O is given as

$$B_C = \frac{\mu_0 I\alpha}{4\pi R}$$

$$\Rightarrow \quad B_C = \frac{\mu_0 I}{4\pi R}(\pi)$$

$$\Rightarrow \quad B_C = \frac{\mu_0 I}{4R} = \frac{4\pi \times 10^{-7} \times 3}{4 \times \frac{\pi}{10}}$$

$$\Rightarrow \quad B_C = 3 \times 10^{-6}\text{ T} = 3\ \mu\text{T}$$

Ans. (D)

Sol. 71 Magnetic flux through a coil of cross section area $\vec{A}$ is calculated as

$$\phi = \vec{B} \cdot \vec{A}$$

Here θ is the angle between $\vec{B}$ and $\vec{A}$ so we have

$$\Rightarrow \quad \phi = BA \cos\theta$$

Most suitable answer is (B) [Otherwise $ABCD$].

Ans. (B)

Sol. 72 Magnetic field due to a wire at different locations in surrounding is given as

$$B = \begin{cases} \dfrac{\mu_0 I r}{\pi a^2} & r \le a\text{; Magnetic field inside cross-section of wire.} \\ \dfrac{\mu_0 I}{\pi r^2} & r \ge a\text{; Magnetic field outside cross-section of wire.} \end{cases}$$

Ans. (C)

Sol. 73 Magnetic field B_C at center of circular coil is given as

$$B_C = \frac{\mu_0 i}{2r}$$

$$\Rightarrow \quad B_C = \frac{4\pi \times 10^{-7}}{2 \times 0.2} \times \sqrt{2}T$$

As two coils are perpendicular to each other, then magnetic field due to each coil will also be perpendicular. Thus, net magnetic field is calculated as

$$B_C\sqrt{2} = \frac{4\pi \times 10^{-7} \times \sqrt{2}}{2 \times 0.2} \times \sqrt{2}T$$

$$B_C = 2\pi \times 10^{-6}\,\text{T}$$

$$\Rightarrow \quad B_C = 200\,\pi \times 10^{-8}\,\text{T}$$

$$\Rightarrow \quad B_C = 2 \times 314 \times 10^{-8}\,\text{T}$$

$$\Rightarrow \quad B_C = 628 \times 10^{-8}\,\text{T}$$

Ans. (628)

Sol. 74 Magnetic intensity at the centre of a long current carrying solenoid.

$$H = \frac{B}{\mu_0} = \frac{\mu_0 ni}{\mu_0} = ni$$

Thus current flowing through solenoid

$$i = \frac{H}{n} = \frac{1.6 \times 10^3}{\left(\frac{8}{10^{-2}}\right)}$$

$$\Rightarrow \quad i = 2\text{A}$$

Ans. (2)

Sol. 75 Figure below shows the coil as described in the question.

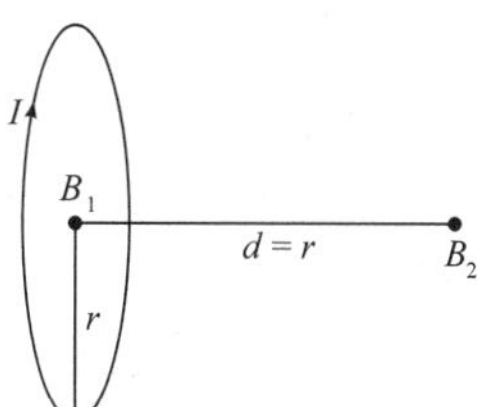

Magnetic field at centre of coil is given as

$$B_1 = \frac{\mu_0 I}{2r}$$

Magnetic field on axis of coil is given as

$$B_2 = \frac{\mu_0 I r^2}{2(r^2 + d^2)^{3/2}}$$

Magnetic field at $d = r$ is calculated as

$$B_2 = \frac{\mu_0 I}{4\sqrt{2}r}$$

$$\Rightarrow \quad \frac{B_1}{B_2} = \frac{\mu_0 I}{2r} \times \frac{4\sqrt{2}r}{\mu_0 I} = \frac{2\sqrt{2}}{1} = \frac{\sqrt{8}}{1}$$

$$\Rightarrow \quad x = 8$$

Ans. (8)

Sol. 76 Magnetic field at O due to straight wires is zero as for axial pointa $Q = 0$ and $B = 0$. Due to semicircular part magnetic field at O is given as

$$\bar{B}_0 = \frac{\mu_0 I \alpha}{4\pi R}$$

At $\alpha = \frac{\pi}{2}$, the magnetic field is calculated as

$$\bar{B}_0 = \frac{\mu_0 I}{8R}$$

$$\Rightarrow \quad B_O = \frac{\mu_0 I}{4R} = \frac{4\pi \times 10^{-7} \times 14}{4 \times 2.2 \times 10^{-2}}$$

$$\Rightarrow \quad B_O = 2 \times 10^{-4}\,\text{T}$$

Ans. (2)

Sol. 77 Magnetic field due to moving charge is given as

$$B = \frac{\mu_0}{4\pi}\frac{qv\sin\theta}{r^2}$$

For the motion on circular path, we have $\theta = \frac{\pi}{2}$

$$B = \frac{\mu_0}{4\pi}\frac{ev\sin(\pi/2)}{r^2}$$

$$\Rightarrow \quad B = \frac{10^{-7} \times 1.6 \times 10^{-19} \times 6.76 \times 10^6}{0.52 \times 0.52 \times 10^{-20}}$$

$$\Rightarrow \quad B = 40\,\text{T}$$

Ans. (40)

18.2 Magnetic Force on Moving Charges

Sol. 1 If B_1 is the magnetic induction at the location of all dl on wire 2 due to wire 1, than force on dl is given as

$$F = B_1 i_2 dl$$

Here $\quad B_1 = \frac{\mu_0 i_1}{4\pi r}(\cos\theta + \cos\theta)$

[We use the expression of $B(r)$ due to a finite length wire]

$$\Rightarrow \quad F = \frac{\mu_0 i_1 i_2 dl}{4\pi r}(2\cos\theta)$$

$$\Rightarrow \quad F = \frac{\mu_0 i_1 i_2 dl}{2\pi r}\cos\theta$$

Ans. (C)

Sol. 2 In uniform magnetic field the radius of circular motion is given as

$$r = mv/qB$$

$\Rightarrow$ Time period of revolution is given as

$$T = \frac{2\pi r}{v} = \frac{2\pi m}{qB}$$

$\Rightarrow$ T is independent of v.

Ans. (A)

Sol. 3

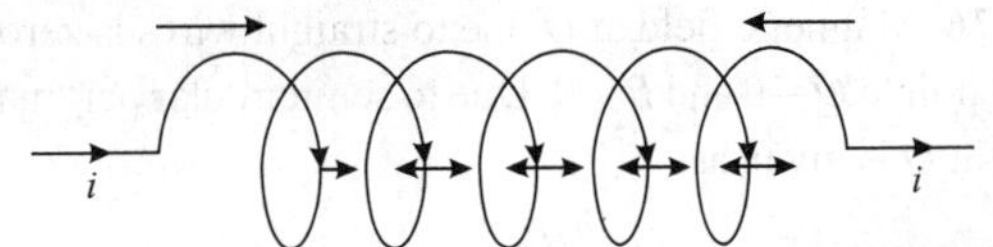

Due to parallel currents in each turn of spring due to same direction of current flow all turns attract each other.

$\Rightarrow$ spring will get compressed. **Ans. (B)**

Sol. 4 When a charge enters in magnetic field normally then it follows a circular path of radius R given as

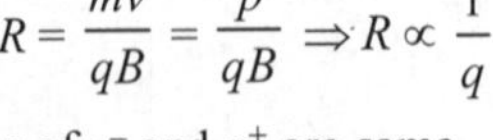

$$R = \frac{mv}{qB} = \frac{p}{qB} \Rightarrow R \propto \frac{1}{q}$$

As charges of e^- and p^+ are same

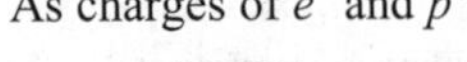

$\Rightarrow \quad R_e = R_p$

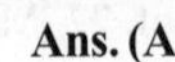

Ans. (A)

Sol. 5

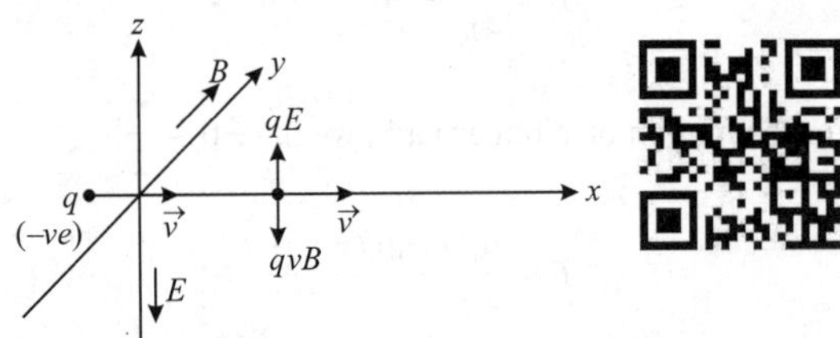

As particle continues motion in x-direction that means electric and magnetic forces on it are balanced.

$$\Rightarrow \quad qE = qvB$$

$$\Rightarrow \quad B = \frac{E}{v} = \frac{10^4}{10} = 10^3 \text{ Wb/m}^2$$ **Ans. (A)**

Sol. 6 Work done in turning a magnetic needle by angle $\theta_1 = 60°$ to $\theta_2 = 60°$ is given as

$$W = MB(\cos\theta_1 - \cos\theta_2)$$

$$W = MB(1 - 1/2) = \frac{1}{2}MB \quad \ldots(1)$$

Torque needed to hold the needle at $\theta_2 = 60°$ to the direction of magnetic field is

$$\tau = MB\sin\theta = MB\left(\frac{\sqrt{3}}{2}\right)$$

$$\Rightarrow \quad \tau = \sqrt{3}W$$ **Ans. (A)**

Sol. 7

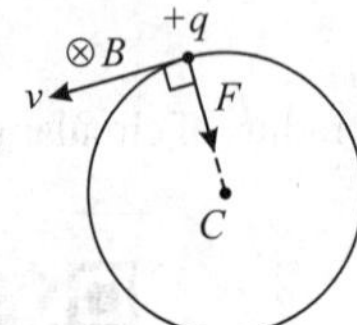

Magnetic force on a moving charge

$$\vec{F} = q(\vec{v} \times \vec{B})$$

as $\vec{F}$ is always $\perp$ to $\vec{v}$

$\Rightarrow$ For a freely moving charge in Magnetic force, work done is always equal to zero. **Ans. (B)**

Sol. 8 Case-I :

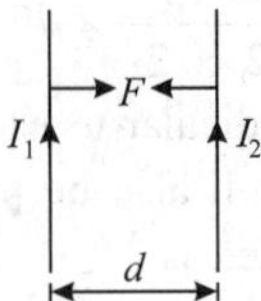

attractive force per unit length between wires is given as

$$F = \frac{\mu_0 I_1 I_2}{4\pi d} \quad \ldots(1)$$

Case-II :

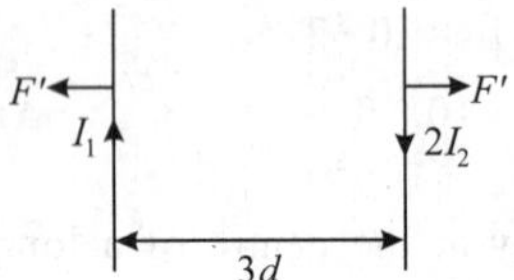

Repulsive force per unit length between wires is given as

$$F' = -\frac{\mu_0 I_1 (2I_2)}{4\pi(3d)} \quad \ldots(2)$$

By (2)/(1) $\Rightarrow \quad \frac{F'}{F} = -\frac{2}{3} \Rightarrow F' = -\frac{2}{3}F$ **Ans. (C)**

Sol. 9 For perpendicular motion of a charge particle in magnetic field, we use

$$qvB = \frac{mv^2}{r}$$

$$\Rightarrow \quad r = \frac{mv}{qB}$$

Time period of one revolution

$$T = \frac{2\pi r}{v} = \frac{2\pi}{v} \times \frac{mv}{qB} = \frac{2\pi m}{qB}$$ **Ans. (D)**

Sol. 10

$B_1 \rightarrow$ magnetic induction due to wire 1 at the location of wire 2.

Force per unit length between the wires is given as

$$F = B_1 il = \left(\frac{\mu_0 i}{2\pi d}\right) il$$

$$\Rightarrow \quad \frac{F}{l} = \frac{\mu_0 i^2}{2\pi d}$$ **Ans. (A)**

Sol. 11

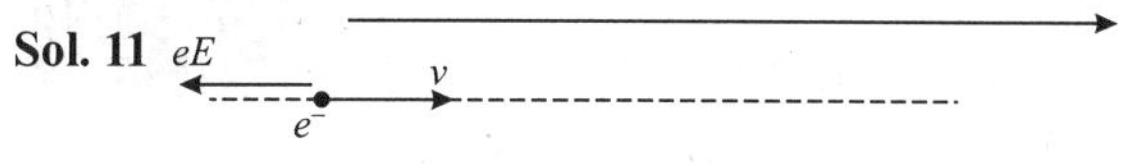

Acceleration of electron

$$a = -\frac{eE}{m}$$

As $\vec{v}$ is in same direction as $\vec{B}$ so no magnetic force will act on electron but electric force due to $\vec{E}$ in direction opposite to $\vec{v}$ so $|\vec{v}|$ will start decreasing. **Ans. (B)**

Sol. 12

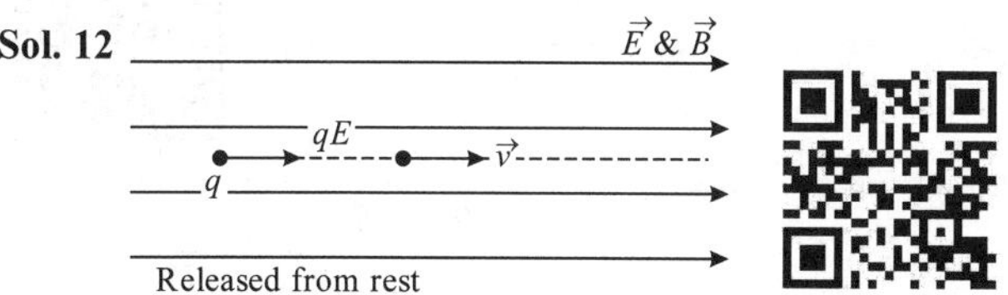

Charge on releasing will start moving due to electric force and it moves along the direction of $\vec{E}$ and $\vec{B}$. At every point the velocity $\vec{v}$ of particle is parallel to magnetic induction $\vec{B}$.

$\Rightarrow$ $F_{magnetic} = 0 \Rightarrow$ path is a straight line **Ans. (B)**

Sol. 13 Whenever a charge moves perpendicular to the direction of magnetic field its path is a circular motion.

$\Rightarrow$ direction of its $\vec{v}$ continuously change and as magnitude force is $\perp$ to motion so no work is done on particle.

$\Rightarrow$ $|\vec{v}| \rightarrow$ constant

$\Rightarrow$ KE $\rightarrow$ constant **Ans. (A)**

Sol. 14

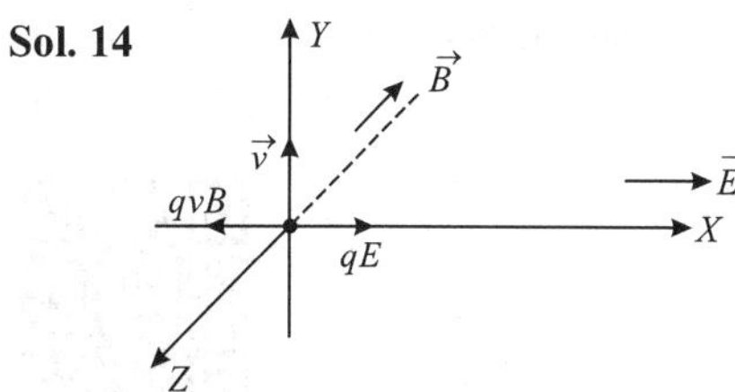

As change comes out without any change in magnitude & direction of $\vec{v}$.

$$\Rightarrow \quad qE = qvB$$

$$\Rightarrow \quad v = \frac{E}{B}$$

Velocity vector will be $\vec{v} = \frac{E}{B}\hat{v}$

Here we can use $\hat{v} = \frac{(\vec{E} \times \vec{B})}{EB}$

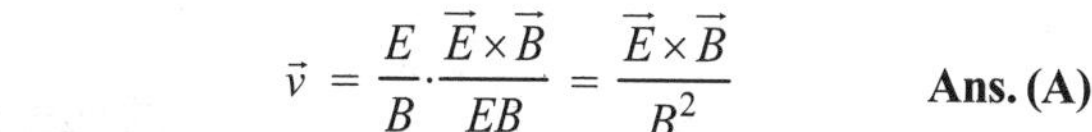

$$\Rightarrow \quad \vec{v} = \frac{E}{B}\cdot\frac{\vec{E}\times\vec{B}}{EB} = \frac{\vec{E}\times\vec{B}}{B^2}$$ **Ans. (A)**

Sol. 15 For uniformly charged & uniform object when rotates then its ratio of magnetic moment to angular momentum remain constant it is given as

$$\frac{\mu}{L} = \frac{q}{2m}$$

For disc, we use $q = \sigma \times \pi R^2$ & angular momentum $L = \left(\frac{1}{2}mr^2\right)\omega$

$$\Rightarrow \quad \mu = \frac{\sigma(\pi R^2)}{2m} \times \frac{1}{2}mR^2\omega$$

$$\Rightarrow \quad \mu = \frac{1}{4}\sigma\pi R^4\omega$$ **Ans. (D)**

Sol. 16 By Lorentz force equation, the force experienced by the charge is given as

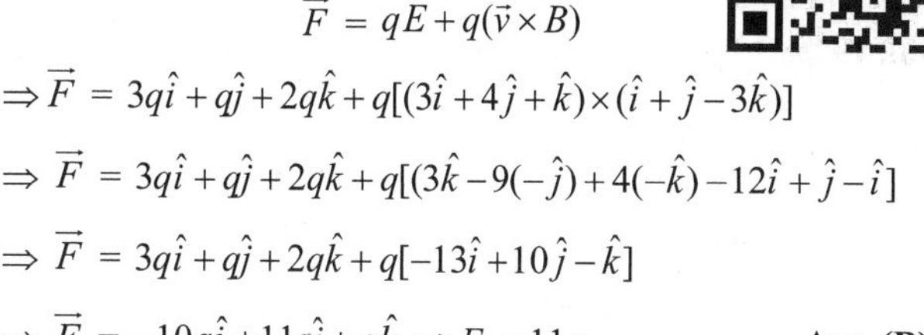

$$\vec{F} = q\vec{E} + q(\vec{v}\times\vec{B})$$

$$\Rightarrow \vec{F} = 3q\hat{i} + q\hat{j} + 2q\hat{k} + q[(3\hat{i} + 4\hat{j} + \hat{k})\times(\hat{i} + \hat{j} - 3\hat{k})]$$

$$\Rightarrow \vec{F} = 3q\hat{i} + q\hat{j} + 2q\hat{k} + q[(3\hat{k} - 9(-\hat{j}) + 4(-\hat{k}) - 12\hat{i} + \hat{j} - \hat{i}]$$

$$\Rightarrow \vec{F} = 3q\hat{i} + q\hat{j} + 2q\hat{k} + q[-13\hat{i} + 10\hat{j} - \hat{k}]$$

$$\Rightarrow \vec{F} = -10q\hat{i} + 11q\hat{j} + q\hat{k} \Rightarrow F_y = 11q$$ **Ans. (B)**

Sol. 17 In magnetic field the radius of circular motion for a charge particle is given as

$$r = \frac{mv}{qB} = \frac{\sqrt{2mk}}{qB}$$

For proton, we use $m = m_p$ & $q = +e \Rightarrow r_p = \frac{\sqrt{2m_pk}}{eB}$

For deutron, we use $m = 2m_p$ & $q = +e$

$$\Rightarrow \quad r_d = \frac{\sqrt{2(2m_p)k}}{eB} = \sqrt{2}r_p$$

For α-particle, we use $m = 4m_p$ & $q = +2e$

$$\Rightarrow \quad r_\alpha = \frac{\sqrt{2(4m_p)k}}{2eB} = r_p$$

$\Rightarrow$ $r_\alpha = r_p < r_d$ **Ans. (C)**

Sol. 18 Initially for equilibrium of wire, we use

$$kx_1 = mg \Rightarrow k \times (0.005) = 0.01 \times 10 \quad \ldots(1)$$

where k = force constant of spring

When wire is subjected to the magnetic field springs further stretch by 0.3 cm.

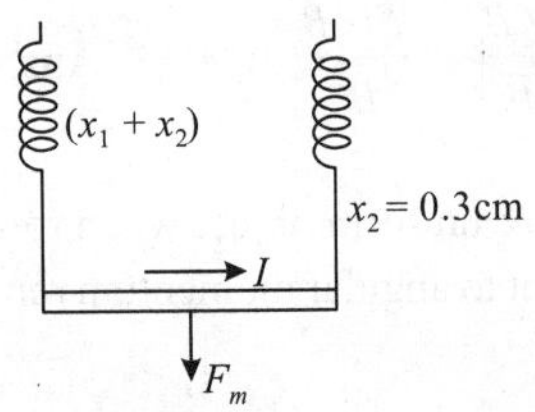

To apply the magnetic force in downward direction the magnetic field must be applied out of the page.

$$F_m = BIl = B \times \left(\frac{24}{12}\right) \times (0.05)$$

The magnetic force is balanced by the spring force due to additional stretch in springs i.e. 0.3 cm.

$\Rightarrow \quad kx_2 = F_m \Rightarrow k \times 0.003 = B \times 2 \times 0.05 \ \ldots(2)$

By (2)/(1) $\Rightarrow \dfrac{B \times 0.1}{0.1} = \dfrac{0.003}{0.005} = \dfrac{3}{5} = 0.6$

$\Rightarrow \quad B = 0.6$ T out of the page **Ans. (A)**

Sol. 19

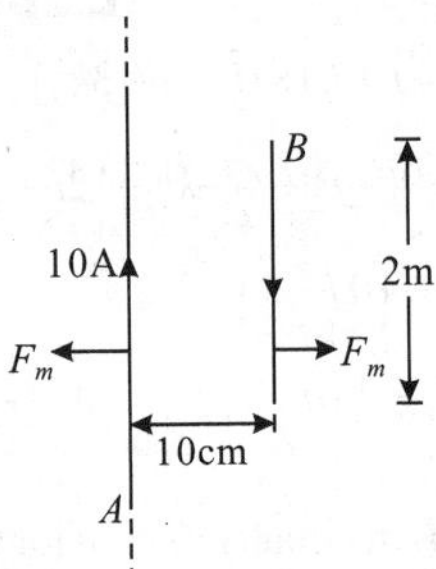

Repulsive force between two opposite and parallel currents is

$$F_m = \frac{\mu_0 I_1 I_2 l}{2\pi d}$$

$$\Rightarrow \quad F_m = \frac{4\pi \times 10^{-7} \times 10 \times 2 \times 2}{2\pi \times 0.1}$$

$\Rightarrow \quad F_m = 8 \times 10^{-5}$ N **Ans. (A)**

Sol. 20 under the influence of crossed magnetic field & electric field the charge particle if pass undeflected.

$\Rightarrow \quad \vec{F_e} = -\vec{F_m}$

$\Rightarrow \quad |\vec{F_e}| = |\vec{F_m}|$

$qE = qvB$

$\Rightarrow \quad v = \dfrac{E}{B} = \dfrac{7.7 \times 10^3}{0.14} = 55 \times 10^3$ m/s

$\Rightarrow \quad v = 55$ km/s **Ans. (C)**

Sol. 21 By Lorentz force equation, the magnetic force on a moving charge is

$$\vec{F} = q(\vec{v} \times \vec{B})$$

$\Rightarrow \quad \vec{F} = 2 \times 10^{-6}[(2\hat{i} + 3\hat{j}) \times 10^6 \times 2\hat{j}]$

$\Rightarrow \quad \vec{F} = 4 \times 10^{-6} \times 10^6[(2\hat{i} + 3\hat{j}) \times \hat{j}]$

$\Rightarrow \quad \vec{F} = 4 \times 2\hat{k} = 8\hat{k}$ N (+z direction) **Ans. (A)**

Sol. 22 In a magnetic field radius of charge particle's circular motion is given as

$$R = \frac{mv}{qB} = \frac{\sqrt{2mk}}{qB}$$

if the particle is accelerated by a potential difference V

$\Rightarrow \quad K = qV$

$\Rightarrow \quad R = \dfrac{\sqrt{2mqV}}{qB}$

for proton $q = e, m = m_p \Rightarrow R_p = \dfrac{1}{B}\dfrac{\sqrt{2m_p eV}}{q}$

for deutron $q = e, m = 2m_p \Rightarrow R_D = \dfrac{1}{B}\dfrac{\sqrt{2(2m_p)eV}}{q} = R$

$\Rightarrow \quad \dfrac{R_p}{R_D} = \dfrac{1}{\sqrt{2}} \Rightarrow R_p = \dfrac{R}{\sqrt{2}}$ **Ans. (B)**

Sol. 23 S-1: In perpendicular motion to a uniform/Non uniform magnetic field as

$$\vec{F}_{\text{magnetic}} = q(\vec{v} \times \vec{B})$$

$\Rightarrow \quad \vec{F}_{\text{magnetic}}$ is $\perp$ to $\vec{v}$ so magnetic force never do work on a moving charge

$\Rightarrow$ KE $\rightarrow$ constant

$\Rightarrow$ S-1 it true

$\vec{F}_{\text{magnetic}}$ is $\perp$ to $\vec{v}$

$\Rightarrow$ **S-2 : it is true and explaining S-1** **Ans. (D)**

Sol. 24

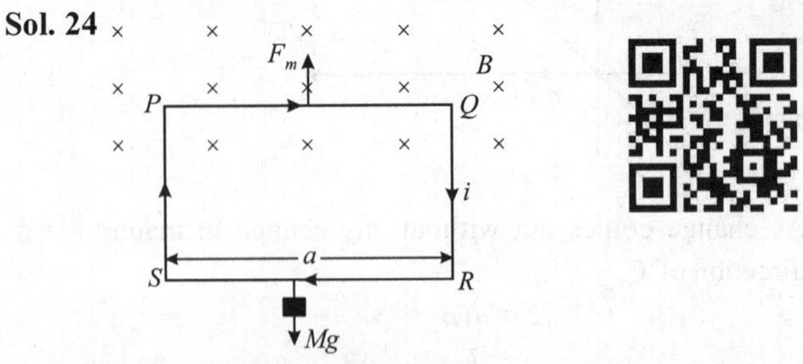

Magnetic force on loop is due to wire PQ, given as

$$F_m = Bia$$

Given that $Bia > mg \Rightarrow$ net force is in upward direction so weight rises up. As net work done by magnetic field is zero and here work is done against gravity. **Ans. (C)**

Sol. 25 As charge is not getting deflected $\Rightarrow$ no net force is acting on it.

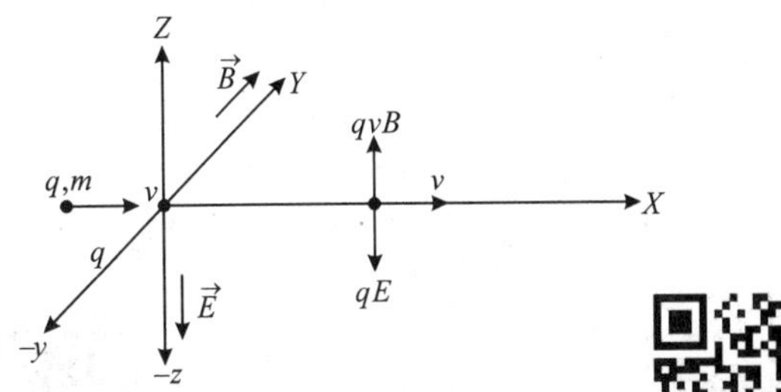

Here we use

$$qE = qvB$$

$$B = \frac{E}{v} = \frac{10^4}{10} = 10^3 \text{ Wb/m}^2 \qquad \textbf{Ans. (C)}$$

Sol. 26 Given that $E_y \propto J_x B_z$

$$\Rightarrow \quad E_y = kJ_xB_z$$

$$\Rightarrow \quad k = \frac{E_y}{J_xB_z}$$

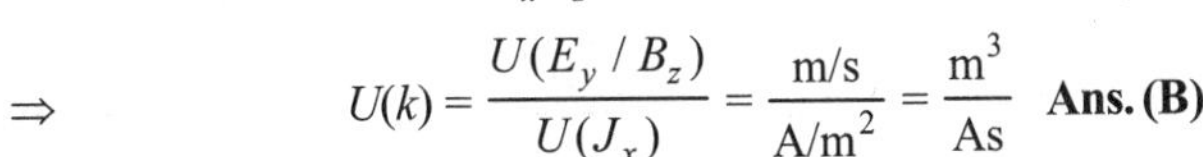

$$\Rightarrow \quad U(k) = \frac{U(E_y / B_z)}{U(J_x)} = \frac{\text{m/s}}{\text{A/m}^2} = \frac{\text{m}^3}{\text{As}} \qquad \textbf{Ans. (B)}$$

Sol. 27 Work done in displacing the wire by an elemental displacement dx is

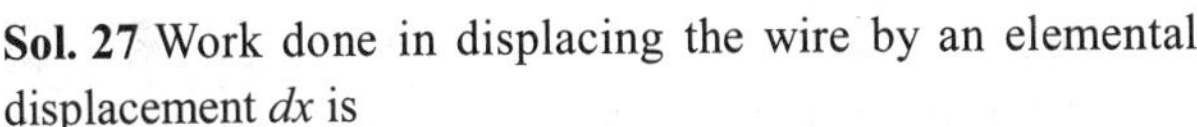

$$dW = BIldx = 3.0 \times 10^{-4} \times e^{-0.2x} \times 10 \times 3 \times dx$$

Total work done in displacing the wire from $x = 0$ to $x = 2$m is

$$W = \int dW = \int_0^{2.0} 9\times10^{-3}.e^{-0.2x}dx$$

$$\Rightarrow \quad W = 9 \times 10^{-3}\left[\frac{e^{-0.2x}}{-0.2}\right]_0^2$$

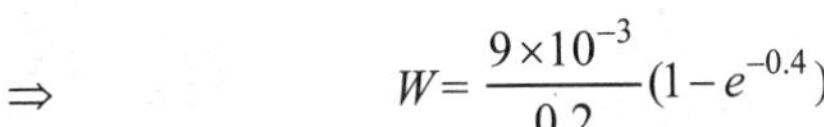

$$\Rightarrow \quad W = \frac{9\times10^{-3}}{0.2}(1-e^{-0.4})$$

$$\Rightarrow \quad W = 4.5 \times 10^{-2}(1-(2.71)^{-0.4}) \text{ J}$$

Power spent is $P = \dfrac{W}{t} = \dfrac{4.5\times10^{-2}\left[1-(2.71)^{-0.4}\right]}{5\times10^{-3}}$

$$\Rightarrow \quad P = 2.96 \text{ W} \qquad \textbf{Ans. (B)}$$

Sol. 28 The force between two parallel current carrying wires is given as

$$F = \frac{\mu_0 I_1 I_2 l}{2\pi d}$$

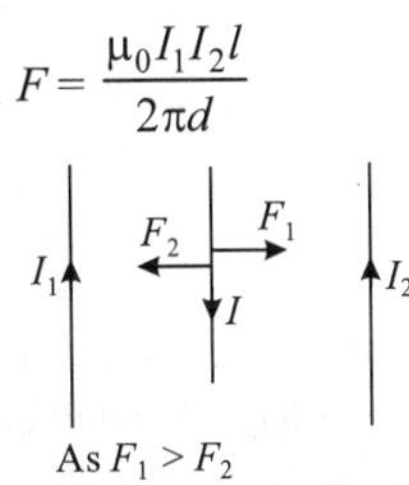

As $F_1 > F_2$

Net force on length l of middle wire is

$$F_{net} = F_1 - F_2 = \frac{\mu_0 I_1 Il}{2\pi(0.03)} - \frac{\mu_0 I_2 Il}{2\pi(0.05)}$$

$$\Rightarrow \quad F_{net} = \frac{\mu_0 Il}{2\pi}\left[\frac{30}{0.03} - \frac{20}{0.05}\right]$$

$$\Rightarrow \quad F_{net} = \frac{4\pi\times10^{-7}\times10\times0.25}{2\pi}\times(1000-400)$$

F_{net} is acting toward right

$$\Rightarrow \quad F_{net} = 2 \times 10^{-7} \times 10 \times 0.25 \times 600$$

$$\Rightarrow \quad F_{net} = 3 \times 10^{-4} \text{ N} \qquad \textbf{Ans. (A)}$$

Sol. 29

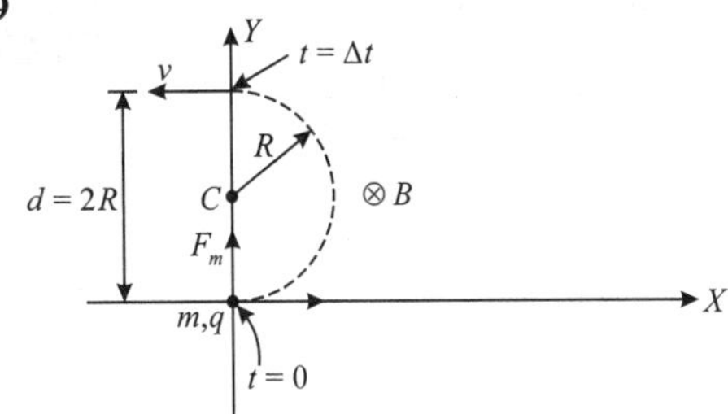

In magnetic field, radius of circular motion of charge is given as

$$R = \frac{mv}{qB} = \frac{d}{2} \qquad \ldots(1)$$

Time required to complete half circle is given as

$$\Delta t = \frac{T}{2} = \frac{1}{2}\left(\frac{2\pi m}{qB}\right) = \frac{\pi m}{qB} \quad \left[\text{As } T = \frac{2\pi m}{qB}\right]$$

from equation-(1) $\Delta t = \pi\left(\dfrac{d}{2v}\right) = \dfrac{\pi d}{2v}$

also from equation-(1) $B = \dfrac{2mv}{qd}$ **Ans. (C)**

Sol. 30 For stable equilibrium $U \rightarrow$ minimum as $U = -\vec{M}\cdot\vec{B}$

so for $\theta = 0°$ it will be in stable equilibrium

and for unstable equilibrium $U \rightarrow$ max $\Rightarrow$ at $\theta = 180°$

it will be in unstable equilibrium.

(a)

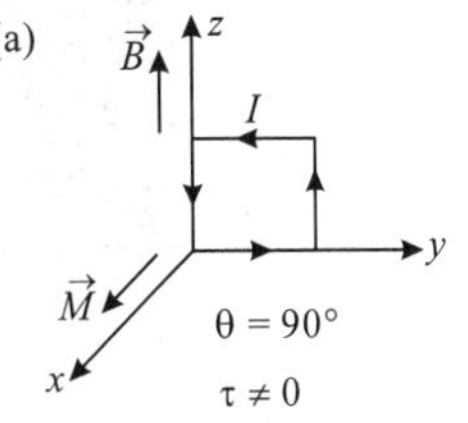

(b)

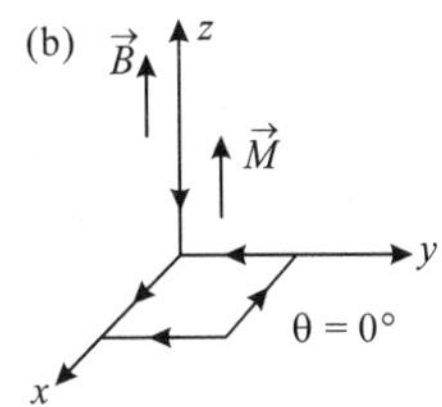

Case of stable eqlibrium

(c)

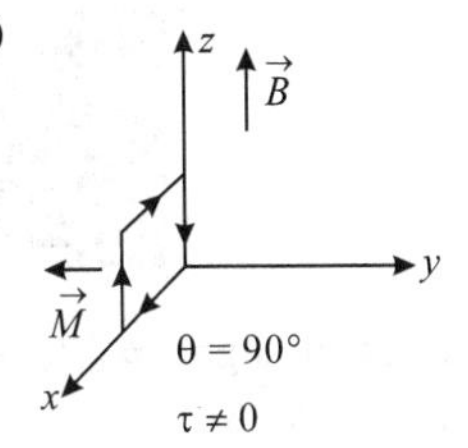

(d)

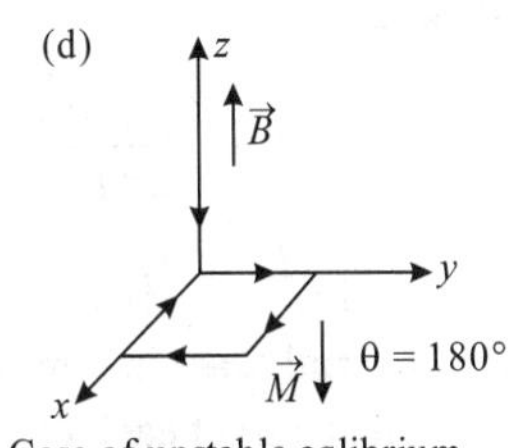

Case of unstable eqlibrium

(b) and (d) are cases of stable and unstable equilibrium respectively. **Ans. (C)**

Sol. 31

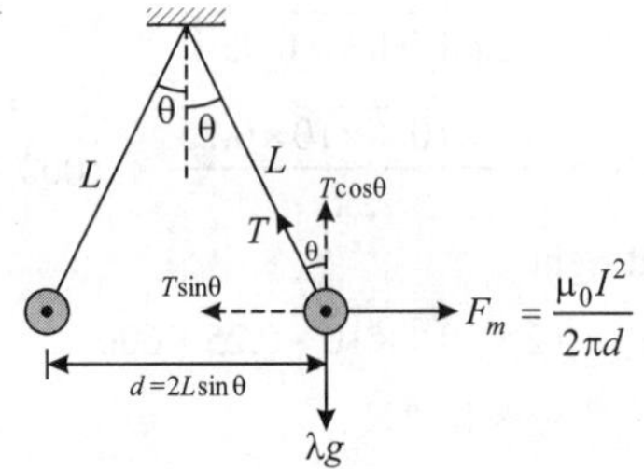

For equilibrium of wires, we have

$$T\sin\theta = \frac{\mu_0 I^2}{2\pi(2L\sin\theta)} \quad \ldots(1)$$

$$T\cos\theta = \lambda g \quad \ldots(2)$$

By (1)/(2) $\Rightarrow$ $$\frac{\sin^2\theta}{\cos\theta} = \frac{\mu_0 I^2}{4\pi L\lambda g}$$

$\Rightarrow$ $$I = 2\sin\theta\sqrt{\frac{\pi\lambda g L}{\mu_0\cos\theta}}$$ **Ans. (B)**

Sol. 32 For solenoid having closely spaced turns these behave like closed loop coils and in uniform magnetic field produced by a solenoid the net force on coils (of other solenoid) is equal to zero.

$\Rightarrow$ $$\vec{F_1} = \vec{F_2} = 0$$ **Ans. (A)**

Sol. 33 $F_m = BIl = 2BIR\sin\theta$ (magnetic force on wire)

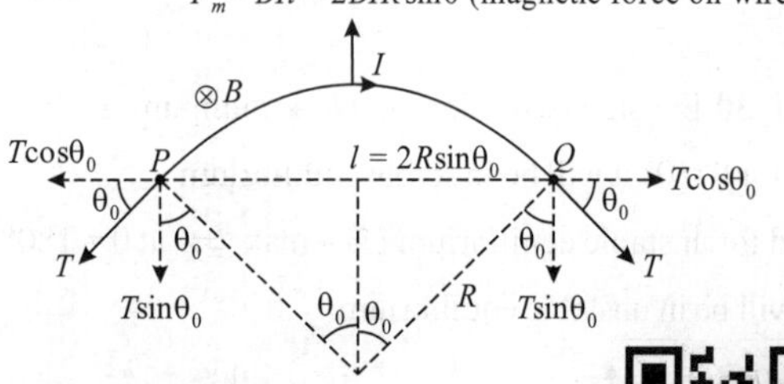

As wire is in equilibrium, we use

$$2T\sin\theta_0 = 2BIR\sin\theta_0$$

$\Rightarrow$ $$T = BIR$$ **Ans. (C)**

Sol. 34 Force between two parallel current carrying wires is given as

$$F = \frac{\mu_0 I_1 I_2 l}{2\pi d}$$

$\Rightarrow$ $$F \propto I_1 I_2$$

$\Rightarrow$ If I_1 and I_2 are in same direction (same sign) $\Rightarrow I_1I_2 \rightarrow$ +ve then wires attract each other $\Rightarrow F \rightarrow$ –ve. And

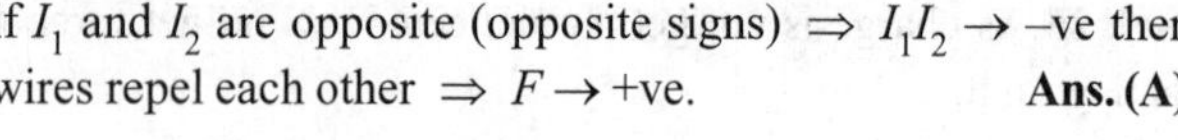

if I_1 and I_2 are opposite (opposite signs) $\Rightarrow I_1I_2 \rightarrow$ –ve then wires repel each other $\Rightarrow F \rightarrow$ +ve. **Ans. (A)**

Sol. 35

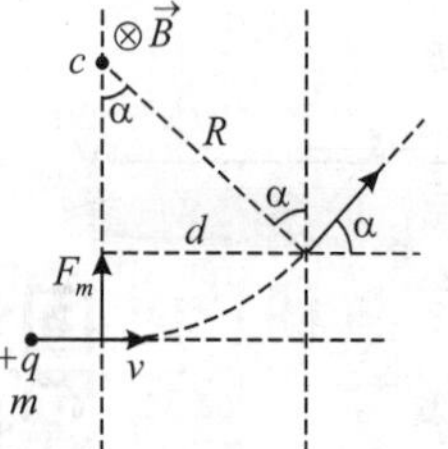

Kinetic energy of proton $K = qV = \frac{1}{2}mv^2$

The radius of proton motion inside the Magnetic Field is

$$R = \frac{mv}{qB} = \frac{\sqrt{2mk}}{qB}$$

from figure $$\sin\alpha = \frac{d}{R} = \frac{qBd}{\sqrt{2mK}} = \frac{qBd}{\sqrt{2mqV}}$$

$\Rightarrow$ $$\sin\alpha = Bd\sqrt{\frac{q}{2mV}}$$ **Ans. (D)**

Sol. 36 A loop with a current is in stable equilibrium in a uniform magnetic field when its PE is minimum, which is given as

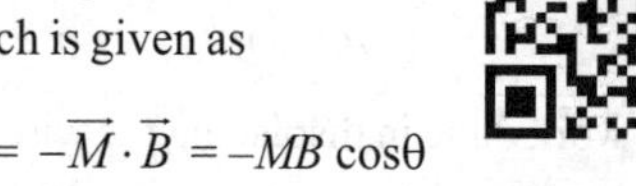
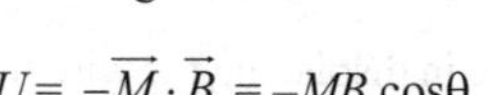

$$U = -\vec{M}\cdot\vec{B} = -MB\cos\theta$$

Stable equilibrium $\Rightarrow \theta = 0°$ when $\vec{M}$ is parallel to $\vec{B}$. **Ans. (C)**

Sol. 37 By Lorentz force equation, force on a moving charge in EM field is given as

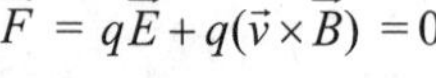

$$\vec{F} = q\vec{E} + q(\vec{v}\times\vec{B}) = 0$$

[As no force is acting on charge]

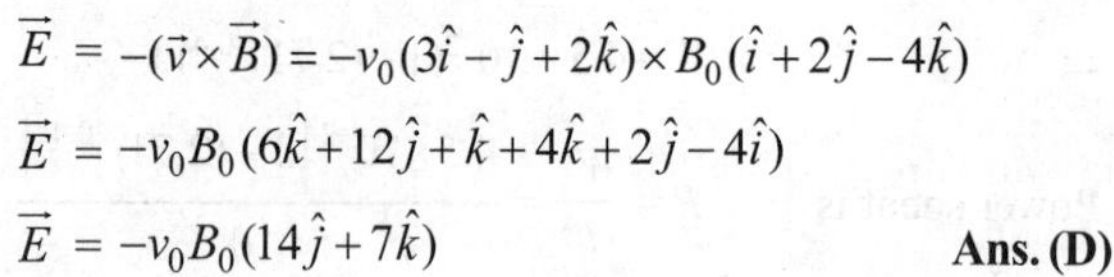

$$\vec{E} = -(\vec{v}\times\vec{B}) = -v_0(3\hat{i} - \hat{j} + 2\hat{k})\times B_0(\hat{i} + 2\hat{j} - 4\hat{k})$$

$$\vec{E} = -v_0B_0(6\hat{k} + 12\hat{j} + \hat{k} + 4\hat{k} + 2\hat{j} - 4\hat{i})$$

$$\vec{E} = -v_0B_0(14\hat{j} + 7\hat{k})$$ **Ans. (D)**

Sol. 38

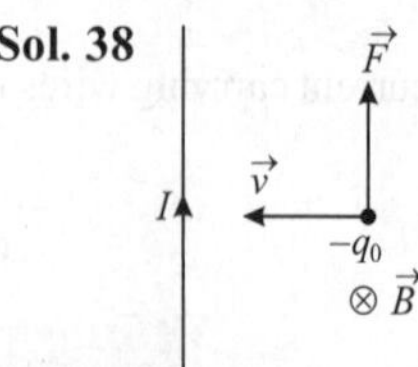

As force $\vec{F}$ and magnetic induction $\vec{B}$ is specified then using Flaming's left hand rule or Right hand Palm Rule we can find the direction of $\vec{v}$ for $-q_0$ charge.

By RHPR fingers point along $\vec{B}$

Palm face vector along force $\vec{F}$

$\Rightarrow$ direction of thumb which is toward right gives the $\vec{v}$ for +ve charge so here as charge is –ve $\Rightarrow$ $\vec{v}$ is leftward. **Ans. (B)**

Sol. 39 Case-I :

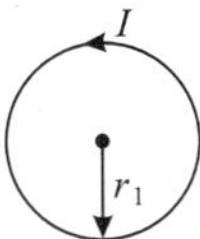

Magnetic moment $m = I\pi r_1^2$

Magnetic induction at centre

$$B_1 = \frac{\mu_0 I}{2r_1} \quad \ldots(1)$$

Case-II : $I \rightarrow$ constant but $m' = 2m$

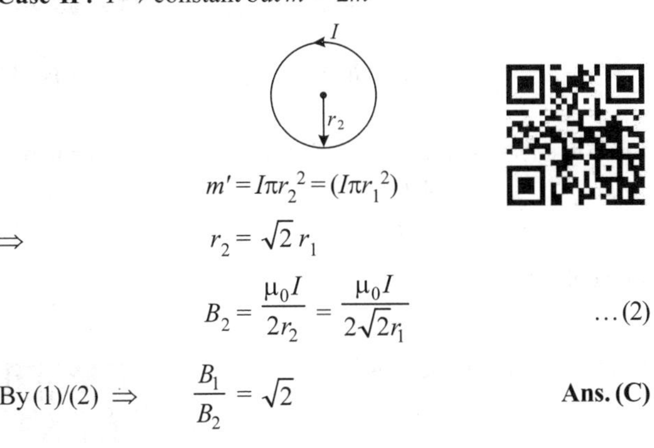

$$m' = I\pi r_2^2 = (I\pi r_1^2)$$

$$\Rightarrow \quad r_2 = \sqrt{2}\, r_1$$

$$B_2 = \frac{\mu_0 I}{2r_2} = \frac{\mu_0 I}{2\sqrt{2} r_1} \quad \ldots(2)$$

By (1)/(2) $\Rightarrow$ $\frac{B_1}{B_2} = \sqrt{2}$ **Ans. (C)**

Sol. 40 In a magnetic field for a charge particle in circular motion, radius is given as

$$r = \frac{mv}{qB} = \frac{\sqrt{2mK}}{qB}$$

for electron $r_e = \frac{\sqrt{2m_e K}}{eB}$

for proton $r_p = \frac{\sqrt{2m_p K}}{eB}$

for α particle $r_\alpha = \frac{\sqrt{2(4m_p)K}}{(2e)B} = \frac{\sqrt{2m_p K}}{eB} = r_p$

As $m_e < m_p$

$\Rightarrow$ $r_e < r_p = r_\alpha$ **Ans. (B)**

Sol. 41

The rotating charged loop behaves like a circular current, which is given as

$$I = \frac{q\omega}{2\pi}$$

$\Rightarrow$ Magnetic moment of this rotating charge is given as

$$\mu = I \times \pi r^2 = \frac{q\omega}{2\pi} \times \pi r^2 = \frac{1}{2} q\omega r^2 \quad \textbf{Ans. (A)}$$

Alternative : We use

$$\frac{\mu}{L} = \frac{q}{2m}$$

$$\Rightarrow \quad \mu = \frac{q}{2m} \times L = \frac{q}{2m} \times mr^2\omega$$

$$\Rightarrow \quad \mu = \frac{1}{2} q\omega r^2 \quad \textbf{Ans. (A)}$$

Sol. 42 Force on loop due to wire is given as

$$|\vec{F}| = m\left|\frac{dB}{dr}\right|$$

Where m is the dipole moment of loop

$$m = i\pi a^2$$

We use

$$\frac{dB}{dx} = \frac{d}{dx}\left(\frac{\mu_0 I}{2\pi x}\right) = -\frac{\mu_0 I}{2\pi x^2}$$

$$\Rightarrow \quad F = \frac{\mu_0 m I}{2\pi x^2} = \frac{\mu_0 i I \pi a^2}{2\pi x^2}$$

$$\Rightarrow \quad F \propto \frac{a^2}{d^2} \quad [\text{As } x = d] \quad \textbf{Ans. (C)}$$

According to Newtons III law same force will be exerted by loop on wire.

Sol. 43 In circular motion under magnetic field, we use

$$\frac{mv^2}{R} = qvB$$

$$mv = qBR \quad \ldots(1)$$

As path becomes straight line by electric field, we use

$$qE = qvB$$

$$\Rightarrow \quad E = vB \quad \ldots(2)$$

From equation-(1) & (2)

$$m = \frac{qB^2R}{E}$$

$$\Rightarrow \quad m = 2.0 \times 10^{-24}\,\text{kg} \quad \textbf{Ans. (A)}$$

Sol. 44 Charge on element of with dx on rod revolving in a circle of radius x is given as

$$dq = \lambda dx$$

its magnetic moment is given as

$$dm = (n)\left(\frac{\rho_0 x}{l}\right) dx \cdot (\pi x^2)$$

Total magnetic moment of rod is

$$M = \int dm = \frac{n\rho_0\pi}{l} \cdot \int_0^l x^3 \cdot dx = \frac{n\rho_0\pi}{l} \cdot \left[\frac{l^4}{4}\right]$$

$$\Rightarrow \quad M = \frac{n\rho_0 \pi l^3}{4}$$ **Ans. (A)**

Sol. 45 Radius of circular motion of electron is

$$r = \frac{\sqrt{2mK}}{eB} = \frac{\sqrt{2meV}}{eB}$$

$$\Rightarrow \quad r = \frac{\sqrt{\frac{2m}{e} \cdot V}}{B} = \frac{\sqrt{\frac{2\times 9.1\times 10^{-31}}{1.6\times 10^{-19}}(500)}}{100\times 10^{-3}}$$

$$\Rightarrow \quad r = \frac{\sqrt{\frac{9.1}{0.16}\times 10^{-10}}}{10^{-1}} = \frac{3}{.4}\times 10^{-4}\ \text{m}$$

$$\Rightarrow \quad r = 7.5 \times 10^{-4}\ \text{m}$$ **Ans. (A)**

Sol. 46

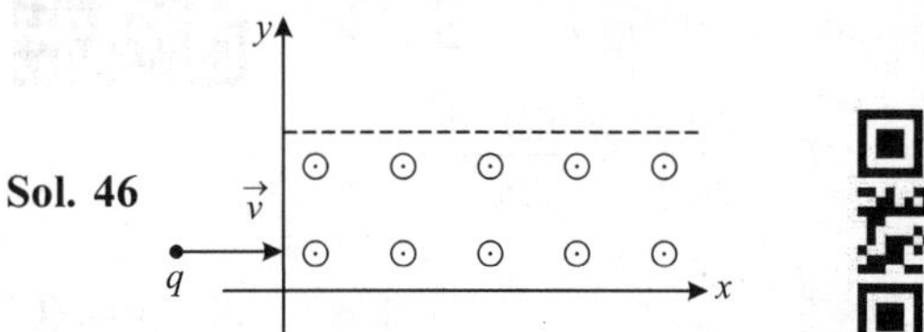

As the initial position of entering particle is not given, acceleration at leaving point cannot be calculated.

Ans. (None)

Sol. 47 Work will only be done against electric force, equal to

$$W = q\vec{E} \cdot \vec{r}$$

$$\Rightarrow \quad W = q(2\hat{i} + 3\hat{j}) \cdot (\hat{i} + \hat{j}) = 5q$$ **Ans. (D)**

Sol. 48 Kinetic energy of a charge q accelerated by a potential difference V is

$$k = qV$$

Radius of circular motion in magnetic field is given as

$$r = \sqrt{\frac{2mk}{q_B}} = \sqrt{\frac{2mkV}{q_B}}$$

$$\Rightarrow \quad r \propto \sqrt{\frac{m}{q}}$$

$$\Rightarrow \quad \frac{r_\rho}{r_\alpha} = \frac{1}{\sqrt{2}}$$ **Ans. (A)**

Sol. 49

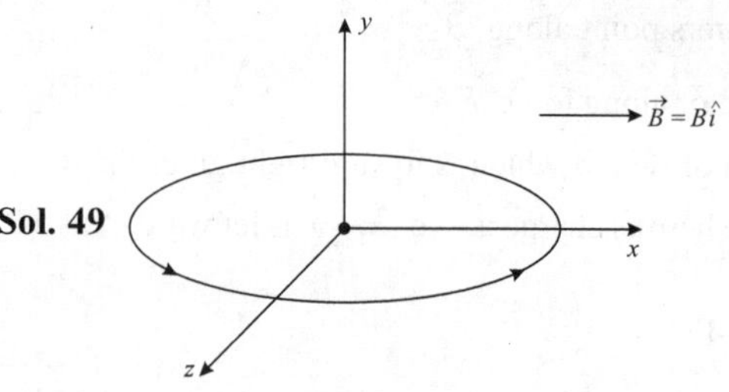

Magnetic moment of coil is given as

$$\overrightarrow{M} = NIA\,\hat{j}$$

$$\vec{\tau} = NI(\pi r^2)\hat{j}$$

Torque on loop (coil)

$$= \overrightarrow{M} \times \overrightarrow{B}$$

$$= NI(\pi r^2)\, B \sin 90° (-\hat{k})$$

$$= NI\pi r^2 B(-\hat{k})$$ **Ans. (A)**

Sol. 50 Given that resitance of galvanometer and shunt are

$$R_g = 50\,\Omega$$

$$R_S = 5000\,\Omega$$

Full deflecttion current is

$$i_g = 4 \times 10^{-3}$$

Thus range of voltmeter is given as

$$V_R = i_g(G + S)$$

$$\Rightarrow \quad V_R = 4 \times 10^{-3}(50 + 5000)\, V$$

$$\Rightarrow \quad V_R = 4 \times 10^{-3}(5050)\, V$$

$$\Rightarrow \quad V_R = 20.2\, V$$ **Ans. (B)**

Sol. 51 Force acting on segments of square frame are shown in figure below

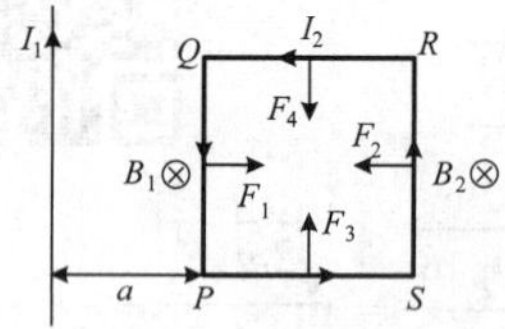

Here F_3 & F_4 cancel each other and force on PQ will be

$$F_1 = I_2 B_1 a$$

$$\Rightarrow \quad F_1 = I_2\left(\frac{\mu_0 I_1}{2\pi a}\right) a$$

$$\Rightarrow \quad F_1 = \frac{\mu_0 I_1 I_2}{2\pi}$$

Force on RS will be

$$F_2 = I_2 B_2 a$$

$$\Rightarrow \quad F_2 = I_2 \frac{\mu_0 I_1}{2\pi(2a)} a$$

$$\Rightarrow \quad F_2 = \frac{\mu_0 I_1 I_2}{4\pi}$$

Net force is given as

$$F_{net} = F_1 - F_2 = \frac{\mu_0 I_1 I_2}{4\pi} \text{ repulsion}$$ **Ans. (B)**

Sol. 52 Torque on coil is given as

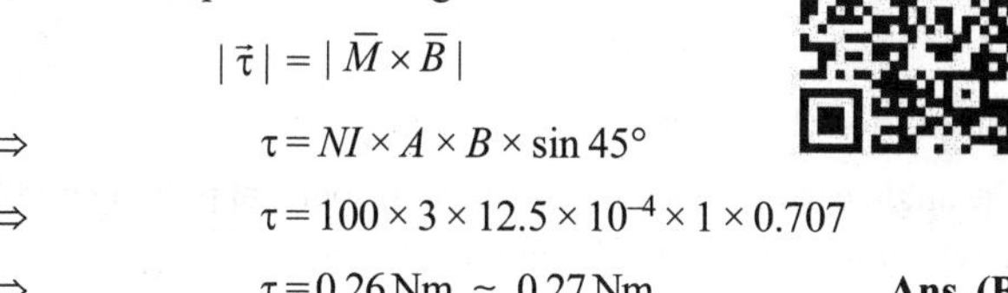

$$|\vec{\tau}| = |\vec{M} \times \vec{B}|$$

$$\Rightarrow \quad \tau = NI \times A \times B \times \sin 45°$$

$$\Rightarrow \quad \tau = 100 \times 3 \times 12.5 \times 10^{-4} \times 1 \times 0.707$$

$$\Rightarrow \quad \tau = 0.26 \text{ Nm} \simeq 0.27 \text{ Nm}$$ **Ans. (B)**

Sol. 53 Torque on coil is given as

$$\tau = \vec{M} \times \vec{B}$$

$$C\theta = i N A B$$

$$\Rightarrow \quad 10^{-6} \times \frac{\pi}{180} = 10^{-3} \times 10^{-4} \times 175 \times B$$

$$\Rightarrow \quad B = 10^{-3} \text{ T}$$ **Ans. (A)**

Sol. 54 Radius of circular motion in magnetic field is given as

$$r = \frac{mv}{qB} = \frac{\sqrt{2mK}}{qB}$$

$$\Rightarrow \quad r_{He} = r_p > r_e$$ **Ans. (C)**

Sol. 55

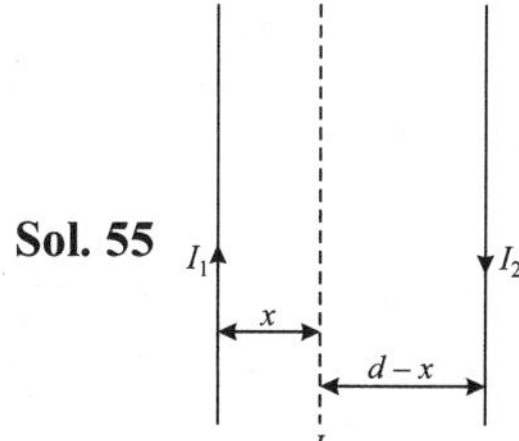

Net force on wire carrying current I per unit length is given as

$$\frac{\mu_0 I_1 I}{2\pi x} + \frac{\mu_0 I_2 I}{2\pi(d-x)} = 0$$

$$\frac{I_1}{x} = \frac{I_2}{x-d}$$

$$\Rightarrow \quad x = \frac{I_1 d}{I_1 - I_2}$$ **Ans. (B)**

Sol. 56 Given that $I_g = 10^{-4}$ A

and if galvanometer resistance is R_g, we use

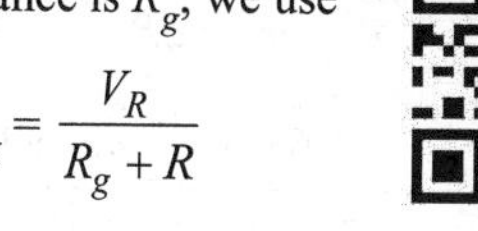

$$I_g = \frac{V_R}{R_g + R}$$

$$10^{-4} = \frac{5}{R_g + 2 \times 10^6}$$

$$\Rightarrow \quad R_g = 5 \times 10^4 - 2 \times 10^6$$
$$= -195 \times 10^4 \,\Omega$$

Practically not possible

None of the options can be chosen. **Ans. (None)**

Sol. 57 When galvanometer is used as an ammeter shunt is used in parallel with galvanometer as shown in figure below

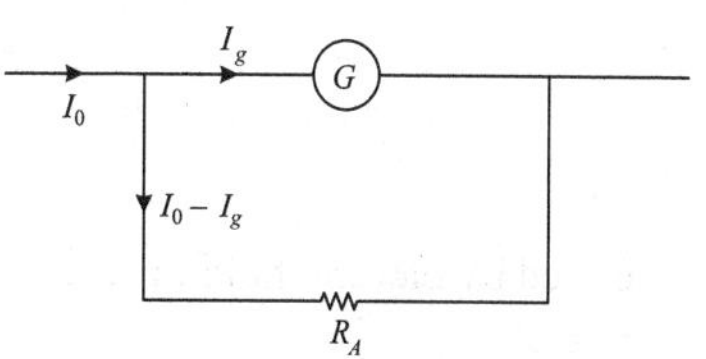

For full deflection, we use

$$I_g G = (I_0 - I_g) R_A$$

$$\Rightarrow \quad R_A = \left(\frac{I_g}{I_0 - I_g}\right) G$$

When galvanometer is used as a voltmeter, resistance is used in series with galvanometer as shown in figure below

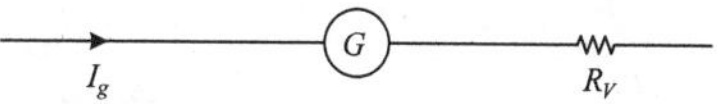

For full deflection, we use

$$I_g(G + R_V) = V = GI_0 \quad (\text{given } V = GI_0)$$

$$\Rightarrow \quad R_V = \frac{(I_0 - I_g)G}{I_g}$$

$$\Rightarrow \quad R_A R_V = G^2$$

and $$\frac{R_A}{R_V} = \left(\frac{I_g}{I_0 - I_g}\right)^2$$ **Ans. (B)**

Sol. 58 Angle of deflection is given as

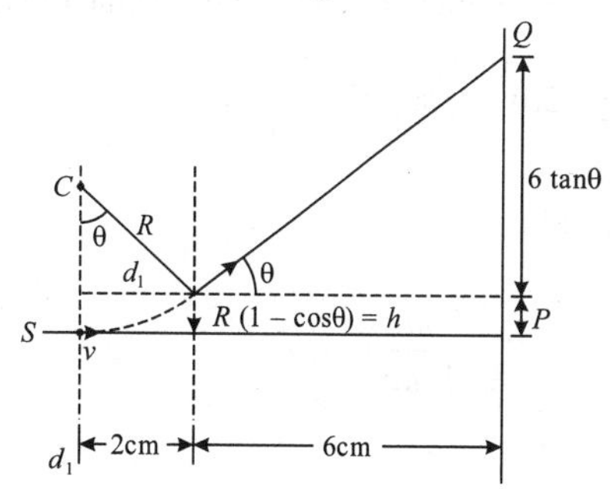

$$\sin\theta = \frac{d_1}{R} = \frac{d_1}{mv/qB} = \frac{qBd_1}{mv} = \frac{qBd_1}{\sqrt{2mE}}$$

$$\Rightarrow \quad \sin\theta = \frac{1.6\times10^{-19}\times1.5\times10^{-3}\times0.02}{\sqrt{2\times9.1\times10^{-31}\times100\times1.6\times10^{-19}}}$$

$$= \frac{4.8\times10^{-24}}{\sqrt{29}\times10^{-24}} = \frac{4.8}{5.4} \simeq 0.89$$

$$\Rightarrow \quad \cos\theta = \sqrt{1-(0.89)^2} = \sqrt{0.2} = 0.45$$

$$\Rightarrow \quad \tan\theta = \frac{0.89}{0.45} \simeq 2$$

$$\Rightarrow \quad d = PQ = 6\tan\theta + h \cong 12 + h \text{ cm}$$ **Ans. (A)**

Sol. 59 We use, after time t

$$(2v_0)^2 = v_0^2 + v_x^2$$

$$\Rightarrow \quad v_x = \sqrt{3}\, v_0$$

As particle is only accelerated by electric field in X-direction, we use

$$\sqrt{3}\, v_0 = 0 + \frac{qE_0}{m}t$$

$$\Rightarrow \quad t = \frac{\sqrt{3}v_0 m}{qE_0}$$ **Ans. (C)**

Sol. 60 Acceleration of protons is

$$a = \frac{qvB}{m}$$

$$\Rightarrow \quad B = \frac{ma}{qv}$$

$$\Rightarrow \quad B = \frac{ma\sqrt{m}}{\sqrt{2k}}$$

$$\Rightarrow \quad B = \frac{m^{3/2}a}{e\sqrt{2k}} = \frac{(1.6\times10^{-27})^{3/2}\times10^{12}}{1.6\times10^{-19}\sqrt{2\times1\times10^6\times1.6\times10^{-19}}}$$

$$\Rightarrow \quad B = 0.71 \text{ mT}$$ **Ans. (D)**

Sol. 61 As work is only done by EF on charge particle, we use

$$qE(2a) = \frac{1}{2}m(2v)^2 - \frac{1}{2}mv^2$$

$$\Rightarrow \quad qE(2a) = \frac{3}{2}mv^2$$

$$\Rightarrow \quad E = \frac{3}{4}\frac{mv^2}{qa}$$

Thus option (A) is correct.

Rate of work done by electric field is

$$P = \vec{F}\cdot\vec{v} = Fv\cos\theta = Fv$$

$$P = qEv$$

$$\Rightarrow \quad P = q\left(\frac{3}{4}\frac{mV^2}{qa}\right)v$$

$$\Rightarrow \quad P = q\left(\frac{3}{4}\frac{mv^3}{qa}\right)$$

$$\Rightarrow \quad P = \frac{3}{4}\frac{mv^3}{a}$$

Thus option (B) is correct.

The angle between electric force and velocity at point Q is 90°, so rate of work done will be zero at Q.

Thus option (C) is correct.

Initial angular momentum about origin is

$$L_i = mva$$

Final angular momentum about origin is

$$L_f = m(2v)(2a)$$

Change in angular momentum $L_f - L_i = 3mva$

Thus option (D) is NOT correct. **Ans. (B)**

Sol. 62 For not to hit the surface of solenoid the radius of electron motion inside must be less than $R/2$, so we use

$$\frac{R}{2} = \frac{mv}{qB}$$

$$\Rightarrow \quad \frac{R}{2} = \frac{mv_{max}}{e(\mu_0 ni)}$$

$$\Rightarrow \quad v_{max} = \frac{e\mu_0 nIR}{2m}$$ **Ans. (B)**

Sol. 63 Figure below shows the given situation

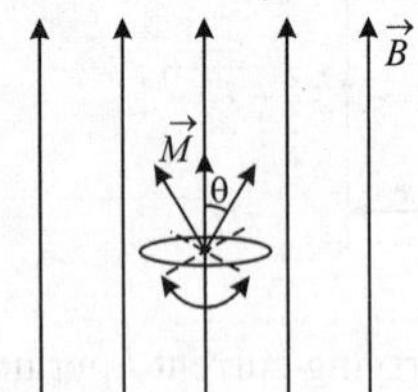

If loop is slightly tilted from equilibrium position (θ = 0°) then restoring torque on it is given as

$$\tau = -MB\sin\theta$$

$$\Rightarrow \quad \tau \simeq -MB\,\theta$$

Thus restoring angular acceleration is given as

$$\alpha = -\left(\frac{MB}{I}\right)\theta$$

Comparing with $\alpha = \omega^2\theta$, we get

$$\Rightarrow \qquad \omega = \sqrt{\frac{MB}{I}} = \sqrt{\frac{(i)(\pi R^2)B}{\left(\frac{mR^2}{2}\right)}}$$

$$\Rightarrow \qquad \omega = \sqrt{\frac{2i\pi B}{m}}$$

Thus time period of oscillations is

$$T = \frac{2\pi}{\omega} = \sqrt{\frac{2\pi m}{iB}}$$ **Ans. (B)**

Sol. 64 Pitch of helix in magnetic field is given as

$$p_H = \frac{2\pi m}{qB} v\cos\theta$$

$$\Rightarrow \qquad p_H = \frac{2(3.14)(1.67\times10^{-27})\times4\times10^5\times\cos60^\circ}{(1.69\times10^{-19})(0.3)}$$

$$\Rightarrow \qquad p_H = 0.04\,\text{m} = 4\,\text{cm}$$ **Ans. (B)**

Sol. 65 Acceleration of particle is given as

$$\vec{a} = \frac{-qE}{m}\hat{j}$$

x & y coordinate of particle at time $t = t$ is

$$x = v_0 t \qquad \ldots(1)$$

$$y = \frac{1}{2}\left(-\frac{qE}{m}\right)t^2 \qquad \ldots(2)$$

From (1) and (2)

$$y = -\frac{qEx^2}{2mv_0^2}$$

At $x = d$, $t = \frac{d}{v}$ $\Rightarrow$ $v_x = v_0$ and $v_y = \frac{qE}{m}\left(\frac{d}{v_0}\right)$

$$\Rightarrow \tan\theta = \frac{v_y}{v_0} = \frac{qEd}{mv_0^2};\quad y_0 = \frac{-qEd^2}{2mv_0^2}$$

Equation of straight line motion of particle after $x = d$ is

$$y = -(x-d)\tan\theta - y_0$$

$$\Rightarrow \qquad y = -x\left(\frac{qEd}{mv_0^2}\right) + \frac{qEd^2}{mv_0^2} - \frac{qEd^2}{2mv_0^2}$$

$$\Rightarrow \qquad y = \frac{qEd}{mv_0^2}\left(\frac{d}{2} - x\right)$$ **Ans. (A)**

Sol. 66 For helical motion of proton, time period is given as

$$T = \frac{2\pi m}{qB}$$

Total time taken by proton is

$$t = 10\,\text{T}$$

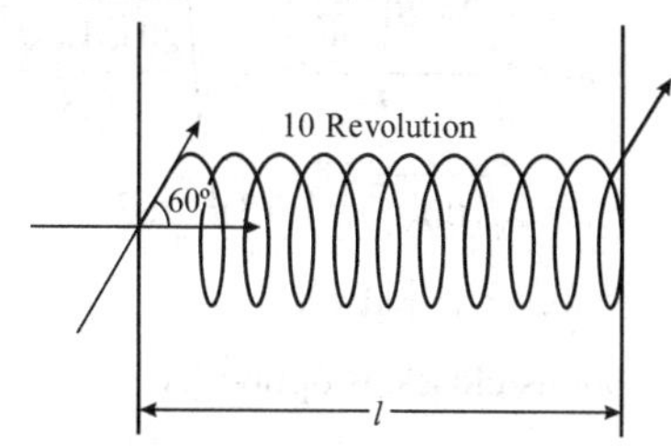

Velocity of proton along axis is

$$v_x = v\cos60^\circ = \frac{v}{2}$$

Thus distance travelled along axis is

$$l = \frac{v}{2}\times t = \frac{v}{2}\times10\times\frac{2\pi m}{qB}$$

$$\Rightarrow \qquad l = 4\times10^5\times10\times\frac{3.14\times1.67\times10^{-27}}{1.6\times10^{-19}\times0.3}\,\text{m}$$

$$\Rightarrow \qquad l = 0.439\,\text{m}$$ **Ans. (C)**

Sol. 67 Magnetic moment of loop *ABCDA* is

$$\overrightarrow{M}_1 = abI\hat{k}$$

Magnetic moment of loop *ADEFA* is

$$\overrightarrow{M}_2 = abI\hat{j}$$

Total magnetic moment of loop *ABCDEAA* is

$$\overrightarrow{M} = \overrightarrow{M}_1 + \overrightarrow{M}_2$$

$$\Rightarrow \qquad \overrightarrow{M} = abI(\hat{k} + \hat{j})$$

$$\Rightarrow \qquad \overrightarrow{M} = \sqrt{2}\,abI\left(\frac{\hat{k}}{\sqrt{2}} + \frac{\hat{j}}{\sqrt{2}}\right)$$ **Ans. (A)**

Sol. 68 Force on particle is given as

$$\vec{F} = q(\vec{v}\times\vec{B})$$

$$\Rightarrow \qquad \vec{F} = 1(2\hat{i} + 3\hat{j} + 4\hat{k})\times(5\hat{i} + 3\hat{j} - 6\hat{k})\times10^{-3}\times10^{-6}\,\text{N}$$

$$\Rightarrow \qquad \vec{F} = (6\hat{k} + 12\hat{j} - 15\hat{k} - 18\hat{i} + 20\hat{j} - 12\hat{i})\times10^{-9}\,\text{N}$$

$$\Rightarrow \qquad \vec{F} = (-30\hat{i} + 32\hat{j} - 9\hat{k})\times10^{-9}\,\text{N}$$ **Ans. (C)**

Sol. 69 Magnetic torque on coil when it is parallel to magnetic field is

$$\tau = BINA = \tau_{ext}$$

$$\Rightarrow \quad B = \frac{\tau_{ext}}{INA} = \frac{1.5}{0.5\times500\times3\times10^{-4}}$$

$$\Rightarrow \quad B = \frac{100}{5} = 20\,\text{T}$$ **Ans. (20)**

Sol. 70 Work done is rotation of coil by 60° is given as

$$\omega = MB(\cos 30^\circ - \cos 90^\circ) = \frac{1}{2}I\omega^2$$

$$\Rightarrow \quad \omega = \sqrt{\frac{2MB\left(\frac{\sqrt{3}}{2}\right)}{I}} = \sqrt{\frac{20\times4\times\sqrt{3}}{0.8}}$$

$$\Rightarrow \quad \omega = \sqrt{100\sqrt{3}} = 10(3)^{1/4}\,\text{rad/s}$$

$$\Rightarrow \quad \omega \simeq 13.16\,\text{rad/s}$$

Out of given options closest is option (A). **Ans. (A)**

Sol. 71 Torque on loop is given as

$$\tau = MB = 4a^2I\times\frac{\mu_0 I}{2\pi b}$$

$$\Rightarrow \quad \tau = \frac{2\mu_0 I^2a^2}{\pi b}$$ **Ans. (A)**

Sol. 72 Magnetic moment of electromagnet is given as

$$M = \mu_r NiA$$

$$\Rightarrow \quad M = \mu_r(nl)iA$$

$$\Rightarrow \quad M = \mu_r niV$$

$$\Rightarrow \quad M = 1000\times10\times100\times0.5\times10^{-3}$$

$$\Rightarrow \quad M = 500\,\text{Am}^2$$ **Ans. (D)**

Sol. 73 Particle will not hit the screen if its radius of circular motion is less then d, so we use

$$R < d$$

$$\Rightarrow \quad \frac{mv}{qB_0} < d$$

$$\Rightarrow \quad v = \frac{qB_0d}{m}$$ **Ans. (B)**

Sol. 74 Force on electron due to electric field is given as

$$F_e = eE = 1.6\times10^{-19}\times300\times100\ \text{(Along} - \text{Y-direction)}$$

For electron to keep moving along X-direction, net force on it must be zero thus magnetic force must balance electric force so act along +Y-direction so for electron moving in +X-direction, magnetic force is given as

$$\vec{F_m} = -e(\vec{v}\times\vec{B})$$

For $\vec{F_m}$ along +Y-direction $\vec{B}$ must be along +Z-direction such that

$$evB = eE$$

$$\Rightarrow \quad B = \frac{E}{v} = \frac{300\times100}{6\times10^6} = 5\times10^{-3}\,\text{T}$$ **Ans. (A)**

Sol. 75 Velocity vector is given as

$$\vec{v} = v_x\hat{i} + v_y\hat{j} \quad \ldots(1)$$

Here force on particle is given as

$$\vec{F} = k(v_y\hat{i} + v_x\hat{j}) \quad \ldots(2)$$

From (1) and (2), if mass of particle is m, we use

$$m = \frac{dv_x}{dt} = kv_y$$

and

$$m\frac{dv_y}{dt} = kv_x$$

$$\Rightarrow \quad \frac{dv_x}{dv_y} = \frac{v_y}{v_x}$$

$$\Rightarrow \quad v_xdv_x - v_ydv_y = 0$$

$$\Rightarrow \quad v_x^2 - v_y^2 = \text{constant}$$

In option (C) $\vec{v}\times\vec{a} = (v_x\hat{i} + v_y\hat{j})\times\frac{k}{m}(v_y\hat{i} + v_x\hat{j})$

$$= (v_x^2 - v_y^2)\times\frac{k}{m}\hat{k} = \text{Constant}$$ **Ans. (C)**

Sol. 76 Radius of circular motion of particle is given as

$$r = \frac{mv}{qB}$$

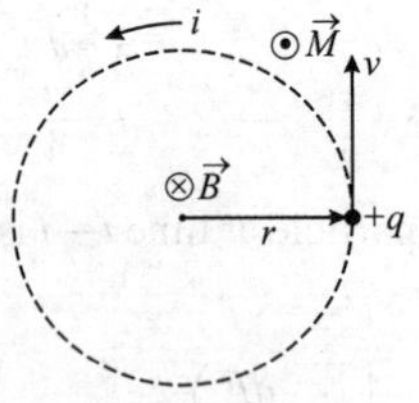

Angular speed of particle is given as

$$\omega = \frac{v}{r} = \frac{qB}{m}$$

Magnetic moment of revolving charge is given as

$$M = i\cdot\pi r^2$$

$$\Rightarrow \quad M = \frac{q\omega}{2\pi}\cdot\pi\left(\frac{mv}{qB}\right)^2 = \frac{1}{2}q\left(\frac{qB}{m}\right)\left(\frac{m^2v^2}{q^2B^2}\right)$$

$$\Rightarrow \quad M = \frac{mv^2}{2B}$$

From figure it is clear that direction of $\overrightarrow{M}$ is opposite to $\vec{B}$ so we write

$$\overrightarrow{M} = -\frac{mv^2}{2B^2}\vec{B}$$ **Ans. (D)**

Sol. 77 Figure shows the forces acting on two side wires of square loop due to the long wire placed at (0, *b*, 0) which is given as

$$F = \frac{\mu_0 I^2 (2a)}{2\pi\sqrt{b^2+a^2}}$$

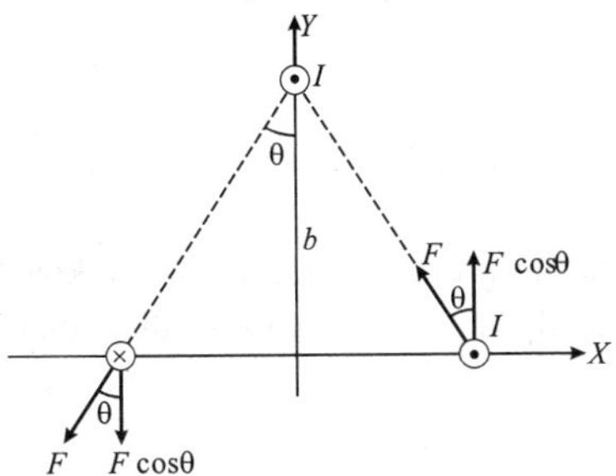

Torque on loop is given as

$$\tau = F\cos\theta \times 2a$$

$$\Rightarrow \quad \tau = \frac{\mu_0 I^2(2a)}{2\pi\sqrt{b^2+a^2}} \times \frac{b}{\sqrt{b^2+a^2}} \times 2a$$

$$\Rightarrow \quad \tau = \frac{2\mu_0 I^2 a^2 b}{\pi(b^2+a^2)} \simeq \frac{2\mu_0 I^2 a^2}{\pi b} (b >> a)$$

For all the option given (A) is most appropriate. **Ans. (A)**

Sol. 78 The magnetic force on a charge moving with some momentum in magnetic field is given as

$$F = q(\vec{v}\times\vec{B}) = \frac{q}{m}(\vec{P}\times\vec{B})$$

$$\Rightarrow \quad F \propto \frac{q}{m}$$

For the three given particles, we can use

$$F_1 : F_2 : F_3 = \frac{q_1}{m_1} : \frac{q_2}{m_2} : \frac{q_3}{m_3} = \frac{e}{m_p} : \frac{e}{2m_p} : \frac{2e}{4m_p}$$

$$\Rightarrow \quad F_1 : F_2 : F_3 = \frac{1}{1} : \frac{1}{2} : \frac{2}{4}$$

$$\Rightarrow \quad F_1 : F_2 : F_3 = 2 : 1 : 1$$

From a given momentum of particle, its speed is given as

$$v \propto \frac{1}{m}$$

$$\Rightarrow \quad v_1 : v_2 : v_3 = \frac{1}{m_p} : \frac{1}{2m_p} : \frac{1}{4m_p}$$

$$\Rightarrow \quad v_1 : v_2 : v_3 = \frac{1}{1} : \frac{1}{2} : \frac{1}{4}$$

$$\Rightarrow \quad v_1 : v_2 : v_3 = 4 : 2 : 1$$ **Ans. (B)**

Sol. 79 The radius of circular motion of a moving charge in magnetic field is given as

$$r = \frac{mv}{qB} = \frac{p}{qB}$$

As $\frac{m_\alpha}{m_p} = 4$, we have

$$\frac{r_p}{r_\alpha} = \frac{p_p}{q_p}\frac{q_\alpha}{p_\alpha} = \frac{2}{1}$$

$$\Rightarrow \quad \frac{p_p}{p_\alpha} = \frac{2q_p}{q_\alpha} = 2\left(\frac{1}{2}\right)$$

$$\Rightarrow \quad \frac{p_p}{p_\alpha} = 1$$

$$\Rightarrow \quad \frac{K_p}{K_\alpha} = \frac{p_p^2}{p_\alpha^2}\frac{m_\alpha}{m_p} = 4 : 1$$ **Ans. (D)**

Sol. 80 In magnetic field the radius of circular motion of charge particle is given as

$$r = \frac{mv}{qB} = \frac{\sqrt{2mK}}{qB}$$

$$\Rightarrow \quad r \propto \frac{\sqrt{m}}{q}$$

As $m_\alpha = 2m_d$ and $q_\alpha = 2q_d$ so ratio of radii of their path is give n as

$$\frac{r_d}{r_\alpha} = \frac{\sqrt{m_d}}{q_d} \times \frac{2q_d}{\sqrt{2m_d}} = \sqrt{2}$$ **Ans. (B)**

Sol. 81 The charge ratio of the ions is given as

$$\frac{Q_1}{Q_2} = \frac{1}{2}$$

The speed ratio is given as

$$\frac{v_1}{v_2} = \frac{2}{3}$$

The radius of circular motion in magnetic field is given as

$$R = \frac{mv}{qB}$$

Thus ratio of radii is given as

$$\Rightarrow \quad \frac{R_1}{R_2} = \frac{v_1}{v_2} \times \frac{Q_2}{Q_1} = \frac{2}{3} \times \frac{2}{1} = \frac{4}{3} \qquad \textbf{Ans. (C)}$$

Sol. 82 If after N revolutions, protons achieve one sixt of the speed of light then we use

$$2\,NeV = \frac{1}{2}mv^2$$

$$\Rightarrow \quad 2N \times 1.6 \times 10^{-19} = \frac{1}{2} \times 1.67 \times 10^{-27} \times (5 \times 10^7)^2$$

$$\Rightarrow \quad N = 543 \qquad \textbf{Ans. (543)}$$

Sol. 83 The radius of circular path in magnetic field is given as

$$r = \frac{P}{qB} = \frac{\sqrt{2mk_e}}{qB}$$

$$\Rightarrow \quad r \propto \frac{\sqrt{m}}{q}$$

$$\Rightarrow \quad \frac{r_1}{r_2} = \frac{3}{4}$$

$$\Rightarrow \quad r_2 > r_1 \qquad \textbf{Ans. (C)}$$

Sol. 84 At $t = 0$, magnetic fields are given as

$$B \text{ at } \left(0, 0, \frac{\pi}{k}\right) = B_0\left(\frac{\hat{i}+\hat{j}}{\sqrt{2}}\right)\cos(\pi)$$

$$\text{and } B \text{ at } \left(0, 0, \frac{3\pi}{k}\right) = B_0\left(\frac{\hat{i}+\hat{j}}{\sqrt{2}}\right)\cos(3\pi)$$

Force on charged particle q_1 moving at a given velocity is calculated as

$$F_1 = q_1\, \vec{V}_1 \times \vec{B}_1$$

$$\Rightarrow \quad F_1 = 4\pi(0.5c\,\hat{i}) \times \left[-B_0\left(\frac{\hat{i}+\hat{j}}{\sqrt{2}}\right)\right] = \frac{4\pi B_0 c}{2\sqrt{2}}(-\hat{k})$$

Force on charged particle q_2 is given as

$$F_2 = q_2\, \vec{V}_2 \times \vec{B}_2$$

$$\Rightarrow \quad F_2 = 2 \times (0.5c\,\hat{i}) \times \left[-B_0\left(\frac{\hat{i}+\hat{j}}{\sqrt{2}}\right)\right]$$

$$\Rightarrow \quad F_2 = \frac{2\pi B_0 c}{2\sqrt{2}}(-\hat{k})$$

$$\Rightarrow \quad \frac{F_1}{F_2} = 2 \qquad \textbf{Ans. (D)}$$

Sol. 85 Magnetic force on a moving charge particle in magnetic field is given as

$$\vec{F} = q(\vec{v} \times \vec{B})$$

As $\vec{F}$ is perpendicular to $\vec{v}$, it does not do any work and no kinetic energy change will occur so its speed and kinetic energy will remain same both (A) and (R) are correct and (R) is correct explanation of (A). **Ans. (A)**

Sol. 86 The radius of circular path of a charge particle in magnetic field is given as

$$R = \frac{\sqrt{2mK}}{qB} \propto \frac{\sqrt{m}}{q}$$

For proton, deuteron and α particle, we use

$$\frac{\sqrt{m}}{e} : \frac{\sqrt{2m}}{e} : \frac{\sqrt{4m}}{2e}$$

$$\Rightarrow \quad 1 : \sqrt{2} : 1 \qquad \textbf{Ans. (D)}$$

Sol. 87 Similar to previous question as the kinetic energy of the proton and alpha particle are same, the ratio of the radii of their circular paths is given as

$$\frac{R_\alpha}{R_P} = \frac{M_\alpha}{M_P} \times \frac{q_P}{q_\alpha}$$

$$\Rightarrow \quad \frac{R_\alpha}{R_P} = \frac{4}{1} \times \frac{1}{2} = 2 \qquad \textbf{Ans. (C)}$$

Sol. 88 For the situation shown in figure below the magnetic field at P due to the two wires is along z direction as field due to wire of 4 A current is more compared to the other one, given as

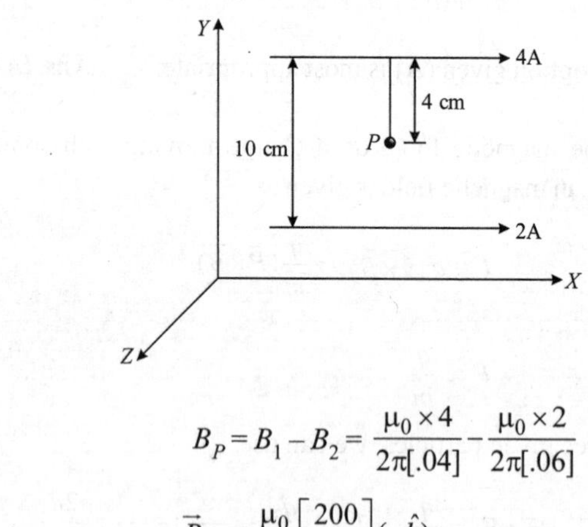

$$B_P = B_1 - B_2 = \frac{\mu_0 \times 4}{2\pi[.04]} - \frac{\mu_0 \times 2}{2\pi[.06]}$$

$$\Rightarrow \quad \vec{B}_P = \frac{\mu_0}{2\pi}\left[\frac{200}{3}\right](-\hat{k})$$

Magnetic force on the charge particle is calculated as

$$\vec{F} = q[\vec{v} \times \vec{B}]$$

$$\Rightarrow \quad \vec{F} = [3\pi]\left[(2\hat{i} + 3\hat{j}) \times \left(\frac{\mu_0}{2\pi}\right)\left(\frac{200}{3}\right) - \hat{k}\right]$$

$$\Rightarrow \quad \vec{F} = 3\pi \times \frac{\mu_0}{2\pi}\left(\frac{200}{3}\right)[2 \times \hat{j} - 3(\hat{i})]$$

$\Rightarrow \quad \vec{F} = (4\pi \times 10^{-7})(100)(-3\hat{i} + 2\hat{j})$

$\Rightarrow \quad \vec{F} = 4\pi \times 10^{-5} \times [-3\hat{i} + 2\hat{j}]$ **Ans. (C)**

Sol. 89 The radius of moving charge in magnetic field is given as

$$R = \frac{mv}{qB} = \frac{\sqrt{2mK}}{eB}$$

For deuteron and proton, we use

$$r_d = \frac{\sqrt{2(2m_p)K}}{eB}$$

and $$r_p = \frac{\sqrt{2m_p K}}{eB}$$

$\Rightarrow \quad \frac{r_d}{r_p} = \frac{\sqrt{2}}{1}$

$\Rightarrow \quad x = 2$ **Ans. (2)**

Sol. 90 For a given kinetic energy, the radius of circular path of the charge particle is given as

$$R = \frac{mv}{qB} = \frac{\sqrt{2mK}}{qB}$$

$$\Rightarrow R = \frac{\sqrt{2\times 24\times 10^{-3}\times 5000\times 1.6\times 10^{-19}}}{\sqrt{6.023\times 10^{23}}\times 1.6\times 10^{-19}\times 0.5}$$

$\Rightarrow R = \frac{61.96}{620.9} = 0.1\text{ m} = 10\text{ cm}$ **Ans. (10)**

Sol. 91 For two parallel wires as shown in figure, attractive force per unit length between wires is given as

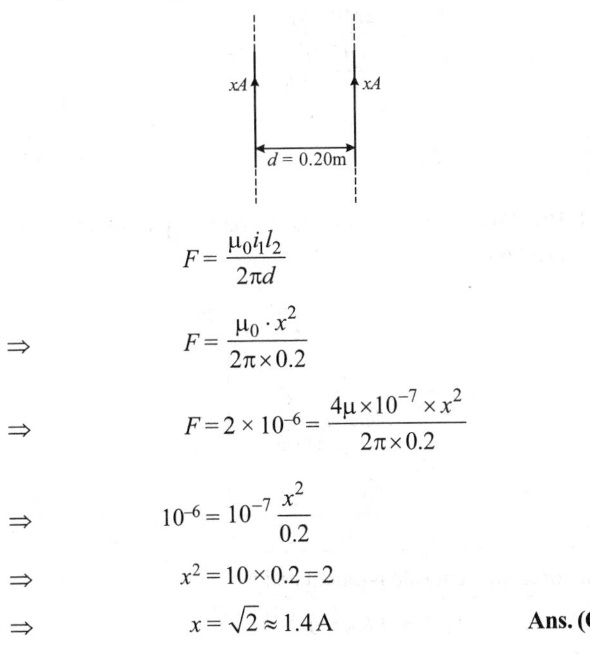

$$F = \frac{\mu_0 i_1 i_2}{2\pi d}$$

$\Rightarrow \quad F = \frac{\mu_0 \cdot x^2}{2\pi\times 0.2}$

$\Rightarrow \quad F = 2\times 10^{-6} = \frac{4\mu\times 10^{-7}\times x^2}{2\pi\times 0.2}$

$\Rightarrow \quad 10^{-6} = 10^{-7}\frac{x^2}{0.2}$

$\Rightarrow \quad x^2 = 10\times 0.2 = 2$

$\Rightarrow \quad x = \sqrt{2} \approx 1.4\text{A}$ **Ans. (C)**

Sol. 92 Radius of charged particle in cyclotron is given as

$$r = \frac{\sqrt{2mK}}{qB}$$

The ratio of new radius to original is calculated as

$\frac{r_n}{r_0} = \sqrt{\frac{K_{new}}{K_{original}}} = \sqrt{4} = 2:1$ **Ans. (C)**

Sol. 93 Electric force as it acts along the direction of field, it can accelerate a charge particle and hence it can change speed and kinetic energy but magnetic force which acts perpendicular to motion of charge cannot change speed and kientic energy of charge particle hence statement-I is correct and statement-II is incorrect. **Ans. (C)**

Sol. 94 Radius of circular path of charges in magnetic field is given as

$$R = \frac{\sqrt{2mK}}{qB}$$

$\Rightarrow \quad q = \frac{\sqrt{2mK}}{RB}$

$\Rightarrow \quad \frac{q_1}{q_2} = \sqrt{\frac{m_1}{m_2}}\times\frac{R_2}{R_1} = \sqrt{\frac{9}{4}}\times\frac{5}{6} = \frac{5}{4}$ **Ans. (B)**

Sol. 95 Magnetic force on a moving charge in magnetic field is given as

$$\vec{F} = q(\vec{v}\times\vec{B})$$

Acceleration of charge is calculated as

$$\vec{a} = \frac{q}{m}(\vec{v}\times\vec{B})$$

Thus $\vec{a}$ & $\vec{B}$ are $\perp$ to each other so we use

$$\vec{a}\cdot\vec{B} = 0$$

$\Rightarrow \quad (\alpha\hat{i} - 4j)\cdot(2\hat{i} + 3\hat{j}) = 0$

$\Rightarrow \quad \alpha(2) + (-4)(3) = 0$

$\Rightarrow \quad \alpha = \frac{12}{2} = 6$ **Ans. (B)**

Sol. 96 The magnetic moment of the loops is calculated as

$$\vec{M} = -i\pi(0.5)^2\hat{k} + i\pi(0.3)^2\hat{k}$$

$\Rightarrow \quad \vec{M} = -7\times\frac{22}{7}\left(\frac{25}{100} - \frac{9}{100}\right)\hat{k}$

$\Rightarrow \quad \vec{M} = -22\left(\frac{16}{100}\right)\hat{k}$

$\Rightarrow \quad \vec{M} = -3.52\hat{k}\text{ Am}^2 = -\frac{7}{2}\hat{k}\text{ Am}^2$ **Ans. (B)**

Sol. 97 The kientic energy of particle is given as

$$E=\frac{1}{2}mv^2$$

$$\Rightarrow \quad 728\,\text{eV}=\frac{1}{2}\times 9.1\times 10^{-31}\times v^2$$

$$\Rightarrow \quad 728\times 1.6\times 10^{-19}=\frac{1}{2}\times 9.1\times 10^{-31}\times v^2$$

$$\Rightarrow \quad v=16\times 10^6\,\text{m/s}$$

For undeflected passing of an electron, we use

$$E=vB$$

$$\Rightarrow \quad E=16\times 10^6\times 12\times 10^{-3}$$

$$\Rightarrow \quad E=192\times 10^3\,\text{V/m}$$

Ans. (A)

Sol. 98 Kinetic energy of electron in cyclotron is given as

$$K=\left[\frac{q^2B^2r_0^2}{2m}\right]$$

$$\Rightarrow \quad K=\frac{(1.6\times10^{-19}\times1\times0.6)^2}{2\times1.6\times10^{-27}\times1.6\times10^{-19}\times10^6}$$

$$\Rightarrow \quad K=18\,\text{MeV}$$

Ans. (B)

Sol. 99 Figure below shows the free body diagram of the rod and all the forces acting on it due to gravity and magnetic field.

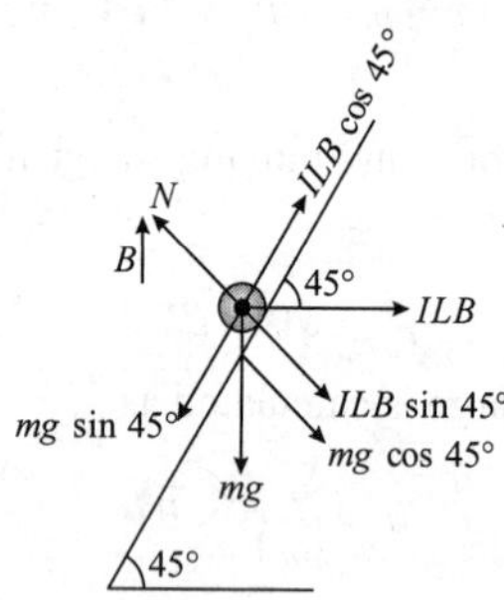

For the rod to be in equilibrium, we use

$$mg\sin 45°=ILB\cos 45°$$

$$\Rightarrow \quad I=\left(\frac{m}{L}\right)\frac{g}{B}$$

$$\Rightarrow \quad I=\frac{(0.45)(10)}{0.15}=30\,\text{A}$$

Ans. (A)

Sol. 100 The magnetic force on segment *CD* is calculated by considering its component perpendicular to magnetic field as effective length on which magnetic force will act, given as

$$F=BIl_{\text{eff}}$$

$$\Rightarrow \quad F=0.5\times(10)\times(5\sin 60\times 10^{-2})$$

$$\Rightarrow \quad F=0.216\,\text{N}$$

Ans. (C)

Sol. 101 Figure below shows the circular coil formed by the wire of given length 314 cm

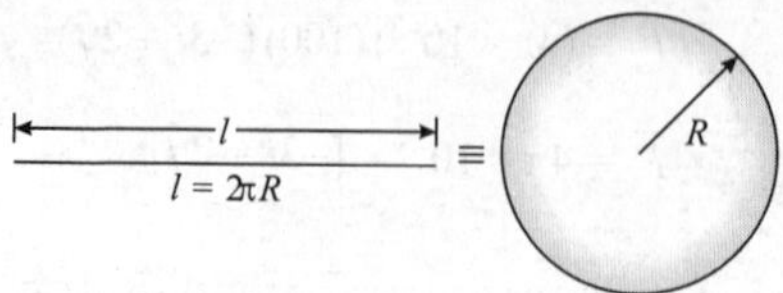

The radius of coil is calculated as

$$\frac{314}{100}=2\pi R$$

$$\Rightarrow \quad R=0.5\,\text{m}$$

The magnetic moment of the coil is given as

$$\Rightarrow \quad \mu=IA$$

$$\Rightarrow \quad \mu=14\times \pi R^2$$

$$\Rightarrow \quad \mu=14\times(3.14)\times\frac{1}{4}$$

$$\Rightarrow \quad \mu=10.99\approx 11.00\,\text{Am}^2$$

Ans. (11)

Sol. 102 As the wires carry current in same direction, the force will be attractive and the force of interaction between two parallel wires is given as

$$F=\frac{\mu_0 I_1 I_2}{2\pi r}l_1$$

$$\Rightarrow \quad F=\frac{4\pi\times10^{-7}\times6\times0.5}{2\pi\times5\times10^{-2}}$$

$$\Rightarrow \quad F=1.2\times 10^{-5}\,\text{N}$$

Ans. (A)

Sol. 103 Force per unit length between two parallel straight current carrying wires is given as

$$F=\frac{\mu_0 i_1 i_2}{2\pi d}$$

$$\Rightarrow \quad \frac{F_1}{F_2}=\frac{\dfrac{\mu_0(10)^2}{2\pi(0.05)}}{\dfrac{\mu_0(20)^2}{2\pi\left(\dfrac{0.05}{2}\right)}}=\frac{1}{8}$$

$$\Rightarrow \quad F_2=8F_1$$

Ans. (A)

Sol. 104 Figure below shows the current loop as described in the question

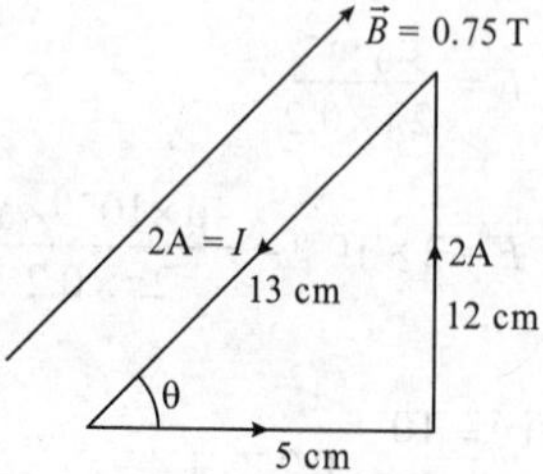

The force on 5 cm side is calculated as

$$|\vec{F}|=ILB\sin\theta$$

$$\Rightarrow \quad |\vec{F}| = (2)(5\times10^{-2})\times\frac{3}{4}\times\frac{12}{13} = \frac{9}{130}\text{ N}$$

$$\Rightarrow \quad x = 9$$ **Ans. (9)**

Sol. 105 At equilibrium position of coil we use restoring torque on coil by the torsional spring is given as

$$\tau = K\theta$$

$$\Rightarrow \quad NiAB = K\theta$$

$$\Rightarrow \quad A = \frac{K\theta}{NiB} = \frac{4\times10^{-5}\times0.05}{200\times10\times10^{-3}\times0.01}$$

$$\Rightarrow \quad A = 10^{-4}\text{ m}^2 = 1\text{cm}^2$$ **Ans. (B)**

Sol. 106 Magnetic induction between two loops is calculated using their flux linkage, given as

$$\phi = Mi$$

Flux through the smaller loop due to magnetic field of larger loop at its centre is given as

$$\phi = (BA)$$

$$\Rightarrow \quad \phi = \pi R^2\left(4\frac{\mu_0}{4\pi}\frac{i}{\left(\frac{L}{2}\right)}\sqrt{2}\right)$$

$$\Rightarrow \quad M = \frac{2\sqrt{2}\mu_0 R^2}{L}$$ **Ans. (C)**

Sol. 107 The magnetic moment of a coil is given as

$$M = NIA$$

Given that

$$M_A = M_B$$

$$\Rightarrow \quad N_A I_A A_A = N_B I_B A_B$$

$$\Rightarrow \quad N_A I_A \pi(0.1)^2 = N_B I_B \pi(0.2)^2$$

$$\Rightarrow \quad N_A I_A = 4N_B I_B$$ **Ans. (C)**

Sol. 108 For equilibrium of loop, we use

$$F_m = mg$$

$$\Rightarrow \quad ILB = mg$$

$$\Rightarrow \quad \left(\frac{V}{R}\right)LB = mg$$

$$\Rightarrow \quad V = \frac{mgR}{LB}$$

$$\Rightarrow \quad V = \frac{(1\times10^{-3}\text{ kg})(10\text{ m/s}^2)(10\Omega)}{(0.1\text{m})(10^3\times10^{-4}\text{T})}$$

$$\Rightarrow \quad V = 10\text{ V}$$ **Ans. (D)**

Sol. 109 Magnetic force per unit length on the wire due to another wire is

$$F = \frac{\mu_0 i_1 i_2}{2\pi r}$$

$$\Rightarrow \quad F \propto I_1 I_2$$

If ammeter reading changes from I to $2I$

$$F_1 : F_{2I} = 1:4$$ **Ans. (B)**

Sol. 110 Workdone in turning the maget is calculated as

$$W = -MB\cos180^\circ\ (-mB\cos0^\circ)$$

$$\Rightarrow \quad W = 2MB = 2\times5\times0.4 = 4\text{J}$$ **Ans. (A)**

Sol. 111 In a moving coil galvanometer curernt sensitivity is given as

Current sensitivity = Voltage sensitivity × R

As current sensitivity is made 1.5 times so for this R also increases 1.5 times as number of turns are increased so new voltage sensitivity is calculated as

$$\text{voltage sensitivity} = \frac{1.5\times\text{current sensitivity}}{1.5\times R}$$

No change in voltage sensitivity. **Ans. (D)**

Sol. 112 The radius of charge particle in magnetic field is calculated as

$$r = \frac{mv}{qB} = \frac{\sqrt{2km}}{qB}$$

$$\Rightarrow \quad m = \frac{r^2q^2B^2}{2k}$$

Figure below shows the motion of charge as described in the question.

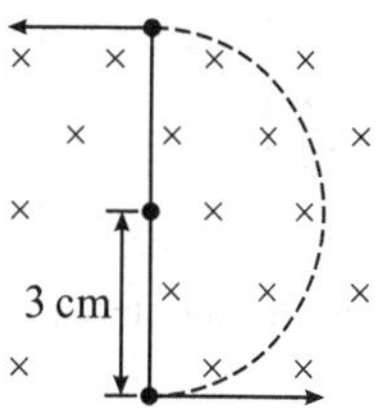

$$\Rightarrow \quad m = \frac{\frac{1}{100}\times\frac{3}{100}\times2\times2\times4\times10^{-3}\times4\times10^{-3}\times10^{-12}}{2\times(100)\times10^{-6}}$$

$$\Rightarrow \quad m = 144\times10^{-18}\text{ kg}$$ **Ans. (144)**

Sol. 113 Magnetic field intensity is given as

$$B = \frac{\pi}{2}\times10^{-3}$$

Kinetic Energy of electron is taken as

$$K = \frac{1}{2}mv^2$$

$$\Rightarrow \quad v = \sqrt{\frac{2KE}{m}}$$

Time period of one revolution is given as

$$T=\frac{2\pi m}{qB}$$

Pitch of helical path is given as

$$P_h = v\cos 60^\circ \times \text{time period of one rotation}$$

$$\Rightarrow \quad P_h = v\cos 60^\circ \times \frac{2\pi m}{eB}$$

$$\Rightarrow \quad P_h = \sqrt{\frac{2\times2\times1.6\times10^{-19}}{1.6\times10^{-27}}}\times\cos 60^\circ \times \frac{2\pi\times1.6\times10^{-27}}{1.6\times10^{-19}\times\frac{\pi}{2}\times10^{-3}}$$

$$\Rightarrow \quad P_h = 2\times10^4\times\frac{1}{2}\times4\times10^{-5}$$

$$\Rightarrow \quad P_h = 4\times10^{-1}\,\text{m} = 40\,\text{cm}$$ **Ans. (40)**

Sol. 114 Due to component of velocity along magnetic field, magnetic force on particle is zero. If this component of velocity is v_1 then v_1 remains unchanged but due to component v_2 magnetic force act in perpendicular direction and causing it to move in circular motion. Thus path is helical with the axis parallel to magnetic field B. **Ans. (C)**

Sol. 115 For a moving coil galvanometer, we have

$$BiNA = k\theta$$

$$\Rightarrow \quad \theta = \left(\frac{BNA}{k}\right)i;$$

Current sensitivity of galvanometer is given as

$$S_i = \frac{BNA}{k}$$

If N is doubled then current sensitivity is doubled. Voltage sensitivity of galvanometer is calculated as

$$B\frac{V}{R}NA = k\theta$$

$$\Rightarrow \quad V = \frac{BNA}{Rk}\theta$$

As N is doubled R is also doubled so, no change in voltage sensitivity. Hence, option (D) is correct. **Ans. (D)**

Sol. 116 We know that current sensitivity and voltage sensitivity

$$I_S = \frac{NBA}{C} \text{ and } V_S = \frac{NBA}{CG}$$

Voltage sensitivity of galvanometer is given as

$$V_S = \frac{I_S}{G}$$

If G (galvanometer resistance) is constant, then $V_S \propto I_S$

Thus percentage change in V_S is also 25%. **Ans. (A)**

Sol. 117 In Solenoid, magnetic field is along the axis of Solenoid. Thus when e^- is moved along axis of Solenoid, force is given as

$$\vec{F} = q(\vec{v}\times\vec{B})$$

As angle between $\vec{v}$ and $\vec{B}$ is 0° so magnetic force is calculated as

$$\vec{F} = 0$$ **Ans. (B)**

Sol. 118 By lorentz magnetic force

$$\vec{F} = -e(\vec{v}\times\vec{B})$$

Force will be along negative y-axis. As magnetic force is perpendicular to velocity, path of electron must be a circle. **Ans. (A)**

Sol. 119 For equilibrium of straight wire with zero tension, we use its weight is balanced by the magnetic force acting on the wire, so we have

$$Mg = IlB$$

Force on conductor in meagnetic field is calculated as

$$F = BIl$$

$$\Rightarrow \quad I = \frac{Mg}{lB} = \frac{40\times10^{-3}\times10}{50\times10^{-2}\times0.4} = 2\text{A}$$ **Ans. (2)**

19.1 Classical and Terrestrial Magnetism

Sol. 1

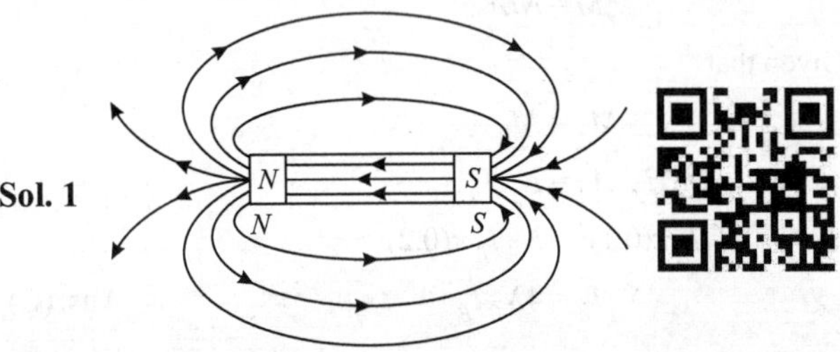

Configuration of magnetic lines in surrounding of a bar magnet. Inside a bar magnet magnetic lines travel from S to N pole. **Ans. (D)**

Sol. 2 The oscillation period of a bar magnet in magnetic field is

$$T = 2\pi\sqrt{\frac{I}{MB}}$$

$$I = \frac{ml^2}{12} \quad \text{and} \quad M = m_s.l$$

If magnet is broken in two equal parts of length $l/2$ then its moment of inertia & magnetic moment will become

$$I' = \frac{(m/2)(l/2)^2}{12} = \frac{I}{8}$$

& $\qquad M' = m_s\left(\frac{l}{2}\right) = \frac{M}{2}$

New time period of oscillation will be

$$\Rightarrow \qquad T' = 2\pi\sqrt{\frac{I'}{M'B}} = 2\pi\sqrt{\frac{I/8}{M/2B}} = \frac{T}{2}$$

$$\Rightarrow \qquad \frac{T'}{T} = \frac{1}{2}$$

Ans. (B)

Sol. 3 In a vibration magnetometer time period of oscillations of a magnet is

$$T = 2\pi\sqrt{\frac{I}{MB}}$$

Here $\qquad I = \frac{ml^2}{12}$ (m = mass)

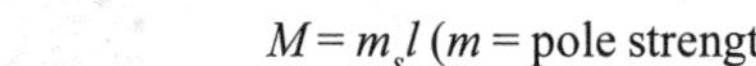

& $\qquad M = m_s l$ (m = pole strength)

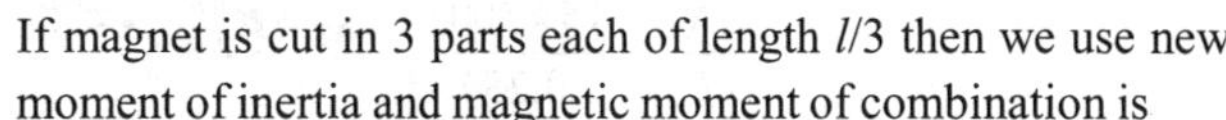

If magnet is cut in 3 parts each of length $l/3$ then we use new moment of inertia and magnetic moment of combination is

$$I' = 3\times\frac{(m/3)(l/3)^2}{12} = \frac{I}{9}$$

and $\quad M' = m_s\frac{l}{3}\times 3 = M$

New time period will be

$$T' = 2\pi\sqrt{\frac{I'}{M'B}} = 2\pi\sqrt{\frac{I/9}{MB}} = \frac{T}{3} = \frac{2}{3}\text{s}$$

Ans. (B)

Sol. 4

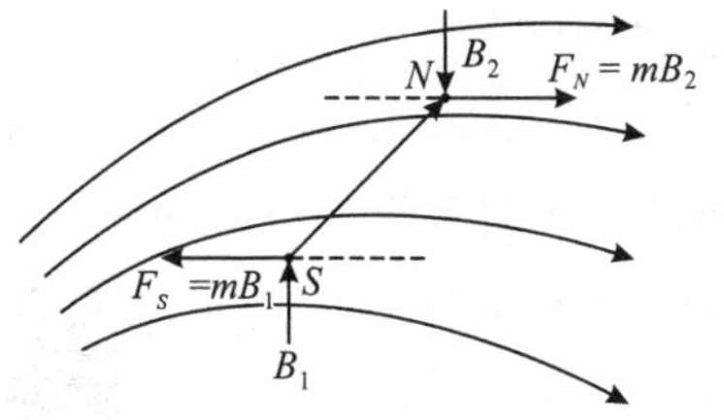

As in non uniform magnetic field $B_1 \neq B_2$

$\Rightarrow$ There will be a net force acting on the needle and as direction F_N and F_S are not along same line.

$\Rightarrow$ There will be a net torque acting as well. **Ans. (C)**

Sol. 5

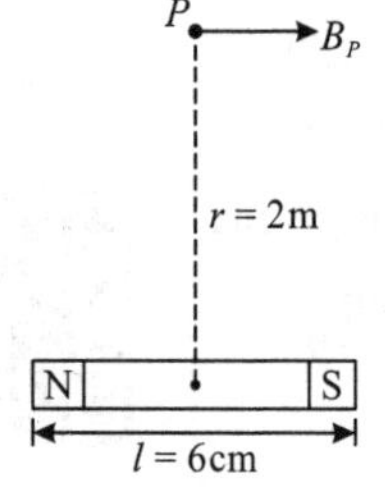

As $r >> l$, here we can use the expression of magnetic induction due to a magnetic dipole, which is given as

For equatorial points $B_P = \frac{kM}{r^3} = \frac{\mu_0}{4\pi}\frac{M}{r^3}$

$$\Rightarrow \qquad B_P = 10^{-7}\times\frac{4}{(2)^3} = 0.5\times 10^{-7}\,\text{T}$$

$$\Rightarrow \qquad B_P = 5\times 10^{-8}\,\text{T}$$

Ans. (C)

Sol. 6

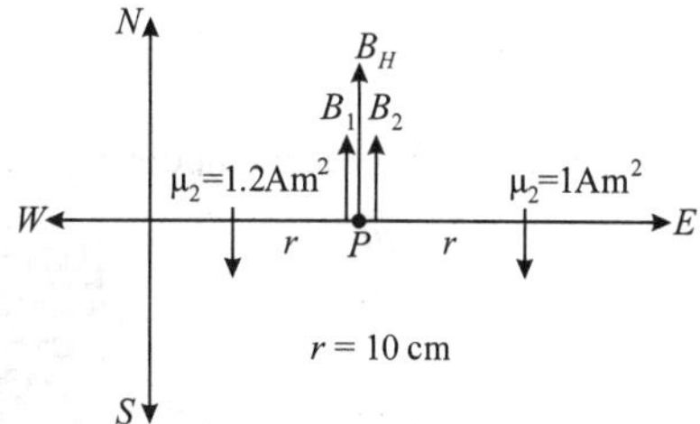

Magnetic moment of the two dipole are μ_1 = 1.20 Am^2 & μ_2 = 1.00 Am^2 and due to these dipoles, magnetic induction at mid point P are given as

$$B_1 = \frac{k\mu_1}{r^3} = \frac{\mu_0}{4\pi}\frac{\mu_1}{r^3} = \frac{10^{-7}\times 1.20}{(0.1)^3}$$

$$\Rightarrow \qquad B_1 = 1.2\times 10^{-4}\,\text{Wb/m}^2$$

and $$B_2 = \frac{k\mu_2}{r^3} = \frac{\mu_0}{4\pi}\frac{\mu_2}{r^3} = \frac{10^{-7}\times 1.00}{(0.1)^3}$$

$$\Rightarrow \qquad B_2 = 1\times 10^{-4}\,\text{Wb/m}^2$$

$\Rightarrow$ Net magnetic induction at P is

$$B_P = B_H + B_1 + B_2$$

$$\Rightarrow \qquad B_P = 3.6\times 10^{-5} + 1.2\times 10^{-4} + 1\times 10^{-4}$$

$$\Rightarrow \qquad B_P = 2.56\times 10^{-4}\,\text{Wb/m}^2$$

Ans. (B)

Sol. 7

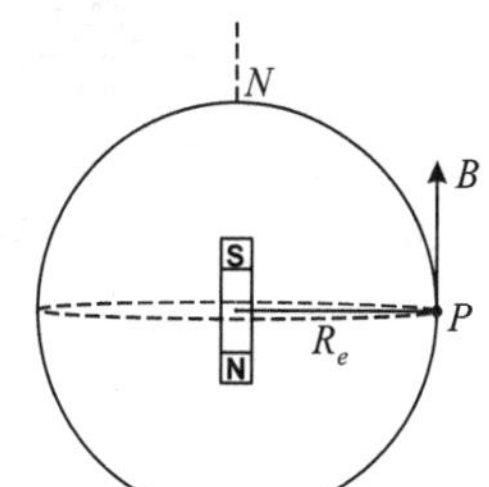

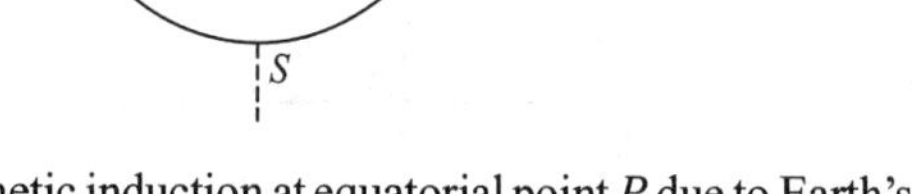

Magnetic induction at equatorial point P due to Earth's magnetic dipole is given as

$$B_P = \frac{kM}{R_e^3}$$

$$\Rightarrow \qquad B_P = \frac{10^{-7}\times 8\times 10^{22}}{(6.4\times 10^6)^3}T$$

$\Rightarrow \quad B_P = 0.302 \times 10^{-4}\,\text{T} = 0.302\,\text{G}$

$1\,\text{G} = 10^{-4}\,\text{T}$ **Ans. (D)**

Sol. 8

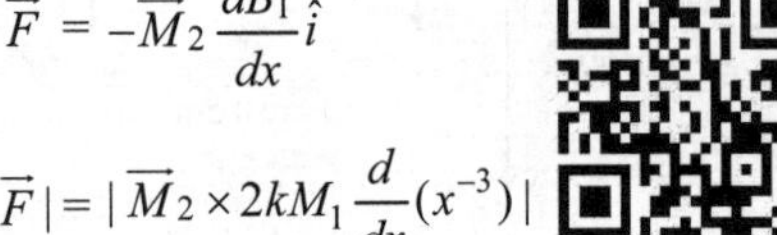

[B_1 is the axial point Magnetic field due to M_1 at a distance x]

Magnetic induction due to M_1 at the location of M_2 is given as

$$B_1 = \frac{2kM_1}{x^3}$$

In non-uniform Magnetic field, force on dipole M_2 is given as

$$\vec{F} = -\vec{M}_2 \frac{dB_1}{dx}\hat{i}$$

In magnitude $\quad |\vec{F}| = |\vec{M}_2 \times 2kM_1 \frac{d}{dx}(x^{-3})|$

$$\Rightarrow \quad |\vec{F}| = \frac{6kM_1M_2}{x^4} \Rightarrow F \propto x^{-4}$$ **Ans. (D)**

Sol. 9

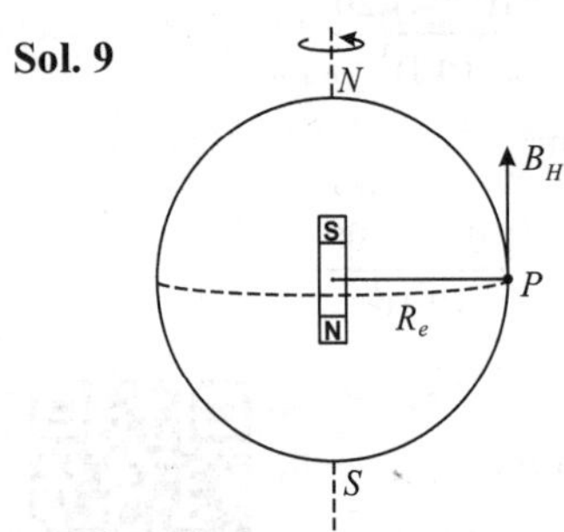

Neglecting declination

For a magnetic dipole at equatorial point, the magnetic induction is given as

$$B_H = \frac{kM}{R_e^3} = 4 \times 10^{-5}$$

$$\Rightarrow \quad M = \frac{4 \times 10^{-5} \times (6.4 \times 10^6)^3}{10^{-7}}$$

$$\Rightarrow \quad M = 1.048 \times 10^{23}\,\text{Am}^2$$ **Ans. (A)**

Sol. 10

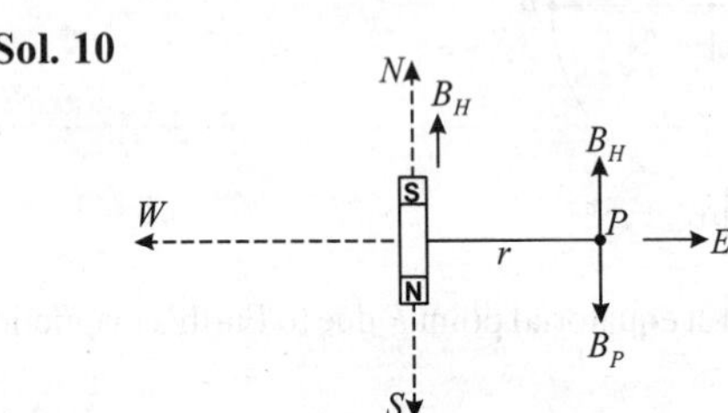

Considering the magnet as a magnetic dipole, magnetic induction at equatorial point P is given as

$$B_P = \frac{KM}{r^3}$$

As at P $\quad B_{\text{net}} = 0$ (null point) Þ $B_P = B_H$

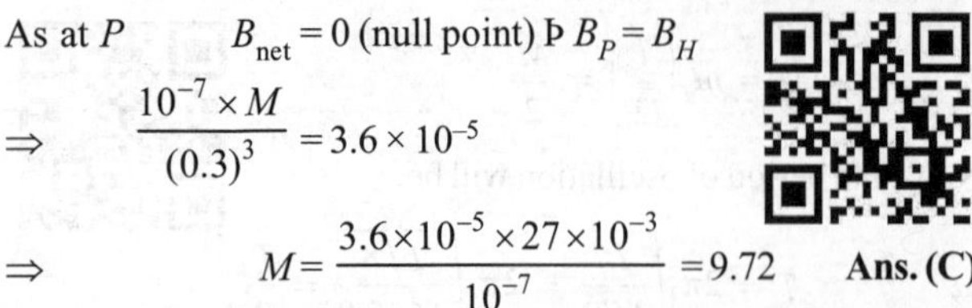

$$\Rightarrow \quad \frac{10^{-7} \times M}{(0.3)^3} = 3.6 \times 10^{-5}$$

$$\Rightarrow \quad M = \frac{3.6 \times 10^{-5} \times 27 \times 10^{-3}}{10^{-7}} = 9.72$$ **Ans. (C)**

Sol. 11

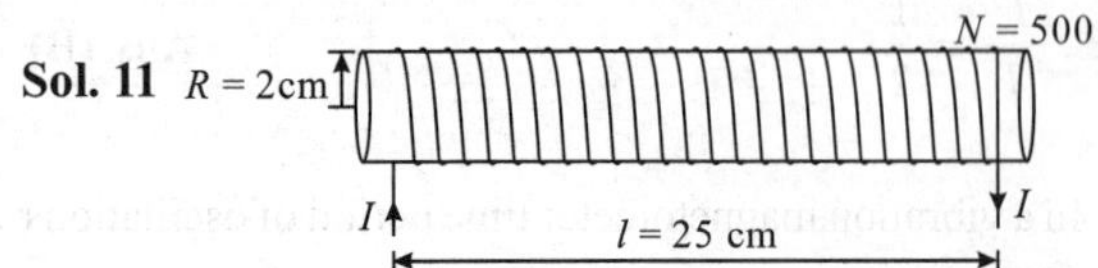

Magnetic moment of solenoid

$$\mu = I(\pi R^2) \times N$$

Volume of solenoid $\quad V = l(\pi R^2)$

Magnetization $\quad |\vec{M}| = \frac{\mu}{V} = \frac{I\pi R^2 N}{l\pi R^2}$

$$= \frac{IN}{l} = \frac{15 \times 500}{0.25}$$

$$\Rightarrow \quad |\vec{M}| = 3 \times 10^4\,\text{A/m}$$ **Ans. (C)**

Sol. 12 Stable equilibrium of a dipole in magnetic field exist when angle θ between its dipole moment and net magnetic induction is zero.

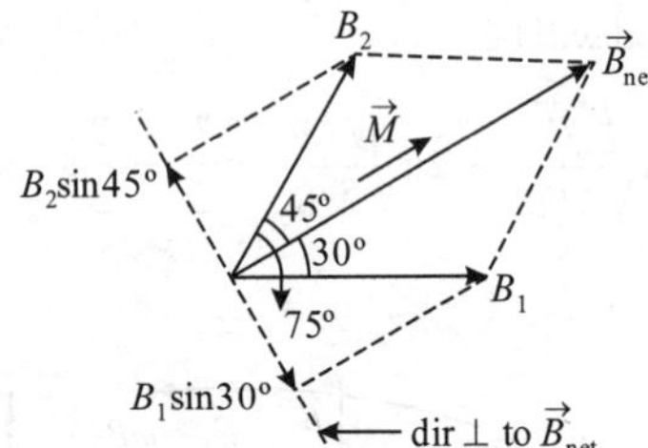

here θ between $\vec{M}$ and $\vec{B}_{\text{net}}$ should be zero. The components of B_1 and B_2 along the direction normal to $\vec{B}_{\text{net}}$ must be equal.

$$\Rightarrow \quad B_1 \sin 30° = B_2 \sin 45°$$

$$15 \times \frac{1}{2} = B_2 \times \frac{1}{\sqrt{2}}$$

$$\Rightarrow \quad B_2 = \frac{15}{\sqrt{2}} = 10.6\,\text{mT}$$ **Ans. (B)**

Sol. 13 For SHM oscillations of magnetic needle must be with small angular amplitude so that restoring torque on it should be given as

$$\tau = MB \sin\theta \approx MB\theta$$

[For small θ we use $\sin\theta \approx \theta$]

⇒ angular acceleration of needle is given as

$$\alpha = -\frac{\tau}{I} = -\frac{MB}{I}\theta$$

[As $\alpha \propto \theta \Rightarrow$ it executes SHM]

Comparing with $\alpha = -\omega^2\theta$, we use $\omega = \sqrt{\frac{MB}{I}}$

Time period of SHM is

$$T = \frac{2\pi}{\omega} = 2\pi\sqrt{\frac{I}{MB}}$$

$$\Rightarrow \quad T = 2\times 3.14\sqrt{\frac{7.5\times 10^{-6}}{6.7\times 10^{-2}\times 0.1}} \approx 0.665\text{s}$$

For 10 oscillations time taken is $= 0.665 \times 10 = 6.66$s **Ans. (C)**

Sol. 14 Potential energy of a magnetic dipole in constant magnetic field is given as

$$U = -\vec{M}\cdot\vec{B}$$

$$\Rightarrow \quad U = -MB\cos\theta$$

Maximum PE is for $\theta = 180°$ when $U_{max} = -MB(-1) = MB$

Minimum PE is for $\theta = 0°$ when $U_{min} = -MB$

Zero PE is for $\theta = 90°$ when $U = -MB(0) = 0$

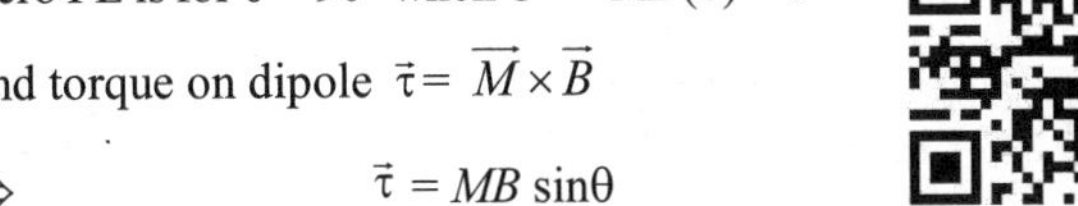

and torque on dipole $\vec{\tau} = \vec{M}\times\vec{B}$

$$\Rightarrow \quad \vec{\tau} = MB\sin\theta$$

$$\Rightarrow \quad \tau = MB\sin 180° = 0 \text{ (Min)}$$

$$\Rightarrow \quad \tau = MB\sin 0 = 0 \text{ (Min)}$$

$$\Rightarrow \quad \tau = MB\sin 90° = MB \text{ (Max)}$$

Ans. (C)

Sol. 15 To keep needle horizontal net torque on it must be zero and net magnetic field at the location of dipole is $\sqrt{2}B$

$$\sqrt{2}B\cdot\mu\sin 45° = F\cdot\frac{l}{2}$$

$$\Rightarrow \quad F = 2mB$$

$$\Rightarrow \quad F = 2\times 1.8\times 18\times 10^{-6}$$

$$\Rightarrow \quad F = 6.48\times 10^{-5}\text{ N}$$

Ans. (B)

Sol. 16 For oscillations of magnetic dipoles, we use

$$T = 2\pi\sqrt{\frac{I}{\mu B}}$$

For hoop, we use

$$T_h = 2\pi\sqrt{\frac{mR^2}{(2\mu)B}}$$

For cylinder, we use

$$T_C = 2\pi\sqrt{\frac{1/2mR^2}{\mu B}} \simeq T_r$$

Ans. (D)

Sol. 17 Magnetic moment of square loop is given as

$$m = NIA = 1\times I\times a^2$$

Here a = side of square

Now, $4a = 2\pi r$

$$r = \frac{2a}{\pi}$$

For circular loop, we use

$$m' = 1\times I\times \pi r^2$$

$$\Rightarrow \quad m' = 1\times I\times\pi\times\left(\frac{2a}{\pi}\right)^2$$

$$m' = \frac{4m}{\pi}$$

Ans. (B)

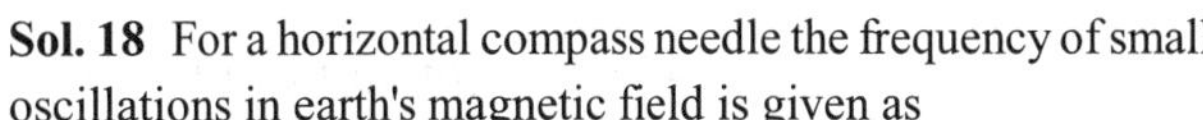

Sol. 18 For a horizontal compass needle the frequency of small oscillations in earth's magnetic field is given as

$$f = \frac{1}{2\pi}\sqrt{\frac{MB_N}{I}} \text{ (horizontal component of earth's field)}$$

When $\theta_D = 45° \rightarrow B_{H_1} = B_1\cos 45°$

$$f_1 = \frac{1}{2\pi}\sqrt{\frac{MB_1\cos 45°}{I}} \quad \ldots(1)$$

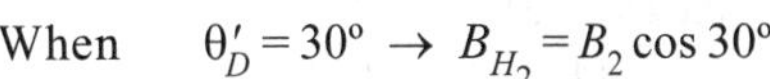

When $\theta'_D = 30° \rightarrow B_{H_2} = B_2\cos 30°$

$$f_2 = \frac{1}{2\pi}\sqrt{\frac{MB_2\cos 30°}{I}} \quad \ldots(2)$$

$$(1)/(2) \Rightarrow \quad \frac{30}{40} = \sqrt{\frac{B_1/\sqrt{2}}{B_2\sqrt{3}/2}}$$

$$\Rightarrow \quad \frac{B_1}{B_2} = \frac{9}{16}\times\sqrt{\frac{3}{2}} \simeq 0.7$$

Ans. (C)

Sol. 19 Stable equilibrium position is at $\theta = 0°$ and unstable equilibrium position is at $\theta = 180°$

Torque on magnet is given as

$$\tau = MB\sin\theta$$

With given values, we use

$$0.018 = M\times 0.06\times(10.5)$$

$$\Rightarrow \quad M = \frac{0.018}{0.06\times 0.5} = 0.6\text{ A-m}^2$$

Work done in rotation of magnet from 0° to 180° is given as

$$W = MB(\cos 0° - \cos 180°)$$

$$\Rightarrow \quad W = 2MB = 2\times 0.6\times 0.06$$

$$\Rightarrow \quad W = 7.2\times 10^{-2}\text{ J}$$

Ans. (D)

Sol. 20 Figure below shows the magnetic field due to the bar magnet at the point P where the field of bar magnet nullifies the horizontal component of earth's field.

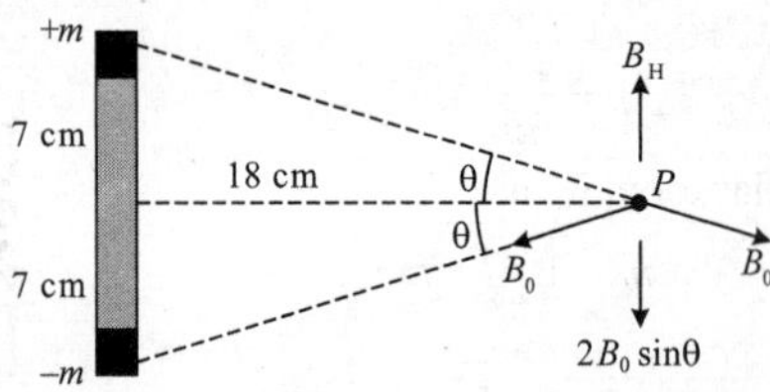

Magnetic induction due to a magnetic pole at point P is given as

$$B_0 = \frac{2\mu_0}{4\pi}\frac{m}{r^2}\times\frac{7}{r} = 0.4\times10^{-4}$$

$$\Rightarrow \quad 2\times10^{-7}\times\frac{m\times7}{(7^2+18^2)^{3/2}}\times10^4 = 0.4\times10^{-4}$$

$$\Rightarrow \quad m = \frac{4\times10^{-2}\times(373)^{3/2}}{14}\text{ A-m}$$

Magnetic moment of the bar magnet is given as

$$M = md = \text{m}\times\frac{14}{100}$$

$$\Rightarrow \quad M = \frac{0.04\times(373)^{3/2}}{14}\times\frac{14}{100}$$

$$\Rightarrow \quad M = 4\times10^{-4}\times7203.82 = 2.88\text{ J/T}$$ **Ans. (C)**

Sol. 21 Apparent dip angle is related to true dip angle by the relation given as

$$\tan\theta_{AD} = \frac{B\sin\phi}{B\cos\phi\cos\alpha}$$

$$\Rightarrow \quad \tan45^\circ = \frac{B\sin\phi}{B\cos\phi\cos30^\circ}$$

$$\Rightarrow \quad \tan\phi = \frac{\sqrt{3}}{2}$$ **Ans. (A)**

Sol. 22 Apparent dip angle is related to true dip angle by the relation given as

$$\tan\theta_{AD} = \frac{B\sin\phi}{B\cos\phi\cos\alpha} = \frac{\tan\phi}{\cos\alpha}$$

Here we can directly state that true dip angle is alware greater than the apparent dip angle. **Ans. (B)**

Sol. 23 Work done in rotation of a magnet in magnetic field is given as

$$W = MB(\cos\theta_1 - \cos\theta_2)$$

We use $\theta_1 = 0^\circ$, $\theta_2 = 60^\circ$ as given in question, so we get

$$W = 2\times10^5\times14\times10^{-5}(1-1/2)$$

$$\Rightarrow \quad W = 14\text{ J}$$ **Ans. (A)**

Sol. 24 Horizontal component of magnetic field at point on earth surface is given as

$$B_H = B\cos\theta$$

$$\Rightarrow \quad B = \frac{B_H}{\cos\theta} = \frac{0.5G}{\cos30^\circ} \Rightarrow \frac{G}{\sqrt{3}}$$ **Ans. (A)**

Sol. 25 The time period of oscillations of a bar magnet in magnetic field is given as

$$T = 2\pi\sqrt{\frac{I}{MB_H}}$$

$$\Rightarrow \quad \frac{T_1}{T_2} = \frac{2\pi\sqrt{\dfrac{I_1}{M_1B_H}}}{2\pi\sqrt{\dfrac{I_2}{M_2B_H}}} = \frac{3}{4}$$

$$\Rightarrow \quad \sqrt{\frac{I_1}{I_2}\times\frac{M_2}{M_1}} = \frac{3}{4}$$

$$\Rightarrow \quad \sqrt{\frac{3}{2}}\times\sqrt{\frac{M_2}{M_1}} = \frac{3}{4}$$

$$\Rightarrow \quad \frac{3}{2}\times\frac{M_2}{M_1} = \frac{9}{16}$$

$$\Rightarrow \quad \frac{M_1}{M_2} = \frac{8}{3}$$ **Ans. (B)**

Sol. 26 Component of horizontal component of earth's magnetic field in the plane of magnet is given as

$$B'_H = B_H\cos\alpha$$

Where α is the angle of magnet plane with magnetic meridian. Writing B_H and B'_H in terms of B_V which is one value at the location of magnet, we have

$$B_V\cot\theta' = B_V\cot\theta\cos\alpha$$

$$\Rightarrow \quad \tan\theta' = \frac{\tan\theta}{\cos\alpha}$$

$$\Rightarrow \quad \sqrt{3} = \frac{\tan\theta}{\frac{1}{\sqrt{2}}}$$

$$\Rightarrow \quad \tan\theta = \sqrt{\frac{3}{2}}$$

$$\Rightarrow \quad \theta = \tan^{-1}\sqrt{\frac{3}{2}}$$ **Ans. (A)**

Sol. 27 Time period of oscillation magnetometer is given as

$$T = 2\pi\sqrt{\frac{I}{B_HM}}$$

$$\Rightarrow \quad T_1 = 3 = 2\pi\sqrt{\frac{I}{(B_P\cos30^\circ)M}}$$

and $T_2 = 6 = 2\pi\sqrt{\frac{I}{(B_Q \cos 60°)M}}$

$\Rightarrow \quad \frac{3}{6} = \sqrt{\frac{1}{\left(B_P \frac{\sqrt{3}}{2}\right)} \times (B_Q/2)}$

$\Rightarrow \quad \frac{3}{6} = \sqrt{\left(\frac{B_Q}{\sqrt{3}B_P}\right)}$

$\Rightarrow \quad \frac{\sqrt{3}}{4} = \frac{B_Q}{B_P}$

$\Rightarrow \quad B_Q : B_P = \sqrt{3} : 4$ **Ans. (A)**

Sol. 28 For a moving coil galvanometer the current passed through galvanometer is related with deflection angle as

$$i = \left(\frac{K}{NAB}\right)\theta$$

Current sensitivity of a galvanometer is given as

$$\frac{d\theta}{di} = \frac{NAB}{K}$$ **Ans. (D)**

Sol. 29 Figure below shows the net magnetic field of earth where dip angle is θ

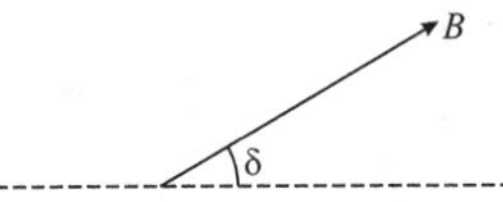

Vertical component of earth's magnetic field is given as

$$B_V = B \sin\theta$$

$\Rightarrow \quad 6 \times 10^{-5} = B\frac{3}{5}$

$\Rightarrow \quad B = 10 \times 10^{-5} = 10^{-4}\,\text{T}$ **Ans. (D)**

Sol. 30 When metal is passing through magnetic field, eddy currents will be induced in the metal body. The motion of body will be opposed by the field of eddy currents, so it will take more time in case of metal body. **Ans. (B)**

Sol. 31 Electromagnets are made of soft iron because it has high permeability and low retentivity. So, both **A** and **R** are correct and **R** is the correct explanation of **A**. **Ans. (B)**

Sol. 32 To demagnetise, magnetic intensity of bar magnet should be equal to induced magnetisation H

$$I = H$$

Given that $I = 2.4 \times 10^3$ A/m

$\Rightarrow \quad 2.4 \times 10^3 = H = ni$

Number of turns per unit length in solenoid is

$$n = \frac{N}{l}$$

$\Rightarrow \quad 2.4 \times 10^3 = \frac{60}{15 \times 10^{-2}} i$

$\Rightarrow \quad i = \frac{2.4 \times 15 \times 10}{60} = \frac{36}{6} = 6\text{A}$ **Ans. (6)**

Sol. 33 After some time force due to weight becomes equal to force due to eddy currents in copper tube. **Ans. (A)**

Sol. 34 As susceptibility of material is given as

$$X_m = 2 \times 10^{-2}$$

Relative permeability of medium is given as

$$\mu_r = 1 + X_m = 1.02$$

Thus magnetic field intensity inside toroid is given as

$\Rightarrow \quad B = \mu_r B_0 = 1.02\, B_0$

Thus percentage increase in magnetic field is given as

$$p = \frac{B - B_0}{B_0} \times 100 = 2\%$$ **Ans. (A)**

Sol. 35 Time period of oscillation for the compass needle is calculated as

$$T = 2\pi\sqrt{\frac{I}{B_H}}$$

$\Rightarrow \quad T \propto \frac{1}{\sqrt{B_H}}$

$\Rightarrow \quad T \propto \frac{1}{\sqrt{B\cos\theta}}$

$\Rightarrow \quad \frac{T_1}{T_2} = \sqrt{\frac{B_2 \cos\theta_2}{B_1 \cos\theta_1}}$

$\Rightarrow \quad \frac{60/20}{60/30} = \sqrt{\frac{B_2 \cos 60°}{B_1 \cos 30°}}$

$\Rightarrow \quad \frac{3}{2} = \sqrt{\frac{B_2}{\sqrt{3}B_1}}$

$\Rightarrow \quad \frac{9}{4} = \frac{B_2}{\sqrt{3}B_1}$

$\Rightarrow \quad \frac{B_1}{B_2} = \frac{4}{9\sqrt{3}} = \frac{4}{\sqrt{243}}$ **Ans. (243)**

19.2 Magnetic Properties of Materials

Sol. 1 Curie temperature is the temperature above which a ferromagnetic material changes to paramagnetic material. **Ans. (A)**

Sol. 2 Materials used for making electromagnets must be quickly magnetized on flow of current & also quickly demagnetize when current is stopped.

$\Rightarrow$ Retentivity and coercivity both must be low. **Ans. (B)**

Sol. 3 Paramagnetic and Diamagnetic materials are weakly magnetized along & opposite to external field.

$\Rightarrow$ N_2 is attracted by external magnet & N_3 is repelled by external magnet weakly Ferromagnetic material strongly magnetized along the direction of external magnetic field.

$\Rightarrow$ N_1 is strongly attracted by magnet. **Ans. (B)**

Sol. 4 For any material always

$$\in_r > 1$$

For a diamagnetic material we use

$$\mu_r < 1$$ **Ans. (B)**

Sol. 5 For demagnetization of magnet we use

$$H_c = \frac{B}{\mu_0} = nI = \frac{N}{l} I$$

$$I = \frac{H_c \times l}{N} = \frac{3 \times 10^3 \times 0.1}{100} = 3\text{ A}$$ **Ans. (C)**

Sol. 6 Diamagnetic $\rightarrow$ repel external magnetic field $\rightarrow$ Bar *B*.

Paramagnetic $\rightarrow$ magnetize in same direction of external Magnetic field $\rightarrow$ Bar *C*.

Ferromagnetic $\rightarrow$ Highly magnetize in the same direction of external magnetic field $\rightarrow$ Bar *A*. **Ans. (B)**

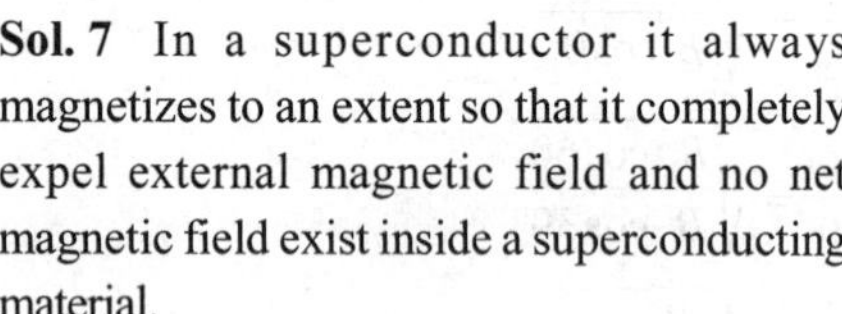

Sol. 7 In a superconductor it always magnetizes to an extent so that it completely expel external magnetic field and no net magnetic field exist inside a superconducting material.

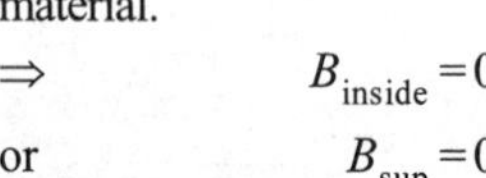

$$\Rightarrow \quad B_{\text{inside}} = 0$$

or $$B_{\text{sup}} = 0$$ **Ans. (B)**

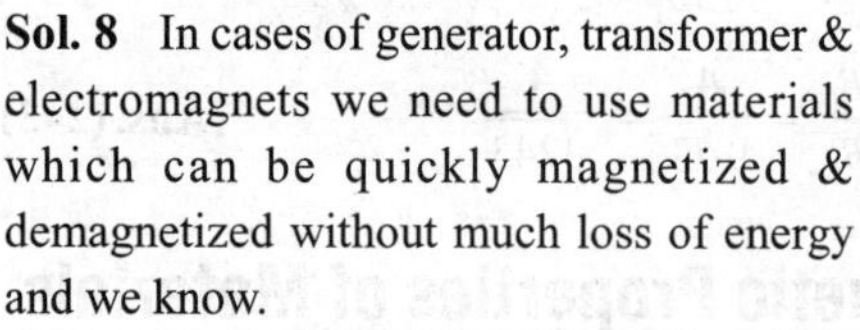

Sol. 8 In cases of generator, transformer & electromagnets we need to use materials which can be quickly magnetized & demagnetized without much loss of energy and we know.

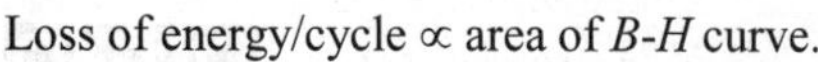

Loss of energy/cycle $\propto$ area of *B-H* curve. **Ans. (D)**

Sol. 9 For demagnetization of ferromagnetic material it is the coercivity which is given as

$$H = 100\text{ A/m}$$

For a solenoid $H = nI$

$$\Rightarrow \quad I = \frac{H}{n} = \frac{(100\text{A/m})}{(10^5\text{ turns/m})} = 10^{-3}\text{ A}$$

$$\Rightarrow \quad I = 1\text{ mA}$$ **Ans. (B)**

Sol. 10 Coercivity of material is given as

$$H_c = \frac{B}{\mu_0}$$

$$\Rightarrow \quad H = ni = \frac{N}{l} i = \frac{100}{0.2} \times 5.2\text{ A/m}$$

$$\Rightarrow \quad H = 2600\text{ A/m}$$ **Ans. (B)**

Sol. 11 Magnetic susceptibility of a material is

$$\chi = \frac{I}{H}$$

Where intensity of magnetization is

$$I = \frac{\text{Magnetic moment}}{\text{Volume}}$$

$$\Rightarrow \quad I = \frac{20 \times 10^{-6}}{10^{-6}} = 20\text{N/m}^2$$

$$\Rightarrow \quad \chi = \frac{20}{60 \times 10^{+3}} = \frac{1}{3} \times 10^{-3}$$

$$\Rightarrow \quad \chi = 0.33 \times 10^{-3} = 3.3 \times 10^{-4}$$ **Ans. (C)**

Sol. 12 We know that susceptibility of a material is given as

$$X \alpha \frac{1}{T_C}$$

$$\Rightarrow \quad \frac{X_1}{X_2} = \frac{T_{C_2}}{T_{C_1}}$$

$$\Rightarrow \quad \frac{2.8 \times 10^{-4}}{X_2} = \frac{300}{350}$$

$$\Rightarrow \quad X_2 = \frac{2.8 \times 350 \times 10^{-4}}{300}$$

$$\Rightarrow \quad X_2 = 3.266 \times 10^{-4}$$ **Ans (C)**

Sol. 13

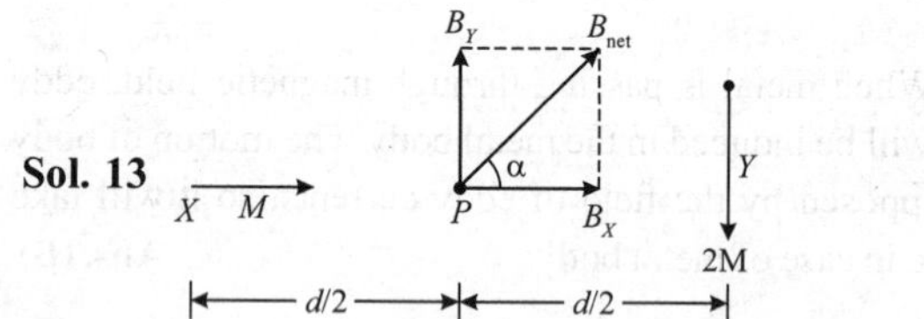

B_X is given as

$$B_X = \frac{2kM_X}{(d/2)^3} = 2\left(\frac{\mu_0}{4\pi}\right) \cdot \frac{M}{(d/2)^3} \qquad (B_X = B_Y)$$

B_Y is given as

$$B_Y = \frac{kM_Y}{(d/2)^3} = \left(\frac{\mu_0}{4\pi}\right) \cdot \frac{2M}{(d/2)^3} \qquad (\alpha = 45°)$$

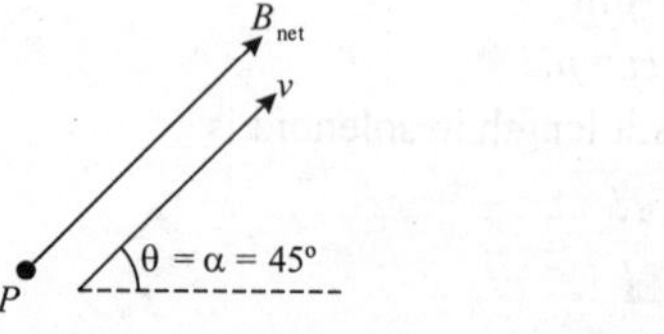

At P, $\vec{B}_{net}$ is $\parallel$ to $\vec{v}$

$\Rightarrow \quad \vec{F}_{mag} = q(\vec{v} \times \vec{B}_{net}) = 0$ **Ans. (D)**

Sol. 14 By definition and from figure, we get

Retentivity = 1.0 T

Coercivity = 50 A/m

Saturation = 1.5 T **Ans. (B)**

Sol. 15 For making of permanent magnets, materials are preferred which have large retentivity and large coercivity. **Ans. (D)**

Sol. 16 The perfect diamagnetic material does not allow external field to enter into it so material inside will not get magnetised. **Ans. (A)**

Sol. 17 By Curies law we use

$$\chi \propto \frac{1}{T}$$

$$\Rightarrow \quad \chi_1 T_1 = \chi_2 T_2$$

$$\Rightarrow \quad \frac{6}{0.4} \times 4 = \frac{I}{0.3} \times 24$$

$$\Rightarrow \quad I = \frac{0.3}{0.4} = 0.75 \text{ A/m}$$ **Ans. (B)**

Sol. 18 Soft ferromagnetic materials are materials which can be magnetised very fast and easily with small magnetic field also and also demagnetised easily by external magnetic field. When external field is applied, the domains experiences a net torque hence change their orientation. **Ans. (B)**

Sol. 19 Domain are the small localized regions of dipoles aligned in same orientation due to coupling of magnetic field by neighbouring dipoles. Out of all the given options, (D) is most appropriate. **Ans. (D)**

Sol. 20 Statements (a) is NOT correct because existance of a charge can be considered as an electric monopole. Statement (b) is NOT correct because magnetic field lines at the ends of a solenoid are not straight but confined as closed loops only. Statement (c) is correct because for a closely wound toroid, magnetic field outside the toroid is zero and all the magnetic field lines are confined within the toroid only. Statement (d) is NOT correct because inside a bar magnet, magnetic lines are considered to be confined within its body only and parallel to each other. Statement (e) is correct because for super conducting materials inside their body the net magnetic field is always zero and they are perfect diamagnetic with relative permeability zero so we have

$$\mu_r = 1 + \chi$$

$$\chi = -1$$ **Ans. (A)**

Sol. 21 Magnetic susceptibility of material is related to its permeability as

$$\mu_r = 1 + \chi$$

$$\Rightarrow \quad \mu_r = 1 + 499 = 500$$

Absolute permeability of the material of rod is given as

$$\mu = \mu_0 \times \mu_r$$

$$\Rightarrow \quad \mu = 500 \times 4\pi \times 10^{-7}$$

$$\Rightarrow \quad \mu = 2\pi \times 10^{-4} \text{ H/m}$$ **Ans. (D)**

Sol. 22 In a ferromagnetic property of a material depends upon temperature and at high temeperature it looses its ferromagnetic properties hence Statement-I is true and Statement-II is flase. **Ans. (D)**

Sol. 23 (m_1 upward, m_2 to the right, B_1 downward)

$$\vec{\tau} = \vec{M}_2 \times \vec{B}_1$$

$$\tau = M_2 B_1 \sin 90° = 1 \times \frac{\mu_0}{4\pi} \frac{M}{(1)^3} 1 = 10^{-7} \text{ N.m}$$

Ans. (1)

Sol. 24 Torque acting on the coil si given as

$$\tau = M \times B$$

$$\Rightarrow \quad \tau = IAB$$

$$\Rightarrow \quad \tau = 0.2 \times \left(\frac{\sqrt{3}a^2}{4}\right) \times 2 \times 10^{-2}$$

$$\Rightarrow \quad \tau = \frac{0.2 \times \sqrt{3} \times 10^{-2} \times 2 \times 10^{-2}}{4} = \sqrt{3} \times 10^{-5} \text{ T}$$

Ans. (3)

Sol. 25 When the coil is in the form of a square having n_1 turns, we have

$$4an_1 = 24a$$

$$\Rightarrow \quad n_1 = 6 \text{ turns}$$

When coil is in form of an equilateral triangle having n_2 turns, we have

$$3an_2 = 24a$$

$$\Rightarrow \quad n_2 = 8 \text{ turns}$$

Magnetic moment of a coil with current i and n turns having area A is given as

$$M = niA$$

$$\Rightarrow \quad \frac{M_1}{M_2} = \frac{n_1 i A_1}{n_2 i A_2} = \frac{6 \times a^2}{8 \times a^2 \times \frac{\sqrt{3}}{4}} = \frac{\sqrt{3}}{1}$$

$$\Rightarrow \quad \frac{M_2}{M_1} = \frac{1}{\sqrt{3}} = \frac{1}{\sqrt{y}}$$

$$\Rightarrow \quad y = 3$$ **Ans. (3)**

Sol. 26 Intensity of magnetisation in the material is given as

$$\vec{M} = \frac{\text{magnetic moment}\,(\vec{m})}{V}$$

$$\Rightarrow \quad \chi \vec{H} = \frac{\vec{m}}{V}$$

$$\Rightarrow \quad \vec{m} = (\mu_r - 1)\,Niv$$

The fractional change in the magnetic moment of core can be calculated as

$$\frac{\Delta m}{m} = \frac{[(750-1)-(500-1)]Niv}{(500-1)Niv} = 250$$

Ans. (250)

Sol. 27 For a diamagnetic material magnetisation and magnetising field are proportional and opposite to each other so option (a) is correct and in diamagnetic materials magnetic suseptibility does not depend on temperature so option (c) is also correct. **Ans. (D)**

Sol. 28 Soft iron is a ferromagnetic material which has high permeability and it can be quickly magnetised or demagnetised hence it has low retentivity. **Ans. (C)**

Sol. 29 According to curie's law, magnetic susceptibility is inversely proportional to temperature for a given external magnetic field, given as

$$\chi = \frac{C}{T}$$

The same is applicable for ferromagnetic material & the relation can be given as

$$\chi = \frac{C}{T - T_C}$$

Here T_C is the Curie Temperature. Hence statement-I is true. Diamagnetism is due to non-cooperative behaviour of orbiting electrons when exposed to external magnetic field hence option statement-II is also correct. **Ans. (A)**

Sol. 30 Susceptibility is related to relative permeability of a material as

$$\mu_r = \frac{\mu}{\mu_0} = 1 + \chi$$

$$\Rightarrow \quad \mu = \mu_0(1+\chi)$$

$$\Rightarrow \quad \mu = 4\pi \times 10^{-7}[1+99]$$

$$\Rightarrow \quad \mu = 4\pi \times 10^{-5}$$

Ans. (C)

Sol. 31 For magnetic susceptibility $\chi = 1.2 \times 10^{-5}$ the material has relative permeability, given as

$$\mu_r = 1 + \chi = 1 + 1.2 \times 10^{-5} = 2.2 \times 10^{-5}$$

Fractional Change in magnetic field is given as

$$f = \frac{\Delta B}{B} = \frac{\mu_0\mu_r ni - \mu_0 ni}{\mu_0 ni} = (\mu_r - 1)$$

$$\Rightarrow \quad f = 1.2 \times 10^{-5}$$

Ans. (A)

Sol. 32 Due to motion of the coil eddy current develops which opposes the motion thus bringing the coil to rest. **Ans. (B)**

Sol. 33 Both Statements are correct as for diamagnetic substances x is negative and so when placed in external magnetic field tend to move from stronger to weaker part. **Ans. (B)**

Sol. 34 When bar magnet is dropped through a metallic pipe, Eddy currents are produced in metallic pipe which opposes the motion. **Ans. (C)**

Sol. 35 Diamagnetic property doesn't depend on temperature, as it is a characteristic property. **Ans. (A)**

20.1 Electromagnetic Induction

Sol. 1

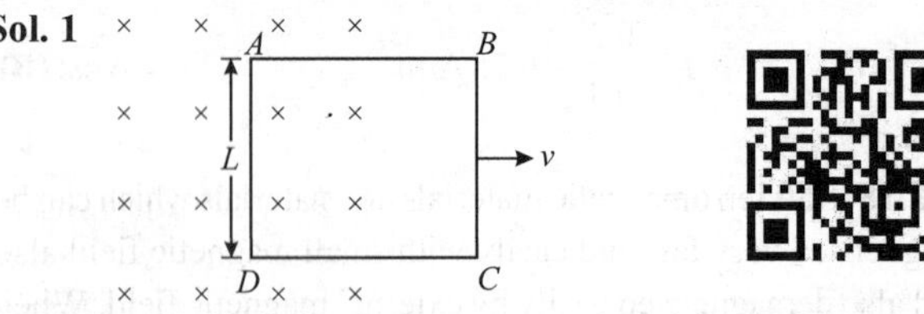

During motion, only wire segment AD will cut magnetic flux and motional emf is induced in it, which is given as

$$e_{AD} = BvL$$

$$v_A - v_D = BvL$$

Ans. (D)

Sol. 2

$\omega = 5\text{rad/s}$, $B_H = 0.2 \times 10^{-4}$, $l = 1\text{m}$

EMF induced across a rotating conductor in uniform and normal magnetic field is given as

$$e = \frac{1}{2}B\omega l^2$$

$$\Rightarrow \quad e = \frac{1}{2} \times 0.2 \times 10^{-4} \times 5 \times (1)^2$$

$$\Rightarrow \quad e = 0.5 \times 10^{-4}\,\text{V} = 50\,\mu\text{V}$$

Ans. (B)

Sol. 3

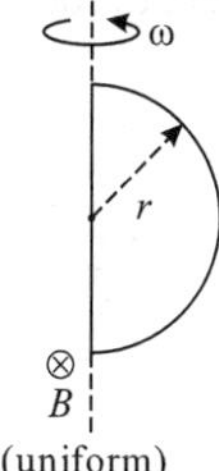

EMF induced in a rotating loop at angular frequency ω is given as

$$e = NBA\sin\omega t = \frac{B\pi\omega r^2}{2}\sin\omega t$$

induced current in loop is $i = e/R$

rate of heat generated (thermal power) in loop is

$$P_{th} = i^2R = \frac{(B\pi\omega r^2)^2}{4R}\sin^2\omega t$$

$$\Rightarrow \text{mean power } \langle P_{th}\rangle = \frac{(B\pi\omega r^2)^2}{4R}\langle\sin^2\omega t\rangle$$

$$\Rightarrow \quad \langle P_{th}\rangle = \frac{(B\pi\omega r^2)^2}{4R}\times\frac{1}{2} = \frac{(B\pi\omega r^2)^2}{8R}$$ **Ans. (B)**

Sol. 4 For a given n turns coil when it is moved such that the flux changes through it is given as

$$\Delta\phi = n(W_2 - W_1)$$

⇒ emf induced in coil is

$$e = -\frac{\Delta\phi}{\Delta t} = -\frac{n(W_2 - W_1)}{t}$$

effective resistance of circuit = $5R$

⇒ Current induced in coil will be

$$i = \frac{e}{5R} = -\frac{n(W_2 - W_1)}{5Rt}$$ **Ans. (B)**

Sol. 5

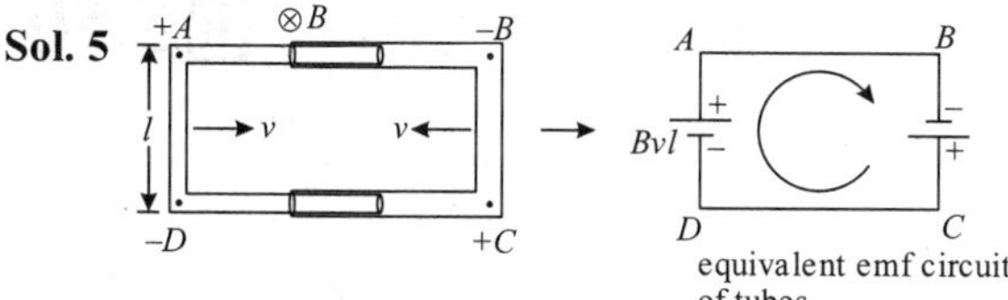

equivalent emf circuit of tubes

As sections AD and BC are moving normal to their lengths so emf is induced only in these sections so motional emf induced is

$$e_{AD} = Bvl \text{ and } e_{BC} = Bvl$$

Total loop emf in circuit

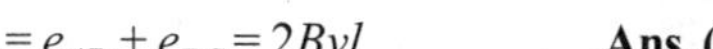

$$e_T = e_{AD} + e_{BC} = 2Bvl$$ **Ans. (D)**

Sol. 6 Motional emf across the conductor PQ is

$$e_{PQ} = Blv$$

⇒ equivalent circuit will be (with emf)

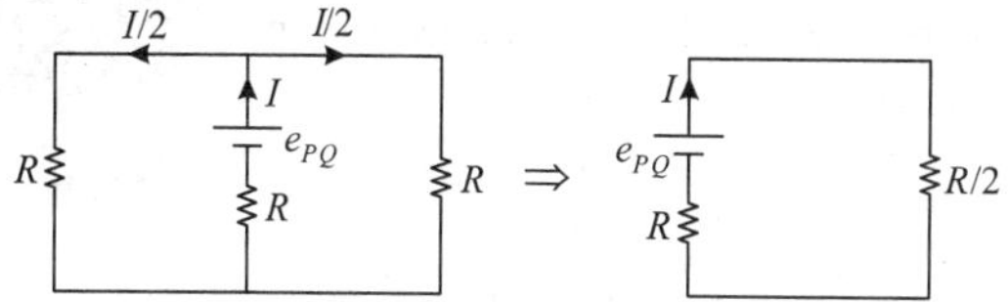

From figure, current supplied by battery is

$$I = \frac{e_{PQ}}{R + \frac{R}{2}} = \frac{Blv}{\frac{3R}{2}} = \frac{2Blv}{3R}$$

Currents through side resistances is

$$I_1 = I_2 = \frac{I}{2} = \frac{Blv}{3R}$$ **Ans. (C)**

Sol. 7 For a straight wire moving perpendicular to the magnetic field, the motional emf is given as

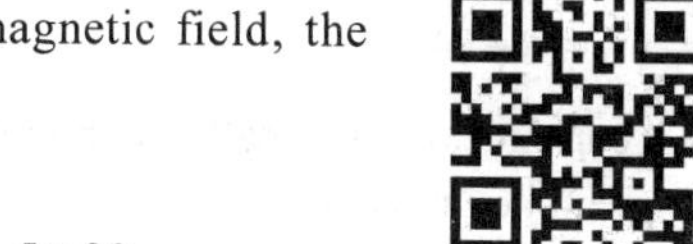

$$e = Bvl$$

$$\Rightarrow \quad e = 0.3\times10^{-4}\times5\times20$$

$$\Rightarrow \quad e = 3\times10^{-3}\text{ V} = 3\text{ mV}$$ **Ans. (B)**

Sol. 8 Here $B_H = 5\times10^{-5}$ T (As 1T = 1 N/A-m)

$l = 2$m [vertical]

$v = 1.5$ m/s

As vertical aerial cuts horizontal component of earth's field normally so motional emf induced in it is given as

$$e = Blv = 5\times10^{-5}\times2\times1.5$$

$$\Rightarrow \quad e = 1.5\times10^{-4} = 0.15\text{ mV}$$ **Ans. (B)**

Sol. 9

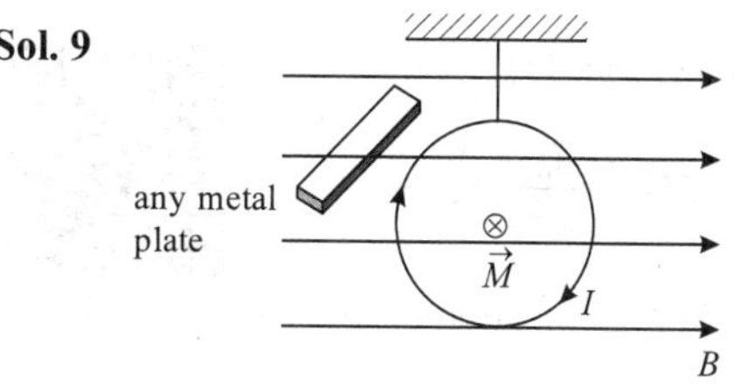

Torque on coil $\vec{\tau} = \vec{M}\times\vec{B}$

$$\tau = MB\sin\theta$$

$$\tau = MB$$

at θ = 0°, coil is in state of stable equilibrium about which it oscillates.

When an aluminium plate (any metal plate) is kept near the coil then due to oscillation of coil $\vec{B}_C$ due to coil will be time varying and eddy currents are induced in plate and that induction opposes its cause which is oscillation of coil so these are damped out. **Ans. (A)**

Sol. 10 When $|\Delta\phi|$ flux is changed through a coil of resistance R, then charge flown through the coil is given as

$$\Delta q_f = \frac{|\Delta\phi|}{R}$$

$$\Rightarrow \quad |\Delta\phi| = \Delta q_f \times R = \text{area under } i\text{-}t \text{ curve} \times R$$

$$\Rightarrow \quad |\Delta\phi| = \frac{1}{2} \times 4 \times 0.1 \times 10 = 2 \text{ Wb}$$ **Ans. (C)**

Sol. 11

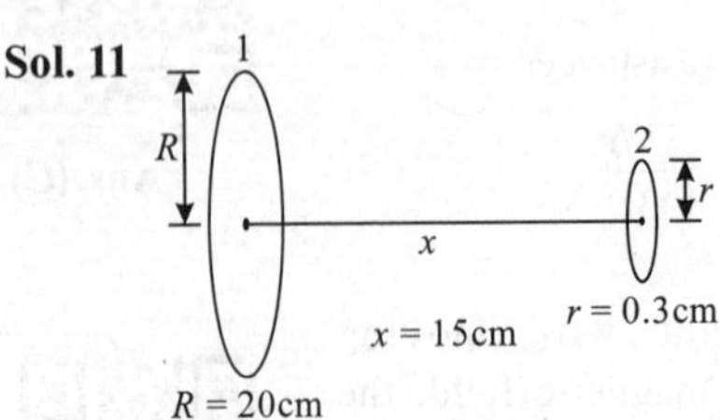

If i_2 current flows in smaller loop then flux linked with bigger loop can be given as

$$\phi_1 = Mi_2 \quad \ldots(1)$$

where $M \rightarrow$ coefficient of mutual inductance between loops 1 and 2.

To calculate M, we supply a current i, through bigger loop & calculate flux linked with smaller loop as

$$\phi_2 = Mi_1 \quad \ldots(2)$$

Magnetic induction at location of loop 2 due to loop 1 is

$$B = \frac{\mu_0 i_1 R^2}{2(R^2 + x^2)^{3/2}}$$

flux through loop -2 is $\phi_2 = B.\pi r^2 = \dfrac{\mu_0 \pi r^2 R^2}{2(R^2 + x^2)^{3/2}} i_1 \quad \ldots(3)$

Comparing equation-(2) and (3) gives

$$M = \frac{\mu_0 \pi r^2 R^2}{2(R^2 + x^2)^{3/2}}$$

Using above value of M in equation-(1), we get

$$\phi_1 = \frac{\mu_0 \pi r^2 R^2}{2(R^2 + x^2)^{3/2}} i_2$$

$$\Rightarrow \quad \phi_1 = \frac{4\pi \times 10^{-7} \times \pi \times (0.3 \times 10^{-2})^2 \times (0.2)^2 \times 2}{2 \times ((0.2)^2 + (0.15)^2)^{3/2}}$$

$$\Rightarrow \quad \phi_1 = \frac{4\pi \times 10 \times 10^{-7} \times 9 \times 10^{-6} \times 4 \times 10^{-2}}{(0.25)^3}$$

$$\Rightarrow \quad \phi_1 = 9.216 \times 10^{-11} \text{ Wb}$$ **Ans. (A)**

Sol. 12 Motional EMF induced in the element of rod as shown is

$$de = B(x\omega)dx$$

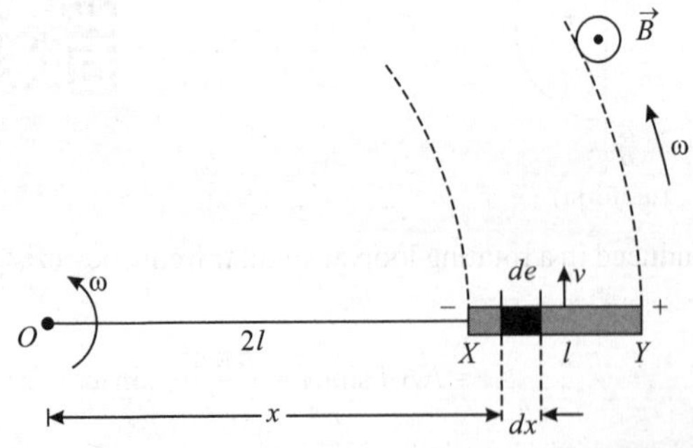

Total EMF induced in rod is

$$e = \int de = B\omega \int_{2l}^{3l} x dx = B\omega \left[\frac{x^2}{2}\right]_{2l}^{3l}$$

$$\Rightarrow \quad e = \frac{1}{2} B\omega l^2 [9 - 4] = \frac{5}{2} B\omega l^2$$ **Ans. (D)**

Sol. 13

The motional emf will be induced in wire segments BC and AD which are given as

$$e_{AD} = B_1 vl = \frac{\mu_0 I}{2\pi x} vl \quad \ldots(1)$$

and

$$e_{BC} = B_2 vl = \frac{\mu_0 I}{2\pi(l + x)} vl \quad \ldots(2)$$

Total emf induced in loop $ABCD$ is given as $(e_{AD} > e_{BC})$

$$e_T = e_{AD} - e_{BC}$$

from (1) and (2)

$$e_T = \frac{\mu_0 Ivl}{2\pi}\left[\frac{1}{x} - \frac{1}{x + l}\right]$$

$$\Rightarrow \quad e_T = \frac{4\pi \times 10^{-7} \times 1 \times 10 \times 0.1}{2\pi}\left[\frac{1}{0.1} - \frac{1}{0.2}\right]$$

$$\Rightarrow \quad e_T = 2 \times 10^{-7} \times \frac{1}{0.2} = 10^{-6} \text{ V}$$

$$\Rightarrow \quad e_T = 1 \,\mu\text{V}$$ **Ans. (B)**

Sol. 14 emf induced in the coil is

$$e = \left|\frac{d\phi_L}{dt}\right|$$

[ϕ_2 - flux linked with all turns of coil]

Here $\phi_L = N \times \phi = N(BA)$

Average emf induced $e = \left|\frac{\Delta\phi_L}{\Delta t}\right| = NA\frac{\Delta B}{\Delta t}$

$\Rightarrow \quad e = 1000 \times 4 \times 10^{-4} \times \frac{10^{-2}}{0.01}$

$\Rightarrow \quad e = 0.4\,\text{V} = 400\,\text{mV}$ **Ans. (A)**

Sol. 15 In a cylindrical region TVMF, the induced EF at a distance r from axis of field is given as

for $r < R \rightarrow \quad E_i = \frac{1}{2}r\frac{dB}{dt} \Rightarrow E \propto r$

for $r < R \rightarrow \quad E_0 = \frac{1}{2}\frac{R^2}{r}\frac{dB}{dt} \Rightarrow E \propto \frac{1}{r}$ **Ans. (A)**

Sol. 16 Magnetic field components at Delhi
As Delhi is in Northern hemisphere, we use

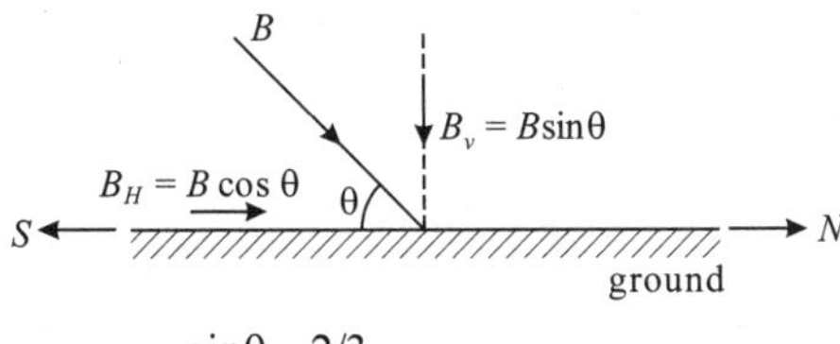

$\sin\theta = 2/3$

$\cos\theta = \sqrt{1-\sin^2\theta} = \frac{\sqrt{5}}{3}$

Flight of plane

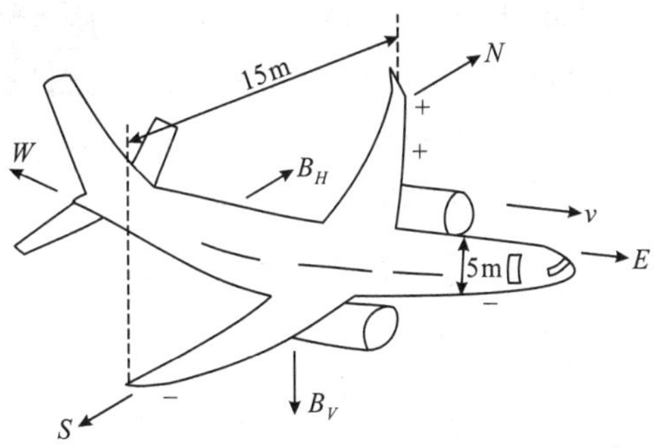

Motional EMF between top and bottom of plane

$e_{TB} = v_B = Blv = B_H hv = B\cos\theta\, hv$

$\Rightarrow \quad v_B = 5 \times 10^{-5}\,T \times \frac{\sqrt{5}}{3} \times 5 \times 240$

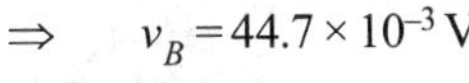

$\Rightarrow \quad v_B = 44.7 \times 10^{-3}\,\text{V}$

Motional EMF between the tips of wings is

$e_{\text{Tips}} = v_w = Blv = B_v wv = B\sin\theta wv$

$\Rightarrow \quad v_w = 5 \times 10^{-5} \times \frac{2}{3} \times 15 \times 240$

$\Rightarrow \quad v_w = 120 \times 10^{-3}\,\text{V}$ **Ans. (D)**

Sol. 17

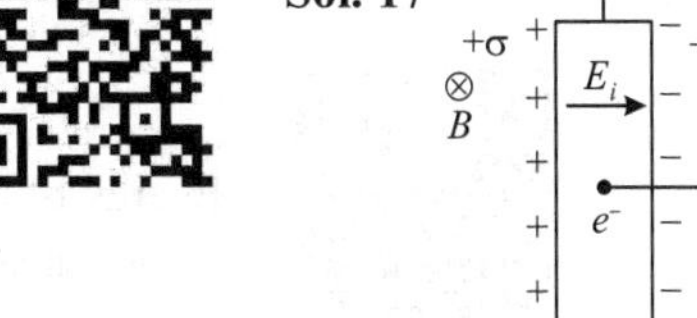

Due to motion of sheet, free electrons in sheet experience force toward right, given as

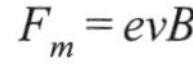

$$F_m = evB$$

Due to magnetic forces the free e^-'s drift and develops internal EF due to induced charges so that net force on free electron should be zero.

$F_e = eE_i \quad e^- \quad F_m$

Where $E_i = \sigma/\epsilon_0$ (due to induced charges on surface of sheet)

we use $\quad eE_i = evB$

$\Rightarrow \quad \frac{\sigma}{\epsilon_0} = vB$

$\Rightarrow \quad \sigma = \epsilon_0 vB$

$\Rightarrow$ On left and right surfaces of sheet induced charges density will be $+\epsilon_0 vB$ and $-\epsilon_0 vB$. **Ans. (B)**

Sol. 18 Magnetic flux through the loop is

$$\phi = B\pi r^2 = \pi r^2 B_0 e^{-t/\tau}$$

loop emf induced in the loop is $e = \left|\frac{d\phi}{dt}\right| = \frac{\pi r^2 B_0}{\tau}e^{-t/\tau}$

induced current in the loop is $i = \frac{e}{R} = \frac{\pi r^2 B_0}{\tau R}e^{-t/\tau}$

Total heat generated in loop is

$$H = \int_0^\infty i^2 R\,dt = \frac{\pi^2 r^4 B_0^2}{\tau^2 R}\int_0^\infty e^{-2t/\tau}dt$$

$\Rightarrow \quad H = \frac{\pi^2 r^4 B_0^2}{\tau^2 R}\left[\frac{e^{-2t/\tau}}{-2/\tau}\right]_0^\infty$

$\Rightarrow \quad H = \frac{\pi^2 r^4 B_0^2}{2\tau R}\left[e^{-2t/\tau}\right]_\infty^0$

$\Rightarrow \quad H = \frac{\pi^2 r^4 B_0^2}{2\tau R}[1-0] = \frac{\pi^2 r^4 B_0^2}{2\tau R}$ **Ans. (B)**

Sol. 19 The charge which flows through a coil of resistance R due to change in magnetic flux $\Delta\phi$ is given as

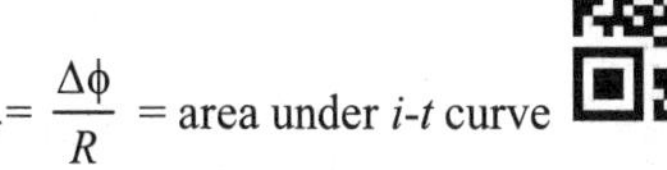

$q_f = \frac{\Delta\phi}{R}$ = area under i-t curve

$\frac{1}{2} \times 10 \times 0.5 = \frac{\Delta\phi}{100}$

$\Rightarrow \quad \Delta\phi = 250\,\text{Wb}$ **Ans. (A)**

Sol. 20

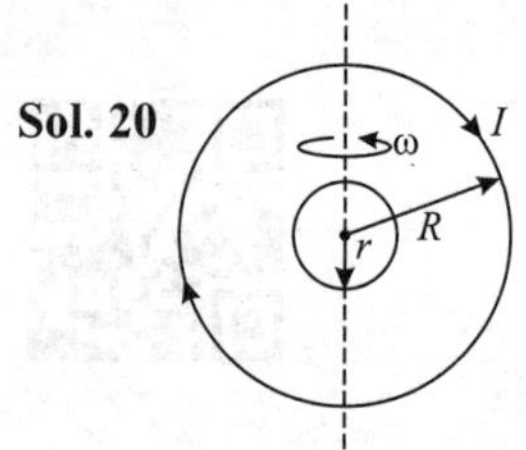

As $r << R$. Magnetic field at centre of larger coil is almost uniform for smaller coil.

The smaller coil is rotating in the magnetic field of larger coil at centre $\left(B_C = \frac{\mu_0 I}{2R}\right)$ so emf induced due to rotation of smaller coil is $e = NBA\omega \sin\omega t$

$$\Rightarrow \quad e = \frac{\mu_0 I}{2R}(\pi r^2)\omega \sin\omega t$$ **Ans. (C)**

Sol. 21

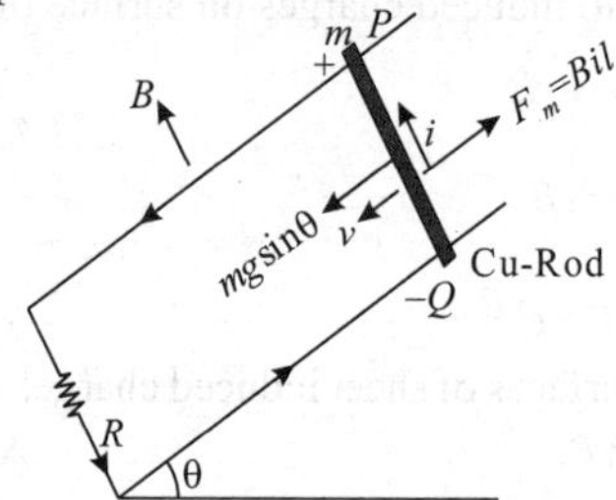

When rod is moving at speed v, the motional emf induced in rod will be

$$e_{PQ} = Bvl$$

$\Rightarrow$ Induced current is

$$i = \frac{e_{PQ}}{R} = \frac{Bvl}{R}$$

Magnetic force on rod due to B is

$$F_m = Bil = \frac{B^2vl^2}{R}$$

At terminal speed $a_{rod} = 0$

$$\Rightarrow \quad mg\sin\theta = F_m$$

$$\Rightarrow \quad mg\sin\theta = \frac{B^2vl^2}{R}$$

$$\Rightarrow \quad v = \frac{mgR\sin\theta}{B^2l^2}$$ **Ans. (B)**

Sol. 22

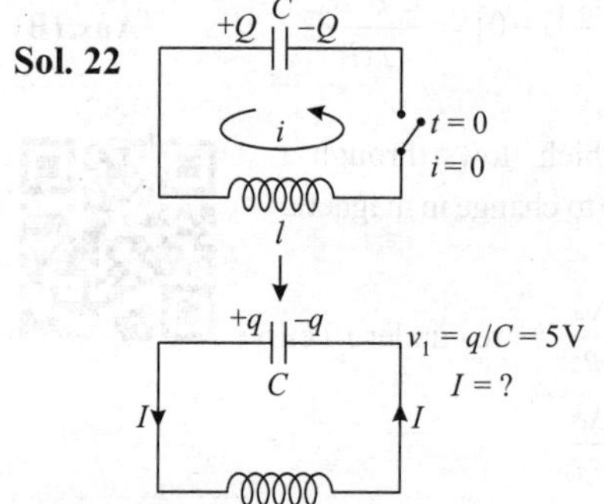

Initial energy in capacitor is

$$E = \frac{1}{2}CV^2 = \frac{1}{2} \times 0.2 \times 10^{-6} \times (10)^2 = 10\ \mu J$$

Energy of circuit (sum of $E_C + E_L$) when potential difference across capacitor is 5 V, is given as

$$E = \frac{1}{2}CV_1^2 + \frac{1}{2}LI^2$$

$$\Rightarrow \quad 10 \times 10^{-6} = \frac{1}{2} \times 0.2 \times 10^{-6} \times (5)^2 + \frac{1}{2} \times 0.5 \times 10^{-3} \times I^2$$

$$\Rightarrow \quad 2.5 \times 10^{-4} I^2 = 7.5 \times 10^{-6} \Rightarrow I^2 = 3 \times 10^{-2}$$

$$\Rightarrow \quad I = \frac{\sqrt{3}}{10}\text{A} = 0.1732\text{ A}$$ **Ans. (A)**

Sol. 23

$$\left.\begin{array}{l}\vec{A} \rightarrow \otimes \\ \vec{B} \rightarrow \otimes\end{array}\right] \Rightarrow \theta = 0^\circ$$

Initial flux through coil $\phi = BAN$

after time t angle between $\vec{A}$ and $\vec{B}$ will be

$$\theta = \omega t$$

$\Rightarrow$ Magnetic flux through coil after time t is given as

$$\phi = (\vec{B} \cdot \vec{A})N = NBA\cos\omega t$$

EMF induced in coil is

$$e = -\frac{d\phi}{dt} = NBA\omega \sin\omega t = e_0 \sin\omega t$$

Here $\quad e_0 = NBA\omega \rightarrow$ maximum emf **Ans. (A)**

Sol. 24 Charge flown through the loop is given as

$$Q = \frac{\Delta\phi}{R} = \frac{1}{10}A(B_f - B_i)$$

$$\Rightarrow \quad Q = \frac{1}{10} \times 3.5 \times 10^{-3}\left(0.4\sin\frac{\pi}{2} - 0\right)$$

$$\Rightarrow \quad Q = \frac{1}{10}(3.5 \times 10^{-3})(0.4 - 0)$$

$$\Rightarrow \quad Q = 1.4 \times 10^{-4} = 0.14\text{ mC}$$ **Ans. (A)**

Sol. 25 Potential difference between two faces perpendicular to x-axis will be the motional emf given as

$$e = Bvl = 12 \text{ mV} \quad \textbf{Ans. (C)}$$

Sol. 26 Work done in reversing direction of current is

$$W = \text{M.B.} (\cos\theta_1 - \cos\theta_2)$$

$$\Rightarrow \quad W = 2 \times 10^{-2} \times 1\cos(0.125)$$

$$\Rightarrow \quad W = 0.019 \text{ J}$$

Closest possible option is (B) **Ans. (B)**

Sol. 27 Motional *emf* induced in wire is

$$e = Bvl$$

$$\Rightarrow \quad e = 0.3 \times 10^{-4} \times 5 \times 10$$

$$\Rightarrow \quad e = 1.5 \times 10^{-3} \text{ V} \quad \textbf{Ans (D)}$$

Sol. 28 For damped oscillation amplitude is given as

$$A = A_0 e^{-t/r}$$

Where $\gamma = \dfrac{2m}{b} \left(b = \dfrac{B^2 l^2}{R} \right)$

If at $t = t_1$, $A = \dfrac{A_0}{e}$

$$\Rightarrow \quad \frac{A_0}{e} = A_0 e^{-t/r}$$

$$\Rightarrow \quad \gamma = \frac{2mR}{B^2 l^2}$$

$$\Rightarrow \quad \gamma = \frac{2 \times 50 \times 10^{-3} \times 10}{(0.1)^2 \times (0.1)^2} = 10^4 \text{ s}$$

Time period of oscillations

$$\Rightarrow \quad T_0 = 2\pi\sqrt{\frac{m}{k}} = 2\pi\sqrt{\frac{50 \times 10^{-3}}{0.5}} = \frac{2\pi}{\sqrt{10}} \text{ s}$$

Number of oscillation of strip in t_1 s are

$$N = \frac{t_1}{T_0} = \frac{10^4}{2\pi\sqrt{10}} \sim 5000 \quad \textbf{Ans. (B)}$$

Sol. 29 Resistance of Wheatstone bridge is

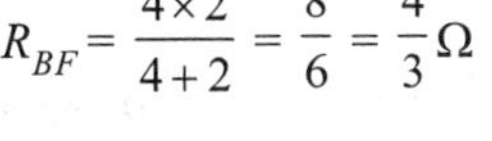

$$R_{BF} = \frac{4 \times 2}{4 + 2} = \frac{8}{6} = \frac{4}{3}\Omega$$

Equal resistance of circuit is

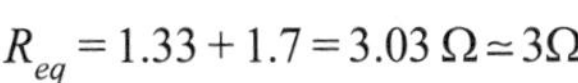

$$R_{eq} = 1.33 + 1.7 = 3.03\ \Omega \simeq 3\Omega$$

EMF induced in L is

$$e = Bvl = 1 \times 0.05 \times 0.01 = 5 \times 10^{-4} \text{ V}$$

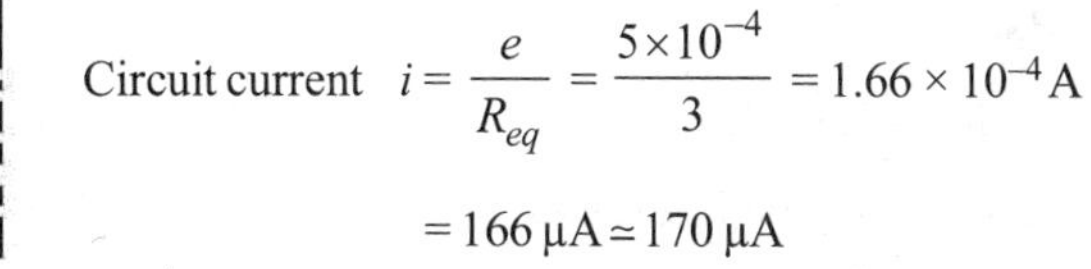

Circuit current $i = \dfrac{e}{R_{eq}} = \dfrac{5 \times 10^{-4}}{3} = 1.66 \times 10^{-4} \text{A}$

$$= 166\ \mu\text{A} \simeq 170\ \mu\text{A} \quad \textbf{Ans. (B)}$$

Sol. 30

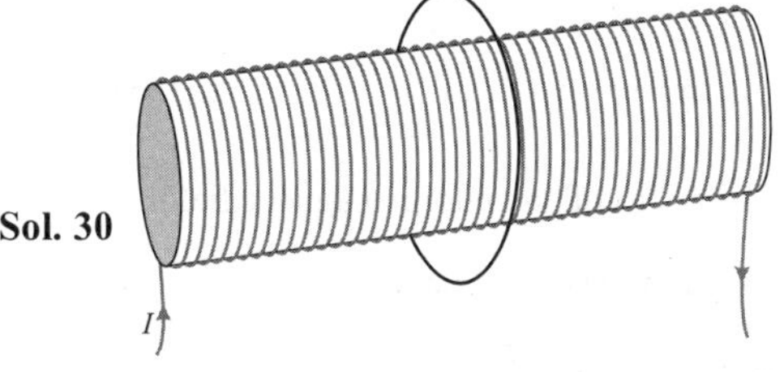

Magnetic flux (ϕ) through ring is

$$\phi = \pi(R)^2 . B$$

$$\Rightarrow \quad \phi = (\pi R^2)(\mu_0 n I)$$

$$\Rightarrow \quad \phi = (\pi R^2 \mu_0 n I_0)(t - t^2)$$

Induced emf is given as

$$e = \frac{d\phi}{dt}$$

$$\Rightarrow \quad e = (\pi R^2 \mu_0 n I_0)(2t - 1)$$

Induced current is given as

$$I_R = \frac{\pi R^2 \mu_0 n I_0 (2t - 1)}{R_{\text{ring}}}$$

In above expressive, we can see that

e and I_R are zero at $t = \dfrac{1}{2} = 0.5$ s

and their sign also changes at $t = 0.5$ s **Ans. (A)**

Sol. 31 Magnetic flux which comes out from the area enclosed by coil is same which goes into the surrounding plane of coil from area outside it. **Ans. (A)**

Sol. 32 Figure shows the loop formed by given corner coordinates

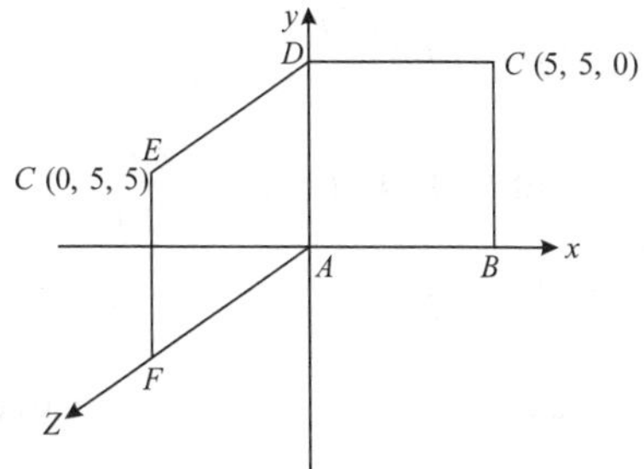

Area vector of loop is given as

$$\vec{A}_{ABCD} = 25\hat{k}$$

$$\vec{A}_{ADEF} = 25\hat{i}$$

$$\vec{A}_{\text{net}} = 25\hat{i} + 25\hat{k}$$

Magnetic flux through loop is given as

$$\phi = \vec{B}\cdot\vec{A}$$

$$\Rightarrow \quad \phi = 25\times 3 + 25\times 4$$

$$\Rightarrow \quad \phi = 175\ Wb$$

Ans. (175)

Sol. 33 For a rotating coil in uniform magnetic field.

Induced emf is given as $|e| = BA\omega \sin \omega t$

emf will be maximum at $\omega t = \frac{\pi}{2}$

$$\Rightarrow \quad \left(\frac{2\pi}{T}\right)t = \frac{\pi}{2}$$

$$\Rightarrow \quad \left(\frac{2\pi}{10}\right)t = \frac{\pi}{2}$$

$$\Rightarrow \quad t = 2.5 \text{ s}$$

and emf will be minimum at $\omega t = \pi$

$$\Rightarrow \quad \left(\frac{2\pi}{10}\right)t = \pi$$

$$\Rightarrow \quad t = 5 \text{ s}$$

Ans. (D)

Sol. 34 For linear reduction in magnetic field, we use

$$\frac{dB}{dt} = \frac{1000-500}{5} = 100 \text{ G/s}$$

$$= 100\times 10^{-4} \text{ T/s}$$

Area of loop is

$$A = 16\times 4 - 4\times 2 = 56 \text{ cm}^2$$

EMF induced in loop is

$$\varepsilon = \frac{dB}{dt}A = 100\times 10^{-4}\times 56\times 10^{-4} = 56\ \mu\text{V}$$

Ans. (C)

Sol. 35 *EMF* induced in a rotating coil in magnetic field is

$$e = BA\omega \sin\omega t$$

Maximum *EMF* is $e_{\max} = BA\omega = BA\left(\frac{2\pi}{T}\right)$

$$\Rightarrow \quad e_{\max} = 3\times 10^{-5}\times 3.14\times (0.1)^2\times \frac{2\times 3.14}{0.4}$$

$$\Rightarrow \quad e_{\max} = \frac{6\times 10}{4}\times 10^{-6}$$

$$\Rightarrow \quad e_{\max} = 15\ \mu\text{V}$$

Ans. (15)

Sol. 36 *emf* induced in rotating loop is

$$e = NBA\omega \sin \omega t$$

Here $\quad N = 1$

$$\Rightarrow \quad e = BA\omega \sin \omega t$$

Average power developed in loop is

$$P_{avg} = \left|\frac{e_{avg}^2}{R}\right| = \frac{B^2A^2\omega^2(\sin^2\omega t)_{avg}}{R}$$

$$\Rightarrow \quad P_{avg} = \frac{B^2A^2\omega^2}{2R} \qquad \left[\text{As } (\sin^2\omega t)_{avg} = \frac{1}{2}\right]$$

$$\Rightarrow \quad P_{avg} = \frac{B^2\omega^2}{2R}(\pi ab)^2 = \frac{\pi^2a^2b^2B^2\omega^2}{2R}$$

Ans. (A)

Sol. 37 Resistance of wire is

$$R = \frac{\rho l}{A} = \frac{1.23\times 10^{-8}\times 30\times 10^{-2}}{\pi(2\times 10^{-3})^2}$$

EMF induced in loop is given as

$$e = \frac{AdB}{dt} = a^2\frac{dB}{dt} = (7.5\times 10^{-2})^2\times 0.032$$

Induced current in loop is

$$i = \frac{e}{R} = \frac{(7.5\times 10^{-2})^2\times 0.032}{1.23\times 10^{-8}\times 30\times 10^{-2}}\times 3.14\times (2\times 10^{-3})^2$$

$$\Rightarrow \quad i \simeq 0.61 \text{ A}$$

Ans. (A)

Sol. 38 At the time of entering of magnet into coil and at the time of leaving the coil due to flux change in opposite direction emf is induced in opposite sense and when magnet is completely inside flux will not change so reading of galvanometer remain zero. Correct match is with option (C). **Ans. (C)**

Sol. 39 Magnetic field at centre of C_2 is

$$B = \frac{\mu_0 I}{2R_2}\cdot N_2$$

flux linked through coil C_1 due to magnetic field of C_2 is

$$\phi = B.\pi R_1^2 N_1 = \frac{\mu_0\pi R_1^2 N_1 N_2}{2R_2}.(5t^2 - 2t + 3)$$

emf induced in C_1 is given as

$$e = \frac{d\phi}{dt} = \frac{\mu_0\pi R_1^2 N_1 N_2}{2R_2}\times(10t - 2)$$

$$\Rightarrow \quad e = \frac{4\pi\times 10^{-7}\times(0.01)^2\times\pi\times 200\times 500}{2\times 0.2}\times(10-2)$$

$$\Rightarrow \quad e = 8\times 10^{-4}\text{ V} = 0.8\ \mu\text{V}$$

Ans. (5.00)

Sol. 40 Emf induced in sliding conductor is given as

$$e = Bvl$$

$\Rightarrow \quad e = \frac{\mu_0 I}{2\pi r} vl$

Induced current in loop is calculated as

$$i = \frac{e}{R} = \frac{\mu_0 I v l}{2\pi r R}$$ **Ans. (B)**

Sol. 41 Motional EMF induced across the plane wings is given as

$$e = Bvl \sin\theta$$

$\Rightarrow \quad e = (2.5 \times 10^{-4}\,\text{T})\left(180 \times \frac{5}{18}\,\text{m/s}\right)(10\,\text{m})\sin 60°$

$\Rightarrow \quad e = 108.25 \times 10^{-3}\,\text{V}$ **Ans. (D)**

Sol. 42 The effective circuit including motional EMF of the moving rod can be shown in figure below.

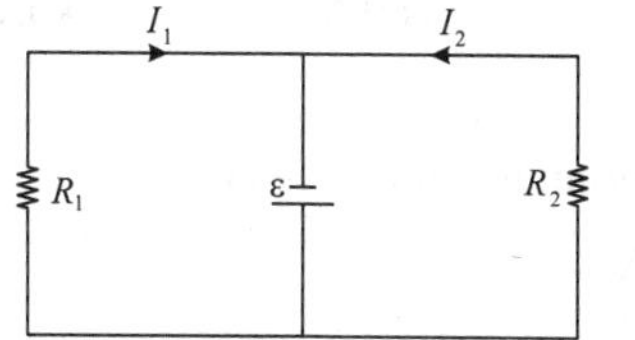

Ans. (C)

Sol. 43 Motional EMF induced in segments parallel to y axes is given as

$$E_1 = \frac{B_0(x+d)}{a} v_0 d$$

$$E_2 = \frac{B_0(x)}{a} v_0 d$$

The equivalent circuit of the loop is shown in below figure.

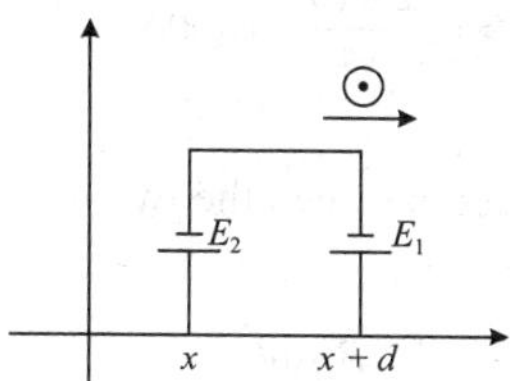

Net loop EMF induced is given as

$$E_{net} = E_1 - E_2$$

$\Rightarrow \quad E_{net} = \frac{B_0 v_0 d^2}{a}$ **Ans. (C)**

Sol. 44 Every elemental part of the wire is pulled by force $Bi(dl)$ and it acts perpendicular to current & magnetic field so it will open up and outward force will be giving it a shape of circle. **Ans. (A)**

Sol. 45 The electric flux passing through the given area is calculated as

$$\phi = \boldsymbol{B}.\boldsymbol{A} = BA\cos 0$$

Area enclosing the flux is given as $A = lvt$, thus flux is given as

$$\phi = Blvt$$

Once rod goes at $a > b$ then flux stops changing and maintained at constant value of Blb. When rod comes back and $x < b$ flux start decreasing so graph of variation of flux with time can be drawn as

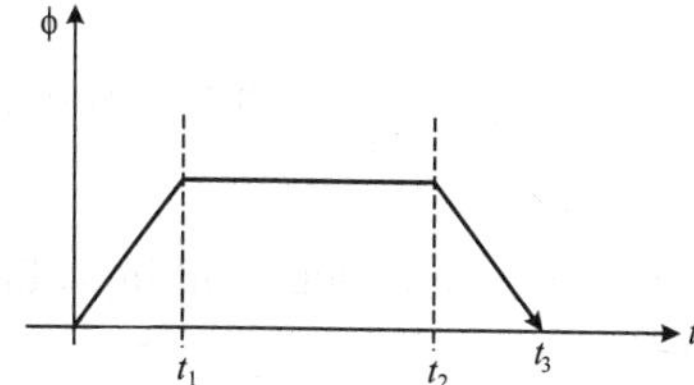

EMF induced can be given as

$$e = \frac{-d\phi}{dt}$$

This can be represented as slope of flux - time curve and *EMF* variation with time can now be plotted as shown in diagram below

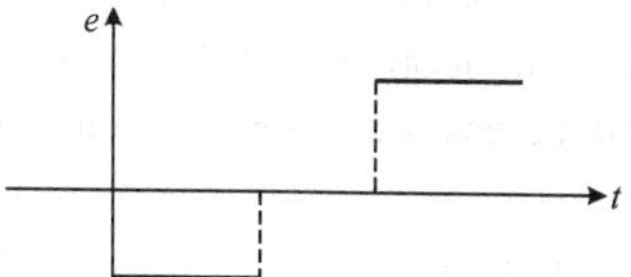

Power dissipated in the circuit is given as e^2/R where resistance is only of rod so resistance of the circuit remain constant thus power is given as

$$P = \frac{B^2 l^2 v^2}{R}$$

Thus power will also remain constant for the duration *EMF* is constant and zero when *EMF* is zero.

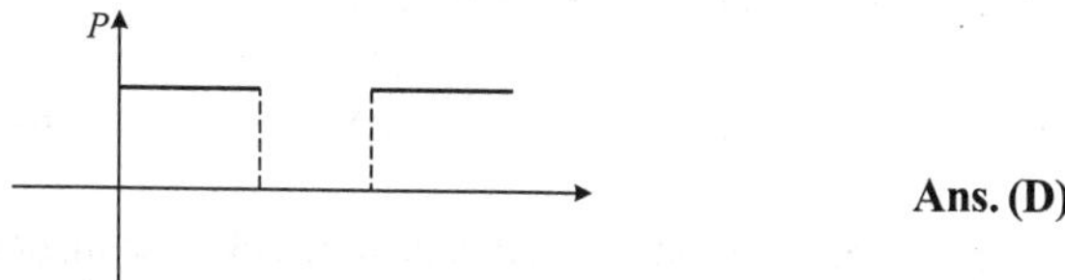

Ans. (D)

Sol. 46 The emf induced in coil can be given as

$$\varepsilon = \left|-A\frac{dB}{dt}\right|$$

$\Rightarrow \quad \varepsilon = \pi(1)^2 \times \frac{4}{\pi} \times \frac{10^{-3}}{100}\,\text{V}$

$\Rightarrow \quad \varepsilon = 4 \times 10^{-5}\,\text{V}$

When $B = 0$, $t = 100$ s, the energy dissipated can be directly calculated as emf in coil is constant, given as

$$H = \frac{\varepsilon^2}{R} \times t = \frac{16 \times 10^{-10}}{2 \times 10^{-6}} \times 100 = 8 \times 10^{-2} = 80\,\text{mJ}$$

Ans. (80)

Sol. 47 EMF induced in the loop is given as

$$\varepsilon = \left|-\frac{d\phi}{dt}\right| = (20t+20)\times 10^{-3}$$

at $t = 5$ s, EMF is calculated as

$$\varepsilon = 120\times 10^{-3}\ \text{V}$$

Current in loop is given as

$$i = \frac{\varepsilon}{R} = \frac{120}{2}\times 10^{-3} = 60\ \text{mA} \qquad \textbf{Ans. (60)}$$

Sol. 48 EMF induced in a rotating coil in uniform magnetic field is given as

$$e = NBA\omega \sin \omega t$$

$$e_{max} = NBA\omega = NB(\pi R^2)\omega$$

$$\Rightarrow \quad e_{max} = 3\times 10^{-2}\times 5\times 20\times \pi \times \frac{64}{10^4}$$

$$\Rightarrow \quad e_{max} = 60\times 10^{-2}\ \text{V} \qquad \textbf{Ans. (60)}$$

Sol. 49 As Magnet is approaches the loop, magnetic flux increases through the loop and rate of increment of flux also increases so emf increases with time. When magnet enters into the loop then at this moment emf is maximum and negative and when magnet is in middle of coil, at this moment emf is zero. Now again flux starts changing however polarity of emf get reversed. When magnet is coming out of the other side from the loop again at this moment emf is maximum and in opposite direction and as it gradually recedes away emf becomes zero. Hence optoin (D) is most appropriate for this situation.

Ans. (D)

Sol. 50 Motional emf induced in the ring when half ring enters in the magnetic field (at t=1s) is given as

$$E = Bv(2R)$$

$$\Rightarrow \quad E = 1\times 1\times 2\times 1 = 2\ \text{V} \qquad \textbf{Ans. (B)}$$

Sol. 51 As the current induced is in anticlock wise manner, it can be due to two reasons : (1) magnetic field is inward and increasing with time or (2) magnetic field is outward and decreasing with time. Hence out of given options, (D) is correct.

Ans. (D)

Sol. 52 Motional emf induced in the loop is given as

$$e = Blv$$

Current in circuit is calculated as

$$i = \frac{Blv}{R_{eq}} = 200\ \text{mA}$$

$$\Rightarrow \quad v = \frac{200\times 10^{-3}\times R_{eq}}{Bl}$$

$$\Rightarrow \quad v = \frac{200\times 10^{-3}\times 5}{5\times (0.2)} = 1\ \text{m/s} \qquad \textbf{Ans. (A)}$$

Sol. 53 Maximum emf induced in a rotating coil in uniform magnetic field is given as

$$e_{max} = BAN\omega$$

$$\Rightarrow \quad e_{max} = 0.07\times 1\times 10^3\times 2\pi$$

$$\Rightarrow \quad e_{max} = 140\pi = 140\times \frac{22}{7} = 440\ \text{V} \qquad \textbf{Ans. (440)}$$

Sol. 54 Magnetic flux through the coil is given as

$$\phi = 5t^3 + 4t^2 + 2t - 5$$

Emf induced in the coil is calculated as

$$|e| = \frac{d\phi}{dt} = 15t^2 + 8t + 2$$

At $t = 2$, we have

$$|e| = 15\times 2^2 + 8\times 2 + 2$$

$$\Rightarrow \quad e = 78\ \text{V}$$

Current in coil is calculated as

$$I = \frac{e}{R} = \frac{78}{5} = 15.60\ \text{A} \qquad \textbf{Ans. (A)}$$

Sol. 55 Average emf induced in coil is given as

$$\langle \varepsilon \rangle = \frac{L\Delta i}{\Delta i}$$

In steady state current in coil at $t = 0$ is given as

$$i_0 = \frac{V}{R} = \frac{20}{10} = 2\text{A}$$

After $t = 100$ μs if current becomes zero, we have induced emf calculated as

$$\langle \varepsilon \rangle = \frac{20\times 10^{-3}\times (2-0)}{100\times 10^{-6}}$$

$$\Rightarrow \quad \langle \varepsilon \rangle = \frac{2\times 10^3}{5} = 400\ \text{V} \qquad \textbf{Ans. (400)}$$

Sol. 56 Emf induced between the two ends of rotating rod is given as

$$e = \frac{B_H \omega l^2}{2}$$

$$\Rightarrow \quad e = \frac{0.2\times 10^{-4}\times 5\times 1}{2} = 0.5\times 10^{-4}$$

$$\Rightarrow \quad e = 50\times 10^{-6}\ \text{V} = 50\ \mu\text{V} \qquad \textbf{Ans. (B)}$$

Sol. 57 Figure below shows the motion of metallic rod as described in question

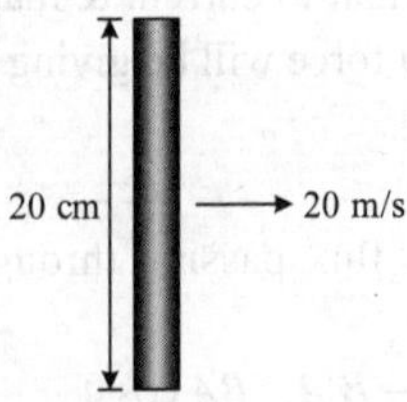

As angle of dip at the location of rod is 45°, the vertical component of earth's magnetic field at this location is calculated as

$$B_H = B_V = 4 \times 10^{-3}\,\text{T}$$

Motional emf induced in rod is given as

$$e = B_V(vl)$$

$$\Rightarrow \quad e = (4 \times 10^{-3})(20)(0.2)$$

$$\Rightarrow \quad e = 16 \times 10^{-3}\,\text{V} = 16\,\text{mV}$$

Ans. (16)

Sol. 58 Power dissipated in the coil due to self induced emf is given as

$$P = \frac{\varepsilon^2}{R} = \frac{\left(NA\frac{dB}{dt}\right)^2 \times A_c}{\rho l}$$

After number of turns are reduced to half and radius is doubled due to which area of coil becomes four times, power dissipated is given as

$$P' = \frac{\left(\frac{NA}{2}\frac{dB}{dt}\right)^2 \times 4A_c}{\rho l/2}$$

$$\Rightarrow \quad P' = 4P$$

Ans. (B)

Sol. 59 Magnetic flux linked with the circuit is given as

$$\phi = 8t^2 - 9t + 5$$

Emf induced in the closed circuit is calculated as

$$e = -\frac{d\phi}{dt} = -(16t - 9)$$

At $t = 0.25$ s, emf is given as

$$e = -[(16 \times 0.25) - 9] = 5\,\text{V}$$

Current in circuit due to induced emf is given as

$$i = \frac{e}{R} = \frac{5}{20}\,\text{A}$$

$$\Rightarrow \quad i = 0.25\,\text{A} = 250\,\text{mA}$$

Ans. (250)

Sol. 60 Flus becomes zero when

$$\phi = \frac{2}{3}(9 - t^2) = 0$$

$$\Rightarrow \quad t = 3\,\text{s}$$

Emf induced in the coil is calculated as

$$e = -\frac{d\phi}{dt} = -\frac{2}{3}(0 - 2t) = \frac{4t}{3}$$

Heat produced in 3 s is calculated as

$$H = \int \frac{e^2}{r}dt = \int_0^3 \frac{16t^2}{9 \times 8}dt = 2\,\text{J}$$

Ans. (2)

Sol. 61 Flux passing through the loop is given as

$$\phi = \vec{B} \cdot \vec{A}$$

$$\Rightarrow \quad \phi = (3t^3\hat{j} + 3t^2\hat{k}) \cdot (\pi(1)^2\hat{k})$$

$$\Rightarrow \quad \phi = 3t^2\pi$$

Induced emf in loop is given as

$$e = \left|\frac{d\phi}{dt}\right| = 6t\pi$$

At $t = 2$, emf is calculated as

$$e = 12\pi\,\text{V}$$

Ans. (12)

Sol. 62 Figure belowshows the R-L growth circuit as described in question.

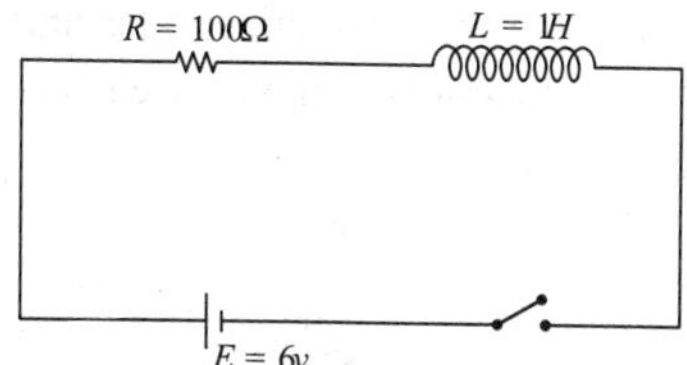

Current in circuit as a function of time is given as

$$i = \frac{E}{R}(1 - e^{-t/\tau})$$

Current becomes half at a time instant given as

$$\frac{E}{2R} = \frac{E}{R}(1 - e^{-t/\tau})$$

$$\Rightarrow \quad t = \tau \ln 2$$

$$\Rightarrow \quad t = \frac{1}{R}\ln 2 = \frac{1}{100}0.693 = 0.00693$$

$$\Rightarrow \quad t = 7\,\text{ms}$$

At time $t = 15$ ms, we have

$$i = \frac{E}{R}(1 - e^{-\frac{15}{10}})$$

$$\Rightarrow \quad i = \frac{1}{100}\left(1 - \frac{1}{4}\right) = \frac{3}{4} \times \frac{6}{100} = 0.045\,\text{A}$$

Energy stored in inductor is given as

$$U = \frac{1}{2}LI^2 = \frac{1}{2} \times 1 \times (0.045)^2 = 0.001\,\text{J}$$

$$\Rightarrow \quad U = 1\,\text{mJ}$$

Ans. (C)

Sol. 63 Energy stored in capacitor in steady state is given as

$$U = \frac{1}{2}CV^2 = \frac{1}{2}500 \times 10^{-6} \times 10^4$$

$$\Rightarrow \quad U = \frac{5}{2}\,\text{J}$$

Current will be maximum when whole energy of capacitor transform into magnetic energy in inductor, given as

$$\frac{1}{2}LI^2 = \frac{5}{2}$$

$$\Rightarrow \quad I = \sqrt{\frac{5}{L}} = \sqrt{\frac{5}{50\times10^{-3}}} = 10\,\text{A} \quad \textbf{Ans. (10)}$$

Sol. 64 Induced emf in the given situation is calculated as

$$e = \frac{d\phi}{dt} = \frac{BA-0}{t}$$

$$\Rightarrow \quad e = \frac{(0.5)(0.01)}{0.5} = 0.01\,\text{V} = 10\,\text{mV} \quad \textbf{Ans. (B)}$$

Sol. 65 Induced emf in case of rotating rod in uniform magnetic field is calculated by integration of motional emf in a small element in the rod as shown in figure below

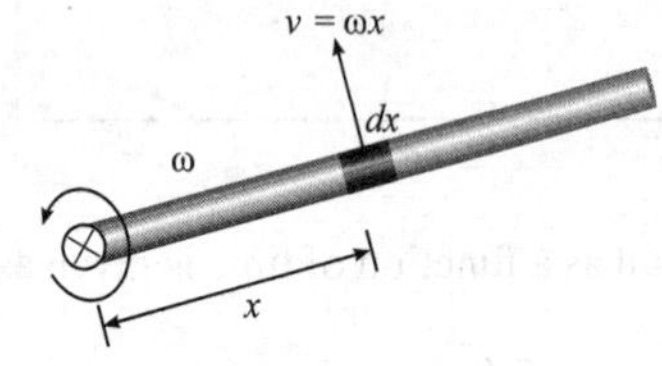

$$\varepsilon = \int d\varepsilon = \int B(\omega x)dx$$

$$\Rightarrow \quad \varepsilon = B\omega\int_0^L x dx = \frac{B\omega L^2}{2} \quad \textbf{Ans. (C)}$$

Sol. 66 The expression of Gauss's Law in electrostatic is given as

$$\phi = \int \vec{E}\cdot d\vec{s} = \frac{q}{\epsilon_0}$$

The expression of Faraday's law is written as

$$\int \vec{E}\cdot d\vec{l} = \frac{-d\phi_B}{dt}$$

The expression of Gauss's law in magnetism is written as

$$\int \vec{B}\cdot d\vec{A} = 0$$

The expression of Ampere's Maxwell law is written as

$$\int \vec{B}\cdot d\vec{l} = \mu_0 i_C + \mu_0\epsilon_0\frac{d\phi_E}{dt}$$

Where i_C is the conduction current and $\epsilon_0\frac{d\phi_E}{dt}$ is the displacement current. **Ans. (A)**

Sol. 67 Motional induced *emf* across the ends of wire is given as

$$e = Bvl$$

$$\Rightarrow \quad e = 2\times8\times1 = 16\,\text{V} \quad \textbf{Ans. (D)}$$

Sol. 68 Induced emf in the loop is calculated as

$$e = \frac{d}{dt}(B\pi r^2)$$

$$\Rightarrow \quad e = 2B\pi r\frac{dr}{dt} = 2\times\pi\times0.1\times0.8\times2\times10^{-2}$$

$$\Rightarrow \quad e = 2\pi\times1.6 = 10.06\,\text{V} \quad \textbf{Ans. (10)}$$

Sol. 69 Average speed of loop which is being pulled out is calculated as

$$v = \frac{0.05}{1} = 0.05\,\text{m/s}$$

Induced current in loop is calculated as

$$i = \frac{Bvl}{R} = \frac{40\times0.05\times0.05}{10} = 0.01\,\text{A}$$

While in motion the magnetic force on loop is given as

$$F = Bil = 40\times0.01\times0.05\,\text{N}$$

$$\Rightarrow \quad F = 0.02\,\text{N}$$

Work done is calculated as

$$W = F\times l = 0.02\times0.05 = 1\times10^{-3}\,\text{J} \quad \textbf{Ans. (B)}$$

Sol. 70 Instantaneous *emf* when the plane of coil is inclined at 60° is calculated as

$$E = NAB\omega\sin\omega t$$

Here ωt is the angle between $\vec{A}$ & $\vec{B}$.

$$\Rightarrow \quad E = 600\times70\times10^{-4}\times0.4\times\frac{100\pi}{6}\times\frac{1}{2}$$

$$\Rightarrow \quad E = 44\,\text{V} \quad \textbf{Ans. (44)}$$

Sol. 71 Figure below shows the situation described in the quesiton.

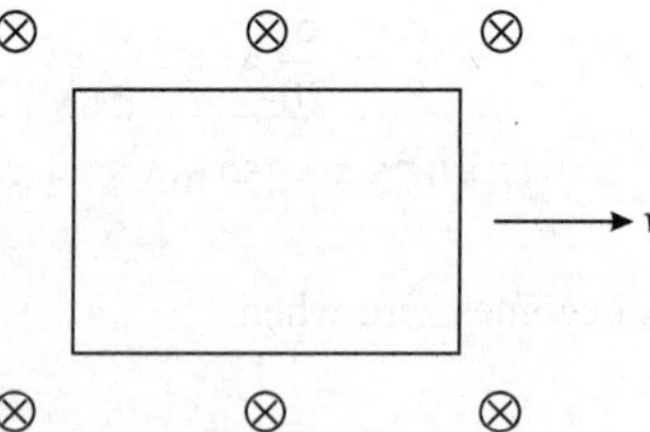

Moving a coil inside a uniform magnetic field either with uniform or non-uniform speed doesn't changes flux, so, no emf is induced a flux passing through coil does not change and rate of change of flux $\frac{d\phi}{dt} = 0$. **Ans. (D)**

Sol. 72 When a current carrying conductor is moved perpendicular to magnetic field, then an emf is induced across the ends of conductor.

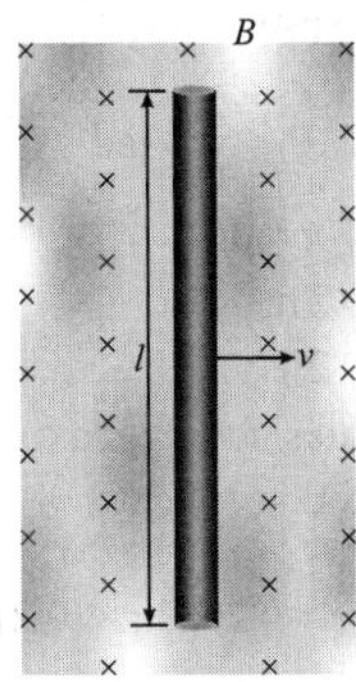

Induced *emf* due to motion of conductor is given as

$$e = Blv$$

$$\Rightarrow \quad 0.08 = 0.4\left(\frac{10}{100}\right)v$$

$$\Rightarrow \quad v = \left(\frac{0.08 \times 10}{0.4}\right)$$

$$\Rightarrow \quad v = 2 \text{ m/s}$$

Ans. (A)

Sol. 73 Force on current carrying conductor in magnetic field

$$F = ilB$$

Induced current in the circuit is given as

$$I = \frac{\varepsilon}{R}$$

Induced emf in the circuit is given as

$$\varepsilon = vBl$$

The force acting on the rod is calculated as

$$F = \left(\frac{\varepsilon}{R}\right)lB = \left(\frac{vBl}{R}\right)lB = \frac{vB^2l^2}{R} = \frac{4}{5} \times \left(\frac{15}{100}\right)^2 \times 1^2$$

$$\Rightarrow \quad F = \frac{4}{5} \times \frac{225}{10^4}$$

$$\Rightarrow \quad F = \frac{180}{10^4} = 0.018 \text{ N}$$

$$\Rightarrow \quad F = 18 \times 10^{-3} \text{ N}$$

Ans. (18)

Sol. 74 Figure below shows the situation described in the question.

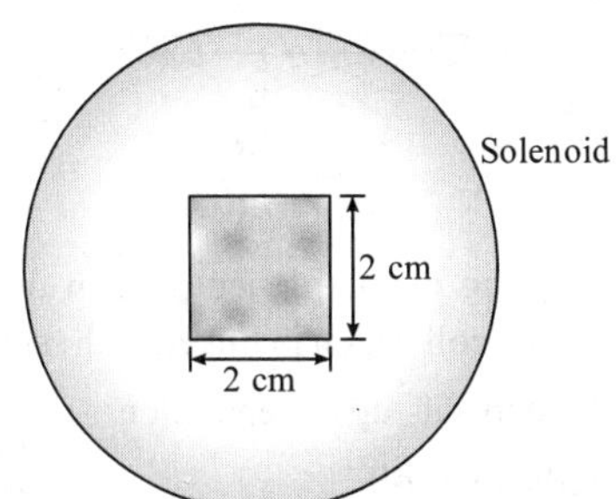

Magnetic induction due to solenoid is given as

$$B_{solenoid} = \mu_0 nI$$

Thus flux through square is calculated as

$$\phi_{square} = \mu_0 nI \times A$$

Induced emf in the loop is calculated as

$$\varepsilon = \frac{d\phi}{dt}$$

$$\Rightarrow \quad \varepsilon = \mu_0 nA \times \frac{dI}{dt}$$

$$\Rightarrow \quad \varepsilon = \mu_0 nA \times I_0 \omega \cos \omega t$$

Amplitude of emf is given as

$$\varepsilon_0 = \mu_0 nA \times I_0 \omega$$

$$\Rightarrow \quad \varepsilon_0 = 4\pi \times 10^{-7} \times \frac{50}{10^{-2}} \times 4 \times 10^{-4} \times 2.5 \times 700$$

$$\Rightarrow \quad \varepsilon_0 = 44 \times 10^{-4} \text{ V}$$

Ans. (44)

Sol. 75 Slope of line is given as

$$m = \tan\theta = \frac{10}{5} = 5$$

$$B = mt$$

$$B = 2t$$

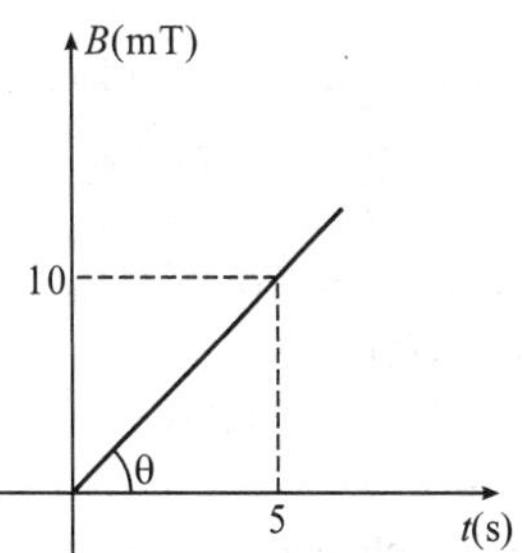

Induced emf

$$\varepsilon = \left|\frac{d\phi}{dt}\right| = \frac{d(BA)}{dt} = \frac{AdB}{dt}$$

$$\Rightarrow \quad \varepsilon = \frac{4d(2t)}{dt} = 4 \times 2 = 8 \text{ mV}$$

Ans. (8)

Sol. 76 Motional emf induced across the faces of cube is

$$\varepsilon = Blv \sin\theta$$

Here $\theta = 90^\circ$ gives

$$\varepsilon = \Delta V$$

$$\Delta V = (v \times B)d$$

$$\Rightarrow \quad \Delta V = (2 \times 1/2)\, 0.15$$

$$\Rightarrow \quad \Delta V = 150 \text{ mV}$$

Ans. (150)

Sol. 77 Figure below shows the circular loop which is considered expanding at a constant rate.

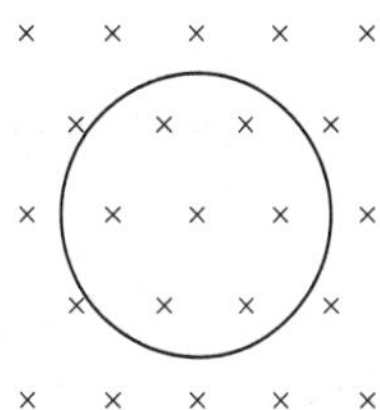

Rate of expansion of radius is given as

$$\frac{dr}{dt} = 10^{-3}\text{ m/s}$$

As $A = \pi r^2$

$$\Rightarrow \quad \frac{dA}{dt} = 2\pi r\frac{dr}{dt}$$

Thus induced emf in the loop is given as

$$\varepsilon = \left|\frac{-d\phi}{dt}\right| = \left|\frac{BdA}{dt}\right|$$

$$\Rightarrow \quad \varepsilon = 0.4\times 2\times\pi\times 2\times 10^{-2}\times 10^{-3}\text{ V}$$

$$\Rightarrow \quad \varepsilon = 16\,\pi\mu\text{V} = 50.24\,\mu\text{V}$$ **Ans. (50)**

Sol. 78 Source of time varying magnetic field may be accelerated or retarded charge which produces varying electric and magnetic fields. An electric field varying linearly with time will not produce time varying magnetic field as current will be constant.

Ans. (A)

Sol. 79 Induced charges can be calculated as

$$\Delta Q = -\frac{\Delta\phi}{R} = -\left(\frac{\phi_2-\phi_1}{R}\right)$$

Flux passing through one coil is given as

$$\phi_1 = NBA$$

When coil rotates by 180°, final flux is given as

$$\phi_2 = -NBA$$

Charge flowing during change of magnetic field is given as

$$\Delta Q = \frac{2NBA}{R} = \frac{2\times100\times1.5\times24\times10^{-4}}{12}$$

$$\Rightarrow \quad \Delta Q = 6\times 10^{-2}$$

$$\Rightarrow \quad \Delta Q = 60\text{ mC}$$ **Ans. (60)**

Sol. 80 When switch is opened in 1 ms then due to change in flux, the emf induced is calculated as

$$emf = -L\frac{dI}{dt}$$

$$\Rightarrow \quad 20 = -L\frac{(0-2)}{10^{-3}}$$

$$\Rightarrow \quad L = 10\text{ mH}$$ **Ans. (D)**

Sol. 81 Figure below shows the rotating rod as described in the question.

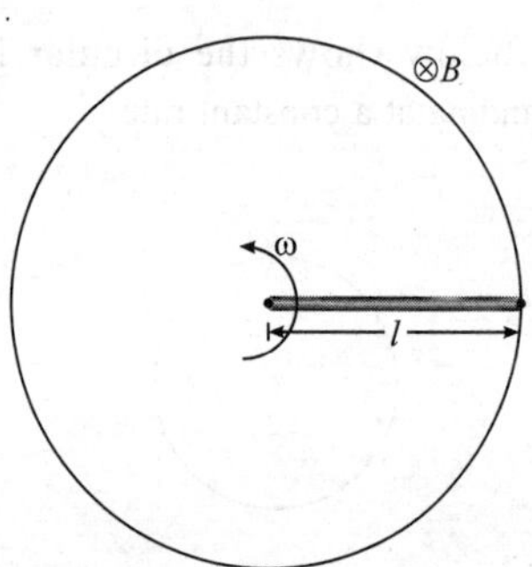

Here ω = 210 rpm so we use

$$\Rightarrow \quad \omega = 210\times\frac{2\pi}{60}\text{ rad/s}$$

$$\Rightarrow \quad \omega = 7\,\pi\text{ rad/s}$$

and $l = 0.2\text{ m}$

and $B = 0.2\text{ T}$

As the metallic rod rotates about an axis normal to rod then *emf* developed across rod is $= B\omega l^2$

$$\Rightarrow \quad e = \frac{1}{2}\times 0.2\times 7\pi\times(0.2)^2$$

$$\Rightarrow \quad e = 88\text{ mV}$$ **Ans. (88)**

20.2 Self and Mutual Induction

Sol. 1 As in given circuit all three inductors are connected in parallel so we use

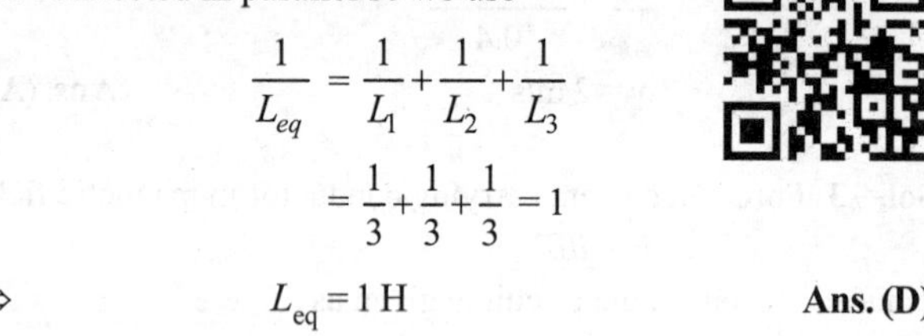

$$\frac{1}{L_{eq}} = \frac{1}{L_1}+\frac{1}{L_2}+\frac{1}{L_3}$$

$$= \frac{1}{3}+\frac{1}{3}+\frac{1}{3} = 1$$

$$\Rightarrow \quad L_{eq} = 1\text{ H}$$ **Ans. (D)**

Sol. 2 If in a coil of self inductance 'L' current changes by di in time dt then magnitude of emf induced in coil is given as

$$e = L\frac{di}{dt}$$

$$\Rightarrow \quad e = L\frac{\Delta i}{\Delta t}$$

$$\Rightarrow \quad 8 = L\times\frac{4A}{0.05\text{s}}$$

$$\Rightarrow \quad L = \frac{8\times0.05}{4}$$

$$\Rightarrow \quad L = 0.1\text{ H}$$ **Ans. (D)**

Sol. 3 If two coils with self induction coefficients L_1 and L_2 are placed to each other than their mutual induction is given as

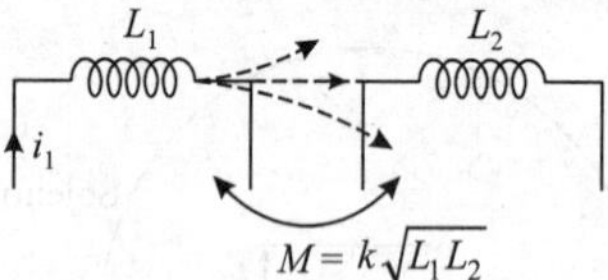

Where $k \rightarrow$ Coupling factor [depends upon the fraction of magnetic flux of first coil which in linked with second coil]

Here k indirectly depends upon separation & orientation of the two coil.

for full flux linkage we use $k = 1$

for No flux linkage we use $k = 0$

Either due to large separation or kept at 90° angle to each other. **Ans. (B)**

Sol. 4

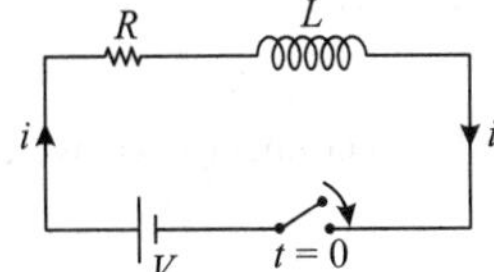

In series *L-R* circuit the current grows according to relation

$$i = \frac{V}{R}(1 - e^{\frac{-Rt}{L}})$$

$V/R \rightarrow$ steady state current

if at $t = t_1$, $i = V/2R$

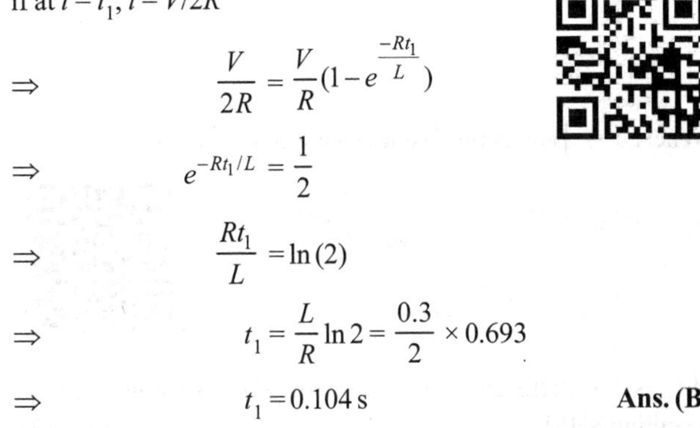

$$\Rightarrow \quad \frac{V}{2R} = \frac{V}{R}(1 - e^{\frac{-Rt_1}{L}})$$

$$\Rightarrow \quad e^{-Rt_1/L} = \frac{1}{2}$$

$$\Rightarrow \quad \frac{Rt_1}{L} = \ln(2)$$

$$\Rightarrow \quad t_1 = \frac{L}{R}\ln 2 = \frac{0.3}{2} \times 0.693$$

$$\Rightarrow \quad t_1 = 0.104\ \text{s}$$

Ans. (B)

Sol. 5 emf induced in a coil due to time varying magnetic flux is given as

$$e = -\frac{d\phi}{dt}$$

$$\Rightarrow \quad e = -\frac{d}{dt}(10t^2 - 50t + 250)$$

$$\Rightarrow \quad e = -[20t - 50]$$

at $t = 3$s, $\quad e = -[20 \times 3 - 50] = -10\ \text{V}$ **Ans. (B)**

Sol. 6 When circuit is in steady state, current through inductor is $\quad I_0 = E/R$

Now battery is disconnected & terminals *A* & *B* are short circuited.

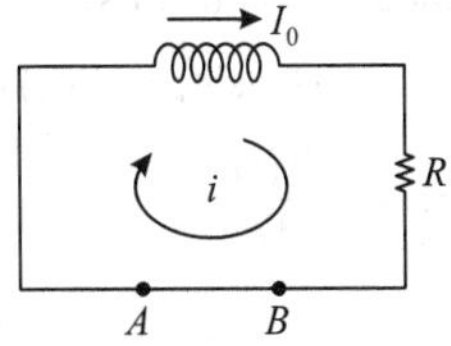

During decay current in circuit is given as

$$i = I_0 e^{-Rt/L} = \frac{E}{R}e^{-Rt/L}$$

$$\Rightarrow \quad i = \frac{100}{100}e^{\frac{-100\times10^{-3}}{100\times10^{-3}}} = e^{-1}$$

$$\Rightarrow \quad i = \frac{1}{e}\ \text{A}$$

Ans. (A)

Sol. 7

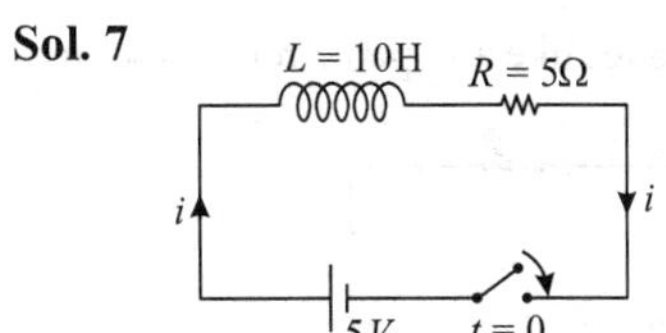

For *R-L* circuit current as a function of time is given as

$$i = \frac{V}{R}(1 - e^{\frac{-Rt}{L}})$$

$$\Rightarrow \quad i = \frac{5}{5}(1 - e^{\frac{-5\times2}{10}})$$

$$\Rightarrow \quad i = 1 - e^{-1}$$

Ans. (D)

Sol. 8

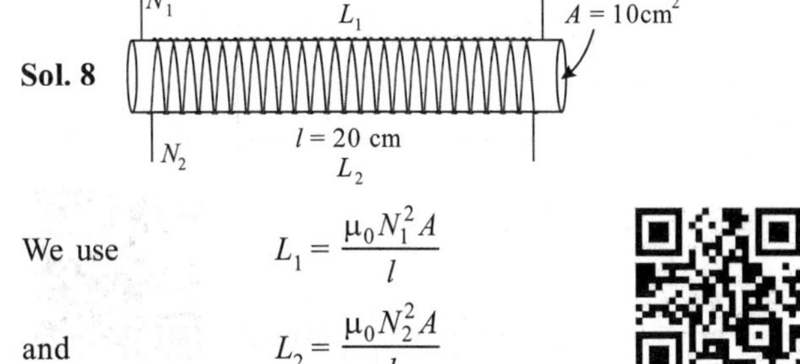

We use $\quad L_1 = \dfrac{\mu_0 N_1^2 A}{l}$

and $\quad L_2 = \dfrac{\mu_0 N_2^2 A}{l}$

As the two solenoids are wound over each other due to complete linkage of flux, the mutual induction between the two is given as $\quad M = k\sqrt{L_1L_2} = \sqrt{L_1L_2}$ [As $k = 1$]

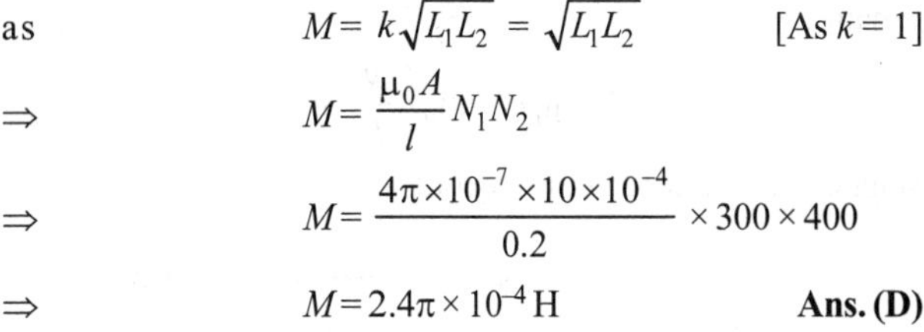

$$\Rightarrow \quad M = \frac{\mu_0 A}{l}N_1N_2$$

$$\Rightarrow \quad M = \frac{4\pi\times10^{-7}\times10\times10^{-4}}{0.2}\times300\times400$$

$$\Rightarrow \quad M = 2.4\pi\times10^{-4}\ \text{H}$$

Ans. (D)

Sol. 9

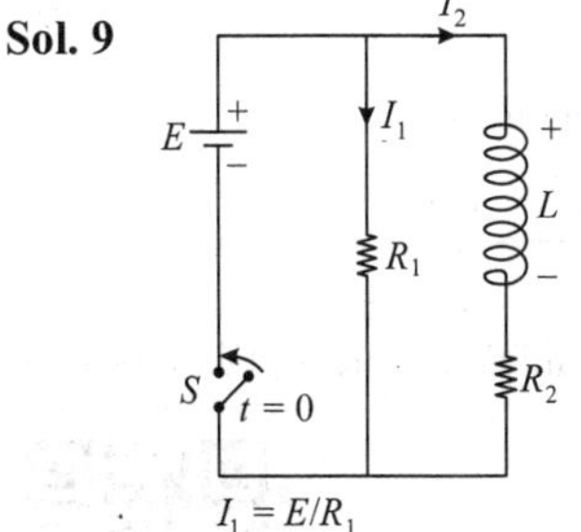

Current I_2 is given as

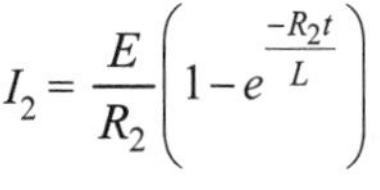

$$I_2 = \frac{E}{R_2}\left(1 - e^{\frac{-R_2 t}{L}}\right)$$

Potential drop across *L* is given as

$$e_L = L\frac{dI_2}{dt} = \frac{LE}{R_2}(e^{\frac{-R_2t}{L}} \times \frac{R_2}{L}) = E\,e^{-R_2t/L}$$

$$\Rightarrow \quad e_L = 12\,e^{\frac{-2t}{0.4}} = 12\,e^{-5t}\ \text{V}$$

Ans. (B)

Sol. 10 At $t = 0$, inductor behaves like an open circuit so we use

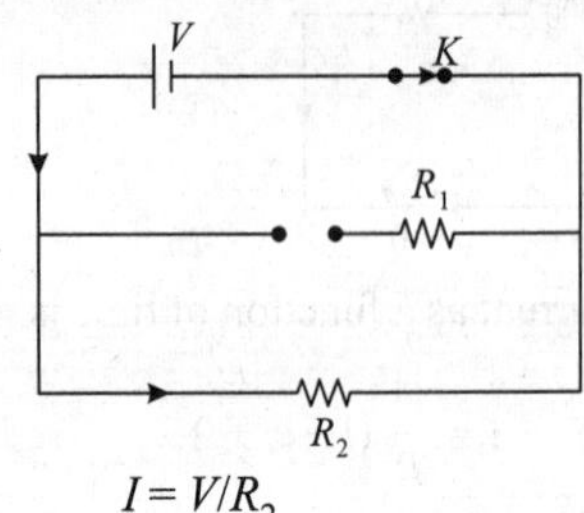

$\Rightarrow$ at $t = 0$, $I = V/R_2$

At $t = \infty$, i.e. in steady state inductor behaves like a straight wire so we use

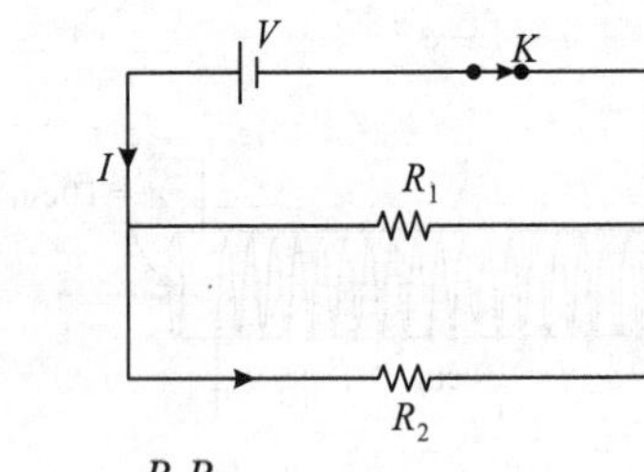

$$R_{eq} = \frac{R_1 R_2}{R_1 + R_2}$$

$$\Rightarrow \quad I = V/R_{eq} = \frac{V(R_1 + R_2)}{R_1 R_2}$$ **Ans. (C)**

Sol. 11 Ideally magnetic induction inside a solenoid everywhere remain constant

$$B = \mu_0 nI = \mu_0 \left(\frac{N}{L}\right) I$$

Magnetic flux linked with solenoid is

$$\phi = NBA = N\left(\frac{\mu_0 NI}{L}\right)\pi r^2 \qquad [\because A = \pi r^2]$$

$$\Rightarrow \quad \phi = \left(\frac{\mu_0 \pi N^2 r^2}{L}\right) I$$

Here self inductance of the solenoid is

$$L_{sol} = \left(\frac{\mu_0 \pi N^2 r^2}{L}\right) I$$

As magnetic induction on the sides of solenoid is less than $\mu_0 nI$

$$\Rightarrow \quad \phi = L_{sol} I$$

Here $\quad L_{sol} < \dfrac{\mu_0 \pi N^2 r^2}{L}$

$\Rightarrow$ **S-1** is true

S-2 is its explanation **Ans. (B)**

Sol. 12

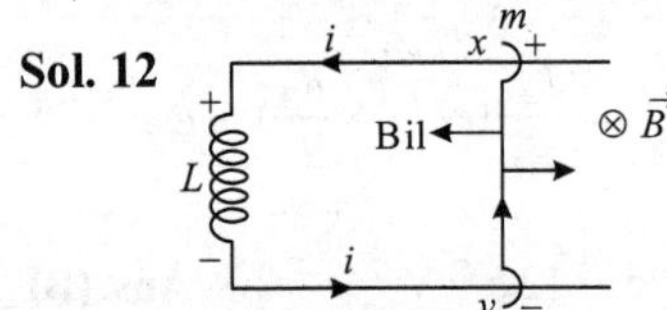

Motional emf induced in connector xy is

$$e_{xy} = Bvl$$

For the inductor as it is in parallel to connector xy, we use

$$Bvl = L\frac{di}{dt} \qquad \ldots(1)$$

If a be the acceleration (–ve) of connector we use

$$a = -\frac{Bil}{m} \qquad \ldots(2)$$

From equation-(1) $\int di = \dfrac{Bl}{L}\int v dt$

$$\Rightarrow \quad i = \frac{Bl}{L}\cdot x$$

[Where $x = \int v dt$ is the displacement of connector]

From equation-(2) $\quad a = -\dfrac{Bl}{m}\left[\dfrac{Bl}{L}\cdot x\right]$

$$\Rightarrow \quad \frac{d^2x}{dt^2} + \frac{B^2 l^2}{mL} x = 0$$

This is the differential equation of SHM so connector is executing SHM. **Ans. (D)**

Sol. 13 As S_2 is kept open so the circuit is simple RC transient and charge on capacitor is given at (time function)

$$q = CV(1 - e^{-t/RC})$$

$$q = CV(1 - e^{-t/\tau})$$

At $t = \tau \rightarrow \quad q = CV(1 - e^{-1})$

At $t = 2\tau \rightarrow \quad q = CV(1 - e^{-2})$ **Ans. (C)**

Sol. 14 for two closely spaced coils, emf induced in one coil due to current variation in other coil is given as

$$V(t) = -M\frac{dI(t)}{dt}$$

[M = mutual inductance between coils]

$\Rightarrow \quad V(t) = -M \times$ slope of $I(t)$ curve

$$\text{slope} = -\frac{V(t)}{M}$$

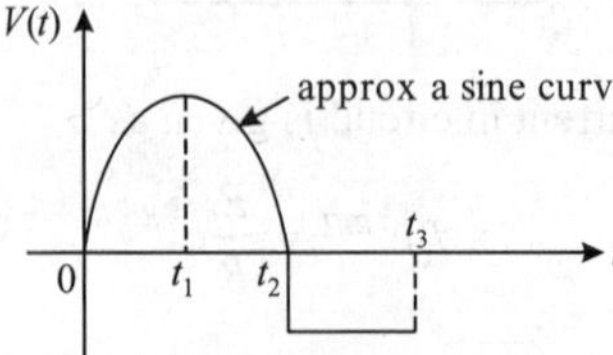

from $\quad$ 0 to $t_1 \rightarrow$ Slope of $I(t)$ curve is increasing (+ve)

t_1 to $t_2 \rightarrow$ Slope of $I(t)$ curve is decreasing (+ve)

t_2 to $t_3 \rightarrow$ Slope of $I(t)$ curve is constant (+ve) as $V(t)$ is –ve

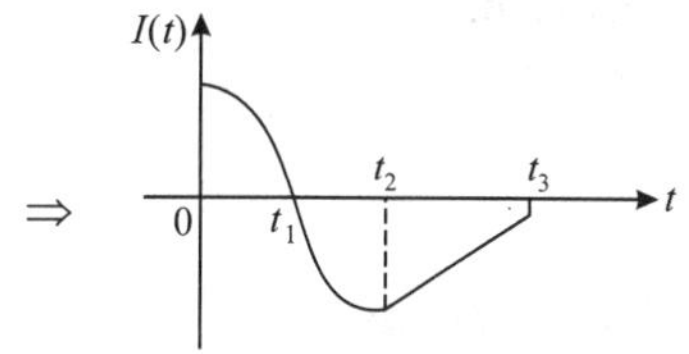

$\Rightarrow$ None of the given options is correct. **Ans. (None)**

Sol. 15 After point C is connected to B then circuit will be

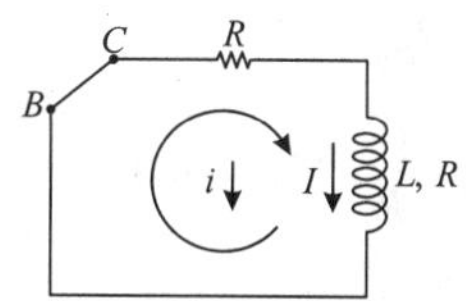

in the given loop we can write

$$V_R + V_L = 0$$

$$\Rightarrow \quad \frac{V_R}{V_L} = -1 \quad \textbf{Ans. (C)}$$

Sol. 16 frequency of LC oscillation is given as

$$f = \frac{1}{2\pi\sqrt{LC}}$$

for $L_1 (> L_2)$ we use $f_1 < f_2$ or $T_1 > T_2$

In case of L_1 maximum charge on capacitor Q_{max} will decrease more slowly. **Ans. (A)**

Sol. 17 When K_1 is closed the steady current in circuit is given as-

$$I_0 = \frac{V}{R} = \frac{15}{150} = 0.1 \text{ A}$$

When at $t = 0$, K_1 is opened and K_2 is closed then inductor current I_0 will start decaying through R and its time function as

$$i = I_0 e^{-Rt/L} = 0.1 \times e^{-150\times10^{-3}/0.03}$$

$$\Rightarrow \quad i = 0.1 \times e^{-5} = \frac{1}{1500} \text{ A} = 0.67 \text{ mA} \quad \textbf{Ans. (D)}$$

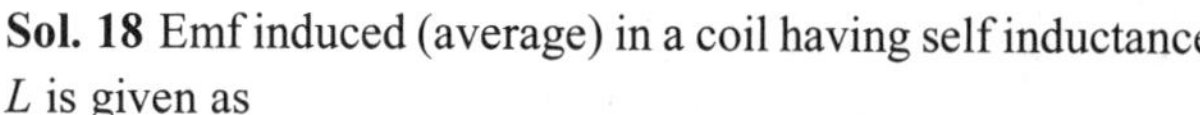

Sol. 18 Emf induced (average) in a coil having self inductance L is given as

$$e = \left| L\frac{\Delta i}{\Delta t} \right|$$

$$\Rightarrow \quad 50 = L\left(\frac{5-2}{0.1}\right)$$

$$\Rightarrow \quad L = \frac{50}{30} = \frac{5}{3} = 1.67 \text{ H} \quad \textbf{Ans. (D)}$$

Sol. 19 In series RC circuit current for charging of capacitor is

$$i_a = \frac{E}{R} e^{-t/RC} \quad \ldots (a)$$

[RC - Transient circuit]

for series LR circuit growth of current is

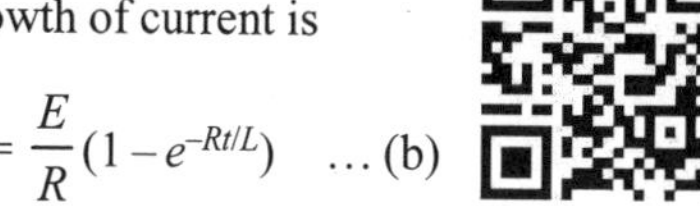

$$i_b = \frac{E}{R}(1 - e^{-Rt/L}) \quad \ldots (b)$$

[RL - Transient circuit] **Ans. (C)**

Sol. 20 Transformer efficiency is given as

$$\eta = \frac{P_{out}}{P_{in}} = \frac{V_S I_S}{V_P I_P}$$

$$\Rightarrow \quad 0.9 = \frac{23 \times I_S}{230 \times 5}$$

$$\Rightarrow \quad I_S = 45 \text{ A} \quad \textbf{Ans. (D)}$$

Sol. 21 Given that

$$L\frac{di}{dt} = 25$$

$$\Rightarrow \quad L \times \frac{15}{1} = 25$$

$$\Rightarrow \quad L = \frac{5}{3} \text{ H}$$

Change in energy of inductance is

$$\Delta E = \frac{1}{2} L (I_2^2 - I_1^2)$$

$$\Rightarrow \quad \Delta E = \frac{1}{2} \times \frac{5}{3} \times (25^2 - 10^2) \text{ J}$$

$$\Rightarrow \quad \Delta E = \frac{5}{6} \times 525 = 437.5 \text{ J} \quad \textbf{Ans. (A)}$$

Sol. 22 From time $t = 0$ to $t = t_0$, current in RL circuit grows and after to current decays.

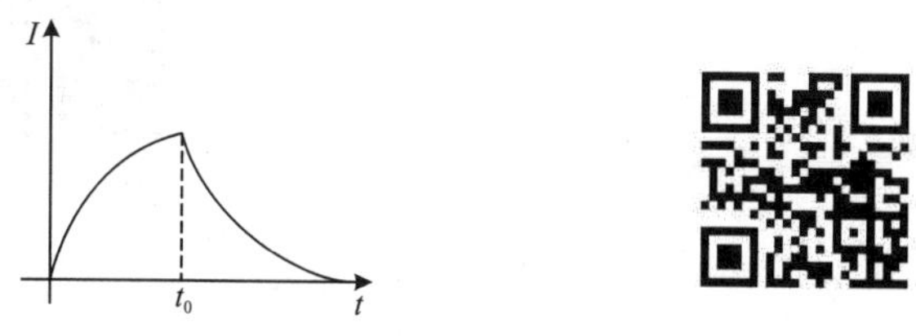

In this case most appropriate option is (B) **Ans. (B)**

Sol. 23 Mutual inductance of the given coils is

$$M = \mu_0 n_1 n_2 \pi r_1^2 l \quad \ldots (1)$$

Self inductance of inner coils is

$$L = \mu_0 n_1^2 \pi r_1^2 l \quad \ldots (2)$$

From (1) and (2)

$$\frac{M}{L} = \frac{n_2}{n_1} \quad \textbf{Ans. (D)}$$

Sol. 24 Self inductance of a solenoid is

$$L = \mu_0 n^2 A l$$

Keeping total length of copper wire same on increasing the frame size to 3 times, the length of solenoid on which terms are wound will become $l/3$ so new self inductance will be

$$L' = \mu_0 n^2 (9A)(l/3) = 3\text{ L}$$ **Ans. (B)**

Sol. 25 Ideal inductor will behave like zero resistance long time after switch is closed so equivalent circuit will be

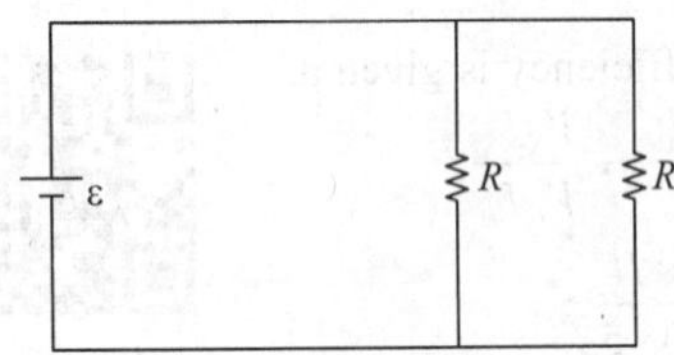

Current through battery is

$$I = \frac{2\varepsilon}{R} = \frac{2 \times 15}{5} = 6\text{ A}$$ **Ans. (A)**

Sol. 26 Flux linked with solenoid is given as

$$\phi = NBA = LI$$

$$\Rightarrow \quad N\mu_0 nI\pi R^2 = LI$$

$$\Rightarrow \quad N\mu_0 \left(\frac{N}{l}\right) I\pi R^2 = LI$$

N & R are to be taken constant, so we use

Self inductance $\quad (L) \propto \dfrac{1}{l} \propto \dfrac{1}{\text{length}}$ **Ans. (B)**

Sol. 27 If mutual induction between coils P and Q is M, we use

$$\phi_Q = MI_P \quad \ldots(1)$$

and $$\phi_P = MI_Q \quad \ldots(2)$$

$$\Rightarrow \quad \frac{\phi_Q}{\phi_P} = \frac{I_P}{I_Q}$$

$$\Rightarrow \quad \frac{10^{-3}}{\phi_P} = \frac{3}{2}$$

$$\Rightarrow \quad \phi_P = \frac{2}{3} \times 10^{-3} = 0.67 \times 10^{-3}\text{ Wb}$$ **Ans. (B)**

Sol. 28 Current in solenoid

$$I_{(t)} = kte^{-\alpha t}$$

Flux through solenoid is

$$\phi_{(t)} = B.A = (\mu_0 n I_{(t)}) \times \pi R^2$$

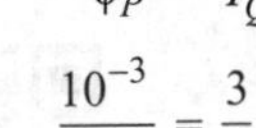

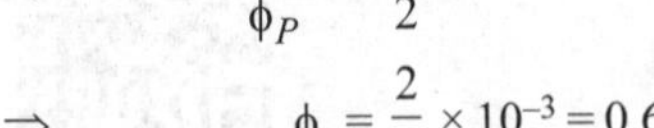

(same flux will pass through coil of radius $2R$)

emf induced in outer coil is

$$e = -\frac{d\phi_{(t)}}{dt}$$

$$\Rightarrow \quad e = -\mu_0 n\pi R^2 k\,[e^{-\alpha t} + t(-\alpha e^{-\alpha t})]$$

$$\Rightarrow \quad e = -Ce^{-\alpha t}(1 - \alpha t)$$

Current induced in coil is

$$I_{coil} = \frac{e}{R_{coil}}$$

$$\Rightarrow \quad I_{coil} = -\frac{C}{R}e^{-\alpha t}(1 - \alpha t)$$

at $\quad t = 0 \rightarrow I_{coil} \rightarrow 0$

at $\quad t = \infty \rightarrow I_{coil} \rightarrow 0$

$\quad I_{coil} \rightarrow -\text{ve}$

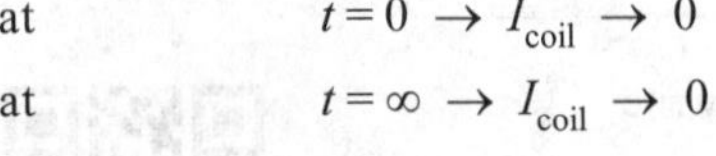

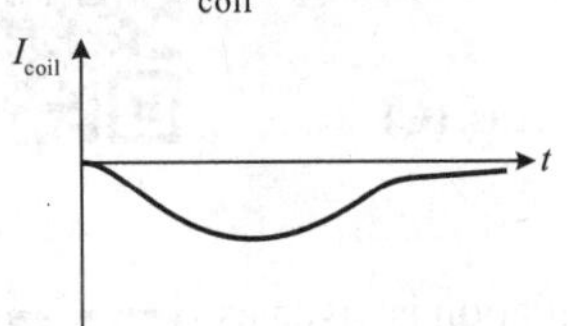

Ans. (B)

Sol. 29

Current in circuit as a function of time is

$$i = i_0(1 - e^{-t/\tau})$$

$$\Rightarrow \quad \frac{80}{100} i_0 = i_0(1 - e^{-t/\tau})$$

$$\Rightarrow \quad 0.8 = 1 - e^{-t/\tau}$$

$$\Rightarrow \quad e^{-t/\tau} = 0.2 = \frac{1}{5}$$

$$\Rightarrow \quad -\frac{t}{\tau} = \ln\left(\frac{1}{5}\right)$$

$$\Rightarrow \quad -\frac{t}{\tau} = -\ln(5)$$

$$\Rightarrow \quad t = \tau.\ln(5)$$

$$\Rightarrow \quad t = \frac{L}{R_{eq}} \cdot \ln(5)$$

$$\Rightarrow \quad t = \frac{10 \times 10^{-3}}{(0.1 + 0.9)} \times 1.6$$

$$\Rightarrow \quad t = 1.6 \times 10^{-2}$$

$$\Rightarrow \quad t = 0.016\text{ s}$$ **Ans. (B)**

Sol. 30 Charge flown through battery is given as

$$q = \int I dt$$

$$\Rightarrow \quad q = \int_0^{L/R} \frac{E}{R}\left[1 - e^{\frac{-Rt}{L}}\right] dt$$

$$\Rightarrow \quad q = \frac{EL}{R^2}\frac{1}{e}$$

$$\Rightarrow \quad q = \frac{EL}{2.7R^2}$$ **Ans. (B)**

Sol. 31 Growth of current in RL circuit is given as

$$i = i_0 (1 - e^{-Rt/L})$$

$$\Rightarrow \quad \frac{i_0}{i} = \frac{1}{1 - e^{-2\times10^4}}$$

$$\Rightarrow \quad \frac{i_0}{i} \simeq 1$$ **Ans. (A)**

Sol. 32 For current growth in LR circuit, we use

$$i = i_0(1 - e^{-Rt/L}) = i_0 (1 - e^{-t/T_C})$$

Charge flow through circuit is given as

$$q = \int_0^{T_C} idt$$

$$\Rightarrow \quad q = \int_0^{T_C} \frac{\varepsilon}{R}(1 - e^{-t/T_C})$$

$$\Rightarrow \quad q = \frac{\varepsilon}{R}\left(t - \frac{e^{-t/T_C}}{-1/T_C}\right)\Bigg|_0^{T_C}$$

$$\Rightarrow \quad q = \frac{e}{R}(T_C - T_C e^{-1}) - \frac{e}{R}(0 + T_C)$$

$$\Rightarrow \quad q = \frac{e}{R} \times T_C e^{-1}$$

$$\Rightarrow \quad q = \frac{\varepsilon}{R} \times \frac{L}{R}\frac{1}{e}$$

$$\Rightarrow \quad q = \frac{\varepsilon L}{eR^2}$$ **Ans. (D)**

Sol. 33 Induced voltage in choke coil is given as

$$V = L\left|\frac{di}{dt}\right|$$

$$\Rightarrow \quad L = \frac{V}{\left|\frac{di}{dt}\right|} = \frac{100}{\frac{0.25}{0.025\times10^{-3}}} = 10 \text{ mH}$$ **Ans. (10)**

Sol. 34 Maximum energy stored in inductor is

$$U_{max} = \frac{1}{2}LI_{max}^2$$

In series L-R circuit growth of current as a function of time is given as

$$i = I_{max}(1 - e^{-Rt/L})$$

When circuit current becomes $\frac{I_{max}}{\sqrt{n}}$ then energy in inductor becomes $\left(\frac{1}{n}\right)$ times maximum value. So we use

$$\frac{I_{max}}{\sqrt{n}} = I_{max}(1 - e^{-Rt/L})$$

$$\Rightarrow \quad e^{-Rt/L} = 1 - \frac{1}{\sqrt{n}} = \frac{\sqrt{n}-1}{\sqrt{n}}$$

$$\Rightarrow \quad -\frac{Rt}{L} = \ln\left(\frac{\sqrt{n}-1}{\sqrt{n}}\right)$$

$$\Rightarrow \quad t = \frac{L}{R}\ln\left(\frac{\sqrt{n}}{\sqrt{n}-1}\right)$$ **Ans. (C)**

Sol. 35 In the given circuit writing equation of potential drop from P to Q, we get

$$V_P - L\frac{di}{dt} - 30 + iR = V_Q$$

$$\Rightarrow \quad V_P - V_Q = 50 \times 10^{-3} \times 10^2 + 30 - 1 \times 2 = 33 \text{ V}$$

NOTE : As current is decreasing in circuit self induced emf in inductor will have high potential end at left side. **Ans. (33)**

Sol. 36 By transformer equation, we use

$$\frac{N_P}{N_S} = \frac{V_P}{V_S}$$

$$\Rightarrow \quad \frac{N_P}{24} = \frac{220}{12}$$

$$\Rightarrow \quad N_P = \frac{220\times24}{12} = 440 \text{ turns}$$ **Ans. (440)**

Sol. 37 Just after the switch is closed, inductor will act like and open circuit or as infinite resistance so the circuit will look like as shown in figure below

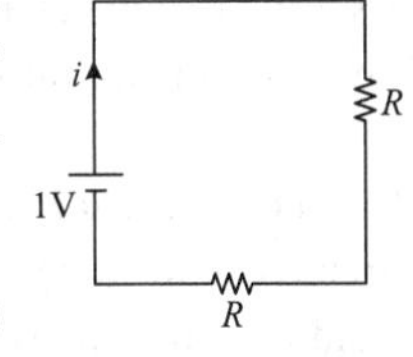

$$i = \frac{9}{R+R} = \frac{9}{4} = 2.25$$ **Ans. (A)**

Sol. 38 In the given circuit, at $t = 0$, inductor act like an open circuit so current through inductor is zero and equivalent resistance of the circuit across battery is given as

$$R_{eq} = (5+1) \,||\, (5+4) = \frac{18}{5}\,\Omega$$

Current through battery can now be calculated as

$$i_1 = \frac{E}{18/5} = \frac{5E}{18}$$

At $t = \infty$ i.e. after a long time, inductor act as a short circuit and it behaves like a straight wire and now the circuit can be drawn

as shown in figure below

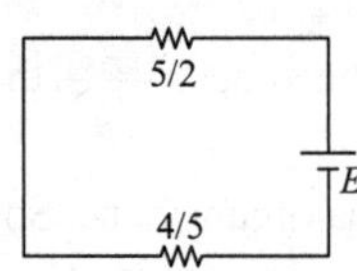

Equivalent resistance of across battery is now given as

$$R_{eq} = (5 \parallel 5) + (4 \parallel 1) = \frac{33}{10}\,\Omega$$

Current through battery can now be calculated as

$$i_2 = \frac{E}{33/10} = \frac{10E}{33}$$ **Ans. (A)**

Sol. 39 From the circuit described in question the EMF of battery remain equal to the potential difference across the inductor, which gives

$$e = \frac{L\,dI}{dt}$$

Current in circuit at $t = 4$ s can be given as

$$\Rightarrow \quad 3\int_0^4 t\,dt = 2\int_0^I dI$$

$$\Rightarrow \quad \frac{3}{2}\times 16 = 2I$$

$$\Rightarrow \quad I = 12$$

Energy stored in coil can now be given as

$$U = \frac{1}{2}LI^2 = \frac{1}{2}\times 2(12)^2 = 144\text{ J}$$ **Ans. (144)**

Sol. 40 Voltage in secondary coil can be calculated from the circuit shown in figure as

$$V_S = \frac{P}{i} = \frac{60}{0.11} = 545.45\text{ V}$$

As primary coil voltage is $V_P = 220$ V so we have $V_S > V_P$ hence this is a step up transformer. **Ans. (C)**

Sol. 41 As the solenoids are identical, current i will be equally divided in the three solenoids B, C and D and the magnetic field inside the solenoid is directly proportional to the current through solenoids so magnetic field inside these three solenoids will be one third of that of A which will be 1 T. **Ans. (D)**

Sol. 42

$$B = 4\times = \frac{\mu_0 i}{4\pi\left(\frac{b}{a}\right)}\sqrt{2} \quad \Rightarrow \quad B = \frac{2\sqrt{2}\mu_0 i}{\pi b}$$

$$\Rightarrow \quad \phi = MI \quad \Rightarrow \quad BA = MI$$

$$\Rightarrow \quad \frac{2\sqrt{2}\mu_0 i}{\pi b}\cdot a^2 = Mi$$

$$\Rightarrow \quad M = \frac{2\sqrt{2}\mu_0 a^2}{\pi b} = \frac{\mu_0}{4\pi}\frac{8\sqrt{2}a^2}{b}$$ **Ans. (D)**

Sol. 43 As inductor energy only depend upon the current passing through it at any instant so initial and final magnetic field energy in inductor is given as

$$U_i = \frac{1}{2}LI_i^2 = 0$$

and

$$U_i = \frac{1}{2}LI_f^2 = \frac{1}{2}\times 2\times(2)^2 = 4\text{ J}$$

Change in energy in inductor is the energy spent during the period, is calculated as

$$\Delta U = U_f - U_i = 4 - 0 = 4\text{ J}$$ **Ans. (4)**

Sol. 44 As shown in figure given in question, current in both the inductor produces magnetic flux in opposite direction so equivalent self inductance of combination is given as

$$L_{eq} = L_1 + L_2 - 2M$$ **Ans. (D)**

Sol. 45 If a current i flows in outer loop, the magnetic field produced by outer loop at its centre is given as

$$B = 4\times\frac{\mu_0 i}{4\pi\times\frac{L}{2}}\times(2\sin 45°) = \frac{2\sqrt{2}\mu_0 i}{\pi L}$$

Mutual inductance of the system is calculated as

$$M = \frac{\text{Flux through inner loop}}{i}$$

$$\Rightarrow \quad M = \frac{2\sqrt{2}\mu_0 l^2}{\pi L}$$ **Ans. (C)**

Sol. 46 In the given circuit in question, just after closing the switch S, inductor behaves like an open circuit so current flows only in the left loop of the circuit. This current is calculated as

$$I = \frac{6}{2+4} = 1\text{ A}$$ **Ans. (1)**

Sol. 47 After a long time after switch is closed, the inductor behaves as a short circuit path. Thus current through cell is calculated as

$$I = \frac{12}{R/3} = 3\text{ A}$$ **Ans. (3)**

Sol. 48 The frequency of LC oscillator is given as

$$\omega_0 = \frac{1}{\sqrt{LC}}$$

Now if inductance is made twice and capacitance is made eight times, new frequency is calculated as

$$\omega = \frac{1}{\sqrt{2L \times 8C}} = \frac{1}{4\sqrt{LC}}$$

$$\Rightarrow \quad \omega = \frac{\omega_0}{4}$$

$$\Rightarrow \quad x = \frac{1}{4}$$

Ans. (A)

Sol. 49 The rate of change of current in the question is given as

$$\frac{dI}{dt} = -1 \text{ A/s}$$

Using KVL equation from A to B, we have

$$V_A - IR - L\frac{dI}{dt} - 12 = V_B$$

$$\Rightarrow \quad V_A - 2 \times 12 - 6(-1) - 12 = V_B$$

$$\Rightarrow \quad V_A - V_B = 36 - 6 = 30 \text{ V}$$

Ans. (30)

Sol. 50 Primary voltage at input of transformer is given as

$$v_p = 12 \times 10^3 \text{ V}$$

Secondary voltage at output of transformer is given as

$$v_s = 120 \text{ V}$$

Power output by transformer is given as

$$P_s = 60 \text{ KW} = v_s \times i_s$$

Thus current at secondary coil is calculated as

$$i_s = \frac{60 \times 10^3}{120} = 5 \times 10^2 \text{ A}$$

So, Resistive load in secondary circuit is given as

$$R_L = \frac{v_s}{i_s} = \frac{120}{5 \times 10^2} = 24 \times 10^{-2} = 240 \times 10^{-3}\ \Omega$$

$$\Rightarrow \quad R_L = 240 \text{ m}\Omega$$

Ans. (240)

Sol. 51 Radius and number of turns in coil 1 and 2 are given as

$r_1 = 1$ cm, $N_1 = 10$

$r_2 = 1000$ cm, $N_2 = 200$

Magnetic flux in coil 1 due to 2 where M is mutual inductance is given as

$$\phi_{12} = MI_2$$

$$\Rightarrow \quad N_2 \vec{B}_2 \cdot N_1 \vec{A}_1 = MI_2$$

$$\Rightarrow \quad N_1 N_2 \frac{\mu_0 I_2}{2r_2} \cdot \pi r_1^2 = MI_2$$

$$\Rightarrow \quad M = \frac{10 \times 200 \times 4\pi \times 10^{-7} \times \pi \times (0.01)^2}{2 \times 10}$$

$$\Rightarrow \quad M = 4 \times 10^{-8} \text{ H}$$

Ans. (4)

Sol. 52 Maximum energy stored in capacitor is same as maximum energy stored in inductor

$$\frac{1}{2}Li_{max}^2 = \frac{1}{2}\frac{Q_{max}^2}{C}$$

Thus maximum current in the circuit is given as

$$i_{max} = \sqrt{\frac{1}{LC}}Q_{max}$$

$$\Rightarrow \quad i_{max} = \frac{2.7 \times 10^{-6}}{\sqrt{75 \times 10^{-3} \times 1.2 \times 10^{-6}}}$$

$$\Rightarrow \quad i_{max} = 9 \text{ mA}$$

Ans. (9)

Sol. 53 Quality factor of a circuit is given as

$$Q = \frac{X_L}{R} = \frac{\omega L}{R}$$

Resonant frequency of resonant circuit is calculated as

$$\omega = \frac{1}{\sqrt{LC}} = \frac{1}{\sqrt{1 \times 6.25 \times 10^{-6}}}$$

$$\Rightarrow \quad \omega = \frac{10^3}{2.5} = 400 \text{ rad/sec}$$

$$\Rightarrow \quad Q\text{-factor} = \frac{400 \times 1}{100} = 4$$

Ans. (4)

Sol. 54 In steady state, current in circuit is given as

2H 4Ω 10V

$$I = \frac{V}{R} = \frac{10}{4}\text{A} = \frac{5}{2}\text{A}$$

Energy stored in the magnetic field is calculated as

$$E = \frac{1}{2}LI^2 = \frac{1}{2} \times 2 \times \left(\frac{5}{2}\right)^2$$

$$\Rightarrow \quad E = 625 \times 10^{-2} \text{ J}$$

Ans. (625)

Sol. 55 Energy stored in inductor at any time is given as

$$E = \frac{1}{2}LI^2$$

Rate of energy storing in inductor is calculated as

$$R_e = \frac{dE}{dt} = LI\frac{dI}{dt}$$

In $R - L$ circuit current as a function of time is given as

$$I = \frac{E}{R}\left(1 - e^{-t\frac{R}{L}}\right)$$

$$\Rightarrow \quad \frac{dI}{dt} = \frac{E}{L}e^{-\frac{rT}{L}}$$

$$\Rightarrow \quad R_e = \frac{dE}{dt} = \frac{E^2}{R}\left(1 - e^{-\frac{tR}{L}}\right)\left(e^{-t\frac{R}{L}}\right)$$

Time at which the rate of power stored will be maximum is calculated when $dR_e/dt = 0$ which gives

$$t = \frac{L}{R\ln 2}$$

$$\Rightarrow \quad R_{max} = \frac{dE}{dt} = \frac{E^2}{R}\left(1 - \frac{1}{2}\right)\times\frac{1}{2}$$

$$\Rightarrow \quad R_{max} = \frac{E^2}{4R} = \frac{E^2}{100} = \frac{E^2}{2\times 50}$$

$$\Rightarrow \quad a = 2, b = 50$$

$$\Rightarrow \quad \frac{b}{a} = 25$$

Ans. (25)

21.1 Alternating Current

Sol. 1 By transformer equation, we have

$$\frac{e_1}{e_2} = \frac{N_1}{N_2} \qquad \ldots(1)$$

and for an ideal transformer we use its power equation for primary and secondary coil is

$$e_1 i_1 = e_2 i_2 \qquad \ldots(2)$$

From (1) and (2) we have

$$\frac{i_2}{i_1} = \frac{N_1}{N_2}$$

$$\Rightarrow \quad i_2 = \frac{N_1}{N_2}i_1 = \frac{140}{280}\times 4$$

$$\Rightarrow \quad i_2 = 2\,\text{A}$$

Ans. (B)

Sol. 2

$$X_L = \omega L\,\underline{|\pi/2}$$

$$R = R\,\underline{|0^\circ}$$

R L

$e = e_0 \sin\omega t$

Impedance triangle

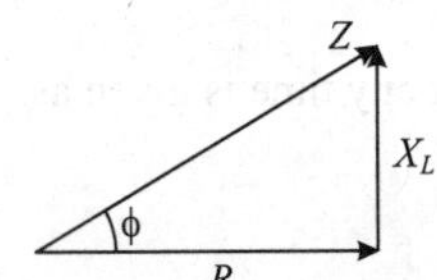

Power factor $\quad \cos\phi = \frac{R}{Z} = \frac{R}{\sqrt{R^2 + \omega^2 L^2}}$ **Ans. (B)**

Sol. 3 The transformer core is laminated to reduce eddy current losses.

Non laminated core :

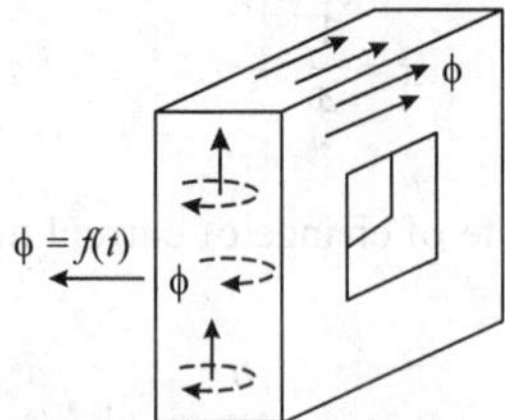

Time varying magnetic flux causes eddy currents to induced over all the cross-section

Laminated core :

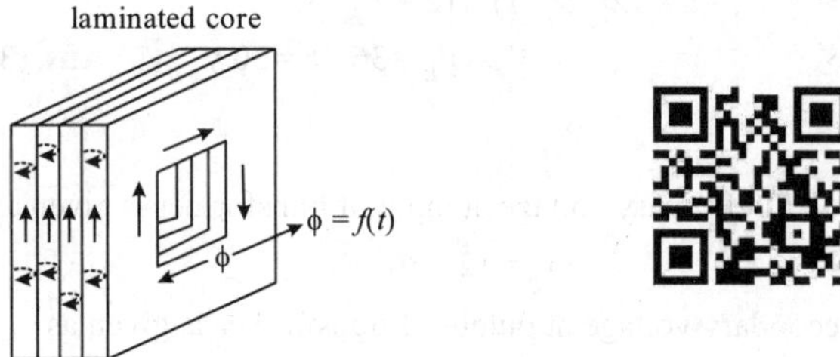

Due to laminated core eddy currents are induced within laminated strips only so losses due these miniature current are reduced. **Ans. (A)**

Sol. 4

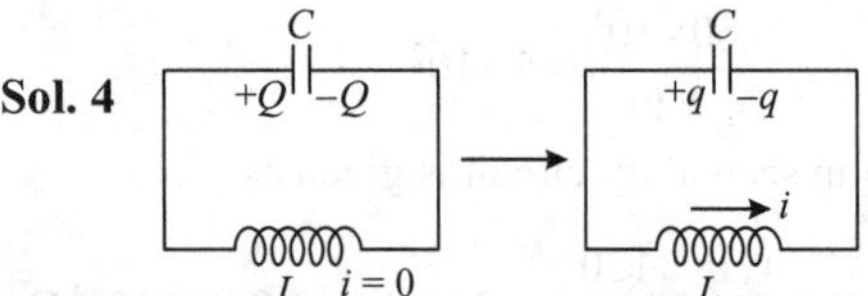

By conservation of energy

$$\frac{Q^2}{2C} = \frac{q^2}{2C} + \frac{1}{2}Li^2$$

When capacitor & inductor have equal amounts of energy stored we use

$$\frac{q^2}{2C} = \frac{1}{2}Li^2$$

$$\frac{Q^2}{2C} = 2\left(\frac{q^2}{2C}\right)$$

$$\Rightarrow \quad q^2 = \frac{Q^2}{2}$$

$$\Rightarrow \quad q = \frac{Q}{\sqrt{2}}$$

Ans. (C)

Sol. 5

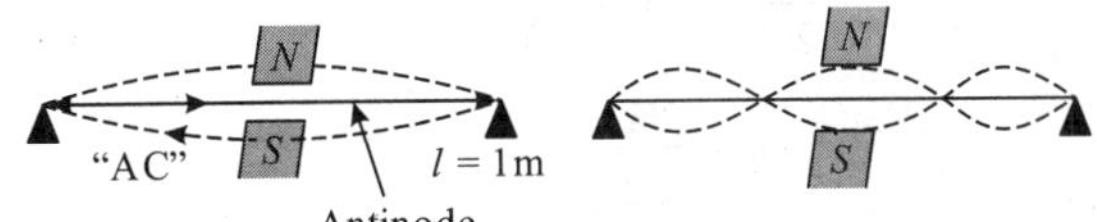

$$T = 10 \times 9.8 \text{ N}$$

$$\mu = 9.8 \times 10^{-3} \text{ kg/m}$$

Here the fundamental frequency of oscillation of wire is given as

$$n_0 = \frac{1}{2l}\sqrt{\frac{T}{\mu}} = \frac{1}{2\times 1}\sqrt{\frac{10\times 9.8}{9.8\times 10^{-3}}}$$

$$\Rightarrow \quad n_0 = \frac{1}{2}\times 100 = 50 \text{ Hz}$$

So when $n = n_0$ or $3n_0$ or $5n_0$ ... the wire will oscillate in resonance.

50 Hz, 150 Hz, 250 Hz.... **Ans. (A)**

Sol. 6 For a given LCR circuit, its resonant frequency is given as

$$f_R = \frac{1}{2\pi\sqrt{LC}}$$

If $C \to 2C$ then to keep f_R same we must use $L \to L/2$

So that product $LC \to$ constant. **Ans. (C)**

Sol. 7

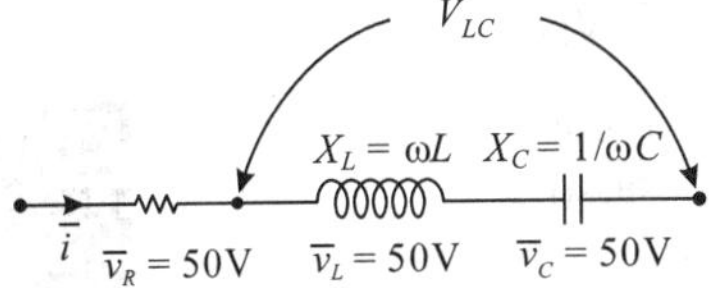

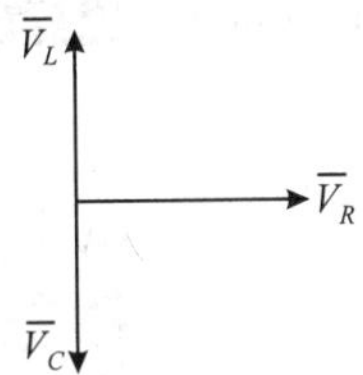

As $\overline{V}_C$ is 90° lagging w.r. to current $\overline{i}$ and $\overline{V}_L$ is 90° ahead w.r. to current $\overline{i}$. So we have $\overline{V}_L$ and $\overline{V}_C$ at phase difference of 180°.

Here $\quad \overline{V}_{LC} = \overline{V}_L + \overline{V}_C = 0$

As $\quad |\overline{V}_L| = |\overline{V}_C|$ **Ans. (D)**

Sol. 8 A *DC* ammeter measures the average value of current passing through it.

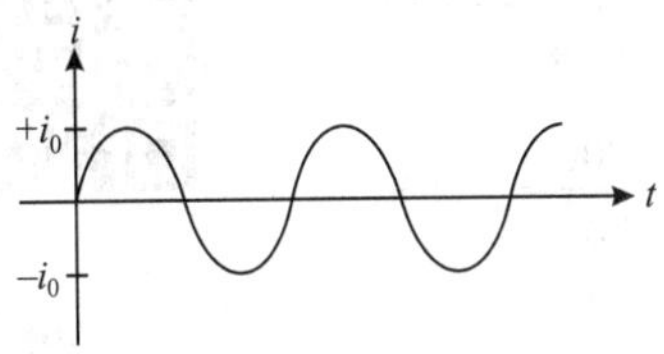

Thus average of a symmetric *AC* is zero. **Ans. (C)**

Sol. 9 Due to an inductor $L \to$ current legs behind emf by $\pi/2$ in phase.

Due to a capacitor $C \to$ current leads emf by $\pi/2$ in phase.

If a resistance R is there in AC circuit then the phase difference between current and emf is between 0 and $\pi/2$.

$\Rightarrow$ Only due to L or C or both L & C phase difference remain $\pi/2$. **Ans. (B)**

Sol. 10 For a given AC circuit power factor is given as

$$\cos\phi = R/Z$$

$$\Rightarrow \quad \cos\phi = \frac{12}{15} = \frac{4}{5} = 0.8$$ **Ans. (A)**

Sol. 11 Power is maximum at resonant frequency given as

$$f_R = \frac{1}{2\pi\sqrt{LC}}$$

$$\Rightarrow \quad C = \frac{1}{4\pi^2 L f_R^2}$$

$$\Rightarrow \quad C = \frac{1}{4\times 10\times 10\times (50)^2} = 10^{-6}\text{ F}$$

$$\Rightarrow \quad C = 1\ \mu\text{F}$$ **Ans. (C)**

Sol. 12 The magnetic flux passing through the coil of area A in magnetic field is given as

$$\phi = \vec{B}\cdot\vec{A}$$

When coil in at an angle $\theta = \omega t$ then total linked flux with coil is

$$\phi_L = N(BA\cos\omega t)$$

emf induced in coil is

$$e = \left|\frac{d\phi_L}{dt}\right| = NBA\sin\omega t = e_0\sin\omega t$$

(AC induced emf in generator)

Maximum emf $\quad e_0 = NBA\omega$ **Ans. (D)**

Sol. 13 $i \to$ $R = 1\text{k}\Omega$, $C = 2\mu\text{F}$, L; X_C, X_L; $V_R = 100$V, $V_L = iX_L$

At resonant frequency in LCR circuit

$$X_L = X_C$$

$$X_L = \frac{1}{\omega C}$$

Circuit current $\quad i = \frac{V_R}{R} = \frac{100}{1000} = 0.1\,\text{A}$

Voltage across inductor is given as

$$V_L = iX_L = i\left(\frac{1}{\omega C}\right) = 0.1 \times \frac{1}{200 \times 2 \times 10^{-6}}$$

$$\Rightarrow \quad V_L = \frac{1}{4 \times 10^{-3}} = \frac{1000}{4} = 250\text{V} \quad \textbf{Ans. (C)}$$

Sol. 14 Average power consumption in an AC circuit is given as

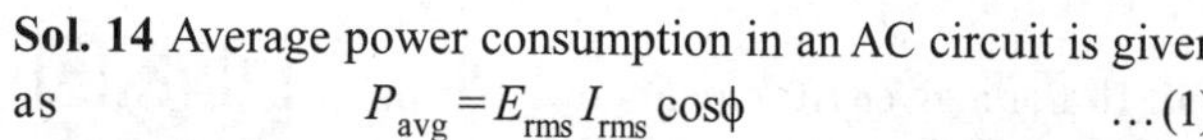

$$P_{avg} = E_{rms} I_{rms} \cos\phi \quad \ldots(1)$$

Here ϕ is the phase difference between emf and current in circuit

Here $\quad E = E_0 \sin\omega t$

and $\quad I = I_0 \sin(\omega t - \pi/2)$

$\Rightarrow \quad \phi = \pi/2$

$\Rightarrow$ From equation-(1)

$$P_{avg} = 0 \quad \textbf{Ans. (B)}$$

Sol. 15 Due to capacitor and inductor separately, power factor of circuit is same as $\phi = 30°$ in both cases (one lagging & one leading)

$\Rightarrow \; X_L = X_C$ so this is the condition of 'Resonance'.

In LCR circuit at resonance, power dissipation is given as

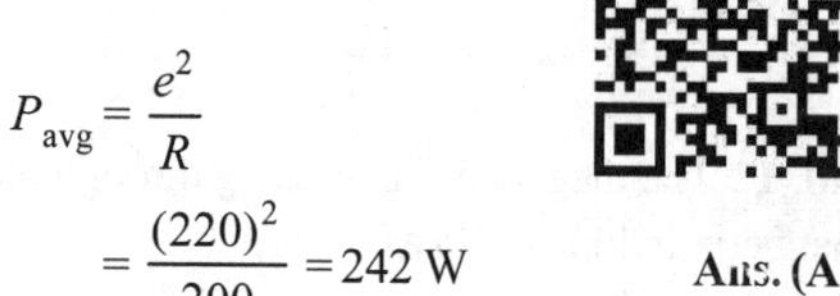

$$P_{avg} = \frac{e^2}{R}$$

$$= \frac{(220)^2}{200} = 242\text{ W} \quad \textbf{Ans. (A)}$$

Sol. 16

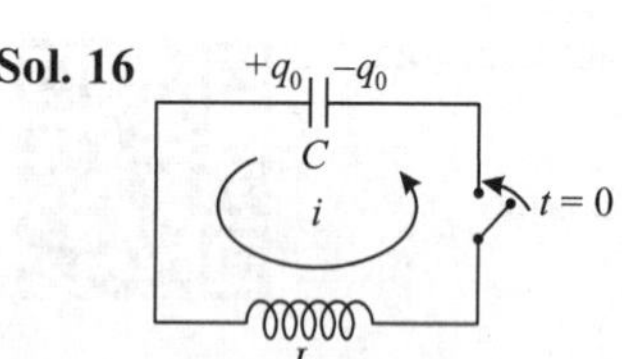

The frequency of LC oscillations is $\omega = \frac{1}{\sqrt{LC}}$

So the charge on capacitor as a function of time can be given as

$$q = q_0 \sin(\omega t + \alpha)$$

at $t = 0,\, q = q_0$

$\Rightarrow \quad \alpha = \pi/2$

$$\Rightarrow \quad q = q_0\cos(\omega t) \quad \ldots(1)$$

If after t, when charge on capacitor is q such that

$$\frac{q_0^2}{2C} = \frac{q^2}{2C} + \frac{1}{2}Li^2 \qquad \left[\text{with } \frac{q^2}{2C} = \frac{1}{2}Li^2\right]$$

$$\Rightarrow \quad \frac{q_0^2}{2C} = 2 \times \frac{q^2}{2C}$$

$$\Rightarrow \quad q = \frac{q_0}{\sqrt{2}}$$

[At this charge on cap $U_{EF} = U_{MF}$]

from equestion-(1) we have

$$q = q_0 \cos\omega t$$

$$\Rightarrow \quad \frac{q_0}{\sqrt{2}} = q_0 \cos(\omega t)$$

$$\Rightarrow \quad \omega t = \frac{\pi}{4}$$

$$\Rightarrow \quad \frac{1}{\sqrt{LC}}t = \frac{\pi}{4}$$

$$\Rightarrow \quad t = \frac{\pi}{4}\sqrt{LC} \quad \textbf{Ans. (D)}$$

Sol. 17 For a given circuit having an inductor L and a capacitor C, the resonant frequency is given as

$$f_R = \frac{1}{2\pi\sqrt{LC}} = 10000\text{ Hz} \quad \ldots(1)$$

When gas of dielectric constant k is filled then $C \rightarrow kC$

$\Rightarrow$ new resonant frequency will be

$$f_R' = \frac{1}{2\pi\sqrt{kLC}} = 9950\text{ Hz} \quad \ldots(2)$$

By (1)/(2) $\Rightarrow \sqrt{k} = \frac{10000}{9950}$

$$\Rightarrow \quad k = \left(\frac{1000}{995}\right)^2 \simeq 1.01 \quad \textbf{Ans. (C)}$$

Sol. 18 We can see in figure that $V_L = V_C$

this happens at resonance only

$\Rightarrow \quad V_R = e = 220\text{ V}$

and circuit current is purely resistivity so

$$i = \frac{e}{R} = \frac{220}{100} = 2.2\text{ A} \quad \textbf{Ans. (D)}$$

Sol. 19 In series LCR circuit at resonance circuit current is maximum so voltage across R is also maximum & it happens at resonant frequency, given as

$$\omega_R = \frac{1}{\sqrt{LC}} = \frac{1}{\sqrt{24 \times 2 \times 10^{-6}}}$$

$$\Rightarrow \quad \omega_R = \frac{1000}{\sqrt{48}} = \frac{1000}{6.9} = 144.5\text{ Hz}$$

Closest option is (B) **Ans. (B)**

Sol. 20

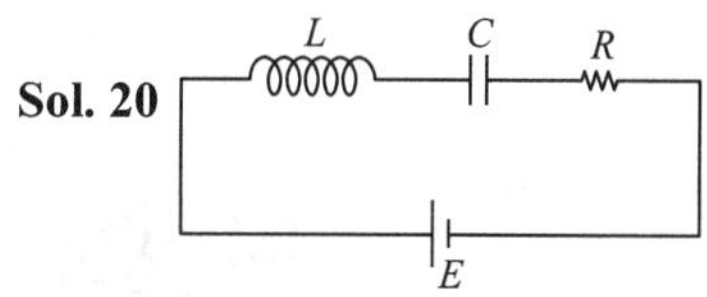

Initially when a DC voltage is applied then in steady state after a long time charge on capacitor is given as

$$q = CE$$

$$\Rightarrow \qquad E = \frac{q}{C} = \frac{10^{-9}}{10^{-11}} = 100\ \text{V}$$

Now DC source is replaced by an AC source of peak value 100 volt.

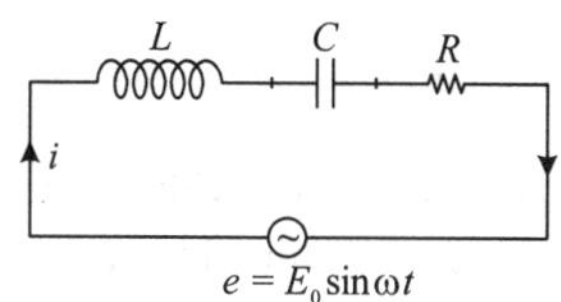

At resonance $\omega = \dfrac{1}{\sqrt{LC}}$ and circuit current will be

$$i = e/R$$

Peak current as $\qquad i_0 = \dfrac{E_0}{R} = \dfrac{100}{100} = 1\text{A}$

Peak voltage across capacitor at resonance is given as

$$V_C = i_0 X_C = 1 \cdot \left(\frac{1}{\omega C}\right)$$

Peak charge on capacitor is

$$q = CV_C = 1 \cdot \left(\frac{1}{\omega C}\right) = \sqrt{LC}$$

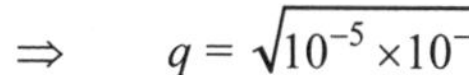

$$\Rightarrow \qquad q = \sqrt{10^{-5} \times 10^{-11}}$$

$$\Rightarrow \qquad q = 10^{-8}\,\text{C}$$

Ans. (D)

Sol. 21 For a series RLC circuit quality factor is given as

$$Q = \frac{\omega_0}{\Delta\omega}$$

Here resonant frequency $\omega_0 \rightarrow$ maximum power in circuit

from plot $\qquad \omega_0 = 5\ \text{kHz}$

and bandwidth $\Delta\omega \rightarrow$ difference of half power frequencies

$$\Delta\omega = \omega_2 - \omega_1$$

From plot we use $\quad \omega_1 \cong 3.6\ \text{kHz}$ and $\omega_2 \cong 6.4\ \text{kHz}$

$\Rightarrow$ Bandwidth $\quad \Delta\omega \cong 6.4 - 3.6 = 2.8\ \text{kHz}$

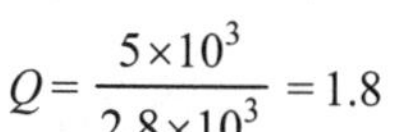

So quality factor of circuit is given as

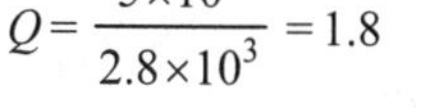

$$Q = \frac{5 \times 10^3}{2.8 \times 10^3} = 1.8$$

Closest option here is (B) **Ans. (B)**

Sol. 22

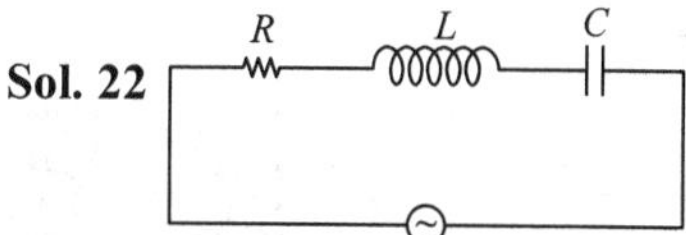

(A) At resonance $X_L = X_C$ so circuit impedance is purely resistive so $\phi = 0°$ and circuit current is in same phase with applied voltage. "It is correct".

(B) At resonant frequency $\omega = \dfrac{1}{\sqrt{LC}}$ we have $\omega L = \dfrac{1}{\sqrt{LC}}$

"It is correct".

(C) At resonance circuit current is $i = e/R$

if R is reduced $\Rightarrow$ i increases.

So voltage across capacitor $V_C = iX_C$ increases with i.

"It is correct".

(D) At resonance circuit impedance is $Z = \sqrt{R^2 + (X_L - X_C)^2}$

As $X_L = X_C \Rightarrow Z$ is minimum.

"It is Not correct". **Ans. (D)**

Sol. 23 Case-I :

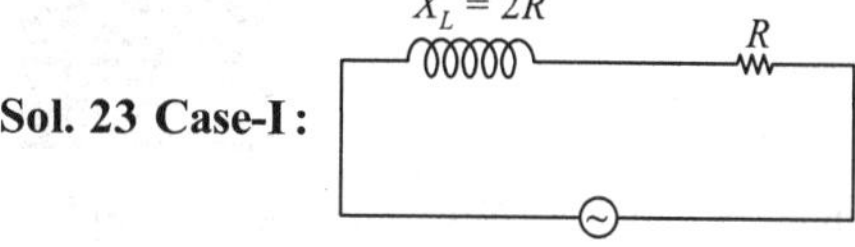

Circuit impedance $\quad Z_1 = \sqrt{R^2 + X_L^2} = \sqrt{5}R$

Power factor $\qquad \cos\phi_1 = \dfrac{R}{Z_1} = \dfrac{1}{\sqrt{5}}$

Case - II :

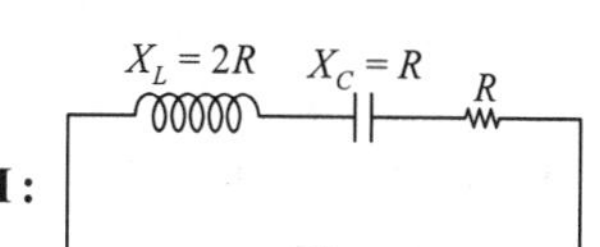

Circuit impedance $\quad Z_2 = \sqrt{R^2 + (X_L - X_C)^2} = \sqrt{2}R$

New power factor, $\cos\phi_2 = \dfrac{R}{Z_2} = \dfrac{1}{\sqrt{2}}$

Ratio of new power factor to old power factor is

$$\frac{\cos\phi_2}{\cos\phi_1} = \frac{1/\sqrt{2}}{1/\sqrt{5}} = \sqrt{\frac{5}{2}}$$

Ans. (D)

Sol. 24

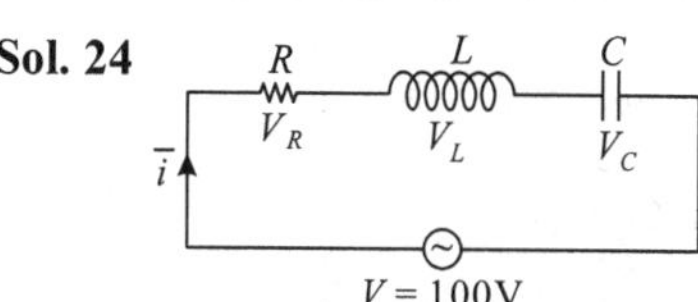

Phasor diagram of voltages is shown below

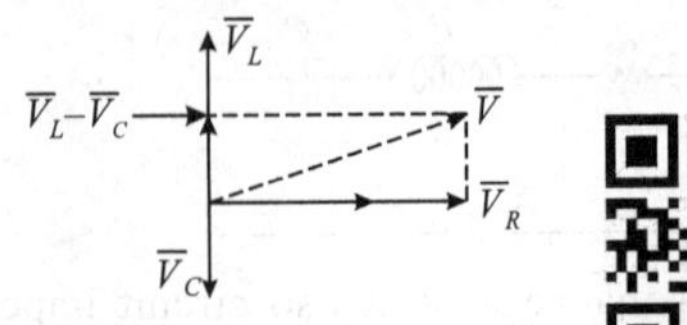

In phasor form $\overline{V} = \overline{V}_R + \overline{V}_L + \overline{V}_C$

in magnitude $V = \sqrt{V_R^2 + (V_L - V_C)^2}$ ...(1)

if $V_L : V_C : V_R = 1 : 2 : 3$

Then we can use in magnitudes $V_L = k$, $V_C = 2k$, $V_R = 3k$

From equation-(1) we have

$$100 = \sqrt{(3k)^2 + (k)^2} = \sqrt{10}k$$

$$\Rightarrow \quad k = 10\sqrt{10}$$

Here we have $V_R = 3k = 30\sqrt{10} \cong 94.5$

Closest option here is option (C) **Ans. (C)**

Sol. 25

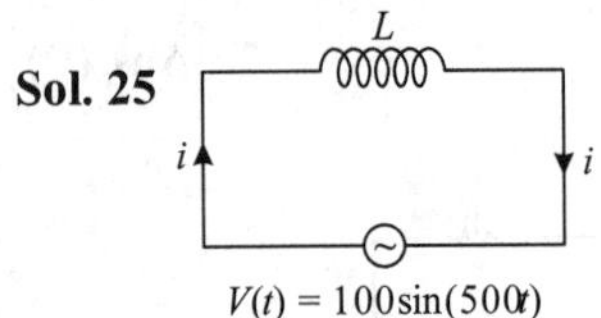

The current through inductor (pure) always lag behind applied emf by 90°

$$\Rightarrow \quad i = i_0 \sin(500t - 90°)$$

Here $i_0 = \dfrac{e_0}{Z} = \dfrac{e_0}{X_L} = \dfrac{e_0}{\omega L} = \dfrac{100}{500 \times 0.02}$

$$\Rightarrow \quad i_0 = \frac{100}{10} = 10\,\text{A}$$

$$\Rightarrow \quad i = 10 \sin(500t - \pi/2)$$

$$\Rightarrow \quad i = -10 \cos(500\,t)$$

Ans. (B)

Sol. 26 As currents & voltages can be measured (RMS value) using Hot wire type meters which use the amount of power dissipation as thermal energy in the meter coil. This does not be affected by change in direction of current.

In devices which work on basis of torque due to magnetic forces will not be useful as due to continuous change in direction of current average torque on coil/magnet = 0. **Ans. (A)**

Sol. 27 Initially current is leading the applied voltage

$$\Rightarrow \quad X_C > X_L \text{ or } \frac{1}{\omega C} > \omega L$$

When C' is connected with C, $X_{CC'} = X_L$

$\Rightarrow \dfrac{1}{\omega C_{eq}} < \dfrac{1}{\omega C}$ so as $C_{eq} > C \Rightarrow C'$ is connected in parallel to C, so we have

$$C_{eq} = C + C'$$

After C' is connected we use for resonance

$$\frac{1}{\omega(C + C')} = \omega L$$

$$\Rightarrow \quad C + C' = \frac{1}{\omega^2 L}$$

$$\Rightarrow \quad C' = \frac{1}{\omega^2 L} - C = \frac{1 - \omega^2 LC}{\omega^2 L}$$

Ans. (C)

Sol. 28 using direct current 10A current flows at 80V emf

$\Rightarrow$ Resistance of arc lamp $R = \dfrac{80}{10} = 8\Omega$

$\Rightarrow$ When arc lamp is connected with an inductor across AC supply then 10A current should flow for lamp to function properly.

$$\Rightarrow \quad i = \frac{220}{Z} \Rightarrow 10 = \frac{220}{Z}$$

$$\Rightarrow \quad Z = 22\,\Omega$$

R L
i = 10A
220V, 50Hz

For a series LR circuit impedance is given as

$$Z = \sqrt{R^2 + \omega^2 L^2}$$

[Here $\omega = 2\pi(50) = 100\pi$ rad/s]

$$\Rightarrow \quad 22 = \sqrt{(8)^2 + (100\pi)^2 L^2}$$

$$\Rightarrow \quad 64 + (100\pi)^2 L^2 = (22)^2$$

$$\Rightarrow \quad (100\pi)^2 L^2 = 484 - 64 = 420$$

$$\Rightarrow \quad L = \frac{\sqrt{420}}{100\pi} = 0.0648\,\text{H} \simeq 0.065\,\text{H}$$

Ans. (D)

Sol. 29 Across a series LR circuit when AC emf is connected, circuit impedance will be

$$Z = \sqrt{R^2 + \omega^2 L^2}$$

and current in circuit is given as

$$i = i_0 \sin(\omega t - \phi)$$

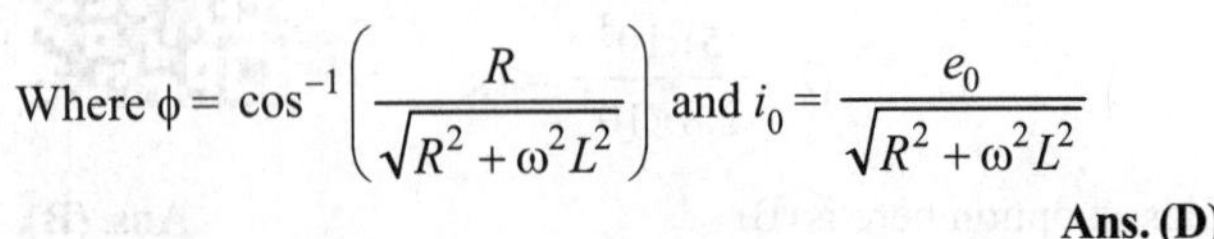

$$\Rightarrow \quad i = \frac{V_0}{\sqrt{R^2 + \omega^2 L^2}} \sin(\omega t - \phi)$$

Where $\phi = \cos^{-1}\left(\dfrac{R}{\sqrt{R^2 + \omega^2 L^2}}\right)$ and $i_0 = \dfrac{e_0}{\sqrt{R^2 + \omega^2 L^2}}$

Ans. (D)

Sol. 30 R, $X_L = \omega L$, $X_C = 1/\omega C$

$$R = 5\Omega$$

$$X_L = \omega L = 320 \times 0.025 = 8\ \Omega$$

$$X_C = \frac{1}{\omega C} = \frac{1}{320 \times 10^{-3}} = 3.1\ \Omega$$

Impedance triangle (phasor form)

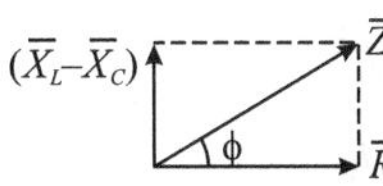

Circuit impedance is given as

$$Z = \sqrt{R^2 + (X_L - X_C)^2}$$

$$Z = \sqrt{(5)^2 + (4.9)^2} = 7\Omega$$

From triangle we have

$$\phi = \tan^{-1}\left(\frac{X_L - X_C}{R}\right) = \tan^{-1}\left(\frac{4.9}{5}\right)$$

$\Rightarrow$ $\phi \simeq \pi/4$ **Ans. (B)**

Sol. 31

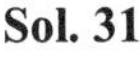

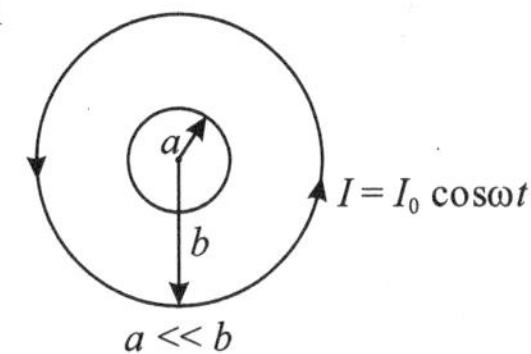

Magnetic flux linked with the smaller loop is

$$\phi = \frac{\mu_0 I}{2b} \times \pi a^2 = \frac{\mu_0 \pi a^2}{2b} I_0 \cos\omega t$$

Emf induced in smaller loop is

$$e = -\frac{d\phi}{dt} = \frac{\mu_0 \pi a^2 I_0}{2b} \omega \sin \omega t$$ **Ans. (A)**

Sol. 32 The quality factor for a series RLC circuit is given as

$$Q = \frac{\omega_R}{\Delta\omega}$$

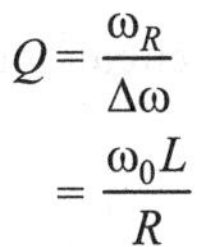

$$= \frac{\omega_0 L}{R}$$ **Ans. (A)**

Sol. 33 Average power consumed by an AC circuit is given as

$$P_{avg} = e_{rms} i_{rms} \cos\phi$$

Here $e_{rms} = \frac{e_0}{\sqrt{2}} = \frac{100}{\sqrt{2}} V$

and $i_{rms} = \frac{i_0}{\sqrt{2}} = \frac{20}{\sqrt{2}} A; \phi = \pi/4$

$\Rightarrow$ $P_{avg} = \frac{100}{\sqrt{2}} \times \frac{20}{\sqrt{2}} \times \frac{1}{\sqrt{2}} = \frac{1000}{\sqrt{2}} W$

Wattless current is given as

$$I_{\omega L} = i_{rms} \sin\phi = \frac{20}{\sqrt{2}} \times \frac{1}{\sqrt{2}} = 10A$$ **Ans.(B)**

Sol. 34 Power equation of a transformer is given as

$$e_1 i_1 \times \eta = e_2 i_2$$

$\Rightarrow$ $2300 \times 5 \times \frac{90}{100} = 230 \times i_2$

$\Rightarrow$ $i_2 = 45A$ **Ans. (C)**

Sol. 35 Capacitive reactance of circuit is given as

$$x_C = \frac{1}{\omega C} = \frac{1}{100\pi \times 120 \times 10^{-6}}$$

$\Rightarrow$ $x_C = 26.52\ \Omega$

Inductive Reactance is

$$x_L = \omega L = 100\pi \times 20 \times 10^{-3} = 2\pi\Omega$$

$\Rightarrow$ $x_C - x_L = 20.24 \approx 20$

Phasor inpedance triangle

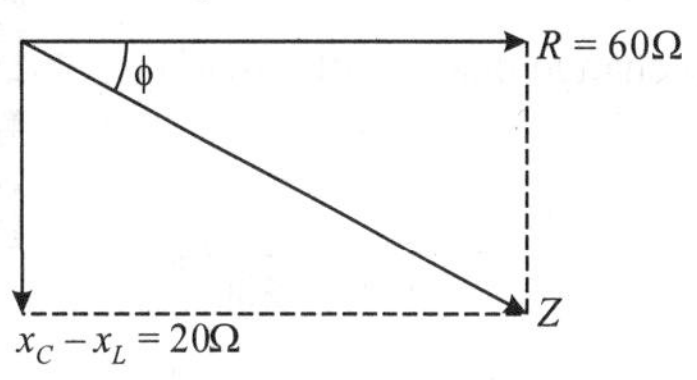

$\Rightarrow$ $Z = \sqrt{R^2 + (x_C - x_L)^2}$

$\Rightarrow$ $Z = 20\sqrt{10}\Omega$

$\Rightarrow$ $\cos\phi = \frac{R}{z} = \frac{3}{\sqrt{10}}$

$\Rightarrow$ $P_{avg} = VI \cos\phi$ where $I = \frac{V}{Z}$

$\Rightarrow$ $P_{avg} = \frac{V^2}{Z} \cos\phi$

$\Rightarrow$ $P_{avg} = 8.64\ W$

Energy spent is

$$Q = P.t = 8.64 \times 60 = 5.18 \times 10^2\ J$$ **Ans. (D)**

Sol. 36 For C-R_2 circuit

$$X_C = \frac{1}{\omega_C} = \frac{2}{10^{-6} \times \sqrt{3} \times 100} = \frac{2 \times 10^4}{\sqrt{3}} \Omega$$

Phase difference of I_2 with AC source emf is given as

$$\tan\theta_1 = \frac{X_C}{R_2} = \frac{2 \times 10^4}{\sqrt{3} \times 20} = \frac{1000}{\sqrt{3}}$$

$\Rightarrow$ θ_1 is close to 90° with current leading emf

For L–R_1 circuit, we use

$$X_L = \omega_L = 100 \times \frac{\sqrt{3}}{10} = 10\sqrt{3}\ \Omega$$

Phase difference of I_1 with AC source emf is given as

$$\tan \theta_2 = \frac{X_L}{R_2} = \frac{10\sqrt{3}}{10} = \sqrt{3}$$

$\Rightarrow$ $\theta_2 = 60°$ with current lagging

Thus phase difference between I_1 and I_2 is given as

$90° + 60° = 150°$ **Ans. (None)**

Sol. 37

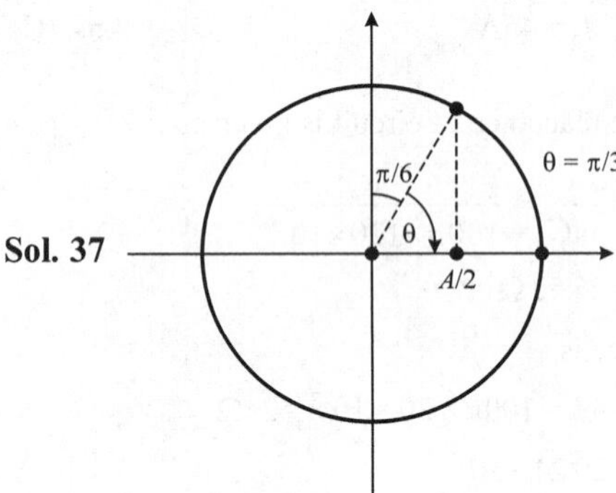

$$V(t) = 220 \sin(100\pi t) \text{ volt}$$

Time taken to change phase angle by $\pi/3$ is

$$t = \frac{\theta}{\omega} = \frac{\frac{\pi}{3}}{100\pi} = \frac{1}{300} \text{ s}$$

$\Rightarrow$ $t = 3.3$ ms **Ans. (C)**

Sol. 38 Given phase difference $\theta = \pi/4$ and $\omega = 100$ rad/s

Phase difference will be 45° when circuit reactance and resitance are equal

$\Rightarrow$ Reactance (X) = Resistance (R) now by checking options

Option (A)

$R = 1000\,\Omega$

and $X_C = \dfrac{1}{10^{-6} \times 100} = 10^4\,\Omega$

Option (B) $R = 10^3\,\Omega$

and $X_L = 10^{-3} \times 100 = 10^{-1}\,\Omega$

Option (C) $R = 10^3\,\Omega$

and $X_L = 10 \times 10^{-3} \times 100 = 1\,\Omega$

Option (D) $R = 10^3\,\Omega$

and 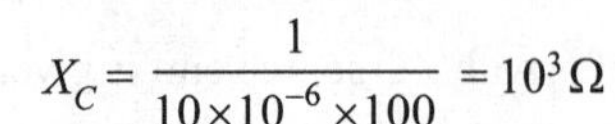 $X_C = \dfrac{1}{10 \times 10^{-6} \times 100} = 10^3\,\Omega$

Option (D) matches the given condition. **Ans. (D)**

Sol. 39 Given that for transformer

$N_P = 300, N_s = 150, P_0 = 2200$ W

$I_s = 10$ A

Output power of transformer is

$$P_0 = V_0 I_0$$

$\Rightarrow$ $2200 = V_0 \times 10$

$\Rightarrow$ $V_0 = 220$ V

Using transformer equation, we have

$$\frac{V_i}{V_0} = \frac{N_P}{N_s}$$

$\Rightarrow$ $V_i = 2 \times 220 = 440$ V

For an ideal transformer

$$V_i I_i = V_O I_O$$

$\Rightarrow$ $I_i = \dfrac{2200}{440} = 5$ A **Ans. (B)**

Sol. 40 For LCR circuit by KVL, we use

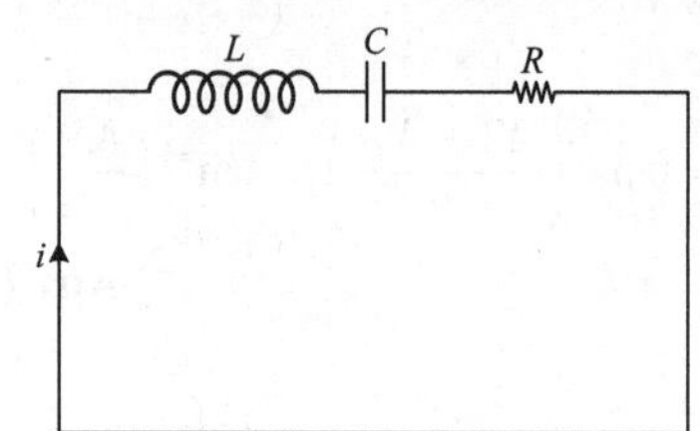

$$-L\frac{di}{dt} - \frac{q}{C} - iR = 0$$

$$L\frac{d^2q}{dt^2} + \frac{1}{C}q + R\frac{dq}{dt} = 0$$

For damped oscillator net force

$$= -kx - bv = ma$$

$$\frac{md^2x}{dt^2} + kx + \frac{bdx}{dt} = 0$$

by comparing; Equivalence is

$L \to m; C \to \dfrac{1}{k}; R \to b$ **Ans. (A)**

Sol. 41 Figure below shows the circuit for situation described in question

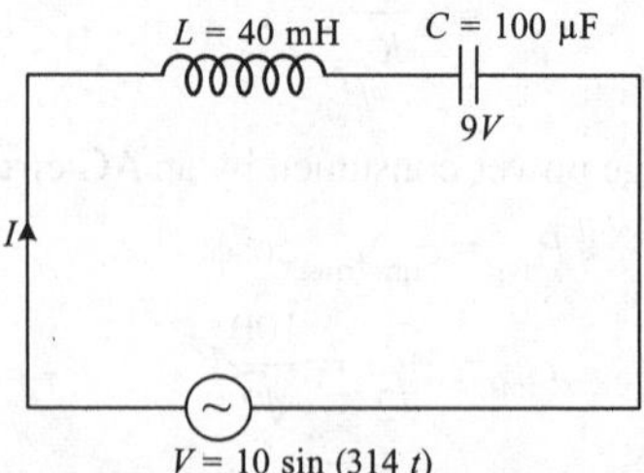

Here reactance of L and C are given as

$$X_L = \omega L = 314 \times 10 \times 10^{-3} = 12.56\,\Omega$$

$$X_C = \frac{1}{\omega C} = \frac{1}{314 \times 100 \times 10^{-6}} = \frac{10^4}{314} = 31.84\,\Omega$$

As $\overline{X_C}$ and $\overline{X_L}$ are in opposite phase we use

$$\overline{Z} = \overline{X_C} - \overline{X_L}$$

$$Z = 31.84 - 12.56 = 19.28\,\Omega$$

Peak current in circuit is given as

$$I = \frac{10}{19.28} = 0.52\,\text{A}$$

As $X_C > X_L$, in this case current will lease voltage by $\pi/2$ so it is given as

$$I = 0.52 \sin\left(314t + \frac{\pi}{2}\right)$$ **Ans. (A)**

Sol. 42 As due to inductance voltage lead current by 45° that means resistance of coil is equal to that of inductive reactance as shown in phasor diagram below

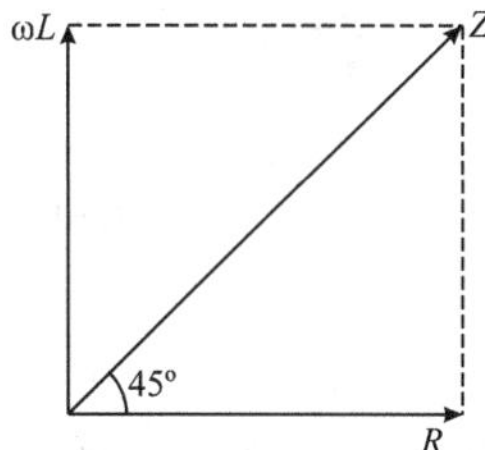

We use

$$R = X_L = \omega L \text{ and } Z = \sqrt{R^2 + \omega^2 L^2}$$

$$\Rightarrow \quad L = \frac{R}{\omega} = \frac{100}{2000\pi} = \frac{1}{20 \times 3.14}$$

$$\Rightarrow \quad L = 1.6 \times 10^{-2}\,\text{H}$$

Most appropriate, among given answers is (A) **Ans. (A)**

Sol. 43 Given circuit impedance is

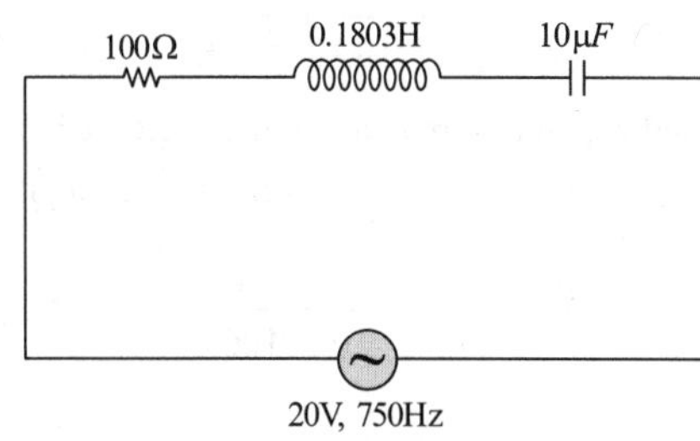

$$Z = \sqrt{R^2 + (X_L - X_C)^2}$$

$$\Rightarrow \quad Z = \sqrt{(100)^2 + \left(2\pi nL - \frac{1}{2\pi nL}\right)^2}$$

$$\Rightarrow \quad Z = \sqrt{(100)^2 + \begin{bmatrix} 2 \times 3.14 \times 750 \times 0.1803 \\ -\dfrac{1}{2 \times 3.14 \times 750 \times 10 \times 10^{-6}} \end{bmatrix}^2}$$

$$\Rightarrow \quad Z \simeq 834\,\Omega$$

Current in circuit is given as

$$i = \frac{e}{Z} = \frac{20}{834} A$$

Thermal power produced in resistance is

$$P_{th} = i^2 R = \left(\frac{20}{834}\right)^2 \times 100\text{W}$$

In time t if temperature of resistance is raised by 10°C, we use

$$\theta = ms\Delta\theta$$

$$\Rightarrow \quad P_{th} t = ms\Delta\theta$$

$$\Rightarrow \quad t = \frac{2 \times 10}{\left(\frac{20}{834}\right)^2 \times 100} = \frac{(834)^2}{20 \times 100}$$

$$\Rightarrow \quad t \cong 348\text{ s}$$ **Ans. (C)**

Sol. 44 Power dissipated in circuit is given as

$$P = \left(\frac{e^2}{Z}\right) \cos\phi$$

$$\Rightarrow \quad 400 = \frac{(250)^2}{Z} \times 0.8$$

$$\Rightarrow \quad Z = 125\,\Omega$$

Capacitance required to change power factor to unity is such that

$$\frac{1}{\omega C} = \omega L = Z \sin\phi$$

$$\Rightarrow \quad C = \frac{1}{\omega Z \sin\phi}$$

$$\Rightarrow \quad C = \frac{1}{2\pi(50) \times 125 \times 0.6}$$

$$\Rightarrow \quad C = \frac{\left(\frac{1}{2500}\right)}{3\pi} F = \frac{n}{3\pi}\,\mu\text{F}$$

$$\Rightarrow \quad n = \frac{10^6}{2500} = 400$$ **Ans. (400)**

Sol. 45 Qality factor of resonance circuit is given as

$$Q = \omega_0 \frac{L}{R} = (2\pi f)\frac{L}{R}$$

$$\Rightarrow \quad Q = 2\pi \times 10^7 \times \frac{2 \times 10^{-4}}{6.28} = 2 \times 10^3 = 2000$$

Ans. (2000)

Sol. 46 (a) Magnetron is a source of microwave frequencies.
(b) Vibration of atoms and molecules causes them to emit infrared frequency.
(c) Radioactive decay of a nucleus causes it to emit radioactive radions including alpha, beta and gamma rays.

(d) Inner shall electrons in heavy elements during transition emits high frequency radiations of the order of X-rays.

Ans. (B)

Sol. 47 At resonance the power of *LCR* circuit is given as

$$P=\frac{V^2}{R}$$

$$\Rightarrow \quad R=\frac{V^2}{P}=\frac{(120)^2}{16}=900\,\Omega$$ **Ans. (900)**

Sol. 48 Current through 60 Ω resistance in the circuit given in question is calculated as

$$i_R=\frac{15}{60}=\frac{1}{4}\text{A}$$

Thus capacitor current is also the same, so we have

$$i_C=\frac{1}{4}\text{A}$$

Potential difference across capacitor is given as

$$V_C=IX_C$$

$$\Rightarrow \quad 10=\frac{1}{4}\times\frac{1}{\omega C}$$

$$\Rightarrow \quad C=\frac{1}{40\omega}=\frac{1}{4000}=250\,\mu\text{F}$$

Now, current through 40 Ω resistance is given as

$$i_{R'}=\frac{20}{40}=\frac{1}{2}\text{A}$$

Thus current through inductor is given as

$$i_L=\frac{1}{2}-\frac{1}{4}=\frac{1}{4}\text{A}$$

Potential difference across inductor is given as

$$V_L=IX_L=\frac{1}{4}\times\omega L$$

$$\Rightarrow \quad 20=\frac{1}{4}\times 100\times L$$

$$\Rightarrow \quad L=0.8\,\text{H}$$ **Ans. (D)**

Sol. 49 As the phase difference between current and voltage is 45°, we can use

$$\tan 45^\circ=\frac{1}{\omega CR}=\frac{\omega L}{R}$$

$$\Rightarrow \quad X_L=X_C$$

At resonance, we have

$$i=\frac{V}{R}=\frac{220}{110}=2\text{A}$$ **Ans. (A)**

Sol. 50 (a) Rectifier is used to convert from *AC* to *DC*.

(b) Stabilizers are used for stable (constant) output voltage when input voltage varies with time.

(c) Transformers are used to Step - up or step - down ac voltages.

(d) Filters are used to remove ripples in the rectified output voltage from a rectifier. **Ans. (A)**

Sol. 51 For the given AC current, RMS value can be calculated by taking square root of average of squares, given as

$$i=i_1\sin\omega t+i_2\sin(\omega t+90)$$

$$i^2=i_1^2\sin^2\omega t+i_2^2\cos^2\omega t+i_1i_2\sin\omega t\cos\omega t$$

$$\langle i^2\rangle=i_1^2\left(\frac{1}{2}\right)+i_2^2\left(\frac{1}{2}\right)+0$$

$$i_{RMS}=\sqrt{\langle i^2\rangle}=\sqrt{\frac{i_1^2+i_2^2}{2}}$$ **Ans. (C)**

Sol. 52 The quality factor of a series *LCR* resonant circuit is given as

$$Q=\frac{X_L}{R}=\frac{\omega L}{R}=\frac{1}{\sqrt{LC}}\frac{L}{R}=\frac{\sqrt{L}}{R\sqrt{C}}$$

When the inductance is increased by two fold and resistance decreased by two fold, the new quality factor becomes

$$Q'=\frac{\sqrt{2L}}{\left(\frac{R}{2}\right)\sqrt{C}}=2\sqrt{2}Q=2\sqrt{2}(100)$$

$$\Rightarrow \quad Q'=282.84$$ **Ans. (283)**

Sol. 53 Impedance of the circuit is given as

$$Z=\sqrt{(X_L-X_C)^2+R^2}$$

Inductive and capacitive reactance of circuit are calculated as

$$X_L=\omega L=100\times 100\times 10^{-3}=10\,\Omega$$

and $$X_C=\frac{1}{\omega C}=\frac{1}{100\times100\times10^{-6}}=10\,\Omega$$

$$\Rightarrow \quad Z=\sqrt{(10-100)^2+R^2}=\sqrt{90^2+120^2}$$

$$\Rightarrow \quad Z=150\,\Omega$$

Peak current in circuit can be calculated as

$$i_{\text{peak}}=\frac{e}{Z}=\frac{30}{150}=\frac{1}{5}=0.2\,\text{A}$$

Resonant frequency of circuit is given by

$$\omega=\frac{1}{\sqrt{LC}}$$

$$\Rightarrow \quad f = \frac{1}{2\pi\sqrt{LC}} = \frac{1}{2\pi\sqrt{100\times10^{-3}\times100\times10^{-6}}}$$

$$\Rightarrow \quad f = \frac{100\sqrt{10}}{2\pi} = \frac{100\pi}{2\pi} = 50\text{ Hz} \qquad \textbf{Ans. (D)}$$

Sol. 54 At resonance power dissipated is given as

$$P = \frac{(V_{\text{rms}})^2}{R}$$

$$P = \frac{(250/\sqrt{2})^2}{8} = 3906.25\text{ W} \approx 4\text{ kW}$$

Ans. (4)

Sol. 55 Hot wire ammeter gives rms value of the current. So it is calculated for the given current as

$$I_{\text{rms}} = \frac{I_0}{\sqrt{2}} = \sqrt{\frac{I_1^2 + I_2^2}{2}} \qquad \textbf{Ans. (B)}$$

Sol. 56 All options can be chosen by definition using phasor analysis for an AC circuit in which an emf is applied across pure resistance, inductor and capacitor. **Ans. (D)**

Sol. 57 Inductive reactance in purely inductive circuit is given as

$$X_L = \omega L$$

Thus circuit current is given as

$$i = \frac{e_0}{\omega L}$$

So on reducing the frequency to half, reactance will get halved and current gets doubled. **Ans. (B)**

Sol. 58 Bandwidth of a series LCR circuit is given as

$$w = R/L$$

$$\Rightarrow \quad w \propto R$$

Thus on increasing resistance, bandwidth will increase. Resonance frequency does not depend on resistance so it will remain same. Quality factor of series LCR circuit is given as

$$Q = \frac{X_L}{R} = \frac{\omega L}{R}$$

Thus on increasing resistnace, quality factor will decrease.

Ans. (A)

Sol. 59 For an ideal AC source, sinusoidal current is given as

$$i = i_0 \cos(\omega t)$$

Here $i = i_0$ at $t = 0$ and $i = \frac{i_0}{\sqrt{2}}$ occurs at $\omega t = \frac{\pi}{4}$ which gives

$$t = \frac{\pi}{4\omega} = \frac{\pi}{4(2\pi f)} = \frac{1}{8f}$$

$$\Rightarrow \quad t = \frac{1}{400} = 2.5\text{ ms} \qquad \textbf{Ans. (A)}$$

Sol. 60 The power factor of the circuit is given as

$$\cos\phi = \frac{R}{Z} \qquad \ldots(1)$$

The circuit impedence in series LCR circuit is given as

$$Z = \sqrt{R^2 + (X_L - X_C)^2} \qquad \ldots(2)$$

$$\Rightarrow \quad Z = \sqrt{6^2 + (10-4)^2}$$

$$\Rightarrow \quad Z = 6\sqrt{2}\ \Omega$$

The power factor is calculated as

$$\cos\phi = \frac{6}{6\sqrt{2}} = \frac{1}{\sqrt{2}} \qquad \textbf{Ans. (C)}$$

Sol. 61 Magnetic energy stored in inductor carrying a current i is given as

$$U = \frac{1}{2}Li^2$$

This energy reaches 25% of its maximum value when current reaches half of maximum value, which gives

$$i = \frac{i_0}{2}$$

For growth of current in LR circuit, current is given as

$$i = i_0(1 - R^{-Rt/L})$$

If at $t = t_1$, $i = \frac{i_0}{2}$, from above expression, we have

$$t_1 = \frac{L}{R}\ln 2 \qquad \textbf{Ans. (C)}$$

Sol. 62 The phasor diagrams of voltage and current in circuit are given as

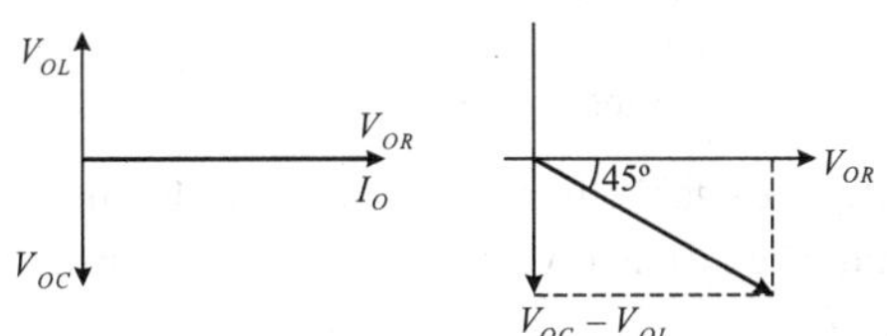

$$\Rightarrow \quad \tan 45^\circ = \frac{V_{0C} - V_{0L}}{V_{0R}} = \frac{X_C - X_L}{R}$$

$$\Rightarrow \quad R = \frac{1}{\omega C} - \omega L$$

$$1\Omega = \frac{1}{300C} - 30\times10^{-3}\times300$$

$$\Rightarrow \quad C = \frac{1}{3}\times10^{-3}\text{ F} \qquad \textbf{Ans. (3)}$$

Sol. 63 The angular frequency of the supply is given as

$$\omega = 2\pi f = 100\,\pi$$

Resistance & reactances of the given circuit are calculated as

$$R = 100\,\Omega$$

$$X_L = \omega L = \frac{100\pi \times 0.5}{1000} = \frac{\pi}{200}\Omega$$

$$X_C = \frac{1}{\omega C} = \frac{10^{12}}{100\pi \times 0.1} = \frac{10^{11}}{\pi}\Omega$$

As $X_C >> X_L$ & $X_C >> R$ so the circuit impedence is dominently capacitive.

Then $$\tan\phi = \frac{|X_L - X_C|}{R} \approx \infty;\ \phi = 90^\circ$$ **Ans. (B)**

Sol. 64 As circuit is given in resonance so we use

$$X_L = X_C$$

$$\Rightarrow \quad Z = R$$

and $i_{rms} = V/Z = V/R$

Power at resonance is given as

$$P = i^2_{rms} R$$

$$\Rightarrow \quad P = \frac{V^2}{R} = \frac{250 \times 250}{5} = 12500\text{ J/s} = 125 \times 10^2\text{ W}$$

Ans. (125)

Sol. 65 In case of a pure capacitive circuit, current leads voltage by an angle $\pi/2$ in phase. **Ans. (D)**

Sol. 66 At resonant frequency in a circuit, emf is in same phase with current and the resonant frequency is given as

$$\omega_R = \frac{1}{\sqrt{LC}}$$

At resonant frequency the capacitive reactance of circuit becomes equal to inductive reactance of circuit and current in circuit becomes maximum, so we use

$$\omega L = \frac{1}{\omega C}$$

If inductive reactance is more current will lag behind emf in phase and when capacitive reactance is more current will lead emf in phase. **Ans. (B)**

Sol. 67 The angular frequency of the AC source is given as

$$\omega = 2\pi f = 100\,\pi\text{ rad/s},$$

The AC emf can be given in standard form as

$$e = e_0 \sin\left(\omega t + \frac{\pi}{2}\right)$$

At rms value of emf, we use

$$\cos\omega t = \frac{1}{\sqrt{2}}$$

$$\Rightarrow \quad \omega t = \frac{\pi}{4}$$

$$\Rightarrow \quad t = \frac{\pi/4}{100\pi} = \frac{1}{400}\text{ s} = \frac{1000}{400}\text{ ms} = 2.5$$ **Ans. (D)**

Sol. 68 Given that per cycle average power dissipation in both the circuits is same, so we have

$$P_1 = P_2$$

$$\left(\frac{V^2}{R}\right)_1 = \left(\frac{V^2}{Z}\right)_2$$

$$\Rightarrow \quad R = Z$$

$$\Rightarrow \quad R = \sqrt{\left(\omega L - \frac{1}{\omega C}\right)^2 + R^2}$$

$$\Rightarrow \quad 5 = \sqrt{\left(\omega L - \frac{1}{\omega C}\right)^2 + R^2}$$

$$\Rightarrow \quad 25 = \left[\omega(0.1) - \frac{1}{\omega(40 \times 10^{-6})}\right]^2 + 25;$$

$$\Rightarrow \quad \omega^2 = (0.1) = \frac{1}{40 \times 10^{-6}}$$

$$\Rightarrow \quad \omega^2 = \frac{1}{4} \times 106$$

$$\Rightarrow \quad \omega = 500\text{ rad/s}$$ **Ans. (500)**

Sol. 69 Angular frequency of the source is given as

$$\omega = 2\pi f = 100\,\pi$$

$$\Rightarrow \quad X_L = \omega L = 100 \times \frac{22}{7} \times 0.07 = 22\,\Omega$$

Circuit impedance can be given as

$$Z = \sqrt{R^2 + x_L^2}$$

$$\Rightarrow \quad Z = \sqrt{(12)^2 + (22)^2} \approx 25.00$$

Circuit current is given as

$$I = \frac{220}{25} = 8.8$$

Phase angle between current and voltage can now be calculated as

$$\tan\phi = 22/12 = 11/6$$

$$\Rightarrow \quad \phi = \tan^{-1}\left(\frac{11}{6}\right)$$ **Ans. (C)**

Sol. 70 Given angular frequncy of source is

$$\omega = 2\pi f = 120\,\pi$$

In resonance condition, we use

$$\omega L = \frac{1}{\omega C}$$

$\Rightarrow \quad L = \frac{1}{\omega^2 C} = \frac{1}{(120\pi)^2 \times 0.1 \times 10^{-6}}$

$\Rightarrow \quad L = 70.3\text{ H}$ **Ans. (D)**

Sol. 71 As the give inductor is storing energy as well as dissipating heat, it acts like a RL circuit shown below.

[Circuit: R and L in series with source E, current 8A]

The energy stored in inductor is given as

$$U = \frac{1}{2}LI^2$$

$$\Rightarrow \quad 64 = \frac{1}{2}.L.(8)^2$$

$$\Rightarrow \quad L = 2\text{ H}$$

Power dissipated in coil resistance is given as

$$P = I^2R$$

$$\Rightarrow \quad 640 = (8)^2.R$$

$$\Rightarrow \quad R = 100$$

Time constant of the circuit is calculated as

$$\tau = \frac{L}{R} = \frac{2}{10} = 0.2$$ **Ans. (D)**

Sol. 72 At resonance of series LCR circuit, we use

$$\omega = 2\pi f = 2\pi \times 50 = 100\pi$$

and $\quad X_C = X_L$

$$\Rightarrow \quad \frac{1}{\omega C} = \omega L$$

$$\Rightarrow \quad \frac{1}{\omega C} = 250\pi$$

$$\Rightarrow \quad C = \frac{1}{100\pi \times 250\pi} = 4\ \mu\text{F}$$ **Ans. (A)**

Sol. 73 The impedence of upper branch in circuit is given as

$$Z_C = \sqrt{\left(\frac{1}{\omega C}\right)^2 + R^2}$$

$$\Rightarrow \quad Z_C = \sqrt{\left(\frac{1}{100 \times 100 \times 10^{-6}}\right)^2 + 100^2}$$

$$\Rightarrow \quad Z_C = \sqrt{100^2 + 100^2} = 100\sqrt{2}\ \Omega$$

The impedence of lower branch in circuit is given as

$$\Rightarrow \quad Z_L = \sqrt{(\omega L)^2 + R^2} = \sqrt{(100 \times 0.5)^2 + 50^2}$$

$$\Rightarrow \quad Z_L = 50\sqrt{2}\ \Omega$$

Current in the two branches of circuit can be calculated as

$$i_L = \frac{200}{Z_L} = \frac{200}{50\sqrt{2}} = \frac{4}{\sqrt{2}} = 2\sqrt{2}\text{ A}$$

$$\Rightarrow \quad i_C = \frac{200}{Z_C} = \frac{200}{100\sqrt{2}} = \sqrt{2}\text{ A}$$

Power factors of the two branches are given as

$$\cos\phi_1 = \frac{100}{100\sqrt{2}} = \frac{1}{\sqrt{2}}, \phi_1 = 45°$$

and $\quad \cos\phi_2 = \frac{50}{50\sqrt{2}} = \frac{1}{\sqrt{2}}, \phi_2 = 45°$

Due to one branch current leads by 45° and due to other branch it lags by 45° so the two currents are at 90° phase difference hence current through source is given as

$$I = \sqrt{I_C^2 + I_L^2} = \sqrt{8+2} = \sqrt{10}A$$

Ans. (None)

Sol. 74 Square of the given current is written as

$$i^2 = 42\sin^2\left(\frac{4t}{T}\right) + 100 + 2\sqrt{42}\sin\left(\frac{4t}{T}\right)$$

Taking mean of each term of squares, gives

$$< i^2 > = \frac{42}{2} + 100$$

Calculating root of mean square gives

$$i_{rms} = \sqrt{< i^2 >} = \sqrt{121} = 11\text{ A}$$ **Ans. (11)**

Sol. 75 Power factor of circuit is given as

$$PF_{initial} = \frac{R}{\sqrt{9R^2 + R^2}} = \frac{1}{\sqrt{10}}$$

Final power factor of circuit is given as

$$PF_{final} = \frac{R}{\sqrt{(3-2)^2 R^2 + R^2}} = \frac{1}{\sqrt{2}}$$

$$\Rightarrow \quad \frac{PF_{final}}{PF_{initial}} = \sqrt{5}$$ **Ans. (1)**

Sol. 76 As $X_L = X_C$ circuit will be in resonance and circuit impedance can be given as

$$Z = \sqrt{(X_L - X_C)^2 + R^2}$$

$$\Rightarrow \quad Z = R$$ **Ans. (B)**

Sol. 77 At very high frequencies i.e. at $\omega \rightarrow \infty$, capacitive reactance can be taken as zero and inductive reactance can be considered as open circuit so the effective circuit can be drawn as shown in figure below

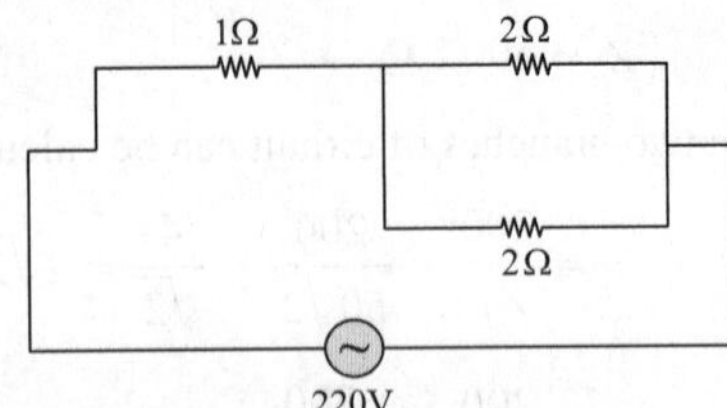

Thus equivalent resistance in this case can be calculated as

$$Z = 1 + \frac{2\times 2}{2+2} = 2\Omega \qquad \textbf{Ans. (2)}$$

Sol. 78 The bandwidth for the given RLC circuit is calculated as

$$\Delta\omega = \frac{R}{L}$$

Quality factor is calculated as

$$Q = \frac{\omega_0}{\Delta\omega} = \omega_0\frac{L}{R}$$

Where resonant frequency is calculated as

$$\omega_0 = \frac{1}{\sqrt{3\times 27\times 10^{-6}}} = \frac{1}{9\times 10^{-3}}$$

$$\Rightarrow \quad \frac{Q}{\Delta\omega} = \frac{\omega_0 \frac{L}{R}}{\frac{R}{L}} = \omega_0\frac{L^2}{R^2} = \sqrt{\frac{1}{LC}}\frac{L^2}{R^2}$$

$$\Rightarrow \quad \frac{Q}{\Delta\omega} = \frac{1}{9\times 10^{-3}} \times \frac{9}{100} = 10\text{ s} \qquad \textbf{Ans. (10)}$$

Sol. 79 Resonance frequency of series LCR circuit when current amplitude is maximum, is given as

$$f = \frac{1}{2\pi\sqrt{LC}}$$

$$\Rightarrow \quad 2000\text{ Hz} = \frac{1}{2\pi\sqrt{L\times 62.5\times 10^{-9}}}$$

$$\Rightarrow \quad L = \frac{1}{4\pi^2\times 2000^2\times 62.5\times 10^{-9}}$$

$$\Rightarrow \quad L = 0.1\text{ H} = 100\text{ mH} \qquad \textbf{Ans. (100)}$$

Sol. 80 Power factor of the series LCR circuit is given as

$$\cos\phi = \frac{R}{Z} = \frac{r}{\sqrt{R^2 + (X_C - X_L)^2}}$$

$$\Rightarrow \quad \cos\phi = \frac{80}{\sqrt{(80)^2 + (60)^2}}$$

$$\Rightarrow \quad \cos\phi = \frac{80}{100} = \frac{8}{10}$$

$$\Rightarrow \quad x = 8 \qquad \textbf{Ans. (8)}$$

Sol. 81 In the first circuit shown in figure below RMS current is calculated as

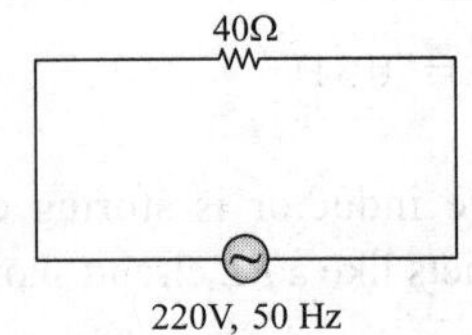

$$I_{\text{rms}} = \frac{220}{40} = 5.5\text{A}$$

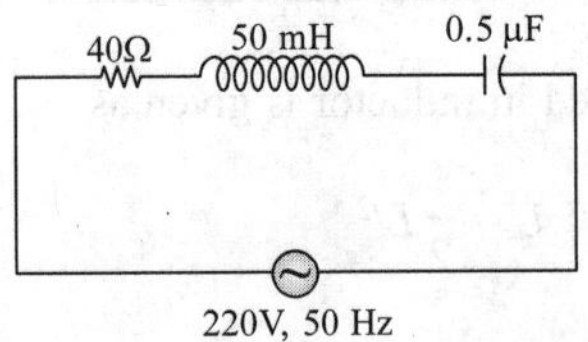

In the above figure in second case at 50 Hz frequency inductive reactance X_L is not equal to capacitive reactance X_C so rms current in second case can never be larger or equal to that compared to first case. **Ans. (A)**

Sol. 82 For maximum current in circuit, we use condition of resonance for which we have

$$\frac{1}{2\pi fC} = 2\pi fL$$

$$\Rightarrow \quad C = \frac{1}{4\pi^2 f^2 L} = \frac{1}{4\times\pi^2\times 49\times 10^6\times 2\times 10^{-6}}$$

$$\Rightarrow \quad C = \frac{1}{3872}F$$

$$\Rightarrow \quad x = 3872 \qquad \textbf{Ans. (3872)}$$

Sol. 83 The power factor of given circuit is written as

$$P = \frac{R}{Z}$$

$$\Rightarrow \quad P_1 = \frac{R}{\sqrt{R^2 + X_L^2}} = \frac{R}{R\sqrt{2}} \quad (\text{as } X_L = R)$$

$$\Rightarrow \quad P_1 = \frac{1}{\sqrt{2}}$$

After adding the capacitor to the circuit, we have

$$P_2 = \frac{R}{\sqrt{R^2 + (X_L - X_L)^2}} = P_2 = 1$$

$$\Rightarrow \quad \frac{P_1}{P_2} = \frac{1}{\sqrt{2}} \qquad \textbf{Ans. (B)}$$

Sol. 84 Impedance of given R-L-C circuit is calculated as

$$z = \sqrt{100^2 + (200-100)^2}$$

$$\Rightarrow \quad z = 100\sqrt{2}\Omega$$

Thus RMS value of current is calculated as

$$i_{rms} = \frac{V_{rms}}{z} = \frac{200\sqrt{2}}{100\sqrt{2}}$$

$$\Rightarrow \quad i_{rms} = 2\,A$$

Ans. (A)

Sol. 85 Maximum value of emf produced due to rotation of coil is

$$\xi_{max} = NAB\omega$$

$$\Rightarrow \quad = 100 \times 14 \times 10^{-2} \times 3 \times \frac{360 \times 2\pi}{60}$$

$$\Rightarrow \quad = 1584\,V$$

Ans. (1584)

Sol. 86 All three parameters given in question have same dimension with same unit ohm therefore $\frac{R}{\sqrt{X_L X_C}}$ is dimensionless. **Ans. (B)**

Sol. 87 As it is specified that current is in same phase with the emf, this indicates that inductive reactance and capacitive reactance are equal, so we use

$$X_L = X_C$$

$$\Rightarrow \quad Z = R = 20\Omega$$

$$\Rightarrow \quad i_{rms} = \frac{220}{20} = 11$$

$$\Rightarrow \quad i_{max} = 11\sqrt{2} = \sqrt{242}$$

Ans. (242)

Sol. 88 Inductive reactance of the circuit is calculated as

$$X_L = L\omega$$

$$\Rightarrow \quad X_L = 5\,mH \times 628$$

$$\Rightarrow \quad X_L = 3.14\,\Omega$$

Ans. (A)

Sol. 89 Impedance of given circuit is calculated as

$$Z = \sqrt{R^2 + (X_L - X_C)^2}$$

$$\Rightarrow \quad Z = \sqrt{80^2 + (100-40)^2}$$

$$\Rightarrow \quad Z = 100\,\Omega$$

The amplitude of current is calculated as

$$i_0 = \frac{V_0}{Z} = \frac{2500}{100} = 25\,A$$

Ans. (25)

Sol. 90 AC generator works on principle of electromagnetic induction, Transformer is based on mutual inductance, Resonance occurs in an AC circuit when effect of capacitance is nullified by the inductance in circuit so both L and C must be present for resonance to occur and sharpness of resonance depends upon the quality factor of the circuit. Hence option (C) is correct. **Ans. (C)**

Sol. 91 To maximize the average rate at which energy is supplied is the situation when power of circuit will be maximum. Thus in LCR circuit maximum power is at the state of resonance when

$$X_L = X_C$$

$$\Rightarrow \quad 79.6 = \frac{1}{\omega C}$$

$$\Rightarrow \quad C = \frac{1}{2\pi \times 50 \times 79.6}$$

$$\Rightarrow \quad C = 40\,\mu F$$

Ans. (40)

Sol. 92 Power will be maximum when circuit impedance is minimum

$$Z = [R^2 + (X_L - X_C)^2]^{1/2}$$

At resonance, $X_L = X_C$

$$\Rightarrow \quad Z_{min} = R$$

Ans. (3)

Sol. 93 Capacitive reactance of circuit is calculated as

$$X_C = \frac{1}{\omega C} = \frac{1}{(2\pi f)C}$$

$$\Rightarrow \quad X_C \propto \frac{1}{f}$$

Thus curve-A represent capacitive circuit. Inductive reactance of the circuit is calculated as

$$X_L = \omega L = (2\pi f)L$$

$$\Rightarrow \quad X_L \propto f$$

Thus curve-B represents inductive circuit. **Ans. (C)**

Sol. 94 Both statements are correct as

$$Z = \sqrt{R^2 + (X_L - X_C)^2} = \sqrt{R^2 + \left(2\pi fL - \frac{1}{2\pi fC}\right)^2}$$

As f increases Z decreases so I increases then at resonance Z minimum so I is maximum then I decreases on fruther increasing f. **Ans. (D)**

Sol. 95 For resonance, $\phi = 0$, hence both inductor & capacitor must be present. Also power factor is zero for pure inductor or pure capacitor hence both the component consume zero average power. **Ans. (A)**

22.1 Electromagnetic Wave

Sol. 1 For an EM wave frequency always remain same and if its wavelength changes from λ to λ' then we use

$$\frac{c}{\lambda} = \frac{v}{\lambda'} \quad \ldots(1)$$

(Where v is wave speed in other medium)

As nothing is specified, we consider medium as non-magnetic ($\mu_r = 1$).

So wave speed in medium is

$$v = \frac{1}{\sqrt{\mu_0 \in_0 \in_r}} = \frac{c}{\sqrt{4.0}} = \frac{c}{2}$$

From (1) $\Rightarrow \frac{c}{\lambda} = \frac{c}{2\lambda'} \Rightarrow \lambda' = \frac{\lambda}{2}$ **Ans. (A)**

Sol. 2 Total power of source = 4×10^3 W

Energy of one photon in eV

$$E = \frac{4\times10^3}{10^{20}} \times \frac{1}{1.6\times10^{-19}} \text{ eV}$$

$\Rightarrow \quad E = 2.5 \times 10^2 = 250$ eV

Wavelength of these photons can be given as

$$\lambda = \frac{hc}{E} = \frac{12431}{250} \text{Å} = 49.72 \text{ Å}$$

As X-Ray wavelength range varies from 0.1Å to 100Å **Ans. (B)**

Sol. 3 Polarization direction of EMF is always along the plane of oscillation of electromagnetic force.

$\Rightarrow \vec{X}$ is parallel to $\vec{E}$

direction of propagation of an EM wave is given by poynting vector which gives direction of flow of energy in the EM wave and it is given as

$$\vec{S} = \frac{1}{\mu_0}(\vec{E} \times \vec{B})$$

Where $\vec{E} \times \vec{B} \rightarrow$ direction of EM waves propagation.

$\Rightarrow \vec{k}$ is parallel to $(\vec{E} \times \vec{B})$ **Ans. (B)**

Sol. 4 The low wavelength part of EM spectrum is

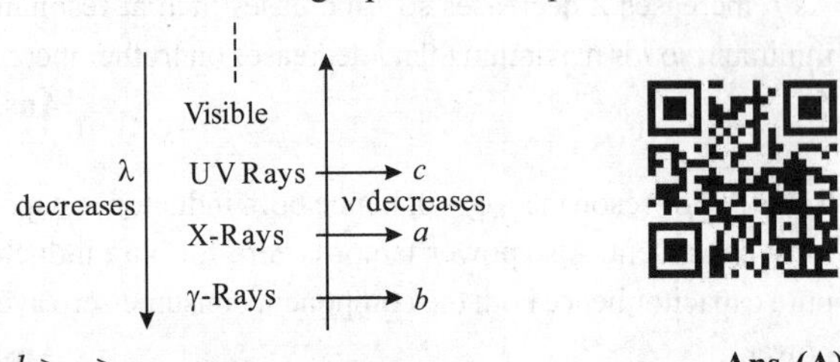

$\Rightarrow b > a > c$ **Ans. (A)**

Sol. 5 Wave propagation in along $\hat{y}$ direction

$\Rightarrow \vec{E} \times \vec{B}$ must be along $\hat{y}$ direction.

$\vec{E} \times \vec{B} \rightarrow$ direction of Poynting vector.

Given that magnetic field is in $+x$ direction i.e. along $\hat{x}$

$\Rightarrow \vec{E}$ must be along $\hat{z}$ direction.

Electric field function is $\vec{E} = E_0\cos(\omega t - \frac{2\pi}{\lambda}y)\,\hat{z}$ **Ans. (C)**

Sol. 6 In EM waves propagation, peak values of electric field E_0 and magnetic field B_0 are related as

$$\frac{E_0}{B_0} = c$$

$\Rightarrow \quad E_0 = B_0 C$

$\Rightarrow \quad E_0 = 20 \times 10^{-9} \times 3 \times 10^8$

$\Rightarrow \quad E_0 = 6$ V/m **Ans. (B)**

Sol. 7 The wavelength of given radiation λ can be given as

$$E = \frac{hc}{\lambda} = 11 \text{ keV}$$

$$\Rightarrow \quad \lambda = \frac{hc}{E} = \frac{6.63\times10^{-34}\times3\times10^8}{11000\times1.6\times10^{-19}}$$

$\Rightarrow \quad \lambda = 11.3 \times 10^{-10}$ m $= 11.3$ Å

As X-rays wavelength range is from 0.1Å to 100Å so this radiation is of X-Rays. **Ans. (A)**

Sol. 8 (A) False - As an EM wave can travel in a medium as well in free space as it does not require a medium for propagation.

(B) False - As EM wave consist of transverse oscillations of EF and MF so these are transverse waves.

(C) False - As EM wave are produced by time variation of EF which can be there by moving charges which accelerates.

(D) True - EM waves carry momentum as well as energy (photons) **Ans. (D)**

Sol. 9 Wave function can be expressed as $f(x - vt)$ by

$$\vec{E} = 4 \times 10^{-7}\vec{E}\left(x - \frac{50t}{4\times10^{-7}}\right)$$

The speed of EM wave in the given medium is given as

$$v = \frac{50}{4\times10^{-7}} \text{ m/s} = 1.25 \times 10^8 \text{ m/s}$$

In terms permittivity & permeability of medium, wave speed can be expressed as

$$v = \frac{1}{\sqrt{\mu \in}} = \frac{1}{\sqrt{\mu_0 \in_0 \in_r}}$$

[As $\mu_r = 1$ for non magnetic medium]

$$\Rightarrow \quad v = \frac{c}{\sqrt{\epsilon_r}}$$

$$\Rightarrow \quad 1.25 \times 10^8 = \frac{3\times10^8}{\sqrt{\epsilon_r}}$$

$$\Rightarrow \quad \sqrt{\epsilon_r} = \frac{3}{1.25} = 2.4$$

$$\Rightarrow \quad \epsilon_r = (2.4)^2 = 5.76$$

Ans. (B)

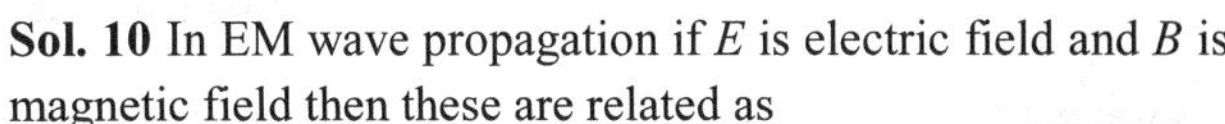

Sol. 10 In EM wave propagation if E is electric field and B is magnetic field then these are related as

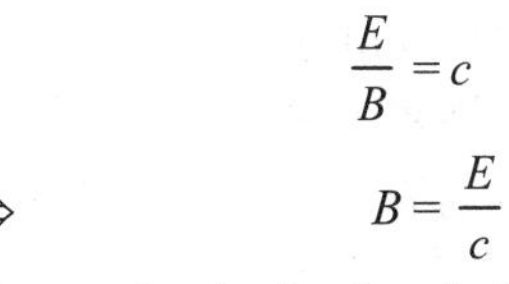

$$\frac{E}{B} = c$$

$$\Rightarrow \quad B = \frac{E}{c}$$

Energy density in electric field wave propagation for electric field and magnetic field are

$$u_E = \frac{1}{2}\epsilon_0 E^2$$

$$u_B = \frac{1}{2}\frac{B^2}{\mu_0} = \frac{1}{2\mu_0}\left(\frac{E}{c}\right)^2 = \frac{1}{2\mu_0}\frac{E^2}{c^2}$$

In free space we use $c = \frac{1}{\sqrt{\mu_0 \epsilon_0}}$

$$\Rightarrow \quad u_B = \frac{1}{2\mu_0}(\mu_0\epsilon_0)E^2 = \frac{1}{2}\epsilon_0 E^2$$

$$\Rightarrow \quad u_B = u_E$$

Ans. (C)

Sol. 11 (1) $\to$ γ-rays ~ 10^{-12} m $\to G$

(2) $\to$ X-rays ~ 10^{-10} m $\to X$

(3) $\to$ UV-rays ~ 10^{-7} m $\to U$

(4) $\to$ Visible ~ 10^{-6} m $\to V$

(5) $\to$ IR-rays ~ 10^{-4} m $\to I$

(6) $\to$ Microwaves ~ 10^{-4} m or more $\to M$

(7) $\to$ Radio waves ~ 10^{-3} m to several km $\to R$

ascending order of wavelength is G, X, U, V, I, M, R **Ans. (C)**

Sol. 12 Average energy density of Electric field in EM waves is given as

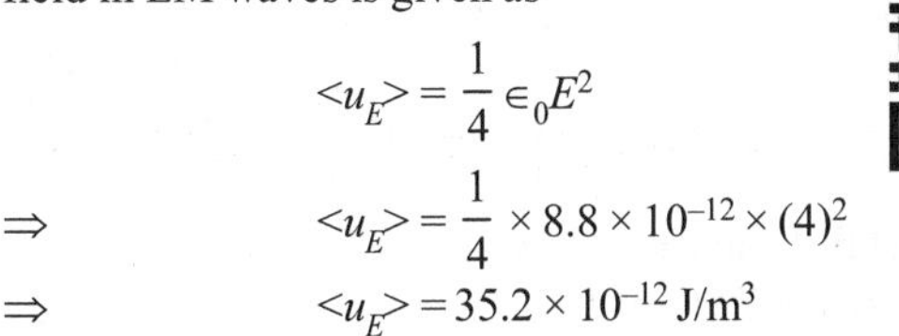

$$\langle u_E \rangle = \frac{1}{4}\epsilon_0 E^2$$

$$\Rightarrow \quad \langle u_E \rangle = \frac{1}{4} \times 8.8 \times 10^{-12} \times (4)^2$$

$$\Rightarrow \quad \langle u_E \rangle = 35.2 \times 10^{-12}\ \text{J/m}^3$$

Ans. (C)

Sol. 13 lamp —— r —— M

$$P_{EM} = 100 \times \frac{3}{100} = 3\text{W}$$

Intensity of EM radiation at $r = 5$m from lamp is given as

$$I = \frac{P}{4\pi r^2} = \frac{3}{4\pi(5)^2} = \frac{3}{100\pi}\ \text{W/m}^2$$

If E_0 is the amplitude of Electric field in wave at point M the we use

$$I = \frac{1}{2}\epsilon_0 E^2 c$$

$$\Rightarrow \quad E_0 = \sqrt{\frac{2I}{\epsilon_0 c}}$$

$$\Rightarrow \quad E_0 = \sqrt{\frac{2\times(3/100\pi)}{\left(\frac{1}{4\pi\times9\times10^9}\right)\times3\times10^8}} = \sqrt{7.2}\ \frac{V}{m}$$

$$\Rightarrow \quad E_0 = 2.68\ \text{V/m}$$

Ans. (B)

Sol. 14 As propagation direction is $+X$ (along $\hat{i}$) so $\vec{B}$ must be along $\hat{j}$ or $\hat{k}$ or in YZ plane.

The amplitude of magnetic field is given as

$$B_0 = \frac{E_0}{c} = \frac{27}{3\times10^8} = 9 \times 10^{-8}\ \text{T}$$

In a propagating wave general equation is written as

$$B = B_0 \sin 2\pi\left(\frac{x}{\lambda} - \frac{t}{T}\right)$$

Average frequency of wave is $\omega = 2\pi\nu = 2\pi(2 \times 10^{14})$ Hz

Wavelength of wave is $\lambda = \frac{c}{\nu} = \frac{3\times10^8}{2\times10^{14}} = 1.5 \times 10^{-6}$ m

$\Rightarrow$ Most appropriate here is option (D), otherwise none is correct. **Ans. (D)**

Sol. 15 For EM wave we use always $\vec{E}\cdot\vec{B} = 0$

and direction of propagation of wave must be along $\vec{E}\times\vec{B}$

here $\vec{E}\times\vec{B}$ must be along $+\hat{k}$ (direction of propagation is z direction).

(A) $(2\hat{i}+3\hat{j}).(\hat{i}+2\hat{j}) = 2+6 = 8 \neq 0$

$\Rightarrow$ Not correct

(B) $(-2\hat{i}-3\hat{j}).(3\hat{i}-2\hat{j}) = -6+6 = 0$

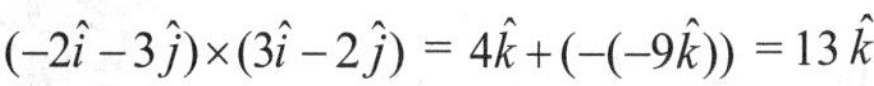

$(-2\hat{i}-3\hat{j})\times(3\hat{i}-2\hat{j}) = 4\hat{k}+(-(-9\hat{k})) = 13\hat{k}$

$\Rightarrow$ Correct.

(C) $(3\hat{i}+4\hat{j}).(4\hat{i}-3\hat{j}) = 12-12 = 0$

$(3\hat{i}+4\hat{j})\times(4\hat{i}-3\hat{j}) = -9\hat{k}-16\hat{k} = -25\hat{k}$ (–z direction)

$\Rightarrow$ Not correct.

(D) $(\hat{i}+2\hat{j})(2\hat{i}-\hat{j}) = 2-2 = 0$

$(\hat{i}+2\hat{j})\times(2\hat{i}-\hat{j}) = -\hat{k}-4\hat{k} = -5\hat{k}$ (–z direction)

$\Rightarrow$ Not correct. **Ans. (B)**

Sol. 16 EM spectrum

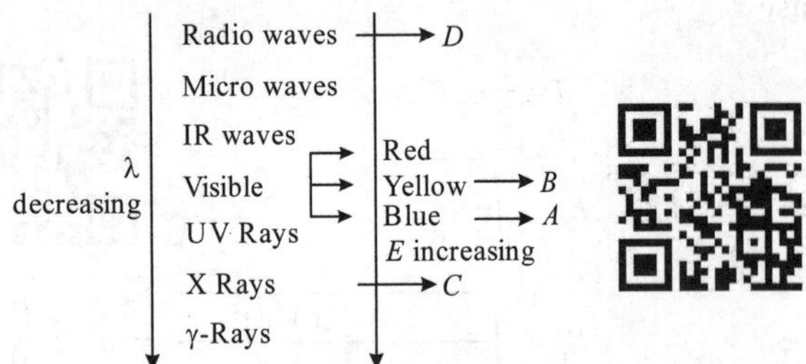

Order of increasing energy, $D \to B \to A \to C$ **Ans. (C)**

Sol. 17

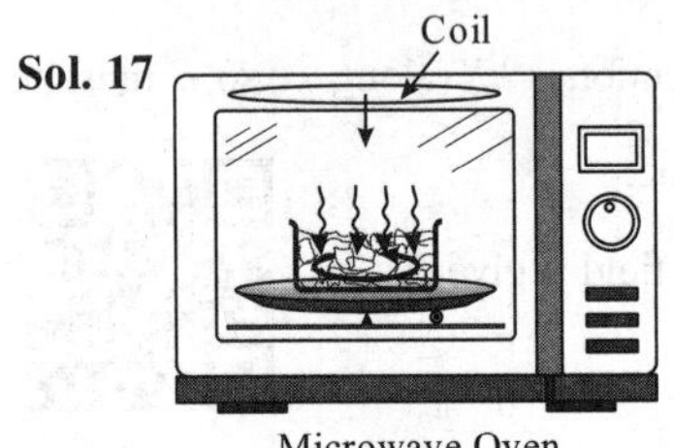

Coil of oven produces time varying magnetic field due to which microwaves are produced.

Due to vibrations in ions (water molecules in ionized state or dipoles) localized heat is produced in substance due to which cooking take place. **Ans. (C)**

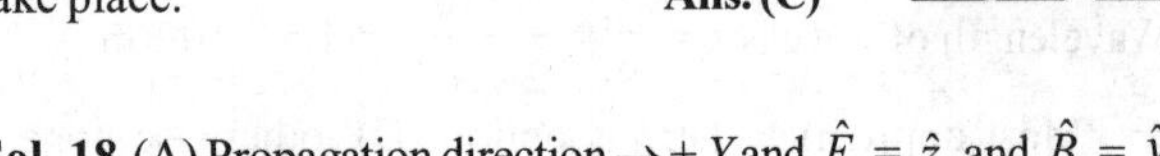

Sol. 18 (A) Propagation direction $\to +Y$ and $\hat{E} = \hat{z}$ and $\hat{B} = \hat{y}$

This is not possible as $\vec{B}$ is along direction of propagation.

$\Rightarrow$ Not correct.

(B) Propagation direction $\to +Y$ and $\hat{E} = \hat{y}$ and $\hat{B} = \hat{z}$

This is not possible as $\vec{E}$ is along direction of propagation.

$\Rightarrow$ Not correct.

(C) Propagation direction $\to +X$ and $\hat{E} = \left(\frac{\hat{y}+\hat{z}}{\sqrt{2}}\right)$ and

$$\hat{B} = \left(\frac{\hat{y}+\hat{z}}{\sqrt{2}}\right)$$

This is not possible as $\vec{E}$ and $\vec{B}$ are parallel $\Rightarrow$ Not correct.

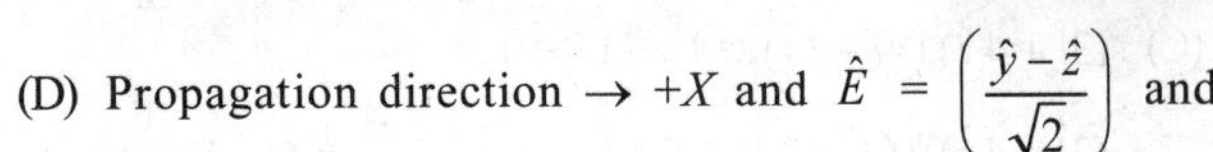

(D) Propagation direction $\to +X$ and $\hat{E} = \left(\frac{\hat{y}-\hat{z}}{\sqrt{2}}\right)$ and

$$\hat{B} = \left(\frac{\hat{y}+\hat{z}}{\sqrt{2}}\right)$$

Here $\vec{E}\cdot\vec{B} = 0$ and both $\vec{E}$ and $\vec{B}$ are $\perp$ to X i.e. in YZ plane and $\vec{E}\times\vec{B}$ direction is along $(\hat{y}-\hat{z})\times(\hat{y}+\hat{z}) = \hat{x}-(-\hat{x}) = 2\hat{x}$ (+x direction). Where $\vec{E}\times\vec{B}$ is direction of energy flow in EM waves. $\Rightarrow$ Correct. **Ans. (D)**

Sol. 19 Electric field amplitude is related to B_0 with speed of EM wave as

$$c = E_0/B_0$$

$$\Rightarrow \quad E_0 = cB_0$$

As wave is travelling along $-X$ direction so vector $\vec{E}\times\vec{B}$ must be along the direction of propagation of EM wave i.e. along $-\hat{i}$ direction. So here Electric field can be given as

$$\vec{E} = E_0 \sin(kx+\omega t)\,\hat{k}$$

$$\Rightarrow \quad \vec{E} = cB_0 \sin(kx+\omega t)\,\hat{k}$$

Ans. (A)

Sol. 20 As we know for an EM wave we use

$$\frac{\partial B}{\partial t} = -\frac{\partial E}{\partial z}$$

$$\Rightarrow \quad \frac{\partial B}{\partial t} = -(-2E_0 k \sin kz \cos\omega t)$$

$$\Rightarrow \quad \partial B = 2E_0 k \sin kz \cos\omega t \partial t$$

$$\Rightarrow \quad \int \partial B = 2E_0 k \sin kz \int \cos\omega t \partial t$$

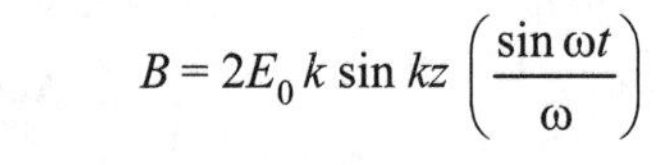

$$\Rightarrow \quad B = 2E_0 k \sin kz \left(\frac{\sin\omega t}{\omega}\right)$$

As we use $\frac{\omega}{k} = c$

$$\Rightarrow \quad B = \frac{2E_0}{c} \sin kz \sin\omega t$$

Ans. (C)

Sol. 21 Wave equation are

$$E_1 = E_{01}\cos\left(\frac{\omega z}{c} - \omega t\right) \quad \ldots(1)$$

[We use $\omega = 2\pi\nu$ and $\omega = kv$ or $v = \omega/k$]

$$E_2 = E_{02}\cos\left(\frac{2\omega z}{c} - \omega t\right) \quad \ldots(2)$$

$$\Rightarrow \quad E_2 = E_{02}\cos\left(\frac{\omega z}{c/2} - \omega t\right) \quad [v = c \text{ in air}]$$

In air speed of EM wave is $v_{air} = c$

In medium speed of EM wave is $v_m = c/2$

The speed of EM wave can be written as

$$v = \frac{1}{\sqrt{\mu \in}} = \frac{1}{\sqrt{\mu_0 \in_0 \mu_r \in_r}}$$

As here medium is non magnetic $\Rightarrow \mu_{r_1} = \mu_{r_2} = 1$

In this case we use

$$\frac{v_{air}}{v_m} = \frac{c}{c/2} = \frac{\frac{1}{\sqrt{\epsilon_{r_1}}}}{\frac{1}{\sqrt{\epsilon_{r_2}}}} = \sqrt{\frac{\epsilon_{r_2}}{\epsilon_{r_1}}}$$

$$\Rightarrow \quad \frac{\epsilon_{r_2}}{\epsilon_{r_1}} = 4 \Rightarrow \frac{\epsilon_{r_1}}{\epsilon_{r_2}} = \frac{1}{4}$$

Ans. (C)

Sol. 22 If $\vec{B}$ is the magnetic field then it should be such that $\vec{E} \times \vec{B}$ is along the direction of propagation of EM wave i.e. parallel to $\left(\frac{\hat{i}+\hat{j}}{\sqrt{2}}\right)$

taking cross product $\hat{k} \times \left(\frac{\hat{i}-\hat{j}}{\sqrt{2}}\right) = \frac{\hat{j}-(-\hat{i})}{\sqrt{2}} = \left(\frac{\hat{i}+\hat{j}}{\sqrt{2}}\right)$

So option (A) and (B) here are resulting $\vec{B}$ such the direction of wave propagation is along the given direction.

here $\vec{r}$ is the wave displacement along direction of propagation

So $\quad \vec{r} = r\hat{r} = r\left(\frac{\hat{i}+\hat{j}}{\sqrt{2}}\right)$

By substituting $\vec{r}$ in (A) and (B) we get

(A) $\frac{E_0}{c}\left(\frac{\hat{i}-\hat{j}}{\sqrt{2}}\right)\cos\left[10^4 \underbrace{\left(\frac{\hat{i}-\hat{j}}{\sqrt{2}}\right)\cdot r\left(\frac{\hat{i}+\hat{j}}{\sqrt{2}}\right)}_{0} - \overset{\omega=2\pi\nu}{\overbrace{3\times10^{12} t}}\right]$

So (A) represent on oscillating field not the travelling field.

(B) $\frac{E_0}{c}\left(\frac{\hat{i}-\hat{j}}{\sqrt{2}}\right)\cos\left[10^4\left(\frac{\hat{i}-\hat{j}}{\sqrt{2}}\right)\cdot r\left(\frac{\hat{i}+\hat{j}}{\sqrt{2}}\right) - 3\times10^{12} t\right]$

$\Rightarrow \quad \frac{E_0}{c}\left(\frac{\hat{i}-\hat{j}}{\sqrt{2}}\right)\cos[10^4 r - 3\times10^{12} t]$ travelling wave equation.

Ans. (B)

Sol. 23 For an EM wave $\frac{E}{B} = c \Rightarrow |\vec{B}| = \frac{E}{c}$

If I is intensity then energy density in EM wave is given as

$$\frac{1}{2}\epsilon_0 E_0{}^2 = \frac{I}{c}$$

$$\Rightarrow \quad E_0 = \sqrt{\frac{2I}{\epsilon_0 c}}$$

We also know that direction of propagation of EM wave is along vector $\vec{E}\times\vec{B}$ which should be along $\hat{j}$ here. **Ans. (C)**

Sol. 24 As z_1 and z_2 are two neighbouring points where EF = 0

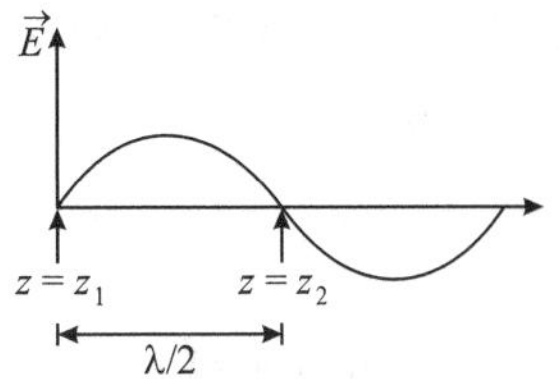

As we have $|(z_2 - z_1)| = \frac{\lambda}{2}$

or $\quad \lambda = c/\nu$

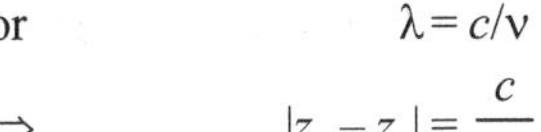

$$\Rightarrow \quad |z_2 - z_1| = \frac{c}{2\nu}$$

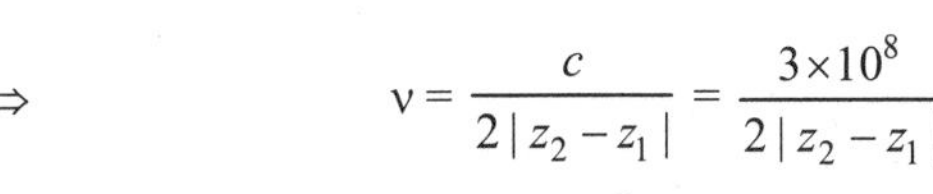

$$\Rightarrow \quad \nu = \frac{c}{2|z_2 - z_1|} = \frac{3\times10^8}{2|z_2 - z_1|}$$

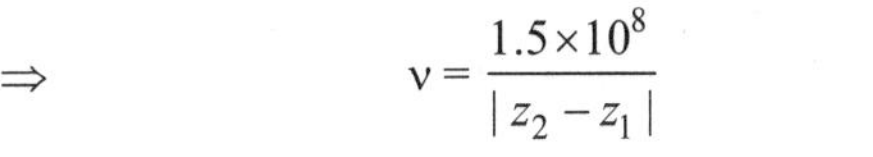

$$\Rightarrow \quad \nu = \frac{1.5\times10^8}{|z_2 - z_1|}$$

Ans. (C)

Sol. 25 In EM wave E, B and C are related as

$$B = \frac{E}{C} = \frac{6.3}{3\times10^8} = 2.1\times10^{-8}\,\text{T}$$

Direction vectors $\hat{E}$, $\hat{B}$ and $\hat{C}$ are related as

$$\Rightarrow \quad \hat{E}\times\hat{B} = \hat{C}$$

$$\Rightarrow \quad \hat{j}\times\hat{B} = \hat{i}$$

$$\Rightarrow \quad \hat{B} = \hat{k}$$

Ans. (C)

Sol. 26 Average energy density of magnetic field,

$u_B = \frac{B_0^2}{2\mu_0}$, B_0 is maximum value of magnetic field.

Average energy density of electric field,

$$u_E = \frac{\epsilon_0 E_0^2}{2}$$

Now, $\quad E_0 = cB_0$, and $c^2 = \frac{1}{\mu_0\epsilon_0}$

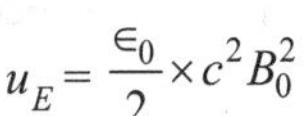

$$u_E = \frac{\epsilon_0}{2}\times c^2 B_0^2$$

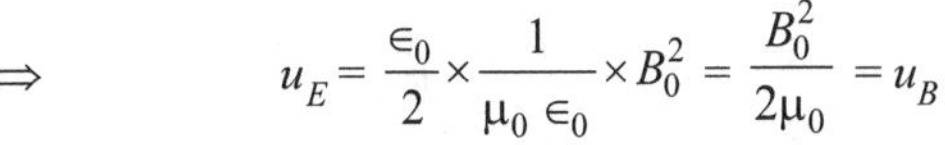

$$\Rightarrow \quad u_E = \frac{\epsilon_0}{2}\times\frac{1}{\mu_0\epsilon_0}\times B_0^2 = \frac{B_0^2}{2\mu_0} = u_B$$

$$\Rightarrow \quad u_E = u_B$$

As energy density of electric & magnetic field is same, energy associated with equal volume will be equal. **Ans. (C)**

Sol. 27 Electric and magnetic field amplitudes are related as

$$E_0 = B_0\times c$$

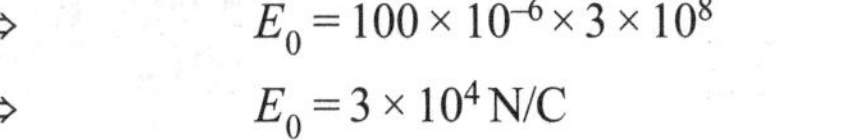

$$\Rightarrow \quad E_0 = 100\times10^{-6}\times3\times10^8$$

$$\Rightarrow \quad E_0 = 3\times10^4\,\text{N/C}$$

Ans. (D)

Sol. 28 The magnetic field magnitude is given as

$$B = \frac{E}{c} = \frac{10}{c}$$

We write EF at $t = 0$ as

$$\vec{E}(x, y) = (10\,\hat{j})\cos[(6\hat{i} + 8\hat{k})\cdot(x\,\hat{i} + z\,\hat{k})]$$

$$= (10\,\hat{j})\cos[\vec{k}\cdot\vec{r}]$$

Here $\vec{k} = 6\hat{i} + 8\hat{k}$ is the direction along which wave propagation

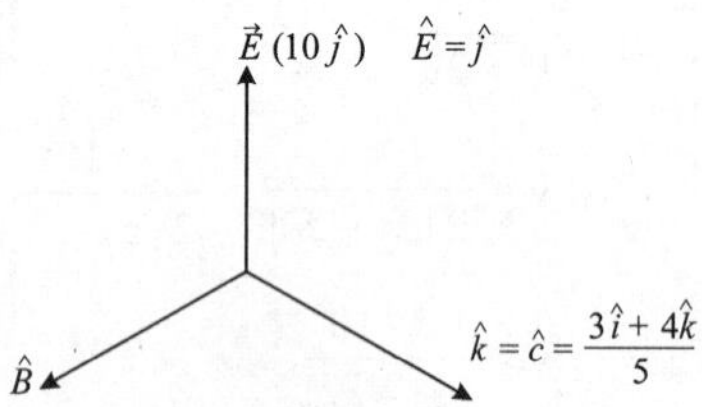

Direction vector of MF is given as

$$\hat{B} = \hat{c}\times\hat{E} = \frac{-4\hat{i} + 3\hat{k}}{5}$$

MF vector is given as

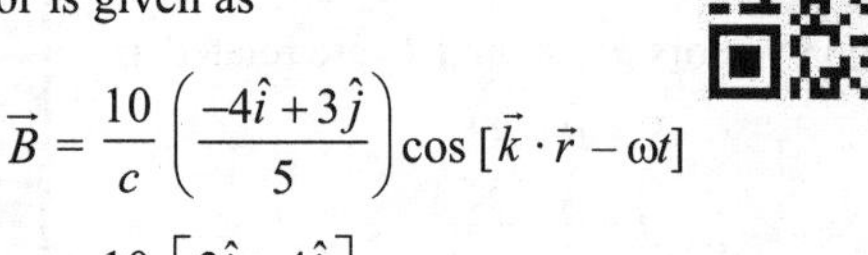

$$\vec{B} = \frac{10}{c}\left(\frac{-4\hat{i} + 3\hat{j}}{5}\right)\cos[\vec{k}\cdot\vec{r} - \omega t]$$

$$\Rightarrow \quad \vec{B} = \frac{10}{c}\left[\frac{3\hat{i} - 4\hat{i}}{5}\right]\cos[6x + 8z - 10ct] \qquad \textbf{Ans. (B)}$$

Sol. 29 Speed of EM wave in free space is

$$c = \frac{1}{\sqrt{\mu_0 \in_0}}$$

In a medium speed is given as

$$v = \frac{1}{\sqrt{k \in_0 \mu_0}} \quad \text{[For transparent medium } \mu_r \approx \mu_0]$$

$$\Rightarrow \quad \frac{c}{v} = \sqrt{k} = n \qquad \text{(refractive index of medium)}$$

EM wave intensity is given as

$$I = \frac{1}{2}\in_0 E_0^2 c = \frac{1}{2}\in_0 kE^2 v$$

$$\Rightarrow \quad E_0^2 c = kE^2 v$$

$$\Rightarrow \quad \frac{E_0^2}{E^2} = \frac{kv}{c} = \frac{n^2}{n}$$

$$\Rightarrow \quad \frac{E_0}{E} = \sqrt{n}$$

For magnetic field we can use

$$\frac{B_0^2 c}{2\mu_0} = \frac{B^2 v}{2\mu_0}$$

$$\Rightarrow \quad \frac{B_0}{B} = \frac{1}{\sqrt{n}} \qquad \textbf{Ans. (B)}$$

Sol. 30 Intensity of electromagnetic wave is given by

$$I = \frac{\text{Power}}{\text{Area}} = \frac{1}{2}\varepsilon_0 E_0^2 c$$

$$\Rightarrow \quad I = \frac{27\times 10^{-3}}{10\times 10^{-6}} = \frac{1}{2}\times 9\times 10^{-12}\times E_0^2 \times 3\times 10^8$$

$$\Rightarrow \quad E_0 = \sqrt{2}\times 10^3 \text{ kV/m}$$

$$\Rightarrow \quad E_0 = 1.4 \text{ kV/m} \quad \textbf{Ans. (C)}$$

Sol. 31 Power of refracted transmitted light is 96% so, we have

$$P_{\text{refracted}} = \frac{96}{100}P_I$$

$$\Rightarrow \quad K_2 A_t^2 = \frac{96}{100}K_1 A_i^2$$

$$\Rightarrow \quad r_2 A_t^2 = \frac{96}{100}r_1 A_i^2$$

$$\Rightarrow \quad A_t^2 = \frac{96}{100}\times\frac{1}{\frac{3}{2}}\times(30)^2 \qquad \text{[Given that } A_i = 30 \text{ V/m]}$$

$$\Rightarrow \quad A_t = \sqrt{\frac{64}{100}\times(30)^2} = 24 \text{ V/m} \qquad \textbf{Ans. (B)}$$

Sol. 32 Intensity of electromagnetic wave is given as

$$I = \varepsilon_0 c E_{\text{rms}}^2$$

$$\text{and} \quad E_{\text{rms}} = cB_{\text{rms}}$$

$$\Rightarrow \quad I = \varepsilon_0 c^3 B_{\text{rms}}^2$$

$$\Rightarrow \quad B_{\text{rms}} = \sqrt{\frac{I}{\in_0 c^3}}$$

$$\Rightarrow \quad B_{\text{rms}} \approx 10^{-4} \qquad \textbf{Ans. (B)}$$

Sol. 33 The direction of propogation of an EM wave is direction of $\vec{E}\times\vec{B}$. Thus from given information

$$\hat{i} = \hat{j}\times\hat{B}$$

$$\Rightarrow \quad \hat{B} = \hat{k}$$

$$c = \frac{E}{B}$$

$$\Rightarrow \quad B = \frac{E}{c} = \frac{6}{3\times 10^8}$$

$$B = 2\times 10^{-8}\text{ T Along } z \text{ direction} \qquad \textbf{Ans. (C)}$$

Sol. 34 If we use that direction of light propagation will be along $\vec{E}\times\vec{B}$. Then (D) option is correct

Detailed solution is as following

magnitude of $E = CB$

$$E = 3\times10^{8}\times1.6\times10^{-6}$$

$$E = 4.8\times10^{2}$$

$\vec{E}$ and $\vec{B}$ are perpendicular to each other

$$\Rightarrow \quad \vec{E}\cdot\vec{B} = 0$$

$\Rightarrow$ either direction of $\vec{E}$ is $\hat{i}-2\hat{j}$ or $-\hat{i}+2\hat{j}$

From given option

Also wave propagation direction is parallel to $\vec{E}\times\vec{B}$ which is $-\hat{k}$

$\Rightarrow$ $\vec{E}$ is along $(-\hat{i}+2\hat{j})$ **Ans. (D)**

Sol. 35 Maximum Electric field is given as

$$E = (B)(c)$$

$$\vec{E}_0 = (3\times10^{-5})c\,(-\hat{j})$$

$$\vec{E}_1 = (2\times10^{-6})c\,(-\hat{i})$$

Maximum force

$$\vec{F}_{0\max} = q\vec{E} = qc(-3\times10^{-5}\hat{j}-2\times10^{-6}\hat{i})$$

$$|\vec{F}_{0\max}| = 10^{-4}\times3\times10^{8}\sqrt{(3\times10^{-5})^2+(2\times10^{-6})^2}$$

$$= 0.9\text{ N}$$

$$F_{rms} = \frac{F_{0\max}}{\sqrt{2}} = 0.63\text{ N (approx)}$$ **Ans. (D)**

Sol. 36 Direction of $\vec{E}\times\vec{B}$ is along the propogaion of EM wave

$$\vec{E}\times\vec{B}\parallel\vec{v} \quad \ldots(1)$$

Given that wave is propagating along positive z-axis and $\vec{E}$ along positive x-axis. Thus from Maxwell's equation we have

$$\vec{\nabla}\times\vec{E} = -\frac{\partial B}{\partial t}$$

$$\Rightarrow \quad \frac{\partial E}{\partial z} = -\frac{\partial B}{dt}$$

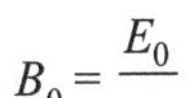

and $$B_0 = \frac{E_0}{c}$$ **Ans. (B)**

Sol. 37 Electric field wave is given as

$$\vec{E} = E_0\hat{n}\sin[\omega t+(6y-8z)]$$

Comparing with equation

$$\vec{E} = E_0\hat{n}\sin[\omega t-\vec{s}\cdot\vec{r}]$$

Where $\vec{s}$ is along direction of propagation and

$$\vec{r} = x\hat{i}+y\hat{i}+z\hat{k}$$

$$\vec{s}\cdot\vec{r} = -6y+8z$$

$$\Rightarrow \quad \vec{s} = -6\hat{i}+8\hat{k}$$

Unit vector along direction of propagation as

$$\hat{s} = \frac{\vec{s}}{|\vec{s}|} = \frac{-6\hat{j}+8\hat{k}}{10} = \frac{-3\hat{j}+4\hat{k}}{5}$$ **Ans. (C)**

Sol. 38 Magnetic field when electromagnetic wave propagates in $+z$ direction so general equation of magnetic field is

$$B = B_0\sin(kz-\omega t)$$

Where $$B_0 = \frac{E_0}{C} = \frac{60}{3\times10^{8}} = 2\times10^{-7}\text{ T}$$

We use $$k = \frac{2\pi}{\lambda} = 0.5\times10^{3}\text{ m}^{-1}$$

and $$\omega = 2\pi f = 1.5\times10^{11}\text{ rad/s}$$ **Ans. (C)**

Sol. 39 Given that $\vec{B}$ is along $\hat{j}$ and wave is travelling in X-direction so using

$$\vec{E}\times\vec{B} = \vec{C} = -\hat{i}$$

$\Rightarrow$ $\vec{E}$ is along $\hat{k}$ and its magnitude is given as

$$E = 3\times10^{-8}\times3\times10^{8} = 9\text{ V/m}$$ **Ans. (A)**

Sol. 40 Force on a charge particle is given as

$$\vec{F} = q(\vec{E}+\vec{v}\times\vec{B})$$

Electric field at location of charge is

$$\vec{E} = E_0\left(\frac{\hat{i}+\hat{j}}{\sqrt{2}}\right)\cos\pi = -E_0\left(\frac{\hat{i}+\hat{j}}{\sqrt{2}}\right)$$

For EM wave, we use

$$\vec{E}\times\vec{B} = \vec{c}$$

$$\Rightarrow \quad +E_0\left(\frac{\hat{i}+\hat{j}}{\sqrt{2}}\right)\times\vec{B} = c\hat{k}$$

$$\Rightarrow \quad \vec{B} = -\left(\frac{\hat{i}-\hat{j}}{\sqrt{2}}\right)\frac{E_0}{c}$$

Thus force on charge is given as

$$\vec{F} = q\left(-E_0\left(\frac{(\hat{i}+\hat{j})}{\sqrt{2}}\right)-v_0\hat{k}\times\left(\frac{(\hat{i}-\hat{j})}{\sqrt{2}}\right)\frac{E_0}{c}\right)$$

$$\Rightarrow \quad \vec{F} \simeq -qE_0\left(\frac{\hat{i}+\hat{j}}{\sqrt{2}}\right) \quad [\text{As } v_0 \ll c]$$

$\Rightarrow$ F is antiparallel to $\dfrac{\hat{i}+\hat{j}}{\sqrt{2}}$ **Ans. (C)**

Sol. 41 For an electromagnetic wave, we use

$$\vec{E} = \vec{B} \times \vec{c}$$

$$\Rightarrow \quad \vec{E} = (5 \times 10^{-8}\,\hat{j}) \times (3 \times 10^{8}\,\hat{k})$$

$$\Rightarrow \quad \vec{E} = 15\hat{i} \text{ V/m}$$

Ans. (B)

Sol. 42 Electric field vector is given as

$$\vec{E}_1 = E_0\,\hat{j}\cos(\omega t - kx)$$

Its corresponding magnetic field will be

$$\vec{B}_1 = \frac{E_0}{c}\,\hat{k}\cos(\omega t - kx)$$

and for second wave corresponding fields are

$$\vec{E}_2 = E_0\,\hat{k}\cos(\omega t - ky)$$

$$\vec{B}_2 = \frac{E_0}{c}\,\hat{i}\cos(\omega t - ky)$$

Net force on charge particle is given as

$$\Rightarrow \vec{F} = q\vec{E}_1 + q\vec{E}_2 + q\vec{v} \times B_1 + q\vec{v} \times B_2$$

$$\Rightarrow \vec{F} = qE_0\,\hat{j} + qE_0\,\hat{k} + q(0.8c\,\hat{j}) \times \left(\frac{E_0}{c}\hat{k}\right) + q(0.8c\,\hat{j}) \times \left(\frac{E_0}{c}\hat{i}\right)$$

$$\Rightarrow \vec{F} = qE_0\,\hat{j} + qE_0\,\hat{k} + 0.8qE_0\,\hat{i} - 0.8qE_0\,\hat{k}$$

$$\Rightarrow \vec{F} = qE_0\,[0.8\,\hat{i} + 1\,\hat{j} + 0.2\,\hat{k}]$$

Ans. (C)

Sol. 43 Direction of polarisation of wave is given as

$$\hat{E} = \hat{k}$$

Direction of propagation

$$\hat{c} = \hat{E} \times \hat{B} = \frac{\hat{i} + \hat{j}}{\sqrt{2}}$$

$$\Rightarrow \quad \hat{B} = \frac{\hat{i} - \hat{j}}{\sqrt{2}}$$

Ans. (A)

Sol. 44 Energy density of electromagnetic waves is given as

$$u_m = \frac{B_0^2}{2\mu_0}$$

$$\Rightarrow \quad B_0 = \sqrt{2\mu_0 u_m} = \sqrt{2 \times 4\pi \times 10^{-7} \times 1.02 \times 10^{-8}}$$

$$\cong \sqrt{256 \times 10^{-16}}$$

$$= 16 \times 10^{-8}\,\text{T} = 160\,\text{nT}$$

Ans. (B)

Sol. 45 Unit vectors along electric and magnetic fields are

$$\hat{E} = \hat{k}$$

$$\hat{B} = \frac{\vec{B}}{|B|} = \frac{2\hat{i} - 2\hat{j}}{2\sqrt{2}}$$

$$\Rightarrow \quad \hat{B} = \frac{1}{\sqrt{2}}(\hat{i} - \hat{j})$$

Direction of wave propagation is given as

$$\hat{c} = \hat{E} \times \hat{B}$$

$$\Rightarrow \quad \hat{c} = \hat{k} \times \left[\frac{1}{\sqrt{2}}(\hat{i} - \hat{j})\right]$$

$$\Rightarrow \quad \hat{c} = \frac{1}{\sqrt{2}}(\hat{k} \times \hat{i} - \hat{k} \times \hat{j})$$

$$\Rightarrow \quad \hat{c} = \frac{1}{\sqrt{2}}(\hat{i} + \hat{j})$$

Ans. (A)

Sol. 46 Direction of speed in an electromagnetic wave is given as

$$\hat{c} = \hat{E} \times \hat{B}$$

(given) $-\hat{j}$ $-\hat{k}$ (analysed) $\hat{i}$ (given)

Amplitudes of electric field and magnetic fields are related as

$$E_0 = cB_0 = 3 \times 10^{8} \times 3 \times 10^{-8} = 9 \text{ V/m}$$

As wave is propagating in $-y$-direction direction of $\vec{E}$ must be along $-\hat{k}$ so we have

$$\vec{E} = -9\sin(200\pi(y + ct))\,\hat{k}$$

Ans. (B)

Sol. 47 At any instance magnetic field magnitude in electromagnetic wave is given as

$$B_0 = \frac{E_0}{c} = E_0\sqrt{\mu_0 \epsilon_0}$$

Direction of wave propagation is given by Poynting vector

$$\vec{S} = \frac{1}{\mu_0}(\vec{E} \times \vec{B})$$

$$\hat{S} = \hat{E} \times \hat{B}$$

$\uparrow$ $\hat{i}$ (given) $\uparrow$ $\hat{j}$ (given) $\downarrow$ $\Rightarrow \hat{k}$ (analysed)

Thus magnetic field is given as

$$B = B_0\cos(\omega t - kx)\,\hat{k}$$

at $t = 0$

$$B = E_0\sqrt{\mu_0 \epsilon_0}\cos kx\hat{k}$$

Ans. (C)

Sol. 48 Electromagnetic spectrum is given as

$\lambda_{\text{increasing}}$ (upward)

Radio waves		few cm to few m
Micro waves		0.01 mm to few cm
IR	~	8000 Å – 10^5 Å
Visible	~	4000 Å – 8000 Å
UV	~	100 Å – 4000 Å
X-Rays	~	0.1 Å – 100 Å

Ans. (C)

Sol. 49 From wave equation, we can see the direction of wave propagation is $+\hat{k}$ so we use

$$\hat{k} = \hat{E} \times \hat{B}$$

Unit vector along $\vec{E}$ is given as

$$\hat{E} = \frac{1}{\sqrt{2}}(\hat{i} + \hat{j})$$

$$\Rightarrow \quad \hat{k} = \frac{1}{2}(\hat{i} + \hat{j}) \times \hat{B}$$

$$\Rightarrow \quad \hat{B} = \frac{1}{\sqrt{2}}(-\hat{i} + \hat{j})$$

Equation of magnetic field wave is given as

$$\vec{B} = \frac{E_0}{C}(-\hat{x} + \hat{y}) \sin(kz - \omega t)$$ **Ans. (B)**

Sol. 50 Magnetic field amplitude in *EM* wave is given as

$$B_0 = \frac{E_0}{C} = \frac{30}{1.5 \times 10^7} \times 5 \times 10^{-2} = 10^{-7}\,\text{T}$$

Maximum force on electron is given as

$$F_{max} = evB_0 = 1.6 \times 10^{-19} \times 0.1 \times 3 \times 10^8 \times 10^{-7}$$

$$\Rightarrow \quad F_{max} = 4.8 \times 10^{-19}\,\text{N}$$ **Ans. (B)**

Sol. 51 *EM* radiation spectrum is given below in decreasing order of wavelength

$\lambda_{decreasing}$ ↓

Radio waves	–	few cm to several meters
Micro waves	–	$10^{-2} - 10^{-5}$ m
IR	–	8000 Å – 10^{-5} m
Visible	–	4000 – 8000 A
UV	–	100 – 4000 Å
X-Rays	–	0.01 – 100 Å
Gama rays	–	10^{-5} to 10^{-3} Å

Ans. (D)

Sol. 52 *Q*-factor of a series *RLC* circuit is given by

$$Q = \frac{1}{R}\sqrt{\frac{L}{C}} = \frac{1}{100}\sqrt{\frac{80 \times 10^{-3}}{2 \times 10^{-6}}}$$

$$\Rightarrow \quad Q = \frac{1}{100}\sqrt{40 \times 10^3}$$

$$\Rightarrow \quad Q = \frac{200}{100} = 2$$ **Ans. (B)**

Sol. 53 Intensity of EM wave is given as

$$I = \epsilon_0 E^2 c$$

$$\Rightarrow \quad E^2 = \frac{I}{\epsilon_0 c}$$

$$\Rightarrow \quad E^2 = \frac{315}{\pi \epsilon_0} \times \frac{1}{3 \times 10^8}$$

$$\Rightarrow \quad E^2 = \frac{315 \times 4 \times 9 \times 10^9}{3 \times 10^8} = 315 \times 120$$

$$\Rightarrow \quad E = \sqrt{315 \times 120} \simeq 194.4$$ **Ans. (194)**

Sol. 54 For given electromagnetic wave electric field amplitude is given as

$$E_0 = cB_0 = 3 \times 10^8 \times 1.2 \times 10^{-7} = 36\,\text{V/m}$$

As wave is travelling in X-direction, we use $\hat{c} = -\hat{i}$

From Poynting vector direction, we use

$$\hat{c} = \hat{E} \times \vec{B}$$

$$-\hat{i} = \hat{E} \times \hat{k}$$

Thus $\hat{E} = -\hat{j}$ so instantaneous electric field is given as

$$\vec{E} = -36 \sin(0.5 \times 10^3 x + 1.5 \times 10^{11} t)\,\hat{j}\ \text{V/m}$$

Ans. (B)

Sol. 55 Frequency of the EM wave given as

$$f = 5\,\text{GHz} = 5 \times 10^9\,\text{Hz}$$

Relative permittivity of medium is $\epsilon_r = 2$

and Relative permeability is also $\mu_r = 2$

Speed of light in a medium in terms of the speed of light in vacuum is given as

$$v = \frac{c}{\sqrt{\mu_r \epsilon_r}} = \frac{3 \times 10^8}{\sqrt{4}} = \frac{30 \times 10^7}{2}\,\text{m/s}$$

$$\Rightarrow \quad v = 15 \times 10^7\,\text{m/s}$$ **Ans. (15)**

Sol. 56 Wavelength in vacuum is given as

$$\lambda = \frac{c}{f} = \frac{3 \times 10^8}{3 \times 10^9} = 0.1\,\text{m}$$

Wavelength in medium is given as

$$\lambda_m = \frac{0.1}{\mu}$$

Where refractive index m of the medium in terms of medium permittivity is given as (considering non-magnetic medium for which relative permeability of medium is taken to be unity)

$$\mu = \sqrt{\varepsilon_r} = \sqrt{2.25} = 1.5$$

$$\Rightarrow \quad \lambda_m = \frac{0.1}{1.5} = \frac{1}{15}m = 6.67\,\text{cm}$$

$$\Rightarrow \quad \lambda_m = 667 \times 10^{-2}\,\text{cm}$$ **Ans. (667)**

Sol. 57 Average intensity of radiation in terms of electric field is given as

$$I = \frac{1}{2} c \epsilon_0 E_0^2$$

Intensity at a distance 10 m from the source of 8 W power with 50% efficiency is given as

$$I = \frac{8}{4\pi\times 10^2}$$

$$\Rightarrow \quad \frac{8}{4\pi\times 10^2} = \frac{1}{2}\times c\times\frac{1}{\mu_0 c^2}\times E_0^2$$

$$\Rightarrow \quad E_0 = \frac{2}{10}\times\sqrt{\frac{\mu_0 c}{\pi}}$$

$$\Rightarrow \quad x = 2$$

Ans. (2)

Sol. 58 Average intensity of radiation in terms of peak electric field is given as

$$I_{avg} = \frac{1}{2}\varepsilon_0 E_0^2 c$$

$$\frac{1.25}{100}\times\frac{1000}{4\pi(2)^2} = \frac{1}{2}\times 8.85\times 10^{-12}\times 3\times 10^8\times E_0^2$$

$$E_0^2 = 187.4$$

$$\Rightarrow \quad E_0 = 13.689 = 136.89\times 10^{-1}\text{ V/m}$$

$$\Rightarrow \quad x = 136.89$$

Rounding off to nearest integer $x = 137$. **Ans. (137)**

Sol. 59 EM wave is travelling towards $+\hat{y}$ direction so electric field is given as

$$\vec{E} = \vec{B}\times\vec{C} = (8\times 10^{-8}\hat{z})\times(3\times 10^8\hat{y})$$

$$\Rightarrow \quad \vec{E} = -24\hat{x}\text{ V/m}$$

Ans. (A)

Sol. 60 In EMW, Average energy density due to electric (U_e) and magnetic (U_m) fields is same. **Ans. (A)**

Sol. 61 The intensity of bulb at a distance 3 m from it is given as

$$I = \frac{100}{4\pi\times 3^2}$$

$$\Rightarrow \quad c\epsilon_0 E^2 = \frac{100}{4\pi\times 3^2}$$

$$\Rightarrow \quad c\epsilon_0\left(\sqrt{\frac{x}{5}}E\right)^2 = \frac{60}{4\pi\times 3^2}$$

$$\Rightarrow \quad \frac{x}{5} = \frac{3}{5}$$

$$\Rightarrow \quad x = 3$$

Ans. (3)

Sol. 62 Conduction current density in seawater is given as

$$J_c = \frac{E}{\rho} = \frac{V}{\rho d}$$

Displacement current density can be given as

$$J_d = \frac{1}{A}\frac{dq}{dt} = \frac{C}{A}\frac{dV_c}{dt} = \frac{\epsilon}{d}\frac{dV_c}{dt}$$

$$\Rightarrow \quad \frac{V_0\sin 2\pi ft}{\rho d} = 10^x\times\frac{80\varepsilon_0}{d}V_0(2\pi f)\cos 2\pi ft$$

$$\Rightarrow \quad \tan\left(2\pi\times\frac{900}{800}\right) = 10^x\times\frac{40}{9\times 10^9}\times 900$$

$$\Rightarrow \quad x = 6$$

Ans. (6)

Sol. 63 Electric field magnitude of EM wave is given as

$$E = Bc = 6\text{ V/m}$$

Direction of propagation of wave is parallel to the vector $(\vec{E}\times\vec{B})$. As the wave is travelling along x-direction and magnetic field is along z-direction hence the electric field vector is given as

$$\vec{E} = 6\hat{j}\text{ V/m}$$

Ans. (C)

Sol. 64 For an electromagnetic wave direction of field vectors is perpendicular to the direction of propagation. As wave is propagating along y-direction, field vectors must have components only in x and z directions. **Ans. (C)**

Sol. 65 Displacement current through capacitor plates can be given as

$$I_d = \varepsilon_0\frac{d\phi_E}{dt}$$

Flux can be calculated as

$$\phi_E = \frac{V(t)A}{d} = \frac{20\sin 100\pi t\cdot 1}{2\times 10^{-3}} = 10^4\sin 100\pi t$$

$$\Rightarrow \quad I_d = \varepsilon_0\frac{d}{dt}(10^4\sin 100\pi t)$$

$$\Rightarrow \quad I_d = 8.85\times 10^{-12}\times 10^4\times 100\pi\cos 10\pi t$$

$$\Rightarrow \quad I_d = 27.79\ \mu\text{A}\cos 100\pi t$$

Ans. (C)

Sol. 66 The direction of propagation of electromagnetic wave in always along the direction of vector $\vec{E}\times\vec{B}$. The unit vector in this direction is given as

$$\hat{n} = \frac{\vec{E}\times\vec{B}}{|\vec{E}\times\vec{B}|}$$

$$\Rightarrow \quad \hat{n} = \frac{E_0\hat{i}\times B_l\hat{k}}{E_0 B_0\sin 90^\circ} = \hat{i}\times\hat{k} = -\hat{j}$$

Ans. (A)

Sol. 67 Average intensity of sunlight is related to the peak electric field in radiation as

$$\frac{1}{2}\varepsilon_0 E_0^2 c = I$$

$$\Rightarrow \quad E_0^2 = \frac{2I}{\varepsilon_0 C}$$

$$\Rightarrow \quad E_0 = \sqrt{\frac{2I}{\varepsilon_0 C}}$$

Peak value of magnetic field can now be given as

$$B_0 = \frac{E_0}{C} = \frac{1}{C}\sqrt{\frac{2I}{\varepsilon_0 C}} = 2.77 \times 10^{-8}\,\text{T} \qquad \textbf{Ans. (C)}$$

Sol. 68 The frequency of the electromagnetic wave is given as

$$2\pi f = 5.3 \times 10^3$$

$$\Rightarrow \quad f = \frac{5.6\times10^3}{2\pi} = \frac{5.6\times10^3}{2\times3.14} = 891.7\,\text{Hz}$$

Wavelength of the electromagnetic wave is given as

$$\lambda = \frac{c}{f} = \frac{3\times10^8}{891.7} = 3.36 \times 10^5\,\text{m}$$

Reflecting wave is in opposite direction considered to have same amplitude is given by the equation

$$E = 3.1\cos[(1.8)z + (5.4 \times 106)t]\,\hat{i}\ \text{N/C}$$

Ans. (B)

Sol. 69 The velocity of light depends upon the electric permittivity & mangetic permeability of the dielectric medium, given as

$$v = \frac{c}{\sqrt{\mu_r \varepsilon_r}} = 3.33 \times 10^7\,\text{m/s} \qquad \textbf{Ans. (A)}$$

Sol. 70 Magnetic field amplitude due to light beam is calculated as

$$B_0 = \frac{E_0}{c} = \frac{800}{3\times10^8}\,\text{V/m}$$

$$\Rightarrow \quad F_m = qvB_0 = \frac{1.6\times10^{-19}\times3\times10^7\times800}{3\times10^8}$$

$$\Rightarrow \quad F_m = 12.8 \times 10^{-18}\,\text{N} \qquad \textbf{Ans. (C)}$$

Sol. 71 Speed of electromagnetic wave is given as

$$v = \frac{\omega}{k} = \frac{10^{10}}{50} = 2 \times 10^8 = \frac{2}{3}c \qquad \textbf{Ans. (A)}$$

Sol. 72 Magnetic field at a point on electromagnetic wave in space **is** given as

$$B = \frac{E}{c} = \frac{6}{3\times10^8} = 2 \times 10^{-8}\,\text{T} \qquad \textbf{Ans. (2)}$$

Sol. 73 Energy density in electromagnetic wave is given as

$$u_e = \frac{1}{2}\in_0 E^2$$

Energy confined in the given volume can be calculated as

$$\text{Energy} = \frac{1}{2}\in_0 E^2 \times \text{volume}$$

$$\Rightarrow \quad 5.5 \times 10^{-12} = \frac{1}{2} \times 8.8 \times 10^{-12} \times 50^2 \times V$$

$$\Rightarrow \quad V = 500\,\text{cm}^3 \qquad \textbf{Ans. (500)}$$

Sol. 74 Wave speed is given as

$$v = \frac{\omega}{x} = \frac{20\times10^{10}}{20\times10^2} = 10^8\,\text{m/s}$$

Refractive index of the wave is given as

$$\mu = \frac{c}{v} = \frac{3\times10^8}{10^8} = 3 = \sqrt{\mu_r \varepsilon_r}$$

$$\Rightarrow \quad \varepsilon_r = 9 \qquad \textbf{Ans. (D)}$$

Sol. 75 Speed of wave in the given medium is given as

$$v = \frac{1}{\sqrt{\mu \in}}$$

Ratio of speed of light in space to that in medium is given as

$$\Rightarrow \quad \frac{c}{v} = \sqrt{\mu_r \in_r} = \sqrt{1.61\times6.44}$$

Ratio of intensity of electric to magnetic field is equal to the speed of light in the medium given as

$$\frac{E}{B} = v = \frac{3\times10^8}{\sqrt{1.61\times6.44}} = 9.32 \times 10^7\,\text{m/s}$$

$$\Rightarrow \quad E = 4.5 \times 10^{-2} \times 4\pi \times 10^{-7} \times 9.32 \times 10^7$$

$$\Rightarrow \quad E = 5.26\,\text{V/m} \qquad \textbf{Ans. (C)}$$

Sol. 76 Intensity of light bulb at a distance r is given as

$$I = \frac{\eta P}{4\pi r^2} = \frac{cB_0^2}{2\mu_0}$$

$$B_0 = \sqrt{\frac{2\mu_0}{4\pi}\frac{\eta P}{c}}\left(\frac{1}{r}\right)$$

$$\Rightarrow \quad B_0 = \frac{1}{4}\sqrt{\frac{2\times10^{-7}\times3.5\times200}{100\times3\times10^8}}$$

$$\Rightarrow \quad B_0 = 1.71 \times 10^{-8}\,\text{T} \qquad \textbf{Ans. (B)}$$

Sol. 77 Intensity of wave is given as

$$I = \frac{1}{2}\varepsilon_0 E_0^2 c$$

$$\Rightarrow \quad I = \frac{1}{2} \times (8.85 \times 10^{-12})(56.5)^2 \times (3 \times 10^8)$$

$$\Rightarrow \quad I = 4.24\,\text{Wm}^{-2} \qquad \textbf{Ans. (B)}$$

Sol. 78 Electromagnetic wave speed in medium is given as

$$v = 2 \times 10^8 \text{ m/s}$$

The refractive index of medium is calculated as

$$\frac{c}{v} = \sqrt{\mu_r \epsilon_r}$$

$$\Rightarrow \quad \frac{9}{4} = 1 \times \epsilon_r$$

$$\Rightarrow \quad \epsilon_r = \frac{9}{4} = 2.25$$ **Ans. (A)**

Sol. 79 The magnetic intensity (in Tesla) is calculated as

$$\vec{B} = \frac{\vec{E}}{c}$$

Electromagnetic wave propagate in the direciton of vector $\vec{E} \times \vec{B}$ so the magnetic field vector magnitude is given by above expression and direction vector is calculated from the direction of propagation which is given as

$$\vec{B} = \frac{301.6}{c} \sin(kz - \omega t)(-\hat{a}_y) + \frac{452.4}{c} \sin(kz - \omega t)(-\hat{a}_x)$$

$$\Rightarrow \quad \vec{H} = \frac{\vec{B}}{\mu_0}$$

$$\Rightarrow \quad \vec{H} = \frac{301.6}{\mu c} \sin(kz - \omega t)(-\hat{a}_y) + \frac{452.4}{\mu c} \sin(kz - \omega t)(-\hat{a}_x)$$

$$\Rightarrow \quad \vec{H} = -0.8 \sin(kz - \omega t)\,\hat{a}_y - 1.2 \sin(kz - \omega t)\,\hat{a}_x$$ **Ans. (C)**

Sol. 80 From electromagnetic wave spectrum the increasing order of wavelengths in the spectrum is shown in figure below.

λ increases ⟶

γ-ray	x-rays	ultra violet	visible	infrared	micro wave	Radio wave

$$\Rightarrow \lambda_{\text{gamma-ray}} < \lambda_{X\text{-ray}} < \lambda_{\text{visible}} < \lambda_{\text{microwave}}$$ **Ans. (B)**

Sol. 81 (a) Ultraviolet rays are used in sterilizing surgical instruments.

(b) Microwaves are high frequency waves used in line of sight communication system used in radars.

(c) Infrared waves causes green house effect in atmosphere as well as in lab experiments.

(d) X-rays are used in studying crystal structures **Ans. (A)**

Sol. 82 Statement-I is true as time varying electric field induced magnetic field and periodic disturbance in either field produces electromagnetic waves. Statement-II is wrong as the velocity of EM wave in a medium is given as

$$v = \frac{1}{\sqrt{\mu\varepsilon}} = \frac{1}{\sqrt{\mu_0\mu_r\varepsilon_0\varepsilon_r}}$$ **Ans. (C)**

Sol. 83 The direction of propagation of electromagetic wave is given by vector $\hat{E} \times \hat{B}$ and magnitude of magnetic field is given as

$$B_0 = \frac{E_0}{c} = \frac{60}{3 \times 10^8} = 2 \times 10^{-7}\,T$$

Angular wave number in medium is given as

$$k = \frac{2\pi}{\lambda} = \frac{\pi}{4} \times 10^3 \text{ m}^{-1}$$

Thus equation of electric field wave is given as

$$E_y = 60 \sin\left[\frac{\pi}{4} \times 10^3 (x - 3 \times 10^8 t)\right] \hat{j} \text{ Vm}^{-1}$$

Similarly the equation of magnetic field wave is given as

$$B_z = 2 \times 10^{-7} \sin\left[\frac{\pi}{4} \times 10^3 (x - 3 \times 10^8 t)\right] \hat{k} T$$

Ans. (B)

Sol. 84 The wave intensity is given as

$$I = \left(\frac{1}{2}\varepsilon_0 E_0^2\right) c$$

$$\Rightarrow \quad E_0 = \sqrt{\frac{2I}{\varepsilon_0 c}} = \sqrt{\frac{2 \times 0.22}{8.85 \times 10^{-12} \times 3 \times 10^8}} = 12.873 \text{ V//m}$$

and $B_0 = \frac{E_0}{c} = \frac{12.873}{3 \times 10^8} = 4.291 \times 10^{-8} = 43 \times 10^{-9}$ T **Ans. (43)**

Sol. 85 For maximum kientic energy we will use higher frequency in the radition, given as

$$f = \frac{9 \times 10^{15}}{2\pi} \text{ Hz}$$

In photoelectric effect maximum kinetic energy of ejected photoelectrons are given as

$$K_{max} = hf - \phi$$

$$\Rightarrow \quad K_{max} = \frac{9 \times 10^{15} \times 4.14 \times 10^{-15}}{2\pi} - 2.50$$

$$\Rightarrow \quad K_{max} = 3.43 \text{ eV}$$

Nearest value in given options is 3.42 eV **Ans. (D)**

Sol. 86 The electric field wave is given as

$$E_y = 540 \sin \pi \times 10^4 (x - ct) \text{ Vm}^{-1}$$

Peak electric field in this wave is

$$E_0 = 540 \text{ Vm}^{-1}$$

Peak magnetic field in this wave is calculated as

$$B_0 = \frac{E_0}{c} = \frac{540}{3 \times 10^8} = 18 \times 10^{-7} \text{ T}$$ **Ans. (A)**

Sol. 87 Amplitude of electric field in wave is calculated as

$$E_0 = cB_0$$

$\Rightarrow$ $$E_0 = (3 \times 10^8)(2 \times 10^{-8})$$

$\Rightarrow$ $$E_0 = 6\ \text{Vm}^{-1}$$

As $\vec{B}$ is along y-axis and from wave equation we can see that wave is travelling along negative x-axis which is the direction of $\hat{E} \times \hat{B}$ vector hence electric field must be along z-axis.

Ans. (C)

Sol. 88 Wave speed is calculated from the given equation of magnetic field as

$$v = \frac{\omega}{k} = \frac{4 \times 10^8}{5} = 8 \times 10^7\ \text{m/s}$$

Amplitude of electric field is calculated as

$$E_0 = vB_0 = 8 \times 10^7 \times 5 \times 10^{-6}$$

$\Rightarrow$ $$E_0 = 4 \times 10^2\ \text{V/m}$$ **Ans. (D)**

Sol. 89 The electric and magnetic force acting on the charge are given as

$$F_E = qE_0$$

and $$F_B = qvB_0$$

$\Rightarrow$ $$\frac{F_E}{F_B} = \frac{E_0}{vB_0} = \frac{c}{v} = \frac{3 \times 10^8}{3 \times 10^7} = 10:1$$ **Ans. (C)**

Sol. 90 (a) is not correct as direction of propagation of wave is perpendicular to the direction of electric and magnetic field.

(b) is correct as average energy in electric and magnetic field are equal in electromagnetic wave.

(c) is not correct as electric and magnetic field in electromagnetic wave are perpendicular to each other.

(d) is correct by definition.

(e) is not correct as the ratio of amplitude of magnetic field to the amplitude of electric field is equal to the reciprocal of speed of light. **Ans. (B)**

Sol. 91 (a) UV rays are used for water purification by killing of bacteria and germs

(b) X-rays are used for diagnosing fracture in bones

(c) Microwaves are used for mobile and radar communication

(d) Infrared waves show less scattering therefore used in foggy days

(a)-(ii), (b)-(i), (c)-(iii), (d)-(iv) **Ans. (B)**

Sol. 92 Magnetic field vector will be in the direction of the vector along the cross product $\hat{K} \times \hat{E}$ and magnitude of magnetic field vector is given as

$$B = \frac{E}{c} = \frac{K}{\omega}E$$

$\Rightarrow$ $$\vec{B} = \frac{1}{\omega}(\vec{K} \times \vec{E})$$ **Ans. (A)**

Sol. 93 The light speed is given in terms of amplitudes of electric and magnetic field as

$$c = \frac{\omega}{k} = \frac{E_0}{B_0}$$ **Ans. (A)**

Sol. 94 Direction of energy flow in an electromagnetic wave is given by Poynting vector, stated as

$$\vec{S} = \frac{1}{\mu_0}(\vec{E} \times \vec{B})$$

Given that the energy flow is along negative z direction and electric field is in positive y direction so using unit vectors in above expression of Poynting vector, we have

$$(-\hat{k}) = (+\hat{j}) \times (\hat{B})$$

Thus magnetic field should be in positive x direction **Ans. (A)**

Sol. 95 Speed of light does not depend on the motion of source as well as intensity of light but depends upon the nature of medium in which it propagates. **Ans. (D)**

Sol. 96 Statement-I is correct as electromagnetic waves are neutral. Statement-II is wrong as the relation in electric and magnetic field amplitude of an em wave is given as

$$E_0 = \sqrt{\frac{1}{\mu_0 \varepsilon_0}}B_0$$ **Ans. (A)**

Sol. 97 Intensity produces by electric field component is

$$\text{Intensity} = \frac{\text{Energy}}{\text{Area} \times \text{time}} = \frac{\text{Power}}{\text{Area}}$$

Where area of the surface energy is crossing is given as

$$\text{Area} = 4\pi \times (5)^2$$

Thus intensity of electric field is calculated as

$$I_{EF} = \frac{1}{2} \times \frac{5}{4\pi \times 5^2}$$

$\Rightarrow$ $$I_{EF} = \frac{1}{40\pi}\ \text{W/m}^2$$ **Ans. (B)**

Sol. 98 Energy of each photon in the radiation is calculated as

$$E = h\nu = \frac{15 \times 10^3}{10^{16}}\ \text{J}$$

Radiation frequency is calculated as

$$\nu = \frac{15 \times 10^{-13}}{6 \times 10^{-34}} = 2.5 \times 10^{21}\ \text{Hz}$$

This falls in range of gamma rays. **Ans. (C)**

Sol. 99 The speed of light in medium is given as

$$v = \frac{c}{\mu}$$

$$\Rightarrow \quad \mu = \frac{c}{v} = \frac{c}{0.2c}$$

$$\Rightarrow \quad \mu = 5$$

$$\Rightarrow \quad \mu = \sqrt{\epsilon_r \mu_r}$$

$$\Rightarrow \quad \epsilon_r = \frac{\mu^2}{\mu_r}$$

$$\Rightarrow \quad \frac{\epsilon_r}{\mu} = \frac{\mu}{\mu_r} = 5$$

Ans. (5)

Sol. 100 (a) Microwaves are used in aircraft navigation.
(b) UV rays are used in lasik eye surgery.
(c) Infra-red rays are used to heal muscular injury and physiotherapy.
(d) X-ray are used in treatment of cancer. **Ans. (C)**

Sol. 101 If $<u_E>$ is average electric energy density and $<u_B>$ is average magnetic energy density then we use

$$<u_E> = <u_B> = \frac{1}{2} <u_{total}>$$

$$\Rightarrow \quad \frac{<u_E>}{<u_{total}>} = \frac{1}{2}$$

Ans. (D)

Sol. 102 Ratio of average electric and magnetic energy density is calculated as

$$\frac{\text{Electric energy density}}{\text{Magnetic energy density}} = \frac{\frac{1}{2} \in E_{rms}^2}{\left(\frac{B_{rms}^2}{2\mu_0}\right)}$$

$$\Rightarrow \quad \frac{u_e}{u_m} = \left(\frac{E_{rms}}{B_{rms}}\right)^2 \cdot \mu_0 \epsilon_0$$

As we know, speed of light is calculated as

$$c = \frac{1}{\mu_0 \epsilon_0}$$

and

$$\frac{E}{B} = c$$

$$\Rightarrow \quad \frac{u_e}{u_m} = \frac{c^2}{c^2} = 1$$

Ans. (A)

Sol. 103 Standard results of average energy densities of electric and magnetic fields given as

$$U_E = \frac{\epsilon_0 E^2}{2} \text{ and } U_B = \frac{B^2}{2\mu_0}$$

Ans. (D)

Sol. 104 Let expression for *EM* wave is given as

$$E = E_0 \sin(\omega t - kx)$$

Energy density of electromangetic wave is given as

$$\Rightarrow \quad u_e = \varepsilon_0 E_0^2 \sin^2(\omega t - kx)$$

$$\Rightarrow \quad u_e = \frac{\varepsilon_0 E_0^2}{2}[1 - \cos(2\omega t - 2kx)]$$

Thus frequency of oscillations of energy in electromagnetic wave is 2ω. **Ans. (C)**

Sol. 105 For electromagnetic wave

$$\frac{E}{B} = c$$

$$\Rightarrow \quad E = Bc$$

$$\Rightarrow \quad E = 6 \times 10^{-7} \times 3 \times 10^8$$

$$\Rightarrow \quad E = 18 \times 10 \text{ V/m}$$

Thus maximum value of electric field is given as

$$E = 180 \text{ V/m}$$

Ans. (B)

Sol. 106 Given that Electric field in electromagnetic wave is

$$\vec{E} = 20 \sin\omega\left(t - \frac{x}{c}\right)\hat{j} \text{ N/C}$$

Average energy density of an electromagnetic wave is given as

$$u_e = \frac{1}{2}\varepsilon_0 E_0^2$$

Energy stored in given volume is calculated as

$$U_e = \left(\frac{1}{2}\epsilon_0 E_0^2\right)(\text{volume})$$

$$\Rightarrow \quad U_e = \frac{1}{2} \times 8.85 \times 10^{-12} \times (20)^2 \times (5 \times 10^{-4}) \text{ J}$$

$$\Rightarrow \quad U_e = 8.85 \times 10^{-13} \text{ J}$$

Ans. (C)

Sol. 107 Given that at a particular space and time, Electric field intensity is given as

$$\vec{E} = 6.6\hat{j}$$

Magnetic intensity magnitude at this point is given as

$$|\vec{B}| = \frac{|\vec{E}|}{c} = 2.2 \times 10^{-8} \text{T}$$

When EM wave propagating in *x*-direction and $\vec{E}$ is *y*-axis, then we have

$$\hat{E} \times \hat{B} = \hat{c}$$

$$\Rightarrow \quad \vec{B} = 2.2 \times 10^{-8}\hat{k} \text{ T}$$

Ans. (B)

Sol. 108 Optical communication is performed in the frequency range of 1 THz to 1000 THz. (Microwave to UV). So, EM waves used for optical communication have shorter wavelength than that of microwaves used in RADAR hence assertion A is false.
As $\nu_{\text{INFRARED}} > \nu_{\text{MICROWAVE}}$
Infrared EM waves are more energetic than microwave hence reason-R is true. **Ans. (A)**

Sol. 109 Based on Maxwell's equations, we have

$$\int \vec{E}\cdot\vec{dl} = -\frac{\partial\phi_B}{\partial t}$$ **Ans. (C)**

Sol. 110 Direction of propagation of electromagnetic wave will be in the direction of $\vec{E}\times\vec{B}$ and as we use

$$\hat{v} = \hat{E}\times\hat{B}$$ **Ans. (B)**

Sol. 111 Increasing order of wave length from electromagnetic spectrum, it is listed below.

X-ray	1 nm to 10^{-3} nm
Ultra Violet	400 nm to 1 nm
Intra red	1 mm to 700 nm
Micro wave	0.1 m to 1 mm

Ans. (B)

23.1 Geometrical Optics-I Reflection of Light

Sol. 1

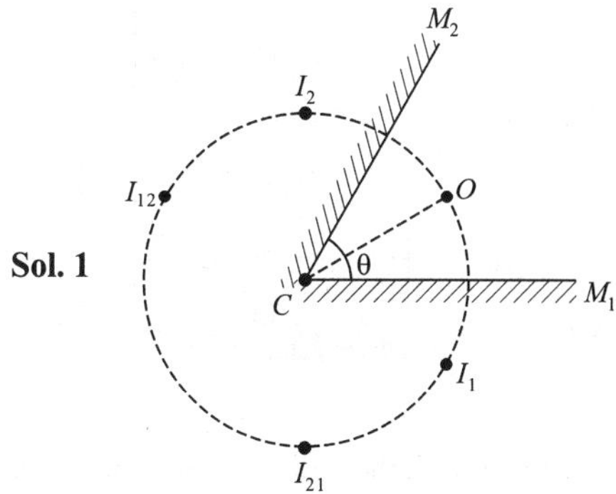

at $\theta = 60°$

$$\frac{180}{60°} = \frac{30}{\text{integer}}$$

If $\frac{180}{\theta} \in I$ then no of images produced are given as

$$N = \frac{360}{\theta} - 1$$

$$N = \frac{360}{60} - 1 = 6 - 1 = 5$$ **Ans. (A)**

Sol. 2 If θ is angle between mirrors that no of images obtained are

$$N = \frac{360}{\theta} - 1 = 3$$

$\Rightarrow$ $$\frac{360}{\theta} = 4$$

$\Rightarrow$ $$\theta = \frac{360}{4} = 90°$$ **Ans. (B)**

Sol. 3

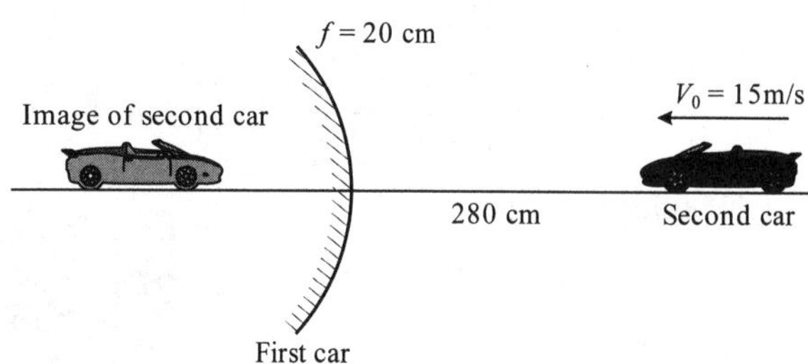

For reflection at convex mirror, we use

$$u = +280\text{ cm}$$

$$f = -20\text{ cm}$$

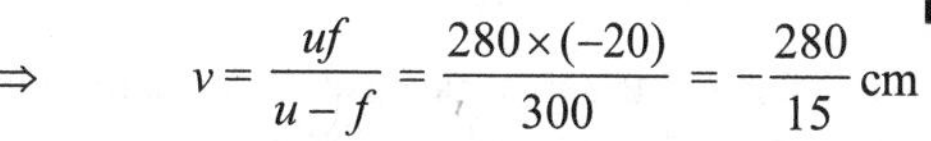

$\Rightarrow$ $$v = \frac{uf}{u-f} = \frac{280\times(-20)}{300} = -\frac{280}{15}\text{ cm}$$

Image velocity in a mirror is given as

$$v_i = m^2\cdot V_0 = \left|\frac{v}{u}\right|^2 \times V_0 = \left(\frac{1}{15}\right)^2 \times 15 \text{ [magnification]}$$

$\Rightarrow$ $$v_i = \frac{1}{15}\text{ m/s}$$ **Ans. (D)**

Sol. 4

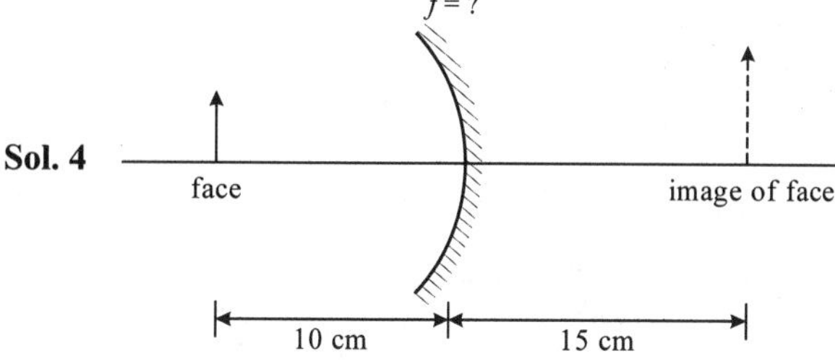

$$R_{\text{mirror}} = 2f$$

For mirror formula we use

$$u = -10\text{ cm}$$

$$V = +15\text{ cm}$$

$\Rightarrow$ $$\frac{1}{f} = \frac{1}{u} + \frac{1}{V}$$

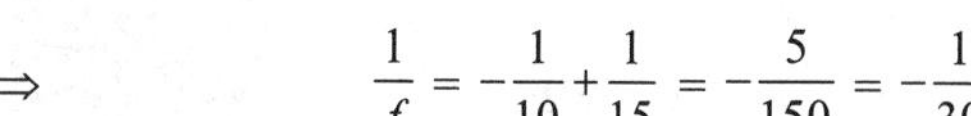

$\Rightarrow$ $$\frac{1}{f} = -\frac{1}{10} + \frac{1}{15} = -\frac{5}{150} = -\frac{1}{30}$$

$\Rightarrow$ $$f = -30\text{ cm}$$

$\Rightarrow$ $$R = 2f = -2\times 30 = -60\text{ cm}$$ **Ans. (C)**

Sol. 5 As object oscillates $\Rightarrow$ image will also oscillate so (a) is correct.

For the two ext positions of object at $u_1 = 8$ cm and $u_2 = 12$ cm the corresponding image location are – (for $f = 5$ cm)

For left position of object

$$u_1 = 8\text{ cm } (+ve)$$

$$f = 5\text{ cm } (+ve)$$

$$\Rightarrow \quad \frac{1}{f} = \frac{1}{v_1} + \frac{1}{u_r}$$

$$\Rightarrow \quad v_1 = \frac{u_1 f}{u_1 - f} = \frac{8 \times 5}{8-5} = +\frac{40}{3} \text{ cm}$$

For right position of object

$$u_2 = 12 \text{ cm}$$

$$f = +5 \text{ cm}$$

$$\Rightarrow \quad \frac{1}{f} = \frac{1}{v_2} + \frac{1}{u_2}$$

$$\Rightarrow \quad v_2 = \frac{u_2 f}{u_2 - f} = \frac{12 \times 5}{12-5} + \frac{60}{7} \text{cm}$$

As v_1 and v_2 are not same so the extreme position in oscillation not symmetric about image mean position which is

$$x_0 = 10 \text{ cm} (2f)$$

$\Rightarrow$ Option (c) is correct

The distance between extreme positions of image is

$$\Delta S = v_1 - v_2$$

$$= \frac{40}{3} - \frac{60}{7} = \frac{280-180}{21} = \frac{100}{21} \text{cm}$$

$\Rightarrow$ Option (d) is correct **Ans. (C)**

Sol. 6

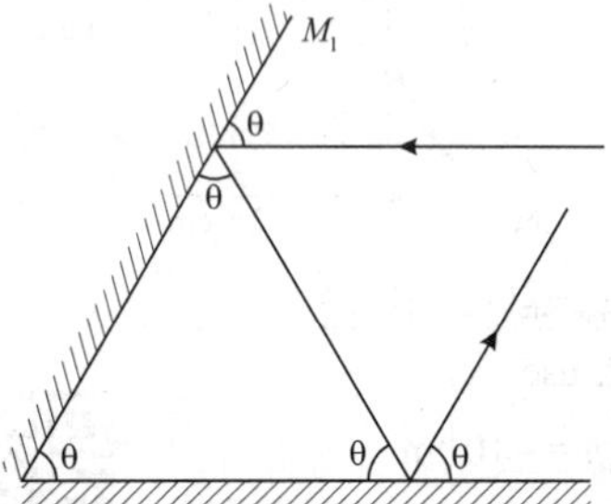

Assuming angles between two mirrors be θ as per geometry, a triangle sum of angles of is

$$3\theta = 180°$$

$$\Rightarrow \quad \theta = 60° \quad \textbf{Ans. (D)}$$

Sol. 7

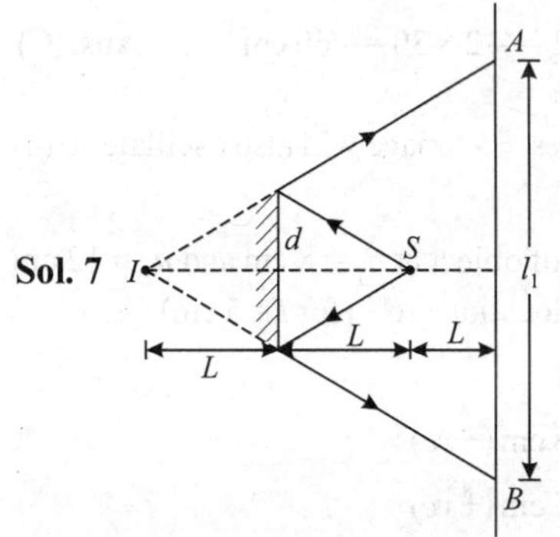

Field of view of image is $l_1 (= AB)$, given as

$$\frac{l_1}{3L} = \frac{d}{L}$$

$$\Rightarrow \quad l_1 = 3d$$ **Ans. (A)**

Sol. 8 Magnification for spherical mirror is given as

$$m = \frac{f}{f-u}$$

$$\Rightarrow \quad 5 = \frac{-40}{-40-u}$$

$$\Rightarrow \quad u = -32 \text{ cm}$$ **Ans. (D)**

Sol. 9 Light from particle P will be reflected at mirror.

So for mirror formula, we use

$$u = -5 \text{ cm},$$

$$f = -\frac{R}{2} = -20 \text{ cm}$$

Using mirror formula, we get

$$\frac{1}{v} + \frac{1}{u} = \frac{1}{f}$$

$$\Rightarrow \quad \frac{1}{v} - \frac{1}{5} = -\frac{1}{20}$$

$$\Rightarrow \quad \frac{1}{v} = \frac{1}{5} - \frac{1}{20} = \frac{3}{20}$$

$$\Rightarrow \quad v = +\frac{20}{3} \text{ cm}$$

This image will act as object for light getting refracted at water surface so, object distance is taken as

$$d = 5 + \frac{20}{3} = \frac{35}{3} \text{ cm}$$

Light rays from this object will refract at plane surface (water level) so after refraction, final image will form at

$$d' = d\left(\frac{\mu_2}{\mu_1}\right)$$

$$\Rightarrow \quad d' = \left(\frac{35}{3}\right)\left(\frac{1}{4/3}\right)$$

$$\Rightarrow \quad d' = \frac{35}{4} = 8.75 \text{cm}$$

$$\Rightarrow \quad d' \approx 8.8 \text{ cm}$$ **Ans. (A)**

Sol. 10 By mirror formula, we have

$$\frac{1}{v} + \frac{1}{u} = \frac{1}{f}$$

At focus $m = \infty$ and $x = f$

At centre $m = -1$ and $x = 2f$

Thus option (B) is most appropriate. **Ans. (B)**

Sol. 11 Using mirror formula, we have

$$f = -4 \text{ cm}$$

$$u = -10 \text{ cm}$$

Substituting values in formula

$$\frac{1}{v} + \frac{1}{u} = \frac{1}{f}$$

$$\Rightarrow \quad \frac{1}{v} + \left(\frac{1}{-10}\right) = \frac{1}{-4}$$

$$\Rightarrow \quad \frac{1}{v} = \frac{1}{10} - \frac{1}{4}$$

$$\Rightarrow \quad \frac{1}{v} = \frac{4-10}{40}$$

$$\Rightarrow \quad v = \frac{40}{-6} \text{ cm}$$

$$\Rightarrow \quad v = \frac{-20}{3} \text{ cm}$$

$$m = \frac{-v}{u} = \frac{-\left(-\frac{20}{3}\right)}{-10} = \frac{-2}{3}$$

$\Rightarrow$ Image will be real, inverted and diminished. **Ans. (D)**

Sol. 12 For the given situation, object and image distances are given as

$$u = 30 \text{ cm}$$

$$v = 10 \text{ cm}$$

Object speed = 9 cm/s

Considering object motion along axis, we use

$$\text{Image speed} = \left|\frac{v}{u}\right|^2 \times \text{Object speed}$$

$$v_{\text{image}} = \left(\frac{1}{3}\right)^2 \times 9 = 1 \text{ cm/s}$$ **Ans. (1)**

Sol. 13 For convex mirror, focus is behind the mirror which is taken as positive side so we use

$$f = +\frac{r}{2}$$ **Ans. (A)**

Sol. 14 As orientation is same image must be virtual and as image is smaller than object, the mirror has to be convex.

Ans. (B)

Sol. 15 The unit vectors along incident ray and reflected ray can be written from the figure shown below as

$$\vec{a} = \sin\theta\hat{i} - \cos\theta\hat{j}$$

$$\vec{b} = \sin\theta\hat{i} + \cos\theta\hat{j}$$

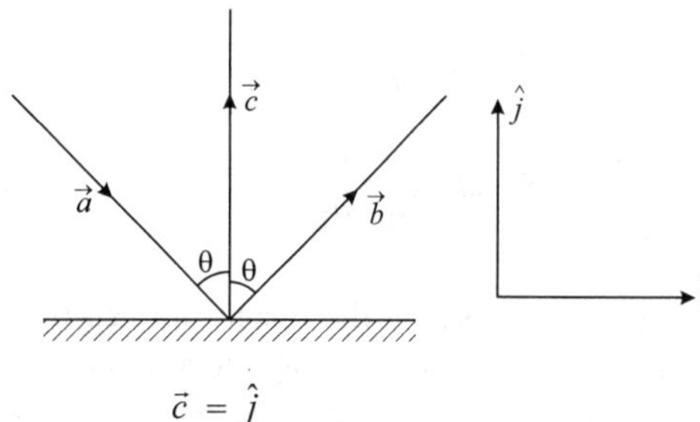

$$\vec{c} = \hat{j}$$

$$\vec{a} - 2(\vec{a}\cdot\vec{c})\vec{c} = \sin\theta\hat{i} + \cos\theta\hat{j}$$

From all the options given (B) is correct. **Ans. (B)**

Sol. 16 The field of view of mirror can be drawn by reflected rays as shown in figure below.

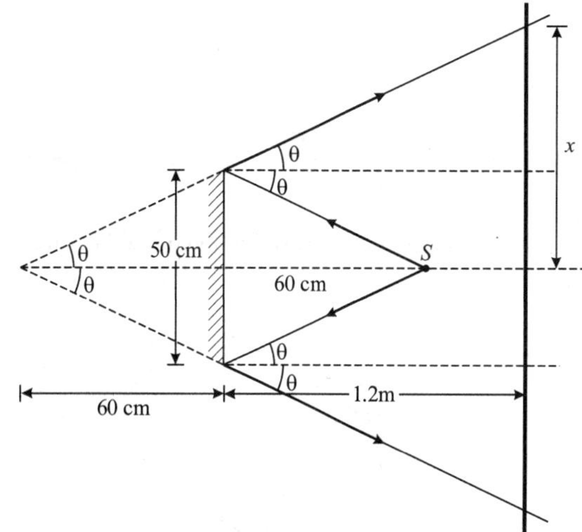

In the above figure, we use

$$\tan\theta = \frac{25}{60} = \frac{x}{180}$$

$$\Rightarrow \quad x = 75 \text{ cm}$$

Thus distance between extreme points is given as

$$d = 2x = 2 \times 75 = 150 \text{ cm}$$ **Ans. (150)**

Sol. 17 Magnification of the mirror, is given a s

$$m = \frac{f}{f-u} = \frac{10}{10-(-190)}$$

$$\Rightarrow \quad m = \frac{10}{200} = \frac{1}{20}$$

Velocity of image in terms of object velocity is given as

$$V_I = -m^2 V_0 = -\left(\frac{1}{20}\right)^2 \times 40$$

$$\Rightarrow \quad V_I = -0.1 \text{ m/s}$$ **Ans. (C)**

Sol. 18 Using Newton's formula, we have

$$xy = f^2$$

$$\Rightarrow \quad (f+d_1)(f-d_2) = f^2$$

$$\Rightarrow \quad f = \frac{d_1 d_2}{d_1 - d_2}$$

$$\Rightarrow \quad R = 2f = \frac{2d_1 d_2}{d_1 - d_2} \qquad \textbf{Ans. (D)}$$

Sol. 19 Figure below shows the images formed by the two mirrors. As mirrors are at right angle total three images are produced but due to direct reflection by the mirrors images I_1 and I_2 are produced.

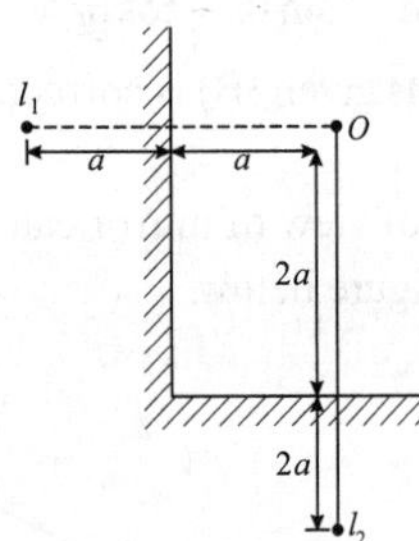

Distance between I_1 and $I_2 = \sqrt{(4a)^2 + (2a)^2} \approx 4.6a$. **Ans. (A)**

Sol. 20 Figure below shows the reflection of light ray from the two mirrors.

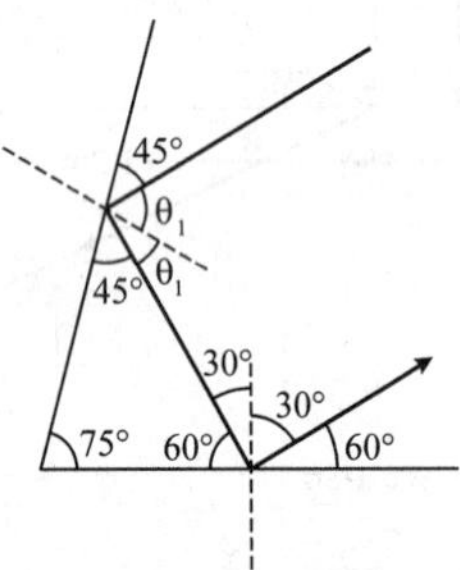

From the above diagram we can calculate that the incident angle of light on mirror M_1 is $\theta_1 = 45°$ so total deviation of light ray at the two reflections is given as

$$\delta_{total} = (180° - 2\theta_1) + (180° - 2 \times 30°)$$

$$= 360° - 2 \times 45° - 2 \times 30°$$

$$\delta_{total} = 210° \qquad \textbf{Ans. (210)}$$

Sol. 21 After 10 s object will be displaced toward mirror by a distance given as

$$x = ut = 2 \times 10 = 20 \text{ cm}$$

Thus object distance from mirror after 10 s is given as

$$u = -80 \text{ cm}$$

and $\quad f = -100$ cm

Using mirror formula, we have

$$\frac{1}{v} + \frac{1}{u} = \frac{1}{f}$$

$$\Rightarrow \quad v = 400 \text{ cm} \qquad \textbf{Ans. (400)}$$

Sol. 22 Plane mirror always forms image which is erected, of same size, laterally inverted and virtual in nature for a real object. **Ans. (A)**

Sol. 23 Using mirror formula we have

$$\frac{1}{v} + \frac{1}{u} = \frac{1}{f}$$

$$\Rightarrow \quad \frac{-1}{120} - \frac{1}{40} = \frac{1}{f}$$

$$\Rightarrow \quad f = -30 \text{ cm}$$

Differentiating mirror formula, we use

$$\frac{-1}{v^2}dv - \frac{-1}{u^2}du = -\frac{1}{f^2}df$$

We also have

$$dv = du = \frac{1}{20} \text{ cm}$$

$$\Rightarrow \quad \frac{\frac{1}{20}}{(120)^2} + \frac{\frac{1}{20}}{(40)^2} = \frac{df}{(30)^2}$$

$$\Rightarrow \quad df = \frac{1}{32} \text{ cm}$$

$$\Rightarrow \quad k = 32 \qquad \textbf{Ans. (32)}$$

Sol. 24 Figure below shows the situation described in the question. To obtain the image of rod we find the image of the ends of rod A and B respectively

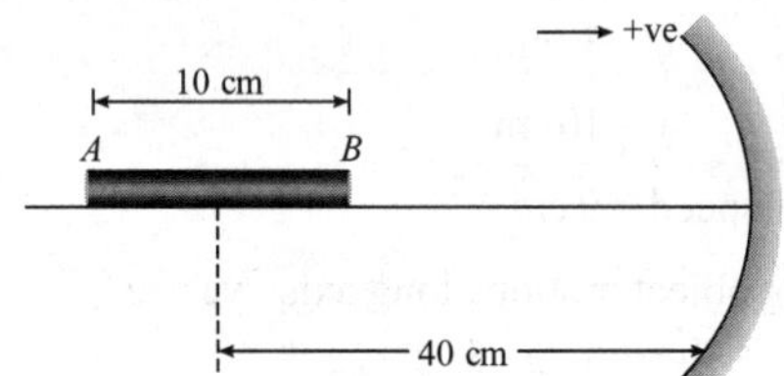

For end A of the rod, we use

$$u_A = -45 \text{ cm},$$

$$f = -20 \text{ cm}$$

$$\Rightarrow \quad v_A = \frac{-45 \times (-20)}{-45 - (-20)} = \frac{-900}{25} = -36 \text{ cm}$$

For end B of the rod, we use

$$u_B = -35 \text{ cm}$$

$$\Rightarrow \quad v_B = \frac{-35 \times (-20)}{-35 - (-20)} = \frac{700}{-15}$$

$$v_A - v_B = \text{length of image}$$

$$\Rightarrow \quad v_A - v_B = \left(-36 + \frac{140}{3}\right) \text{ cm}$$

$$\Rightarrow \quad v_A - v_B = \frac{-108 + 140}{3} \text{ cm}$$

$$\Rightarrow \quad v_A - v_B = \frac{32}{3} \text{ cm}$$

$$\Rightarrow \quad x = 32 \qquad \textbf{Ans. (32)}$$

Sol. 25 By mirror formula, for the object placed at 15 cm from the mirror, we have

$$\frac{1}{v}+\frac{1}{u}=\frac{1}{f}$$

$$\Rightarrow \quad \frac{1}{v_1}+\frac{1}{-15}=\frac{1}{(-20)}$$

$$\Rightarrow \quad \frac{1}{v_1}=-\frac{1}{20}+\frac{1}{15}$$

$$\Rightarrow \quad \frac{1}{v_1}=\frac{-3+4}{60}$$

$$\Rightarrow \quad v_1=60\,\text{cm}$$

For the object placed at 25 cm, by mirror formula we have

$$\frac{1}{v_2}+\frac{1}{(-25)}=\frac{1}{(-20)}$$

$$\Rightarrow \quad \frac{1}{v_2}=\frac{-1}{20}+\frac{1}{25}$$

$$\Rightarrow \quad =\frac{-5+4}{100}=\frac{-1}{100}$$

$$\Rightarrow \quad v_2=-100\,\text{cm}$$

Thus distance between images formed is given as

$$d=60+100=160\,\text{cm}$$ **Ans. (C)**

Sol. 26 Figure below shows the image formation by the two mirrors for the point object.

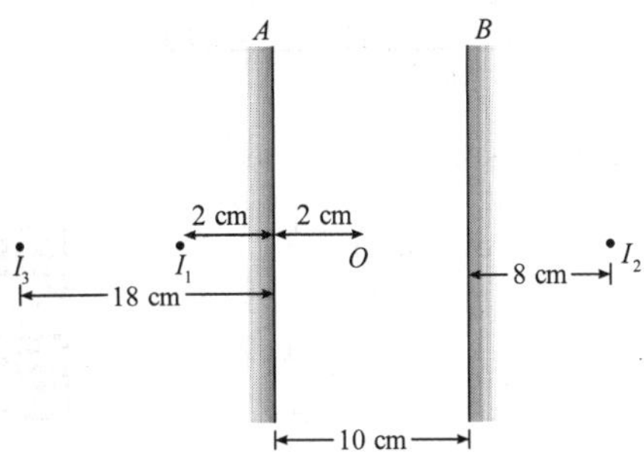

In plane mirror distance of image is same as distance of object and image in one mirror behave as object for second mirror so the desired image is at 18 cm from A. **Ans. (18)**

Sol. 27 Initial situation of mirror and object position is shown in figure below.

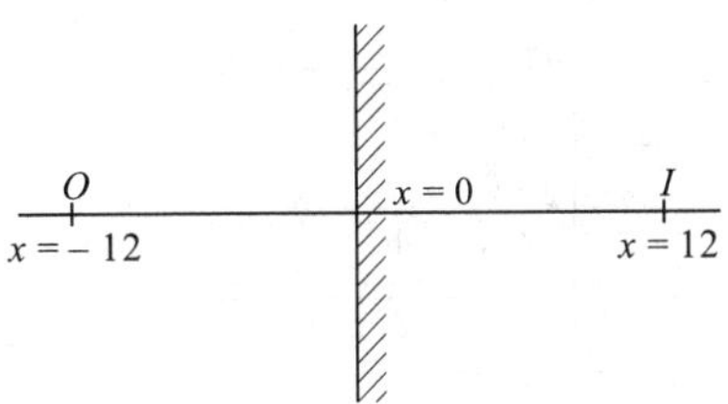

When mirror is moved by 4 cm towards stationary object, the situation is shown in figure below.

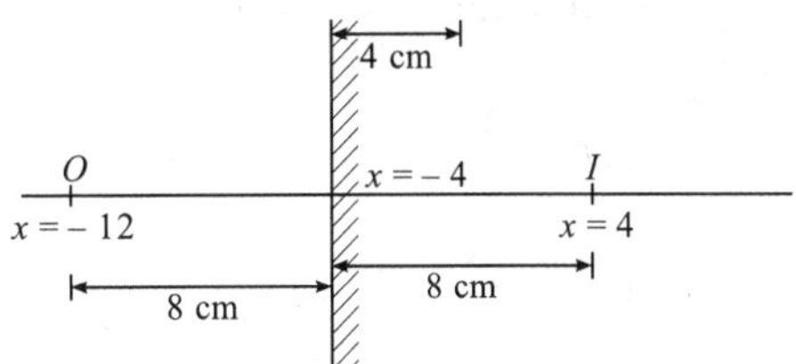

In this case image will be shifted 8 cm towards mirror.

Ans. (B)

Sol. 28 Figure below shows the situation described in the question

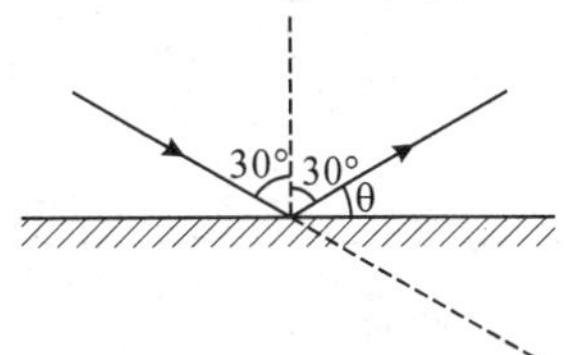

Angle of deviation of ray after reflection

$$\delta=180°-2\theta$$

$$\Rightarrow \quad \delta=180°-60°=120°$$ **Ans. (B)**

23.2 Geometrical Optics-II Refraction of Light

Sol. 1

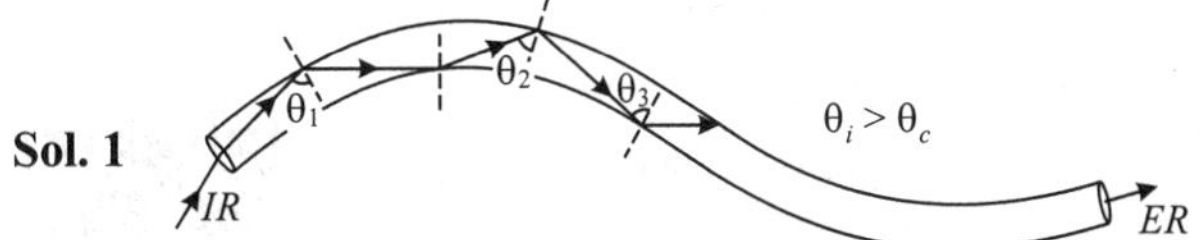

In an optical fibre at every point of inner surface light is internally reflected so light comes out only from the other end of fibre. Total internal reflection. **Ans. (A)**

Sol. 2 As light ray is internally reflected

$$\Rightarrow \quad 45° > \theta_c = \sin^{-1}\left(\frac{1}{n}\right)$$

$$\Rightarrow \quad \frac{1}{n} < \sin 45°$$

$$\Rightarrow \quad n > \sqrt{2}$$ **Ans. (B)**

Sol. 3

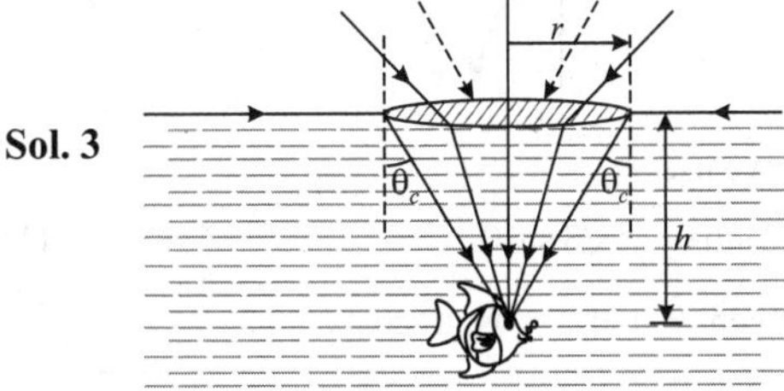

$\theta_c \rightarrow$ critical angle of water air interface

$$\theta_c = \sin^{-1}\left(\frac{1}{\mu_\omega}\right) = \sin^{-1}\left(\frac{3}{4}\right)$$

$$\Rightarrow \qquad \sin\theta_c = \frac{3}{4}$$

$$\Rightarrow \qquad \tan\theta_c = \frac{3}{\sqrt{7}}$$

In the figure shown, we have

$$\tan\theta_c = \frac{r}{h}$$

$$\Rightarrow \qquad r = h\tan\theta_c$$

$$\Rightarrow \qquad r = 12\times\frac{3}{\sqrt{7}} = \frac{36}{\sqrt{7}}\text{ cm}$$

Ans. (B)

Sol. 4

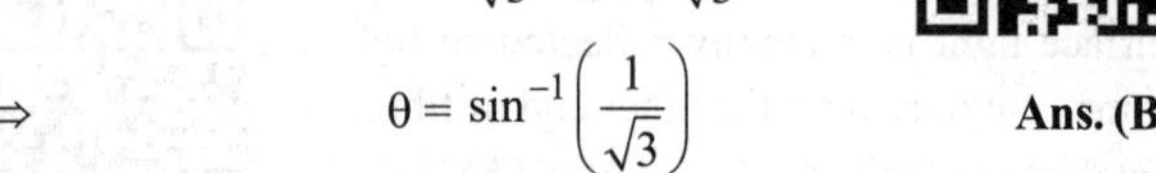

If light ray to graze along the wall of rod

$$\Rightarrow \qquad 90° - r = \theta_c = \sin^{-1}\left(\frac{1}{\mu}\right) = \sin^{-1}\left(\frac{\sqrt{3}}{2}\right) = 60°$$

$$\Rightarrow \qquad r = 90° - 60° = 30°$$

By Snell's law at point A we have

$$\Rightarrow \qquad \sin\theta = \mu\sin r$$

$$= \frac{2}{\sqrt{3}}\times\frac{1}{2} = \frac{1}{\sqrt{3}}$$

$$\Rightarrow \qquad \theta = \sin^{-1}\left(\frac{1}{\sqrt{3}}\right)$$

Ans. (B)

Sol. 5

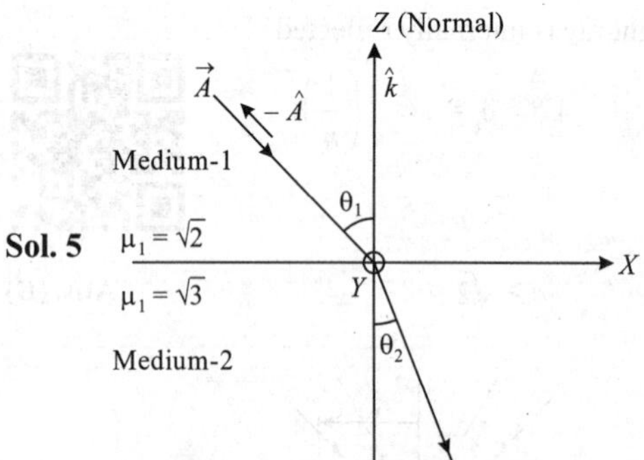

By Snell's law, we use

$$\sqrt{2}\sin\theta_1 = \sqrt{3}\sin\theta_2 \qquad \ldots(1)$$

The angle between $\hat{k}$ and $-\hat{A}$ is given as

$$\cos\theta_1 = -\hat{A}\cdot\hat{k} \qquad \ldots(2)$$

Unit vector along $\vec{A}$ is given as

$$\hat{A} = \frac{\vec{A}}{|\vec{A}|} = \frac{6\sqrt{3}\hat{i} + 8\sqrt{3}\hat{j} - 10\hat{k}}{\sqrt{(6\sqrt{3})^2 + (8\sqrt{3})^2 + (10)^2}}$$

$$\Rightarrow \qquad \hat{A} = \frac{1}{20}(6\sqrt{3}i + 8\sqrt{3}\hat{j} - 10\hat{k})$$

$$\Rightarrow \qquad \hat{A} = \frac{3\sqrt{3}}{10}\hat{i} + \frac{4\sqrt{3}}{10}\hat{j} - \frac{1}{2}\hat{k}$$

From equation (2), we have

$$\cos\theta_1 = -\left(\frac{3\sqrt{3}}{10}\hat{i} + \frac{4\sqrt{3}}{10}\hat{j} - \frac{1}{2}\hat{k}\right)\cdot\hat{k} = \frac{1}{2}$$

$$\Rightarrow \qquad \theta_1 = 60°$$

Now from equation-(1), we have

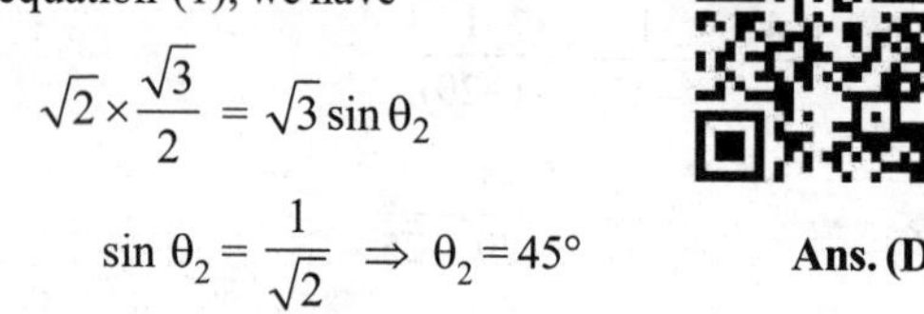

$$\sqrt{2}\times\frac{\sqrt{3}}{2} = \sqrt{3}\sin\theta_2$$

$$\Rightarrow \qquad \sin\theta_2 = \frac{1}{\sqrt{2}} \Rightarrow \theta_2 = 45°$$

Ans. (D)

Sol. 6

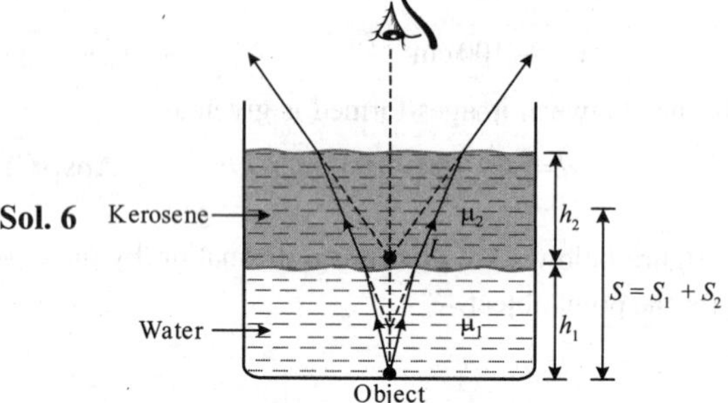

Object shift along normal is given as

$$S_1 = h_1\left(1 - \frac{1}{\mu_1}\right)$$

$$S_2 = h_2\left(1 - \frac{1}{\mu_2}\right)$$

Total shift of object as seen normally is given as

$$S = h_1\left(1 - \frac{1}{\mu_2}\right) + h_2\left(1 - \frac{1}{\mu_2}\right)$$

Ans. (C)

Sol. 7

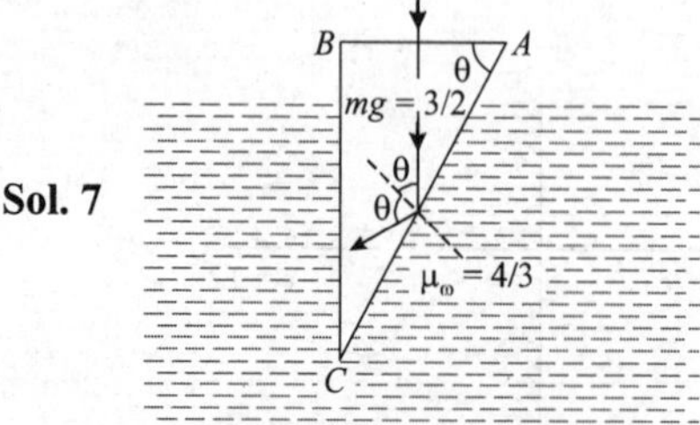

Light will be totally reflected at face AC to reach face BC if

$$\theta > \theta_c$$

$[\theta_c \rightarrow$ critical angle glass-water interface$]$

$\Rightarrow$ $\sin\theta > \sin\theta_c$

$\Rightarrow$ $\sin\theta > \dfrac{1}{{}_\omega\mu_g} = \dfrac{\mu_\omega}{\mu g} = \dfrac{4/3}{3/2}$

$\Rightarrow$ $\sin\theta > \dfrac{8}{9}$ **Ans. (C)**

Sol. 8

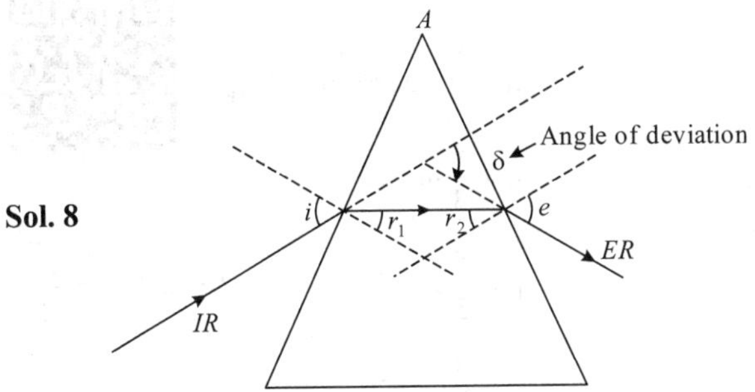

$$\delta = i + e - A$$

When $i = e \rightarrow \delta$ is minimum

$\Rightarrow$ $r_1 = r_2 = A/2$

and light passes through prism symmetrically

$$\sin i = \mu \sin r_1$$

$$\sin e = \mu \sin r_2 = \mu \sin (A - r_1)$$

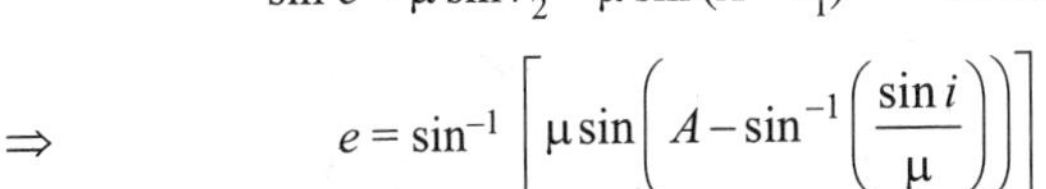

$\Rightarrow$ $e = \sin^{-1}\left[\mu \sin\left(A - \sin^{-1}\left(\dfrac{\sin i}{\mu}\right)\right)\right]$

$\Rightarrow$ $\delta = i + \sin^{-1}\left[\mu \sin^{-1}\left(A \sin^{-1}\left[\dfrac{\sin i}{\mu}\right]\right)\right] - A$

This is a non linear relation. **Ans. (C)**

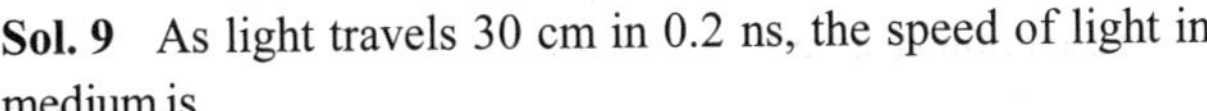

Sol. 9 As light travels 30 cm in 0.2 ns, the speed of light in medium is

$$v = \frac{3\times10^{-2}}{0.2\times10^{-9}} = 1.5\times10^8 \text{ m/s}$$

Refractive index of medium is

$$\mu = \frac{c}{v} = \frac{3\times10^8}{1.5\times10^8} = 2$$

Critical angle for the medium is

$$\theta_c = \sin^{-1}\left(\frac{1}{\mu}\right) = \sin^{-1}\left(\frac{1}{2}\right) = 30°$$

For all incidence angle $\theta > 30°$, light will suffer TIR

In A & B only for case (B), light will suffer TIR **Ans. (B)**

Sol. 10

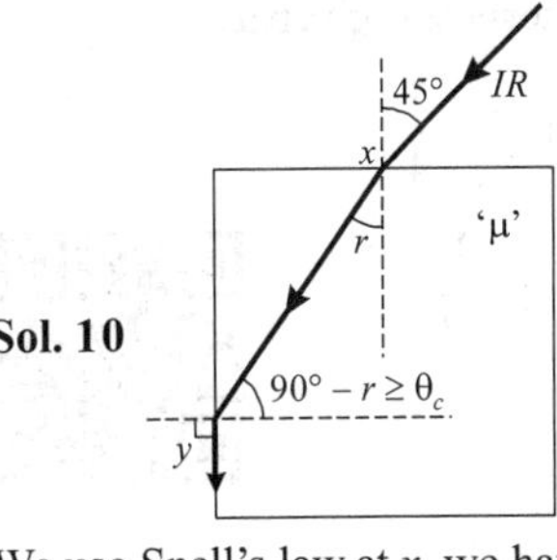

We use Snell's law at x, we have

$$\sin 45° = \mu \sin r \quad \ldots(1)$$

At point y for TIR here $90° - r = \theta_c$

$\Rightarrow$ $\mu \sin(90° - r) = \sin 90°$

$\Rightarrow$ $\mu \cos r = 1 \quad \ldots(2)$

$(1)^2 + (2)^2 \Rightarrow \mu^2 \sin^2 r + \mu^2 \cos^2 r = 1 + \dfrac{1}{2} = \dfrac{3}{2}$

$\Rightarrow$ $\mu^2 = \dfrac{3}{2} \Rightarrow \mu = \sqrt{\dfrac{3}{2}}$ **Ans. (D)**

Sol. 11

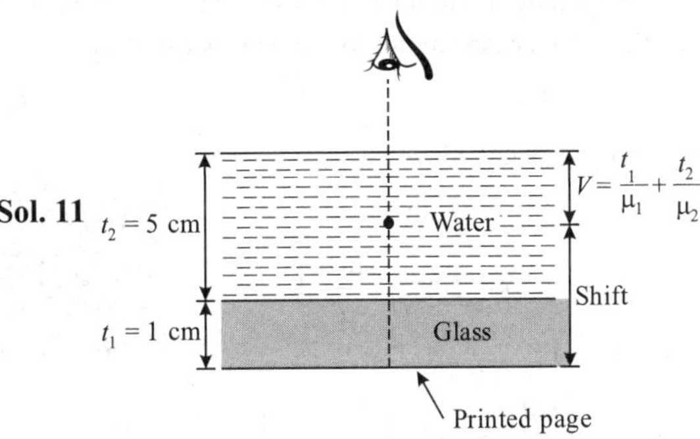

Apparent shift of printed page due to glass and water is given as

$$S = t_1\left(1 - \frac{1}{\mu_1}\right) + t_2\left(1 - \frac{1}{\mu_2}\right)$$

$\Rightarrow$ $S = 1\left(1 - \dfrac{1}{3/2}\right) + 5\left(1 - \dfrac{1}{4/3}\right)$

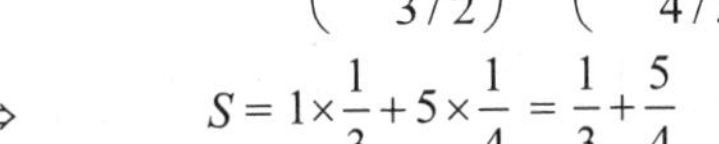

$\Rightarrow$ $S = 1\times\dfrac{1}{3} + 5\times\dfrac{1}{4} = \dfrac{1}{3} + \dfrac{5}{4}$

$\Rightarrow$ $S = \dfrac{19}{12}$ cm

$\Rightarrow$ $S \simeq 1.58$ cm **Ans. (None)**

Sol. 12

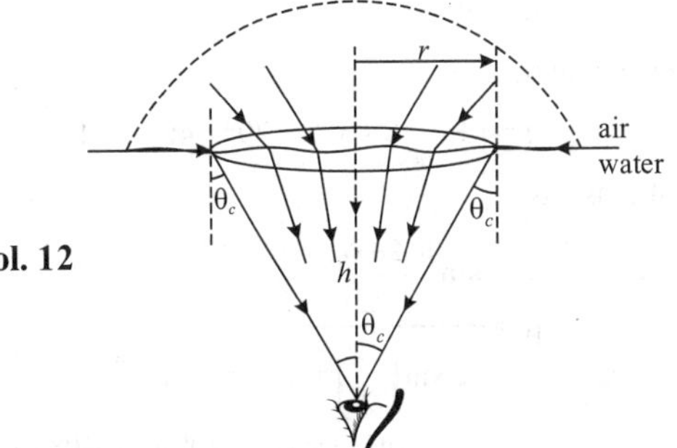

The critical angle at air-water interface is given as

$$\theta_c = \sin^{-1}\left(\frac{1}{\mu_\omega}\right) = \sin^{-1}\left(\frac{3}{4}\right)$$

$$\Rightarrow \quad \sin\theta_c = \frac{3}{4} \quad \Rightarrow \quad \tan\theta_c = \frac{3}{\sqrt{7}}$$

$$\tan\theta_c = \frac{r}{h}$$

$$\Rightarrow \quad r = h\tan\theta_c$$

$$\Rightarrow \quad h = h\left(\frac{3}{\sqrt{7}}\right) = \frac{15\times 3}{\sqrt{3}} \text{ cm}$$

Ans. (D)

Sol. 13

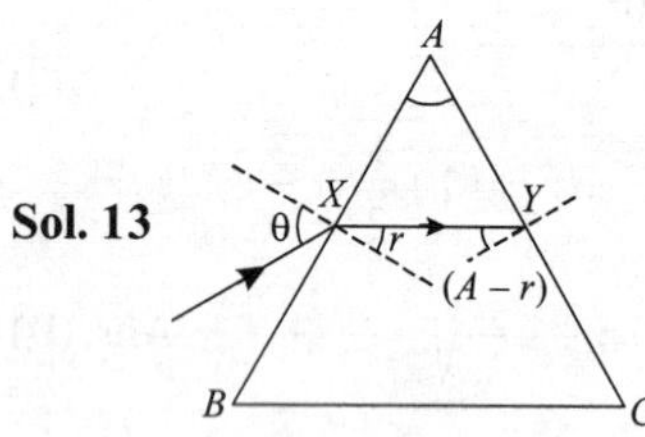

If $(A-r) > \theta_c$ then light will not get not get transmitted through force AC. So in this case we use for transmission of light

$$A - r < \theta_c \quad \ldots(1)$$

At point x, we use Snell's law

$$\sin\theta = \mu\sin r$$

$$\Rightarrow \quad r = \sin^{-1}\left(\frac{\sin\theta}{\mu}\right)$$

From equation-(1)

$$A - \sin^{-1}\left(\frac{\sin\theta}{\mu}\right) < \sin^{-1}\left(\frac{1}{\mu}\right) \quad \left[\text{As } \theta_c = \sin^{-1}\left(\frac{1}{\mu}\right)\right]$$

$$\Rightarrow \quad \sin^{-1}\left(\frac{\sin\theta}{\mu}\right) > A - \sin^{-1}\left(\frac{1}{\mu}\right)$$

$$\Rightarrow \quad \frac{\sin\theta}{\mu} > \sin\left(A - \sin^{-1}\left(\frac{1}{\mu}\right)\right)$$

$$\Rightarrow \quad \theta > \sin^{-1}\left[\mu\sin\left(A - \sin^{-1}\left(\frac{1}{\mu}\right)\right)\right]$$

Ans. (A)

Sol. 14 We are given with

$$\left.\begin{array}{l} i = 35^\circ \\ \delta = 40^\circ \\ e = 79^\circ \end{array}\right] \rightarrow \quad \delta = i + e - A$$

Prism angle is given as

$$A = i + e - \delta = 35^\circ + 79^\circ - 40^\circ = 74^\circ$$

We use RI of glass as

$$\mu = \frac{\sin\left(\frac{A+\delta_m}{2}\right)}{\sin\left(\frac{A}{2}\right)}$$

[we use $\delta_m < 40^\circ$ as at 40° $i \neq e$]

$$\Rightarrow \quad \mu = \frac{\sin\left(\frac{A+\delta_m}{2}\right)}{\sin(37^\circ)} \quad [\text{As } \sin 37^\circ = 3/5]$$

$$\Rightarrow \quad \mu = \frac{\sin\left(\frac{A+\delta_m}{2}\right)}{3/5}$$

$$\Rightarrow \quad \mu = \frac{5}{3}\sin(37^\circ + \theta)$$

Where $\theta = \frac{\delta_m}{2} < 20^\circ$

$$\Rightarrow \quad \mu = \frac{5}{3}\sin[37 + (< 20^\circ)]$$

$$\Rightarrow \quad \mu = \frac{5}{3}\sin[< 57^\circ]$$

$$\Rightarrow \quad \mu < \frac{5}{3}\sin 57^\circ \quad \left[\text{As } \sin 60^\circ = \frac{\sqrt{3}}{2}\right]$$

$$\Rightarrow \quad \mu < \frac{5}{3}\times 1.67$$

$$\Rightarrow \quad \mu < 1.45$$

Here out of given options closest possible maximum volume can be 1.5.

Ans. (A)

Sol. 15

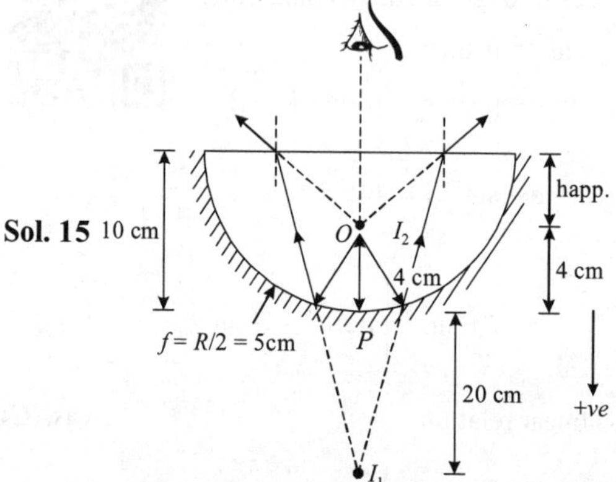

For reflection at spherical mirror we use

$$u = -4 \text{ cm}$$

$$f = -5 \text{ cm}$$

So from mirror formula, we have

$$\frac{1}{f} = \frac{1}{u} + \frac{1}{v}$$

$$\Rightarrow \quad \frac{1}{v} = \frac{1}{f} - \frac{1}{u} = -\frac{1}{5} + \frac{1}{4} = +\frac{1}{20}$$

$$\Rightarrow \quad v = +20 \text{ cm}$$

Due to refraction from plane surface of glass to air final image is seen at

$$h_{ap} = \frac{h}{\mu} = \frac{30}{1.5} = 20 \text{ cm}$$

Ans. (B)

Sol. 16 We are given that

$$n_{12} = \frac{n_1}{n_2} = \frac{1}{\sin\theta_c} \qquad (n_1 > n_2)$$

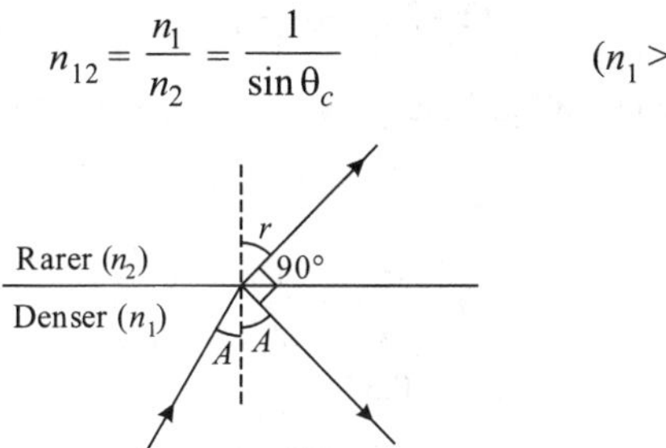

By Snells' law we have

$$n_1 \sin A = n_2 \sin r$$

$$\frac{\sin A}{\sin\theta_c} = \sin(90° - A) = \cos A$$

$$\Rightarrow \quad \tan A = \sin\theta_c$$

$$\Rightarrow \quad A = \tan^{-1}(\sin\theta_c)$$

Ans. (D)

Sol. 17

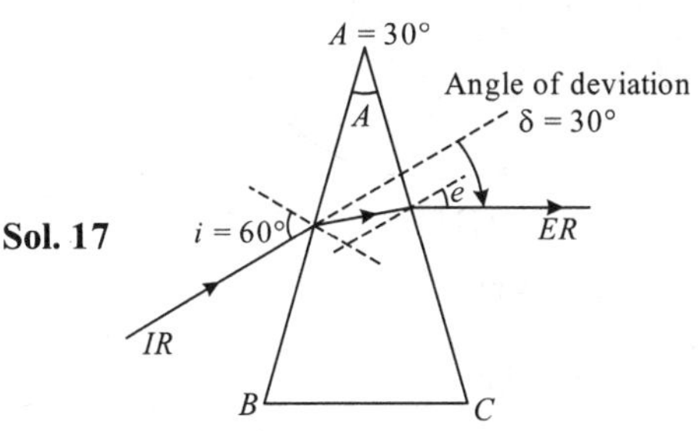

For a trihedral prism we know

$$\delta = i + e - A$$

$$\Rightarrow \quad e = \delta + A - i = 30° + 30° - 60° = 0°$$

$\Rightarrow$ Emergent ray comes out of prism perpendicular to face AC.

Ans. (C)

Sol. 18 Using refraction formula at spherical surface, we have

$$\frac{\mu_2}{v} - \frac{\mu_1}{u} = \frac{\mu_2 - \mu_1}{R}$$

$R = 7.8$ mm

$\mu = 1 \quad \mu = 1.34$

$$\Rightarrow \quad \frac{1.34}{v} - \frac{1}{\infty} = \frac{1.34 - 1}{7.8}$$

$$\Rightarrow \quad v = 30.7\,\text{mm}$$

Ans. (C)

Sol. 19 For minimum deviations by a prism, we use

$$i = e$$

and

$$r_1 = r_2 = \frac{A}{2} = 30°$$

by Snell's law, we use

$$\mu_{air} \sin i = \mu_p \sin r$$

$$\Rightarrow \quad 1 \times \sin i = \sqrt{3} \times \frac{1}{2} = \frac{\sqrt{3}}{2}$$

$$\Rightarrow \quad i = 60°$$

Ans. (D)

Sol. 20 By Snell's law, we use

$$1.\sin 40° = 1.31 \sin\theta_2$$

$$\Rightarrow \quad \sin\theta_2 = \frac{0.64}{1.31} = \frac{64}{131}$$

$$\Rightarrow \quad \tan\theta_2 = \frac{64}{\sqrt{(131)^2 - (64)^2}} = \frac{64}{\sqrt{13065}} \simeq 0.49 = \frac{d}{x}$$

$$\Rightarrow \quad x = \frac{d}{0.49}$$

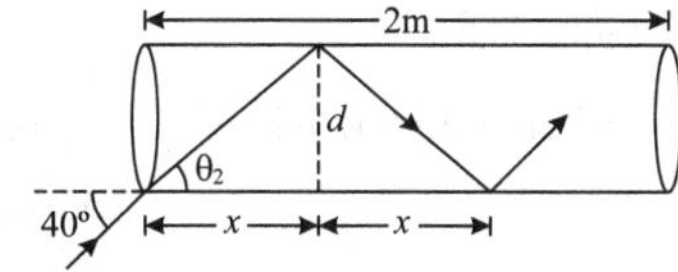

Total number of reflections in fiber are

$$N = \frac{2 \times 0.49}{20 \times 10^{-6}} = 55991$$

Closes to 55000

Ans. (A)

Sol. 21 From Snell's law

$$1.\sin 60° = \mu \sin 30°$$

$$\Rightarrow \quad \mu = \sqrt{3}$$

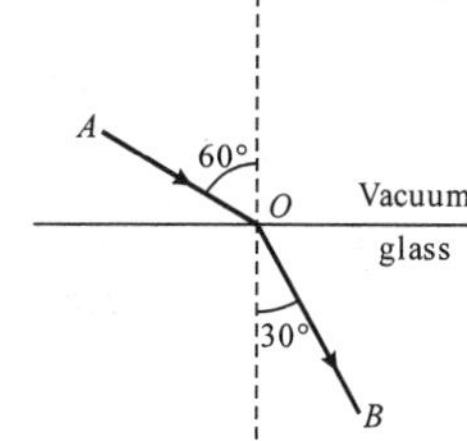

Optical path length of light ray from A to B is

$$OA = \mu(OB)$$

$$= \frac{a}{\cos 60°} + \sqrt{3}\frac{b}{\cos 30°}$$

$$= 2a + 2b$$

Ans. (A)

Sol. 22

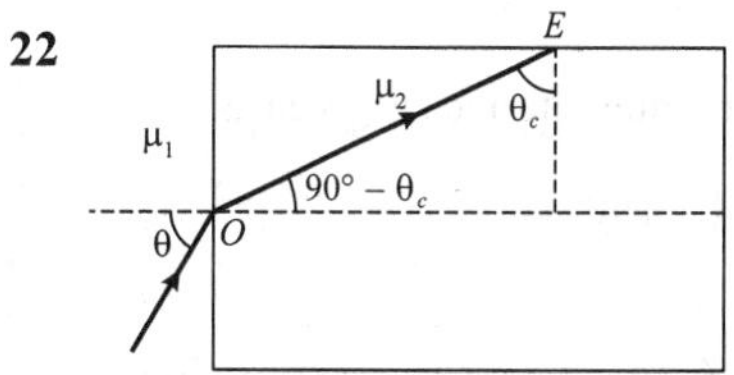

For total internal reflection at point E, we use

$$\sin\theta_c = \frac{\mu_1}{\mu_2}$$

For refraction at point O, we use

$$\mu_1 \sin\theta = \mu_2 \sin(90^\circ - \theta_c)$$

$$\Rightarrow \quad \sin\theta = \frac{\mu_2\sqrt{1-\frac{\mu_1^2}{\mu_2^2}}}{\mu_1}$$

$$\Rightarrow \quad \theta = \sin^{-1}\sqrt{\frac{\mu_2^2-\mu_1^2}{\mu_1^2}}$$

For total internal reflection to take place at point E, we use

$$\theta < \sin^{-1}\sqrt{\frac{\mu_2^2}{\mu_1^2}-1}$$ **Ans. (B)**

Sol. 23 Critical angle for a medium is given as

$$\sin\theta_C = \frac{1}{\mu} = \frac{1}{\sqrt{3\times 4/3}}$$

[where Refractive index $\mu = \sqrt{\mu_0\varepsilon_0}$]

$$\theta_C = 30^\circ$$ **Ans. (D)**

Sol. 24 Figure shows the situation described in question

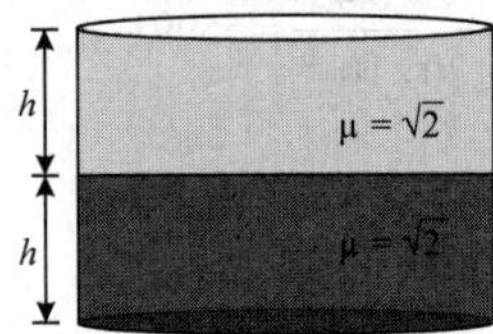

For normal viewing, apparent depth of vessel is given as

$$h_{app} = \frac{h}{\mu_1} + \frac{h}{\mu_2}$$

$$\Rightarrow \quad h_{app} = \frac{h}{\sqrt{2}} + \frac{h}{2\sqrt{2}}$$

$$\Rightarrow \quad h_{app} = \frac{3}{2\sqrt{2}}h = \frac{3}{4}\sqrt{2}\,h$$ **Ans. (B)**

Sol. 25

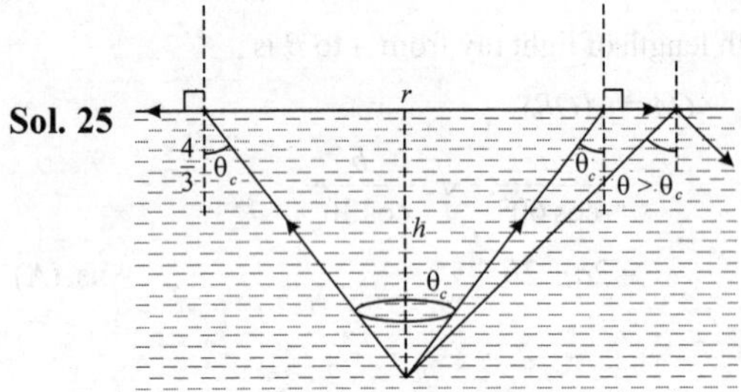

Critical angle for air-water interface is given as

$$\frac{4}{3}\sin\theta_c = 1 \sin 90^\circ$$

$$\sin\theta_c = \frac{3}{4}$$

Solid angle of cone from which light escapes out from water surface is

$$\Omega = 2\pi(1-\cos\theta)$$

$$\Rightarrow \quad \Omega = 2\pi\left(1-\frac{\sqrt{7}}{4}\right)$$

Fraction of light energy which escapes out from water is

$$P_{esc} = \frac{P}{4\pi}\times\Omega$$

$$\Rightarrow \quad f = \frac{P_{esc}}{P} = \frac{1}{2}\left(1-\frac{\sqrt{7}}{4}\right)$$

$$\Rightarrow \quad f = \frac{0.33}{2} = 0.17$$

$\Rightarrow$ % fraction = 17% **Ans. (A)**

Sol. 26

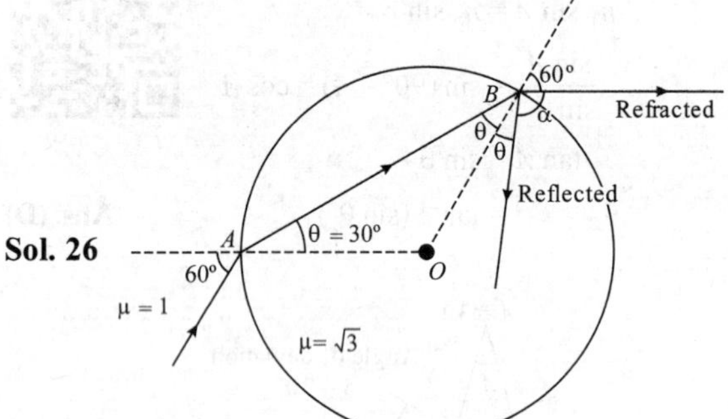

Using Snell's law at point A, we have

$$1\times\sin 60^\circ = \sqrt{3}\times\sin\theta$$

$$\Rightarrow \quad \frac{\sqrt{3}}{2} = \sqrt{3}\times\sin\theta$$

$$\Rightarrow \quad \sin\theta = \frac{1}{2}$$

$$\Rightarrow \quad \theta = 30^\circ$$

At point B, we have

$$\theta + 60^\circ + \alpha = 180^\circ$$

$$\Rightarrow \quad 30^\circ + 60^\circ + \alpha = 180^\circ$$

$$\Rightarrow \quad \alpha = 90^\circ$$ **Ans. (90)**

Sol. 27 From figure, we have

$$\tan r = \frac{15}{30} = \frac{1}{2}$$

$$\Rightarrow \quad \sin r = \frac{1}{\sqrt{5}}$$

By smell's law, we have

$$\mu \sin r = \sin 45^\circ$$

$$\Rightarrow \quad \mu = \frac{\frac{1}{\sqrt{2}}}{\frac{1}{\sqrt{5}}} = \sqrt{\frac{5}{2}} = 1.581$$

Given value of

$$\mu = \frac{N}{100} = 1.581$$

$$\Rightarrow \quad N = 158.1 \qquad \textbf{Ans. (158)}$$

Sol. 28 Angle of minimum deviation for small angled prism is given as

$$\delta_{\min} = (\mu - 1)A$$

$$= (1.5 - 1)1$$

$$= 0.5^\circ$$

$$\delta_{\min} = \frac{N}{10} = 0.5^\circ$$

$$\Rightarrow \quad N = 5 \qquad \textbf{Ans. (5)}$$

Sol. 29 By lens makers formula, we use

$$P = (\mu - 1)\left[\frac{2}{R}\right] \qquad \ldots(1)$$

For another plano convex lens with power $1.5P$, we use

$$1.5P = (\mu - 1)\left[\frac{1}{R'}\right] \qquad \ldots(2)$$

From (1) and (2), we have

$$1.5 = \frac{1/R'}{2/R}$$

$$\Rightarrow \quad R' = \frac{R}{2 \times 1.5} = \frac{R}{3} \qquad \textbf{Ans. (D)}$$

Sol. 30 By lens makers formula, we have

$$\frac{1}{f} = \left[\frac{\mu_L}{\mu_S} - 1\right]\left[\frac{1}{R_1} - \frac{1}{R_2}\right]$$

If $\mu_L = \mu_S$, it gives $f = \infty$. **Ans. (B)**

Sol. 31 Using the construction diagram below for the lens, the radius of lens can be related with its thickness and radius of curvature as

$$R^2 = r^2 + (R - t)^2$$

$$\Rightarrow \quad R^2 = r^2 + R^2 + t^2 - 2Rt \qquad \ldots(1)$$

For small thickness neglecting t^2 in equation-(1), we get

$$R = \frac{r^2}{2t}$$

Using lens maker's formula, we have

$$\frac{1}{f} = (\mu - 1)\left(\frac{1}{R} - \frac{1}{\infty}\right) = \frac{\mu - 1}{R}$$

$$\Rightarrow \quad f = \frac{R}{\mu - 1} = \frac{r^2}{2t(\mu - 1)}$$

$$\Rightarrow \quad f = \frac{(3 \times 10^{-2})^2}{2 \times 3 \times 10^{-3} \times \left(\frac{3}{2} - 1\right)}$$

$$\Rightarrow \quad f = \frac{9 \times 10^{-4}}{6 \times 10^{-3} \times 1} \times 2 = 0.3 \text{ m} = 30 \text{ cm} \qquad \textbf{Ans. (D)}$$

Sol. 32 As wavelength of light in medium is given as

$$\lambda_m = \frac{\lambda_a}{\mu}$$

Thus refractive index of the medium is given as

$$\mu = \frac{3}{2}$$

Using refraction formula for the surface, we have

$$\frac{\mu}{v} - \frac{1}{u} = \frac{\mu - 1}{R}$$

Given that image is located at 10 cm behind the surface and it is 2/3 of the object distance so object distance can be taken as 15 cm in from of the lens. Substituting $v = +10$ cm and $u = -15$ cm in above formula, we have

$$\Rightarrow \quad \frac{3}{2 \times 10} + \frac{1}{15} = \frac{\frac{3}{2} - 1}{R}$$

$$\Rightarrow \quad R = \frac{30}{13} \text{ cm}$$

$$\Rightarrow \quad x = 30 \qquad \textbf{Ans. (30)}$$

Sol. 33 In the situation described in question we consider that the given right angled prism is an isoceles one, so the angles at which light will incident on face PR is 45° as shown below.

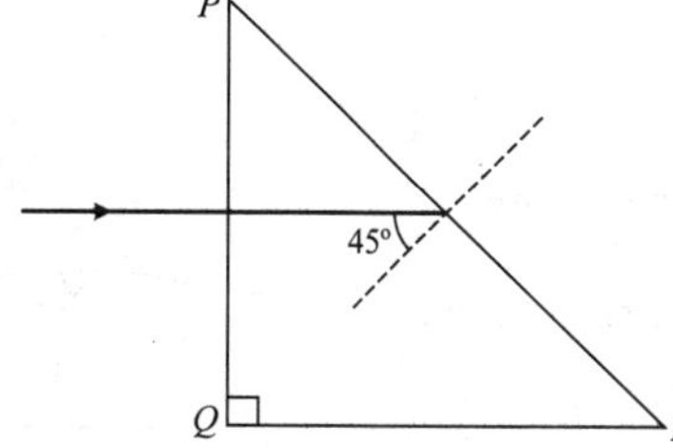

Critical angle θ_C for the three light rays is given as

$$\theta_C = \sin^{-1}\left(\frac{1}{\mu}\right)$$

If $\theta_C \geq 45°$ for any ray then this light ray will pass through the prism.

$$(\theta_C)_{Red} = \sin^{-1}\left(\frac{1}{1.27}\right) = 51.94°$$

Thus red light will pass

$$(\theta_C)_{Green} = \sin^{-1}\left(\frac{1}{1.42}\right) = 44.76°$$

Thus green light will not pass

$$(\theta_C)_{Blue} = \sin^{-1}\left(\frac{1}{1.49}\right) = 42.15°$$

Thus blue light will not pass. So only red light will pass through the prism face *PR*. **Ans. (B)**

Sol. 34 Using refraction formula for image formation, we have

$$\frac{n_2}{v} - \frac{n_1}{u} = \frac{n_2 - n_1}{R}$$

$$\Rightarrow \quad \frac{1.4}{v} - \frac{1.25}{-40} = \frac{1.4 - 1.25}{(-25)}$$

$$\Rightarrow \quad v = -37.58 \text{ cm}$$ **Ans. (D)**

Sol. 35 In figure, we can see that

$$i = r \;\&\; r' = 90 - r$$

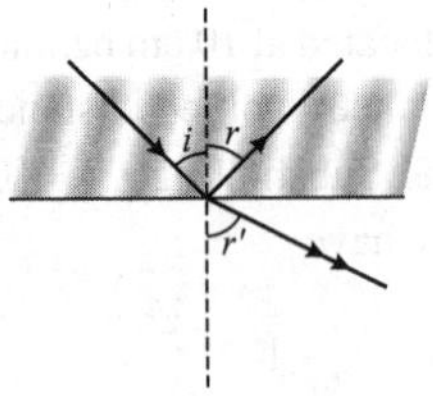

Critical angle is given as

$$\theta_C = \sin^{-1}\left(\frac{1}{\mu}\right) = \sin^{-1}\left(\frac{\sin i}{\sin r'}\right)$$

$$\Rightarrow \quad \theta_C = \sin^{-1}\left(\frac{\sin i}{\sin r}\right) = \sin^{-1}(\tan r)$$ **Ans. (A)**

Sol. 36 The situation described in question is shown in figure below. In this state the Prism angle is given as

$$A = \frac{i}{2} + \frac{i}{2} = i$$

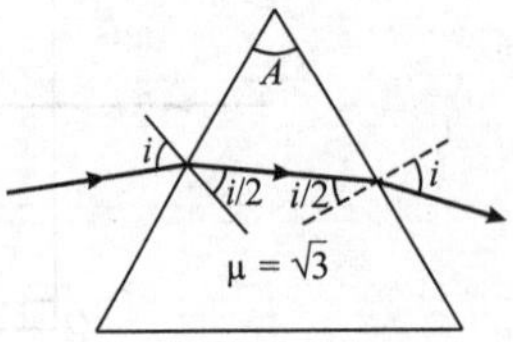

From Snell's law we use

$$1 \sin i = \mu \sin\frac{i}{2}$$

$$\Rightarrow \quad 2\sin\frac{i}{2}\cos\frac{i}{2} = \sqrt{3}\sin\frac{i}{2}$$

$$\Rightarrow \quad \cos\frac{i}{2} = \frac{\sqrt{3}}{2}$$

$$\Rightarrow \quad \frac{i}{2} = 30°$$

$$\Rightarrow \quad i = 60°$$

$$\Rightarrow \quad A = i = 60°$$ **Ans. (60)**

Sol. 37 Critical angle for diamond-air interface is given as

$$\theta_C = \left(\frac{1}{2.42}\right) = 24.4°$$

Given Incident angle 30° is more than critical angle so laser will suffer total internal reflection and will not get refracted. **Ans. (C)**

Sol. 38 By Snell's law, we use

$$1 \times \sin\theta = \frac{4}{3}\sin(90 - C)$$

$$\Rightarrow \quad \sin\theta = \frac{4}{3}\cos C$$

For critical angle, we use

$$\sin C = \frac{1}{\mu} = \frac{5}{4}$$

$$\Rightarrow \quad \cos C = \frac{\sqrt{7}}{4}$$

$$\Rightarrow \quad \sin\theta = \frac{4}{3} \times \frac{\sqrt{7}}{4} = \frac{\sqrt{7}}{3}$$

For Total Internal Reflection, we have

$$\sin\theta \leq \frac{\sqrt{7}}{3} \text{ as } \theta = \sin^{-1}\frac{\sqrt{7}}{3}$$ **Ans. (C)**

Sol. 39 Angle of deviation for the light ray passing through the prism first decreases with increase in angle of incidence from 90° as for grazing incidence and grazing emergence the deviation is maximum and when light ray passes through the prism symmetrically, it is minimum thus option (C) only matches with this analysis. **Ans. (C)**

Sol. 40 As it is specified that in initial siutation, image coincides with the object this indecases that Image oby lens must be on centre of curvature of mirror. Hence, distance between object and image after removing mirror = 12 + 8 + 30 = 50 cm. **Ans. (50)**

Sol. 41 For first refraction, using lens formula, we have

$$\frac{1}{v} - \frac{1}{u} = \frac{1}{f}$$

$$\Rightarrow \quad \frac{1}{v} - \frac{1}{-30} = \frac{1}{10}$$

$$\Rightarrow \quad \frac{1}{v} = \frac{1}{10} - \frac{1}{30}$$

$$\Rightarrow \quad \frac{1}{v} = \frac{3-1}{30}; v = 15 \text{ cm}$$

For second refraction we use $u = +10$ cm and again by using lens formula, we have

$$\frac{1}{v} - \frac{1}{10} = \frac{1}{-10}$$

$$\Rightarrow \quad \frac{1}{v} = \frac{-1}{10} + \frac{1}{10} \quad v = \infty$$

For third refraction as light rays incident parallel to principal axis, final image is formed at focal point of the third lens so we have $v = +30$ cm from third lens. **Ans. (B)**

Sol. 42 By lens maker's formula for plano convex and plano concave lens, we have

$$\frac{1}{f_1} = (\mu_1 - 1)\left[\frac{1}{R}\right]$$

and

$$\frac{1}{f_1} = (\mu_2 - 1)\left[\frac{1}{-R}\right]$$

$$\Rightarrow \quad \frac{1}{f_{eq}} = \frac{1}{f_1} + \frac{1}{f_2} = \frac{(\mu_1 - 1)}{R} - \frac{(\mu_2 - 1)}{R}$$

$$\Rightarrow \quad \frac{1}{f_{eq}} = \frac{(\mu_1 - \mu_2)}{R}$$

$$\Rightarrow \quad \frac{R}{f_{eq}} = \mu_1 - \mu_2$$

Ans. (D)

Sol. 43 de-Broglie wavelength of electron is given as

$$\lambda_D = \frac{h}{\sqrt{2m\left(\frac{3}{2}kT\right)}}$$

$$\Rightarrow \quad \lambda_D = \frac{h}{\sqrt{3mkT}}$$

$$\Rightarrow \quad \lambda_D = \frac{6.6 \times 10^{-34}}{\sqrt{3 \times 9 \times 10^{-31} \times 1.38 \times 10^{-23} \times 300}}$$

$$\Rightarrow \quad \lambda_D = \frac{6.6 \times 10^{-34}}{105.72 \times 10^{-27}} = 62 \text{ Å} = 6.2 \text{ nm}$$

Ans. (A)

Sol. 44 Height of water as observed by observer is given as

$$h' = \frac{H}{\mu_w} = \frac{H}{(4/3)} = \frac{3H}{4}$$

Height of air as observed by observer = 17.5 – H

Acoording to the given condition, we use

$$\frac{3H}{4} = 17.5 - H$$

$$\Rightarrow \quad H = 10 \text{ cm}$$

Ans. (B)

Sol. 45 Lateral shift due to refraction by a glass slab of thickness t is given as

$$l = t \sin i \left[1 - \frac{\cos i}{\sqrt{\mu^2 - \sin^2 i}}\right]$$

$$\Rightarrow \quad 4\sqrt{3} = t \sin 60° \left[1 - \frac{\cos 60°}{\sqrt{3 - \frac{3}{4}}}\right]$$

$$\Rightarrow \quad t = 12 \text{cm}$$

Ans. (12)

Sol. 46 For total internal reflection, we use

$$i > \theta_C$$

$$\Rightarrow \quad \sin i > \sin \theta_C$$

$$\Rightarrow \quad \sin i > \frac{\mu_R}{\mu_D} \qquad \ldots(1)$$

Refractive index of medium in terms of relative permeability and dielectric constant is given as

$$\mu = \sqrt{\mu_r \in_r}$$

$$\Rightarrow \quad \frac{\mu_R}{\mu_D} = \frac{\sqrt{1 \times 1}}{\sqrt{4 \times 1}} = \frac{1}{2}$$

From equation-(1), we have

$$\sin i > \frac{1}{2}$$

$$\Rightarrow \quad i > 30°,$$

Thus from given options in question, we use

$$i = 60°$$

Ans. (D)

Sol. 47 Refractive index of a media is given as

$$v = \frac{c}{\mu}$$

$$\Rightarrow \quad v_B = \frac{3 \times 10^8}{1.47} = 2.04 \times 10^8 = 20.4 \times 10^7 \text{ m/s}$$

Given that the difference in speed of light in the two media is

$$v_A - v_B = 2.6 \times 10^7 \text{ m/s}$$

$$\Rightarrow \quad v_A = (20.4 + 2.6) \times 10^7 = 23 \times 10^7 \text{ m/s}$$

$$\Rightarrow \quad \frac{\mu_B}{\mu_A} = \frac{v_A}{v_B} = \frac{23 \times 10^7}{20.4 \times 10^7} = 1.13$$

Ans. (C)

Sol. 48 Critical angle for the given situation as shown in figure below is calculated as

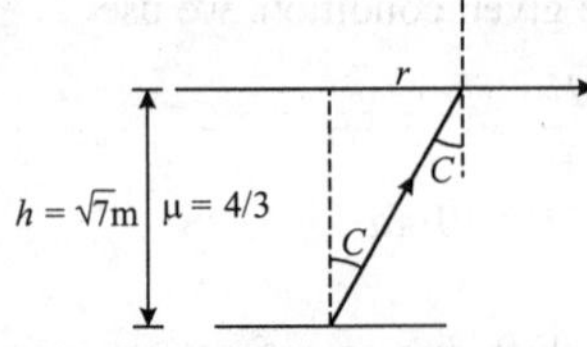

$$\tan C = \frac{r}{h}$$

$$\Rightarrow \quad r = h \tan C$$

$$\Rightarrow \quad \sin C = \frac{1}{\mu} = \frac{3}{4}$$

$$\Rightarrow \quad \tan C = \frac{3}{\sqrt{7}}$$

$$\Rightarrow \quad r = \sqrt{7} \times \frac{3}{\sqrt{7}} = 3$$

Area of circular surface on water through which light will emerge out is calculated as

$$A = \pi r^2 = 9\,\pi\text{m}^3$$ **Ans. (9)**

Sol. 49 Given that

$$i = 2r$$

Using Snell's law at the surface we have

$$\sin i \times n_1 = \sin r \times n_2$$

$$\Rightarrow \quad \sin i \times 1 = \sin\frac{i}{2} \times \sqrt{2n}$$

$$\Rightarrow \quad \frac{\sin i}{\sin\frac{i}{2}} = \sqrt{2n}$$

$$\Rightarrow \quad \frac{2\sin\frac{i}{2}\cos\frac{i}{2}}{\sin\frac{i}{2}} = \sqrt{2n}$$

$$\Rightarrow \quad \cos\frac{i}{2} = \sqrt{\frac{n}{2}}$$

$$\Rightarrow \quad \frac{i}{2} = \cos^{-1}\left(\sqrt{\frac{n}{2}}\right)$$

$$\Rightarrow \quad i = 2\cos^{-1}\left(\sqrt{\frac{n}{2}}\right)$$ **Ans. (D)**

Sol. 50 In refraction frequency never changes but getting into deser medium speed decreases and correspondingly wavelength also decreases. **Ans. (C)**

Sol. 51 Refractive index of material of the prism is given as

$$\mu = \frac{\sin\left(\frac{a+\delta_m}{2}\right)}{\sin\frac{A}{2}}$$

Given that $\mu = \cot\frac{A}{2}$ so we have

$$\sin\left(\frac{A+\delta_m}{2}\right) = \cos\frac{A}{2}$$

$$\Rightarrow \quad \delta_m = 180 - 2A$$ **Ans. (A)**

Sol. 52 Critical angle of light is given as

$$\sin i_c = \frac{n_r}{n_d} = \frac{v_d}{v_r} = \frac{1.5\times10^{10}}{2\times10^{10}}$$

$$\Rightarrow \quad \sin i_c = \frac{3}{4}$$

$$\Rightarrow \quad i_c = \sin^{-1}\left(\frac{3}{4}\right)$$

For total internal reflection $\theta > i_c$

$$\Rightarrow \quad \theta > \sin^{-1}\left(\frac{3}{4}\right)$$ **Ans. (D)**

Sol. 53 Given that the ratio of refractive indices of media is

$$\frac{\mu_A}{\mu_B} = \frac{c/V_A}{c/V_B} = \frac{V_B}{V_A} = \frac{1}{2}$$

If the thickness of material is d the difference in time taken to cross the material by light is given as

$$\frac{d}{v_B} - \frac{d}{v_A} = 5 \times 10^{-10}\text{s}$$

$$\Rightarrow \quad d = \frac{5\times10^{-10} \times v_A v_B}{v_A - v_B}$$

As $v_A = 2v_B$ we have

$$d = 5 \times 10^{-10} \times 2v_B$$

or $$d = 5 \times 10^{-10} \times v_A$$ **Ans. (A)**

Sol. 54 Ratio of refractive indices is given as

$$\frac{\mu_2}{\mu_{air}} = \frac{c}{v_2}$$

$$\Rightarrow \quad \frac{\sqrt{\mu_{r_2}\varepsilon_{r_2}}}{(1)} = \frac{c}{v_2}$$

$$\Rightarrow \quad \sqrt{(1)(9)} = \frac{c}{v_2}$$

$$\Rightarrow \quad v_2 = \frac{c}{3} = 1.0 \times 10^8\ \text{ms}^{-1}$$ **Ans. (A)**

Sol. 55 The refraction of light ray from medium 1 to medium 2 is shown in figure below and the vector along incident ray is given as

$$\vec{A} = 4\sqrt{3}\hat{i} - 3\sqrt{3}\hat{j} - 5\hat{k}$$

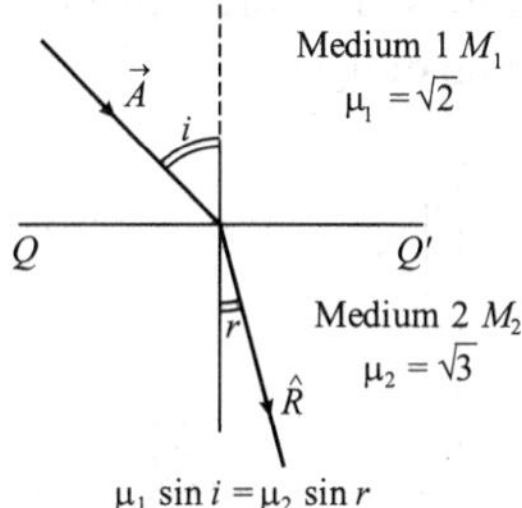

As incident vector A makes i angle with normal along z-axis & vector along refracted ray R makesan angle r with the normal. The incidence angle i is given as

$$i = \cos^{-1}\left(\frac{A_z}{A}\right)$$

$$\Rightarrow \quad i = \cos^{-1}\left(\frac{5}{\sqrt{(4\sqrt{3})^2 + (3\sqrt{3})^2 + 5^2}}\right)$$

$$\Rightarrow \quad i = \cos^{-1}\left(\frac{5}{10}\right)$$

$$\Rightarrow \quad i = 60°$$

Using Snell's law, we have

$$\sqrt{2}\sin 60° = \sqrt{3} \times \sin r$$

$$\Rightarrow \quad r = 45°$$

Difference between i & r is calculated as

$$\delta = 60° - 45° = 15°$$

Ans. (15)

Sol. 56 Situation described in question is shown by the refraction of light ray in figure below from air to the given medium.

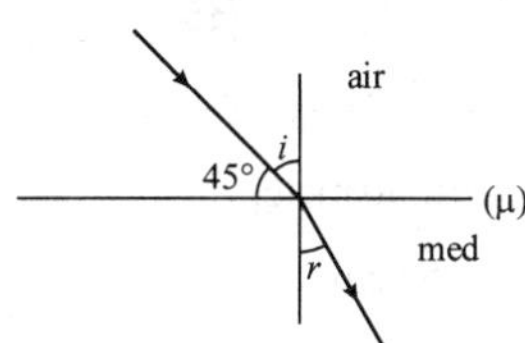

Deviation angle of light is given as

$$\delta = i - r$$

$$\Rightarrow \quad 15° = 45 - r$$

$$\Rightarrow \quad r = 30°$$

Using Snell's law we have

$$n_1 \sin i = n_2 \sin r$$

$$\Rightarrow \quad 1 \sin 45° = \mu \sin 30°$$

$$\Rightarrow \quad \frac{1}{\sqrt{2}} = \mu\frac{1}{2}$$

$$\Rightarrow \quad \mu = \sqrt{2} = 1.414$$

Ans. (C)

Sol. 57 Figure below shows the situation described in the question.

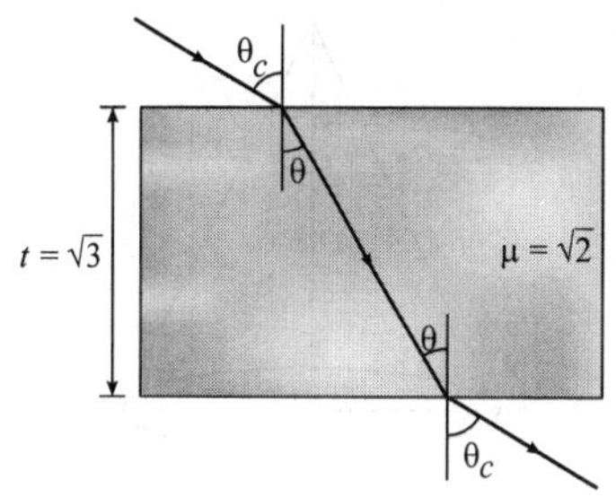

For the given refractive index, critical angle θ_c is given as

$$\sin \theta_c = \frac{1}{\sqrt{2}}$$

$$\Rightarrow \quad \theta_c = 45°$$

Using Snell's law, we have

$$\sin \theta_c = \mu \sin \theta$$

$$\Rightarrow \quad \frac{1}{\sqrt{2}} = \sqrt{2}\sin\theta$$

$$\Rightarrow \quad \theta = 30°$$

Lateral displacement of light ray is calculated as

$$x = t \sin(i - r) \sec r$$

$$\Rightarrow \quad x = \sqrt{3}\,\sin(45° - 30°)\sec 30°$$

$$\Rightarrow \quad x = \sqrt{3}\,(0.26)\left(\frac{2}{\sqrt{3}}\right)$$

$$\Rightarrow \quad x = 0.52\text{ cm} = 52 \times 10^{-2}\text{ cm}$$

Ans. (52)

Sol. 58 For dispersion without deviation, we use

$$\delta_1 = \delta_2$$

For thin prism deviation $\delta = (\mu - 1)A$

$$\Rightarrow \quad 6°(1.54 - 1) = A(1.72 - 1)$$

$$\Rightarrow \quad A = \frac{6° \times 0.54}{0.72} = \frac{18°}{4} = 4.5°$$

Ans. (D)

Sol. 59 Shift of object is calculated as

$$\text{Shift} = \left(d - \frac{d}{\mu}\right) = 30\text{ cm}$$

$$\Rightarrow \quad d\left[1 - \frac{1}{\frac{5}{3}}\right] = 30$$

$$\Rightarrow \quad d = \frac{30 \times 5}{2} = 75\text{ cm}$$

Ans. (A)

Sol. 60 Figure below shows the revolving particle as described in question.

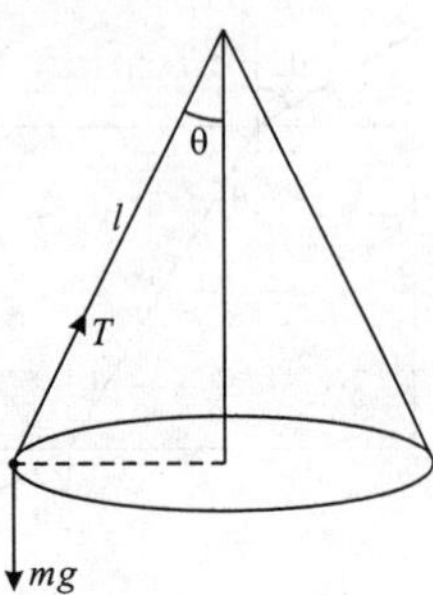

Using free body diagram of stone in rotating frame of reference, we have

$$T\cos\theta = mg$$

and $$T\sin\theta = \frac{mv^2}{l^2\sin\theta}$$

$$\Rightarrow \quad \cos\theta = \frac{mg}{T} \quad \ldots(1)$$

$$\Rightarrow \quad \sin^2\theta = \frac{mv^2}{Tl^2} \quad \ldots(2)$$

From equations-(1) & (2), we have

$$1 = \left(\frac{mg}{T}\right)^2 + \frac{mv^2}{Tl^2}$$

$$\Rightarrow \quad 1 = \left(\frac{10}{400}\right)^2 + \frac{v^2}{400}$$

$$\Rightarrow \quad v^2 = 399.75$$

$$\Rightarrow \quad v = 20 \text{ m/s}$$

Ans. (A)

Sol. 61

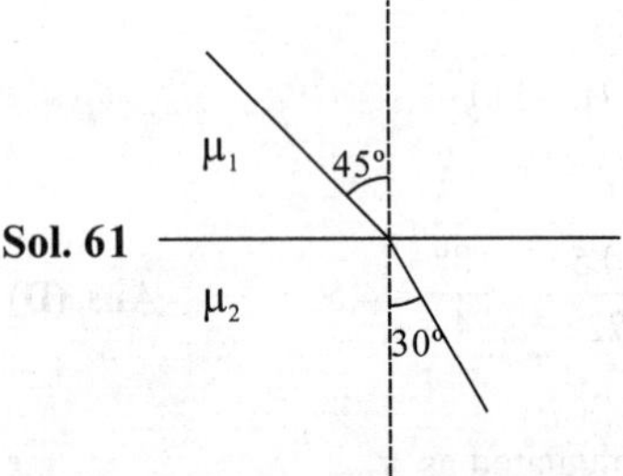

Applying Snell's law at air-medium surface, we have

$$\mu_1 \sin 45° = \mu_2 \sin 30°$$

$$\Rightarrow \quad \frac{\mu_1}{\mu_2} = \frac{1}{\sqrt{2}}$$

$$\Rightarrow \quad \frac{\mu_1}{\mu_2} = \frac{v_2}{v_1} = \frac{f\lambda_2}{f\lambda_1}$$

For monochromatic light as frequency remains constant even after the change in medium.

$$\Rightarrow \quad \frac{\mu_1}{\mu_2} = \frac{\lambda_2}{\lambda_1} = \frac{1}{\sqrt{2}}$$

$$\Rightarrow \quad \lambda_2 = \frac{\lambda_1}{\sqrt{2}}$$

Ans. (B)

Sol. 62 Figure below shows the situation described in the question for refraction of light.

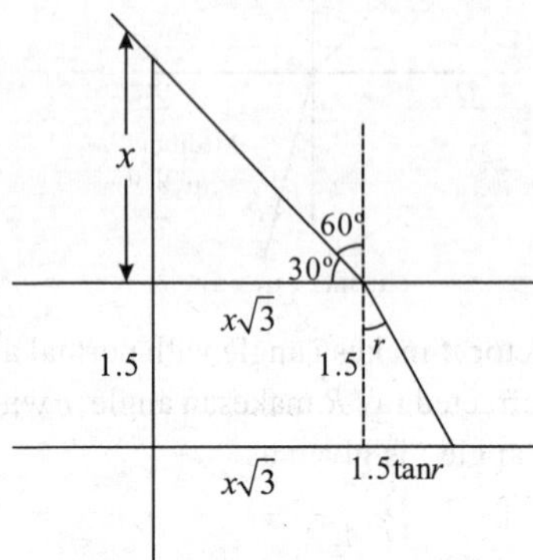

By Snell's law at air-water interface

$$\sin 60° = \frac{4}{3}\sin r$$

$$\Rightarrow \quad \sin r = \frac{3\sqrt{3}}{8}$$

$$\Rightarrow \quad \tan r = \frac{3\sqrt{3}}{\sqrt{37}}$$

From the diagram, when light is incident at 30° then length of shadow of pole within water can be calculated by the equation written as

$$x\sqrt{3} + 1.5\tan r = 2.15$$

$$\Rightarrow \quad x\sqrt{3} = 2.15 - 1.5 \times \frac{3\sqrt{3}}{\sqrt{37}}$$

$$\Rightarrow \quad x = \frac{2.15}{\sqrt{3}} - \frac{1.5\times 3}{\sqrt{37}}$$

$$\Rightarrow \quad x = 1.241 - 0.739 = 0.502$$

$$\Rightarrow \quad x \approx 0.50 \text{ meter}$$

$$\Rightarrow \quad x = 50 \text{ cm}$$

Ans. (50)

Sol. 63 Figure below shows the situation described in the question.

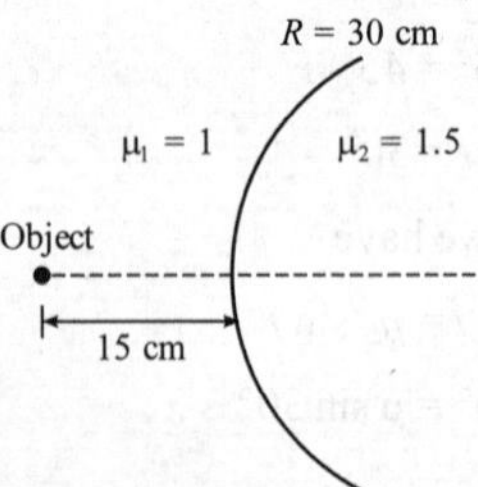

By refraction at the spherical surface, we use

$$\frac{\mu_2}{v} - \frac{\mu_1}{u} = \frac{\mu_2 - \mu_1}{R}$$

$$\Rightarrow \quad \frac{1.5}{v} - \frac{1}{-15} = \frac{1.5-1}{30} = \frac{1}{60}$$

$$\Rightarrow \quad \frac{1.5}{v}+\frac{1}{15}=\frac{1}{60}$$

$$\Rightarrow \quad \frac{1.5}{v}=\frac{1}{60}-\frac{1}{15}=\frac{-1}{20}$$

$$\Rightarrow \quad \frac{1.5}{v}=-\frac{1}{20}$$

$$\Rightarrow \quad v=-30 \text{ cm}$$

Ans. (30)

Sol. 64 If i_C is the critical angle for denser-rarer interface then by
Snell's law, we have

$$\frac{v}{C}=\frac{1}{\mu}=\sin i_C=\sin 45°=\frac{1}{\sqrt{2}}$$

Thus speed of light in denser medium is given as

$$\Rightarrow \quad v=\frac{C}{\sqrt{2}}=\frac{3\times10^8}{\sqrt{2}} \text{ m/s}=2.12\times10^8 \text{ m/s}$$

Ans. (B)

Sol. 65 By lens maker formula, we have

$$\frac{1}{F}=P=(1.8-1)\left(\frac{1}{20}+\frac{1}{20}\right)$$

When lens is immersed in a liquid of refractive index 1.5, new poer is given as

$$P'=\left(\frac{1.8}{1.5}-1\right)\left(\frac{1}{20}+\frac{1}{20}\right)$$

$$\Rightarrow \quad \frac{P}{P'}=\frac{0.8}{1.2-1}=4$$

Ans. (4)

Sol. 66 Figure below shows the image formation after refraction and reflection in the system given.

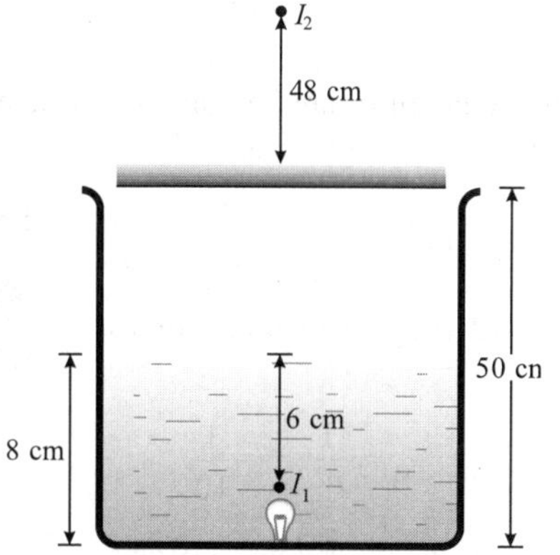

Shift of bulb due to refraction at water surface is calculated as

$$S=t\left(1-\frac{1}{\mu}\right)=8\left(1-\frac{1}{4/3}\right)$$

$$\Rightarrow \quad S=8\left(1-\frac{3}{4}\right)=8\times\frac{1}{4}=2 \text{ cm}$$

Thus light rays from bulb will fall on mirror such that these will appear to be coming from a distance of 48 cm so at the same distance behind the mirror image of bulb will be obtained. Thus distance between O and I_2 is calculated as

$$d=48+50=98 \text{ cm}$$

Ans. (98)

Sol. 67 Due to refraction, apparent distance of bubble from on side is 12 cm so we use

$$d_{\text{apparent}}=\frac{d_{\text{actual}}}{\mu_{\text{rel}}}$$

$$\Rightarrow \quad 12=\frac{x}{\mu} \qquad \ldots(1)$$

When viewed from opposite side distance of bubble is calculated as

$$4=\frac{24-x}{\mu} \qquad \ldots(2)$$

On solving we get $\mu = 1.5$.

Ans. (B)

Sol. 68 Apparent depth of vessel when viewed from above will be if

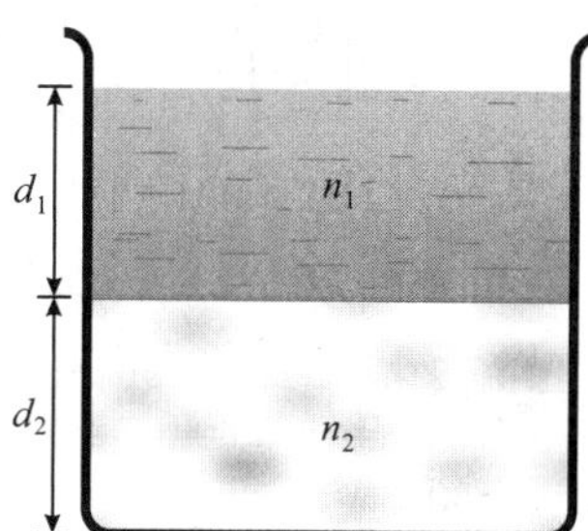

as $d_1 = d_2 = \frac{d}{2}$, apparent depth is given as

$$d_{\text{app}}=\frac{d_1}{n_1}+\frac{d_2}{n_2}$$

$$\Rightarrow \quad d_{\text{app}}=\frac{d}{2}\left[\frac{n_1+n_2}{n_1 n_2}\right]$$

Ans. (B)

Sol. 69 Position of bird w.r.t. fish can be calculated from the figure shown below in which bird is taken as an object and fish as an observer

$$y_{net} = y + \mu y'$$

$$\Rightarrow \quad \frac{dy_{net}}{dt} = \frac{dy}{dt} + \frac{\mu dy'}{dt}$$

$$\Rightarrow \quad 12 = 8 + \frac{4}{3}(v)$$

$$\Rightarrow \quad v = (12-8) \times \frac{3}{4}$$

$$\Rightarrow \quad v = 3 \text{ m/s}$$

Ans. (3)

Sol. 70 Refraction index of a transparent liquid filled in equilateral hollow prism is given as

$$\mu = \frac{\sin\left(\frac{D_{min}+A}{2}\right)}{\sin\left(\frac{A}{2}\right)}$$

$$\angle A = 60°$$

$$\Rightarrow \quad \sqrt{2} = \frac{\sin\frac{(D_{min}+60)}{2}}{\sin\left(\frac{60}{2}\right)}$$

$$\Rightarrow \quad \frac{1}{\sqrt{2}} = \sin\left(\frac{D_{min}+60}{2}\right)$$

$$\Rightarrow \quad \frac{D_{min}+60}{2} = 45$$

$$\Rightarrow \quad D_{min} = 30$$

Ans. (30)

23.3 Geometrical Optics-III Thin Lenses

Sol. 1

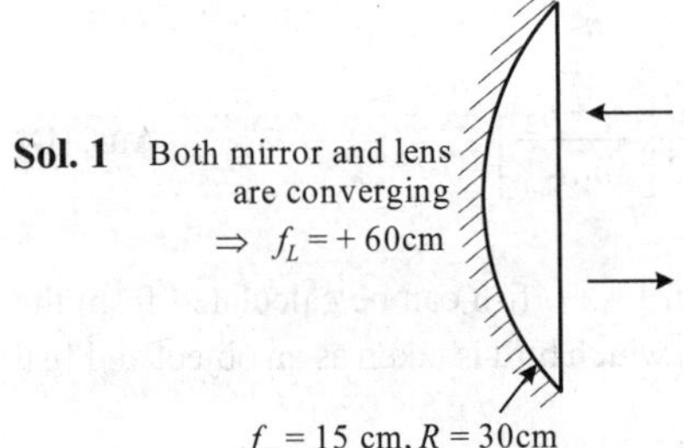

This system acts like a combination of a concave mirror and a plano convex lens

Focal length of mirror is

$$f_m = \frac{R}{2} = 15 \text{ cm}$$

Focal length of lens is

$$f_L = \frac{R}{\mu-1} = \frac{30}{0.5} = 60 \text{ cm}$$

Equivalent focal length of combination is given as

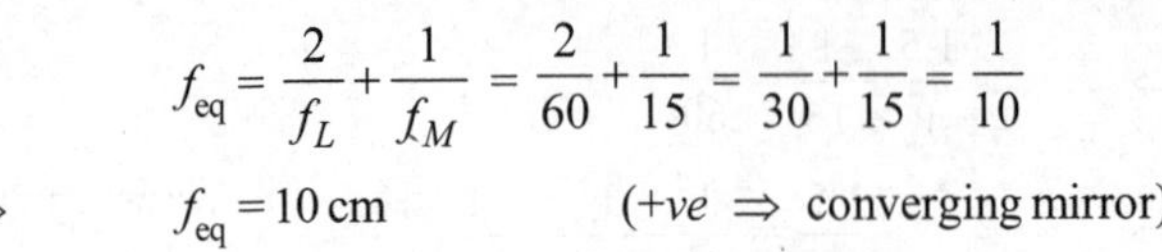

$$f_{eq} = \frac{2}{f_L} + \frac{1}{f_M} = \frac{2}{60} + \frac{1}{15} = \frac{1}{30} + \frac{1}{15} = \frac{1}{10}$$

$$\Rightarrow \quad f_{eq} = 10 \text{ cm} \qquad (+ve \Rightarrow \text{converging mirror})$$

This system acts like a concave mirror of focal length 10 cm

A real image of size of object is produced when object is placed at a distance

$$2f_e = 20 \text{ cm}$$

Ans. (A)

Sol. 2 Given that $P_{air} = -5$D [$-ve$ Diverging lens]

⇒ focal length of lens in air

$$f_{air} = \frac{1}{5} = 0.2 \text{ m} = 20 \text{ cm}$$

By lens makers formula we have

$$P = (\mu-1)\left[\frac{1}{R_1} - \frac{1}{R_2}\right]$$

$$-\frac{5}{100} = (1.5-1)\left[\frac{1}{R_1} - \frac{1}{R_2}\right] \qquad \ldots(1)$$

If focal length of lens in liquid medium is f_L then we use

$$\frac{1}{f_L} = \left(\frac{\mu_g}{\mu_L} - 1\right)\left(\frac{1}{R_1} - \frac{1}{R_2}\right)$$

$$\Rightarrow \quad \frac{1}{f_L} = \left(\frac{1.5}{1.6} - 1\right)\left[\frac{1}{R_1} - \frac{1}{R_2}\right] \qquad \ldots(2)$$

$$\frac{(1)}{(2)} \Rightarrow \quad -\frac{f_L}{20} = -\frac{0.5}{0.1} \times 1.6$$

$$\Rightarrow \quad f_L = 5 \times 20 \times 1.6 = 160 \text{ cm} = 1.6 \text{ m}$$

Power lens in liquid medium

$$P_{liq} = \frac{\mu_L}{f_L} = \frac{1.6}{1.6} = +1D$$

Ans. (A)

Sol. 3 When thin lenses are kept in contact then equivalent power of combination is given as

$$P_{eq} = P_1 + P_2$$

$$= -15D + 5D = 10D$$

equivalent focal length of combination is given as

$$f_{eq} = \frac{1}{P_{eq}} = -\frac{1}{10}$$

$$= -0.1 \text{ m} = -10 \text{ cm}$$

Ans. (B)

Sol. 4 Object and image distances from a lens are related by "Lens formula", given as

$$\frac{1}{v} - \frac{1}{u} = \frac{1}{f} \qquad [u \to \text{object distance from lens}]$$

[$v \to$ image distance from lens]

$$\Rightarrow \qquad v = \frac{uf}{u-f}$$

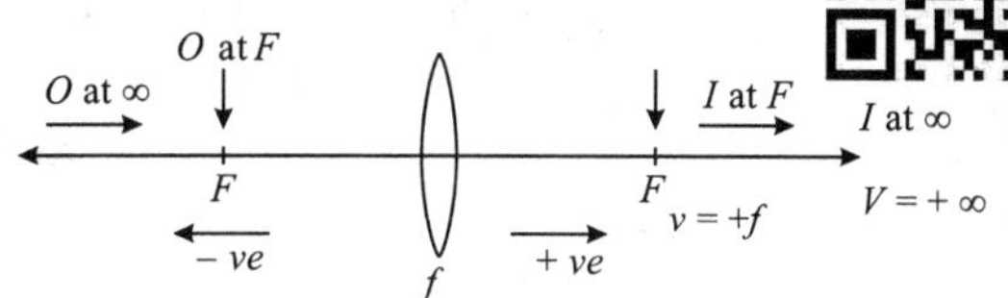

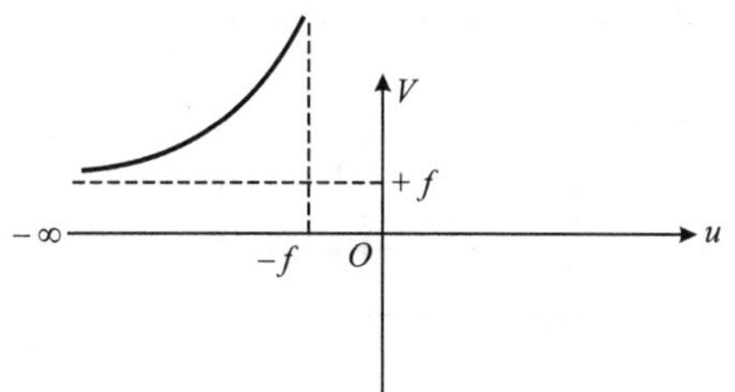

Ans. (C)

Sol. 5 For a convex lens the curve between | u | and | v | is shown as

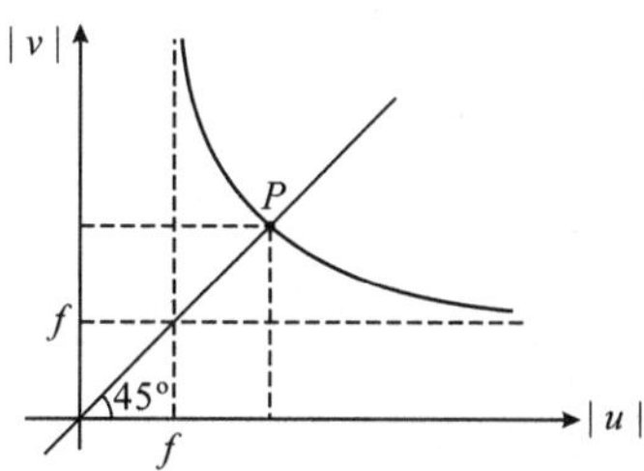

P is graph is a point where object distance and image distance are equal and in case of a convex lens for a real image

$$|u| = |v| = 2f$$

$\Rightarrow \ P \to (2f, 2f)$ **Ans. (D)**

Sol. 6 Case-I :

O ——— 240 cm ——— (lens f) ——— I (Film); u, v = 12cm

Using lens formula, we have

$$u = -240 \text{ cm}$$

$$v = +12 \text{ cm}$$

$$f = ?$$

$$\frac{1}{v} - \frac{1}{u} = \frac{1}{f}$$

$$\Rightarrow \qquad \frac{1}{12} + \frac{1}{240} = \frac{1}{f}$$

$$\Rightarrow \qquad f = \frac{12 \times 240}{252} = \frac{240}{21} \text{ cm}$$

Case-II :

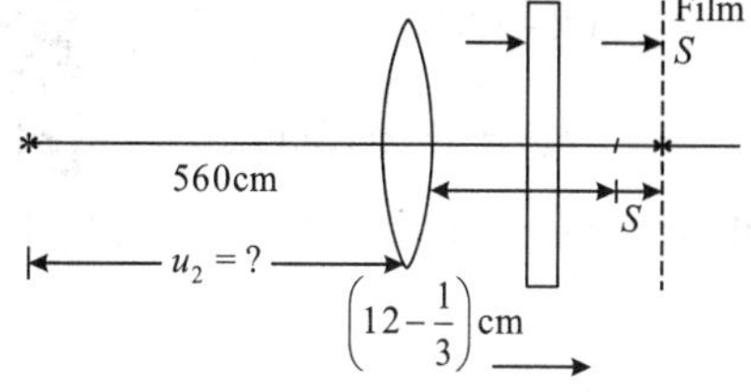

Due to glass plate image produced by lens will be shifted to right by S were

$$S = t\left(1 - \frac{1}{\mu}\right)$$

Shift $$S = 1\left(1 - \frac{1}{3/2}\right) = 1 - \frac{2}{3} = \frac{1}{3} \text{ cm}$$

Lens has to produce the final image at a distance $\left(12 - \frac{1}{3}\right)$cm so as to get sharp image on screen.

For case-II

$$v_2 = +\left(12 - \frac{1}{3}\right) = +\frac{35}{3} \text{ cm}$$

$$f = +\frac{240}{21} \text{ cm}$$

object distance $u_2 = ?$

By lens formula

$$\frac{1}{v} - \frac{1}{u} = \frac{1}{F}$$

$$\frac{3}{35} - \frac{1}{u} = \frac{21}{240}$$

$$\Rightarrow \qquad \frac{1}{u} = \frac{3}{35} - \frac{21}{240}$$

$$\Rightarrow \qquad u = -560 \text{ cm}$$

Ans. (A)

Sol. 7 Refractive index of lens material is

$$\mu = \frac{c}{v} = \frac{3 \times 10^8}{2 \times 10^8} = 1.5$$

From the given information, we have (by figure)

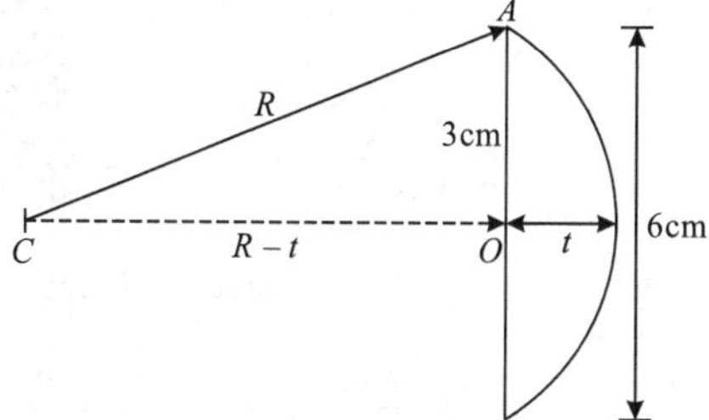

Very thin lines with thickness $t = 3$ mm

In ΔCAO, we use

$$R^2 = (3)^2 + (R - 0.3)^2$$

$$\Rightarrow \quad R^2 = 9 + R^2 + 0.09 - 0.6R$$

$$\Rightarrow \quad 0.6R = 9.09$$

$$\Rightarrow \quad R = \frac{90.9}{6} \simeq 15 \text{ cm}$$

Focal length of a thin lens ($t << R$) is given by lens makes formula as

$$f = \frac{R}{\mu - 1} = \frac{15}{1.5 - 1} = 30 \text{ cm}$$

Ans. (C)

Sol. 8

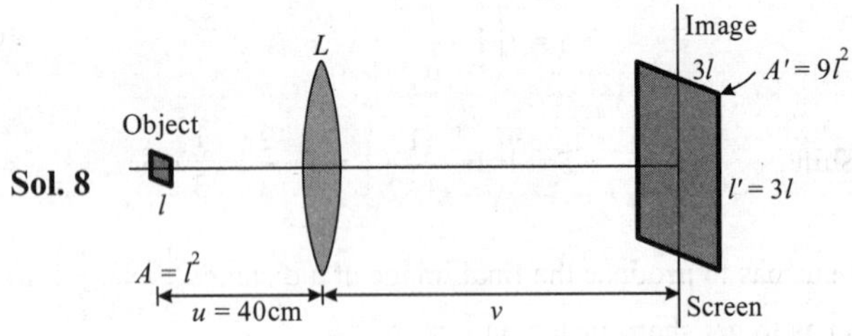

As image size in 3 times the object size

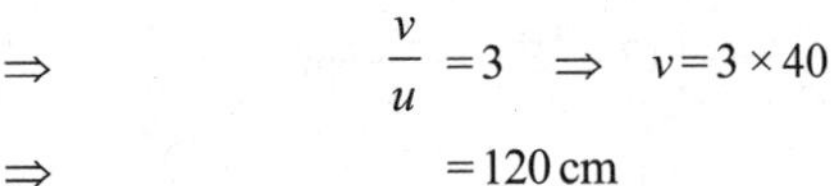

$$\Rightarrow \quad \frac{v}{u} = 3 \Rightarrow v = 3 \times 40$$

$$\Rightarrow \quad = 120 \text{ cm}$$

By lens formula we have

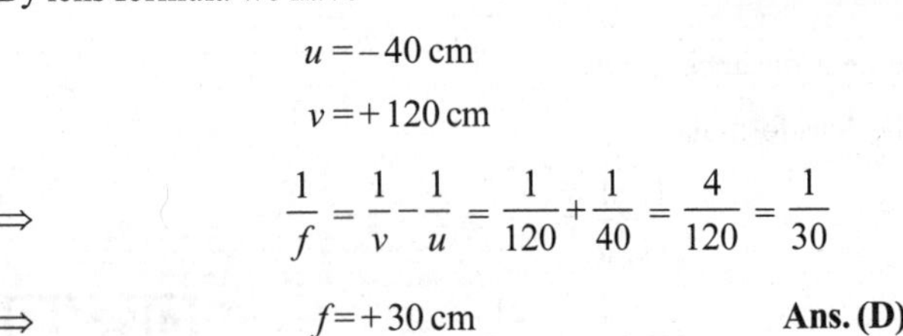

$$u = -40 \text{ cm}$$

$$v = +120 \text{ cm}$$

$$\Rightarrow \quad \frac{1}{f} = \frac{1}{v} - \frac{1}{u} = \frac{1}{120} + \frac{1}{40} = \frac{4}{120} = \frac{1}{30}$$

$$\Rightarrow \quad f = +30 \text{ cm}$$

Ans. (D)

Sol. 9 By lens makes formula

$$\frac{1}{f} = (\mu - 1)\underbrace{\left[\frac{1}{R_1} - \frac{1}{R_2}\right]}_{C}$$

$$\Rightarrow \quad f = \frac{1}{C(\mu - 1)} = \frac{1}{C(3/2 - 1)} = \frac{2}{C}$$

When the lens is submerged in a medium of RI μ_m then we use

$$\frac{1}{f} = \left(\frac{\mu}{\mu_m} - 1\right)C$$

$$\Rightarrow \quad f = \frac{\mu_m}{C(\mu - \mu_m)}$$

For medium having RI $\mu_1 = 4/3$, we have focal length

$$f_1 = \frac{4/3}{C\left(\frac{3}{2} - \frac{4}{3}\right)} = \frac{4/3}{C\left(\frac{1}{6}\right)} = \frac{8}{C}$$

For medium having RI $\mu_2 = 5/3$, we have focal length

$$f_2 = \frac{5/3}{C\left(\frac{3}{2} - \frac{5}{3}\right)} = \frac{5/3}{C\left(-\frac{1}{6}\right)} = -\frac{10}{C}$$

Ans. (B)

Sol. 10

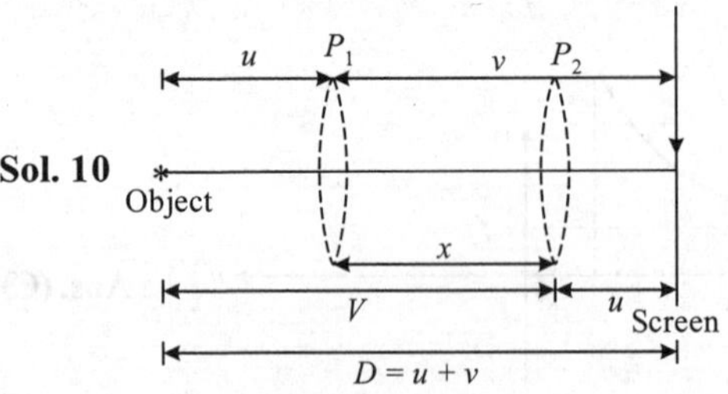

$$I_1 = \frac{v}{u} S_0$$

$$I_2 = \frac{u}{v} S_0$$

The lens displacement is given as

$$x = v - u = 10 \text{ cm} \quad \ldots(1)$$

$$\frac{I_1}{I_2} = \frac{v^2}{u^2} = \frac{3}{2} \quad \ldots(2)$$

from (1) and (2), we use

$$v - \sqrt{\frac{2}{3}}v = 10$$

$$\Rightarrow \quad v = \frac{\sqrt{3}(10)}{\sqrt{3} - \sqrt{2}}$$

$$\Rightarrow \quad u = \sqrt{\frac{2}{3}}(10) \Rightarrow u = \frac{\sqrt{2}(10)}{\sqrt{3} - \sqrt{2}}$$

Separation between object and screen is given as

$$D = u + v$$

$$\Rightarrow \quad D = 10\left(\frac{\sqrt{3} + \sqrt{2}}{\sqrt{3} - \sqrt{2}}\right) \text{cm}$$

$$\Rightarrow \quad D = 10\left(\frac{1.732 + 1.414}{1.732 - 1.414}\right) = 10 \times \frac{3.146}{0.318}$$

$$\Rightarrow \quad D = 98.9 \text{ cm}$$

Ans. (D)

Sol. 11 Method-1 : As we know a lens is submerged in a medium of RI greater then that of lens material then behaviour of lens is changed i.e. convex → converging

Here as $\mu_1 > \mu$, $\Rightarrow$ Concave lens converges parallel light beam.

Ans. (A)

Method-2 : Using lens makers formula

$$\frac{1}{f} = ({}_m\mu_L - 1)\left[\frac{1}{R_1} - \frac{1}{R_2}\right]$$

Here say $|R_1| = |R_2|$ (we consider an equiconcave lens) and

$$_m\mu_L = \frac{\mu}{\mu_1}$$

$$R_1 \rightarrow -ve\,(-R)$$

$$R_2 \rightarrow +ve\,(+R)$$

$$\Rightarrow \quad \frac{1}{f} = \underbrace{\left(\frac{\mu}{\mu_1}-1\right)}_{-ve}\underbrace{\left[-\frac{1}{R}-\frac{1}{R}\right]}_{-ve}$$

As $(\mu < \mu_1)$

$\Rightarrow$ $f \rightarrow +ve$

$\Rightarrow$ Lens behave as a converging lens **Ans. (A)**

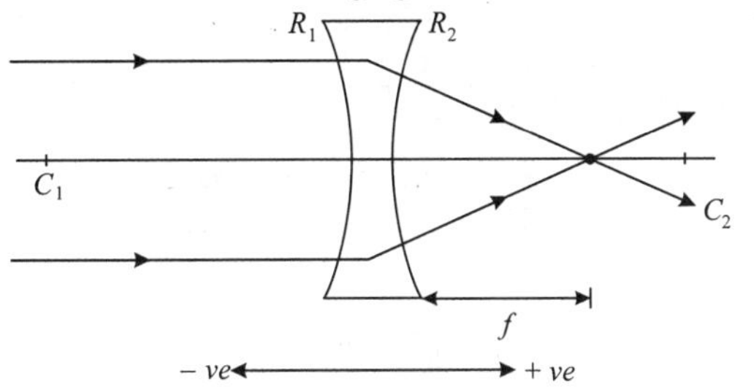

Sol. 12 The lens mirror combination behaves like a spherical mirror of which focal length f_m is given as

$$f_{\text{plain}} = \infty$$

$$f_{\text{lens}} = f$$

$$\frac{1}{f_m} = \frac{2}{f_{\text{lens}}} + \frac{1}{f_{\text{plain}}} = \frac{2}{f}$$

$$\Rightarrow \quad f_m = \frac{f}{2}$$

Using mirror formula we have

$$u = +a$$

$$V = +a/3$$

$$\Rightarrow \quad \frac{1}{f_m} = \frac{1}{u} + \frac{1}{V} = \frac{1}{a} + \frac{3}{a} = \frac{4}{a}$$

$$\Rightarrow \quad \frac{a}{4} = \frac{f}{2}$$

$$\Rightarrow \quad a = 2f$$

Ans. (D)

Sol. 13

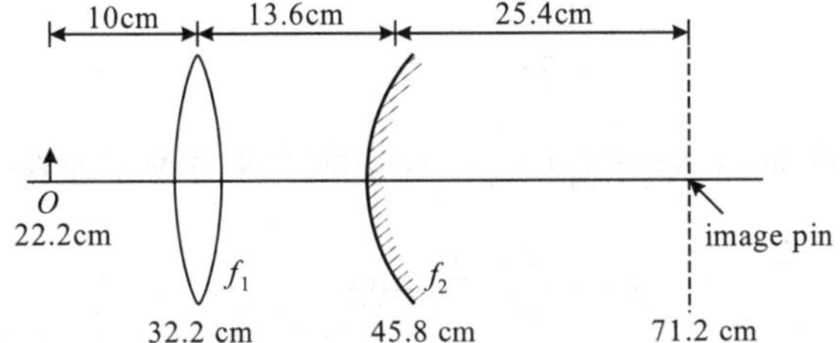

Option (C) and (D) are not possible as $f_2 \neq 25.4$ cm.

In option (A) and (B) $f_2 < 25.4$ cm $\Rightarrow$ Image produced by lens must be on right side of mirror

$\Rightarrow$ for lens $v > 13.6$ cm $\Rightarrow$ Real enlarged image produced by lens

$$\Rightarrow \quad f_1 < 10\text{ cm}$$

As to produce enlarged real image ($|v| > |u|$) object must be located at

$$|f_1| < u < |2f_1|$$

$\Rightarrow$ Only possible option is (A) where $f_1 = 7.8$

& here (B) is also not possible as $f_1 > 10$ cm **Ans. (A)**

Sol. 14 First refraction at convex lens L_1

$u = -60$ cm lens

$f = +30$ cm formula

$$\frac{1}{f} = \frac{1}{v} - \frac{1}{u}$$

$$\Rightarrow \quad \frac{1}{v} = \frac{1}{30} - \frac{1}{60} = \frac{1}{60}$$

$$\Rightarrow \quad v = +60\text{ cm}$$

Second refraction at concave lens L_2 for which object is image, produced by L_1

$u = +40$ cm lens

$f = -120$ cm formula

$$\frac{1}{f} = \frac{1}{v} - \frac{1}{u}$$

$$\Rightarrow \quad \frac{1}{v} = -\frac{1}{120} + \frac{1}{40} = \frac{1}{60}$$

$$\Rightarrow \quad v = +60\text{ cm}$$

Third refraction at L_2 again for which I_3 is the object

$$u = -40\text{ cm}$$

$$f = -120\text{ cm}$$

Using lens formula

$$\frac{1}{f} = \frac{1}{v} - \frac{1}{u}$$

$$\Rightarrow \quad \frac{1}{v} = -\frac{1}{120} - \frac{1}{40} = -\frac{1}{30}$$

$$\Rightarrow \quad v = -30\text{ cm}$$

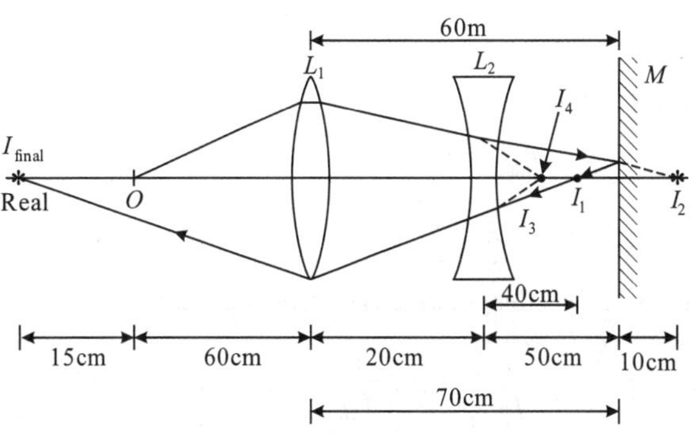

For the refraction at L_1 again for which I_4 is the object

$u = -50$ cm lens

$f = +30$ cm formula

$$\frac{1}{v} = \frac{1}{f} + \frac{1}{u} = \frac{1}{30} - \frac{1}{50} = \frac{5-3}{150} = \frac{1}{75}$$

$\Rightarrow$ $v = 75$ cm

If whole combination ($L_1 + L_2 + M$) is used once then image produced I_3 is located at a distance 60 cm from the convex lens. So most appropriate answer here is option (A). **Ans. (A)**

Sol. 15

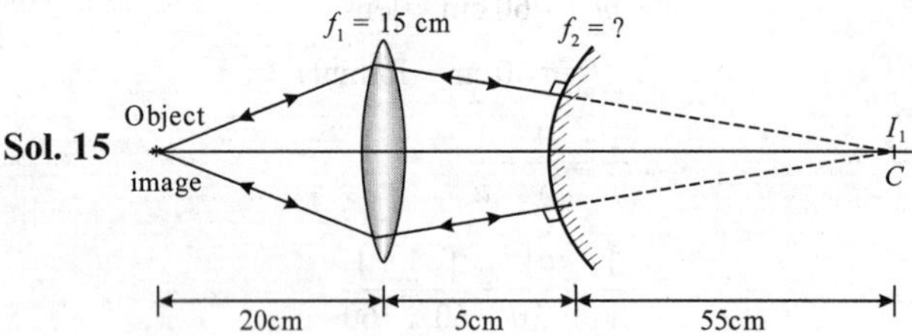

For refraction through lens we use lens formula for which we use

$u = -20$ cm

$f_1 = +15$ cm

$$\Rightarrow \quad \frac{1}{f} = \frac{1}{v} - \frac{1}{u}$$

$$\Rightarrow \quad \frac{1}{v} = \frac{1}{15} - \frac{1}{20} = \frac{1}{60}$$

$\Rightarrow$ $v = 60$ cm

If image is produced on object then it can happen light rays retrace the path of incident ray after reflection or IR at mirror falls normally. For this I_1 must be at centre of curvature of mirror

$\Rightarrow$ $R = 55$ cm

$$\Rightarrow \quad f_2 = \frac{R}{2} = 27.5 \text{ cm}$$ **Ans. (A)**

Sol. 16 If μ be the refractive index of lens and R be its radius of curvature then focal length of a thin plano convex lens is given by lens makers formula as

$$f = \frac{R}{(\mu-1)} \quad \left[\text{As } \frac{1}{f} = (\mu-1)\left[\frac{1}{\underset{R}{R_1}} - \frac{1}{\underset{\infty}{R_2}}\right]\right]$$

Case-I : (Figure-A)

$f_{\text{mirror}} = \infty$

$$f_L = \frac{R}{\mu-1}$$

Effective focal length of lens system is given as

$$\frac{1}{f_{eq}} = \frac{2}{f_L} + \frac{1}{f_M}$$

$$\Rightarrow \quad f_{\text{eq}} = \frac{f_L}{2} = \frac{R}{2(\mu-1)}$$

$$\Rightarrow \quad \frac{R}{2(\mu-1)} = 28 \text{ cm} \qquad \ldots(1)$$

Case-II : (Figure-B)

$$f_M = \frac{R}{2}$$

$$f_L = \frac{R}{2(\mu-1)}$$

Equivalent focal length of lens system is given as

$$\frac{1}{f'_{eq}} = \frac{2}{f_C} + \frac{1}{f_M}$$

$$\Rightarrow \quad \frac{1}{f'_{eq}} = \frac{2(\mu-1)}{R} + \frac{2}{R} = \frac{24}{R}$$

$$\Rightarrow \quad f'_{eq} = \frac{R}{2\mu} = 10 \text{ cm} \qquad \ldots(2)$$

We use $\frac{(1)}{(2)}$

$$\Rightarrow \quad \frac{\mu}{\mu-1} = \frac{28}{10} = 2.8$$

$$\Rightarrow \quad \mu = 2.8\mu - 2.8$$

$$\Rightarrow \quad \mu = \frac{2.8}{1.8} = \frac{14}{9} = 1.55$$ **Ans. (B)**

Sol. 17

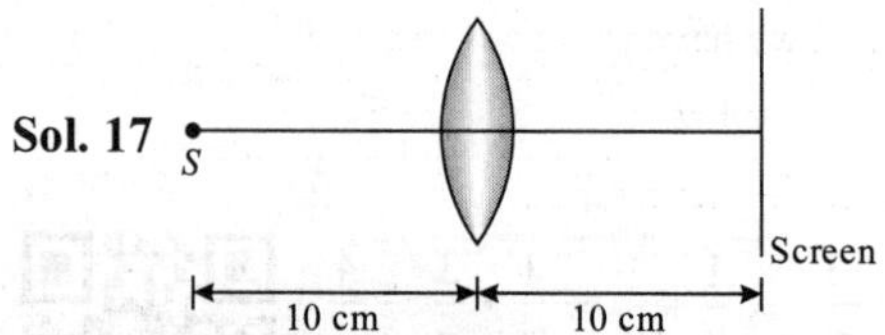

By lens formula, we have

$$\frac{1}{v} - \frac{1}{u} = \frac{1}{f}$$

$$\Rightarrow \quad \frac{1}{10} - \frac{1}{-10} = \frac{1}{f}$$

$\Rightarrow$ $f = 5$ cm

Shift due to slab $= t\left(1 - \frac{1}{\mu}\right)$ in the direction of incident ray

$$\Rightarrow \quad S = 1.5\left(1 - \frac{2}{3}\right) = 0.5$$

Again using lens formula, we have

$$\frac{1}{v} - \frac{1}{-9.5} = \frac{1}{5}$$

$$\Rightarrow \quad \frac{1}{v} = \frac{1}{5} - \frac{2}{19} = \frac{9}{95}$$

$$\Rightarrow \quad v = \frac{95}{9} = 10.55 \text{ cm} \quad \textbf{Ans. (A)}$$

Sol. 18 By lens maker's formula, we have

$$\frac{1}{2f_2} = \frac{1}{f_1} = (\mu_1 - 1)\left(\frac{1}{\infty} - \frac{1}{-R}\right)$$

and $$\frac{1}{f_2} = (\mu_2 - 1)\left(\frac{1}{-R} - \frac{1}{\infty}\right)$$

$$\Rightarrow \quad \frac{(\mu_1 - 1)}{R} = \frac{(\mu_2 - 1)}{2R}$$

$$2\mu_2 - \mu_1 = 1$$

Ans. (B)

Sol. 19 From lens formula, we have

$$\frac{1}{v} - \frac{1}{u} = \frac{1}{f}$$

$$\frac{1}{v} - \frac{1}{(-20)} = \frac{1}{(.3)} = \frac{10}{3}$$

$$\frac{1}{v} = \frac{10}{3} - \frac{1}{20}$$

$$\frac{1}{v} = \frac{197}{60}; v = \frac{60}{197} \text{m}$$

Lateral magnification by lens is

$$m = \left(\frac{v}{u}\right) = \frac{\left(\frac{60}{197}\right)}{20} = \frac{3}{197}$$

Velocity of image w.r.t. to lens is given by

$$v_I = m^2 v_O$$

Direction of velocity of image is same as that of object

As $v_O = 5$ m/s

$$\Rightarrow \quad v_I = \left(\frac{60 \times 1}{197 \times 20}\right)^2 (5)$$

$$\Rightarrow \quad v_I = 1.16 \times 10^{-3} \text{ m/s toward lens}$$

Ans. (C)

Sol. 20 For first lens, using lens formula, we have

$$\frac{1}{v} - \frac{1}{-20} = \frac{1}{5}$$

$$v = \frac{20}{3} \text{ cm}$$

For second lens, using lens formula, we have

$$u_2 = \frac{20}{3} - 2 = \frac{14}{3} \quad \text{(object distance)}$$

$$\Rightarrow \quad \frac{1}{v_2} - \frac{1}{\frac{14}{3}} = \frac{1}{-5}$$

$$\Rightarrow \quad v_2 = 70 \text{ cm}$$

Ans. (D)

Sol. 21 Using lens marker's formula, new focal length of lens will be given as

$$\frac{1}{f} = (\mu_{rel} - 1)\left(\frac{1}{R_1} - \frac{1}{R_2}\right)$$

As focal length of lens will change hence image disappears from the screen. **Ans. (A)**

Sol. 22

Focal length of this combination is given as

$$\frac{1}{f} = \frac{1}{f_1} + \frac{1}{f_2} = \frac{1-\mu_1}{R} + \frac{\mu_2 - 1}{R} = \frac{\mu_2 - \mu_1}{R}$$

Ans. (C)

Sol. 23

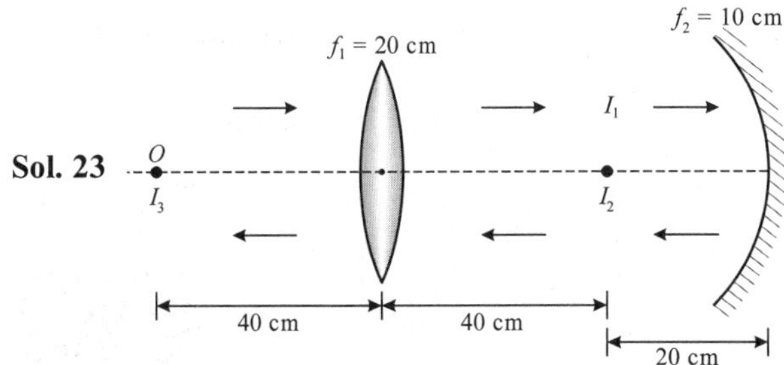

After all refractions and reflection final image is produced at object i.e. 40 cm from lens and of same size.

If I_2 is considered final image after one refraction and one reflection. Then **Ans. (None)**

Safe and most appropriate.

Sol. 24

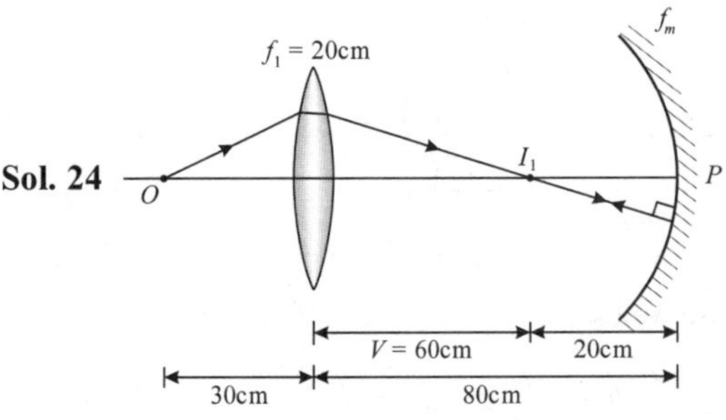

By lens formula we use

$$u = -30 \text{ cm}$$

$$f = +0\,20 \text{ cm}$$

$$\Rightarrow \quad \frac{1}{v} + \frac{1}{30} = \frac{1}{20}$$

$$\Rightarrow \quad V = +60 \text{ cm}$$

I_1 will be producing image at same position in cancave mirror if it is centre of curvature

$$\Rightarrow \quad R_m = 20 \text{ cm}$$

$$f_m = \frac{R_m}{2} = 10 \text{ cm}$$

A cancave mirror produces a virtual image for object to be located from P to F.

$$u_{max} = f_m = 10 \text{ cm}$$ **Ans. (B)**

Sol. 25 Initially we use

$$f_L = \frac{R}{2(\mu - 1)} R$$

We also use $\frac{1}{f_{eq}} = \frac{1}{f_L} + \frac{1}{\infty} + \frac{1}{f_L} = \frac{2}{f_L}$

$$\Rightarrow \quad f_L = 2f_{eq} = 18 \text{ cm} = R$$

After insertion of liquid, the focal length of plano concave lens is $\quad f_{L_2} = \frac{R}{\mu_1 - 1} = \frac{18}{\mu_1 - 1}$ (–ve) as diverging lens

equal focal length for the combination now is

$$\frac{1}{f'_{eq}} = \frac{2}{f_L} + \frac{2}{f_{L_2}} \quad \text{[here we use } 2f'_{eq} = 27 \text{ cm]}$$

$$\frac{1}{27/2} = \frac{2}{18} - \frac{2(\mu_1 - 1)}{18}$$

$$\Rightarrow \quad -\frac{\mu_1 - 1}{18} = \frac{1}{27} - \frac{1}{18} = \frac{2-3}{54} = -\frac{1}{54}$$

$$\Rightarrow \quad \mu_1 - 1 = \frac{1}{3}$$

$$\Rightarrow \quad \mu_1 = 1 + \frac{1}{3} = \frac{4}{3}$$ **Ans. (B)**

Sol. 26

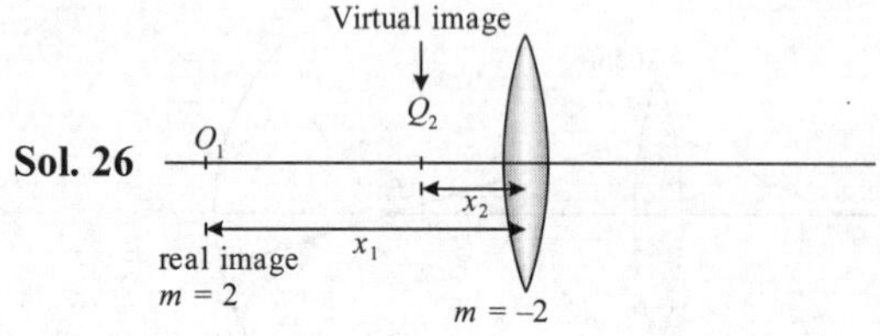

In first case we use

$$m = \frac{f}{x_1 - f} = 2$$

$$\Rightarrow \quad f = 2x_1 - 2f$$

$$\Rightarrow \quad x_1 = \frac{3f}{2} \quad \ldots(1)$$

In second case we use

$$m = \frac{f}{x_2 - f} = -2$$

$$\Rightarrow \quad f = -2x_2 + 2f$$

$$\Rightarrow \quad x_2 = \frac{f}{2} \quad \ldots(2)$$

$$(1)/(2) \quad \Rightarrow \quad \frac{x_1}{x_2} = \frac{3}{1}$$ **Ans. (D)**

Sol. 27 By lens maker's formulas, focal lengths of the two lenses are

$$f_1 = \frac{R}{\mu_1 - 1} \text{ and } f_2 = -\frac{R}{\mu_2 - 1}$$

For combination, we use

$$\frac{1}{f_{eq}} = \frac{1}{f_1} + \frac{1}{f_2} = \frac{\mu_1 - 1}{R} - \frac{\mu_2 - 1}{R} = \frac{\mu_1 - \mu_2}{R}$$

$$\Rightarrow \quad f_{eq} = \frac{R}{\mu_1 - \mu_2}$$ **Ans. (D)**

Sol. 28 By lens formula, we have

$$\frac{1}{v} - \frac{1}{u} = \frac{1}{f}$$

$$\Rightarrow \quad 1 - \frac{v}{u} = \frac{v}{f}$$

$$\Rightarrow \quad 1 - m = \frac{v}{f}$$

$$\Rightarrow \quad m = 1 - \frac{v}{f}$$

At $\quad v = a$,

$$m_1 = 1 - \frac{a}{f}$$

At $\quad v = a + b$,

$$m_2 = 1 - \frac{a+b}{f}$$

$$m_2 - m_1 = c = \left(1 - \frac{a+b}{f}\right) - \left(1 - \frac{a}{f}\right)$$

In magnitude $\quad c = \frac{b}{f}$

$$f = \frac{b}{c}$$ **Ans. (D)**

Sol. 29 Using Lens Maker's Formula, we have

$$\frac{1}{f} = \left(\frac{\mu_2}{\mu_1} - 1\right)\left(\frac{1}{R_1} - \frac{1}{R_2}\right)$$

In air, we use

$$\frac{1}{f}=\left(\frac{1.5}{1}-1\right)\left(\frac{1}{R_1}-\frac{1}{R_2}\right) \quad \ldots(1)$$

In liquid, we use

$$\frac{1}{f_1}=\left(\frac{1.5}{1.42}-1\right)\left(\frac{1}{R_1}-\frac{1}{R_2}\right) \quad \ldots(2)$$

$$\Rightarrow \quad \frac{f_1}{f}=\frac{0.5}{0.056}=8.93\approx 9$$ **Ans. (C)**

Sol. 30 By use of Lens-Maker's formula, we have

$$\frac{1}{f}=(\mu-1)\left(\frac{1}{R_1}-\frac{1}{R_2}\right)$$

$$\Rightarrow \quad \frac{1}{f}=(1.5-1)\left(\frac{1}{30}-0\right)$$

$$\Rightarrow \quad f=60\text{ cm}$$ **Ans. (60)**

Sol. 31 Using displacement method on optical bench we have

$$f=\frac{D^2-d^2}{4D}$$

$$\Rightarrow \quad f=\frac{100^2-40^2}{4(100)}=21\text{ cm}$$

Power of lens is given as

$$P=\frac{1}{f}=\frac{100}{21}D=\left(\frac{N}{100}\right)_D$$

$$\Rightarrow \quad N=\frac{10000}{21}\approx 476$$ **Ans. (476)**

Sol. 32 By lens formula, we have

$$\frac{1}{v}-\frac{1}{u}=\frac{1}{f}$$

For $v=u$ (both positive)

This happens only when object is at pole

For + *ve* u, *ve* is given as

$$\frac{1}{v}=\frac{1}{f}+\frac{1}{u}$$

Thus $|v|<|u|$ so the $u-v$ curve must lie below $u=v$ reference line and at $u\to\infty$, $v=f$ thus option (D) is correct. **Ans. (D)**

Sol. 33 For lens refraction, we use its image produced is I_1 at position v from lens, given as

$$u=-1\text{ m}$$

$$f=+0.5\text{ m}$$

$$\Rightarrow \quad v=\frac{uf}{u+f}=\frac{-1\times 0.5}{-0.5}=1\text{ m}$$

Image I_1 produced by convex lens above will act like object for plane mirror so in mirror its image I_2 is produced at a distance 1 m behind the mirror. (Refer ray diagram given below)

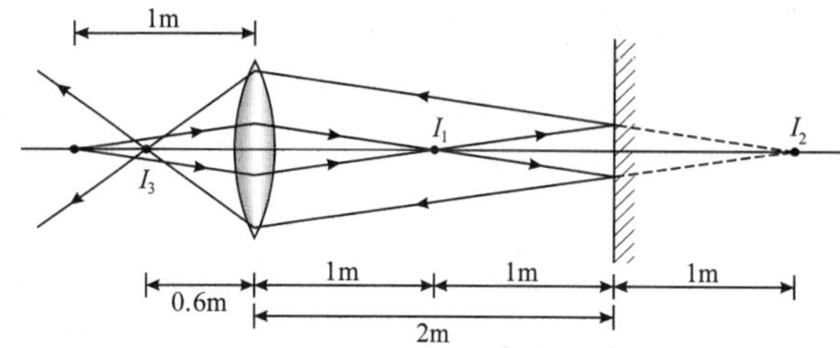

For lens this image I_2 will act as object so again for lens refraction we use

$$u=+3\text{ m}$$

$$f=-0.5\text{ m}$$

$$\Rightarrow \quad v=\frac{uf}{u+f}=\frac{3\times -0.5}{2.5}=-\frac{3}{5}=-0.6\text{ m}$$

This final image is produced to the left of lens at 0.6 m and real. So correct answer is option-(C) **Ans. (C)**

Sol. 34 By lens formula, we have

$$\frac{1}{f}=\frac{1}{v}-\frac{1}{u}$$

Using $u=-f$ and $f=-f$, we have

$$\frac{1}{-f}=\frac{1}{v}-\frac{1}{-f}$$

$$\Rightarrow \quad \frac{-1}{f}=\frac{1}{f}-\frac{1}{v}$$

$$\Rightarrow \quad \frac{-2}{f}=\frac{1}{v}$$

$$\Rightarrow \quad v=-\frac{f}{2}$$

Magnification is given as

$$m=v/u=\frac{-f}{2(-f)}=\frac{1}{2}$$ **Ans. (A)**

Sol. 35 Focal length of the biconvex lenses is given as

$$\frac{1}{f_1}=\frac{1}{15}=\left(\frac{3}{2}-1\right)\left[\frac{2}{R}\right]$$

$$\Rightarrow \quad \frac{1}{R}=\frac{1}{15}$$

If f_{eq} is the equivalent focal length of the combination, it is given as

$$\frac{1}{f_{eq}}=\frac{1}{f_1}+\frac{1}{f_2}+\frac{1}{f_3}$$

$$\Rightarrow \quad \frac{1}{f_{eq}} = \frac{1}{15} + \left(\frac{5}{4} - 1\right)\left[\frac{-2}{R}\right] + \frac{1}{15}$$

$$\Rightarrow \quad \frac{1}{f_{eq}} = \frac{1}{15} - \frac{1}{30} + \frac{1}{15}$$

$$\Rightarrow \quad \frac{1}{f_{eq}} = \frac{2-1+2}{30}$$

$$\Rightarrow \quad \frac{1}{f_{eq}} = \frac{3}{30} = \frac{1}{10}$$

$$\Rightarrow \quad f_{eq} = 10\,\text{cm}$$ **Ans. (10)**

Sol. 36 For the pieces cut as shown in figure the power of each part is shown in the figure below.

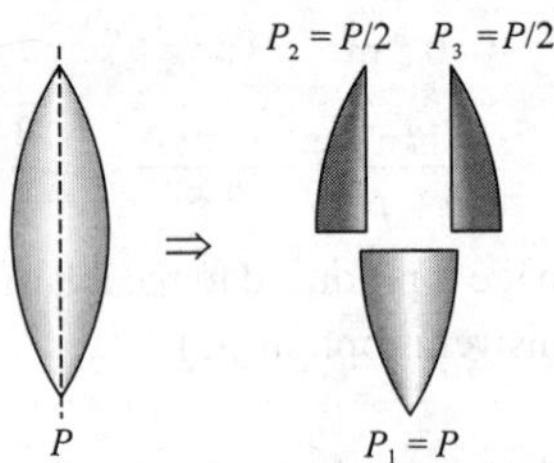

If a lens is cut by a plane along principal axis, every part of lens has same focal length so the lower part has same power as that of the original lens and when a lens is cut by a plane perpendicular to principal axis symmetrically then focal length of each part is double than the original lens so lens power is reduced to half **Ans. (A)**

Sol. 37 Figure below shows the refraction of light as described in the question

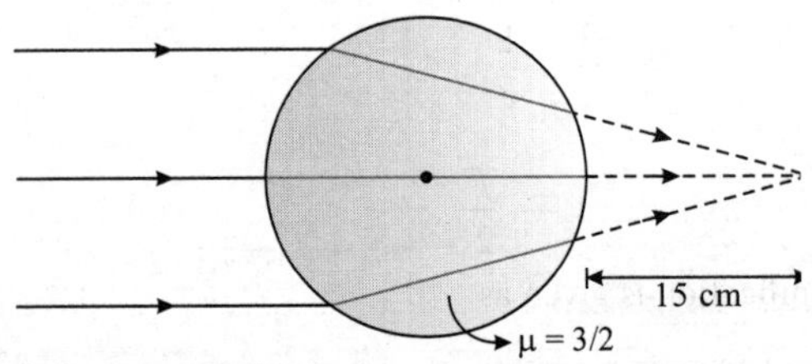

Using refraction formula for the left surface of sphere, we have

$$\frac{\mu_2}{v} - \frac{\mu_1}{u} = \frac{\mu_2 - \mu_1}{R}$$

$$\Rightarrow \quad \frac{\frac{3}{2}}{v} - \frac{1}{\infty} = \frac{\frac{3}{2} - 1}{15}$$

$$\Rightarrow \quad \frac{3}{2v} = \frac{1}{30}$$

$$\Rightarrow \quad v = 45\,\text{cm}$$

Thus after first refraction from the left surface of sphere, light rays will converge at a distance 45 cm from this surface. These light rays will fall again on the right surface of the lens for which we consider virtual object at a distance of 15 cm as shown in figure. Using refraction formula at right surface of lens, now we use

$$\frac{\mu_2}{v} - \frac{\mu_1}{u} = \frac{\mu_2 - \mu_1}{R}$$

$$\Rightarrow \quad \frac{1}{v} - \frac{\frac{3}{2}}{15} = \frac{1 - \frac{3}{2}}{-15}$$

$$\Rightarrow \quad \frac{1}{v} - \frac{1}{10} = \frac{1}{30}$$

$$\Rightarrow \quad \frac{1}{v} = \frac{1}{10} + \frac{1}{30} = \frac{4}{30}$$

$$\Rightarrow \quad v = 7.5\,\text{cm}$$

Thus distance of final image (Converging point of light rays) from centre of sphere is calculated as

$$l = 15 + 7.5 = 22.5\,\text{cm}$$

$$\Rightarrow \quad l = 225\,\text{mm}$$ **Ans. (225)**

Sol. 38 Figure below shows the situation described in question initially.

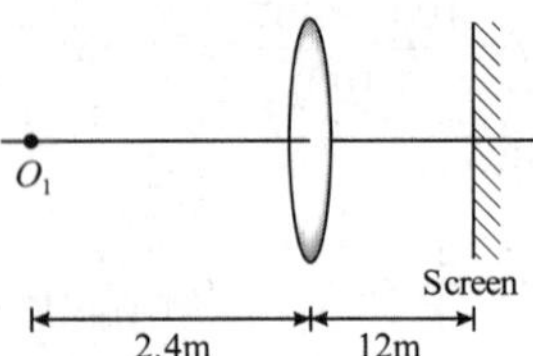

Using lens formula, the focal length of lens is calculated as

$$\frac{1}{0.12} + \frac{1}{2.4} = \frac{1}{f}$$

$$\Rightarrow \quad \frac{1}{f} = \frac{210}{24}\,\text{m}^{-1}$$

After placing the glass slab, shift of image due to glass slab is calculated asas

$$\Delta x = t\left(1 - \frac{1}{\mu}\right) = \frac{1}{3}\,\text{cm}$$

Now distance of final image from lens is calculated as

$$v = 12 - \frac{1}{3} = \frac{35}{3}\,\text{cm} = 0.12\,\text{m}$$

Using lens formula again, we have

$$\frac{1}{0.12} + \frac{1}{u} = \frac{1}{f} = \frac{210}{24}$$

$$\Rightarrow \quad u = -5.6\,\text{m}$$

Thus shift of object to produce image on screen is calculated as

$$\Delta u = 5.6 - 2.4 = 3.2\,\text{m}$$ **Ans. (B)**

Sol. 39 Figure below shows the situation described in the question.

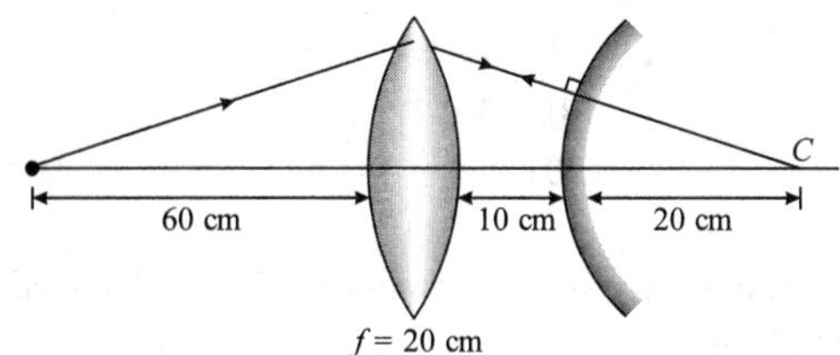

As image is produced on object itself, the light rays must fall on mirror normally so that reflected rays will back trace their own path. Thus for lens, we use lens formula as

$$\frac{1}{v}-\frac{1}{u}=\frac{1}{f}$$

$$\Rightarrow \quad \frac{1}{v}-\frac{1}{(-60)}=\frac{1}{20}$$

$$\Rightarrow \quad \frac{1}{v}+\frac{1}{60}=\frac{1}{20}$$

$$v=30\text{ cm}$$

For final image to be formed on the object itself, this image is to be produced at the centre of curvature of mirror so we have

$$R=30-10=20\text{ cm}$$

Thus focal length of mirror is given as

$$f=R/2=10\text{ cm}$$ **Ans. (10)**

Sol. 40 For point B as shown in graph below, we have

$$\frac{1}{u}=-0.10\text{ cm}^{-1}$$

and $$\frac{1}{v}=0$$

$\Rightarrow$ $$u=-10\text{ cm}$$

and $$v=\infty$$

$\Rightarrow$ $$f=10\text{ cm}$$

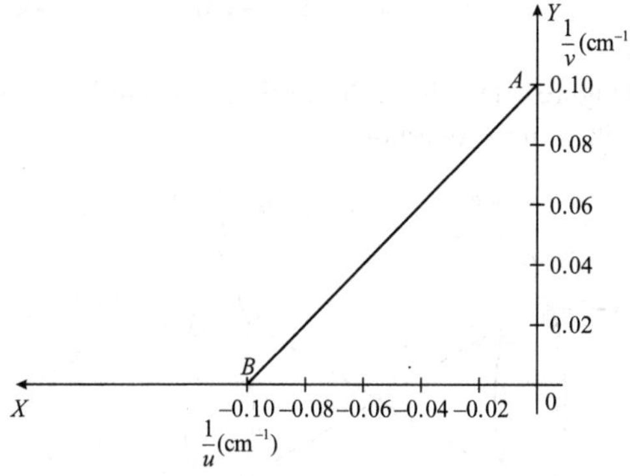

Using lens maker's formula, we have

$$\frac{1}{10}=(1.5-1)\left(\frac{2}{R}\right)=\frac{1}{R}$$

$\Rightarrow$ $$R=10\text{ cm}$$ **Ans. (10)**

Sol. 41 Figure below shows the situation for grazing emergence for the light ray

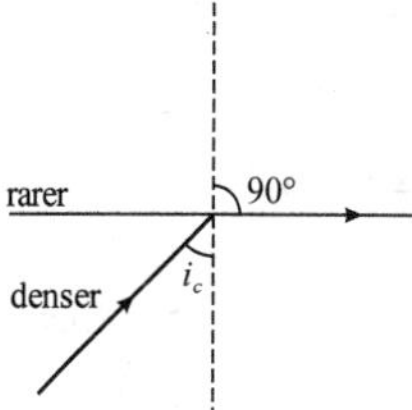

Using Snell's law we have

$$n_d \sin i_c = n_r \sin 90°$$

$$\Rightarrow \quad \sin i_c=\frac{n_r}{n_d}=\frac{v_d}{v_r}$$

$$\Rightarrow \quad \sin i_c=\frac{1.5\times10^8}{2\times10^8}=\frac{1.5}{2}$$

$$\Rightarrow \quad \sin i_c=\frac{3}{4}$$

$$\Rightarrow \quad \tan i_c=\frac{3}{\sqrt{4^2-3^2}}\Rightarrow\frac{3}{\sqrt{7}}$$

$$\Rightarrow \quad i_c=\tan^{-1}\left(\frac{3}{\sqrt{7}}\right)$$ **Ans. (A)**

Sol. 42 For the grazing emergence of light at the face AC, we use Snell's law as

$$1.5\sin 60°=n\times\sin 90°$$

$$\Rightarrow \quad 1.5\times\frac{\sqrt{3}}{2}=n=\frac{\sqrt{x}}{4}$$

$$\Rightarrow \quad 3\sqrt{3}=\sqrt{x}$$

$$\Rightarrow \quad x=27$$ **Ans. (27)**

Sol. 43 Average deviation produced by the combination of prisms is given as

$$\delta=A(\mu_y-1)-A'(\mu_y'-1)$$

$$\Rightarrow \quad \delta=6(1.5-1)-5(1.55-1)$$

$$\Rightarrow \quad \delta=\frac{1}{4}$$ **Ans. (4)**

Sol. 44 Power of a lens is a medium is given as

$$P=\frac{\mu_m}{f}=(\mu_L-\mu_m)\left(\frac{1}{R_1}-\frac{1}{R_2}\right)$$

Here μ_L is refractive index of lens and μ_m is that of surrounding medium

$$1.25=(1.5-\mu_m)\left(\frac{1}{0.2}+\frac{1}{0.4}\right)$$

$$\Rightarrow \quad \frac{1.25\times0.08}{0.6}=(1.5-\mu_m)$$

$$\Rightarrow \quad \mu_m=\frac{4}{3}$$ **Ans. (B)**

Sol. 45 By Newton's formula, in which all distances of object and image are measured from focal point of the lens, we use

$$vu = f^2$$

$$\Rightarrow \quad f^2 = 225$$

$$\Rightarrow \quad f = 15 \text{ cm}$$ **Ans. (15)**

Sol. 46 By lens makers formula the focal length of first lens is given as

$$\frac{1}{f_1} = (1.75-1)\left(-\frac{1}{30}\right)$$

$$\Rightarrow \quad f_1 = -40 \text{ cm}$$

For second lens, we use

$$\frac{1}{f_2} = (1.75-1)\left(\frac{1}{30}\right)$$

$$\Rightarrow \quad f_2 = 40 \text{ cm}$$

Image from L_1 will be virtual and produced on the left of L_1 at focal length 40 cm. So the object for L_2 will be 80 cm from L_2 which is located at its $2f$. Thus final image is produced by second lens at 80 cm from L_2 on the right of it so we have

$$x = 120$$ **Ans. (120)**

Sol. 47 Focal length of the biconvex lens immersed in water is calculated as

$$\frac{1}{f_m} = \left(\frac{\mu_g}{\mu_{H_2O}} - 1\right)\left(\frac{2}{R}\right)$$

$$\Rightarrow \quad \frac{1}{f_m} = \frac{1}{8}\left(\frac{2}{R}\right)$$

$$\Rightarrow \quad \frac{1}{f_m} = \frac{1}{4f_{air}}$$

$$\Rightarrow \quad f_m = 4f_{air} = 72 \text{ cm}$$

Thus change in focal length is calculated as

$$\Delta f = 72 - 18 = 54 \text{ cm}$$ **Ans. (54)**

Sol. 48 Figure below shows the refraction of light rays from object through the given lens. This shows that the image is produced by lens at 15 cm from the lens

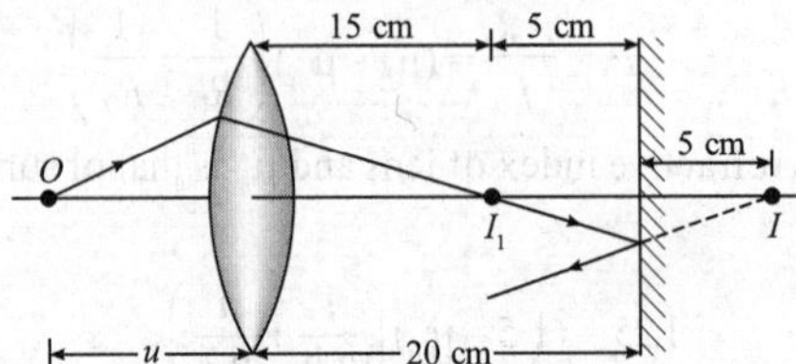

Using lens formula for the lens, we have

$$f = 10 \text{ cm}$$

$$\Rightarrow \quad \frac{1}{v} - \frac{1}{u} = \frac{1}{f}$$

$$\Rightarrow \quad \frac{1}{15} - \frac{1}{-u} = \frac{1}{10}$$

$$\Rightarrow \quad \frac{1}{u} = \frac{1}{10} - \frac{1}{15}$$

$$\Rightarrow \quad u = 30 \text{ cm}$$ **Ans. (30)**

Sol. 49 From I[st] lens image formation can be calculated using lens formula as

$$\frac{1}{v} + \frac{1}{6} = \frac{1}{24}$$

$$\Rightarrow \quad \frac{1}{v} = \frac{1}{24} - \frac{1}{6} = -\frac{1}{8}$$

$$\Rightarrow \quad v = -8 \text{ cm}$$

From IInd lens successive image formation can be calculated by using lens formula as

$$\frac{1}{v} + \frac{1}{18} = \frac{1}{9}$$

$$\Rightarrow \quad \frac{1}{v} = \frac{1}{18}$$

$$\Rightarrow \quad v = 18 \text{ cm}$$

Final ray diagram of the image formation is shown in figure below.

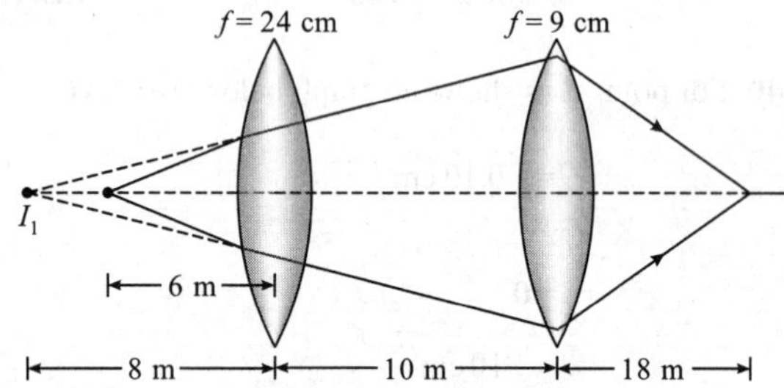

Thus distance between object and its image can be calculated as

$$d = 6 + 10 + 18 = 34 \text{ cm}$$ **Ans. (34)**

Sol. 50 Figure below shows the ray diagram of image formation as described in the question.

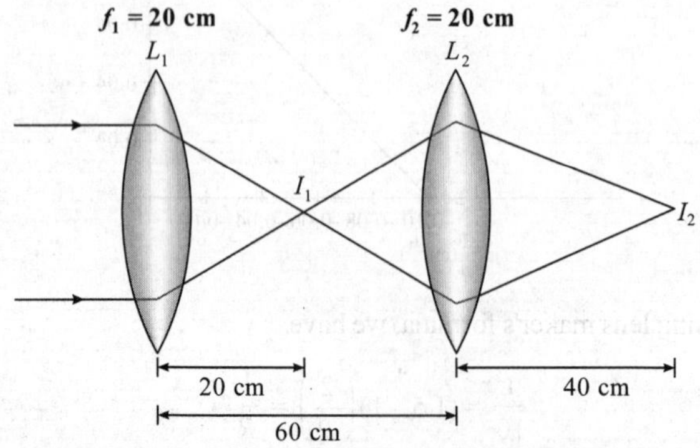

Here 1st refraction in L_1 for formation of I_1, using lens formula we have

$$\frac{1}{v}-\frac{1}{u}=\frac{1}{f}$$

As light rays are parallel so we use $u=-\infty$

$$\Rightarrow \qquad \frac{1}{v}-\frac{1}{\infty}=\frac{1}{f}$$

$$\Rightarrow \qquad v=f$$

For 2nd refraction in L_2 we use $I_1\rightarrow$ object and $I_2\rightarrow$ image so we have

$$u=-40 \text{ cm}$$

and $$f=20 \text{ cm}$$

Using lens formula, we have

$$\frac{1}{v}-\frac{1}{u}=\frac{1}{f}$$

$$\Rightarrow \qquad \frac{1}{v}-\frac{1}{(-40)}=\frac{1}{20}$$

$$\Rightarrow \qquad \frac{1}{v}=\frac{1}{20}-\frac{1}{40}=\frac{6-3}{120}$$

$$\Rightarrow \qquad \frac{1}{v}=\frac{3}{120}=\frac{1}{40}$$

$$\Rightarrow \qquad v=40 \text{ cm}$$

Thus distance final image from first lens is 100 cm. **Ans. (100)**

Sol. 51 Figure below shows the lens cut in two identical parts along the dotted line as described in question.

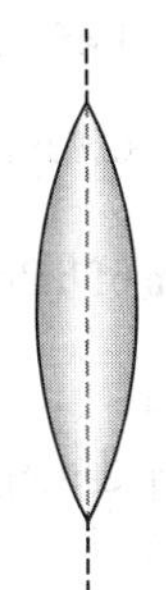

If power of each part is P_1, then we use

$$P_1+P_1=P=\frac{1}{f}$$

$$\Rightarrow \qquad 2P_1=\frac{1}{0.1}=10$$

$$\Rightarrow \qquad P_1=5\text{ D}$$ **Ans. (5)**

23.4 Dispersion of Light

Sol. 1 For a prism of refracting angle A, the refractive index 'μ' and minimum deviation angle D' are related as

$$\mu=\frac{\sin\left(\frac{A+D}{2}\right)}{\sin\left(\frac{a}{2}\right)}$$

As μ increases for a given prism $\Rightarrow$ D increases

here $\mu_R=1.520\rightarrow D_1$

and $\mu_B=1.525\rightarrow D_2$

As $\mu_B>\mu_R$

$\Rightarrow$ $D_2>D_1$ **Ans. (A)**

Sol. 2 We know that refractive index of a material, is related to light wavelength as

$$\mu=A+\frac{B}{\lambda^2}$$

So for blue light $\lambda_B<\lambda_R$

$\Rightarrow$ $\mu_B>\mu_R$

By lens makers formula, we use

$$\frac{1}{f}=(\mu-1)\left[\frac{1}{R_1}-\frac{1}{R_2}\right]$$

As $\mu_B>\mu_R$

$\Rightarrow$ $f_B<f_R$ **Ans. (B)**

Sol. 3

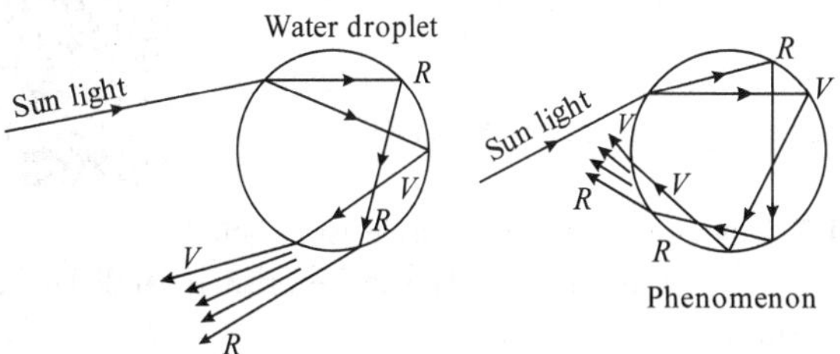

When light enters & emerges out from drop $\rightarrow$ refraction

At inner surface of drop $\rightarrow$ Internal reflection

inside the drop $\rightarrow$ dispersion.

Here internal reflection is not total internal reflection but for the given options most appropriate is option (A) only. **Ans. (A)**

Sol. 4

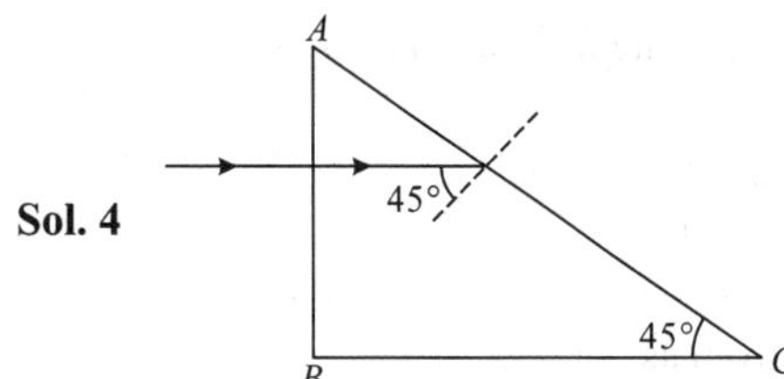

For *TIR* at face *AC*

$$\theta_C<45^\circ$$

$$\Rightarrow \qquad \sin\theta_C<\frac{1}{\sqrt{2}}$$

$$\Rightarrow \qquad \frac{1}{\mu}<\frac{1}{\sqrt{2}} \Rightarrow \mu>\sqrt{2}$$

Those colours which have $\mu > 1.414$ will suffer total internal reflection at face AC and will not come out.

$\Rightarrow$ Green & Blue colour light will not come out. **Ans. (D)**

Sol. 5 For doublet lenses at a separation d and with focal lengths f_1 and f_2 the equivalent focal length is given as

$$\frac{1}{f_{eq}} = \frac{1}{f_1} + \frac{1}{f_2} - \frac{d}{f_1 f_2}$$

Here $f_{eq} = 10\text{ cm}$

and $d = 2\text{ cm}$

$$\Rightarrow \quad \frac{1}{10} = \frac{1}{f_1} + \frac{1}{f_2} - \frac{2}{f_1 f_2} = \left(\frac{f_2 + f_1 - 2}{f_1 f_2}\right)$$

For option (A) $f_1 = 18\text{ cm}$ & $f_2 = 20\text{ cm}$

$$\Rightarrow \quad \frac{1}{10} = \frac{20 + 18 - 2}{18 \times 20} = \frac{36}{18 \times 20} = \frac{1}{10}$$

$\Rightarrow$ LHS = RHS **Ans. (A)**

Sol. 6 As for thin prism, minimum deviation is

$$D_m = (\mu - 1)\text{A}$$

So on increasing the wavelength, μ decreases & hence D_m decreases. **Ans. (B)**

Sol. 7 Deviation is minimum in a prism when :

$i = e$, $r_1 = r_2$ and refracted ray inside prism is parallel to base of prism.

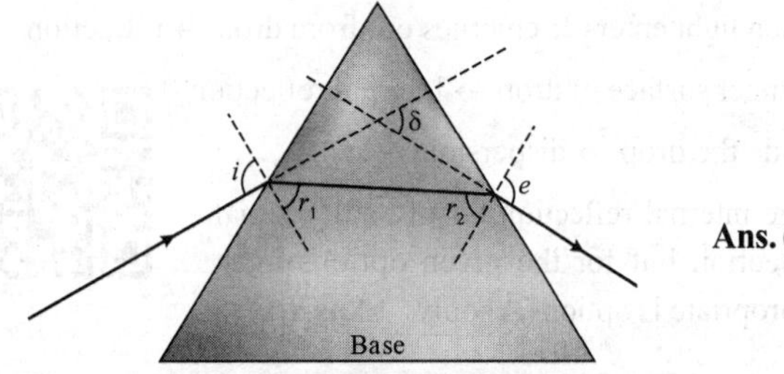

Ans. (A)

Sol. 8 For Achromatic combination of prisms, we use

$$\theta_{net} = 0$$

$$\Rightarrow \quad \theta_1 - \theta_2 = 0$$

$$\Rightarrow \quad \theta_1 = \theta_2$$

$$\Rightarrow \quad \omega_1 \delta_1 = \omega_2 \delta_2$$

Total deviation is given as

$$\delta_{net} = \delta_1 - \delta_2 = 2^\circ$$

$$\Rightarrow \quad \delta_1 - \frac{\omega_1 \delta_1}{\omega_2} = 2^\circ$$

$$\Rightarrow \quad \delta_1\left(1 - \frac{\omega_1}{\omega_2}\right) = 2^\circ$$

$$\Rightarrow \quad \delta_1\left(1 - \frac{2}{3}\right) = 2^\circ$$

$$\Rightarrow \quad \delta_1 = 6^\circ$$

For small angle prism, deviation of light is given as

$$\delta_1 = (\mu_1 - 1) A_1$$

$$\Rightarrow \quad 6^\circ = (1.5 - 1) A_1$$

$$\Rightarrow \quad A_1 = 12^\circ$$

Ans. (12)

Sol. 9 The refractive index of prism is given as

$$\mu = \frac{\sin\left(\frac{A + \delta_m}{2}\right)}{\sin\left(\frac{A}{2}\right)}$$

$$\Rightarrow \quad \mu = \frac{\sin A}{\sin A/2}$$

$$\Rightarrow \quad \mu = 2\cos\frac{A}{2}$$

$$\Rightarrow \quad A = 2\cos^{-1}\left(\frac{\mu}{2}\right)$$

Ans. (D)

Sol. 10 For without deviation passing at interface, we must have same refractive index on the two sides of the interface, so we use

$$n_1 = n_2$$

$$\Rightarrow \quad 1.2 + \frac{1.8 \times 10^{-14}}{\lambda^2} = 1.45 + \frac{1.8 \times 10^{-14}}{\lambda^2}$$

$$\Rightarrow \quad \frac{(10.8 - 1.8) \times 10^{-14}}{\lambda^2} = 1.45 - 1.2 = 0.25$$

$$\Rightarrow \quad \lambda = 600\text{ nm}$$

Ans. (600)

Sol. 11 Minimum deviation of light through the prism is given as

$$\delta_{min} = 2i - A$$

As per given condition in question, we use

$$A = 2i - A$$

Refractive index of prism in terms of minimum deviation angle is given as

$$\mu = \frac{\sin\left(\frac{\delta_{min} + A}{2}\right)}{\sin\left(\frac{A}{2}\right)} = \frac{\sin(A)}{\sin\left(\frac{A}{2}\right)} = \frac{\sin 60^\circ}{\sin 30^\circ}$$

$$\Rightarrow \quad \mu = \sqrt{3}$$

Velocity of light in prism is given as

$$v = \frac{c}{\mu} = \frac{3 \times 10^8}{\sqrt{3}}\text{ m/s}$$

Time taken by light to cover the distance AP is given as

$$t = \frac{AP}{V} = \frac{5\sqrt{3} \times 10^{-3}}{\left(\frac{3 \times 10^8}{\sqrt{3}}\right)} = 5 \times 10^{-10}\text{ s}$$

Ans. (5)

Sol. 12 In primary rainbow, red colour is at top and violet is at bottom and intensity of secondary rainbow is less in comparison to primary rainbow. **Ans. (A)**

Sol. 13 Due to different colours the refractive index of lens material which is different for different colours, focal points for each colour is different thats why instead of a single focal point, for white light blurred image is formed at focus. This distortion in focus due to multiple wavelengths in incident light is called chromatic aberration. **Ans. (B)**

23.5 Optical Instruments

Sol. 1 Resolving power of an astronomical telescope is given as

$$R_\lambda = \frac{D}{1.22\lambda}$$

$D \rightarrow$ Aperture diameter of objective lens

$\lambda \rightarrow$ Wavelength of light

An astronomical telescope is use to observe distant objects and heavenly bodies so resolution required is high so keep high resolving power D must be kept large. **Ans. (B)**

Sol. 2 For an optical instrument we know that

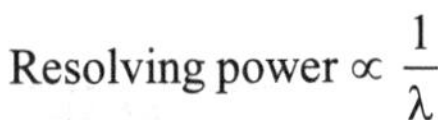
Resolving power $\propto \frac{1}{\lambda}$

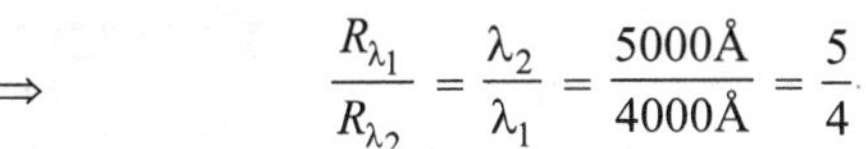
$$\Rightarrow \quad \frac{R_{\lambda_1}}{R_{\lambda_2}} = \frac{\lambda_2}{\lambda_1} = \frac{5000\text{Å}}{4000\text{Å}} = \frac{5}{4}$$ **Ans. (D)**

Sol. 3 As object is placed in front of objective lens between F and $2F$ points $\Rightarrow$ image produced is real and enlarged.

Ans. (C)

Sol. 4 Angular width of resolution

$$\theta = \frac{1.22\lambda}{D} = \frac{d}{x}$$

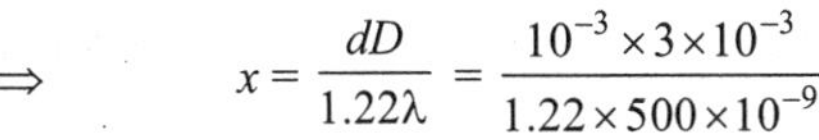
$$\Rightarrow \quad x = \frac{dD}{1.22\lambda} = \frac{10^{-3} \times 3 \times 10^{-3}}{1.22 \times 500 \times 10^{-9}}$$

$$\Rightarrow \quad x = \frac{3}{1.22 \times 0.5} = 4.92\,\text{m} = 5\,\text{m}$$ **Ans. (A)**

Sol. 5 For a travelling microscope readings are taken from meter scale on the vertical mount and vernier scale on the microscope.

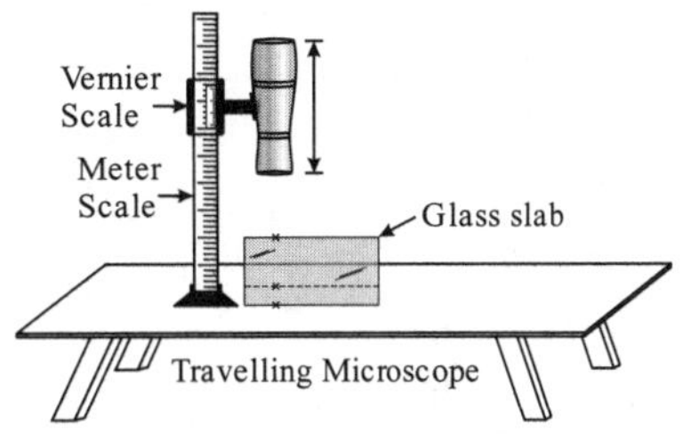

Ans. (A)

Sol. 6 For normal adjustment (to obtain final image at ∞) the image due to objective lens must be obtained at focal point of eyepiece lens as shown, lens formula for objective

$$f_e = +20\,\text{mm}$$

$$u_e = -25\,\text{mm}$$

$$\Rightarrow \quad \frac{1}{f_e} = \frac{1}{v_e} - \frac{1}{u_e}$$

$$\Rightarrow \quad \frac{1}{v_e} = \frac{1}{f_e} + \frac{1}{u_e} = \frac{1}{20} - \frac{1}{25} = \frac{1}{100}$$

$$\Rightarrow \quad v_e = 100\,\text{mm}$$

lens separation $= f_e + v_e$

$= 20\,\text{mm} + 100\,\text{mm} = 120\,\text{mm}$ **Ans. (C)**

Sol. 7 Limit of resolution of given telescope is

$$\theta_L = \frac{1.22\lambda}{D} = \frac{1.22 \times 5.5 \times 10^{-7}}{3 \times 10^{-2}}$$

$$\Rightarrow \quad \theta_L = 2.33 \times 10^{-5}\,\text{rad}$$

The angle subtended by the given wire mesh on telescope is

$$\theta = \frac{x}{L} = \frac{2 \times 10^{-3}}{80} = 2.5 \times 10^{-5}\,\text{rad}$$

As $\theta > \theta_L$ so the mesh spacing will be clearly seen. **Ans. (A)**

Sol. 8 Lens formula for objective lens

$$u_0 = -200\,\text{cm}$$

$$f_0 = +50\,\text{cm}$$

$$\Rightarrow \quad \frac{1}{f_0} = \frac{1}{v_0} - \frac{1}{u_0}$$

$$\Rightarrow \quad \frac{1}{50} = \frac{1}{v_0} + \frac{1}{200}$$

$$\Rightarrow \quad \frac{V}{v_0} = \frac{1}{50} - \frac{1}{200} = \frac{3}{200}$$

$$\Rightarrow \quad v_0 = \frac{200}{3}\,\text{cm}$$

Lens formula for eyepiece lens

$$u_e = -u_e$$

$$v_e = -D = -25\,\text{cm}$$

$$\Rightarrow \quad \frac{1}{f_e} = \frac{1}{v_e} - \frac{1}{u_e}$$

$$\Rightarrow \quad \frac{1}{5} = -\frac{1}{25} + \frac{1}{u_e}$$

$$\Rightarrow \quad \frac{1}{u_e} = \frac{1}{5} + \frac{1}{25} = \frac{6}{25}$$

$$u_e = \frac{25}{6}\text{ cm}$$

Magnification produced by objective lens

$$M_0 = \frac{v_0}{u_0} = \frac{+200/3}{-200} = -\frac{1}{3}$$

Magnification produced by objective lens

$$M_e = \frac{v_e}{u_e} = \frac{-25}{-25/6} = +6$$

Total magnification by telescope is

$$M_T = M_0 \times M_e = -\frac{1}{3} \times 6 = -2 \qquad \textbf{Ans. (D)}$$

Sol. 9 S-1: True as reflecting telescopes are easier to construct due to handling of spherical mirrors. As only one side of mirror is used so on other side the mirror can be mounted where in case of lenses mounting can only be done at thin edges which is practically difficult. Thats why very large size telescopes are made by using mirrors instead of lenses

⇒ **S-2 :** It is true and explaining S-1. **Ans. (C)**

Sol. 10 For standard adjustment (image at D) of a telescope the magnifying power is given as

$$M = \frac{f_0}{f_e}\left(1 - \frac{f_e}{D}\right)$$

$$\Rightarrow \quad M = \frac{30}{3}\left(1 - \frac{3}{25}\right)$$

$$\Rightarrow \quad M = \frac{30}{3} \times \frac{22}{25} = \frac{44}{5} = 8.8 \qquad \textbf{Ans. (D)}$$

Sol. 11 For normal adjustment (final image at ∞) the magnifying power of microscope is given as

$$M = \left|\frac{V_0}{u_0} \cdot \frac{D}{f_e}\right| \qquad \ldots(1)$$

by using lens formula for objective lens, we have

$$\frac{1}{f_0} = \frac{1}{v_0} - \frac{1}{u_0}$$

$$\Rightarrow \quad \frac{1}{1.2} = \frac{1}{v_0} + \frac{1}{1.25}$$

$$\Rightarrow \quad \frac{1}{v_0} = \frac{1}{1.2} - \frac{1}{1.25} = \frac{1}{30}$$

$$\Rightarrow \quad v_0 = 30\text{ cm}$$

From-(1), we have

$$M = \frac{30}{1.25} \times \frac{25}{3} = 200 \qquad \textbf{Ans. (A)}$$

Sol. 12 $D = 0.25 \times 2$

$$\Rightarrow \quad D = 0.5\text{ cm}$$

We use for limit of resolution we have

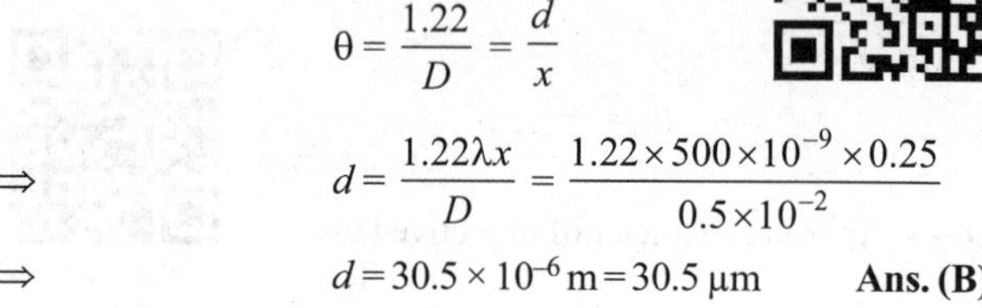

$$\theta = \frac{1.22}{D} = \frac{d}{x}$$

$$\Rightarrow \quad d = \frac{1.22\lambda x}{D} = \frac{1.22 \times 500 \times 10^{-9} \times 0.25}{0.5 \times 10^{-2}}$$

$$\Rightarrow \quad d = 30.5 \times 10^{-6}\text{ m} = 30.5\ \mu\text{m} \qquad \textbf{Ans. (B)}$$

Sol. 13 Magnifying power of telescope

$$M = \frac{\theta}{\alpha} = \frac{\text{Angle subtanded by image}}{\text{Angle subtanded by object}}$$

For normal setting we consider final image at ∞

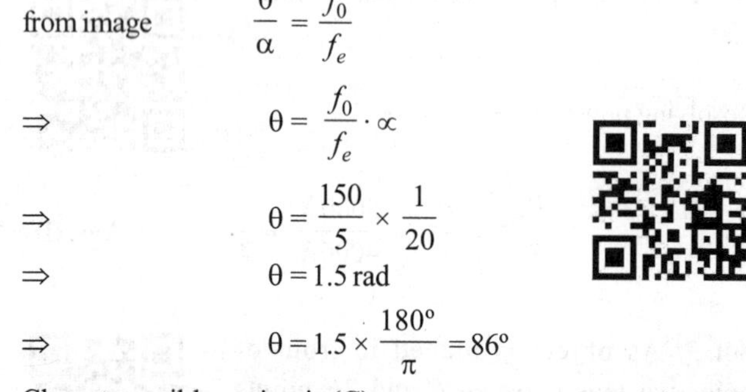

from image $\dfrac{\theta}{\alpha} = \dfrac{f_0}{f_e}$

$$\Rightarrow \quad \theta = \frac{f_0}{f_e} \cdot \alpha$$

$$\Rightarrow \quad \theta = \frac{150}{5} \times \frac{1}{20}$$

$$\Rightarrow \quad \theta = 1.5\text{ rad}$$

$$\Rightarrow \quad \theta = 1.5 \times \frac{180^\circ}{\pi} = 86^\circ$$

Closest possible option is (C). **Ans. (C)**

Sol. 14 As $m = \dfrac{\text{Image size}}{\text{object size}} = 20$

Angular magnification for normal condition

Angular size

Image size (Apparent size observer is scaling)
= 20 × object size **Ans. (C)**

Sol. 15 Step-I : A reading of bottom surface = 'a' without slab

Step-II : A reading of bottom surface with the slab = 'b'

Step-III : A reading of top surface of slab = 'c'

⇒ Thickness of slab $t = a - c$ and we use

$$(a - b) = (a - c)\left[1 - \frac{1}{\mu}\right]$$

$$\Rightarrow \quad 1-\left(\frac{a-b}{a-c}\right)=\frac{1}{\mu}$$

$$\Rightarrow \quad \mu=\left(\frac{a-c}{b-c}\right)$$

Thus readings *a*, *b* and *c* are required. **Ans. (C)**

Sol. 16 Limit of resolution of telescope is given as

$$\theta=\frac{1.22\lambda}{D}$$

$$\Rightarrow \quad \theta=\frac{1.22\times 500\times 10^{-9}}{200\times 10^{-2}}$$

$$\Rightarrow \quad \theta=305\times 10^{-9}\text{ rad}$$

Ans. (A)

Sol. 17 Limit of resolution for objective of telescope is

$$\theta=\frac{1.22\lambda}{d}$$

$$\Rightarrow \quad \theta=\frac{1.22\times 600\times 10^{-9}}{250\times 10^{-2}}$$

$$\Rightarrow \quad \theta=2.9\times 10^{-7}\text{ rad}$$

Ans. (C)

Sol. 18 Numerical aperature of the microscope is given as

$$NA=\frac{0.61\lambda}{d}$$

Where d = minimum sparation between two points to be seen as distinct

$$\Rightarrow \quad d=\frac{0.61\lambda}{NA}=\frac{(0.61)\times(5000\times 10\text{m}^{-10})}{1.25}$$

$$\Rightarrow \quad d=2.4\times 10^{-7}\text{ m}$$

$$\Rightarrow \quad d=0.24\ \mu\text{m}$$

Ans. (A)

Sol. 19 Magnification of compound microscope is

$$m=\frac{LD}{f_e\times f_0}=\frac{150\times 250}{f_e\times 25}=375$$

$$\Rightarrow \quad f_e=20\text{ mm}$$

Ans. (A)

Sol. 20 Tube length is taken as

$$L=f_0+f_e=60\text{ cm}$$

and

$$M=\frac{f_0}{f_e}=5$$

$$\Rightarrow \quad f_0=5f_e$$

$$\Rightarrow \quad 6f_e=60\text{ cm}$$

$$\Rightarrow \quad f_e=10\text{ cm}$$

Ans. (D)

Sol. 21 If the distance between objects on moon is l and D is diameter of telescope and d = distance between earth and moon, then we use

$$\theta=\frac{1.22\lambda}{D}$$

$$\Rightarrow \quad \frac{l}{d}=\frac{1.22\lambda}{D}$$

$$\Rightarrow \quad l=\frac{1.22\times(5500\times 10^{-10})\times(4\times 10^{8})}{5}=53.68\text{ m}$$

Here most appropriate option is (C) **Ans. (C)**

Sol. 22 If object is placed at a distance x from objective, its image is produced at for objective

$$\frac{1}{v}+\frac{1}{x}=1$$

$$\Rightarrow \quad v=\frac{x}{x-1}$$

and

$$m_0=\left|\frac{v}{u}\right|=\frac{1}{x-1}$$

For eye-piece

$$\frac{1}{-25}-\frac{1}{-\left(20-\frac{x}{x-1}\right)}=\frac{1}{f_e}$$

and

$$m_e=\frac{25}{20-\frac{x}{x-1}}$$

Given that

$$\left(\frac{1}{x-1}\right)\times\left(\frac{25}{20-\frac{x}{x-1}}\right)=100$$

Solving we get

$$x=\frac{81}{76}\text{ cm}$$

For eye piece lens formula, we use

$$-\frac{1}{25}-\frac{1}{20-\frac{81/76}{81/76-1}}=\frac{1}{f_e}$$

Solving we get

$$f_e\cong 4.5\text{ cm}$$

Answer can be 4 or 5. **Ans. (5)**

Sol. 23 Given focal lengths of objective and eyepiece are

$$f_0=1\text{ cm}$$

$$f_e=5\text{ cm}$$

$$l=10\text{ cm (Tube length)}$$

For final image at infinity (for least strain on eye), image produced by objective must be at focal point of eyepiece i.e. 5 cm from it. For objective we use

$$f_0=+1\text{ cm}$$

$V_0 = +5$ cm

$$\Rightarrow \quad u = \frac{V_0 f_0}{V_0 - f_0} = \frac{5\times1}{4} = 1.25 \text{ cm} = \frac{N}{40}$$

$$\Rightarrow \quad N = 50$$

Ans. (50)

Sol. 24 In case of a simple microscope the line of vision of object and image is same for eye so assertion is correct and magnification of simple microscope is also greater than one so both A and R are true but R is not the correct explanation of A. **Ans. (C)**

Sol. 25 For a person if far away objects are not clear or blurry then problem is Myopia. When objects are looking distorted then it is due to Astigmatism. **Ans. (B)**

Sol. 26 Resolving power of a microscope is reated to wavelength used as

$$RP \propto \frac{1}{\lambda}$$

de-Broglie wavelength of the protons is given as

$$\lambda = \frac{h}{P} = \frac{h}{mv}$$

$$\Rightarrow \quad RP \propto \frac{mv}{h}$$

$$\Rightarrow \quad RP \propto m$$

Mass of proton is 1837 times that of an electron. **Ans. (A)**

Sol. 27 Magnification of microscopic lens is given as

$$m = 1 + \frac{D}{f_0}$$

$$\Rightarrow \quad 6 = 1 + \frac{25}{f_0}$$

$$\Rightarrow \quad f_0 = 5 \text{ cm}$$

Magnification of compound microscope when image is formed at infinity, is given as

$$m = \frac{l}{f_0} \times \frac{D}{f_e}$$

$$\Rightarrow \quad 12 = \frac{60}{5} \times \frac{25}{f_e}$$

$$f_e = 25 \text{ cm}$$

Ans. (25)

Sol. 28 Resolving power of the telescope is given as

$$RP = \frac{d}{1.22\lambda} = \frac{24.4\times10^{-2}}{1.22\times2440\times10^{-10}}$$

$$\Rightarrow \quad RP = 8.2\times10^5$$

Ans. (C)

Sol. 29 Resolving power of the microscope in air is given as

$$(\text{R.P})_{\text{air}} = \frac{2\sin\theta}{1.22\lambda}$$

Resolving power of the microscope in oil is given as

$$(\text{R.P})_{\text{oil}} = \frac{2\sin\theta}{1.22\lambda_{\text{oil}}} = \frac{2\sin\theta\times\mu}{1.22\lambda}$$

$$\Rightarrow \quad (\text{R.P})_{\text{oil}} = (\text{R.P})_{\text{air}} \times 2$$

Ans. (C)

Sol. 30 Tube length of telescope for normal adjustment is given as

$$f_0 + f_e = 30$$

Magnification of telescope for normal adjustment is given as

$$m = \frac{f_0}{f_e}$$

$$\Rightarrow \quad 2 = \frac{f_0}{f_e}$$

$$\Rightarrow \quad f_0 = 2f_e$$

$$\Rightarrow \quad f_0 + \frac{f_0}{2} = 30$$

$$\Rightarrow \quad f_0 = 20 \text{ cm}$$

Ans. (A)

Sol. 31 Using lens formula for reading glasses, we use

$$\frac{1}{v} - \frac{1}{u} = \frac{1}{f}$$

$$P = 2D = 2 \text{ m}^{-1}$$

$$\Rightarrow \quad \frac{1}{f} = \frac{2}{100} \text{ cm}^{-1}$$

For object placed at 25 cm to be seen clearly, we use

$$\Rightarrow \quad \frac{1}{v} - \left(-\frac{1}{25}\right) = \frac{2}{100}$$

$$\Rightarrow \quad \frac{1}{v} = \frac{1}{50} - \frac{1}{25}$$

$$\Rightarrow \quad v = -50 \text{ cm}$$

Thus the person is able to see the clear image at 50 cm which is the least distance of distinct vision for him. **Ans. (D)**

Sol. 32 Figure below shows the cross sectional view of the reflecting telescope.

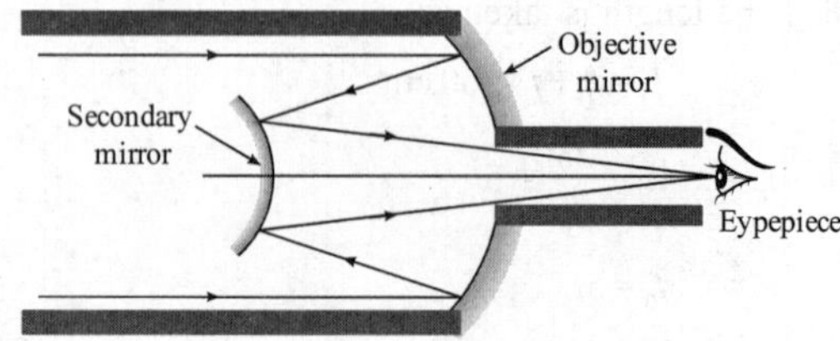

It has advantage of a large focal length in a short telescope. Secondary mirror concentrates light into a beam. **Ans. (D)**

24.1 Interference of Light

Sol. 1 Interference take place only when two or more coherent waves superpose

Coherence → The waves have exactly same frequency and constant (invarient in time) phase difference. **Ans. (D)**

Sol. 2

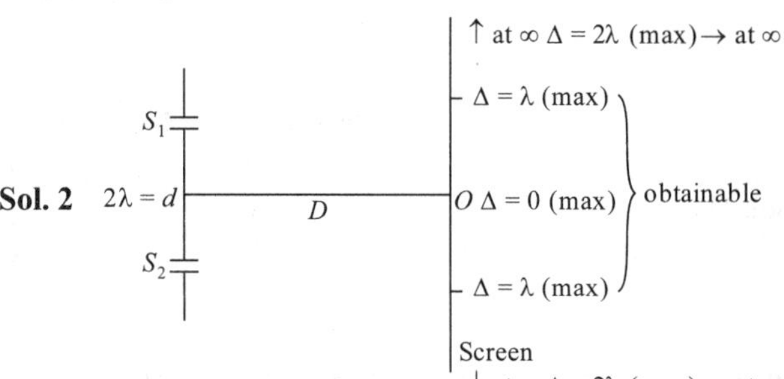

A maxima is obtained whenever path difference in waves from S_1 and S_2 is a multiple of λ, here out of 5 maxima only 3 can be obtained on screen in *YDSE*. **Ans. (C)**

Sol. 3 **Case-I :** If slits are considered as sources in *YDSE* for a source of light

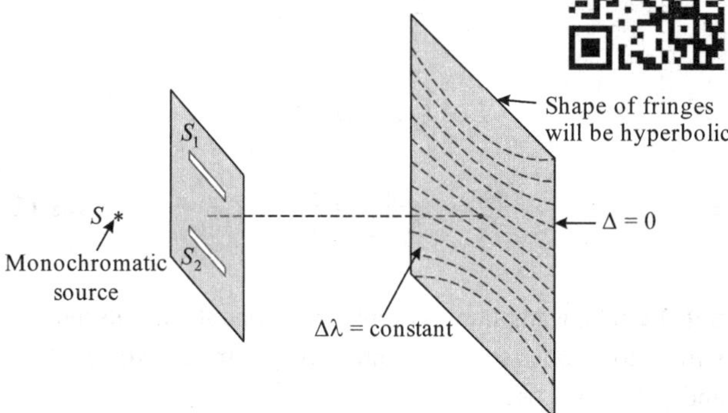

Because the locus of all points on screen which have path difference from S through S_1 and S_2 is constant is a hyperbola.

Case-II : If slits are considered as sources in *YDSE* for a line source

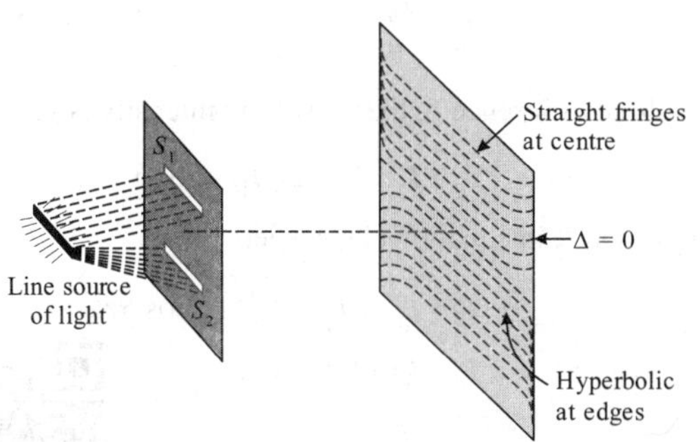

Here most appropriate is option (A). **Ans. (A)**

Sol. 4

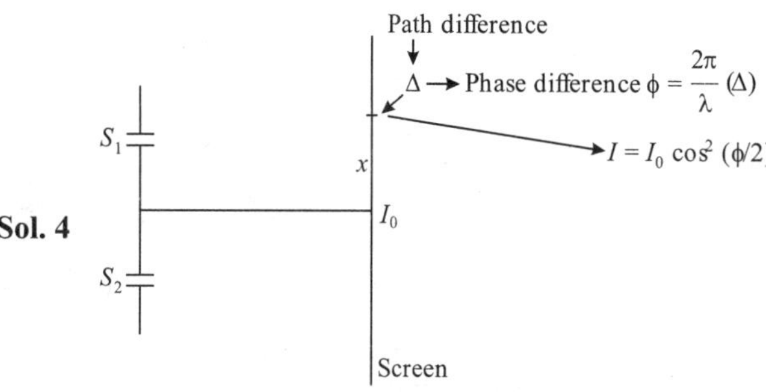

On either side of central maxima on screen the light intensity is given as

$$I = I_0 \cos^2(\phi/2) \qquad \ldots(1)$$

Where ϕ → phase difference between waves coming from S_1 and S_2 to that point for a point having path difference $\lambda/6$, then phase difference is

$$\phi = \frac{2\pi}{\lambda}(\Delta) = \frac{2\pi}{\lambda}\times\frac{\lambda}{6} = \frac{\pi}{3}$$

⇒ From equation-(1)

$$I = I_0\cos^2\left(\frac{\lambda}{6}\right) = I_0\left(\frac{\sqrt{3}}{2}\right)^2 = \frac{3I_0}{4}$$

$$\Rightarrow \quad \frac{I}{I_0} = \frac{3}{4}$$ **Ans. (D)**

Sol. 5 For a given wavelength of light used in *YDSE*, the distance of nth bright fringe from central maxima on screen is given as

$$x_{nB} = \frac{n\lambda D}{d}$$

[where D → distance between slit plane and screen]

[d → slit separation]

For known wavelength (λ_1) and unknown wavelength (λ_2) we use

$$\frac{3\lambda_1 D}{d} = \frac{4\lambda_2 D}{d}$$

$$\Rightarrow \quad \lambda_2 = \frac{3\lambda_1}{4} = \frac{3}{4}\times 590\text{ nm}$$

$$\Rightarrow \quad \lambda_2 = 442.5\text{ nm}$$ **Ans. (A)**

Sol. 6 Initially → beam is cylindrical

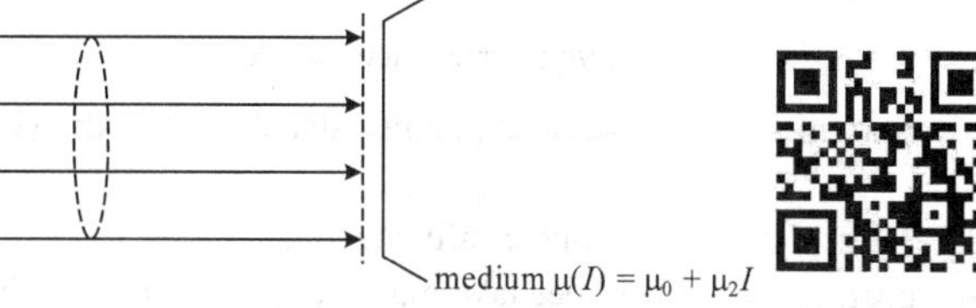

As light travel at same speed at all points of beam so wavefront will be planer. **Ans. (A)**

Sol. 7

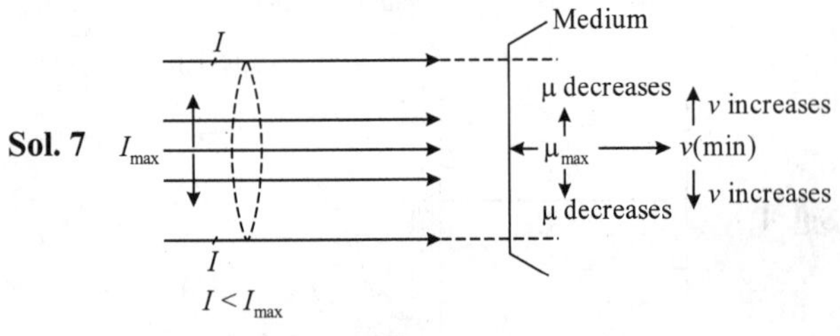

$$\mu(I) = \mu_0 + \mu_2 I \qquad \ldots(1)$$

As on the two all sides of axis of beam as intensity of light decreases so from equation-(1) refractive index also decreases

$$\mu = \frac{\text{Speed of light in vaccum}}{\text{Speed in medium}} = \frac{c}{v}$$

$$\Rightarrow \quad v = \frac{c}{\mu} \quad \textbf{Ans. (B)}$$

Sol. 8

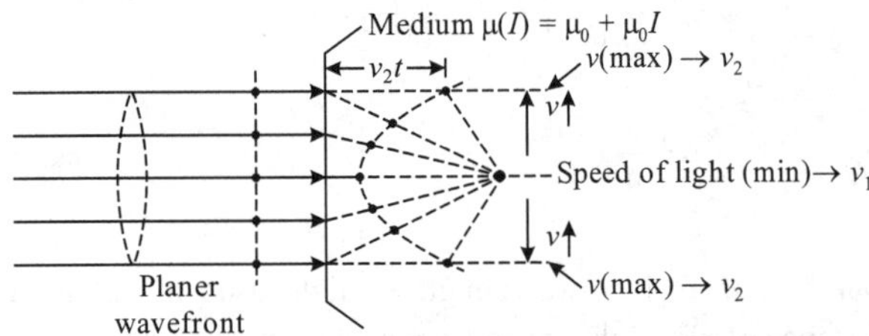

Concave wavefront after beam enters in medium

⇒ beam converges to a point. **Ans. (C)**

Sol. 9

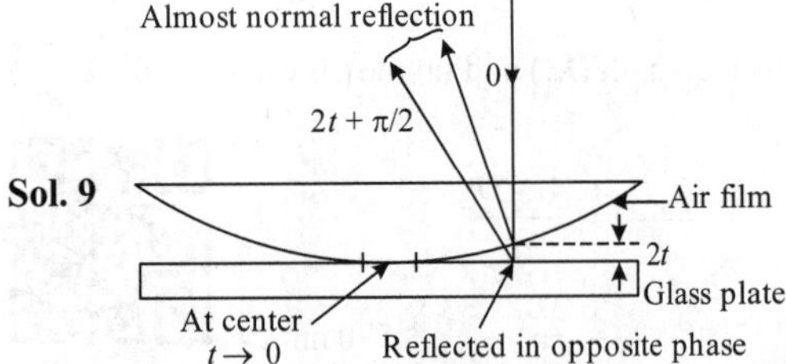

Statement-1 : It is true because a wave whenever reflected from a denser maximum suffers a phase change of π rad or a path delay of λ/2

⇒ $\Delta = \lambda/2 \to$ destructive interference ⇒ "dark"

⇒ **Statement-2 :** is true & explaining statement-1. **Ans. (D)**

Sol. 10 In first case as sources are coherent

So constructive interference take place at screen centre so the resulting intensity will be given as

$$I_1 = I_{max} = (\sqrt{I_1} + \sqrt{I_2})^2 = 4I_0 \qquad \ldots(1)$$

In second case as slits are in coherent sources so no interference take place and intensity will be

$$I_2 = I_R = I_1 + I_2 = 2I_0 \quad [I_0 = kA] \qquad \ldots(2)$$

Intensity due to each slit on screen

$$I_1 = I_2 = I_0$$

$$\frac{I_1}{I_2} = 2 \quad \textbf{Ans. (B)}$$

Sol. 11

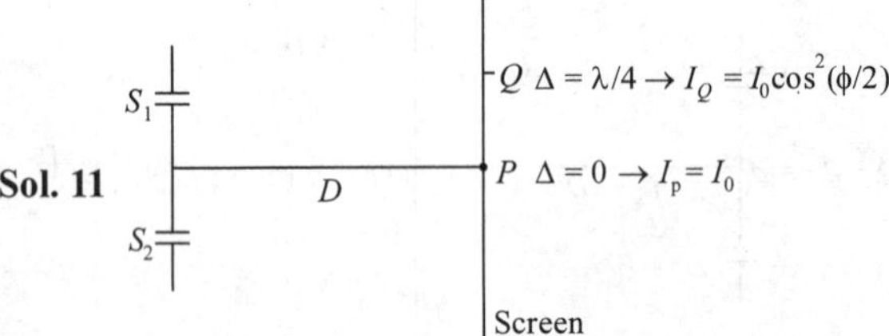

Intensity of light at a point on screen where phase difference between waves coming from S_1 and S_2 is & can be given as

$$I = I_0 \cos^2(\phi/2) \qquad \ldots(1)$$

At point P as $\quad \Delta = 0 \ \Rightarrow \ \phi = 0$

So $\quad I_P = I_0$

and at point Q as

$$\Delta = \lambda/4$$

$$\Rightarrow \quad \phi = \frac{2\pi}{\lambda} \cdot \Delta = \frac{2\pi}{\lambda}\left(\frac{\lambda}{4}\right) = \frac{\pi}{2}$$

$$\Rightarrow \quad I_Q = I_0 \cos^2(\pi/4) = I_0\left(\frac{1}{\sqrt{2}}\right)^2 = \frac{I_0}{2}$$

$$\Rightarrow \quad \frac{I_P}{I_Q} = \frac{I_0}{I_0/2} = \frac{2}{1} \qquad \textbf{Ans. (B)}$$

Sol. 12 If light amplitude is double from one slit then its intensity will be four times so if light intensities from the two slits are I_1 and I_2 then we use

If $I_1 = I_0 \ \Rightarrow \ I_2 = 4I_0$ (from wider slit)

Maximum intensity on screen is given as

$$I_m = (\sqrt{I_1} + \sqrt{I_2})^2 = (\sqrt{I_0} + \sqrt{4I_0})^2 = 9I_0$$

$$\Rightarrow \quad I_0 = \frac{I_m}{9}$$

For a phase difference ϕ, then resulting intensity is given as

$$I_R = I_1 + I_2 + 2\sqrt{I_1 I_2}\cos\phi \qquad \ldots(1)$$

Using I_1 and I_2 in equation (1), we get

$$I_R = I_0 + 4I_0 + 2\sqrt{4I_0^2}\cos\phi$$

$$\Rightarrow \quad I_R = I_0(5 + 4\cos\phi)$$

$$\Rightarrow \quad I_R = \frac{I_m}{9}(5 + 4\cos\phi)$$

$$\Rightarrow \qquad I_R = \frac{I_m}{9}\,(5+4\,(2\cos^2\phi/2-1))$$

$$\Rightarrow \qquad I_R = \frac{I_m}{9}\,(1+8\cos^2\phi/2) \qquad \textbf{Ans. (A)}$$

Sol. 13

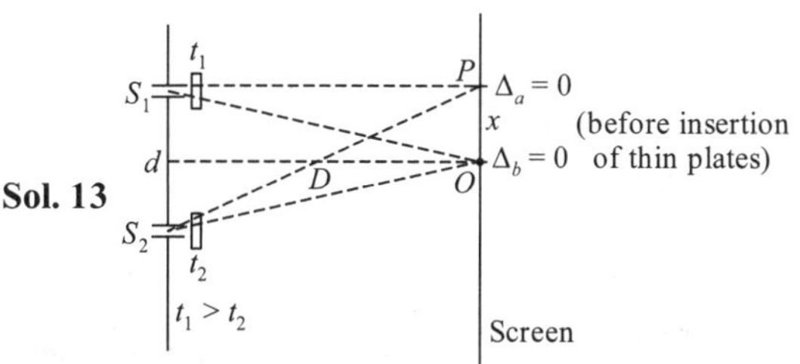

After insertion of thin glass plates in path of beams from S_1 and S_2 as shown, path difference at screen center becomes

$$\Delta_a = t_1(\mu-1) - \mu_2(\mu-1)$$

Say after insertion of plates at point P (above O) the physical path difference $\left(\frac{dx}{D}\right)$ will balance the optical path difference due to plates so finally effective path difference at P is zero at point P we use

$$\Delta_{ap} = \frac{dx}{D} - \Delta_a = 0$$

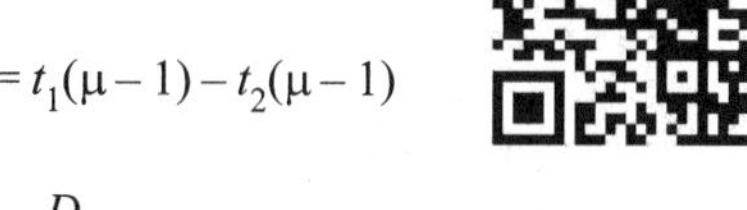

$$\Rightarrow \qquad \frac{dx}{D} = t_1(\mu-1) - t_2(\mu-1)$$

$$\Rightarrow \qquad x = \frac{D}{d}(\mu-1)(t_1-t_2) \qquad \ldots(1)$$

In *YDSE* fringe width is given as

$$\beta = \frac{\lambda D}{d} \quad \Rightarrow \quad \frac{D}{d} = \frac{\beta}{\lambda}$$

$$\Rightarrow \text{ From (1)} \qquad x = \frac{\beta}{\lambda}(\mu-1)(t_1-t_2) \qquad \textbf{Ans. (C)}$$

Sol. 14

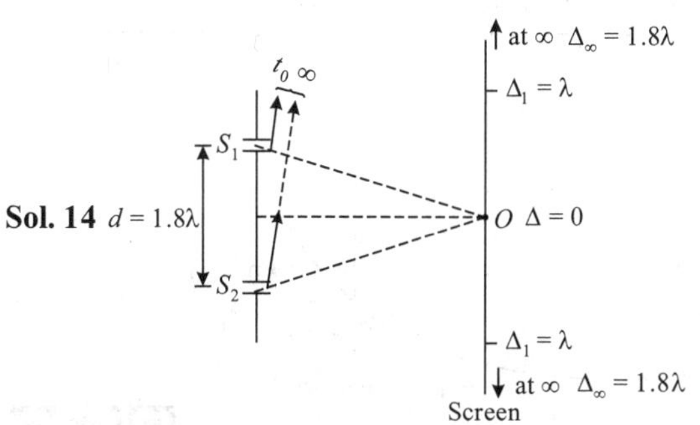

At ∞ on screen path difference in waves coming from states S_1 and S_2 is

$$\Delta_\infty = d = 1.8\lambda$$

between screen centre at ∞ somewhere path difference will be $\Delta_1 = \lambda$ on both sides of screen centre.

So at $\Delta = 0$ and $\Delta_1 = \lambda$ (two sides)

Constructive interference of light takes place

$\Rightarrow$ No of maxima = 3 **Ans. (B)**

Sol. 15 Intensity of light on *YDSE* screen due to a slit is proportional is slit width so we consider light intensities due to the two slits on screen are I_0 and $25\,I_0$.

The ratio of maximum to minimum intensities in interference pattern is given as

$$\frac{I_{max}}{I_{min}} = \left(\frac{\sqrt{I_1}+\sqrt{I_2}}{\sqrt{I_2}-\sqrt{I_1}}\right)^2$$

$$\Rightarrow \qquad \frac{I_{max}}{I_{min}} = \left(\frac{5+1}{5-1}\right)^2 = \left(\frac{6}{4}\right)^2 = \frac{9}{4} \qquad \textbf{Ans. (C)}$$

Sol. 16

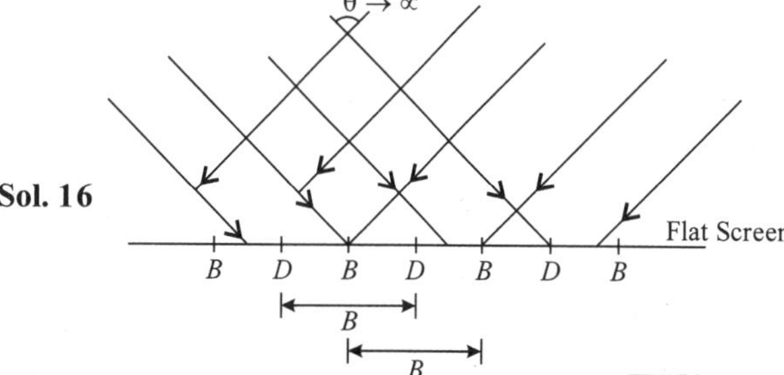

Fringe width for interference of parallel beams on a surface is given as

$$B = \frac{\lambda}{\alpha} \qquad \textbf{Ans. (B)}$$

Sol. 17

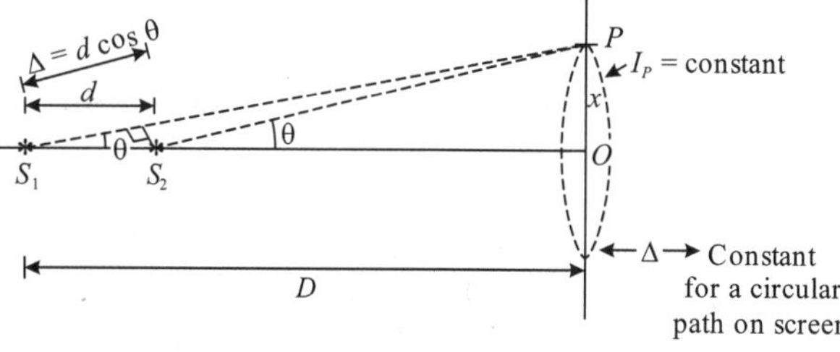

here $d \ll D$

On screen the locus of points where θ is a constant and at all these points path difference

$\Delta = d\cos\theta$ is a constant, will be a circle.

$\Rightarrow$ Fringes obtained are circular with centre at point O. **Ans. (D)**

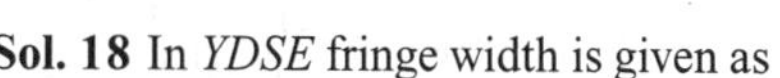

Sol. 18 In *YDSE* fringe width is given as

$$\beta = \frac{\lambda D}{d}$$

$$\Rightarrow \qquad \beta \propto \lambda$$

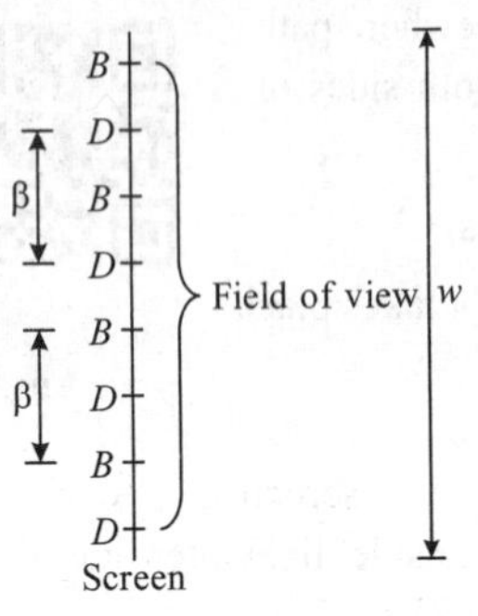

$$\text{No of fringes} = \frac{\text{field of view}}{\beta} = \frac{wd}{\lambda D}$$

$\Rightarrow$ No of fringes $\propto \frac{1}{\lambda}$

$\Rightarrow$ Statement-1 is true & statement-2 is also true and explaining statement-1. **Ans. (A)**

Sol. 19 For interference maxima of wavelength λ_1, the path difference we use bright fringe

$$\Delta = m\lambda_1 \quad \ldots(1)$$

For interference minima (Dark fringe) of wavelength λ_2, the path difference we use

$$\Delta = (2n+1)\frac{\lambda_2}{2} \quad \ldots(2)$$

At one point on screen if bright fringe of λ_1 coincides with dark fringe of λ_2 then we use

$$\Rightarrow \quad \Delta = m\lambda_1 = (2n+1)\frac{\lambda_2}{2}$$

$$\Rightarrow \quad \frac{m}{2n+1} = \frac{\lambda_2}{2\lambda_1} = \frac{600}{2\times 500} = \frac{3}{5}$$

For least distance from central maxima where these fringes coincide will be corresponding if

$$m = 3$$

and $$2n+1 = 5$$

$$\Rightarrow \quad \Delta = 3\lambda_1 = 3\times 500 = 1500\,\text{nm}$$ **Ans. (D)**

20.

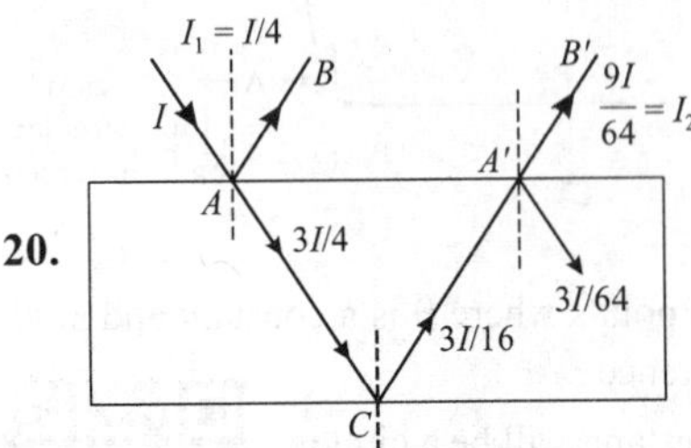

When rays AB and $A'B'$ interfere then the ratio of maximum to minimum intensity in interference pattern is given as

$$\frac{I_{max}}{I_{min}} = \left(\frac{\sqrt{I_1}+\sqrt{I_2}}{\sqrt{I_1}-\sqrt{I_2}}\right)^2 = \left(\frac{\sqrt{\frac{1}{4}}+\sqrt{\frac{9}{64}}}{\sqrt{\frac{1}{4}}-\sqrt{\frac{9}{64}}}\right)^2$$

$$\Rightarrow \quad \frac{I_{max}}{I_{min}} = \left(\frac{\frac{1}{2}+\frac{3}{8}}{\frac{1}{2}-\frac{3}{8}}\right)^2$$

$$\Rightarrow \quad \frac{I_{max}}{I_{min}} = \left(\frac{7}{1}\right)^2 = \frac{49}{1}$$ **Ans. (A)**

Sol. 21 For n coherent waves when, interfere constructively then resulting intensity is given as

$$I_{max} = n^2 I_0 \quad \ldots(1)$$

for n non-coherent waves when superpose then resulting intensity at the point of superposition is given as

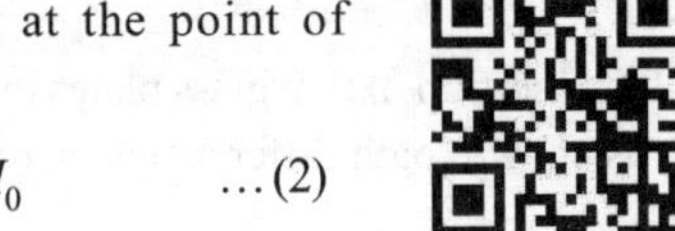

$$I'_{max} = nI_0 \quad \ldots(2)$$

$$\Rightarrow \quad \frac{I_{max}}{I_{max}} = \frac{n^2 I_0}{nI_0} = n$$ **Ans. (D)**

Sol. 22 If I_1 and I_2 are interfering intensities then maxima & minima intensities are given as

$$I_{max} = (\sqrt{I_1}+\sqrt{I_2})^2$$

and $$I_{min} = (\sqrt{I_1}-\sqrt{I_2})^2$$

$$\Rightarrow \quad \frac{I_{max}}{I_{min}} = \left(\frac{4+3}{4-1}\right)^2 = \left(\frac{7}{1}\right)^2 = \frac{49}{1}$$ **Ans. (D)**

Sol. 23

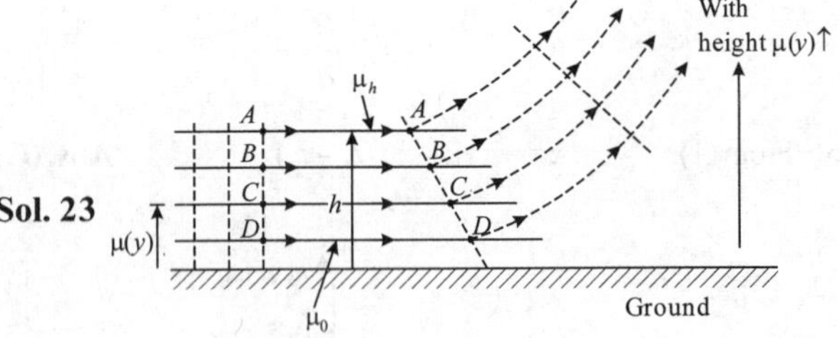

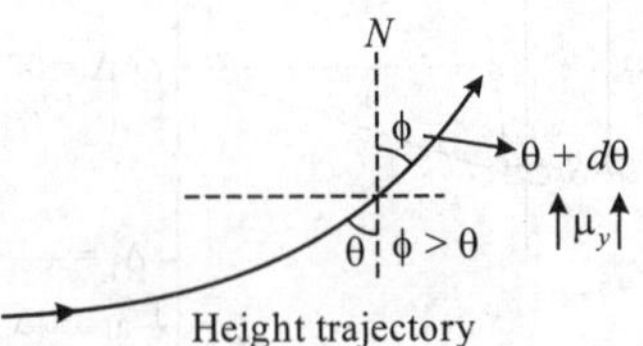

$\mu_h > \mu_0 \Rightarrow v$ (speed of light) decreases with height so the speed of light at some height is lesser then compared to ground. **Ans. (D)**

Sol. 24

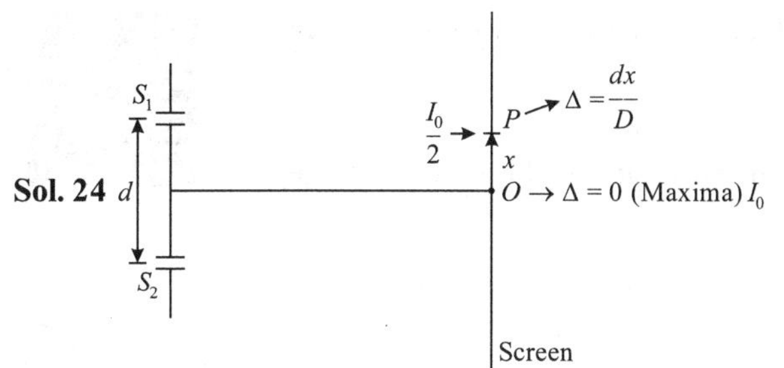

In *YDSE* fringe width is given as

$$\beta = \frac{\lambda D}{d} \quad \ldots(1)$$

The path difference between waves reaching point P at a distance x from screen centre is

$$\Delta_P = \frac{dx}{D}$$

$\Rightarrow$ Phase difference at P

$$\phi_P = \frac{2\pi}{\lambda}\cdot\Delta_P = \frac{2\pi}{\lambda}\cdot\frac{dx}{D} \quad \ldots(2)$$

If phase difference at point P is ϕ_P

$\Rightarrow$ Resulting light intensity at P is given as

$$I_P = I_0 \cos^2\left(\frac{\phi_P}{2}\right) = \frac{I_0}{2}$$

$$\Rightarrow \quad \cos^2\left(\frac{\phi_P}{2}\right) = \frac{1}{\sqrt{2}}$$

$$\Rightarrow \quad \frac{\phi_P}{2} = \frac{\pi}{4} \quad \text{or} \quad \phi_P = \frac{\pi}{2}$$

From equation-(2) we have

$$\phi_P = \frac{2\pi}{\lambda}\frac{dx}{D} = \frac{\pi}{2}$$

$$\Rightarrow \quad x = \frac{D\lambda}{4d} = \frac{\beta}{4}$$ **Ans. (C)**

Sol. 25 In *YDSE* the distance of nth bright fringe from central maximum is given as

$$x_{nB} = \frac{n\lambda D}{d}$$

Where $D = 1.5\text{m}$

$d = 0.5 \times 10^{-3}\text{ m}$

If n^{th} bright fringe of λ, coincides with n_2^{th} bright fringe of λ_2 then we use

$$x_{n_1B} = x_{n_2B}$$

$$\Rightarrow \quad \frac{n_1\lambda_1 D}{d} = \frac{n_2\lambda_2 D}{d}$$

$$\Rightarrow \quad \frac{n_1}{n_2} = \frac{\lambda_2}{\lambda_1} = \frac{520}{560} = \frac{4}{5}$$

Last distance from central maxima of coinciding bright fringe is

$$x_{n_1B} = \frac{4\times 650\times 10^{-9}\times 1.5}{0.5\times 10^{-3}}$$

$n_1 = 4$

$n_2 = 5$

$$\Rightarrow \quad x_{n_1B} = 7.8\times 10^{-3}\text{m}$$ **Ans. (D)**

Sol. 26 Ratio of maximum to minimum intensities of interference is given as

$$\frac{I_{max}}{I_{min}} = 16$$

If I_1 and I_2 are intensities of compoment waves, we have

$$\Rightarrow \quad \frac{I_{max}}{I_{min}} = \frac{\sqrt{I_1}+\sqrt{I_2}}{\sqrt{I_1}-\sqrt{I_2}} = 4$$

$$\Rightarrow \quad \sqrt{\frac{I_1}{I_2}} = \frac{5}{3}$$

$$\Rightarrow \quad \frac{I_1}{I_2} = \frac{25}{9}$$ **Ans. (B)**

Sol. 27

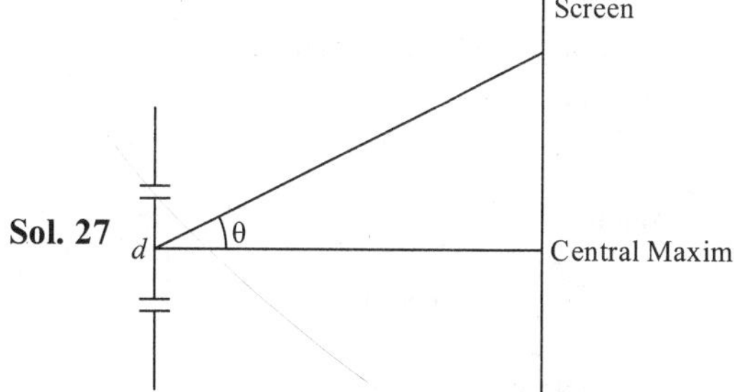

Path difference for angular position θ is

$$d\sin\theta = n\lambda$$

$$0.32\times 10^{-3}\sin(30°) = n\times 500\times 10^{-9}$$

$$n = 320$$

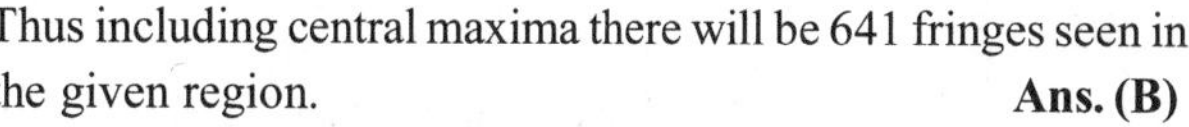

Thus including central maxima there will be 641 fringes seen in the given region. **Ans. (B)**

Sol. 28 Path difference

$$\Delta = d\sin\theta \approx d\theta$$

$$\Rightarrow \quad \Delta = 0.1\times\frac{1}{40}\text{mm} = 2500\text{ nm}$$

On bright fringe, path difference must be integral multiple of λ.

$\Rightarrow \quad 2500 = n\lambda_1 = m\lambda_2$

$\Rightarrow \quad \lambda_1 = 625,$

$\lambda_2 = 500$ (from $m = 5$ and $n = 4$) **Ans. (B)**

Sol. 29 For first minima in front of one slit, path difference at point P is

$$\sqrt{5}d - 2d = \frac{\lambda}{2}$$

$$\Rightarrow \quad d = \frac{\lambda}{2(\sqrt{5}-2)} \quad \textbf{Ans. (D)}$$

Sol. 30 Given that path difference at the point

$$\Delta x = \frac{\lambda}{8}$$

Phase difference at this point is

$$\Delta\phi = \frac{(2\pi)}{\lambda} \times \frac{\lambda}{8} = \frac{\pi}{4}$$

Intensity at this point is

$$I = I_0 \cos^2\left(\frac{\Delta\phi}{2}\right) = I_0 \cos^2\left(\frac{\pi}{8}\right)$$

$$\Rightarrow \quad \frac{I}{I_0} = \cos^2\left(\frac{\pi}{8}\right) = \frac{1+\cos\left(\frac{\pi}{4}\right)}{2}$$

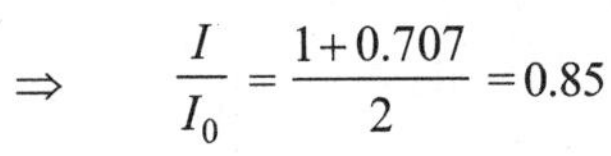

$$\Rightarrow \quad \frac{I}{I_0} = \frac{1+0.707}{2} = 0.85 \quad \textbf{Ans. (C)}$$

Sol. 31 For diffraction, location of 1st minima

$$y_1 = \frac{D\lambda}{b} = 0.2469\,D\lambda$$

location of 2nd minima is

$$y_2 = \frac{2D\lambda}{b} = 0.4938 D\lambda$$

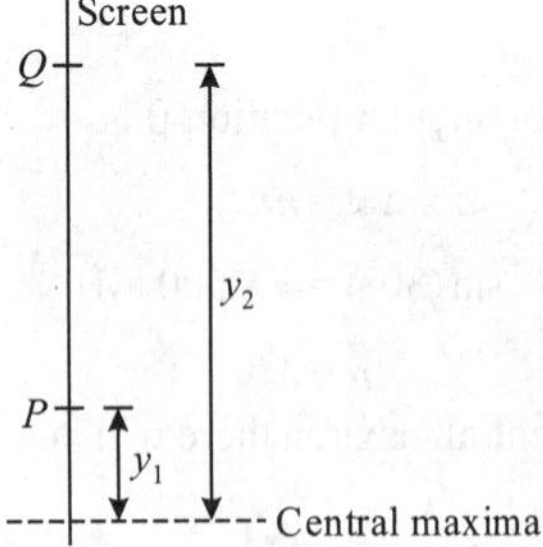

Now for interference

Path difference at point P on screen is

$$\frac{dy}{D} = 4.8\,\lambda$$

Path difference at point Q on screen is

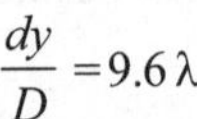

$$\frac{dy}{D} = 9.6\,\lambda$$

So orders of maxima in between P & Q is are 5, 6, 7, 8, 9

Thus 5 bright fringes are present between P & Q. **Ans. (D)**

Sol. 32 Given that ratio of amplitudes is

$$\frac{a_1}{a_2} = \frac{1}{3}$$

Ratio of intensities of the waves can be given as

$$\frac{I_1}{I_2} = \left(\frac{a_1}{a_2}\right)^2 = \frac{1}{9}$$

Ratio of maximum to minimum intensities is given as

$$\frac{I_{max}}{I_{min}} = \left(\frac{\sqrt{I_1}+\sqrt{I_2}}{\sqrt{I_1}-\sqrt{I_2}}\right)^2 = \left(\frac{1+3}{1-3}\right)^2 = 4 \quad \textbf{Ans. (A)}$$

Sol. 33 If we introduce a thin sheet in path of light from one of the slits then path difference introduced is

$$\Delta = t(\mu - 1)$$

If fringe pattern displaces by x, we use

$$\Delta = \frac{xd}{D}\left(\frac{\lambda D}{d}\right) = n\lambda$$

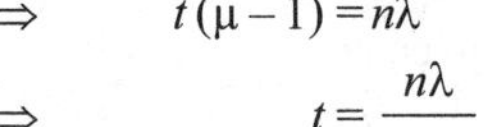

$$\Rightarrow \quad t(\mu-1) = n\lambda$$

$$\Rightarrow \quad t = \frac{n\lambda}{\mu - 1}$$

Ans. (None)

Sol. 34 If I_1 and I_2 are light intensities from the two slits, we use

$$I_2 = I_0$$

$$\Rightarrow \quad I_1 = 4I_0$$

We use $$\frac{I_{max}}{I_{min}} = \left[\frac{\sqrt{I_1}+\sqrt{I_2}}{\sqrt{I_1}-\sqrt{I_2}}\right]^2$$

$$\Rightarrow \quad \frac{I_{max}}{I_{min}} = \left[\frac{2\sqrt{I_0}+\sqrt{I_0}}{2\sqrt{I_0}-\sqrt{I_0}}\right]^2 = \frac{9}{1} \quad \textbf{Ans. (B)}$$

Sol. 35

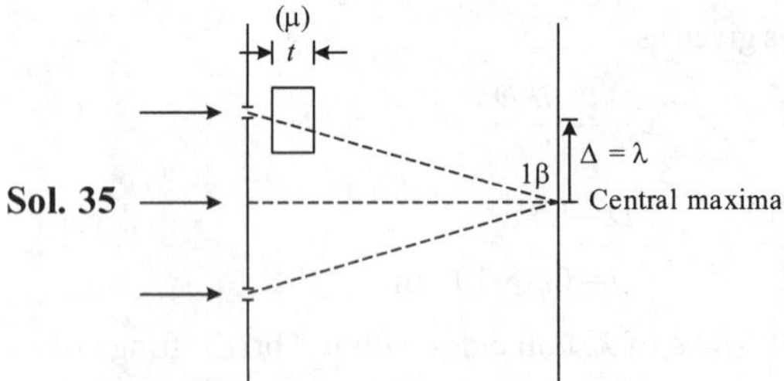

Path difference due to introducing a thin film is given as

$$\Delta = t(\mu - 1)$$

As shift of central maxima is by one fringe width

$$\Rightarrow \quad \Delta = \lambda$$

$$t(\mu - 1) = \lambda$$

$$\Rightarrow \quad t = \frac{\lambda}{\mu - 1} \quad \textbf{Ans. (D)}$$

Sol. 36 Fringe width in YDSE is given as,

$$\beta = \frac{D\lambda}{d}$$

$$\Rightarrow \quad \beta = \frac{1.5 \times 589 \times 10^{-9}}{0.15 \times 10^{-3}} \text{ m}$$

$$\Rightarrow \quad \beta = 5.9 \times 10^{-3} \text{ m}$$

$$\Rightarrow \quad \beta = 5.9 \text{ mm}$$

Ans. (B)

Sol. 37 Intensity of light on YDSE screen is given as

$$I = I_0 \cos^2\left(\frac{\phi}{2}\right)$$

$$\frac{I}{I_0} = \cos^2\left(\frac{\phi}{2}\right)$$

Phase difference for the given point is

$$\phi = \frac{2\pi}{\lambda} \times \frac{\lambda}{8} = \frac{\pi}{4}$$

$$\Rightarrow \quad \frac{I}{I_0} = \cos^2\left(\frac{\pi}{8}\right)$$

$$\Rightarrow \quad \frac{I}{I_0} = 0.853$$

Ans. (D)

Sol. 38 If amplitude of each wave is considered as A then for resulting amplitude, we use the figure below

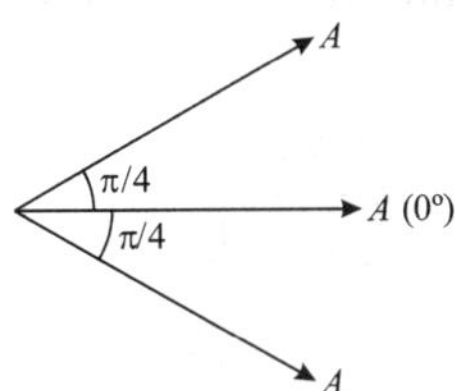

Resulting amplitude is given as

$$R = A + 2\left(\frac{A}{\sqrt{2}}\right) = A(\sqrt{2} + 1)$$

As intensity is proportional to square of amplitude, we use

$$I_0 = kA^2 \text{ and } I_R = kR^2$$

$$\Rightarrow \quad \frac{I_R}{I_0} = (\sqrt{2} + 1)^2$$

$$\Rightarrow \quad I_R \cong 5.8\, I_0$$

Ans. (A)

Sol. 39 The length of portion on screen used for 15 fringes is

$$l = 15B = 15\frac{\lambda_0 D}{d}$$

When wavelength changes to λ, 10 fringes are observed on same position of screen, so we use

$$l = 10\frac{\lambda D}{d} = \frac{15\lambda_0 D}{d}$$

$$\Rightarrow \quad l = \frac{15\lambda_0}{10} = \frac{3}{2} \times 500 = 750 \text{ nm}$$

Ans. (750)

Sol. 40 Distance of nth bright fringe from centre is

$$Y = \frac{nD\lambda}{d}$$

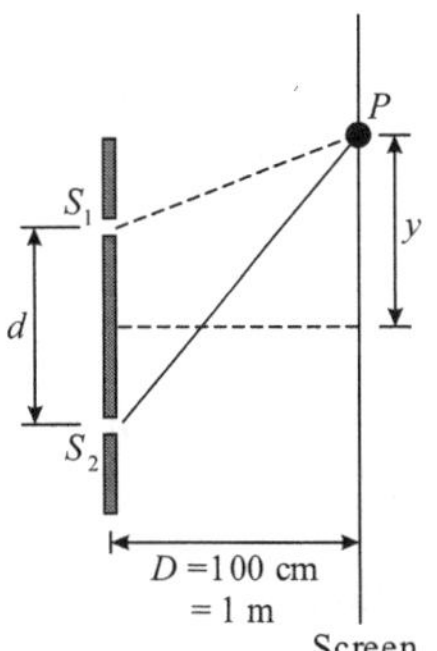

$$\Rightarrow \quad n = \frac{Yd}{D\lambda} = \frac{1.27 \times 10^{-3} \times 10^{-3}}{1 \times 632.8 \times 10^{-9}} = 2$$

For bright fringe path difference is given as

$$\Delta x = n\lambda$$

$$\Rightarrow \quad \Delta x = 2 \times 632.8 \text{ nm}$$

$$\Rightarrow \quad \Delta x = 1265.6 \text{ nm}$$

$$\Rightarrow \quad \Delta x = 1.27\ \mu\text{m}$$

Ans. (A)

Sol. 41 If N fringes are there in screen segment of length l and fringe width is β, we use

$$N\beta = l$$

$$\Rightarrow \quad N\left(\frac{\lambda D}{d}\right) = l$$

For first wavelength $\lambda_1 = 700$ nm, we use

$$\frac{N_1\lambda_1 D}{d} = l$$

For second wavelength $\lambda_2 = 400$ nm, we use

$$\frac{N_2\lambda_2 D}{d} = l$$

$$\Rightarrow \quad N_1\lambda_1 = N_2\lambda_2$$

$$\Rightarrow \quad 16 \times 700 = N_2 \times 400$$

$$\Rightarrow \quad N_2 = 28$$

Ans. (A)

Sol. 42 In YDSE setup given data is

$$d = 0.05 \times 10^{-3} \text{ m}$$

and

$$\lambda = 5 \times 10^{-7} \text{ m}$$

Angular fringe width is given as

$$\theta = \frac{\beta}{D} = \frac{\lambda D/d}{D} = \frac{\lambda}{d} \text{ rad}$$

In degrees it is given as

$$\theta = \frac{\lambda}{d} \times \frac{180}{\pi} \text{ degree}$$

$$\Rightarrow \qquad \theta = \frac{5 \times 10^{-7} \times 180}{5 \times 10^{-5} \times 3.14} \simeq 0.57^\circ \qquad \textbf{Ans. (D)}$$

Sol. 43 Optical path difference between two waves is given as

$$\Delta x = n_1 L_1 - n_2 L_2$$

Phase difference is given as

$$\Delta\phi = \frac{2\pi}{\lambda}\Delta k = \frac{2\pi}{\lambda}(n_1 L_1 - n_2 L_2) \qquad \textbf{Ans. (A)}$$

Sol. 44 At initial position

$$S_1 L = \sqrt{2^2 + \left(\frac{3}{2}\right)^2} = \frac{5}{2}\text{m}$$

and $\quad S_2 L = 2\text{ m}$

Path difference at initial position is given as

$$\Delta = S_1 L - S_2 L = 2.5 - 2 = 0.5\text{ m}$$

Here $\quad \Delta = \frac{\lambda}{2}$

Thus intensity is minimum. When L moves away to new position L', next maximum is obtained when

$$\Delta = \lambda = 1\text{ m}$$

$$\Rightarrow \qquad d = S_1 L' = S_2 L' + 1 = 3\text{ m} \qquad \textbf{Ans. (B)}$$

Sol. 45 As P is 90° ahead in phase so path lead of P w.r. to Q is

$$\Delta = \frac{\lambda}{2\pi} \times \frac{\pi}{2} = \frac{\lambda}{4} = \frac{20}{4} = 5\text{ m}$$

For point A, path difference in waves from P and Q is

$$\Delta_A = 5 - 5 = 0$$

Thus at A intensity is maximum I_0

For point B, path difference in waves from P and Q is

$$\Delta = 5\text{ m}$$

$$\Rightarrow \text{ Phase difference } \phi = \frac{2\pi}{\lambda}\Delta = \frac{2\pi}{20} \times 5 = \frac{\pi}{2}$$

Thus intensity at B is

$$I_B = I_0 \cos^2\frac{\phi}{2} = I_0 \cos^2\frac{\phi}{4} = \frac{I_0}{2}$$

For point C, path difference in waves from P and Q is

$$\Delta = 5 + 5 = 10\text{ m}$$

As $\Delta = \frac{\lambda}{2}$ so intensity will be minimum at C due to distractive interference, so we have

$$\Rightarrow \quad I_A : I_B : I_C = I_0 : \frac{I_0}{2} : 0 = 2:1:0 \qquad \textbf{Ans. (D)}$$

Sol. 46 The point on screen where path difference is λ is the point of bright fringe so at a general point resulting light intensity is given as

$$I_R = K \cdot \cos^2\frac{\phi}{2}$$

At the point where path difference is $\frac{\lambda}{6}$, phase difference is given as

$$\phi = \frac{2\pi}{\lambda} \times \frac{\lambda}{6} = \frac{\pi}{3}$$

Thus intensity at this point is given as

$$\Rightarrow \qquad I = \frac{3K}{4} = \frac{nK}{12}$$

$$\Rightarrow \qquad n = 9 \qquad \textbf{Ans. (9)}$$

Sol. 47 Amplitude of light wave is directly proportional to the width of slit in YDSE setup so we use

$$A_2 = 3A_1$$

$$\Rightarrow \qquad \frac{I_{max}}{I_{min}} = \left(\frac{\sqrt{I_1} + \sqrt{I_2}}{|\sqrt{I_1} - \sqrt{I_2}|}\right)^2$$

As light intensity is directly proportional to square of amplitude, we use

$$\frac{I_{max}}{I_{min}} = \left(\frac{A_1 + A_2}{|A_1 - A_2|}\right)^2$$

$$\Rightarrow \qquad \frac{I_{max}}{I_{min}} = \left(\frac{A_1 + 3A_1}{|A_1 - 3A_1|}\right)^2$$

$$\Rightarrow \qquad \frac{I_{max}}{I_{min}} = \left(\frac{4A_1}{2A_1}\right)^2 = 4:1 \qquad \textbf{Ans. (B)}$$

Sol. 48 Fringe width in YDSE setup is given as

$$\beta = \frac{\lambda . D}{d}$$

As $\lambda_R > \lambda_V \Rightarrow \beta_R > \beta_V$

Thus fringe pattern will shrink on changing color from red to violet. **Ans. (D)**

Sol. 49 The ratio of component light intensities is given as

$$\frac{I_1}{I_2} = 2x$$

Maximum and minimum intensities due to interference are given as

$$I_{max} = (\sqrt{I_1} + \sqrt{I_2})^2$$

and $$I_{min} = (\sqrt{I_1} - \sqrt{I_2})^2$$

$$\Rightarrow \frac{I_{max} - I_{min}}{I_{max} + I_{min}} = \frac{2\sqrt{I_1 I_2}}{I_1 + I_2}$$

$$\Rightarrow \frac{I_{max} - I_{min}}{I_{max} + I_{min}} = \frac{2\sqrt{I_1 / I_2}}{\frac{I_1}{I_2} + 1} = \frac{2\sqrt{2x}}{2x+1}$$ **Ans. (A)**

Sol. 50 Fringe width in YDSE pattern is given as

$$\beta = \frac{\lambda D}{d} = \frac{500 \times 10^{-9} \times 1}{2 \times 10^{-3}}$$

$$\Rightarrow \beta = \frac{5}{2} \times 10^{-4}\,\text{m} = 2.5 \times 10^{-1}\,\text{mm}$$

$$\Rightarrow \beta = 0.25\,\text{mm}$$ **Ans. (A)**

Sol. 51 Fringe withd in YDSE setup is given as

$$\beta = \frac{\lambda D}{d}$$

$$\Rightarrow \lambda = \frac{\beta d}{D} = \frac{6 \times 10^{-3} \times 10^{-3}}{10}$$

$$\Rightarrow \lambda = 6 \times 10^{-7}\,\text{m} = 600 \times 10^{-9}\,\text{m} = 600\,\text{nm}$$

Ans. (600)

Sol. 52 In *YDSE* setup, fringe width in interference pattern is given as

$$\beta = \frac{D\lambda}{d}$$

$$\Rightarrow \beta = \frac{D\lambda}{(d_0 + a_0 \sin wt)}$$

The difference between the largest and smallest fringe width can be calculated as

$$\beta_{max} - \beta_{min} = \frac{D\lambda}{d_0 - a_0} - \frac{D\lambda}{d_0 + a_0}$$

$$\Rightarrow \beta_{max} - \beta_{min} = D\lambda\left[\frac{d_0 + a_0 - d_0 + a_0}{d_0^2 + a_0^2}\right] = \frac{2\lambda D a_0}{(d_0^2 - a_0^2)}$$

Ans. (A)

Sol. 53 Distance of first red and violet fringe from central white fringe is given as

$$y_1 = \frac{\lambda_r D}{d}$$

and $$y_2 = \frac{\lambda_v D}{d}$$

$$\Rightarrow \lambda_r - \lambda_v = (y_1 - y_2)\frac{d}{D} = (3.5 - 2)\,\text{mm} \times \frac{0.3\text{mm}}{1.5\text{m}} = 300\,\text{nm}$$

Ans. (300)

Sol. 54 As $I_1 = I_2 = I_0$, the maximum intensity at the point of interference is given as

$$I_{max} = (\sqrt{I_1} + \sqrt{I_2})^2 = 4I_1$$ **Ans. (C)**

Sol. 55 In YDSE setup intensity of light is proportional to the slit width. The ratio of maximum to minimum intensity is given as

$$\frac{I_{max}}{I_{min}} = \frac{(\sqrt{I_1} + \sqrt{I_2})^2}{(\sqrt{I_1} - \sqrt{I_2})^2} = \left[\frac{\sqrt{9I} + \sqrt{I}}{\sqrt{9I} - \sqrt{I}}\right]^2$$

$$\Rightarrow \frac{I_{max}}{I_{min}} = \left[\frac{(3+1)\sqrt{I}}{(3-1)\sqrt{I}}\right]^2 = 4 = \frac{x}{4}$$

$$\Rightarrow x = 16$$ **Ans. (16)**

Sol. 56 For the slit width a, distance between slit and screen L and wavelength of light λ we use for first maximum

$$a \sin\theta = \frac{3\lambda}{2}$$

If y is the distance of first maxima from screen centre then we use

$$\theta = \frac{y}{L} = \frac{3\lambda}{2a}$$

$$\Rightarrow y_1 = \frac{3\lambda_1 L}{2a} \qquad \lambda_2 = 655\,\text{nm}$$

$$\Rightarrow y_2 = \frac{3\lambda_2 L}{2a} \qquad \lambda_1 = 650\,\text{nm}$$

Separation between the two first maxima is calculated as

$$\Delta y = y_2 - y_1 = \frac{2(\lambda_2 - \lambda_1)}{2a}L$$

$$\Rightarrow \Delta y = \frac{3(655 - 650)}{2 \times 0.5 \times 10^{-3}} \times 2 \times 10^{-9}$$

$$\Rightarrow \Delta y = \frac{3 \times 5 \times 2}{1 \times 10^{-3}} \times 10^{-9}$$

$$\Rightarrow \Delta y = 3 \times 10^{-5}$$ **Ans. (3)**

Sol. 57 Amplitude ratio of light beams is calculated as

$$\frac{A_1}{A_2} = \sqrt{\frac{I_1}{I_2}} = \sqrt{\frac{9}{4}} = \frac{3}{2}$$

The ratio of intensity of maxima and minima i.e. contrast ratio is calculated as

$$\gamma = \left(\frac{\sqrt{I_1} + \sqrt{I_2}}{\sqrt{I_1} - \sqrt{I_2}}\right)^2 = 5^2 = 25$$ **Ans. (D)**

Sol. 58 Resulting intensity at P is given as

$$I_P = I + 9I + 2\sqrt{I \times 9I}\ \cos\frac{\pi}{2}$$

$$\Rightarrow \quad I_P = 10I$$

Resulting intensity at Q is given as

$$I_Q = I + 9I + 2\sqrt{I \times 9I}\ \cos\pi$$

$$\Rightarrow \quad I_Q = 10I - 6I = 4I$$

Difference in intensities at P and Q is given as

$$I_P - I_Q = 10I - 4I = 6I$$

Ans. (B)

Sol. 59 Ratio of intensitites of the two light sources is given as

$$\frac{I_1}{I_2} = 4$$

$$\Rightarrow \quad \frac{I_{max}}{I_{min}} = \left[\frac{\sqrt{I_1} + \sqrt{I_2}}{\sqrt{I_1} - \sqrt{I_2}}\right]^2$$

$$\Rightarrow \quad \frac{I_{max}}{I_{min}} = \left[\frac{2\sqrt{I_1} + \sqrt{I_2}}{2\sqrt{I_1} - \sqrt{I_2}}\right]^2$$

$$\Rightarrow \quad \frac{I_{max}}{I_{min}} = 9$$

$$\Rightarrow \quad \frac{I_{max} + I_{min}}{I_{max} - I_{min}} = \frac{10}{8}$$

$$\Rightarrow \quad \frac{5}{x} = \frac{10}{8}$$

$$\Rightarrow \quad x = 4$$

Ans. (D)

Sol. 60 As first dark fringe is formed in front of one of the slits so we use $x = d/2$. The path difference at this point on screen is given as

$$\Delta = \frac{dx}{D} = \frac{\lambda}{2}$$

$$\Rightarrow \quad \lambda = \frac{d^2}{D}$$

$$\Rightarrow \quad \lambda = 450 \times 10^{-9}\,\text{m}$$

Ans. (450)

Sol. 61 Figure below shows the situation described in question.

Path difference at O (Screen Center) will be $n\lambda$ as at screen centre a maxima is there so we use

$$(\mu - 1)t = n\lambda$$

$$\Rightarrow \quad (1.5 - 1)x\lambda = n\lambda$$

$$\Rightarrow \quad x = 2n$$

For $n = 1$ it gives $x = 2$

Ans. (B)

Sol. 62 When system is immersed in a medium, the new fringe width will become

$$\beta_m = \frac{\beta_{air}}{\mu} = \frac{0.35 \times 5}{7} = 0.25$$

$$\Rightarrow \quad \frac{1}{\alpha} = \frac{25}{100}$$

$$\Rightarrow \quad \alpha = 4$$

Ans. (4)

Sol. 63 Fringe width in YDSE setup is given as

$$\beta = \frac{D\lambda}{d}$$

At $\lambda_1 = 5000$ Å, we have

$$\beta_1 = \frac{D}{d}(5000 \times 10^{-10}) = 5 \times 10^{-4}\,\text{m} \quad \ldots(1)$$

And for $\lambda_2 = 6000$ Å, we have

$$\beta_2 = \frac{D}{(2d)}(6000 \times 10^{-10}) = x \quad \ldots(2)$$

Dividing equation-(2) & (1) gives

$$\frac{\beta_2}{\beta_1} = \frac{3000 \times 10^{-10}}{5000 \times 10^{-10}} = \frac{x}{5 \times 10^{-4}}$$

$$\Rightarrow \quad x = 3 \times 10^{-4}\,\text{m or } 0.3\,\text{mm}$$

Ans. (D)

Sol. 64 In YDSE setup fringe width is given as

$$\beta = \frac{\lambda D}{d}$$

Change in fringe width is calculated as

$$\Delta\beta = \frac{\lambda}{d}\Delta D$$

$$\Rightarrow \quad \lambda = \frac{\Delta\beta \cdot d}{\Delta D}$$

$$\Rightarrow \quad \lambda = \frac{3 \times 10^{-5} \times 1 \times 10^{-3}}{5 \times 10^{-2}}$$

$$\Rightarrow \quad \lambda = 60 \times 10^{-8} = 600 \times 10^{-9}\,\text{m}$$

$$\Rightarrow \quad \lambda = 600\,\text{nm}$$

Ans. (600)

Sol. 65 In a medium light wavelength changed to

$$\lambda_{med} = \frac{\lambda_{vacuum}}{\mu}$$

Fringe width in YDSE setup is given as

$$\beta = \frac{\lambda D}{d}$$

$$\Rightarrow \qquad \beta_{med} = \frac{\beta_{vacuum}}{\mu} = \frac{12}{\frac{4}{3}} = 9 \text{ mm} \qquad \textbf{Ans. (B)}$$

Sol. 66 Phase differences between the beams at points A and B are given as

$$\phi_A = \frac{\pi}{2}$$

and $$\phi_B = \frac{\pi}{3}$$

Resulting intensity of light at point A is given as

$$I_A = I + 4I + 2\sqrt{I}\sqrt{4I}\cos\left(\frac{\pi}{2}\right)$$

$$\Rightarrow \qquad I_A = 5I + 4I(0) = 5I$$

Resulting intensity of light at point B is given as

$$I_B = I + 4I + 2\sqrt{I}\sqrt{4I}\cos(60°)$$

$$\Rightarrow \qquad I_B = 5I + 4I \times \frac{1}{2} = 7I$$

Difference in light intensities at the two points is given as

$$I_B - I_A = 7I - 5I = 2I \qquad \textbf{Ans. (2)}$$

Sol. 67 Intensity ratio of lights is given as

$$\frac{I_1}{I_2} = \frac{1}{4}$$

$$\Rightarrow \qquad I_2 = 4I_1$$

Maximum and minimum intensities of light are given as

$$I_{max} = I_1 + 4I_1 + 2\sqrt{I_1 4I_1} = 9I_1$$

and $$I_{min} = I_1 + 4I_1 - 2\sqrt{I_1 4I_1} = I_1$$

$$\Rightarrow \qquad \frac{9I_1 + I_1}{9I_1 - I_1} = \frac{10}{8} = \frac{5}{4} = \frac{2\alpha + 1}{\beta + 1}$$

$$\Rightarrow \qquad \alpha = 2$$

and $$\beta = 1$$

$$\Rightarrow \qquad \frac{\alpha}{\beta} = \frac{2}{1} = 2 \qquad \textbf{Ans. (B)}$$

Sol. 68 In YDSE setup the fringe width is directly proportional to light wavelength, so we use

$$\beta \propto \lambda$$

Given that $$\lambda_2 = \frac{9}{8}\lambda_1$$

$$\Rightarrow \qquad \lambda_2 = \frac{9}{8} \times 560 = 630 \text{ nm} \qquad \textbf{Ans. (630)}$$

Sol. 69 Resulting light intensity at phase difference ϕ is given as

$$I_{net} = I_1 + I_2 + 2\sqrt{I_1}\sqrt{I_2}\cos\phi$$

At point A, we have I_{max} and at point B we have I_{min} so we use

$$I_{max} = (\sqrt{I_1} + \sqrt{I_2})^2 = (\sqrt{9I} + \sqrt{4I})^2 = 25I$$

$$I_{min} = (\sqrt{I_1} - \sqrt{I_2})^2 = (\sqrt{9I} - \sqrt{4I})^2 = I$$

$$\Rightarrow \qquad I_{max} - I_{min} = 25I - I = 24I \qquad \textbf{Ans. (24)}$$

Sol. 70 Radiation power of bulb is given as

$$P' = 10\% \text{ of } 110 \text{ W}$$

$$P' = \frac{10}{100} \times 110 \text{ W} = 11 \text{ W}$$

Change in intensities of the bulb radiation at distances 1 m and 5 m is calculated as

$$I_1 - I_2 = \frac{P'}{4\pi r_1^2} - \frac{P'}{4\pi r_2^2}$$

$$\Rightarrow \qquad I_1 - I_2 = \frac{11}{4\pi}\left[\frac{1}{1} - \frac{1}{25}\right]$$

$$\Rightarrow \qquad I_1 - I_2 = \frac{11}{4\pi} \times \frac{24}{25}$$

$$\Rightarrow \qquad I_1 - I_2 = \frac{264}{\pi} \times 10^{-2} = 84 \times 10^{-2} \text{ W/m}^2 \qquad \textbf{Ans. (84)}$$

Sol. 71 In Young's double slit experiment, the distance of nth bright fringe from screen centre is given as

$$y_{n\text{th}} = \frac{n\lambda D}{d}$$

$$\Rightarrow \qquad \frac{5 \times 600 \times 10^{-9} \times 1}{d} = 5 \times 10^{-2}$$

$$\Rightarrow \qquad d = \frac{5 \times 600 \times 10^{-9} \times 1}{5 \times 10^{-2}} = 60 \times 10^{-6} \text{ m}$$

$$\Rightarrow \qquad d = 60\,\mu\text{m} \qquad \textbf{Ans. (A)}$$

Sol. 72 If first minimum is obtained at point P, the condition of path difference is given as

$$A_2P - A_1P = \frac{\lambda}{2}$$

$$\Rightarrow \qquad \sqrt{D^2 + a^2} - D = \frac{\lambda}{2}$$

$$\Rightarrow \qquad D\left(1 + \frac{a^2}{D^2}\right)^{1/2} - D = \frac{\lambda}{2}$$

$$\Rightarrow \qquad D\left(1 + \frac{1}{2} \times \frac{a^2}{D^2}\right) - D = \frac{\lambda}{2}$$

$$\Rightarrow \qquad \frac{a^2}{2D} = \frac{\lambda}{2}$$

$\Rightarrow \quad a = \sqrt{\lambda . D}$

$\Rightarrow \quad a = \sqrt{800 \times 10^{-6} \times 50}$

$\Rightarrow \quad a = 0.2 \text{ nm}$ **Ans. (C)**

Sol. 73 If central bright fringe is shifted by a distance y as shown in figure below, the path difference at point P is calculated as given below

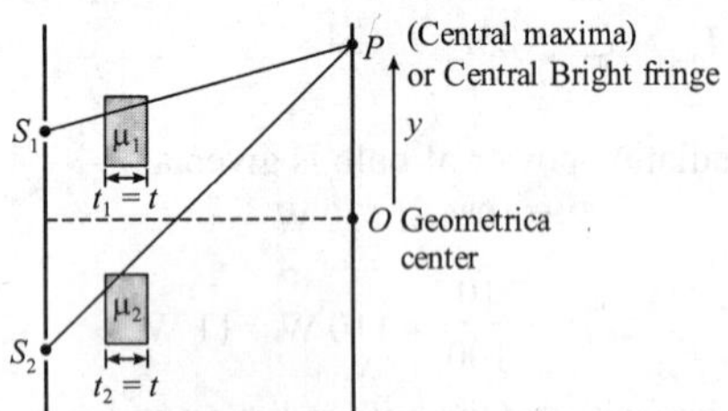

Considering path difference at P to be Δx, we use

$$\Delta x = (\mu_2 - \mu_1)t$$

$$\Rightarrow \quad \Delta x = (1.55 - 1.51)\, 0.1 \text{ mm}$$

$$\Rightarrow \quad \Delta x = 0.04 \times 10^{-4}$$

$$\Rightarrow \quad \Delta x = 4 \times 10^{-6} = 4 \text{ μm}$$

The distance y is now given as

$$y = \frac{\Delta x D}{d} = 4 \times 10^{-6} \frac{D}{d}$$

$$\text{Fringe width} = \frac{\lambda D}{d} \times 4 \times 10^{-6} \text{ m} \frac{D}{d} = 4 \text{ μm} \frac{D}{d}$$

Number of bright fringes which will cross the central point of screen is calculated as

$$N = \frac{y}{\beta}$$

$$\Rightarrow \quad N = \frac{4 \times 10^{-6} D/d}{4 \times 10^{-7} D/d} = 10$$ **Ans. (10)**

Sol. 74 Resultant intensity by interference is

$$I = 4I_0 \cos^2\left(\frac{\Delta\phi}{2}\right)$$

$$\text{Phase difference} = \frac{2\pi}{\lambda} \times \text{Path difference}$$

$$\Delta\phi = \frac{2\pi}{\lambda} \times \Delta x$$

we use $\quad \Delta\phi_1 = \frac{2\pi}{\lambda} \times \frac{\lambda}{4}$

$$\Rightarrow \quad \Delta\phi_1 = \frac{\pi}{2}$$

and $\quad \Delta\phi_2 = \frac{2\pi}{\lambda} \times \frac{\lambda}{3}$

$$\Rightarrow \quad \Delta\phi_2 = \frac{2\pi}{3}$$

We use

$$I_1 = 4I_0 \cos^2\left(\frac{\pi}{4}\right) = 2I_0$$

and $\quad I_2 = 4I_0 \cos^2\left(\frac{\pi}{3}\right) = I_0$

$$\Rightarrow \quad \frac{I_1 + I_2}{I_0} = 3$$ **Ans. (3)**

Sol. 75 Fringe width when light of 800 nm and 600 nm are used to obtain interference pattern

$$\omega_1 = \frac{\lambda_1 D}{d} = 16 \text{ mm}$$

& $\quad \omega_2 = \frac{\lambda_2 D}{d} = 12 \text{ mm}$

Thus first common distance where the bright fringes will overlap is given by lowest common factor of above two fringe widths, given as

$$l = 48 \text{ mm}$$

Thus at 48 mm distance both bright fringes will be found.

Ans. (48)

Sol. 76 Displacement equation of two interfering wave is given as

$$y_2 = 5(\sin \omega t + \sqrt{3} \cos \omega t)$$

$$\Rightarrow \quad y_2 = 10 \sin\left(\omega t + \frac{\pi}{3}\right)$$

Thus the phase difference between the waves is 0 thus resulting amplitude of the wave is calculated as

$$A = A_1 + A_2 = 20 \text{ cm}$$ **Ans. (20)**

Sol. 77 When glass slab inserted infront of one slit then the shift of central fringe is given as

$$\Delta x = \frac{t(\mu - 1)}{\lambda} \beta_0$$

$$\Rightarrow \quad \Delta x = \frac{10 \times 10^{-6}(1.2 - 1)}{5 \times 10^{-7}} \beta_0$$

$$\Rightarrow \quad \Delta x = \frac{10^{-5} \times 0.2}{5 \times 10^{-7}} = 4\beta_0 = x\beta_0$$

$$\Rightarrow \quad x = 4$$ **Ans. (4)**

Sol. 78 As medium changes, optical path changes, but frequency remain constant. The phase difference is given as

$$\Delta\phi = \frac{2\pi}{\lambda}(\Delta x)$$

Here, $\Delta\phi$ is phase difference and Δx is path difference

$$\Rightarrow \quad \frac{\Delta x}{\lambda} = \frac{\Delta\phi}{2\pi}$$

Hence, phase difference changes. **Ans. (C)**

Sol. 79 Fringe width in YDSE is given as

$$\beta = \frac{\lambda D}{d}$$

If n^{th} bright fringe of λ_1 match with m^{th} bright fringe of λ_2, we use

$$\Rightarrow \quad n\beta_1 = m\beta_2$$

$$\Rightarrow \quad n\lambda_1 = m\lambda_2$$

$$\Rightarrow \quad \frac{n}{m} = \frac{\lambda_2}{\lambda_1} = \frac{5500}{7000}$$

$$\Rightarrow \quad \frac{n}{m} = \frac{11}{14}$$

Distance where bright fringe will match each other is calculated as

$$\Rightarrow \quad x = n\beta_1 = \frac{11 \times 7000Å \times 150 \text{ cm}}{0.25 \text{ cm}}$$

$$\Rightarrow \quad x = 462 \times 10^{-5} \text{m}$$ **Ans. (462)**

Sol. 80 Fringe width β for light of wavelength λ is given as

$$\beta = \frac{D\lambda}{d}$$

$$\Rightarrow \quad \beta \propto \lambda$$

$$\Rightarrow \quad \frac{\beta_2}{\beta_1} = \frac{\lambda_2}{\lambda_1}$$

Given that $\beta_1 = 2$ mm, so we have

$$\Rightarrow \quad \frac{\beta_2}{2\text{mm}} = \frac{600 \text{ nm}}{400\text{nm}} = \frac{3}{2}$$

$$\Rightarrow \quad \beta_2 = 3 \text{ mm}$$ **Ans. (C)**

Sol 81 .Resulting intensity at point of intereference P when phase difference is $\frac{\pi}{3}$ is given as

$$I_{RP} = I_1 + I_2 = 2\sqrt{I_1 I_2} \cos\phi$$

$$I_{RP} = I_0 + I_0 + 2I_0 \cos\frac{\pi}{3}$$

$$\Rightarrow \quad I_{RP} = 2I_0 + 2I_0 \times \frac{1}{2} = 3I_0$$

Intensity when phase difference is $\frac{\pi}{2}$ at point of interference Q is given as

$$I_{RQ} = I_0 + I_0 + 2I_0 \cos 90° = 2I_0$$

Ratio of intensities at P and Q is given as

$$\Rightarrow \quad \frac{I_P}{I_Q} = \frac{3}{2}$$ **Ans. (C)**

Sol. 82 For lowest wavelength of Lyman series, we use

$$\frac{1}{\lambda} = RZ^2\left[\frac{1}{1^2} - \frac{1}{\infty^2}\right] = RZ^2$$

For lowest wavelength of Balmer series, we use

$$\frac{1}{\lambda'} = RZ^2\left[\frac{1}{2^2} - \frac{1}{\infty^2}\right] = \frac{RZ^2}{4}$$

$$\Rightarrow \quad \lambda' = \frac{4}{RZ^2} = 4 \times 917$$

$$\Rightarrow \quad \lambda' = 3668 \text{ Å}$$ **Ans. (3668)**

Sol. 83 Given that the amplitude ratio is

$$\frac{A_1}{A_2} = \frac{2}{1}$$

Thus ratio of maximum to minimum intensity in interference pattern is given as

$$\frac{I_{max}}{I_{min}} = \left(\frac{A_1 + A_2}{A_1 + A_2}\right)^2 = \frac{9}{1} = 9 : 1$$ **Ans. (D)**

Sol. 84 As for first diffraction minima, we use

$$a \sin\theta = \lambda$$

$$\Rightarrow \quad a \sin 30° = 600 \times 10^{-9}$$

$$\Rightarrow \quad a = 1200 \times 10^{-9} \text{m}$$

$$\Rightarrow \quad a = 1.2 \text{ μm}$$ **Ans. (B)**

24.2 Diffraction of Light

Sol. 1

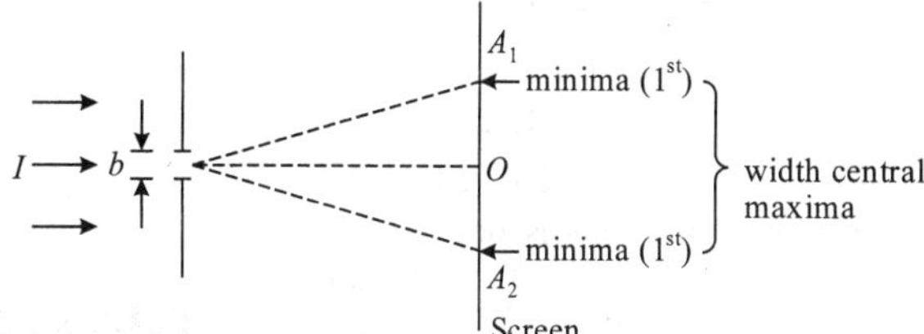

Intensity of central maxima on screen.

$$I_0 = \frac{\text{Power incident on central max}}{\text{area of screen } (A_1 A_2)}$$

If slit width is doubled then power incident on screen will get doubled and we use $\theta = \frac{\lambda}{b}$ (for central maxima)

So for double slit width we have angular separation $\rightarrow \theta'$

$$= \frac{\lambda}{2b} = \frac{\theta}{2}$$

$\Rightarrow$ On doubling slit width the area of central maxima will reduce to half

So from equation we use

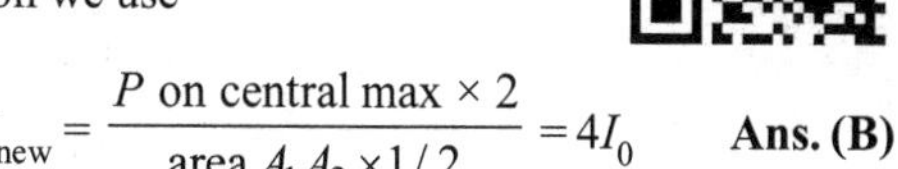

$$I_{new} = \frac{P \text{ on central max} \times 2}{\text{area } A_1 A_2 \times 1/2} = 4I_0$$ **Ans. (B)**

Sol. 2 If single slit diffraction for n^{th} minima on screen we use

$$b \sin \theta = n\lambda \qquad [b \to \text{width of slit}]$$

For first minima, we use

$$n = 1$$

$$\Rightarrow \quad b \sin \theta = \lambda$$

$$\Rightarrow \quad b = \frac{\lambda}{\sin\theta} = \frac{5\times10^{-7}}{0.5} = 10\times10^{-7}\,\text{m}$$

$$\Rightarrow \quad b = 10\times10^{-5}\,\text{cm}$$

Ans. (C)

Sol. 3 **Statement-1 :** As we know wavelength of radio waves is more then that of microwaves so the diffraction possibility is more for radio waves because diffraction is more in case of wavelength of the order of obstacles in surrounding or in path of waves.

⇒ **Statement-1 :** is true

Statement-2 : As both are *EM* waves so for

$$\lambda_R > \lambda_\mu$$

$$\Rightarrow \quad \nu_R < \nu_\mu$$

⇒ **Statement-2 :** is false **Ans. (C)**

Sol. 4 The minimum separation between two lines which can be just resolved by a microscope is given as

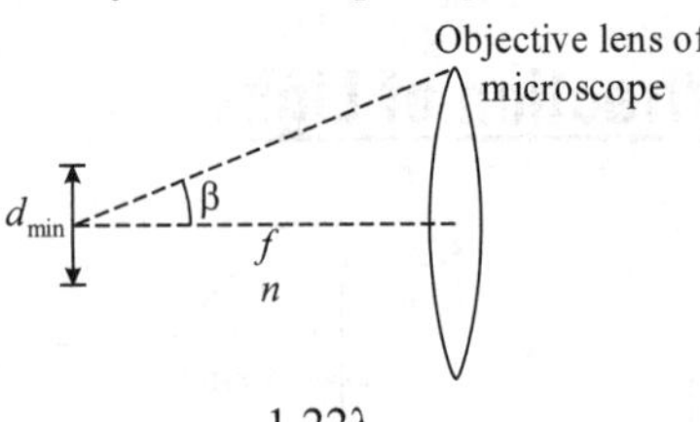

$$d_{min} = \frac{1.22\lambda}{2n\sin\beta}$$

Resolving power of microscope

$$R = \frac{1}{d_{min}} = \frac{2n\sin\beta}{1.22\lambda}$$

$$\sin 2\beta = 2\sin\beta\cos\beta$$

& for small β, we use

$$\cos\beta \simeq 1$$

$$\Rightarrow \quad R \propto n\sin 2\beta$$

Ans. (C)

Sol. 5

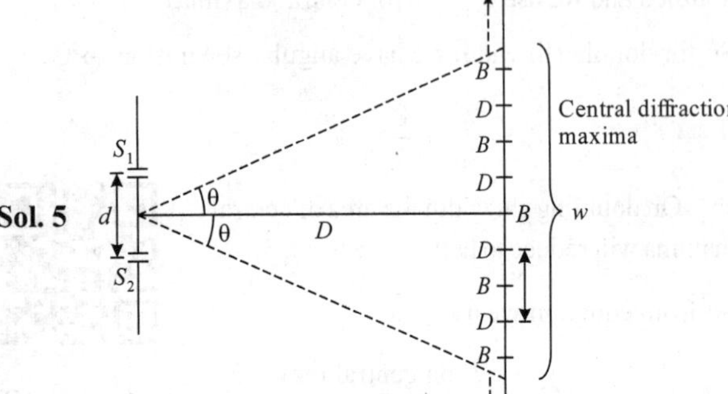

$$d = 6.1b$$

$b \to$ slit width

As interference of significant intensity of tight from the two slits will take place in the region of central diffraction maxima as shown in figure

Fringe width of *YDSE* pattern is

$$\beta = \frac{\lambda D}{d} \qquad \ldots(1)$$

Angular width of central diffraction maxima is given as

$$2\theta = \frac{2\lambda}{b}$$

Width of central diffraction maxima is

$$w = 2\theta D = \frac{2\lambda D}{b} \qquad \ldots(2)$$

No of bright fringes in width of central maxima is

$$N = \frac{w}{\beta} = \frac{2\lambda D}{b}\times\frac{d}{\lambda D}$$

We use $d = 1b$

$$\Rightarrow \quad N = \frac{2\times1.1b}{b} = 12.2$$

⇒ Complete 12 bright fringes are obtained in width of central maxima. **Ans. (C)**

Sol. 6 In a single slit diffraction pattern if θ is angular position of a point on screen from central maxima then we have

For a wavelength λ_1 (red light) at the position of first minimum we use

$$b \sin\theta = \lambda_1 \qquad \ldots(1)$$

For one there wavelength λ_1 at the same angle θ for first maximum, we use

$$b \sin\theta = \frac{3\lambda_2}{2} \qquad \ldots(2)$$

We use from (1) and (2)

$$\lambda_1 = \frac{3\lambda_2}{2}$$

$$\Rightarrow \quad \lambda_2 = \frac{2\lambda_1}{3} = \frac{2\times6600}{3}$$

$$= 4400\,\text{Å}$$

Ans. (B)

Sol. 7

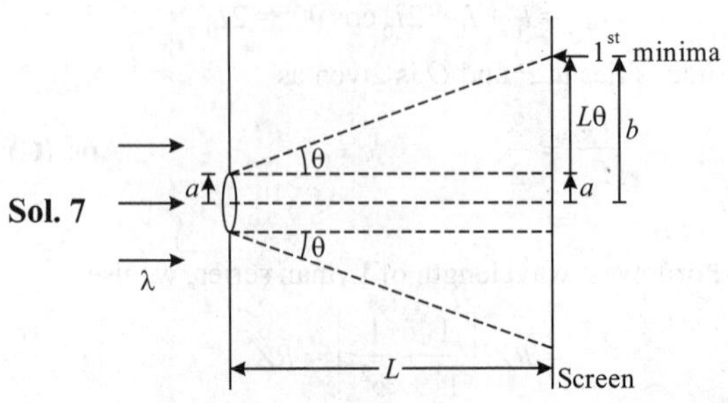

The angular position of first diffraction minima on either side of the hole is

$$a \sin\theta = \lambda \qquad [n=1]$$

$$\sin\theta = \frac{\lambda}{a}$$

For small θ we use

$$\theta = \frac{\lambda}{a} \qquad [\sin\theta \simeq \theta]$$

The radius of spot can be given as

$$b = L\theta + a$$

Spread due to differaction = Geometric spread

$$\Rightarrow \quad b = \frac{L\lambda}{a} + a$$

b is minimum when

$$\frac{db}{da} = 0 = -\frac{L\lambda}{a^2} + 1$$

$$\Rightarrow \quad a = \sqrt{L\lambda}$$

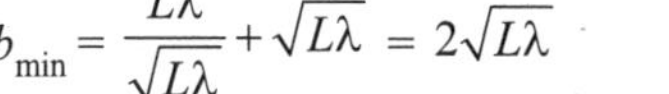
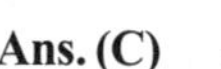

So $$b_{min} = \frac{L\lambda}{\sqrt{L\lambda}} + \sqrt{L\lambda} = 2\sqrt{L\lambda}$$ **Ans. (C)**

Sol. 8

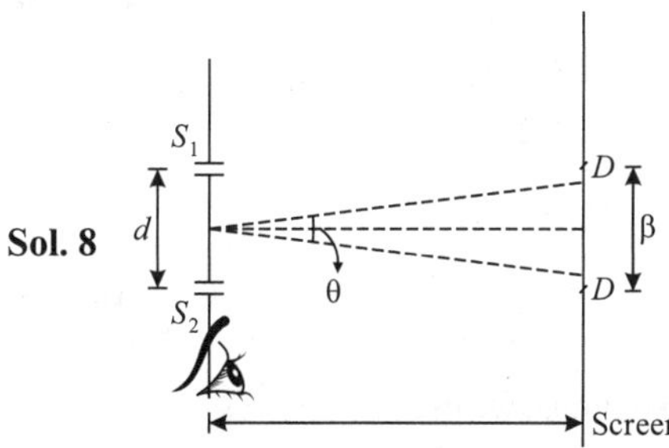

Angular fringe width in *YDSE* is given as

$$\theta = \frac{\beta}{D} = \frac{(\lambda D/d)}{D} = \frac{\lambda}{d}$$

Fringes will not be clearly seen by observer if $\theta < \theta_R$ (resolution limit of eye)

The fringes will just disappear at $d = d_0$

⇒ We use $\theta = \theta_R$ at $d = d_0$

$$\frac{\lambda}{d_0} = \frac{1}{60} \times \frac{\pi}{180}$$

$$\Rightarrow \quad d_0 = \frac{60 \times 180 \times 600 \times 10^{-9}}{3.14}$$

$$\Rightarrow \quad d_0 = 2.06 \times 10^{-3}\,\text{m}$$

$$\Rightarrow \quad d_0 = 2.06\,\text{mm}$$ **Ans. (C)**

Sol. 9 The angular resolution limit of a telescope is given as

$$\theta_R = \frac{1.22\lambda}{D}$$

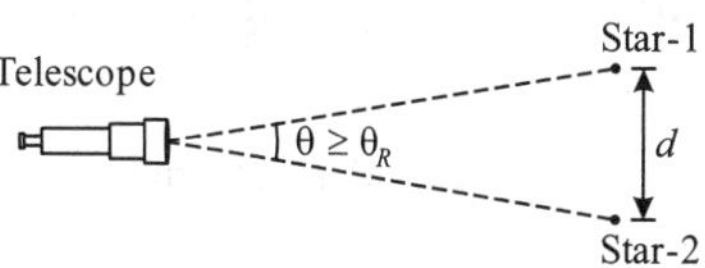

The two starts will be just resolved by the telescope when

$$L = 10 \text{ light yr}$$

$$\theta = \frac{d}{L}$$

$$\theta = \theta_R$$

$$\Rightarrow \quad \frac{d}{L} = \frac{1.22\lambda}{D}$$

$$\Rightarrow \quad d = \frac{1.22\lambda}{D}$$

$$\Rightarrow \quad d = \frac{1.22 \times 600 \times 10^{-9} \times 10 \times 9.46 \times 10^{15}}{0.3}$$

$$\Rightarrow \quad d = 2.308 \times 10^{11}\,\text{m}$$

$$\Rightarrow \quad d = 2.308 \times 10^{8}\,\text{km}$$ **Ans. (A)**

Sol. 10

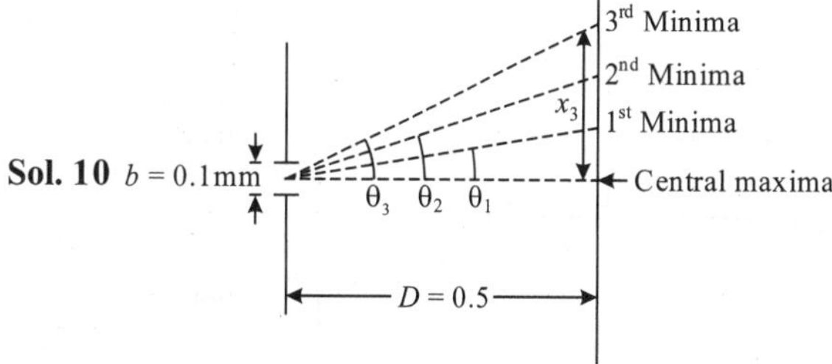

The angular separation of nth minima from screen centre (central maxima) is given as

$$b \sin\theta_n = n\lambda$$

from 3rd dark band ($n = 3$) we have

$$\theta_3 \simeq \sin\theta_3 = \frac{3\lambda}{b} \qquad [\text{as } \theta \text{ is } v \text{ small}]$$

distance of 3rd minima from O is

$$x_3 = D\theta_3 = 0.5 \times \frac{3 \times 6000 \times 10^{-10}}{0.1 \times 10^{-3}}$$

$$\Rightarrow \quad x_3 = 9 \times 10^{-3}\,\text{m} = 9\,\text{mm}$$ **Ans. (B)**

Sol. 11

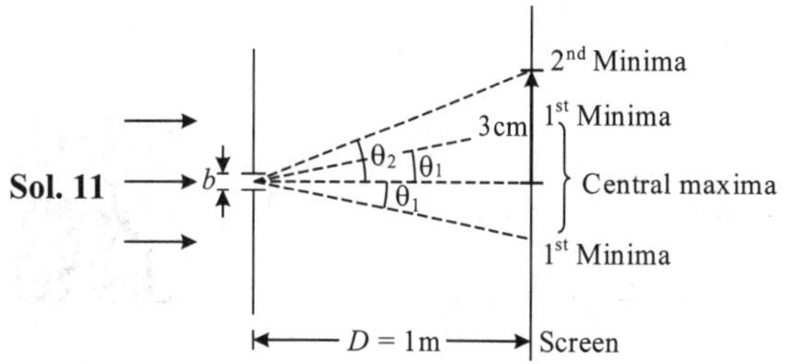

For first minima in diffraction pattern we use

$$b \sin \theta = n\lambda \qquad \ldots (1)\,[n=1]$$

The angular position of second minimum from screen centre is given as

$$\tan \theta_2 \simeq \sin \theta_2 \simeq \theta_2 = \frac{3\text{cm}}{100\text{cm}}$$

$$\Rightarrow \qquad \theta_2 = \frac{3\text{cm}}{1\text{cm}} = 0.03$$

From equation (1), for 2nd minima, we use

$$b \sin \theta_2 = 2\lambda$$

$$\Rightarrow \qquad \theta_2 = \frac{2\lambda}{b} = 0.03$$

Angular width of central maxima in diffraction pattern is given as

$$2\theta_1 = \frac{2\lambda}{b} = 0.03$$

Width of central maxima is

$$w = 2\theta_1 D = 0.03 \times 1\text{ m} = 0.03\text{ m} = 3\text{ cm}$$

Ans. (B)

Sol. 12

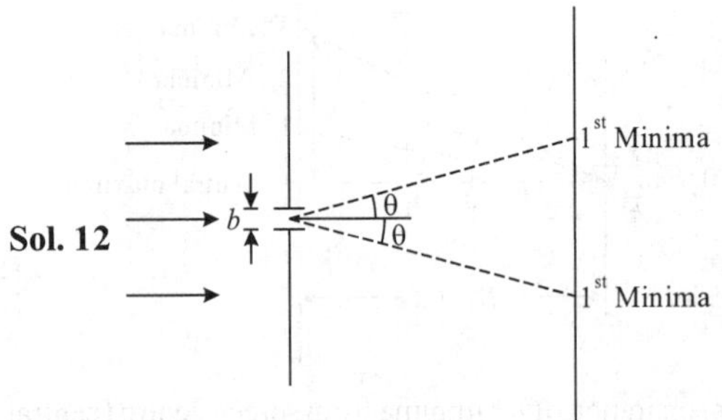

Angular position of first minima

$$b \sin \theta = \lambda$$

$$\Rightarrow \qquad \sin \theta = \frac{\lambda}{b}$$

So wavelength of light used

$$\lambda = b \sin 30° = \frac{b}{2}$$

given $\quad 2\theta = 60°$

$$\Rightarrow \qquad \theta = 30°$$

If similar two slits are used in *YDSE* at a separation '*d*' then we have

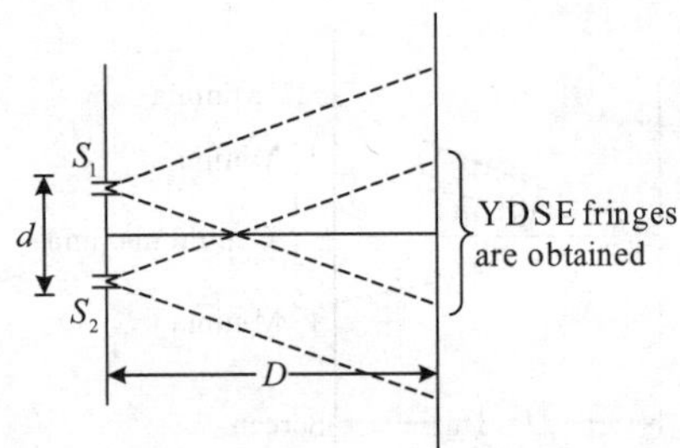

Fringe width in *YDSE* interference pattern is given as

$$b = \frac{\lambda D}{d} = \frac{bD}{2d}$$

$\Rightarrow$ Slit separation

$$d = \frac{bD}{2\beta} = \frac{10^{-6} \times 0.5}{2 \times 0.01} = 25 \times 10^{-6}\text{m}$$

$$\Rightarrow \qquad d = 25\ \mu\text{m}$$

Ans. (A)

Sol. 13

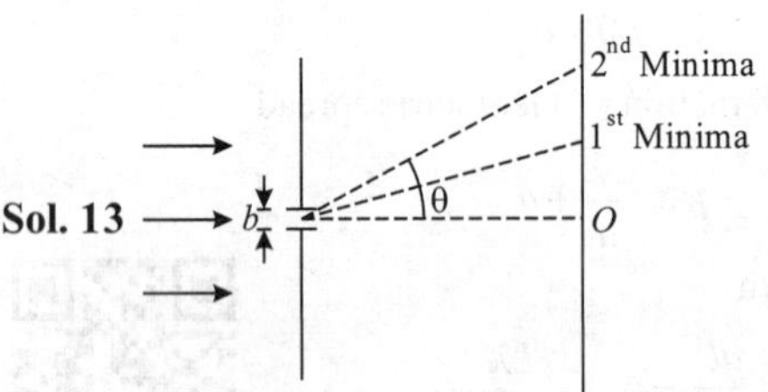

If at angular position θ from central maxima we get n^{th} minima then we use

$$b \sin \theta = n\lambda$$

So for 2nd minima we use $n = 2$

$$\Rightarrow \qquad b \sin \theta = 2\lambda$$

$$\Rightarrow \qquad \sin \theta = \frac{2\lambda}{b} = \frac{2 \times 550 \times 10^{-9}}{22 \times 10^{-7}} = 50 \times 10^{-2}$$

$$\Rightarrow \qquad \sin \theta = 0.5$$

$$\Rightarrow \qquad \theta = \frac{\pi}{6}$$

Ans. (D)

Sol. 14 For second minima, we use

$$b \sin \theta_2 = 2\lambda$$

$$\Rightarrow \qquad \sin 60° = \frac{2\lambda}{b}$$

$$\Rightarrow \qquad \frac{\lambda}{b} = \sqrt{\frac{3}{4}}$$

Now for first minima, we use

$$\sin \theta_1 = \frac{\lambda}{b} = \frac{\sqrt{3}}{4}$$

$$\Rightarrow \qquad \theta_1 \simeq 25°$$

Ans. (D)

Sol. 15 For n^{th} diffraction minima, we use

$$b \sin \theta = n\lambda$$

$$\Rightarrow \qquad \sin \theta = \frac{n\lambda}{b} \leq 1 \qquad \text{(for maximum limit)}$$

$$\Rightarrow \qquad n \leq \frac{b}{\lambda} = \frac{6 \times 10^{-5}}{6 \times 10^{-7}} = 100$$

Thus maximum number of minima on either side are 99 so total minima on screen are

$$99 \times 2 = 198$$ **Ans. (198)**

Sol. 16 First intensity maximum occurs at an angle θ, given as

$$2d \sin \theta = \lambda = \frac{h}{\sqrt{2mE}}$$

$$\Rightarrow \quad E = \frac{h^2}{8md^2 \sin^2 \theta} \times \frac{1}{e} eV$$

$$\Rightarrow \quad E = \frac{(6.64 \times 10^{-34})^2}{8 \times 9.1 \times 10^{-31} \times (10^{-10})^2 \times \frac{3}{4} \times 1.6 \times 10^{-19}}$$

$$\Rightarrow \quad E \cong 50.47$$ **Ans. (50)**

Sol. 17 For a pin hole of diameter a, the angular width of disc produced is given as

$$\sin \theta = \frac{m\lambda}{a}$$

When a increases, θ decreases due to which the area of disc decreases so intensity will increases. **Ans. (A)**

Sol. 18 As the size of object compared to wavelength is given as

$$\frac{a}{\lambda} = \frac{1}{100}$$

For reflection size of obstacle must be very high compared to wavelength, for diffraction size should be of the order of wavelength and here as the size of object is $\frac{\lambda}{100}$, which is much smaller than wavelength, so the phenomenon of scattering will dominate. **Ans. (D)**

Sol. 19 Resolving power of a compound microscope is given as

$$RP = \frac{2\mu \sin \theta}{1.22\lambda}$$

Hence increasing refractive index increases the resolving power. **Ans. (B)**

24.3 Polarization of Light

Sol. 1 Interference, reflection and diffraction are phenomenon which can be seen for any type of waves but polarization is based upon restricting the plane of oscillation of $\vec{E}$ in EM waves only in one plane & plane of oscillations are analyzed only in transverse waves. **Ans. (A)**

Sol. 2 Using Brewster's law, we have reflected light is completely polarized when IR incident at such an angle for which reflected and refracted ray are mutually perpendicular

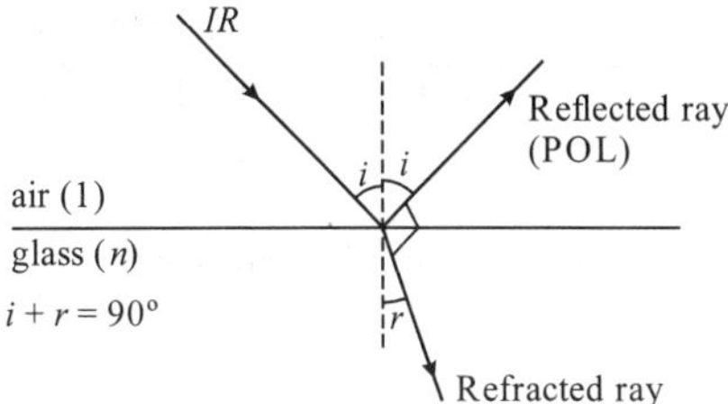

By Snell's law

$$\sin i = n \sin r = n \sin (90 - i)$$

$$\sin i = n \cos i$$

$$\Rightarrow \quad \tan i = n$$

$$\Rightarrow \quad i = \tan^{-1} (n)$$ **Ans. (D)**

Sol. 3

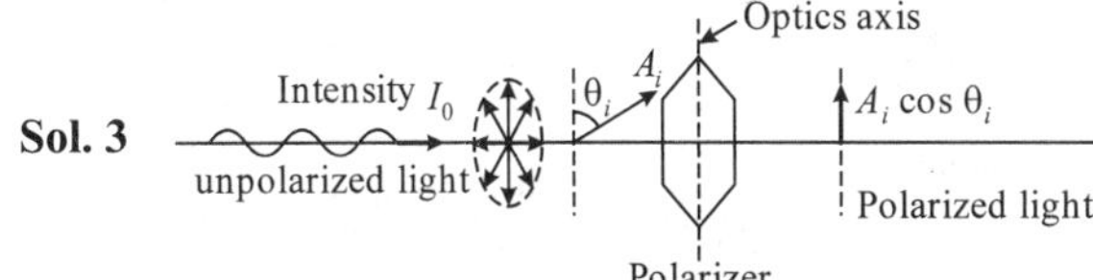

If a specific oscillation is considered before incidence on polarizer which has oscillations at an angle θ to optic axis as shown then in polarized light $A \cos \theta$ component will be transmitted so total intensity of polarized light can be given as

$$I_0 = \Sigma k A_i^2 = kA^2$$

$$I_P = k\Sigma(A_i \cos \theta_i)^2$$

$$I_P = kA_i^2 (\Sigma \cos^2 \theta_i) \text{ (average)}$$

$$I_P = kA_i^2 \left(\frac{1}{2}\right) = \frac{1}{2} kA^2 = \frac{1}{2} I_0$$

Here half incident light intensity does not get transmitted. **Ans. (A)**

Sol. 4 When sunlight is scattered by air particles in atmosphere

$$\text{Scattering} \propto \frac{1}{\lambda^4}$$

$\Rightarrow$ λ is least for blue/violet light so sky appears blue.

When light is seen at a specific region of atmosphere then polarized light due to scattering also enters the observer's eye so through polarizer (analyzer) light intensity varies on rotating it. **Ans. (C)**

Sol. 5

$2I_0$ Unpolarized light — A — I_0 Polarized light — B — I_0 Transmitted light

Final situation

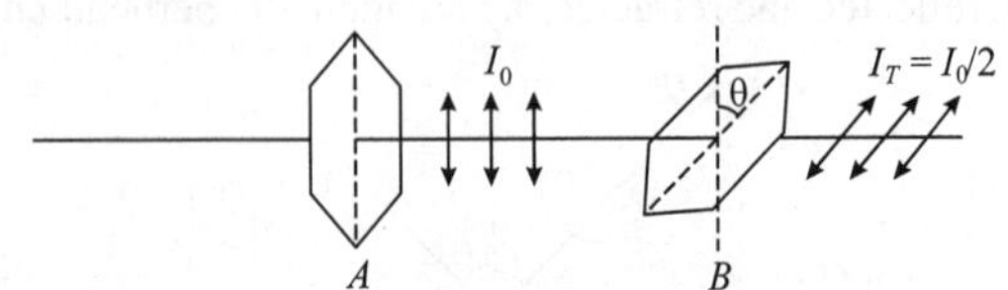

By Malus law the intensity of transmitted beam is

$$I_T = I_0 \cos^2 \theta$$

$$\Rightarrow \quad \frac{I_0}{2} = I_0 \cos^2\theta$$

$$\Rightarrow \quad \cos \theta = \frac{1}{\sqrt{2}}$$

$$\Rightarrow \quad \theta = 45°, 135°, 225°, 315°$$

Ans. (A)

Sol. 6

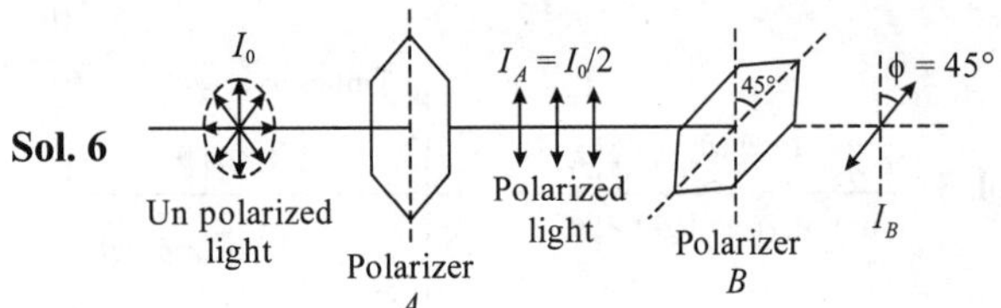

Intensity of light emerging from polarizer B is given by Malus law as

$$I_B = I_A \cos^2 \phi = \frac{I_0}{2} \cos^2(45°)$$

$$\Rightarrow \quad I_B = \frac{I_0}{2} \times \left(\frac{1}{\sqrt{2}}\right)^2 = \frac{I_0}{4}$$

Ans. (C)

7.

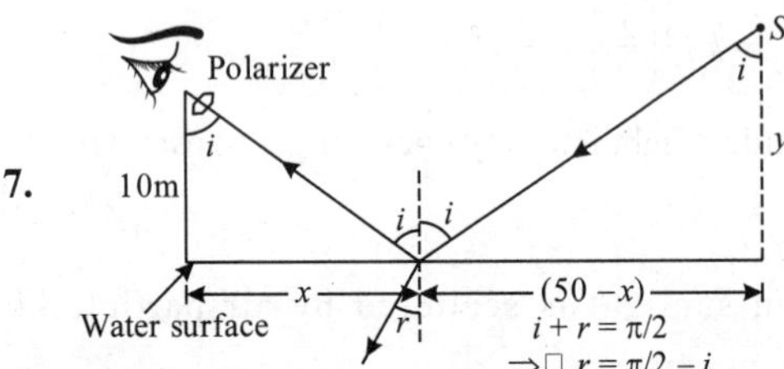

We use Snell's law at the point of reflection

$$\sin i = \mu_\omega \sin r$$

$$\Rightarrow \quad \sin i = \mu_\omega \sin\left(\frac{\pi}{2} - 1\right)$$

$$\Rightarrow \quad \tan i = \mu_\omega = \frac{4}{3}$$

In the figure shown we use

$$\tan i = \frac{x}{10} = \frac{50-x}{y} = \frac{4}{3} \quad \ldots(1)$$

$$\Rightarrow \quad y = \frac{10(50-x)}{x}$$

and $x = \frac{40}{3} m \simeq 13.33\,\text{m}$

$$\Rightarrow \quad y = \frac{10\left(50 - \frac{40}{3}\right)}{\frac{40}{3}} = \frac{3}{4} \times \frac{110}{3} = 27.5\,\text{m}$$

Ans. (B)

Sol. 8

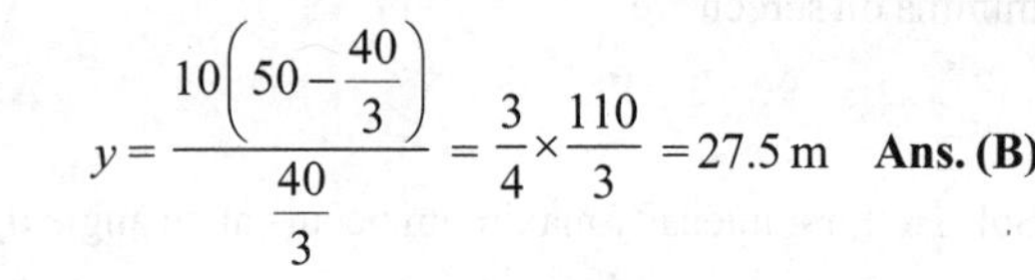

Here the polaroid axis after rotation by 30° from polarization plane of light A it is at an angle 60° from the polarization plane of light B.

By Malus law the transmitting intensities of lights A and B are

$$I_{AT} = I_A \cos^2 30°$$

and $$I_{BT} = I_B \cos^2 60°$$

Given that $$I_{AT} = I_{BT}$$

$$\Rightarrow \quad I_A \cos^2 30° = I_B \cos^2 60°$$

$$\Rightarrow \quad \frac{I_A}{I_B} = \frac{\cos^2 60°}{\cos^2 30°} = \frac{\left(\frac{1}{2}\right)^2}{\left(\frac{\sqrt{3}}{2}\right)^2} = \frac{1}{3}$$

Ans. (D)

Sol. 9

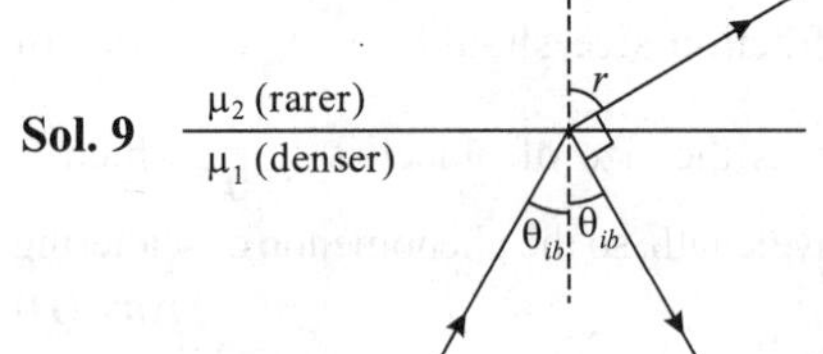

For Brewster's angle of incidence, we use

$$r = \frac{\pi}{2} - \theta_{ib}$$

By Snell's law, we have

$$\mu_1 \sin \theta_{ib} = \mu_2 \sin r = \mu_2 \sin\left(\frac{\pi}{2} - \theta_{ib}\right)$$

We use $$\sin \theta_{ic} = \frac{\mu_2}{\mu_1} \quad \ldots(1)$$

$$\Rightarrow \quad \mu_1 \sin \theta_{ib} = \mu_2 \cos \theta_{ib}$$

We also use $$\frac{\sin \theta_{ic}}{\sin \theta_{ic}} = 1.28 \quad \ldots(2)$$

$$\Rightarrow \quad \frac{\mu_2}{\mu_1} = \frac{\sin \theta_{ib}}{\cos \theta_{ib}} \quad \ldots(3)$$

From (1), (2) and (3) we get

$$\sin \theta_{ic} = \frac{\sin \theta_{ib}}{\cos \theta_{ib}}$$

$$\Rightarrow \quad \cos \theta_{ib} = \frac{1}{1.28}$$

So from equation-(3), we have

$$\frac{\mu_2}{\mu_1} = \frac{\sin\theta_{ib}}{\cos\theta_{ib}} = \frac{\sqrt{1-\cos^2\theta_{ib}}}{\cos\theta_{ib}}$$

$$\Rightarrow \quad \frac{\mu_2}{\mu_1} = \sqrt{\left(\frac{1}{\cos\theta_{ib}}\right)^2 - 1} = \sqrt{(1.28)^2 - 1}$$

$$\Rightarrow \quad \frac{\mu_2}{\mu_1} = \sqrt{0.64} = 0.8$$

Ans. (C)

Sol. 10

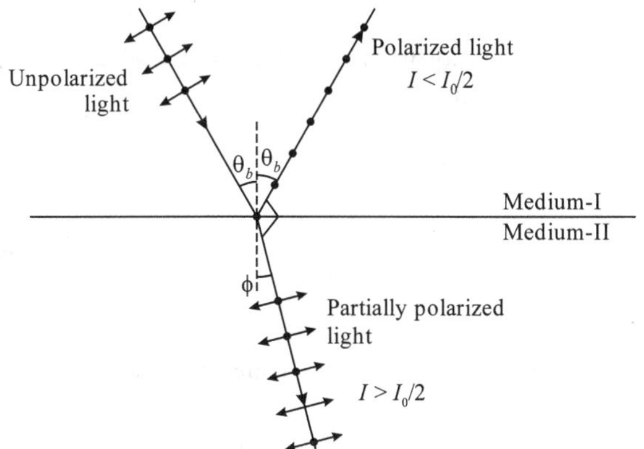

As transmitted light is partially polarized and it also contains the oscillation in the plane parallel to boundary which are there is reflected light.

So intensity of reflected light

$$I_{\text{reflected}} < I_0/2$$

Ans. (A)

Sol. 11

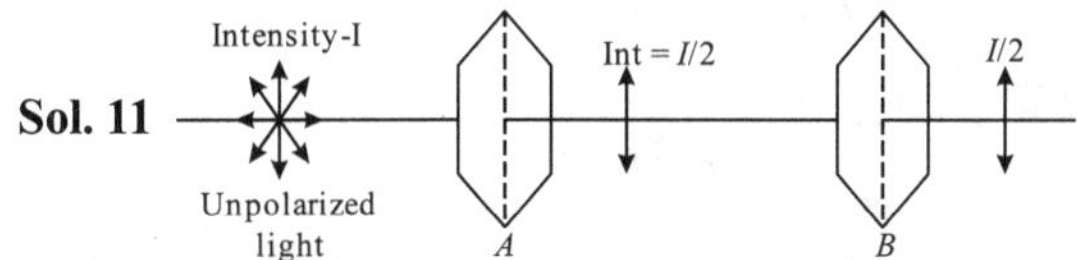

Initially when A and B are present then light intensity does not change on passing through $B \Rightarrow$ Axis of B is parallel to axis of A.

After inserting polarizer C, intensity after B becomes $I/8$

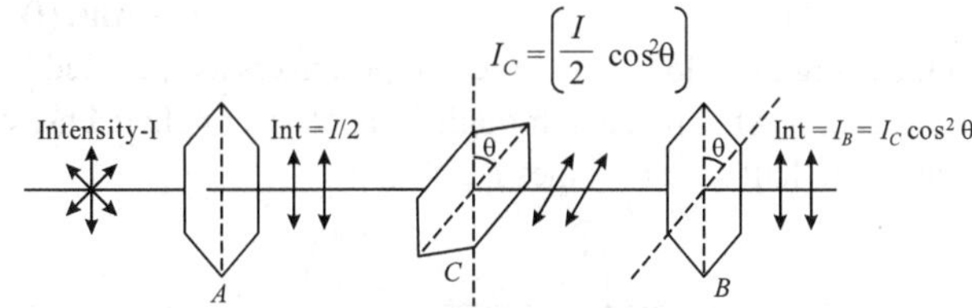

By Malus law $\quad I_C = \frac{I}{2}\cos^2\theta$

and $\quad I_B = I_C\cos^2\theta = \frac{I}{2}\cos^4\theta$

Given that finally $\quad I_B = \frac{I}{8}$

$$\Rightarrow \quad \frac{I}{8} = \frac{I}{2}\cos^4\theta$$

$$\Rightarrow \quad \cos^4\theta = \frac{1}{4}$$

$$\Rightarrow \quad \cos\theta = \frac{1}{\sqrt{2}}$$

$$\Rightarrow \quad \theta = 45°$$

Ans. (C)

Sol. 12

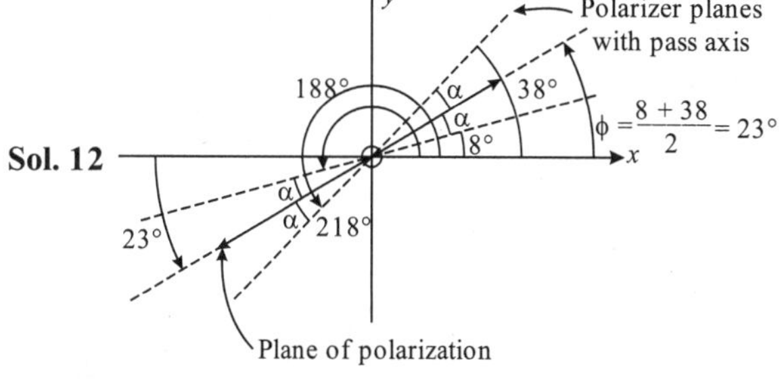

As the light intensity after passing through polarizer is same for the given pass axis (polarization axis)

$\Rightarrow$ Plane of polarization must be at same angle with the given direction of pass axis.

$$I_T = I\cos^2\alpha$$

Values of $\quad \phi = 23°$

or $\quad 180 + 23 = 203°$

Ans. (A)

Sol. 13 Case-I :

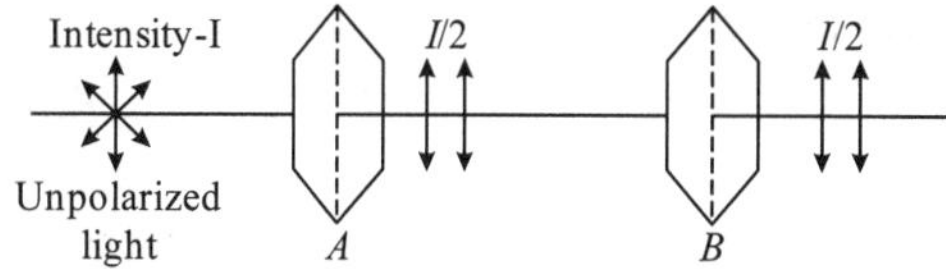

As light intensity does not change after passing through polarizer B

$\Rightarrow$ Axis of B is parallel to axis of A

Case-II :

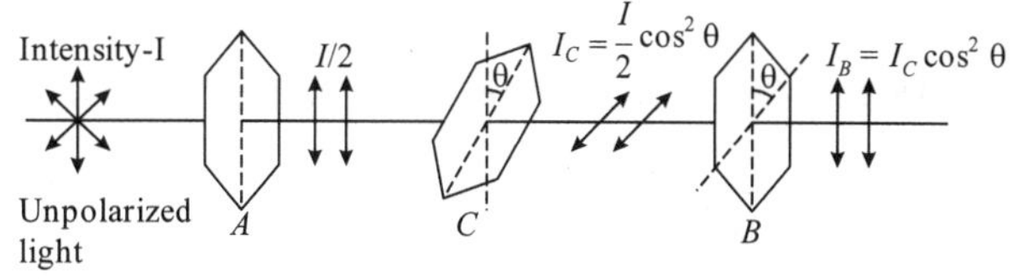

By Malus law have

$$I_C = \frac{I}{2}\cos^2\theta$$

and $\quad I_B = I_C\cos^2\theta = \frac{I}{2}\cos^4\theta$

Given that $I_B = \frac{I}{3}$

$\Rightarrow \quad \frac{I}{3} = \frac{I}{2}\cos^4\theta$

$\Rightarrow \quad \cos\theta = \left(\frac{2}{3}\right)^{1/4}$ **Ans. (A)**

Sol. 14 For light to never be reflected as polarized, incident angle on liquid glass interface must be less then brewster's angle so we use

$$\theta_{CL} < i_b$$

$\Rightarrow \quad \sin\theta_c < \sin i_b$

Here we use $\left[\tan i_b = \frac{1.5}{\mu}\right]$

$\Rightarrow \quad \sin i_b = \frac{1.5}{\sqrt{\mu^2 + (1.5)^2}}$

$\Rightarrow \quad \frac{1}{\mu} < \frac{1.5}{\sqrt{\mu^2 + (1.5)^2}}$

$\Rightarrow \quad \sqrt{\mu^2 + (1.5)^2} < 1.5 \times \mu$

$\Rightarrow \quad \mu^2 + (1.5)^2 < (\mu \times 1.5)^2$

$\Rightarrow \quad \mu > \frac{3}{\sqrt{5}}$ **Ans. (A)**

Sol. 15 Since unpolarised light falls on P_1, intensity of light transmitted from P_1 is given as

$$I_1 = I_0/2$$

Pass axis of P_2 is at an angle of 30° with P_1 so intensity of light transmitted from P_2 is given by Malus law as

$$I_2 = \frac{I_0}{2}\cos^2 30° = \frac{3I_0}{8}$$

Pass axis of P_3 is at an angle of 60° with P_2 so intensity of light transmitted from P_3 is given as

$$I_3 = \frac{3I_0}{8}\cos^2 60° = \frac{3I_0}{32}$$

$\Rightarrow \quad \left(\frac{I_0}{I}\right) = \frac{32}{3} = 10.67$ **Ans. (D)**

Sol. 16 Initially 10% light comes out so we use

$$\frac{I_0}{10} = \frac{I_0}{2} \times \cos^2\theta$$

$\Rightarrow \quad \cos\theta = \frac{1}{\sqrt{5}}$

$\Rightarrow \quad \theta = 63.44°$

Thus rotation angle is $\theta_R = 90 - 63.44° = 26.56°$

Out of given options, closest is (A) **Ans. (A)**

Sol. 17 Total power of light incident on polarizer is

$$P = IA = 3.3 \times 3 \times 10^{-4} = 9.9 \times 10^{-4}\,\text{W}$$

Average light intensity passes through polarizer is $I/2$ so we use amount of light energy passing through it in one revolution is

$$E = \frac{I}{2} \times A \times t = \frac{1}{2} \times 9.9 \times 10^{-4} \times \frac{2\pi}{31.4}$$

$\Rightarrow \quad E = 4.95 \times 10^{-4} \times 0.2$

$\Rightarrow \quad E = 0.99 \times 10^{-4}$

$\Rightarrow \quad E \simeq 1 \times 10^{-4}\,\text{J}$ **Ans. (C)**

Sol. 18 The intensity of emerging light from analyzer after it is rotated by 30° is given as

$$I = \frac{I_0}{2}\cos^2 30°$$

$\Rightarrow \quad I = 100\left(\frac{\sqrt{3}}{2}\right)^2 = 100 \times \frac{3}{4} = 75$ **Ans. (75)**

Sol. 19 The situation described in question is shown in figure below

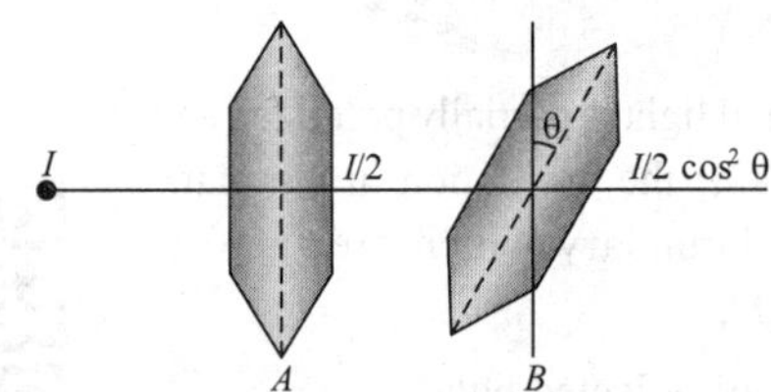

The intensity of light after polarizer A becomes $I/2$ then after passing through polarizer B it is given by Malus's law as

$$I' = \frac{I}{2}\cos^2\theta$$

$\Rightarrow \quad \frac{3I}{8} = \frac{I}{2}\cos^2\theta$

$\Rightarrow \quad \theta = 30°$ **Ans. (30)**

Sol. 20 As stated in the question that in the incident light electric field vectors are completely removed so there will be no reflection of light and there will be total transmission of light. **Ans. (D)**

Sol. 21 Figure below shows the two polarizers as specified in the question. After passing through first polarizer light intensity reduced to half so it becomes I_0

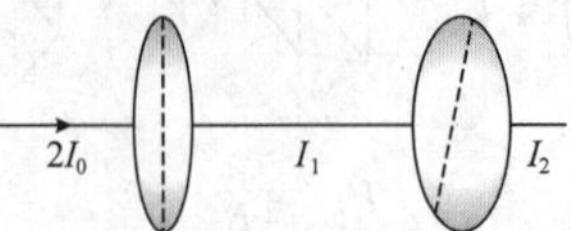

After passing through second polarizer the light intensity is given by Malus law, stated as

$$I_2 = I_1\cos^2 30°$$

$$\Rightarrow \quad I_2 = I_0 \cdot \frac{3}{4} = \frac{3I_0}{4} \quad \textbf{Ans. (C)}$$

Sol. 22 At Brewster's angle the reflected and refracted ray are perpendicular as shown in figure below.

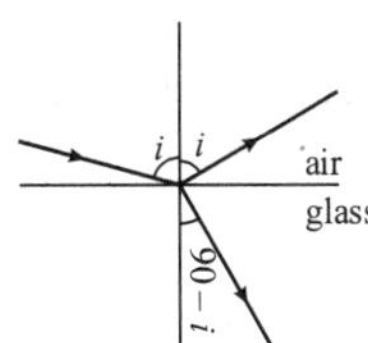

When light goes from air to glass, we use

$$\mu_a \sin i_1 = \mu_g \sin(90 - i_1)$$

$$\Rightarrow \quad \tan i_1 = \frac{\mu_g}{\mu_a}$$

When going from glass to air, we use

$$\tan i_2 = \frac{\mu_a}{\mu_g} = \cot i_1$$

$$\Rightarrow \quad i_2 = \frac{\pi}{2} - i_1 \quad \textbf{Ans. (B)}$$

Sol. 23 After passing through the first polaroid P_1 the light intensity will be reduced to half and it get polarised along the axis of polaroid. Next P_2 and P_3 will reduce the light intensity by $\cos^2(60°)$ and $\cos^2(30°)$ times respectively according to Malus Law. Final intensity of light coming out is given as

$$I = \frac{256}{2} \times \frac{1}{4} \times \left(\frac{\sqrt{3}}{2}\right)^2 = \frac{256 \times 3}{2 \times 4 \times 4}$$

$$\Rightarrow \quad I = 24 \quad \textbf{Ans. (24)}$$

Sol. 24 Refractive indices of the two media are calculated as

$$\mu_1 = \sqrt{2.8 \times 1} = \sqrt{2.8}$$

$$\mu_2 = \sqrt{6.8 \times 1} = \sqrt{6.8}$$

As the reflected and refracted rays are perpendicular to each other, using Snell's law, we have

$$\mu_1 \sin i = \mu_2 \cos i$$

$$\Rightarrow \quad \tan i = \frac{\mu_2}{\mu_1} = \sqrt{\frac{6.8}{2.8}}$$

$$\Rightarrow \quad \tan i = \left(\frac{2.8 + 4}{2.8}\right)^{1/2}$$

$$\Rightarrow \quad i = \tan^{-1}\left(1 + \frac{10}{7}\right)^{1/2}$$

$$\Rightarrow \quad \theta = 7 \quad \textbf{Ans. (7)}$$

Sol. 25 Intensity of unpolarized light after pssing through the polariser A is reduced to half, given as

$$I_A = \frac{I_0}{2}$$

Next polaroid C is placed with its polarization axis at 45° to that of A so the intensity of light coming out of C is given by Malus law as

$$I_C = \frac{I_0}{2} \cos^2(45°) = \frac{I_0}{4}$$

Similarly the light intensity after passing through the polaroid B is given as

$$I_B = I_C \cos^2(45°) = \frac{I_0}{8} \quad \textbf{Ans. (C)}$$

Sol. 26 After passing through first sheet, the light intensity becomes

$$I_1 = \frac{I}{2}$$

After passing through second sheet, light intensity will become

$$I_2 = I_1 \cos^2(45°) = \frac{I}{4}$$

After passing through n^{th} sheet, it can be written as

$$I_n = \frac{I}{2^n} = \frac{I}{64}$$

$$\Rightarrow \quad n = 6 \quad \textbf{Ans. (B)}$$

Sol. 27 Intensity of unpolarised light is given as

$$I_0 = 32 \text{ W/m}^2$$

Angle between transmission axes of p_1 and p_2 is θ, then angle between p_2 and p_3 is $(90 - \theta)$

If I_0 is incident intensity then from p_1 intensity coming out is given as

$$I_1 = \frac{I_0}{2}$$

By Malus law the inetensity of light coming out from p_2 is given as

$$I_2 = I_1 \cos^2\theta = \frac{I_0}{2}\cos^2\theta$$

Again using Malus law the intensity of light coming out from p_3 is given as

$$I_3 = I_2 \cos^2(90 - \theta) = \frac{I_0}{2}\cos^2\theta \sin^2\theta$$

$$\Rightarrow \quad I_3 = \frac{I_0 \sin^2(2\theta)}{8}$$

$$\Rightarrow \quad \sin^2 2\theta = \frac{8I_3}{I_0} = \frac{8 \times 3}{32} = \frac{3}{4}$$

$$\Rightarrow \quad \sin 2\theta = \frac{\sqrt{3}}{2}$$

$$\Rightarrow \quad 2\theta = 60°$$

$$\Rightarrow \quad \theta = 30° \text{ or } 60° \quad \textbf{Ans. (30, 60)}$$

25.1 Photoelectric Effect

Sol. 1 As we know work function ϕ and threshold wavelength λ_{th} are related as

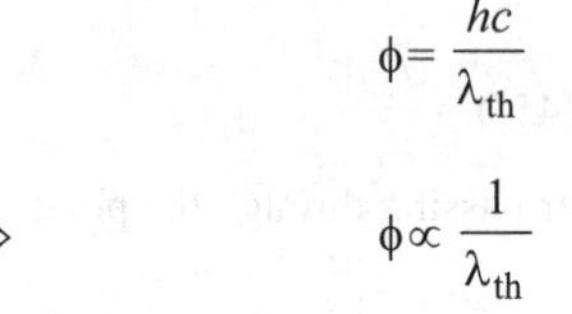

$$\phi = \frac{hc}{\lambda_{th}}$$

$$\Rightarrow \quad \phi \propto \frac{1}{\lambda_{th}}$$

We use $\dfrac{\phi_{Na}}{\phi_{Cu}} = \dfrac{\lambda_{Cu}}{\lambda_{Na}} = \dfrac{2.3}{4.5}$

$$\Rightarrow \quad \frac{\lambda_{Na}}{\lambda_{Cu}} = \frac{45}{23} \simeq \frac{2}{1}$$ **Ans. (C)**

Sol. 2 For photoelectric effect of given light frequencies on photocathods (work function = ϕ) we use

$$hf_1 = \phi + \frac{1}{2} mv_1^2 \quad \ldots(1)$$

and
$$hf_2 = \phi + \frac{1}{2} mv_2^2 \quad \ldots(2)$$

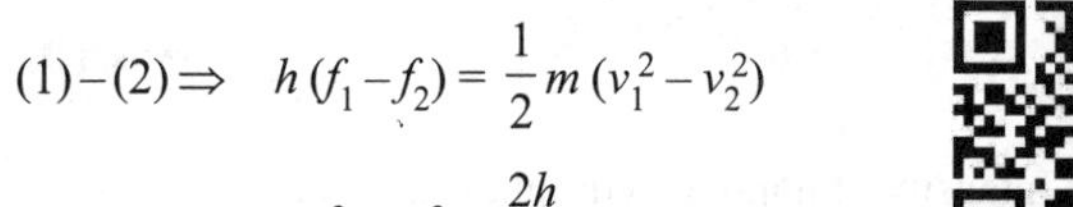

$$(1)-(2) \Rightarrow \quad h(f_1 - f_2) = \frac{1}{2} m (v_1^2 - v_2^2)$$

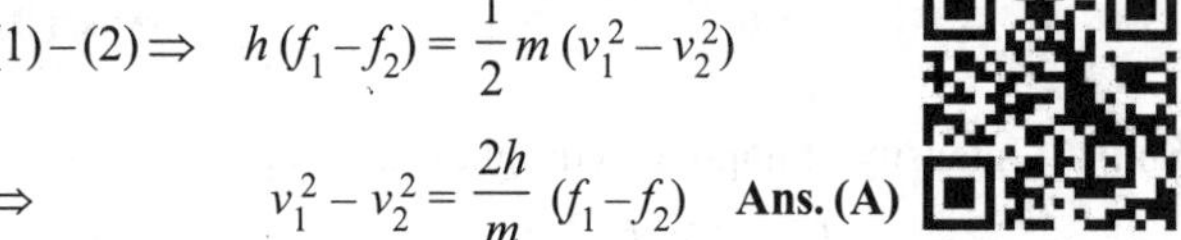

$$\Rightarrow \quad v_1^2 - v_2^2 = \frac{2h}{m}(f_1 - f_2)$$ **Ans. (A)**

Sol. 3 The largest wavelength (minimum energy) which can cause photoelectric emission is the threshold wavelength and it is given as :

$$\frac{hc}{\lambda_{th}} = \phi$$

$$\Rightarrow \quad \lambda_{th} = \frac{hc}{\phi} = \frac{6.63 \times 10^{-34} \times 3 \times 10^{8}}{4 \times 1.6 \times 10^{-19}}$$

$$\Rightarrow \quad \lambda_{th} = \frac{12431}{4} \text{ Å} \simeq 310.8 \text{ nm}$$ **Ans. (C)**

Sol. 4 In photo electric effect for a radiation frequency ν ($\nu > \nu_{th}$) when falls on a metal with work function ϕ then these are related to the $KE_{(max)}$ of ejected electron as

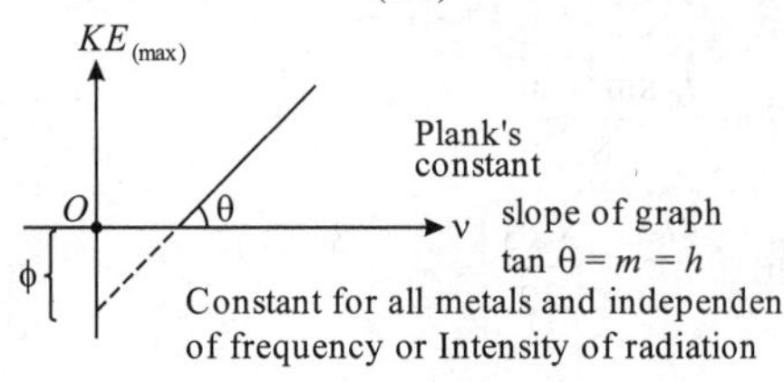

$$h\nu = \phi + KE_{max}$$

$$\Rightarrow \quad KE_{max} = h\nu - \phi$$

$$y = mx + c$$ **Ans. (D)**

Sol. 5

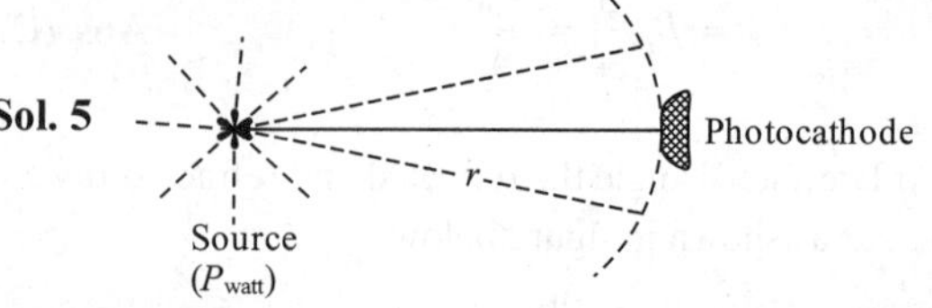

Intensity of light due to source at the location of photocathode is

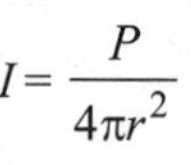

$$I = \frac{P}{4\pi r^2}$$

Here $I \propto \dfrac{1}{r^2}$

So when distance r is changed from 1 m to 0.5 m so intensity will become 4 times.

$\Rightarrow$ Number of ejected electrons will also increase by a factor of 4. **Ans. (B)**

Sol. 6

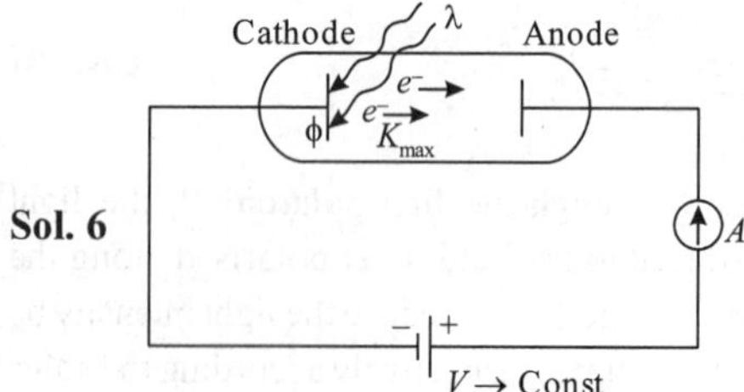

At cathode we use

$$\frac{hc}{\lambda} = \phi = K_{max}$$

As $\lambda \uparrow$ the energy $\dfrac{hc}{\lambda} \downarrow \Rightarrow K_{max} \downarrow$

As $K_{max} \downarrow \Rightarrow$ less electrons will reach anode $\Rightarrow$ current I decreases. **Ans. (B)**

Sol. 7 As threshold frequency gives the work function of a metal surface

$$\Rightarrow \quad \phi = h\nu_{th} = 6.2 \text{ eV}$$

Stopping potential for a metal surface gives the maximum kinetic energy for ejected photoelectrons

$$\Rightarrow \quad KE_{max} = eV_0 = 5 \text{ eV}$$

Energy of incident radiation

$$E_i = \phi = KE_{max} = 6.2 + 5 = 11.2 \text{ eV}$$

Wavelength λ of incident radiation is given as

$$\lambda = \frac{hc}{E_i} = \frac{12431}{11.2} \text{ Å} \simeq 1109.9 \text{ Å}$$

As UV region wavelength exist from ~ 100 Å to 4000 Å **Ans. (A)**

Sol. 8 The energy of incident light photon is taken as

$$E = \frac{hc}{\lambda} = \frac{12431}{4000}\text{ eV} = 3.10\text{ eV}$$

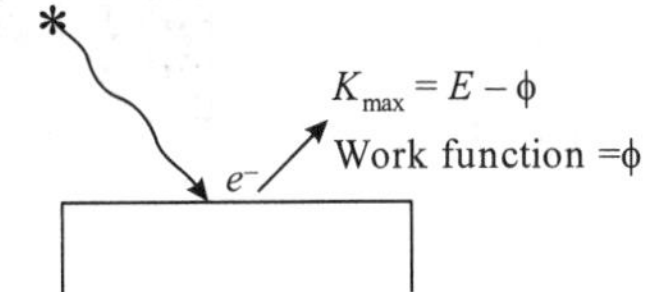

Here we use $KE_{max} = E - \phi$

$$1.68 = 3.10 - \phi$$

$\Rightarrow$ $\phi = 3.10 - 1.68 = 1.42$ eV **Ans. (D)**

Sol. 9 In Statement-1 we use

$$E_i = h\nu = \phi + KE_{max} \quad \ldots(1)$$

and stopping potential

$$V_0 = \frac{KE_{max}}{e}$$

When UV light is replaced by X-rays. Then we use $E_{i_2} > E_{i_1}$

$\Rightarrow$ $KE_{max_2} > KE_{max_1}$

from equation-(1) as ϕ is a constant KE_{max} increases with E_i

$\Rightarrow$ $V_{0_2} > V_{0_1}$

$\Rightarrow$ Statement-1 is True

In Statement-2 we have

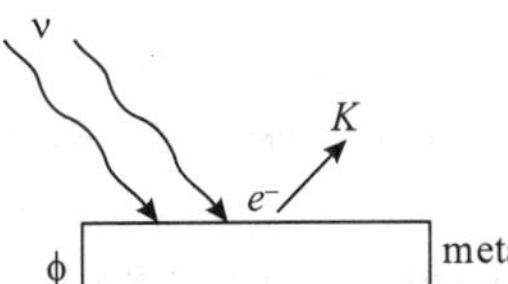

electrons in the process of ejection from metal surface looses some energy in collisions and other phenomenon so ejected KE is less than KE_{max}.

$\Rightarrow$ Statement-2 is False. **Ans. (A)**

Sol. 10 For Statement-1 we know in photoelectric effect

$$h\nu = \phi + KE_{max} \quad \ldots(1)$$

$$\Rightarrow \quad h\nu = \phi + eV_0 \quad \ldots(2)$$

Due to doubling the value of frequency ν, KE_{max} and V_0 will not get doubled.

$\Rightarrow$ Statement-1 is false and Statement-2 is true as by equation (1) and (2) it is clear that KE_{max} and V_0 are linearly dependent on frequency ν. **Ans. (B)**

Sol. 11 Photocurrent $\propto$ photon flux of incident light (Intensity of light)

and $KE_{max} = h\nu - \phi$

if $\nu \leftarrow 2\nu \Rightarrow$ $KE'_{max} = 2h\nu - \phi$

$\Rightarrow$ $KE'_{max} > 2KE_{max}$

$\Rightarrow$ Statement-1 is False

For Statement-2

As $KE_{max} = h\nu - \phi$

linear function

$\Rightarrow$ Statement-2 is True **Ans. (B)**

Sol. 12 (A) Threshold frequency $\nu_{th} = \frac{W}{h}$. So to emit electrons $\nu > \nu_{th}$

$\Rightarrow$ (A) is correct

(B) As there is no time leg in absorption of a photon and emission of electron

$\Rightarrow$ (B) is correct

(C) By photo electric effect equation max kinetic energy of ejected electrons is given as

$$KE_{max} = h\nu - W$$

$\Rightarrow$ (C) is wrong

(D) As KE_{max} depends only on incident frequency

$\Rightarrow$ (D) is correct **Ans. (C)**

Sol. 13

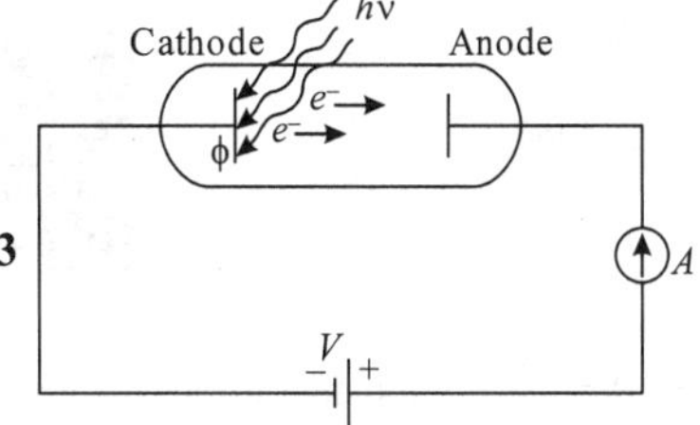

For photoelectric effect at cathode we use

$$h\nu = \frac{hc}{\lambda} = \phi + KE_{max}$$

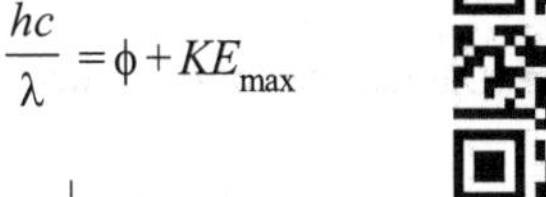

As $\lambda \uparrow \Rightarrow \frac{hc}{\lambda} \downarrow$ and $KE_{max} \downarrow$

At lesser KE of ejected electrons the circuit current reduces. **Ans. (D)**

Sol. 14

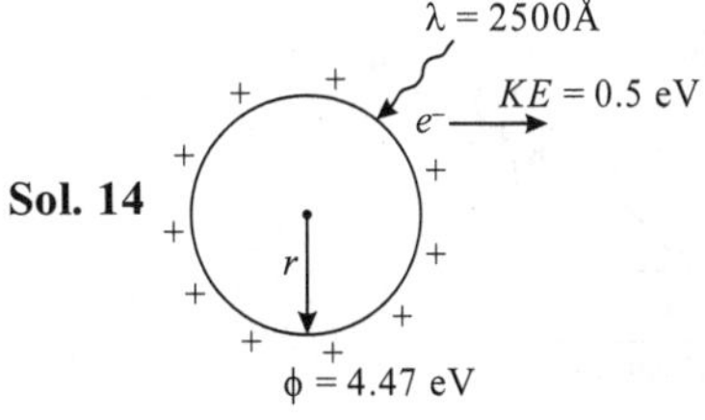

$$r = 1\text{ cm} = 10^{-2}\text{ m}$$

Energy of incident photon is

$$E_i = \frac{hc}{\lambda}$$

$$\Rightarrow \quad E_i = \frac{12431}{2500} = 4.97\text{ eV}$$

Maximum kinetic energy of ejected e^- from ball is

$$KE_{max} = E_i - \phi = 4.97 - 4.47 = 0.5\text{ eV}$$

As positive charge gets accumulated on ball due to ejection of electrons, this positive charge will attract back the ejected electrons, these electrons will not escape from ball if ball's potential will rise to 0.5 V.

If q is the charge acquired by ball upto this time, then we use

$$V = \frac{kq}{r} = \frac{1}{4\pi\varepsilon_0}\cdot\frac{q}{r}$$

$$\Rightarrow \quad 0.5 = \frac{9\times10^9 \times q}{10^{-2}}$$

$$\Rightarrow \quad q = \frac{5\times10^{-3}}{9\times10^9} = 5.5\times10^{-13}\text{ C}$$ **Ans. (A)**

Sol. 15 By Photo electric effect equation we use

$$\frac{hc}{\lambda} = \phi + eV_0$$

$$\Rightarrow \quad V_0 = \left(\frac{hc}{e}\right)\frac{1}{\lambda} - \frac{\phi}{e}$$

Comparing with $\quad y = mx + c$

Slope of V_0 vs $\frac{1}{\lambda}$ curve is

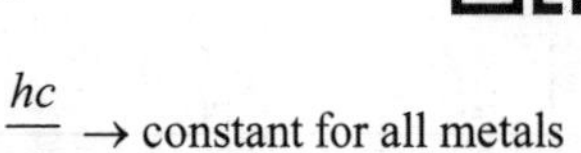

$$\Rightarrow \quad \tan\theta = \frac{hc}{e} \rightarrow \text{constant for all metals}$$

As in graph shown slope for A and B is not same. **Ans. (D)**

Sol. 16 The radiation energy emitted in $3 \rightarrow 2$ transition is given as

$$\Delta E = E_3 - E_2 = (-1.51\text{ eV}) - (-3.4\text{ eV})$$

$$\Rightarrow \quad \Delta E = 1.89\text{ eV}$$

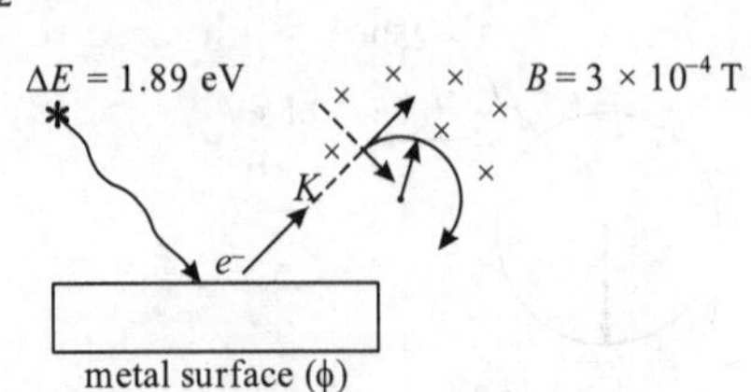

In magnetic field the radius of circular motion is given as

$$r = \frac{mv}{qB}$$

$$\Rightarrow \quad p = mv = qBr$$

The maximum kinetic energy of ejected electron is given as

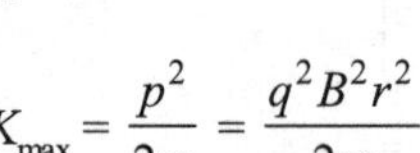

$$K_{max} = \frac{p^2}{2m} = \frac{q^2B^2r^2}{2m}$$

$$\Rightarrow \quad K_{max} = \frac{(1.6\times10^{-19})^2\times(3\times10^{-4})^2\times(10\times10^{-3})^2}{2\times9.1\times10^{-31}\times(1.6\times10^{-19})}\, eV$$

$$\Rightarrow \quad K_{max} = 0.8\text{ eV}$$

For photoelectric effect at metal surface we use

$$h\nu = \phi + KE_{max}$$

$$\Rightarrow \quad \phi = h\nu - KE_{max} = 1.89 - 0.8 = 1.09\text{ eV}$$

Ans. (B)

Sol. 17 Energy of each photon in the two radiations are

$$E_1 = \frac{hc}{\lambda_1} = \frac{12431}{4972}\text{ eV} \simeq 2.5\text{ eV}$$

and $$E_2 = \frac{hc}{\lambda_2} = \frac{12431}{6216}\text{ eV} \simeq 1.99\text{ eV} (<2.3\text{ eV})$$

So second radiation is not capable of ejecting electrons from the metal. The power incident on 1 cm^2 of metal surface in first radiation is

$$P = IA = 1.8\times10^{-3}\times10^{-4} = 1.8\times10^{-7}\text{ W}$$

Number of photons incident on metal/second are

$$N = \frac{P\lambda}{hc}$$

Number of electrons ejected in 2 second are

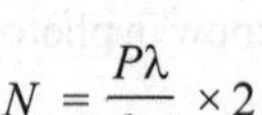

$$N_e = \frac{P\lambda}{hc}\times2$$

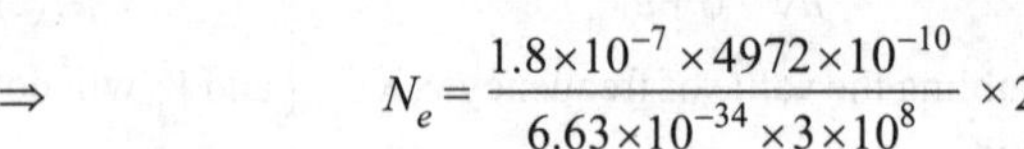

$$\Rightarrow \quad N_e = \frac{1.8\times10^{-7}\times4972\times10^{-10}}{6.63\times10^{-34}\times3\times10^8}\times2$$

$$\Rightarrow \quad N_e = 9\times10^{11}$$ **Ans. (B)**

Sol. 18 If ϕ be the work function of metal used, then we have

$$\frac{1}{2}mv^2 = \frac{hc}{\lambda} - \phi$$

$$\Rightarrow \quad v = \sqrt{\frac{2hc}{\lambda m} - \frac{2\phi}{m}} \qquad \ldots(1)$$

If λ is changed to $\frac{3\lambda}{4}$ so now the speed of fastest electron emitted now will be

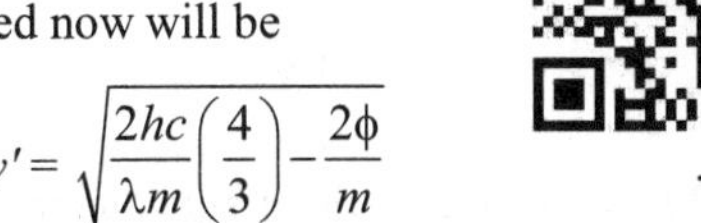

$$v' = \sqrt{\frac{2hc}{\lambda m}\left(\frac{4}{3}\right) - \frac{2\phi}{m}} \quad \ldots(2)$$

From equation-(1) $\sqrt{\frac{4}{3}}\,v = \sqrt{\frac{2hc}{\lambda m}\left(\frac{4}{3}\right) - \frac{2\phi}{m}\left(\frac{4}{3}\right)} \quad \ldots(3)$

from equation-(2) and (3) comparing, we get

$$v' > \sqrt{\frac{4}{3}}\,v \qquad \textbf{Ans. (A)}$$

Sol. 19 Using photo electric effect equation we have

$$\frac{hc}{\lambda_1} = \phi + eV \quad \ldots(1)$$

$$\frac{hc}{\lambda_2} = \phi + 3\,eV \quad \ldots(2)$$

$(1) \times 3 - (2)$ gives

$$\Rightarrow \frac{3hc}{\lambda_1} - \frac{hc}{\lambda_2} = 2\phi \Rightarrow \phi = \frac{3hc}{2\lambda_1} - \frac{hc}{2\lambda_2}$$

When wavelength λ_3 is used

$$\frac{hc}{\lambda_3} = \phi + eV_0$$

$$\Rightarrow V_0 = \frac{1}{e}\left(\frac{hc}{\lambda_3} - \phi\right) = \frac{hc}{e}\left(\frac{1}{\lambda_3} - \frac{3}{2\lambda_1} + \frac{1}{2\lambda_2}\right) \qquad \textbf{Ans. (D)}$$

Sol. 20 We use photo electric effect equation which gives

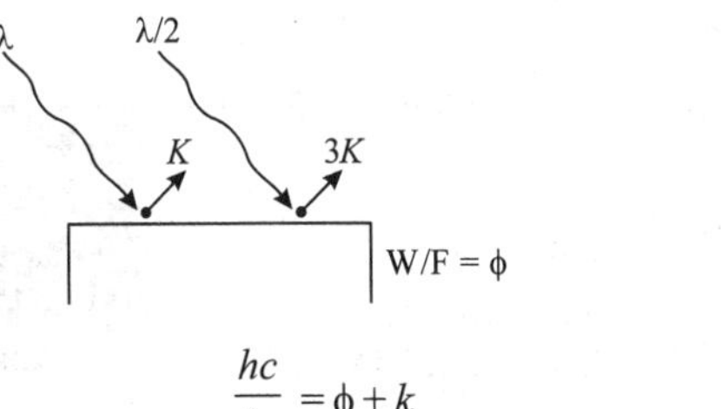

$$\frac{hc}{\lambda} = \phi + k \quad \ldots(1)$$

$$\frac{hc}{(\lambda/2)} = \phi + 3k \quad \ldots(2)$$

Using $(1) \times 3 - (2) \Rightarrow \frac{hc}{\lambda} = 2\phi \Rightarrow \phi = \frac{hc}{2\lambda}$ **Ans. (A)**

Sol. 21 By photo electric effect equation we use

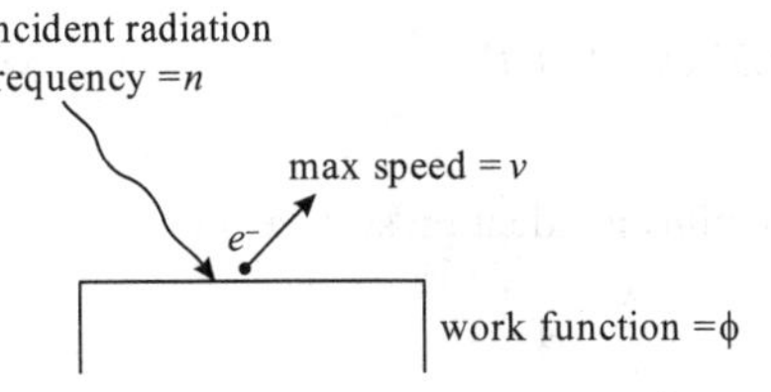

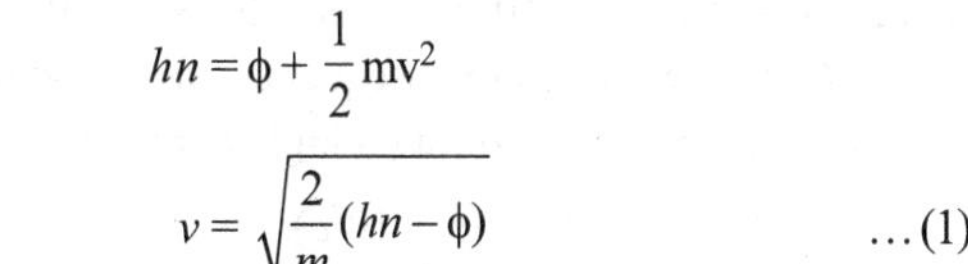

$$hn = \phi + \frac{1}{2}mv^2$$

$$\Rightarrow v = \sqrt{\frac{2}{m}(hn - \phi)} \quad \ldots(1)$$

If incident radiation frequency is increased to $3n \Rightarrow$ The speed of ejected fastest electron is given as

$$\Rightarrow v' = \sqrt{\frac{2}{m}(3hn - \phi)}$$

$$\Rightarrow v' > \sqrt{3}\,v \qquad \textbf{Ans. (C)}$$

Sol. 22 $\delta t = 60\text{ms}$ $\quad \lambda = 6 \times 10^{-7}$ m $\quad P_{watt}$

Pulse energy $E = \rho\delta t$

Energy of one photon in pulse $E_{ph} = \frac{hc}{\lambda}$

Number of photons in light pulse are

$$N = \frac{E}{E_{ph}} = \frac{E\lambda}{hc} = \frac{\rho\delta t\lambda}{hc}$$

$$\Rightarrow N = \frac{0.5 \times 10^3 \times 60 \times 10^{-3} \times 6 \times 10^{-7}}{6.62 \times 10^{-34} \times 3 \times 10^8}$$

$$\Rightarrow N \simeq \frac{10^{-7} \times 10}{10^{-34} \times 10^8} = 10^{20} \qquad \textbf{Ans. (A)}$$

Sol. 23 We use for velocity of ejected photoelectrons

$$\frac{hc}{\lambda_1} = \phi + \frac{1}{2}m(2v)^2 \quad \ldots(1)$$

and $$\frac{hc}{\lambda_2} = \phi + \frac{1}{2}mv^2 \quad \ldots(2)$$

$$\frac{(1)}{(2)} \Rightarrow \frac{\frac{hc}{\lambda_1} - \phi}{\frac{hc}{\lambda_2} - \phi} = 4$$

$$\Rightarrow \frac{hc}{\lambda_1} - \phi = \frac{4hc}{\lambda_2} - 4\phi$$

$$\Rightarrow \frac{4hc}{\lambda_2} - \frac{hc}{\lambda_1} = 3\phi$$

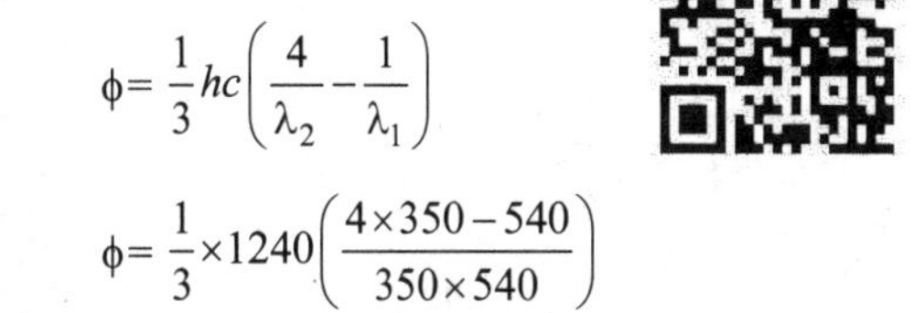

$$\Rightarrow \phi = \frac{1}{3}hc\left(\frac{4}{\lambda_2} - \frac{1}{\lambda_1}\right)$$

$$\Rightarrow \phi = \frac{1}{3} \times 1240\left(\frac{4 \times 350 - 540}{350 \times 540}\right)$$

$$\Rightarrow \phi = 1.8\,\text{eV} \qquad \textbf{Ans. (A)}$$

Sol. 24 $B = B_0 \sin(\pi \times 10^7 c)t + B_0 \sin(2\pi \times 10^7 c)t$ since there are two *EM* waves with different frequency, to get maximum kinetic energy we take the photon with higher frequency

$$B_1 = B_0 \sin(\pi \times 10^7 c)t$$

$$\nu_1 = \frac{10^7}{2} \times c$$

$$B_2 = B_0 \sin(2\pi \times 10^7 c)t$$

$$\nu_2 = 10^7 c$$

Where c is speed of light

As $\nu_2 > \nu_1$ so *KE* of photoelectron will be maximum for photon of higher energy $\nu_2 = 10^7 c$ Hz

We use

$$h\nu = \phi + KE_{max} \quad \ldots(1)$$

Energy of photon incident is

$$E_{ph} = h\nu = 6.6 \times 10^{-34} \times 10^7 \times 3 \times 10^9$$

$$\Rightarrow \quad E_{ph} = 6.6 \times 3 \times 10^{-19}\,\text{J}$$

$$\Rightarrow \quad E_{ph} = \frac{6.6 \times 3 \times 10^{-19}}{1.6 \times 10^{-19}}\,\text{eV} = 12.375\,\text{eV}$$

from (1)

$$KE_{max} = E_{ph} - \phi$$

$$\Rightarrow \quad KE_{max} = 12.375 - 4.7$$

$$= 7.675\,\text{eV} \approx 7.7\,\text{eV}$$ **Ans. (A)**

Sol. 25 Energy incident on plate per second is

$$E = IA$$

Number of electrons emitted per second are

$$n = \frac{E \times 0.1}{10\text{eV}} = \frac{16 \times 10^{-3} \times 1 \times 10^{-4} \times 0.1}{10 \times 1.6 \times 10^{-19}}$$

$$= 10^{11}\ \text{electrons/sec}$$

Maximum *KE* of ejected electrons is

$$KE = 10 - 5 = 5\,\text{eV}$$ **Ans. (D)**

Sol. 26 For photoelectric effect, we use

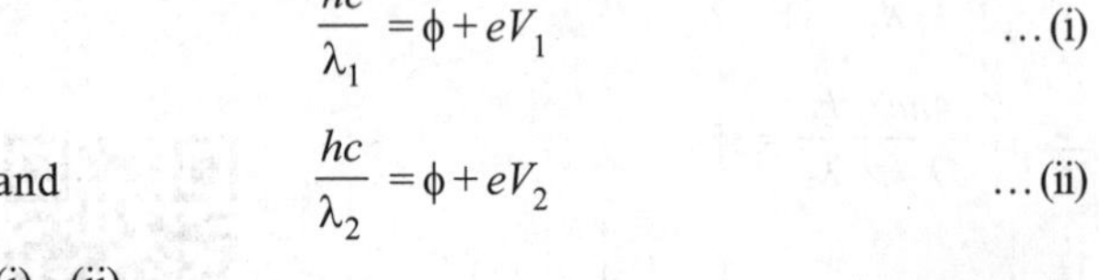

$$\frac{hc}{\lambda_1} = \phi + eV_1 \quad \ldots(\text{i})$$

and $$\frac{hc}{\lambda_2} = \phi + eV_2 \quad \ldots(\text{ii})$$

(i) − (ii)

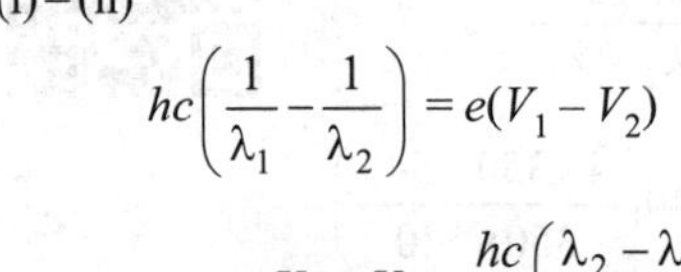

$$hc\left(\frac{1}{\lambda_1} - \frac{1}{\lambda_2}\right) = e(V_1 - V_2)$$

$$\Rightarrow \quad V_1 - V_2 = \frac{hc}{e}\left(\frac{\lambda_2 - \lambda_1}{\lambda_1 - \lambda_2}\right)$$

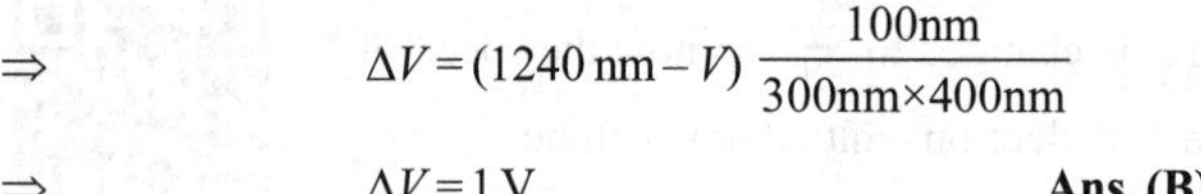

$$\Rightarrow \quad \Delta V = (1240\,\text{nm} - V)\frac{100\text{nm}}{300\text{nm} \times 400\text{nm}}$$

$$\Rightarrow \quad \Delta V = 1\,\text{V}$$ **Ans. (B)**

Sol. 27 By photoelectric effect equation, we use

$$h_\nu = \phi + \frac{eV_0}{2} \quad \ldots(1)$$

and $$\frac{h\nu}{2} = \phi + eV_0 \quad \ldots(2)$$

Above equations cannot holds so given data is incorrect.

Ans. (None)

Sol. 28 Frequency of the incident light is

$$\nu = \frac{\omega}{2\pi} = \frac{2\pi \times 6 \times 10^{14}}{2\pi} = 6 \times 10^{14}\,\text{Hz}$$

Light wavelength is

$$\lambda = \frac{c}{\nu} = \frac{3 \times 10^8}{6 \times 10^{14}} = 5000\,\text{Å}$$

Photo energy $E = \frac{12375}{5000} = 2.475\,\text{eV}$

KE (max) of ejected photoelectrons is

$$KE_{max} = E - \phi = 2.475 - 2 = 0.475\,\text{eV}$$

Stopping potential

$$V = \frac{KE_{max}}{e} = 0.475\,\text{V} \simeq 0.48\,\text{V}$$ **Ans. (A)**

Sol. 29 Max *KE* of ejected photoelectrons is given as

$$K_{max} = \frac{hc}{\lambda} - \phi$$

$$\Rightarrow \quad K_{max} = \frac{hc}{\lambda} - \frac{hc}{\lambda_0} = \frac{hc}{260\text{ nm}} - \frac{hc}{380\text{ nm}}$$

$$\Rightarrow \quad K_{max} = 1237\left[\frac{1}{260} - \frac{1}{380}\right]$$

$$= 1237\left[\frac{380 - 260}{3800 \times 260}\right]\text{eV} \simeq 1.5\,\text{eV}$$ **Ans. (A)**

Sol. 30 V_0 vs ν graph starts at threshold frequency so from graph, we have

$$\Rightarrow \quad \nu_{th} = 4 \times 10^{14}\,\text{Hz}$$

So work function for this metal is given as

$$\phi = h\nu_{th} = \frac{6.63 \times 10^{-34} \times 4 \times 10^{14}}{1.6 \times 10^{-19}}\,\text{eV}$$

$$\Rightarrow \quad \phi = 1.657\,\text{eV} \simeq 1.66\,\text{eV}$$ **Ans. (C)**

Sol. 31 Total photons incident on surface are

$$n = \frac{I\lambda}{hc} \times A$$

$$\Rightarrow \quad n = \frac{6.4\times10^{-5}\times1}{\frac{1240}{310}\times1.6\times10^{-19}} = \frac{6.4\times10^{-5}\times310\times1}{1240\times1.6\times10^{-19}}$$

$$\Rightarrow \quad n = 10^{14} \text{ photons/sec}$$

Number of electrons emitted from surface per second are

$$N_e = 10^{14}\times10^{-3} = 10^{11} \text{ per/sec}$$

$$\Rightarrow \quad x = 11$$

Ans. (11)

Sol. 32 Given that $\lambda_B = 2\lambda_A$

$$\Rightarrow \quad \frac{h}{\sqrt{2T_B m}} = \frac{2h}{\sqrt{2T_A m}}$$

$$\Rightarrow \quad T_A = 4T_B \quad \ldots(1)$$

Also given that

$$T_B = (T_A - 1.5)\,\text{eV} \quad \ldots(2)$$

$$\Rightarrow \quad 3T_B = 1.5\,\text{eV}$$

$$\Rightarrow \quad T_B = 0.5\,\text{eV}$$

and $\quad 0.5\,\text{eV} = 4.5\,\text{eV} - \phi_B$

$$\Rightarrow \quad \phi_B = 4\,\text{eV}$$

Ans. (C)

Sol. 33 Using dimensions of given physical quantities, we have

$$[V_0] = [h]^x [c]^y [G]^z [A]^w$$

$$\Rightarrow \quad \frac{ML^2T^{-2}}{AT} = (ML^2T^{-1})^x (LT^{-1})^y (M^{-1}L^3T^{-2})^z A^w$$

Comparing dimensions at LHS and RHS, we have

$\Rightarrow \quad w = -1, x = 0, z = 1$ and $y = 5$

$$\Rightarrow \quad [V_0] = h^0 c^5 G^{-1} A^{-1}$$

So Bonus. **Ans. (Bonus)**

Sol. 34 If the metal surface work function is ϕ then maximum kinetic energy of ejected electrons will be

$$KE_{max} = \frac{hc}{\lambda} - \phi$$

In magnetic field, radius of circular motion will be

$$R = \frac{mv}{qB} = \frac{\sqrt{2mKE_{max}}}{qB} = \frac{\sqrt{2m\left(\frac{hc}{\lambda} - \phi\right)}}{qB}$$

$$\Rightarrow \quad \frac{R^2q^2B^2}{2m} = \frac{hc}{\lambda} - \phi$$

$$\Rightarrow \quad \phi = \frac{hc}{\lambda} - \frac{R_{max}^2 q^2 B^2}{2m} = 1.0899\,\text{eV} \approx 1.1\text{eV}$$

Ans. (B)

Sol. 35 We use for given data

$$\frac{hc}{\lambda} = \frac{hc}{\lambda_{th}} + eV \quad \ldots(1)$$

and $$\frac{hc}{3\lambda} = \frac{hc}{\lambda_{th}} + \frac{eV}{4} \quad \ldots(2)$$

$4\times(2) - (1)$ gives

$$\frac{1}{3}\frac{hc}{\lambda} = \frac{3hc}{\lambda_{th}}$$

$$\Rightarrow \quad \lambda_{th} = 9\lambda$$

$$\Rightarrow \quad n = 9$$

Ans. (9)

Sol. 36 By photoelectric effect equation, we use

$$\frac{hc}{\lambda} = \phi + k_m$$

We use $$\frac{hc}{500\text{ nm}} = \phi + k_m \quad \ldots(1)$$

and $$\frac{hc}{200\text{ nm}} = \phi + 3k_m \quad \ldots(2)$$

from (1) and (2) we get

$$3 = \frac{\frac{hc}{200\text{ nm}} - \phi}{\frac{hc}{500\text{ nm}} - \phi}$$

We use $\quad hc = 1243.1\,\text{eV} - \text{nm}$

$$\Rightarrow \quad 3\times\frac{1243.1}{500} - 3\phi = \frac{1243.1}{200} - \phi$$

$$\Rightarrow \quad 2\phi = 1243.1\left[\frac{3}{500} - \frac{1}{200}\right]$$

$$\Rightarrow \quad \phi = \frac{1243.1}{1000}[6-5]$$

$$\Rightarrow \quad \phi = \frac{1.243}{2} = 0.612\,\text{eV}$$

Ans. (A)

Sol. 37 From the given graph, we use

$$hv = \phi + eV_0$$

At point B, $\quad V_0 = 0,$

$$v = 5.5\times10^{14}\,\text{Hz}$$

$$\Rightarrow \quad h(5.5\times10^{14}) = \phi$$

$$\Rightarrow \quad \phi = \frac{6.62\times10^{-34}\times5.5\times10^{14}}{1.6\times10^{-19}}$$

$$\Rightarrow \quad \phi = 2.27\,\text{eV}$$

Ans. (A)

Sol. 38 With variation in intensity we consider change in photon flux at constant frequency so the graph does not change. **Ans. (C)**

Sol. 39 For emitted photoelectrons, we use

$$E_1 = \phi + k_1$$

and $$E_2 = \phi + k_2$$

As $$\frac{v_1}{v_2} = 2$$

$$\Rightarrow \quad \frac{k_1}{k_2} = \frac{v_1^2}{v_2^2} = 4$$

Given that $$\frac{k_1}{k_2} = 4 = \frac{4-\phi}{2.5-\phi}$$

$$\Rightarrow \quad 10 - 4\phi = 4 - \phi$$

$$\Rightarrow \quad \phi = 2\,\text{eV}$$ **Ans. (2)**

Sol. 40 By photoelectric effect equation, we have

$$\Rightarrow \quad \frac{hc}{\lambda} = \phi + eV_s$$

$$\Rightarrow \quad \frac{1240}{491} = \phi + 0.71 \quad \ldots(1)$$

$$\Rightarrow \quad \frac{1240}{\lambda} = \phi + 1.43 \quad \ldots(2)$$

$$\Rightarrow \quad \lambda = 382\,\text{nm}$$ **Ans. (A)**

Sol. 41 By photoelectric effect equation, we have

$$KE_{max} = h\nu - \phi$$

$$\Rightarrow \quad \frac{1}{2}mv^2 = h\nu - \phi$$

$$\Rightarrow \quad v = \sqrt{\frac{2(h\nu - \phi)}{m}}$$

If ν_1 and ν_2 are the frequencies of the incident radiations, it is given that

$$h\nu_1 = 2\phi$$

and $$h\nu_2 = 10\phi$$

$$\Rightarrow \quad \frac{v_1}{v_2} = \sqrt{\frac{h\nu_1 - \phi}{h\nu_2 - \phi}}$$

$$\Rightarrow \quad \frac{v_1}{v_2} = \sqrt{\frac{2\phi - \phi}{10\phi - \phi}} = \frac{1}{3}$$ **Ans. (1)**

Sol. 42 Stopping potential changes linearly with frequency of incident radiation and it does not depend on any other listed physical quantitites. **Ans. (D)**

Sol. 43 Force due to radiation pressure on the surface is calculated as

$$F = \frac{IA}{C}$$

$$\Rightarrow \quad I = \frac{FC}{A} = \frac{2.5\times10^{-6}\times3\times10^{8}}{30} = 25\,\text{W/cm}^2$$ **Ans. (25)**

Sol. 44 By photoelectric effect equation in the two cases, we use

$$\frac{1}{2}mv_1^2 = hf_1 - \phi$$

and $$\frac{1}{2}mv_2^2 = hf_2 - \phi$$

$$\Rightarrow \quad v_1^2 - v_2^2 = \frac{2h}{m}[f_1 - f_2]$$ **Ans. (A)**

Sol. 45 Energy of radiation from hydrogen atom is given as

$$E_p = 13.6\left[\frac{1}{R_1^2} - \frac{1}{R_2^2}\right]\text{eV} = 13.6\left[\frac{1}{4} - \frac{1}{9}\right]\text{eV}$$

$$\Rightarrow \quad E_p = 1.89\,\text{eV}$$

Speed of electrons moving in circular path in magnetic field is given as

$$v = \frac{RqB}{m}$$

$$\Rightarrow \quad v = \frac{7\times10^{-3}\times1.6\times10^{-19}\times5\times10^{-4}}{9.1\times10^{-31}} = 6.15\times10^5$$

Maximum kinetic energy of electrons is given as

$$KE_{max} = \frac{1}{2}mv^2$$

$$\Rightarrow \quad KE_{max} = \frac{1}{2}\times\frac{9.1\times10^{-31}\times(6.15\times10^5)}{1.6\times10^{-19}}\,\text{eV} = 1.075\,\text{eV}$$

For Gold plate, we use

$$\phi = E_p - KE_{max}$$

$$\Rightarrow \quad \phi = 1.89 - 1.075;\ \phi = 0.82\,\text{eV}$$ **Ans. (A)**

Sol. 46 By photoelectric effect equation, we have

$$KE = h\nu - \phi$$

$$\Rightarrow \quad eV = \frac{hc}{\lambda} - \phi$$

For first case when stopping potential is $3V_0$, we have

$$e(3V_0) = \frac{hc}{\lambda} - \phi \quad \ldots(1)$$

For second case when stopping potential is $2V_0$, we have

$$e(2V_0) = \frac{hc}{2\lambda} - \phi \quad \ldots(2)$$

From equations-(1) and (2), we use

$$\phi = \frac{hc}{4\lambda} = \frac{hc}{\lambda_{th}}$$

$$\Rightarrow \quad \frac{hc}{4\lambda} = \frac{hc}{\lambda_{th}}$$

$$\Rightarrow \quad \lambda_{th} = 4\lambda$$ **Ans. (4)**

Sol. 47 By photoelectric effect equation, we have

$$KE = h\nu - \phi$$

$$\Rightarrow \quad eV = \frac{hc}{\lambda} - \phi$$

For first case, we use

$$4.8 = \frac{hc}{\lambda_0} - \phi \quad \ldots(1)$$

For second case, we use

$$1.6 = \frac{hc}{2\lambda_0} - \phi \quad \ldots(2)$$

From equation (1) and (2)

$$\phi = \frac{hc}{4\lambda_0}$$

For threshold wavelength λ_{th} we have

$$\phi = \frac{hc}{\lambda_{th}}$$

$$\Rightarrow \quad \frac{hc}{4\lambda_0} = \frac{hc}{\lambda_{th}}$$

$$\Rightarrow \quad \lambda_{th} = 4\lambda_0$$ **Ans. (B)**

Sol. 48 Using photoelectric effect equation, we have

$$\frac{hc}{\lambda} = \phi + KE_{max}$$

$$\Rightarrow \quad \frac{1240}{500} = 1.25 + KE_{max}$$

$$\Rightarrow \quad KE_{max} = 1.23\text{ eV}$$

Radius of circular motion of electron in magnetic field is given as

$$R = \frac{mv}{qB} = \frac{\sqrt{2mKE}}{qB}$$

$$\Rightarrow \quad B = \frac{\sqrt{2mKE}}{qR}$$

$$\Rightarrow \quad B = \frac{\sqrt{2\times 9.1\times 10^{-31}\times 1.23\times 1.6\times 10^{-19}}}{1.6\times 10^{-19}\times 30\times 10^{-2}}$$

$$\Rightarrow \quad B = 0.125\times 10^{-4} = 125\times 10^{-7}$$ **Ans. (125)**

Sol. 49 Initial kinetic energy of electron is given as 3 eV. After the proton captures the electron, hydrogen atom in second excited state is formed i.e. in $n = 3$. In this state final energy of electron is – 1.51 eV. Thus the energy taken out by photon is given as

$$E_P = 3 + 1.51 = 4.51\text{ eV}$$

Work function of photosensititve metal is given as

$$\phi = \frac{12431}{4000} = 3.1\text{ eV}$$

By photoelectric effect equation, the maximum kinetic energy of ejected electron is given as

$$K_{max} = E_P - \phi$$

$$\Rightarrow \quad K_{max} = 4.51 - \frac{12431}{4000}$$

$$\Rightarrow \quad K_{max} = 4.51 - 3.1 = 1.41\text{ eV}$$ **Ans. (B)**

Sol. 50 Photoelectric effect equation is written as

$$eV_0 = \frac{hc}{\lambda} - \phi$$

For the two cases, we have

$$eV_1 = \frac{hc}{280nm} - \phi$$

and $$eV_2 = \frac{hc}{400\text{nm}} - \phi$$

$$\Rightarrow \quad e(V_1 - V_2) = 1240\left(\frac{120}{280\times 400}\right) = 1.3\text{ eV}$$ **Ans. (C)**

Sol. 51 The amplitude of electric field

$$E_0 = 200\text{ V/m}$$

Intensity of wave is given as

$$I = \frac{1}{2}e_0E_0^2.c$$

Radiation pressure due to wave is given as

$$P = \frac{2I}{c} = \left(\frac{2}{c}\right)\left(\frac{1}{2}\varepsilon_0E_0^2c\right) = e_0E_0^2$$

$$\Rightarrow \quad P = 8.85\times 10^{-12}\times 200^2$$

$$\Rightarrow \quad P = 8.85\times 10^{-8}\times 4 = \frac{354}{10^9}$$ **Ans. (1)**

Sol. 52 The number of photons incident per second on a surface depends on the intensity of light as

$$N = \frac{IA}{h\nu}$$

Thus with increase in light intensity, number of photons increases but it does not change the kinetic energy of photoelectrons as frequency is not changing. **Ans. (A)**

Sol. 53 From photoelectric effect equation, we have

$$eV_s = \frac{hc}{\lambda} - \phi$$

$$\Rightarrow \quad e(V'-V)=hc\left(\frac{1}{\lambda'}-\frac{1}{\lambda}\right)$$

$$\Rightarrow \quad V'=\frac{hc}{e}\left(\frac{1}{\lambda'}-\frac{1}{\lambda}\right)+V$$

$$\Rightarrow \quad V'=1240\left(\frac{1}{474}-\frac{1}{670}\right)+0.48=1.25\,\text{eV}$$

Ans. (C)

Sol. 54 Using conservation of energy, we have

$$K_e = E_{H\text{-atom}} + E_{\text{photon}}$$

$$\Rightarrow \quad 2.6=\frac{-13.6}{4}+h\nu$$

$$\Rightarrow \quad h\nu=6\,\text{eV}$$

$$\Rightarrow \quad \nu=\frac{6\times1.6\times10^{-19}}{6.626\times10^{-34}}$$

$$\Rightarrow \quad \nu=1.45\times10^{15}\,\text{Hz}=1.45\times10^{9}\,\text{MHz}$$

Ans. (B)

Sol. 55 Using photoelectric effect equation, we have

$$h\nu=h\nu_{\text{th}}+\frac{1}{2}mv^2$$

It is given that $\nu=2\nu_{\text{th}}$ so we have

$$2h\nu_{\text{th}}=h\nu_{\text{th}}+\frac{1}{2}mv_1^2 \qquad \ldots(1)$$

Now frequency is changed to $\nu=5\nu_{\text{th}}$ so we have

$$5h\nu_{\text{th}}=h\nu_{\text{th}}+\frac{1}{2}mv_2^2 \qquad \ldots(2)$$

$$\Rightarrow \quad \frac{\frac{1}{2}mv_1^2}{\frac{1}{2}mv_2^2}=\frac{h\nu_{\text{th}}}{4h\nu_{\text{th}}}$$

$$\Rightarrow \quad \left(\frac{v_1}{v_2}\right)^2=\frac{1}{4}$$

$$\Rightarrow \quad v_2=2v_1$$

Ans. (2)

Sol. 56 The ratio of maximum speed of photoeelctrons is given as

$$\frac{v_1}{v_2}=\sqrt{\frac{3.8-0.6}{1.4-0.6}}=\sqrt{\frac{3.2}{0.8}}=2$$

Ans. (B)

Sol. 57 A small change in minority charge carriers produces high value of reverse bias current so its detection is sensitive and immediate. **Ans. (D)**

Sol. 58 Radius of circular path followed by charged particle in magnetic field is given as

$$R=\frac{\sqrt{2Km}}{qB}$$

$$\Rightarrow \quad K=\frac{(qBR)^2}{2m}$$

$$\Rightarrow \quad K=\frac{(1.6\times10^{-19}\times2\times10^{-3}\times2\times10^{-3})^2}{2\times9.1\times10^{-31}}$$

$$\Rightarrow \quad K=\frac{(6.4)^2}{2\times9.1}\times\frac{10^{-50}}{10^{-31}}$$

$$\Rightarrow \quad K=2.25\times10^{-19}\,\text{J}$$

$$\Rightarrow \quad K=\frac{2.25\times10^{-19}}{1.6\times10^{-18}}\,\text{eV}=1.40\,\text{eV}$$

Energy of incident photon is given as

$$E=\frac{12400}{4500}=2.76\,\text{eV}$$

Thus work function of metal surface is calculated as

$$\phi=E-K=(2.76-1.40)\,\text{eV}$$

$$\Rightarrow \quad \phi=1.36\,\text{eV}$$

Ans. (A)

Sol. 59 Energy of incident photon is calculated as

$$E=\frac{12431}{6630}=1.84\,\text{eV}$$

By photoelectric effect equation, we have

$$E=\phi+eV_0$$

$$\phi=E-eV_0$$

$$\Rightarrow \quad \phi=(1.87-0.42)=1.45\,\text{eV}$$

Work function is related to threshold frequency as

$$\phi=h\nu_0$$

$$\Rightarrow \quad 1.45\times1.6\times10^{-19}=6.63\times10^{-34}\times\nu_0$$

$$\Rightarrow \quad \nu_0=0.35\times10^{15}$$

$$\Rightarrow \quad \nu_0=35\times10^{13}\,\text{s}^{-1}$$

$$\Rightarrow \quad x=35$$

Ans. (35)

Sol. 60 By photoelectric effect equation, we have

$$\frac{hc}{\lambda_1}-\phi=K_1$$

and

$$\frac{hc}{\lambda_2}-\phi=K_2$$

Given that $\lambda_1=3\lambda_2$ so we have

$$3K_1=\frac{3hc}{\lambda_1}-3\phi$$

$$\Rightarrow \quad 3K_1=\frac{hc}{\lambda_2}-3\phi$$

$$\Rightarrow \quad 3K_1=K_2-2\phi$$

$$\Rightarrow \quad 3K_1<K_2$$

$$\Rightarrow \quad K_1<\frac{K_2}{3}$$

Ans. (B)

Sol. 61 To eject a free electron from metal surface minimum energy required, is equal to the work function of that metal hence Assertion **A**, is correct

If incident energy of radiation is equal to work function then it ejects a photoelectron with almost zero kinetic energy hence reason **R**, is also correct, but **R** is not the correct explanation of A. **Ans. (B)**

Sol. 62 From photoelectric effect equation the maximum kinetic energy of ejected electron from 800 nm wavelength is given as

$$K_1 = \frac{1230}{800} - \phi \quad \ldots(1)$$

When light of wavelength 500 nm is used, we have

$$K_2 = 2K_1 = \frac{1230}{500} - \phi \quad \ldots(2)$$

Eliminating K_1 from equations-(1) and (2) we get

$$0 = \frac{1230}{500} - \frac{1230}{400} + \phi$$

$$\Rightarrow \quad \phi = 0.615 \text{ eV}$$ **Ans. (C)**

Sol. 63 By photoelectric effect equaiton maximum kinetic energy of ejected photoelectrons is given as

$$\frac{1}{2}mv_{max}^2 = hf - \phi$$

Hence option (a) is correct.

When light source is displaced away from the metal surface, light intensity decreases and due to this the saturation current decreases hence option (b) is NOT correct.

Power of LED increases the intensity not the frequency so it does not affect the maximum kinetic energy of ejected electrons hence option (c) is NOT correct.

Immediate emission of electron can be explained by particle nature of light not by wave nature hence option (d) is NOT correct. Due to same reason option (e) is correct. **Ans. (B)**

Sol. 64 If work function of a metal surface is ϕ, then for the two streams of photons we use

$$\frac{1}{2}mv_1^2 = 5\phi - \phi$$

and

$$\frac{1}{2}mv_2^2 = 10\phi - \phi$$

$$\Rightarrow \quad \frac{v_1}{v_2} = \frac{2}{3}$$ **Ans. (C)**

Sol. 65 By photoelectric effect equation, we use

$$E = \frac{hc}{\lambda} - \phi \quad \ldots(1)$$

and

$$2E = \frac{hc}{\lambda'} - \phi \quad \ldots(2)$$

Subtracting equation-(1) from (2), gives

$$E = hc\left(\frac{1}{\lambda'} - \frac{1}{\lambda}\right)$$

$$\Rightarrow \quad \lambda' = \frac{hc\lambda}{E\lambda + hc}$$ **Ans. (B)**

Sol. 66 (a) is NOT correct as stopping potential depends on frequency of incident light and work function of metal.

(b) is correct as saturation photocurrent is directly proportional to the intensity of incident light

(c) is NOT correct as maximum kinetic energy depends on frequency of light

(d) is NOT correct as photoelectric effect is explained using particle nature of light not the wave nature. **Ans. (C)**

Sol. 67 (a) Dimensions of Planck's constant is calculated as

$$[hv] = [E]$$

$$[h] = [\frac{E}{v}] = \frac{M^1L^2T^{-2}}{T^{-1}} = M^1L^2T^{-1}$$

Matches with option (iii) in list-II.

(b) Dimensions of potential is calculated as

$$[E] = [qV]$$

$$[V] = [\frac{E}{q}] = \frac{M^1L^2T^{-2}}{A^1T^1} = M^1\,L^2\,T^{-3}\,A^{-1}$$

Matches with option (iv) in list-II.

(c) Dimensions of work fuction is that of energy, given as

$$[\phi] = M^1\,L^2\,T^{-2}$$

Matches with option (i) in list-II.

(d) Dimensions of Momentum is calculated as

$$[p] = [F.t]$$

$$[p] = M^1\,L^1\,T^{-2}\,T^1 = M^1\,L^1\,T^{-1}$$

Matches with option (ii) in list-II. **Ans. (B)**

Sol. 68 Stopping potential in photoelectric effect is given as

$$V_0 = \frac{KE_{max}}{e}$$

$$\Rightarrow \quad V_0 = \frac{h\nu - \phi}{e}$$

Stopping potential does not depend on intensity or power of light used, it only depends on frequency or wavelength of incident light and work function of metal. Thus both statements-I and II are correct. **Ans. (D)**

Sol. 69 For photoelectric emission to start incident wavelength should be less than 5500 Å which corresponds to visible light so wavelength lesser than visible is of ultra violet radiation hence options of lamp in (c) and (d) are correct. **Ans. (D)**

Sol. 70 Energy of the radiation is calculated as

$$\phi = \frac{hc}{\lambda} = \frac{1240}{350}\,\text{eV} = 3.54\,\text{eV}$$

Thus only metal B will emit photoelectron. **Ans. (D)**

Sol. 71 By Einstein's photoelectric equation, we have

$$K_{max} = hf - hf_0$$

When light of frequency $2f_0$ is incident on the metal plate, we have

$$\frac{1}{2}mv_1^2 = 2\,hf_0 - hf_0 = hf_0 \quad \ldots(1)$$

For $f = 5f_0$ we have

$$\frac{1}{2}mv_2^2 = 5\,hf_0 - hf_0 = 4hf_0 \quad \ldots(2)$$

From equations-(1) and (2), we get

$$\Rightarrow \quad \frac{v_1}{v_2} = \frac{1}{2}$$ **Ans. (A)**

Sol. 72 By Einstein Photoelectric equation, we have

$$eV_s = K_{max}$$

$$\Rightarrow \quad V_s = \left(\frac{h}{e}\right)f + \left(\frac{-\phi}{e}\right)$$

Slope of curve is independent of nature of metal, so we use

Slope $(V_s)^{\text{Gold}}$ = Slope $(V_s)^{\text{Aluminium}}$ **Ans. (C)**

Sol. 73 Light intensity is related to photo current and other factors as listed below

Intensity of light $\propto$ number of photons
Intensity of light $\propto$ no. of photo electrons
Intensity of light $\propto$ photo current

Thus we can say when intensity of light increases, photocurrent increases hence (a) is correct. By Einstein photoelectric equation, we have

$$K_{max} = h\nu - \phi$$

Thus K_{max} depends on frequency hence (c) is correct. **Ans. (A)**

Sol. 74 By Einstein photo electric equation

$$eV_0 = h\nu - \phi$$

Initially $V_0 = 0$

$$\Rightarrow \quad 0 = h\nu - \phi$$
$$\Rightarrow \quad \phi = h\nu$$
$$\Rightarrow \quad \phi = 6.6 \times 10^{-34} \times 5 \times 10^{14}$$
$$\Rightarrow \quad \phi = 33 \times 10^{-20}\,\text{J}$$

Thus work function for the surface in eV is calculated as

$$\phi = \frac{33 \times 10^{-20}}{1.6 \times 10^{-19}} = 2.07\,\text{eV}$$ **Ans. (C)**

Sol. 75 From the equation of photoelectric effect we have

$$eV_0 = \frac{hc}{\lambda} - \phi_0 = \frac{hc}{\lambda} - \frac{hc}{\lambda_0}$$

and $$\frac{eV_0}{4} = \frac{hc}{2\lambda} = \frac{hc}{\lambda_0}$$

$$\Rightarrow \quad \frac{1}{4}\left(\frac{hc}{\lambda} - \frac{hc}{\lambda_0}\right) = \frac{hc}{2\lambda} - \frac{hc}{\lambda_0}$$

$$\Rightarrow \quad \frac{1}{\lambda_0} - \frac{1}{4\lambda_0} = \frac{1}{2\lambda} - \frac{1}{4\lambda}$$

$$\Rightarrow \quad \frac{3}{4\lambda_0} = \frac{1}{4\lambda}$$

$$\Rightarrow \quad \lambda_0 = 3\lambda$$ **Ans. (D)**

Sol. 76 We known that energy of a photon is given as

$$E = \frac{hc}{\lambda}$$

$$\Rightarrow \quad \lambda = \frac{hc}{E} = \frac{12400}{E(ev)}$$

Thus wavelength of A is calculated as

$$\lambda_A = \left(\frac{1242}{9}\right) = 138\,\text{nm}$$

and wavelength of B is calculated as

$$\lambda_B = \left(\frac{1242}{4.5}\right) = 276\,\text{nm}$$

Thus difference between threshold wavelengths

$$\lambda_B - \lambda_A = 138\,\text{nm}$$ **Ans. (B)**

Sol. 77 *UV* rays have maximum frequency hence are most effective for emission of electrons from a metallic surface. By Einstein photoelectric equation

$$KE_{max.} = hf - hf_0$$

Hence statement-I is true and statement-II is false. **Ans. (A)**

25.2 Matter Waves and X-Rays

Sol. 1

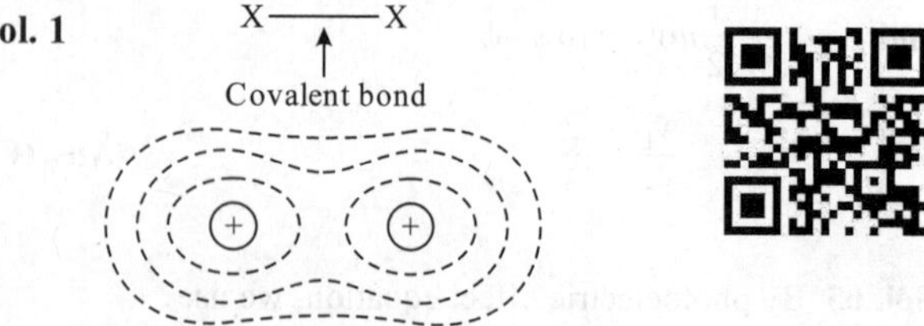

Overlapping region of electron cloud due to Van der waals force between electrons and Nu.

This is a region between two atoms where probability density of existence of electrons now exist. **Ans. (A)**

Sol. 2

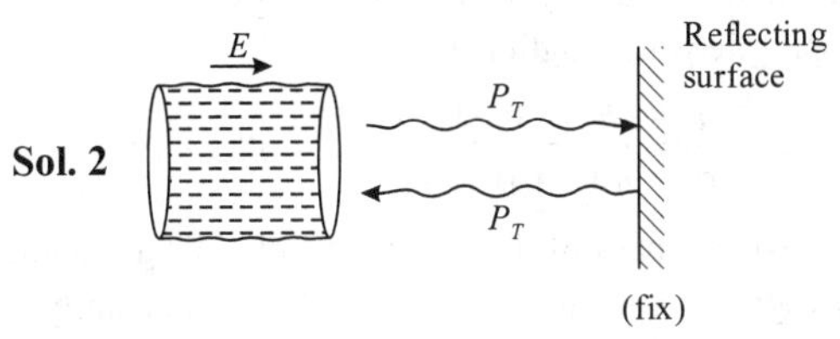

If radiation wavelength is λ, then no of photons in the radiation are $N = \frac{E}{hc/\lambda} = \frac{E\lambda}{hc}$

Each photon in radiation has momentum

$$p = \frac{h}{\lambda}$$

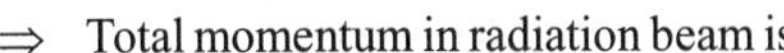

$\Rightarrow$ Total momentum in radiation beam is

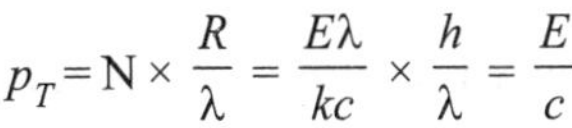

$$p_T = \text{N} \times \frac{R}{\lambda} = \frac{E\lambda}{kc} \times \frac{h}{\lambda} = \frac{E}{c}$$

Total momentum change due to reflection is momentum transferred to surface

$$\Delta p_T = p_T - (-p_T) = 2p_T = \frac{2E}{c}$$ **Ans. (B)**

Sol. 3 The momentum of a moving free electron at KE 'K' is given as

$$p = \sqrt{2mK} \quad [\text{As } k = \frac{1}{2}mv^2]$$

De-Broglie wavelength of moving electron can be given as

$$\lambda = \frac{h}{p} = \frac{h}{\sqrt{2mK}} \qquad \ldots(1)$$

If K is doubled to $2K$ then new De-Broglie wavelength is given as

$$\lambda' = \frac{h}{\sqrt{2m(2K)}} = \frac{\lambda}{\sqrt{2}}$$ **Ans. (C)**

Sol. 4 As a photon having wavelength λ has momentum associated with it is given by concept of matter wave

$$p = \frac{h}{\lambda} \qquad \ldots(1)$$

If ν be the photon frequency we use

$$\lambda = \frac{c}{\nu}$$

from equation-(1) $p = \frac{h}{c/\nu} = \frac{h\nu}{c}$ **Ans. (D)**

Sol. 5 Path difference in the electron beams which are diffracted from points O and O' are

$$\Delta = 2d\cos i$$

For constructive interference in the two beams we use

$$\Delta = 2d\cos i = n\lambda \quad (n \in \text{I})$$

Wavelength of electron wave is given as

$$\lambda = \frac{h}{mv} = \frac{h}{\sqrt{2mK}} = \frac{h}{\sqrt{2meV}} \quad [\text{As } K = eV]$$

Here we have

$$2d\cos i = \frac{nh}{\sqrt{2meV}}$$

$$\Rightarrow \quad V = \frac{n^2h^2}{8med^2\cos^2 i}$$

For least value of V, we take $n = 1$

$$\Rightarrow \quad V = \frac{h^2}{8med^2\cos^2 i}$$

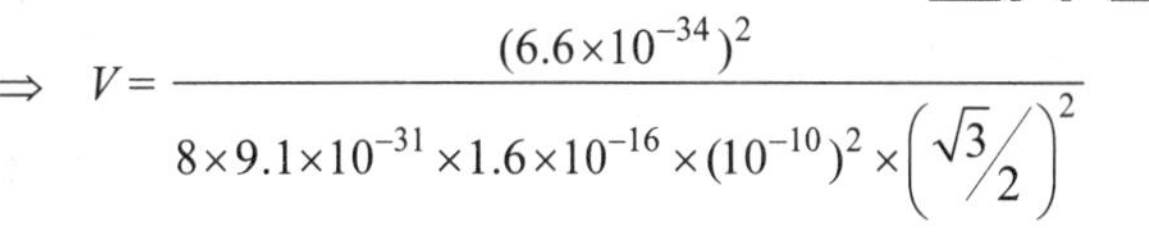

$$\Rightarrow \quad V = \frac{(6.6\times10^{-34})^2}{8\times9.1\times10^{-31}\times1.6\times10^{-16}\times(10^{-10})^2\times\left(\sqrt{3}/2\right)^2}$$

$\Rightarrow \quad V \simeq 50\,\text{V}$ **Ans. (B)**

Sol. 6 Path difference in waves AOA' and $BO'B'$ after reflection from atomic layers is given as

$$\Delta = 2d\cos i$$

For constructive interference at angle i we use

$$\Delta = 2d\cos i = n\lambda_{dB} \quad [n \in I]$$ **Ans. (B)**

Sol. 7 **Statement-1 :** It is true as in Davisson-Germer experiment interference of electrons considering their de-Broglie wavelength was experimentally verified.

Statement-2 : It is true and explaining S-1 also. **Ans. (D)**

Sol. 8 The de-Broglie wavelength for a particle with charge q and mass m which is accelerated at potential difference V is given as

$$\lambda = \frac{h}{\rho} = \frac{h}{\sqrt{2mK}} = \frac{h}{\sqrt{2mqV}} \quad [K = qV]$$

For e^- we use $m = m_e, q = e, V = V$

$$\Rightarrow \quad \lambda_e = \frac{h}{\sqrt{2m_e eV}} \qquad \ldots(1)$$

for a proton we use $m = m_p, q = e, V = 4V$

$$\Rightarrow \quad \lambda_p = \frac{h}{\sqrt{2m_p e(4V)}} = \frac{h}{2\sqrt{2m_p eV}} \qquad \ldots(2)$$

So we have from (1) and (2)

$$\frac{\lambda_e}{\lambda_p} = \frac{h/\sqrt{2m_eeV}}{h/2\sqrt{2m_peV}}$$

$$\Rightarrow \quad \frac{\lambda_e}{\lambda_p} = 2\sqrt{\frac{m_p}{m_e}}$$

Ans. (D)

Sol. 9 **(a)** Davisson and Germer experiment → related to diffraction of electron as a wave. →(i)

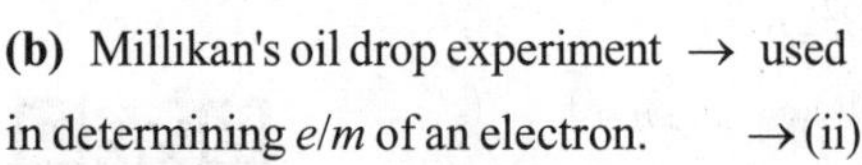

(b) Millikan's oil drop experiment → used in determining *e*/*m* of an electron. →(ii)

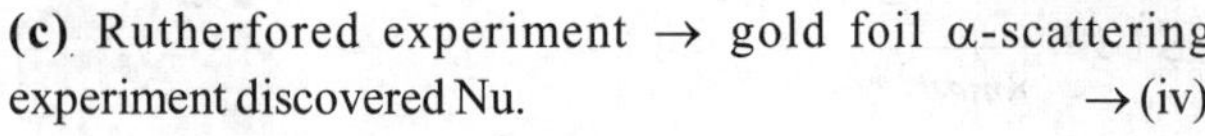

(c) Rutherfored experiment → gold foil α-scattering experiment discovered Nu. →(iv)

(d) Franck-Hertz experiment → energy level quantization. →(iii)

Ans. (B)

Sol. 10

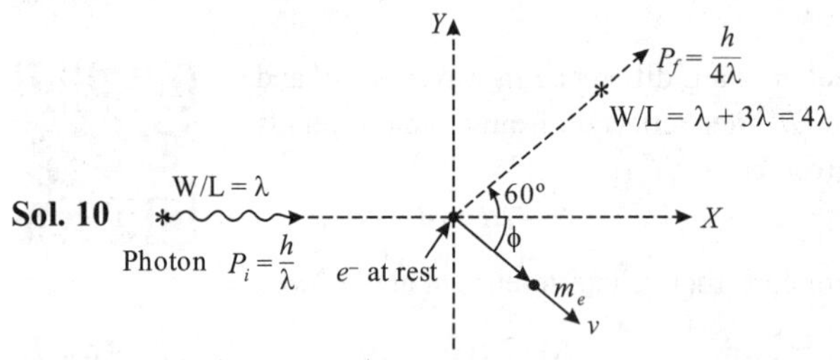

By conservation of linear momentum in above scattering, we use

$$\text{X-dir} \rightarrow \frac{h}{\lambda} = \frac{h}{4\lambda}\cos 60^\circ + m_e v\cos\phi \quad \ldots(1)$$

$$\text{Y-dir} \rightarrow \frac{h}{4\lambda}\sin 60^\circ = m_e v\sin\phi \quad \ldots(2)$$

From (1) and (2) we have

$$m_e v\cos\phi = \frac{h}{\lambda} - \frac{h}{8\lambda} = \frac{7h}{8\lambda} \quad \ldots(1)$$

$$m_e v\sin\phi = \frac{h}{4\lambda}\left(\frac{\sqrt{3}}{2}\right) \quad \ldots(2)$$

$$(2)(1) \Rightarrow \tan\phi = \frac{\sqrt{3}}{7}$$

$$\Rightarrow \quad \tan\phi = \frac{\sqrt{3}}{7} \simeq 0.25$$ **Ans. (C)**

Sol. 11 For a dust particle if we consider its mass

$$m \sim 1\mu g = 10^{-6} \times 10^{-3}\,\text{kg}$$
$$= 10^{-9}\,\text{kg}$$

If we consider it to be moving at 1 cm/s ~ 0.01 m/s

Then its de-Broglie wavelength is given as

$$\lambda = \frac{h}{\rho} = \frac{h}{mv} = \frac{6.63\times10^{-34}}{10^{-9}\times10^{-2}} \simeq 6.63 \times 10^{23}\,\text{m}$$

In such cases where mass of a moving particle is significant than its de-Broglie wavelength is so low that experimentally it cannot be determined. **Ans. (D)**

Sol. 12 For list-I

(A) Frank-Hertz Experiment → (b)→(ii)

it has given the discrete spectral lines for H or any atom.

It was the first practical experiment which analyzed quantum behaviour of radiation emitted from an atom.

(B) Photo electric effect → (b)→(i)

Photon energy is absorbed by electron to eject from a metal surface.

(C) Davisson Germer Experiment → (c)→(iii)

Scattering and interference of electrons in a specific direction by considering de-Broglie wavelength of moving electrons.

Ans. (C)

Sol. 13 For circular motion in magnetic field we use

$$qvB = \frac{mv^2}{r} \quad \ldots(1)$$

By Bohr model using quantization of angular momentum, we have

$$mvr = \frac{nh}{2\pi} \quad \ldots(2)$$

$$\Rightarrow \quad r = \frac{nh}{2\pi mv}$$

$$\Rightarrow \quad mv^2 = \frac{qBnh}{2\pi m}$$

from (1) and (2) we use

$$mv = qBr$$

$$\Rightarrow \quad mv = qB\left(\frac{nh}{2\pi mv}\right)$$

TE of charge particle is

$$E = \frac{1}{2}mv^2 = \frac{qBnh}{4\pi m}$$ **Ans. (C)**

Sol. 14 de-Broglie wavelength of an electron in n^{th} level is given as

$$\lambda = \frac{h}{\rho_n} = \frac{h}{mv_n} \qquad \ldots(1)$$

As we know $\quad v_n = \dfrac{2\pi kZe^2}{nh}$

$$\Rightarrow \quad v_n \to \frac{1}{n} \qquad \ldots(2)$$

From (1) and (2) we use

$$\lambda \propto n$$

$\Rightarrow$ For $n = 4$ we use

$$\lambda_{n=4} = 4\lambda_{n=1} \text{ (for ground state)}$$

Ans. (B)

Sol. 15 KE of an electron accelerated by a voltage V is

$$K = eV$$

de-Broglie wavelength of this e^- is given as

$$\lambda = \frac{h}{P} = \frac{h}{\sqrt{2mK}}$$

$$\Rightarrow \quad \lambda = \frac{6.6\times10^{-34}}{\sqrt{2\times9.1\times10^{-31}\times1.6\times10^{-19}\times50}}$$

$$\cong 1.7\times10^{-10}\,\text{m} = 1.7\,\text{Å}$$

Ans. (C)

Sol. 16 After passing through slit the first minima angular position can be given as

$$d\sin\theta = \lambda \qquad \ldots(1)$$

Where λ we consider de-Broglie wavelength of incident electron out of which it is given that one electron acquires momentum p_y in y-div so for this electron we use

$$\lambda = \frac{h}{|P_y|} \qquad \ldots(2)$$

From equation-(1) and (2) we use for all these electrons which gain momentum p_y in y-direction (one electron is given in question)

$$d\sin\theta = \frac{h}{|P_y|}$$

$$\Rightarrow \quad \frac{h}{|P_y|\,d} = \sin\theta$$

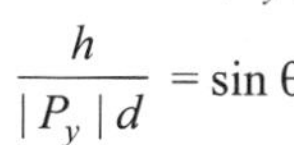
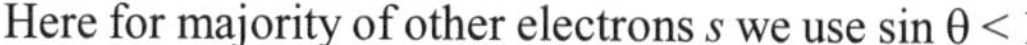

Here for majority of other electrons s we use $\sin\theta < 1$

$$\Rightarrow \quad \frac{h}{|P_y|\,d} < 1$$

$$\Rightarrow \quad |P_y|\,d > h$$

Ans. (A)

Sol. 17

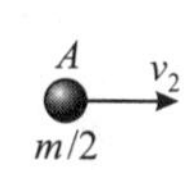

For head-on electric collision we use

$$v_1 = \left(\frac{m_1 - m_2}{m_1 + m_2}\right)u_1 + \left(\frac{2m_2}{m_1 + m_2}\right)u_2$$

$$\Rightarrow \quad v_1 = \frac{m/2}{3m/2}\cdot v + 0 = \frac{v}{3}$$

$$\Rightarrow \quad v_2 = \left(\frac{m_1 - m_1}{m_1 + m_2}\right)u_2 + \left(\frac{2m_1}{m_1 + m_2}\right)u_1$$

$$\Rightarrow \quad v_2 = 0 + \frac{2m}{3m/2}\cdot v = \frac{4v}{3}$$

After collision de-Broglie wavelength of A and B are

$$\Rightarrow \quad \lambda_A = \frac{h}{mv_1} \text{ and } l_B = \frac{h}{\left(\frac{m}{2}\right)v_2} = \frac{3h}{2mv}$$

$$\Rightarrow \quad \frac{\lambda_A}{\lambda_B} = \frac{3h/mv}{3h/2mv} = 2$$

Ans. (D)

Sol. 18 For continuous X-Ray the cut of wavelength λ_{min} is related to the voltage applied in discharge tube V as

$$\lambda_{min} = \frac{hc}{eV} = \frac{12431}{V}\,\text{Å}$$

Taking log on both sides

$$\log\lambda_{min} = \log\frac{12431}{V} = \log(12431) - \log V$$

$$\Rightarrow \quad \log\lambda_{min} = -\log V + \log(12431)$$

Ans. (C)

Sol. 19 It is given that de-Broglie wavelength for proton and α-particle are equal

$$\Rightarrow \quad \lambda_p = \lambda_\alpha$$

$$\Rightarrow \quad \frac{h}{m_\rho v_\rho} = \frac{h}{m_\alpha v_\alpha}$$

$$\Rightarrow \quad \frac{v_\rho}{v_\alpha} = \frac{m_\alpha}{m_\rho} = \frac{4m_\rho}{m_\rho} = \frac{4}{1} \quad [\text{As } m_\alpha = 4m_\rho]$$

Ans. (C)

Sol. 20

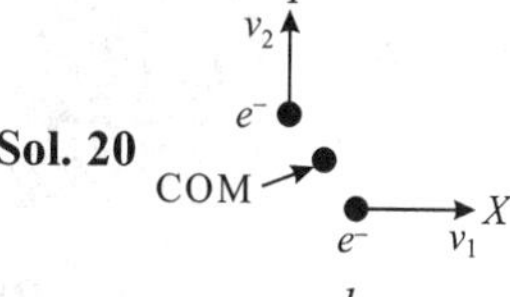

As $\vec{v}_1 = v_1\hat{i} = \dfrac{h}{m\lambda_1}\hat{i}$

$$\vec{v}_2 = v_2\hat{i} = \frac{h}{m\lambda_2}\hat{j}$$

We also use

$$\lambda_1 = \frac{h}{mv_1} \text{ and } \lambda_2 = \frac{h}{mv_2}$$

Vector speed of COM of this two electron system is given as

$$\vec{V_C} = \frac{m\vec{v}_1 + m\vec{v}_2}{m+m} = \frac{1}{2}(\vec{v}_1 + \vec{v}_2)$$

$$\Rightarrow \quad \vec{V_C} = \frac{1}{2}\frac{h}{m}\left(\frac{1}{\lambda_1}\hat{i} + \frac{1}{\lambda_2}\hat{j}\right)$$

If reference frame of COM, velocity of first electron travelling in X-direction is given as

$$\vec{V}_{1C} = \vec{v}_1 - \vec{v}_2$$

$$\Rightarrow \quad \vec{V}_{1C} = \frac{h}{m\lambda_1}\hat{i} - \left(\frac{h}{2m\lambda_1}\hat{i} + \frac{h}{2m\lambda_2}\hat{j}\right)$$

$$\Rightarrow \quad \vec{V}_{1C} = \frac{h}{2m\lambda_1}\hat{i} - \frac{h}{2m\lambda_2}\hat{j}$$

$$\Rightarrow \quad |\vec{V}_{1C}| = \frac{h}{2m}\sqrt{\frac{1}{\lambda_1^2} + \frac{1}{\lambda_2^2}}$$

[speed of first electron in frame of COM]

de-Broglie wavelength of first electron in COM frame is given as

$$\lambda_{1C} = \frac{h}{mv_{1C}} = \frac{h}{m} \times \frac{1}{\frac{h}{2m}\sqrt{\frac{1}{\lambda^2} + \frac{1}{\lambda_2^2}}}$$

$$\Rightarrow \quad \lambda_{1C} = \frac{2\lambda_1\lambda_2}{\sqrt{\lambda_1^2 + \lambda_2^2}}$$ **Ans. (C)**

Sol. 21 By Bohr's II Postulate we have

$$mvr = \frac{nh}{2\pi} \Rightarrow mv = \frac{nh}{2\pi r}$$

de-Broglie wavelength of electron in second excited state ($n = 3$), we use

$$\lambda = \frac{h}{mv} = \frac{2\pi h}{n} = \frac{2\pi}{n}(0.529\, n^2) \text{ [as } z = 1]$$

$$\Rightarrow \quad \lambda = 1.06\,\pi n \Rightarrow \lambda \propto n$$

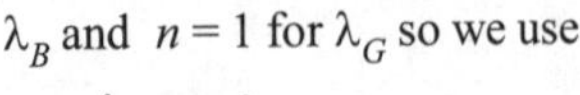
As $n = 3$ for λ_B and $n = 1$ for λ_G so we use

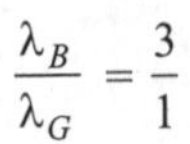
$$\frac{\lambda_B}{\lambda_G} = \frac{3}{1}$$

$$\Rightarrow \quad \lambda_B = 3\lambda_G$$ **Ans. (D)**

Sol. 22 deBroglie wavelength of electrons is

$$\lambda = \frac{h}{p} \qquad [\lambda = 7.5 \times 10^{-12}]$$

$$\Rightarrow \quad p = \frac{h}{\lambda}$$

Kinetic energy of electrons is

$$KE = \frac{p^2}{2m} = \frac{(h/\lambda)^2}{2m} = \frac{\left\{\frac{6.6\times10^{-34}}{7.5\times10^{-12}}\right\}^2}{2\times9.1\times10^{-31}}\text{J}$$

$$\Rightarrow \quad KE = 25\,\text{keV}$$ **Ans. (C)**

Sol. 23 Given that de-Broglie wavelength of electron is

$$\lambda_e = \frac{h}{mv} = 10^{-3}\left(\frac{3\times10^8}{6\times10^{14}}\right)$$

$$\Rightarrow \quad v = \frac{6.63\times10^{-34}\times6\times10^{14}}{9.1\times10^{-31}\times3\times10^5}$$

$$\Rightarrow \quad v = 1.45 \times 10^6 \text{ m/s}$$ **Ans. (A)**

Sol. 24 De-Broglie wavelength of an accelerated charge particle is given as

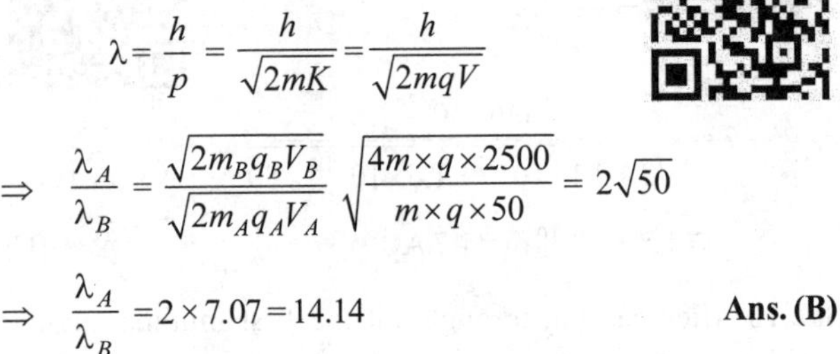
$$\lambda = \frac{h}{p} = \frac{h}{\sqrt{2mK}} = \frac{h}{\sqrt{2mqV}}$$

$$\Rightarrow \quad \frac{\lambda_A}{\lambda_B} = \frac{\sqrt{2m_Bq_BV_B}}{\sqrt{2m_Aq_AV_A}}\sqrt{\frac{4m\times q\times2500}{m\times q\times50}} = 2\sqrt{50}$$

$$\Rightarrow \quad \frac{\lambda_A}{\lambda_B} = 2\times7.07 = 14.14$$ **Ans. (B)**

Sol. 25 Momentum of the two particle are given as

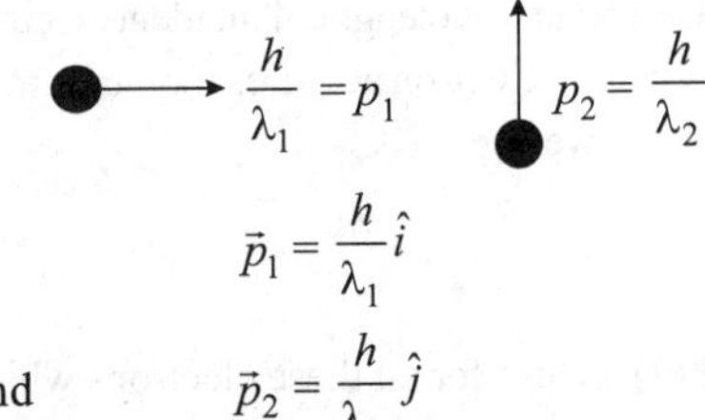

$$\vec{p}_1 = \frac{h}{\lambda_1}\hat{i}$$

and $$\vec{p}_2 = \frac{h}{\lambda_2}\hat{j}$$

Using momentum conservation before and after collision we have

$$\vec{p} = \vec{p}_1 + \vec{p}_2$$

$$\vec{p} = \frac{h}{\lambda_1}\hat{i} + \frac{h}{\lambda_2}\hat{j}$$

$$|\vec{p}| = \sqrt{\left(\frac{h}{\lambda_1}\right)^2 + \left(\frac{h}{\lambda_2}\right)^2}$$

de-Broglie wavelength of final particle λ is given as

$$\frac{h}{\lambda} = \sqrt{\left(\frac{h}{\lambda_1}\right)^2 + \left(\frac{h}{\lambda_2}\right)^2}$$

$$\Rightarrow \quad \frac{1}{\lambda^2} = \frac{1}{\lambda_1^2} + \frac{1}{\lambda_2^2}$$ **Ans. (D)**

Sol. 26 A (v_0), $\lambda_A = \dfrac{h}{mv_0}$; C ($v/2$, $m/2$), B (v, $m/2$)

By conservation of momentum

$$mv_0 = \frac{m}{2}v - \frac{mv}{4} = \frac{mv}{4}$$

$\Rightarrow$ $$v = 4v_0$$

De-Broglie wavelength of B is

$$\lambda_B = \frac{2h}{mv} = \frac{2h}{4mv_0} = \frac{\lambda_A}{2}$$

De-Broglie wavelength of C is

$$\lambda_C = \frac{4h}{mv} = \frac{4h}{4mv_0} = \lambda_A \qquad \textbf{Ans. (D)}$$

Sol. 27 (x) → ← (y) → (P) →

By momentum conservation before and after collision, we have

$$p_x - p_y = p_P$$

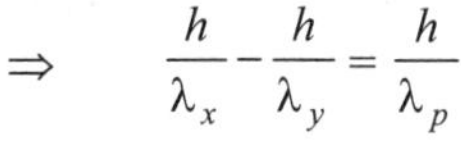

$\Rightarrow$ $$\frac{h}{\lambda_x} - \frac{h}{\lambda_y} = \frac{h}{\lambda_p}$$

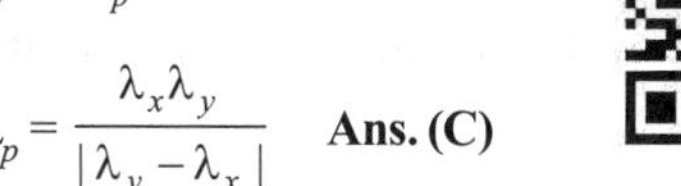

$\Rightarrow$ $$\lambda_p = \frac{\lambda_x \lambda_y}{|\lambda_y - \lambda_x|} \qquad \textbf{Ans. (C)}$$

Sol. 28

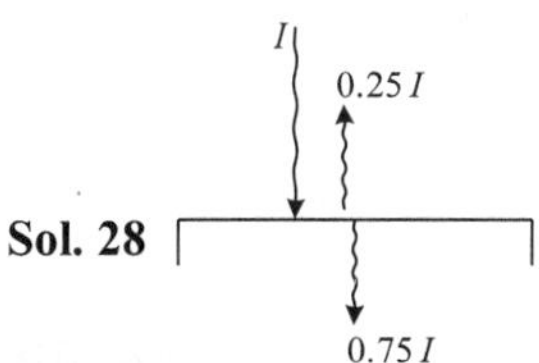

Pressure on surface due to incident, reflected and absorbed radiation is

$$p = 0.25\left[\frac{2I}{c}\right] + 0.75\left[\frac{I}{c}\right]$$

$$= 1.25\frac{I}{c} = 1.25 \times \frac{50}{3\times10^8} = 20.83 \times 10^{-8}\,\text{N} \qquad \textbf{Ans. (D)}$$

Sol. 29 Radiation pressure on surface is

$$p = \frac{I}{c}$$

Momentum transferred in time Δt is

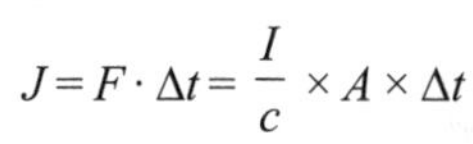

Impulse $$J = F\cdot\Delta t = \frac{I}{c} \times A \times \Delta t$$

$$= \frac{25\times10^4}{3\times10^8} \times 25 \times 10^{-4} \times 40 \times 60$$

$$= 5 \times 10^{-3}\,\text{N-s} \qquad \textbf{Ans. (A)}$$

Sol. 30 From an energy source of power P_{watt}, number of photons emitted/sec are

$$n = \frac{P\lambda}{hc}$$

$$= \frac{2\times10^{-3}\times500\times10^{-9}}{6.6\times10^{-34}\times3\times10^8} \simeq 1.5 \times 10^{16}\,\text{ph/s} \qquad \textbf{Ans. (B)}$$

Sol. 31 Photon energy is given as

$$E = \frac{hc}{\lambda_{\text{photon}}} \qquad \ldots(1)$$

and $$\lambda_{\text{electron}} = \frac{h}{\sqrt{2mE}} \qquad \ldots(2)$$

From (1) and (2) we have

$\Rightarrow$ $$\frac{\lambda_{\text{electron}}}{\lambda_{\text{photon}}} = \frac{1}{c}\left(\frac{E}{2m}\right)^{1/2} \qquad \textbf{Ans. (B)}$$

Sol. 32 Using de-Broglie hypothesis, wavelength of electron is given as

$$\lambda = \frac{h}{mv}$$

$\Rightarrow$ $$\lambda_0 = \frac{h}{m\sqrt{2}v_0} \qquad [\text{As } v = \sqrt{2}v_0] \qquad \ldots(1)$$

After time t, we use

$$\lambda' = \frac{h}{\sqrt{v_0^2 + v_0^2 + \left(\dfrac{eE_0 t}{m}\right)^2}}$$

$\Rightarrow$ $$\lambda' = \frac{h}{m\sqrt{2v_0^2 + \dfrac{e^2E_0^2t^2}{m^2}}} \qquad \ldots(2)$$

From (1) and (2), we use

$\Rightarrow$ $$\lambda' = \frac{\lambda_0}{\sqrt{1 + \dfrac{e^2E_0^2t^2}{2m^2v_0^2}}} \qquad \textbf{Ans. (C)}$$

Sol. 33 de-Broglie wavelength of particle of kinetic energy E is given as

$$\lambda = \frac{h}{\sqrt{2mE}} \qquad \ldots(1)$$

and $$\frac{\lambda}{2} = \frac{h}{\sqrt{2m(E+\Delta E)}} \qquad \ldots(2)$$

From (1) and (2), we have

$$\frac{E+\Delta E}{E} = 4$$

$\Rightarrow$ $$\Delta E = 3E \qquad \textbf{Ans. (C)}$$

Sol. 34 Acceleration of electron due to electric field is

$$a=\frac{eE}{m}$$

Speed of electron after time t is given as

$$v=u+at=\left(\frac{eE}{m}\right)t$$

de Broglie wavelength of electron is

$$\lambda=\frac{h}{mv}=\frac{h}{eEt}$$

$$\Rightarrow \quad \frac{d\lambda}{dt}=-\frac{h}{|e|Et^2} \qquad \textbf{Ans. (A)}$$

Sol. 35 deBroglie wavelength of particle is given as

$$\lambda_d=\frac{h}{p}=\frac{h}{mv}$$

$$\Rightarrow \quad \lambda_{dp}=\frac{h}{m(5v)} \qquad \ldots(1)$$

$$\Rightarrow \quad \lambda_{de}=\frac{h}{m_e v} \qquad \ldots(2)$$

Given that

$$\frac{\lambda_{dp}}{\lambda_{de}}=1.878\times10^{-4}$$

$$\Rightarrow \quad m=\frac{m_e}{5\times1.878\times10^{-4}}$$

$$\Rightarrow \quad m=\frac{9.1\times10^{-31}}{5\times1.878\times10^{-4}}$$

$$\Rightarrow \quad m=9.7\times10^{-28}\text{ kg} \qquad \textbf{Ans. (D)}$$

Sol. 36 Number of photons emitted per second are given as

$$N_1=\frac{P\lambda_1}{hc}$$

and $$N_2=\frac{P\lambda_2}{hc}$$

$$\Rightarrow \quad \frac{N_1}{N_2}=\frac{\lambda_1}{\lambda_2}=\frac{1}{500} \qquad \textbf{Ans. (A)}$$

Sol. 37 Before Collision

m/2 (A) → V_0 m/3 (B)

After Collision

A → V_A B → V_B

By momentum conservation, we use

$$\frac{m}{2}V_0=\frac{m}{2}V_A+\frac{m}{3}V_B$$

$$\Rightarrow \quad V_0=V_A+\frac{2V_B}{3} \qquad \ldots(1)$$

As collision is elastic

$$V_0=V_B-V_A \qquad \ldots(2)$$

Solving (1) & (2), we get

$$V_A=\frac{V_0}{5} \text{ and } V_B=\frac{6V_0}{5}$$

De-Broglie wavelength of A before collision

$$\lambda_{DA}=\frac{2h}{mV_0}$$

De-Broglie wavelength A after collision

$$\lambda'_{DA}=\frac{h}{mV_A}=\frac{10h}{mV_0}$$

$$\Rightarrow \quad \Delta\lambda=\lambda'_{DA}-\lambda_{DA}=\frac{8h}{mv_0}=4\lambda_{DA}=4\lambda_0 \qquad \textbf{Ans. (A)}$$

Sol. 38 For a particle of kinetic energy E, de-Broglie wavelength is given as

$$\lambda=\frac{h}{p}=\frac{h}{\sqrt{2mE}}$$

$$\Rightarrow \quad \lambda\propto\frac{1}{\sqrt{m}}$$

As $\quad m_{He^{++}}>m_p>m_e$

$$\Rightarrow \quad \lambda_{He^{++}}<\lambda_p<\lambda_e \qquad \textbf{Ans. (D)}$$

Sol. 39 RMS speed of molecules is given as

$$v_{rms}=\sqrt{\frac{3RT}{M}}=\sqrt{\frac{3KT}{m}}$$

Here $M=m\,N_{AV}$ where m is mass of one molecule and $R=k{\cdot}N_{AV}$

de-Broglie wavelength of molecules is given as

$$\lambda=\frac{h}{mv}$$

$$\Rightarrow \quad \lambda=\frac{h}{m\sqrt{\frac{3KT}{m}}}$$

$$\Rightarrow \quad \lambda=\frac{h}{\sqrt{3KTm}}$$

$$\Rightarrow \quad \lambda=\frac{6.63\times10^{-34}}{\sqrt{3\times1.38\times10^{-23}\times400\times\left(\frac{28\times10^{-3}}{6.023\times10^{-23}}\right)}}$$

$$\Rightarrow \qquad \lambda = \frac{6.63 \times 10^{-11}}{2.77} = 2.39 \times 10^{-11}\,\text{m}$$

$$\Rightarrow \qquad \lambda \simeq 0.24\,\text{Å}$$ **Ans. (B)**

Sol. 40 If linear momentum of photons are equal then it can be considered that wavelengths are also equal because wavelength is inversely proportional to momentum so Statement-I is true. On decreasing wavelength, momentum and energy of photon increases so Statement-II is false. **Ans. (C)**

Sol. 41 The shortest wavelength of the photon produced in X-Ray tube can be given as

$$\lambda_{\min} = \frac{1240}{\Delta V}\,(\text{nm})$$

$$\Rightarrow \qquad \lambda_{\min} = \frac{1240}{1.24 \times 10^{6}} = 10^{-3}\,\text{nm}$$ **Ans. (C)**

Sol. 42 de-Broglie wavelength of a moving particle is given as

$$\lambda = \frac{h}{mv}$$

Given that $\lambda_p = \lambda_\alpha$

$$\Rightarrow \qquad m_p v_p = m_\alpha v_\alpha$$

$$\Rightarrow \qquad m_p v_p = 4m_p v_\alpha \qquad (\text{As } m_\alpha = 4m_p)$$

$$\Rightarrow \qquad \frac{v_p}{v_\alpha} = 4$$ **Ans. (D)**

Sol. 43 de-Broglie wavelength of particles can be given as

$$\lambda = \frac{h}{p} = \frac{h}{\sqrt{2mqV}}$$

$$\Rightarrow \qquad \frac{\lambda_p}{\lambda_\alpha} = \sqrt{\frac{m_\alpha q_\alpha}{m_p q_p}} = \sqrt{\frac{4m_p \times 2e}{m_p \times e}}$$

$$\Rightarrow \qquad \frac{\lambda_p}{\lambda_\alpha} = \sqrt{8} = 2\sqrt{2} == 2 \times 1.4 = 2.8$$ **Ans. (C)**

Sol. 44 The ratio of de-Broglie wavelengths of electron and proton is given as

$$\frac{\lambda_e}{\lambda_p} = \frac{\dfrac{h}{m_e v}}{\dfrac{h}{m_p v}} = 1836$$ **Ans. (C)**

Sol. 45 Considering energy of X-Ray beam when equated to the energy of the given mass, we get

$$\frac{hc}{\lambda} = mc^2$$

$$\Rightarrow \qquad m = \frac{h}{c\lambda}$$ **Ans. (10)**

Sol. 46 Resolving power $\propto \frac{1}{\lambda}$

As the wavelength of electron is much less than visible light, its resolving power will be much more. **Ans. (D)**

Sol. 47

(ΔE) Releases when photon going from $n = 5$ to

$$n = \Delta E = (13.6 - 0.54)\,\text{eV} = 13.06\,\text{eV}$$

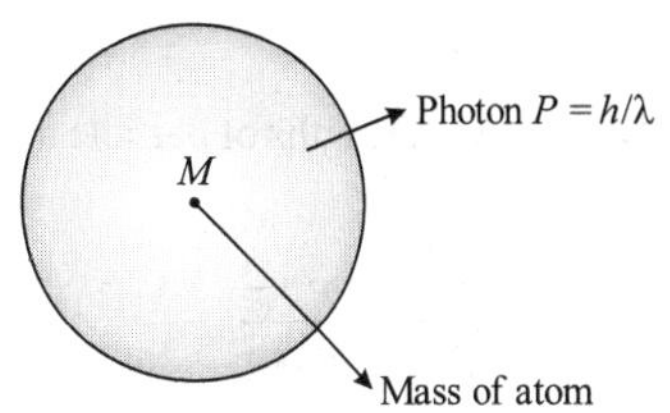

$$P_i = P_f \quad \text{(By linear momentum conservation)}$$

$$0 = \frac{h}{\lambda} - Mv = V_{\text{Recoil}} = \frac{h}{\lambda M} \qquad \ldots\text{(i)}$$

$$\& \qquad \Delta E = \frac{hc}{\lambda} = \frac{hc}{\lambda M} \times M$$

$$\Rightarrow \quad McV_{\text{Recoil}}$$

$$V_{\text{Recoil}} = \frac{\Delta E}{Mc} = \frac{13.06 \times 1.6 \times 10^{-19}}{1.67 \times 10^{-27} \times 3 \times 10^{8}}$$

$$= 4.17\,\text{m/sec}$$ **Ans. (C)**

Sol. 48 The de-Broglie wavelength of a charge particle is given as

$$\lambda = \frac{h}{mv} = \frac{h}{\sqrt{2mK}} = \frac{h}{\sqrt{2mqV}}$$

$$\Rightarrow \qquad \frac{\lambda_1}{\lambda_2} = \sqrt{\frac{m_2}{m_1}}$$

$$\Rightarrow \qquad \frac{\lambda_e}{\lambda_p} = \sqrt{\frac{m_p}{m_e}} = \sqrt{1831.4} = 42.79$$ **Ans. (D)**

Sol. 49 Red light and blue light have different wavelength and different frequency in space as both travel at same speed. **Ans. (A)**

Sol. 50 de-Broglie wavelength of electron with kinetic energy

E is given as

$$\lambda_1 = \frac{h}{\sqrt{2mE}}$$

Wavelength of photon is given as

$$\lambda_2 = \frac{hc}{E}$$

$$\Rightarrow \quad \frac{\lambda_1}{\lambda_2} = \frac{1}{c}\left(\frac{E}{2m}\right)^{1/2}$$ **Ans. (B)**

Sol. 51 de-Broglie wavelength for the particle travelling at momentum p is given as

$$\lambda = \frac{h}{p}$$

The ratio of de-Broglie wavelengths of particle to that of electron is given as

$$\frac{\lambda_p}{\lambda_e} = \frac{p_e}{p_p} = \frac{m_e v_e}{m_p v_p}$$

$$\Rightarrow \quad 2 = \frac{m_e}{m_p}\left(\frac{v_e}{4v_e}\right)$$

$$\Rightarrow \quad m_p = \frac{m_e}{8}$$ **Ans. (D)**

Sol. 52 De-broglie wavelength of electron can be given as

$$\lambda = \frac{h}{P}$$

$$\Rightarrow \quad P = \frac{h}{\lambda}$$

Kinetic energy of electron in the tube can be given as

$$\Rightarrow \quad E = \frac{P^2}{2m} = \frac{h^2}{2m\lambda^2}$$

Cut-off wavelength of emitted X-Ray can be calculated as

$$E = \frac{hc}{\lambda}$$

$$\Rightarrow \quad \frac{h^2}{2m\lambda^2} = \frac{hc}{\lambda}$$

$$\Rightarrow \quad \lambda = \frac{2mc\lambda^2}{h}$$ **Ans. (D)**

Sol. 53 de-Broglie wavelength associated with the charged particle accelerated through a given potential difference is given as

$$\lambda = \frac{h}{p} = \frac{h}{\sqrt{2mqV}}$$

As it is given that potential difference is same for both electron and proton, so we have

$$\Rightarrow \quad \frac{\lambda_e}{\lambda_p} = \sqrt{\frac{m_p}{m_e}}$$ **Ans. (D)**

Sol. 54 de-Broglie wavelengths of the particles is given as

$$\lambda = \frac{h}{P}$$

As momentum is same for both the particles, we use

$$\frac{\lambda_{2m}}{\lambda_m} = \frac{1}{1}$$ **Ans. (C)**

Sol. 55 The de-Broglie wavelength of a particle with kinetic energy K is given as

$$\lambda = \frac{h}{\sqrt{2mK}}$$

$$\Rightarrow \quad \lambda \propto \frac{1}{\sqrt{m}}$$

As $m_\alpha > m_p > m_e$,

$$\Rightarrow \quad \lambda_e > \lambda_p > \lambda_\alpha$$ **Ans. (D)**

Sol. 56 de-Broglie wavelength of the particle is given as

$$\lambda = \frac{h}{p}$$

Thus kinetic energy of photon is related to its momentum as

$$K_p = mc^2 = pc \quad \ldots(1)$$

Kinetic energy of electron is related to its momentum as

$$K_e = \frac{1}{2}mv^2 = \frac{pv}{2} \quad \ldots(2)$$

$$\Rightarrow \quad \frac{K_e}{K_p} = \frac{pv/2}{pc} = \frac{v}{2c}$$ **Ans. (C)**

Sol. 57 Particle deBroglie wavelength is given as

$$\lambda_{\text{particle}} = \frac{h}{p} = \frac{h}{9.1\times10^{-31}\times10^{6}}$$

Wavelength of photon is given as

$$\lambda_{\text{photon}} = \frac{h}{p} = \frac{h}{10^{-27}}$$

$$\Rightarrow \quad \lambda_{\text{photon}}/\lambda_{\text{particle}} = 910$$ **Ans. (910)**

Sol. 58 From the given data, we have

$$K_\alpha = \frac{hc}{\lambda} = 0.071 \times 10^{-9} = 17.49 \text{ eV}$$

Energy difference between L and K lines can be given as

$$E_K - E_L = 17.49 \text{ eV}$$

$$\Rightarrow \quad E_L = 27.5 - 17.49 \approx 10 \text{ eV} \qquad \textbf{Ans. (10)}$$

Sol. 59 Momentum of a particle is inversely proportional to its de-Broglie wavelength so its energy is related to wavelength as

$$E \propto \frac{1}{\lambda^2}$$

$$\Rightarrow \quad E = \frac{k}{\lambda^2}$$

If de-Broglie wavlength reduces to 75% of initial value then final energy will become

$$E_1 = \frac{16k}{9\lambda^2},$$

Thus extra energy required will be

$$\Rightarrow \quad \Delta E = \frac{7}{9}E \qquad \textbf{Ans. (B)}$$

Sol. 60 From photoelectric effect equation, we have

$$eV_s = \frac{hc}{\lambda} - \phi$$

$$\Rightarrow \quad e(V' - V) = hc\left(\frac{1}{\lambda'} - \frac{1}{\lambda}\right)$$

$$\Rightarrow \quad V' = \frac{hc}{e}\left(\frac{1}{\lambda'} - \frac{1}{\lambda}\right) + V$$

$$\Rightarrow \quad V' = 1240\left(\frac{1}{474} - \frac{1}{670}\right) + 0.48 = 1.25 \text{ eV}$$

Ans. (C)

Sol. 61 de-Broglie wavelength of particles is given as

$$\lambda_{\text{deb}} = \frac{h}{p} = \frac{h}{\sqrt{2mK}}$$

If λ is same then P will be same so we have

$$K \propto 1/m$$

As $\quad m_p > m_e$

$$\Rightarrow \quad K_p < K_e \qquad \textbf{Ans. (A)}$$

Sol. 62 In Davisson-Germer experiment the electrons exhibit diffraction which proves that electrons have wave nature hence both statement are correct. **Ans. (A)**

Sol. 63 DeBroglie wavelength of protonk, neutron, electron and α particle are given by

$$\lambda = \frac{h}{\sqrt{2Em}}$$

$$\Rightarrow \quad \lambda \propto \frac{h}{\sqrt{m}}$$

$$\Rightarrow \quad \lambda_e > \lambda_p > \lambda_n > \lambda_\alpha \qquad \textbf{Ans. (B)}$$

Sol. 64 Given that wavelength of electron and photon are same, so we have

$$\lambda_e = \lambda_{\text{photon}}$$

$$\Rightarrow \quad \frac{h}{mv} = \frac{h}{P_{\text{photon}}}$$

$$\Rightarrow \quad P_{\text{photon}} = mv$$

$$\Rightarrow \quad \frac{E_e}{E_{ph}} = \frac{\frac{1}{2}mv^2}{\frac{hc}{\lambda}} = \frac{1}{2}\frac{mv}{P_{ph}C} \times v = \frac{v}{2C} \qquad \textbf{Ans. (B)}$$

Sol. 65 Kinetic energy of a particle having momentum p is given as

$$K = \frac{p^2}{2m}$$

DeBroaglie wavelengths of these particles is calculated as

$$\lambda = \frac{h}{p}$$

$$\Rightarrow \quad \lambda \alpha \frac{1}{p}$$

$$\Rightarrow \quad \lambda \alpha \frac{1}{\sqrt{m}}$$

$$\Rightarrow \quad \frac{\lambda_\alpha}{\lambda_{C_{12}}} = \frac{\sqrt{3}}{1} \qquad \textbf{Ans. (B)}$$

Sol. 66 DeBroglei wavelength of a particle is given as

$$\lambda_e = \frac{h}{\sqrt{2mK}}$$

Also for photon, energy is given as

$$E = \frac{hc}{\lambda_p}$$

$$\Rightarrow \qquad \lambda_e = \frac{h\sqrt{\lambda_p}}{\sqrt{2mhc}}$$

$$\Rightarrow \qquad \lambda_p \propto \lambda e^2$$

Ans. (A)

Sol. 67 Total kinetic energy gained by a charged particle accelerated by a potential V is qV

$$K = qV$$

Body momentum gained p is written as

$$\frac{p^2}{2m} = qV$$

$$\Rightarrow \qquad p = \sqrt{2mqV}$$

$$\Rightarrow \qquad p = \frac{h}{\lambda}$$

de-Broglie wavelength of the particle is calculated as

$$\lambda = \frac{h}{\sqrt{2mqV}}$$

$$\Rightarrow \qquad \frac{\lambda_p}{\lambda_d} = \sqrt{\frac{m_d V_d}{m_p V_p}}$$

$$\Rightarrow \qquad \frac{1}{\sqrt{2}} = \sqrt{\frac{2V_d}{V_p}}$$

$$\Rightarrow \qquad \frac{V_p}{V_d} = 4$$

Ans. (D)

Sol. 68 Wavelength of incident beam $\lambda = 900 \times 10^{-9}$ m
Intensity of incident beam $= I$ W/m^2
No. of photons crossing per unit time are given as

$$n = \frac{E_{\text{net}}}{E_{\text{single photon}}} = \frac{IA\lambda}{hc}$$

$$\Rightarrow \quad n = \frac{(100)(1\times10^{-4})(900\times10^{-9})}{6.62\times10^{-34}\times3\times10^{8}} = 4.5\times10^{16}$$

Ans. (B)

Sol. 69 de-Broglie wavelength of each part is given as

$$\lambda = \frac{h}{P}$$

By conservation of linear momentum the two parts have same momentum in opposite direction hence both will have same de broglie wavelength. **Ans. (C)**

Sol. 70 Due to the electric field $-E_0\hat{i}$ the velocity of electron after time t is calculated as

$$v = v_0 + \frac{eE_0 t}{m}$$

The de-Broaglie wavelength of electron after time t is calculated as

$$\lambda = \frac{h}{mv} = \frac{h}{m\left(v_0 + \frac{eE_0}{m}t\right)}$$

$$\Rightarrow \qquad \lambda' = \frac{h}{mv_0\left(1 + \frac{eE_0}{mv_0}t\right)}$$

$$\Rightarrow \qquad \lambda' = \frac{\lambda_0}{\left(1 + \frac{eE_0}{mv_0}t\right)}$$

Ans. (D)

Sol. 71 de-Broglie wavelength of any moving particle is given as

$$\lambda = \frac{h}{mv}$$

$$\Rightarrow \qquad \lambda = \frac{h}{\sqrt{2m(K\cdot E)}}$$

$$\Rightarrow \qquad h = \frac{h}{\sqrt{2mqV}}$$

Substituting the values of m and q gives

$$\lambda = \frac{1.22}{\sqrt{V}} nm$$

Ans. (D)

Sol. 72 Force due to radiation incident on a surface is given as

$$F = \frac{IA}{c}$$

$$\Rightarrow \qquad 7.2\times10^{-9} = \frac{I}{c}\times36\times10^{-4}$$

$$\Rightarrow \qquad I = \frac{7.2\times10^{-9}\times3\times10^{8}}{36\times10^{-9}\times10}$$

$$\Rightarrow \qquad I = \frac{6\times10^{-1}}{10^{-3}}$$

$$\Rightarrow \qquad I = 6\times10^{2}\ \text{W/m}^2 = 0.06\ \text{W/cm}^2$$

Ans. (D)

Sol. 73 De-Broglie wavelength of a particle having kinetic energy K is given as

$$\lambda_D = \frac{h}{p} = \frac{h}{\sqrt{2mK}}$$

$$\Rightarrow \qquad \lambda \propto \frac{1}{\sqrt{m}}$$

As we have $m_\alpha > m_p > m_e$ so this gives

$$\lambda_e > \lambda_p > \lambda_\alpha$$ **Ans. (B)**

Sol. 74 When an electron beam is accelerated through potential difference V, then gain in kinetic energy is given as

$$K = eV$$

de-Broglie wavelength of electron can be given as

$$\lambda = \frac{h}{p} = \frac{h}{\sqrt{2meV}}$$

$$\Rightarrow \qquad \lambda \alpha \frac{1}{\sqrt{V}}$$

$$\Rightarrow \qquad \frac{\lambda}{\lambda_0} = \sqrt{\frac{20}{40}}$$

$$\Rightarrow \qquad \lambda = \frac{\lambda_0}{\sqrt{2}}$$ **Ans. (D)**

Sol. 75 The ratio of de-Broglie wavelength of α-particle and the proton as mentioned in the question is calculated as

$$\frac{\lambda_\alpha}{\lambda_p} = \frac{\dfrac{h}{\sqrt{2m_\alpha q_\alpha V}}}{\dfrac{h}{\sqrt{2m_p q_p V}}}$$

$$\Rightarrow \qquad \frac{\lambda_\alpha}{\lambda_p} = \sqrt{\frac{1}{8}}$$

$$\Rightarrow \qquad m = 8$$ **Ans. (C)**

Sol. 76 Momentum of photon is given as

$$\text{Momentum} = \frac{\text{Energy}}{c}$$

$$p = \frac{\text{Power} \times \text{time}}{c}$$

$$\Rightarrow \qquad p = \frac{(20\times10^{-3}\,\text{w})(300\times10^{-9}\,\text{s})}{3\times10^8\,\text{m/s}}$$

$$\Rightarrow \qquad p = 2\times10^{-17}\,\text{kg}-\text{m/s}$$ **Ans. (B)**

Sol. 77 Figure below shows the situation described in the question. We consider an elemental strip on the surface of the hemisphere as shown in figure below.

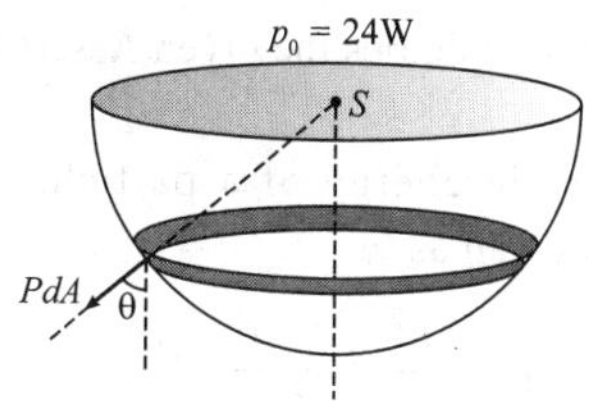

Due to the power of source, the force acting on the strip in vertically downward direction is calculated as

$$dF = PdA\cos\theta$$

Total force on the hemisphere is calculated as

$$F = \int PdA\cos\theta$$

$$\Rightarrow \qquad F = \frac{2I}{C}\int dA\cos\theta = \frac{2I}{C}\pi R^2 = 2\frac{P_0}{4\pi R^2}\cdot\frac{\pi R^2}{C}$$

$$\Rightarrow \qquad F = \frac{P_0}{2C} = \frac{24}{2\times3\times10^8} = 4\times10^{-8}\,\text{N}$$ **Ans. (4)**

Sol. 78 By momentum-kinetic energy relation, we have

$$KE = \frac{P^2}{2m}$$

De-Broglie wavelength of electron is given as

$$P = \frac{h}{\lambda}$$

When electron is accelerated through potential difference of V_1, gain in kinetic energy is calculated as

$$K_1 = eV_1$$

$$\Rightarrow \qquad eV_1 = \frac{\left(\dfrac{h}{\lambda}\right)^2}{2m}$$

When electron is accelerated through potential difference of V_2, we use

$$K_2 = eV_2$$

$$\Rightarrow \qquad eV_2 = \frac{\left(\dfrac{h}{1.5\lambda}\right)^2}{2m}$$

The ratio of potential difference is calculated as

$$\frac{V_1}{V_2} = (1.5)^2 = \frac{9}{4}$$ **Ans. (B)**

Sol. 79 A high speed electron beam acts like a wave because of high energy and momentum, its wave nature will be significant hence Assertion-A is correct. With the Davisson Germer experiment it was well evident that electron beam accelerated at high potential difference behave like a wave hence Reason-R is also correct and it explaines the given Assertion. **Ans. (D)**

Sol. 80 The kinetic energy of a particle with de-Broglie wavelength λ is given as

$$K=\frac{p^2}{2m}=\frac{h^2}{2m\lambda^2}$$

Ratio of kinetic energies of proton and alpha particle given in the question are calculated as

$$\frac{KE_p}{KE_\alpha}=\frac{m_\alpha}{m_p}=4:1$$ **Ans. (B)**

Sol. 81 Table below shows the deBroglie wavelenth calculation for the three particles mentioned in question.

	Electron	**Alpha**	**Proton**
Mass :	$\frac{m}{1840}$	$4m$	m
Charge :	e	$2e$	e
Kinetic : energy	$4K$	$2K$	K
$\lambda=\frac{h}{\sqrt{2mK}}$	$\frac{h}{\sqrt{2\cdot\frac{m}{1840}\cdot 4K}}$	$\frac{h}{\sqrt{2\cdot 4m\cdot 2K}}$	$\frac{h}{\sqrt{2mK}}$

Comparing wavelengths of electron, α-Particle and proton we have

$$\lambda_\alpha<\lambda_p<\lambda e$$ **Ans. (C)**

Sol. 82 De-Broglie wavelength of particles is calculated as

$$\lambda=\frac{h}{mv}$$

Given that de-Broglie wavelength of proton and electron are same so we use

$$\lambda_P=\lambda_e$$

$$\Rightarrow \quad m_P v_P=m_e v_e$$

$$\Rightarrow \quad p_P=p_e$$

Thus the ratio of momentum is 1. **Ans. (D)**

Sol. 83 X-rays are emitted when target metal is bombarded with high energy electron. **Ans. (B)**

Sol. 84 From kinetic theory of gases, RMS speed of gas molecules is given as

$$v_{RMS}=\sqrt{\frac{3k_BT}{m}}$$

$$\Rightarrow \quad v_{RMS}\propto\sqrt{T}$$

Thus de-Broglie wavelength is calculated as

$$\frac{h}{mv_{RMS}}=\lambda$$

$$\Rightarrow \quad \lambda\propto\frac{1}{\sqrt{T}}$$

$$\Rightarrow \quad \frac{\lambda_2}{\lambda_1}=\sqrt{\frac{T_1}{T_2}}=\sqrt{\frac{300}{600}}=\frac{1}{\sqrt{2}}$$

$$\Rightarrow \quad \lambda_2=\frac{\lambda_1}{\sqrt{2}}$$ **Ans. (A)**

Sol. 85 Using de-Broglie wavelength, we have

$$\lambda=\frac{h}{P}=\frac{h}{\sqrt{2mK}}$$

As we know, $p^2=2mK$, we use

$$\lambda\propto\frac{1}{\sqrt{m}}$$

The ratio of de-Broglie wavelength of proton and electron is given as

$$\Rightarrow \quad \frac{\lambda_P}{\lambda_e}=\sqrt{\frac{m_e}{m_P}}=1:43$$ **Ans. (A)**

Sol. 86 De-Broglie wavelength of particles which are accelerated by a potential difference are given as

$$\lambda=\frac{h}{mv}=\frac{h}{\sqrt{2mK}}=\frac{h}{\sqrt{2mq\Delta V}}$$

Ratio of wavelength for α-particle and proton is given as

$$\frac{\lambda_\alpha}{\lambda_p}=\sqrt{\frac{m_pV_pq_p}{m_\alpha V_\alpha q_\alpha}}$$

$$\Rightarrow \quad \frac{\lambda_\alpha}{\lambda_p}=\sqrt{\frac{1\times2\times1}{4\times4\times2}}=\frac{1}{4}$$

$$\Rightarrow \quad \lambda_p:\lambda_\alpha=4:1$$ **Ans. (A)**

Sol. 87 De-Broglie wavelength of electron is given as

$$\lambda=\frac{h}{\sqrt{2mE}}$$

When kinetic energy of electron becomes $\frac{E}{4}$ then de-Broglie wavelength will be given as

$$\lambda'=\frac{h}{\sqrt{2m\left(\frac{E}{4}\right)}}=\frac{2h}{\sqrt{2mE}}=2\lambda$$ **Ans. (C)**

26.1 Atomic Structure

Sol. 1 Ionization energy (to remove on e^-) from n^{th} state of a hydrogenic atom is given as

$$IP_n = \frac{13.6Z^2}{n^2}$$

here from Li^{+2} ($Z = 3$) and $n = 2$ we use

$$IP_2 = \frac{13.6 \times 3^2}{2^2} = \frac{13.6 \times 9}{4}$$

$$\Rightarrow \quad IP_2 = 3.4 \times 9 = 30.6 \text{ eV}$$ **Ans. (A)**

Sol. 2 As H and D have slight variation in spectrum, it is mainly because the nuclei of these two have different masses so if we account for this, we need to use reduced mass of electron while revolving around nucleus at centre of orbit. Here reduced mass of electron are different so energies of different energy levels will be slightly different. **Ans. (C)**

Sol. 3 Ionization of H-atom from $n = 1$ to ∞ requires 13.6 eV So from n^{th} state of H-atom ionization requires the energy

$$IP_n = \frac{13.6Z^2}{n^2} \text{ eV}$$

$$\Rightarrow \quad IP_2 = \frac{13.6 \times 1^2}{2^2} = \frac{13.6}{4} = 3.4 \text{ eV}$$ **Ans. (C)**

Sol. 4

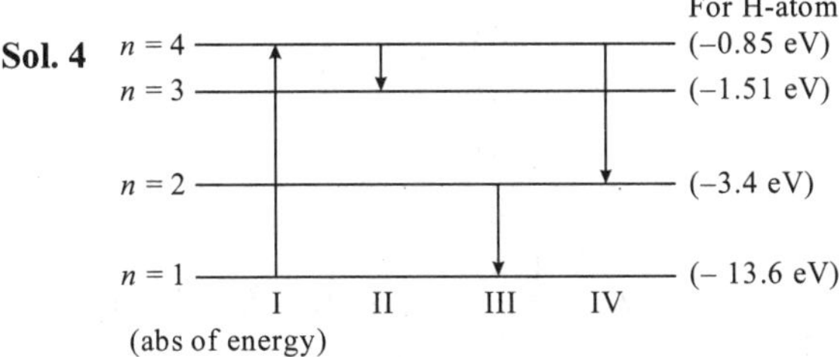

$$E_{\text{II}} = (-0.95 \text{ eV}) - (-1.51 \text{ eV}) = 0.66 \text{ eV}$$

$$E_{\text{III}} = (-3.4 \text{ eV}) - (-13.6 \text{ eV}) = 10.2 \text{ eV}$$

$$E_{\text{IV}} = (-0.85 \text{ eV}) - (-3.4 \text{ eV}) = 2.55 \text{ eV}$$

Ans. (A)

Sol. 5 Options (A) and (D) are absorbing photons

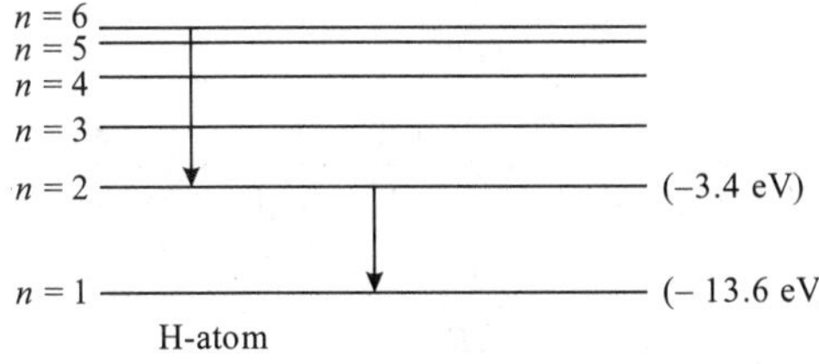

$$E_{6 \to 2} < 3.4 \text{ eV}$$

$$E_{2 \to 1} = (-3.4 \text{ eV}) - (-13.6 \text{ eV})$$

$$= 10.2 \text{ eV}$$ **Ans. (C)**

Sol. 6 In Bohr model for n^{th} orbit we can use for circular motion of electron

$$\frac{mv^2}{r} = \frac{k}{r}$$

$$\Rightarrow \quad mv^2 = k$$

$$v = \sqrt{\frac{k}{m}} \quad \text{(independent of } n\text{)}$$

kE of electron is $T_n = \frac{1}{2} mv_n^2 = \frac{1}{2} k$ (independent of n)

By Bohr's quantization postulate we have

$$mvr = \frac{nh}{2\pi}$$

$$\Rightarrow \quad r_n = \frac{nh}{2\pi mv}$$

$$\Rightarrow \quad r_n \propto n$$ **Ans. (B)**

Sol. 7 As $\lambda_{\text{IR}} > \lambda_{\text{UV}}$

$$\Rightarrow \quad \Delta E_{\text{IR}} < \Delta E_{\text{UV}}$$

For H-atom

In H-atom IR radiation will be corresponding to a transition which emits energy less than $4 \to 3$ transition

(A) $E_{4 \to 2} > E_{4 \to 3}$

(B) $E_{5 \to 4} < E_{4 \to 3}$

(C) $E_{2 \to 1} > E_{4 \to 3}$

(D) $E_{3 \to 2} > E_{4 \to 3}$

Ans. (B)

Sol. 8 Energy involved in transition from n_1 to n_2 Bohr orbit is given as

$$\Delta E = 13.6 Z^2 \left(\frac{1}{n_1^2} - \frac{1}{n_2^2}\right) \text{ eV}$$

$$\Rightarrow \quad \Delta E = 13.6 \times (3)^2 \times \left[\frac{1}{1^2} - \frac{1}{3^2}\right] \text{ eV}$$

$$\Rightarrow \quad \Delta E = 13.6 \times 8 \times \frac{8}{9} = 108.8 \text{ eV}$$ **Ans. (A)**

Sol. 9

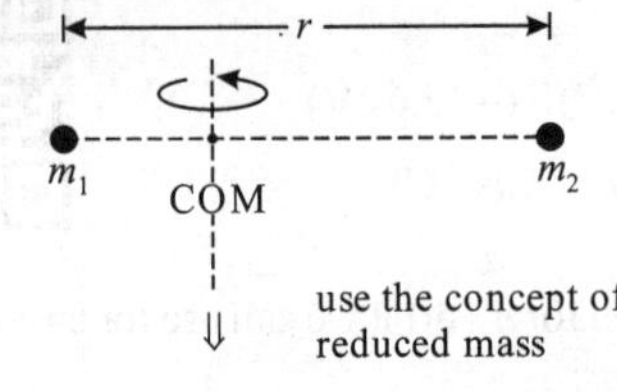

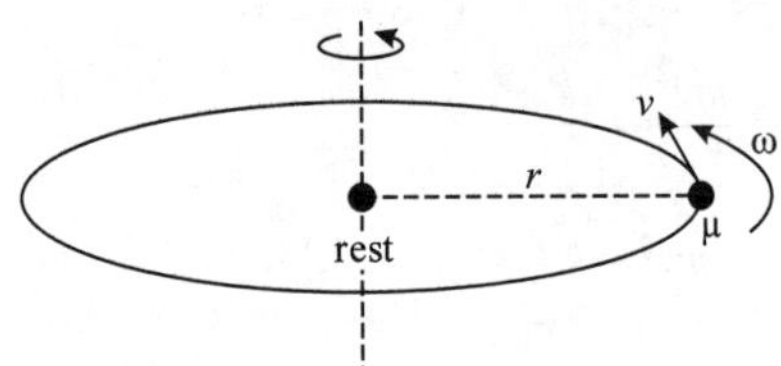

Rotational KE of system is given as

$$E=\frac{1}{2}I\omega^2=\frac{1}{2}(\mu r^2)\left(\frac{v}{r}\right)^2=\frac{1}{2}\mu v^2 \ldots (1)$$

Where $\left(\mu=\dfrac{m_1m_2}{m_1+m_2}\right)$

Using Bohr's quantization rule, we have

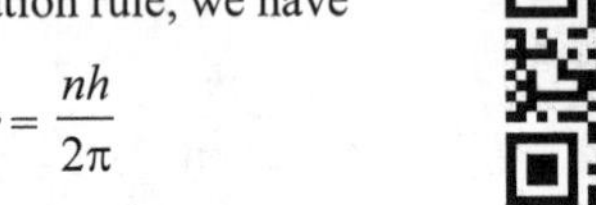

$$\mu vr=\frac{nh}{2\pi}$$

$$\Rightarrow \quad v=\frac{nh}{2\pi\mu r} \quad \ldots (2)$$

From equation (1) and (2) we have

$$E_{RE}=\frac{1}{2}\mu\left(\frac{nh}{2\pi\mu r}\right)^2$$

$$\Rightarrow \quad E_{RE}=\frac{1}{2}\mu\left(\frac{n^2h^2}{4\pi^2\mu^2r^2}\right)$$

$$\Rightarrow \quad E_{RE}=\frac{n^2h^2}{8\pi^2\left(\dfrac{m_1m_2}{m_1+m_2}\right)r^2}=\frac{(m_1+m_2)n^2h^2}{8\pi^2m_1m_2r^2}$$

Most appropriate among given options is option (A). **Ans. (A)**

Sol. 10 If an atom (H-atom) the electron is excited to n^{th} state than no of spectral lines in emission spectrum are

$$N={}^nC_2=\frac{n(n-1)}{2}$$

$$\Rightarrow \quad N=\frac{4\times 3}{2}=6 \qquad \textbf{Ans. (A)}$$

Sol. 11 By Rydberg's formula we use

$$\frac{1}{\lambda}=RZ^2\left[\frac{1}{n_1^2}-\frac{1}{n_2^2}\right]$$

[We use $n_1=n$ and $n_2=n+1$]

$$\Rightarrow \quad \frac{1}{\lambda}=RZ^2\left[\frac{1}{n^2}-\frac{1}{(n+1)^2}\right]$$

$$\Rightarrow \quad \frac{1}{\lambda}=RZ^2\left[\frac{(n+1)^2-n^2}{n^2(n+1)^2}\right]$$

$$\Rightarrow \quad \frac{1}{\lambda}=R\left[\frac{2n+1}{n^2(n+1)^2}\right]$$

$$\Rightarrow \quad \lambda=\frac{n^2(n+1)^2}{(2n+1)R}$$

For $n >> 1$

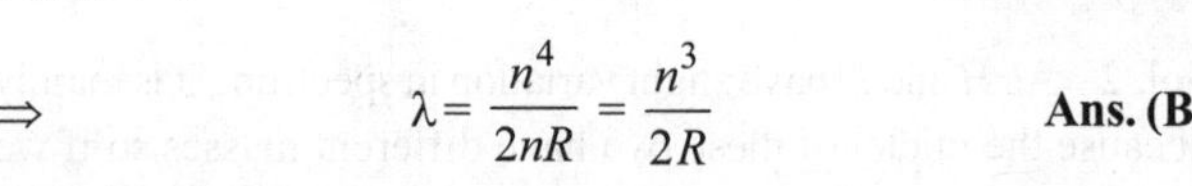

$$\Rightarrow \quad \lambda=\frac{n^4}{2nR}=\frac{n^3}{2R} \qquad \textbf{Ans. (B)}$$

Sol. 12

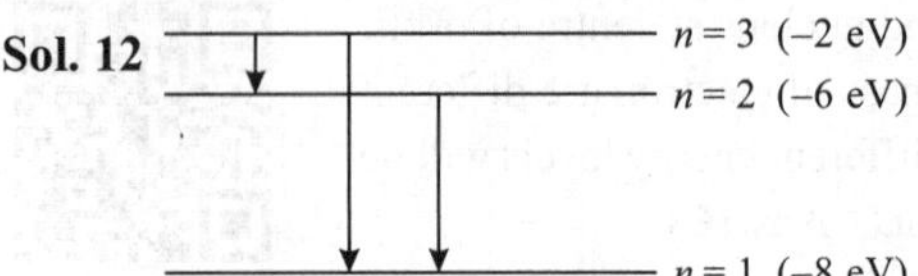

For the three energy levels there will be 3 emission lines in spectrum corresponding to energies

$$\Delta E_{32}=(-2)-(-6)=4\text{ eV}$$

$$\Delta E_{31}=(-2)-(-8)=6\text{ eV}$$

$$\Delta E_{21}=(-6)-(-8)=2\text{ eV}$$

Wavelength emitted corresponding to the energies are

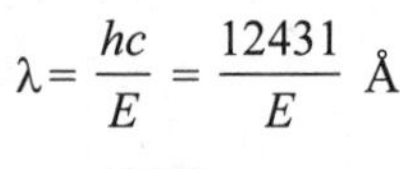

$$\lambda=\frac{hc}{E}=\frac{12431}{E}\text{ Å}$$

$$\lambda_{32}=\frac{12431}{4}\text{ Å}=3108\text{ Å}\simeq 310\text{ nm}$$

$$\lambda_{31}=\frac{12431}{6}\text{ Å}=2072\text{ Å}\simeq 207\text{ nm}$$

$$\lambda_{21}=\frac{12431}{2}\text{ Å}=6215.5\text{ Å}\simeq 621\text{ nm}$$

$\Rightarrow$ 465 nm is not present in the spectrum **Ans. (B)**

Sol. 13 For Li^{+2} Z = 3 and by Rydberg's formula these wavelengths are given as :

$$\underset{n_2}{2}\longrightarrow\underset{n_1}{1} \quad \frac{1}{\lambda_{21}}=R(3)^2\left[\frac{1}{1^2}-\frac{1}{2^2}\right]=9R\times\frac{3}{4}$$

$$\Rightarrow \quad \lambda_{21}=\frac{4}{27R}$$

$$\underset{n_2}{3}\longrightarrow\underset{n_1}{1} \quad \frac{1}{\lambda_{31}}=R(3)^2\left[\frac{1}{1^2}-\frac{1}{3^2}\right]=9R\times\frac{8}{9}$$

$$\Rightarrow \quad \lambda_{31}=\frac{9}{72R}$$

$$3 \underset{n_2 \quad\quad n_1}{\longrightarrow} 2 \quad \frac{1}{\lambda_{32}} = R(3)^2\left[\frac{1}{2^2}-\frac{1}{3^2}\right] = 9R \times \frac{5}{36}$$

$$\Rightarrow \quad \lambda_{32} = \frac{36}{45R}$$

$$\Rightarrow \quad \frac{\lambda_{32}}{\lambda_{31}} = \frac{36/45R}{9/42R} = \frac{32}{5} = 6.4$$

$$\frac{\lambda_{21}}{\lambda_{31}} = \frac{4/27R}{9/72R} = \frac{32}{27} = 1.2$$

Ans. (C)

Sol. 14 Speed of electron in n^{th} orbit is given as

$$v_n = \frac{2\pi ke^2}{nh} \quad (\text{for } z = 1)$$

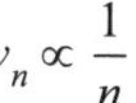

$$\Rightarrow \quad v_n \propto \frac{1}{n}$$

Graph-B shows the correct case **Ans. (A)**

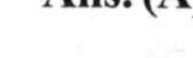

Sol. 15 As α particles are positively charged so these must be repelled away from the nucleus so here option (C) is not possible.

Ans. (C)

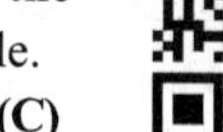

Sol. 16 IP of Li^{+2} which is a hydrogenic atom can be given by Bohr model as energy required to remove third electron from Li^{+2}

$$\Rightarrow \quad IP = 13.6\,Z^2 = 13.6\,(3)^2 = 13.6 \times 9 = 122.4\text{ eV}$$

Total energy required to remove all 3 electrons from Li atom is

$$E = 5.4 + 75.6\text{ eV} + 122.4\text{ eV}$$

$$\Rightarrow \quad E = 203.4\text{ eV}$$ **Ans. (C)**

Sol. 17 If ν be the frequency of radiation emitted, we use

$$\Delta E = (IP)\,Z^2\left[\frac{1}{n_1^2}-\frac{1}{n_2^2}\right]$$

(For a transition $n_2 \rightarrow n_1$)

$$\Rightarrow \quad \Delta E = h\nu = (IP)\,Z^2\left[\frac{1}{(n-1)^2}-\frac{1}{n^2}\right]$$

$$\Rightarrow \quad \nu = \frac{(IP)Z^2}{h}\left[\frac{2n-1}{n^2(n-1)^2}\right]$$

Here $2n >> 1$ and $n >> 1$

$$\Rightarrow \quad \nu = \frac{(IP)(Z)^2}{h}\left(\frac{2n}{n^4}\right) = \frac{2IP(Z^2)}{h}\left(\frac{1}{n^3}\right)$$

$$\Rightarrow \quad \nu \propto \frac{1}{n^3}$$

Ans. (D)

Sol. 18 Equivalent current due to e^- motion in n^{th} orbit is given as

$$I_n = \frac{e\omega_n}{2\pi} = \frac{ev_n}{2\pi r_n}$$

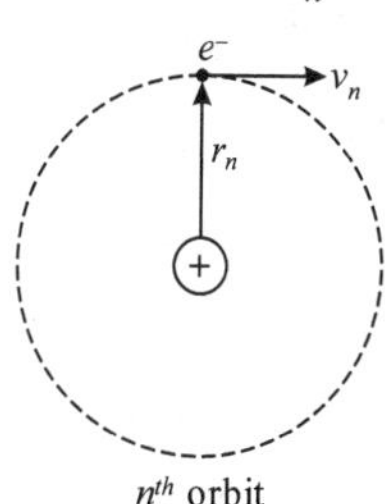

So magnetic moment due to electron current in n^{th} orbit is

$$\mu_n = I_n \times \pi r_n^2 = \frac{ev_n}{2\pi r_n} \times \pi r_n^2$$

$$\mu_n = \frac{1}{2}\,ev_nr_n \quad \ldots(1)$$

For n^{th} orbit (for $Z = 1$)

$$v_n = \frac{2\pi Ke^2}{nh}$$

and $$r_n = \frac{n^2h^2}{4\pi^2Ke^2m}$$

From equation-(1) we use

$$\mu_n = \frac{1}{2}e\left(\frac{2\pi Ke^2}{nh}\right) \times \left(\frac{n^2h^2}{4\pi^2Ka^2m}\right)$$

$$\Rightarrow \quad \mu_n = \frac{enh}{4\pi m}$$

Out of given options most appropriate is option (C).
But here for n^{th} excited state $n \leftarrow n + 1$. **Ans. (C)**

Sol. 19

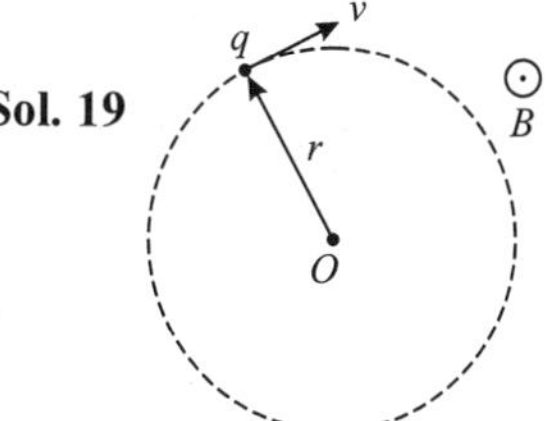

We use $$2\pi r = n\lambda = n\left(\frac{h}{mv}\right)$$

$$\Rightarrow \quad 2\pi r = \frac{nh}{mv} \quad \ldots(1)$$

For circular motion in magnetic field, we have

$$qvB = \frac{mv^2}{r}$$

$$\Rightarrow \quad mv = qBr \quad \ldots(2)$$

From (1) and (2) we use

$$2\pi r = \frac{nh}{qBr}$$

$$\Rightarrow \quad 2\pi qBr^2 = nh$$

$$\Rightarrow \quad r = \sqrt{n}$$

Ans. (C)

Sol. 20 e^- ⟶ H-atom, $E = 12.5$ eV

We know the excitation energies of H-atom are :

$$\Delta E_{1\to 2} = 10.2\,\text{eV}$$

$$\Delta E_{1\to 3} = 12.09\,\text{eV}$$

$$\Delta E_{1\to 4} = 12.75\,\text{eV}$$

By the collisions with incident electron, the H-atom can get excited max to $n = 3$ level

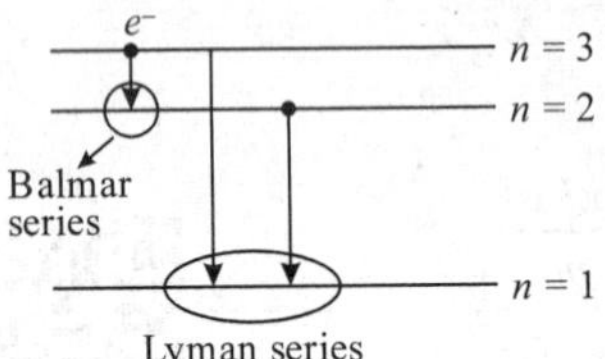

Ans. (A)

Sol. 21 For a revolving electron in speed v_n and radius r_n for n^{th} orbit, we use :

$$F = \frac{mv_n^2}{r_n}$$

$$\Rightarrow \quad \frac{e^2}{4\pi E_0}\left(\frac{1}{r_n^2} + \frac{\beta}{r_n^3}\right) = \frac{mv_n^2}{r_n} \qquad \ldots(1)$$

By using Bohr's quantization postulate we use

$$\Rightarrow \quad mv_n r_n = \frac{nh}{2\pi} \qquad \ldots(2)$$

$$\Rightarrow \quad v_n = \frac{nh}{2\pi m r_n}$$

From equation (1) we have

$$\frac{e^2}{4\pi\varepsilon_0}\left[\frac{1}{r_n} + \frac{\beta}{r_n^2}\right] = m\left[\frac{nh}{2\pi m r_n}\right]^2$$

We multiply r_n^2 in LHS and RHS gives :

$$\frac{e^2}{4\pi\varepsilon_0}[r_n + \beta] = \frac{n^2h^2}{4\pi^2 m}$$

$$\Rightarrow \quad r_n = n^2\frac{h^2\varepsilon_0}{e^2\pi m} - \beta$$

$$\Rightarrow \quad r_n = n^2 a_0 - \beta$$

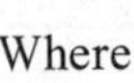

Where $a_0 = \dfrac{h^2\varepsilon_0}{e^2\pi m}$ **Ans. (C)**

Sol. 22 For a given transition $n_2 \to n_1$, the wavelength emitted can be given as :

$$\frac{1}{\lambda} = RZ^2\left[\frac{1}{n_1^2} - \frac{1}{n_2^2}\right]$$

As $n_2 = 2$ and $n_1 = 1$ for all

$$\Rightarrow \quad \lambda \propto \frac{1}{Z^2}$$

For ${}_1H^1$ and ${}_1H^2$ we take $Z = 1$

$$\Rightarrow \quad \lambda_1 \cong \lambda_2$$

and for $({}_2He^4)^+$ $Z = 2$

$$\Rightarrow \quad \lambda_3 = \frac{\lambda_1}{4}$$

and for $({}_3Li^6)^{+2}$ $Z = 3$

$$\Rightarrow \quad \lambda_4 = \frac{\lambda_1}{9}$$

$$\Rightarrow \quad \lambda_1 = \lambda_2 = 4\lambda_3 = 9\lambda_4$$

Ans. (C)

Sol. 23 For a given hydrogenic atom (Li^{+2}) the ionization energy for an electron in n^{th} state is given as :

$$IP_n = 13.6\frac{Z^2}{n^2}$$

Here $Z = 3$ and for first excited state we use $n = 2$

$$\Rightarrow \quad IP_{n=2} = 13.6 \times \frac{(3)^2}{(2)^2}$$

$$= \frac{13.6 \times 9}{4} = 30.3\,\text{eV}$$

Ans. (B)

Sol. 24 KE, PE and TE of electron in a hydrogenic atom according to Bohr model is given as :

$$TE = -\frac{KZe^2}{2r_n};\ KE = \frac{KZe^2}{2r_n};\ PE = -\frac{KZe^2}{r_n}$$

When electron makes a transition from excited state to ground state r_n decreases.

As KE is positive

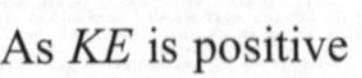

$\Rightarrow$ $KE\uparrow$ (+ve); $PE\downarrow$ (–ve); $TE\downarrow$ (–ve) **Ans. (A)**

Sol. 25 m neutron ⟶ - - - - m H-atom (rest) - - - - $2m$ ⟶ v_f

By conservation of momentum if inelastic collision occurs

$$mv = 2mv_f$$

$$\Rightarrow \quad v_f = \frac{v}{2}$$

Loss in KE $\quad \Delta E_2 = \frac{1}{2}mv^2 - \frac{1}{2}(2m)\left(\frac{v}{2}\right)^2 = \frac{1}{4}mv^2$

For this collision to happen ΔE_2 must be minimum 10.2 eV (Minimum energy which can be absorbed by H-atom in excitation from $n = 1$ to $n = 2$.)

$$\Rightarrow \quad \frac{1}{4}mv^2 = 10.2\,\text{eV}$$

$$\Rightarrow \quad K_{in} = 20.4\,\text{eV}$$

Ans. (A)

Sol. 26 (H-atom) $n = 2 \downarrow (*)$ $n = 1$ $\quad \Delta E = 13.6\left[\frac{1}{1^2} - \frac{1}{2^2}\right] = 10.2\,\text{eV}$ $\longrightarrow$ (Li^{+2} atom) $\quad \infty \uparrow$ n^{th} state

Energy required to completely remove the electron of Li^{+2} in n^{th} state is

$$\Delta E_{IP(n)} = \frac{13.6Z^2}{n^2} = \frac{13.6(3)^2}{n^2}$$

$$\Rightarrow \quad 10.2 \geq \frac{13.6 \times 9}{n^2}$$

$$\Rightarrow \quad n^2 \geq \frac{13.6 \times 9}{10.2}$$

$$\Rightarrow \quad n^2 \geq 12$$

$$\Rightarrow \quad n \geq 2\sqrt{3} \cong 3.5$$

Ans. (B)

Sol. 27 For first transition, we use

$$\frac{hc}{\lambda_1} = -E - (-2E) = E$$

$$\Rightarrow \quad \lambda_1 = \frac{hc}{E} \qquad \ldots(1)$$

For second transition, we use

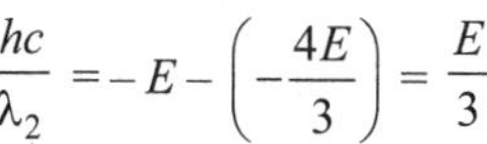

$$\frac{hc}{\lambda_2} = -E - \left(-\frac{4E}{3}\right) = \frac{E}{3}$$

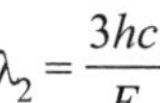

$$\Rightarrow \quad \lambda_2 = \frac{3hc}{E} \qquad \ldots(2)$$

From equation-(1) and (2)

$$\frac{\lambda_1}{\lambda_2} = \frac{hc/E}{3hc/E} = \frac{1}{3}$$

Ans. (B)

Sol. 28 Due to motion of electron in n^{th} orbit, the magnetic induction at centre is given by Biot Savart's law as

$$B = \frac{\mu_0}{4\pi}\frac{qv\sin\theta}{r^2} = \frac{\mu_0}{4\pi}\cdot\frac{ev_n}{r_n^2}$$

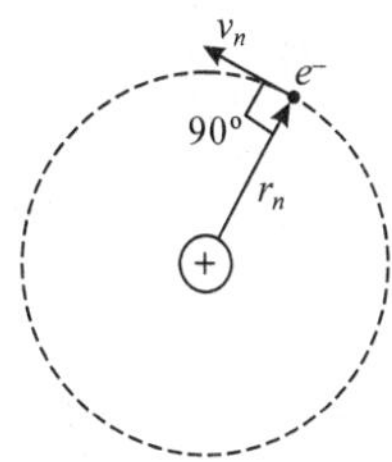

As for n^{th} orbit – for $Z = 1$

$$v_n = \frac{2\pi ke^2}{nh}$$

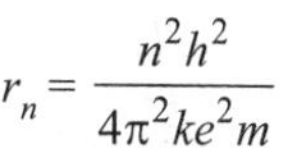

$$r_n = \frac{n^2h^2}{4\pi^2ke^2m}$$

$$\Rightarrow \quad B_n = \frac{\mu_0}{4\pi}e \times \frac{2\pi ke^2}{nh} \times \left(\frac{4\pi^2ke^2m}{n^2h^2}\right)^2$$

$$\Rightarrow \quad B_n \propto n^{-5}$$

Ans. (B)

Sol. 29 From Bohr's II Postulate we use

$$mvr = \frac{h}{2\pi}\,[n = 1] \qquad \ldots(1)$$

Acceleration of electron in first orbit of H-atom is :

$$a = \frac{v^2}{r} \qquad \ldots(2)$$

From equation-(1) we use

$$v = \frac{h}{2\pi mr}$$

From equaton-(2) we have

$$a = \left(\frac{h}{2\pi mr}\right)^2 \times \frac{1}{r} = \frac{h^2}{4\pi^2m^2r^3}$$

Ans. (C)

Sol. 30 Series limit of Lyman series is corresponding to transition $\infty \to 1$

Here by using Rydberg's formula, we have

$$\nu_L = \frac{c}{\lambda_{\infty\to1}} = c\left[RZ^2\left(\frac{1}{1^2} - \frac{1}{\infty^2}\right)\right]$$

$$\Rightarrow \quad \nu_L = RcZ^2$$

For Pfund series limit is corresponding to transition $\infty \to 5$ So we use

$$\nu_F = \frac{c}{\lambda_{\infty\to5}} = c\left[RZ^2\left(\frac{1}{5^2} - \frac{1}{\infty^2}\right)\right]$$

$$\Rightarrow \quad \nu_F = RcZ^2\left(\frac{1}{25}\right) = \frac{\nu_L}{25}$$

Ans. (D)

Sol. 31 For transition from n^{th} state to ground state emitted wavelength is given as

$$\frac{1}{\Lambda_n} = RZ^2\left[\frac{1}{1^2}-\frac{1}{n^2}\right]$$

As $n >> 1$
$$\Lambda_n = \frac{1}{RZ^2}\left(1-\frac{1}{n^2}\right)^{-1}$$

$$\Lambda_n = \frac{1}{RZ^2}\left(1+\frac{1}{n^2}\right) \quad \ldots(1)$$

[As $\frac{1}{n^2} << 1$ we use $(1+x)^b \simeq (1+xb)$]

For n^{th} state de-Broglie wavelength of e^- is given as :

$$\lambda_n = \frac{h}{p_n} = \frac{h}{mv}$$

By angular momentum quantization postulate of Bohr model, we use

$$mvr = \frac{nh}{2\pi}$$

$$\Rightarrow \quad \lambda_n = \frac{h}{mv}$$

$$= \frac{2\pi}{n} \times (0.529 \times n^2) \simeq 1.06\,\pi n \quad \ldots(2)$$

From equation-(1) we have

$$\Lambda_n = \frac{1}{RZ^2} + \frac{1}{RZ^2 n^2}$$

$$\Rightarrow \quad \Lambda_n = \frac{1}{RZ^2} + \frac{(1.06\pi)^2}{RZ^2}\cdot\frac{1}{\lambda_n^2}$$

$$\Rightarrow \quad \Lambda_n = A + \frac{B}{\lambda_n^2}$$

Ans. (A)

Sol. 32 Radius of nth orbit for Muonic atom is

$$r_\mu = \frac{n^2h^2}{4\pi^2 kZe^2 m_\mu} \simeq \frac{n^2h^2}{4\pi^2 kZe^2 m_e} \times \frac{1}{200} = \frac{r_e}{200}$$

[As $m_\mu = 200\, m_e$]

$\Rightarrow$ Option (a) is correct.

Speed of muon in n^{th} orbit can be given as

$$v_\mu = \frac{2\pi kZe^2}{nh}$$

[It is independent of mass of orbiting particle]

$$\Rightarrow \quad v_\mu = v_e$$

$\Rightarrow$ Option (b) is NOT correct.

The ionization energy in Bohr model is given as

$$E_\mu = \frac{2\pi^2k^2Z^2e^4m_\mu}{n^2h^2}$$

$$\simeq \frac{2\pi^2k^2Z^2e^4m_e}{n^2h^2} \times 200 = 200\,E_e$$

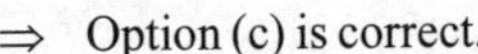

$\Rightarrow$ Option (c) is correct.

Momentum of muon in atom is given as :

$$P_\mu = m_\mu v_\mu = 200\, m_e \cdot v_e = 200\, P_e$$

$\Rightarrow$ Option (d) is correct. **Ans. (D)**

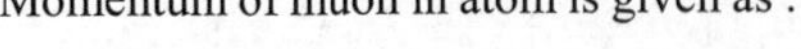

Sol. 33 He $\xrightarrow{E_1}$ He$^+$ $\xrightarrow{E_2}$ He^{+2}

Total energy required to completely ionize He atom is $(E_1 + E_2)$. As He$^+$ is a hydrogenic atom (single electron system) so E_2 in the *IP* for He$^+$ which is given by Bohr model as

$$E_2 = 13.6\, Z^2 = 13.6 \times (2)^2 = 13.6 \times 4 = 54.4 \text{ eV}$$

It is given that

$$E_2 = 2.2\, E_1$$

$$\Rightarrow \quad E_1 = \frac{54.4}{2.2} \times \frac{272}{11} = 24.72 \text{ eV}$$

Total required energy is $E_1 + E_2 = 54.4 + 24.72 = 79.12$ eV

Ans. (B)

Sol. 34 Energy of photon $= \frac{12500}{980} = 12.75$ eV

So electron will excite to $n = 4$ state

As radius of nth orbit $r_n \propto n^2$ so the radius of atom will be $16a_0$. **Ans. (D)**

Sol. 35 For $M \rightarrow L$ shell transition, we use

$$\frac{1}{\lambda} = K\left(\frac{1}{2^2}-\frac{1}{3^2}\right) = \frac{K\times 5}{36} \quad \ldots(1)$$

For $N \rightarrow L$ shell transition, we use

$$\frac{1}{\lambda'} = K\left(\frac{1}{2^2}-\frac{1}{4^2}\right) = \frac{K\times 3}{16} \quad \ldots(2)$$

From (1) and (2)

$$\Rightarrow \quad \lambda' = \frac{20}{27}\lambda$$

Ans. (C)

Sol. 36 For the given circular orbit, we use

$$F = \frac{dU}{dr} = kr = \frac{mv^2}{r} \quad \ldots(1)$$

and
$$mvr = \frac{nh}{2\pi} \quad \ldots(2)$$

$$\Rightarrow \quad r^2 \propto n$$

$$\Rightarrow \quad r \propto \sqrt{n}$$

Energy of particle in n^{th} energy level is

$$E = \frac{1}{2}kr^2 + \frac{1}{2}mv^2$$

$$\Rightarrow \quad E = \frac{1}{2}kr^2 + \frac{1}{2}kr^2 = kr^2$$

$$\Rightarrow \quad E \propto n$$

Ans. (D)

Sol. 37 The energy lost by electrons in excitation of mercury atoms is

$$\Delta E = 5.6 - 0.7 = 4.9\text{ eV}$$

Minimum wavelength (maximum energy) of photons emitted by mercury atoms in de-excitation is given as

$$\lambda = \frac{hc}{\Delta E} = \frac{1240}{4.9} = 253\text{ nm}$$

Ans. (C)

Sol. 38 Energy released for transition $n = 2$ to $n = 1$ of hydrogen atom is given as

$$E = 13.6\,Z^2\left(\frac{1}{n_1^2} - \frac{1}{n_2^2}\right)$$

For $Z = 1, n_1 = 1, n_2 = 2$, we get

$$E = 13.6 \times 1 \times \left(\frac{1}{1^2} - \frac{1}{2^2}\right)$$

$$\Rightarrow \quad E = 13.6 \times \frac{3}{4}\text{ eV}$$

For He^+ ion $z = 2$

(A) $n = 1$ to $n = 4$

$$E = 13.6 \times 2^2 \times \left(\frac{1}{1^2} - \frac{1}{4^2}\right) = 13.6 \times \frac{15}{4}\text{ eV}$$

(B) $n = 2$ to $n = 4$

$$E = 13.6 \times 2^2 \times \left(\frac{1}{2^2} - \frac{1}{4^2}\right) = 13.6 \times \frac{3}{4}\text{ eV}$$

(C) $n = 2$ to $n = 5$

$$E = 13.6 \times 2^2 \times \left(\frac{1}{2^2} - \frac{1}{5^2}\right) = 13.6 \times \frac{21}{25}\text{ eV}$$

(D) $n = 2$ to $n = 3$

$$E = 13.6 \times 2^2 \times \left(\frac{1}{2^2} - \frac{1}{3^2}\right) = 13.6 \times \frac{5}{9}\text{ eV}$$

Energy required for transition of He^+ for $n = 2$ to $n = 4$ matches exactly with energy released in transition of H for $n = 2$ to $n = 1$

Ans. (B)

Sol. 39 Using Rydberg's formula for first line of Balmer series, we have

$$\frac{1}{660} = R\left(\frac{1}{2^2} - \frac{1}{3^2}\right) = \frac{5R}{36} \qquad \ldots(1)$$

For second line of Balmer series, we have

$$\frac{1}{\lambda} = R\left(\frac{1}{2^2} - \frac{1}{4^2}\right) = \frac{3R}{16} \qquad \ldots(2)$$

From equations-(1) and (2), we use

$$\frac{\lambda}{660} = \frac{5 \times 16}{36 \times 3}$$

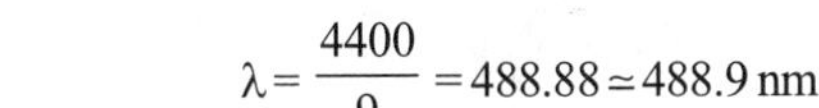

$$\Rightarrow \quad \lambda = \frac{4400}{9} = 488.88 \simeq 488.9\text{ nm}$$

Ans. (C)

Sol. 40 Energy levels in Hydrogen like atom are given as

$$E = -13.6\,\frac{Z^2}{n^2}\text{ eV}$$

As He^+ is 1st excited state, we use

$$z = 2, n = 2$$

$$\Rightarrow \quad E = -13.6\text{ eV}$$

As total energy of He^+ in 1st excited state is –13.6 eV, its ionisation energy will be +13.6 eV. **Ans. (B)**

Sol. 41 For 6 spectral lines emission, excitation must be upto $n = 4$

$\Rightarrow$ Energy required for transition from $n = 1$ to $n = 4$ is

$$\Delta E = E_4 - E_1$$

$$= (-0.85\,z^2) - (-13.6\,z^2)\text{ eV}$$

$$= 12.75\,z^2\text{ eV}$$

Wavelength of radiation of this energy

$$12.75 \times 9 = \frac{hc}{\lambda}$$

or $$\lambda = \frac{hc}{9 \times 12.75e} = \frac{12431}{9 \times 12.75}\text{ Å} \simeq 108\text{ Å}$$

Ans. (C)

Sol. 42

108.5 nm

i

30.4 nm

$n = 1$

$$\Delta E_1 = \frac{1240}{108.5}\text{ eV} = 11.73\text{ eV}$$

$$\Delta E_2 = \frac{1240}{30.4}\text{ eV} = 40.79\text{ eV}$$

$$= 52.22\text{ eV}$$

$$E_n - E_1 = 52.22\text{ eV}$$

For He^+, we use $z = 2$

$$\left(-\frac{13.6z^2}{n^2}\right) - (-13.6\,z^2) = 52.22\text{ eV}$$

$$54.4 - \frac{54.4}{n^2} = 52.22\text{ eV}$$

$$\Rightarrow \quad 54.4 - 52.22 = \frac{54.4}{n^2}$$

$$\Rightarrow \quad n^2 = \frac{54.4}{2.18} = 24.95 \simeq 25$$

$$\Rightarrow \quad n = 5$$

Ans. (A)

Sol. 43 By Rydberg's formula

$$\frac{1}{\lambda} = R\left(\frac{1}{n_1^2} - \frac{1}{n_2^2}\right)$$

For $n=4 \rightarrow n=3$ transition

$$\frac{1}{\lambda_1} = R\left[\frac{1}{3^2} - \frac{1}{4^2}\right] = \frac{16-9}{144}R$$

$$\Rightarrow \qquad \lambda_1 = \frac{144}{7R} \qquad \ldots(1)$$

For $n=3 \rightarrow n=2$ transition

$$\frac{1}{\lambda_2} = R\left[\frac{1}{2^2} - \frac{1}{3^2}\right] = \frac{9-4}{36}R$$

$$\Rightarrow \qquad \lambda_2 = \frac{36}{5R} \qquad \ldots(2)$$

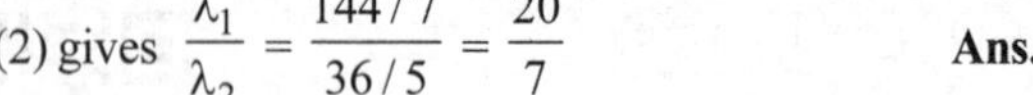

(1)/(2) gives $\frac{\lambda_1}{\lambda_2} = \frac{144/7}{36/5} = \frac{20}{7}$ **Ans. (D)**

Sol. 44 De-Broglie wavelength is given as

$$\lambda = \frac{h}{mv} = \frac{hr}{mvr}$$

By Bohr's II Postulate, we use

$$mvr = \frac{hr}{2\pi}$$

$$\Rightarrow \qquad \lambda = \frac{hr}{(hr/2\pi)} = \frac{2\pi r}{n} \qquad [n=3]$$

$$\Rightarrow \qquad \lambda = \frac{2\times 3.14\times 4.65\times 10^{-10}}{3}$$

$$\Rightarrow \qquad \lambda \simeq 9.7\,\text{Å}$$ **Ans. (C)**

Sol. 45 Time period of revolution of electron in n^{th} orbit is given as

$$T \propto \frac{n^3}{Z^2}$$

$$\Rightarrow \qquad \frac{T_2}{T_1} = \frac{(2)^3}{(1)^3} = 8$$

$$\Rightarrow \qquad T_2 = 8\times 1.6\times 10^{-16}\,\text{s}$$

Frequency in $n=2$ orbit is

$$f_2 = \frac{1}{T_2} = \frac{10^{16}}{8\times 1.6}$$

$$\approx 7.8\times 10^{14}\,\text{Hz}$$ **Ans. (C)**

Sol. 46 For scattering experiment we use number of α particles at a specific angle are inversely proportional to fourth power of sine of scattering angle. Thus most appropriate option is (C).

Ans. (C)

Sol. 47 Using Rydberg's formula, we use

$$\frac{1}{\lambda} = R_H\left(\frac{1}{2^2} - \frac{1}{n_2^2}\right) \qquad \text{[For Balmer Series]}$$

For first and second line of Balmer Series, we have

$$\frac{\lambda_2}{\lambda_1} = \frac{\left(\frac{1}{2^2} - \frac{1}{3^2}\right)}{\left(\frac{1}{2^2} - \frac{1}{4^2}\right)}$$

$$\Rightarrow \qquad \frac{\lambda_2}{6561} = \frac{5/36}{3/16}$$

$$\Rightarrow \qquad \lambda_2 = \frac{20}{27}\times 6561$$

$$\Rightarrow \qquad \lambda_2 = 4860\,\text{Å}$$

$$\Rightarrow \qquad \lambda_2 = 486\,\text{nm}$$ **Ans. (486)**

Sol. 48 Ionization energy of H-like ion with atomic number z is given as

$$IP = 13.6\,Z^2 = 9\times 13.6\,\text{eV}$$

$$\Rightarrow \qquad Z = 3$$

Using Rydberg formula, we use

$$\frac{1}{\lambda} = RZ^2\left(\frac{1}{1^2} - \frac{1}{3^2}\right) = 1.09\times 10^7\times 9\times \frac{8}{9}$$

$$\Rightarrow \qquad \lambda = 11.4\,\text{nm}$$ **Ans. (D)**

Sol. 49 For very high value of n, transition between adjoining orbits causes emission of radiation of frequency equal to revolution frequency of electron in n^{th} orbit which is given as

$$f_n = \frac{v_n}{2\pi r_n} = \frac{\frac{2\pi kZe^2}{nh}}{\frac{n^2h^2}{2\pi kZe^2 m}}$$

$$\Rightarrow \qquad f_n \propto \frac{Z^2}{n^3}$$

$$\Rightarrow \qquad f_n \propto \frac{1}{n^3}$$ **Ans. (B)**

Sol. 50 For lyman series first & last time wavelengths are given as

$$\frac{1}{\lambda_1} = R\left[\frac{1}{1^2} - \frac{1}{2^2}\right] = \frac{3R}{4}$$

$$\frac{1}{\lambda_L} = R\left[\frac{1}{1^2} - \frac{1}{\infty^2}\right] = R$$

$$\Delta\lambda_L = \lambda_1 - \lambda_L = \frac{4}{3R} - \frac{1}{R} = \frac{1}{3R} = 304\,\text{Å}$$

For paschen series we use

$$\frac{1}{\lambda_1} = R\left[\frac{1}{3^2} - \frac{1}{4^2}\right] = \frac{7R}{144}$$

$$\Rightarrow \quad \frac{1}{\lambda_L} = R\left[\frac{1}{3^2} - \frac{1}{\infty^2}\right] = \frac{R}{9}$$

$$\Rightarrow \quad \Delta\lambda_p = \lambda_1 - \lambda_L = \frac{144}{7R} - \frac{9}{R} = \frac{81}{7R}$$

$$\Rightarrow \quad \Delta\lambda_p = \frac{81}{7} \times 304 \times 3 = 10553.14\,\text{Å}$$ **Ans. (10553)**

Sol. 51 If initial KE of particle is E_1 and KE of H-atom after collision is E_2 we use

$$E_1 - 10.2\,\text{eV} = E_2$$

By conservation of momentum we use

$$\sqrt{2m_P E_1} = \sqrt{2m_H E_2}$$

$$m_P E_1 = m_H(E_1 - 10.2)$$

$$\Rightarrow \quad E_1 = \frac{10.2 \times m_H}{m_H - m_P} eV$$

$$\Rightarrow \quad E_1 = \frac{10.2 \times 10^9}{10^9 - 2 \times 10^8} eV$$

$$\Rightarrow \quad E_1 = \frac{10.2}{0.8} = 12.75\, eV = \frac{N}{4}$$

$$\Rightarrow \quad N = 51$$ **Ans. (51)**

Sol. 52 Series limit of Lymen series is corresponding to the transision from infinite to $n = 1$ which is denoted by A. Third member of Balmer series is corresponding to $n = 5$ to $n = 2$ which is denoted by B. Second member of Paschen series is corresponding to $n = 5$ to $n = 3$ which is denoted by C. **Ans. (A)**

Sol. 53 Out of the given options, maximum energy difference is there in transition from $n = 2$ to $n = 1$. **Ans. (A)**

Sol. 54 For the given transition energy difference is 10.2 eV and corresponding wavelength is given as

$$\lambda = \frac{12431}{10.2} = 1218.72\,\text{A} = 121.8\,\text{nm}$$ **Ans. (A)**

Sol. 55 By Rydberg's formula, we have

$$\frac{1}{\lambda_1} = R\left[\frac{1}{1^2} - \frac{1}{4^2}\right]$$

$$\frac{1}{\lambda_2} = R\left[\frac{1}{3^2} - \frac{1}{4^2}\right]$$

$$\Rightarrow \quad \frac{\lambda_1}{\lambda_2} = \frac{\left[\frac{1}{9} - \frac{1}{16}\right]}{\left[1 - \frac{1}{16}\right]} = \frac{7}{9 \times 15} = \frac{7}{135}$$ **Ans. (D)**

Sol. 56 For 1st line of Balmer series, its wavelength is given as

$$\frac{1}{\lambda_1} = Rz^2\left(\frac{1}{2^2} - \frac{1}{3^2}\right)$$

$$\Rightarrow \quad \frac{1}{\lambda_1} = Rz^2 \frac{5}{36} \qquad \ldots(1)$$

For 3rd line of Balmer series, its wavelength is given as

$$\frac{1}{\lambda_3} = Rz^2\left(\frac{1}{2^2} - \frac{1}{5^2}\right)$$

$$\Rightarrow \quad \frac{1}{\lambda_3} = Rz^2 \frac{21}{100} \qquad \ldots(2)$$

From equations-(1) and (2), we have

$$\frac{\lambda_1}{\lambda_3} = \frac{21}{100} \times \frac{36}{5} = 1.512 = 15.12 \times 10^{-1}$$

$$\Rightarrow \quad x \approx 15$$ **Ans. (15)**

Sol. 57 Velocity of electron in n^{th} energy level of hydrogen atom is given as

$$v = \frac{2\pi k Z e^2}{nh}$$

$$\Rightarrow \quad v \propto \frac{1}{n}$$ **Ans. (B)**

Sol. 58 Energy of hydrogen atom in ground state is given as

$$E = -13.6\, Z^2/n^2$$

For $Z = 1$ & $n = 1$, we have

$$E = -13.6 \times \frac{1^2}{1^2} = -13.6\,\text{eV}$$

For single ionized carbon atom having same energy, we use

$$E = -13.6 \frac{Z^2}{n^2} = -13.6\,\text{eV}$$

$$\Rightarrow \quad n^2 = 6^2$$

$$\Rightarrow \quad n = 6$$ **Ans. (B)**

Sol. 59 As a fact it is studied from hydrogen spectrum that Lyman series lies in UV region, Balmer series lies in visible region and and next 3 lies in infrared region. **Ans. (D)**

Sol. 60 For hydrogen like atoms ionization potential is related to mass of revolving particle as

$$E \propto m$$

Thus ionization potential of muon, we use

$$E_\mu = 13.6 \times \frac{(\text{Mass}_\mu) eV}{(\text{Mass}_e)}$$

$$\Rightarrow \quad E_\mu = 13.6 \times 207\,\text{eV} = 2815.2\,\text{eV}$$ **Ans. (B)**

Sol. 61 Energy band gap of the PN junction can be calculated by using the light wavelength which triggers the electron-hole pair generation, given as

$$E=\frac{hc}{\lambda}=\frac{1224-\text{nm}}{e\times 612\text{nm}}=2\text{ eV}$$ **Ans. (C)**

Sol. 62 Average current due to electron revolution is given as

$$i=\frac{q}{t}=\frac{q}{2\pi r/v}$$

$$\Rightarrow \quad i=\frac{1.6\times10^{-19}\times2.2\times10^{6}}{2\pi\times0.5\times10^{-10}}$$

$$\Rightarrow \quad i=\frac{1.6\times10^{-19}\times2.2\times10^{6}}{2\times\frac{22}{7}\times0.5\times10^{-10}}$$

$$\Rightarrow \quad i=112\times10^{-2}\text{mA}$$ **Ans. (112)**

Sol. 63 Statement-(1) is correct as atom of each element has various fixed energy levels so thats why emit characteristics spectrum
Statement-(2) is correct, as Bohr's model shows the motion of electrons revolution in well define orbits called energy levels
Statement-(3) is wrong, as density nucleus almost remain constant and that is the basis of defining fermi radius.
Statement-(4) is wrong, as neutrons can decay in proton, electron and an antineutrino
Statement-(5) is Correct, as unstable nuclei shows radioactivity.
Ans. (A)

Sol. 64 If the thickness of air column is taken as l, difference in number of waves through air and vacuum columns are given as

$$n_a-n_v=1$$

$$\Rightarrow \quad l\left(\frac{1}{\lambda_a}-\frac{1}{\lambda_v}\right)=1$$

$$\Rightarrow \quad l\left(\frac{1.0003}{6000}-\frac{1}{6000}\right)=1$$

$$\Rightarrow \quad l=2\text{mm}$$ **Ans. (2)**

Sol. 65 Photon energy radiated from hydrogen like ion for transion of electron between orbits, we have

$$h\upsilon=E_0Z^2\left[\frac{1}{n_1^2}-\frac{1}{n_2^2}\right]$$

$$\Rightarrow \quad \upsilon\ \alpha\left[\frac{1}{n_1^2}-\frac{1}{n_2^2}\right]$$

$$\Rightarrow \quad \frac{2.19\times10^{15}}{\upsilon}=\frac{\left[\frac{1}{1}-\frac{1}{9}\right]}{\left[\frac{1}{1}-\frac{1}{4}\right]}$$

$$\Rightarrow \quad \upsilon=2.19\times10^{15}\frac{3}{4}\times\frac{9}{8}=2.46\times10^{15}\text{ Hz}$$

Ans. (B)

Sol. 66 Number of spectral lines emitted when atoms of hydrogen gas are excited to nth energy level, are given as

$$N={}^6C_2=15$$ **Ans. (15)**

Sol. 67 According to Rutherford's model electrons revolves around nucleus in circular orbits. Thus electrons are always have acceleration due to circular motion. An accelerating change emits electromagnetic radiation thus electrons should also loose energy and finally should collapse in the nucleus. **Ans. (C)**

Sol. 68 At higher energy level as n increases hence potential as well as total energy increases because both negative and kinetic energy decreases as it is positive. **Ans. (B)**

Sol. 69 Speed of electron in Bohr model depends upon n and Z as

$$v\propto\frac{Z}{n}$$

As for both atoms n is same, we use

$$\frac{v_{\text{He}^+}}{v_{\text{H}}}=\frac{Z_{\text{He}^+}}{Z_{\text{H}}}=\frac{2}{1}$$ **Ans. (D)**

Sol. 70 10.2 eV is the energy of transition from $n=1$ to $n=2$ so electron transits to second energy level so its angular momentum changes from initial value

$$L_i=\frac{h}{2\pi}\times1$$

To final value

$$L_F=\frac{2h}{2\pi}$$

Thus change in angular momentum is calculated as

$$\Delta L=L_F-L_i=\frac{h}{2\pi}=\frac{6.6\times10^{-34}}{2\times\frac{22}{7}}\text{ Js}$$

$$\Rightarrow \quad \Delta L=1.05\times10^{-34}\text{ Js}$$ **Ans. (B)**

Sol. 71 Wavelength of photon required for lithium ion to excite from first to third orbit is calculated as

$$\frac{1}{\lambda}=RZ^2\left(\frac{1}{n_1^2}-\frac{1}{n_2^2}\right)$$

Using $n_1=1$ and $n_2=3$ gives

$$\frac{1}{\lambda}=R(9)\left(\frac{1}{1}-\frac{1}{9}\right)=8R$$

$$\Rightarrow \quad \lambda=\frac{1}{8R}=114\times10^{-10}\text{ m}$$ **Ans. (114)**

Sol. 72 When electron transits from lower energy level to higher energy level, energy is absorbed so statement-I is NOT correct. When electron transits from higher energy level to lower energy level, energy of emitted radiation photon is given as

$$\Delta E = E_2 - E_1$$

$$\Rightarrow \quad hf = E_2 - E_1$$

$$\Rightarrow \quad f = \frac{E_2 - E_1}{h}$$

Thus statement-II is NOT correct. **Ans. (D)**

Sol. 73 For scattering of α particle from a gold nucleus, impact parameter is related to scattering angles as

$$\sqrt{d} \propto \cot\frac{\theta}{2}$$

$$\Rightarrow \quad \cot^2 30° = x\cot^2 45°$$

$$\Rightarrow \quad x = 3$$

Ans. (3)

Sol. 74 Angular momentum of electron in n^{th} orbit is integral multiple of $\frac{h}{2\pi}$ so we have

$$mvr = \frac{nh}{2\pi}$$

Thus linear momentum is given as

$$mv = \frac{nh}{2\pi r}$$

Ans. (A)

Sol. 75 Ratio of magnetic moment and angular momentum of revolving electron is given as

$$\frac{\mu}{L} = \frac{q}{2m}$$

$$\Rightarrow \quad \vec{\mu} = -\frac{e}{2m}\vec{L}$$

Ans. (B)

Sol. 76 Frequency of electron revolution is given as

$$f = \frac{1}{T} = \frac{eB}{2\pi m}$$

$$\Rightarrow \quad f = \frac{1.6\times10^{-19}\times10^{-4}}{2\pi\times9\times10^{-31}}$$

$$\Rightarrow \quad f = 2.8\times10^6\text{ Hz}$$

Ans. (C)

Sol. 77 For electron transition from state n to $n = 1$, the emitted wavelength λ is given as

$E_n = \frac{-\text{Rch}}{n^2}(1)$ ———— $n = n$

$E_{\text{photon}} = E_n - E_1$

$E_1 = \frac{-\text{Rch}}{(1)^2}(1)$ ———— $n = 1$

$$E_n - E_1 = \frac{hc}{\lambda}$$

$$\Rightarrow \quad \frac{-\text{Rch}}{(n)^2} + \frac{\text{Rch}}{1} = \frac{hc}{\lambda}$$

$$\Rightarrow \quad \frac{-R}{n^2} + R = \frac{1}{\lambda}$$

$$\Rightarrow \quad R - \frac{1}{\lambda} = \frac{R}{n^2}$$

$$\Rightarrow \quad \frac{\lambda R - 1}{\lambda} = \frac{R}{n^2}$$

$$\Rightarrow \quad n^2 = \frac{\lambda R}{\lambda R - 1}$$

$$\Rightarrow \quad n = \sqrt{\frac{\lambda R}{\lambda R - 1}}$$

Ans. (B)

Sol. 78 For the given transitions in the hydrogen atom, ratio of energies is written as

$$\frac{13.6\left(\frac{1}{2^2} - \frac{1}{3^2}\right)}{13.6\left(\frac{1}{2^2} - 0\right)} = \frac{x}{x+4}$$

$$\Rightarrow \quad \frac{\frac{1}{4} - \frac{1}{9}}{\frac{1}{4}} = \frac{x}{x+4}$$

$$\Rightarrow \quad \frac{5}{9} = \frac{x}{x+4}$$

$$\Rightarrow \quad 5x + 20 = 9x$$

$$\Rightarrow \quad 4x = 20$$

$$\Rightarrow \quad x = 5$$

Ans. (5)

Sol. 79 For first line of Lyman series, we use

$$\frac{1}{\lambda} = R\left(1 - \frac{1}{4}\right) = R\left(\frac{3}{4}\right)$$

$$\Rightarrow \quad \lambda = \frac{4}{3R} \qquad \ldots(1)$$

For 3rd line of Paschen series, we use

$$\frac{1}{\lambda_3} = R\left(\frac{1}{2^2} - \frac{1}{6^2}\right) = \frac{R}{9}\times\frac{3}{4} \qquad \ldots(2)$$

For 2nd line of Balmer series, we use

$$\frac{1}{\lambda_2} = R\left(\frac{1}{2^2} - \frac{1}{4^2}\right) = \frac{R}{4}\times\frac{3}{4} \qquad \ldots(3)$$

From equations-(1), (2) and (3) we have

$$d\lambda = \lambda_3 - \lambda_2 = \frac{12}{R} - \frac{16}{3R} = \frac{20}{3R}$$

$\Rightarrow \quad a\left(\frac{4}{3R}\right)=\frac{20}{3R}$

$\Rightarrow \quad a=5$ **Ans. (5)**

Sol. 80 Energy of electron in n[th] energy level is given as

$$E_n=-\frac{13.6}{n^2}\text{ eV}$$

Thus the desired ratio is calculated as

$$\frac{E_2-E_1}{E_\infty-E_1}=\frac{13.6\left(1-\frac{1}{4}\right)}{13.6}=\frac{3}{4}$$ **Ans. (A)**

Sol. 81 Wavelength of emitted photon in transition is given as

$$\frac{hc}{\lambda}=\left[1-\frac{1}{16}\right](13.6\text{ eV})$$

$\Rightarrow \quad \lambda=94.1\text{ nm}$ **Ans. (A)**

Sol. 82 Wavelength of radiation emitted for transition of electron from $n=3$ to $n=2$ is given as

$$\frac{hc}{\lambda_0}=13.6\left(\frac{1}{2^2}-\frac{1}{3^2}\right)\text{eV} \qquad \ldots(1)$$

For transition from third excited state ($n=4$) to $n=2$ the emitted wavelength is given as

$$\frac{hc}{(20\lambda_0/x)}=13.6\left(\frac{1}{2^2}-\frac{1}{4^2}\right)\text{eV} \qquad \ldots(2)$$

From equations-(1) and (2), we have

$$\frac{x}{20}=\frac{\frac{1}{2^2}-\frac{1}{4^2}}{\frac{1}{2^2}-\frac{1}{3^2}}$$

$\Rightarrow \quad x=27$ **Ans. (27)**

Sol. 83 Energy corresponding to the given wavelength is calculated as

$$\Delta E=\frac{12412}{1241}=10\text{ eV}$$

This energy is emitted in transition *D*. **Ans. (D)**

Sol. 84 $\quad v_n\propto\frac{Z}{n}$

For $Z=1$, we use

$\Rightarrow \quad v_n\propto\frac{1}{n}$

$\Rightarrow \quad \frac{v_3}{v_7}=\frac{7}{3}$

$\Rightarrow \quad v_3=\frac{7}{3}v_7$

$\Rightarrow \quad v_3=\frac{7}{3}\times3.6\times10^6\text{ m/s}$

$\Rightarrow \quad v_3=8.4\times10^6\text{ m/s}$ **Ans. (D)**

Sol. 85 For electron transion between energy levels, wave length emitted is given as

$$\frac{1}{\lambda}=RZ^2\left[\frac{1}{n_1^2}-\frac{1}{n_2^2}\right]$$

For transition between levels 3 to 1, we have

$$\frac{1}{\lambda_1}=RZ^2\left[\frac{1}{1^2}-\frac{1}{3^2}\right]=\frac{8}{9}RZ^2 \qquad \ldots(1)$$

For transition between levels 2 to 1, we have

$$\frac{1}{\lambda_2}=RZ^2\left[\frac{1}{1^2}-\frac{1}{2^2}\right]=\frac{3}{4}RZ^2 \qquad \ldots(2)$$

Dividing equation (1) by (2) gives

$$\frac{\lambda_2}{\lambda_1}=\frac{8}{9}\times\frac{4}{3}=\frac{32}{27}$$

$\Rightarrow \quad \frac{\lambda_1}{\lambda_2}=\frac{27}{32}$ **Ans. (27)**

Sol. 86 Radius of electron in n^{th} orbit

$$r\propto\frac{n^2}{Z}$$

So ratio of e^- in 2nd orbit to 3rd orbit

$$\frac{r_{2\text{nd}}}{r_{3\text{rd}}}=\left(\frac{n_2}{n_3}\right)^2$$

$\Rightarrow \quad \frac{R}{r_{3\text{rd}}}=\left(\frac{2}{3}\right)^2$

$\Rightarrow \quad r_{3d}=\frac{9}{4}R$

$\Rightarrow \quad =2.25R$ **Ans. (B)**

Sol. 87 Binding energy of ground state electron in a hydrogen atom is 13.6 eV. Energy required to remove electron from second excited state of Li^{+2} is calculated below.

For $n=3$, $Z=3$, we have

$$E_{\text{Li+2}}=13.6\times\frac{Z^2}{n^2}$$

$\Rightarrow \quad E_{\text{Li+2}}=13.6\times\frac{(3)^2}{(3)^2}$

$\Rightarrow \quad E_{\text{Li+2}}=13.6\text{ eV}$

$\Rightarrow \quad 136\times10^{-1}\text{ eV}=x\times10^{-1}\text{ eV}$

$\Rightarrow \quad x=136$ **Ans. (136)**

Sol. 88 Ground state energy of electron in hydrogen atom is –13.6 eV. After absorbing radiation of energy 12.75 eV it gets excited to n^{th} state, for which energy is calculated as

$$\frac{-13.6\text{eV}}{n^2} = -13.6 + 12.75$$

$$\Rightarrow \quad \frac{-13.6\text{eV}}{n^2} = -0.85$$

$$\Rightarrow \quad n = \sqrt{16}$$

$$\Rightarrow \quad n = 4$$

Angular momentum of electron in n^{th} state of hydrogen atom is calculated as

$$L = \frac{nh}{2\pi} = \frac{4h}{2\pi} = \frac{2h}{\pi}$$

$$\Rightarrow \quad L = \frac{2}{\pi} \times 4.14 \times 10^{-15}$$

$$\Rightarrow \quad L = \frac{828 \times 10^{-17}}{\pi} \text{ eVs}$$ **Ans. (828)**

Sol. 89 Energy released in e^- transition from 4th energy state to 2nd energy state

$$\Delta E = 13.6 Z^2\left[\frac{1}{2^2} - \frac{1}{4^2}\right] \text{eV}$$

$$\Rightarrow \quad = 13.6 \times (4)^2\left(\frac{1}{4} - \frac{1}{16}\right) \text{eV}$$

$$\Rightarrow \quad = 13.6\,[4 - 1]\,\text{eV}$$

$$\Rightarrow \quad = 13.6 \times 3 = 40.8\,\text{eV}$$ **Ans. (D)**

Sol. 90 Energy difference in hydrogen atom is given as

$$\Delta E = \frac{hc}{\lambda}$$

$$\Rightarrow \quad \lambda \propto \frac{1}{\Delta E}$$

For shortest wavelength, energy gap should be maximum. So, correct choice is transition from $n = 3$ to $n = 1$. **Ans. (B)**

Sol. 91 Radius of nth orbit in H-like atom (Li^{+2}) is given as

$$r_n = r_0 \frac{n^2}{z}$$

As $n = 5$, we use

$$\Rightarrow \quad r_n = 0.51 \times \frac{25}{3} \text{Å}$$

$$\Rightarrow \quad r_n = 4.25 \times 10^{-10}\,\text{m}$$

$$\Rightarrow \quad r_n = 425 \times 10^{-12}\,\text{m}$$ **Ans. (425)**

Sol. 92 Potential energy of particle of mass m moving on circular path of radius r is calculated as

$$U = \frac{1}{2} m\omega^2 r^2$$

Force on particle is calculated as

$$F = -\frac{dU}{dr} = -m\omega^2 r$$

This force provides the centripetal force for circular motion

$$m\omega^2 r = \frac{mv^2}{r}$$

$$\Rightarrow \quad v = \omega r \qquad \text{...(i)}$$

By conservation of angular momentum, we use

$$mvr = \frac{nh}{2\pi} \qquad \text{...(ii)}$$

From equations-(i) and (ii), we have

$$m\omega r^2 = \frac{nh}{2\pi}$$

$$\Rightarrow \quad r \propto \sqrt{n}$$ **Ans. (A)**

Sol. 93 Binding energy of system = $\frac{ke^2}{2r}$ joule taken as

$$\frac{ke^2}{2r} = 12.8\,\text{eV}$$

$$\Rightarrow \quad \frac{9 \times 10^9 \times (1.6 \times 10^{-19})^2}{2r} = 12.8 \times 1.6 \times 10^{-19}$$

Thus orbital radius of electron is calculated as

$$\Rightarrow \quad r = \frac{9 \times 10^9 \times 1.6 \times 10^{-19}}{12.8 \times 2}$$

$$\Rightarrow \quad r = \frac{9 \times 10^{-10}}{16}\,\text{m}$$

$$\Rightarrow \quad x = 16$$ **Ans. (16)**

Sol. 94 Wavelength emitted from a Hydrogen like atom is calculated by Rydberg's formula given as

$$\frac{1}{\lambda} = R\left[\frac{1}{n_1^2} - \frac{1}{n_2^2}\right]$$

For Balmer series, $n_1 = 2$ so we have

$$\frac{1}{\lambda} = R\left[\frac{1}{4} - \frac{1}{n_2^2}\right]$$

For H_α line $n_2 = 3$ & for H_β line $n_2 = 4$ so we have

$$\Rightarrow \quad \frac{1}{\lambda_{H_\alpha}} = R\left[\frac{1}{4} - \frac{1}{9}\right] = \frac{5R}{36} \qquad \text{...(1)}$$

$$\Rightarrow \quad \frac{1}{\lambda_{H_\beta}} = R\left[\frac{1}{4} - \frac{1}{16}\right] = \frac{3R}{16} \qquad \text{...(2)}$$

For ratio of wavelength of H_α and H_β lines, we use

$$\Rightarrow \frac{\frac{1}{\lambda_{H_\alpha}}}{\frac{1}{\lambda_{H_\beta}}} = \frac{\frac{5R}{36}}{\frac{3R}{16}}$$

$$\Rightarrow \frac{\lambda_{H_\alpha}}{\lambda_{H_\beta}} = \frac{27}{20} = \frac{x}{20}$$

$$\Rightarrow x = 27$$ **Ans. (27)**

Sol. 95 Angular momentum of electron in Bohr's orbit is given as

$$L = mvr$$

Radius of nth orbit is given as

$$r_n \propto n^2$$

Velocity of electron in nth orbit is given as

$$v_n \propto \frac{1}{n}$$

$$\Rightarrow L \propto n$$

Angular momentum of electron in nth orbit is also given as

$$L = \frac{nh}{2\pi}$$

For $n = 1$, we have

$$L_1 = L = \frac{1 \cdot h}{2\pi}$$

For $n = 2$, we sue

$$L_2 = 2[L] = 2L$$

$$\Rightarrow L_2 = \frac{2h}{2\pi}$$

Thus change in angular momentum is calculated as

$$\Delta L = L_2 - L_1 = 2L - L = L$$ **Ans. (C)**

Sol. 96 For 6 wavelength emission, we have

$$6 = {}^4C_2$$

$$\Rightarrow n_2 = 4 \text{ (Third excited state)}$$

Photon energy required to excite the atom to 4th energy level is calculated as

$$hv = E_4 - E_1$$

Frequency of incident light for this is calculated as

$$v = 13.6\left(\frac{1}{1^2} - \frac{1}{4^2}\right) \times \frac{1}{4.25 \times 10^{-15}}$$

$$\Rightarrow v = 3 \times 10^{15}\,\text{Hz}$$ **Ans. (3)**

Sol. 97 Energy in n^{th} orbit of H-like atoms

$$E_n = \frac{-13.6Z^2}{n^2}$$

For He, $Z = 2$ and for $n = 2$ which is first excited state, we have

$$E_2 = \frac{-13.6 \times 4}{4} = -13.6\,\text{eV}$$ **Ans. (C)**

Sol. 98 According to Bohr's postulates, an electron makes jump to higher energy orbital if it absorbs a photon of energy equal to difference between the energies of an excited state and the ground state. Assuming that collided electron takes energy equal to 10.2 eV or 12.09 eV from incoming electron beam (some part lost due to collision). The maximum excited state is $n = 3$.

So, number of spectral lines is $\frac{3(3-1)}{2} = 3$. **Ans. (C)**

Sol. 99 For H-like atoms, the radius of orbits is given as

$$r \propto \frac{n^2}{Z}$$

$$\Rightarrow \frac{r_{He^+}}{r_{Be^{3+}}} = \frac{2^2 \times 4}{2 \times 4 \times 4} = \frac{1}{2}$$ **Ans. (2)**

Sol. 100 The net energy absorbed by the atom in the absorption process is calculated as

$$E = E_1 - E_2 = \frac{hc}{\lambda_1} - \frac{hc}{\lambda_2} = hc\left(\frac{1}{\lambda_1} - \frac{1}{\lambda_2}\right)$$

$$\Rightarrow E = 6.6 \times 10^{-34} \times 3 \times 10^8 \left(\frac{1}{500 \times 10^{-9}} - \frac{1}{600 \times 10^{-9}}\right)$$

$$\Rightarrow E = 6.6 \times 10^{-20}\,\text{J}$$

$$\Rightarrow E = \frac{6.6 \times 10^{-20}}{1.6 \times 10^{-19}}\,\text{eV} = 4.125 \times 10^{-1}\,\text{eV}$$

$$\Rightarrow E = 4125 \times 10^{-4}\,\text{eV}$$ **Ans. (4125)**

Sol. 101 Figure below shows the corresponding transitions.

n = 4 —— C
λ_2
n = 3 —— B
λ_1
n = 2 —— A

As wave number is given as

$$\frac{1}{\lambda} = RZ^2\left[\frac{1}{n_1^2} - \frac{1}{n_2^2}\right]$$

From $n = 3$ to $n = 2$ transition we use

$$\Rightarrow \frac{1}{\lambda_1} = R(1)^2\left[\frac{1}{(2)^2} - \frac{1}{(3)^2}\right] = R\left(\frac{5}{36}\right) \quad \ldots(1)$$

From $n = 4$ to $n = 3$ transition we use

$$\text{and} \quad \frac{1}{\lambda_2} = R(1)^2\left[\frac{1}{(3)^2} - \frac{1}{(4)^2}\right] = R\left(\frac{7}{144}\right) \quad \ldots(2)$$

(ii) ÷ (i) gives

$$\frac{\lambda_1}{\lambda_2} = \frac{7/144}{5/36} = \frac{7}{20} = \frac{7}{4\times 5}$$

$\Rightarrow \quad n = 5$ **Ans. (5)**

27.1 Radioactivity

Sol. 1 Only charged particles moving in field experience magnetic force and get deflected

$\Rightarrow$ At the instant emission can contain

(i) e^- (– ve)

(ii) protons (+ ve)

(iii) He^{+2} (+ ve)

Will get deflected in MF **Ans. (A)**

Sol. 2 By Radioactive Decay Law we have

$$N = N_0\, 2^{-t/T}$$

at $T = 5$ years and $t = 15$ years

$$\Rightarrow \quad N = N_0\,(2)^{-15/5}$$

$$\Rightarrow \quad N = \frac{N_0}{2^3} = \frac{N_0}{8}$$ **Ans. (A)**

Sol. 3 Using Radioactive decay law, we have

$$A_c = A_{co}\, e^{-\lambda t}$$

here at $t = 0$, $A_{co} = 5000$ dis/min

at $t = 5$ min, $A_c = 1250$ dis/mim

$$\Rightarrow \quad 1250 = 5000\, e^{-\lambda(s)}$$

$$\Rightarrow \quad e^{5\lambda} = \frac{5000}{1250} = 4$$

$$\Rightarrow \quad 5\lambda = \ln(4) = 2\ln(2)$$

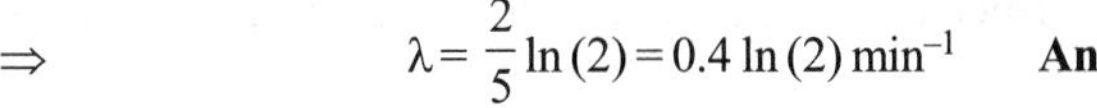

$$\Rightarrow \quad \lambda = \frac{2}{5}\ln(2) = 0.4\ln(2)\ \text{min}^{-1}$$ **Ans. (A)**

Sol. 4 Least wavelength $\Rightarrow$ maximum energy here γ-rays have highest energy

So it will be having least wavelength among other option

"fact based question" **Ans. (A)**

Sol. 5

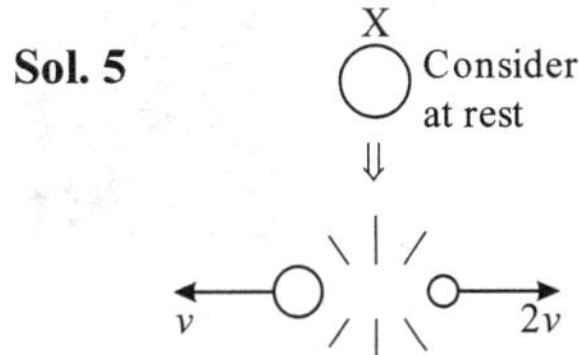

By conservation of linear momentum

$$m_1(2v) = m_2 v$$

$$m_2 = 2m_1$$

$\Rightarrow \quad A_2 = 2A_1 \quad A_1, A_2$ are molar atomic masses

By concept of fermi radius, the nuclear radius is given as

$$R = R_0\,(A)^{1/3}$$

$$\Rightarrow \quad R_1 = R_0\,(m_1)^{1/3}$$

and $$R_2 = R_0\,(2m_1)^{1/3}$$

$$\Rightarrow \quad \frac{R_1}{R_2} = \frac{1}{2^{1/3}}$$ **Ans. (D)**

Sol. 6 We use formula of fermi radius of a Nu, given as

$$R = R_0\,(A)^{1/3}$$

For ^{27}Al we use

$$3.6\times 10^{-15} = R_0\,(27)^{1/3} \qquad \ldots(1)$$

For ^{125}Te we use

$$R = R_0\,(125)^{1/3} \qquad \ldots(2)$$

(2) and (1) gives $\rightarrow$ $$\frac{R}{3.6\times 10^{-15}} = \left(\frac{125}{27}\right)^{1/3} = \frac{5}{3}$$

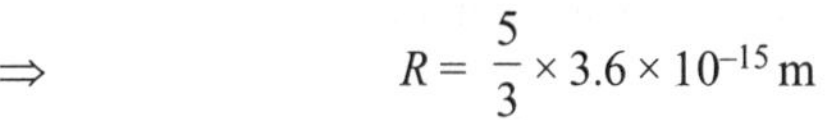

$$\Rightarrow \quad R = \frac{5}{3}\times 3.6\times 10^{-15}\ \text{m}$$

$$= 6\times 10^{-15}\ \text{m} = 6\ \text{fermi}$$ **Ans. (A)**

Sol. 7 Using RA decay equation we have

$$N = N_0\,(2)^{-t/T}$$

At $t = 15$ min, $N = \frac{N_0}{8}$ as $\frac{7}{8}N_0$ decay in times 15 min

$$\Rightarrow \quad \frac{N_0}{8} = N_0\,(2)^{-15/T}$$

$$\Rightarrow \quad 2^{-3} = 2^{-15/T}$$

$$\Rightarrow \quad \frac{15}{T} = 3$$

$$\Rightarrow \quad T = \frac{15}{3} = 5\ \text{min}$$ **Ans. (C)**

Sol. 8 Half life of X = Mean life of Y

If λ_X and λ_Y are decay constants of elements X and Y, we have

$$\frac{\ln Z}{\lambda_X} = \frac{1}{\lambda_Y}$$

$$\Rightarrow \quad \lambda_X = (\ln 2)\lambda_Y = 0.696\,\lambda_Y$$

$$\Rightarrow \quad \lambda_X < \lambda_Y$$

$\Rightarrow$ For same N_0 we use $\lambda_X N_0 < \lambda_Y N_0$ **Ans. (B)**

Sol. 9 From RA decay equation we have

$$N = N_0\,(2)^{-t/T} \quad [T = 20\ \text{min}]$$

at $t = t_2$ we use $N = \dfrac{N_0}{3}$ (as $\dfrac{2}{3}N_0$ has decayed)

$$\Rightarrow \frac{N_0}{3} = N_0(2)^{-t_2/20} \Rightarrow \frac{t_2}{20} = \log_2(3)$$

$$\Rightarrow t_2 = 20\log_2(3) \Rightarrow \log_2(3) = \frac{t_2}{20}$$

at $t = t_1$ we use $N = \dfrac{2N_0}{3}$ (as $\dfrac{N_0}{3}$ has decayed)

$$\Rightarrow \frac{2N_0}{3} = N_0(2)^{-t_1/20}$$

$$\Rightarrow \frac{t_1}{20} = \log_2\left(\frac{3}{2}\right) = \log_2(3) - 1$$

$$\Rightarrow \frac{t_1}{20} = \frac{t_2}{20} - 1 \Rightarrow t_2 - t_1 = 20 \text{ min}$$ **Ans. (A)**

Sol. 10 By RA decay equation we have $N = N_0(2)^{-t/T}$

For $A \rightarrow N_A = N_0(2)^{-80/20} = \dfrac{N_0}{2^4} = \dfrac{N_0}{16}$

Number of nuclei of A decayed in 80 min are

$$N_{AD} = N_0 - N_A = N_0 - \frac{N_0}{16} = \frac{15N_0}{16}$$

For $B \rightarrow N_B = N_0(2)^{-80/20} = \dfrac{N_0}{2^2} = \dfrac{N_0}{4}$

Number of nuclei of B decayed in 80 min are

$$N_{BD} = N_0 - N_B = N_0 - \frac{N_0}{4} = \frac{3N_0}{4}$$

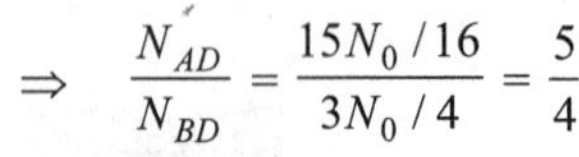

$$\Rightarrow \frac{N_{AD}}{N_{BD}} = \frac{15N_0/16}{3N_0/4} = \frac{5}{4}$$ **Ans. (D)**

Sol. 11 Using radioactive decay equation we have

$$A_C = A_{C_0}(2)^{-t/T}$$

at

$t = 0$, $A_{C_0} = 1600$ counts/s

$t = 8$s, $A_C = 100$ counts/s

$$\Rightarrow 100 = 1600(2)^{-8/T}$$

$$\Rightarrow 2^{-8/T} = 2^4 \Rightarrow \frac{8}{T} = 4 \Rightarrow T = 2\text{ s}$$

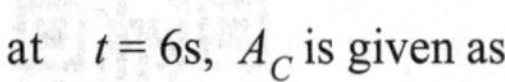

at $t = 6$s, A_C is given as

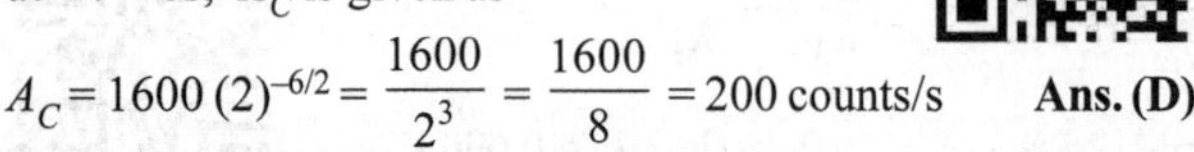

$$A_C = 1600(2)^{-6/2} = \frac{1600}{2^3} = \frac{1600}{8} = 200 \text{ counts/s}$$ **Ans. (D)**

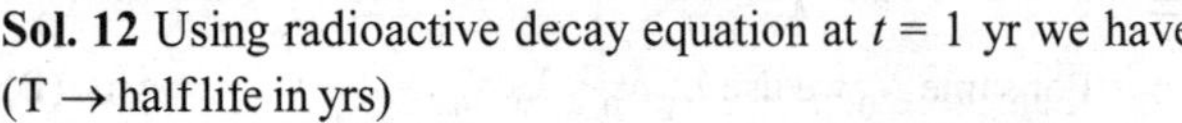

Sol. 12 Using radioactive decay equation at $t = 1$ yr we have (T $\rightarrow$ half life in yrs)

$$N_1 = N_0(2)^{-1/T} \quad [N_0 = 10^{20} \text{ atoms}]$$

at $t = 2$ yr $\quad N_2 = N_0(2)^{-2/T}$

at $t = 3$ yr $\quad N_3 = N_0(2)^{-3/T}$

N_0 of α-particles emitted in 3rd yr

$$\Delta N_3 = N_2 - N_3 = N_0((2)^{-2/T} - (2)^{-3/T}) \quad \ldots(1)$$

N_0 of α-particles emitted in 2nd yr

$$\Delta N_2 = N_1 - N_2 = N_0((2)^{-1/T} - (2)^{-2/T}) \quad \ldots(2)$$

N_0 of α-particles emitted in 1st yr

$$\Delta N_1 = N_0 - N_1 = N_0(1 - (2)^{-1/T}) \quad \ldots(3)$$

From (1) and (2) we have

$$\frac{\Delta N_3}{\Delta N_2} = \frac{N_0(2^{-2/T} - 2^{-3/T})}{N_0(2^{-1/T} - 2^{-2/T})} = 0.3$$

$$\Rightarrow 2^{-2/T} - 2^{-3/T} = 0.3\,2^{-1/T} - 0.3\,2^{-2/T}$$

$$\Rightarrow 1.3\,2^{-2/T} - 2^{-3/T} = -0.3\,2^{-1/T} = 0$$

$$\Rightarrow 1.3\,2^{-1/T} - 2^{-2/T} = -0.3 = 0$$ [multiplied the equation with $2^{1/T}$]

Using $2^{-1/T} = x \Rightarrow x^2 - 1.3x + 0.3 = 0$

$$\Rightarrow x = \frac{1.3 \pm \sqrt{1.69 - 1.2}}{2} = \frac{1.3 \pm 0.7}{2} = 1 \text{ or } 0.3$$

Here we use $2^{-1/T} = 0.3$

From equation-(1) we have

$$\Delta N_1 = N_0(1 - 2^{-1/T}) = N_0(1 - 0.3) = 0.7\,N_0$$

$$\Rightarrow \Delta N_1 = 0.7 \times 10^{20} = 7 \times 10^{19}$$ **Ans. (D)**

Sol. 13 (A) ✓ $A_C = \lambda N$ independent of any extra factor fission

(B) × Nuclear forces are independent of charge

(C) × Nuclei having same number of neutrons are called isotones

(D) × For photons momentum $P = \dfrac{h}{\lambda} \rightarrow$

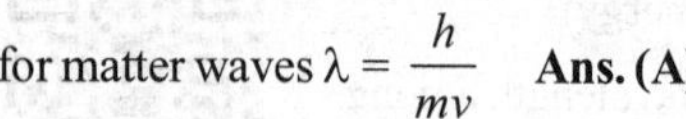

for matter waves $\lambda = \dfrac{h}{mv}$ **Ans. (A)**

Sol. 14 In case of simultaneous decay modes of a radioactive element equivalent decay constant is given as

$$\lambda_{eq} = \lambda_\alpha + \lambda_\beta \; [\lambda_\alpha = \frac{\ln_2}{T_\alpha} \text{ and } \lambda_\beta = \frac{\ln_2}{T_\beta}]$$

Average half life of element is

$$T = \frac{\ln_2}{\lambda_{eq}} = \frac{\ln_2}{\lambda_\alpha + \lambda_\beta}$$

$$\Rightarrow T = \frac{\ln_2}{\dfrac{\ln_2}{T_\alpha} + \dfrac{\ln_2}{T_\beta}} = \frac{T_\alpha T_\beta}{T_\alpha + T_\beta}$$ **Ans. (C)**

Sol. 15 Decay rate of a radioactive element is given as $A_C = \lambda N$. Given that half life of A = mean life of B

$$\Rightarrow \quad \frac{\ln 2}{\lambda_A} = \frac{1}{\lambda_B}$$

$$\Rightarrow \quad \lambda_A = \ln(2) \cdot \lambda_B = 0.693\,\lambda_B$$

$$\Rightarrow \quad \lambda_B > \lambda_A$$ so B will decay faster than A **Ans. (D)**

Sol. 16 If a radioactive nuclei element is produced at a constant rate r nuclei/s than the rate of accumulation of these nuclei is given as

$$\frac{dN}{dt} = r - \lambda N$$

$$\Rightarrow \quad \int_0^N \frac{dN}{r-\lambda N} = \int_0^t dt$$

$$\Rightarrow \quad -\frac{1}{\lambda}\left[\ln(r-\lambda N)\right]_0^N = t$$

$$\Rightarrow \quad \ln\left(\frac{r-\lambda N}{r}\right) = -\lambda t$$

$$\Rightarrow \quad \ln \frac{r-\lambda N}{r} = e^{-\lambda t}$$

$$\Rightarrow \quad N = \frac{r}{\lambda}(1-e^{-\lambda t}) \qquad \ldots(1)$$

From equation-(1) $N = 50$ at time t gives

$$50 = \frac{1000}{0.5}(1-e^{-0.5t})$$

$$\Rightarrow \quad 50 = 200(1-e^{-0.5t})$$

$$\Rightarrow \quad \frac{1}{4} = 1 - e^{-0.5t}$$

$$\Rightarrow \quad e^{0.5t} = \frac{4}{3}$$

$$\Rightarrow \quad 0.5t = \ln\left(\frac{4}{3}\right)$$

$$\Rightarrow \quad t = 2\ln\left(\frac{4}{3}\right) \text{ s}$$ **Ans. (B)**

Sol. 17 Activity at $t = 0$ (living animal) is $A_{C_0} = 16$ dpm/gm If animal died t yrs ago then activity now is $A_C = 12$ dpm/gm using radioactive decay equation we have

$$A_C = A_{C_0}(2)^{-t/T}$$

$$\Rightarrow \quad 12 = 16(2)^{-t/T}$$

$$\Rightarrow \quad 2^{-t/5760} = \frac{4}{3}$$

$$\Rightarrow \quad \frac{t}{5760} \times \log 2 = 2\log 2 - \log 3$$

$$\frac{t}{5760} \times 0.301 = 0.602 - 0.477 = 0.125$$

$$\Rightarrow \quad t = \frac{0.125 \times 5760}{0.301} \simeq 2392 \text{ yrs}$$ **Ans. (B)**

Sol. 18 By radioactive decay equation, we have

$$A_C = A_{C_0}(2)^{-t/T} \qquad \ldots(1)$$

From freshly cut tree at $t = 0$ activity is $A_{C_0} = 20$ dpm

If a tree was cut t yrs ago its activity is $A_C = 2$ dpm

from equation-(1) $\quad 2 = 20(2)^{-t/5730}$

$$2^{t/5730} = 10$$

$$\Rightarrow \quad t = \frac{\log 10}{\log 2} \times 5730$$

$$\Rightarrow \quad t = \frac{5730}{0.301} = 19036.55 \text{ yrs.}$$ **Ans. (C)**

Sol. 19 Number of Na^{24} nuclei in beginning (at $t = 0$) are

$$N_0 = \frac{1\times10^{-3}\times6.023\times10^{23}}{24\times10^{-3}}$$

$$\Rightarrow \quad N_0 = \frac{6.023\times10^{23}}{24} \cong 0.25\times10^{23}$$

by radioactive decay equation number of Na^{24} nuclei left at $t = 7.5$ hr are

$$N = N_0(2)^{-t/T} = N_0(2)^{-7.5/15} = \frac{N_0}{\sqrt{2}}$$

Number of decays in 7.5 hr are

= Number of β emissions

$$N_\beta = N_0 - \frac{N_0}{\sqrt{2}} = N_0\left(1-\frac{1}{\sqrt{2}}\right)$$

$$\Rightarrow \quad N_\beta = N_0(1-0.707) \simeq 0.3\,N_0$$

$$\Rightarrow \quad N_\beta = 0.3\times0.25\times10^{23} = 7.5\times10^{21}$$ **Ans. (B)**

Sol. 20 After time t, number of undecayed nuclei of A is given by radioactive decay equation as

$$N_A = N_0(2)^{-t/T} \qquad \ldots(1)$$

[where $N_0 \to$ number of nuclei of A at $t = 0$]

Number of nuclei of B at time t are

$$N_B = N_0 - N_A = N_0(1-(2)^{-t/T}) \qquad \ldots(2)$$

$$(2)\div(1) \Rightarrow \quad \frac{N_B}{N_A} = \frac{N_0[1-(2)^{-t/T}]}{N_0(2)^{-t/T}} = 0.3$$

$$\Rightarrow \quad 1-(2)^{-t/T} = 0.3(2)^{-t/T}$$

$$\Rightarrow \quad 1.3(2)^{-t/T} = 1$$

$$\Rightarrow \quad 2^{t/T} = 1.3$$

$$\Rightarrow \quad t = \frac{T\log(1.3)}{\log(2)}$$ **Ans. (D)**

Sol. 21 At $t = 0$, Activity of sample is $A_{C_0} = 0.8\ \mu Ci$

$$= 0.8 \times 10^{-6} \times 3.7 \times 10^{10} \text{ dps}$$

At $t = 10$hrs activity/cm^3 is given as AC/cm^3 = 300 dpm

$$= \frac{300}{60} = 5 \text{ dps}$$

If V is the volume of animal's blood then total activity at $t = 10$ hr is $A_C = 5V$ dps

Using radioactive decay equation, we have

$$A_C = A_{C_0} e^{-\lambda t}$$

$$5V = 0.8 \times 3.7 \times 10^4 \times 0.84$$

$$V = \frac{0.8 \times 3.7 \times 10^4 \times 0.84}{5} \text{ cm}^2$$

$$\Rightarrow \quad V = 4.97 \times 10^{-3} \text{ m}^3 \simeq 5 \text{ ltr}$$

Ans. (D)

Sol. 22 Activity of a sample

$$A_C = \lambda N$$

If S_1 has N_0 nuclei then S_2 will be having $\frac{N_0}{2}$ nuclei at that instant, so we use

$$A_{CS_1} = \lambda_1 N_0 \qquad \ldots(1)$$

and

$$A_{CS_2} = \lambda_2 \left(\frac{N_0}{2}\right) \qquad \ldots(2)$$

$$(1) \div (2) \Rightarrow \quad \frac{5}{10} = \frac{\lambda_1}{\lambda_2 / 2}$$

$$\Rightarrow \quad \frac{\lambda_1}{\lambda_2} = \frac{1}{4} = \frac{T_2}{T_1}$$

$$\Rightarrow \quad T_1 = 4T_2$$

So half life of S_1 is 4 times that of S_2 **Ans. (D)**

Sol. 23 For RA material A if there ae $2N_0$ atoms then, we use

$$A_{C_A} = \lambda_A (2N_0) \qquad \ldots(1)$$

For material B, number of atoms are N_0 so, we use

$$A_{C_B} = \lambda_B (N_0) \qquad \ldots(2)$$

$$(1)/(2) \quad \frac{A_{CA}}{A_{CB}} = \frac{2\lambda_A}{\lambda_B} = \frac{10}{20}$$

$$\Rightarrow \quad 4\lambda_A = \lambda_B$$

If half lives of A and B are T_A and T_B

$$\Rightarrow \quad \frac{T_A}{T_B} = \frac{\lambda_B}{\lambda_A} = 4$$

Ans. (A)

Sol. 24 Activities of A and B after time t are given as

$$R_A = A_{COA} e^{-\lambda_A t} \quad \left[\text{Here } \lambda_A = \frac{\ln(2)}{T_A} = \frac{\ln(2)}{\ln(2)} = 1\right]$$

$$R_B = A_{COB} e^{-\lambda_B t}$$

$$\Rightarrow \quad \frac{R_B}{R_A} = \frac{A_{COB} e^{-\lambda_B t}}{A_{COA} e^{-\lambda_A t}} = e^{-(\lambda_B - \lambda_A)t} = e^{-3t}$$

$$\Rightarrow \quad \lambda_B - \lambda_A = 3$$

$$\Rightarrow \quad \lambda_B = 3 + 1 = 4$$

$$\Rightarrow \quad T_B = \frac{\ln(2)}{\lambda_B} = \frac{\ln(2)}{4}$$

Ans. (C)

Sol. 25 By radioactive decay law, we use

$$A = A_0 (2)^{-t/T}$$

at $t = 0$, $A_O = 1600$ cps

at $t = 8$s, $A_C = 100$ cps

$$\Rightarrow \quad 100 = 1600\,(2)^{-8/T} \text{ half life}$$

$$\Rightarrow \quad 2^{8/T} = 2^4$$

$$\Rightarrow \quad \frac{8}{T} = 4$$

$$\Rightarrow \quad T = 2\text{s}$$

At $t = 6$s, we use

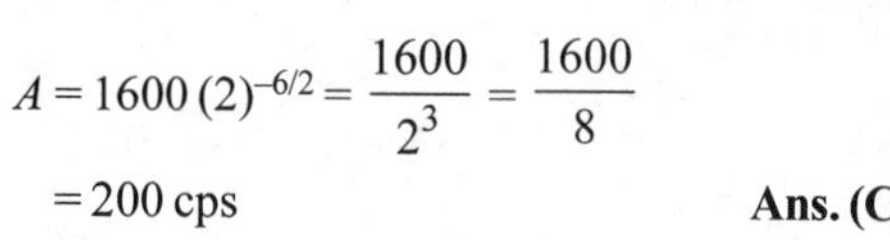

$$A = 1600\,(2)^{-6/2} = \frac{1600}{2^3} = \frac{1600}{8}$$

$$= 200 \text{ cps}$$

Ans. (C)

Sol. 26 If we write nulcear equation for α and β decay

$${}^{232}_{90}\text{Th} \longrightarrow {}^{208}_{78}Y + 6\,{}^{4}_{2}\alpha$$

$${}^{208}_{78}Y \longrightarrow {}^{208}_{82}X + 4\,{}^{0}_{-1}\beta$$

$$Z = 82 \text{ and } A = 208$$

Ans. (D)

Sol. 27 By using RA decay law, we have

For $A \leftarrow N = N_0 e^{-\lambda t} \rightarrow$ For A

$$N_A = N_0 e^{-10\lambda t} \qquad \ldots(1)$$

$$N_B = N_0 e^{-\lambda t} \qquad \ldots(2)$$

$$(1)/(2) \Rightarrow \quad \frac{N_A}{N_B} = e^{-9\lambda t} = \frac{1}{e}$$

$$\Rightarrow \quad 9\lambda t = 1$$

$$\Rightarrow \quad t = \frac{1}{9\lambda}$$

Ans. (B)

Sol. 28 By RA decay equation for A and B, number of nuclei after time t are

$$N_A = N_0 e^{-5\lambda t}$$

and

$$N_B = N_0 e^{-\lambda t}$$

$$\Rightarrow \quad \frac{N_A}{N_B} = \frac{e^{-5\lambda t}}{e^{-\lambda t}} = \frac{1}{e^2}$$

$$\Rightarrow \quad e^{-4\lambda t} = e^{-2}$$

$$\Rightarrow \quad 4\lambda t = 2$$

$$\Rightarrow \quad t = \frac{1}{2\lambda}$$

Ans. (C)

Sol. 29 By radioactive decay equation, we use

For $A \rightarrow \quad N_A = N_0\, 2^{-t/T_A}$

Number of decayed nuclei

$$N_{DA} = N_0 - N_A$$

$$= N_0 - N_0\, 2^{-6} = N_0\left(1 - \frac{1}{64}\right) = \frac{63}{64}N_0$$

For $B \rightarrow \quad N_B = N_0\, 2^{-t/T_B}$

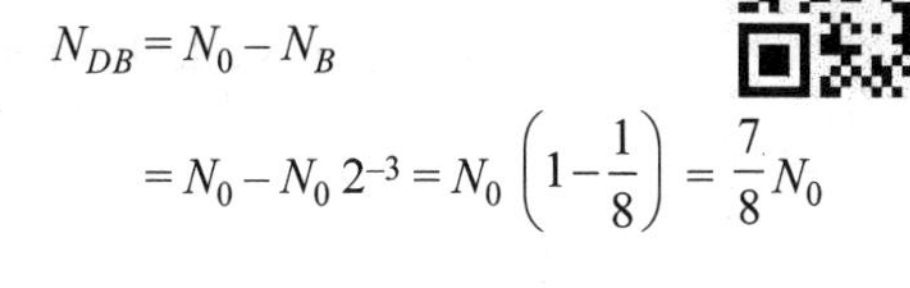

Number of decayed nuclei

$$N_{DB} = N_0 - N_B$$

$$= N_0 - N_0\, 2^{-3} = N_0\left(1 - \frac{1}{8}\right) = \frac{7}{8}N_0$$

$$\frac{N_{DA}}{N_{DB}} = \frac{63/64 N_0}{7/8 N_0} = \frac{9}{8}$$

Ans. (A)

Sol. 30 From radioactive decay law, we use

$$A = A_0 2^{-t/T}$$

$$\Rightarrow \quad 500 = 700\, 2^{-t/T}$$

$$\Rightarrow \quad 0.7 \simeq 2^{-t/T}$$

$$\Rightarrow \quad 2^{-1/2} \simeq 2^{-t/T}$$

$$\Rightarrow \quad T \simeq 2t = 2 \times 30 = 60 \text{ m}$$

Ans. (D)

Sol. 31 By radiative decay equation, we use

$$N = N_0 e^{-\lambda t}$$

Given that $\quad \frac{N}{N_0} = \frac{9}{16} = e^{-\lambda t}$

After time $t/2$, fraction remaining is

$$\left(\frac{N}{N_0}\right)_{t/2} = e^{-\lambda t/2} = \sqrt{e^{-\lambda t}} = \sqrt{\frac{9}{16}} = \frac{3}{4}$$

Ans. (A)

Sol. 32 By radioactive decay equation, we have

$$R = R_0\, e^{-\lambda t}$$

$$\Rightarrow \quad \ln R = \ln R_0 - \lambda t$$

For element A we use

$$\lambda_A = \frac{6}{10}$$

$$\Rightarrow \quad T_A = \frac{10}{6}\ln 2$$

For element B we use

$$\lambda_B = \frac{6}{5}$$

$$\Rightarrow \quad T_B = \frac{5\ln 2}{6}$$

for element C we use

$$\lambda_C = \frac{2}{5}$$

$$\Rightarrow \quad T_C = \frac{5\ln 2}{2}$$

$$\Rightarrow \quad T_A : T_B : T_C = \frac{10}{6} : \frac{5}{6} : \frac{15}{6} :: 2:1:3$$

Ans. (C)

Sol. 33 For simultaneous decay modes of nucleus, effective decay constant is given as

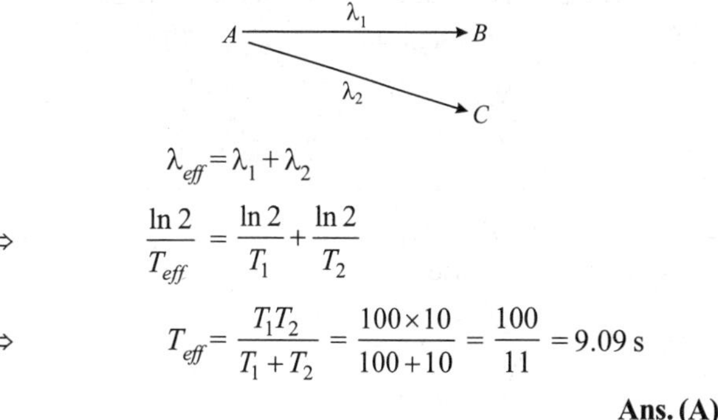

$$\lambda_{eff} = \lambda_1 + \lambda_2$$

$$\Rightarrow \quad \frac{\ln 2}{T_{eff}} = \frac{\ln 2}{T_1} + \frac{\ln 2}{T_2}$$

$$\Rightarrow \quad T_{eff} = \frac{T_1 T_2}{T_1 + T_2} = \frac{100 \times 10}{100 + 10} = \frac{100}{11} = 9.09 \text{ s}$$

Ans. (A)

Sol. 34 If half life of X is t, than that of Y will be $2t$. After three half lives of Y i.e. after a time $6t$ the number of nuclei of both becomes equal so we use

$$N_1' = N_2'$$

$$\Rightarrow \quad N_1 e^{-\lambda_1 6t} = N_2 e^{-\lambda_2 6t}$$

$$\Rightarrow \quad \frac{N_1}{N_2} = e^{(\lambda_1 - \lambda_2)6t} = e^{\ln 2\left(\frac{1}{t} - \frac{1}{2t}\right) \times 6t} = e^{(\ln 2) \times 3}$$

$$\Rightarrow \quad \frac{N_1}{N_2} = e^{\ln 8} = 8$$

Ans. (A)

Sol. 35 If initial activity of the sample be A_0 then by radioactive decay law, after time t_1 and t_2, we have

$$A = A_0\, e^{-\lambda t_1} \quad \ldots(1)$$

and $\quad \frac{A}{5} = A_0 e^{-\lambda t_2} \quad \ldots(2)$

From above equations-(1) and (2), we have

$$5 = e^{\lambda(t_2 - t_1)}$$

$$\Rightarrow \quad \lambda = \frac{\ln 5}{t_2 - t_1} = \frac{1}{\tau}$$

$$\Rightarrow \quad \tau = \frac{t_2 - t_1}{\ln 5}$$ **Ans. (A)**

Sol. 36 The activity is given as

$$A = \lambda N \qquad \left[t_{1/2} = \frac{\ln 2}{\lambda}\right]$$

where $N = nN_A$

$$\Rightarrow \quad N = \left(\frac{1.5 \times 10^{-3}}{198}\right) N_A$$

$$\Rightarrow \quad A = \left(\frac{\ln 2}{t_{1/2}}\right) N$$

As 1 Ci = 3.7×10^{10} Bq, calculating for activity we get

$$A = 365 \text{ Ci}$$

Thus closes matching option is (B). **Ans. (B)**

Sol. 37 By radioactive deay law, for first case we have

$$N_1 = N_0 e^{-\lambda t_1}$$

$$\Rightarrow \quad \frac{N_1}{N_0} = e^{-\lambda t_1}$$

$$\Rightarrow \quad 0.67 = e^{-\lambda t_1}$$

$$\Rightarrow \quad \ln(0.67) = -\lambda t_1 \qquad \ldots(1)$$

For second case, we use

$$N_2 = N_0 e^{-\lambda t_2}$$

$$\Rightarrow \quad \frac{N_2}{N_0} = e^{-\lambda t_2}$$

$$\Rightarrow \quad 0.33 = e^{-\lambda t_2}$$

$$\Rightarrow \quad \ln(0.33) = -\lambda t_2 \qquad \ldots(2)$$

From equations-(1) annd (2), we have

$$\ln(0.67) - \ln(0.33) = \lambda t_1 - \lambda t_2$$

$$\Rightarrow \quad \lambda(t_1 - t_2) = \ln\left(\frac{0.67}{0.33}\right)$$

$$\Rightarrow \quad \lambda(t_1 - t_2) \cong \ln 2$$

$$\Rightarrow \quad t_1 - t_2 = \frac{\ln 2}{\lambda} = t_{1/2} = 20 \text{ min}$$ **Ans. (B)**

Sol. 38 For independent decay processes in a radioactive sample, the effective decay constant is given as

$$\lambda_{\text{eff}} = \lambda_1 + \lambda_2$$

$$\Rightarrow \quad \frac{1}{T_{1/2}} = \frac{1}{T_{1/2}^{(1)}} + \frac{1}{T_{1/2}^{(2)}}$$

$$\Rightarrow \quad T_{1/2} = \frac{T_{1/2}^{(1)} T_{1/2}^{(2)}}{T_{1/2}^{(1)} + T_{1/2}^{(2)}}$$ **Ans. (C)**

Sol. 39 For simultaneous decay modes of the element X, we can write the total decay rate of the element as

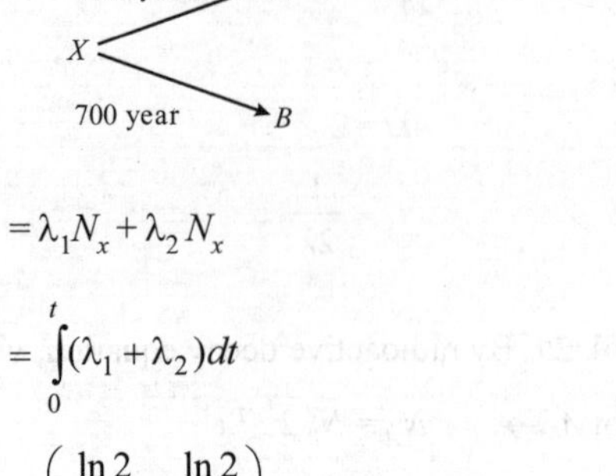

$$-\frac{dN_x}{dt} = \lambda_1 N_x + \lambda_2 N_x$$

$$\Rightarrow \quad \int_N^{N/3} \frac{dN_x}{N_x} = \int_0^t (\lambda_1 + \lambda_2) dt$$

$$\Rightarrow \quad \ln 3 = \left(\frac{\ln 2}{1400} + \frac{\ln 2}{700}\right) t$$

$$\Rightarrow \quad t = \frac{\ln 3}{\ln 2} \times \frac{1400}{3} = 740 \text{ year}$$ **Ans. (C)**

Sol. 40 By radioactive decay law, the activity of substance after time t is given as

$$R = R_0 e^{-\lambda t}$$

$$\ln R = \ln R_0^{-\lambda t}$$

Slope of curve from the given graph is calculated as

$$-\lambda = \frac{-6}{40}$$

$$\Rightarrow \quad \lambda = \frac{3}{20}$$

$$\Rightarrow \quad t_{1/2} = \frac{\ln 2}{\lambda} = \frac{\ln 2}{3} \times 20 = \frac{20}{3} \ln 2 = 4.62 \text{ s}$$ **Ans. (A)**

Sol. 41 As activity is decayed to one sixteenth of its initial value means total time passed is four half lives. Thus the half life of the substance is 80/4 = 20 days. **Ans. (3)**

Sol. 42 By Radioactive decay law, we have

$$N = N_0 e^{-\lambda t}$$

$$\Rightarrow \quad \frac{3N_0}{4} = N_0 e^{-\lambda t_1}$$

$$\Rightarrow \quad t_1 = \frac{\ln\left(\frac{4}{3}\right)}{\lambda}$$

$$\Rightarrow \quad t_2 = \frac{\ln 2}{\lambda}$$

$$\Rightarrow \quad t_2 - t_1 = \frac{\ln 2}{\lambda} - \frac{\ln\left(\frac{4}{3}\right)}{\lambda} = \frac{\ln\left(\frac{3}{2}\right)}{\lambda}$$ **Ans. (A)**

Sol. 43 Number of nuclei in 2mg of gold are calculated as

$$N = \frac{m}{M} . N_A = \frac{2 \times 10^{-3}}{198} \times 6.02 \times 10^{23}$$

Activity of the substance is given as

$$A_0 = \lambda N = \frac{0.693}{3\times 27\times 60\times 60} \times \frac{2\times 10^{-3}}{198} \times 6.02\times 10^{23}$$

$\Rightarrow$ $A_0 = 1618 \times 10^{-8} \times 10^{18} = 16.18 \times 10^{12}$ **Ans. (D)**

Sol. 44 By radioactive decay law, we use

$$A = A_0 e^{-\lambda t}$$

After half life, we have

$$A/2 = A_0 e^{-\lambda T}$$

$$\Rightarrow \quad \frac{1}{2} = e^{-\lambda T} \quad \ldots(1)$$

Given that half life is 30 years, so we use

$$1/8 = e^{-\lambda 30} \quad \ldots(2)$$

Solving (1) and (2)

$$e^{-3\lambda T} = e^{-\lambda 30}$$

$\Rightarrow$ $T = 10$ Yrs **Ans. (10)**

Sol. 45 By radioactive decay law, we have

$$N = N_0 e^{-\lambda t}$$

Number of nuclei decayed upto time t are given as

$$N_d = N - N_0$$

$$\Rightarrow \quad N_d = N_0(1 - e^{-\lambda t})$$

The fraction f is now given as

$$f = \frac{N_d}{N_0}(1 - e^{-\lambda t})$$

$\Rightarrow$ $\dfrac{df}{dt} = \lambda e^{-\lambda t}$ **Ans. (A)**

Sol. 46 During discharging of capacitor, charge on capacitor is given

$$q = q_0 e^{\frac{-t}{RC}} \quad \ldots(1)$$

$$A = A_0 e^{-\lambda t} \quad \ldots(2)$$

Dividing equations-(1) & (2), we get

$$\frac{q}{A} = \frac{q_0}{A_0}\frac{e^{\frac{-t}{RC}}}{e^{-\lambda t}}$$

$$\Rightarrow \quad -\lambda t = -\frac{-t}{RC}$$

$$\Rightarrow \quad RC = 1/\lambda$$

$\Rightarrow$ $R = \dfrac{30\times 10^{-3}}{200\times 10^{-6}} = 150\,\Omega$ **Ans. (150)**

Sol. 47 By radioactive decay law for element A we have

$$N_A = N_{0A} e^{-\lambda t}$$

Rate at which nuclie of B are being produced is given as

$$\frac{dN_B}{dt} = \lambda N_A - \lambda N_B$$

$$\Rightarrow \quad \frac{dN_B}{dt} = \lambda N_{0A} e^{-\lambda t} - \lambda N_B$$

$$\Rightarrow \quad \frac{dN_B}{dt} + \lambda N_B = \lambda N_{0A} e^{-\lambda t}$$

$$\Rightarrow \quad N_B = N_{0B}(1 - 2\lambda t e^{-\lambda t})$$

$$\Rightarrow \quad \frac{N_B(t)}{N_B(0)} = 1 - 2\lambda t e^{-\lambda t}$$

Out of given option, best is matching with option (A).

Ans. (A)

Sol. 48 By radioactive decay law, we have

$$N = N_0 e^{-\left(\frac{\ln 2}{T}\right)t}$$

$\Rightarrow$ $N = 10^{10} e^{-0.35} = 7 \times 10^9$ **Ans. (B)**

Sol. 49 Equation below shows the disintegration in the radioactive series from elements A to element C.

$$A \xrightarrow{\lambda_1} B \xrightarrow{\lambda_2} C$$

Accumulation rate of nuclei of B is given by the below expression

$$\frac{dN_B}{dt} = \lambda_1 N_1 - \lambda_2 N_2$$

As time passes decay rate of A decreases and decay rate of B increases as quantity of B is increasing. In such situation after a period of time when the quantity of B becomes maximum it will be in radioactive equilibrium and after this point quantity of B starts decreasing. Hence out of given option only (D) is most appropriate. **Ans. (D)**

Sol. 50 From the given condition, we have

$$\frac{0.693}{\lambda_x} = \frac{1}{\lambda_y}$$

$$\Rightarrow \quad \lambda_x = 0.693\,\lambda_y$$

$\Rightarrow$ $\lambda_x < \lambda_y$ **Ans. (D)**

Sol. 51 By radioactive decay law, we use

$$N_t = N_0 (0.5)^{\frac{t}{t_{1/2}}}$$

$$N_t = \frac{m}{M} \times N_A (0.5)^{\frac{t}{t_{1/2}}}$$

For the two substances, ratio of their number of nuclei are calculated as

$$\frac{N_1}{N_2} = \frac{M_2}{M_1}(0.5)^{t\left[\frac{1}{T_A} - \frac{1}{T_B}\right]}$$

$$\Rightarrow \quad \frac{N_1}{N_2} = 2(0.5)^{16\times\frac{1}{8}}$$

$$\Rightarrow \quad \frac{N_1}{N_2} = \frac{2}{4}$$

$$\Rightarrow \quad \frac{x}{100} = \frac{1}{2}$$

$$\Rightarrow \quad x = 50$$

Ans. (50)

Sol. 52 Expression of Fermi radii of a nucleus is given as

$$R = R_0 A^{1/3}$$

$$\Rightarrow \quad \ln\frac{R}{R_0} = \frac{1}{3}\ln A$$

Thus graph will be a straight line as shown below

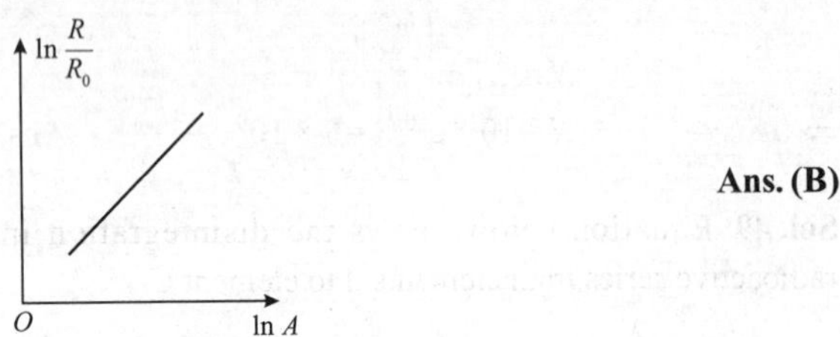

Ans. (B)

Sol. 53 For multiple decay modes of a radioactive nucleus, we use

$$\lambda_{eq} = \lambda_1 + \lambda_2$$

$$\Rightarrow \quad \frac{\ln 2}{T_{eq}} = \frac{\ln 2}{T_1} + \frac{\ln 2}{T_2}$$

$$\Rightarrow \quad T_{eq} = \frac{T_1 T_2}{T_1 + T_2}$$

$$\Rightarrow \quad T_{eq} = \frac{3\times 4.5}{3+4.5} = \frac{3\times 4.5}{7.5} = \frac{3\times 3}{5}$$

$$\Rightarrow \quad T_{eq} = 1.8 \text{ hr}$$

Ans. (D)

Sol. 54 (a) is NOT correct as radioactivity is independent from physical and chemical properties of an element.
(b) is correct according to radioactive decay law.
(c) is correct according to mathematical expression of decay law.
(d) is NOT correct as the product of decay constant and half life time of a radioactive element is a constant equal to ln(2).

Ans. (C)

Sol. 55 By radioactive decay law, we use

$$A = A_0 (2)^{-t/T}$$

$$\Rightarrow \quad 2\times 10^{-5} = 2.56\times 10^{-3}(2)^{-t/T}$$

$$\Rightarrow \quad (2)^{-5/T} = \frac{1}{128} = (2)^{-7}$$

$$\Rightarrow \quad T = 7\times 5 = 35 \text{ day}$$

Ans. (B)

Sol. 56 By radioactive decay law, we use

$$N = N_0 (2)^{-t/T}$$

$$\Rightarrow \quad \frac{N}{N_0} = \frac{1}{16} = \left(\frac{1}{2}\right)^4$$

$$\Rightarrow \quad t = 4\times 5 = 20 \text{ years}$$

Ans. (20)

Sol. 57 By radioactive decay law, we use

$$A = A_0 e^{-\lambda t}$$

$$\Rightarrow \quad 2250 = 4250 e^{-\lambda(10)}$$

$$\Rightarrow \quad \lambda(10) = \ln\left(\frac{4250}{2250}\right)$$

$$\Rightarrow \quad \lambda = 0.063 \text{ min}^{-1}$$

Ans. (C)

Sol. 58 By radioactive decay law, we use

$$A = A_0 e^{-\lambda t}$$

$$\Rightarrow \quad -\lambda t = \ln\left(\frac{A}{A_0}\right)$$

$$\Rightarrow \quad -\frac{\ln 2}{t_{1/2}}\times 30 = \ln\left(\frac{1}{16}\right)$$

$$\Rightarrow \quad -\frac{\ln 2}{t_{1/2}}\times 30 = -4\ln 2$$

$$\Rightarrow \quad t_{1/2} = \frac{30}{4} = 7.5 \text{ years}$$

Ans. (C)

Sol. 59 Using radioactive decaly law, we have

$$A = A_0 e^{-\lambda t}$$

$$\Rightarrow \quad 5\times 10^{-6} = 6.4\times 10^{-4} e^{-\lambda t}$$

$$\Rightarrow \quad \frac{5}{6.4}\times 10^{-2} = e^{-\lambda t}$$

$$\Rightarrow \quad 7.8\times 10^{-3} = e^{-\lambda t}$$

$$\Rightarrow \quad \log(7.8\times 10^{-3}) = -\lambda t \ln e$$

$$\Rightarrow \quad \ln(7.8\times 10^{-3}) = -\frac{\lambda n2}{5}\cdot t$$

$$\Rightarrow \quad \frac{5\times 4.853}{0.693} = t$$

$$\Rightarrow \quad t = 35 \text{ days}$$

Ans. (D)

Sol. 60 7/8 of its original mass is disintegrates which means its 1/8 part of original mass left undecayed so by radioactive decay law, we use

$$\left(\frac{1}{2}\right)^3 N_0 = N_0 (2)^{-t/T}$$

$$t = 180 \text{ days}$$

Ans. (C)

Sol. 61 By radioactive decay law, we have

$$A = A_0(2)^{-t/T}$$

$$\Rightarrow \quad \frac{A_0}{64} = A_0(2)^{-t/T}$$

$$\Rightarrow \quad t = 6T = 6 \times 2{\cdot}5 = 15 \text{ hours}$$ **Ans. (15)**

Sol. 62 7/8 of its original mass is disintegrates which means its 1/8 part of original mass left undecayed so by radioactive decay law, we use

$$\left(\frac{1}{2}\right)^3 N_0 = N_0 (2)^{-t/T}$$

$$\Rightarrow \quad 3T = 15 \text{ min}$$

$$\Rightarrow \quad T = 5 \text{ min}$$ **Ans. (A)**

Sol. 63 By radioactive decay law, we have

$$N = N_0 e^{-\lambda t}$$

$$\Rightarrow \quad \frac{N_B}{N_A} = \frac{e^{-\lambda_2 t}}{e^{-\lambda_1 t}} = e^{-\lambda_2 t} \cdot e^{\lambda_1 t}$$

$$\Rightarrow \quad e^1 = e^{(\lambda_1 - \lambda_2)t}$$

$$\Rightarrow \quad (\lambda_1 - \lambda_2)t = 1$$

$$\Rightarrow \quad t = \frac{1}{\lambda_1 - \lambda_2} = \frac{1}{25\lambda - 16\lambda} = \frac{1}{9\lambda}$$ **Ans. (9)**

Sol. 64 By radioactive decay law, we use

$$\frac{N}{N_0} = \left(\frac{1}{2}\right)^{t/t\frac{1}{2}} = \left(\frac{1}{2}\right)^{\frac{90}{30}}$$

$$\frac{N}{N_0} = \left(\frac{1}{2}\right)^3 = \frac{1}{8}$$ **Ans. (A)**

Sol. 65 By conservation of charge, we use

$$P = 92 - 2 - 2 + 1 - 1 - 1$$

$$P = 92 - 5 = 87$$ **Ans. (87)**

Sol. 66 Number of moles of substances A and B are

$$N_{0A} = \frac{320}{16} = 20 \text{ moles}$$

$$N_{0B} = \frac{320}{32} = 10 \text{ moles}$$

After time $t = 2$ days, we have

$$N_A = \frac{N_{0A}}{(2)^{2/1}} = \frac{20}{4} = 5$$

and $$N_B = \frac{N_{0B}}{(2)^{2/5}} = \frac{10}{2^4} = 0.625$$

Total number of moles after 2 days present are $N = 5.625$ thus total number of atoms present after 2 days are

$$N = 5.625 \times 6.023 \times 10^{23}$$

$$\Rightarrow \quad N = 3.38 \times 10^{24}$$ **Ans. (A)**

Sol. 67 Radioactive decay equation gives

$$\frac{dN}{dt} = -\lambda N$$

For the first process, we have

$$\frac{dN_1}{dt} = -\lambda_1 N \qquad \ldots(1)$$

For the second process, we have

$$\frac{dN_2}{dt} = -\lambda_2 N \qquad \ldots(2)$$

From equations-(1) and (2) net radioactive decay

$$\frac{dN}{dt} = -(\lambda_1 + \lambda_2)N$$

$$\Rightarrow \quad \lambda_{eq} = \lambda_1 + \lambda_2$$

So effective half life on nucleus is calculated as

$$\Rightarrow \quad \frac{1}{t_{1/2}} = \frac{1}{300} + \frac{1}{30} = \frac{11}{300}$$

$$\Rightarrow \quad t_{1/2} = \frac{300}{11} \text{ s}$$ **Ans. (300)**

Sol. 68 Radioactive material after n half lives becomes

$$N = N_0 \left(\frac{1}{2}\right)^n$$

Radioactive material is reduced to $\frac{1}{8}$ th of its original amount so we use

$$\frac{N_0}{8} = N_0 \left(\frac{1}{2}\right)^n$$

$$n = 3$$

$$\Rightarrow \quad 3 \text{ half lives} = 3 \text{ days}$$

$$\Rightarrow \quad 5 \text{ days} = 5 \text{ half life}$$

After 5 days, material left is calculated as

$$N = N_0 \left(\frac{1}{2}\right)^n$$

$$\Rightarrow \quad 8 \times 10^{-3} = N_0 \left(\frac{1}{2}\right)^5$$

$$\Rightarrow \quad N_0 = 256 \times 10^{-3} \text{ kg}$$

$$\Rightarrow \quad N_0 = 256 \text{ g}$$ **Ans. (D)**

Sol. 69 Given that decay constant for radio active nuclide is

$$\lambda = 1.5 \times 10^{-5} \text{ s}^{-1}$$

$$\text{No. of mole} = \frac{1\times10^{-6}}{60} = \frac{10^{-7}}{6}$$

$$\text{No. of atoms} = \text{no. of moles} \times N_A$$

$$\Rightarrow \quad N_0 = \frac{10^{-7}}{6} \times 6 \times 10^{23} = 10^{16}$$

Activity of substance is calculated as

$$A = \lambda N_0 e^{-\lambda t}$$

At $t = 0, A = A_0 = N_0\lambda$

$$\Rightarrow \quad A_0 = 1.5 \times 10^{-5} \times 10^{16} = 15 \times 10^{10}\ \text{Bq}$$

Ans. (15)

Sol. 70 As $\frac{7}{8}$ th part is disintegrated so remaining substance is $\frac{1}{8}$ of the original mass. This happens after three half lives as shown in decay series below

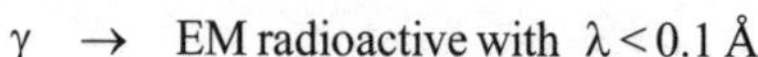

$$\Rightarrow \quad t_{7/8} = 3T$$

Ans. (A)

Sol. 71 Given that half life of A is same as average life of B

$$T_{1/2}(A) = T_{av}(B)$$

Half life is given as

$$T_{1/2} = \frac{\ln 2}{\lambda_A}$$

$$\frac{\ln 2}{\lambda_A} = \frac{1}{\lambda_B}$$

$$\Rightarrow \quad \lambda_A = \lambda_B l2$$

Ans. (C)

Sol. 72 Energy released in given alpha decay is calculated as

$$\Delta E = (\Delta m)_{\text{amu}} \times 931.5\ \text{MeV}$$

$$\Rightarrow \quad \Delta E = (m_u - m_{\text{Th}} - m_{\text{He}})_{\text{amu}} \times 931.5\ \text{MeV}$$

$$\Rightarrow \quad \Delta E = 0.0044 \times 931.5\ \text{MeV} = 4.0986\ \text{MeV}$$

Ans. (4)

Sol. 73 15 year is equal to 3 half lives in which number of active nuclei will reduce to $(1/2)^3$ so we have

$$\text{Number of active nuclei} = \frac{N_0}{8}$$

Thus number of atoms decayed in 15 years is given as

$$N_d = N - \frac{N_0}{8} = \frac{7N_0}{8}$$

Ans. (C)

27.2 Nuclear Physics

Sol. 1 Fusion will start when average thermal energy of particles in gaseous state will be equal to the PE of the fusing particles

Average thermal energy $E = \frac{3}{2}kT$

$\Rightarrow$ here $\frac{3}{2}kT \geq 7.7 \times 10^{-14}$ for fusion to start

$$\Rightarrow \quad T \geq \frac{7.7\times10^{-14}}{1.5\times1.38\times10^{-23}} \simeq 3.7 \times 10^{9}$$

Ans. (D)

Sol. 2 We know that in general radio active decay causes α, β and γ emissions

$\alpha \rightarrow He^{+2}$

$\beta \rightarrow e^-(\bar{\nu})$ or $e^+(\nu)$ where $\bar{\nu}$, ν are neutrinos

$\gamma \rightarrow$ EM radioactive with $\lambda < 0.1$ Å

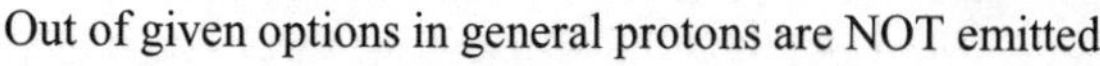

Out of given options in general protons are NOT emitted.

Ans. (A)

Sol. 3 In the decay sequence parent nucleus ($Z = 92$) is emitting

8 α particles $\Rightarrow$ this will decrease Z by $8 \times 2 = 16$ (–ve)

4 β^- particles $\Rightarrow$ this will decrease Z by $4 \times 1 = 4$ (+ve)

2 β^+ particles $\Rightarrow$ this will decrease Z by $2 \times 1 = 2$ (–ve)

Final Z of the resulting end product is

$$Z_{\text{end product}} = 92 - 16 + 4 - 2 = 92 - 14 = 78$$

Ans. (B)

Sol. 4 By conservation of linear momentum we use

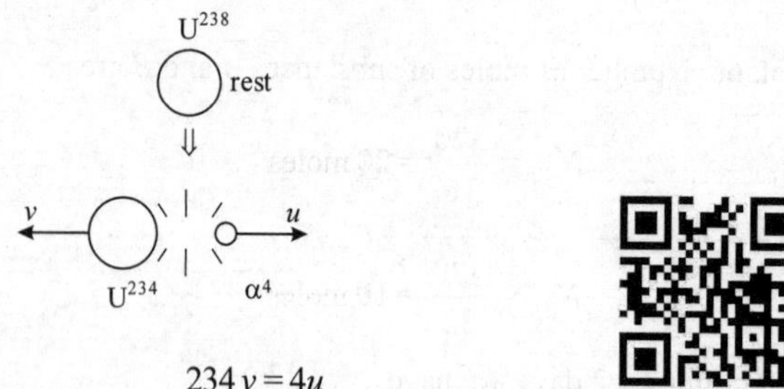

$$234\,v = 4u$$

$$\Rightarrow \quad v = \frac{4u}{234}$$

Ans. (C)

Sol. 5 At the point of closest approach the initial KE of α-particle gets transformed into the electrical PE of (α – U)

system, so we use

+2e $v = 0$
α-particle + 92e
r
fixed U-nucleus
Z = 92

$$KE_i = \frac{Kq_1q_2}{r} \quad [q_1 = +92e \ \& \ q_2 = +2e]$$

$$\Rightarrow \quad r = \frac{9\times10^9\times92\times2\times1.6\times10^{-19}}{5\times10^6}$$

$$\Rightarrow \quad r = 5.3\times10^{-14}\,\text{m}$$

$$\Rightarrow \quad r = 5.3\times10^{-12}\,\text{cm}$$

Ans. (C)

Sol. 6 The reaction of nuclear fusion is

$$_1H^2 + {}_1H^2 \longrightarrow {}_2H^2 + \Delta E$$

Here the Q-value of nuclear reaction (energy released) is given as

$$\Delta E = 4\,(BE_N)_{_2He^4} - 2\times2\times(BE_N)_{_1H^2}$$

$$\Rightarrow \Delta E = 4\times7 - 4\times1.1 = 28 - 4.4 = 23.6\,\text{MeV}$$ **Ans. (C)**

Sol. 7 $X(n, \alpha) \rightarrow {}^7_3\text{Li}$ can be written as

$${}^1_0n + {}^A_ZX \longrightarrow {}^7_3\text{Li} + {}^4_2\alpha$$

In above reaction conserving charge and mass number gives

$$1 + A = 7 + 4 \rightarrow A = 11 - 1 = 10$$

$$0 + Z = 3 + 2 \rightarrow Z = 5$$

Element X is ${}^{10}_5\text{B}$ **Ans. (B)**

Sol. 8 We use the γ-ray intensity at a penetration depth x, is gives as

$$I = I_0e^{-\mu x}$$

[$\mu \rightarrow$ absorption coefficient of lead for γ-rays]

$$I_0 = I$$

and $I = \frac{I}{8}$ at $x = 36$ mm

$$\Rightarrow \quad \frac{I}{8} = Ie^{-\mu(36)}$$

$$\Rightarrow \quad 36\mu = \ln 8 = 3\ln 2$$

$$\Rightarrow \quad \mu = \frac{3\ln 2}{36} = \frac{\ln 2}{12}\,\text{mm}^{-1}$$

In next case at $I = \frac{I}{2}$ we use

$$x = r$$

$$\Rightarrow \quad \frac{I}{2} = Ie^{-\mu r}$$

$$\Rightarrow \quad \mu r = \ln(2)$$

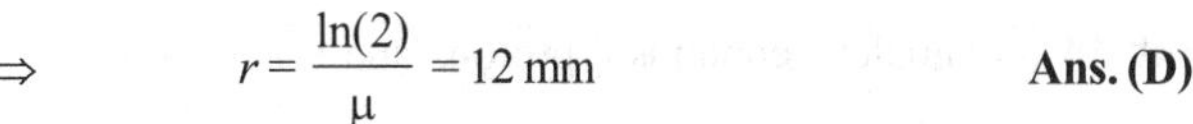

$$\Rightarrow \quad r = \frac{\ln(2)}{\mu} = 12\,\text{mm}$$ **Ans. (D)**

Sol. 9 Fact based

emitted β-particles from a radioactive source have widely spreaded energy spectrum with speeds ranging 0.01c to 0.99c. Option (C) is most appropriate matching option. **Ans. (C)**

Sol. 10 The nuclear reaction of above given case is written as

$${}^7_3\text{Li} + {}^1_1\text{p} \longrightarrow {}^8_4\text{Be} + a$$

In every nuclear reaction the obtained product can be found in excited state which can further transit to ground state and emit gamma radiation.

$${}^7_3\text{Li} + {}^1_1\text{p} \longrightarrow {}^8_4\text{Be}^*$$

$${}^8_4\text{Be}^* \longrightarrow {}^8_4\text{Be} + \gamma$$

$$\Rightarrow \quad a = \gamma$$ **Ans. (C)**

Sol. 11 At the point of closest approach whole KE transform into electric PE

v $v_f = 0$
α(+2e) +Ze
r_m

$$\Rightarrow \quad \frac{1}{2}mv^2 = \frac{Kq_1q_2}{r_m} = \frac{K(2e)(Ze)}{r_m}$$

$$\Rightarrow \quad r_m = \frac{4KZe^2}{mv^2}$$

$$\Rightarrow \quad r_m \propto \frac{1}{m}$$

Ans. (B)

Sol. 12 In the given nuclear reaction

$$p + {}^7_3\text{Li} \longrightarrow 2\,{}^4_2\text{He}$$

As some initial energy of proton is required so by conservation of mass-energy in this equation we use

$$E_p + E_{Li} = 2E_{He}$$

$$\Rightarrow \quad E_p + 7\times5.60 = 2\times4\times7.06$$

$$\Rightarrow \quad E_p = 56.48 - 392 = 17.28\,\text{MeV}$$ **Ans. (B)**

Sol. 13 A gamma ray emission takes place only from on excited Nu when it transit to ground state

$${}^A_ZX^* \longrightarrow {}^A_ZX + \gamma\text{-photon}$$

Nucleus in excited state — Nucleus in ground state

As Z and A remain same is γ-emission so no change in proton and neutron number. **Ans. (B)**

Sol. 14 $^{17}_{8}O$ nucleus contains 8 protons and 9 neutrons

The mass defect in formation of nucleus $^{17}_{8}O$ is

$$\Delta m = 8M_P + 9M_n - M_O$$

energy released at the time of formation $^{17}_{8}O$ nucleus is

$$\Delta E = \Delta Mc^2 = [8M_P + 9M_n - M_O]c^2$$ **Ans. (B)**

Sol. 15

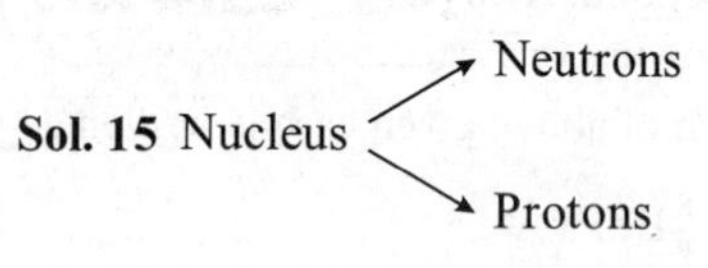

Ans. (B)

Sol. 16 In case of fission of heavy nuclei or fusion of light Nu overall $(BE)_N$ increases and energy is released

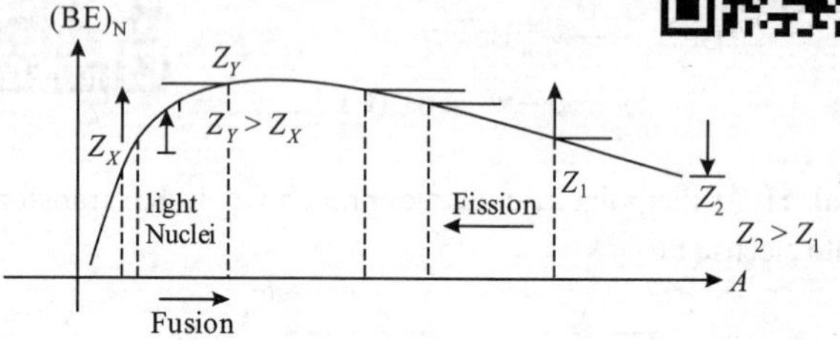

⇒ From above graph we can state that **S-1** is true and **S-2** is false. **Ans. (D)**

Sol. 17 In any reaction if overall BE per nucleon increases

⇒ Energy is released

(i) $A + B \rightarrow C + E$

As $E_A, E_B < E_C$ ⇒ energy is released ⇒ E is +ve

(ii) $C \rightarrow A + B + E$

As $E_C > E_A, E_B$ ⇒ energy is absorbed ⇒ E is –ve

(iii) $D + E \rightarrow F + E$

As $E_F < E_D, E_E$ ⇒ energy is absorbed ⇒ E is –ve

(iv) $F \rightarrow D + E + E$

As $E_D, E_E > E_F$ ⇒ energy is released ⇒ E is +ve

Ans. (C)

Sol. 18 The effective nuclear reaction is

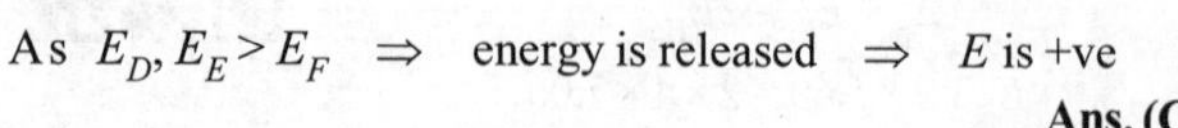

$$^{A}_{Z}X \longrightarrow {}^{A_1}_{Z_1}Y + 3\,^{4}_{2}\alpha + 2F_e$$

By conservation of charge, we use

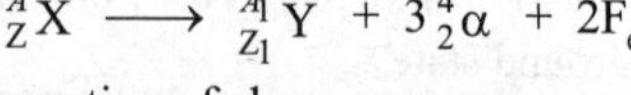

$$Z = Z_1 + 6 + 2 \Rightarrow Z_1 = Z - 8$$

By conservation mass, we use

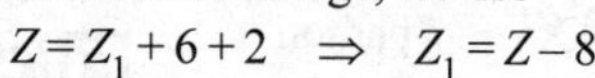

$$A = A_1 + 12 \Rightarrow A_1 = A - 12$$

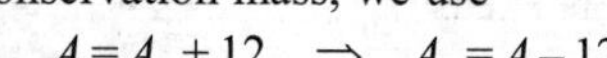

In Y ratio of neutrons to protons is $\dfrac{N_n}{N_P} = \dfrac{A_1 - Z_1}{Z_1} = \dfrac{A - Z - 4}{Z - 8}$

Ans. (C)

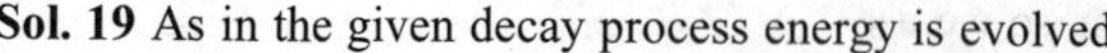

Sol. 19 As in the given decay process energy is evolved

$M + \Delta m$ (rest) ⇓ $v \leftarrow \frac{M}{2}$, $\frac{M}{2} \rightarrow v$

⇒ Final BE per nucleon must be less than initial BE per nucleon

⇒ $$E_2 < E_1$$ **Ans. (C)**

Sol. 20 Energy evolved = Δmc^2

due to released energy the daughter nuclei gain KE so we use

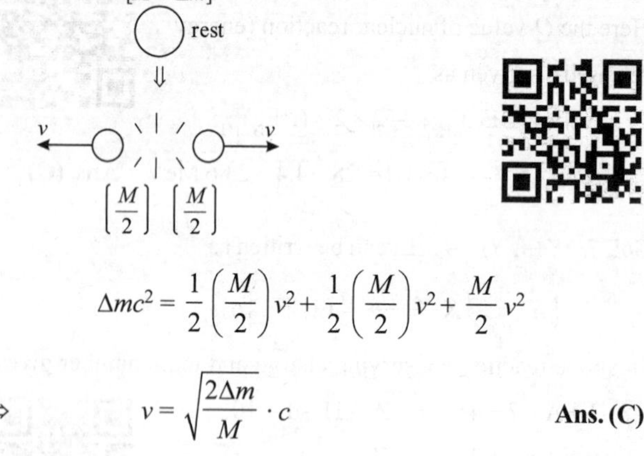

$$\Delta mc^2 = \frac{1}{2}\left(\frac{M}{2}\right)v^2 + \frac{1}{2}\left(\frac{M}{2}\right)v^2 + \frac{M}{2}v^2$$

⇒ $$v = \sqrt{\frac{2\Delta m}{M}} \cdot c$$ **Ans. (C)**

Sol. 21 $0 \le E_\beta \le E_1 - E_2$

$$\underset{E_1}{(X)} \longrightarrow \underset{E_2}{(Y)} + \beta^-$$

β particles are emitted with varying energy in range from O to $E_1 - E_2$. This is because the energy different is carried by another particle called nutrino.

S-1: It is correct

S-2: By neutrino hypothesis energy and momentum conservation was considered to be valid so it explains S-1. **Ans. (C)**

Sol. 22 As $P_i = 0 \Rightarrow P_f = 0$

⇒ $$P_{m_1} = P_{sm_1}$$

$P_i = 0$, $O \longrightarrow$ neutron, M ⇓ $v_2 \leftarrow (\lambda')\ 5m_1$, $m_1\ (\lambda) \rightarrow v_1$

We use de-Broglie wavelength of m_1 is λ

$$\Rightarrow \qquad \lambda = \frac{h}{P_{m_1}}$$

For de-Broglie wavelength of $5m_1$ we use

$$\lambda' = \frac{h}{P_{m_2}} = \frac{h}{P_{m_1}} = \lambda$$ **Ans. (D)**

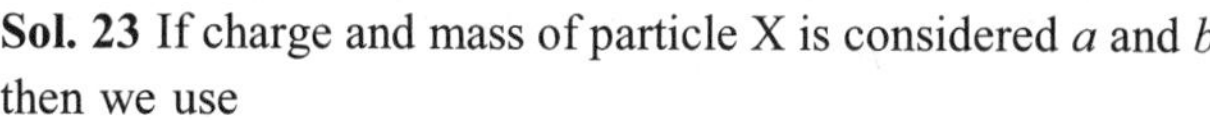

Sol. 23 If charge and mass of particle X is considered a and b then we use

By conservation of mass $\rightarrow$ $235 + 1 = 141 + 92 + 3b$

$$\Rightarrow \qquad 3b = 236 - 233 = 3$$

$$\Rightarrow \qquad b = 1$$

By conservation of charge $\rightarrow$ $92 + 0 = 56 + 36 + 3a$

$$\Rightarrow \qquad 3a = 92 - 92 = 0$$

$$\Rightarrow \qquad a = 0$$

$\Rightarrow \; {}^{b}_{a}x \rightarrow {}^{1}_{0}x \; \Rightarrow \; {}^{1}_{0}n$ **Ans. (C)**

Sol. 24 The nuclear reaction is

$${}^{2}_{1}\text{H} + {}^{2}_{1}\text{H} \longrightarrow {}^{4}_{2}\text{He} + \Delta E$$

The released energy is equal to difference in total binding energies of product in this reaction.

Total BE of the reaction $E_R = 4 \times 1.1 = 4.4$ MeV

Total BE of the reaction $E_P = 7 \times 4 = 28$ MeV

Energy released (Q-value) $\Delta E = E_P - E_R = 28 - 4.4 = 23.6$ MeV **Ans. (C)**

Sol. 25 Mass m produces energy $E = mc^2$

Power of reactor $P = 10^9\ W$

Mass of fuel consumed per second is $m' = \dfrac{P}{c^2} = \dfrac{10^9}{(3\times10^8)^2}$ kg

Mass of fuel consumed per hour is $m = m' \times 3600$

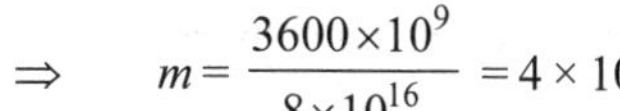

$$\Rightarrow \qquad m = \frac{3600\times10^9}{8\times10^{16}} = 4\times10^{-5}\ \text{kg}$$

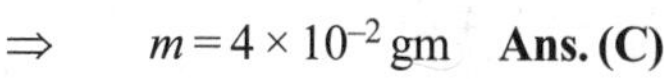

$\Rightarrow \qquad m = 4\times10^{-2}$ gm **Ans. (C)**

Sol. 26 We consider an element X

$X^* \longrightarrow X + E_r \qquad (\lambda_N \sim 0.1\ \text{Å or less} \approx 0.01\text{Å})$

(E_r = Energy of the order of some MeV)

For atom X when its e^- transit from $n = 2$ to $n = 1$

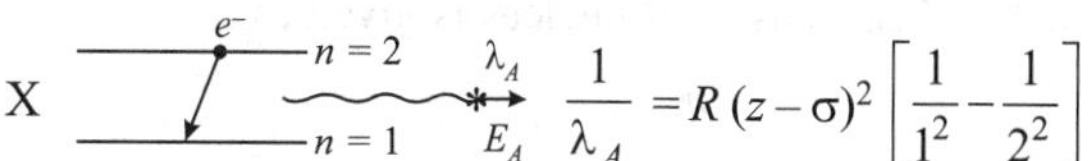

$$\frac{1}{\lambda_A} = R(z-\sigma)^2\left[\frac{1}{1^2} - \frac{1}{2^2}\right]$$

Energy is of the order of few eV only.

For photons we can use

$$E_r = \frac{hc}{\lambda_N} \text{ and } E_A = \frac{hc}{\lambda_A}$$

$$\Rightarrow \frac{\lambda_N}{\lambda_A} = \frac{E_A}{E_r} = \frac{\text{order of } eV}{\text{order of } MeV} \simeq \frac{1}{10^6} = 10^{-6}$$ **Ans. (A)**

Sol. 27 Formula for fermi radius is $R = R_0(A)^{1/3}$

rest

$\Downarrow$

$8v \leftarrow m_1 \qquad m_2 \rightarrow 27v$

By conservation of linear momentum we use

$$m_1(8v) = m_2(27v)$$

$$8m_1 = 27m_2$$

If R_1 and R_2 are the radii of nuclei of masses m_1 and m_2, we have

$$\frac{R_1}{R_2} = \left(\frac{m_1}{m_2}\right)^{1/3}$$

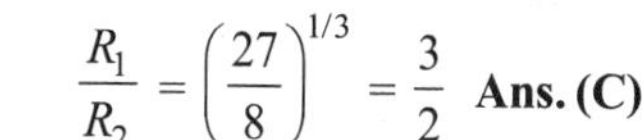

$$\Rightarrow \qquad \frac{R_1}{R_2} = \left(\frac{27}{8}\right)^{1/3} = \frac{3}{2}$$ **Ans. (C)**

Sol. 28 For the given nuclear reaction energy released is

$$E_B = (BE)_{\text{product}} - (BE)_{\text{reactant}}$$

$$= 2\times7.07\times4 + 7.86\times12 - 8.03\times20$$

$$= -9.72 \text{ MeV}$$

Thus minimum 9.72 MeV or more energy to be supplied. Thus option (C) is most appropriate **Ans. (C)**

Sol. 29 Mass densities of all nuclei are same so their ratio is 1 **Ans. (A)**

Sol. 30 Power output is energy released per second. Total energy released in 30 days is given as

$$E = 200\text{ MeV} \times \left[\frac{2\times6.03\times10^{23}}{235\times10^{-3}}\right]$$

Power output of reactor is given as

$$\Rightarrow \qquad P = \frac{E}{T} = \frac{200\times10^6\times1.6\times10^{-19}}{30\times24\times86400} \times \frac{2\times6.023\times10^{23}}{235\times10^{-3}}$$

$$\cong 63.2\times10^6\ \text{W}$$ **Ans. (B)**

Sol. 31 Mass density of nucleus is given as

$$\rho = \frac{m}{V}$$

$$\Rightarrow \quad \rho = \frac{m}{V} = \frac{A}{\frac{4}{3}\pi R^3} = \frac{3}{4\pi r^3}$$

$$\Rightarrow \quad \rho = 2.3 \times 10^{17}\ \text{kg/m}^3$$ **Ans. (C)**

Sol. 32 Binding energy of $^{120}_{50}S_n$ given as

$$\Delta E = \Delta mc^2$$

where $\Delta m = 50(1.00783) + 70(1.00867) - 119.902199$

$$\Rightarrow \quad \Delta E = \Delta m \times 931 = 1020.56\ \text{MeV}$$

BE per nucleon is given as

$$(BE)_n = \frac{1020.56}{120} = 8.5\ \text{MeV}$$ **Ans. (A)**

Sol. 33 Mass defect of proton capture reaction is given as

$$^7_3\text{Li} + ^1_1\text{H} \rightarrow 2^4_2\text{He}$$

$$\Delta m = m_{\text{Li}} + m_{\text{H}} - 2M_{\text{He}}$$

$$\Rightarrow \quad \Delta m = [(7.016 + 1.0079) - 2 \times 4.0026] = 0.0187\ u$$

Energy released by 20 gm of Li is given as

$$E = \frac{\Delta mc^2}{m_{\text{Li}}} \times 20$$

$$\Rightarrow \quad E = \frac{0.0187 \times 1.6 \times 10^{-19} \times 10^9}{7.016 \times 1.6 \times 10^{-24}} \times 20\ \text{J}$$

$$\Rightarrow \quad E = 5.33 \times 10^{12}\ \text{J}$$

$$\Rightarrow \quad E = \frac{5.33 \times 10^{12}}{1000 \times 3600} = 1.48 \times 10^6\ \text{kWh}$$ **Ans. (B)**

Sol. 34 Only in option-(A) masses of reactants is more than that of product so this is only possible reaction. **Ans. (A)**

Sol. 35 It is possible only inside the nucleus for a proton to decay into a neutron otherwise it is not possible as mass of a free proton is less than that of a free neutron. **Ans. (B)**

Sol. 36 When nucleus emits the photon it also recoils, say with a recoil speed v to conserve momentum. By conservation of linear momentum, we use

$$Mv = \frac{h}{\lambda} = \left(\frac{h\nu}{c}\right)$$

Total loss of energy of nucleus is given as

$$E = \frac{1}{2}Mv^2 + h\nu = \frac{1}{2}\frac{p^2}{M} + h\nu = \frac{1}{2M}\left(\frac{h\nu}{c}\right)^2 + h\nu$$

$$\Rightarrow \quad E = h\nu\left(1 + \frac{h\nu}{2Mc^2}\right)$$ **Ans. (B)**

Sol. 37 As already explained in Physics Galaxy Vol. 4 about the kinetic energy of α-particle as

$$k_\alpha = \left(\frac{A-4}{A}\right)Q = 5.38\ \text{MeV}$$ **Ans. (C)**

Sol. 38 Binding energy of Al nucleus is given as

$$BE = \Delta mC^2$$

Mass defect is calculated as

$$\Delta m = [13 \times 1.00726 + 14 \times 1.00866 - 27.18846]$$

$$\Delta m = 0.02716 = 27.16 \times 10^{-3}\ u \approx 27\ u$$

$$\Rightarrow \quad BE = 27 \times 10^{-3}\ \text{J}$$ **Ans. (27)**

Sol. 39 Q value of the given fission is calculated by difference in binding energies of the product and reactants which is calculated as

$$Q = (\text{B·E})_{\text{P}} - (\text{B·E})_{\text{R}}$$

$$Q = (105 + 115)(6.4) - (220)(5.6)$$

$$Q = 176\ \text{MeV}$$ **Ans. (D)**

Sol. 40 The balanced nuclear reaction for the given decay is written below.

$$^{238}_{92}\text{U} \rightarrow \boxed{8}\ ^4_2\text{He} + \boxed{6}\ ^0_{-1}\text{e} + ^{206}_{82}\text{Pb}$$

Thus in this reaction 8α particles and 6β particles are emitted. **Ans. (D)**

Sol. 41 A projectile particle p fired on an element nucleus X at rest. If product nucleus Y is formed and residual particle b is emitted, the nuclear reaction is written as

$$p + X \rightarrow Y + b$$

Q value of this nuclear reaction is calculated as

$$Q = K_Y + K_b - K_p$$

$$\Rightarrow \quad Q + K_p = K_Y + K_b$$

$$\Rightarrow \quad Q + K_p > 0$$ **Ans. (D)**

Sol. 42 For the given nuclear reaction using conservation of charge and mass gives

$$A = 182 - 4 - 4 = 174$$

$$Z = 74 - 2 + 1 - 2 = 71$$ **Ans. (A)**

Sol. 43 Radius of nucleus is given as

$$R = R_0 A^{1/3}$$

$$\text{Density of nucleus} = \frac{\text{Mass of nucleus}}{\text{volume of nucleus}}$$

$$\Rightarrow \quad \rho = \frac{m \times A}{\frac{4}{3}\pi R^3}$$

$$\Rightarrow \quad \rho = \frac{m \times A}{\frac{4}{3}\pi R_0^3 A}$$

$$\rho \propto \text{Å}$$

Hence density of nucleus is independent of mass number.

Ans. (C)

Sol. 44 Energy released in the given process = Binding energy of product-Binding energy of reactants

$$= 7.6 \times 4 - (1.1 \times 2) \times 2$$
$$= 30.4 - 4.4$$
$$= 26 \text{ MeV}$$

Ans. (26)

Sol. 45 $R \propto A^{1/3}$

$$V = \frac{4}{3}\pi R^3 \propto A$$

Mass $\propto A$

So density is independent of A. **Ans. (B)**

Sol. 46 In the given equation, writing charges numbers and mass numbers at every step, the equation is given as written below

$${}^{218}_{84}A \xrightarrow{\alpha} {}^{214}_{82}A_1 \xrightarrow{\beta^-} {}^{214}_{83}A_2 \xrightarrow{\gamma} {}^{214}_{83}A_3$$

$${}^{214}_{83}A_3 \xrightarrow{\alpha} {}^{210}_{81}A_4 \xrightarrow{\beta^+} {}^{210}_{80}A_5 \xrightarrow{\gamma} {}^{210}_{80}A_6$$

Ans. (C)

Sol. 47 Density of a nucleus is given as

$$\rho = \frac{\text{mass of nuclei}}{\text{volume of nuclei}}$$

$$\Rightarrow \quad \rho = \frac{1.6 \times 10^{-27} A}{\frac{4}{3}\pi(1.5 \times 10^{-15})^3 A}$$

$$\Rightarrow \quad \rho = \frac{1.6 \times 10^{-27}}{14.14 \times 10^{-45}} = 0.113 \times 10^{18}$$

The required ratio is calculated as

$$\frac{\rho}{\rho_w} = 11.31 \times 10^{13}$$

Ans. (11)

Sol. 48 Total number of atoms in 120 g or half mole of pure element is given as

$$\text{No. of atoms} = \frac{1}{2} \times N_A$$

Total energy released by all these atoms is calculated as

$$E = \frac{1}{2} \times 6 \times 10^{23} \times 200$$

$$E = 6 \times 10^{25} \text{ MeV}$$

Ans. (6)

Sol. 49 Nuclear density is independent of mass number hence both nuclei have same density. **Ans. (C)**

Sol. 50 Figure below shows the nuclear parts after disintegration of nucleus

$v_1 \leftarrow (m_1) \quad (m_2) \rightarrow v_2$

Given velocity ratios of the two fragments, we have

$$\frac{v_1}{v_2} = \frac{3}{2}$$

By conservation of momentum we can write

$$m_1 v_1 = m_2 v_2$$

$$\Rightarrow \quad \frac{m_1}{m_2} = \frac{2}{3}$$

Since, Nuclear mass density is constant, we use

$$\frac{m_1}{\frac{4}{3}\pi r_1^3} = \frac{m_2}{\frac{4}{3}\pi r_2^3}$$

$$\Rightarrow \quad \left(\frac{r_1}{r_2}\right)^3 = \frac{m_1}{m_2}$$

$$\Rightarrow \quad \frac{r_1}{r_2} = \left(\frac{2}{3}\right)^{1/3}$$

$$\Rightarrow \quad x = 2$$

Ans. (2)

Sol. 51 Nuclear density is independent of **A** so Assertion is false. **Ans. (B)**

Sol. 52 As neutron has more rest mass than proton so it can decay into a proton and release energy but for the decay of a proton into neutron external energy needs to be suppoed to induce the reaction. **Ans. (D)**

Sol. 53 A Klystron valve or a magnetron is source of microwaves. Gamma rays are produced in a radioactive decay of a nucleus. Radio waves are produced by rapid acceleration and deceleration of electrons in antennas and X-rays are produced by inner shell electron transition in a heavy atom.

Ans. (B)

Sol. 54 Binding energy of Helium is calculated as

$$\Delta E = (2m_p + 2m_N - m_{He})c^2$$
$$= 28.4 \text{ MeV}$$

Ans. (B)

Sol. 55 For mass number A = 34, total binding energy is calculated as

$$\Delta E = 1.2 \times 34 = 40.8 \text{ MeV}$$

For mass number $B = 26$, total binding energy is calculated as

$$\Delta E = 1.8 \times 26 = 46.8 \text{ MeV}$$

Difference in binding energy is calculated as

$$\Delta BE = 6 \text{ MeV}$$ **Ans. (6)**

Sol. 56 Surface energy per nucleon is given as

$$u_s \propto \frac{r^2}{A} \propto \frac{A^{2/3}}{A} \propto \frac{1}{A^{1/3}}$$

Thus (1) is incorrect. Contribution to binding energy by coulombic force is calculated as

$$\Rightarrow \quad b_c = \frac{-a_2 Z(Z-1)}{A^{1/3}}$$

Thus (2) is incorrect. Volume of nucleus is proportional to the mass so we have

Volume energy $\propto A$

Thus (3) is correct. For (4), if we consider only surface energy contribution then option is correct.

For (5) only a few interactions contribute to surface energy hence not correct. **Ans. (A)**

Sol. 57 Initial binding energy is given as

$$E_1 = 242 \times 7.6 \text{ MeV}$$

Final binding energy is calculated as

$$E_2 = 121 \times 8.1 \text{ MeV} + 121 \times 8.1 \text{ MeV}$$

$$\Rightarrow \quad E_2 = 242 \times 8.1 \text{ MeV}$$

Total gain in binding energy is calculated as

$$\Delta E = 242(8.1 - 7.6) = 121 \text{ MeV}$$ **Ans. (121)**

Sol. 58 As linear momentum of system must be conserved so we have

$$m_1 v_1 - m_2 v_2 = 0$$

$$\Rightarrow \quad \frac{v_1}{v_2} = \frac{m_2}{m_1}$$

As nuclear density remain constant for all nuclei, mass of a nucleus is proportional to the volume of nucleus, so we have

$$m \propto \frac{4}{3}\pi r^3$$

$$\Rightarrow \quad \frac{m_2}{m_1} \propto \left(\frac{r_2}{r_1}\right)^3$$

$$\Rightarrow \quad \frac{m_2}{m_1} = \left[\frac{(2)^{1/3}}{1}\right]^3$$

$$\Rightarrow \quad \frac{m_2}{m_1} = \frac{2}{1}$$

$$\Rightarrow \quad \frac{v_1}{v_2} = \frac{2}{1}$$

$$\Rightarrow \quad n = 2$$ **Ans. (2)**

Sol. 59 Amount of energy released in the given reaction can be calculated as

$$Q = (m_A - m_B - m_D) \times 931.5 \text{ MeV}$$

$$\Rightarrow \quad Q = (238.05079 \times 234.04363 - 4.00260) \times 931.5$$

$$\Rightarrow \quad Q = 4.25 \text{ MeV}$$ **Ans. (D)**

Sol. 60 Binding energy per nucleon is almost same for nuclei of mass number ranging 30 to 170. Nuclear force is very short range force. **Ans. (D)**

28.1 Semiconductor

Sol. 1 Energy band gap

in case of metals $E \sim 0$ eV [here V.B. and C.B. are overlapped]

in case of Semiconductor $E < 2 - 2.5$ eV

in case of insulators $E > 5 - 6$ eV **Ans. (C)**

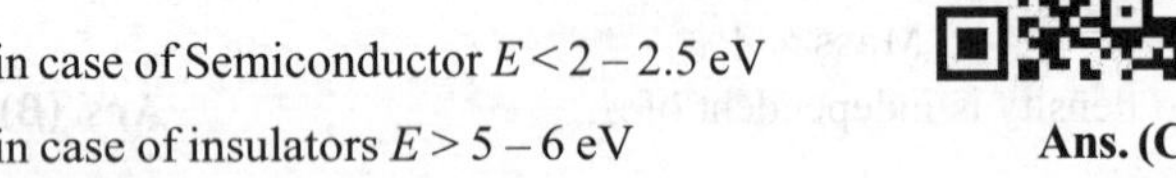

Sol. 2 For conductors on increasing temperature mobility of charge carriers decrease and concentration of free charge carriers does not change.

$$\rho_c = \frac{1}{\sigma} = \frac{1}{n_e e \mu_e}$$

$\Rightarrow$ on increasing T, ρ decreases

For semiconductors on increasing temperature the concentration of charge carriers increase due to more e-h pair generation at high temperature

$$\rho_{sc} = \frac{1}{\sigma_{sc}} = \frac{1}{n_e e \mu_e + n_h e \mu_h}$$

$\Rightarrow$ on increasing T, ρ decreases.

Effect of temperature in a semi-conductor on n_e and n_h is high compared to μ_e and μ_h. **Ans. (C)**

Sol. 3 As band gap of Si is 1.14 eV. This energy can be obtained at room temperature by thermal agitation so at room temperature it conducts electricity.

But at $T = 0$ K temperature no energy will be available for generation of e-h pairs

$\Rightarrow$ No free charge carriers will be there is Si lattice so Si behave like an insulator at 0 K temperature. **Ans. (C)**

Sol. 4 As "Copper" is a metal $\Rightarrow$ its resistance increases with rise in temperature, here as temperature is reduced, its resistance "decreases". And as "Germanium" in a semiconductor

$\Rightarrow$ its resistance decreases with rise in temperature here as temperature is reduced, its resistance "increases". **Ans. (C)**

Sol. 5 With temperature increase due to thermal agitation more e - h pairs are generated in a Semi-conductor due to which concentration of charge carriers in electricity conduction increases. **Ans. (B)**

Sol. 6

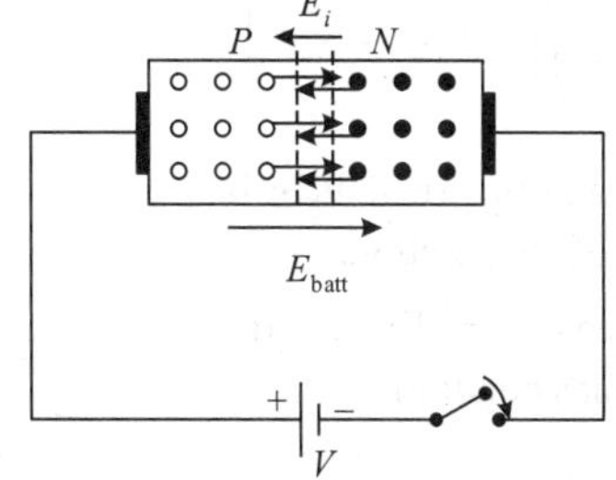

$E_i \rightarrow$ Electric field due to P and N type semi-conductor doping ions.

As external electric field due to battery overcomes the barrier field so barrier height reduces and diffusion of electrons and holes start in depletion layer so it becomes narrower. **Ans. (A)**

Sol. 7 Conductivity increases $\Rightarrow$ more e - h pair generation starts in Semi-conductor.

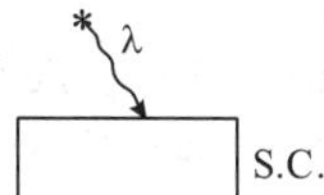

Energy band gap of Semi-conductor is given as

$$\Delta E_g = \frac{hc}{e\lambda} \text{eV}$$

$$\Delta E_g = \frac{12431}{(24800\text{Å})} \text{eV} \approx 0.5 \text{ eV} \quad \textbf{Ans. (D)}$$

Sol. 8 The currents in the Semiconductor due to electrons and holes can be given as

$$I_e = n_e eA\, v_{d_e}$$

$$I_h = n_e eA\, v_{d_h}$$

here $A \rightarrow$ cross section area of Semi-conductor substrate.

$$\Rightarrow \quad \frac{I_e}{I_h} = \frac{n_e v_{d_e}}{n_n v_{d_h}}$$

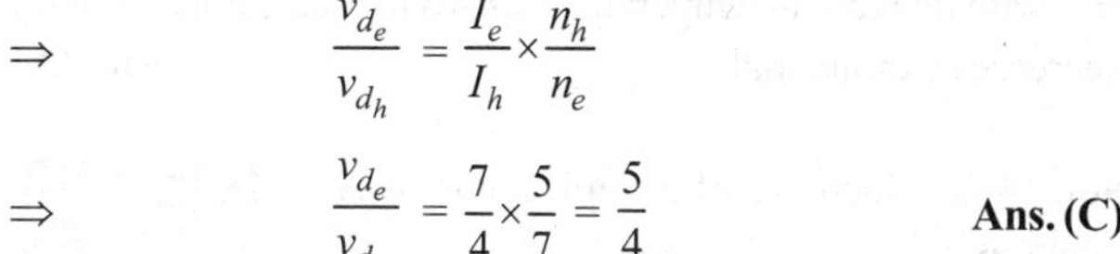

$$\Rightarrow \quad \frac{v_{d_e}}{v_{d_h}} = \frac{I_e}{I_h} \times \frac{n_h}{n_e}$$

$$\Rightarrow \quad \frac{v_{d_e}}{v_{d_h}} = \frac{7}{4} \times \frac{5}{7} = \frac{5}{4}$$

Ans. (C)

Sol. 9 If conductivity increases with temperature $\Rightarrow$ the substance is a semiconductor (Si or Ge or any other)

Bonding $\rightarrow$ Covalent bonds **Ans. (B)**

Sol. 10 If density of a material increases, the energy levels in atoms become more prominent bands or E_c and E_v increase due to influence by neighbouring atoms.

On decreasing lattice constant E_c and E_v will increase and E_g decreases.

$\Rightarrow$ If lattice constant increase $\Rightarrow E_c$ and E_v will decrease $\Rightarrow E_g$ will increase $\Rightarrow$ density decrease. **Ans. (B)**

Sol. 11 The band energy gap in C, Si and Ge are

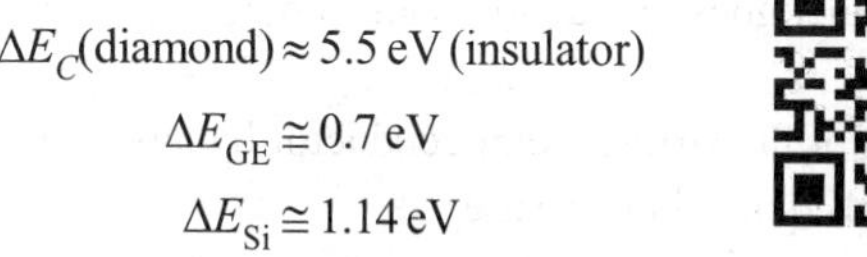

$$\Delta E_C(\text{diamond}) \approx 5.5 \text{ eV (insulator)}$$

$$\Delta E_{GE} \cong 0.7 \text{ eV}$$

$$\Delta E_{Si} \cong 1.14 \text{ eV}$$

ΔE_{Ge} and $\Delta E_{Si} < 2$ eV so these gain energy to excite electrons from valence band to conduction band at room temperature

Ans. (A)

Sol. 12 Statement-1 : For a semiconductor as temperature rises, its conductivity increases $\Rightarrow$ Resistance decreases

$\Rightarrow$ its α is negative $\Rightarrow$ True

Statement-2 : Conductivity of a semiconductor is given as

$$\sigma = n_e e\mu_e + n_h e\mu_h$$

[n_e, $n_h \rightarrow e^-$ and hole concentration]

As temperature rises more e - h pair generation take place

$\Rightarrow$ n_e, n_h increases $\Rightarrow \sigma \uparrow$ **Ans. (D)**

Sol. 13 As Cu is a metal we use its resistivity

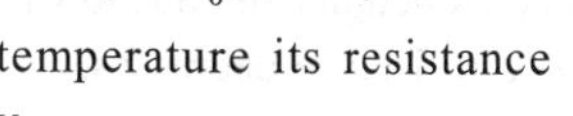

$$\rho = \rho_0(1 + \alpha T)$$

For Cu with temperature its resistance increase linearly.

For intrinsic Si the carrier concentration is given as

$$n_i = n_0 e^{-\Delta E_g / kT}$$

the conductivity of a substance $\sigma \propto n_i$

$\Rightarrow$ with increase in temperature resistivity of a Semiconductor decreases exponentially. **Ans. (A)**

Sol. 14 Conductivity of a semiconductor is given as

$$\sigma = n_e\mu_e e + n_n\mu_h e$$

$$\Rightarrow \sigma = (n_e\mu_e + n_n\mu_h)e$$

$$\Rightarrow \sigma = [5\times 10^{18}\times 2.0 + 5\times 10^{19}\times 0.01]\times 1.6\times 10^{-19}$$

$$\Rightarrow \sigma = [10^{19} + 0.05\times 10^{19}]\times 1.6\times 10^{-19}$$

$$\Rightarrow \sigma = 1.05\times 1.6 = 1.68\,(\Omega\text{-m})^{-1}$$ **Ans. (A)**

Sol. 15 Resistivity of a semiconductor is given as

$$\rho = \frac{1}{\sigma} = \frac{1}{ne\mu_e}$$

$$\Rightarrow \qquad \rho = \frac{1}{10^{19}\times 1.6\times 10^{-19}\times 1.6}$$

$$\Rightarrow \qquad \rho = 0.4\,\Omega\text{m}$$ **Ans. (B)**

Sol. 16 By fundamental theory it is evident that on increasing doping concentration the fermi level of p type semiconductor goes down due to acceptor impurity and that of n type semiconduct goes up due to donor impurity. **Ans. (A)**

Sol. 17 For in extrinsic semi-conductor by law of mass action in thermal equilibrium, we use

$$n_e n_h = n_i^2$$

$$\Rightarrow \quad n_e = \frac{n_i^2}{n_h} = \frac{(1.5\times 10^{16})^2}{4.5\times 10^{22}} = 5\times 10^9\,\text{m}^{-3}$$ **Ans. (5)**

Sol. 18 Collector current in circuit is given as

$$I_C = \frac{0.6}{1000}$$

Using $\beta = \dfrac{I_C}{I_B}$, we have

$$\Rightarrow \qquad I_B = \frac{0.6}{1000\times 24} = 25\mu\text{A}$$ **Ans. (25)**

Sol. 19 Pentavalent material have excess free electrons so free electron density increases but overall semiconductor always remain neutral. **Ans. (D)**

Sol. 20 For a semiconductor as temperature increases, more electrons excite to conduction band due to which conductivity increases and hence resistance decreases. **Ans. (C)**

Sol. 21 Resistivity of a conductor is given as

$$\rho = \frac{m}{ne^2\tau}$$

With rise in temperature, number density (n) of electrons and holes increases for semiconductors as m, e, τ are constant.

$$\Rightarrow \qquad \rho \propto \frac{1}{n}$$

$$\Rightarrow \qquad \rho \propto \frac{1}{T}$$ **Ans. (B)**

28.2 PN Junction and Diodes

Sol. 1

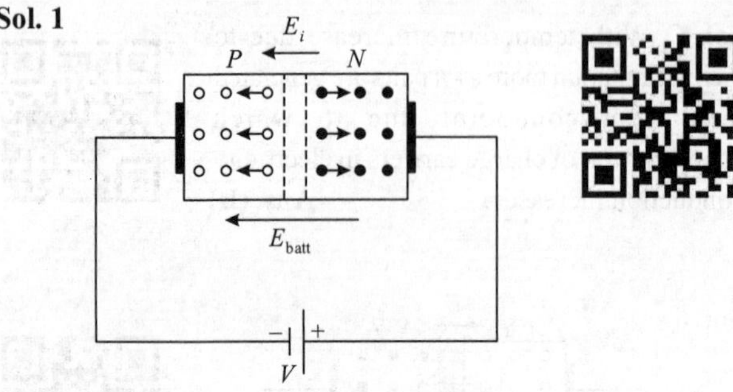

In Reverse Bias mode the electric field by battery adds up to the internal Electric field of the junction.

So barrier height increases $\Rightarrow$ Electric field in depletion layer increase in Reverse Bias condition. **Ans. (C)**

Sol. 2 As "Copper" is a metal $\Rightarrow$ its resistance increases with rise in temperature. Here as temperature is reduced $\Rightarrow$ its resistance "decreases".

And As "Germanium" is a Semiconductor $\Rightarrow$ its resistance decreases with rise in temperature.

Here as temperature is reduced $\Rightarrow$ its resistance "increases". **Ans. (C)**

Sol. 3 In above circuit for the applied battery D_2 is Forward biased and D_1 is reverse biased so the equivalent circuit for this case is

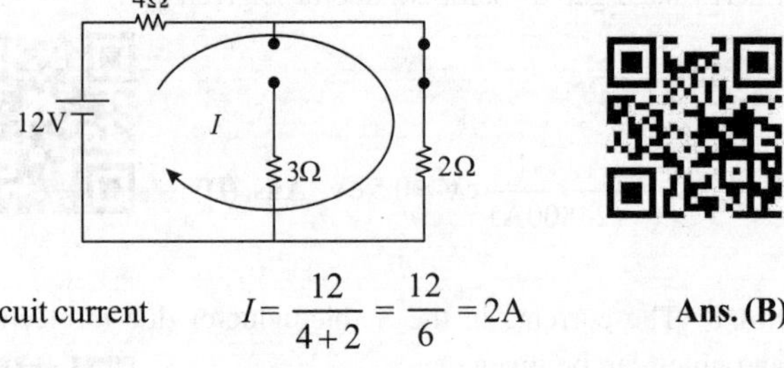

Circuit current $\qquad I = \dfrac{12}{4+2} = \dfrac{12}{6} = 2\text{A}$ **Ans. (B)**

Sol. 4 In Reverse Bias condition

P side of diode $\rightarrow$ low potential

N side of diode $\rightarrow$ high potential

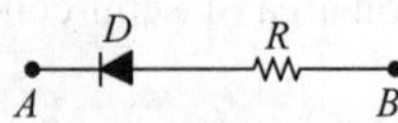
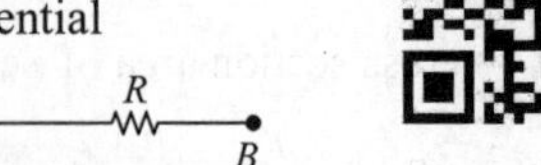

For Reverse Bias we use $V_A > V_B$ **Ans. (D)**

Sol. 5

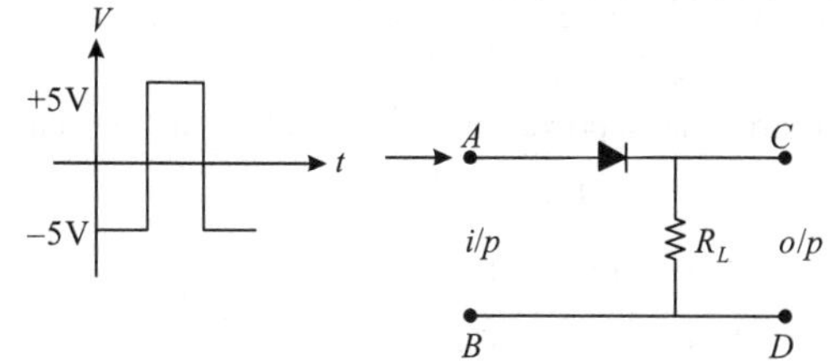

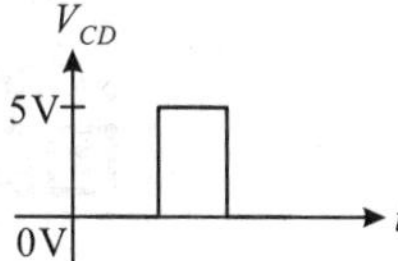

Current will pass through diode only when it is forward biased or when input voltage at A is more than 0 V so the current through R will only be there for $V_A = 0$ to 5 V. **Ans. (A)**

Sol. 6 The given circuit is using "D" as a half wave rectifier. **Ans. (B)**

Sol. 7 LED of difference band gaps emit different coloured visible radiations because the band gap is of the order of energy of visible light.

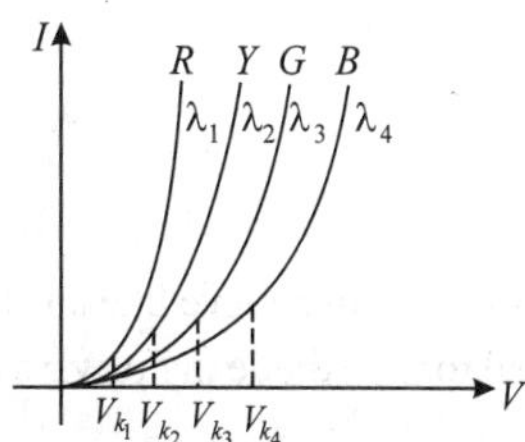

Knee point voltage increases as band gap increase ⇒ for higher ΔE_g (for lower λ) V_K is more.

Here $\Delta E_{G_1} < \Delta E_{G_2} < \Delta E_{G_3} < \Delta E_{G_4}$

$\Rightarrow \lambda_1 > \lambda_2 > \lambda_3 > \lambda_4$ **Ans. (A)**

Sol. 8 Here in circuit we can see due to 6V battery diodes D_1 and D_2 are forward biased and D_3 is reverse biased so equivalent circuit of this case is

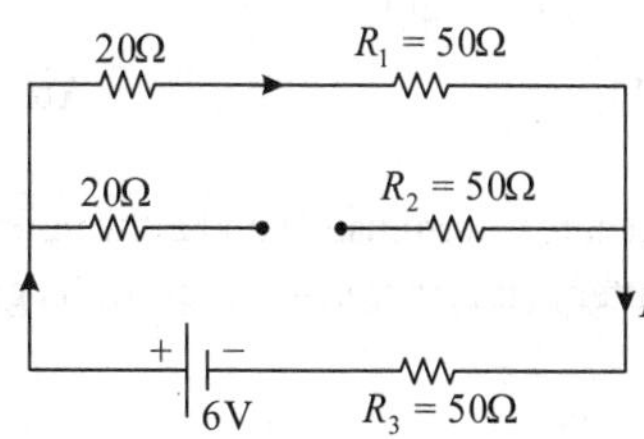

Current through R_3 (in outer loop) is given as

$$I = \frac{6}{120} = \frac{1}{20} = 50 \text{ mA}$$ **Ans. (A)**

Sol. 9 In Forward Bias connection P side of diode is connected to high potential and N side is connected to low potential.

A B

For Forward Bias $V_A > V_B$ **Ans. (A)**

Sol. 10 Equivalent circuit of given situation is shown below

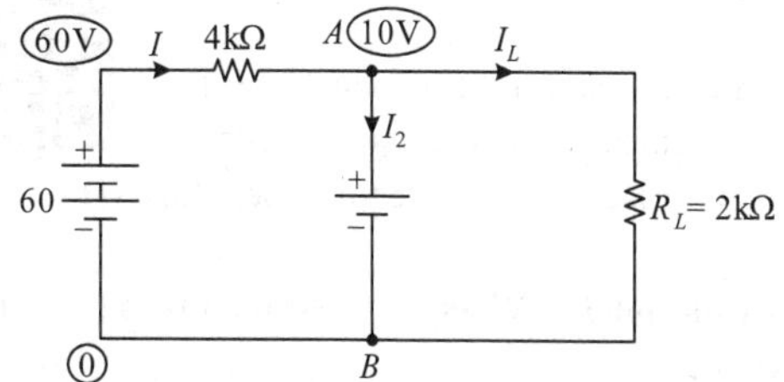

Current I in 4 kΩ resistance is

$$I = \frac{60-10}{4000} = \frac{5}{400} = 12.5 \text{ mA}$$

Current in 2 kΩ resistance is

$$I_L = \frac{10-0}{2000} = 5 \text{ mA}$$

and $$I_Z = I - I_L$$

$$= 12.5 - 5 = 7.5 \text{ mA}$$ **Ans. (C)**

Sol. 11 Light radiation (visible) range is form ~ 4000 Å to 7000 Å for average wavelength of about 6000 Å the corresponding energy should be

$$E \text{ (in eV)} = \frac{hc}{\lambda e} = \frac{12431}{\lambda(\text{in Å})}$$

$$= \frac{12431}{6000} \approx 2.07 \text{ eV}$$ **Ans. (D)**

Sol. 12

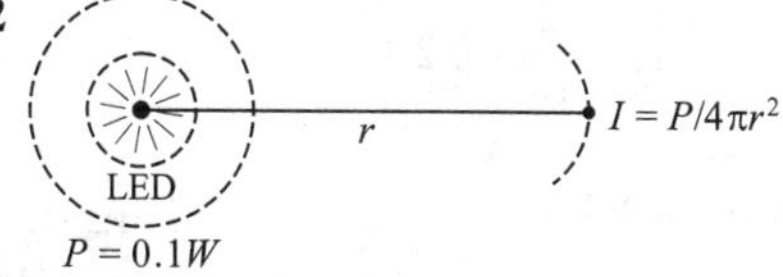

For an EM wave the energy density in light beam is given as

$$u_e = \frac{I}{c} = \frac{1}{2} \epsilon_0 E^2$$

$$\Rightarrow \quad E = \sqrt{\frac{2I}{c\epsilon_0}} = \sqrt{\frac{2P}{4\pi\epsilon_0 cr^2}}$$

$$\Rightarrow \quad E = \sqrt{\frac{2 \times 0.1 \times 9 \times 10^9}{3 \times 10^8 \times (1)^2}} = \sqrt{6}$$

$$\Rightarrow \quad E = 2.45 \text{ V/m}$$ **Ans. (D)**

Sol. 13

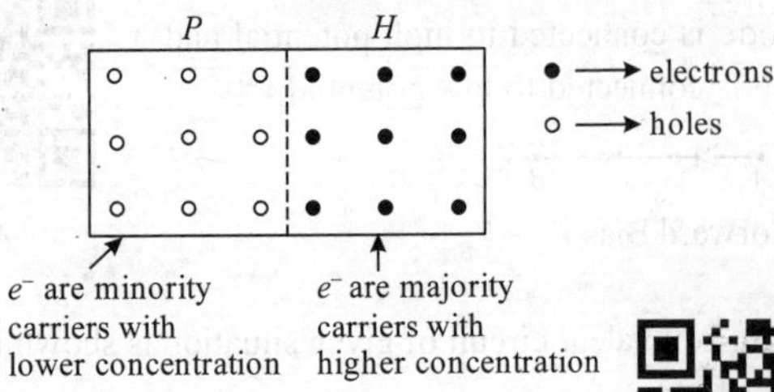

As in n-side concentration of electrons is higher compared to p-side so electron diffusion take place from n side to p side. **Ans. (D)**

Sol. 14 Situation-1 : When +ve terminal of battery is at A, circuit will be

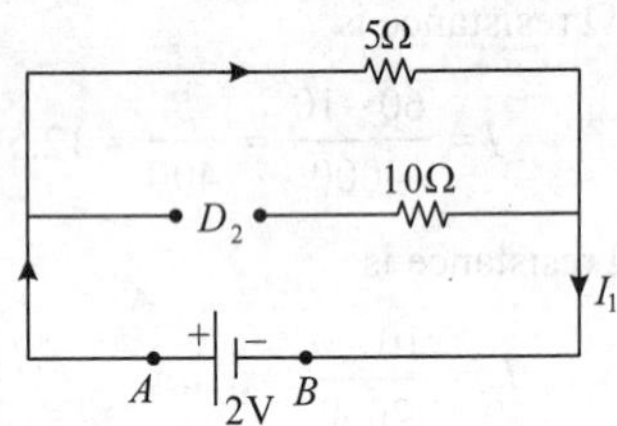

Current through battery is

$$I_1 = \frac{2}{5} = 0.4\,\text{A}$$

Situation-2 : When +ve terminal of battery is at B, circuit will be

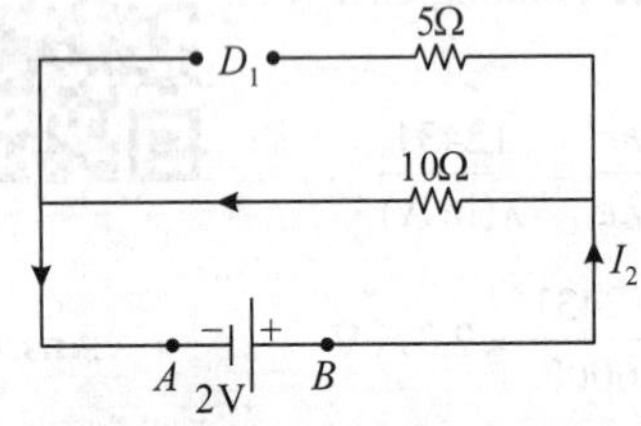

Current through battery is

$$I_2 = \frac{2}{10} = 0.2\,\text{A}$$ **Ans. (A)**

Sol. 15 (i) → Normal diode

(ii) → Zener diode

(iii) → Photo cell/Solar cell (Reverse Bias)

(iv) → Light dependent resistance (Forward Bias) **Ans. (C)**

Sol. 16

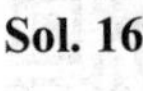

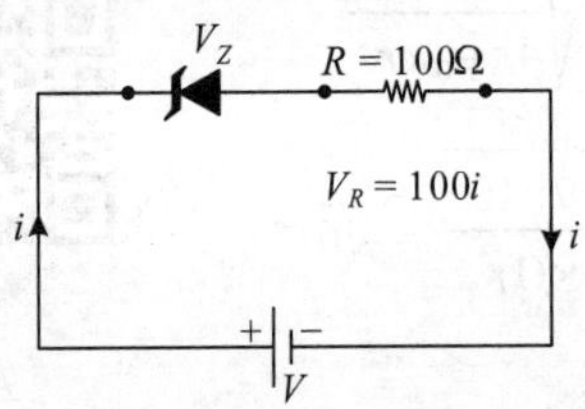

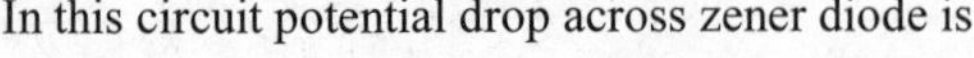

In this circuit potential drop across zener diode is

$$V_Z = V - 100\,i$$

Zener diode has a power dissipation rating of 1W (max)

$$\Rightarrow \quad V_z i = 1$$

$$\Rightarrow \quad (V - 100i)i = 1$$

$$\Rightarrow \quad Vi - 100i^2 = 1$$

$$\Rightarrow \quad 100i^2 - Vi + 1 = 0$$

For real values of i we must have

$$D \geq 0$$

$$\Rightarrow \quad V^2 - 400 \geq 0$$

$$\Rightarrow \quad V \geq 20\text{ volt}$$ **Ans. (B)**

Sol. 17 Forward bias resistance is given as

$$r_{\text{FB}} = \left.\frac{\Delta V}{\Delta I}\right|_{FB} = \frac{0.8 - 0.7}{(20-10)\times 10^{-3}}$$

$$\Rightarrow \quad r_{\text{FB}} = \frac{0.1}{10\times 10^{-3}} = 10\,\Omega$$

Reverse bias resistance is given as

$$r_{\text{RB}} = \left.\frac{\Delta V}{\Delta I}\right|_{RB} = \frac{10-0}{(1-0)\times 10^{-6}} = 10^7\,\Omega$$

$$\Rightarrow \quad \frac{r_{\text{FB}}}{r_{\text{RB}}} = \frac{10}{10^7} = 10^{-6}$$ **Ans. (B)**

Sol. 18 As silicon is forward biased, it will behave like an emf of 0.7 V which is the barrier potential for a Si diode in Forward bias. So equivalent circuit here is

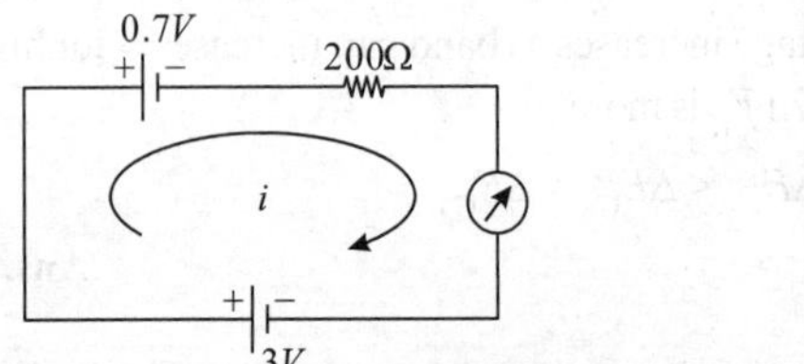

Current in circuit is

$$I = \frac{3-0.7}{200} = \frac{2.3}{200} = 0.00115\,\text{A}$$

$$\Rightarrow \quad I = 11.5\text{ mA}$$ **Ans. (C)**

Sol. 19 In the given circuit across terminals A and B the potential difference due to 15 V battery (if Zener is not present) is given as

$$V_{AB} = \frac{15}{2000}\times 1500 = \frac{225}{20} = 11.25\,\text{V}$$

So here $V_{AB} > V_Z \Rightarrow$ Zener will operate in breakdown region and maintain V_Z across terminals A and B.

Under the state of breakdown the equilibrium circuit will be

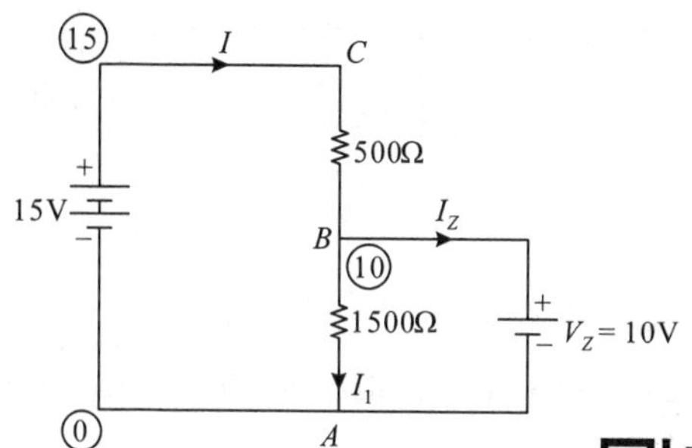

Current in 1500 Ω resistance is

$$I_1 = \frac{10}{1500} = \frac{1}{150}\text{A}$$

Current in 500 Ω resistance is

$$I = \frac{15-10}{500} = \frac{5}{500} = \frac{1}{100}\text{A}$$

Current through zener diode is

$$I_Z = I - I_1 = \frac{1}{100} - \frac{1}{150} = \frac{3-2}{300}$$

$$\Rightarrow \quad I_Z = \frac{1}{300} = 3.33\text{ mA} \qquad \textbf{Ans. (B)}$$

Sol. 20 Initially *Ge* is forward biased so current will pass through *Ge* diode with a drop of 0.3V so *Si* will not be forward biased.

If "*Ge*" is reversed then current will flow through "*Si*" diode hence an effective drop of $(0.7-0.3) = 0.4$V is observed. **Ans. (C)**

Sol. 21 Assuming zener diode doesn't breakdown, current in circuit

$$= \frac{120}{15000} = 8\text{ mA}$$

$\Rightarrow$ The voltage drop across diode = 80 V > 50 V.

$\Rightarrow$ The diode will breakdown and maintain 50 V across 10 kΩ resistance as shown below

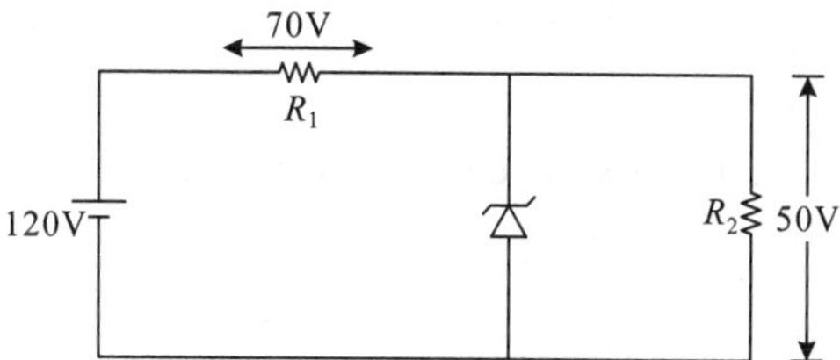

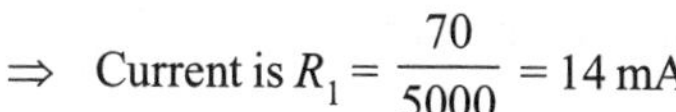

$\Rightarrow$ Current is $R_1 = \dfrac{70}{5000} = 14$ mA

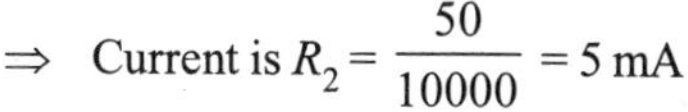

$\Rightarrow$ Current is $R_2 = \dfrac{50}{10000} = 5$ mA

$\Rightarrow$ Current through diode = 9 mA **Ans. (D)**

Sol. 22 As voltage across zener diode must be less than 10V so it will not work in breakdown region, & its resistance will be infinite & current through it = 0. **Ans. (D)**

Sol. 23 Here diode D_2 is reverse biased and D_1 is forward biased so current flows in outer loop only, given as

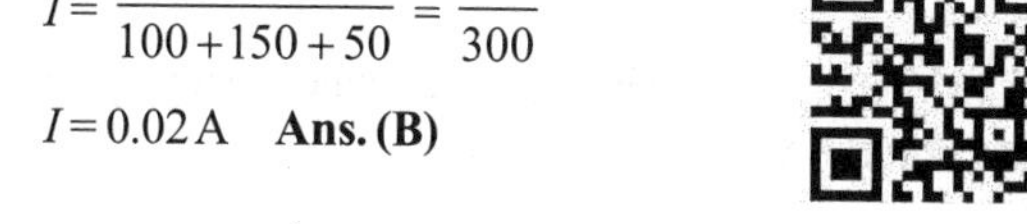

$$I = \frac{6}{100+150+50} = \frac{6}{300}$$

$\Rightarrow \quad I = 0.02$ A **Ans. (B)**

Sol. 24

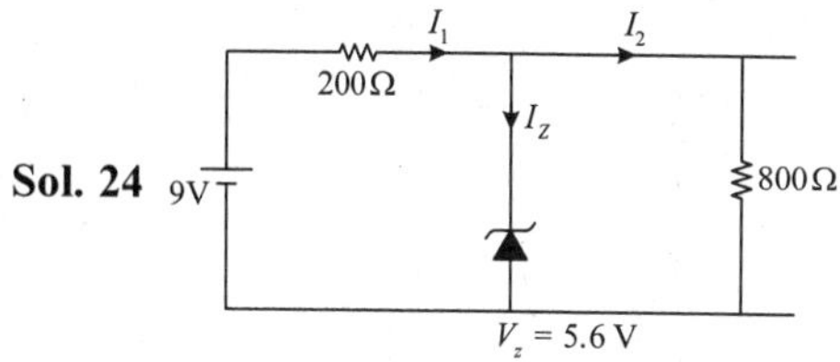

In first loop

$$9V = 200\,I_1 + 5.6$$

$$\Rightarrow \quad I_1 = \frac{3.4}{200} = 17\text{ mA}$$

In second loop

$$I_2 = \frac{5.6}{800} = 7\text{ mA}$$

Here $\quad I_Z = I_1 - I_2$

$\Rightarrow \quad I_Z = 17 - 7 = 10$ mA **Ans. (C)**

Sol. 25 At $\quad V = 8$V

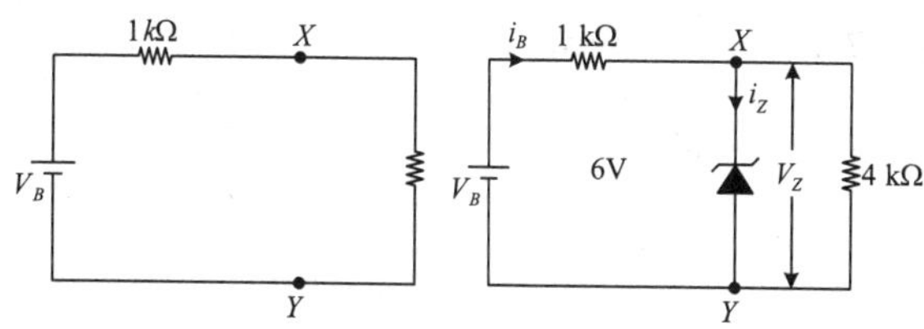

$$V_{XY} = \frac{V_B}{5} \times 4 = \frac{32}{5} = 6.4\text{ V} > \frac{V_Z}{6\text{V}}$$

At $V = 16$ V

$$V_{XY} = \frac{16 \times 4}{5} = \frac{64}{5} = 12.8\text{ V} > 6\text{ V}$$

In breakdown state

$$V_{XY} = 6\text{ V}$$

$$\Rightarrow \quad i_L = \frac{6}{4\text{k}} = 1.5 \times 10^{-3}\text{A}$$

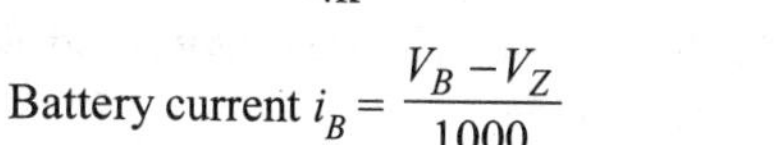

$\Rightarrow$ Battery current $i_B = \dfrac{V_B - V_Z}{1000}$

At 8 V $\rightarrow$ $i_B = \dfrac{8-6}{1000} = 2\times 10^{-3}\,\text{A}$

$$i_Z = i_B - i_L = (2-1.5)\times 10^{-3} = 0.5\,\text{mA}$$

At 16 V $\rightarrow$ $i_B' = \dfrac{16-6}{1000} = 10\times 10^{-3}\,\text{A}$

$$i_Z = i_B' - i_L = (10\ \ 1.5)\times 10^{-3} = 8.5\,\text{mA}$$ **Ans. (B)**

Sol. 26 Current I_2 will be maximum when input voltage is also maximum at 16 V

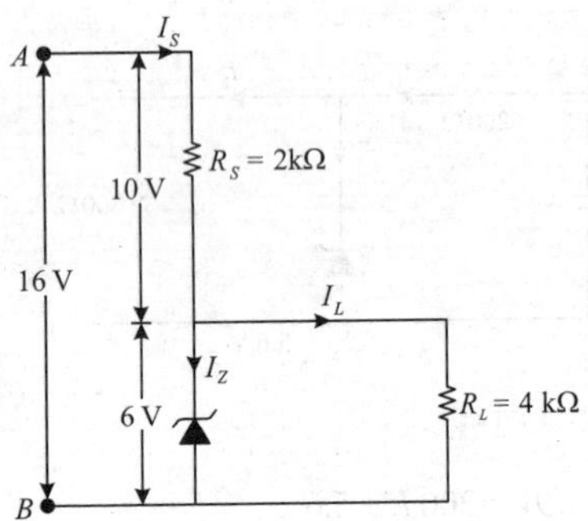

At voltage across R_L is maintained at $6V$

$$\Rightarrow \quad I_L = \frac{6}{4000} = 1.5\,\text{mA}$$

Current through R_S is

$$I_S = \frac{10}{2000} = 5\,\text{mA}$$

$\Rightarrow$ Zoner current

$$I_Z = I_S - I_L = 5 - 1.5 = 3.5\,\text{mA}$$ **Ans. (B)**

Sol. 27 As diode is forword based, we use circuit as

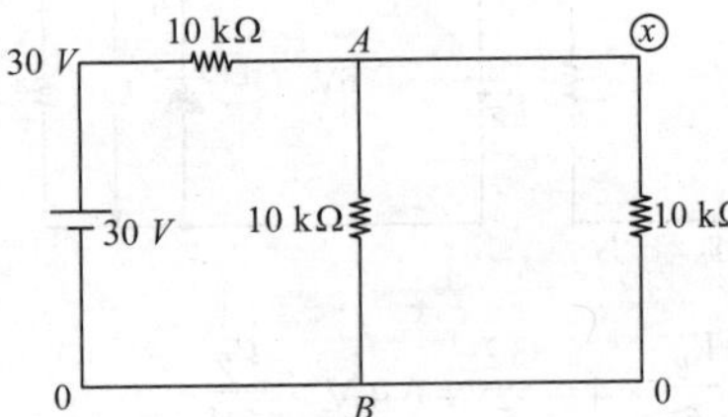

Using KCL equation for x, we have

$$\frac{30-x}{10} + \frac{0-x}{10} + \frac{0-x}{10} = 0$$

$$3 = \frac{3x}{10}$$

$$x = 10\text{V}$$ **Ans. (B)**

Sol. 28 In the figure shown we can if diode D_1 is forward biased potential of point A will become

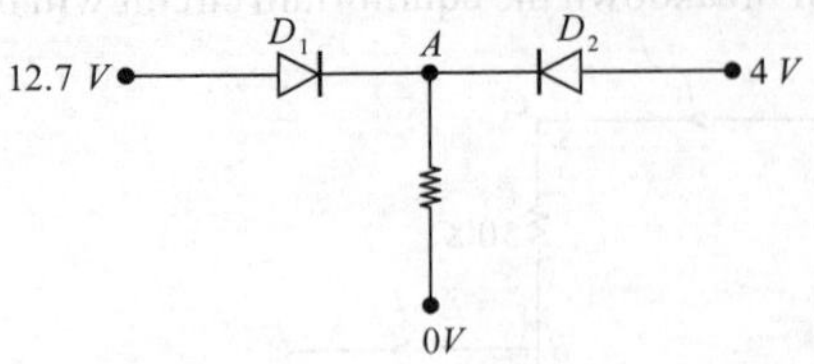

$$V_A = 12.7 - 0.7 = 12\text{V}$$

and diode D_2 will be reverse biased and behave like open circuit. **Ans. (12)**

Sol. 29 In given circuit both the diodes will be reverse biased so effective circuit is shown below

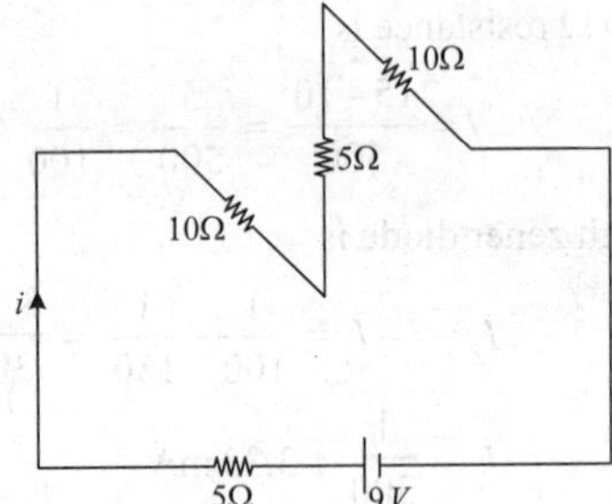

Thus current in circuit is given as

$$i = \frac{9}{(5+10+5+10)} = \frac{9}{30}\,\text{A}$$ **Ans. (C)**

Sol. 30 Diode in circuit of capacitor A will remain reverse biased so no current flows. Thus charge on capacitor A will remain constant at

$$Q_A = CV$$

Discharge current flows in circuit of capacitor B so charge capacitor is given as

$$\Rightarrow \quad q = CVe^{-\frac{t}{RC}}$$

at $t = CR$, charge is given as

$$Q_B = CVe^{-1} = \frac{CV}{e}$$ **Ans. (A)**

Sol. 31 For the given circuit potential difference across 400 Ω resistance is 4V, so current in circuit is

$$I = \frac{4}{400} = \frac{1}{100}A$$

Power dissipated in each diode is given as

$$P_d = \text{VI}$$

$$\Rightarrow \quad P_d = 4\times\frac{1}{100}W$$

$$\Rightarrow \quad P_d = 40\times 10^{-3}\,\text{mW}$$ **Ans. (40)**

Sol. 32 Circuit of described situation is shown below

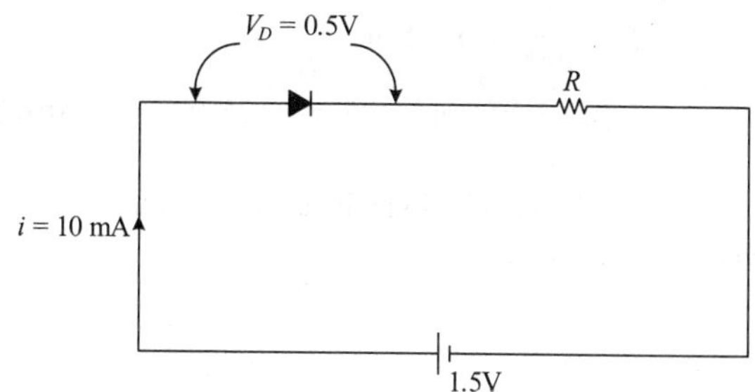

Using KVL equation in loop gives

$$1.5 - 0.5 - iR = 0$$

$$\Rightarrow \quad R = \frac{1}{i} \times \frac{1}{10 \times 10^{-3}} = 100\,\Omega$$ **Ans. (A)**

Sol. 33 Energy band gap is given as

$$\Delta E = \frac{hc}{\lambda} = \frac{19.89 \times 10^{-26}}{400 \times 10^{-9} \times 1.6 \times 10^{-19}}$$

$$\Rightarrow \quad \Delta E = 3.107\,\text{eV}$$ **Ans. (C)**

Sol. 34 As there are two zener diodes in opposite sense in both cycles one will breakdown and cut off voltage at 6 V so option (B) is correct. **Ans. (B)**

Sol. 35 For photodiode on increasing reverse voltage, current increases then saturation arises at a specific intensity of light falling on it.

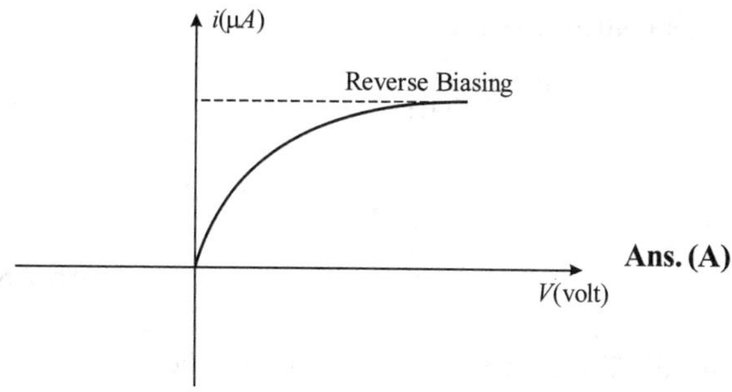

Ans. (A)

Sol. 36 When input voltage increases from 0, breakdown in 4V zener take place when $V_{in} = 4V$ but when $V_{in} > 4V$ then also V_{out} increases because of potential drop across 100Ω resistance and when $V_{in} = 6V$ or more breakdown in 6V zener occurs and V_{out} remain constant at 6V now on lord. **Ans. (D)**

Sol. 37 During $t = 0$ to $t = 5$ μC charge on capacitor grows and becomes constant for $t = 5$ μC to $t = 10$ μC then again grows with time so most appropriate option is (A) **Ans. (A)**

Sol. 38 In the figure shown current through 2 kΩ resistance can be directly given by zener voltage as

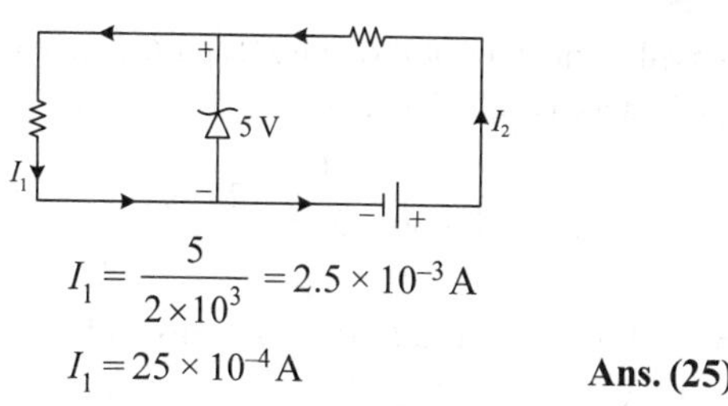

$$I_1 = \frac{5}{2 \times 10^3} = 2.5 \times 10^{-3}\,\text{A}$$

$$I_1 = 25 \times 10^{-4}\,\text{A}$$ **Ans. (25)**

Sol. 39 Zener diode is constructed with high doping concentration and have a thin depletion layer. **Ans. (B)**

Sol. 40 Back to back connections of diodes will not act like a transistor because its working require different dopin concentrations in the emitter, base and collector along with different sizes of these regions so Statement-I is false. For a transistor the current amplification factor is defined as

$$\beta = \frac{i_c}{i_b}$$

Thus Statement-II is true. **Ans. (B)**

Sol. 41 After connection of battery as described in question, circuit is shown in figure below

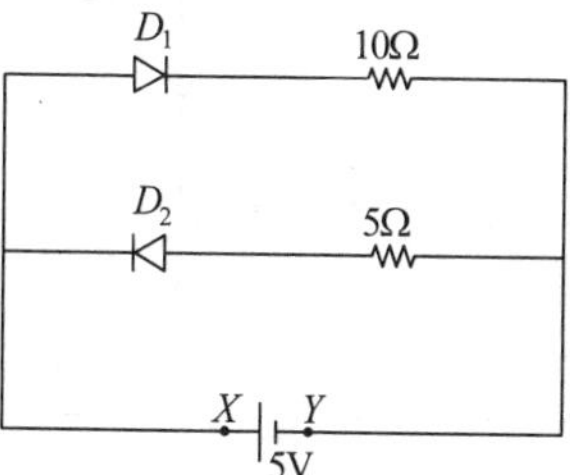

In this only D_1 will be forward biased and conduct and for silicon diode, potential drop across D_1 will be 0.7 V so current is given as

$$I = \frac{5 - 0.7}{10} = 0.43\,\text{A}$$ **Ans. (C)**

Sol. 42 From the energy gap given, the wavelength of emitted light due to transition of electrons in depletion layer can be calculated as

$$\lambda = \frac{12431}{1.9} \times 10^{-10} = 654\,\text{nm}$$

The color corresponding to this wavelength is red. **Ans. (C)**

Sol. 43 After connection of battery as described in question, circuit is shown in figure below

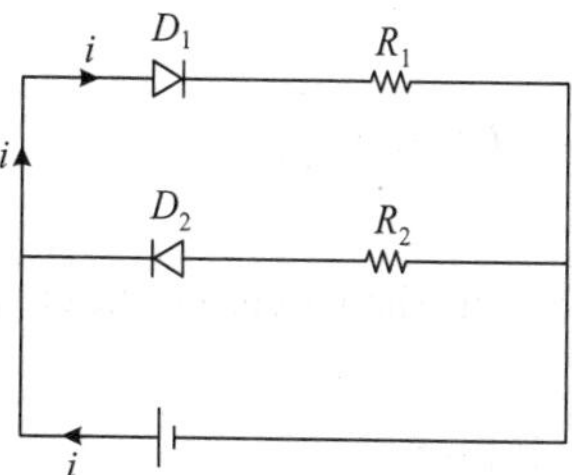

In this circuit D_1 will be forward biased and D_2 will be revers

biased so no current through D_2 and R_2. Here current through the battery is calculated as

$$i = \frac{6}{120+130+50} = \frac{6}{300} = 20\text{ mA} \quad \textbf{Ans. (20)}$$

Sol. 44 Under breakdown conditions current through 4000 Ω resistance can be given as

$$i = \frac{60}{4000}\text{ A}$$

Current through 5000 Ω resistance is given as

$$i_1 = \frac{30}{5000}\text{ A}$$

Current through the diode can now be calculated as

$$i_z = i - i_1 = \frac{60}{4000} - \frac{30}{5000} = \frac{9}{1000} = 9\text{ mA} \quad \textbf{Ans. (9)}$$

Sol. 45 Figure below shows the current distibution in the given circuit.

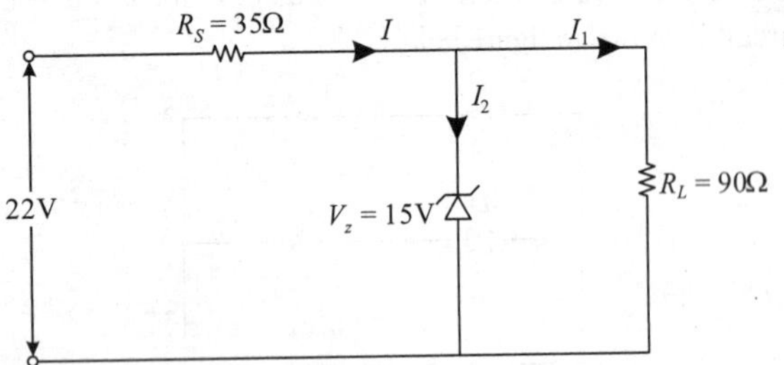

Voltage across R_S is calculated as

$$V_S = 22 - 15 = 7\text{ V}$$

Current through R_S is calculated as

$$I = \frac{7}{35} = \frac{1}{5}\text{ A}$$

Current through 90Ω resistance is calculated as

$$I_2 = \frac{15}{90} = \frac{1}{6}\text{ A}$$

Current through the zener diode is calculated as

$$I_1 = \frac{1}{5} - \frac{1}{6} = \frac{1}{30}\text{ A}$$

Power dissipated across zener diode is calculated as

$$P = VI$$

$$\Rightarrow \quad P = 15 \times \frac{1}{30} = 0.5 = 5 \times 10^{-1}\text{ W} \quad \textbf{Ans. (5)}$$

Sol. 46 In the circuit shown current through battery can be calculated as

$$I = \frac{100-50}{1000} = 50\text{ mA}$$

Current through 2000Ω resistance is calculated as

$$I' = \frac{50}{2000} = 0.025\text{ mA}$$

$$\Rightarrow \quad I_z = I - I' = 50\text{ mA} - 25\text{ mA} = 25\text{ mA} \quad \textbf{Ans. (B)}$$

Sol. 47 For a PN junction diode,dynamic resistance in forward biasing is given as

$$r_D = \frac{\Delta V}{\Delta I}$$

$$\Rightarrow \quad r_D = \frac{0.75-0.65}{(5-1)\times 10^{-3}} = 25 \quad \textbf{Ans. (25)}$$

Sol. 48 For maximum potential drop across the diode through potential divider, the voltage across protective resistance will be 20 – 8 = 12 V and the effective resistance across the zener diode can be given as

$$R_D = \frac{V_Z^2}{P} = \frac{64}{0.5} = 128\ \Omega$$

So the current through the branch is given as

$$i = \frac{20}{128 + R_P}$$

$$\Rightarrow \quad 12 = \frac{20}{128 + R_P} \times R_P$$

$$\Rightarrow \quad 1536 = 8R_P$$

$$\Rightarrow \quad R_P = 192\ \Omega \quad \textbf{Ans. (192)}$$

Sol. 49 From the circuit given in question, as current through zener is 90 mA, we use

$$\frac{50-5}{R} \geq 90\text{ mA}$$

$$\Rightarrow \quad R \leq \frac{45}{90} \times 10^3$$

$$\Rightarrow \quad R \leq 500\ \Omega \quad \textbf{Ans. (500)}$$

Sol. 50 The currents in the various branches

$$i_1 = \frac{14}{1} = 14\text{ mA}$$

$$i_3 = \frac{10}{5} = 2\text{ mA}$$

$$i_2 = i_1 - i_3 = 12\text{ mA}$$

Power consume in zerer diode is given as

$$\text{Pavg} = V \times i = 0.12\text{ W} \quad \textbf{Ans. (120)}$$

Sol. 51 Power consumption in diode can be written as

$$P = Vi$$

$$\Rightarrow \quad i = \frac{P}{V} = \frac{2}{10} = 0.2\text{ A}$$

For $V_S = 14$ V, the voltage drop across R_S can be given as

$$V_R = 14 - 10 = 4\text{ V}$$

By Ohm's law for R_S, we have

$$V = iR_S$$

$$\Rightarrow \quad R_S = \frac{4}{0.2\times10^{-2}} = 20\ \Omega$$ **Ans. (20)**

Sol. 52 In the given circuit current will flow through the diode when it is in forward biased state. The potential difference across the diode $\Delta V = (10\sin\omega t - 3)$ V and ΔV across the diode will be zero, when diode is in reverse biased state. For this case the correct waveform is drawn in figure below.

Ans. (D)

Sol. 53 A capacitor has low reactance in AC circuit but act as an open circuit for dc so it short circuits the pulsating voltage and allows direct current to pass hence statement-I is true. Similarly an inductor has high reactance in AC circuit but act as short circuit for dc so it is connected in series to allow dc to pass through it and blocks ac hence statement-II is also true. **Ans. (C)**

Sol. 54 Equivalent resistance of the given circuit with forward biased diodes to be considered short circuited and reverse biased diods as open circuit, is calculated as

$$R_{eq} = \frac{160}{2} + 20$$

$$\Rightarrow \quad R_{eq} = 80 + 20$$

$$\Rightarrow \quad R_{eq} = 100$$

$$\Rightarrow \quad i = \frac{200}{100} = 2\text{A}$$ **Ans. (A)**

Sol. 55 Current I_L is caused by zener voltage, given as

$$I_L = \frac{5}{1000} = 5\text{ mA}$$ **Ans. (5)**

Sol. 56 Dynamic resistance of a pn junction diode is given as

$$r_d = \frac{\Delta V}{\Delta i}$$

$$\Rightarrow \quad \frac{R_1}{R_2} = \frac{\Delta v_1}{\Delta v_2}\frac{\Delta i_2}{\Delta i_1} = \frac{0.1}{0.2}\times\frac{50}{5} = 5$$ **Ans. (B)**

Sol. 57 In the given circuit diode D_1 and D_3 are forward biased and D_2 is reversed biased. Figure below shows the equivalent circuit for current distribution.

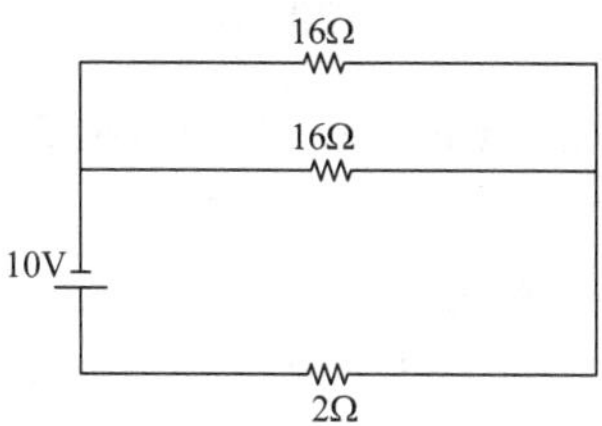

Equivalent resistance across battery here is 10 Ω. The current is calculated by Ohm's law as

$$V = IR_{net}$$

$$\Rightarrow \quad 10 = I\times 10$$

$$\Rightarrow \quad I = 1\text{A}$$ **Ans. (1)**

Sol. 58 In the circuit shown in figure diode D_1 is forward biased and D_2 is reverse biased and the potential drop across D_1 is 0.6V thus equivalent circuit for current flow is shown in figure below.

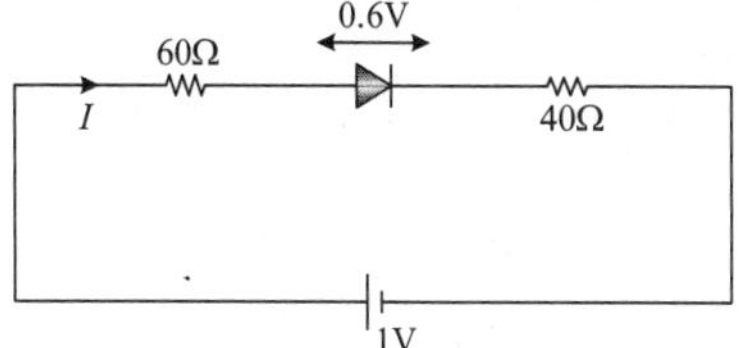

Writing KVL equation for the loop gives

$$1 - I(60) - 0.6 - I(40) = 0$$

$$\Rightarrow \quad \frac{0.4}{100} = I$$

$$\Rightarrow \quad I = 4\text{ mA}$$ **Ans. (4)**

Sol. 59 In forward bias diode conducts and in reverse bias it does not conducts so it conducts current only in one direction. **Ans. (B)**

Sol. 60 In first half cycle of voltage diode D_1 is forward biased and D_2 is reverse biased and in second half cycle it is opposite for the two diodes. The output voltage will be same as the voltage across teh forward biased diode which is equal to 0.6 V if the input voltage is more than this and if it is less than 0.6 V then V_{out} is same as V_{in}. **Ans. (D)**

Sol. 61 The circuit below shows the breakdown situation of zener diode and in this state the voltage across the 100 Ω resistance will be 2 V.

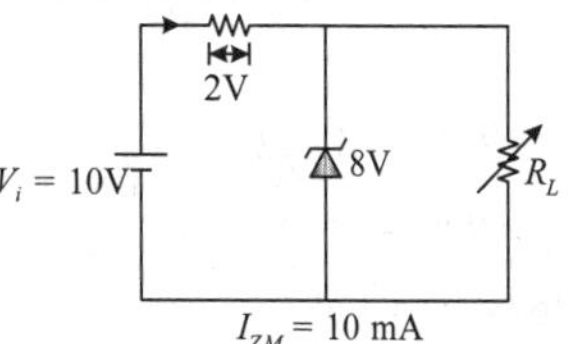

Thus current through 100 Ω resistance is given as

$$I = \frac{2}{100} = 20 \text{ mA}$$

When zener current is maximum at breakdown $I_L = 10$ mA, current through load resistance is calculated as

$$I = I_Z + I_L$$

$$\Rightarrow \quad I_L = I - I_Z = 20 - 10 = 10 \text{ mA}$$

Voltage across load resistance is the zener voltage so we use

$$V_L = I_L R_L$$

$$\Rightarrow \quad 8 = 10 \times 10^{-3} \times R_{L_{max}}$$

$$\Rightarrow \quad R_{L_{max}} = \frac{4}{5} \times 10^3 \, \Omega$$

$$\Rightarrow \quad R_{L_{max}} = 800 \, \Omega$$

In no breakdown state zener current is $I_Z = 0$ and full current of 100 Ω resistance will pass through load resistance

$$I_L = 20 \text{ mA}$$

$$V_L = I_L R_L$$

$$\Rightarrow \quad R_{L_{max}} = \frac{8}{20} \times 10^3 \, \Omega$$

$$\Rightarrow \quad R_{L_{min}} = 400 \, \Omega$$

$$\Rightarrow \quad \frac{R_{L_{max}}}{R_{L_{min}}} = \frac{800}{400} = 2$$

Ans. (2)

Sol. 62 While crossing the barrier, the change in kinetic energy of electron is equal to the work done on it, given as

$$eV = K_f - K_i$$

$$\Rightarrow \quad \frac{1}{2}mv^2 - \frac{1}{2}mu^2 = -1.6 - 10^{-19} \times 0.4$$

$$\Rightarrow \quad \frac{1}{2} 9 \times 10^{-31} (v^2 - u^2) = -0.64 \times 10^{-19}$$

$$\Rightarrow \quad u^2 - v^2 = \frac{2 \times 0.64 \times 10^{12}}{9}$$

$$\Rightarrow \quad v^2 = \left(36 - \frac{128}{9}\right) \times 10^{10}$$

$$\Rightarrow \quad v = \frac{14}{3} \times 10^5 \text{ m/s}$$

$$\Rightarrow \quad x = 14$$

Ans. (14)

Sol. 63 Energy band gap to produce light of wavelength λ is given as

$$E_g = \frac{hc}{\lambda} = \frac{1242}{\lambda(nm)} = \frac{1242}{400} = 3.105 \text{ eV}$$

Answer is rounded off to 3 eV. **Ans. (3)**

Sol. 64 As terminal A is at higher potential, first diode will be forward biased and second one will be reverse biased and will not be conducting so equivalent circuit is as shown in figure below.

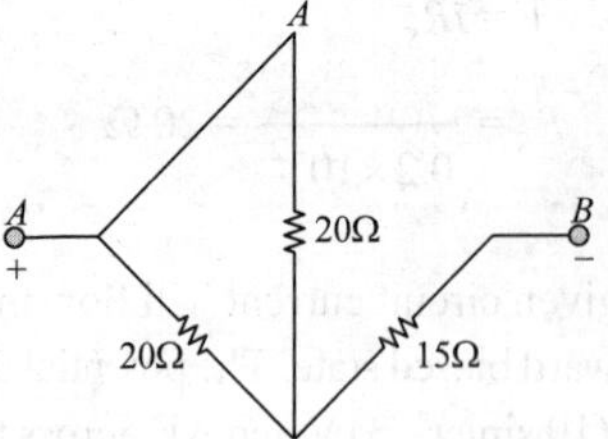

The forward biased diode will conduct while the reverse biased will not

$$R_{AB} = 10 + 15 = 25 \, \Omega$$

Ans. (25)

Sol. 65 In the given circuit the current in 4000Ω resistance is calculated as

$$I = \frac{(120 - 60)V}{4000\Omega} = 0.015\text{A}$$

Current in load resistance is due to zener voltage and it is equal to 60/10000 = 0.006 A. Thus current through zener diode is calculated as

$$I_Z = I - I_L$$

$$\Rightarrow \quad I_Z = 0.015 - 0.006 = 0.009\text{A}$$

$$\Rightarrow \quad I_Z = 9 \text{ mA}$$

Ans. (9)

Sol. 66 Solar cell V–I Characteristics is shown in figure option (B) as solar cells produce maximum current at zero voltage and no current at maximum voltage. As current is produced by the electric field of junction so current is negative. **Ans. (B)**

Sol. 67 Electric field in the depletion layer of pn junction is given as

$$E = \frac{V}{d}$$

$$\Rightarrow \quad E = \frac{0.6V}{6 \times 10^{-6} m} = 1 \times 10^5 \text{ V/m}$$

$$\Rightarrow \quad E = 1 \times 10^5 \text{ N/C}$$

Ans. (1)

Sol. 68 In the given circuit using KVL equation in left loop gives

$$\varepsilon - IR - V_z = 0$$

$$\Rightarrow \quad 20 - IR - 6 = 0$$

$$\Rightarrow \quad IR = 12$$

$$\Rightarrow \quad 25 \times 10^{-3} R = 12$$

$$\Rightarrow \quad R = \frac{12}{25 \times 10^{-3}} = 480 \, \Omega$$

Ans. (480)

Sol. 69 Photodiodes are operated in reverse bias as fractional change in current due to light is easier to detect because of high sensitivity in reverse bias hence Assertion is correct. Reason is also correct by definition but not explaining the Assertion. **Ans. (B)**

Sol. 70 As already analysed that photodiodes are operated in reverse bias condition so Assertion-A is false. For *P-N* junction current in forward bias (for $V \geq V_0$) is always greater than current in reverse bias (for $V \leq V_Z$) so Reason-R is true. **Ans. (C)**

Sol. 71 *LED* works in forward biasing and light energy maybe slightly less or equal to band gap. **Ans. (C)**

Sol. 72 Statement-I is correct because when *P-N* junction is formed, an electric field is produced form *N*-side to *P*-side due to which barrier potential get developed across the depletion layer & majority charge carrier can not flow through the junction due to barrier potential so current is zero unless we apply forward bias voltage hence statement-II is false. **Ans. (D)**

Sol. 73 For an intrinsic semiconductor the fermi level is between conduction and valance band whereas for *n* type semiconductor due to donor impurity fermi level is near valance band and for *p* type semiconductor due to acceptor impurity atoms, fermi level is near conduction band. For metals fermi level is inside the conduction band due to plenty of free electrons available at room temperature. **Ans. (C)**

Sol. 74 Works as voltage regulator in reverse bias and as simple *p-n*-junction in forward bias. **Ans. (A)**

Sol. 75 In forward biased condition, diffusion of majority charge carriers takes place from *p*-side to *n*-side which constitute the diffusion current. **Ans. (C)**

Sol. 76 Figure below shows the potentials distributed in the given circuit of zener diode.

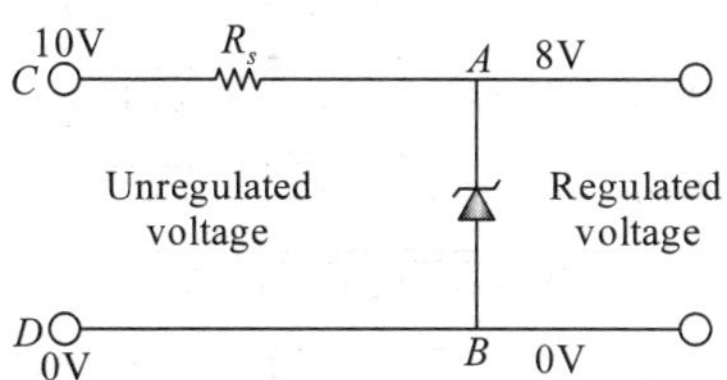

Current through zener diode is calculated as

$$i = \frac{P}{V} = \frac{1.6W}{8V} = 0.2\text{ A}$$

$$\Rightarrow \quad V_C - V_A = 10 - 8 = 2\text{ V}$$

$$\Rightarrow \quad R = \frac{V_C - V_A}{i} = \frac{2}{0.2} = 10\ \Omega$$

A zener diode can regulate only if input voltage is greater than zener breakdown voltage the range of input voltage should be 8 to 10 V so that output voltage remains constant = 8 V **Ans. (C)**

Sol. 77 Considering forward biased diodes, simplified circuit is drawn below.

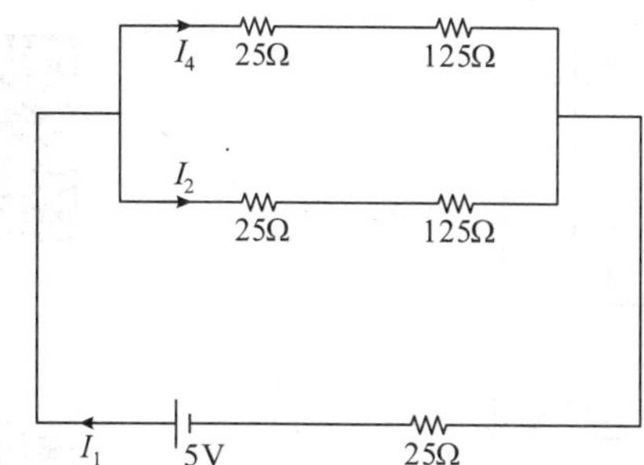

Equivalent resistance of this circuit is given as

$$R_{\text{equivalent}} = \frac{150 \times 150}{300} + 25 = 100\ \Omega$$

Current in circuit is calculated as

$$I_1 = \frac{5}{10} = 0.05\text{ A}$$

Current is equally divided as resistance of both branches are same

$$I_2 = I_4 = \frac{0.05}{2} = 0.025\text{ A}$$

$$\Rightarrow \quad \frac{I_1}{I_2} = 2$$ **Ans. (D)**

28.3 Logic Gates

Sol. 1

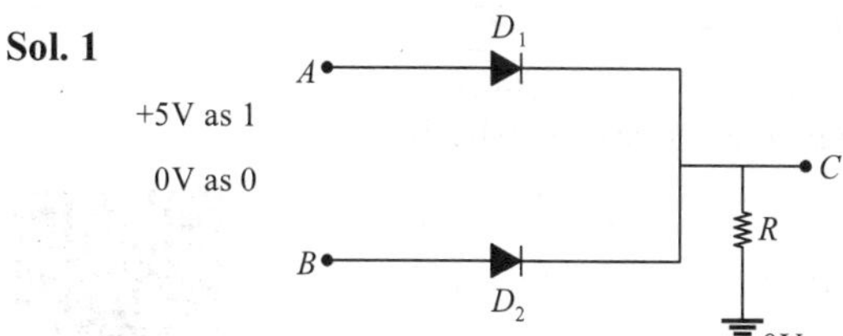

if current passes through $R \Rightarrow C$ is at high potential state (1)

Truth table of given circuit

A	B	C	
0	0	0	→ D_1, D_2 are RB
0	1	1	→ D_1 is RB & D_2 is FB
1	0	1	→ D_1 is FB & D_2 is RB
1	1	1	→ D_1, D_2 are FB

⇒ This is Truth table of OR gate. **Ans. (D)**

Sol. 2

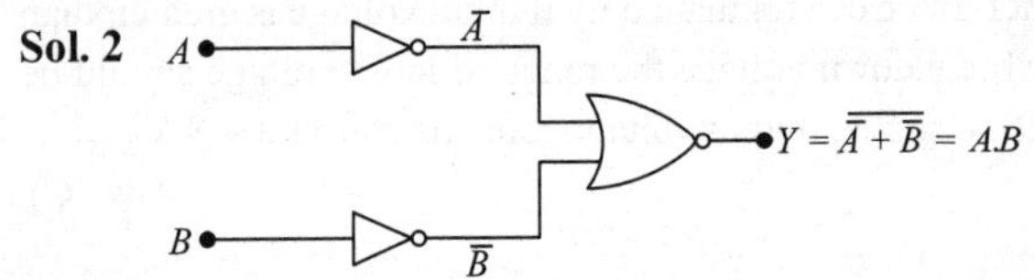

This logic circuit is equivalent to AND gate. The truth table used will be

A	B	Y
0	0	0
0	1	0
1	0	0
1	1	1

Ans. (D)

Sol. 3

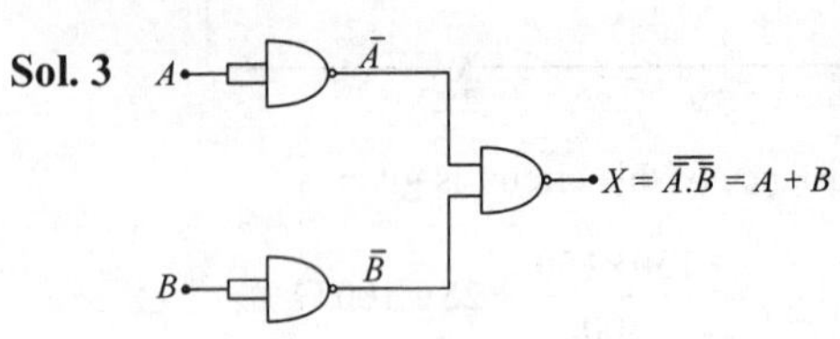

looking at output, this combination represents an OR gate.

Ans. (A)

Sol. 4

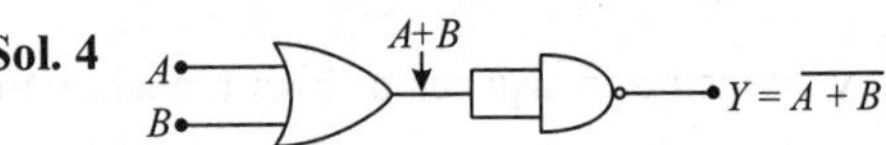

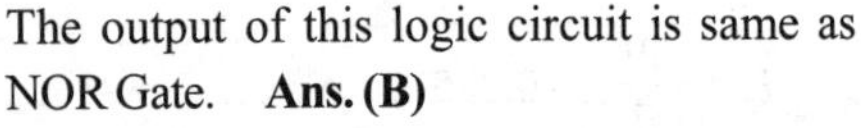

The output of this logic circuit is same as NOR Gate. **Ans. (B)**

Sol. 5

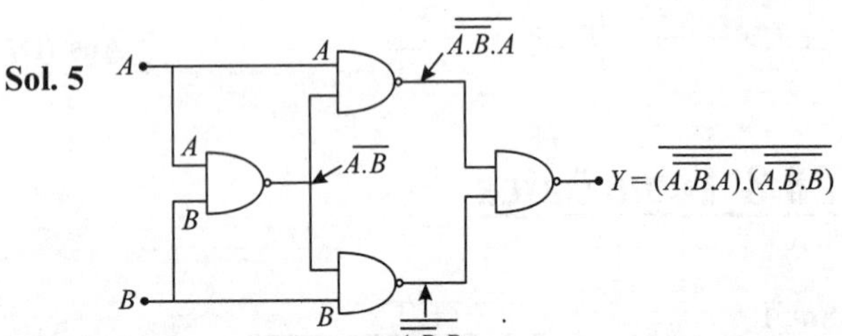

The simplified Boolean expression is

$$Y = \overline{\overline{A \cdot B} \cdot A} + \overline{\overline{A \cdot B} \cdot B}$$

for $A = 0, B = 0 \Rightarrow Y = 1.0 + 1.0 = 0 + 0 = 0$

for $A = 0, B = 1 \Rightarrow Y = 1.0 + 1.1 = 0 + 1 = 1$

for $A = 1, B = 0 \Rightarrow Y = 1.1 + 1.0 = 1 + 0 = 1$

for $A = 1, B = 1 \Rightarrow Y = 0.1 + 0.1 = 0 + 0 = 0$ **Ans. (A)**

Sol. 6 In the given circuit either of switches A or B is closed will make the lamp glow.

Inputs → if ON state of a switch is taken as '1' and OFF state is taken as '0'

output → lamp glow → '1'

lamp off → '0'

Truth table

A	B	Y
0	0	0
0	1	1
1	0	1
1	1	1

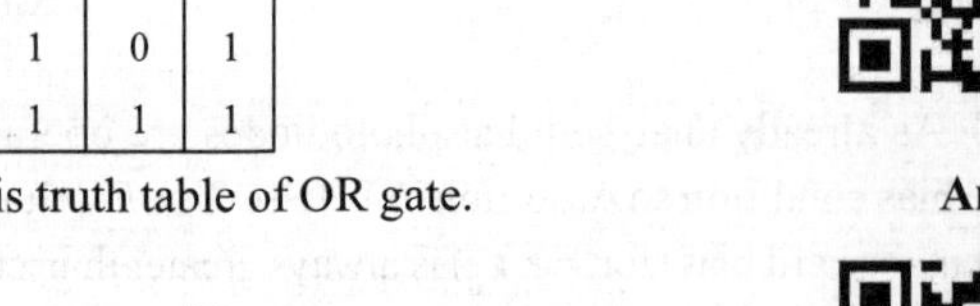

This is truth table of OR gate. **Ans. (B)**

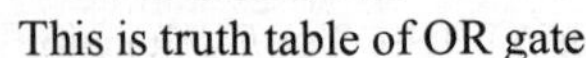

Sol. 7

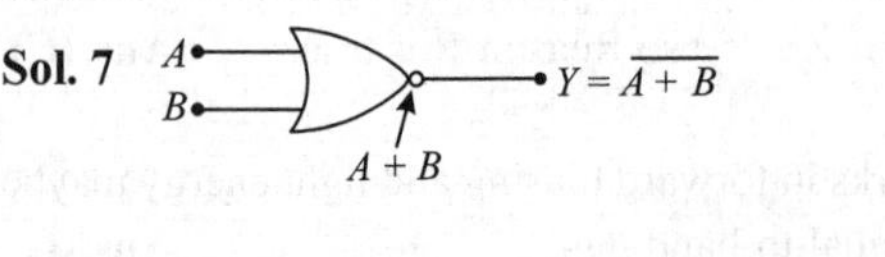

Ans. (A)

Sol. 8 Here the output of given system of gates is given as

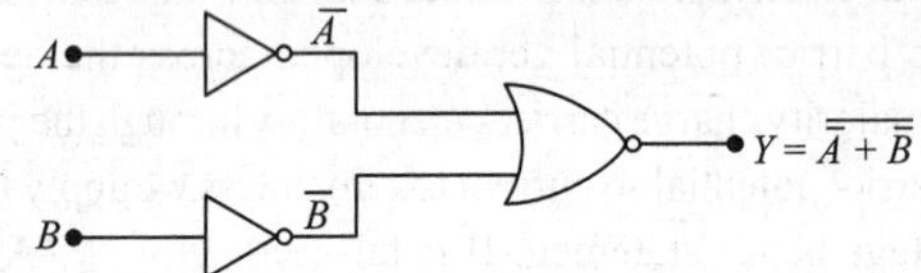

Truth table

A	B	Y
0	0	0
0	1	0
1	0	0
1	1	1

This is the truth table for AND gate. **Ans. (C)**

Sol. 9

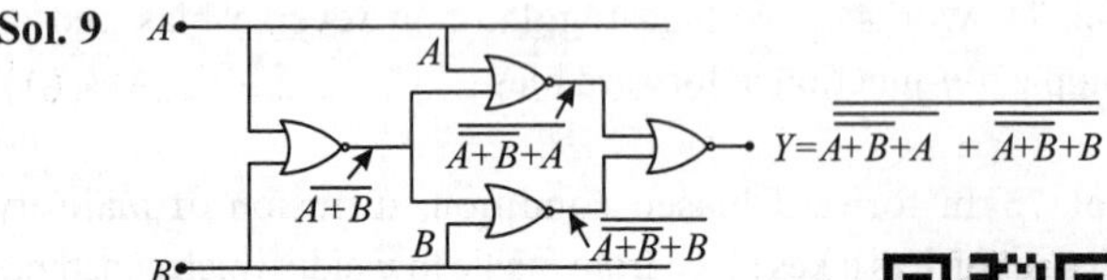

To make the truth table of this system we use

$A = 0, 0, 1, 1$

$B = 0, 1, 0, 1$

$$Y = \overline{\overline{\overline{A+B}+A} + \overline{\overline{A+B}+B}}$$

At $A = 0, B = 0 \Rightarrow$ $Y = \overline{\overline{\overline{1+0}+1} + \overline{\overline{1+0}}} = \overline{0+0} = 1$

At $A = 0, B = 1 \Rightarrow$ $Y = \overline{\overline{\overline{0+0}+0} + \overline{\overline{0+1}}} = \overline{1+0} = 0$

At $A = 1, B = 0 \Rightarrow$ $Y = \overline{\overline{\overline{0+1}+0} + \overline{\overline{0+0}}} = \overline{0+1} = 0$

At $A = 1, B = 1 \Rightarrow$ $Y = \overline{\overline{\overline{0+1}+0} + \overline{\overline{0+1}}} = \overline{0+0} = 1$ **Ans. (A)**

Sol. 10 (A) $C = \overline{A \cdot B + \bar{A}}$ (B) $C = (A + B) \cdot B$

(A)

A	B	C
0	0	0
0	1	0
1	0	1
1	1	0

(B)

A	B	C
0	0	1
0	1	0
1	0	1
1	1	1

(C) $C=(A \cdot B)+\overline{B}$

A	B	C
0	0	0
0	1	1
1	0	0
1	1	1

(D) $C=\overline{A.B}$

A	B	C
0	0	1
0	1	1
1	0	1
1	1	0

Ans. (A)

Sol. 11 In the given circuit if any input A or B will be at high potential +5 V (state-1) then transistor will be in ON state and C will get connected to earth (0 V) so C will be at stable 0. Only when A & B are 0 then C will be at high position (state-1)

Truth table of this circuit will be

A	B	Y
0	0	1
0	1	0
1	0	0
1	1	0

This is Truth table of NAND gate. **Ans. (A)**

Sol. 12

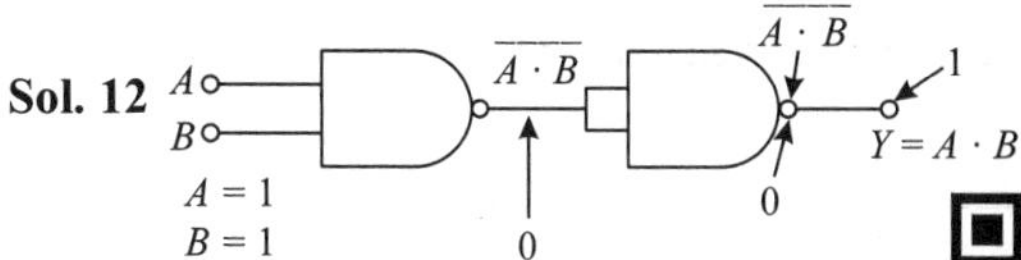

$Y=A.B$ is expression for AND gate.

Ans. (A)

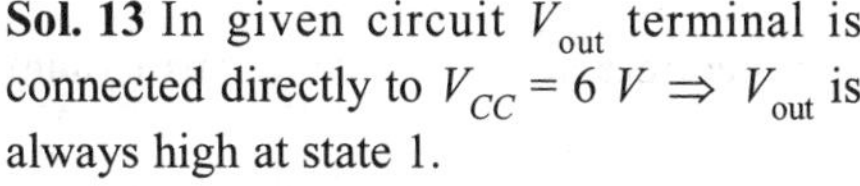

Sol. 13 In given circuit V_{out} terminal is connected directly to $V_{CC}=6\ V \Rightarrow V_{out}$ is always high at state 1.

So the output is at logic 1 irrespective of input states.

So here it is Not representing any logic gate.

$\Rightarrow$ None of the option is correct. **Ans. (None)**

Sol. 14 For A and B both zero, output Y is zero and if either of A or B is 1, output Y is 1.

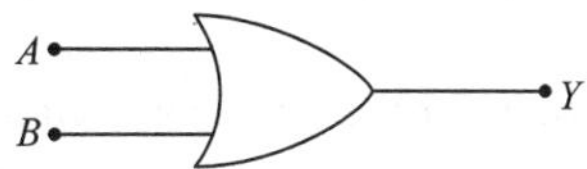

Ans. (A)

Sol. 15 Truth table for circuit

	a	b	c	Y
(A)	0	0	1	0
(B)	1	0	0	0
(C)	1	0	1	1
(D)	0	1	0	0

Boolean expression for the output of this gated circuit is

$$Y=(A+B)C$$

Ans. (C)

Sol. 16

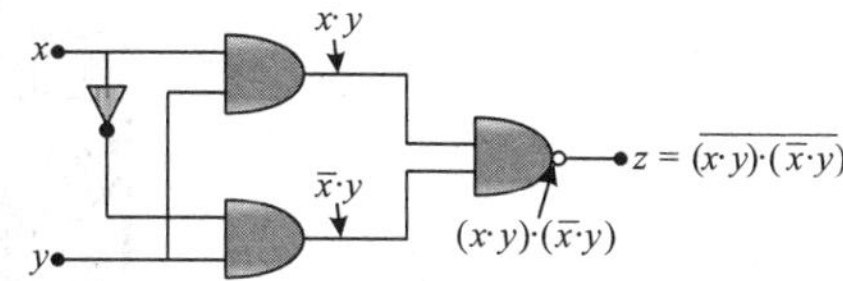

We calculate z for cases

$x=0, 0, 1, 1$

$y=0, 1, 0, 1$

output $\qquad z=\overline{(x.y).(\overline{x}.y)}$

At $x=0, y=0 \Rightarrow \quad z=\overline{0 \cdot 0}=1$

At $x=0, y=1 \Rightarrow \quad z=\overline{0 \cdot 1}=1$

At $x=1, y=0 \Rightarrow \quad z=\overline{0 \cdot 0}=1$

At $x=1, y=1 \Rightarrow \quad z=\overline{1 \cdot 0}=1$

Ans. (C)

Sol. 17 $\qquad P=\overline{X}+Y$

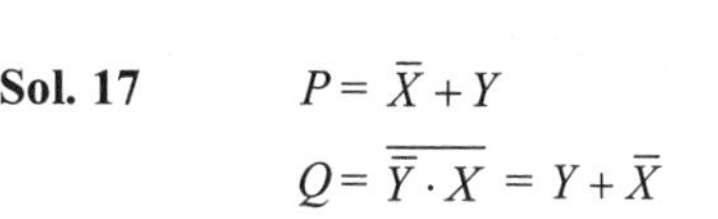

$$Q=\overline{\overline{Y} \cdot X}=Y+\overline{X}$$

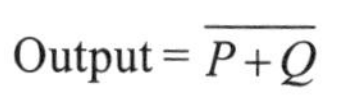

$$\text{Output}=\overline{P+Q}$$

To make output '1' $P+Q$ must be '0'

So, $Y=0$ and $X=1$ **Ans. (D)**

Sol. 18 The boolean output of given circuit is shown in figure below

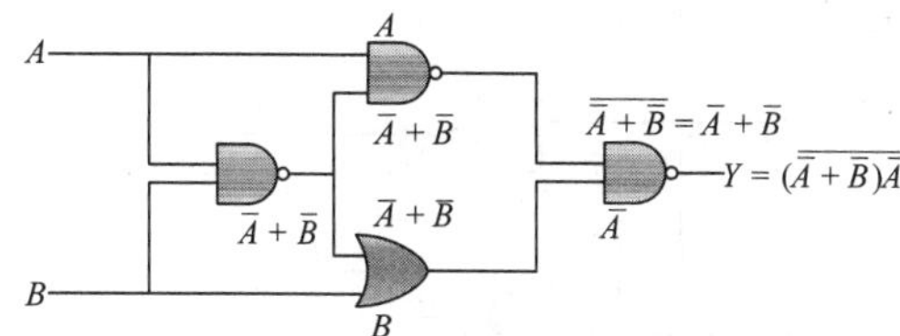

The output is simplified as

$$Y=\overline{(\overline{A}+\overline{B})\overline{A}}$$

$\Rightarrow \qquad Y=\overline{\overline{A}+\overline{A}B}$

$\Rightarrow \qquad Y=A(\overline{\overline{A}B})$

$\Rightarrow \qquad Y=A(A+\overline{B})$

$\Rightarrow \qquad Y=A+A\overline{B}=A\overline{B}$ **Ans. (B)**

Sol. 19 Truth table of the given logic circuit is written as

A	B	Equivalent
0	0	0
0	1	1
1	0	1
1	1	1

This is the table for 'OR' gate **Ans. (A)**

Sol. 20

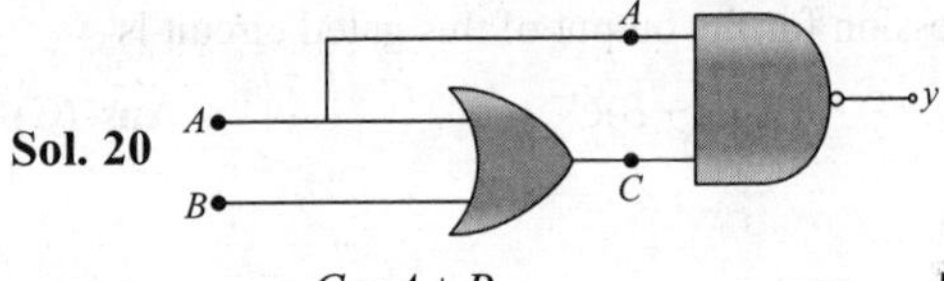

$$C = A + B$$

and $$y = \overline{A.C}$$

Truth table for this circuit is given as

A	B	$C = (A+B)$	$A.C.$	$y = \overline{A.C}$
0	0	0	0	1
0	1	1	0	1
1	0	1	1	0
1	1	1	1	0

Ans. (A)

Sol. 21 Reversible operation is only possible in one input, one output systems so only option is (B). **Ans. (B)**

Sol. 22 For the given circuit boolean expression of output Y is given as

$$Y = \overline{\overline{AB}.A}$$

$$\Rightarrow \quad Y = \overline{\overline{AB}} + \overline{A}$$

$$\Rightarrow \quad Y = 0 + 0$$

$$\Rightarrow \quad Y = 0$$

Ans. (B)

Sol. 23 In below figure intermediate outputs of gates are mentioned as C, D and E

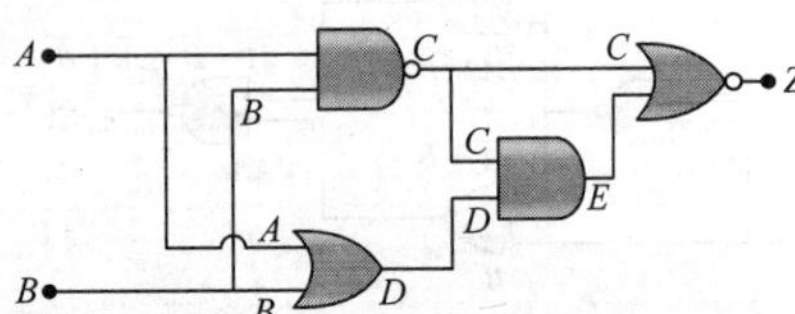

Relation between output z and inputs A and B are deduced as

$$Z = (\overline{C + E})$$

$$\Rightarrow \quad Z = (\overline{C + CD})$$

$$\Rightarrow \quad Z = (\overline{C(1+D)}) = \overline{C}$$

$$\Rightarrow \quad Z = \overline{\overline{AB}} = AB$$

Thus truth table for given circuit is made as shown in figure

A	B	Z
0	0	0
0	1	0
1	0	0
1	1	1

Ans. (C)

Sol. 24 The truth table of given circuit is shown below

A	B	C	
0	0	0	0
1	0	0	0
0	1	0	0
0	0	1	0
1	1	0	0
1	0	1	0
0	1	1	0
1	1	1	1

This corresponds to AND gate. **Ans. (A)**

Sol. 25 Truth table of given logic circuit is given as

A	B	Y
0	0	0
0	1	1
1	0	1
1	1	1

For the given input wavelength correct output wavelength is given as

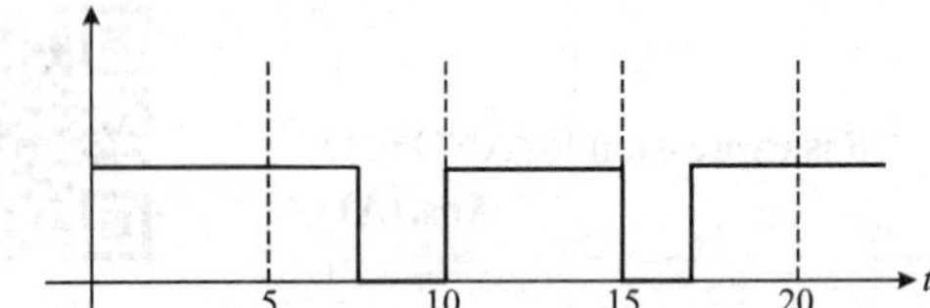

None of given option is correct. **Ans. (None)**

Sol. 26 The Boolean expression for output of the given logic circuit can be written as

$$\overline{A + \overline{B}} = C$$

Simplifying the expression, we get

$$C = \overline{A} \cdot B$$

This expression is realized by the circuit shown in option (D). **Ans. (D)**

Sol. 27 The boolean expression for the given logic circuit is written as

$$y = \overline{(A\overline{B} + \overline{A}B)}$$

$$\Rightarrow \quad y = \overline{A\overline{B}} \cdot \overline{\overline{A}B}$$

$$\Rightarrow \quad y = (\overline{A} + B) \cdot (A + \overline{B})$$

$$\Rightarrow \quad y = \overline{A}.A + \overline{A}.\overline{B} + A.B + B.\overline{B}$$

$$\Rightarrow \quad y = AB + \overline{A}\,\overline{B}$$

$$\Rightarrow \quad Y = A(\overline{AB}) + B(\overline{AB})$$

$$\Rightarrow \quad Y = A(\overline{A} + \overline{B}) + B(\overline{A} + \overline{B})$$

$$\Rightarrow \quad Y = 0 + A\overline{B} + B\overline{A} + 0$$

This represent XOR gate. **Ans. (D)**

Sol. 48 In transistor, emitter collector and base have different doping levels and collector is the thickest while base is thinnest segment. **Ans. (C)**

Sol. 49 Circuit is closed when neither A nor B is closed so current flows in the lamp for $A = 0$ and $B = 0$ and when either or both of A & B is closed we get current bypass through the switch due to short circuiting. Hence this is equivalent to "NOR" gate. **Ans. (B)**

Sol. 50 Truth table for the logic circuit is shown below

A	B	Y
0	0	0
0	1	1
1	0	1
1	1	1

According to truth table, resultant graph is shown below

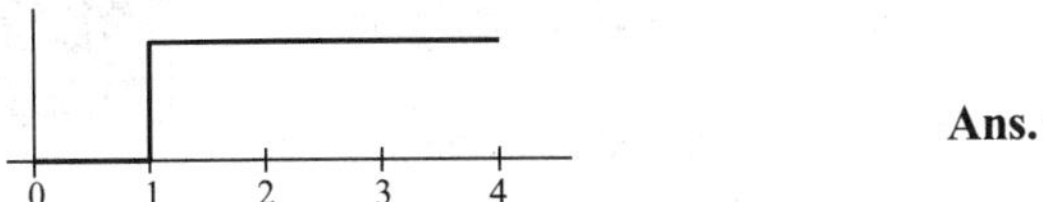

Ans. (D)

Sol. 51 Considering the transistor in saturation mode, we use

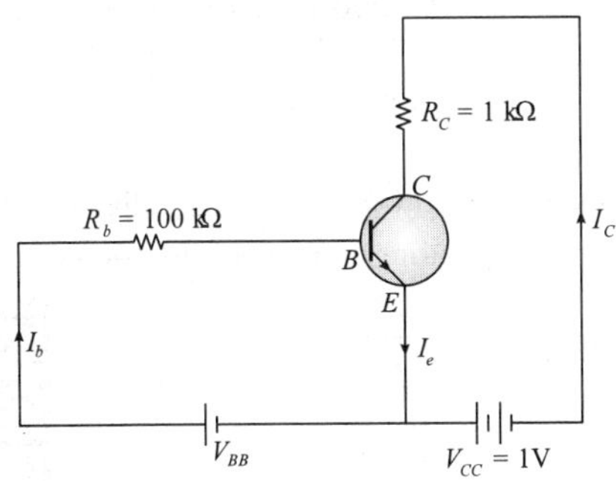

$$V_{CE} = 0$$

Using KVL gives

$$-I_c R_c + V_{CC} = 0$$

$$\Rightarrow \quad I_c = \frac{V_{CC}}{R_C}$$

$$\Rightarrow \quad I_c = 10^{-3}\text{ A}$$

Current gain is given as

$$\beta = \frac{I_c}{I_b}$$

$$\Rightarrow \quad I_b = \frac{10^{-3}}{100} = 10^{-5}\text{ A}$$

$$\Rightarrow \quad I_b = 10\ \mu\text{A}$$

Ans. (D)

Sol. 52 Figure below shows the ouputs of individula gates in the circuit given in question.

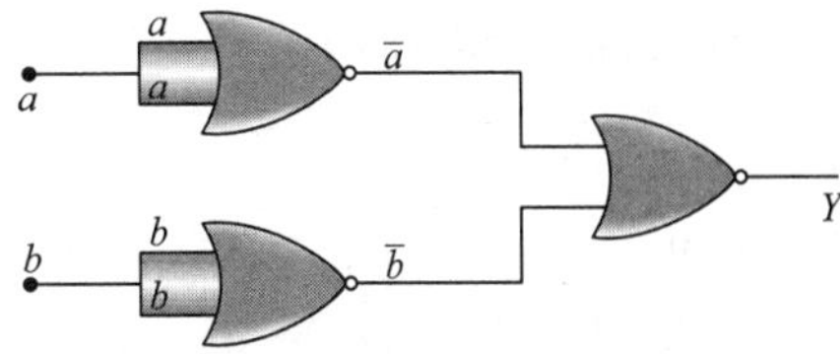

Final output of the circuit is given as

$$Y = \overline{\overline{a} + \overline{b}} = a \cdot b$$

The truth table for the given circuit can be drawn as shown in figure below.

a	b	output
0	0	0
0	1	0
1	0	0
1	1	1

Hence this is equivalent to AND gate. **Ans. (A)**

Sol. 53 Diagram below shows the terminal outputs of all gates in circuit.

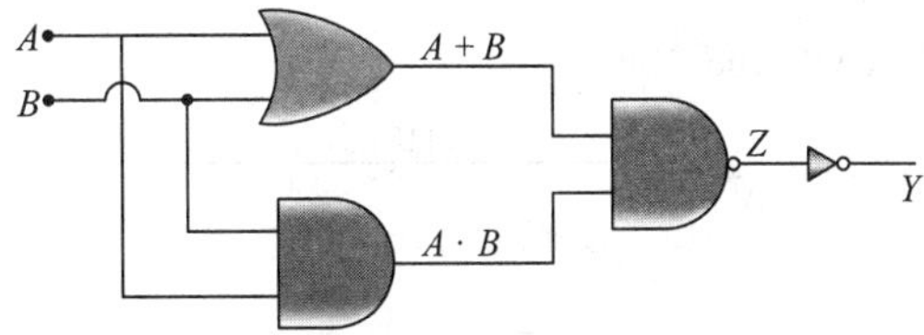

The final output of the circuit is given as

$$Z = \overline{(A+B)\cdot(A\cdot B)}$$

$$\Rightarrow \quad Y = \overline{Z} = (A+B)\cdot(A\cdot B)$$

$$\Rightarrow \quad Y = A \cdot B$$

Thus it is an AND Gate. **Ans. (A)**

Sol. 54 Current gain in common emitter transistor is given as

$$\beta = \frac{\Delta I_c}{\Delta I_B} = \frac{16\text{ mA} - 5\text{ mA}}{200\ \mu\text{A} - 100\ \mu\text{A}} = \frac{11\text{ mA}}{100\ \mu\text{A}} = 110$$ **Ans. (A)**

Sol. 55 By demorgan's theorem final output of the given logic circuit is written as

$$Y = \overline{\overline{A}\cdot B} = A + \overline{B}$$

Thus for given input answer will be option (A). **Ans. (A)**

Sol. 56 Power gain in common emitter configuration is calculated as

$$\Rightarrow \quad A_v A_I = B\frac{R_C}{R_B} \cdot B = B_2 \frac{R_C}{R_B}$$

$$\Rightarrow \quad A_v A_I = \left(\frac{(20-10)\times10^{-3}}{(200-100)\times10^{-6}}\right) \times \frac{1\times10^3}{10\times10^3} = 10^3$$

Hence $x = 3$ **Ans. (3)**

Sol. 57 Truth table for NAND gate is

A	B	$Y = \overline{A \cdot B}$
0	0	1
0	1	1
1	0	1
1	1	0

On the basis of given input A and B the truth table is

A	B	Y
1	1	0
0	0	1
0	1	1
1	0	1
1	1	0
0	0	1
0	1	1

Ans.(A)

29.1 Communication Systems

Sol. 1 (A) True

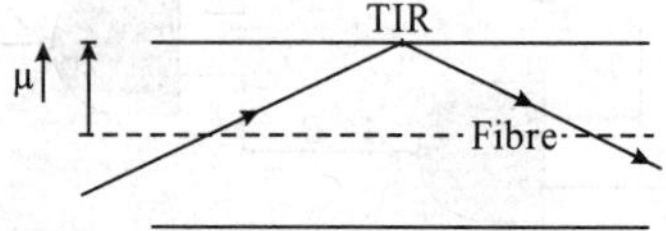

(B) Not true

As only by internal reflection light passes in optical fibre so outside interference is not possible.

(C) True

As absorption or scattering of light in OF is negligible.

(D) True

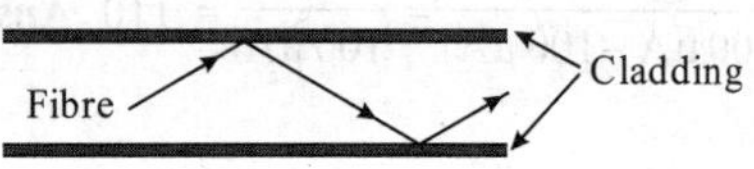

Ans. (B)

Sol. 2 **Statement-1 :** Due to energy absorption by Earth in 'Ground waves' the range of these is less so in sky wave range in relativity vary large as these are reflected from ionospheric layers.

$\Rightarrow$ True

Due to Change in state of ionization in ionosphere with temperature and other physical conditions the signals in sky waves are less stable.

$\Rightarrow$ **Statement-2 :** It is correct & explaining Statement-1 also. **Ans. (A)**

Sol. 3 $h = 500$ m

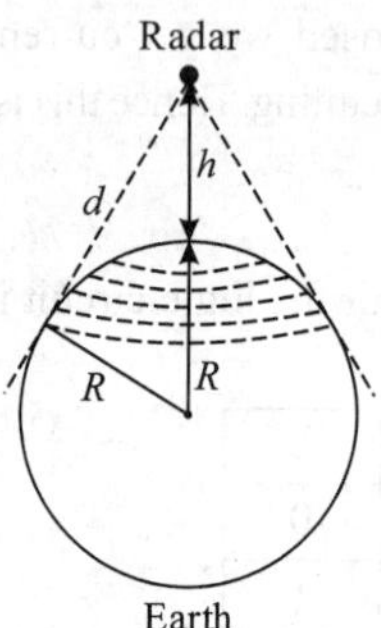

If d is the maximum distance from radar upto which objects can be detected on earth surface, we use

$$(h+R)^2 = d^2 + R^2$$

$$\Rightarrow \quad d = \sqrt{(h+R)^2 - R^2}$$

$$\Rightarrow \quad d = \sqrt{h^2 + 2hR}$$

$$\Rightarrow \quad d = \sqrt{(500)^2 + 2\times500\times6400\times10^3}$$

$$\Rightarrow \quad d = 500\sqrt{1 + 2\times6.4\times\frac{10^6}{500}}$$

$$\Rightarrow \quad d = 80000\text{ m}$$ **Ans. (A)**

Sol. 4 If a source of power P watt emits radiation of wavelength λ then number of photons emitted/s from source are

$$N = \frac{P}{hc/\lambda} = \frac{P\lambda}{hc}$$

$$\Rightarrow \quad N = \frac{10\times10^3\times500}{6.62\times10^{-34}\times3\times10^8}$$

$$\Rightarrow \quad N = \frac{5}{6.62\times3} \times 10^{34-8+6} = 0.25\times10^{32}$$

$$\Rightarrow \quad N = 2.5\times10^{31}$$ **Ans. (B)**

Sol. 5 Wavelength of EM waves transmitted is

$$\lambda = \frac{c}{\nu} = \frac{3\times10^8}{830\times10^3} \simeq 361\text{ m}$$

The electric field and magnetic field in EM waves are related as

$$\frac{E}{B} = c$$

$$\Rightarrow \quad E = Bc = 4.82 \times 10^{-11} \times 3 \times 10^{8}$$

$$\Rightarrow \quad E = 14.46 \times 10^{-3}$$

$$\Rightarrow \quad E = 0.01446 \text{ V/m}$$

Ans. (D)

Sol. 6 Statement-1 : True

As in sky wave propagation we use frequency upto 30 MHz. (short waves) and long distance communication is possible due to reflection by ionospheric layers.

Statement-2 : Not correct

In ionosphere → plasma. (free ions) The frequency which are above 30 MHz can penetrate the atmosphere → No total Internal reflection occurs because with frequency Refractive Index of ionosphere changes. **Ans. (A)**

Sol. 7 (A) In sky wave signals ionization status of ionosphere changes with time due to thermal agitation change in day, night, seasons & almost every hour. (Correct)

(B) frequency which are more than 30 MHz can penetrate atmospheric layers in ionosphere. (Correct)

(C) ionosphere penetration occurs at $f > 30$ MHz (Not correct)

(D) In night ionospheric ionization is more stable due to lesser temperature variation. (Correct) **Ans. (C)**

Sol. 8 In LOS mode the maximum distance of clear communication is given as

$$d_{max} = \sqrt{2Rh_T} + \sqrt{2Rh_R}$$

$$\Rightarrow \quad d_{max} = \sqrt{2 \times 6400 \times 10^3 \times 32} + \sqrt{2 \times 6400 \times 10^3 \times 50}$$

$$\Rightarrow \quad d_{max} = \sqrt{64 \times 64 \times 10^5} + \sqrt{6400 \times 10^5} = (64 + 80) \times \sqrt{10^5}$$

$$\Rightarrow \quad d_{max} = 144 \times \sqrt{10} \times 10^2$$

$$\Rightarrow \quad d_{max} = 144 \times 3.16 \times 100$$

$$\Rightarrow \quad d_{max} \simeq 45.5 \text{ km}$$

Ans. (B)

Sol. 9 For sky wave propagation of radio waves, frequency range is

$$\Delta f_{SW} \sim 2 \text{ MHz} \leftrightarrow 30 \text{ MHz}$$

Among options given only option (B) is in this range.

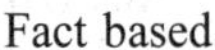

Fact based. **Ans. (B)**

Sol. 10 for sky wave propagation radio waves, frequency range is

$$\Delta f_{SW} \sim 2 \text{ MHz} \leftrightarrow 30 \text{ MHz}$$

Among the given options only option (B) is in range of this.

Fact based. **Ans. (B)**

Sol. 11 For resonating condition in antena for transmission of EM wave the antenna length should be of the order of λ.

Power transmission in EM waves is inversely proportional to λ^2

$$\Rightarrow \quad P_{eff} \propto \frac{1}{\lambda^2}$$

$$\Rightarrow \quad P_{eff} = K\left(\frac{1}{\lambda}\right)^2$$

Ans. (A)

Sol. 12 Maximum distance upto which signal can be broadcasted is

$$d_{max} = \sqrt{2Rh_T} + \sqrt{2Rh_R}$$

$$\Rightarrow \quad d_{max} = \sqrt{2 \times 6.4 \times 10^6}\left[\sqrt{140} + \sqrt{40}\right]$$

$$\Rightarrow \quad d_{max} = 65 \text{ km}$$

Ans. (D)

Sol. 13 As range of transmission on ground is

$$R \propto \sqrt{h}$$

⇒ To double the range h should be made 4 times. **Ans. (B)**

Sol. 14 Range of transmission in LOS communication is given as

$$\text{Range} = \sqrt{2Rh_T} + \sqrt{2Rh_R}$$

$$\Rightarrow \quad 50 \times 10^3 = \sqrt{2 \times 6400 \times 10^3 \times h_T} + \sqrt{2 \times 6400 \times 10^3 \times 70}$$

$$\Rightarrow \quad h_T = 32 \text{ m}$$

Ans. (C)

Sol. 15 The given question is based on conceptual facts. Optical fibre communication uses infrared light for signal transmission, radar uses radio waves, sonar uses ultrasound and mobile phones use microwaves for their signal transmission. **Ans. (D)**

Sol. 16 As it is given that sound level decreases by 5dB every km so sound level decreased in 20 km will be 100 dB. So we use

$$L_2 - L_1 = 10\log_{10}\frac{I_2}{I_1}$$

$$\Rightarrow \quad -100 = 10\log_{10}\frac{I_2}{I_1}$$

$$\Rightarrow \quad \frac{I_1}{I_2} = 10^{10}$$

$$\Rightarrow \quad I_2 = 10^{-10} I_1$$

$$\Rightarrow \quad P_2 = 10^{-10} P_1$$

$$\Rightarrow \quad P_2 = 10^{-10}\,\text{W}$$

$$\Rightarrow \quad P_1 = 10^{-10} \times 100 = 10^{-8}\,\text{W}$$

$$\Rightarrow \quad x = 8$$ **Ans. (8)**

Sol. 17 Resonant frequency of the circuit is given as

$$\omega_0 = \frac{1}{\sqrt{LC}}$$

$$\Rightarrow \quad 2\pi f_0 = \frac{1}{\sqrt{LC}}$$

$$\Rightarrow \quad 2\pi \frac{c}{\lambda} = \frac{1}{\sqrt{LC}}$$

$$\Rightarrow \quad 4\pi^2 \frac{c^2}{\lambda^2} = \frac{1}{LC}$$

$$\Rightarrow \quad \frac{4\times 10\times (3\times 10^8)^2}{(960)^2} = \frac{1}{L\times 2.56\times 10^{-6}}$$

$$\Rightarrow \quad \frac{1}{L} = \frac{4\times 10\times 9\times 10^{16}\times 2.56\times 10^{-6}}{960\times 960}$$

$$\Rightarrow \quad L = 10\times 10^{-8}\,\text{H}$$ **Ans. (10)**

Sol. 18 Signal Bandwidth is for AM broadcast is given as
$BW = 2 \times$ maximum frequency at modulating signal

$$= 2\times 5\text{kHz}$$

$$= 10\,\text{kHz}$$

Number of stations accommodated using this bandwidth in 90kHz total bandwidth are calculated as

$$N = \frac{90}{10} = 9$$ **Ans. (9)**

Sol. 19 Length of antenna tower is taken one fourth of the wavelength of transmission signal so we use

$$L = 25\text{m} = \frac{\lambda}{4}$$

$$\Rightarrow \quad \lambda = 100\,\text{m}$$ **Ans. (D)**

Sol. 20 For line of sight communication the distance between antennas is given as

$$D = 2\sqrt{2Rh}$$

$$\Rightarrow \quad h = \frac{D^2}{8R} = \frac{45^2}{8\times 6400}\,\text{km} \cong 39.55\,\text{m}$$ **Ans. (B)**

Sol. 21 Ground distance between antennas is given as

$$d = \sqrt{2Rh}$$

where $A = \pi d^2$, so we have

$$A = \pi 2Rh = 3.14\times 2\times 6400\times \frac{30}{1000}$$

$$\Rightarrow \quad A = 1205.76\,\text{km}^2 \sim 1206\,\text{km}^2$$ **Ans. (1206)**

Sol. 22 Bandwidth of the modulated signal is given as

$$w = 2f_m$$

Angular frequency of the signal is given as

$$\omega_m = 1.57\times 10^8 = 2\pi f_m$$

$$\Rightarrow \quad w = 2f_m = \frac{10^8}{2}\,\text{Hz} = 50\,\text{MHz}$$ **Ans. (D)**

Sol. 23 In order of atmospheric layers from bottom we have Troposphere, stralospherre, Mesospehre, Thermosphere.
Ans. (A)

Sol. 24 Antenna range in first case is calculated as

$$R_1 = \sqrt{2Rh}$$

Antenna range in nsecond case is calculated as

$$R_2 = \sqrt{2Rh} + \sqrt{2Rh'}$$

Where $h = 20\,\text{m}$ & $h' = 5\,\text{m}$

Percentage increase in range can be given as

$$\Delta = \frac{R_2 - R_1}{R_1}\times 100\%$$

$$\Rightarrow \quad \Delta = \frac{\sqrt{2Rh'}}{\sqrt{2Rh}}\times 100\%$$

$$\Rightarrow \quad \Delta = \frac{\sqrt{5}}{\sqrt{20}}\times 100\% = 50\%$$ **Ans. (50)**

Sol. 25 Transmitting distance is given as

$$d = \sqrt{2Rh}$$

$$\Rightarrow \quad h = \frac{d^2}{2R} = \frac{150\times 150\times 10^6}{2\times 6.5\times 10^6} = 1730.7 \approx 1731\,\text{m}$$

Population covered can be given as

$$p = 2\pi Rh \times \text{Population density}$$

$$\Rightarrow \quad p = 2\pi\times 6.5\times 10^6\times 1730.7\times 2000\times 10^{-6} \approx 1413\times 10^5$$

Ans. (A)

Sol. 26 Band width of the modulated signal is given as

$$BW = 2f_m = 2\times 10^5\,\text{Hz} = 200\,\text{kHz}$$ **Ans. (B)**

Sol. 27 From the given data, we can use

$$\frac{a}{10}=\frac{b}{10}=\frac{\mu A_c}{2}$$

$$\Rightarrow \quad \frac{a}{b}=1$$ **Ans. (1)**

Sol. 28 Range between transmitting and receiving antenna is given as

$$R=\sqrt{2Rh_1}+\sqrt{2Rh_2}=57.28\text{ km}$$ **Ans. (B)**

Sol. 29 Range on ground by line of sight is given as

$$R=\sqrt{2Rh_1}+\sqrt{2Rh_2}$$

$$\Rightarrow \quad R=\sqrt{2\times 6400\times 320}+\sqrt{2\times 6400\times 2000}$$

$$\Rightarrow \quad R\approx 224\text{ km}$$ **Ans. (224)**

Sol. 30 For better range and directivity height of antenna should be proportional to the wavelength of signal to be transmitted. The higher the antenna, better is the range and for and optimum antenna its height should be equal to its wavelength or longer than that so option (A) is most appropriate out of all given options. **Ans. (A)**

Sol. 31 Sum of heights of transmitting and receiving antennas is given as

$$h_r+h_t=160$$

Range of LOS communication is given as

$$R=\sqrt{2R}(\sqrt{h_r}+\sqrt{h_t})$$

$$\Rightarrow \quad R=\sqrt{2R}(\sqrt{h_r}+\sqrt{160-h_r})$$

For maximum range of LOS communication, we have

$$\frac{dR}{dh_r}=0=\sqrt{2}\,R\left(\frac{1}{2\sqrt{h_r}}-\frac{1}{2\sqrt{160-h_r}}\right)=0$$

$$\Rightarrow \quad h_r=h_t=80\text{ m}$$

$$\Rightarrow \quad R=\sqrt{2R}(\sqrt{80}+\sqrt{80})$$

$$\Rightarrow \quad R=2\sqrt{2\times 6400\times 10^3\times 80}=64\text{ km}$$

Ans. (64)

Sol. 32 Total bandwidth in terms of number of channels and modulating frequency can be calculated as

Band width = 2 × n × highest modulating frequency

$$\Rightarrow \quad 6\text{ MHz}=2\times n\times 6\text{ KHz}$$

$$\Rightarrow \quad n=\frac{6\times 10^6}{2\times 6\times 10^3}=500$$ **Ans. (500)**

Sol. 33

$$C'=\frac{C}{\sqrt{\mu_r\varepsilon_r}}=\frac{3\times 10^8}{\sqrt{6.25}}=\frac{3\times 10^8}{2.5}$$

$$f\lambda=\frac{3}{2.5}\times 10^8\text{ s}$$

$$\Rightarrow \quad f(5\times 10^{-3}\times 4)=1.25\times 10^8$$

$$f=6\text{ GHz}$$

So $f\approx 6$ **Ans. (6)**

Sol. 34 Optical fibre frequency range is 1 THz to 1000 THz.
Ans. (B)

Sol. 35

$$=[[\overline{A+A}]+[\overline{B+B}]]$$

$$Y=\overline{\bar{A}+\bar{B}}$$ (*D'*MORGAN LAW)

$$Y=AB$$ **Ans. (B)**

Sol. 36

$$LOS=\sqrt{2Rh_T}+\sqrt{2Rh_R}$$

$$=\sqrt{2R}(\sqrt{h_T}+\sqrt{h_R})$$

$$=\sqrt{2\times 64\times 10^5}\,(\sqrt{25}+\sqrt{49})$$

$$=192\sqrt{5}\times 10^2\text{ m}$$

$$K=192$$ **Ans. (192)**

Sol. 37 (a) For low frequency or high wavelength size of antenna required is high.

(b) EPR is low for longer wavelength.

(c) Yes we want to avoid mixing up signals transmitted by different transmitter simultaneously.

(d) Low frequency signals sent to long distance by superimposing with high frequency. **Ans. (C)**

Sol. 38

$$C=\frac{E_0}{B_0}=\frac{2.25}{1.5\times 10^{-8}}=1.5\times 10^8\text{ ms}^{-1}$$

$$t=\frac{6\times 10^3}{1.5\times 10^8}=4\times 10^{-5}\text{ s}$$ **Ans. (B)**

Sol. 39 Theory **Ans. (C)**

Sol. 40 For minimum impedance

$$X_L=X_C$$

$$\Rightarrow \quad \omega L=\frac{1}{\omega C}$$

$$\Rightarrow \quad L=\frac{1}{\omega^2 C}=10^{-1}\text{ H}$$

$$=100\text{ mH}$$ **Ans. (100)**

Sol. 41 Frequency at

$$1000\,\text{nm} = \frac{3\times10}{1000\times10^{-9}}$$

$\Rightarrow\ 3\times10^{14}$ Hz

Available for channel band width

$$= \frac{2}{100}\times3\times10^{14}$$

$\Rightarrow\ 6\times10^{12}$ Hz

Bandwidth for 1 channel = 8000 Hz

$$\Rightarrow \text{No. of channel} = \frac{6\times10^{12}}{8\times10^{3}} \Rightarrow \frac{600}{8}\times10^{7}$$

$$= 75\times10^{7}$$ **Ans. (B)**

Sol. 42 Ranged

$$d=\sqrt{2Rh}$$

$$d_2=2d_1$$

$$\sqrt{2Rh_2}=2\sqrt{2Rh_1}$$

$$h_2=4h_1=500\text{ m}$$

$$\Delta h=500\text{ m}-125\text{ m}=375\text{ m}$$ **Ans. (C)**

Sol. 43

$$d=\sqrt{2Rh}$$

$$d=\sqrt{2\times6400\times h\times10^{-3}}\ (h\text{ in }m)$$

$$\text{Area}=\pi d^2$$

$$=(\pi\times2\times6400\times h\times10^{-3})\text{ km}^2$$

$$6.03\times100000=100\times\pi\times2\times6400\times10^{-3}h$$

$$h=\frac{6.03\times10^{5}}{10\times\pi\times128}$$

$$h=150\text{ m}$$ **Ans. (150)**

Sol. 44 Given : Frequency

$$f_1=6\text{ MHz}$$

Frequency $f_2=10$ MHz

$$\lambda_1=\frac{c}{f_1}$$

$$\lambda_2=\frac{c}{f_2}$$

Wavelength bandwidth $=\lambda_2-\lambda_1=20$ m. **Ans. (B)**

Sol. 45

Let d be range

$$d^2=(h+R)^2-R^2$$

$$=h^2+R^2+2RH-R^2$$

$$d^2=h^2+2Rh$$

as $R>>>>h$

then $d\approx\sqrt{2Rh}$ ...(1)

Now, if coverage is to be increased 3 times

$$3d=\sqrt{2Rh'} \quad ...(2)$$

Divide 2 and 1 $\dfrac{3d}{d}=\sqrt{\dfrac{2Rh'}{2Rh}}$

$$9=\frac{h'}{h}$$

$$9h=h'$$

If $h=100$ m then tower of height 900 m is required. **Ans. (D)**

Sol. 46 The atmospheric layers in the given list are in order of Troposphere then D part of Stratosphere then E part of stratosphere then F_2 part of thermosphere . **Ans. (A)**

Sol. 47 Between two antennas, maximum line of sight distance is given as

$$d_M=\sqrt{2Rh_T}+\sqrt{2R.h_R}$$

$$\Rightarrow\quad d_M=2\times\sqrt{2\times6.4\times10^{6}\times80}=64\text{ km}$$ **Ans. (D)**

Sol. 48 (a) Loss of signal strength while propagating through a medium is called attenuation.

(b) Transducers converts one form of energy into another at transmitting and receiving ends of signal.

(c) Demodulation is the process of extraction of base signal from the carrier wave at receiving end.

(d) Repeater is the combination of receiver and transmitter and used to reduce attenuation for long range communication. **Ans. (D)**

Sol. 49 In amplitude modulation frequency of carrier wave remains unchanged. Thus minimum size of antenna should be $\frac{1}{4}$th of wavelength. **Ans. (D)**

Sol. 50 Range of transmission is given as

$$d=\sqrt{2Rh}$$

When height is increased by 21 percent, we have

$$h_2=h_1+\left(h_1\times\frac{21}{100}\right)=1.21\,h_1$$

$$\Rightarrow\quad d_2=\sqrt{2Rh_2}=\sqrt{2R(1.21)h_1}=1.1\sqrt{2Rh_1}$$

$$\Rightarrow\quad d_2=1.1\,d_1$$

Percentage increase in range is calculated as

$$\Rightarrow \qquad p = \frac{d_2 - d_1}{d_1} \times 100 = \left(\frac{d_2}{d_1} - 1\right) \times 100$$

$$\Rightarrow \qquad p = (1.1 - 1) \times 100 = 10\% \qquad \textbf{Ans. (C)}$$

Sol. 51 Height of transmitting antenna is 98 m.

$$h_T = 98\text{ m}, h_R = 0,$$

$$R = 6400\text{ km}$$

Range of transmitting antenna

$$d = \sqrt{2h_T \cdot R} + \sqrt{2h_R \cdot R}$$

$$\Rightarrow \qquad d = \sqrt{2 \times 98 \times 6400 \times 10^3} + 0 = \frac{112}{\sqrt{10}}\text{ km}$$

Area of communication is calculated as

$$S = \pi d^2$$

$$\Rightarrow \qquad S = 3.14 \times \frac{112^2}{10} = 3942\text{ km}^2 \qquad \textbf{Ans. (B)}$$

Sol. 52 Power radiated from a linear antenna of length l is given as

$$P \propto \left(\frac{l}{\lambda}\right)^2$$

Here l is length of antenna and λ is wavelength of wave.

Ans. (D)

Sol. 53 Line of sight distance from antenna to receive signal is given as

$$d_r = \sqrt{2h_r R}$$

Height of antenna required is calculated as

$$h_r = \frac{d_r^2}{2R}$$

$$\Rightarrow \qquad h_r = \frac{(4km)^2}{2(6400km)} = \left(\frac{1}{800}\right)\text{km}$$

$$= 1.25\text{ m} \qquad \textbf{Ans. (A)}$$

Sol. 54 The uplink frequency band for satellite communication is 5.925 to 6.422 MHz. **Ans. (B)**

Sol. 55 The atmospheric layers are at specific height above the earth's surface. F_1 layer is in range from 170 to 190km. D-layer is from 60 - 80 km. Troposphere is the lowest layer of atmosphere upto around 10km range and E-layer is above D layer at around 100 km height. **Ans. (D)**

Sol. 56 Minimum length of antenna should be $\frac{\lambda}{4}$. **Ans. (B)**

Sol. 57 Maximum line of sight distance

$$d_{\text{max.}} = \sqrt{2Rh_t} + \sqrt{2Rh_r}$$

$$\Rightarrow \qquad d_{\text{max}} = \sqrt{2 \times 64 \times 10^5 \times 180} + \sqrt{2 \times 64 \times 10^5 \times 245}$$

$$\Rightarrow \qquad d_{\text{max}} = \{(8 \times 6 \times 10^3) + (8 \times 7 \times 10^3)\}\text{ m}$$

$$\Rightarrow \qquad d_{\text{max}} = (48 + 56)\text{ km}$$

$$\Rightarrow \qquad d_{\text{max}} = 104\text{ km} \qquad \textbf{Ans. (D)}$$

29.2 Modulation and Demodulation

Sol. 1 Modulation $\rightarrow$ using high frequency carrier frequency for long distance communication of modulation signals because of noise in direct transmission of these signals.

(A) $\rightarrow$ Speed of EM wave remain same even after modulation. Time lag between T_x and R_x is

$$t = \frac{\text{distance}}{c} \rightarrow \text{remain same (Not correct)}$$

(B) $\rightarrow$ After modulation frequency increases $\Rightarrow$ wavelength decreases $\Rightarrow$ size of antenna required reduce. (Correct)

(C) $\rightarrow$ $\Delta\omega = 2\omega_m$ and fractional bandwidth $= \frac{\Delta\omega}{\omega_c} = \frac{2\omega_m}{\omega_c} \downarrow$ (Correct)

(D) $\rightarrow$ As fractional band width on spectrum decreases $\Rightarrow$ selectivity increases. (Correct) **Ans. (A)**

Sol. 2 $E = E_c \cos\omega_c t + E_m \cos\omega_m t \cos\omega_c t$

$$\Rightarrow \quad E = E_c \cos\omega_c t + \frac{E_m}{2}\cos(\omega_c - \omega_m)t + \frac{E_m}{2}\cos(\omega_c + \omega_m)t$$

where $\omega_c \rightarrow$ Carrier frequency

$(\omega_c - \omega_m) \rightarrow$ lower side band frequency

$(\omega_c + \omega_m) \rightarrow$ upper side band frequency

Ans. (B)

Sol. 3 Noise $\rightarrow$ unwanted signal which is caught from atmosphere and distort the transmitting signal.

It is low frequency (or higher wavelengths)

To avoid (easily extract) noise upto maximum limit the signal frequency is to be kept high that's why modulation is needed.

Tolerance of modulated signal against noise (Noise tolerance) is high for a high frequency wave (short wave). **Ans. (B)**

Sol. 4 During de-modulation the maximum frequency which can be detected is given as

$$f_{\text{max}} = \frac{1}{2\pi\mu RC}$$

$$\Rightarrow f_{max} = \frac{1}{2 \times 3.14 \times 0.6 \times 100 \times 10^3 \times 250 \times 10^{-12}}$$

$$\Rightarrow f_{max} = \frac{10^8}{6.28 \times 6 \times 250} = \frac{1000 \times 10^5}{6.28 \times 6 \times 250}$$

$\Rightarrow f_{max} \simeq 10^4\text{ Hz} \simeq 10\text{ kHZ}$ **Ans. (B)**

Sol. 5

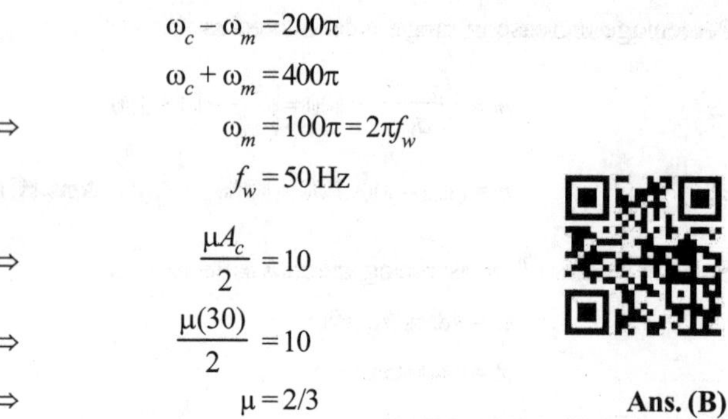

Where $R = A_c + A_m \sin\omega_m t$

$\Rightarrow$ Output as modulated wave is given as

$$C_m(t) = (A_c + A_m \sin\omega_m t)\sin\omega_c t$$

$$= A(1 + \sin\omega_m t)\sin\omega_c t$$

Here modulation index $\mu = \frac{A_m}{A_c} = \frac{A}{A} = 1$ **Ans. (C)**

Sol. 6 When a modulating signal of frequency ω_m is modulated on a carrier wave of frequency ω_c then the resulting modulated signal will have three frequencies

(1) Carrier frequency $\rightarrow \omega_c \rightarrow 2000$ kHz

(2) lower side band frequency $\rightarrow \omega_c \rightarrow 1995$ kHz

(2) upper side band frequency $\rightarrow \omega_c + \omega_m \rightarrow 2005$ kHz **Ans. (A)**

Sol. 7 Carrier wave is used for long distance propagation of signal using modulation.

here modulating signal $\rightarrow$ audio signal

if carrier wave is $y_c(t) = A_c \sin\omega_c t$

in AM, amplitude of carrier wave is varied in proportion to amplitude of MW (MW = audio signal modulating wave)

in FM, frequency of carrier wave is varied in proportion to amplitude of MW. **Ans. (C)**

Sol. 8 The modulated signal is

$$C_m(t) = 30 \sin 300\,\pi t + 10 \cos 200\,\pi t - 10 \cos 400\,\pi t \quad \ldots(1)$$

If is of the form

$$C_m(t) = A_c \sin\omega_c t + \frac{\mu A_c}{2}\cos(\omega_c - \omega_m)t - \frac{\mu A_c}{2}\cos(\omega_c + \omega_m)t \ldots(2)$$

Comparing-(1) and (2) we get

$$A_c = 30$$

$$\Rightarrow \quad \omega_c = 300\pi = 2\pi f_c$$

$$\Rightarrow \quad f_c = 150\text{ Hz}$$

$$\omega_c - \omega_m = 200\pi$$

$$\omega_c + \omega_m = 400\pi$$

$$\Rightarrow \quad \omega_m = 100\pi = 2\pi f_w$$

$$f_w = 50\text{ Hz}$$

$$\Rightarrow \quad \frac{\mu A_c}{2} = 10$$

$$\Rightarrow \quad \frac{\mu(30)}{2} = 10$$

$\Rightarrow \quad \mu = 2/3$ **Ans. (B)**

Sol. 9 Human speech signal range ~ 200 Hz $\leftrightarrow$ 2700 Hz

$$(\Delta f_{HS}) = 2500\text{ Hz}$$

Music signal range ~ 10200 Hz $\leftrightarrow$ 15200 Hz

$$\Delta f_{MS} = 5000\text{ Hz}$$

In case both the signals are transmitted together than overall frequency range. 200 Hz $\leftrightarrow$ 15200 Hz

The frequency range for both signals will be $\Delta f_{both} = 15000$Hz

$$\frac{\Delta\omega_{both}}{\Delta\omega_{MS}} = \frac{\Delta f_{both}}{\Delta f_{MS}} = \frac{15000}{2500} = 6$$ **Ans. (C)**

Sol. 10 If carrier frequency is ω_c than side band frequencies as

lower side band $\omega_1 = \omega_c - \omega_m$

upper side band $\omega_2 = \omega_c + \omega_m$

in transmission of modulated signal all frequencies between $\omega_1 \leftrightarrow \omega_2$ are transmitted.

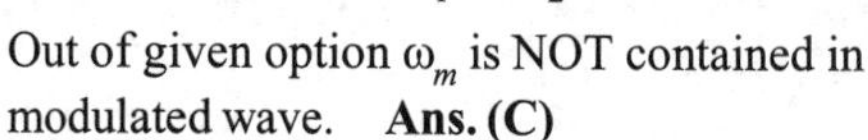

Out of given option ω_m is NOT contained in modulated wave. **Ans. (C)**

Sol. 11 Modulation index

$$m = \frac{V_{0m}}{V_{0c}} = \frac{5}{25} = \frac{1}{5} = 0.2$$

Modulation frequency $f_m = 20$ kHz

Carrier frequency $f_c = 1.2\text{ MHz} = 1200\text{ kHz}$

Side band frequencies as

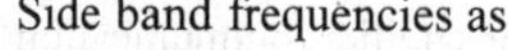

$$f_{LSB} = f_c - f_m = 1200 - 20 = 1180\text{ kHz}$$

and

$$f_{USB} = f_c + f_m = 1200 + 20 = 1220\text{ kHz}$$

Ans. (D)

Sol. 12 Modulation Index

$$m = \frac{V_{0m}}{V_{0c}}$$

$$\Rightarrow \quad V_{0m} = m \times V_{0c}$$

$\Rightarrow \quad V_{0m} = 0.8 \times 14 = 11.2\text{ V}$ **Ans. (A)**

Sol. 13 Total transmission band width available is

$$= \frac{10}{100} \times 10 \times 10^9 \text{ Hz} = 10^9 \text{ Hz}$$

Each channel requires and individual Band width = 5×10^3 Hz

If the total Band width can accommodate N channels then we use

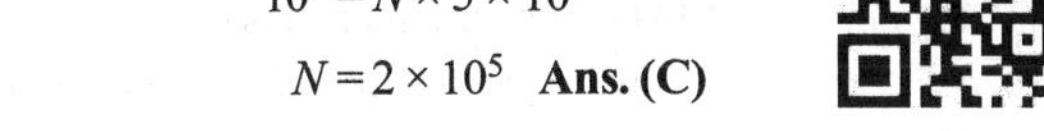

$$10^9 = N \times 5 \times 10^3$$

$$\Rightarrow \quad N = 2 \times 10^5 \quad \textbf{Ans. (C)}$$

Sol. 14 The frequency of carrier wave is given as

$$f_c = \frac{1}{2\pi\sqrt{LC}}$$

$$\Rightarrow \quad f_c = \frac{1}{2\pi\sqrt{49 \times 10^{-6} \times 2.5 \times 10^{-9}}}$$

$$\Rightarrow \quad f_c = \frac{1}{2\pi \times 7 \times 5 \times 10^{-8}} = \frac{10^8}{70\pi}$$

$$f_c = 454.54 \text{ kHz}$$

The frequency range used for transmission of this signal is

$$\Delta f = 2f_m$$

$$f_1 \leftrightarrow f_2 = f_c \pm f_m$$

$$\Rightarrow \quad f_1 = f_c - f_m = 454.54 - 12 = 442.54 \text{ kHz}$$

$$\Rightarrow \quad f_2 = f_c + f_m = 454.54 + 12 = 466.54 \text{ kHz} \qquad \textbf{Ans. (C)}$$

Sol. 15 Total bandwidth available for transmission is

$$\Delta BW = 300 \times 10^3 \text{ Hz}$$

Band width of each channel is $CBW = 2f_m = 2 \times 15 \times 10^3$

$$\Rightarrow \quad CBW = 30 \times 10^3 \text{ Hz}$$

If N are the number of channels accommodated for transmission, we have

$$300 \times 10^3 = N \times 30 \times 10^3$$

$$\Rightarrow \quad N = 10 \quad \textbf{Ans. (B)}$$

Sol. 16 Frequency of source is given as

$$f = \frac{3 \times 10^8}{8 \times 10^{-7}} = \frac{30}{8} \times 10^{14} \text{ Hz}$$

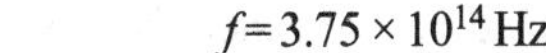

$$\Rightarrow \quad f = 3.75 \times 10^{14} \text{ Hz}$$

$$1\% \text{ of } f = 0.0375 \times 10^{14} \text{ Hz}$$

$$\Rightarrow \quad \Delta f = 3.75 \times 10^{12} \text{ Hz} = 3.75 \times 10^6 \text{ MHz}$$

$\Rightarrow$ Number of channels

$$= \frac{3.75 \times 10^6}{6} = 6.25 \times 10^5 \qquad \textbf{Ans. (D)}$$

Sol. 17 $f_{\text{carrier}} = \frac{250}{0.1}$ kHz

$\Rightarrow$ Range of signal = 2250 Hz to 2750 Hz

In given options; for 2000 kHz

$$f_{\text{mod}} = 200 \text{ Hz}$$

So its range is 1800 Hz to 2200 Hz

Which is not overlapping with the existing range. **Ans. (B)**

Sol. 18 The given signal can be written as

$$V(t) = 10 \sin B + \frac{3}{2}[2 \cos A \sin B]$$

$$\Rightarrow \quad V(t) = 10 \sin B + 10 + \frac{3}{2}[\sin(A + B) - \sin(A - B)]$$

$$\Rightarrow \quad V(t) = 10 \sin B (5.5 \times 10^5 t) + \frac{3}{2}[\sin(57.2 \times 10^4 t)]$$

$$- \sin(52.8 \times 10^4 t)$$

Here $\omega_1 = 57.2 \times 10^4 = 2\pi f_1$

$$\Rightarrow \quad f_1 = \frac{57.2 \times 10^4}{2 \times \left(\frac{22}{7}\right)} = 9.1 \times 10^4$$

$$\Rightarrow \quad f_1 \simeq 91 \text{ kHz}$$

and $\omega_2 = 52.8 \times 10^4 t = 2\pi f_2$

$$\Rightarrow \quad f_2 = \frac{52.8 \times 10^4}{2 \times \left(\frac{22}{7}\right)}$$

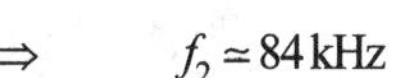

$$\Rightarrow \quad f_2 \simeq 84 \text{ kHz} \qquad \textbf{Ans. (C)}$$

Sol. 19 Analysis of graph shows

(1) Amplitude varies between
$8 - 10\,V$ or 9 ± 1

(2) Two time period are
100 μs (modulating wave) & 8 μs (carrier wave)

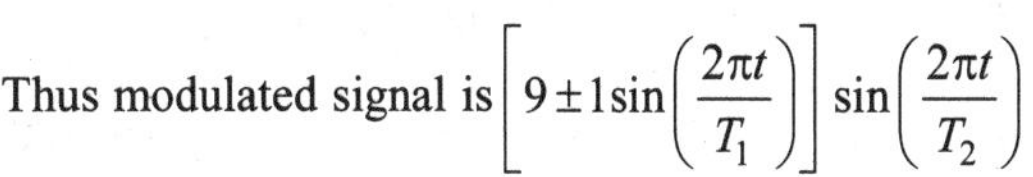

Thus modulated signal is $\left[9 \pm 1 \sin\left(\frac{2\pi t}{T_1}\right)\right] \sin\left(\frac{2\pi t}{T_2}\right)$

$$= [9 \pm 1 \sin(2\pi \times 10^4 t)] \sin(2.5\pi \times 10^5 t) \qquad \textbf{Ans. (D)}$$

Sol. 20 Given that by modulation carrier wave amplitude changes maximum upto 160 V, so we use

$$A_m + A_c = 160$$

$$\Rightarrow \quad A_m + 100 = 160$$

$$\Rightarrow \quad A_m = 60$$

Modulation index is

$$\mu = \frac{A_m}{A_c} = \frac{60}{100}$$

$\Rightarrow \quad \mu = 0.6$ **Ans. (A)**

Sol. 21 To minimise attenuation, wavelength of carrier waves is close to 1500 nm **Ans. (D)**

Sol. 22 AM signal contains one carrier frequency and two (ω_0) side band frequencies

$$USB = \omega_0 + \omega$$

$$LSB = \omega_0 - \omega$$ **Ans. (B)**

Sol. 23 The physical size of antenna of reciver and transmitter both directly proportional to the wavelength of carrier wave so both are inversely proportional to carrier frequency. **Ans. (C)**

Sol. 24 Modulation inxex

$$= \frac{V_{0m}}{V_{0c}} = \frac{100}{400} = 0.25$$

USB frequency $= f_c + f_m$

LSB frequency $= f_c - f_m$

$USB - LSB = 2f_m = 2 \times 100 \times 10^6 = 2 \times 10^8$ Hz **Ans. (D)**

Sol. 25 Modulation index

$$m = \frac{A_m}{A_c} = \frac{2}{4} = 0.5$$

Carrier frequency

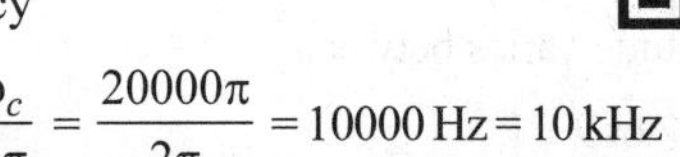

$$f_c = \frac{\omega_c}{2\pi} = \frac{20000\pi}{2\pi} = 10000\text{ Hz} = 10\text{ kHz}$$

Modulating frequency

$$f_m = \frac{\omega_m}{2\pi} = \frac{2000\pi}{2\pi} = 1000\text{ Hz} = 1\text{ kHz}$$

LSB frequency $= f_c - f_m = 10 - 1 = 9$ kHz **Ans. (A)**

Sol. 26 Modulated signed voltage is given as

$$V_m = 5\,(1 + 0.6\cos(6280t)\sin(2\pi \times 10^4 t)$$

$$V_m = 5\sin(2\pi \times 10^4 t) + 3\cos(6280t)\sin(2\pi \times 10^4 t)$$

In above wave maximum amplitude is

$$V_{max} = 5 + 3 = 8\text{ V}$$

$$V_{min} = 5 - 3 = 2\text{ V}$$ **Ans. (None)**

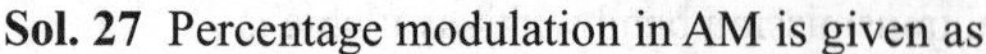

Sol. 27 Percentage modulation in AM is given as

$$p = \frac{Am}{Ac} \times 100$$

$$\Rightarrow \quad p = \frac{20}{80} \times 100 = 25\%$$ **Ans. (25)**

Sol. 28 Band width requirement for FM is twice the modulation frequency i.e. $2f_m = 4$ kHz thus Statement-I is correct. Sideband frequencies are $= f_c \pm f_m$ which are (1000 ± 2) kHz or 998 kHz & 1002 kHz. Thus Statement-II is also correct. **Ans. (A)**

Sol. 29 The wavelength of signal in air will be at carrier frequency only so it is given as

$$\lambda = \frac{c}{f_c}$$ **Ans. (D)**

Sol. 30 Modulation index for the modulated wave is given as

$$m = \frac{A_{max} - A_{min}}{A_{max} + A_{min}}$$

$$m = \frac{16-8}{16+8} = \frac{8}{24} = \frac{1}{3} = 0.33$$

$$\Rightarrow \quad x \times 10^{-2} = 0.33$$

$$\Rightarrow \quad x = 33$$ **Ans. (33)**

Sol. 31 Maximum voltage in the transmitting signal can be given as

$$V_{max} = A_C + A_m$$

$$\Rightarrow \quad 200 = 160 + A_m$$

$$\Rightarrow \quad A_m = 40\text{ V}$$ **Ans. (40)**

Sol. 32 Modulation index of the wave is calculated as

$$\mu = \frac{A_m}{A_c} = \frac{20}{20} = 1$$ **Ans. (1)**

Sol. 33 Modulation index of AM wave is given as

$$\mu = \frac{A_{max} - A_{min}}{A_{max} + A_{min}}$$

$$\mu = \frac{12-3}{12+3} = 0.6$$

$$\Rightarrow \quad x = 1$$ **Ans. (1)**

Sol. 34 The amplitude modelated wave is generally represented as

$$C_m = Ac\left[1 + \frac{A_m}{A_C}\sin 2\pi f_m t\right]$$

Comparing with the given signal, we have

$$2\pi f_m = 12560$$

$$\Rightarrow \quad f_m = 2000\text{ Hz} = 2\text{ KHz}$$ **Ans. (2)**

Sol. 35 Maximum and minimum amplitude of modulated signal are given as

$$A_{\max} = A_c + A_m = 250 + 150 = 400$$

and $$A_{\min} = A_c - A_m = 250 - 150 = 100$$

$$\Rightarrow \quad \frac{A_{\min}}{A_{\max}} = \frac{100}{400} = \frac{1}{4} = \frac{50}{200}$$

$$\Rightarrow \quad x = 200$$

Ans. (200)

Sol. 36 $f_c = 3.5$ GHz

$f_m = 3.5$ MHz

Side band frequencies are $f_e - f_m$ & $f_c + f_m$. Which are almost f_c

$$\lambda = \frac{c}{f_c}$$

Minimum length of antenna

$$= \frac{c}{f_c 4} = \frac{\lambda}{4} = \frac{3\times10^8}{3.5\times10^9\times4} = 21.4 \text{ mm}$$

Ans. (C)

Sol. 37 In amplitude modulation the amplitude of high frequency carrier wave is varied in accordance with message signal. **Ans. (C)**

Sol. 38 $y(t) = 40 \sin(10 \times 10^6 \pi t)$

$x(t) = 20 \sin(1000 \pi t)$

$\Rightarrow \quad \omega_c = 10^7 \pi$

$\omega_m = 10^3 \pi$

$A_C = 40$

$A_m = 20$

Equation of modulated wave

$$= (A_C + A_m \sin \omega_m t) \sin \omega_c t$$

$$= A_c\left(1 + \frac{A_m}{A_c} \sin \omega_m t\right) \sin \omega_c t$$

$$= A_c(1 + \mu \sin \omega_m t) \sin \omega_c t,$$

$$\mu = \frac{A_m}{A_c}$$

$$= A_c \sin \omega_c t + \frac{\mu A_c}{2}$$

$$[\cos(\omega_c - \omega_m)t - \cos(\omega_c + \omega_m)t]$$

Amplitude of minimum frequency

$$= \frac{\mu A_c}{2} = \frac{A_m}{A_c} \times \frac{A_c}{2} = \frac{A_m}{2} = 10$$

Ans. (D)

Sol. 39 Bandwidth $= 2f_m$

$= 2 \times 10^4$ Hz $= 20 \times 10^3$ Hz

$= 20$ kHz **Ans. (C)**

Sol. 40 Modulation index

$$= \frac{V_{\max} - V_{\min}}{V_{\max} + V_{\min}} \times 100\%$$

$$= \frac{6-2}{6+2} \times 100\% = 50\%$$

Ans. (D)

Sol. 41 $V_{\max} = 60$

$V_{\min} = 20$

$$\% \text{ modulation} = \left(\frac{V_{\max} - V_{\min}}{V_{\max} + V_{\min}}\right) 100$$

$$\Rightarrow \left(\frac{60-20}{60+20}\right) 100 \Rightarrow \left(\frac{40}{80}\right) 100$$

$\Rightarrow$ 50% **Ans. (B)**

Sol. 42 Modulation Index

$$\mu = \frac{A_m}{A_C} = \frac{1}{5} = 0.2$$

A_C = amp. of carrier wave. **Ans. (A)**

Sol. 43 $$\mu = \frac{A_m}{A_c}$$

$$\mu \le 1$$

to avoid distortion because $\mu > 1$ will result in interference between career frequency & message frequency. **Ans. (A)**

Sol. 44 Given

FM broadcast

Modulating frequency $= 20$ k Hz $= f$

$$\text{Deviation ratio} = \frac{\text{frequency deviation}}{\text{modulating frequency}} = \frac{\Delta f}{f}$$

$\Rightarrow$ frequency deviation $- \Delta f = f \times 10$

$= 20$ kHz $\times 10 = 200$ kHz

$\Rightarrow$ Bandwidth $= 2(f + \Delta f)$

$= 2(20 + 200)$ kHz

$= 440$ kHz **Ans. (D)**

Sol. 45 Modulation index :

$$m = \frac{A_m}{A_c}$$

Given $2A_m = 8$

$A_m + A_c = 9$

$\Rightarrow \quad A_c = 5$

$$\Rightarrow \quad m = \frac{4}{5} = 0.8$$

Ans. (A)

Sol. 46 Frequencies present in output of square law device

$2f_c, f_c+f_m, f_c, f_c-f_m, 2f_m, f_m$

After passing through band bass filte

f_c+f_m, f_c, f_c-f_m

$$\text{Band width} = 2f_m$$

$$= \frac{2\omega_m}{2\pi} = \frac{6.28\times10^6}{3.14}$$

$$= 2\text{ MHz}$$ **Ans. (2)**

Sol. 47 Modulation index of a signal is given as

$$\mu = \frac{\text{Amplitude of modulating signal}}{\text{Amplitude of carrier wave}}$$

$$\Rightarrow \quad \mu = \frac{1}{2}$$ **Ans. (D)**

Sol. 48 (a) AM Broadcast frequency is in range 540 – 1600 KHz

(b) FM Broadcast frequency is in range 88 – 108 MHz

(c) Television broadcast frequency is in range 54 – 890 MHz

(d) Satellite communication frequency is in range 3.7 – 4.2 GHz

Ans. (A)

Sol. 49 For the given signal frequency of 5 kHz and carrier wave frequency 2 MHz. The resultant signal will have band with of frequency given as

$$\Delta f = [(f_c+f_m)-(f_c-f_m)]$$

$$\Rightarrow \quad \Delta f = [(2000+5)-(2000-5)]\text{ kHz}$$

$$\Rightarrow \quad \Delta f = 10\text{ kHz}$$ **Ans. (C)**

Sol. 50 Modulation index of the given AM wave is written as

$$\mu = \frac{A_{\max} - A_{\min}}{A_{\max} + A_{\min}}$$

$$\Rightarrow \quad \mu = \frac{14-6}{14+6} = 0.4$$ **Ans. (B)**

Sol. 51 From the given data, we have

$$A_c + A_m = 120$$

$$A_c - A_m = 80$$

Solving above equation, we get

$$A_c = 100$$

and $$A_m = 20$$

Modulation index is given as

$$\mu = \frac{20}{100} = \frac{1}{5}$$

Amplitude of each sideband is calculated as

$$\mu = A_c \times \frac{\mu}{2}$$

$$\Rightarrow \quad \mu = 100 \times \frac{1}{10} = 10\text{ V}$$ **Ans. (B)**

Sol. 52 In the given carrier wave, frequency is calculated as

$$v_C = \frac{100\pi}{2\pi} = 500\text{ Hz}$$

Modulating wave frequency is given as

$$v_m = \frac{4\pi}{2\pi} = 2\text{ Hz}$$

Side band and carrier frequencies are given as

$$v_C - v_m,\ v_C,\ v_C + v_m$$

$$= 498\text{ Hz}. 500\text{ Hz}, 502\text{ Hz}.$$ **Ans. (B)**

Sol. 53 *FM* broadcast range is 88 MHz to 108 MHz. **Ans. (B)**

Sol. 54 Modulation Index for a communication system is given as

$$\mu = \frac{A_m}{A_c}$$

When modulating signal having amplitude of xV is superimposed with carrier signal of amplitude YV, the modulation index is calculated as

$$\Rightarrow \quad \mu_1 = \frac{X}{Y}$$

When modulating signal of xV is superimposed with carrier signal of $2YV$

$$\mu_2 = \frac{X}{2Y}$$

So ratio of μ_1 and μ_2

$$\frac{\mu_1}{\mu_2} = \frac{2}{1}$$ **Ans. (C)**

Sol. 55 Modulation index of wave is given as

$$\mu = \frac{A_m}{A_c} = 0.6$$

Minimum amplitude of modulated wave is given as

$$A_{min} = A_c - A_m = 3$$

$$\Rightarrow \quad A_c - 0.6A_c = 3$$

$$\Rightarrow \quad 0.4A_c = 3$$

$$\Rightarrow \quad A_c = \frac{3}{0.4} = \frac{15}{2} = 7.5\text{ V}$$

$$\Rightarrow \quad A_m = 0.6A_c = 4.5\text{ V}$$

Maximum amplitude is given as

$$A_{mzx} = A_c + A_m$$

$$\Rightarrow \quad A_{mzx} = 7.5 + 4.5 = 12\text{ V}$$ **Ans. (B)**

Sol. 56 Given amplitudes of career and modulating wave are

$$A_c = 15\,\text{V}$$

and $$A_m = 3\,\text{V}$$

Maximum amplitude of modulated wave is given as

$$A_{\text{max.}} = A_c + A_m = 15 + 3 = 18\,\text{V}$$

Minimum amplitude of modulated wave is given as

$$A_{\text{min}} = A_c + A_m = 15 + 3 = 12\,\text{V}$$

The ratio of maximum to minimum amplitude is given as

$$\frac{A_c + A_m}{A_c - A_m} = \frac{18}{12} = \frac{3}{2}$$ **Ans. (B)**

Sol. 57 Bandwidth $= 2f_m$

$\Rightarrow$ $$\Delta f = 2 \times 3\,\text{kHz}$$

$\Rightarrow$ $$\Delta f = 6\,\text{kHz}$$ **Ans. (D)**

Sol. 58 Equation of Carrier wave is given as

$$c(t) = 15\sin(1000\,\pi t)$$

Frequency of career wave is calculated as

$$f_i = \frac{\omega_c}{2\pi} = \frac{1000\pi}{2\pi} = 500\,\text{Hz}$$

Equation of modulated wave is given as

$$m(t) = 10\sin(4\,\pi t)$$

Frequency of modulated wave is calculated as

$$f_m = \frac{\omega_m}{2\pi} = \frac{4\pi}{2\pi} = 2\,\text{Hz}$$

Frequencies contained in resultant amplitude modulated wave are calculated as

$$f_{AM} = (500 - 2)\,\text{Hz},\ 500\,\text{Hz and } (500 + 2)\,\text{Hz}.$$

Thus correct ans is (D). **Ans. (D)**

30.1 Errors & Experiments

Sol. 1 Least count of main scale = 1mm

$$\text{least count of circular scale} = \frac{\text{Pitch}}{\text{No. of CSD}} = \frac{0.5\,\text{mm}}{50} = 0.01\text{mm}$$

[As 2 turns of CS gives 1mm displacement on MS]

Diameter of wire = MS + CS − ZE

$$= 3\,\text{mm} + 35 \times 0.01\,\text{mm} - (-0.03\,\text{mm})$$

$$= 3 + 0.35 + 0.03 = 3.38\,\text{mm}$$ **Ans. (D)**

Sol. 2 Momentum of body is given as

$$p = mv = \underbrace{3.513}_{4SD} \times \underbrace{5.00}_{3SD}$$

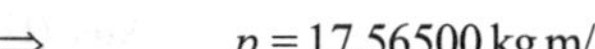

$\Rightarrow$ $$p = 17.56500\,\text{kg m/s}$$

Rounding off to 3SD, we get

$$p = 17.6\,\text{kg m/s}$$ **Ans. (A)**

Sol. 3 Given that 29 MSD = 30 VSD

$\Rightarrow$ $$1\text{VSD} = \frac{29}{30}\text{MSD}$$ [1 MSD = 0.5°]

Least count of Instrument

$$\text{LC} = 1\,\text{MSD} - 1\,\text{VSD}$$

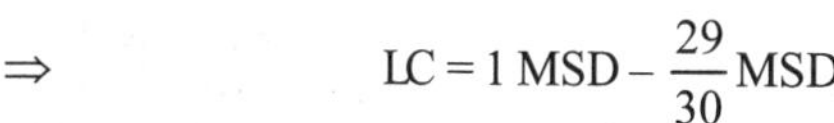

$\Rightarrow$ $$\text{LC} = 1\,\text{MSD} - \frac{29}{30}\text{MSD}$$

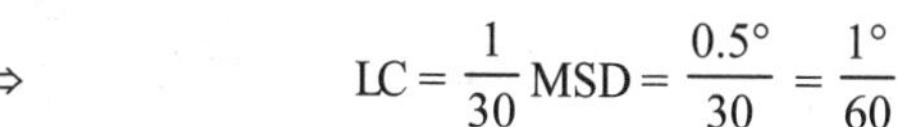

$\Rightarrow$ $$\text{LC} = \frac{1}{30}\text{MSD} = \frac{0.5^\circ}{30} = \frac{1^\circ}{60}$$

$\Rightarrow$ $$\text{LC} = 1'\,(\text{one minute})$$ **Ans. (D)**

Sol. 4

	No of SD
1st figure $\underset{12\ 345}{23.023}$	5
2nd figure $\underset{1}{0.0003}$	1
2nd figure $\underset{1\ 2}{2.1 \times 10^{-3}}$	2

Ans. (A)

Sol. 5 Least count of circular scale

$$= \frac{\text{Pitch}}{\text{no. of division}}$$

$$= \frac{1\,\text{mm}}{100} = 0.01\,\text{mm}$$

$$= 0.001\,\text{cm}$$

Diameter of wire = MS + CS

$$= 0 + 52 \times 0.001 = 0.052\,\text{cm}$$ **Ans. (A)**

Sol. 6 Least count of Main Scale = 0.5°

$$\text{least count of Vernier Scale} = \frac{1}{30} \times 0.5^\circ = \frac{1^\circ}{60}$$

$$\text{Prism Angle} = 58.5\,(\text{MS}) + 9(\text{VS}) \times \frac{1}{60}$$

$$= 58.5^\circ + 0.15^\circ = 58.65^\circ$$ **Ans. (C)**

Sol. 7 By Ohm's law we use $V = IR$

$\Rightarrow$ $$R = V/I$$

% (max) error in measurement of R is

$$100 \times \frac{dR}{R} = \left(\frac{dV}{V} + \frac{dI}{I}\right) \times 100$$

$\Rightarrow$ $$\frac{dR}{R} \times 100 = 3 + 3 = 6\%$$ **Ans. (A)**

Sol. 8 (A) 5.3200 cm $\rightarrow$ LC = 0.0001 cm

(B) 5.3 cm $\rightarrow$ LC = 0.1 cm

(C) 5.32 cm $\rightarrow$ LC = 0.01 cm

(D) 5.320 cm $\rightarrow$ LC = 0.001 cm **Ans. (D)**

Sol. 9 Least count = 1 MSD – 1 VSD

If we consider a and b are 1 MSD and 1 VSD lengths then we use

$$Na = (N+1)b \Rightarrow b = \frac{Na}{N+1}$$

$$LC = a - \frac{Na}{N+1} = \frac{a}{N+1}$$ **Ans. (D)**

Sol. 10 The length measured is 3.50 cm.

least count $\rightarrow$ 0.01 cm

(A) Meter scale LC = 0.1 cm

(B) Vernier Calliper LC = $\frac{1}{10}$ mm = 0.1 mm = 0.01 cm

(C) Screw Gauge LC = $\frac{1}{100}$ mm = 0.01 mm = 0.001 cm

(D) Screw Gauge LC = $\frac{1}{50}$ mm = 0.02 mm = 0.002 cm **Ans. (B)**

Sol. 11 The current is given as

$$I = (e^{1000V/T} - 1) \text{ mA} \rightarrow \text{at } I = 5 \text{ mA we have}$$

$$e^{1000V/T} = 6 \text{ mA}$$

$$dI = \frac{1000}{T} e^{1000V/T} dV \text{ mA}$$

$$\Rightarrow \quad dI = \frac{1000}{300} \times 6 \times 0.01 \times 10^{-3}$$

$$\Rightarrow \quad dI = 2 \times 10^{-4} \text{ A} = 0.2 \text{ mA}$$ **Ans. (A)**

Sol. 12 For a simple pendulum time period is given as

$$T = 2\pi\sqrt{\frac{L}{g}}$$

$$\Rightarrow \quad g = \frac{4\pi^2 L}{T^2}$$

Maximum relative error in measurement of g is given as

$$\frac{dg}{g} = \frac{dL}{L} + 2\frac{dT}{T}$$

$$\% \frac{dg}{g} \times 100 = \left(\frac{0.1}{20} + 2 \times \frac{1}{90}\right) \times 100$$

$$= (0.005 + 0.0222) \times 100 = 2.72\%$$ **Ans. (B)**

Sol. 13 If λ be the potential gradient on potentiometer wire than we use $V_s = \lambda l_1$

$$\Rightarrow \quad 1.1 = \lambda(440) \Rightarrow \lambda = \frac{1.1}{440} = \frac{1}{400} \text{ V/cm}$$

If V_R be the potential difference across the resistance we use for balancing length $l_2 = 220$ cm

$$V_R = \lambda l_2$$

$$\Rightarrow \quad V_R = \frac{1}{400} \times 220 = \frac{22}{40} = 0.55 \text{ V}$$

As reading of voltmeter is 0.5 V so error in reading is

$$\varepsilon = 0.5 - 0.55 = -0.05 \text{ V}$$ **Ans. (D)**

Sol. 14 (a) Rotation period of earth, $T = 24$ Hz $= 86400$ s $\sim 10^5$ s

(b) Revolution period of earth, $T = 365 \times 86400 \sim 10^7$ s

(c) Period of light wave ($\lambda \rightarrow 4000 - 8000$ Å)

$$T = \frac{\lambda}{c} = \frac{6\times10^{-7}}{3\times10^{8}} = 2 \times 10^{-15} \sim 10^{-15} \text{ s}$$

(d) Period of sound wave ($\nu \rightarrow 20 - 20000$ Hz)

$$T = \frac{1}{\nu} = \frac{1}{1000} = 10^{-3} \text{ s}$$ **Ans. (A)**

Sol. 15 Using the expression of time period of simple pendulum, g is calculated as

$$g = \frac{4\pi^2 L}{T^2}$$

% relative error in measurement of g is given as

$$\frac{dg}{g} \times 100 = \frac{dL}{L} \times 100 + 2\frac{dT}{T} \times 100$$

$$\Rightarrow \quad \frac{dg}{g} \times 100 = \frac{0.01}{20} \times 100 + 2 \times \frac{1}{90} \times 100$$

$$= (0.005 + 0.0222) \times 100$$

$$= 0.0272 \times 100 = 2.72\%$$ **Ans. (D)**

Sol. 16 Least count of VS is $LC_V = \frac{0.1}{10} = 0.01$ cm

The diameter readings are

(1) $0.5 + 8 \times 0.01 - (-0.03) = 0.61$ cm R_1

(2) $0.5 + 4 \times 0.01 - (-0.03) = 0.57$ cm R_2

(3) $0.5 + 6 \times 0.01 - (-0.03) = 0.59$ cm R_3

mean diameter of ball is

$$D = \frac{R_1 + R_2 + R_3}{3} = \frac{0.61 + 0.57 + 0.59}{3}$$

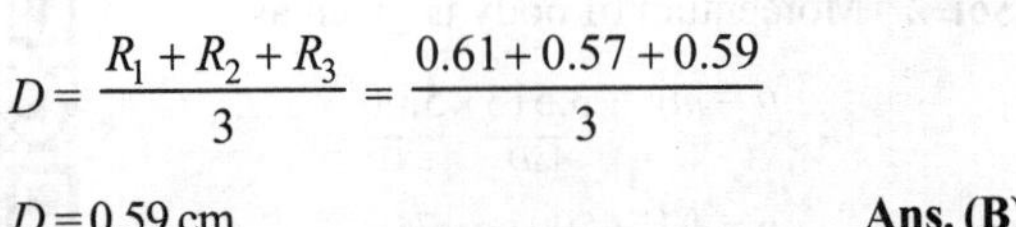

$$\Rightarrow \quad D = 0.59 \text{ cm}$$ **Ans. (B)**

Sol. 17 Mean value of measurement

$$T_m = \frac{T_1 + T_2 + T_3 + T_4}{4} = \frac{90+91+95+92}{4}$$

$$\Rightarrow \quad T_m = 92 \text{ s}$$

Deviation values of measurements from mean value is

$$\Delta T_1 = |T_1 - T_m| = 2\text{s}$$

$$\Delta T_2 = |T_2 - T_m| = 1\text{s}$$

$$\Delta T_3 = |T_3 - T_m| = 3\text{s}$$

$$\Delta T_4 = |T_4 - T_m| = 0\text{s}$$

Mean deviation of the data set from mean value is

$$\Delta T_m = \frac{\Delta T_1 + \Delta T_2 + \Delta T_3 + \Delta T_4}{4}$$

$$\Delta T_m = \frac{2+1+3+0}{4} = 1.5 \text{ s}$$

mean time measurement

$$T_m \pm \Delta T_m = 92 \pm 1.5 \text{ s}$$ **Ans. (C)**

Sol. 18 Main scale least count (1 division) = 0.5 mm

Circular scale lest count $= \frac{0.5}{50}$ mm $= 0.01$ mm

zero error = 5 division on the –ve side of scale

$= 5 \times 0.01 \text{ mm} = 0.05 \text{ mm}$

Thickness of Al scale Reading = 0.5 (MS) mm + 25 (CS) × 0.01

$= 0.5 + 0.25 = 0.75$ mm

As sheet thickness = 0.75 – (–0.05) = 0.80 mm **Ans. (D)**

Sol. 19 For the given relation of surface tension, the relative % error in its measurement is given as

$$\frac{dT}{T} = \frac{dr}{r} + \frac{dh}{h} = \frac{dD}{D} + \frac{dh}{h} \quad \ldots(1)$$

In measurement of diameter and water rise the least count of readings are

$$dr = \frac{dD}{2} = \frac{0.01\times10^{-2}}{2}\text{m}$$

and $dh = 0.01 \times 10^{-2}$ m

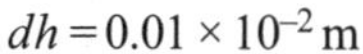

from equation-(1) we use

$$\% \text{ error} = \frac{dT}{T} \times 100 = \left(\frac{dD}{D} + \frac{dh}{h}\right) \times 100$$

$$\Rightarrow \quad \frac{dT}{T} \times 100 = \left(\frac{0.01\times10^{-2}}{1.25\times10^{-2}} + \frac{0.01\times10^{-2}}{1.45\times10^{-2}}\right) \times 100$$

$$\Rightarrow \quad \frac{dT}{T} \times 100 = \frac{1}{1.25} + \frac{1}{1.45} = 0.8 + 0.69\%$$

$= 1.49\%$ **Ans. (D)**

Sol. 20 From the given expression the measurement of P gives its relative maximum % error as

$$\frac{dP}{P} = \frac{1}{2}\left(\frac{da}{a}\right) + 2\left(\frac{db}{b}\right) + 3\left(\frac{dc}{c}\right) + 4\left(\frac{dd}{d}\right)$$

$$\Rightarrow \quad \% \text{ error in } P = \frac{1}{2} \times 2 + 2 \times 1 + 3 \times 3 + 4 \times 5$$

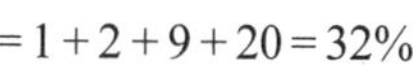

$= 1 + 2 + 9 + 20 = 32\%$

Ans. (C)

Sol. 21 Expression for A is

$$A = \frac{P^3 Q^3}{\sqrt{R}S}$$

Maximum relative % error in A is given as

$$\frac{dA}{A} \times 100\% = \left(3\frac{dP}{P} + 2\frac{dQ}{Q} + \frac{1}{2}\frac{dR}{R} + \frac{dS}{S}\right) \times 100\%$$

$$\Rightarrow \quad \% \text{ error in } A = 3 \times 0.5 + 2 \times 1 + \frac{1}{2} \times 3 + 1.5$$

$= 1.5 + 2 + 1.5 + 1.5 = 6.5\%$ **Ans. (D)**

Sol. 22 5 revolution displacement of 0.25 cm

$$\Rightarrow \quad 1 \text{ main scale division} = \frac{0.25}{5} = 0.05 \text{ cm}$$

$$\text{least count of Screw gauge} = \frac{1 \text{ main scale division}}{\text{no. of circular scale division}}$$

$$= \frac{0.05}{100} = 5 \times 10^{-4} \text{ cm}$$

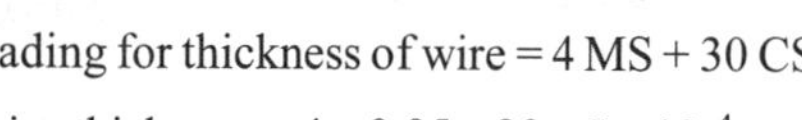

Reading for thickness of wire = 4 MS + 30 CS

wire thickness $= 4 \times 0.05 + 30 \times 5 \times 10^{-4}$ cm

$= 0.2 + 0.015 = 0.215$ cm **Ans. (D)**

Sol. 23 If R is radius of a sphere then

its surface area $S = 4\pi R^2$

$$\Rightarrow \quad \frac{dS}{S} = \frac{2dR}{R} = \alpha$$

$$\Rightarrow \quad \frac{dR}{R} = \frac{\alpha}{2}$$

its volume $V = \frac{4}{3}\pi R^3$

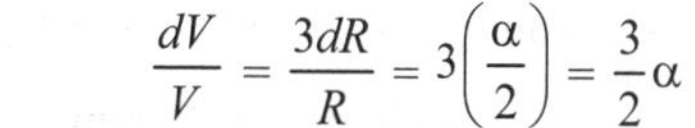

$$\Rightarrow \quad \frac{dV}{V} = \frac{3dR}{R} = 3\left(\frac{\alpha}{2}\right) = \frac{3}{2}\alpha$$ **Ans. (C)**

Sol. 24 Using Kepler's law we have

$$T^2 = \frac{4\pi^2}{GM} R^3$$

taking log on both sides and differentiating, gives

$$2\ln T = \ln\left(\frac{4\pi^2}{GM}\cdot R^3\right)$$

$$\Rightarrow \quad 2\ln T = \ln\left(\frac{4\pi^2 R^3}{G}\right) - \ln M$$

$$\Rightarrow \quad 2\frac{dT}{T} = \frac{dM}{M}$$

$$\Rightarrow \quad \frac{dM}{M} = 2\times 10^{-2}$$ **Ans. (C)**

Sol. 25 Using mass and length of a cube, its density is given as

$$\rho = \frac{\text{Mass}}{\text{Volume}} = \frac{M}{L^3}$$

relative error (maximum) in density calculation is given as

$$\frac{\Delta\rho}{\rho} = \frac{\Delta M}{M} + 3\frac{\Delta L}{L}$$

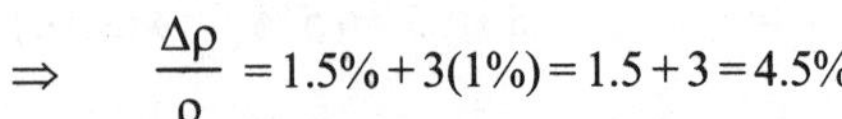

$$\Rightarrow \quad \frac{\Delta\rho}{\rho} = 1.5\% + 3(1\%) = 1.5 + 3 = 4.5\%$$ **Ans. (C)**

Sol. 26 Resistance of a conductor is given as

$$R = \frac{\rho l}{A} \text{ and volume } (V) = Al$$

$$R = \frac{\rho l^2}{V}$$

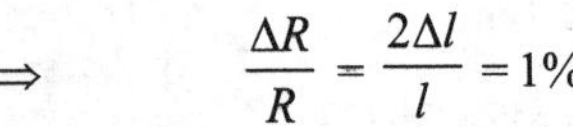

$$\Rightarrow \quad \frac{\Delta R}{R} = \frac{2\Delta l}{l} = 1\%$$ **Ans. (C)**

Sol. 27 Least count of screw gauge is given as

$$LC = \frac{\text{Pitch}}{\text{No. of division}}$$

$$\Rightarrow \quad LC = 0.5\times 10^{-2}\text{ mm}$$

Zero error is given as

$$z = 3\times 0.5\times 10^{-2}\text{ mm}(+ve)$$

$$\Rightarrow \quad z = 1.5\times 10^{-2}\text{ mm} = 0.015\text{ mm}$$

Thickness of sheet measured by screw gauge is

$$t = MSR + CSR - ZE \; (+ve \text{ zero error})$$

$$\Rightarrow \quad t = 5.5\text{ mm} + (48\times 0.5\times 10^{-2}) - 0.015$$

$$\Rightarrow \quad t = 5.5 + 0.24 - 0.015 = 5.725\text{ mm}$$ **Ans. (B)**

Sol. 28 For volume of cylinder, relative error is given as

$$\frac{\Delta V}{V} = 2\frac{\Delta d}{d} + \frac{\Delta h}{h} = 2\left(\frac{0.1}{12.6}\right) + \frac{0.1}{34.2} \simeq 0.016$$

$$V = 34.2\times\frac{\pi}{4}\times(12.6)^2$$

$$\Rightarrow \quad V = 4262.23\text{ cm}^3$$

and $\Delta V = 0.016\times 4262.23 = 68.19\text{ cm}^3$

Volume in 3 SD is given as

$$V = 4260 \pm 68\text{ cm}^3$$

Out of given options most appropriate is (A). **Ans. (A)**

Sol. 29 Least count of a screw gauge is given as

$$\text{Least count} = \frac{\text{Pitch}}{\text{Number of division on circular scale}}$$

$$\Rightarrow \quad 5\times 10^{-6} = \frac{10^{-3}}{N}$$

$$\Rightarrow \quad N = 200$$ **Ans. (C)**

Sol. 30 Time of one oscillation measured is

$$T = \frac{30\text{ sec}}{20}$$

Maximum possible error in measurement (least count) is

$$\Delta T = \frac{1}{20}\text{ sec}$$

Length of pendulum is given as

$$L = 55\text{ cm}$$

$$\Delta L = 1\text{mm} = 0.1\text{ cm}$$

Value of g is calculated as

$$g = \frac{4\pi^2 L}{T^2}$$

Percentage error in measurement of g is

$$\frac{\Delta g}{g}\times 100\% = \left(\frac{\Delta L}{L} + \frac{2\Delta T}{T}\right)100\%$$

$$= \left(\frac{0.1}{55} + \frac{2\left(\frac{1}{20}\right)}{\frac{30}{20}}\right)100\% \simeq 6.8\%$$ **Ans. (D)**

Sol. 31 Density of cube is given as

$$\rho = \frac{m}{v} \qquad [v = l^3]$$

Percentage error or fractional error in density measurement is

$$\frac{\Delta\rho}{\rho} = \frac{\Delta m}{m} + 3\frac{\Delta l}{l}$$

$$\Rightarrow \quad \Delta\rho = \left[\frac{0.10}{10.00} + 3\left(\frac{0.01}{0.10}\right)\right]\times\frac{10.00}{(0.10)^3}\text{ kg/m}^3$$

$\Rightarrow \quad \Delta\rho = [0.01 + 0.3] \times 10^4\ kg/m^3$

$\Rightarrow \quad \Delta\rho = 0.31 \times 10^4\ kg/m^3$ **Ans. (None)**

Sol. 32 Total Area of seven squares is

$$S = A_1 + A_2 + \ldots A_7$$

$$\Rightarrow \quad S = A + A + \ldots 7\text{ times}$$

$$\Rightarrow \quad S = 37.03\ cm^2$$

Addition of 7 terms all having 2 terms beyond decimal, so final answer must have 2 terms beyond decimal (as per rules of significant digits.) **Ans. (C)**

Sol. 33 Time period of a simple pendulum is given as

$$T = 2\pi\sqrt{\frac{l}{g}}$$

$$\Rightarrow \quad g = \frac{4\pi^2 l}{T^2}$$

Fractional error is measurement of g is

$$\frac{\Delta g}{g} = \frac{\Delta l}{l} + \frac{2\Delta T}{T}$$

$$\Rightarrow \quad \frac{\Delta g}{g} = \frac{0.1}{25} + \frac{2\times 1}{50} = 0.044$$

Thus percentage error or accuracy is given as

$$\frac{\Delta g}{g} = 4.4\%$$ **Ans. (C)**

Sol. 34 As by on six rotation, reading of main scale changes by 3mm, so main scale least count is 0.5 mm. Number of division on circular scale are 50 so least count of the screw gauge will be

$$LC = \frac{0.5}{50}\ mm = 0.001\ cm$$ **Ans. (B)**

Sol. 35 For the given set of physical quantities we use

$$A_1 + B_1 + C_1 = 24.36 + 0.0724 + 256.2$$
$$= 280.6324$$
$$= 280.6$$

$$A_2 + B_2 + C_2 = 24.44 + 16.082 + 240.2$$
$$= 280.722$$
$$= 280.7$$

$$A_3 + B_3 + C_3 = 25.2 + 19.2812 + 236.183$$
$$= 280.6642$$
$$= 280.7$$

$$A_4 + B_4 + C_4 = 25 + 236.191 + 19.5$$
$$= 280.691$$
$$= 281$$

$$A_4 + B_4 + C_4 > A_3 + B_3 + C_3 = A_2 + B_2 + C_2 > A_1 + B_1 + C_1$$

Out of given options none is correct. **Ans. (Bonus)**

Sol. 36 Zero error in scale is

$$Z = 0 + 0.01 \times 7 = 0.07\ cm$$

Reading of vernier caliper for length of cylinder is

$$R = 3.1 + 0.01 \times 4 - 0.07 = 3.07\ cm$$ **Ans. (D)**

Sol. 37 Pitch of screw gauge $p = 0.1$ cm

No of divisions on circular scale

$$N = 50$$

least count of screw gauge is

$$L = \frac{p}{N} = 0.002\ cm$$

The result of measurement by this screw gauge must be in multiples of L, so here result would be 2.124 cm. **Ans. (D)**

Sol. 38 Percentage error in measurement of z is given as

$$\frac{\Delta z}{z} = \frac{2\Delta a}{a} + \frac{2}{3}\frac{\Delta b}{b} + \frac{1}{2}\frac{\Delta c}{c} + \frac{3\Delta d}{d}$$

$$\Rightarrow \quad \frac{z}{\Delta z} = 2 \times 2 + \frac{2}{3} \times 15 + \frac{1}{2} \times 4 + 3 \times 2.5\%$$

$$\Rightarrow \quad \frac{\Delta z}{z} = 14.5\%$$ **Ans. (B)**

Sol. 39 Pitch of circular scale = 0.5 mm

Number of divisions of CS = 50

$$\text{Least count} = \frac{0.5}{50}\ mm$$
$$= 0.01\ mm$$
$$= 0.01 \times 10^{-3}\ m$$
$$= 10 \times 10^{-6}\ m = 40\ \mu m$$

Zero error $z = 4 \times 0.01 = 0.04$ mm (+ve) **Ans. (B)**

Sol. 40 Fractional error in density measurement is given as

$$\frac{d\rho}{\rho} = \frac{dm}{m} + \frac{3dD}{D} \qquad \left[\text{As } \rho = \frac{m}{\frac{4}{3}\pi\left(\frac{D}{2}\right)^3}\right]$$

We use

$$\frac{d\rho}{\rho} = \times 100 = 6 + 3 \times 1.5 = 10.5\% = \frac{x}{100}$$

$$\Rightarrow \quad x = 1050$$ **Ans. (1050)**

Sol. 41 As all readings are given in 3 significant digits, final results must be in 3 significant digits only so option (C) can be only correct answer. **Ans. (C)**

Sol. 42 The least count of screw guage given is calculated as

$$LC = \frac{1\text{mm}}{100} = 0.01\text{mm}$$

Zero error in the given screw guage is given as

$$z = +8 \times LC = +0.08 \text{ mm}$$

True reading for diameter of wire is given as

$$TR = (1 \text{ mm} + 72 \times LC) - z$$

$$\Rightarrow \quad TR = (1 \text{ mm} + 72 \times 0.01 \text{ mm}) - 0.08 \text{ mm}$$

$$\Rightarrow \quad TR = 1.72 \text{ mm} - 0.08 \text{ mm} = 1.64 \text{ mm}$$

Radius of wire is given as

$$r = \frac{1.64}{2} = 0.82 \text{ mm}$$ **Ans. (D)**

Sol. 43 As volume of wire remains constant while stretching so we use

$$A_0 l_0 = A_1 l_1$$

$$\Rightarrow \quad A_1 = \frac{l_0 A_0}{l_1}$$

Resistance of wire can be given as

$$R = \frac{\rho l}{A}$$

If R_0 and R_1 are the initial and final resistances of wire, we use

$$\frac{R_0}{R_1} = \frac{l_0 / A_0}{l_1 / A_1}$$

$$\Rightarrow \quad \frac{1}{R_1} = \frac{l_0}{A_0}\left(\frac{l_0 A_0}{l_1 \times l_1}\right)$$

$$\Rightarrow \quad R_1 = \frac{l_1^2}{l_0^2} = 1.5625\,\Omega$$

Thus percentage change in resistance is given as

$$p = \frac{R_1 - R_0}{R_0} \times 100\%$$

$$\Rightarrow \quad p = \frac{1.5625 - 1}{1} \times 100\% = 56.25\%$$ **Ans. (D)**

Sol. 44 The percentage error in value of resistance R is given as

$$\frac{\Delta R}{R} \times 100 = \frac{\Delta V}{V} \times 100 + \frac{\Delta I}{I} \times 100$$

$$\frac{\Delta R}{R} \times 100 = \frac{2}{50} \times 100 + \frac{0.2}{20} \times 100$$

$$\frac{\Delta R}{R} \times 100 = 4 + 1 = 5\%$$ **Ans. (5)**

Sol. 45 Positive zero error is to be subtracted from the final reading. As main scale reading is 8.5 cm and vernier coincidence is 6, so final reading of vernier scale reading is given as

$$\text{VSR} = 6 \times 0.01 = 0.06 \text{ cm}$$

Thus final reading is calculated as

$$\text{MSR} + \text{VSR} - \text{ZE} = 8.5 + 0.06 - 0.02 = 8.54 \text{ cm}$$ **Ans. (B)**

Sol. 46 By the expression of time period of a simple pendulum, value of g is given as

$$g = \frac{4\pi^2 l}{T^2}$$

Fractional error in value of g is given as

$$\frac{\Delta g}{g} = \frac{\Delta l}{l} + 2\frac{\Delta T}{T} = \frac{0.1}{10} + 2\left(\frac{1}{\frac{200}{0.5}}\right)$$

$$\Rightarrow \quad \frac{\Delta g}{g} = \frac{1}{100} + \frac{1}{50}$$

Percentage error in value of g is

$$\Rightarrow \quad \frac{\Delta g}{g} \times 100 = 3\%$$ **Ans. (B)**

Sol. 47 Volume of sphere is expressed as

$$V = \frac{4}{3}\pi r^3$$

Fractional error in volume can be written as

$$\frac{dV}{V} = 3\frac{dr}{r}$$

Percentage error in volume is given as

$$\frac{dV}{V} \times 100\% = \frac{3 \times 0.85}{7.5} \times 100\% = 34\%$$ **Ans. (34)**

Sol. 48 Percentage error in calculation of g can be written as

$$\frac{\Delta g}{g} = \left(\frac{\Delta l}{l} + \frac{2\Delta T}{T}\right) \times 100\%$$ **Ans. (1)**

Sol. 49 The difference between the two student reading for any set of experiment should be equal to approximately the difference between zero error in the two instruments. From the figure, it is calculated as

$$= 5 - (-8) = 13 \text{ divisions}$$ **Ans. (13)**

Sol. 50 The magnetic permeability of the medium is related to susceptibility of medium is given as

$$\mu_r = 1 + \chi$$

$$\Delta\mu_r = \Delta\chi$$

As $B \propto \mu r$

$\Rightarrow$ $B = k\mu_r$ (Here k is a constant)

Percentage change in magnetic field is given as

$$\frac{\Delta B}{B} \times 100 = \frac{k(\Delta\mu_r)}{k\mu_r} \times 100 = \frac{2.2\times10^{-5}}{1} \times 100 = 0.022$$

$\Rightarrow$ $x = 22$ **Ans. (22)**

Sol. 51 Pitch of screw gauge is the distance travelled on main scale in one rotation which is 1 mm thus least count is given as

$$LC = 1\text{ mm}/50 = 0.02\text{ mm}$$

Thus Statement-1 is wrong. Statement-2 is right **Ans. (B)**

Sol. 52 Percentage error in calculation of y is given as

$$\%y = 2(\%m) + 4(\%r) + x(\%g) + \frac{3}{2}(\%l)$$

$\Rightarrow$ $18 = 2 \times 1 + 4 \times 0.5 + x(p) + \frac{3}{2} \times 4$

$\Rightarrow$ $18 = 2 + 2 + x(p) + 6$

$\Rightarrow$ $x(p) = 8$

Out of given options only option (C) can be correct. **Ans. (C)**

Sol. 53 Time period of simple pendulum is given as

$$T = 2\pi\sqrt{\frac{l}{g}}$$

For percentage error calculation, we have

$$\frac{\Delta g}{g} + \frac{2\Delta T}{T} = \frac{\Delta L}{L}$$

$\Rightarrow$ $4 + 2 \times 3 = 10\%$

$\Rightarrow$ $E = \frac{1}{2}m\omega^2A^2 = \frac{1}{2}m\frac{g}{l}A^2$

$\Rightarrow$ $E = \frac{\Delta g}{g} + \frac{\Delta L}{L} = 4 + 10 = 14\%$ **Ans. (14)**

Sol. 54 For the expression of Z, maximum error is calculated as

$$Z = \frac{A^2B^3}{C^4}$$

$\Rightarrow$ $\frac{\Delta Z}{Z} = \frac{2\Delta A}{A} + \frac{3\Delta B}{B} + \frac{4\Delta C}{C}$ **Ans. (C)**

Sol. 55 For the expression of z, maximum percentage error is calculated as

$$z = a^2 x^3 y^{1/2}$$

$\Rightarrow$ $\frac{\Delta z}{z} = \frac{2\Delta a}{a} + \frac{3\Delta x}{x} + \frac{1}{2}\frac{\Delta y}{y}$

Here a is constant so we have

$$\frac{\Delta z}{z} \times 100 = 3(4\%) + \frac{1}{2}(12\%) = 18\%$$ **Ans. (18)**

Sol. 56 For the given vernier callipers, given that

$$20\ MSD = 1\text{ cm}$$

$\Rightarrow$ $1\ MSD = \frac{1}{20}\text{ cm}$

and $10\ VSD = 9\ MSD$

$\Rightarrow$ $1\ VSD = \frac{9}{10}\ MSD$

$\Rightarrow$ $1\ VSD = \frac{9}{10} \times \frac{1}{20} = \frac{9}{200}\text{ cm}$

Vernier constant or least count of vernier callipers is given as

$$VC = 1\ MSD - 1\ VSD$$

$\Rightarrow$ $VC = \frac{1}{20}\text{ cm} - \frac{9}{200}\text{ cm}$

$\Rightarrow$ $VC = \frac{1}{200} \times 10\text{ mm} = 5 \times 10^{-2}\text{ mm}$ **Ans. (5)**

Sol. 57 For the given vernier scale, given that

$$50\ VSD = 49\ MSD$$

$\Rightarrow$ $1\ VSD = \frac{49}{50}\ MSD$

Least count of given scale is calculated as

$$LC = 1\ MSD - 1\ VSD$$

$\Rightarrow$ $LC = \left(1 - \frac{49}{50}\right)MSD = \frac{1}{50}\ MSD$

$\Rightarrow$ $LC = \frac{1}{50 \times 40}\text{ cm}$

$\Rightarrow$ $LC = \frac{1}{2000}\text{ cm} = \frac{1}{2} \times 10^{-5}\text{ m}$

$\Rightarrow$ $LC = 5 \times 10^{-6}\text{ m}$ **Ans. (5)**

Sol. 58 Density of the silver wire is written as

$$\rho = \frac{m}{V} = \frac{m}{\pi r^2 l}$$

Fractional error in density is calculated as

$$\frac{\Delta\rho}{\rho}=\frac{\Delta m}{m}+\frac{2\Delta r}{r}+\frac{\Delta l}{l}$$

$$\Rightarrow \quad \frac{\Delta\rho}{\rho}=\frac{0.006}{0.6}+\frac{2\times 0.005}{0.5}+\frac{0.04}{4}$$

Percentage error is calculated as

$$100\times\frac{\Delta\rho}{\rho}=4\times 10^{-2}\times 100$$

$$\Rightarrow \quad \frac{\Delta\rho}{\rho}\times 100=4\%$$

Ans. (A)

Sol. 59 The average value of wire thickness t is calculated as

$$t=\frac{1.22\text{mm}+1.23\text{mm}+1.19\text{mm}+1.20\text{mm}}{4}$$

$$\Rightarrow \quad t=1.21\text{ mm}$$

Average deviation from average value of t is calculated as

$$\Delta t=\frac{0.01+0.02+0.02+0.01}{4}=\frac{0.06}{4}=0.015$$

Percentage error is calculated as

$$\Delta t(\text{in }\%)=\frac{0.015}{1.21}\times 100$$

$$\Rightarrow \quad x=150$$

Ans. (150)

Sol. 60 Measured diameter of sphere is written as

$$d=MSR+VSR\times VC$$

$$\Rightarrow \quad d=1.7+0.01\times 5=1.75\text{ cm}$$

Corrected diameter is calculated as

$$d_c=\text{Measured}-\text{Error}$$

$$\Rightarrow \quad d_c=1.75-(-0.05)$$

$$\Rightarrow \quad d_c=1.80\text{ cm}=180\times 10^{-2}\text{ cm}$$

Ans. (180)

Sol. 61 Heat dissipation is calculated in time t as

$$E_H=I^2R\times t$$

Percentage error in calculation of heat is given as

$$\frac{\Delta E}{E}\times 100=\frac{2\Delta I}{I}\times 100+\frac{\Delta R}{R}\times 100+\frac{\Delta T}{T}\times 100$$

$$\Rightarrow \quad \frac{\Delta E}{E}\times 100=2\times 2+1+3=8\%$$

Ans. (D)

Sol. 62 Least count of the given screw guage is

$$\text{LC}=\frac{P}{N}=\frac{0.5\text{mm}}{50}=0.01\text{ nm}$$

Length of wire is already given as constant at 6.8 cm. The diameter of wire is measured as

$$d=1.5\text{ mm}+7\times\text{LC}$$

$$\Rightarrow \quad d=1.5\text{ mm}+7\times .01=1.57\text{ mm}$$

Curved surface area of wire is calculated as

$$S=\pi Dl$$

$$\Rightarrow \quad S=3.14\times 6.8\times 1.57\times 10^{-1}\text{ cm}^2$$

$$\Rightarrow \quad S=3.352\text{ cm}^2$$

Rounding off to two significant digits gives

$$S=3.4\text{ cm}^2$$

Ans. (B)

Sol. 63 For the given scale, we use

$$1\text{ MSD}=1\text{ mm}$$

and

$$9\text{ MSD}=10\text{ VSD}$$

$$\Rightarrow \quad 1\text{ VSD}=0.9\text{ MSD}=0.9\text{ mm}$$

Least count of vernier caliper is calculated as

$$LC=1-0.9=0.1\text{ mm}=0.01\text{ cm}$$

Zero error in scale is given as

$$z=-(10-4)\times 0.1\text{ mm}=-0.6\text{ mm}$$

Reading of scale is given as

$$R=\text{MSR}+\text{VSR}-z$$

$$\Rightarrow \quad R=3\text{cm}+6\times 0.01-[-0.06]$$

$$\Rightarrow \quad R=3+0.06+0.06=3.12\text{ cm}$$

Nearest given answer in the options is 3.10. **Ans. (C)**

Sol. 64 Dimensional formula for Torque is written as

$$[\tau]=[\text{ML}^2\text{T}^{-2}]$$

Percentage error in torque calcluation is given as

$$\Delta\tau\text{ (in }\%)=(\Delta\text{M in }\%)+2(\Delta\text{L in }\%)+2(\Delta\text{T in }\%)$$

$$\Rightarrow \quad \Delta\tau\text{ (in }\%)=25\%$$

Ans. (B)

Sol. 65 For the given vernier scale we have

$$10\text{ VSD}=9\text{ MSD}$$

$$\Rightarrow \quad 1\text{ VSD}=.1\text{ mm}=.01\text{ cm}$$

Zero error in scale is given as

$$z=+0.4\text{ mm}=0.04\text{ cm}$$

Diameter reading is calculated as

$$d=4.1\text{ cm}+6\times .01-0.04\text{ cm}=4.12\text{ cm}$$

$$\Rightarrow \quad d=412\times 10^{-2}\text{ cm}$$

Ans. (412)

Sol. 66 Time period of oscillations of pendulum is given as

$$T = 2\pi\sqrt{\frac{l}{g}}$$

$$\Rightarrow \quad g = \frac{1}{4\pi^2}\frac{T^2}{l}$$

Fractional error in calculation of g is given as

$$\frac{\Delta g}{g} = \frac{2\Delta T}{T} + \frac{\Delta l}{l}$$

$$\Rightarrow \quad \frac{\Delta g}{g} = 2\cdot\frac{1}{100\times 0.5} + \frac{1}{100} = \frac{5}{100}$$

Percentage error in calculation of g is given as

$$\frac{\Delta g}{g}\times 100 = 5$$

Ans. (5)

Sol. 67 For the travelling microscope, we have

$$1\,\text{MSD} = \frac{1}{20}\text{cm}$$

and $$1\,\text{VSD} = \frac{24}{25}\,\text{MSD} = \frac{24}{25}\times\frac{1}{20}\text{ cm}$$

Least count of scale is calculated as

$$LC = \text{MSD} - \text{VSD} = \frac{1}{20}\left(1-\frac{24}{25}\right)\text{cm}$$

$$\Rightarrow \quad LC = \frac{1}{20}\times\frac{1}{25} = \frac{1}{500}\text{cm} = 0.002\text{ cm}$$ **Ans. (C)**

Sol. 68 For the given scale of screw guage, we have

$$MSR = 2.5\text{ mm}$$

and $$CSR = 45\times\frac{0.5}{50}\text{ mm} = 0.45\text{ mm}$$

Diameter of wire is given as

$$d = MSR + CSR - \text{zero error}$$

$$\Rightarrow \quad d = 2.5 + 0.45 - (-0.03) = 2.98\text{ mm}$$ **Ans. (C)**

Sol. 69 The refractive index of glass slab is calculated using concept of apparent depth as

$$\mu = \frac{h}{h'} = \frac{5.25}{5.00} \qquad ...(1)$$

Least count of travelling microscope is calculated as

$$\text{LC} = \frac{1}{20}\text{ cm} - \frac{49}{50}\cdot\frac{1}{20}\text{cm}$$

$$\Rightarrow \quad \text{LC} = \frac{1}{50}\times\frac{1}{20}\text{ cm} = 0.01\text{ mm}$$

Taking log on both sides of equation (1) gives

$$\ln\mu = \ln h - \ln h'$$

Differentiating above equation gives

$$\frac{d\mu}{\mu} = \frac{dh}{h} - \frac{dh'}{h'}$$

$$\Rightarrow \quad d\mu = \left[\frac{0.01}{5.25} + \frac{0.01}{5.00}\right]\frac{5.25}{5.00} = 0.0041$$

$$\Rightarrow \quad d\mu = \frac{41}{10}\times 10^{-3}$$

$$\Rightarrow \quad x = 41$$ **Ans. (41)**

Sol. 70 For the screw gauge, least count is calculated as

$$LC = \frac{\text{Pitch}}{\text{No. of circular divisions}}$$

$$\Rightarrow \quad LC = \frac{0.5\text{mm}}{100}$$

$$\Rightarrow \quad LC = 5\times 10^{-3}\text{ mm}$$

Error in the given state of question is calculated as

$$\text{Positive Error} = MSR + CSR\,(LC)$$
$$= 0\text{ mm} + 6(5\times 10^{-3}\text{ mm})$$

Diameter of wire is calculated as

$$D = MSR + CSR\,(LC) - \text{Positive zero error}$$
$$= 4\times 0.5\text{ mm} + (46(5\times 10^{-3})) - 6(5\times 10^{-3})\text{ mm}$$
$$= 2\text{ mm} + 40\times 5\times 10^{-3}\text{ mm}$$
$$= 2.2\text{ mm}$$ **Ans. (220)**

Sol. 71 When two resistances R_1 and R_2 are connected in parallel, equivalent resistance is calculated as

$$\frac{1}{R} = \frac{1}{R_1} + \frac{1}{R_2} \qquad ...(1)$$

$$\Rightarrow \quad R = \frac{R_1R_2}{R_1+R_2} = \frac{10\times 15}{10+15} = 6\Omega$$

Differentiating equation-(1) on both sides, we get

$$\frac{\Delta R}{R^2} = \frac{\Delta R_1}{R_1^2} + \frac{\Delta R_2}{R_2^2}$$

$$\Rightarrow \quad \frac{\Delta R}{R} = \left(\frac{\Delta R_1}{R_1^2} + \frac{\Delta R_2}{R_2^2}\right)R$$

$$\Rightarrow \quad \frac{\Delta R}{R} = \left(\frac{0.5}{100} + \frac{0.5}{225}\right)6$$

$$\Rightarrow \quad \frac{\Delta R}{R} = \left(\frac{6\times 0.5}{25}\right)\left(\frac{1}{4}+\frac{1}{9}\right) = \frac{13}{300}$$

Percentage error in measurement of equivalent resistance is given as

$$\frac{\Delta R}{R}\times 100=\frac{13}{3}=4.33\% \qquad \textbf{Ans. (C)}$$

Sol. 72 Initial resistance of the wire is written as

$$R=\rho\frac{l}{A}$$

When length is increased by 20% by stretching, new length becomes $1.2l$ and also the cross-section area is reduced by 4%, so it becomes 0.96 A. The new resistance can be written as

$$R'=\rho\frac{1.2l}{0.96\text{A}}=1.25\rho\frac{l}{A}=1.25\,\text{R}$$

Percentage change in resistance is calculated as

$$p=\frac{1.25R-R}{R}\times 100=25\% \qquad \textbf{Ans. (25)}$$

Sol. 73 Least count of scale = 0.2 cm

Distance of object from optical centre is calculated as

$$u=(100\pm 0.2)-(80\pm 0.2)=(20\pm 0.4)\,\text{cm}$$

Distance of image from optical centre is calculated as

$$v=(180\pm 0.2)-(100\pm 0.2)=(80\pm 0.4)\,\text{cm}$$

From lens formula, we have

$$\frac{1}{v}-\frac{1}{u}=\frac{1}{f}$$

$$\Rightarrow \quad \frac{1}{f}=\frac{1}{80}-\frac{1}{-20}$$

$$\Rightarrow \quad f=16\,\text{cm}$$

Differentiating both sides of lens formula for maximum percentage error gives

$$\frac{\Delta v}{v^2}+\frac{\Delta u}{u^2}=\frac{\Delta f}{f^2}$$

Percentage error in finding focal length is given as

$$\Rightarrow \quad \frac{\Delta f}{f}\times 100=\left(\frac{\Delta v}{v^2}+\frac{\Delta u}{u^2}\right)\times f\times 100$$

$$\Rightarrow \quad \frac{\Delta f}{f}\times 100=\left(\frac{0.4}{400}+\frac{0.4}{6400}\right)\times 16\times 100$$

$$\Rightarrow \quad \frac{\Delta f}{f}\times 100=1.70 \qquad \textbf{Ans. (C)}$$

Sol. 74 Density of wire is calculated as

$$\text{Density}=\frac{\text{Mass}}{\text{Volume}}$$

$$\Rightarrow \quad \rho=\frac{m}{\pi r^2 l}$$

For maximum error by differentiation, it gives

$$\Rightarrow \quad \left|\frac{d\rho}{\rho}\right|_{\max}=2\left|\frac{dr}{r}\right|+\left|\frac{dl}{l}\right|$$

$$\Rightarrow \quad \left|\frac{d\rho}{\rho}\right|_{\max}=\frac{0.01}{0.4}+\frac{2(0.03)}{6}+\frac{0.04}{8}$$

Percentage error in density is calculated as

$$p=\left(\frac{d\rho}{\rho}\right)\times 100\%$$

$$\Rightarrow \quad p=(2.5+1+0.5)\%=4\% \qquad \textbf{Ans. (C)}$$

Sol. 75 Percentage error in measurement of the physical quantity is calculated as

$$\frac{\Delta P}{P}\times 100\%=\left(2\frac{\Delta a}{a}+3\frac{\Delta b}{b}+\frac{\Delta c}{c}+\frac{1}{2}\frac{\Delta d}{d}\right)\times 100\%$$

$$\Rightarrow \quad \frac{\Delta P}{P}\times 100\%=2(1\%)+3(2\%)+3\%+\frac{1}{2}\times 4\%=13\%$$

Ans. (A)

Sol. 76 Least count of the vernier is given as

$$LC=0.1\,\text{mm}$$

Zero Error= $6\times LC=0.6$ mm

Reading= $MSR+VSR\times LC-$ Zero Error

Here MSR is main scale reading VSR is vernier scale reading, using values, the diameter of bob is measured as

$$d=[32\,\text{mm}+(0.1)4\,\text{mm}]-0.6\,\text{mm}$$

$$\Rightarrow \quad d=31.8\,\text{mm}$$

$$\Rightarrow \quad d=3.18\,\text{cm} \qquad \textbf{Ans. (A)}$$

* * * * *